Understanding Microbial Biofilms

Understanding Microbial Biofilms

Fundamentals to Applications

Edited by

Surajit Das
Department of Life Science, National Institute of Technology, Rourkela, Odisha, India

Neelam Amit Kungwani
Gujarat Biotechnology University, Gandhinagar, Gujarat, India

ELSEVIER

ACADEMIC PRESS
An imprint of Elsevier

Academic Press is an imprint of Elsevier
125 London Wall, London EC2Y 5AS, United Kingdom
525 B Street, Suite 1650, San Diego, CA 92101, United States
50 Hampshire Street, 5th Floor, Cambridge, MA 02139, United States
The Boulevard, Langford Lane, Kidlington, Oxford OX5 1GB, United Kingdom

ISBN 978-0-323-99977-9

For information on all Academic Press publications
visit our website at https://www.elsevier.com/books-and-journals

Publisher: Stacy Masucci
Acquisitions Editor: Kattie Washington
Editorial Project Manager: Susan E. Ikeda
Production Project Manager: Maria Bernard
Cover Designer: Christian J. Bilbow

Typeset by STRAIVE, India

Contents

Section E
Biofilms and pathogenesis

Section F
Application of microbial biofilms

Contributors

Numbers in parentheses indicate the pages on which the authors' contributions begin.

Mohd Adil (427), Department of Plant, Food and Environmental Sciences, Faculty of Agriculture, Dalhousie University, Truro, NS, Canada

Iqbal Ahmad (59,315,587), Department of Agricultural Microbiology, Faculty of Agricultural Sciences, Aligarh Muslim University, Aligarh, Uttar Pradesh, India

Sanjeev Anand (415), Midwest Dairy Research Center, Department of Dairy and Food Science, South Dakota State University, Brookings, SD, United States

Firoz Ahmad Ansari (59), Department of Agricultural Microbiology, Faculty of Agricultural Sciences, Aligarh Muslim University, Aligarh, Uttar Pradesh, India

Yaser Arafath (291), Department of Microbiology, School of Life Sciences, Pondicherry University, Kalapet, Puducherry, India

Marco Artini (81), Department of Public Health and Infectious Diseases, Sapienza University of Rome, Rome, Italy

Ana Azevedo (665), LEPABE—Laboratory for Process Engineering, Environment, Biotechnology and Energy; ALiCE – Associate Laboratory in Chemical Engineering, Faculty of Engineering, University of Porto, Porto, Portugal

Sivaraj Barath (373), Department of Biotechnology, Alagappa University, Karaikudi, Tamil Nadu, India

Samsudeen Malik Basha (629), Department of Biotechnology, Alagappa University, Karaikudi, Tamil Nadu, India

Justyna Bauza-Kaszewska (125), Department of Microbiology and Food Technology, Jan and Jędrzej Śniadecki University of Technology in Bydgoszcz, Bydgoszcz, Poland

Akshaya Kumar Behera (147), Algal Biotechnology and Molecular Systematic Laboratory, Post Graduate Department of Botany, Berhampur University, Berhampur, Odisha, India

Chhandashree Behera (147), Algal Biotechnology and Molecular Systematic Laboratory, Post Graduate Department of Botany, Berhampur University, Berhampur, Odisha, India

Assirbad Behura (461), Laboratory of Mycobacterial Immunology, Department of Life Science, National Institute of Technology, Rourkela, Odisha, India

Sudipa Bhadra (699), Department of Biotechnology, National Institute of Technology Warangal, Warangal, Telangana, India

Prajna Paramita Bhuyan (147), Department of Botany, Maharaja Sriram Chandra Bhanja Deo University, Baripada, Odisha, India

Níura Madalena Bila (521), School of Pharmaceutical Science, Universidade Estadual Paulista, Araraquara, SP, Brazil

Damon C. Brown (173), Department of Biological Sciences, University of Calgary, Calgary, AB, Canada

Maria Pia Busnelli (683), Chemistry Area, National University General Sarmiento, Buenos Aires, Argentina

Fábio M. Carvalho (665), LEPABE—Laboratory for Process Engineering, Environment, Biotechnology and Energy; ALiCE – Associate Laboratory in Chemical Engineering, Faculty of Engineering, University of Porto, Porto, Portugal

Shruti Chatterjee (247), Applied Phycology and Biotechnology Division, CSIR—Central Salt and Marine Chemicals Institute, Bhavnagar, Gujarat, India

Rui Chen (91,275), Biofilm Centre, 5D Health Protection, Group, Liverpool, United Kingdom

Christudoss Abisha Christy (629), Department of Biotechnology, Alagappa University, Karaikudi, Tamil Nadu, India

Caroline Barcelos Costa-Orlandi (521), Department of Clinical Analysis, School of Pharmaceutical Sciences, São Paulo State University (UNESP), Araraquara, SP, Brazil

Loredana Cristiano (81), Department of Clinical Medicine, Public Health, Life and Environmental Sciences, University of Aquila, L'Aquila, Italy

Ana Carolina Moreira da Silva Pires (521), Food and Nutrition, UNESP, Araraquara, São Paulo, Brazil

Maria Alejandra Daniel (683), Chemistry Area, National University General Sarmiento, Buenos Aires, Argentina

Kannan Dheerka Darsha (629), Department of Biotechnology, Alagappa University, Karaikudi, Tamil Nadu, India

Mousumi Das (461), Laboratory of Mycobacterial Immunology, Department of Life Science, National Institute of Technology, Rourkela, Odisha, India

Samanta de Matos Silva (521), Sao Paulo State University, São Paulo, Brazil

Purbajyoti Deka (475), Department of Biotechnology, Mizoram University, Aizawl, Mizoram, India

Rohan Dhiman (461), Laboratory of Mycobacterial Immunology, Department of Life Science, National Institute of Technology, Rourkela, Odisha, India

Singh Dikshant (373), Department of Biotechnology, Alagappa University, Karaikudi, Tamil Nadu, India

Orlando Donfrancesco (81), Department of Anatomy, Histology, Forensic Medicine and Orthopaedics, Sapienza University of Rome, Rome, Italy

Aifa Fathima (291), Department of Microbiology, School of Life Sciences, Pondicherry University, Kalapet, Puducherry, India

Lauren Foulkes (91,275), Microbiology, Biofilm Centre, Liverpool, United Kingdom

Ana Marisa Fusco-Almeida (521), School of Pharmaceutical Sciences, São Paulo State University (UNESP), Araraquara, Brazil

Gashahun Gezahegne (315), Department of Agribusiness and Value Chain Management, College of Agriculture and Forestry, Mettu University, Bedelle, Ethiopia

Anindya Sundar Ghosh (357), Molecular Microbiology Laboratory, Department of Biotechnology, Indian Institute of Technology Kharagpur, Kharagpur, West Bengal, India

Lisa M. Gieg (173), Department of Biological Sciences, University of Calgary, Calgary, AB, Canada

Luciana C. Gomes (665), LEPABE—Laboratory for Process Engineering, Environment, Biotechnology and Energy; ALiCE – Associate Laboratory in Chemical Engineering, Faculty of Engineering, University of Porto, Porto, Portugal

Vinoj Gopalakrishnan (165), Department of Chemical and Biological Engineering; Data Driven Material Discovery Center for Bioengineering Innovation, South Dakota School of Mines and Technology, Rapid City, SD, United States

Eugenia Gospodarek-Komkowska (125), Department of Microbiology, Nicolaus Copernicus University in Toruń, Ludwik Rydygier Collegium Medicum in Bydgoszcz, Bydgoszcz, Poland

Katarzyna Grudlewska-Buda (125), Department of Microbiology, Nicolaus Copernicus University in Toruń, Ludwik Rydygier Collegium Medicum in Bydgoszcz, Bydgoszcz, Poland

Anshu Gupta (699), University School of Environment Management, Guru Gobind Singh Indraprastha University, New Delhi, India

Soumya Haldar (247), Analytical Discipline and Centralized Instrument Facility, CSIR—Central Salt and Marine Chemicals Institute, Bhavnagar, Gujarat, India

Ekramul Haque (247), Analytical Discipline and Centralized Instrument Facility, CSIR—Central Salt and Marine Chemicals Institute, Bhavnagar, Gujarat, India

Singhal Hardik (373), Department of Biotechnology, Alagappa University, Karaikudi, Tamil Nadu, India

Saqib Hassan (291), Department of Microbiology, School of Life Sciences, Pondicherry University, Kalapet, Puducherry; Division of Non-Communicable Diseases, Indian Council of Medical Research (ICMR), New Delhi, India

Fohad Mabood Husain (315), Department of Food Science and Nutrition, King Saud University, Riyadh, Saudi Arabia

Julfequar Hussain (247), Department of Biotechnology, School of Life Sciences, Pondicherry University, Kalapet, Puducherry, India

Mrutyunjay Jena (147), Algal Biotechnology and Molecular Systematic Laboratory, Post Graduate Department of Botany, Berhampur University, Berhampur, Odisha, India

P. Jishma (71), School of Biosciences, Mahatma Gandhi University, Kottayam, Kerala, India

Karthikeyan Kandaswamy (335), Department of Biotechnology, Alagappa University, Karaikudi, Tamil Nadu, India

Ashwin Kannan (41), Department of Chemical Engineering, Visvesvaraya National Institute of Technology, Nagpur, Maharashtra, India

Thirupathi Kasthuri (335,373,629), Department of Biotechnology, Alagappa University, Karaikudi, Tamil Nadu, India

Anubha Kaushik (699), University School of Environment Management, Guru Gobind Singh Indraprastha University, New Delhi, India

Vaishali Kaushik (209), Department of Biochemistry, Central University of Rajasthan, Ajmer, Rajasthan, India

R. Kavi Bharathi (605), Biofilm Biology Laboratory, Anusandhan Kendra-2, Centre for Research on Infectious Diseases (CRID), School of Chemical and Biotechnology, SASTRA Deemed to be University, Thanjavur, Tamil Nadu, India

Asad U. Khan (427), Interdisciplinary Biotechnology Unit, Aligarh Muslim University, Aligarh, Uttar Pradesh, India

Atif Khan (187), Biofouling and Biofilm processes Section, Water & Steam Chemistry Division, Bhabha Atomic Research Centre Facilities, Kalpakkam, Tamil Nadu; Homi Bhabha National Institute, Mumbai, Maharastra, India

Javed Ahamad Khan (315), Department of Natural Resource Management, Mettu University, Bedele Campus, Mettu, Ethiopia

Rosina Khan (427), Department of Plant, Food and Environmental Sciences, Faculty of Agriculture, Dalhousie University, Truro, NS, Canada

George Segal Kiran (291), Department of Food Science and Technology, School of Life Sciences, Pondicherry University, Kalapet, Puducherry, India

Ashish Kumar (461), Laboratory of Mycobacterial Immunology, Department of Life Science, National Institute of Technology, Rourkela, Odisha, India

Joanna Kwiecińska-Piróg (125), Department of Microbiology, Nicolaus Copernicus University in Toruń, Ludwik Rydygier Collegium Medicum in Bydgoszcz, Bydgoszcz, Poland

Irene C. Lazzarini Behrmann (683), Chemistry Area, National University General Sarmiento, Buenos Aires, Argentina

Swarupa Mallick (447), Laboratory of Infection Immunology, Department of Life Science, National Institute of Technology, Rourkela, Odisha, India

Nambiraman Malligarjunan (485), Department of Biotechnology, Alagappa University, Karaikudi, Tamil Nadu, India

Amiya Kumar Mandal (147), Algal Biotechnology and Molecular Systematic Laboratory, Post Graduate Department of Botany, Berhampur University, Berhampur, Odisha, India

Gajanan Mane (261), National Centre for Microbial Resources (NCMR), NCCS, NCCS Complex, S.P. Pune University, Pune, Maharashtra, India

Caroline Maria Marcos (521), School of Pharmaceutical Sciences, São Paulo State University (UNESP), Araraquara, Brazil

Kaila Petronila Medina-Alarcón (521), Department of Clinical Analysis, School of Pharmaceutical Sciences, São Paulo State University (UNESP), Araraquara, SP, Brazil

Maria José Soares Mendes-Giannini (521), Clinical Analysis, School of Pharmacy-UNESP, Araraquara, Sao Paulo, Brazil

Filipe J.M. Mergulhão (665), LEPABE—Laboratory for Process Engineering, Environment, Biotechnology and Energy; ALiCE – Associate Laboratory in Chemical Engineering, Faculty of Engineering, University of Porto, Porto, Portugal

Abtar Mishra (461), Laboratory of Mycobacterial Immunology, Department of Life Science, National Institute of Technology, Rourkela, Odisha, India

Amit Mishra (461), Cellular and Molecular Neurobiology Unit, Indian Institute of Technology Jodhpur, Jodhpur, Rajasthan, India

Neha Mishra (447), Laboratory of Infection Immunology, Department of Life Science, National Institute of Technology, Rourkela, Odisha, India

Nishu Mittal (569), Institute of Biosciences and Technology, Shri Ramswaroop Memorial University, Barabanki, Uttar Pradesh, India

Andrea M. Monroy (683), Chemistry Area, National University General Sarmiento; CONICET, Buenos Aires, Argentina

Hari Prasath Nagaiah (335,485,629), Department of Biotechnology, Alagappa University, Karaikudi, Tamil Nadu, India

Lincoln Naik (461), Laboratory of Mycobacterial Immunology, Department of Life Science, National Institute of Technology, Rourkela, Odisha, India

Muruganandam Nandhakumar (373), Department of Biotechnology, Alagappa University, Karaikudi, Tamil Nadu, India

Dev Kiran Nayak (461), Laboratory of Mycobacterial Immunology, Department of Life Science, National Institute of Technology, Rourkela, Odisha, India

Rabindra Nayak (147), Algal Biotechnology and Molecular Systematic Laboratory, Post Graduate Department of Botany, Berhampur University, Berhampur, Odisha, India

Vidya Devi Negi (447), Laboratory of Infection Immunology, Department of Life Science, National Institute of Technology, Rourkela, Odisha, India

Stefania Nottola (81), Department of Anatomy, Histology, Forensic Medicine and Orthopaedics, Sapienza University of Rome, Rome, Italy

Gloria N. Okpala (173), Department of Biological Sciences, University of Calgary, Calgary, AB, Canada

Lipsa Panda (395), Laboratory No. 6, Nosocomial Pathogen Biology Laboratory, Department of Biotechnology, Pondicherry University, Kalapet, Puducherry, India

Shunmugiah Karutha Pandian (3,335,373,485,629), Department of Biotechnology, Alagappa University, Karaikudi, Tamil Nadu, India

Rosanna Papa (81), Department of Public Health and Infectious Diseases, Sapienza University of Rome, Rome, Italy

Sudhamayee Parida (147), Algal Biotechnology and Molecular Systematic Laboratory, Post Graduate Department of Botany, Berhampur University, Berhampur, Odisha, India

Salina Patel (447,461), Laboratory of Mycobacterial Immunology; Laboratory of Infection Immunology, Department of Life Science, National Institute of Technology, Rourkela, Odisha, India

Milind S. Patole (261), National Centre for Microbial Resources (NCMR), NCCS, NCCS Complex, S.P. Pune University, Pune, Maharashtra, India

Srimanta Patra (147), Cancer and Cell Death Laboratory, Department of Life Science, National Institute of Technology Rourkela, Rourkela, Odisha, India

Steven L. Percival (91,275), R&D, 5D Health Protection, Group, Liverpool, United Kingdom

Satish T. Pote (261), National Centre for Cell Science, NCCS Complex, S.P. Pune University, Pune, Maharashtra, India

Akhilesh A. Potnis (23), Molecular Biology Division, Bhabha Atomic Research Centre, Mumbai, Maharashtra, India

Biswajita Pradhan (147), Algal Biotechnology and Molecular Systematic Laboratory, Post Graduate Department of Botany, Berhampur University, Berhampur, Odisha, India; Department of Biotechnology, Sangmyung University, Seoul, South Korea

Jagannath Pradhan (447), Laboratory of Infection Immunology, Department of Life Science, National Institute of Technology, Rourkela, Odisha, India

Jasmin Pradhan (447), Laboratory of Infection Immunology, Department of Life Science, National Institute of Technology, Rourkela, Odisha, India

K. Prashanth (395,509), Laboratory No. 6, Nosocomial Pathogen Biology Laboratory, Department of Biotechnology, Pondicherry University, Kalapet, Puducherry, India

R. Prathiviraj (291), Department of Microbiology, School of Life Sciences, Pondicherry University, Kalapet, Puducherry, India

Arumugam Priya (335,485), Department of Biotechnology, Alagappa University, Karaikudi, Tamil Nadu, India

Luciano Procópio (109), Industrial Microbiology and Bioremediation Department, Federal University of Rio de Janeiro (UFRJ), Caxias, Rio de Janeiro, Brazil

E.K. Radhakrishnan (71), School of Biosciences, Mahatma Gandhi University, Kottayam, Kerala, India

Prashanth S. Raghavan (23), Molecular Biology Division, Bhabha Atomic Research Centre, Mumbai, Maharashtra, India

Hema Rajaram (23), Molecular Biology Division, Bhabha Atomic Research Centre; Homi Bhabha National Institute, Training School Complex, Mumbai, Maharashtra, India

Satish Kumar Rajasekharan (649), School of Chemical Engineering, Yeungnam University, Gyeongsan, South Korea

Meora Rajeev (3), Department of Biological Sciences, Inha University, Incheon, Republic of Korea

Augustus Akshaya Rani (629), Department of Biotechnology, Alagappa University, Karaikudi, Tamil Nadu, India

Ankit Rao (699), University School of Environment Management, Guru Gobind Singh Indraprastha University, New Delhi, India

Toleti Subba Rao (187,713), School of Arts & Sciences, Sai University, Chennai, Tamil Nadu; Biofouling and Biofilm processes Section, Water & Steam Chemistry Division, Bhabha Atomic Research Centre Facilities, Kalpakkam, Tamil Nadu; Homi Bhabha National Institute, Mumbai, Maharastra, India

Sumit Kumar Rastogi (357), Molecular Microbiology Laboratory, Department of Biotechnology, Indian Institute of Technology Kharagpur, Kharagpur, West Bengal, India

Shailabh Rauniyar (165), Department of Chemical and Biological Engineering; Data Driven Material Discovery Center for Bioengineering Innovation, South Dakota School of Mines and Technology, Rapid City, SD, United States

Michela Relucenti (81), Department of Anatomy, Histology, Forensic Medicine and Orthopaedics, Sapienza University of Rome, Rome, Italy

Mohd Sajjad Ahmad Khan (587), Department of Basic Sciences, Deanship of Preparatory Year and Supporting

Studies, Imam Abdulrahman Bin Faisal University, Dammam, Saudi Arabia

Samreen (315,587), Department of Agricultural Microbiology, Faculty of Agricultural Sciences, Aligarh Muslim University, Aligarh, Uttar Pradesh, India

Rajesh Kumar Sani (165), Department of Chemical and Biological Engineering; Data Driven Material Discovery Center for Bioengineering Innovation; BuG ReMeDEE Consortium, South Dakota School of Mines and Technology, Rapid City, SD, United States

Ajit Ramesh Sawant (395,509), Laboratory No. 6, Nosocomial Pathogen Biology Laboratory, Department of Biotechnology, Pondicherry University, Kalapet, Puducherry, India

Priya Saxena (165), Department of Chemical and Biological Engineering; Data Driven Material Discovery Center for Bioengineering Innovation, South Dakota School of Mines and Technology, Rapid City, SD, United States

Laura Selan (81), Department of Public Health and Infectious Diseases, Sapienza University of Rome, Rome, Italy

Joseph Selvin (291), Department of Microbiology, School of Life Sciences, Pondicherry University, Kalapet, Puducherry, India

Surajbhan Sevda (699), Department of Biotechnology, National Institute of Technology Warangal, Warangal, Telangana, India

Sriram Shankar (247), Department of Biotechnology, School of Life Sciences, Pondicherry University, Kalapet, Puducherry, India

Rohit Sharma (261), School of Sciences, SAGE University, Bhopal, Madhya Pradesh, India

Shraddha Sharma (209,261), Department of Biochemistry, Central University of Rajasthan, Ajmer, Rajasthan; Centre Instrumental Facility for Flow Cytometry (CIFF), National Centre for Biological Sciences, Bengaluru, Karnataka, India

Moshe Shemesh (649), Volcani Institute, Food Science, Agricultural Research Organization, Rishon LeZion, Israel; Biology Department, Northeastern University, Boston, MA, United States

Bhim Pratap Singh (475), Department of Agriculture and Environmental Sciences (AES), National Institute of Food Technology Entrepreneurship and Management (NIFTEM), Sonepat, Haryana, India

Dhananjay Singh (569), Department of Bioengineering and Biosciences, Integral University, Lucknow, Uttar Pradesh, India

Garima Singh (475), Department of Botany (Life-Science), Mizoram University (A Central University), Pachhunga University College Campus, Aizawl, Mizoram, India

Prashant Kumar Singh (475), Department of Biotechnology (Life-Science), Mizoram University (A Central University), Pachhunga University College Campus, Aizawl, Mizoram, India

Ram Nageena Singh (165), Department of Chemical and Biological Engineering; Data Driven Material Discovery Center for Bioengineering Innovation, South Dakota School of Mines and Technology, Rapid City, SD, United States

Shiv Mohan Singh (261), Department of Botany, Banaras Hindu University (BHU), Ajagara, Varanasi, Uttar Pradesh, India

Krzysztof Skowron (125), Department of Microbiology, Nicolaus Copernicus University in Toruń, Ludwik Rydygier Collegium Medicum in Bydgoszcz, Bydgoszcz, Poland

Jae Jun Song (475), Department of Otorhinolaryngology—Head and Neck Surgery, Korea University College of Medicine, Guro Hospital, Seoul, South Korea

C.S. Srinandan (605), Biofilm Biology Laboratory, Anusandhan Kendra-2, Centre for Research on Infectious Diseases (CRID), School of Chemical and Biotechnology, SASTRA Deemed to be University, Thanjavur, Tamil Nadu, India

T.J. Sushmitha (3), Department of Biotechnology, Alagappa University, Karaikudi, Tamil Nadu, India

Marcus Swann (91,275), 5D Health Protection Group Ltd., Centre of Excellence in Biofilm Science and Technologies (CEBST), Liverpool, United Kingdom

Payal Thakur (165), Department of Chemical and Biological Engineering; Data Driven Material Discovery Center for Bioengineering Innovation, South Dakota School of Mines and Technology, Rapid City, SD, United States

Vihang Thite (605), Department of Pharmacology, Sun Pharmaceutical Industries Limited, Vadodara, Gujarat, India

Vishvanath Tiwari (209), Department of Biochemistry, Central University of Rajasthan, Ajmer, Rajasthan, India

Abhilash Kumar Tripathi (165), Department of Chemical and Biological Engineering; Data Driven Material Discovery Center for Bioengineering Innovation, South Dakota School of Mines and Technology, Rapid City, SD, United States

Raymond J. Turner (173), Department of Biological Sciences, University of Calgary, Calgary, AB, Canada

Shipra Varshney (699), University School of Environment Management, Guru Gobind Singh Indraprastha University, New Delhi, India

Carolina Orlando Vaso (521), School of Pharmaceutical Science, Universidade Estadual Paulista, Araraquara, SP, Brazil

Pratishtha Verma (415), Midwest Dairy Research Center, Department of Dairy and Food Science, South Dakota State University, Brookings, SD, United States

Diana L. Vullo (683), Chemistry Area, National University General Sarmiento; CONICET, Buenos Aires, Argentina

Ewa Wałecka-Zacharska (125), Department of Food Hygiene and Consumer Health, Wrocław University of Environmental and Life Sciences, Wrocław, Poland

Fergus Watson (91,275), Operations, 5D Health Protection Group Ltd., Liverpool, United Kingdom

Natalia Wiktorczyk-Kapischke (125), Department of Microbiology, Nicolaus Copernicus University in Toruń, Ludwik Rydygier Collegium Medicum in Bydgoszcz, Bydgoszcz, Poland

Mukesh Kumar Yadav (475), Department of Biotechnology (Life-Science), Mizoram University (A Central University), Pachhunga University College Campus, Aizawl, Mizoram, India

Alicja Zimmer (91,275), Getinge Applikon, Liverpool, England, United Kingdom

Ruth Zomuansangi (475), Department of Life-Science, Mizoram University (A Central University), Pachhunga University College Campus, Aizawl, Mizoram, India

Zothanpuia (475), Department of Biotechnology (Life-Science), Mizoram University (A Central University), Pachhunga University College Campus, Aizawl, Mizoram, India

Preface

Microorganisms are considered unicellular creatures that can be effectively studied as experimental model. The unicellular bacterium has always been considered by researchers as a source of biological machinery to provide fundamental insight into regulatory machinery of life. However, the existence of these unicellular organisms is beyond their individual identity. They not only live with organisms of different species, but they have social life and community. The social life and community of microbes may sound strange, but in nature they do prefer to live in community. One such well-organized community of microbes is known as biofilm. The existence of these microbial communities has been discovered and rediscovered by several researchers. However, the "Father of biofilms" and "The King of Slime" Late Dr. John William (Bill) Costertron was the one who introduced these social communities of microbes to the world through his benchmark paper "How Bacteria Stick" [Costerton JW, Geesey GG, Cheng KJ. (1978) Scientific American 238: 86–95]. Biofilm, a sessile microbial community, is enclosed in a self-sustaining matrix known as EPS, the "dark matter" of biofilms. We humans, being social animals, use inorganic compounds to make our buildings for living; these microbes used their own synthesized cementing material rich in biomolecules to build the city known as "city of microbes." Thus, it would not be mistaken if we refer biofilms as the tiny microbial buildings.

Most of the microorganisms, particularly bacteria, live as biofilm at one or other phase of their life under natural conditions. The distribution of biofilms ranges from extreme environment to clinical settings. On the one end, they are responsible for many infectious diseases. On the other hand, they are responsible for the degradation of many pollutants in nature. To get rid of biofilms and associated problems, understanding the mechanisms may provide a way to utilize these tiny buildings as valuable sources. Thus, this book brings together contributions from pioneers across the globe and from diverse areas that have facilitated the progress in biofilm studies. The focus of the book is to bring forth both fundamentals and applications of microbial biofilms to a distinct platform. The chapters have been written by experts who have been working on different aspects of microbial biofilms. The book covers six foremost aspects of microbial biofilm. This book is an editorial artwork of our research work on bacterial community and communication from different environments. We have put substantial efforts in gathering researchers and covering the past and recent advancements on biofilm research. The book will serve as an ideal reference for the academician and researchers who wish to understand the microbial buildings and its architecture in a better way.

Surajit Das
Neelam Amit Kungwani

Section A

Introduction to biofilms

Chapter 1

Marine biofilms: Bacterial diversity and dynamics

T.J. Sushmitha[a], Meora Rajeev[b], and Shunmugiah Karutha Pandian[a]

[a]*Department of Biotechnology, Alagappa University, Karaikudi, Tamil Nadu, India,* [b]*Department of Biological Sciences, Inha University, Incheon, Republic of Korea*

1. Introduction

Marine environment is the intricate and expanded ecosystem on the Earth planet, occupying about 70% of the total surface (Kennedy et al., 2010). This environment is dominated by a wide range of microorganisms including bacteria, archaea, unicellular fungus, and protists. It is estimated that the oceans contain 3.6×10^{29} microbial cells. These microbes play crucial roles in biogeochemical cycles and are responsible for 98% of the primary carbon production (Sogin et al., 2006). Marine environment provides habitats for a diverse range of uncharacterized bacterial communities that collaborate with the surrounding environment to perform a variety of ecologically vital tasks such as nutrient recycling, bioremediation, and organic waste decomposition (Wang, Liu, Zheng, Zhu, & Wang, 2013). It has been stated that coastal systems are among the most productive ecosystems in the marine environments, providing more than US$ 14 trillion ecosystem products (e.g., food and raw materials) every year (Harley et al., 2006).

Among several other microorganisms, bacteria contribute a major portion to the marine environment. Bacteria can survive either as a planktonic or as an adherent form in the marine environment. Planktonic cells are generally free-floating and constitute a larger fraction of the total community (Milici et al., 2016). On the other hand, several bacterial assemblages are capable of colonizing various natural and artificial surfaces immersed in the water, leading to the formation of a complex and dynamic community termed biofilms. Biofilms are the unsolicited colonization of microorganisms on solid surfaces within a self-produced and embedded polymeric matrix called extracellular polymeric substances (EPSs) (Bellou, Garcia, Colijn, & Herndl, 2020; Doghri et al., 2015). Biofilm formation is known to be the fundamental survival strategy for a majority of microbes as it provides several advantages including higher environmental stability, increased access to nutrients, and enhanced cellular interactions (multicellular life) (Costerton, Lewandowski, Caldwell, Korber, & Lappin-Scott, 1995; Dang & Lovell, 2016). Marine bacterial biofilms, also regarded as "microfouling," contribute significantly to the fitness of the marine ecosystems through their important roles in biogeochemical cycling (Fuhrman, Cram, & Needham, 2015; Hawley et al., 2017), by providing chemical cues for the appropriate settlement of macrofoulers (Hadfield, 2011) and their interaction with them. Although they play a significant role in the marine ecosystem, bacterial biofilm formation on the man-made structures causes huge economic loss (Rajeev, Sushmitha, Toleti, & Pandian, 2019). For example, obstruction to flow rate of water and heat transfer efficiency across heat exchangers, mechanical blockage (Choi, Noh, Yu, & Kang, 2010), and microbiologically influenced corrosion (MIC) of structural components (Rao, Kora, Chandramohan, Panigrahi, & Narasimhan, 2009; Saravanan, Nancharaiah, Venugopalan, Rao, & Jayachandran, 2006).

In addition to this, the establishment of early-stage biofilms on any immersed surfaces is considered to be a key factor driving the colonization and recruitment of higher multicellular eukaryotes (e.g., crustaceans, mussels, oysters, tube worms, etc.) to form a complex biofouling community, known as macrofouling (Sushmitha, Rajeev, Sriyutha Murthy, et al., 2021). Therefore, understanding the diversity, community composition, and dynamics of marine biofilms is considered an essential factor not only to decipher the interactions between bacterial communities and macrofoulers but also to develop an effective antifouling strategy.

With this background, this chapter discusses bacterial assemblages responsible for biofilm formation in a marine environment. In addition, readers will walk through the detrimental consequences and plausible applications of marine biofilms.

Understanding Microbial Biofilms. https://doi.org/10.1016/B978-0-323-99977-9.00032-6

2. Biofilm development in the marine environment

Once immersed in the seawater, surfaces are rapidly covered with the layer of organic and inorganic molecules, forming a conditioning film. The conditioning film mostly occurs within seconds to minutes, wherein proteins and glycoproteins are dominantly adsorbed onto the surface followed by polysaccharides, lipids, humic acid, and many other biomolecules. The established conditioning film affects the physicochemical properties of the immersed substrata such as increasing roughness and hydrophilicity. In this way, conditioning film develops the desired surface and acts as solid support for nutrient entrapment, which in turn facilitates microbial adhesion. The initial establishment of microbial attachment to the immersed surface in seawater is favored either by microbial swimming or by diffusive transport of the seawater. Among all the primary colonizers, bacteria contribute to the most, followed by diatoms, algae, and lower eukaryotes. **Field-emission scanning electron microscope** (FESEM) images depicting biofilms formation on acrylic coupons in the vicinity of Kudankulam, Laccadive Sea, are shown in Fig. 1.

2.1 Phase I—Reversible adhesion of bacteria

The microbial approach to the conditioned surface is a crucial stage in biofilm development. Depending on whether the bacteria are motile or transported by aqueous phase forces, this process can be active or passive. The physicochemical properties of the bacterial cell surface determine their attachment on the surfaces. Bacteria possess colloidal dimensions and a net negative charge at pH levels found in natural waters. Despite the fact that the bacterial surfaces are charged, they approach a surface by van der Waals forces. During this stage, desorption may occur due to the release of reversibly adsorbed cells caused by fluid shear forces or a hostile environment (Rao, 2015; Rao, Kesavamoorthy, Rao, & Nair, 1997).

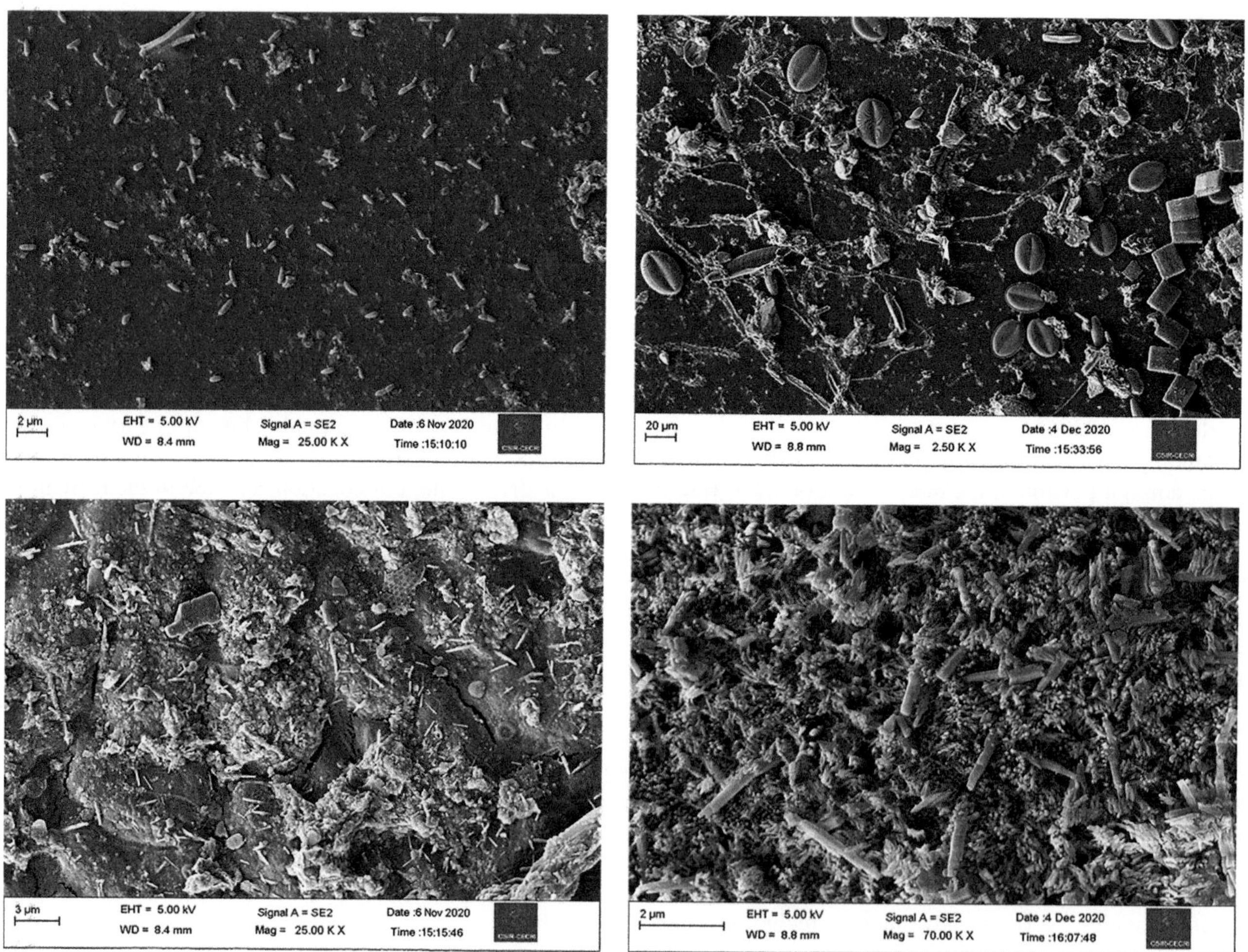

FIG. 1 (FESEM) images of early-stage marine biofilms formed on acrylic coupons in the southern coastal seawater of India.

2.2 Phase II—Irreversible adsorption

The irreversible attachment of the microbe to a substratum occurs at this important stage in biofilm growth. Although repelling forces prohibit the bacterial cells to make direct contact with the surface, they can nevertheless cling by developing surface appendages. Flagella, fimbriae, and exopolysaccharide fibrils can easily pass through the energy barrier that allows robust surface adherence. Cells can hover at this location and engage in dipoleedipole, ionedipole, and hydrophobic interactions, among other short-range interactions. Polymeric fibrils operate as a link between the bacterium and the surface, cementing the bond in an irreversible way (Sutherland, 1997).

Bacterial communities can detect and respond to environmental cues on the surface of the cell, and they can actively begin surface adhesion by changing gene expression, which causes changes in physiology, cell surface chemistry, and behavior. Microorganisms respond differently to different chemical cues, allowing them to settle efficiently onto the surface. The process of species sorting has been identified as a fundamental mechanism in the determination of initial surface colonizers. Species sorting is a process that occurs when local abiotic and biotic environmental circumstances select bacteria from a pool of species to establish a community (Zhang et al., 2014). Adsorption of biomolecules varies with the surface having different substratum physicochemical properties, resulting in the selection of different primary colonizers (Dang & Lovell, 2000). However, the effect of species sorting on the surface type remains controversial; for example, few studies have concluded that different substrate types harbored different primary colonizers (Meier, Tsaloglou, Mowlem, Keevil, & Connelly, 2013; Zhang et al., 2014), while other studies didn't observe any influence of surface types on the primary bacterial colonization (Bellou, Papathanassiou, Dobretsov, Lykousis, & Colijn, 2012; Sushmitha, Rajeev, Sriyutha Murthy, et al., 2021).

The variation in the colonization of microorganisms is mostly observed between biotic and abiotic surfaces, wherein the colonization on biotic surfaces is mediated by a specific process such as ligand-receptor interactions, secretion of specific proteins (secretome) (Chagnot, Zorgani, Astruc, & Desvaux, 2013), whereas colonization on abiotic surfaces is nonspecific, facilitated by pili, holdfast, flagella, bacterial secreted adhesins (Chan, Fakra, Emerson, Fleming, & Edwards, 2011; Mandlik, Swierczynski, Das, & Ton-That, 2008; Proft & Baker, 2009). The colonized bacterial community grows, multiplies, and secretes various EPSs. This bacterial consortium shielded with self-produced EPS is called biofilm. EPS helps the bacterial community to cement the cells to the surface and protect them from environmental fluctuations. The exopolymer gel is not a passive matrix; it responds to external stimuli in physical, chemical, and electrical ways. The biofilm's structural basis is provided by the exopolymer generated by bacteria. These biofilms contain various bacterial populations that have multispecies interactions, and the developed biofilms provide the residing microorganisms a favorable environment to have functional interdependent relationships. Unlike free-floating microorganisms (planktonic cells), biofilms maintain a broad spectrum of organisms with diverse cooperation, communication, and competition. This enables the microfouling organisms to have a favorable mode of survival even in extreme environments (Hibbing, Fuqua, Parsek, & Peterson, 2010).

There are two major questions related to biofilms in marine ecology: (i) which are the primary colonizing bacterial groups and (ii) which are the dominant bacterial groups? As a response to the former question, a number of studies (Antunes et al., 2020; Dang & Lovell, 2000, 2002) have observed *Alphaproteobacteria* (members of *Rhodobacteraceae*) as primary colonizers, while the other few studies (Briand et al., 2017; Pollet et al., 2018; Zhang et al., 2014) identified members of *Gammaproteobacteria* as primary colonizers. A recent study by Sushmitha, Rajeev, Sriyutha Murthy, et al. (2021) observed a clear shift in the biofilm-forming bacterial community in the southern coastal region of India. This study found that the proportion of *Alteromonadales* was much greater on succession days 1 and 9, whereas *Vibrionales* dominated the biofilm-forming microbiota on succession days 3 and 12. In comparison to previous succession days, a wide number of bacterial orders were found on succession days 6 and 15. *Oceanospirillales* was the third most prevalent bacterial order discovered on days 6 and 15 of succession. Studies delineating changes in pioneering biofilm-forming bacterial dynamics are highly essential as they determine the further succession and growth of the community.

2.3 Phase III—Microbial biofilm formation

Regardless of the system geometry or ecology, biofilm communities predominate all nutrient-sufficient aquatic systems. Microbial cell-cell interactions between the initial colonizing bacteria and additional microorganisms with differing nutritional requirements are required for the biofilm to develop further. Biofilms have a high level of functional homogeneity in their activities, which is dependent on the biofilm's structural integrity. Biofilms, often also known as "quasitissues," mimic tissue in their physiological cooperativity and use primitive homeostasis to protect themselves against changes in the bulk environment conditions (Costerton et al., 1995). Variations in the spatial structure of microbial cells in biofilms are also present. Biofilm structural variability is a result of its growth features and metabolic activities. Physiological synergism is a key component in determining the shape of a biofilm's structure, as well as the establishment of microcolony interactions that lead to mature biofilms. Mixed microbial communities are known for their physiological congruity and heterogeneity. The biofilm eventually enters a plateau phase, where it has reached a certain thickness and metabolic capacity and can now serve as a foundation for biofouling succession, allowing higher organisms to colonize. The established biofilms support the attachment of spores and larvae of macrofoulers such as macroalgae, bryozoans, barnacles, molluscs, polychaete, coelenterates, and tunicates by providing required nutrients and food sources. These spores and larvae further grow and form a complex biotic community called biofouling. Eventually, marine biofilms lead to biofouling, which is a prevalent problem in marine systems (Sushmitha, Rajeev, Sriyutha Murthy, et al., 2021).

2.4 Phase IV—Biofilm detachment and dispersal

The biofilm-forming bacteria detach, scatter, and colonize new habitats in order to propagate. The urge to disseminate starts with bacterial proliferation in biofilms. Biofilm growth enhances surface roughness while simultaneously providing protection from shear pressures, as well as increased surface area and convective mass movement at the surface. Surface roughness is produced by the sessile accretions of microbial cells, which increases turbulence and mass transport at the colonized surface (Donlan & Costerton, 2002).

3. Bacterial competition and cooperation in shaping the diversity and dynamics of marine biofilm

The marine environment harbors a huge spectrum of bacterial species capable of forming biofilms. These bacterial communities compete and coexist with one another for space and nutrients to form a biofilm with specific structures and functions. Research targeting interspecies competition and cooperation strategies has shown that there are diverse mechanisms through which bacterial communities can compete and coexist with other bacterial species or other organisms in the same pool. The existence of competition and cooperation among diverse bacterial communities in the biofilm allows not only the transfer of nutrients but also exchange of extracellular enzymes, membrane vesicles, intraspecies, and interspecies signaling molecules that facilitate bacterial interactions. These interactions among the bacterial communities often result in a change in the physiology and function of the whole biofilm-forming community (Sadiq et al., 2021). Major interactions occurring among the biofilm-forming bacterial communities are depicted in Fig. 2.

Besides aiding in early colonization on surfaces, signaling molecules produced by the bacterial communities also help in the recruitment of surface-associated secondary microorganisms from the surrounding planktonic phase. The recruitment of secondary microorganisms is also facilitated by microbial coaggregation that is initiated from a specific cell surface adhesin-receptor interaction among the participating microorganisms. Coaggregation is the best-known example of bacterial cooperation within multispecies biofilm that leads to an increase in biomass of overall biofilm. It draws diverse microorganisms into a close association, which enhances effective cell-cell signaling, contact-dependent gene expression, genetic exchange, and importantly protection from extreme conditions. Bacterial coaggregation and exchange of signaling molecules (quorum sensing) result in the formation of complex multispecies bacterial biofilm. Research on coaggregation and bacterial signaling has been vastly studied in freshwater and wastewater environments, and their implications for marine biofilms are clear. For example, in activated sludge flocs and biofilms, ammonia-oxidizing bacteria (AOB) and nitrite-oxidizing bacteria (NOB) create symbiotic relationships via coaggregation and signal transfer to carry out nitrification (Almstrand, Daims, Persson, Sörensson, & Hermansson, 2013; Gruber-Dorninger et al., 2015).

Rather than cooperation, surface-colonizing bacterial communities also possess various mechanisms of competition. Bacteria compete for nutrients and space and try to hinder the growth of other bacteria by producing toxic substances such as bacteriocins in their biofilm consortia. For example, *Pseudoalteromonas tunicate* produces an antibacterial protein (AlpP) that hinders the growth of other microorganisms isolated from the same marine environment (Rao, Webb, & Kjelleberg, 2005). These competitive interactions often lead to the selection of variants that are better suited to colonize the surface. Competition also gets initiated from the cooperative behavior of the bacterial communities, when an individual bacterial population gets benefits. In this situation, cooperative interactions leading to the beneficiary of all residing bacterial community breaks and social cheaters emerge (Hibbing et al., 2010). Pure culture studies have observed the rise of social cheaters in various bacterial species. For example, bacterial species with mutations in *lasR* were found to proliferate faster when *Pseudomonas aeruginosa* cells were cultured with conditions requiring quorum-sensing regulated extracellular proteases. These social cheaters take advantage of the protease activity of enzymes secreted by their neighbors without having to spend the energy required to generate and secrete the enzymes. In marine environments, most of the social cheaters arise within the biofilm. For instance, most of the species in the natural environment are quorum-sensing-deficient, and they cheat by utilizing the benefits

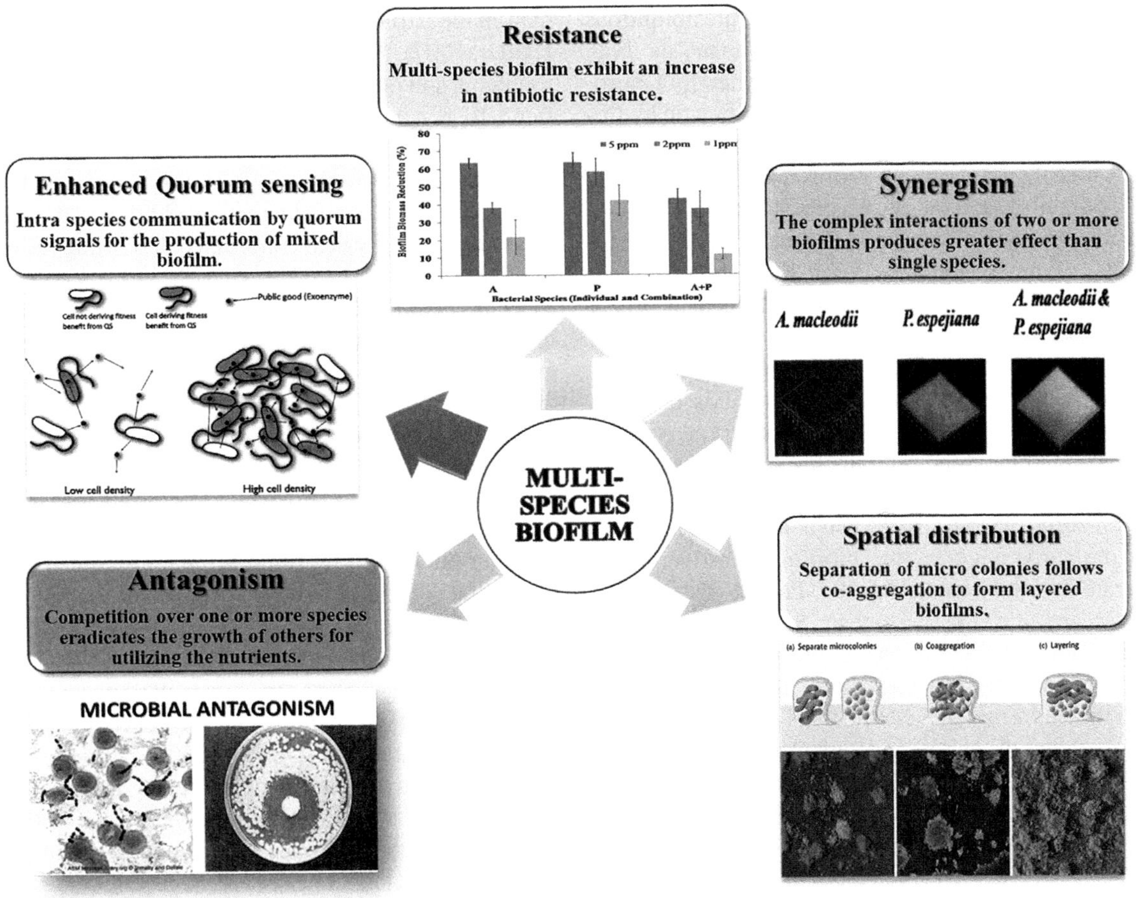

FIG. 2 Multispecies biofilm interactions and their characteristics.

of quorum-sensing-mediated outcomes such as EPS production. They enjoy the benefits of EPS such as trapping nutrients and protection from the extreme environment.

Apart from enjoying the benefits produced by cobacteria, some competition strategies are also strictly antagonistic. The antagonistic nature is more common in particle-associated biofilm-forming marine bacteria than the free-floating. For example, most members of *Roseobacter* clade produce antibiotics such as indigoidine, tryptanthrin, and tropodithietic acid (Dang & Lovell, 2016; Wagner-Döbler & Biebl, 2006) through quorum sensing. Antibiotic production is 10 times more likely to be observed in surface-associated *Roseobacter* clade than their planktonic counterpart (Long & Azam, 2001). There are few other marine genera such as *Bdellovibrio* and its like organisms (BALOs), belonging to the class *Deltaproteobacteria* and *Micavibrio* belonging to *Alphaproteobacteria*, known as obligate predators of other marine microorganisms (Sockett, 2009). Living in environments such as biofilms also provides these organisms the required protection and resources, which enhances their survival. Biofilm-associated bacterial communities also use some developed competitive behavior such as the use of contact-dependent growth inhibition that secrets toxins with the activity of RNase, DNase, and membrane-pore-forming activities toward the target cell. Furthermore, for the microorganisms to survive in their consortia, it becomes mandatory for the microorganisms to benefit their cooperators, tackle their competitors, and maintain interactions with the associated communities (Dang & Lovell, 2002).

4. Microbial diversity of marine biofilms

As discussed in the previous sections, marine bacterial communities have two major lifestyles, free-living and surface-associated. Most of these organisms prefer either one lifestyle or switch between two depending on the environmental conditions. Most of the bacterial populations belonging to SAR11 and SAR86 classes are free-living, while members of *Alphaproteobacteria* (namely *Rhodobacteraceae*), *Gammaproteobacteria* (*Alteromonadaceae* and *Vibrionaceae*), and *Bacteroidetes* (*Flavobacteria*) are surface-associated. Overall, biofilm-forming bacterial communities are vastly diverse than the planktonic counterpart, and major of this diversity comes from the *Roseobacter* and *Bacteroidetes* clades.

Members belonging to *Roseobacter* clade are ubiquitous in the marine environment and are dominant as both biofilm-forming and planktonic. They are generally heterotrophic bacteria that can metabolize various aromatic hydrocarbons and many recalcitrant organic substances, and they also take advantage of quick proliferation even in the presence of small quantities of organic substances such as amino acids and simple sugars (Buchan, González, & Moran, 2005; Wagner-Döbler & Biebl, 2006). Free-floating members of this clade are associated with the fecal pellets of zooplankton and other higher organisms and surfaces of algae, diatoms, and dinoflagellates. The genome of *Roseobacter* harbors several genes encoding chemotaxis, motility, chemoreceptor proteins, c-di-GMP signaling proteins, and quorum-sensing regulators, which play a significant role in bacterial adhesion, proliferation, and survival in the biofilm consortia. Many studies have found their dominancy in the early-stage marine biofilms, mostly formed in temperate coastal waters of the Pacific Ocean, the Atlantic Ocean, and Mediterranean Sea (Briand et al., 2012, 2017; Dang & Lovell, 2002).

Other dominant bacterial community of marine biofilms belongs to the *Bacteroidetes*. Members of *Bacteroidetes* are known for their diversity and adaptation to surface-associated lifestyles. Research on *Bacteroidetes* has shown that this is the second dominant marine phylum found abundant in organic-particle-rich coastal waters and is most prone to the surface-associated lifestyle (Fernández-Gomez et al., 2013; Gómez-Pereira et al., 2012). The members of *Bacteroidetes* possess gliding motility than flagella-mediated. Hence, rather than sensing a surface and attaching, they tend to aggregate, enlarge, and agglomerate with other surfaces such as detritus or phytoplankton, wherein they aid in the attachment to surfaces. Once attached, their gliding movement enhances the exposure to surfaces. Further, phylum *Bacteroidetes* also possess advantages such as the production of rhodopsin pigments for light energy harvesting (Campbell, Waidner, Cottrell, & Kirchman, 2008; José et al., 2003). This helps in attaining the required energy directly from light and makes them independent from other associated bacterial communities. Other bacterial taxa found abundant in marine biofilms are members of *Alteromonadaceae*, *Vibrionaceae* (Rajeev et al., 2019; Sushmitha, Rajeev, Sriyutha Murthy, et al., 2021), and *Flavobacteria* (Pollet et al., 2018). Geographical origin, surface type, and seasonality determine the type of bacteria and their colonization.

5. Factors influencing the establishment of marine biofilms

Understanding the parameters that drive bacterial colonization on surfaces is critical for preventing fouling on man-made structures (Qian, Lau, Dahms, Dobretsov, & Harder, 2007). As marine biofilms grow in a variety of environments and on both natural and artificial surfaces, it's difficult to generalize how their structure and physiological activities are influenced. The succession of marine biofilm communities is an evolving process impacted by a wide range of environmental factors, in addition to the substratum and prevailing physical conditions on biofilms. Biofilm formation and features are known to be influenced by physical aspects of surfaces such as surface wettability, surface hydrophobicity, critical surface tension, and surface molecular topography (Lee, Nam, Kim, Lee, & Lee, 2008). Temperature, radiation, pH, dissolved oxygen, CO_2, and nutrition availability, among other factors, can influence the rate of bacterial colonization, as well as the composition and ecophysiology of biofilm-forming species. Physical adsorption processes dominate biomass and polymer buildup in biofilms, which are regulated by environmental processes such as nutrient recycling (Moss, Nocker, Lepo, & Snyder, 2006).

Environmental conditions change seasonally or spatially, affecting planktonic populations and bacterial attachment, which might affect the formation of a biofilm microbial community (Donlan, 2002). Variations in environmental nutrient levels may cause the production of marine biofilms during community formation and modify the community makeup (McDougald, Rice, Barraud, Steinberg, & Kjelleberg, 2012). A recent study found that a drop in the diversity of microorganisms responsible for biofilm development resulted from a decrease in mineral nutrient availability to biofilm photoautotrophs, compromising both the stability and function of marine biofilms (Lawes, Neilan, Brown, Clark, & Johnston, 2016). Environmental factors influence the precise settings and surfaces where biofilms grow in diverse ways. Similarly, the colonizing microorganisms' lifestyle can influence biofilm diversity and the microbial community's responses to environmental stimuli. Biofilms that grow on ship hulls are typically more intricate and dynamic than those that grow on other types of artificial surfaces, and they are influenced by factors such as the surface's dynamic circumstances (Zargiel & Swain, 2014). When compared with biofilm communities growing on other surfaces, microfouling communities forming on these surfaces are subject to selective physical pressures from the environment and are more protected against variables such as insulation, oxidative stress, predation, and shearing (Matz et al., 2008). These external conditions change the water column's physiochemical and biological dynamics, as well as the biofilm's community structure (Patil & Anil, 2005). The maximum age of the biofilm was associated with higher bacterial densities. It has been argued that as the biofilm matures, its composition becomes more comparable to the local microbial population (Hall & Pepe-Ranney, 2015).

However, biofilm-forming bacterial communities show higher variations when compared with their peripheral planktonic fraction in their early developmental stage.

6. Bacterial protection from harsh environmental conditions

Bacterial survival in the marine environment is highly complicated due to its heterogeneity. Brownian diffusion allows a nonmotile bacterium with an average cell diameter of 0.4–2 m to ingest nutrients from the diffusion boundary layer and allows the cells to explore around 45 pL of seawater in 10 min (Stocker, 2012). Swimming bacteria may receive 0.5 μL of freshwater every 10 min, while a chemotactic velocity of 10 μm/s allows the cell to travel 6 mm in the same time. Bacterial species are transported via sea currents, and their chemotaxis (sensing) only works when a surface is available near to them. Bacterial cells in seawater are commonly present as multicellular structures such as cell aggregates and clusters that benefit them by providing improved nutrient acquisition and protection from predators. These clusters also benefit them to potentially invade any surfaces. Surface-associated bacteria are metabolically more active than their planktonic counterpart. Bacteria produce EPS, which absorb, transport solutes and microorganisms while also acting as a protective cover for cells, especially in harsh settings. The organization provided by the EPS aids in (i) nutrient and toxic compound entrapment, (ii) extracellular enzyme activity and stabilization by buffering pH and salinity fluctuations, (iii) quorum sensing, (iv) genetic information exchange, (v) physical anchorage, and (vi) predation protection (De Carvalho & Fernandes, 2010; Decho & Gutierrez, 2017).

EPSs are made up of polysaccharides, proteins, and in some cases, lipids, nucleic acids, and other biopolymers (Flemming & Wingender, 2010). Additionally, sulfates, phosphates, and uronic acids present in EPS are ionizable, which makes bacterial cells interact with cations such as Fe^{2+}, Cu^{2+}, and Zn^{2+} to increase the concentration of metal ions to survive oligotrophic circumstances (van der Merwe et al., 2009).

The high survival rate of microorganisms in intertidal zones is one of the greatest instances of biofilm protection against acute stress induced by UV exposure, brief dehydration, competition, and limited access to nutrients (Ortega-Morales, Chan-Bacab, De la Rosa, & Camacho-Chab, 2010). Cells employ the ability to maintain enough fluidity of the cellular membrane by modifying the fatty acid profile of the phospholipids in these situations, where cells face abrupt cold/hot temperature, high/low salinity, dried/immersed circumstances, UV exposure, and pH variations. Several strains, including *Shewanella colwelliana* and *Vibrio splendidus*, have been identified from anoxic intertidal sediments that produce polyunsaturated fatty acids (Freese, Rütters, Köster, Rullkötter, & Sass, 2009). Additionally, microbes in biofilms and mats produce mycosporine-like amino acids and carotenoid pigments that protect them against UV radiation, free radicals, salt, pH, temperature, radioactive chemicals, and reactive oxygen species. Biofilms defend against UV radiation by limiting radiation penetration via the biofilm matrix. This shield might have also protected against the early Earth's severe and variable temperature, pH, and UV radiation (Minaev, 2007).

7. Traditional and modern methods to study biofilm-forming community

Every member of the biofilm-forming bacterial community serves a distinct purpose in determining the structure and dynamics of trophic web networks and controlling the environment's biogeochemical cycle as equal to their planktonic fraction. Understanding the composition and pattern of microbial community structure is a critical aspect for determining their function. A better understanding of the entire biofilm-forming community, their individual members, and how they interact with the surrounding environment is considered essential, which has been attained using several approaches. The traditional culture-dependent methods have been constantly refined to characterize bacterial diversity in varied settings. However, these methods support merely a 0.1%–1% fraction of the total community and thus fail to provide a comprehensive information on the majority of microorganisms. Later advancement in culture-independent molecular methods has enabled the identification of not only culturable bacteria but also hitherto unculturable bacterial communities. These methods use direct extraction of metagenomic DNA from environmental samples and thus circumvent the limitations associated with culture-dependent methods. The diversity assessment methods through culture-independent approach include PCR-based methods such as clone library construction, denaturing gradient gel electrophoresis (DGGE), terminal-restriction fragment length polymorphism (T-RFLP), automated ribosomal intergenic spacer analysis (ARISA), and amplified ribosomal DNA restriction analysis (ARDRA) (DeSantis et al., 2007; Fisher & Triplett, 1999; Liu, Marsh, Cheng, & Forney, 1997; Muyzer, De Waal, & Uitterlinden, 1993) and non-PCR based methods such as Fluorescence in situ Hybridization (FISH) (Amann, Ludwig, & Schleifer, 1995; Rajeev, Sushmitha, & Pandian, 2020).

The common methodological steps involved in all culture-independent methods include direct extraction of whole community nucleic acids (metagenomic DNA) from biofilm samples, amplification of molecular marker gene from mixed nucleic acids population, and finally, the analysis (depending on the method) and sequencing of the amplified gene products. Though culture-independent techniques were successfully employed in several previous studies to determine the community composition of marine biofilms (Briand et al., 2012; Zhang, Choi, Dionysiou, Sorial, & Oerther, 2006), these methods are hampered with various limitations (e.g., cumbersome and erroneous). The bottlenecks associated with traditional fingerprinting methods were surmounted by the development of high-throughput next-generation sequencing (NGS) technologies with capabilities to sequence millions of DNA fragments in a single instrumental run.

The advent of NGS techniques allowed the scientific community to access the bacterial community at a higher resolution and thereby provided a rapid advancement in marine ecology (Rajeev, Sushmitha, Aravindraja, Toleti, & Pandian, 2021; Sushmitha, Rajeev, & Pandian, 2020). In respect to marine biofilms, NGS techniques provide a wide range of applications and are therefore successfully applied in several studies pertaining to marine biofilms. For example, pioneer biofilm-forming bacterial community and their succession pattern on artificial surfaces (Pollet et al., 2018; Sushmitha, Rajeev, Sriyutha Murthy, et al., 2021), bacterial community variations between biofilm-forming and peripheral seawater (Rajeev et al., 2019; Zhang et al., 2014), genes/pathways involved in EPS biosynthesis of potent biofilm-forming bacterial isolates (Rajeev, Sushmitha, Toleti, & Pandian, 2021; Sushmitha, Rajeev, Toleti, & Pandian, 2021), and influence of biofilm community on biocorrosion (Moura et al., 2018) were investigated.

Recent advancement in NGS techniques, in particular metabarcoding, has further advanced marine biofilm studies. Metabarcoding employs high-throughput sequencing of several marker genes (e.g., 18S rRNA gene, cytochrome oxidase I gene) concurrently to unveil the diversity and structure of eukaryotic microfouling assemblages (Taberlet, Coissac, Pompanon, Brochmann, & Willerslev, 2012). In addition, to improve the overall assessment method, metabarcoding enables the taxonomic resolution at a finer level, that is, identification of organisms at genus/species level. An increasing number of recent studies have utilized this approach to characterize marine biofilms (Azevedo et al., 2020; Zaiko et al., 2016). A representative list of important studies related to the bacterial community analysis of marine biofilms and their variations with respect to environmental factors is given in Table 1.

TABLE 1 Comprehensive list of important studies pertaining to the bacterial community composition analysis of marine biofilms and the methodology used.

Objective	Geographic location	Methodology	Inference	References
Exploration of pioneering surface-biofilm bacteria on various polydimethyl-siloxane (PDMS)-coated surfaces	Xiaoshi Island, China	Ion torrent sequencing	*Proteobacteria* and *Cyanobacteria* dominated the PDMS coated surfaces	Sun, Lang, Yan, Wang, and Zhang (2020)
To determine the variations in the bacterial community composition of biofilm due to internal tides	The coast of Fintas, Kuwait	Illumina sequencing	The bacterial biomass decreased significantly with depth. *Alphaproteobacteria* and *Bacteroidetes* dominated the upper surface of seawater.	Al Senafi et al. (2020)
To analyze the biofilm community structure and functional profile of the intertidal ecosystem	Pearl River Estuary, China	Illumina sequencing	The bacterial diversity decreased along the salinity gradient. Functions including amino acid and carbohydrate metabolism and membrane transport were altered in the estuarine environment.	Mai, Peng, and Lai (2020)
To analyze the primary and secondary colonizers in the Arabian Gulf	Muscat, Omen	Illumina sequencing	*Alphaproteobacteria* and *Flavobacteriia* were identified as primary colonizers as they decreased with time, while *Actinobacteria* and *Planctomycetia* were identified as secondary colonizers as they increased with time.	Abed, Al Fahdi, and Muthukrishnan (2019)

TABLE 1 Comprehensive list of important studies pertaining to the bacterial community composition analysis of marine biofilms and the methodology used—cont'd

Objective	Geographic location	Methodology	Inference	References
To determine the influence of Zn and Cu_2O-based coating of steel surfaces on micro and macrofouling	Patos Lagoon, Brazil	Flow cytometry and scanning electron microscopy	The coatings of coupons did not affect the microfouling development, while macrofouling was affected by coupon orientation.	Agostini, Macedo, Muxagata, and Pinho (2019)
Assessment of planktonic and biofilm communities	Bay of Bengal	Ion Torrent Sequencing	Planktonic and biofilm communities are influenced by the dominant community in photic regions and rare-species in the deep sea.	Angelova, Ellis, Wijesekera, and Vora (2019)
How biofilm community associated with plastics in marine environment differs from community associated with other polymers?	Northern Adriatic, Croatia	Scanning electron microscopy and Illumina sequencing	Two groups of bacterial community were observed. One community was common to all polymers that included *Flavobacteriaceae, Rhodobacteraceae, Planctomycetaceae* and *Phyllobacteriaceae* and the others were unique to specific plastic.	Pinto, Langer, Hüffer, Hofmann, and Herndl (2019)
To monitor the microfouling development on submerged nylon nets	Red Sea coast of Saudi Arabia	Culture-dependent and scanning electron microscope analyses	Protein was observed to be the major component in biofilm with bacteria and diatoms being the primary colonizers. Significant seasonal variations in the biofilm community were also observed.	Balqadi, Salama, and Satheesh (2018)
To ascertain the role of *Flavobacteriia* in prokaryotic community succession	Toulon Bay, France	Flow cytometry and Illumina sequencing	Bacterial communities of biofilms exhibited higher temporal dynamics with *Gammaproteobacteria* dominating the first few hours, followed by *Alphaproteobacteria* and *Flavobacteriia*.	Pollet et al. (2018)
To determine the impact of various artificial surfaces on biofouling community	Auckland Westhaven Marina, New Zealand	Metabarcoding	Copper exposed panels exhibited a low complexity in biofouling	von Ammon et al. (2018)
Comparison of bacterial biofilm community composition of groomed and ungroomed fouling release ship hull coatings	Port Canaveral, Florida	Illumina sequencing	Biofilm community on groomed surfaces was dominated by *Gammaproteobacteria* and *Alphaproteobacteria*, while ungroomed surfaces were dominated by *Phyllobacteriaceae, Xenococcaceae*, and *Rhodobacteraceae*.	Hunsucker et al. (2018)
To investigate the microfouling bacterial community in a freshwater distribution system	Madras Atomic Power Station (MAPS), East coast, India	Culture-dependent and DGGE analysis	Proteobacteria, Cyanobacteria, Bacteroidetes, Actinobacteria and Firmicutes were dominant phyla.	Palaniappan and Toleti (2016)

Continued

TABLE 1 Comprehensive list of important studies pertaining to the bacterial community composition analysis of marine biofilms and the methodology used—cont'd

Objective	Geographic location	Methodology	Inference	References
Differentiate the taxonomic composition of attached and free-living bacteria	Northwest Mediterranean Sea	454 Pyro-sequencing	*Alphaproteobacteria* dominated the free-living bacterial community and *Gammaproteobacteria* dominated the attached community.	Crespo, Pommier, Fernández-Gómez, and Pedrós-Alió (2013)
Characterization of bacterial diversity in the cooling water system (CWS) of a fast breeder test reactor (FBTR) with specific emphasize on sulfate reducing bacteria (SRB)	Madras Atomic Power Station (MAPS), Kalpakkam, East coast of India	Culture-dependent and group-specific PCR-DGGE	The predominant presence of three SRB groups viz. group I (*Desulfotomaculum*), group 5 (*Desulfonema, Desulfosarcina, Desulfococcus*) and group 6 (*Desulfovibrio, Desulfomicrobium*) was observed.	Balamurugan, Hiren Joshi, and Rao (2011)
Comparative analyses of marine biofilm formed on plexiglass panels in an ambient and modified environment to decipher the influence of temperature	Madras Atomic Power Station (MAPS), Kalpakkam, East coast of India	Culture-dependent analysis of nitrate-reducing bacteria, ammonia-oxidizing bacteria and heterotrophic bacteria	The biofilm progression was highly influenced by nutrients content and temperature. The biofilms developed at the temperature-impacted station was comparatively thick and consist of diverse microorganisms.	Rao et al. (2009)
Evaluation of bacterial community composition in the initial stage of biofilm formed on artificial surfaces	Sacheon harbor, Korean Peninsula	T-RFLP and 16S rRNA gene clone library	Changes were dramatically observed between 9 and 24 h. *Gammaproteobacteria* class was predominant during 0–9 h, while the ratio of *Alphaproteobacteria* increased >2-folds during 24–36 h.	Lee et al. (2008)
Examination of the bacterial community structure of biofilms formed on stainless steel and polycarbonate surfaces	Delaware Bay, USA	16S rDNA library, FISH and DGGE	Community structure differed between the free-living and attached community. *Alphaproteobacteria* dominated the attached community.	Jones, Cottrell, Kirchman, and Dexter (2007)
How site-specific variations occur in the biofilm community established on artificial surfaces	Antarctic Ocean	FISH and DGGE	Biofilm community was dominated by *Gammaproteobacteria, Bacteroidetes, Verrucomicrobia,* and *Planctomycetes.*	Webster and Negri (2006)
Fouling of manganese-oxidizing bacteria on artificial surfaces	Tuticorin harbor waters, India	Culture-dependent techniques	Biofilm development was poor on copper surface. Gram-positive bacteria were more abundant than Gram-negative bacteria on surfaces. Bacillus species, a manganese-oxidizing bacterium was common on all surfaces.	Palanichamy, Maruthamuthu, Manickam, and Rajendran (2002)

TABLE 1 Comprehensive list of important studies pertaining to the bacterial community composition analysis of marine biofilms and the methodology used—cont'd

Objective	Geographic location	Methodology	Inference	References
To investigate the impact of substratum color on the formation of micro- and macro-fouling communities	Muscat, Sea of Oman	Amplicon sequencing	The bacterial densities on both black and white colored substrata were similar, except for the day 10, when the black surface consisted of a higher abundance than white surfaces.	Dobretsov, Abed, and Voolstra (2013)
Culture-dependent and -independent analysis and appraisal of early stage biofilm	Kudankulam, Laccadive sea	Culture-dependent, DGGE and Illumina sequencing	*Vibrionaceae* and *Alteromonadaceae* are key components in early stage of biofilm	Rajeev et al. (2019)
Bacterial community structure and temporal structure of early-stage biofilms	Kudankulam, Laccadive sea	FESEM and Illumina sequencing	*Proteobacteria* decreased over biofilm succession days, while *Bacteroidetes* increased, suggesting their role as initial and late colonizers.	Sushmitha, Rajeev, Sriyutha Murthy, et al. (2021)
Monitoring biofouling communities using a combined morphological and metabarcoding approach	Portuguese port	Metabarcoding using Illumina sequencing	Increase in the colonization of *cyanobacteria*, green algae, and red algae during the sampled period with the replacement of less abundant groups	Azevedo et al. (2020)

8. Economic implications of marine biofilms

8.1 Biofouling

Biofouling is the unwanted deposition of micro- and macro-organisms with other elements of natural waterways that compromise industrial equipment performance and operational efficiency. Biofouling is widely recognized as a severe problem that has an impact on the technoeconomics of industrial units around the world (Rajeev, Sushmitha, Prasath, Toleti, & Pandian, 2020). The different stages of biofouling succession are given in Fig. 3.

8.2 Biofouling in ship hulls

Bacteria and other microbes form multispecies biofilms. Microfouling is caused by bacteria and microalgae, which allow larger organisms such as algae, mussels, and barnacles to adhere, causing macrofouling. They stick to any surface that comes into contact with seawater, such as pipes and ship hulls. The expenses of biofilms adhering to ship hulls are primarily due to increased fuel consumption, greater frictional drag, though hull cleaning, coating, and painting expenditures may also be significant. In 2010, the US Navy estimated that the total cost of hull fouling for the entire mid-sized naval surface ship class DDG-51 was 56 million dollars a year, and the total cost of hull fouling for the entire US Navy surface fleet was 180–260 million dollars (Schultz, Bendick, Holm, & Hertel, 2011). Microorganisms contribute to microfouling and play an essential part in surface colonization, despite the fact that organisms such as mussels and barnacles are responsible for macrofouling.

Bacteria can also quickly colonize pipes and heat exchangers, promoting the settlement of invertebrates' larvae (Railkin, 1998). Biofilms generally stimulate larval and spore settlement of macrofoulers, while negative and neutral effects have also been documented (Salta, Wharton, Blache, Stokes, & Briand, 2013). However, discrepancies in experimental methodologies utilized in the few research dedicated to assessing the effect of biofilms on the macrofouling settlement could explain the inconsistent results. Nonetheless, both micro and macrofoulers appear to be influenced by local environmental and hydrodynamic factors. Surprisingly, containerships that sail at fast speeds and stop in ports for brief periods of time appear to be less colonized than other ships (Davidson, Brown, Sytsma, & Ruiz, 2009). In research including 22 containerships with submerged surface areas greater than $7000\,m^2$ in San

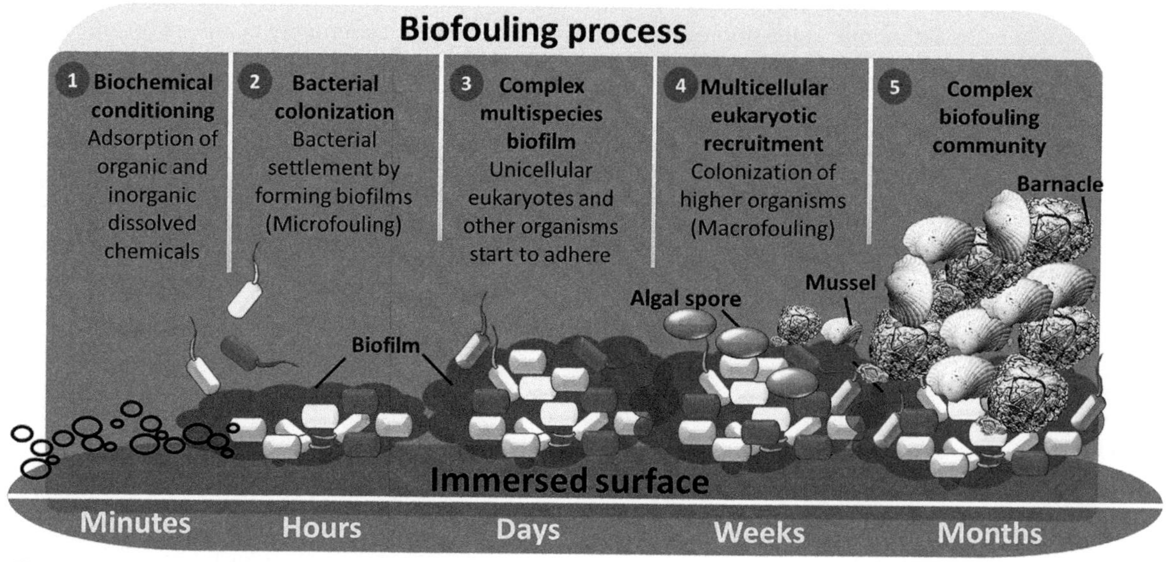

FIG. 3 The growth stages of the marine biofouling community.

Francisco Bay, fewer than 1% of the exposed hull was colonized, with the exception of one vessel, which had approximately 90% of the hull covered (Davidson et al., 2009). The latter ship was considerably smaller than its contemporaries, with a shorter journey range and slower speeds. Green algae and barnacles, as well as hydroids, bryozoans, bivalves, and ascidians, were discovered to be mostly positioned in sheltered places such as rudders, stern tubes, and intake gratings.

The ability of biofilms to affect the qualities of the surface to which they have adhered is one of their most essential characteristics. However, it has been shown that various species colonize hydrophobic and hydrophilic surfaces with distinct morphologies and cell-cell interactions and that their early colonization behavior may not be adequate to predict eventual biofilm growth. Natural films containing organic matter, particularly biofilms, formed on seawater-immersed surfaces can alter the adhesion force interfacial energy: whereas strong hydrophobic interactions can be diminished, weaker forces on hydrophilic surfaces can be strengthened (Johnson & Azetsu-Scott, 1995).

8.3 Biofouling in desalination plants

One of the most pressing issues that humanity will face in the near future is the provision of safe drinking water to an expanding global population. Saline water and wastewater processed by reverse osmosis are used in one of the main technologies for producing freshwater. Water flows across a membrane (such as cellulose acetate, triacetate, or diacetate) that rejects dissolved chemicals in the feeding water in this process. If these solutes reversibly or irreversibly adsorb to the membrane, fouling or biofouling occurs, depending on the chemical composition of the molecules adsorbed. Biofouling is the most important issue because it reduces water flow, raises pressure drops in membrane modules, reduces salt rejection, and causes membrane biodegradation and failure.

Samples from five desalination plants in various regions of the world were collected for a study to determine the microorganisms responsible for membrane biofouling in seawater reverse osmosis (Zhang et al., 2011). The findings revealed that bacterial communities from membranes were not the same, but that several dominant species were frequent, whereas bacterial populations in source seawater were highly diverse and based on location and season.

Autopsies of 500 membrane elements revealed that biological/organic and particulate/colloidal materials were the main causes of membrane fouling in 60% of the membranes, inorganic fouling in 22% of the membranes, and metal fouling in 10%. Fouling on the spacer was found in 30% of membranes with biofilms, decreased flow values were found in 37% of membranes, and poor salt rejection was found in 47% of membranes. Membranes with biofilms, on the other hand, suffer minimal physical integrity defects, implying that cleaning could restore their performance. Membrane scaling and metal fouling result in permanent damage. *Proteobacteria* dominated bacterial populations in reverse osmosis membranes in a desalination facility, according to 16S rRNA gene metabarcoding, with *Alphaproteobacteria* detected in badly fouled membranes. The community composition and diversity of biofilm communities produced by high-throughput DNA sequencing technology are, however, reliant on the DNA extraction method used.

8.4 Biofouling in power plants

The higher value of freshwater and the high need for cooling water have promoted the construction of power plants along the coast, where seawater is readily supplied at low costs. Biofouling of heat exchanger units, on the other hand, is a significant issue. Biofouling reduces heat transfer efficiency in heat exchangers. Iron oxide has a thermal conductivity of $2.9\,W\,m^{-1}\,K^{-1}$, whereas biofilm has a thermal conductivity of $0.6\,W\,m^{-1}\,K^{-1}$ (Flynn, 2009), making the latter even more insulating. The fouling of power plant cooling systems demands both operational and plant shutdowns.

Furthermore, biofouling produces scale and corrosion because the cells adjacent to the tube wall have limited access to oxygen as the biofilm accumulates. In a process known as microbiologically influenced corrosion, bacteria such as sulfate-reducing strains produce compounds that attack the metal. Cleaning, fluid treatment, part replacement, and output loss due to heat exchanger fouling are estimated to cost 0.25% of the GDP of industrialized countries, according to studies conducted in the 1980s and early 1990s (Müller-Steinhagen, Malayeri, & Watkinson, 2005). Repairing heat exchangers and boilers is expected to cost roughly 15% of a process plant's maintenance expenditures, with fouling accounting for about half of that (Ibrahim, 2012). The major economic implication of marine biofilms is provided in Table 2.

8.5 Biocorrosion

NACE International, The Worldwide Corrosion Authority, projected the global cost of corrosion to be around 2.5 trillion dollars in 2016. Corrosion has been discovered to have a significant impact in the oil and gas industry, storage, and transportation, as well as the drinking and wastewater industries. Microbes play a vital role in oil bioremediation; however, microbial contamination of oil and natural gas facilities is undesirable because they can metabolize hydrocarbons, change sulfur content, and affect oil density and viscosity.

The anaerobic conditions common in the oil industry, together with the abundance of microorganism substrates (e.g., hydrocarbons and organosulfur compounds), stimulate the formation of biofilms that cause microbially influenced corrosion. Aerobic microbes such as *Pseudomonas* sp. SS304, on the other hand, can cause corrosion. Bacteria made up more than 98% of the microbial population assessed by qPCR in samples from three anaerobic biofilms collected inside a badly corroded steel pipe in an offshore oil facility in the Gulf of Mexico, but archaea species were also found (Vigneron et al., 2016). Bacteria were mostly represented by the *Desulfovibrio* genus. The biofilms' major population consisted of sulfate-, elemental sulfur-, and iron-reducing bacteria, which were likely to blame for the significant corrosion rate and corrosive compounds found. Bacteria that produce acetic acid, such as *Acetobacter* spp. and *Gluconacetobacter* spp., have been found to cause corrosion in tanks containing fuel-grade ethanol (Williamson, Jain, Mishra, Olson, & Spear, 2015), while hydrogen-sulfide-producing strains of the genus *Acidithiobacillus* have been found to corrode concrete sewers.

Nonetheless, several bacterial strains with corrosion-protecting properties have been identified: *Vibrio neocaledonicus* reduced ASTM A36 steel corrosion by 64-fold (Moradi, Song, & Tao, 2015); and *Pseudomonas* S9 and *Serratia marcescens* EF190 reduced corrosion of ASTM A619 carbon steel under aerobic conditions (Pedersen & Hermansson, 1989). The elimination of toxic substances, restriction of the development of corrosion-causing

TABLE 2 Economic implications of marine biofilms in maritime industries.

Industry	Problem	Estimated cost	References
Aquaculture	Fouling of both stock and infrastructures, Biofouling on fish cages and shellfish sites	1.5–3 billion US dollars per year	Fitridge, Dempster, Guenther, and De Nys (2012)
Heat exchanger in power plants	Decreased heat transfer and induced corrosion	c. 7.5% of maintenance costs of a process plant	Ibrahim (2012)
Oil and gas industry	Microbial influenced corrosion	20%–30% of corrosion-related costs	Skovhus, Enning, & Lee (2017)
Maritime transport	Increased drag Fouling of ballast water with nonindigenous species	35%–50% increased fuel consumption	Schultz et al. (2011)
Water desalination	Fouling of ballast water with nonindigenous species	5%–20% of operational costs for cleaning; cost of biofouling c. 30% of total operating costs	Maddah and Chogle (2017)

microorganisms by their noncorrosive counterparts, and generation of a protective layer by noncorrosive microbes overproducing EPS are three putative strategies for inhibiting biofilm-induced metal corrosion (Zuo, 2007).

9. Advantages of marine biofilms

Despite several negative impacts, bacterial biofilms may offer beneficial roles in a number of areas, which have been well reviewed (Muhammad et al., 2020). For instance, biofilms have been used as biofertilizers as well as controlling agents against phytopathogens in agriculture (Bogino, Oliva, Sorroche, & Giordano, 2013). Biofilm forming and antagonists producing capability of *Bacillus subtilis*, a dominant rhizobacterium, is recognized as an efficient strategy to prevent the plant from bacterial and fungal pathogens (Bais, Fall, & Vivanco, 2004). Similarly, microbial biofilms control the inorganic nutrients load in the wastewater treatment plants and thus support a sustainable wastewater treatment practice (Guzzon, Bohn, Diociaiuti, & Albertano, 2008). In fact, several approaches have been adapted to facilitate the attachment of nitrifying bacteria on membrane surfaces, which in turn promote the nitrification (oxidation of ammonia to nitrite and subsequently to nitrate) in wastewater treatment plants (Hibiya, Tsuneda, & Hirata, 2000). Furthermore, bio-electrochemical systems and microbial fuel cells utilize biofilms to transform organic waste into electrical energy (Bajracharya et al., 2016). Though there have been several uses of biofilms, marine biofilms are mostly applied in bioremediation.

Rapid industrialization, urbanization, and other anthropogenic activities that occurred during the past few decades have caused contamination of several pollutants such as heavy metals, plastics, antibiotics, pesticides, xenobiotics, petroleum hydrocarbons, and polycyclic aromatic hydrocarbons (PAHs) in marine environments. These contaminants mostly originate from anthropogenic activities and ultimately exert deleterious impacts on living organisms including microbes, when their concentrations exceed beyond definite threshold levels (Gillan, Danis, Pernet, Joly, & Dubois, 2005). Marine microbes are considered a suitable biomonitoring tool to ascertain the health status of a marine environment as they respond quickly to environmental perturbations (Rajeev, Sushmitha, Aravindraja, et al., 2021; Sun, Dafforn, Brown, & Johnston, 2012). In order to mitigate the distribution of these hazardous materials in the environment, remediation is emerging as a potential method that involves the conversion of toxic hazardous materials into less toxic substances or complete degradation (Van Dillewijn, Nojiri, Van Der Meer, & Wood, 2009). Remediation processes mediated by living organisms and their derivatives (bioremediation) are preferred over the conventional chemical-based methods due to the fact that bioremediation processes are relatively effective, environmental-friendly, and technoeconomically doable (Singh, Paul, & Jain, 2006).

Compared with the planktonic (free-living) counterpart, marine bacteria isolated from biofilms and capable to form robust biofilms are considered a suitable candidate for bioremediation of various organic and inorganic compounds as they possess strong catabolic pathways to decompose these compounds (Van Dillewijn et al., 2009). Biofilm-mediated remediation methods are generally considered more efficient owing to the easy bioavailability of pollutants to the functioning microorganisms. Moreover, bacterial cells residing in a biofilm are often encased in a self-produced matrix of polymers that provide a solid defense against environmental hazards (Mangwani, Kumari, & Das, 2016). By contrast, the bioremediation efficiency of planktonic cells is proven to be less due to their increased sensitivity toward changes in nutrient content, predation, exposure to a high concentration of pollutants, and other environmental stressors such as changes in pH, temperature, and salt content. Considering the overall durability, the importance of facilitating biofilms in a natural environment to enhance the removal of various pollutants such as pesticides, heavy metals, dyes, explosives, radioactive substances, and pharmaceutical products in various environments has recently gained considerable attention (Edwards & Kjellerup, 2013).

Marine biofilm-mediated bioremediations are considered potential toward the removal of various contaminants. For example, marine bacterial consortium composed of five biofilm-forming marine bacterial strains such as *Stenotrophomonas acidaminiphila*, *Alcaligenes faecalis*, *Pseudomonas mendocina*, *Pseudomonas aeruginosa*, and *Pseudomonas pseudoalcaligenes* showed potent bioremediation for PAHs in soil microcosm (Mangwani et al., 2016). Similarly, biofilm-forming bacterial isolates, obtained from coastal sediments of India, showed promising uranium sequestering characteristics (Manobala, Shukla, Rao, & Kumar, 2019). A recent fascinating study by Bôto et al. (2021) described the potential of native marine microbes in the Iberian Peninsula NW coast for bioremediation of oil spills. Although the biofilm-forming potential of major hydrocarbon-degrading bacterial genera (e.g., *Alcanivorax*, *Oleibacter*, *Acinetobacter*, *Marinobacter*, *Rhodococcus*, *Pseudomonas*, *Flavobacterium*, and *Thalassospira*) has not been specified in this study. Moreover, at the time of accidents in the oil industries, for example, the Exxon Valdez and the BP Deepwater Horizon oil spills, biofertilizers were used to enhance the population of naturally occurring oil-degrading bacteria and thus to establish a successful bioremediation approach (Atlas & Hazen, 2011; de Carvalho, 2018).

10. Conclusion

Despite the extensive research on microfouling communities in various marine environments, studies delineating macrofouling communities at the genus and species level using modern sequencing technologies are rather rare. Several methods are being practiced by the different maritime industries to control microbial aggregation and biofilm formation. However, the majority of these strategies are successful only in a particular environment and/or pose detrimental impacts on the environment. The quest for novel, effective, broad-range, and environmental-friendly strategies will be long-lasting. Secondly, a large number of research studies describing detrimental impacts of biofilms on various settings such as agriculture, water distribution system, medical, food processing industries, and maritime industries have been conducted so far. However, the beneficial roles of marine biofilms are higher than assumed previously. Considering the genetic potential and success of biofilm-forming bacteria in various environments, future research efforts may expand the application of biofilm-forming marine bacteria in mitigating environmental hazards and thus in the restoration of contaminated ecosystems. Furthermore, exploiting modern technologies such as shotgun metagenomics, metaproteomics, and metabolomics to unveil the fundamental cellular pathways involved in the bioremediation of hazardous compounds is another exciting area, warranting instant scientific attention.

References

Abed, R. M., Al Fahdi, D., & Muthukrishnan, T. (2019). Short-term succession of marine microbial fouling communities and the identification of primary and secondary colonizers. *Biofouling*, *35*(5), 526–540.

Agostini, V. O., Macedo, A. J., Muxagata, E., & Pinho, G. L. L. (2019). Surface coatings select their micro and macrofouling communities differently on steel. *Environmental Pollution*, *254*, 113086.

Al Senafi, F., Abed, R. M., Muthukrishnan, T., Anis, A., Al Rawahi, A. N., & Mahmoud, H. (2020). Development and diversity of bacterial biofilms in response to internal tides, a case study off the coast of Kuwait. *Frontiers in Marine Science*, *7*, 21.

Almstrand, R., Daims, H., Persson, F., Sörensson, F., & Hermansson, M. (2013). New methods for analysis of spatial distribution and coaggregation of microbial populations in complex biofilms. *Applied and Environmental Microbiology*, *79*(19), 5978–5987.

Amann, R. I., Ludwig, W., & Schleifer, K. H. (1995). Phylogenetic identification and in situ detection of individual microbial cells without cultivation. *Microbiological Reviews*, *59*(1), 143–169.

Angelova, A. G., Ellis, G. A., Wijesekera, H. W., & Vora, G. J. (2019). Microbial composition and variability of natural marine planktonic and biofouling communities from the Bay of Bengal. *Frontiers in Microbiology*, *10*, 2738.

Antunes, J. T., Sousa, A. G., Azevedo, J., Rego, A., Leão, P. N., & Vasconcelos, V. (2020). Distinct temporal succession of bacterial communities in early marine biofilms in a Portuguese Atlantic Port. *Frontiers in Microbiology*, *11*, 1938.

Atlas, R. M., & Hazen, T. C. (2011). Oil biodegradation and bioremediation: A tale of the two worst spills in US history. *Environmental Science and Technology*, *45*, 6709–6715.

Azevedo, J., Antunes, J. T., Machado, A. M., Vasconcelos, V., Leão, P. N., & Froufe, E. (2020). Monitoring of biofouling communities in a Portuguese port using a combined morphological and metabarcoding approach. *Scientific Reports*, *10*(1), 1–15.

Bais, H. P., Fall, R., & Vivanco, J. M. (2004). Biocontrol of *Bacillus subtilis* against infection of Arabidopsis roots by *Pseudomonas syringae* is facilitated by biofilm formation and surfactin production. *Plant Physiology*, *134*(1), 307–319.

Bajracharya, S., Sharma, M., Mohanakrishna, G., Benneton, X. D., Strik, D. P., Sarma, P. M., et al. (2016). An overview on emerging bioelectrochemical systems (BESs): Technology for sustainable electricity, waste remediation, resource recovery, chemical production and beyond. *Renewable Energy*, *98*, 153–170.

Balamurugan, P., Hiren Joshi, M., & Rao, T. S. (2011). Microbial fouling community analysis of the cooling water system of a nuclear test reactor with emphasis on sulphate reducing bacteria. *Biofouling*, *27* (9), 967–978.

Balqadi, A. A., Salama, A. J., & Satheesh, S. (2018). Microfouling development on artificial substrates deployed in the central Red Sea. *Oceanologia*, *60*(2), 219–231.

Bellou, N., Garcia, J. A. L., Colijn, F., & Herndl, G. J. (2020). Seasonality combined with the orientation of surfaces influences the microbial community structure of biofilms in the deep Mediterranean Sea. *Deep Sea Research Part II: Topical Studies in Oceanography*, *171*, 104703.

Bellou, N., Papathanassiou, E., Dobretsov, S., Lykousis, V., & Colijn, F. (2012). The effect of substratum type, orientation and depth on the development of bacterial deep-sea biofilm communities grown on artificial substrata deployed in the Eastern Mediterranean. *Biofouling*, *28*(2), 199–213.

Bogino, P. C., Oliva, M. D. L. M., Sorroche, F. G., & Giordano, W. (2013). The role of bacterial biofilms and surface components in plant-bacterial associations. *International Journal of Molecular Sciences*, *14*(8), 15838–15859.

Bôto, M. L., Magalhães, C., Perdigão, R., Alexandrino, D. A., Fernandes, J. P., Bernabeu, A. M., et al. (2021). Harnessing the potential of native microbial communities for bioremediation of oil spills in the Iberian Peninsula NW coast. *Frontiers in Microbiology*, *12*, 879.

Briand, J. F., Barani, A., Garnier, C., Réhel, K., Urvois, F., LePoupon, C., et al. (2017). Spatio-temporal variations of marine biofilm communities colonizing artificial substrata including antifouling coatings in contrasted French coastal environments. *Microbial Ecology*, *74*(3), 585–598.

Briand, J. F., Djeridi, I., Jamet, D., Coupé, S., Bressy, C., Molmeret, M., et al. (2012). Pioneer marine biofilms on artificial surfaces including antifouling coatings immersed in two contrasting French Mediterranean coast sites. *Biofouling*, *28*(5), 453–463.

Buchan, A., González, J. M., & Moran, M. A. (2005). Overview of the marine Roseobacter lineage. *Applied and Environmental Microbiology*, *71*(10), 5665–5677.

Campbell, B. J., Waidner, L. A., Cottrell, M. T., & Kirchman, D. L. (2008). Abundant proteorhodopsin genes in the North Atlantic Ocean. *Environmental Microbiology*, *10*(1), 99–109.

Chagnot, C., Zorgani, M. A., Astruc, T., & Desvaux, M. (2013). Proteinaceous determinants of surface colonization in bacteria: Bacterial adhesion and biofilm formation from a protein secretion perspective. *Frontiers in Microbiology*, *4*, 303.

Chan, C. S., Fakra, S. C., Emerson, D., Fleming, E. J., & Edwards, K. J. (2011). Lithotrophic iron-oxidizing bacteria produce organic stalks to control mineral growth: Implications for biosignature formation. *The ISME Journal*, *5*(4), 717–727.

Choi, D. H., Noh, J. H., Yu, O. H., & Kang, Y. S. (2010). Bacterial diversity in biofilms formed on condenser tube surfaces in a nuclear power plant. *Biofouling*, *26*(8), 953–959.

Costerton, J. W., Lewandowski, Z., Caldwell, D. E., Korber, D. R., & Lappin-Scott, H. M. (1995). Microbial biofilms. *Annual Review of Microbiology*, *49*(1), 711–745.

Crespo, B. G., Pommier, T., Fernández-Gómez, B., & Pedrós-Alió, C. (2013). Taxonomic composition of the particle-attached and free-living bacterial assemblages in the Northwest Mediterranean Sea analyzed by pyrosequencing of the 16S rRNA. *Microbiologyopen*, *2*(4), 541–552.

Dang, H., & Lovell, C. R. (2000). Bacterial primary colonization and early succession on surfaces in marine waters as determined by amplified rRNA gene restriction analysis and sequence analysis of 16S rRNA genes. *Applied and Environmental Microbiology*, *66*(2), 467–475.

Dang, H., & Lovell, C. R. (2002). Numerical dominance and phylotype diversity of marine Rhodobacter species during early colonization of submerged surfaces in coastal marine waters as determined by 16S ribosomal DNA sequence analysis and fluorescence in situ hybridization. *Applied and Environmental Microbiology*, *68*(2), 496–504.

Dang, H., & Lovell, C. R. (2016). Microbial surface colonization and biofilm development in marine environments. *Microbiology and Molecular Biology Reviews*, *80*(1), 91–138.

Davidson, I. C., Brown, C. W., Sytsma, M. D., & Ruiz, G. M. (2009). The role of containerships as transfer mechanisms of marine biofouling species. *Biofouling*, *25*(7), 645–655.

de Carvalho, C. C. (2018). Marine biofilms: A successful microbial strategy with economic implications. *Frontiers in Marine Science*, *5*, 126.

De Carvalho, C. C., & Fernandes, P. (2010). Production of metabolites as bacterial responses to the marine environment. *Marine Drugs*, *8*(3), 705–727.

Decho, A. W., & Gutierrez, T. (2017). Microbial extracellular polymeric substances (EPSs) in ocean systems. *Frontiers in Microbiology*, *8*, 922.

DeSantis, T. Z., Brodie, E. L., Moberg, J. P., Zubieta, I. X., Piceno, Y. M., & Andersen, G. L. (2007). High-density universal 16S rRNA microarray analysis reveals broader diversity than typical clone library when sampling the environment. *Microbial Ecology*, *53*(3), 371–383.

Dobretsov, S., Abed, R. M., & Voolstra, C. R. (2013). The effect of surface colour on the formation of marine micro and macrofouling communities. *Biofouling*, *29*(6), 617–627.

Doghri, I., Rodrigues, S., Bazire, A., Dufour, A., Akbar, D., Sopena, V., et al. (2015). Marine bacteria from the French Atlantic coast displaying high forming-biofilm abilities and different biofilm 3D architectures. *BMC Microbiology*, *15*(1), 1–10.

Donlan, R. M. (2002). Biofilms: Microbial life on surfaces. *Emerging Infectious Diseases*, *8*(9), 881.

Donlan, R. M., & Costerton, J. W. (2002). Biofilms: Survival mechanisms of clinically relevant microorganisms. *Clinical Microbiology Reviews*, *15*(2), 167–193.

Edwards, S. J., & Kjellerup, B. V. (2013). Applications of biofilms in bioremediation and biotransformation of persistent organic pollutants, pharmaceuticals/personal care products, and heavy metals. *Applied Microbiology and Biotechnology*, *97*(23), 9909–9921.

Fernández-Gomez, B., Richter, M., Schüler, M., Pinhassi, J., Acinas, S. G., González, J. M., et al. (2013). Ecology of marine Bacteroidetes: A comparative genomics approach. *The ISME Journal*, *7*(5), 1026–1037.

Fisher, M. M., & Triplett, E. W. (1999). Automated approach for ribosomal intergenic spacer analysis of microbial diversity and its application to freshwater bacterial communities. *Applied and Environmental Microbiology*, *65*(10), 4630–4636.

Fitridge, I., Dempster, T., Guenther, J., & De Nys, R. (2012). The impact and control of biofouling in marine aquaculture: A review. *Biofouling*, *28*(7), 649–669.

Flemming, H. C., & Wingender, J. (2010). The biofilm matrix. *Nature Reviews Microbiology*, *8*(9), 623–633.

Flynn, D. (2009). *Nalco water handbook*. McGraw-Hill Education.

Freese, E., Rütters, H., Köster, J., Rullkötter, J., & Sass, H. (2009). Gammaproteobacteria as a possible source of eicosapentaenoic acid in anoxic intertidal sediments. *Microbial Ecology*, *57*(3), 444–454.

Fuhrman, J. A., Cram, J. A., & Needham, D. M. (2015). Marine microbial community dynamics and their ecological interpretation. *Nature Reviews Microbiology*, *13*(3), 133–146.

Gillan, D. C., Danis, B., Pernet, P., Joly, G., & Dubois, P. (2005). Structure of sediment-associated microbial communities along a heavy-metal contamination gradient in the marine environment. *Applied and Environmental Microbiology*, *71*(2), 679–690.

Gómez-Pereira, P. R., Schüler, M., Fuchs, B. M., Bennke, C., Teeling, H., Waldmann, J., et al. (2012). Genomic content of uncultured Bacteroidetes from contrasting oceanic provinces in the North Atlantic Ocean. *Environmental Microbiology*, *14*(1), 52–66.

Gruber-Dorninger, C., Pester, M., Kitzinger, K., Savio, D. F., Loy, A., Rattei, T., et al. (2015). Functionally relevant diversity of closely related Nitrospira in activated sludge. *The ISME Journal*, *9*(3), 643–655.

Guzzon, A., Bohn, A., Diociaiuti, M., & Albertano, P. (2008). Cultured phototrophic biofilms for phosphorus removal in wastewater treatment. *Water Research*, *42*(16), 4357–4367.

Hadfield, M. G. (2011). Biofilms and marine invertebrate larvae: What bacteria produce that larvae use to choose settlement sites. *Annual Review of Marine Science*, *3*, 453–470.

Hall, E., & Pepe-Ranney, C. (2015). The effect of carbon subsidies on marine planktonic niche partitioning and recruitment during biofilm assembly. *Frontiers in Microbiology*, *6*, 703.

Harley, C. D., Randall Hughes, A., Hultgren, K. M., Miner, B. G., Sorte, C. J., Thornber, C. S., et al. (2006). The impacts of climate change in coastal marine systems. *Ecology Letters*, *9*(2), 228–241.

Hawley, A. K., Nobu, M. K., Wright, J. J., Durno, W. E., Morgan-Lang, C., Sage, B., et al. (2017). Diverse Marinimicrobia bacteria may mediate coupled biogeochemical cycles along eco-thermodynamic gradients. *Nature Communications*, *8*(1), 1–10.

Hibbing, M. E., Fuqua, C., Parsek, M. R., & Peterson, S. B. (2010). Bacterial competition: Surviving and thriving in the microbial jungle. *Nature Reviews Microbiology*, *8*(1), 15–25.

Hibiya, K., Tsuneda, S., & Hirata, A. (2000). Formation and characteristics of nitrifying biofilm on a membrane modified with positively-charged polymer chains. *Colloids and Surfaces B: Biointerfaces*, *18*(2), 105–112.

Hunsucker, K. Z., Vora, G. J., Hunsucker, J. T., Gardner, H., Leary, D. H., Kim, S., et al. (2018). Biofilm community structure and the associated drag penalties of a groomed fouling release ship hull coating. *Biofouling*, *34*(2), 162–172.

Ibrahim, H. A. (2012). Fouling in heat exchangers. In V. N. Katsikis (Ed.), *MATLAB—A fundamental tool for scientific computing and engineering applications* (pp. 57–96). Croatia: Intech.

Johnson, B. D., & Azetsu-Scott, K. (1995). Adhesion force and the character of surfaces immersed in seawater. *Limnology and Oceanography*, *40*(4), 802–808.

Jones, P. R., Cottrell, M. T., Kirchman, D. L., & Dexter, S. C. (2007). Bacterial community structure of biofilms on artificial surfaces in an estuary. *Microbial Ecology*, *53*(1), 153–162.

José, R., Christianson, L. M., Béja, O., Suzuki, M. T., Karl, D. M., Heidelberg, J., et al. (2003). Proteorhodopsin genes are distributed among divergent marine bacterial taxa. *Proceedings of the National Academy of Sciences*, *100*(22), 12830–12835.

Kennedy, J., Flemer, B., Jackson, S. A., Lejon, D. P., Morrissey, J. P., O'gara, F., et al. (2010). Marine metagenomics: New tools for the study and exploitation of marine microbial metabolism. *Marine Drugs*, *8*(3), 608–628.

Lawes, J. C., Neilan, B. A., Brown, M. V., Clark, G. F., & Johnston, E. L. (2016). Elevated nutrients change bacterial community composition and connectivity: High throughput sequencing of young marine biofilms. *Biofouling*, *32*(1), 57–69.

Lee, J. W., Nam, J. H., Kim, Y. H., Lee, K. H., & Lee, D. H. (2008). Bacterial communities in the initial stage of marine biofilm formation on artificial surfaces. *The Journal of Microbiology*, *46*(2), 174–182.

Liu, W. T., Marsh, T. L., Cheng, H., & Forney, L. J. (1997). Characterization of microbial diversity by determining terminal restriction fragment length polymorphisms of genes encoding 16S rRNA. *Applied and Environmental Microbiology*, *63*(11), 4516–4522.

Long, R. A., & Azam, F. (2001). Antagonistic interactions among marine pelagic bacteria. *Applied and Environmental Microbiology*, *67*(11), 4975–4983.

Maddah, H., & Chogle, A. (2017). Biofouling in reverse osmosis: Phenomena, monitoring, controlling and remediation. *Applied Water Science*, *7*(6), 2637–2651.

Mai, Y. Z., Peng, S. Y., & Lai, Z. N. (2020). Structural and functional diversity of biofilm bacterial communities along the Pearl River Estuary, South China. *Regional Studies in Marine Science*, *33*, 100926.

Mandlik, A., Swierczynski, A., Das, A., & Ton-That, H. (2008). Pili in Gram-positive bacteria: Assembly, involvement in colonization and biofilm development. *Trends in Microbiology*, *16*(1), 33–40.

Mangwani, N., Kumari, S., & Das, S. (2016). Bacterial biofilms and quorum sensing: Fidelity in bioremediation technology. *Biotechnology and Genetic Engineering Reviews*, *32*(1–2), 43–73.

Manobala, T., Shukla, S. K., Rao, T. S., & Kumar, M. D. (2019). Uranium sequestration by biofilm-forming bacteria isolated from marine sediment collected from Southern coastal region of India. *International Biodeterioration & Biodegradation*, *145*, 104809.

Matz, C., Webb, J. S., Schupp, P. J., Phang, S. Y., Penesyan, A., Egan, S., et al. (2008). Marine biofilm bacteria evade eukaryotic predation by targeted chemical defense. *PLoS One*, *3*(7), e2744.

McDougald, D., Rice, S. A., Barraud, N., Steinberg, P. D., & Kjelleberg, S. (2012). Should we stay or should we go: Mechanisms and ecological consequences for biofilm dispersal. *Nature Reviews Microbiology*, *10* (1), 39–50.

Meier, A., Tsaloglou, N. M., Mowlem, M. C., Keevil, C. W., & Connelly, D. P. (2013). Hyperbaric biofilms on engineering surfaces formed in the deep sea. *Biofouling*, *29*(9), 1029–1042.

Milici, M., Tomasch, J., Wos-Oxley, M. L., Wang, H., Jáuregui, R., & Camarinha-Silva, A. (2016). Low diversity of planktonic bacteria in the tropical ocean. *Scientific Reports*, *6*(1), 1–9.

Minaev, B. F. (2007). Electronic mechanisms of activation of molecular oxygen. *Russian Chemical Reviews*, *76*(11), 989.

Moradi, M., Song, Z., & Tao, X. (2015). Introducing a novel bacterium, *Vibrio neocaledonicus* sp., with the highest corrosion inhibition efficiency. *Electrochemistry Communications*, *51*, 64–68.

Moss, J. A., Nocker, A., Lepo, J. E., & Snyder, R. A. (2006). Stability and change in estuarine biofilm bacterial community diversity. *Applied and Environmental Microbiology*, *72*(9), 5679–5688.

Moura, V., Ribeiro, I., Moriggi, P., Capão, A., Salles, C., Bitati, S., et al. (2018). The influence of surface microbial diversity and succession on microbiologically influenced corrosion of steel in a simulated marine environment. *Archives of Microbiology*, *200*(10), 1447–1456.

Muhammad, M. H., Idris, A. L., Fan, X., Guo, Y., Yu, Y., Jin, X., et al. (2020). Beyond risk: Bacterial biofilms and their regulating approaches. *Frontiers in Microbiology*, *11*, 928.

Müller-Steinhagen, H., Malayeri, M. R., & Watkinson, A. P. (2005). Fouling of heat exchangers-new approaches to solve an old problem. *Heat Transfer Engineering*, *26*(1), 1–4.

Muyzer, G., De Waal, E. C., & Uitterlinden, A. (1993). Profiling of complex microbial populations by denaturing gradient gel electrophoresis analysis of polymerase chain reaction-amplified genes coding for 16S rRNA. *Applied and Environmental Microbiology*, *59*(3), 695–700.

Ortega-Morales, B. O., Chan-Bacab, M. J., De la Rosa, S. D. C., & Camacho-Chab, J. C. (2010). Valuable processes and products from marine intertidal microbial communities. *Current Opinion in Biotechnology*, *21*(3), 346–352.

Palaniappan, B., & Toleti, S. R. (2016). Characterization of microfouling and corrosive bacterial community of a firewater distribution system. *Journal of Bioscience and Bioengineering*, *121*(4), 435–441.

Palanichamy, S., Maruthamuthu, S., Manickam, S. T., & Rajendran, A. (2002). Microfouling of manganese-oxidizing bacteria in Tuticorin harbour waters. *Current Science*, 865–869.

Patil, J. S., & Anil, A. C. (2005). Biofilm diatom community structure: Influence of temporal and substratum variability. *Biofouling*, *21*(3–4), 189–206.

Pedersen, A., & Hermansson, M. (1989). The effects on metal corrosion by *Serratia marcescens* and a *Pseudomonas* sp. *Biofouling*, *1*(4), 313–322.

Pinto, M., Langer, T. M., Hüffer, T., Hofmann, T., & Herndl, G. J. (2019). The composition of bacterial communities associated with plastic biofilms differs between different polymers and stages of biofilm succession. *PLoS One*, *14*(6), e0217165.

Pollet, T., Berdjeb, L., Garnier, C., Durrieu, G., Le Poupon, C., Misson, B., et al. (2018). Prokaryotic community successions and interactions in marine biofilms: The key role of *Flavobacteriia*. *FEMS Microbiology Ecology*, *94*(6), fiy083.

Proft, T., & Baker, E. N. (2009). Pili in Gram-negative and Gram-positive bacteria—Structure, assembly and their role in disease. *Cellular and Molecular Life Sciences*, *66*(4), 613–635.

Qian, P. Y., Lau, S. C., Dahms, H. U., Dobretsov, S., & Harder, T. (2007). Marine biofilms as mediators of colonization by marine macroorganisms: Implications for antifouling and aquaculture. *Marine Biotechnology*, *9*(4), 399–410.

Railkin, A. I. (1998). The pattern of recovery of disturbed microbial communities inhabiting hard substrates. *Hydrobiologia*, *385*(1), 47–57.

Rajeev, M., Sushmitha, T. J., Aravindraja, C., Toleti, S. R., & Pandian, S. K. (2021). Exploring the impacts of heavy metals on spatial variations of sediment-associated bacterial communities. *Ecotoxicology and Environmental Safety*, *209*, 111808.

Rajeev, M., Sushmitha, T. J., & Pandian, S. K. (2020). Culture-dependent and-independent strategies in bacterial diversity appraisal. In *Microbial systematics* (pp. 1–27). CRC Press.

Rajeev, M., Sushmitha, T. J., Prasath, K. G., Toleti, S. R., & Pandian, S. K. (2020). Systematic assessment of chlorine tolerance mechanism in a potent biofilm-forming marine bacterium *Halomonas boliviensis*. *International Biodeterioration & Biodegradation*, *151*, 104967.

Rajeev, M., Sushmitha, T. J., Toleti, S. R., & Pandian, S. K. (2019). Culture dependent and independent analysis and appraisal of early stage biofilm-forming bacterial community composition in the Southern coastal seawater of India. *Science of the Total Environment*, *666*, 308–320.

Rajeev, M., Sushmitha, T. J., Toleti, S. R., & Pandian, S. K. (2021). Draft genome sequencing of *Pseudoalteromonas tetraodonis* strain kknpp56, a potent biofilm-forming bacterium isolated from early-stage marine biofilm. *Microbiology Resource Announcements*, *10*(38), e00605–e00621.

Rao, T. S. (2015). Biofouling in industrial water systems. In *Mineral scales and deposits* (pp. 123–140). Elsevier.

Rao, T. S., Kesavamoorthy, R., Rao, C. B., & Nair, K. V. K. (1997). Influence of flow on ordering characteristics of a bacterial biofilm. *Current Science*, 69–74.

Rao, T. S., Kora, A. J., Chandramohan, P., Panigrahi, B. S., & Narasimhan, S. V. (2009). Biofouling and microbial corrosion problem in the thermo-fluid heat exchanger and cooling water system of a nuclear test reactor. *Biofouling*, *25*(7), 581–591.

Rao, D., Webb, J. S., & Kjelleberg, S. (2005). Competitive interactions in mixed-species biofilms containing the marine bacterium *Pseudoalteromonas tunicata*. *Applied and Environmental Microbiology*, *71*(4), 1729–1736.

Sadiq, F. A., Burmølle, M., Heyndrickx, M., Flint, S., Lu, W., Chen, W., et al. (2021). Community-wide changes reflecting bacterial interspecific interactions in multispecies biofilms. *Critical Reviews in Microbiology*, *47*(3), 338–358.

Salta, M., Wharton, J. A., Blache, Y., Stokes, K. R., & Briand, J. F. (2013). Marine biofilms on artificial surfaces: Structure and dynamics. *Environmental Microbiology*, *15*(11), 2879–2893.

Saravanan, P., Nancharaiah, Y. V., Venugopalan, V. P., Rao, T. S., & Jayachandran, S. (2006). Biofilm formation by *Pseudoalteromonas ruthenica* and its removal by chlorine. *Biofouling*, *22*(6), 371–381.

Schultz, M. P., Bendick, J. A., Holm, E. R., & Hertel, W. M. (2011). Economic impact of biofouling on a naval surface ship. *Biofouling*, *27*(1), 87–98.

Singh, R., Paul, D., & Jain, R. K. (2006). Biofilms: Implications in bioremediation. *Trends in Microbiology*, *14*(9), 389–397.

Skovhus, T. L., Enning, D., & Lee, J. S. (Eds.). (2017). *Microbiologically influenced corrosion in the upstream oil and gas industry* CRC Press.

Sockett, R. E. (2009). Predatory lifestyle of *Bdellovibrio bacteriovorus*. *Annual Review of Microbiology*, *63*, 523–539.

Sogin, M. L., Morrison, H. G., Huber, J. A., Welch, D. M., Huse, S. M., Neal, P. R., et al. (2006). Microbial diversity in the deep sea and the underexplored "rare biosphere". *Proceedings of the National Academy of Sciences*, *103*(32), 12115–12120.

Stocker, R. (2012). Marine microbes see a sea of gradients. *Science*, *338*(6107), 628–633.

Sun, M. Y., Dafforn, K. A., Brown, M. V., & Johnston, E. L. (2012). Bacterial communities are sensitive indicators of contaminant stress. *Marine Pollution Bulletin*, *64*(5), 1029–1038.

Sun, Y., Lang, Y., Yan, Z., Wang, L., & Zhang, Z. (2020). High-throughput sequencing analysis of marine pioneer surface-biofilm bacteria communities on different PDMS-based coatings. *Colloids and Surfaces B: Biointerfaces*, *185*, 110538.

Sushmitha, T. J., Rajeev, M., & Pandian, S. K. (2020). Next generation sequencing advances and applications in the world of bacterial diversity. In *Microbial systematics* (pp. 178–209). CRC Press.

Sushmitha, T. J., Rajeev, M., Sriyutha Murthy, P., Ganesh, S., Toleti, S. R., & Karutha Pandian, S. (2021). Bacterial community structure of early-stage biofilms is dictated by temporal succession rather than substrate types in the southern coastal seawater of India. *PLoS One*, *16*(9), e0257961.

Sushmitha, T. J., Rajeev, M., Toleti, S. R., & Pandian, S. K. (2021). Complete genome sequence of *Halomonas boliviensis* strain kknpp38, a chlorine-resistant bacterium isolated from the early-stage marine biofilm. *Marine Genomics*, 100890.

Sutherland, I. W. (1997). Microbial exopolysaccharides-structural subtleties and their consequences. *Pure and Applied Chemistry*, *69*(9), 1911–1918.

Taberlet, P., Coissac, E., Pompanon, F., Brochmann, C., & Willerslev, E. (2012). Towards next-generation biodiversity assessment using DNA metabarcoding. *Molecular Ecology*, *21*(8), 2045–2050.

van der Merwe, P., Lannuzel, D., Nichols, C. M., Meiners, K., Heil, P., & Norman, L. (2009). Biogeochemical observations during the winter–spring transition in East Antarctic sea ice: Evidence of iron and exopolysaccharide controls. *Marine Chemistry*, *115*(3–4), 163–175.

Van Dillewijn, P., Nojiri, H., Van Der Meer, J. R., & Wood, T. K. (2009). Bioremediation, a broad perspective. *Microbial Biotechnology*, *2*(2), 125.

Vigneron, A., Alsop, E. B., Chambers, B., Lomans, B. P., Head, I. M., & Tsesmetzis, N. (2016). Complementary microorganisms in highly corrosive biofilms from an offshore oil production facility. *Applied and Environmental Microbiology*, *82*(8), 2545–2554.

von Ammon, U., Wood, S. A., Laroche, O., Zaiko, A., Tait, L., Lavery, S., et al. (2018). The impact of artificial surfaces on marine bacterial and eukaryotic biofouling assemblages: A high-throughput sequencing analysis. *Marine Environmental Research*, *133*, 57–66.

Wagner-Döbler, I., & Biebl, H. (2006). Environmental biology of the marine Roseobacter lineage. *Annual Review of Microbiology*, *60*, 255–280.

Wang, L., Liu, L., Zheng, B., Zhu, Y., & Wang, X. (2013). Analysis of the bacterial community in the two typical intertidal sediments of Bohai Bay, China by pyrosequencing. *Marine Pollution Bulletin*, *72*(1), 181–187.

Webster, N. S., & Negri, A. P. (2006). Site-specific variation in Antarctic marine biofilms established on artificial surfaces. *Environmental Microbiology*, *8*(7), 1177–1190.

Williamson, C. H., Jain, L. A., Mishra, B., Olson, D. L., & Spear, J. R. (2015). Microbially influenced corrosion communities associated with fuel-grade ethanol environments. *Applied Microbiology and Biotechnology*, *99*(16), 6945–6957.

Zaiko, A., Schimanski, K., Pochon, X., Hopkins, G. A., Goldstien, S., Floerl, O., et al. (2016). Metabarcoding improves detection of eukaryotes from early biofouling communities: Implications for pest monitoring and pathway management. *Biofouling*, *32*(6), 671–684.

Zargiel, K. A., & Swain, G. W. (2014). Static vs dynamic settlement and adhesion of diatoms to ship hull coatings. *Biofouling*, *30*(1), 115–129.

Zhang, K., Choi, H., Dionysiou, D. D., Sorial, G. A., & Oerther, D. B. (2006). Identifying pioneer bacterial species responsible for biofouling membrane bioreactors. *Environmental Microbiology*, *8*(3), 433–440.

Zhang, M., Jiang, S., Tanuwidjaja, D., Voutchkov, N., Hoek, E. M., & Cai, B. (2011). Composition and variability of biofouling organisms in seawater reverse osmosis desalination plants. *Applied and Environmental Microbiology*, *77*(13), 4390–4398.

Zhang, W. P., Wang, Y., Tian, R. M., Bougouffa, S., Yang, B., Cao, H. L., et al. (2014). Species sorting during biofilm assembly by artificial substrates deployed in a cold seep system. *Scientific Reports*, *4*(1), 1–7.

Zuo, R. (2007). Biofilms: Strategies for metal corrosion inhibition employing microorganisms. *Applied Microbiology and Biotechnology*, *76*(6), 1245–1253.

Chapter 2

Cyanobacterial biofilms: Perspectives from origin to applications

Prashanth S. Raghavan[a], Akhilesh A. Potnis[a], and Hema Rajaram[a,b]

[a] *Molecular Biology Division, Bhabha Atomic Research Centre, Mumbai, Maharashtra, India,* [b] *Homi Bhabha National Institute, Training School Complex, Mumbai, Maharashtra, India*

1. Introduction

Cyanobacteria are the most ancient microbes on Earth (Brock, 1973), inhabiting every nook and corner and withstanding several extreme environmental stresses during their existence (Albertano, 2012; Castenholz, 1973). This does not leave a doubt, that the first biofilms, which would have appeared on earth, on soil, water, or other surfaces, would have invariably been those of cyanobacteria along with the other coexisting microbes. Hence, it is imperative to talk about cyanobacterial biofilms in a book devoted to biofilms. In general, biofilms are microbial communities that are encapsulated in a matrix that is produced by the microbes themselves (Schatz et al., 2013). These biofilm structures can be found in a variety of environments (Albertano, 2012; Joh & Lee, 2012; Rana, Kour, Yadav, Yadav, & Saxena, 2020) and serve various purposes including providing fodder for molluscans due to the presence of high nutritional value (Nagarkar, Williams, Subramanian, & Saha, 2004). Cyanobacterial biofilms are one of the most widely distributed microbial systems ranging from aquatic ecosystems to terrestrial ecosystems. This chapter covers nature and role of cyanobacterial biofilms pertaining to ecology, composition, formation, genetics, molecular biology, and applications.

2. Cyanobacterial biofilms

Cyanobacteria have been known to inhabit earth for over ~2.5 billion years (Schopf, 1975). They are among the earliest inhabitants of earth with very simple requirements of water, CO_2, inorganic compounds, and light for growth (Brock, 1973; Schopf, 1975). On account of being photosynthetic and nitrogen fixers, they are the primary source of food in several ecosystems and also responsible for the oxygenation of earth (Brock, 1973; Schopf, 1975). In spite of being photosynthetic in nature, some cyanobacteria are known to survive long periods of complete darkness (Fay, 1965), thus enabling them to survive extreme conditions as well. In several habitats or bare surfaces with little or no nutritional value, cyanobacteria are often the first to colonize and perform important functions of ecosystems and recycle nutrient elements (Whitton, 1992). Such nutritionally bare or impoverished substrates include volcanic ash, rocks, and sands of desert (Dor & Danin, 1996; Jaag, 1945). They also exist in symbiotic associations across the Plant Kingdom (Rai, 1990).

Cyanobacteria also carry out oxygen-labile nitrogen fixation, in addition to oxygenic photosynthesis (Stewart, 1980). In order to overcome the inhibition of nitrogen fixation by oxygen, cyanobacteria have evolved intelligent mechanisms by separation of the two processes by the means of either spatial or temporal separation (Robson & Postgate, 1980). In filamentous cyanobacteria such as in *Anabaena* and *Nostoc*, the two processes are physically separated by compartmentalizing nitrogen fixation in specialized cells called heterocysts while oxygen-evolving photosynthesis occurs in the vegetative cells (Haselkorn, 1978). As this physical compartmentalization is not possible in unicellular cyanobacteria, temporal separation is achieved by carrying out photosynthesis during day time and nitrogen fixation during the night in unicellular cyanobacteria (Mitsui et al., 1986).

Biofilm architecture provides a more stable and enhanced endurance against harsh environments in comparison to planktonic cell formation (Baqué, Scalzi, Rabbow, Rettberg, & Billi, 2013) and can comprise monospecies or multispecies microorganisms (Barranguet, Charantoni, Plans, & Admiraal, 2000; Rana et al., 2020; Thapa, Bharti, & Prasanna, 2017). Monospecies biofilm of cyanobacterium *Nostoc muscorum* comprising filamentous cells shows tightly knit structure as visualized by light microscopy (Fig. 1A) and atomic force microscopy (Fig. 1B).

In nature, cyanobacteria are found as part of multispecies biofilms comprising other bacterial populations.

Understanding Microbial Biofilms. https://doi.org/10.1016/B978-0-323-99977-9.00041-7

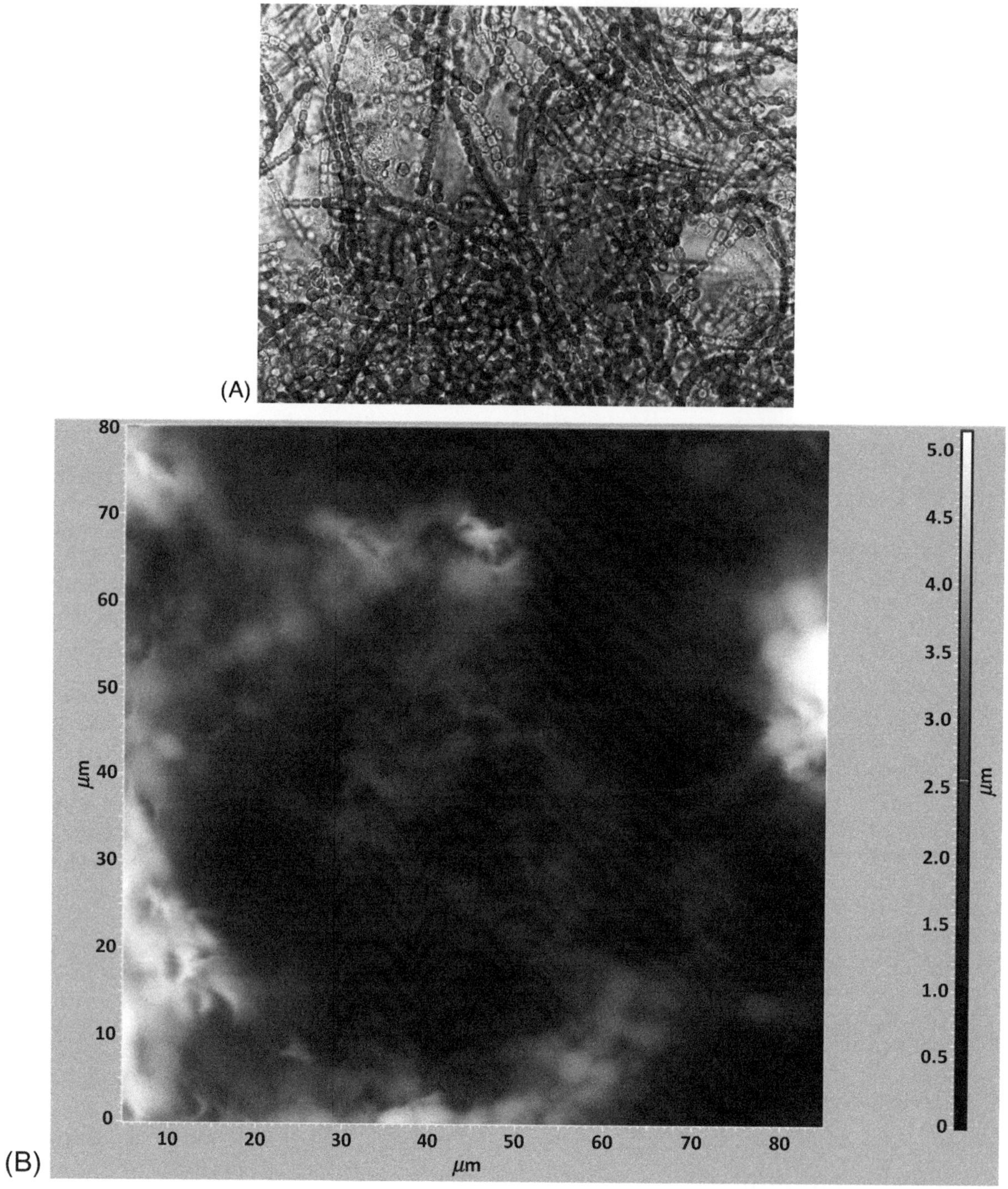

FIG. 1 (A) Light microscope image of *Nostoc muscorum* biofilm (400×). (B) Atomic force microscope image of *N. muscorum* biofilm.

In case of multispecies biofilms comprising both phototrophs and heterotrophs, the upper layer is quite often composed of phototrophs, which include oxygenic phototrophs (such as cyanobacteria) as well as/or anoxygenic such as green and purple sulfur bacteria (Rana et al., 2020; Thapa et al., 2017). Microbial mats are predominated by some groups of organisms such as cyanobacteria, purple sulfur bacteria, etc. (Rana et al., 2020). Cyanobacteria and algae perform the role of primary producers and are thus the force driving functioning and existence in most microbial mats (Van Gemerden, 1993). Nitrogen-fixing cyanobacteria capable of oxygen-evolving photosynthesis are the ones capable of colonizing sand flats that are nutrient-poor and hostile to support other organisms (Van Gemerden, 1993). Their gliding motility helps them move vertically and overcome sand deposition (Van Gemerden, 1993). The nutrient requirements for growth and survival of other organisms in a microbial mat are directly dependent on the excretion, decomposition products of these cyanobacteria (Van Gemerden, 1993). Those comprising cyanobacterial species have been observed to be prevalent in a variety of ecological niches including roots of plants to monuments (Ahmed, Stal, & Hasnain, 2014; Albertano, 2012; Garcia-Meza, Barranguet, & Admiraal, 2005; Gorbushina & Broughton, 2009; Samad & Adhikary, 2008; Schatz et al., 2013), etc.

Cyanobacteria are capable of transforming from singular cell to cellular aggregates/biofilms upon stimulation by environmental factors (Li, Fang, Ye, Zhou, & Li, 2021) and achieve the complex process of biofilm formation through a series of steps (Rana et al., 2020). Studies on succession pattern in biofilm formation in a thermal vent showed the presence of initial colonization of unicellular cyanobacterial species such as *Synechococcus* and gradually leading to mixture of unicellular as well as filamentous species of cyanobacteria eventually forming a complex microbial biofilm (Boomer, Noll, Geesey, & Dutton, 2009).

2.1 Cyanobacterial biofilm distribution and natural diversity

Cyanobacterial biofilms have been known to inhabit wide-ranging ecological habitats (Albertano, 2012). These consist of rocky shores, caves, walls of ancient monuments, deserts, marine environments, also epidermis of dead dolphin (Brown et al., 2021; Keshari & Adhikary, 2013; Mehta & Shah, 2021; Nagarkar et al., 2004; Ramirez, Hernandez-Marine, Novelo, & Roldán, 2010). Microphagous grazers such as intertidal mollusks utilize these cyanobacterial biofilms as an important food source due to their high nutritional content (Nagarkar et al., 2004). Phototrophic biofilms present in cold sulfur spring with biofilms comprising cyanobacterial species exhibit capability of sulfide-based anoxygenic photosynthesis in addition to resistance to sulfide effect on oxygenic photosynthesis (Camacho, Rochera, Silvestre, Vicente, & Hahn, 2005). Phospholipid fatty acid analysis on biofilms from ancient limestone Mayan monuments in Uxmal (Yucatan, Mexico) alluded that cyanobacteria may be the main dominant species in terms of biomass in most of the biofilms characterized from that region (Ortega-Morales, Narvaez-Zapata, Schmalenberger, Sosa-López, & Tebbe, 2004). Similar study on Mayan monuments of Rio Bec style revealed major component of the biofilms in patina formation on these monuments comprising green algae *Trentepholia* and coccoid cyanobacteria *Gloeocapsa* (Ortega-Morales et al., 2013). These biofilms influence the environment that they inhabit by causing bioerosion involving bio-decay and transformation of the rock substratum on which they inhabit (Albertano, 2012; Gorbushina, 2007; Urzi, 2004). Biological activity has been cited to be responsible for deterioration of 20%–30% of stone (Macedo, Miller, Dionísio, & Saiz-Jimenez, 2009; Wakefield & Jones, 1998).

Cyanobacterial biofilms are capable of influencing the rock substratum that they grow on through biogeochemical or biogeophysical mode of biodeterioration (Macedo et al., 2009). Biogeochemical degradation is dependent on the mineral chemistry, while biophysical deterioration is dependent on the shape and porosity of the stone (Macedo et al., 2009; Warscheid & Braams, 2000). Repeated cycles of moistening and drying of cyanobacteria exert large force resulting in rock grain loosening (Macedo et al., 2009; Sáiz-Jiménez, 1999). Studies on biofilms formed on walls of churches in Porto Alegre, RS, Brazil, showed predominance of cyanobacteria of the order *Pleurocapsales* and *Chroococcales* (Crispim, Gaylarde, & Gaylarde, 2004). Cyanobacteria belonging to the *Pleurocapsa* genera have been shown to possess biogeochemical capability of boring stone by secreting a liquid that can digest carbonate from the apical cells (Crispim et al., 2004; Mao-Che et al., 1996). Unicellular cyanobacteria *Synechocystis* and *Gloeocapsa* are also capable of stone-boring activity in addition to filamentous cyanobacteria belonging to *Schizothrix*, *Stigonema*, *Scytonema*, and *Mastiglocladus* found at the site (Crispim et al., 2004). The increase in pH from 8 to 10.5 during photosynthesis tends to increase solubility of SiO_2 in rock formations causing loosening/exfoliation of rock formations (Büdel, 1999). Thus, cyanobacterial biofilm formation influences the ecology and physical characteristics of environment in which they grow. The reason for this ability to grow in hard environments is because cyanobacterial biofilms are highly adapted to extreme conditions, and concomitant growth of cyanobacterial population of *Scytonema* in biofilm community was also observed to be instrumental as a protective layer for the microbial community against high UV irradiation (Crispim, Gaylarde, & Gaylarde, 2003). Scytonemin has been observed to be one of the most important UV protection substances that is present in cyanobacterial sheaths (Büdel, 1999; Proteau, Gerwick, Garcia-Pichel, & Castenholz, 1993). *Scytonema millei* capable of tolerating surface temperature of almost 60°C as well as desiccation was observed to be the dominant cyanobacterial species in the biofilms that colonized a heritage monument in Shantiniketan, India (Keshari, Das, & Adhikary, 2021). Induction of carbohydrate and SOD (superoxide dismutase) activity upon exposure to stress was observed in this strain, and this seemed to be one of the important strategies for stress tolerance (Keshari et al., 2021).

Over 57 taxa of cyanobacteria were isolated from surfaces of monuments, and buildings from various regions of India with four cyanobacterial species *Chroococcidiopsis kashayi*, *Scytonema geitleri*, *Pseudophormidium indicum*, and *Plectonema puteale* were observed throughout the year (Samad & Adhikary, 2008). Molecular-biology-based studies have thrown more light on the population structure of microbial biofilms in nature and the nature of cyanobacterial components in these biofilms. Molecular phylogenetic analysis utilizing 16S rRNA gene sequencing of 24 cyanobacterial species isolated from monuments and temples in India along with those from other parts of the world showed that species from tropical region formed separate cluster from those isolated from temperate region (Keshari & Adhikary, 2014). Filamentous cyanobacteria such as *Scytonema*, *Microcoleus*, and *Leptolyngbya*, as well as *Lyngbya*, have been found in biofilms that predominate in high-moisture areas in the sandstone monuments of Angkor wat complex while major colonizer of the area was *Chroococcidiopsis* (Gaylarde, Rodríguez, Navarro-Noya, & Ortega-Morales, 2012). More recently, a study on microbial biofilms from sandstone in two sites from China using 16S and 18S rRNA sequence analysis showed the microbial population with large presence of sequences attributed to cyanobacteria (Duan et al., 2021).

2.2 Phototroph-heterotroph interaction in biofilm

Microbial mats vary in different niches, as it is dependent on the interactions between the microbes, and hence, understanding these interactions is very important. While cyanobacteria are a primary producer, being an oxygenic photosynthetic organism, the consortia it creates around it would be dependent on the autotroph-heterotroph and heterotroph-heterotroph interactions (Cole et al., 2014). Unicyanobacterial consortium (UCC), a symbiotic system where a cyanobacterium serves the role of a primary autotroph providing the heterotrophs with carbon and oxygen that are fixed photosynthetically, is a perfect model system for understanding these interactions (Cole et al., 2014). The heterotrophs in turn provide key metabolites for the promotion of cyanobacterial growth (Cole et al., 2014; Paerl & Pinckney, 1996; Paerl, Pinckney, & Steppe, 2000). The two cultured UCCs from Hot Lake, Washington, were analyzed for metabolic function and community structure of the assembling phototrophic biofilm over a period of 28 days (Cole et al., 2014). Between the two consortia designated as UCC-a and UCC-o, the latter produced higher nonproteinaceous EPS matrix significantly more than UCC-a (Cole et al., 2014). The Chl *a* and total protein contents of both the consortia increased concurrently till day 17 followed by plateauing of Chl *a*, the protein content continued increasing indicative of increase in heterotrophs over autotrophs or a decrease in growth of autotrophs (Cole et al., 2014).

Further studies revealed differences in the localization of heterotrophs in the two biofilm consortia. In UCC-a, the cyanobacteria were colocalized vertically with heterotrophs at all the time points, while in UCC-o, cyanobacteria were dominant on the top surface followed by heterotrophs in the lower part (Cole et al., 2014). The presence of cyanobacteria in the upper layers could be due to the phototaxis behavior of *Phormidium* sp. OSCR (in UCC-o), which is absent in *Phormidesmis priestleyi* str ANA (in UCC-a) (Cole et al., 2014). Such studies need to be carried out extensively in varied niches, which could throw light on the evolution of biofilm formation as well as close associations of specific microbes. Biofilm architecture and the extent of cell collisions in the biofilm microbial population have been shown to influence the extent of transfer of genetic material (plasmid-based) through conjugation (Wuertz, Okabe, & Hausner, 2004), and this could contribute to the horizontal gene transfer across communities during the evolution under harsh conditions.

3. Factors influencing cyanobacterial biofilm formation

Biofilm formation being a natural process is expected to be dependent on host of factors. This includes physical, chemical forces, surface interaction, genetics, biochemical parameters, environmental factors, to name the major ones (Agostoni, Koestler, Waters, Williams, & Montgomery, 2013; Albertano, 2012; Allen, Habimana, & Casey, 2018; Faria et al., 2020; Li et al., 2021; Parnasa et al., 2016; Zhang et al., 2020). The following subsections give a detailed overview on the role of these factors.

3.1 Physical factors

Physical factors are major influencers of cyanobacterial biofilm formation and have real-time consequences on growth, application, and utilization of these biofilms. A hallmark of formation of cyanobacterial biofilm is that it can thrive on any kind of environment/surface with adequate presence of moisture and light (Joh & Lee, 2012). These factors along with high temperatures account for and promote the colonization and biofilm formation on ancient monuments (Crispim et al., 2003). A sustained biofilm formation is additionally dependent on the cues provided by the cyanobacterial cells, which participate in the initial colonization. These cues, which include surface hydrophobicity of cell as well as the substratum (Faria et al., 2020; Fattom & Shilo, 1984), along with other physical factors such as surface properties of the substrate, nutrient composition (Barros, Gonçalves, & Simões, 2019), nutrient availability such as that of inorganic nitrogen (Stauch-White, Srinivasan, Kuo-Dahab, Park, & Butler, 2017), and the incident light color/wavelength (Conradi et al., 2019), influence the nature and stability of the biofilm. Biofilm formation can also be induced under laboratory conditions, wherein several physical factors such as application of hydrodynamic conditions and shape of flask can influence exopolysaccharide generation and biofilm formation (David, Bühler, & Schmid, 2015; Veerabadhran, Chakraborty, Mitra, Karmakar, & Mukherjee, 2018) and also possibly influenced through repeated passaging/culturing as visualized in *Synechococcus elongatus* PCC7942 (Yang et al., 2018). The importance of light and physical forces in biofilm formation is explained below:

3.1.1 Irradiance

Light is particularly important for cyanobacterial biofilm formation as its growth was found to be arrested at irradiances below 60 μmol photons m^{-2} s^{-1} (Di Pippo et al., 2012). Even the wavelength of the irradiance possesses a role in the extent of support to formation of biofilm, as in the case of growth of cyanobacteria *Acaryochloris marina* biofilms being more pronounced when grown under near-infrared wavelength in comparison to visible light irradiance (Behrendt et al., 2012), and also the far-red light supported superior and improved biofilm formation compared with white light (Hernández-Prieto, Li, Postier, Blankenship, & Chen, 2018). Within the visible light spectra, it was shown that the combination of blue, green, and red wavelengths prevented growth of phototrophic biofilm (Bruno & Valle, 2017). The influence of light on biofilm formation in cyanobacteria is due to its photosynthetic nature. It has been shown that phototrophic biofilms get acclimated to variations in irradiances by bringing variations to photosynthetic rate, efficiency, and pigment content (Guzzon, Congestri, & Albertano, 2005). Adsorption of cyanobacterial biofilm on glass has been shown to be improved by the presence of high blue to green light ratio (Agostoni, Waters, & Montgomery, 2016; Conradi et al., 2019; Enomoto, Narikawa, & Ikeuchi, 2015). Color of the surface also plays a key aspect in formation of biofilm with white and red color promoting maximal biofilm-forming filaments, while blue and black surfaces promoted oxidative stress and EPS formation (Gambino et al., 2019). These contribute significantly to bioreceptivity of the surfaces on which biofilm formation occurs (Gambino et al., 2019). It has also been hypothesized based on experimental observations that filamentous cyanobacteria could surpass unicellular cyanobacteria in terms of competing for light in phototrophic biofilms due to their effective migration toward optimal light (Roeselers, van Loosdrecht, & Muyzer, 2007) and could thus outnumber them in the consortium of microbial mats.

While the irradiance is important for regulating formation of biofilm, the growth temperature also plays an equally important role. Using cultured phototrophic biofilm consisting of cyanobacteria *Synechocystis* and green alga *Chlorococcum* sp. grown with combination of two temperatures [20°C and 30°C] and irradiance [60 (LC60) or 120 (LC120) μmol photons m^{-2} s^{-1}]. At 20°C, the biofilm formation exhibited a lag phase of 13 and 10 days and an active growth phase on 25th and 20th day, respectively at low and high irradiance (Di Pippo et al., 2012). While at 30°C, the lag phase with LC60 was 18 days, and at LC120 it was 15 days. Among all these conditions, well-defined biofilms and mature communities were observed at LC120 and 30°C, the biofilm detachment was higher at 20°C irrespective of the irradiance (Di Pippo et al., 2012). Additionally, the difference was also observed in the biomass and the abundance of the species, with Chl *a* being higher in LC60 cultures than LC120 at both the temperatures, higher effective quantum yield for LC120 over LC60 at both the growth temperatures, and higher phototrophic biomass and biovolume at a temperature of 30°C compared with 20°C. The increase in biomass accumulation linked with enhanced EPS production at higher temperatures and light intensities resulted in stable mature biofilm (Di Pippo et al., 2012). Among the two species tested, the *Chlorococcum* sp. dominated the biofilm growing at 20°C, while at 30°C, almost 99% of the biofilm comprised *Synechocystis* sp. (Di Pippo et al., 2012).

3.1.2 Shear forces and hydrophobicity

Physical forces such as shear forces are known to influence not just on the structure of the biofilm but also on its composition, exopolysaccharide content, metabolism as well as genetics and changes at the molecular level (Moreira, Simões, Melo, & Mergulhão, 2015; Romeu et al., 2019). However, the shear rate had negligible influence on the initial adhesion of the cells to the substratum (Romeu et al., 2019). This was reflected by the similar nature of the development of the biofilm across exposure to varied shear forces in the initial 2 weeks of biofilm growth, and shear force influence was more prominent in the biofilm maturation phase (Romeu et al., 2019). Influence of shear forces was higher than that of surface hydrophobicity as no significant variation in wet weight or chlorophyll *a*, used as a measure of cyanobacterial biofilm formation on varied surfaces, was observed in the biofilm under influence of similar shear force (Romeu et al., 2019).

Cell-to-cell collision has been shown to give rise to reticulate structure formation that determines effective biofilm formation in cyanobacteria (Shepard & Sumner, 2010), and cell-to-cell interaction forms a more crucial role in biofilm formation than physical forces (Papadopoulos, Economou, Tekerlekopoulou, & Vayenas, 2021). Cell surface hydrophobicity has been shown to be affected by the presence of divalent cations in the medium (Fattom & Shilo, 1984). The hydrophobicity of the biofilms of cyanobacteria *Synechococcus* was higher than that of the planktonic cells (Jiang & Ohtaguchi, 2006).

Pseudochlorococcum was observed to have faster growth when grown on hydrophilic surface (Ji et al., 2014; Wang, Liu, & Liu, 2017). Subsequently, it was noticed that more hydrophilic surfaces such as glass hosted better cyanobacterial biofilm growth than those that were not too hydrophilic, and cyanobacterial cells, which were hydrophilic in nature, formed less effective biofilms on glass (Faria et al., 2021). However, there are also contradictory reports that show no influence of substratum surface properties and hydrophobicity for cell adhesion (Irving & Allen, 2011; Katsikogianni & Missirlis, 2004; Talluri, Winter, & Salem, 2020). While the hydrophobicity of the surface on which the biofilm grows is a key factor in determining biofilm growth, hydrophilicity/hydrophobicity of the cyanobacterial cells is equally important in determining effective biofilm formation. It is proposed that surface chemical functional groups in conditional biofilms play a more important role than surface hydrophobicity and roughness (Talluri et al., 2020). Studies on the surface properties of six species of algae demonstrated the presence of hydrophobic groups capable of electron donation (Wang et al., 2017; Xia, Li, & Song, 2016). The combination of roles of surface hydrophobicity, hydrodynamic conditions, and shear forces could influence differently on various cyanobacterial species with respect to biofilm formation (Faria et al., 2020, 2021). Surprisingly, the biofilm architecture was not greatly affected by geochemical or physical characteristic of the substratum, as shown by near similar architecture of biofilms of the cyanobacteria *Oculatella subterranea*, isolated from archeological park sites irrespective of the nature of substratum on which it formed (Del Mondo, Pinto, Carbone, Pollio, & De Natale, 2018). Romeu et al. (2020) observed that more differences could be noticed in proteomic profile of biofilms when grown under varied hydrodynamic conditions than when grown under different surfaces.

3.2 Molecular mechanisms and genetic factors

Influence of physical factors on biofilm is manifested and modulated via various molecular mechanisms, which include signaling molecules such as cyclic di-guanosine monophosphate (c-di-GMP) (Chrismas, Barker, Anesio, & Sánchez-Baracaldo, 2016), and has been linked to the ability of certain species of *Nostocales* to form biofilms (Agostoni et al., 2013). The photoreceptor protein Cph2 has been shown to promote flocculation of cyanobacterial cells by modulating the cAMP and c-di-GMP levels in the cells (Conradi et al., 2019; Wallner, Pedroza, Voigt, Kaever, & Wilde, 2020). The Cph2 system in *Synechocystis* possesses a phytochrome system capable of red/far-red interconversion at the N-terminal, EAL domain capable of degrading c-di-GMP, and green-blue absorbing cyanobacteriochrome at the C-terminal that in turn is linked to a GGDEF domain (Savakis et al., 2012; Schuergers, Mullineaux, & Wilde, 2017; Wu & Lagarias, 2000). In the presence of excess blue light, increased production of c-di-GMP is activated by GGDEF domain, and this can cause inhibition of motility (Savakis et al., 2012; Schuergers et al., 2017). The change in c-di-GMP levels as a function of light quality varied with different cyanobacterial species (Agostoni et al., 2013). While the blue light caused maximal synthesis of the c-di-GMP in *Synechocystis* PCC6803, it was white light and red light that induced maximal synthesis of c-di-GMP in *Fremyella diplosiphon* (Agostoni et al., 2013). Synthesis of this signaling molecule also regulated by Wsp pathway, and almost complete gene cassette encoding the Wsp chemosensory pathway has been observed in the cyanobacteria *Phormidesmis priestleyi* (Chrismas et al., 2016). Other modes of regulation of c-di-GMP levels identified were a phosphodiesterase activity induced by blue light in *S. elongatus* (Cao, Livoti, Losi, & Gärtner, 2010), diguanylate cyclase in *Synechocystis*, which increased c-di-GMP levels inside the cell and promoted formation of biofilm (Agostoni et al., 2016) and in response to salt stress in *Nostoc* PCC7120 (Agostoni et al., 2018). Expression of GTP-dependent diguanylate cyclase activity was observed in at least six random proteins containing GGDEF domains across a diverse species of bacteria, namely cyanobacteria, *Deinococcus-Thermus*, Thermotogae, Spirochaetes, and α- and γ-Proteobacteria (Römling, Gomelsky, & Galperin, 2005). c-di-GMP has been observed to interact with proteins such as cellulose synthase possibly as a monomer as well as via its PilZ domain that are involved in cellular aggregation as observed in *Thermosynechococcus elongatus* (Enomoto, Okuda, & Ikeuchi, 2018). In the cyanobacterium *Microcystis aeruginosa*, comparative genomic analysis of the genomes indicated significant presence of genes encoding PilZ domain containing proteins across the genome that could interact with c-di-GMP and promote exopolysaccharide synthesis and biofilm formation (Chen et al., 2020).

Small proteins and secretory molecules have been demonstrated to play a vital role in biofilm formation (Nagar & Schwarz, 2015). One of the unicellular cyanobacteria, *Synechococcus elongatus* does not form biofilms naturally, which was attributed to the phenomenon of self-repression of biofilm formation (Nagar & Schwarz, 2015). Impairment of protein secretion through the inactivation of Synpcc7942_2071, which is homologous to a protein E belonging to type II secretion/type IV pilus assembly system, allowed biofilm formation in this organism suggesting the role of secretory molecules in biofilm formation (Schatz et al., 2013), through altering the exo-proteome milieu and disrupting pilli formation (Nagar et al., 2017). Four small proteins, designated as EbFG1–4 (enable biofilm formation with GG motif), were identified in the *Synechococcus* mutant, which was able to form biofilms (Parnasa et al., 2016). The transport of these molecules was achieved through the protein named PteB (peptidase transport essential for biofilm) (Parnasa et al., 2016). The addition of extracellular fluids from the wild-type cells inhibited the biofilm formation of the mutant (Parnasa et al., 2016). In *Synechocystis* PCC 6803, it was shown that mutants that did not have cell surface structures such as S-layer and type IV pili are not capable of forming biofilms (Allen, Rittmann, & Curtiss III, 2019).

Exopolysaccharide synthesis has been closely associated with and facilitates biofilm formation in various cyanobacteria (Gupta, Snehi, & Singh, 2017; Rana et al., 2020). An important group of genes found to be important for biofilm formation were the glycosyl transferases. Increased expression of genes encoding glycosyl transferases was observed under far red-light illumination in the cyanobacteria *Acaryochloris marina*, a condition that stimulates its change from a planktonic state observed under white light influence to biofilm like immobile state in the presence of far red light (Hernández-Prieto et al., 2018). On the other hand, in *Synechocystis*, inactivation of Ogt, a homolog of glycosyl transferase, was shown to cause robust biofilm formation, and this protein was found to be involved in the glycosylation of pili (Suban, Sendersky, Golden, & Schwarz, 2022). Mutational analysis of three putative glycosyl transferases in *Synechocystis* 6803 promoted biofilm formation, and the cells possessed lower hydrophobicity and lower surface negative charge (Li et al., 2021). Bioinformatic analysis of the genome of *Nostoc* PCC7120 threw across several genes with probable involvement in EPS biosynthesis, including over 81 genes coding for glycosyl transferase (Potnis, Raghavan, & Rajaram, 2021). Whole genome sequencing of a biofilm forming filamentous cyanobacteria *Jacksonvillea* sp. ISTCYN1 revealed the presence of several biofilm forming and EPS biosynthesis genes (Mishra, Gupta, Kumari, Pal, & Thakur, 2021). Cellulose synthases have also been observed to play a vital role in cellular aggregation in the cyanobacteria *Thermosynechococcus vulcanus* (Kawano et al., 2011). Inactivation of three genes showing similarity to cellulose synthase revealed that one of the ORFs *Tvtll0007* was important for low temperature and light-induced cellular aggregation and cellulose accumulation that are in turn important for biofilm formation (Kawano et al., 2011).

Lipopeptides are important molecules characterized in bacteria that play a very important part in biofilm formation and gene clusters encoding lipopeptide synthetic pathways have been observed in biofilm forming cyanobacterial species as well (Galica, Hrouzek, & Mareš, 2017). Disruption of genes encoding ABC transporter permease, putative C-3 methyl transferase, and ABC transporter ATP-binding subunit resulted in increased adherence of cells to glass surface resulting in a simple cellular monolayer unlike complex biofilm structure suggesting the possible role of such genes in biofilm formation (Fisher, Allen, Luo, & Curtiss III, 2013). Knockout of two arginine decarboxylases involved in the synthesis of spermidine resulted in biofilm formation in the cyanobacteria *Synechocystis* even under nonstress conditions (Kera et al., 2018). Among the regulatory proteins, histidine kinases (Hik), which play a critical role in signal processing in cyanobacteria, have also been implicated in biofilm formation, with the knockout strains of Hik36 and Hik43 in *Synechocystis* PCC 6803 unable to form biofilms under salt stress (Kera et al., 2020). However, lot more studies need to be carried out to identify different proteins involved in regulating the formation of biofilms across the cyanobacterial species. These studies show how the presence of close interaction between physical and molecular factors that can influence biofilm formation.

4. Impact of abiotic factors such as temperature, nutrients, and environmental conditions on cyanobacterial mats/biofilms

4.1 Effect of warming on light- and dark-grown microbial biofilms

The ambient temperature for growth of cyanobacteria in natural habitats is also extraordinarily diverse. With global warming increasing by the year, it is not only expected to affect the flora and fauna but also the microbial community, which could be free-living, symbiotic, or as microbial mats/biofilms (Romaní et al., 2014). Cyanobacterial species has been documented in hot springs, namely *Synechococcus* at temperatures greater than 90°C and *Oscillatoria terebriformis* growing at approx. 54°C (Castenholz, 1968, 1973), mountain streams (Kann, 1988), polar lakes of the Arctic, and *Nostoc* sp., *Synechococcus* sp., and *Phormidium frigidum* in Antarctic (Wynn-Williams, 1990) as well as snow and ice (Kol, 1968; Laamanen, 1996). Increase in primary production, community respiration, and rate of denitrification are some of the well-known consequences of temperature on biofilm and metabolism in biofilms (Boulêtreau, Salvo, Lyautey, Mastrorillo, & Garabetian, 2012; Perkins et al., 2012; Romaní et al., 2014). The microbial mats/biofilms have existed for ages, but the changing climate is expected to have an effect on the community composition and stability of the mats (Lear, Anderson, Smith, Boxen, & Lewis, 2008; Romaní et al., 2014). Temperature along with light can cause variation in interactions within different members of the biofilm due to the controls exerted by these two factors on the growth of photoautotrophs such as cyanobacteria and algae, which are the primary producers in the biofilm (Romaní et al., 2014). Influence of light and temperature on structure of microbial community in biofilms has been studied in detail by Romaní et al. (2014). Increase in autotrophic component of microbial mats increases the heterotrophic dependency on organic carbon generated by algae and cyanobacteria (Neely & Wetzel, 1995; Romaní et al., 2014; Ziegler, Lyon, & Townsend, 2009). Change in temperature and light-dark cycle also influences the diversity of the biofilm as studied in biomass accrual (algal, bacterial, ciliate, rotifier), 12 h light-dark cycle (in comparison to 24 h dark), and a 2°C increase in temperature had a positive impact on the chlorophyll *a* accrual of the biofilm (Romaní et al., 2014). Prokaryotes increased with light and were adversely affected by increase in temperature clearly showing that negative impact of temperature was more pronounced in the absence of light (Romaní et al., 2014). Activity of various extracellular enzymes involved in the utilization of carbon, nitrogen, and phosphorous organic compounds was not affected by an elevation in temperature of 2°C; however, β-glucosidase activity was lower at the end of the biofilm progression, and activity of leucine-aminopeptidase was stimulated by higher temperature in the presence of light and retarded in dark (Romaní et al., 2014). A significant decrease in the photosynthesis activity of the mature biofilms was observed due to warming (Romaní et al., 2014). Due to the increased temperature, the contribution of cyanobacteria and diatoms in the photosynthesis activity was significantly higher than that of green algae (Romaní et al., 2014).

Variations in dissolved organic matter in terms of quality as well as quantity are prevalent across different water bodies and thereby influence the microbial population and biomass (Bott, Kaplan, & Kuserk, 1984; Volk, Volk, & Kaplan, 1997; Warren, Wales, Davis, & Doudoroff, 1964; Ziegler et al., 2009). Under sufficient light conditions, autotrophic bacteria thrive and tend to produce thick biofilms, and algal biomass has been reported to reach almost 60%–90% of the total biomass of autotrophic biofilms (Romaní, 2010; Romaní et al., 2014). This generates photoautotrophically synthesized organic carbon, thereby facilitating a bacterial population structure of more diversity dependent on this nutrient availability involving species such as *Alphaproteobacteria*, *Betaproteobacteria* and *Gammaproteobacteria*, *Gemmatimonadetes*, and cyanobacteria (Romaní et al., 2014; Watanabe et al., 2008; Ylla, Borrego, Romaní, & Sabater, 2009). However, in the absence of light, thinner biofilms are observed largely dominated by *Actinobacteria* and *Betaproteobacteria* and substantial amounts of photosensitive ammonia-oxidizing archaea reflecting a lower diversity (Church, Wai, Karl, & DeLong, 2010; Guenet, Danger, Abbadie, & Lacroix, 2010; Merbt et al., 2012; Romaní et al., 2004, 2014; Ylla et al., 2009).

4.2 Effect of phosphate and sulfide availability on cultured biofilms

Studies have shown the degree of complex physiology and metabolism present in cyanobacterial microbial mats with respect to influence of elements such as phosphates and sulfides in their external environment. Effect of the availability of phosphate on the structure and physiology of community of cultured biofilms was studied using the phototrophic biofilms, comprising primarily of *Phormidium*, *Oscillatoria*, *Lyngbya*, *Chroococcus*, species and *Scenedesmus*, *Westella*, *Carteria*, and also species from the genera *Oocystis*, *Chlorella*, *Stigeoclonium*, isolated from Nanhu Lake, Wuhan, China (Li et al., 2016). While low levels of phosphate inhibited the growth of algae in terms of photosynthetic ability and biomass, in cultured biofilms, it also resulted in a rapid and continuous increase in alkaline phosphatase levels within 4 days, unlike a steady and low level of alkaline phosphatase activity observed in biofilms grown in phosphate-rich media. The alkaline phosphatase activity was found to be much higher in the extracellular medium under phosphate-limiting conditions compared with those with medium or high levels of phosphate in the growth media (Li et al., 2016). Cyanobacteria showed a more vigorous competitive edge over other algae in terms of dominating the biofilms under phosphate-limiting conditions (Li et al., 2016).

The brownish freshwater filamentous motile cyanobacterium *Oscillatoria boryana* is known to inhabit warm to hot springs in several parts of the world and also tolerate high levels of sulfides (Castenholz, 1976; Geitler, 1932; Gomont, 1982). The behavioral versatility of *O. boryana* has been extensively studied in terms of the influence of sulfide-rich environment on the photosynthetic ability of cyanobacteria, which revealed the complex adaptive mechanism used by these organisms to survive in such extreme environments (Castenholz, Jørgensen, D'Amelio, & Bauld, 1991). *O. boryana*, in spite of being present as dense populations of motile trichomes, forms a microbial mat up to a temperature of about 50°C and above pH 6.0 in a sulfide-rich environment (Castenholz et al., 1991). Light intensity and sulfide concentration were important reasons for the motility of *O. boryana*, which indicates acclimatization to changing microbial mat environment (Castenholz et al., 1991).

This was also observed for *O. terebriformis*, wherein sulfide concentration was found to be responsible for slowing or stopping the downward motility and resulted in formation of microbial mat-like layered structure (Richardson & Castenholz, 1987). Positive phototaxis resulting in movement toward the surface in decreasing light intensity was observed in *O. boryana* (Richardson & Castenholz, 1987) similar to few other cyanobacteria (Castenholz, 1982). As observed in the case of *Microcoleus chthonoplastes* (Cohen, Jørgensen, Revsbech, & Poplawski, 1986),

O. boryana consumes sulfide under illuminated conditions and exclusively employs anoxygenic photosynthesis at sulfide concentrations above 0.05–0.1 mM in the absence of DCMU and in anaerobic conditions in the presence of DCMU (Castenholz et al., 1991). Also incubation under high illumination greater than 200 W m^{-2} acted as an immediate trigger and fast tracked the recovery of oxygenic photosynthesis post sulfide inhibition in *O. boryana*, which was not observed under lower incident light of 50 or 150 W m^{-2} (Castenholz et al., 1991). In comparison to lower concentrations, recovery of photosystem II post exposure to high concentrations of sulfide was much slower in *O. amphigranulata* (Castenholz et al., 1991; Garcia-Pichel & Castenholz, 1990). However, exposure to very high irradiance also negatively impacted oxygenic photosynthesis and the presence of sulfide, however, seemed to mitigate the photooxidative damage (Castenholz et al., 1991). The migration of *O. boryana* to the lower layers in response to very high light intensity toward tolerable lower light intensity seems to be a response seeking a protection from the possible inducement of photooxidative damage by high light exposure in addition to anerobic and sulfide-containing environment (Castenholz et al., 1991). Thus, *O. boryana* appears to be a highly versatile organism by virtue of its ability to move upward or downward in order to occupy a position that is optimum in light intensity for carrying out photosynthesis or prevent photoinhibition, the presence of capability to move from oxygenic photosynthesis to anoxygenic mode of photosynthesis and/or carry out both together (Castenholz et al., 1991). Further studies on similar mechanisms that may be present in other cyanobacterial species residing in mineral/metal-rich environment, unsuitable pH and temperature would help account for the varied microbial mats found under diverse environmental conditions, particularly with those having cyanobacteria or algae as one of the major constituents.

4.3 Role of environmental conditions

Environmental influences on species diversity of biofilms have been observed, and among these, pollution levels have been observed to have a significant influence on the cyanobacterial biofilm species diversity. A decline in the species diversity along a pollution gradient was observed in Guadarrama river in Spain (Loza, Berrendero, Perona, & Mateo, 2013). Species such as *Nostoc punctiforme*, *Tolypothrix* cf. *tenuis*, *Chamaesiphon investiens*, *Nostoc piscinale*, *Chroococcus minor*, *Leptolyngbya tenuis*, and *Leptolyngbya nostocorum* predominated waters containing low level of nutrients while *Cyanobium* sp., *Phormidium autumnale*, *Chamaesiphon polymorphus*, *Leptolyngbya boryana*, *Aphanocapsa* cf., *Aphanocapsa* cf. *rivularis*, and *Phormidium* sp., dominated polluted waters (Loza et al., 2013). A third group of species consisting of *Nostoc carneum*, *Pseudanabaena catenata*, and *Aphanocapsa muscicola* were present in downstream and upstream sites (Loza et al., 2013). The other important factor that seems to influence the species diversity of cyanobacteria is the nature of the substratum on which the colonization happens (Albertano, 2012; Tomaselli, Lamenti, Bosco, & Tiano, 2000). Calcareous substrates showed a predominance of *Scytonema*, *Myxosarcina*, *Chroococcus*, *Pleurocapsa* along with *Apatococcus* that belonged to Phylum Chlorophyta on calcareous substrates while the *Nostoc* sp. was more predominant on artificial substratum (Tomaselli et al., 2000). Climatic conditions further influence species diversity of cyanobacteria in natural biofilms wherein species diversity of biofilms characterized from stone monuments in India, a tropical environment varied distinctly from those characterized from temperate region albeit the substratum for growth of these species, was similar (Keshari & Adhikary, 2013). Presence of varied species depends on factors such as temperature, moisture availability, desiccation, humidity, and other environmental conditions (Samad & Adhikary, 2008).

Impact of light on the distribution of species shows variations of even species such as in sites in Bozana caves in Serbia where low light regions were dominated by coccoid cyanobacterial forms while in well-lit condition, *Nostocales* members were observed (Popović et al., 2015). Similarly cyanobacterial biofilms were largely limited to entrance of caves where there was a presence of natural light in a study conducted in caves in Serbia (Nikolić et al., 2020). Multiple types of phototrophic biofilms with different compositions of cyanobacteria and other bacteria, fungi, lichens have also been observed in caves depending on the environmental factors such as availability of light, nutrients, humidity, etc. (Roldán & Hernández-Mariné, 2009). In addition to these external environmental conditions that influence the diversity of biofilms, the cyanobacterial biofilms such as those of *Phormidium* are very thick and cohesive, and these create an internal environment inside with respect to pH, heavy metal concentrations, nutrients, dissolved oxygen, etc., in the biofilms, which also influence the microbial structure of these biofilms (Echenique-Subiabre et al., 2018; Wood, Miller, Sims, & Takemoto, 2015). Studies on submerged cyanobacterial biofilms in reefs in Zanzibar revealed a large presence of cyanobacterial biofilms ranging from 5% to 56% of surface cover, and cyanobacterial biofilms present in the depth were better adapted to low light photosynthesis in comparison to those in shallow depths (Lugomela, Söderbäck, & Björk, 2005). Thus, it is evident that cyanobacterial biofilms and phototrophic biofilms containing cyanobacteria play a very critical role in determining the ecology as well as the ecological succession of the ecosystem in which they are present.

5. Application and utilization of cyanobacterial biofilms

5.1 Agriculture and biofuel

Cyanobacteria hold a major potential in terms of utilizing their physiological and biochemical process outputs for eco-friendly and potential benefits (Miranda et al., 2017; Rana et al., 2020). Plant phyllosphere region has been documented to possess cyanobacterial biofilm-like growth, and this helps in nitrogen fixation for crops such as rice (Rana et al., 2020; Schwartz, 1975; Thapa et al., 2017) as well as playing a role in biocontrol (Triveni et al., 2015). Leguminous plants such as chick pea showed better nodule formation when inoculated with cyanobacteria *Mesorhizobium ciceri*–based biofilm biofertilizer (Prasanna et al., 2017). Cyanobacteria-based biofilm biofertilizer showed high hydrolytic enzyme activity and fungicidal activity with maximal fungicidal activity displayed by cyanobacterium-bacillus biofilms and cyanobacterial-fungal biofilms (Prasanna et al., 2013). Microalgal cultivation based on biofilm-based cultivation is very advantageous due to the fact it requires very less water and low cost of harvesting (Lan, Wu, Yang, Zhang, & Hu, 2017). Cyanobacteria can be used as effective matrices for utilizing agriculturally important microbes (Prasanna, Pattnaik, Sugitha, Nain, & Saxena, 2011). Mixed biofilm cultures consisting of *Anabaena torulosa* as matrix and other bacteria such as *Pseudomonas striata*, *Serratia marcescens*, *Azotobacter chroococcum*, and *Mesorhizobium ciceri* showed agriculturally important chemical processes such as generation of IAA, phosphate solubilization, and fungicidal activity (Prasanna et al., 2011).

Biofertilizers prepared as biofilms utilizing cyanobacterial matrix showed better persistence and longer nitrogen-fixing capabilities and higher P uptake by plants than that seen in corresponding comparative chemical fertilizer applied (Swarnalakshmi et al., 2013). Improvement in soil nitrogen content was also observed when applied to wheat crop as a biofertilizer (Swarnalakshmi et al., 2013). Algal biofilm-based biofertilizers also gave promising results in terms of increased yield in soybean (Prasanna et al., 2013). Application of algal biofilm-based biofertilizers also showed increased accumulation of Zn in the flag leaf of maize plants in addition to the higher activity of enzymes such as PAL, PPO, and peroxidases in the roots of the plants (Prasanna et al., 2015). Soil with application of algal biofilm-based biofertilizer showed higher concentration of soil elements such as zinc and iron (Manjunath et al., 2016). Application of cyanobacterial biofilm-based biofertilizers using two different techniques as dry formulation or soil drench methods showed considerable improvement in both micro and macro nutrients in the soil, and significant improvements in plant physiological parameters were observed in terms of chlorophyll, carotenoid pigments, plant height, and leaf number, etc. (Prasanna et al., 2018). Application of cyanobacterial-based biofertilizer in hydroponics resulted in biofilm formation on the roots of chrysanthemum plants, which showed improved chlorophyll and root biomass in the presence of biofertilizers (Bharti, Prasanna, Kumar, Kumar, & Nain, 2019).

In addition to acting as biofertilizers, microalgal biofilms have also been studied for their utilities in biofuel/biodiesel generation. Three cyanobacterial isolates from biofilms in field sites in Italy were grown as batch cultures as well as semicontinuous-flow methods capable of biofilm growth. Fatty acid profile of the biofilms was predominantly palmitic acid (about >50%) and lesser amount of other saturated fatty acids and monounsaturated fatty acids (Bruno et al., 2012). The composition of fatty acid seen in these isolates was suited for generation of good-quality biodiesel production (Bruno et al., 2012). Lipid production in algal biofilms has been observed to be enhanced by nutrient starvation (Schnurr & Allen, 2015). Gismondi, Pippo, Bruno, Antonaroli, and Congestri (2016) reported that in addition to removal of phosphate, the cyanobacterial biofilm generating strains could generate lipids that were suited for production of biofuel. Thus, it is evident from abovementioned studies that cyanobacterial biofilms serve primary roles in agriculture not only in terms of improving soil nutrition and serving a beneficial role for crops but also for generation of useful products.

5.2 Wastewater treatment and bioremediation

Algal biofilms are efficient systems to be recruited for treatment of wastewater due to their energy-efficient ability to absorb important elements such as nitrogen and phosphorous from wastewater (Miranda et al., 2017). Phototrophic biofilms have also been shown to be effective in removing petroleum hydrocarbons (Chavan & Mukherji, 2008). Biofilms-based wastewater treatment plants are more economical than open pond photobioreactors in that they can be easily scraped away, reducing the costs of harvesting of cyanobacterial biomass and dewatering (Papadopoulos et al., 2021). Phototrophic biofilms comprising cyanobacterial species such as *Phormidium*, *Pseudanabaena*, and *Synechocystis* showed high uptake of phosphate from wastewater (Guzzon, Bohn, Diociaiuti, & Albertano, 2008). Phototrophic biofilms showed accumulation of phosphate in between thylakoids and widespread across cytoplasm of the *Phormidium* sp. (Guzzon et al., 2008).

Microalgal biofilms cultivated in photobioreactors have been effectively shown to remove phosphate from wastewater (Sukačová, Trtílek, & Rataj, 2015). The microalgal biofilm could remove close to 97% of the phosphate from wastewater with an average phosphate uptake of $0.16\,g^{-2}\,d^{-1}$ (Sukačová et al., 2015). The phosphorous removal rates varied from 36% to 41% under natural light illumination with a 12h day/night photoperiod (Sukačová et al., 2015). Similar study on another *Leptolyngbya* sp. showed removal of metals and organic contaminants and correspondingly with a good fatty acid profile for biodiesel production (Singh & Thakur, 2015). Rai, Kumar, Pandey, Yadav, and Gaur (2016) studied the phosphate removal capability of four cyanobacterial biofilms predominated by *Phormidium* and *Oscillatoria* grown on polyester mesh discs. All four biofilms were capable of removing phosphate with the biofilm dominated by *P. fragile* performed the best among all the four biofilms (Rai et al., 2016). Cyanobacterial biofilms based on the heterocystous cyanobacteria *Calothrix* sp. showed from 6% to 43% removal of phosphorous that was varying across the strains used to cultivate the biofilm (Gismondi et al., 2016).

In addition to the removal of phosphate, the biofilms could also remove nitrate from the growth medium (Rai et al., 2016). Biofilm photobioreactor utilizing *Leptolyngbya* containing biofilm showed decrease in concentration of phosphate, ammonia, nitrite, COD, and nitrate from brewery wastewater (Papadopoulos et al., 2021). Biofilm-based integrated photobioelectrochemical (IPB) systems have been developed where in addition to removal of contaminants, bioenergy generation can be carried out (Xiao et al., 2015). In this IPB system, in addition to other microbial species, cyanobacteria such as *Leptolyngbya* and *Merismopedia glauca* were a part of this system. This system could achieve almost 100% removal of ammonium nitrogen and 90% of organic compounds (Xiao et al., 2015). In addition, bioenergy of $0.089\,kWh\,m^{-3}$ comprising direct electric energy of $0.055\,kWh\,m^{-3}$ and $0.034\,kWh\,m^{-3}$ indirect electric energy could be generated (Xiao et al., 2015).

In addition to phosphate and nitrogen, the ability to remove sulfate from acid mine drainage utilizing revolving algal biofilm reactor with removal capacity of up to 46% along with a removal rate of $0.56\,g\,L^{-1}$ has also been demonstrated (Zhou, Sheng, Zhao, Gross, & Wen, 2018). Microalgal-based biofilm bioreactor was also capable of removing suspended solids and nutrients from petroleum refinery wastewater (Hodges, Fica, Wanlass, Van Darlin, & Sims, 2017). Cyanobacterial biofilms have been also known to take up and cumulate insecticides from aquatic systems (Kour et al., 2020).

In addition to organic constituents, removal of pesticides, heavy metal by algal biofilms has also been demonstrated (da Costa & de França, 2003; Husien, Labena, El-Belely, Mahmoud, & Hamouda, 2020; Raghavan, Potnis, Bhattacharyya, Salaskar, & Rajaram, 2020, Velu, Cirés, Brinkman, & Heimann, 2020). Cd was effectively cumulated by the cyanobacteria *Spirulina* with the metal accumulating in different layers on the surface of the cyanobacteria (da Costa & de França, 2003). Biofilm system formed from axenic cultures of *N. muscorum* was also shown to adsorb Cd(II) at varying pH, concentrations, and volumes (Raghavan et al., 2020). The *N. muscorum* biofilm revealed a Langmuir adsorption of Cd(II) and capable of adsorbing Cd(II) from very low Cd(II) concentrations to high concentration of almost 100 ppm Cd(II). This biofilm system also could work across a wide initial pH range from 5 to 9 (Raghavan et al., 2020). The robustness of this system was reflected in the fact that the system could tolerate the harsh constituent of wastewater (Raghavan et al., 2020). Husien et al. (2020) showed effective chromium removal by *Nostoc* sp. biofilm with over 86.498% chromium removal after 7 days of exposure to chromium. Removal of multiple metals by *Tolypothrix*-based biofilm was reported from simulated ash dam wastewater, and the biofilm exhibited over 90% removal of Al, Sr, and Zn, ~70%–80% removal of Cu and Fe, ~65%–75% As (Velu et al., 2020). Increase in removal of Mo was observed from ~58.9%–63.6% (in absence of CO_2 fertilization) to ~98.9% when fertilized with carbon dioxide and converse was true in case of Ni where metal removal decreased from ~79%–80% removal rate to ~7.5% in the presence of carbon dioxide fertilization, while removal of other heavy metals was largely unaffected by the absence or presence of carbon dioxide (Velu et al., 2020).

6. Conclusion and future perspectives

Cyanobacteria, by virtue of their ancient existence and exposure to multiple stresses both biotic and abiotic, developed a useful strategy of forming biofilms among each other and with other bacteria and even higher organisms. While several physical, chemical, environmental, and biological factors influenced the biofilm formation, because of their photosynthetic nature, the cyanobacteria controlled the biofilm formation as being the primary provider of food to other microbes and also formed the top layer of biofilms.

The cyanobacterial biofilms exhibit versatility in terms of evolution, widespread distribution, and resource utilization, which has widely enhanced their scope of application in varied fields. The inputs on the genetic mechanism and interactions of different cyanobacteria for biofilm formation under varied conditions can be well utilized to generate strains in the laboratory for greater use in the field of bioremediation from soil and water, thereby contributing to a green clean environment, and also enhance its utility as a biofertilizer and a weedicide to decrease the dependence on chemical fertilizers and weedicide/pesticides and help turn toward organic farming with lower economic inputs and higher gains. A recent review article on cyanobacterial EPS synthesis genes and their contribution to biofilm formation (Potnis et al., 2021) also throws open newer avenues to synthesize designed cyanobacteria as well as EPS for multiple applications. Thus, we conclude to say that cyanobacterial biofilms are not restricted to their ancient origin but hold a lot of promise for the future as well, and their utility will keep evolving with time.

References

Agostoni, M., Koestler, B. J., Waters, C. M., Williams, B. L., & Montgomery, B. L. (2013). Occurrence of cyclic di-GMP-modulating output domains in cyanobacteria: An illuminating perspective. *MBio*, *4*(4), e00451-13.

Agostoni, M., Logan-Jackson, A. R., Heinz, E. R., Severin, G. B., Bruger, E. L., Waters, C. M., et al. (2018). Homeostasis of second messenger cyclic-di-AMP is critical for cyanobacterial fitness and acclimation to abiotic stress. *Frontiers in Microbiology*, *9*, 1121.

Agostoni, M., Waters, C. M., & Montgomery, B. L. (2016). Regulation of biofilm formation and cellular buoyancy through modulating intracellular cyclic di-GMP levels in engineered cyanobacteria. *Biotechnology and Bioengineering*, *113*(2), 311–319.

Ahmed, M., Stal, L. J., & Hasnain, S. (2014). Biofilm formation and indole-3-acetic acid production by two rhizospheric unicellular cyanobacteria. *Journal of Microbiology and Biotechnology*, *24*(8), 1015–1025.

Albertano, P. (2012). Cyanobacterial biofilms in monuments and caves. In *Ecology of cyanobacteria II* (pp. 317–343). Dordrecht: Springer.

Allen, A., Habimana, O., & Casey, E. (2018). The effects of extrinsic factors on the structural and mechanical properties of *Pseudomonas fluorescens* biofilms: A combined study of nutrient concentrations and shear conditions. *Colloids and Surfaces B: Biointerfaces*, *165*, 127–134.

Allen, R., Rittmann, B. E., & Curtiss, R., III. (2019). Axenic biofilm formation and aggregation by *Synechocystis* sp. strain PCC 6803 are induced by changes in nutrient concentration and require cell surface structures. *Applied and Environmental Microbiology*, *85*(7), e02192-18.

Baqué, M., Scalzi, G., Rabbow, E., Rettberg, P., & Billi, D. (2013). Biofilm and planktonic lifestyles differently support the resistance of the desert cyanobacterium *Chroococcidiopsis* under space and Martian simulations. *Origins of Life and Evolution of Biospheres*, *43*(4), 377–389.

Barranguet, C., Charantoni, E., Plans, M., & Admiraal, W. (2000). Short-term response of monospecific and natural algal biofilms to copper exposure. *European Journal of Phycology*, *35*(4), 397–406.

Barros, A. C., Gonçalves, A. L., & Simões, M. (2019). Microalgal/cyanobacterial biofilm formation on selected surfaces: The effects of surface physicochemical properties and culture media composition. *Journal of Applied Phycology*, *31*(1), 375–387.

Behrendt, L., Schrameyer, V., Qvortrup, K., Lundin, L., Sørensen, S. J., Larkum, A. W., et al. (2012). Biofilm growth and near-infrared radiation-driven photosynthesis of the chlorophyll d-containing cyanobacterium *Acaryochloris marina*. *Applied and Environmental Microbiology*, *78*(11), 3896–3904.

Bharti, A., Prasanna, R., Kumar, G., Kumar, A., & Nain, L. (2019). Co-cultivation of cyanobacteria for raising nursery of chrysanthemum using a hydroponic system. *Journal of Applied Phycology*, *31*(6), 3625–3635.

Boomer, S. M., Noll, K. L., Geesey, G. G., & Dutton, B. E. (2009). Formation of multilayered photosynthetic biofilms in an alkaline thermal spring in Yellowstone National Park, Wyoming. *Applied and Environmental Microbiology*, *75*(8), 2464–2475.

Bott, T. L., Kaplan, L. A., & Kuserk, F. T. (1984). Benthic bacterial biomass supported by streamwater dissolved organic matter. *Microbial Ecology*, *10*(4), 335–344.

Boulêtreau, S., Salvo, E., Lyautey, E., Mastrorillo, S., & Garabetian, F. (2012). Temperature dependence of denitrification in phototrophic river biofilms. *Science of the Total Environment*, *416*, 323–328.

Brock, T. D. (1973). Evolutionary and ecological aspects of the cyanophytes. In N. G. Carr, & B. A. Whitton (Eds.), *The biology of the blue-green algae* (pp. 487–500). Oxford: Blackwell Scientific Publications.

Brown, A. O., Romanis, C. S., Dvořák, P., Foss, A. J., Gibson, Q. A., Villanueva, C. D., et al. (2021). A new species of cryptic cyanobacteria isolated from the epidermis of a bottlenose dolphin and as a bioaerosol. *Phycologia*, 1–16.

Bruno, L., Di Pippo, F., Antonaroli, S., Gismondi, A., Valentini, C., & Albertano, P. (2012). Characterization of biofilm-forming cyanobacteria for biomass and lipid production. *Journal of Applied Microbiology*, *113*(5), 1052–1064.

Bruno, L., & Valle, V. (2017). Effect of white and monochromatic lights on cyanobacteria and biofilms from Roman Catacombs. *International Biodeterioration & Biodegradation*, *123*, 286–295.

Büdel, B. (1999). Ecology and diversity of rock-inhabiting cyanobacteria in tropical regions. *European Journal of Phycology*, *34*(4), 361–370.

Camacho, A., Rochera, C., Silvestre, J. J., Vicente, E., & Hahn, M. W. (2005). Spatial dominance and inorganic carbon assimilation by conspicuous autotrophic biofilms in a physical and chemical gradient of a cold sulfurous spring: The role of differential ecological strategies. *Microbial Ecology*, *50*(2), 172–184.

Cao, Z., Livoti, E., Losi, A., & Gärtner, W. (2010). A blue light-inducible phosphodiesterase activity in the cyanobacterium *Synechococcus elongatus*. *Photochemistry and Photobiology*, *86*(3), 606–611.

Castenholz, R. W. (1968). The behavior of *Oscillatoria terebriformis* in hot springs. *Journal of Phycology*, *4*(2), 132–139.

Castenholz, R. W. (1973). The possible photosynthetic use of sulphide by the filamentous phototrophic bacteria of hot springs. *Limnology and Oceanography*, *18*(6), 863–876.

Castenholz, R. W. (1976). The effect of sulphide on the bluegreen algae of hot springs. I. New Zealand and Iceland 1. *Journal of Phycology*, *12*(1), 54–68.

Castenholz, R. W. (1982). Motility and taxes. In N. G. Carr, & B. A. Whitton (Eds.), *The biology of cyanobacteria* (pp. 413–440). Blackwell Scientific Publications.

Castenholz, R. W., Jørgensen, B. B., D'Amelio, E., & Bauld, J. (1991). Photosynthetic and behavioral versatility of the cyanobacterium *Oscillatoria boryana* in a sulphide-rich microbial mat. *FEMS Microbiology Letters*, *86*(1), 43–57.

Chavan, A., & Mukherji, S. (2008). Treatment of hydrocarbon-rich wastewater using oil degrading bacteria and phototrophic microorganisms in rotating biological contactor: Effect of N:P ratio. *Journal of Hazardous Materials*, *154*(1–3), 63–72.

Chen, M., Xu, C. Y., Wang, X., Ren, C. Y., Ding, J., & Li, L. (2020). Comparative genomics analysis of c-di-GMP metabolism and regulation in *Microcystis aeruginosa*. *BMC Genomics*, *21*(1), 1–17.

Chrismas, N. A., Barker, G., Anesio, A. M., & Sánchez-Baracaldo, P. (2016). Genomic mechanisms for cold tolerance and production of exopolysaccharides in the Arctic cyanobacterium *Phormidesmis priestleyi* BC1401. *BMC Genomics*, *17*(1), 1–14.

Church, M. J., Wai, B., Karl, D. M., & DeLong, E. F. (2010). Abundances of crenarchaealamo A genes and transcripts in the Pacific Ocean. *Environmental Microbiology*, *12*(3), 679–688.

Cohen, Y., Jørgensen, B. B., Revsbech, N. P., & Poplawski, R. (1986). Adaptation to hydrogen sulphide of oxygenic and anoxygenic photosynthesis among cyanobacteria. *Applied and Environmental Microbiology*, *51*(2), 398–407.

Cole, J. K., Hutchison, J. R., Renslow, R. S., Kim, Y. M., Chrisler, W. B., Engelmann, H. E., et al. (2014). Phototrophic biofilm assembly in microbial-mat-derived unicyanobacterial consortia: Model systems for the study of autotroph-heterotroph interactions. *Frontiers in Microbiology*, *5*, 109.

Conradi, F. D., Zhou, R. Q., Oeser, S., Schuergers, N., Wilde, A., & Mullineaux, C. W. (2019). Factors controlling floc formation and structure in the cyanobacterium *Synechocystis* sp. strain PCC 6803. *Journal of Bacteriology*, *201*, e00344-19–19.

Crispim, C. A., Gaylarde, P. M., & Gaylarde, C. C. (2003). Algal and cyanobacterial biofilms on calcareous historic buildings. *Current Microbiology*, *46*(2), 0079–0082.

Crispim, C. A., Gaylarde, C. C., & Gaylarde, P. M. (2004). Biofilms on church walls in Porto Alegre, RS, Brazil, with special attention to cyanobacteria. *International Biodeterioration & Biodegradation*, *54*(2–3), 121–124.

da Costa, A. C. A., & de França, F. P. (2003). Cadmium interaction with microalgal cells, cyanobacterial cells, and seaweeds; toxicology and biotechnological potential for wastewater treatment. *Marine Biotechnology*, *5*(2), 149–156.

David, C., Bühler, K., & Schmid, A. (2015). Stabilization of single species Synechocystis biofilms by cultivation under segmented flow. *Journal of Industrial Microbiology and Biotechnology*, *42* (7), 1083–1089.

Del Mondo, A., Pinto, G., Carbone, D. A., Pollio, A., & De Natale, A. (2018). Biofilm architecture on different substrates of an *Oculatella subterranea* (Cyanobacteria) strain isolated from Pompeii archaeological site (Italy). *Environmental Science and Pollution Research*, *25*(26), 26079–26089.

Di Pippo, F., Ellwood, N. T. W., Guzzon, A., Siliato, L., Micheletti, E., De Philippis, R., et al. (2012). Effect of light and temperature on biomass, photosynthesis and capsular polysaccharides in cultured phototrophic biofilms. *Journal of Applied Phycology*, *24*(2), 211–220.

Dor, I., & Danin, A. (1996). Cyanobacterial desert crusts in the Dead Sea Valley, Israel map: 1. *Algological Studies/Archiv für Hydrobiologie*, 197–206. Supplement Volumes.

Duan, Y., Wu, F., He, D., Gu, J. D., Feng, H., Chen, T., et al. (2021). Bacterial and fungal communities in the sandstone biofilms of two famous Buddhist grottoes in China. *International Biodeterioration & Biodegradation*, *163*, 105267.

Echenique-Subiabre, I., Zancarini, A., Heath, M. W., Wood, S. A., Quiblier, C., & Humbert, J. F. (2018). Multiple processes acting from local to large geographical scales shape bacterial communities associated with *Phormidium* (cyanobacteria) biofilms in French and New Zealand rivers. *Scientific Reports*, *8*(1), 1–12.

Enomoto, G., Narikawa, R., & Ikeuchi, M. (2015). Three cyanobacteriochromes work together to form a light color-sensitive input system for c-di-GMP signaling of cell aggregation. *Proceedings of the National Academy of Sciences*, *112*(26), 8082–8087.

Enomoto, G., Okuda, Y., & Ikeuchi, M. (2018). Tlr1612 is the major repressor of cell aggregation in the light-color-dependent c-di-GMP signaling network of *Thermosynechococcus vulcanus*. *Scientific Reports*, *8*(1), 1–10.

Faria, S. I., Teixeira-Santos, R., Romeu, M. J., Morais, J., Jong, E. D., Sjollema, J., et al. (2021). Unveiling the antifouling performance of different marine surfaces and their effect on the development and structure of cyanobacterial biofilms. *Microorganisms*, *9*(5), 1102.

Faria, S. I., Teixeira-Santos, R., Romeu, M. J., Morais, J., Vasconcelos, V., & Mergulhão, F. J. (2020). The relative importance of shear forces and surface hydrophobicity on biofilm formation by coccoid cyanobacteria. *Polymers*, *12*(3), 653.

Fattom, A., & Shilo, M. (1984). Hydrophobicity as an adhesion mechanism of benthic cyanobacteria. *Applied and Environmental Microbiology*, *47*(1), 135–143.

Fay, P. (1965). Heterotrophy and nitrogen fixation in *Chlorogloea fritschii*. *Microbiology*, *39*(1), 11–20.

Fisher, M. L., Allen, R., Luo, Y., & Curtiss, R., III. (2013). Export of extracellular polysaccharides modulates adherence of the cyanobacterium *Synechocystis*. *PLoS One*, *8*(9), e74514.

Galica, T., Hrouzek, P., & Mareš, J. (2017). Genome mining reveals high incidence of putative lipopeptide biosynthesis NRPS/PKS clusters containing fatty acyl-AMP ligase genes in biofilm-forming cyanobacteria. *Journal of Phycology*, *53*(5), 985–998.

Gambino, M., Sanmartín, P., Longoni, M., Villa, F., Mitchell, R., & Cappitelli, F. (2019). Surface colour: An overlooked aspect in the study of cyanobacterial biofilm formation. *Science of the Total Environment*, *659*, 342–353.

Garcia-Meza, J. V., Barranguet, C., & Admiraal, W. (2005). Biofilm formation by algae as a mechanism for surviving on mine tailing. *Environmental Toxicology and Chemistry*, *3*, 575–581.

Garcia-Pichel, F., & Castenholz, R. W. (1990). Comparative anoxygenic photosynthetic capacity in 7 strains of a thermophilic cyanobacterium. *Archives of Microbiology*, *153*(4), 344–351.

Gaylarde, C. C., Rodríguez, C. H., Navarro-Noya, Y. E., & Ortega-Morales, B. O. (2012). Microbial biofilms on the sandstone monuments of the Angkor Wat complex, Cambodia. *Current Microbiology*, *64*(2), 85–92.

Geitler, L. (1932). *Cyanophyceae in rabenhorst kryptogamen-flora von deutschland, osterreich und der schweiz XIV* (pp. 1–1196). Leipzig, Germany: Akademische Verlagsgesellschaft.

Gismondi, A., Pippo, F. D., Bruno, L., Antonaroli, S., & Congestri, R. (2016). Phosphorus removal coupled to bioenergy production by three cyanobacterial isolates in a biofilm dynamic growth system. *International Journal of Phytoremediation*, *18*(9), 869–876.

Gomont, M. (1982). Monographie des Oscillatoriées (Nostocacéeshomocystées). *Annales des Sciences Naturellesserie 7 Botanique*, *16*, 91–264.

Gorbushina, A. A. (2007). Life on the rocks. *Environmental Microbiology*, *9*(7), 1613–1631.

Gorbushina, A. A., & Broughton, W. J. (2009). Microbiology of the atmosphere-rock interface: How biological interactions and physical stresses modulate a sophisticated microbial ecosystem. *Annual Review of Microbiology*, *63*, 431–450.

Guenet, B., Danger, M., Abbadie, L., & Lacroix, G. (2010). Priming effect: Bridging the gap between terrestrial and aquatic ecology. *Ecology*, *91* (10), 2850–2861.

Gupta, G., Snehi, S. K., & Singh, V. (2017). Role of PGPR in biofilm formations and its importance in plant health. *Biofilms Plant Soil Health*, *27*, 27–40.

Guzzon, A., Bohn, A., Diociaiuti, M., & Albertano, P. (2008). Cultured phototrophic biofilms for phosphorus removal in wastewater treatment. *Water Research, 42*(16), 4357–4367.

Guzzon, A., Congestri, R., & Albertano, P. (2005). Light-induced changes in photosynthesis and structure of cyanobacterial cultured biofilms from an Italian wastewater treatment plant. *Algological Studies/Archiv für Hydrobiologie*, 223–238. Supplement Volumes.

Haselkorn, R. (1978). Heterocysts. *Annual Review of Plant Physiology, 29* (1), 319–344.

Hernández-Prieto, M. A., Li, Y., Postier, B. L., Blankenship, R. E., & Chen, M. (2018). Far-red light promotes biofilm formation in the cyanobacterium *Acaryochloris marina. Environmental Microbiology, 20*(2), 535–545.

Hodges, A., Fica, Z., Wanlass, J., Van Darlin, J., & Sims, R. (2017). Nutrient and suspended solids removal from petrochemical wastewater via microalgal biofilm cultivation. *Chemosphere, 174*, 46–48.

Husien, S., Labena, A., El-Belely, E., Mahmoud, H., & Hamouda, A. (2020). Application of *Nostoc* sp. for hexavalent chromium [Cr (VI)] removal: Planktonic and biofilm. *International Journal of Environmental Analytical Chemistry*, 1–22.

Irving, T. E., & Allen, D. G. (2011). Species and material considerations in the formation and development of microalgal biofilms. *Applied Microbiology and Biotechnology, 92*(2), 283–294.

Jaag, O. (1945). *Investigations into the vegetation and biology of the algae of the naked Rock in the Alps, in the Jura and in the Swiss plateau. Cryptogam flora of the Switzerland. Vol. IX, Issue 3*. Bern: Kommissionsverlag Buchdruckerei Büchler and Co.

Ji, B., Zhang, W., Zhang, N., Wang, J., Lutzu, G. A., & Liu, T. (2014). Biofilm cultivation of the oleaginous microalgae *Pseudochlorococcum* sp. *Bioprocess and Biosystems Engineering, 37*(7), 1369–1375.

Jiang, G., & Ohtaguchi, K. (2006). Kinetics of biofilm formation of a cyanobacterium on bioreactor surfaces. *Journal of Chemical Engineering of Japan, 39*(4), 453–460.

Joh, G., & Lee, J. (2012). Cyanobacterial biofilms on sedimentation basins in a water treatment plant in South Korea. *Journal of Applied Phycology, 24*(2), 285–293.

Kann, E. (1988). On the autocology of benthic cyanophytes in pure European lakes and rivers. *Algological Studies/Archiv für Hydrobiologie, 50–53*, 473–495. Suppl. 80.

Katsikogianni, M., & Missirlis, Y. F. (2004). Concise review of mechanisms of bacterial adhesion to biomaterials and of techniques used in estimating bacteria-material interactions. *European Cells & Materials, 8*(3), 37–57.

Kawano, Y., Saotome, T., Ochiai, Y., Katayama, M., Narikawa, R., & Ikeuchi, M. (2011). Cellulose accumulation and a cellulose synthase gene are responsible for cell aggregation in the cyanobacterium *Thermosynechococcus vulcanus* RKN. *Plant and Cell Physiology, 52*(6), 957–966.

Kera, K., Nagayama, T., Nanatani, K., Saeki-Yamoto, C., Tominaga, A., Souma, S., et al. (2018). Reduction of spermidine content resulting from inactivation of two arginine decarboxylases increases biofilm formation in *Synechocystis* sp. strain PCC 6803. *Journal of Bacteriology, 200*(9), e00664-17.

Kera, K., Yoshizawa, Y., Shigehara, T., Nagayama, T., Tsujii, M., Tochigi, S., et al. (2020). Hik36–Hik43 and Rre6 act as a two-component regulatory system to control cell aggregation in *Synechocystis* sp. PCC6803. *Scientific Reports, 10*(1), 1–9.

Keshari, N., & Adhikary, S. P. (2013). Characterization of cyanobacteria isolated from biofilms on stone monuments at Santiniketan, India. *Biofouling, 29*(5), 525–536.

Keshari, N., & Adhikary, S. P. (2014). Diversity of cyanobacteria on stone monuments and building facades of India and their phylogenetic analysis. *International Biodeterioration & Biodegradation, 90*, 45–51.

Keshari, N., Das, S. K., & Adhikary, S. P. (2021). Colonization and survival of a stress tolerant cyanobacterium on a heritage monument of Santiniketan, India. *International Biodeterioration & Biodegradation, 164*, 105294.

Kol, E. (1968). Algae from the Antarctica. *Annales Historico-Naturales Musei Nationalis Hungarici, 60*, 71–77.

Kour, D., Rana, K. L., Kaur, T., Yadav, N., Yadav, A. N., Rastegari, A. A., et al. (2020). Microbial biofilms: Functional annotation and potential applications in agriculture and allied sectors. In *New and future developments in microbial biotechnology and bioengineering: Microbial biofilms* (pp. 283–301). Elsevier.

Laamanen, M. (1996). Cyanoprokaryotes in the Baltic Sea ice and winter plankton. *Algological Studies/Archiv für Hydrobiologie*, 423–433. Supplement Volumes.

Lan, S., Wu, L., Yang, H., Zhang, D., & Hu, C. (2017). A new biofilm based microalgal cultivation approach on shifting sand surface for desert cyanobacterium *Microcoleus vaginatus. Bioresource Technology, 238*, 602–608.

Lear, G., Anderson, M. J., Smith, J. P., Boxen, K., & Lewis, G. D. (2008). Spatial and temporal heterogeneity of the bacterial communities in stream epilithic biofilms. *FEMS Microbiology Ecology, 65*(3), 463–473.

Li, J., Fang, D., Ye, R., Zhou, C., & Li, P. (2021). The released polysaccharide inhibits cell aggregation and biofilm formation in the cyanobacterium *Synechocystis* sp. PCC 6803. *European Journal of Phycology, 56*(2), 119–128.

Li, S., Wang, C., Qin, H., Li, Y., Zheng, J., Peng, C., et al. (2016). Influence of phosphorus availability on the community structure and physiology of cultured biofilms. *Journal of Environmental Sciences, 42*, 19–31.

Loza, V., Berrendero, E., Perona, E., & Mateo, P. (2013). Polyphasic characterization of benthic cyanobacterial diversity from biofilms of the Guadarrama river (Spain): Morphological, molecular, and ecological approaches 1. *Journal of Phycology, 49*(2), 282–297.

Lugomela, C., Söderbäck, E., & Björk, M. (2005). Photosynthesis rates in cyanobacteria-dominated sub-tidal biofilms near Zanzibar, Tanzania. *Estuarine, Coastal and Shelf Science, 63*(3), 439–446.

Macedo, M. F., Miller, A. Z., Dionísio, A., & Saiz-Jimenez, C. (2009). Biodiversity of cyanobacteria and green algae on monuments in the Mediterranean Basin: An overview. *Microbiology, 155*(11), 3476–3490.

Manjunath, M., Kanchan, A., Ranjan, K., Venkatachalam, S., Prasanna, R., Ramakrishnan, B., et al. (2016). Beneficial cyanobacteria and eubacteria synergistically enhance bioavailability of soil nutrients and yield of okra. *Heliyon, 2*(2), e00066.

Mao-Che, L. M., Le Campion-Alsumard, T., Boury-Esnault, N., Payri, C., Golubic, S., & Bézac, C. (1996). Biodegradation of shells of the black pearl oyster, *Pinctada margaritifera* var. *cumingii*, by microborers and sponges of French Polynesia. *Marine Biology, 126*(3), 509–519.

Mehta, D., & Shah, D. (2021). Cyanobacteria and microalgae growing on monuments of UNESCO World Heritage site Champaner Pavagadh, India: Biofilms and their exopolysaccharide composition. *Archives of Microbiology*, 1–9.

Merbt, S. N., Stahl, D. A., Casamayor, E. O., Martí, E., Nicol, G. W., & Prosser, J. I. (2012). Differential photoinhibition of bacterial and archaeal ammonia oxidation. *FEMS Microbiology Letters*, *327*(1), 41–46.

Miranda, M., Noyma, N., Pacheco, F. S., de Magalhães, L., Pinto, E., Santos, S., et al. (2017). The efficiency of combined coagulant and ballast to remove harmful cyanobacterial blooms in a tropical shallow system. *Harmful Algae*, *65*, 27–39.

Mishra, A., Gupta, J., Kumari, T., Pal, R., & Thakur, I. S. (2021). Unravelling the attributes of novel cyanobacteria *Jacksonvillea* sp. ISTCYN1 by draft genome sequencing. *Bioresource Technology*, *337*, 125473.

Mitsui, A., Kumazawa, S., Takahashi, A., Ikemoto, H., Cao, S., & Arai, T. (1986). Strategy by which nitrogen-fixing unicellular cyanobacteria grow photoautotrophically. *Nature*, *323*(6090), 720–722.

Moreira, J. M. R., Simões, M., Melo, L. F., & Mergulhão, F. J. (2015). The combined effects of shear stress and mass transfer on the balance between biofilm and suspended cell dynamics. *Desalination and Water Treatment*, *53*(12), 3348–3354.

Nagar, E., & Schwarz, R. (2015). To be or not to be planktonic? Self-inhibition of biofilm development. *Environmental Microbiology*, *17*(5), 1477–1486.

Nagar, E., Zilberman, S., Sendersky, E., Simkovsky, R., Shimoni, E., Gershtein, D., et al. (2017). Type 4 pili are dispensable for biofilm development in the cyanobacterium *Synechococcus elongatus*. *Environmental Microbiology*, *19*(7), 2862–2872.

Nagarkar, S., Williams, G. A., Subramanian, G., & Saha, S. K. (2004). Cyanobacteria-dominated biofilms: A high quality food resource for intertidal grazers. In *Asian pacific phycology in the 21st century: Prospects and challenges* (pp. 89–95). Dordrecht: Springer.

Neely, R. K., & Wetzel, R. G. (1995). Simultaneous use of 14 C and 3 H to determine autotrophic production and bacterial protein production in periphyton. *Microbial Ecology*, *30*(3), 227–237.

Nikolić, N., Zarubica, N., Gavrilović, B., Predojević, D., Trbojević, I., Subakov-Simić, G., et al. (2020). Lampenflora and the entrance biofilm in two show caves: Comparison of microbial community, environmental, and biofilm parameters. *Journal of Cave and Karst Studies*, *82*(2), 69–81.

Ortega-Morales, B. O., Gaylarde, C., Anaya-Hernandez, A., Chan-Bacab, M. J., De la Rosa-García, S. C., Arano-Recio, D., et al. (2013). Orientation affects Trentepohlia-dominated biofilms on Mayan monuments of the Rio Bec style. *International Biodeterioration & Biodegradation*, *84*, 351–356.

Ortega-Morales, B. O., Narvaez-Zapata, J. A., Schmalenberger, A., Sosa-López, A., & Tebbe, C. C. (2004). Biofilms fouling ancient limestone Mayan monuments in Uxmal, Mexico: A cultivation-independent analysis. *Biofilms*, *1*(2), 79–90.

Paerl, H. W., & Pinckney, J. L. (1996). A mini-review of microbial consortia: Their roles in aquatic production and biogeochemical cycling. *Microbial Ecology*, *31*(3), 225–247.

Paerl, H. W., Pinckney, J. L., & Steppe, T. F. (2000). Cyanobacterial–bacterial mat consortia: Examining the functional unit of microbial survival and growth in extreme environments. *Environmental Microbiology*, *2*(1), 11–26.

Papadopoulos, K. P., Economou, C. N., Tekerlekopoulou, A. G., & Vayenas, D. V. (2021). A Cyanobacteria-based biofilm system for advanced brewery wastewater treatment. *Applied Sciences*, *11*(1), 174.

Parnasa, R., Nagar, E., Sendersky, E., Reich, Z., Simkovsky, R., Golden, S., et al. (2016). Small secreted proteins enable biofilm development in the cyanobacterium *Synechococcus elongatus*. *Scientific Reports*, *6*(1), 1–10.

Perkins, D. M., Yvon-Durocher, G., Demars, B. O., Reiss, J., Pichler, D. E., Friberg, N., et al. (2012). Consistent temperature dependence of respiration across ecosystems contrasting in thermal history. *Global Change Biology*, *18*(4), 1300–1311.

Popović, S., Subakov Simić, G., Stupar, M., Unković, N., Predojević, D., Jovanović, J., et al. (2015). Cyanobacteria, algae and microfungi present in biofilm from Božana Cave (Serbia). *International Journal of Speleology*, *44*(2), 4.

Potnis, A. A., Raghavan, P. S., & Rajaram, H. (2021). Overview on cyanobacterial exopolysaccharides and biofilms: Role in bioremediation. *Reviews in Environmental Science and Bio/Technology*, *20*(3), 781–794.

Prasanna, R., Bidyarani, N., Babu, S., Hossain, F., Shivay, Y. S., & Nain, L. (2015). Cyanobacterial inoculation elicits plant defense response and enhanced Zn mobilization in maize hybrids. *Cogent Food & Agriculture*, *1*(1), 998507.

Prasanna, R., Kumar, A., Babu, S., Chawla, G., Chaudhary, V., Singh, S., et al. (2013). Deciphering the biochemical spectrum of novel cyanobacterium-based biofilms for use as inoculants. *Biological Agriculture & Horticulture*, *29*(3), 145–158.

Prasanna, R., Pattnaik, S., Sugitha, T. K., Nain, L., & Saxena, A. K. (2011). Development of cyanobacterium-based biofilms and their in vitro evaluation for agriculturally useful traits. *Folia Microbiologica*, *56*(1), 49–58.

Prasanna, R., Ramakrishnan, B., Simranjit, K., Ranjan, K., Kanchan, A., Hossain, F., et al. (2017). Cyanobacterial and rhizobial inoculation modulates the plant physiological attributes and nodule microbial communities of chickpea. *Archives of Microbiology*, *199*(9), 1311–1323.

Prasanna, R., Saxena, G., Singh, B., Ranjan, K., Buddhadeo, R., Velmourougane, K., et al. (2018). Mode of application influences the biofertilizing efficacy of cyanobacterial biofilm formulations in chrysanthemum varieties under protected cultivation. *Open Agriculture*, *3*(1), 478–489.

Proteau, P. J., Gerwick, W. H., Garcia-Pichel, F., & Castenholz, R. (1993). The structure of scytonemin, an ultraviolet sunscreen pigment from the sheaths of cyanobacteria. *Experientia*, *49*(9), 825–829.

Raghavan, P. S., Potnis, A. A., Bhattacharyya, K., Salaskar, D. A., & Rajaram, H. (2020). Axenic cyanobacterial (*Nostoc muscorum*) biofilm as a platform for Cd (II) sequestration from aqueous solutions. *Algal Research*, *46*, 101778.

Rai, A. N. (1990). *CRC handbook of symbiotic cynobacteria* (pp. 1–8). CRC Press Inc.

Rai, J., Kumar, D., Pandey, L. K., Yadav, A., & Gaur, J. P. (2016). Potential of cyanobacterial biofilms in phosphate removal and biomass production. *Journal of Environmental Management*, *177*, 138–144.

Ramirez, M., Hernandez-Marine, M., Novelo, E., & Roldán, M. (2010). Cyanobacteria-containing biofilms from a Mayan monument in Palenque, Mexico. *Biofouling*, *26*(4), 399–409.

Rana, K. L., Kour, D., Yadav, A. N., Yadav, N., & Saxena, A. K. (2020). Agriculturally important microbial biofilms: Biodiversity, ecological significances, and biotechnological applications. In *New and future developments in microbial biotechnology and bioengineering: Microbial biofilms* (pp. 221–265). Elsevier.

Richardson, L. L., & Castenholz, R. W. (1987). Diel vertical movements of the cyanobacterium *Oscillatoria terebriformis* in a sulphide-rich hot spring microbial mat. *Applied and Environmental Microbiology*, *53*(9), 2142–2150.

Robson, R. L., & Postgate, J. R. (1980). Oxygen and hydrogen in biological nitrogen fixation. *Annual Reviews in Microbiology*, *34*(1), 183–207.

Roeselers, G., van Loosdrecht, M. C., & Muyzer, G. (2007). Heterotrophic pioneers facilitate phototrophic biofilm development. *Microbial Ecology*, *54*(3), 578–585.

Roldán, M., & Hernández-Mariné, M. (2009). Exploring the secrets of the three-dimensional architecture of phototrophic biofilms in caves. *International Journal of Speleology*, *38*(1).

Romaní, A. M. (2010). Freshwater biofilms. In J. C. Thomason, & S. Dürr (Eds.), *Biofouling* (pp. 137–153). Oxford: Wiley-Blackwell.

Romaní, A. M., Borrego, C. M., Díaz-Villanueva, V., Freixa, A., Gich, F., & Ylla, I. (2014). Shifts in microbial community structure and function in light-and dark-grown biofilms driven by warming. *Environmental Microbiology*, *16*(8), 2550–2567.

Romaní, A. M., Guasch, H., Munoz, I., Ruana, J., Vilalta, E., Schwartz, T., et al. (2004). Biofilm structure and function and possible implications for riverine DOC dynamics. *Microbial Ecology*, *47*(4), 316–328.

Romeu, M. J., Alves, P., Morais, J., Miranda, J. M., de Jong, E. D., Sjollema, J., et al. (2019). Biofilm formation behaviour of marine filamentous cyanobacterial strains in controlled hydrodynamic conditions. *Environmental Microbiology*, *21*(11), 4411–4424.

Romeu, M. J. L., Domínguez-Pérez, D., Almeida, D., Morais, J., Campos, A., Vasconcelos, V., et al. (2020). Characterization of planktonic and biofilm cells from two filamentous cyanobacteria using a shotgun proteomic approach. *Biofouling*, *36*(6), 631–645.

Römling, U., Gomelsky, M., & Galperin, M. Y. (2005). C-di-GMP: The dawning of a novel bacterial signalling system. *Molecular Microbiology*, *57*(3), 629–639.

Sáiz-Jiménez, C. (1999). Biogeochemistry of weathering processes in monuments. *Geomicrobiology Journal*, *16*(1), 27–37.

Samad, L. K., & Adhikary, S. P. (2008). Diversity of micro-algae and cyanobacteria on building facades and monuments in India. *Algae*, *23*(2), 91–114.

Savakis, P., De Causmaecker, S., Angerer, V., Ruppert, U., Anders, K., Essen, L. O., et al. (2012). Light-induced alteration of c-di-GMP level controls motility of *Synechocystis* sp. PCC 6803. *Molecular Microbiology*, *85*(2), 239–251.

Schatz, D., Nagar, E., Sendersky, E., Parnasa, R., Zilberman, S., Carmeli, S., et al. (2013). Self-suppression of biofilm formation in the cyanobacterium *Synechococcus elongatus*. *Environmental Microbiology*, *15*(6), 1786–1794.

Schnurr, P. J., & Allen, D. G. (2015). Factors affecting algae biofilm growth and lipid production: A review. *Renewable and Sustainable Energy Reviews*, *52*, 418–429.

Schopf, J. W. (1975). Precambrian paleobiology: Problems and perspectives. *Annual Review of Earth and Planetary Sciences*, *3*(1), 213–249.

Schuergers, N., Mullineaux, C. W., & Wilde, A. (2017). Cyanobacteria in motion. *Current Opinion in Plant Biology*, *37*, 109–115.

Schwartz, W. (1975). GE Fogg, WDP Stewart, P. Fay And AE Walsby, The Blue-Green Algae. VII, 459 S., 170 Abb. und 8 farbige Abb., 33 Tab., 1 Farbtafel. London-New York 1973: Academic Press, £ 8.50. *Zeitschrift für allgemeine Mikrobiologie*, *15*(1), 68. 68.

Shepard, R. N., & Sumner, D. Y. (2010). Undirected motility of filamentous cyanobacteria produces reticulate mats. *Geobiology*, *8*(3), 179–190.

Singh, J., & Thakur, I. S. (2015). Evaluation of cyanobacterial endolith *Leptolyngbya* sp. ISTCY101, for integrated wastewater treatment and biodiesel production: A toxicological perspective. *Algal Research*, *11*, 294–303.

Stauch-White, K., Srinivasan, V. N., Kuo-Dahab, W. C., Park, C., & Butler, C. S. (2017). The role of inorganic nitrogen in successful formation of granular biofilms for wastewater treatment that support cyanobacteria and bacteria. *AMB Express*, *7*(1), 1–10.

Stewart, W. D. P. (1980). Some aspects of structure and function in n fixing cyanobacteria. *Annual Reviews in Microbiology*, *34*(1), 497–536.

Suban, S., Sendersky, E., Golden, S. S., & Schwarz, R. (2022). Impairment of a cyanobacterial glycosyltransferase that modifies a pilin results in biofilm development. *Environmental Microbiology Reports*, *14*, 218–229. bio Rxiv.

Sukačová, K., Trtílek, M., & Rataj, T. (2015). Phosphorus removal using a microalgal biofilm in a new biofilm photobioreactor for tertiary wastewater treatment. *Water Research*, *71*, 55–63.

Swarnalakshmi, K., Prasanna, R., Kumar, A., Pattnaik, S., Chakravarty, K., Shivay, Y. S., et al. (2013). Evaluating the influence of novel cyanobacterial biofilmed biofertilizers on soil fertility and plant nutrition in wheat. *European Journal of Soil Biology*, *55*, 107–116.

Talluri, S. N., Winter, R. M., & Salem, D. R. (2020). Conditioning film formation and its influence on the initial adhesion and biofilm formation by a cyanobacterium on photobioreactor materials. *Biofouling*, *36*(2), 183–199.

Thapa, S., Bharti, A., & Prasanna, R. (2017). Algal biofilms and their biotechnological significance. In *Algal green chemistry* (pp. 285–303). Elsevier.

Tomaselli, L., Lamenti, G., Bosco, M., & Tiano, P. (2000). Biodiversity of photosynthetic micro-organisms dwelling on stone monuments. *International Biodeterioration & Biodegradation*, *46*(3), 251–258.

Triveni, S., Prasanna, R., Kumar, A., Bidyarani, N., Singh, R., & Saxena, A. K. (2015). Evaluating the promise of Trichoderma and Anabaena based biofilms as multifunctional agents in *Macrophomina phaseolina*-infected cotton crop. *Biocontrol Science and Technology*, *25*(6), 656–670.

Urzi, C. (2004). Microbial deterioration of rocks and marble monuments of the Mediterranean basin: A review. *Corrosion Reviews*, *22*(5–6), 441–458.

Van Gemerden, H. (1993). Microbial mats: A joint venture. *Marine Geology*, *113*(1–2), 3–25.

Veerabadhran, M., Chakraborty, S., Mitra, S., Karmakar, S., & Mukherjee, J. (2018). Effects of flask configuration on biofilm growth and metabolites of intertidal Cyanobacteria isolated from a mangrove forest. *Journal of Applied Microbiology*, *125*(1), 190–202.

Velu, C., Cirés, S., Brinkman, D. L., & Heimann, K. (2020). Bioproduct potential of outdoor cultures of Tolypothrix sp.: Effect of carbon dioxide and metal-rich wastewater. *Frontiers in Bioengineering and Biotechnology*, *8*, 51.

Volk, C. J., Volk, C. B., & Kaplan, L. A. (1997). Chemical composition of biodegradable dissolved organic matter in streamwater. *Limnology and Oceanography*, *42*(1), 39–44.

Wakefield, R. D., & Jones, M. S. (1998). An introduction to stone colonizing micro-organisms and biodeterioration of building stone. *Quarterly Journal of Engineering Geology and Hydrogeology*, *31*(4), 301–313.

Wallner, T., Pedroza, L., Voigt, K., Kaever, V., & Wilde, A. (2020). The cyanobacterial phytochrome 2 regulates the expression of motility-related genes through the second messenger cyclic di-GMP. *Photochemical & Photobiological Sciences*, *19*(5), 631–643.

Wang, J., Liu, W., & Liu, T. (2017). Biofilm based attached cultivation technology for microalgal biorefineries—A review. *Bioresource Technology*, *244*, 1245–1253.

Warren, C. E., Wales, J. H., Davis, G. E., & Doudoroff, P. (1964). Trout production in an experimental stream enriched with sucrose. *The Journal of Wildlife Management*, 617–660.

Warscheid, T., & Braams, J. (2000). Biodeterioration of stone: A review. *International Biodeterioration & Biodegradation*, *46*(4), 343–368.

Watanabe, K., Imase, M., Aoyagi, H., Ohmura, N., Saiki, H., & Tanaka, H. (2008). Development of a novel artificial medium based on utilization of algal photosynthetic metabolites by symbiotic heterotrophs. *Journal of Applied Microbiology*, *105*(3), 741–751.

Whitton, B. A. (1992). Diversity, ecology, and taxonomy of the cyanobacteria. In *Photosynthetic prokaryotes* (pp. 1–51). Boston, MA: Springer.

Wood, J. L., Miller, C. D., Sims, R. C., & Takemoto, J. Y. (2015). Biomass and phycocyanin production from cyanobacteria dominated biofilm reactors cultured using oilfield and natural gas extraction produced water. *Algal Research*, *11*, 165–168.

Wu, S. H., & Lagarias, J. C. (2000). Defining the bilinlyase domain: Lessons from the extended phytochrome superfamily. *Biochemistry*, *39*(44), 13487–13495.

Wuertz, S., Okabe, S., & Hausner, M. (2004). Microbial communities and their interactions in biofilm systems: An overview. *Water Science and Technology*, *49*(11–12), 327–336.

Wynn-Williams, D. D. (1990). Ecological aspects of Antarctic microbiology. *Advances in Microbial Ecology*, 71–146.

Xia, L., Li, H., & Song, S. (2016). Cell surface characterization of some oleaginous green algae. *Journal of Applied Phycology*, *28*(4), 2323–2332.

Xiao, L., Young, E. B., Grothjan, J. J., Lyon, S., Zhang, H., & He, Z. (2015). Wastewater treatment and microbial communities in an integrated photo-bioelectrochemical system affected by different wastewater algal inocula. *Algal Research*, *12*, 446–454.

Yang, Y., Lam, V., Adomako, M., Simkovsky, R., Jakob, A., Rockwell, N. C., et al. (2018). Phototaxis in a wild isolate of the cyanobacterium *Synechococcus elongatus*. *Proceedings of the National Academy of Sciences*, *115*(52), E12378–E12387.

Ylla, I., Borrego, C., Romaní, A. M., & Sabater, S. (2009). Availability of glucose and light modulates the structure and function of a microbial biofilm. *FEMS Microbiology Ecology*, *69*(1), 27–42.

Zhang, N., Chang, Y. G., Tseng, R., Ovchinnikov, S., Schwarz, R., & LiWang, A. (2020). Solution NMR structure of Se0862, a highly conserved cyanobacterial protein involved in biofilm formation. *Protein Science*, *29*(11), 2274–2280.

Zhou, H., Sheng, Y., Zhao, X., Gross, M., & Wen, Z. (2018). Treatment of acidic sulfate-containing wastewater using revolving algae biofilm reactors: Sulfur removal performance and microbial community characterization. *Bioresource Technology*, *264*, 24–34.

Ziegler, S. E., Lyon, D. R., & Townsend, S. L. (2009). Carbon release and cycling within epilithic biofilms in two contrasting headwater streams. *Aquatic Microbial Ecology*, *55*(3), 285–300.

Chapter 3

From understanding bacterial interactions to developing bactericidal surfaces

Ashwin Kannan
Department of Chemical Engineering, Visvesvaraya National Institute of Technology, Nagpur, Maharashtra, India

1. Introduction

Adhesion of bacterial cells to solid surfaces results in the formation of organized multicellular structures called biofilms (Dunne, 2002; Vadillo-Rodríguez, Guerra-García-Mora, Perera-Costa, Gónzalez-Martín and Fernández-Calderón, 2018). Biofilms are tough to eradicate, as once formed, the bacterial cells in the biofilm exhibit an enhanced tolerance to the effect of antibacterial agents (Colvin et al., 2011; Flemming, Neu, & Wozniak, 2007; Flemming & Wingender, 2010; Kannan et al., 2014; Subbiahdoss, Kuijer, Grijpma, van der Mei, & Busscher, 2009). In the industrial context, formation of biofilms has been implicated in microbiologically influenced corrosion (MIC) (Li, Jia, Al-Mahamedh, Xu, & Gu, 2016; Liu, Dong, Liu, & Yin, 2010; Ogawa et al., 2020; Procópio, 2019; Qi, Chen, & Zhang, 2016; Teng, Guan, & Zhu, 2008; Wu, Liu, Jin, Xiu, & Sun, 2016; Xu et al., 2017). MIC has been implicated in the biocorrossion of biomedical implants (Jia, Yang, Xu, & Gu, 2017), biocorrossion of pipelines in the oil and gas industry (Liu et al., 2016), accelerated low water corrosion (Ramírez et al., 2016) and biofouling of cooling water systems (Jia, Li, Al-Mahamedh, & Gu, 2017). In the medical context, colonization of biomedical devices by bacterial cells and subsequent formation of biofilms (thereby causing serious infections) have been reported (Costerton, Stewart, & Greenberg, 1999; Elving, van der Mei, Busscher, van Weissenbruch, & Albers, 2003; Gristina, 1987; Hall-Stoodley, Costerton, & Stoodley, 2004; Kamal, Pfaller, Rempe, & Jebson, 1991; Klug, Wallet, Kacet, & Courcol, 2003; Padera, 2006; Petrini et al., 2006; Tunney et al., 2007). In light of the above, to mitigate the serious consequences of biofouling, it is desirable to prevent initial bacterial colonization over treatment of biofilms (Costerton et al., 1999; Palmer, Flint, & Brooks, 2007; Vadillo-Rodríguez et al., 2018). To achieve this, it is imperative that the factors that influence adhesion of bacterial cells to surfaces are studied and understood. They include surface properties of both the cell and the substrate (Bos, Van der Mei, & Busscher, 1999; Mitik-Dineva et al., 2009; Papenburg, Rodrigues, Wessling, & Stamatialis, 2010) surface chemistry and surface topography of the substrate (Webb et al., 2013).

The effect of surface chemistry and surface physical properties on bacterial adhesion has been extensively investigated in the last few decades. While several studies have been able to successfully explain the effect of the aforementioned properties on bacterial adhesion to surfaces, discrepancies between experimental observations and theoretical expectations have also been frequently observed. In this chapter, (some of) the progress made in understanding the effect of (material) surface properties on bacterial adhesion (to surfaces) with the help of analytical techniques (with a focus on microscopy) and how this has led to the development of synthetic biomimetic bactericidal (yet bioactive) surfaces is discussed.

2. Effect of surface chemistry on bacterial interaction: Applications of analytical techniques

Bacterial adhesion as a function of hydrophobicity has been reported by a number of research groups (Costerton, Lewandowski, Caldwell, Korber, & Lappin-Scott, 1995; Kitano et al., 1996; Morra and Cassinelli, 1998; van Loosdrecht, Lyklema, Norde, & Zehnder, 1989). Earlier, contradictory results (on the effect of hydrophobicity) were reported due to heterogeneity in the bacterial world and lack of adequate experimental setup (Bayer, 1982; Bos et al., 1999; Olofsson, Zita, & Hermansson, 1998). Bos, van der Mei, Gold, and Busscher (2000) studied the influence of hydrophobicity (of substrate) on adhesion of *Streptococcus sobrinus* in a parallel flow chamber using phase-contrast microscopy. This was performed under controlled hydrodynamic conditions to minimize biological and experimental variations. Micro-patterned surfaces on glass with a

Understanding Microbial Biofilms. https://doi.org/10.1016/B978-0-323-99977-9.00007-7

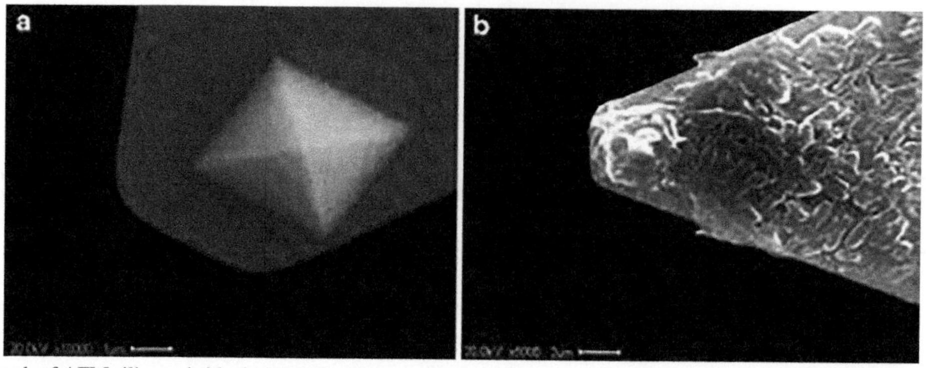

FIG. 1 SEM micrograph of AFM silicon nitride tip (A) before being coated and (B) after being coated with *E. coli* JM 109 bacterial cells. *(Reproduced with permission from Cao, T., Tang, H., Liang, X., Wang, A., Auner, G. W., Salley, S. O., et al. (2006). Nanoscale investigation on adhesion of* E. coli *to surface modified silicone using atomic force microscopy.* Biotechnology and Bioengineering, *94(1), 167–76. https://doi.org/10.1002/bit.20841. PMID: 16538682 and with permission from John Wiley and Sons Copyright (2006).)*

chemical pattern consisting of 10 μm wide hydrophobic lines of dimethyldichlorosilane (DDS) with 20 μm hydrophilic spacings in between were fabricated using lithography (Healy et al., 1996). Water contact angle measurements revealed that DDS coatings on glass resulted in a hydrophobic surface. Bacterial adhesion experiments revealed that there was preferential retention of bacteria on the hydrophilic spacings close to the interface between the hydrophilic and hydrophobic regions. After exposure to high shear forces, Bos et al. (2000) observed that a large number of bacteria adhered close to the interface between hydrophilic and hydrophobic regions. This they suggested was due the most optimal interaction between the heterogeneous bacterial cell and the substratum surface.

Bacterial contamination of biomedical devices causes infections, which may lead to acute bacterial meningitis, pneumonia, or even bacteremia (Barnett & Stephens, 1997; Cao et al., 2006). In this context, modification of synthetic materials (by surface heparization) has been considered to reduce the incidence of infection (Arnander, Bagger-Sjöbäck, Frebelius, Larsson, & Swedenborg, 1987; Davenas, Thevenard, Philippe, & Arnaud, 2002; Sefton, 1989). Different microscopic and quantitative adhesion assays have been used to study bacterial adhesion to surface of substrates (Bowen, Fenton, Lovitt, & Wright, 2002; Kockro et al., 2000; Livni, Yuhas, Ashkenazi, & Michowiz, 2004; Simhi, van der Mei, Ron, Rosenberg, & Busscher, 2000; Zhao et al., 2004). Cao et al. (2006) studied the effect of (substrate) hydrophobicity and surface groups on bacterial adhesion to modified silicone surfaces. Silicone is widely used as a material for medical applications because of its flexibility, fatigue resistance, high degree of chemical inertness, thermal stability, and nontoxicity. For this purpose, they prepared hydrophobic (silicone, octadecyltrichlorosilane (OTS)/silicone and fluoroalkylsilane (FAS)/silicone) and hydrophilic (heparin/OTS/silicone and hyaluronan/OTS/silicone) surfaces. The wettability of the surfaces (a measure of its hydrophobicity/hydrophilicity) was characterized using contact angles measured with DI water and Tris solution. Force spectroscopy (FS) was used to quantify the interaction between silicon nitride probes coated with *Escherichia coli* (JM 109) cells and silicone/modified silicone surfaces (Fig. 1).

Attachment of *E. coli* cells to the probe and their integrity were confirmed using scanning electron microscopy (SEM). While the "approach curve" provides information on long-range physicochemical interactions between the (modified) probe and substrate, the "retraction curve" provides information on the "pull-off" force (force required to break the interaction between the probe and substrate) and "pull-off" distance (Abu-Lail & Camesano, 2002, 2003; Cao et al., 2006; Wang, Palmer, Schwartz, & Razatos, 2004). From the approach and retraction curves generated between the modified probe and the different substrates, they noted that FAS/silicone showed the lowest pull-off force (with the *E. coli* cells). They also observed that the pull-off forces for silicone, heparin/OTS/silicone, and hyaluronan/OTS/silicone were comparable. Bacterial adhesion on the different surfaces was also evaluated using SEM. SEM images (of bacterial adhesion on the different surfaces) revealed that hyaluronan/OTS/silicone and heparin/OTS/silicone were the surfaces with highest number of *E. coli* cells. The investigations of Cao et al. (2006) further revealed that while FAS/silicone and OTS/silicone showed potential for minimizing bacterial adhesion, heparin/OTS/silicone and hyaluronan/OTS/silicone facilitated bacterial adhesion. These results also helped them conclude that hydrophobicity alone does not determine *E. coli* adhesion to substrates and that the nature of functional groups on the surface of substrate plays important role in the interaction between *E. coli* cells and substrates.

Polymeric biomaterials used in biomedical devices are mainly based on polylactic acid, polyglycolic acid, and their copolymers (Mauclaire, Brombacher, Bünger, & Zinn, 2010). Poly([R]-3-hydroxyalkanoate) (PHA) is one of the (biodegradable and biocompatible) polymers, which has been reported to have potential for applications in the medical field (Mauclaire et al., 2010). Mauclaire and his coworkers investigated the influence of PHA material properties on adhesion of *Staphylococcus aureus* and *E. coli* cells using epifluorescence microscopy (EM). Results of their study indicated that surface energy of the bacterial cell wall and presence of impurities in the polymer coating influenced the adhesion of bacterial cells.

Pranzetti et al. (2012) studied the initial stages of bacterial adhesion using surface plasmon resonance (SPR). SPR can detect molecular interactions on a metal surface in real time in a label-free environment (Pranzetti et al., 2012). Pranzetti et al. (2012) used self-assembled monolayers (SAMs) of organosulfur compounds on gold substrates to study the effect of charge and wettability on bacterial adhesion. These substrates were characterized using contact angle measurement and ellipsometry. SPR was then used to investigate the effect of aforementioned factors on bacterial adhesion as a function of SAM type. They reported that electrostatic interactions played an important role in the initial bacterial adhesion process and that the physicochemical characteristics of the substrates had a major effect on bacterial adhesion of both *Cobetia marina* and *Marinobacter hydrocarbonoclasticus*. This was supported by the higher bacterial adhesion on charged SAMs than in neutral SAMs. They also reported that conditioning by EPS reduced the effect of the surface properties on bacterial adhesion.

Investigations into the factors affecting adhesion of *Helicobacter pylori* to plumbing materials were carried out because of their presence in biofilms associated with drinking water. This is because *H. pylori* is one of the most common infectious agents and the issue of antimicrobial resistance has resulted in partial failure of standard therapeutic treatments with antibiotics (Mégraud, 2004a, 2004b; Parreira et al., 2011). Azevedo, Pacheco, Keevil, and Vieira (2006) investigated the effect of different environmental conditions on the adhesion of *H. pylori* cells to stainless steel 304 coupons. Computational studies revealed that high shear stresses had a negative effect on the adhesion of *H. pylori* to stainless steel and polypropylene. Investigations (using EM) on the effect of temperature and inoculation concentration revealed that they had no effect on adhesion to substrate. They also observed a lack of formation of three-dimensional structures. The aforementioned results helped them conclude that the presence of the pathogen in well water could be related to the ability of *H. pylori* to integrate into biofilms under conditions of low shear stress (Azevedo, Pacheco, et al., 2006). In continuation of this work, Azevedo, Pinto, Reis, Vieira, and Keevil (2006) studied the effect of different substrates on the adhesion of *H. pylori* cells, their cell morphology, and their ability to form three-dimensional structures. Studies on adhesion pattern and changes to cell morphology were done using EM and SEM. They observed that all abiotic surfaces were rapidly colonized by the bacterial cells with cell morphology depending on the substrate with spiral morphology being associated with nonpolymeric substrate. In addition, large three-dimensional structures associated with the formation of biofilm were mostly observed on the surface of copper substrates. However, further studies on these structures revealed that most of *H. pylori* cells attached to copper or stainless steel had a compromised cell membrane after only 48 h. Parreira et al. (2011) investigated the effect of surface chemistry on the adhesion, viability, and morphology of *Helicobacter pylori* using (hydrophilic and hydrophobic) SAMs of alkanethiols on gold. Studies on bacterial adhesion, viability, and morphology were conducted using FM and SEM. AFM (intermittent contact mode) was used to characterize the fabricated SAMs. Zeta potentials of the different SAMs were determined using a commercial electrokinetic analyzer and the water contact angle of the different *H. pylori* strains was also determined. Parreira et al. (2011) reported that bacterial adhesion was highest on the hydrophobic SAMs (except for 17875/Leb strain) with the nonfouling EG4-SAM preventing bacterial adhesion. They also observed the spiral morphology of *H. pylori* only on the hydrophobic SAMs and on the control for 2 h, after which it changed to coccoid. From the results of their experiments, Parreira et al. (2011) concluded that different *H. pylori* strains exhibit different adsorption patterns and morphologies due to the differences in surface chemistry and functional groups that are exposed.

Xu et al. (2013) quantitatively measured the adhesion between *E. coli* ZK1056 biofilm cells and different substrates with distinct and well-characterized surface chemistries using AFM. For this, *E. coli* biofilms were grown on polylysine-coated AFM cantilevers. This nonpathogenic strain was chosen because of its ability to form biofilms on a wide range of substrates and hence, can serve as a model system to evaluate the behavior of pathogenic *E. coli* strains. Ellipsometry and contact angle measurements were used to characterize mica, silicon, and silanized silicon surfaces. Adhesion measurements (using AFM) revealed that after only 0–10 s of contact time, the biofilms adhered strongly to fluorinated, amine-terminated and mica and weakly to antifouling PEG monolayers. This revealed the ability of the cells to adapt to the wide variation in hydrophobicity and charge of the different substrates. Xu and his coworkers also demonstrated the critical role played by type I pili in adhesion by this strain to the different surfaces.

3. Effect of surface physical characteristics on bacterial adhesion: Applications of analytical techniques

Metals are generally used to fabricate biomedical implants because of their excellent mechanical properties (Katti, 2004; Talha, Ma, Kumar, Lin, & Singh, 2019; Virtanen et al., 2008). Titanium and its alloys (in particular) are the materials of choice for fabrication of biomedical implants because of their biocompatibility, chemical stability, and excellent mechanical properties (Brunette, Tengvall, Textor, & Thomsen, 2001; Estrin, Kasper, Diederichs, & Lapovok, 2009; Geetha, Singh, Asokamani, & Gogia, 2009; Hannig & Hannig, 2010; Kim et al., 2008; Liu, Chu, & Ding, 2004; Mendonça, Mendonça, Aragao, & Cooper, 2008; Valiev et al., 2008).

One of the most important process in determining the success of an orthopedic implant is osseointegration. Osseointegration is defined as the formation of a direct structural and functional connection between the living bone and the surface of implant (Lorenzetti et al., 2014). Parameters such as physicochemical composition, wettability, surface charge, stiffness, and surface topography of the biomaterial's superficial zone (which is in direct contact with fluids and corporal tissues) determine the osseointegration of the implant (Lorenzetti et al., 2014). Hence, to improve osseointegration, it is imperative that the surface of implant is modified (and optimized) using suitable techniques. However, such biomedical devices (with engineered surfaces) are susceptible to biofouling, which ultimately leads to their failure (Rizzello et al., 2011). This is because once bacterial cells colonize the biomaterial surface, their gene expression changes and growth rates are altered. As a result, defense mechanisms of the host are no longer able to remove them from the body (Gorth et al., 2012). In addition, antibiotic-resistant strains cannot be treated by antibiotic therapy after their adhesion to the prosthetic surface. This results in the development of a biofilm, which is responsible for many chronic infections and prevents proper integration of the implant to the surrounding tissue (Gorth et al., 2012). Treatment of such implant-related infections requires extensive repeat surgery, implant extrication, extended duration of systemic antibiotic therapy, bone loss, and substantial cost, suffering, and disability (Gorth et al., 2012). Although antibiotic-loaded implants have demonstrated superior curative effects, their potential toxicity and increased bacterial drug resistance due to the slow-release doses have increased risks associated with surgery (Zhang, Zhang, & Li, 2013). In light of the above, it has been considered that the most effective way to prevent failure of implants would be to minimize the initial bacterial adhesion. To achieve this, a proper understanding of the effect of surface topography on bacterial adhesion is necessary.

The effect of surface roughness on bacterial adhesion to titanium surfaces was reported by Truong et al. (2010). They investigated the effect of nanoscale roughness of ultrafine grained titanium (fabricated using equal channel angular pressing (ECAP) technique) on the attachment of *S. aureus* and *Pseudomonas aeruginosa* cells. To understand the role of surface conditioning (due to grain refinement) on bacterial proliferation, they used techniques such as AFM, CLSM, optical profilometry (OP), X-ray photoelectron spectrometry (XPS) and high-resolution SEM.

AFM and OM were used to characterize surface topography, and the data obtained indicated that the peaks were more uniform and higher on ECAP processed titanium when compared with as-received titanium substrates, thereby reflecting the ultrafine grain size of titanium substrate. Contact angle measurements showed that ECAP titanium was (slightly) more hydrophobic when compared with as-received titanium. XPS was used to determine the chemical composition of the two titanium substrates, and the results indicated that titanium and oxygen were the two most abundant elements. SEM and CLSM were used to evaluate the attachment of both *S. aureus* and *P. aeruginosa* cells to ECAP processed titanium and as-received titanium surfaces. The investigations revealed that both the bacterial cells displayed a distinct preference for ECAP processed titanium (even though the chemistry and wettability of both the as-received and the ECAP modified titanium were similar). This led them to suggest that the surface architecture of ECAP processed titanium could have driven the attachment of (spherical) *S. aureus* cells (to its surface). The aforementioned results were in excellent agreement with other reports on the responses of bacterial cells to changes in surface topography of glass and polymer surfaces (Ivanova et al., 2008; Mitik-Dineva et al., 2009). These investigations helped confirm the hypothesis that substrate topography and morphology play a crucial role in the attachment of bacterial cells to surfaces (Ivanova et al., 2008; Mitik-Dineva et al., 2009; Truong et al., 2010).

The effect of surface chemistry (as one of the factors influencing bacterial cell attachment to a surface) has been extensively studied and (relatively) well understood (Fletcher & Loeb, 1979; Jacobs, Lafolie, Herry, & Debroux, 2007; Mauclaire et al., 2010; Oliveira, Song, Alves, & Mano, 2011; Statz, Barron, and Messersmith, 2008) when compared with the effect of surface topography (Díaz, Schilardi, Salvarezza, & Fernández Lorenzo de Mele, 2007; Emerson & Camesano, 2004; Fadeeva et al., 2011; Ivanova et al., 2010; Mitik-Dineva et al., 2009; Neoh, Hu, Zheng, & Kang, 2012; Puckett, Taylor, Raimondo, & Webster, 2010; Whitehead, Colligon, & Verran, 2005). This is because some of the earlier studies used only average and root mean square (RMS) roughness (to characterize surface topography) in order to determine the effect of surface topography on bacterial cell attachment (Boulangé-Petermann, Rault, & Bellon-Fontaine, 1997; Medilanski, Kaufmann, Wick, Wanner, & Harms, 2002; Mitik-Dineva et al., 2008). These roughness parameters have been reported to be inadequate and inconsistent in characterizing the sample surface (Anselme et al., 2010; Medilanski et al., 2002; Neoh et al., 2012; Whitehead et al., 2005) as average and RMS roughness parameters describe the deviations about a mean profile line (in only one dimension) without taking into account the shape or spatial distribution of peaks. As a consequence, two surfaces with very different surface structures may have identical roughness parameters (Brune, Hellborg, Whitlow, & Hunderi, 1997; Kannan et al., 2014; Thomas & Rosén, 2000; Webb et al.; 2012). To overcome this shortcoming, Webb et al. (2013), in an extension of the work done by Truong et al. (2010), used nine different surface roughness parameters to comprehensively characterize the sample surface. In addition to average and RMS roughness, they also calculated a range of roughness parameters that reflected the spatial arrangement and the shape of the structures on the sample surface. Webb et al. (2013) investigated the attachment of *P. aeruginosa* and *S. aureus* on three different (subnanometrically smooth) titanium substrates. Three different titanium substrates were fabricated—chemical mechanically polished titanium, ECAP processed titanium, and magnetron-sputtered titanium thin films onto clean silicon wafers. AFM images were resolved in tapping mode, and nine individual roughness parameters were calculated (from the AFM images) for each of the titanium substrates. AFM analysis indicated that the surfaces of all three substrates were nanometrically smooth. AFM analysis further revealed that the (titanium) surface fabricated by sputtering was the roughest (over a scan area of 1 μm × 1 μm) and had the most even distribution of peaks and valleys. Bacterial attachment to the (fabricated) titanium surfaces was studied using Autodesk Maya software. Studies on bacterial attachment revealed that the attachment of spherical *S. aureus* cells was more favorable (to the three titanium surfaces) when compared with the attachment of rod shaped *P. aeruginosa* cells. This, they suggested, was favored by the rougher titanium surfaces that had an even distribution of peaks and valleys (surface features that are not reflected in traditionally used roughness parameters but reflected in other less used roughness parameters such as skewness, developed surface area ratio, and autocorrelation length).

The interaction between biomedical implants and the physiological environment is controlled by the use of nanostructured biomaterials (Singh et al., 2011). As soon as the implants are inserted into the body, they are covered with a layer of adsorbed protein (Talha et al., 2019). This layer of protein plays a role in cell adhesion (Singh et al., 2011; Talha et al., 2019). In this context, there have been a few reports describing the role of nanoscale morphology on protein adsorption (Cai, Bossert, & Jandt, 2006; Han, Sethuraman, Kane, & Belfort, 2003; Rechendorff, Hovgaard, Foss, Zhdanov, & Besenbacher, 2006; Riedel, Müller, & Wintermantel, 2001; Roach, Farrar, & Perry, 2006; Röcker, Pötzl, Zhang, Parak, & Nienhaus, 2009) with very different results. This was due to the lack of suitable techniques to study protein adsorption on nanostructured surfaces. Scopelliti et al. (2010) overcame this by using supersonic cluster beam deposition (SCBD) to produce samples of nanostructured titanium substrates with different surface roughness. They used CM and AFM to study the amount of adsorbed protein and to characterize the structure of the adsorbed layer. As a consequence, they were able to show that adsorption of proteins depends significantly on surface nanotopography.

Singh et al. (2011) extended the work of Scopelliti et al. (2010) to quantitatively evaluate the influence of surface properties (of titanium substrates) on protein adsorption as well as adhesion of bacterial cells. The results of this study, they reasoned, could help in the design of implant surfaces that would be able to promote osseointegration as well as inhibit bacterial colonization and subsequent biofilm formation (Stevens & George, 2005). Singh et al. (2011) prepared titanium substrates of varying roughness using SCBD. This technique was selected because nanostructured surfaces could be fabricated with quantitative control over both surface morphology and surface chemistry (Barborini et al., 2003; Kholmanov et al., 2003; Podestà et al., 2009). Also, the morphological features produced using this technique were reported to resemble the nanostructures found in the extracellular matrix (ECM) (Abrams, Goodman, Nealey, Franco, & Murphy, 2000; Carbone et al., 2006), thereby making them "model" surfaces to study protein adsorption and cell response (Belicchi et al., 2010; Scopelliti et al., 2010).

The surface morphology of surfaces (with varying roughness) was resolved and quantitatively characterized using AFM (Fig. 2). The surface chemical characteristics were determined using X-ray photoelectron spectroscopy (XPS), contact angle, and surface energy measurements. (Singh et al., 2011) observed that the surfaces were highly porous (RMS value between 16 nm and 32 nm) with a granular structure with no change in their wetting character. The effect of different nanostructured samples on the amount of adsorbed protein, adhesion on *E. coli* and *S. aureus* cells (Figs. 3 and 4, respectively) and subsequent biofilm formation was evaluated using CLSM.

Singh et al. (2011) reported that bacterial adhesion and subsequent biofilm formation did not strictly increase with surface roughness of the sample. They observed that bacterial adhesion initially increased with surface roughness (at low corrugations) after which it decreased. They also noted that accumulation of proteins on the rougher surfaces downplayed bacterial adhesion and biofilm formation. This, they suggested, was due to the formation of a thick (passivating) layer of proteins, which flattened the (rough) surface, thereby reducing the interaction of bacteria with the nanostructured surface. As a consequence of this "flattening effect" of the protein layer, the effect of nanostructures on microbial cell adhesion was greatly reduced. These results led them to suggest that surface roughness is not the only morphological parameter that influences bacteria

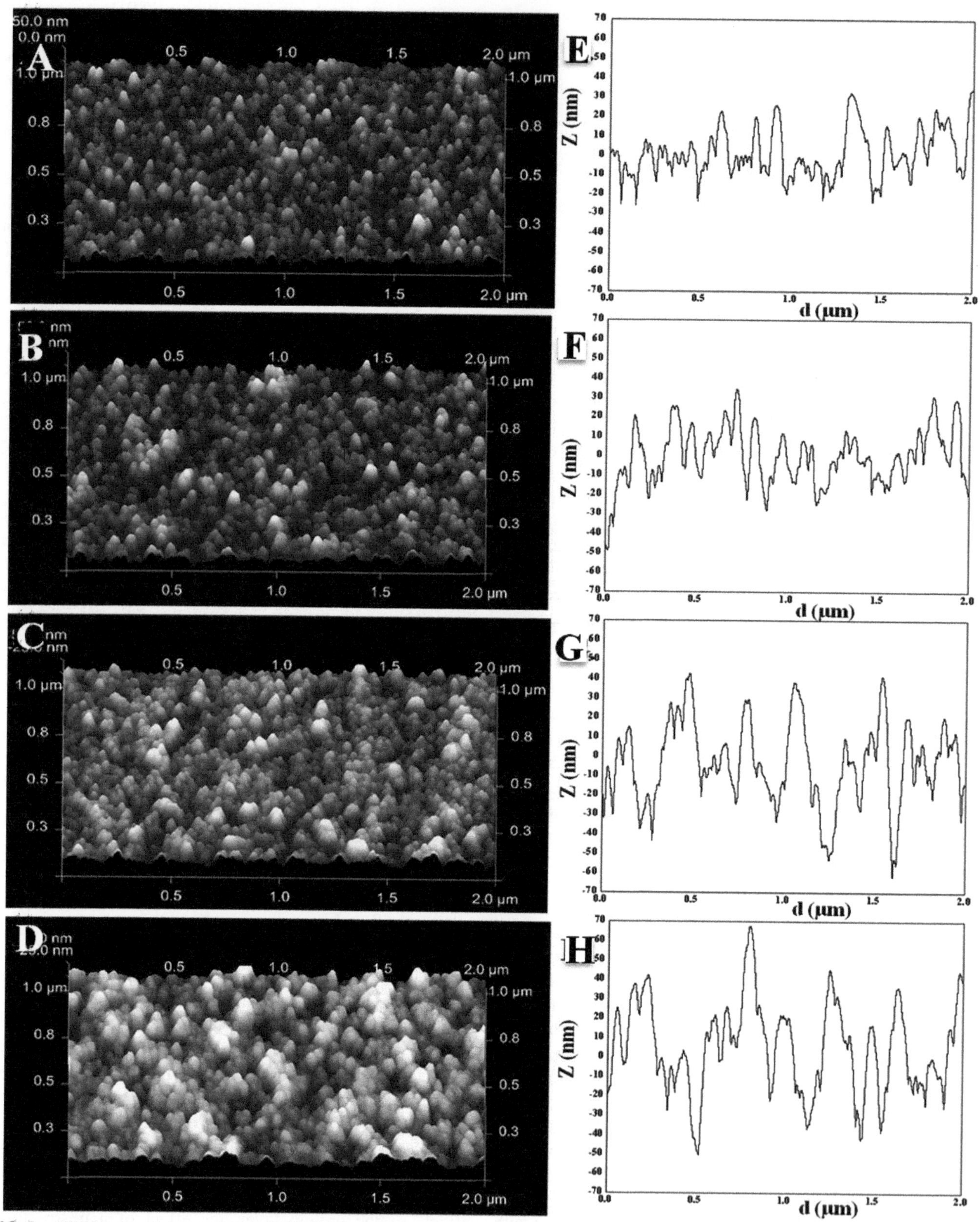

FIG. 2 AFM images of nanostructured titanium substrates resolved over a scan area of 2 μm × 1 μm. (A–D) Representative height images (in 3D) of nanostructured titanium surfaces films with increasing thickness (50, 100, 200, and 300 nm); (E–H) Representative surface profiles. *(Reproduced with permission from Singh, A. V., Vyas, V., Patil, R., Sharma, V., Scopelliti, P.E., Bongiorno, G., et al. (2011). Quantitative characterization of the influence of the nanoscale morphology of nanostructured surfaces on bacterial adhesion and biofilm formation. PLoS ONE, 6(9), e25029. https://doi.org/10.1371/journal.pone.0025029.)*

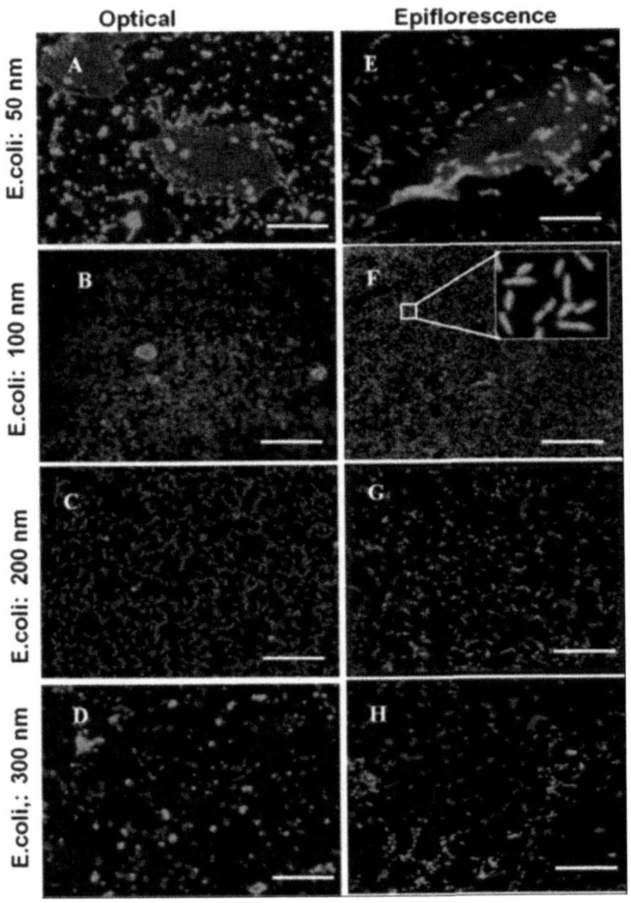

FIG. 3 Optical and epifluorescence micrographs of live/dead *E. coli* cells. (A–D) Optical microscopy images. (E–H) CLSM epifluorescence images of cells stained with BacLight Live *(green, light gray in print version)*/Dead *(red, dark gray in print version)* staining kit. *(Reproduced with permission from Singh, A. V., Vyas, V., Patil, R., Sharma, V., Scopelliti, P.E., Bongiorno, G., et al. (2011). Quantitative characterization of the influence of the nanoscale morphology of nanostructured surfaces on bacterial adhesion and biofilm formation. PLoS ONE, 6(9), e25029. https://doi.org/10.1371/journal.pone.0025029.)*

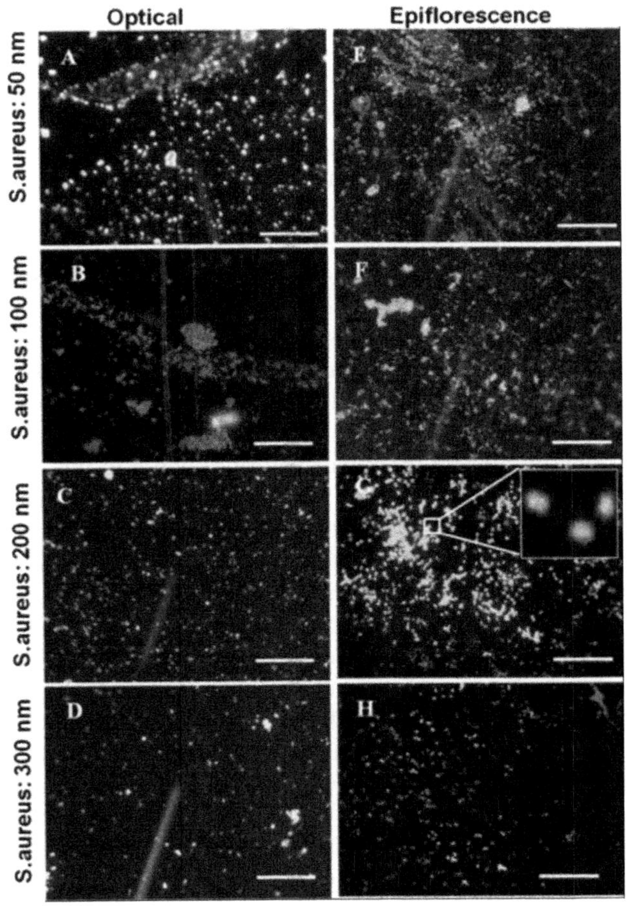

FIG. 4 Optical and epifluorescence micrographs of live/dead *S. aureus* cells. (A–D) Optical microscopic images. (E–H) CLSM epifluorescence images of cells stained with BacLight Live/Dead staining kit. *(Reproduced with permission from Singh, A. V., Vyas, V., Patil, R., Sharma, V., Scopelliti, P.E., Bongiorno, G., et al. (2011). Quantitative characterization of the influence of the nanoscale morphology of nanostructured surfaces on bacterial adhesion and biofilm formation. PLoS ONE, 6(9), e25029. https://doi.org/10.1371/journal.pone.0025029.)*

adhesion and that it is possible to optimize the surface of titanium implants to promote osseointegration while simultaneously inhibiting bacterial colonization.

Initially bacterial cells were considered to be inert colloidal micro particles (while interacting with a surface) (Anselme et al., 2010; Dorobantu, Bhattacharjee, Foght, & Gray, 2009; Liu, Strauss, & Camesano, 2008; Mandlik, Swierczynski, Das, & Ton-That, 2008; Ong, Razatos, Georgiou, & Sharma, 1999; O'Toole & Kolter, 1998; Razatos, Ong, Sharma, & Georgiou, 1998; Riedewald, 2006; Scardino, Harvey, & De Nys, 2006; Shellenberger & Logan, 2002; Sheng, Ting, & Pehkonen, 2007), but studies have shown that physicochemical properties (of a substrate) can regulate biological responses (Jefferson, 2004; Mitragotri & Lahann, 2009; Popham & Young, 2003; Rizzello et al., 2011; Thanassi & Hultgren, 2000; Verma et al., 2008). Rizzello et al. (2011) investigated the effect of nanostructured surfaces on the proteomic profile, expression of regulatory and structural genes, and the morphology of *E. coli*. For this, nanostructured gold substrates were fabricated using an electroless plating technique—spontaneous galvanic displacement reaction (SGDR). AFM analysis of the nanostructured substrates revealed the presence of high protrusions distributed uniformly across the surface. The fabricated surfaces had surface roughness of ranging from 30 to 100 nm. Flat gold substrates (fabricated by thermal evaporation onto premodified aminopropyltriethoxysilane glass slides) and glass substrates with roughness less than 1 nm were used as reference substrates. The surface characteristics of the substrates were evaluated using water contact angle measurements. The data confirmed that the surfaces were hydrophilic in nature, which (they suggested) was due to the adsorption of proteins on the surface of substrates. Energy-dispersive X-ray analysis of the nanostructured surfaces further revealed the absence of contaminants. *E. coli* was grown on the different gold substrates (varying in their surface roughness) and substrates taken for reference to study the response of *E. coli* cells to the change in surface topography. The cells were monitored for any changes in morphology in response to the (physical) stimuli using AFM and SEM analysis. A clear change in the morphology of *E. coli* cells (grown on substrates with roughness from 20 nm to 100 nm), namely the absence of type-1 fimbriae, was observed. These organelles enable the cell to target and colonize specific host tissues (Klemm & Schembri, 2004). Their key role in promoting the formation of biofilm, enhancing the virulence of *E. coli*, enhancing the inflammatory response to infection has been reported (Connell et al., 1996; Klemm, Vejborg, & Hancock, 2010; Pouttu et al., 1999; Schembri, Kjærgaard, & Klemm, 2003). This biological response to nanotopography was confirmed using SEM analysis. This led (Rizzello et al., 2011) to conclude that surface features of the order of 0.01% in terms of the nanostructure/bacterium volume ratio are sufficient to elicit a biological response in bacterial cells. They also reported that the cells (initially grown on nanostructured surfaces) recovered the ability to express fimbriae when incubated on flat surfaces, thereby indicating that the variation in expression of fimbriae was a transient phenomenon. To further evaluate the effect of surface topography on the biological response of bacterial cells, (Rizzello et al., 2011) investigated the genomic and proteomic profile of *E. coli*. Proteomic profile studied using 2D-DIGE revealed a difference in expression of proteins involved in transcription regulation, transport of oligopeptides and amino acids, energy metabolism, stress response, and synthesis and modification of macromolecules. Genomic profile revealed that nanostructured surfaces resulted in a loss of expression of virulence factors.

4. Development of antimicrobial surfaces: Applications of analytical techniques

Advances in science combined with an increase in life expectancy of an aging population have led to an exponential increase in the use of biomedical devices. Associated with this development is the serious concern of device-associated infections (DAIs) as a consequence of formation of biofilms. The nature of biofilms and the increased tolerance to the effect of antimicrobial agents exhibited by cells in the biofilm matrix have dictated the need for a multi-target approach. Considering this, different approaches such as preventing microbial attachment and/or killing the microbe upon contact with the surface have been proposed for the development of antifouling surfaces (Ganjian et al., 2019). Since the surfaces that prevent the attachment of bacterial cells also tend to have a negative effect on the osseointegration of the implant, they are considered to be of little use (Ganjian et al., 2019). Hence, researchers have focused their efforts on the fabrication of noncytotoxic bactericidal surfaces that selectively kill bacterial cells while at the same time promoting osseointegration. The aforementioned bactericidal surfaces employ a chemical or physical killing mechanism (Ganjian et al., 2019). The bactericidal surfaces employing chemical methods are tuned to deliver antimicrobial agents. However, the downside to this technique is the possible development of bacterial resistance to the delivered antimicrobial agents (Elbourne, Crawford, & Ivanova, 2017; Enright et al., 2002). Another significant drawback of this technique is the introduction of potentially toxic agents into the body that can cause tissue toxicity. Hence, as an alternative, engineering surfaces (to obtain high-aspect-ratio surface topographies) that exhibit bactericidal properties as a consequence of physicomechanical interactions between the nanostructures (of substrate) and the microbial cells have been considered. A number of researchers have taken inspiration from naturally occurring bactericidal surfaces to develop next-generation surfaces (that exhibit bactericidal activity due to the presence of physical structures) for use in medical devices.

Ivanova's group was among the first to report the bactericidal effect of the surface of Cicada wings. Cicada wings have been reported to be mostly hydrophobic with antireflective, antiwetting, and self-cleaning properties (Feng & Jiang, 2006; Stoddart, Cadusch, Boyce, Erasmus, & Comins, 2006; Sun, Feng, Gao, & Jiang, 2005; Watson, Myhra, Cribb, & Watson, 2008). To confirm whether there is a direct relationship between the self-cleaning and antibiofouling properties of Cicada wings, Ivanova et al. (2012) investigated the adhesion behavior of *P. aeruginosa* cells on cicada wing surfaces. Results of SEM and CLSM analysis revealed that the surface of Cicada wing was not effective at preventing the adhesion of *P. aeruginosa* cells. However, results of SEM analysis helped confirm the change in the morphology of cells when in contact with the surface of Cicada wing. This was attributed to the penetration of cells by the nanopillars (present on the surface), which resulted in the spillage of cellular components. CLSM analysis revealed that the aforementioned cells were dead. Further analysis involving the use of SEM and CLSM revealed that most of the cells were killed within 5 min. AFM analysis revealed the slow downward movement (approximately 200 nm) of the cells into the surface of the wing followed by a sudden short downward displacement, which resulted in cell rupture. To further confirm the bactericidal mechanism involved, Cicada wings were sputtered with gold, which resulted in a 10 nm thick film. This resulted in a less hydrophobic surface with AFM analysis revealing no change to the surface morphology of gold-coated surface of wing (when compared with native surface of wing).

SEM and CLSM analysis revealed that the bactericidal activity of gold-coated surface was preserved in spite of the significant change to the surface chemistry of wings. These results helped Ivanova et al. (2012) to conclude that the physical structures on the surface of wing were primarily responsible for the bactericidal effect on *P. aeruginosa* cells. These nanostructures were reported to be 200nm in height, 100nm in diameter at the base, and 60nm in diameter at the cap, with an interpillar spacing of 170nm (Ivanova et al., 2012). Ivanova et al. (2012) also reported that though the surface of Cicada wing was not antifouling, the formation of biofilm was prevented because it was highly bactericidal. These results led them to suggest that the nanostructures imparted bactericidal and hydrophobic properties to the surface. A theoretical study performed by Pogodin et al. (2013) supported the aforementioned findings. In an extension of the work by Ivanova et al. (2012), Hasan et al. (2013) conducted a wide spectrum bacterial study on the bactericidal activity of Cicada wings. SEM and CLSM analysis revealed that *P. fluorescens*, *E. coli*, *B. catarrhalis*, and *P. aeruginosa* cells that attached to the surface of the wing were deformed and killed in the same way as reported by Ivanova et al. (2012). These results helped Hasan et al. (2013) conclude that the bactericidal activity exhibited by Cicada wings were unaffected by the morphology of the organism. However, SEM and CLSM analysis of Gram-positive species such as *B. subtilis*, *S. aureus*, and *P. maritimus* revealed that all Gram-positive cells were unaffected by the nanostructures on Cicada wing. Hasan et al. (2013) concluded that the bactericidal activity of Cicada wing was independent of bacterial morphology.

Kelleher et al. (2016) evaluated the bactericidal activity of three different Cicada wings on *P. fluorescens* cells. CLSM and SEM analysis revealed that Cicada wings of *Megapomponia intermedia* and *Cryptotympana aguila* exhibited strong bactericidal properties. This, they suggested, was due to the increase in bactericidal activity as with decrease in nano-pillar pitch and diameter. Xue, Liu, Guo, Zhang, and Li (2015) explained these results with a theoretical model that proposed that shorter sharp nanopillars cause a greater strain on bacterial membranes than taller nanopillars.

With the work of Ivanova et al. (2012) as basis, development of synthetic biomimetic antimicrobial surfaces was pioneered by Ivanova et al. (2013). They investigated the bactericidal activity of black silicon on both Gram-positive and Gram-negative bacteria. For this, nanostructures with high aspect ratio were fabricated on silicon substrates (black silicon) using reactive-ion etching technique. The resulting structures mimicked the surface nanofeatures found on the wings of *Diplacodes bipunctata* species of dragon fly (Ivanova et al., 2013). SEM analysis revealed the presence of hierarchical nanostructures on both the surfaces (black silicon and *D. bipunctata* wings) form structures through the formation of clusters of adjacent nanoprotrusions. However, the nanostructures were taller, sharper, more discretely arranged and displayed a lower degree of clustering on black silicon. CLSM analysis and results of viability assays confirmed the bactericidal effect of both surfaces against both Gram-negative and Gram-positive bacteria with a kill rate of 450,000 cells min^{-1} cm^{-2} over the first 3h, which further declined to ~50,000 cells min^{-1} cm^{-2} for *S. aureus* (Ivanova et al., 2013). The bactericidal efficacy of black silicon was made evident when the kill rate was taken in context of minimum infective dose (*P. aeruginosa*—10^3 cells, *S. aureus*—10^5 cells, and *Bacillus cereus*—10^5 cells) (Ivanova et al., 2013). This study established the potential bactericidal properties of mechano-responsive surfaces that incorporate biomimetic analogs and also highlighted the need for development and optimization of such surfaces for their use in novel next-generation bactericidal devices. Wang et al. (2016) showed that such bactericidal surfaces are not limited for use in medical devices alone by incorporating black silicon in a microfluidic device to purify contaminated water. Viability assays, SEM and CLSM analysis confirmed 99% killing efficiency of the device. This represented a chemical-free method for cleaning water contaminated by bacteria.

Dickson, Liang, Rodriguez, Vollereaux, and Yee (2015) investigated the bactericidal effect of nanopillars of different periodicities on *E. coli* cells. For this, they fabricated biomimetic analogs of nanostructures found on Cicada wings on polymethyl methacralate films (PMMA) using nanoimprint lithography. The surface topography of fabricated surfaces was examined using SEM and AFM. After incubation of *E. coli* cells on the fabricated PMMA surfaces, SEM and AFM analysis revealed that *E. coli* cells were deformed. Composite fluorescence microscopy images confirmed that a higher number of cells were dead on fabricated PMMA than on flat films (Dickson et al., 2015). They also reported that smaller, closely spaced nanostructures were more bactericidal than larger and more widely spaced nanostructures. Further analysis of cell orientation data revealed that that the minimum threshold for optimal nanopillar spacing is between 130 and 380nm. The results from this study suggest that biomimetic analogs could be transferred to other materials to make effective antifouling medical devices. These findings were supported by the work of Kelleher et al. (2016). Fisher et al. (2016) were the first to report on the bactericidal effects of biomimetic diamond nanocone-patterned surfaces. These surfaces were biomimetic analogs of the surface of Cicada wings and were fabricated using microwave plasma chemical vapor deposition, followed by bias-assisted reactive ion etching. Two structurally distinct nanocone surfaces were produced and their surface topography was characterized using SEM. Raman spectroscopy was used to evaluate the phase composition of the nanocones formed, and the results indicated that they retained the microstructure of the microcrystalline diamond film (Fisher et al., 2016). SEM and FM analysis revealed bactericidal ability of both the nanostructured surfaces and that less dense, high-aspect-ratio nonuniform nanocone surfaces were more bactericidal when compared with uniform, highly dense nanocone surfaces (Fisher et al., 2016). Xue et al. (2015) explained these results with a mathematical model by suggesting that the antibacterial properties of a surface are enhanced when the nanostructures are less dense and sharper.

A number of researchers have fabricated biomimetic analogs on titanium substrates to produce bactericidal surfaces to be used as implants. Bhadra et al. (2015) fabricated titanium surfaces with nanowire arrays using chemical hydrothermal treatment followed by high-temperature treatment. The resulting surface architecture resembled the bactericidal nanopatterns found on dragonfly wings. An optical profilometer was used to characterize surface topography, and SEM was used to visualize the orientation of nanowire arrays. Wettability analysis revealed that the fabricated titanium surfaces were more hydrophobic when compared with the as-received titanium surfaces.

SEM and CM were used to study the attachment of *S. aureus* and *P. aeruginosa* to the two surfaces (Fig. 5), and the bactericidal activity of the fabricated titanium surface was confirmed using CM. Bhadra and coworkers reported that the nanowire arrays present on the fabricated titanium surface rendered them effective (to a greater extent) against *P. aeruginosa* cells. This, they suggested,

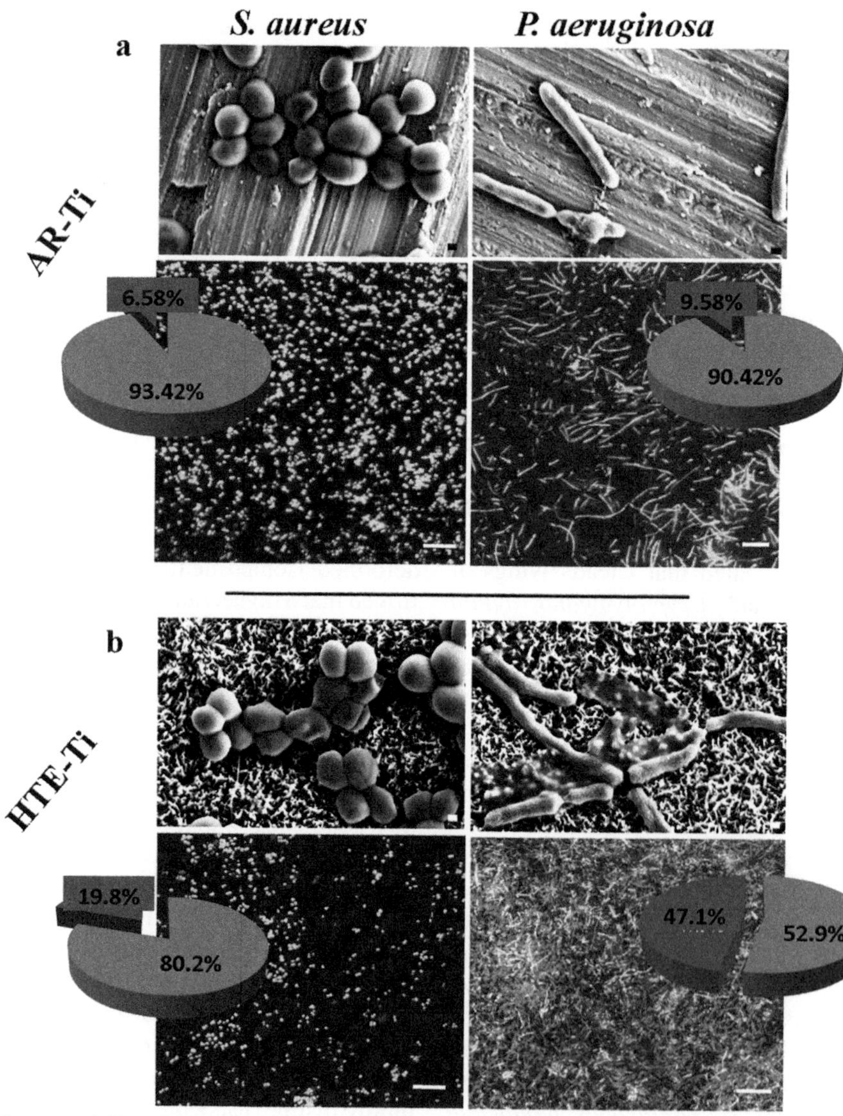

FIG. 5 Representative *S. aureus* (left) and *P. aeruginosa* (right) attachment patterns on the as-received titanium (AR-Ti) and hydrothermally etched titanium (HTE-Ti) surfaces after an 18 h incubation period. SEM images represent the attachment pattern. CSLM images reveal the viable (stained *green, gray in print version*) and the nonviable cells (stained *red, dark gray in print version*). The bactericidal activity is given in the individual pie charts. *(Reproduced with permission from Bhadra, C., Khanh Truong, V., Pham, V., et al. (2015). Antibacterial titanium nano-patterned arrays inspired by dragonfly wings.* Scientific Reports, 5, *16817. https://doi.org/10.1038/srep16817.)*

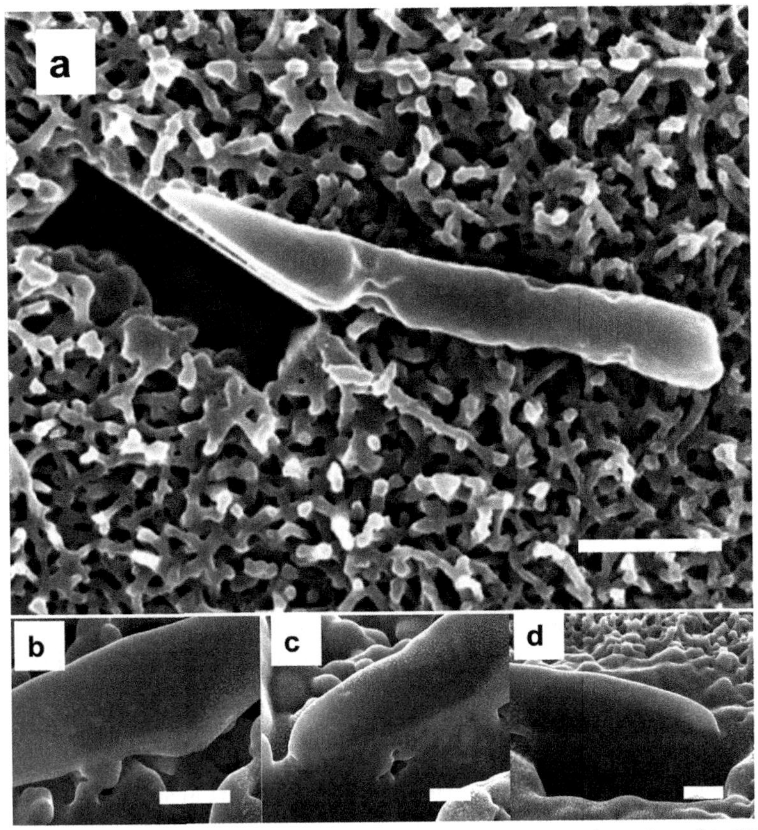

FIG. 6 Interaction between the membrane of the *P. aeruginosa* cells and the nanowires present on the surface of the HTE-Ti as visualized by FIB-SEM. (A) Top view of a *P. aeruginosa* cell on the HTE-Ti surface (scale bar 1 μm). (B–D) Engulfment of *P. aeruginosa* cell membrane into the nanowires on the HTE-Ti surface. *(Reproduced with permission from Bhadra, C., Khanh Truong, V., Pham, V. et al. (2015). Antibacterial titanium nano-patterned arrays inspired by dragonfly wings.* Scientific Reports, 5, *16817. https://doi.org/10.1038/srep16817.)*

was due to the higher susceptibility of the cell walls of *P. aeruginosa* to morphological deform (when compared with *S. aureus* cells).

They confirmed the morphological deformation of *P. aeruginosa* cell wall using focused ion beam scanning electron microscopy (Fig. 6). Bhadra et al. (2015) also suggested that a modification of parameters such as the heat treatment temperature and the concentration of the working solution could help in the fabrication of more effective bactericidal surfaces. Wandiyanto et al. (2019) investigated the bactericidal effect of nanostructured titanium surfaces against methicillin- and gentamicin-susceptible (MSSA) and resistant (MRSA) *Staphylococcus aureus* strains. Their hypothesis was that nanostructured surfaces will be effective against both the MSSA and MRSA *S. aureus* strains, as long as their cell wall structures and rigidity are not significantly different. *S. aureus* is a species that is highly adaptable and as a consequence, develops antibiotic resistance quickly. It has also been reported that methicillin-resistant *S. aureus* strains were isolated only 2 years after introduction of the antibiotics (Wandiyanto et al., 2019). Surfaces with nanostructures were fabricated by hydrothermal etching of titanium substrates. AFM and optical profilometry were used to characterize the topography of nanostructured titanium surfaces. Using SEM and CLSM, the researchers reported that the nanostructured titanium surfaces were highly bactericidal to both MSSA and MRSA strains. They concluded that nanostructures have the potential to overcome multidrug-resistant bacteria as the physical method of bacterial cell rupturing employs nonselective bactericidal activity.

There have also been reports that have assessed both bactericidal and specific bioactive properties of a titanium implant surface with nanowires. This is necessary as nanostructured implant surfaces are not of use if the cells of host cannot proliferate after coming into contact with such high-aspect-ratio surfaces. Tsimbouri et al. (2016) studied the bactericidal and osteogenic properties of different types of nanowires hydrothermally grown on titanium surfaces. SEM analysis of the fabricated surfaces revealed the effect of time (of hydrothermal treatment) on surface topography of titanium surfaces (Tsimbouri et al., 2016). SEM images revealed the presence of compact, homogeneous brush-like spiky features at 2–2.5 h and presence of intertwined structures at 3 h. Bactericidal studies involving the use of viability assays and CLSM revealed a kill rate of 30% after

1 h and 58% after 18 h toward *P. aeruginosa* cells by titanium nanowire substrates fabricated over a 2-h period. SEM images of punctured cells (after 18 h of attachment) confirmed the aforementioned results. In vitro studies with human osteoblast/osteoclast cocultures further demonstrated the osteogenic properties of the aforementioned titanium substrates (without an unwanted osteoclastic response) (Tsimbouri et al., 2016). This is very important as an optimal implant surface should show reduced risk of infection along with increased bone contact. Tsimbouri et al. (2016) suggested that the fabrication of such implants with innate bactericidal properties would lead to their use without high risk of infection. Diu et al. (2014) investigated the effect of cicada wing inspired nanowire topographies on bacterial colonization and proliferation of human osteoblast-like cells. The topographies were engineered using an alkaline hydrothermal process as a function of time (Diu et al., 2014). This resulted in the generation of brush type (dense coverage of spike-like structures) and niche-type structures on the titanium substrates (longer structures that intertwined at their tips with the formation of dispersed niches or pockets). SEM analysis revealed that these structures peaked at c. 3 μm and were an order of magnitude taller than the nanostructures on Cicada wings and black silicon (Diu et al., 2014). Their rationale for fabricating taller nanopillars was that taller spikes have the ability to pierce bacterial cells and hence mitigate bacterial adhesion. Fluorescence micrographs of *P. aeruginosa* on the titanium substrates revealed that the nanopillars were strongly bactericidal. However, fluorescence micrographs also revealed negligible bactericidal activity against *S. aureus* even though piercing of the cells by the nanopillars was evident. Further analysis (involving the use of FM) with Gram-negative *E. coli* and *K. pneumonia*, as well as Gram-positive *B. subtilis* and *E. faecalis*, revealed extensive bactericidal activity against motile cells (Diu et al., 2014). Fluorescence micrographs and scanning electron micrographs of osteoblast-like cells on the fabricated titanium substrates indicated that cell proliferation via distinguishable proliferation was supported.

A number of studies have reported the bactericidal potential of a nanostructured surface over a short time frame (Bhadra et al., 2015; Elliott, Wiggins, & Dua, 2021; Ferraris et al., 2019; Wandiyanto et al., 2020). Though bioinspired surfaces have been shown to be highly bactericidal, there are very few reports on the bactericidal potential of such surfaces over a clinically relevant time frame. This is important because biofilms are known to evolve over days and weeks (Bright et al., 2022). Bright et al. (2022) investigated the long-term potential of bioinspired surfaces (with high-aspect-ratio nanostructures) to resist biofilm formation as reduced bacterial viability (during the early stages) or delayed formation of biofilm on the surface of implant greatly improves the integration of host tissue (Bright et al., 2021). They evaluated the bactericidal efficacy of nanostructured titanium substrates of *S. aureus* and *P. aeruginosa* over a time period of 21 days. The nanostructures were fabricated by hydrothermal treatment of etched titanium (HTE). SEM and AFM analysis of the fabricated surfaces revealed the presence of nano-spikes with high aspect ratio. CLSM images revealed a significant reduction in live bacteria on the nanostructured surfaces (when compared with the control surfaces) throughout the time course for both bacteria (Bright et al., 2021). Further analysis revealed a significantly lower biovolume (on the nanostructured surfaces) with the bacterial cells killed not just on the surface but also at locations above the surface. SEM analysis revealed fragmented morphology of *S. aureus* and *P. aeruginosa* cells, thereby confirming contact killing of bacterial cells. Also, lack of mature and organized biofilm structures on the nanostructured surfaces was also confirmed using SEM analysis. These results led Bright et al. (2021) to suggest that the use of such (bactericidal) nanostructured implants may lead to reduced occurrence of implant-associated infections.

5. Conclusions

Advancements in analytical techniques have helped in visualizing and understanding bacterial interactions with surfaces. This has led to the development of synthetic bactericidal surfaces. These surfaces are biomimetic analogs of naturally occurring surfaces. Bacterial cells cannot develop resistance because of mechano bactericidal activity of high-aspect-ratio nanostructures on such surfaces. Recent research has also proved the biocompatibility of the aforementioned bactericidal analogs. In this context, optimization of the surface topography for maximum bactericidal activity, tuning of fabrication techniques, and scaling-up production of such nanomaterials to meet the ever-rising demand for prosthetic devices are considered the final frontiers.

Acknowledgment

The author would like to thank the Department of Science and Technology for their financial support (EMR/2017/004730). The author would also like to thank Mr. Kamlesh Pandey for his assistance.

References

Abrams, G. A., Goodman, S. L., Nealey, P. F., Franco, M., & Murphy, C. J. (2000). Nanoscale topography of the basement membrane underlying the corneal epithelium of the rhesus macaque. *Cell and Tissue Research*, *299*(1), 39–46.

Abu-Lail, N. I., & Camesano, T. A. (2002). Elasticity of *Pseudomonas putida* KT2442 surface polymers probed with single-molecule force microscopy. *Langmuir*, *18*(10), 4071–4081.

Abu-Lail, N. I., & Camesano, T. A. (2003). Role of lipopolysaccharides in the adhesion, retention, and transport of *Escherichia coli* JM109. *Environmental Science & Technology*, *37*(10), 2173–2183.

Anselme, K., Davidson, P., Popa, A. M., Giazzon, M., Liley, M., & Ploux, L. (2010). The interaction of cells and bacteria with surfaces structured at the nanometre scale. *Acta Biomaterialia*, *6*(10), 3824–3846.

Arnander, C., Bagger-Sjöbäck, D., Frebelius, S., Larsson, R., & Swedenborg, J. (1987). Long-term stability in vivo of a thromboresistant heparinized surface. *Biomaterials*, *8*(6), 496–499.

Azevedo, N. F., Pacheco, A. P., Keevil, C. W., & Vieira, M. J. (2006). Adhesion of water stressed *Helicobacter pylori* to abiotic surfaces. *Journal of Applied Microbiology*, *101*(3), 718–724.

Azevedo, N. F., Pinto, A. R., Reis, N. M., Vieira, M. J., & Keevil, C. W. (2006). Shear stress, temperature, and inoculation concentration influence the adhesion of water-stressed *Helicobacter pylori* to stainless steel 304 and polypropylene. *Applied and Environmental Microbiology*, *72*(4), 2936–2941.

Barborini, E., Kholmanov, I. N., Conti, A. M., Piseri, P., Vinati, S., Milani, P., et al. (2003). Supersonic cluster beam deposition of nanostructured titania. *The European Physical Journal D-Atomic, Molecular, Optical and Plasma Physics*, *24*(1), 277–282.

Barnett, B. J., & Stephens, D. S. (1997). Urinary tract infection: An overview. *The American Journal of the Medical Sciences*, *314*(4), 245–249.

Bayer, R. E. (1982). Conditioning surfaces to suit the biomedical environment. *Journal of Biomedical Engineering*, *140*, 257–271.

Belicchi, M., Erratico, S., Razini, P., Meregalli, M., Cattaneo, A., Jacchetti, E., et al. (2010). Ex vivo expansion of human circulating myogenic progenitors on cluster-assembled nanostructured TiO2. *Biomaterials*, *31*(20), 5385–5396.

Bhadra, C. M., Khanh Truong, V., Pham, V. T., Al Kobaisi, M., Seniutinas, G., Wang, J. Y., et al. (2015). Antibacterial titanium nano-patterned arrays inspired by dragonfly wings. *Scientific Reports*, *5*(1), 1–12.

Bos, R., Van der Mei, H. C., & Busscher, H. J. (1999). Physico-chemistry of initial microbial adhesive interactions—its mechanisms and methods for study. *FEMS Microbiology Reviews*, *23*(2), 179–230.

Bos, R, van der Mei, HC, Gold, J, & Busscher, HJ. (2000). Retention of bacteria on a substratum surface with micro-patterned hydrophobicity. *FEMS Microbiology Letters*, *189*(2), 311–315. https://doi.org/10.1111/j.1574-6968.2000.tb09249.x.

Boulangé-Petermann, L., Rault, J., & Bellon-Fontaine, M. N. (1997). Adhesion of *Streptococcus thermophilus* to stainless steel with different surface topography and roughness. *Biofouling*, *11*(3), 201–216.

Bowen, W. R., Fenton, A. S., Lovitt, R. W., & Wright, C. J. (2002). The measurement of *Bacillus mycoides* spore adhesion using atomic force microscopy, simple counting methods, and a spinning disk technique. *Biotechnology and Bioengineering*, *79*(2), 170–179.

Bright, R., Fernandes, D., Wood, J., Palms, D., Burzava, A., Ninan, N., et al. (2022). Long-term antibacterial properties of a nanostructured titanium alloy surface: An in vitro study. *Materials Today Bio*, *13*, 100176. https://doi.org/10.1016/j.mtbio.2021.100176.

Brune, D., Hellborg, R., Whitlow, H. J., & Hunderi, O. (1997). *Surface characterization*. Scandinavian Science Publication.

Brunette, D. M., Tengvall, P., Textor, M., & Thomsen, P. (2001). *Titanium in medicine: Material science, surface science, engineering, biological responses and medical applications* (p. 232). Berlin: Springer.

Cai, K., Bossert, J., & Jandt, K. D. (2006). Does the nanometre scale topography of titanium influence protein adsorption and cell proliferation? *Colloids and Surfaces B: Biointerfaces*, *49*(2), 136–144.

Cao, T., Tang, H., Liang, X., Wang, A., Auner, G. W., & Salley, S. O. (2006). Nanoscale investigation on adhesion of *E. coli* to surface modified silicone using atomic force microscopy. *Biotechnology and Bioengineering*, *94*(1), 167–176.

Carbone, R., Marangi, I., Zanardi, A., Giorgetti, L., Chierici, E., Berlanda, G., et al. (2006). Biocompatibility of cluster-assembled nanostructured TiO2 with primary and cancer cells. *Biomaterials*, *27*(17), 3221–3229.

Colvin, K. M., Gordon, V. D., Murakami, K., Borlee, B. R., Wozniak, D. J., Wong, G. C., et al. (2011). The pel polysaccharide can serve a structural and protective role in the biofilm matrix of *Pseudomonas aeruginosa*. *PLoS Pathogens*, *7*(1), e1001264. https://doi.org/10.1371/journal.ppat.1001264. 21298031. PMCID:PMC3029257.

Connell, I., Agace, W., Klemm, P., Schembri, M., Mărild, S., & Svanborg, C. (1996). Type 1 fimbrial expression enhances *Escherichia coli* virulence for the urinary tract. *Proceedings of the National Academy of Sciences*, *93*(18), 9827–9832.

Costerton, J. W., Lewandowski, Z., Caldwell, D. E., Korber, D. R., & Lappin-Scott, H. M. (1995). Microbial biofilms. *Annual Review of Microbiology*, *49*(1), 711–745.

Costerton, J. W., Stewart, P. S., & Greenberg, E. P. (1999). Bacterial biofilms: A common cause of persistent infections. *Science*, *284*(5418), 1318–1322.

Davenas, J., Thevenard, P., Philippe, F., & Arnaud, M. N. (2002). Surface implantation treatments to prevent infection complications in short term devices. *Biomolecular Engineering*, *19*(2–6), 263–268.

Díaz, C., Schilardi, P. L., Salvarezza, R. C., & Fernández Lorenzo de Mele, M. (2007). Nano/microscale order affects the early stages of biofilm formation on metal surfaces. *Langmuir*, *23*(22), 11206–11210.

Dickson, M. N., Liang, E. I., Rodriguez, L. A., Vollereaux, N., & Yee, A. F. (2015). Nanopatterned polymer surfaces with bactericidal properties. *Biointerphases*, *10*(2), 021010.

Diu, T., Faruqui, N., Sjöström, T., Lamarre, B., Jenkinson, H. F., Su, B., et al. (2014). Cicada-inspired cell-instructive nanopatterned arrays. *Scientific Reports*, *4*(1), 1–7. https://doi.org/10.1038/srep07122.

Dorobantu, L. S., Bhattacharjee, S., Foght, J. M., & Gray, M. R. (2009). Analysis of force interactions between AFM tips and hydrophobic bacteria using DLVO theory. *Langmuir*, *25*(12), 6968–6976.

Dunne, W. M., Jr. (2002). Bacterial adhesion: Seen any good biofilms lately? *Clinical Microbiology Reviews*, *15*(2), 155–166.

Elbourne, A., Crawford, R. J., & Ivanova, E. P. (2017). Nano-structured antimicrobial surfaces: From nature to synthetic analogues. *Journal of Colloid and Interface Science*, *508*, 603–616.

Elliott, D. T., Wiggins, R. J., & Dua, R. (2021). Bioinspired antibacterial surface for orthopedic and dental implants. *Journal of Biomedical Materials Research Part B*, *109*(7), 973–981.

Elving, G. J., van der Mei, H., Busscher, H., van Weissenbruch, R., & Albers, F. (2003). Influence of different combinations of bacteria and yeasts in voice prosthesis biofilms on air flow resistance. *Antonie Van Leeuwenhoek*, *83*(1), 45–55.

Emerson, R. J., IV, & Camesano, T. A. (2004). Nanoscale investigation of pathogenic microbial adhesion to a biomaterial. *Applied and Environmental Microbiology*, *70*(10), 6012–6022.

Enright, M. C., Robinson, D. A., Randle, G., Feil, E. J., Grundmann, H., & Spratt, B. G. (2002). The evolutionary history of methicillin-resistant *Staphylococcus aureus* (MRSA). *Proceedings of the National Academy of Sciences*, *99*(11), 7687–7692.

Estrin, Y., Kasper, C., Diederichs, S., & Lapovok, R. (2009). Accelerated growth of preosteoblastic cells on ultrafine grained titanium. *Journal of Biomedical Materials Research Part A: An Official Journal of The Society for Biomaterials, The Japanese Society for Biomaterials, and The Australian Society for Biomaterials and the Korean Society for Biomaterials*, *90*(4), 1239–1242.

Fadeeva, E., Truong, V. K., Stiesch, M., Chichkov, B. N., Crawford, R. J., Wang, J., et al. (2011). Bacterial retention on superhydrophobic titanium surfaces fabricated by femtosecond laser ablation. *Langmuir*, *27*(6), 3012–3019.

Feng, X. J., & Jiang, L. (2006). Design and creation of superwetting/antiwetting surfaces. *Advanced Materials*, *18*(23), 3063–3078.

Ferraris, S., Cochis, A., Cazzola, M., Tortello, M., Scalia, A., Spriano, S., et al. (2019). Cytocompatible and anti-bacterial adhesion nanotextured titanium oxide layer on titanium surfaces for dental and orthopedic implants. *Frontiers in Bioengineering and Biotechnology*, *7*(103), 1–12.

Fisher, L. E., Yang, Y., Yuen, M. F., Zhang, W., Nobbs, A. H., & Su, B. (2016). Bactericidal activity of biomimetic diamond nanocone surfaces. *Biointerphases*, *11*(1), 011014.

Flemming, H. C., Neu, T. R., & Wozniak, D. J. (2007). The EPS matrix: The "house of biofilm cells". *Journal of Bacteriology*, *189*(22), 7945–7947.

Flemming, H. C., & Wingender, J. (2010). The biofilm matrix. *Nature Reviews Microbiology*, *8*(9), 623–633.

Fletcher, M., & Loeb, G. (1979). Influence of substratum characteristics on the attachment of a marine pseudomonad to solid surfaces. *Applied and Environmental Microbiology*, *37*(1), 67–72.

Ganjian, M., Modaresifar, K., Ligeon, M. R., Kunkels, L. B., Tümer, N., Angeloni, L., et al. (2019). Nature helps: Toward bioinspired bactericidal nanopatterns. *Advanced Materials Interfaces*, *6*(16), 1900640. https://doi.org/10.1002/admi.201900640.

Geetha, M., Singh, A. K., Asokamani, R., & Gogia, A. K. (2009). Ti based biomaterials, the ultimate choice for orthopaedic implants—A review. *Progress in Materials Science*, *54*(3), 397–425.

Gorth, D. J., Puckett, S., Ercan, B., Webster, T. J., Rahaman, M., & Bal, B. S. (2012). Decreased bacteria activity on Si3N4 surfaces compared with PEEK or titanium. *International Journal of Nanomedicine*, *7*, 4829–4840.

Gristina, A. G. (1987). Biomaterial-centered infection: Microbial adhesion versus tissue integration. *Science*, *237*(4822), 1588–1595.

Hall-Stoodley, L., Costerton, J. W., & Stoodley, P. (2004). Bacterial biofilms: From the natural environment to infectious diseases. *Nature Reviews Microbiology*, *2*(2), 95–108.

Han, M., Sethuraman, A., Kane, R. S., & Belfort, G. (2003). Nanometer-scale roughness having little effect on the amount or structure of adsorbed protein. *Langmuir*, *19*(23), 9868–9872.

Hannig, M., & Hannig, C. (2010). Nanomaterials in preventive dentistry. *Nature Nanotechnology*, *5*(8), 565–569.

Hasan, J., Webb, H. K., Truong, V. K., Pogodin, S., Baulin, V. A., Watson, G. S., et al. (2013). Selective bactericidal activity of nanopatterned superhydrophobic cicada *Psaltoda claripennis* wing surfaces. *Applied Microbiology and Biotechnology*, *97*(20), 9257–9262. https://doi.org/10.1007/s00253-012-4628-5.

Healy, K. E., Thomas, C. H., Rezania, A., Kim, J. E., McKeown, P. J., Lom, B., et al. (1996). Kinetics of bone cell organization and mineralization on materials with patterned surface chemistry. *Biomaterials*, *17*(2), 195–208.

Ivanova, EP, Truong, VK, Wang, JY, Berndt, CC, Jones, RT, Yusuf, … Crawford, RJ. (2010). Impact of nanoscale roughness of titanium thin film surfaces on bacterial retention. *Langmuir*, *26*(3), 1973–1982. https://doi.org/10.1021/la902623c.

Ivanova, E. P., Hasan, J., Webb, H. K., Gervinskas, G., Juodkazis, S., Truong, V. K., et al. (2013). Bactericidal activity of black silicon. *Nature Communications*, *4*(1), 1–7. https://doi.org/10.1038/ncomms3838.

Ivanova, E. P., Hasan, J., Webb, H. K., Truong, V. K., Watson, G. S., Watson, J. A., et al. (2012). Natural bactericidal surfaces: Mechanical rupture of *Pseudomonas aeruginosa* cells by cicada wings. *Small*, *8* (16), 2489–2494.

Ivanova, E. P., Mitik-Dineva, N., Wang, J., Pham, D. K., Wright, J. P., Nicolau, D. V., et al. (2008). *Staleya guttiformis* attachment on poly (tert-butylmethacrylate) polymeric surfaces. *Micron*, *39*(8), 1197–1204.

Jacobs, A., Lafolie, F., Herry, J. M., & Debroux, M. (2007). Kinetic adhesion of bacterial cells to sand: Cell surface properties and adhesion rate. *Colloids and Surfaces B: Biointerfaces*, *59*(1), 35–45.

Jefferson, K. K. (2004). What drives bacteria to produce a biofilm? *FEMS Microbiology Letters*, *236*(2), 163–173. https://doi.org/10.1111/j.1574-6968.2004.tb09643.x.

Jia, R., Li, Y., Al-Mahamedh, H. H., & Gu, T. (2017). Enhanced biocide treatments with D-amino acid mixtures against a biofilm consortium from a water cooling tower. *Frontiers in Microbiology*, *8*, 1538. https://doi.org/10.3389/fmicb.2017.01538.

Jia, R., Yang, D., Xu, D., & Gu, T. (2017). Mitigation of a nitrate reducing *Pseudomonas aeruginosa* biofilm and anaerobic biocorrosion using ciprofloxacin enhanced by D-tyrosine. *Scientific Reports*, *7*(1), 1–11. https://doi.org/10.1038/s41598-017-07312-7.

Kamal, G. D., Pfaller, M. A., Rempe, L. E., & Jebson, P. J. (1991). Reduced intravascular catheter infection by antibiotic bonding: A prospective, randomized, controlled trial. *JAMA*, *265*(18), 2364–2368.

Kannan, A., Karumanchi, S. L., Krishna, V., Thiruvengadam, K., Ramalingam, S., & Gautam, P. (2014). Nanoscale investigation on Pseudomonas aeruginosa biofilm formed on porous silicon using atomic force microscopy. *Scanning*, *36*(5), 551–553.

Katti, K. S. (2004). Biomaterials in total joint replacement. *Colloids and Surfaces B: Biointerfaces*, *39*(3), 133–142.

Kelleher, S. M., Habimana, O., Lawler, J., O'reilly, B., Daniels, S., Casey, E., et al. (2016). Cicada wing surface topography: An investigation into the bactericidal properties of nanostructural features. *ACS Applied Materials & Interfaces*, *8*(24), 14966–14974.

Kholmanov, I. N., Barborini, E., Vinati, S., Piseri, P., Podesta, A., Ducati, C., et al. (2003). The influence of the precursor clusters on the structural and morphological evolution of nanostructured TiO2 under thermal annealing. *Nanotechnology*, *14*(11), 1168.

Kim, T. N., Balakrishnan, A., Lee, B. C., Kim, W. S., Dvorankova, B., Smetana, K., et al. (2008). In vitro fibroblast response to ultra fine grained titanium produced by a severe plastic deformation process. *Journal of Materials Science: Materials in Medicine*, *19*(2), 553–557.

Kitano, T., Yutani, Y., Shimazu, A., Yano, I., Ohashi, H., & Yamano, Y. (1996). The role of physicochemical properties of biomaterials and bacterial cell adhesion in vitro. *The International Journal of Artificial Organs*, *19*(6), 353–358.

Klemm, P., & Schembri, M. (2004). Type 1 fimbriae, curli, and antigen 43: Adhesion, colonization, and biofilm formation. *EcoSal Plus*, *1* (1). https://doi.org/10.1128/ecosalplus.8.3.2.6.

Klemm, P., Vejborg, R. M., & Hancock, V. (2010). Prevention of bacterial adhesion. *Applied Microbiology and Biotechnology*, *88*(2), 451–459.

Klug, D., Wallet, F., Kacet, S., & Courcol, R. J. (2003). Involvement of adherence and adhesion *Staphylococcus epidermidis* genes in pacemaker lead-associated infections. *Journal of Clinical Microbiology*, *41*(7), 3348–3350.

Kockro, R. A., Hampl, J. A., Jansen, B., Peters, G., Scheihing, M., Giacomelli, R., et al. (2000). Use of scanning electron microscopy to investigate the prophylactic efficacy of rifampin-impregnated CSF shunt catheters. *Journal of Medical Microbiology*, *49*(5), 441–450.

Li, Y., Jia, R., Al-Mahamedh, H. H., Xu, D., & Gu, T. (2016). Enhanced biocide mitigation of field biofilm consortia by a mixture of D-amino acids. *Frontiers in Microbiology*, *7*, 896. https://doi.org/10.3389/fmicb.2016.00896.

Liu, X., Chu, P. K., & Ding, C. (2004). Surface modification of titanium, titanium alloys, and related materials for biomedical applications. *Materials Science and Engineering: R: Reports*, *47*(3–4), 49–121.

Liu, T., Dong, L., Liu, T., & Yin, Y. (2010). Investigations on reducing microbiologically-influenced corrosion of aluminum by using super-hydrophobic surfaces. *Electrochimica Acta*, *55*(18), 5281–5285.

Liu, H., Gu, T., Zhang, G., Wang, W., Dong, S., Cheng, Y., et al. (2016). Corrosion inhibition of carbon steel in CO2-containing oilfield produced water in the presence of iron-oxidizing bacteria and inhibitors. *Corrosion Science*, *105*, 149–160. https://doi.org/10.1016/j.corsci.2016.01.012.

Liu, Y., Strauss, J., & Camesano, T. A. (2008). Adhesion forces between *Staphylococcus epidermidis* and surfaces bearing self-assembled monolayers in the presence of model proteins. *Biomaterials*, *29*(33), 4374–4382.

Livni, G., Yuhas, Y., Ashkenazi, S., & Michowiz, S. (2004). In vitro bacterial adherence to ventriculoperitoneal shunts. *Pediatric Neurosurgery*, *40*(2), 64–69.

Lorenzetti, M., Pellicer, E., Sort, J., Baró, M. D., Kovač, J., Novak, S., et al. (2014). Improvement to the corrosion resistance of Ti-based implants using hydrothermally synthesized nanostructured anatase coatings. *Materials*, *7*(1), 180–194.

Mandlik, A., Swierczynski, A., Das, A., & Ton-That, H. (2008). Pili in Gram-positive bacteria: Assembly, involvement in colonization and biofilm development. *Trends in Microbiology*, *16*(1), 33–40.

Morra, M., & Cassinelli, C. (1998). Bacterial adhesion to polymer surfaces: A critical review of surface thermodynamic approaches. *Journal of Biomaterials Science, Polymer Edition*, *9*(1), 55–74.

Mauclaire, L., Brombacher, E. B. J. D., Bünger, J. D., & Zinn, M. (2010). Factors controlling bacterial attachment and biofilm formation on medium-chain-length polyhydroxyalkanoates (mcl-PHAs). *Colloids and Surfaces B: Biointerfaces*, *76*(1), 104–111.

Medilanski, E., Kaufmann, K., Wick, L. Y., Wanner, O., & Harms, H. (2002). Influence of the surface topography of stainless steel on bacterial adhesion. *Biofouling*, *18*(3), 193–203.

Mégraud, F. (2004a). Basis for the management of drug-resistant *Helicobacter pylori* infection. *Drugs*, *64*(17), 1893–1904.

Mégraud, F. (2004b). *H pylori* antibiotic resistance: Prevalence, importance, and advances in testing. *Gut*, *53*(9), 1374–1384.

Mendonça, G., Mendonça, D. B., Aragao, F. J., & Cooper, L. F. (2008). Advancing dental implant surface technology—from micron- to nanotopography. *Biomaterials*, *29*(28), 3822–3835.

Mitik-Dineva, N., Wang, J., Mocanasu, R. C., Stoddart, P. R., Crawford, R. J., & Ivanova, E. P. (2008). Impact of nano-topography on bacterial attachment. *Biotechnology Journal: Healthcare Nutrition Technology*, *3*(4), 536–544.

Mitik-Dineva, N., Wang, J., Truong, V. K., Stoddart, P., Malherbe, F., Crawford, R. J., et al. (2009). *Escherichia coli*, *Pseudomonas aeruginosa*, and *Staphylococcus aureus* attachment patterns on glass surfaces with nanoscale roughness. *Current Microbiology*, *58*(3), 268–273.

Mitragotri, S., & Lahann, J. (2009). Physical approaches to biomaterial design. *Nature Materials*, *8*(1), 15–23. https://doi.org/10.1038/nmat2344.

Neoh, K. G., Hu, X., Zheng, D., & Kang, E. T. (2012). Balancing osteoblast functions and bacterial adhesion on functionalized titanium surfaces. *Biomaterials*, *33*(10), 2813–2822.

Ogawa, A., Takakura, K., Hirai, N., Kanematsu, H., Kuroda, D., Kougo, T., et al. (2020). Biofilm formation plays a crucial rule in the initial step of carbon steel corrosion in air and water environments. *Materials*, *13*(4), 923. https://doi.org/10.3390/ma13040923.

Olofsson, A. C., Zita, A., & Hermansson, M. (1998). Floc stability and adhesion of green-fluorescent-protein-marked bacteria to flocs in activated sludge. *Microbiology*, *144*(2), 519–528.

Oliveira, S. M., Song, W., Alves, N. M., & Mano, J. F. (2011). Chemical modification of bioinspired superhydrophobic polystyrene surfaces to control cell attachment/proliferation. *Soft Matter*, *7*(19), 8932–8941.

Ong, Y. L., Razatos, A., Georgiou, G., & Sharma, M. M. (1999). Adhesion forces between *E. coli* bacteria and biomaterial surfaces. *Langmuir*, *15*(8), 2719–2725.

O'Toole, G. A., & Kolter, R. (1998). Flagellar and twitching motility are necessary for *Pseudomonas aeruginosa* biofilm development. *Molecular Microbiology*, *30*(2), 295–304.

Padera, R. F. (2006). Infection in ventricular assist devices: The role of biofilm. *Cardiovascular Pathology*, *15*(5), 264–270.

Palmer, J., Flint, S., & Brooks, J. (2007). Bacterial cell attachment, the beginning of a biofilm. *Journal of Industrial Microbiology and Biotechnology*, *34*(9), 577–588.

Papenburg, B. J., Rodrigues, E. D., Wessling, M., & Stamatialis, D. (2010). Insights into the role of material surface topography and wettability on cell-material interactions. *Soft Matter*, *6*(18), 4377–4388.

Parreira, P., Magalhães, A., Gonçalves, I. C., Gomes, J., Vidal, R., Reis, C. A., et al. (2011). Effect of surface chemistry on bacterial adhesion, viability, and morphology. *Journal of Biomedical Materials Research Part A*, *99*(3), 344–353. https://doi.org/10.1002/jbm.a.33178.

Petrini, P., Arciola, C. R., Pezzali, I., Bozzini, S., Montanaro, L., Tanzi, M. C., et al. (2006). Antibacterial activity of zinc modified titanium oxide surface. *The International Journal of Artificial Organs*, *29*(4), 434–442.

Podestà, A., Bongiorno, G., Scopelliti, P. E., Bovio, S., Milani, P., Semprebon, C., et al. (2009). Cluster-assembled nanostructured titanium oxide films with tailored wettability. *The Journal of Physical Chemistry C*, *113*(42), 18264–18269.

Pogodin, S., Hasan, J., Baulin, V. A., Webb, H. K., Truong, V. K., Nguyen, T. H. P., et al. (2013). Biophysical model of bacterial cell interactions with nanopatterned cicada wing surfaces. *Biophysical Journal*, *104*(4), 835–840.

Popham, D. L., & Young, K. D. (2003). Role of penicillin-binding proteins in bacterial cell morphogenesis. *Current Opinion in Microbiology*, *6*, 594–599.

Pouttu, R., Puustinen, T., Virkola, R., Hacker, J., Klemm, P., & Korhonen, T. K. (1999). Amino acid residue Ala-62 in the FimH fimbrial adhesin is critical for the adhesiveness of meningitis-associated *Escherichia coli* to collagens. *Molecular Microbiology*, *31*(6), 1747–1757.

Pranzetti, A., Salaün, S., Mieszkin, S., Callow, M. E., Callow, J. A., Preece, J. A., et al. (2012). Model organic surfaces to probe marine bacterial adhesion kinetics by surface plasmon resonance. *Advanced*

Functional Materials, *22*(17), 3672–3681. https://doi.org/10.1002/adfm.201103067.

Procópio, L. (2019). The role of biofilms in the corrosion of steel in marine environments. *World Journal of Microbiology and Biotechnology*, *35* (5), 1–8. https://doi.org/10.1007/s11274-019-2647-4.

Puckett, S. D., Taylor, E., Raimondo, T., & Webster, T. J. (2010). The relationship between the nanostructure of titanium surfaces and bacterial attachment. *Biomaterials*, *31*(4), 706–713.

Qi, Z., Chen, L., & Zhang, W. (2016). Comparison of transcriptional heterogeneity of eight genes between batch *Desulfovibrio vulgaris* biofilm and planktonic culture at a single-cell level. *Frontiers in Microbiology*, *7*, 597. https://doi.org/10.3389/fmicb.2016.00597.

Ramírez, G. A., Hoffman, C. L., Lee, M. D., Lesniewski, R. A., Barco, R. A., Garber, A., et al. (2016). Assessing marine microbial induced corrosion at Santa Catalina Island, California. *Frontiers in Microbiology*, *7*, 1679. https://doi.org/10.3389/fmicb.2016.01679.

Razatos, A., Ong, Y. L., Sharma, M. M., & Georgiou, G. (1998). Molecular determinants of bacterial adhesion monitored by atomic force microscopy. *Proceedings of the National Academy of Sciences*, *95* (19), 11059–11064.

Rechendorff, K., Hovgaard, M. B., Foss, M., Zhdanov, V. P., & Besenbacher, F. (2006). Enhancement of protein adsorption induced by surface roughness. *Langmuir*, *22*(26), 10885–10888.

Riedewald, F. (2006). Bacterial adhesion to surfaces: The influence of surface roughness. *PDA Journal of Pharmaceutical Science and Technology*, *60*(3), 164–171.

Riedel, M., Müller, B., & Wintermantel, E. (2001). Protein adsorption and monocyte activation on germanium nanopyramids. *Biomaterials*, *22* (16), 2307–2316.

Rizzello, L., Sorce, B., Sabella, S., Vecchio, G., Galeone, A., Brunetti, V., et al. (2011). Impact of nanoscale topography on genomics and proteomics of adherent bacteria. *ACS Nano*, *5*(3), 1865–1876. https://doi.org/10.1021/nn102692m.

Roach, P., Farrar, D., & Perry, C. C. (2006). Surface tailoring for controlled protein adsorption: Effect of topography at the nanometer scale and chemistry. *Journal of the American Chemical Society*, *128*(12), 3939–3945.

Röcker, C., Pötzl, M., Zhang, F., Parak, W. J., & Nienhaus, G. U. (2009). A quantitative fluorescence study of protein monolayer formation on colloidal nanoparticles. *Nature Nanotechnology*, *4*(9), 577–580.

Scardino, A. J., Harvey, E., & De Nys, R. (2006). Testing attachment point theory: Diatom attachment on microtextured polyimide biomimics. *Biofouling*, *22*(1), 55–60.

Schembri, M. A., Kjærgaard, K., & Klemm, P. (2003). Global gene expression in *Escherichia coli* biofilms. *Molecular Microbiology*, *48*(1), 253–267.

Scopelliti, P. E., Borgonovo, A., Indrieri, M., Giorgetti, L., Bongiorno, G., Carbone, R., et al. (2010). The effect of surface nanometre-scale morphology on protein adsorption. *PLoS One*, *5*(7), e11862. https://doi.org/10.1371/journal.pone.0011862.

Sefton, M. V. (1989). Blood, guts and chemical engineering. *The Canadian Journal of Chemical Engineering*, *67*(5), 705–712.

Shellenberger, K., & Logan, B. E. (2002). Effect of molecular scale roughness of glass beads on colloidal and bacterial deposition. *Environmental Science & Technology*, *36*(2), 184–189.

Sheng, X., Ting, Y. P., & Pehkonen, S. O. (2007). Force measurements of bacterial adhesion on metals using a cell probe atomic force microscope. *Journal of Colloid and Interface Science*, *310*(2), 661–669.

Simhi, E., van der Mei, H. C., Ron, E. Z., Rosenberg, E., & Busscher, H. J. (2000). Effect of the adhesive antibiotic TA on adhesion and initial growth of *E. coli* on silicone rubber. *FEMS Microbiology Letters*, *192*(1), 97–100.

Singh, A. V., Vyas, V., Patil, R., Scopelliti., P. E., … Milani, P. (2011). Quantitative characterization of the influence of the nanoscale morphology of nanostructured surfaces on bacterial adhesion and biofilm formation. *PLoS One*, *6*(9), e25029. https://doi.org/10.1371/journal.pone.0025029.

Statz, A. R., Barron, A. E., & Messersmith, P. B. (2008). Protein, cell and bacterial fouling resistance of polypeptoid-modified surfaces: Effect of side-chain chemistry. *Soft Matter*, *4*(1), 131–139.

Stevens, M. M., & George, J. H. (2005). Exploring and engineering the cell surface interface. *Science*, *310*(5751), 1135–1138. https://doi.org/10.1126/science.1106587. 16293749.

Stoddart, P. R., Cadusch, P. J., Boyce, T. M., Erasmus, R. M., & Comins, J. D. (2006). Optical properties of chitin: Surface-enhanced Raman scattering substrates based on antireflection structures on cicada wings. *Nanotechnology*, *17*(3), 680.

Subbiahdoss, G., Kuijer, R., Grijpma, D. W., van der Mei, H. C., & Busscher, H. J. (2009). Microbial biofilm growth vs. tissue integration: "the race for the surface" experimentally studied. *Acta Biomaterialia*, *5*(5), 1399–1404.

Sun, T., Feng, L., Gao, X., & Jiang, L. (2005). Bioinspired surfaces with special wettability. *Accounts of Chemical Research*, *38*(8), 644–652.

Talha, M., Ma, Y., Kumar, P., Lin, Y., & Singh, A. (2019). Role of protein adsorption in the bio corrosion of metallic implants—A review. *Colloids and Surfaces B: Biointerfaces*, *176*, 494–506. https://doi.org/10.1016/j.colsurfb.2019.01.038.

Teng, F., Guan, Y. T., & Zhu, W. P. (2008). Effect of biofilm on cast iron pipe corrosion in drinking water distribution system: Corrosion scales characterization and microbial community structure investigation. *Corrosion Science*, *50*(10), 2816–2823.

Thanassi, G., & Hultgren, S. J. (2000). Multiple pathways allow protein secretion across the bacterial outer membrane. *Current Opinion in Cell Biology*, *12*(4), 420–430.

Thomas, T. R., & Rosén, B. G. (2000). Determination of the optimum sampling interval for rough contact mechanics. *Tribology International*, *33* (9), 601–610.

Truong, V. K., Lapovok, R., Estrin, Y. S., Rundell, S., Wang, J. Y., Fluke, C. J., et al. (2010). The influence of nano-scale surface roughness on bacterial adhesion to ultrafine-grained titanium. *Biomaterials*, *31* (13), 3674–3683. https://doi.org/10.1016/j.biomaterials.2010.01.071. Epub 2010 Feb 16 20163851.

Tsimbouri, P. M., Fisher, L., Holloway, N., Sjostrom, T., Nobbs, A. H., Meek, R. D., et al. (2016). Osteogenic and bactericidal surfaces from hydrothermal titania nanowires on titanium substrates. *Scientific Reports*, *6*(1), 1–12. https://doi.org/10.1038/srep36857.

Tunney, M. M., Dunne, N., Einarsson, G., McDowell, A., Kerr, A., & Patrick, S. (2007). Biofilm formation by bacteria isolated from retrieved failed prosthetic hip implants in an in vitro model of hip arthroplasty antibiotic prophylaxis. *Journal of Orthopaedic Research*, *25*(1), 2–10.

Vadillo-Rodríguez, V., Guerra-García-Mora, A. I., Perera-Costa, D., González-Martín, M. L., & Fernández-Calderón, M. C. (2018). Bacterial response to spatially organized microtopographic surface patterns with nanometer scale roughness. *Colloids and Surfaces B: Biointerfaces*,

169, 340–347. https://doi.org/10.1016/j.colsurfb.2018.05.038. Epub 2018 May 17 29800909.

Valiev, R. Z., Semenova, I. P., Latysh, V. V., Rack, H., Lowe, T. C., Petruzelka, J., et al. (2008). Nanostructured titanium for biomedical applications. *Advanced Engineering Materials*, *10*(8), B15–B17.

van Loosdrecht, M., Lyklema, J., Norde, W., & Zehnder, A. J. (1989). Bacterial adhesion: A physicochemical approach. *Microbial Ecology*, *17*(1), 1–15.

Verma, A., Uzun, O., Hu, Y., Hu, Y., Han, H. S., Watson, N., et al. (2008). Surface-structure-regulated cell-membrane penetration by monolayer-protected nanoparticles. *Nature Materials*, *7*(7), 588–595. https://doi.org/10.1038/nmat2202.

Virtanen, S., Milošev, I., Gomez-Barrena, E., Trebše, R., Salo, J., & Konttinen, Y. T. (2008). Special modes of corrosion under physiological and simulated physiological conditions. *Acta Biomaterialia*, *4*(3), 468–476.

Wandiyanto, J. V., Cheeseman, S., Truong, V. K., Al Kobaisi, M., Bizet, C., Juodkazis, S., et al. (2019). Outsmarting superbugs: Bactericidal activity of nanostructured titanium surfaces against methicillin-and gentamicin-resistant *Staphylococcus aureus* ATCC 33592. *Journal of Materials Chemistry B*, *7*(28), 4424–4431.

Wandiyanto, J. V., Tamanna, T., Linklater, D. P., Truong, V. K., Al Kobaisi, M., Baulin, V. A., et al. (2020). Tunable morphological changes of asymmetric titanium nanosheets with bactericidal properties. *Journal of Colloid and Interface Science*, *560*, 572–580.

Wang, X., Bhadra, C. M., Dang, T. H. Y., Buividas, R., Wang, J., Crawford, R. J., et al. (2016). A bactericidal microfluidic device constructed using nano-textured black silicon. *RSC Advances*, *6*(31), 26300–26306.

Wang, M. S., Palmer, L. B., Schwartz, J. D., & Razatos, A. (2004). Evaluating protein attraction and adhesion to biomaterials with the atomic force microscope. *Langmuir*, *20*(18), 7753–7759.

Watson, G. S., Myhra, S., Cribb, B. W., & Watson, J. A. (2008). Putative functions and functional efficiency of ordered cuticular nanoarrays on insect wings. *Biophysical Journal*, *94*(8), 3352–3360.

Webb, H. K., Boshkovikj, V., Fluke, C. J., Truong, V. K., Hasan, J., Baulin, V. A., et al. (2013). Bacterial attachment on sub-nanometrically smooth titanium substrata. *Biofouling*, *29*(2), 163–170. https://doi.org/10.1080/08927014.2012.757697.

Webb, H. K., Truong, V. K., Hasan, J., Fluke, C., Crawford, R. J., & Ivanova, E. P. (2012). Roughness parameters for standard description of surface nanoarchitecture. *Scanning*, *34*(4), 257–263.

Whitehead, K. A., Colligon, J., & Verran, J. (2005). Retention of microbial cells in substratum surface features of micrometer and sub-micrometer dimensions. *Colloids and Surfaces B: Biointerfaces*, *41*(2–3), 129–138.

Wu, S., Liu, G., Jin, W., Xiu, P., & Sun, C. (2016). Antibiofilm and anti-infection of a marine bacterial exopolysaccharide against *Pseudomonas aeruginosa*. *Frontiers in Microbiology*, *7*, 102. https://doi.org/10.3389/fmicb.2016.00102.

Xu, H., Murdaugh, A. E., Chen, W., Aidala, K. E., Ferguson, M. A., Spain, E. M., et al. (2013). Characterizing pilus-mediated adhesion of biofilm-forming *E. coli* to chemically diverse surfaces using atomic force microscopy. *Langmuir*, *29*(9), 3000–3011. https://doi.org/10.1021/la304745s. Epub 2013 Feb 19 23421314. PMCID: PMC3590879.

Xu, D., Xia, J., Zhou, E., Zhang, D., Li, H., Yang, C., et al. (2017). Accelerated corrosion of 2205 duplex stainless steel caused by marine aerobic *Pseudomonas aeruginosa* biofilm. *Bioelectrochemistry*, *113*, 1–8. https://doi.org/10.1016/j.bioelechem.2016.08.001.

Xue, F., Liu, J., Guo, L., Zhang, L., & Li, Q. (2015). Theoretical study on the bactericidal nature of nanopatterned surfaces. *Journal of Theoretical Biology*, *385*, 1–7.

Zhang, P., Zhang, Z., & Li, W. (2013). Antibacterial coating incorporating silver nanoparticles by microarc oxidation and ion implantation. *Journal of Nanomaterials*, *2013*.

Zhao, X., Hilliard, L. R., Mechery, S. J., Wang, Y., Bagwe, R. P., Jin, S., et al. (2004). A rapid bioassay for single bacterial cell quantitation using bioconjugated nanoparticles. *Proceedings of the National Academy of Sciences*, *101*(42), 15027–15032.

Further reading

Ghilini, F., Pissinis, D. E., Miñán, A., Schilardi, P. L., & Diaz, C. (2019). How functionalized surfaces can inhibit bacterial adhesion and viability. *ACS Biomaterials Science & Engineering*, *5*(10), 4920–4936.

Ivanova, E. P., Linklater, D. P., Werner, M., Baulin, V. A., Xu, X., Vrancken, N., et al. (2020). The multi-faceted mechano-bactericidal mechanism of nanostructured surfaces. *Proceedings of the National Academy of Sciences*, *117*(23), 12598–12605. https://doi.org/10.1073/pnas.1916680117. Epub 2020 May 26 32457154. PMCID: PMC7293705.

Ivanova, E. P., Truong, V. K., Webb, H. K., Baulin, V. A., Wang, J. Y., Mohammodi, N., et al. (2011). Differential attraction and repulsion of *Staphylococcus aureus* and *Pseudomonas aeruginosa* on molecularly smooth titanium films. *Scientific Reports*, *1*(1), 1–8.

Karthik, L., Kumar, G., Keswani, T., Bhattacharyya, A., Chandar, S. S., & Bhaskara Rao, K. V. (2014). Protease inhibitors from marine actinobacteria as a potential source for antimalarial compound. *PLoS One*, *9*(3), e90972.

Katsikogianni, M., & Missirlis, Y. F. (2004). Concise review of mechanisms of bacterial adhesion to biomaterials and of techniques used in estimating bacteria-material interactions. *European Cells & Materials*, *8*(3), 37–57.

Lee, S. W., Phillips, K. S., Gu, H., Kazemzadeh-Narbat, M., & Ren, D. (2021). How microbes read the map: Effects of implant topography on bacterial adhesion and biofilm formation. *Biomaterials*, *268*, 120595.

Linklater, D. P., Nguyen, H. K. D., Bhadra, C. M., Juodkazis, S., & Ivanova, E. P. (2017). Influence of nanoscale topology on bactericidal efficiency of black silicon surfaces. *Nanotechnology*, *28*(24), 245301.

Lundqvist, M., Stigler, J., Elia, G., Lynch, I., Cedervall, T., & Dawson, K. A. (2008). Nanoparticle size and surface properties determine the protein corona with possible implications for biological impacts. *Proceedings of the National Academy of Sciences*, *105*(38), 14265–14270.

Tai, W., He, L., Zhang, X., Pu, J., Voronin, D., Jiang, S., et al. (2020). Characterization of the receptor-binding domain (RBD) of 2019 novel coronavirus: Implication for development of RBD protein as a viral attachment inhibitor and vaccine. *Cellular & Molecular Immunology*, *17*(6), 613–620.

Chapter 4

Microbial biofilms: Unravel their potential for agricultural applications under agro-ecosystem

Firoz Ahmad Ansari and Iqbal Ahmad
Department of Agricultural Microbiology, Faculty of Agricultural Sciences, Aligarh Muslim University, Aligarh, Uttar Pradesh, India

1. Introduction

Approximately more than 90% microbes especially bacteria are predicted to prevail in self-producing matrix of biofilms under natural agro-ecosystem. The soil environment harbors a multitude of microbiota in micro-aggregate to form complex and compact soil structures (Flemming & Wuertz, 2019; Gutiérrez Castorena et al., 2016; Kuzyakov & Blagodatskaya, 2015). The rising evidences indicate that the sessile (biofilm) growth style is also the prevalent mode of lifestyle for soil microbes. Biofilm formation in the soil comprises single and multispecies interaction and establishes to inhabit a number of interfaces and substrata, namely plant roots, pore places, and mineral surfaces (Ansari, Jabeen, & Ahmad, 2021; Zhang et al., 2014). At the time of biofilm formation, extra-polymeric substances (EPSs) excrete and design a protective layer of biofilm matrix, which mediate surface (solid and semisolid) attachment and retaining corrosiveness of biofilm growth (Costa, Raaijmakers, & Kuramae, 2018; Flemming & Wingender, 2010). Bacterial biofilms in soil ecosystem also provide a kind of competitive advantages over free living bacterial growth including improved fitness against dehydration and prominent prospects for horizontal gene transfer in their surroundings. Also, soil biofilm formation positively impacts the soil physicochemical attributes by maintaining and enhancing soil moisture through its aggregation (Cai et al., 2019; Lennon & Lehmkuhl, 2016; Moyano, Manzoni, & Chenu, 2013). The EPS secreted during biofilm growth creates network or meshwork-like connection among soil's minerals and fortifies the essential cohesion of soil micro-aggregates for longer period of time. Consequently, soil biofilms enhance the survivability of beneficial microbes under different environmental challenges and play an advantageous role in influencing microenvironment of soil (Lin et al., 2016; Muhammad et al., 2020).

Bacterial biofilm developments are established in different niches (rhizosphere and rhizoplane) in agro-ecosystem, which effectively compete with other microbes and colonize efficiently their host. The biofilm existence and effective rhizosphere colonization provide the benefits to microbe's metabolic activities and therefore improve plants and soil health in agro-ecosystems (Ansari & Ahmad, 2019; Burmolle, Kjoller, & Sorensen, 2012; Flemming & Wuertz, 2019). Multispecies biofilms naturally occur in the rhizosphere and perform a key activity in the enhancement of solubilization and acquisition of soil nutrients, plant growth stimulation, and tolerance against adverse impact of environmental and biological constraints (Ahmad et al., 2017; Ansari, Ahmad, & Pichtel, 2019; Fierer, 2017). Understanding of such agriculturally indispensable bacterial biofilm in better details delivers additional understandings for their consequences on growth and bioprotection of plants, enhancing plants nutrition and soil health improvement. The bacterial population in the rhizosphere, as well as their multipartite interaction in biofilms, is expected to have a considerable impact on plant metabolism (Ahmad et al., 2017; Cai et al., 2019; Ren, Madsen, Sørensen, & Burmølle, 2015). Nonetheless, there is a requirement to more and in-depth exploration about role and mechanisms involved in interaction and their significance in unraveling the solicitation of these beneficial bacterial biofilms in enhanced agricultural productivity under sustainable agro-ecosystem. Thus, this chapter highlights the occurrence of bacterial biofilms in rhizospheric niches, interaction to plants roots, and their prospective applications in plants and soil health management under natural settings.

Understanding Microbial Biofilms. https://doi.org/10.1016/B978-0-323-99977-9.00039-9

2. Bacterial biofilms in soil system

Micro-communities are the ancient form of life in the soil ecosystem, which are responsible for determining the processes and direction of the earth's evolution and maintaining agricultural production. Occurrence of microbial biofilms in any kind of soil is known as a key driver, which regulates numerous essential biogeochemical processes. In the natural soil systems, microbes directly involve to maintain soil fertility, nutrient cycling, and storage of organic carbons, and therefore, it creates direct or indirect impact on plants and soil health in ecosystems (Ahmad et al., 2017; Cai et al., 2019; Dodd et al., 2017). Microbial populations in natural agro-ecosystem frequently occur as surface adhered or attached structures called biofilms. Bacterial biofilms secret EPSs and colonize carbon-rich soil surfaces including clay minerals and plant roots. Soil bacteria constituting aggregates of multicellular structure are termed soil biofilms and unsaturated biofilms as a result of limited water availability along defining environmental characteristic in the native ecosystem (Fierer, 2017; Pandit, Adholeya, Cahill, Brau, & Kochar, 2020). Bacterial biofilms in the soil are characterized according to their complexity, which arrays from simple to multifaceted, tower to flat, featureless to multilayer structures, and have a range of physiological features (Cai et al., 2019; O'Toole & Wong, 2016). Rhizospheric soil is highly active region for microbe's richness and functions generally due to secretion as well as deposition of root exudates. In the rhizosphere, bacteria not only fight for nutrient but also for space and existence. Therefore, bacterial communities migrate from bulk soil to rhizosphere through signaling processes and colonize rhizosphere effectively (Jung, Hong, Park, Kim, & Shin, 2018; Pandit et al., 2020). After bacterial-surface adherence, a well-coordinated structure forms, which differs from simple to mushroom-like appearance. The bacterial biofilm morphology in the rhizosphere is determined by different factors such as oxygen gradient, carbon sources, and flow velocity (Duvernoy et al., 2018). In the beginning, individual bacteria attach to the surfaces and are transformed into microcolonies after few cycles of division. The morphogenesis of microcolony involves cell rearrangement, mechanical coupling in cell elongation forces, fraction, and steric interaction, and adhesion or attachment (Flemming et al., 2016; Nunan, 2017; Rillig, Muller, & Lehmann, 2017), Initially, proliferation of bacterial cells takes place and further constitutes microcolony in a single layer and swiftly transforms into biofilm-like structure on the solid-liquid interfaces. Fig. 1 depicts a clear typical representation and illustrates all of the phases of biofilm growth in the rhizosphere.

The interactions and effective rhizobacterial colonization on the different parts of root surfaces are primarily reliant on adherence of cells and microcolonies development (Ansari & Ahmad, 2019; Cai et al., 2019; Ren et al., 2015). There are numerous studies regarding beneficial associations between rhizobacteria and plant roots have been broadly investigated in the context of microbe-plant interaction. Biofilm development on plants roots can also be triggered by some signals (nutrients, water availability, and bacteria-bacteria associations). Root exudates act key influential functions in biofilm development on multilocation root sites. The zone of root cap and cell division are important location for bacterial colonization in comparison to matured root hairs and root zone where less bacterial colonization occurs (Jung et al., 2018; Pandit et al., 2020; Vardharajula, Ali, Grover, Reddy, & Venkateswarlu, 2011). Roots of many agricultural crops (wheat, mustard, and rice) contain a complex microbial consortium in biofilm mode of life that improved growth and productivity of plants. The chemical constituents of root exudates, namely sugars, organic acids, amino acids, fatty acids, sterols, proteins, etc., play a direct impact on micro-communities in the rhizosphere (Saleh, Sharma, Seguin, & Jabaji, 2020; Yuan et al., 2015). It has evidenced that the microbial protein (TasA) and exopolysachharides help in attachment, biofilm, and colonization in the *B. subtilis* on the surface of mustard roots (Hassani, Durán, & Hacquard, 2018). It was also revealed in other investigations that carbohydrates (pectin, arabinogalactan, and xylan) of the root exudates were enhanced during early stage of biofilm development (Poole, 2017). A special attention is giving for bacterial biofilms and their interaction only for clinically relevance and biotechnological applications (Kenny & Balskus, 2018; Magana et al., 2018; Perez-Garcia, Lear, & Singhal, 2016), whereas the significance of soil bacterial biofilms and their interaction has mostly been overlooked. In field of soil microbiology, research in several domains is required to create the groundwork for investigating bacterial biofilms in soil and to accelerate this area into a new era for future researchers. The beneficial soil bacterial biofilms with their associated plant roots are depicted in Table 1.

3. Bacterial interaction in biofilms under soil ecosystem

Bacterial cells in biofilm's structure are well known to connect with one another through the process of quorum sensing. This kind of association in biofilm state is engaged in a variety of social behaviors as a result of their close proximity, which led to the coining of the anthropomorphic term "sociomicrobiology." The term sociomicrobiology elucidates the connection between quorum sensing and biofilms. Generally, biofilms cells are exposed to substantially greater local cell densities in comparison to planktonic or free-floating cells. Biofilm cells are exposed to higher levels of secondary metabolites, metabolic by-products, and other expelled or secreted microbial substances as a result of this

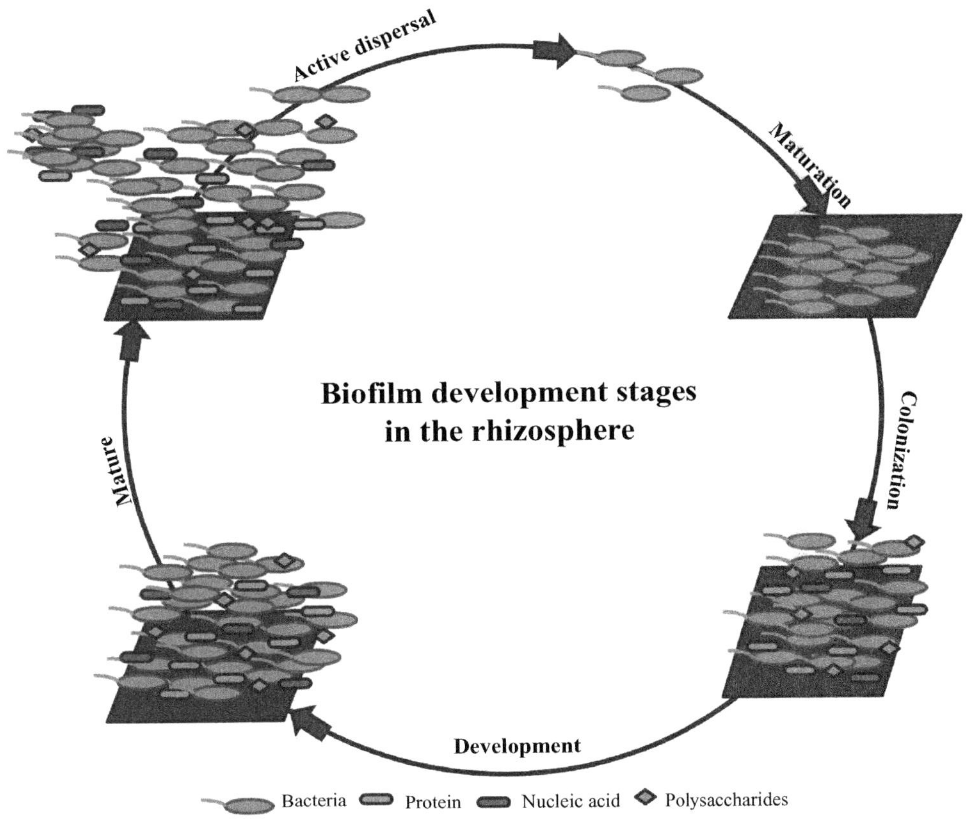

FIG. 1 Model of bacterial biofilm development and their various stages in the rhizosphere.

TABLE 1 Association of beneficial bacterial biofilm with plant roots.

Bacterial isolates	Host plants (site)	Functional attributes	References
B. subtilis	Arabidopsis (rhizoplane)	Biocontrol	Dwivedi, Khare, Chaturvedi, and Singh (2017)
P. entomophila	Wheat (rhizosphere and rhizoplane)	Plant growth stimulation	Ansari and Ahmad (2018)
B. pumilus	Wheat (rhizosphere and rhizoplane)	Plant growth stimulation and stress mitigation	Ansari et al. (2019)
P. azotoformans	Wheat (rhizosphere and rhizoplane)	Plant growth stimulation and stress mitigation	Ansari et al. (2021)
P. fluorescens	Wheat (rhizosphere and rhizoplane)	Synergistic growth enhances the plant-growth-promoting attributes	Ansari and Ahmad (2019)
P. putida	Chickpea (roots)	Synergistic growth enhances the plant-growth-promoting attributes	Kumar et al. (2016)
A. brasilense	Sorghum (rhizosphere)	Enhance NO and indole-3-acetic acid	Koul, Tripathi, Adholeya, and Kochar (2015)
A. brasilense	Wheat (rhizosphere)	Nitrogen fixation	Souza et al. (2014)
B. amyloliquefaciens and *B. polymyxa*	Tomato (rhizosphere)	Biocontrol	Nihorimbere et al. (2012)
P. putida	Chickpea (rhizosphere)	Plant growth promoting activities	Srivastava et al. (2008)

cellular interaction (Parsek & Greenberg, 2005; Ren et al., 2015). These social activities of the bacteria encompass inter and intraspecies-specific associations, and these interactions enhance the exchange of signal molecules and metabolic products (Ansari & Ahmad, 2019; Burmølle, Ren, Bjarnsholt, & Sørensen, 2014; Ren et al., 2015). Owing to the microbial diversity in biofilm communities, associations in biofilm state are constantly dynamic and diverse, which that plays imperative roles in development and functions of biofilms in soil system.

3.1 Cooperative interactions

In natural environments, microbes do not survive alone and can associate and cooperate with each other to form biofilms. The development of biofilms is known to conducive to protect against various hostile environments, which involve that microbes develop multispecies biofilms with their cooperation. Multispecies biofilm provide a strong protection against hostile living situations. For instance, the development of dual species biofilm in between *P. putida* SB5 and *Chryseobacterium* sp. SB9 is cooperative and strain-specific kind of biofilm reported by Burmølle, Hansen, and Sørensen (2007). Another investigation has also revealed significant compatible and synergistic interactions in four bacterial species biofilm comprising *Xanthomonas retroflexus*, *Stenotrophomonas rhizophila*, *Paenibacillus amylolyticus*, and *Microbacterium oxydans* with *X. retroflexus* performing a prevailing function, and other isolates act as an additional functions in multispecies biofilm development (Cai et al., 2019; Ren et al., 2015).

Soil bacteria in biofilm mode reside in close contact with each other, which is essential for cell-to-cell metabolite exchange. In natural soil ecosystem, bacteria typically utilize specific metabolites comprising amino acids synthesized by other bacterial strains for their growth through cross-feeding process (Pande et al., 2016). A large number of soil micro-communities in biofilm state are found auxotrophic, and >98% of microbes sequenced thus far lack important mechanisms or key regulatory genes for amino acids production implying that metabolite exchange between various species in biofilms is essential for existence (Zengler & Zaramela, 2018). For instance, metabolic cooperation may occur when ammonia (NH_3) is provided to NH_3 oxidizing microbes via nitrite (NO_2) oxidizers (*Nitrospira moscoviensis*) in the soil ecosystem. Both the bacterial candidates reveal cooperative association in soils (Flemming et al., 2016). An enhancing tolerance of biofilms against several antibiotics and surfactants is also a result of cooperative as well as mutual associations (Ansari, Jafri, Ahmad, & Abulreesh, 2017; Harrison, Ceri, & Turner, 2007; Lee et al., 2014; Olsen, 2015). Furthermore, bacteria can cooperate together to degrade contaminants such as herbicide linuron and 2,4-dichlorophenoxyacetic acid under soil ecosystems (Breugelmans et al., 2008; Flemming, 2010). Generally, co-metabolism in bacterial biofilms under soil system is thought to be important for nutrient cycling, organic matter turnover, as well as contaminant degradation.

3.2 Inter and intraspecies competition

Cooperative interactions appear to be the most important ecological associations among biofilm forming microbes; however, competition also exists in the soil ecosystem. On the one side, individual or populations of the microbes may strive for advantages in terms of resource utilization and space occupancy. Diversely, soil biofilm's microenvironment is still susceptible to exterior disruption, which can abolish the stabilization, therefore resulting in community's structure shuffling, which we called as postdisruption biofilms competition (Ansari & Ahmad, 2019; Ren et al., 2015). In a soil biofilm, the exploitation of different niches is a community consequence, which is the result of time- and space-constrained evolution. Most of the natural habitats harbor amazing collection of diverse microbial communities. Bacteria within these communities compete with their nearby residents for nutrients, space, and other important resources. Various experiments with pure and mixed bacterial cultures exhibited numerous active pathways by which bacteria may kill or impair other micro-communities for their existence. A growing amount of experimental as well as theoretical population research suggests that associations within and across bacterial population can have a significant effect on the consequences of natural competition (Cai et al., 2018; Hibbing, Fuqua, Parsek, & Peterson, 2010; Pandit et al., 2020). For instance, methanogenic archaea and sulfate-reducing bacteria compete for substrates such as hydrogen, acetate, and organic matters in the anaerobic biofilm development on the sewer pipes' surfaces. The sulfate-reducing soil bacteria were dominant at the initial phase of biofilm development but as the stage of fully matured biofilm, methanogenic colonizers became more adapted and eventually replaced sulfate reducers (Auguet, Pijuan, Batista, Borrego, & Gutierrez, 2015). The competition pressure is particularly very high for nutrients between phylogenetically identical strains under natural and soil agro-ecosystem (Hawlena, Bashey, & Lively, 2010; Madsen et al., 2016; Patin, Duncan, Dorrestein, & Jensen, 2016; Schoustra, Dench, Dali, Aaron, & Kassen, 2012). In recent past, researchers studied a combinatorial interaction of 64 isolates of soil's Streptomyces strains discovered a mutualism association between strains, which means that if one strain promotes or inhibits the growth of a second strain, the second strain is likely to have the same effect on the first strain (Madsen et al., 2016; Vetsigian, Jajoo, & Kishony, 2011).

This reciprocity is boosted expressively in the isolates recovered from the same soil samples, implying that it is a feature of coexisting micro-communities. Another investigation reveals that biofilm development appears to be a common adaptive feature for microbes to stable and everlasting coexistence in natural soil (Madsen et al., 2016). An experiment with 67 bacterial combinations revealed that antagonistic interactions are more likely to occur among closely related species, with more metabolically identical strains having a higher chance of antagonistic interactions (Russel, Røder, Madsen, Burmølle, & Sørensen, 2017). Considering the aforementioned aspects, it is evident that both inter- and intraspecific bacterial diversities are critical indicators for understanding as well as forecasting the activities of biofilms in agro-ecosystems.

A bacterial taxon can use as a variety of strategies to acquire a competitive advantage, including (i) restricting the growth and development of other bacterial taxa through nutrients competition via siderophore synthesis and (ii) killing the other bacterial competitors via the production of broad-spectrum antibiotics in their surroundings (Gao, Zhang, Wu, Huang, & Cai, 2019; Hibbing et al., 2010). For instance, bacteria synthesize antimicrobial agents including antibiotics, surfactants, and bacteriocins to restrict the growth and multiplication of neighboring species, which results in cellular competition among each other. Likewise, biocontrol bacteria (*Bacillus* spp. and *Pseudomonas* spp.) synthesizes a special kind of antibacterial agent (2,4-diacetylphloroglucinol) for the suppression of growth and multiplication of phytopathogens (Raaijmakers & Mazzola, 2012). Some pathogenic bacteria such as *P. aeruginosa* have a Type VI secretory system through which virulent proteins transport into surrounding cells, and these virulent proteins play a deleterious role in killing and hindering the growth of cells (Basler, Ho, & Mekalanos, 2013; Nadell, Drescher, & Foster, 2016).

Because of the enormous variety and rapid coevolution of microorganisms in natural habitats, high order of intra and intertaxa associations emerge as fundamental features of ecological networks. The majority of current research examines the association between biodiversity and their ecosystem's functioning through studying multispecies combinations made up of taxa selected randomly (Ansari & Ahmad, 2019; Foster & Bell, 2012). The productivity rises quickly and considerably as more taxa are added, especially for ecologically native taxa, which could explain the observed increase in total community size as taxonomic diversity increases (Ansari & Ahmad, 2019; Freilich et al., 2011; Gravel et al., 2011; Pandit et al., 2020).

4. Significance of biofilms in agroecosystem

The plant growth promotion by PGPR is governed by effective root colonization to the host plant. These bacterial biofilms essentially depend on root exudates for food and nutrition. They are major plant-driven factors, which triggering biofilm formation. Table 2 highlights the several pathways of rhizosphere colonization in PGPR that have been studied thus far. Soil matrix, root tip, rhizoplane, and rhizosphere colonization are chief bacterial colonization sites. Chemotaxix, motility, and cell adhesion to

TABLE 2 Current studies describing the various mechanisms of PGPR to effectively colonize plant's rhizosphere and rhizoplane.

Bacterial colonization sites	Mechanisms	Key profits of functions in rhizosphere	References
Root tip and soil matrix colonization	Biofilm development	Nitrogenase activity, IAA production, phosphate solubilization, siderophore production, ammonia production, and higher resistance to adverse environmental are higher in PGPR biofilm than the planktonic cells.	Wang, Xu, Elmerich, and Ma (2017), Ansari et al. (2019), Banerjee, Sarkar, Cuadros-Orellana, and Bandopadhyay (2019)
Rhizoplane and rhizosphere colonization	Chemotaxis and motility	Chemical attractants of beneficial bacteria in the rhizosphere	Olanrewaju, Ayangbenro, Glick, and Babalola (2019), Phour, Sehrawat, Sindhu, and Glick (2020)
Endophytic and rhizoplane colonization	Chemotaxis and motility	Successful colonization will only be achieved if preceded by the detection (chemotaxis) of root exudates, and movement (motility) of the microorganisms toward the plant roots.	Böhm, Hurek, and Reinhold-Hurek (2007), Martínez-Cámara, Montejano-Ramírez, Moreno-Hagelsieb, Santoyo, and Valencia-Cantero (2020), Fernández-Llamosas, Díaz, and Carmona (2021)

Continued

TABLE 2 Current studies describing the various mechanisms of PGPR to effectively colonize plant's rhizosphere and rhizoplane—cont'd

Bacterial colonization sites	Mechanisms	Key profits of functions in rhizosphere	References
Soil matrix, root tip, rhizoplane and rhizosphere colonization	Solubilization and acquisition of nutrients	Siderophores are ferric ion specific chelators, and they possess antimicrobial properties. Phosphate solubilizing microorganisms mediate bioavailable soil P to plants. Benefit the host plant mainly by fixing atmospheric nitrogen	Hider and Kong (2010), Glick (2012), Jog, Pandya, Nareshkumar, and Rajkumar (2014), Dimkpa (2016), Heil, Vereecken, and Brüggemann (2016), Shin et al. (2016), Khatoon et al. (2020)
Root and rhizosphere colonization	Suppression and evasion of plant immune system	Suppression refers to bypassing the plant immune system, through the use of effector proteins. Evasion consists, primarily, of preventing the activation of the plant immune system.	Stringlis, Zhang, Pieterse, Bolton, and de Jonge (2018), Yu, Pieterse, Bakker, and Berendsen (2019)
Rhizosphere colonization	Volatile organic compounds	Increase the biosynthesis of secondary metabolites, improve the antioxidant status in some plants grown under salt stress, and inhibit pathogenic microbes	Martínez-Cámara et al. (2020)
Rhizosphere colonization	Antioxidant properties	PGPRs maintain high antioxidant enzyme activity under stress conditions.	(Ansari et al., 2019, 2021)

the surface are major mechanisms involved in the colonization. Biofilm formation is one of them. As a result of these associations, plant growth gets affected in direct or indirect ways. The direct benefit includes enhanced nitrogenase activity, nitrogen fixation, phosphate solubilization, siderophore production, and synthesis of various phytohormones at rhizoplane and rhizosphere. Indirectly it affects the plant defense mechanism and synthesis of secondary metabolites (Ansari et al., 2019; Banerjee et al., 2019; Böhm et al., 2007; Fernández-Llamosas et al., 2021; Martínez-Cámara et al., 2020; Wang et al., 2017).

Plant-associated biofilms can interact with plants in diverse ways. The interaction can have negative effects such as biofilm forming phytopathogenes, and positive effects like PGPB and biocontrol organisms. Thus, microbes can have commensal, mutualistic, and pathogenic interactions with plants. These organisms can adhere to different parts of the plant such as leaves, stems, roots, seeds, and the vascular system. Overall plant microbiome has direct impact on plant health. Broadly biofilms are affecting the agroecosystem by contributing toward soil health or crop health or both. The multispecies biofilms present in soil are important for ecological balance and fertility of soil (Ahmad et al., 2017).

4.1 Biofilms in soil health

The biofilms regulate soil fertility, soil carbon storage, and nutrient cycling; all of these have a direct or indirect influence on soil health in agro-ecosystem (Ahmad et al., 2017; Ansari & Ahmad, 2019; Fierer, 2017). Biofilm mode of lifestyle of the bacteria offers a significant fitness benefits over their planktonic growth state in the soil ecosystem (Ansari & Ahmad, 2019; Ansari et al., 2019; Kragh et al., 2016). Biofilms development in the plant rhizosphere is a hot spot for interkingdom interactions, which influence microbial aggregation as well (Hassani et al., 2018). The biofilm-based biofertilizers can have substantial potential benefits in the sector of agriculture. These environment-friendly alternatives have the ability to address the shortcomings of traditional chemical-based and synthetic fertilizers. Biofilm development in the rhizosphere soil may also play a crucial role in improving microbial biomass along with water stability, which may improve the response of root exudation under environmental constraints (Kasim, Gaafar, Abou-Ali, Omar, & Hewait, 2016). These biofilm formers could be used as potent biofertilizers directly in the soil to upraise the crops growth, development, and productivity in eco-friendly way (Ansari & Ahmad, 2018; Hettiarachchi et al., 2014). Bacterial interactions through key taxonomic groups within the biofilms also display advantageous effect on aggregation as well as fertility of soil (Lehmann, Zheng, & Rillig, 2017). Additionally, dry mass of biofilms contains 80% EPS, which can provide several beneficial roles including protecting microbes against abiotic and biotic stress factors in soil ecosystem (Or, Phutane, & Dechesne, 2007). For instance, extrapolymeric substances of bacterial biofilms can retain 15–20 times more water to its own weight, and therefore, it expressively improves agricultural soil's water-retaining capacity (Adessi, de Carvalho, De Philippis, Branquinho, & da Silva, 2018; Chenu & Roberson, 1996). Biofilm EPSs also add significantly in the development and constancy of soil aggregation for better crop productivity. The volume of this soil aggregates is mostly dependent on the features of EPS and physicochemical properties of soil (Bezzate et al., 2000; Büks & Kaupenjohann, 2016; Chenu & Cosentino, 2011).

The biofilms cover a limited soil area, and these areas become hotspot for ecologically essential biogeochemical reactions (Kuzyakov & Blagodatskaya, 2015). For instance, soil biofilms of *B. japonicum* SEMIA5019 enhanced the mineralization of organic phosphorus and nitrogen and exhibited maximum nitrogenase activity even under higher concentration of NO_3 in comparison to planktonic cells of similar bacterium in the soil (Seneviratne & Jayasinghearachchi, 2005). Soil biofilm biomass either in live or dead form also plays an important function in maintaining soil health by balancing nutrients pool as microbial turnover. Microbial turnover is in biofilm mode determined by microbe's cell death and cell production. The dead cells of microbes may either attach/adhere to the different soil particles and enter into the soil organic carbon (SOC) pool or be metabolized and utilized by living micro-communities (Hagerty et al., 2014). Soil microbes mediate the cycling and mobilization of carbon, phosphorus, nitrogen, sulfur, iron, and other key nutrients, and therefore, a key aim of microbes-to-ecosystems model has recognized the exact relations among microbial abundance and biogeochemical cycling. Subsequently, enhanced microbial turnover can improve respiration per unit of soil microbial biomass carbon (Kivlin & Hawkes, 2020). By forming a nutrient pool in the soil system, the biofilms in the soil can sustain and meet the requirements of important nutrients for plants and microbial flora, potentially leading to soil fertility. Yet, in the context of soil biofilms, the possible linkage between soil micro-communities turnover and their linked functions remains a greater challenge for young researchers in the current era.

Bacterial biofilms are recalcitrant to hostile environments and might provide protection to microbes against extreme temperature, pH, excess salinity, antibiotics, UV radiations, poor nutrients, high pressure, and high cold by acting as protective layer in the agro-ecosystem (Ansari & Ahmad, 2019; Yin, Wang, Liu, & He, 2019). The application and benefits of biofilm-based crop and soil health management under natural agro-ecosystem in sustainable manner are illustrated in Fig. 2.

4.2 Biofilms in crop health

Plant cells do not have mobile defense and adaptable immune system unlike human cells. Nevertheless, they have evolved a complex innate immune system that can detect prospective invading microbes over the use of transmembrane pattern recognition receptors (PRRs). This receptor system responds to microbe-associated molecular patterns (MAMPs), and these molecules are shared by an extensive array of micro-communities (Jones & Dangl, 2006; Santoyo, Urtis-Flores, Loeza-Lara, Orozco-Mosqueda, & Glick, 2021). Plants' immune systems also recognize beneficial microbes as a potential threat because they have immunogenic MAMPs that are strikingly similar to pathogenic microbes (Stringlis et al., 2018). Beneficial micro-communities have been displayed to trigger the plants' immune system during their initial interaction with roots, which are identified by PRRs of plants (Danhorn & Fuqua, 2007; Yu et al., 2019). For instance, the biofilm forming bacterium *P. simiae* strain (WCS417) stimulated the immune response in roots of *Arabidopsis* by recognizing the expression of MAMP-responsive genes. Likewise, biofilm producing *B. japonicum* activated the defense genes expression at the initial stage of their association with cells of root hairs (Libault et al., 2010). Though, beneficial biofilm forming microbes have also been reported for their ability to bypass or repress the defensive system in order to accomplish successful and effective rhizospheric and endophytic colonization (Santoyo, Moreno-Hagelsieb, del Carmen Orozco-Mosqueda, & Glick, 2016; Stringlis et al., 2018). The primary goal of evasion is to prevent the plant immune system from becoming activated. There are few of the mechanisms utilized by numerous PGPRs, which are discussed in the following text.

The most significant structural protein (Flagellin) is found in the bacterial flagella, which is essentially required for the bacterial motility (Rossez, Wolfson, Holmes, Gally, & Holden, 2015). The PRRs of *Arabidopsis* FLS2 usually identify this protein and bind to immunogenic epitope fgl22, which is conserved sequence of 22 amino acids located at the N terminus of the protein structure. After this binding, plants activate their immune systems as well (Sun et al., 2013).

Nevertheless, biofilm forming *Sinorhizobium meliloti* has major changes in this highly conserved region, which is adequate to prevent Arabidopsis FLS2 immune activation (Felix, Duran, Volko, & Boller, 1999). Afterward, another study revealed that the purified flagellin from *Mesorhizobium loti* (symbiont) did not activate the immunological response of *Lotus japonicus* plants. While the epitope (fgl22) from *P. aeruginosa* accelerated the synthesis of ethylene and the stimulation of defense-linked genes among additional molecules from the same plants (Lopez-Gomez, Sandal, Stougaard, & Boller, 2012). A likewise observation was also revealed that the grapevine FLS2 differentially recognized the epitope (flg22) of the PGPR *B. phytofirmans* from the pathogenic bacteria *X. campestris* and *P. aeruginosa*, since grapevine demonstrated a significant diminished immune response to *B. phytofirmans* when compared with reactions to the

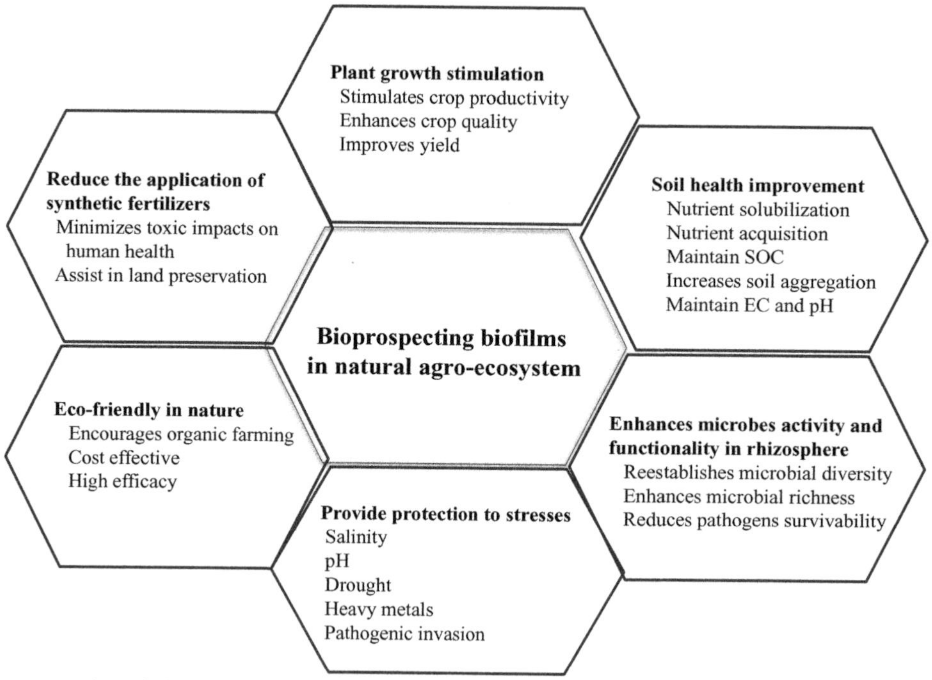

FIG. 2 A pictorial representation of the bioprospecting bacterial biofilms in crop and soil health management under natural agro-ecosystem in sustainable manner.

pathogenic bacteria's epitope (Trdá et al., 2014). Reduced flagella synthesis is one of the most imperative mechanisms for bacterial evasion of plant immunity to protect plants' health. For instance, synthesis of flagella in *Pseudomonas* spp. is controlled by a switch molecule (cyclic di-GMP), which controls the shifting from planktonic to sessile mode of growth state. As a result, large amounts of cyclic-di-GMP block the synthesis of flagellin, which prevents the production of ROS in plant system (Pfeilmeier, Caly, & Malone, 2016).

To suppress the plant immune system, beneficial microbes use a range of effector proteins, including the T3SS system, which is present in the many bacterial genomes, including the *Pseudomonas* and *Rhizobium* genera (Stringlis, Zamioudis, Berendsen, Bakker, & Pieterse, 2019). It was discovered that the inoculation of *Arabidopsis thaliana* with *P. simiae* strain (WCS417) reduced more than half of the MAMP-induced transcriptional responses. This type of reduction in MAMPs was mostly reported in the case of *Pseudomonas* spp. According to the authors of this study, this could lead to establishment of a positive association among beneficial bacteria and plant roots (Yu et al., 2019). So far, these findings recommend that the beneficial microbial biofilms in the rhizosphere also interfere with the plant immune system's signals by synthesizing suppressive effector molecules. Despite all these things, considering the paucity of study on the beneficial bacterial biofilms in crop and soil health, there is still a lot of ground to cover in this interesting area of research.

5. Conclusion and future perspective

In this chapter, few key functions that beneficial bacterial biofilms perform in order to compete, survive, inhabit, colonize, and establish themselves efficiently in the plants' rhizoplane and rhizosphere to maintain soil and crop health have been deliberated. Application of competitive biofilm producing PGPR in open field tests has proven to be a valuable tool for reducing or eliminating pesticide use in agricultural productivity. The biofilm forming PGPR used different mechanisms comprising nutrient uptake, recognition of chemical signals, rhizosphere colonization, motility, and soil aggregation. Furthermore, the synthesis of antimicrobial chemicals benefits PGPR in competing with other rhizospheric microbes. Though, numerous of these mechanisms are also used by phyto-pathogenic microbes, therefore isolating and selecting strong biofilm formers are critical to achieve better outcomes in the natural agro-ecosystems. Furthermore, these PGPRs must have the ability to enhance crop growth and development after effective and successful colonization in the rhizosphere. Certain rhizospheric isolates also associate with the plant, penetrating root tissues and colonizing other plant's parts such as fruits leaves and stems, where they might also employ protective or stimulatory role against nonrhizospheric phytopathogens. Bacterial biofilms use a variety of mechanisms in the rhizosphere and rhizoplane to effective colonization, which are still being researched. The certain attributes such as biofilm development have been demonstrated to be critical for establishing and adhering to plant roots and rhizosphere colonization and existence for longer duration. If this attribute is imperative, it is important to ensure that the experimental settings under it are functioning; this is because of various abiotic and biotic factors that inflect PGPR's beneficial functionality. Currently, high-throughput sequence technologies such as metagenomics and studies of diversity's ITSs or ribosomal genes or DGGE patterns have revealed the influence of biofilm forming PGPR inoculation on the native microbiota in the plants' rhizosphere (Compant, Samad, Faist, & Sessitsch, 2019; Kumar & Dubey, 2020). These approaches could also be used to display if colonizing PGPR has an indirect beneficial effect on the plant, such as enhancing growth-stimulating taxa or reducing potential pathogenic microbes. Though, it is necessary to isolate and characterize these microbial strains and endorse their useful properties to avoid negative crop results.

Finally, because just a few strategies have been proposed, it would be advantageous to build novel screening methods for important functions in biofilm forming PGPR to colonize the rhizosphere effectively. Additionally, it is critical to establish the precise functions of few mechanisms for stimulating plant growth and development in rhizosphere colonization; for instance, few volatile compounds have been demonstrated to have antibiotic properties in the laboratory. Filling in the gaps in our understanding of PGPR's rhizosphere colonization mechanisms would make it easier to choose superior isolates as bioinoculants under natural agro-ecosystem.

References

Adessi, A., de Carvalho, R. C., De Philippis, R., Branquinho, C., & da Silva, J. M. (2018). Microbial extracellular polymeric substances improve water retention in dryland biological soil crusts. *Soil Biology & Biochemistry*, *116*, 67–69.

Ahmad, I., Khan, M. S., Altaf, M. M., Qais, F. A., Ansari, F. A., & Rumbaugh, K. P. (2017). Biofilms: An overview of their significance in plant and soil health. In I. Ahmad, & F. M. Husain (Eds.), *Biofilms in plant and soil health* (pp. 1–25). Chichester: John Wiley & Sons, Ltd.

Ansari, F. A., & Ahmad, I. (2018). Biofilm development, plant growth promoting traits and rhizosphere colonization by *Pseudomonas entomophila* FAP1: A promising PGPR. *Advances in Microbiology*, *8* (03), 235.

Ansari, F. A., & Ahmad, I. (2019). Fluorescent *Pseudomonas*-FAP2 and *Bacillus licheniformis* interact positively in biofilm mode enhancing plant growth and photosynthetic attributes. *Scientific Reports*, *9*(1), 1–12.

Ansari, F. A., Ahmad, I., & Pichtel, J. (2019). Growth stimulation and alleviation of salinity stress to wheat by the biofilm forming *Bacillus pumilus* strain FAB10. *Applied Soil Ecology*, *143*, 45–54.

Ansari, F. A., Jabeen, M., & Ahmad, I. (2021). *Pseudomonas azotoformans* FAP5, a novel biofilm-forming PGPR strain, alleviates drought stress in wheat plant. *International Journal of Environmental Science and Technology*, 1–16.

Ansari, F. A., Jafri, H., Ahmad, I., & Abulreesh, H. H. (2017). Factors affecting biofilm formation in in vitro and in the rhizosphere. In *Biofilms in plant and soil health* (pp. 275–290). Wiley.

Auguet, O., Pijuan, M., Batista, J., Borrego, C. M., & Gutierrez, O. (2015). Changes in microbial biofilm communities during colonization of sewer systems. *Applied and Environmental Microbiology*, *81*, 7271–7280.

Banerjee, A., Sarkar, S., Cuadros-Orellana, S., & Bandopadhyay, R. (2019). Exopolysaccharides and biofilms in mitigating salinity stress: The biotechnological potential of halophilic and soil-inhabiting PGPR microorganisms. In *Microorganisms in saline environments: Strategies and functions* (pp. 133–153). Cham: Springer.

Basler, M., Ho, B. T., & Mekalanos, J. J. (2013). Tit-for-tat: Type VI secretion system counterattack during bacterial cell-cell interactions. *Cell*, *152*, 884–894.

Bezzate, S., Aymerich, S., Chambert, R., Czarnes, S., Berge, O., & Heulin, T. (2000). Disruption of the *Paenibacillus polymyxa* levansucrase gene impairs its ability to aggregate soil in the wheat rhizosphere. *Environmental Microbiology*, *2*, 333–342.

Böhm, M., Hurek, T., & Reinhold-Hurek, B. (2007). Twitching motility is essential for endophytic rice colonization by the N_2-fixing endophyte *Azoarcus* sp. strain BH72. *Molecular Plant Microbe Interactions*, *20* (5), 526–533.

Breugelmans, P., Barken, K. B., Tolker-Nielsen, T., Hofkens, J., Dejonghe, W., & Springael, D. (2008). Architecture and spatial organization in a triple-species bacterial biofilm synergistically degrading the phenylurea herbicide linuron. *FEMS Microbiology Ecology*, *64*, 271–282.

Büks, F., & Kaupenjohann, M. (2016). Enzymatic biofilm digestion in soil aggregates facilitates the release of particulate organic matter by sonication. *Soil (Göttingen)*, *2*, 499–509.

Burmølle, M., Hansen, L. H., & Sørensen, S. J. (2007). Establishment and early succession of a multispecies biofilm composed of soil bacteria. *Microbial Ecology*, *54*, 352–362.

Burmolle, M., Kjoller, A., & Sorensen, S. J. (2012). An invisible workforce: Biofilms in the soil. In L. Gavin, & L. Gillian (Eds.), *Microbial Biofilm: Current Research and Applications* (pp. 61–71). Caister Academic Press.

Burmølle, M., Ren, D., Bjarnsholt, T., & Sørensen, S. J. (2014). Interactions in multispecies biofilms: Do they actually matter? *Trends in Microbiology*, *22*, 84–91.

Cai, P., Liu, X., Ji, D., Yang, S., Walker, S. L., Wu, Y., et al. (2018). Impact of soil clay minerals on growth, biofilm formation, and virulence gene expression of *Escherichia coli* O157:H7. *Environmental Pollution*, *243*, 953–960.

Cai, P., Sun, X., Wu, Y., Gao, C., Mortimer, M., Holden, P. A., et al. (2019). Soil biofilms: Microbial interactions, challenges, and advanced techniques for ex-situ characterization. *Soil Ecology Letters*, *1*(3), 85–93.

Chenu, C., & Cosentino, D. (2011). Microbial regulation of soil structural dynamics. In K. Ritz, & I. Young (Eds.), *The architecture and biology and soils: Life in inner space* (pp. 37–70). London: CAB International.

Chenu, C., & Roberson, E. B. (1996). Diffusion of glucose in microbial extracellular polysaccharide as affected by water potential. *Soil Biology & Biochemistry*, *28*, 877–884.

Compant, S., Samad, A., Faist, H., & Sessitsch, A. (2019). A review on the plant microbiome: Ecology, functions, and emerging trends in microbial application. *Journal of Advanced Research, 19*, 29–37.

Costa, O. Y. A., Raaijmakers, J. M., & Kuramae, E. E. (2018). Microbial extracellular polymeric substances: Ecological function and impact on soil aggregation. *Frontiers in Microbiology, 9*, 1636.

Danhorn, T., & Fuqua, C. (2007). Biofilm formation by plant-associated bacteria. *Annual Review Microbiology, 61*, 401–422.

Dimkpa, C. (2016). Microbial siderophores: Production, detection and application in agriculture and environment. *Endocytobiosis Cell Research, 27*, 7–16.

Dodd, M. S., Papineau, D., Grenne, T., Slack, J. F., Rittner, M., Pirajno, F., et al. (2017). Evidence for early life in Earth's oldest hydrothermal vent precipitates. *Nature, 543*, 60–64.

Duvernoy, M. C., Mora, T., Ardré, M., Croquette, V., Bensimon, D., Quilliet, C., et al. (2018). Asymmetric adhesion of rod-shaped bacteria controls microcolony morphogenesis. *Nature Communications, 9*, 1120.

Dwivedi, D., Khare, M., Chaturvedi, H., & Singh, V. (2017). Plant pathogenic bacteria: Role of quorum sensing and biofilm in disease development. In I. Ahmad, & F. M. Husain (Eds.), *Biofilms in plant and soil health* (pp. 387–407). Hoboken, NJ: Wiley-Blackwell.

Felix, G., Duran, J. D., Volko, S., & Boller, T. (1999). Plants have a sensitive perception system for the most conserved domain of bacterial flagellin. *The Plant Journal, 18*(3), 265–276.

Fernández-Llamosas, H., Díaz, E., & Carmona, M. (2021). Motility, adhesion and c-di-GMP influence the endophytic colonization of rice by *Azoarcus* sp. CIB. *Microorganisms, 9*(3), 554.

Fierer, N. (2017). Embracing the unknown: Disentangling the complexities of the soil microbiome. *Nature Reviews Microbiology, 15*, 579–590.

Flemming, H. C. (2010). Biodeterioration of synthetic materials—A brief review. *Materials and Corrosion, 61*, 986–992.

Flemming, H. C., & Wingender, J. (2010). The biofilm matrix. *Nature Review Microbiology, 8*, 623.

Flemming, H. C., Wingender, J., Szewzyk, U., Steinberg, P., Rice, S. A., & Kjelleberg, S. (2016). Biofilms: An emergent form of bacterial life. *Nature Reviews Microbiology, 14*, 563–575.

Flemming, H. C., & Wuertz, S. (2019). Bacteria and archaea on Earth and their abundance in biofilms. *Nature Review Microbiology, 17*, 247–260.

Foster, K. R., & Bell, T. (2012). Competition, not cooperation, dominates interactions among culturable microbial species. *Current Biology, 22*, 1845–1850.

Freilich, S., Zarecki, R., Eilam, O., Segal, E. S., Henry, C. S., Kupiec, M., et al. (2011). Competitive and cooperative metabolic interactions in bacterial communities. *Nature Communications, 2*, 589.

Gao, C. H., Zhang, M., Wu, Y., Huang, Q., & Cai, P. (2019). Divergent influence to a pathogen invader by resident bacteria with different social interactions. *Microbial Ecology, 77*, 76–86.

Glick, B. R. (2012). Plant growth-promoting bacteria: Mechanisms and applications. *Scientifica*, 963401.

Gravel, D., Bell, T., Barbera, C., Bouvier, T., Pommier, T., Venail, P., et al. (2011). Experimental niche evolution alters the strength of the diversity–productivity relationship. *Nature, 469*, 89–92.

Gutiérrez Castorena, E. V., Gutiérrez-Castorena, M. D. C., González Vargas, T., Cajuste Bontemps, L., Delgadillo Martínez, J., Suástegui Méndez, E., et al. (2016). Micromapping of microbial hotspots and biofilms from different crops using digital image mosaics of soil thin sections. *Geoderma, 279*, 11–21.

Hagerty, S. B., Van Groenigen, K. J., Allison, S. D., Hungate, B. A., Schwartz, E., Koch, G. W., et al. (2014). Accelerated microbial turnover but constant growth efficiency with warming in soil. *Nature Climate Change, 4*(10), 903–906.

Harrison, J. J., Ceri, H., & Turner, R. J. (2007). Multimetal resistance and tolerance in microbial biofilms. *Nature Reviews Microbiology, 5*, 928–938.

Hassani, M. A., Durán, P., & Hacquard, S. (2018). Microbial interactions within the plant holobiont. *Microbiome, 6*(1), 1–17.

Hawlena, H., Bashey, F., & Lively, C. M. (2010). The evolution of spite: Population structure and bacteriocin-mediated antagonism in two natural populations of *Xenorhabdus bacteria*. *Evolution, 64*, 3198–3204.

Heil, J., Vereecken, H., & Brüggemann, N. (2016). A review of chemical reactions of nitrification intermediates and their role in nitrogen cycling and nitrogen trace gas formation in soil. *European Journal of Soil Science, 67*(1), 23–39.

Hettiarachchi, R. P., Dharmakeerthi, R. S., Jayakody, A. N., Seneviratne, G., de Silva, E., Gunathilake, T., et al. (2014). Effectiveness of fungal bacterial interactions as biofilmed biofertilizers on enhancement of root growth of Hevea seedlings. *Journal of Environmental Professionals Sri Lanka, 3*, 25–40.

Hibbing, M. E., Fuqua, C., Parsek, M. R., & Peterson, S. B. (2010). Bacterial competition: Surviving and thriving in the microbial jungle. *Nature Reviews Microbiology, 8*, 15–25.

Hider, R. C., & Kong, X. (2010). Chemistry and biology of siderophores. *Natural Product Reports, 27*(5), 637–657.

Jog, R., Pandya, M., Nareshkumar, G., & Rajkumar, S. (2014). Mechanism of phosphate solubilization and antifungal activity of Streptomyces spp. isolated from wheat roots and rhizosphere and their application in improving plant growth. *Microbiology, 160*(4), 778–788.

Jones, J. D., & Dangl, J. L. (2006). The plant immune system. *Nature, 444* (7117), 323–329.

Jung, B. K., Hong, S. J., Park, G. S., Kim, M. C., & Shin, J. H. (2018). Isolation of *Burkholderia cepacia* JBK9 with plant growth-promoting activity while producing pyrrolnitrin antagonistic to plant fungal diseases. *Applied Biological Chemistry, 61*, 173–180.

Kasim, W. A., Gaafar, R. M., Abou-Ali, R. M., Omar, M. N., & Hewait, H. M. (2016). Effect of biofilm forming plant growth promoting rhizobacteria on salinity tolerance in barley. *Annals of Agricultural Sciences, 61*, 217–227.

Kenny, D. J., & Balskus, E. P. (2018). Engineering chemical interactions in microbial communities. *Chemical Society Reviews, 47*, 1705–1729.

Khatoon, Z., Huang, S., Rafique, M., Fakhar, A., Kamran, M. A., & Santoyo, G. (2020). Unlocking the potential of plant growth-promoting rhizobacteria on soil health and the sustainability of agricultural systems. *Journal of Environmental Management, 273*, 111118.

Kivlin, S. N., & Hawkes, C. V. (2020). Spatial and temporal turnover of soil microbial communities is not linked to function in a primary tropical forest. *Ecology, 101*(4), e02985.

Koul, V., Tripathi, C., Adholeya, A., & Kochar, M. (2015). Nitric oxide metabolism and indole acetic acid biosynthesis cross-talk in *Azospirillum brasilense* SM. *Research in Microbiology, 166*, 174–185.

Kragh, K. N., Hutchison, J. B., Melaugh, G., Rodesney, C., Roberts, A. E. L., Irie, Y., et al. (2016). Role of multicellular aggregates in biofilm formation. *Molecular Biology, 7*. e00237–e16.

Kumar, A., & Dubey, A. (2020). Rhizosphere microbiome: Engineering bacterial competitiveness for enhancing crop production. *Journal of Advanced Research, 24*, 337–352.

Kumar, M., Mishra, S., Dixit, V., Kumar, M., Agarwal, L., Chauhan, P. S., et al. (2016). Synergistic effect of *Pseudomonas putida* and *Bacillus amyloliquefaciens* ameliorates drought stress in chickpea (*Cicer arietinum* L.). *Plant Signaling Behaviour*, *11*, e1071004.

Kuzyakov, Y., & Blagodatskaya, E. (2015). Microbial hotspots and hot moments in soil: Concept & review. *Soil Biology & Biochemistry*, *83*, 184–199.

Lee, K. W. K., Periasamy, S., Mukherjee, M., Xie, C., Kjelleberg, S., & Rice, S. A. (2014). Biofilm development and enhanced stress resistance of a model, mixed-species community biofilm. *ISME Journal*, *8*, 894–907.

Lehmann, A., Zheng, W., & Rillig, M. C. (2017). Soil biota contributions to soil aggregation. *Nature Ecology & Evolution*, *1*, 1828–1835.

Lennon, J. T., & Lehmkuhl, B. K. (2016). A trait-based approach to bacterial biofilms in soil. *Environmental Microbiology*, *18*, 2732–2742.

Libault, M., Farmer, A., Brechenmacher, L., Drnevich, J., Langley, R. J., Bilgin, D. D., et al. (2010). Complete transcriptome of the soybean root hair cell, a single-cell model, and its alteration in response to *Bradyrhizobium japonicum* infection. *Plant Physiology*, *152*(2), 541–552.

Lin, D., Ma, W., Jin, Z., Wang, Y., Huang, Q., & Cai, P. (2016). Interactions of EPS with soil minerals: A combination study by ITC and CLSM. *Colloids and Surfaces B: Biointerfaces*, *138*, 10–16.

Lopez-Gomez, M., Sandal, N., Stougaard, J., & Boller, T. (2012). Interplay of flg22-induced defence responses and nodulation in *Lotus japonicus*. *Journal of Experimental Botany*, *63*(1), 393–401.

Madsen, J. S., Røder, H. L., Russel, J., Sørensen, H., Burmølle, M., & Sørensen, S. J. (2016). Coexistence facilitates interspecific biofilm formation in complex microbial communities. *Environmental Microbiology*, *18*, 2565–2574.

Magana, M., Sereti, C., Ioannidis, A., Mitchell, C. A., Ball, A. R., Magiorkinis, E., et al. (2018). Options and limitations in clinical investigation of bacterial biofilms. *Clinical Microbiology Reviews*, *31*. e00084–e16.

Martínez-Cámara, R., Montejano-Ramírez, V., Moreno-Hagelsieb, G., Santoyo, G., & Valencia-Cantero, E. (2020). The volatile organic compound dimethylhexadecylamine affects bacterial growth and swarming motility of bacteria. *Folia Microbiologica*, *65*(3), 523–532.

Moyano, F. E., Manzoni, S., & Chenu, C. (2013). Responses of soil heterotrophic respiration to moisture availability: An exploration of processes and models. *Soil Biology and Biochemistry*, *59*, 72–85.

Muhammad, M. H., Idris, A. L., Fan, X., Guo, Y., Yu, Y., Jin, X., et al. (2020). Beyond risk: Bacterial biofilms and their regulating approaches. *Frontiers in Microbiology*, *11*, 928.

Nadell, C. D., Drescher, K., & Foster, K. R. (2016). Spatial structure, cooperation and competition in biofilms. *Nature Reviews Microbiology*, *14*, 589–600.

Nihorimbere, V., Cawoy, H., Seyer, A., Brunelle, A., Thonart, P., & Ongena, M. (2012). Impact of rhizosphere factors on cyclic lipopeptide signature from the plant beneficial strain *Bacillus amyloliquefaciens* S499. *FEMS Microbiology Ecology*, *79*, 176–191.

Nunan, N. (2017). The microbial habitat in soil: Scale, heterogeneity and functional consequences. *Journal of Plant Nutrition and Soil Science*, *180*, 425–429.

O'Toole, G. A., & Wong, G. C. (2016). Sensational biofilms: Surface sensing in bacteria. *Current Opinion in Microbiology*, *30*, 139–146.

Olanrewaju, O. S., Ayangbenro, A. S., Glick, B. R., & Babalola, O. O. (2019). Plant health: Feedback effect of root exudates-rhizobiome interactions. *Applied Microbiology and Biotechnology*, *103*(3), 1155–1166.

Olsen, I. (2015). Biofilm-specific antibiotic tolerance and resistance. *European Journal of Clinical Microbiology & Infectious Diseases*, *34*, 877–886.

Or, D., Phutane, S., & Dechesne, A. (2007). Extracellular polymeric substances affecting pore-scale hydrologic conditions for bacterial activity in unsaturated soils. *Vadose Zone Journal*, *6*, 298–305.

Pande, S., Kaftan, F., Lang, S., Svatoš, A., Germerodt, S., & Kost, C. (2016). Privatization of cooperative benefits stabilizes mutualistic cross-feeding interactions in spatially structured environments. *ISME Journal*, *10*, 1413–1423.

Pandit, A., Adholeya, A., Cahill, D., Brau, L., & Kochar, M. (2020). Microbial biofilms in nature: Unlocking their potential for agricultural applications. *Journal of Applied Microbiology*, *129*(2), 199–211.

Parsek, M. R., & Greenberg, E. P. (2005). Sociomicrobiology: The connections between quorum sensing and biofilms. *Trends in Microbiology*, *13*(1), 27–33.

Patin, N. V., Duncan, K. R., Dorrestein, P. C., & Jensen, P. R. (2016). Competitive strategies differentiate closely related species of marine actinobacteria. *ISME Journal*, *10*, 478–490.

Perez-Garcia, O., Lear, G., & Singhal, N. (2016). Metabolic network modeling of microbial interactions in natural and engineered environmental systems. *Frontiers in Microbiology*, *7*, 673.

Pfeilmeier, S., Caly, D. L., & Malone, J. G. (2016). Bacterial pathogenesis of plants: Future challenges from a microbial perspective: Challenges in bacterial molecular plant pathology. *Molecular Plant Pathology*, *17* (8), 1298–1313.

Phour, M., Sehrawat, A., Sindhu, S. S., & Glick, B. R. (2020). Interkingdom signaling in plant-rhizomicrobiome interactions for sustainable agriculture. *Microbiological Research*, 126589.

Poole, P. (2017). Shining a light on the dark world of plant root–microbe interactions. *Proceedings of the National Academy of Sciences of the United States of America*, *114*, 4281–4283.

Raaijmakers, J. M., & Mazzola, M. (2012). Diversity and natural functions of antibiotics produced by beneficial and plant pathogenic bacteria. *Annual Review of Phytopathology*, *50*, 403–424.

Ren, D., Madsen, J. S., Sørensen, S. J., & Burmølle, M. (2015). High prevalence of biofilm synergy among bacterial soil isolates in cocultures indicates bacterial interspecific cooperation. *ISME Journal*, *9*, 81–89.

Rillig, M. C., Muller, L. A., & Lehmann, A. (2017). Soil aggregates as massively concurrent evolutionary incubators. *ISME Journal*, *11*, 1943–1948.

Rossez, Y., Wolfson, E. B., Holmes, A., Gally, D. L., & Holden, N. J. (2015). Bacterial flagella: Twist and stick, or dodge across the kingdoms. *PLoS Pathogens*, *11*(1), e1004483.

Russel, J., Røder, H. L., Madsen, J. S., Burmølle, M., & Sørensen, S. J. (2017). Antagonism correlates with metabolic similarity in diverse bacteria. *Proceedings of the National Academy of Sciences of the United States of America*, *114*, 10684–10688.

Saleh, D., Sharma, M., Seguin, P., & Jabaji, S. (2020). Organic acids and root exudates of *Brachypodium distachyon*: Effects on chemotaxis and biofilm formation of endophytic bacteria. *Canadian Journal of Microbiology*, *66*(10), 562–575.

Santoyo, G., Moreno-Hagelsieb, G., del Carmen Orozco-Mosqueda, M., & Glick, B. R. (2016). Plant growth-promoting bacterial endophytes. *Microbiological Research*, *183*, 92–99.

Santoyo, G., Urtis-Flores, C. A., Loeza-Lara, P. D., Orozco-Mosqueda, M., & Glick, B. R. (2021). Rhizosphere colonization determinants by plant growth-promoting rhizobacteria (PGPR). *Biology*, *10*(6), 475.

Schoustra, S. E., Dench, J., Dali, R., Aaron, S. D., & Kassen, R. (2012). Antagonistic interactions peak at intermediate genetic distance in clinical and laboratory strains of *Pseudomonas aeruginosa*. *BMC Microbiology*, *12*, 40.

Seneviratne, G., & Jayasinghearachchi, H. S. (2005). A rhizobial biofilm with nitrogenase activity alters nutrient availability in a soil. *Soil Biology & Biochemistry*, *37*, 1975–1978.

Shin, W., Islam, R., Benson, A., Joe, M. M., Kim, K., Gopal, S., et al. (2016). Role of diazotrophic bacteria in biological nitrogen fixation and plant growth improvement. *Korean Journal Soil Science and Fertility*, *49*, 17–29.

Souza, E. M., Chubatsu, L. S., Huergo, L. F., Monteiro, R., Camilios-Neto, D., Wassem, R., et al. (2014). Use of nitrogen-fixing bacteria to improve agricultural productivity. *BMC Proceeding*, *8*.

Srivastava, S., Yadav, A., Seem, K., Mishra, S., Chaudhary, V., & Nautiyal, C. S. (2008). Effect of high temperature on *Pseudomonas putida* NBRI0987 biofilm formation and expression of stress sigma factor RpoS. *Current Microbiology*, *56*, 453–457.

Stringlis, I. A., Zamioudis, C., Berendsen, R. L., Bakker, P. A., & Pieterse, C. M. (2019). Type III secretion system of beneficial rhizobacteria *Pseudomonas simiae* WCS417 and *Pseudomonas defensor* WCS374. *Frontiers in Microbiology*, *10*, 1631.

Stringlis, I. A., Zhang, H., Pieterse, C. M., Bolton, M. D., & de Jonge, R. (2018). Microbial small molecules—Weapons of plant subversion. *Natural Product Reports*, *35*(5), 410–433.

Sun, Y., Li, L., Macho, A. P., Han, Z., Hu, Z., Zipfel, C., et al. (2013). Structural basis for flg22-induced activation of the *Arabidopsis* FLS2-BAK1 immune complex. *Science*, *342*(6158), 624–628.

Trdá, L., Fernandez, O., Boutrot, F., Héloir, M. C., Kelloniemi, J., Daire, X., et al. (2014). The grapevine flagellin receptor Vv FLS 2 differentially recognizes flagellin-derived epitopes from the endophytic growth-promoting bacterium *Burkholderia phytofirmans* and plant pathogenic bacteria. *New Phytologist*, *201*(4), 1371–1384.

Vardharajula, S., Ali, S., Grover, M., Reddy, G., & Venkateswarlu, B. (2011). Drought-tolerant plant growth promoting *Bacillus* spp. effect on growth, osmolytes, and antioxidant status of maize under drought stress. *Journal of Plant Interaction*, *6*, 1–14.

Vetsigian, K., Jajoo, R., & Kishony, R. (2011). Structure and evolution of Streptomyces interaction networks in soil and in silico. *PLoS Biology*, *9*, e1001184.

Wang, D., Xu, A., Elmerich, C., & Ma, L. Z. (2017). Biofilm formation enables free-living nitrogen-fixing rhizobacteria to fix nitrogen under aerobic conditions. *The ISME Journal*, *11*(7), 1602–1613.

Yin, W., Wang, Y., Liu, L., & He, J. (2019). Biofilms: The microbial "protective clothing" in extreme environments. *International Journal of Molecular Sciences*, *20*(14), 3423.

Yu, K., Pieterse, C. M., Bakker, P. A., & Berendsen, R. L. (2019). Beneficial microbes going underground of root immunity. *Plant Cell & Environment*, *42*(10), 2860–2870.

Yuan, J., Zhang, N., Huang, Q., Raza, W., Li, R., Vivanco, J. M., et al. (2015). Organic acids from root exudates of banana help root colonization of PGPR strain *Bacillus amyloliquefaciens* NJN-6. *Scientific Reports*, *5*(1), 1–8.

Zengler, K., & Zaramela, L. S. (2018). The social network of microorganisms-how auxotrophies shape complex communities. *Nature Reviews Microbiology*, *16*, 383–390.

Zhang, N., Wang, D., Liu, Y., Li, S., Shen, Q., & Zhang, R. (2014). Effects of different plant root exudates and their organic acid components on chemotaxis, biofilm formation and colonization by beneficial rhizosphere-associated bacterial strains. *Plant and Soil*, *374*, 689–700.

Chapter 5

Chemotaxis and rhizobacterial biofilm formation in plant-microbe interaction

P. Jishma and E.K. Radhakrishnan
School of Biosciences, Mahatma Gandhi University, Kottayam, Kerala, India

1. Introduction to microbial interaction with plants

In natural and agricultural ecosystems, various types of microorganisms, mainly bacteria and fungi, are inhabited on both aboveground and belowground of plants as a mutualistic interaction (Dolatabadian, 2020; Wu, Bernard, Andersen, & Chen, 2009). The beneficial interactions stimulate nutrient transfer and make available mineral nutrients and fixed nitrogen to plants, augmentation of plant growth through phytohormones production by direct mechanisms, inhibition of pathogenic microorganisms, and alleviation of stress through indirect mechanisms (Ribeiro & Cardoso, 2012). The microbial association on the plants can be grouped into epiphytes that habitat on the surface, endophytes that are present within the plant tissues, phyllospheric microbes that locate on surface of leaves, and rhizospheric microbes that grow in the soil, which is influenced by the plant roots. Among these associations, rhizomicrobiome has most significant role in the growth and development stages of plant and its resistance to biotic and abiotic stresses (Vishwakarma et al., 2020).

The rhizosphere is composed of various types of chemical compounds, biomolecules, and different species of microorganisms, which include both beneficial and harmful to the plants (Hayat, Ali, Amara, Khalid, & Ahmed, 2010; Mendes, Garbeva, & Raaijmakers, 2013). The microorganisms present in the soil have a vital role in shaping the rhizosphere, as these can produce a variety of compounds and released into the rhizosphere region. These compounds determine the microbial composition of rhizosphere, because many of these compounds have antimicrobial property and inhibit the growth of various microorganisms present in the soil. Those surviving rhizobacteria are mostly beneficial to plants and are attracted toward the plant root in response to the compounds released by the roots (Compant, Samad, Faist, & Sessitsch, 2019).

Rhizobia and mycorrhizal fungi are two significant associations in the rhizosphere with plant roots. These rhizospheric microbes produce various types of signals, including volatile organic compounds, hormones and hormone mimics, carbohydrate- and protein-based signals (Plett & Martin, 2018). In leguminous plant roots, the symbiotic association of rhizobia begins with detection of signal molecules such as plant-derived flavonoids and produces a lipo-chitooligosaccharide signal (Nod factor). The plants recognize Nod factor, resulting in activation of following symbiotic reactions that lead to rhizobial infection and nodule organogenesis (Shimoda et al., 2020). In the arbuscular mycorrhizal association, communication starts by exchanging signal molecules between plant root and fungi. The spore germination and hyphae growth are stimulated by the strigolactone released by the plant. In response, factors including lipo-chitooligosaccharides and chitooligosaccharides are released by mycorrhizal fungi and stimulate the signaling pathway of the symbiosis association in the plant root (Oldroyd, 2013). The present chapter provides a detailed outlook on the rhizobacterial chemotaxis and biofilm formation in beneficial plant-microbe interaction.

2. Chemotaxis: Underground communication of rhizobacteria with plant roots

Plant-rhizobacterial interaction has a significant role in the growth and health of plants. Plant growth promoting rhizobacterial (PGPR) genera such as *Pseudomonas*, *Rhizobium*, *Azospirillum*, *Bacillus*, and *Burkholderia* are widely applied in agricultural field to augment plant growth, crop yield, and control soil-borne diseases (Lugtenberg & Kamilova, 2009). Efficient and successful root colonization of such rhizobacteria is essential for their active functioning and stimulation of plant beneficial effects. The microbial activity in the rhizosphere is influenced and regulated by chemical compounds released by plant roots known as root exudates or rhizodeposits that form a chemical gradient in the rhizosphere (Egamberdieva et al., 2007). Rhizobacteria

Understanding Microbial Biofilms. https://doi.org/10.1016/B978-0-323-99977-9.00005-3

have to move toward the rhizodeposits to survive and succeed in the rhizosphere, and this movement of rhizobacteria is chemotaxis that determines the structure and architecture of the microbial community and chemical composition in the rhizosphere (Feng et al., 2021).

Chemotaxis is considered as the communication of rhizobacteria with plants through chemical signals that gives an advantage to motile flagellated soil bacteria in the plant root colonization, which is a primary requirement for the establishment of useful interactions (Gao et al., 2020). This chemical signaling facilitates motile soil bacteria to recognize and act in response to gradients of chemical compounds released by plant roots. As a result, the rhizobacteria can actively migrate toward plant roots and is thus significant for active colonization on root surface (Scharf, Hynes, & Alexandre, 2016). The previous studies on analysis of complete genome sequences of various rhizobacterial species have reported the occurrence of multiple chemotaxis systems and a variety of chemoreceptors. Also, most of the soil bacteria are motile, and various chemotaxis-encoding genes are found in their genetic material that makes them capable for chemotaxis (Meier, Muschler, & Scharf, 2007; Phillips, Joseph, & Maxwell, 1992; Webb, Hildreth, Helm, Scharf, & Voordouw, 2014).

3. Role of root exudates in chemotaxis

Root exudates or rhizodeposits are very complex mixtures of chemoattractants released by plant roots into the surroundings. These chemoeffectors include amino acids, organic acids, flavonoids, sugars, alcohols, phenolics, polyamines, fatty acids, purines, phytohormones, terpenoids, benzoxazinoids, mucilage (polysaccharides), and proteins (Bais, Weir, Perry, Gilroy, & Vivanco, 2006; Compton & Scharf, 2021). Organic acids or carboxylates and amino acids are the commonly found molecules in the root exudates. The mono- and di-carboxylates can act as carbon sources and are weak attractants than the amino acids (Moe, 2013; Webb et al., 2014). The flavonoids can stimulate the expression of symbiotic genes in rhizobia (Abdel-Lateif, Bogusz, & Hocher, 2014), and the phenolic compounds include phenylpropanoid derivatives of aromatic acids, which are present in plant metabolites and precursors to flavonoids can also have a vital role in chemotaxis (Kape, Parniske, & Werner, 1991). Sugars or saccharides are some of the common carbon sources in the rhizodeposits. Gluconate and glucose in addition to structural components of pectins such as arabinose and xylose also serve as chemoattractants for rhizobia. Chemotaxis toward these compounds facilitates rhizobacterial recruitment and colonization that leads to the establishment of rhizobacteria-root association (Compton & Scharf, 2021; Jishma & Radhakrishnan, 2020).

The composition of exudates varies with plant species, soil nutrient content, and physiological features of soil. All the compounds present in the rhizodeposit released by a specific plant species cannot act as attractants for specific bacterial species. Chemotactic responses toward root exudates are started by ligand sensing at chemoreceptors or methyl-accepting chemotaxis proteins (MCPs) present in the rhizobacteria and are very significant for colonization on root surface and the plant beneficial functions of PGPR (Allard-Massicotte et al., 2016). Sometime, only one compound can be recognized by the chemoreceptors and facilitates chemotaxis and colonization (Table 1). Also, plant-specific root exudates exhibit the specific selection of rhizobacterial species. Certain compounds may be found in different forms, such as isomers or optical rotation, which may lead to difference in chemotactic response of a strain. For example, *Pseudomonas fluorescens* Pf0-1 exhibited very strong chemotactic response to L-malate, but could not recognize D-malate (Oku et al., 2014).

4. Recognition of chemoattractants and mechanism of chemotaxis

Beneficial plant-rhizobacterial interaction involves different steps such as recognition of chemoattractant in rhizodeposits by the rhizobacteria at chemoreceptors or MCPs, chemotaxis, root colonization, and biofilm formation. This successful recruitment of PGPR on plant root is essential for stimulation of their plant beneficial functions resulting in plant growth enhancement and disease resistance (Fig. 1) (Allard-Massicotte et al., 2016; Feng et al., 2021; Scharf et al., 2016).

Every species or strain of rhizobacteria carries several chemoreceptors or MCPs that can detect a wide array of chemical attractants present in the root exudates of respective host plants (Compton & Scharf, 2021). The sensing mechanism begins through communication of the chemoeffector with the MCP ligand-binding domain (LBD) that is usually located in the extracytoplasmic space. A chemosensory array is formed by transmembrane MCPs, the histidine kinase CheA, and the coupling protein CheW (Matilla, Martin-Mora, & Krell, 2020; Parkinson, Hazelbauer, & Falke, 2015). MCPs detect chemoattractant directly or through periplasmic binding proteins. The receptors transduce this binding stimulus across the cellular membrane to regulate the autophosphorylation activity of the histidine kinase CheA. CheA is a five-domain (P1–P5) enzyme, and each domain has different functions. The P1 domain has the substrate histidine that turns into phosphorylated, P2 docks the response regulator proteins CheY and CheB, P3 dimerizes the protein, P4 binds Mg^{+2}-ATP and catalyzes phosphoryl transfer to the histidine substrate on P1, and P5 couples CheA to the other

TABLE 1 Chemoattractants produced by the plants and rhizobacterial chemotaxis.

Chemoattractant	Plant species	Attracted rhizobacteria	Reference
Citric acids	*Cucumis sativus* (cucumber)	*Bacillus amyloliquefaciens* SQR9	Zhang et al. (2013)
Fumaric acid	*Musaceae* (banana)	*Bacillus subtilis* N11	Zhang et al. (2013)
Malic acid and Citric acid	*Citrullus lanatus* (watermelon)	*Paenibacillus polymyxa* SQR-21	Ling, Raza, Ma, Huang, and Shen (2011)
Benzoxazinoids	*Zea mays* (maize)	*Pseudomonas putida* KT2440	Yang, Neal, Ahmad, Gordon-Weeks, and Ton (2012)
Malic acid	*Arabidopsis thaliana*	*Bacillus subtilis* FB17	Rudrappa, Czymmek, Pare, and Bais (2008)
Malic acid	*Musaceae* (banana)	*Bacillus amyloliquefaciens* NJN-6	Yuan et al. (2015)
Malic acid and Citric acid	*Solanum lycopersicum* (tomato)	*Pseudomonas fluorescens* WCS365	De Weert et al. (2002)
Isoliquiritigenin	*Glycine max* (soybean)	*Bradyrhizobium japonicum*	Kape, Parniske, Brandt, and Werner (1992)
Glycine betaine, trigonelline, and choline	*Medicago sativa* (alfalfa)	*Sinorhizobium meliloti*	Compton, Hildreth, Helm, Scharf, and Stock (2018)
L-Malate, succinate, and fumarate	*Solanum lycopersicum* (tomato)	*Pseudomonas fluorescens* Pf0-1	Oku, Komatsu, Nakashimada, Tajima, and Kato (2014)
Malic acid	*Solanum lycopersicum* (tomato)	*Bacillus subtilis*	Sasse, Martinoia, and Northen (2018)

array components by binding both CheW and the chemoreceptors (Greenswag, Muok, Li, & Crane, 2015; Swanson, Bourret, & Simon, 1993).

CheA begins an intracellular phosphorelay after its activation that eventually controls flagella rotation. A previous study in *Escherichia coli* by Parkinson et al. (2015) has been reported that an enhancement in repellent or reduction in attractant in the surroundings enhances CheA autophosphorylation (Parkinson et al., 2015). The phosphorylated CheA contributes phosphate from phosphoryl-histidine to an aspartate residue on the response regulator protein CheY. Phosphorylated CheY (CheY-P) directly binds the flagellar rotor protein FliM to support rotation of the flagella in clockwise direction, encouraging the flagellar bundle to disperse, and cell tumbling (Flack & Parkinson, 2018; Sourjik & Wingreen, 2012; Wadhams & Armitage, 2004). When CheA is deactivated by attractant, a decline in CheY-P causes rotation of the rotor in counterclockwise direction, flagellar filaments turn to bundle, leading to smooth straight swimming. The rhizobacteria follow gradients of attractants and repellents by alternating between tumbling and smooth swimming. In addition to this, two enzymes, the methyltransferase CheR and the methylesterase CheB, make a slower adaptation response that provides the tracking of chemoattractant gradients. Also, CheR and CheB modify the methylation of glutamate residues on the receptors and thereby regulate affinity of receptor ligand and modulation of CheA activity (Fig. 2) (Feng et al., 2021; Muok, Briegel, & Crane, 2020).

Many PGPR including *Pseudomonas putida*, *P. fluorescens*, *Bacillus subtilis*, *Bacillus amyloliquefaciens*, and *Sinorhizobium meliloti* harbor multiple MCPs, and their corresponding ligands that facilitate chemotactic responses toward various root exudates (Fernandez, Matilla, Ortega, & Krell, 2017; Glekas et al., 2012; Oku et al., 2014; Webb et al., 2017). Chemoreceptors such as McpA for glucose, alpha-D-ethylglucoside, McpB for asparagine, aspartic acid, glutamine, histidine (Hanlon & Ordal, 1994; Tohidifar et al., 2020), and McpC for proline, threonine, glycine, serine, valine, alanine, tyrosine, isoleucine, tryptophan, phenylalanine, leucine, asparagine, lysine, glutamine, methionine have been identified in *B. subtilis* OI1085 (Glekas et al., 2012). The MCPs present in the *S. meliloti* MV II-1 consist of McpU that can detect proline (Webb et al., 2017), McpV that can recognize propionate, acetate, glycolate, pyruvate, acetoacetate, formate, butyrate, L-lactate, glyoxylate, methyl pyruvate, α-hydroxybutyrate, α-ketobutyrate (Compton et al., 2018; Meier et al., 2007), and McpX that can identify choline, glycine betaine, stachydrine, trigonelline, choline, betonicine, proline betaine as

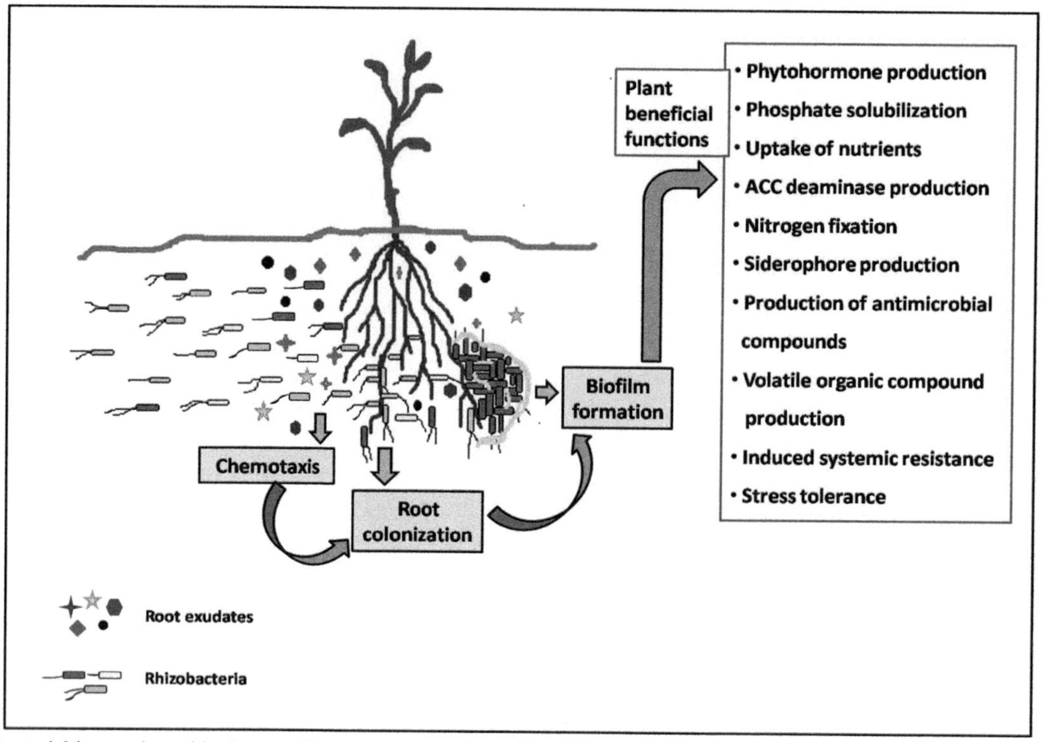

FIG. 1 Rhizobacterial interaction with plants and its beneficial functions for plant growth.

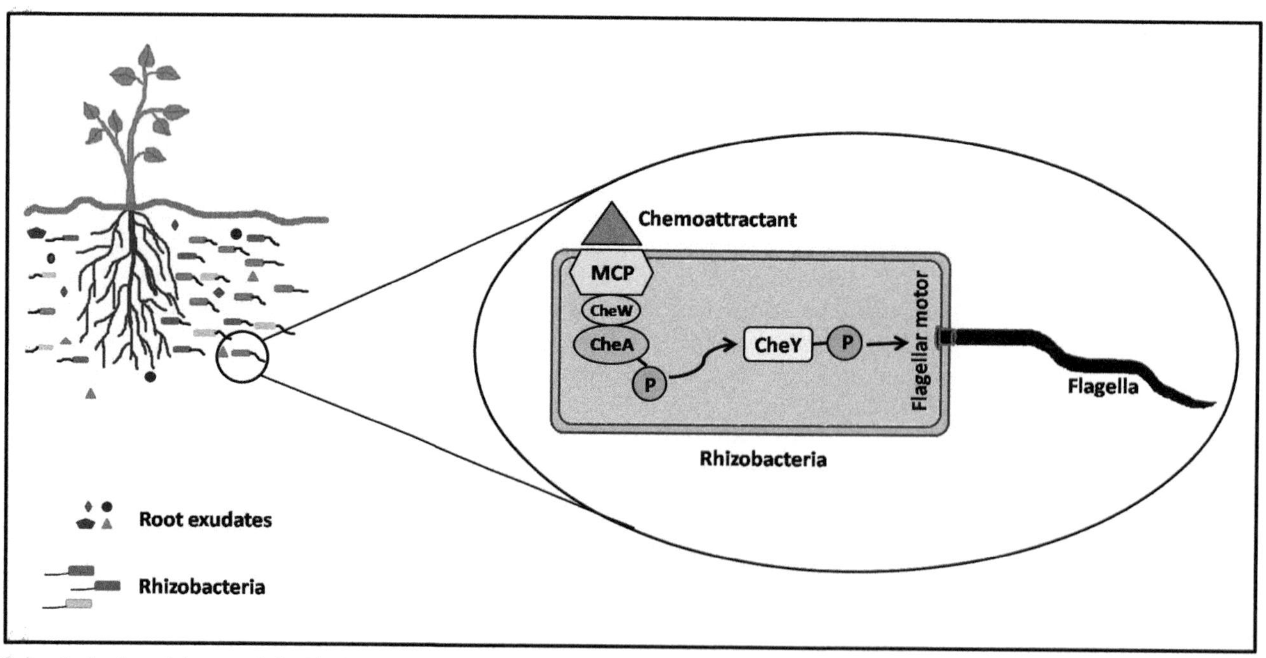

FIG. 2 Activation of chemotaxis in rhizobacteria toward the chemoattractant present in the root exudates. *(Modified from Feng, H., Fu, R., Hou, X., Lv, Y., Zhang, N., Liu, Y., et al. (2021). Chemotaxis of beneficial rhizobacteria to root exudates: The first step towards root–microbe rhizosphere interactions.* International Journal of Molecular Sciences, *22(13), 6655. https://doi.org/10.3390/ijms22136655.)*

chemoattractants (Shrestha et al., 2018; Webb et al., 2017). CtaA, CtaB, CtaC, and McpT were identified as chemoreceptors in *P. fluorescens* Pf0-1. Among these, CtaA and CtaB can sense alanine, arginine, asparagine, cystine, glycine, histidine, isoleucine, lysine, methionine, phenylalanine, serine, threonine, tyrosine, and valine. In addition to these, CtaA can also detect leucine and proline and CtaB can identify glutamine and glutamic acid also. CtaC can act as receptor for arginine, cystine, glycine, methionine, and threonine and McpT for L-malate and succinate (Oku et al., 2014).

5. Colonization and biofilm formation of rhizobacteria on plant root

Plants trigger an induced systemic resistance response after chemotaxis and attachment of rhizobacteria on the root surface. This response is induced in spite of the beneficial or pathogenic nature of the bacterial species (Soto, Dominguez-Ferreras, Perez-Mendoza, Sanjuan, & Olivares, 2009). This signaling activates defense response such as induction of reactive oxygen species (ROS) (Nanda, Andrio, Marino, Pauly, & Dunand, 2010; Pieterse et al., 2014). Antioxidant systems such as production of catalase and superoxide dismutase enzymes are present in many rhizobacterial species to resist the ROS activity. These enzymes help to maintain ROS at equilibrium in the cells. The rhizobacterial species with strong resistance to ROS are capable of being colonized successfully and establish beneficial association with plants (Amaya-Gomez, Porcel, Mesa-Garriga, Gomez-Alvarez, & Zhou, 2020).

For the active functioning of the rhizobacterial-plant association, biofilm formation is regarded as a remarkable trait (Amaya-Gomez et al., 2020). Biofilms are bacterial cells enclosed tightly together within a self-produced extracellular matrix (ECM), which is typically formed by extracellular polymeric substance (EPS). Generally the core components of EPS are exopolysaccharides, extracellular DNA (eDNA), proteins, organic matter, nucleic acid, enzyme, lipids, membrane vesicles, and bacterial refractory compounds, which vary with diverse rhizobacterial species (Angus & Hirsch, 2013; Flemming & Wingender, 2010; Jamal et al., 2018). Each component in the ECM has a particular function in the formation of biofilm and its stability (Fong et al., 2015). Exopolysaccharides have a significant role in the formation and organization of bacterial biofilm. Thus, polysaccharides are grouped as aggregative, protective, and architectural component, which facilitate adhesion of bacterial cells on the surface, act as diffusion barrier where it protects the cells from antibiotic penetration, and regulate biofilm formation and its structure (Limoli et al., 2015).

The most studied biofilm-forming bacteria are *Pseudomonas aeruginosa* and *B. subtilis*. The biofilm produced by *P. aeruginosa* consist of three polysaccharides, which are Pel, alginate, and Psl, and these provide architecture to the biofilm matrix. Only a subset of *P. aeruginosa* strains can produce alginate, and most environmental isolates exude either Pel or Psl (Franklin, Nivens, Weadge, & Howell, 2011). The ECM of *B. subtilis* consists of an exopolysaccharide (EPS) and the proteins TasA and BslA (Toyofuku et al., 2016). The EPS has a role as adhesive element of the bacterial cells at the interface of cell and surface, which facilitates biofilm attachment. The BslA acts as hydrophobin, which produces a thin external hydrophobic layer and provides hydrophobic property to the biofilm. The TasA protein forms fibers resistant to unfavorable physicochemical conditions and gives structural stability to the biofilm (Camara-Almiron et al., 2020).

The biofilm formation of rhizobacteria consists of different stages. The motile bacteria having pili or flagella move and primarily adhere to the root surface with the help of adhesions enzyme. Motility can facilitate the bacteria to move against the hydrophobic forces that repel them from surfaces. The successfully adhered bacteria start to multiply, spread around the primary site, and form microcolonies. These microcolonies secrete different exopolysaccharides and form an external matrix. Finally, the microcolonies enclosed within the exopolymer matrix start to detach from the matrix. This process may repeat at different sites on the plant root for their active functioning (Zboralski & Filion, 2020).

5.1 Role of quorum sensing in biofilm development

The rhizobacterial colonization is regulated in a population-density-dependent means by quorum sensing (QS) (Whitehead, Barnard, Slater, Simpson, & Salmond, 2001). QS is a communication mechanism between bacterial cells, and by this, rhizobacteria regulate gene expression in response to cell density (Miller & Bassler, 2001). This process involves the production and release of extracellular signaling molecules known as autoinducers, and the bacterial cells detect and act in response to these autoinducers. The autoinducers accumulate in the environment with the increase in bacterial population density (Papenfort & Bassler, 2016), and bacterial cells sense this signal to track changes in their cell density and regulate gene expression (Rutherford & Bassler, 2012).

Gram-positive and Gram-negative bacteria have different types of QS systems. Gram-positive bacteria use autoinducing peptides (AIPs) as signaling molecules (Jung et al., 2020). These bacteria release the AIPs into the surroundings. At high cell density, the extracellular concentration of the AIP increases and attaches to a cognate membrane-bound histidine kinase receptor. This binding stimulates kinase activity of the receptor and gets autophosphorylated and transfers phosphate to a cognate cytoplasmic response regulator. The phosphorylated response regulator induces transcription of the genes in the QS regulon. In certain conditions, AIPs are moved back into the cell cytoplasm and interact with transcription factors to alter its activity and gene expression changes (Rutherford & Bassler, 2012).

Gram-negative bacteria use acyl-homoserine lactones (AHLs) as autoinducers for cell-to-cell communication. Many bacteria possess LuxI-type synthases that use (*S*)-adenosylmethionine (SAM) and acylated acyl carrier proteins or acyl-coA thioesters as substrates to produce signals with different acyl tails (Papenfort & Bassler, 2016; Wei, Perez, Ng, Semmelhack, & Bassler, 2011; Whitehead et al., 2001). AIs produced by the bacterial cells diffuse across the inner and outer membranes. At high cell density, the AIs become sufficiently elevate and bind with the cytoplasmic receptors or transcription factors. These activated transcription factors modulate the genes expression in the QS regulon (Rutherford & Bassler, 2012).

6. Significance of rhizobacterial chemotaxis and biofilm formation

Chemotaxis and biofilm formation are most important processes involved in the beneficial association of rhizobacteria with plants. This interaction is advantageous not only to plants, but also to the rhizobacteria. The rhizosphere is an environment enriched with various nutrients and root exudates, which can act as carbon sources and can be utilized by the PGPR to multiply, move, compete, and colonize on the root (Santoyo, Urtis-Flores, Loeza-Lara, Orozco-Mosqueda, & Glick, 2021). Biofilm formed by rhizobacteria acts as a protective barrier from adverse conditions and environmental stresses. Bacteria within a biofilm are more resistant to ROS, variable conditions in soil, to antimicrobial compounds produced by other soil microorganisms, and to predation (Batista et al., 2018). In addition to this, biofilm protects the bacterial cells from harmful effect of osmotic shock, UV radiation, change in pH, and desiccation (Koo & Yamada, 2016). The external matrix can function as a hydrated barrier between the bacterial cells and outer environment and thus facilitates accumulation of organic and inorganic compounds, protection against dehydration and penetration of toxic elements into the biofilm (Flemming et al., 2016).

The biofilm formation has been found to enhance the activity level of PGPR. Elevated plant growth promoting traits such as nitrogenase activity, indole acetic acid (IAA) production, phosphate solubilization, siderophore production, and ammonia production can be observed in biofilm PGPR than the planktonic PGPR (Mah et al., 2003). IAA has a key role in apical dominance, division, and cell differentiation, photosynthesis, seed germination, biosynthesis of metabolites, development of roots, and stress resistance in plants (Maheshwari, Dheeman, & Agarwal, 2015). Phosphate solubilization by rhizobacteria stimulates cell division, photosynthesis, respiration, root system development, biosynthesis of macromolecules, stem strength, crop maturity and production, and quality improvement of many fruits, vegetables, and grain crops (Hakim et al., 2021; Sharma, Sayyed, Trivedi, & Gobi, 2013). Nitrogen-fixing rhizobacteria make available the nitrogen in the form of ammonia to the plants, which is essential for the synthesis of nucleic acid, proteins, and enzymes (Beeckman, Motte, & Beeckman, 2018). Also, biofilm PGPR can produce high level of antimicrobial compounds than the planktonic PGPR that can suppress the growth and establishment of soil-borne phytopathogens (Pandin, Le Coq, Canette, Aymerich, & Briandet, 2017).

The exudates released by plant roots attract the rhizobacteria toward the root and in response to this process, the rhizobacteria enhance plant growth and disease resistance. Leguminous plants secrete flavonoid compounds through the roots and are sensed by rhizobia, which in turn produce nodulation factors (Nod factors) that activate the nodule formation on the roots. The rhizobia colonize within the nodules and fix atmospheric nitrogen and make it accessible to the plant in the form of ammonia and thereby enhance plant growth and development (Dixon & Kahn, 2004). Malic acid exuded by tomato (*Solanum lycopersicum*) plant can act as chemoattractant for *B. subtilis* and promote colonization of the bacteria on the root. In response to this, *B. subtilis* produce antimicrobial compounds such as bacilomycin that suppress the growth of phytopathogens and enhance disease resistance of the plant (Sasse et al., 2018).

7. Conclusion

The microorganisms associated with plants are considered as second genome of the plants, which have significant role in plant growth promotion and disease resistance. In the plant-microbe interaction, rhizobacteria play a significant role to augment plant growth. The rhizosphere is the most complex microhabitat, consisting of a plant root signaling network, soil with diverse nutrients, and various microbial communities. The plant root secretes exudates into the rhizosphere that act as chemoattractant for certain rhizobacteria and bind with the receptors present on the bacterial cells. This leads to the chemotactic movement of rhizobacterial species and successful colonization on plant roots. This further facilitates biofilm formation that protects the bacterial cells from adverse stress conditions in the rhizosphere and stimulates high level of plant beneficial traits such as indole acetic acid production, phosphate solubilization, nitrogen fixation, production of siderophore, hydrogen cyanide, ammonia, volatile organic compounds, and antimicrobial compounds. This leads to the augmentation of plant growth and disease resistance.

References

Abdel-Lateif, K., Bogusz, D., & Hocher, V. (2014). The role of flavonoids in the establishment of plant roots endosymbioses with arbuscular mycorrhiza fungi, rhizobia and Frankia bacteria. *Plant Signaling & Behavior*, *7*(6), 636–641. https://doi.org/10.4161/psb.20039.

Allard-Massicotte, R., Tessier, L., Lecuyer, F., Lakshmanan, V., Lucier, J. F., Garneau, D., et al. (2016). *Bacillus subtilis* early colonization of *Arabidopsis thaliana* roots involves multiple chemotaxis receptors. *MBio*, *7*(6). https://doi.org/10.1128/mBio.01664-16.

Amaya-Gomez, C. V., Porcel, M., Mesa-Garriga, L., Gomez-Alvarez, M. I., & Zhou, N.-Y. (2020). A framework for the selection of plant growth-promoting rhizobacteria based on bacterial competence mechanisms. *Applied and Environmental Microbiology*, *86*(14), 1–13. https://doi.org/10.1128/aem.00760-20.

Angus, A. A., & Hirsch, A. M. (2013). Biofilm formation in the rhizosphere: Multispecies interactions and implications for plant growth. In F. J. D. Bruijn (Ed.) (Vols. 1 & 2). *Molecular microbial ecology of the rhizosphere* (pp. 701–712). New York: Wiley.

Bais, H. P., Weir, T. L., Perry, L. G., Gilroy, S., & Vivanco, J. M. (2006). The role of root exudates in rhizosphere interactions with plants and other organisms. *Annual Review of Plant Biology*, *57*(1), 233–266. https://doi.org/10.1146/annurev.arplant.57.032905.105159.

Batista, B. D., Lacava, P. T., Ferrari, A., Teixeira-Silva, N. S., Bonatelli, M. L., Tsui, S., et al. (2018). Screening of tropically derived, multi-trait plant growth-promoting rhizobacteria and evaluation of corn and soybean colonization ability. *Microbiological Research*, *206*, 33–42. https://doi.org/10.1016/j.micres.2017.09.007.

Beeckman, F., Motte, H., & Beeckman, T. (2018). Nitrification in agricultural soils: Impact, actors and mitigation. *Current Opinion in Biotechnology*, *50*, 166–173. https://doi.org/10.1016/j.copbio.2018.01.014.

Camara-Almiron, J., Navarro, Y., Diaz-Martinez, L., Magno-Perez-Bryan, M. C., Molina-Santiago, C., Pearson, J. R., et al. (2020). Dual functionality of the amyloid protein TasA in *Bacillus* physiology and fitness on the phylloplane. *Nature Communications*, *11*(1). https://doi.org/10.1038/s41467-020-15758-z.

Compant, S., Samad, A., Faist, H., & Sessitsch, A. (2019). A review on the plant microbiome: Ecology, functions, and emerging trends in microbial application. *Journal of Advanced Research*, *19*, 29–37. https://doi.org/10.1016/j.jare.2019.03.004.

Compton, K. K., Hildreth, S. B., Helm, R. F., Scharf, B. E., & Stock, A. M. (2018). *Sinorhizobium meliloti* chemoreceptor McpV senses short-chain carboxylates via direct binding. *Journal of Bacteriology*, *200* (23). https://doi.org/10.1128/jb.00519-18.

Compton, K. K., & Scharf, B. E. (2021). Rhizobial chemoattractants, the taste and preferences of legume symbionts. *Frontiers in Plant Science*, *12*. https://doi.org/10.3389/fpls.2021.686465.

De Weert, S., Vermeiren, H., Mulders, I. H., Kuiper, I., Hendrickx, N., Bloemberg, G. V., et al. (2002). Flagella-driven chemotaxis towards exudate components is an important trait for tomato root colonization by *Pseudomonas fluorescens*. *Molecular Plant-Microbe Interactions*, *15*(11), 1173–1180. https://doi.org/10.1094/MPMI.2002.15.11.1173.

Dixon, R., & Kahn, D. (2004). Genetic regulation of biological nitrogen fixation. *Nature Reviews Microbiology*, *2*(8), 621–631. https://doi.org/10.1038/nrmicro954.

Dolatabadian, A. (2020). Plant–microbe interaction. *Biology*, *10*(1), 15. https://doi.org/10.3390/biology10010015.

Egamberdieva, D., Kamilova, F., Validov, S., Gafurova, L., Kucharova, Z., & Lugtenberg, B. (2007). High incidence of plant growth-stimulating bacteria associated with the rhizosphere of wheat grown on salinated soil in Uzbekistan. *Environmental Microbiology*, *10*(1), 1–9. https://doi.org/10.1111/j.1462-2920.2007.01424.x.

Feng, H., Fu, R., Hou, X., Lv, Y., Zhang, N., Liu, Y., et al. (2021). Chemotaxis of beneficial rhizobacteria to root exudates: The first step towards root–microbe rhizosphere interactions. *International Journal of Molecular Sciences*, *22*(13), 6655. https://doi.org/10.3390/ijms22136655.

Fernandez, M., Matilla, M. A., Ortega, A., & Krell, T. (2017). Metabolic value chemoattractants are preferentially recognized at broad ligand range chemoreceptor of *Pseudomonas putida* KT2440. *Frontiers in Microbiology*, *8*, 990. https://doi.org/10.3389/fmicb.2017.00990.

Flack, C. E., & Parkinson, J. S. (2018). A zipped-helix cap potentiates HAMP domain control of chemoreceptor signaling. *Proceedings of the National Academy of Sciences*, *115*(15), E3519–E3528. https://doi.org/10.1073/pnas.1721554115.

Flemming, H.-C., & Wingender, J. (2010). The biofilm matrix. *Nature Reviews Microbiology*, *8*(9), 623–633. https://doi.org/10.1038/nrmicro2415.

Flemming, H. C., Wingender, J., Szewzyk, U., Steinberg, P., Rice, S. A., & Kjelleberg, S. (2016). Biofilms: An emergent form of bacterial life. *Nature Reviews Microbiology*, *14*(9), 563–575. https://doi.org/10.1038/nrmicro.2016.94.

Fong, J. N. C., Yildiz, F. H., Ghannoum, M., Parsek, M., Whiteley, M., & Mukherjee, P. (2015). Biofilm matrix proteins. *Microbiology Spectrum*, *3*(2). https://doi.org/10.1128/microbiolspec.MB-0004-2014.

Franklin, M. J., Nivens, D. E., Weadge, J. T., & Howell, P. L. (2011). Biosynthesis of the *Pseudomonas aeruginosa* extracellular polysaccharides, alginate, Pel, and Psl. *Frontiers in Microbiology*, *2*, 167. https://doi.org/10.3389/fmicb.2011.00167.

Gao, X., Li, T., Liu, W., Zhang, Y., Shang, D., Gao, Y., et al. (2020). Enhancing the 1-aminocyclopropane-1-carboxylate metabolic rate of *Pseudomonas* sp. UW4 intensifies chemotactic rhizocompetence. *Microorganisms*, *8*(1), 71. https://doi.org/10.3390/microorganisms8010071.

Glekas, G. D., Mulhern, B. J., Kroc, A., Duelfer, K. A., Lei, V., Rao, C. V., et al. (2012). The *Bacillus subtilis* chemoreceptor McpC senses multiple ligands using two discrete mechanisms. *The Journal of Biological Chemistry*, *287*(47), 39412–39418. https://doi.org/10.1074/jbc.M112.413518.

Greenswag, A. R., Muok, A., Li, X., & Crane, B. R. (2015). Conformational transitions that enable histidine kinase autophosphorylation and receptor array integration. *Journal of Molecular Biology*, *427*(24), 3890–3907.

Hakim, S., Naqqash, T., Nawaz, M. S., Laraib, I., Siddique, M. J., Zia, R., et al. (2021). Rhizosphere engineering with plant growth-promoting microorganisms for agriculture and ecological sustainability. *Frontiers in Sustainable Food Systems*, *5*. https://doi.org/10.3389/fsufs.2021.617157.

Hanlon, D. W., & Ordal, G. W. (1994). Cloning and characterization of genes encoding methyl-accepting chemotaxis proteins in *Bacillus subtilis*. *The Journal of Biological Chemistry*, *269*, 14038–14046.

Hayat, R., Ali, S., Amara, U., Khalid, R., & Ahmed, I. (2010). Soil beneficial bacteria and their role in plant growth promotion: A review. *Annals of Microbiology*, *60*(4), 579–598. https://doi.org/10.1007/s13213-010-0117-1.

Jamal, M., Ahmad, W., Andleeb, S., Jalil, F., Imran, M., Nawaz, M. A., et al. (2018). Bacterial biofilm and associated infections. *Journal of the Chinese Medical Association*, *81*(1), 7–11. https://doi.org/10.1016/j.jcma.2017.07.012.

Jishma, P., & Radhakrishnan, E. K. (2020). Rhizobacterial biofilm and plant growth promoting trait enhancement by organic acids and sugars. *Biofouling*, *36*(8), 990–999. https://doi.org/10.1080/08927014.2020.1832219.

Jung, B. K., Ibal, J. C., Pham, H. Q., Kim, M.-C., Park, G.-S., Hong, S.-J., et al. (2020). Quorum sensing system affects the plant growth promotion traits of *Serratia fonticola* GS2. *Frontiers in Microbiology*, *11*, 536865. https://doi.org/10.3389/fmicb.2020.536865.

Kape, R., Parniske, M., Brandt, S., & Werner, D. (1992). Isoliquiritigenin, a strong nod gene- and glyceollin resistance-inducing flavonoid from soybean root exudate. *Applied and Environmental Microbiology*, *58* (5), 1705–1710. https://doi.org/10.1128/aem.58.5.1705-1710.1992.

Kape, R., Parniske, M., & Werner, D. (1991). Chemotaxis and nod gene activity of *Bradyrhizobium japonicum* in response to hydroxycinnamic acids and isoflavonoids. *Applied and Environmental Microbiology*, *57* (1), 316–319. https://doi.org/10.1128/aem.57.1.316-319.1991.

Koo, H., & Yamada, K. M. (2016). Dynamic cell-matrix interactions modulate microbial biofilm and tissue 3D microenvironments. *Current Opinion in Cell Biology*, *42*, 102–112. https://doi.org/10.1016/j.ceb.2016.05.005.

Limoli, D. H., Jones, C. J., Wozniak, D. J., Ghannoum, M., Parsek, M., Whiteley, M., et al. (2015). Bacterial extracellular polysaccharides in biofilm formation and function. *Microbiology Spectrum*, *3* (3). https://doi.org/10.1128/microbiolspec.MB-0011-2014.

Ling, N., Raza, W., Ma, J., Huang, Q., & Shen, Q. (2011). Identification and role of organic acids in watermelon root exudates for recruiting *Paenibacillus polymyxa* SQR-21 in the rhizosphere. *European Journal of Soil Biology*, *47*(6), 374–379. https://doi.org/10.1016/j.ejsobi.2011.08.009.

Lugtenberg, B., & Kamilova, F. (2009). Plant-growth-promoting rhizobacteria. *Annual Review of Microbiology*, *63*(1), 541–556. https://doi.org/10.1146/annurev.micro.62.081307.162918.

Mah, T.-F., Pitts, B., Pellock, B., Walker, G. C., Stewart, P. S., & O'Toole, G. A. (2003). A genetic basis for *Pseudomonas aeruginosa* biofilm antibiotic resistance. *Nature*, *426*(6964), 306–310. https://doi.org/10.1038/nature02122.

Maheshwari, D. K., Dheeman, S., & Agarwal, M. (2015). Phytohormone-producing PGPR for sustainable agriculture. In D. Maheshwari (Ed.), *Vol. 12. Bacterial metabolites in sustainable agroecosystem* (pp. 159–182). Springer.

Matilla, M. A., Martin-Mora, D., & Krell, T. (2020). The use of isothermal titration calorimetry to unravel chemotactic signalling mechanisms. *Environmental Microbiology*, *22*(8), 3005–3019. https://doi.org/10.1111/1462-2920.15035.

Meier, V. M., Muschler, P., & Scharf, B. E. (2007). Functional analysis of nine putative chemoreceptor proteins in *Sinorhizobium meliloti*. *Journal of Bacteriology*, *189*(5), 1816–1826. https://doi.org/10.1128/jb.00883-06.

Mendes, R., Garbeva, P., & Raaijmakers, J. M. (2013). The rhizosphere microbiome: Significance of plant beneficial, plant pathogenic, and human pathogenic microorganisms. *FEMS Microbiology Reviews*, *37*(5), 634–663. https://doi.org/10.1111/1574-6976.12028.

Miller, M. B., & Bassler, B. L. (2001). Quorum sensing in bacteria. *Annual Reviews in Microbiology*, *55*(1), 165–199.

Moe, L. A. (2013). Amino acids in the rhizosphere: From plants to microbes. *American Journal of Botany*, *100*(9), 1692–1705. https://doi.org/10.3732/ajb.1300033.

Muok, A. R., Briegel, A., & Crane, B. R. (2020). Regulation of the chemotaxis histidine kinase CheA: A structural perspective. *Biochimica et Biophysica Acta (BBA)—Biomembranes*, *1862*(1), 183030. https://doi.org/10.1016/j.bbamem.2019.183030.

Nanda, A. K., Andrio, E., Marino, D., Pauly, N., & Dunand, C. (2010). Reactive oxygen species during plant-microorganism early interactions. *Journal of Integrative Plant Biology*, *52*(2), 195–204. https://doi.org/10.1111/j.1744-7909.2010.00933.x.

Oku, S., Komatsu, A., Nakashimada, Y., Tajima, T., & Kato, J. (2014). Identification of *Pseudomonas fluorescens* chemotaxis sensory proteins for malate, succinate, and fumarate, and their involvement in root colonization. *Microbes and Environments*, *29*(4), 413–419. https://doi.org/10.1264/jsme2.ME14128.

Oldroyd, G. E. D. (2013). Speak, friend, and enter: Signalling systems that promote beneficial symbiotic associations in plants. *Nature Reviews Microbiology*, *11*(4), 252–263. https://doi.org/10.1038/nrmicro2990.

Pandin, C., Le Coq, D., Canette, A., Aymerich, S., & Briandet, R. (2017). Should the biofilm mode of life be taken into consideration for microbial biocontrol agents? *Microbial Biotechnology*, *10*(4), 719–734. https://doi.org/10.1111/1751-7915.12693.

Papenfort, K., & Bassler, B. L. (2016). Quorum sensing signal–response systems in Gram-negative bacteria. *Nature Reviews Microbiology*, *14*(9), 576–588. https://doi.org/10.1038/nrmicro.2016.89.

Parkinson, J. S., Hazelbauer, G. L., & Falke, J. J. (2015). Signaling and sensory adaptation in *Escherichia coli* chemoreceptors: 2015 update. *Trends in Microbiology*, *23*(5), 257–266. https://doi.org/10.1016/j.tim.2015.03.003.

Phillips, D. A., Joseph, C. M., & Maxwell, C. A. (1992). Trigonelline and stachydrine released from Alfalfa seeds activate NodD2 protein in *Rhizobium meliloti*. *Plant Physiology*, *99*(4), 1526–1531. https://doi.org/10.1104/pp.99.4.1526.

Pieterse, C. M. J., Zamioudis, C., Berendsen, R. L., Weller, D. M., Van Wees, S. C. M., & Bakker, P. A. H. M. (2014). Induced systemic resistance by beneficial microbes. *Annual Review of Phytopathology*, *52*(1), 347–375. https://doi.org/10.1146/annurev-phyto-082712-102340.

Plett, J. M., & Martin, F. M. (2018). Know your enemy, embrace your friend: Using omics to understand how plants respond differently to pathogenic and mutualistic microorganisms. *The Plant Journal*, *93* (4), 729–746. https://doi.org/10.1111/tpj.13802.

Ribeiro, C. M., & Cardoso, E. J. B. N. (2012). Isolation, selection and characterization of root-associated growth promoting bacteria in Brazil Pine (*Araucaria angustifolia*). *Microbiological Research*, *167*(2), 69–78. https://doi.org/10.1016/j.micres.2011.03.003.

Rudrappa, T., Czymmek, K. J., Pare, P. W., & Bais, H. P. (2008). Root-secreted malic acid recruits beneficial soil bacteria. *Plant Physiology*, *148*(3), 1547–1556. https://doi.org/10.1104/pp.108.127613.

Rutherford, S. T., & Bassler, B. L. (2012). Bacterial quorum sensing: Its role in virulence and possibilities for its control. *Cold Spring Harbor Perspectives in Medicine*, *2*(11), a012427. https://doi.org/10.1101/cshperspect.a012427.

Santoyo, G., Urtis-Flores, C. A., Loeza-Lara, P. D., Orozco-Mosqueda, M. D. C., & Glick, B. R. (2021). Rhizosphere colonization determinants by plant growth-promoting rhizobacteria (PGPR). *Biology*, *10* (6), 475. https://doi.org/10.3390/biology10060475.

Sasse, J., Martinoia, E., & Northen, T. (2018). Feed your friends: Do plant exudates shape the root microbiome? *Trends in Plant Science*, *23*(1), 25–41. https://doi.org/10.1016/j.tplants.2017.09.003.

Scharf, B. E., Hynes, M. F., & Alexandre, G. M. (2016). Chemotaxis signaling systems in model beneficial plant–bacteria associations. *Plant Molecular Biology*, *90*(6), 549–559. https://doi.org/10.1007/s11103-016-0432-4.

Sharma, S. B., Sayyed, R. Z., Trivedi, M. H., & Gobi, T. A. (2013). Phosphate solubilizing microbes: Sustainable approach for managing phosphorus deficiency in agricultural soils. *SpringerPlus*, *2*(1), 587. https://doi.org/10.1186/2193-1801-2-587.

Shimoda, Y., Nishigaya, Y., Yamaya-Ito, H., Inagaki, N., Umehara, Y., Hirakawa, H., et al. (2020). The rhizobial autotransporter determines the symbiotic nitrogen fixation activity of *Lotus japonicus* in a host-specific manner. *Proceedings of the National Academy of Sciences*, *117*(3), 1806–1815. https://doi.org/10.1073/pnas.1913349117.

Shrestha, M., Compton, K. K., Mancl, J. M., Webb, B. A., Brown, A. M., Scharf, B. E., et al. (2018). Structure of the sensory domain of McpX from *Sinorhizobium meliloti*, the first known bacterial chemotactic sensor for quaternary ammonium compounds. *Biochemical Journal*, *475*(24), 3949–3962. https://doi.org/10.1042/bcj20180769.

Soto, M. J., Dominguez-Ferreras, A., Perez-Mendoza, D., Sanjuan, J., & Olivares, J. (2009). Mutualism versus pathogenesis: The give-and-take in plant-bacteria interactions. *Cellular Microbiology*, *11*(3), 381–388. https://doi.org/10.1111/j.1462-5822.2009.01282.x.

Sourjik, V., & Wingreen, N. S. (2012). Responding to chemical gradients: Bacterial chemotaxis. *Current Opinion in Cell Biology*, *24*(2), 262–268. https://doi.org/10.1016/j.ceb.2011.11.008.

Swanson, R. V., Bourret, R. B., & Simon, M. I. (1993). Intermolecular complementation of the kinase activity of CheA. *Molecular Microbiology*, *8*(3), 435–441. https://doi.org/10.1111/j.1365-2958.1993.tb01588.x.

Tohidifar, P., Bodhankar, G. A., Pei, S., Cassidy, C. K., Walukiewicz, H. E., Ordal, G. W., et al. (2020). The unconventional cytoplasmic sensing mechanism for ethanol chemotaxis in *Bacillus subtilis*. *MBio*, *11*(5). https://doi.org/10.1128/mBio.02177-20. e02177-02120.

Toyofuku, M., Inaba, T., Kiyokawa, T., Obana, N., Yawata, Y., & Nomura, N. (2016). Environmental factors that shape biofilm formation. *Bioscience, Biotechnology, and Biochemistry*, *80*(1), 7–12. https://doi.org/10.1080/09168451.2015.1058701.

Vishwakarma, K., Kumar, N., Shandilya, C., Mohapatra, S., Bhayana, S., & Varma, A. (2020). Revisiting plant–microbe interactions and microbial consortia application for enhancing sustainable agriculture: A review. *Frontiers in Microbiology*, *11*. https://doi.org/10.3389/fmicb.2020.560406.

Wadhams, G. H., & Armitage, J. P. (2004). Making sense of it all: Bacterial chemotaxis. *Nature Reviews Molecular Cell Biology*, *5*(12), 1024–1037. https://doi.org/10.1038/nrm1524.

Webb, B. A., Hildreth, S., Helm, R. F., Scharf, B. E., & Voordouw, G. (2014). *Sinorhizobium meliloti* chemoreceptor McpU mediates chemotaxis toward host plant exudates through direct proline sensing. *Applied and Environmental Microbiology*, *80*(11), 3404–3415. https://doi.org/10.1128/aem.00115-14.

Webb, B. A., Karl Compton, K., Castañeda Saldaña, R., Arapov, T. D., Keith Ray, W., Helm, R. F., et al. (2017). *Sinorhizobium meliloti* chemotaxis to quaternary ammonium compounds is mediated by the chemoreceptor McpX. *Molecular Microbiology*, *103*(2), 333–346. https://doi.org/10.1111/mmi.13561.

Wei, Y., Perez, L. J., Ng, W.-L., Semmelhack, M. F., & Bassler, B. L. (2011). Mechanism of *Vibrio cholerae* autoinducer-1 biosynthesis. *ACS Chemical Biology*, *6*(4), 356–365. https://doi.org/10.1021/cb1003652.

Whitehead, N. A., Barnard, A. M., Slater, H., Simpson, N. J., & Salmond, G. P. (2001). Quorum-sensing in Gram-negative bacteria. *FEMS Microbiology Reviews*, *25*(4), 365–404. https://doi.org/10.1111/j.1574-6976.2001.tb00583.x.

Wu, C. H., Bernard, S. M., Andersen, G. L., & Chen, W. (2009). Developing microbe-plant interactions for applications in plant-growth promotion and disease control, production of useful compounds, remediation and carbon sequestration. *Microbial Biotechnology*, *2*(4), 428–440. https://doi.org/10.1111/j.1751-7915.2009.00109.x.

Yang, C.-H., Neal, A. L., Ahmad, S., Gordon-Weeks, R., & Ton, J. (2012). Benzoxazinoids in root exudates of maize attract *Pseudomonas putida* to the rhizosphere. *PLoS One*, *7*(4), e35498. https://doi.org/10.1371/journal.pone.0035498.

Yuan, J., Zhang, N., Huang, Q., Raza, W., Li, R., Vivanco, J. M., et al. (2015). Organic acids from root exudates of banana help root colonization of PGPR strain *Bacillus amyloliquefaciens* NJN-6. *Scientific Reports*, *5*, 13438. https://doi.org/10.1038/srep13438.

Zboralski, A., & Filion, M. (2020). Genetic factors involved in rhizosphere colonization by phytobeneficial *Pseudomonas* spp. *Computational and Structural Biotechnology Journal*, *18*, 3539–3554. https://doi.org/10.1016/j.csbj.2020.11.025.

Zhang, N., Wang, D., Liu, Y., Li, S., Shen, Q., & Zhang, R. (2013). Effects of different plant root exudates and their organic acid components on chemotaxis, biofilm formation and colonization by beneficial rhizosphere-associated bacterial strains. *Plant and Soil*, *374*(1–2), 689–700. https://doi.org/10.1007/s11104-013-1915-6.

Chapter 6

Variable pressure SEM and conventional high vacuum SEM protocols for biofilm imaging

Michela Relucenti[a], Orlando Donfrancesco[a], Loredana Cristiano[b], Stefania Nottola[a], Marco Artini[c], Rosanna Papa[c], and Laura Selan[c]

[a]*Department of Anatomy, Histology, Forensic Medicine and Orthopaedics, Sapienza University of Rome, Rome, Italy,* [b]*Department of Clinical Medicine, Public Health, Life and Environmental Sciences, University of Aquila, L'Aquila, Italy,* [c]*Department of Public Health and Infectious Diseases, Sapienza University of Rome, Rome, Italy*

1. Introduction

The term "Biofilm" was coined in 1978 by William J. Costerton; he used this word to specifically mark the surface-attached microbial communities (Costerton, Geesey, & Cheng, 1978). In later years, he implemented the meaning of the term, introducing the concept of the role of the host organism and that of the 3D organization of biofilm components. Currently, biofilm is considered a complex community of microorganisms that live attached to a surface (or interface) and is embedded in a self-secreted extracellular matrix (EPS). EPS is species-specific in its three-dimensional arrangement (Di Martino, 2018a) and its chemical composition (Costerton, Lewandowski, Caldwell, Korber, & Lappin-Scott, 1995) (mainly polysaccharides, proteins, nucleic acid, lipids). EPS composition is so complex that, recently, the term "matrixome" was introduced to underline how it is important, for the peculiar characteristics of biofilm behavior, the coexistence in the EPS of biomolecules diverse in their chemical composition and functional roles (Karygianni, Ren, Koo, & Thurnheer, 2020). The complexity of EPS supports the complex cooperative behavior (Kolenbrander, 2000) shown by single bacterial cells in the biofilm population, which act like units of a multicellular organism (Shapiro, 1998). Biofilms cause numerous infectious diseases in humans (up to 75%), such as implant-related infections and/or tissue-associated infections, as stated by the National Institutes of Health (NIH) (Yadav, Song, Singh, & Vidal, 2020). Recent data concerning UE and EEA (Union and European Economic Area) healthcare-associated infection related to biofilms accounts for 8.9 million cases/year (Suetens et al., 2018). The real problem consists not only of the number of these types of infections but especially the fact that very often they are recurrent and resistant to treatments (Breidenstein, De La Fuente-Nunez, & Hancock, 2011; Vestby, Grønseth, Simm, & Nesse, 2020). This characteristic is due to EPS; it protects the resident bacterial cells from the host immune response and/or antimicrobial drug effects (Li, Xiao, Wang, & Hao, 2020). This is why recently the design of new antibiofilm drugs and the screening of natural chemicals with antibiofilm activity are of crucial importance to find new molecules that will be effective in minimizing and eradicating biofilm-related infections. Morphological investigations play an important role in improving the knowledge of the intimate details of the biofilms' microstructure and ultrastructure. EPS 3-D architecture, cell population aspect, and position in the biofilm mass, together with the identification of cell-cell relations, are all parameters that can be visualized and analyzed by morphological techniques. Several imaging modalities have been used in biofilm investigation, scanning electron microscopy (SEM) is a morphological technique that images sample surface using an electron beam moving in a raster mode, under high vacuum conditions. This technique is useful for biofilm study having a wide range of magnifications (20 × to 30,000 ×), it is powerful for the investigation at a high resolution (from 50 to 100 nm) of biofilm surface structure, providing information on sample shape, size, and localization. The large depth of field of SEM micrographs provides a 3-D appearance, useful for understanding the surface structure of the sample. During the past decade, SEM has been a preferred method for biofilm imaging, allowing investigation of its spatial structure and detecting the presence of EPS (Hung et al., 2013; Rodrigues, Bañobre-López, Espiña, Rivas, & Azeredo, 2013). Recently also a quantitative SEM approach for both bacterial and fungal biofilms was undertaken using specific image analysis software (Mountains Map 8, Digital

Understanding Microbial Biofilms. https://doi.org/10.1016/B978-0-323-99977-9.00019-3

Surf, France). Digital SEM images were acquired, a specific region of interest was extracted, and biofilm 3D structure was reconstructed and analyzed (Relucenti et al., 2021; Vrenna et al., 2021) in a paper focusing on drug treatments' effects on biofilm morphology (Papa et al., 2021; Relucenti et al., 2021; Vrenna et al., 2021). SEM has some downsides in the imaging of highly hydrated samples, so recently another SEM technique, the variable-pressure scanning electron microscopy (VP-SEM), is emerging as a new possible option in morphological biofilm analysis.

In this paper, two ultrastructural techniques, SEM and the innovative VP-SEM, and their application in biofilm studies, are discussed.

2. Conventional scanning electron microscopy (SEM) applied to biofilm imaging

To accurately describe biofilm morphology, high magnification and high-resolution images are needed, conventional SEM can satisfy this requirement (Bossù, Matassa, et al., 2020; Hung et al., 2013; Rodrigues et al., 2013). The undisputed advantages of SEM consist of the combined ability to image with a wide range of magnifications (from 20× to 30,000×) coupled with high resolution and depth of field. The conventional SEM protocol consists of the steps summarized in Table 1.

This standard protocol can be modified to adjust it according to the specific characteristics of the highly hydrated extracellular matrices (Relucenti et al., 2019, 2020) that are easily damageable during preparation procedures. Biofilm EPS can be preserved from disruption during critical point drying (damage is due to ethanol flow) (Alhede et al., 2012; Timp & Matsudaira, 2008), performing drying procedures in ascending hexamethyldisilazane (HDMS) series (De Araujo et al., 2003; Di Bonaventura et al., 2008; Karcz, Bernas, Nowak, Talik, & Woznica, 2011). Complete dehydration and metal coating allowed samples a long-time resistance in the condition of high vacuum and high voltage (20kV), but conventional SEM has some downsides when it is applied in biofilm observation.

TABLE 1 Conventional SEM protocol for biofilm, detailed steps description.

Steps	Conventional SEM protocol
Fixation	Glutaraldehyde 2.5% in PB 0.1 M pH7.4 at least 48h
Washing	10min for two times in PB 0.1 M pH7.4
Postfixation	OsO_4 2% in 1h
Washing	10min x2 times in PB 0.1 M pH7.4
Dehydration	Ascending ethanol series (30%–50%–70%–95%–100%)
Drying	Ascending Hexamethyldisilazane series or Critical point Drying
Pt sputter coating	15mA, 2min
Operating conditions	15–20kV, high vacuum

Streptococcus mutans biofilm was used to demonstrate the pros and cons of conventional SEM (Figs. 1 and 2). At low magnification (Fig. 1A), bacterial cells appear settled onto the surface of EPS. Increasing the magnification (5000×, Fig. 1B) single cells are visible, they are spherical, and they are resting on the EPS matrix, not embedded in. At high magnifications (10,000×, 20,000× Fig. 2A and B), bacterial cells show an irregular shape, and their surface appears devoid of any EPS residue ("naked"). In turn, EPS seems like a dense and shrunken mass with a rough surface. Fig. 2B evidences the separation between bacterial cells and EPS, which seems to be a substrate, rather than an environment in which bacterial cells are embedded, as demonstrated by numerous studies (Davies et al., 1998; Di Martino, 2018b; Karygianni et al., 2020; Kolenbrander, 2000; Shapiro, 1998). To overcome conventional SEM and FESEM limitations, and obtain pictures corresponding to the actual biofilm ultrastructure, highly tailored protocols or alternative SEM modalities such as VP-SEM had to be used.

3. SEM tailored protocol for biofilm

In the paper of Bossù, Selan, et al. (2020), *Scardovia wiggsiae* biofilm was characterized for the first time with SEM. To observe the actual biofilm ultrastructure (preserving its high hydration degree) at high magnification, a new and tailored protocol was built up. In this procedure (summarized in Table 2), the fixation with glutaraldehyde was followed by postfixation with osmium tetroxide (OsO_4) and ruthenium red (RR) 50%–50%, then two stEPS were performed: the impregnation with tannic acid (TA) and the ionic liquid (IL, Hitachi Hilem 1000, Hitachi, Japan) drop casting (substitute for sputter coating). This new procedure was named OsO_4-RR-TA-IL protocol.

This procedure was established to maintain the advantages of the conventional SEM protocol (image quality, magnification, resolution, and long resistance under electron beam) with other vantages as a faster timing (the protocol has few stEPS); a minimal amount of sample loss; the preservation of the actual sample structure (avoiding alcohol dehydration and critical point drying procedures).

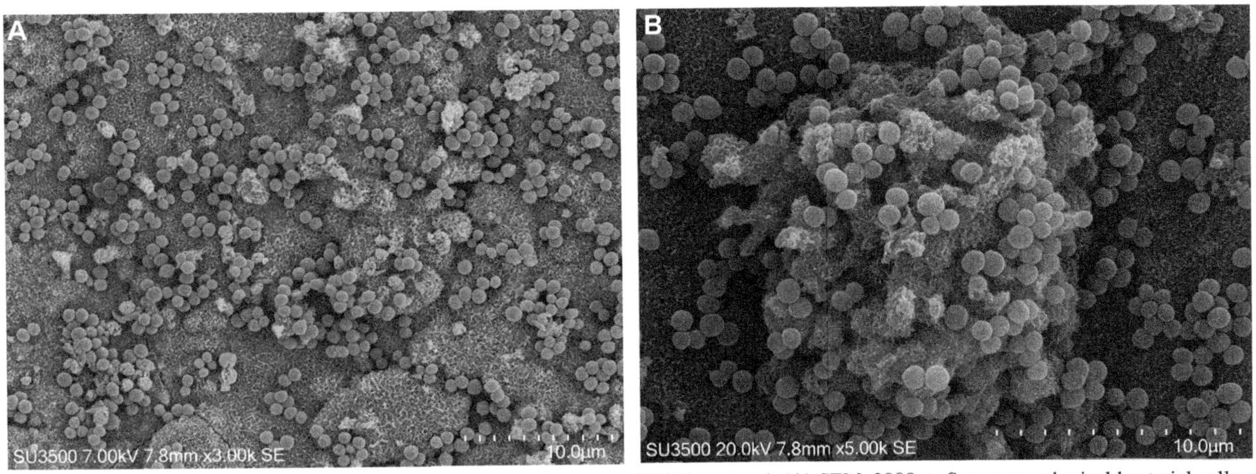

FIG. 1 *S. mutans* prepared by conventional scanning electron microscope (SEM) protocol. (A) SEM, 3000×. *S. mutans* spherical bacterial cells were scattered on the biofilm matrix compact surface. (B) SEM, 5000×. Increasing magnification, *S. mutans* spherical bacterial cells appeared clustered in small groups on the surface of a rough and dense extracellular matrix (EPS). *(From Relucenti, M., Familiari, G., Donfrancesco, O., Taurino, M., Li, X., Chen, R., et al. (2021). Microscopy methods for biofilm imaging: Focus on SEM and VP-SEM pros and cons.* Biology, 10*(1), 51. MDPI AG. Retrieved from https://doi.org/10.3390/biology10010051.)*

High magnifications and high-resolution images obtained under these conditions allowed ultrastructural characterization of *S. wiggsiae* without artifacts.

The effectiveness of OsO_4-RR-TA-IL is due to the specific properties of every single reagent. RR was used for the first time by Luft (1964), to visualize extracellular matrices with electron microscopy and was then used on several bacterial biofilms: *Staphylococcus aureus* (Fassel & Edmiston, 1999a, 1999b; Fassel, Schaller, & Remsen, 1992), *Pseudomonas* spp. (Carr & Toner, 1982), *Enterococcus faecalis*, and *Klebsiella pneumoniae* (Erlandsen, Kristich, Dunny, & Wells, 2004). RR it is a polycationic dye, its cationic properties enable EPS polysaccharides preservation creating electrostatic or ionic links with EPS components. Those links stabilize EPS and avoid sample shrinking (Chang & Rittmann, 1986; Erlandsen et al., 2004; Fassel & Edmiston, 1999a, 1999b; Fulcher et al., 2001; Priester et al., 2007; Vitkov, Hannig, et al., 2005; Vitkov, Hermann, et al., 2005; Vitkov, Krautgartner, & Hannig, 2005). Specifically, RR binding to polysaccharides stabilizes their chain in the extracellular matrix network, avoiding their loss during conventional SEM procedures (Chang & Rittmann, 1986).

TA forms complexes that link to proteins and carbohydrates (Murakami et al., 1983), it reacts also with osmium tetroxide and increases lipid retention. The creation of

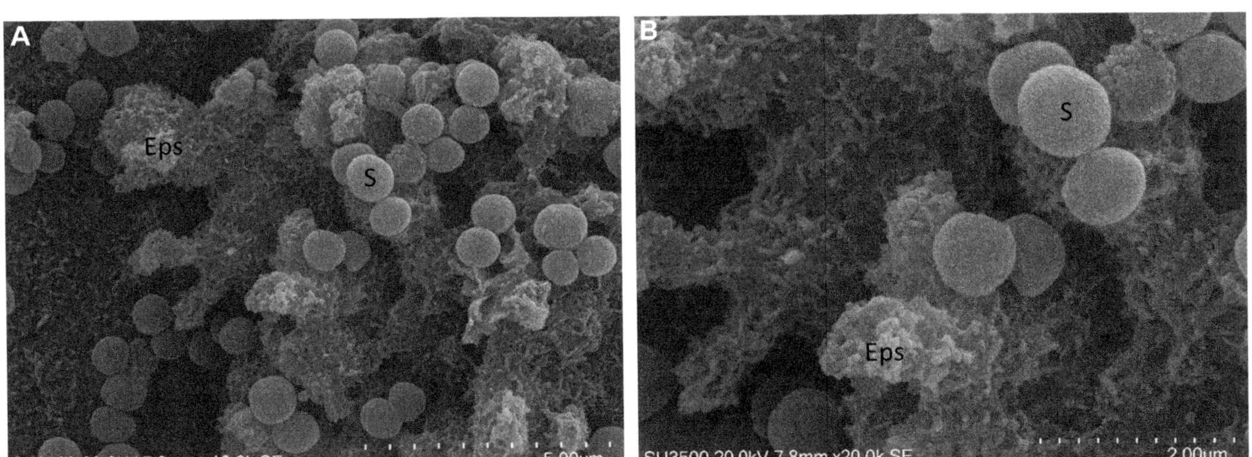

FIG. 2 *S. mutans* prepared by conventional SEM procedure. (A) SEM, 10,000×. EPS forms a canalicular system of compact trabeculae with a spiny surface. Bacterial cells, S, are adherent to the EPS surface. *EPS*, extracellular polymeric substance; (B) SEM, 20,000×. Bacterial cells appear irregular, EPS micro-canalicular system is not developed, only superficial holes are visible. Bacterial cells lay down on the EPS surface, and they appear naked, without a matrix covering. Bacterial cells are sometimes fragmented or indented; EPS showed a compact aspect due to the collapse of its fine structure. Bacterial cells, uncovered by the matrix, rest on the EPS surface. *EPS*, extracellular polymeric substance; *S*, *S. mutans*. *(From Bossù, M., Matassa, R., Relucenti, M., Iaculli, F., Salucci, A., Di Giorgio, G., et al. (2020). Morpho-chemical observations of human deciduous teeth enamel in response to biomimetic toothpastes treatment.* Materials, 13, *1803.)*

TABLE 2 OsO_4-RR-TA-IL protocol detailed stEPS description.

	OsO_4-RR-TA-IL protocol
StEPS	
Fixation	Glutaraldehyde 2.5% in PB 0.1 M pH7.4 at least 48h
Washing	10min x2 times in PB 0.1 M pH7.4
Postfixation	OsO_4 2%+RR 0.2% 1:1 solution, 1h
Washing	10min x2 times in PB 0.1 M pH7.4
Impregnation	Tannic Acid 1% in d H_2O 30min
Dehydration	–
Drying	–
Pt sputter coating	–
Operating conditions	15–20kV, high vacuum

multiple links between ECM components and bacterial cells surface produces biofilm hardening (Murakami, 1973, 1974, 1976; Murakami, Yamamoto, Itoshima, & Irino, 1977) which enhances its resistance to mechanical damage during SEM preparation procedures. An additional advantage of this method is that it renders the sample itself conductive (not only its surface, as it happens with sputter coating) and enhances the image contrast without charging phenomena, allowing the observation of its subsurface structures (i.e., bacteria are visible even if they are completely immersed in EPS).

The last reagent used in OsO_4-RR-TA-IL protocol is an ionic liquid (Hitachi Hilem1000, Hitachi, Japan), it is a salt, fluid at ambient conditions, which resists evaporation under SEM high-vacuum conditions (having irrelevant vapor pressure) and allows SEM investigation of samples without their dehydration (Brodusch, Waters, Demers, & Gauvin, 2014; Tsuda et al., 2011). It has high electronic conductivity so that it can replace the sample surface metal coating (Arimoto, Sugimura, Kageyama, Torimoto, & Kuwabata, 2008; Brodusch et al., 2014; Imashuku, Kawakami, Ze, & Kawai, 2012).

The overall effect of OsO_4-RR-TA-IL protocol on *S. mutans* biofilm ultrastructure observed by SEM (high-vacuum, high-voltage conditions) is shown in Figs. 3 and 4.

The biofilm surface appears compact in some areas and spongy in other ones (Fig. 3A and B). The spongy aspect is due to the presence of holes that are the superficial openings of a well-developed and intricate microchannels system (Fig. 4A and B). At high magnifications (20,000 ×), the biofilm topography is perfectly preserved and EPS components appear structured in an intricate 3D network in which no sign of deformation is apparent (Fig. 4A and B). The shape of the *S. mutans* cells was smooth and spherical. Bacterial cells emerged from EPS in some areas (Fig. 3B), while in others, they were partially embedded in it.

As visible in Figs. 3 and 4, the appearance of *S. mutans* biofilm treated with the OsO_4-RR-TA-IL is different from that observed after the conventional SEM protocol. Data from other imaging techniques, from molecular and behavioral analysis of biofilm, correlate better than that obtained with conventional SEM, with the structure suggested from

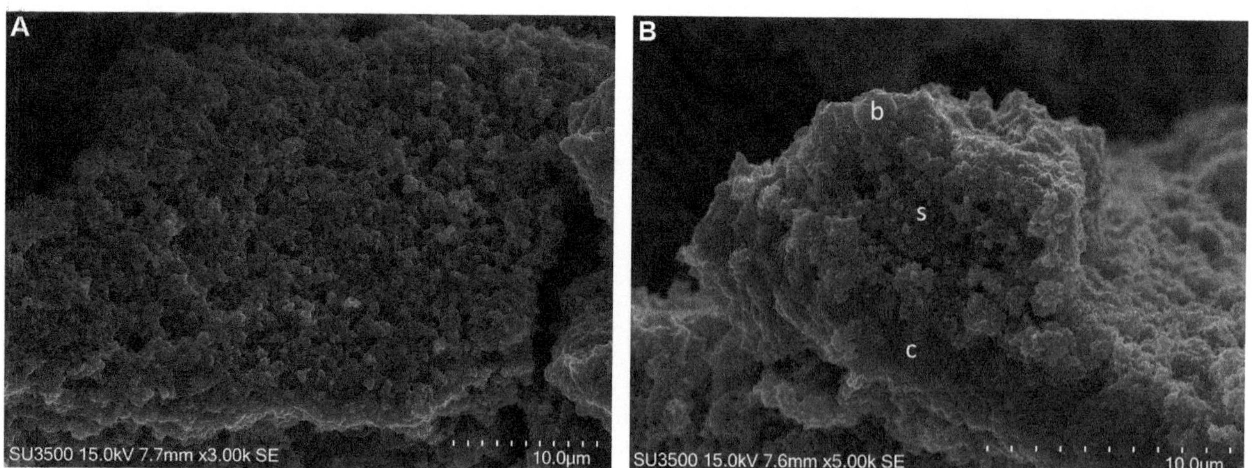

FIG. 3 *S. mutans* prepared by the OsO_4-RR-TA-IL procedure. (A) 3000 ×. The biofilm surface shows a spongy appearance. (B) 5000 ×. The biofilm surface shows compact (c) spongy (s) aspect, a single bacterial cell (b) is partially embedded in EPS. *(From Relucenti, M., Familiari, G., Donfrancesco, O., Taurino, M., Li, X., Chen, R., et al. (2021). Microscopy methods for biofilm imaging: Focus on SEM and VP-SEM pros and cons.* Biology, *10(1), 51. MDPI AG. Retrieved from https://doi.org/10.3390/biology10010051.)*

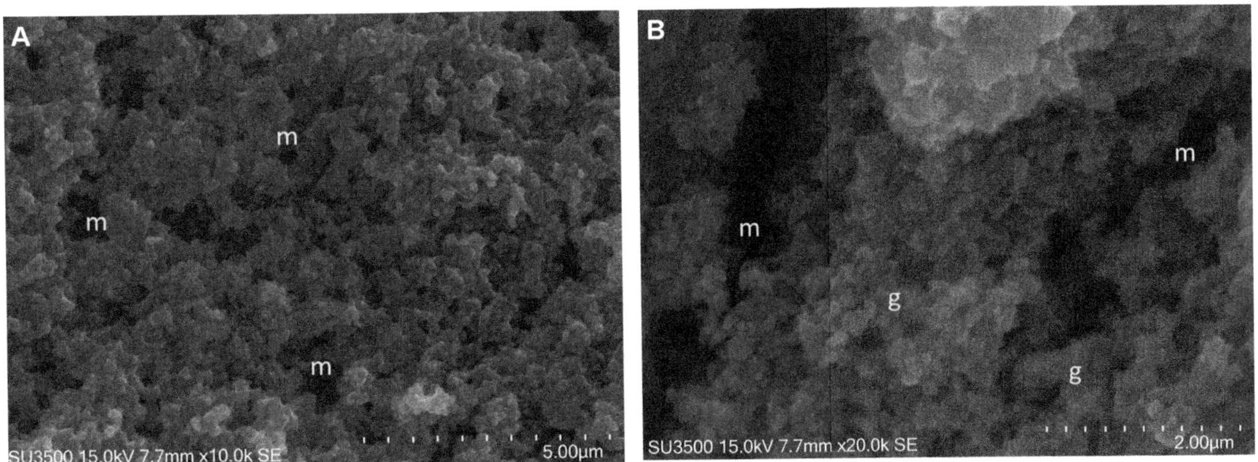

FIG. 4 *S. mutans* prepared by the OsO_4-RR-TA-IL method. (A) At high magnification, the spongy aspect of EPS appears due to the openings of a well-developed micro-canalicular system, m, 10,000. (B). High-voltage, high-magnification, and a high-resolution image of the microcanalicular system, m. EPS appears as a fully hydrated and spongy structure formed by globular aggregates, g, no collapsed structures are visible, 20,000 ×. *(From Relucenti, M., Familiari, G., Donfrancesco, O., Taurino, M., Li, X., Chen, R., et al. (2021). Microscopy methods for biofilm imaging: Focus on SEM and VP-SEM pros and cons.* Biology, *10(1), 51. MDPI AG. Retrieved from https://doi.org/10.3390/biology10010051.)*

images obtained with OsO_4-RR-TA-IL protocol, in which bacterial cells appear embedded in the matrix and not settled on it, and the presence of a micro-canalicular system in EPS is evident.

4. Variable-pressure scanning electron microscopy (VP-SEM) in biofilm studies

VP-SEM allows high-resolution imaging of fully hydrated samples (fixed in glutaraldehyde watery solution and post-fixed OsO_4 watery solution) avoiding dehydration stEPS, as described in Table 3 (Bossù et al., 2019; Bossù, Selan, et al., 2020; Redler et al., 2019).

The experimental conditions of low vacuum, and absence of metal coating, allow sample observation without artifacts due to dehydration procedure or high vacuum, and biofilm loss during preparation is negligible. Images of *S. mutans* biofilm observed at VP-SEM are presented in Figs. 5 and 6.

The images in Fig. 5A and B clearly show the intricate micro-canalicular system peculiar of *S. mutans* biofilm, which appears well-preserved in its topography. EPS appears in some areas compact, in others spongy. Bacterial cells are regularly shaped and embedded, partially or totally in EPS.

In Fig. 6, the images at higher magnifications show, on the bacterial cells' surface, the fine graininess of freshly secreted EPS components. Good quality and very informative pictures were obtained up to 10k.

The maintenance of the sample in its natural hydration state and the negligible sample loss during the preparation procedure are very useful characteristics in studies on drug effects on biofilm when biofilm has to be evaluated before and after treatment. In this kind of studies it is mandatory to maintain the sample unaltered as possible to ascribe morphological changes only to the drug effect, and not to the preparation procedure, and very low sample loss is needed to compare biofilm surface before and after treatment.

An example of VP-SEM imaging of a biofilm subjected to different treatments is present in the paper of Papa et al. (2020). In this study, the effect of organic extract from cell-free supernatant from *Psychrobacter* sp. TAE2020 on *Pseudomonas aeruginosa* P23 was evaluated. The control

TABLE 3 OsO_4-RR-TA-IL protocol.

OsO_4-RR-TA-IL protocol	
StEPS	
Fixation	Glutaraldehyde 2.5% in PB 0.1 M pH7.4 at least 48 h
Washing	10 min x2 times in PB 0.1 M pH7.4
Postfixation	OsO_4 2% 1 h
Washing	10 min x2 times in PB 0.1 M pH7.4
Impregnation	–
Dehydration	–
Drying	–
Pt sputter coating	–
Operating conditions	5–10 kV, low vacuum 20–30 Pa

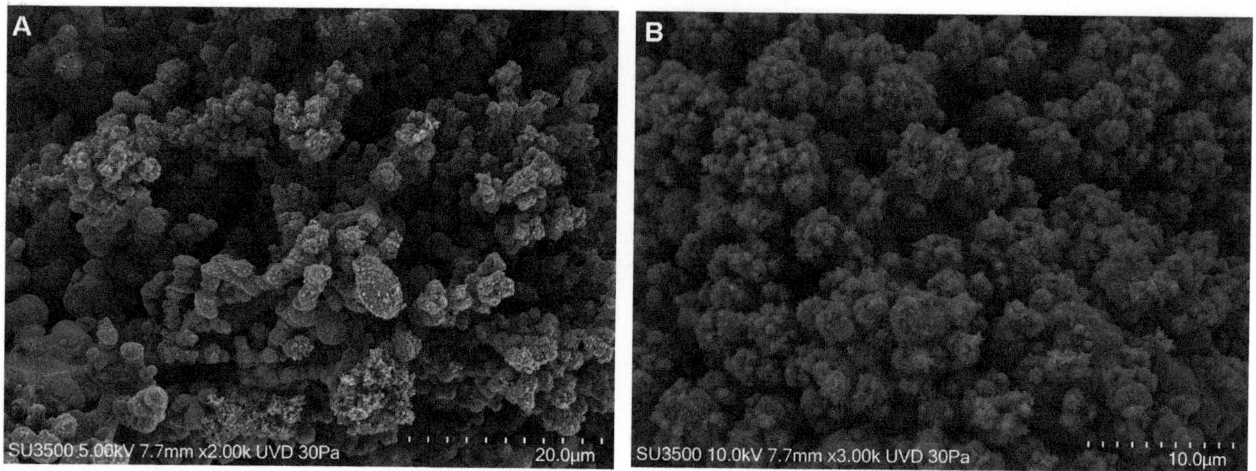

FIG. 5 *S. mutans* prepared by conventional VP-SEM procedure. (A) VP-SEM 2000×. An intricate micro-canalicular system develops among bacterial towers. (B) VP-SEM 3000×. Hydration preservation confers biofilm a spongy aspect, without shrinking or any sign of collapse. *(From Relucenti, M., Familiari, G., Donfrancesco, O., Taurino, M., Li, X., Chen, R., et al. (2021). Microscopy methods for biofilm imaging: Focus on SEM and VP-SEM pros and cons.* Biology, 10*(1), 51. MDPI AG. Retrieved from https://doi.org/10.3390/biology10010051.)*

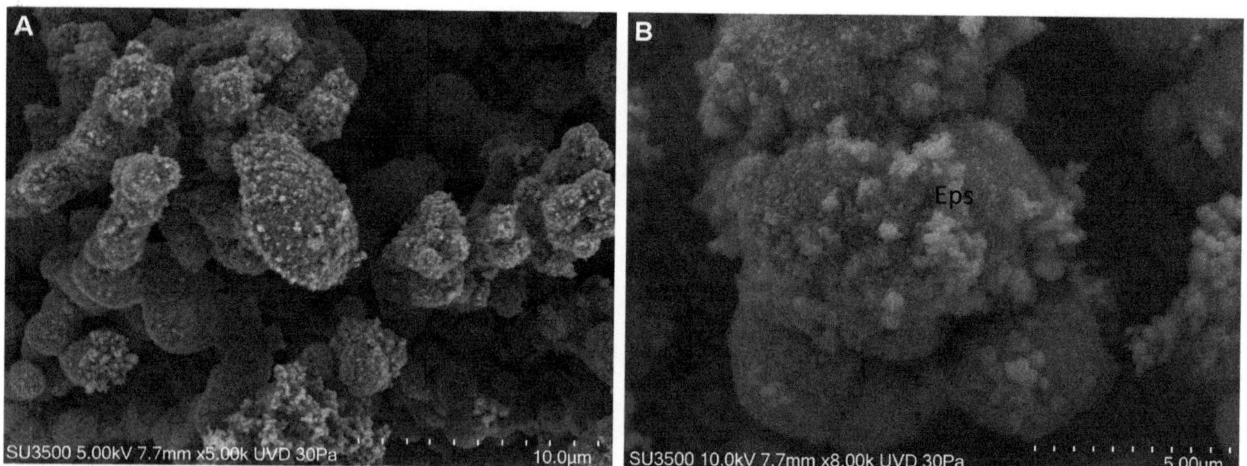

FIG. 6 *S. mutans* prepared by conventional variable-pressure SEM (VP-SEM) procedure. (A) VP-SEM 5000×. Bacterial towers are lined by a superficial granulation representing EPS secretion. (B) VP-SEM 8000×. On *S. mutans* cells surface is visible the fine graininess of freshly secreted EPS components. *(From Relucenti, M., Familiari, G., Donfrancesco, O., Taurino, M., Li, X., Chen, R., et al. (2021). Microscopy methods for biofilm imaging: Focus on SEM and VP-SEM pros and cons.* Biology, 10*(1), 51. MDPI AG. Retrieved from https://doi.org/10.3390/biology10010051.)*

sample shows a rough surface aspect (Fig. 7A), a few superficial openings of the micro-channel system are visible on the sample, which has a dense and compact appearance. Observation at higher magnification (Fig. 7B) shows some bacterial cells emerging from the matrix surface. The sample treated with SN_TAE2020 extract exhibits a different topography. The biofilm surface presents very large ladders (Fig. 7C), and the dense matrix is modified into a spongy and meshed structure. At higher magnifications, the biofilm appears to be formed by a loose and stratified network of EPS trabeculae, in which bacterial cells are more or less embedded (Fig. 7D).

VP-SEM biofilm observation presents advantages but also the following downsides: at higher magnifications, the overall image quality drops significantly due to signal-to-noise ratio lowering; the resolution of VP-SEM images was lower than that of SEM ones, and the sample resistance time at an operating condition of vacuum 30 Pa and electron beam of 5–6 kV was about 1 h before the appearance of cracking phenomena.

After having illustrated the results obtained with different protocols and techniques of SEM in the observation of bacterial biofilms, we can make an assessment of what are the advantages and disadvantages of each method.

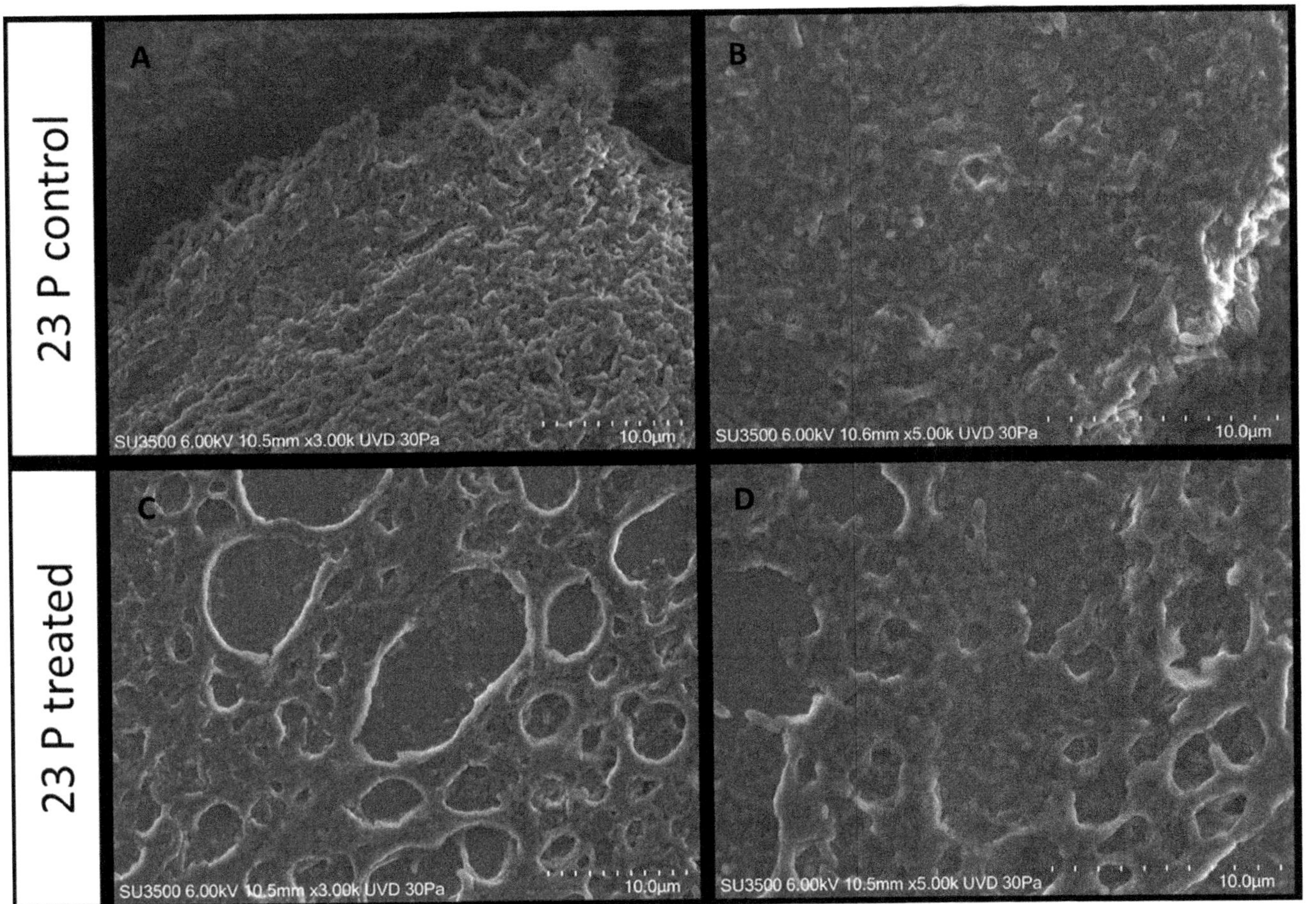

FIG. 7 *P. aeruginosa* 23 P untreated (A, B) and treated (C, D) with SN_TAE2020.VP-SEM (A) 3000×, biofilm appears as a dense and rough-surfaced structure; (B) 5000×, bacterial cells are visible on the matrix surface and entrapped in the matrix bulk; (C) 3000×, the treatment affects the biofilm surface, and large ladders disrupt its surface; (D) 5000×, image at higher magnification shows an irregularly stratified network of extracellular polymeric substances in which bacterial cells are more or less entrapped. *(From Papa, R., Vrenna, G., D'Angelo, C., Casillo, A., Relucenti, M., Donfrancesco, O., et al. (2021). Anti-virulence activity of the cell-free supernatant of the Antarctic bacterium* Psychrobacter *sp. TAE2020 against* Pseudomonas aeruginosa *clinical isolates from cystic fibrosis patients.* Antibiotics, 10*(8), 944. MDPI AG.)*

Each protocol has each protocol has advantages and drawbacks, to evaluate them it is useful to summarize in a table (Table 4) all the parameters we considered, together with the overall informative value given by image magnification and resolution.

5. Conclusion

Based on the findings of our previous works cited above, if high magnification and long-time observation are required (as in the case of new biofilm species characterization), the best protocol in terms of resolution and achievable magnification is the OsO_4-RR-TA-IL for the following reasons: (1) it is a fast procedure (timing comparable with that of VP-SEM protocol, despite having more stEPS than VP-SEM protocol); (2) modest sample loss during preparation procedures; (3) sample hardening, induced by the conductive staining protocol, increased the resistance of EPS to mechanical damage throughout protocol stEPS and allows subsurface observation; (4) lacking dehydration and drying stEPS preserves high EPS water content and EPS actual 3-D structure; (5) high-quality images at magnifications up to 30k are achieved; and (6) sample resists under operating conditions for a long time (even for 2h in high vacuum conditions). OsO_4-RR-TA-IL protocol combines the advantages of SEM and VP-SEM, giving the possibility of achieving images of high quality, at high magnification, and resolution (typical of conventional SEM) faster, with the protocol being made up by few stEPS, having minimal sample loss, and actual sample structure preservation (advantages of VP-SEM).

If, however, drug effects on biofilm morphology have to be evaluated, or if a low amount of sample is available and magnification up to 8000× is sufficient, VP-SEM has to be preferred. It is fast and has a few-step protocol that ensures very minimal sample loss.

TABLE 4 Different parameters to be considered in evaluating the most suitable technique to be used in biofilm study.

Parameters	Conventional SEM	VP-SEM	OsO_4-RR-TA-IL
Procedure time	2 days	1 h and 30 min	2 h and 10 min
Sample loss	StEPS produce sample loss of about 60%	About 20%	About 20%
Dehydration and drying	Yes	None	None
Pt sputter coating	Yes	None	Replaced by IL
Resistance in vacuum	Excellent, it is possible to observe for hours	Good for 1 h	Excellent, it is possible to observe for hours
Operating conditions	15–20 kV, high vacuum	5–10 kV, 30 Pa	15–20 kV, high vacuum
Magnification	Up to 40k	Up to 10k	Up to 40k
Image quality	Excellent up to 30k	Good up to 8k	Excellent up to 30k

References

Alhede, M., Qvortrup, K., Liebrechts, R., Høiby, N., Givskov, M., & Bjarnsholt, T. (2012). Combination of microscopic techniques reveals a comprehensive visual impression of biofilm structure and composition. *FEMS Immunology and Medical Microbiology*, *65*, 335–342.

Arimoto, S., Sugimura, M., Kageyama, H., Torimoto, T., & Kuwabata, S. (2008). Development of new techniques for scanning electron microscope observation using ionic liquid. *Electrochimica Acta*, *53*, 6228–6234.

Bossù, M., Matassa, R., Relucenti, M., Iaculli, F., Salucci, A., Di Giorgio, G., et al. (2020). Morpho-chemical observations of human deciduous teeth enamel in response to biomimetic toothpastes treatment. *Materials*, *13*, 1803.

Bossù, M., Saccucci, M., Salucci, A., Di Giorgio, G., Bruni, E., Uccelletti, D., et al. (2019). Enamel remineralization and repair results of biomimetic hydroxyapatite toothpaste on deciduous teeth: An effective option to fluoride toothpaste. *Journal of Nanbiotechnology*, *17*, 17.

Bossù, M., Selan, L., Artini, M., Relucenti, M., Familiari, G., Papa, R., et al. (2020). Characterization of *Scardovia wiggsiae* biofilm by original scanning electron microscopy protocol. *Microorganisms*, *8*(6), 807. MDPI AG. Retrieved from https://doi.org/10.3390/microorganisms8060807.

Breidenstein, E. B., De La Fuente-Nunez, C., & Hancock, R. E. (2011). *Pseudomonas aeruginosa*: All roads lead to resistance. *Trends in Microbiology*, *19*, 419–426.

Brodusch, N., Waters, K., Demers, H., & Gauvin, R. (2014). Ionic liquid-based observation technique for nonconductive materials in the scanning electron microscope: Application to the characterization of a rare earth ore. *Microscopy Research and Technique*, *77*, 225–235.

Carr, K. E., & Toner, P. G. (1982). *Cell structure. An introduction to biomedical electron microscopy* (3rd ed.). Churchill Livingstone.

Chang, H. T., & Rittmann, B. E. (1986). Biofilm loss during sample preparation for scanning electron microscopy. *Water Research*, *20*, 1451–1456.

Costerton, J. W., Geesey, G. G., & Cheng, K. J. (1978). How bacteria stick. *Scientific American*, *238*(1), 86–95.

Costerton, J. W., Lewandowski, Z., Caldwell, D. E., Korber, D. R., & Lappin-Scott, H. M. (1995). Microbial biofilms. *Annual Review of Microbiology*, *49*, 711–745.

Davies, D. G., Parsek, M. R., Pearson, J. P., Iglewski, B. H., Costerton, J. W., & Greenberg, E. P. (1998). The involvement of cell-to-cell signals in the development of a bacterial biofilm. *Science (New York, N.Y.)*, *280*(5361), 295–298.

De Araujo, J. C., Téran, F. C., Oliveira, R. A., Nour, E. A. A., Montenegro, M. A. P., Campos, J. R., et al. (2003). Comparison of hexamethyldisilazane and critical point drying treatments for SEM analysis of anaerobic biofilms and granular sludge. *QJM: An International Journal of Medicine*, *52*, 429–433.

Di Bonaventura, G., Piccolomini, R., Paludi, D., D'Orio, V., Vergara, A., Conter, M., et al. (2008). Influence of temperature on biofilm formation by *Listeria monocytogenes* on various food-contact surfaces: Relationship with motility and cell surface hydrophobicity. *Journal of Applied Microbiology*, *104*, 1552–1561.

Di Martino, P. (2018a). Extracellular polymeric substances, a key element in understanding biofilm phenotype. *AIMS Microbiology*, *4*, 274–288.

Di Martino, P. (2018b). Bacterial adherence: Much more than a bond. *AIMS Microbiology*, *4*(3), 563–566.

Erlandsen, S. L., Kristich, C. J., Dunny, G. M., & Wells, C. L. (2004). High-resolution visualization of the microbial glycocalyx with low-voltage scanning electron microscopy: Dependence on cationic dyes. *The Journal of Histochemistry and Cytochemistry*, *52*, 1427–1435.

Fassel, T. A., & Edmiston, C. E., Jr. (1999a). Bacterial biofilms: Strategies for preparing glycocalyx for electron microscopy. *Methods in Enzymology*, *310*, 194–203.

Fassel, T. A., & Edmiston, C. E., Jr. (1999b). Ruthenium red and the bacterial glycocalyx. *Biotechnic & Histochemistry*, *74*, 194–212.

Fassel, T. A., Schaller, M. J., & Remsen, C. C. (1992). Comparison of alcian blue and ruthenium red effects on preservation of outer envelope ultrastructure in methanotrophic bacteria. *Microscopy Research and Technique*, *20*, 87–94.

Fulcher, T. P., Dart, J. K., McLaughlin-Borlace, L., Howes, R., Matheson, M., & Cree, I. (2001). Demonstration of biofilm in infectious crystalline keratopathy using ruthenium red and electron microscopy. *Ophthalmology*, *108*, 1088–1092.

Hung, C., Zhou, Y., Pinkner, J. S., Dodson, K. W., Crowley, J. R., Heuser, J., et al. (2013). *Escherichia coli* biofilms have an organized and complex extracellular matrix structure. *MBio*, *4*(5), e00645-13.-https://doi.org/10.1128/mBio.00645-13.

Imashuku, S., Kawakami, T., Ze, L., & Kawai, J. (2012). Possibility of scanning electron microscope observation and energy dispersive X-ray analysis in microscale region of insulating samples using diluted ionic liquid. *Microscopy and Microanalysis*, *18*, 365–370.

Karcz, J., Bernas, T., Nowak, A., Talik, E., & Woznica, A. (2011). Application of lyophilization to prepare the nitrifying bacterial biofilm for imaging with scanning electron microscopy. *Scanning*, *34*, 26–36.

Karygianni, L., Ren, Z., Koo, H., & Thurnheer, T. (2020). Biofilm matrixome: Extracellular components in structured microbial communities. *Trends in Microbiology*, *28*, 668–681.

Kolenbrander, P. E. (2000). Oral microbial communities: Biofilms, interactions, and genetic systems. *Annual Review of Microbiology*, *54*, 413–437.

Li, Y., Xiao, P., Wang, Y., & Hao, Y. (2020). Mechanisms and control measures of mature biofilm resistance to antimicrobial agents in the clinical context. *ACS Omega*, *5*, 22684–22690.

Luft, J. H. (1964). Electron microscopy of cell extraneous coats as revealed by ruthenium red staining. *The Journal of Cell Biology*, *23*, 54A–55A.

Murakami, T. (1973). A metal impregnation method of biological specimens for SEM. *Archivum Histologicum Japonicum*, *35*, 323–326.

Murakami, T. (1974). A revised tannin-osmium method for non-coated SEM specimens. *Archivum Histologicum Japonicum*, *36*, 189–193.

Murakami, T. (1976). Puncture perfusion of small tissue pieces for SEM. *Archivum Histologicum Japonicum*, *39*, 99–103.

Murakami, T., Iida, N., Taguchi, T., Ohtani, O., Kikuta, A., Ohtsuka, A., et al. (1983). Conductive staining of biological specimens for scanning electron microscopy with special reference to ligand-mediated osmium impregnation. *Scanning Electron Microscopy*, *1*, 235–246.

Murakami, T., Yamamoto, K., Itoshima, T., & Irino, S. (1977). Modified tannin osmium conductive staining method for non-coated SEM specimens. Its application to microdissection SEM of the spleen. *Archivum Histologicum Japonicum*, *40*, 35–40.

Papa, R., Garzoli, S., Vrenna, G., Sabatino, M., Sapienza, F., Relucenti, M., et al. (2020). Essential oils biofilm modulation activity, chemical and machine learning analysis. Application on *Staphylococcus aureus* isolates from cystic fibrosis patients. *International Journal of Molecular Sciences*, *21*, 9258.

Papa, R., Vrenna, G., D'Angelo, C., Casillo, A., Relucenti, M., Donfrancesco, O., et al. (2021). Anti-virulence activity of the cell-free supernatant of the Antarctic bacterium *Psychrobacter* sp. TAE2020 against *Pseudomonas aeruginosa* clinical isolates from cystic fibrosis patients. *Antibiotics*, *10*(8), 944. MDPI AG.

Priester, J. H., Horst, A. M., Van de Werfhorst, L. C., Saleta, J. L., Mertes, L. A., & Holden, P. A. (2007). Enhanced visualization of microbial biofilms by staining and environmental scanning electron microscopy. *Journal of Microbiological Methods*, *68*(3), 577–587.

Redler, A., Miglietta, S., Monaco, E., Matassa, R., Relucenti, M., Daggett, M., et al. (2019). Ultrastructural assessment of the anterolateral ligament. *Orthopaedic Journal of Sports Medicine*, *7*. 2325967119887920.

Relucenti, M., Familiari, G., Donfrancesco, O., Taurino, M., Li, X., Chen, R., et al. (2021). Microscopy methods for biofilm imaging: Focus on SEM and VP-SEM pros and cons. *Biology*, *10*(1), 51. MDPI AG. Retrieved from https://doi.org/10.3390/biology10010051.

Relucenti, M., Miglietta, S., Bove, G., Donfrancesco, O., Battaglione, E., Familiari, P., et al. (2020). SEM BSE 3D image analysis of human incus bone affected by cholesteatoma ascribes to osteoclasts the bone erosion and VpSEMdEDX analysis reveals new bone formation. *Scanning*, *2020*, 1–9.

Relucenti, M., Miglietta, S., Covelli, E., Familiari, P., Battaglione, E., Familiari, G., et al. (2019). Ciliated cell observation by SEM on the surface of human incudo-malleolar-joint articular cartilage: Are they a new chondrocyte phenotype? *Acta Oto-Laryngologica*, *139*, 439–443.

Rodrigues, D., Bañobre-López, M., Espiña, B., Rivas, J., & Azeredo, J. (2013). Effect of magnetic hyperthermia on the structure of biofilm and cellular viability of a food spoilage bacterium. *Biofouling*, *29*(10), 1225–1232. https://doi.org/10.1080/08927014.2013.834893.

Shapiro, J. A. (1998). Thinking about bacterial populations as multicellular organisms. *Annual Review of Microbiology*, *52*, 81–104.

Suetens, C., Latour, K., Kärki, T., Ricchizzi, E., Kinross, P., Moro, M. L., … Monnet, D. L. (2018). Prevalence of healthcare-associated infections, estimated incidence and composite antimicrobial resistance index in acute care hospitals and long-term care facilities: results from two European point prevalence surveys, 2016 to 2017. *Euro Surveillance: Bulletin Europeen sur les maladies transmissibles = European Communicable Disease Bulletin*, *23*(46), 1800516. https://doi.org/10.2807/1560-7917.ES.2018.23.46.1800516.

Timp, W., & Matsudaira, P. (2008). Chapter 14 Electron microscopy of hydrated samples. *Methods in Cell Biology*, *89*, 391–407.

Tsuda, T., Nemoto, N., Kawakami, K., Mochizuki, E., Kishida, S., Tajiri, T., et al. (2011). SEM observation of wet biological specimens pretreated with room-temperature ionic liquid. *ChemBioChem*, *12*, 2547–2550.

Vestby, L. K., Grønseth, T., Simm, R., & Nesse, L. L. (2020). Bacterial biofilm and its role in the pathogenesis of disease. *Antibiotics*, *9*, 59.

Vitkov, L., Hannig, M., Krautgartner, W. D., Herrmann, M., Fuchs, K., Klappacher, M., et al. (2005). Ex vivo gingival-biofilm consortia. *Letters in Applied Microbiology*, *41*, 404–411.

Vitkov, L., Hermann, A., Krautgartner, W. D., Herrmann, M., Fuchs, K., Klappacher, M., et al. (2005). Chlorhexidine-induced ultrastructural alterations in oral biofilm. *Microscopy Research and Technique*, *68*, 85–89.

Vitkov, L., Krautgartner, W. D., & Hannig, M. (2005). Bacterial internalization in periodontitis. *Oral Microbiology and Immunology*, *20*, 317–321.

Vrenna, G., Artini, M., Ragno, R., Relucenti, M., Fiscarelli, E. V., Tuccio Guarna Assanti, V., et al. (2021). Anti-virulence properties of *Coridothymus capitatus* essential oil against *Pseudomonas aeruginosa* clinical isolates from cystic fibrosis patients. *Microorganisms*, *9*(11), 2257. MDPI AG. Retrieved from https://doi.org/10.3390/microorganisms9112257.

Yadav, M. K., Song, J.-J., Singh, B. P., & Vidal, J. E. (2020). Microbial biofilms and human disease: A concise review. In M. K. Yadav, & B. P. Singh (Eds.), *New and future developments in microbial-l biotechnology and bioengineering; current research and future trends in microbial biofilms* (pp. 1–13). Amsterdam: Elsevier (Chapter 1).

Chapter 7

Aspects of biofilms on medical devices

Fergus Watson[a], Marcus Swann[b], Alicja Zimmer[c], Rui Chen[d], Lauren Foulkes[e], and Steven L. Percival[f]

[a]*Operations, 5D Health Protection Group Ltd., Liverpool, United Kingdom,* [b]*5D Health Protection Group Ltd., Centre of Excellence in Biofilm Science and Technologies (CEBST), Liverpool, United Kingdom,* [c]*Getinge Applikon, Liverpool, England, United Kingdom,* [d]*Biofilm Centre, 5D Health Protection, Group, Liverpool, United Kingdom,* [e]*Microbiology, Biofilm Centre, Liverpool, United Kingdom,* [f]*R&D, 5D Health Protection, Group, Liverpool, United Kingdom*

1. Introduction

Indwelling medical devices, such as urinary catheters, orthopedic implants, and endotracheal tubes are commonly used in modern medical practice to support the function of many vital organs (Shunmugaperumal, 2010). Although medical devices are an important aspect of modern medicine, improving prognosis and quality of life of patients, they are susceptible to microbial colonization and infection, with at least half of healthcare-associated infections (HAIs) being linked to their use (Gilmore & Carson, 2015; VanEpps & Younger, 2016).

Medical-device-related infections are difficult to treat, and recurrence is frequent. The fundamental reason for this is due to the adherence of bacteria to indwelling medical device surfaces and the formation of biofilms, which are associated with considerable morbidity and mortality and thus pose a significant impact on healthcare delivery (Donlan, 2008). According to the National Health Institute (NIH), biofilms are responsible for up to 80% of microbial infections in humans (Jamal et al., 2018; Sharma, Misba, & Khan, 2019). Biofilms can be formed by both Gram-positive and Gram-negative bacteria on medical equipment; however, *Enterococcus faecalis*, *Staphylococcus aureus*, *Staphylococcus epidermidis*, *Streptococcus viridans*, *Escherichia coli*, *Klebsiella pneumoniae*, *Proteus mirabilis*, and *Pseudomonas aeruginosa* are the most prevalent (Chen, Yu, & Sun, 2013). Specifically, *S. aureus* and *S. epidermidis* are thought to be responsible for 40%–50% of prosthetic heart valve infections, 50%–70% of catheter biofilm infections, and 87% of bloodstream infections (Agarwal, Singh, & Jain, 2010; Crowe, Sculco, & Kahn, 2003). Upon the formation of a biofilm on a medical device, the eradication of microorganisms becomes exceedingly difficult, and the expense can be significant as a result of the recurrent need for extended hospitalization, surgery, and extensive antibiotic therapy (Crowe et al., 2003).

Microorganisms in biofilms differ from their planktonic counterparts in terms of growth rates, structural properties, microenvironments, and resistance to host immune responses and antimicrobials as bacteria in mature biofilms over 7 days old are 500–5000 times more resistant to antimicrobial treatments (Khoury, Lam, Ellis, & Costerton, 1992; Wi & Patel, 2018). This is largely due to the formation of a matrix of extracellular polymeric substances (EPS), which is secreted by attached cells and encases the bacterial aggregates (Percival, Malone, Mayer, Salisbury, & Gregory Schultz, 2018; Zhang, Li, Yu, & Wang, 2020) (Fig. 1). Polysaccharides, extracellular DNA, and proteins make up the majority of EPS, which aid in the protection of microorganisms from external dangers such as immune system components and antimicrobial treatments (Percival, Thomas, & Williams, 2010).

Biofilm formation is enabled through quorum sensing (QS), which is cell-to-cell communication among bacterial cells that utilizes autoinducers (Ding et al., 2011; Waters & Bassler, 2005). This enables bacteria to alter gene expression through conjugation, transformation, and transduction, secrete virulence factors, and increase antimicrobial resistance.

Once the biofilm develops on the surface of medical devices, they are resistant to the host's endogenous defenses and are typically treated with a combination of antibiotic therapy and tissue debridement (Donlan & Costerton, 2002; Khatoon, McTiernan, Suuronen, Mah, & Alarcon, 2018). However, the administration of combination antibiotic therapy has previously been linked to the development of antibiotic-resistant bacteria strains, thus highlighting the need for novel approaches to managing biofilm formation (Jamal et al., 2018; Khatoon et al., 2018).

The purpose of this review is to establish the role of biofilm formation in medical-device-related infections, the challenges in treatment and diagnosis, and future prospects in prevention, diagnosis, and treatment of these biofilms.

2. Biofilm formation on medical devices

A medical device is a product or equipment intended for a medical purpose and can range in complexity; from tongue depressors, lavatory pans, and absorbent sutures to more

Understanding Microbial Biofilms. https://doi.org/10.1016/B978-0-323-99977-9.00020-X

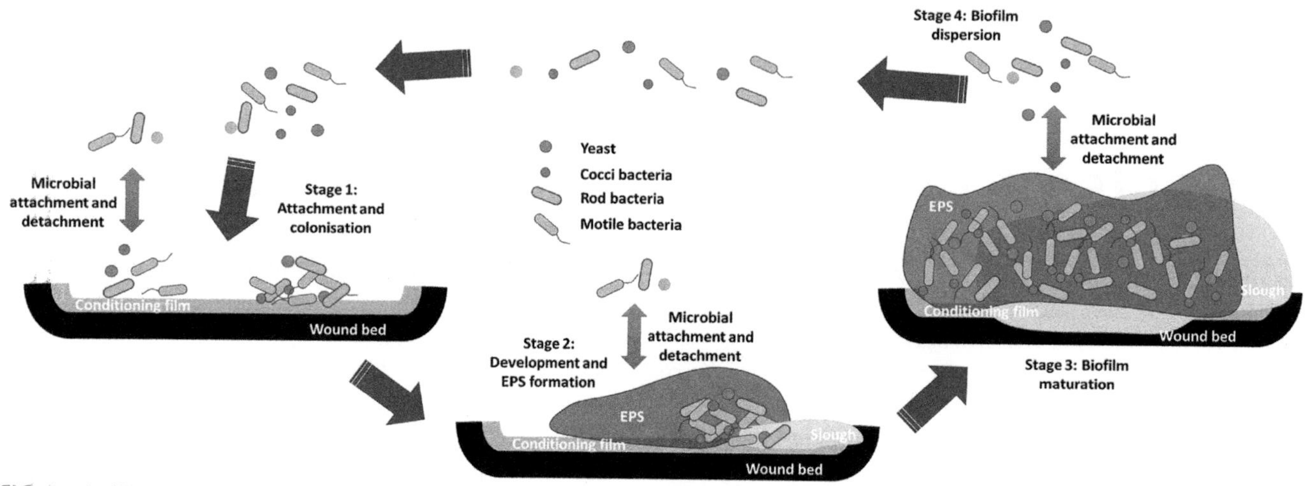

FIG. 1 An illustration of the stages for biofilm formation in a wound bed. Slough is a by-product of the wound's inflammatory response and comprises fibrin, leucocytes, dead and living cells.

advance technologies such as ultrasound machines, computed tomography scanners, and catheters. For implantable medical devices such as intravenous catheters, upon contact with the patient's skin, bacteria will readily adhere and begin to form a community of microorganisms known as a biofilm. Infections associated with the device materials are frequent due to the patient's own endogenous flora and, among various reasons, can result in an increased number of revised procedures. For example, revised prosthetic joint implant surgery in which the original implant has become a source of infection occurs in two stages: initially the contaminated device and surroundings are removed followed by often extensive antibiotic therapy; and then an additional round of surgery occurs to situate the replacement device (Gbejuade, Lovering, & Webb, 2015; Vickery, Hu, Jacombs, Bradshaw, & Deva, 2013). This must occur irrespective of the inherent risks of surgery, to ensure the biofilm doesn't continue to function as a reservoir for pathogens and infections.

It is believed the majority of bacterial strains are capable of forming biofilms. Up to 65% of all hospital-acquired infections (HAIs) are linked to biofilm formation, and therefore, it is no surprise significant research has gone into understanding their formation and association on medical devices (Vega & Gore, 2014).

A variety of both synthetic and natural materials are incorporated in the production of medical devices including polyethylene, silicone rubber, nylon, stainless steel, titanium, and many more. These materials are selected based on their biocompatibility and long-term stability and are generally considered nontoxic, inert, and do not have any adverse biological response during contact with the human body. In fact, some materials are chosen for their advantageous biocompatibility with the host's biological system by promoting a synergistic interaction (Festas, Ramos, & Davim, 2020). As an example, biodegradable polymers comprising naturally occurring polysaccharides and proteins can be extensively used in sutures and adhesives for surgical procedures for many years (Sáenz, Rivera, Brostow, & Castaño, 1999).

Biofilm formation occurs through a series of processes including attachment, development and maturation, and finally, dispersal. Upon contact and/or insertion of a medical device, a conditioning film of macromolecules begins to form comprising proteins and glycoproteins such as fibronectin, fibrinogen, albumin, and immunoglobulins providing an ideal substrate upon which bacteria, and other microorganisms, can adhere (Bryers, 2008). The attachment of planktonic bacteria is initially thought to be reversible, in which the physical interaction is instantaneous or transient and the forces between the bacteria and surface are weak (Gupta, Sarkar, Das, Bhattacharjee, & Tribedi, 2016). This can also be influenced by the appendages of the cell such as flagella or pili, allowing the microorganisms to actively seek out the conditioning film and thus the medical device (Floyd, Eberly, & Hadjifrangiskou, 2017). The adherence forces at this stage are mostly governed by adhesive protein and electrostatic, hydrophobic, and steric interactions (Pavithra & Doble, 2008). The physical attraction between the microbe and substrate is commonly known as the Brownian motion or Van der Waals attraction. These attractive forces are categorized according to the distance from the target surface; for example, long-range interactions (>50 nm) are nonspecific and occur between the microbe and the substrate surface; and short-range interactions (<5 nm) are specific bonds that may arise between the microbe and surface such as hydrogen bonds and polarized connections. Once the attractive forces begin to outweigh the repulsive forces, the attachment phase becomes irreversible and the microbe is then immobilized and trapped at the surface (Stoica, Chifiriuc, Rapa, & Lazăr, 2017). Bacteria cell wall proteins referred to as adhesins regulate this attachment and promote the formation of biofilms; *S. epidermidis*, for example, readily binds to polymer surfaces via an autolysin (At1E) found on the surface of the cell (Veerachamy, Yarlagadda, Manivasagam, & Yarlagadda, 2014). These proteins have been shown once in sufficiently high concentrations can stimulate a behavioral response and interaction between substrate-bound bacteria promoting the production of extracellular polymeric substances (EPS) and enhancing the aggregation of microorganisms to the surface (Cerca, Pier, Vilanova, Oliveira, & Azeredo, 2005; Flemming & Wingender, 2010; Harris, Tosatti, Wieland, Textor, & Richards, 2004). Following attachment, the biofilm begins to develop in structural and organizational complexity, through the secretion of diffusible signaling molecules in a process known as quorum sensing. These molecules are specific peptides and allow cells to communicate and regulate the production of EPS such as polysaccharide adhesins, proteins, and nucleic acids (Hobley, Harkins, MacPhee, & Stanley-Wall, 2015). In doing so, the initial microbes stimulate further adhesion of other microbes, which in turn contribute to quorum sensory activity and thus a self-promoting cycle (Van Houdt & Michiels, 2005). In some instances of *Staphylococcus* biofilms, large surface-associated proteins called clumping factors are produced and used to further mediate the formation of biofilms. These have become targets for antibiofilm technologies particularly in catheters where chelating agents are used (Abraham & Jefferson, 2012).

During this phase, bacteria within the biofilm produce a variety of extracellular substances, which facilitate the structuring and organizational features within the matrix. For example, glycocalyx, a carbohydrate-lipid-protein complex, is present in all biofilms to aid the attachment to solid substrates (Rodis, Tsapadikou, Potsios, & Xaplanteri, 2021). The absorbance properties of Glycocalyx have been shown to improve the survival of microbes during desiccation in unfavorable environmental conditions, as well as providing protection against antimicrobials by readily quenching these agents and thus limiting their penetration within the microbial community (Abebe, 2020; Saïdi et al., 2020). As the biofilm develops, and matures, the EPS traps microbes of both the same and different species, which in turn form a chemical connection and synergistic relationships (Rao, 2020).

Polymicrobial biofilms are present in nearly all microbial infections and contribute significantly to their persistence as seen in chronic wounds (Percival et al., 2012). For example, the use of fluorescence microscopy by numerous research groups has allowed for the visualization of *Pseudomonas* sp., *Staphylococcus* sp., *Streptococcus* sp., and *Micrococcus* sp. within biofilms, forming on both the surface of reconstituted human epidermis models and in clinical biopsies from chronic wound patients (Malic et al., 2009; Perry-O'Keefe et al., 2001). In this instance, the group used fluorescent nucleic acid stains in combination with confocal laser scanning microscopy in a technique referred to as fluorescence in situ hybridization. In addition to bacterial multispecies biofilms, the significant impact of bacterial and fungal polymicrobial infections has become more widely recognized for biofilm-associated disease on indwelling medical devices such as central venous catheters (CVCs). It is increasingly common for *Candida albicans* and *S. aureus*, commensal microorganisms of the human body, to cohabit these infection sites. On a number of occasions both species have been found to express an enhanced adhesion toward each other, and thus an affinity to form biofilms (Peters et al., 2012; Peters, Ward, Rane, Lee, & Noverr, 2013). The *Staphylococci* adhere to the hyphae of the fungus through a series of membrane proteins on the surface of the *C. albicans*. The presence of these adhesins would indicate a probable symbiotic relationship between the two organisms (El-Azizi, Starks, & Khardori, 2004). The close proximity of this augments the exchange of effector molecules necessary for quorum sensory activity. The relationship between these species can be both symbiotic and parasitic depending on their metabolic activities and requirements and can dynamically alter an infection microenvironment and thus the progression of diseases. In 2014, Falsetta and colleagues demonstrated *C. albicans* when coexisting with *Streptococcus mutans* enhanced the expression of virulence genes and contributed to formation of dense EPS matrices. The severity of the infection was greatly increased as a result. Other virulence factors such as antimicrobial susceptibility, upregulation of endotoxins, and impairment of host immune cells can result from multispecies biofilm infections (Armbruster et al., 2010; Duan, Dammel, Stein, Rabin, & Surette, 2003; Korgaonkar, Trivedi, Rumbaugh, & Whiteley, 2013; Ryan et al., 2008; Sibley et al., 2008).

The final stage in biofilm formation is dispersal, triggered by a number of stress-inducing factors including nutrient depletion, overpopulation/crowding, exposure to chemical stresses and physical shear forces. Dispersion of the biofilm is a critical stage in the life cycle of these communities by allowing the microorganisms to migrate toward more favorable conditions and thus away from unfavorable conditions. Once detached from the biofilm, the microbes migrate within the human body and are the common cause for chronic infections in a process often referred to as metastatic seeding (Gupta et al., 2016; Veerachamy et al., 2014). Maturation of the biofilm readily causes a limitation in resources and a build of toxic metabolic by-products causing either single cells or clusters of cells to slough off the surface. While there are many trigger mechanisms for dispersal, the majority are centered upon the dissolution of the EPS and the role of a messenger molecule called cyclic dimeric guanosine monophosphate (c-di-GMP) (Abdel-Aziz & Aeron, 2014). The majority of factors, both environmental and chemical, that are linked to induced dispersion are associated with c-di-GMP regulation (Ha & O'Toole, 2015). Generally, high intracellular levels of c-di-GMP promote EPS production by bacteria and therefore the formation of biofilms. In comparison, decreased levels of c-di-GMP inhibit this production while inducing dispersal of the bacteria back into their planktonic state (Hengge, 2009; Römling, Gomelsky, & Galperin, 2005). Many studies have shown by interrupting the production of c-di-GMP and depleting its content levels within cells, established biofilms can be effectively dispersed. A recent example demonstrated that fluorinated compounds, often used as an intermediate in pharmacology or cosmetics industries, in combination with targeted antibiotics enhanced the inactivation of *P. aeruginosa* biofilm in medical-device-associated infections in mice (Andersen et al., 2021). Once detached, many bacteria revert to their planktonic state of susceptibility to antimicrobials, antibiotics, and the hosts own immune system, and thus, dissolution of the biofilm is a key area of research for treating medical biofilms (Fleming & Rumbaugh, 2017).

3. Biofilm models

Biofilm models are designed to improve our understanding of their formation, function, resistances, and regulatory capacity. Buckingham-Meyer, Goeres, and Hamilton (2007) once demonstrated, through a series of biofilm efficacy studies, the importance of correctly identifying key characteristics of the environment from which the biofilm originates. Failure to do so can provide false interruption of results and thus inaccurate translation during in vitro to in situ application. Key characteristics to consider include the following:

- Substrate materials—type, texture, chemical properties, mechanical properties
- Growth conditions—pH, temperature, time, nutrients, conditioning film
- Microbiome—species, phenotype, virulence, pathogenicity
- Host cells—cell type, immune response

Nonetheless, the development of biofilm models has significantly contributed to the continual development of novel antimicrobial and antibiofilm therapies. The development of the *P. aeruginosa* ATCC 15442 Centres for Disease Control and Prevention (CDC) biofilm reactor has enabled vast amounts of research to be carried out in order to investigate the formation of biofilms on medical-device-related materials such as titanium and zirconium (Hadi, Vickery, Deva, & Charlton, 2010; Veerachamy et al., 2014). This particular model enables confluent growth under high shear conditions with a continuous flow of nutrients, providing a platform to demonstrate the efficacy and disruptive performance of antibiofilm surfaces for indwelling device materials (Artini et al., 2017; Francolini, Vuotto, Piozzi, & Donelli, 2017). In a study by Cai, Wu, Xi, and Meyerhoff (2012), an antibiofilm surface using a nitric oxide releasing film on silicone rubber exhibited an 98% reduction in biomass for *S. aureus* and almost complete reduction (99.9%) in *E. coli* biofilms. Alternative models such as the 96-well plate microfluidic device, comprising microchannels in a microtiter plate and a peg lid, are designed for higher-throughput analysis of biofilm growth under differing conditions; often utilized for measuring minimum biofilm eradication concentrations of common antimicrobial compounds (Doll, Jongsthaphongpun, Stumpp, Winkel, & Stiesch, 2016; Olson, Ceri, Morck, Buret, & Read, 2002; Ravi, Aslam, & Veeraraghavan, 2019). Magacho, Pinto, Souza, Pereira, and Ferreira-Strixino (2020) recently highlighted the impact of modern photodynamic therapy, which relies on the activation of photosensitive substances to produce reactive oxygen species to elicit cell damage, on clinically relevant Gram-negative bacteria (e.g., *E. coli*, *K. pneumoniae*, and *Enterobacter aerogenes*). Using viability assays and culture analysis, the results indicated significant reductions in bacterial viability and metabolic activity.

In addition to the advancement of more sophisticated biofilm models, vast amounts of research have been targeting the development of new techniques for the visualization and quantification of biofilm growth on both abiotic and biotic surfaces. Such techniques utilize the use of stains for the labeling of specific targets within the matrix and EPS such as: Alexa-488-labeled Aleuria aurantia lectins for glycoconjugates; Sypro Orange or Sypro Ruby for proteins, and Syto-9 and propidium iodide for nucleic acid (Almatroudi et al., 2015; Kragh, Alhede, Kvich, & Bjarnsholt, 2019; Welch, Cai, & Strømme, 2012) (Fig. 2). For example, Shin and Eom (2019) utilized confocal laser scanning microscopy to highlight the dissemination of EPS components (proteins, carbohydrates, nucleic acids) in an *S. aureus* and *C. albicans* biofilm. The group demonstrated significant reduction in biomass and cell density upon exposure to the fungicide zerumbone. While these models provide valuable data, the vast majority remain focused on single or dual species, whereas in situ biofilms are likely to contain a variety of other species all with generating unique synergies and micro-niches.

4. Medical-device-related infections

Although medical devices are an integral aspect of medicine, 60%–70% of nosocomial infections are associated with their use (Yasir, Willcox, & Dutta, 2018). Medical-device-related infections (MDRI) are associated with high mortality and morbidity and incur significant healthcare

FIG. 2 An epifluorescence micrograph of a biofilm model generated using a drip flow reactor. *(Modified from Watson, F., Keevil, C. W., Wilks, S. A., & Chewins, J. (2018). Modelling vaporised hydrogen peroxide efficacy against mono-species biofilms.* Scientific Reports, *8(1), 1–7.)*

costs. The infection typically occurs as a result of microorganism contamination from the patient's skin, mucous membrane, and even from the hands of the clinician during insertion (von Eiff, Jansen, Kohnen, & Becker, 2005). One of the main causative agents associated with MDRI is the formation of biofilms on the surface of the implantable device.

4.1 Central venous catheters

CVCs are inserted through a large, central vein allowing for the administration of several medicines and fluids and allow access for sampling blood. However, CVCs are at high risk of microbial colonization and biofilm formation, thus often lead to catheter-related bloodstream infection (CRBSI) with approximately 70% of all nosocomial bloodstream infections occurring in patients with CVCs (Gominet, Compain, Beloin, & Lebeaux, 2017). The location of the biofilm often depends on the length of catheterization. Extraluminal biofilms are thought to be the leading cause of CRBSI during the first week of catheterization (Percival, Suleman, Vuotto, & Donelli, 2015). However, catheters that have been in place for more than 30 days revealed signs of mainly luminal colonization and biofilm development (Percival et al., 2015; Raad et al., 1993).

Biofilm development on CVCs was first discovered in 1982, due to an outbreak of *S. epidermidis* bloodstream infection (Christensen, Simpson, Bisno, & Beachey, 1982). The microbial colonization of the outer lumen of the catheter typically occurs due to the close proximity of the CVC to the skin, enabling the skin flora to contaminate the device surface (Murga, Miller, & Donlan, 2001). Several studies have suggested that excessive cutaneous colonization is linked to CVC-related bloodstream infections (Safdar & Maki, 2004).

4.2 Endotracheal tube

Endotracheal intubation (ETT) is a life-saving procedure, which enables mechanical ventilation in patients with breathing difficulties. ETT causes the cough reflex to be disrupted, allowing tracheobronchial secretions and mucus to accumulate. This increases the risk of infection as it provides pathogenic bacteria with a direct route to the lower respiratory tract (Cairns et al., 2011; Pneumatikos, Dragoumanis, & Bouros, 2009). The tube, which consists of polyvinyl chloride, is at high risk of rapid microbial colonization and adhesion, resulting in the formation of biofilms.

Biofilm formation on endotracheal tube (ETT) surface occurs rapidly after insertion and is considered to be a significant source of bacterial inoculation of the lungs (Pneumatikos et al., 2009). There are several bacteria that have been found to colonize ETTs and develop in the form of a biofilm, such as the multidrug-resistant MRSA as well as Gram-negative bacilli including *K. pneumoniae*, *E. coli*, *P. aeruginosa*, and *Acinetobacter* spp. (Bauer et al., 2002; Percival et al., 2015; Ramirez, Ferrer, & Torres, 2007). The development of biofilms on ETTs has previously been linked to ventilator-associated pneumonia (VAP), with an incidence of 28% among mechanically ventilated patients (Ruffell & Adamcova, 2008; Wilson, Gray, Karakiozis, & Thomas, 2012). VAP significantly increases healthcare costs due to prolonged hospitalization and has a high mortality rate of 75% (Cairns et al., 2011).

4.3 Urinary catheters

Urinary catheters consist of flexible tubes, which are inserted through the urethra and connected to a drainage pouch, allowing for the urinary bladder to be emptied. Catheters are one of the most commonly used medical devices, with 15%–25% of patients receiving indwelling urinary catheters for urine retention during hospitalization (Warren, 2001). The use of indwelling urinary catheters can often lead to catheter-associated urinary tract infection (CAUTI), with 80% of nosocomial urinary tract infections and 36% of all healthcare infections being associated with their use (Greene, Marx, & Oriola, 2008; Nicolle, 2014).

The formation of a biofilm on urinary catheter surfaces is the most relevant cause of bacteriuria (Stickler, 2008). Microorganisms that are most commonly responsible for the biofilm formation on urinary catheters are Gram-negative bacteria such as *E. coli*, *S. epidermidis*, *P. aeruginosa* (Niveditha, Pramodhini, Umadevi, Kumar, & Stephen, 2012). The deposition of organic molecules from urine components, such as proteins and electrolytes, facilitates the development of biofilms in indwelling urinary catheters (Trautner & Darouiche, 2004).

It has also been observed that the production of urease by certain bacteria increases urinary pH, which leads to the formation of crystalline biofilms within the catheter (Stickler, 2008). Damage to the bladder and urethral epithelia may occur as the crystalline biofilms can develop on the outer surface of the catheter, around the balloon and the catheter tip (Percival 2015; Stickler, 2008). The flow of urine through the catheter may also be impaired as these crystalline biofilms are able to obstruct the catheter lumen (Percival, Sabbuba, Kite, & Stickler, 2009; Stickler, 2008).

4.4 Orthopedic devices

Orthopedic devices are used in the prevention and treatment of a variety of musculoskeletal pathologies. However, due to the disruption of the skin barrier upon insertion of the device, there is a risk of microbial contamination, which can lead to infection. The emergence of complex infections following orthopedic implant surgery is a cause for concern. Implant infections have a substantial effect on patients' quality of life and incur significant costs for healthcare services worldwide (Gibb & Hadjiargyrou, 2021).

Biofilms are heavily associated with orthopedic-device-related infections and osteomyelitis (Gristina, Oga, Webb, & Hobgood, 1985; Matthews, Berendt, McNally, & Byren, 2009). Biofilms found on prosthetic joints can be either mono or polymicrobial and are most often caused by *S. aureus*, beta-hemolytic *Streptococci*, and aerobic Gram-negative rods (Berbari et al., 1998). The growth of bacterial biofilms in the joint space is also considered to be responsible for around 65% of prosthetic joint infections, which commonly cause the device to fail, necessitating implant replacement and frequently resulting in chronic and/or recurrent illness (Arciola, Campoccia, Ehrlich, & Montanaro, 2015; Gbejuade et al., 2015). In fact, prosthetic joint infections are responsible for 15% of all hip revision and 25% of all knee revision surgeries (Bozic et al., 2009; Gbejuade et al., 2015).

5. Diagnosis of medical-device-related biofilms

Although a variety of methods have been developed to aid the detection of biofilms on medical devices, accurate diagnosis remains a challenge. Diagnostic strategies for device-associated infections with microbial biofilms typically involve investigating the surface of the device. Although sampling the surface of the device may sometimes be required, it largely depends on the type of indwelling medical device as to what procedure is best suited. In order to directly sample the surface, surgery is required, which can often cause complications for the patient. For example, with a joint prosthesis, a swab sample would not be suitable due to the high rate of false-negative results as microorganisms are exceedingly difficult to detach from the surface sample (Bjerkan, Witsø, & Bergh, 2009; Tunney et al., 1998). However, a preoperative culture of aspirated synovial fluid may be more accurate (Gomez & Patel, 2011).

Although swab cultures offer a fast, cost-effective approach, they are generally not recommended as negative results are not always confirmatory due to slow-growth rates of microbes in biofilms (Li, Mendis, Trigui, Oliver, & Faucher, 2014). These can result in a viable but nonculturable state of the microbes, thus making diagnosis difficult as they are undetectable with standard growth media. Bacteria can develop in multispecies biofilms, which require specific growth factors, thus further preventing accurate detection (Hall-Stoodley et al., 2012). However, bacterial recovery rates can be increased considerably through the use of staining. This approach supports the visualization of the biofilm on the sample area rather than utilizing randomized sampling and is therefore more accurate.

Sonication may also be considered to further enhance the sensitivity of cultured samples as it displaces microorganisms embedded in the biofilm. This method has been found to be more effective in improving sensitivity of microbiological tests, than utilizing physical removal via surgical blade (Bjerkan et al., 2009; Bozic & Ries, 2005).

Molecular methods may provide a better approach to diagnostics in MDRIs than standard microbiological methods as they offer better detection and identification of bacteria. Polymerase chain reaction (PCR) is one of the most widely used methods as it is fast and offers high sensitivity with the ability to detect small amounts of microorganisms in sonicated fluid. However, there are some limitations to this method, which may compromise the specificity of this approach. As such, when clinical materials are analyzed directly, some 16S rRNA gene PCR primers cross-react with human DNA, thus reducing specificity (Huys et al., 2008).

Fluorescence in situ hybridization (FISH) or nucleic acid staining is also a commonly used molecular method for the identification, visualization, and quantification of microbes in samples (Lukumbuzya, Schmid, Pjevac, & Daims, 2019). FISH offers high sensitivity and specificity as well as enabling the localization, visualization of targeted bacteria inside complex microenvironments including multispecies biofilms (Almstrand, Daims, Persson, Sörensson, & Hermansson, 2013; Amann, 1995; Mark Welch, Rossetti, Rieken, Dewhirst, & Borisy, 2016). Regular PCR alone is susceptible to contamination; however, when paired with FISH, it provides highly effective differentiation between contaminant bacterial cells and actual infection

(Aistleitner et al., 2018; Eichinger, Kikhney, Moter, Wießner, & Eichinger, 2019; Frickmann et al., 2017).

6. Current treatments

The treatment of biofilm-related infections on implanted medical devices has thus far depended on conventional antimicrobial or surgical intervention. Treatment of implant-associated infections currently depends highly on conventional antimicrobial therapies, involving the administration of high-dose antibiotics, posing a significant threat to the patients' health due to the selective pressure toward antibiotic-resistant strains (Sunarintyas, 2016). The alternative if symptoms persist is to carry out costly and risky surgical replacement of the contaminated medical device. However, reinfection of the new devices is extremely likely (Fux, Stoodley, Hall-Stoodley, & Costerton, 2003). The combination of the increase in antimicrobial resistance toward the high-dose antibiotics and the likelihood for reinfection following replacement has meant treatment of contaminated medical devices poses a significant issue for treating patients (Gupta et al., 2016). Due to the associated morbidity and mortality rates that arise from acute and chronic infections worldwide, there is an ever-increasing pressure for alternative treatment and control methods. As a result, vast amounts of research are currently being carried out in order to investigate cutting-edge strategies to prevent and treat the formation of biofilm-based infections on medical devices (Khatoon et al., 2018). At present, the main strategies and areas of investigation addressing biofilm-associated infections include that of prevention, diagnosis, and treatment. Some areas of prevention have included surface modification, antibacterial coating, and biofilm disruption, as it is believed it will prove easier, safer, and more cost-effective than treating a biofilm that has already formed. Areas of research for diagnosis have included that of biosensors for early detection, advanced imaging, and theranostic nanoparticles. Treatment areas being investigated include that of antimicrobial PDT, bioacoustics, electric, and magnetic fields, and the dissolution of EPS (Khatoon et al., 2018).

6.1 Prevention

6.1.1 Antibacterial coatings

The coating of implanted medical devices with antibacterial and antibiofilm complexes has been one of the most promising approaches for the future treatment and prevention of biofilm-associated infections (Veerachamy et al., 2014). Inhibiting the initial attachment of planktonic cells to the medical device surface ensures that no formation of a biofilm is observed. Numerous strategies are under investigation and have been designed to produce antibiofilm coatings, utilizing both natural and synthetic materials (Khatoon et al., 2018).

The addition of antimicrobials to the surface of medical devices, such as catheters, has been the focus of much research and can be addressed in three ways: the agent can be applied as a thin film on the surface of the catheter, ionically bound to the surface, or bound to the surface within a polymer matrix (Shintani, 2004). A number of factors influence the effectiveness of catheters, for example, treated with an antiinfective agent; the solubility, the hydrophilicity, and the affinity to surrounding tissue are all factors that affect the release and duration of antiinfective activity (Shintani, 2004). Preliminary research into antimicrobial-impregnated devices, such as that demonstrated by Darouiche et al. (1999), compared the effectiveness of minocycline/rifampicin and chlorhexidine/silver sulfadiazine-impregnated CVCs in terms of their action against catheter colonization and bloodstream infection. It was observed that catheters impregnated with minocycline/rifampicin were associated with a lower rate of infection than catheters impregnated with chlorhexidine/silver sulfadiazine. However, the chlorhexidine/silver sulfadiazine-impregnated catheters were only treated with the antimicrobials on the external surface, whereas the comparator catheters were treated with minocycline/rifampicin on both the luminal and external surfaces. In addition, the concentration and availability of the antimicrobials used in the minocycline/rifampicin catheters were considerably higher than those in the chlorhexidine/silver sulfadiazine-treated catheters (Darouiche et al., 1999; Yousif, Jamal, & Raad, 2015).

Nanomaterials, nanofilms, nanocoatings, and nanostructures surfaces are being widely studied for the implementation into medical devices surfaces (Veerachamy et al., 2014). The antibacterial properties of silver have been acknowledged for some time, and nanosilver coatings have been implemented into several devices including that of catheters, heart valves, and wound dressings, being shown to be efficacious (Barnes & Cooper, 2014). Silver nanoparticles have a unique mechanism for eradicating bacterial cells, involving binding to the cell wall and therefore causing membrane disruption (Veerachamy et al., 2014) and the accumulation of peroxides that oxidize the cell walls (Ahumada, Bohne, Oake, & Alarcon, 2018), attacking the respiratory chain of the bacterial cells, and cell disruption via hydroxyl radicals and other reactive oxygen species. It has also been demonstrated that silver nanoparticles also have antibacterial and antibiofilm properties that result in synergistic effects with particular antibiotics (Alarcon, Griffith, & Udekwu, 2015). However, it is reported that nanosilver applications are most effective when applied for short-term purposes rather than long term due to the extent of their biocompatibility being unknown, therefore making them most viable for the treatment of surgical site infections (An & Friedman, 1998).

Conventional approaches mainly consist of the deposition of metallic silver on the device polymer surface. In this regard, silver-impregnated ETTs delay microbial colonization in animal studies (Olson, Harmon, & Kollef, 2002), and this effect has been replicated in human studies. Furthermore, a randomized control study reported that patients intubated for 24 h or longer with a silver-coated ETT indeed had significantly lower colonization rates than control groups and confirmed that reduction in microbial bacterial colonization and biofilm formation could lower the incidence of VAP (Kollef et al., 2008). Interesting results in this field are also produced by directly incorporating silver ions in polymeric substrates, in order to obtain medical devices refractory to microbial colonization. In fact, preclinical studies have shown that the presence of silver-impregnated ETTs can have a positive effect in reducing colonization and biofilm development of *P. aeruginosa* in an in vitro model of the early pathogenesis of VAP (Rello et al., 2010). Despite these promising in vitro results, larger randomized controlled trials, such as that of Pickard and colleagues, demonstrated that silver-alloy-coated catheters were not effective in reducing the incidence of CAUTI, casting doubt over the routine use of antimicrobial-treated catheters (Pickard et al., 2012).

Other examples of coatings that have been implemented in medical devices include that of hydroxyapatite coatings, which have been widely implemented due to their ability to be altered by surface adsorbed antibiotics by immersion in antibiotic solutions, thus producing antibiotic hydroxyapatite-based coatings (Veerachamy et al., 2014).

Other strategies that have been investigated to reduce the initial adhesion of microorganism to catheters both in vitro and in vivo include that of hydrophilic polymers such as hyaluronic acid, hydrogels, and heparin coatings (Francolini & Donelli, 2010). While there are a variety of silane coatings, nanoplasma trimethyl silane coatings specifically have been shown to be most effective on hydrophilic surfaces and stainless steel to *prevent S. epidermidis* biofilms (Gupta et al., 2016). In order to address long-term antifouling, some biomaterials have also been treated with ceramics such as calcium phosphate and other biodegradable polymers, which can serve as an effective coating (Chen et al., 2013). Surfaces have also been functionalized with antiadhesive high-density polymers that present steric repulsion in order to make bacterial attachment more difficult, serving as another potential coating (Neoh, Wang, & Kang, 2015).

6.1.2 Surface modification of biomaterials

Surface modification of biomaterials used in indwelling medical devices has also been another area of prevention that has been widely investigated. Surface modification of the biomaterial is performed by utilizing the nonadhesive properties that are part of the material itself. With numerous techniques, including that of matrix-assisted pulsed laser evaporation, research has been directed into modifying biomaterial surfaces to prevent initial attachment of microorganisms (Khatoon et al., 2018). Properties of the medical device biomaterials such as surface roughness, surface area, surface energy, and hydrophilicity can enhance or lessen protein absorption and microbial attachment (Moriarty, Zaat, & Busscher, 2012). Delaviz, Finer, and Santerre (2014) demonstrated that increasing the material's stiffness has previously demonstrated an increase in bacterial adhesion for *S. epidermidis* and *E. coli* species, therefore such correlation can help modify and optimize the surface in order to prevent adhesion on biomaterial surfaces in order to prevent biofilm prevention (Kannan, 2015).

The modification of a medical device surface with a hydrogel is one way in which biocompatibility can be achieved; a hydrogel is a hydrophilic polymer that has a capacity to absorb large quantities of water and therefore can promote a soft surface that helps minimize microbial colonization (Shintani, 2004). A number of studies, both in vivo and in vitro, have demonstrated the effectiveness of hydrogel and silver-hydrogel catheters through reduced microbial colonization (Ahearn et al., 2000; Bull, Chilton, Gould, & Sutton, 1991; Lai & Fontecchio, 2002).

More recently, modifications to ETT surfaces at the nanoscale level have also been researched. A study by Machado and coworkers explored the effect of nanomodified ETT on *S. aureus* biofilm formation (Machado, Tarquinio, & Webster, 2012). This study showcased the creation of a textured, nanomodified surface using a *Rhizopus arrhizus* lipase that was able to enzymically degrade the polyvinylchloride material of the ETT. The nanomodified ETT was exposed to a constant flow of *S. aureus* medium and incorporated in an airway model. The results showed significantly reduced colony-forming units for bacteria in the nanomodified ETT when compared with the untreated control. In addition, there was an increase in protein absorption by the nanomodified ETT, which the authors hypothesized may prevent the colonization and formation of biofilms (Machado et al., 2012).

6.1.3 Disruption of biofilm formation and antibiotic enhancers

Due to major role played by QS systems in both biofilm development and microbial virulence, much attention has been focused on QS regulators in the formation and resilience of biofilm-based medical-device-related infections (Percival et al., 2015). One strategy to prevent QS is to inhibit the signaling molecules involved from binding to their receptors on the bacterial cell surface, which can be achieved by using analog of the signal molecules that compete for the binding sites. Hence, the use of molecules that interfere with QS is a promising method to prevent biofilm formations (Francolini & Donelli, 2010). In turn, if the formation of aggregates is prevented or the EPS is dissolved, the exposed bacterial cells can be susceptible to therapies once again (Bjarnsholt, 2013.)

Some examples of this have been observed, with Vuong et al demonstrating that the *S. epidermidis* QS regulator agr has been shown to be involved in biofilm detachment, and an isogenic agr mutant showed increased biofilm development. By disabling the agr QS regulator, *S. epidermidis* seems to generate an increased capacity for biofilm development on medical devices (Vuong, Kocianova, Yao, Carmody, & Otto, 2004). The aim of QS inhibitors is to enhance the susceptibility of the biofilm to antimicrobials (Bjarnsholt & Givskov, 2008). Three main pharmacological targets of QS systems include the signal generator, the QS molecule itself, and the signal receptor (Rasmussen & Givskov, 2006). Christensen and colleagues showed that *P. aeruginosa* biofilms in an in vivo mouse model could be disrupted by the use of the antibiotic tobramycin and several QS molecules, including furanone and horseradish juice extract. Synergy was seen between both treatments, and the presence of QS inhibitor molecules increased the susceptibility of the *P. aeruginosa* biofilm to tobramycin (Christensen et al., 2012).

Similarly, Qin et al. demonstrated that *S. epidermidis* biofilm formation could be inhibited on polystyrene, glass, and mica surfaces if treated with two benzoate derivatives. In addition, it was demonstrated that two carboxamide derivatives could inhibit initial adhesion and cell division by reacting with the bacterial cells. However, it was found that these compounds did not affect preestablished biofilms (Qin et al., 2009).

6.2 Diagnosis

Currently, there is no gold standard detection method available for the diagnosis of a biofilm within an implanted medical device. The use of traditional culture methods to determine colonization is not indicative of biofilm growth (Hall-Stoodley et al., 2012). Furthermore, negative results from swab samples may not necessarily imply the absence of an infection but could possibly be due to the slow growth rate within a biofilm of species that cannot be detected within the usual detection range (Lindsay & Von Holy, 2006).

To improve the diagnosis of device-related infections, methods such as the sonication of suspected infected implants may improve culture positivity (Achermann, Vogt, Leunig, Wüst, & Trampuz, 2010). In addition to this, more sophisticated molecular methods of identification, including PCR and fluorescence in situ hybridization, are now being used to identify bacteria in complex biological samples and are proving to be a more accurate means of detection. Several publications indicate differences between culture and molecular diagnostic methods. For example, in otitis media with effusion, pathogens are identified around 25%–30% of the time by culture methods and 80%–100% of the time by culture-independent molecular methods (Hall-Stoodley, Costerton, & Stoodley, 2004).

6.2.1 Biosensors for early detection

The use of biosensors and hyperspectral imaging methods for the early diagnosis of biofilm-related infections is also an active field of research for the diagnosis, prevention, and control of biofilm growth on medical device material surfaces such as stainless steel, titanium, and titanium alloys (Kim, Kim, & Hwang, 2012). Current diagnostic tools include blood tests for increased leukocytes, scanning electron microscopy (SEM), C-T scans, MRIs, atomic force microscopy (AFM), and colony-forming unit (CFU) counting (Ribeiro, Monteiro, & Ferraz, 2012); however, biofilm-associated infections are still not effectively diagnosed. New diagnosis and quantification techniques that have been investigated include improved CFU counting, light microscopy, SEM, confocal scanning laser microscopy, AFM, Fourier transform infrared spectroscopy, radiolabeling, contact angle measurements, and qPCR (Ribeiro et al., 2012). Research into impedance microbiology, a method that can detect the presence of microorganisms in samples, has been used to develop an interdigitated microelectrode sensor system that showed low frequencies (10–100 Hz) to be more sensitive for *S. epidermidis* detection (Paredes, Becerro, & Arana, 2014). The use of surface-enhanced Raman spectroscopy (SERS) for the early detection of biofilm-associated infection has also been widely researched for molecular sensing. Raman spectra usually contain many sharp peaks that correspond to specific molecular vibrational frequencies that provide detection of the presence of specific molecules in a sample (Paredes et al., 2014).

6.3 Treatment

Currently, treatment of biofilm infection requires highly sensitive and well-penetrating antibiotics to ensure a sufficient concentration of effective antibiotic for medical-device-associated biofilm infections. In general, macrolides, lincosamides, tetracyclines, rifamycins, quinolones, fusidic acid, nitroimidazole, sulfonamides, and oxazolidinones penetrate better in tissues and cells than betalactam (including penicillins, cephalosporins, and carbapenems), aminoglycosides, glycopeptide, and polymyxin. Biofilm infection may lead to inflammation, resulting in faster metabolism and significant consumption of oxygen locally or systemically. Therefore, if oxygen supply does not meet the demand, glycolysis will be activated leading to acidosis, therefore the antibiotics will potentially be affected by pH values. As Laub and Schneider have previously demonstrated that the presence of low pH may decrease the effects of β-lactam antibiotics while increasing the effects of rifamycin. Hence, antibiotic treatment and correction of acid-base balance disorders may potentially be important for the treatment of biofilm-associated infections (Wu, Moser, Wang, Høiby, & Song, 2015).

6.3.1 Bioacoustics effect

Research has been demonstrated that ultrasonication (500kHz) in combination with antibiotics increases the transport of antibiotics across biofilms, a process termed the bioacoustic effect (Vergidis & Patel, 2012). Freebairn et al. (2013) demonstrated that the use of a bioacoustic effect also effectively bypasses the conditioning film, therefore preventing the surface adhesion of microorganisms including *E. coli* and *P. aeruginosa*. In a study by Ensing et al. (2005), it was demonstrated that ultrasound waves in combination with gentamycin entrapped in bone cements were able to prevent 70% of biofilm formation in a rabbit model. Research by Kopel, Degtyar, and Banin (2011) also observed that on an indwelling catheter, which transmitted low-frequency surface acoustic waves, it was able to eradicate more than 85% of *E. coli*, *S. epidermidis*, and *P. aeruginosa* biofilms when applied with gentamicin. Another study reported on the use of low-frequency vibration therapy in successfully assisting tobramycin in killing *P. aeruginosa* biofilms at subminimal inhibitory concentrations (Bandara, Harb, Kolacny, Martins, & Smyth, 2014). These applications of these technologies may be used as efficient noninvasive alternatives for treating implant-associated infections that have shown antibiofilm effects when combined with antibiotics.

6.3.2 Antimicrobial photodynamic therapies

Photodynamic therapy is based on the production of reactive oxygen species when light-absorbing compounds (photosensitizers) react with light and oxygen. When this technique is used to eradicate bacterial cells, it is termed antimicrobial photodynamic therapy. Various light sources have been used in in vitro and in vivo studies, such as yttrium aluminum garnet (YAG) lasers (Saffarpour et al., 2016), potassium yttrium tungstate (KYW) lasers, femtosecond (Cunha et al., 2016) and near-infrared lasers, demonstrating the potential of photodynamic therapy to treat and control biofilm-based infections of medical devices. The use of photodynamic therapy has been shown to have an antimicrobial effect on antibiotic-resistant *P. aeruginosa* and clinical MRSA biofilms grown on ETTs (Percival, Mccarty, Hunt, & Woods, 2014). A methylene-blue-based photosensitizer in the lumen of the infected ETT before exposing the ETT to light from a fiber optic diffuser at 644nm wavelength, resulting in a 99.9% reduction in polymicrobial biofilm growth. A wide range of light in the electromagnetic spectrum has been shown to be useful in destroying bacteria, including the higher-energy UVC and blue light, when combined with antibiotics (Yin et al., 2013). However, despite these noninvasive technologies seeming to be effective, further research is required in their cytotoxicity in particular.

7. Future prospects

Biofilms are of particular importance in the control of healthcare-associated and other infections. This is not only due to their aptitude to act as a safe haven for those microorganisms that are of public health significance, but also due to their inherent tolerance of and "resistance" to traditional antimicrobial treatment (Percival et al., 2015). The diagnosing of a biofilm infection represents an area of grave concern, with identification significantly easier on abiotic surfaces following removal of a medical device than on biotic surfaces. For example, evidencing a biofilm within a chronic wound is complex and presently there is no "gold standard" (Percival et al., 2014). With this in mind, focus upon diagnostic methods that incorporate routine microbiological procedures with more sophisticated methods that are low-cost, reliable, and easily incorporated into routine clinical laboratory testing should be considered. However, the initial prevention of biofilm formation on medical devices represents the most important aspect in reference to tackling biofilm-associated infections. This can be achieved by a number of techniques, in particular, devices that incorporate antimicrobial coatings and surface alterations or pharmacological inhibitors of QS molecules provide promising opportunities. Consequently, for biofilm-associated infections of medical devices, there is still a major need to further develop the understanding of microbial interaction with biotic and abiotic surfaces and of how the adverse environment of the host affects microbial survival, proliferation, and recalcitrance (Percival et al., 2015). Therefore, future research should aim to delve deeper into our understanding of microbial biofilms and their interactions with biotic and abiotic surfaces and further developing possible control strategies such as the use of antimicrobial-treated medical devices for biofilm prevention. An ideal indwelling medical device would possess surfaces that are similar to those of a healthy human, limiting bacterial adhesion and thus preventing infection (Percival et al., 2015). To achieve biocompatibility, the surface of the medical device should be smooth and uniform to allow the growth of healthy tissue and evasion of invading pathogens. In addition to the use of antiinfective agents, consideration of the surface physicochemical properties of the medical devices is also key and can help to overcome the limitations associated with medical devices pretreated with antimicrobials (Percival et al., 2015).

The use of probiotics to inhibit biofilm formation has attracted interest in the medical world in recent years and may potentially demonstrate itself to be a novel area of treatment for biofilm-associated infections of medical devices going forward. However, the current knowledge is limited and has not yet been properly reviewed (Carvalho, Teixeira-Santos, Mergulhão, & Gomes, 2021). Probiotics are able to competitively attach to medical equipment, thus preventing pathogenic microbes from attaching and forming biofilms (Barzegari et al., 2020). Through utilizing a variety of mechanisms, probiotics inhibit bacterial adhesion to surfaces, prevent QS, biofilm formation, and the survival of the biofilm microorganisms, eventually leading to biofilm removal (Barzegari et al., 2020). Many metabolites with antimicrobial and antibiofilm properties such as antimicrobial peptides, biosurfactants, and hydrogen peroxide are known to be produced by probiotics and are often released into the surrounding medium to help aid pathogenic biofilm eradication (Bermudez-Brito, Plaza-Diaz, Muñoz-Quezada, Gómez-Llorente, & Gil, 2012).

References

Abdel-Aziz, S. M., & Aeron, A. (2014). Bacterial biofilm: Dispersal and inhibition strategies. *Scholarena Journal of Biotechnology*, *1*(1), 105.

Abebe, G. M. (2020). The role of bacterial biofilm in antibiotic resistance and food contamination. *International Journal of Microbiology*, *2020*.

Abraham, N. M., & Jefferson, K. K. (2012). *Staphylococcus aureus* clumping factor B mediates biofilm formation in the absence of calcium. *Microbiology*, *158*(Pt 6), 1504.

Achermann, Y., Vogt, M., Leunig, M., Wüst, J., & Trampuz, A. (2010). Improved diagnosis of periprosthetic joint infection by multiplex PCR of sonication fluid from removed implants. *Journal of Clinical Microbiology*, *48*(4), 1208–1214.

Agarwal, A., Singh, K. P., & Jain, A. (2010). Medical significance and management of staphylococcal biofilm. *FEMS Immunology and Medical Microbiology*, *58*, 147–160.

Ahearn, D. G., Grace, D. T., Jennings, M. J., Borazjani, R. N., Boles, K. J., Rose, L. J., et al. (2000). Effects of hydrogel/silver coatings on in vitro adhesion to catheters of bacteria associated with urinary tract infections. *Current Microbiology*, *41*(2), 120–125.

Ahumada, M., Bohne, C., Oake, J., & Alarcon, E. I. (2018). Protein capped nanosilver free radical oxidation: Role of biomolecule capping on nanoparticle colloidal stability and protein oxidation. *Chemical Communications*, *54*(37), 4724–4727.

Aistleitner, K., Jeske, R., Wölfel, R., Wießner, A., Kikhney, J., Moter, A., et al. (2018). Detection of *Coxiella burnetii* in heart valve sections by fluorescence in situ hybridization. *Journal of Medical Microbiology*, *67*(4), 537–542.

Alarcon, E. I., Griffith, M., & Udekwu, K. I. (2015). *Silver nanoparticle applications*. Springer International Publishing.

Almatroudi, A., Hu, H., Deva, A., Gosbell, I. B., Jacombs, A., Jensen, S. O., et al. (2015). A new dry-surface biofilm model: An essential tool for efficacy testing of hospital surface decontamination procedures. *Journal of Microbiological Methods*, *117*, 171–176.

Almstrand, R., Daims, H., Persson, F., Sörensson, F., & Hermansson, M. (2013). New methods for analysis of spatial distribution and coaggregation of microbial populations in complex biofilms. *Applied and Environmental Microbiology*, *79*(19), 5978–5987.

Amann, R. I. (1995). Fluorescently labelled, rRNA-targeted oligonucleotide probes in the study of microbial ecology. *Molecular Ecology*, *4*(5), 543–554.

An, Y. H., & Friedman, R. J. (1998). Concise review of mechanisms of bacterial adhesion to biomaterial surfaces. *Journal of Biomedical Materials Research*, *43*(3), 338–348.

Andersen, J. B., Hultqvist, L. D., Jansen, C. U., Jakobsen, T. H., Nilsson, M., Rybtke, M., et al. (2021). Identification of small molecules that interfere with c-di-GMP signaling and induce dispersal of *Pseudomonas aeruginosa* biofilms. *NPJ Biofilms Microbiomes*, *7*, 59. https://doi.org/10.1038/s41522-021-00225-4.

Arciola, C. R., Campoccia, D., Ehrlich, G. D., & Montanaro, L. (2015). Biofilm-based implant infections in orthopaedics. In *Vol. 1. Biofilm-based healthcare-associated infections* (pp. 29–46). Springer-Verlag Berlin Heidelberg.

Armbruster, C. E., Hong, W., Pang, B., Weimer, K. E., Juneau, R. A., Turner, J., et al. (2010). Indirect pathogenicity of *Haemophilus influenzae* and *Moraxella catarrhalis* in polymicrobial otitis media occurs via interspecies quorum signaling. *MBio*, *1*(3), e00102–e00110.

Artini, M., Cicatiello, P., Ricciardelli, A., Papa, R., Selan, L., Dardano, P., et al. (2017). Hydrophobin coating prevents *Staphylococcus epidermidis* biofilm formation on different surfaces. *Biofouling*, *33*(7), 601–611.

Bandara, H. M. H. N., Harb, A., Kolacny, D., Martins, P., & Smyth, H. D. C. (2014). Sound waves effectively assist tobramycin in elimination of *Pseudomonas aeruginosa* biofilms in vitro. *AAPS PharmSciTech*, *15*(6), 1644–1654.

Barnes, L., & Cooper, I. (Eds.). (2014). *Biomaterials and medical device-associated infections* Elsevier.

Barzegari, A., Kheyrolahzadeh, K., Hosseiniyan Khatibi, S. M., Sharifi, S., Memar, M. Y., & Zununi Vahed, S. (2020). The battle of probiotics and their derivatives against biofilms. *Infection and Drug Resistance*, *13*, 659–672.

Bauer, T. T., Torres, A., Ferrer, R., Heyer, C. M., Schultze-Werninghaus, C., & Rasche, K. (2002). Biofilm formation in endotracheal tubes. Association between pneumonia and the persistence of pathogens. *Monaldi Archives for Chest Disease*, *57*(1), 84–87.

Berbari, E. F., Hanssen, A. D., Duffy, M. C., Steckelberg, J. M., Ilstrup, D. M., Harmsen, W. S., et al. (1998). Risk factors for prosthetic joint infection: Case-control study. *Clinical Infectious Diseases*, *27*(5), 1247–1254.

Bermudez-Brito, M., Plaza-Diaz, J., Muñoz-Quezada, S., Gómez-Llorente, C., & Gil, A. (2012). Probiotic mechanisms of action. *Annals of Nutrition and Metabolism*, *61*, 160–174.

Bjarnsholt, T. (2013). The role of bacterial biofilms in chronic infections. *APMIS*, *121*, 1–58.

Bjarnsholt, T., & Givskov, M. (2008). Quorum sensing inhibitory drugs as next generation antimicrobials: Worth the effort? *Current Infectious Disease Reports*, *10*(1), 22–28.

Bjerkan, G., Witsø, E., & Bergh, K. (2009). Sonication is superior to scraping for retrieval of bacteria in biofilm on titanium and steel surfaces in vitro. *Acta Orthopaedica*, *80*(2), 245–250.

Bozic, K. J., Kurtz, S. M., Lau, E., Ong, K., Vail, T. P., & Berry, D. J. (2009). The epidemiology of revision total hip arthroplasty in the United States. *Journal of Bone and Joint Surgery*, *91*(1), 128–133.

Bozic, K. J., & Ries, M. D. (2005). The impact of infection after total hip arthroplasty on hospital and surgeon resource utilization. *The Journal of Bone and Joint Surgery*, *87*(8), 1746–1751.

Bryers, J. D. (2008). Medical biofilms. *Biotechnology and Bioengineering*, *100*(1), 1–18.

Buckingham-Meyer, K., Goeres, D. M., & Hamilton, M. A. (2007). Comparative evaluation of biofilm disinfectant efficacy tests. *Journal of Microbiological Methods*, *70*(2), 236–244.

Bull, E., Chilton, C. P., Gould, C. A. L., & Sutton, T. M. (1991). Single-blind, randomised, parallel group study of the Bard Biocath catheter and a silicone elastomer coated catheter. *British Journal of Urology*, *68*(4), 394–399.

Cai, W., Wu, J., Xi, C., & Meyerhoff, M. E. (2012). Diazeniumdiolate-doped poly (lactic-co-glycolic acid)-based nitric oxide releasing films as antibiofilm coatings. *Biomaterials*, *33*(32), 7933–7944.

Cairns, S., Thomas, J. G., Hooper, S. J., Wise, M. P., Frost, P. J., Wilson, M. J., et al. (2011). Molecular analysis of microbial communities in endotracheal tube biofilms. *PLoS One*, *6*, e14759.

Carvalho, F. M., Teixeira-Santos, R., Mergulhão, F. J., & Gomes, L. C. (2021). The use of probiotics to fight biofilms in medical devices: A systematic review and meta-analysis. *Microorganisms*, *9*(1), 27.

Cerca, N., Pier, G. B., Vilanova, M., Oliveira, R., & Azeredo, J. (2005). Quantitative analysis of adhesion and biofilm formation on hydrophilic and hydrophobic surfaces of clinical isolates of *Staphylococcus epidermidis*. *Research in Microbiology*, *156*(4), 506–514.

Chen, M., Yu, Q., & Sun, H. (2013). Novel strategies for the prevention and treatment of biofilm related infections. *International Journal of Molecular Sciences*, *14*(9), 18488–18501.

Christensen, G. D., Simpson, W. A., Bisno, A. L., & Beachey, E. H. (1982). Adherence of slime-producing strains of *Staphylococcus epidermidis* to smooth surfaces. *Infection and Immunity*, *37*(1), 318–326.

Christensen, L. D., Van Gennip, M., Jakobsen, T. H., Alhede, M., Hougen, H. P., Høiby, N., et al. (2012). Synergistic antibacterial efficacy of early combination treatment with tobramycin and quorum-sensing inhibitors against *Pseudomonas aeruginosa* in an intraperitoneal foreign-body infection mouse model. *Journal of Antimicrobial Chemotherapy*, *67*(5), 1198–1206.

Crowe, J. F., Sculco, T. P., & Kahn, B. (2003). Revision total hip arthroplasty: Hospital cost and reimbursement analysis. *Clinical Orthopaedics and Related Research*, *413*, 175–182.

Cunha, A., Elie, A. M., Plawinski, L., Serro, A. P., do Rego, A. M. B., Almeida, A., et al. (2016). Femtosecond laser surface texturing of titanium as a method to reduce the adhesion of *Staphylococcus aureus* and biofilm formation. *Applied Surface Science*, *360*, 485–493.

Darouiche, R. O., Raad, I. I., Heard, S. O., Thornby, J. I., Wenker, O. C., Gabrielli, A., et al. (1999). A comparison of two antimicrobial-impregnated central venous catheters. *New England Journal of Medicine*, *340*(1), 1–8.

Delaviz, Y., Finer, Y., & Santerre, J. P. (2014). Biodegradation of resin composites and adhesives by oral bacteria and saliva: A rationale for new material designs that consider the clinical environment and treatment challenges. *Dental Materials*, *30*(1), 16–32.

Ding, X., Yin, B., Qian, L., Zeng, Z., Yang, Z., Li, H., et al. (2011). Screening for novel quorum-sensing inhibitors to interfere with the formation of *Pseudomonas aeruginosa* biofilm. *Journal of Medical Microbiology*, *60*(12), 1827–1834.

Doll, K., Jongsthaphongpun, K. L., Stumpp, N. S., Winkel, A., & Stiesch, M. (2016). Quantifying implant-associated biofilms: Comparison of microscopic, microbiologic and biochemical methods. *Journal of Microbiological Methods*, *130*, 61–68.

Donlan, R. M. (2008). Biofilms on central venous catheters: Is eradication possible? *Current Topics in Microbiology and Immunology*, *322*, 133–161.

Donlan, R. M., & Costerton, J. W. (2002). Biofilms: Survival mechanisms of clinically relevant microorganisms. *Clinical Microbiology Reviews*, *15*(2), 167–193.

Duan, K., Dammel, C., Stein, J., Rabin, H., & Surette, M. G. (2003). Modulation of *Pseudomonas aeruginosa* gene expression by host microflora through interspecies communication. *Molecular Microbiology*, *50*(5), 1477–1491.

Eichinger, S., Kikhney, J., Moter, A., Wießner, A., & Eichinger, W. B. (2019). Fluorescence in situ hybridization for identification and visualization of microorganisms in infected heart valve tissue as addition to standard diagnostic tests improves diagnosis of endocarditis. *Interactive Cardiovascular and Thoracic Surgery*, *29*(5), 678–684.

El-Azizi, M. A., Starks, S. E., & Khardori, N. (2004). Interactions of *Candida albicans* with other Candida spp. and bacteria in the biofilms. *Journal of Applied Microbiology*, *96*, 1067–1073.

Ensing, G. T., Roeder, B. L., Nelson, J. L., Van Horn, J. R., Van Der Mei, H. C., Busscher, H. J., et al. (2005). Effect of pulsed ultrasound in combination with gentamicin on bacterial viability in biofilms on bone cements in vivo. *Journal of Applied Microbiology*, *99*(3), 443–448.

Festas, A. J., Ramos, A., & Davim, J. P. (2020). Medical devices biomaterials—A review. *Proceedings of the Institution of Mechanical Engineers, Part L: Journal of Materials: Design and Applications*, *234* (1), 218–228.

Fleming, D., & Rumbaugh, K. P. (2017). Approaches to dispersing medical biofilms. *Microorganisms*, *5*(2), 15.

Flemming, H. C., & Wingender, J. (2010). The biofilm matrix. *Nature Reviews Microbiology*, *8*(9), 623–633.

Floyd, K. A., Eberly, A. R., & Hadjifrangiskou, M. (2017). Adhesion of bacteria to surfaces and biofilm formation on medical devices. In *Biofilms and implantable medical devices* (pp. 47–95). Woodhead Publishing.

Francolini, I., & Donelli, G. (2010). Prevention and control of biofilm-based medical-device-related infections. *FEMS Immunology and Medical Microbiology*, *59*(3), 227–238.

Francolini, I., Vuotto, C., Piozzi, A., & Donelli, G. (2017). Antifouling and antimicrobial biomaterials: An overview. *APMIS*, *125*(4), 392–417.

Freebairn, D., Linton, D., Harkin-Jones, E., Jones, D. S., Gilmore, B. F., & Gorman, S. P. (2013). Electrical methods of controlling bacterial adhesion and biofilm on device surfaces. *Expert Review of Medical Devices*, *10*(1), 85–103.

Frickmann, H., Zautner, A. E., Moter, A., Kikhney, J., Hagen, R. M., Stender, H., et al. (2017). Fluorescence in situ hybridization (FISH) in the microbiological diagnostic routine laboratory: A review. *Critical Reviews in Microbiology*, *43*(3), 263–293.

Fux, C. A., Stoodley, P., Hall-Stoodley, L., & Costerton, J. W. (2003). Bacterial biofilms: A diagnostic and therapeutic challenge. *Expert Review of Anti-Infective Therapy*, *1*(4), 667–683.

Gbejuade, H. O., Lovering, A. M., & Webb, J. C. (2015). The role of microbial biofilms in prosthetic joint infections: A review. *Acta Orthopaedica*, *86*(2), 147–158.

Gibb, B. P., & Hadjiargyrou, M. (2021). Bacteriophage therapy for bone and joint infections. *The Bone & Joint Journal*, *103-B*(2), 234–244.

Gilmore, B. F., & Carson, L. (2015). Bioactive biomaterials for controlling biofilms. In L Barnes, & I Cooper (Eds.), *Biomaterials and medical device-associated infections* (pp. 163–183). Elsevier.

Gomez, E., & Patel, R. (2011). Laboratory diagnosis of prosthetic joint infection, Part I. *Clinical Microbiology Newsletter*, *33*(8), 55–60.

Gominet, M., Compain, F., Beloin, C., & Lebeaux, D. (2017). Central venous catheters and biofilms: Where do we stand in 2017? *APMIS Journal of Pathology, Microbiology and Immunology*, *125*(4), 365–375.

Greene, L., Marx, J., & Oriola, J. (2008). *Guide to the elimination of catheter-associated urinary tract infections (CAUTIs): Developing and applying facility-based prevention interventions in acute and long-term care setting*. Centres for Disease Control and Prevention.

Gristina, A. G., Oga, M., Webb, L. X., & Hobgood, C. D. (1985). Adherent bacterial colonization in the pathogenesis of osteomyelitis. *Science*, *228*(4702), 990–993.

Gupta, P., Sarkar, S., Das, B., Bhattacharjee, S., & Tribedi, P. (2016). Biofilm, pathogenesis and prevention—A journey to break the wall: A review. *Archives of Microbiology*, *198*(1), 1–15.

Ha, D. G., & O'Toole, G. A. (2015). c-di-GMP and its effects on biofilm formation and dispersion: A *Pseudomonas aeruginosa* review. *Microbiology Spectrum*, *3*(2). 3-2.

Hadi, R., Vickery, K., Deva, A., & Charlton, T. (2010). Biofilm removal by medical device cleaners: Comparison of two bioreactor detection assays. *Journal of Hospital Infection*, *74*(2), 160–167.

Hall-Stoodley, L., Costerton, J. W., & Stoodley, P. (2004). Bacterial biofilms: From the natural environment to infectious diseases. *Nature Reviews Microbiology*, *2*(2), 95–108.

Hall-Stoodley, L., Stoodley, P., Kathju, S., Høiby, N., Moser, C., William Costerton, J., et al. (2012). Towards diagnostic guidelines for biofilm-associated infections. *FEMS Immunology and Medical Microbiology*, *65*(2), 127–145.

Harris, L. G., Tosatti, S., Wieland, M., Textor, M., & Richards, R. G. (2004). *Staphylococcus aureus* adhesion to titanium oxide surfaces coated with non-functionalized and peptide-functionalized poly (L-lysine)-grafted-poly (ethylene glycol) copolymers. *Biomaterials*, *25* (18), 4135–4148.

Hengge, R. (2009). Principles of c-di-GMP signalling in bacteria. *Nature Reviews Microbiology*, *7*(4), 263–273.

Hobley, L., Harkins, C., MacPhee, C. E., & Stanley-Wall, N. R. (2015). Giving structure to the biofilm matrix: An overview of individual strategies and emerging common themes. *FEMS Microbiology Reviews*, *39* (5), 649–669.

Huys, G., Vanhoutte, T., Joossens, M., Mahious, A. S., De Brandt, E., Vermeire, S., et al. (2008). Coamplification of eukaryotic DNA with 16S rRNA gene-based PCR primers: Possible consequences for population fingerprinting of complex microbial communities. *Current Microbiology*, *56*(6), 553–557.

Jamal, M., Ahmad, W., Andleeb, S., Jalil, F., Imran, M., Nawaz, M. A., et al. (2018). Bacterial biofilm and associated infections. *Journal of the Chinese Medical Association*, *81*(1), 7–11.

Kannan, M. B. (2015). Biodegradable polymeric coatings for surface modification of magnesium-based biomaterials. In T. S Narayanan, Song Park, & Min-Ho Lee (Eds.), *Vol. 2. Surface modification of magnesium and its alloys for biomedical applications* (pp. 355–376). Elsevier.

Khatoon, Z., McTiernan, C. D., Suuronen, E. J., Mah, T. F., & Alarcon, E. I. (2018). Bacterial biofilm formation on implantable devices and approaches to its treatment and prevention. *Heliyon*, *4*(12), e01067.

Khoury, A. E., Lam, K., Ellis, B., & Costerton, J. W. (1992). Prevention and control of bacterial infections associated with medical devices. *American Society for Artificial Internal Organs*, *38*(3), 174–178.

Kim, D. H., Kim, M. S., & Hwang, J. (2012). Monitoring of biofilm formation on different material surfaces of medical devices using hyperspectral imaging method. In *Vol. 8215. Design and quality for biomedical technologies V* (p. 821507). International Society for Optics and Photonics.

Kollef, M. H., Afessa, B., Anzueto, A., Veremakis, C., Kerr, K. M., Margolis, B. D., et al. (2008). Silver-coated endotracheal tubes and incidence of ventilator-associated pneumonia: The NASCENT randomized trial. *JAMA*, *300*(7), 805–813.

Kopel, M., Degtyar, E., & Banin, E. (2011). Surface acoustic waves increase the susceptibility of *Pseudomonas aeruginosa* biofilms to antibiotic treatment. *Biofouling*, *27*(7), 701–711.

Korgaonkar, A., Trivedi, U., Rumbaugh, K. P., & Whiteley, M. (2013). Community surveillance enhances *Pseudomonas aeruginosa* virulence during polymicrobial infection. *Proceedings of the National Academy of Sciences*, *110*(3), 1059–1064.

Kragh, K. N., Alhede, M., Kvich, L., & Bjarnsholt, T. (2019). Into the well—A close look at the complex structures of a microtiter biofilm and the crystal violet assay. *Biofilm*, *1*, 100006.

Lai, K. K., & Fontecchio, S. A. (2002). Use of silver-hydrogel urinary catheters on the incidence of catheter-associated urinary tract infections in hospitalized patients. *American Journal of Infection Control*, *30*(4), 221–225.

Li, L., Mendis, N., Trigui, H., Oliver, J. D., & Faucher, S. P. (2014). The importance of the viable but non-culturable state in human bacterial pathogens. *Frontiers in Microbiology*, *5*, 258.

Lindsay, D., & Von Holy, A. (2006). Bacterial biofilms within the clinical setting: What healthcare professionals should know. *Journal of Hospital Infection*, *64*(4), 313–325.

Lukumbuzya, M., Schmid, M., Pjevac, P., & Daims, H. (2019). A multicolor fluorescence in situ hybridization approach using an extended set of fluorophores to visualize microorganisms. *Frontiers in Microbiology*, *10*, 1383.

Machado, M. C., Tarquinio, K. M., & Webster, T. J. (2012). Decreased *Staphylococcus aureus* biofilm formation on nanomodified endotracheal tubes: A dynamic airway model. *International Journal of Nanomedicine*, *7*, 3741.

Magacho, C. C., Pinto, J. G., Souza, B. M. N., Pereira, A. H. C., & Ferreira-Strixino, J. (2020). Comparison of photodynamic therapy with methylene blue associated with ceftriaxone in gram-negative bacteria; an in vitro study. *Photodiagnosis and Photodynamic Therapy*, *30*, 101691.

Malic, S., Hill, K. E., Hayes, A., Percival, S. L., Thomas, D. W., & Williams, D. W. (2009). Detection and identification of specific bacteria in wound biofilms using peptide nucleic acid fluorescent in situ hybridization (PNA FISH). *Microbiology*, *155*(8), 2603–2611.

Mark Welch, J. L., Rossetti, B. J., Rieken, C. W., Dewhirst, F. E., & Borisy, G. G. (2016). Biogeography of a human oral microbiome at the micron scale. *Proceedings of the National Academy of Sciences of the United States of America*, *113*(6), E791–E800. https://doi.org/10.1073/pnas.1522149113.

Matthews, P. C., Berendt, A. R., McNally, M. A., & Byren, I. (2009). Diagnosis and management of prosthetic joint infection. *BMJ*, *338*.

Moriarty, T. F., Zaat, S. A., & Busscher, H. J. (Eds.). (2012). *Biomaterials associated infection: Immunological aspects and antimicrobial strategies* Springer Science & Business Media.

Murga, R., Miller, J. M., & Donlan, R. M. (2001). Biofilm formation by gram-negative bacteria on central venous catheter connectors: Effect of conditioning films in a laboratory model. *Journal of Clinical Microbiology, 39*(6), 2294–2297.

Neoh, K. G., Wang, R., & Kang, E. T. (2015). Surface nanoengineering for combating biomaterials infections. In *Biomaterials and medical device–associated infections* (pp. 133–161). Elsevier.

Nicolle, L. E. (2014). Catheter associated urinary tract infections. *Antimicrobial Resistance and Infection Control, 3*, 23.

Niveditha, S., Pramodhini, S., Umadevi, S., Kumar, S., & Stephen, S. (2012). The isolation and biofilm formation of uropathogens in the patients with catheter associated urinary tract infections (CAUTI). *Journal of Clinical and Diagnostic Research, 6*, 1478–1482.

Olson, M. E., Ceri, H., Morck, D. W., Buret, A. G., & Read, R. R. (2002). Biofilm bacteria: Formation and comparative susceptibility to antibiotics. *Canadian Journal of Veterinary Research, 66*(2), 86.

Olson, M. E., Harmon, B. G., & Kollef, M. H. (2002). Silver-coated endotracheal tubes associated with reduced bacterial burden in the lungs of mechanically ventilated dogs. *Chest, 121*(3), 863–870.

Paredes, J., Becerro, S., & Arana, S. (2014). Label-free interdigitated microelectrode based biosensors for bacterial biofilm growth monitoring using Petri dishes. *Journal of Microbiological Methods, 100*, 77–83.

Pavithra, D., & Doble, M. (2008). Biofilm formation, bacterial adhesion and host response on polymeric implants—Issues and prevention. *Biomedical Materials, 3*(3), 034003.

Percival, S. L., Hill, K. E., Williams, D. W., Hooper, S. J., Thomas, D. W., & Costerton, J. W. (2012). A review of the scientific evidence for biofilms in wounds. *Wound Repair and Regeneration, 20*(5), 647–657.

Percival, S. L., Mccarty, S., Hunt, J. A., & Woods, E. J. (2014). The effects of pH on wound healing, biofilms, and antimicrobial efficacy. *Wound Repair and Regeneration, 22*, 174–186.

Percival, S. L., Sabbuba, N. A., Kite, P., & Stickler, D. J. (2009). The effect of EDTA instillations on the rate of development of encrustation and biofilms in Foley catheters. *Urological Research, 37*(4), 205–209.

Percival, S. L., Suleman, L., Vuotto, C., & Donelli, G. (2015). Healthcare-associated infections, medical devices and biofilms: Risk, tolerance and control. *Journal of Medical Microbiology, 64*(4), 323–334.

Percival, S. L., Thomas, J. G., & Williams, D. W. (2010). The world of microbiology and biofilmology. In S. L. Percival, J. G. Thomas, & D. W. Williams (Eds.), *Microbiology of wounds* (pp. 1–58). CRC Press.

Percival, S. L., Malone, M., Mayer, D., Salisbury, A. M., & Gregory Schultz, G. S. (2018). Role of anaerobes in polymicrobial communities and biofilms complicating diabetic foot ulcers. *International Wound Journal, 15*, 776–782.

Perry-O'Keefe, H., Stender, H., Broomer, A., Oliveira, K., Coull, J., & Hyldig-Nielsen, J. J. (2001). Filter-based PNA in situ hybridization for rapid detection, identification and enumeration of specific microorganisms. *Journal of Applied Microbiology, 90*(2), 180–189.

Peters, B. M., Ovchinnikova, E. S., Krom, B. P., Schlecht, L. M., Zhou, H., Hoyer, L. L., et al. (2012). *Staphylococcus aureus* adherence to *Candida albicans* hyphae is mediated by the hyphal adhesin Als3p. *Microbiology, 158*, 2975–2986.

Peters, B. M., Ward, R. M., Rane, H. S., Lee, S. A., & Noverr, M. C. (2013). Efficacy of ethanol against *Candida albicans* and *Staphylococcus aureus* polymicrobial biofilms. *Antimicrobial Agents and Chemotherapy, 57*, 74–82.

Pickard, R., Lam, T., MacLennan, G., Starr, K., Kilonzo, M., McPherson, G., et al. (2012). Antimicrobial catheters for reduction of symptomatic urinary tract infection in adults requiring short-term catheterisation in hospital: A multicentre randomised controlled trial. *The Lancet, 380* (9857), 1927–1935.

Pneumatikos, I. A., Dragoumanis, C. K., & Bouros, D. E. (2009). Ventilator-associated pneumonia or endotracheal tube-associated pneumonia? An approach to the pathogenesis and preventive strategies emphasizing the importance of endotracheal tube. *Anesthesiology, 110* (3), 673–680.

Qin, Z., Zhang, J., Hu, Y., Chi, Q., Mortensen, N. P., Qu, D., et al. (2009). Organic compounds inhibiting *S. epidermidis* adhesion and biofilm formation. *Ultramicroscopy, 109*(8), 881–888.

Raad, I., Costerton, W., Sabharwal, U., Sadlowski, M., Anaissie, E., & Bodey, G. P. (1993). Ultrastructural analysis of indwelling vascular catheters: A quantitative relationship between luminal colonization and duration placement. *Journal of Infectious Diseases, 168*, 400–407.

Ramirez, P., Ferrer, M., & Torres, A. (2007). Prevention measures for ventilator-associated pneumonia: A new focus on the endotracheal tube. *Current Opinion in Infectious Diseases, 20*(2), 190–197.

Rao, T. S. (2020). Bacterial biofilms and implant infections: A perspective. *Archives of Orthopaedics, 1*(4).

Rasmussen, T. B., & Givskov, M. (2006). Quorum-sensing inhibitors as anti-pathogenic drugs. *International Journal of Medical Microbiology, 296*(2–3), 149–161.

Ravi, N. S., Aslam, R. F., & Veeraraghavan, B. (2019). *A new method for determination of minimum biofilm eradication concentration for accurate antimicrobial therapy in Acinetobacter baumannii* (pp. 61–67). New York, NY: Humana Press.

Rello, J., Afessa, B., Anzueto, A., Arroliga, A. C., Olson, M. E., Restrepo, M. I., et al. (2010). Activity of a silver-coated endotracheal tube in preclinical models of ventilator-associated pneumonia and a study after extubation. *Critical Care Medicine, 38*(4), 1135–1140.

Ribeiro, M., Monteiro, F. J., & Ferraz, M. P. (2012). Infection of orthopedic implants with emphasis on bacterial adhesion process and techniques used in studying bacterial-material interactions. *Biomatter, 2*(4), 176–194.

Rodis, N., Tsapadikou, V. K., Potsios, C., & Xaplanteri, P. (2021). A review on resistance mechanisms in bacterial biofilm formations. In *Vol. 3. Recent developments in medicine and medical research* (pp. 7–18). BP International.

Römling, U., Gomelsky, M., & Galperin, M. Y. (2005). C-di-GMP: The dawning of a novel bacterial signalling system. *Molecular Microbiology, 57*(3), 629–639.

Ruffell, A., & Adamcova, L. (2008). Ventilator-associated pneumonia: Prevention is better than cure. *Nursing in Critical Care, 13*(1), 44–53.

Ryan, R. P., Fouhy, Y., Garcia, B. F., Watt, S. A., Niehaus, K., Yang, L., et al. (2008). Interspecies signalling via the *Stenotrophomonas maltophilia* diffusible signal factor influences biofilm formation and polymyxin tolerance in *Pseudomonas aeruginosa*. *Molecular Microbiology, 68*(1), 75–86.

Sáenz, A., Rivera, E., Brostow, W., & Castaño, V. M. (1999). Ceramic biomaterials: An introductory overview. *Journal of Materials Education, 21*(5/6), 267–276.

Safdar, N., & Maki, D. G. (2004). The pathogenesis of catheter-related bloodstream infection with noncuffed short-term central venous catheters. *Intensive Care Medicine, 30*, 62–67.

Saffarpour, A., Fekrazad, R., Heibati, M. N., Bahador, A., Saffarpour, A., Rokn, A. R., et al. (2016). Bactericidal effect of erbium-doped yttrium aluminum garnet laser and photodynamic therapy on *Aggregatibacter actinomycetemcomitans* biofilm on implant surface. *International Journal of Oral and Maxillofacial Implants*, *31*(3).

Saïdi, F., Jolivet, N. Y., Lemon, D. J., Nakamura, A., Garza, A. G., Veyrier, F. J., et al. (2020). Bacterial glycocalyx integrity drives multicellular swarm biofilm dynamism. *bioRxiv*.

Sharma, D., Misba, L., & Khan, A. U. (2019). Antibiotics versus biofilm: An emerging battleground in microbial communities. *Antimicrobial Resistance and Infection Control*, *8*, 76.

Shin, D. S., & Eom, Y. B. (2019). Efficacy of zerumbone against dual-species biofilms of *Candida albicans* and *Staphylococcus aureus*. *Microbial Pathogenesis*, *137*, 103768.

Shintani, H. (2004). Modification of medical device surface to attain anti-infection. *Biomaterials and Artificial Organs*, *18*(1), 1–8.

Shunmugaperumal, T. (2010). Microbial colonization of medical devices and novel preventive strategies. *Recent Patents on Drug Delivery & Formulation*, *4*(2), 153–173.

Sibley, C. D., Duan, K., Fischer, C., Parkins, M. D., Storey, D. G., Rabin, H. R., et al. (2008). Discerning the complexity of community interactions using a Drosophila model of polymicrobial infections. *PLoS Pathogens*, *4*(10), e1000184.

Stickler, D. J. (2008). Bacterial biofilms in patients with indwelling urinary catheters. *Nature Clinical Practice Urology*, *5*(11), 598–608.

Stoica, P., Chifiriuc, M. C., Rapa, M., & Lazăr, V. (2017). Overview of biofilm-related problems in medical devices. In *Biofilms and implantable medical devices* (pp. 3–23). Woodhead Publishing.

Sunarintyas, S. (2016). Bioadhesion of biomaterials. In *Biomaterials and medical devices* (pp. 103–125). Cham: Springer.

Trautner, B. W., & Darouiche, R. O. (2004). Role of biofilm in catheter-associated urinary tract infection. *American Journal of Infection Control*, *32*(3), 177–183.

Tunney, M. M., Patrick, S., Gorman, S. P., Nixon, J. R., Anderson, N., Davis, R. I., et al. (1998). Improved detection of infection in hip replacements: A currently underestimated problem. *The Journal of Bone and Joint Surgery. British Volume*, *80*(4), 568–572.

Van Houdt, R., & Michiels, C. W. (2005). Role of bacterial cell surface structures in *Escherichia coli* biofilm formation. *Research in Microbiology*, *156*(5–6), 626–633.

VanEpps, J. S., & Younger, J. G. (2016). Implantable device-related infection. *Shock (Augusta, Ga.)*, *46*(6), 597–608.

Veerachamy, S., Yarlagadda, T., Manivasagam, G., & Yarlagadda, P. K. (2014). Bacterial adherence and biofilm formation on medical implants: A review. *Proceedings of the Institution of Mechanical Engineers, Part H: Journal of Engineering in Medicine*, *228*(10), 1083–1099.

Vega, N. M., & Gore, J. (2014). Collective antibiotic resistance: Mechanisms and implications. *Current Opinion in Microbiology*, *21*, 28–34.

Vergidis, P., & Patel, R. (2012). Novel approaches to the diagnosis, prevention, and treatment of medical device-associated infections. *Infectious Disease Clinics*, *26*(1), 173–186.

Vickery, K., Hu, H., Jacombs, A. S., Bradshaw, D. A., & Deva, A. K. (2013). A review of bacterial biofilms and their role in device-associated infection. *Healthcare Infection*, *18*(2), 61–66.

von Eiff, C., Jansen, B., Kohnen, W., & Becker, K. (2005). Infections associated with medical devices. *Drugs*, *65*, 179–214.

Vuong, C., Kocianova, S., Yao, Y., Carmody, A. B., & Otto, M. (2004). Increased colonization of indwelling medical devices by quorum-sensing mutants of *Staphylococcus epidermidis* in vivo. *The Journal of Infectious Diseases*, *190*(8), 1498–1505.

Warren, J. W. (2001). Catheter-associated urinary tract infections. *International Journal of Antimicrobial Agents*, *17*(4), 299–303.

Waters, C. M., & Bassler, B. L. (2005). Quorum sensing: Cell-to-cell communication in bacteria. *Annual Review of Cell and Developmental Biology*, *21*, 319–346.

Welch, K., Cai, Y., & Strømme, M. (2012). A method for quantitative determination of biofilm viability. *Journal of Functional Biomaterials*, *3*(2), 418–431.

Wi, Y. M., & Patel, R. (2018). Understanding biofilms and novel approaches to the diagnosis, prevention, and treatment of medical device-associated infections. *Infectious Disease Clinics of North America*, *32*(4), 915–929.

Wilson, A., Gray, D., Karakiozis, J., & Thomas, J. (2012). Advanced endotracheal tube biofilm stage, not duration of intubation, is related to pneumonia. *Journal of Trauma and Acute Care Surgery*, *72*(4), 916–923.

Wu, H., Moser, C., Wang, H. Z., Høiby, N., & Song, Z. J. (2015). Strategies for combating bacterial biofilm infections. *International Journal of Oral Science*, *7*(1), 1–7.

Yasir, M., Willcox, M. D. P., & Dutta, D. (2018). Action of antimicrobial peptides against bacterial biofilms. *Materials*, *11*(12), 2468.

Yin, R., Dai, T., Avci, P., Jorge, A. E. S., de Melo, W. C., Vecchio, D., et al. (2013). Light based anti-infectives: Ultraviolet C irradiation, photodynamic therapy, blue light, and beyond. *Current Opinion in Pharmacology*, *13*(5), 731–762.

Yousif, A., Jamal, M. A., & Raad, I. (2015). Biofilm-based central line-associated bloodstream infections. In *Biofilm-based healthcare-associated infections* (pp. 157–179). Springer.

Zhang, K., Li, X., Yu, C., & Wang, Y. (2020). Promising therapeutic strategies against microbial biofilm challenges. *Frontiers in Cellular and Infection Microbiology*, *10*, 359.

Section B

Biofilms in nature

Chapter 8

Metagenomic analysis of electroactive microorganisms in corrosion: Impact of the corrosive biofilms in the oil industry

Luciano Procópio
Industrial Microbiology and Bioremediation Department, Federal University of Rio de Janeiro (UFRJ), Caxias, Rio de Janeiro, Brazil

1. Introduction

The corrosion of metal alloys is a problem that affects several productive sectors, with financial costs of around 4 billion dollars, which represent 3.3% of the world GDP (Baorong et al., 2017). Losses related to corrosion impact critical structural sectors, such as pipelines of offshore oil-gas, ship hulls, water cooling systems, aviation fuel tanks, sewer, and drinking water distribution systems (Okabe, Odagiri, Ito, & Satoh, 2007; Rajasekar, Anandkumar, Maruthamuthu, Ting, & Rahman, 2010; Rauch et al., 2006). However, this value does not take into account the economic losses related to lost productivity, personnel training, research and development, and preventive maintenance, which makes this total increase considerably (Kato, Yumoto, & Kamagata, 2015). Such costs and problems related to corrosion make an effort to employ innumerable methods to prevent or mitigate those in different types of structures. The use of organic and inorganic coatings, plastics, and paints, aiming to prevent direct contact with the metallic surface with the corrosive environment, is still the most used today (Price & Figueira, 2017). Cathodic protection techniques, which consist of creating a galvanic cell, thus acting as a "sacrificial metal," are also widely used, especially in offshore structures, pipelines, and oil tankers (Baeckmann, 1997). Adding these corrosion control costs at the initial amount, the total is believed to reach about $121 billion worldwide (Koch, Brongers, Thompson, Virmani, & Payer, 2018). Although all these sectors mentioned are considerably impacted by corrosion, the oil industry is undoubtedly the main sector of corrosion costs among various industries due to pipelines and downhole applications (Khan, Qurashi, Badeghaish, Noui-Mehidi, & Aziz, 2020).

Metal corrosion is a spontaneous process in which electrons are transferred from an electron donor site, called "anodic site," to an electron acceptor, called "cathodic site." These reactions happen concurrently and are dependent on each other. Frequently, the transfer of electrons from the anodic site to the cathodic site will induce the dissolution of the anodic site or corrosion itself (Gadd, 2004; Hamilton, 2003). However, a considerable part of the corrosion processes is attributed to microbiologically influenced corrosion (MIC). This statement can be corroborated by surveys that indicate that about 20% of corrosion in oil pipelines is directly attributed to MIC (Javaherdashti, 2016). MIC or biocorrosion is characterized by the interaction between microorganisms, in the form of biofilm structures, the corroded metal surface, and the surrounding environment (Procópio, 2019). The action of microorganisms present in biofilms can significantly affect corrosion kinetics, which can accelerate corrosion rates from 100 to 1000 times (Rajala et al., 2015). This is because, inside it, metabolic activities alter conditions such as the availability of dissolved oxygen, pH levels, and the production of organic acids and complexing agents. These factors influence the electrochemical potential current states in a way that influences the states, as well as corrosion kinetics (Beech & Sunner, 2004; Videla & Herrera, 2005). Studies on the identification of microorganisms involved in biocorrosion point to the participation of fungi, such as *Aspergillus*, *Chrysosporium*, and *Arthrinium*, and algae, for example, *Cymbella*, *Bleakeleya*, and *Nitzschia* genera (De Messano, Sathler, Reznik, & Coutinho, 2009; Lugauskas et al., 2009; Mitbavkar & Anil, 2000). However, is indisputable the action of bacteria is preponderant in corrosion, with numerous representatives of different taxonomic groups. Among the species associated with corrosion highlight the species of sulfur-reducing bacteria (SRB), nitrate-reducing bacteria (NRB), iron-oxidizing bacteria (IOB), and acid-producing bacteria groups (Andrews et al., 2013; Enning & Garrelfs, 2014; Iino et al., 2015; Peccia, Marchand, Silverstein, & Hernandez, 2000). In recent years, members of Phylum Archaea have drawn

Understanding Microbial Biofilms. https://doi.org/10.1016/B978-0-323-99977-9.00037-5

the attention in biocorrosion assessments, especially in environmental anaerobic conditions, when methanogenic archaea are described as active participants in the corrosion of metals alongside SRB (Dinh et al., 2004; Larsen et al., 2010).

The relationship between microorganisms present in biofilm and their role in the deterioration of metal surfaces is a challenge in biocorrosion studies, mainly in the analysis of samples from microbial communities in situ. In the last decades, advances in DNA sequencing techniques have allowed providing information from sets of genes present in a given community, such as the sequencing of the 16S rRNA gene. This approach allows to establish a taxonomy assignment to sequenced genes present in the environmental sample under study. Furthermore, it was possible to define correlations among the microbial community and environmental factors (Lian et al., 2019; Shin, Lee, Park, Moon, & Hyun, 2017). However, concerns and challenges related to the use of metagenomics reside in the complexity of the studied habitats. Fluctuations in temperatures, pH, oxygen levels, interactions between microorganisms are factors to be considered when sampling the analyzed community (Goh et al., 2019; Riesenfeld, Schloss, & Handelsman, 2004). In addition, low or high diversity of the microbial community also tends to influence the results obtained. This is because extreme environments, such as high pressure on the deep ocean floor, high soil salinity, favor certain microbes with metabolism adapted to these conditions. On the other hand, highly diverse environments such as oceans, soils, and sediments are more complex and abundant in the genomic repertoire present (Konstantinidis & Tiedje, 2005; White et al., 2016). Both described situations reflect when choosing the metagenomic strategy adopted, e.g., obtaining eDNA (problems during cell lysis, subsequent DNA degradation), and low representativeness of the present community (Wesolowska-Andersen et al., 2014).

Another important aspect in biocorrosion is the active participation of microorganisms capable of using electrodes as electron acceptors (called electrogens) or electron donors (called electrotrophs), in a process called extracellular electron transfer (EET) mechanism (Gralnick & Newman, 2007; Richter, Schicklberger, & Gescher, 2012). A wide variety of microorganisms are described as capable of exchanging electrons with electrodes, which contributes to a wide range of practical applications, such as bioremediation, biofuel production, energy production, biosensorization, and a leading role in inducing corrosion on metal surfaces (Lovley, 2012; Nealson & Rowe, 2016). With the advances in large-scale sequencing techniques, there has been an expansion of the understanding of the direct participation of microorganisms in metal corrosion. These developments demonstrate the importance of the mechanisms employed by electroactive species often described in MIC. Based on these questions, this chapter focuses on the recent progress of metagenomic research on electroactive microbes in corrosive biofilms on metal surfaces.

2. Electroactive microorganisms

The conservation of energy through respiration is an indispensable activity for all living organisms, including microorganisms. A characteristic that makes microorganisms inhabitants of different habitats is their ability to use different inorganic compounds as electron donors. Some microbial groups have the ability to accept or donate the electrons needed to convert energy from extracellular solid compounds, including graphite and metal electrode, in a process called "extracellular electron transfer (EET)" (Bond & Lovley, 2003; Gregory, Bond, & Lovley, 2004). Currently, it is known that innumerable bacteria and archaea are capable of growing pure cultures or in microbial pools using electrons of compounds such as nitrate, sulfate, and different metals (El-Naggar et al., 2010). These microorganisms can transfer electrons to the anode through different paths: (i) direct contact of the cytochromes of the outer membrane on the cell surface; (ii) through endogenous (riboflavin and ferredoxin) and exogenous mediators (ferrocyanide); and (iii) through conductive pili or flagellum that can achieve electron transfer in long range (Fig. 1) (Myers & Myers, 2001; Pirbadian et al., 2014; von Canstein, Ogawa, Shimizu, & Lloyd, 2008). The initial reports on EET came in 1988 with studies directed by two metal-reducing bacterial dissimilatory, the *Shewanella* and *Geobacter* genera (Lovley & Phillips, 1988; Myers & Nealson, 1988). These studies showed that EET is mainly composed of cytochromes, which are present as electron ducts between intracellular catabolic reactions and extracellular conductive materials (Kasai, Kouzuma, Nojiri, & Watanabe, 2015; Kouzuma, Kasai, Hirose, & Watanabe, 2015). Currently, in biofilm samples, other representatives are commonly described, such as members of the Geobacteraceae Family, which is a group of Deltaproteobacteria that gains energy for growth by reducing ferric iron (Fe[III]) minerals (Nealson & Rowe, 2016).

The hypothesis that established the relationship between corrosion of metals and electroactive microorganisms was initially raised by Dinh et al. (2004). In this study, it was demonstrated that some species of SRB and methanogens obtained electrons from metallic iron in a more direct way than via the free hydrogen produced by abiotic corrosion. Other experiments in laboratory conditions confirmed the cathodic effect of some species of the SRB group, in addition to describing the role of iron-reducing bacteria (IRB) and iron oxidizers (IOB) in addition to SRB and methanogens (Kato, 2016; Uchiyama, Ito, Mori, Tsurumaru, & Harayama, 2010; Venzlaff et al., 2013). Furthermore, bioenergetics has described that electron

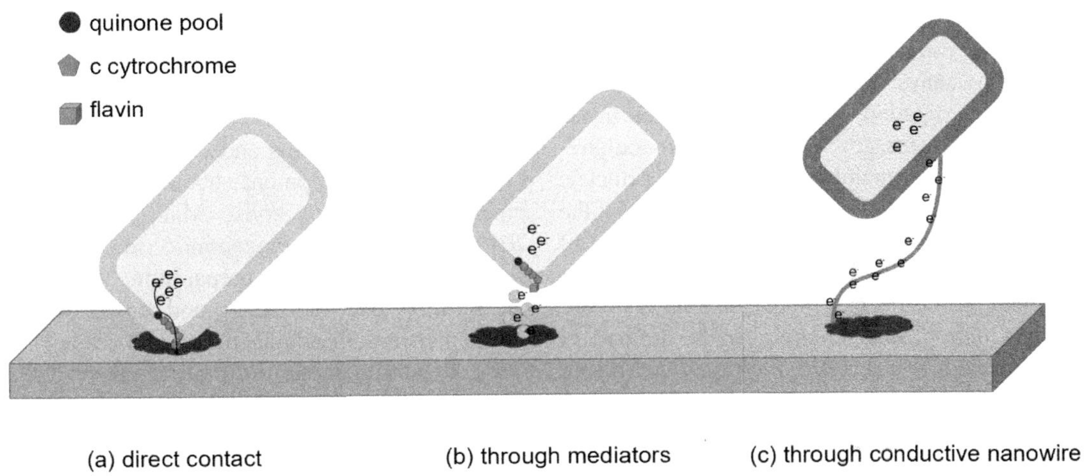

FIG. 1 Probable mechanisms are described in the transfer of electrons by electroactive microorganisms. (A) Transfer of electrons through direct contact between the cell and the extracellular surface, with the participation of active redox proteins, such as flavin *(in red, light gray in print version)*, type C cytochromes *(in blue, gray in print version)*, and the pool of quinones *(in purple, dark gray in print version)*. (B) Electron transfer by reducing soluble electron launchers synthesized by electroactive microorganisms, where oxidized transport molecules on the outer surface of the cell are reduced, and reduced transport molecules donate electrons to the outer surfaces. (C) Electron transport is mediated by a conductor via electrically conductive pili synthesized by the microorganism, the short range allows the transfer of electrons from the cell mediated by extracellular cytochromes.

transport into the cell, through the cell wall and membrane, is thermodynamically favorable, despite that the mechanism of transmembrane transport has not been explained. Subsequently, it was demonstrated that the species SRB *Desulfovibrio vulgaris* induced corrosion of carbon steel using electrons donated by lactate (Xu & Gu, 2014). MIC investigations using the methanogen *Methanosarcina barkeri* showed that under conditions of deprivation of common energy sources for this, such as methanol and H_2, there was a considerable acceleration in corrosion rates (Tan, Goh, & Blackwood, 2017). Similarly, the NRB *P. aeruginosa* (PAO1) showed acceleration in the corrosion of C1018 carbon steel in the absence of nutritional carbon sources (Jia, Yang, Xu, Xu, & Gu, 2017).

In the mechanism by direct contact of the outer membrane, electrons are transferred employing redox proteins establishing a direct connection with the intracellular repeating system (Fig. 1A). This system has been widely studied in iron-reducing and -oxidizing bacteria (IRB and IOB, respectively), including *Geobacter sulfurreducens*, *Shewanella oneidensis*, and *Acidithiobacillus ferrooxidans* (Castelle et al., 2008; Weber, Achenbach, & Coates, 2006). However, differences occur with respect to the details of the specific proteins involved between the different species, especially between *Geobacter* and *Shewanella*. In both models, the similarity occurs during the electron transport of the dehydrogenase to a pool of quinones embedded in the cytoplasmic membrane, which are then transferred to type C cytochromes and, finally, to a reductase in the outer membrane (Stams et al., 2006). The second mechanism proposed in the EET would be for indirect transfer by means of diffusible mediators (Fig. 1B) (Watanabe, Manefield, Lee, & Kouzuma, 2009). This occurs in certain microorganisms that reduce or oxidize intracellular electron mediators, after which they diffuse out of the cell, where on solid surfaces they can donate or accept electrons, lastly, the oxidized or reduced mediators return to the cells, and after being recycled, they are again used as in the next electron transfer cycle (Marsili et al., 2008; Rabaey & Rozendal, 2010). The mechanism of electron transfer by pili during MIC was observed by Xu, Zhang, and Cheng (2007), when they described that the deprivation of organic carbon source in SRB induced the transfer of electrons from the steel to the cell cytoplasm for the reduction of reaction sulfate (Fig. 1C). The pili structures appear to behave like microbial nanowires associated with cytochromes. The distance between the cell and the surfaces established by the extension of the nanowires should also be considered, which is quite short, less than 3 nm, although there are still no reliable studies on this distance.

3. Metagenomic approaches in studies of MIC

In the last decades, traditional approaches to the identification of microorganisms were based on PCR amplification of the 16S rRNA gene and subsequently sequenced by the Sanguer method. It was mandatory to grow and maintain the microbial isolates in laboratory conditions (Narihiro & Kamagata, 2013; Nichols, 2007). With the advent of large-scale sequencing techniques, and later methodologies that allowed sequencing from environmental samples, a new stage of discoveries about microbial diversity began (Procópio, Macrae, van Elsas, & Seldin, 2013). Nowadays, a considerable number of approaches in the investigation of microbes involved in the corrosion of metallic surfaces are based on culture-independent methodologies, mainly in the analysis of nucleic acids, to evaluate the structure, function, and dynamics of the microbial communities. Not surprisingly, our knowledge of MIC participants has been enhanced by high-throughput sequencing techniques, especially metagenomic methodology (Procópio, 2020a). Metagenomic can be defined as the comprehensive sequence analysis of total DNA isolated from environmental samples. This approach dramatically reshaped the field of microbial ecology and biocorrosion showed to the realization that in most ecosystems, the existing microorganisms outnumber those that are accessible through cultivation by orders of magnitude (Gutleben et al., 2018; Marchesi & Ravel, 2015).

The first challenge for obtaining satisfactory results in metagenomic methodologies is the environmental sample. In fact, when dealing with corrosive biofilms, the most varied environments and conditions can be described, such as oil pipeline biofilms and water collection system pipelines, offshore equipment submerged in deep waters, gas pipelines buried in soils (Maia, Capão, & Procópio, 2019). Then, the metagenomic sequencing involves a crucial stage of library preparation, which is commonly the limiting step to obtain good final results. Depending on the platform, biological material is needed in small quantities, about 10 nanograms, up to expressive amounts of up to 10 micrograms of eDNA. A subsequent step will produce fragments ranging from 150 to 800 bp, which are later incorporated into the fragments, in a process called Tagmentation (Fig. 2A) (Hennig et al., 2018). Certain platforms, especially the second generation, use a fragments methodology, which helps in the development of a strong detectable signal

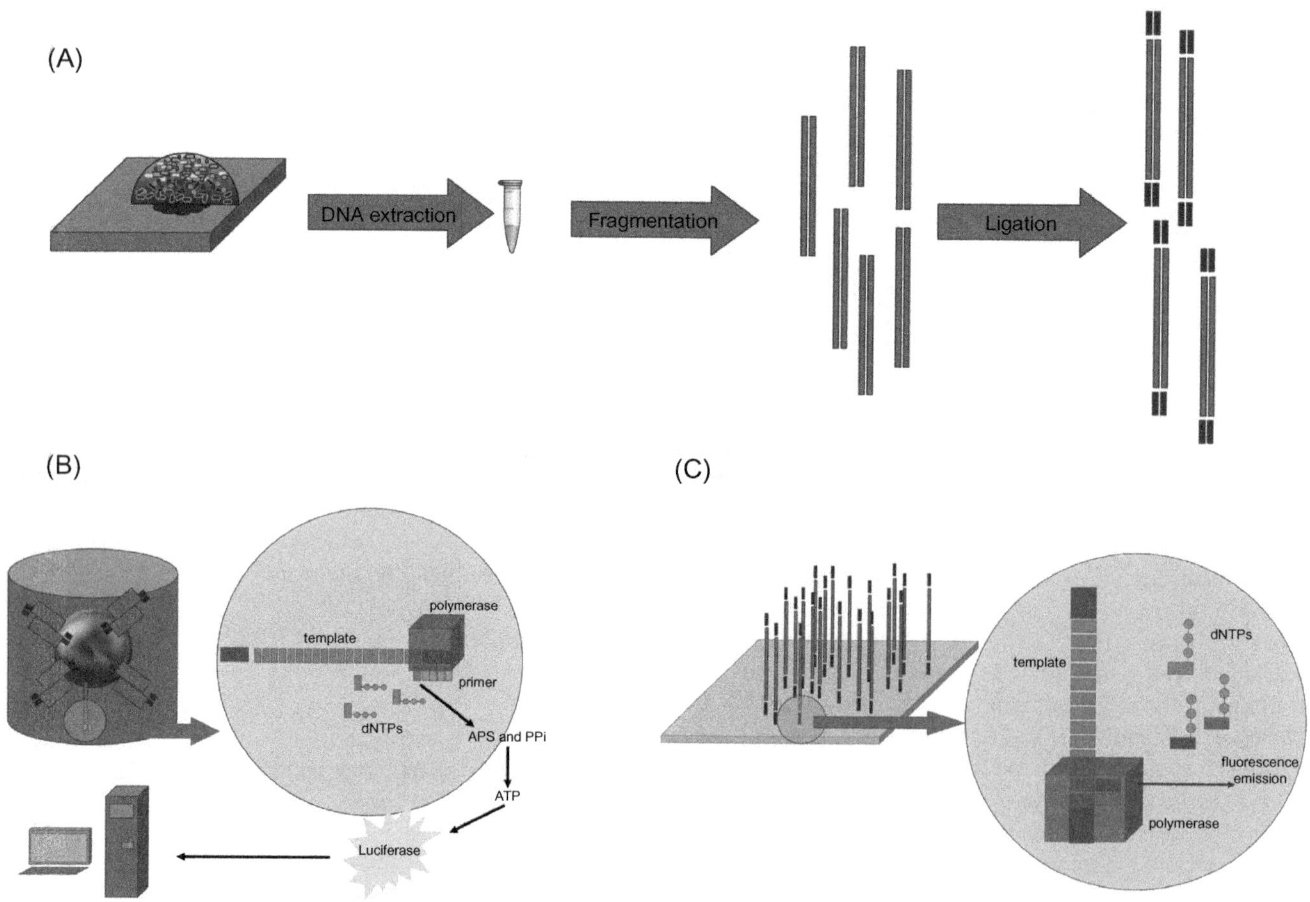

FIG. 2 Illustrative diagrams describing the sequencing steps of biofilm corrosion. (A) Obtaining samples of metallic surfaces, sequenced by DNA extraction, fragmentation into pieces of 150–800 bp, and connection of adapters. (B) On 454 platforms, the fragmented DNA previously attached to adapters is attached to magnetic spheres for emulsion PCR amplification. Each granule containing copies of a single DNA molecule is packaged in wells. After the addition of the necessary reagents, amplifications occur from the synthesis of the complementary DNA strand, the incorporated nucleotides are sequentially identified by the detection of the light resulting from the release of the pyrophosphate by the enzyme luciferase. (C) Illumina sequencing involves linking DNA molecules with adapters. Bridge amplification and clustering of single molecules are followed by the amplification of the template, which for each nucleotide incorporated a corresponding wavelength of light is emitted and converted into a DNA sequence.

in the next step of the sequencing (Shao et al., 2011). The sequencing itself will vary depending on the platform, the most used being 454 from Roche and MiSeq and HiSeq from Illumina. For example, Illumina sequencing employs the reversible termination (RT) approach, which consists of a fluorescent marker with different emission wavelengths for each nucleotide (Berglund, Kiialainen, & Syvänen, 2011; Reuter, Spacek, & Snyder, 2015). The sequencing by detection of hydrogen ions used by the 454 instead of using the terminators presents an approach that allows to measure the pH variation induced by the release of the hydrogen ion during the amplification of the DNA (Fig. 2B). Illumina sequencing employs the reversible termination approach, where the RT nucleotide is marked by fluorescence, which has different emission wavelengths for each nucleotide and with 3′-OH protected groups. At Illumina, HiSeq and MiSeq use four-channel sequencing, where each base is detected by an individual image, thus reducing the time it takes to capture the image, the number of cycles, and the costs (Fig. 2C) (Berglund et al., 2011). The third generation of sequencing known as NGS technology or single real-time molecule (SMRT) has the important advantage of eliminating the PCR bias introduced by clonal amplification, used in the technologies described above. Although still incipient, the use of third-generation technologies, there is a promise that cost reduction and sequencing time for platforms that offer these methodologies. NGS technologies use the sequencing of a single strand of DNA, instead of sequencing a model amplified by cloning, thereby reducing the use of biochemical steps and reagents throughout the process, in addition to leading the nanoscale sequencing process. The main platforms are Helicos' Genetics Analysis System, Pacific Biosciences Single Molecular Real-Time and Sequencing of Nanopore Oxford (Greninger et al., 2015; Thompson & Steinmann, 2010).

Each of the sequencing technologies described above has its advantages and limitations. Nevertheless, a common limitation to all technologies lies in the small size of the generated sequences called reads, usually between 30 and 300 nucleotides, despite the 454 GS Junior platform producing larger reads with sizes over 500 bases. The limited size of these sequences poses a problem during the association of the reads. All platforms produce different kinds of errors during the base incorporations, such as substitution, insertions, deletions, and AT bias, whereas the third-generation sequencing, which employs single-molecule real-time sequencing, shows high error levels. These techniques described above also allowed the study of genomes and genes of noncultivable microorganisms directly from the environment, in the technique called metagenomics. From the shotgun sequencing methods of environmental samples, it is possible to describe the microbial composition present in a taxonomic approach from marker genes identification. For example, microbial amplicons of the 16S rRNA genes enabled the description of the taxonomic profile and functional profiling of complex systems of microbial populations present in different samples from natural environmental sites, buildings, and the human body (Turnbaugh & Gordon, 2008; Wicker et al., 2006). However, the use of metagenomic sequencing techniques also presents limitations. The sequencing of a sample will present a cross section of that community at a given time only, regardless of the dynamics that occur over time. Another concern in the use of metagenomics is in the fact that the results on phylogeny are at the species level, while much of the biogeochemical or physiological processes occur at the level of strains. Finally, the number of sequenced samples may not be a faithful representation of the populations present in the environment in question.

4. Corrosive biofilms on metal surfaces

A fundamental condition for the occurrence of corrosion influenced by microorganisms is the presence of a biofilm established at the metal-environment interface. This is because the biofilm structure creates the appropriate conditions so that electrochemical reactions can be favored (Little & Lee, 2014). The biofilm is a complex aggregate of extracellular polymeric substance (EPS) produced by different microorganisms present inside, which is attached to biotic and abiotic surfaces (Flemming et al., 2016; Vert et al., 2012). The biofilms found in natural environments present a microbial community that varies in its composition and the representation of species according to the time and conditions around them. Unfavorable conditions such as temperature, presence of biocides, shear, starvation are common driving forces for the formation of biofilms (Salta, Wharton, Blache, Stokes, & Briand, 2013). Another important aspect to be considered is the benefits that a cooperative relationship between species, such as defensive compounds, and metabolite exchange, since metabolites secreted or produced by a certain species can be useful for other microbial groups (Daddi Oubekka, Briandet, Fontaine-Aupart, & Steenkeste, 2012). However, the relationships within the biofilm are not always favorable to all its members. Competitions for access to nutrients, intraspecies predation, and overlapping populations due to changes in their interior, such as changes in pH and oxygen levels, are normal factors during the maturation of the structure (Serra & Hengge, 2014). These changes may reflect on the role of the biofilm established on the surface, such as a structure capable of influencing the corrosion of metallic surfaces (Garcia & Procópio, 2020). Allied to all this, the nature of the metallic surface also influences the formation of biofilms. The adhesion of surrounding cells also depends on the chemical composition and roughness of the metal. Porous metallic surfaces with more irregular aspects, with the presence of scratches and cracks, and weld-joint constitute more favorable environments for the adhesion of colonizing cells (Liduino, Lutterbach, & Sérvulo, 2018).

The formation of biofilms on metallic surfaces constitutes a diverse and dynamic process, with different levels of development and characteristics inherent to the surface to which it has adhered (Beloin & McDougald, 2021) (Fig. 3). Although biofilms are structures present in different ecosystems, such as sediments, soils, and subsoils, most models describe their formation in aquatic environments. At first, planktonic cells adhere to the surface forming microcolonies, which initiate EPS synthesis, which also promotes cell-cell cohesion, facilitating microbial aggregation (Fig. 3A and B) (Costerton et al., 1987). As these structures are built concurrently, channels are formed allowing the passage of nutrients and cellular waste. The biofilm formed has multilayers with a distribution of different microbial populations, which interact with each other through quorum sensing, through signals such as nitric oxide and cyclic-di-GMP (Fig. 3C and D) (Jenal, Reinders, & Lori, 2017; Rice, McDougald, Kumar, & Kjelleberg, 2005). In addition, the composition of its extracellular matrix plays a role in the capture and environmental resources, such as nutrients, gases, metal ions, and inorganic particles. These elements are present in the aqueous phase or associated with the surface to which the biofilm is adhered. The nutrient capture mechanism employs the properties of passive sorption existing in the EPS matrix-like sponge structure. The most common metals present in the biofilm matrix are calcium, iron, and manganese (Grumbein, Opitz, & Lieleg, 2014; Okabe, Yasuda, & Watanabe, 1997). They contribute to the stabilization of the structure, preventing the deterioration of the extracellular matrix, in addition to being electrically conductive particles, thus supporting intraspecies electron transfer (Oppenheimer-Shaanan et al., 2016). The biofilm-metal interface is established with the layer in contact with the surface and played an important role in the corrosive process (Fig. 3D). This is because the physical-chemical properties of the matrix confer spatial organization, which allows the creation of a dynamic of synergistic interactions through cell-cell communication, which reflects in the formation of favorable conditions to induce localized corrosion (Flemming et al., 2016).

The importance of biofilms in corrosion processes, whether in induction, acceleration, or even inhibition, can be evidenced by the numerous studies that attempt to establish their relationship with the present microbial community. In fact, efforts to elucidate the role of different microbial species in corrosion are a challenge, especially in conditions in situ or in laboratory microcosms that try to simulate the environment of interest (Marty et al., 2014). These surveys of the communities comprising these structures attempt to describe their composition, dynamics, and physiology. Furthermore, studies also show that the maintenance of the structure constitutes a decisive factor in the corrosion process, and in biofilms present in situ, it does not depend only on the species of bacteria directly related to the corrosion. Metagenomic studies show that bacteria present in natural environments colonize metal surfaces in a succession of different populations throughout the development of the biofilm. Colonizing bacteria called "pioneers" have an essential performance in the initial development of biofilms, and metagenomic surveys on bacterial succession in marine environments have shown that the phyla Proteobacteria and Bacteroidetes are predominant in samples obtained from metallic surfaces (Capão, Moreira-Filho, Garcia, Bitati, & Procópio, 2020; Lanneluc et al., 2015; McBeth & Emerson, 2016; Moura et al., 2018). The presence of representatives of the phylum Proteobacteria is easily explained due to its metabolic versatility, which makes this group the largest bacterial phylum and the most widely distributed and abundant in marine environments (McBeth, Little, Ray, Farrar, & Emerson,

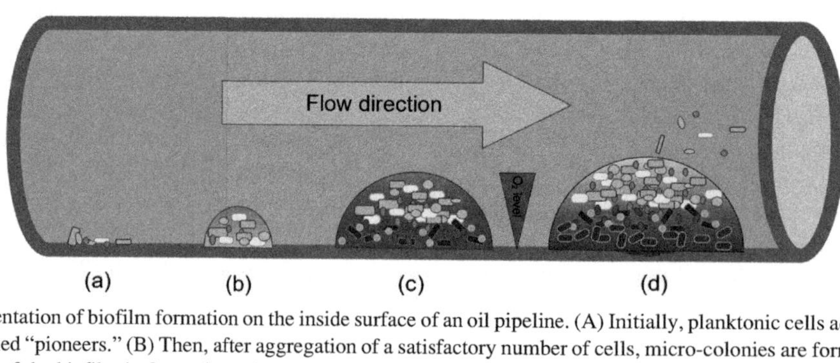

FIG. 3 Schematic representation of biofilm formation on the inside surface of an oil pipeline. (A) Initially, planktonic cells adhere to the metal surface of the duct, these cells are called "pioneers." (B) Then, after aggregation of a satisfactory number of cells, micro-colonies are formed, which secrete EPS that will compose the structure of the biofilm in formation. (C) As the final structure of the biofilm matures, internal micro-environments are created, with a difference in pH and oxygenation. (D) In a corrosive biofilm, these conditions allow the establishment of late anaerobe microbes that directly participate in the corrosion of the metallic surfaces in contact. Also in this phase, the detachment of cells occurs, usually from the most superficial layers.

2011). Notwithstanding any of the specimens involved in metallic corrosion, they play a role in the construction of biofilm structures, especially members of the Alphaproteobacteria and Gammaproteobacteria classes (Li, Duan, et al., 2017; Vigneron et al., 2016). In addition, countless species belonging to the phylum Proteobacteria are associated with the sulfur cycle, which is determinate in corrosive processes (Lanneluc et al., 2015). In metallic surfaces immersed in seawater, they are colonized in a short time by the ubiquitous group Bacteroidetes (Kirchman, 2002; Moura et al., 2018). In fact, species of this phylum are widely associated with the construction and maintenance of corrosive biofilms (Dang et al., 2011; Doghri et al., 2015). Species of Bacteroidetes synthesize innumerable constituent molecules of the extracellular matrix, such as adhesive exopolysaccharides, adhesion proteins, and hydrolases, which allow to maintain the viable structure and create favorable environments for corrosion (Edwards et al., 2010). This colonization process is important for the establishment and maintenance of its structure and the role it will have in the metal. Depending on the internal environmental conditions and the chemical composition of the mature biofilm, it will be favorable for the presence of electroactive microorganisms directly involved in the corrosion of the metal.

5. Metagenomic studies of electroactive bacteria

5.1 Sulfate and nitrate reducers

In recent years, numerous metagenomic surveys have described the presence of electroactive microbial species in samples of corrosive biofilms. The participation of different electroactive microorganisms has been increasingly accepted in corrosive processes in different electrochemical conditions and metals. However, the SRB and methanogens groups are still the most widely distributed in metagenomic analysis, being commonly detected where sulfate is present (Blackwood, 2020). Due to their metabolic versatility, which allows the use of several electron acceptors and donors, SRBs inhabit a variety of different environmental conditions. As such, SRBs are ubiquitous in many natural and designed environments, such as in the offshore oil industry, anaerobic wastewater, agro-industrial treatment systems, and water distribution systems, where they can cause considerable economic losses through induced corrosion. In natural environments, metagenomic research has detected SRB in marine sediments, hydrothermal vents, volcanic mud, and different niches with extreme pH and temperature conditions (Enning & Garrelfs, 2014). Most SRB species are free-living; however, when associated with corrosive biofilms, they are commonly described alongside aerobic IOB, and methanotrophic archaea, the latter being especially in oil installations (Boetius et al., 2000; Chamritski, Burns, Webster, & Laycock, 2004; Leloup et al., 2007; Xu et al., 2007). A possible explanation for this would be the fact that the hydrocarbons abundant in these systems usually serve as a source of organic carbons for the growth of SRBs. The relationship between the presence of SRBs and the corrosion of equipment in the oil industry has been established many decades ago (Barton & Fauque, 2009). However, with the advent of large-scale sequencing techniques, many studies using 16S rRNA sequencing have described the participation of SRBs in corrosive biofilms, as well as identifying the relationships with other microbial groups in these structures, such as NRB, and methanogens. In fact, numerous studies on microbial communities involved in the corrosion of oil facilities, such as pipelines and water injection systems, describe the predominance of specimens in this group. The presence of sulfate and nitrate, which are abundant in oil fields alongside hydrocarbons dissolved in the oil, provides the source of carbon and energy necessary for the growth and permanence of SRB (Gittel, Kofoed, Sørensen, Ingvorsen, & Schramm, 2012; Vigneron, Head, & Tsesmetzis, 2018).

It is possible to roughly divide the main representatives of sulfate reducers involved in the corrosion into two groups, according to the temperature: (1) a group normally described in an environment with low temperatures, with emphasis on the genera *Desulfovibrio*, *Desulfovermiculus*, *Desulfomicrobium*, *Desulfotomaculum*, *Desulfobulbus*, and *Desulfotignum* (Bødtker, Lysnes, Torsvik, Bjørnestad, & Sunde, 2009; Callbeck et al., 2011; Grigoryan et al., 2008; Kumaraswamy, Ebert, Gray, Fedorak, & Foght, 2011; Neria-González, Wang, Ramírez, Romero, & Hernández-Rodríguez, 2006; Shartau et al., 2010); (2) a group detected at high temperatures, the genera *Archaeoglobus*, *Thermodesulfobacterium*, *Desulfotomaculum*, and *Desulfacinum* being the main representatives (Duncan et al., 2009; Gittel, Sørensen, Skovhus, Ingvorsen, & Schramm, 2009; Kaster, Bonaunet, Berland, Kjeilen-Eilertsen, & Brakstad, 2009) (Table 1). Many of the species of these genera, known for using hydrogen and other organic compounds for growth, are commonly found in metagenomic surveys of corrosive biofilms in equipment used in the oil industry. A study on microbiological succession during corrosion of carbon steel in the presence of different hydrocarbons as electron donors demonstrated that even in the absence of sulfate, there was the persistence of the genera *Desulfovibrio* and *Desulfobulbus* in the communities analyzed, and a concomitant increase in the severity of general and localized corrosion occurred with the addition of sulfate (Liang, Davidova, Hirano, Duncan, & Suflita, 2019). The analysis of the microbial communities of samples of water production samples with different levels of oil, pH, and temperatures showed the presence of different genera of the SRB group, such as *Desulfotignum*, *Desulfuromonas*, and *Desulfovibrio* potentially active in

TABLE 1 Description of the main representatives of electroactive microbes involved in metal corrosion.

Group	Environmental condition	Growth temperature	Genre	References
Sulfate reducers	Aquatic environments, moderately halophilic environmental, anaerobic conditions	Moderate	*Desulfovibrio, Desulfovermiculus, Desulfomicrobium, Desulfotomaculum, Desulfobulbus, Desulfotignum,* and *Desulfovirga*	Procópio (2020a, 2020b), Procópio et al. (2013), Rabaey and Rozendal (2010), Rajala et al. (2015), Rajasekar et al. (2010)
		Elevate	*Archeoglobus, Thermodesulfobacterium, Desulfotomaculum,* and *Desulfacinum*	Rauch et al. (2006), Ren et al. (2015), Reuter et al. (2015)
Methanogens	Anaerobic or low-oxygen conditions, found in aquatic sediments, thriving in alkaline conditions	Moderate	*Methanobacterium, Methanosaeta, Methanolobus, Methanocalculus,* and *Methanosarcina*	Nealson and Rowe (2016), Salgar-Chaparro, Darwin, Kaksonen, and Machuca (2020), Serra and Hengge (2014), Usher, Kaksonena, and MacLeod (2014), Venzlaff et al. (2013)
		Elevate	*Methanococcus, Methanosaeta,* and *Methanothermobacter*	Neria-González et al. (2006), Videla and Herrera (2005), Callbeck et al. (2011)
Other electroactive corrosive bacteria	Facultatively anaerobic. Some species live in extreme environments, such as temperature and pressure		*Shewanella, Pseudomonas,* and *Geobacter*	Myers and Myers (2001), Nealson and Rowe (2016), Shartau et al. (2010), Tian et al. (2017), Vigneron et al. (2018), Villahermosa, Corzo, Garcia-Robledo, González, and Papaspyrou (2016), Wang et al. (2020), Watanabe et al. (2009), Weber et al. (2006), Wicker et al. (2006), Xu et al. (2007)

the process of metallic corrosion (Lahme, Mand, Longwell, Smith, & Enning, 2021; Li, Duan, et al., 2017; Wang et al., 2020). Investigations conducted on structures of bacterial and archaea communities present in offshore oil production facilities commonly indicate the dominant presence of species of the *Desulfovibrio* and *Desulfobulbus* genera alongside methanogenic archaea in the microbial communities of biofilms (Lahme et al., 2021; Staniszewska et al., 2019). The presence of SBRs in natural gas pipelines has also been found in metagenomic studies. These surveys showed the presence of numerous members of the Order Desulfovibrionales, e.g., *Desulfosporosinus*, *Desulfotomaculum*, and *Desulfovibrio* (Su, Tang, Peng, Gao, & Han, 2020) (Table 1). Despite the information about SRBs at high temperatures is still incipient, recent studies have pointed out the presence of members of this group in biofilms present in equipment used in oil exploration and recovery, which constantly suffer from corrosion failures (Duncan et al., 2009). In these temperature conditions, the genera *Archeoglobus*, *Thermodesulforhabdus*, and *Desulfotomaculum* are the most commonly described in conditions with the highest temperatures (Guan et al., 2014; Li, Yang, et al., 2017; Tian et al., 2017) (Table 1).

Stainless steel includes different types of alloys, nickel, chromium, molybdenum, and ruthenium. Although its surface has greater resistance compared with other metals, duplex stainless steel is reported to be susceptible to MIC (Hsu, Chen, Lo, & Lee, 2019; Li et al., 2015; Lopes, Perrin, & Féron, 2007; Prithiraj, Otunniyi, Osifo, & van der Merwe, 2019). In a study of the characterization of biofilms grown in stainless steel used in urban wastewater systems, microbiological succession shows an initial predominance of IOB, which are later supplanted by SRBs, e.g., *Desulfovirga*, *Desulfovibrio*, *Desulfuromusa*, *Desulfococcus*, and *Desulfosarcina*, when oxygen levels decrease (Ziadi, El-Bassi, Bousselmi, & Akrout, 2020) (Table 1).

The use of stainless steel in petroleum equipment was also shown to be susceptible to corrosion by electroactive bacteria of the SRB group. Analysis of the microbial communities in wastewater treatment systems for reinjection oilfields described the presence of different SRB species, mainly *Desulfuromonas*, *Desulfovibrio*, *Desulfotignum*, alongside NRB (Wang et al., 2020) (Table 1). In addition, corrosion studies influenced by microbial communities in stainless steel pipelines used to transport brackish water pointed to the massive presence of *Desulfomicrobium* and *Desulfocapsa*, alongside methanogenic archaea species, after the addition of sodium bisulfite to the system (An et al., 2016; Park et al., 2011).

In secondary oil recovery process, the produced water reinjection (PWRI) is often employed in reservoirs to improve oil and gas recovery (Bødtker et al., 2008). This process leads to increased corrosion and acidification of reservoirs due to the production of hydrogen sulfide (H_2S) by sulfidogenic microorganisms (Garcia-de-Lomas et al., 2007). One of the approaches used to mitigate biogenic acidification is the injection of nitrate into the reservoir mixed with PWRI (Drønen et al., 2014). However, in recent years, studies have reported that the addition of nitrate has led to corrosion of petroleum facilities, which may be due to the increased population of NRB that can out-compete SRB members for carbon sources. Members of this metabolic group are more widely distributed phylogenetically, when compared with the SRB group. Recent metagenomic studies have described representatives of the NRB group in the Phylum Proteobacteria, Bacteriodetes, and Firmicutes (Villahermosa et al., 2016). Genera of the Epsilonproteobacteria Class, such as the genera *Arcobacter*, *Sulfurimonas*, *Sulfuricurvum*, and *Sulfurospirillum*, are commonly detected in oil exploration facilities and have been identified in corrosive biofilms when nitrate is added to the PWRI (Drønen et al., 2014).

5.2 Methanogens involved in metal corrosion

The relationship between corrosion of methanogenic archaea has been confirmed by numerous studies that demonstrate the emission of methane with the concomitant loss of mass of the metal in experiments conducted under laboratory conditions (Vigneron et al., 2018). In fact, methane production by these microorganisms during corrosion is dominant in anaerobic environments, especially when depleted of sulfate sources (Mori, Tsurumaru, & Harayama, 2010; Uchiyama et al., 2010; Zhang et al., 2019). Despite this relation about sulfate, the metagenomic surveys of corrosion-associated communities are generally associated with members of SRBs and acetogens, especially in environments such as offshore oil production facilities, production water of oil reservoirs, and metal sheet piles. Recently, numerous sequences of the biofilm communities present in oil facilities have described the genera *Methanococcus*, *Methanobacterium*, *Methanosaeta*, *Methanolobus*, *Methanocalculus*, *Methanosarcina* as commonly associated with corrosion (Lahme et al., 2021; Mand et al., 2016; Okoro et al., 2014; Vigneron et al., 2016) (Table 1). In addition, some of these cited genera, for example, *Methanobacterium*, can alter their metabolism to reduce dissimilatory elemental sulfur, producing high concentrations of hydrogen sulfide, thereby contributing to corrosion (Stetter & Gaag, 1983). It is plenty common for oil facilities to be exposed to temperature gradients, which influences both the community present in biofilms and the levels of severity in corrosion influenced by microorganisms (Okoro et al., 2014). For example, a survey of these communities indicates that under ambient temperature conditions, around 30°C, there is a predominance of methanogenic archaea from the *Methanosaeta*, *Methanolobus*, and *Methanobacterium* genera. However, at higher temperatures, above 50°C, the genus *Methanothermobacter* was found predominantly in the samples, and although the presence of methanogens at higher temperatures is uncommon, the *Methanosaeta* and *Methanothermobacter* genera were identified at 63°C (Conlette, Emmanuel, & Chijoke, 2016; Liang, Grizzle, Duncan, McInerney, & Suflita, 2014) (Table 1). The use of sodium bisulfite (SBS) is widely used as an oxygen scavenger, thus reducing corrosion in pipelines that transport brackish groundwater used in bitumen production. Metagenomic studies carried out show the induced corrosion influenced by methanogenic arches, both in brackish waters and in the ducts that conduct them, being the main representatives of *Methanolobus* and *Methanocalculus* genera (An et al., 2016; Park et al., 2011).

5.3 Other electroactive corrosive bacteria

The ability to grow in aerobic and anaerobic environments, combined with its ability to use a wide variety of electrons, allows the *Shewanella* genus to inhabit corrosive biofilms present in different niches and environmental conditions. Furthermore, it is not uncommon for metagenomic sequencing to identify specimens of the genus *Shewanella* in considerable representativeness among members of the community directly involved in the corrosion of metals. Under aerobic conditions, the genus *Shewanella* was identified in biofilms grown on the surfaces of different types of stainless steel (Capão et al., 2020; Ziadi et al., 2020) (Table 1). Under anaerobic conditions, the participation of *Shewanella* in the corrosion of gas and oil ducts seems to be associated with the presence of hydrocarbons, which can be used as a carbon source (Bonifay et al., 2017; Salgar-Chaparro et al., 2020; Staniszewska et al., 2019). The *Pseudomonas* genus is commonly described in oil and gas facilities, where the presence of hydrocarbons provides nutrients for its growth (Patel, Homaei, Patil, &

Daverey, 2019; Procópio, 2020b). Analyses of biofilms involved in the corrosion of metal surfaces identify different species of *Pseudomonas* genera, such as *P. aeruginosa*, *P. putida*, and *P. stutzeri*, normally associated with the presence of the *Shewanella* genus (Jia, Yang, Xu, & Gu, 2017; Kouzuma, Kato, & Watanabe, 2015; Procópio, 2020b) (Table 1). However, the presence of *Pseudomonas* in biofilms grown on metal surfaces in different environments, e.g., metallic soda, stainless steel, water distribution systems, is also reported by metagenomic studies (Javed, Neil, Stoddart, & Wade, 2016; Ren et al., 2015). The electroactive bacterium *Geobacter*, described as the most important current-producing bacterium, it is normally found in biofilms grown at a pH close to neutral and freshwater (Lin, Yu, Chen, Liu, & Zhang, 2013). In metagenomic surveys in communities of wastewater distribution systems, the genus *Geobacter* has been identified in biofilms associated with rust in both stainless steel and cast iron, suggesting its important role in the biogeochemical cycle of iron and sulfur (Zhang et al., 2018) (Table 1).

6. Conclusions

The participation of multispecies in the formation of biofilms and their role in the corrosion of metallic surfaces have attracted the attention of researchers in the field of biocorrosion. Recent advances in DNA sequencing of environmental samples have allowed the description of participants from microbial communities present on different surfaces and environmental conditions. Although EET mechanisms in biofilms and their role in MIC are still an issue for discussion, metagenomics has commonly described the presence of species from SRB and NRB groups, alongside metagenomic archaea in corrosive biofilms. EET has a wide range of potential applications for understanding MIC, thus a better understanding of the ecophysiological relationships of electroactive microorganisms will be essential for the future development of biocorrosion mitigation strategies in metals.

Funding

Not applicable.

Conflicts of interest/competing interests

The authors declare no competing interests.

Ethics approval

Not applicable.

Consent to participate

Not applicable.

Consent for publication

Not applicable.

Availability of data and material (data transparency)

Not applicable.

Code availability (software application or custom code)

Not applicable.

References

An, D., Dong, X., An, A., Park, H. S., Strous, M., & Voordouw, G. (2016). Metagenomic analysis indicates Epsilonproteobacteria as a potential cause of microbial corrosion in pipelines injected with bisulfite. *Frontiers in Microbiology*, *7*, 7–28. https://doi.org/10.3389/fmicb.2016.00028.

Andrews, S., Norton, I., Salunkhe, A. S., Goodluck, H., Aly, W. S. M., Mourad-Agha, H., et al. (2013). Control of iron metabolism in bacteria. In L. Banci (Ed.), *Metallomics and the cell* (pp. 203–239). Switzerland: Springer.

Baeckmann, W. V. (1997). *Handbook of cathodic corrosion protection*. Massachusetts: Elsevier.

Baorong, H., Xiaogang, L., Xiumin, M., Cuiwei, D., Dawei, D., Meng, Z., et al. (2017). The cost of corrosion in China. *npj Materials Degradation*, *1*, 4. https://doi.org/10.1038/s41529-017-0005-2.

Barton, L. L., & Fauque, G. D. (2009). Biochemistry, physiology and biotechnology of sulfate-reducing bacteria. *Advances in Applied Microbiology*, *68*, 41–98. https://doi.org/10.1016/S0065-2164(09)01202-7.

Beech, I. B., & Sunner, J. (2004). Biocorrosion: Towards understanding interactions between biofilms and metals. *Current Opinion in Biotechnology*, *15*, 181–186. https://doi.org/10.1016/S0958-1669(04)00063-1.

Beloin, C., & McDougald, D. (2021). Speciality grand challenge for "biofilms". *Frontiers in Cellular and Infection Microbiology*, *11*, 632429. https://doi.org/10.3389/fcimb.2021.632429.

Berglund, E. C., Kiialainen, A., & Syvänen, A. C. (2011). Next-generation sequencing technologies and applications for human genetic history and forensics. *Investigative Genetics*, *2*, 23. https://doi.org/10.1186/2041-2223-2-23.

Blackwood, D. J. (2020). An alectrochemist perspective of microbiologically influenced corrosion. *Corrosion and Materials Degradation*, *1*, 59–76. https://doi.org/10.3390/cmd1010005.

Bødtker, G., Lysnes, K., Torsvik, T., Bjørnestad, E.Ø., & Sunde, E. (2009). Microbial analysis of backflowed injection water from a nitrate-treated North Sea oil reservoir. *Journal of Industrial Microbiology & Biotechnology*, *36*, 439–450. https://doi.org/10.1007/s10295-008-0515-6.

Bødtker, G., Thorstenson, T., Lillebø, B. L., Thorbjørnsen, B. E., Ulvøen, R. H., Sunde, E., et al. (2008). The effect of long-term nitrate treatment on SRB activity, corrosion rate and bacterial community composition in offshore water injection systems. *Journal of Industrial Microbiology & Biotechnology*, *35*(12), 1625–1636. https://doi.org/10.1007/s10295-008-0406-x.

Boetius, A., Ravenschlag, K., Schubert, C. J., Rickert, D., Widdel, F., Gieseke, A., et al. (2000). A marine microbial consortium apparently mediating anaerobic oxidation of methane. *Nature*, *407*, 623–626. https://doi.org/10.1038/35036572.

Bond, D. R., & Lovley, D. R. (2003). Electricity production by *Geobacter sulfurreducens* attached to electrodes. *Applied and Environmental Microbiology*, *69*, 1548–1555.

Bonifay, V., Wawrik, B., Sunner, J., Snodgrass, E. C., Aydin, E., Duncan, K. E., et al. (2017). Metabolomic and metagenomic analysis of two crude oil production pipelines experiencing differential rates of corrosion. *Frontiers in Microbiology*, *31*, 8–99. https://doi.org/10.3389/fmicb.2017.00099.

Callbeck, C. M., Dong, X., Chatterjee, I., Agrawal, A., Caffrey, S. M., Sensen, C. W., et al. (2011). Microbial community succession in a bioreactor modeling a souring low-temperature oil reservoir subjected to nitrate injection. *Applied Microbiology and Biotechnology*, *91*, 799–810. https://doi.org/10.1007/s00253-011-3287-2.

Capão, A., Moreira-Filho, P., Garcia, M., Bitati, S., & Procópio, L. (2020). Marine bacterial community analysis on 316L stainless steel coupons by Illumina MiSeq sequencing. *Biotechnology Letters*, *42*, 1431–1448. https://doi.org/10.1007/s10529-020-02927-9.

Castelle, C., Guiral, M., Malarte, G., Ledgham, F., Leroy, G., Brugna, M., et al. (2008). A new iron-oxidizing/O_2-reducing supercomplex spanning both inner and outer membranes, isolated from the extreme acidophile *Acidithiobacillus ferrooxidans*. *The Journal of Biological Chemistry*, *283*, 25803–25811. https://doi.org/10.1074/jbc.M802496200.

Chamritski, I. G., Burns, G. R., Webster, B. J., & Laycock, N. J. (2004). Effect of iron-oxidizing bacteria on pitting of stainless steel. *Corrosion*, *60*, 658–669.

Conlette, O. C., Emmanuel, N. E., & Chijoke, O. G. (2016). Methanogen population of an oil production skimmer pit and the effects of environmental factors and substrate availability on methanogenesis and corrosion rates. *Microbial Ecology*, *72*, 175–184. https://doi.org/10.1007/s00248-016-0764-2.

Costerton, J. W., Cheng, K. J., Geesey, G. G., Ladd, T. I., Nickel, J. C., Dasgupta, M., et al. (1987). Bacterial biofilms in nature and disease. *Annual Review of Microbiology*, *41*, 435–464. https://doi.org/10.1146/annurev.mi.41.100187.002251.

Daddi Oubekka, S., Briandet, R., Fontaine-Aupart, M. P., & Steenkeste, K. (2012). Correlative time-resolved fluorescence microscopy to assess antibiotic diffusion-reaction in biofilms. *Antimicrobial Agents and Chemotherapy*, *56*, 3349–3358. https://doi.org/10.1128/AAC.00216-12.

Dang, H., Chen, R., Wang, L., Shao, S., Dai, L., Ye, Y., et al. (2011). Molecular characterization of putative biocorroding microbiota with a novel niche detection of Epsilon- and Zetaproteobacteria in Pacific Ocean coastal seawaters. *Environmental Microbiology*, *13*, 3059–3074. https://doi.org/10.1111/j.1462-2920.2011.02583.x.

De Messano, L. R. V., Sathler, L., Reznik, L. Y., & Coutinho, R. (2009). The effect of biofouling on localized corrosion of the stainless steels N08904 and UNS S32760. *International Biodeterioration and Biodegradation*, *63*, 607–614. https://doi.org/10.1016/j.ibiod.2009.04.006.

Dinh, H. T., Kuever, J., Mussmann, M., Hassel, A. W., Stratmann, M., & Widdel, F. (2004). Iron corrosion by novel anaerobic microorganisms. *Nature*, *427*, 829–832. https://doi.org/10.1038/nature02321.

Doghri, I., Rodrigues, S., Bazire, A., Dufour, A., Akbar, D., Sopena, V., et al. (2015). Marine bacteria from the French Atlantic coast displaying high forming-biofilm abilities and different biofilm 3D architectures. *BMC Microbiology*, *15*, 231. https://doi.org/10.1186/s12866-015-0568-4.

Drønen, K., Roalkvam, I., Beeder, J., Torsvik, T., Steen, I. H., Skauge, A., et al. (2014). Modeling of heavy nitrate corrosion in anaerobe aquifer injection water biofilm: A case study in a flow rig. *Environmental Science & Technology*, *48*(15), 8627–8635. https://doi.org/10.1021/es500839u.

Duncan, K. E., Gieg, L. M., Parisi, V. A., Tanner, R. S., Tringe, S. G., Bristow, J., et al. (2009). Biocorrosive thermophilic microbial communities in Alaskan North Slope oil facilities. *Environmental Science & Technology*, *43*, 7977–7984. https://doi.org/10.1021/es9013932.

Edwards, J. L., Smith, D. L., Connolly, J., McDonald, J. E., Cox, M. J., Joint, I., et al. (2010). Identification of carbohydrate metabolism genes in the metagenome of a marine biofilm community shown to be dominated by Gammaproteobacteria and Bacteroidetes. *Genes*, *1*, 371–384. https://doi.org/10.3390/genes1030371.

El-Naggar, M. Y., Wanger, G., Leung, K. M., Yuzvinsky, T. D., Southam, G., Yang, J., et al. (2010). Electrical transport along bacterial nanowires from *Shewanella oneidensis* MR-1. *Proceedings of the National Academy of Sciences of the United States of America*, *107*, 18127–18131. https://doi.org/10.1073/pnas.1004880107.

Enning, D., & Garrelfs, J. (2014). Corrosion of iron by sulfate-reducing bacteria: New views of an old problem. *Applied and Environmental Microbiology*, *80*, 1226–1236. https://doi.org/10.1128/AEM.02848-13.

Flemming, H. C., Wingender, J., Szewzyk, U., Steinberg, P., Rice, S. A., & Kjelleberg, S. (2016). Biofilms: An emergent form of bacterial life. *Nature Reviews Microbiology*, *14*, 563–575. https://doi.org/10.1038/nrmicro.2016.94.

Gadd, G. M. (2004). Microbial influence on metal mobility and application for bioremediation. *Geoderma*, *122*, 109–119. https://doi.org/10.1016/j.geoderma.2004.01.002.

Garcia, M., & Procópio, L. (2020). Distinct profiles in microbial diversity on carbon steel and different welds in simulated marine microcosm. *Current Microbiology*, *77*, 967–978. https://doi.org/10.1007/s00284-020-01898-4.

Garcia-de-Lomas, J., Corzo, A., Carmen Portillo, M., Gonzalez, J. M., Andrades, J. A., Saiz-Jimenez, C., et al. (2007). Nitrate stimulation of indigenous nitrate-reducing, sulfide-oxidising bacterial community in wastewater anaerobic biofilms. *Water Research*, *41*(14), 3121–3131. https://doi.org/10.1016/j.watres.2007.04.004.

Gittel, A., Kofoed, M. V., Sørensen, K. B., Ingvorsen, K., & Schramm, A. (2012). Succession of Deferribacteres and Epsilonproteobacteria through a nitrate-treated high-temperature oil production facility. *Systematic and Applied Microbiology*, *35*, 165–174. https://doi.org/10.1016/j.syapm.2012.01.003.

Gittel, A., Sørensen, K. B., Skovhus, T. L., Ingvorsen, K., & Schramm, A. (2009). Prokaryotic community structure and sulfate reducer activity in water from high-temperature oil reservoirs with and without nitrate treatment. *Applied and Environmental Microbiology*, *75*, 7086–7096. https://doi.org/10.1128/AEM.01123-09.

Goh, K. M., Shahar, S., Chan, K. G., Chong, C. S., Amran, S. I., Sani, M. H., et al. (2019). Current status and potential applications of underexplored prokaryotes. *Microorganisms*, *7*, 468. https://doi.org/10.3390/microorganisms7100468.

Gralnick, J. A., & Newman, D. K. (2007). Extracellular respiration. *Molecular Microbiology*, *65*, 1–11. https://doi.org/10.1111/j.1365-2958.2007.05778.x.

Gregory, K. B., Bond, D. R., & Lovley, D. R. (2004). Graphite electrodes as electron donors for anaerobic respiration. *Environmental Microbiology*, *6*, 596–604.

Greninger, A. L., Naccache, S. N., Federman, S., Yu, G., Mbala, P., Bres, V., et al. (2015). Rapid metagenomic identification of viral pathogens in clinical samples by real-time nanopore sequencing analysis. *Genome Medicine*, *29*(7), 99. https://doi.org/10.1186/s13073-015-0220-9.

Grigoryan, A. A., Cornish, S. L., Buziak, B., Lin, S., Cavallaro, A., Arensdorf, J. J., et al. (2008). Competitive oxidation of volatile fatty acids by sulfate- and nitrate-reducing bacteria from an oil field in Argentina. *Applied and Environmental Microbiology*, *74*, 4324–4335. https://doi.org/10.1128/AEM.00419-08.

Grumbein, S., Opitz, M., & Lieleg, O. (2014). Selected metal ions protect *Bacillus subtilis* biofilms from erosion. *Metallomics*, *6*, 1441–1450. https://doi.org/10.1039/c4mt00049h.

Guan, J., Zhang, B. L., Mbadinga, S. M., Liu, J. F., Gu, J. D., & Mu, B. Z. (2014). Functional genes (*dsr*) approach reveals similar sulphidogenic prokaryotes diversity but different structure in saline waters from corroding high temperature petroleum reservoirs. *Applied Microbiology and Biotechnology*, *98*, 1871–1882. https://doi.org/10.1007/s00253-013-5152-y.

Gutleben, J., Chaib De Mares, M., van Elsas, J. D., Smidt, H., Overmann, J., & Sipkema, D. (2018). The multi-omics promise in context: From sequence to microbial isolate. *Critical Reviews in Microbiology*, *44*, 212–229. https://doi.org/10.1080/1040841X.2017.1332003.

Hamilton, W. A. (2003). Microbially influenced corrosion as a model system for the study of metal microbe interactions: A unifying electron transfer hypothesis. *Biofouling*, *19*, 65–76.

Hennig, B. P., Velten, L., Racke, I., Tu, C. S., Thoms, M., Rybin, V., et al. (2018). Large-scale low-cost NGS library preparation using a robust Tn5 purification and tagmentation protocol. *G3 (Bethesda)*, *8*, 79–89. https://doi.org/10.1534/g3.117.300257.

Hsu, C. W., Chen, T. E., Lo, K. Y., & Lee, Y. L. (2019). Inhibitive properties of benzyldimethyldodecylammonium chloride on microbial corrosion of 304 stainless steel in a *Desulfovibrio desulfuricans*-inoculated medium. *Materials (Basel)*, *12*, 307. https://doi.org/10.3390/ma12020307.

Iino, T., Ito, K., Wakai, S., Tsurumaru, H., Ohkuma, M., & Harayama, S. (2015). Iron corrosion induced by nonhydrogenotrophic nitrate-reducing *Prolixibacter* sp. strain MIC1-1. *Applied and Environmental Microbiology*, *81*, 1839–1846. https://doi.org/10.1128/AEM.03741-14.

Javaherdashti, R. (2016). *Microbiologically influenced corrosion: An engineering insight*. London: Springer.

Javed, M. A., Neil, W. C., Stoddart, P. R., & Wade, S. A. (2016). Influence of carbon steel grade on the initial attachment of bacteria and microbiologically influenced corrosion. *Biofouling*, *32*, 109–122. https://doi.org/10.1080/08927014.2015.

Jenal, U., Reinders, A., & Lori, C. (2017). Cyclic di-GMP: Second messenger extraordinaire. *Nature Reviews Microbiology*, *15*, 271–284. https://doi.org/10.1038/nrmicro.2016.190.

Jia, R., Yang, D., Xu, D., & Gu, T. (2017). Electron transfer mediators accelerated the microbiologically influence corrosion against carbon steel by nitrate reducing *Pseudomonas aeruginosa* biofilm. *Bioelectrochemistry*, *118*, 38–46. https://doi.org/10.1016/j.bioelechem.2017.06.013.

Jia, R., Yang, D., Xu, J., Xu, D., & Gu, T. (2017). Microbiologically influenced corrosion of C1018 carbon steel by nitrate reducing *Pseudomonas aeruginosa* biofilm under organic carbon starvation. *Corrosion Science*, *127*, 1–9. https://doi.org/10.1016/j.corsci.2017.08.007.

Kasai, T., Kouzuma, A., Nojiri, H., & Watanabe, K. (2015). Transcriptional mechanisms for differential expression of outer membrane cytochrome genes omcA and mtrC in *Shewanella oneidensis* MR-1. *BMC Microbiology*, *12*, 68. https://doi.org/10.1186/s12866-015-0406-8.

Kaster, K. M., Bonaunet, K., Berland, H., Kjeilen-Eilertsen, G., & Brakstad, O. G. (2009). Characterisation of culture-independent and -dependent microbial communities in a high-temperature offshore chalk petroleum reservoir. *Antonie Van Leeuwenhoek*, *96*, 423–439. https://doi.org/10.1007/s10482-009-9356-1.

Kato, S. (2016). Microbial extracellular electron transfer and its relevance to iron corrosion. *Microbial Biotechnology*, *9*, 141–148. https://doi.org/10.1111/1751-7915.12340.

Kato, S., Yumoto, I., & Kamagata, Y. (2015). Isolation of acetogenic bacteria that induce biocorrosion by utilizing metallic iron as the sole electron donor. *Applied and Environmental Microbiology*, *81*, 67–73. https://doi.org/10.1128/AEM.02767-14.

Khan, A., Qurashi, A., Badeghaish, W., Noui-Mehidi, M. N., & Aziz, M. A. (2020). Frontiers and challenges in electrochemical corrosion monitoring; Surface and downhole applications. *Sensors (Basel)*, *20*, 6583. https://doi.org/10.3390/s20226583.

Kirchman, D. (2002). The ecology of Cytophaga–Flavobacteria in aquatic environments. *FEMS Microbiology Ecology*, *39*, 91–100. https://doi.org/10.1016/s0168-6496(01)00206-9.

Koch, G. H., Brongers, M. P. H., Thompson, N. G., Virmani, Y. P., & Payer, J. H. (2018). *Corrosion cost and preventive strategies in the United States*. NACE International.

Konstantinidis, K. T., & Tiedje, J. M. (2005). Genomic insights that advance the species definition for prokaryotes. *Proceedings of the National Academy of Sciences of the United States of America*, *102*, 2567–2572. https://doi.org/10.1073/pnas.0409727102.

Kouzuma, A., Kasai, T., Hirose, A., & Watanabe, K. (2015). Catabolic and regulatory systems in *Shewanella oneidensis* MR-1 involved in electricity generation in microbial fuel cells. *Frontiers in Microbiology*, *6*, 609. https://doi.org/10.3389/fmicb.2015.00609.

Kouzuma, A., Kato, S., & Watanabe, K. (2015). Microbial interspecies interactions: Recent findings in syntrophic consortia. *Frontiers in Microbiology*, *6*, 477. https://doi.org/10.3389/fmicb.2015.00477.

Kumaraswamy, R., Ebert, S., Gray, M. R., Fedorak, P. M., & Foght, J. M. (2011). Molecular- and cultivation-based analyses of microbial communities in oil field water and in microcosms amended with nitrate to control H_2S production. *Applied Microbiology and Biotechnology*, *89*, 2027–2038. https://doi.org/10.1007/s00253-010-2974-8.

Lahme, S., Mand, J., Longwell, J., Smith, R., & Enning, D. (2021). Severe corrosion of carbon steel in oil field produced water can be linked to methanogenic Archaea containing a special type of [NiFe] hydrogenase. *Applied and Environmental Microbiology*, *87*, e01819–e01820. https://doi.org/10.1128/AEM.01819-20.

Lanneluc, I., Langumier, M., Sabot, R., Jeannin, M., Refait, P., & Sablé, S. (2015). On the bacterial communities associated with the corrosion product layer during the early stages of marine corrosion of carbon steel. *International Biodeterioration and Biodegradation*, *99*, 55–65. https://doi.org/10.1016/j.ibiod.2015.01.003.

Larsen, J., Rasmussen, K., Pedersen, H. S., Sørensen, K. B., Lundgaard, T., & Skovhus, T. L. (2010). Consortia of MIC bacteria and Archaea causing pitting corrosion in top side oil production facilities. In *NACE-international corrosion conference series* (pp. 14–18).

Leloup, J., Loy, A., Knab, N. J., Borowski, C., Wagner, M., & Jørgensen, B. B. (2007). Diversity and abundance of sulfate-reducing microorganisms in the sulfate and methane zones of a marine sediment, Black Sea. *Environmental Microbiology*, *9*, 131–142. https://doi.org/10.1111/j.1462-2920.2006.01122.x.

Li, X., Duan, J., Xiao, H., Li, Y., Liu, H., Guan, F., et al. (2017). Analysis of bacterial community composition of corroded steel immersed in Sanya and Xiamen seawaters in China via method of Illumina MiSeq sequencing. *Frontiers in Microbiology*, *8*, 1737. https://doi.org/10.3389/fmicb.2017.01737.

Li, H., Xu, D., Li, Y., Feng, H., Liu, Z., Li, X., et al. (2015). Extracellular electron transfer is a bottleneck in the microbiologically influenced corrosion of C1018 carbon steel by the biofilm of sulfate-reducing bacterium *Desulfovibrio vulgaris*. *PLoS One*, *10*, e0136183. https://doi.org/10.1371/journal.pone.0136183.

Li, X. X., Yang, T., Mbadinga, S. M., Liu, J. F., Yang, S. Z., Gu, J. D., et al. (2017). Responses of microbial community composition to temperature gradient and carbon steel corrosion in production water of petroleum reservoir. *Frontiers in Microbiology*, *8*, 2379. https://doi.org/10.3389/fmicb.2017.02379.

Lian, T., Mu, Y., Jin, J., Ma, Q., Cheng, Y., Cai, Z., et al. (2019). Impact of intercropping on the coupling between soil microbial community structure, activity, and nutrient-use efficiencies. *PeerJ*, *7*, e6412. https://doi.org/10.7717/peerj.6412.

Liang, R., Davidova, I., Hirano, S. I., Duncan, K. E., & Suflita, J. M. (2019). Community succession in an anaerobic long-chain paraffin-degrading consortium and impact on chemical and electrical microbially influenced iron corrosion. *FEMS Microbiology Ecology*, *95*, fiz111. https://doi.org/10.1093/femsec/fiz111.

Liang, R., Grizzle, R. S., Duncan, K. E., McInerney, M. J., & Suflita, J. M. (2014). Roles of thermophilic thiosulfate-reducing bacteria and methanogenic archaea in the biocorrosion of oil pipelines. *Frontiers in Microbiology*, *5*(5), 89. https://doi.org/10.3389/fmicb.2014.00089.

Liduino, V. S., Lutterbach, M. T. S., & Sérvulo, E. F. C. (2018). Biofilm activity on corrosion of API 5L X65 steel weld bead. *Colloids and Surfaces B: Biointerfaces*, *172*, 43–50. https://doi.org/10.1016/j.colsurfb.2018.08.026.

Lin, W., Yu, Z., Chen, X., Liu, R., & Zhang, H. (2013). Molecular characterization of natural biofilms from household taps with different materials: PVC, stainless steel, and cast iron in drinking water distribution system. *Applied Microbiology and Biotechnology*, *97*, 8393–8401. https://doi.org/10.1007/s00253-012-4557-3.

Little, B. J., & Lee, J. S. (2014). Microbiologically influenced corrosion: An update. *International Materials Review*, *59*, 384–393. https://doi.org/10.1038/nrmicro.2016.94.

Lopes, F. A., Perrin, S., & Féron, D. (2007). A dual-electrochemical cell to study the biocorrosion of stainless steel. *Water Science and Technology*, *55*, 499–504. https://doi.org/10.2166/wst.2007.296.

Lovley, D. R. (2012). Electromicrobiology. *Annual Review of Microbiology*, *66*, 391–409. https://doi.org/10.1146/annurev-micro-092611-150104.

Lovley, D. R., & Phillips, E. J. (1988). Novel mode of microbial energy metabolism: Organic carbon oxidation coupled to dissimilatory reduction of iron or manganese. *Applied and Environmental Microbiology*, *54*, 1472–1480.

Lugauskas, A., Prosyčevas, I., Ramanauskas, R., Grigucevičienė, A., Selskienė, A., & Pakštas, V. (2009). The influence of micromycetes on the corrosion behaviour of metals (steel, al) under conditions of the environment polluted with organic substances. *Materials Science*, *15*, 224–235.

Maia, M., Capão, A. S. V., & Procópio, L. (2019). Biosurfactant produced by oil-degrading *Pseudomonas putida* AM-b1 strain with potential for microbial enhanced oil recovery. *Bioremediation Journal*, 302–310. https://doi.org/10.1080/10889868.2019.1669527.

Mand, J., Park, H. S., Okoro, C., Lomans, B. P., Smith, S., Chiejina, L., et al. (2016). Microbial methane production associated with carbon steel corrosion in a Nigerian oil field. *Frontiers in Microbiology*, *6*, 1538. https://doi.org/10.3389/fmicb.2015.01538.

Marchesi, J. R., & Ravel, J. (2015). The vocabulary of microbiome research: A proposal. *Microbiome*, *3*, 31. https://doi.org/10.1186/s40168-015-0094-5.

Marsili, E., Baron, D. B., Shikhare, I. D., Coursolle, D., Gralnick, J. A., & Bond, D. R. (2008). *Shewanella* secretes flavins that mediate extracellular electron transfer. *Proceedings of the National Academy of Sciences of the United States of America*, *105*, 3968–3973. https://doi.org/10.1073/pnas.0710525105.

Marty, F., Gueuné, H., Malard, E., Sánchez-Amaya, J. M., Sjögren, L., Abbas, B., et al. (2014). Identification of key factors in accelerated low water corrosion through experimental simulation of tidal conditions: Influence of stimulated indigenous microbiota. *Biofouling*, *30*, 281–297. https://doi.org/10.1080/08927014.2013.864758.

McBeth, J. M., & Emerson, D. (2016). In situ microbial community succession on mild steel in estuarine and marine environments: Exploring the role of iron-oxidizing bacteria. *Frontiers in Microbiology*, *7*, 767. https://doi.org/10.3389/fmicb.2016.00767.

McBeth, J. M., Little, B. J., Ray, R. I., Farrar, K. M., & Emerson, D. (2011). Neutrophilic iron-oxidizing "zetaproteobacteria" and mild steel corrosion in nearshore marine environments. *Applied and Environmental Microbiology*, *77*, 1405–1412. https://doi.org/10.1128/AEM.02095-10.

Mitbavkar, S., & Anil, A. C. (2000). Diatom colonization on stainless steel panels in estuarine waters of Goa, west coast of India. *Indian Journal of Geo-Marine Sciences*, *29*, 273–276.

Mori, K., Tsurumaru, H., & Harayama, S. (2010). Iron corrosion activity of anaerobic hydrogen-consuming microorganisms isolated from oil facilities. *Journal of Bioscience and Bioengineering*, *110*, 426–430. https://doi.org/10.1016/j.jbiosc.2010.04.012.

Moura, V., Ribeiro, I., Moriggi, P., Capão, A., Salles, C., Bitati, S., et al. (2018). The influence of surface microbial diversity and succession on microbiologically influenced corrosion of steel in a simulated marine environment. *Archives of Microbiology*, *200*, 1447–1456. https://doi.org/10.1007/s00203-018-1559-2.

Myers, J. M., & Myers, C. R. (2001). Role for outer membrane cytochromes OmcA and OmcB of *Shewanella putrefaciens* MR-1 in reduction of manganese dioxide. *Applied and Environmental Microbiology*, *67*, 260–269.

Myers, C. R., & Nealson, K. H. (1988). Bacterial manganese reduction and growth with manganese oxide as the sole electron acceptor. *Science*, *240*, 1319–1321.

Narihiro, T., & Kamagata, Y. (2013). Cultivating yet-to-be cultivated microbes: The challenge continues. *Microbes and Environments*, *28*, 163–165. https://doi.org/10.1264/jsme2.ME2802rh.

Nealson, K. H., & Rowe, A. R. (2016). Electromicrobiology: Realities, grand challenges, goals and predictions. *Microbial Biotechnology*, *9*, 595–600. https://doi.org/10.1111/1751-7915.12400.

Neria-González, I., Wang, E. T., Ramírez, F., Romero, J. M., & Hernández-Rodríguez, C. (2006). Characterization of bacterial community associated to biofilms of corroded oil pipelines from the southeast of Mexico. *Anaerobe*, *12*, 122–133. https://doi.org/10.1016/j.anaerobe.2006.02.001.

Nichols, D. (2007). Cultivation gives context to the microbial ecologist. *FEMS Microbiology Ecology*, *60*, 351–357.

Okabe, S., Odagiri, M., Ito, T., & Satoh, H. (2007). Succession of sulfur-oxidizing bacteria in the microbial community on corroding concrete in sewer systems. *Applied and Environmental Microbiology*, *73*, 971–980. https://doi.org/10.1128/AEM.02054-06.

Okabe, S., Yasuda, T., & Watanabe, Y. (1997). Uptake and release of inert fluorescence particles by mixed population biofilms. *Biotechnology and Bioengineering*, *53*, 459–469. https://doi.org/10.1002/(SICI)1097-0290.

Okoro, C., Smith, S., Chiejina, L., Lumactud, R., An, D., Park, H. S., et al. (2014). Comparison of microbial communities involved in souring and corrosion in offshore and onshore oil production facilities in Nigeria. *Journal of Industrial Microbiology & Biotechnology*, *41*, 665–678. https://doi.org/10.1007/s10295-014-1401-z.

Oppenheimer-Shaanan, Y., Sibony-Nevo, O., Bloom-Ackermann, Z., Suissa, R., Steinberg, N., Kartvelishvily, E., et al. (2016). Spatio-temporal assembly of functional mineral scaffolds within microbial biofilms. *NPJ Biofilms and Microbiomes*, *2*, 15031. https://doi.org/10.1038/npjbiofilms.2015.31.

Park, H. S., Chatterjee, I., Dong, X., Wang, S. H., Sensen, C. W., Caffrey, S. M., et al. (2011). Effect of sodium bisulfite injection on the microbial community composition in a brackish-water-transporting pipeline. *Applied and Environmental Microbiology*, *77*, 6908–6917. https://doi.org/10.1128/AEM.05891-11.

Patel, S., Homaei, A., Patil, S., & Daverey, A. (2019). Microbial biosurfactants for oil spill remediation: Pitfalls and potentials. *Applied Microbiology and Biotechnology*, *103*, 27–37. https://doi.org/10.1007/s00253-018-9434-2.

Peccia, J., Marchand, E. A., Silverstein, J., & Hernandez, M. (2000). Development and application of small-subunit rRNA probes for assessment of selected *Thiobacillus* species and members of the genus *Acidiphilium*. *Applied and Environmental Microbiology*, *66*, 3065–3072. https://doi.org/10.1128/AEM.66.7.3065-3072.2000.

Pirbadian, S., Barchinger, S. E., Leung, K. M., Byun, H. S., Jangir, Y., Bouhenni, R. A., et al. (2014). *Shewanella oneidensis* MR-1 nanowires are outer membrane and periplasmic extensions of the extracellular electron transport components. *Proceedings of the National Academy of Sciences of the United States of America*, *111*, 12883–12888. https://doi.org/10.1073/pnas.1410551111.

Price, S. J., & Figueira, R. B. (2017). Corrosion protection systems and fatigue corrosion in offshore wind structures: Current status and future perspectives. *Coatings*, *7*, 25. https://doi.org/10.3390/coatings7020025.

Prithiraj, A., Otunniyi, I. O., Osifo, P., & van der Merwe, J. (2019). Corrosion behaviour of stainless and carbon steels exposed to sulphate—Reducing bacteria from industrial heat exchangers. *Engineering Failure Analysis*, *104*, 977–986. https://doi.org/10.1016/j.engfailanal.2019.06.042.

Procópio, L. (2019). The role of biofilms in the corrosion of steel in marine environments. *World Journal of Microbiology and Biotechnology*, *35*, 73. https://doi.org/10.1007/s11274-019-2647-4.

Procópio, L. (2020a). The era of 'omics' technologies in the study of microbiologically influenced corrosion. *Biotechnology Letters*, *42*, 341–356. https://doi.org/10.1007/s10529-019-02789-w.

Procópio, L. (2020b). Changes in microbial community in the presence of oil and chemical dispersant and their effects on the corrosion of API 5L steel coupons in a marine-simulated microcosm. *Applied Microbiology and Biotechnology*, *104*, 6397–6411. https://doi.org/10.1007/s00253-020-10688-8.

Procópio, L., Macrae, A., van Elsas, J. D., & Seldin, L. (2013). The putative α/β-hydrolases of *Dietzia cinnamea* P4 strain as potential enzymes for biocatalytic applications. *Antonie Van Leeuwenhoek*, *103*(3), 635–646. https://doi.org/10.1007/s10482-012-9847-3.

Rabaey, K., & Rozendal, R. A. (2010). Microbial electrosynthesis—Revisiting the electrical route for microbial production. *Nature Reviews Microbiology*, *8*, 706–716. https://doi.org/10.1038/nrmicro2422.

Rajala, P., Carpén, L., Vepsäläinen, M., Raulio, M., Sohlberg, E., & Bomberg, M. (2015). Microbially induced corrosion of carbon steel in deep groundwater environment. *Frontiers in Microbiology*, *6*, 647. https://doi.org/10.3389/fmicb.2015.00647.

Rajasekar, A., Anandkumar, B., Maruthamuthu, S., Ting, Y. P., & Rahman, P. K. (2010). Characterization of corrosive bacterial consortia isolated from petroleum-product-transporting pipelines. *Applied Microbiology and Biotechnology*, *85*, 1175–1188. https://doi.org/10.1007/s00253-009-2289-9.

Rauch, M. E., Graef, H. W., Rozenzhak, S. M., Jones, S. E., Bleckmann, C. A., Kruger, R. L., et al. (2006). Characterization of microbial contamination in United States Air Force aviation fuel tanks. *Journal of Industrial Microbiology & Biotechnology*, *33*, 29–36. https://doi.org/10.1007/s10295-005-0023-x.

Ren, H., Wang, W., Liu, Y., Liu, S., Lou, L., Cheng, D., et al. (2015). Pyrosequencing analysis of bacterial communities in biofilms from different pipe materials in a city drinking water distribution system of East China. *Applied Microbiology and Biotechnology*, *99*, 10713–11024. https://doi.org/10.1007/s00253-015-6885-6.

Reuter, J. A., Spacek, D. V., & Snyder, M. P. (2015). High-throughput sequencing technologies. *Molecular Cell*, *58*, 586–597. https://doi.org/10.1016/j.molcel.2015.05.004.

Rice, S. A., McDougald, D., Kumar, N., & Kjelleberg, S. (2005). The use of quorum-sensing blockers as therapeutic agents for the control of biofilm-associated infections. *Current Opinion in Investigational Drugs*, *6*, 178–184.

Richter, K., Schicklberger, M., & Gescher, J. (2012). Dissimilatory reduction of extracellular electron acceptors in anaerobic respiration. *Applied and Environmental Microbiology*, *78*, 913–921. https://doi.org/10.1128/AEM.06803-11.

Riesenfeld, C. S., Schloss, P. D., & Handelsman, J. (2004). Metagenomics: Genomic analysis of microbial communities. *Annual Review of Genetics*, *38*, 525–552. https://doi.org/10.1146/annurev.genet.38.072902.091216.

Salgar-Chaparro, S. J., Darwin, A., Kaksonen, A. H., & Machuca, L. L. (2020). Carbon steel corrosion by bacteria from failed seal rings at an offshore facility. *Scientific Reports*, *10*, 12287. https://doi.org/10.1038/s41598-020-69292-5.

Salta, M., Wharton, J. A., Blache, Y., Stokes, K. R., & Briand, J. F. (2013). Marine biofilms on artificial surfaces: Structure and dynamics. *Environmental Microbiology*, *11*, 2879–2893. https://doi.org/10.1111/1462-2920.

Serra, D. O., & Hengge, R. (2014). Stress responses go three dimensional-the spatial order of physiological differentiation in bacterial macrocolony biofilms. *Environmental Microbiology*, *16*, 1455–1471. https://doi.org/10.1111/1462-2920.

Shao, K., Ding, W., Wang, F., Li, H., Ma, D., & Wang, H. (2011). Emulsion PCR: A high efficient way of PCR amplification of random DNA libraries in aptamer selection. *PLoS One*, *6*, e24910. https://doi.org/10.1371/journal.pone.0024910.

Shartau, S. L., Yurkiw, M., Lin, S., Grigoryan, A. A., Lambo, A., Park, H. S., et al. (2010). Ammonium concentrations in produced waters from a mesothermic oil field subjected to nitrate injection decrease through formation of denitrifying biomass and anammox activity. *Applied and Environmental Microbiology*, *76*, 4977–4987. https://doi.org/10.1128/AEM.00596-10.

Shin, D., Lee, Y., Park, J., Moon, H. S., & Hyun, S. P. (2017). Soil microbial community responses to acid exposure and neutralization treatment. *Journal of Environmental Management*, *204*, 383–393. https://doi.org/10.1016/j.jenvman.2017.09.014.

Stams, A. J., de Bok, F. A., Plugge, C. M., van Eekert, M. H., Dolfing, J., & Schraa, G. (2006). Exocellular electron transfer in anaerobic microbial communities. *Environmental Microbiology*, *8*, 371–382. https://doi.org/10.1111/j.1462-2920.2006.00989.x.

Staniszewska, A., Kunicka-Styczyńska, A., Otlewska, A., Gawor, J., Gromadka, R., Żuchniewicz, K., et al. (2019). High-throughput sequencing approach in analysis of microbial communities colonizing natural gas pipelines. *Microbiologyopen*, *8*, e00806. https://doi.org/10.1002/mbo3.806.

Stetter, K. O., & Gaag, G. (1983). Reduction of molecular sulphur by methanogenic bacteria. *Nature*, *305*, 309–311. https://doi.org/10.1038/305309a0.

Su, H., Tang, R., Peng, X., Gao, A., & Han, Y. (2020). Corrosion behavior and mechanism of carbon steel influenced by interior deposit microflora of an in-service pipeline. *Bioelectrochemistry*, *132*, 107406. https://doi.org/10.1016/j.bioelechem.2019.107406.

Tan, J. L., Goh, P. C., & Blackwood, D. J. (2017). Influence of H_2S-producing chemical species in culture medium and energy source starvation on carbon steel corrosion caused by methanogens. *Corrosion Science*, *119*, 102–111. https://doi.org/10.1016/j.corsci.2017.02.014.

Thompson, J. F., & Steinmann, K. E. (2010). Single molecule sequencing with a HeliScope genetic analysis system. *Current Protocols in Molecular Biology*, *7*, 10. https://doi.org/10.1002/0471142727.mb0710s92.

Tian, H., Gao, P., Chen, Z., Li, Y., Li, Y., Wang, Y., et al. (2017). Compositions and abundances of sulfate-reducing and sulfur-oxidizing microorganisms in water-flooded petroleum reservoirs with different temperatures in China. *Frontiers in Microbiology*, *8*, 143. https://doi.org/10.3389/fmicb.2017.00143.

Turnbaugh, P. J., & Gordon, J. L. (2008). An invitation to the marriage of metagenomics and metabolomics. *Cell*, *134*, 708–713. https://doi.org/10.1016/j.cell.2008.08.025.

Uchiyama, T., Ito, K., Mori, K., Tsurumaru, H., & Harayama, S. (2010). Iron-corroding methanogen isolated from a crude-oil storage tank. *Applied and Environmental Microbiology*, *76*, 1783–1788. https://doi.org/10.1128/AEM.00668-09.

Usher, K. M., Kaksonena, A. H., & MacLeod, I. D. (2014). Marine rust tubercles harbour iron corroding archaea and sulphate reducing bacteria. *Corrosion Science*, *83*, 189–197. https://doi.org/10.1016/j.corsci.2014.02.014.

Venzlaff, H., Enning, D., Srinivasan, J., Mayrhofer, K. J. J., Hassel, A. W., Widdel, F., et al. (2013). Accelerated cathodic reaction in microbial corrosion of iron due to direct electron uptake by sulfate-reducing bacteria. *Corrosion Science*, *66*, 88–96. https://doi.org/10.1016/j.corsci.2012.09.006.

Vert, M., Doi, Y., Hellwich, K. H., Hess, M., Hodge, P., Przemyslaw, K., et al. (2012). Terminology for biorelated polymers and applications. *Pure and Applied Chemistry*, *84*, 377–410. https://doi.org/10.1351/PAC-REC-10-12-04.

Videla, H. A., & Herrera, L. K. (2005). Microbiologically influenced corrosion: Looking to the future. *International Microbiology*, *8*, 169–180.

Vigneron, A., Alsop, E. B., Chambers, B., Lomans, B. P., Head, I. M., & Tsesmetzis, N. (2016). Complementary microorganisms in highly corrosive biofilms from an offshore oil production facility. *Applied and Environmental Microbiology*, *82*, 2545–2554. https://doi.org/10.1128/AEM.03842-15.

Vigneron, A., Head, I. M., & Tsesmetzis, N. (2018). Damage to offshore production facilities by corrosive microbial biofilms. *Applied Microbiology and Biotechnology*, *102*, 2525–2533. https://doi.org/10.1007/s00253-018-8808-9.

Villahermosa, D., Corzo, A., Garcia-Robledo, E., González, J. M., & Papaspyrou, S. (2016). Kinetics of indigenous nitrate reducing sulfide oxidizing activity in microaerophilic wastewater biofilms. *PLoS One*, *11*(2), e0149096. https://doi.org/10.1371/journal.pone.0149096.

von Canstein, H., Ogawa, J., Shimizu, S., & Lloyd, J. R. (2008). Secretion of flavins by *Shewanella* species and their role in extracellular electron transfer. *Applied and Environmental Microbiology*, *74*, 615–623. https://doi.org/10.1128/AEM.01387-07.

Wang, B., Liu, W., Cai, W., Li, J., Yang, L., Li, X., et al. (2020). Reinjection oilfield wastewater treatment using bioelectrochemical system and consequent corrosive community evolution on pipe material. *Journal of Bioscience and Bioengineering*, *129*, 199–205. https://doi.org/10.1016/j.jbiosc.2019.09.001.

Watanabe, K., Manefield, M., Lee, M., & Kouzuma, A. (2009). Electron shuttles in biotechnology. *Current Opinion in Biotechnology*, *20*, 633–641. https://doi.org/10.1016/j.copbio.2009.09.006.

Weber, K. A., Achenbach, L. A., & Coates, J. D. (2006). Microorganisms pumping iron: Anaerobic microbial iron oxidation and reduction. *Nature Reviews Microbiology*, *4*, 752–764. https://doi.org/10.1038/nrmicro1490.

Wesolowska-Andersen, A., Bahl, M. I., Carvalho, V., Kristiansen, K., Sicheritz-Ponten, T., Gupta, R., et al. (2014). Choice of bacterial DNA extraction method from fecal material influences community structure as evaluated by metagenomic analysis. *Microbiome*, *2*, 19. https://doi.org/10.1186/2049-2618-2-19.

White, R. A., Bottos, E. M., Roy Chowdhury, T., Zucker, J. D., Brislawn, C. J., Nicora, C. D., et al. (2016). Moleculo long-read sequencing facilitates assembly and genomic binning from complex soil metagenomes. *mSystems*, *1*, e00045-16. https://doi.org/10.1128/mSystems.00045-16.

Wicker, T., Schlagenhauf, E., Graner, A., Close, T. J., Keller, B., & Stein, N. (2006). 454 sequencing put to the test using the complex genome of barley. *BMC Genomics*, *7*, 275. https://doi.org/10.1186/1471-2164-7-275.

Xu, D., & Gu, T. (2014). Carbon source starvation triggered more aggressive corrosion against carbon steel by the *Desulfovibrio vulgaris* biofilm. *International Biodeterioration and Biodegradation, 91*, 74–81.

Xu, C. M., Zhang, Y. H., & Cheng, G. X. (2007). Localized corrosion behavior of 316L stainless steel in the presence of sulfate- reducing and iron-oxidizing bacteria. *Materials Science and Engineering A, 443*, 235. https://doi.org/10.1016/j.msea.2006.08.110.

Zhang, G., Li, B., Liu, J., Luan, M., Yue, L., Jiang, X. T., et al. (2018). The bacterial community significantly promotes cast iron corrosion in reclaimed wastewater distribution systems. *Microbiome, 6*, 222. https://doi.org/10.1186/s40168-018-0610-5.

Zhang, Y., Ma, Y., Duan, J., Li, X., Wang, J., & Hou, B. (2019). Analysis of marine microbial communities colonizing various metallic materials and rust layers. *Biofouling, 35*, 429–442. https://doi.org/10.1080/08927014.2019.1610881.

Ziadi, I., El-Bassi, L., Bousselmi, L., & Akrout, H. (2020). Characterization of the biofilm grown on 304L stainless steel in urban wastewaters: Extracellular polymeric substances (EPS) and bacterial consortia. *Biofouling, 36*, 977–989. https://doi.org/10.1080/08927014.2020.1836163.

Further reading

Chang, Y. W., Fragkopoulos, A. A., Marquez, S. M., Kim, H. D., Angelini, T. E., & Fernández-Nieves, A. (2015). Biofilm formation in geometries with different surface curvature and oxygen availability. *New Journal of Physics, 17*, 033017. https://doi.org/10.1088/1367-2630/17/3/033017.

Davidova, I. A., Duncan, K. E., Choi, O. K., & Suflita, J. M. (2006). *Desulfoglaeba alkanexedens* gen. nov., sp. nov., an n-alkane-degrading, sulfate-reducing bacterium. *International Journal of Systematic and Evolutionary Microbiology, 56*, 2737–2742. https://doi.org/10.1099/ijs.0.64398-0.

Davis, J. R. (2006). *Corrosion of weldments* (1st ed.). Ohio: ASM International.

Forster, S. C. (2017). Illuminating microbial diversity. *Nature Reviews Microbiology, 15*, 578. https://doi.org/10.1038/nrmicro.2017.106.

Kojima, T., Takei, Y., Ohtsuka, M., Kawarasaki, Y., Yamane, T., & Nakano, H. (2005). PCR amplification from single DNA molecules on magnetic beads in emulsion: Application for high-throughput screening of transcription factor targets. *Nucleic Acids Research, 33*, e150. https://doi.org/10.1093/nar/gni143.

Lane, D. J., Pace, B., Olsen, G. J., Stahl, D. A., Sogint, M. L., & Pace, N. R. (1985). Rapid determination of 16S ribosomal RNA sequences for phylogenetic analyses. *Proceedings of the National Academy of Sciences of the United States of America, 82*, 6955–6959.

Li, X. X., Liu, J. F., Zhou, L., Mbadinga, S. M., Yang, S. Z., Gu, J. D., et al. (2017). Diversity and composition of sulfate-reducing microbial communities based on genomic DNA and RNA transcription in production water of high temperature and corrosive oil reservoir. *Frontiers in Microbiology, 8*, 1011. https://doi.org/10.3389/fmicb.2017.01011.

Liang, R., Davidova, I. A., Marks, C. R., Stamps, B. W., Harriman, B. H., Stevenson, B. S., et al. (2016). Metabolic capability of a predominant *Halanaerobium* sp. in hydraulically fractured gas wells and its implication in pipeline corrosion. *Frontiers in Microbiology, 7*, 988. https://doi.org/10.3389/fmicb.2016.00988.

Mand, J., & Enning, D. (2021). Oil field microorganisms cause highly localized corrosion on chemically inhibited carbon steel. *Microbial Biotechnology, 14*, 171–185. https://doi.org/10.1111/1751-7915.13644.

Newman, D. K., & Kolter, R. (2000). A role for excreted quinones in extracellular electron transfer. *Nature, 405*, 94–97. https://doi.org/10.1038/35011098.

Skovhus, T. L., Eckert, R. B., & Rodrigues, E. (2017). Management and control of microbiologically influenced corrosion (MIC) in the oil and gas industry—Overview and a North Sea case study. *Journal of Biotechnology, 20*(256), 31–45. https://doi.org/10.1016/j.jbiotec.2017.07.003.

Xu, D., Li, Y., Song, F., & Gu, T. (2013). Laboratory investigation of microbiologically influenced corrosion of C1018 carbon steel by nitrate reducing bacterium *Bacillus licheniformis*. *Corrosion Science, 77*, 385–390. https://doi.org/10.1016/j.corsci.2013.07.044.

Chapter 9

Biofilms in dairy industry

Krzysztof Skowron[a], Justyna Bauza-Kaszewska[b], Katarzyna Grudlewska-Buda[a], Natalia Wiktorczyk-Kapischke[a], Joanna Kwiecińska-Piróg[a], Ewa Wałecka-Zacharska[c], and Eugenia Gospodarek-Komkowska[a]

[a]*Department of Microbiology, Nicolaus Copernicus University in Toruń, Ludwik Rydygier Collegium Medicum in Bydgoszcz, Bydgoszcz, Poland,* [b]*Department of Microbiology and Food Technology, Jan and Jędrzej Śniadecki University of Technology in Bydgoszcz, Bydgoszcz, Poland,* [c]*Department of Food Hygiene and Consumer Health, Wrocław University of Environmental and Life Sciences, Wrocław, Poland*

1. Introduction

The emergence of biofilm in production spaces, regardless of the specificity of a given industries, carries a number of adverse consequences. In the case of a negative action of microorganisms in biofilm on any of the aspects of production, actions are taken to prevent its emergence or leading to its elimination. In some situations, however, the creation of biofilm can bring positive effects and be deliberately used, for example, for biological plant protection, corrosion prevention, wastewater treatment, or soil bioremediation (Morikawa, 2006; Muhammad et al., 2020).

The formation of biofilms by microorganisms accompanying food production is of particular importance. The presence of pathogenic bacteria in them creates a direct threat to consumers and broadly understood public health. In turn, microbes spoiling the food cause changes to the organoleptic properties of products, leading to the loss of their commercial value (Teh et al., 2014; Wagner et al., 2020). Direct contact with the consumer also imposes some restrictions on disinfection procedures applied in the plant. The used disinfectants on the one hand can not pose a danger to consumers, and on the other hand, they should effectively limit the presence of harmful microbes in the production space. In the case of biofilm, achieving such a compromise is extremely difficult, due to its increased, compared with planktonic forms, resistance to chemicals (Galié, García-Gutiérrez, Miguélez, Villar, & Lombó, 2018; Shi & Zhu, 2009).

The most important factors affecting the creation of biofilm are associated with the individual properties of a given microorganism, the trait suites, and the specificity of the environment in which biofilm arises (Van Houdt & Michiels, 2010). In the case of the dairy industry, the diversity and richness of microorganisms related to raw milk and processed products, materials used in production lines in direct contact with food, and production conditions make this branch of the food industry extremely vulnerable to the formation of biofilms and the resulting consequences.

A positive role of dairy products on human diet is commonly known and well documented. Their suitable supply in childhood and during puberty regulates the economy with mineral ingredients in the body reducing the effects of their shortages at later stages of life (Givens, 2020). Cow milk, which is the dominant raw material and product in dairy industry, composes of: water (87%), solids, or fats, proteins (casein and whey protein), lactose and various minerals (calcium, potassium, magnesium, phosphorus, sodium, iodine), and vitamins (ryboflavin, vitamin D, vitamin A, niacin) (Marangoni et al., 2019; Ryu et al., 2021). The benefits of drinking milk are usually associated with the nutritional value and biological properties of proteins contained in them. A special role that can be performed in the correct feed scheme determines the presence of nine amino acids necessary for humans. The fat content in cow milk varies from 3.3 to 6.4 g/100 g; however, depending on the consumer requirements, it may be subject to modifications. The final composition of milk is varied and depends on the factors such as the species and breed of animal, its life environment, or the way of nutrition (Burke et al., 2018; Davoodi et al., 2016; Gantner, Mijic, Baban, Skrtic, & Turalija, 2015; Roy, Ye, Moughan, & Singh, 2020).

The range of food produced from milk is very wide, and their attractiveness from both high nutritional value and organoleptic values. The most popular dairy products known and widely available include cream, cheese, butter, yoghurts, ice cream, and used as food components such as powdered milk, whey protein concentrates, and caseinates. In many places around the world, on a smaller scale, various types of specific dairy products, enjoying popularity and recognition of local consumers, are also produced (Bremer, Flint, Brooks, & Palmer, 2015; Burke et al., 2018). Fermented dairy products contain lactic fermentation bacteria, including probiotic, whose beneficial effects on various aspects of human health are widely known. Many of these products, including cheese, constitute a valuable calcium source for people with lactose intolerance (Silanikove, Leitner, & Merin, 2015). The ability to reduce

Understanding Microbial Biofilms. https://doi.org/10.1016/B978-0-323-99977-9.00023-5

fat content, e.g., in yoghurt, cheese, and hard cheese, contributes to high interest in such products among people using a diet with a reduced calorie content.

Forecasts of specialists regarding the development of the Dairy Industry Development take into account many factors. There is significant importance in this matter that emerging nutritional trends among consumers, even those of short term, which are the result of a temporary fashion for a given assortment of food products. It can be assumed that interest in vegan products, alternative to obtained from milk and its products, can reduce their consumption (Burke et al., 2018). On the other hand, shifting consumer preferences from dairy meat, as a main source of protein, can be treated as an incentive to increase the meaning of dairy industry. The growing consumer awareness of the nutritional value of milk and dairy products and the impact of their components on the proper functioning of the body contributed to the growth of their popularity during the Covid-19 pandemic (KBV Research, 2021). Certain controversy related to the dairy industry, on the other hand, the share of cattle breeding, necessary for milk production, in general greenhouse gas emissions (Burke et al., 2018).

2. Microorganisms in milk and dairy products

Milk and dairy products are generally included in the category of food easily spoiled (milk and cream) or semismoking (cheese, butter and yogurt) (Bremer et al., 2015). Rich in nutrients, the composition of milk and products produced from it, similar to the neutral pH and the presence of growth stimulants, e.g., orotic acid, pyrimidine precursor conducive to the development of microorganisms, makes them a great ground for the growth of many groups of microorganisms, including those with very high requirements of food (Mogha, Shah, Prajapati, & Chaudhari, 2014). The number of bacteria in 1 mL of raw milk may exceed 10^5–10^6 CFU, and their source is most often an external environment, animal udders, and equipment used for milking (Efimochkina et al., 2019; Hayes & Boor, 2001; Revelli, Sbodio, & Tercero, 2004). The most numerous group of microorganisms in raw milk are lactic fermentation bacteria, mainly *Lactococcus*, *Streptococcus*, *Lactobacillus*, *Leuconostoc*, and *Enterococcus* sp. and psychotrophic bacteria from types *Pseudomonas*, *Acinetobacter*, and *Aeromonas* (Quigley et al., 2013; Seale, Bremer, Flint, Brooks, & Palmer, 2015). The presence of *Lactobacillus* in milk after subjecting it to the pasteurization process results in their high number of primary or secondary impurities after heat treatment (Henning, 2001). Contamination after pasteurization is most often the effect of insuf ficient hygiene of personnel or contamination by direct contact with the equipment surfaces used for production and premises (Quigley et al., 2013).

The importance of microorganisms isolated from milk and products produced may be positive or negative, and their presence sometimes unacceptable (pathogens) or on the contrary—necessary to obtain a specific type of food, with specific, typical organoleptic properties (Lactic Acid Bacteria, LAB).

2.1 Lactic acid bacteria (LAB)

LAB are metabolically and physiologically related group of Gram-positive, catalase-negative, nonspore-forming (except for *Sporolactobacillus*) cocci and bacilli. They are anaerobes or aerotolerants that produce lactic acid through fermentative respiration (Björkroth & Koort, 2011). One of the criteria for LAB classification is the molar composition of DNA bases below 50% G+C, which phylogenetically includes these bacteria in the so-called *Clostridium* branch (De Filippis, Pasolli, & Ercolini, 2020). The LAB species associated with dairy products are primarily of the genera: *Lactococcus* and *Lactobacillus*, *Enterococcus*, *Leuconostoc*, *Pediococcus*, and *Streptococcus*. These bacteria have GRAS (Generally Recognized as Safe) status, and many of them have probiotic properties resulting from the production of bioactive metabolites such as bacteriocins, hydrogen peroxide, vitamins, organic acids, bioactive peptides, enzymes, and exopolysaccharides. The most important of the probiotic species are *Lactobacillus acidophilus*, *L. gasseri*, *L. helveticus*, *L. johnsonii*, *L. (para) casei*, *L. reuteri*, *L. plantarum*, *L. rhamnosus*, and *L. fermentum* and belonging to the genus *Bifidobacterium*: *Bifidobacterium bifidum*, *B. longum*, *B. animalis*, and *B. breve* (Linares et al., 2017; Widyastuti, *Rohmatussolihat*, & *Febrisiantosa*, 2014). Presence of LAB and their metabolites in fermented dairy products (yogurt, kefir) is considered to be the key factor determining the health-promoting properties of these products (Stiemsma, Nakamura, Nguyen, & Michels, 2020). They also improve the digestibility of proteins, inhibit the release of fatty acids by populating the human digestive tract, they support the proper functioning of the gut microbiome (García-Cano et al., 2019). Moreover, thanks to its activity enzymatic, LAB determine the organoleptic properties of the fermented dairy products—their specific taste, aromas, and texture as a result hydrolysis of protein, fat, and carbohydrates (García-Cano et al., 2019; Teh, Flint, & Bremer, 2015; Widyastuti et al., 2014). Their main metabolite, lactic acid, acts as a preservative food by lowering the pH to a value that prevents the growth of bacteria pathogenic and putrefactive (e.g., *Pseudomonas* spp.) (Stiemsma et al., 2020).

LAB inhabit the environment rich in nutrients, are isolated from the soil, plant surfaces (fruit, vegetables), but also the genital tract and the digestive tract of animals and humans (Miranda et al., 2021; Yu, Leveau, & Marco, 2020). The penetration of microorganisms into the originally sterile milk takes place in the animal's organism. LAB are the dominant group of bacteria isolated from raw milk, and the abundance of genera *Lactococcus*, *Streptococcus*, *Lactobacillus*, *Leuconostoc*, and *Enterococcus* spp. varies within limits of 10^1–10^4 CFU mL^{-1}(Quigley et al., 2013). In many products, however, LAB are additionally introduced into the raw material in the form of special industrial vaccines, the purpose of which is to effectively conduct lactic acid fermentation and obtain organoleptic characteristics appropriate for a given type of product. Depending on the product, combinations of different types and species of microorganisms, usually strains, are used in starter cultures *Lactococcus lactis*, *S. thermophilus*, *Leuconostoc* sp., and *Lactobacillus* sp. (Garneau & Moineau, 2011; Laranjo, Potes, & Elias, 2019). The starter cultures used in the production of fermented milk include *L.brevis*, *L. casei*, *L. acidophilus*, *L. bulgaricus*, *B. longum*, *B. animalis*, *Streptococcus thermophilus*, and *Lactococcus lactis*, yogurt—*L. acidophilus*, *B lactis*, *L. bulgaricus*, *S. thermophilus*, and *L. casei*. In turn, among the species isolated with camembert cheese, there were *S. thermophilus*, *Lactobacillus* spp., *L. lactis*, *Leuconostoc*, *Hafnia alvei*, and yeast from genus *Geotrichum* (Stiemsma et al., 2020).

In order to obtain high-quality cheese, the pasteurized milk is inoculated with an average amount of starter cultures of 10^7/mL. In practice, the production of 1 ton of cheese requires the activity of 10^{14} bacterial cells (Garneau & Moineau, 2011). As the cheese matures, the number of starter bacteria cells decreases and the number of nonground bacteria increases (Henning, 2001). In the production and maturation of cheese, in addition to starter bacteria (mainly *Lactococcus*), intentionally added to the milk during production, microorganisms that get into the milk or cheese curd, among others as environmental contamination, also play an important role. The types and strains of nonchartered bacteria in cheese also depend on their initial number in the milk used for production and the ability to form a biofilm in the production environment (Henning, 2001). These include, in addition to the typical LAB (*Lactobacillus*, *Pediococcus*, *Enterococcus*, and *Leuconostoc*), also bacteria of the genera *Micrococcus* and *Staphylococcus*. Their proteolytic and lipolytic activity contributes to the maturation of the cheese (Ogier, Son, Gruss, Tailliez, & Delacroix-Buchet, 2002). Nonchartered lactic acid bacteria (NSLAB) in the first stage of cheese maturation may be the dominant group of microorganisms, and their number from the initial stage reaching 2–3 log CFU g^{-1} even increases to 8 log CFU g^{-1} (Agarwal, Sharma, Swanson, Yüksel, & Clark, 2006). Unfortunately, the results of some studies indicate that NSLAB strains may also be responsible for the unfavorable quality and commercial value of ripening cheeses (Weber, Liedtke, Plattes, & Lipski, 2019).

2.2 Pathogenic microorganisms in dairy products

Pathogenic microorganisms in raw milk enter it from various sources, which are elements of the primary (farm) or proper food production environment (Parkar, Flint, Teh, Brooks, & Bremer, 2015). Many pathogens inhabit the skin and other parts of the body, especially the udder of an infected animal and its secretions, from where they can pass into the milk. Subclinical mastitis (MS), which is common in dairy herds, seems to be a particular problem. The result of this productive disease is a decrease in the milk yield of cows and changes in the composition of the milk (Adriaens et al., 2021). Microbes isolated from animals with MS include *S. aureus*, *non-aureus* staphylococci, *Streptococcus agalactiae*, environmental streptococci, Gram-negative bacteria, *Corynebacterium* spp., and yeast. Despite the absence of an inflammatory reaction and changes in the gland, the appearance of pathogens triggers the body's response in the form of an increase in the number of somatic cells in milk (Goncalves et al., 2020). In the absence of typical disease symptoms, the correct diagnosis of the infection is difficult. Predicting behavioral changes that are early heralds of the disease can be based on the analysis of interactions in the immune system—central nervous system (Caplen & Held, 2021).

Another source of contamination is direct or indirect contact with contaminated elements of the dairy farm's environment, such as litter, fodder, water, or environmental surfaces, mainly floors and walls. For indirect contact, pathogen vectors can be air, farm workers, or cleaning systems (Parkar et al., 2015; Quigley et al., 2013; Simoes, Simoes, & Vieira, 2010). In order to ensure microbiological safety, the dairy industry uses technological processes that eliminate pathogens from milk and dairy products. The most common use for this purpose is the pasteurization process. Unpasteurized milk and dairy products are 150 times more likely than pasteurized products to cause foodborne illness and cause 13 times more hospitalizations (Burke et al., 2018). Contaminated raw milk, as a source of biofilms created by pathogens on elements of production lines, may in turn cause secondary contamination of dairy products, even those previously pasteurized (Simoes et al., 2010). Some pathogenic bacteria are able to survive and grow in the production environment after pasteurization (Oliver, Jayarao, & Almeida, 2005).

Among pathogenic bacteria found in both raw milk and dairy products, the most important are species of the genera *Salmonella*, *Escherichia coli*, *Listeria monocytogenes*, *S. aureus*, and *Campylobacter jejuni* (Table 1). The presence of *Mycobacterium bovis*, which is also dangerous for humans, is a problem in some parts of the world (Bolaños, Paula, Guerra, Franco, & Ribeiro, 2017). Mold fungi metabolites, mainly aflatoxins, found in milk, also have a highly toxic effect on human health (Bellio et al.,

TABLE 1 Selected studies on the outbreaks related to milk and dairy products consumption in different countries from 2014 to 2021.

Pathogen		Year	Country	Product	Number of cases	Hospitalizations/ fatalities	Reference
Escherichia coli (STEC) serotypes	O103:H2	2017	Germany	Raw milk	39 confirmed cases	4 hospitalizations	Mylius et al. (2018)
	O26	2019	France	Soft raw milk cheeses	16 confirmed cases	14 hospitalizations (HUS)	Jones et al. (2019)
	O121	2018	Canada	Raw milk Gouda-like cheese	7 confirmed cases	No hospitalizations	Boyd et al. (2021)
Salmonella Newport ST118		2018	France	Raw milk cheese	147 cases (133 confirmed, 13 probable, 1 possible)	13 hospitalizations	Robinson et al. (2020)
Salmonella Agona		2017	France	Infant milk products	37 confirmed cases	18 hospitalizations	Jourdan-da Silva et al. (2018)
Listeria monocytogenes		2020–21	United States	Soft cheeses—queso fresco	13 cases	12 hospitalizations 1 death (2 pregnancy losses 1 premature birth)	Centers for Disease Control and Prevention CDC (2021)
		2017	United States	Soft raw milk cheese	8 cases	8 hospitalizations 2 deaths	Centers for Disease Control and Prevention CDC (2017)
		2015–16	Canada	Pasteurized chocolate milk	34 cases	32 hospitalizations 4 deaths	Hanson et al. (2019)
		2018	Switzerland	Pasteurized soft cheese	34 confirmed cases	10 deaths	Nüesch-Inderbinen et al. (2021)
Campylobacter jejuni		2014	United States	Raw milk	99 cases (59 confirmed and 40 probable)	10 hospitalizations 1 death	Davis et al. (2016)
Staphylococcus aureus		2014	Switzerland	Soft raw milk cheese	14 cases exhibiting clinical signs of staphylococcal intoxication	1 person sought medical care and treated	Johler et al. (2015)

2016; Puga-Torres, Salazar, Cachiguango, Cisneros, & Gómez-Bravo, 2020).

2.2.1 *Escherichia coli*

Bacteria from the coli group, to which the *E. coli* species belongs, colonize the large intestines of humans and animals, hence their presence in raw milk and dairy products most often results from fecal contamination during the milking process or improper hygienic practices (Anderson et al., 2011; Chaleshtori, Arani, Aghadavod, Naseri, & Chaleshtori, 2017). Cows infected with *E. coli* can excrete up to 10^8 CFU/mL, but not all strains are pathogenic (Hayes & Boor, 2001). The most dangerous are the verotoxigenic Shiga toxin-producing *E. coli* (STEC) strains especially serotype 0157:H7 (Chaleshtori et al., 2017; Teh et al.,

2015). Dairy cattle are considered the main reservoir of these bacteria. They cause diarrhea, but also hemorrhagic colitis and hemolytic uremic syndrome (HUS), and their infectious dose is only 10 cells (Garbaj et al., 2016; Weimer, 2001). In the studies of Rosario et al. (2021), STEC isolates were detected in 4% of the samples of pasteurized dairy products (cream, cheese). STEC O174:H2 and O116: H48 serotypes were detected in approx. 9% of the tested samples of a semihard raw milk cheese (Cortimiglia, Borney, Bassi, & Cocconcelli, 2021)

2.2.2 *Salmonella* spp.

Salmonella spp., similar to *E. coli*, belonging to the *Enterobacteriaceae* family and the order Enterobacterales are responsible for food poisoning many times caused by the consumption of contaminated food, including milk and dairy products. The strains most often isolated from food belong to the serotypes *Salmonella* Enteritidis, *S.* Dublin, and *S.* Typhimurium (Anderson et al., 2011; Castañeda-Salazar, Pulido-Villamarín, Ángel-Rodríguez, Zafra-Alba, & Oliver-Espinosa, 2021; Lobacz & Zulewska, 2021). The number of *Salmonella* bacteria excreted daily in the feces of sick animals can be as high as 10^{14} CFU (Castañeda-Salazar et al., 2021). Elimination of these pathogens from raw milk is difficult due to their presence in the feces of both sick and asymptomatic *Salmonella* carriers. This limits the possibility of separating them from the rest of the herd and inhibits the spread of pathogenic bacteria in the environment of the production farm (Teh et al., 2015).

2.2.3 *Listeria monocytogenes*

The particular risk resulting from poisoning with food contaminated with *L. monocytogenes* is related mainly to the accompanying high mortality, reaching even several dozen percent. The most severe cases of listeriosis concern pregnant women, newborns, and infants, as well as elderly people treated with immunosuppressants (Hunt et al., 2012; Mansouri-Najand, Kianpour, Sami, & Jajarmi, 2015; Simoes et al., 2010). Also, the resistance of these bacteria to adverse environmental conditions, including high salinity, extreme pH, and especially low temperature, contributes to the increased risk of *Listeria* spp. in milk and dairy products. The psychotrophic nature of *L. monocytogenes* allows replication in chilled, ready-to-eat foods that have been contaminated during processing and packaging (Simoes et al., 2010). The source of these bacteria in raw milk is primarily the production environment, less often the organism of a sick animal (Hunt et al., 2012; Mansouri-Najand et al., 2015; Oliver et al., 2005). The factors determining the degree of milk contamination by *L. monocytogenes* include the farm management system, feed quality, herd size, geographical location, season, and animal housing hygiene. Despite the sensitivity of *L. monocytogenes* to the effects of typical temperatures used in pasteurization processes, inadequate temperature and procedural errors may lead to postpasteurization of contamination (Teh et al., 2015).

2.2.4 *Staphylococcus aureus*

S. aureus is a pathogen that enters raw milk from the udder of dairy cows affected by mastitis caused by these bacteria (Mekonnen et al., 2018). The direct cause of food poisoning by food products contaminated with *S. aureus* is produced during their growth of enterotoxins. The presence of these substances in raw milk is usually directly correlated with the high number of bacteria themselves, and the action of high temperature destroying their cells does not deactivate the thermostable enterotoxin (Dai et al., 2019). Contamination of finished products is the result of the use of low-quality raw materials or noncompliance with hygiene standards by employees (Palmer, Seale, & Flint, 2015).

2.2.5 *Campylobacter jejuni*

C. jejuni is a mesophilic and microaerophilic species, which, despite being sensitive to desiccation and the presence of oxygen, has developed survival mechanisms that ensure its protection against atmospheric oxygen, heat shock, low pH, or osmotic stress, and the possibility of growth in conditions of nutrient deficiency. These properties allow it to survive under stress and colonize the infected host (Jaakkonen, Kivistö, Aarnio, Kalekivi, & Hakkinen, 2020). *C. jejuni* is the leading cause of acute bacterial gastroenteritis in humans worldwide. Although poisoning occurs most often from the consumption of contaminated poultry meat, the pathogen is also common in cattle, and the consumption of raw cow's milk has been involved in several epidemics of campylobacteriosis (Anderson et al., 2011). *C. jejuni* has been detected in raw and improperly pasteurized milk and cheese (Teh et al., 2015).

2.3 Spoilage microorganisms in dairy industry

Microorganisms responsible for the spoilage of milk and milk products, through their metabolic activity, can lead to a significant reduction or complete loss of their nutritional value, sensory properties, and commercial usefulness (Teh et al., 2015). Their widespread presence in the production environment makes it virtually impossible to eliminate them from raw milk (Perin, Pereira, Bersot, & Nero, 2018). The group of microorganisms of greatest importance in terms of the quality and organoleptic characteristics of dairy products are psychrotrophs, which have developed a number of adaptive mechanisms that allow them to develop metabolic activity at cooling temperatures. The most important psychrotrophic bacteria causing spoilage of milk and dairy products are genera *Pseudomonas*, *Bacillus*, and

Lactobacillus (Anderson et al., 2011; de Oliveira, Favarin, Luchese, & McIntosh, 2015; Teh et al., 2015). They produce lipases and proteases already during the initial processing of raw milk, but thanks to the resistance of these enzymes to the temperature used during pasteurization, they can also reduce the value of processed products (Ribeiro Júnior et al., 2018; Simoes et al., 2010). Organoleptic changes resulting from the action of enzymes hydrolyzing milk proteins, mainly casein, may consist of coagulation and gelation of UHT milk. Bittering of cheeses results from the appearance of peptides from the decomposition of casein, while rancidity is caused by the action of lipases, which, by hydrolyzing triglycerides, release free fatty acids—butyric and caproic acids responsible for the unpleasant taste of milk and dairy products (Perin et al., 2018; Ribeiro Júnior et al., 2018).

A particular problem for many branches of the food industry, including the dairy industry, is the contamination of raw materials and products with bacteria that produce heat-resistant spores. Spore bacteria are common in the dairy farm environment where they have been detected in feed, manure, soil, and washing water. The source of contamination of fresh milk is most often the teats of cows that have been in contact with the soil or litter. The concentration of spores in 1 mL of fresh milk can even exceed 5000 (André, Vallaeys, & Planchon, 2017; Teh et al., 2015; Warren, 2001).

Species of special importance in the dairy industry are *Bacillus cereus*, *B. licheniformis*, *B. coagulans*, *B. subtilis*, *Geobacillus stearothermophilus*, and *Anoxybacillus flavithermus* (Gopal et al., 2015). The last two, as typical thermophiles, pose a particular threat in the production of highly processed food, e.g., powdered milk (Lücking, Stoeckel, Atamer, Hinrichs, & Ehling-Schulz, 2013). On the other hand, in refrigeration conditions, spoilage of pasteurized dairy products can be caused by a psychrotrophic species of *Paenibacillus* that produces a large variety, which grows there in the absence of other bacteria, e.g., *Pseudomonas* (Ranieri et al., 2012).

3. Biofilms in the dairy industry

Research on the role of biofilms in food production most often focuses on their production by harmful or pathogenic species and the accompanying problems with the effective elimination of these specific structures in disinfection processes. However, the specificity of the dairy industry is that the importance of biofilms produced by microorganisms accompanying the processes of milk processing and dairy production can be both negative and positive.

As in other branches of the food industry, biofilms formed by microorganisms can be the cause of spoilage of raw materials and finished products and the source of the spread of dangerous human and animal pathogens (Bremer et al., 2015). Even those produced by microorganisms that are neutral in terms of health risk or product quality changes can become a habitat and shelter for pathogenic bacteria, e.g., *L. monocytogenes* (Weber et al., 2019). In addition to reducing the safety, durability, as well as the nutritional and organoleptic value of dairy products, biofilms are also a technological problem, contributing to the reduction of milk and liquid products flow in blocked pipes, shortening the operation time of individual installations, corrosion of stainless steel components, or reduction of heat transfer through plate heat exchangers (Kumari & Sarkar, 2016).

The process of cell adhesion to various surfaces, considered necessary and crucial for biofilm formation, and the formation of the proper biofilm were observed at every stage in the dairy industry (Teh et al., 2014). Biofilms produced by bacteria in the udder of cows with mastitis are a source of contamination of raw milk and reduce the effectiveness of antibiotics used to eliminate these pathogens from animal organisms (Rudenko et al., 2021; Schönborn & Krömker, 2016). The presence of biofilms has been reported on the surfaces of teat cups on dairy farms, in pipe bends, conveyor belts, floors, or plate heat exchangers (Sharma & Anand, 2002; Teh et al., 2014). The places favorable for their formation are, in particular, corners, slots, gaskets, valves, and joints of individual elements of the production line (Simoes et al., 2010).

Bremer et al. (2015) introduced the division of biofilms produced by microorganisms in the dairy industry into two categories: process, quickly formed on the surfaces of direct contact with the product and subject to the selective pressure of factors typical for a given processing stage, and environmental, the formation of which takes up to several weeks, and are created in the specific, niche conditions of hard-to-reach places and around the sewage system.

Both the ability of bacteria to adhere and the formation of biofilm depend on many factors that can both favor and inhibit the speed and intensity of these processes (Teh et al., 2014; Weber et al., 2019). The presence in the culture medium of various substrates, similar to substances naturally present in the environment of milk and dairy production, may stimulate the formation of biofilms by the microorganisms inhabiting them. Dancer, Mah, and Kang (2009) found that the nitrogen source influences the biofilm formation in skim milk by *Cronobacter* spp. to a greater extent than the carbohydrate source. In the case of *Staphylococcus uberis* strains not showing the potential for biofilm formation, an intensification of biofilm production was observed with the addition of a small amount of milk. Since this species is isolated from cows suffering from mastitis, milk is treated as an optimal and natural growth medium for it. The main inducers of biofilm growth in their case were α and β-casein (Varhimo et al., 2011). The presence of milk proteins in the culture medium also favored the adhesion and biofilm formation by the psychrotrophic strain of *P. fluorescens* (Aswathanarayan & Vittal, 2014). On the other hand, Thiran et al. (2018) observed that the *Pseudomonas* spp. strains isolated from raw milk showed a higher ability to form biofilm in culture in the classic TSB growth medium than in a medium with a composition similar to milk.

Milk, dairy products, and dairy by-products (skimmed milk, buttermilk) contain organic compounds such as proteins or fats, the molecules of which in direct contact with working surfaces create on them so-called conditioning layer. Its formation largely determines the adhesion of bacteria to the surface and then the formation of biofilm by them (Dat et al., 2014; Gomes & Mergulhão, 2017). When analyzing the influence of conditioning, both intentional and accidental, on the development of biofilm, it is necessary to take into account the material from which the contact surface with the product is made (stainless steel, glass, rubber, plastic) and the microtopography of this surface, taking into account the presence of cracks, gaps, and scratches (Mogha et al., 2014). The results of most studies suggest that the effectiveness of cell adhesion is the highest on rough and hydrophobic surfaces; however, the existence of such a relationship is not clearly confirmed. It seems that the differences in the reported findings result from the degree of roughness specific for a given type of material as well as the intensity and depth of scratches or microcracks resulting from intensive mechanical cleaning or progressive natural wear (Palmer, Flint, & Brooks, 2007; Palmer et al., 2015; Simoes et al., 2010). Mariani et al. (2007), while studying the biofilm formed by microorganisms on the wooden surface of the shelves for the storage of ripened cheese, confirmed their higher abundance in cracks and pores in the wood.

The discrepancies also concern the influence of the conditioning layer on the efficiency of cell adhesion and biofilm formation. Milk or its ingredients, such as casein or β-lactoglobulin, generally reduce the effectiveness of bacteria sticking to surfaces used as materials for the construction of production line elements and equipment and for the packaging of dairy products. Robitaille, Choinière, Ells, Deschênes, and Mafu (2014) observed that conditioning with milk proteins (whey proteins isolate, β-lactoglobulin, bovine serum albumin) inhibited the adhesion of *L. monocytogenes* on the polystyrene surface, with this inhibitory effect being transient. The synergistic effect of the studied proteins and salts on the adhesion of bacterial cells to abiotic surfaces was minimal. In turn, Hamadi, Asserne, Bensouda, Mabrounki, and Latrache (2014) confirmed the ability of *S. aureus* to adhere and form biofilm on the surface of stainless steel conditioned with various types of milk: UHT, skimmed UHT, and semi-skimmed UHT. The strongest adhesion occurred with the application of milk with the lowest fat content, while the weakest with the use of semiskimmed milk. In the studies by Dat et al. (2014), the skim milk, buttermilk, and butter serum solutions used for the conditioning of stainless steel effectively formed layers of different thickness on its surface. Buttermilk and butter serum inhibited the adherence of lactic acid bacteria of the *Lactococcus*, *Leuconostoc*, and *Lactobacillus* genera more strongly than skim milk. Moreover, the use of buttermilk and butter serum for surface conditioning was more effective than skim milk in the situation of long-term (over 180 min) exposure of the surface to contact with bacteria. Conditioning of stainless steel with SMP skim milk proteins (1% and 10%) reduced the adhesion of vegetative cells and thermoresistant spores of *Bacillus* spp. strains isolated from dairy products and the dairy industry environment by Parkar, Flint, Palmer, and Brooks (2001). The use of 10% SMP resulted in greater adhesion of surviving forms than in the case of 1% SMP, but similar relationships did not apply to vegetative forms. Skimmed milk and its proteins (α-casein, β-casein, κ-casein, and α-lactalbumin) reduce the adhesion of *S. aureus* and *L. monocytogenes* to the stainless steel treated with these compounds (Barnes, Lo, Adams, & Chamberlain, 1999). The key factor causing the weakening of the adherence of bacteria to the work surfaces covered with organic compounds is probably the competition of proteins from the processed product and bacterial proteins for the binding site (Palmer et al., 2015). Optimized abiotic surface conditioning technology using milk proteins can help to effectively control the adhesion and biofilm processes on work surfaces in the dairy industry (Parkar et al., 2001).

Biofilms produced by microorganisms essential for the dairy industry can be mono-, bi-, or multispecies. Galié et al. (2018) report that the formation of a biofilm involving more than one species of microorganisms may increase its adhesion to the inhabited surface. Moreover, such a structure is more resistant to disinfection. Colonization of *Pseudomonas fluorescens* biofilm by *L. monocytogenes* cells protected them against disinfectants and drying. Increased adhesion efficiency of *L. monocytogenes* was also observed in the presence of previously prepared biofilm of *P. fluorescens* (Pang & Yuk, 2019). In the study by van der Veen and Abee (2011), the mixed biofilm formed by *L. monocytogenes* and *L. plantarum* ensured a higher resistance of these bacteria cells to disinfecting agents than in monospecies biofilms. The relationships within the multispecies biofilm formed by thermophilic spore bacteria on the surface of stainless steel immersed in skim milk were of various characters. While a synergistic effect was noted between *B. licheniformis* and *Anoxybacillus flavithermus*, a competitive effect was already observed in the case of *B. licheniformis* and *Geobacillus stearothermophilus* (Wang, Jin, He, & Yuan, 2021). In turn, in the studies by Alonso, Harada, and Kabuki (2020), the antagonist substance produced by *B. cereus* inhibited the adhesion of *L. monocytogenes* cells. However, in the presence of the

Bacillus strain that did not produce this active substance, the relationships within the bispecies biofilm were cooperative. The interactions between the microorganisms that make up bi- or multispecies biofilms depend on the substrates used (BHI, whey protein, and skim milk) (Alonso & Kabuki, 2019).

The risk associated with the negative impact of biofilms produced by microorganisms accompanying dairy production on the quality and safety of food increases in the event of the emergence of antibiotic-resistant strains. The cause of this phenomenon may be, for example, the permanent exposure of microorganisms to the presence of antibiotics or other substances with antibacterial activity in the environment in subinhibitory concentrations. It was also found that the formation of a multidrug-resistant microorganism population is favored by a specific microenvironment biofilm, which creates conditions that facilitate gene exchange between microorganisms. It can occur both within one species and between different species that make up the multispecies biofilm (Bhardwaj et al., 2021). As a result of combining the ability of microorganisms to form a biofilm with antibiotic resistance, the risk of persistent persistence of undesirable microorganisms in the production environment increases (Marti et al., 2017). Conclusions from the study regarding the potential correlation between antibiotic resistance and cell adhesion and the formation of permanent biofilms are not unequivocal. In the studies of Lin et al. (2019), *S. aureus* strains isolated from dairy showed multidrug-resistance (MDR), but their ability to form biofilm was low. According to Phophi, Petzer, and Qekwana (2019), among coagulase-negative *Staphylococcus* (CoNS) isolated from milk samples from cows with subclinical mastitis, 42% of formers biofilms were MDR. Bissong and Ateba (2020) observed the lack of a clear relationship between the ability to form biofilm and antibiotic resistance in the case of *S. aureus* isolated from milk. The tendency to multiresistance among *Streptococcus* sp. biofilms that threaten the safety of dairy product consumers was confirmed in turn by Ali (2020). Studies on the characteristics of enterococci isolated from cheeses produced in various regions of the Balkans proved that resistance to ciprofloxacin in strains capable of forming biofilm was correlated with the place of origin of the products (Popović et al., 2018). The strong ability to create biofilm on the surface of stainless steel under the conditions of the dairy industry combined with multidrug resistance was confirmed in a heat-resistant strain of *E. coli* isolated from raw milk cheese (Marti et al., 2017). Resistance to many antibiotics was observed in pathogenic strains of *L. monocytogenes* isolated from cow's milk producing strong biofilms on the surface of polystyrene and glass (Osman et al., 2016).

The ability of microorganisms to form biofilms in the dairy industry, however, also has a positive effect and allows, among other things, to obtain local dairy products with specific organoleptic properties. The biofilms of the surface of wooden containers (vats) used for the production of ripening cheeses are, in addition to raw materials, the main source of LAB necessary for their traditional production. In addition, LAB characterized by the ability to adhere and form biofilm can play a control role in preventing the appearance of pathogens and limiting their presence on production surfaces. The mechanisms that enable this type of activity include competition for the colonization site and the production of bacteriocin-like inhibitory substances (BLIS) (Scatassa et al., 2015; Wallis, Krömker, & Paduch, 2019). EPS, a component of the LAB biofilm, may inhibit the formation of biofilms by some pathogens (Salas-Jara, Ilabaca, Vega, & García, 2016). These topics will be described in more detail in the following chapters.

4. Biofilm by lactic acid bacteria

The production of biofilms by LAB, due to their natural presence in raw milk already at the stage of its milking and the key role they play in the production of individual types of dairy products, is of particular importance for the dairy industry. When analyzing the effects of the appearance of LAB biofilms in the production space, one should also take into account the differentiation of the effects of their impact on the quality of the obtained product, which may sometimes deteriorate due to their activity.

Studies on the ability of LAB, including *Lactobacillus* sp. to create biofilms on various abiotic surfaces, prove that not all strains have this property (Salas-Jara et al., 2016). Sun et al. (2020) report that most LAB strains isolated from fermented dairy products produced biofilm on abiotic surfaces. In the studies by Bujňáková and Kmeť (2012), *Lactobacillus* strains isolated from dairy products showed a high ability to form biofilm, which, combined with the inhibitory effect on intestinal pathogens, including *Salmonella* spp., proves the possibility of their use as probiotic starter cultures. Tested by Kubota, Senda, Nomura, Tokuda, and Uchiyama (2008), *Lactobacillus* spp. strains responsible for food spoilage formed morphologically diverse biofilms on the glass surface.

In the case of *L. brevis* and *L. fructivorans*, it was observed that some of the cells in the biofilm structure were longer than in the planktonic form. Studies with the use of *Enterococcus hirae*, *S. uberis*, *L. lactis* subsp. *lactis*, and *L. fermentum* showed that their ability to form biofilm in vitro was changed by various physical and chemical factors. *S. uberis* was the only one of the tested isolates that did not form a biofilm. All other strains showed higher biofilm formation efficiency at lower pH values of the growth medium. In turn, the salt concentration not exceeding 8% stimulated the production of a stronger biofilm in all tested strains, except for *E. hirae*. Inhibitory influence on its formation by *L. lactis* subsp. *lactis* had a neutral and alkaline medium

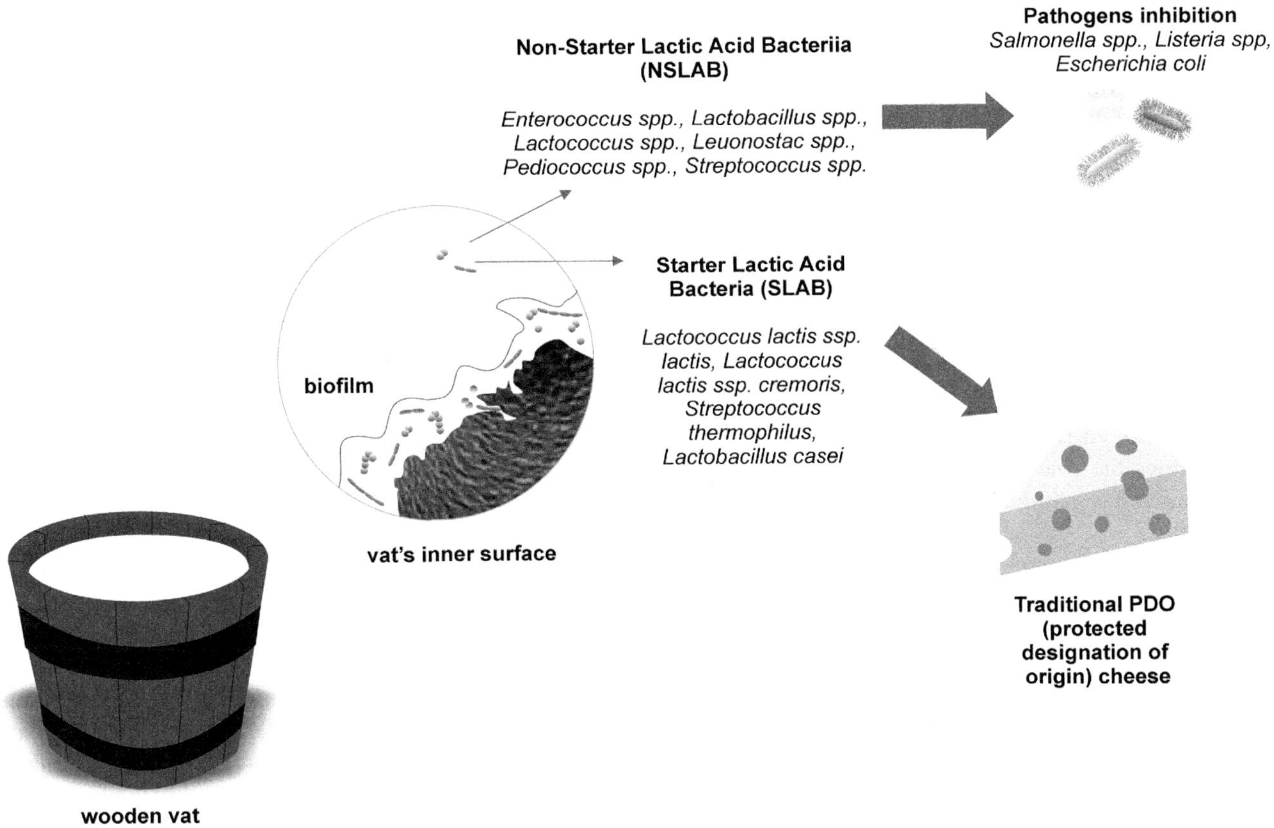

FIG. 1 LAB biofilm in production of PDO (protected designation of origin) cheese.

and an 8% NaCl concentration (Muruzovic, Mladenovic, & Comic, 2018).

A special type of biofilm typical of the dairy industry is that produced by LAB on the surface of wooden cheese vats (Fig. 1). They are the main source of specific LAB strains necessary to obtain the original flavor and aroma characteristic only for these products.

Examples of such products are the famous cheeses from the protected designation of origin (PDO) originating in Italy and France, such as Roquefort, Pecorino, Comté, Reblochon, and Beaufort. The best known types of vats for the production of these cheeses are *tina* and *gerle*. Using them for cheese production eliminates the need to add starter cultures to raw milk, and the bacteria responsible for fermentation come from the raw milk or the surface of the vat (Didienne et al., 2012; Gaglio, Todaro, & Settanni, 2021; Lortal, Licitra, & Valence, 2014). The biofilm created on the *tina* vats used in the production of Sicilian Ragusano cheese consists mainly of LAB, mainly *S. thermophilus*, thermophilic *Lactobacillus* spp., and *Lactococcus*. Research by Lortal et al. (2009) proved that after 10 min of contact with the surface of the vat, from 10^4 to 10^6 CFU/mL of *S.thermophilus* and 10^4–10^5 CFU/mL of thermophilic lactobacilli are released into the milk. LAB of the genus *Lactobacillus* and *Leuconostoc* dominate in the biofilm on the surface of the *gerle* vat, their numbers reaching the order of 6 and 5.2 $\log \text{cm}^{-2}$, respectively. Moreover, Gram-negative bacteria, yeasts, and molds were isolated from it (Lortal et al., 2014). The experience of Cruciata et al. (2018) proved that LAB biofilms produced on the surface of a vat for the production of traditional Italian cheeses prevented the formation of own biofilms by *Salmonella* spp., *L. monocytogenes*, *E. coli*, and coagulase-positive staphylococci. The predominant species in the LAB biofilm were *L. fermentum* and *L. lactis*. The suggested reason for the effective inhibition of pathogen adhesion on the ladle surface by LAB biofilms was the production of specific antibiotic metabolites—bacteriocins and organic acids that lower the pH.

LAB biofilm formation may be a problem when the bacteria that produce them adversely affect the quality of the obtained product. Nonchartered lactic acid bacteria (NSLAB), which very often contribute to the desired and product-specific organoleptic properties, are also the cause of cheese defects. These include, first of all, the formation of gases and different or unpleasant taste and smell from the traditional one. In the case of Cheddar cheeses, a disadvantage is also the formation of calcium lactate crystals on the surface of the cheese, which are harmless but undesirable from the point of view of both producers and consumers. Agarwal et al. (2006) claim that the process of creating these structures was related to the activity of NSLAB, especially *L. curvatus*, the presence of which was detected in the biofilms created by these bacteria at the Cheddar cheese production facility. According to Somers, Johnson, and Wong (2001), the biofilm produced by *L. curvatus* on the surface of vats used at the initial stage of cheese production was the cause of its contamination and the formation of calcium lactate crystals on the surface. Weber et al. (2019) believe that the appearance of certain NSLAB species in biofilms in dairy plants after their presence on the surface of milking equipment in dairy farms demonstrates the role that biofilm plays in the spread of these, sometimes harmful, microorganisms along the chain production.

The effectiveness of the method of reducing the risk associated with the presence of pathogenic microorganisms based on the prevention of their biofilms formation or replacement of existing ones by those created by LAB has not been confirmed by Jara et al. (2020). In the study, mature biofilms of *L. fermentum* MP26 and *L. salivarius* were used as substrates that potentially inhibit the adhesion of *L. monocytogenes* on the glass surface. However, even despite the lack of the confirmed inhibitory effect of lactobacilli on the adhesion of *L. monocytogenes*, a different type of relationship was observed between these bacteria, which could increase the safety in food production. It has been shown that the spread of pathogen cells in the production environment can be limited by immobilizing them in the internal layers of biofilm created jointly with LAB dual species. In the studies of Wallis et al. (2019) aimed at determining the ability of LAB to replace pathogenic biofilms produced by *S. aureus* and *S. xylosus* strains responsible for mastitis, such a feature was confirmed in two out of five tested lactobacilli strains—reference *L. rhamnosus* ATCC 7469 and isolated from milk *L. plantarum*.

5. Production of biofilms by pathogenic microorganisms

The richness of nutrients that characterize milk and its products undoubtedly contributes to the development of a very wide spectrum of microorganisms, including those responsible for diseases that are more or less serious for consumers. In the case of most of the previously described pathogenic microorganisms isolated from raw milk, processed milk, and dairy products, the ability to produce biofilm on biotic or abiotic surfaces of dairy farms and milk processing plants has been confirmed. The effects of this phenomenon include, among others, persistent persistence and spread of pathogen cells in the production space and the associated increased risk of food contamination, including food already processed. These problems, combined with the difficult disinfection of biofilm-covered production areas, constitute a significant stimulus to undertake research on the mechanisms that determine the effectiveness of pathogens to create biofilms and how to prevent its formation.

As suggested by Teh et al. (2015), the appearance of *Salmonella* spp. biofilms in the environment of farms and dairy plants may cause more frequent cases of salmonellosis among consumers of products manufactured there. Inadequately washed and disinfected raw milk storage tanks may be a potential source of contamination of raw milk by cells released from biofilm. In studies on the ability to produce biofilm on stainless steel and polypropylene by *S. aureus*, *Cronobacter sakazakii*, and *S.* Typhimurium strains isolated from raw milk and dairy products, the number of *Salmonella* recovered count exceeded the value of 7 $\log_{10}$ cm^{-2} and was approx. 2 $\log_{10}$ cm^{-2} higher than other strains (Lamas et al., 2018). In a study by Bhardwaj et al. (2021), *E. coli* isolates from milk and dairy products showed poor to moderate biofilm formation potential. However, it is suspected that the biofilms produced by these bacteria may be the cause of persistent intramammary *E. coli* infections in dairy cows. Concerns are also raised by the possibility of transferring to *E. coli* strains of thermoresistance, which, together with their ability to form biofilm, would ensure the possibility of survival and contamination of food products both by harmless to humans, but causing early blowing of cheese saprotrophs, and pathogenic *E. coli* serotypes, especially STEC (Marti et al., 2017).

5.1 *Staphylococcus aureus* biofilm

Considering the role of methicillin-resistant *S. aureus* (MRSA) strains in nosocomial infections, research aimed at limiting their presence and the ability to form biofilm concerns the surfaces most often used in medical practice. Meanwhile, the results of experiments aimed at determining the frequency of staphylococci occurrence in the environment of dairy farms and dairy processing plants prove the legitimacy of their conduct, pointing to a real problem related to, inter alia, their potential in the field of biofilm production. The appearance and long-term maintenance of *S. aureus* in the dairy industry are important because, apart from serious human poisoning as a result of eating food contaminated with these bacteria and their toxins, this pathogen is also a frequent cause of mastitis in cows, also in subclinical form. This causes a decrease in milk yield and the resulting economic losses in milk production, making sick animals a reservoir of pathogenic microorganisms dangerous for humans (Lee et al., 2014; Palmer et al., 2015).

The frequency of *S. aureus* occurrence in farms and dairy plants varies (Palmer et al., 2015). Rola, Czubkowska, Korpysa-Dzirba, and Osek (2016) confirmed the presence of *S. aureus* in 50% of samples from raw milk, semifinished and finished products as well as swabs from dairy farms producing cheese from cow's milk in Poland. In the studies of Jørgensen, Mørk, and Rørvik (2005), where *S. aureus* was isolated from animals, workers, equipment, the environment, and the final product—cheese, the dominance of one pulsotype was found, identified in most of the farm test sites. Such a phenomenon may indicate the long-term maintenance of a specific strain in the production environment, which may result, inter alia, from its ability to produce biofilm.

The production of biofilm by *S. aureus* in vivo occurs much faster than in vitro, where this process often requires modification of the atmosphere (anaerobiosis) or supplementation of the growth medium. The increase in the amount of biofilm produced has been observed under the influence of stressors such as reduced availability of nutrients, thermal stress, or the effects of subinhibitory concentrations of ethanol, salt, and antibiotics (Arciola, Campoccia, Ravaioli, & Montanaro, 2015; Fabres-Klein, Caizer Santos, Contelli Klein, Nunes de Souza, & de Oliveira Barros Ribon, 2015). By stimulating biofilm production by *S. aureus* in environments characterized by poor composition and enzyme accumulation, biofilm production takes on the function of a mechanism that allows cells to survive under stressful conditions (Seale et al., 2015). This is confirmed by the results of Alonso's and Kabuki's experiments (Alonso & Kabuki, 2019), in which *S. aureus* was the only microorganism that did not form a single-species biofilm on stainless steel within 24 h, which was considered the effect of too rich composition of the culture media used for the research.

The results of the research prove that not all strains of staphylococci isolated from dairy farms are able to produce biofilm. It depends, among other things, on the specifics of the strain and the properties of the potentially colonized surface. The study of 31 *S. aureus* isolates from 849 samples collected in Brazilian dairy farms from milk, milking machine surfaces, and farm workers showed that almost half (45.2%) of the pulsotypes produced biofilm on the polystyrene surface, while 41.9% and 38.7% on stainless steel and rubber, respectively. However, none of the pulsotypes had the ability to form biofilms on silicone surfaces (Lee et al., 2014). The Fabres-Klein et al. (2015) study proved that among 54 *S. aureus* isolates collected from cows with subclinical mastitis belonging to dairy herds, the majority (47) had the ability to form biofilm. Among the mastitis-inducing *S. aureus* strains, over 50% had a low and 25% high biofilm potential when cultured on TSBg (Trypticase soy broth supplemented with glucose). After the use of milk as a growth medium, 6% of the strains did not show the ability to form biofilm, while half showed moderate or high potential in this area. Milk whey had an inhibitory effect on the intensity of biofilm formation by the tested *S. aureus* strains (Cáceres et al., 2021).

S. aureus genotype B (GTB) strains isolated from milk and dairy products, the high incidence of which in the herd suggests higher infectivity and virulence compared with other genotypes, also produced more biofilm. The *S. aureus* GTB dairy isolate had a higher biofilm production index (BPI) value than the *S. aureus* PR 281 (not GTB) strain isolated from poultry, with a confirmed high biofilm production potential (Thiran et al., 2018). *S. aureus* GTB under static conditions produced a stronger biofilm than most other genotypes, including methicillin-resistant. On the other hand, under dynamic conditions, more similar to those prevailing during production, the biofilm formation process was less intense (Thiran, 2018).

The analysis of the properties of the *S. aureus* JDFM SA01 strain isolated from the milk filter used on the dairy farm showed its higher biofilm formation capacity compared with the reference *S. aureus* strain RN4220. Taking into account the fact that the milk filter is in direct contact with the milking machine, the presence of a biofilm-forming pathogenic *S. aureus* strain on its surface may pose a risk of continuous contamination of various places and equipment in the plant (milking machines, milking pipes, tanks), leading to contamination of raw milk (Ryu et al., 2021).

5.2 *Listeria monocytogenes* biofilm

The presence of *L. monocytogenes* in dairy products is most often the result of cross-contamination due to contact with contaminated production surfaces (Fagerlund, Langsrud, & Møretrø, 2021). A particular problem is the emergence of persistent *L. monocytogenes* strains, characterized by the ability to stay in a specific place in the production environment for a long time—from months to many years (Magalhães et al., 2017; Taylor & Stasiewicz, 2019). Long-term environmental pollution throughout the cheese production facility was identified as a potential cause of the listeriosis epidemic caused by *L. monocytogenes* 4b ST6 CT7448 in Switzerland in 2018–20 (Nüesch-Inderbinen et al., 2021). The ability to form a biofilm is sometimes considered one of the characteristics of persistent strains and responsible for their survival in the production environment (Latorre et al., 2010; Magalhães et al., 2017; Osman et al., 2016; Taylor & Stasiewicz, 2019). The opinions of scientists on the existence and real meaning of this relationship, however, vary. Fagerlund et al. (2021) argue that the ability of *L. monocytogenes* to form mono-species biofilms is not related to their ability to survive in food processing plants. Similar conclusions were also drawn by other authors based on the results of research on *L. monocytogenes* strains of isolated samples of cheese, brine, and the environment of cheese processing plants (Costa, Bertolotti, Brito, & Civera, 2016; In Lee et al., 2017). According to Magalhães et al. (2017), discrepancies in these results may result from differences between strains and experimental conditions, as well as ambiguous criteria for classifying a strain as persistent or sporadic, formulated individually for the needs of individual studies.

The ability of *L. monocytogenes* to form biofilm on various abiotic surfaces has been confirmed many times (Costa, Lourenco, Civera, & Brito, 2018; Koo, Ndahetuye, O'Bryan, Ricke, & Crandall, 2014; Latorre et al., 2010; Maury et al., 2019; Weber et al., 2019). The research proves the ability to form biofilm structures on surfaces from materials used in dairy for the construction of vessels, tanks, and elements of production and processing lines. The most frequently tested is stainless steel, which, thanks to its resistance to the use of aggressive mechanical and chemical methods of cleaning and disinfection, has many applications in the dairy industry (Skowron et al., 2019). The results of Skowron, Wiktorczyk, et al. (2019) showed that the ability of *L. monocytogenes* isolated from blue cheese to form biofilm on stainless steel was higher in comparison to *L. innocua* and *L. welshimeri.* Alonso and Kabuki (2019) observed the formation of single species biofilm by *L. monocytogenes* on the surface of stainless steel (4.36 log CFU cm^{-2}) in whey protein. However, the same strain in the same medium did not produce multispecies biofilm. In a study by In Lee et al. (2017), *L. monocytogenes* isolates formed biofilm on both stainless steel and polystyrene. According to Skowron et al. (2019), the strains derived from milk were characterized by a higher efficiency of biofilm formation on the rubber surface, compared with clinical isolates from blood. Moreover, a high positive correlation was observed between the biofilm formation efficiency and the number of virulence genes of *L. monocytogenes* strains from milk. Also, Osman et al. (2016) showed a strong potential of the 1/2a and 4a and 4c serotypes coding for virulence genes (*plcB*, *actA*, *hlyA*, and *iap*) for biofilm formation on polystyrene and glass. In turn, studies by In Lee et al. (2017) did not confirm the relationship between the studied serotypes of *L. monocytogenes* 4b, 1/2b, and 1/2c, and their ability to produce biofilm. Confirming or denying the relationship between individual serotypes and their potential for biofilm formation is of particular importance in the event of emergence in the production space of serotypes 1/2a, 1/2b, and 4b isolated from food and responsible for the majority of listeriosis outbreaks in humans.

6. Production of biofilm by spoilage bacteria in milk and dairy products

6.1 *Pseudomonas* spp.

Pseudomonas spp. are often isolated from dairy products and the environment of farms and dairy plants. The processes of milk spoilage caused by them during storage in psychotropic conditions are the result of the production of lipases and extracellular proteases. The species most often isolated in dairy processing plants is *P. fluorescens*, which has the ability to produce a specific blue pigment (Rossi et al., 2018). This species, showing high activity of extracellular protease, lipase, and lecithinase, is the main cause of spoilage of processed dairy products (Rossi et al., 2016). Among the factors that pose an additional problem related to the presence of *P. fluorescens* is also their resistance to heat treatment, short generation time, and the ability to form biofilms. The production of persistent biofilms by *Pseudomonas* causes their enzymatic activity to increase compared with cells in the planktonic form, which contributes to the faster deterioration of food contaminated with these bacteria. This is confirmed by the results of the studies by Aswathanarayan and Vittal (2014), in which high metabolic activity of the biofilm produced at low temperatures by the psychrotrophic *P. fluorescens* PSD4 strain was observed. The intensity of this process, however, depended on factors such as the availability of nutrients. The presence of milk proteins (TSC medium) and citrate as a carbon source in the medium favored the adhesion process and biofilm formation, in contrast to glucose contained in the medium.

According to Rossi et al. (2018), *Pseudomonas* spp. strains isolated from milk, dairy products, and the environment of a dairy plant mostly produced biofilm, although differences between individual strains were observed. In the case of the investigated *P. fluorescens* isolates, the ability to form biofilm was positively correlated with the production of blue dye. Biofilm formation, apart from species variability and the composition of the medium, also depends on the conditions and time of incubation. In general, higher intensity of biofilm formation by *Pseudomonas* spp. is observed at the temperature corresponding to the optimum temperature for the growth of a given species. Lowering the temperature is associated with the extension of the time necessary for the production of biofilm (Thiran et al., 2018). The comparative studies of 64 strains of *P. fluorescens* conducted by Rossi et al. (2016) confirmed this ability in 57 strains, with high variability between the tested temperatures of 10 and 30°C. The results of these experiments indicate a greater importance of temperature than of the cultivation time for the efficiency of biofilm production. The assessment of the ability of *P. fluorescens* to produce biofilm in conditions simulating those present in a dairy plant proved the possibility of its production on both polystyrene and stainless steel (Maggio et al., 2021). They also showed that the surfaces inhabited by the biofilm previously produced by *Pseudomonas* facilitated the adherence of pathogenic *L. monocytogenes* to them. It has been proved that *P. fluorescens*, by producing a dense biofilm characterized by a large amount of EPS, creates a protective layer for *L. monocytogenes*, which itself produces matrix in a much smaller amount. The incorporation of *Listeria* cells into the biofilm produced by *Pseudomonas* spp. led to an increase in its density by stimulating the production of EPS (Puga, Dahdouh, San Jose, & Orgaz, 2018).

6.2 Biofilm formation by spore formers

Spore-forming microorganisms, thanks to the high resistance to temperature of the spores they produce, preventing their elimination in classic pasteurization processes, are a huge challenge in every branch of the food industry, including dairy. The problem with the appearance of undesirable species, both from the point of view of product quality and pathogenic, becomes even more serious when they show the ability to form biofilms on the surfaces in contact with milk or dairy products. The matrix of biofilms produced by vegetative forms, which mainly includes EPS, can serve as a protective environment in which thermoresistant endospores temporarily stay, mature, and are gradually released into the environment, leading to troublesome contamination of both products and elements of production lines (Ostrov, Paz, & Shemesh, 2019; Radmehr, Zaferanloo, Tran, Beale, & Palombo, 2020).

Spores produced by various strains of *Bacillus* spp., derived from dairy products and the production space of dairy plants, showed a higher tendency to adhere than vegetative forms on the surface of stainless steel (Parkar et al., 2001). According to Jindal et al. (2016), commonly found in dairy processing, thermophilic spore bacteria are able to form a biofilm on the surface of stainless steel. The number of *G. stearothermophilus*, *B. licheniformis*, and *B. sporothermodurans* cells reisolated from biofilm produced after 72h incubation in nonfat dry milk (NFDM) was log 5.10, log 5.13, and log 3.09 CFU cm^{-2}, respectively. Also in the studies by Alonso, de Oliveira Morais, and Kabuki (2021), *G. stearothermophilus* turned out to be the strongest biofilm producer, compared with other strains of sporulating bacteria, also isolated from UHT milk—*B. sporothermodurans* and *B. cereus*. However, the intensity of biofilm formation within species varied and some isolates did not show this ability at all. According to Sadiq et al. (2017), among the species associated with the dairy industry that make up the biofilm, *G. thermoleovorans*, *B. licheniformis*, the group of *G. stearothermophilus* and *Anoxybacillus flavithermus* are the most important. The number of cells recovered from the biofilms they produced on stainless steel in the presence of skim milk was log 4.5; 4.8; 5.2; and 5.3 CFU cm^{-2}, respectively. The analysis of the specificity of biofilms produced by strains of spore bacteria isolated from milk and milk concentrate production lines showed that thermophiles of the genus *Anoxybacillus* and *Geobacillus* form these structures on the liquid-air interface rather than on submerged surfaces. These studies also proved that *G. thermoglucosidasius* was not able to grow in a medium enriched with milk (Zhao et al., 2013). Ostrov et al. (2019) found phenotypic differences between the surface of biophymes produced by dairy-associated strains of *Bacillus* spp. and the nondairy strain of *B. subtilis*, it was also observed that the cells of the former were surrounded by a greater amount of EPS. The strains from the dairy industry environment were characterized by higher resistance to disinfection processes.

The ability to create biofilm was also confirmed in the case of *B. cereus*, which is responsible for both spoilage of dairy products and poisoning in humans (Kumari & Sarkar, 2016). Among 41 *B. cereus* isolates obtained from raw and pasteurized milk, more than 50% formed biofilm, however, only 4.9% were classified as strong biofilm formers. More biofilm producers were isolated from pasteurized milk than raw (Radmehr et al., 2020). In a study by Kumari and Sarkar (2014), over 70% of *B. cereus* isolates isolated from various dairy products in India were able to form a biofilm at 4°C. Contrary to most *Bacillus* spp., the production of biofilm by *B. cereus* is more intense in whole milk than in diluted milk, probably due to the high content of surfactants in the former, which facilitate cell adhesion on work surfaces (Kumari & Sarkar, 2016).

7. Prevention and control of biofilm in the dairy industry

Increasing the safety of food produced in dairy plants by limiting the harmful effects of biofilm can take place in several ways. Both among scientists and producers dealing with the problem of biofilm in everyday practice, there is a belief about the key role of preventive measures, preventing or significantly reducing the appearance of these problematic and specific biological structures in the environment of farms and dairy plants (Thiran, 2018).

The inhibition of biofilm formation on the surfaces of the production space may be the result of thoughtful planning and construction of the production line, limiting, for example, the laminate, biofilm-promoting, flow of liquid products in working installations, and assuming the use of appropriate, nonpromoting cell adhesion, and formation of bacterial biofilm, construction materials (Thiran, 2018). Some possibilities are also related to the conditioning of working surfaces with agents that prevent the adhesion process of microbes. In the studies of Jindal et al. (2016), among several types of substances used as coatings reducing bacterial adhesion and biofilm formation, Ni-P-PTFE (nickel-phosphorus-polytetrafluoroethylene) showed the highest effectiveness. In turn, Friedlander, Nir, Reches, and Shemesh (2019) observed a reduction in adhesion and biofilm production by *P. aeruginosa* and *B. licheniformis* on a peptide-based stainless steel surface. The surface modified in this way did not lose its properties also in contact with raw milk. However, the use of the method of surface conditioning with chemical agents is associated with certain concerns related, inter alia, to the risk of the release of ingredients of these coatings into food and their impact on consumer safety, the length of time of the actual exposure of bioactive substances to cells, but also the possibility of microorganisms becoming resistant to their cells' action (Bremer et al., 2015).

The reduction of surface contamination, especially by organic compounds, directly reduces the risk of microbial contamination. The residues of raw materials and dairy products rich in proteins and fats can, on the one hand, be a source of nutrients for microorganisms, and on the other hand, have a protective function and contribute to increasing the resistance of bacteria to stress factors, including drying out. An optimal cleaning program using highly effective formulations can remove unnecessary residues and sediments from milk and processed dairy products from the treated surfaces, reducing the degree of contamination and the risk of biofilm formation (Kuda, Nakano, Takahashi, & Kimura, 2016; Simoes et al., 2010).

If the presence of biofilm in the production space of the dairy industry is confirmed, cleaning and disinfection of the surfaces covered by it are carried out together as part of the cleaning-in-place (CIP) strategy. The combination of these procedures seems necessary, because for effective disinfection, including the elimination of biofilm, it is necessary to at least partially damage or degrade the biofilm matrix, which protects the cells of the bacteria producing them from the negative influence of external factors, which is ensured by the precleaning process. The classic CIP scheme is based on the alternating or combined action of high temperature, acid or alkaline agents, and disinfectants (Guerrero-Navarro et al., 2021; Mazaheri, Cervantes-Huamán, Bermúdez-Capdevila, Ripolles-Avila, & Rodríguez-Jerez, 2021; Ostrov et al., 2019). The application of NaOH or HNO_3 is mainly aimed at removing organic impurities from the surface, and in order to increase the effectiveness of their operation, in addition to the main active substances, compounds of the nature of surfactants, emulsifiers, and chelators are also added to the composition of commercial preparations. In the disinfection process, preceded by rinsing out the residues of the previously used agents and contaminants, ultimately aimed at the elimination of unwanted microorganism cells, the most commonly used are hydrogen peroxide, sodium hypochlorite, ozone, and peracetic acid (Coughlan, Cotter, Hill, & Alvarez-Ordóñez, 2016). Standard CIP parameters used in the dairy industry, i.e., 1% concentration of sodium hydroxide, 0.8%–1% nitric acid at process temperatures of 60–70°C, turn out to be ineffective in eliminating biofilms on work surfaces, similarly to the traditionally used, ineffective disinfection programs (Kumari & Sarkar, 2016; Simoes et al., 2010; Weber et al., 2019). The combination of increased concentrations of NaOH (2%) and HNO_3 (1.8%) and temperature (75°C) shows a higher efficiency in inactivating microorganisms, but does not remove all biofilm residues from the treated surfaces, which confirms the need to include the mechanical cleaning stage in the full CIP procedure (Knight, 2015).

The unsatisfactory effects in terms of the elimination of biofilms may result directly from their increased, compared with the planktonic form, resistance to chemicals. Studies on the antibiotic activity of the agents used in the subsequent stages of CIP against microorganisms isolated from the production environment of a dairy plant (*Bacillus*, *Enterococcus*, *Streptococcus*, *Staphylococcus*, *Micrococcus*, *Aeromonas*, *Corynebacterium*, *Pseudomonas*, *Klebsiella*, and *Escherichia* cells) have proven a lower effectiveness of sanitizing their biophylls compared with planktonic cells (Anand & Singh, 2013). Higher than the planktonic form, resistance to 11% acetic acid was also demonstrated by the biofilm produced by the LAB *L. plantarum* strain M606 (Kubota et al., 2008). The low sensitivity of microorganisms inhabiting the surfaces of the production space in the form of a biofilm to the action of disinfectants is attributed to many factors and mechanisms, which, however, despite many scientific analyses, are not yet fully explained (Li, Xiao, Wang, & Hao, 2020; Mah & O'Toole, 2001; Ulusoy & Chirkena, 2019). Another problem may be the practical use of disinfectants with the effectiveness confirmed by the results of research on microorganisms in the planktonic form or under conditions insufficiently simulating the conditions of food production. Improper application and dosage of disinfectants may, in turn, lead to the gradual resistance of microorganisms to their effects (Coughlan et al., 2016). For these reasons, as well as in connection with the search for solutions that are more economically effective and less harmful to the natural environment, modifications are made to the concentrations and types of disinfectants used, temperature values, and frequency of individual activities. In order to increase the effectiveness of CIP, preparations based on enzymes, e.g., proteolytic, cellulolytic, or DNAses, are also included in the protocols (Delhalle et al., 2020; Marchand et al., 2012). Their function is primarily the degradation of carbohydrates, lipids, and DNA, which are part of the biofilm matrix, and the destruction of connections between cells, leading to a reduction in the integrity of the biofilm structure (Coughlan et al., 2016). The use of enzymes may be a solution to the problem related to the presence of biofilm on the surfaces of dairy membranes. Membrane filtration techniques are used, inter alia, to improve the microbiological quality of milk and whey or the production of protein concentrates. The specific structure of membranes and their sensitivity to acids and bases do not allow the use of standard CIP methods for this purpose and require an individual approach to the problem (Tang, Flint, Bennett, & Brooks, 2010). Due to the high degree of contamination of membranes with milk or whey components, the use of enzyme preparations seems to be a reasonable solution in this case. They can simultaneously destructively affect the cells of microorganisms that make up the biofilm and effectively penetrate (especially with the addition of surfactants) and disintegrate the biofilm matrix (Anand, Singh, Avadhanula, & Marka, 2014; Chamberland, Lessard, Doyen, Labrie, & Pouliot, 2017).

Parallel to the optimization of the already functioning CIP programs, attempts are being made more and more often by scientists and producers in the dairy industry to develop new, alternative methods of eliminating biofilm or preventing its formation. Among the unconventional methods of limiting the intensity of biofilm formation by microorganisms in the dairy industry, the use of bacteriophages, which enable the lysis of microbial cells in the biofilm and the destruction of the EPS structure, is of great interest. Their use in the dairy industry is of particular importance due to the high specificity of phages toward the host, excluding their influence on the microorganisms of starter cultures (Coughlan et al., 2016). Preparations based on the activity of microorganisms with competitive and antagonistic properties against undesirable biofilm producers have also shown effective action. As such features are exhibited by many LAB, which play a key role in dairy production, the possibility of their additional use for biological control of adhesion and destruction of biofilms seems attractive and justified. Gavrilova et al. (2019) showed that *L. plantarum* and *L. fermentum* strains isolated from fermented plant material, in addition to the confirmed probiotic potential, effectively eliminated biofilms formed by *E. coli*, *S. aureus* and *P. aeruginosa*. Such properties were not demonstrated by the reference probiotic strain *L. plantarum* 8PA3. On the other hand, Wallis et al. (2019) found that LAB strains isolated from milk and dairy farm environment showed the ability to eliminate biofilms formed by the reference strain of *S. aureus* and *S. xylosus* and *S. epidermidis* isolated from the udder of cows with mastitis. In some cases, the pathogen biofilm was replaced with the biofilm produced by the tested LAB. The results of many studies confirm that inhibition of the production and elimination of biofilm does not require the presence of cells of beneficial bacterial strains and is also possible with the use of metabolites synthesized by them. In a study by Singh et al. (2018), cell-free supernatants (CFS) obtained from LAB cultures from goat's milk reduced metabolic activity and inhibited biofilm formation by *P. aeruginosa* and *S. aureus*. Bioactive bacterial substances can also act as biosurfactants that prevent cell adhesion, chelates that inhibit the growth of microbial cells by limiting the availability of iron, or quorum sensing inhibitors (Coughlan et al., 2016; Simoes et al., 2010).

8. Conclusion

The presence of microorganisms in the production spaces of the dairy industry plants is a necessary condition for its functioning. Microorganisms introduced into products in the form of starter cultures provide them with appropriate quality, health-promoting values, and specific organoleptic properties. The formation of biofilms by LAB on food contact surfaces can prevent the growth of pathogens from contaminated raw materials or the external environment. In some cases, the spontaneous and uncontrolled development of NSLAB makes it possible to obtain local dairy products, giving them a unique and typical only for them taste, smell, and texture. The biofilms that are problematic for the dairy industry arise as a result of the activity of microorganisms, the presence of which is undesirable for various reasons. Due to the high content of nutrients in raw materials and dairy products, their residues accumulating on the surfaces of production lines are quickly and easily colonized by all kinds of microorganisms, including spoilage and pathogens. The first group adversely affects the quality and organoleptic characteristics of dairy products through the production of enzymes and by-metabolites. The second, which includes pathogens as dangerous to humans as *L. monocytogenes*, STEC, *Salmonella* sp., or *S. aureus*, poses a direct threat to the health of consumers of food contaminated with them. The appearance of representatives of each of these groups of microorganisms in the form of a biofilm and the subsequent gradual migration of cells from it may contaminate the entire production environment, including the food produced. This obviously reduces the shelf life of the product and its health safety.

The characteristics of raw materials and dairy products as well as technological solutions typical for the dairy industry used within production lines make it an excellent environment for the development of biofilms. In addition, the construction of some elements of these lines prevents the effective elimination of biofilms inhabiting their surfaces, due to the difficult access to them disinfectants. These difficulties, combined with the high resistance to stress factors typical of the biofilm structure and the limitations of disinfectants used in the dairy industry, make the formation of biofilms in the dairy industry one of the most serious problems related to its functioning. Therefore, it seems that the main challenge for both the dairy industry and scientists cooperating with it should be the development of effective CIP systems, limiting the creation of new and eliminating the already existing biofilms in dairy production plants.

References

Adriaens, I., Van Den Brulle, I., Geerinckx, K., D'Anvers, L., De Vliegher, S., & Aernouts, B. (2021). Milk losses linked to mastitis treatments at dairy farms with automatic milking systems. *Preventive Veterinary Medicine*, *194*, 105420. https://doi.org/10.1016/j.prevetmed.2021.105420.

Agarwal, S., Sharma, K., Swanson, B. G., Yüksel, G. U., & Clark, S. (2006). Nonstarter lactic acid bacteria biofilms and calcium lactate crystals in Cheddar cheese. *Journal of Dairy Science*, *89*(5), 1452–1466. https://doi.org/10.3168/jds.S0022-0302(06)72213-5.

Ali, F. F. (2020). Detection of biofilm formation and antibiotics resistance for *Streptococcus* spp. isolated from some dairy products in Diwanyah City of Iraq. *Al-Mustansiriyah Journal of Science*, *31*(4), 28–35. https://doi.org/10.23851/MJS.V31I4.893.

Alonso, V. P. P., & Kabuki, D. Y. (2019). Formation and dispersal of biofilms in dairy substrates. *International Journal of Dairy Technology*, *72*, 472–478. https://doi.org/10.1111/1471-0307.12587.

Alonso, V., de Oliveira Morais, J., & Kabuki, D. Y. (2021). Incidence of *Bacillus cereus*, *Bacillus sporothermodurans* and *Geobacillus stearothermophilus* in ultra-high temperature milk and biofilm formation capacity of isolates. *International Journal of Food Microbiology*, *354*, 109318. https://doi.org/10.1016/j.ijfoodmicro.2021.109318.

Alonso, V. P. P., Harada, A. M. M., & Kabuki, D. Y. (2020). Competitive and/or cooperative interactions of *Listeria monocytogenes* with *Bacillus cereus* in dual-species biofilm formation. *Frontiers in Microbiology*, *11*, 1–10. https://doi.org/10.3389/fmicb.2020.00177.

Anand, S., & Singh, D. (2013). Resistance of the constitutive microflora of biofilms formed on whey reverse-osmosis membranes to individual cleaning steps of a typical clean-in-place protocol. *Journal of Dairy Science*, *96*(10), 6213–6222. https://doi.org/10.3168/jds.2013-7012.

Anand, S., Singh, D., Avadhanula, M., & Marka, S. (2014). Development and control of bacterial biofilms on dairy processing membranes. *Comprehensive Reviews in Food Science and Food Safety*, *13*, 18–33. https://doi.org/10.1111/1541-4337.12048.

Anderson, M., Hinds, P., Hurditt, S., Miller, P., McGrowder, D., & Alexander-Lindo, R. (2011). The microbial content of unexpired pasteurized milk from selected supermarkets in a developing country. *Asian Pacific Journal of Tropical Biomedicine*, *1*(3), 205–211. https://doi.org/10.1016/S2221-1691(11)60028-2.

André, S., Vallaeys, T., & Planchon, S. (2017). Spore-forming bacteria responsible for food spoilage. *Research in Microbiology*, *168*(4), 379–387. https://doi.org/10.1016/j.resmic.2016.10.00.

Arciola, C. R., Campoccia, D., Ravaioli, S., & Montanaro, L. (2015). Polysaccharide intercellular adhesin in biofilm: Structural and regulatory aspects. *Frontiers in Cellular and Infection Microbiology*, *5*, 7. https://doi.org/10.3389/fcimb.2015.00007.

Aswathanarayan, J. B., & Vittal, R. R. (2014). Attachment and biofilm formation of *Pseudomonas fluorescens* PSD4 isolated from a dairy processing line. *Food Science and Biotechnology*, *23*, 1903–1910. https://doi.org/10.1007/s10068-014-0260-8.

Barnes, L. M., Lo, M. F., Adams, M. R., & Chamberlain, A. H. (1999). Effect of milk proteins on adhesion of bacteria to stainless steel surfaces. *Applied and Environmental Microbiology*, *65*(10), 4543–4548. https://doi.org/10.1128/AEM.65.10.4543-4548.1999.

Bellio, A., Bianchi, D. M., Gramaglia, M., Loria, A., Nucera, D., Gallina, S., et al. (2016). Aflatoxin M1 in cow's milk: Method validation for milk sampled in Northern Italy. *Toxins*, *8*(3), 57. https://doi.org/10.3390/toxins8030057.

Bhardwaj, D. K., Taneja, N. K., Dp, S., Chakotiya, A. S., Patel, P., Taneja, P., et al. (2021). Phenotypic and genotypic characterization of biofilm forming, antimicrobial resistant, pathogenic *Escherichia coli* isolated from Indian dairy and meat products. *International Journal of Food Microbiology*, *336*, 108899. https://doi.org/10.1016/j.ijfoodmicro.2020.108899.

Bissong, M., & Ateba, C. N. (2020). Genotypic and phenotypic evaluation of biofilm production and antimicrobial resistance in *Staphylococcus aureus* isolated from milk, North West Province, South Africa. *Antibiotics (Basel, Switzerland)*, *9*(4), 156. https://doi.org/10.3390/antibiotics9040156.

Björkroth, J., & Koort, J. (2011). Lactic acid bacteria. Taxonomy and biodiversity. In J. W. Fuguay, P. F. Fox, & P. L. H. McSweeney (Eds.), *Encyclopedia of dairy sciences* (2nd ed., pp. 45–48). Elsevier. https://doi.org/10.1016/B978-0-12-374407-4.00255-7.

Bolaños, C., Paula, C. L., Guerra, S. T., Franco, M., & Ribeiro, M. G. (2017). Diagnosis of mycobacteria in bovine milk: An overview. *Revista do Instituto de Medicina Tropical de Sao Paulo*, *59*, e40. https://doi.org/10.1590/S1678-9946201759040.

Boyd, E., Trmcic, A., Taylor, M., Shyng, S., Hasselback, P., Man, S., et al. (2021). *Escherichia coli* O121 outbreak associated with raw milk Gouda-like cheese in British Columbia, Canada, 2018. *Canada Communicable Disease Report*, *47*(1), 11–16. https://doi.org/10.14745/ccdr.v47i01a03.

Bremer, P., Flint, S., Brooks, J., & Palmer, J. (2015). Introduction to biofilms—Definition and basic concepts. In K. H. Teh, S. Flint, J. Brooks, & G. Knight (Eds.), *Biofilms in the dairy industry* (pp. 1–16). John Wiley & Sons Ltd. https://doi.org/10.1002/9781118876282.

Bujňáková, D., & Kmeť, V. (2012). Functional properties of Lactobacillus strains isolated from dairy products. *Folia Microbiologica*, *57*, 263–267. https://doi.org/10.1007/s12223-012-0121-x.

Burke, N., Zacharski, K. A., Southern, M., Hogan, P., Ryan, M. P., & Adley, C. C. (2018). The dairy industry: Process, monitoring, standards, and quality. In A. V. Diaz, & R. M. Garcia-Gimeno (Eds.), *Descriptive food science* IntechOpen. https://doi.org/10.5772/intechopen.80398.

Cáceres, M. E., Ledesma, M. M., Serratab, A. L., Vay, C., Sordellia, D. O., Giacomodonato, M. N., et al. (2021). Growth conditions affect biofilms of *Staphylococcus aureus* producing mastitis: Contribution of MALDI-TOF-MS to strain characterization. *Current Research in Microbial Sciences*, *2*, 100073. https://doi.org/10.1016/j.crmicr.2021.100073.

Caplen, G., & Held, S. (2021). Changes in social and feeding behaviors, activity, and salivary serum amyloid A in cows with subclinical mastitis. *Journal of Dairy Science*, *104*(10), 10991–11008. https://doi.org/10.3168/jds.2020-20047.

Castañeda-Salazar, R., Pulido-Villamarín, A.d. P., Ángel-Rodríguez, G. L., Zafra-Alba, C. A., & Oliver-Espinosa, O. J. (2021). Isolation and identification of *Salmonella* spp. in raw milk from dairy herds in Colombia. *Brazilian Journal of Veterinary Research and Animal Science*, *58*, e172805. https://doi.org/10.11606/issn.1678-4456.bjvras.2021.172805.

Centers for Disease Control and Prevention CDC. (2017). Available at: https://www.cdc.gov/listeria/outbreaks/soft-cheese-03-17/index.html (Accessed 7 December 2021).

Centers for Disease Control and Prevention CDC. (2021). Available at: https://www.cdc.gov/listeria/outbreaks/hispanic-soft-cheese-02-21/index.html (Accessed 7 December 2021).

Chaleshtori, F. S., Arani, N. M., Aghadavod, E., Naseri, A., & Chaleshtori, R. S. (2017). Molecular characterization of *Escherichia coli* recovered from traditional milk products in Kashan, Iran. *Veterinary World*, *10* (10), 1264–1268. https://doi.org/10.14202/vetworld.2017.1264-1268.

Chamberland, J., Lessard, M. H., Doyen, A., Labrie, S., & Pouliot, Y. (2017). Biofouling of ultrafiltration membrane by dairy fluids: Characterization of pioneer colonizer bacteria using a DNA metabarcoding approach. *Journal of Dairy Science*, *100*(2), 981–990. https://doi.org/10.3168/jds.2016-11829.

Cortimiglia, C., Borney, M. F., Bassi, D., & Cocconcelli, P. S. (2021). Genomic investigation of virulence potential in Shiga toxin *Escherichia coli* (STEC) strains from a semi-hard raw milk cheese. *Frontiers in Microbiology*, *11*, 629189. https://doi.org/10.3389/fmicb.2020.629189.

Costa, A., Bertolotti, L., Brito, L., & Civera, T. (2016). Biofilm formation and disinfectant susceptibility of persistent and nonpersistent *Listeria monocytogenes* isolates from Gorgonzola cheese processing plants. *Foodborne Pathogens and Disease*, *13*(11), 602–609. https://doi.org/10.1089/fpd.2016.2154.

Costa, A., Lourenco, A., Civera, T., & Brito, L. (2018). *Listeria innocua* and *Listeria monocytogenes* strains from dairy plants behave similarly in biofilm sanitizer testing. *LWT*, *92*, 477–483. https://doi.org/10.1016/j.lwt.2018.02.073.

Coughlan, L. M., Cotter, P. D., Hill, C., & Alvarez-Ordóñez, A. (2016). New weapons to fight old enemies: Novel strategies for the (bio) control of bacterial biofilms in the food industry. *Frontiers in Microbiology*, *7*, 1641. https://doi.org/10.3389/fmicb.2016.01641.

Cruciata, M., Gaglio, R., Scatassa, M. L., Sala, G., Cardamone, C., Palmeri, M., et al. (2018). Formation and characterization of early bacterial biofilms on different wood typologies applied in dairy production. *Applied and Environmental Microbiology*, *84*(4), e02107–e02117. https://doi.org/10.1128/AEM.02107-17.

Dai, J., Wu, S., Huang, J., Wu, Q., Zhang, F., Zhang, J., et al. (2019). Prevalence and characterization of *Staphylococcus aureus* isolated from pasteurized milk in China. *Frontiers in Microbiology*, *10*, 641. https://doi.org/10.3389/fmicb.2019.00641.

Dancer, G. I., Mah, J. H., & Kang, D. H. (2009). Influences of milk components on biofilm formation of *Cronobacter* spp. (*Enterobacter sakazakii*). *Letters in Applied Microbiology*, *48*(6), 718–725. https://doi.org/10.1111/j.1472-765X.2009.02601.x.

Dat, N. M., Manh, L. D., Hamanaka, D., Hung, D. V., Tanaka, F., & Uchino, T. (2014). Surface conditioning of stainless steel coupons with skim milk, buttermilk, and butter serum solutions and its effect on bacterial adherence. *Food Control*, *42*, 94–100. https://doi.org/10.1016/j.foodcont.2014.01.040.

Davis, K. R., Dunn, A. C., Burnett, C., McCullough, L., Dimond, M., Wagner, J., et al. (2016). *Campylobacter jejuni* infections associated with raw milk consumption—Utah, 2014. *MMWR Morbidity and Mortality Weekly Report*, *65*, 301–305. https://doi.org/10.15585/mmwr.mm6512a1.

Davoodi, S. H., Shahbazi, R., Esmaeili, S., Sohrabvandi, S., Mortazavian, A., Jazayeri, S., et al. (2016). Health-related aspects of milk proteins. *Iranian Journal of Pharmaceutical Research: IJPR*, *15*(3), 573–591.

De Filippis, F., Pasolli, E., & Ercolini, D. (2020). The food-gut axis: Lactic acid bacteria and their link to food, the gut microbiome and human health. *FEMS Microbiology Reviews*, *44*(4), 454–489. https://doi.org/10.1093/femsre/fuaa015.

de Oliveira, G. B., Favarin, L., Luchese, R. H., & McIntosh, D. (2015). Psychrotrophic bacteria in milk: How much do we really know? *Brazilian Journal of Microbiology*, *46*(2), 313–321. https://doi.org/10.1590/S1517-838246220130963.

Delhalle, L., Taminiau, B., Fastrez, S., Fall, A., Ballesteros, M., Burteau, S., et al. (2020). Evaluation of enzymatic cleaning on food processing installations and food products bacterial microflora. *Frontiers in Microbiology*, *11*, 1827. https://doi.org/10.3389/fmicb.2020.01827.

Didienne, R., Defargues, C., Callon, C., Meylheuc, T., Hulin, S., & Montel, M. C. (2012). Characteristics of microbial biofilm on wooden vats ('gerles') in PDO Salers cheese. *International Journal of Food Microbiology*, *156*(2), 91–101. https://doi.org/10.1016/j.ijfoodmicro.2012.03.007.

Efimochkina, N. R., Stetsenko, V. V., Markova, Y. M., Bykova, I. B., Pichugina, T. V., Polyanina, A. S., et al. (2019). The study of the raw milk contamination by bacteria of the genus *Campylobacter* using traditional microbiological methods and quantitative PCR assay. *Voprosy Pitaniia*, *88*(5), 17–23. https://doi.org/10.24411/0042-8833-2019-10049.

Fabres-Klein, M. H., Caizer Santos, M. J., Contelli Klein, R., Nunes de Souza, G., & de Oliveira Barros Ribon, A. (2015). An association between milk and slime increases biofilm production by bovine *Staphylococcus aureus*. *BMC Veterinary Research*, *11*, 3. https://doi.org/10.1186/s12917-015-0319-7.

Fagerlund, A., Langsrud, S., & Møretrø, T. (2021). Microbial diversity and ecology of biofilms in food industry environments associated with *Listeria monocytogenes* persistence. *Current Opinion in Food Science*, *37*, 171–178. https://doi.org/10.1016/j.cofs.2020.10.015.

Friedlander, A., Nir, S., Reches, M., & Shemesh, M. (2019). Preventing biofilm formation by dairy-associated bacteria using peptide-coated surfaces. *Frontiers in Microbiology*, *10*, 1405. https://doi.org/10.3389/fmicb.2019.01405.

Gaglio, R., Todaro, M., & Settanni, L. (2021). Improvement of raw milk cheese hygiene through the selection of starter and non-starter lactic acid bacteria: The successful case of PDO pecorino siciliano cheese. *International Journal of Environmental Research and Public Health*, *18*(4), 1834. https://doi.org/10.3390/ijerph18041834.

Galié, S., García-Gutiérrez, C., Miguélez, E. M., Villar, C. J., & Lombó, F. (2018). Biofilms in the food industry: Health aspects and control methods. *Frontiers in Microbiology*, *9*, 898. https://doi.org/10.3389/fmicb.2018.00898.

Gantner, V., Mijic, P., Baban, M., Skrtic, Z., & Turalija, A. (2015). The overall and fat composition of milk of various species. *Mljekarstvo*, *65*, 223–231. https://doi.org/10.15567/mljekarstvo.2015.0401.

Garbaj, A. M., Awad, E. M., Azwai, S. M., Abolghait, S. K., Naas, H. T., Moawad, A. A., et al. (2016). Enterohemorrhagic *Escherichia coli* O157 in milk and dairy products from Libya: Isolation and molecular identification by partial sequencing of 16S rDNA. *Veterinary World*, *9*(11), 1184–1189. https://doi.org/10.14202/vetworld.2016.1184-1189.

García-Cano, I., Rocha-Mendoza, D., Ortega-Anaya, J., Wang, K., Kosmerl, E., & Jiménez-Flores, R. (2019). Lactic acid bacteria isolated from dairy products as potential producers of lipolytic, proteolytic and antibacterial proteins. *Applied Microbiology and Biotechnology*, *103*(13), 5243–5257. https://doi.org/10.1007/s00253-019-09844-6.

Garneau, J. E., & Moineau, S. (2011). Bacteriophages of lactic acid bacteria and their impact on milk fermentations. *Microbial Cell Factories*, *10*, S20. https://doi.org/10.1186/1475-2859-10-S1-S20.

Gavrilova, E., Anisimova, E., Gabdelkhadieva, A., Nikitina, E., Vafina, A., Yarullina, D., et al. (2019). Newly isolated lactic acid bacteria from silage targeting biofilms of foodborne pathogens during milk fermentation. *BMC Microbiology*, *19*, 248. https://doi.org/10.1186/s12866-019-1618-0.

Givens, D. I. (2020). MILK Symposium review: The importance of milk and dairy foods in the diets of infants, adolescents, pregnant women, adults, and the elderly. *Journal of Dairy Science*, *103*(11), 9681–9699. https://doi.org/10.3168/jds.2020-18296.

Gomes, L. C., & Mergulhão, F. J. (2017). SEM analysis of surface impact on biofilm antibiotic treatment. *Scanning*, *2017*. https://doi.org/10.1155/2017/2960194, 2960194.

Goncalves, J. L., Kamphuis, C., Vernooij, H., Araujo, J. P., Grenfell, R. C., Juliano, L., et al. (2020). Pathogen effects on milk yield and composition in chronic subclinical mastitis in dairy cows. *The Veterinary Journal*, *262*, 105473. https://doi.org/10.1016/j.tvjl.2020.105473.

Gopal, N., Hill, C., Ross, P. R., Beresford, T. P., Fenelon, M. A., & Cotter, P. D. (2015). The prevalence and control of Bacillus and related spore-forming bacteria in the dairy industry. *Frontiers in Microbiology*, *6*, 1418. https://doi.org/10.3389/fmicb.2015.01418.

Guerrero-Navarro, A., Rios Castillo, A. G., Ripolles-Avila, C., Zamora, A., Hascoët, A. S., Felipe, X., et al. (2021). Effectiveness of enzymatic treatment for reducing dairy fouling at pilot-plant scale under real cleaning conditions. *LWT*, *154*, 112634. https://doi.org/10.1016/j.lwt.2021.112634.

Hamadi, F., Asserne, S., Bensouda, S., Mabrounki, M., & Latrache, H. (2014). Adhesion of *Staphylococcus aureus* on stainless steel treated with three types of milk. *Food Control*, *38*, 104–108. https://doi.org/10.1016/j.foodcont.2013.10.006.

Hanson, H., Whitfield, Y., Lee, C., Badiani, T., Minielly, C., Fenik, J., et al. (2019). *Listeria monocytogenes* associated with pasteurized chocolate milk, Ontario, Canada. *Emerging Infectious Diseases*, *25*(3), 581–584. https://doi.org/10.3201/eid2503.180742.

Hayes, M. C., & Boor, K. (2001). Raw milk and fluid milk products. In E. H. Marth, & J. L. Steele (Eds.), *Applied dairy microbiology* (pp. 59–76). Marcel Dekker Inc. https://doi.org/10.1201/9781482294606.

Henning, D. R. (2001). Fermented by-products. In E. H. Marth, & J. L. Steele (Eds.), *Applied dairy microbiology* (pp. 385–396). Marcel Dekker Inc. https://doi.org/10.1201/9781482294606.

Hunt, K., Drummond, N., Murphy, M., Butler, F., Buckley, J., & Jordan, K. (2012). A case of bovine raw milk contamination with *Listeria monocytogenes*. *Irish Veterinary Journal*, *65*(1), 13. https://doi.org/10.1186/2046-0481-65-13.

In Lee, S. H., Barancelli, G. V., de Camargo, T. M., Corassin, C. H., Rosim, R. E., da Cruz, A. G., et al. (2017). Biofilm-producing ability of *Listeria monocytogenes* isolates from Brazilian cheese processing plants. *Food Research International (Ottawa, Ont.)*, *91*, 88–91. https://doi.org/10.1016/j.foodres.2016.11.039.

Jaakkonen, A., Kivistö, R., Aarnio, M., Kalekivi, J., & Hakkinen, M. (2020). Persistent contamination of raw milk by *Campylobacter jejuni* ST-883. *PLoS One*, *15*(4), e0231810. https://doi.org/10.1371/journal.pone.0231810.

Jara, J., Perez-Ramos, A., del Solar, G., Rodriguez, J. M., Fernandez, L., & Orgaz, B. (2020). Role of *Lactobacillus* biofilms in *Listeria monocytogenes* adhesion to glass surfaces. *International Journal of Food Microbiology*, *334*, 108804. https://doi.org/10.1016/j.ijfoodmicro.2020.108804.

Jindal, S., Anand, S., Huang, K., Goddard, J., Metzger, L., & Amamcharla, J. (2016). Evaluation of modified stainless steel surfaces targeted to reduce biofilm formation by common milk sporeformers. *Journal of Dairy Science*, *99*(12), 9502–9513. https://doi.org/10.3168/jds.2016-11395.

Johler, S., Weder, D., Bridy, C., Huguenin, M. C., Robert, L., Hummerjohann, J., et al. (2015). Outbreak of staphylococcal food poisoning among children and staff at a Swiss boarding school due to soft cheese made from raw milk. *Journal of Dairy Science*, *98*(5), 2944–2948. https://doi.org/10.3168/jds.2014-9123.

Jones, G., Lefèvre, S., Donguy, M. P., Nisavanh, A., Terpant, G., Fougère, E., et al. (2019). Outbreak of Shiga toxin-producing *Escherichia coli* (STEC) O26 paediatric haemolytic uraemic syndrome (HUS) cases associated with the consumption of soft raw cow's milk cheeses, France, March to May 2019. *Euro Surveillance: Bulletin Europeen Sur Les Maladies Transmissibles = European Communicable Disease Bulletin*, *24*(22), 1900305. https://doi.org/10.2807/1560-7917.ES.2019.24.22.1900305.

Jørgensen, H. J., Mørk, T., & Rørvik, L. M. (2005). The occurrence of *Staphylococcus aureus* on a farm with small-scale production of raw milk cheese. *Journal of Dairy Science*, *88*(11), 3810–3817. https://doi.org/10.3168/jds.S0022-0302(05)73066-6.

Jourdan-da Silva, N., Fabre, L., Robinson, E., Fournet, N., Nisavanh, A., Bruyand, M., et al. (2018). Ongoing nationwide outbreak of *Salmonella* Agona associated with internationally distributed infant milk products, France, December 2017. *Euro Surveillance: Bulletin Europeen Sur Les Maladies Transmissibles = European Communicable Disease Bulletin*, *23*(2), 17–00852. https://doi.org/10.2807/1560-7917.ES.2018.23.2.17-00852.

KBV Research. (2021, May). *Dairy products market size*. Available at: https://www.kbvresearch.com/dairy-products-market/table-of-content/ (Accessed 7 December 2021).

Knight, G. (2015). Biofilm control in dairy manufacturing plants. In K. H. Teh, S. Flint, J. Brooks, & G. Knight (Eds.), *Biofilms in the dairy industry* (pp. 229–251). John Wiley & Sons Ltd. https://doi.org/10.1002/9781118876282.

Koo, O. K., Ndahetuye, J. B., O'Bryan, C. A., Ricke, S. C., & Crandall, P. G. (2014). Influence of *Listeria innocua* on the attachment of *Listeria monocytogenes* to stainless steel and aluminum surfaces. *Food Control*, *39*, 135–138. https://doi.org/10.1016/j.foodcont.2013.11.008.

Kubota, H., Senda, S., Nomura, N., Tokuda, H., & Uchiyama, H. (2008). Biofilm formation by lactic acid bacteria and resistance to environmental stress. *Journal of Bioscience and Bioengineering*, *106*(4), 381–386. https://doi.org/10.1263/jbb.106.381.

Kuda, T., Nakano, A., Takahashi, H., & Kimura, B. (2016). Effect of the quantities of food residues on the desiccation resistance of spoilage lactic acid bacteria adhered to a stainless steel surface. *Food Control*, *68*, 40–44. https://doi.org/10.1016/j.foodcont.2016.03.026.

Kumari, S., & Sarkar, P. K. (2014). Prevalence and characterization of *Bacillus cereus* group from various marketed dairy products in India. *Dairy Science and Technology*, *94*, 483–497. https://doi.org/10.1007/s13594-014-0174-5.

Kumari, S., & Sarkar, P. K. (2016). *Bacillus cereus* hazard and control in industrial dairy processing environment. *Food Control*, *69*, 20–29. https://doi.org/10.1016/j.foodcont.2016.04.012.

Lamas, A., Regal, P., Vazquez, B., Miranda, J. M., Cepeda, A., & Franco, C. M. (2018). Influence of milk, chicken residues and oxygen levels on biofilm formation on stainless steel, gene expression and small RNAs in *Salmonella enterica*. *Food Control*, *90*, 1–9. https://doi.org/10.1016/j.foodcont.2018.02.023.

Laranjo, M., Potes, M. E., & Elias, M. (2019). Role of starter cultures on the safety of fermented meat products. *Frontiers in Microbiology*, *10*, 853. https://doi.org/10.3389/fmicb.2019.00853.

Latorre, A. A., Van Kessel, J. S., Karns, J. S., Zurakowski, M. J., Pradhan, A. K., Boor, K. J., et al. (2010). Biofilm in milking equipment on a dairy farm as a potential source of bulk tank milk contamination with *Listeria monocytogenes*. *Journal of Dairy Science*, *93*(6), 2792–2802. https://doi.org/10.3168/jds.2009-2717.

Lee, S. H., Mangolin, B. L., Gonçalves, J. L., Neeff, D. V., Silva, M. P., Cruz, A. G., et al. (2014). Biofilm-producing ability of *Staphylococcus aureus* isolates from Brazilian dairy farms. *Journal of Dairy Science*, *97*(3), 1812–1816. https://doi.org/10.3168/jds.2013-7387.

Li, Y., Xiao, P., Wang, Y., & Hao, Y. (2020). Mechanisms and control measures of mature biofilm resistance to antimicrobial agents in the clinical context. *ACS Omega*, *5*(36), 22684–22690. https://doi.org/10.1021/acsomega.0c02294.

Lin, Q., Sun, H., Yao, K., Cai, J., Ren, Y., & Chi, Y. (2019). The prevalence, antibiotic resistance and biofilm formation of *Staphylococcus aureus* in bulk ready-to-eat foods. *Biomolecules*, *9*(10), 524. https://doi.org/10.3390/biom9100524.

Linares, D. M., Gómez, C., Renes, E., Fresno, J. M., Tornadijo, M. E., Ross, R. P., et al. (2017). Lactic acid bacteria and Bifidobacteria with potential to design natural biofunctional health-promoting dairy foods. *Frontiers in Microbiology*, *8*, 846. https://doi.org/10.3389/fmicb.2017.00846.

Lobacz, A., & Zulewska, J. (2021). Fate of Salmonella spp. in the fresh soft raw milk cheese during storage at different temperatures. *Microorganisms*, *9*(5), 938. https://doi.org/10.3390/microorganisms9050938.

Lortal, S., Di Blasi, A., Madec, M. N., Pediliggieri, C., Tuminello, L., Tanguy, G., et al. (2009). Tina wooden vat biofilm: A safe and highly efficient lactic acid bacteria delivering system in PDO Ragusano cheese making. *International Journal of Food Microbiology*, *132*(1), 1–8. https://doi.org/10.1016/j.ijfoodmicro.2009.02.026.

Lortal, S., Licitra, G., & Valence, F. (2014). Wooden tools: Reservoirs of microbial biodiversity in traditional cheesemaking. *Microbiology Spectrum*, *2*(1). https://doi.org/10.1128/microbiolspec.CM-0008-2012. CM–2012.

Lücking, G., Stoeckel, M., Atamer, Z., Hinrichs, J., & Ehling-Schulz, M. (2013). Characterization of aerobic spore-forming bacteria associated with industrial dairy processing environments and product spoilage. *International Journal of Food Microbiology*, *166*(2), 270–279. https://doi.org/10.1016/j.ijfoodmicro.2013.07.004.

Magalhães, R., Ferreira, V., Biscottini, G., Brandão, T. R. S., Almeida, G., & Teixeira, P. (2017). Biofilm formation by persistent and non-persistent *Listeria monocytogenes* strains on abiotic surfaces. *Acta Alimentaria*, *46*(1), 43–50. https://doi.org/10.1556/066.2017.46.1.6.

Maggio, F., Rossi, C., Chaves-López, C., Serio, A., Valbonetti, L., Pomilio, F., et al. (2021). Interactions between *L. monocytogenes* and *P. fluorescens* in dual-species biofilms under simulated dairy processing conditions. *Foods (Basel, Switzerland)*, *10*(1), 176. https://doi.org/10.3390/foods10010176.

Mah, T. F., & O'Toole, G. A. (2001). Mechanisms of biofilm resistance to antimicrobial agents. *Trends in Microbiology*, *9*(1), 34–39. https://doi.org/10.1016/s0966-842x(00)01913-2.

Mansouri-Najand, L., Kianpour, M., Sami, M., & Jajarmi, M. (2015). Prevalence of *Listeria monocytogenes* in raw milk in Kerman, Iran. *Veterinary Research Forum: An International Quarterly Journal*, *6*(3), 223–226.

Marangoni, F., Pellegrino, L., Verduci, E., Ghiselli, A., Bernabei, R., Calvani, R., et al. (2019). Cow's milk consumption and health: A health professional's guide. *Journal of the American College of Nutrition*, *38*(3), 197–208. https://doi.org/10.1080/07315724.2018.1491016.

Marchand, S., De Block, J., De Jonghe, V., Coorevits, A., Heyndrickx, M., & Herman, L. (2012). Biofilm formation in milk production and processing environments: Influence on milk quality and safety. *Comprehensive Reviews in Food Science and Food Safety*, *11*(2), 133–147. https://doi.org/10.1111/j.1541-4337.2011.00183.x.

Mariani, C., Briandet, R., Chamba, J. F., Notz, E., Carnet-Pantiez, A., Eyoug, R. N., et al. (2007). Biofilm ecology of wooden shelves used in ripening the French raw milk smear cheese Reblochon de Savoie. *Journal of Dairy Science*, *90*(4), 1653–1661. https://doi.org/10.3168/jds.2006-190.

Marti, R., Schmid, M., Kulli, S., Schneeberger, K., Naskova, J., Knøchel, S., et al. (2017). Biofilm formation potential of heat-resistant *Escherichia coli* dairy isolates and the complete genome of multidrug-resistant, heat-resistant strain FAM21845. *Applied and Environmental Microbiology*, *83*(15), e00628-17. https://doi.org/10.1128/AEM.00628-17.

Maury, M. M., Bracq-Dieye, H., Huang, L., Valès, G., Lavina, M., Thouvenot, P., et al. (2019). Hypervirulent *Listeria monocytogenes* clones' adaption to mammalian gut accounts for their association with dairy products. *Nature Communications*, *10*. https://doi.org/10.1038/s41467-019-10380-0.

Mazaheri, T., Cervantes-Huamán, B., Bermúdez-Capdevila, M., Ripolles-Avila, C., & Rodríguez-Jerez, J. J. (2021). *Listeria monocytogenes* biofilms in the food industry: Is the current hygiene program sufficient to combat the persistence of the pathogen? *Microorganisms*, *9*(1), 181. https://doi.org/10.3390/microorganisms9010181.

Mekonnen, S. A., Lam, T., Hoekstra, J., Rutten, V., Tessema, T. S., Broens, E. M., et al. (2018). Characterization of *Staphylococcus aureus* isolated from milk samples of dairy cows in small holder farms of North-Western Ethiopia. *BMC Veterinary Research*, *14*(1), 246. https://doi.org/10.1186/s12917-018-1558-1.

Miranda, C., Contente, D., Igrejas, G., Câmara, S. P. A., Dapkevicius, M.d. L. E., & Poeta, P. (2021). Role of exposure to lactic acid bacteria from foods of animal origin in human health. *Foods*, *10*(9), 2092. https://doi.org/10.3390/foods10092092.

Mogha, K. V., Shah, N., Prajapati, J. B., & Chaudhari, A. R. (2014). Biofilm—A threat to dairy industry. *Indian Journal of Dairy Science*, *67*(6), 459–466. https://doi.org/10.5146/IJDS.V67I6.39317.G19823.

Morikawa, M. (2006). Beneficial biofilm formation by industrial bacteria *Bacillus subtilis* and related species. *Journal of Bioscience and Bioengineering*, *101*(1), 1–8. https://doi.org/10.1263/jbb.101.1.

Muhammad, M. H., Idris, A. L., Fan, X., Guo, Y., Yu, Y., Jin, X., et al. (2020). Beyond risk: Bacterial biofilms and their regulating approaches. *Frontiers in Microbiology*, *11*, 928. https://doi.org/10.3389/fmicb.2020.00928.

Muruzovic, M. Z., Mladenovic, K. G., & Comic, L. R. (2018). In vitro evaluation of resistance to environmental stress by planktonic and biofilm form of lactic acid bacteria isolated from traditionally made cheese from Serbia. *Food Bioscience*, *23*, 54–59. https://doi.org/10.1016/j.fbio.2018.03.005.

Mylius, M., Dreesman, J., Pulz, M., Pallasch, G., Beyrer, K., Claußen, K., et al. (2018). Shiga toxin-producing *Escherichia coli* O103:H2 outbreak in Germany after school trip to Austria due to raw cow milk, 2017—The important role of international collaboration for outbreak investigations. *International Journal of Medical Microbiology: IJMM*, *308*(5), 539–544. https://doi.org/10.1016/j.ijmm.2018.05.005.

Nüesch-Inderbinen, M., Bloemberg, G. V., Müller, A., Stevens, M., Cernela, N., Kollöffel, B., et al. (2021). Listeriosis caused by persistence of *Listeria monocytogenes* serotype 4b sequence type 6 in cheese production environment. *Emerging Infectious Diseases*, *27*(1), 284–288. https://doi.org/10.3201/eid2701.203266.

Ogier, J. C., Son, O., Gruss, A., Tailliez, P., & Delacroix-Buchet, A. (2002). Identification of the bacterial microflora in dairy products by temporal temperature gradient gel electrophoresis. *Applied and Environmental Microbiology*, *68*(8), 3691–3701. https://doi.org/10.1128/AEM.68.8.3691-3701.2002.

Oliver, S. P., Jayarao, B. M., & Almeida, R. A. (2005). Foodborne pathogens in milk and the dairy farm environment: Food safety and public health implications. *Foodborne Pathogens and Disease*, *2*(2), 115–129. https://doi.org/10.1089/fpd.2005.2.115.

Osman, K. M., Samir, A., Abo-Shama, U. H., Mohamed, E. H., Orabi, A., & Zolnikov, T. (2016). Determination of virulence and antibiotic resistance pattern of biofilm producing *Listeria* species isolated from retail raw milk. *BMC Microbiology*, *16*, 263. https://doi.org/10.1186/s12866-016-0880-7.

Ostrov, I., Paz, T., & Shemesh, M. (2019). Robust biofilm-forming *Bacillus* isolates from the dairy environment demonstrate an enhanced resistance to cleaning-in-place procedures. *Foods (Basel, Switzerland)*, *8*(4), 134. https://doi.org/10.3390/foods8040134.

Palmer, J., Flint, S., & Brooks, J. (2007). Bacterial cell attachment, the beginning of a biofilm. *Journal of Industrial Microbiology & Biotechnology*, *34*(9), 577–588. https://doi.org/10.1007/s10295-007-0234-4.

Palmer, J., Seale, B., & Flint, S. (2015). Significance of bacterial attachment: A focus on the food industry. In K. H. Teh, S. Flint, J. Brooks, & G. Knight (Eds.), *Biofilms in the dairy industry* (pp. 17–35). John Wiley & Sons Ltd. https://doi.org/10.1002/9781118876282.

Pang, X., & Yuk, H. G. (2019). Effects of the colonization sequence of *Listeria monocytogenes* and *Pseudomonas fluorescens* on survival of biofilm cells under food-related stresses and transfer to salmon. *Food Microbiology*, *82*, 142–150. https://doi.org/10.1016/j.fm.2019.02.002.

Parkar, S. G., Flint, S. H., Palmer, J. S., & Brooks, J. D. (2001). Factors influencing attachment of thermophilic bacilli to stainless steel. *Journal of Applied Microbiology*, *90*(6), 901–908. https://doi.org/10.1046/j.1365-2672.2001.01323.x.

Parkar, S., Flint, S., Teh, K. H., Brooks, J., & Bremer, P. (2015). Pathogen contamination in dairy manufacturing environments. In K. H. Teh, S. Flint, J. Brooks, & G. Knight (Eds.), *Biofilms in the dairy industry* (pp. 154–188). John Wiley & Sons Ltd. https://doi.org/10.1002/9781118876282.

Perin, L. M., Pereira, J. G., Bersot, L. S., & Nero, L. A. (2018). The microbiology of raw milk. In L. A. Nero, & A. F. De Carvalho (Eds.), *Raw milk: Balance between hazards and benefits* (pp. 45–64). Academic Press. https://doi.org/10.1016/C2016-0-00282-6.

Phophi, L., Petzer, I. M., & Qekwana, D. N. (2019). Antimicrobial resistance patterns and biofilm formation of coagulase-negative *Staphylococcus* species isolated from subclinical mastitis cow milk samples submitted to the Onderstepoort Milk laboratory. *BMC Veterinary Research*, *15*, 420. https://doi.org/10.1186/s12917-019-2175-3.

Popović, N., Dinić, M., Tolinački, M., Mihajlović, S., Terzić-Vidojević, A., Bojić, S., et al. (2018). New insight into biofilm formation ability, the presence of virulence genes and probiotic potential of *Enterococcus* sp. dairy isolates. *Frontiers in Microbiology*, *9*, 78. https://doi.org/10.3389/fmicb.2018.00078.

Puga, C. H., Dahdouh, E., San Jose, C., & Orgaz, B. (2018). *Listeria monocytogenes* colonizes *Pseudomonas fluorescens* biofilms and induces matrix over-production. *Frontiers in Microbiology*, *9*, 1706. https://doi.org/10.3389/fmicb.2018.01706.

Puga-Torres, B., Salazar, D., Cachiguango, M., Cisneros, G., & Gómez-Bravo, C. (2020). Determination of aflatoxin M1 in raw milk from different provinces of Ecuador. *Toxins*, *12*(8), 498. https://doi.org/10.3390/toxins12080498.

Quigley, L., O'Sullivan, O., Stanton, C., Beresford, T. P., Ross, R. P., Fitzgerald, G. F., et al. (2013). The complex microbiota of raw milk. *FEMS Microbiology Reviews*, *37*(5), 664–698. https://doi.org/10.1111/1574-6976.12030.

Radmehr, B., Zaferanloo, B., Tran, T., Beale, D. J., & Palombo, E. A. (2020). Prevalence and characteristics of *Bacillus cereus* group isolated from raw and pasteurised milk. *Current Microbiology*, *77*(10), 3065–3075. https://doi.org/10.1007/s00284-020-02129-6.

Ranieri, M. L., Ivy, R. A., Mitchell, W. R., Call, E., Masiello, S. N., Wiedmann, M., et al. (2012). Real-time PCR detection of *Paenibacillus* spp. in raw milk to predict shelf life performance of pasteurized fluid milk products. *Applied and Environmental Microbiology*, *78*(16), 5855–5863. https://doi.org/10.1128/AEM.01361-12.

Revelli, G. R., Sbodio, O. A., & Tercero, E. J. (2004). Recuento de bacterias totales en leche cruda de tambos que caracterizan la zona noroeste de Santa Fe y sur de Santiago del Estero [Total bacterial count in raw milk from the dairy farms that characterize the zone northwest of Santa Fe and south of Santiago del Estero]. *Revista Argentina de Microbiologia*, *36*(3), 145–149.

Ribeiro Júnior, J. C., de Oliveira, A. M., Silva, F. G., Tamanini, R., de Oliveira, A., & Beloti, V. (2018). The main spoilage-related psychrotrophic bacteria in refrigerated raw milk. *Journal of Dairy Science*, *101*(1), 75–83. https://doi.org/10.3168/jds.2017-13069.

Robinson, E., Travanut, M., Fabre, L., Larréché, S., Ramelli, L., Pascal, L., et al. (2020). Outbreak of *Salmonella* Newport associated with internationally distributed raw goats' milk cheese, France, 2018. *Epidemiology and Infection*, *148*, e180. https://doi.org/10.1017/S0950268820000904.

Robitaille, G., Choinière, S., Ells, T., Deschènes, L., & Mafu, A. A. (2014). Attachment of *Listeria innocua* to polystyrene: Effects of ionic strength and conditioning films from culture media and milk proteins. *Journal of Food Protection*, *77*(3), 427–434. https://doi.org/10.4315/0362-028X.JFP-13-353.

Rola, J. G., Czubkowska, A., Korpysa-Dzirba, W., & Osek, J. (2016). Occurrence of *Staphylococcus aureus* on farms with small scale production of raw milk cheeses in Poland. *Toxins*, *8*(3), 62. https://doi.org/10.3390/toxins8030062.

Rosario, A., Castro, V. S., Santos, L. F., Lisboa, R. C., Vallim, D. C., Silva, M., et al. (2021). Shiga toxin-producing *Escherichia coli* isolated from pasteurized dairy products from Bahia, Brazil. *Journal of Dairy Science*, *104*(6), 6535–6547. https://doi.org/10.3168/jds.2020-19511.

Rossi, C., Chaves-López, C., Serio, A., Goffredo, E., Goga, B. T., & Paparella, A. (2016). Influence of incubation conditions on biofilm formation by *Pseudomonas fluorescens* isolated from dairy products and dairy manufacturing plants. *Italian Journal of Food Safety*, *5*(3), 5793. https://doi.org/10.4081/ijfs.2016.5793.

Rossi, C., Serio, A., Chaves-López, C., Anniballi, F., Auricchio, B., Goffredo, E., et al. (2018). Biofilm formation, pigment production and motility in *Pseudomonas* spp. isolated from the dairy industry. *Food Control*, *86*, 241–248. https://doi.org/10.1016/j.foodcont.2017.11.018.

Roy, D., Ye, A., Moughan, P. J., & Singh, H. (2020). Composition, structure, and digestive dynamics of milk from different species—A review. *Frontiers in Nutrition*, *7*, 577759. https://doi.org/10.3389/fnut.2020.577759.

Rudenko, P., Sachivkina, N., Vatnikov, Y., Shabunin, S., Engashev, S., Kontsevaya, S., et al. (2021). Role of microorganisms isolated from cows with mastitis in Moscow region in biofilm formation. *Veterinary World*, *14*(1), 40–48. https://doi.org/10.14202/vetworld.2021.40-48.

Ryu, S., Shin, M., Yun, B., Lee, W., Choi, H., Kang, M., et al. (2021). Bacterial quality, prevalence of pathogens, and molecular characterization of biofilm-producing *Staphylococcus aureus* from Korean dairy farm environments. *Animals: An Open Access Journal From MDPI*, *11*(5), 1306. https://doi.org/10.3390/ani11051306.

Sadiq, F. A., Flint, S., Yuan, L., Li, Y., Liu, T., & He, G. (2017). Propensity for biofilm formation by aerobic mesophilic and thermophilic spore forming bacteria isolated from Chinese milk powders. *International Journal of Food Microbiology*, *262*, 89–98. https://doi.org/10.1016/j.ijfoodmicro.2017.09.015.

Salas-Jara, M. J., Ilabaca, A., Vega, M., & García, A. (2016). Biofilm forming *Lactobacillus*: New challenges for the development of probiotics. *Microorganisms*, *4*(3), 35. https://doi.org/10.3390/microorganisms4030035.

Scatassa, M. L., Gaglio, R., Macaluso, G., Francesca, N., Randazzo, W., Cardamone, C., et al. (2015). Transfer, composition and technological characterization of the lactic acid bacterial populations of the wooden vats used to produce traditional stretched cheeses. *Food Microbiology*, *52*, 31–41. https://doi.org/10.1016/j.fm.2015.06.008.

Schönborn, S., & Krömker, V. (2016). Detection of the biofilm component polysaccharide intercellular adhesin in *Staphylococcus aureus* infected cow udders. *Veterinary Microbiology*, *196*, 126–128. https://doi.org/10.1016/j.vetmic.2016.10.023.

Seale, B., Bremer, P., Flint, S., Brooks, J., & Palmer, J. (2015). Overview of the problems resulting from biofilm contamination in the dairy industry. In K. H. Teh, S. Flint, J. Brooks, & G. Knight (Eds.), *Biofilms in the dairy industry* (pp. 49–64). John Wiley & Sons Ltd. https://doi.org/10.1002/9781118876282.

Sharma, M., & Anand, S. K. (2002). Characterization of constitutive microflora of biofilms in dairy processing lines. *Food Microbiology*, *19*(6), 627–636. https://doi.org/10.1006/fmic.2002.0472.

Shi, X., & Zhu, X. (2009). Biofilm formation and food safety in food industries. *Trends in Food Science and Technology*, *20*(9), 407–413. https://doi.org/10.1016/j.foodcont.2012.12.001.

Silanikove, N., Leitner, G., & Merin, U. (2015). The interrelationships between lactose intolerance and the modern dairy industry: Global perspectives in evolutional and historical backgrounds. *Nutrients*, *7*(9), 7312–7331. https://doi.org/10.3390/nu7095340.

Simoes, M., Simoes, L. C., & Vieira, M. J. (2010). A review of current and emergent biofilm control strategies. *LWT - Food Science and Technology*, *43*(4), 573–583. https://doi.org/10.1016/j.lwt.2009.12.008.

Singh, N., Sharma, C., Gulhane, R. D., Rokana, N., Singh, B. P., Puniya, A. K., et al. (2018). Inhibitory effects of lactobacilli of goat's milk origin against growth and biofilm formation by pathogens: An in vitro study. *Food Bioscience*, *22*, 129–138. https://doi.org/10.1016/j.fbio.2018.02.001.

Skowron, K., Wiktorczyk, N., Grudlewska, K., Kwiecińska-Piróg, J., Wałecka-Zacharska, E., Paluszak, Z., et al. (2019). Drug-susceptibility, biofilm-forming ability and biofilm survival on stainless steel of *Listeria* spp. strains isolated from cheese. *International Journal of Food Microbiology*, *296*, 75–82. https://doi.org/10.1016/j.ijfoodmicro.2019.02.021.

Skowron, K., Wałecka-Zacharska, E., Grudlewska, K., Wiktorczyk, N., Kaczmarek, A., Gryń, G., et al. (2019). Characteristics of *Listeria monocytogenes* strains isolated from milk and humans and the possibility of milk-borne strains transmission. *Polish Journal of Microbiology*, *68*(3), 353–369. https://doi.org/10.33073/pjm-2019-038.

Somers, E. B., Johnson, M. E., & Wong, A. C. (2001). Biofilm formation and contamination of cheese by nonstarter lactic acid bacteria in the dairy environment. *Journal of Dairy Science*, *84*(9), 1926–1936. https://doi.org/10.3168/jds.S0022-0302(01)74634-6.

Stiemsma, L. T., Nakamura, R. E., Nguyen, J. G., & Michels, K. B. (2020). Does consumption of fermented foods modify the human gut microbiota? *Journal of Nutrition*, *150*(7), 1680–1692. https://doi.org/10.1093/jn/nxaa077.

Sun, L., Zhang, Y., Guo, X., Zhang, L., Zhang, W., Man, C., et al. (2020). Characterization and transcriptomic basis of biofilm formation by *Lactobacillus plantarum* J26 isolated from traditional fermented dairy products. *LWT*, *125*, 109333. https://doi.org/10.1016/j.lwt.2020.109333.

Tang, X, Flint, SH, Bennett, RJ, & Brooks, JD. (2010). The efficacy of different cleaners and sanitisers in cleaning biofilms on UF membranes used in the dairy industry. *Journal of Membrane Science*, *352*, 71–75. https://doi.org/10.1016/j.memsci.2010.01.063.

Taylor, A. J., & Stasiewicz, M. J. (2019). Persistent and sporadic *Listeria monocytogenes* strains do not differ when growing at 37°C, in planktonic state, under different food associated stresses or energy sources. *BMC Microbiology*, *19*, 257. https://doi.org/10.1186/s12866-019-1631-3.

Teh, K. H., Flint, S., Palmer, J., Andrewes, P., Bremer, P., & Lindsay, D. (2014). Biofilm—An unrecognised source of spoilage enzymes in dairy products? *International Dairy Journal*, 3432–3440. https://doi.org/10.1016/j.idairyj.2013.07.002.

Teh, K. H., Flint, S., & Bremer, P. (2015). Raw milk quality influenced by biofilms and the effect of biofilm growth on dairy product quality. In K. H. Teh, S. Flint, J. Brooks, & G. Knight (Eds.), *Biofilms in the dairy industry* (pp. 65–98). John Wiley & Sons Ltd. https://doi.org/10.1002/9781118876282.

Thiran, E. (2018). *Role of biofilms in bacterial persistence in dairy environment* (Doctoral dissertation No. 25065). Zürich: Eidgenössische Technische Hochschule ETH. https://doi.org/10.3929/ethz-b-000281595.

Thiran, E., Di Ciccio, P. A., Graber, H. U., Zanardi, E., Ianieri, A., & Hummerjohann, J. (2018). Biofilm formation of *Staphylococcus aureus* dairy isolates representing different genotypes. *Journal of Dairy Science*, *101*(2), 1000–1012. https://doi.org/10.3168/jds.2017-13696.

Ulusoy, B. H., & Chirkena, K. (2019). Two perspectives of *Listeria monocytogenes* hazards in dairy products: The prevalence and the antibiotic resistance. *Food Quality and Safety*, *3*(4), 233–241. https://doi.org/10.1093/fqsafe/fyz035.

van der Veen, S., & Abee, T. (2011). Mixed species biofilms of *Listeria monocytogenes* and *Lactobacillus plantarum* show enhanced resistance to benzalkonium chloride and peracetic acid. *International Journal of Food Microbiology*, *144*(3), 421–431. https://doi.org/10.1016/j.ijfoodmicro.2010.10.029.

Van Houdt, R., & Michiels, C. W. (2010). Biofilm formation and the food industry, a focus on the bacterial outer surface. *Journal of Applied Microbiology*, *109*(4), 1117–1131. https://doi.org/10.1111/j.1365-2672.2010.04756.x.

Varhimo, E., Varmanen, P., Fallarero, A., Skogman, M., Pyörälä, S., Iivanainen, A., et al. (2011). Alpha- and β-casein components of host milk induce biofilm formation in the mastitis bacterium *Streptococcus uberis*. *Veterinary Microbiology*, *149*(3–4), 381–389. https://doi.org/10.1016/j.vetmic.2010.11.010.

Wagner, E. M., Pracser, N., Thalguter, S., Fischel, K., Rammer, N., Pospíšilová, L., et al. (2020). Identification of biofilm hotspots in a meat processing environment: Detection of spoilage bacteria in multi-species biofilms. *International Journal of Food Microbiology*, *328*, 108668. https://doi.org/10.1016/j.ijfoodmicro.2020.108668.

Wallis, J. K., Krömker, V., & Paduch, J. H. (2019). Biofilm challenge: Lactic acid bacteria isolated from bovine udders versus staphylococci. *Foods (Basel, Switzerland)*, *8*(2), 79. https://doi.org/10.3390/foods8020079.

Wang, N., Jin, Y., He, G., & Yuan, L. (2021). Development of multi-species biofilm formed by thermophilic bacteria on stainless steel immerged in skimmed milk. *Food Research International*, *150*, 110754. https://doi.org/10.1016/j.foodres.2021.110754.

Warren, S. C., Jr. (2001). Concentrated and dry milks and wheys. In E. H. Marth, & J. Steele (Eds.), *Applied dairy microbiology* (pp. 77–92). Marcel Dekker Inc. https://doi.org/10.1201/9781482294606.

Weber, M., Liedtke, J., Plattes, S., & Lipski, A. (2019). Bacterial community composition of biofilms in milking machines of two dairy farms assessed by a combination of culture-dependent and—Independent methods. *PLoS One*, *14*(9), e0222238. https://doi.org/10.1371/journal.pone.0222238.

Weimer, P. J. (2001). Microbiology of the dairy animal. In E. H. Marth, & J. L. Steele (Eds.), *Applied dairy microbiology* (pp. 1–58). Marcel Dekker Inc. https://doi.org/10.1201/9781482294606.

Widyastuti, Y., Rohmatussolihat, & Febrisiantosa, A. (2014). The role of lactic acid bacteria in milk fermentation. *Food of Nutrition and Science*, *5*, 425–434. https://doi.org/10.4236/fns.2014.54051.

Yu, A. O., Leveau, J., & Marco, M. L. (2020). Abundance, diversity and plant-specific adaptations of plant-associated lactic acid bacteria. *Environmental Microbiology Reports*, *12*(1), 16–29. https://doi.org/10.1111/1758-2229.12794.

Zhao, Y., Caspers, M. P., Metselaar, K. I., de Boer, P., Roeselers, G., Moezelaar, R., et al. (2013). Abiotic and microbiotic factors controlling biofilm formation by thermophilic sporeformers. *Applied and Environmental Microbiology*, *79*(18), 5652–5660. https://doi.org/10.1128/AEM.00949-13.

Chapter 10

Microalgal biofilm and their prospective application for wastewater treatment and biofuel production

Rabindra Nayak[a], Biswajita Pradhan[a,b], Srimanta Patra[c], Prajna Paramita Bhuyan[d], Chhandashree Behera[a], Sudhamayee Parida[a], Akshaya Kumar Behera[a], Amiya Kumar Mandal[a], and Mrutyunjay Jena[a]

[a]*Algal Biotechnology and Molecular Systematic Laboratory, Post Graduate Department of Botany, Berhampur University, Berhampur, Odisha, India,* [b]*Department of Biotechnology, Sangmyung University, Seoul, South Korea,* [c]*Cancer and Cell Death Laboratory, Department of Life Science, National Institute of Technology Rourkela, Rourkela, Odisha, India,* [d]*Department of Botany, Maharaja Sriram Chandra Bhanja Deo University, Baripada, Odisha, India*

1. Introduction

Algae are the primary photosynthetic microorganism found in both marine and freshwater, maintains the ecosystem balance as a primary producer for the aquatic food chain. Over these properties, it maintains ecological balance to give a maximum amount of O_2 (Behera, Dash, Pradhan, Jena, & Adhikary, 2020; Behera et al., 2021; Colman, Huertas, Bhatti, & Dason, 2002; Dash, Pradhan, Behera, & Jena, 2020; Dash, Pradhan, Behera, Nayak, & Jena, 2021; Maharana, Pradhan, Jena, & Misra, 2019; Mohanty et al., 2020; Pradhan, Maharana, Bhakta, & Jena, 2021; Smith & Piedrahita, 1988). Furthermore, algae are chlorophyll-containing single to multicellular creatures that lack real roots, stalks, and leaves (Bellinger & Sigee, 2015; Pradhan et al., 2022, 2020). Nature bears a diverse group of algae such as micro, macro, aerial, subaerial, Aquatica, and the small leafy structure found in *Chara* sp. The microalgae and several groups of algae have more economic value, food value, and therapeutic effects (Pradhan, Patra, Behera, et al., 2021; Pradhan, Patra, Behera, et al., 2020; Pradhan, Patra, Dash, et al., 2021; Pradhan, Patra, Nayak, et al., 2020; Pulz & Gross, 2004; Suganya, Varman, Masjuki, & Renganathan, 2016). They may be toxic or nontoxic. When compared with energy crops, microalgae have demonstrated substantial benefits in generating biofuels (Schenk et al., 2008). Aside from their rapid growth rates and high lipid/triacylglycerol yields, microalgae can thrive in wastewaters. They efficiently eliminate primary nutrients, heavy metals, and micropollutants. Moreover, they do not compete with crops for arable land. Nevertheless, the high expenses of collecting the algae from the water remain significant obstacles to the industrial utilization of microalgae for biofuel manufacture (Christenson & Sims, 2011). In addition, they sustain by utilizing autotrophic, mixotrophic, or heterotrophic metabolic methods (Barone et al., 2020; Perez-Garcia & Bashan, 2015). Since the 1950s, many microalgae species have been exploited to produce valuable bioactives. Protection of the environment and green energy substitute mandates have sparked a revival in microalgae research in recent decades. Moreover, microalgal biomass as a feedstock for biodiesel production and other valuable products in bioengineering, such as the food, cosmetics, aquaculture, and pharmaceutical sectors, and the recovery of components such as phosphorus and iodine has also been exploited (Harun, Singh, Forde, & Danquah, 2010; Tang et al., 2020). The adaptability character of microalgae to adjust their metabolism in different cultural conditions provides the diverse used in biotechnology, which is known as a living cell factory. On the other hand, microalgae provide eco-friendly bioproducts, cost-effective and profitable in wastewater treatment as well as heavy and light metal removal (Pradhan, Patra, Dash, et al., 2020b; Pradhan, Patra, Maharana, et al., 2021; Pradhan, Patra, Nayak, et al., 2021). Furthermore, compared with terrestrial crops, it can use polluted municipal or industrial wastewater as a medium also requires lower nutrients (Cai, Park, & Li, 2013). Despite these benefits, the production of microalgae for biofuel has been slow to scale up since it is not economically competitive with other fuels (Choudhary, Malik, & Pant, 2017; Milano et al., 2016). Lower growth rates due to photoinhibition, contamination with less productive species, poor harvesting efficiency, mixing/transportation expenses, and a high water need are some of the causes (Kim, Moon, Kwak, Lee, & Chang, 2018; Park et al., 2012). Significant research is being

Understanding Microbial Biofilms. https://doi.org/10.1016/B978-0-323-99977-9.00021-1

conducted to improve the cultivation of microalgae in suspension, both by increasing biomass concentration and by improving process and bioreactor designs, particularly more efficient biomass recovery technologies (Wollmann et al., 2019). Furthermore, there is a lot of interest in looking into a completely new approach: the potential of the connected culture of algae as biofilm. Additionally, when combined with processes such as wastewater treatment, bioproduct manufacture from microalgae can be more ecologically friendly, cost effective, and profitable (Ravindran et al., 2016) (Fig. 1). On the other hand, to solve the crisis of necessity, they constructed a biofilm based on the qualities of algae.

A biofilm defined that an immobilized cell is a live cell that is prevented from migrating freely from its initial position to all areas of an aqueous phase of a system by natural or artificial causes (Tampion & Tampion, 1987). Algae such as green algae, diatoms, and cyanobacteria species are more potent among the microorganisms, which creates natural biofilm in the ecosystem. Furthermore, microalgae are inextricably linked to a vast microbial population that coats their surface. Because this microenvironment is so diverse in terms of biodiversity and food availability, living there is complicated and competitive (Lage & Graça, 2016; Lewandowski & Evans, 2000). On the other hand, microalgal biofilm are also known as phototrophic biofilms, which use extracellular polymeric molecules to spontaneously adhere to a surface in the presence of light, moisture, and nutrition (EPS) (Kesaano & Sims, 2014; Mahto, Kumari, & Das, 2021; Mangwani, Shukla, Kumari, Rao, & Das, 2014). Microalgal biofilms are a major source of autochthonous carbon for other trophic levels in natural marine or freshwater habitats, and they influence nutrient recycling at the ecosystem level (Bengtsson, Wagner, Schwab, Urich, & Battin, 2018; Kanavillil, Thorn, & Kurissery, 2012). A similar principle may be used in wastewater treatment, where microalgal biofilms are used for nutrient recycling (removal of nutrients or heavy metals) as well as carbon and oxygen supplies to boost bacterial activity (Dalrymple et al., 2013; Samal, Kar, & Trivedi, 2019). Likewise, utilizing a photobioreactor to create an artificial algal biofilm plays a significant part in meeting the criterion (Tao et al., 2017). Therefore, the importance of this chapter is centered on the treatment of wastewater using microalgal biofilm and the resolution of the biodiesel dilemma. Given the rise of this modern biotechnology, it will be advantageous for microalgal biofilm research groups to pursue pathways that promote not only the sustainability of the planned system but also the valorization of microalgal metabolite production. On the other hand, by enhancing the interaction between microalgae and bacteria, effective water purification and a large quantity of biodiesel will be achieved in the future.

2. Microalgal biofilm

A biofilm is a collection of microorganisms attached to a permanently submerged surface and connected with an organic matrix (Gómez-Ramírez et al., 2019; Pandey, Bharti, & Kumar, 2014). Biofilm are microbial communities that thrive on solid surfaces and are encapsulated in an extracellular polymeric matrix (EPS) (Kesaano & Sims, 2014; Strieth, Ulber, & Muffler, 2018). Polysaccharides make up most of these EPS, although they also contain proteins, nucleic acids, and lipids (Flemming & Wingender, 2001). EPS plays a vital part in the attachment of microorganisms to a substratum in addition to their involvement in biofilm cohesiveness. Furthermore, EPS can shield organisms from grazers, aid in trace element binding, and serve as a water storage system. A biofilm can build on any substance that is exposed to a certain amount of moisture, and it is generally apparent as a slimy coating. Biofilm may thus be found in all natural environments, including aquatic and soil ecosystems, plant tissues, and animal tissues (Varnam & Evans, 2000). Biofilm is also found in filters and other porous materials, reservoirs, plumbing systems, pipelines, ship hulls, heat exchangers, and separation membranes, among other technological

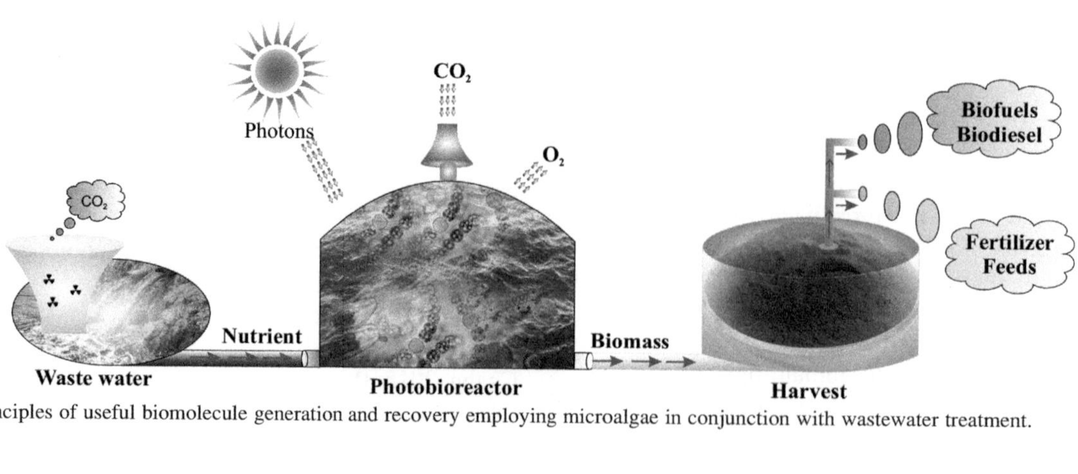

FIG. 1 Principles of useful biomolecule generation and recovery employing microalgae in conjunction with wastewater treatment.

systems (Flemming, 2002). Based on mobilization, biofilm can be divided into two categories, i.e., active and passive (Moreno-Garrido, 2008). If there is adequate light, the biofilm will also include microalgae and green in color (Garnham, Codd, & Gadd, 1992).

Microalgae, also known as microphytes, are unicellular organisms ranging from macro to microscopic. They are found in marine and freshwater waters as single cells, filaments, or aggregates (Thapa, Bharti, & Prasanna, 2017). Microalgae play an essential part in ecological succession. Microalgal biofilm may grow on any surface with enough light and moisture to support them. The creatures in these films are both autotrophic and heterotrophic (Ebeling, Timmons, & Bisogni, 2006). Microalgal biofilm, like bacterial biofilm, may adapt to changes in the environment, maintain colonies on a surface, and disassociate as a single colony or in clumps from the surface. Despite their widespread presence in nature, microalgal biofilm has mainly been researched for its harmful impacts on buildings to regulate or prevent their spread, primarily owing to safety and economic considerations. As a result, these films can include many microorganisms, including diatoms, cyanobacteria, bacteria, and protozoa (Galie, García-Gutiérrez, Miguélez, Villar, & Lombó, 2018). Oxygenic phototrophs such as cyanobacteria and diatoms and merged layers of anoxygenic phototrophic bacteria such as green and purple sulfur bacteria occupy the top layer of microbial mats. Biofilm formed by members of Cyanophyta and Chlorophyta are usually found on terrestrial surfaces (Hallmann et al., 2013; Thapa et al., 2017). Most green biofilm algae are unicellular, colonial, or filamentous (Bellinger & Sigee, 2015; Crispim, Gaylarde, & Gaylarde, 2003). In one research, many algae species, primarily the green algae *Ulothrix* (Orandi, Lewis, & Moheimani, 2012), *Stigeoclonium* (Kim et al., 2015), *Chlamydomonas* (Shen et al., 2019), and the blue-green alga *Oscillatoria*, created an algal biofilm (Schumacher & Sekoulov, 2002). The summary of microalgal biofilm is enumerated in Table 1.

3. The life cycle of algal biofilm

In natural aquatic and aero-terrestrial ecosystems, microalgal colonization of hard surfaces is a typical occurrence with ecological and industrial implications. Initial attachment (adsorption), irreversible attachment via EPS generation (consolidation), early development (colonization), and maturity of biofilm architecture are the four phases of microbial mature biofilm formation (Moreno Osorio, Pollio, Frunzo, Lens, & Esposito, 2021; Stoodley, Sauer, Davies, & Costerton, 2002) (Fig. 2).

3.1 Adsorption

The cell solid and cell to cell interactions, which are based on the physicochemical features of the contacting surfaces, such as the thermodynamic balance of interaction energies, affect the initial adherence of microbial cells to solid materials (Fig. 2.I). Cells can shift toward the conditioned surface in various ways at first: gravity, advection, or actively moving via motile processes. Most marine algae's attachment is assumed to be aided by their specific reproductive propagules, which allow first contact with a solid surface. Furthermore, it could depend on the comparison to the bulk aquatic environment; the concentration of glycoproteins, proteins, cations, and organic compounds near the surface can create a nutritional zone for microbes (Berner, Heimann, & Sheehan, 2015; Moreno Osorio et al., 2021; Qureshi, Annous, Ezeji, Karcher, & Maddox, 2005).

3.2 Consolidation

Diatoms, unicellular or colonial algae, are thought to be the initial colonizers in freshwater (Romaní et al., 2008). The spontaneous, often reversible, adsorption of organic and inorganic aqueous molecules to a solid surface is followed by a nonreversible secondary attachment of cells via EPS formation. By providing a diffusive barrier to biofilm cells, it protects them from any poisonous substances that may harm them, as well as phagocytes and bacteriocides (Qureshi et al., 2005). Surface charge and hydrophobicity of cell walls and membranes can be adjusted by forming surface structures such as pili, curli, fimbriae, and flagella, or by regulating EPS production, all of which may improve adhesion to a habitable solid surface, establishing a stronger binding of the algal cells forming colonies (Fig. 2.II) (Hori & Matsumoto, 2010; Lasaro et al., 2009; Moreno Osorio et al., 2021).

3.3 Colonization

Because EPS facilitates cell to cell contact, the microcolonies create more EPS, which helps to keep the developing biofilm on the surface and thicken it during the growth phase (Toyofuku et al., 2016). Secondary organisms can colonize the biofilm, acting as a sink for other particulate matter in the environment (Lee et al., 2005). In response to shear stresses, the biofilm architecture evolves during this phase. Heterogeneous matrices containing clusters of cells and polymers, randomly dispersed within a three-dimensional (3D) architecture and surrounded by water channels that occasionally reach from the bottom to the top of the biofilm, are the primary components of this structure (Fig. 2.III) (Moreno Osorio et al., 2021; Petroff et al., 2011; Toyofuku et al., 2016).

3.4 Maturation

During this stage of development, the cells in the biofilm consume nutrients primarily for the synthesis of EPS rather than for biomass increase. Cells must emigrate from the EPS

TABLE 1 An overview of algal biofilm grown on various substrates.

Phylum	Algal species	Culture system	Substratum	References
Cyanobacteria	*Anabaena variabilis*	Porous substrate Bioreactor	Glass fiber filter	Murphy and Berberoglu (2014)
	Chroococcus minutus	Petridish biofilm	Titanium	Gómez-Ramírez et al. (2019)
	Spirulina platensis	Porous substrate photobioreactor	Electrostatic flocking cloth (EFC)	Zhang et al. (2015)
Bacillariophyta	*Amphora coffeaeformis*	Petri-dish biofilm	Polyvinylidene fluoride commercial membrane	Tong and Derek (2021)
	Cylindrotheca fusiformis	Single layer vertical plate attached photobioreactor	Cellulose acetate/nitrate membrane	Schultze et al. (2015)
	Navicula sp.	Curtain biofilm in aquarium	Plastic net	Gómez-Ramírez et al. (2019)
	Nitzschia palea	Glass	Flat plate photobioreactor	Schnurr, Espie, and Allen (2013)
	Nitzschia amphibia	Petridish biofilm	Titanium	Sekar, Venugopalan, Satpathy, Nair, and Rao (2004)
	Phaeodactylum tricornutum	Twin layer PBR	Filter paper	Naumann, Çebi, Podola, and Melkonian (2013)
Chlorophyta	*Acutodesmus obliquus*	Attached cultivation system	Cellulose acetate/nitrate filter	Ji et al. (2014)
	Botryococcus braunii	Multiple layer	Filter paper	Wang, Liu, and Liu (2015)
		Attached cultivation reactor	Cellulose acetate/nitrate filter	Cheng et al. (2013)
		Multilayers photobioreactor	Polyethylene foam, Glass fiber reinforced plastic Polyethylene foam	Shen, Zhang, Xu, and Lin (2015)
		Single layer vertical plate attached photobioreactor	Cellulose acetate/nitrate membrane	Liu et al. (2013)
		Algal biofilm photobioreactor	Concrete	Ozkan, Kinney, Katz, and Berberoglu (2012)
	Chlorella sp.	Rocking attached cultivation system	Polystyrene	Johnson and Wen (2010)
		Filtration photobioreactor	Membrane	Zhang, Fung, and Ng (2014)
	Chlorella pyrenoidosa	Attach cultivation system	Filter paper	Cheng, Wang, Liu, and Liu (2017)
	Chlorella sorokiniana	Algadisk (rotating biological contractor)	Stainless steel mesh	Blanken et al. (2014)
	Chlorella vulgaris	Ceramic substrate photobioreactor	Ceramic membrane	Dalirian, Najafabadi, and Movahedirad (2021)
		Petridish biofilm	Titanium	Sekar et al. (2004)
		Algal biofilm photobioreactor	Flexible fiber bundles	Gao et al. (2015)
		Walnut shell based self permeating biofilm reactor (WSPBR)	Walnut shell	Zou, Xu, Chang, Qu, and Li (2021)
		Rotating flat plate photobioreactor	PVC	Melo, Fernandes, Caetano, and Borges (2018)

TABLE 1 An overview of algal biofilm grown on various substrates—cont'd

Phylum	Algal species	Culture system	Substratum	References
	Chlorococcum sp.	Rocking cultivation reactor	Glass fiber reinforced plastic	Shen, Chen, Chen, and Xu (2014)
	Halochlorella rubescens	Twin layer PBR	Polycarbonate membranes	Schultze et al. (2015)
		Twin layer PBR	Nylon filter	Vickers (2017)
	Scenedesmus vacuolatus	Petridish biofilm	Cotton, Jute fabrics	Carbone, Gargano, Pinto, De Natale, and Pollio (2017)
	Scenedesmus dimorphus	Biofilm cultivation reactor	Cellulose acetate/nitrate membrane	Toninelli, Wang, Liu, Wu, and Liu (2016)
	Scenedesmus obliquus	Flat plate photobioreactor	Glass	Schnurr et al. (2013)
	Scenedesmus obliquus	Walnut shell based self permeating biofilm reactor (WSPBR)	Walnut shell	Zou, Xu, et al. (2021)
	Scenedesmus rubescens	Rectangular reactor vessel	Cork, Silicone rubber, Denim, Sponge towel, Stainless steel, Plexiglass	Tsavatopoulou and Manariotis (2020)
	Scenedesmus sp.	Flask	Loofah sponge	Deantes-Espinosa et al. (2019)
	Tetraselmis suecica	Twin layer PBR	Filter paper	Naumann et al. (2013)
	Trentepohlia aurea	Filter in test tube	Chromatography filter	Abe, Matsumura, Imamaki, and Hirano (2003)
Haptophyta	*Isochrysis* sp.	Tube-type-twin-layer PBR, Twin-layer photobioreactor	Plain printing paper	Naumann et al. (2013)
Ochrophyta	*Nannochloropsis oculata*	Attached cultivation system	Glass fiber-reinforced plastic	Shen et al. (2014)
	Nannochloropsis sp.	Twin layer PBR	Filter paper	Naumann et al. (2013)
Mixed culture	Cyanobacterial strains (dominated By *Phormidium* or *Oscillatoria*)	Biofilm system	Polyester mesh discs	Rai, Kumar, Pandey, Yadav, and Gaur (2016)
	Scenedesmus obliquus, Chlorella vulgaris, Oscillatoria tenuis	Flat plate algal biofilm photobioreactor	Rice husk, Pine sawdust, Oak sawdust, Sugarcane bagasse	Zhang et al. (2017)
	Nannochloropsis sp. *Tetraselmis suecica Phaeodactylum tricornutum*	Tube type twin layer PBR	Plain printing paper	Naumann et al. (2013)
	Chlorella and *Phormidium*	Algal biofilm reactor (ABR)	Nonwoven spun bond fabric	Choudhary, Prajapati, Kumar, Malik, and Pant (2017)
	Anabaena augstumalis, Calothrix sp., *Nostoc* sp.	Continuous flow lane incubator	Polycarbonate slides	Di Pippo et al. (2013), Gismondi, Pippo, Bruno, Antonaroli, and Congestri (2016)

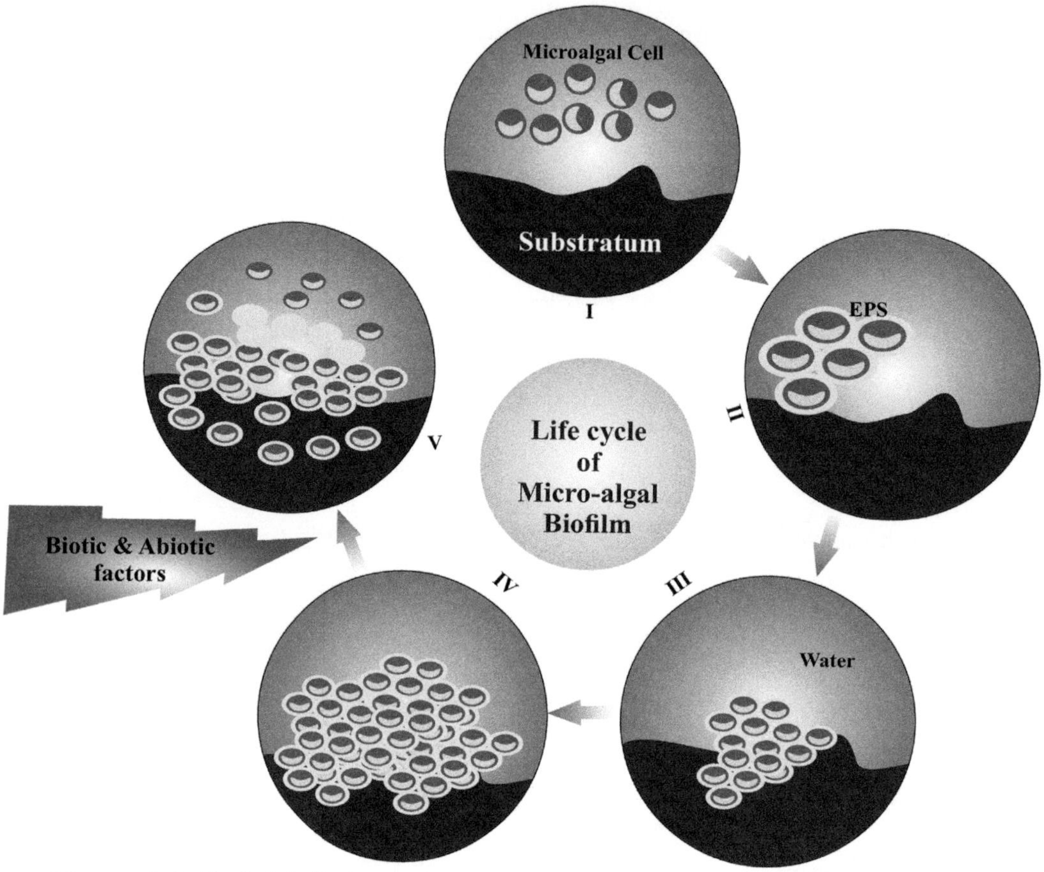

FIG. 2 Developmental stages of a biofilm include (I) adsorption/initial attachment, (II) consolidation/irreversible attachment, (III) colonization/growth, (IV) maturation, and (V) detachment. EPS stands for Extracellular Polymeric Substances.

matrix when nutrients become limited. In nutritional depletion, biofilm associated cells can produce enzymes that can break down the EPS matrix (Qureshi et al., 2005). The build-up of cellular and polymeric biomass is inextricably tied to biofilm growth (Fig. 2.IV).

Biofilm detachment, or the loss of tiny or more significant portions of biofilm biomass, has long been recognized as a feature of the biofilm life cycle. In addition to environmental variables such as cell separation due to famine or nutritional insufficiency and physiological changes such as abrasion, shear stress, sloughing, and grazing, biotic variables such as microbial metabolic activity and microbial gene expression influence biofilm separation (Celmer, Oleszkiewicz, & Cicek, 2008; Garny, Horn, & Neu, 2008; Telgmann, Horn, & Morgenroth, 2004; Woodcock & Sloan, 2017) (Fig. 2.V). In forming and maintaining a healthy biofilm community, a balance between biofilm attachment, growth, and detachment is critical.

4. Substratum for microalgal biofilm

Microalgal biofilm, made up of cyanobacteria and green microalgae, are found in practically every photic aquatic habitat. Microalgal biofilm may evolve on any surface or carrier material with enough moisture and light (Goffredo et al., 2017). The wide range of materials that have been successfully investigated as carriers for microalgal biofilm emphasize this point, many different plastics have been tested, including polyvinylchloride (PVC), polyethylene (PE), polyurethane (PUR), polymethyl methacrylate (PMMA), polystyrene, polycarbonate, and polyamide (nylon), various natural materials, including old reed stems, green reeds, bamboo pipe, granite, andesite; and many other materials, including borosilicate glass, cardboard, loofah sponge, fiber, etc. (Bouabidi, El-Naas, & Zhang, 2019; Mantzorou & Ververidis, 2019). Biofilms have the unique ability to develop and be functionally regulated by gradients of energy sources and chemical products. The requirement to identify two main subtypes of "algal biofilm," as previously defined, is significant in the coupled culture. The first of those is a biofilm made up of only one microalga species (axenic culture). The connected culture for this biofilm is started with a concentrated algal biomass inoculum deposited onto the carrier medium. Biofouling is associated with the second form of biofilm. It may begin with axenic microalgal biomass, but as the biofilm grows, allochthonous bacteria, fungus, and cyanobacteria settle inside, resulting in a complex microbial community. As a result, Mantzorou and Ververidis (2019) defined a biofilm as a consortium of microbes immersed in EPS that create a complex structure on solid surfaces. The colonization of the carrier material, which is commonly diatoms and green algae, is the first step in forming a microalgal biofilm. The community progresses into filamentous green algae and cyanobacteria as the biomass rises exponentially at first, then linearly (Liu, Danneels, Vanormelingen, & Vyverman, 2016; Sekar, Nair, Rao, & Venugopalan, 2002). Following that, biofilm growth slows until it equals the losses caused by respiration, cell death, parasitism, illness, grazing, and sloughing at a specific biofilm thickness (Wu, 2016). At this point, the losses exceed the biomass increase. External forces, such as shear, can cause biofilm detachment if they are greater than the internal strength of the matrix that holds the biofilm together. As a result, biofilm detachment occurs when a biofilm's internal strength is reduced, or external shear pressures are raised (Celmer et al., 2008; Picioreanu, Blauert, Horn, & Wagner, 2018; Wijeyekoon, Mino, Satoh, & Matsuo, 2004).

5. Microalgal biofilm and biomass production

Microalgae have recently piqued the scientific community's interest due to their numerous potentials in wastewater treatment and the pharmaceutical sector. However, the currently used growing methods are still outrageous. Low biomass productivities, difficulties in biomass harvesting and recovery, high installation and running costs, high water needs, and other bottlenecks plague traditional growing systems where microalgae are suspended in liquid media. Although microalgal biofilms are known to be a nuisance due to surface fouling, they have emerged as a cutting edge method for growing microalgae adhered to a solid surface (Vo et al., 2019). In comparison to traditional growing techniques, this methodology appears to be favorable. Low biomass productivities, water management, and biomass recovery are all issues that conventional culture techniques have. Microalgal biofilm systems potentially overcome these issues. However, recent renewed interest in microalgal biofilm has been driven by the need for wastewater remediation strategies for nutrient control, alternative biofuel feedstock, and effective, low-cost biomass harvesting techniques (Kesaano & Sims, 2014).

Algal biofilm technology has lately been touted as a viable approach for producing algal biomass as a biofuel feedstock. However, the present carrier materials for forming algal biofilm are challenging to grow at a cheap cost or unendurable. The commercialization of biofilm technology for algal biomass generation necessitates the development of innovative and low-cost biofilm carriers with high biomass production capabilities.

5.1 Photobioreactor for production of microalgal biofilm

A diverse variety of microalgae activities also addresses the significant commercialization potential of microalgae based goods, including biofuels, nutraceuticals, cosmetics, medicines, animal and aquacultural feeds, and so on (Moreno Osorio et al., 2021). The cultivation of microalgae on a big scale and at a cheap cost is one of the significant technologies that support the growth of the microalgae industry. This microalgae cultivation method is related to the design of open or closed culture systems and photobioreactors and the discovery of operating parameters that lead to the maximum growth performance of the target microalgae (Singh & Sharma, 2012).

Bioreactors are closed systems that function outdoors or inside and are injected with a specific organism to maintain a clean culture. Closed cultivation systems offer better control of contamination and cell physiology than open systems, leading to higher growth and quality of the harvested product, but manufacturing costs increase (Mata, Martins, & Caetano, 2010). Furthermore, solar receivers are how photobioreactors function. Any light that is not used in photosynthesis and is absorbed by the photobioreactor walls or the culture is transformed into thermal energy, which can cause the culture temperature to rise, leading to culture failure even intolerant species. At the same time, closed photobioreactors are flexible systems that can be optimized according to the biological and physiological characteristics of the microalgal species involved. The stirred tank reactor (typically with internal illumination), the vertical column reactor, the horizontal tubular reactor, and the flat plate reactor are the kinds of PBR previously solar driven hydrogen production in green algae in the literature (Koller, 2015). Stirred tank reactors have a lot of back mixing and have a lot of light penetration through the culture. In the case of the vertical column reactor, artificial illumination is necessary due to the reactor's orientation; a high degree of turbulence is also required to provide adequate mixing and light dark cycling of the culture. Furthermore, it shows that two advantages of vertical column reactors are high mass transfer rates and effective culture mixing with reduced shear stress on the algae (Mirón et al., 2003). The straight, coiled, or looped transparent tubing is put out in a specific geometric pattern to enhance the light collection in tubular photobioreactors. It may be composed of horizontal, helical, conical, and a-shape reactor geometries, as well as materials ranging from glass capillaries to plastic bags (Acién et al., 2017). It depicts that *Chlorella* culture has been commercially successful using horizontal tubular plastic bag reactors (Carvalho, Meireles, & Malcata, 2006). Likewise, flat plate reactors are operationally flexible since they may be run in batch or continuous modes. Outdoor flat plate reactors are slanted at an angle equivalent to the mean solar irradiation angle.

In contrast, artificially lighted flat plate reactors are generally vertical, and the irradiation is incident on one of the big reactor surfaces (Kunjapur & Eldridge, 2010).

The mathematical representation of photosynthetic processes and its application to bioreactor design have become extremely important due to the importance of photobioreactor design in industrial biochemical production and, in particular, in matters related to the sustainability of human activities on the Earth, such as wastewater recovery, carbon dioxide recycling, and biofuel production (Bees & Croze, 2014; Satyanarayana, Mariano, & Vargas, 2011).

6. Application of microalgal biofilm

As mentioned above, algae have diversified applications in biotechnology and industries. Microalgal biofilms are extremely important in the environment, industry, agriculture, and human health, both in terms of use and annoyance (Zou, Zeng, Li, Liu, & Lu, 2021). Most early biofilm studies were uncontrolled in aquatic bodies or artificial structures. Furthermore, biofilm development is a medical plague. Still, agriculture may be a formidable candidate for integrated nutrition and disease management or soil structure improvement since it aids in better and effective colonization in soil and around roots, allowing a connection between soil and plant (Moënne-Loccoz, Mavingui, Combes, Normand, & Steinberg, 2015). Additionally, they can be used in aqueous treatment progressions for an assortment of purposes, the reduction of both chemical and biochemical oxygen demand, including the removal of coliform bacteria, the removal of heavy metals, and the removal of phosphorous and nitrogen, as they can further convert them to more acquiescent forms (Mishra, Chowdhary, & Bharagava, 2019). Microalgae manufacture of high additional benefit molecules enables forms of value addition, such as health food and quality feed, carbohydrates, polyunsaturated fatty acids, and pigments, to penetrate their markets and achieve economic competitiveness. In addition to their high macronutrient content, microalgae can collect and express secondary metabolites under stressful situations. This paves the way for synthesizing tailored bioactive chemicals using biofilm culture systems to penetrate the cosmetics and pharmaceutical sectors eventually. Wound healing, antifungal, anticancer, and antibacterial activities are among the biomedical uses of algal-derived nanoparticles. However, it shows great potential for biotechnological interventions in wastewater treatment, bioremediation, and biofouling, as well as biofuels, biopolymers, and bioactive metabolite synthesis (Thapa et al., 2017) (Fig. 3). They're being studied more and more as biofertilizers, as well as for their involvement in the production of biological soil crusts and the mitigation of desertification (Tirkey & Adhikary, 2005). Table 2 shows the role of microalgal biofilm in wastewater treatment and biodiesel generation.

6.1 Microalgal biofilm used in wastewater treatment

The atmosphere and natural streams of the earth have become dumps for agricultural and industrial pollutants as human populations have grown. The majority of remediation strategies used in the previous half century have failed. Surface water eutrophication has been a major environmental concern for over half a century. Surface waters are eutrophic in many US watersheds, while coastal water bodies such as the Chesapeake Bay and the Gulf of Mexico are becoming more hypoxic (Malone & Newton, 2020). The primary goal of wastewater treatment is to obtain clean water. It entails a trio of physical, chemical, and biological sciences, with biological activity essential to the process. Despite the availability of various innovative technologies for wastewater treatment, finding a dependable, cost effective, and environmentally acceptable solution remains a challenge (Raheem et al., 2018). Microbes' inherent power to clean water by degrading contaminants (such as industrial waste, pesticides, heavy metals, and so on) and removing them has enormous potential (Dixit et al., 2015).

After removing organic waste by heterotrophic bacteria during secondary treatment, microalgae are among the possible microflora that may thrive even in low quality water enhanced with inorganic substances found in wastewater (Thapa et al., 2017). As a result, tertiary treatment operations are carried out biologically, with microalgae replacing dangerous chemicals in the process. Their photosynthetic metabolism style decreases BOD with O_2 release, eliminates heavy metals, and provides high value products in the form of clean energy sources, nutraceuticals, and fertilizers, among other things (Abdelaziz, Leite, & Hallenbeck, 2013). The algal turf scrubber (ATS) is a designed device for pulsing wastewaters over sloping surfaces with naturally seeded filamentous algae attached. This treatment has been proven for tertiary sewage, agricultural wastes, streams, and extensive aquaculture systems, with rates ranging from 40 million to 80 million liters per day. In addition, attached growth/fixed film systems have been used in wastewater treatment since 1893, when trickling filters were first introduced in England (Kesaano & Sims, 2014). On the other hand, wastewater treatment using fixed film systems is accomplished by fostering the establishment of the desired consortia of microorganisms capable of metabolizing the contaminants of concern (Mohan, 2009). Because flocculation is not required and biofilms or microalgal mats on the sidewalls may be quickly scraped and removed, aggregated biomass recovery is less expensive (Thapa et al., 2017).

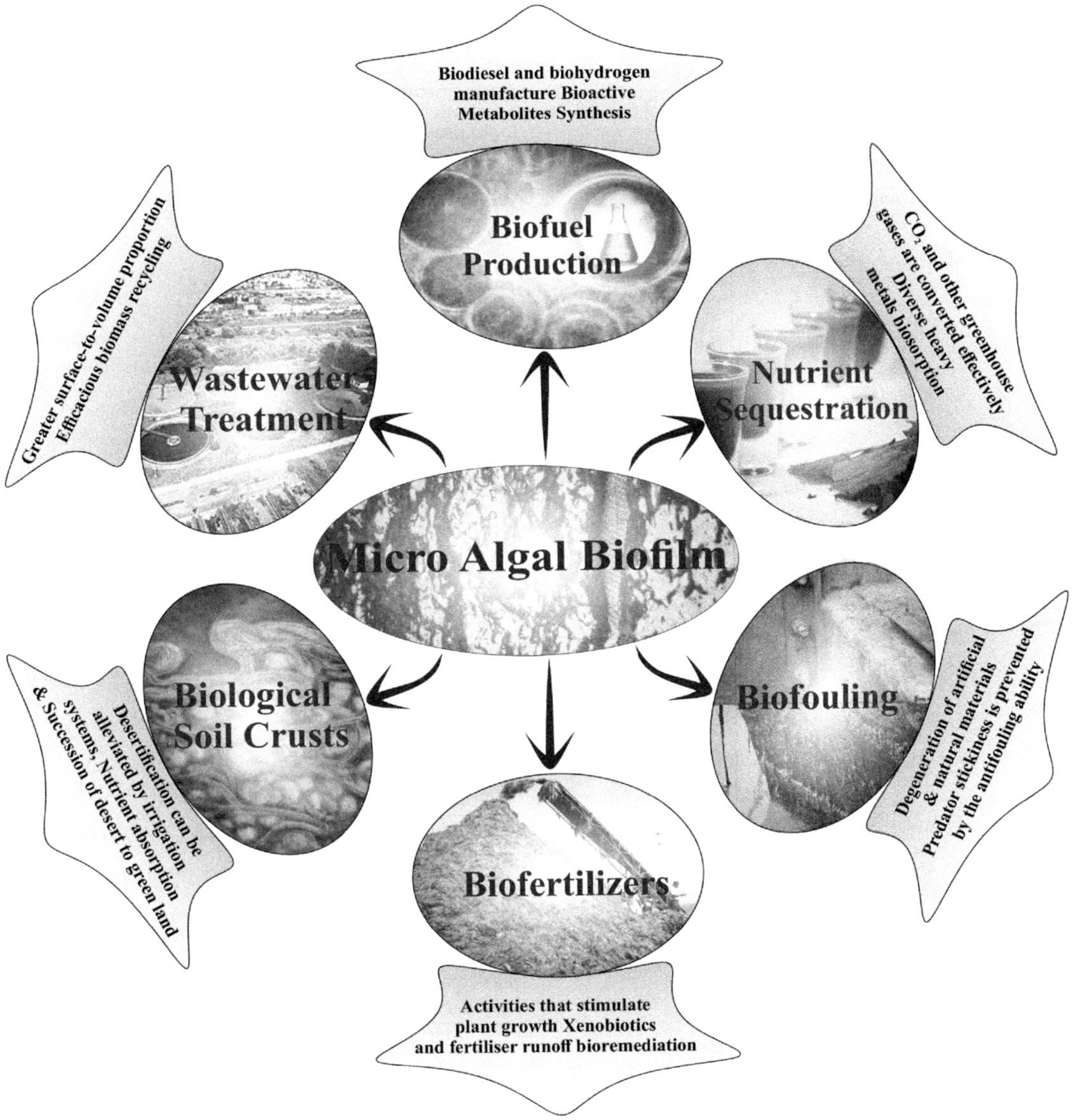

FIG. 3 Summary of the uses of algal biofilm.

The effluents created by home and industrial operations constitute a significant source of contamination of receiving water bodies regularly, putting a strain on water quality management. Pathogenic bacteria, phosphate and nitrogen, hydrocarbons, heavy metals, endocrine disruptors, and organic waste are all examples of pollution (Akpor, Otohinoyi, Olaolu, & Aderiye, 2014). Wastewater treatment plants may exploit microalgae's capacity to consume nitrogen (N) and phosphorous (P) to meet tight effluent nutrient limitations, and the collected biomass might be used to make bioproducts (Menger-Krug, Niederste-Hollenberg, Hillenbrand, & Hiessl, 2012).

Depending on the source, raw wastewater contains enough nutrients for algae development. However, nutrient addition may be required when algal biofilms are used for tertiary wastewater treatment (effluent polishing) to maximize carbon, nitrogen, and phosphorus (C:N:P) required for growth. Microalgae use both direct and indirect ways of nutrient remediation. They use the luxury uptake route to directly take up inorganic nitrogen sources such as nitrate or nitrite and phosphate and store them in polyphosphate granules, similar to cyanobacteria. In comparison to macroalgae, microalgae can effectively absorb nitrogenous chemicals, and they commonly collect phosphorus in the form of phosphate. Ammonia is removed from the solution indirectly owing to an elevation in pH caused by photon absorption by the cell, which converts NH_4^+ ions into NH_3 gas. The cyanobacterium *Phormidium bohneri* has been found to have a higher efficacy of indirect ammonium removal (Whitton, 2016). For many years, microalgae have been used in algal ponds to treat municipal wastewater on a small scale. By absorbing these nutrients into their biomass,

TABLE 2 Microalgal biofilm used in wastewater treatment and biodiesel generation.

Algal species	Bioreactor/culture type	Application	References
Nitzschia palea	Horizontal biofilm reactor	Lipid production	Schnurr et al. (2013)
Botryococcus braunii	Multilayer photobioreactor	Wastewater treatment (Manure wastewater)	Shen et al. (2015)
	Photobioreactor	Biodiesel production	Rokicka, Zieliński, Dudek, and Dębowski (2021)
Chlorella vulgaris	Twin-layer photobioreactor	Municipal and synthetic secondary wastewater	Shi, Podola, and Melkonian (2007)
	Polymethyl methacrylate flow cell with coupons	Wastewater treatment (Municipal wastewater)	Irving and Allen (2011)
	Algal biofilm membrane photobioreactor	Wastewater treatment (Simulated secondary effluent)	Gao et al. (2015)
	Photobioreactor	Biodiesel production	Rokicka et al. (2021)
Chlorella sorokiniana	Tubular photobioreactor, flat-plate photobioreactor, algal turf reactor shallow open pond	Wastewater treatment (Mineral salt medium)	Muñoz, Köllner, and Guieysse (2009)
Chlorella sp.	Polystyrene rocker system	Dairy manure	Johnson and Wen (2010)
	Rocker cultivation system	Biodiesel production	Johnson and Wen (2010)
	Rotating biological contactors	Generation of biomass for biodiesel applications	Sims and Christenson (2011)
Scenedesmus obliquus	Algem lab-scale photobioreactor	Secondary wastewater	Whitton et al. (2018)
	Polymethyl methacrylate flow cell with coupons	Wastewater treatment (Municipal wastewater)	Irving and Allen (2011)
	Horizontal biofilm reactor	Lipid production	Schnurr et al. (2013)
Scenedesmus rubescens	Twin layer photobioreactor	Municipal and synthetic secondary wastewater	Shi et al. (2007)
	Cork, sponge towel, denim, plexiglass, stainless steel, and silicone rubber	Purify wastewater and at the same time to be able to produce biofuel	Tsavatopoulou and Manariotis (2020)
Scenedesmus sp.	Polymethyl methacrylate reactor with cylindrical polypropylene fiber bundle carriers	Wastewater treatment (Municipal wastewater)	He and Xue (2010)
Trentepohlia aurea	Filter paper	Synthetic wastewater	Abe et al. (2003)
Microalgal consortia	Flow cell	Wastewater treatment (Domestic sewage)	Boelee, Temmink, Janssen, Buisman, and Wijffels (2011)

microalgae extract inorganic N and P from wastewater as ammonium (NH_4^+) and phosphate (PO_4^{3-}), and from (nitrified) wastewater effluent as nitrate (NO_3^-) and PO_4^{3-}. Proteins and nucleic acids are synthesized using assimilated nitrogen, with P also being used in the latter. Phospholipids are also made from P, and luxurious absorption of P, i.e., the internal storage of PO_4^{3-} as polyphosphates, can occur. PO_4^{3-} can be eliminated by precipitation in addition to assimilation. When microalgae absorb more NO_3^- and carbon from the atmosphere than can be supplied by absorption, the pH of the water rises. At high pH, dissolved PO_4^{3-} can precipitate with cations such Ca^{2+} and Mg^{2+}, which are commonly found in urban wastewater. *Chlorella* and *Scenedesmus* are two microalgae studied for nutrient removal in wastewaters, but *Nannochloris*, *B. braunii*, and the cyanobacteria *Phormidium* and *Spirulina* have also been studied (Baglieri et al., 2016; Johnson & Wen, 2009, 2010).

Furthermore, *Trentepohlia aurea* biofilms might be a beneficial tool for removing nitrate and phosphate from water (Abe et al., 2003). When a produced biofilter was dampened with medium and incubated between 5 and 50 mol photons m^{-2} s^{-1} under low light intensity (10 mol photons m^{-2} s^{-1}), the efficiency of inorganic compound removal from the water was enhanced without the destruction of chlorophylls in the cells. However, most microalgal wastewater treatment systems do not use a monoculture of microalgae, and the microalgae mentioned above can be found with a variety of other species (Abe, Takahashi, & Hirano, 2008; Luo, Le-Clech, & Henderson, 2020; Xiong, Kurade, & Jeon, 2017).

Microalgal biofilm are a potential method for promoting microbial cell adhesion in an exopolysaccharide matrix with water, resulting in excess biomass that may be readily separated at a low cost and used to make biofuels or bioactive compounds. Furthermore, heavy metals and radioactive elements can be removed from these microalgal biofilm communities by intracellular absorption, surface adsorption, or sulfide or phosphate precipitates (Thapa et al., 2017). They are involved in the intracellular or surface adsorption of specific metal ions (Cd, Pb, Hg, As, Cr, Fe, etc.). As demonstrated by immobilized *Anabaena doliolum*, immobilized cyanobacteria have more metal removal capacity than their free living counterparts (Lau, Matsui, & Abdullah, 2015; Pradhan, Patra, Dash, et al., 2020a). In a country like India, where most of the water requirement is met by monsoon fed rainwater, which is unpredictable in the current global climate change scenario, treated wastewater through algal biofilms can meet the growing demand for irrigation water for agriculture (requires 80% of total water). Cu and Fe uptake were 45% and 23% greater in immobilized *A. doliolum*, respectively than in free living cells. Additionally, mucilage sheaths obtained from the cyanobacteria *Microcystis aeruginosa* and *Aphanothece halophytica* have a significant affinity for heavy metal ions such as copper, lead, and zinc, according to De Philippis and Micheletti (2017). The increased pH inside photosynthetically active biofilms may promote metal removal via precipitation in addition to biosorption and bioaccumulation (Roeselers, Van Loosdrecht, & Muyzer, 2008).

The most significant greenhouse gas emissions in history were recorded in the recent decade, negatively impacting global climate change and threatening fair and sustainable development across the continent. According to the fifth IPCC (Intergovernmental Panel on Climate Change) report 2014, anthropogenic activities have increased emissions of greenhouse gases such as CO_2 over the last three decades, and CO_2 concentration as a greenhouse gas has increased to around 78%, resulting in a 0.19 m increase in the sea (Intergovernmental Panel on Climate Change, 2014). On the other hand, biological sequestration, which involves biological agents such as plants and microorganisms fetching carbon from the atmosphere through their metabolic activity and acting as a carbon sink in various forms such as forest bodies, grassland, peat bogs, and microbial biomass, is one of the valuable mitigation approaches. In addition to this, biofilm made of microalgae can be effective. Furthermore, CO_2 is converted to biomass from its emission source in an engineered system such as a photobioreactor, producing bioenergy with carbon dioxide collection and storage (Rinanti, Dewi, Kardena, & Astuti, 2016).

Companies and businesses are now exploiting clean water in various ways at the laboratory, bench, and pilot scale, but large-scale systems for controlled algal cultivation and harvesting for wastewater treatment and subsequent bioproduct processing are absent (Christenson & Sims, 2011). More research and development of large scale production and harvesting methods for biofuels and bioproducts are needed, especially for less well studied but promising technologies such as connected algal biofilm cultures.

6.2 The microalgal biofilm used in biofuel production

The global energy crisis and rising greenhouse gas emissions fueled the hunt for alternative and ecologically benign renewable energy sources. Microalgae biofuel has been highlighted as one of the primary renewable energy sources for sustainable development, having the potential to replace fossil based fuels, according to life cycle analysis. The critical problems associated with oil crops and lignocellulose based biofuels were not present in microalgae biofuel. Algae-based biofuels are technically and commercially viable, cost competitive, and require no new lands or water. They also help to reduce CO_2 levels in the atmosphere. However, due to low biomass concentration and expensive downstream procedures, commercial production of microalgae biodiesel is currently not practical. Upstream and downstream processes are involved in developing microalgae biomass and biofuel generation. Different growing strategies are used in the upstream phase to enhance biomass quality and quantity, while the downstream stage focuses on harvesting technologies and long term biofuel generation. In addition to internal lipids and proteins, thorough biomass characterization investigations indicated the presence of extracellular polymeric material (EPS) comprising polysaccharides, proteins, nucleic acids, and phospholipids (Mata et al., 2010; Williams & Laurens, 2010). Microalgal biofilm have a significant potential for conversion into biodiesel, bioethanol, biogas, or biooil because of their diverse composition (Pulz & Gross, 2004). Most microalgal biofilm research focuses on lipid synthesis and its subsequent application in biodiesel generation.

Chlorella vulgaris, which has high lipid content, has been recognized as a significant species in the scaling-up of algal biodiesel production (Chisti, 2013; Griffiths, van Hille, & Harrison, 2012). In addition, the rocker culture method uses dairy manure wastewater as the growth medium to cultivate *Chlorella* sp. as a biofuel feedstock (Johnson & Wen, 2010). The attached culture had a more significant total fatty acid (TFA) concentration (9% w/w) than the suspended culture (8.9% w/w), according to the findings. Furthermore, the neutral lipid productivities of *Scenedesmus obliquus* and *Nitzschia palea* were assessed using a horizontal biofilm reactor and revealed that increase in neutral lipid concentrations during food deprivation (Schnurr et al., 2013).

Cyanobacteria have received less attention than microalgae in biofilm based lipid extraction research, making them an intriguing choice for biofuel feedstock despite their excellent commercial culture and different features. Rai et al. (2016) assessed biofilm-forming cyanobacteria for application in biodiesel production in this context. Palmitic acid was discovered as the primary fatty acid (Gara-Ali, Zili, Hosni, Ouada, & Ben-Mahrez, 2021), with maximum lipid content of 18% (*w*/w). The authors determined that the high amount of saturated fatty acids found in cyanobacterial isolates, together with monounsaturated fatty acids, provided the best fuel quality (Bruno et al., 2012).

6.2.1 Improved microalgal biofuel production through molecular strategies

Because of its unicellular nature, genetic engineering may be used to manipulate metabolic pathways in microalgae. The primary goal of using genetic engineering to increase biomass and biodiesel production in microalgae is to improve biomass and biodiesel production. Until recently, development in microalgal genetic engineering was incredibly sluggish. Genetic engineering technology is considerably aided by the availability of microalgae genome sequences (Sinha, 2021). Genome sequencing projects for several microalgae species have been completed to date and sequencing projects for other microalgae species such as *Fragilariopsis cylindrus*, *Pseudo-nitzschia*, *Thalassiosira rotula*, *Botryococcus braunii*, *C. vulgaris*, *Dunaliella salina*, *Galdieria sulphuraria*, and *Porphyra purpurea* (Radakovits, Jinkerson, Darzins, & Posewitz, 2010). Furthermore, green, red, and brown algae, diatoms, euglenoids, and dinoflagellates, have advanced techniques, and 30 microalgae strains have been successfully changed to the furthest. Most of the transformation studies were carried out on the model green alga *Chlamydomonas reinhardtii* at both the nuclear and chloroplast levels (Tran & Kaldenhoff, 2020). On the other hand, blocking energy rich compounds (e.g., starch) produces metabolic pathways to decrease lipid catabolism, that is, elimination of fatty acid oxidation that consumes TAGs; modification of lipid characteristics; direct biological synthesis of fatty acids; and secretion of TAGs, free fatty acids, alkane, and wax esters directly into the medium are some of the current molecular strategies required to improve microalgae biodiesel production (Wase, Black, & DiRusso, 2018).

6.2.2 Biofuel production interactions with bacterial biofilms

Microalgae and bacteria develop a symbiotic connection by forming a "Phycosphere, "similar to the "rhizosphere" between plants and bacteria (Fig. 4). Microalgae create extracellular materials to establish a matrix like structure on their surfaces, fostering and facilitating bacterial biofilm formation. In the unicellular microalgae *Chlamydomonas reinhardtii*, Hu, Xiao, Liao, Leng, and Lu (2021) discovered microalgae-bacteria interactions. Furthermore, symbiotic bacterial interactions with microalgae are accomplished by the exchange of metabolites or elements such as dissolved organic matter (DOM), vitamin (Vit), iron (Fe), sulfur (S), nitrogen (N), and phosphorus (P). The stored carbohydrates in microalgal cells have the potential to create hydrogen, while the conversion of photosynthate to triglyceride has the potential to make diesel fuels. Yeast or anaerobic bacteria can ferment the leftover biomass of microalgae to produce ethanol or biogas (Yao et al., 2019). The majority of molecular signals, genes, and enzymes involved in microalgal–bacteria symbiosis are yet unknown. On the other hand, the synthesis of hydrogen by microalgae is dependent on the activity of a hydrogenase enzyme that is extremely sensitive to oxygen (Alavi, Miller, Erlandson, Schneider, & Belas, 2001). As a result, rigorous anaerobic conditions are required for microalgae to produce hydrogen efficiently. Some bacteria such as *Brevundimonas* sp., *Rhodococcus* sp., and *Leifsonia* sp. were discovered to

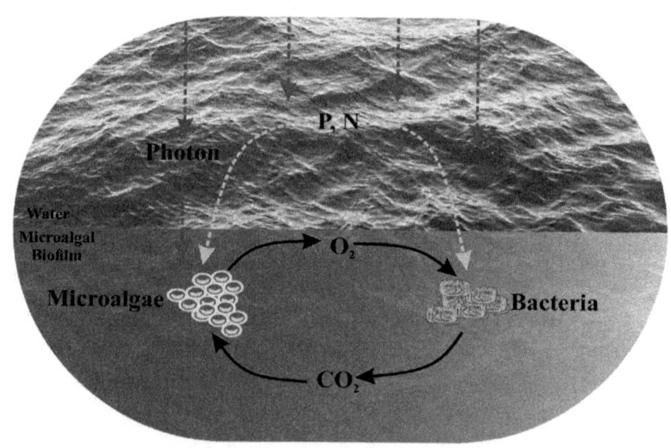

FIG. 4 The symbiotic microbial biofilm reactor is depicted schematically. Microalgae use light and CO_2 to produce biomass and O_2 that bacteria can use.

boost hydrogen generation in the microalga *Chlamydomonas*. This is because effective bacterial respiration eliminates oxygen, which is required for the activation of a Fe-dependent hydrogenase in *Chlamydomonas* (Lakatos et al., 2014). In addition, when a hydrogenase-deficient *Escherichia coli* was used as an artificial symbiotic bacteria to *Chlamydomonas*, the maximum hydrogen output was obtained (Lakatos et al., 2014). In general, triacylglycerol and other lipids generated by microalgae may be transesterified into fatty acid methyl esters and are of great interest as biofuel precursors. In comparison to axenic growth, coculturing the green microalga *Auxenochlorella protothecoides* with *E. coli* resulted in a two to sixfold increase in algal growth and a doubling in neutral lipid content, suggesting a symbiotic connection between the partners (Higgins & Vander Gheynst, 2014). A further investigation found that *E. coli*'s supply of thiamine derivatives and degradation products to *A. protothecoides* had a beneficial impact (Higgins et al., 2016). Because of its high lipid content and rapid development, the marine microalga *Tetraselmis striata* was chosen as a candidate for biodiesel production (Chisti, 2007). *Pelagibaca bermudensis* and *Stappia* sp., both isolated from *T. striata* bulk culture, displayed growth promoting activities on this microalga. These microalgae symbionts might be employed in the manufacture of industrial biodiesel (Park et al., 2017; Yao et al., 2019).

7. Conclusion and future prospective

Microalgae can be a significant and long term renewable energy source that can fulfill world demand. Despite the availability of fresh and creative biofilm designs with high productivity potential and easy harvesting capabilities, there is a significant difference between published research owing to variability and lack of standardization. Most of the described systems are only evaluated with a limited range of environmental factors and species, limiting their applicability. This also restricts our ability to comprehend and identify the fundamental drivers of algal biomass generation and subsequent conversion to biofuels. Furthermore, there is a lack of variety in growing settings and external applications in the substantial study on wastewater cleanup employing biofilms. Engineering adaptable systems that should be evaluated under more realistic environments are required to integrate algal biofilm into wastewater treatment processes successfully. Using biomass for various applications necessitates research into the correlation of wastewater properties from multiple sources with biofilm development, species composition, nutrient removal trends, and biochemical composition. Furthermore, using genetic engineering techniques to manipulate microalgae metabolic pathway is a viable method for increasing biomass and biofuel output. The use of genetic engineering technologies is very significant in manufacturing high quality products at low costs. In the generation of microalgae biomass and biofuels, biotic interaction with bacterial biofilms is also crucial. As a result, more research into innovative upstream and downstream technologies will help commercially produce microalgae-based biofuels.

Acknowledgments

The authors are thankful to Berhampur University for providing the necessary facilities.

Conflicts of interest

The authors have no conflicts of interest to disclose.

References

Abdelaziz, A. E., Leite, G. B., & Hallenbeck, P. C. (2013). Addressing the challenges for sustainable production of algal biofuels: I. Algal strains and nutrient supply. *Environmental Technology*, *34*, 1783–1805.

Abe, K., Matsumura, I., Imamaki, A., & Hirano, M. (2003). Removal of inorganic nitrogen sources from water by the algal biofilm of the aerial microalga *Trentepohlia aurea*. *World Journal of Microbiology and Biotechnology*, *19*, 325–328.

Abe, K., Takahashi, E., & Hirano, M. (2008). Development of laboratory-scale photobioreactor for water purification by use of a biofilter composed of the aerial microalga *Trentepohlia aurea* (Chlorophyta). *Journal of Applied Phycology*, *20*, 283–288.

Acién, F., Molina, E., Reis, A., Torzillo, G., Zittelli, G., Sepúlveda, C., et al. (2017). Photobioreactors for the production of microalgae. In C. Gonzalez-Fernandez, & R. Muñoz (Eds.), *Microalgae-based biofuels and bioproducts* (pp. 1–44). Elsevier.

Akpor, O. B., Otohinoyi, D., Olaolu, D., & Aderiye, B. (2014). Pollutants in wastewater effluents: Impacts and remediation processes. *International Journal of Environmental Research and Earth Science*, *3*, 050–059.

Alavi, M., Miller, T., Erlandson, K., Schneider, R., & Belas, R. (2001). Bacterial community associated with *Pfiesteria*-like dinoflagellate cultures. *Environmental Microbiology*, *3*, 380–396.

Baglieri, A., Sidella, S., Barone, V., Fragalà, F., Silkina, A., Nègre, M., et al. (2016). Cultivating *Chlorella vulgaris* and *Scenedesmus quadricauda* microalgae to degrade inorganic compounds and pesticides in water. *Environmental Science and Pollution Research*, *23*, 18165–18174.

Barone, R., De Napoli, L., Mayol, L., Paolucci, M., Volpe, M. G., D'Elia, L., et al. (2020). Autotrophic and heterotrophic growth conditions modify biomolecole production in the microalga *Galdieria sulphuraria* (cyanidiophyceae, rhodophyta). *Marine Drugs*, *18*, 169.

Bees, M. A., & Croze, O. A. (2014). Mathematics for streamlined biofuel production from unicellular algae. *Biofuels*, *5*, 53–65.

Behera, C., Dash, S. R., Pradhan, B., Jena, M., & Adhikary, S. P. (2020). Algal diversity of Ansupa lake, Odisha, India. *Nelumbo*, *62*, 207–220.

Behera, C., Pradhan, B., Panda, R., Nayak, R., Nayak, S., & Jena, M. (2021). Algal diversity of Saltpans, Huma (Ganjam), India. *Journal of the Indian Botanical Society*, *101*, 107–120.

Bellinger, E. G., & Sigee, D. C. (2015). *Freshwater algae: Identification, enumeration and use as bioindicators*. John Wiley & Sons.

Bengtsson, M. M., Wagner, K., Schwab, C., Urich, T., & Battin, T. J. (2018). Light availability impacts structure and function of phototrophic stream biofilms across domains and trophic levels. *Molecular Ecology*, *27*, 2913–2925.

Berner, F., Heimann, K., & Sheehan, M. (2015). Microalgal biofilms for biomass production. *Journal of Applied Phycology*, *27*, 1793–1804.

Blanken, W., Janssen, M., Cuaresma, M., Libor, Z., Bhaiji, T., & Wijffels, R. (2014). Biofilm growth of *Chlorella sorokiniana* in a rotating biological contactor based photobioreactor. *Biotechnology and Bioengineering*, *111*, 2436–2445.

Boelee, N., Temmink, H., Janssen, M., Buisman, C., & Wijffels, R. (2011). Nitrogen and phosphorus removal from municipal wastewater effluent using microalgal biofilms. *Water Research*, *45*, 5925–5933.

Bouabidi, Z. B., El-Naas, M. H., & Zhang, Z. (2019). Immobilization of microbial cells for the biotreatment of wastewater: A review. *Environmental Chemistry Letters*, *17*, 241–257.

Bruno, L., Di Pippo, F., Antonaroli, S., Gismondi, A., Valentini, C., & Albertano, P. (2012). Characterization of biofilm-forming cyanobacteria for biomass and lipid production. *Journal of Applied Microbiology*, *113*, 1052–1064.

Cai, T., Park, S. Y., & Li, Y. (2013). Nutrient recovery from wastewater streams by microalgae: Status and prospects. *Renewable and Sustainable Energy Reviews*, *19*, 360–369.

Carbone, D. A., Gargano, I., Pinto, G., De Natale, A., & Pollio, A. (2017). Evaluating microalgae attachment to surfaces: A first approach towards a laboratory integrated assessment. *Chemical Engineering Transactions*, *57*.

Carvalho, A. P., Meireles, L. A., & Malcata, F. X. (2006). Microalgal reactors: A review of enclosed system designs and performances. *Biotechnology Progress*, *22*, 1490–1506.

Celmer, D., Oleszkiewicz, J., & Cicek, N. (2008). Impact of shear force on the biofilm structure and performance of a membrane biofilm reactor for tertiary hydrogen-driven denitrification of municipal wastewater. *Water Research*, *42*, 3057–3065.

Cheng, P., Ji, B., Gao, L., Zhang, W., Wang, J., & Liu, T. (2013). The growth, lipid and hydrocarbon production of *Botryococcus braunii* with attached cultivation. *Bioresource Technology*, *138*, 95–100.

Cheng, P., Wang, Y., Liu, T., & Liu, D. (2017). Biofilm attached cultivation of *Chlorella pyrenoidosa* is a developed system for swine wastewater treatment and lipid production. *Frontiers in Plant Science*, *8*, 1594.

Chisti, Y. (2007). Biodiesel from microalgae. *Biotechnology Advances*, *25*, 294–306.

Chisti, Y. (2013). Constraints to commercialization of algal fuels. *Journal of Biotechnology*, *167*, 201–214.

Choudhary, P., Malik, A., & Pant, K. K. (2017). Algal biofilm systems: An answer to algal biofuel dilemma. In *Algal biofuels* (pp. 77–96). Springer.

Choudhary, P., Prajapati, S. K., Kumar, P., Malik, A., & Pant, K. K. (2017). Development and performance evaluation of an algal biofilm reactor for treatment of multiple wastewaters and characterization of biomass for diverse applications. *Bioresource Technology*, *224*, 276–284.

Christenson, L., & Sims, R. (2011). Production and harvesting of microalgae for wastewater treatment, biofuels, and bioproducts. *Biotechnology Advances*, *29*, 686–702.

Colman, B., Huertas, I. E., Bhatti, S., & Dason, J. S. (2002). The diversity of inorganic carbon acquisition mechanisms in eukaryotic microalgae. *Functional Plant Biology*, *29*, 261–270.

Crispim, C. A., Gaylarde, P. M., & Gaylarde, C. C. (2003). Algal and cyanobacterial biofilms on calcareous historic buildings. *Current Microbiology*, *46*, 0079–0082.

Dalirian, N., Najafabadi, H. A., & Movahedirad, S. (2021). Surface attached cultivation and filtration of microalgal biofilm in a ceramic substrate photobioreactaor. *Algal Research*, *55*, 102239.

Dalrymple, O. K., Halfhide, T., Udom, I., Gilles, B., Wolan, J., Zhang, Q., et al. (2013). Wastewater use in algae production for generation of renewable resources: A review and preliminary results. *Aquatic Biosystems*, *9*, 1–11.

Dash, S., Pradhan, B., Behera, C., & Jena, M. (2020). Algal diversity of Kanjiahata Lake, Nandankanan, Odisha, India. *Journal of the Indian Botanical Society*, *99*, 11–24.

Dash, S., Pradhan, B., Behera, C., Nayak, R., & Jena, M. (2021). Algal flora of Tampara Lake, Chhatrapur, Odisha, India. *Journal of the Indian Botanical Society*, *101*, 1–15.

De Philippis, R., & Micheletti, E. (2017). Heavy metal removal with exopolysaccharide-producing cyanobacteria. In *handbook of advanced industrial and hazardous wastes management* (pp. 931–964). CRC Press.

Deantes-Espinosa, V. M., Zhang, T.-Y., Wang, X.-X., Wu, Y., Dao, G.-H., & Hu, H.-Y. (2019). Attached cultivation of *Scenedesmus* sp. LX1 on selected solids and the effect of surface properties on attachment. *Frontiers of Environmental Science & Engineering*, *13*, 1–9.

Di Pippo, F., Ellwood, N. T., Gismondi, A., Bruno, L., Rossi, F., Magni, P., et al. (2013). Characterization of exopolysaccharides produced by seven biofilm-forming cyanobacterial strains for biotechnological applications. *Journal of Applied Phycology*, *25*, 1697–1708.

Dixit, R., Malaviya, D., Pandiyan, K., Singh, U. B., Sahu, A., Shukla, R., et al. (2015). Bioremediation of heavy metals from soil and aquatic environment: An overview of principles and criteria of fundamental processes. *Sustainability*, *7*, 2189–2212.

Ebeling, J. M., Timmons, M. B., & Bisogni, J. (2006). Engineering analysis of the stoichiometry of photoautotrophic, autotrophic, and heterotrophic removal of ammonia–nitrogen in aquaculture systems. *Aquaculture*, *257*, 346–358.

Flemming, H.-C. (2002). Biofouling in water systems—Cases, causes and countermeasures. *Applied Microbiology and Biotechnology*, *59*, 629–640.

Flemming, H.-C., & Wingender, J. (2001). Relevance of microbial extracellular polymeric substances (EPSs)—Part I: Structural and ecological aspects. *Water Science and Technology*, *43*, 1–8.

Galie, S., García-Gutiérrez, C., Miguélez, E. M., Villar, C. J., & Lombó, F. (2018). Biofilms in the food industry: Health aspects and control methods. *Frontiers in Microbiology*, *9*, 898.

Gao, F., Yang, Z.-H., Li, C., Zeng, G.-M., Ma, D.-H., & Zhou, L. (2015). A novel algal biofilm membrane photobioreactor for attached microalgae growth and nutrients removal from secondary effluent. *Bioresource Technology*, *179*, 8–12.

Gara-Ali, M., Zili, F., Hosni, K., Ouada, H. B., & Ben-Mahrez, K. (2021). Lipophilic extracts of the thermophilic cyanobacterium *Leptolyngbya* sp. and chlorophyte *Graesiella* sp. and their potential use as food and anticancer agents. *Algal Research*, *60*, 102511.

Garnham, G. W., Codd, G. A., & Gadd, G. M. (1992). Accumulation of cobalt, zinc and manganese by the estuarine green microalga *Chlorella salina* immobilized in alginate microbeads. *Environmental Science & Technology*, *26*, 1764–1770.

Garny, K., Horn, H., & Neu, T. R. (2008). Interaction between biofilm development, structure and detachment in rotating annular reactors. *Bioprocess and Biosystems Engineering*, *31*, 619–629.

Gismondi, A., Pippo, F. D., Bruno, L., Antonaroli, S., & Congestri, R. (2016). Phosphorus removal coupled to bioenergy production by three cyanobacterial isolates in a biofilm dynamic growth system. *International Journal of Phytoremediation*, *18*, 869–876.

Goffredo, G. B., Accoroni, S., Totti, C., Romagnoli, T., Valentini, L., & Munafò, P. (2017). Titanium dioxide based nanotreatments to inhibit microalgal fouling on building stone surfaces. *Building and Environment*, *112*, 209–222.

Gómez-Ramírez, A. L., Enriquez-Ocaña, L. F., Miranda-Baeza, A., Cordero Esquivel, B., López-Elías, J. A., & Martínez-Córdova, L. R. (2019). Biofilm-forming capacity of two benthic microalgae, *Navicula incerta* and *Navicula* sp., on three substrates (Naviculales: Naviculaceae). *Revista de Biología Tropical*, *67*, 599–607.

Griffiths, M. J., van Hille, R. P., & Harrison, S. T. (2012). Lipid productivity, settling potential and fatty acid profile of 11 microalgal species grown under nitrogen replete and limited conditions. *Journal of Applied Phycology*, *24*, 989–1001.

Hallmann, C., Stannek, L., Fritzlar, D., Hause-Reitner, D., Friedl, T., & Hoppert, M. (2013). Molecular diversity of phototrophic biofilms on building stone. *FEMS Microbiology Ecology*, *84*, 355–372.

Harun, R., Singh, M., Forde, G. M., & Danquah, M. K. (2010). Bioprocess engineering of microalgae to produce a variety of consumer products. *Renewable and Sustainable Energy Reviews*, *14*, 1037–1047.

He, S., & Xue, G. (2010). Algal-based immobilization process to treat the effluent from a secondary wastewater treatment plant (WWTP). *Journal of Hazardous Materials*, *178*, 895–899.

Higgins, B. T., Gennity, I., Samra, S., Kind, T., Fiehn, O., & Vander Gheynst, J. S. (2016). Cofactor symbiosis for enhanced algal growth, biofuel production, and wastewater treatment. *Algal Research*, *17*, 308–315.

Higgins, B. T., & Vander Gheynst, J. S. (2014). Effects of *Escherichia coli* on mixotrophic growth of *Chlorella minutissima* and production of biofuel precursors. *PLoS One*, *9*, e96807.

Hori, K., & Matsumoto, S. (2010). Bacterial adhesion: From mechanism to control. *Biochemical Engineering Journal*, *48*, 424–434.

Hu, Y., Xiao, Y., Liao, K., Leng, Y., & Lu, Q. (2021). Development of microalgal biofilm for wastewater remediation: From mechanism to practical application. *Journal of Chemical Technology & Biotechnology*, *96*, 2993–3008.

Intergovernmental Panel on Climate Change. (2014). *AR5 climate change 2014: Mitigation of climate change*.

Irving, T. E., & Allen, D. G. (2011). Species and material considerations in the formation and development of microalgal biofilms. *Applied Microbiology and Biotechnology*, *92*, 283–294.

Ji, C., Wang, J., Zhang, W., Liu, J., Wang, H., Gao, L., et al. (2014). An applicable nitrogen supply strategy for attached cultivation of *Aucutodesmus obliquus*. *Journal of Applied Phycology*, *26*, 173–180.

Johnson, M. B., & Wen, Z. (2009). Production of biodiesel fuel from the microalga Schizochytrium limacinum by direct transesterification of algal biomass. *Energy & Fuels*, *23*, 5179–5183.

Johnson, M. B., & Wen, Z. (2010). Development of an attached microalgal growth system for biofuel production. *Applied Microbiology and Biotechnology*, *85*, 525–534.

Kanavillil, N., Thorn, M., & Kurissery, S. (2012). Characterization of natural biofilms in temperate inland waters. *Journal of Great Lakes Research*, *38*, 429–438.

Kesaano, M., & Sims, R. C. (2014). Algal biofilm based technology for wastewater treatment. *Algal Research*, *5*, 231–240.

Kim, B.-H., Kim, D.-H., Choi, J.-W., Kang, Z., Cho, D.-H., Kim, J.-Y., et al. (2015). Polypropylene bundle attached multilayered stigeoclonium biofilms cultivated in untreated sewage generate high biomass and lipid productivity. *Journal of Microbiology and Biotechnology*, *25*, 1547–1554.

Kim, S., Moon, M., Kwak, M., Lee, B., & Chang, Y. K. (2018). Statistical optimization of light intensity and CO 2 concentration for lipid production derived from attached cultivation of green microalga *Ettlia* sp. *Scientific Reports*, *8*, 1–13.

Koller, M. (2015). Design of closed photobioreactors for algal cultivation. In *Algal biorefineries* (pp. 133–186). Springer.

Kunjapur, A. M., & Eldridge, R. B. (2010). Photobioreactor design for commercial biofuel production from microalgae. *Industrial & Engineering Chemistry Research*, *49*, 3516–3526.

Lage, O. M., & Graça, A. P. (2016). Biofilms: An extra coat on macroalgae. In *Algae—Organisms for imminent biotechnology* IntechOpen.

Lakatos, G., Deák, Z., Vass, I., Rétfalvi, T., Rozgonyi, S., Rákhely, G., et al. (2014). Bacterial symbionts enhance photo-fermentative hydrogen evolution of *Chlamydomonas* algae. *Green Chemistry*, *16*, 4716–4727.

Lasaro, M. A., Salinger, N., Zhang, J., Wang, Y., Zhong, Z., Goulian, M., et al. (2009). F1C fimbriae play an important role in biofilm formation and intestinal colonization by the *Escherichia coli* commensal strain Nissle 1917. *Applied and Environmental Microbiology*, *75*, 246–251.

Lau, N.-S., Matsui, M., & Abdullah, A. A.-A. (2015). Cyanobacteria: Photoautotrophic microbial factories for the sustainable synthesis of industrial products. *BioMed Research International*, *2015*.

Lee, B., Haagensen, J. A., Ciofu, O., Andersen, J. B., Høiby, N., & Molin, S. (2005). Heterogeneity of biofilms formed by nonmucoid *Pseudomonas aeruginosa* isolates from patients with cystic fibrosis. *Journal of Clinical Microbiology*, *43*, 5247–5255.

Lewandowski, Z., & Evans, L. (2000). Structure and function of biofilms. In L. V. Evans (Ed.), *Vol. 1. Biofilms: Recent advances in their study and control* (pp. 21–38). London: CRC Press.

Liu, J., Danneels, B., Vanormelingen, P., & Vyverman, W. (2016). Nutrient removal from horticultural wastewater by benthic filamentous algae *Klebsormidium* sp., *Stigeoclonium* spp. and their communities: From laboratory flask to outdoor Algal Turf Scrubber (ATS). *Water Research*, *92*, 61–68.

Liu, T., Wang, J., Hu, Q., Cheng, P., Ji, B., Liu, J., et al. (2013). Attached cultivation technology of microalgae for efficient biomass feedstock production. *Bioresource Technology*, *127*, 216–222.

Luo, Y., Le-Clech, P., & Henderson, R. K. (2020). Characterisation of microalgae-based monocultures and mixed cultures for biomass production and wastewater treatment. *Algal Research*, *49*, 101963.

Maharana, S., Pradhan, B., Jena, M., & Misra, M. K. (2019). Diversity of phytoplankton in Chilika Lagoon, Odisha, India. *Environment and Ecology*, *37*, 737–746.

Mahto, K. U., Kumari, S., & Das, S. (2021). Unraveling the complex regulatory networks in biofilm formation in bacteria and relevance of biofilms in environmental remediation. *Critical Reviews in Biochemistry and Molecular Biology*, 1–28.

Malone, T. C., & Newton, A. (2020). The globalization of cultural eutrophication in the coastal ocean: Causes and consequences. *Frontiers in Marine Science*, *7*.

Mangwani, N., Shukla, S. K., Kumari, S., Rao, T. S., & Das, S. (2014). Characterization of *Stenotrophomonas acidaminiphila* NCW-702 biofilm for implication in the degradation of polycyclic aromatic hydrocarbons. *Journal of Applied Microbiology*, *117*, 1012–1024.

Mantzorou, A., & Ververidis, F. (2019). Microalgal biofilms: A further step over current microalgal cultivation techniques. *Science of the Total Environment*, *651*, 3187–3201.

Mata, T. M., Martins, A. A., & Caetano, N. S. (2010). Microalgae for biodiesel production and other applications: A review. *Renewable and Sustainable Energy Reviews*, *14*, 217–232.

Melo, M., Fernandes, S., Caetano, N., & Borges, M. T. (2018). *Chlorella vulgaris* (SAG 211-12) biofilm formation capacity and proposal of a rotating flat plate photobioreactor for more sustainable biomass production. *Journal of Applied Phycology*, *30*, 887–899.

Menger-Krug, E., Niederste-Hollenberg, J., Hillenbrand, T., & Hiessl, H. (2012). Integration of microalgae systems at municipal wastewater treatment plants: Implications for energy and emission balances. *Environmental Science & Technology*, *46*, 11505–11514.

Milano, J., Ong, H. C., Masjuki, H., Chong, W., Lam, M. K., Loh, P. K., et al. (2016). Microalgae biofuels as an alternative to fossil fuel for power generation. *Renewable and Sustainable Energy Reviews*, *58*, 180–197.

Mirón, A. S., Garcıa, M. C. C., Gómez, A. C., Camacho, F. G., Grima, E. M., & Chisti, Y. (2003). Shear stress tolerance and biochemical characterization of *Phaeodactylum tricornutum* in quasi steady-state continuous culture in outdoor photobioreactors. *Biochemical Engineering Journal*, *16*, 287–297.

Mishra, S., Chowdhary, P., & Bharagava, R. N. (2019). Conventional methods for the removal of industrial pollutants, their merits and demerits. In *Emerging and eco-friendly approaches for waste management* (pp. 1–31). Springer.

Moënne-Loccoz, Y., Mavingui, P., Combes, C., Normand, P., & Steinberg, C. (2015). Microorganisms and biotic interactions. In *Environmental microbiology: Fundamentals and applications* (pp. 395–444). Springer.

Mohan, S. V. (2009). Harnessing of biohydrogen from wastewater treatment using mixed fermentative consortia: Process evaluation towards optimization. *International Journal of Hydrogen Energy*, *34*, 7460–7474.

Mohanty, S., Pradhan, B., Patra, S., Behera, C., Nayak, R., & Jena, M. (2020). Screening for nutritive bioactive compounds in some algal strains isolated from coastal Odisha. *Journal of Advanced Plant Sciences*, *10*, 1–8.

Moreno Osorio, J. H., Pollio, A., Frunzo, L., Lens, P. N. L., & Esposito, G. (2021). A review of microalgal biofilm technologies: Definition, applications, settings and analysis. *Frontiers in Chemical Engineering*, *3*.

Moreno-Garrido, I. (2008). Microalgae immobilization: Current techniques and uses. *Bioresource Technology*, *99*, 3949–3964.

Muñoz, R., Köllner, C., & Guieysse, B. (2009). Biofilm photobioreactors for the treatment of industrial wastewaters. *Journal of Hazardous Materials*, *161*, 29–34.

Murphy, T. E., & Berberoglu, H. (2014). Flux balancing of light and nutrients in a biofilm photobioreactor for maximizing photosynthetic productivity. *Biotechnology Progress*, *30*, 348–359.

Naumann, T., Çebi, Z., Podola, B., & Melkonian, M. (2013). Growing microalgae as aquaculture feeds on twin-layers: A novel solid-state photobioreactor. *Journal of Applied Phycology*, *25*, 1413–1420.

Orandi, S., Lewis, D., & Moheimani, N. (2012). Biofilm establishment and heavy metal removal capacity of an indigenous mining algal-microbial consortium in a photo-rotating biological contactor. *Journal of Industrial Microbiology and Biotechnology*, *39*, 1321–1331.

Ozkan, A., Kinney, K., Katz, L., & Berberoglu, H. (2012). Reduction of water and energy requirement of algae cultivation using an algae biofilm photobioreactor. *Bioresource Technology*, *114*, 542–548.

Pandey, P., Bharti, V., & Kumar, K. (2014). Biofilm in aquaculture production. *African Journal of Microbiology Research*, *8*, 1434–1443.

Park, J., Park, B. S., Wang, P., Patidar, S. K., Kim, J. H., Kim, S.-H., et al. (2017). Phycospheric native bacteria *Pelagibaca bermudensis* and *Stappia* sp. ameliorate biomass productivity of *Tetraselmis striata* (KCTC1432BP) in co-cultivation system through mutualistic interaction. *Frontiers in Plant Science*, *8*, 289.

Park, K. C., Whitney, C., McNichol, J. C., Dickinson, K. E., MacQuarrie, S., Skrupski, B. P., et al. (2012). Mixotrophic and photoautotrophic cultivation of 14 microalgae isolates from Saskatchewan, Canada: Potential applications for wastewater remediation for biofuel production. *Journal of Applied Phycology*, *24*, 339–348.

Perez-Garcia, O., & Bashan, Y. (2015). Microalgal heterotrophic and mixotrophic culturing for bio-refining: From metabolic routes to techno-economics. In A. Prokop, R. K. Bajpai, & M. E. Zappi (Eds.), *Algal biorefineries* (pp. 61–131). Springer.

Petroff, A. P., Wu, T.-D., Liang, B., Mui, J., Guerquin-Kern, J.-L., Vali, H., et al. (2011). Reaction–diffusion model of nutrient uptake in a biofilm: Theory and experiment. *Journal of Theoretical Biology*, *289*, 90–95.

Picioreanu, C., Blauert, F., Horn, H., & Wagner, M. (2018). Determination of mechanical properties of biofilms by modelling the deformation measured using optical coherence tomography. *Water Research*, *145*, 588–598.

Pradhan, B., Bhuyan, P. P., Patra, S., Nayak, R., Behera, P. K., Behera, C., et al. (2022). Beneficial effects of seaweeds and seaweed-derived bioactive compounds: Current evidence and future prospective. *Biocatalysis and Agricultural Biotechnology*, *39*, 102242.

Pradhan, B., Maharana, S., Bhakta, S., & Jena, M. (2021). Marine phytoplankton diversity of Odisha coast, India with special reference to new record of diatoms and dinoflagellates. *Vegetos*. https://doi.org/10.1007/s42535-021-00301-2.

Pradhan, B., Nayak, R., Patra, S., Jit, B. P., Ragusa, A., & Jena, M. (2020). Bioactive metabolites from marine algae as potent pharmacophores against oxidative stress-associated human diseases: A comprehensive review. *Molecules*, *26*(1), 37. https://doi.org/10.3390/molecules26010037.

Pradhan, B., Patra, S., Behera, C., Nayak, R., Jit, B. P., Ragusa, A., & Jena, M. (2021). Preliminary investigation of the antioxidant, anti-diabetic, and anti-inflammatory activity of *Enteromorpha intestinalis* extracts. *Molecules*, *26*(4), 1171. https://doi.org/10.3390/molecules26041171.

Pradhan, B., Patra, S., Behera, C., Nayak, R., Patil, S., Bhutia, S. K., & Jena, M. (2020). *Enteromorpha compressa* extract induces anticancer activity through apoptosis and autophagy in oral cancer. *Molecular Biology Reports*, *47*, 9567–9578.

Pradhan, B., Patra, S., Dash, S. R., Maharana, S., Behera, C., & Jena, M. (2020a). Antioxidant responses against aluminum metal stress in *Geitlerinema amphibium*. *SN Applied Sciences*, *2*, 1–8.

Pradhan, B., Patra, S., Dash, S. R., Maharana, S., Behera, C., & Jena, M. (2020b). Antioxidant responses against aluminum metal stress in *Geitlerinema amphibium*. *SN Applied Sciences*, *2*, 800.

Pradhan, B., Patra, S., Dash, S. R., Nayak, R., Behera, C., & Jena, M. (2021). Evaluation of the anti-bacterial activity of methanolic extract of *Chlorella vulgaris* Beyerinck [Beijerinck] with special reference to antioxidant modulation. *Future Journal of Pharmaceutical Sciences*, *7*, 17.

Pradhan, B., Patra, S., Maharana, S., Behera, C., Dash, S. R., & Jena, M. (2021). Demarcating antioxidant response against aluminum induced oxidative stress in *Westiellopsis prolifica* Janet 1941.

International Journal of Phytoremediation, *23*(3), 238–251. https://doi.org/10.1080/15226514.2020.1807906.

Pradhan, B., Patra, S., Nayak, R., Behera, C., Dash, S. R., Nayak, S., et al. (2020). Multifunctional role of fucoidan, sulfated polysaccharides in human health and disease: A journey under the sea in pursuit of potent therapeutic agents. *International Journal of Biological Macromolecules*, *164*, 4263–4278.

Pradhan, B., Patra, S., Nayak, R., Swain, S. S., Jit, B. P., Behera, C., et al. (2021). *Low-dose priming of gamma radiation enhanced cadmium tolerance in Chlamydomonas Reinhardtii by modulating physio-biochemical pathways.*

Pulz, O., & Gross, W. (2004). Valuable products from biotechnology of microalgae. *Applied Microbiology and Biotechnology*, *65*, 635–648.

Qureshi, N., Annous, B. A., Ezeji, T. C., Karcher, P., & Maddox, I. S. (2005). Biofilm reactors for industrial bioconversion processes: Employing potential of enhanced reaction rates. *Microbial Cell Factories*, *4*, 1–21.

Radakovits, R., Jinkerson, R. E., Darzins, A., & Posewitz, M. C. (2010). Genetic engineering of algae for enhanced biofuel production. *Eukaryotic Cell*, *9*, 486–501.

Raheem, A., Sikarwar, V. S., He, J., Dastyar, W., Dionysiou, D. D., Wang, W., et al. (2018). Opportunities and challenges in sustainable treatment and resource reuse of sewage sludge: A review. *Chemical Engineering Journal*, *337*, 616–641.

Rai, J., Kumar, D., Pandey, L. K., Yadav, A., & Gaur, J. (2016). Potential of cyanobacterial biofilms in phosphate removal and biomass production. *Journal of Environmental Management*, *177*, 138–144.

Ravindran, B., Gupta, S. K., Cho, W.-M., Kim, J. K., Lee, S. R., Jeong, K.-H., et al. (2016). Microalgae potential and multiple roles—Current progress and future prospects—An overview. *Sustainability*, *8*, 1215.

Rinanti, A., Dewi, K., Kardena, E., & Astuti, D. I. (2016). Biotechnology carbon capture and storage by microalgae to enhance CO2 removal efficiency in closed-system photobioreactor. In N. Thajuddin, & D. Dhanasekaran (Eds.), *Algae—Organisms for imminent biotechnology* (pp. 105–109). IntechOpen.

Roeselers, G., Van Loosdrecht, M. C., & Muyzer, G. (2008). Phototrophic biofilms and their potential applications. *Journal of Applied Phycology*, *20*, 227–235.

Rokicka, M., Zieliński, M., Dudek, M., & Dębowski, M. (2021). Effects of ultrasonic and microwave pretreatment on lipid extraction of microalgae and methane production from the residual extracted biomass. *BioEnergy Research*, *14*, 752–760.

Romaní, A. M., Fund, K., Artigas, J., Schwartz, T., Sabater, S., & Obst, U. (2008). Relevance of polymeric matrix enzymes during biofilm formation. *Microbial Ecology*, *56*, 427–436.

Samal, K., Kar, S., & Trivedi, S. (2019). Ecological floating bed (EFB) for decontamination of polluted water bodies: Design, mechanism and performance. *Journal of Environmental Management*, *251*, 109550.

Satyanarayana, K., Mariano, A., & Vargas, J. (2011). A review on microalgae, a versatile source for sustainable energy and materials. *International Journal of Energy Research*, *35*, 291–311.

Schenk, P. M., Thomas-Hall, S. R., Stephens, E., Marx, U. C., Mussgnug, J. H., Posten, C., et al. (2008). Second generation biofuels: High-efficiency microalgae for biodiesel production. *Bioenergy Research*, *1*, 20–43.

Schnurr, P. J., Espie, G. S., & Allen, D. G. (2013). Algae biofilm growth and the potential to stimulate lipid accumulation through nutrient starvation. *Bioresource Technology*, *136*, 337–344.

Schultze, L. K., Simon, M.-V., Li, T., Langenbach, D., Podola, B., & Melkonian, M. (2015). High light and carbon dioxide optimize surface productivity in a twin-layer biofilm photobioreactor. *Algal Research*, *8*, 37–44.

Schumacher, G., & Sekoulov, I. (2002). Polishing of secondary effluent by an algal biofilm process. *Water Science and Technology*, *46*, 83–90.

Sekar, R., Nair, K., Rao, V., & Venugopalan, V. (2002). Nutrient dynamics and successional changes in a lentic freshwater biofilm. *Freshwater Biology*, *47*, 1893–1907.

Sekar, R., Venugopalan, V., Satpathy, K., Nair, K., & Rao, V. (2004). Laboratory studies on adhesion of microalgae to hard substrates. In *Asian pacific phycology in the 21st century: Prospects and challenges* (pp. 109–116). Springer.

Shen, Y., Chen, C., Chen, W., & Xu, X. (2014). Attached culture of Nannochloropsis oculata for lipid production. *Bioprocess and Biosystems Engineering*, *37*, 1743–1748.

Shen, Y., Yu, T., Xie, Y., Chen, J., Ho, S.-H., Wang, Y., et al. (2019). Attached culture of *Chlamydomonas* sp. JSC4 for biofilm production and TN/TP/Cu (II) removal. *Biochemical Engineering Journal*, *141*, 1–9.

Shen, Y., Zhang, H., Xu, X., & Lin, X. (2015). Biofilm formation and lipid accumulation of attached culture of *Botryococcus braunii*. *Bioprocess and Biosystems Engineering*, *38*, 481–488.

Shi, J., Podola, B., & Melkonian, M. (2007). Removal of nitrogen and phosphorus from wastewater using microalgae immobilized on twin layers: An experimental study. *Journal of Applied Phycology*, *19*, 417–423.

Sims, R., & Christenson, L. (2011). Production and harvesting of microalgae for wastewater treatment, biofuels and subproducts. *Biotechnology Advances*, *29*, 686–702.

Singh, R., & Sharma, S. (2012). Development of suitable photobioreactor for algae production—A review. *Renewable and Sustainable Energy Reviews*, *16*, 2347–2353.

Sinha, R. (2021). Metabolic engineering of Cyanobacteria & micro algae for enhanced production of bio fuels & high value products. *Annals of the Romanian Society for Cell Biology*, 21160–21179.

Smith, D. W., & Piedrahita, R. H. (1988). The relation between phytoplankton and dissolved oxygen in fish ponds. *Aquaculture*, *68*, 249–265.

Stoodley, P., Sauer, K., Davies, D. G., & Costerton, J. W. (2002). Biofilms as complex differentiated communities. *Annual Reviews in Microbiology*, *56*, 187–209.

Strieth, D., Ulber, R., & Muffler, K. (2018). Application of phototrophic biofilms: From fundamentals to processes. *Bioprocess and Biosystems Engineering*, *41*, 295–312.

Suganya, T., Varman, M., Masjuki, H., & Renganathan, S. (2016). Macroalgae and microalgae as a potential source for commercial applications along with biofuels production: A biorefinery approach. *Renewable and Sustainable Energy Reviews*, *55*, 909–941.

Tampion, J., & Tampion, M. (1987). *Immobilized cells: Principles and applications.* Cambridge University Press.

Tang, D. Y. Y., Khoo, K. S., Chew, K. W., Tao, Y., Ho, S.-H., & Show, P. L. (2020). Potential utilization of bioproducts from microalgae for the quality enhancement of natural products. *Bioresource Technology*, *304*, 122997.

Tao, Q., Gao, F., Qian, C.-Y., Guo, X.-Z., Zheng, Z., & Yang, Z.-H. (2017). Enhanced biomass/biofuel production and nutrient removal in an algal biofilm airlift photobioreactor. *Algal Research*, *21*, 9–15.

Telgmann, U., Horn, H., & Morgenroth, E. (2004). Influence of growth history on sloughing and erosion from biofilms. *Water Research*, *38*, 3671–3684.

Thapa, S., Bharti, A., & Prasanna, R. (2017). Algal biofilms and their biotechnological significance. In *Algal green chemistry* (pp. 285–303). Elsevier.

Tirkey, J., & Adhikary, S. P. (2005). Cyanobacteria in biological soil crusts of India. *Current Science*, *89*, 515–521.

Tong, C., & Derek, C. (2021). Biofilm formation of benthic diatoms on commercial polyvinylidene fluoride membrane. *Algal Research*, *55*, 102260.

Toninelli, A. E., Wang, J., Liu, M., Wu, H., & Liu, T. (2016). *Scenedesmus dimorphus* biofilm: Photoefficiency and biomass production under intermittent lighting. *Scientific Reports*, *6*, 1–10.

Toyofuku, M., Inaba, T., Kiyokawa, T., Obana, N., Yawata, Y., & Nomura, N. (2016). Environmental factors that shape biofilm formation. *Bioscience, Biotechnology, and Biochemistry*, *80*, 7–12.

Tran, N. T., & Kaldenhoff, R. (2020). Achievements and challenges of genetic engineering of the model green alga *Chlamydomonas reinhardtii*. *Algal Research*, *50*, 101986.

Tsavatopoulou, V. D., & Manariotis, I. D. (2020). The effect of surface properties on the formation of *Scenedesmus rubescens* biofilm. *Algal Research*, *52*, 102095.

Varnam, A., & Evans, M. (2000). *Environmental microbiology*. CRC Press.

Vickers, N. J. (2017). Animal communication: When i'm calling you, will you answer too? *Current Biology*, *27*, R713–R715.

Vo, H. N. P., Ngo, H. H., Guo, W., Nguyen, T. M. H., Liu, Y., Liu, Y., et al. (2019). A critical review on designs and applications of microalgae-based photobioreactors for pollutants treatment. *Science of the Total Environment*, *651*, 1549–1568.

Wang, J., Liu, J., & Liu, T. (2015). The difference in effective light penetration may explain the superiority in photosynthetic efficiency of attached cultivation over the conventional open pond for microalgae. *Biotechnology for Biofuels*, *8*, 1–12.

Wase, N., Black, P., & DiRusso, C. (2018). Innovations in improving lipid production: Algal chemical genetics. *Progress in Lipid Research*, *71*, 101–123.

Whitton, R. L. (2016). *Algae reactors for wastewater treatment*. England: Cranfield University.

Whitton, R., Santinelli, M., Pidou, M., Ometto, F., Henderson, R., Roddick, F., et al. (2018). Tertiary nutrient removal from wastewater by immobilised microalgae: Impact of wastewater nutrient characteristics and hydraulic retention time (HRT). *H2Open Journal*, *1*, 12–25.

Wijeyekoon, S., Mino, T., Satoh, H., & Matsuo, T. (2004). Effects of substrate loading rate on biofilm structure. *Water Research*, *38*, 2479–2488.

Williams, P. J. L. B., & Laurens, L. M. (2010). Microalgae as biodiesel & biomass feedstocks: Review & analysis of the biochemistry, energetics & economics. *Energy & Environmental Science*, *3*, 554–590.

Wollmann, F., Dietze, S., Ackermann, J. U., Bley, T., Walther, T., Steingroewer, J., et al. (2019). Microalgae wastewater treatment: Biological and technological approaches. *Engineering in Life Sciences*, *19*, 860–871.

Woodcock, S., & Sloan, W. T. (2017). Biofilm community succession: A neutral perspective. *Microbiology*, *163*, 664–668.

Wu, Y. (2016). *Periphyton: Functions and application in environmental remediation*. Elsevier.

Xiong, J.-Q., Kurade, M. B., & Jeon, B.-H. (2017). Ecotoxicological effects of enrofloxacin and its removal by monoculture of microalgal species and their consortium. *Environmental Pollution*, *226*, 486–493.

Yao, S., Lyu, S., An, Y., Lu, J., Gjermansen, C., & Schramm, A. (2019). Microalgae–bacteria symbiosis in microalgal growth and biofuel production: A review. *Journal of Applied Microbiology*, *126*, 359–368.

Zhang, L., Chen, L., Wang, J., Chen, Y., Gao, X., Zhang, Z., et al. (2015). Attached cultivation for improving the biomass productivity of *Spirulina platensis*. *Bioresource Technology*, *181*, 136–142.

Zhang, D., Fung, K. Y., & Ng, K. M. (2014). Novel filtration photobioreactor for efficient biomass production. *Industrial & Engineering Chemistry Research*, *53*, 12927–12934.

Zhang, Q., Liu, C., Li, Y., Yu, Z., Chen, Z., Ye, T., et al. (2017). Cultivation of algal biofilm using different lignocellulosic materials as carriers. *Biotechnology for Biofuels*, *10*, 1–16.

Zou, X., Xu, K., Chang, W., Qu, Y., & Li, Y. (2021). A novel microalgal biofilm reactor using walnut shell as substratum for microalgae biofilm cultivation and lipid accumulation. *Renewable Energy*, *175*, 676–685.

Zou, Y., Zeng, Q., Li, H., Liu, H., & Lu, Q. (2021). Emerging technologies of algae-based wastewater remediation for bio-fertilizer production: A promising pathway to sustainable agriculture. *Journal of Chemical Technology & Biotechnology*, *96*, 551–563.

Chapter 11

Biochemical and molecular mechanisms of sulfate-reducing bacterial biofilms

Vinoj Gopalakrishnan[a,b], Ram Nageena Singh[a,b], Abhilash Kumar Tripathi[a,b], Shailabh Rauniyar[a,b], Priya Saxena[a,b], Payal Thakur[a,b], and Rajesh Kumar Sani[a,b,c]

[a]*Department of Chemical and Biological Engineering, South Dakota School of Mines and Technology, Rapid City, SD, United States,* [b]*Data Driven Material Discovery Center for Bioengineering Innovation, South Dakota School of Mines and Technology, Rapid City, SD, United States,* [c]*BuG ReMeDEE Consortium, South Dakota School of Mines and Technology, Rapid City, SD, United States*

1. Introduction

The development of biofilms is a survival strategy employed by microorganisms to protect them from changing or challenging/harsh environments. Biofilm development is an issue of rising importance in community, and different bacterial models are currently being investigated to better understand the molecular strategies that bacteria go through to develop biofilms. Biofilm environments can maintain water, sorb nutrients, and keep against instant variations in salinity, osmolarity, pH, nutrient availability, and redox. Biofilm formation is a complex multistep development determined on flagellar function, motility, extracellular polymeric substances (EPS) production, and cellular metabolic state. Sulfate reducing bacteria (SRB) are capable of overcoming the oxygen stress in the environment by forming biofilm (Liu, Meng, Li, Gu, & Liu, 2019). The reduced oxygen diffusion through the biofilm results in anoxic zones, which in turn promotes SRB growth (Scarascia, Wang, & Hong, 2016). SRB are a distinct group of nonpathogenic, anaerobic bacteria and utilize dissimilatory sulfate reduction (DSR) to generate energy by sulfate reduction to sulfide (Tripathi et al., 2021). DSR acts an important role in SRB by providing energy for the creation of adenosine triphosphate (ATP) (Fig. 1). SRB oxidize organic compounds (e.g., lactate, propionate, and butyrate) or hydrogen and use sulfate as a terminal electron acceptor and provide energy for growth (Kushkevych et al., 2018). Most of the enzymes that are associated with this process are localized in the cytoplasm and periplasm of the bacterial cells.

SRB are of primary concern because they accelerate metallic corrosion under mild conditions, including neutral pH, ambient temperatures, and absence of oxygen, both in natural and engineered aquatic systems. SRB are often responsible for microbiologically influenced corrosion (MIC), accelerating corrosion reactions, or shifting corrosion mechanisms. It has been reported that SRB biofilm cells have differences in their gene expressions, for example, two pyruvate:ferredoxin oxidoreductases [e.g., *oor* (upregulated) and *por* (downregulated)]were overexpressed as compared with planktonic cells. The expression of pyruvate formate lyase was unchanged but formate dehydrogenase expression (*hybA* transcript and protein, alpha subunit) was increased. At the same time, cytochrome c553 (electron transfer partner for formate dehydrogenase) was found to increase in *Desulfovibrio vulgaris* Hildenborough (DVH) (Sebban-Kreuzer, Blackledge, Dolla, Marion, & Guerlesquin, 1998). It has also been reported that SRB biofilm has 26-fold higher formate-dependent (dehydrogenase) activity than planktonic cells and assumed that it was because of formate cycling and their role in electron flow.

SRB biofilms have been found in underground pipelines, oil and gas fields and are known to cause MIC. The presence of biofilm on metal surfaces enhances the frequency of corrosion by up to 10,000 times, compared with the planktonic state. Microbial attachment to the surfaces involves a complex mechanism that influences interactions between bacterial surface and physicochemical proprieties of the substratum. In a biological system, most effective hydrophobic interactions facilitate adhesion to metal surfaces (Dantam, Subbaraman, & Jones, 2020). The development of SRB biofilm on a metal surface can impact the kinetics of anodic and cathodic reactions, leading to an acceleration of MIC. SRB can directly uptake electrons released from iron oxidation, which travel to the cell's cytoplasm or periplasm through extracellular electron transfer. During starvation, SRB uptake electrons from carbon steel and activate corrosion (Li et al., 2020). Moreover, adhesion on surfaces (biofilm formation) allows direct contact between outer membrane proteins or electroconductive

Understanding Microbial Biofilms. https://doi.org/10.1016/B978-0-323-99977-9.00016-8

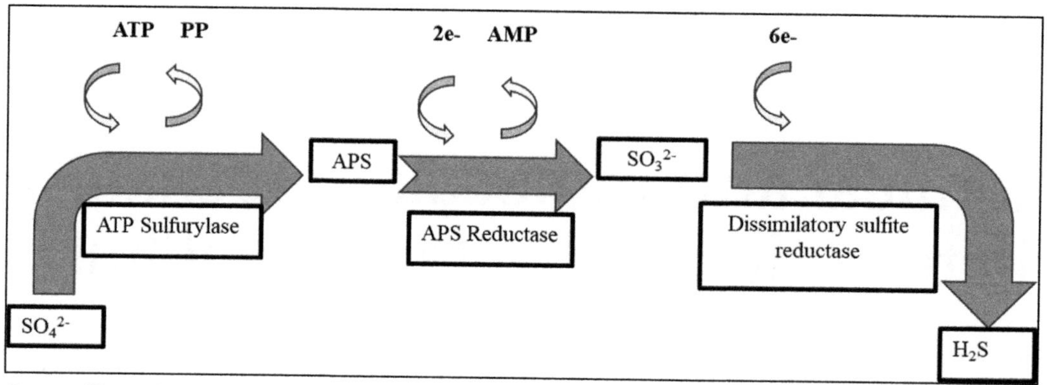

FIG. 1 Dissimilatory sulfate reduction in sulfate-reducing bacteria. *AMP*, adenosine monophosphate; *APS*, adenosine phosphosulfate; *ATP*, adenosine triphosphate; *PPi*, pyrophosphate.

nanowires and the metal surface (Xu, Li, & Gu, 2016). We describe here external factors as well as regulatory pathways that contribute to the biochemical and molecular mechanisms of SRB biofilms.

2. Exopolysaccharide

Biofilm formation is a complex multistep process dependent on flagellar motility, EPSs production, and cellular metabolic state. Bacterial biofilms are cells embedded in a dense extracellular matrix, generally comprising organic polymers such as polysaccharides, proteins, lipids, and extracellular DNA, which support bacteria to irreversibly bind to surfaces (Wettstadt, 2020). EPSs not only play significant roles in the development of biofilms but also participate in the transformation of cells from reversible to irreversible attachment and promote the sessile cells to form mature biofilms and enhance the structure of biofilms. Studies have shown that several genes are associated with EPS production in *D. alaskensis* G20 such as two genes Dde_3587 and Dde_3588 encode for OmpA domain proteins and three genes of heptosyl transferase family (Dde_0565, Dde_1714, and Dde_2022), several glycosyltransferases (Dde_0337 and Dde_0426), a glycerol-3-phosphate dehydrogenase (Dde_3178), and a UDP gene (Dde_2187) (Williamson et al., 2018). Besides, lipid fraction has an influence on the hydrophobic properties of EPS (Ma & Wood, 2009) and may act as a stimulus to bacterial adhesion to the substratum.

The EPS production relies on the biofilm growth phase where maximum EPS production has been observed in the biofilm maturation stage. Large quantities of EPS produced caused the creation of a compact and uniform biofilm structure and a role as a survival strategy. In fact, EPS matrix is formed in response to the important requirements of bacteria according to their stage of growth in the biofilm. Consequently, the production of EPS resulted in the initiation of a stable structure to protect the biofilm from the harsh environment (Wilkins, Hall-Stoodley, Allan, & Faust, 2014). The EPSs are the most important agents responsible for the structural and functional integrity and for the maintenance of biological and the physicochemical properties of the biofilm structure. Furthermore, the EPSs play a defending role for the bacteria against the external stress and manage as energy reservoirs during starvation condition (Gómez-Suárez et al., 2002).

D. vulgaris consists of EPS that includes glucose (major carbohydrate), mannose, and galactose (Zhu et al., 2018). Most of the SRB produce excess amount of EPS to form the noncovalent binding between the cells and inhibit the access of oxygen and promote the protective biofilm matrix (Ziadi, El-Bassi, Bousselmi, & Akrout, 2020). The primary bacterial attachment development is induced by the adsorption of EPS to the metal surface, which would adjust the physicochemical characteristics of the metal. EPS, which is the most essential factor of SRB biofilm architecture, and influences corrosion processes by complexing agents with metal ions (Dong, Liu, & Liu, 2011). Recent genomic studies have shown that expression of SRB biofilm associated cells regularly demonstrates high levels of gene expression heterogeneity related to EPS and histidine kinases implicated in biofilm formation as well as hydrogenases and cytochromes. Indeed, stress-specific responses genes include hydroxylamine reductase (*hcp* gene) in nitrite stress, Na^+/H^+ antiporter gene (*nhaC2*) in alkalinity stress, and genes involved in exopolysaccharide biosynthesis were upregulated in SRB biofilms during growth on a steel surface (Zhang, Culley, Nie, & Scholten, 2007).

EPSs contain functional groups, such as carboxyl, phosphoric, amine, and hydroxyl groups, which play a crucial role in the biosorption of heavy metals (Yue, Li, Li, Chen, & Wang, 2015). The corrosive activity of SRB is often associated with bacterial biofilms formed on the metal surfaces (Dinh et al., 2004; Dong et al., 2011). EPS is one of the relevant causes for both the corrosion and corrosion inhibition of metals. The EPS of *D. vulgaris* could accelerate the corrosion of metal surfaces, while those of *D. alaskensis* impeded corrosion due to the exclusion of oxygen from the biofilm (Liu et al., 2019; Stadler, Wei, Fürbeth, Grooters, & Kuklinski, 2010). The metabolic product (H_2S) of SRB biofilm has been known to induce metal corrosion. SRB reduce sulfate ion (SO_4^{-2}) and generate H_2S or HS^-, which interacts with Fe to generate iron sulfides (Zhang, Wen, & Cao, 2011). SRB corrosion had been initiated by direct electron exchange between metal surface and cells.

3. Signaling pathways

SRB use multiple signaling pathways to monitor and respond exactly to reconstruct surroundings. Quorum sensing (QS) and cyclic di-GMP (c-di-GMP) signaling are two chemicals signaling structures that regulate this shift. QS permits SRB to recognize the population density and species composition of the surrounding bacterial consortium through the secretion and detection of chemical signals called autoinducers. Moreover, studies have demonstrated that Gram-negative bacteria, in the high-cell-density QS state, biofilm formation expression were repressed. c-di-GMP is a nearly ubiquitous bacterial second messenger that induces biofilm formation. In contrast to QS, c-di-GMP activates the expression of genes necessary for biofilm formation. QS and c-di-GMP signaling are described below.

3.1 Quorum sensing (QS)

To form a community and an organized structure, bacteria communicate using broad range of biochemical agents. The ability of communication enables bacteria to start a social structure and develop biofilm. Many bacterial groups can form this specialized community with high complexity and plasticity that help them for better adaptability for diverse growth conditions. Biofilm formation is a kind of developmental cycle regulated by cell-cell communication events known as quorum sensing (QS). QS, the secretion and detection of a diffusible molecule or extracellular signaling molecules by the population, can be in charge of bacterial community to response to the stress, for instance, the production of EPS, cell surface hydrophobicity, and antibiotic resistance (Fuqua & Greenberg, 2002). Bacteria in biofilm structure respond as a population, which promotes cell-cell interactions, transmission of a certain QS signals, whereas the same mechanism is limited in the planktonic counterpart. QS system allows bacteria to regulate their gene expression according to their cell density. QS signal mainly consists of acyl-homoserine lactones (AHLs), autoinducing peptides (AIPs), and autoinducer-2 (AI-2) that have been implicated in the initiation and control of biofilm formation (Eickhoff & Bassler, 2018). When the bacterial secretion reaches the maximum limit of AHLs molecules in the environment, molecules will diffuse the cell membrane and bind to the signaling transcriptional regulators, thus promoting specific gene upregulation (Zeng, Wang, Yu, & Huang, 2017). In SRB QS system, Lux-I type AHL synthase circuit has been considered as the signals producer. With more studies, QS signals (*LuxS*, *LuxQ*, and *LuxO*) are gaining more and more attention due to their regulatory roles in bacterial growth and biofilm formation. High abundance of diverse AHLs (C6-AHL, oxo-C6-AHL, C8-AHL, C10-AHL, and C12-AHL) was also detected from SRB biofilms (Decho et al., 2009; Kawaguchi, Yung, Norman, & Decho, 2008; Montgomery, Charlesworth, LeBard, Visscher, & Burns, 2013).

AHLs with long chains of alkyl groups are more stable than short-chain AHLs, further long-chain alkyl AHLs were formed by SRB to accelerate sulfide oxidation (Montgomery et al., 2013). In SRB, LuxR and LuxS were found in QS-based gene homologous, further analysis of the genomes in the NCBI database also found that proteins homologous to LuxS AI-2 synthase and the AI-2 receptors (i.e., *LuxP* and *LuxQ*) are present in many *Desulfovibrio* species (Brady, Chao, & Clardy, 2004; Kimura, 2014; Williamson et al., 2005). *LuxO* system, which is responsible for phosphorylation cascade regulating the expression of the QS associated genes, was also observed in SRB. NCBI data QS systems of *Vibrio harveyi* showed similar signaling system as SRB, with potential homologous proteins related to *LuxS*, *LuxP*, *LuxQ*, and *LuxO* of *D. hydrothermalis* and *LuxS*, *LuxQ*, and *LuxO* of *Desulfovibrio salexigens* (Fig. 2). SRB orphan receptors might be able to sense and respond to signal produced by other bacteria or even environmentally modified autoinducers (Decho, Norman, & Visscher, 2010). The complexation of Fe^{2+} ions by polysaccharides and proteins configures the QS detection system of bacteria. In SRB, autoinducer-2 influences establishment and growth of the biofilm by regulating the synthesis and secretion of EPS, as similar reported of influence of AHL on *Pseudomonas aeruginosa* (Li, Wu, & Zhang, 2021).

Genetic factors engaged in the sulfate reduction pathway including sulfate adenylyl transferase (*sat*), adenylsulfate reductase (*apsAB*), and dissimilatory sulfite reductase (*dsrABC*) were altered by the inhibition or the stimulation of the QS system. It was observed that obstructing the QS system of *D. vulgaris* inhibits both biological and QS regulated physiological traits including plasmid transfer, surface polysaccharide production anaerobic respiration, cell activity, biofilm formation, and biocorrosion of carbon steel (Scarascia et al., 2020).

3.2 Cyclic-di-GMP

In the environment, many inter and intraspecies interactions are mediated by small-molecule signals. Signaling systems that use second messenger signaling are critical to existence in different environments and provide a competitive

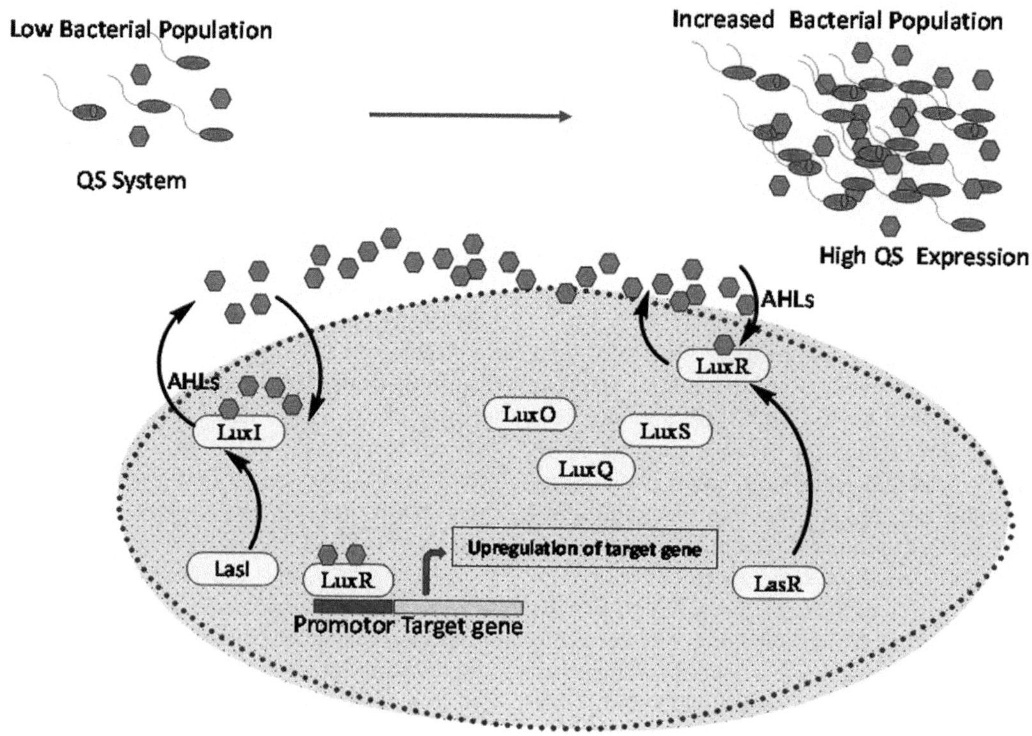

FIG. 2 Quorum sensing signaling in *Desulfovibrio vulgaris* Hildenborough (DVH). Acyl homoserine lactones (AHLs) are autoinducers used by DVH to communicate. The enzyme LuxI synthesizes the AHLs, and the latter can diffuse freely through the membrane. The autoinducers are recognized by the receptors LuxS, LuxPQ, LuxPO, and LuxR. AHL binds to its receptor and triggers the expression of genes involved in motility, stress, and biofilm formation.

advantage to biological systems under different environmental constraint. Bis-(3′-5′)-cyclic dimeric guanosine monophosphate (c-di-GMP) has been identified as a global second messenger in bacteria. Cyclic-di-GMP modulating response regulators are involved in the molecular decision ranging from attachment, motility, impacts biofilm formation, QS to numerous other developmental, pathogenic, and virulence responses (Rajeev et al., 2014). c-di-GMP-modulating response regulators are part of two-component signaling in the SRB (Heidelberg et al., 2004). SRB have a diverse signaling mechanism regulating biofilm formation for a variety of stress responses (De León et al., 2017). In SRB, two diguanylate cyclases (DGCs) as part of two-component signaling show a significant role in the biofilm pattern even though the role has downregulated in planktonic growth (Rajeev et al., 2014). There have been various studies supporting the biofilm formation in the SRB, but only few studies have been reported on signaling pathways. Further investigation is required to understand SRB signaling pathways on the microbial community structure and their interactions.

4. Flagellum

Flagellar motility is an important part of nutrient acquisition, biofilm formation, and prevails over the challenging environment (McCarter, 2006). The architecture of *P. aeruginosa* and *E. coli* mature biofilm also appears to be influenced by the presence or absence of flagella (Sharma et al., 2016). The relation between flagella biosynthesis and biofilm formation has also been reported in *D. alaskensis* G20 (Krumholz et al., 2015). *D. alaskensis* G20-deficient genes necessary for glycosyltransferase, the pilus assembly protein, and the flagellar biosynthesis proteins were unable to form biofilm and cell-to-cell interactions (Krumholz et al., 2015).Two-component signaling includes the regulation of cell motility (flagella and pili), exopolysaccharide production in most of Gram-negative bacteria (Rajeev et al., 2011).

5. Chemotaxis

Chemotaxis is an important activity performed by flagella. Chemosensory-like pathway regulates flagellar biosynthesis gene expression in SRB (Stolyar et al., 2007; Wilkins et al., 2014). *cheA3* is the terminal gene in an operon that determines various chemotaxis genes and genes with other putative functions that have a role in motility (Ray et al., 2014). Sigma^{70} transcription factor has been predicted to modulate genes involved in the formation of the flagellum as well as *cheA3*.

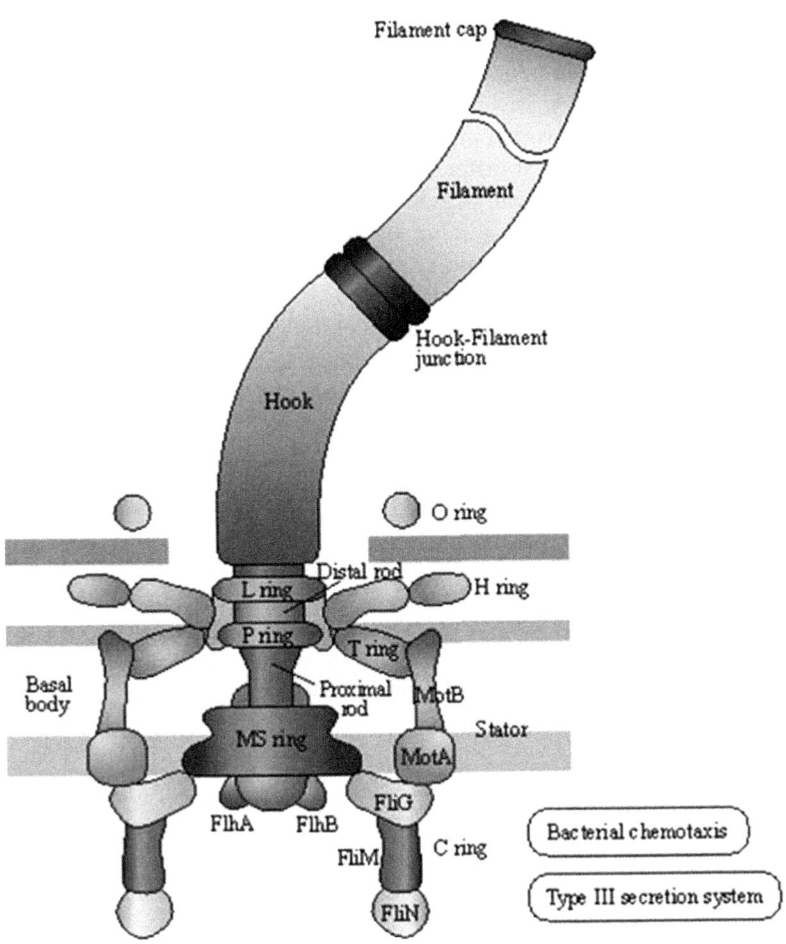

FIG. 3 Structure of flagellum of SRB residing within the cell wall and membranes. C ring, *FliM* and *FliN*, interact with the signaling molecule of the chemotaxis network, *CheY*, to control the direction of rotation.

gene (Novichkov et al., 2009; Ray et al., 2014). In SRB, *flhA*, *tadC*, and glycosyltransferase mutants showed fragile biofilm structure and decrease in EPS production when compared with wild-type biofilms (Krumholz et al., 2015) (Fig. 3). *D. alaskensis*, which were unable to produce flagellar biosynthetic protein *flhB*, were nonmotile and were unable to form biofilm (Williamson et al., 2018).

6. Functional categories of genes and proteins in SRB biofilm cells

6.1 Electron transfer

The participation of metals, either intra or extracellular, for the microbial metabolism is dependent on the metal capacity to act as electron donor or acceptor. In the SRB, expression of genes involved with energy conversion and carbon flow within the biofilm is most important contributor to the distinctiveness of the biofilm formation. When compared with planktonic cells, biofilm cells have different mRNA and protein expression associated to the pyruvate to acetate and formate metabolic nodes. Two annotated pyruvate: ferredoxin oxidoreductases gene (*oor* and *por*), which convert pyruvate to acetyl-CoA, reported altered expression in biofilm cells where gene *oor* was upregulated and gene *por* was downregulated compared with planktonic counterpart. The cytochrome c553, which is an electron transfer partner for the formate dehydrogenase, was upregulated in an SRB *D. vulgaris* (Sebban-Kreuzer et al., 1998). Therefore, it suggests that formate dehydrogenase highly expressed in biofilm cells might imply higher efficiency of formate cycling compared with planktonic cells (Clark et al., 2012). In SRB interconnection between biofilm formation and extracellular electron transfer could play an important factor in accelerating the MIC.

6.2 Stress response

Biofilm formation involves stress response regulators, and different stress conditions trigger changes in the expression of genes that help microbes to survive. The altered expression of the stress response regulators depends upon the environments under which biofilm formation has triggered and the genetic background of the microbe (Jefferson, 2004). Microbes have established a range of resistance mechanisms to counteract different stress responses such as media, nutrient, pH, oxygen radicals, osmotic stress, temperature, heavy metal, and metal surfaces. Main adaptive strategies are proton pumping by ATP synthase, increase of metabolic acid production, modification of cell surface, and upregulation of antiporters (Padan, Bibi, Ito, & Krulwich, 2005). Several regulators for stress response system were observed in SRB. In challenging environments, various chaperone proteins, including *uspA* and *hsp20* genes, had increased transcripts within the biofilm compared with planktonic cells (Stolyar et al., 2007).

D. vulgaris biofilms showed increased transcripts for the stress response proteins such as superoxide dismutase and alkyl hydroperoxide reductase. An increase in expression of three more genes alkyl hydroperoxide reductase *ahpC* (DVU2247), *msrB* (DVU0576), and *ahpC*/TSA (thiol-specific antioxidant) was also reported in *D. vulgaris* Hildenborough (Clark et al., 2012). Oxidative stress is another major phenomenon in SRB biofilms, which induces a defense mechanism and results in expression of genes to regulate and counteract the stress. In *D. vulgaris* Hildenborough, it is reported that two genes rubrerythrin (*rbr*) and thioredoxin reductase (*trxB*) were upregulated in response to oxidative stress, suggesting a key role in the oxidative damage control (Zhou et al., 2010). The upregulation of *trxB* suggests that thiol-specific redox system might be involved in the oxidative stress response in SRB. Other proteins (the [NiFeSe] hydrogenase, formate dehydrogenase(s), and the *Hmc* membrane complex) were reported for increased expression, which is associated with periplasmic hydrogenase-dependent mechanism during oxidative stress in DVH.

6.3 Iron acquisition

Iron uptake by SRB is necessary for maintaining an optimal concentration of cellular biomass and enzymatic activity. Iron plays significant roles within biofilm formation. Various organisms use iron as a signal to produce dense biofilms (Hindré, Brüggemann, Buchrieser, & Héchard, 2008; Zhang et al., 2007). Therefore, to maintain appropriate iron concentration in different layer levels of the *D. vulgaris* biofilm, higher gene expression heterogeneity of *Fur* gene (involved in ferrous iron transport) was reported (Qi, Chen, & Zhang, 2016). Iron storage protein bacterioferritin (DVU1397) and ferritin storage protein (DVU1568) are important components of iron management in SRB biofilm cells (Zhang et al., 2007).

7. Conclusion

SRB are unique and important group of anaerobic bacteria found at diverse places and form biofilms. Biofilm formation is a complex process, and several mechanisms are involved at different levels to sustain the bacterial growth and survival. These include EPS formation, flagellar motility, QS system, signaling molecule expression, stress response regulators, and iron uptake. The whole genome analysis of SRB—*D. alaskensis* G20 revealed that still large number of genes are unknown or had hypothetical functions. There is a huge scope to perform gene mutation and knock-out studies to characterize these genes and identify the pathways and mechanisms involved in SRB biofilm formation.

Acknowledgment

We gratefully acknowledge support from the National Science Foundation (Awards #1849206, #1920954, and #1736255) and the Department of Chemical and Biological Engineering at the South Dakota School of Mines and Technology.

References

Brady, S. F., Chao, C. J., & Clardy, J. (2004). Long-chain N-acyltyrosine synthases from environmental DNA. *Applied and Environmental Microbiology*, *70*(11), 6865–6870. https://doi.org/10.1128/AEM.70.11.6865-6870.2004.

Clark, M. E., He, Z., Redding, A. M., Joachimiak, M. P., Keasling, J. D., Zhou, J. Z., et al. (2012). Transcriptomic and proteomic analyses of *Desulfovibrio vulgaris* biofilms: Carbon and energy flow contribute to the distinct biofilm growth state. *BMC Genomics*, *13*(1). https://doi.org/10.1186/1471-2164-13-138.

Dantam, J., Subbaraman, L. N., & Jones, L. (2020). Adhesion of *Pseudomonas aeruginosa*, *Achromobacter xylosoxidans*, *Delftia acidovorans*, *Stenotrophomonas maltophilia* to contact lenses under the influence of an artificial tear solution. *Biofouling*, *36*(1), 32–43. https://doi.org/10.1080/08927014.2019.1710832.

De León, K. B., Zane, G. M., Trotter, V. V., Krantz, G. P., Arkin, A. P., Butland, G. P., et al. (2017). Unintended laboratory-driven evolution reveals genetic requirements for biofilm formation by *Desulfovibrio vulgaris* Hildenborough. *MBio*, *8*(5), 1–16. https://doi.org/10.1128/mBio.01696-17.

Decho, A. W., Norman, R. S., & Visscher, P. T. (2010). Quorum sensing in natural environments: Emerging views from microbial mats. *Trends in Microbiology*, *18*(2), 73–80. https://doi.org/10.1016/j.tim.2009.12.008.

Decho, A. W., Visscher, P. T., Ferry, J., Kawaguchi, T., He, L., Przekop, K. M., et al. (2009). Autoinducers extracted from microbial mats reveal a surprising diversity of N-acylhomoserine lactones (AHLs) and abundance changes that may relate to diel pH. *Environmental Microbiology*, *11*(2), 409–420. https://doi.org/10.1111/j.1462-2920.2008.01780.x.

Dinh, H. T., Kuever, J., Mußmann, M., Hassel, A. W., Stratmann, M., & Widdel, F. (2004). Iron corrosion by novel anaerobic microorganisms. *Nature*, *427*(6977), 829–832. https://doi.org/10.1038/nature02321.

Dong, Z. H., Liu, T., & Liu, H. F. (2011). Influence of EPS isolated from thermophilic sulphate-reducing bacteria on carbon steel corrosion. *Biofouling*, *27*(5), 487–495. https://doi.org/10.1080/08927014.2011.584369.

Eickhoff, M. J., & Bassler, B. L. (2018). SnapShot: Bacterial quorum sensing. *Cell*, *174*(5), 1328–1328.e1. https://doi.org/10.1016/j.cell.2018.08.003.

Fuqua, C., & Greenberg, E. P. (2002). Listening in on bacteria: Acyl-homoserine lactone signalling. *Nature Reviews Molecular Cell Biology*, *3*(9), 685–695. https://doi.org/10.1038/nrm907.

Gómez-Suárez, C., Pasma, J., van der Borden, A. J., Wingender, J., Flemming, H. C., Busscher, H. J., et al. (2002). Influence of extracellular polymeric substances on deposition and redeposition of *Pseudomonas aeruginosa* to surfaces. *Microbiology*, *148*(4), 1161–1169. https://doi.org/10.1099/00221287-148-4-1161.

Heidelberg, J. F., Seshadri, R., Haveman, S. A., Hemme, C. L., Paulsen, I. T., Kolonay, J. F., et al. (2004). The genome sequence of the anaerobic, sulfate-reducing bacterium *Desulfovibrio vulgaris* Hildenborough. *Nature Biotechnology*, *22*(5), 554–559. https://doi.org/10.1038/nbt959.

Hindré, T., Brüggemann, H., Buchrieser, C., & Héchard, Y. (2008). Transcriptional profiling of *Legionella pneumophila* biofilm cells and the influence of iron on biofilm formation. *Microbiology*, *154*(1), 30–41. https://doi.org/10.1099/mic.0.2007/008698-0.

Jefferson, K. K. (2004). What drives bacteria to produce a biofilm? *FEMS Microbiology Letters*, *236*(2), 163–173. https://doi.org/10.1016/j.femsle.2004.06.005.

Kawaguchi, T., Yung, P. C., Norman, R. S., & Decho, A. W. (2008). Rapid screening of quorum-sensing signal N-acyl homoserine lactones by an in vitro cell-free assay. *Applied and Environmental Microbiology*, *74*(12), 3667–3671. https://doi.org/10.1128/AEM.02869-07.

Kimura, N. (2014). Metagenomic approaches to understanding phylogenetic diversity in quorum sensing. *Virulence*, *5*(3), 433–442. https://doi.org/10.4161/viru.27850.

Krumholz, L. R., Bradstock, P., Sheik, C. S., Diao, Y., Gazioglu, O., Gorby, Y., et al. (2015). Syntrophic growth of *Desulfovibrio alaskensis* requires genes for H2 and formate metabolism as well as those for flagellum and biofilm formation. *Applied and Environmental Microbiology*, *81*(7), 2339–2348. https://doi.org/10.1128/AEM.03358-14.

Kushkevych, I., Vítězová, M., Vítěz, T., Kováč, J., Kaucká, P., Jesionek, W., et al. (2018). A new combination of substrates: Biogas production and diversity of the methanogenic microorganisms. *Open Life Sciences*, *13*(1), 119–128. https://doi.org/10.1515/biol-2018-0017.

Li, Y., Feng, S., Liu, H., Tian, X., Xia, Y., Li, M., et al. (2020). Bacterial distribution in SRB biofilm affects MIC pitting of carbon steel studied using FIB-SEM. *Corrosion Science*, *167*(May 2019). https://doi.org/10.1016/j.corsci.2020.108512.

Li, E., Wu, J., & Zhang, D. (2021). Exogenous autoinducer-2 inhibits biofilm development of *Desulfovibrio* sp. Huiquan2017. *World Journal of Microbiology and Biotechnology*, *37*(7), 1–8. https://doi.org/10.1007/s11274-021-03071-w.

Liu, H., Meng, G., Li, W., Gu, T., & Liu, H. (2019). Microbiologically influenced corrosion of carbon steel beneath a deposit in CO2-saturated formation water containing *Desulfotomaculum nigrificans*. *Frontiers in Microbiology*, *10*(Jun), 1–13. https://doi.org/10.3389/fmicb.2019.01298.

Ma, Q., & Wood, T. K. (2009). OmpA influences Escherichia coli biofilm formation by repressing cellulose production through the CpxRA two-component system. *Environmental Microbiology*, *11*(10), 2735–2746. https://doi.org/10.1111/j.1462-2920.2009.02000.x.

McCarter, L. L. (2006). Regulation of flagella. *Current Opinion in Microbiology*, *9*(2), 180–186. https://doi.org/10.1016/j.mib.2006.02.001.

Montgomery, K., Charlesworth, J. C., LeBard, R., Visscher, P. T., & Burns, B. P. (2013). Quorum sensing in extreme environments. *Life*, *3*(1), 131–148. https://doi.org/10.3390/life3010131.

Novichkov, P. S., Laikova, O. N., Novichkova, E. S., Gelfand, M. S., Arkin, A. P., Dubchak, I., et al. (2009). RegPrecise: A database of curated genomic inferences of transcriptional regulatory interactions in prokaryotes. *Nucleic Acids Research*, *38*(Suppl. 1), 111–118. https://doi.org/10.1093/nar/gkp894.

Padan, E., Bibi, E., Ito, M., & Krulwich, T. A. (2005). Alkaline pH homeostasis in bacteria: New insights. *Biochimica et Biophysica Acta - Biomembranes*, *1717*(2), 67–88. https://doi.org/10.1016/j.bbamem.2005.09.010.

Qi, Z., Chen, L., & Zhang, W. (2016). Comparison of transcriptional heterogeneity of eight genes between batch *Desulfovibrio vulgaris* biofilm and planktonic culture at a single-cell level. *Frontiers in Microbiology*, *7*(Apr), 1–14. https://doi.org/10.3389/fmicb.2016.00597.

Rajeev, L., Luning, E. G., Altenburg, S., Zane, G. M., Baidoo, E. E. K., Catena, M., et al. (2014). Identification of a cyclic-di-GMP-modulating response regulator that impacts biofilm formation in a model sulfate reducing bacterium. *Frontiers in Microbiology*, *5*(July), 1–13. https://doi.org/10.3389/fmicb.2014.00382.

Rajeev, L., Luning, E. G., Dehal, P. S., Price, M. N., Arkin, A. P., & Mukhopadhyay, A. (2011). Systematic mapping of two component response regulators to gene targets in a model sulfate reducing bacterium. *Genome Biology*, *12*(10). https://doi.org/10.1186/gb-2011-12-10-r99.

Ray, J., Keller, K. L., Catena, M., Juba, T. R., Zemla, M., Rajeev, L., et al. (2014). Exploring the role of CheA3 in *Desulfovibrio vulgaris* Hildenborough motility. *Frontiers in Microbiology*, *5*(Mar), 1–9. https://doi.org/10.3389/fmicb.2014.00077.

Scarascia, G., Lehmann, R., Machuca, L. L., Morris, C., Cheng, K. Y., Kaksonen, A., et al. (2020). Effect of quorum sensing on the ability of *Desulfovibrio vulgaris* to form biofilms and to biocorrode carbon steel in saline conditions. *Applied and Environmental Microbiology*, *86*(1). https://doi.org/10.1128/AEM.01664-19.

Scarascia, G., Wang, T., & Hong, P. Y. (2016). Quorum sensing and the use of quorum quenchers as natural biocides to inhibit sulfate-reducing bacteria. *Antibiotics*, *5*(4). https://doi.org/10.3390/antibiotics5040039.

Sebban-Kreuzer, C., Blackledge, M., Dolla, A., Marion, D., & Guerlesquin, F. (1998). Tyrosine 64 of cytochrome c553 is required for electron exchange with formate dehydrogenase in *Desulfovibrio vulgaris* Hildenborough. *Biochemistry*, *37*(23), 8331–8340. https://doi.org/10.1021/bi980142u.

Sharma, G., Sharma, S., Sharma, P., Chandola, D., Dang, S., Gupta, S., et al. (2016). *Escherichia coli* biofilm: Development and therapeutic strategies. *Journal of Applied Microbiology*, *121*(2), 309–319. https://doi.org/10.1111/jam.13078.

Stadler, R., Wei, L., Fürbeth, W., Grooters, M., & Kuklinski, A. (2010). Influence of bacterial exopolymers on cell adhesion of *Desulfovibrio vulgaris* on high alloyed steel: Corrosion inhibition by extracellular polymeric substances (EPS). *Materials and Corrosion*, *61*(12), 1008–1016. https://doi.org/10.1002/maco.201005819.

Stolyar, S., He, Q., Joachimiak, M. P., He, Z., Zamin, K. Y., Borglin, S. E., et al. (2007). Response of *Desulfovibrio vulgaris* to alkaline stress. *Journal of Bacteriology*, *189*(24), 8944–8952. https://doi.org/10.1128/JB.00284-07.

Tripathi, A. K., Thakur, P., Saxena, P., Rauniyar, S., Gopalakrishnan, V., Singh, R. N., et al. (2021). Gene sets and mechanisms of sulfate-reducing bacteria biofilm formation and quorum sensing with impact on corrosion. *Frontiers in Microbiology*, *12*(October), 1–18. https://doi.org/10.3389/fmicb.2021.754140.

Wettstadt, S. (2020). Breaking free from home: Biofilm dispersal by a glycosidase from *Desulfovibrio vulgaris*. *Environmental Microbiology*, *22*(2), 557–558. https://doi.org/10.1111/1462-2920.14883.

Wilkins, M., Hall-Stoodley, L., Allan, R. N., & Faust, S. N. (2014). New approaches to the treatment of biofilm-related infections. *Journal of Infection*, *69*(S1), S47–S52. https://doi.org/10.1016/j.jinf.2014.07.014.

Williamson, L. L., Borlee, B. R., Schloss, P. D., Guan, C., Allen, H. K., & Handelsman, J. (2005). Intracellular screen to identify metagenomic clones that induce or inhibit a quorum-sensing biosensor. *Applied and Environmental Microbiology*, *71*(10), 6335–6344. https://doi.org/10.1128/AEM.71.10.6335-6344.2005.

Williamson, A. J., Carlson, H. K., Kuehl, J. V., Huang, L. L., Iavarone, A. T., Deutschbauer, A., et al. (2018). Dissimilatory sulfate reduction under high pressure by *Desulfovibrio alaskensis* G20. *Frontiers in Microbiology*, *9*(Jul), 1–10. https://doi.org/10.3389/fmicb.2018.01465.

Xu, D., Li, Y., & Gu, T. (2016). Mechanistic modeling of biocorrosion caused by biofilms of sulfate reducing bacteria and acid producing bacteria. *Bioelectrochemistry*, *110*, 52–58. https://doi.org/10.1016/j.bioelechem.2016.03.003.

Yue, Z. B., Li, Q., Li, C.c., Chen, T.h., & Wang, J. (2015). Component analysis and heavy metal adsorption ability of extracellular polymeric substances (EPS) from sulfate reducing bacteria. *Bioresource Technology*, *194*, 399–402. https://doi.org/10.1016/j.biortech.2015.07.042.

Zeng, Y., Wang, Y., Yu, Z., & Huang, Y. (2017). Hypersensitive response of plasmid-encoded AHL synthase gene to lifestyle and nutrient by *Ensifer adhaerens* X097. *Frontiers in Microbiology*, *8*(Jun), 1–11. https://doi.org/10.3389/fmicb.2017.01160.

Zhang, W., Culley, D. E., Nie, L., & Scholten, J. C. M. (2007). Comparative transcriptome analysis of *Desulfovibrio vulgaris* grown in planktonic culture and mature biofilm on a steel surface. *Applied Microbiology and Biotechnology*, *76*(2), 447–457. https://doi.org/10.1007/s00253-007-1014-9.

Zhang, C., Wen, F., & Cao, Y. (2011). Progress in research of corrosion and protection by sulfate-reducing bacteria. *Procedia Environmental Sciences*, *10*(Part B), 1177–1182. https://doi.org/10.1016/j.proenv.2011.09.188.

Zhou, A., He, Z., Redding-Johanson, A. M., Mukhopadhyay, A., Hemme, C. L., Joachimiak, M. P., et al. (2010). Hydrogen peroxide-induced oxidative stress responses in *Desulfovibrio vulgaris* Hildenborough. *Environmental Microbiology*, *12*(10), 2645–2657. https://doi.org/10.1111/j.1462-2920.2010.02234.x.

Zhu, L., Poosarla, V. G., Song, S., Wood, T. L., Miller, D. S., Yin, B., et al. (2018). Glycoside hydrolase DisH from *Desulfovibrio vulgaris* degrades the N-acetylgalactosamine component of diverse biofilms. *Environmental Microbiology*, *20*(6), 2026–2037. https://doi.org/10.1111/1462-2920.14064.

Ziadi, I., El-Bassi, L., Bousselmi, L., & Akrout, H. (2020). Characterization of the biofilm grown on 304L stainless steel in urban wastewaters: Extracellular polymeric substances (EPS) and bacterial consortia. *Biofouling*, *36*(8), 977–989. https://doi.org/10.1080/08927014.2020.1836163.

Chapter 12

Biofilms and their role in corrosion in marine environments

Damon C. Brown, Gloria N. Okpala, Lisa M. Gieg, and Raymond J. Turner
Department of Biological Sciences, University of Calgary, Calgary, AB, Canada

1. Introduction

We now recognize the concept of the microbiome, the community of organisms that are associated with humans living practically symbiotically (Cho & Blaser, 2012). But the concept of a microbiome should also be considered for a given ecological niche. Microbiomes are not only associated with certain biological organs and areas (oral), but also niches such as soils, aquatic and marine systems, drinking water distribution systems, plant surfaces, rhizospheres, and even smaller organisms such as fungi can have unique bacteria associated with them (Berg, Rybakova, Fischer, et al., 2020). One often overlooks abiotic and various anthropogenic surfaces that can collect associated microbial communities. In the case of marine environments, these include ship halls, moorings, underwater cables, mining equipment (offshore oil exploration), pipelines, and floating/suspended pollutants such as plastics.

Depending on the local sea temperature and salinity, marine microorganisms found as single, free-living organisms referred to as the planktonic fraction can be quite small. However, regardless of the marine conditions, microorganisms are abundantly found assembled in groups or communities that are most often attached to surfaces; this form is referred to as a biofilm. As a surface-attached version of microbial life, we see biofilms featured in every ecosystem as their sessile microbiome as they influence the environment, geochemistry, industrial processes, and animal and human well-being. These biofilms can become quite thick (hundreds of μm) and complex in structure and species composition that can become visible on various surfaces (e.g., as a green scum, slime, or film). Biofilms can comprise a complex mixture of algae, cyanobacteria, heterotrophic microorganisms, and detritus (dead organic material) that is attached to submerged surfaces in most aquatic and marine ecosystems and is referred to as a periphyton. As the periphyton matures, one will observe small eukaryotic organisms (such amoeba and paramecium) grazing and feeding on the microbial mat, which attracts higher-level organisms including protozoans, snails, fish, reptiles, birds, and mammals. Thus, the periphyton becomes the archetype of the bottom of the food chain.

It is also worthwhile to consider how microorganisms contribute to our planet's biogeochemistry. They participate in the global carbon, nitrogen, sulfur, and phosphorus cycles. In fact, bacteria are crucial to water distribution and climate as their metabolism produces volatile compounds (e.g., dimethyl sulfate, sulfite, dimethyl ammonium) that can form aerosols that serve as nucleation sites for water molecules in the atmosphere to seed cloud formation (Malfatti, Lee, Tinta, et al., 2019). Recognizing this widespread and integral biogeochemistry allows one to appreciate how microbial metabolism leads to corrosive properties on anthropogenically produced marine infrastructure. Microorganisms coevolved with the planet's geochemistry and have become well suited to the wide variety of marine micro and macro niche environments. Many microorganisms are able to survive in oxic as well as anoxic environments, at extremes of pH (e.g., pH 1.5–10), temperature (as low as -14 and $>100°C$), pressure, ionic strength, redox potential, and utilize extremely varied carbon and energy sources.

To control corrosion and biofouling from microbial biofilms or extensive periphytons, industry applies a variety of antimicrobial coatings and biocides. Yet such treatments are only transiently effective due to the character of biofilms to show high tolerance against antimicrobial agents (Jolivet-Gougeon & Bonnaure-Mallet, 2014; Yan & Bassler, 2019). The tolerance to antimicrobials in part comes from biofilm physiology, which is remarkably distinct from the planktonic form. Upon forming a biofilm, approximately 30% of the transcriptome is downregulated and a different 30% is upregulated to give about a 60% difference between planktonic and biofilm lifestyles. Additionally, upon surface attachment, genes are expressed that initially leads to exopolysaccharide (EPS) production and then to a complex extracellular matrix—a mixture of EPS, proteins, eDNA, eRNA, lipids, and various signaling molecules

Understanding Microbial Biofilms. https://doi.org/10.1016/B978-0-323-99977-9.00036-3

and metabolites—resulting in a mature biofilm (Antunes et al., 2020; Flemming & Wingender, 2010). This matrix provides diffusional barrier protection of the biofilm cells against antimicrobials. Other processes that lead to enhanced antimicrobial tolerance are traits of enhanced horizontal gene transfer, enhanced species variants, and increased persister cell numbers. Nearly all environments contain multiple coexisting species of prokaryotes and microbial eukaryotes. Emerging evidence suggests that biofilm stress resistance may be dictated by the numbers and types of microbial species in the community. Signaling molecules (quorum sensing) and other metabolites are used in interspecies communication allowing the biofilm community to mount a collective defense against a biocide challenge. For more detail of such processes, see the review by Hall and Mah (2017).

2. Problems with marine biofilms

In marine environments, the impact of biofilms is most often observed as biofouling. Biofouling is the unwelcome, visible accumulation of the periphyton on surfaces such as marine vessel hulls, pipework, etc., that is not only unsightly, but can decrease performance. However, most papers/reviews do not consider the role of microbial activity but consider biofouling to be the result of the mature periphyton of marine algae and animals (Fitridge, Dempster, Guenther, & de Nys, 2012). This is an easy oversight as such biofouling issues are not considered until the periphyton is robustly visible by the human eye. Unfortunately, this overlooks the initial stage where bacteria can rapidly attach on a pristine metal surface. Typically, within hours of immersion, a conditioning film of complex biomolecules from sea life begins to cover the surface as a molecular monolayer, which acts as an anchor for bacterial cell attachment (Dang & Lovell, 2016; Fischer, Friedrichs, & Lachnit, 2014). Microfouling is the initial step of biofouling, causing bacteria to start communities of scattered microcolony biofilm structures. As the microbial biofilm proliferates, the biofilm recruits algae and fungi and secretes excessive amounts of EPS such that the biofilm forms a thick slime that can reach thicknesses of 500 μm (Callow & Callow, 2002). It is at this stage that the higher-order life forms are attracted considering the microbial base layer for nutrients. Macrofouling follows suit, where larger invertebrates and corals adhere, which can grow along with or overtake the microfouling community. This macrofouling effect is what will subsequently cause performance issues such as reducing hull speed of marine vessels.

Beyond biofouling issues of marine biofilms, it is now recognized that the microbial community of a biofilm can facilitate corrosion processes. Below follows a description of microbiologically influenced corrosion (MIC) processes and the species of bacteria and archaea associated with this process. The topic of biofilms and marine corrosion has been explored in recent literature, and the reader is directed to several timely reviews (Caruso, 2020; Dang & Lovell, 2016; Ma et al., 2020; Procopio, 2019).

3. Mechanisms of microbiologically influenced corrosion (MIC)

MIC can be broadly defined as the interaction between microorganisms and a metal substrate resulting in the solubilization of the metal substrate. This may be a direct result of microbial metabolism where the metal substrate is used as the electron donor or acceptor or indirectly through chemical reactions with metabolic products such as sulfide or hydrogen gas. These two different types of MIC have been named electrical MIC (eMIC) and chemical (cMIC) (Enning et al., 2012). The corrosion processes are summarized in Fig. 1.

3.1 Chemical microbiologically influenced corrosion

The process of cMIC is an indirect corrosion mechanism resulting from otherwise normal microbial metabolism. This process may be due to the action of several different types of microorganisms (Sharma & Voordouw, 2016); however, sulfate reducers are the most notorious types of organisms involved in MIC. Sulfate reducers can comprise both SRB and sulfate-reducing archaea (SRA) and are thus also referred to as sulfate-reducing prokaryotes (SRPs).

SRPs are the best studied organisms capable of cMIC due to the aggressive corrosive nature of the sulfur-containing metabolites produced. These types of anaerobic microorganisms use sulfate as their electron acceptor, producing sulfide as the respiratory end product. In some cases, sulfur and/or polysulfides may also form depending on the types of SRP present. Elemental sulfur is rarely produced; however, as it is an intermediary reduced state meaning it is typically further reduced to sulfide. As SRPs are anaerobic microorganisms, they will most often be found deep into the structure or at the base of biofilms, often in direct contact with the metal substrate to allow aerobic organisms to consume oxygen creating an anaerobic micro niche for SRP to survive. Additionally, in marine environments where sulfate is abundant (e.g., 25–30 mM), growth of SRP is only limited by other nutrients, not by a carbon source or electron acceptor (Canfield & Farquhar, 2009). Sulfide molecules interact with metal to remove the electrons from the cathode (Ewing, 1955). Ferrous ions that dissolve in the water phase can then react with the produced sulfide molecules and CO_2 to produce iron sulfides and iron carbonates that settle on the surface of the metal. These

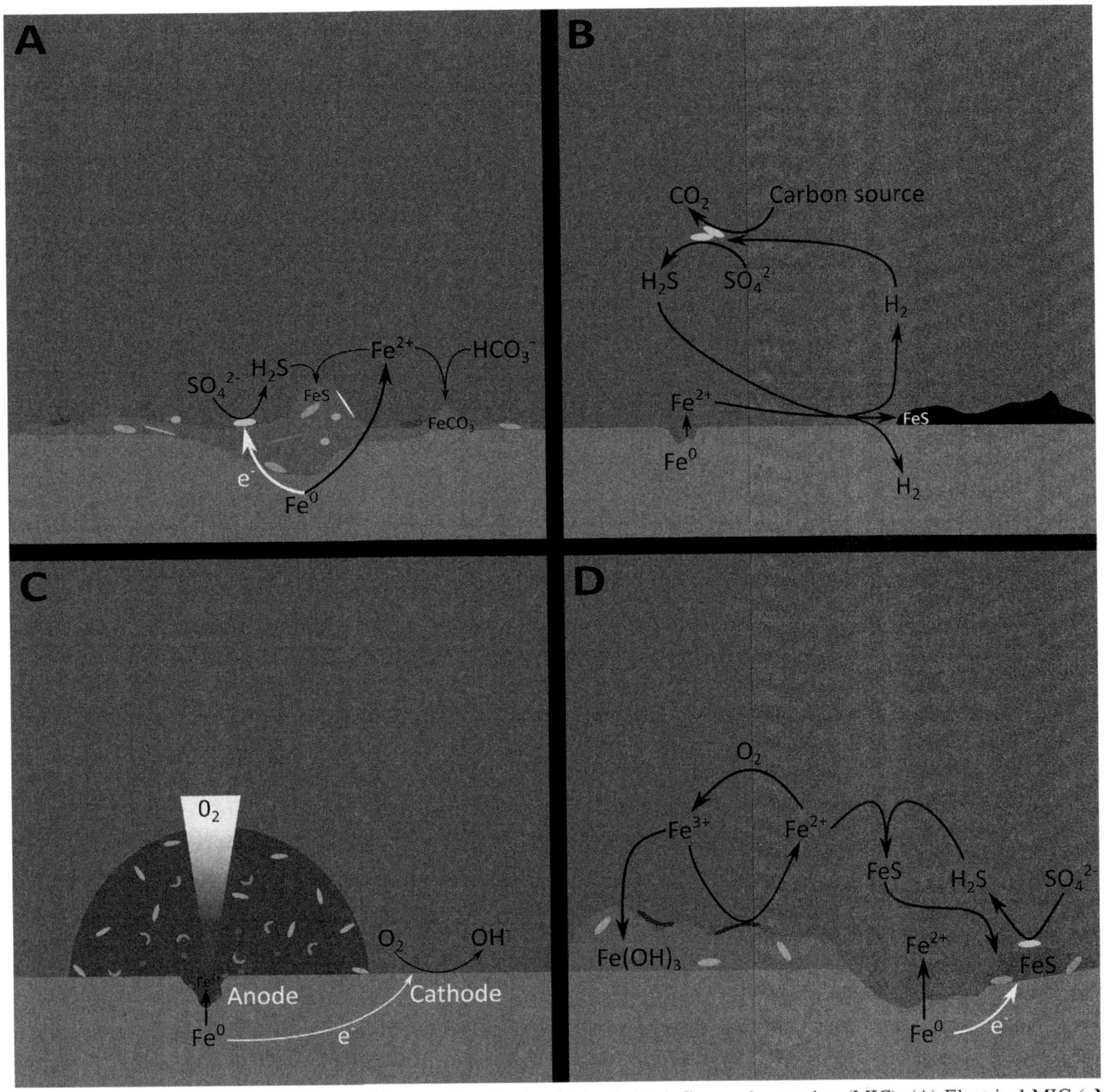

FIG. 1 Simplified and unbalanced representation of four modes of microbiologically influenced corrosion (MIC). (A) Electrical MIC (eMIC), where sulfate-reducing bacteria (SRB) directly remove the electrons from the iron to use in the reduction of sulfate, releasing ferrous ions, which can precipitate with sulfide or bicarbonate, creating a conductive-scale sediment. (B) Chemical MIC (cMIC), where sulfide produced by SRB chemically reacts with ferrous ions to precipitate as ferrous sulfide. (C) Concentration gradient, where mixed species within a biofilm coexist and the aerobic microbes rapidly reduce the oxygen present creating an anoxic micro niche at the base of the biofilm. This creates an anodic location with respect to exposed iron surrounding the biofilm where electrons from the anodic region are used to reduce oxygen, completing the electrochemical circuit allowing release of ferrous ions beneath the biofilms. (D) Accelerated corrosion where iron-reducing bacteria convert ferric ions to ferrous ions, which are free to react with sulfide produced by SRB resulting in a brittle FeS crust, which can easily be removed, further exposing more iron surface for continued corrosion.

products are electrically conductive and thereby increase the size of the cathode leading to increased corrosion rates (Enning et al., 2012). Acid-producing bacteria (e.g., fermentative microorganisms) have also been associated with corrosion due to their acidic metabolic end products. These bacterial waste products are weak organic acids (such as acetic acid) and become concentrated within the biofilm that can produce localized pH values that are magnitudes different than the surrounding environment. In dental biofilms (plaque), pH within the biofilm can be dropped to 4.5–5.5 compared with the bulk pH of 7 because of *Streptococcus mutans* producing organic acids through fermentation (Hwang et al., 2016; Xiao, Klein, Falsetta, et al., 2012). The weak acids become sources of protons, which remove electrons from the cathodic region of the metal resulting in corrosion.

The production of H_2 gas can also be damaging to metal infrastructure. The H_2 gas can penetrate the metal surface and accumulate in pockets creating bubbles within the metal and causing metal embrittlement (Thomas, Edyvean, & Brook, 1988). However, at room temperature, this is not considered a risk due to very high dissociation constant of

H_2 (429.5 kJ/mol) and very low diffusion coefficient ($1.5 \times 10^8 cm^2/s$) (Woodtli & Kieselbach, 2000).

3.2 Electrical microbiologically influenced corrosion

Microorganisms can also affect corrosion directly through a process known as electrical microbiologically influenced corrosion (eMIC). To accomplish this, microorganisms interact directly with the metal substrate, stripping electrons away or depositing electrons as part of their metabolism. This process is known as extracellular electron transfer (EET), which presumably conducts electrons from insoluble iron along c-type cytochromes to the inner membrane where it is used to reduce the oxidized electron acceptors. The arrangement of the c-type cytochromes takes three forms: direct contact between the cell and the metal surface via biological nanowires, which are extended lipid and/or protein structures extending several times the length of the cell, or through soluble, external electron shuttles (Li, Xu, Chen, et al., 2018). SRP and nitrate-reducing bacteria (NRB) species are capable of EET, removing electrons directly from a metal substrate for use in their metabolism (Beese-Vasbender, Nayak, Erbe, Stratmann, & Mayrhofer, 2015; Jia, Yang, Xu, Xu, & Gu, 2017; Venzlaff et al., 2013; Xu, Li, Song, & Gu, 2013). In most cases, the preferred metal is iron in the form of different steel types. It should be noted that EET is promoted under carbon starvation and turned off when carbon sources are present (Sherar et al., 2011).

There are many questions remaining regarding the EET process at the molecular level. It has not been conclusively demonstrated whether the c-type cytochromes directly contact the metal or if an electron shuttle such as flavin mononucleotide is required for the interaction between the metal and cytochromes (Edwards, White, Norman, et al., 2015). The addition of riboflavin and flavin mononucleotide has been observed to increase corrosion rates by SRP and NRB (Jia et al., 2017; Li et al., 2015; Zhang, Xu, Li, Yang, & Gu, 2015).

Although SRPs receive most of the focus of MIC-related studies, research is showing other species previously considered innocuous are capable of contributing to MIC (e.g., the NRB *Pseudomonas aeruginosa*, typically thought of as simply a biofilm pioneer and not a threat for MIC directly). An additional potential mechanism for these innocuous organisms to be indirectly corrosive may simply be via the formation of concentration cells on the surface of metals. In concentration cells, key chemicals such as oxygen are excluded from the surface of the metal beneath the biofilm due to microbial consumption. This creates a difference in electrical potential on the surface of the metal surrounding the biofilm with respect to under the biofilm. The surrounding metal then becomes cathodic, completing the electrochemical corrosion cell (Li et al., 2018). While oxygen gradient is the simplest example resulting from a concentration of metabolizing cells, the accumulation of other compounds such as metal ions can become trapped in a biofilm and concentration on the surface of the metal. This type of accumulation may not be directly linked to the microbial metabolism at all and just a passive accumulation in a biofilm, regardless of whether the cells in the biofilm are active or not. Some methanogens are also capable of eMIC under marine conditions (Deutzmann, Sahin, & Spormann, 2015; Lahme, Mand, Longwell, Smith, & Enning, 2021; Tsurumaru, Ito, Mori, et al., 2018).

4. Taxa involved in marine MIC

Surface-attached microorganisms found on marine structures contribute to the process of degradation of these structures as well as in biogeochemical cycles. There are generally two groups of microbes that should be considered when looking at the role of biofilms in marine corrosion. The first group constitutes those that are directly implicated in the corrosion mechanism observed (Emerson, 2018; Garrison & Field, 2020), while the second group constitutes those that facilitate biofilm formation and structural stability in MIC active sites (Garrison & Field, 2020). As mentioned, marine biofilms are complex assemblages of microorganisms that are attached to inorganic or organic substrates and are generally composed of diverse numbers and species of bacteria/archaea, fungi, algae, protozoans, diatoms, and detritus (de Carvalho, 2018). Of these groups of microbes, heterotrophic bacteria are crucial for initializing biofilm formation as they are typically the first to adhere to submerged surfaces. Once they adhere, they alter the properties of the surfaces onto which they are attached and begin to produce EPS and other matrix components starting with biofilm microcolonies. The EPS controls the adhesion of other microorganisms to the surface substratum (Caruso, 2020). This matrix facilitates the adhesion of other species of microbes to the surface and to join the initiated biofilm. Several factors affect the colonization and adhesion of microbes to surfaces of marine structures. Surface colonization by marine microorganisms is controlled by environmental conditions such as temperature, salinity, water depth, nutrient availability, and organic matter cycling (Antunes, Leão, & Vasconcelos, 2019; Dang & Lovell, 2016). Additionally, surface colonization and species diversity of a biofilm are dependent on the dispersal and aggregation of organisms within given microenvironments.

Proteobacteria, the most abundant group of heterotrophic bacteria in marine environments, are among the major drivers of corrosion in the marine environment. Due to an abundance of these prolific biofilm formers among members of this phylum, they have consistently been identified using genomics as the earliest colonizers of metal surfaces submerged in marine environment (Dang & Lovell,

2016; Ma et al., 2020; Zhang, Ding, & Li, 2019). Members of the marine *Roseobacter* clade (MRC) found in the {alpha}-3 subclass of the *Alphaproteobacteria* in the family *Rhodobacteraceae* (Antunes et al., 2020; Dang & Lovell, 2016) and *Alteromonas* (*Gammaproteobacteria*) (Briand, Barani, Garnier, et al., 2017; Pollet, Berdjeb, Garnier, et al., 2018; Sushmitha et al., 2021) are notable early colonizers. Both MRC and *Alteromonas* have been reported as the primary biofilm colonizers within the first 9–72 h after exposure of pristine surfaces to the marine environment (Salta, Wharton, Blache, Stokes, & Briand, 2013). Members of MRC group can thrive under low levels of nutrient in their environment due to their symbiotic association with marine algae (Pohlner, Degenhardt, von Hoyningen-Huene, et al., 2017; Tang, Yang, Lin, et al., 2016), possess vast metabolic flexibility (Geng & Belas, 2010; Ma et al., 2020), and have a well-conditioned quorum sensing regulatory system that enables them to react to changes in their environment and control their interaction with algae (Ramanan, Kim, Cho, Oh, & Kim, 2016). This symbiotic association of *Roseobacter* lineage/clade with algae enables them to be major contributors to the cycling of carbon and sulfur in marine environments (Ramanan et al., 2016). Depending on the type of surface, other minor early colonizers identified are *Acidobacteria*, *Actinobacteria*, *Chlroo-flexi*, *Cyanobacteria*, *Firmicutes*, *Planctomyces*, *Beta-*, *Delta-*, *and Epsilon-proteobacteria*, and *Verrucomicrobia*.

Table 1 provides examples of the types of microbes associated with marine biofilms that have been found to be associated with different corrosion events.

5. Corrosion of steel materials

Marine biofilms cause significant damages to steel structures in marine environments due to corrosion. Corrosion of marine infrastructure results in increased maintenance costs, safety risks, and huge economic losses (Ma et al., 2020; Zhang et al., 2015). Currently the annual global cost of corrosion is estimated to be about $4 trillion USD (Li et al., 2015). Microorganisms have been implicated in causing approximately 20% of the total corrosion of steel structures in marine environments (Little & Lee, 2007). The demand for steel for use in building marine platforms has increased. Due to corrosion threat by marine biofilms to steel structures, different types of steel and steel alloys are used in the construction of marine infrastructure. For many applications, duplex stainless steel (DSS) is now used as an alternative to stainless steel because of its higher corrosion resistance and mechanical properties (Li, Zhou, Zhang, et al., 2016). Even with these properties, DSS structures in marine environments are still subject to localized pitting corrosion (Little and Lee, 2014). Other stainless steel types used for the construction of long-lasting structures in marine environments are hyper duplex SS (HDSS) and high alloy super austenitic SS (SASS), both of which are composed of high percentage of Cr, Mo, and Ni (Li et al., 2016; Ma et al., 2020).

How, then, do biofilms result in the corrosion of steel? The activities of these early surface colonizers, many of whom are noncorrosive, serve the function of creating a conducive environment for other species to join the community through nutrient cycling, biofilm building, maintenance of biofilm stability, and community function (Garrison & Field, 2020). Overtime, the microenvironment within the biofilm particularly at the metal surface becomes oxygen-deficient and anoxic, thus driving the eventual corrosion of the surface by specific members of the biofilm community (Ma et al., 2020). Owing to the abundance of the electron acceptor sulfate in marine environments, the biocorrosion of steel structures is driven by bacteria involved in the cycling of sulfur and iron, which serves as a substrate and energy source for some bacterial groups such as SRP, iron-reducing bacteria (IRB), iron-oxidizing bacteria (IOB), and sulfur/sulfide-oxidizing bacteria (SOB) (Lanneluc et al., 2015; Little & Lee, 2014; Ma et al., 2020; McBeth & Emerson, 2016; Mugge, Brock, & Salerno, 2019). The collective physiology of these types of microorganisms leads to cMIC of steel-containing materials.

Of the SRPs often identified, members of the genus *Desulfovibrio*, *Desulfotomaculum*, and *Desulfobacter* were reported by Li et al. (2017) as accounting for ~46% of the total bacterial population from the inner rust layer closest to the corroded steel plate. Zhang, Ma, Zhang, et al. (2019), on the other hand, while examining the distribution of SRB in the outer, middle, and inner rust layers of carbon steel, identified that *Desulfotomaculum* dominated the outer rust layer, the obligate extremely alkaliphilic anaerobe *Desulfonatronum* in the middle rust layer, while *Desulfobacter* and *Desulfovibrio* present in the inner layer. Also seen across these different rust layers were methanogens *Methanococcus* and *Methanothermococcus*. Some marine *Methanococcus* species are now well known to be involved in carbon steel corrosion (Deutzmann et al., 2015; Lahme et al., 2021; Tsurumaru et al., 2018).

6. Potential for copper use to control marine corrosion of materials

Copper and copper-containing alloys are used extensively in the construction of marine infrastructure such as heat exchangers, offshore structures, aquaculture facilities, and piping systems and in electrical cables. Beneficial properties of copper include strong resistance to corrosion and macrofouling, ease of mechanical workability, and good electrical and thermal conductivity (Little, Wagner, & Mansfeld, 1997). Copper and its alloys can reduce microbial colonization because of the release copper ions from the

TABLE 1 Marine biofilms formed on different surfaces submerged in marine environments potentially involved in corrosion.

Dominant taxa	Analysis	Surfaces	Sampling sites	Time periods	Salinity	Temp (°C)	References
Bacillus spp.	16S rRNA gene sequencing	Polyvinyl chloride; Glass; Mild steel; 304-Stainless steel; Cu-Ni alloys; Brass; Copper.	Southern India	7 days	31.4 ppt	29.8	Krishnan, Dahms, Seeni, et al. (2017)
Proteobacteria (*Zeta-proteobacteria* and *Beta-proteobacteria*) and *Bacteroidetes*	High-throughput amplicon sequencing of the SSU rRNA gene Single cell genomics analysis	Mild carbon steel coupons (C1018)	Great Salt Bay West Boothbay Harbor	3 to 43 days	30.2%	19–21	McBeth and Emerson (2016)
Inner biofilms Algae: *Stramenopiles, Bacteroidetes: Flavobacteriaceae, Gamma-proteobacteria, Alphaproteobacteria: Rhodobacteriales* **Outer biofilms** *Cyanobacteria, Bacteroidetes, Gamma-proteobacteria*	16S rRNA gene sequencing	Mild steel coupons	Gulf of Guinea	8 weeks			Celikkol-Aydin et al. (2016)
Desulfovibrio, Desulfomicrobium, Clostridium, Ruminococcus	16S rRNA gene sequencing	Mild steel	Victoria Harbor, Hong Kong	90 days			(Zhang & Fang, 2001)
Epsilonbacteria (Sulfurimonas autotrophica, *Sulfurospirillum arachonense*) *Alpha-proteobacteria* (*Roseobacter, Phaeobacter arcticus*, and *Sulfitobacter mediterraneus*)	TTGE and 16S rRNA gene sequencing	Carbon steel coupons	Atlantic Coast, France	1 week to 8 weeks			Lanneluc et al. (2015)
Bacteroidetes and *Cyanobacteria*	16S T-RFLP	Titanium, aluminum, limestone, glass and shale	Ionian Sea	155 days	38.76 ppt	13.4	. (Bellou, Colijn, & Papathanassiou, 2011)
Alpha-proteobacteria, Gamma-proteobacteria, Bacteroidetes	16S T-RFLP	Stainless steel covered with Cu-based AF coating		4 weeks	28.9%	29.4	Chen, Maki, Rittschof, and Teo (2013)
Firmicutes, Proteobacteria and *Bacteroidetes*		Steel plates	Hongtang Bay, South China Sea	6 months to 8 years			Li, Duan, Xiao, et al. (2017)

Gamma-proteobacteria: Vibrionaceae, Alteromonadaceae, and *Pseudoalteromonadaceae*		Steel panels	Laccadive Sea	72 h	38 g/L	30	Rajeev, Sushmitha, Toleti, et al. (2019)
Proteobacteria and *Bacteroidetes*		Polystyrene Petri, zinc, aluminum, poly(etherether-ketone), poly-tetrafluoroethylene, poly(vinyl chloride), stainless steel, and titanium	South Atlantic, Red Sea, Hong Kong Seawater, Yung Shu O Bay, East China Sea and South China Sea	3, 12, and 30 days			Zhang, Ma, Zhang, et al. (2019)
Alpha-proteobacteria and *Cyanobacteria, Planctomycetes*	16S rRNA gene sequencing and metagenomics	Zinc-coated aluminum wire panels	Central Rea Sea	30 days			Ding, Zhang, Alikunhi, et al. (2019)
Rust biofilm	16S rRNA gene sequencing	Carbon steel (Q235)	Qingdao marine station, China	1–2 weeks	32	13.5	Duan et al. (2008)
Proteobacteria, Bacteroidetes, Firmicutes, and *Verrucomicrobia* **30 days**: *Alcanivorax, Oceanospirillum,* and *Shewanella* **60 days**: *Flexibacter, Pseudoalteromonas* **90 days:** *Alteromonas, Bacteriovorax,* and *Nautella*	16S rRNA gene sequencing, high throughput sequencing	Stainless steel	Guanabara Bay, Rio de Janeiro	30, 60, 90 days			Capao, Moreira-Filho, Garcia, et al. (2020)
Carbon steel biofilm *Pseudomonas, Candidatus Arthromitus, Desulfovibrio* and *Labrenzia, Thaumarchaeota—Candidatus Nitrosopumilus* **Carbon steel rust layers** Inner layer: *Desulfovibrio, Desulfobacter* and *Desulfotomaculum, Methanothermococcus* Middle layer: *Desulfonatronum* Outer layer: *Desulfotomaculum, Methanococcus* **Copper alloy biofilm** *Lactobacillus, Helicobacter, Pseudomonas, Oceanicella* **Aluminum alloy biofilm** *Bacillus, Ruegeria,* and *Candidatus Arthromitus*	16S rRNA gene sequencing, high throughput sequencing	Biofilms and rust layers from carbon steel (Q235), Copper alloy-99.9% Cu (Q235), Aluminum alloy–99.6% (1060)	Hongtang Bay, South China Sea	30 months	3.5%	25.1	Zhang, Ma, Zhang, et al. (2019)

Continued

TABLE 1 Marine biofilms formed on different surfaces submerged in marine environments potentially involved in corrosion—cont'd

Dominant taxa	Analysis	Surfaces	Sampling sites	Time periods	Salinity	Temp (°C)	References
Proteobacteria-Rhodobacteraceae, Bacteroidetes—Flavobacteriales, Cyanobacteria	16S rRNA amplicon based sequencing	Stainless steel (316L) Austenitic chrome nickel steel	Leixoes Port, northern Portugal	30 days in 2 contrasting seasons	31.3	17.1	Antunes et al. (2020)
Copper alloy *Proteobacteria: Woeseia, Hyphomonas* *Cyanobacteria: Pleurocapsa, Leptolyngbya* **Aluminum alloy** *Proteobacteria: Woeseia, Candidatus Pelagibacter* *Cyanobacteria: Pleurocapsa, Leptolyngbya* *Chloroflexi: Ardenticatena* *Thaumarchaeota: Nitrosopumilus*	Metagenomics	Copper alloy (T2) and aluminum alloy (1060)	Sanya, South Sea China	44 months		25.1	Zhang, Ma, Zhang, et al. (2019)
Zetaproteobacteria (*Mariprofundus*), *Gammaproteobacteria* (*Sulfurimonas*) *Rhodobacterales* (*Roseobacter*) *Alteromonadales* (*Colwelliaceae*)	16S rRNA gene sequencing Metagenomics	Carbon steel disks	Anona, Ewing Bank, Halo, Mica and U-166 shipwrecks in Gulf of Mexico	16 weeks			Mugge, Lee, Brown, and Hamdan (2019)

metal surface, which facilitate a Fenton-type reaction that leads to the production of toxic reactive oxygen species (ROS) in microbial cells (Harrison, Ceri, & Turner, 2007; Lemire, Harrison, & Turner, 2013; Mathews, Hans, Mücklich, & Solioz, 2013) that damage the cell's DNA, protein, and lipids. Thus, copper can be used in copper-based antifouling coatings to prevent fouling or corrosion of existing structures. However, microbes can still form biofilms on copper-containing surfaces, although perhaps not as robustly as on carbon steel surfaces. Studies have shown that copper ions affect biofilms by reducing EPS production (Chari, Felix, Davoodbasha, Ali, & Nooruddin, 2017; Gomes, Simões, & Simões, 2020; Lin, Wang, Zhao, Chen, & Chen, 2020). Gomes et al. (2020) observed that the protein and EPS content produced by biofilms formed on copper alloys containing 57% and 96% copper were very low compared with those formed on stainless steel. Lou, Lin, Xu, et al. (2016) found that exposure of *P.aeruginosa* biofilms on copper-bearing 2205 DSS in artificial seawater resulted in cell death and overall reduction in biofilm thickness. The presence of copper on surfaces also slows down the settlement of macrofouling species such as barnacles and mussels (Little et al., 1997). Overall diversity of microflora on copper-associated surfaces is low, and biofilms that do form on copper-associated surfaces do so because they harbor certain stress, transport, and copper resistance systems. Zhang et al. (2019), using metagenomic sequencing, reported an enrichment of genes for heavy metal transport systems such as RNA polymerase sigma factors (*rpoE*), TP-binding cassette (ABC) transport system permease protein (ABC.CD.P), and the copper-resistance genes *cus* (Cu-sensing), *cop*, and *pco* in copper-associated biofilms growing on copper alloy (99.9% CU) immersed in coastal seawater for 44 months. The dominant copper-tolerant communities within these biofilms were affiliated to *Alpha-*, *Beta-*, and *Gamma-proteobacteria*, *Bacteroidetes*, and *Cyanobacteria*. In a similar study by Zhang et al. (2019), the dominant genus found after an exposure period of 30 months of copper alloy was *Lactobacillus*

Although eMIC from biofilms on copper-containing surfaces is possible, the strong antimicrobial activity of copper ions leaves it challenging for a robust biofilm to form for significant biofouling and corrosion to occur. However, copper lacks the important material properties for construction that steel has. Additionally, copper is considerably more expensive leading it to be quite cost-prohibitive.

7. Other microbial control approaches in marine systems

Biocides are chemicals that are frequently used in some industries (such as the oil and gas industry) to help prevent detrimental microbial activities such as souring and MIC. These chemicals (such as glutaraldehyde, quaternary phosphonium or ammonium compounds, or isothiazolone, to name a few) are designed to interfere with critical microbial cell functionality, such as by disrupting cell membranes or inhibiting protein synthesis (Morris & van der Kraan, 2017) resulting in a decrease of microbial numbers in a system. However, these types of chemicals are often applied in enclosed environments (such as oil well tubulars or pipeline systems) and thus may not work as effectively in more open marine environments.

Methods for controlling MIC in marine structures depend on where the corrosion is happening. To control MIC in marine environments, two main methods can be applied: (1) the use of chemical inhibitors or (2) the application of anticorrosion coatings eliminates the formation of biofilms on the surfaces (Ou, Tran, & Lin, 2018; San Keskin, Yaylaci, Durgun, Deniz, & Nazir, 2020). These chemicals are often very toxic and could result in environmental pollution. Antifouling paints/coatings formulated with tin and copper such as tributyl-tin (TBT) have been used with success (Gipperth, 2009). However, due to the toxic nature TBT, its usage in antifouling coatings has been banned. Currently, new nontoxic, eco-friendly coatings are being explored. Examples include silicone-based coatings such as poly(dimethylsiloxane), fluorine-based coatings such as poly(tetrafluoro ethylene) (de Carvalho, 2018; Nurioglu, Esteves, & de With, 2015), and graphene coatings (Parra et al., 2015).

There is currently an increased interest in use of combined biocides and corrosion inhibitors (multifunctional inhibitors) to achieve longer-lasting protection of steel structures in marine environments. Such multifunctional inhibitor compounds have been shown to have enhanced solubility and antimicrobial effects (Table 2). Some examples of multifunctional inhibitors include cetrimonium nalidixate (combination of cetrimonium bromide and nalidixic acid) (Seter, Thomson, Chong, MacFarlane, & Forsyth, 2013). Recently, Tuck et al. (2021) reported that a multifunctional organic corrosion inhibitor, a quaternary ammonium carboxylate compound (CTA-4OHcinn) composed of organic 4-hydroxy-cinnamate anion and hexadecyl trimethyl ammonium cation, was able to control attachment of marine bacteria on the surface of carbon steel (Table 2).

Conductive polymers are another group of compounds reported to be effective in controlling biocorrosion. Several factors such as high environmental stability, high conductivity, and distinctive redox mechanism make them potentially useful (Li & Ning, 2019). Conductive polymers that have been applied for corrosion prevention in aluminum, mild steel, stainless steels, copper, and copper alloys include polypyrrole (PPY), polyaniline (PANI), and polythiophene (PBT) (Guo & Facchetti, 2020). Improved anticorrosion efficacy of conductive polymers with immobilized silver nanoparticles has been reported (Guo & Facchetti, 2020). Silver nanoparticles coupled with bithiophene were shown to inhibit SRB-induced MIC on copper surfaces (Videla &

TABLE 2 Mode of action of some biocides and corrosion inhibitors for mitigating MIC in marine environments.

Chemical type	Structure	Mode of action	Microbes targeted	Effective dose	Reference
Composite Biocide: Thymol and Benzyldimethyldodecylammonium chloride (BDMDAC)	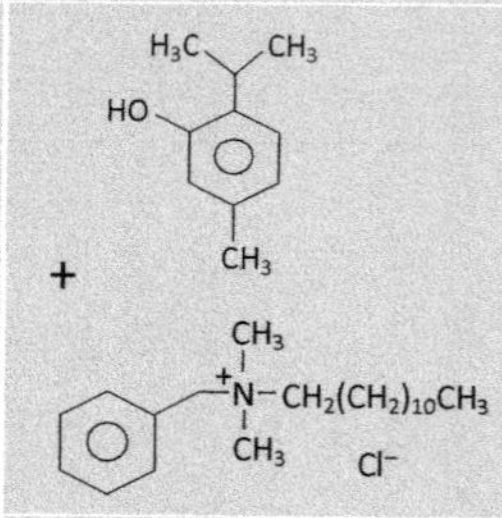	Reduce adhesion of biofilms to 304 SS metal surface	*Shewanella* and *Vibrio*	Thymol + BDMDAC 60 mg/L + 20 mg/L	Chang et al. (2021)
Green Inhibitor: Cetrimonium *trans*-4-hydroxy cinnamide (CTA-4OHcinn)	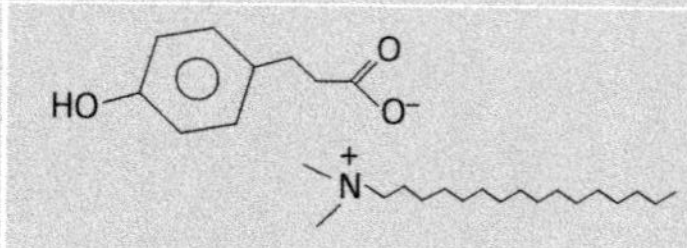	Inhibits cell attachment, eliminates biofilm formation, and reduces the viability of mature biofilms	*Shewanella chilikensis, Pseudomonas balearica,* and *Klebsiella pneumoniae*	10 mM for 4 h	Tuck et al. (2021), Tuck et al. (2022)
Green Inhibitor: Lanthanum *trans*-4-Hydroxycinnamate + Cetrimonium *trans*-4-hydroxycinnamate	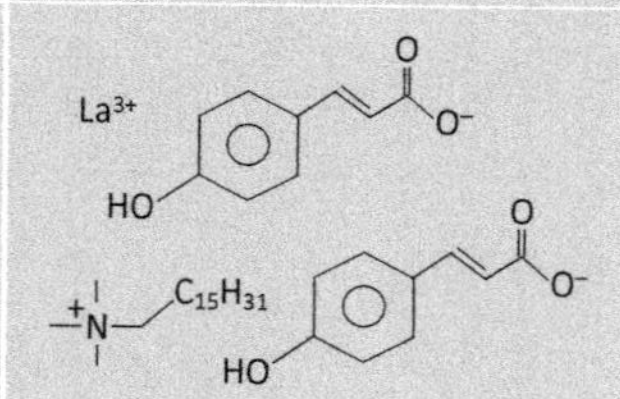	Reduce cell density		La + Cet 1.1 mM:1.1 mM	Catubig et al. (2021)

Herrera, 2009). Although these types of studies have shown numerous types of chemicals that can be effective in preventing detrimental microorganisms, additional, longer-term research is needed to evaluate their efficacy in marine environments.

8. Conclusion

Overall, biofilms are common in the marine environment. As such, these surface-attached microbial communities can attract higher life forms that can result in biofouling (micro- and macrofouling), as well contribute to the corrosion of infrastructure in marine systems. Molecular microbiological approaches have helped to identify many types of microorganisms that may be involved in corrosion of marine infrastructure (e.g., Table 1). Further, some materials such as stainless steels and copper have shown promise for reducing corrosive biofilms, but their higher costs can be prohibitive for widespread use. Ongoing research to identify corrosive microorganisms found in diverse marine biofilms will allow for improved understanding of how to potentially mitigate costly marine infrastructure damage.

References

Antunes, J., Leão, P., & Vasconcelos, V. (2019). Marine biofilms: Diversity of communities and of chemical cues. *Environmental Microbiology Reports*, *11*, 287–305.

Antunes, J. T., Sousa, A. G. G., Azevedo, J., Rego, A., Leão, P. N., & Vasconcelos, V. (2020). Distinct temporal succession of bacterial communities in early marine biofilms in a Portuguese Atlantic Port. *Frontiers in Microbiology*, *11*, 1938.

Beese-Vasbender, P. F., Nayak, S., Erbe, A., Stratmann, M., & Mayrhofer, K. J. J. (2015). Electrochemical characterization of direct electron uptake in electrical microbially influenced corrosion of iron by the lithoautotrophic SRB *Desulfopila corrodens* strain IS4. *Electrochimica Acta*, *167*, 321–329.

Bellou, N., Colijn, F., & Papathanassiou, E. (2011). Experimental settlement study in the Eastern Mediterranean deep sea (Ionian Sea). *Nuclear Instruments and Methods in Physics Research Section A: Accelerators, Spectrometers, Detectors and Associated Equipment*, *626*, S102–S105.

Berg, G., Rybakova, D., Fischer, D., et al. (2020). Microbiome definition re-visited: Old concepts and new challenges. *Microbiome*, *8*, 103.

Briand, J. F., Barani, A., Garnier, C., et al. (2017). Spatio-temporal variations of marine biofilm communities colonizing artificial substrata including antifouling coatings in contrasted French coastal environments. *Microbial Ecology*, *74*(3), 585–598.

Callow, M. E., & Callow, J. E. (2002). Marine biofouling: A sticky problem. *Biologist*, *49*(1), 10–14.

Canfield, D. E., & Farquhar, J. (2009). Animal evolution, bioturbation, and the sulfate concentration of the oceans. *PNAS*, *106*(20), 8123–8127.

Capao, A., Moreira-Filho, P., Garcia, M., et al. (2020). Marine bacterial community analysis on 316L stainless steel coupons by Illumina MiSeq sequencing. *Biotechnology Letters*, *42*, 1431–1448.

Caruso, G. (2020). Microbial colonization in marine environments: Overview of current knowledge and emerging research topics. *Journal of Marine Science and Engineering*, *8*, 78.

Catubig, R., Michalczyk, A., Neil, W., McAdam, G., Forsyth, J., Ghorbani, M., et al. (2021). Multifunctional inhibitor mixture for reducing bacteria growth and corrosion on marine grade steel. *ChemRxiv*, *1*, 1–22.

Celikkol-Aydin, S., Gaylarde, C., Lee, T., Melchers, R. E., Witt, D. L., & Beech, I. B. (2016). 16S rRNA gene profiling of planktonic and biofilm microbial populations in the Gulf of Guinea using Illumina NGS. *Marine Environmental Research*, *122*, 105–112.

Chang, S. Y., Huang, S. Y., Chu, Y. R., Jian, S. Y., Lo, K. Y., & Lee, Y. L. (2021). Antimicrobial and anticorrosion activity of a novel composite biocide against mixed bacterial strains in Taiwanese marine environments. *Materials (Basel)*, *14*(20), 6156.

Chari, N., Felix, L., Davoodbasha, M., Ali, A. S., & Nooruddin, T. (2017). In vitro and in vivo antibiofilm effect of copper nanoparticles against aquaculture pathogens. *Biocatalysis and Agricultural Biotechnology*, *10*, 336–341.

Chen, C., Maki, J. S., Rittschof, D., & Teo, S. L. (2013). Early marine bacterial biofilm on a copper-base antifouling paint. *International Biodeterioration and Biodegradation*, *83*, 71–76.

Cho, I., & Blaser, M. (2012). The human microbiome: At the interface of health and disease. *Nature Reviews Genetics*, *13*, 260–270.

Dang, H., & Lovell, C. R. (2016). Microbial surface colonization and biofilm development in marine environments. *Microbiology and Molecular Biology Reviews*, *80*, 91–138.

de Carvalho, C. C. C. R. (2018). Marine biofilms: A successful microbial strategy with economic implications. *Frontiers in Marine Science*, *5*, 126.

Deutzmann, J. S., Sahin, M., & Spormann, A. M. (2015). Extracellular enzymes facilitate electron uptake in biocorrosion and bioelectrosynthesis. *mBio*, *6*, e00496-15.

Ding, W., Zhang, W., Alikunhi, N. M., et al. (2019). Metagenomic analysis of zinc surface–associated marine biofilms. *Microbial Ecology*, *77*, 406–416.

Duan, J., Wu, S., Zhang, X., Huang, G., Du, M., & Hou, B. (2008). Corrosion of carbon steel influenced by anaerobic biofilm in natural seawater. *Electrochimica Acta*, *54*(1), 22–28.

Edwards, M. J., White, G. F., Norman, M., et al. (2015). Redox linked flavin sites in extracellular decaheme proteins involved in microbe-mineral electron transfer. *Scientific Reports*, *5*, 11677.

Emerson, D. (2018). The role of iron-oxidizing bacteria in biocorrosion: A review. *Biofouling*, *34*, 989–1000.

Enning, D., Venzlaff, H., Garrelfs, J., Dinh, H. T., Meyer, V., & Mayrhofer, K. (2012). Marine sulfate-reducing bacteria cause serious corrosion of iron under electroconductive biogenic mineral crust. *Environmental Microbiology*, *14*, 1772–1787.

Ewing, S. P. (1955). Electrochemical studies of the hydrogen sulfide corrosion mechanism. *Corrosion*, *11*, 51–55.

Fischer, M., Friedrichs, G., & Lachnit, T. (2014). Fluorescence-based quasi continuous and in situ monitoring of biofilm formation dynamics in natural marine environments. *Applied and Environmental Microbiology*, *80*(12), 3721–3728.

Fitridge, I., Dempster, T., Guenther, J., & de Nys, R. (2012). The impact and control of biofouling in marine aquaculture: A review. *Biofouling*, *28*(7), 649–669.

Flemming, H. C., & Wingender, J. (2010). The biofilm matrix. *Nature Reviews Microbiology*, *8*, 623–633.

Garrison, C. E., & Field, E. K. (2020). Introducing a "core steel microbiome" and community functional analysis associated with microbially influenced corrosion. *FEMS Microbiology Ecology*, *97*(1), fiaa237.

Geng, H., & Belas, R. (2010). Molecular mechanisms underlying *Roseobacter*-phytoplankton symbioses. *Current Opinion in Biotechnology*, *21*(3), 332–338.

Gipperth, L. (2009). The legal design of the international and European Union ban on tributyltin antifouling paint: Direct and indirect effects. *Journal of Environmental Management*, *90*, S86–S95.

Gomes, I. B., Simões, M., & Simões, L. C. (2020). Copper surfaces in biofilm control. *Nanomaterials (Basel)*, *10*(12), 2491.

Guo, X, & Facchetti, A. (2020). The journey of conducting polymers from discovery to application. *Nature Materials*, *19*, 922–928.

Hall, C. W., & Mah, T. F. (2017). Molecular mechanisms of biofilm-based antibiotic resistance and tolerance in pathogenic bacteria. *FEMS Microbiology Reviews*, *41*, 276–301.

Harrison, J. J., Ceri, H, & Turner, R. J. (2007). Multimetal resistance and tolerance in microbial biofilms. *Nature Reviews Microbiology*, *5*, 928–938.

Hwang, G., Liu, Y., Kim, D., Sun, V., Aviles-Reyes, A., & Kajfasz, J. K. (2016). Simultaneous spatiotemporal mapping of in situ pH and bacterial activity within an intact 3D microcolony structure. *Scientific Reports*, *6*, 1–11.

Jia, R., Yang, D., Xu, J., Xu, D., & Gu, T. (2017). Microbiologically influenced corrosion of C1018 carbon steel by nitrate reducing *Pseudomonas aeruginosa* biofilm under organic carbon starvation. *Corrosion Science*, *127*, 1–9.

Jolivet-Gougeon, A., & Bonnaure-Mallet, M. (2014). Biofilms as a mechanism of bacterial resistance. *Drug Discovery Today: Technologies*, *11*, 49–56.

Krishnan, M., Dahms, H. U., Seeni, P., et al. (2017). Multi metal assessment on biofilm formation in offshore environment. *Materials Science & Engineering, C: Materials for Biological Applications*, *73*, 743–755.

Lahme, S., Mand, J., Longwell, J., Smith, R., & Enning, D. (2021). Severe corrosion of carbon steel in oil field produced water can be linked to methanogenic archaea containing a special type of [NiFe] hydrogenase. *Applied and Environmental Microbiology*, *87*(3), e01819–e01820.

Lanneluc, I., Langumier, M., Sabot, R., Jeannin, M., Refait, P., & Sable, S. (2015). On bacterial communities associated with the corrosion product layer during the early stages of marine corrosion of carbon steel. *International Biodeterioration and Biodegradation*, *99*, 55–65.

Lemire, J., Harrison, J. J., & Turner, R. J. (2013). Antimicrobial activity of metals: Mechanisms, molecular targets and applications. *Nature Reviews Microbiology*, *11*, 371–384.

Li, X., Duan, J., Xiao, H., et al. (2017). Analysis of bacterial community composition of corroded steel immersed in Sanya and Xiamen seawaters in China via method of Illumina MiSeq sequencing. *Frontiers in Microbiology*, *8*, 1737.

Li, Y., & Ning, C. (2019). Latest research progress of marine microbiological corrosion and bio-fouling, and new approaches of marine anti-corrosion and anti-fouling. *Bioactive Materials*, *4*, 189–195.

Li, Y., Xu, D., Chen, C., et al. (2018). Anaerobic microbiologically influenced corrosion mechanisms interpreted using bioenergetics and bioelectrochemistry: A review. *Journal of Materials Science & Technology*, *34*, 1713–1718.

Li, X., Zhang, D., Liu, Z., Li, Z., Du, C., & Dong, C. (2015). Material science: Share corrosion data. *Nature*, *527*(7579), 441–442.

Li, H., Zhou, E., Zhang, D., et al. (2016). Microbiologically influenced corrosion of 2707 hyper-duplex stainless steel by marine *Pseudomonas aeruginosa* biofilm. *Scientific Reports*, *6*, 20190.

Lin, H., Wang, C., Zhao, H., Chen, G., & Chen, X. (2020). A subcellular level study of copper speciation reveals the synergistic mechanism of microbial cells and EPS involved in copper binding in bacterial biofilms. *Environmental Pollution*, *263*, 114485.

Little, B. J., & Lee, J. S. (2007). Microbiologically influenced corrosion. *Wiley Series in Corrosion* (p. 279). John Wiley and Sons.

Little, B. J., & Lee, J. S. (2014). Microbiologically influence corrosion: An update. *International Materials Review*, *59*, 384–393.

Little, B. J., Wagner, P. A., & Mansfeld, F. (1997). Microbiologically influenced corrosion. *NACE*, *5*, 13–15.

Lou, Y., Lin, L., Xu, D., et al. (2016). Antibacterial ability of a novel Cu-bearing 2205 duplex stainless steel against *Pseudomonas aeruginosa* biofilm in artificial seawater. *International Biodeterioration and Biodegradation*, *110*, 199–205.

Ma, Y., Zhang, Y., Zhang, R., Guan, F., Hou, B., & Duan, J. (2020). Microbiologically influenced corrosion of marine steels within the interaction between steel and biofilms: A brief view. *Applied Microbiology and Biotechnology*, *104*(2), 515–525.

Malfatti, F., Lee, C., Tinta, T., et al. (2019). Detection of active microbial enzymes in nascent sea spray aerosol: Implications for atmospheric chemistry and climate. *Environmental Science & Technology Letters*, *6*, 171–177.

Mathews, S., Hans, M., Mücklich, F., & Solioz, M. (2013). Contact killing of bacteria on copper is suppressed if bacterial-metal contact is prevented and is induced on iron by copper ions. *Applied and Environmental Microbiology*, *79*, 2605–2611.

McBeth, J. M., & Emerson, D. (2016). In situ microbial community succession on mild steel in estuarine and marine environments.: Exploring the role of iron-oxiding bacteria. *Frontiers in Microbiology*, *7*, 767.

Morris, B. E. L., & van der Kraan, G. M. (2017). Application of biocides and chemical treatments to both combat microorganisms and reduce (bio)corrosion. In T. L. Skovhus, D. Enning, & J. S. Lee (Eds.), *Microbiologically influenced corrosion in the upstream oil and gas industry* (pp. 229–253). Boca Raton, FL: CRC Press.

Mugge, R. L., Brock, M. L., & Salerno, J. L. (2019). Deep-Sea biofilms, historic shipwreck preservation and the Deepwater Horizon Spill. *Frontiers in Marine Science*, *6*, 48.

Mugge, R. L., Lee, J. S., Brown, T. T., & Hamdan, L. J. (2019). Marine biofilm bacterial community response and carbon steel loss following Deepwater Horizon spill contaminant exposure. *Biofouling*, *35*(8), 870–882.

Nurioglu, A. G., Esteves, A. C. C., & de With, G. (2015). Non-toxic, non-biocide-release antifouling coatings based on molecular structure design for marine applications. *Journal of Materials Chemistry B*, *3*, 6547–6570.

Ou, H. H., Tran, Q. T. P., & Lin, P. H. (2018). A synergistic effect between gluconate and molybdate on corrosion inhibition of recirculating cooling water systems. *Corrosion Science*, *133*, 231–239.

Parra, C., Dorta, F., Jimenez, E., Henríquez, R., Ramírez, C., Rojas, R., et al. (2015). A nanomolecular approach to decrease adhesion of biofouling-producing bacteria to graphene-coated material. *Journal of Nanbiotechnology*, *13*, 82.

Pohlner, M., Degenhardt, J., von Hoyningen-Huene, A. J. E., et al. (2017). The biogeographical distribution of benthic *Roseobacter* group members along a Pacific Transect is structured by nutrient availability within the sediments and primary production in different oceanic provinces. *Frontiers in Microbiology*, *8*, 2550.

Pollet, T., Berdjeb, L., Garnier, C., et al. (2018). Prokaryotic community successions and interactions in marine biofilms: The key role of *Flavobacteriia*. *FEMS Microbiology Ecology*, *94*(6), fiy083.

Procopio, L. (2019). The role of biofilms in the corrosion of steel in marine environments. *World Journal of Microbiology and Biotechnology*, *35*, 73.

Rajeev, M., Sushmitha, T. J., Toleti, S. R., et al. (2019). Culture dependent and independent analysis and appraisal of early-stage biofilm-forming bacterial community composition in the Southern coastal seawater of India. *Science of the Total Environment*, *666*, 308–320.

Ramanan, R., Kim, B. H., Cho, D. H., Oh, H. M., & Kim, H. S. (2016). Algae-bacteria interactions: Evolution, ecology and emerging applications. *Biotechnology Advances*, *34*, 14–29.

Salta, M., Wharton, J. A., Blache, Y., Stokes, K. R., & Briand, J. F. (2013). Marine biofilms on artificial surfaces: Structure and dynamics. *Environmental Microbiology*, *15*(11), 2879–2893.

San Keskin, N. O., Yaylaci, E., Durgun, S. G., Deniz, F., & Nazir, N. (2020). Anticorrosive properties of green silver nanoparticles to prevent microbiologically influenced corrosion on copper in the marine environment. *Journal of Marine Science and Application*, *20*, 10–20.

Seter, M., Thomson, M. J., Chong, A., MacFarlane, D. R., & Forsyth, M. (2013). Cetrimonium nalidixate as a multifunctional inhibitor to combat biofilm formation and microbiologically influenced corrosion. *Australian Journal of Chemistry*, *66*, 921–929.

Sharma, M., & Voordouw, G. (2016). MIC detection and assessment—A holistic approach. In T. L. Skovhus, D. Enning, & J. S. Lee (Eds.), *Microbiologically influenced corrosion in the upstream oil and gas industry* (pp. 177–212). Boca Raton, FL: CRC Press.

Sherar, B. W. A., Power, I. M., Keech, P. G., Mitlin, S., Southam, G., & Shoesmith, D. W. (2011). Characterizing the effect of carbon steel exposure in sulfide containing solutions to microbially induced corrosion. *Corrosion Science*, *53*, 955–960.

Sushmitha, T. J., Rajeev, M., Sriyutha, M. P., Ganesh, S., Toleti, S. R., & Karutha, P. S. (2021). Bacterial community structure of early-stage biofilms is dictated by temporal succession rather than substrate types in the southern coastal seawater of India. *PLoS ONE*, *16*(9), e0257961.

Tang, K., Yang, Y., Lin, D., et al. (2016). Genomic, physiologic, and proteomic insights into metabolic versatility in *Roseobacter* clade bacteria isolated from deep-sea water. *Scientific Reports*, *6*, 35528.

Thomas, C. J., Edyvean, R. G. J., & Brook, R. (1988). Biologically enhanced corrosion fatigue. *Biofouling*, *1*(1), 65–77.

Tsurumaru, H., Ito, N., Mori, K., et al. (2018). An extracellular [NiFe] hydrogenase mediating iron corrosion is encoded in a genetically unstable genomic island in *Methanococcus maripaludis*. *Scientific Reports*, *8*, 15149.

Tuck, B., Leinecker, N., Watkin, E., Somers, A., Forsyth, M., & Machuca, L. L. (2022). Efficiency of a novel multifunctional corrosion inhibitor against biofilms developed on carbon steel. *Frontiers in Bioengineering and Biotechnology*, *10*, 803559.

Tuck, B., Watkin, E., Forsyth, M., Somers, A., Ghorbani, M., & Machuca, L. L. (2021). Evaluation of a novel, multi-functional inhibitor compound for prevention of biofilm formation on carbon steel in marine environments. *Scientific Reports*, *11*, 15697.

Venzlaff, H., Enning, D., Srinivasan, J., Mayrhofer, K. J. J., Hassel, A. W., Widdel, F., et al. (2013). Accelerated cathodic reaction in microbial corrosion of iron due to direct electron uptake by sulfate-reducing bacteria. *Corrosion Science*, *66*, 88–96.

Videla, H. A., & Herrera, L. K. (2009). Understanding microbial inhibition of corrosion. A comprehensive overview. *International Biodeterioration & Biodegradation*, *63*, 896–900.

Woodtli, J., & Kieselbach, R. (2000). Damage due to hydrogen embrittlement and stress corrosion cracking. *Engineering Failure Analysis*, *7*, 427–450.

Xiao, J., Klein, M. I., Falsetta, M. L., et al. (2012). The exopolysaccharide matrix modulates the interaction between 3D architecture and virulence of a mixed-species oral biofilm. *PLoS Pathogens*, *8*, e1002623.

Xu, D., Li, Y., Song, F., & Gu, T. (2013). Laboratory investigation of microbiologically influenced corrosion of C1018 carbon steel by nitrate reducing bacterium *Bacillus licheniformis*. *Corrosion Science*, *77*, 385–390.

Yan, J., & Bassler, B. L. (2019). Surviving as a community: Antibiotic tolerance and persistence in bacterial biofilms. *Cell Host & Microbe*, *26*, 15–21.

Zhang, P., Xu, D., Li, Y., Yang, K., & Gu, T. (2015). Electron mediators accelerate the microbiologically influenced corrosion of 304 stainless steel by the *Desulfovibrio vulgaris* biofilm. *Bioelectrochemistry*, *101*, 14–21.

Zhang, T, & Fang, H. H. P. (2001). Phylogenetic diversity of a SRB-rich marine biofilm. *Applied Microbiology Biotechnology*, *57*, 437–440.

Zhang, W., Ding, W., Li, Y., et al. (2019). Marine biofilms constitute a bank of hidden microbial diversity and functional potential. *Nature Communications*, *10*(1), 517.

Zhang, Y., Ma, Y., Duan, J., Li, X., Wang, J., & Hou, B. (2019). Analysis of marine microbial communities colonizing various metallic materials and rust layers. *Biofouling*, *35*(4), 429–442.

Zhang, Y., Ma, Y., Zhang, R., et al. (2019). Metagenomic resolution of functional diversity in copper surface-associated marine biofilms. *Frontiers in Microbiology*, *10*, 2863.

Further reading

Jaume, J., Marques, M. F. J., Delia, M. L., et al. (2022). Surface modification of 5083 aluminum-magnesium induced by marine microorganisms. *Corrosion Science*, *194*, 1–14.

Kessler, R. W., Weiss, A., Kuegler, S., Hermes, C., & Wichard, T. (2018). Macroalgal–bacterial interactions: Role of dimethylsulfoniopropionate in microbial gardening by *Ulva* (Chlorophyta). *Molecular Ecology*, *27*, 1808–1819.

Malmstrom, R. R., Kiene, R. P., & Kirchman, D. L. (2004). Identification and enumeration of bacteria assimilating dimethylsulfoniopropionate (DMSP) in the North Atlantic and Gulf of Mexico. *Limnology and Oceanography*, *49*, 597–606.

Pei, P., Aslam, M., Du, H., Liang, H., Wang, H., Liu, X., et al. (2021). Environmental factors shape the epiphytic bacterial communities of *Gracilariopsis lemaneiformis*. *Scientific Reports*, *11*(1), 8671.

San, N. O., Nazır, H., & Dönmez, G. (2014). Microbially influenced corrosion and inhibition of nickel–zinc and nickel–copper coatings by *Pseudomonas aeruginosa*. *Corrosion Science*, *79*, 177–183.

Chapter 13

Natural biofilms: Structure, development, and habitats

Atif Khan[a,b] and Toleti Subba Rao[a,b,c]

[a]*Biofouling and Biofilm processes Section, Water & Steam Chemistry Division, Bhabha Atomic Research Centre Facilities, Kalpakkam, Tamil Nadu, India,* [b]*Homi Bhabha National Institute, Mumbai, Maharastra, India,* [c]*School of Arts & Sciences, Sai University, Chennai, Tamil Nadu, India*

1. Introduction

Microorganisms recurrently dwelling a diverse form of a dense and diverse community called biofilms, which are mostly attached to the surface and encased in the polysaccharide matrix (Hobley, Harkins, MacPhee, & Stanley-Wall, 2015). In the mid of 20th century, Claude Zobell has found that aquatic bacteria are numerously present on the surface of sample carrying containers than in the suspended forms (Zobell, 1943). From then, they used several techniques and stains and found out that biofilms are not just the assemblage of cells on the surface but a complex and dynamic biological system. Biofilms are considered the most successful and diverse form of life on this planet (Stoodley, Sauer, Davies, & Costerton, 2002). Cells of the biofilms exhibit a social phenotype where they interact among each other to influence their evolutionary fitness. This fitness comes through the secretion of digestive enzymes, surface adhesins, siderophores, and signaling molecules (Absalon, Dellen, & Watnick, 2011; Allison, 2005; Cordero, Ventouras, DeLong, & Polz, 2012; West, Diggle, Buckling, Gardner, & Griffin, 2007). Cells inside the biofilms have advantages over planktonic forms by the provision of protection from external threats and division of labor inside the biofilm. All the organisms including humans are harbored by bacteria able to form biofilms and may be associated with persistent infections including plants and animals. Biofilms in the environment are responsible for biofouling and corrosion of the submerged surfaces along with contamination of drinking water (Flemming, Wingender, & Szewzyk, 2011; Wingender & Flemming, 2011). Biofilms communities are continuously emerging novel structures, activities, and patterns leading to the development of synergistic interactions among cells of biofilms to promote gene exchange and enhanced tolerance to antimicrobials (Flemming & Wingender, 2010). An increasing number of evidences suggest that plasmids and biofilm structure and function are tangled through numerous interactions, which point to their role in bacterial evolution. The matrix of the biofilm confers organizational structure to the biofilm in the form of hydration, resource capture, and digestive capacity ultimately deriving highly diverse and synergistic interactions.

Biofilms in marine or aquatic habitats are dominated by a mixed community of bacteria along with diatoms surrounded by extracellular polymeric substance (EPS). Marine biofilms act as a preferred substrate for the settlement of algae and invertebrate larvae, which later influences the development of biofouling and corrosion processes. Under natural conditions, interspecies interaction within a biofilm for nutrients and physical resources signifies the presence of multiple species in a biofilm and is called a multispecies biofilm. The multispecies biofilm is considered the most dominant form of biofilm in natural environments as well as in living hosts as in the case of cystic fibrosis (Høiby, Ciofu, & Bjarnsholt, 2010) (Fig. 1).

2. Structure of biofilms

Biofilms are a ubiquitous type of natural ancient adaptation acquired by the bacteria and most prevalently distributed in about 65% of the total infections (Hall-Stoodley, Costerton, & Stoodley, 2004). They become more problematic when they develop resistance toward antibiotics and bypass antibiotic therapies, which are most commonly used in clinical settings (Kolter, 2010). Bacteria changed to biofilm style around 3.25 billion years ago, which were proved beneficial to seepage the harsh environment of the primitive earth (Hall-Stoodley et al., 2004). Biofilm mode of life has activated certain molecular pathways, which ensued in the occurrence of expedient biofilms making them escape from hostile environments in the current environmental circumstances. When the bacteria sense any critical condition, they initiate their attachment to the surface followed by the

Understanding Microbial Biofilms. https://doi.org/10.1016/B978-0-323-99977-9.00008-9

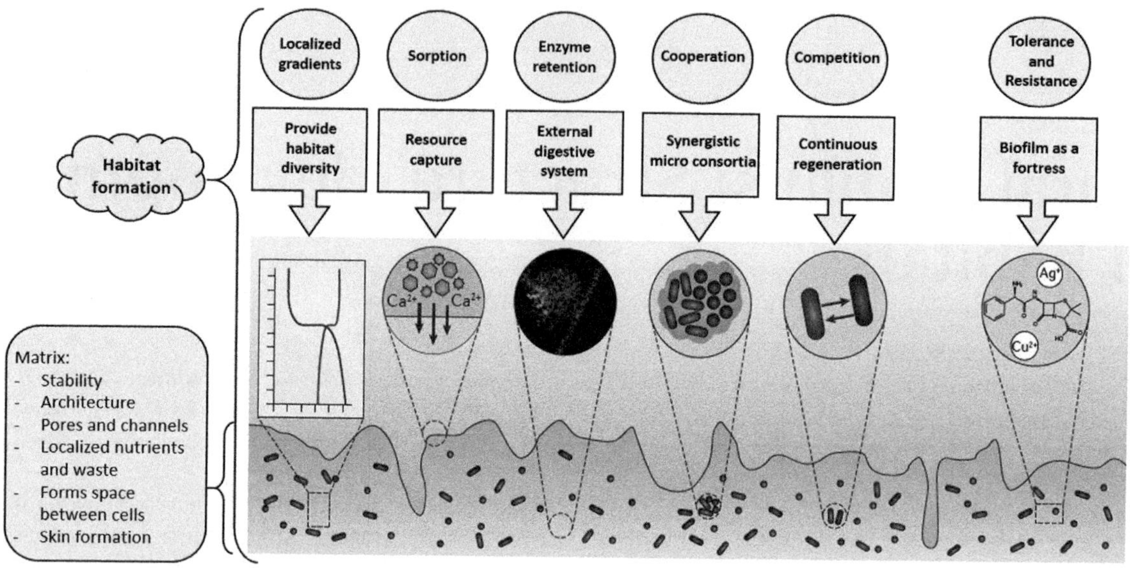

FIG. 1 Formation of habitats in the biofilms. *(Modified from Flemming, H.-C., Wingender, J., Szewzyk, U., Steinberg, P., Rice, S. A., & Kjelleberg, S. (2016). Biofilms: An emergent form of bacterial life. Nature Reviews Microbiology, 14(9), 563–575.)*

development of biofilm to enable greater environmental sustainability. Cells inside any mature biofilm tend to produce certain compounds that enable the dispersal of biofilm cells, to form a new biofilm on a new substratum (Kolodkin-Gal et al., 2010). Apart from some of the universal characteristics of biofilms, every other biofilm community is considered unique (Tolker-Nielsen & Molin, 2000). The structure of biofilms is highly heterogeneous, with multiple different bacterial species encased in the EPS matrix and separated from each other by water channels (Quan et al., 2021). These water channels are involved in the diffusion of nutrients, oxygen, and even biofilm inhibitory compounds. The composition of EPS (polysaccharides, proteins, and nucleic acids) among different bacterial species varies with the composition and hence their physical and chemical properties. Some of them are neutral macromolecules, but most of them are polyanionic or ketal-linked pyruvate. Nutrients are confined inside the matrix for its consumption by bacteria along with water retained by forming H-bond with hydrophilic polysaccharides (Flemming & Wingender, 2010). The production of EPS is affected by nutritional status and of the surrounding medium, availability of carbon, nitrogen, potassium, or phosphate. There has been a substantial resemblance between the polysaccharides formed by the planktonic cells and those produced by biofilms, with some of the polysaccharides produced as a result of the stress response (Sutherland, 2001). The disparity in the polysaccharide content also depends on the microbial species composition and physiological state of biofilm. EPS contributes to the properties of the biofilm by binding water and providing mechanical strength to endure shear forces. These structural constituents escalate to a highly hydrated and robust biofilm structure that keeps the bacteria in close vicinity for cell–cell communication and DNA interchange (Koo, Xiao, Klein, & Jeon, 2010). This structural integrity also safeguards the resident bacteria and biofilm from desiccation, antimicrobial agents, predation, and other damaging factors (Mai-Prochnow et al., 2008). It is found that O-acetylation of alginate endorses structural advancement in mucoid strains of *P. aeruginosa* (Nivens, Ohman, Williams, & Franklin, 2001). Also, overexpression of alginate encourages structural complexity in wild-type nonmucoid strains of *P. aeruginosa*. Some reports proposed the requirement of EPS and colonic acid for the development of complex structures and disruption of the same leads to reduced mechanical complexity as well as augmented vulnerability to antimicrobial agents (Watnick, Lauriano, Klose, Croal, & Kolter, 2001) (Fig. 2).

The structure of biofilm, i.e., thickness is also affected by the type of inhabitant microorganism. Single species biofilm could be thinner compared with dual or multispecies biofilm as a result of one species enhancing the stability of the other (Jones, Roth, & Sanders, 1969). Studies have found the existence of persister cells inside the biofilms that are metabolically sluggish and nondividing. They are found to be the same as the rest of the cells in biofilm but are highly tolerant to antibiotics and play an important role in reestablishing the biofilm after the dissolution of unfavorable conditions (Lewis, 2008). Currently, there is a growing number of techniques that can be used to study the structure of biofilms, for example, TEM, AFM, CLSM, but all of them are associated with some problems of interpretation of images produced by them (Wimpenny, Manz, & Szewzyk, 2000).

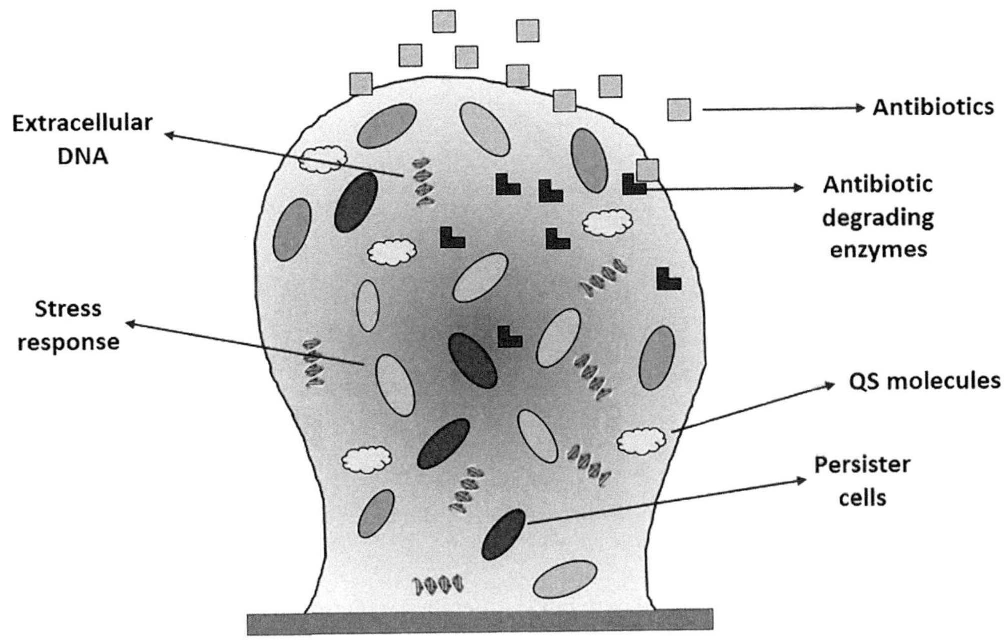

FIG. 2 Structure of biofilm.

The structure of biofilm is also predisposed by the interaction of nonmicrobial constituents of environmental origin as found in the biofilms formed in the native heart valves containing platelets, fibrin, and EPS (Lauten et al., 2021). Similarly, biofilms in the urinary catheters can hydrolyze urea to produce ammonia through the activity of the enzyme urease resulting in the precipitation of minerals such as calcium phosphate (hydroxyapatite) and magnesium ammonium phosphate (Espinosa-Ortiz, Eisner, Lange, & Gerlach, 2019). Minerals such as calcium carbonate, iron oxide, and soil particles also accumulate in the biofilms of industrial water systems.

3. Development of biofilms

Biofilm is an assembly of microbial cells irreversibly attached with a solid surface and enclosed in the material of polysaccharide matrix. The components of the matrix may contain clay, silt, or mineral particles, which may vary with the surrounding environment of the biofilm. The interface between a solid and liquid is considered an ideal environment for the attachment and growth of microorganisms. But, many other properties of the substratum play a critical role in the development of biofilm when any substratum is submerged in an aqueous medium. The roughness and physicochemical properties of any surface increase the attachment and colonization of microorganisms (Zheng et al., 2021). Some studies confirmed it by finding the rapid microbial attachment to the nonpolar hydrophobic surface compared with hydrophilic surfaces (Kimkes & Heinemann, 2020). The formation of conditioning film after submerging any surface in an aqueous medium is a critical step in microbial attachment to the surface (Bhagwat, O'Connor, Grainge, & Palanisami, 2021). There are several types of conditioning film produced by hosts such as those of sugar, blood, urine, or saliva that may influence the attachment of bacteria to any biomaterials transplanted inside a human body. The hydrodynamic property of an aqueous medium also plays an important role in the development of biofilm. Under low linear velocity, an association of the cell with the surface depends on cell size and motility, and under higher velocities, cells tend to settle faster on the surface until the velocity is sufficient enough to detach the cells from the surface (Krsmanovic et al., 2021). Physical and chemical properties of the aqueous medium such as pH, temperature, nutrient levels play role in bacterial attachment and development of biofilm. A study found that increasing the concentration of cations promoted bacterial settlement on the surface by reducing the repulsive forces between negatively charged cells and glass surfaces (de Kerchove & Elimelech, 2008). Cell surface characteristics of bacterial-cell-like presence of fimbriae or production of EPS affect the surface attachment of microbial cells. Fimbriae being possessed with hydrophobic amino acids contribute to cell surface hydrophobicity and attachment by overcoming electrostatic repulsion between cell and surface. Similarly, flagella also promote bacterial attachment to the surface even against the water flow compared with nonmotile bacterial cells (Berne, Ellison, Ducret, & Brun, 2018). All these processes lead to successful attachment of the bacterial cells to the favorable surface and produce EPS to form a biofilm-like structure.

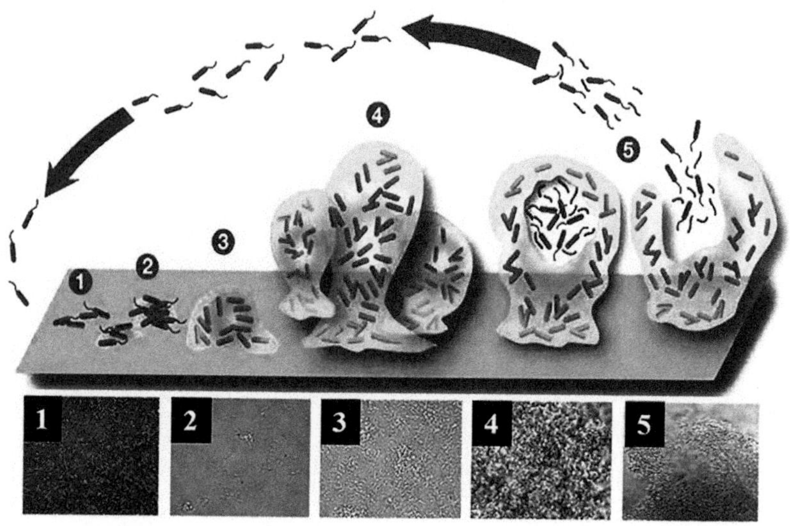

FIG. 3 Developmental stages in biofilm formation. *(Reproduced with permission from Verderosa, A. D., Totsika, M., & Fairfull-Smith, K. E. (2019). Bacterial biofilm eradication agents: A current review. Frontiers in Chemistry, 7, 824.)*

The occurrence of uronic acids or ketal-linked pyruvates in the EPS permits binding with divalent cations such as calcium and magnesium, which cross-link with EPS and deliver binding force in a developed biofilm (Donlan, 2002). The architecture of biofilm changes continuously due to external and internal processes occurring in the immediate environment as described in the interactions of *P. aeruginosa* and *P. putida* in the flow cell system (Tolker-Nielsen & Molin, 2000). The development of biofilm is divided into multiple steps: irreversible attachment, maturation, cell-to-cell communication, genetic regulation, and detachment. The conversion of a weak reversible attachment to a strong irreversible attachment is governed by the extension and retraction of type IV pili to form microcolonies (Piepenbrink & Sundberg, 2016). Production of EPS by biofilm-resident bacteria is the differential feature from the bacteria that are simply attached to the substratum. The production of alginate, best characterized EPS, is governed by a variety of environmental factors. This includes alginate promoters, *algC*, and *algD*, which are activated under conditions of nitrogen limitation, membrane perturbation, and high osmolarity (Hay, Wang, Moradali, Rehman, & Rehm, 2014). Studies have indicated that alginate production is an early event in the biofilm formation that prepares the structural and mechanical framework for the maturation of biofilm (Hay, Rehman, Moradali, Wang, & Rehm, 2013). Biofilm maturation, the next step in biofilm development, results in the formation of channels, pores, and redistribution of bacteria inside the biofilm (Quan et al., 2021). The proteome analysis in a mature biofilm showed that 300 out of 800 proteins were detectable in mature biofilm, which was not detectable in planktonic bacteria (Sauer, Camper, Ehrlich, Costerton, & Davies, 2002). Another study with DNA microarray technology showed that 70 genes including those of translation, membrane transport, and metabolism were found to be upregulated in mature biofilms (Whiteley et al., 2001) (Fig. 3).

Bacteria communicate with each other through signaling molecules produced by members of their species for density-dependent growth. In the biofilm mode of life, signaling molecules called autoinducers have been detected in natural biofilms, which suggest density-dependent gene regulation (McLean, Whiteley, Stickler, & Fuqua, 1997). A study showed that deletion of genes responsible for the production of autoinducers leads to an 80% reduction in thickness of biofilms compared with their wild-type counterparts (Davies et al., 1998). But when the deletion mutants were grown in media containing autoinducers, they formed biofilms indistinguishable from wild types confirming the role of autoinducers in the development of the complex architecture of mature biofilms. These results suggest that intercellular signaling mechanisms are complex circuits of processes under intercellular and extracellular controls. Programmed lysis and cell death play an important role in the development of biofilms by releasing the nutrient and genomic DNA, which improves biofilm adhesion and stability (Bayles, 2007). A study performed on inhibition of *P. aeruginosa* biofilm using DNase I revealed that released DNA is derived from the genome of *P. aeruginosa* and occupies an ordered pattern in the biofilm (Allesen-Holm et al., 2006). This concluded that released DNA is an integral part of the biofilm matrix and supports the notion that biofilm cell lysis contributes to the structural stability of biofilm.

4. Interactions in biofilms

Biofilms provide a unique mode of life to the bacteria by making them survive under hostile environments by increasing resistance toward disinfectants and antimicrobials (Mah & O'Toole, 2001). Most of the studies are performed on single-species biofilm, but the complexity inside the biofilm increases with multispecies biofilms. This complexity could be at the level of the cell-to-cell communication and competition for physical and chemical resources. The development of biofilm relies highly on the interactions happening between the EPS encased bacteria inside a multispecies biofilm. One of the important modes of such type of communication is quorum sensing, which induces the neighboring cells inside the biofilms by using autoinducers (Pena et al., 2019). A list of quorum sensing systems and their controlled social trait is described in Table 1. The outcome of this interaction includes biofilm resistance to stress, altered EPS composition, improved utilization of nutrients, and distribution of biofilm community. They also get benefitted from division of labor, collective and cooperative activities with their neighbors in the biofilm (Kuramitsu, He, Lux, Anderson, & Shi, 2007). The interaction between the biofilm bacteria in a multispecies biofilm exhibits synergistic and mutualistic behavior that facilitates biofilm formation. For example, the formation of mixed biofilm by *Actinomyces naeslundii* and *Streptococcus oralis* is dependent on the production of AI-2 (Autoinducer 2) by *S. oralis* higher than *A. naeslundii* (Rickard et al., 2006). Autoinducer peptides are another group of signaling molecules utilized by Gram-positive bacteria. These peptides mediate interspecies communication and manipulate quorum sensing signals to affect their biofilm formation by inhibiting the concentration of extracellular signaling molecules (Wang, Deutch, Hong, & Kuramitsu, 2011) (Fig. 4).

The diversity and heterogeneity in biofilms impact the communication ability as well as communication distance in a mixed-species biofilm. An interesting study found that AHLs produced by an individual cell are sufficient enough to communicate the adjacent even if they are separated by a dense population (Gantner et al., 2006). Another study on communication range found that signaling occurs within and not across the cell clusters. Both the studies hypothesized that aggregation of the cells has a greater impact on the accumulation of signals rather than population density (Hense et al., 2007). Species inside the biofilm either compete for nutrients or cooperate to better utilize the nutrients and survive harsh environments giving a perfect example of synergism and antagonism. These interactions include coexistence, cooperation, competition, growth inhibition, and resistance to host defense and antimicrobial agents (Elias & Banin, 2012). Another level of interaction between different species during the establishment of multispecies biofilm results in changes in gene expression. This could be due to synergistic, antagonistic, or any other interaction that arose from environmental cues, which ultimately leads to architecture and shaping the community of multispecies biofilm (Luppens et al., 2008). Sometimes the interaction between two species in the same biofilm community results in the evolution of a species variant (Hansen et al., 2007). All the above interaction suggests that interaction inside the biofilm should be taken into account before designing experiments or therapies.

Interspecies interactions were significantly identified for the first time in the biofilms of the oral cavity and also from various environments such as soil, seawater, and dairy farms (Kolenbrander, Palmer, Periasamy, & Jakubovics, 2010; Schwering, Song, Louie, Turner, & Ceri, 2013). These interactions (also referred to as synergistic interactions) in the multispecies biofilm sometimes lead to enhancement of a

TABLE 1 **Bacterial quorum sensing systems and their controlled traits.**

Microorganism	Major signal molecules	Regulatory system	Benefits
Vibrio harveyi	HAI-1, CAI-1 AI-2	LuxLM/LuxN LuxP/LuxQ	Bioluminescence emission, symbiosis
Staphylococcus aureus	AIP-I, AIP-II, AIP-II, AIP-IV	AgrC/AgrA	Biofilm formation, virulence factors
Myxococcus xanthus	A-signal C-signal	SasSRN	Fruiting body formation or Sporulation
Bacillus subtilis	ComX CSF (PhrC) PhrA, -E, -F, -K, -H	ComP/ComA	Rap proteins Competence, sporulation, biofilm formation, antibiotic production,
Pseudomonas aeruginosa	3O-C12-HSL C4-HSL	LasI/LasR RhlI/RhlR OscR (orphan)	Structured biofilm formation, Virulence factors.

Modified from Li, Y.-H., & Tian, X. (2012). Quorum sensing and bacterial social interactions in biofilms. *Sensors, 12*(3), 2519–2538.

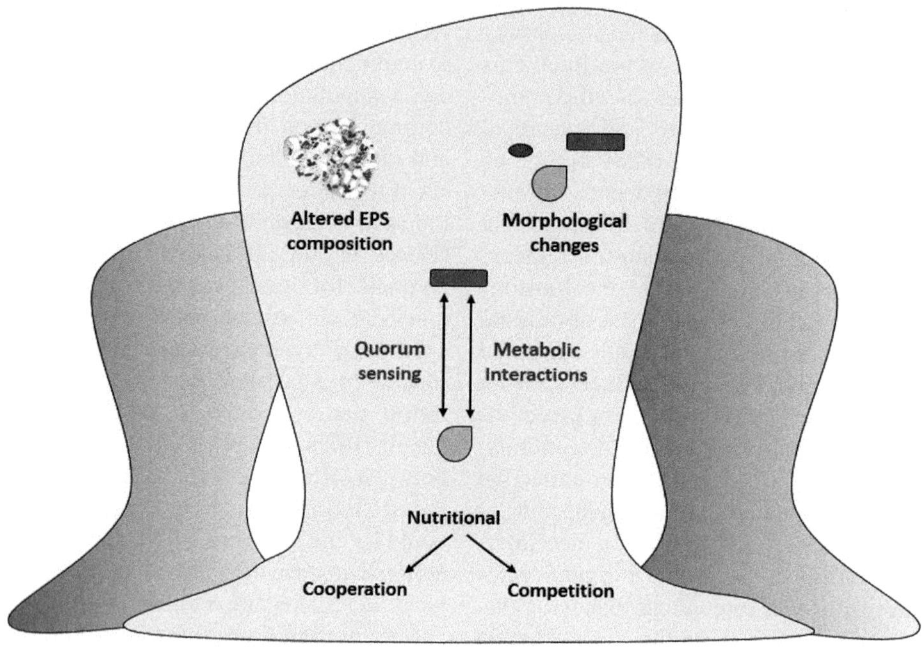

FIG. 4 Individual and social interactions in biofilm community. *(Reproduced with permission from Elias, S., & Banin, E. (2012). Multi-species biofilms: Living with friendly neighbors. FEMS Microbiology Reviews, 36(5), 990–1004.)*

protective effect or function of the biofilm, which cannot be extrapolated from a monospecies biofilm. The biofilm-associated protection is induced by a concert of multiple factors such as changed biofilm structure and gene expression, virulence factors, and reduced growth rates (Mah & O'Toole, 2001). A study found that different marine isolates when cocultured produced 167% more biofilm biomass when compared with monospecies biofilm (Burmølle et al., 2006). The presence of *S. aureus* and *C. albicans* in most of the infections displays enhanced virulence and antibiotic resistance due to antibiotic resistance provided by *S. aureus* and protection against immune evasion and oxidative stress (Harriott & Noverr, 2009). Biofilms are also involved in many chronic infections where they play a significant role in boosting virulence. Autoinducers produced by separate microcolonies may reach a certain distance and signals from thousands of such microcolonies may even go further, which concludes that different species in a microcolony affect each other in certain ways (Fazli et al., 2009). Biofilms are also known to be involved in the degradation of a variety of organic compounds. The enzymes involved in the degradation of the substrate lie in the close vicinity to the cells producing it so that the product formed is utilized by the nearby cells (Flemming & Wingender, 2010). The structure of the biofilm matrix absorbs substrate in such a way that inhabits the existence of aerobes as well as anaerobes for an efficient degradation (Burmølle, Kjøller, & Sørensen, 2012). It is also observed that the addition of emulsifier-producing bacterial species enhances the process of degradation and adds a new product to the degrading consortia (Ron & Rosenberg, 2002). Several conflicting pieces of evidence arise from the role of plasmid-mediated horizontal gene transfer (HGT) on the formation of biofilm (Reisner, Höller, Molin, & Zechner, 2006; Røder, Hansen, Sørensen, & Burmølle, 2013). The multispecies biofilm community facilitates the availability of plasmid recipients to facilitate the production of novel genotypic combinations mediated by HGT. Another mode of HGT in biofilm is transformation due to the availability of extracellular DNA as a source of new information for the members of the biofilm community (Antonova & Hammer, 2011). This signifies biofilms serve as a hotspot of novel genetic recombinations due to the presence of a rich source of genetic materials and their recipients assisted by HGT.

5. Biofilms in marine habitats

Biofilms in the aquatic habitats, prominently the marine biofilms, are often termed as microfouling due to the presence of communities of mixed microorganisms surrounded by EPSs. In aquatic ecosystems, the number of microorganisms on the surface vastly outnumbers those in the suspension stage indicating that biofilm is a more dominant form of lifestyle (Flemming & Wingender, 2010). Bacterial colonization of the surface followed by biofilm development provided them some important advantages such as availability of inorganic micronutrients, organic carbon, energy sources, electron donors, and acceptors (Hall-Stoodley et al., 2004). Biofilm-forming microorganisms, abundantly present in the seawater, colonize any natural or artificial structure dipped in the sea to form a highly complex and dynamic 3D surface structure (Huggett, Nedved, & Hadfield, 2009). Natural as well as artificial structures are continuously colonized by biofilm-forming bacteria leading to macrofouling colonization along with a huge economic loss (Schultz, Bendick, Holm, & Hertel, 2011). The community composition of the marine biofilms is directly affected by the biogeochemical, physical, seasonal, and geographical variations where some of them have a direct impact on the structural development of biofilm (Briand et al., 2012). A diverse form of deep-sea biofilm communities was described and found that biofilm diversity increases with depth and does not affect by the type of substratum (Bellou, Papathanassiou, Dobretsov, Lykousis, & Colijn, 2012). It is also found that known human pathogens such as *E. coli* and *V. cholera* were also found in the marine biofilms giving an indication of pathogen dissemination in the marine ecosystem (Inbakandan, Murthy, Venkatesan, & Khan, 2010). Biofilm community composition decreases with increased shear stress leading to reduced biofilm maturation and maintenance in early developmental stages.

Once a surface is submerged in seawater, it is quickly covered by the proteins and glycoproteins, which act as a conditioning film for the attachment of microbial cells. Following the formation of conditioning film, bacteria attach and form microcolonies, which further mature to form a biofilm community (Flemming et al., 2016). The bacterial cells in the biofilm produce EPS, which forms a highly hydrated matrix of biofilm. The formation of biofilm alters the physicochemical properties of the surface and influences colonization of diatoms, cyanobacteria, and microalgae, all of which lead to macrofouling of the surface (Dang & Lovell, 2016). The final step includes the attachment of protozoans and invertebrate larvae, which culminates in the formation of a mature biofilm. Diatoms colonizing the mature biofilms get benefitted from the ready availability of nutrients, protection from UV, toxic compounds, pathogens, and protozoan grazing (Matz & Kjelleberg, 2005). Multiple surface-related factors promote bacterial colonization on the surface, such as wettability, hydrophobicity, surface tension, surface topography. Some other environmental parameters such as temperature, pH, nutrient availability, dissolved oxygen, and radiation also play an important role in the composition and ecophysiology of biofilm. A study found that decreasing the nutrient availability reduced the diversity of biofilm-forming microorganisms indirectly affecting the biofilm stability and function (Lawes, Neilan, Brown, Clark, & Johnston, 2016). It is also found that as the biofilm ages, its bacterial diversity becomes progressively similar to that of the local bacterial community (Hall & Pepe-Ranney, 2015). Microorganisms are the primary colonizers of the surface that later forms biofilms, they are responsible for the settlement of barnacle and mussel larvae to introduce macrofouling (Salta, Wharton, Blache, Stokes, & Briand, 2013). Biofilms are known to influence the settlement of micro and macrofouling algae or their spores in a positive, negative, or neutral manner. Studies have found out that single-species biofilms of certain bacterial species such as strains of *Vibrio* and *Shewanella* promote algal settlement while species of *Pseudoalteromonas* have strong antialgal activity for algal species of *Ulva*, *Polysiphonia*, and some diatoms (Bernbom, Ng, Kjelleberg, Harder, & Gram, 2011; Patel, Callow, Joint, & Callow, 2003). On the other hand, bacterial biofilms formed by strains of *Halomonas* and *Phaeobacter* show no effect on the settlement of *Ulva* sp. and diatoms. Microorganisms are also known for the property of bioremediation of xenobiotic compounds, and thus, they are present in oil and natural gas facilities. The anaerobic conditions along with the availability of nutrient substrates favor the formation of biofilm and microbial-induced corrosion (Bakar et al., 2017). The corroded steel pipe in the offshore facility of the Gulf of Mexico represented more than 98% of the population in the biofilms from the genus *Desulfovibrio*, sulfate, and iron-reducing bacteria (Vigneron et al., 2016).

The surface is considered as the hotspot of microbial activities due to a comparable distinction between surface-associated and free-living microorganisms. Microorganisms attached to the surface have a distinct morphology, abundance, composition, dynamics, and functions (Caron, Davis, Madin, & Sieburth, 1982; Grossart, 2010). The association of microorganisms in the form of biofilms on the surface leads to intense competition for resources, nutrients as well as a deteriorating condition due to the accumulation of metabolic wastes. Viruses are also found to be concentrated with marine particles, which infect microorganisms inside the biofilms by the rapid production of viral particles due to high cell density (Tan, Dahl, & Middelboe, 2015). Biofilms associated with the surface tend to absorb toxic organic compounds and heavy metals (Hao, Li, Kappler, & Obst, 2013). Surface-associated bacterial communities participate in multiple carbon cycling processes such as photosynthesis, methane production, and oxidation, heterotrophic respiration, and many more (Orcutt, Sylvan, Knab, & Edwards, 2011). They are also involved in N cycling processes such as nitrification, denitrification, and anaerobic ammonium oxidation (Klawonn, Bonaglia, Brüchert, & Ploug, 2015). Surface or particle-associated marine bacteria also play role in iron cycling processes such as siderophore-mediated iron oxidation, iron redox reaction, and iron transport (Li et al., 2014).

6. Biofilm in extreme habitats

Biofilms are aggregations of microbial cells encapsulated in EPSs produced by the underlying cells. Biofilms act as a protective shelter for extreme environmental conditions such as temperature, ultraviolet radiation, pH, salinity, and pressure by acting like a protective layer for encapsulated cells. These extreme environments were previously considered inhabitable for life, but after huge explorations, they are found to contain a huge diversity of microorganisms called extremophiles (Wakai, 2019). These extremophiles could be acidophiles, alkaliphiles, halophiles, thermophiles, piezophiles, etc., with each category having its survival mechanism including biofilm formation (Flemming et al., 2016). The presence of biofilms in deep-sea volcanogenic sulfide deposits, hydrothermally influenced vents, deep-sea vents, and hot springs suggests that biofilms can protect the underlying microorganisms from extreme environments (Lawes et al., 2016; Reysenbach & Cady, 2001; Westall et al., 2001). Additionally, resistant biofilms in the extreme environment make the biofilm cells capable of exchanging information more conveniently with the neighboring cells, making it a self-protective mechanism for the growth of microorganisms.

Ultraviolet light from the sun has UV-A, B, and C where UV-A damages DNA by generating ROS and inducing single-strand breaks, UV-B is absorbed by nucleotides to mutate nucleotides while UV-C photoproducts. Biofilm formed by *Listeria monocytogenes* protects them from UV-C radiation isolated from the salmon smokehouse, while those formed by *Deinococcus geothermalis* were more tolerant to UV compared with planktonic cells (Makarova et al., 2007). High microbial diversity in the biofilm of radioactive hydrothermal springs confirms the biofilm adaptation to the high radioactivity environment (Enyedi et al., 2019). An extreme temperature has a significant influence on the biofilm formation by enabling the microorganisms to become more resistant to temperature stress. Biofilms resist the high temperature in the external environment to ensure a suitable interior environment for growth and reproduction. Similarly, when it is low temperature in the external environment, biofilms stabilize the internal environment by preventing the freezing of cells and ensuring their ability to survive. A study found that an incubation temperature of 65°C is more effective in the formation of biofilms in the family *Bacillaceae* compared with 55°C for other thermophilic bacterial genera (Coleri Cihan, Karaca, Ozel, & Kilic, 2017). A study found that four- to fivefold increase in biofilm formation was observed in *Sulfolobus acidocaldarius* and *Sulfolobus solfataricus* when their temperature was changed to 60°C (Koerdt, Gödeke, Berger, Thormann, & Albers, 2010). Similarly, temperature below 10°C leads to a reduction in planktonic bacteria but not in the bacteria on the surface in the form of biofilms. A study also found that *Bacteriovorax* survives 50% longer in biofilms at the temperature of 5°C, and EPS in the biofilms of bacteria from Antarctic regions (*Winogradskyella* and *Shewanella*) forms stable emulsions to protect the underlying cells from extremes of temperatures (Caruso et al., 2018; Williams, Turng, & Kelley, 2009) (Fig. 5).

Acidophiles and alkaliphiles are the bacteria that thrive in acidic and alkaline environments, and the biofilms formed by them help resist the damaging effects of extreme

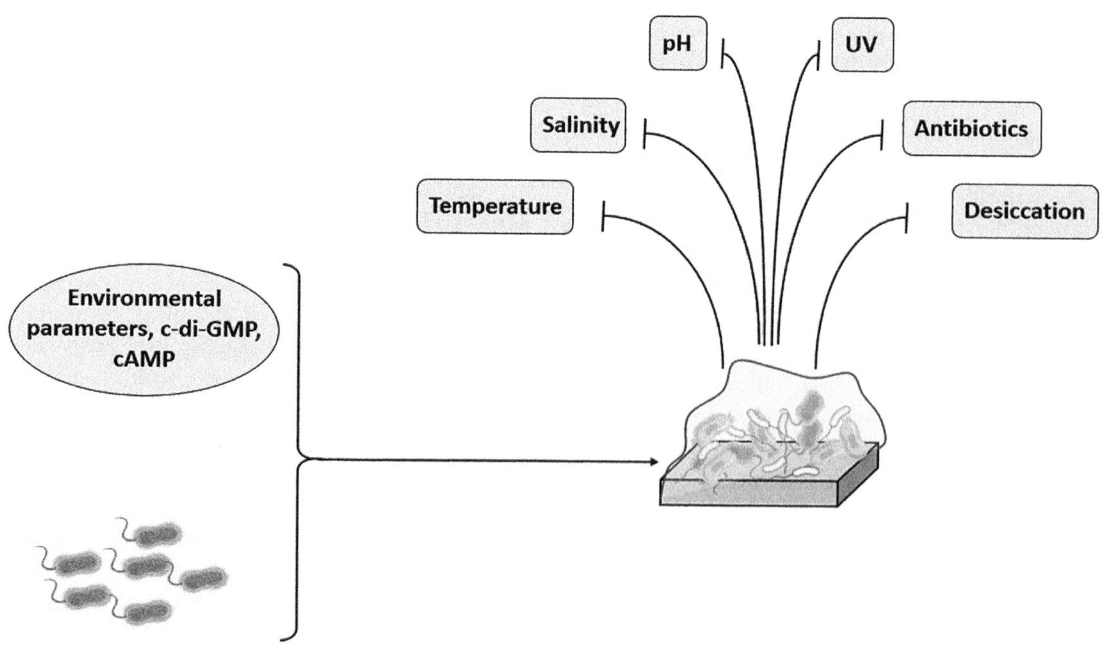

FIG. 5 Schematic showing the adaptability of biofilms in extreme habitats.

pH. Acidic conditions are known to promote the dissolution of heavy metals making the environment more toxic. This combination of acidic pH and toxic environment alters the EPS composition that plays an important role in the adaptation of microorganisms in such environments. EPS is also involved not only in capturing the heavy metals, but also the trace elements as found in the case prevalence of specific sugars such as inositol polyphosphate and methylglucose containing polysaccharides (Hawkins et al., 1993; Shah, Ray, Garg, & Madamwar, 2000). Similarly, under alkaline conditions, the thick layer of EPS provides shelter to alkalophilic communities by allowing them to grow at an external pH of 11.0–11.5 and maintaining an internal pH of 10.4–10.7. This demonstrates biofilms as a survival strategy in the hyperalkaline environments to a larger extent (Charles et al., 2017). Microorganisms encased inside the biofilm are protected from alkaline conditions much better than that of the planktonic microorganisms as found in this study (Chávez de Paz, Bergenholtz, Dahlén, & Svensäter, 2007). But the exact mechanism of the acid and alkali tolerance in biofilm, and whether the biofilm is involved in the development of such tolerance, still needs to be studied.

Microorganisms are also present in the high-salinity environment where most of the other organisms die due to higher osmotic pressure (Kimata-Kino, Ikeda, Kurosawa, & Toda, 2011). These microorganisms escape these extreme salt conditions by forming biofilms as found in the case of *Halomonas stenophila*, which can form biofilms in the presence of 5% salt (Amjres et al., 2015). Biofilms formed by *Kocuria flava* and *Bacillus vietnamensis* produce a huge amount of EPS, which can effectively absorb metals and develop intrinsic resistance. An investigation on their salt resistance mechanism revealed that bacteria stresses with saline conditions maintain cellular activity and produce EPS to continue the formation of biofilms (Kim & Chong, 2017). A study on the composition of EPS with varying salt concentrations concluded with an increase in the protein and polysaccharide content of soluble, loose, and tightly bound EPS (Zhao et al., 2016). EPS being a gel-like matrix is linked to the absorption of various metals and promotion of salt stress under saline conditions. Similar to halophiles, piezophiles are microorganisms that live in high-pressure environments, especially in the deep sea, but they are not well studied for their ability to form biofilms under such environments. Some of the studies suggest that increased pressure upregulates the expression of outer membrane proteins (Simonato et al., 2006). Some other studies have also shown that microorganisms inside the biofilm are more resistant to high hydrostatic pressure compared with planktonic microorganisms (Masanta, Hinz, & Zautner, 2015). Biofilms also protect the underlying cells from high mechanical pressure as found in the comparison study performed between EPS producing and nonproducing *Staphylococcus aureus*, and found EPS producers are more resistant to mechanical pressure due to sudden increase in polysaccharide content (Hou, Veeregowda, van de Belt-Gritter, Busscher, & van der Mei, 2018).

7. Biofilm in humans

For many centuries, humans suffered from life-threatening attacks of bacterial infections by planktonic cells of different bacterial pathogens. We have successfully encountered them by vaccines or antibiotics, but these organisms being successful in surviving in the environment can continuously infect humans. One of their strategies of continuous infection is biofilms, in which they are protected in the environment and also impart resistance toward drugs and antibiotics, which were meant to destroy them. They use different mechanisms to develop resistance toward antibiotics: delayed penetration of antibiotics, altered growth rate, and certain physiological changes (Donlan & Costerton, 2002). Biofilms may develop on inert surfaces such as dead tissue or medical devices and fragments of dead bone. Biofilms are involved in several human-related infections such as native valve endocarditis, otitis media, chronic bacterial prostatitis, periodontitis, and cystic fibrosis that will be discussed further. Apart from being directly involved in infection, biofilms are also formed on several medical devices implanted inside the body. Some of the devices such as prosthetic heart valves, central venous catheters, urinary catheters, IUDs, contact lenses, and central unit water lines (Fig. 6).

(a) **Native valve endocarditis (NVE):** It is a condition resulting from the interaction of vascular endothelial tissues such as aortic, bicuspid, or tricuspid valve with bacteria circulating in the bloodstream (Chambers & Bayer, 2020). Most of the cases of infection are caused by Streptococci (56%) and Staphylococci (25%) by gaining access to the bloodstream by the oropharynx or GI tract (Chamat-Hedemand et al., 2020). Damaged endothelium produces fibronectin that leads to the accumulation of platelets, fibrins, and RBCs at the point of injury as found in the case of nonbacterial thrombotic endocarditis (NBTE). Some of the bacterial species such as *Staphylococcus aureus* and *Streptococcus* display fibronectin receptor that binds to fibronectin molecules at the site of injury (Speziale, Arciola, & Pietrocola, 2019). Some of the Streptococci also produce dextrans that promote thrombus adherence in the case of NBTE. Biofilms on the native heart valves damage the valve tissue, and the release of cells or cell clumps into the bloodstream occurs in the case of NVE biofilms. The mechanism adopted by these bacterial species to form biofilms varies with changing

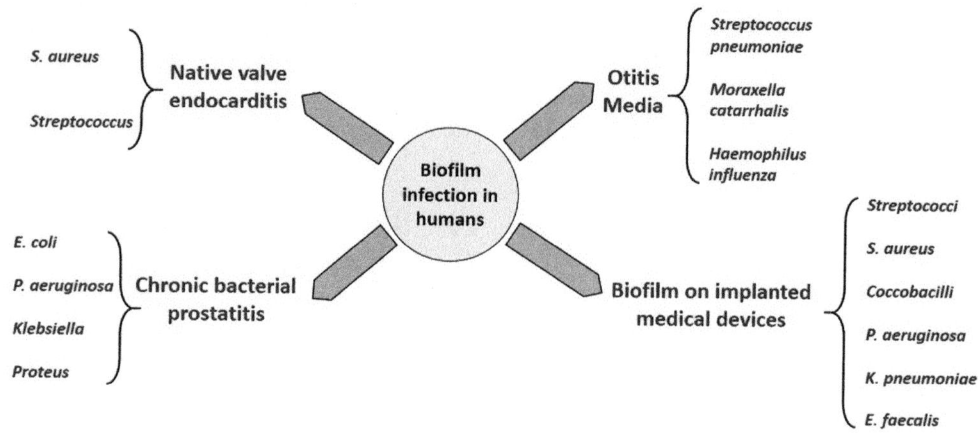

FIG. 6 Biofilm infection in humans.

environmental conditions keeping some of the biofilm elements conserved.

(b) Otitis media: Otitis media is a disease involving inflammation of the mucoperiosteal lining in the middle ear primarily caused by *Streptococcus pneumoniae*, *Moraxella catarrhalis*, and *Haemophilus influenza* and other organisms (Ngo, Massa, Thornton, & Cripps, 2016). Under conditions of chronic infections with highly viscous fluid, tympanostomy tubes are implanted to prevent hearing loss. These tubes are subjected to contamination and biofilm formation by bacterial species such as *Pseudomonas aeruginosa*, *Staphylococcus aureus*, and *Staphylococcus epidermidis* present in the surrounding environment (Ojano-Dirain, Silva, & Antonelli, 2013). Biofilms are known to be more resistant to antibiotics and in the case of biofilm in otitis media, less penetration of antibiotics is observed in the middle ear fluid. These biofilms are known to be the major contributor to the pathogenesis of otitis media and related infections. A large proportion of the biofilm community formed by these pathogenic strains undergoes in VBNC state where traditional techniques fail to detect them, and they also escape defense mechanisms and several antimicrobials.

(c) Chronic bacterial prostatitis: This is the condition of bacterial infection in the prostate gland caused by ascending of bacteria from the urethra or reflux of infected urine (Lipsky, Byren, & Hoey, 2010). Initial infection can be controlled by antibiotics, but their long-term persistence from microcolonies and biofilms that adhere to the epithelial cells of ducts. Major organisms isolated from these infections include *E. coli*, *P. aeruginosa*, *Klebsiella*, *Proteus*, and many others (Skerk et al., 2004). The reason for treatment failure is the local environment surrounding these organisms and their ability to form biofilms resulting in their altered metabolic activity and resistance toward antimicrobials (Liaqat et al., 2019). It is also known that if bacteria persist from acute prostate inflammations, they can form microcolonies or biofilms attached to the epithelium of ductal systems. The formation of biofilms is suspected to persuade an immunologically distinct microenvironment that induces microorganisms to crystallize, calcify, and form calculi that are now considered a very common condition in prostatitis (Bartoletti et al., 2007).

(d) Biofilm on implanted medical devices: Biofilms on different medical devices have been studied in the last several years using basic microbiological tools and techniques such as viable culturing and scanning electron microscopy. Some of the important devices where biofilms are known are prosthetic heart valves where biofilm forms as a result of tissue damage during implantation followed by accumulation of platelets and fibrins. The organisms responsible for these biofilms include *Streptococci*, *S. aureus*, *coccobacilli*, or fungi (Palraj, Knoll, Baddour, & Wilson, 2015). Central nervous catheters caused by *S. aureus*, *P. aeruginosa*, *K. pneumoniae*, *E. faecalis*, and others with an infection rate of 3%–5% pose a greater risk than any other device-related infections (Gahlot, Nigam, Kumar, Yadav, & Anupurba, 2014). These catheters being in the bloodstream are coated with plasma, platelets, and serum proteins that act as a conditioning film to adhere and form biofilms. Hard contact lenses of the eye movement with each blink and oxygen-containing tears beneath the lens promote the adhesion and biofilm formation by *P. aeruginosa* on the surface of the lens (Raksha, Gangashettappa, Shantala, Nandan, & Sinha, 2020). Depending on the lens formulation. This adhesion is found to be further promoted by immunoglobulins, lysozymes, lactoferrins, and mucins. Some other medical devices prone to biofilm infection include urinary catheters, intrauterine devices, and dental unit waterlines (Srivastava, Chandra, & Kumar, 2019).

8. Biofilms on animal surfaces

The outer surface of the marine organisms acts as an interface between the host and biofilm but is often threatened by other organisms through fouling. This type of nontrophic interaction between the host and the growing organism is known as epibiosis (Wahl, 1989). Under these circumstances, if any outer skin is unprotected, it will be covered by micro and macrofoulers within a short time leading to heavyweight, friction, interrupted gas exchange, altered smell. In the case of marine epibiotic biofilms, the constituents are not comprehensively described due to less accessibility of culture techniques and failure of recent molecular techniques to identify all organisms in the biofilm. Biofilm bacteria, compared with planktonic bacteria, are more densely packed, show higher enzyme activity and lateral gene transfer. The presence of biofilm alters the physical and chemical properties of the substratum and can recruit other organisms such as algae, diatoms, larvae, and others (Hadfield, 2011). Epibiosis is determined by the physical and chemical property of the host surface as well as interaction among the settlers to form a specific community (Steinberg, Rice, Campbell, McDougald, & Harder, 2011). All the marine organisms exhibit epibiotic microbial biofilms on their surface (except colonial didemnid ascidians) of variable density and composition. Epibiotic biofilms play role in the replacement of the host epidermis as an interface between host and environment with epibiotic biofilm (Lage & Graça, 2016). Cells inside the biofilm interact, exchange information and metabolites, grow, and disperse cells to form new biofilms on other surfaces. Biofilms on the surface of marine organisms are mostly dominated by prokaryotes while eukaryotes such as diatoms, protozoa, and fungi are found in lower abundance (Webster & Taylor, 2012). The ratio of bacteria:diatoms:flagellates in biofilms is 640:4:1, which shows that every macroorganism is found to maintain a more or less stable relationship with prokaryotes (Railkin, 2003). For example, the density of epibiotic bacteria *Pseudoalteromonas tunicata* is three to four times lower on a eukaryotic host compared with an inert surface (Skovhus, Ramsing, Holmström, Kjelleberg, & Dahllöf, 2004). Reduction in biofilm density compared with undefended surface and their taxonomic composition designates a dynamic or submissive role of host in recruiting epibiotic bacteria.

Algae being uni to multicellular, phylogenetically and morphologically diverse groups with varying lengths are also subjected to bacterial settlement on their surface. Unicellular algae are subjected to bacterial settlement in the form of structured bacterial communities such as biofilms while multicellular algae are covered by diverse microbial communities that include bacteria, fungi, and protists (Follows & Dutkiewicz, 2011; Kohlmeyer & Volkmann-Kohlmeyer, 2003). Algal tissue represents a rich source of organic nutrients that acts as a cue for some bacteria that are capable of its rapid metabolization and start the process of biofilm formation (Goecke, Labes, Wiese, & Imhoff, 2010). There are pieces of evidence showing that biofilm formed on the algal surface differs from that formed in the surrounding environment or nearby undefended surface (Burke, Thomas, Lewis, Steinberg, & Kjelleberg, 2011). At higher trophic levels, Proteobacteria and Bacteroitedes dominate the surface of microalgae in all aquatic environments (Sapp, Wichels, Wiltshire, & Gerdts, 2007). Other phyla such as Verrucomicrobia, Cyanobacteria, and Planctomycetes are recently identified colonizers on the surface of macroalgae. Some recent reports confirm that different macroalgal species in the same environment are associated with different bacterial communities, while the same species in different environments have highly similar bacterial communities (Lachnit, Blümel, Imhoff, & Wahl, 2009; Trias et al., 2012). The epibacterial community composition on the macroalgal surface is affected by spatial and temporal shifts in the aquatic ecosystem as well as age, senescence, and disease associated with the host (Seyedsayamdost, Carr, Kolter, & Clardy, 2011) (Fig. 7).

Studies with the internal or external association of microorganisms with the animal surface are very limited compared with endosymbiotic-associated microorganisms. Invertebrates and vertebrates stay relatively free from macrofouling, but microbial association on their surface has been observed (Dobretsov, Dahms, & Qian, 2006). Specificity in the association of these microbes with animal surface since they are different than those present in the water column or other types of substrata in the neighboring environment (Webster & Taylor, 2012). Epibiotic bacterial

FIG. 7 Host tissue surface of *Fucus vesiculosus* (algae) with unobstructed and biofilm-covered regions. *(Reproduced with permission from Wahl, M., Goecke, F., Labes, A., Dobretsov, S., & Weinberger, F. (2012). The second skin: Ecological role of epibiotic biofilms on marine organisms. Frontiers in Microbiology, 3, 292.)*

communities show changes along with the temporal variation in the environment, but some bacteria remain associated with particular marine organisms and not present in the water columns or with other organisms (Sharp, Eam, Faulkner, & Haygood, 2007). A study also suggested that the interaction of juvenile organisms and epibiotic bacteria plays an important role in the host-specific assemblage of bacteria. The density of epibiotic bacteria varies from the individual organism to species, habitats, regions, or even seasons. For example, bacterial density on sponge surface varies from sterile (60 cells/cm^2) to strongly reduced (4×10^4 cells/cm^2) to normally fouled (7×10^6 cells/cm^2). When the host surface is unsuitable for settlement of bacteria due to altered physiological condition of the epidermis, bacterial diversity inside the epibiotic biofilms on the host surface reduces compared with their neighboring bacteria in a particular habitat.

Bacteria in the epibiotic biofilms sometimes penetrate the host tissue by overcoming the cell membrane and develop an endosymbiotic relationship dependence with their host (Hollants, Decleyre, Leliaert, De Clerck, & Willems, 2011). This type of relationship is observed in major metazoan groups representing their flexibility and evolutionary success. They transfer from one to other hosts in multiple ways such as horizontal transfer, vertical transfer, or by reinfecting a new generation of the host. Epibiotic biofilm is involved in the horizontal and reinfecting mode of transmission (as found in *Vibrio fischeri*) where a likely contact of biofilm is involved. In the absence of their host, the cultivation of these microbial consortia is hindered by technical difficulties elucidating the biological roles of their hosts (Visick & Ruby, 2006).

9. Biofilm in plants

Plants tissues such as leaves, roots, and seeds are usually colonized by bacteria. Each plant tissue being distinct in its physical and chemical attributes brings new challenges for microbial colonists to successful occupation. Association between plants and microbes is defined by the physical interactions between them that result in multicellular assemblies in the form of cell aggregates, microcolonies, and cell clusters. Water availability and saturation level play an important role in biofilm structure (van de Mortel & Halverson, 2004), and tissues such as leaves are comparatively dry compared with roots and soil-borne seeds and thus are more prone to bacterial interactions and biofilm formation. Bacteria deposit on the plant tissues either by active deposition (chemotaxis, motility) or by passive deposition (wind and rain splash). Motility offers important colonization criteria in the case of competitive root colonization and seed association (Turnbull, Morgan, Whipps, & Saunders, 2001). The rhizosphere contains exudates such as vitamins, organic acids, and amino acids that are released by the plants as a result of their metabolism or cell lysis that act as a substrate for microbial growth (Knights, Jorrin, Haskett, & Poole, 2021). But unlike the rhizosphere, the leaf surface has a highly heterogeneous distribution of nutrients originating from wounds or epidermal leakage. Under these conditions, bacteria follow two major strategies for their growth and survival: tolerance strategy to resist tolerance to environmental stress and avoidance strategy for seeking a protected area to avoid stress (Kumar, Singh, Gaurav, Srivastava, & Verma, 2020). Properties of plant surface and bacterial properties also play an important role in the attachment of bacteria to plants (Yaron & Römling, 2014). Biofilm formation on plant surfaces by *S. enterica* and *E. coli* is a survival strategy adopted by bacteria to withstand harsh environmental conditions (Franz et al., 2007). The initial adhesion of bacteria to plant surface is governed by reversible, weak, and nonspecific binding followed by a strong and irreversible attachment (Muhammad et al., 2020). Bacterial biofilms can be divided based on the environment they are colonizing (as described in Table 2): commensals and pathogens colonize the rhizosphere and surfaces, endophytes and vascular pathogens colonize xylem and phloem and find their way to seeds and sprouts (Danhorn & Fuqua, 2007). The development of biofilm also contributes to enhanced virulence in pathogenic bacteria blocking xylem vessels, resistance to antimicrobial compounds, and colonization of specific habitats (Mansfield et al., 2012). Danhorn and Fuqua described a detailed mechanism of biofilm formation in plants involving regulatory mechanisms, motility, and chemotaxis, the role of adhesins and biofilm matrix components, and the role of different metabolites affecting biofilms (Danhorn & Fuqua, 2007). In general, plant-bacterial interactions depend upon bacterial adhesion, cell-cell interactions, and plant colonization that ultimately depends upon multiple environmental, genetic, and structural factors (Bogino, de las Mercedes Oliva, Sorroche, & Giordano, 2013) (Fig. 8 and Table 3).

The rhizospheric zone of soil is supposed to be rich in nutrients and can support 10^6–10^9 bacteria per gram of soil (Mendes, Garbeva, & Raaijmakers, 2013). Rhizosphere-associated *Pseudomonads* are used as a biocontrol agent due to their plant-growth-promoting ability. They are reported to either colonize the root surface in the form of small biofilms on epidermal fissures or dense confluent biofilms on the root surface (Bloemberg, Wijfjes, Lamers, Stuurman, & Lugtenberg, 2000; Walker et al., 2004). Similarly, *Azospirillum brasilense* colonizes root elongation zones and root hairs in the form of biofilms (Assmus et al., 1995). *Bacillus cereus* represents surface-associated biofilms along with their biocontrol ability. Polysaccharides play an important role in biofilm formation as well as their interaction with plant roots and appendages. For example, *A. tumefaciens* produces cellulose to facilitate normal root adherence and exceptionally dense biofilms (Augimeri,

TABLE 2 Functional role of biofilm in EPS plant-associated beneficial and pathogenic bacteria.

S. no.	Bacteria/plant association	Outcome
1.	*B. japonicum* symbiosis with *Glycine max*	Biofilm formation on both inert and biotic surfaces. Roles during the early stages of interaction with the host plant (initial attachment of rhizobia to root epidermal cells).
2.	*M. tianshanense* symbiosis with *Glycyrrhiza uralensis*	Involved in biofilm formation and successful establishment of symbiosis.
3.	*E. amylovora* plant pathogen	Pathogenicity factor required for biofilm formation. Virulence factor also contributes to biofilm formation.
4.	*R. leguminosarum* symbiosis with *Trifolium, Pisum, Vicia,* and *Phaseolus* spp.	Development of a structured biofilm required for infection and nodulation.
5.	*X. campestris, X. axonopodis* plant pathogens	Essential for microcolony formation. Formation of structured biofilms on abiotic surfaces and in infected plants.

Modified from Bogino, P. C., de las Mercedes Oliva, M., Sorroche, F. G., & Giordano, W. (2013). The role of bacterial biofilms and surface components in plant-bacterial associations. *International Journal of Molecular Sciences, 14*(8), 15838–15859.

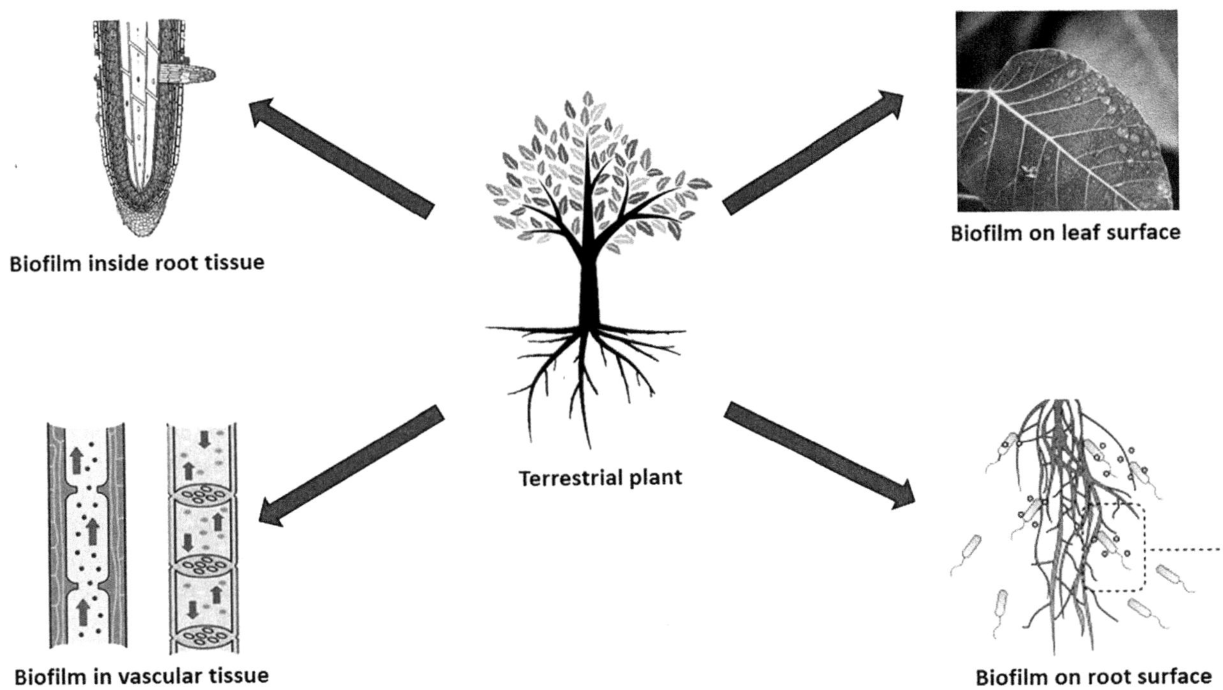

FIG. 8 Microbial biofilm formation on different parts of terrestrial plants.

Varley, & Strap, 2015). Along with the genotypic and physiological characteristic of any particular strain, environmental factors such as pH, mineral, water content, and physiological status of the plant determine the sustainability of bacteria in the rhizosphere (Yan et al., 2021).

Bacterial adherence with seeds acts as an initial inoculum that influences rhizosphere colonization, and sometimes they also act as a common source of human infections. Some studies found nonpathogenic actinobacteria in wheat tissues that were derived from interior colonization of surface-sterilized seeds (Coombs & Franco, 2003). Bacteria attached to seeds act as an initial inoculum for colonization of rhizosphere while endophytic bacteria of seeds transfer to multiple plant generations. Sometimes, animal pathogens coincidently come in contact with seeds (*Salmonella* in cantaloupe melons, *E. coli* in mung bean) due to contaminated irrigation water or processing facilities (Aruscavage, Lee, Miller, & LeJeune, 2006). This leads to their surface attachment and

TABLE 3 **Beneficial bacterial biofilm associated with plant roots.**

S. no.	Bacterial species and associated plant	Beneficial trait
1.	*Pseudomonas entomophila* (rhizoplane of wheat)	Tolerance to abiotic stress
2.	*Bacillus subtilis* (rhizosphere of Arabidopsis)	Biocontrol
3.	*Pseudomonas putida* (rhizosphere of chickpea)	Plant-growth-promoting activities
4.	*Paenibacillus polymyxa* (rhizosphere of wheat)	Protection against abiotic and pathogen stress

Modified from Pandit, A., Adholeya, A., Cahill, D., Brau, L., & Kochar, M. (2020). Microbial biofilms in nature: Unlocking their potential for agricultural applications. *Journal of Applied Microbiology, 129*(2), 199–211.

biofilm formation enabling them to withstand washing and antibacterial treatments.

Biofilms formed by vascular pathogens are dependent upon the insect vectors or open wounds for dissemination. The phloem of the plant being rich in nutrients is not an ideal habitat for bacteria (except the genera *Spiroplasma* and *Phytoplasma* that reside in phloem) due to higher osmotic pressure (Bové, Renaudin, Saillard, Foissac, & Garnier, 2003). Thus, bacteria indirectly reside in the gut of phloem sap-feeding insects where they transmit by membrane proteins. Pathogens colonizing the xylem are more widespread compared with phloem and have pronounced economic importance. Biofilm cells of *Xylella fastidiosa* responsible for Pierce's disease of grapevine reside in the foregut tissue of insects (Newman, Almeida, Purcell, & Lindow, 2003). Colonization of xylem depends upon the *rpf* quorum sensing system and diffusible signal factors that govern insect-specific adhesion factors (Newman, Almeida, Purcell, & Lindow, 2004). *Xanthomonas campestris* causes black rot on cruciferous plants by accessing their vascular system through wounded root tissues (Vicente & Holub, 2013). On the other hand, *Pantoea stewartii* causes Stewart's wilt disease in maize in which bacteria residing in the xylem produce huge exopolysaccharides as a function of the cell density regulatory system (Carlier, Burbank, & von Bodman, 2009).

10. Bacteriophages and biofilms

The ability to form biofilms is now considered omnipresent for all the bacterial species present in the ecosystem. Biofilms represent resistance toward biocides and antibiotics due to the presence of extracellular matrix and metabolically inactive persister cells. The presence of different nutrient and metabolic conditions creates distinct zones within the biofilm. Biofilms also use quorum sensing to communicate with the surrounding community rather than single independent cells (Li & Tian, 2012). Due to their resistance to antibiotics, biofilms are involved in multiple bacterial infections of clinical origin. Bacteriophages are naturally occurring viruses that specifically infect bacteria independent of their antibiotic and biocidal resistance. Bacteriophages either insert their genome into the bacterial genome and sustain a lysogenic lifestyle or destroy their host by undergoing lytic mode of infection. Interaction of bacteriophages with biofilm is different than that of chemical antibiotics due to the presence of certain enzymes in the phage genome that can degrade the EPS of biofilms (Ferriol-González & Domingo-Calap, 2020). Once the phages reach inside the biofilm and find their suitable host, they infect and replicate within the biofilm killing all the biofilm cells leading to lysis of biofilm. A study found that bacteriophages can grow inside the *P. aeruginosa* biofilms in the early stages of biofilm growth (Chegini et al., 2020). While this strain can block the activity of antibiotics, bacteriophages are found to kill the bacteria under situations where antibiotics are unable to do so. However, filamentous phages of *P. aeruginosa* (Pf phages) are known to promote biofilm formation in cystic fibrosis. The matrix of *P. aeruginosa* assembles into liquid crystals by entropic interactions between polymers and Pf phages that increase antibiotic tolerance and desiccation in biofilms (Secor et al., 2015) (Fig. 9).

Interaction of phages with their bacterial hosts becomes very limited when it happens inside biofilms in natural environments. This leads to the reduced dissolution of particulates and solutes, heterogeneity in nutritional access, and physiological state. Some studies have found a considerable difference between phage host interaction in biofilms and mixed liquid cultures (Pires, Melo, & Azeredo, 2021). A relative increase in the number of phage-resistant cells protects susceptible cells from phage infection, and thus, an array of susceptible and resistant phenotypes is established in biofilm (Simmons et al., 2020). This also confirms that the phage-susceptible phenotype is protected from phage exposure if they are growing next to the phage-resistant phenotype. Studies have found out that biofilm matrix did not protect the underlying cells from phage infection, possibly due to the presence of insufficient polymer on cell aggregates or diffused matrix to allow easy penetration of phages (Doolittle, Cooney, & Caldwell, 1995). The multiplicity of infection (MOI) for phages remains low at the starting of infection but may increase by the lysis of planktonic cells that will infect biofilms. While in the case of multispecies biofilms, the physiological state and availability of receptors on the host cell surface impact the outcome of

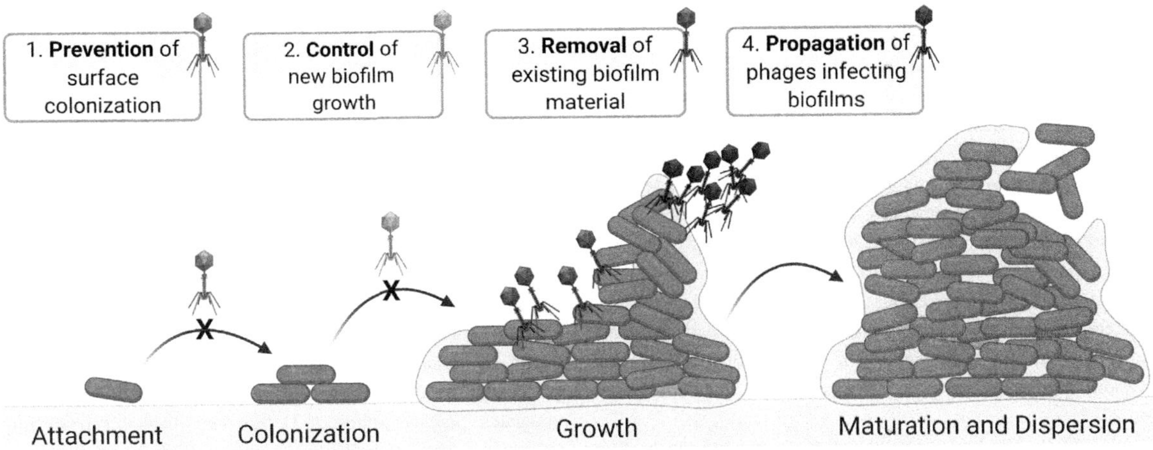

FIG. 9 Different aspects of biofilm development and phage-biofilm interactions. *(From Abedon, S. T., Danis-Wlodarczyk, K. M., Wozniak, D. J., & Sullivan, M. B. (2021). Improving phage-biofilm in vitro experimentation. Viruses, 13(6), 1175.)*

phage infection. Bacterial coaggregation observed in the biofilm communities from different habitats and tight cell-cell binding results in the blocking of phage receptors (Rickard, Gilbert, High, Kolenbrander, & Handley, 2003). In both cases, bacteriophages can be engineered to express biofilm degrading enzymes to kill bacterial cells inside the matrix (Lu & Collins, 2007).

11. Conclusion

Biofilms are microbial communities and encased in a self-secreted EPS that occupied all the ecological niches including extreme environments. Based on the molecular and microscopic techniques, it is found that the development of biofilm is a well-regulated process that results in the formation of complex communities. Some studies have found that EPS molecules contain biopolymers that impart physical properties to the matrix, along with retaining distinct molecules. Also, EPS of biofilm contains water channels that provide oxygen, nutrients, and other metabolites for the maintenance of biofilm. The formation of the biofilm community is also regulated by the integration of internal and external signals by quorum sensing molecules, determining density and type of neighbors and behaviors associated with morphological changes. The biofilm characteristics, such as age and composition, determine their susceptibility to antimicrobial treatments. Apart from the ecosystem, they have started colonizing living organisms including algae, plants, animals, and even humans. In humans, they have developed an extreme tolerance to antimicrobials leading to life-threatening conditions. While in marine habitats, biofilm on the surface is categorized by bacterial diversity in the immediate environment and their ability to compete with other organisms to form biofilm. Marine biofilms are highly dynamic and change over time and interact with distinct environmental parameters. Some of the marine biofilms are involved in the release of chemical cues that induce larval settlement and metamorphosis. Conclusively, bacteriophages play a dual role by promoting as well as inhibiting the biofilm formation in different natural environments. Many other aspects of environmental biofilms and their interaction with their surroundings need to be studied to develop a better understanding.

References

Absalon, C., Dellen, K. V., & Watnick, P. I. (2011). A communal bacterial adhesin anchors biofilm and bystander cells to surfaces. *PLoS Pathogens*, *7*(8), e1002210.

Allesen-Holm, M., Barken, K. B., Yang, L., Klausen, M., Webb, J. S., Kjelleberg, S., et al. (2006). A characterization of DNA release in *Pseudomonas aeruginosa* cultures and biofilms. *Molecular Microbiology*, *59*(4), 1114–1128.

Allison, S. D. (2005). Cheaters, diffusion and nutrients constrain decomposition by microbial enzymes in spatially structured environments. *Ecology Letters*, *8*(6), 626–635.

Amjres, H., Béjar, V., Quesada, E., Carranza, D., Abrini, J., Sinquin, C., et al. (2015). Characterization of haloglycan, an exopolysaccharide produced by *Halomonas stenophila* HK30. *International Journal of Biological Macromolecules*, *72*, 117–124.

Antonova, E. S., & Hammer, B. K. (2011). Quorum-sensing autoinducer molecules produced by members of a multispecies biofilm promote horizontal gene transfer to *Vibrio cholerae*. *FEMS Microbiology Letters*, *322*(1), 68–76.

Aruscavage, D., Lee, K., Miller, S., & LeJeune, J. T. (2006). Interactions affecting the proliferation and control of human pathogens on edible plants. *Journal of Food Science*, *71*(8), R89–R99.

Assmus, B., Hutzler, P., Kirchhof, G., Amann, R., Lawrence, J. R., & Hartmann, A. (1995). In situ localization of *Azospirillum brasilense* in the rhizosphere of wheat with fluorescently labeled, rRNA-targeted oligonucleotide probes and scanning confocal laser microscopy. *Applied and Environmental Microbiology*, *61*(3), 1013–1019.

Augimeri, R. V., Varley, A. J., & Strap, J. L. (2015). Establishing a role for bacterial cellulose in environmental interactions: Lessons learned from diverse biofilm-producing proteobacteria. *Front Microbiol.* [cited 2021 Nov 3]; Available from: https://www.frontiersin.org/articles/10.3389/fmicb.2015.01282/full.

Bakar, A. A., Ali, M. F., Noor, N., Yahaya, N., Ismail, M., & Abdullah, A. (2017). Bio-corrosion of carbon steel by sulfate reducing bacteria consortium in oil and gas pipelines. *Journal of Mechanical Engineering and Sciences*, *11*(2), 2592–2600.

Bartoletti, R., Cai, T., Mondaini, N., Dinelli, N., Pinzi, N., Pavone, C., et al. (2007). Prevalence, incidence estimation, risk factors and characterization of chronic prostatitis/chronic pelvic pain syndrome in urological hospital outpatients in Italy: Results of a multicenter case-control observational study. *The Journal of Urology*, *178*(6), 2411–2415. discussion 2415.

Bayles, K. W. (2007). The biological role of death and lysis in biofilm development. *Nature Reviews Microbiology*, *5*(9), 721–726.

Bellou, N., Papathanassiou, E., Dobretsov, S., Lykousis, V., & Colijn, F. (2012). The effect of substratum type, orientation and depth on the development of bacterial deep-sea biofilm communities grown on artificial substrata deployed in the Eastern Mediterranean. *Biofouling*, *28*(2), 199–213.

Bernbom, N., Ng, Y. Y., Kjelleberg, S., Harder, T., & Gram, L. (2011). Marine bacteria from Danish coastal waters show antifouling activity against the marine fouling bacterium *Pseudoalteromonas* sp. strain S91 and zoospores of the green alga *Ulva australis* independent of bacteriocidal activity. *Applied and Environmental Microbiology*, *77*(24), 8557–8567.

Berne, C., Ellison, C. K., Ducret, A., & Brun, Y. V. (2018). Bacterial adhesion at the single-cell level. *Nature Reviews Microbiology*, *16*(10), 616–627.

Bhagwat, G., O'Connor, W., Grainge, I., & Palanisami, T. (2021). Understanding the fundamental basis for biofilm formation on plastic surfaces: Role of conditioning films. *Frontiers in Microbiology.* [cited 2021 Sep 14]; Available from: https://www.frontiersin.org/articles/10.3389/fmicb.2021.687118/full.

Bloemberg, G. V., Wijfjes, A. H., Lamers, G. E., Stuurman, N., & Lugtenberg, B. J. (2000). Simultaneous imaging of *Pseudomonas fluorescens* WCS365 populations expressing three different autofluorescent proteins in the rhizosphere: New perspectives for studying microbial communities. *Molecular Plant-Microbe Interactions*, *13*(11), 1170–1176.

Bogino, P. C., de las Mercedes Oliva, M., Sorroche, F. G., & Giordano, W. (2013). The role of bacterial biofilms and surface components in plant-bacterial associations. *International Journal of Molecular Sciences*, *14*(8), 15838–15859.

Bové, J. M., Renaudin, J., Saillard, C., Foissac, X., & Garnier, M. (2003). *Spiroplasma citri*, a plant pathogenic molligute: Relationships with its two hosts, the plant and the leafhopper vector. *Annual Review of Phytopathology*, *41*, 483–500.

Briand, J.-F., Djeridi, I., Jamet, D., Coupé, S., Bressy, C., Molmeret, M., et al. (2012). Pioneer marine biofilms on artificial surfaces including antifouling coatings immersed in two contrasting French Mediterranean coast sites. *Biofouling*, *28*(5), 453–463.

Burke, C., Thomas, T., Lewis, M., Steinberg, P., & Kjelleberg, S. (2011). Composition, uniqueness and variability of the epiphytic bacterial community of the green alga *Ulva australis*. *The ISME Journal*, *5* (4), 590–600.

Burmølle, M., Kjøller, A., & Sørensen, S. J. (2012). An invisible workforce: Biofilms in the soil. *Microbial Biofilms: Current Research and Applications*, 61–71.

Burmølle, M., Webb, J. S., Rao, D., Hansen, L. H., Sørensen, S. J., & Kjelleberg, S. (2006). Enhanced biofilm formation and increased resistance to antimicrobial agents and bacterial invasion are caused by synergistic interactions in multispecies biofilms. *Applied and Environmental Microbiology*, *72*(6), 3916–3923.

Carlier, A., Burbank, L., & von Bodman, S. B. (2009). Identification and characterization of three novel EsaI/EsaR quorum-sensing controlled stewartan exopolysaccharide biosynthetic genes in *Pantoea stewartii* ssp. *stewartii*. *Molecular Microbiology*, *74*(4), 903–913.

Caron, D. A., Davis, P. G., Madin, L. P., & Sieburth, J. M. (1982). Heterotrophic bacteria and bacterivorous protozoa in oceanic macroaggregates. *Science.* [cited 2021 Sep 29]; Available from: https://www.science.org/doi/abs/10.1126/science.218.4574.795.

Caruso, C., Rizzo, C., Mangano, S., Poli, A., Di Donato, P., Finore, I., et al. (2018). Production and biotechnological potential of extracellular polymeric substances from sponge-associated Antarctic bacteria. *Applied and Environmental Microbiology*, *84*(4).

Chamat-Hedemand, S., Dahl, A., Østergaard, L., Arpi, M., Fosbøl, E., Boel, J., et al. (2020). Prevalence of infective endocarditis in streptococcal bloodstream infections is dependent on streptococcal species. *Circulation*, *142*(8), 720–730.

Chambers, H. F., & Bayer, A. S. (2020). Native-valve infective endocarditis. *The New England Journal of Medicine*, *383*(6), 567–576.

Charles, C. J., Rout, S. P., Patel, K. A., Akbar, S., Laws, A. P., Jackson, B. R., et al. (2017). Floc formation reduces the pH stress experienced by microorganisms living in alkaline environments. *Applied and Environmental Microbiology*, *83*(6).

Chávez de Paz, L. E., Bergenholtz, G., Dahlén, G., & Svensäter, G. (2007). Response to alkaline stress by root canal bacteria in biofilms. *International Endodontic Journal*, *40*(5), 344–355.

Chegini, Z., Khoshbayan, A., Taati Moghadam, M., Farahani, I., Jazireian, P., & Shariati, A. (2020). Bacteriophage therapy against *Pseudomonas aeruginosa* biofilms: A review. *Annals of Clinical Microbiology and Antimicrobials*, *19*(1), 45.

Coleri Cihan, A., Karaca, B., Ozel, B. P., & Kilic, T. (2017). Determination of the biofilm production capacities and characteristics of members belonging to Bacillaceae family. *World Journal of Microbiology and Biotechnology*, *33*(6), 118.

Coombs, J. T., & Franco, C. M. M. (2003). Isolation and identification of actinobacteria from surface-sterilized wheat roots. *Applied and Environmental Microbiology*, *69*(9), 5603–5608.

Cordero, O. X., Ventouras, L.-A., DeLong, E. F., & Polz, M. F. (2012). Public good dynamics drive evolution of iron acquisition strategies in natural bacterioplankton populations. *Proceedings of the National Academy of Sciences of the United States of America*, *109*(49), 20059–20064.

Dang, H., & Lovell, C. R. (2016). Microbial surface colonization and biofilm development in marine environments. *Microbiology and Molecular Biology Reviews: MMBR*, *80*(1), 91–138.

Danhorn, T., & Fuqua, C. (2007). Biofilm formation by plant-associated bacteria. *Annual Review of Microbiology*, *61*, 401–422.

Davies, D. G., Parsek, M. R., Pearson, J. P., Iglewski, B. H., Costerton, J. W., & Greenberg, E. P. (1998). The involvement of cell-to-cell signals in the development of a bacterial biofilm. *Science*, *280* (5361), 295–298.

de Kerchove, A. J., & Elimelech, M. (2008). Calcium and magnesium cations enhance the adhesion of motile and nonmotile *Pseudomonas aeruginosa* on alginate films. *Langmuir*, *24*(7), 3392–3399.

Dobretsov, S., Dahms, H.-U., & Qian, P.-Y. (2006). Inhibition of biofouling by marine microorganisms and their metabolites. *Biofouling*, *22*(1–2), 43–54.

Donlan, R. M. (2002). Biofilms: Microbial life on surfaces. *Emerging Infectious Diseases*, *8*(9), 881–890.

Donlan, R. M., & Costerton, J. W. (2002). Biofilms: Survival mechanisms of clinically relevant microorganisms. *Clinical Microbiology Reviews*, *15*(2), 167–193.

Doolittle, M. M., Cooney, J. J., & Caldwell, D. E. (1995). Lytic infection of *Escherichia coli* biofilms by bacteriophage T4. *Canadian Journal of Microbiology*, *41*(1), 12–18.

Elias, S., & Banin, E. (2012). Multi-species biofilms: Living with friendly neighbors. *FEMS Microbiology Reviews*, *36*(5), 990–1004.

Enyedi, N. T., Anda, D., Borsodi, A. K., Szabó, A., Pál, S. E., Óvári, M., et al. (2019). Radioactive environment adapted bacterial communities constituting the biofilms of hydrothermal spring caves (Budapest, Hungary). *Journal of Environmental Radioactivity*, *203*, 8–17.

Espinosa-Ortiz, E. J., Eisner, B. H., Lange, D., & Gerlach, R. (2019). Current insights into the mechanisms and management of infection stones. *Nature Reviews. Urology*, *16*(1), 35–53.

Fazli, M., Bjarnsholt, T., Kirketerp-Møller, K., Jørgensen, B., Andersen, A. S., Krogfelt, K. A., et al. (2009). Nonrandom distribution of *Pseudomonas aeruginosa* and *Staphylococcus aureus* in chronic wounds. *Journal of Clinical Microbiology*, *47*(12), 4084–4089.

Ferriol-González, C., & Domingo-Calap, P. (2020). Phages for biofilm removal. *Antibiotics (Basel, Switzerland)*, *9*(5).

Flemming, H.-C., & Wingender, J. (2010). The biofilm matrix. *Nature Reviews Microbiology*, *8*(9), 623–633.

Flemming, H.-C., Wingender, J., & Szewzyk, U. (Eds.). (2011). *Biofilm highlights*. Berlin Heidelberg: Springer-Verlag. [cited 2021 Jul 9]. (Springer Series on Biofilms). Available from: https://www.springer.com/gp/book/9783642199394.

Flemming, H.-C., Wingender, J., Szewzyk, U., Steinberg, P., Rice, S. A., & Kjelleberg, S. (2016). Biofilms: An emergent form of bacterial life. *Nature Reviews Microbiology*, *14*(9), 563–575.

Follows, M. J., & Dutkiewicz, S. (2011). Modeling diverse communities of marine microbes. *Annual Review of Marine Science*, *3*, 427–451.

Franz, E., Visser, A. A., Van Diepeningen, A. D., Klerks, M. M., Termorshuizen, A. J., & van Bruggen, A. H. C. (2007). Quantification of contamination of lettuce by GFP-expressing *Escherichia coli* O157:H7 and *Salmonella enterica* serovar Typhimurium. *Food Microbiology*, *24*(1), 106–112.

Gahlot, R., Nigam, C., Kumar, V., Yadav, G., & Anupurba, S. (2014). Catheter-related bloodstream infections. *International Journal of Critical Illness and Injury Science*, *4*(2), 162–167.

Gantner, S., Schmid, M., Dürr, C., Schuhegger, R., Steidle, A., Hutzler, P., et al. (2006). In situ quantitation of the spatial scale of calling distances and population density-independent N-acylhomoserine lactone-mediated communication by rhizobacteria colonized on plant roots. *FEMS Microbiology Ecology*, *56*(2), 188–194.

Goecke, F., Labes, A., Wiese, J., & Imhoff, J. (2010). Chemical interactions between marine macroalgae and bacteria. *Marine Ecology Progress Series*, *409*, 267–299.

Grossart, H.-P. (2010). Ecological consequences of bacterioplankton lifestyles: Changes in concepts are needed. *Environmental Microbiology Reports*, *2*(6), 706–714.

Hadfield, M. G. (2011). Biofilms and marine invertebrate larvae: What bacteria produce that larvae use to choose settlement sites. *Annual Review of Marine Science*, *3*, 453–470.

Hall, E., & Pepe-Ranney, C. (2015). The effect of carbon subsidies on marine planktonic niche partitioning and recruitment during biofilm assembly. *Frontiers in Microbiology*. [cited 2021 Sep 22]; Available from: https://www.frontiersin.org/articles/10.3389/fmicb.2015.00703/full.

Hall-Stoodley, L., Costerton, J. W., & Stoodley, P. (2004). Bacterial biofilms: From the natural environment to infectious diseases. *Nature Reviews Microbiology*, *2*(2), 95–108.

Hansen, S. K., Haagensen, J. A. J., Gjermansen, M., Jørgensen, T. M., Tolker-Nielsen, T., & Molin, S. (2007). Characterization of a *Pseudomonas putida* rough variant evolved in a mixed-species biofilm with *Acinetobacter* sp. strain C6. *Journal of Bacteriology*, *189*(13), 4932–4943.

Hao, L., Li, J., Kappler, A., & Obst, M. (2013). Mapping of heavy metal ion sorption to cell-extracellular polymeric substance-mineral aggregates by using metal-selective fluorescent probes and confocal laser scanning microscopy. *Applied and Environmental Microbiology*, *79*(21), 6524–6534.

Harriott, M. M., & Noverr, M. C. (2009). Candida albicans and *Staphylococcus aureus* form polymicrobial biofilms: Effects on antimicrobial resistance. *Antimicrobial Agents and Chemotherapy*, *53*(9), 3914–3922.

Hawkins, P. T., Poyner, D. R., Jackson, T. R., Letcher, A. J., Lander, D. A., & Irvine, R. F. (1993). Inhibition of iron-catalysed hydroxyl radical formation by inositol polyphosphates: A possible physiological function for myo-inositol hexakisphosphate. *The Biochemical Journal*, *294*(3), 929–934.

Hay, I. D., Rehman, Z. U., Moradali, M. F., Wang, Y., & Rehm, B. H. A. (2013). Microbial alginate production, modification and its applications. *Microbial Biotechnology*, *6*(6), 637–650.

Hay, I. D., Wang, Y., Moradali, M. F., Rehman, Z. U., & Rehm, B. H. A. (2014). Genetics and regulation of bacterial alginate production. *Environmental Microbiology*, *16*(10), 2997–3011.

Hense, B. A., Kuttler, C., Müller, J., Rothballer, M., Hartmann, A., & Kreft, J.-U. (2007). Does efficiency sensing unify diffusion and quorum sensing? *Nature Reviews Microbiology*, *5*(3), 230–239.

Hobley, L., Harkins, C., MacPhee, C. E., & Stanley-Wall, N. R. (2015). Giving structure to the biofilm matrix: An overview of individual strategies and emerging common themes. *FEMS Microbiology Reviews*, *39*(5). [cited 2021 Jul 9]; Available from: https://pubmed.ncbi.nlm.nih.gov/25907113/.

Høiby, N., Ciofu, O., & Bjarnsholt, T. (2010). Pseudomonas aeruginosa biofilms in cystic fibrosis. *Future Microbiology*, *5*(11), 1663–1674.

Hollants, J., Decleyre, H., Leliaert, F., De Clerck, O., & Willems, A. (2011). Life without a cell membrane: Challenging the specificity of bacterial endophytes within Bryopsis (Bryopsidales, Chlorophyta). *BMC Microbiology*, *11*, 255.

Hou, J., Veeregowda, D. H., van de Belt-Gritter, B., Busscher, H. J., & van der Mei, H. C. (2018). Extracellular polymeric matrix production and relaxation under fluid shear and mechanical pressure in *Staphylococcus aureus* biofilms. McBain AJ, editor *Applied and Environmental Microbiology*, *84*(1). [cited 2021 Oct 11];; Available from: https://journals.asm.org/doi/10.1128/AEM.01516-17.

Huggett, M. J., Nedved, B. T., & Hadfield, M. G. (2009). Effects of initial surface wettability on biofilm formation and subsequent settlement of *Hydroides elegans*. *Biofouling*, *25*(5), 387–399.

Inbakandan, D., Murthy, P. S., Venkatesan, R., & Khan, S. A. (2010). 16S rDNA sequence analysis of culturable marine biofilm forming bacteria from a ship's hull. *Biofouling*, *26*(8), 893–899.

Jones, H. C., Roth, I. L., & Sanders, W. M. (1969). Electron microscopic study of a slime layer. *Journal of Bacteriology*, *99*(1), 316–325.

Kim, L. H., & Chong, T. H. (2017). Physiological responses of salinity-stressed *Vibrio* sp. and the effect on the biofilm formation on a nanofiltration membrane. *Environmental Science & Technology*, *51*(3), 1249–1258.

Kimata-Kino, N., Ikeda, S., Kurosawa, N., & Toda, T. (2011). Saline adaptation of granules in mesophilic UASB reactors. *International Biodeterioration and Biodegradation*, *65*(1), 65–72.

Kimkes, T. E. P., & Heinemann, M. (2020). How bacteria recognise and respond to surface contact. *FEMS Microbiology Reviews*, *44*(1), 106–122.

Klawonn, I., Bonaglia, S., Brüchert, V., & Ploug, H. (2015). Aerobic and anaerobic nitrogen transformation processes in N2-fixing cyanobacterial aggregates. *The ISME Journal*, *9*(6), 1456–1466.

Knights, H. E., Jorrin, B., Haskett, T. L., & Poole, P. S. (2021). Deciphering bacterial mechanisms of root colonization. *Environmental Microbiology Reports*, *13*(4), 428–444.

Koerdt, A., Gödeke, J., Berger, J., Thormann, K. M., & Albers, S.-V. (2010). Crenarchaeal biofilm formation under extreme conditions. *PLoS One*, *5*(11), e14104.

Kohlmeyer, J., & Volkmann-Kohlmeyer, B. (2003). Marine ascomycetes from algae and animal hosts. *Botanica Marina*, *46*(3), 285–306.

Kolenbrander, P. E., Palmer, R. J., Periasamy, S., & Jakubovics, N. S. (2010). Oral multispecies biofilm development and the key role of cell–cell distance. *Nature Reviews Microbiology*, *8*(7), 471–480.

Kolodkin-Gal, I., Romero, D., Cao, S., Clardy, J., Kolter, R., & Losick, R. (2010). d-Amino acids trigger biofilm disassembly. *Science*. [cited 2021 Sep 13]; Available from: https://www.science.org/doi/abs/10.1126/science.1188628.

Kolter, R. (2010). Biofilms in lab and nature: A molecular geneticist's voyage to microbial ecology. *International Microbiology*, *13*(1), 1–7.

Koo, H., Xiao, J., Klein, M. I., & Jeon, J. G. (2010). Exopolysaccharides produced by *Streptococcus mutans* glucosyltransferases modulate the establishment of microcolonies within multispecies biofilms. *Journal of Bacteriology*, *192*(12), 3024–3032.

Krsmanovic, M., Biswas, D., Ali, H., Kumar, A., Ghosh, R., & Dickerson, A. K. (2021). Hydrodynamics and surface properties influence biofilm proliferation. *Advances in Colloid and Interface Science*, *288*, 102336.

Kumar, A., Singh, S., Gaurav, A. K., Srivastava, S., & Verma, J. P. (2020). Plant growth-promoting bacteria: Biological tools for the mitigation of salinity stress in plants. *Frontiers in Microbiology*. [cited 2021 Nov 1]; Available from: https://www.frontiersin.org/articles/10.3389/fmicb.2020.01216/full.

Kuramitsu, H. K., He, X., Lux, R., Anderson, M. H., & Shi, W. (2007). Interspecies interactions within oral microbial communities. *Microbiology and Molecular Biology Reviews: MMBR*, *71*(4), 653–670.

Lachnit, T., Blümel, M., Imhoff, J. F., & Wahl, M. (2009). *Specific epibacterial communities on macroalgae: Phylogeny matters more than habitat*. [cited 2021 Oct 27]; Available from: http://oceanrep.geomar.de/4516/1/2009_Lachnit.pdf.

Lage, O. M., & Graça, A. P. (2016). Biofilms: An extra coat on macroalgae. In *Algae—Organisms for imminent biotechnology* IntechOpen. [cited 2021 Oct 26]; Available from: https://www.intechopen.com/chapters/50582.

Lauten, A., Martinović, M., Kursawe, L., Kikhney, J., Affeld, K., Kertzscher, U., et al. (2021). Bacterial biofilms in infective endocarditis: An in vitro model to investigate emerging technologies of antimicrobial cardiovascular device coatings. *Clinical Research in Cardiology*, *110*(3), 323–331.

Lawes, J. C., Neilan, B. A., Brown, M. V., Clark, G. F., & Johnston, E. L. (2016). Elevated nutrients change bacterial community composition and connectivity: High throughput sequencing of young marine biofilms. *Biofouling*, *32*(1), 57–69.

Lewis, K. (2008). Multidrug tolerance of biofilms and persister cells. *Current Topics in Microbiology and Immunology*, *322*, 107–131.

Li, Y.-H., & Tian, X. (2012). Quorum sensing and bacterial social interactions in biofilms. *Sensors*, *12*(3), 2519–2538.

Li, M., Toner, B. M., Baker, B. J., Breier, J. A., Sheik, C. S., & Dick, G. J. (2014). Microbial iron uptake as a mechanism for dispersing iron from deep-sea hydrothermal vents. *Nature Communications*, *5*(1), 3192.

Liaqat, I., Liaqat, M., Ali, S., Ali, N. M., Haneef, U., & Mirza, S. A. (2019). Biofilm formation, maturation and prevention: A review. *Journal of Bacteriology & Mycology*, *6*(1), 1092.

Lipsky, B. A., Byren, I., & Hoey, C. T. (2010). Treatment of bacterial prostatitis. *Clinical Infectious Diseases*, *50*(12), 1641–1652.

Lu, T. K., & Collins, J. J. (2007). Dispersing biofilms with engineered enzymatic bacteriophage. *Proceedings of the National Academy of Sciences*, *104*(27), 11197–11202.

Luppens, S. B. I., Kara, D., Bandounas, L., Jonker, M. J., Wittink, F. R. A., Bruning, O., et al. (2008). Effect of *Veillonella parvula* on the antimicrobial resistance and gene expression of *Streptococcus mutans* grown in a dual-species biofilm. *Oral Microbiology and Immunology*, *23*(3), 183–189.

Mah, T.-F. C., & O'Toole, G. A. (2001). Mechanisms of biofilm resistance to antimicrobial agents. *Trends in Microbiology*, *9*(1), 34–39.

Mai-Prochnow, A., Lucas-Elio, P., Egan, S., Thomas, T., Webb, J. S., Sanchez-Amat, A., et al. (2008). Hydrogen peroxide linked to lysine oxidase activity facilitates biofilm differentiation and dispersal in several gram-negative bacteria. *Journal of Bacteriology*, *190*(15), 5493–5501.

Makarova, K. S., Omelchenko, M. V., Gaidamakova, E. K., Matrosova, V. Y., Vasilenko, A., Zhai, M., et al. (2007). *Deinococcus geothermalis*: The pool of extreme radiation resistance genes shrinks. *PLoS One*, *2*(9), e955.

Mansfield, J., Genin, S., Magori, S., Citovsky, V., Sriariyanum, M., Ronald, P., et al. (2012). Top 10 plant pathogenic bacteria in molecular plant pathology. *Molecular Plant Pathology*, *13*(6), 614–629.

Masanta, W. O., Hinz, R., & Zautner, A. E. (2015). Infectious causes of cholesteatoma and treatment of infected ossicles prior to reimplantation by hydrostatic high-pressure inactivation. *BioMed Research International*, *2015*, e761259.

Matz, C., & Kjelleberg, S. (2005). Off the hook—How bacteria survive protozoan grazing. *Trends in Microbiology*, *13*(7), 302–307.

McLean, R. J., Whiteley, M., Stickler, D. J., & Fuqua, W. C. (1997). Evidence of autoinducer activity in naturally occurring biofilms. *FEMS Microbiology Letters*, *154*(2), 259–263.

Mendes, R., Garbeva, P., & Raaijmakers, J. M. (2013). The rhizosphere microbiome: Significance of plant beneficial, plant pathogenic, and human pathogenic microorganisms. *FEMS Microbiology Reviews*, *37*(5), 634–663.

Muhammad, M. H., Idris, A. L., Fan, X., Guo, Y., Yu, Y., Jin, X., et al. (2020). Beyond risk: Bacterial biofilms and their regulating approaches. *Frontiers in Microbiology*. [cited 2021 Nov 2]; Available from: https://www.frontiersin.org/articles/10.3389/fmicb.2020.00928/full.

Newman, K. L., Almeida, R. P. P., Purcell, A. H., & Lindow, S. E. (2003). Use of a green fluorescent strain for analysis of *Xylella fastidiosa* colonization of *Vitis vinifera*. *Applied and Environmental Microbiology*, *69*(12), 7319–7327.

Newman, K. L., Almeida, R. P. P., Purcell, A. H., & Lindow, S. E. (2004). Cell-cell signaling controls *Xylella fastidiosa* interactions with both insects and plants. *Proceedings of the National Academy of Sciences*, *101*(6), 1737–1742.

Ngo, C. C., Massa, H. M., Thornton, R. B., & Cripps, A. W. (2016). Predominant bacteria detected from the middle ear fluid of children experiencing otitis media: A systematic review. *PLoS One*, *11*(3), e0150949.

Nivens, D. E., Ohman, D. E., Williams, J., & Franklin, M. J. (2001). Role of alginate and its O acetylation in formation of *Pseudomonas aeruginosa* microcolonies and biofilms. *Journal of Bacteriology*, *183*(3), 1047–1057.

Ojano-Dirain, C. P., Silva, R. C., & Antonelli, P. J. (2013). Biofilm formation on coated silicone tympanostomy tubes. *International Journal of Pediatric Otorhinolaryngology*, *77*(2), 223–227.

Orcutt, B. N., Sylvan, J. B., Knab, N. J., & Edwards, K. J. (2011). Microbial ecology of the dark ocean above, at, and below the seafloor. *Microbiology and Molecular Biology Reviews: MMBR*, *75*(2), 361–422.

Palraj, R., Knoll, B. M., Baddour, L. M., & Wilson, W. R. (2015). 83—Prosthetic valve endocarditis. In J. E. Bennett, R. Dolin, & M. J. Blaser (Eds.), *Mandell, Douglas, and Bennett's principles and practice of infectious diseases* (8th ed., pp. 1029–1040.e4). Philadelphia: W.B. Saunders. [cited 2021 Oct 25]; Available from: https://www.sciencedirect.com/science/article/pii/B9781455748013000837.

Patel, P., Callow, M. E., Joint, I., & Callow, J. A. (2003). Specificity in the settlement—Modifying response of bacterial biofilms towards zoospores of the marine alga *Enteromorpha*. *Environmental Microbiology*, *5*(5), 338–349.

Pena, R. T., Blasco, L., Ambroa, A., González-Pedrajo, B., Fernández-García, L., López, M., et al. (2019). Relationship between quorum sensing and secretion systems. *Frontiers in Microbiology*. [cited 2021 Sep 20]; Available from: https://www.frontiersin.org/articles/10.3389/fmicb.2019.01100/full.

Piepenbrink, K. H., & Sundberg, E. J. (2016). Motility and adhesion through type IV pili in Gram-positive bacteria. *Biochemical Society Transactions*, *44*(6), 1659–1666.

Pires, D. P., Melo, L. D. R., & Azeredo, J. (2021). Understanding the complex phage-host interactions in biofilm communities. *Annual Review of Virology*, *8*(1), 73–94.

Quan, K., Hou, J., Zhang, Z., Ren, Y., Peterson, B. W., Flemming, H.-C., et al. (2021). Water in bacterial biofilms: Pores and channels, storage and transport functions. *Critical Reviews in Microbiology*, 1–20.

Railkin, A. I. (2003). *Marine biofouling: Colonization processes and defenses* (p. 320). Boca Raton, FL: CRC Press.

Raksha, L., Gangashettappa, N., Shantala, G. B., Nandan, B. R., & Sinha, D. (2020). Study of biofilm formation in bacterial isolates from contact lens wearers. *Indian Journal of Ophthalmology*, *68*(1), 23–28.

Reisner, A., Höller, B. M., Molin, S., & Zechner, E. L. (2006). Synergistic effects in mixed *Escherichia coli* biofilms: Conjugative plasmid transfer drives biofilm expansion. *Journal of Bacteriology*, *188*(10), 3582–3588.

Reysenbach, A.-L., & Cady, S. L. (2001). Microbiology of ancient and modern hydrothermal systems. *Trends in Microbiology*, *9*(2), 79–86.

Rickard, A. H., Gilbert, P., High, N. J., Kolenbrander, P. E., & Handley, P. S. (2003). Bacterial coaggregation: An integral process in the development of multi-species biofilms. *Trends in Microbiology*, *11*(2), 94–100.

Rickard, A. H., Palmer, R. J., Blehert, D. S., Campagna, S. R., Semmelhack, M. F., Egland, P. G., et al. (2006). Autoinducer 2: A concentration-dependent signal for mutualistic bacterial biofilm growth. *Molecular Microbiology*, *60*(6), 1446–1456.

Røder, H. L., Hansen, L. H., Sørensen, S. J., & Burmølle, M. (2013). The impact of the conjugative IncP-1 plasmid pKJK5 on multispecies biofilm formation is dependent on the plasmid host. *FEMS Microbiology Letters*, *344*(2), 186–192.

Ron, E. Z., & Rosenberg, E. (2002). Biosurfactants and oil bioremediation. *Current Opinion in Biotechnology*, *13*(3), 249–252.

Salta, M., Wharton, J. A., Blache, Y., Stokes, K. R., & Briand, J.-F. (2013). Marine biofilms on artificial surfaces: Structure and dynamics. *Environmental Microbiology*, *15*(11), 2879–2893.

Sapp, M., Wichels, A., Wiltshire, K. H., & Gerdts, G. (2007). Bacterial community dynamics during the winter–spring transition in the North Sea. *FEMS Microbiology Ecology*, *59*(3), 622–637.

Sauer, K., Camper, A. K., Ehrlich, G. D., Costerton, J. W., & Davies, D. G. (2002). *Pseudomonas aeruginosa* displays multiple phenotypes during development as a biofilm. *Journal of Bacteriology*, *184*(4), 1140–1154.

Schultz, M. P., Bendick, J. A., Holm, E. R., & Hertel, W. M. (2011). Economic impact of biofouling on a naval surface ship. *Biofouling*, *27*(1), 87–98.

Schwering, M., Song, J., Louie, M., Turner, R. J., & Ceri, H. (2013). Multi-species biofilms defined from drinking water microorganisms provide increased protection against chlorine disinfection. *Biofouling*, *29*(8), 917–928.

Secor, P. R., Sweere, J. M., Michaels, L. A., Malkovskiy, A. V., Lazzareschi, D., Katznelson, E., et al. (2015). Filamentous bacteriophage promote biofilm assembly and function. *Cell Host & Microbe*, *18*(5), 549–559.

Seyedsayamdost, M. R., Carr, G., Kolter, R., & Clardy, J. (2011). Roseobacticides: Small molecule modulators of an algal-bacterial symbiosis. *Journal of the American Chemical Society*, *133*(45), 18343–18349.

Shah, V., Ray, A., Garg, N., & Madamwar, D. (2000). Characterization of the extracellular polysaccharide produced by a marine cyanobacterium, *Cyanothece* sp. ATCC 51142, and its exploitation toward metal removal from solutions. *Current Microbiology*, *40*(4), 274–278.

Sharp, K. H., Eam, B., Faulkner, D. J., & Haygood, M. G. (2007). Vertical transmission of diverse microbes in the tropical sponge *Corticium* sp. *Applied and Environmental Microbiology*, *73*(2), 622–629.

Simmons, E. L., Bond, M. C., Koskella, B., Drescher, K., Bucci, V., & Nadell, C. D. (2020). Biofilm structure promotes coexistence of phage-resistant and phage-susceptible bacteria. *mSystems*. [cited 2021 Nov 10]; Available from: https://journals.asm.org/doi/abs/10.1128/mSystems.00877-19.

Simonato, F., Campanaro, S., Lauro, F. M., Vezzi, A., D'Angelo, M., Vitulo, N., et al. (2006). Piezophilic adaptation: A genomic point of view. *Journal of Biotechnology*, *126*(1), 11–25.

Skerk, V., Krhen, I., Schonwald, S., Cajic, V., Markovinovic, L., Roglic, S., et al. (2004). The role of unusual pathogens in prostatitis syndrome. *International Journal of Antimicrobial Agents*, *24*(Suppl. 1), S53–S56.

Skovhus, T. L., Ramsing, N. B., Holmström, C., Kjelleberg, S., & Dahllöf, I. (2004). Real-time quantitative PCR for assessment of abundance of Pseudoalteromonas species in marine samples. *Applied and Environmental Microbiology*, *70*(4), 2373–2382.

Speziale, P., Arciola, C. R., & Pietrocola, G. (2019). Fibronectin and its role in human infective diseases. *Cells*, *8*(12). [cited 2021 Oct 18]; Available from: https://www.ncbi.nlm.nih.gov/pmc/articles/PMC6952806/.

Srivastava, A., Chandra, N., & Kumar, S. (2019). The role of biofilms in medical devices and implants. In S. Kumar, N. Chandra, L. Singh, M. Z. Hashmi, & A. Varma (Eds.), *Biofilms in human diseases: Treatment and control* (pp. 151–165). Cham: Springer International Publishing. [cited 2021 Oct 25]; Available from: https://doi.org/10.1007/978-3-030-30757-8_11.

Steinberg, P. D., Rice, S. A., Campbell, A. H., McDougald, D., & Harder, T. (2011). Interfaces between bacterial and eukaryotic "neuroecology". *Integrative and Comparative Biology*, *51*(5), 794–806.

Stoodley, P., Sauer, K., Davies, D. G., & Costerton, J. W. (2002). Biofilms as complex differentiated communities. *Annual Review of Microbiology*, *56*, 187–209.

Sutherland, I. (2001). Biofilm exopolysaccharides: A strong and sticky framework. *Microbiology (Reading, England)*, *147*(Pt 1), 3–9.

Tan, D., Dahl, A., & Middelboe, M. (2015). Vibriophages differentially influence biofilm formation by *Vibrio anguillarum* strains. *Applied and Environmental Microbiology*, *81*(13), 4489–4497.

Tolker-Nielsen, T., & Molin, S. (2000). Spatial organization of microbial biofilm communities. *Microbial Ecology*, *40*(2), 75–84.

Trias, R., García-Lledó, A., Sánchez, N., López-Jurado, J. L., Hallin, S., & Bañeras, L. (2012). Abundance and composition of epiphytic bacterial and archaeal ammonia oxidizers of marine red and brown macroalgae. *Applied and Environmental Microbiology*, *78*(2), 318–325.

Turnbull, G. A., Morgan, J. A. W., Whipps, J. M., & Saunders, J. R. (2001). The role of bacterial motility in the survival and spread of *Pseudomonas fluorescens* in soil and in the attachment and colonisation of wheat roots. *FEMS Microbiology Ecology*, *36*(1), 21–31.

van de Mortel, M., & Halverson, L. J. (2004). Cell envelope components contributing to biofilm growth and survival of *Pseudomonas putida* in low-water-content habitats. *Molecular Microbiology*, *52*(3), 735–750.

Vicente, J. G., & Holub, E. B. (2013). *Xanthomonas campestris* pv. *campestris* (cause of black rot of crucifers) in the genomic era is still a worldwide threat to brassica crops. *Molecular Plant Pathology*, *14*(1), 2–18.

Vigneron, A., Alsop, E. B., Chambers, B., Lomans, B., Head, I., & Tsesmetzis, N. (2016). Complementary microorganisms in highly corrosive biofilms from an offshore oil production facility. *Applied and Environmental Microbiology*, *82*(8), 2545–2554.

Visick, K. L., & Ruby, E. G. (2006). *Vibrio fischeri* and its host: It takes two to tango. *Current Opinion in Microbiology*, *9*(6), 632–638.

Wahl, H. M. (1989). Marine epibiosis. I. Fouling and antifouling: Some basic aspects. *Marine Ecology Progress Series*. [cited 2021 Oct 26]; Available from: https://www.scienceopen.com/document?vid=af03b4b1-ade3-4218-9b4f-8626e9c93eaa.

Wakai, S. (2019). Biochemical and thermodynamic analyses of energy conversion in extremophiles. *Bioscience, Biotechnology, and Biochemistry*, *83*(1), 49–64.

Walker, T. S., Bais, H. P., Déziel, E., Schweizer, H. P., Rahme, L. G., Fall, R., et al. (2004). *Pseudomonas aeruginosa*-plant root interactions. Pathogenicity, biofilm formation, and root exudation. *Plant Physiology*, *134*(1), 320–331.

Wang, B.-Y., Deutch, A., Hong, J., & Kuramitsu, H. K. (2011). Proteases of an early colonizer can hinder *Streptococcus mutans* colonization in vitro. *Journal of Dental Research*, *90*(4), 501–505.

Watnick, P. I., Lauriano, C. M., Klose, K. E., Croal, L., & Kolter, R. (2001). The absence of a flagellum leads to altered colony morphology, biofilm development and virulence in *Vibrio cholerae* O139. *Molecular Microbiology*, *39*(2), 223–235.

Webster, N. S., & Taylor, M. W. (2012). Marine sponges and their microbial symbionts: Love and other relationships. *Environmental Microbiology*, *14*(2), 335–346.

West, S. A., Diggle, S. P., Buckling, A., Gardner, A., & Griffin, A. S. (2007). The social lives of microbes. *Annual Review of Ecology, Evolution, and Systematics*, *38*(1), 53–77.

Westall, F., de Wit, M. J., Dann, J., van der Gaast, S., de Ronde, C. E. J., & Gerneke, D. (2001). Early Archean fossil bacteria and biofilms in hydrothermally-influenced sediments from the Barberton greenstone belt, South Africa. *Precambrian Research*, *106*(1), 93–116.

Whiteley, M., Bangera, M. G., Bumgarner, R. E., Parsek, M. R., Teitzel, G. M., Lory, S., et al. (2001). Gene expression in *Pseudomonas aeruginosa* biofilms. *Nature*, *413*(6858), 860–864.

Williams, H. N., Turng, B.-F., & Kelley, J. I. (2009). Survival response of *Bacteriovorax* in surface biofilm versus suspension when stressed by extremes in environmental conditions. *Microbial Ecology*, *58*(3), 474–484.

Wimpenny, J., Manz, W., & Szewzyk, U. (2000). Heterogeneity in biofilms. *FEMS Microbiology Reviews*, *24*(5), 661–671.

Wingender, J., & Flemming, H.-C. (2011). Biofilms in drinking water and their role as reservoir for pathogens. *International Journal of Hygiene and Environmental Health*, *214*(6), 417–423.

Yan, Y., Li, B., Huang, Z., Zhang, H., Wu, X., Farooq, T. H., et al. (2021). Characteristics and driving factors of rhizosphere bacterial communities of Chinese fir provenances. *Forests*, *12*(10), 1362.

Yaron, S., & Römling, U. (2014). Biofilm formation by enteric pathogens and its role in plant colonization and persistence. *Microbial Biotechnology*, *7*(6), 496–516.

Zhao, L., She, Z., Jin, C., Yang, S., Guo, L., Zhao, Y., et al. (2016). Characteristics of extracellular polymeric substances from sludge and biofilm in a simultaneous nitrification and denitrification system under high salinity stress. *Bioprocess and Biosystems Engineering*, *39*(9), 1375–1389.

Zheng, S., Bawazir, M., Dhall, A., Kim, H.-E., He, L., Heo, J., et al. (2021). Implication of surface properties, bacterial motility, and hydrodynamic conditions on bacterial surface sensing and their initial adhesion. *Frontiers in Bioengineering and Biotechnology*. [cited 2021 Sep 14]; Available from: https://www.frontiersin.org/articles/10.3389/fbioe.2021.643722/full.

Zobell, C. E. (1943). The effect of solid surfaces upon bacterial activity. *Journal of Bacteriology*, *46*(1), 39–56.

Section C

Biofilm lifestyle of various microorganisms and its control

Chapter 14

Role of biofilms in hospital-acquired infections (HAIs)

Shraddha Sharma, Vaishali Kaushik, and Vishvanath Tiwari
Department of Biochemistry, Central University of Rajasthan, Ajmer, Rajasthan, India

1. Introduction

Hospital-acquired infections (HAIs) are the fourth driving reason for sickness in industrialized nations and address one of the major medical problems (Guggenbichler, Assadian, Boeswald, & Kramer, 2011). Long-term care facilities (LTCFs) are a heterogeneous gathering of foundations, going from rehabilitation focuses to those giving persistent consideration to the sick and old. These clinical settings provide protection to all types of related contaminations and it is yet to be standardized completely (High et al., 2009). The host susceptibility adds critical variables to the transmission, also securing HAIs (Moro, Jans, Cookson, & Fabry, 2010). The meaning of HAI utilized by the World Health Organization is "a contamination happening in a patient in an emergency clinic or other health care centre in whom the disease was absent or brooding at the hour of confirmation" (Benenson, 1995; Ducel, Haxhe, Tanner, & Zumofen, 1979; World Health Organization, 2002). The HAIs are viewed as those diseases growing over 2 days after the affirmation. It has been assessed that over 60% of HAIs are biofilm-related diseases (Archibald & Gaynes, 1997; Fridkin, Welbel, & Weinstein, 1997; Goldmann & Huskins, 1997; Huebner et al., 1994) with a yearly expense of \$1.5 billion in the United Kingdom alone (Allen, 2000). Further developing WaSH (water, sanitation and hygiene) across India's public medical care area and supporting administrations among overhauled offices for 1 year would cost US\$354 million in capital expenses and US\$289 million in repetitive expenses according to the supplier point of view. The most expensive intercessions were those on water (US\$238 million), linen reprocessing (US\$112 million) and sterilization (US\$104 million), while the most unexorbitant were mediations close by cleanliness (US\$52 million), clinical gadget reprocessing (US\$56 million) and ecological surface cleaning (US\$80 million). Altogether, interests in provincial health care facilities (HCFs) would represent 64.4% of complete expenses, of which 52.3% would go towards essential well-being habitats (Tseng, Joshi, Shrivastava, & Klein, 2020). In the current review, the pace of HAI was found as 21.59%, the death pace of all patients was 26.13%. The death pace of patients who experienced HAI was viewed as higher than patients without HAI (60.60%, 18.18%, separately). It was seen that the advancement of HAI is a significant element that expands the death rate. In investigations directed in the writing, it was seen that experiencing HAI expands the death rate (Yardim, Yeşildağ, Çelik, Yildiz, & Dogan, 2021).

Three investigations revealed the qualified sterilization costs. Kumar, Jithesh, and Gupta (2015) reported costs of sewage plant activity allotted to one polytrauma and one particular neuro-medical procedure emergency unit in a level-IV ER in India at USD 0.15 and USD 0.09 each bed-day, separately. However, the study did not determine included costs. Liu, Zhou, Chen, and Zheng (2010) reported capital equipment and working expenses of biomembrane reactors at four distinct emergency clinics in China going from USD 0.23 to 0.62 per cubic meter of sewage. Riewpaiboon, Malaroje, and Kongsawatt (2007) costed disinfection for a region medical clinic in Thailand at USD 14,766 yearly, with district hospitals in the locale averaging 30–60 beds and no extra office characteristics described. Costs were not plainly portrayed yet may bar support, as "building maintenance," was cost as a different classification (Anderson et al., 2021).

Because of the insertion of medical devices in human body, there are various common HAI caused such as urinary tract infections (UTIs), respiratory infections, and gastrointestinal infections. These are mainly affecting delicate tissues, e.g., skin (Smith, Seip, Schaefer, & Bell-Dixon, 2000). As is notable, most HAIs are preventable through hand cleanliness at the ideal opportunity and in the correct way. As indicated by the new WHO Rules on Hand Hygiene in Health Care (Buković et al., 2021; Pittet et al., 2009), the individual-to-individual transmission through the hands of medical service laborers seems, by all accounts, to be the best approach to spread HAIs (Chang et al., 2021; McDonald et al., 2021). It is additionally notable that

Understanding Microbial Biofilms. https://doi.org/10.1016/B978-0-323-99977-9.00001-6

multidrug resistant (MDR) microorganisms frequently emerge from antimicrobial misuse, wrong medication choice, or improper antimicrobial treatment length (Hujer, Bethel, Hujer, & Bonomo, 2004; van Buul et al., 2012).

When patients or natural surfaces of the setting have been colonized by resistant strains, they can endure, quickly disperse, and get endemic. In this way, antimicrobial stewardship comprising of better clinical administration of patients by advancing the ideal medication routine, portion, course of organization, and span of treatment is suggested in the latest infection control rules (Pope, Dellit, Owens, & Hooton, 2009). Microbial life inside these multicellular networks encased in a hydrated framework of extracellular polymeric substances (EPS) provides organisms with more noteworthy security from resistant reactions and compound assaults, simpler quality exchange among cells, and higher obstruction to antimicrobials (Donelli & Vuotto, 2014).

Biofilms have been embroiled in a huge number of intense and ongoing diseases, and inhabiting clinical gadgets. Biofilms are collective bacterial cells which together form a consortium forming extracellular polysaccharide matrix (EPS), which aids in adherence to biotic/abiotic surfaces (Solano, Echeverz, & Lasa, 2014). Biofilm development includes adherence of planktonic cells and maturation and dispersal of bacterial planktonic cells (Preda & Săndulescu, 2019) (Fig. 1). Biofilm provides the foundation of bacterial communities on the tissues/surfaces of clinical gadgets accepted to restrain the viability of infection treatment, sidestep have guard systems and work with bacterial correspondence, prompting articulation of harmfulness determinants. Notably, development of bacteria as biofilm associated with the contaminations can affect the efficacy of antibiotic treatments, being less helpless as therapeutic agents (Patel, 2005). Among the bacterial species as often as possible isolated from catheter-associated urinary tract infections (CAUTIs) (Gordon & Jones, 2003), the most researched for their part in biofilm arrangement are *Escherichia coli*, *Klebsiella pneumoniae*,

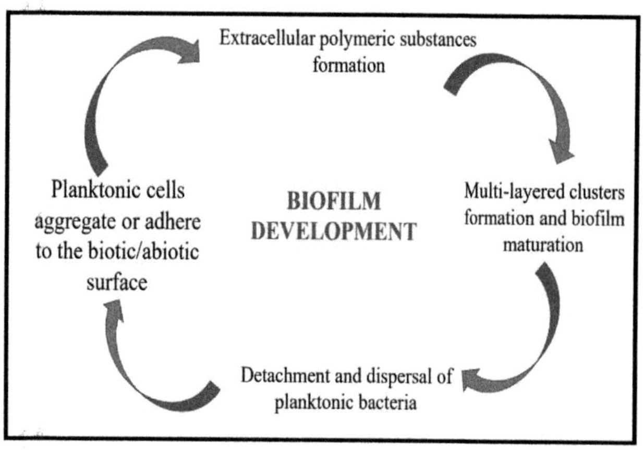

FIG. 1 Steps involved in biofilm formation (Preda & Săndulescu, 2019).

Proteus mirabilis, and, latest, the arising *Acinetobacter baumannii*. Lately, these species have recorded an expanding spread at the emergency clinic level and significant changes in their antimicrobial opposition patterns (Paterson & Lipman, 2007), exacerbated by the notably diminished vulnerability of biofilms to antimicrobial (Mah, 2012).

The Gram-negative, facultative anaerobic, *Proteus mirabilis* is the second most frequent etiological specialist of CAUTIs after *E. coli*. It is nothing but countless destructiveness factors, including fimbriae, flagellar motility, case antigen, lipopolysaccharide (LPS), struvite and apatite *gem* development, urease and hemolysin creation, adhesiveness, and amassing capacity, the last empowering *Proteus* to colonize and endure (Hola, Peroutkova, & Ruzicka, 2012; Jacobsen, Stickler, Mobley, & Shirtliff, 2008; Rather, 2005; Rocha, Pelayo, & Elias, 2007; Rózalski, Sidorczyk, & Kotełko, 1997).

Till now, just a single investigation is accessible on the impact of various classes of antibiotics on various periods of biofilm development in *K. pneumoniae*. Some antibiotics were piperacillin, amikacin sulfate, and ciprofloxacin hydrochloride, having a place with β-lactam, aminoglycoside, and fluoroquinolone classes separately. Of the antibiotics tried, amikacin had all the earmarks of being the best in the eradication of *K. pneumoniae* youthful biofilm, the higher creation of EPS making development of biofilm less penetrable to therapeutics agents (Singla, Harjai, & Chhibber, 2013).

An expanding number of *A. baumannii* clinical segregates have been accounted for worldwide as multidrug-resistance microorganisms; that is, impervious to at least three of the accompanying antibiotic classes: β-lactams, including penicillin, cephalosporin, and monobactam, carbapenems, fluoroquinolones, and aminoglycosides. On the other hand, pan-drug-resistance *A. baumannii* strains, characterized as impervious to all accessible antibiotic choices, are frequently separated at the clinic level (Grosso, Quinteira, & Peixe, 2010; Nicasio, Kuti, & Nicolau, 2008). Information on the vulnerability of *A. baumannii* biofilm to colistin reports that, the full bactericidal impact was observed when its concentration arrived at 8 μg/mL, a worth around 50 times higher than the MIC (0.125 μg/mL). This polymyxin was more powerful in development than youthful biofilm (Cai, Wang, Liang, & An, 2009).

Basically, 70% of all nosocomial bloodstream infections happen in patients who have central venous access devices (CVADs) (Gominet, Compain, Beloin, & Lebeaux, 2017; Ryder, 2005). This remembers patients for serious consideration, patients with malignant growth, and patients who need hemodialysis or total parenteral nutrition (TPN) (Lai et al., 2006; Schiffer et al., 2013; Zhang et al., 2010). There is proof that colonized CVADs can be significant transporters of methicillin-resistant *Staphylococcus aureus* (MRSA) and other multidrug-resistant microscopic organisms in networks and medical clinics around the world (O'Grady et al., 2011; Ryder, 2005).

1.1 Hospital-acquired pneumonia

Pulmonary complexities, including different types of pneumonia, are normal in the postoperative period and influence a significant part of patients following a medical procedure with general anesthesia. Ventilator-related pneumonia is the second most common clinic procured contamination (Arozullah, Khuri, Henderson, & Daley, 2001; Richards, Edwards, Culver, & Gaynes, 1999). Ventilator-related pneumonia is a main source of death in sick patients of intensive care unit (ICU), with assessed predominance paces of 10%–65% and death paces of 25%–60%, contingent upon the examination, the patient populace, and the clinical conditions included (Craven, Barber, Steger, & Montecalvo, 1990; Craven & Steger, 1995; Donald, Kathleen, & Thomas, 1991; Kollef, 1995, 2004; Vincent & Bensoussan, 1995). Ventilator-related pneumonia and different types of medical clinic related pneumonia are factors for mortality in hospitalized patients, independent of the seriousness and kind of fundamental disease (Fagon et al., 1993). A scene of medical clinic developed pneumonia adds 5–6 days to the length of emergency clinic stay and a large number of dollars in cost to clinical consideration (Kollef, 1995, 2004; Vincent & Bensoussan, 1995). The danger of creating ventilator-related pneumonia in the clinical and careful emergency unit from 5 to 21 for every 1000 ventilator days (Lynch & Lama, 2000). The danger of ventilator-related pneumonia, or different types of clinic procured pneumonia, is significantly expanded by sedation and medical procedure. In one examination completed by the U.S. Habitats for Disease Control and Prevention, 75% of patients with bacterial pneumonia became tainted after a medical procedure. Various danger factors have significant implications for the perioperative period (Cook, De Jonghe, Brochard, & Brun-Buisson, 1998). Of main concern is the relationship of pneumonia with the position of an endotracheal tube and mechanical ventilation. The occurrence of respiratory contaminations in patients requiring an endotracheal tube is generous and the danger of procuring ventilator-related pneumonia increments by as much as 1%–5% each hospital day (Cook et al., 1998).

The endotracheal tube gives a latent, nonshedding surface to which microorganisms hold fast and develop to shape biofilms, from which microbes are shed and aspirated into the lower airway. Moreover, the endotracheal tube incites mechanical abrasion, bothering of the respiratory mucosa, weakness of typical laryngeal capacity, and expanded sedation, all of which lead to an expanded danger of desire of upper respiratory parcel emissions. Bacterial floor, as well as the gastric substance, may likewise be significant in the pathogenesis of postoperative pneumonia (Donald et al., 1991; Kollef, 2001). Even though ventilators alone are not viewed as a significant wellspring of bacterial spread, breathing circuits can turn out to be vigorously defiled with microorganisms from the patients' oropharynx and windpipe (Craven, Connolly, Lichtenberg, Primeau, & McCabe, 1982).

Microscopic organisms have a least of three proteins that direct copper homeostasis: (i) a copper chaperone, which sequesters and courses cytoplasmic copper; (ii) copper-exporting ATPase which pumps copper from the cytoplasmic layer; and (iii) copper-responsive controllers, which direct the articulation of these proteins. Different cycles, for example, intra- and extracellular sequestration, enzymatic detoxification and dynamic efflux keep up with the copper balance in bacterial cells (Bondarczuk & Piotrowska-Seget, 2013). On account of copper, planktonic structures are more sensitive than the biofilms. The presence of these EPS viz., proteins, extracellular DNA (eDNA), polysaccharides that makes the biofilm microbes unique (and that is more resistant to antimicrobials) from their planktonic structures (Flemming et al., 2016). *Pseudomonas aeruginosa* PAO1 is an opportunistic microbe and one of the infamous biofilm shaping microorganisms ensnared in infections like cystic fibrosis, bacterial endocarditis, human and veterinary wounds, and otitis media (Percival, Suleman, Vuotto, & Donelli, 2015). According to the Centre for Disease Control and Prevention (CDC) (2013), it is a causal specialist in the majority of the medical clinic gained contaminations expanding the medical care-associated costs. A few investigations completed worldwide show that hospitalizations lead to nosocomial diseases, which add to longer medical clinic stays, mortality, and morbidity (El Moussaoui et al., 2015).

According to ongoing investigation in India, *P. aeruginosa* is the main source of bacteremia, pneumonia, urinary plot infections, and skin and delicate tissue infections in fundamentally sick and people with compromised immune frameworks (Pragasam, Veeraraghavan, Nalini, Anandan, & Kaye, 2018). Improper/experimental organization of antibiotics regularly prompts antimicrobial resistance in the microorganism. As per the World Health Organization (Tacconelli et al., 2018), carbapenem-resistant *P. aeruginosa* is recorded as a basic microbe. *Pseudomonas* biofilms increment the danger of clinical device-associated diseases (Khatoon, McTiernan, Suuronen, Mah, & Alarcon, 2018; Percival et al., 2015). Diverse systems that advance the variation, endurance, and persistence in different conditions and at various phases of pathogenesis happen in *P. aeruginosa* (Moradali, Ghods, & Rehm, 2017). With an enormous genome (~5–7 Mbp), it has an arsenal of subatomic instruments and ecological pressure prompts the declaration of a totally unique set of qualities explicit to the planktonic and the biofilm structure (Gellatly & Hancock, 2013; Krahn et al., 2012; Moradali et al., 2017). *P. aeruginosa* is exceptionally impervious to a few antibiotics (Poole, 2017; Ryan et al., 2014) including metal particles like Cu^{2+}. In an investigation in *P. aeruginosa*, a populace of cells was presented to raise copper levels for a brief time frame (demonstrating copper shock), and another populace was adjusted to high copper contaminations. The transcriptional profiles of planktonic cells presented to two separate copper pressure conditions (at copper contaminations 10 mM, given as $CuSO_4$) uncovered that two obstruction nodulation division efflux frameworks, viz., a P-type ATPase, and a two-segment controller were engaged with copper tolerance (Teitzel et al., 2006).

In another examination, *P. aeruginosa* (PAO1) strain were associated with their endurance on copper-containing medium (enhanced with copper salt, viz., $CuCl_2$) or copper compound surfaces. The investigation likewise brings up to the way that solubilized copper is not the sole reason for harmfulness but rather surface properties of copper (if there should be an occurrence of copper cast combinations, addressing copper in its "mass" structure) caused killing of the *P. aeruginosa* cells (Elguindi, Wagner, & Rensing, 2009). Copper nanoparticles hindered the efflux pumps while CuO nanoparticles at fixations 700 μg/mL made oxidative killing of planktonic *P. aeruginosa* cells (Ulloa-Ogaz & Piñón-Castillo, 2017).

In a new report performed utilizing RNA-sequencing, critical upregulation of qualities engaged with copper opposition, resistance-nodulation-division (RND), P-type ATPase efflux, and cation dispersion facilitator carriers was seen in planktonic *P. aeruginosa* PAO1 cells presented to 1–50 mg/L CuO nanoparticles (Guo et al., 2017). This load of studies show expanded resilience of planktonic *P. aeruginosa* PAO1 to copper. The clinical relevance and relative simplicity of biofilm arrangement arise an occurrence of *P. aeruginosa*. The way that reviews separating between the commitments of "copper particles" and "nanoparticles" in *P. aeruginosa* PAO1 biofilm are lacking, provoked them to attempt the current investigation. In light of the literature they recognized a bunch of agent qualities associated with biofilm arrangement, metabolic cycles, majority detecting and endeavored at recognizing "ions" versus "particles-" explicit impacts of copper in preformed *P. aeruginosa* biofilms. Quantitative constant PCR (qRT-PCR) permits concurrent checking of the declaration of a prechosen set of qualities and henceforth they utilized it as an apparatus for this similar investigation (Singh, Paknikar, & Rajwade, 2019).

Copper nanoparticles are being investigated as a powerful antibiotic in clinical settings to defeat contaminations due to biofilm shaping microorganisms. Gram-negative microbes are accounted for to be impervious to copper. Since nanoparticles address novel materials with properties not quite the same as their mass partners, they embraced an examination of the particles-explicit impacts by considering the statement of biofilm explicit qualities in light of copper (Singh et al., 2019). With *P. aeruginosa* biofilms as a model and with the restricted arrangement of qualities researched, the study demonstrated that CuNPs were to be sure better than Cu^{2+} in light of the fact that they were more harmful. CuNPs, for example "nano" copper, caused unmistakable reactions, viz. decline in articulation of qualities liable for EPS, influence film capacities and harm the cell surface. Thus, by utilizing CuNPs, it can accomplished for better killing at lower fixations. Our discoveries are in backing of the utilization of nanoparticles in clinical settings for checking nosocomial diseases (Singh et al., 2019).

2. Sites of infection

The principal nosocomial diseases brought about by biofilms are bacteremia, UTI, and pneumonia. Among sick patients in the United States, 95% of UTIs are catheter-related, 87% of bacteremia cases have an inhabiting vascular gadget, and 86% of cases of pneumonia are ventilator related (Richards et al., 1999), featuring the significance of an "unfamiliar body" to give a steady concentration to bacterial grip and increase (Ebrey, Hamilton, Cairns, & Lappin-Scott, 2004).

2.1 Bacteremia

In emergency clinics in England, the frequency of hospital-gained bacteremia is 3.5 cases/1000 confirmations or 0.6 bacteremia/1000 patient days. In ICUs, the rate increments to 9.0 cases/1000 patient days (Hongsuwan et al., 2014). More than 40% of bacteremia cases in clinics in Britain are brought about by *Staphylococci*: 24% by *Staphylococcus aureus* (a generally equivalent split between MRSA and methicillin-defenseless *S. aureus*) and, 17% by CoNS. A further 8.6% of bacteremia are brought about by enterococci, and one report concerns 50 of the 495 scenes where the seclude was impervious to vancomycin. In the United States, 300,000 instances of catheter-related bacteremia were anticipated for 2001 (Hanna, Darouiche, & Raad, 2001). *Staphylococcus lugdunensis* in clinical practice acts more like *S. aureus*. Practically, one of a kind in its destructiveness among CoNS, having the capacity to cause native value endocarditis (NVE), and biofilm contamination, *S. lugdunensis* has likewise been separated from a ventriculoperitoneal shunt in which it was framing biofilm contamination (Sandoe & Longshaw, 2001).

2.2 Urinary tract infection

UTIs represent 30% of all nosocomial infections, a large portion of which are related to urinary catheters (Dixon, 2002). Of all patients with a urinary catheter, 30% will create bacteriuria Furthermore, up to 6% of those will create a clinically critical UTI (Ebrey et al., 2004). *E. coli* is a Gram's-negative, facultative anaerobic, rod-shaped bacterium, causing more than 30% of nosocomially procured UTIs and is the most regularly detached microorganism from the urine of patients influenced by CAUTIs (Bouza et al., 2001; Kahlmeter, 2000; Tinelli et al., 2012). The danger of UTI has been calculated to increment by 5% for every day that the catheter is in situ (Dixon, 2002). Eliminating and supplanting urinary catheters are regularly went with bacteremia created by organic entities passing through the damaged urethral mucosa (Ebrey et al., 2004). *A. baumannii* is known to form biofilm in clinical settings and viably colonize the surfaces of urinary catheters. The long constancy in clinical conditions (Espinal, Martí, & Vila, 2012) and the rising therapeutics obstruction announced for these species (Navon-Venezia, Ben-Ami, & Carmeli, 2005) could be identified with their capacity to frame biofilm (Lee et al., 2008; Rodriguez-Bano et al., 2008; Vidal et al., 1996). A few highlights of this microorganism have been related to biofilm arrangement, such as a pilus assembly framework (Tomaras, Dorsey, Edelmann, & Actis, 2003), a homolog to the staphylococcal Bap engaged with intercellular adhesion inside the developed biofilm (Loehfelm, Luke, & Campagnari, 2008), the external layer protein OmpA required for the development of powerful biofilms on polystyrene (Gaddy, Tomaras, & Actis, 2009), a poly-β-(1-6)-*N*-acetyl glucosamine (Choi, Slamti, Avci, Pier, & Maira-Litrán, 2009), and an auto inducer synthase, which is engaged with quorum-sensing and needed for typical biofilm development (Niu, Clemmer, Bonomo, & Rather, 2008).

The creation and practical job of type 3 fimbriae were found to impact biofilm development of *E. coli* (Ong et al., 2008). Few considers have so far looked at changed *E. coli* phylogenetic gatherings engaged with UTIs with the presence of virulence factors, biofilm-shaping capacity, and antimicrobial opposition (Ejrnaes et al., 2011; Novais et al., 2012). In specific, Ejrnaes and colleagues have attempted to discover a relationship between biofilm formations and constancy of uropathogenic *E. coli* strains segregated from local area obtained lower UTIs in women (Ejrnaes et al., 2011). They have mentioned, as a team with Portuguese associates, on the absence of a positive connection between biofilm creation capacities of various *E. coli* clonal bunches having a place with phylogenetic group D and their high contagiousness among various settings. Momentarily, biofilm creation appears not to be straightforwardly identified with the epidemiological achievement of these strains for the most part disengaged from urine (Novais et al., 2013). A few elements have been accounted for to be related to *E. coli* biofilm improvement, including curly, flagella, type 1, and F9 fimbriae (Reisner, Haagensen, Schembri, Zechner, & Molin, 2003; Ulett, Mabbett, Fung, Webb, & Schembri, 2007) and antigen 43 (Ejrnaes et al., 2011; Klemm, Hjerrild, Gjermansen, & Schembri, 2004).

2.3 Pneumonia

Nosocomial pneumonia is known as an infection that occurs after 48 h of exposure. It happens in up to 1% of all patients, representing 40% of ICU contaminations (Dixon, 2002). Nosocomial pneumonia tragically stays a successive occasion for which suitable antitoxin treatment is key to further developing results. Doctors should pick an early and suitable exact treatment, putting together their choice with respect to the security profile and conceivable incidental effects (Ceccato, Di Giannatale, Nogas, & Torres, 2021). Nosocomial pneumonia essentially affects mortality, morbidity, and medical care costs, particularly when the ensnared microbes are multidrug-resistant ones (Xu, Pérez-Torres, Fragkou, Zahar, & Koulenti, 2021). The frequency of ventilator-related pneumonia ranges from 7% to 40%, has truly been predominantly Gram's-negative, and might be mono-or polymicrobial, neither having more noteworthy mortality than the other (Combes et al., 2002). The image is evolving, notwithstanding, with increasing quantities of patients with MRSA ventilator-associated pneumonia and an attendant higher mortality if not analyzed early enough (Ebrey et al., 2004). Lung infection, in the case of cystic fibrosis, is known to be accelerated due to biofilm mode of growing pathogen (Martin, Waters, & Grasemann, 2021). Its main cause is mutation in the cystic fibrosis transmembrane conductance regulator (CFTR) gene; inflammation and chronic bacterial infection being the main characteristics (Martin et al., 2021). Microbes like *Stenotrophomonas maltophilia* and *P. aeruginosa* are known to be involved in cystic fibrosis lung infections (Harrington, Sweeney, & Harrison, 2020; Pompilio, Savini, Fiscarelli, Gherardi, & Di Bonaventura, 2020).

2.4 Surgical wound infections/pressure ulcers/ wound infections

While not biofilm contamination essentially, surgical injury infections can give a core interest from which microorganisms may spread hematogenously, to seed and shape biofilm infections at different locales, e.g., local and prosthetic heart valves, prosthetic joints, and inhabiting vascular gadgets (Wilson & Gross, 2022). Most biofilm-related diseases happen in the clinical inhabiting gadgets, e.g., dialysis catheters in renal medication, "long queues" in oncology, prosthetic joints in muscular surgery, blood vessel and venous catheters (central venous catheter, CVC), and the plenty of lines, urinary catheters, pressure ulcers, wound infections, and tracheal tubes found in ICU patients (Ebrey et al., 2004; Ielapi et al., 2020; Versey et al., 2021). Wound infections are serious concerns as delayed wound healing may lead to diabetic ulcers, surgical site infections, chronic injuries, and pressure ulcers (Evelhoch, 2020; Versey et al., 2021). Pressure ulcers, also known as bed sores, are sores of skin, along with its surrounding tissues and bones caused due to long exposed compression force (Gomes & Furtado, 2022). *Staphylococcus aureus* biofilms are commonly found associated with pressure ulcers (Fayolle & Morsli, 2021).

3. Pathogenesis of attachment

Adherence of the organic entity to a gadget is governed by three chief factors: the host's reaction to gadget inclusion, the intrinsic ability of the life form to stick to surfaces, and the constitution of the gadget material and surface qualities (Raad, Darouiche, Hachem, Mansouri, & Bodey, 1996). At the point when a catheter or other gadget is embedded in the body, it is perceived as immunologically unfamiliar and a reaction happens. At first, a thrombin sleeve, wealthy in fibrin and fibronectin, is conformed to the gadget and shields any implanted organisms from additional safe reaction (Raad et al., 1996). Practically, all pathogenic Gram-positive microorganism have a group of adhesins known as MSCRAMMs (microbial surface components recognizing adhesive matrix molecules) (Navarre & Schneewind, 1994; Patti & Höök, 1994), which work with restricting to different human or organism tissues, serum proteins that may cover inhabiting gadgets and polypeptides of the extracellular network. Specifically, the MSCRAMMs of *S. aureus* (which are essentially integrated during the remarkable development stage (Lowy, 1998) comprise of (I) protein A, which ties to the Fc parcel of immunoglobulins, accordingly forestalling their opsonization and phagocytosis; (ii) fibrinogen-binding proteins ("bunching factor") A and B, which tie to fibrinogen, advancing bacterial bond to thrombi on heart valves; and (iii) fibronectin-restricting proteins (FnbA and FnbB), which tie to fibronectin and can instigate attack and apoptosis of contaminated epithelial cells (Jönsson, Signäs, Müller, & Lindberg, 1991; Menzies & Kourteva, 2000). Optional connection of *S. aureus* is facilitated by the creation of polysaccharide intercellular adhesin (poly-*N*-succinyl-β-1-6-glucosamine), coded for by the *icaADBC* operon, and is engaged with intercellular total (Cucarella et al., 2002), a substance delivered utilizing the same system as biofilm-shaping strains of CoNS (Mack et al., 1994) [it has been shown that creation of polysaccharide intercellular adhesin is upregulated under anaerobic conditions (Cramton, Ulrich, Götz, & Döring, 2001)]. In any case, in CoNS, the essential connection as proposed in Mack's two-stage model (Mack et al., 1994; Mack, Siemssen, & Laufs, 1992) is interceded by a capsular polysaccharide adhesin. The commitment of host-tissue ligands on a superficial level has been additionally showing that *S. aureus* binds more promptly to surfaces of vascular catheters that had been eliminated from patients than to unused gadgets (Weinstein & Darouiche, 2001).

Moreover, in both the case, also type 1 and 3 fimbriae, along with the commitment of a sort 2 majority detecting regulatory framework, have been demonstrated to be included in the capacity of *K. pneumoniae* to frame biofilm (Balestrino, Haagensen, Rich, & Forestier, 2005; Di Martino, Cafferini, Joly, & Darfeuille-Michaud, 2003; Dzul et al., 2011; Stahlhut et al., 2009; Stahlhut, Struve, Krogfelt, & Reisner, 2012). Additionally, signature-tagged mutagenesis (STM) showed that, past the extracellular parts (case and fimbriae), the qualities ORF-4 (wza homologous, external layer protein for the vehicle of polysaccharides), wbbM (combination of D-galactan I), wzm (an ABC-2 layer family engaged with the LPS transport pathway), and *pgaA* (encoding poly-β-1,6-*N*-acetyl-D-glucosamine) are additionally included in *K. pneumoniae* biofilm development (Balestrino, Ghigo, Charbonnel, Haagensen, & Forestier, 2008; Boddicker, Anderson, Jagnow, & Clegg, 2006).

Connection systems of enterococci are ineffectively observed, yet enterococcal surface protein has been embroiled by Nathan et al. (2001), who discovered esp. communicated in approximately 40% of enterococcal bacteremia and endocarditis detaches, contrasted, and not exactly 3% of fecal detaches (Nathan et al., 2001). Enterococcal total substance works with bond to numerous eukaryotic cells, including refined porcine renal rounded cells (Chow et al., 1993). Further adhesins have likewise been depicted in *Enterococcus faecalis* that tie to human bladder carcinoma cells (Shiono & Ike, 1999) and Girardi heart cells (Guzman, Pruzzo, LiPira, & Calegari, 1989; Guzman, Pruzzo, Platé, Guardati, & Calegari, 1991). Weinstein and Darouiche (2001) recognized five significant standards to consider corresponding to bacterial adherence and the gadget surface: (i) unique microbes may cling diversely to the equivalent gadget material; (ii) similar microbes may cling diversely to various gadget materials; (iii) similar microscopic organisms may follow in an unexpected way to a similar gadget material kept under different conditions, remembering the mode for which the gadget is kept (hydrophobic versus hydrophilic medium), kind of stream (dynamic versus fixed) and temperature; (iv) in vitro hindrance of bacterial colonization of the gadget doesn't guarantee antiinfective efficacy in vivo; and (v) the clinical advantage of a specific surface-changing methodology may change starting with one application then onto the next (Ebrey et al., 2004; Weinstein & Darouiche, 2001).

The pathogenicity of *K. pneumoniae* has been credited to various virulence factors, including antiphagocytic capsule, LPS, membrane transporters, and siderophores (Hsieh et al., 2012; Podschun, Fischer, & Ullmann, 1992; Podschun & Ullmann, 1992; Simoons-Smit, Verweij-van Vught, &

MacLaren, 1986). On the other hand, its persistence is by all accounts related with the capacity to adhere to explicit epithelial tissues or on the other hand abiotic surfaces, and this cycle, mindful for the initial step of biofilm development (Kuchma & O'Toole, 2000), has been accounted for to be intervened by type 1 and 3 fimbriae (Hornick, Allen, Horn, & Clegg, 1992; Würker, Beuth, Ko, Przondo-Mordarska, & Pulverer, 1990).

4. Diagnosis of biofilm infection

Research center affirmation of a suspected biofilm contamination on a prosthetic gadget regularly requires evacuation of the gadget. For strategies based on microscopy and additional culture. Maki et al. proposed a culture technique that elaborate moving the external surface of a vascular catheter straightforwardly onto an agar plate, albeit this won't recognize intraluminal living beings (Maki, 1977). Endoluminal brushes have been proposed as a method for gathering intraluminal development (Goudarzi, Navidinia, Khadembashi, & Rasouli, 2021). Microscopy has been performed utilizing auramine staining and immunofluorescence to picture microbes present on the inhabiting gadget; confocal laser microscopy has been effective, despite the fact that past the method for some routine symptomatic research centers. Late subatomic strategies have included PCR enhancement and recognition of the *ica* operon to decide if a staphylococcal disengage is equipped for creating polysaccharide intercellular adhesin (Arciola, Collamati, Donati, & Montanaro, 2001) and discovery of the bacterial 16S rRNA quality (Tunney et al., 1999). The increasing frequency of *eno* and *icaD* genes hints toward the role of these genes as being synergistically functional in the initial and progressive bacterial colonization and biofilm formation (Haghi Ghahremanloi Olia, Ghahremani, Ahmadi, & Sharifi, 2020).

5. Factors affecting biofilm formation within the clinical and hospital setting

Connection of microbes to surfaces is a complex measure affected by a few components including supplement accessibility, surface science, and have reactions (Fig. 2) (Donlan & Costerton, 2002). Every one of these will be managed thus yet space constraints forbid conversations on other factors like hydrodynamics, oxygen availability, or impact of salt fixation on biofilm development (Ebrey et al., 2004; Wilson & Gross, 2022). Deferred wound mending can cause huge issues for stable and maturing people just as those living with co-horrible conditions like diabetes, cardiovascular illness, and malignant growth. These defer increment in patient's danger for contamination and, in serious cases, can bring about the arrangement of persistent, nonrecuperating ulcers (e.g., diabetic foot ulcers, careful site diseases, pressure ulcers, and venous leg ulcers) (Versey et al., 2021).

5.1 Nutrients

Supplement accessibility assumes a basic part in the development of biofilms. While this sessile development permits microbes to continue in a low nutrient climate with an eased back metabolic and replication rate starting endurance and colonization of surfaces are impacted by supplement supply. The idea of the liquids controlled through a clinical gadget may upgrade or repress improvement of biofilm: *Candida* spp. are particularly connected with lipid-containing arrangements and some Gram-negative microorganism (e.g., *P. aeruginosa*, *Enterobacter* spp.) have been displayed to persevere in low supplement liquids, for example, some intravenous liquids, more energetically than Gram-positive microorganisms (e.g., *Staphylococcus epidermidis*, *S. aureus*) (Donlan, 2001a). Regularly utilized liquids fluctuate in nutrient organization from extremely poor (0.9% saline) to exceptionally rich (Intralipid) and organic entities would be expected to use these supplements. Utilizing a catheter model, Donlan et al. demonstrated that the development capability of liquids rich in supplements upheld upward of 108 cells/ml (Donlan, Murga, & Carson, 1999). This development rate would effectively be distinguished by the turbidity of the liquid, yet more slow development rates in low supplements might be ignored. Many diseases because of *S. epidermidis* appear to emerge precipitously, just distinguished months to years following the joining, having persevered undetected in a biofilm on the joint surface for quite a while, perhaps since the implantation. A mouse model was utilized to research in skin joint biofilms (Bergamini, Corpus, Brittian, Peyton, & Cheadle, 1994). The results showed an abatement in *S. epidermidis* graft-associated biofilm that could be secluded from the twisted after longer spans of contamination. In this investigation, the scientists trait this to low supplements encompassing the biofilm and to have invulnerable reaction (Bergamini et al., 1994). In vivo, microbes are probably going to encounter a variance in nourishing components dissimilar to that utilized in research center media. Increasing salt concentrations restrict the growth of *Salmonella enterica* biofilms, rather in case the pH is increased in low salinity condition of less than 4% NaCl concentration biofilms growth is normal (Iliadis, Daskalopoulou, Simões, & Giaouris, 2018). Iron, a fundamental component for bacterial development, is a decent example of a factor that can be restricting inside the have if hard to find (Evans, Brown, & Gilbert, 1994; Gedefie et al., 2021).

Iron sequestration is an intrinsic host protection in which have iron-restricting proteins, for example, transferrin, binds free iron, making it unavailable to microorganisms. Investigations of iron consumption in pathogenic oral

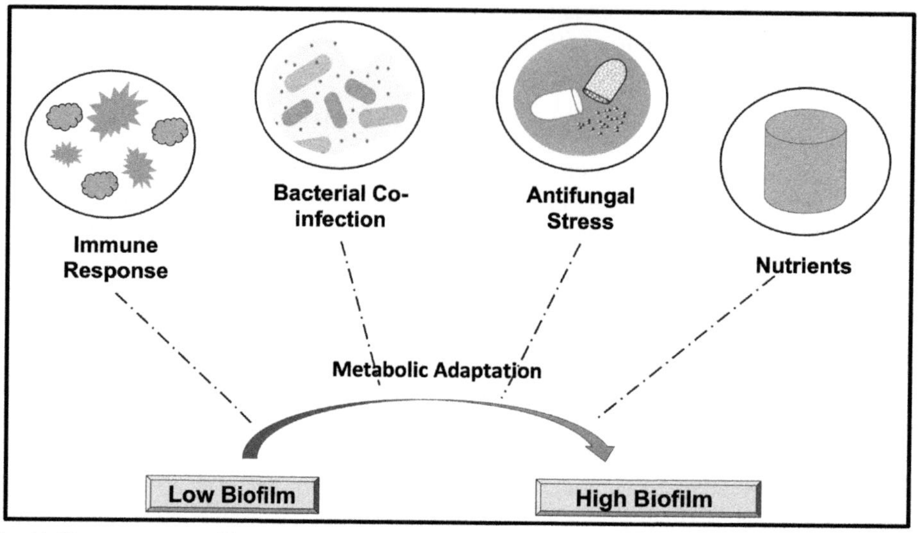

FIG. 2 Factors affecting biofilm arrangement. There are various improvements that can prompt biofilm formation including the immune reaction, antifungal stress, and bacterial determined metabolites. Environmental stressors can likewise invigorate biofilm arrangement, and these incorporate the availability of supplements, temperature, and pH. *Dotted lines* address various variables that partner with biofilm development.

biofilm have shown expanded record and additionally discharge of various explicit proteins, like those engaged with regulation and transport (Fletcher, Nair, Ward, Henderson, & Wilson, 2001). Additionally, overabundance or restricting iron has been answered to control a few qualities significant in protection against oxidative pressure (Bollinger, Hassett, Iglewski, Costerton, & McDermott, 2001) and destructiveness factors (Baldassarri et al., 2001; Bjorn, Sokol, & Iglewski, 1979) in *P. aeruginosa*. The study conducted by Çam and Brinkmeyer shows that there exists an evident correlation between iron concentration and biofilm production in strains of *Vibrio vulnificus* (Çam & Brinkmeyer, 2020). According to Kotay et al. (2020), mostly protein rich nutrients are supportive for biofilm formation ability in bacterial strains (Kotay et al., 2020). Besides proteinaceous liquids administered to patients, consider the impact of regularly endorsed pharmaceutical specialists in supporting biofilm development. Catecholamine chemicals, like adrenaline, dobutamine, dopamine, isoprenaline, and noradrenaline, have been displayed to invigorate development of Gram-positive and Gram-negative species inside intravascular catheters (Freestone, Haigh, Williams, & Lyte, 1999; Lyte & Ernst, 1992). Albeit these compounds are thought of "nonantibiotic" specialists, exploratory proof has shown that they may expand the danger of bacteremia by making iron accessible to entrepreneurial pathogens. Since these specialists are ordinarily utilized in the ICU, this may go some approach to clarifying the presence of catheter-related HAI. Andrew et al. announced more noteworthy development in media enhanced with these specialists than in media alone and suggested that the primary characteristic of these chemicals is the key (Andrew et al., 2001). It was recommended that these medications could shape complexes with ferric iron because of their remarkable structure in this way furnishing the microorganisms with a method for getting iron (Ebrey et al., 2004; Wilson & Gross, 2022).

5.2 Host response

Bacterial biofilm networks are complicatedly intended for every types of microorganisms so they can effectively sidestep the host immune responses (both inborn and adaptive insusceptible reactions) and keep on getting by inside the host (Ghosh & Ray, 2021). Resistant reaction is for the most part considered as good for the host; in any case, this is not generally the situation. It has been suggested that tissue encompassing a biofilm could itself be harmed by insusceptible environment generated as a reaction to contamination (Hoiby et al., 1995). Besides, consider that clinically, biofilms endure in the host as an endurance instrument instead of exclusively as a microorganism, creating destructiveness factors (Stewart & Costerton, 2001).

In the lungs of cystic fibrosis (CF) patients, there are abnormally high chloride concentrations because of an imperfect CF transmembrane controller chloride siphon. It has been recommended that adjusted ionic strength could influence the host reaction in that neutrophils are flawed in phagocytic killing and apoptosis of these resistant protection cells is higher than typical (Lyczak, Cannon, & Pier, 2002). Also, the high convergence of circulating antibodies to *P. aeruginosa* in the CF persistent has been displayed to correspond with broad aspiratory tissue harm from aggravation (Costerton, Stewart, & Greenberg, 1999). In vivo study on a hare model have demonstrated how host factors are in some cases inefficient in disposing of gadget-related disease. In this examination, biofilm present on a peritoneal gadget embedded into an immunized hare could not be annihilated (Ward, Olson, Lam, & Costerton, 1992).

5.3 pH

A biofilm method of development has been advantageous in enduring unfriendly conditions such as changes in pH (Rowbury, 1995). Additionally, these progressions in pH have been shown to altogether adjust biofilm structure and propose that this might be a method of ensuring the mobility inside the biofilm. Stoodley et al. showed that pH shifts somewhere in the range of 3 and 10 caused reversible extension or on the other hand withdrawal in biofilm thickness (Stoodley, deBeer, & Lappin-Scott, 1997). This reversible impact was ascribed to hydrogen bond development and changes in aversion between charged carboxylate bunches commonly found in the extracellular polymeric network of gram-negative microbes at a low pH. The study done by Fulaz et al. focused on the assessment of generation of pH within the *Pseudomonas fluorescens* biofilms via use of Fluorescent Mesoporous Silica Nanosensors (Fulaz et al., 2019). The biofilms showed distinct acidic pH regions and the lowest pH observed within the core of microcolonies was 5.1. Changes in pH have additionally been displayed to correlate with diminished biofilm-framing limit, adherence to epithelial cells, and freaks defective in tryptophan's creation in *E. coli* (Di Martino, Merieau, Phillips, Orange, & Hulen, 2002). This study uncovered changes in the bacterial environment during connection to both biotic and abiotic surfaces and recommended that the tryptophan's structure assumed a part of pH tweak. These outcomes affirmed before work by McCabe and Horn (1994) that demonstrated pH changes were quite unique when contrasting follower and no adherent strains of *E. coli* (Ebrey et al., 2004; Versey et al., 2021). To target the *S. aureus* biofilms causing periprosthetic joint infection, a cationic amphipathic peptide (WLBU2) was engineered by Mandell et al. (2020). The ability of WLBU2 was observed to be more effective in treating biofilms in lesser acidic pH, i.e., alkaline pH supported the efficacy of the amphipathic peptide to target biofilms (Mandell et al., 2020). In dental caries, *Streptococcus mutans*, *S. sobrinus*, and *Lactobacillus acidophilus* biofilms have been observed as high surviving at a low pH, and the metabolic activity was observed to be maximum at pH6–6.5, also the biofilm mass was seen highest at pH7–7.5 (Schultze, Maldonado, Lussi, Sculean, & Eick, 2021).

5.4 Temperature

Vibrio vulnificus infections causing gastroenteritis and septicemia are related with biofilm formation. The effects of temperature were analyzed upon biofilm survival by growing biofilms at various temperatures, higher than the environmental strains, like 24°C, 30°C, and 37°C. The results showcased that the biofilms were more prominently surviving at temperature of 24°C (Çam & Brinkmeyer, 2020).

5.5 Surfaces

Microorganisms attach to various surfaces in hospital setting and colonize forming biofilms. The surfaces can be the furniture, beds, instruments, of hospital rooms and ICU's, and even the medical indwelling devices. Biofilms associate with wet/damp surfaces more commonly (Otter et al., 2015). Biofilms also have the ability to survive for long durations on dessicated surfaces and are known to be less sensitive to biocides, antibiotics, and physical stress (Otter et al., 2015).

6. Device-related infections

Colonization of clinical gadgets by microorganisms is answerable for an enormous extent of nosocomial diseases, particularly in critically sick patients (Weinstein & Darouiche, 2001). In particular, CVC's are of the main sources of nosocomial contamination and essential scenes of bacteremia (Maki, 1992), and, urinary catheters are the most habitually colonized gadgets, though as a rule with less genuine sequelae. Table 1 summarizes the biofilm-related infections and their details.

Colonization of the inhabiting gadget might be an introduction to both contamination and malfunction of the gadget, with significant clinical and financial ramifications (Weinstein & Darouiche, 2001). Clinically, biofilm diseases are portrayed by repeating side effects notwithstanding rehashed antimicrobial treatment: ordinary treatment takes out just the free-living (planktonic) cells, leaving the appended (sessile) cells to recreate inside the biofilm, possibly allowed to proceed with dissemination at the suspension of treatment (Ebrey et al., 2004).

The scope of inhabiting gadgets utilized in current medication is enormous, each having differing affinity to colonization and a more noteworthy or on the other hand lesser potential for going about as a repository of genuine contamination. Albeit more than 30,000,000 urinary catheters are utilized in the United States consistently and up to 30% of patients have contaminations, mortality from a catheter-related UTI is exceptionally low (not exactly 5%). In the United States 85,000 mechanical heart valves are fitted per annum, with a low pace of contamination (1%–3%); however, at the point when they do become contaminated,

TABLE 1 Summary of the natural history of classical and biofilm-related infections (Ebrey et al., 2004).

	Acute infection	Chronic infection	Biofilm infection			
			Nondevice associated		Device associated	
Infection caused	Abscess	Tuberculosis	Dental caries	Cystic fibrosis pneumonia	Urinary catheter	Prosthetic heart valve
Organism involved	*S. aureus*	*M. tuberculosis*	*S. mutans*	*P. aeruginosa*	*P. mirabilis*, MRSA	*S. epidermidis*
Onset	Rapid	Slow to develop but long-lasting	Slow	Rapid	Rapid	Slow
Course	Days	Months to years	Years	Years	Days	May take weeks to become apparent
Characteristics	Rapid resolution with drainage and appropriate antibiotics	Failure of host to clear intracellular pathogen. Reactivation of latent infection with immunocompromization	Uncertain onset, ecological imbalance	Deterioration in lung function of compromised host	Ability of organism to move over device. Underlying disease	Risk of secondary infection from emboli
Treatment to cure infection	± Short term antibiotic	Long-term antibiotic, supportive	Physical, preventive (diet/ hygiene)	Long-term antibiotic, physiotherapy	Device removal, ± antibiotic	Antibiotics, surgical removal

the concomitant death rate is high (Weinstein & Darouiche, 2001). HAIs are a typical issue, complicated by expanding microbial antibiotic resistance and determination notwithstanding efforts to build up prescribed procedures for the avoidance of HAIs. In a huge forthcoming companion study, Lambert and partners tracked down that 7% of patients introduced to an intensive care unit related pneumonia and 4% created blood stream diseases (Lambert et al., 2011). The most widely recognized segregates were *E. coli*, *S. aureus*, *A. baumannii*, and *P. aeruginosa*. Antibiotic resistance was observed in 11% of the *E. coli*, 34% of the *S. aureus*, 22% of the *P. aeruginosa*, and 75% of the *A. baumannii* samples (Lambert et al., 2011).

6.1 Infected pressure ulcers

Confined to bed patients moved from general hospitals to LTCFs are regularly influenced by pressure ulcers. Indeed, occurrence rates are accounted for to reach up to 38% for medical clinics, 24% for LTCFs, and 17% for home consideration (Lyder, 2003). Pressure ulcers require a brief and viable treatment, especially to keep away from wound-related contaminations and the foundation of microbial colonization that defers wound healing (Garcia & Thomas, 2006; Livesley & Chow, 2002). Except this, infected pressure ulcers can bring about profound delicate tissue contaminations that, when not sufficiently treated, may cause osteomyelitis and secondary bacteremia related with a half death rate (Garcia & Thomas, 2006; Livesley & Chow, 2002; Smith et al., 2000).

As indicated by an examination performed on patients conceded to their emergency clinic in Rome from 2007 to 2009, an enormous number of various antibiotic-resistant microorganisms were gathered from pressure ulcers at stages III and IV utilizing the irrigation-aspiration technique (Ehrenkranz, Alfonso, & Nerenberg, 1990). The most often-times separated species were MRSA (32%), *P. aeruginosa* (21%), *E. coli* (23%), and *Proteus* spp. (17%) (Donelli & Vuotto, 2014). Their outcomes agree with the high rates of MRSA carriage, going between 8.8% and 23%, announced in the writing for other LTCFs (Bradley et al., 1991; Cretnik et al., 2005; Manzur & Pujol, 2008; Vovko et al., 2005).

The deficiency of the inborn hindrance capacity of the skin facilitates the colonization of wounds by microorganisms and advances biofilm improvement. Bacterial biofilms are applied not just in the disappointment of wound mending yet in addition in the occurrence of persistent aggravation by the EPS matrix-mediated avoidance of the immune reaction and enlistment of a nonhealing inflammatory stage. Direct proof of biofilm presence in injuries, particularly in constant ones, has been given by James and colleagues, who researched biopsies from intense and persistent injuries by light microscopy and SEM (James et al., 2008). Results exhibited a very lower generality of biofilm in intense injuries in correlation with persistent injuries (6% vs 60%) also, the polymicrobial idea of persistent injury biofilms, comprised by a few animal varieties, including *Staphylococcus* (65%), *Enterococcus* (62%), *Pseudomonas* (35%), *Proteus* (24%), *Escherichia* (14%), *Klebsiella* (5%), and *Acinetobacter* (5%). Pressure ulcers of grade III and IV can likewise be affected by Gram-positive anaerobes, *Finegoldia* spp., *Clostridium* spp., and *Veillonella* spp., and these species must have been showing to have the option to develop a single- or multispecies biofilm (Donelli, Vuotto, Cardines, & Mastrantonio, 2012).

The recuperation of biofilm from an infected pressure ulcer addresses a troublesome testing since a completely acceptable inspecting technique has not, however, been characterized. Indeed, there is no experimental information on the unwavering quality of the noninvasive aspiration-irrigation procedure (Ehrenkranz et al., 1990) thought about with the utilization of swabs or tissue biopsies (Percival et al., 2012). Regardless, SEM has been utilized to notice the presence of biofilms inside injury tissues, while confocal laser scanning microscopy has given data on the tridimensional formation of biofilm and EPS framework by turning to a reach of staining and molecular procedures, including FISH (Fazli et al., 2009; Kirketerp-Møller et al., 2008; Malic et al., 2009).

6.2 Central venous catheters (CVCs)

Central venous catheters (CVCs) are utilized to manage liquids and blood supply in fundamentally sick patients, and colonization of the catheter by life forms in a biofilm lattice can be exhibited in practically all cases (Raad, 1998; Raad et al., 1993). Septicemia, vascular impediment, and abscess development are the most continuous dangerous complications related with such gadgets (Donlan et al., 2001; Raad, 1998) and 87% of septicemias in basically sick patients begin from inhabiting vascular catheters (Richards et al., 1999). Essentially to the uropathogenic *E. coli* strains, strains segregated from CAUTIs express explicit O antigen and capsule (K) types and show an enormous number of virulence factors, including poisons and adhesins, the best-described ones being P and type 1 fimbria, both associated with adherence to catheters (Källenius et al., 1981; Klemm & Schembri, 2000; Mobley, Chippendale, Tenney, Hull, & Warren, 1987).

Besides the clinical sepsis, fringe central line blood stream colonization, due to the microbe present inside catheters in the form of biofilm, may frequently be sterilized (Matsumura & Simor, 1998). Of the 20 million intravascular catheters utilized yearly in the United States (Raad & Bodey, 1992), more than 7 million are CVCs which, with a minimalistically assessed infection pace of 5%, represent about 350,000 instances of catheter-related septicemia per annum (Hanna et al., 2001), costing approximately $9000 per scene (Saint, Veenstra, & Lipsky, 2000), despite the fact that assessments of contamination rates differ between 4% and 14% (Gil, Kruse, Thill-Baharozian, & Carlson, 1989; Michel, McMichan, & Bachy, 1979; Powell, Kudsk, Kulich, Mandelbaum, & Fabri, 1988; Prager & Silva, 1984). Yearly vascular catheter utilization rates in the United Kingdom have not been distributed; however, Christine, Térésa, Joseph, George, and Alain (1994) assessed that an ICU patient would have six inhabiting vascular catheters embedded during their visit. This is most likely a think little of with parenteral taking care of lines, pressure-observing gadgets, and dialysis being progressively utilized since that time (Christine et al., 1994). Numerous examinations have looked to decide regardless of whether the danger of septicemia is more noteworthy when utilizing triple-lumen CVCs than single-lumen CVCs. Raad et al. (1996) audited a few examinations and discovered the danger with multilumen catheters was more prominent however a new metaanalysis (Dezfulian, Lavelle, Nallamothu, Kaufman, & Saint, 2003) presumed that "at the point when investigations were included, multilumen catheters were found to be not related with a higher danger of catheter-related circulatory system disease." Triple-lumen catheters are bound to be embedded into those patients that have a more extreme underlying ailment, requiring numerous mixtures, and who might along these lines be more inclined to creating catheter-related septicemia (Gil et al., 1989; Raad et al., 1996).

Catheter-colonizing organic entities often arises from the skin inclusion site and catheter center and less often from either hematogenous cultivating of the catheter or sullied infusates (Raad et al., 1996). However, as of late, it has been proposed that CoNS may emerge from the oropharynx or gut of neutropenic patients with CoNS septicemia (Horn et al., 2000).

The degree and area of catheter colonization relies particularly upon the term of the catheter's utilization. A transient catheter (10 days) is probably going to have a more prominent colonizing load on the outer surface than in the lumen. Notwithstanding, a drawn out catheter (30 days) is liable to have twofold higher colonization on the inward surface than outside (Hanna et al., 2001). The idea of the liquids administered through the gadget may upgrade or repress advancement of biofilm on the inward surfaces (Anderson, Highsmith, & Holland, 1986; Donlan et al., 1999; Failla, Benedict, & Weinberg, 1975; Maki & Martin, 1975; Maki, Mermel, Bennett, & Brachman, 1998). *Candida* spp. are particularly connected with lipid-containing arrangements, for example, those found altogether parenteral sustenance. Henceforth, *Candida*-line contaminations and candidemia are especially connected with absolute parenteral sustenance lines, particularly patients in the ICU or who have pancreatitis, the last mentioned requiring intravenous taking care of for quite a long time (Ebrey et al., 2004; Versey et al., 2021).

6.3 Urinary catheters

Urinary Foley catheters are rounded latex or silicone gadgets embedded urethra to empower seepage of urine from the bladder. They readily attain microbes on their surface (Donlan, 2001b), accomplishing 10 to half colonization inside 7 days of inclusion (Stickler, 1996) and colonization is widespread by 30 days (Kunin & McCormack, 1966). Moreover, work by Jones et al. showed that MRSA biofilms in urinary catheters accomplish a profundity of five to six cells inside 2 h (Mathai, Jones, & Pfaller, 2001). In vitro, microbe, and clinical investigations showed that catheter-related UTIs are rising (Nickel, Downey, & Costerton, 1992; Nickel, Grant, Lam, Olson, & Costerton, 1991).

These trial models showed that thick biofilms create around tainted seepage spouts, expanding proximally into the waste sack and along these lines into the catheter. The bacterial biofilm rises by possibly (i) quickly separating bacterial cells spreading along the intraluminal catheter surface inside the extracellular polymeric network of the biofilm, and additionally and (ii) planktonic cells exhibiting saltatory development. Saltatory development includes planktonic cells "leap-frogging" only in front of the biofilm, potentially helped by the disturbance coming about because of when the urine stream meets the progressing biofilm front. A few living beings ensnared in catheter-related UTI, including all individuals from the family Proteus, constitutively produce urease (Stickler, Ganderton, King, Nettleton, & Winters, 1993; Stickler, Morris, Moreno, & Sabbuba, 1998).

Urease hydrolyzes urea within the sight of water to frame alkali as well as ammonium carbonate. Smelling salts raises the pH of the urine from 5.5 to a soluble pH, which can inactivate supplement and inclines for the development of a few life forms. Also, the basic pH makes precipitation of Ca^{+2} and Mg^{+2} structure stones, or calculi. The stones may frame in the catheter and become fused in the biofilm, causing encrustation (Stickler, 1996). As the follower bacterial and mineral total expands, the naturally visible biofilm will upset urinary stream, obstructing the catheter outlet and narrowing the measurement of the lumen. This "debris," practically speaking, requires numerous catheter changes and endeavors to "flush" out, which would itself be able to produce bacteremia (Ebrey et al., 2004).

UTIs are known to add to the dreariness of patients recuperated in LTCFs regardless of whether asymptomatic bacteriuria is extremely successive, as affirmed by a few commonness's examines acted in various nations, reporting that 15%–30% of urine tests from men and, 25%–50% from ladies were culture positive for at least one uropathogen (Nicolle, 1997).

Sample of urinary catheters eliminated from patients following a few days of persistence and noticed by scanning electron microscopy (SEM) frequently shows the multispecies idea of these biofilms, in some cases adding valuable data on the species included not generally recognized by culture-subordinate methods. If a catheter stays in situ for an adequate number of days to permit biofilm development, simply the way of life of a urine test gathered from the siphoned patient could be utilized to distinguish the bacterial composition of the biofilm created in the catheter lumen, keeping away from the need to examine the catheter itself. In any case, it should be stressed that this approach could lead to false adverse outcomes (Donelli & Vuotto, 2014).

The ongoing comorbid (more than one illness) conditions requiring patient admittance to the LTCF are broadly perceived as the principal causative variables of bacteriuria. Indeed, cerebrovascular mishaps just as degenerative neurological sicknesses, including Alzheimer's and Parkinson's illness, are related to a hindered purging of the bladder (neurogenic bladder) and related ureteric reflux, with both these marvels adding to bacteriuria (Nicolle & Yoshikawa, 2000).

Around 80% of UTIs in LTCFs are expected to the short- or long haul inclusion of a urinary catheter. In clinics for neuromotor rehabilitation, a drawn-out bladder catheterization is needed for most patients (Nicolle, 2012).

There are many reasons for CAUTIs:

(i) Movement of microorganisms from the urethra to the bladder following catheter inclusion. Notwithstanding, when the urinary plot is not influenced by explicit illnesses, microorganisms are expeditiously cleared by the safeguard component of bladder mucosa, the most solid being addressed by polymorphonuclear leukocytes.

(ii) Bladder defilement as an outcome of the passage of microorganisms through the thin liquid layer present at the urethral mucosa-catheter interface. This defilement course is supported by catheter developments and includes intestinal microorganisms colonizing the patient perineum that can climb the urethra after catheter inclusion.

(iii) Bacterial relocation through the catheter lumen as an outcome of defilement of the drainage pack framework. Around 30% of CAUTIs are assessed to be related with this intraluminal infection course, while the initial two extraluminal infection courses represent 70% of CAUTIs. The included microorganisms are generally obtained from outer sources, principally by means of the hands of clinical staff, with this source coming to up to 15% of diseases in gatherings of hospitalized patients (Mangal, Pho, Arcia, & Carter, 2021).

Patients with a urinary catheter set up for more than 30 days are considered to have a long haul, or on the other hand constant, catheter (Kalsi, Arya, Wilson, & Mundy, 2003). Obviously, the more extended urinary catheters stay set up, the more prominent the inclination of microscopic organisms to colonize their surfaces and develop as biofilm. It has been over and again detailed that two to five different microorganisms are typically recognizable in biofilm-based CAUTIs, the most well-known ones being Gram-negative *Enterobacteriaceae*, counting *E. coli*, *K. pneumoniae*, and *Citrobacter freundii*, just as urease-producing microbes, like *Proteus mirabilis*, *Morganella morganii*, and *Providencia stuartii*. Other Gram-negative species frequently happening in CAUTIs are *Pseudomonas aeruginosa* and *Acinetobacter baumannii*. Gram-positive microorganisms, like coagulase-negative *staphylococci*, *enterococci*, and bunch B *streptococci*, are additionally regularly segregated. Yeast might be at times confined from patients treated with rehashed courses of antimicrobials, *Candida albicans* being the most well-known, trailed by *Candida glabrata* and *Candida tropicalis* (Stickler, 1996; Warren, 1997).

Concerning the cycle of biofilm development, the addition of a urinary catheter in a patient is before long followed by a bond to the gadget surfaces of host proteins, shaping a "molding layer" ready to advance microbial adherence and biofilm development (Chenoweth & Saint, 2013; Jacobsen et al., 2008). Given the huge job of biofilm in the pathogenesis, the treatment also, the anticipation of CAUTIs (Hooton et al., 2010; Stickler, 2008), appropriate consideration has been as of late centered around the clinical ramifications of microbial sessile development. Indeed, the foundation of bacterial biofilm networks on urinary catheters is accepted to repress the viability of antitoxin treatment, sidestep have guard systems and work with bacterial correspondence, prompting the expression of virulence determinants (Davey & O'toole, 2000).

6.4 Catheter-related bloodstream infections (CRBSIs)

The early exchange of seriously injured patients from intensive care units to LTCFs has given a new unique to the rate of CRBSIs, for the most part, brought about by the improvement of a microbial biofilm on both the inward and external surfaces of vascular catheters. Indeed, extraluminal biofilm is the significant source of bloodstream contaminations in intravascular catheters (Ielapi et al., 2020) inside the main 7-day stretch of catheterization while, in both short and long-term catheters, the intraluminal biofilm turns into the significant reason following a multiweek of implantation. The arrangement of biofilm is the consequence of a nearby collaboration among microorganisms, gadget surfaces, and host proteins and emerge from two consecutive occasions: skin colonization around the catheter inclusion site by bacteria as well as organisms and their relocation along the catheter, where organisms join to its polymeric surfaces and begin to create as biofilm (Donelli et al., 2001; Nishikawa, Takasu, Morita, Tsumori, & Sakamoto, 2010; Raad, 1994).

In intravascular catheter-related diseases, the most elaborate microorganisms are coagulase-negative *Staphylococci*, representing approximately 40%, *Staphylococcus aureus* (~20%), and organisms, especially *Candida* species (~10%). The staying 30% are addressed by other Gram-positive (*Enterococcus faecalis*) and Gram-negative (*P. aeruginosa*, *E. coli*, etc.) microbes (O'Grady et al., 2002). The notable wonders of encrustation and impediment of catheters are identified with the foundation of catheter colonization by urease-producing microorganisms, ready to create smelling salts from urea and hoist the urine pH. As the pH esteem increments, long rectangular stones made of magnesium ammonium phosphate (struvite) and hydroxylated calcium phosphate microcrystals (apatite) encourage, leading to the turn of events of a microcrystalline layer, which is then quickly covered by a thick poly-microbial biofilm comprising of *P. mirabilis*, *E. coli*, *and P. aeruginosa* (Jacobsen & Shirtliff, 2011; Stickler, 2008).

Biofilm advancement causes the easing back of urine flow inside the catheter (Stickler, 2008), and different kinds of Foley catheters being influenced by this problem. Presently, there are no compelling techniques accessible for its control. Indeed, experiments in a research center model created by Stickler and Morgan have shown that even antimicrobial catheters, for example, those covered with silver or impregnated with nitrofurazone, are quickly encrusted and hindered by biofilm-framing, urease-creating microbes, especially *P. mirabilis* (Stickler, 2008). Perceptions on encrusted silver catheters eliminated from a patient have affirmed that the glasslike layer shapes additionally in vivo (Stickler, 2008).

Aside from the essential job of urinary pH in the grip of *P. mirabilis* to catheter surfaces, surface abnormalities of the Foley catheter eyeholes advance colonization and crystallization and lead to device blockage, as uncovered by SEM examinations (Stickler, Young, Jones, Sabbuba, & Morris, 2003). The Gram-negative, rod molded, encapsulated, and nonmotile *K. pneumoniae* is today considered the causative specialist of around 10%–20% of CAUTIs (Niveditha, Pramodhini, Umadevi, Kumar, & Stephen, 2012; Tinelli et al., 2012). Lately, extraordinary consideration has been done on the expanded occurrence of extended spectrum β-lactamases and carbapenemase producing *K. pneumonia*, not just when all is said in done emergency clinic yet in addition in rehabilitation places (Baraniak et al., 2013).

S. epidermidis, comprising a relevant segment of the typical human skin flora and mucosal epithelia, is viewed as a wholesome commensal occupant of these tissues, regardless of whether it can be effectively presented as a foreign substance specialist during the careful implantation of an intravascular catheter (Otto, 2008), implying that it is the causative specialist most oftentimes segregated from CRBSIs. Aside from the ubiquity of coagulase-negative *Staphylococci*, *S. aureus* likewise has a preeminent part as a causative specialist of CRBSIs. The development, lately, of methicillin-resistant *S. aureus* (MRSA) in everyday emergency clinics, especially in intensive care units, has drawn highlight regarding its spread in LTCFs. Truth be told, in the system of the MOSAR study, information was provided on the generality of MRSA and related hazard factors for colonization in four European rehabilitation centers (Bilavsky et al., 2012). Of a sum of 1204 patients, 105 (8.7%) were positive for MRSA furthermore, among them, 74.3% had been embedded with an invasive clinical gadget in the past month. Since the mid-1980s, electron microscopy perceptions showed that multifaceted bacterial networks of *Staphylococci* were encased in a slimy lattice (Baldassarri, Simpson, Donelli, & Christensen, 1993) and clung to intravascular catheters eliminated from patients (Peters, Locci, & Pulverer, 1982).

Atomic science examinations added more detail to the various steps of biofilm development in both *S. epidermidis* and *S. aureus*. In *S. aureus*, and especially in MRSA strains, the introductory connection to have extracellular matrix proteins is interceded by microbial surface components perceiving adhesive matrix molecules, like FnBPs (Pozzi et al., 2012). In contrast in *S. epidermidis*, and the majority of the methicillin-resistant *S. aureus* strains, the creation of the exopolysaccharide poly-β-1,6-*N*-acetyl-D-glucosamine is needed for polysaccharide intercellular adhesion-dependent biofilm development and encoded by the extra intercellular attachment operon (Beenken et al., 2004; Cramton, Gerke, Schnell, Nichols, & Götz, 1999; Foster & Höök, 1998; Mack et al., 1996; Pozzi et al., 2012; Rohde, Frankenberger, Zähringer, & Mack, 2010). Poly-β-1,6-*N*-acetyl-D-glucosamine has been utilized as an objective in an antibiofilm procedure based on the protein dispersin B (Fig. 3), which can disrupt the framework of *S. epidermidis*, in this manner advancing antitoxin dispersion into the biofilm and afterward its killing action (Donelli et al., 2007). *S. aureus* Bap is another surface protein discovered to be basic for biofilm development in 5% of 350 *S. aureus* isolates from bovine mastitis, regardless of whether not found in each of the 75 *S. aureus* human strains analyzed (Cucarella et al., 2001). Up until now, Bap has never been accounted for in *S. aureus* human clinical isolates.

Furthermore, the staphylococcal accessory regulator (*sarA*) and accessory gene regulator (*agr*) have been implicated in biofilm formation: the first induces attachment and early biofilm formation (Beenken, Blevins, & Smeltzer, 2003) and the latter leads to the upregulation of several virulence factors and supports seeding dispersal in mature biofilms (Boles & Horswill, 2008; Chan, Coyle, & Williams, 2004).

In a new report on the subatomic epidemiology of *S. epidermidis*, a bunch of 33 separates causing CRBSI in hospitalized patients has been contrasted and a bunch of 33 commensal separates, exhibiting that strains from CRBSI patients were more impervious to most of the tested antibiotics and more regularly is positive than the commensal ones, the larger part in this manner being solid biofilm producers (Cherifi et al., 2013).

Enterococcus spp. are progressively considered arising causative specialists of intravascular CRBSIs. Indeed, a critical expansion in nosocomial contaminations brought about by *Enterococci* has been seen in numerous countries (Huycke, Sahm, & Gilmore, 1998), this bacterial species being the fourth most normal reason for bloodstream contaminations in Europe (Fluit, Schmitz, & Verhoef, 2001).

Clinic procured enterococcal bacteremia related to central venous catheters have expanded drastically since the mid-1980s (Patterson et al., 1995; Shlaes, Levy, & Wolinsky, 1981). Everywhere in the world, the capacity of enterococci to frame a biofilm on the polymeric surfaces of intravascular catheters, along with their characteristic protection from a few antimicrobial, brought about the expanding spread of MDR enterococcal strains in clinical settings. As the biofilm develops and gets fully grown, single or grouped microorganisms isolate from the highest point of the biofilm and, returning their planktonic method of development, spread to uncolonized regions, so rehashing the biofilm improvement cycle on new surfaces. Some adherence factors are involved in enterococcal colonization of intravascular catheters are Efa A protein (Lowe, Lambert, & Smith, 1995); as surface protein (Clewell, 1993; Joyanes, Pascual, Martinez-Martinez, Hevia, & Perea, 2000); Carbohydrates of the cell surface (Guzman et al., 1991); Ace (adhesins of collagen from enterococci) protein acting as a collagen-binding adhesins (Rich et al., 1999).

Besides, the cell surface protein (Esp) has been exhibited to assume a crucial part in enterococcal biofilm arrangement (Toledo-Arana et al., 2001), its succession being like the *S. aureus* surface protein Bap, basic in biofilm arrangement (Cucarella et al., 2001).

6.5 Prosthetic heart valves

Mechanical heart valves, while somewhat low in numbers and recurrence of disease, have a high inferable mortality when they become contaminated. Prosthetic valve endocarditis accounts for 7%–25% of all instances of infective endocarditis in the created world (Mylonakis & Calderwood, 2001). Biofilm improvement is more normal on the tissue encompassing the prosthesis and on the

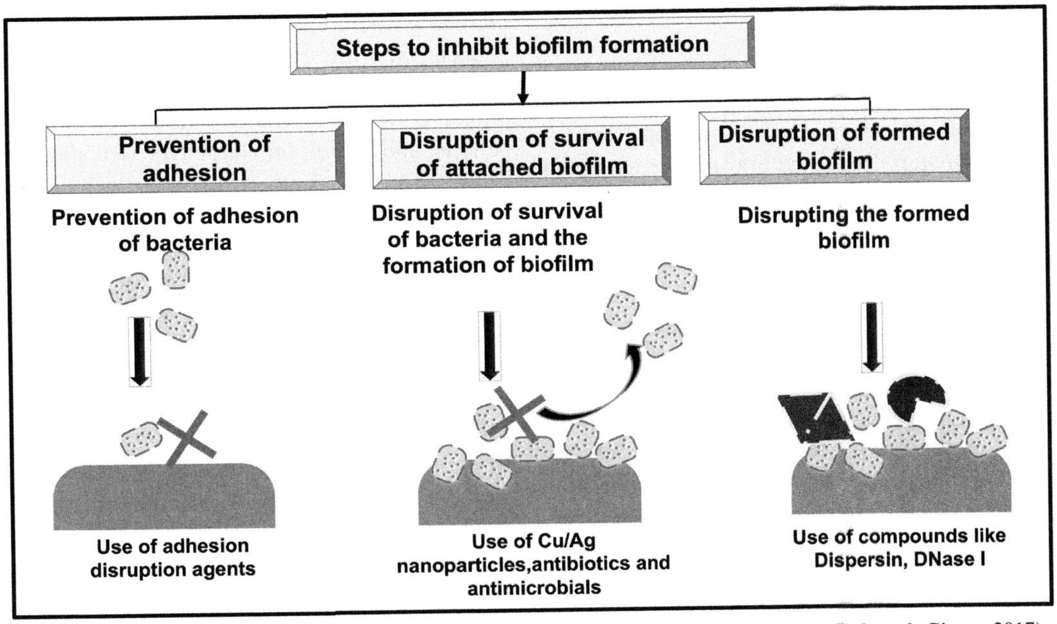

FIG. 3 Prevention strategies of biofilm formation by use of different strategies on implant surfaces (Berhe, Tefera, & Gizaw, 2017).

sewing sleeve texture (Carrel, Nguyen, Kipfer, & Althaus, 1998; Illingworth, Tweden, Schroeder, & Cameron, 1998) than on the prosthetic valve itself (Karchmer & Gibbons, 1994). Endeavors to lessen biofilm improvement have included (likewise with vascular catheters) silver covering the sleeve texture (Illingworth et al., 1998). As with biofilm contaminations in other inhabiting prostheses, not many are settled by antitoxin treatment alone (Waldvogel & Bisno, 2000). Surgical intercession for expulsion is regularly necessary to forestall the genuine results of embolization and valve burst. Helpless join and patient endurance likewise happened in patients with contaminated vascular unions, ordinarily requiring surgical evacuation, and substitution (Alsulaimani, 2021; Wilcox, 1994).

6.6 Orthopedic devices

Outer obsession gadgets for open breaks, being presented to the climate, are more liable to become colonized and afterward the focal point of disease than shut break (internal) fixators. Interior fixators are almost certain thusly to become contaminated than joint prostheses (Weinstein & Darouiche, 2001). Examination of microbes utilizing CoNS biofilms on treated steel screws accomplished a 90% achievement rate while treating with antibiotics alone (Isiklar, Darouiche, Landon, & Beck, 1996). Other work to decide the infiltration of antimicrobials into biofilms on tempered steel muscular nuts neglected to annihilate the contamination (Darouiche et al., 1994). It has for some time been realized that diseases of prosthetic joints are regularly difficult to fix by antitoxin treatment alone. Fitzgerald et al. (1977) detailed just a 20% achievement rate in saving the joint by traditionalist administration, utilizing antibiotic and careful waste. This has been generously worked on in later a long time, in spite of the fact that "modification" medical procedure has its own orderly dangers of grimness and mortality and a far higher danger of resulting contamination (Ebrey et al., 2004; Versey et al., 2021).

7. Prevention strategies

7.1 Hand hygiene

As the hands of medical service workers are the most continuous vehicle of HAIs, hand cleanliness (hand washing and hand sanitization) is the essential preventive measure. Cross-contamination by means of the hands of medical care work force is assessed to be answerable for 20%–40% of HAIs (Ellingson et al., 2014). Subsequently, the cleaning of hands, both earlier and following to patient contact, is imperative in limiting the spread of bacteria (Mathur, 2011). Thorough hand washing with satisfactory amounts of water and cleanser eliminates over 90% of shallow, greenery including all or most impurities. Killing all transient verdure with all toxins inside a brief time frame (a couple seconds) requires sterile hand sanitization utilizing liquor or alcoholic arrangements. Hand sterilization with alcohol does not substitute the requirement for ordinary and exhaustive hand washing with cleanser and water (van Roekel, Reinhard, & Grimmelikhuijsen, 2021; Widmer, 2000).

7.2 Cleaning

Perhaps the most fundamental measures for the support of cleanliness, and one that is especially significant in the medical clinic climate, is cleaning. Cleaning is the evacuation of unfamiliar material (like natural matter and soil) from objects utilizing cleanser or enzymatic items. Medical clinic surfaces are cleaned hourly, day by day or week by week in the event that the surfaces show up messy, there are spillages, or after tolerant release. In the United Kingdom, routine cleaning is performed physically by utilizing gear like mops, wipes, materials, cans and brushes, and vacuum cleaners. Basic surfaces, for example, as often as possible contacts surfaces should be cleaned altogether more routinely (Smith, 2008). Cleansers and cleansers go about as solubilizing specialists and have no antimicrobial action, and in this way, the viability of the cleaning interaction is totally subject to the 60 mechanical cycle. Careful cleaning will eliminate over 90% of microorganisms. Notwithstanding, thoughtless and shallow cleaning is substantially less viable and could even have an adverse consequence, by scattering the microorganisms over a more noteworthy surface and expanding the chance that they may debase other objects (Rutala & Weber, 2008).

For instance, a *P. aeruginosa* episode was accounted for in a hematology-oncology unit because of tainting of the surface cleaning gear when nongermicidal cleaning arrangements were utilized to disinfect the patients' current circumstance rather than disinfectants (Hardy, Oppenheim, Gossain, Gao, & Hawkey, 2006). Therefore, the CDC suggests the utilization of dye and other substance sanitizers like aldehydes, alcohols, and quaternary ammonium mixtures to dispose of microorganisms (Rutala & Weber, 2008).

7.3 Disinfection

Sterilization known to the utilization of substance specialists on lifeless surfaces to annihilate or repress microorganisms, except for high quantities of bacterial spores. There is no great sanitizer and the best trade-off will rely upon various variables including:

(i) The number and area of microorganisms
(ii) The intrinsic obstruction of the microorganisms
(iii) The focus and intensity of the sanitizers
(iv) Duration of exposure to disinfectant
(v) Nature of pollutant (natural or inorganic matter)
(vi) Presence of biofilms (these can be impervious to sanitizers and harder to eliminate)
(vii) Chemical and actual components (for example, temperature, pH, stickiness, and water hardness) (Jing et al., 2020).

A scope of biocides (substance specialists that inactivate microorganisms) are accessible for use as sterilizers including alcohols (fundamentally ethanol and isopropanol), chlorine, and chlorine compounds, quaternary ammonium mixtures and phenolic, formaldehyde, iodophors, peracetic corrosive, hydrogen peroxide, glutaraldehyde, and *ortho*-phthalaldehyde (Rutala & Weber, 2021).

7.4 Sterilization

Sterilization alludes to a physical or compound interaction that totally obliterates or eliminates all microbial life, including spores (McDonnell & Russell, 1999). By definition, it impacts a decrease in the quantity of microorganisms by a factor of more than 10^6 (for example >99.9999%). Sanitization can be accomplished by both physical and compound means. Actual techniques depend on the activity of warmth (autoclaving, dry warm, or wet warm sanitization), on illumination, or on mechanical division by filtration. Compound means incorporate gas disinfection with ethylene oxide, hydrogen peroxide, peracetic corrosive, or different gases and submersion in a sanitizer arrangement with disinfecting properties (for example, glutaraldehyde) (McDonnell & Russell, 1999; Rutala & Weber, 2008).

7.5 Behaviors and practices affecting cleaning and hospital-acquired infection

Viability of cleaning and adherence to contamination control conventions is not reliant just on the strength of the cleaning arrangements alone yet in addition the common wellbeing society in medical services offices. Notwithstanding the danger of openness to irresistible specialists, consistence to disease control, cleanliness, and personal protection equipment (PPE) conventions are regularly very low and are credited to various elements (Owusu, 2020).

To begin with, there is an overall absence of value preparing for clinic cleaning staff and other medical care faculty with regard to methods of HAI transmission, appropriate hand cleanliness, and PPE usage (Larson & Liverman, 2011). In one examination at an emergency unit, a 9-month time span, the securing of vancomycin-resistant *Enterococcus* (VRE) dropped 3-overlap after a time of instructive mediation to work on natural cleaning and a time of multimodal hand cleanliness intervention (Hayden et al., 2006).

Also, for medical clinic cleaning staff, obstructions experienced while preforming their work has a sway on cleaning rehearses. One report recorded not having adequate opportunity to perform every day cleaning and release cleaning appropriately just as interferences to help with different errands as difficulties to accomplishing ideal cleaning and sanitization in hospitals (Bernstein et al., 2016).

Institutional perspectives and strategies toward security, cleanliness, and contamination control are significant in developing a solid well-being society. Hierarchical obligation to disease control is reflected in the accessibility of and interest in preparing and supplemental classes, peer adherence to cleanliness and disease control arrangements, authoritative help for staff, patient security and well-being—advancing cleanliness even to the detriment of productivity, positive support of individual consistence conduct, and negative support of resistance. Gershon et al. (1995) discovered that representatives who see a solid organization wide obligation to wellbeing have been discovered to be more than 2.5 occasions bound to follow to security conventions than the individuals who need such perceptions (Gershon et al., 1995; Larson & Liverman, 2011).

In this way, distinguishing main drivers of rebelliousness and planning intercessions to address these issues guarantees viable and predictable practices that lessen HAI transmission and keep procurement of HAIs low (Owusu, 2020).

7.6 Brass alloys: Copper-bottomed solutions against hospital-acquired infections

Brasses are made up of copper and zinc and come under alloys, but bronze is made up of copper and tin. The percentage of zinc alloy can differ from 5% to 40% (Choucri et al., 2019). The zinc percentage less than 35% is known as copper rich and forms a face-centered cubic (fcc), and single solution copper matrix known as alpha brass (Choucri et al., 2019; Laws et al., 2015). And, a solution having more than 35% of Zn, and ordered body-centered-cubic (bcc) phase, known as beta brass. The beta phase is stronger than the alpha phase (Choucri et al., 2019; Laws et al., 2015).

According to the layman view, high Zn-containing brasses are named as yellow brasses and high copper amount containing brasses are named as red brasses. To improve the corrosion-resistant and technological properties, high entropy alloys (HEA) are formed using various chemical elements added to brass like aluminum, silicium or lead, nickel, and manganese (Laws et al., 2015; Nagase, Shibata, Matsumuro, Takemura, & Semboshi, 2019).

7.6.1 Brass alloys to reduce hospital-acquired infections (Fig. 4)

The primary legitimate effort to assess a potential decrease in the occurrence of HAIs utilizing copper-containing combinations were studied by Salgado et al. (2013). The creators enrolled a critical decrease in HAIs and additionally colonization by MRSA and VRE in ICUs in a forthcoming, aim to-treat concentrate in a room containing six items made of copper-based combination versus rooms with control objects. There was no impact of length of stay on the occurrence of HAI colonization in either kind of room. A huge connection between natural weight and occurrence of HAI was also seen (Salgado et al., 2013).

Restrictions were all things considered highlighted, for example, the expulsion of some copper-containing objects from patient rooms or the presentation of such articles in charge rooms. Additionally, the investigation was not twofold dazed, yet such an examination configuration would be hard to accomplish due to the unmistakable parts of copper compounds and their control partners. Alterations close by hygiene schedules following the presentation of copper-containing objects and the beginning of the examination, thus relieving the impact of this putative meddling element. Inquiries regarding the chance of specific detailing and the natural believability of the discoveries were likewise raised (Harbarth, Maiwald, & Dancer, 2013).

Another planned examination on the conceivable decrease of HAIs occurrence was held in Chile (Rivero, Brenner, & Nercelles, 2014) in a setup of emergency unit. Notwithstanding a slower pace of catheter-related bacteremia in copper-fitted rooms, no huge distinction in HAI worldwide occurrence was found between rooms fitted with copper and control rooms. Notwithstanding, a few constraints have been advanced by the actual creators, for example, the set number of included patients contrasted with the one conjecture to get a huge factual outcome or the way that diseases not related with intrusive gadgets were not considered. Others can be added, for example, a base length of stay to be remembered for the investigation of 24 h when HAIs is typically diagnosed after a 48 h stay. Additionally, extraordinary cleaning conventions utilized for copper-based gadgets and their partners (Dauvergne & Mullié, 2021).

7.7 Copper alloy surfaces kill bacteria and reduce hospital-acquired infections

There is clinical evidence that shows that copper alloy surfaces are a good source to reduce microbial burden (Schmidt et al., 2012) and the health risk of HAIs (Salgado et al., 2013). After observation with infection control policies and hand washing, the solid copper alloy touch surfaces usage provide a second line of defense against health-threatening microbes. The enrolment of copper alloys with the U.S. Environmental Protection Act (EPA) grants disclosing well-being claims against six illness-causing microbes (Michels & Anderson, 2008). By killing microbes, copper alloys may offer passive insurance between routine cleaning and sterilization (Michels & Anderson, 2008).

Notwithstanding the microbes covered by EPA registration, the viability of copper compounds against a scope of significant microorganisms, including infections and parasites, is very much portrayed. Handwashing, cleaning, and

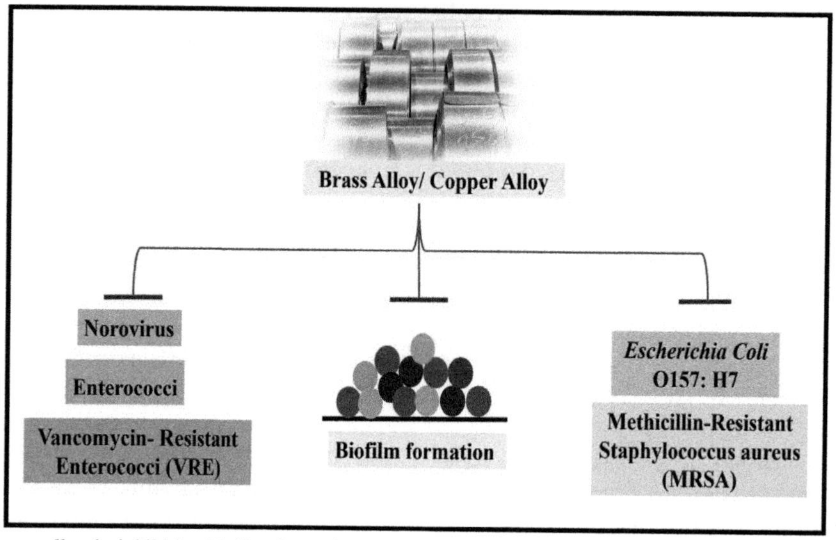

FIG. 4 Action of brass/copper alloy in inhibiting biofilm formation.

sanitizing are basic in relieving the transmission of organisms. In any case, these arrangements depend on human conduct. The natural antimicrobial action of metallic copper compounds is a passive framework that requires no human interference and its activity is constant as opposed to occasional (Michels, Keevil, Salgado, & Schmidt, 2015).

A new clinical preliminary showed that the restricted position of copper alloy surfaces inside the fabricated climate brought about a normal 83% decrease in the bacterial weight (Schmidt et al., 2012) and a 58% decrease in the frequency of HAI (Salgado et al., 2013). The determination and sending of parts with contact surfaces produced using copper compounds may decrease HAIs, lower treatment expenses, and save lives (Table 2) (Michels et al., 2015).

Copper nanoparticles (CuNPs) are a viable antimicrobial specialist next just to silver, filling in as a station for the sustained however constant arrival of copper particles. Copper-and copper-containing nanomaterials are not difficult to fabricate and are less expensive in contrast with silver. Like silver nanoparticles, CuNPs cause contact killing by means of explicit activities, for example, film interruption, and DNA harm because of copper flood into the cells, oxidative harm, removal of iron from Fe-S bunch compounds, and so forth (Chen et al., 2014; Gao et al., 2014; Henry, 1943; Macomber & Imlay, 2009; Tian et al., 2012; Vincent et al., 2018). It is seen that the powerful antibacterial grouping of copper nanoparticles is profoundly considered. For instance, the MIC of CuNP was in the scope of 1.1–1000μg/mL (Kaweeteerawat et al., 2015; Khurana & Chudasama, 2018; Pourbeyram, Bayrami, & Dadkhah, 2017).

7.7.1 *Methicillin-resistant* Staphylococcus aureus *(MRSA)*

Staphylococci are omnipresent and effortlessly moved between people, or to objects where they are therefore gained by medical care workers or patients and promptly moved to other people. *S. aureus* is a segment of numerous individuals' skin flora. Even though anybody can be tainted by *Staph*, a few people are more prone. At the point when these microscopic organisms discover a weakness, particularly an open injury, or when an immune-compromised person is uncovered, a lethal disease may happen (Tierno, 2002). Among antibiotic-resistant bacteria, MRSA is of specific concern. MRSA diseases cause around 126,000 hospitalizations every year in the United States (Kuehnert et al., 2005), although numerous medical clinics have effectively decreased the rates lately (Jain et al., 2011). Copper's viability against MRSA has been exhibited in a few independent investigations (Gould et al., 2009; Mehtar, Wiid, & Todorov, 2008; Michels, Noyce, & Keevil, 2009; Noyce, Michels, & Keevil, 2006a; Weaver et al., 2010).

For this situation, a 99.9% copper composite and stainless steel were immunized with 700,000 colony-forming units for every square in, a sum of multiple times in a 24h period, with no irregular cleaning. The grouping of every inoculum is somewhere around one significant degree higher than the surface levels estimated on noncopper surfaces in the ICUs of three US emergency clinics, which went from a few hundred to a few large numbers of CFUs (Attaway et al., 2012; Schmidt et al., 2012, 2013).

However, the stainless steel coupons continued to not only harbor viable MRSA but also showed substantial and increasing concentrations of MRSA after each inoculation. Copper alloys can continue to kill MRSA, even after repeated inoculations, while bacteria resident on the stainless steel continues to survive and grow. It has been reported (Gould et al., 2009) that three of the five clinically important MRSA strains tested on pure copper were killed within 60min, and the remaining two strains were killed within 80–100 min. Furthermore, these authors (Gould

TABLE 2 **Biofilm-targeting compounds/molecules and their effects.**

Biofilm targeting compound/ molecule	Effect of compound/molecule on biofilm	References
CuNP	Contact killing by film interruption, DNA/oxidative damage/ harm, removal of iron-sulfur bunch because of copper flooding into cells.	Macomber and Imlay (2009), Henry (1943), Gao et al. (2014), Tian et al. (2012), Vincent, Duval, Hartemann, and Engels-Deutsch (2018), Chen et al. (2014)
Pure copper (Cu)	60% Methicillin-resistant *Staphylococcus aureus* (MRSA) killed within 60 min remaining 40% killed within 80–100 min. 100% community-acquired Methicillin-sensitive *Staphylococcus aureus* (CA-MSSA) killed within 80 min.	Gould et al. (2009)
Compounds containing 90% or more copper (Cu)	Vancomycin-resistant *Enterococci* (VRE) (*Enterococcus faecalis* and *Enterococcus faecium*) killing within 1 h.	Gould et al. (2009)
Cu-Ni-Zn combination (C75200, 65% Cu) and Cu-Ni-Zn+Metal combination (C28000, 60% Cu)	Killing of more than one billion CFU of *Escherichia coli* within 15 min.	Santo, Taudte, Nies, and Grass (2008), Michels et al. (2015)
Compounds containing more than 70% Cu	Critical decrease in *Clostridium difficile* (vegetative spores) in 6 h. Total killing of *Clostridium difficile* (stronger spores) between 24 and 48 h.	Weaver, Michels, and Keevil (2008), Michels et al. (2015)
Cu test/Cu amalgams	Can latent influenza/inactivate 75% of influenza A (H1N1) in 1 h and practically 100% after 6 h	Weaver, Noyce, Michels, and Keevil (2010)
	Can latent norovirus and can assist with controlling its transmission	Michels et al. (2015)
99.9% of Cu surfaces	No recuperation of irresistible infections of Norovirus after 30 min exposure.	Warnes and Keevil (2013)
Chitosans (CS) (Polysaccharides)	Countered biofilms shaped by *Klebsiella pneumoniae* and *Escherichia coli.*	Campana, Biondo, Mastrotto, Baffone, and Casettari (2018)

et al., 2009) also reported similar efficacy against the community-acquired methicillin-sensitive *S. aureus*. All five tested strains were killed within 80 min. These findings are significant because the number of cases of CA-MSSA continues to increase, as healthcare-associated MRSA has spread to nonhealthcare settings in the community. It is important to note that most laboratory efficacy tests utilize very high initial inoculation levels, which are typically greater than one million CFU/in.^{2}, to demonstrate efficacy. Alloys containing lower levels of copper, including brass and bronze, were tested against MRSA and also exhibited greater than 99.9% kill within 2 h when challenged with these high inoculation levels. It has also been shown that lowering the initial bacterial load to levels closer to those found in the clinical setting resulted in a shorter time to kill MRSA. This has practical significance because, in the typical clinical setting, copper alloys kill most of the bacteria in a matter of minutes rather than hours (Michels et al., 2015).

7.7.2 Vancomycin-resistant enterococci (VRE)

Around 33% of enterococcal diseases in ICUs in the United States are brought about by VRE (Warnes, Green, Michels, & Keevil, 2010). VRE is spread by contact between patients, medical services workers, and natural surfaces (Drees et al., 2008). It was accounted (Warnes et al., 2010) for that strains of vancomycin-resistant *Enterococcus faecalis* and *Enterococcus faecium* were killed by compounds containing 90% or more copper in 1 h, when beginning inoculations were more prominent than 1000,000 CFU/cm^2, however, made due for half a month on treated steel. Different specialists (Gould et al., 2009) revealed that around 10 million CFU/cm^2 of five clinical strains of VRE were killed by a pure copper alloy inside 1 h (Michels et al., 2015).

7.7.3 Escherichia coli *O157:H7*

This virulent strain of *E. coli* is answerable for various food reviews, diseases, and deaths. Episodes are usually connected with half-cooked meat (Noyce, Michels, & Keevil, 2006b), albeit polluted drinking water can correspondingly bring about disease. Exposure may outcome in gentle to extreme vomiting, stomach cramps, and kidney failure as a result of the hemolytic uremic disorder. Notwithstanding, there are numerous strains of *E. coli*, and some are found in clinics (Jarvis & Martone, 1992). More than 30 years prior, another specialist (Kuhn, 1983) covered an understudy nurses' project led in a clinic that exhibited that door handles made of metal, a copper alloy, showed inadequate development of *E. coli*, while the stainless steel counterparts were intensely colonized (Michels et al., 2015).

In a lab study, this specialist (Kuhn, 1983) likewise tracked down that little microscopic organisms were available on copper and metal strips vaccinated with *E. coli*, while heavy contamination persevered on aluminum and stainless steel samples. Later, Noyce et al. (2006b) inoculated different cast copper combinations with *E. coli* O157: H7. The entirety of the copper composites killed the life form on occasions going from 45 min to 6 h. A few free investigations have affirmed that copper alloys viably kill different strains of *E. coli* (Gould et al., 2009; Michels & Anderson, 2008; Noyce et al., 2006b; Santo et al., 2008; Warnes, Highmore, & Keevil, 2012; Wilks, Michels, & Keevil, 2005).

A technique named a "dry inoculation strategy" was created (Santo et al., 2008). This strategy better stimulates this present reality circumstance when people contact surfaces by restricting the volume, however, not the convergence of a marker microorganism applied to a surface. A volume equivalent to 1 μL containing a known convergence of the bacterium under evaluation is applied to a surface utilizing a swab application. The fluid rapidly evaporates putting the organism in direct contact with the microbicidal copper surface. This strategy was worked on copper surfaces. This technique was rearranged and made more reproducible for VRE (Warnes et al., 2010) and hence *E. coli* O157: H7 (Warnes & Keevil, 2013; Warnes et al., 2012), by adding 1 μL straightforwardly to the surface and spreading with the pipette tip, permitting drying in short order. At the point when these dry inoculum strategies are utilized, the time needed for a considerable decrease in reasonability of the applied population of microorganisms is extraordinarily diminished, when contrasted with the chance to death saw in the generally utilized wet technique for immunization utilized in lab exploration and EPA viability testing. The dry inoculum strategy was additionally used to challenge a copper-nickel-zinc combination (C75200, 65% Cu) and a metal [C28000, 60% Cu (Santo et al., 2008)]. Both alloys brought about the demise of more than one billion CFUs of *E. coli* within 15 min (Michels et al., 2015).

7.7.4 Clostridium difficile

This anaerobic spore-forming bacterium is exceptionally strong and frequently burdens patients taking expansive range antibiotics. At the point when focused, vegetative *C. difficile* reacts by shaping spores to endure. It was accounted (Kramer, Schwebke, & Kampf, 2006) for that *C. difficile* spores can make due on evaporating surfaces for upto 5 months. Also, *C. difficile* spores are not killed by all medical clinic-grade sanitizers. It has been seen by different specialists (Weaver et al., 2008) that it is simpler for copper combinations to kill vegetative *C. difficile* than its spores (Michels et al., 2015).

It was discovered (Weaver et al., 2008) that exposure to compounds containing more noteworthy than 70% copper brought about a critical decrease in both vegetative *C. difficile* and its spores in 6 h, and a total kill of the stronger *C. difficile* spores somewhere in the range of 24 and 48 h, while stainless steel showed no decrease even after 168 h of exposure (Michels et al., 2015).

Vegetative *C. difficile* died quicker, albeit both vegetative microorganisms and spores can be shed in contaminated feces. Although 24 h may recommend an insufficient kill rate, viability showed by copper composites is possibly critical, taking into account that *C. difficile* spores are particularly difficult to inactivate and can grow even following quite a while of exposure to surrounding oxygen on lifeless surfaces in the fabricated climate (Kramer et al., 2006). Different specialists (Wheeldon et al., 2008) used a germinant to improve copper's viability against *C. difficile* spores. A germinant is a synthetic that induces the change of spores to more vulnerable vegetative cells. By instigating the developing spore express, these analysts detailed more noteworthy than a 99% decrease of more than 1000,000 CFUs/cm^2 inside 3 h of openness to an unadulterated copper compound against *C. difficile* spores, however, this may not be useful in medical services setting (Michels et al., 2015).

7.7.5 *Influenza A*

This infection is answerable for various hospitalizations. And, hospitalized patients with sensitive immune frameworks are helpless to influenza disease. Immunization of medical care workers can lessen the pace of nosocomial diseases among patients (Salgado, Giannetta, Hayden, & Farr,

2004). Influenza A will be a viral microbe that causes huge mortality and morbidity in the old and different gatherings at high danger (Noyce, Michels, & Keevil, 2007). These scientists likewise revealed that copper tests inactivated 75% of influenza A (H1N1) in 1 h and practically 100% after 6 h. Viruses are known as committed parasites. Consequently, they cannot finish their life cycle without exploiting appropriate hosts and along these lines are not viewed as alive. They contain a bunch of guidelines that, when brought into the host, can create pathobiological outcomes. Nonetheless, copper amalgams can latent influenza A (Noyce et al., 2007) and along these lines may fundamentally diminish its pathobiological outcomes (Michels et al., 2015).

7.7.6 Norovirus

Norovirus is infectious, causing 267 million instances of gastroenteritis yearly around the world, bringing about 5–8,000,000 passing (Warnes & Keevil, 2013). Around 23 million are tainted in the United States every year. It is moved by hand-to-hand contact, contacting ecological surfaces, and ingesting infected food. Outbreaks do happen in medical clinics and are additionally now and again provided details regarding journey ships. A few components clarify why norovirus is so infectious. It has an exceptionally little distance across comparative with other infections, and as not many as the possible reason of disease. Contaminated people likewise shed norovirus for as long as 3 weeks after side effects stop. Norovirus has a defensive shell or capsid. In any case, a newly weakened (1:10) arrangement of family bleach (5.25%), adequate to produce the last grouping of 5000 sections for every millions of accessible chlorine is powerful and inactivates norovirus. As of now, neither immunization nor a successful treatment is accessible (Michels et al., 2015).

Human norovirus cannot at present be refined in the lab, which has collapsed the improvement of viable mediations or control measures. Examination on a feline surrogate, a respiratory microorganism, has met with restricted achievement. A murine norovirus, MNV-1, a nearer proxy since it is a gastrointestinal microbe, was mercifully given by Professor Herbert W. Virgin VI of Washington University. It was seen (Warnes, Summersgill, & Keevil, 2015) that the infectivity of MNV-1 was lost upon exposure to copper and copper composites, however, held on hardened steel. These specialists could not recuperate irresistible infection after just 30 min of openness to copper (99.9% Cu) and 60 min to copper-nickel (90% Cu) surfaces, when utilizing the wet inoculum strategy (20 μL volume) to reproduce moderate drying vomitus. In the dry inoculum strategy, which reproduces hand-to-hand or hand-to-protest move, when a 1 μL volume containing norovirus was put onto test surfaces (Warnes & Keevil, 2013), the inactivation was discovered to be 5 min for both copper (99.9% Cu) and copper-nickel (90% Cu) surfaces. They likewise showed that contact of the infection with copper surfaces destroyed their RNA genome. Along these lines, it was seen (Warnes et al., 2015) that capsid respectability was compromised after interacting with copper composites. Along these lines, copper amalgams can latent norovirus and can assist with controlling its transmission (Michels et al., 2015).

7.7.7 Enterococci

The movement of enterococci from the gut into the circulatory system starts with gut colonization, then, at that point total substance helping enterococci to cross sound intestinal epithelial (Olmsted, Dunny, Erlandsen, & Wells, 1994; Wells et al., 2000), prompting unconstrained bacteremia (Edmond et al., 1995). Gut carriage of resistant enterococci need not be related to a clinic stay or earlier antibiotic treatment, since patients are frequently conceded from the local area previously colonized with VRE. In a medical clinic with no recorded instances of disease with VRE, 3.5% of inpatients were found to have gut colonization (Gordts, Van Landuyt, Ieven, Vandamme, & Goossens, 1995). Such carriage has demonstrated hard to kill by antibiotics (Hachem & Raad, 2002). The extent of HAI brought about by VRE in the United Kingdom is higher than in the rest of Europe, however, not exactly that in the United States. Colonization of patients by VRE has been demonstrated to be an inclining factor for resulting clinical disease shows an examining electron micrograph of a VRE biofilm. Many seclude of VRE from patients clinically contaminated in episodes are clonal, showing nosocomial transmission.

Author examination developed a new molecule that helps to prevent the sticky biomolecules (proteins, sugars, and so on) from sticking to surfaces, and therefore decreases biofilm development (Maity, Nir, Zada, & Reches, 2014). To shape this "nonstick" covering that will prevent the connection of organic particles to a surface, they copied the science of Teflon. Teflon is a nonstick material utilized in the creation of a frying pan. This atom created is comprised of amino acids. Amino acids are the structure blocks of proteins. Authors picked amino acids since they are not harmful to people, they are not difficult to orchestrate in enormous amounts, they can be put away in the fridge for quite a long time, and, most strangely, because they can self-assemble. Self-assembly is a cycle by which atoms spontaneously structure an arranged construction, with no requirement for external energy (Reches, 2018).

All in all, these amino acids can communicate with one another and interface like Lego building blocks. To follow our self-amassing nonstick particles to the surface, they utilized another amino acid, one that fills in as the principal segment of mussel's adhesive proteins (MAPs). Mussels can join firmly to rocks in the ocean and hang on during elevated and low tide. Mussels follow utilizing strings made of MAPs that are rich in the amino acid L-3,4-dihydroxyphenylalanine (DOPA). This amino acid empowers MAPs to adhere to any surface metal, glass, and plastic (Reches, 2018).

They synthesized nonstick particles in the lab, as a powder that resembled powdered sugar. They dissolved this powder in a fluid and sprayed it on various surfaces. They then, at that point played out a few estimations to demonstrate that the particles, in reality, self-assembled on a superficial level. To test if the covering preventing bacterial bond, they incubated the covered surface in an arrangement that contained 1,000,000 bacterial cells and their necessary supplements. They generally tried two sorts of surfaces: one with the covering (the experiment) and the other without the covering (the control) (Reches, 2018).

7.7.8 Antibiotic coating of devices

Impressive work has been attempted to examine the presentation of covered gadgets, particularly vascular and urinary catheters covered with antimicrobial substances, to take out, or basically diminish, the rate at which colonization happens. The presentation of different antimicrobial specialists has been explored, including dicloxacillin, clindamycin, fusidic corrosive, and chlorhexidine (Sherertz, Carruth, Hampton, Byron, & Solomon, 1993; Sherertz, Forman, & Solomon, 1989). Novel utilization of tridodecylmethyl-ammonium chloride as a holding cationic surfactant to tie cefazolin to catheter surfaces created a close to sevenfold diminish in the rate of catheter colonization (Kamal, Pfaller, Rempe, & Jebson, 1991). In spite of this praiseworthy outcome, some vibe the utilization of cephalosporins might be considered a particular pressing factor empowering the development of safe organic entities like VRE and MRSA, the two organic entities innately impervious to cephalosporins and known to frame biofilms. As an option in contrast to remedial antibiotics catheter coatings, different germ-killers, for example, chlorhexidine and silver sulfadiazine, have been utilized (Maki, Alvarado, & Ringer, 1991; Maki et al., 1988), who announced up to a fourfold abatement in colonization, albeit later work has not been ready to repeat their discoveries (Ciresi, Albrecht, Volkers, & Scholten, 1996; Pemberton et al., 1996). Later work has zeroed in on covering venous catheters with a blend of rifampicin and minocycline. This blend showed a protection from colonization better than that acquired by the chlorhexidine/silver sulfadiazine blend and was not related with development of any antimicrobial safe organisms (Francolini & Donelli, 2010; Raad et al., 1996). While the cefazolin covering has provoked concerns with respect to the rise of antibiotic-resistant organic entities, it appears to be judicious to keep away from cephalosporins (alongside other valuable broadspectrum specialists, for example, glycopeptides, lactams, quinolones, and aminoglycoside antiinfection agents) to prophylactically cover vascular catheters. Nonetheless, sterile specialists like chlorhexidine and silver sulfadiazine, or then again less normally utilized antimicrobial specialists for example, fusidic corrosive, novobiocin, minocycline, rifampin, and polymyxin, might be thought of appropriate as they would not deter observational, fruitful expansive range recommending. When clinically demonstrated, vancomycin is normally the first-line exact antimicrobial used to treat patients with CoNS in a solitary blood culture taken through the catheter. In numerous such cases, the CoNS are thought to be an accidental by-product of intraluminal or center point colonization of the catheter instead of an essential bacteremia. CVCs in which both the inner and outer surfaces are covered with antimicrobial specialists are less inclined to luminal colonization and henceforth might bring about a lower recurrence of positive colonizations with coagulase-negative staphylococci. This, thus, may diminish the pointless use of vancomycin and subsequently may diminish the hazard of rise of vancomycin-resistant antimicrobials (Raad, 1997).

The grip of microorganisms to surfaces, trailed by cell division and colonization, is considered as biofilm foundation (Fuquay, McSweeney, & Fox, 2011). Biofilms presumably address the main component of microorganism connection and colonization in nature since they are coordinated in complex local area structure that holds fast to an inactive or living surface (Costerton et al., 1999). The significant element of the biofilm is oneself created EPS comprising of polysaccharides, nucleic acids (extracellular DNA), and proteins, which shields the microorganisms from antibacterial/antimicrobials just as from human immune system parts (Costerton et al., 1999). Surely, microscopic organisms inserted in biofilms lessen the adequacy of polymorphonuclear neutrophils also, macrophages, and, within the sight of these cells, biofilm arrangement is improved and segments of host invulnerable cells are acclimatized into the EPS grid (Watters, Fleming, Bishop, & Rumbaugh, 2016). In such manner, the relationship among biofilms and clinical device-related contaminations has been perceived in a wide scope of polymeric clinical gadgets, including urinary catheters (Percival et al., 2015). Planktonic microscopic organisms can be wiped out by antimicrobial treatment, while microscopic organisms inside biofilms are for the most part less helpless to the chemotherapeutic treatment and whenever it has been finished up, the biofilm can again shed planktonic cells, bringing about repetitive intense disease (Anderson & O'toole, 2008). Breaking this pattern of infection is definitely not a trivial issue, and it regularly requires the expulsion of the defiled gadget to dispense with the bacterial biofilm (Foxman, 2014).

In any case, this training is just a transitory arrangement as the supplanting with a new gadget just gives another reasonable surface to biofilm arrangement. Likewise, antibiotics utilized for the medicines of bacterial diseases have frequently demonstrated to be insufficient against microorganisms developing as biofilms, and until this point, no antimicrobial has been created for pragmatic use against biofilm contaminations (Wolfmeier, Pletzer, Mansour, & Hancock, 2018). The catheter-associated urinary tract infections (CAUTIs) address an extremely normal issue in the clinical setting and the expenses identified with these possibly genuine complexities are significant (Meddings et al., 2012).

The microorganisms most much of the time related with CAUTIs are Gram-positive bacteria, for example, *S. aureus* and *S. epidermidis*, and Gram's-negative microorganisms including *E. coli*, *K. pneumoniae*, *P. mirabilis*, and *P. aeruginosa* (Frank, Wilson, St. Amand, & Pace, 2009; Niveditha et al., 2012). A few endeavors have been made to limit CAUTI, including the presentation of entirely shut waste frameworks, prophylactic treatment with fundamental antiinfection agents or covering the catheter surface with a covering to forestall bacterial attachment and biofilm development (Gordon & Jones, 2003; Niël-Weise, van den Broek, da Silva, & Silva, 2012; Regev-Shoshani, Ko, Crowe, & Av-Gay, 2011).

Next to the normal utilized antimicrobial specialists, new mixtures can address a novel field. Of examination for biofilm avoidance and control. Among these, chitosan's have very much perceived antimicrobial action as of late inspected by Perinelli et al. (2018) that underlined additionally their movement in reducing or forestalling bacterial grip and biofilm arrangement. Chitosans (CS) are inferred through the fractional deacetylation of chitin, making a group of polysaccharide comprising of β-[1–4] connected glucosamine (2 amino 2 deoxy D glucose) and *N*-acetyl glucosamine (2 acetamido 2 deoxy D glucose) units. The presence of the amino bunches in the design gives a cationic nature to the polysaccharide which permits, at low pH, the association with adversely charged compounds. Regardless of their insolubility in water at pH higher than 6, CS have exceptionally intriguing properties (for example, biocompatibility, mucoadhesion, film shaping capacity), which makes them adept competitors as antimicrobial option in contrast to customary medicines (Alves & Mano, 2008; Raafat & Sahl, 2009; Ruel-Gariépy & Leroux, 2006).

The point of this exploration was to examine the counter biofilm impact of two chitosans with various atomic loads against biofilms shaped by *K. pneumoniae* and *E. coli* clinical confines capable of urinary tract infections. The exploratory plan incorporated the development of microbial biofilms on silicone urinary catheters pieces in artificial urine medium (AUM) in the presence or absence of high and low subatomic weight chitosan and the resulting assurance of the polysaccharides viability in diminishing bacterial re-development after antimicrobial treatment (Campana et al., 2018).

The microbial biofilm arrangement on urinary catheters surface is the serious issue in the treatment of CAUTIs (Chavant, Gaillard-Martinie, Talon, Hébraud, & Bernardi, 2007; Salvarci, Koroglu, & Erayman, 2016; Schumm & Lam, 2008), which address almost 80% of the nosocomial contaminations overall (Lo, Lange, & Chew, 2014). Different approaches have been utilized to forestall CAUTI, for example, a superior treatment of catheters, imaginative urinary catheter coatings, improving catheter plan, and underscoring momentary use (Chenoweth & Saint, 2013; Singha, Locklin, & Handa, 2017). Among these, the utilization of options against ineffective catheters is as yet viewed as potentially the best answer for this applicable clinical issue. For this reason, a few combinations have been utilized like antibiotics, antimicrobial specialists (for example chlorhexidine and triclosan), antimicrobial peptides (AMPs), proteins, and finally normal mixtures. In light of their past experience identified with the capacity of chitosan to repress the grip of a few uro-microbes on Foley catheters (Campana et al., 2018), in this investigation they pointed toward growing further the exploration testing of chitosan answers for their biofilm action on a similar model. For this examination, they have chosen to utilize AUM-CSHMWC and AUMCSLMWC arrangements just at pH 5.0 to impersonate the urinary pH and, besides, since chitosan movement is emphatically improved in acidic condition (Kong, Chen, Xing, & Park, 2010). In light of our information, they can express that biofilm development of the inspected clinical strains (*K. pneumoniae* and *E. coli*) on urinary catheters pieces happens after 48 h and 72 h of hatching at 37°C, while after 24 h just a simple grip was recently noticed (Campana et al., 2018).

In reality, both the inspected species resulted sensitive to CS in a period subordinate way, while the molecular weight did not assume a vital part in the antibiofilm movement as a comparable decrease in biofilm arrangement after 48 and 72 h was noticed in any case the MW of the CS (Campana et al., 2018). A clarification of these discoveries could be the comparable level of deacetylation (DD) of the analyzed chitosan tests ($\geq$90%) that gives to the mixtures practically identical antimicrobial action. In the writing, the job of chitosan MW and DD has been broadly discussed and inverse outcomes were accounted for (Chen & Zhao, 2012; Mellegård, Strand, Christensen, Granum, & Hardy, 2011; Younes, Sellimi, Rinaudo, Jellouli, & Nasri, 2014); however, it is for the most part acknowledged that the principle antimicrobial impact of chitosan is because of the capacity of its charged amine gatherings to bound onto the adversely charge surface of bacterial cell layer, causing alteration of the cell penetrability, osmotic harm with efflux of particles and proteins from the cytoplasm outside the microorganism (Lu, Zhang, & Han, 2004).

Another fascinating movement displayed by CS arrangements in the present investigation was the viability of the antimicrobial medicines in nearly totally decreasing the bacterial re-development on the analyzed clinical gadgets; to be sure, low upsides of log CFU/catheter were recuperated from 48 and 72h treated catheters for every one of the strains, with the solitary exemption of *K. pneumoniae* 6/4 (N 2 log CFU/catheter) (Campana et al., 2018).

These perceptions are in line with the recently analyzed CS against attachment action on catheters after 24h of brooding with AUM-CS (Campana et al., 2018) and focused on the possibility over the long run of these mixtures. With respect to the inspected uro-microbes, it tends to be seen that they all have antimicrobial protection from somewhere around two distinct antimicrobial classes; among the tried strains, *E. coli* 10/5 is considered a multidrug-obstruction strain (3 classes of antiinfection) showing an aggregate of 13 protections (Campana et al., 2018).

This element results especially significant as the increasing opposition of microbes dependable of urinary plot diseases to therapeutic agents, including *Klebsiella* spp. *E. coli*, is a principle worry in clinical practices (Mazzariol, Bazaj, & Cornaglia, 2017). Notwithstanding, in examination, CS arrangements had the option to lessen biofilm development of both the inspected species regardless the particular antimicrobial opposition of each single strain, opening another wide future utilization of CS in this field (Campana et al., 2018).

Especially, the consolidation of these polysaccharides into the biomaterials utilized in catheter improvement can produce a catheter surface that could repress bacterial attachment and consequently biofilm arrangement. In this viewpoint, further examinations are continuous to comprehend the connection between CS and catheters surface (Campana et al., 2018). In the United Kingdom, HAIs expand medical clinic stays by a normal of 2.5 occasions and cost an extra US$4400 per patient (Plowman, 2000). Prior information for the United States shows that treatment of every scene of catheter-related septicemia in basically sick patients cost the US $28,690 and expanded medical clinic stays in concentrated consideration units (ICUs) by 6.5days (Raad et al., 1996). Moreover, focal venous catheter-related septicemia alone costs the US medical care framework an overabundance of US $11.5 billion every year, a moderate gauge that does exclude the lesser frequencies of sepsis-related with little fringe venous or blood vessel catheters (Raad & Bodey, 1992).

Information assembled by the US-based National Nosocomial Infection Surveillance Scheme from ICUs (Fridkin & Gaynes, 1999) showed that Gram-positive microorganisms were liable for three of the four most successive reasons for HAI; staphylococci (e.g., coagulase-negative staphylococci [CoNS] and *S. aureus*, including methicillin-resistant *S. aureus* [MRSA]) and enterococci (both vancomycin-susceptible, and -resistant strains [VSE and VRE]) by and large involved almost 34% of the individual organic entities recorded. Gram-negative living beings as a wellspring of HAI have effectively been broadly announced in this specific situation (Blot, Vandewoude, De Bacquer, & Colardyn, 2002; Fujita, Lilly, & Ayliffe, 1982; Ledingham, 1988; Raymond et al., 2003; Rose & Babcock, 1975; Stoddart, 1974; Toltzis & Blumer, 1995).

Of specific clinical interest, living beings causing HAI to keep on creating protection from antimicrobial treatment (Weinstein & Darouiche, 2001). The most of the late accessible information show that three kinds of numerous antitoxin safe Gram's-positive cocci (methicillin-resistant CoNS, MRSA, and VRE) represent the greater part of the nosocomial diseases in US ICUs. Organic entities ensured and sustained in a biofilm are more than 1000 overlay more impervious to therapeutic agents than free-living (or planktonic) microbes (Gilbert, Das, & Foley, 1997).

Staphylococci once excused as innocuous skin verdure, CoNS are ensnared in 48%–67% of focal sensory system shunt intricacies and influence 50%–70% of catheter-related contaminations, 40 to half of prosthetic heart valve diseases, and 20 to half of the prosthetic joint diseases (O'Gara & Humphreys, 2001). The prevalence of CoNS among the biofilm-framing *Staphylococci* in causing contaminations has provoked the proposition of another term, "ongoing polymer associated diseases" (Götz, 2002). A new survey (O'Gara & Humphreys, 2001) examined the strains of CoNS answerable for numerous contaminations in singular units and showed that while there was no connection between the CoNS disconnects liable for different diseases sometimes, singular clones were pervasive in others. CoNS are a significant reason for the endogenous disease of intravascular catheters, either by following along their outside surface or by entering through the catheter center during control by clinical staff (Raad et al., 1996). Edmond et al. (1995) reasoned that most neutropenic patients in their flare-up of VRE bacteremia became contaminated or colonized in the gastrointestinal tract using the nosocomial transmission. The movement of gut microscopic organisms across the stripped intestinal mucosa is normal in neutropenic patients. In this way, once in the circulatory system, colonization of inhabiting gadgets could happen. *S. aureus*, paying little heed to its weakness to methicillin, stays the most well-known specialist of careful site diseases in the United States and Europe (Dixon, 2002). The numerous protections regularly showed by MRSA lessen the remedial choices accessible. The new rise of vancomycin-resistant MRSA from patients getting renal dialysis will additionally lessen restorative choices for clinicians (Hiramatsu et al., 1997). Mathai et al. (2001) exhibited the capacity of MRSA to shape biofilms in a urinary catheter model, and its protection from 10mg of vancomycin for every litter, regardless of seeming powerless by traditional in vitro techniques (Ebrey et al., 2004).

8. Conclusion

Biofilm-related contaminations will definitely increase with the quantities of inhabiting gadgets utilized except if more powerful procedures can be conceived to forestall their colonization by microorganisms. Such procedures could incorporate novel uses of therapeutic agents and more prophylactic antimicrobial use with innate dangers of producing resistant pathogens. A comprehension of the actual elements that influence biofilm development inside the body is fundamental in accomplishing this point. The morbidity and mortality related with the colonization and resulting contamination must firmly focus on this medical service suppliers and makers of such devices.

Strong biofilm makers coming from patients hospitalized in ICUs are frequently more vulnerable to drugs than feeble biofilm makers. Strain classification dependent on capacity to shape biofilms loses significant data. To be conceivable, various variables can to be utilized to manage relapse model. Portrayal of the connection between biofilm development and drug susceptibility should consider about different elements, for example, the wards from which the strains begin. The more prominent level of delicate strains among biofilm makers can be eradicated by the way that, microscopic organisms that accomplish resistance through biofilm formation do not have to create or keep up with the mechanisms answerable for obstruction of planktonic cells. Copper nanoparticles are being investigated as a viable antimicrobial in clinical settings to curb contaminations due to biofilm framing microorganisms. Gram-negative microbes are accounted for to be impervious to copper.

HAIs represent a genuine risk to patients and clinical staff and result in expanded mortality and morbidity to those tainted. This is particularly so, because of the spread of AMR bringing about HAIs that are lethargic to the medicines regularly used to treat them. Contaminated items, instruments, and surfaces fill in as bacterial repositories and are the main drivers of HAIs, supporting the requirement for antibacterial surfaces that can diminish or forestall bacterial colonization. In this manner, compelling antibacterial surfaces can possibly decrease the frequency of HAIs by upsetting the pattern of transmission of microscopic organisms from surfaces to patients or medical personnel.

The new pattern in the early permission to LTCFs of seriously harmed patients moved from general clinics has given another dynamic to the rate of medical services related contaminations, including biofilm-based diseases identified with the embed of urinary and intravascular catheters, and the beginning of pressure ulcers. Catheter-related urinary tract diseases lead in the majority of the overviews on LTCFs, roughly 80% of urinary tract contaminations in these settings because of the short- or long duration of a urinary catheter. Besides, the implantation of intravascular catheters is regularly liable for catheter-related circulation system contaminations brought about by the advancement of an intraluminal biofilm. Pressure ulcers, every now and again happening in bedridden patients conceded to LTCFs, are additionally helpless to contamination by biofilm-developing high-impact and anaerobic microorganisms. The biofilm on the injury the primary justification for postponed healing in patients.

Many hospital-acquired infections are related with biofilm diseases of implantable clinical gadgets like intravascular catheters. Inside biofilms, microscopic organisms are essentially less helpless to therapeutic agents and host safeguards, making biofilm contaminations hard to analyze and treat, and regularly requiring evacuation of the tainted embed.

Acknowledgments

Dr. Vishvanath Tiwari would like to thank the SERB (EMR/2017/001854), India for providing funding. Ms. Shraddha would like to thank SERB (EMR/2017/001854), India for providing JRF.

Author contributions

Conceived and design the chapter: Vishvanath Tiwari; literature survey: Shraddha Sharma; analyzed the data: Vishvanath Tiwari; wrote the manuscript: Shraddha Sharma, Vaishali Kaushik.

Competing interests

The authors have declared that no competing interests exist.

References

Allen, G. (2000). Maximizing nurses' advocacy role to improve patient outcomes. *AORN Journal*, *71*(5), 1038–1050.

Alsulaimani, F. F. F. (2021). *Use of affimers for targeting S. aureus biofilms*. University of Leeds. https://etheses.whiterose.ac.uk/29377.

Alves, N. M., & Mano, J. F. (2008). Chitosan derivatives obtained by chemical modifications for biomedical and environmental applications. *International Journal of Biological Macromolecules*, *43*(5), 401–414.

Anderson, D. M., Cronk, R., Fejfar, D., Pak, E., Cawley, M., & Bartram, J. (2021). Safe healthcare facilities: A systematic review on the costs of establishing and maintaining environmental health in facilities in low-and middle-income countries. *International Journal of Environmental Research and Public Health*, *18*(2), 817.

Anderson, R. L., Highsmith, A. K., & Holland, B. W. (1986). Comparison of the standard pour plate procedure and the ATP and Limulus amebocyte lysate procedures for the detection of microbial contamination in intravenous fluids. *Journal of Clinical Microbiology*, *23*(3), 465–468.

Anderson, G. G., & O'Toole, G. A. (2008). Innate and induced resistance mechanisms of bacterial biofilms. In T. Romeo (Ed.), *Bacterial biofilms. Current topics in microbiology and immunology* (Vol. 322, pp. 85–105). Berlin, Heidelberg: Springer.

Andrew, L. N., Somkiet, T., Alice, D., David, M., Brent, M. P., & Gill, G. G. (2001). Iron sulfides and sulfur species produced at hematite surfaces in the presence of sulfate-reducing bacteria. Associate editor: D. J. Wesolowski *Geochimica et Cosmochimica Acta*, *65*(2), 223–235.

Archibald, L. K., & Gaynes, R. P. (1997). Hospital-acquired infections in the United States: The importance of interhospital comparisons. *Infectious Disease Clinics of North America*, *11*(2), 245–255.

Arciola, C. R., Collamati, S., Donati, E., & Montanaro, L. (2001). A rapid PCR method for the detection of slime-producing strains of *Staphylococcus epidermidis* and *S. aureus* in periprosthesis infections. *Diagnostic Molecular Pathology*, *10*(2), 130–137.

Arozullah, A. M., Khuri, S. F., Henderson, W. G., & Daley, J. (2001). Development and validation of a multifactorial risk index for predicting postoperative pneumonia after major noncardiac surgery. *Annals of Internal Medicine*, *135*(10), 847–857.

Attaway, H. H., III, Fairey, S., Steed, L. L., Salgado, C. D., Michels, H. T., & Schmidt, M. G. (2012). Intrinsic bacterial burden associated with intensive care unit hospital beds: Effects of disinfection on population recovery and mitigation of potential infection risk. *American Journal of Infection Control*, *40*(10), 907–912.

Baldassarri, L., Cecchini, R., Bertuccini, L., Ammendolia, M. G., Iosi, F., Arciola, C. R., et al. (2001). *Enterococcus* spp. produces slime and survives in rat peritoneal macrophages. *Medical Microbiology and Immunology*, *190*(3), 113–120.

Baldassarri, L., Simpson, W., Donelli, G., & Christensen, G. (1993). Variable fixation of staphylococcal slime by different histochemical fixatives. *European Journal of Clinical Microbiology and Infectious Diseases*, *12*(11), 866–868.

Balestrino, D., Ghigo, J. M., Charbonnel, N., Haagensen, J. A., & Forestier, C. (2008). The characterization of functions involved in the establishment and maturation of *Klebsiella pneumoniae* in vitro biofilm reveals dual roles for surface exopolysaccharides. *Environmental Microbiology*, *10*(3), 685–701.

Balestrino, D., Haagensen, J. A., Rich, C., & Forestier, C. (2005). Characterization of type 2 quorum sensing in *Klebsiella pneumoniae* and relationship with biofilm formation. *Journal of Bacteriology*, *187*(8), 2870–2880.

Baraniak, A., Izdebski, R., Fiett, J., Sadowy, E., Adler, A., Kazma, M., et al. (2013). Comparative population analysis of *Klebsiella pneumoniae* strains with extended-spectrum β-lactamases colonizing patients in rehabilitation centers in four countries. *Antimicrobial Agents and Chemotherapy*, *57*(4), 1992–1997.

Beenken, K. E., Blevins, J. S., & Smeltzer, M. S. (2003). Mutation of sarA in *Staphylococcus aureus* limits biofilm formation. *Infection and Immunity*, *71*(7), 4206–4211.

Beenken, K. E., Dunman, P. M., McAleese, F., Macapagal, D., Murphy, E., Projan, S. J., et al. (2004). Global gene expression in *Staphylococcus aureus* biofilms. *Journal of Bacteriology*, *186*(14), 4665–4684.

Benenson, A. S. (1995). *Control of communicable diseases manual*. The Association.

Bergamini, T. M., Corpus, R. A., Jr., Brittian, K. R., Peyton, J. C., & Cheadle, W. G. (1994). The natural history of bacterial biofilm graft infection. *Journal of Surgical Research*, *56*(5), 393–396.

Berhe, N., Tefera, Y., & Gizaw, T. (2017). International Journal of Advanced Research in Biological Sciences Review on biofilm formation and its control options. *Biological Science*. https://doi.org/10.13140/RG.2.2.16800.81921.

Bernstein, D. A., Salsgiver, E., Simon, M. S., Greendyke, W., Eiras, D. P., Ito, M., et al. (2016). Understanding barriers to optimal cleaning and disinfection in hospitals: A knowledge, attitudes, and practices survey of environmental services workers. *Infection Control & Hospital Epidemiology*, *37*(12), 1492–1495.

Bilavsky, E., Lerman, Y., Rabinovich, A., Salomon, J., Lawrence, C., Rossini, A., et al. (2012). Carriage of methicillin-resistant *Staphylococcus aureus* on admission to European rehabilitation centres—A prospective study. *Clinical Microbiology and Infection*, *18*(6), E164–E169.

Bjorn, M. J., Sokol, P. A., & Iglewski, B. H. (1979). Influence of iron on yields of extracellular products in *Pseudomonas aeruginosa* cultures. *Journal of Bacteriology*, *138*(1), 193–200.

Blot, S., Vandewoude, K., De Bacquer, D., & Colardyn, F. (2002). Nosocomial bacteremia caused by antibiotic-resistant gram-negative bacteria in critically ill patients: Clinical outcome and length of hospitalization. *Clinical Infectious Diseases*, *34*(12), 1600–1606.

Boddicker, J. D., Anderson, R. A., Jagnow, J., & Clegg, S. (2006). Signature-tagged mutagenesis of *Klebsiella pneumoniae* to identify genes that influence biofilm formation on extracellular matrix material. *Infection and Immunity*, *74*(8), 4590–4597.

Boles, B. R., & Horswill, A. R. (2008). Agr-mediated dispersal of *Staphylococcus aureus* biofilms. *PLoS Pathogens*, *4*(4), e1000052.

Bollinger, N., Hassett, D. J., Iglewski, B. H., Costerton, J. W., & McDermott, T. R. (2001). Gene expression in *Pseudomonas aeruginosa*: Evidence of iron override effects on quorum sensing and biofilm-specific gene regulation. *Journal of Bacteriology*, *183*(6), 1990–1996.

Bondarczuk, K., & Piotrowska-Seget, Z. (2013). Molecular basis of active copper resistance mechanisms in Gram-negative bacteria. *Cell Biology and Toxicology*, *29*(6), 397–405.

Bouza, E., San Juan, R., Munoz, P., Voss, A., Kluytmans, J., & on behalf of the Co-operative Group of the European Study Group on Nosocomial Infections (ESGNI). (2001). A European perspective on nosocomial urinary tract infections I. Report on the microbiology workload, etiology and antimicrobial susceptibility (ESGNI–003 study). *Clinical Microbiology and Infection*, *7*(10), 523–531.

Bradley, S. F., Terpenning, M. S., Ramsey, M. A., Zarins, L. T., Jorgensen, K. A., Sottile, W. S., et al. (1991). Methicillin-resistant *Staphylococcus aureus*: Colonization and infection in a long-term care facility. *Annals of Internal Medicine*, *115*(6), 417–422.

Buković, E., Kurtović, B., Rotim, C., Svirčević, V., Friganović, A., & Važanić, D. (2021). Compliance with Hand Hygiene Among Healthcare workers in preventing healthcare associated infections—A systematic review. *Journal of Applied Health Sciences [Internet]*, *7*(1), 57–69.

Cai, Y., Wang, R., Liang, B. B., & An, M. M. (2009). In-vitro bactericidal activity of colistin against biofilm-associated *Pseudomonas aeruginosa* and *Acinetobacter baumannii*. *The Journal of Hospital Infection*, *72*(4), 368–370.

Çam, S., & Brinkmeyer, R. (2020). The effects of temperature, pH, and iron on biofilm formation by clinical versus environmental strains of *Vibrio vulnificus*. *Folia Microbiologia (Praha)*, *65*(3), 557–566.

Campana, R., Biondo, F., Mastrotto, F., Baffone, W., & Casettari, L. (2018). Chitosans as new tools against biofilms formation on the surface of silicone urinary catheters. *International Journal of Biological Macromolecules*, *118*, 2193–2200.

Carrel, T., Nguyen, T., Kipfer, B., & Althaus, U. (1998). Definitive cure of recurrent prosthetic endocarditis using silver-coated St. Jude Medical heart valves: A preliminary case report. *The Journal of Heart Valve Disease*, *7*(5), 531–533.

Ceccato, A., Di Giannatale, P., Nogas, S., & Torres, A. (2021). Safety considerations of current drug treatment strategies for nosocomial pneumonia. *Expert Opinion on Drug Safety*, *20*(2), 181–190.

Centre for Disease Control and Prevention (CDC). (2013). https://www.cdc.gov.

Chan, W. C., Coyle, B. J., & Williams, P. (2004). Virulence regulation and quorum sensing in staphylococcal infections: Competitive AgrC antagonists as quorum sensing inhibitors. *Journal of Medicinal Chemistry*, *47*(19), 4633–4641.

Chang, N.-C. N., Reisinger, H. S., Schweizer, M. L., Jones, I., Chrischilles, E., Chorazy, M., et al. (2021). Hand hygiene compliance at critical points of care. *Clinical Infectious Diseases*, *72*(5), 814–820.

Chavant, P., Gaillard-Martinie, B., Talon, R., Hébraud, M., & Bernardi, T. (2007). A new device for rapid evaluation of biofilm formation potential by bacteria. *Journal of Microbiological Methods*, *68*(3), 605–612.

Chen, C.-W., Hsu, C.-Y., Lai, S.-M., Syu, W.-J., Wang, T.-Y., & Lai, P.-S. (2014). Metal nanobullets for multidrug resistant bacteria and biofilms. *Advanced Drug Delivery Reviews*, *78*, 88–104.

Chen, J. L., & Zhao, Y. (2012). Effect of molecular weight, acid, and plasticizer on the physicochemical and antibacterial properties of β-chitosan based films. *Journal of Food Science*, *77*(5), E127–E136.

Chenoweth, C., & Saint, S. (2013). Preventing catheter-associated urinary tract infections in the intensive care unit. *Critical Care Clinics*, *29*(1), 19–32.

Cherifi, S., Byl, B., Deplano, A., Nonhoff, C., Denis, O., & Hallin, M. (2013). Comparative epidemiology of *Staphylococcus epidermidis* isolates from patients with catheter-related bacteremia and from healthy volunteers. *Journal of Clinical Microbiology*, *51*(5), 1541–1547.

Choi, A. H., Slamti, L., Avci, F. Y., Pier, G. B., & Maira-Litrán, T. (2009). The pgaABCD locus of *Acinetobacter baumannii* encodes the production of poly-β-1-6-N-acetylglucosamine, which is critical for biofilm formation. *Journal of Bacteriology*, *191*(19), 5953–5963.

Choucri, J., Zanotto, F., Grassi, V., Balbo, A., Ebn Touhami, M., Mansouri, I., et al. (2019). Corrosion behavior of different brass alloys for drinking water distribution systems. *Metals*, *9*(6), 649.

Chow, J., Thal, L., Perri, M., Vazquez, J. A., Donabedian, S., Clewell, D., et al. (1993). Plasmid-associated hemolysin and aggregation substance production contribute to virulence in experimental enterococcal endocarditis. *Antimicrobial Agents and Chemotherapy*, *37*(11), 2474–2477.

Christine, M., Térésa, P., Joseph, F., George, W., & Alain, E. (1994). A method to perform direct transcutaneous intrathecal injection in rats. *Journal of Pharmacological and Toxicological Methods*, *32*(4), 197–200.

Ciresi, D. L., Albrecht, R. M., Volkers, P. A., & Scholten, D. J. (1996). Failure of antiseptic bonding to prevent central venous catheter-related infection and sepsis. *The American Surgeon*, *62*(8), 641–646.

Clewell, D. B. (1993). Bacterial sex pheromone-induced plasmid transfer. *Cell*, *73*(1), 9–12. https://doi.org/10.1016/0092-8674(93)90153-h.

Combes, A., Figliolini, C., Trouillet, J.-L., Kassis, N., Wolff, M., Gibert, C., et al. (2002). Incidence and outcome of polymicrobial ventilator-associated pneumonia. *Chest*, *121*(5), 1618–1623.

Cook, D., De Jonghe, B., Brochard, L., & Brun-Buisson, C. (1998). Influence of airway management on ventilator-associated pneumonia: Evidence from randomized trials. *JAMA*, *279*(10), 781–787.

Costerton, J. W., Stewart, P. S., & Greenberg, E. P. (1999). Bacterial biofilms: A common cause of persistent infections. *Science*, *284* (5418), 1318–1322.

Cramton, S. E., Gerke, C., Schnell, N. F., Nichols, W. W., & Götz, F. (1999). The intercellular adhesion (ica) locus is present in *Staphylococcus aureus* and is required for biofilm formation. *Infection and Immunity*, *67*(10), 5427–5433.

Cramton, S. E., Ulrich, M., Götz, F., & Döring, G. (2001). Anaerobic conditions induce expression of polysaccharide intercellular adhesin in *Staphylococcus aureus* and *Staphylococcus epidermidis*. *Infection and Immunity*, *69*(6), 4079–4085.

Craven, D. E., Barber, T. W., Steger, K. A., & Montecalvo, M. A. (1990). Nosocomial pneumonia in the 1990s: Update of epidemiology and risk factors. *Seminars in Respiratory Infections*, *5*(3), 157–172.

Craven, D. E., Connolly, M. G., Jr., Lichtenberg, D. A., Primeau, P. J., & McCabe, W. R. (1982). Contamination of mechanical ventilators with tubing changes every 24 or 48 hours. *The New England Journal of Medicine*, *306*(25), 1505–1509.

Craven, D. E., & Steger, K. A. (1995). Epidemiology of nosocomial pneumonia. New perspectives on an old disease. *Chest*, *108*(2 Suppl), 1s–16s.

Cretnik, T. Z., Vovko, P., Retelj, M., Jutersek, B., Harlander, T., Kolman, J., et al. (2005). Prevalence and nosocomial spread of methicillin-resistant *Staphylococcus aureus* in a long-term-care facility in Slovenia. *Infection Control & Hospital Epidemiology*, *26*(2), 184–190.

Cucarella, C., Solano, C., Valle, J., Amorena, B., Lasa, Í., & Penadés, J. R. (2001). Bap, a *Staphylococcus aureus* surface protein involved in biofilm formation. *Journal of Bacteriology*, *183*(9), 2888–2896.

Cucarella, C., Tormo, M.Á., Knecht, E., Amorena, B., Lasa, Í., Foster, T. J., et al. (2002). Expression of the biofilm-associated protein interferes with host protein receptors of *Staphylococcus aureus* and alters the infective process. *Infection and Immunity*, *70*(6), 3180–3186.

Darouiche, R. O., Dhir, A., Miller, A. J., Landon, G. C., Raad, I. I., & Musher, D. M. (1994). Vancomycin penetration into biofilm covering infected prostheses and effect on bacteria. *Journal of Infectious Diseases*, *170*(3), 720–723.

Dauvergne, E., & Mullié, C. (2021). Brass alloys: Copper-bottomed solutions against hospital-acquired infections? *Antibiotics*, *10*, 286.

Davey, M. E., & O'toole, G. A. (2000). Microbial biofilms: From ecology to molecular genetics. *Microbiology and Molecular Biology Reviews*, *64*(4), 847–867.

Dezfulian, C., Lavelle, J., Nallamothu, B. K., Kaufman, S. R., & Saint, S. (2003). Rates of infection for single-lumen versus multilumen central venous catheters: A meta-analysis. *Critical Care Medicine*, *31*(9), 2385–2390.

Di Martino, P., Cafferini, N., Joly, B., & Darfeuille-Michaud, A. (2003). *Klebsiella pneumoniae* type 3 pili facilitate adherence and biofilm formation on abiotic surfaces. *Research in Microbiology*, *154*(1), 9–16.

Di Martino, P., Merieau, A., Phillips, R., Orange, N., & Hulen, C. (2002). Isolation of an *Escherichia coli* strain mutant unable to form biofilm on polystyrene and to adhere to human pneumocyte cells: Involvement of tryptophanase. *Canadian Journal of Microbiology*, *48*(2), 132–137.

Dixon, G. (2002). Sources of surgical infection. *Surgery (Oxford)*, *20*(8), 179–185.

Donald, E. C., Kathleen, A. S., & Thomas, W. B. (1991). Preventing nosocomial pneumonia: State of the art and perspectives for the 1990s. *The American Journal of Medicine*, *91*(3, Suppl. 2), S44–S53.

Donelli, G., De Paoli, P., Fadda, G., Marone, P., Nicoletti, G., Varaldo, P., et al. (2001). A multicenter study on central venous catheter-associated infections in Italy. *Journal of Chemotherapy*, *13*(sup4), 251–262.

Donelli, G., Francolini, I., Romoli, D., Guaglianone, E., Piozzi, A., Ragunath, C., et al. (2007). Synergistic activity of dispersin B and cefamandole nafate in inhibition of staphylococcal biofilm growth on polyurethanes. *Antimicrobial Agents and Chemotherapy*, *51*(8), 2733–2740.

Donelli, G., & Vuotto, C. (2014). Biofilm-based infections in long-term care facilities. *Future Microbiology*, *9*(2), 175–188.

Donelli, G., Vuotto, C., Cardines, R., & Mastrantonio, P. (2012). Biofilm-growing intestinal anaerobic bacteria. *FEMS Immunology & Medical Microbiology*, *65*(2), 318–325.

Donlan, R. M. (2001a). Biofilm formation: A clinically relevant microbiological process. *Clinical Infectious Diseases*, *33*(8), 1387–1392.

Donlan, R. M. (2001b). Biofilms and device-associated infections. *Emerging Infectious Diseases*, *7*(2), 277.

Donlan, R. M., & Costerton, J. W. (2002). Biofilms: Survival mechanisms of clinically relevant microorganisms. *Clinical Microbiology Reviews*, *15*(2), 167–193.

Donlan, R. M., Murga, R., Bell, M., Toscano, C. M., Carr, J. H., Novicki, T. J., et al. (2001). Protocol for detection of biofilms on needleless connectors attached to central venous catheters. *Journal of Clinical Microbiology*, *39*(2), 750–753.

Donlan, R., Murga, R., & Carson, L. (1999). Growing biofilms in intravenous fluids. In *Biofilms, the good, the bad, and the ugly. Presented at the fourth meeting of the Biofilm Club.*

Drees, M., Snydman, D. R., Schmid, C. H., Barefoot, L., Hansjosten, K., Vue, P. M., et al. (2008). Prior environmental contamination increases the risk of acquisition of vancomycin-resistant enterococci. *Clinical Infectious Diseases*, *46*(5), 678–685.

Ducel, G., Haxhe, J., Tanner, F., & Zumofen, M. (1979). *Practical guide to the prevention of hospital-acquired infections.* Geneva: World Health Organization.

Dzul, S. P., Thornton, M. M., Hohne, D. N., Stewart, E. J., Shah, A. A., Bortz, D. M., et al. (2011). Contribution of the *Klebsiella pneumoniae* capsule to bacterial aggregate and biofilm microstructures. *Applied and Environmental Microbiology*, *77*(5), 1777–1782.

Ebrey, R., Hamilton, M. S., Cairns, G., & Lappin-Scott, H. M. (2004). Biofilms and hospital-acquired infections. In *Microbial biofilms* (pp. 294–313). American Society of Microbiology.

Edmond, M. B., Ober, J. F., Weinbaum, D. L., Pfaller, M. A., Hwang, T., Sanford, M. D., et al. (1995). Vancomycin-resistant *Enterococcus faecium* bacteremia: Risk factors for infection. *Clinical Infectious Diseases*, *20*(5), 1126–1133.

Ehrenkranz, N., Alfonso, B., & Nerenberg, D. (1990). Irrigation-aspiration for culturing draining decubitus ulcers: Correlation of bacteriological findings with a clinical inflammatory scoring index. *Journal of Clinical Microbiology*, *28*(11), 2389–2393.

Ejrnaes, K., Stegger, M., Reisner, A., Ferry, S., Monsen, T., Holm, S. E., et al. (2011). Characteristics of *Escherichia coli* causing persistence or relapse of urinary tract infections: Phylogenetic groups, virulence factors and biofilm formation. *Virulence*, *2*(6), 528–537.

El Moussaoui, N., Ramli, I., El Amraoui, M., Abdou, A., Znati, K., Ismaili, N., et al. (2015). Primary anetoderma revealing autoimmune disease. *Annales de Dermatologie et de Vénéréologie*, *142*(11), 716–718.

Elguindi, J., Wagner, J., & Rensing, C. (2009). Genes involved in copper resistance influence survival of *Pseudomonas aeruginosa* on copper surfaces. *Journal of Applied Microbiology*, *106*(5), 1448–1455.

Ellingson, K., Haas, J. P., Aiello, A. E., Kusek, L., Maragakis, L. L., Olmsted, R. N., et al. (2014). Strategies to prevent healthcare-associated infections through hand hygiene. *Infection Control & Hospital Epidemiology*, *35*(8), 937–960.

Espinal, P., Martí, S., & Vila, J. (2012). Effect of biofilm formation on the survival of *Acinetobacter baumannii* on dry surfaces. *Journal of Hospital Infection*, *80*(1), 56–60.

Evans, E., Brown, M. R., & Gilbert, P. (1994). Iron chelator, exopolysaccharide and protease production in *Staphylococcus epidermidis*: A comparative study of the effects of specific growth rate in biofilm and planktonic culture. *Microbiology*, *140*(1), 153–157.

Evelhoch, S. R. (2020). Biofilm and chronic nonhealing wound infections. *The Surgical Clinics of North America*, *100*(4), 727–732.

Fagon, J. Y., Chastre, J., Hance, A. J., Montravers, P., Novara, A., & Gibert, C. (1993). Nosocomial pneumonia in ventilated patients: A cohort study evaluating attributable mortality and hospital stay. *The American Journal of Medicine*, *94*(3), 281–288.

Failla, M. L., Benedict, C., & Weinberg, E. (1975). Bacterial and fungal growth in total parenteral nutrition solutions. *Antonie Van Leeuwenhoek*, *41*(1), 319–328.

Fayolle, M., & Morsli, M. (2021). The persistence of *Staphylococcus aureus* in pressure ulcers: A colonising role. *Genes*, *12*(12).

Fazli, M., Bjarnsholt, T., Kirketerp-Møller, K., Jørgensen, B., Andersen, A. S., Krogfelt, K. A., et al. (2009). Nonrandom distribution of *Pseudomonas aeruginosa* and *Staphylococcus aureus* in chronic wounds. *Journal of Clinical Microbiology*, *47*(12), 4084–4089.

Fitzgerald, R. H., Nolan, D. R., Ilstrup, D. M., Van Scoy, R. E., Washington, J. A., & Coventry, M. B. (1977). Deep wound sepsis following total hip arthroplasty. *The Journal of Bone and Joint Surgery. American Volume*, *59*(7), 847–855.

Flemming, H. C., Wingender, J., Szewzyk, U., Steinberg, P., Rice, S. A., & Kjelleberg, S. (2016). Biofilms: An emergent form of bacterial life. *Nature Reviews Microbiology*, *14*(9), 563–575.

Fletcher, J. M., Nair, S. P., Ward, J. M., Henderson, B., & Wilson, M. (2001). Analysis of the effect of changing environmental conditions on the expression patterns of exported surface-associated proteins of the oral pathogen *Actinobacillus actinomycetemcomitans*. *Microbial Pathogenesis*, *30*(6), 359–368.

Fluit, A., Schmitz, F.-J., & Verhoef, J. (2001). Frequency of isolation of pathogens from bloodstream, nosocomial pneumonia, skin and soft tissue, and urinary tract infections occurring in European patients. *European Journal of Clinical Microbiology and Infectious Diseases*, *20*(3), 188–191.

Foster, T. J., & Höök, M. (1998). Surface protein adhesins of *Staphylococcus aureus*. *Trends in Microbiology*, *6*(12), 484–488.

Foxman, B. (2014). Urinary tract infection syndromes: Occurrence, recurrence, bacteriology, risk factors, and disease burden. *Infectious Disease Clinics*, *28*(1), 1–13.

Francolini, I., & Donelli, G. (2010). Prevention and control of biofilm-based medical-device-related infections. *FEMS Immunology & Medical Microbiology*, *59*(3), 227–238.

Frank, D. N., Wilson, S. S., St. Amand, A. L., & Pace, N. R. (2009). Culture-independent microbiological analysis of foley urinary catheter biofilms. *PLoS One*, *4*(11), e7811.

Freestone, P. P., Haigh, R. D., Williams, P. H., & Lyte, M. (1999). Stimulation of bacterial growth by heat-stable, norepinephrine-induced autoinducers. *FEMS Microbiology Letters*, *172*(1), 53–60.

Fridkin, S. K., & Gaynes, R. P. (1999). Antimicrobial resistance in intensive care units. *Clinics in Chest Medicine*, *20*(2), 303–316.

Fridkin, S. K., Welbel, S. F., & Weinstein, R. A. (1997). Magnitude and prevention of nosocomial infections in the intensive care unit. *Infectious Disease Clinics of North America*, *11*(2), 479–496.

Fujita, K., Lilly, H., & Ayliffe, G. (1982). Spread of resistant Gram-negative bacilli in a burns unit. *Journal of Hospital Infection*, *3*(1), 29–37.

Fulaz, S., Hiebner, D., Barros, C. H. N., Devlin, H., Vitale, S., Quinn, L., et al. (2019). Ratiometric imaging of the in situ pH distribution of biofilms by use of fluorescent mesoporous silica nanosensors. *ACS Applied Materials & Interfaces*, *11*(36), 32679–32688.

Fuquay, J. W., McSweeney, P. L., & Fox, P. F. (2011). *Encyclopedia of dairy sciences*. Academic Press.

Gaddy, J. A., Tomaras, A. P., & Actis, L. A. (2009). The *Acinetobacter baumannii* 19606 OmpA protein plays a role in biofilm formation on abiotic surfaces and in the interaction of this pathogen with eukaryotic cells. *Infection and Immunity*, *77*(8), 3150–3160.

Gao, X., Liu, X., Xu, J., Xue, C., Xue, Y., & Wang, Y. (2014). Dietary trimethylamine N-oxide exacerbates impaired glucose tolerance in mice fed a high fat diet. *Journal of Bioscience and Bioengineering*, *118*(4), 476–481.

Garcia, A. D., & Thomas, D. R. (2006). Assessment and management of chronic pressure ulcers in the elderly. *Medical Clinics*, *90*(5), 925–944.

Gedefie, A., Demsis, W., Ashagrie, M., Kassa, Y., Tesfaye, M., Tilahun, M., et al. (2021). *Acinetobacter baumannii* biofilm formation and its role in disease pathogenesis: A review. *Infection and Drug Resistance*, *14*, 3711.

Gellatly, S. L., & Hancock, R. E. (2013). *Pseudomonas aeruginosa*: New insights into pathogenesis and host defenses. *Pathogens and Disease*, *67*(3), 159–173.

Gershon, R. R., Vlahov, D., Felknor, S. A., Vesley, D., Johnson, P. C., Delcios, G. L., et al. (1995). Compliance with universal precautions among health care workers at three regional hospitals. *American Journal of Infection Control*, *23*(4), 225–236.

Ghosh, S., & Ray, R. R. (2021). Immune response to biofilm. In *Biofilm-mediated diseases: Causes and controls* (pp. 127–155). Springer.

Gil, R. T., Kruse, J. A., Thill-Baharozian, M. C., & Carlson, R. W. (1989). Triple-vs single-lumen central venous catheters: A prospective study in a critically ill population. *Archives of Internal Medicine*, *149*(5), 1139–1143.

Gilbert, P., Das, J., & Foley, I. (1997). Biofilm susceptibility to antimicrobials. *Advances in Dental Research*, *11*(1), 160–167.

Goldmann, D. A., & Huskins, W. C. (1997). Control of nosocomial antimicrobial-resistant bacteria: A strategic priority for hospitals worldwide. *Clinical Infectious Diseases*, *24*(Suppl._1), S139–S145.

Gomes, F., & Furtado, G. E. (2022). The skin microbiome of infected pressure ulcers: A review and implications for health professionals. *European Journal of Clinical Investigation*, *52*(1), e13688.

Gominet, M., Compain, F., Beloin, C., & Lebeaux, D. (2017). Central venous catheters and biofilms: Where do we stand in 2017? *APMIS*, *125*(4), 365–375.

Gordon, K. A., & Jones, R. N. (2003). Susceptibility patterns of orally administered antimicrobials among urinary tract infection pathogens from hospitalized patients in North America: Comparison report to Europe and Latin America. Results from the SENTRY Antimicrobial Surveillance Program (2000). *Diagnostic Microbiology and Infectious Disease*, *45*(4), 295–301.

Gordts, B., Van Landuyt, H., Ieven, M., Vandamme, P., & Goossens, H. (1995). Vancomycin-resistant enterococci colonizing the intestinal tracts of hospitalized patients. *Journal of Clinical Microbiology*, *33* (11), 2842–2846.

Götz, F. (2002). *Staphylococcus* and biofilms. *Molecular Microbiology*, *43* (6), 1367–1378.

Goudarzi, M., Navidinia, M., Khadembashi, N., & Rasouli, R. (2021). Biofilm matrix formation in human: Clinical significance, diagnostic techniques, and therapeutic drugs. *Archives of Clinical Infectious Diseases*, *16*(3).

Gould, S. W., Fielder, M. D., Kelly, A. F., Morgan, M., Kenny, J., & Naughton, D. P. (2009). The antimicrobial properties of copper surfaces against a range of important nosocomial pathogens. *Annals of Microbiology*, *59*(1), 151–156.

Grosso, F., Quinteira, S., & Peixe, L. (2010). Emergence of an extreme-drug-resistant (XDR) *Acinetobacter baumannii* carrying blaOXA-23 in a patient with acute necrohaemorrhagic pancreatitis. *The Journal of Hospital Infection*, *75*(1), 82–83.

Guggenbichler, J. P., Assadian, O., Boeswald, M., & Kramer, A. (2011). Incidence and clinical implication of nosocomial infections associated with implantable biomaterials—Catheters, ventilator-associated pneumonia, urinary tract infections. *GMS Krankenhhyg Interdiszip*, *6*(1), Doc18.

Guo, J., Gao, S. H., Lu, J., Bond, P. L., Verstraete, W., & Yuan, Z. (2017). Copper oxide nanoparticles induce lysogenic bacteriophage and metal-resistance genes in *Pseudomonas aeruginosa* PAO1. *ACS Applied Materials & Interfaces*, *9*(27), 22298–22307.

Guzman, C., Pruzzo, C., LiPira, G., & Calegari, L. (1989). Role of adherence in pathogenesis of *Enterococcus faecalis* urinary tract infection and endocarditis. *Infection and Immunity*, *57*(6), 1834–1838.

Guzman, C. A., Pruzzo, C., Platé, M., Guardati, M. C., & Calegari, L. (1991). Serum dependent expression of *Enterococcus faecalis* adhesins involved in the colonization of heart cells. *Microbial Pathogenesis*, *11*(6), 399–409.

Hachem, R., & Raad, I. (2002). Failure of oral antimicrobial agents in eradicating gastrointestinal colonization with vancomycin-resistant enterococci. *Infection Control & Hospital Epidemiology*, *23*(1), 43–44.

Haghi Ghahremanloi Olia, A., Ghahremani, M., Ahmadi, A., & Sharifi, Y. (2020). Comparison of biofilm production and virulence gene distribution among community- and hospital-acquired *Staphylococcus aureus* isolates from northwestern Iran. *Infection, Genetics and Evolution*, *81*, 104262.

Hanna, H., Darouiche, R., & Raad, I. (2001). New approaches for prevention of intravascular catheter-related infections. *Infections in Medicine*, *18*(1), 38–48.

Harbarth, S., Maiwald, M., & Dancer, S. J. (2013). The environment and healthcare-acquired infections: Why accurate reporting and evaluation of biological plausibility are important. *Infection Control & Hospital Epidemiology*, *34*(9), 996–997.

Hardy, K. J., Oppenheim, B. A., Gossain, S., Gao, F., & Hawkey, P. M. (2006). A study of the relationship between environmental contamination with methicillin-resistant *Staphylococcus aureus* (MRSA) and patients' acquisition of MRSA. *Infection Control & Hospital Epidemiology*, *27*(2), 127–132.

Harrington, N. E., Sweeney, E., & Harrison, F. (2020). Building a better biofilm—Formation of in vivo-like biofilm structures by *Pseudomonas aeruginosa* in a porcine model of cystic fibrosis lung infection. *Biofilms, 2*, 100024.

Hayden, M. K., Bonten, M. J., Blom, D. W., Lyle, E. A., van de Vijver, D. A., & Weinstein, R. A. (2006). Reduction in acquisition of vancomycin-resistant enterococcus after enforcement of routine environmental cleaning measures. *Clinical Infectious Diseases, 42*(11), 1552–1560.

Henry, R. J. (1943). The mode of action of sulfonamides. *Bacteriological Reviews, 7*(4), 175–262.

High, K. P., Bradley, S. F., Gravenstein, S., Mehr, D. R., Quagliarello, V. J., Richards, C., et al. (2009). Clinical practice guideline for the evaluation of fever and infection in older adult residents of long-term care facilities: 2008 update by the Infectious Diseases Society of America. *Clinical Infectious Diseases, 48*(2), 149–171.

Hiramatsu, K., Aritaka, N., Hanaki, H., Kawasaki, S., Hosoda, Y., Hori, S., et al. (1997). Dissemination in Japanese hospitals of strains of *Staphylococcus aureus* heterogeneously resistant to vancomycin. *The Lancet, 350*(9092), 1670–1673.

Hoiby, N., Fomsgaard, A., Jensen, E. T., Johansen, H., Kronborg, G., Pedersen, S., et al. (1995). *The immune response to bacterial biofilms*. Cambridge: Cambridge University Press.

Hola, V., Peroutkova, T., & Ruzicka, F. (2012). Virulence factors in *Proteus* bacteria from biofilm communities of catheter-associated urinary tract infections. *FEMS Immunology and Medical Microbiology, 65*(2), 343–349.

Hongsuwan, M., Srisamang, P., Kanoksil, M., Luangasanatip, N., Jatapai, A., Day, N. P., et al. (2014). Increasing incidence of hospital-acquired and healthcare-associated bacteremia in northeast Thailand: A multicenter surveillance study. *PLoS One, 9*(10), e109324.

Hooton, T. M., Bradley, S. F., Cardenas, D. D., Colgan, R., Geerlings, S. E., Rice, J. C., et al. (2010). Diagnosis, prevention, and treatment of catheter-associated urinary tract infection in adults: 2009 International Clinical Practice Guidelines from the Infectious Diseases Society of America. *Clinical Infectious Diseases, 50*(5), 625–663.

Horn, D. L., Morrison, D. C., Opal, S. M., Silverstein, R., Visvanathan, K., & Zabriskie, J. B. (2000). What are the microbial components implicated in the pathogenesis of sepsis? Report on a symposium. *Clinical Infectious Diseases, 31*(4), 851–858.

Hornick, D. B., Allen, B. L., Horn, M. A., & Clegg, S. (1992). Adherence to respiratory epithelia by recombinant *Escherichia coli* expressing *Klebsiella pneumoniae* type 3 fimbrial gene products. *Infection and Immunity, 60*(4), 1577–1588.

Hsieh, P.-F., Lin, T.-L., Yang, F.-L., Wu, M.-C., Pan, Y.-J., Wu, S.-H., et al. (2012). Lipopolysaccharide O1 antigen contributes to the virulence in *Klebsiella pneumoniae* causing pyogenic liver abscess. *PLoS One, 7* (3), e33155.

Huebner, J., Pier, G. B., Maslow, J. N., Muller, E., Shiro, H., Parent, M., et al. (1994). Endemic nosocomial transmission of *Staphylococcus epidermidis* bacteremia isolates in a neonatal intensive care unit over 1O years. *Journal of Infectious Diseases, 169*(3), 526–531.

Hujer, A. M., Bethel, C. R., Hujer, K. M., & Bonomo, R. A. (2004). Antibiotic resistance in the institutionalized elderly. *Clinics in Laboratory Medicine, 24*(2), 343–361.

Huycke, M. M., Sahm, D. F., & Gilmore, M. S. (1998). Multiple-drug resistant enterococci: The nature of the problem and an agenda for the future. *Emerging Infectious Diseases, 4*(2), 239.

Ielapi, N., Nicoletti, E., Lorè, C., Guasticchi, G., Avenoso, T., Barbetta, A., et al. (2020). The role of biofilm in central venous catheter related bloodstream infections: Evidence-based nursing and review of the literature. *Reviews on Recent Clinical Trials, 15*(1), 22–27.

Iliadis, I., Daskalopoulou, A., Simões, M., & Giaouris, E. (2018). Integrated combined effects of temperature, pH and sodium chloride concentration on biofilm formation by *Salmonella enterica* ser. Enteritidis and Typhimurium under low nutrient food-related conditions. *Food Research International, 107*, 10–18.

Illingworth, B. L., Tweden, K., Schroeder, R. F., & Cameron, J. D. (1998). In vivo efficacy of silver-coated (Silzone) infection-resistant polyester fabric against a biofilm-producing bacteria, *Staphylococcus epidermidis*. *The Journal of Heart Valve Disease, 7*(5), 524–530.

Isiklar, Z. U., Darouiche, R. O., Landon, G. C., & Beck, T. (1996). Efficacy of antibiotics alone for orthopaedic device related infections. *Clinical Orthopaedics and Related Research®, 332*, 184–189.

Jacobsen, S. M., & Shirtliff, M. E. (2011). *Proteus mirabilis* biofilms and -catheter-associated urinary tract infections. *Virulence, 2*(5), 460–465.

Jacobsen, S. M., Stickler, D. J., Mobley, H. L., & Shirtliff, M. E. (2008). Complicated catheter-associated urinary tract infections due to *Escherichia coli* and *Proteus mirabilis*. *Clinical Microbiology Reviews, 21*(1), 26–59.

Jain, R., Kralovic, S. M., Evans, M. E., Ambrose, M., Simbartl, L. A., Obrosky, D. S., et al. (2011). Veterans Affairs initiative to prevent methicillin-resistant *Staphylococcus aureus* infections. *New England Journal of Medicine, 364*(15), 1419–1430.

James, G. A., Swogger, E., Wolcott, R., Pulcini, E.d., Secor, P., Sestrich, J., et al. (2008). Biofilms in chronic wounds. *Wound Repair and Regeneration, 16*(1), 37–44.

Jarvis, W. R., & Martone, W. J. (1992). Predominant pathogens in hospital infections. *Journal of Antimicrobial Chemotherapy, 29*(suppl_A), 19–24.

Jing, J. L. J., Pei Yi, T., Bose, R. J., McCarthy, J. R., Tharmalingam, N., & Madheswaran, T. (2020). Hand sanitizers: A review on formulation aspects, adverse effects, and regulations. *International Journal of Environmental Research and Public Health, 17*(9), 3326.

Jönsson, K., Signäs, C., Müller, H. P., & Lindberg, M. (1991). Two different genes encode fibronectin binding proteins in *Staphylococcus aureus*: The complete nucleotide sequence and characterization of the second gene. *European Journal of Biochemistry, 202*(3), 1041–1048.

Joyanes, P., Pascual, A., Martinez-Martinez, L., Hevia, A., & Perea, E. (2000). In vitro adherence of *Enterococcus faecalis* and *Enterococcus faecium* to urinary catheters. *European Journal of Clinical Microbiology and Infectious Diseases, 19*(2), 124–127.

Kahlmeter, G. (2000). The ECO• SENS Project: A prospective, multinational, multicentre epidemiological survey of the prevalence and antimicrobial susceptibility of urinary tract pathogens—Interim report. *Journal of Antimicrobial Chemotherapy, 46*(suppl_1), 15–22.

Källenius, G., Svenson, S., Hultberg, H., Möllby, R., Helin, I., Cedergren, B., et al. (1981). Occurrence of P-fimbriated *Escherichia coli* in urinary tract infections. *The Lancet, 318*(8260–8261), 1369–1372.

Kalsi, J., Arya, M., Wilson, P., & Mundy, A. (2003). Hospital-acquired urinary tract infection. *International Journal of Clinical Practice, 57* (5), 388–391.

Kamal, G. D., Pfaller, M. A., Rempe, L. E., & Jebson, P. J. (1991). Reduced intravascular catheter infection by antibiotic bonding: A prospective, randomized, controlled trial. *JAMA, 265*(18), 2364–2368.

Karchmer, A. W., & Gibbons, G. (1994). Infections of prosthetic heart valves and vascular grafts. In *Infections associated with indwelling medical devices* (2nd ed., p. 213). Washington, DC: American Society for Microbiology.

Kaweeteerawat, C., Chang, C. H., Roy, K. R., Liu, R., Li, R., Toso, D., et al. (2015). Cu nanoparticles have different impacts in *Escherichia coli* and *Lactobacillus brevis* than their microsized and ionic analogues. *ACS Nano*, *9*(7), 7215–7225.

Khatoon, Z., McTiernan, C. D., Suuronen, E. J., Mah, T. F., & Alarcon, E. I. (2018). Bacterial biofilm formation on implantable devices and approaches to its treatment and prevention. *Heliyon*, *4*(12), e01067.

Khurana, C., & Chudasama, B. (2018). Nanoantibiotics: Strategic assets in the fight against drug-resistant superbugs. *International Journal of Nanomedicine*, *13*(T-NANO 2014 Abstracts), 3.

Kirketerp-Møller, K., Jensen, P.Ø., Fazli, M., Madsen, K. G., Pedersen, J., Moser, C., et al. (2008). Distribution, organization, and ecology of bacteria in chronic wounds. *Journal of Clinical Microbiology*, *46*(8), 2717–2722.

Klemm, P., Hjerrild, L., Gjermansen, M., & Schembri, M. A. (2004). Structure-function analysis of the self-recognizing antigen 43 autotransporter protein from *Escherichia coli*. *Molecular Microbiology*, *51*(1), 283–296.

Klemm, P., & Schembri, M. A. (2000). Bacterial adhesins: Function and structure. *International Journal of Medical Microbiology*, *290*(1), 27–35.

Kollef, M. H. (1995). The identification of ICU-specific outcome predictors: A comparison of medical, surgical, and cardiothoracic ICUs from a single institution. *Heart & Lung*, *24*(1), 60–66.

Kollef, M. H. (2001). Hospital-acquired pneumonia and de-escalation of antimicrobial treatment. *Critical Care Medicine*, *29*(7), 1473–1475.

Kollef, M. H. (2004). Prevention of hospital-associated pneumonia and ventilator-associated pneumonia. *Critical Care Medicine*, *32*(6), 1396–1405.

Kong, M., Chen, X. G., Xing, K., & Park, H. J. (2010). Antimicrobial properties of chitosan and mode of action: A state of the art review. *International Journal of Food Microbiology*, *144*(1), 51–63.

Kotay, S. M., Parikh, H. I., Barry, K., Gweon, H. S., Guilford, W., Carroll, J., et al. (2020). Nutrients influence the dynamics of *Klebsiella pneumoniae* carbapenemase producing enterobacterales in transplanted hospital sinks. *Water Research*, *176*, 115707.

Krahn, T., Gilmour, C., Tilak, J., Fraud, S., Kerr, N., Lau, C. H., et al. (2012). Determinants of intrinsic aminoglycoside resistance in *Pseudomonas aeruginosa*. *Antimicrobial Agents and Chemotherapy*, *56*(11), 5591–5602.

Kramer, A., Schwebke, I., & Kampf, G. (2006). How long do nosocomial pathogens persist on inanimate surfaces? A systematic review. *BMC Infectious Diseases*, *6*(1), 1–8.

Kuchma, S. L., & O'Toole, G. A. (2000). Surface-induced and biofilm-induced changes in gene expression. *Current Opinion in Biotechnology*, *11*(5), 429–433.

Kuehnert, M. J., Hill, H. A., Kupronis, B. A., Tokars, J. I., Solomon, S. L., & Jernigan, D. B. (2005). Methicillin-resistant—*Staphylococcus aureus* hospitalizations, United States. *Emerging Infectious Diseases*, *11* (6), 468.

Kuhn, P. J. (1983). Doorknobs: A source of nosocomial infection. *Diagnostic Medicine*, *6*(8), 62–63.

Kumar, P., Jithesh, V., & Gupta, S. K. (2015). Does a single specialty intensive care unit make better business sense than a multi-specialty intensive care unit? A costing study in a trauma center in India. *Saudi Journal of Anaesthesia*, *9*(2), 189.

Kunin, C. M., & McCormack, R. C. (1966). Prevention of catheter-induced urinary-tract infections by sterile closed drainage. *New England Journal of Medicine*, *274*(21), 1155–1161.

Lai, C. H., Wong, W. W., Chin, C., Huang, C. K., Lin, H. H., Chen, W. F., et al. (2006). Central venous catheter-related *Stenotrophomonas maltophilia* bacteraemia and associated relapsing bacteraemia in haematology and oncology patients. *Clinical Microbiology and Infection*, *12*(10), 986–991.

Lambert, M.-L., Suetens, C., Savey, A., Palomar, M., Hiesmayr, M., Morales, I., et al. (2011). Clinical outcomes of health-care-associated infections and antimicrobial resistance in patients admitted to European intensive-care units: A cohort study. *The Lancet Infectious Diseases*, *11*(1), 30–38.

Larson, E. L., & Liverman, C. T. (Eds.). (2011). *Institute of Medicine (US) Committee on personal protective equipment for healthcare personnel to prevent transmission of pandemic influenza and other viral respiratory infections: Current research issues. Preventing transmission of pandemic influenza and other viral respiratory diseases: Personal protective equipment for healthcare personnel: Update 2010*. Washington, DC: National Academies Press.

Laws, K. J., Crosby, C., Sridhar, A., Conway, P., Koloadin, L. S., Zhao, M., et al. (2015). High entropy brasses and bronzes—Microstructure, phase evolution and properties. *Journal of Alloys and Compounds*, *650*, 949–961.

Ledingham, I. (1988). Prevention and management of infections with gram-negative bacteria in intensive care units. *British Journal of Clinical Practice. Supplement*, *57*, 45.

Lee, H.-W., Koh, Y., Kim, J., Lee, J.-C., Lee, Y.-C., Seol, S.-Y., et al. (2008). Capacity of multidrug-resistant clinical isolates of *Acinetobacter baumannii* to form biofilm and adhere to epithelial cell surfaces. *Clinical Microbiology and Infection*, *14*(1), 49–54.

Liu, Q., Zhou, Y., Chen, L., & Zheng, X. (2010). Application of MBR for hospital wastewater treatment in China. *Desalination*, *250*(2), 605–608.

Livesley, N. J., & Chow, A. W. (2002). Infected pressure ulcers in elderly individuals. *Clinical Infectious Diseases*, 1390–1396.

Lo, J., Lange, D., & Chew, B. H. (2014). Ureteral stents and foley catheters-associated urinary tract infections: The role of coatings and materials in infection prevention. *Antibiotics*, *3*(1), 87–97.

Loehfelm, T. W., Luke, N. R., & Campagnari, A. A. (2008). Identification and characterization of an *Acinetobacter baumannii* biofilm-associated protein. *Journal of Bacteriology*, *190*(3), 1036–1044.

Lowe, A., Lambert, P., & Smith, A. (1995). Cloning of an *Enterococcus faecalis* endocarditis antigen: Homology with adhesins from some oral streptococci. *Infection and Immunity*, *63*(2), 703–706.

Lowy, F. D. (1998). *Staphylococcus aureus* infections. *New England Journal of Medicine*, *339*(8), 520–532.

Lu, X., Zhang, C., & Han, Y. (2004). Low-density polyethylene superhydrophobic surface by control of its crystallization behavior. *Macromolecular Rapid Communications*, *25*(18), 1606–1610.

Lyczak, J. B., Cannon, C. L., & Pier, G. B. (2002). Lung infections associated with cystic fibrosis. *Clinical Microbiology Reviews*, *15*(2), 194–222.

Lyder, C. H. (2003). Pressure ulcer prevention and management. *JAMA*, *289*(2), 223–226.

Lynch, J., & Lama, V. (2000). Diagnosis and therapy of nosocomial ventilator associated pneumonia. *AFC*, *4*(1), 19–26.

Lyte, M., & Ernst, S. (1992). Catecholamine induced growth of gram negative bacteria. *Life Sciences*, *50*(3), 203–212.

Mack, D., Fischer, W., Krokotsch, A., Leopold, K., Hartmann, R., Egge, H., et al. (1996). The intercellular adhesin involved in biofilm accumulation of *Staphylococcus epidermidis* is a linear beta-1, 6-linked glucosaminoglycan: Purification and structural analysis. *Journal of Bacteriology, 178*(1), 175–183.

Mack, D., Nedelmann, M., Krokotsch, A., Schwarzkopf, A., Heesemann, J., & Laufs, R. (1994). Characterization of transposon mutants of biofilm-producing *Staphylococcus epidermidis* impaired in the accumulative phase of biofilm production: Genetic identification of a hexosamine-containing polysaccharide intercellular adhesin. *Infection and Immunity, 62*(8), 3244–3253.

Mack, D., Siemssen, N., & Laufs, R. (1992). Parallel induction by glucose of adherence and a polysaccharide antigen specific for plastic-adherent *Staphylococcus epidermidis*: Evidence for functional relation to intercellular adhesion. *Infection and Immunity, 60*(5), 2048–2057.

Macomber, L., & Imlay, J. A. (2009). The iron-sulfur clusters of dehydratases are primary intracellular targets of copper toxicity. *Proceedings of the National Academy of Sciences, 106*(20), 8344–8349.

Mah, T. F. (2012). Biofilm-specific antibiotic resistance. *Future Microbiology, 7*(9), 1061–1072.

Maity, S., Nir, S., Zada, T., & Reches, M. (2014). Self-assembly of a tripeptide into a functional coating that resists fouling. *Chemical Communications, 50*(76), 11154–11157.

Maki, D. G. (1977). Preventing infection in intravenous therapy. *Anesthesia and Analgesia, 56*(1), 141–153.

Maki, D. G. (1992). Infections due to infusion therapy. In *Hospital infections* (pp. 849–898). Little Brown.

Maki, D., Alvarado, C., & Ringer, M. (1991). Prospective randomised trial of povidone-iodine, alcohol, and chlorhexidine for prevention of infection associated with central venous and arterial catheters. *The Lancet, 338*(8763), 339–343.

Maki, D. G., Cobb, L., Garman, J. K., Shapiro, J. M., Ringer, M., & Helgerson, R. B. (1988). An attachable silver-impregnated cuff for prevention of infection with central venous catheters: A prospective randomized multicenter trial. *The American Journal of Medicine, 85* (3), 307–314.

Maki, D. G., & Martin, W. T. (1975). Nationwide epidemic of septicemia caused by contaminated infusion products. IV. Growth of microbial pathogens in fluids for intravenous infusion. *Journal of Infectious Diseases, 131*(3), 267–272.

Maki, D., Mermel, L., Bennett, J., & Brachman, P. (1998). Infections due to infusion therapy. In *Vol. 44. Hospital infections*. Philadelphia: Lippincott-Raven.

Malic, S., Hill, K. E., Hayes, A., Percival, S. L., Thomas, D. W., & Williams, D. W. (2009). Detection and identification of specific bacteria in wound biofilms using peptide nucleic acid fluorescent in situ hybridization (PNA FISH). *Microbiology, 155*(8), 2603–2611.

Mandell, J. B., et al. (2020). Direct antimicrobial activity of cationic amphipathic peptide WLBU2 against *Staphylococcus aureus* biofilms is enhanced in physiologic buffered saline. *Journal of Orthopaedic Research, 38*(12), 2657–2663.

Mangal, S., Pho, A., Arcia, A., & Carter, E. (2021). Patient and family engagement in catheter-associated urinary tract infection (CAUTI) prevention: A systematic review. *The Joint Commission Journal on Quality and Patient Safety, 47*(9), 591–603.

Manzur, A., & Pujol, M. (2008). Impact and control of methicillin-resistant *Staphylococcus aureus* (MRSA) in long-term care facilities. *Revista Española de Geriatría y Gerontología, 43*(4), 235–238.

Martin, I., Waters, V., & Grasemann, H. (2021). Approaches to targeting bacterial biofilms in cystic fibrosis airways. *International Journal of Molecular Sciences, 22*(4).

Mathai, D., Jones, R. N., & Pfaller, M. A. (2001). Epidemiology and frequency of resistance among pathogens causing urinary tract infections in 1,510 hospitalized patients: A report from the SENTRY Antimicrobial Surveillance Program (North America). *Diagnostic Microbiology and Infectious Disease, 40*(3), 129–136.

Mathur, P. (2011). Hand hygiene: Back to the basics of infection control. *The Indian Journal of Medical Research, 134*(5), 611.

Matsumura, S., & Simor, A. E. (1998). Treatment of endocarditis due to vancomycin-resistant *Enterococcus faecium* with quinupristin/dalfopristin, doxycycline, and rifampin: A synergistic drug combination. *Clinical Infectious Diseases*, 1554–1556.

Mazzariol, A., Bazaj, A., & Cornaglia, G. (2017). Multi-drug-resistant Gram-negative bacteria causing urinary tract infections: A review. *Journal of Chemotherapy, 29*(sup1), 2–9.

McCabe, B. J., & Horn, G. (1994). Learning-related changes in Fos-like immunoreactivity in the chick forebrain after imprinting. *Proceedings of the National Academy of Sciences, 91*(24), 11417–11421.

McDonald, M. V., Brickner, C., Russell, D., Dowding, D., Larson, E. L., Trifilio, M., et al. (2021). Observation of hand hygiene practices in home health care. *Journal of the American Medical Directors Association, 22*(5), 1029–1034.

McDonnell, G., & Russell, A. D. (1999). Antiseptics and disinfectants: Activity, action, and resistance. *Clinical Microbiology Reviews, 12* (1), 147–179.

Meddings, J. A., Reichert, H., Rogers, M. A., Saint, S., Stephansky, J., & McMahon, L. F., Jr. (2012). Effect of nonpayment for hospital-acquired, catheter-associated urinary tract infection: A statewide analysis. *Annals of Internal Medicine, 157*(5), 305–312.

Mehtar, S., Wiid, I., & Todorov, S. (2008). The antimicrobial activity of copper and copper alloys against nosocomial pathogens and *Mycobacterium tuberculosis* isolated from healthcare facilities in the Western Cape: An in-vitro study. *Journal of Hospital Infection, 68* (1), 45–51.

Mellegård, H., Strand, S., Christensen, B., Granum, P., & Hardy, S. (2011). Antibacterial activity of chemically defined chitosans: Influence of molecular weight, degree of acetylation and test organism. *International Journal of Food Microbiology, 148*(1), 48–54.

Menzies, B. E., & Kourteva, I. (2000). *Staphylococcus aureus* α-toxin induces apoptosis in endothelial cells. *FEMS Immunology & Medical Microbiology, 29*(1), 39–45.

Michel, L., McMichan, J. C., & Bachy, J.-L. (1979). Microbial colonization of indwelling central venous catheters: Statistical evaluation of potential contaminating factors. *The American Journal of Surgery, 137*(6), 745–748.

Michels, H. T., & Anderson, D. G. (2008). Antimicrobial regulatory efficacy testing of solid copper alloy surfaces in the USA. *Metal ions in Biology and Medicine, 10*(2), 185–190.

Michels, H. T., Keevil, C. W., Salgado, C. D., & Schmidt, M. G. (2015). From laboratory research to a clinical trial: Copper alloy surfaces kill bacteria and reduce hospital-acquired infections. *HERD: Health Environments Research & Design Journal, 9*(1), 64–79.

Michels, H., Noyce, J., & Keevil, C. W. (2009). Effects of temperature and humidity on the efficacy of methicillin-resistant *Staphylococcus aureus* challenged antimicrobial materials containing silver and copper. *Letters in Applied Microbiology, 49*(2), 191–195.

Mobley, H., Chippendale, G., Tenney, J., Hull, R., & Warren, J. (1987). Expression of type 1 fimbriae may be required for persistence of *Escherichia coli* in the catheterized urinary tract. *Journal of Clinical Microbiology*, *25*(12), 2253–2257.

Moradali, M. F., Ghods, S., & Rehm, B. H. (2017). *Pseudomonas aeruginosa* lifestyle: A paradigm for adaptation, survival, and persistence. *Frontiers in Cellular and Infection Microbiology*, *7*, 39.

Moro, M. L., Jans, B., Cookson, B., & Fabry, J. (2010). The burden of healthcare-associated infections in European long-term care facilities. *Infection Control and Hospital Epidemiology*, *31*(Suppl. 1), S59–S62.

Mylonakis, E., & Calderwood, S. B. (2001). Infective endocarditis in adults. *New England Journal of Medicine*, *345*(18), 1318–1330.

Nagase, T., Shibata, A., Matsumuro, M., Takemura, M., & Semboshi, S. (2019). Alloy design and fabrication of ingots in Cu-Zn-Mn-Ni-Sn high-entropy and Cu-Zn-Mn-Ni medium-entropy brasses. *Materials & Design*, *181*, 107900.

Nathan, S., Lockatell, C. V., Arto, S. B., Drachenberg, C., Michael, S. G., & David, E. J. (2001). Role of *Enterococcus faecalis* surface protein Esp in the pathogenesis of ascending urinary tract infection. *Infection and Immunity*, *69*(7), 4366–4372.

Navarre, W. W., & Schneewind, O. (1994). Proteolytic cleavage and cell wall anchoring at the LPXTG motif of surface proteins in Gram-positive bacteria. *Molecular Microbiology*, *14*(1), 115–121.

Navon-Venezia, S., Ben-Ami, R., & Carmeli, Y. (2005). Update on *Pseudomonas aeruginosa* and *Acinetobacter baumannii* infections in the healthcare setting. *Current Opinion in Infectious Diseases*, *18*(4), 306–313.

Nicasio, A. M., Kuti, J. L., & Nicolau, D. P. (2008). The current state of multidrug-resistant gram-negative bacilli in North America. *Pharmacotherapy*, *28*(2), 235–249.

Nickel, J. C., Downey, J., & Costerton, J. (1992). Movement of *Pseudomonas aeruginosa* along catheter surfaces A mechanism in pathogenesis of catheter-associated infection. *Urology*, *39*(1), 93–98.

Nickel, J. C., Grant, S. K., Lam, K., Olson, M. E., & Costerton, J. W. (1991). Bacteriologically stressed animalmodel of new closed catheter drainage system with microbicidal outlet tube. *Urology*, *38*(3), 280–289.

Nicolle, L. E. (1997). Asymptomatic bacteriuria in the elderly. *Infectious Disease Clinics of North America*, *11*(3), 647–662.

Nicolle, L. E. (2012). Urinary catheter-associated infections. *Infectious Disease Clinics of North America*, *26*(1), 13–27.

Nicolle, L. E., & Yoshikawa, T. T. (2000). Urinary tract infection in long-term-care facility residents. *Clinical Infectious Diseases*, *31*(3), 757–761.

Niël-Weise, B. S., van den Broek, P. J., da Silva, E. M., & Silva, L. A. (2012). Urinary catheter policies for long-term bladder drainage. *Cochrane Database of Systematic Reviews*, *8*.

Nishikawa, K., Takasu, A., Morita, K., Tsumori, H., & Sakamoto, T. (2010). Deposits on the intraluminal surface and bacterial growth in central venous catheters. *Journal of Hospital Infection*, *75*(1), 19–22.

Niu, C., Clemmer, K. M., Bonomo, R. A., & Rather, P. N. (2008). Isolation and characterization of an autoinducer synthase from *Acinetobacter baumannii*. *Journal of Bacteriology*, *190*(9), 3386–3392.

Niveditha, S., Pramodhini, S., Umadevi, S., Kumar, S., & Stephen, S. (2012). The isolation and the biofilm formation of uropathogens in the patients with catheter associated urinary tract infections (UTIs). *Journal of Clinical and Diagnostic Research: JCDR*, *6*(9), 1478.

Novais, Â., Pires, J., Ferreira, H., Costa, L., Montenegro, C., Vuotto, C., et al. (2012). Characterization of globally spread *Escherichia coli* ST131 isolates (1991 to 2010). *Antimicrobial Agents and Chemotherapy*, *56*(7), 3973–3976.

Novais, Â., Vuotto, C., Pires, J., Montenegro, C., Donelli, G., Coque, T. M., et al. (2013). Diversity and biofilm-production ability among isolates of *Escherichia coli* phylogroup D belonging to ST69, ST393 and ST405 clonal groups. *BMC Microbiology*, *13*(1), 1–9.

Noyce, J., Michels, H., & Keevil, C. (2006a). Potential use of copper surfaces to reduce survival of epidemic meticillin-resistant *Staphylococcus aureus* in the healthcare environment. *Journal of Hospital Infection*, *63*(3), 289–297.

Noyce, J., Michels, H., & Keevil, C. (2006b). Use of copper cast alloys to control *Escherichia coli* O157 cross-contamination during food processing. *Applied and Environmental Microbiology*, *72*(6), 4239–4244.

Noyce, J., Michels, H., & Keevil, C. (2007). Inactivation of influenza A virus on copper versus stainless steel surfaces. *Applied and Environmental Microbiology*, *73*(8), 2748–2750.

O'Gara, J. P., & Humphreys, H. (2001). *Staphylococcus epidermidis* biofilms: Importance and implications. *Journal of Medical Microbiology*, *50*(7), 582–587.

O'Grady, N. P., Alexander, M., Burns, L. A., Dellinger, E. P., Garland, J., Heard, S. O., et al. (2011). Guidelines for the prevention of intravascular catheter-related infections. *Clinical Infectious Diseases*, *52*(9), e162–e193.

O'Grady, N. P., Alexander, M., Dellinger, E. P., Gerberding, J. L., Heard, S. O., Maki, D. G., et al. (2002). Guidelines for the prevention of intravascular catheter-related infections. *American Journal of Infection Control*, *30*(8), 476–489.

Olmsted, S. B., Dunny, G. M., Erlandsen, S. L., & Wells, C. L. (1994). A plasmid-encoded surface protein on *Enterococcus faecalis* augments its internalization by cultured intestinal epithelial cells. *Journal of Infectious Diseases*, *170*(6), 1549–1556.

Ong, C.-L. Y., Ulett, G. C., Mabbett, A. N., Beatson, S. A., Webb, R. I., Monaghan, W., et al. (2008). Identification of type 3 fimbriae in uropathogenic *Escherichia coli* reveals a role in biofilm formation. *Journal of Bacteriology*, *190*(3), 1054–1063.

Otter, J. A., Vickery, K., Walker, J. T., deLancey Pulcini, E., Stoodley, P., Goldenberg, S. D., et al. (2015). Surface-attached cells, biofilms and biocide susceptibility: Implications for hospital cleaning and disinfection. *The Journal of Hospital Infection*, *89*(1), 16–27.

Otto, M. (2008). Staphylococcal biofilms. *Bacterial biofilms*, 207–228.

Owusu, E. G. (2020). *Light-activated antimicrobial surfaces containing quantum dots for the prevention of hospital-acquired infections*. UCL (University College London).

Patel, R. (2005). Biofilms and antimicrobial resistance. *Clinical Orthopaedics and Related Research*, *437*, 41–47.

Paterson, D. L., & Lipman, J. (2007). Returning to the pre-antibiotic era in the critically ill: The XDR problem. *Critical Care Medicine*, *35*(7), 1789–1791.

Patterson, J. E., Sweeney, A. H., Simms, M., Carley, N., Mangi, R., Sabetta, J., et al. (1995). An analysis of 110 serious enterococcal infections. Epidemiology, antibiotic susceptibility, and outcome. *Medicine*, *74*(4), 191–200.

Patti, J. M., & Höök, M. (1994). Microbial adhesins recognizing extracellular matrix macromolecules. *Current Opinion in Cell Biology*, *6*(5), 752–758.

Pemberton, L. B., Ross, V., Cuddy, P., Kremer, H., Fessler, T., & McGurk, E. (1996). No difference in catheter sepsis between standard and antiseptic central venous catheters: A prospective randomized trial. *Archives of Surgery*, *131*(9), 986–989.

Percival, S. L., Hill, K. E., Williams, D. W., Hooper, S. J., Thomas, D. W., & Costerton, J. W. (2012). A review of the scientific evidence for biofilms in wounds. *Wound Repair and Regeneration*, *20*(5), 647–657.

Percival, S. L., Suleman, L., Vuotto, C., & Donelli, G. (2015). Healthcare-associated infections, medical devices and biofilms: Risk, tolerance and control. *Journal of Medical Microbiology*, *64*(Pt 4), 323–334.

Perinelli, D. R., Fagioli, L., Campana, R., Lam, J. K. W., Baffone, W., Palmieri, G. F., et al. (2018). Chitosan-based nanosystems and their exploited antimicrobial activity. *European Journal of Pharmaceutical Sciences*, *117*, 8–20.

Peters, G., Locci, R., & Pulverer, G. (1982). Adherence and growth of coagulase-negative staphylococci on surfaces of intravenous catheters. *Journal of Infectious Diseases*, *146*(4), 479–482.

Pittet, D., Allegranzi, B., Boyce, J., & World Health Organization World Alliance for Patient Safety First Global Patient Safety Challenge Core Group of Experts. (2009). The World Health Organization guidelines on hand hygiene in health care and their consensus recommendations. *Infection Control & Hospital Epidemiology*, *30*(7), 611–622.

Plowman, R. (2000). The socioeconomic burden of hospital acquired infection. *Eurosurveillance*, *5*(4), 49–50.

Podschun, R., Fischer, A., & Ullmann, U. (1992). Siderophore production of *Klebsiella* species isolated from different sources. *Zentralblatt für Bakteriologie*, *276*(4), 481–486.

Podschun, R., & Ullmann, U. (1992). Klebsiella capsular type K7 in relation to toxicity, susceptibility to phagocytosis and resistance to serum. *Journal of Medical Microbiology*, *36*(4), 250–254.

Pompilio, A., Savini, V., Fiscarelli, E., Gherardi, G., & Di Bonaventura, G. (2020). Clonal diversity, biofilm formation, and antimicrobial resistance among *Stenotrophomonas maltophilia* strains from cystic fibrosis and non-cystic fibrosis patients. *Antibiotics*, *9*(1).

Poole, K. (2017). At the nexus of antibiotics and metals: The impact of Cu and Zn on antibiotic activity and resistance. *Trends in Microbiology*, *25* (10), 820–832.

Pope, S. D., Dellit, T. H., Owens, R. C., & Hooton, T. M. (2009). Results of survey on implementation of Infectious Diseases Society of America and Society for Healthcare Epidemiology of America guidelines for developing an institutional program to enhance antimicrobial stewardship. *Infection Control and Hospital Epidemiology*, *30*(1), 97–98.

Pourbeyram, S., Bayrami, R., & Dadkhah, H. (2017). Green synthesis and characterization of ultrafine copper oxide reduced graphene oxide (CuO/rGO) nanocomposite. *Colloids and Surfaces A: Physicochemical and Engineering Aspects*, *529*, 73–79.

Powell, C., Kudsk, K. A., Kulich, P. A., Mandelbaum, J. A., & Fabri, P. J. (1988). Effect of frequent guidewire changes on triple-lumen catheter sepsis. *Journal of Parenteral and Enteral Nutrition*, *12*(5), 462–464.

Pozzi, C., Waters, E. M., Rudkin, J. K., Schaeffer, C. R., Lohan, A. J., Tong, P., et al. (2012). Methicillin resistance alters the biofilm phenotype and attenuates virulence in *Staphylococcus aureus* device-associated infections. *PLoS Pathogens*, *8*(4), e1002626.

Pragasam, A. K., Veeraraghavan, B., Nalini, E., Anandan, S., & Kaye, K. S. (2018). An update on antimicrobial resistance and the role of newer antimicrobial agents for *Pseudomonas aeruginosa*. *Indian Journal of Medical Microbiology*, *36*(3), 303–316.

Prager, R., & Silva, J., Jr. (1984). Colonization of central venous catheters. *Southern Medical Journal*, *77*(4), 458–461.

Preda, V. G., & Săndulescu, O. (2019). Communication is the key: Biofilms, quorum sensing, formation and prevention. *Discoveries (Craiova)*, *7*(3), e100.

Raad, I. I. (1994). The pathogenesis and prevention of central venous catheter-related infections. *Middle East Journal of Anaesthesiology*, *12*(4), 381–403.

Raad, I. I. (1997). Vascular catheters impregnated with antimicrobial agents present knowledge and future direction. *Infection Control & Hospital Epidemiology*, *18*(4), 227–229.

Raad, I. (1998). Intravascular-catheter-related infections. *The Lancet*, *351* (9106), 893–898. https://doi.org/10.1016/S0140-6736(97)10006-X.

Raad, I. I., & Bodey, G. P. (1992). Infectious complications of indwelling vascular catheters. *Clinical Infectious Diseases*, *15*(2), 197–208.

Raad, I., Costerton, W., Sabharwal, U., Sadlowski, M., Anaissie, E., & Bodey, G. P. (1993). Ultrastructural analysis of indwelling vascular catheters: A quantitative relationship between luminal colonization and duration of placement. *Journal of Infectious Diseases*, *168*(2), 400–407.

Raad, I., Darouiche, R., Hachem, R., Mansouri, M., & Bodey, G. P. (1996). The broad-spectrum activity and efficacy of catheters coated with minocycline and rifampin. *Journal of Infectious Diseases*, *173*(2), 418–424.

Raafat, D., & Sahl, H. G. (2009). Chitosan and its antimicrobial potential—A critical literature survey. *Microbial Biotechnology*, *2*(2), 186–201.

Rather, P. N. (2005). Swarmer cell differentiation in *Proteus mirabilis*. *Environmental Microbiology*, *7*(8), 1065–1073.

Raymond, D. P., Pelletier, S. J., Crabtree, T. D., Evans, H. L., Pruett, T. L., & Sawyer, R. G. (2003). Impact of antibiotic-resistant Gram-negative bacilli infections on outcome in hospitalized patients. *Critical Care Medicine*, *31*(4), 1035–1041.

Reches, M. (2018). Fighting bacteria: How can we prevent hospital-acquired infections? *Age*, *11*, 12.

Regev-Shoshani, G., Ko, M., Crowe, A., & Av-Gay, Y. (2011). Comparative efficacy of commercially available and emerging antimicrobial urinary catheters against bacteriuria caused by *E. coli* in vitro. *Urology*, *78*(2), 334–339.

Reisner, A., Haagensen, J. A., Schembri, M. A., Zechner, E. L., & Molin, S. (2003). Development and maturation of *Escherichia coli* K-12 biofilms. *Molecular Microbiology*, *48*(4), 933–946.

Rich, R. L., Kreikemeyer, B., Owens, R. T., LaBrenz, S., Narayana, S. V., Weinstock, G. M., et al. (1999). Ace is a collagen-binding MSCRAMM from *Enterococcus faecalis*. *Journal of Biological Chemistry*, *274*(38), 26939–26945.

Richards, M. J., Edwards, J. R., Culver, D. H., & Gaynes, R. P. (1999). Nosocomial infections in medical intensive care units in the United States. National Nosocomial Infections Surveillance System. *Critical Care Medicine*, *27*(5), 887–892.

Riewpaiboon, A., Malaroje, S., & Kongsawatt, S. (2007). Effect of costing methods on unit cost of hospital medical services. *Tropical Medicine & International Health*, *12*(4), 554–563.

Rivero, P., Brenner, P., & Nercelles, P. (2014). Impact of copper in the reduction of hospital-acquired infections, mortality and antimicrobial costs in the Adult Intensive Care Unit. *Revista chilena de infectologia: Organo oficial de la Sociedad Chilena de Infectologia*, *31*(3), 274–279.

Rocha, S. P., Pelayo, J. S., & Elias, W. P. (2007). Fimbriae of uropathogenic *Proteus mirabilis*. *FEMS Immunology and Medical Microbiology*, *51* (1), 1–7.

Rodriguez-Bano, J., Marti, S., Soto, S., Fernández-Cuenca, F., Cisneros, J. M., Pachón, J., et al. (2008). Biofilm formation in *Acinetobacter baumannii*: Associated features and clinical implications. *Clinical Microbiology and Infection*, *14*(3), 276–278.

Rohde, H., Frankenberger, S., Zähringer, U., & Mack, D. (2010). Structure, function and contribution of polysaccharide intercellular adhesin (PIA) to *Staphylococcus epidermidis* biofilm formation and pathogenesis of biomaterial-associated infections. *European Journal of Cell Biology*, *89*(1), 103–111.

Rose, H. D., & Babcock, J. B. (1975). Colonization of intensive care unit patients with gram-negative bacilli. *American Journal of Epidemiology*, *101*(6), 495–501.

Rowbury, R. (1995). An assessment of environmental factors influencing acid tolerance and sensitivity in *Escherichia coli*, *Salmonella* spp. and other enterobacteria. *Letters in Applied Microbiology*, *20*(6), 333–337.

Rózalski, A., Sidorczyk, Z., & Kotełko, K. (1997). Potential virulence factors of *Proteus* bacilli. *Microbiology and Molecular Biology Reviews*, *61*(1), 65–89.

Ruel-Gariépy, E., & Leroux, J.-C. (2006). Chitosan: A natural polycation with multiple applications. In *Polysaccharides for drug delivery and pharmaceutical applications*. https://doi.org/10.1021/bk-2006-0934.ch012.

Rutala, W. A., & Weber, D. J. (2008). *Guideline for disinfection and sterilization in healthcare facilities, 2008*.

Rutala, W. A., & Weber, D. J. (2021). Disinfection and sterilization in health care facilities: An overview and current issues. *Infectious Disease Clinics*, *35*(3), 575–607.

Ryan, A., Kaplan, E., Nebel, J. C., Polycarpou, E., Crescente, V., Lowe, E., et al. (2014). Identification of NAD(P)H quinone oxidoreductase activity in azoreductases from *P. aeruginosa*: Azoreductases and NAD(P)H quinone oxidoreductases belong to the same FMN-dependent superfamily of enzymes. *PLoS One*, *9*(6), e98551.

Ryder, M. (2005). Catheter-related infections: It's all about biofilm. *Topics in Advanced Practice Nursing*, *5*.

Saint, S., Veenstra, D. L., & Lipsky, B. A. (2000). The clinical and economic consequences of nosocomial central venous catheter-related infection: Are antimicrobial catheters useful? *Infection Control & Hospital Epidemiology*, *21*(6), 375–380.

Salgado, C. D., Giannetta, E. T., Hayden, F. G., & Farr, B. M. (2004). Preventing nosocomial influenza by improving the vaccine acceptance rate of clinicians. *Infection Control & Hospital Epidemiology*, *25* (11), 923–928.

Salgado, C. D., Sepkowitz, K. A., John, J. F., Cantey, J. R., Attaway, H. H., Freeman, K. D., et al. (2013). Copper surfaces reduce the rate of healthcare-acquired infections in the intensive care unit. *Infection Control & Hospital Epidemiology*, *34*(5), 479–486.

Salvarci, A., Koroglu, M., & Erayman, B. (2016). Investigation of antibacterial activity and biofilm formation of silicones coated with minocycline-rifampicin, silver nitrate, and nitrofurantoin for short-term utilization in in vitro urinary system models. *Urology*, *88*, 66–75.

Sandoe, J., & Longshaw, C. (2001). Ventriculoperitoneal shunt infection caused by *Staphylococcus lugdunensis*. *Clinical Microbiology and Infection*, *7*(7), 385–387.

Santo, C. E., Taudte, N., Nies, D. H., & Grass, G. (2008). Contribution of copper ion resistance to survival of *Escherichia coli* on metallic copper surfaces. *Applied and Environmental Microbiology*, *74*(4), 977–986.

Schiffer, C. A., Mangu, P. B., Wade, J. C., Camp-Sorrell, D., Cope, D. G., El-Rayes, B. F., et al. (2013). Central venous catheter care for the patient with cancer: American Society of Clinical Oncology clinical practice guideline. *Journal of Clinical Oncology*, *31*(10), 1357–1370.

Schmidt, M. G., Attaway, H. H., Fairey, S. E., Steed, L. L., Michels, H. T., & Salgado, C. D. (2013). Copper continuously limits the concentration of bacteria resident on bed rails within the intensive care unit. *Infection Control & Hospital Epidemiology*, *34*(5), 530–533.

Schmidt, M. G., Attaway, H. H., Sharpe, P. A., John, J., Jr., Sepkowitz, K. A., Morgan, A., et al. (2012). Sustained reduction of microbial burden on common hospital surfaces through introduction of copper. *Journal of Clinical Microbiology*, *50*(7), 2217–2223.

Schultze, L. B., Maldonado, A., Lussi, A., Sculean, A., & Eick, S. (2021). The impact of the pH value on biofilm formation. *Monographs in Oral Science*, *29*, 19–29.

Schumm, K., & Lam, T. B. L. (2008). Types of urethral catheters for management of short-term voiding problems in hospitalized adults: A short version Cochrane review. *Neurourology and Urodynamics: Official Journal of the International Continence Society*, *27*(8), 738–746.

Sherertz, R. J., Carruth, W. A., Hampton, A., Byron, M. P., & Solomon, D. D. (1993). Efficacy of antibiotic-coated catheters in preventing subcutaneous *Staphylococcus aureus* infection in rabbits. *Journal of Infectious Diseases*, *167*(1), 98–106.

Sherertz, R., Forman, D., & Solomon, D. (1989). Efficacy of dicloxacillin-coated polyurethane catheters in preventing subcutaneous *Staphylococcus aureus* infection in mice. *Antimicrobial Agents and Chemotherapy*, *33*(8), 1174–1178.

Shiono, A., & Ike, Y. (1999). Isolation of *Enterococcus faecalis* clinical isolates that efficiently adhere to human bladder carcinoma T24 cells and inhibition of adhesion by fibronectin and trypsin treatment. *Infection and Immunity*, *67*(4), 1585–1592.

Shlaes, D. M., Levy, J., & Wolinsky, E. (1981). Enterococcal bacteremia without endocarditis. *Archives of Internal Medicine*, *141*(5), 578–581.

Simoons-Smit, A., Verweij-van Vught, A., & MacLaren, D. (1986). The role of K antigens as virulence factors in *Klebsiella*. *Journal of Medical Microbiology*, *21*(2), 133–137.

Singh, N., Paknikar, K. M., & Rajwade, J. (2019). Gene expression is influenced due to 'nano' and 'ionic' copper in pre-formed *Pseudomonas aeruginosa* biofilms. *Environmental Research*, *175*, 367–375.

Singha, P., Locklin, J., & Handa, H. (2017). A review of the recent advances in antimicrobial coatings for urinary catheters. *Acta Biomaterialia*, *50*, 20–40.

Singla, S., Harjai, K., & Chhibber, S. (2013). Susceptibility of different phases of biofilm of *Klebsiella pneumoniae* to three different antibiotics. *Journal of Antibiotics (Tokyo)*, *66*(2), 61–66.

Smith, V. S. (2008). *Clean: A history of personal hygiene and purity*. Oxford University Press.

Smith, P. W., Seip, C. W., Schaefer, S. C., & Bell-Dixon, C. (2000). Microbiologic survey of long-term care facilities. *American Journal of Infection Control*, *28*(1), 8–13.

Solano, C., Echeverz, M., & Lasa, I. (2014). Biofilm dispersion and quorum sensing. *Current Opinion in Microbiology*, *18*, 96–104.

Stahlhut, S. G., Chattopadhyay, S., Struve, C., Weissman, S. J., Aprikian, P., Libby, S. J., et al. (2009). Population variability of the FimH type 1 fimbrial adhesin in *Klebsiella pneumoniae*. *Journal of Bacteriology*, *191*(6), 1941–1950.

Stahlhut, S. G., Struve, C., Krogfelt, K. A., & Reisner, A. (2012). Biofilm formation of *Klebsiella pneumoniae* on urethral catheters requires either type 1 or type 3 fimbriae. *FEMS Immunology & Medical Microbiology*, *65*(2), 350–359.

Stewart, P. S., & Costerton, J. W. (2001). Antibiotic resistance of bacteria in biofilms. *The Lancet*, *358*(9276), 135–138.

Stickler, D. (1996). Bacterial biofilms and the encrustation of urethral catheters. *Biofouling*, *9*(4), 293–305.

Stickler, D. J. (2008). Bacterial biofilms in patients with indwelling urinary catheters. *Nature Clinical Practice Urology*, *5*(11), 598–608.

Stickler, D., Ganderton, L., King, J., Nettleton, J., & Winters, C. (1993). *Proteus mirabilis* biofilms and the encrustation of urethral catheters. *Urological Research*, *21*(6), 407–411.

Stickler, D., Morris, N., Moreno, M.-C., & Sabbuba, N. (1998). Studies on the formation of crystalline bacterial biofilms on urethral catheters. *European Journal of Clinical Microbiology and Infectious Diseases*, *17*(9), 649–652.

Stickler, D., Young, R., Jones, G., Sabbuba, N., & Morris, N. (2003). Why are Foley catheters so vulnerable to encrustation and blockage by crystalline bacterial biofilm? *Urological Research*, *31*(5), 306–311.

Stoddart, J. (1974). Gram-negative infections in the ICU. *Critical Care Medicine*, *2*(1), 17–22.

Stoodley, P., deBeer, D., & Lappin-Scott, H. M. (1997). Influence of electric fields and pH on biofilm structure as related to the bioelectric effect. *Antimicrobial Agents and Chemotherapy*, *41*(9), 1876–1879.

Tacconelli, E., Carrara, E., Savoldi, A., Harbarth, S., Mendelson, M., Monnet, D. L., et al. (2018). Discovery, research, and development of new antibiotics: The WHO priority list of antibiotic-resistant bacteria and tuberculosis. *The Lancet Infectious Diseases*, *18*(3), 318–327.

Teitzel, G. M., Geddie, A., De Long, S. K., Kirisits, M. J., Whiteley, M., & Parsek, M. R. (2006). Survival and growth in the presence of elevated copper: Transcriptional profiling of copper-stressed *Pseudomonas aeruginosa*. *Journal of Bacteriology*, *188*(20), 7242–7256.

Tian, W.-X., Yu, S., Ibrahim, M., Almonaofy, A. W., He, L., Hui, Q., et al. (2012). Copper as an antimicrobial agent against opportunistic pathogenic and multidrug resistant Enterobacter bacteria. *Journal of Microbiology*, *50*(4), 586–593.

Tierno, P. M. (2002). *The secret life of germs: Observations and lessons from a microbe hunter*. Simon and Schuster.

Tinelli, M., Cataldo, M. A., Mantengoli, E., Cadeddu, C., Cunietti, E., Luzzaro, F., et al. (2012). Epidemiology and genetic characteristics of extended-spectrum β-lactamase-producing Gram-negative bacteria causing urinary tract infections in long-term care facilities. *Journal of Antimicrobial Chemotherapy*, *67*(12), 2982–2987.

Toledo-Arana, A., Valle, J., Solano, C., Arrizubieta, M.a. J., Cucarella, C., Lamata, M., et al. (2001). The enterococcal surface protein, Esp, is involved in *Enterococcus faecalis* biofilm formation. *Applied and Environmental Microbiology*, *67*(10), 4538–4545.

Toltzis, P., & Blumer, J. L. (1995). Antibiotic-resistant gram-negative bacteria in the critical care setting. *Pediatric Clinics of North America*, *42* (3), 687–702.

Tomaras, A. P., Dorsey, C. W., Edelmann, R. E., & Actis, L. A. (2003). Attachment to and biofilm formation on abiotic surfaces by *Acinetobacter baumannii*: Involvement of a novel chaperone-usher pili assembly system. *Microbiology*, *149*(12), 3473–3484.

Tseng, K. K., Joshi, J., Shrivastava, S., & Klein, E. (2020). Estimating the cost of interventions to improve water, sanitation and hygiene in healthcare facilities across India. *BMJ Global Health*, *5*(12), e003045.

Tunney, M. M., Patrick, S., Curran, M. D., Ramage, G., Hanna, D., Nixon, J. R., et al. (1999). Detection of prosthetic hip infection at revision arthroplasty by immunofluorescence microscopy and PCR amplification of the bacterial 16S rRNA gene. *Journal of Clinical Microbiology*, *37*(10), 3281–3290.

Ulett, G. C., Mabbett, A. N., Fung, K. C., Webb, R. I., & Schembri, M. A. (2007). The role of F9 fimbriae of uropathogenic *Escherichia coli* in biofilm formation. *Microbiology*, *153*(7), 2321–2331.

Ulloa-Ogaz, A. L., & Piñón-Castillo, H. A. (2017). Oxidative damage to *Pseudomonas aeruginosa* ATCC 27833 and *Staphylococcus aureus* ATCC 24213 induced by CuO-NPs. *Environmental Science and Pollution Research*, *24*(27), 22048–22060.

van Buul, L. W., van der Steen, J. T., Veenhuizen, R. B., Achterberg, W. P., Schellevis, F. G., Essink, R. T., et al. (2012). Antibiotic use and resistance in long term care facilities. *Journal of the American Medical Directors Association*, *13*(6). 568.e561–513.

van Roekel, H., Reinhard, J., & Grimmelikhuijsen, S. (2021). Improving hand hygiene in hospitals: Comparing the effect of a nudge and a boost on protocol compliance. *Behavioural Public Policy*, 1–23.

Versey, Z., da Cruz Nizer, W. S., Russell, E., Zigic, S., DeZeeuw, K. G., Marek, J. E., et al. (2021). Biofilm-innate immune interface: Contribution to chronic wound formation. *Frontiers in Immunology*, *12*, 648554.

Vidal, R., Dominguez, M., Urrutia, H., Bello, H., Gonzalez, G., Garcia, A., et al. (1996). Biofilm formation by *Acinetobacter baumannii*. *Microbios*, *346*(86), 49–58.

Vincent, F., & Bensoussan, T. A. (1995). Pneumocystis carinii pneumonia: A major complication of immunosuppressive therapy in patients with Wegener's granulomatosis. *American Journal of Respiratory and Critical Care Medicine*, *152*(6 Pt 1), 2202.

Vincent, M., Duval, R. E., Hartemann, P., & Engels-Deutsch, M. (2018). Contact killing and antimicrobial properties of copper. *Journal of Applied Microbiology*, *124*(5), 1032–1046.

Vovko, P., Retelj, M., Cretnik, T. Z., Jutersek, B., Harlander, T., Kolman, J., et al. (2005). Risk factors for colonization with methicillin-resistant *Staphylococcus aureus* in a long-term-care facility in Slovenia. *Infection Control & Hospital Epidemiology*, *26*(2), 191–195.

Waldvogel, F. A., & Bisno, A. L. (2000). *Infections associated with indwelling medical devices*. Washington, DC: ASM Press.

Ward, K., Olson, M., Lam, K., & Costerton, J. (1992). Mechanism of persistent infection associated with peritoneal implants. *Journal of Medical Microbiology*, *36*(6), 406–413.

Warnes, S., Green, S., Michels, H., & Keevil, C. (2010). Biocidal efficacy of copper alloys against pathogenic enterococci involves degradation of genomic and plasmid DNAs. *Applied and Environmental Microbiology*, *76*(16), 5390–5401.

Warnes, S. L., Highmore, C. J., & Keevil, C. W. (2012). Horizontal transfer of antibiotic resistance genes on abiotic touch surfaces: Implications for public health. *MBio*, *3*(6), e00489-00412.

Warnes, S. L., & Keevil, C. W. (2013). Inactivation of norovirus on dry copper alloy surfaces. *PLoS One*, *8*(9), e75017.

Warnes, S. L., Summersgill, E. N., & Keevil, C. W. (2015). Inactivation of murine norovirus on a range of copper alloy surfaces is accompanied by loss of capsid integrity. *Applied and Environmental Microbiology*, *81*(3), 1085–1091.

Warren, J. W. (1997). Catheter-associated urinary tract infections. *Infectious Disease Clinics of North America*, *11*(3), 609–622.

Watters, C., Fleming, D., Bishop, D., & Rumbaugh, K. (2016). Host responses to biofilm. *Progress in Molecular Biology and Translational Science*, *142*, 193–239.

Weaver, L., Michels, H., & Keevil, C. (2008). Survival of *Clostridium difficile* on copper and steel: Futuristic options for hospital hygiene. *Journal of Hospital Infection*, *68*(2), 145–151.

Weaver, L., Noyce, J., Michels, H., & Keevil, C. (2010). Potential action of copper surfaces on meticillin-resistant *Staphylococcus aureus*. *Journal of Applied Microbiology*, *109*(6), 2200–2205.

Weinstein, R. A., & Darouiche, R. O. (2001). Device-associated infections: A macroproblem that starts with microadherence. *Clinical Infectious Diseases*, *33*(9), 1567–1572.

Wells, C. L., Moore, E. A., Hoag, J. A., Hirt, H., Dunny, G. M., & Erlandsen, S. L. (2000). Inducible expression of *Enterococcus faecalis* aggregation substance surface protein facilitates bacterial internalization by cultured enterocytes. *Infection and Immunity*, *68*(12), 7190–7194.

Wheeldon, L., Worthington, T., Lambert, P. A., Hilton, A., Lowden, C., & Elliott, T. S. (2008). Antimicrobial efficacy of copper surfaces against spores and vegetative cells of *Clostridium difficile*: The germination theory. *Journal of Antimicrobial Chemotherapy*, *62*(3), 522–525.

Widmer, A. F. (2000). Replace hand washing with use of a waterless alcohol hand rub? *Clinical Infectious Diseases*, *31*(1), 136–143.

Wilcox, M. H. (1994). Treating medical device associated infections. In *Bacterial biofilms and their control in medicine and industry* (pp. 155–158). Cardiff: BioLine Press.

Wilks, S., Michels, H., & Keevil, C. (2005). The survival of *Escherichia coli* O157 on a range of metal surfaces. *International Journal of Food Microbiology*, *105*(3), 445–454.

Wilson, S. K., & Gross, M. S. (2022). Biofilm and penile prosthesis infections in the era of coated implants: 2021 update. *International Journal of Impotence Research*, *34*(5), 411–415.

Wolfmeier, H., Pletzer, D., Mansour, S. C., & Hancock, R. E. (2018). New perspectives in biofilm eradication. *ACS Infectious Diseases*, *4*(2), 93–106.

World Health Organization. (2002). *Prevention of hospital-acquired infections: a practical guide/editors: G. Ducel, J. Fabry and L. Nicolle, 2nd. ed.* World Health Organization.

Würker, M., Beuth, J., Ko, H., Przondo-Mordarska, A., & Pulverer, G. (1990). Type of fimbriation determines adherence of *Klebsiella* bacteria to human epithelial cells. *Zentralblatt für Bakteriologie*, *274*(2), 239–245.

Xu, E., Pérez-Torres, D., Fragkou, P. C., Zahar, J.-R., & Koulenti, D. (2021). Nosocomial pneumonia in the era of multidrug-resistance: Updates in diagnosis and management. *Microorganisms*, *9*(3), 534.

Yardim, A., Yeşildağ, K., Çelik, H., Yildiz, O., & Dogan, M. (2021). The relationship between mortality and hospital-acquired infections in patients followed-up with neurological complaints in the third level intensive care unit. *New Trends in Medicine Sciences*, *2*(1), 24–30.

Younes, I., Sellimi, S., Rinaudo, M., Jellouli, K., & Nasri, M. (2014). Influence of acetylation degree and molecular weight of homogeneous chitosans on antibacterial and antifungal activities. *International Journal of Food Microbiology*, *185*, 57–63.

Zhang, L., Sriprakash, K. S., McMillan, D., Gowardman, J. R., Patel, B., & Rickard, C. M. (2010). Microbiological pattern of arterial catheters in the intensive care unit. *BMC Microbiology*, *10*, 266.

Chapter 15

Implication of *Vibrio* biofilms in human and seafood sector

Ekramul Haque[b], Julfequar Hussain[c], Sriram Shankar[c], Soumya Haldar[b], and Shruti Chatterjee[a]

[a]*Applied Phycology and Biotechnology Division, CSIR—Central Salt and Marine Chemicals Institute, Bhavnagar, Gujarat, India,* [b]*Analytical Discipline and Centralized Instrument Facility, CSIR—Central Salt and Marine Chemicals Institute, Bhavnagar, Gujarat, India,* [c]*Department of Biotechnology, School of Life Sciences, Pondicherry University, Kalapet, Puducherry, India*

1. Introduction

Vibrio is a genus of bacteria that consists of more than 70 species; among them, at least 12 are a potential threat to humans (Farmer, Janda, & Birkhead, 2003; Thompson, 2004). They are comma-shaped Gram-negative bacteria that are frequently found to colonize gastrointestinal region of many marine organisms. Often these bacteria are transmitted to humans by the consumption of undercooked seafood and upon contact with seawater through a small cut or wound on their skin (Baker-Austin et al., 2018). Upon infection, they can cause vomiting, nausea, diarrhea and may even be fatal if remain untreated. Bacterial biofilms are associated with several diseases in humans, for example, they are frequently implicated in colitis, periodontitis, urethritis, prostatitis, keratitis osteomyelitis, and conjunctivitis, to name a few (Marsh, 2010). Apart from causing diseases, they are particularly of great concern in hospital settings as they can readily form biofilms on medical devices such as abdominal drains, catheters, contact lenses, prosthetic joints, cardiac pacemaker valves, and intrauterine contraceptive devices (Del Pozo, 2018). A study estimates that, in the United States alone, the expenditure for treatment of device-related biofilm infection is above 11 billion dollars. *Vibrio* biofilms affect the seafood sector as they often cause a variety of diseases in shrimps, fish, oysters, and shellfish and cause a huge loss to the aquaculture industry. Visible symptoms of vibriosis include skin ulcers, black or brownish cuticular lesions, increase in musculature and corneal opacity, appendage tip melanization, tail rot and pale gills, etc. (Chen, 1992; Marudhupandi, Kumar, Prakash, Balamurugan, & Dhayanithi, 2017; Toranzo, Magariños, & Romalde, 2005). Due to the excessive usage of antibiotics, over a period of time, several *Vibrio* spp. have acquired resistance against them. The mechanism by which they evade these antibiotics has been discussed in detail. Several virulence genes that are responsible for the formation of biofilm and their hyperinfectivity phenomenon have been elucidated. Finally, the strategies employed to combat the menace of *Vibrio* biofilm has been elaborated, some of these strategies include the usage of bacteriophages, chemotherapeutic agents, quorum-sensing (QS) disruptors, immune priming and biogenic compounds, etc.

2. Biofilms

Biofilms are nothing but the aggregation of microbes on biotic and abiotic surfaces, these microbes secrete extra polymeric substances such as lipids, proteins, and polysaccharides that make up the extracellular slimy matrix of the biofilm, which consists of carbohydrates, DNA, fibrous proteins, etc. The benefits of biofilms are numerous, they confer adhesion, cellular communication, antibiotic and detergent resistance, protection from dehydration, UV radiation, and shear stress, etc. In case of a biofilm formation inside a host body, they also provide protection from phagocytes and help in evading the host immune response (Santos et al., 2018). Bacterial antibiotic resistance develops and persists primarily through mutations in the bacterial DNA or through horizontal gene transfer (consisting of biocide/antibiotic-resistant genes) processes such as transformation (uptake of free DNA from the environment), conjugation (transfer of genetic material between two bacteria in direct contact), and transduction (foreign DNA transfer mediated by bacteriophage) (Sun, Jeannot, Xiao, & Knapp, 2019).

While physical and chemical treatments have been employed for controlling bacterial biofilms, their effectiveness is restricted due to toxicity they cause (Donato & Zani, 2010). Controlling of biofilms is a twofold challenge, on the one hand, excessive usage of biocides/antibiotics and chemicals in the seafood sector poses a health risk to humans when they are consumed while on the other hand, usage of these biocides in reduced or subinhibitory

Understanding Microbial Biofilms. https://doi.org/10.1016/B978-0-323-99977-9.00038-7

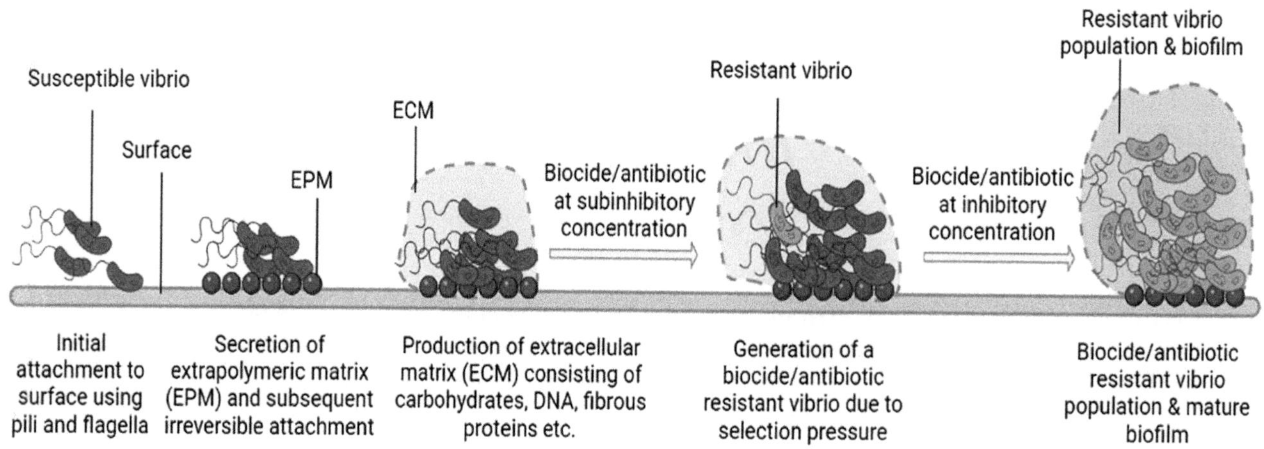

FIG. 1 Schematic diagram of a hypothetical model of *Vibrio* biofilm formation and simultaneous generation of biocide/antibiotic resistance through mutation.

concentrations selectively enhance the biocide resistant properties of the bacteria and biofilms. This is an example of positive selection pressure giving rise to resistant bacteria and requires special emphasis for the development of robust antibiofilm agents required in the seafood sector and food processing industries. A schematic diagram depicting the mechanism of formation of *Vibrio* biofilm and simultaneous generation of biocide/antibiotic resistance is illustrated in Fig. 1.

3. Biofilm formation

Vibrio can transit between a biofilm state and a free-swimming planktonic state. The biofilm life cycle consists of three stages: cell attachment, biofilm maturation, and dispersal (Bridges, Fei, & Bassler, 2020). Under favorable conditions, individual cells undergo a series of extensive metabolic, phenotypic, and physiological changes upon encountering a surface that causes the switch from planktonic to sessile state.

Pili and flagella play important roles in the formation of biofilms through attachment and anchoring to host surface. In *Vibrio* species, flagellar motility has been identified as a significant virulence component. The polar flagellum plays an essential role in *V. cholerae* movement, adhesion, and colonization. The flagellar protein FlgT helps to maintain the flagellar apparatus near the cell pole (Martinez, Jude, Kirn, Skorupski, & Taylor, 2010). The mobility of *V. cholerae* flagella may influence irreversible attachment and microcolony development (Utada et al., 2014). In case of *V. parahaemolyticus*, polar flagellar protein contributes to cell attachment and colonization. In addition, stress circumstances have been reported to influence the flagellar assembly route and adhesion ability of *V. alginolyticus* (Hongprayoon, 1993; Wang, Huang, et al., 2015). Through adherence and biofilm development on host surfaces, functional flagella systems affect the pathogenicity of pathogenic *V. parahaemolyticus* (Merino, Shaw, & Tomás, 2006). *Vibrio* initial attachment, adhesion, biofilm formation, and horizontal gene transfer are all facilitated by the two type-IV pili (IVa and IVb) (Paranjpye & Strom, 2005). Pathogenic *Vibrio* species possess the *msh*A and *pil*A genes that help in colonization. The *pil*A has been implicated in the development of biofilms on abiotic surfaces as well as on living cells. For example, *pil*A gene is expressed when *V. vulnificus* forms biofilms on epithelial cells and also when *V. cholerae* colonizes the gastrointestinal system in humans (Hang et al., 2003; Paranjpye & Strom, 2005). In addition to pili and flagella, transcription factors and membrane proteins also play significant role in the development of biofilms, for instance, the transcriptional activator. ToxR regulates the outer membrane protein of *V. cholerae*, which promotes colonization by regulating virulence factors. It has sticky characteristics that attach to laminin, collagen, and fibronectin and aid in their pathogenesis (Sperandio, Giron, Silveira, & Kaper, 1995). The membrane protein VpadF is a key adhesion factor for *V. parahaemolyticus* and controls both the attachment of the pathogen to the epithelial cells and their dissemination in the afflicted tissues (Goo et al., 2006).

4. Genes involved in biofilm formation

In a study, it was found that the transcript abundance of genes associated with the formation of biofilm matrix such as the major biofilm regulators, *Vibrio* polysaccharide (*vps*) genes, namely *vps*R and *vps*T, was greater in biofilm-grown cells, whereas genes involved in flagellar mobility had a lower transcript abundance. Various transcriptional activators are reported as VpsR, VpsT, AphA, and repressors are HapR, H-NS, which regulate the biofilm formation (Conner, Teschler, Jones, & Yildiz, 2016). VpsR mainly regulates biofilm formation by activating Vps genes, matrix protein genes, other genes encoding part of the type II secretion system needed for matrix protein secretion (Conner et al., 2016). VpsR and VpsT are known to bind c-di-GMP in vitro, and the level of c-di-GMP regulates VpsR activation of VpsT and AphA, a positive upstream regulator of virulence factors (Zamorano-Sánchez, Fong, Kilic, Erill, & Yildiz, 2015). VpsT directly senses c-di-GMP and inversely regulates ECM production and motility (Krasteva et al., 2010). VpsT contains an N-terminal receiver (REC) domain and a C-terminal helix-turn-helix (HTH) domain, which mediates DNA binding (Krasteva et al., 2010). The c-di-GMP binding motif in VpsT consists of four conserved residue W[F/L/M][T/S]R sequence, and VpsT undergoes a change in oligomerization upon binding of c-di-GMP (Krasteva et al., 2010).

In addition, a number of virulence-related genes, including the important virulence regulators such as *tox*R, *tox*T, *tox*S, *tcp*A, *tcp*H, *tcp*P, and *ctx*A, as well as the essential virulence factors controlled by them, had higher message abundance in biofilm-grown cells. The major virulence genes that are differentially expressed in *V. cholerae* during biofilm formation have been listed in Table 1. Toxin-coregulated pilus (TCP) is a type IV pilus that is essential for intestinal colonization, and cholera toxin (CTX) is a secreted enterotoxin responsible for cholera's watery diarrhea (Dutta, Panse, & Kulkarni, 1959; Taylor, Miller, Furlong, & Mekalanos, 1987). Moreover, TCP is also a receptor for the CTXΦ, a lysogenic bacteriophage that carries the CTX gene cluster and may infect nonpathogenic *Vibrio* species and impart virulence (Waldor & Mekalanos, 1996). Furthermore, single-strain colonization studies using biofilm-grown wild-type *V. cholerae* or biofilm-grown Δ*tcp*A. *V. cholerae* revealed a mammoth decrease in intestinal colonization, with Δ*tcp*A abundance in the gut being reduced by 6000-fold as compared with the wild type. Thus, it is clear that *tcp*A is one of the major virulence genes responsible for the biofilm hyperinfectivity phenotype in *Vibrio*.

TABLE 1 Virulence gene involved in biofilm formation.

Differentially expressed genes during biofilm formation in *V. cholerae*	
Upregulated genes	**Downregulated genes**
aldA, tagD, tagE, tcpA, tcpB, tcpI, tcpH, tcpP, tcpQ, tcpT, tcpF, tcpJ, toxT, toxR, acfA, acfB, acfC, ctxA, ctxB, vpsA, vpsB, vpsC, vpsD, vpsL, vpsM, vpsR, vpsT, vpsU, rbmA, rbmC, rbmF, bap1, flaB, flaC, flaD, motX, fliE, flrB, flrC	*mshC, mshD, mshE, mshG, mshM, mshN, vpsP, cheD, cheR, cheV-4, cheW, cheW-1, fihB, fliP, fliQ, fliR, flrA, flaL, flgB, flgC, flgD, flgG, flgH, flgI, flgJ, flgK, flgL, flgO, flgP*

5. Signaling cascade involved in biofilm formation

Cyclic-di-GMP (c-di-GMP) has been identified as intracellular signaling molecule, which coordinates the transition between motility and sessility and vice versa. It is synthesized from two molecules of GTP by enzymes called diguanylate cyclases (DGCs), which have GGDEF domains, and is degraded by phosphodiesterases containing EAL or HD-GYP domains into pGpG and/or GMP (Conner, Zamorano-Sánchez, Park, Sondermann, & Yildiz, 2017). Usually, high c-di-GMP concentration promotes expression of extracellular matrix and biofilm formation while a low c-di-GMP concentration leads to altered motility upon changes in flagellar or pili function or production (Liu, Beyhan, Lim, Linington, & Yildiz, 2010). DCG and PDE genes have been identified, such as *cdg*G, *cdg*D, *cdg*H, *roc*S, *cdg*J, *cdg*K, *cdg*L, which are involved in regulating motility. *Cdg*J is predicted to be a PDE while *cdg*K and *cdg*L are predicted to be DGCs (Liu et al., 2010). The *pil*Z domain containing c-di-GMP receptor proteins functions in controlling motility. Most receptors use either Arg or Asp/Glu amino acid residues to bring c-di-GMP monomers, dimers, or tetramers with stacked guanine bases. The *V. cholerae* genome codes for five proteins, which are thought to have *pil*Z domains, i.e., PlzA–PlzE. Major components of *V. cholerae* biofilm are *Vibrio* polysaccharide (VPSs) encoded by *vps* genes, which are organized into *vps*I and *vps*II coding regions separated by an intergenic region. Genes located in the *vps* intergenic region, namely r*mb*B–*rmb*F (rugosity and biofilm structure modulators B–F), have been found to encode proteins involved in biofilm matrix production, structure, and stability. C-di-GMP is known to regulate both initial attachment and matrix production.

Biofilm formation is stimulated by surface attachment, which is regulated by the MshA pilus assembled by the polymerization of MshA subunits (Conner et al., 2017). MshE, an AAA+ATPase, forms a hexameric ring and triggers the assembly of MshA pilus upon binding c-di-GMP. The process involves MSHA biogenesis proteins MshG, MshL, and the major pilin protein MshA. A schematic diagram represents the assembly of the MshA pilus.

Various transcriptional activators are VpsR, VpsT, AphA, and repressors are HapR, H-NS (Conner et al., 2016). The *vps*R mainly regulates biofilm formation by activating Vps genes, matrix protein genes, genes encoding part of the type II secretion system needed for matrix protein secretion (Conner et al., 2016). VpsR and VpsT are known to bind c-di-GMP in vitro, and the level of c-di-GMP regulates VpsR activation of VpsT and AphA (Zamorano-Sánchez et al., 2015). VpsT directly senses c-di-GMP and inversely regulates ECM production and motility (Krasteva et al., 2010). VpsT contains an N-terminal receiver (REC) domain and a C-terminal HTH domain, which mediates DNA binding (Krasteva et al., 2010). The c-di-GMP binding motif in VpsT consists of four conserved residue W[F/L/M][T/S]R sequence, and VpsT undergoes a change in oligomerization upon binding of c-di-GMP (Krasteva et al., 2010).

6. Factors affecting *Vibrio* biofilms

The main factors that affect *Vibrio* biofilm were recognized as temperature regime, hydrodynamic conditions, and water quality. Other factors are likely substratum, medium, physical and chemical properties of surfaces. The substrate physicochemical properties such as stiffness, mechanical stability, elasticity, and diffusion too regulate physiology and biofilm production of bacteria (Renner & Weibel, 2011). C-di-GMP levels can be regulated by environmental parameters through sensory domains present in *V. cholerae* DGCs and PDEs. *V. cholerae* experiences temperature fluctuations within the aquatic environment or during transition from human host and aquatic bodies. It was seen that reduced temperature during growth resulted in increased biofilm formation in *V. cholerae* mediated by a surge in c-di-GMP levels. Apart from this extracellular polymeric material production, surface charges, swimming speed are also important for biofilm formation (Characklis, 1990; Marshall, Stout, & Mitchell, 1971).

7. Resistance against antibiotics and drugs

Bacteria usually have a variety of ways for displaying resistance to a broad range of antibiotics. *Vibrio* species displays resistance against drugs and antibiotic by employing chromosomal mutations, efflux pumps, and horizontal gene transfer methods. It exports drugs by pumping the drugs outward via efflux pumps. As example *V. cholerae* has multidrug efflux pumps that enable it to export a wide variety of antibiotics, detergents, and chemicals, etc., thereby conferring resistance against biocides and antimicrobials (Paulsen, Brown, & Skurray, 1996). *V. cholerae* possess multidrug resistance efflux pumps such as the NorM, MATE family, and RND efflux systems, which protect them from quinolones, aminoglycosides, tetracycline, doxorubicin, and Triton X-100, etc. (Begum et al., 2005; Bina, Provenzano, Nguyen, & Bina, 2008; Kitaoka, Miyata, Unterweger, & Pukatzki, 2011; Singh, Haldar, Mandal, & Kundu, 2006). Apart from the efflux pumps, *V. cholerae* has been reported to acquire increased quinolone resistance (such as norfloxacin and ciprofloxacin) through mutation in its *gyr*A and *par*C genes due to positive selection pressure over a period of 6 years while quinolones were being used for the treatment of cholera (Baranwal, Dey, Ramamurthy, Nair, & Kundu, 2002; Kim et al., 2010). Biofilms also provide protection against antimicrobials such as chlorine. It has been reported that in biofilms with thickness around 150–200 mm, chlorine at a concentration of 25 mg L^{-1} is unable to penetrate beyond 100 mm.

8. Vibriosis in human and aquaculture

Vibrios are naturally found in estuarine and marine environments (Baker-Austin, Trinanes, Gonzalez-Escalona, & Martinez-Urtaza, 2017). The *Vibrio* genomes are divided into two chromosomes, which are driven by recombination and horizontal gene transfer (Baker-Austin et al., 2018). *Vibrios* can contaminate seafood and are responsible for dramatic increase of seafood-borne infections (Bonnin-Jusserand et al., 2019). *Vibrio* spp. infection is usually caused by consumption of contaminated water or raw, uncooked contaminated sea food and produces symptoms such as fever, bloody diarrhea (Altekruse et al., 2000). Diseases caused by *Vibrio* spp. can be divided into two groups: cholera and noncholera infections. Cholera caused by the bacteria O1 and O139 *V. cholerae* is characterized by watery diarrhea and rapid dehydration, which can lead to hypotonic shock and death within 12 h of first symptom if left untreated (Conner et al., 2016). Since the most frequent cause of death is dehydration, fluid replacement therapy is fundamental in cholera treatment programs. In mild to moderate dehydration, Oral Rehydration Therapy (ORS) containing water, salt, and glucose can be administered but in case of severe dehydration when ORS cannot be ingested by mouth, treatment consists of passing ORS fluid down an inserted nasogastric tube or by repleting fluids and electrolytes intravenously (Chowdhury, Khan, Faruque, & Ryan, 2010). Noncholera infections are caused by non-O1 and non-O139 *V. cholerae*, *V. vulnificus*, and *V. parahaemolyticus*. *V. vulnificus* infection varies from gastrointestinal illness including vomiting, fever, diarrhea to septicemia. *V. parahaemolyticus* infections can lead to gastroenteritis, symptoms including, fever, nausea, chills, headache, abdominal cramps, and diarrhea. The major *Vibrio* spp. that are pathogenic to humans along with the diseases caused by them have been listed in Table 2.

Vibriosis is ubiquitous in aquaculture system and is responsible for mortality of cultured aquaculture organisms around the globe (Chen, Liu, & Lee, 2000). Several species including *V. vulnificus*, *V. alginolyticus*, *V. harveyi*, *V. splendidus*, *V. parahaemolyticus*, *V. campbellii*, and

TABLE 2 *Vibrio* **sp. causing infection to humans.**

Pathogen	Disease caused in humans	Reference
V. cholerae	Cholera, diarrhea, and dysentery	Glenn Morris (2003)
V. parahaemolyticus	Acute gastroenteritis, diarrhea	Cai, Li, Thompson, Li, and Han (2007), DePaola, Nordstrom, Bowers, Wells, and Cook (2003)
V. vulnificus	Wound infections, Septicemia (sepsis)	Drake, DePaola, and Jaykus (2007), Li et al. (2006)
V. alginolyticus	Ear infection, soft tissue, and wound infections	(Lee, Yu, Yang, Liu, & Chen, 1996)
V. fluvialis	Gastroenteritis, wound infections, hemorrhagic cellulitis, and cerebritis	Tall et al. (2003)
V. furnissii	Infantile diarrhea (Child)	Sung, Hsu, Chen, Ting, and Chao (2001)
V. harveyi	Wound infections	Baker-Austin et al. (2018)

V. anguillarum contributes majorly toward the morbidity and mortality of aquacultured organisms, especially associated with shrimp diseases (Haldar et al., 2011; Haldar, Maharajan, et al., 2009; Haldar, Neogi, et al., 2009). For instance, *Vibrio harveyi* causes luminescent vibriosis and is considered the most important causative agent for the mass mortality in *Penaeus monodon* (Aguirre-Guzmán et al., 2010; Austin, Pride, & Rhodie, 2003; Caipang & Aguana, 2010). Also, *V. vulnificus*, *V. salmonicida*, and *V. anguillarum* cause severe morbidity and mortality in many fish species. Chemotactic behavior of bacteria helps in penetrating the surface tissue of fishes, thereby causing infection leading to release of extracellular products including protease, hemolysin, and deployment of iron sequestration. Edible fish, *Sparus aurata* (moribund seabream), isolated from a hatchery situated in Malta has been observed to have blood clots and mucus secretion due to hemolysin activity of *V. harveyi* infection (Haldar, Maharajan, et al., 2009). Other common symptoms of disease in fishes caused by *Vibrio* spp. include anemia, liquid in the air bladder, intestinal necrosis, petechial hemorrhages in the muscle wall, and ascitic fluid. Even shrimps get infected by *Vibrio* spp., and their route of infection is through feed, hepatopancreas, and gill and colonizes the host tissues by breaching epithelial cells (Martin-Laurent et al., 2001). Although it is understood that the main cause of infection in aquaculture is due to *Vibrio* spp., a simple and reliable animal model for investigating the virulence potential and mechanism of causing infection is lacking and is at developmental stage. In recent years, *Artemia nauplii* is being used as a model organism as it can be cultured in gnotobiotic condition (Haldar et al., 2011). Bacterial colonization and interactions with host are difficult to study in a complex organism, and several techniques such as usage of cell lines observation under scanning electron microscope have been employed (Lavilla-Pitogo, Baticados, Cruz-Lacierda, & Leobert, 1990; Olsson, Westerdahl, Conway, & Kjelleberg, 1992). However, fixation step and sample preparations do not guarantee the confirmation. Thus, the colonization potential in vivo is studied by staining the pathogenic *Vibrio* spp. with fluorescent dye that is considered as an appropriate method. For instance, in a recent study, *V. campbellii* isolates previously considered as nonpathogenic have been established to possess pathogenic potential (Haldar et al., 2011). The major *Vibrio* spp. that are pathogenic to marine organisms along with the antibiofilm agents that can be used to control their pathogenicity have been elucidated in Table 3.

9. Seafood sector and economic losses

Aquatic animals, such as mollusks, crustaceans, and fishes, are associated with distinct species of bacteria belonging to the genus *Vibrio* (Beaz-Hidalgo, Balboa, Romalde, & Figueras, 2010; Cuéllar-Anjel, Corteel, Galli, Alday-Sanz, & Hasson, 2014; Mohamad et al., 2019). A few of them cause infection leading to vibriosis. The term includes both local and general septicemic infections (Egidius, 1987). Crustaceans are infected particularly during larval production stage and can have devastating effects on their population (Goulden, Hall, Bourne, Pereg, & Høj, 2012).

As all crustaceans are susceptible to infection caused by *Vibrio* spp., substantial production losses are common due to *Vibrio* spp. outbreaks and other diseases such as acute hepatopancreatic necrosis disease (AHPND) are a huge concern for shrimp farm operators. For instance, an estimate of about 44 million dollar has been reported between 2010 and 2016 in Thailand, China, Vietnam, Mexico, and Malaysia. Other vibriosis diseases include shell disease syndrome, tail necrosis, red body disease, limp lobster disease, and summer syndrome (Cawthorn, 2011; Cuéllar-Anjel et al., 2014; Jayasree et al., 2006; Mancuso et al., 2010; Radhakrishnan & Kizhakudan, 2019; Soto-Rodríguez, Armenta, & Gomez-Gil, 2006; Sullivan & Neigel, 2018).

TABLE 3 *Vibrio* **sp. causing infection to marine species along with their antibiofilm agents.**

Pathogen	Species affected (marine)	Disease caused	Antibiofilm agent	Reference
V. cholerae	Ayu (*Plecoglossus altivelis*), Indian prawn (*Penaeus indicus*), Tiger prawn (*Penaeus monodon*), Blue crab (*Callinectes sapidus*),Oysters, Shell fish	Bacteremia	Auromomycin, Chromophore, Benzimidazole, Benzo oxazines, Diguanylate, Cyclase Enzymes	Yamanoi, Muroga, and Takahashi (1980), Haldar, Chatterjee, Asakura, Vijayakumaran, and Yamasaki, (2007), Peach (2013), Sambanthamoorthy et al. (2011), Warner, Cheng, Yildiz, and Linington (2015)
V. parahaemolyticus	Whiteleg shrimp (*Penaeus vannamei*), American lobster (*Homarus americanus*), Milk fish (*Chanos chanos*), Sea snail (*Haliotis supertexta*), Oyster (*Crassostrea rhizophorae*)	Septic Hepatopancreatic Necrosis Disease, Lobster hatchery vibriosis, shell disease, Zoea II Syndrome	Coenzyme Q0 (CoQ0), *Bacillus pumilus*, Copper Nanoparticles (CuNPs), Bacteriophages	Cao, An, Zheng, and He (2015), Baker-Austin et al. (2018), Jayasree, Janakiram, and Madhavi (2006), Sousa, Vieira, Menezes, Reis, and Hofer (2004), Zheng et al. (2020), Nithya, Devi, and Karutha Pandian (2011), Chari, Felix, Davoodbasha, Ali, and Nooruddin (2017), Jun et al. (2014)
V. vulnificus	European eel (*Anguilla anguilla*), Japanese eel (*Anguilla japonica*), Tiger prawn (*Penaeus monodon*), Ovate pompano (*Trachinotus ovatus*)	Septic Hepatopancreatic Necrosis Disease, shell disease, bacteremia	Curcumin, *Synechococcus* sp.	Tison, Nishibuchi, Greenwood, and Seidler (1982), Høi, Dalsgaard, DePaola, Siebeling, and Dalsgaard (1998), Baker-Austin et al. (2018), Jayasree et al. (2006), Na et al. (2011), Santhakumari et al. (2016)
V. alginolyticus	Cobia (*Rachycentron canadum*) Silver Pompano (*Trachinotus blochii*) Indian Pompano (*Trachinotus mookalee*) Orange spotted Grouper (*Epinephelus coioides*), Seabream (*Sparus aurata*), Black sea bream (*Spondyliosoma cantharus*), Long-legged spiny lobster (*Panulirus longipes*),Ornate rock lobster (*Panulirus ornatus*), Mud spiny lobster (*Panulirus polyphagus*)	Red body disease, Zoea II Syndrome, Bacteremia	*Bacillus pumilus*, pVa-21 (Bacteriophage), Copper Nanoparticles (CuNPs)	Baker-Austin et al. (2018), Radhakrishnan and Kizhakudan (2019), Nithya et al. (2011)), Kim et al. (2019), Chari et al. (2017)
V. fluvialis	American lobster (*Homarus americanus*), Shell fish	Limp lobster disease	Polymyxin B-capped silver nanoparticles	Tall et al. (2003), Cawthorn (2011)
V. harveyi	Gazimi crab (*Portunus pelagicus*), Flower crab (*Portunus trituberculatus*)	Crab hatchery vibriosis, Zoea II Syndrome	Citrus limonoids, 4-fluoro-DPD, Isolimonic acid, *Piper betle*, Polymethoxyflavones from *Citrus miaray*	Talpur et al. (2011), Vikram, Jesudhasan, Jayaprakasha, Pillai, and Patil (2011), Kadirvel et al. (2014), Rabin et al. (2015), Srinivasan, Santhakumari, and Ravi (2017)

In general, global aquaculture production is severely affected by vibriosis and results in major economic losses and poses a risk to global food security (Cuéllar-Anjel et al., 2014; Talpur et al., 2011).

10. Control of *Vibrio* biofilm

10.1 Bacteriophages

The infections caused by *Vibrio* spp. significantly affect humans in various forms that include decrease in crop production, clinical infection, and aquaculture. Management and preventive measures for controlling *Vibrio* infections depend greatly on antibiotics as a result of which the emergence of multidrug-resistant pathogens in environment is rapid and posing a severe threat to humankind. Thus, the need to find an alternative to antibiotics for treating bacterial infections is of great importance. One such alternative makes use of viruses that infect bacteria (Bacteriophages) in combating multidrug resistance.

It is an interesting fact that bacteriophages have been used worldwide since 1929, even before the discovery of broad spectrum antibiotics and their application as therapeutic agent against microbial infections (Tan, Chan, & Lee, 2014).*V. cholerae* was the first bacterium to be tested against bacteriophage wherein the in vitro activity of phage was reported to be higher than in vivo (Adams, 1959). There was a great reduction in the application of bacteriophages after the advent of antibiotics (Tan et al., 2014); nevertheless, phage research and development were actively carried out in Poland and former Soviet Union (Sulakvelidze, Alavidze, & Morris Jr, 2001). In recent years, due to the emergence of multidrug-resistant bacteria, bacteriophage therapy has been rejuvenated.

"Luminous vibriosis disease" caused by *V. campbellii*, *V. harveyi*, and *V. parahemolyticus* results in 50%–100% of *Vibrio* infection in human and mortality rate among shrimps (Chatterjee & Haldar, 2012; Letchumanan, Chan, & Lee, 2014; Tan, Chan, Lee, & Goh, 2016; Wang, Chen, et al., 2015). The control of luminous vibriosis disease has been proven to be effective using bacteriophages, implying the phages' potential as a biocontrol agent (Jassim & Limoges, 2014; Karunasagar, Shivu, Girisha, Krohne, & Karunasagar, 2007; Vinod et al., 2006). Few studies have highlighted the importance of lytic cholera phage as a prophylactic agent (Marčuk et al., 1971; Monsur et al., 1970). Also, a study using rabbit model demonstrated the efficacy of five lytic vibriophage cocktail against *V. cholerae* O1El Tor biotype Ogawa MAK 757 (ATCC 51352). Shedding of bacteria was reduced significantly ($P < .01$) after oral administration of vibriophage cocktail followed by oral bacterial administration suggesting a suitable alternative to antibiotic treatment against *Vibrio* infection (Jaiswal, Koley, Ghosh, Palit, & Sarkar, 2013). Moreover, US Food and Drug Administration (FDA) has approved the commercial usage of phage cocktail ListShield against *Listeria monocytogenes* that confirms the applicability of phages for human application and consumption (Housby & Mann, 2009).

10.2 Chemotherapeutics agents

Chemotherapeutics is employed in aquaculture for preventing and controlling vibriosis, but their usage is banned and restricted in Europe and North America due their negative effects, which include environmental contamination, bacterial resistance, and residues in animal tissue (Sutili & Gressler, 2021). Although, few countries in Asia still utilize antimicrobials as prophylactic agents. For instance, oxytetracycline ($10\,mg\,L^{-1}$), chloramphenicol ($8–10\,mg\,L^{-1}$), nifurpirinol ($1\,mg\,L^{-1}$), ciprofloxacin ($4–8\,mg\,L^{-1}$) (i.e., nitrofuran, $2–5\,mg\,L^{-1}$) are used against *P. monodon vibriosis* in Bangladesh (Aftabuddin, Kader, Kamal, & Zafar, 2009). Moreover, for the control of shell disease in larvae production, antimicrobials, namely malachite green ($0.0075\,mg\,L^{-1}$), and formalin ($25\,mg\,L^{-1}$), were used. Also, antimicrobials such as nitrofurans ($10\,mg\,L^{-1}$) and erythromycin were used for controlling shrimp hatchery vibriosis and oxolinic acid ($2\,g\,kg^{-1}$ feed) for combating general shrimp vibriosis. In Singapore and Malaysia, benzalkonium chloride ($1–1.25\,mg\,L^{-1}$) and sulfonamides ($20\,mg\,L^{-1}$) are predominantly used for controlling vibriosis in shrimp hatcheries, and calcium hypochlorite is employed for prophylactic treatment ($0.08\,mg\,L^{-1}$) or disinfection ($20–30\,mg\,L^{-1}$) in shrimp ponds (Shariff, Bond, & Johnson, 2000). Antimicrobials such as rifampicin ($0.1\,mg\,L^{-1}$), chloramphenicol ($1\,mg\,L^{-1}$), furazolidone ($0.5–1\,mg\,L^{-1}$), and oxytetracycline ($1–2\,mg\,L^{-1}$) are widely used in shrimp hatcheries in the Philippines for combating luminous vibriosis (Cruz-Lacierda, De la Peña, & Lumanlan-Mayo, 2000). Kim, Seid, McElhiney, and Tlusty (2018) documented the role of formalin and malachite green in treating lobster shell disease. *Homarus americanus* infected with shell disease was treated with malachite green ($5\,mg\,L^{-1}$) or formalin ($100\,mg\,L^{-1}$) for 1 h, two times a week, and was observed for the formation of lesions. Interestingly, both chemicals were effective in reducing the formation of new lesions and the expansion of preexisting ones. However, Bourne et al. (2004) reported that antibiotic treatment and formalin has no effect in reducing *Vibrio* spp. on the surface of *Panulirus ornatus* females and eggs when oxytetracycline ($8\,mg\,L^{-1}$), streptomycin ($20\,mg\,L^{-1}$), formalin ($25\,mg\,L^{-1}$), and erythromycin ($9\,mg\,L^{-1}$) were administered.

Fu et al., 2016 reported that combinatorial use of trimethoprim and sulfamethoxazole ($50\,mg\,kg^{-1}$) orally was effective in controlling *V. parahemolyticus* infection in

Portunus trituberculatus. Few reports have highlighted the detrimental effects of chemotherapeutics for both environment and animal health. For instance, Austin (2017) demonstrated that malachite green is toxic, teratogenic, mutagenic, and carcinogenic. Also, the Food and Agriculture Organization (FAO) banned the addition of malachite green in aquafeed and restricted their usage only to ornamental animals (Kim et al., 2018). Similarly, the European Commission (Regulation (EU) 37/2010, EU 2010) added the chemical to the prohibited chemical list. Few other chemotherapeutics have been banned by EU legislation, which include nitrofurans (as furazolidone) and chloramphenicol, whereas few other drugs have been accepted at a permissible level, which include erythromycin (200 $\mu g\,kg^{-1}$, muscle), benzalkonium chloride (0.05%, excipient use only), oxytetracycline (100 $\mu g\,kg^{-1}$, muscle), oxolinic acid (100 $\mu g\,kg^{-1}$, muscle), sulfonamides (100 $\mu g\,kg^{-1}$, muscle), and ciprofloxacin (100 $\mu g\,kg^{-1}$, muscle).

10.3 Quorum sensing (QS) disruptors

Bacterial cells communicate with each other and respond by secreting chemicals called autoinducers that increase as a function of cell density. Individual cells communicate with neighbor cells before initiating a cooperative behavior such as biofilm formation and virulence (Defoirdt, 2018). Thus, disrupting quorum sensing (QS) mechanism is potentially considered as a preventive and therapeutic measure for controlling infectious diseases. For instance, QS disrupting compounds or quorum quenchers/quorum interfering drugs are widely accepted to be antipathogenic but not antibacterial, also known as antivirulent therapy. These compounds (either natural or synthetic) attenuate virulence of pathogens including *Vibrio* but not essentially interfering with the microbial growth (Defoirdt, Boon, Sorgeloos, Verstraete, & Bossier, 2007). For example, a study reported the efficacy of copper sulfate (40 $mg\,L^{-1}$) as a potent QS disrupting compound that inhibits QS bioluminescence regulation by interfering with luxD genes in *V. harveyi* without affecting the bacterial growth (Nakayama, Nomura, & Matsumura, 2007). Even live organisms such as *Bacillus* sp. NFMC-C have been reported to disrupt QS mechanism of *V. campbellii* by decreasing bioluminescence and degrading autoinducer acylhomoserine lactone molecules (Pande et al., 2015). The survival of larvae of giant river prawn improved significantly, from 42% in control to 67% in treated, after infecting them with *V. campbellii* (Pande et al., 2015). However, quorum quenching technology still requires intensified research and development focusing on their interactions with nonpathogens, host, aquatic pathogens, and environmental impact. Thus, further research and studies will possibly highlight their impact and applicability in commercial aquatic farms (Zhao, Chen, Quan, & Fan, 2015).

10.4 Immune priming

Innate memory also called as immune priming confers memory-based responses that allow specific and effective immune response similar to the mechanism of vaccine and are commonly observed in decapods. For instance, crustaceans including crayfish and shrimp possess specific memory that can last as long as 30 days (Chang, Kumar, Ng, & Wang, 2018). The administration is either through oral or by immersion, and the dosage is species-dependent. Although, injection through cephalothorax or intramuscular is another possible mode of administration, but is not preferred due to its technical and economical constraints.

Inactivated or subunit vaccines of *Vibrio* are available, and the animals are generally vaccinated during early stages of development especially at larval or postlarval stages (Amatul-Samahah et al., 2020). The commercially available vaccines include AquaVac Vibromax (Schering-Plow Animal Health Corporation, United Kingdom), an inactivated vaccine used as an enrichment in live feed for *Artemia nauplii* (Wongtavatchai, López-Dóriga, & Francis, 2010), and Vibrogen-S (AquaHealth, Asia Ltd), an inactivated vaccine promoted as injection-based vaccine effective against *V. parahemolyticus* infection in shrimp hatcheries (Tonguthai, 2000). Another study reported the efficacy of immune priming against *V. parahemolyticus* infection in *Scylla paramamosain* (mud crab). In this study, the animals that received formalin inactivated *V. parahemolyticus* (2×105 $cell\,g^{-1}$ dosage, hemocoel injection) had a significant survival rate (75.5%) after infection with live *V. parahemolyticus* (Yang et al., 2020). Although qualitative results on efficacy have been reported, deciphering the molecular details of immune priming by elucidating the mechanism of vaccine interaction with host immune system for proper understanding of immune priming requires extensive research. For instance, few questions regarding the unspecific protection as a result of cross immune reactions and ineffectiveness of immune priming against all potential pathogens have been raised (Chang et al., 2018; Pope et al., 2011). Although immune priming is in developmental stage, it is considered as a promising prophylactic and therapeutic agent in addressing vibriosis in crustacean farming.

10.5 Biogenic compounds

Biologically active compounds and natural products extracted from animals, plants, and microorganisms have been proven to be effective against vibriosis. They are predominantly secondary metabolites and have been studied for their applicability in pharmaceutical, nutraceuticals, and cosmetics. Many reports have been documented regarding the extraction and application of these biogenic compounds in aquafeeds as well as therapeutic agents. Examples include the application of ethanolic extract from the seeds of *Rhodomyrtus tomentosa* (3.5 or 7% concentration, coated feed pellets) increases the survival percentage of *Penaeus vannamei* against acute hepatopancreatic necrosis disease (AHPND) caused by *V. parahemolyticus*. The ethanolic extract contains resveratrol, hydrolyzable tannins, and piceatannol that have been demonstrated to possess antibacterial activities, which can boost the immune status of shrimp (Dang, Nguyen, Hoang, Lai, & Nguyen, 2019).

Also, ginger extract has been documented to improve the survival rate up to 60% as compared with control groups of *P. vannamei* against *V. parahemolyticus* infection (Soowannayan et al., 2019). Furthermore, the ethanolic extracts obtained from *Sargassum* spp. (150 or 200 mg L^{-1}, immersion treatment) have been reported to boost the immune system of *P. vannamei* and have been considered as a promising candidate for controlling vibriosis (Mulyadi & Iba, 2020). Félix et al. (2020) have reported the application of ethanolic extracted obtained from *Asparagopsis armata* in increasing the survival rate of *P. vannamei* infected with *V. parahemolyticus* by 50%. Although, bioactive agents and natural products have promising application for improving and treating diseases. Their commercialization remains unclear and the molecular aspects of these compounds such as lethal doses, pharmacological activity, synergistic interaction with other drugs, toxicity, and biodegradability are yet to be understood.

10.6 Algal technology

Green water technology (GWT) utilizes microalgae on site for larval rearing and is commonly practiced among shrimp farmers. GWT employs microalgal population that has the ability to potentially control bacterial disease outbreaks by interfering with QS mechanism (Natrah et al., 2011). The marine green algae *Tetraselmis suecica* are commonly given as live larval supplement or feed and can also potentially control *Vibrio* spp. For instance, 30,000 algal cells mL^{-1} is considered as effective dose for larval rearing (Regunathan & Wesley, 2004). Also, supplementation of *Platymonas helgolandica*, green algae to *P. vannamei*, has been documented to increase resistance toward *V. parahemolyticus* at a concentration of 1×10^5 cell mL^{-1}(Ge et al., 2017). Additionally, GWT can act as immunostimulatory, improvement of water quality, inorganic and organic nutrient competition (Becker, 2013; Charoonnart, Purton, & Saksmerprome, 2018).

10.7 Probiotic, prebiotics, and synbiotics

Prebiotics, synbiotics, and probiotics are considered as functional food supplements that are proven to promote healthy microbiota primarily involved in the maintenance of gut homeostasis (Pandey, Naik, & Vakil, 2015). However, future investigations are required to study the impact of microbiota present in hepatopancreas, gills, and surface of the skin. In general, probiotic preparation includes live beneficial microbes that determine the microbiota of the host by competing with pathogens and aid in maintaining homeostasis. Prebiotics are predominantly carbohydrates that are not metabolized by our body but are metabolized by the beneficial microbes present in our gut, thus prebiotics help in promoting probiotic microbes in our body system. Synbiotics comprises both probiotics and prebiotics and has been extensively described by Schrezenmeir and de Vrese (2001) and Lauzon, Dimitroglou, Merrifield, Ringø, and Davies (2014). So far, scientific community has achieved success at the laboratory scale; however, scaling-up for commercialization requires keen analysis and extensive experimentation. For instance, synbiotics, probiotics, and prebiotics have greater applicability in crustacean production.

11. Concluding remarks and future perspectives

The diversified and ubiquitous genus *Vibrio* is nearly impossible to exclude from an aquaculture setting, and majority of them are opportunistic pathogens. They have the ability to transfer their antibiotic resistance gene to the surrounding microbial pathogens. *Vibrio* species are considered to be the main source of food contamination as they are capable of forming biofilm in the various settings in the food and dairy industry. Day by day, the consumption of seafood is increasing across the globe. Therefore, they can pose a huge threat to the human health. Thus, control and management of *Vibrio* spp. population and their infections are highly essential. Prevention plays a key role to avoid onset of vibriosis outbreak and their emergence. Thus, an effective prevention plan must include a comprehensive disease management plan that mandates use of robust stock animals, maintaining high water quality, immune resistance, and nutritional importance. In addition, substantial improvement in the form of immune priming, phage therapy, and QS disruption, validation of efficacy, and biosafety at the farm level is also necessary. To ensure global food security, aquaculture is considered as a sustainable solution; however, vibriosis is a big hurdle in our goal of sustainability. Scientific communities must address this issue with proper understanding on how vibriosis is established in a shrimp or aquaculture system, thereby minimizing their impacts. Vibriosis has been studied and well understood in majority of hosts, but there are certain knowledge gaps in studying crabs and lobsters. A detailed understanding is expected in near future so as to secure food security and further the productivity of the aquaculture industry.

CSIR-CSMCRI communication no. 246/2021.

References

Adams, M. H. (1959). *Bacteriophages* (pp. 450–454). New York, NY: Interscience Publishers. Inc.

Aftabuddin, S., Kader, A., Kamal, A. M., & Zafar, M. (2009). Present status on the use of antibiotics and chemicals in shrimp hatcheries and grow-out ponds and their environmental implications in Bangladesh. *Aquaculture, Aquarium, Conservation & Legislation, 2*(4), 369–379.

Aguirre-Guzmán, G., Sánchez-Martínez, J. G., Pérez-Castañeda, R., Palacios-Monzón, A., Trujillo-Rodríguez, T., & De La Cruz-Hernández, N. I. (2010). Pathogenicity and infection route of *Vibrio parahaemolyticus* in American white shrimp, *Litopenaeus vannamei. Journal of the World Aquaculture Society, 41*(3), 464–470.

Altekruse, S. F., Bishop, R. D., Baldy, L. M., Thompson, S. G., Wilson, S. A., Ray, B. J., et al. (2000). Vibrio gastroenteritis in the US Gulf of Mexico region: The role of raw oysters. *Epidemiology and Infection, 124*(3), 489–495.

Amatul-Samahah, M. A., Omar, W. H. H. W., Ikhsan, N. F. M., Azmai, M. N. A., Zamri-Saad, M., & Ina-Salwany, M. Y. (2020). Vaccination trials against vibriosis in shrimp: A review. *Aquaculture Reports, 18*, 100471.

Austin, B. (2017). Antibiotics and disinfectants. In *Diagnosis and control of diseases of fish and shellfish* (pp. 263–278). John Wiley & Sons Ltd.

Austin, B., Pride, A. C., & Rhodie, G. A. (2003). Association of a bacteriophage with virulence in *Vibrio harveyi. Journal of Fish Diseases, 26* (1), 55–58.

Baker-Austin, C., Oliver, J. D., Alam, M., Ali, A., Waldor, M. K., Qadri, F., et al. (2018). *Vibrio* spp. infections. *Nature Reviews Disease Primers, 4* (1), 1–19.

Baker-Austin, C., Trinanes, J., Gonzalez-Escalona, N., & Martinez-Urtaza, J. (2017). Non-cholera vibrios: The microbial barometer of climate change. *Trends in Microbiology, 25*(1), 76–84.

Baranwal, S., Dey, K., Ramamurthy, T., Nair, G. B., & Kundu, M. (2002). Role of active efflux in association with target gene mutations in fluoroquinolone resistance in clinical isolates of *Vibrio cholerae. Antimicrobial Agents and Chemotherapy, 46*(8), 2676–2678.

Beaz-Hidalgo, R., Balboa, S., Romalde, J. L., & Figueras, M. J. (2010). Diversity and pathogenecity of *Vibrio* species in cultured bivalve molluscs. *Environmental Microbiology Reports, 2*(1), 34–43.

Becker, E. W. (2013). Microalgae for aquaculture: Nutritional aspects. In *Handbook of microalgal culture: Applied phycology and biotechnology* (pp. 671–691). John Wiley & Sons Ltd.

Begum, A., Rahman, M. M., Ogawa, W., Mizushima, T., Kuroda, T., & Tsuchiya, T. (2005). Gene cloning and characterization of four MATE family multidrug efflux pumps from *Vibrio cholerae* non-O1. *Microbiology and Immunology, 49*(11), 949–957.

Bina, X. R., Provenzano, D., Nguyen, N., & Bina, J. E. (2008). *Vibrio cholerae* RND family efflux systems are required for antimicrobial resistance, optimal virulence factor production, and colonization of the infant mouse small intestine. *Infection and Immunity, 76*(8), 3595–3605.

Bonnin-Jusserand, M., Copin, S., Le Bris, C., Brauge, T., Gay, M., Brisabois, A., et al. (2019). *Vibrio* species involved in seafood-borne outbreaks (*Vibrio cholerae, V. parahaemolyticus* and *V. vulnificus*): Review of microbiological versus recent molecular detection methods in seafood products. *Critical Reviews in Food Science and Nutrition, 59*(4), 597–610.

Bourne, D. G., Young, N., Webster, N., Payne, M., Salmon, M., Demel, S., et al. (2004). Microbial community dynamics in a larval aquaculture system of the tropical rock lobster, *Panulirus ornatus. Aquaculture, 242*(1–4), 31–51.

Bridges, A. A., Fei, C., & Bassler, B. L. (2020). Identification of signaling pathways, matrix-digestion enzymes, and motility components controlling *Vibrio cholerae* biofilm dispersal. *Proceedings of the National Academy of Sciences, 117*(51), 32639–32647.

Cai, J., Li, J., Thompson, K. D., Li, C., & Han, H. (2007). Isolation and characterization of pathogenic Vibrio parahaemolyticus from diseased post-larvae of abalone Haliotis diversicolor supertexta. *Journal of Basic Microbiology, 47*(1), 84–86.

Caipang, C. M. A., & Aguana, M. P. N. (2010). Rapid diagnosis of vibriosis and white spot syndrome (WSS) in the culture of shrimp, *Penaeus monodon* in Philippines. *Veterinary Research Communications, 34* (7), 597–605.

Cao, H., An, J., Zheng, W., & He, S. (2015). *Vibrio cholerae* pathogen from the freshwater-cultured whiteleg shrimp *Penaeus vannamei* and control with Bdellovibrio bacteriovorus. *Journal of Invertebrate Pathology, 130*, 13–20.

Cawthorn, R. J. (2011). Diseases of American lobsters (*Homarus americanus*): A review. *Journal of Invertebrate Pathology, 106*(1), 71–78.

Chang, Y. H., Kumar, R., Ng, T. H., & Wang, H. C. (2018). What vaccination studies tell us about immunological memory within the innate immune system of cultured shrimp and crayfish. *Developmental & Comparative Immunology, 80*, 53–66.

Characklis, W. G. (1990). Biofilm processes. In W. G. Characklis, & K. C. Marshall (Eds.), *Biofilms* (pp. 195–231). New York, NY: John Wiley& Sons.

Chari, N., Felix, L., Davoodbasha, M., Ali, A. S., & Nooruddin, T. (2017). In vitro and in vivo antibiofilm effect of copper nanoparticles against aquaculture pathogens. *Biocatalysis and Agricultural Biotechnology, 10*, 336–341.

Charoonnart, P., Purton, S., & Saksmerprome, V. (2018). Applications of microalgal biotechnology for disease control in aquaculture. *Biology, 7* (2), 24.

Chatterjee, S., & Haldar, S. (2012). Vibrio related diseases in aquaculture and development of rapid and accurate identification methods. *Journal of Marine Science Research and Development S, 1*(1), 1–7.

Chen, D. (1992). An overview of the disease situation, diagnostic techniques, treatments and preventatives used on shrimp farms in China. In *Diseases of cultured penaeid shrimp in Asia and the Unites States* (pp. 47–55). Hawaii: The Oceanic Institute.

Chen, F. R., Liu, P. C., & Lee, K. K. (2000). Lethal attribute of serine protease secreted by *Vibrio alginolyticus* strains in kuruma prawn *Penaeus japonicus. Zeitschrift für Naturforschung. Section C, 55*(1–2), 94–99.

Chowdhury, F., Khan, A. I., Faruque, A. S. G., & Ryan, E. T. (2010). Severe, acute watery diarrhea in an adult. *PLoS Neglected Tropical Diseases, 4*(11), e898.

Conner, J. G., Teschler, J. K., Jones, C. J., & Yildiz, F. H. (2016). Staying alive: *Vibrio cholerae*'s cycle of environmental survival, transmission, and dissemination. In *Virulence mechanisms of bacterial pathogens* (pp. 593–633). Microbiology Spectrum.

Conner, J. G., Zamorano-Sánchez, D., Park, J. H., Sondermann, H., & Yildiz, F. H. (2017). The ins and outs of cyclic di-GMP signaling in *Vibrio cholerae. Current Opinion in Microbiology, 36*, 20–29.

Cruz-Lacierda, E. R., De la Peña, L., & Lumanlan-Mayo, S. (2000). The use of chemicals in aquaculture in the Philippines. In *Use of chemicals in aquaculture in Asia: Proceedings of the meeting on the use of chemicals in aquaculture in Asia 20–22 May 1996* (pp. 155–184).

Tigbauan, Iloilo, Philippines: Aquaculture Department, Southeast Asian Fisheries Development Center.

Cuéllar-Anjel, J., Corteel, M., Galli, L., Alday-Sanz, V., & Hasson, K. W. (2014). Principal shrimp infectious diseases, diagnosis and management. In V. Alday-Sanz (Ed.), *The shrimp book* (pp. 517–622). UK: Nottingham University Press.

Dang, L. T., Nguyen, H. T., Hoang, H. H., Lai, H. N., & Nguyen, H. T. (2019). Efficacy of rose myrtle *Rhodomyrtus tomentosa* seed extract against acute Hepatopancreatic necrosis disease in Pacific whiteleg shrimp *Penaeus vannamei*. *Journal of Aquatic Animal Health*, *31*(4), 311–319.

Defoirdt, T. (2018). Quorum-sensing systems as targets for antivirulence therapy. *Trends in Microbiology*, *26*(4), 313–328.

Defoirdt, T., Boon, N., Sorgeloos, P., Verstraete, W., & Bossier, P. (2007). Alternatives to antibiotics to control bacterial infections: Luminescent vibriosis in aquaculture as an example. *Trends in Biotechnology*, *25* (10), 472–479.

Del Pozo, J. L. (2018). Biofilm-related disease. *Expert Review of Anti-Infective Therapy*, *16*(1), 51–65.

DePaola, A., Nordstrom, J. L., Bowers, J. C., Wells, J. G., & Cook, D. W. (2003). Seasonal abundance of total and pathogenic Vibrio parahaemolyticus in Alabama oysters. *Applied and Environmental Microbiology*, *69*(3), 1521–1526.

Donato, F., & Zani, C. (2010). Chronic exposure to organochlorine compounds and health effects in adults: Cancer, non-Hodgkin lymphoma. Review of literature. *Annali di Igiene: Medicina Preventiva e di Comunita*, *22*(4), 357–367.

Drake, S. L., DePaola, A., & Jaykus, L. A. (2007). An overview of Vibrio vulnificus and Vibrio parahaemolyticus. *Comprehensive Reviews in Food Science and Food Safety*, *6*(4), 120–144.

Dutta, N. K., Panse, M. V., & Kulkarni, D. R. (1959). Role of cholera toxin in experimental cholera. *Journal of Bacteriology*, *78*(4), 594–595.

Egidius, E. (1987). Vibriosis: Pathogenicity and pathology. A review. *Aquaculture*, *67*(1–2), 15–28.

Farmer, J. J., Janda, J., & Birkhead, K. (2003). *Vibrio: Manual of clinical microbiology*. ASM Press, Washington.

Félix, R., Félix, C., Januário, A. P., Carmona, A. M., Baptista, T., Gonçalves, R. A., et al. (2020). Tailoring shrimp aquafeed to tackle Acute Hepatopancreatic Necrosis Disease by inclusion of industry-friendly seaweed extracts. *Aquaculture*, *529*, 735661.

Fu, G., Peng, J., Wang, Y., Zhao, S., Fang, W., Hu, K., et al. (2016). Pharmacokinetics and pharmacodynamics of sulfamethoxazole and trimethoprim in swimming crabs (*Portunus trituberculatus*) and in vitro antibacterial activity against *Vibrio*: PK/PD of SMZ-TMP in crabs and antibacterial activity against *Vibrio*. *Environmental Toxicology and Pharmacology*, *46*, 45–54.

Ge, H., Li, J., Chen, P., Chang, Z., Shen, M., & Zhao, F. (2017). Cultivation of green algae *Platymonas helgolandica* in rearing water enhances the growth performance and resistance of *Litopenaeus vannamei* against *Vibrio parahaemolyticus* infection. *Aquaculture International*, *25*(3), 1279–1290.

Glenn Morris, J. (2003). Cholera and other types of vibriosis: A story of human pandemics and oysters on the half shell. *Clinical Infectious Diseases*, *37*(2), 272–280.

Goo, S. Y., Lee, H. J., Kim, W. H., Han, K. L., Park, D. K., Lee, H. J., et al. (2006). Identification of OmpU of *Vibrio vulnificus* as a fibronectin-binding protein and its role in bacterial pathogenesis. *Infection and Immunity*, *74*(10), 5586–5594.

Goulden, E. F., Hall, M. R., Bourne, D. G., Pereg, L. L., & Høj, L. (2012). Pathogenicity and infection cycle of *Vibrio owensii* in larviculture of the ornate spiny lobster (*Panulirus ornatus*). *Applied and Environmental Microbiology*, *78*(8), 2841–2849.

Haldar, S., Chatterjee, S., Asakura, M., Vijayakumaran, M., & Yamasaki, S. (2007). Isolation of Vibrio parahaemolyticus and Vibrio cholerae (Non-O1 and O139) from moribund shrimp (Penaeus monodon) and experimental challenge study against post larvae and juveniles. *Annals of Microbiology*, *57*(1), 55–60.

Haldar, S., Chatterjee, S., Sugimoto, N., Das, S., Chowdhury, N., Hinenoya, A., et al. (2011). Identification of *Vibrio campbellii* isolated from diseased farm-shrimps from south India and establishment of its pathogenic potential in an *Artemia* model. *Microbiology*, *157*(1), 179–188.

Haldar, S., Maharajan, A., Chatterjee, S., Hunter, S. A., Chowdhury, N., Hinenoya, A., et al. (2009). Identification of *Vibrio harveyi* as a causative bacterium for a tail rot disease of sea bream *Sparus aurata* from research hatchery in Malta. *Microbiological Research*, *165*(8), 639–648.

Haldar, S., Neogi, S. B., Kogure, K., Chatterjee, S., Chowdhury, N., Hinenoya, A., et al. (2009). Development of a haemolysin gene-based multiplex PCR for simultaneous detection of *Vibrio campbellii*, *Vibrio harveyi* and *Vibrio parahaemolyticus*. *Letters in Applied Microbiology*, *50*, 146–152.

Hang, L., John, M., Asaduzzaman, M., Bridges, E. A., Vanderspurt, C., Kirn, T. J., et al. (2003). Use of in vivo-induced antigen technology (IVIAT) to identify genes uniquely expressed during human infection with *Vibrio cholerae*. *Proceedings of the National Academy of Sciences*, *100*(14), 8508–8513.

Høi, L., Dalsgaard, I., DePaola, A., Siebeling, R. J., & Dalsgaard, A. (1998). Heterogeneity among isolates of *Vibrio vulnificus* recovered from eels (*Anguilla anguilla*) in Denmark. *Applied and Environmental Microbiology*, *64*(12), 4676–4682.

Hongprayoon, R. (1993). *The role of flagellar proteins in adhesion of Vibrio parahaemolyticus, and isolation of the corresponding gene (s)*. Louisiana State University and Agricultural & Mechanical College.

Housby, J. N., & Mann, N. H. (2009). Phage therapy. *Drug Discovery Today*, *14*(11–12), 536–540.

Jaiswal, A., Koley, H., Ghosh, A., Palit, A., & Sarkar, B. (2013). Efficacy of cocktail phage therapy in treating *Vibrio cholerae* infection in rabbit model. *Microbes and Infection*, *15*(2), 152–156.

Jassim, S. A., & Limoges, R. G. (2014). Natural solution to antibiotic resistance: Bacteriophages 'The Living Drugs'. *World Journal of Microbiology and Biotechnology*, *30*(8), 2153–2170.

Jayasree, L., Janakiram, P., & Madhavi, R. (2006). Characterization of *Vibrio* spp. associated with diseased shrimp from culture ponds of Andhra Pradesh (India). *Journal of the World Aquaculture Society*, *37*(4), 523–532.

Jun, J. W., Shin, T. H., Kim, J. H., Shin, S. P., Han, J. E., Heo, G. J., et al. (2014). Bacteriophage therapy of a *Vibrio parahaemolyticus* infection caused by a multiple-antibiotic–resistant O3: K6 pandemic clinical strain. *The Journal of Infectious Diseases*, *210*(1), 72–78.

Kadirvel, M., Fanimarvasti, F., Forbes, S., McBain, A., Gardiner, J. M., Brown, G. D., et al. (2014). Inhibition of quorum sensing and biofilm formation in *Vibrio harveyi* by 4-fluoro-DPD; a novel potent inhibitor of AI-2 signalling. *Chemical Communications*, *50*(39), 5000–5002.

Karunasagar, I., Shivu, M. M., Girisha, S. K., Krohne, G., & Karunasagar, I. (2007). Biocontrol of pathogens in shrimp hatcheries using bacteriophages. *Aquaculture*, *268*(1–4), 288–292.

Kim, S. G., Jun, J. W., Giri, S. S., Yun, S., Kim, H. J., Kim, S. W., et al. (2019). Isolation and characterisation of pVa-21, a giant bacteriophage with anti-biofilm potential against *Vibrio alginolyticus*. *Scientific Reports*, *9*(1), 1–10.

Kim, A., Seid, C., McElhiney, A., & Tlusty, M. (2018). Treatment of a laboratory-based model of shell disease in hatchery raised American lobsters (*Homarus americanus*). *Bulletin of Marine Science*, *94*(3), 923–943.

Kim, H. B., Wang, M., Ahmed, S., Park, C. H., LaRocque, R. C., Faruque, A. S., et al. (2010). Transferable quinolone resistance in *Vibrio cholerae*. *Antimicrobial Agents and Chemotherapy*, *54*(2), 799–803.

Kitaoka, M., Miyata, S. T., Unterweger, D., & Pukatzki, S. (2011). Antibiotic resistance mechanisms of *Vibrio cholerae*. *Journal of Medical Microbiology*, *60*(4), 397–407.

Krasteva, P. V., Fong, J. C., Shikuma, N. J., Beyhan, S., Navarro, M. V., Yildiz, F. H., et al. (2010). *Vibrio cholerae* VpsT regulates matrix production and motility by directly sensing cyclic di-GMP. *Science*, *327* (5967), 866–868.

Lauzon, H. L., Dimitroglou, A., Merrifield, D. L., Ringø, E., & Davies, S. J. (2014). Probiotics and prebiotics: Concepts, definitions and history. In *Aquaculture nutrition: Gut health, probiotics and prebiotics* (pp. 169–184). John Wiley & Sons, Ltd.

Lavilla-Pitogo, C. R., Baticados, M. C. L., Cruz-Lacierda, E. R., & Leobert, D. (1990). Occurrence of luminous bacterial disease of *Penaeus monodon* larvae in the Philippines. *Aquaculture*, *91*(1–2), 1–13.

Lee, K. K., Yu, S. R., Yang, T. I., Liu, P. C., & Chen, F. R. (1996). Isolation and characterization of Vibrio alginolyticus isolated from diseased kuruma prawn, Penaeus japonicus. *Letters in Applied Microbiology*, *22*(2), 111–114.

Letchumanan, V., Chan, K. G., & Lee, L. H. (2014). *Vibrio parahaemolyticus*: A review on the pathogenesis, prevalence, and advance molecular identification techniques. *Frontiers in Microbiology*, *5*, 705.

Li, J., Tan, B., Mai, K., Ai, Q., Zhang, W., Xu, W., & Ma, H. (2006). Comparative study between probiotic bacterium Arthrobacter XE-7 and chloramphenicol on protection of Penaeus chinensis post-larvae from pathogenic vibrios. *Aquaculture*, *253*(1–4), 140–147.

Liu, X., Beyhan, S., Lim, B., Linington, R. G., & Yildiz, F. H. (2010). Identification and characterization of a phosphodiesterase that inversely regulates motility and biofilm formation in *Vibrio cholerae*. *Journal of Bacteriology*, *192*(18), 4541–4552.

Mancuso, M., Costanzo, M. T., Maricchiolo, G., Gristina, M., Zaccone, R., Cuccu, D., et al. (2010). Characterization of chitinolytic bacteria and histological aspects of Shell Disease Syndrome in European spiny lobsters (*Palinurus elephas*) (Fabricius 1787). *Journal of Invertebrate Pathology*, *104*(3), 242–244.

Marčuk, L. M., Nikiforov, V. N., Ščerbak, J. F., Levitov, T. A., Kotljarova, R. I., Naumšina, M. S., et al. (1971). Clinical studies of the use of bacteriophage in the treatment of cholera. *Bulletin of the World Health Organization*, *45*(1), 77.

Marsh, P. D. (2010). Microbiology of dental plaque biofilms and their role in oral health and caries. *Dental Clinics of North America*, *54*(3), 441–454.

Marshall, K., Stout, R., & Mitchell, R. (1971). Mechanism of the initial events in the sorption of marine bacteria to surfaces. *Journal of General Microbiology*, *68*(3), 337–348. https://doi.org/10.1099/00221287-68-3-337.

Martinez, R. M., Jude, B. A., Kirn, T. J., Skorupski, K., & Taylor, R. K. (2010). Role of FlgT in anchoring the flagellum of *Vibrio cholerae*. *Journal of Bacteriology*, *192*(8), 2085–2092.

Martin-Laurent, F., Philippot, L., Hallet, S., Chaussod, R., Germon, J. C., Soulas, G., et al. (2001). DNA extraction from soils: Old bias for new microbial diversity analysis methods. *Applied and Environmental Microbiology*, *67*(5), 2354–2359.

Marudhupandi, T., Kumar, T. T. A., Prakash, S., Balamurugan, J., & Dhayanithi, N. B. (2017). *Vibrio parahaemolyticus* a causative bacterium for tail rot disease in ornamental fish, *Amphiprion sebae*. *Aquaculture Reports*, *8*, 39–44.

Merino, S., Shaw, J. G., & Tomás, J. M. (2006). Bacterial lateral flagella: An inducible flagella system. *FEMS Microbiology Letters*, *263*(2), 127–135.

Mohamad, N., Amal, M. N. A., Yasin, I. S. M., Saad, M. Z., Nasruddin, N. S., Al-saari, N., et al. (2019). Vibriosis in cultured marine fishes: A review. *Aquaculture*, *512*, 734289.

Monsur, K. A., Rahman, M. A., Huq, F., Islam, M. N., Northrup, R. S., & Hirschhorn, N. (1970). Effect of massive doses of bacteriophage on excretion of vibrios, duration of diarrhoea and output of stools in acute cases of cholera. *Bulletin of the World Health Organization*, *42*(5), 723.

Mulyadi, I. N., & Iba, W. (2020). Efficacy of seaweed (*Sargassum* sp.) extract to prevent vibriosis in white shrimp (*Litopenaeus vannamei*) juvenile. *International Journal of Zoological Research*, *16*, 1–11. Research article.

Na, H. S., Cha, M. H., Oh, D. R., Cho, C. W., Rhee, J. H., & Kim, Y. R. (2011). Protective mechanism of curcumin against *Vibrio vulnificus* infection. *FEMS Immunology and Medical Microbiology*, *63*(3), 355–362.

Nakayama, T., Nomura, N., & Matsumura, M. (2007). The effect of copper concentration on the virulence of pathogenic *Vibrio harveyi*. *Journal of Applied Microbiology*, *102*(5), 1300–1306.

Natrah, F. M. I., Kenmegne, M. M., Wiyoto, W., Sorgeloos, P., Bossier, P., & Defoirdt, T. (2011). Effects of micro-algae commonly used in aquaculture on acyl-homoserine lactone quorum sensing. *Aquaculture*, *317* (1–4), 53–57.

Nithya, C., Devi, M. G., & Karutha Pandian, S. (2011). A novel compound from the marine bacterium *Bacillus pumilus* S6-15 inhibits biofilm formation in Gram-positive and Gram-negative species. *Biofouling*, *27*(5), 519–528.

Olsson, J. C., Westerdahl, A. L. L. A. N., Conway, P. L., & Kjelleberg, S. T. A. F. F. A. N. (1992). Intestinal colonization potential of turbot (*Scophthalmus maximus*)-and dab (*Limanda limanda*)-associated bacteria with inhibitory effects against *Vibrio anguillarum*. *Applied and Environmental Microbiology*, *58*(2), 551–556.

Pande, G. S. J., Natrah, F. M. I., Flandez, A. V. B., Kumar, U., Niu, Y., Bossier, P., et al. (2015). Isolation of AHL-degrading bacteria from micro-algal cultures and their impact on algal growth and on virulence of *Vibrio campbellii* to prawn larvae. *Applied Microbiology and Biotechnology*, *99*(24), 10805–10813.

Pandey, K. R., Naik, S. R., & Vakil, B. V. (2015). Probiotics, prebiotics and synbiotics—A review. *Journal of Food Science and Technology*, *52* (12), 7577–7587.

Paranjpye, R. N., & Strom, M. S. (2005). A *Vibrio vulnificus* type IV pilin contributes to biofilm formation, adherence to epithelial cells, and virulence. *Infection and Immunity*, *73*(3), 1411–1422.

Paulsen, I. T., Brown, M. H., & Skurray, R. A. (1996). Proton-dependent multidrug efflux systems. *Microbiological Reviews*, *60*(4), 575–608.

Peach, K. C. (2013). Image-based screening in Vibrio cholerae: Platform development, image analysis, and small molecule biofilm modulators (Doctoral dissertation, UC Santa Cruz).

Pope, E. C., Powell, A., Roberts, E. C., Shields, R. J., Wardle, R., & Rowley, A. F. (2011). Enhanced cellular immunity in shrimp (*Litopenaeus vannamei*) after 'vaccination'. *PLoS One*, *6*(6), e20960.

Rabin, N., Zheng, Y., Opoku-Temeng, C., Du, Y., Bonsu, E., & Sintim, H. O. (2015). Agents that inhibit bacterial biofilm formation. *Future Medicinal Chemistry*, *7*(5), 647–671.

Radhakrishnan, E. V., & Kizhakudan, J. K. (2019). Health management in lobster aquaculture. In *Lobsters: Biology, fisheries and aquaculture* (pp. 571–601). Singapore: Springer.

Regunathan, C., & Wesley, S. G. (2004). Control of *Vibrio* spp. in shrimp hatcheries using the green algae *Tetraselmis suecica*. *Asian Fisheries Science*, *17*(1/2), 147–158.

Renner, L. D., & Weibel, D. B. (2011). Physicochemical regulation of biofilm formation. *MRS Bulletin*, *36*(5), 347–355. https://doi.org/10.1557/mrs.2011.65.

Sambanthamoorthy, K., Gokhale, A. A., Lao, W., Parashar, V., Neiditch, M. B., Semmelhack, M. F., & Waters, C. M. (2011). Identification of a novel benzimidazole that inhibits bacterial biofilm formation in a broad-spectrum manner. *Antimicrobial Agents and Chemotherapy*, *55*(9), 4369–4378.

Santhakumari, S., Kannappan, A., Pandian, S. K., Thajuddin, N., Rajendran, R. B., & Ravi, A. V. (2016). Inhibitory effect of marine cyanobacterial extract on biofilm formation and virulence factor production of bacterial pathogens causing vibriosis in aquaculture. *Journal of Applied Phycology*, *28*(1), 313–324.

Santos, A. L. S. D., Galdino, A. C. M., Mello, T. P. D., Ramos, L. D. S., Branquinha, M. H., Bolognese, A. M., et al. (2018). What are the advantages of living in a community? A microbial biofilm perspective! *Memórias do Instituto Oswaldo Cruz*, *113*(9).

Schrezenmeir, J., & de Vrese, M. (2001). Probiotics, prebiotics, and synbiotics—Approaching a definition. *The American Journal of Clinical Nutrition*, *73*(2), 361s–364s.

Shariff, Z. M., Bond, J. T., & Johnson, N. E. (2000). Nutrition and educational achievement of urban primary schoolchildren in Malaysia. *Asia Pacific Journal of Clinical Nutrition*, *9*(4), 264–273.

Singh, A. K., Haldar, R., Mandal, D., & Kundu, M. (2006). Analysis of the topology of *Vibrio cholerae* NorM and identification of amino acid residues involved in norfloxacin resistance. *Antimicrobial Agents and Chemotherapy*, *50*(11), 3717–3723.

Soowannayan, C., Boonmee, S., Puckcharoen, S., Anatamsombat, T., Yatip, P., Ng, W. K., et al. (2019). Ginger and its component shogaol inhibit *Vibrio* biofilm formation in vitro and orally protect shrimp against acute hepatopancreatic necrosis disease (AHPND). *Aquaculture*, *504*, 139–147.

Soto-Rodríguez, S., Armenta, M., & Gomez-Gil, B. (2006). Effects of enrofloxacin and florfenicol on survival and bacterial population in an experimental infection with luminescent *Vibrio campbellii* in shrimp larvae of *Litopenaeus vannamei*. *Aquaculture*, *255*(1–4), 48–54.

Sousa, O. V. D., Vieira, R. H. S. D. F., Menezes, F. G. R. D., Reis, C. M. F. D., & Hofer, E. (2004). Detection of *Vibrio parahaemolyticus* and *Vibrio cholerae* in oyster, *Crassostrea rhizophorae*, collected from a natural nursery in the Cocó river estuary, Fortaleza, Ceará, Brazil. *Revista do Instituto de Medicina Tropical de São Paulo*, *46*, 59–62.

Sperandio, V., Giron, J. A., Silveira, W. D., & Kaper, J. B. (1995). The OmpU outer membrane protein, a potential adherence factor of *Vibrio cholerae*. *Infection and Immunity*, *63*(11), 4433–4438.

Srinivasan, R., Santhakumari, S., & Ravi, A. V. (2017). Inávitro antibiofilm efficacy of *Piper betle* against quorum sensing mediated biofilm formation of luminescent *Vibrio harveyi*. *Microbial Pathogenesis*, *110*, 232–239.

Sulakvelidze, A., Alavidze, Z., & Morris, J. G., Jr. (2001). Bacteriophage therapy. *Antimicrobial Agents and Chemotherapy*, *45*(3), 649–659.

Sullivan, T. J., & Neigel, J. E. (2018). Effects of temperature and salinity on prevalence and intensity of infection of blue crabs, *Callinectes sapidus*, by *Vibrio cholerae*, *V. parahaemolyticus*, and *V. vulnificus* in Louisiana. *Journal of Invertebrate Pathology*, *151*, 82–90.

Sun, D., Jeannot, K., Xiao, Y., & Knapp, C. W. (2019). Horizontal gene transfer mediated bacterial antibiotic resistance. *Frontiers in Microbiology*, *10*, 1933.

Sung, H. H., Hsu, S. F., Chen, C. K., Ting, Y. Y., & Chao, W. L. (2001). Relationships between disease outbreak in cultured tiger shrimp (Penaeus monodon) and the composition of Vibrio communities in pond water and shrimp hepatopancreas during cultivation. *Aquaculture*, *192*(2–4), 101–110.

Sutili, F. J., & Gressler, L. G. (2021). Antimicrobial agents. In *Aquaculture pharmacology* (pp. 131–168). Academic Press.

Tall, B. D., Fall, S., Pereira, M. R., Ramos-Valle, M., Curtis, S. K., Kothary, M. H., & Bayer, R. C. (2003). Characterization of Vibrio fluvialis-like strains implicated in limp lobster disease. *Applied and Environmental Microbiology*, *69*(12), 7435–7446.

Talpur, A. D., Memon, A. J., Khan, M. I., Ikhwanuddin, M., Daniel, M. D., & Abol Munafi, A. B. (2011). A novel of gut pathogenic bacteria of blue swimming crab *Portunus pelagicus* (Linneaus, 1758) and pathogenicity of *Vibrio harveyi* a transmission agent in larval culture under hatchery conditions. *Research Journal of Applied Sciences*, *6*(2), 116–127.

Tan, T. L. H., Chan, K. G., & Lee, L. H. (2014). Application of bacteriophage in biocontrol of major foodborne bacterial pathogens. *Journal of Molecular Biology and Molecular Imaging*, *1*(9).

Tan, L. T. H., Chan, K. G., Lee, L. H., & Goh, B. H. (2016). *Streptomyces* bacteria as potential probiotics in aquaculture. *Frontiers in Microbiology*, *7*, 79.

Taylor, R. K., Miller, V. L., Furlong, D. B., & Mekalanos, J. J. (1987). Use of phoA gene fusions to identify a pilus colonization factor coordinately regulated with cholera toxin. *Proceedings of the National Academy of Sciences*, *84*(9), 2833–2837.

Thompson, F., Iida, T., & Swings, J. (2004). Biodiversity of vibrios. *Microbiology and Molecular Biology Reviews*, *68*(3), 403–431.

Tison, D. L., Nishibuchi, M., Greenwood, J. D., & Seidler, R. J. (1982). *Vibrio vulnificus* biogroup 2: New biogroup pathogenic for eels. *Applied and Environmental Microbiology*, *44*(3), 640–646.

Tonguthai, K. (2000). The use of chemicals in aquaculture in Thailand. In *Use of chemicals in aquaculture in Asia: Proceedings of the meeting on the use of chemicals in aquaculture in Asia 20–22 May 1996* (pp. 207–220). Tigbauan, Iloilo, Philippines: Aquaculture Department, Southeast Asian Fisheries Development Center.

Toranzo, A. E., Magariños, B., & Romalde, J. L. (2005). A review of the main bacterial fish diseases in mariculture systems. *Aquaculture*, *246*(1–4), 37–61.

Utada, A. S., Bennett, R. R., Fong, J. C., Gibiansky, M. L., Yildiz, F. H., Golestanian, R., et al. (2014). *Vibrio cholerae* use pili and flagella synergistically to effect motility switching and conditional surface attachment. *Nature Communications*, *5*(1), 1–8.

Vikram, A., Jesudhasan, P. R., Jayaprakasha, G. K., Pillai, S. D., & Patil, B. S. (2011). Citrus limonoids interfere with *Vibrio harveyi* cell–cell signalling and biofilm formation by modulating the response regulator LuxO. *Microbiology*, *157*(1), 99–110.

Vinod, M. G., Shivu, M. M., Umesha, K. R., Rajeeva, B. C., Krohne, G., Karunasagar, I., et al. (2006). Isolation of *Vibrio harveyi* bacteriophage with a potential for biocontrol of luminous vibriosis in hatchery environments. *Aquaculture*, *255*(1–4), 117–124.

Waldor, M. K., & Mekalanos, J. J. (1996). Lysogenic conversion by a filamentous phage encoding cholera toxin. *Science*, *272*(5270), 1910–1914.

Wang, L., Chen, Y., Huang, H., Huang, Z., Chen, H., & Shao, Z. (2015). Isolation and identification of *Vibrio campbellii* as a bacterial pathogen for luminous vibriosis of *Litopenaeus vannamei*. *Aquaculture Research*, *46*(2), 395–404.

Wang, L., Huang, L., Su, Y., Qin, Y., Kong, W., Ma, Y., et al. (2015). Involvement of the flagellar assembly pathway in *Vibrio alginolyticus* adhesion under environmental stresses. *Frontiers in Cellular and Infection Microbiology*, *5*, 59.

Warner, C. J., Cheng, A. T., Yildiz, F. H., & Linington, R. G. (2015). Development of benzo [1, 4] oxazines as biofilm inhibitors and dispersal agents against Vibrio cholerae. *Chemical Communications*, *51*(7), 1305–1308.

Wongtavatchai, J., López-Dóriga, M. V., & Francis, M. J. (2010). Effect of AquaVacTM VibromaxTM on size and health of post larva stage of Pacific White shrimp *Litopenaeus vannamei* and Black Tiger shrimp *Penaeus monodon*. *Aquaculture*, *308*(3–4), 75–81.

Yamanoi, H., Muroga, K., & Takahashi, S. (1980). Physiological characteristics and pathogenicity of NAG vibrio isolated from diseased ayu. *Fish Pathology*, *15*(2), 69–73.

Yang, W., Tran, N. T., Zhu, C., Zhang, M., Yao, D., Aweya, J. J., et al. (2020). Enhanced immune responses and protection against the secondary infection in mud crab (*Scylla paramamosain*) primed with formalin-killed *Vibrio parahemolyticus*. *Aquaculture*, *529*, 735671.

Zamorano-Sánchez, D., Fong, J. C., Kilic, S., Erill, I., & Yildiz, F. H. (2015). Identification and characterization of VpsR and VpsT binding sites in *Vibrio cholerae*. *Journal of Bacteriology*, *197*(7), 1221–1235.

Zhao, J., Chen, M., Quan, C. S., & Fan, S. D. (2015). Mechanisms of quorum sensing and strategies for quorum sensing disruption in aquaculture pathogens. *Journal of Fish Diseases*, *38*(9), 771–786.

Zheng, X., Guo, J., Rao, H., Guo, D., Huang, Y., Xu, Y., et al. (2020). Antibacterial and antibiofilm activity of coenzyme Q0 against *Vibrio parahaemolyticus*. *Food Control*, *109*, 106955.

Further reading

Aagesen, A. M., Phuvasate, S., Su, Y. C., & Häse, C. C. (2013). Persistence of *Vibrio parahaemolyticus* in the Pacific oyster, *Crassostrea gigas*, is a multifactorial process involving pili and flagella but not type III secretion systems or phase variation. *Applied and Environmental Microbiology*, *79*(10), 3303–3305.

Austin, B., Austin, D., Sutherland, R., Thompson, F., & Swings, J. (2005). Pathogenicity of vibrios to rainbow trout (*Oncorhynchus mykiss*, Walbaum) and *Artemia* nauplii. *Environmental Microbiology*, *7*(9), 1488–1495.

Bondad-Reantaso, M. G., McGladdery, S. E., East, I., & Subasinghe, R. P. (2001). *Asia diagnostic guide to aquatic animal diseases*. Rome (Italy): FAO.

Briet, A., Helsens, N., Delannoy, S., Debuiche, S., Brisabois, A., Midelet, G., et al. (2018). NDM-1-producing *Vibrio parahaemolyticus* isolated from imported seafood. *Journal of Antimicrobial Chemotherapy*, *73*(9), 2578–2579.

Charles, R. C., & Ryan, E. T. (2011). Cholera in the 21st century. *Current Opinion in Infectious Diseases*, *24*(5), 472–477.

Chou, S. H., & Galperin, M. Y. (2016). Diversity of cyclic di-GMP-binding proteins and mechanisms. *Journal of Bacteriology*, *198*(1), 32–46.

Cruz-Lacierda, E. R., De la Peña, L., & Lumanlan-Mayo, S. (2000). The use of chemicals in aquaculture in the Philippines. In *Use of chemicals in aquaculture in Asia: Proceedings of the meeting on the use of chemicals in aquaculture in Asia 20-22 May 1996, Tigbauan, Iloilo, Philippines* (pp. 155–184). Aquaculture Department, Southeast Asian Fisheries Development Center.

De Schryver, P., Defoirdt, T., & Sorgeloos, P. (2014). Early mortality syndrome outbreaks: A microbial management issue in shrimp farming? *PLoS Pathogens*, *10*(4), e1003919.

Deysine, M. (2004). Infections associated with surgical implants. *The New England Journal of Medicine*, *351*(2), 193–195.

Dibartola, A. C., Swearingen, M. C., Granger, J. F., Stoodley, P., & Dusane, D. H. (2017). Biofilms in orthopedic infections: A review of laboratory methods. *APMIS*, *125*(4), 418–428.

Fong, J. C., & Yildiz, F. H. (2007). The rbmBCDEF gene cluster modulates development of rugose colony morphology and biofilm formation in *Vibrio cholerae*. *Journal of Bacteriology*, *189*(6), 2319–2330.

Li, G., & Wang, M. Y. (2020). The role of *Vibrio vulnificus* virulence factors and regulators in its infection-induced sepsis. *Folia Microbiologica*, *65*(2), 265–274.

Li, X., Zhou, Y., Jiang, Q., Yang, H., Pi, D., Liu, X., et al. (2019). Virulence properties of *Vibrio vulnificus* isolated from diseased zoea of freshness shrimp *Macrobrachium rosenbergii*. *Microbial Pathogenesis*, *127*, 166–171.

Malone, M., Goeres, D. M., Gosbell, I., Vickery, K., Jensen, S., & Stoodley, P. (2017). Approaches to biofilm-associated infections: The need for standardized and relevant biofilm methods for clinical applications. *Expert Review of Anti-Infective Therapy*, *15*(2), 147–156.

Tang, K. F., Bondad-Reantaso, M. G., & Arthur, J. R. (2019). *Shrimp infectious myonecrosis strategy manual*. FAO Fisheries and Aquaculture Circular, (C1187), I-43.

Townsley, L., & Yildiz, F. H. (2015). Temperature affects c-di-GMP signalling and biofilm formation in *Vibrio cholerae*. *Environmental Microbiology*, *17*(11), 4290–4305.

Valentini, M., & Filloux, A. (2016). Biofilms and cyclic di-GMP (c-di-GMP) signaling: Lessons from *Pseudomonas aeruginosa* and other bacteria. *Journal of Biological Chemistry*, *291*(24), 12547–12555.

World Health Organization [WHO]. (2006). WHO guidelines for the safe use of wastewater, excreta and grey water. In J. Bartram, & R. Carr (Eds.), *Vol. 2. Wastewater use in agriculture*. Geneva: World Health Organization.

World Health Organization [WHO]. (2008). Cholera: Global surveillance summary. *Weekly Epidemiological Record*, *84*, 309–324.

Chapter 16

Candida: Biofilm formation and antifungal resistance

Shraddha Sharma[e], Shiv Mohan Singh[b], Gajanan Mane[c], Satish T. Pote[d], Milind S. Patole[c], and Rohit Sharma[a]

[a]*School of Sciences, SAGE University, Bhopal, Madhya Pradesh, India,* [b]*Department of Botany, Banaras Hindu University (BHU), Ajagara, Varanasi, Uttar Pradesh, India,* [c]*National Centre for Microbial Resources (NCMR), NCCS, NCCS Complex, S.P. Pune University, Pune, Maharashtra, India,* [d]*National Centre for Cell Science, NCCS Complex, S.P. Pune University, Pune, Maharashtra, India,* [e]*Centre Instrumental Facility for Flow Cytometry (CIFF), National Centre for Biological Sciences, Bengaluru, Karnataka, India*

1. Introduction

There is an increase in the fungal infections over past several years including infections caused by *Candida*. During the COVID-19 pandemic also, it has been a cause of concern in some countries. In general also, the fungal infections are a serious problem due to hospital-associated multidrug-resistant strains. The change in climate, i.e., increase in temperature is also a factor due to which many environmental strains have now a human or animal pathogen. This is because the fungal tolerance range of many environmental isolate has increased and is now adaptive to human body conditions. Another factor is the increased number of immune-deficiency patients. With the increase in organ transplant surgeries, virus infections, and autoimmune diseases, there is high risk of fungal infections. These infections had been neglected in the past as most of them were not life-threatening. However, with the increase in the frequency of infections and drug resistance, the rate of mortality has increased a lot. In recent times, some of the pathogens have become serious and fatal pathogens due to high drug resistance and high mortality rates such as *Candida auris* (Muñoz et al., 2018). Similarly, *Candida parapsilosis* complex species, which mainly causes nosocomial catheter-related infection in immune-compromised patients, is also pathogen of concern (Modiri et al., 2019). The available drugs for the treatment of fungal infections including those caused by genus *Candida* are less. They are mostly azoles (fluconazole, itraconazole, ketoconazole, miconazole, and clotrimazole), polyene (amphoterecin B and nystatin), and echinocandins (caspofungin, micafungin, and anidulafungin), etc. The capacity of a strain to form biofilm is one of the important characteristics of its pathogenicity. These biofilms are known to protect them from antifungal drugs and host immune response. The characteristic feature of biofilm formed by any species of *Candida* depends on the ability of the organism to produce extracellular polymeric substances and dimorphism. Species of *Candida*, namely *Candida tropicalis*, *Candida albicans*, *Candida glabrata*, *C. parapsilosis*, etc., are known to form biofilm. Various factors that affect the virulence factor of pathogenic yeast are biofilm formation, adhesion, filamentation, and extracellular polymeric substances production. Historical evolution in the yeast genomics has paved the way to improve the biological knowledge of the organism (Fig. 1). Formation of biofilms on the implanted medical devices such as central venous catheters is also an important factor, which may require implant replacement (d'Enfert & Janbon, 2016). The chapter discusses the various aspects of biofilm formation and antifungal susceptibility in the genus *Candida*. The recent advances that happened in the field of genomics are also discussed.

2. *Candida* and human disease

The genus *Candida* belongs to phylum *Ascomycota*. Species of *Candida* are part of normal human microflora contributing a major portion of the total microbial population. In general, *Candida* does not cause any pathological infections in our body until there is a change or disturbance in the local population or reduced immunity. Candidiasis is one of the common diseases caused by species of *Candida*. They not only cause infections of skin but also infect blood stream causing several invasive and noninvasive diseases (Sharma & Jangid, 2016). As yeast is known to be dimorphic, the genus *Candida* is also found in their mycelial stage and yeast form (Fig. 2). The mycelial forms help in the penetration of the yeast into the skin. Once inside the body due to the temperature (37°C), the yeast converts to yeast form due to quorum sensing effect. Although *C. albicans*

Understanding Microbial Biofilms. https://doi.org/10.1016/B978-0-323-99977-9.00024-7

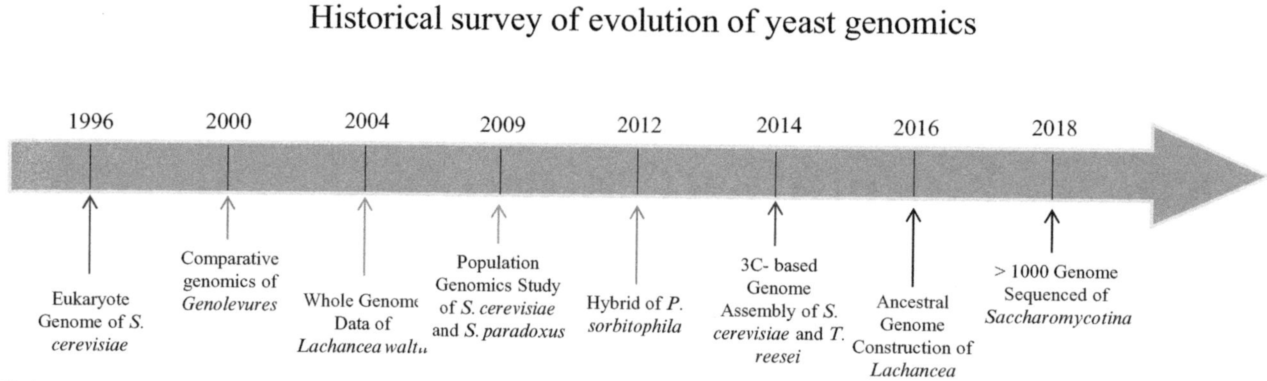

FIG. 1 Important discoveries and turning points in yeast genomics. *(Modified from Fairhead, C., Fischer, G., Liti, G., Neuvéglise, C., & Schacherer, J. (2019). André Goffeau's imprinting on second generation yeast "genomologists".* Yeast, *36(4), 167–175.)*

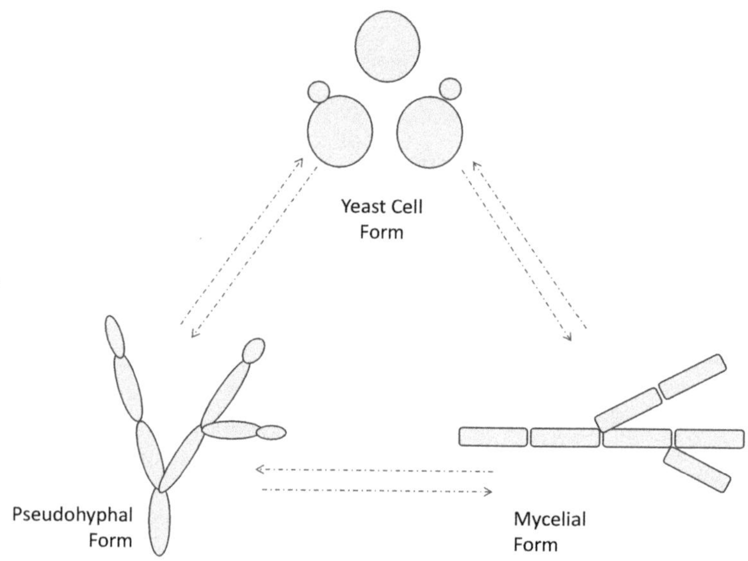

FIG. 2 The species of *Candida* show morphological transitions between yeast form, pseudohyphal and mycelial form.

is the most common species, others such as *C. glabrata*, *C. parapsilosis*, and *C. tropicalis* are also common. *Candida auris* has also emerged as an important pathogen with mostly multidrug-resistant strains. Pote et al. (2020) also reported species of *Candida* as the most frequent pathogen, which also included several new environmental isolates as pathogens. Different kinds of *Candida* infections that are common are mucosal infections, vaginal irritations, skin infections, systemic candidiasis, chronic mucocutaneous candidiasis, and organ transplant *Candida* infections.

3. *Candida* biofilm formation

Biofilms are primarily coordinated microbial networks connected to the surfaces of embedded gadgets exemplified inside a strong defensive extracellular grid (Spormann, Thormann, Saville, Shukla, & Entcheva, 2004). All *Candida* species portray contrasts as far as their biofilm development, structure, changes in the morphology extracellular matrix (ECM), and potential to oppose antifungal drugs (Cavalheiro & Teixeira, 2018). National Nosocomial Infection Survey (NNIS) showed that *Candida* spp. were the fourth most normal reason for nosocomial circulation system diseases during the 1990s. Nonetheless, in additional ongoing examinations, it has been found that *Candida* spp. are the third most incessant nosocomial circulation system disconnects (Lohse, Gulati, Johnson, & Nobile, 2018; Perlroth, Choi, & Spellberg, 2007). Moreover, *Candida* is the third standard disease of central line-related blood system infections (CLABSIs) and the second most unmistakable justification for catheter-related urinary lot diseases in the United States (Lockhart, 2014). Subsequently, the frequency of scattered candidiasis, which comprehensively alludes to the mucosal, cutaneous, also well-established organ disease brought about by *Candida* class, has expanded 15–20-overlap over the most recent 20 years (Lockhart, 2014; Pappas, Lionakis, Arendrup, Ostrosky-Zeichner, & Kullberg, 2018).

It has been accounted for that foundational candidiasis, brought about by *C. albicans*, is one of the significant reasons for death in nosocomial just as deftly appalling contagious disease in patients due to its capacity to frame a biofilm, which diminishes the helplessness of contagious cells toward drug treatment. Thus, it has been accounted for every year that up to half of fundamental candidiasis grown-up patients and roughly 30% of the youthful populace pass on because of candidiasis correlated with biofilms. Besides, there is an assessment of 100,000 passings in light of the disease started by biofilm development. Roughly, \$6500 million is spent each year in the United States toward its treatment causing a high monetary way on the state (Gulati & Nobile, 2016; Silva, Rodrigues, Araújo, Rodrigues, & Henriques, 2017). The five most inescapable species, *C. albicans*, *C. glabrata*, *C. parapsilosis*, *C. tropicalis*, and *Candida krusei*, are the purpose for 92%–95% of Intensive Care Unit (ICU) cases in the United States, despite the fact that dissemination of each of the five species may shift (Kojic & Darouiche, 2004; Ten Cate, Klis, Pereira-Cenci, Crielaard, & De Groot, 2009). As indicated by National Institute of Health (NIH) reports, pathogenic biofilms account for over 80% of every microbial disease (Atriwal et al., 2021; Lohse et al., 2018; Perlroth et al., 2007; Ramage, Martínez, & López-Ribot, 2006; Tsui, Kong, & Jabra-Rizk, 2016), of which 75% cases are of vaginal yeast contamination in ladies once or at least a couple of times in the course of their life (Perlroth et al., 2007; Yapar, 2014).

Candida albicans is a commensal parasite, asymptomatically related with typical microflora of the host; however, it becomes obtrusive and harmful when changed over to hyphal structure covered by an extracellular polymeric substance (EPS), which is a dimorphic nature of *C. albicans* (Gulati & Nobile, 2016). It is the most cliche parasitic microorganism in people, causing infection running from obvious mucosal to deadly scattered circulation system contaminations, which make an impressive commitment of more than 40% of death rates (Lohse et al., 2018). The three phases of the advancement of *C. albicans* contain bond of the yeast cells on the clinical gadgets (beginning phase), separation of the yeast cells to hyphal cells (middle stage), and an expansion in the framework, which is the development stage (Alim, Sircaik, & Panwar, 2018). The adult biphasic construction of *C. albicans* is advanced by the glue hydrophobic nature of inhabiting gadgets with the appropriate developing climate. Biomedical gadgets embedded during transplantation catheter are the ideal ones as they give sustenance like glucose from the discharged items (Alim et al., 2018; Ramage et al., 2006). *Candida albicans* colonizes the skin, gastrointestinal mucosa as well as genital mucosa of 30%–70% of solid people. Hence, under ordinary conditions, the growth doesn't cause any critical infection (Ramage, Robertson, & Williams, 2014).

Besides, around 15% of sepsis cases obtained in a clinic because of clinical gadgets happen because of the dominating parasites *Candida* (Lohse et al., 2018). As indicated by some new reports, 63.5% of patients out of 224 patients experiencing septic shock showed a positive outcome for *Candida* and were viewed as dead. In this way, it is the main source of demise in emergency clinic acquired illnesses (Ramage et al., 2014). In India, only 6%–18% of instances of Candidemia are accounted for. A review displays 53 rehashed episodes of Candidemia in 48 patients in 1.5 years. Various species made their commitment, as 45% of contaminations were because of *C. tropicalis*, while 23% of cases by *C. albicans* and the rest 32% by other *Candida* spp. (Kothari & Sagar, 2009). Western nation's pattern on candidemia contrasts from India as *C. albicans* is connected with over half of all candidemia frequencies (Kaur, Goyal, Dhakad, Bhalla, & Kumar, 2014).

4. Factors affecting biofilm

To examine the impact of starches on candida bond or biofilm arrangement, past laborers have prerefined yeast cells with sugar supplemented media for up to 18h (McCourtie & Douglas, 1984; Samaranayake & MacFarlane, 1982). The subsequent bond or biofilm studies were directed in without sugar media as the carbon source may influence the development pace of yeasts during the tests and jumble the resultant information. However, in an organic milieu, for example, the oral hole, sugars are consistently present, and their impact on both the cell attachment and ensuing biofilm development should be evaluated. Consequently, media with sugar supplementation were utilized all through our bond and biofilm measures (with the exception of the bond period of the biofilm tests). Our information showed that galactose (500mM)-developed yeasts displayed fundamentally lower levels of both attachment and biofilm development than glucose (100mM)-developed contagious cells, in concurrence with the past work utilizing the equivalent exploratory protocol (Nikawa, Nishimura, Hamada, Kumagai, & Samaranayake, 1997). Yet, a few early investigations have shown that galactose-developed yeasts are definitely more disciple than glucose-developed counterparts (Hawser & Douglas, 1994; Samaranayake & MacFarlane, 1982). However, it is imperative that the contagious cells were prerefined up to 18h in various sugars preceding attachment or biofilm tests in the last examinations. In addition, these investigations were acted in static media, which were not renewed as in the current review, and exhaustion of supplements might have changed biofilm development. Development profiles of planktonic yeast cells in glucose and galactose were additionally analyzed in the current review, and previous inspired a higher development rate. The differential development pace of glucose- and galactose-developed yeasts just as the delayed exploratory period contrasted with previous studies is possible with explanations behind the noticed more serious level of bond and biofilm creation in glucose-enhanced media. On the principal day of biofilm arrangement, glucose-grown biofilms grew more quickly than the partners filled in galactose. After day 1 and until day 3, the development rate in galactose was quicker than that of glucose. Past investigations have demonstrated that galactose might modify the peripheral surface parts of yeast wall coming about in expanded union of fibrillar mannoprotein adhesins (Douglas, 1992; McCourtie & Douglas, 1985; Samaranayake & MacFarlane, 1980, 1982). These extra adhesins are thought to work with contagious attachment and resulting biofilm formation likewise, contrasted and cells filled in different sugars, galactose-developed yeasts seem to show the most extensive level of hydrophobicity and in this manner bound to total in aqueous media (Hawser, Baillie, & Douglas, 1998; Hobden, Teevan, Jones, & O'Shea, 1995; McCourtie & Douglas, 1985). These might be explanations behind the noticed quick advancement of biofilms in galactose contrasted and glucose. It is notable that a few salivary proteins, such as human fibronectin and individuals from the fundamental proline-rich protein family, can go about as receptors for *C. albicans* (O'Sullivan, Cannon, Sullivan, & Jenkinson, 1997; Skerl, Segal, Sreevalsan, & Scheld, 1984). On the one hand, these proteins, when immobilized on a surface, may potentially advance *Candida* grip and ensuing biofilm development by going about as receptors free of charge living planktonic yeasts while on the other, they may at the same time block restricting destinations initially present on the substratum (Jin, Samaranayake, Samaranayake, & Yip, 2004).

Barely any investigations on the inhibitory impact of berberine on biofilms of *C. albicans* were accounted for, essentially on the morphological change of *C. albicans* cells treated with berberine hydrochloride (BH), the inhibitory impact of BH on practicality and engineering and the helplessness of sessile cells of *C. albicans* biofilms (Da Silva et al., 2016; Xie, Liu, & Zhou, 2020; Zorić et al., 2017). It has been observed that BH has great antifungal action against *C. albicans* biofilm arrangement by lessening the thickness of the biofilm, killing the cells in the biofilm and annihilating the whole construction of the biofilm. The consequences of 2,3-bis (2-methoxy-4-nitro-5-sulfophenyl)-2*H*-tetrazolium-5-carboxanilide (XTT) test showed that 128 and 32 μg/mL BH could altogether repress the cell action of ATCC10231 biofilm ($P < .05$). The after-effects of Confocal Laser Scanning Microscopy (CLSM) demonstrated that BH can obliterate the entire design of the biofilm, hinder the development of hyphae, and essentially decline the thickness of the biofilm.

The development interaction of biofilm is complicated and impacted by many variables, and the contagious change from the yeast state to the mycelial province of *C. albicans* assumes a significant part. Lower groupings of alizarin (2 μg/mL) and Sanguinarine (0.8 μg/mL) have been accounted for that restrain the hyphal advancement of *C. albicans* (Manoharan, Lee, Kim, & Lee, 2017; Zhong et al., 2017). Similarly, Zorić et al. (2017) observed that MIC of BH against *C. albicans* was 17.75 μg/mL, the cytotoxic impact of BH was focus subordinate, and repressed the morphological change of *C. albicans* cells into filamentous morphology in sub-MIC fixations. It was observed that BH showed a similar action at 128 and 32 μg/mL. Additionally, BH at 32 μg/mL annihilated the construction of biofilm (at 128 μg/mL totally disturbed). The information recommends that BH might stifle *C. albicans* dimorphic changing to restrain the advancement of biofilm.

The contagious change from the yeast state to the mycelial province of *C. albicans* was impacted by an assortment of administrative elements, and the component may connect with the downstream four vital qualities of the aggregate exchanged Ras1-cAMP-EFG1 flagging pathway (Manoharan et al., 2017). At the point when the body is impacted by natural factors, for example, serum, which enacts the whole flagging pathway and upmanages EFG1 articulation; EFG1 articulation can quickly initiate the statement of a progression of hyphal development explicit qualities, accordingly advancing the arrangement of biofilm (Berkow & Lockhart, 2017). ECE1 and HWP1 are hyphae-shaping qualities, which are not communicated in the yeast state and are bounteously communicated in the mycelial state, and their appearance is controlled by the EFG1 record factor (Berkow & Lockhart, 2017). Furthermore, EFG1 can likewise decidedly direct the outflow of the cell-grip-related quality ALS1. ALS1 is exceptionally communicated during biofilm arrangement, and the deficiency of ALS1 quality can cause genuine imperfections in biofilm (Manoharan et al., 2017).

The study showed that BH inhibitory affects the arrangement of *C. albicans* biofilms. The system might be connected with downguideline of the declaration of hyphae arrangement, which has likely incentive for clinical application. The limit of this study was they just noticed the change of *C. albicans* cells into filamentous morphology, which is just a significant stage in biofilm development. The consequences for layer ergosterol blend and prompted film penetrability increment require further review. Contrasted and the utilization of ATCC standard strains and BH alone, it will be important to utilize clinically safe strains and synergistic opposition of BH and fluconazole for additional examination (Huang et al., 2020).

Biofilm arrangement is a harmfulness variable of *C. albicans*, which is a pioneering parasitic microbe. In the report, *B. rotunda* displayed hostility to biofilm arrangement impact on *C. albicans*. Pinostrobin and pinocembrin, which are significant constituents of the concentrate, perceptibly repressed the biofilm advancement stage. The concentrate and pinocembrin diminished the cell surface hydrophobicity and the degree of ALS3 mRNA. Hence, the *B. rotunda* separate repressed biofilm arrangement of *C. albicans* by decreasing cell surface hydrophobicity and ALS3 mRNA articulation. Pinocembrin strongly affected ALS3 mRNA articulation than pinostrobin. Just pinocembrin essentially diminished the ACT1 mRNA articulation (Kanchanapiboon et al., 2020). In general, mitochondrial disability in *C. albicans* leads to inhibition of hyphal formation. Biofilm formation is highly linked up with *Candida* infections. A regulatory core complex in *Candida* is responsible for biofilm formation. Mitochondria in *C. albicans* are significant supporters of energy creation, harmfulness, drug opposition, and cell divider trustworthiness; in this way, the respiratory chain is an alluring objective for the advancement of medications against human contagious microbes. At present, various respiratory chain inhibitors have shown extensive antifungal movement. Chitosan is a biodegradable and nontoxic normal polysaccharide, and expansive range antifungal instruments could focus on the cell surface and mitochondria and apply a few intracellular activities against human microbes invest it with significant business potential. These elements demonstrate that chitosan is a promising atom for future antifungal medication improvement either alone or in blend with current antifungal medications (Ke, Liao, & Lin, 2021).

Proteases may assume a few unique parts in *C. albicans* biofilm development, a thought upheld by the way that proteases are powerfully communicated over the span of *C. albicans* biofilm development (Fox et al., 2015; Lohse, Gulati, Craik, Johnson, & Nobile, 2020; Nailis et al., 2010). For instance, Sap5 and Sap6, two emitted aspartyl proteases that are profoundly upregulated at specific phases of biofilm arrangement, are known to intervene bond of *C. albicans* cells to surfaces and conceivably of *C. albicans* cells to each other (Kumar, Saraswat, Tati, & Edgerton, 2015; Winter et al., 2016). Proteases may likewise add to the breakdown and securing of supplements, the handling of atoms significant for biofilm arrangement (e.g., bond particles), majority detecting, and additionally extracellular network creation all through biofilm arrangement and upkeep. Albeit the contribution of emitted proteases in biofilm development is a moderately new idea, there is some point of reference for this thought in bacterial biofilms, where extracellular proteases were viewed as involved in the handling of bond particles during biofilm arrangement of *Staphylococcus* species (Koziel & Potempa, 2013; Martínez-García, Rodríguez-Martínez, Cancino-Diaz, & Cancino-Diaz, 2018; Paharik et al., 2017).

5. Recent advances to inhibit *C. albicans* biofilm formation

All through this article, we have seen various procedures and techniques to battle antifungal medication opposition. Nonetheless, later every one of the endeavors and endeavors we are as yet stuck at a point where biofilm development is hard to treat with both nonantifungal and antifungal medications as a result of the helpless infiltration what's more nonparticularity of the medications. Thusly, to address this issue, specialists are dealing with tracking down ways of expanding the entrance of medications into the extracellular grid of biofilm. In the course of the most recent couple of years, distinctive metal nanoparticles have arisen as forthcoming possibility to treat microbial diseases relating to their solid potential as antimicrobial specialists. In this setting, distinctive examination bunches have started the investigation to distinguish the antibiofilm movement of various nanomaterials against *C. albicans* (Vera-González & Shukla, 2020). The utilization of silver as an antimicrobial specialist traverses a century.

Silver was utilized as a sanitizer by Greeks and Romans to clean their water and food supplies. Silver was moreover utilized in old occasions to treat consumes and wounds (Adhya et al., 2015; Alexander et al., 2009; Barillo & Marx, 2014). Concentrating on silver nanoparticles against biofilms has acquired huge acknowledgment. Silver nanoparticles apply an inhibitory impact against contagious biofilms by harming cell divider generally by twisting and interruption of the external surface of the parasitic cell divider (Lara et al., 2015). In resulting tests, AgNPs when used to functionalize clinical and natural surfaces, illustrated biofilm restraint (>50%) at generally low focuses (2.3–0.28 ppm) (Lara et al., 2020). Blend treatment of Fluconazole with silver nanoparticles has likewise shown a significant decline in the MIC (Longhi et al., 2015). The protection from fluconazole (FLC) in *C. albicans* is unequivocally connected with overexpression of qualities encoding efflux siphons or lanosterol 14α-demethylase (Sagatova et al., 2016). Thusly, with the interruption of the cell divider and cytoplasmic layer of *C. albicans* by AgNPs, these silver particles cause an increment in responsive oxygen species and hydroxyl revolutionary creation, which can likewise add to cell layer harm (Hwang, Hwang, Choi, Kim, & Lee, 2012; Monteiro et al., 2012).

As of late, scientists have recognized bismuth oxide (BiO_3) nanoparticles with intense antimycotic action against *C. albicans* development. Additionally, when these nanoparticles were contrasted with economically accessible antifungal medications such as chlorhexidine, nystatin, they totally destroyed the biofilm arrangement (Hernandez-Delgadillo et al., 2013). In another review, magnesium oxide nanoparticles (MgO NP) were utilized to examine their antifungal and antibiofilm property against *C. albicans*. The review uncovered that MgO nanoparticles successfully restrained *C. albicans* biofilm development. Moreover, attachment tests showed that MgO NPs stifled the starter attachment of *Candida albicans* (Kong et al., 2020). In another review, when gold nanoparticles (AuNP) were tried on the biofilms shaped because of *C. albicans*, these particles showed powerful inhibitory movement. Furthermore, these AuNPs expanded the host insusceptible reaction movement against these pathogenic life forms (Nani et al., 2019; Yu et al., 2016). From past investigations, chitosan and its nanoparticle (ChNPs) were at that point known for their antifungal property (Gondim et al., 2018; Mohamed et al., 2017; Yien, Zin, Sarwar, & Katas, 2012). Along these lines, to set up its antibiofilm movement, ChNPs were explored, and it was found that ChNPs fundamentally restrained the biofilm arrangement causing extremely less changes in the acrylic tar surface (Gondim et al., 2018). Moreover, one more review directed with selenium nanoparticles (Se-NP) showed solid restraint on *Candida* biofilm development as these particles out contend sulfur in organic cycle because of their comparable compound properties. Likewise, these nanoparticles go into the cell to contract and upset the external layer construction of *C. albicans* cell (Guisbiers et al., 2017).

The titanium oxide nanoparticles (Ti-NP) when investigated for their antifungal property showed solid antifungal movement against planktonic type of *C. albicans*. They likewise portrayed powerful potential against biofilm arrangement (Haghighi, Mohammadi, Mohammadi, Hosseinkhani, & Shipour, 2013). Likewise, zinc oxide nanoparticles (ZnO-NPs) utilized on *Candida* biofilm showed a reduction in the biofilm arrangement by about 62%–85% at differed centralization of 125 ppm–250 ppm separately. The ZnO-NPs proposed to repress hyphae arrangement in *C. albicans* by creation of responsive oxygen species in a portion subordinate way (Golipour, Habibipour, & Moradihaghgou, 2019; Hosseini, Ghaemi, & Koohsar, 2018; Mahamuni-Badiger et al., 2020). Furthermore, another review including copper oxide nanoparticles (Cu_2ONP what's more CuONP) showed strong inhibitory activity against biofilm arrangement by *C. albicans*. The review uncovered that copper oxide nanoparticles oblige the yeast for hyphae change. Moreover, CuONP evoked the responsive oxygen species while Cu_2ONP articulated the film harm in *C. albicans*. In examination to the two kinds of copper oxide nanoparticles, CuONP was steadier and portrayed better antifungal movement in contrast with Cu_2ONP (Padmavathi et al., 2020; Rasool, Sah, & Hemalatha, 2021). In another study, iron-oxide nanoparticles (Fe_3O_4 NPs) were explored on *C. albicans* biofilm, which showed 87.2%–100% restraint based on their molecule size, which was 100 ppm–200 pm separately (Seddighi, Salari, & Izadi, 2017). The nitric oxide nanoparticles (NO-NP) showed solid potential as an antifungal specialist by upsetting the extracellular grid and biofilm arrangement on the outer layer of biomaterials. It was observed that NO-NPs additionally decline the metabolic movement of *Candida* cells both in vitro and in vivo (Ahmadi et al., 2016).

6. *Candida* biofilms and *Candida*-bacterial interactions

Adherence and total of *Candida* with abiotic surfaces, bacteria, or different organisms rely upon surface responses. Thus, it is vital to concentrate on the cell-divider properties of *Candida* spp. The cell mass of *C. albicans* comprises four significant parts: (i) mannoproteins, which represent around 40% of the cell-divider biomass; (ii) β-1,3-glucan, the significant pressure bearing polysaccharide of the divider; (iii) β-1,6-glucan, a water-dissolvable part without a customary structure that interconnects mannoproteins to β-1,3-glucan and chitin chains; and (iv) a limited quantity of chitin (a direct pressure bearing polysaccharide comprising of β-1,4-connected *N*-acetylglucosaminebuildups). β-1,3-Glucan structures a three-layered, hydrogen-bond-settled, organization of fibrils that run corresponding to the cell surface (Ten Cate et al., 2009). Chitin bound to β-1,3-glucan further fortifies this organization. The atomic structures of the yeast and hyphal mass of *C. albicans* are comparable. In any case, the synthesis of its cell divider, especially its protein structure, is variable and emphatically depends on ecological conditions (Nett et al., 2007; Sosinska et al., 2008; Walker et al., 2008; Yin, De Groot, De Koster, & Klis, 2008). The cell mass of *C. albicans* is basically bilayered, with an inside skeletal layer that is encircled by a layer of fibrillar proteins that radiate outward (Kapteyn et al., 2000; Tokunaga, Kusamichi, & Koike, 1986) (Fig. 3). The majority of the cell-divider proteins (CWPs) are GPI proteins (GPI, glycosylphosphatidylinositol). They for the most part comprise a cell-surface-uncovered amino-terminal effector area that is upheld by a spacer space; also, they have carboxy-terminal highlights, including a leftover of a GPI anchor that makes covalent connection the cell-divider β-1,3-glucan network (Klis, Boorsma, & De Groot, 2006). The combination of β-1,3-glucan is hindered by the new echinocandin class of medications (anidulafungin, caspofungin, and micafungin). They are fungicidal and dynamic against a wide spectrum of growths (Cappelletty & Eiselstein-McKitrick, 2007).

Candida biofilm cells are for the most part significantly more impervious to antifungal specialists than are their planktonic partners. Regardless of its clear clinical pertinence, the atomic components underlying this peculiarity are not completely perceived (d'Enfert, 2006; Mukherjee, Zhou, Munyon, & Ghannoum, 2005; Ramage, Saville, Thomas, & López-Ribot, 2005). Obviously, biofilm opposition is a complex multifactorial peculiarity. Multidrug efflux siphons (ABC carriers) have been recommended to contribute to the expanded obstruction of cells in the beginning phases of biofilm development (Mukherjee, Chandra, Kuhn, & Ghannoum, 2003; Ramage, Bachmann, Patterson, Wickes, & López-Ribot, 2002). Hindrance of dispersion of antifungals by the extracellular biofilm network and the presence of "antifungal persister" cells may likewise

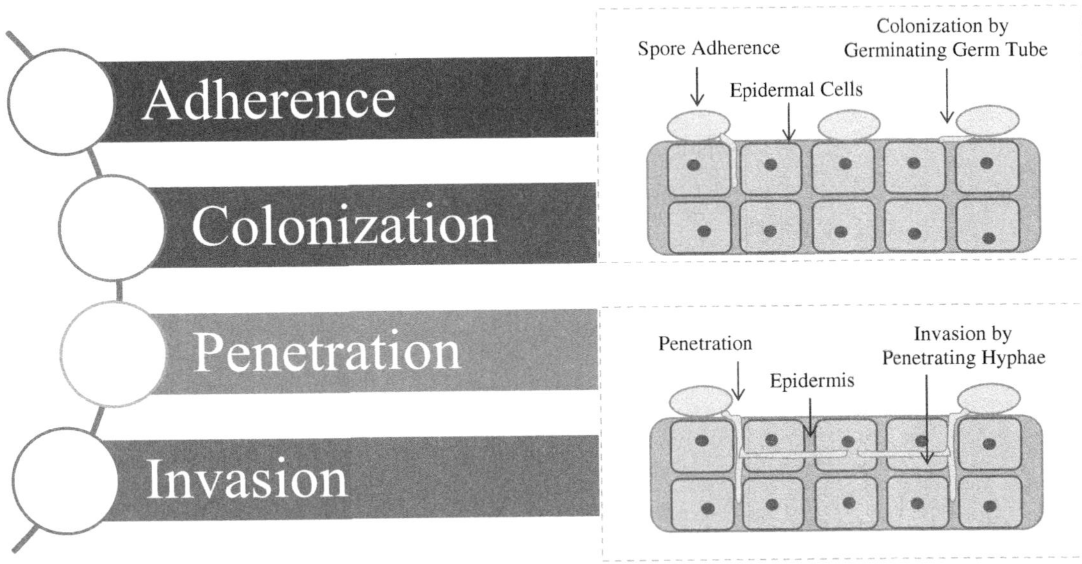

FIG. 3 Schematic representation of different steps in the invasion of human tissue by *Candida*.

be important (Al-Dhaheri & Douglas, 2008; Al-Fattani & Douglas, 2006; LaFleur, Kumamoto, & Lewis, 2006). As of late, a possible job for β-1,3-glucan in antifungal obstruction was hypothesized after it was shown that cell dividers from biofilm cells could tie antifungals better than planktonic variants, and that exogenous β-1,3-glucan decreased the action of fluconazole against planktonic *C. albicans* cells (Nett et al., 2007). Another contributing variable may be the job of cross resistance. It has been habitually seen that gentle types of stress might get ready cells for ensuing (expanded) stress conditions of an alternate sort (Kara, Luppens, & ten Cate, 2006). Biofilm cells live in a supplement poor, hypoxic climate, and it is possible that the imperfect development conditions in biofilms may bring about expanded resilience to different types of stress, including antimycotics-related cell stress.

Since *C. albicans* and other *Candida* spp. share their (oral) climate with numerous bacterial species, the inquiry emerges regardless of whether bacterial-contagious collaborations can influence biofilm formation, morphology, and destructiveness of *Candida*. This is for sure the case. *C. albicans* can tie to the oral organisms *Streptococcus gordonii*, *Streptococcus oralis*, and *Streptococcus sanguinis*, bringing about co-accumulation. This connection is intervened by streptococcal cell-divider polysaccharides and cell-surface proteins and by at this point obscure *Candida* adhesins and is promoted by explicit salivary proteins (O'Sullivan, Jenkinson, & Cannon, 2000). Not just direct actual cell-cell interactions can influence biofilm development and destructiveness, yet additionally roundabout associations interceded by discharged metabolic results also by extracellular flagging particles, for example, "majority sensing" atoms (Wargo & Hogan, 2006). Connections between and among various host microorganisms in the oral hole might effetely affect surface colonization and biofilm development. For instance, rivalry between distinctive *Candida* species (*C. albicans* and *C. krusei*) and among *Candida* and microbes has been accounted for, with an additional balancing impact by spit (Thein, Samaranayake, & Samaranayake, 2007). A negative relationship between *Porphyromonas gingivalis* and *C. albicans* biomass in a biofilm brought about hypothesis on a potential restraint of *Candida* colonization applied by this periodontal microorganism in the gingival crevicular region (Thein, Samaranayake, & Samaranayake, 2006). It was additionally shown that morphological advances, and in this way harmfulness, can be affected by the presence of oral microorganisms in a *Candida* biofilm (Fig. 4) (Pereira-Cenci et al., 2008). The job of the majority detecting atoms emitted by *C. albicans* has been contemplated in impressive detail. *C. albicans* utilizes somewhere around two majority detecting atoms, tyrosol and farnesol. Under conditions tolerant for microbe tube development, tyrosol animates hyphal development in *C. albicans*, while farnesol restrains the change from yeast to hyphal development (Chen, Fujita, Feng, Clardy, & Fink, 2004; Hornby et al., 2001). As of late, it was found that *C. albicans* additionally secretes dodecanol, which likewise restrains the yeast-to-hypha change (Martins et al., 2007). Significantly, culture supernatants from mature *C. albicans* biofilms hinder filamentous development by planktonic *C. albicans* cells, which may demonstrate that farnesol or potentially dodecanol is delivered in situ in biofilms (Ramage et al., 2002). Further, farnesol has been displayed to connect with the cell layers of a few bacterial animal groups, including those that are available in multispecies oral biofilms. Accordingly, bacterial development, digestion, and polysaccharide arrangement are impacted (Koo et al., 2003). Studies on

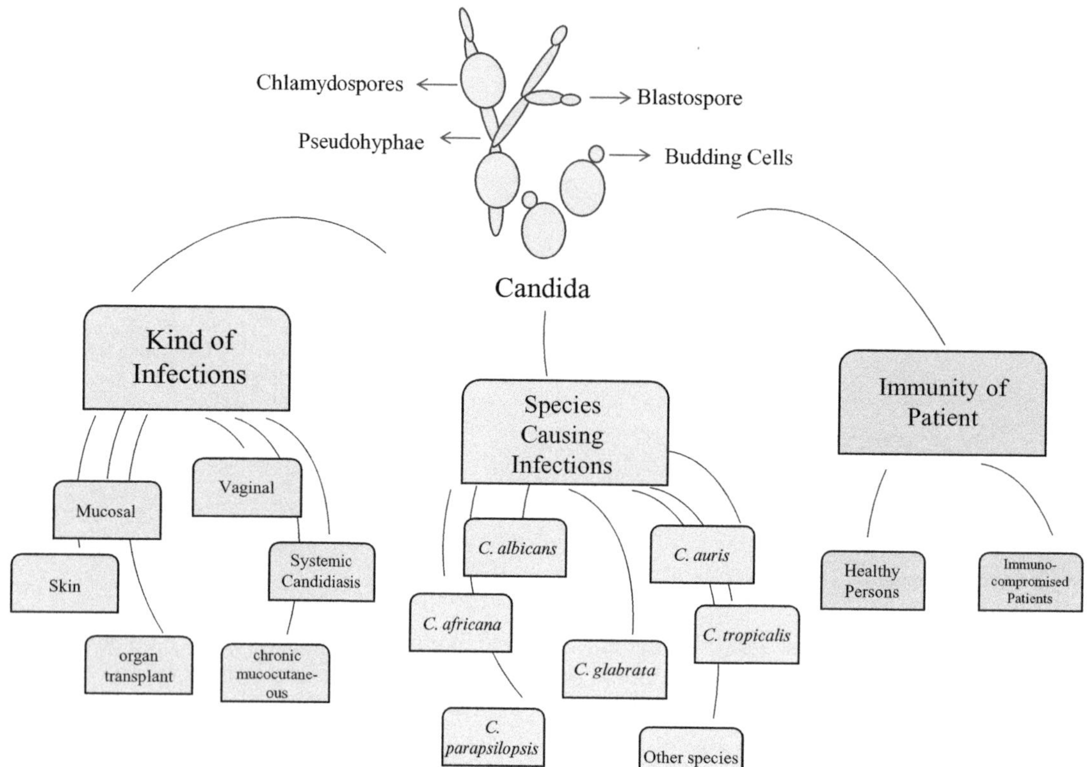

FIG. 4 Schematic representation of features of species of *Candida* in causing kind of infections and patients it infects.

the impacts of farnesol on *Staphylococcus aureus* (biofilms) validated the impact on bacterial cell films, in any case, more critically, showed sensitization of *S. aureus* (and *Escherichia coli*) to different antimicrobials upon preopenness to farnesol (Brehm-Stecher & Johnson, 2003; Inoue et al., 2004). The way that farnesol inhibits biofilm development by *Staphylococci* (Jabra-Rizk, Shirtliff, James, & Meiller, 2006) demonstrates that it has incredible potential in the control of blended *Candida*-microscopic organisms biofilms.

Candida signals influence microorganisms, yet similarly bacterial quorum-sensing atoms can affect *C. albicans*. The bacterial majority detecting signals 3-oxo-C12 homoserine lactone andcis-11-methyl-2-dodecanoic corrosive both regulate filamentation, furthermore in this way destructiveness, in *C. albicans* (Hogan, Vik, & Kolter, 2004; Wang et al., 2004). For instance, the flagging particle homoserine lactone assumes a part in the medicinally significant communications between *C. albicans* and *Pseudomonas aeruginosa*, which separately or together structure biofilms on clinical gadgets, and increment each other's opposition toward antimicrobials (Hogan & Kolter, 2002; Pierce, 2005; Wargo & Hogan, 2006). An illustration of a significantly more intricate cross-flagging that might happen in a blended biofilm is the perception that farnesol diminishes the development of the (Pseudomonas) flagging atom quinolone in societies of *P. aeruginosa* (Cugini et al., 2007). The quinolone signal in *P. aeruginosa* directs harmfulness factors such as the amalgamation of pyocyanin, a redox-dynamic phenazine compound that hinders development of *C. albicans*. This complex substance communication might advance *C. albicans* endurance.

7. Genomics of biofilm formation and antifungal resistance

Biofilm is one of the most important virulence factors of species of *Candida* both in terms of antifungal resistance and protection from host immune. The biofilms help the yeast to colonize both living and nonliving surfaces with equal ease such as skin surface, venous catheters, intracardiac prosthetic devices, and other implanted devices (Modiri et al., 2019). As discussed above, the production and formation of biofilm are a complex but well-coordinated process of yeasts cells. The biofilm-related genes in species of *Candida*, namely cell wall regulator1 (BCR1), beta-1,3-glucan synthase catalytic subunit 1 (FKS1), and enhanced filamentous growth protein 1 (EFG1) together form the biofilm regulatory network genes (Araújo, Henriques, & Silva, 2017; Ding et al., 2011; Holland et al., 2014). The development of resistance against available antifungal drugs by several strains of *Candida* has become a global problem. In-depth studies are required to look into the point mutations in genes involved in drug resistance. Advanced techniques of whole genome are an important tool to study the same. The mutations in the genes responsible for antifungal resistance can only be studied or monitored on regular basis. Khalaf, Fattouh, Medvecky, and Hrabak (2021) showed by whole-genome data that the increased resistance of an isolate of *C. albicans* CA77 is due to the increased ergosterol content and point mutations in genes of ergosterol biosynthesis pathway. Studies on the expression of genes responsible for adherence, growth, and biofilm formation, the use of various antifungal extracts of plants are studied using such techniques. Kim and Kim (2021) studied the effect of *Hedera rhombea* extract on the growth and biofilm formation of *C. tropicalis*, *C. parapsilosis*, and *C. glabrata*. The molecular mechanism of biofilm formation and inhibition by extracts of *H. rhombea* was elucidated by looking into the expression of hyphal-associated gene, extracellular matrix gene, and other pathways.

Whole-genome data are used to monitor the antifungal resistance pattern among populations across globe. Surveillance at the local as well as global level is important to trace the antifungal resistance pattern prevalent among various strains. The whole-genome sequencing provides a good opportunity for the same. It helps in understanding the mechanism of resistance in prevalent clinical isolates, which can help in drug discovery. Moreover, it may help to compare the clinical isolates with those of environmental isolates. Since the molecular mechanism of acquired resistance is complex and depends on more than one factor, whole-genome sequence data helps in understanding those multiple factors. It would be otherwise difficult to elucidate experimentally and more time-consuming. McTaggart, Cabrera, Cronin, and Kus (2020) showed that resistance to antifungal agent such as azole was probably due to homozygous ERG3 mutation and a previously undescribed chromosomal-duplication event (ERG11 and TAC1 orthologs) in *C. albicans* and *C. tropicalis*, respectively.

8. Conclusion

Multiple factors are responsible for the formation of biofilm by species of *Candida*. The factors include adhesion, changes in cell morphology and exo-polysaccharide secretion. Apart from this, the potential of strain, surface to be adhered, etc., also have effect on the characteristic of biofilm. From the clinical and treatment point of view, research has to be concentrated on the biofilm formation and its relation to antifungal resistance of *Candida* to available drugs. The recent data of whole-genome sequence have helped a lot in understanding many concepts and phenomenon. This has helped in developing new therapies and selection of suitable combination of drug treatment. Despite so much work, there is a need to study species-wise mechanism of biofilm formation, increased pathogenesis, and persistent infections.

Acknowledgment

RS thanks Department of Biotechnology, New Delhi, for the extramural grant no. BT/PR25490/NER/95/1220/2017 dated 28.06.2018. GM thanks DBT, New Delhi, for JRF fellowship.

References

Adhya, A., Bain, J., Ray, O., Hazra, A., Adhikari, S., Dutta, G., et al. (2015). Healing of burn wounds by topical treatment: A randomized controlled comparison between silver sulfadiazine and nano-crystalline silver. *Journal of Basic and Clinical Pharmacy*, *6*, 29–34.

Ahmadi, M. S., Lee, H. H., Sanchez, D. A., Friedman, A. J., Tar, M. T., Davies, K. P., et al. (2016). Sustained nitric oxide-releasing nanoparticles induce cell death in *Candida albicans* yeast and hyphal cells, preventing biofilm formation in vitro and in a rodent central venous catheter model. *Antimicrobial Agents and Chemotherapy*, *60*(4), 2185–2194.

Al-Dhaheri, R. S., & Douglas, L. J. (2008). Absence of amphotericin B-tolerant persister cells in biofilms of some *Candida* species. *Antimicrobial Agents and Chemotherapy*, *52*, 1884–1887.

Alexander, K. D., Hampton, M. J., Zhang, S., Dhawan, A., Xu, H., & Lopez, R. (2009). A high-throughput method for controlled hot-spot fabrication in SERS-active gold nanoparticle dimer arrays. *Journal of Raman Spectroscopy: An International Journal for Original Work in all Aspects of Raman Spectroscopy, Including Higher Order Processes, and also Brillouin and Rayleigh Scattering*, *40*(12), 2171–2175.

Al-Fattani, M. A., & Douglas, L. J. (2006). Biofilm matrix of *Candida albicans* and *Candida tropicalis*: Chemical composition and role in drug resistance. *Journal of Medical Microbiology*, *55*, 999–1008.

Alim, D., Sircaik, S., & Panwar, S. L. (2018). The significance of lipids to biofilm formation in *Candida albicans*: An emerging perspective. *Journal of Fungi*, *4*, 140. https://doi.org/10.3390/jof4040140.

Araújo, D., Henriques, M., & Silva, S. (2017). Portrait of *Candida* species biofilm regulatory network genes. *Trends in Microbiology*, *25*(1), 62–75.

Atriwal, T., Azeem, K., Husain, F. M., Hussain, A., Khan, M. N., Alajmi, M. F., et al. (2021). Mechanistic understanding of *Candida albicans* biofilm formation and approaches for its inhibition. *Frontiers in Microbiology*, *12*, 932.

Barillo, D. J., & Marx, D. E. (2014). Silver in medicine: A brief history BC 335 to present. *Burns*, *40*, S3–S8.

Berkow, E. L., & Lockhart, S. R. (2017). Fluconazole resistance in *Candida* species: A current perspective. *Infection and Drug Resistance*, *10*, 237–245. https://doi.org/10.2147/IDR.S118892.

Brehm-Stecher, B. F., & Johnson, E. A. (2003). Sensitization of *Staphylococcus aureus* and *Escherichia coli* to antibiotics by the sesquiterpenoids nerolidol, farnesol, bisabolol, and apritone. *Antimicrobial Agents and Chemotherapy*, *47*, 3357–3360.

Cappelletty, D., & Eiselstein-McKitrick, K. (2007). The echinocandins. *Pharmacotherapy*, *27*, 369–388.

Ten Cate, J. M., Klis, F. M., Pereira-Cenci, T., Crielaard, W., & De Groot, P. W. (2009). Molecular and cellular mechanisms that lead to *Candida* biofilm formation. *Journal of Dental Research*, *88*(2), 105–115. https://doi.org/10.1177/0022034508329273. 19278980.

Cavalheiro, M., & Teixeira, M. C. (2018). *Candida* biofilms: Threats, challenges, and promising strategies. *Frontiers in Medicine*, *5*, 28. https://doi.org/10.3389/fmed.2018.00028.

Chen, H., Fujita, M., Feng, Q., Clardy, J., & Fink, G. R. (2004). Tyrosol is a quorum-sensing molecule in *Candida albicans*. *Proceedings of the National Academy of Sciences of the United States of America*, *101*, 5048–5052.

Cugini, C., Calfee, M. W., Farrow, J. M., 3rd, Morales, D. K., Pesci, E. C., & Hogan, D. A. (2007). Farnesol, a common sesquiterpene, inhibits PQS production in *Pseudomonas aeruginosa*. *Molecular Microbiology*, *65*, 896–906.

d'Enfert, C. (2006). Biofilms and their role in the resistance of pathogenic *Candida* to antifungal agents. *Current Drug Targets*, *7*, 465–470.

d'Enfert, C., & Janbon, G. (2016). Biofilm formation in *Candida glabrata*: What have we learnt from functional genomics approaches? *FEMS Yeast Research*, *16*(1).

Da Silva, A. R., de Andrade Neto, J. B., da Silva, C. R., de Sousa, C. R., Silva, R. A. C., Freitas, D. D., et al. (2016). Berberine antifungal activity in fluconazole-resistant pathogenic yeasts: Action mechanism evaluated by flow cytometry and biofilm growth inhibition in *Candida* spp. *Antimicrobial Agents and Chemotherapy*, *60*, 3551–3557. https://doi.org/10.1128/AAC.01846-15.

Ding, C., Vidanes, G. M., Maguire, S. L., Guida, A., Synnott, J. M., Andes, D. R., et al. (2011). Conserved and divergent roles of Bcr1 and CFEM proteins in *Candida parapsilosis* and *Candida albicans*. *PLoS One*, *6* (12), e28151.

Douglas, L. J. (1992). Mannoprotein adhesins of *Candida albicans*. In J. E. Bennett, R. J. Hay, & P. K. Peterson (Eds.), *New strategies in fungal disease* (pp. 34–50). Edinburgh: Churchill Livingstone.

Fox, E. P., Bui, C. K., Nett, J. E., Hartooni, N., Mui, M. C., Andes, D. R., et al. (2015). An expanded regulatory network temporally controls *Candida albicans* biofilm formation. *Molecular Microbiology*, *96*, 1226–1239. https://doi.org/10.1111/mmi.13002.

Golipour, F., Habibipour, R., & Moradihaghgou, L. (2019). Investigating effects of superparamagnetic iron oxide nanoparticles on *Candida albicans* biofilm formation. *Medical Laboratory Journal*, *13*(6), 44–50.

Gondim, B. L. C., Castellano, L. R. C., de Castro, R. D., Machado, G., Carlo, H. L., Valença, A. M. G., et al. (2018). Effect of chitosan nanoparticles on the inhibition of *Candida* spp. biofilm on denture base surface. *Archives of Oral Biology*, *94*, 99–107.

Guisbiers, G., Lara, H. H., Mendoza-Cruz, R., Naranjo, G., Vincent, B. A., Peralta, X. G., et al. (2017). Inhibition of *Candida albicans* biofilm by pure selenium nanoparticles synthesized by pulsed laser ablation in liquids. *Nanomedicine: Nanotechnology, Biology and Medicine*, *13*(3), 1095–1103.

Gulati, M., & Nobile, C. J. (2016). *Candida albicans* biofilms: Development, regulation, and molecular mechanisms. *Microbes and Infection*, *18*, 310–321. https://doi.org/10.1016/j.micinf.2016.01.002.

Haghighi, F., Mohammadi, S. R., Mohammadi, P., Hosseinkhani, S., & Shipour, R. (2013). Antifungal activity of TiO_2 nanoparticles and EDTA on *Candida albicans* biofilms. *Infection, Epidemiology and Microbiology*, *1*(1), 33–38.

Hawser, S. P., Baillie, G. S., & Douglas, L. J. (1998). Production of extracellular matrix by *Candida albicans* biofilms. *Journal of Medical Microbiology*, *47*, 253–256.

Hawser, S. P., & Douglas, L. J. (1994). Biofilm formation by *Candida* species on the surface of catheter materials in vitro. *Infection and Immunity*, *62*, 915–921.

Hernandez-Delgadillo, R., Velasco-Arias, D., Martinez-Sanmiguel, J. J., Diaz, D., Zumeta-Dube, I., Arevalo-Nino, K., et al. (2013). Bismuth oxide aqueous colloidal nanoparticles inhibit *Candida albicans* growth and biofilm formation. *International Journal of Nanomedicine*, *8*, 1645–1652.

Hobden, C., Teevan, C., Jones, L., & O'Shea, P. (1995). Hydrophobic properties of the cell surface of *Candida albicans*: A role in aggregation. *Microbiology*, *141*, 1875–1881.

Hogan, D. A., & Kolter, R. (2002). *Pseudomonas-Candida* interactions: An ecological role for virulence factors. *Science*, *296*, 2229–2232.

Hogan, D. A., Vik, A., & Kolter, R. (2004). A *Pseudomonas aeruginosa* quorum-sensing molecule influences *Candida albicans* morphology. *Molecular Microbiology*, *54*, 1212–1223.

Holland, L. M., Schröder, M. S., Turner, S. A., Taff, H., Andes, D., Grózer, Z., et al. (2014). Comparative phenotypic analysis of the major fungal pathogens *Candida parapsilosis* and *Candida albicans*. *PLoS Pathogens*, *10*(9), e1004365.

Hornby, J. M., Jensen, E. C., Lisec, A. D., Tasto, J. J., Jahnke, B., Shoemaker, R., et al. (2001). Quorum sensing in the dimorphic fungus *Candida albicans* is mediated by farnesol. *Applied and Environmental Microbiology*, *67*, 2982–2992.

Hosseini, S. S., Ghaemi, E., & Koohsar, F. (2018). Influence of ZnO nanoparticles on *Candida albicans* isolates biofilm formed on the urinary catheter. *Iranian Journal of Microbiology*, *10*(6), 424.

Huang, X., Zheng, M., Yi, Y., Patel, A., Song, Z., & Li, Y. (2020). Inhibition of berberine hydrochloride on *Candida albicans* biofilm formation. *Biotechnology Letters*, *42*(11), 2263–2269.

Hwang, I.-S., Hwang, J. H., Choi, H., Kim, K.-J., & Lee, D. G. (2012). Synergistic effects between silver nanoparticles and antibiotics and the mechanisms involved. *Journal of Medical Microbiology*, *61*(12), 1719–1726.

Inoue, Y., Shiraishi, A., Hada, T., Hirose, K., Hamashima, H., & Shimada, J. (2004). The antibacterial effects of terpene alcohols on Staphylococcus aureus and their mode of action. *FEMS Microbiology Letters*, *237*, 325–331.

Jabra-Rizk, M. A., Shirtliff, M., James, C., & Meiller, T. (2006). Effect of farnesol on *Candida dubliniensis* biofilm formation and fluconazole resistance. *FEMS Yeast Research*, *6*, 1063–1073.

Jin, Y., Samaranayake, L. P., Samaranayake, Y., & Yip, H. K. (2004). Biofilm formation of *Candida albicans* is variably affected by saliva and dietary sugars. *Archives of Oral Biology*, *49*(10), 789–798.

Kanchanapiboon, J., Kongsa, U., Pattamadilok, D., Kamponchaidet, S., Wachisunthon, D., Poonsatha, S., et al. (2020). *Boesenbergia rotunda* extract inhibits *Candida albicans* biofilm formation by pinostrobin and pinocembrin. *Journal of Ethnopharmacology*, *261*, 113193.

Kapteyn, J. C., Hoyer, L. L., Hecht, J. E., Muller, W. H., Andel, A., Verkleij, A. J., et al. (2000). The cell wall architecture of *Candida albicans* wild-type cells and cell wall-defective mutants. *Molecular Microbiology*, *35*, 601–611.

Kara, D., Luppens, S. B., & ten Cate, J. M. (2006). Differences between single- and dual-species biofilms of *Streptococcus mutans* and *Veillonella parvula* in growth, acidogenicity and susceptibility to chlorhexidine. *European Journal of Oral Sciences*, *114*, 58–63.

Kaur, R., Goyal, R., Dhakad, M. S., Bhalla, P., & Kumar, R. (2014). Epidemiology and virulence determinants including biofilm profile of *Candida* infections in an ICU in a tertiary hospital in India. *The Journal of Mycology*, *27*, 171–172.

Ke, C. L., Liao, Y. T., & Lin, C. H. (2021). MSS2 maintains mitochondrial function and is required for chitosan resistance, invasive growth, biofilm formation and virulence in *Candida albicans*. *Virulence, 12* (1), 281–297.

Khalaf, R. A., Fattouh, N., Medvecky, M., & Hrabak, J. (2021). Whole genome sequencing of a clinical drug resistant *Candida albicans* isolate reveals known and novel mutations in genes involved in resistance acquisition mechanisms. *Journal of Medical Microbiology, 70*(4).

Kim, D., & Kim, K. Y. (2021). *Hedera rhombea* inhibits the biofilm formation of *Candida*, thereby increases the susceptibility to antifungal agent, and reduces infection. *PLoS One, 16*(10), e0258108.

Klis, F. M., Boorsma, A., & De Groot, P. W. J. (2006). Cell wall construction in *Saccharomyces cerevisiae*. *Yeast, 23*, 185–202.

Kojic, E. M., & Darouiche, R. O. (2004). *Candida* infections of medical devices. *Clinical Microbiology Reviews, 17*, 255–267. https://doi.org/10.1128/CMR.17.2.255-267.2004.

Kong, F., Wang, J., Han, R., Ji, S., Yue, J., Wang, Y., et al. (2020). Antifungal activity of magnesium oxide nanoparticles: Effect on the growth and key virulence factors of *Candida albicans*. *Mycopathologia, 185*(3), 485–494.

Koo, H., Hayacibara, M. F., Schobel, B. D., Cury, J. A., Rosalen, P. L., Park, Y. K., et al. (2003). Inhibition of *Streptococcus mutans* biofilm accumulation and polysaccharide production by apigenin and tt-farnesol. *The Journal of Antimicrobial Chemotherapy, 52*, 782–789.

Kothari, A., & Sagar, V. (2009). Epidemiology of *Candida* bloodstream infections in a tertiary care institute in India. *Indian Journal of Medical Microbiology, 27*, 171. https://doi.org/10.4103/0255-0857.49440.

Koziel, J., & Potempa, J. (2013). Protease-armed bacteria in the skin. *Cell and Tissue Research, 351*, 325–337. https://doi.org/10.1007/s00441-012-1355-2.

Kumar, R., Saraswat, D., Tati, S., & Edgerton, M. (2015). Novel aggregation properties of *Candida albicans* secreted aspartyl proteinase Sap6 mediate virulence in oral candidiasis. *Infection and Immunity, 83*, 2614–2626. https://doi.org/10.1128/iai.00282-15.

LaFleur, M. D., Kumamoto, C. A., & Lewis, K. (2006). *Candida albicans* biofilms produce antifungal-tolerant persister cells. *Antimicrobial Agents and Chemotherapy, 50*, 3839–3846.

Lara, H. H., Romero-Urbina, D. G., Pierce, C., Lopez-Ribot, J. L., Arellano-Jiménez, M. J., & Jose-Yacaman, M. (2015). Effect of silver nanoparticles on *Candida albicans* biofilms: An ultrastructural study. *Journal of Nanobiotechnology, 13*(1), 1–12.

Lara, P., Palma-Florez, S., Salas-Huenuleo, E., Polakovicova, I., Guerrero, S., Lobos-Gonzalez, L., et al. (2020). Gold nanoparticle based double-labeling of melanoma extracellular vesicles to determine the specificity of uptake by cells and preferential accumulation in small metastatic lung tumors. *Journal of Nanobiotechnology, 18*(1), 1–17.

Lockhart, S. R. (2014). Current epidemiology of *Candida* infection. *Clinical Microbiology Newsletter, 36*, 131–136. https://doi.org/10.1016/j.clinmicnews.2014.08.001.

Lohse, M. B., Gulati, M., Craik, C. S., Johnson, A. D., & Nobile, C. J. (2020). Combination of antifungal drugs and protease inhibitors prevent *Candida albicans* biofilm formation and disrupt mature biofilms. *Frontiers in Microbiology, 11*, 1027.

Lohse, M. B., Gulati, M., Johnson, A. D., & Nobile, C. J. (2018). Development and regulation of single- and multi-species *Candida albicans* biofilms. *Nature Reviews Microbiology, 16*, 19–31. https://doi.org/10.1038/nrmicro.2017.107.

Longhi, C., Santos, J. P., Morey, A. T., Marcato, P. D., Duran, N., Pinge-Filho, P., et al. (2015). Combination of fluconazole with silver nanoparticles produced by *Fusarium oxysporum* improves antifungal effect against planktonic cells and biofilm of drug-resistant *Candida albicans*. *Sabouraudia, 54*(4), 428–432.

Mahamuni-Badiger, P. P., Patil, P. M., Badiger, M. V., Patel, P. R., Thorat-Gadgil, B. S., Pandit, A., et al. (2020). Biofilm formation to inhibition: Role of zinc oxide-based nanoparticles. *Materials Science and Engineering: C, 108*, 110319.

Manoharan, R. K., Lee, J. H., Kim, Y. G., & Lee, J. (2017). Alizarin and chrysazin inhibit biofilm and hyphal formation by *Candida albicans*. *Frontiers in Cellular and Infection Microbiology, 7*, 447. https://doi.org/10.3389/fcimb.2017.00447.

Martínez-García, S., Rodríguez-Martínez, S., Cancino-Diaz, M. E., & Cancino-Diaz, J. C. (2018). Extracellular proteases of *Staphylococcus epidermidis*: Roles as virulence factors and their participation in biofilm. *APMIS, 126*, 177–185. https://doi.org/10.1111/apm.12805.

Martins, M., Henriques, M., Azeredo, J., Rocha, S. M., Coimbra, M. A., & Oliveira, R. (2007). Morphogenesis control in *Candida albicans* and *Candida dubliniensis* through signaling molecules produced by planktonic and biofilm cells. *Eukaryotic Cell, 6*, 2429–2436.

McCourtie, J., & Douglas, L. J. (1984). Relationship between cell surface composition, adherence, and virulence of *Candida albicans*. *Infection and Immunity, 45*, 6–12.

McCourtie, J., & Douglas, L. J. (1985). Extracellular polymer of *Candida albicans*: Isolation, analysis and role in adhesion. *Journal of General Microbiology, 131*, 495–503.

McTaggart, L. R., Cabrera, A., Cronin, K., & Kus, J. V. (2020). Antifungal susceptibility of clinical yeast isolates from a large Canadian reference laboratory and application of whole-genome sequence analysis to elucidate mechanisms of acquired resistance. *Antimicrobial Agents and Chemotherapy, 64*(9), e00402–e00420.

Modiri, M., Hashemi, S. J., Ghazvini, R. D., Khodavaisy, S., Ahmadi, A., Ghaffari, M., et al. (2019). Antifungal susceptibility pattern and biofilm-related genes expression in planktonic and biofilm cells of *Candida parapsilosis* species complex. *Current Medical Mycology, 5*(4), 35.

Mohamed, A. A., Fouda, A., Elgamal, M. S., EL-Din Hassan, S., Shaheen, T. I., & Salem, S. S. (2017). Enhancing of cotton fabric antibacterial properties by silver nanoparticles synthesized by new Egyptian strain *Fusarium keratoplasticum* A1-3. *Egyptian Journal of Chemistry, 60* (Conference Issue), 63–71. (*The 8th international conference of the textile research division (ICTRD 2017). Cairo, Egypt: National Research Centre*).

Monteiro, D. R., Silva, S., Negri, M., Gorup, L. F., De Camargo, E. R., Oliveira, R., et al. (2012). Silver nanoparticles: Influence of stabilizing agent and diameter on antifungal activity against *Candida albicans* and *Candida glabrata* biofilms. *Letters in Applied Microbiology, 54* (5), 383–391.

Mukherjee, P. K., Chandra, J., Kuhn, D. M., & Ghannoum, M. A. (2003). Mechanism of fluconazole resistance in *Candida albicans* biofilms: Phase-specific role of efflux pumps and membrane sterols. *Infection and Immunity, 71*, 4333–4340.

Mukherjee, P. K., Zhou, G., Munyon, R., & Ghannoum, M. A. (2005). *Candida* biofilm: A well-designed protected environment. *Medical Mycology, 43*, 191–208.

Muñoz, J. F., Gade, L., Chow, N. A., Loparev, V. N., Juieng, P., Berkow, E. L., et al. (2018). Genomic insights into multidrug-resistance, mating and virulence in *Candida auris* and related emerging species. *Nature Communications*, *9*(1), 1–13.

Nailis, H., Kucharíková, S., Ricicová, M., Van Dijck, P., Deforce, D., Nelis, H., et al. (2010). Real-time PCR expression profiling of genes encoding potential virulence factors in *Candida albicans* biofilms: Identification of model-dependent and -independent gene expression. *BMC Microbiology*, *10*, 114. https://doi.org/10.1186/1471-2180-10-114.

Nani, B. D., Sardi, J. C. O., Lazarini, J. G., Silva, D. R., Massariolli, A. P., Cunha, T. M., et al. (2019). Anti-inflammatory and anti-*Candida* effects of Brazilian organic propolis, a promising source of bioactive molecules and functional food. *Journal of Agricultural and Food Chemistry*, *68*(10), 2861–2871.

Nett, J., Lincoln, L., Marchillo, K., Massey, R., Holoyda, K., Hoff, B., et al. (2007). Putative role of β-1,3 glucans in *Candida albicans* biofilm resistance. *Antimicrobial Agents and Chemotherapy*, *51*, 510–520.

Nikawa, H., Nishimura, H., Hamada, T., Kumagai, H., & Samaranayake, L. P. (1997). Effects of dietary sugars and, saliva and serum on *Candida* biofilm formation on acrylic surfaces. *Mycopathologia*, *139*, 87–91.

O'Sullivan, J. M., Cannon, R. D., Sullivan, P. A., & Jenkinson, H. F. (1997). Identification of salivary basic proline-rich proteins as receptors for *Candida albicans* adhesion. *Microbiolgy*, *143*, 341–348.

O'Sullivan, J. M., Jenkinson, H. F., & Cannon, R. D. (2000). Adhesion of *Candida albicans* to oral streptococci is promoted by selective adsorption of salivary proteins to the streptococcal cell surface. *Microbiology*, *146*(Pt 1), 41–48.

Padmavathi, A. R., Das, A., Priya, A., Sushmitha, T. J., Pandian, S. K., & Toleti, S. R. (2020). Impediment to growth and yeast-to-hyphae transition in *Candida albicans* by copper oxide nanoparticles. *Biofouling*, *36*(1), 56–72.

Paharik, A. E., Kotasinska, M., Both, A., Hoang, T.-M. N., Büttner, H., Roy, P., et al. (2017). The metalloprotease SepA governs processing of accumulation-associated protein and shapes intercellular adhesive surface properties in *Staphylococcus epidermidis*. *Molecular Microbiology*, *103*, 860–874. https://doi.org/10.1111/mmi.13594.

Pappas, P. G., Lionakis, M. S., Arendrup, M. C., Ostrosky-Zeichner, L., & Kullberg, B. J. (2018). Invasive candidiasis. *Nature Reviews Disease Primers*, *4*, 18026. https://doi.org/10.1038/nrdp.2018.26.

Pereira-Cenci, T., Deng, D. M., Kraneveld, E. A., Manders, E. M., Del BelCury, A. A., Ten Cate, J. M., et al. (2008). The effect of *Streptococcus mutans* and *Candida glabrata* on *Candida albicans* biofilms formed on different surfaces. *Archives of Oral Biology*, *53*, 755–764.

Perlroth, J., Choi, B., & Spellberg, B. (2007). Nosocomial fungal infections: Epidemiology, diagnosis, and treatment. *Medical Mycology*, *45*, 321–346. https://doi.org/10.1080/13693780701218689.

Pierce, G. E. (2005). *Pseudomonas aeruginosa*, *Candida albicans*, and device-related nosocomial infections: Implications, trends, and potential approaches for control. *Journal of Industrial Microbiology & Biotechnology*, *32*, 309–318.

Pote, S. T., Sonawane, M. S., Rahi, P., Shah, S. R., Shouche, Y. S., Patole, M. S., et al. (2020). Distribution of pathogenic yeasts in different clinical samples: Their identification, antifungal susceptibility pattern, and cell invasion assays. *Infection and Drug Resistance*, *13*, 1133.

Ramage, G., Bachmann, S., Patterson, T. F., Wickes, B. L., & López-Ribot, J. L. (2002). Investigation of multidrug efflux pumps in relation to fluconazole resistance in *Candida albicans* biofilms. *The Journal of Antimicrobial Chemotherapy*, *49*, 973–980.

Ramage, G., Martínez, J. P., & López-Ribot, J. L. (2006). *Candida* biofilms on implanted biomaterials: A clinically significant problem. *FEMS Yeast Research*, *6*, 979–986. https://doi.org/10.1111/j.1567-1364.2006.00117.x.

Ramage, G., Robertson, S. N., & Williams, C. (2014). Strength in numbers: Antifungal strategies against fungal biofilms. *International Journal of Antimicrobial Agents*, *43*(2), 114–120.

Ramage, G., Saville, S. P., Thomas, D. P., & López-Ribot, J. L. (2005). *Candida* biofilms: An update. *Eukaryotic Cell*, *4*, 633–638.

Rasool, U., Sah, S. K., & Hemalatha, S. (2021). Effect of biosynthesized copper nanoparticles (CUNPS) on the growth and biofilm formation of fluconazole-resistant *Candida albicans*. *Journal of Microbiology, Biotechnology and Food Sciences*, *2021*, 21–24.

Sagatova, A. A., Keniya, M. V., Wilson, R. K., Sabherwal, M., Tyndall, J. D. A., & Monk, B. C. (2016). Triazole resistance mediated by mutations of a conserved active site tyrosine in fungal lanosterol 14α-demethylase. *Scientific Reports*, *6*(1), 1–11.

Samaranayake, L. P., & MacFarlane, T. W. (1980). An in vitro study of the adherence of *Candida albicans* to acrylic surfaces. *Archives of Oral Biology*, *25*, 603–609.

Samaranayake, L. P., & MacFarlane, T. W. (1982). The effect of dietary carbohydrates on the in vitro adhesion of *Candida albicans* to epithelial cells. *Journal of Medical Microbiology*, *15*, 511–517.

Seddighi, N. S., Salari, S., & Izadi, A. R. (2017). Evaluation of antifungal effect of iron-oxide nanoparticles against different *Candida* species. *IET Nanobiotechnology*, *11*(7), 883–888.

Sharma, R., & Jangid, K. (2016). Role of Quorum Sensing in fungal morphogenesis and pathogenesis. In *Fungal metabolites* (pp. 1–28). Cham, Switzerland: Springer International Publishing.

Silva, S., Rodrigues, C. F., Araújo, D., Rodrigues, M. E., & Henriques, M. (2017). *Candida* species biofilms' antifungal resistance. *Journal of Fungi*, *3*, 8. https://doi.org/10.3390/jof3010008.

Skerl, K. G., Segal, E., Sreevalsan, T., & Scheld, W. M. (1984). In vitro binding of *Candida albicans* yeast cells to human fibronectin. *Canadian Journal of Microbiology*, *30*, 221–227.

Sosinska, G. J., De Groot, P. W. J., Teixeira de Mattos, M. J., Dekker, H. L., De Koster, C. G., Hellingwerf, K. J., et al. (2008). Hypoxic conditions and iron restriction affect the cell wall proteome of *Candida albicans* grown under vagina-simulative conditions. *Microbiology*, *154*, 510–520.

Spormann, A. M., Thormann, K., Saville, R., Shukla, S., & Entcheva, P. (2004). Microbial biofilms. In *Nanoscale Technology in Biological Systems* (pp. 341–357). https://doi.org/10.1201/9780203500224.

Thein, Z. M., Samaranayake, Y. H., & Samaranayake, L. P. (2006). Effect of oral bacteria on growth and survival of *Candida albicans* biofilms. *Archives of Oral Biology*, *51*, 672–680.

Thein, Z. M., Samaranayake, Y. H., & Samaranayake, L. P. (2007). Characteristics of dual species *Candida* biofilms on denture acrylic surfaces. *Archives of Oral Biology*, *52*, 1200–1208.

Tokunaga, M., Kusamichi, M., & Koike, H. (1986). Ultrastructure of outermost layer of cell wall in *Candida albicans* observed by rapid-freezing technique. *Journal of Electron Microscopy*, *35*, 237–246.

Tsui, C., Kong, E. F., & Jabra-Rizk, M. A. (2016). Pathogenesis of *Candida albicans* biofilm. *Pathogens and Disease*, *74*, ftw018. https://doi.org/10.1093/femspd/ftw018.

Vera-González, N, & Shukla, A. (2020). Advances in biomaterials for the prevention and disruption of *Candida* biofilms. *Frontiers in Microbiology*, *11*, 538602. https://doi.org/10.3389/fmicb.2020.538602.

Walker, L. A., Munro, C. A., de Bruijn, I., Lenardon, M. D., McKinnon, A., & Gow, N. A. (2008). Stimulation of chitin synthesis rescues *Candida albicans* from echinocandins. *PLoS Pathogens*, *4*, e1000040.

Wang, L. H., He, Y., Gao, Y., Wu, J. E., Dong, Y. H., He, C., et al. (2004). A bacterial cell-cell communication signal with cross-kingdom structural analogues. *Molecular Microbiology*, *51*, 903–912.

Wargo, M. J., & Hogan, D. A. (2006). Fungal-bacterial interactions: A mixed bag of mingling microbes. *Current Opinion in Microbiology*, *9*, 359–364.

Winter, M. B., Salcedo, E. C., Lohse, M. B., Hartooni, N., Gulati, M., Sanchez, H., et al. (2016). Global identification of biofilm-specific proteolysis in *Candida albicans*. *MBio*, *7*, e01514–e01516. https://doi.org/10.1128/mBio.01514-16.

Xie, Y., Liu, X., & Zhou, P. (2020). In vitro antifungal effects of berberine against *Candida* spp. In planktonic and biofilm conditions. *Drug Design, Development and Therapy*, *14*, 87–101. https://doi.org/10.2147/DDDT.S230857.

Yapar, N. (2014). Epidemiology and risk factors for invasive candidiasis. *Therapeutics and Clinical Risk Management*, *10*, 95–105. https://doi.org/10.2147/TCRM.S40160.

Yien, L., Zin, N. M., Sarwar, A., & Katas, H. (2012). Antifungal activity of chitosan nanoparticles and correlation with their physical properties. *International Journal of Biomaterials*, *2012*.

Yin, Q. Y., De Groot, P. W. J., De Koster, C. G., & Klis, F. M. (2008). Mass spectrometry-based proteomics of fungal wall glycoproteins. *Trends in Microbiology*, *16*, 20–26.

Yu, S., Li, W., Liu, X., Che, J., Wu, Y., & Lu, J. (2016). Distinct expression levels of ALS, LIP, and SAP genes in *Candida tropicalis* with diverse virulent activities. *Frontiers in Microbiology*, *7*, 1175.

Zhong, H., Hu, D., Hu, G., Su, J., Bi, S., Zhang, Z., et al. (2017). Activity of Sanguinarine against *Candida albicans* biofilms. *Antimicrobial Agents and Chemotherapy*, *61*, e02259. https://doi.org/10.1128/AAC.02259-16.

Zorić, N., Kosalec, I., Tomić, S., Bobnjarić, I., Jug, M., Vlainić, T., et al. (2017). Membrane of *Candida albicans* as a target of berberine. *BMC Complementary and Alternative Medicine*, *17*(1), 268. https://doi.org/10.1186/s12906-017-1773-5.

Chapter 17

Removal and control of biofilms in wounds

Fergus Watson[a], Alicja Zimmer[b], Rui Chen[c], Marcus Swann[d], Lauren Foulkes[e], and Steven L. Percival[f]
[a]*Operations, 5D Health Protection Group Ltd., Liverpool, United Kingdom,* [b]*Getinge Applikon, Liverpool, England, United Kingdom,* [c]*Biofilm Centre, 5D Health Protection, Group, Liverpool, United Kingdom,* [d]*5D Health Protection Group Ltd., Centre of Excellence in Biofilm Science and Technologies (CEBST), Liverpool, United Kingdom,* [e]*Microbiology, Biofilm Centre, Liverpool, United Kingdom,* [f]*R&D, 5D Health Protection, Group, Liverpool, United Kingdom*

1. Introduction

Unlike acute wound infections, where microorganisms are assumed to exist predominately in the planktonic form, over 75% of chronic wounds have been found to support the presence of biofilms, where microbes are embedded in extracellular polymeric substances (EPS), comprising polysaccharides, proteins, extracellular DNA (eDNA), and lipids. EPS not only protects microbes from being engulfed and killed by neutrophils and macrophages, but it also acts as a diffusion barrier to antimicrobial agents by impeding their penetration and reaching therapeutic concentrations (Attinger & Wolcott, 2012). While most acute wounds heal through the "normal" healing cascade of events, chronic wounds are stalled possibly due predominately to the presence of biofilm.

Biofilm-based wound care on chronic wounds has been brought up by researchers and clinicians in recent publications and best practice statements (Bjarnsholt, Eberlein, Malone, & Schultz, 2017; Ochsner, Reiser, Fiechter, & Witholt, 1995; Swanson, Haesler, Angel, & Sussman, 2016). In these studies, they suggested that nonhealing chronic wounds, which failed to respond to standard care, should be assumed to house biofilms and that treatments should be focused toward the biofilms in the wounds, for example, wound bed, slough, wound dressing. This also includes wound cleaning, wound bed preparation, removal of biofilm, and prevention biofilm regrowth by the application of an appropriate antimicrobial dressing to keep the numbers of microorganisms reduced (Percival & Suleman, 2015).

2. Biofilm in wounds

Chronic wounds are often classified as such if they do not heal within a period of approximately 6 weeks (Cullell-Dalmau, Otero-Viñas, Ferrer-Solà, Sureda-Vidal, & Manzo, 2021); however, this time period has been cited to be as long as 12 weeks. Such wounds may include venous leg ulcers, pressure ulcers, and diabetic foot ulcers (Percival, 2017). Chronic wounds often do not progress past the inflammatory phase and are susceptible to a number of biochemical and physiological defects that can impair the healing process. Biofilm formation and wound chronicity are often linked as biofilms often lend themselves to the development of the chronic nature of the wound, delaying the healing process. Both acute and chronic wounds are treated by using a process described by Schultz et al., known as TIME (Schultz et al., 2003). The process involves removal of nonviable tissue from around the wound by debridement (T), then infection and inflammation are minimized with the use of antibiotics and antiinflammatories (I). Then, moisture imbalance within the wound is addressed and corrected using appropriate wound dressings (M). Finally, epithelialization and formation of new tissue are encouraged using the application of specific techniques and therapies such as growth factors (E) (Schultz et al., 2003). Although great care and diligence are put into the treatment of acute and chronic wounds, the treatment of the more chronic wounds often fails, and one of the major reasons for this failure is that acute infections can lead to the formation of biofilms (Bjarnsholt, 2013).

Chronic and acute wounds differ in both physiology and microbial flora. Acute wounds that are commonly treated with antibiotics are usually caused by a small diverse group of microorganisms. Chronic wounds are often deeper and polymicrobial in nature. Common microbial genera and species found within the chronic wound bed often include but are not limited to *Pseudomonas aeruginosa*, *Bacteroides* spp., *Enterococcus* spp., and *Staphylococcus* spp. These bacterial species are also all found as part of normal microbial flora of the body. Furthermore, the more chronic a wound, the more likely it is to find a greater number of anaerobic species due the low tissue oxygen level. This has been found particularly in pressure ulcers (Wolcott et al., 2010). Common anaerobic microbial species found in the chronic wound bed have included *Enterobacter cloacae*, *Peptostreptococcus magnus*, *Corynebacterium striatum*, and *Helcococcus kunzii*.

Chronic wound infections are not limited to bacterial species as a number of fungal species have also been

Understanding Microbial Biofilms. https://doi.org/10.1016/B978-0-323-99977-9.00006-5

implicated in wound bed biofilm formation. The most prevalent species is *Candida albicans* (Kalan et al., 2016). In more recent times, multidrug-resistant *Candida auris* has been identified as a global health threat due to multiple reasons: multidrug resistance is common within *C. auris* species, and some isolates are resistant to all three classes of antifungal drugs (e.g., azoles, polyenes, and echinocandins) (Lockhart et al., 2017). *C. auris* is also often misidentified by clinical laboratories as it presents in a similar way to other *Candida* species, meaning many cases of *C. auris* infection can go undetected. *C. auris* is also readily transmissible between patients in healthcare settings and is known to cause healthcare-associated outbreaks. Colonization of patients, especially on skin, can persist for weeks in a healthcare environment, possibly indefinitely (Forsberg et al., 2019).

The treatment of biofilms in chronic wounds can be difficult as they are highly recalcitrant to the more common antimicrobials unlike their planktonic counterparts (Spoering & Lewis, 2001). Traditional disinfectants, although effective, are cytotoxic, and therefore, their usage within a chronic wound setting is often discouraged unless used at effective therapeutic levels. More common therapies are longer-term and often utilize silver, iodine, and benzalkonium-chloride-based dressings and wound gels (Salisbury et al., 2018; Weigelt, McNamara, Sanchez, Hirt, & Kirsner, 2021; Wolcott et al., 2020). As previously mentioned, most anti-biofilm agents are cytotoxic, which leads to treatment being more intensive and dressing rotation becoming more frequent.

The study of biofilms within wounds has largely been performed on in vitro models, but these are reported to vary greatly from their in vivo counterparts. Biofilms grown in vitro often form large, single 3D structures and exhibit classic mushroom-like multicellular structures. In vivo grown biofilms on the other hand are smaller, don't exhibit the aforementioned mushroom-like structure, and tend to form many smaller aggregates segregated by host tissue. These aggregates do not necessarily have to be attached to a surface, whereas most in vitro models require the biofilm to be adhered to a surface for testing (Weigelt et al., 2021).

Biofilm infections can be split into two categories: infections associated with medical devices, e.g., central venous catheters, pacemakers, contact lenses, etc., and direct biofilm infections, e.g., cystic fibrosis, otitis media (OM), etc. (Lee & Yoon, 2017). Cystic fibrosis patients are well known to suffer from biofilm infections, mainly consisting of the bacterium *P. aeruginosa*; a water-borne bacterial species that is known to cause advanced chronic infections in compromised hosts. The infections that occur within the lungs of the patient can be difficult to treat. Antibiotic therapy relieves symptoms in short term but then fails to cure the ongoing infection caused by the biofilm (Costerton, Stewart, & Greenberg, 1999). Biofilms within the lungs of cystic fibrosis patients often consist of small aggregates encapsulated in EPS, and inflammation is induced by recruiting polymorphonuclear leukocytes (Lee & Yoon, 2017).

The condition OM is a complication of Eustachian tube dysfunction that occurs during viral infection of the upper respiratory tract. The three main bacterial species that are implicated in biofilm formation in OM patients are non-typeable *Haemophilus influenzae*, *Streptococcus pneumoniae*, and *Moraxella catarrhalis*, which all readily form biofilms both in vivo and in vitro (Bakaletz, 2007). When viewed inside an animal host, OM-related biofilms are viable, well-organized, and highly structured (Bakaletz, 2007). If left untreated, OM infections can spread to structures in the aural cavity, brain, and meninges. Often OM infections will be treated with antibiotics, but can also escalate to the need for surgery if the infection becomes too advanced for antibiotics and antimicrobial prophylactics to deal with (Lisby-Sutch, Nemec-Dwyer, Deeter, & Gaur, 1990).

3. Importance of biofilm to wound infections and management

As mentioned previously, wound healing occurs through a process of several overlapping steps including hemostasis, inflammation, cell proliferation, and tissue remodeling. Disruption in these steps can prevent wound healing and can lead to the formation of chronic wounds. Microbial colonization of wounds and the formation of biofilm have demonstrated to have a significant role in delayed wound healing (Attinger & Wolcott, 2012; Honorato-Sampaio, Guedes, Lima, & Borges, 2014; Omar, Wright, Schultz, Burrell, & Nadworny, 2017; Percival, Mayer, Malone, Swanson, & Schultz, 2017; Percival, Malone, Mayer, Salisbury, & Schultz, 2018). The presence of biofilms have been shown to prolong and arrest wound healing at the inflammatory phase directly or indirectly (Wolcott et al., 2010). On the one side, biofilms not only provide a physical shield for microbial cells to reduce the contact with immune cells and antimicrobial agents, but also produce chemical agents to kill or malfunction polymorphonuclear leukocytes (PMNs) and macrophages. For examples, *P. aeruginosa* biofilm produces cytotoxic agents, such as a quorum sensing molecule rhamnolipids (Ochsner et al., 1995; Pearson, Pesci, & Iglewski, 1997), to rapidly necrose and fend off PMNs (Jensen et al., 2007; Jensen, Givskov, Bjarnsholt, & Moser, 2010), and malfunction macrophages by distortion of the membrane to inhibit bacterial binding and phagocytosis (Mcclure & Schiller, 1992). One the other side, the persistence of microbial biofilm in chronic wound recruits excessive PMNs and macrophages and arrests the wound healing phase at a prolonged or unresolved inflammatory stage. Zhao et al. (2010) observed that the expressions of proinflammatory cytokines such as IL-1b, IL-6, and MMP-10 increased significantly by *P. aeruginosa* biofilm challenged in the diabetic mice model compared with without biofilm where no significant increase in TNF-a, TGF-b, and VEGF expressions was evident. Edwards and Harding (2004) observed that *S. aureus* biofilm initiated a sustained low-grade inflammatory response, which leads to unresolved inflammation and delayed wound healing. Other studies by Hanke, Heim, Angle, Sanderson, and Kielian (2013) and Thurlow et al. (2011) demonstrated that *S. aureus* biofilms were capable of modulating the immune response by stimulating macrophages toward an M2 instead of M1 response, attenuating inflammation and promoting fibrosis. Seth et al. (2012) demonstrated that polymicrobial biofilms increased the expression of IL-1b and TNF-a by a comparative study on single-species and polymicrobial wound biofilm containing *P. aeruginosa* and *S. aureus* using an in vivo rabbit ear model. Compared with each mono-species infection, Interestingly, only the components in *P. aeruginosa* biofilm EPS, such as lipopolysaccharides (LPS), deoxyribonucleic acids (DNA) (without bacterial cells) can activate and polarize PMNs and macrophages into a more aggressive proinflammatory subtype, known as a hyperinflammatory phenotype. PMNs and macrophages treated with LPS, DNA, and EPS secreted higher levels of TNF-a, IL-6, and IL12 p40, which is similar to that of N1/M1-type (Ciszek-Lenda et al., 2019).

Most chronic wounds are arrested at the inflammatory stage by the presence of microbial biofilms, which affects subsequent epithelialization, granulation tissue formation, angiogenesis, and tissue remodeling and leads to delayed or nonhealing wounds. Proliferative epidermis and deficient vascularization were observed in *P. aeruginosa* biofilm-challenged wounds in histology analysis. Gurjala et al. (2011) demonstrated that a sustained low-grade inflammatory response created by *S. aureus* biofilm impairs epithelial migration and granulation tissue formation. In the above polymicrobial biofilm model, Seth et al. (2012) also demonstrated upgradation in proinflammatory cytokines by polymicrobial biofilms, which increased the epithelial gaps in the wound and delayed wound healing. Zhao et al. (2012) observed that the healing time of *P. aeruginosa* biofilm challenged on wounds in diabetic mice increased approximately to 6 weeks, compared with 4 weeks for nonbiofilm-challenged normal wound.

4. Strategies of biofilm management in wounds

A consensus guideline by Schultz et al. (2017) claimed that the presence of biofilm should be assumed in all chronic nonhealing wounds and a biofilm-based treatment should be multifaceted with clinicians taking a step-down approach (Bjarnsholt et al., 2017). In vitro and clinic evidence all suggested that systemic antibiotics and topical antiseptics are of limited effect in biofilm management in chronic wound healing. Bacterial cells within biofilms have been shown to be up to 1000 times more tolerant to antimicrobial agents than the planktonic cells (Davies, 2003), due to the limited penetration of antibiotics and antiseptics through the EPS barrier and the limited metabolic activity of microbial cells embedded in the biofilm for antimicrobials to disrupt. Hence, different strategies are required for biofilm management in chronic wound care. Despite a growing consensus that biofilms are a significant barrier to chronic wound healing, there are no standard guidelines for antibiofilm strategies in wound care at present. Here, we would like to discuss three overlapping stages in antibiofilm strategies for biofilm-based wound care, which include prevention, removal, and killing (Percival, 2017).

4.1 Prevention

Prevention is always better than cure. All the open wounds, except the wounds that are created under aseptic conditions, are inevitably contaminated by microbes. The first stage in biofilm management is preventing the formation of a biofilm within the wound bed by preventing microbes entering the wound and attaching on to a biotic or abiotic surface. These include the clinician complying with standard procedures during the surgery, such as use of sterile gloves, cross-contamination prevention, use of disinfectants and antiseptics for skin preparation before a surgical procedure, and the use of prophylactic antibiotics for high-risk patients. To prevent wound contamination postsurgery, appropriate choices of wound dressings have an important role to play in helping to manage the wound's microbial bioburden, cross-contamination, infection risk, and supporting the patient's ability to heal the wound. Wound dressings that are used for acute wounds must be able to absorb exudate, reduce pain and discomfort for the patient, allow visual inspection of the wound, protect and support the formation of newly formed tissue, support a moist wound environment, and immobilize or sequester microbes and prevent their dissemination into the wound bed. For cost and safety reasons, the simple conventional dressings (such as basic wound contact or gauze dressings) should be the first choice for noninfected acute wounds. For chronic wounds, such as diabetic foot ulcers, pressure ulcers, venous leg ulcers, and arterial leg ulcers, advanced antimicrobial wound dressings should be considered. The advanced dressings (e.g., alginate, film, foam, hydrocolloid, and hydrogel dressings) for chronic wounds provide the optimal environment for wound healing by their special physical and chemical properties, which help in retaining moisture and absorbing exudate. For local wound infections, some dressings with antimicrobial properties can be used to reduce the level of bacteria at the wound surface.

For examples, dressings containing iodine or silver can be used to treat clinically infected wounds. Medical-grade honey has been used in clinical settings due to its antimicrobial and antiinflammatory properties. Dressing selection should be made after careful clinical assessment of the person's wound, their clinical condition and underlying pathologies, and their personal experience and preferences. Factors that should be considered in dressing selection include wound conditions, such as exudate, odor, infection and wound stage; dressing properties, such as absorption ability and adhesiveness; and patient experience, with regard to irritation, preference, frequency of dressing changes, pain and ease of operation. Of course, the cost-effectiveness also should be considered in dressing selection (Cullum et al., 2016). It is also important that the patient is discharged from hospital as quickly as possible following a surgical procedure to avoid microbial attachment and infection (Percival, 2017).

4.2 Removal

In most chronic wounds, microbial colonization has already occurred, which means the wound healing process has been impeded by the biofilm. Wound bed eschar and slough serve as a reservoir for microorganisms, damaged or devitalized cells, and inflammatory chemokines, which work as harmful agents that initiate persistent inflammatory response. Hence, the prerequisite of chronic wound healing is the successful resolution of inflammation by the neutralization and elimination of the harmful agents residing in the eschar and slough (Chen, Salisbury, & Percival, 2020; Percival & Suleman, 2015; Shi et al., 2009).

Debridement, which is the removal of dead, damaged, and infected tissue within eschar and slough, is often a treatment used in the management of chronic wounds. The main methods of debridement include biological debridement using sterile maggots to digest dead tissue and pathogens; chemical/enzymatic debridement using enzymes to chemically liquefy necrotic tissue, autolytic debridement using the patients' own enzymes with a hydrogel or hydrocolloids to moisten and liquefy necrotic tissue; mechanical debridement using a certain amount of force to physically remove devitalized tissue nonselectively; and surgical sharp and conservative sharp debridement using a scalpel, pair of scissors, or forceps to remove devitalized tissue, usually by a surgical practitioner (Table 1). Sometimes, an array of debridement methods is applied in complex chronic wound care to reduce risk of infection and ameliorate wound healing (Percival & Suleman, 2015; Schultz et al., 2003).

TABLE 1 Types of debridement.

Type	Advantages	Disadvantages
Sharp (Surgical)—Removing nonviable tissue through surgical instruments	• Immediate results • Targeted/selective • Used to remove large amounts of necrotic tissue and eschar	• Required trained and skilled clinician • Potential for complications • Often causes bleeding and pain
Biological—Maggot therapy	• Targeted/selective • Maggots exhibit an antimicrobial function	• Not applicable to all cases
Mechanical—Removing nonviable tissue through nonselective and physical interactions	• Easily and commonly performed by healthcare professionals • Often faster than autolytic and chemical debridement	• Nonselective • May cause damage to or remove healthy tissue surrounding the wound
Chemical—Use of proteolytic enzymes to remove nonviable tissue	• Useful in smaller wounds that may lack infection • Easily performed by clinicians • Selective—only targeting necrotic tissue and not harming viable tissue	• Causes irritation to surround tissue • Process is time-consuming • Effectiveness is reduced in dry environments
Autolytic—The natural digestion of the nonviable tissue by body's leukocytes	• Natural • Targeted/selective • Generally painless	• Process is time-consuming. • Effectiveness is reduced in infected wounds

Among these methods, autolytic debridement causes relatively little pain; it is versatile and selective and requires minimal skill/training resulting in it being the most commonly performed method for wound debridement. However, the process of autolytic debridement is slow and delayed by malnutrition, aging, and chronic diseases, which may increase the potential of infection (McCarty, Cochrane, Clegg, & Percival, 2012). Various wound dressings, such as hydrocolloids, hydrogels, transparent films, and honey, were designed and applied to enhance natural autolytic debridement as well as to provide an ideal environment for wound healing by rehydrating, softening, and even melting the hard eschar and slough (Percival & Suleman, 2015; Schultz, Chin, Moldawer, & Diegelmann, 2011; Schultz et al., 2003). Surfactant containing dressings, including solutions and gels, allow wound exudate and debris to soften, solubilize, and be more easily rinsed off with water during dressing changes, leaving a cleaner wound (Percival et al., 2018). This results in less tissue damage from the cleaning process, which aids the body's own repair process.

4.3 Treatments to kill the microbes

Microbial biofilms are inherently characterized by their tolerance to antimicrobial treatment and notably within the wound environment, their tolerance to the host's immune response. In spite of this, broad-spectrum antibiotic therapy is widely revered and an effective control measure for most microbial infections. Early and aggressive use of antibiotic treatment on newly formed biofilms has been shown to be more effective compared with a more developed or mature biofilm; however, most clinical biofilm-related infections are considered the latter, and therefore, alternative therapies are necessary in the absence of appropriate early diagnosis of wound infections (Hengzhuang, Wu, Ciofu, Song, & Høiby, 2011). The following subsections describe some unconventional antimicrobial strategies for combatting bacterial infections and thus biofilm formation within wounds.

4.3.1 Phage therapy

Bacteriophages prey on bacteria, by injecting their DNA into the cells, multiplying within before being released by lysis causing irreparable damage. Alternatively, as a result of the phage infection, depolymerase and alginase enzymes are produced within the bacterial cells disrupting the structural components of a biofilm. This type of virus has had considerable attention since its antimicrobial usage was discovered in the early 1900s and has been successfully used without any adverse effects in clinical trials to treat acute forms of *P. aeruginosa* and *S. aureus* wounds (Jault et al., 2019; Merabishvili et al., 2017). Naturally, this has led to biofilm research, whereby the bacteriophages' specificity toward pathogenic bacteria can be utilized to ensure the therapy does not harm the commensal flora and due to the inherently enhanced cell density of biofilms phage therapy in theory would exhibit improved efficiencies during treatment (Harper et al., 2014; Moye et al., 2018). In contrast, the matrix of EPS and variety in metabolic state of cells within the biofilm can hinder the progress of phage infection; and studies show that the complexity of biofilm (i.e., structure, bacterial strain, and composition) has a significant impact, thus making it difficult to conclusively state the performance of phage within this environment (Harper et al., 2014; Pires, Melo, Vilas Boas, Sillankorva, & Azeredo, 2017).

Bacteriophages therapy has potential use in wound care treatment as an up and coming antibiofilm agent (Donlan, 2009). However, there are many hurdles to overcome, many of which are shared by other targeted antimicrobial agents, these include: limited penetration through the biofilm matrix, whereby phage molecules are readily quenched by proteins and polysaccharides excreted by the biofilm cells (Rakhuba, Kolomiets, Dey, & Novik, 2010); differences in metabolic activity, the physiological state of biofilm cells can drastically influence phage performance whereby more active cells produce greater results, and studies have suggested that the survival response of cells in nutrient-depleted conditions such as biofilms has a reduced effect on phage infection efficacy, or quorum sensory activity, which sees that the communication molecules released by biofilm cells reduce the number of phage receptors on cells inhibiting attachment (Tan, Svenningsen, & Middelboe, 2015).

In more recent studies, a new approach of using multiple phage strains with an array of target receptors in a single treatment is referred to as a phage cocktail. By broadening their range of activity, the therapy reduces the likelihood of phage-resistance developing and increases performance particularly against multispecies biofilms (Chan, Abedon, & Loc-Carrillo, 2013). Chronic wound models comprising *Pseudomonas* spp., *Acinetobacter* spp., and *Staphylococcus* spp. when exposed to a phage "cocktail" demonstrated significant reductions in bacterial bioburden with the potential to promote the wound healing process. There are several reports to support the safe use of bacteriophages in clinical trials most notably in burn wound patients suffering with suspected *P. aeruginosa* infections (Jault et al., 2019; Kortright, Chan, Koff, & Turner, 2019). While successful continued research is required on fully formed biofilms, which present the greatest challenge for this therapy and the regulatory hurdles that would need to be overcome for exploitation commercially.

4.3.2 Probiotic therapeutics

The microbial ecology of chronic wounds is highly complex, and as a result, it is difficult to determine the "true" pathogenic strain or strains that are responsible for delay wound healing and infection. Previous pathology studies have shown that alterations in the skin microbiome could hinder the healing process in chronic wounds and therefore promote the potential use of topical probiotic wound treatment to support the host's natural microbiome (Vuotto, Longo, & Donelli, 2014; Wong, Martindale, Longaker, & Gurtner, 2013). Probiotics have successfully proven therapeutic effects in many pathological situations including inflammatory bowel diseases, viral gastroenteritis, and antibiotic-associated diarrhea (Chapman, Plosker, & Figgitt, 2006; Guandalini, Cernat, & Moscoso, 2015; Johnston, Goldenberg, Vandvik, Sun, & Guyatt, 2011). In principle, probiotic therapy utilizes nonpathogenic bacteria to combat and outcompete the proliferation of pathogenic species while supporting the endogenous flora; in this instance, promoting the restoration of the commensal microbiota at the wound site. By restoring balance of production of antiinflammatory signals, physiological state of host cells, and production of antimicrobial compounds, the levels of pathogenic bioburden can be reduced, ultimately promoting the wound healing process (Scales & Huffnagle, 2013). This is often referred to as recalibrating the wound's immune response.

Several studies, using *Lactobacillus* spp., have been able to show antibiofilm formation effects against *Pseudomonas* spp. and *Staphylococcus* spp. with clinical relevance. For example, the studies by Lopes et al. (2017) and Varma, Nisha, Dinesh, Kumar, and Biswas (2011) demonstrated the strains capacity to adhere to keratin and prevented the adhesion of potential skin-associated pathogens; this would owe itself to be a viable postsurgical treatment for preventing the biofilm formation in wound sites. Furthermore, probiotics using *Lactobacillus* spp. have been known to influence the migration, proliferation, and thus recruitment of keratinocyte for re-epithelialization, a key stage in the healing process (Mohammedsaeed, Cruickshank, McBain, & O'Neill, 2015). Probiotics are not an uncommon feature both in domestic and in clinical settings nor the concept of treating bacteria with more bacteria and, with the ongoing debate of antibiotic stewardship, may prove promising for wound biofilm management.

4.3.3 Blue light therapy

The wound healing process is a complex multistage process, and the use of low-intensity light has been revealed to enhance specific repair stages (Percival, Francolini, & Donelli, 2015). Light, both infrared and visible spectrum, has the ability to promote processes such as granulation tissue formation and collagen production within wounds (Adamskaya et al., 2011). In the late 2010s, visible light was assessed for its antimicrobial properties, whereby photodynamic inactivation of cells occurs as a consequence of photoexcitation of microbial polyphyrins. This resulted in the release of toxic reactive oxygen species and subsequent cell damage as a result. In a study by Halstead et al. (2016), the group was able to demonstrate an antibiofilm aspect to this technology and achieve significant reductions in mature biofilms using clinically relevant strains such as *Acinetobacter baumannii* and *Klebsiella pneumoniae*. Other applications for light therapy include the use of photosensitive dyes, which similarly upon exposure can induce the production of reactive oxygen species and thus cell death.

There is limited clinical data to support the implementation of this therapy into wound treatment at this current moment in time. Existing data such as preclinical studies on *P. aeruginosa* and *A. baumannii* biofilms in infected burns of mice show that this form of treatment was only found to be effect when administered soon after the initial infection began to occur (Wang et al., 2016). Additional, responses by Gram-positive and Gram-negative bacteria vary, making this type of approach less effective for multispecies chronic wounds. While this should be considered, the noninvasive approach of light therapy can be readily combined with other chronic wound treatments to reduce the biofilm burden within wounds across a range of pathogenic organisms.

5. Current developments for biofilm management in wounds

5.1 Addressing the challenges of biofilms in wounds

The high prevalence of biofilms in chronic wounds and the significant impact of these wounds on the long-term prospects and care for patients, as well as the concomitant impact on the health care system, have spurred a large body of research and developments in the area. However, the significant inherent challenges mean that the field is evolving, and there are still many uncertainties. The challenges stem from the wide range of wounds and underlying medical conditions, which may become infected and establish a biofilm, and these multiple factors make the most appropriate treatment highly wound-specific. Wounds have a range of origins, caused by burns, diabetic or pressure ulcers, surgical procedures, or trauma. These are impacted by physiological conditions such as general health and nutritional status of patient and localized blood flow at the wound site. There are also a range of bacterial and fungal species that may make up what is often a multispecies biofilm within the wound (Hall-Stoodley et al., 2012). This, together with the typically microscopic nature of biofilms, also makes diagnosis difficult. Furthermore, the very nature of wound care means that ongoing treatment is often community-based and wounds are only assessed periodically when the dressings are changed. This means that if a wound progresses from acute to chronic state, the healthcare professionals treating the patient on a day-to-day or week-to-week basis may be doing so without access to the infrastructure and diagnostic facilities available within a hospital environment.

Current focus for improving the management of biofilms in wounds spans all areas of care, with developments in identification and diagnostics, enhanced treatments with increased antimicrobial and biofilm efficacy, all with the aim to reduce discomfort for the patient and minimize damage to underlying healthy or healing tissue. It remains the case however that optimal protocols for treating wounds and biofilms within wounds are not yet fully established.

5.2 Biofilm identification and wound monitoring

Over the last decade, the growing appreciation of the importance of biofilms in wounds has led a number of authors in the scientific literature providing guidance for the identification and treatment of biofilms in wounds (Schultz et al., 2017). However, a widespread lack of understanding of biofilms in wounds or diagnostic tests to aid with their identification and treatment has also been addressed by a number of expert groups who have sought to establish a consensus and set of common criteria for wound assessment and treatment, but controversy and clinical applicability prevail (Høiby et al., 2015).

While most if not all chronic wounds contain biofilms, they cannot be visualized by the naked eye due to their size and the presence of slough or wound eschar and exudate, which also will house biofilms. The most reliable diagnostic techniques for medically related biofilms, proposed by the European Society of Clinical Microbiology and Infectious Diseases (ESCMID) biofilm study group (ESGB), are microscopy-based such as scanning electron microscopy (SEM) and confocal laser scanning microscopy (CLSM) (Høiby et al., 2015). However, these are highly specialized techniques and not widely available in clinical settings and would require biopsy samples from the patient, which would make this clinically unethical in many circumstances.

Sample collection for diagnosis is also complicated as the biofilm does not typically form as a continuous layer on the wound surface, but is dispersed aggregated patch, which can be missed in discrete sampling. Indeed, microbial colonies need not be present on the surface of the wound, but may also form deeper within the wound tissue, which cannot not be probed using swabs. This means that the most rigorous methods for establishing the presence of biofilms involve taking curettage samples, punch biopsies, and other samples from sharp debridement containing wound tissue (Singer & Clark, 1999).

The lack of availability and impracticality of rigorous diagnostic testing for biofilms means that currently the most useful method relies on clinical indicators for the presence of biofilm, which have been proposed by a number of expert groups (Percival, 2017; Percival & Suleman, 2015), the most commonly listed of which are:

- Shiny, slimy, or gelatinous wound surface
- Exudate and malodour
- Persistent slough or necrotic tissue
- Low level erythema or chronic inflammation
- Recalcitrance to or failure of treatment with antibiotics or antiseptics
- Delayed healing despite appropriate therapy
- Repeated cycles of infection

In practice, current guidance suggests that clinicians should assume that all chronic wounds that exhibit delayed healing, despite optimal standards of care, should be regarded as having biofilm present (Schultz et al., 2017) and should be treated as such.

A number of studies have shown that treatment of wounds using a targeted biofilm approach has improved the clinical outcomes (Dowd, Wolcott, Kennedy, Jones, & Cox, 2011; Wolcott, 2015). Nonetheless, approaches that combine diagnostics with antibiofilm treatment have also been shown to be effective. For example, Mori et al. (2019) used wound blotting as a point-of-care diagnostic method to guide ultrasonic debridement and were able to show a faster rate of healing for the treatment group compared with the control group receiving standard treatment.

Current developments in wound management systems (WMSs) and in particular with the inclusion of wound imaging as a means of tracking the progression of wound healing provide one way of capturing and formalizing clinical observations that may be indicative of biofilms (Rosemary Hill & Rennie, 2018); this represents an emerging diagnostic methodology. WMS has the potential to address several of the challenges for treatment of biofilms in wounds. The systems should be able to record and flag up the clinical markers for biofilm, both from analysis of the captured images and through prompts to the carer, clearly to track the rate of healing and recognize a slow healing wound or deviation from a normal healing process to a more recalcitrant one. It would then be able to provide the clinician with guidance on the most appropriate range of treatment options and potentially easy access to specialist input and advice.

5.3 Methods for debridement

The initial method for removal of biofilms, slough, necrotic, or devitalized tissue is debridement or desloughing. The established methods for wound debridement while effective have significant drawbacks, which are being addressed with current developments. Sharp or surgical debridement is fast, effective, and may be the best way to remove macroscopic amounts of necrotic tissue but is also painful and may damage underlying viable tissue. Autolytic debridement is gentle, but slow, and may allow time for bacteria to attach and form biofilms and infection to become established if not used in conjunction with antimicrobial agents. The use of sterile maggots for cleaning wounds is gentle and highly selective to necrotic tissue and eschar, but necessarily involves careful management of the maggots.

One alternative method to sharp debridement is hydrosurgical debridement. This uses a high-pressure jet of water to cut through the affected tissue and slice layers from the wound surface, while at the same time keeping the wound irrigated and flushing away any dislodged material (Hirokawa et al., 2019). The study on a rat wound model suggested that hydrosurgical debridement was easier and quicker to implement and caused minimal damage to health tissue, which led to reduced inflammation and fibrotic tissue and resulted in earlier wound healing. A clinical trial suggested that it could also reduce the incidence of readmission due to surgical site infection (James et al., 2021).

Another emerging method for debridement and cleaning wounds is the use of ultrasound (Dhar & Han, 2020). The mechanical vibration induced by the acoustic waves causes the structures within the biofilm to loosen and become more easily broken up and dislodged from the wound surface. The method has been shown to be effective for the treatment of biofilms in conjunction with point-of-care biofilm diagnostic testing. The ultrasound method has also developed further, with the introduction of microbubbles to the treatment surface (Butcher & Pinnuck, 2013). The presence of microbubbles in the ultrasound field has a magnifying effect on the physical shear and displacement forces, which causes a physical loosening and disruption of the underlying structures. Another combined approach is to use ultrasound in conjunction with antibiotics (Cai, Wang, Liu, Wang, & Xia, 2017; Ennis et al., 2005), as the former has been shown to sensitize bacteria within a biofilm to the antibiotics, overcoming some of the increased resistance that the biofilm environment provides. This is also likely to be the case for treatment of the biofilms with surfactants and other molecules that target the integrity of the biofilm matrix or the structure of microbial membranes.

5.4 Antibiofilm agents

The ongoing problem of antimicrobial resistance and the challenges of addressing the intractability of biofilms in wounds have led to the development of a wide range of novel antibiofilm and antimicrobial agents (Kadam, Shai, Shahane, & Kaushik, 2019). These range from antimicrobial peptides (Thapa, Diep, & Tønnesen, 2020), engineered nanoparticles designed to aid with drug delivery and release to nanometal structures, which can be intrinsically antimicrobial, act slowly to release bactericidal metal ions or act as a mediator for photothermal (Zhang, Tan, Wu, Zhang, & Liao, 2021) or photodynamic therapy (Shanmugapriya, Kim, & Kang, 2020).

Another class of antibiofilm agents are quorum sensing inhibitors (QSI) (Anju, Dyavaiah, & Siddhardha, 2021; Das & Singh, 2018). These compounds interfere with the formation of biofilms by disrupting communication via quorum sensing between bacteria within a colony, which is required to regulate the genes that code for various virulence factors and for the cooperative progression of the biofilm to a mature state in which individual organisms are metabolically less active and significantly more resistant to treatment. While QSIs may be synthetic compounds, there has been extensive research into QSIs derived from natural sources as well as quorum quenching enzymes, which degrade the signaling molecules (Rémy et al., 2018). The combination of QSIs with other antimicrobial or antibiotic agents has been found to be particularly beneficial as the effect of the inhibitors is to reduce resistance and maintain the bacterial susceptibility to other treatments (Simonetti et al., 2016).

5.5 Antibiofilm wound dressings

Recently, a wide range of wound dressings and treatments have been developed to address the need to treat biofilms in wounds, and several dressings have been launched onto the market (Table 2). While antibiofilm dressings are not generally intended to replace the need for debridement, the debridement process will nonetheless often leave

TABLE 2 Different types of wound dressings.

Dressing type	Type of application
Alginate	Fibrous dressing enables easy insertion into the wound cavity. Used mostly on heavily exuding or bleed wounds.
Hydrogel	Amorphous gel for dry as well as sloughy wounds. The dressing increases the moisture levels within the wound.
Semipermeable films	Used for lacerations, abrasions and other superficial wounds. Acts as a microbial barrier to prevent infection and is not intended for use with exudates.
Foams	Highly absorbent dressing intended for use on heavily exuding wounds. Thicker foam dressings provide protective cushioning effect to wounds.
Antimicrobial	Used on infected or colonized wounds. Exhibits a broad spectrum of nonselective antimicrobial action.
Deodorizing	Often comprised of activated charcoal to absorb and reduce odor from wounds.
Nonadherent	Designed to prevent adhesion to dry wound tissue and reduce the risk of tissue damage and patient discomfort upon removal.

residual biofilm on the wound surface, and biofilms can also rapidly reform. This residual or reforming biofilm needs to be dealt with to ensure optimal progression of wound healing. Antibiofilm dressings need to manage the level of exudate in a wound and typically consist of a combination of active components, which can act to breakdown and reduce the bioburden in a wound as well as control the potential growth of biofilm within the wound dressing. For example, the dressing may contain: highly absorbent polymer materials to remove exudate from the wound, the ability to gel and conform to the wound surface, antimicrobials such as silver or iodine, surfactants to reduce bacterial adhesion or disrupt bacterial membranes, or metal ion chelators to destabilize the ECM within the biofilm (Salisbury et al., 2018).

Another approach is the development of dressing materials that strongly bind bacterial membranes. This aims to capture and immobilize bacteria within the dressing structure and so prevent their proliferation and dissemination back into the wound, as well as to adhere to bacteria or biofilm on the wound surface and remove them when the dressing is removed. One such example is dressings coated with dialkyl carbamoyl chloride (DACC) (Cooper & Jenkins, 2016).

5.6 Other antibiofilm treatments

5.6.1 Cold atmospheric plasma

Atmospheric or cold plasma (CAP) is a technique where alternating high voltages are used in a device to excite ambient gas molecules and cause them to break down and form a plasma. The plasma formed is a highly reactive, partially ionized, gas, which is nonetheless close to room temperature. A low level of ultraviolet light is also emitted and the reactive components of the plasma produce reactive oxygen species (ROS) and reactive nitrogen species (RNS). A CAP can be formed as a thin layer close to the generating electrode, via dielectric barrier discharge, or in the form of a jet, which is pumped out of a plasma pen device. The nature of the plasma produced can be tailored and conditions controlled to produce a plasma that is at the same time viricidal and bactericidal with low cytotoxicity (Shimizu, Lachner, & Zimmermann, 2017). The reactive hydroxyl radicals formed damage the prokaryotic cell membranes and intracellular macromolecules such as DNA and cause cell death. The effect on eukaryotic cells is somewhat different, in that the oxidative stress induced by the plasma can stimulate the cells' survival mechanism, the release of antimicrobial peptides such as defensins, production of proinflammatory cytokines, and the stimulation of cell recovery (Kleineidam, Nokhbehsaim, Deschner, & Wahl, 2019). The combination of the UV light and plasma can have a synergistic antimicrobial effect, and CAP has also been shown to be effective against biofilms (Ziuzina, Boehm, Patil, Cullen, & Bourke, 2015). The use of CAP was shown to enhance the healing of full-depth wounds in a mouse model and a number of small-scale clinical trials have also been undertaken (Jung et al., 2021), with one on patients with venous leg ulcers showing greater reduction of the wound area compared with the control group over the period of the study (Brehmer et al., 2015). Larger-scale trials are required to fully validate the technique. Portable CAP devices are already commercially available for wound treatment.

5.6.2 Electrochemistry and electroceuticals

Several methods for enhanced wound healing and combatting biofilms have used electrochemical approaches. While not yet extensively developed, electrochemical methods can act in a number of different ways. A dressing printed with separated silver and zinc islands, which produces local electrical fields in the presence of conductive wound exudate dressings, was shown to disrupt biofilms of *P. aeruginosa* and reduce the activity of quorum sensing genes (Banerjee, 2014). Electric fields can encourage the migration of keratinocytes within a wound, can stimulate the production of superoxide radicals reactive oxygen species. The potential for electrochemical dressing is to enable the metered delivery of antimicrobial agents on demand. One electrochemically driven printed silver wound dressing has been shown to produce silver ions and hypochlorous acid, both of which show efficacy against biofilms (Dusane et al., 2019). The approach was also used to successfully treat a persistent polymicrobial infections in a canine subject (Heald et al., 2020). Electrochemical scaffolds have also been used to generate low levels of hydrogen peroxide within wound dressings (Sultana et al., 2015) and shown to significantly reduce *A. baumannii* biofilm that was grown on a porcine explant.

5.7 Treating the wound environment and stimulating the body's response

Another approach to combatting the development of biofilms in wounds is by stimulating or modifying the wound environment to promote healing and reduce the likelihood of biofilm formation. This encompasses variables such as control of temperature, pH, light, humidity, pressure, or oxygen level as well as enhancing the body's natural healing processes and mechanisms for fighting bacterial infection or reducing the damage caused to cells caused by the body's inflammatory response with the use of antiinflammatory treatments (Paes Leme & da Silva, 2021).

One method that has become well established for the treatment nonhealing wounds is negative pressure wound therapy (NPWT). This technique applies suction to the outside of a wound to draw subcutaneous fluid through the wound bed. This improves nutrient and oxygen supply to the wound and helps remove exudate and slough from the wound surface. NPWT has also been shown to reduce bacterial load in wounds when combined with antiseptic irrigation (Ludolph et al., 2018). Combination with acetic acid instillation has also shown promising results against biofilms in recalcitrant wounds (Jeong et al., 2015).

5.7.1 Oxygen therapies

One particular aspect of a wound that favors poor healing and the formation of biofilms is poor oxygen supply to the wound tissue and can be a consequence of the patient's comorbidities. This has multiple effects on the wound healing process as large amount of oxygen is used during the inflammation and proliferation stages and is required for successful angiogenesis and epithelialization. Poor oxygen supply favors the growth of anaerobic bacteria, which are often present within biofilms where the oxygen levels can be low, which reduces the metabolic activity of aerobes making them more resistant to antibiotics. A compounding effect of low oxygen levels is to reduce the activity of a range of antimicrobial and antibiotic molecules that rely on the generation of oxygen radicals for their function (Lobritz et al., 2015); these include common antimicrobials such as nano-particulate silver. One approach to address these issues is to use oxygen therapies, such as hyperbaric oxygen therapy (HBOT) or topical oxygen therapy (TOT) (Bishop, 2019). In HBOT, the patient is typically enclosed in a chamber with elevated oxygen levels for a period of time. This is repeated several times a day for a few days. Whereas for TOT, oxygen is supplied in a variety of different forms locally to the wound, which enable the patient to be more mobile (Gottrup, 2017; Gottrup et al., 2017). TOT may take the form of continuous delivery of nonpressurized oxygen, oxygen release from a dressing or gel, or oxygen transfer materials such as hemoglobin spray (Hunt, Elg, & Percival, 2018). Oxygen therapies have been shown to be effective at killing anaerobic bacteria and enhance leucocyte function. It has also been shown to upregulate angiogenic, endothelial, and fibroblast growth factors, stimulate the growth of new blood vessels, enhance wound bed granulation, and collagen tissue formation and lead to wound closure. This approach has also been used to enhance the activity of antimicrobials in a combined therapeutic approach (Arenbergerova et al., 2013; Frykberg, 2021; Hu et al., 2020; Jensen et al., 2019).

5.7.2 Vaccines

As antibiotic resistance increases, vaccination has been explored as an additional method for combatting bacterial infection associated with biofilm formation. The approach has been shown to be effective in reducing infections in patients with cystic fibrosis with a vaccine against *P. aeruginosa* flagella and vaccination against biofilm-specific antigens has been proposed as an approach to improve treatment for tuberculosis (Döring et al., 2007; Flores-Valdez, 2016). For the treatment of wounds, the importance of specifically targeting the formation of biofilm has been recognized, for example, in *Acinetobacter*-related infections, which are associated with infections in war wounds and from natural disasters as well in *S. aureus*–related infections, which are prevalent in a high proportion in chronic wounds (Harro et al., 2010). Prevention of biofilm formation or microbial adhesion can be achieved with multivalent vaccines against components of the biofilm matrix, which is made up of proteins, polysaccharides, and extracellular DNA (eDNA), e.g., fibronectin-binding protein, clumping factor A, or collagen adhesin (Parastan, Kargar, Solhjoo, & Kafilzadeh, 2020). Members of the bacterial eDNA-binding proteins, DNABII, have been shown to be promising targets for antibiofilm antibodies or vaccines and can be effective against established biofilms (D'Andrea & Lau, 2020). DNABII proteins are an integral part of the biofilm structure, common to all bacterial species tested by the authors, and have been found to have a high degree of sequence conservation, making them good candidates for species-independent antibiofilm treatments. It has been shown that anti-DNABII antibodies are able to sequester the proteins from biofilms, causing the biofilm structure to collapse, bacteria to detach and become susceptible to antibiotic clearance. Peptides for the eDNA-binding tip of the DNABII have also been effective against a range of biofilm models of chronic human diseases in rodents, such as OM caused by nontypeable *Haemophilus influenzae* (NTHi) in chinchillas. Immunization based on these peptides also significantly delayed onset of disease with a high degree of efficacy, 85% (Novotny, Goodman, & Bakaletz, 2020). However, for completely clearing infection, combined approaches, or vaccines that target both planktonic and biofilm species, will be required.

5.7.3 Human-derived treatments

Mesenchymal stem cells (MSCs) have received a lot of attention for their potential to treat a wide variety of diseases, including for the treatment of chronic wounds (Hu, Borrelli, Lorenz, Longaker, & Wan, 2018). MSCs are self-renewing multipotent cells, which occur naturally in the body, in bone marrow, adipose tissue, or placenta and have been shown to reduce inflammation, increase angiogenesis, improve granulation, enhance reepithelialization, and speed wound closure with reduced scarring. MSCs display multiple modes of action in the healing process and in combatting infections. While they have the ability to transdifferentiate into skin cells, MSCs can act directly against bacteria and biofilms as well as indirectly by modulating the body's immune and healing response via their paracrine action. In response to biofilms, MSCs secrete antimicrobial peptides, such as cathelicidin, defensins, and lipocalins, which act on the bacterial surface and against intracellular targets (Alcayaga-Miranda, Cuenca, & Khoury, 2017; Krasnodembskaya et al., 2010; Mezey & Nemeth, 2015). Media used to culture MSC has been shown to inhibit formation of *P. aeruginosa* biofilms and also has been observed to attach to the surface of and engulf *S. aureus* biofilms (Wood et al., 2018). MSCs are reported to have efficacy against a range of pathogens relevant to wounds (Harman, Yang, He, & Van de Walle, 2017). The effect of MSCs on the body's healing response is due to excretion of a wide range of paracrine factors, growth factors, antiinflammatory cytokines, and chemokines (Huang, Gou, Da, Zhang, & Xie, 2020). The combination of MSCs and antibiotics has been shown to be particularly effective in a mouse model. MSCs have been administered intravenously, injected locally to the wound as well as transplanted onto the wound surface, and clinical studies are under way (Johnson et al., 2017). While there are still questions relating to the mode of delivery and optimized methods, this represents a promising approach for the future treatment of chronic wounds.

An alternative method of increasing the level of growth factors in chronic wounds has been by using platelet concentrates or platelet-rich plasma. More recently, an additive-free method has been developed to extract fibrin, platelets, and autologous polymorphonuclear leucocytes by centrifugation from a small volume of the patient's blood. The leucopatch method, which requires no other additives, produces discs of the material, which is used to cover the wound before dressing and is changed weekly (Jørgensen, Karlsmark, Vogensen, Haase, & Lundquist, 2011). Controlled trials on recalcitrant wounds showed significant improvement in the rate of healing (Game et al., 2018). Evaluation of the mechanism of action was able to show that it was in part due to activity of the leucocytes within the patch. Polymorphonuclear neutrophils from the leucopatch showed a significant respiratory burst, bacterial-induced migration, as well as active phagocytosis and killing of *P. aeruginosa* (Thomsen et al., 2016).

6. Future prospects

The wide-ranging developments in the treatments for chronic wounds and biofilms and plethora of new technologies hold up the promise of significant improvements in patient care with reductions in the time required for wounds to heal and reduce patient discomfort. The growing understanding of biofilms in wounds is likely to lead to ever more effective treatments, and this is expected to lead to incremental improvements in wound care. Careful assessment of clinical relevance of new technologies is required and effective comparisons against established methods undertaken, so that the most effective techniques are taken forward. Combinations of methods that target different aspects or stages of the biofilm evolution have been shown to be most effective, and refining this is likely to lead to significant improvements in performance. In order to realize this full potential, the most appropriate treatment or combinations of treatment need to be established and delivered to the patient at the right time, within the context of the patient's healing process.

The development of wound monitoring systems is expected to help select the best treatment approach for a particular wound, and clinicians can look forward to a time when difficult-to-heal wounds are identified before they become chronic or infected. For this to happen, however, accurate diagnosis of the wound condition is required, and it is clear that the better an individual wound is understood, the more likely the best treatment can be offered. In order to achieve this, better diagnostics and methods for monitoring wounds are required. Ideally, wound management systems integrated with sensors would be able to provide real-time assessment of a wound, without requiring the dressing to be removed (Bruinink, 2018). Such a system would allow proactive intervention in the case of bacterial proliferation or a wound stopping from progressing at the anticipated rate. On the other hand, dressings and the wound bed could be left undisturbed if a dressing change or other intervention was not indicated.

Academic research into sensors for wounds is being investigated, and some interesting concepts have been developed (Dargaville et al., 2013). Milne, Connolly, Al Hamad, and Seoudi (2014) presented sensors for moisture, pH, and wound matrix metalloproteinases (MMP) enzyme activity. Another concept for a theragnostic sensor uses a sensor for wound temperature coupled with a UV responsive antimicrobial hydrogel to administer antimicrobial agents in response to increased wound temperature. Indicator dressings that change color in response to the presence of bacterial toxins have also been demonstrated (Zhou et al., 2011). Significant developments have been made in wearable sensors for sport and lifestyle as well as health monitoring, with several papers detailing sensors for metabolic as well as physiological markers (Sempionatto et al., 2021), some of which would also be applicable to wounds (Brown, Ashley, & Koh, 2018).

References

Adamskaya, N., Dungel, P., Mittermayr, R., Hartinger, J., Feichtinger, G., Wassermann, K., et al. (2011). Light therapy by blue LED improves wound healing in an excision model in rats. *Injury*, *42*(9), 917–921.

Alcayaga-Miranda, F., Cuenca, J., & Khoury, M. (2017). Antimicrobial activity of mesenchymal stem cells: Current status and new perspectives of antimicrobial peptide-based therapies. *Frontiers in Immunology*, *8*, 339.

Anju, V. T., Dyavaiah, M., & Siddhardha, B. (2021). Quorum sensing as a therapeutic target in the treatment of chronic wound infections. In *Wound healing research* (pp. 643–659). Singapore: Springer.

Arenbergerova, M., Engels, P., Gkalpakiotis, S., Dubská, Z., Arenberger, P., & Arenbergerova, M. (2013). Effect of topical haemoglobin on healing in patients with venous leg ulcers. *EWMA Journal, 13*(2), 25–30.

Attinger, C., & Wolcott, R. (2012). Clinically addressing biofilm in chronic wounds. *Advances in Wound Care (New Rochelle)*, 127–132.

Bakaletz, L. O. (2007). Bacterial biofilms in otitis media. *The Pediatric Infectious Disease Journal, 26*(10), S17–S19.

Banerjee, J. (2014). Electroceutical wound dressing disrupts biofilm. *PLoS One*.

Bishop, A. (2019). Hyperbaric oxygen therapy for problem wounds: An update. *Wounds UK, 15*(4).

Bjarnsholt, T. (2013). The role of bacterial biofilms in chronic infections. *APMIS, 121*, 1–58.

Bjarnsholt, T., Eberlein, T., Malone, M., & Schultz, G. (2017). Management of wound biofilm. *Wounds International, 8*(2).

Brehmer, F., Haenssle, H. A., Daeschlein, G., Ahmed, R., Pfeiffer, S., Görlitz, A., et al. (2015). Alleviation of chronic venous leg ulcers with a hand-held dielectric barrier discharge plasma generator (PlasmaDerm® VU-2010): Results of a monocentric, two-armed, open, prospective, randomized and controlled trial (NCT 01415622). *Journal of the European Academy of Dermatology and Venereology, 29*(1), 148–155.

Brown, M. S., Ashley, B., & Koh, A. (2018). Wearable technology for chronic wound monitoring: Current dressings, advancements, and future prospects. *Frontiers in Bioengineering and Biotechnology, 6*, 47.

Bruinink, A. (2018). Biosensor-bearing wound dressings for continuous monitoring of hard-to-heal wounds: Now and next. *Biosensors and Bioelectronics Open Access, BBOA-117*, 100017. https://doi.org/10.29011/BBOA-117.

Butcher, G., & Pinnuck, L. (2013). Wound bed preparation: Ultrasonic-assisted debridement. *British Journal of Nursing, 22*(Suppl. 4), S36–S43.

Cai, Y., Wang, J., Liu, X., Wang, R., & Xia, L. (2017). A review of the combination therapy of low frequency ultrasound with antibiotics. *BioMed Research International, 2017*.

Chan, B. K., Abedon, S. T., & Loc-Carrillo, C. (2013). Phage cocktails and the future of phage therapy. *Future Microbiology, 8*, 769–783.

Chapman, T. M., Plosker, G. L., & Figgitt, D. P. (2006). VSL# 3 probiotic mixture. *Drugs, 66*(10), 1371–1387.

Chen, R., Salisbury, A., & Percival, S. (2020). A comparative study on the cellular viability and debridement efficiency of antimicrobial-based wound dressings. *International Wound Journal, 17*, 73–82.

Ciszek-Lenda, M., Strus, M., Walczwska, M., Majka, G., Machul-Zwirbla, A., Mikolajczyk, D., et al. (2019). *Pseudomonas aeruginosa* biofilm is a potent inducer of phagocyte hyperinflammation. *Inflammation Research, 68*, 397–413.

Cooper, R., & Jenkins, L. (2016). Binding of two bacterial biofilms to dialkyl carbamoyl chloride (DACC)-coated dressings in vitro. *Journal of Wound Care, 25*(2), 76–82.

Costerton, J. W., Stewart, P. S., & Greenberg, E. P. (1999). Bacterial biofilms: A common cause of persistent infections. *Science, 284* (5418), 1318–1322.

Cullell-Dalmau, M., Otero-Viñas, M., Ferrer-Solà, M., Sureda-Vidal, H., & Manzo, C. (2021). A toolkit for the quantitative evaluation of chronic wounds evolution for early detection of non-healing wounds. *Journal of Tissue Viability, 30*(2), 161–167.

Cullum, N., Davies, M., Harrison, M., Howers, M. L., Murphy, S., & Walters, B. (2016, March 30). *Chronic wounds: Advanced wound dressings and antimicrobial dressings*. Retrieved from nice.org.uk. www.nice.org.uk/guidance/esmpb2.

D'Andrea, M. M., & Lau, G. W. (2020). DNABII targeting antibodies as vaccines against biofilm diseases. *EBioMedicine, 58*, 102921.

Dargaville, T. R., Farrugia, B. L., Broadbent, J. A., Pace, S., Upton, Z., & Voelcker, N. H. (2013). Sensors and imaging for wound healing: A review. *Biosensors and Bioelectronics, 41*, 30–42.

Das, L., & Singh, Y. (2018). Quorum sensing inhibition: A target for treating chronic wounds. In *Biotechnological applications of quorum sensing inhibitors* (pp. 111–126). Singapore: Springer.

Davies, D. (2003). Understanding biofilm resistance to antibacterial agents. *Nature Reviews Drug Discovery, 2*, 114–122.

Dhar, Y., & Han, Y. (2020). Current developments in biofilm treatments: Wound and implant infections. *Engineered Regeneration, 1*, 64–75.

Donlan, R. M. (2009). Preventing biofilms of clinically relevant organisms using bacteriophage. *Trends in Microbiology, 17*, 66–72.

Döring, G., Meisner, C., Stern, M., & Flagella Vaccine Trial Study Group. (2007). A double-blind randomized placebo-controlled phase III study of a *Pseudomonas aeruginosa* flagella vaccine in cystic fibrosis patients. *Proceedings of the National Academy of Sciences, 104*(26), 11020–11025.

Dowd, S. E., Wolcott, R. D., Kennedy, J., Jones, C., & Cox, S. B. (2011). Molecular diagnostics and personalised medicine in wound care: Assessment of outcomes. *Journal of Wound Care, 20*(5), 232–239.

Dusane, D. H., Lochab, V., Jones, T., Peters, C. W., Sindeldecker, D., Das, A., et al. (2019). Electroceutical treatment of *Pseudomonas aeruginosa* biofilms. *Scientific Reports, 9*(1), 1–13.

Edwards, R., & Harding, K. (2004). Bacteria and wound healing. *Current Opinion in Infectious Diseases, 17*, 91–96.

Ennis, W. J., Formann, P., Mozen, N., Massey, J., Conner-Kerr, T., & Meneses, P. (2005). Ultrasound therapy for recalcitrant diabetic foot ulcers: Results of a randomized, double-blind, controlled, multicenter study. *Ostomy Wound Management, 51*(8), 24.

Flores-Valdez, M. A. (2016). Vaccines directed against microorganisms or their products present during biofilm lifestyle: Can we make a translation as a broad biological model to tuberculosis? *Frontiers in Microbiology, 7*, 14.

Forsberg, K., Woodworth, K., Walters, M., Berkow, E. L., Jackson, B., Chiller, T., et al. (2019). *Candida auris*: The recent emergence of a multidrug-resistant fungal pathogen. *Medical Mycology, 57*(1), 1–12.

Frykberg, R. G. (2021). Topical wound oxygen therapy in the treatment of chronic diabetic foot ulcers. *Medicina, 57*(9), 917.

Game, F., Jeffcoate, W., Tarnow, L., Jacobsen, J. L., Whitham, D. J., Harrison, E. F., et al. (2018). LeucoPatch system for the management of hard-to-heal diabetic foot ulcers in the UK, Denmark, and Sweden: An observer-masked, randomised controlled trial. *The Lancet Diabetes & Endocrinology, 6*(11), 870–878.

Gottrup, F. (2017). Oxygen therapies for wound healing: EWMA findings and recommendations. *Wounds International, 2017*(8), 4.

Gottrup, F., Dissemond, J., Baines, C., Frykberg, R., Jensen, P.Ø., Kot, J., et al. (2017). Use of oxygen therapies in wound healing: Focus on topical and hyperbaric oxygen treatment. *Journal of Wound Care*, *26* (Suppl. 5), S1–S43.

Guandalini, S., Cernat, E., & Moscoso, D. (2015). Prebiotics and probiotics in irritable bowel syndrome and inflammatory bowel disease in children. *Beneficial Microbes*, *6*(2), 209–217.

Gurjala, A., Geringer, M., Seth, A., Hong, S., Smeltzer, M., Galiano, R., et al. (2011). Development of a novel, highly quantitative in vivo model for the study of biofilm-impaired cutaneous wound healing. *Wound Repair and Regeneration*, *19*, 2159–2168.

Hall-Stoodley, L., Stoodley, P., Kathju, S., Høiby, N., Moser, C., William Costerton, J., et al. (2012). Towards diagnostic guidelines for biofilm-associated infections. *FEMS Immunology & Medical Microbiology*, *65* (2), 127–145.

Halstead, F. D., Thwaite, J. E., Burt, R., Laws, T. R., Raguse, M., Moeller, R., et al. (2016). Antibacterial activity of blue light against nosocomial wound pathogens growing planktonically and as mature biofilms. *Applied and Environmental Microbiology*, *82*(13), 4006–4016.

Hanke, M., Heim, C., Angle, A., Sanderson, S., & Kielian, T. (2013). Targeting macrophage activation for the prevention and treatment of *Staphylococcus aureus* biofilm infections. *Journal of Immunology*, *190*, 2159–2168.

Harman, R. M., Yang, S., He, M. K., & Van de Walle, G. R. (2017). Antimicrobial peptides secreted by equine mesenchymal stromal cells inhibit the growth of bacteria commonly found in skin wounds. *Stem Cell Research & Therapy*, *8*(1), 1–14.

Harper, D., Parracho, H., Walker, J., Sharp, R., Hughes, G., Werthén, M., et al. (2014). Bacteriophages and biofilms. *Antibiotics*, *3*, 270–284.

Harro, J. M., Peters, B. M., O'May, G. A., Archer, N., Kerns, P., Prabhakara, R., et al. (2010). Vaccine development in Staphylococcus aureus: Taking the biofilm phenotype into consideration. *FEMS Immunology & Medical Microbiology*, *59*(3), 306–323.

Heald, R., Bennett, M., Subramaniam, V. V., Dusane, D., Lochab, V., Sundaram, P. M., et al. (2020). Printed electroceutical dressings for the inhibition of biofilms and treatment of chronic wounds. *Journal of Microelectromechanical Systems*, *29*(5), 918–923.

Hengzhuang, W., Wu, H., Ciofu, O., Song, Z., & Høiby, N. (2011). Pharmacokinetics/pharmacodynamics of colistin and imipenem on mucoid and nonmucoid *Pseudomonas aeruginosa* biofilms. *Antimicrobial Agents and Chemotherapy*, *55*(9), 4469–4474.

Hirokawa, E., Sato, T., Fujino, T., Gotoh, Y., Yokogawa, H., & Ichioka, S. (2019). Hydrosurgical debridement as an approach to wound healing: An animal thermal burn model. *Journal of Wound Care*, *28* (5), 304–311.

Høiby, N., Bjarnsholt, T., Moser, C., Bassi, G. L., Coenye, T., Donelli, G., et al. (2015). ESCMID guideline for the diagnosis and treatment of biofilm infections 2014. *Clinical Microbiology and Infection*, *21*, S1–S25.

Honorato-Sampaio, K., Guedes, A., Lima, V., & Borges, E. (2014). Bacterial biofilm in chronic venous ulcer. *The Brazilian Journal of Infectious Diseases*, *18*(3), 350–351.

Hu, M. S., Borrelli, M. R., Lorenz, H. P., Longaker, M. T., & Wan, D. C. (2018). Mesenchymal stromal cells and cutaneous wound healing: A comprehensive review of the background, role, and therapeutic potential. *Stem Cells International*, *2018*.

Hu, D., Zou, L., Yu, W., Jia, F., Han, H., Yao, K., et al. (2020). Relief of biofilm hypoxia using an oxygen nanocarrier: A new paradigm for enhanced antibiotic therapy. *Advanced Science*, *7*(12), 2000398.

Huang, Y. Z., Gou, M., Da, L. C., Zhang, W. Q., & Xie, H. Q. (2020). Mesenchymal stem cells for chronic wound healing: Current status of preclinical and clinical studies. *Tissue Engineering Part B: Reviews*, *26* (6), 555–570.

Hunt, S. D., Elg, F., & Percival, S. L. (2018). Assessment of clinical effectiveness of haemoglobin spray as adjunctive therapy in the treatment of sloughy wounds. *Journal of Wound Care*, *27*(4), 210–219.

James, C. V., Patel, M., Ilonzo, N., Wallace, K., Lee, J., Chan, M., et al. (2021). Hydrosurgical debridement use associated with decreased surgical site-related readmissions: A retrospective analysis. *Wounds*, *33* (6), 139–142.

Jault, P., Leclerc, T., Jennes, S., Pirnay, J. P., Que, Y. A., Resch, G., et al. (2019). Efficacy and tolerability of a cocktail of bacteriophages to treat burn wounds infected by *Pseudomonas aeruginosa* (PhagoBurn): A randomised, controlled, double-blind phase 1/2 trial. *The Lancet Infectious Diseases*, *19*(1), 35–45.

Jensen, P., Bjarnsholt, T., Phipps, R., Rasmussen, T., Calum, H., Christoffersen, L., et al. (2007). Rapid necrotic killing of polymorphonuclear leukocytes is caused by quorum-sensing-controlled production of rhamnolipid by *Pseudomonas aeruginosa*. *Microbiology*, *153*, 1329–1338.

Jensen, P., Givskov, M., Bjarnsholt, T., & Moser, C. (2010). The immune system vs. *Pseudomonas aeruginosa* biofilms. *FEMS Immunology and Medical Microbiology*, *59*, 292–305.

Jensen, P.Ø., Møller, S. A., Lerche, C. J., Moser, C., Bjarnsholt, T., Ciofu, O., et al. (2019). Improving antibiotic treatment of bacterial biofilm by hyperbaric oxygen therapy: Not just hot air. *Biofilm*, *1*, 100008.

Jeong, H. S., Lee, B. H., Lee, H. K., Kim, H. S., Moon, M. S., & Suh, I. S. (2015). Negative pressure wound therapy of chronically infected wounds using 1% acetic acid irrigation. *Archives of Plastic Surgery*, *42*(1), 59.

Johnson, V., Webb, T., Norman, A., Coy, J., Kurihara, J., Regan, D., et al. (2017). Activated mesenchymal stem cells interact with antibiotics and host innate immune responses to control chronic bacterial infections. *Scientific Reports*, *7*(1), 1–18.

Johnston, B. C., Goldenberg, J. Z., Vandvik, P. O., Sun, X., & Guyatt, G. H. (2011). Probiotics for the prevention of pediatric antibiotic-associated diarrhea. *Cochrane Database of Systematic Reviews*, (11), CD004827.

Jørgensen, B., Karlsmark, T., Vogensen, H., Haase, L., & Lundquist, R. (2011). A pilot study to evaluate the safety and clinical performance of Leucopatch, an autologous, additive-free, platelet-rich fibrin for the treatment of recalcitrant chronic wounds. *The International Journal of Lower Extremity Wounds*, *10*(4), 218–223.

Jung, J. M., Yoon, H. K., Jung, C. J., Jo, S. Y., Hwang, S. G., Lee, H. J., et al. (2021). Cold plasma treatment promotes full-thickness healing of skin wounds in murine models. *The International Journal of Lower Extremity Wounds*. 15347346211002144.

Kadam, S., Shai, S., Shahane, A., & Kaushik, K. S. (2019). Recent advances in non-conventional antimicrobial approaches for chronic wound biofilms: Have we found the 'chink in the armor'? *Biomedicines*, *7*(2), 35.

Kalan, L., Loesche, M., Hodkinson, B. P., Heilmann, K., Ruthel, G., Gardner, S. E., et al. (2016). Redefining the chronic-wound microbiome: Fungal communities are prevalent, dynamic, and associated with delayed healing. *MBio*, *7*(5), e01058-16.

Kleineidam, B., Nokhbehsaim, M., Deschner, J., & Wahl, G. (2019). Effect of cold plasma on periodontal wound healing—An in vitro study. *Clinical Oral Investigations*, *23*(4), 1941–1950.

Kortright, K. E., Chan, B. K., Koff, J. L., & Turner, P. E. (2019). Phage therapy: A renewed approach to combat antibiotic-resistant bacteria. *Cell Host & Microbe*, *25*(2), 219–232.

Krasnodembskaya, A., Song, Y., Fang, X., Gupta, N., Serikov, V., Lee, J. W., et al. (2010). Antibacterial effect of human mesenchymal stem cells is mediated in part from secretion of the antimicrobial peptide LL-37. *Stem Cells*, *28*(12), 2229–2238.

Lee, K., & Yoon, S. S. (2017). Pseudomonas aeruginosa biofilm, a programmed bacterial life for fitness. *Journal of Microbiology and Biotechnology*, *27*(6), 1053–1064.

Lisby-Sutch, S. M., Nemec-Dwyer, M. A., Deeter, R. G., & Gaur, S. M. (1990). Therapy of otitis media. *Clinical Pharmacy*, *9*(1), 15–34.

Lobritz, M. A., Belenky, P., Porter, C. B., Gutierrez, A., Yang, J. H., Schwarz, E. G., et al. (2015). Antibiotic efficacy is linked to bacterial cellular respiration. *Proceedings of the National Academy of Sciences*, *112*(27), 8173–8180.

Lockhart, S. R., Etienne, K. A., Vallabhaneni, S., Farooqi, J., Chowdhary, A., Govender, N. P., et al. (2017). Simultaneous emergence of multidrug-resistant *Candida auris* on 3 continents confirmed by whole-genome sequencing and epidemiological analyses. *Clinical Infectious Diseases*, *64*(2), 134–140.

Lopes, E. G., Moreira, D. A., Gullón, P., Gullón, B., Cardelle-Cobas, A., & Tavaria, F. K. (2017). Topical application of probiotics in skin: Adhesion, antimicrobial and antibiofilm in vitro assays. *Journal of Applied Microbiology*, *122*(2), 450–461.

Ludolph, I., Fried, F. W., Kneppe, K., Arkudas, A., Schmitz, M., & Horch, R. E. (2018). Negative pressure wound treatment with computer-controlled irrigation/instillation decreases bacterial load in contaminated wounds and facilitates wound closure. *International Wound Journal*, *15*(6), 978–984.

McCarty, S., Cochrane, C., Clegg, P., & Percival, S. L. (2012). The role of endogenous and exogenous enzymes in chronic wounds: A focus on the implications of aberrant levels of both host and bacterial proteases in wound healing. *Wound Repair and Regeneration*, *20*, 125–136.

Mcclure, C., & Schiller, N. (1992). Effects of *Pseudomonas aeruginosa* rhamnolipids on human monocyte-derived macrophages. *Journal of Leukocyte Biology*, *51*, 97–102.

Merabishvili, M., Monserez, R., Van Belleghem, J., Rose, T., Jennes, S., De Vos, D., et al. (2017). Stability of bacteriophages in burn wound care products. *PLoS ONE*, *12*, e0182121.

Mezey, É., & Nemeth, K. (2015). Mesenchymal stem cells and infectious diseases: Smarter than drugs. *Immunology Letters*, *168*(2), 208–214.

Milne, S. D., Connolly, P., Al Hamad, H., & Seoudi, I. (2014 August). Development of wearable sensors for tailored patient wound care. In *2014 36th annual international conference of the IEEE engineering in medicine and biology society* (pp. 618–621). IEEE.

Mohammedsaeed, W., Cruickshank, S., McBain, A. J., & O'Neill, C. A. (2015). *Lactobacillus rhamnosus* GG lysate increases re-epithelialization of keratinocyte scratch assays by promoting migration. *Scientific Reports*, *5*, 16147.

Mori, Y., Nakagami, G., Kitamura, A., Minematsu, T., Kinoshita, M., Suga, H., et al. (2019). Effectiveness of biofilm-based wound care system on wound healing in chronic wounds. *Wound Repair and Regeneration*, *27*(5), 540–547.

Moye, Z., Woolston, J., Sulakvelidze, A., Moye, Z. D., Woolston, J., & Sulakvelidze, A. (2018). Bacteriophage applications for food production and processing. *Viruses*, *10*, 205.

Novotny, L. A., Goodman, S. D., & Bakaletz, L. O. (2020). Targeting a bacterial DNABII protein with a chimeric peptide immunogen or humanised monoclonal antibody to prevent or treat recalcitrant biofilm-mediated infections. *EBioMedicine*, *59*, 102867.

Ochsner, U., Reiser, J., Fiechter, A., & Witholt, B. (1995). Production of *Pseudomonas aeruginosa* rhamnolipid biosurfactants in heterologous hosts. *Applied and Environmental Microbiology*, *61*, 3503–3506.

Omar, A., Wright, J., Schultz, G., Burrell, R., & Nadworny, P. (2017). Microbial biofilms and chronic wounds. *Microorganisms*, *5*(1), 9.

Paes Leme, R. C., & da Silva, R. B. (2021). Antimicrobial activity of non-steroidal anti-inflammatory drugs on biofilm: Current evidence and potential for drug repurposing. *Frontiers in Microbiology*, *12*, 707629. https://doi.org/10.3389/fmicb.2021.707629.

Parastan, R., Kargar, M., Solhjoo, K., & Kafilzadeh, F. (2020). *Staphylococcus aureus* biofilms: Structures, antibiotic resistance, inhibition, and vaccines. *Gene Reports*, *20*, 100739.

Pearson, J., Pesci, E., & Iglewski, B. (1997). Roles of *Pseudomonas aeruginosa* las and rhl quorum-sensing systems in control of elastase and rhamnolipid biosynthesis genes. *Journal of Bacteriology*, *179*, 5756–5767.

Percival, S. L. (2017). Importance of biofilm formation in surgical infection. *BJS*, *104*, e85–e94.

Percival, S. L., Francolini, I., & Donelli, G. (2015). Low-level laser therapy as an antimicrobial and antibiofilm technology and its relevance to wound healing. *Future Microbiology*, *10*(2), 255–272.

Percival, S. L., Malone, M., Mayer, D., Salisbury, A., & Schultz, G. (2018). Role of anaerobes in polymicrobial communities and biofilms complicating diabetic foot ulcers. *International Wound Journal*, *15*(2), 776–782.

Percival, S. L., Mayer, D., Malone, M., Swanson, T., & Schultz, G. (2017). Surfactants and their role in wound cleansing and biofilm management. *Journal of Wound Care*, *26*(11), 680.

Percival, S. L., & Suleman, L. (2015). Slough and biofilm: Removal of barriers to wound healing by desloughing. *Journal of Wound Care*, *24*(11), 498–510.

Pires, D. P., Melo, L. D. R., Vilas Boas, D., Sillankorva, S., & Azeredo, J. (2017). Phage therapy as an alternative or complementary strategy to prevent and control biofilm-related infections. *Current Opinion in Microbiology*, *39*, 48–56.

Rakhuba, D. V., Kolomiets, E. I., Dey, E. S., & Novik, G. I. (2010). Bacteriophage receptors, mechanisms of phage adsorption and penetration into host cell. *Polish Journal of Microbiology*, *59*, 145–155. https://doi.org/10.33073/pjm-2010-023.

Rémy, B., Mion, S., Plener, L., Elias, M., Chabrière, E., & Daudé, D. (2018). Interference in bacterial quorum sensing: A biopharmaceutical perspective. *Frontiers in Pharmacology*, *9*, 203.

Rosemary Hill, B. S. N., & Rennie, M. Y. (2018). Using bacterial fluorescence imaging and antimicrobial stewardship to guide wound management practices: A case series. *Ostomy Wound Management*, *64* (8), 18–28.

Salisbury, A. M., Woo, K., Sarkar, S., Schultz, G., Malone, M., Mayer, D. O., et al. (2018). Tolerance of biofilms to antimicrobials and significance to antibiotic resistance in wounds. *Surgical Technology International*, *33*, 59–66.

Scales, B. S., & Huffnagle, G. B. (2013). The microbiome in wound repair and tissue fibrosis. *The Journal of Pathology*, *229*(2), 323–331.

Schultz, G., Bjarnsholt, T., James, A. G., Leaper, J. D., McBain, J. A., Malone, M., et al. (2017). Consensus guidelines for the identification and treatment of biofilms in chronic nonhealing wounds. *Wound Repair and Regeneration*, *25*, 744–757. https://doi.org/10.1111/wrr.12590.

Schultz, G., Chin, G., Moldawer, L., & Diegelmann, R. (2011). Principles of wound healing. In I. R. Fitridge, & M. Thompson (Eds.), *Mechanisms of vascular disease: A reference book for vascular specialists* (pp. 423–450). The University of Adelaide, Barr Smith Press.

Schultz, G., Sibbald, R., Falanga, V., Ayello, E., Dowsett, C., Harding, K., et al. (2003). Wound bed preparation: A systematic approach to wound management. *Wound Repair and Regeneration*, *11*(Suppl. 1), S1–28.

Sempionatto, J. R., Lin, M., Yin, L., Pei, K., Sonsa-ard, T., de Loyola Silva, A. N., et al. (2021). An epidermal patch for the simultaneous monitoring of haemodynamic and metabolic biomarkers. *Nature Biomedical Engineering*, *5*(7), 737–748.

Seth, A. K., Geringer, M. R., Hong, S. J., Leung, K. P., Galiano, R. D., & Mustoe, T. A. (2012). Comparative analysis of single-species and polybacterial wound biofilms using a quantitative, in vivo, rabbit ear model. *PLoS One*, *7*(8), e42897. https://doi.org/10.1371/journal.pone.0042897.

Shanmugapriya, K., Kim, H., & Kang, H. W. (2020). Fucoidan-loaded hydrogels facilitates wound healing using photodynamic therapy by in vitro and in vivo evaluation. *Carbohydrate Polymers*, *247*, 116624.

Shi, L., Ermis, R., Lam, K., Cowart, J., Attar, P., & Aust, D. (2009). Study on the debridement efficacy of formulated enzymatic wound debriding agents by in vitro assessment using artificial wound eschar and by an in vivo pig model. *Wound Repair and Regeneration*, *17*, 853–862.

Shimizu, T., Lachner, V., & Zimmermann, J. L. (2017). Surface microdischarge plasma for disinfection. *Plasma Medicine*, *7*(2).

Simonetti, O., Cirioni, O., Cacciatore, I., Baldassarre, L., Orlando, F., Pierpaoli, E., et al. (2016). Efficacy of the quorum sensing inhibitor FS10 alone and in combination with tigecycline in an animal model of staphylococcal infected wound. *PLoS One*, *11*(6), e0151956.

Singer, A. J., & Clark, R. A. (1999). Cutaneous wound healing. *New England Journal of Medicine*, *341*(10), 738–746.

Spoering, A. L., & Lewis, K. I. M. (2001). Biofilms and planktonic cells of *Pseudomonas aeruginosa* have similar resistance to killing by antimicrobials. *Journal of Bacteriology*, *183*(23), 6746–6751.

Sultana, S. T., Atci, E., Babauta, J. T., Mohamed Falghoush, A., Snekvik, K. R., Call, D. R., et al. (2015). Electrochemical scaffold generates localized, low concentration of hydrogen peroxide that inhibits bacterial pathogens and biofilms. *Scientific Reports*, *5*(1), 1–10.

Swanson, T., Haesler, E., Angel, D., & Sussman, G. (2016). IWII wound infection in clinical practice consensus document 2016 update. *Wound Practice & Research: Journal of the Australian Wound Management Association*, *24*(4), 194–198.

Tan, D., Svenningsen, S. L., & Middelboe, M. (2015). Quorum sensing determines the choice of antiphage defense strategy in *Vibrio anguillarum*. *mBio*, *6*(3), e00627.

Thapa, R. K., Diep, D. B., & Tønnesen, H. H. (2020). Topical antimicrobial peptide formulations for wound healing: Current developments and future prospects. *Acta Biomaterialia*, *103*, 52–67.

Thomsen, K., Trøstrup, H., Christophersen, L., Lundquist, R., Høiby, N., & Moser, C. (2016). The phagocytic fitness of leucopatches may impact the healing of chronic wounds. *Clinical & Experimental Immunology*, *184*(3), 368–377.

Thurlow, L., Hanke, M., Fritz, T., Angle, A., Aldrich, A., Williams, S., et al. (2011). *Staphylococcus aureus* biofilms prevent macrophage phagocytosis and attenuate inflammation in vivo. *Journal of Immunology*, *186*, 6585–6596.

Varma, P., Nisha, N., Dinesh, K. R., Kumar, A. V., & Biswas, R. (2011). Anti-infective properties of Lactobacillus fermentum against *Staphylococcus aureus* and *Pseudomonas aeruginosa*. *Microbial Physiology*, *20*(3), 137–143.

Vuotto, C., Longo, F., & Donelli, G. (2014). Probiotics to counteract biofilm-associated infections: Promising and conflicting data. *International Journal of Oral Science*, *6*(4), 189–194.

Wang, G. Q., Li, T. T., Li, Z. R., Zhang, L. C., Zhang, L. H., Han, L., et al. (2016). Effect of negative pressure on proliferation, virulence factor secretion, biofilm formation, and virulence-regulated gene expression of *Pseudomonas aeruginosa* in vitro. *BioMed Research International*, *2016*, 1–7. https://doi.org/10.1155/2016/7986234.

Weigelt, M. A., McNamara, S. A., Sanchez, D., Hirt, P. A., & Kirsner, R. S. (2021). Evidence-based review of antibiofilm agents for wound care. *Advances in Wound Care*, *10*(1), 13–23.

Wolcott, R. (2015). Disrupting the biofilm matrix improves wound healing outcomes. *Journal of Wound Care*, *24*(8), 366–371.

Wolcott, R. D., Cook, R. G., Johnson, E., Jones, C. E., Kennedy, J. P., Simman, R., et al. (2020). A review of iodine-based compounds, with a focus on biofilms: Results of an expert panel. *Journal of Wound Care*, *29*(Suppl. 7), S38–S43.

Wolcott, R., Rhoads, D., Bennett, M., Wolcott, B., Gogokhia, L., Costerton, J., et al. (2010). Chronic wounds and the medical biofilm paradigm. *Journal of Wound Care*, *19*. 45–46, 48–50, 52–53.

Wong, V. W., Martindale, R. G., Longaker, M. T., & Gurtner, G. C. (2013). From germ theory to germ therapy: Skin microbiota, chronic wounds, and probiotics. *Plastic and Reconstructive Surgery*, *132*(5), 854e–861e.

Wood, C. R., Al Dhahri, D., Al Delfi, I., Pickles, N. A., Sammons, R. L., Worthington, T., et al. (2018). Human adipose tissue-derived mesenchymal stem/stromal cells adhere to and inhibit the growth of *Staphylococcus aureus* and *Pseudomonas aeruginosa*. *Journal of Medical Microbiology*, *67*(12), 1789–1795.

Zhang, X., Tan, B., Wu, Y., Zhang, M., & Liao, J. (2021). A review on hydrogels with photothermal effect in wound healing and bone tissue engineering. *Polymers*, *13*(13), 2100.

Zhao, G., Hochwalt, P., Usui, M., Underwood, R., Singh, P., James, G., et al. (2010). Delayed wound healing in diabetic (db/db) mice with *Pseudomonas aeruginosa* biofilm challenge: A model for the study of chronic wounds. *Wound Repair and Regeneration*, *18*, 467–477.

Zhao, G., Usui, M., Underwood, R., Singh, P., James, G., Stewart, P., et al. (2012). Time course study of delayed wound healing in a biofilm-challenged diabetic mouse model. *Wound Repair and Regeneration*, *20*, 342–352.

Zhou, J., Tun, T. N., Hong, S. H., Mercer-Chalmers, J. D., Laabei, M., Young, A. E., et al. (2011). Development of a prototype wound dressing technology which can detect and report colonization by pathogenic bacteria. *Biosensors and Bioelectronics*, *30*(1), 67–72.

Ziuzina, D., Boehm, D., Patil, S., Cullen, P. J., & Bourke, P. (2015). Cold plasma inactivation of bacterial biofilms and reduction of quorum sensing regulated virulence factors. *PLoS One*, *10*(9), e0138209.

Chapter 18

Microbial biofilms: A persisting public health challenge

Aifa Fathima[a], Yaser Arafath[a], Saqib Hassan[a,b], R. Prathiviraj[a], George Segal Kiran[c], and Joseph Selvin[a]

[a]*Department of Microbiology, School of Life Sciences, Pondicherry University, Kalapet, Puducherry, India,* [b]*Division of Non-Communicable Diseases, Indian Council of Medical Research (ICMR), New Delhi, India,* [c]*Department of Food Science and Technology, School of Life Sciences, Pondicherry University, Kalapet, Puducherry, India*

1. Introduction

Many bacteria and fungi can form biofilms, which allow them to live in communities and adhere to the surfaces. The formation of biofilm begins from motile stage to sessile stage with the formation of extracellular matrix. More than 70% of the bacteria in the planet forms biofilm (Flemming & Wuertz, 2019). Biofilm-forming bacteria that are pathogenic in nature can affect food industries causing major health risk for the public. Biofilms are thought to be responsible for 80% of all microbial infections in humans (Davies, 2003). Biofilm formation is an important virulence mechanism in many bacterial pathogens (Verderosa, Totsika, & Fairfull-Smith, 2019). In medical implants, biofilm infection causes major concern. Microbial biofilm formation is a common occurrence in implanted medical devices, and it can cause severe illness and sometimes it may be life threatening (Habash & Reid, 1999). Commonly used medical implants are heart valves, intravascular catheters, orthopedic implants, pacemakers, urinary catheters, and stents are commonly used to save lives. But, when these get colonized by the bacterial biofilms, they can give rise to a significant health risk (Divya, Kiran, Hassan, & Selvin, 2019; Francolini & Donelli, 2010).

The bacterial biofilm forms a matrix consisting of polysaccharides, secreted proteins, and DNAs (extracellular) on the surface area where they gets attached (Tremblay, Lévesque, Segers, & Jacques, 2013). A biofilm can be made up of a single microbial species or a consortium of algae, archaea, bacteria, filamentous fungi, protozoa, and yeast that form strong bonds with one another and with biotic or abiotic surfaces (Muhammad et al., 2020). The microorganisms have the ability to from biofilm with same and different species. Biofilm formation appears to be a survival mechanism that gives microorganisms more options than planktonic cells, such as more remarkable strength to survive in oligotrophic environments, greater access to nutritional resources, improved survival to biocides, increased productivity and interactions, and better environmental stability (Dang & Lovell, 2016; Røder et al., 2018). The formation of biofilm helps the bacteria to survive under harsh environmental conditions.

The biofilms of pathogenic microorganisms are a primary medical concern because they are challenging to treat. Most antimicrobial treatments have been developed and tested on microorganisms that live in a planktonic (free-living) state. As a result, antimicrobial treatments are frequently ineffective against pathogenic biofilms (Lebeaux, Ghigo, & Beloin, 2014), which can be a thousand times more resistant to antimicrobials. Vaginitis, colitis, conjunctivitis, gingivitis, urethritis, and otitis are just a few of the diseases that have been attributed to biofilm infections (Verderosa et al., 2019). This chapter focusses on a detailed study of biofilm formation, the complexity of biofilms, and different diseases caused by the biofilm formation.

2. Biofilm formation and maturation

Biofilm is formed when bacteria excrete a slimy, glue-like substance that adheres to surfaces in moist environments. Biofilms can be formed almost on all surfaces including metals, plastics, medical implant materials, and even on plants and human tissues. Biofilm formation consists of several steps including different bacterial species (Al-Ahmad et al., 2010). Environmental stresses like UV radiation, desiccation, limited nutrition, extreme pH, thermal stress, high salinity, high pressure, and antimicrobial treatments cause bacteria to develop biofilms. Biofilm production is thought to begin with a reversible attachment of bacteria to a surface, followed by an irreversible adhesion, which is frequently helped by adhesive structures possessed by the bacteria and short-range

Understanding Microbial Biofilms. https://doi.org/10.1016/B978-0-323-99977-9.00004-1

interactions. The synthesis of EPS (exopolysaccharides) helps the bacteria in reversible attachment. Bacterial biofilms secrete EPS, various polysaccharides, proteins, fatty acids, and various nucleic acids (Khatoon, McTiernan, Suuronen, Mah, & Alarcon, 2018). The external polysaccharides forms 80% of the biofilm. Finally, bacterial cells can break off from the mature biofilm and colonize new sites in the environment (Muhammad et al., 2020). There are five significant steps involving biofilm formation, which are demonstrated in Fig. 1 and discussed below.

2.1 Reversible attachment

The first stage in the biofilm formation is attachment of bacterial cells to a surface. Pathogenic bacterial cells are opportunistic; consequently, when planktonic cells come in contact with a conditioning film (a thin layer of proteins), they cling to it by physical forces or bacterial appendages like pili or flagella (Khatoon et al., 2018). It all starts with a positive interaction between a few planktonic cells and the substrate surfaces. Brownian motion, sedimentation, or convection must carry the bacteria to the target site (Palmer, Flint, & Brooks, 2007). The interaction between the cell surfaces and the conditioned surface is determined by the net total of repulsive or attractive forces generated. If the attractive forces exceed the repulsive forces, the bacteria will adhere to the surface and vice versa (Carniello, Peterson, van der Mei, & Busscher, 2018). A conditioning film is a thin layer of proteins like collagen, fibronectin, fibrinogen, laminin, thrombospondin, vitronectin, von Willebrand factor, and polysaccharides that form on an implanted medical device or living tissues (Shunmugaperumal, 2010; Veerachamy, Yarlagadda, Manivasagam, & Yarlagadda, 2014). Because of the weak connections between the bacteria and the surface, this stage is called reversible attachment because the first interaction can be temporary and reversible. The ability, rate, and extent of bacterial adhesion to an implant surface are determined by the material composition, temperature, pressure, and the bacterial cell's surface features (Gupta, Sarkar, Das, Bhattacharjee, & Tribedi, 2016; Veerachamy et al., 2014). When bacteria adhere to a surface, cohesion occurs when bacteria attach (Gupta et al., 2016). Hydrophobic, steric, and Vander Waal forces are essential factors in biofilm adhesion to materials (Khatoon et al., 2018).

The bacterial attachment can take place in both biological and inert surfaces. Every substance that comes in touch with a bacterial solution is considered a biofilm

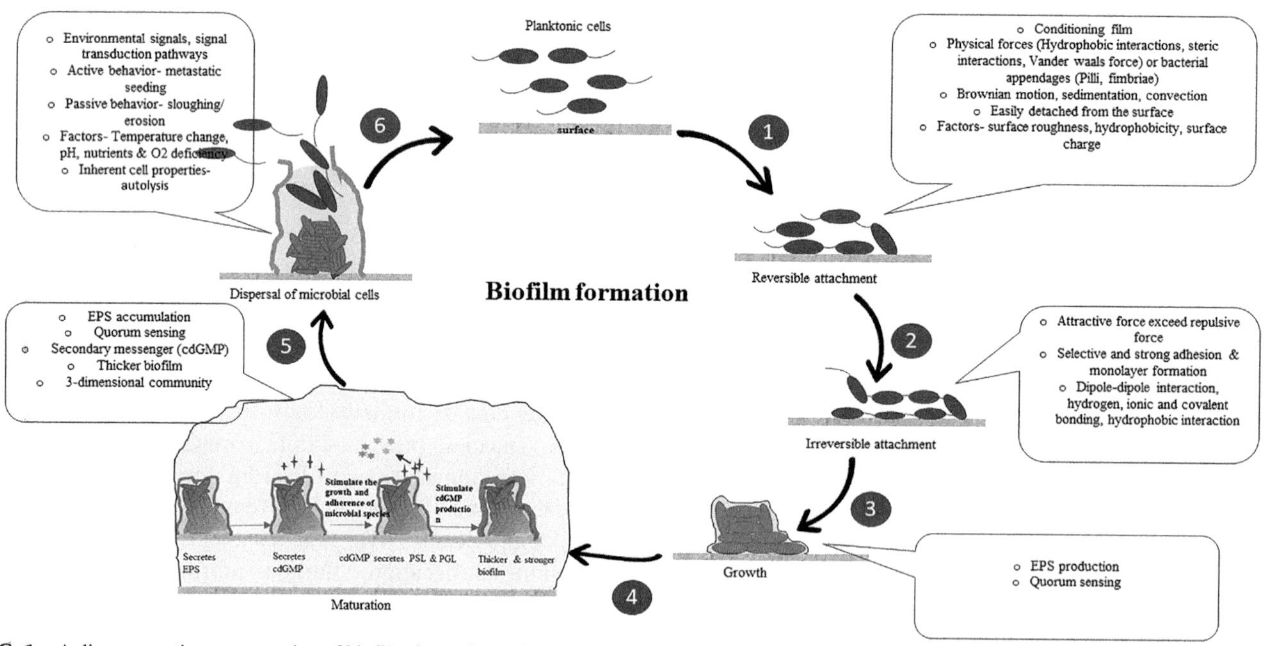

FIG. 1 A diagrammatic representation of biofilm formation—(1) Individual planktonic cells migrate and adhere to a surface; (2) cells adhere to substrate reversibly via physical forces or bacterial appendages; (3) adherent cells encased in small amounts of exopolymeric material become irreversibly attached to the surface, resulting in cell aggregation and matrix formation; (4) the biofilm begins to grow by developing microcolonies and water channel architecture while also becoming significantly more layered; (5) the fully mature biofilm reaches its maximum cell density and is now considered a three-dimensional community; (6) as the biofilm matures, it releases microcolonies of cells from the main community, which are free to migrate to new surfaces, spreading the infection to new areas.

growth substrate (Donlan, 2002). Surface hydrophobicity, roughness, charge, and presence of conditioning films are all physicochemical features of a substratum surface that might affect bacterial attachment (Muhammad et al., 2020).

2.2 Irreversible attachment

When attractive forces exceed repulsive forces, cells that are reversibly attached tend to be immobilized, whereas cells that are irreversibly attached shows stronger adhesion and monolayer formation (Gupta et al., 2016). Short-range interactions like dipole-dipole interactions, hydrophobic interactions, hydrogen bonding, ionic and covalent bonding, and along with the participation of bacterial structural adhesions, resulting in irreversible attachment (Bos, Van der Mei, & Busscher, 1999). Bacteria have a variety of adhesins on their surfaces that project outward from the cell surface to the extracellular environment (Berne, Ducret, Hardy, & Brun, 2015). The adhesive structures of the bacterial biofilm comprise of flagella, pili/fimbriae, and nonfimbrial adhesions facilitating the biofilm formation. These surface organelles assist bacterial cells in making initial physical contact with substrates (Muhammad et al., 2020).

2.3 Growth

Irreversible adhesion is facilitated by EPS production, which is regulated by the resident bacterial cell's quorum sensing (QS). Bacteria produce and secrete EPSs, essential components of the extracellular matrix in biofilms. EPS can mediate bacterial cohesion and biofilm adhesion to surfaces (Costa, Raaijmakers, & Kuramae, 2018; Fahs, Quilès, Jamal, Humbert, & Francius, 2014). EPS mediates intercellular signaling molecule like cyclic dimeric guanosine monophosphate (c-dGMP), which is found in almost all bacterial species. The growth and adhesion of bacteria are stimulated by this signaling process (Khatoon et al., 2018).

C-dGMP aids in the synthesis of matrix components such as polysaccharides and proteins, e.g., in *P. aeruginosa*, c-dGMP facilitates the secretion of various polysaccharides such as pentasaccharides (PSL), glucose-rich polysaccharides (PEL), and alginate. PSL and PEL function as signal molecules to further stimulate c-dGMP production, resulting in higher levels of c-dGMP that help to biofilm thicker and more robust (Khatoon et al., 2018).

2.3.1 Extracellular polymeric substances (EPS)

EPS that surrounds the cells in biofilm is made up of lipids, nucleic acids, i.e., extracellular DNA, proteins, and polysaccharides (Annous, Fratamico, & Smith, 2009). These substances aid in biofilm adhesion to the surface, nutrition trapping, structural support, and biofilm protection from host immune responses and antimicrobial treatments. Besides these functions, the EPS is responsible for preserving the biofilm cells together by facilitating cell to cell communication, QS, and enabling the exchange of genetic material via horizontal gene transfer (Hausner & Wuertz, 1999).

The development of biofilm significantly dependent on the presence of polysaccharides in a bacterial cell. Polysaccharides present in bacteria are generally neutral or polyanionic. The anionic property of the polysaccharides is due to the presence of the uronic acid or ketal-linked pyruvate. It is believed to make it easier for divalent cations like magnesium and calcium to bind together, which is essential for cross-linking polymer strands and increasing binding force in a developed biofilm (Donlan, 2002). However, in gram-positive bacteria like *Staphylococci*, the EPS is mostly cationic (Donlan, 2002). Proteinaceous structures like pili and fimbriae are also abundant in the EPS matrix. Each bacterial species has it unique protein that helps in the EPS production. In *Staphylococci*, proteases, nucleases, teichoic acids, and soluble phenol modules enhance the production of EPS and biofilm formation. But in *Streptococci*, glucan binding proteins (Gbps) are responsible for EPS growth (Khatoon et al., 2018). Furthermore, DNA is a component of the EPS matrix, which serves as an intercellular connector. Lipids found in the matrix are also crucial for *Thiobacillus ferrooxidans* attachment to surfaces.

2.4 Maturation

The adhered cells secrete autoinducer (AI) signals and help in the growth and maturation of the cells through which leads to the expression of biofilm-specific genes (Gupta et al., 2016). Signaling cues that aid in virulence and gene regulation corresponds to increased autoinducer production (Gu, Hou, Yongyat, De Tore, & Ren, 2013). At this point, bacteria begin secreting the EPS surrounding them, stabilizing the biofilm network and protecting them from antibacterial agents (Gupta et al., 2016).

When the intensity of autoinducers (AIs) exceeds a particular threshold, the genetic machinery of EPS, such as the 15-gene-long epsA-O cluster involved in biofilm formation in *Bacillus subtilis*, gets activated. Using the AIs signals, bacteria continue to grow within the embedded EPS matrix, leading to the creation of microcolonies and biofilm maturation (Muhammad et al., 2020). Changes in gene expression are induced following the formation of microcolonies and the accumulation of EPS. The products of these genes are used to produce EPS, which serves as a biological "glue" between embedded bacterial cells. Following the formation of the matrix, water-filled channels act as circulatory systems, transporting nutrients to cell communities and removing waste products. Microcolonies have a mushroom-shaped multicellular structure, according to structural analysis. As the production of bacterial surface structures is inhibited and the gene expression pattern changes during the maturation process, motility within microcolonies is restricted. As the production of bacterial surface structures is inhibited during the maturation process, motility within microcolonies is limited, and the gene expression pattern of sessile cells differs significantly from that of planktonic cells.

Furthermore, QS allows bacteria of the same or different species to communicate by secreting and detecting AIs. Bacteria use these signaling molecules to aggregate and control gene expression in response to population density changes. In this case, AIs play a critical role in maintaining existing biofilms (Muhammad et al., 2020).

2.5 Dispersal

The last step in biofilm formation is the bacterial detachment process or dispersal. The dispersed bacteria spreads from one part to the other in response to specific conditions. Ecological signals, signal transduction pathways, and effectors play a role in dispersal (Kaplan, 2010). However, bacterial dispersal mechanisms differ. It comprises of three stages, i.e., cell detachment from microcolonies, cell movement to a new substrate, and cell adhesion to the new substrate (Muhammad et al., 2020).

In most cases, a biofilm is made up of two layers. The lower part of the biofilm consists of microbial cells once they are released it spreads to the surrounding surfaces, which leads to the development of chronic infections and other serious health problems, and this process is described as metastatic seeding (Khatoon et al., 2018). In this active state, biofilm cells initiate detachment in response to environmental changes such as antimicrobial stress, matrix-degrading enzymes, and nutrient deprivation, or passive behaviors mediated by external forces such as shear forces (Muhammad et al., 2020). During active dispersal, genes involved in cell motility (flagella synthesis and EPS breakdown) are commonly increased, whereas genes associated in EPS production (polysaccharide synthesis), adhesion, and fimbriae synthesis are frequently downregulated (Kostakioti, Hadjifrangiskou, & Hultgren, 2013). Inhibiting the c-dGMP pathway is another effective way to disperse biofilm because lower intracellular c-dGMP levels inhibit or enhance biofilm development (Kaplan, 2010). Biofilm dispersal can also be aided by environmental factors such as temperature changes, pH, nutrients, and oxygen deficiency (Kostakioti et al., 2013).

Resources become limited as the biofilm matures, and the concentration of toxic products tends to increase. As a result, the cells disperse to other parts of the host's body including medical implants, i.e., obtain nutrition, eliminate stress-inducing conditions and waste. The bacterial cell disperse either as a single cell or as clumps. This process is triggered by low oxygen levels (in aerobic biofilms) or nutrient starvation. Small molecules like fatty acid DSF (cis-11-methyl-2-dodecenoic acid) are stimulated due to starvation, causing autophosphorylation and activation of c-dGMP phosphodiesterase, which degrades c-dGMP. It leads to shear forces to tear clusters apart or release planktonic cells, which dissolve a portion of the EPS. While gene regulation pathways are one mechanism involved in the dissolution of EPS, others are (Khatoon et al., 2018). For example, the bacterial cells within the biofilm produce saccharolytic enzymes, which break down the biofilm's stabilizing polysaccharides, allowing the surface bacteria to escape (Gupta et al., 2016). After being released, the bacteria either form new biofilms in another part of the body or float freely on the surface by upregulating the expression of flagellar proteins, which aid motility (Gupta et al., 2016; Veerachamy et al., 2014). The dissolution of EPS is a central area because it may hold the key to tearing biofilms and preventing their spread (Khatoon et al., 2018). Biofilm dispersal is aided by a variety of physicochemical and inherent cell properties like autolysis (Muhammad et al., 2020).

3. Infections associated with biofilms

Human infections are frequently caused by biofilms (Costerton, Stewart, & Greenberg, 1999). Biofilms, which can be life-threatening colonizers of biomedical devices, are responsible for the majority of the hospital-acquired infections. Pathogen colonization on implanted medical devices like respirators, catheters, prosthetic heart valves, and orthopedic implants is linked to a large proportion of nosocomial infections. 95% of urinary tract infections (UTIs) are connected to a urinary catheter, 80% of pneumonias are related to mechanical ventilation, and 87% of bloodstream infections are linked to intravascular devices. The bacteria *Staphylococci* and *Enterococci* are responsible for the majority of hospital borne diseases. The human bacterial biofilm flora of the skin, upper respiratory tract, lower gastrointestinal tract, and urogenital tract contains these two bacterial genera as commensal residents. As a result, they are among the organisms most likely to colonize implanted medical devices (Ebrey, Hamilton, Cairns, & Lappin-Scott, 2004). The most common bacteria linked to device-associated infections are Coagulase-negative staphylococci (CoNS) *Staphylococcus epidermidis*, *Staphylococcus lugdunensis*, *Staphylococcus haemolyticus*, and *Staphylococcus aureus* (Otto, 2008). CoNS can cause endogenous infection of intravascular catheters by tracking along their external surface or entering through the catheter hub during healthcare worker manipulation. *S. epidermidis* infections are more chronic and less acute, whereas *S. aureus* infections are more acute due to their ability to elicit a more robust immune response from the host (Toutain, Caiazza,

& O'Toole, 2004). Despite their low virulence, *enterococci* have been linked to severe nosocomial infections in recent years. Almost all of the infections they cause are usually food to be associated with hospitalized patients with severely weakened immune systems. Two major bacterial species cause most biofilm infections: *Enterococcus faecalis* 80%–90% of enterococcal infections, and *Enterococcus faecium*, which causes 10%–15% (Mohamed & Huang, 2007). *Enterococci* are one of the most common causes of nosocomial infections, especially in intensive care units, where enterococci can be passed from one patient to the next through the hands of medical personnel. The urinary tract and soft tissues adjacent to the intestinal flora, where enterococci live, are the most common sites of infection. Patients who have had extensive abdominal surgery, have indwelling devices or have undergone abdominal procedures are the most vulnerable. Antibiotic resistance, particularly resistance against vancomycin, has become a significant issue (Zaph, 2010). Biofilm infections have been linked to pathogens associated with chronic infections such as periodontitis, cystic fibrosis pneumonia, chronic (UTIs), chronic otitis media, and chronic wound infections. In contrast to acute infections primarily caused by planktonic growth, the planktonic phenotype usually only exists transiently and as a minor population in chronic bacterial infections. Because chronic and acute infections are different in fundamentally, so that different treatment methods are used to cure biofilm infections effectively. Treatment of chronic biofilm infections is complicated because microbial biofilms are a protected mode of growth that allows cells to survive in hostile environments (Dufour, Leung, & Lévesque, 2010).These characteristics are frequently cited as reasons for bacterial persistence in the human body for extended periods (Dufour et al., 2010).

4. Molecular mechanism of biofilm formation

Microorganisms were ubiquitously present in nature, Most of microbes (99%) get adhered to any surface by means of biofilm formation which involves multi species microorganisms (Davey & O'Toole, 2000). The bacterial biofilm comprises of 15% bacterial cells and 85% of matrix material, the arrangements between the matrix material determines the complexity and the functionality of the biofilm (Kokare, Chakraborty, Khopade, & Mahadik, 2009). It is a dynamic process which can develop into a biofilm life cycle, During the process of biofilm formation certain biochemical components are produced by some of the microorganisms which makes them the early colonizers and helps them to get adhere to the surface very firmly (Kolenbrander, Palmer, Periasamy, & Jakubovics, 2010). The exopolysaccharide plays a major role in the structural organization of the biofilm, consisting of polysaccharides, proteins (undefined), and some extracellular DNA.

The first exopolysaccharide that was isolated from *Pseudomonas aeruginosa* is alginate, which was seen in mucoid strains isolated from Cystic Fibrosis patients (Doggett, 1969). The effective formation of the biofilm depends on the proper coordination that comes from the external polysaccharide interaction between bacteria. The in vitro analysis of the biofilm shows the presence of various proteins, exopolysaccharides, eDNA, and other polymeric substances, which leads to the production of bacterial agglomerations which has no role in the formation of biofilm rather acts as a protective agents against several antibiotics and host defense system (Mai, Seow, Pier, McCormack, & Thong, 1993).

4.1 *Pseudomonas*

P. aeruginosa is Gram-negative bacteria which belongs to the member of γ-proteobacteria, there are 12 members in the group and *P. aeruginosa* is one among them (Todar, 2008). It is a free-living organism which is responsible for causing about 20% of the infections in hospitalized patients, mostly seen in the immunosuppressed patients. They can survive in less nutrient conditions and have higher degree of tolerance to several physical factors (Fazeli et al., 2012). The environmental strain of *P. aeruginosa* has two different exopolysaccharides *psl* containing (D-mannose, D-glucose, and L-rhamnose) that helps in interacting with other cells and facilitates biofilm formation (Ma et al., 2009). The *psl* operon consists of 15 co-transcribed genes (*pslA* to *pslO*) which help in the synthesis of the Psl, the *pslD* gene present in the PslD has major role in biofilm formation (Campisano, Schroeder, Schemionek, Overhage, & Rehm, 2006). *P. aeruginosa* forms pellicle in liquid media (standing liquid media) by the help of *pel* (operon) which is made up of seven genes (*pelA* to *pelG*) (Friedman & Kolter, 2004; Ghafoor, Jordens, & Rehm, 2013). The *pel* operon is cellulose-sensitive which has galactosamine and glucosamine sugars which are partially acetylated (1-4 linked) (Jennings et al., 2015).

Elements like adhesins, flagella, and pili (type IV) help in the formation of biofilm, which gets attached to different parts of the cell surface (Cai et al., 2015). T4P pili has functions like colonization, twitching motility, bacteriophage infection, and natural DNA transformation. Pilin aggregates and forms polymeric fibers (Hazes, Sastry, Hayakawa, Read, & Irvin, 2000).

4.2 *Escherichia coli*

Escherichia coli is a gram-negative facultative anaerobic bacterium, which serves as a model organism in industrial applications (Lee, 1996). *E. coli* is a commensal that colonizes the intestine of the warm-blooded organisms. The biofilm formed by these bacteria causes intestinal infections where the development of biofilm takes place through the communication by signaling pathways QS attaining in the release of the autoinducers (AI) helps in strengthening the biofilm formation and helps in the maturation process (Sturbelle et al., 2015). The pathogenicity varies with strains, and the enteropathogenic strains are of two types intestinal pathogenic *E. coli* (InPEC) and extraintestinal pathogenic *E. coli* (ExPEC). The diseases such as enteric syndromes and hemorrhage are associated with InPEC due to various toxins. Whereas the ExPEC is associated with (UTIs), neonatal sepsis, and meningitis (Vogeleer, Tremblay, Mafu, Jacques, & Harel, 2014). The urinary tract infection caused by *E. coli* is mostly of catheter-associated urinary tract infection (CAUTI) (Reisner et al., 2014).

The biofilm of *E. coli* contains β-1,6-*N*-acetyl-D-glucosamine polymer (PGA), cellulose, and colanic acid apart from these the lipopolysaccharides and capsules forms the major role in the biofilm formation in *E. coli*. It also helps in the stabilization of the biofilm structure. The operon *pgaABCD* of *E. coli* encodes for PgaC glycosyltransferase protein, which helps in the synthesis of PGA polymer. The presence of cellulose in the biofilm makes it more rigid. The presence of colanic acid protects it against the stressful environmental conditions (Vogeleer et al., 2014). There are other components such as lipopolysaccharides (LPS O antigen), capsular polysaccharides (polysaccharide K antigen) that have indirect effect on the cell to cell interaction making it the strong (Beloin, Roux, & Ghigo, 2008). Certain proteins do not require any accessory proteins for their translocation into outer membrane are called auto-transporter proteins (Beloin et al., 2008). The virulent strain *E. coli* has two transporter proteins AidA and TibA that helps in biofilm formation (Vogeleer et al., 2014).

The autoinducers varies with different bacterial species that governs the expression of genes (Daniels, Vanderleyden, & Michiels, 2004). The thickness of the biofilm can be of hundred microns that makes it harder for antibiotic treatment due to exopolymers (Larson et al., 2003). The external capsule of *E. coli* is strong that protects it from the environmental stresses (Larson et al., 2003). The mutated strain of *E. coli* has rapA gene, which altered the expression of 22 genes (biofilm) and removal of the gene *yhcQ* (multidrug resistance pump) lowers the biofilm resistance between biofilm and *E. coli* (Lynch et al., 2007). During the stress condition, the *E. coli* expresses stress resistance genes like Hfq protein helps in biofilm formation under harsh environments like urinary tract, the gene *YcfR/BhsA* shows resistance to acid, heat, peroxide and cadmium. The gene *ymgB* secretes AriR protein (regulator of acid resistance influenced by indole) helps in the survival of *E. coli* in low pH (Wood, 2009).The sigma factor S encoded by *rpoS* gene helps in the regulation of stress response, where the structural and regulatory genes such as curli, cellulose, matrix components, and other enzymes involving in the formation of biofilm depend on sigma factor (Serra & Hengge, 2014). The cell surface of *E. coli* has specialized amyloid fibers that helps in the cell to cell interaction and in biofilm formation.

E. coli strain K-12 needs amino acids supplement for the formation of biofilm and *E. coli* O157:H7 forms biofilm under low nutrient conditions (Dewanti & Wong, 1995; Pratt & Kolter, 1998). *E. coli* has an outer membrane protein antigen 43 (coded by agn43 gene) which helps the *E. coli* cells to aggregate in liquid media during stationary phase, when this gene is deleted the binding ability of the bacterium toward PVC surface is retained, whereas the density of the amount of biofilm formation is reduced upon testing with crystal violet stain (Henderson, Meehan, & Owen, 1997). The *agn43* gene encodes for two protein subunits α- and β, β subunit is an integral protein that is present in the outer membrane and the α subunit is present on the cell surface. Finally, the Ag43-Ag43 interaction helps in the bacterial cell aggregation (Hasman, Chakraborty, & Klemm, 1999). This α subunit forms a L-shaped protein structure through hydrogen bonding and salt bridging thus stabilizing the protein dimers between bacteria. The arrangement looks like head to tail conformation leading to protein "velcro" formation (cells aggregation) (Schembri, Hjerrild, Gjermansen, & Klemm, 2003). The antigen 43 also found to be expressed in *P. fluorescens* helps in the formation of biofilm and also in aggregation of bacterial cells, when *E. coli* and *P. fluorescens* expresses agn43 they co-aggregate and effectively form mixed biofilm (Kjærgaard, Schembri, Hasman, & Klemm, 2000).

4.3 *Staphylococcus aureus*

S. aureus colonizes skin and mucosal membrane of the human body. It acts as both commensal and as an opportunistic pathogen causing infections like skin infections, bacteremia, and severe pneumonia (Gillet et al., 2002). The clinical strain *S. aureus* forms biofilm on the catheter and other medical implants, where it resists the antimicrobial therapy and evades host immune system (Otto, 2008). Biofilm-associated food-borne infections are public health concern (Srey, Jahid, & Ha, 2013). The spectrum of virulence factor varies with different genetical make of the specified strain (Jarraud et al., 2002). The *agr* (QS system) upregulates the expression of δ-toxin, which helps in the biofilm development and adhesion in biofilm formation along with polysaccharide intercellular adhesin (PIA) (Jiang, Jin, & Sun, 2018). The Staphylococcal protein (*spa*) facilitates adhesion steps in biofilm formation (Lasa & Penadés, 2006).

S. aureus uses electrostatic force and hydrophobic interactions to get attach toward abiotic surfaces where the polystyrene surface undergoes change for better biofilm development (Kennedy & O'Gara, 2004). The presence of teichoic acid on the cell surface also helps in the formation of biofilm on polystyrene and glass surface (Gross, Cramton, Götz, & Peschel, 2001). The presence of surface proteins molecules known as microbial surface components recognizing adhesive matrix molecules (MSCRAMMs) which expresses different binding capacity toward different components like fibrinogen, collagen, and cytokeratin (Speziale et al., 2009). The MSCRAMMs complex proteins includes FnBPA and FnBPB, which are fibronectin-binding proteins, repeated proteins from serine—aspartate like SdrC, SdrD, and SdrE, clumping factors, adhesin form collagen, Protein A, sensitive protein of plasmin, iron-regulated surface determinants like IsdA, IsdB, IsdC, and IsdH and bone sialoprotein (Bbp) helps in the biofilm network formation (Vazquez et al., 2011).

The Agr QS mechanism in *S. aureus* regulates phenol-soluble modulin (PSM), which helps in the biofilm adherence (Periasamy et al., 2012). The δ-toxin prevents hydrophobic interaction between the bacterial cell and the polymeric surface, whereas AtlA helps in the fibronectin attachment and assists in early biofilm development (Bose, Lehman, Fey, & Bayles, 2012). The *Staphylococcal* proteins like FnBPs, ClfB, and SdrC help in the accumulation and attachment of the biofilm (Speziale, Pietrocola, Foster, & Geoghegan, 2014). There are other associated proteins that help in the biofilm attachment (Aap) *S. epidermidis* accumulation-associated protein, SasG protein from *S. aureus* (Geoghegan et al., 2010). Recent studies show that cell wall-associated proteins have nothing to do with multiplication, *S. aureus* uses cytoplasmic proteins during biofilm formation and the formation of matrix completely strain dependent (Foulston, Elsholz, DeFrancesco, & Losick, 2014). There are two operons associated with the diverse colony of *S. aureus cidABC* and *lrgAB* where the fast growing colony shows increased *lrgAB* expression and delayed *cidABC* expression and vice-versa in other colony types (Moormeier et al., 2013). The genes *ackA* and *pta* help *S. aureus* in converting the pyruvate to acetate making a shift in the biofilm formation. The initiation of the dispersal of *S. aureus* biofilm takes place under the Agr controlled QS (Vuong, Saenz, Götz, & Otto, 2000). The Agr QS system depends on the cell signaling molecules, cell density, and auto-inducers like AIP (auto-inducing peptide) which is and octapeptide pheromone once attaining a threshold level it activates histidine kinase and AgrC. This AgrC activates AgrA by phosphorylation leading to *agr* operon activation (P3 promoter region). The RNA responsible for the virulence factor expression and biofilm associated genes is released (Novick & Geisinger, 2008).

4.4 *Klebsiella pneumoniae*

Klebsiella pneumoniae is a capsulated bacterium that is present in the mucosa, water bodies, and vegetative soil, whereas in human body they colonize gastrointestinal tract and oropharynx from where it enters to the other parts of the human body causing infections like pneumonia, bacteremia, and ventilator-associated pneumonia. They cause severe infection in immune compromised patients like bladder neuropathy or diabetes mellitus (Seifi et al., 2016). *K. pneumoniae* has the capacity to form biofilm that consists of proteins, exopolysaccharides, DNA, and lipopeptides, it also has virulence factors like capsule polysaccharide, fimbriae (type 1 and 3), and they use these virulence factors for escaping the host immune system (Li, Zhao, Liu, Chen, & Zhou, 2014). In 2005, about 28% *K. pneumoniae* isolates extended spectrum betalactamase (ESBLs) were reported to be the causal agent for hospital-acquired infections in Soeradji Tirtonegoro hospital (Parathon et al., 2017). The genes that are responsible for the formation of biofilm in *K. pneumoniae rmp*A (Capsular gene) present in K1/K2 virulent type. The genes *fim*H and *mrk*D responsible for the type 1 and 3 fimbriae formation, and the gene *uge* is responsible for the synthesis of uridine diphosphate galacturonate 4-epimerase (Balestrino, Ghigo, Charbonnel, Haagensen, & Forestier, 2008). The gene *kfu* is responsible for the regulation of the iron uptake by *K. pneumoniae* (Ma, Fang, Lee, Shun, & Wang, 2005). The synthesis of poly-β-1,6-*N*-acetyl-D-glucosamine adhesin is regulated by type 2 QS and ABCD operon, which helps in the cell-to-cell communication and adhesion process too (Chen et al., 2014).

There are more than 300 types of ESBL variants have been reported and the genes that are responsible for the regulation of β-lactamases such as bla_{SHV}, bla_{TEM}, and $bla_{CTX\text{-}M}$ (Paterson & Bonomo, 2005). *K. pneumoniae* forms a thick outer layer (extracellular biofilm) for attachment and also for protecting against antibiotic effect (Vuotto, Longo, Balice, Donelli, & Varaldo, 2014). In one of the studies, it was observed that among the 40 isolated carbapenem resistant *K. pneumoniae* from blood samples (13) showed highest degree of biofilm formation (Diago-Navarro et al., 2014). The antibiotic resistance of a mature biofilm is 1000 times than that of the planktonic bacteria making it harder to eliminate and treat with antibiotics (Lebeaux et al., 2014). In a catheterized bladder model *K. pneumoniae* was reported to expresses higher expression of type 1 and type 3 fimbriae with higher biofilm formation (Stahlhut, Struve, Krogfelt, & Reisner, 2012).The protein MrkA forms the basic subunit for type 3 fimbriae, the protein MrkD has adhesive properties helps the bacteria to gets attached toward the surfaces like Polyethylene (PE), Polyvinyl chloride (PVC) (Chung, 2016).

5. Components of biofilms

5.1 Curli

The Curli fibers form the major component of *E. coli* biofilm matrix, this helps the bacterium to form biofilm outside the host (curli-dependent) at 30°C (Römling, Bian, Hammar, Sierralta, & Normark, 1998). There are two operons encoded for curli fibers *csgBAC* and *csgDEFG*. The operon *csgBAC* encodes for structural fibers like CsgA and CsgB and periplasmic protein CsgC, whereas the *csgDEFG* operon encodes CsgD which controls the production of cellulose and curli along with accessory proteins and CsgE and CsgF (Hobley, Harkins, MacPhee, & Stanley-Wall, 2015). The fiber contains CsgB helps in the polymerization of the fiber formation (minority), and CsgA forms the major part of the fiber resulting in the formation of β-sheet structures (Hammar, Bian, & Normark, 1996). Dyes such as Congo Red and Thioflavin T (ThT) interact with these fibers, and the polymerized form of these fibers is highly resistant to any denaturation methods (Collinson, Emödy, Müller, Trust, & Kay, 1991). The CsgA has three regions signal peptide, N-terminal region and C-terminal domain (core of the β-sheet-rich fiber) (Collinson, Parker, Hodges, & Kay, 1999).

The β-strand comprises of five repeating units the first and the fifth are responsible for the fiber formation. When the first N-terminal repeat is deleted (in vitro), the CsgA shows less fiber formation and when the fifth C-terminal is deleted the fiber formation completely stops (Wang, Hammer, & Chapman, 2008). The purified form of curli fibers from *E. coli* is rich in *β-sheet* (Chapman et al., 2002). The CsgB varies from CsgA in its fifth C-terminal, which is comparatively less conserved than the other four along with that they have the positively charged amino acid. The deletion of fourth or fifth repeat resulted in the loss of outer membrane and CsgA polymerization (Hammer, Schmidt, & Chapman, 2007). The outer membrane of *E. coli* is formed of CsgB in the absence of CsgA short fibers that are expressed on the *E. coli* cell surface. The curli biogenesis and functional studies have been performed on *E. coli* and *Salmonella* through bioinformatics approach few other bacteria like *Proteobacteria*, *Thermodesulfobacteria*, and *Halanaerobium* have also been studied (Dueholm, Albertsen, Otzen, & Nielsen, 2012). The operon *csgEFG* is found in most of the bacteria with CsgA/B homologs with/without CsgC and CsgD.

5.2 Cellulose

The most abundant organic polymer in nature is cellulose, it has linear glucose chain with 1-4 beta linkage which has a protective nature during biofilm development (Solano et al., 2002). The gene cluster responsible for the cellulose production is conserved in *Salmonella typhimurium* and *E. coli* (Zogaj, Nimtz, Rohde, Bokranz, & Römling, 2001). The proteins responsible for cellulose synthesis are encoded in *yhjR-bcsQABZC* and *bcsEFG* in *E. coli*. The cellulase synthase complex has two subunits which are encoded by proteins BcsA and BcsB helps in the conversion of UDP glucose to cellulose (Omadjela et al., 2013). Sometimes over production of cellulose has negative effect on the biofilm formation as in *E. coli* strain MG1655 (Gualdi et al., 2008). The over expression of the cellulose may interfere with the curli-mediated adhesion by masking them with excessive cellulose, thus this over expression makes *E. coli* resistant to desiccation. The resistance to desiccation can be lowered by deleting the curli subunit CsgA, this curli and cellulose protects the bacteria form external damage (Gualdi et al., 2008). This network protects the bacteria from *Caenorhabditis elegans* and by the predatory bacterium *Myxococcus xanthus* (DePas et al., 2014).

5.3 Flagellum

Bacterial flagellum has major role in the formation of biofilm matrix, the *E. coli* K12 strain W3110 with point mutation at *bcsQ* gene results in the cellulose defect. The colony formed by the strain (macrocolony) has three distinct regions outer smooth zone, middle concentric ring shaped, and the inner most area has wrinkled structure. The development of the biofilm depends on the production of the flagella and its ability to rotate, any mutation in the flagellar filament protein *fliC* and motor protein *motA* resulting in nonringed pattern colony. In mutant strains, the genes responsible for flagella synthesis are *flhDC* and *fliA* when these are deleted the secondary cyclic-di-GMP increases. The increase in the cyclic-di-GMP tends to increase the curli production due to the absence of flagellar filament (Pesavento et al., 2008). The genes that encode flagellar proteins are *fleQ*, *fleR*, *fleS*, *fliA*, *flgM*, and *fleN* they are separately formed in the chromosome (Marsin & Forterre, 1999).

5.4 Alginate

Alginate protects the bacterium form external stress and improves the surface adhesion capacity, it is made up of β-1-4 glycosidic linked α-L-guluronic acid and β-D-mannuronic acid (Hay, Remminghorst, & Rehm, 2009). The genes responsible for alginate gets transcribed (12-gene cluster) after the attachment of the bacteria to any surface leading to biofilm formation. Primarily, it is made up of D-mannuronic acid homopolymer, which later on transforms into periplasm (Wiens, Vasil, Schurr, & Vasil, 2014). In the biofilm of *P. aeruginosa*, different types of alginate and hydrolytic enzymes are secreted which protect the bacteria from antibiotic attacks (Tielen et al., 2013). The

Pseudomonas strain affecting the lungs in the clinical patients shows development of mucoid phenotype from nonmucoid phenotype due to the production of alginate (Hay et al., 2009). This alginate helps *Pseudomonas* from phagocytic attack, and alginate targets macrophage released free radicals (Aguilar, Vlamakis, Losick, & Kolter, 2007). Over production of alginate leads to the well-structured biofilm formation showing higher degree of resistance to antibiotics (tobramycin), it can coexist with *S. aureus* and cause clinical infections. The alginate interferes with the production of siderophores and rhamnolipids that kills *S. aureus* affecting the antimicrobial activity (Limoli et al., 2017).

During the favorable conditions, *P. aeruginosa* produces alginate lyase *algL* that divides the alginate into smaller oligosaccharides releasing it from its attaching ability on any surface (dispersal) and allow the bacterium to occupy new surface (Boyd & Chakrabarty, 1995).

5.5 eDNA

The major function of *eDNA* is to strengthen the biofilm structure at the time of nutrient deficiency and also helps in extension of biofilm, resistance against antibiotics, and also in connecting microcolonies (Wang et al., 2015). The expression of pyocyanin (phenazine molecules) also triggers the release of *eDNA*. The mutant strain of *P. aeruginosa* having defective gene on the expression of phenazine molecules produces less *eDNA* when compared to nonmutant strain of *P. aeruginosa* (Das & Manefield, 2012). The molecule pyocyanin and *eDNA* show increased cellular surface hydrophobicity and nonspecific interactions. There is a 40% decrease in the biofilm production due to *eDNA* removal and 80% reduction of biofilm due to phenazine deficiency (Das, Ibugo, Klare, & Manefield, 2016).

There are two ways in which the *eDNA* is produced in *P. aeruginosa* one is by non-QS way which releases very less *eDNA* and during late log phase (QS dependent) which releases abundant *eDNA* (Montanaro et al., 2011). The biofilm of *P. aeruginosa* is mainly made up of *psl and eDNA* which combines together and forms network scaffold assisting the bacteria in attachment and growth (Wang et al., 2015).

The stabilization and cross binding of biofilm between bacteria are facilitated by the *eDNA* (negatively charged) binding with Ca^{2+} leading to the biofilm formation and also it binds to antimicrobials and aminoglycosides (Chiang et al., 2013). The presence of *eDNA* in the biofilm lowers the pH and acidifies the biofilm, at lower pH two factors are activated *PhoPQ* (virulence regulating factor) and *PmrAB* (aminoglycosides resistance factor) (Wilton, Charron-Mazenod, Moore, & Lewenza, 2016). Gram-negative bacteria has Type VI secretion system (T6SS) helping in the translocation of proteins and virulence factor. The regulatory process is carried out by *eDNA*, which chelates the cations present in the outer membrane allowing the bacteria to over grown in any microbial communities (Wilton et al., 2016).

5.6 Proteins

There are major components like adhesins, flagella, and type IV pili, which helps in the formation of biofilm, they gets attached to (top and bottom) various part of the epithelial surface of the cells (Cai et al., 2015). Different protein components occupy different parts of the epithelial cells Type IV pili protein fibers (apical region), flagella (basal region) (Bouvet et al., 2010). There are smaller structural proteins called Pilins, which has protein interaction and transmembrane domain (Giltner, Nguyen, & Burrows, 2012). Pilins are formed from protein subunit *pilA*, which is secreted by pilQ and transported out of the cell to form fimbrial strands (Laverty, Gorman, & Gilmore, 2014). The Type IV pili helps in colonization, formation of biofilm, infection by bacteriophage, the pilin can form ordered layers by the help of polymeric fibers (Hazes et al., 2000). There are two types of pilus components present in the membrane complexes are *pilM* and *pilN* and at the N-terminal region of the subcomplex has *pilMNOP* which connects the inner and the outer pat of the membrane (McCallum et al., 2016).

Flagella is made up of protein subunit called flagellin which is of two types a and b, flagellin is a globular protein which is recognized by the immune system (innate) (Bucior, Pielage, & Engel, 2012). In *P. aeruginosa* flagella forms an important virulence factor which is also used for developing vaccine (Campodónico et al., 2010). The chemotaxis in *P. aeruginosa* is similar to other microorganisms. The continuous rotation of the flagella (Clockwise and Counter clock wise) forms a chemoattractant gradient (Cai, Li, Ouyang, Luo, & Gordon, 2016). In the present chapter we also revealed the vital genes and protein are involved in bioflim communication between bacterial communities. Thus, a protein-protein interaction networks (PPINs) is a study to identify the essential (hub) genes/proteins involved in their respective metabolism (Kiran et al., 2021; Prathiviraj, Berchmans, & Chellapandi, 2019; Prathiviraj & Chellapandi, 2020). The draft PPINs was constructed from STRING v11.5 server (Szklarczyk et al., 2019) based on Gene Ontology terms and KEGG pathways using the keywords "*pilin*, *flagellin*, *chemotaxis*" against *E. coli*, *P. aeruginosa*, and *S. aureus*. The final network was designed and validates using CytoScape v3.9.0 (Shannon et al., 2003) software. From the result of PPINs, we found that some essential genes play a significant role in flagellar assembly process between bacterial domains (Fig. 2). A detailed biological and molecular function was represented in Fig. 3A and B.

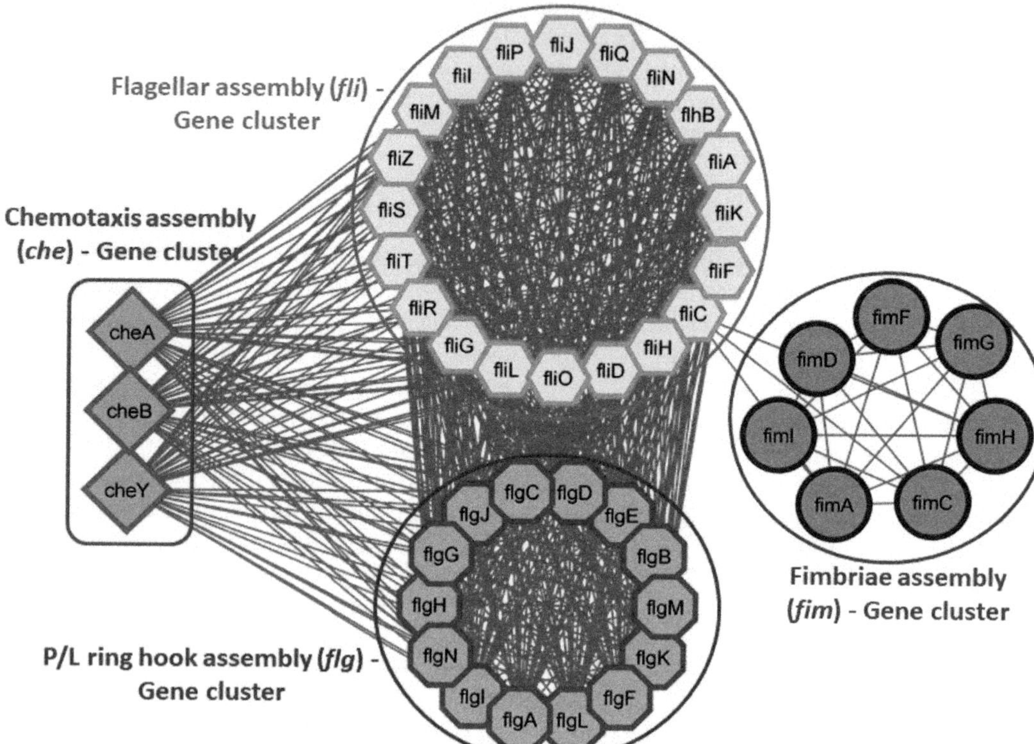

FIG. 2 Construction of PPINs using the gene clusters involved in flagellar assembly process in bacterial domain. The network consists of 44 nodes and 689 edges. The network was validated by clustering coefficient (0.982). The *P*-Value of protein-protein interaction enrichments was found to be $<1.0e^{-16}$.

In *E. coli*, the cruli fibers and flagella are distinctively seen. The upper side of the biofilm shows an extracellular mesh with small ovoid colonies. Along with this the cellulose-producing isolates, *E. coli* K12 W3110 shows large flat and tissue-like appearance in the edge of the colony and the cells are coated with smooth coatings (cellulose and curli). The cellulose stops both flagellar rotation and turns off the flagellar gene transcription (Zorraquino et al., 2013). The *E. coli* strain UTI89 forms floating pellicle with two distinctive layer upper layer with dense fibrous materials (basket like) and base with less fibers material (McCrate, Zhou, Reichhardt, & Cegelski, 2013).The NMR (nuclear magnetic resonance) analysis of the biofilm showed that the presence of curli fibers in comparison to cellulose was 85%:15% (McCrate et al., 2013).

In *Vibrio cholera*, the biofilm matrix contains Bap1 (biofilm-associated protein 1), RbmA (rugosity and biofilm modulator A), and RbmC (rugosity and biofilm modulator C), which is secreted by N-terminal signal peptide and is then released in to the biofilm matrix by (T-IISS) type two secretion system (Johnson et al., 2014). The presence of Bap1 helps in the identification of different genes that are expressed between the biofilm and the bacterial cell, it is also involved in the antibiotic resistance mechanism (Duperthuy et al., 2013). The RbmC is made up of 957 amino acids and shows 70% similarity to Bap, at the time of biofilm formation these two are functionally inert (Absalon, Van Dellen, & Watnick, 2011). The RbmA is made up of 271 amino acid and is detergent resistant and provides rugosity to the biofilm. It is secreted and released extracellularly by the Type II secretion system (Johnson et al., 2014).

B. subtilis is an endospore forming Gram-positive bacilli. By 1870s, by Burton-Sanderson and Ferdinand Cohn officially reported the formation of biofilm by this bacterium (Cohn, 1875). The external matrix is helpful in producing resistant spores on the surface of the biofilm (Veening, Hamoen, & Kuipers, 2005). The biofilm matrix contains polysaccharides of higher molecular weight and extracellular proteins TasA and TapA. The production of BslA (biofilm surface layer protein A) helps in the maturation of the biofilm along with other proteins (Ostrowski, Mehert, Prescott, Kiley, & Stanley-Wall, 2011). The BslA is hydrophobic in nature and forms wild type biofilm around the bacterial community (Hobley et al., 2013). The biofilm fiber TasA is dependent on TapA (253 amino-acid protein), TapA is present in the cell wall (Romero, Vlamakis, Losick, & Kolter, 2011).

In *S. aureus*, there are special proteins in the biofilm matrix known as cell wall anchored proteins (CWAP). They are recognized by the presence of "LPXTG" motif at the C-terminal region. The enzyme Sortase A (SrtA) cleaves

(a)

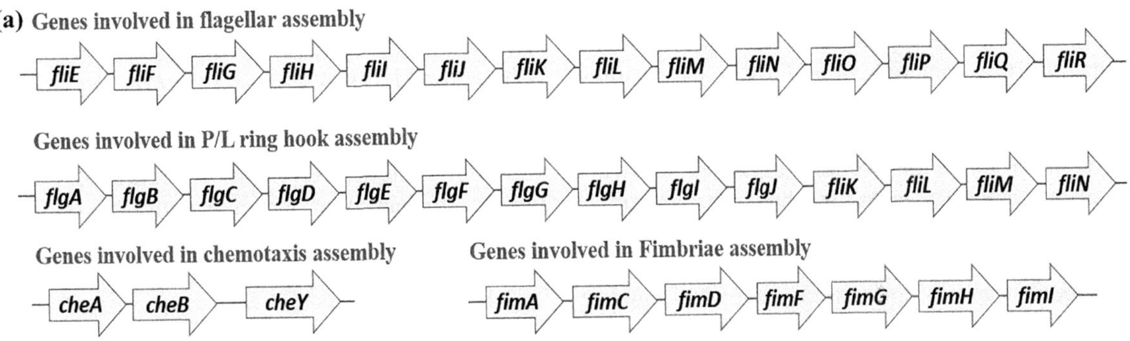

(b)

Genes involved in flagellar assembly		**Genes involved in P/L ring hook assembly**	
Gene	**Biological role**	**Gene**	**Biological role**
FliA	RNA polymerase sigma factor for flagellar operon	*FlgA*	Assembly protein for flagellar basal-body periplasmic p ring
FliC	Flagellar filament structural protein (flagellin)	*FlgB, FlgC, FlgF, FlgG*	Flagellar component of cell-proximal portion of basal-body rod
FliD	Flagellar hook-associated protein 2	*FlgD, FlgE*	Flagellar biosynthesis, initiation of hook assembly
FliF	Flagellar basal-body MS-ring and collar protein	*FlgH, FlgI*	Flagellar protein of basal-body outer-membrane l ring
FliG, FliM-N	Flagellar motor switching and energizing component	*FlgJ*	Flagellar rod assembly protein and murein hydrolase
FliH	Negative regulator of *FliI* ATPase activity	*FlgK*	Flagellar hook-filament junction protein 1
FliI	Flagellum-specific ATP synthase	*FlgL*	Flagellar biosynthesis; hook-filament junction protein
FliJ	Flagellar biosynthesis protein	*FlgM*	Negative regulator of flagellin synthesis
FliK	Flagellar hook-length control protein	*FlgN*	Flagellar biosynthesis protein
FliL-R	Flagellar biosynthesis protein	**Genes involved in chemotaxis assembly**	
FliS	Flagellar secretion chaperone	**Gene**	**Biological role**
FliT	Flagellar synthesis and assembly chaperone	*CheA, CheY*	chemotaxis family sensor kinase
FlhB	Flagellin export apparatus, substrate specificity protein	*CheB*	Involved in chemotaxis
Genes involved in Fimbriae assembly			
Gene	**Biological role**	**Gene**	**Biological role**
FimA	Major type 1 subunit fimbrin (pilin)	*FimI*	Fimbrial protein
FimG, FimH	Involved in regulation of length and mediation of adhesion of type 1 fimbriae	*FimC*	Type 1 fimbriae periplasmic chaperone
FimD, FimF	Fimbrial usher outer membrane porin protein		

FIG. 3 (A and B) Molecular function and biological process of flagellar assembly proteins found in bacterial domain. The functional annotation was performed based on Gene Ontology terms.

the proteins and helps in attachment to the peptidoglycan layer of the cell wall (Ton-That, Liu, Mazmanian, Faull, & Schneewind, 1999). There are 24 different CWA proteins found in *S. aureus*, and the expression of these can vary with strain and cultural conditions (Foster, Geoghegan, Ganesh, & Höök, 2014). The phenol soluble modulins (PSM) help in the development and stabilization of the biofilm primarily discovered in *S. epidermidis* (highly conserved) (Cheung, Chatterjee, & Otto, 2014).

5.7 Quorum sensing

In small bacterial colonies, the bacterial cell envelope is essential for intercellular signaling and communication between neighboring cells (Khatoon et al., 2018). QS is a bacterial communication system in which the bacteria produce and secrete small signaling molecules called autoinducers, which allow them to express specific genes at specified population densities (Kiran, Hassan, Sajayan, & Selvin, 2017; Schachter, 2003). When the bacterial population achieves a high density, QS affects major bacterial activities like pathogenicity, antibiotic resistance, biofilm formation, and bioluminescence (Hassan, Ahmad, Bashir, Kiran, & Selvin, 2019). The communication among cell is a complicated regulatory process that keeps the thickness of biofilm cells from becoming unsustainable (Nadell, Xavier, Levin, & Foster, 2008). Autoinducers that are signaling molecules are used in QS. Autoinducers are produced continuously by bacterial cells, and as cell density rises, so does the level of autoinducers. The QS takes place when a critical concentration of autoinducers is achieved at a specific cell density. Autoinducer receptor binding causes the repression or activation of several target genes. This QS process allows bacteria to respond in a coordinated manner that benefits the entire bacterial community by maintaining optimal biofilm size and coordinating virulence phenotypes (Verderosa et al., 2019). This collective response allows the biofilm to act more like a multicellular organism, allowing the bacterial community to adapt to changing conditions. Controlling population density is not the only advantage of QS. QS has been shown to help spread beneficial mutations throughout the biofilm colony, improve nutrient access, and contribute to antibiotic tolerance (Hannan et al., 2010).

Because disruption of the QS system can inhibit the growth of bacteria in the EPS, it has become a popular research topic (Puiu et al., 2017). It has been proposed that manipulating the underlying pathways of QS can cause pre-established biofilms to disassemble through a phenomenon known as quorum quenching (QQ), thus serving as a pathway for the development of potential treatments (Oppenheimer-Shaanan, Steinberg, & Kolodkin-Gal, 2013). Furthermore, QQ has been shown to increase biofilm susceptibility to antibiotics, as evidenced by the QS inhibitors cinnamaldehyde and baicalin hydrate, which reduced biofilm resistance to tobramycin in *P. aeruginosa* and *Burkholderia cepacia* (Brackman, Cos, Maes, Nelis, & Coenye, 2011).

The formation of biofilm depends upon chemical substances called autoinducers that are secreted by the microorganisms, such process is termed as QS. After reaching an optimum level, they either activates or repress the target genes (transcriptional regulators) giving out a coordinated response (Annous et al., 2009). It can activate either same species or interspecies of microorganisms (Annous et al., 2009). QS helps the bacteria to modulate its activities that cannot be attain as a single organism (Li & Tian, 2012). During the time of infection, QS process plays an important role by inducing phenotypic changes (González & Keshavan, 2006). The development of acute and chronic infection is dependent on the QS gene expression (10%) especially in *P. aeruginosa*. QS genes are mostly responsible for the expression of the virulence factors, biofilm formation, and antibiotic resistance and governs metabolic pathway (Moradali, Ghods, & Rehm, 2017).

Some of the microorganisms lose their ability to quorum sense during chronic infection. In *P. aeruginosa*, there are genes that help in the maturation and persistence of the biofilm. The mutant strains of *P. aeruginosa* encoded genes for various functions (*ΔlasI ΔrhlI* and *ΔlasR ΔrhlR*) shows under developed biofilm structure and very sensitive to antibiotic attacks (Nelson, D'Amours, Sproule-Willoughby, Morck, & Ceri, 2009). The rhamnolipid encoded pathways *rhl* and the formation of *N*-butanoyl-L-homoserine lactone (C4-HSL) unique to mucoid population at the chronic stage facilitate the release of biosurfactants and alginates (Bjarnsholt et al., 2010). The rhamnolipids supports the interior of the biofilm with nutrients and water (Chrzanowski, Ławniczak, & Czaczyk, 2012). The pigment pyocyanin combines with *eDNA* and increases the physiochemical nature of the biofilm which helps in the aggregation of bacterial cells (Das et al., 2015). *P. aeruginosa* QS system has interconnected pathways of *las*, *pqs*, *iqs*, and *rhl*, whereas *las* and *rhl* are the most important among them (Lee & Zhang, 2015).

The autoinducers *C4-HSL and 3-oxo-C12-HSL* were controlled by *RhlI* and *lasI* system. There are two more signaling molecule *Pseudomonas* quinolone signal (PQS) and 2-heptyl-3-hydroxy-4(1*H*)-quinolone also produced by *P. aeruginosa* (De Kievit, 2009).

6. Biofilm infections

More than 65% of the bacterial infections are caused by biofilm formations, which involve both device and nondevice-related infections, implants on breast, prosthetic joints 2%, heart valves, pacemakers 4%, also in 40% ventricular-assisted devices (Darouiche, 2004). Bacterium like *Streptococci*, *Staphylococci* notable gram-negative bacteria with fungi causes native valve endocarditis (Kokare et al., 2009). The common entry of the microbes is through gastrointestinal tract, urinary tract, and oropharynx, and they target the damaged part of the valve resulting in nonbacterial thrombotic endocarditis (NBTE) forming a thrombus (platelets, red blood cells, and fibrin). The preferred choice of treatment for biofilm infections is by high dosage of tobramycin and ciprofloxacin during the primary stage (Høiby, Frederiksen, & Pressler, 2005). The penetration power of these antibiotics is less and ineffective so drug carriers are used for slow release against biofilm development (Høiby et al., 2005). By combining shock wave and antibiotic treatment, the implant-associated biofilm infection can be overcome (Gnanadhas et al., 2015) (Fig. 4).

6.1 Implant related infections

The most commonly affected medical implants are contact lenses, catheters (dialysis and venous), mechanical heart valves, and prosthetics (joint and pacemakers). The biofilm composition depends on the microbial community (Donlan, 2002). The microorganisms that form biofilm on the contact lenses are *E. coli*, *P. aeruginosa*, *S. aureus*, *S. epidermidis*, *Candida* species, *Serratia*, and *Proteus*, etc. the stability of the biofilm depends on the substrate and electrolyte present in it (Donlan, 2002). In central venous catheter, the nature of the biofilm gets disturbed depending on the nature of the fluid administered, where *S. epidermidis* and *S. aureus* show lesser growth when compared to *P. aeruginosa*,

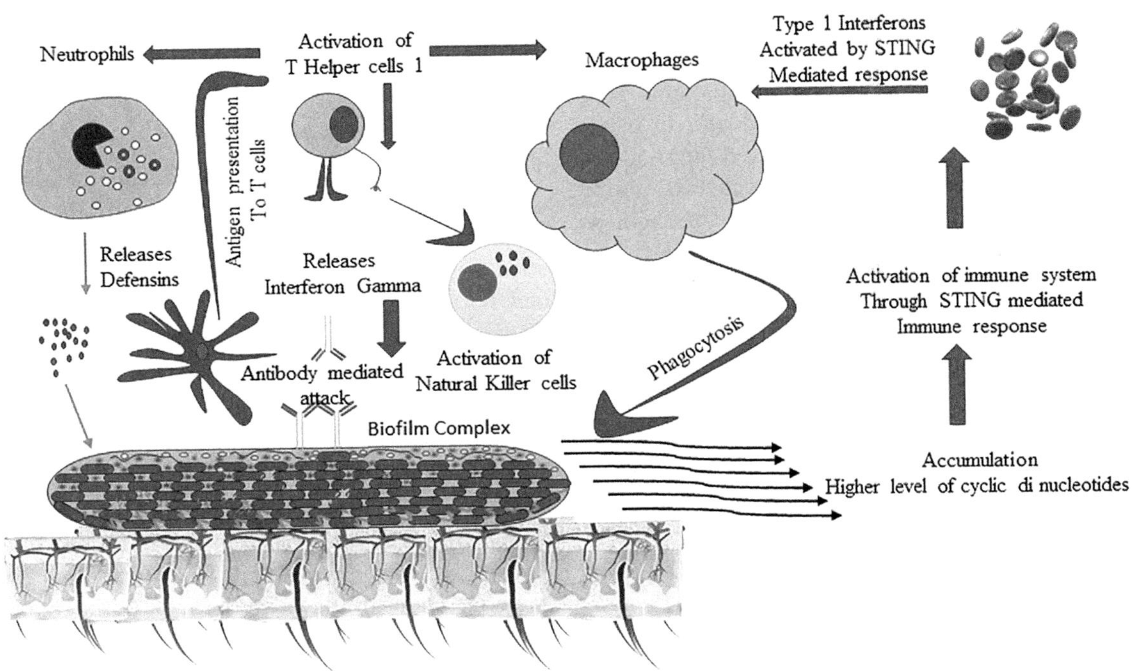

FIG. 4 The secretion and release of cyclic dinucleotides (C-di-NMPs) by biofilm-forming bacteria activates the STING-associated immune response. Dendritic cell presents the antigen to the T helper cells. Type 1 Interferons get activated and in turn activates T Helper cell 1, macrophages, and neutrophils. T helper cell 1 activates NK cells, which secretes interferon gamma. This activates the neutrophils to produce defensins, production if antibodies, facilitation of phagocytes.

Enterobacter species, and *Klebsiella* species (Maki & Mermel, 1998).

The microorganisms that cause prosthetic valve endocarditis (PVE) are *Streptococcus* species, *S. aureus*, *S. epidermidis*, *gram-negative Bacillus*, *Enterococcus*, *and Candida* spp. The common mode of entry of these microbes is by skin and medical implants (Donlan, 2002). Urinary catheters are produced using latex or silicone. The most common biofilm-forming organism are *E. coli*, *E. faecalis*, *S. epidermidis*, *P. aeruginosa*, *Proteus mirabilis*, *K. pneumoniae*, and other gram-negative bacteria (Donlan, 2002). The disease associated with the formation of biofilm is termed as chronic polymer-associated infection (CPAI) (Götz, 2002). The production and formation of the slime (EPS) by external polysaccharide is the characteristic of *Staphylococcal* colonization. In medical devices, the most common biofilm formers are *S. aureus* and *S. epidermidis* (Römling et al., 2014). The *S. aureus* infection is mostly accompanied gram-negative bacteria such as *Acinetobacter baumannii*, *E. coli*, *K. pneumoniae*, and *P. aeruginosa*, multidrug resistant commonly seen in CAUTI (Niveditha, Pramodhini, Umadevi, Kumar, & Stephen, 2012).

6.2 Nonimplant infections

6.2.1 Periodontitis

The oral cavity is home to a variety of biofilms. Dental cavities and periodontitis are caused by the multimicrobial biofilm that is present in the tooth plaque (Morse et al., 2018). The gingival crevice fluid (GCF) supports the growth of plaque in the subgingival crevice. Proteolytic enzymes are created after the oral biofilm forms, causing direct harm to soft and hard tissues while also interfering with the host's defense mechanisms (Bramhachari, Ahmed, Selvin, & Hassan, 2019; Donlan & Costerton, 2002). The disease that affects the gums and soft tissue of the mouth is termed as Periodontitis that results from poor oral hygiene (Kokare, Kadam, Mahadik, & Chopade, 2007). *P. aerobicus* and *Fusobacterium nucleatum* commonly cause periodontitis and also occupies the mucosal surface of the mouth (Lamont & Jenkinson, 1998). These microbes colonize on the surface of the teeth then enters the mucosal cells resulting in the plaque formation which then gets mineralized with calcium and phosphate ions developing calculus (Overman, 2000).

6.2.2 Cystic fibrosis

The disease that affects the lungs due to autosomal recessive genetic disorder. The protein cystic fibrosis transmembrane conductance regulator protein (CFTR) is affected and releases heavy mucous into the lungs. The biofilm formed by *P. aeruginosa* affects the lungs and secretes HSL, which can be seen in patients' sputum (Singh et al., 2000). The lower respiratory part in CF (cystic fibrosis) patients are found to be colonized by *S. aureus and Haemophilus influenzae*. By the time of adolescence, the CF patients have *P.*

aeruginosa infection (Koch & Hoiby, 1993). The growth and attachment of *P. aeruginosa* is supported by the acylated receptors that are present on the epithelial cells (Koch & Hoiby, 1993). First, *P. aeruginosa* thrives inside the biofilm within the cystic fibrosis-infected lungs (Potts, Roggli, & Spock, 1995). The QS compound HSL shows signals that are consistent when compared to *P. aeruginosa* that were grown planktonically (Singh et al., 2000). Secondly, the mucoid variant of the *P. aeruginosa* which secretes larger quantity of alginate, the presence of mucoid represents the chronic infection. Phenotypical changes of *P. aeruginosa* happens in the CF patients lungs (Mathee et al., 1999).

P. aeruginosa targets the undifferentiated epithelial cells and adhere between αvβ1 integrins and the fibronectin RGD (Arg-Gly-Asp) receptor also with an outer membrane protein. They possess various types of adhesins and get attached toward various receptors (Ofek, Hasty, Doyle, & Ofek, 2003). The QS molecule 3-oxo-C12-HSL in *P. aeruginosa* induces inflammatory cytokines like IL-1, IL-6, and interferon (INF)-γ and certain inflammatory proteins present in the macrophage (Smith, Harris, Phipps, & Iglewski, 2002). These compounds worsen the pathological conditions of the lungs. The anaerobic environmental condition increases toxic nitrogen metabolites along with 40-fold increase in the porin and OprF in the outer membrane (Yoon et al., 2002).

6.2.3 Osteomyelitis

The disease that affects the bones are termed as osteomyelitis, which causes disease in 2 out of 10,000 individuals. Trauma is the major way for the entry of the bacteria causing infection to the bones. The invading microbes are surrounded by the leukocytes present in the blood stream. They secrete an enzyme that digests/degrades the intact bone and infects the metaphysics (Kumar, Abbas Abul, Fausto, & Mitchell, 2007). The common bacterial isolated from the infectious part is *S. aureus*. The fibrin receptors get attached to the fibrinogen present in the bone and forms biofilm. Some of the other substances that *S. aureus* gets attached toward are collagen, laminin, and fibronectin (Kumar et al., 2007).

6.2.4 Wound infections

Wounds are the resultant from the loss of dead skin, which is moist and nutrient dense supports microbial colonization. The presence of certain bacteria like *P. aeruginosa* leads to ulcer and prolongs healing process (Gjødsbøl et al., 2006). The biofilm formed by *P. aeruginosa* at the wound site is more deeply located inside the wound than the biofilm of *S. aureus* thus arresting the healing progress (Fazli et al., 2009). This keeps the wound in constant chronic inflammatory stage with high level of metalloproteases and cytokines (Bjarnsholt et al., 2008).

6.2.5 Rhinosinusitis

The inflammation of the sinus part is called as rhinosinusitis with common symptoms like irritation in the nose, mucus discharge, and plugged nose with pain. The common causative organisms for inflammatory sinus are *S. pneumoniae* and *H. influenza* (Overman, 2000). In chronic infection, *S. pneumoniae*, *S. aureus* (50% on the nasal surface), *H. influenza*, and *Moraxella catarrhalis* (Foreman, Singhal, Psaltis, & Wormald, 2010).

6.2.6 Infective endocarditis

The major causative agent for the infective endocarditis is *Streptococcus* species (50% of the infection) followed by *Streptococcus* species (Donlan & Costerton, 2002). The bacterial endocarditis comprises of fibrin, platelets, and bacterial cells, which gets adhered toward the damaged epithelium of the heart valve. The platelet and fibrin forms clot that blocks the implanted valves and grafts leading to endocarditis. The bacteria present within the clot is less active. The *fap1* gene (peritrichous fimbriae) that is responsible for the formation of biofilm on plastic, which was produced by *Streptococcus parasanguis* (colonizer of the tooth) (Froeliger & Fives-Taylor, 2001). The mutant strain of *S. aureus* (defective fibronectin) has lower binding capacity toward damaged heart valves (Kuypers & Proctor, 1989). The fibrinogen-binding protein Clumping factor A (ClfA) produced by *S. aureus* assists in attaching to plastic surface and as a virulence factor (Moreillon et al., 1995). In vivo studies done on animal model (rabbit) with *E. coli* as causative agent required antibiotic concentration that is 200 times of serum bactericidal concentration (Joly et al., 1987).

6.3 Immunological response

Innate immunity and adaptive immune system are involved in defense mechanism against biofilm infection, but the mechanism of action is not well understood (Jesaitis et al., 2003). Compounds like human cationic host defense peptide LL-37 and lactoferrin are present on the surface of the polymorpho neutrophils highly inhibit the formation of biofilm by *P. aeruginosa* in vitro. The presence of lactoferrin induces binding to iron and helps in causing twitching motility (Singh, Parsek, Greenberg, & Welsh, 2002).

The human host defense peptide LL-37 (cationic) interferes with surface motility, QS system, and attachment phase of *P. aeruginosa* (Overhage et al., 2008). The biofilm-forming bacteria counter acts by secreting protease enzyme that destroys lactoferrin and LL-37 (Britigan, Hayek, Doebbeling, & Fick Jr, 1993). In cystic fibrosis lungs oxidative burst takes place due to increased glucose uptake by the polymorpho neutrophils. Releasing high amount of L-lactate in sputum of chronic lung infection sputum (Jensen, Givskov, Bjarnsholt, & Moser, 2010). The immature biofilms are weak and highly susceptible to immune attack. PMNs can clear biofilm formed by *S. aureus* by phagocytosis (Günther et al., 2009). Polymorpho neutrophils decrease the biofilm density, degranulate lactoferrin and elastase, finally with the release of DNA (Günther et al., 2009). Macrophages attack biofilms (*S. aureus*) that are lesser in density, size, and with less complexity. The host immune system secretes antibody (Immunoglobulin G) assisting in opsonization of *S. aureus* biofilm by PMNs by increasing oxygen radical production (Stroh et al., 2011). The major role of biofilm is to protect the bacteria from antibody-mediated phagocytosis (Stroh et al., 2011). The macrophages that are present near to the biofilm matrix are found dead when compared to the cells that are present on the top of the biofilm. This phenomenon is known as "layering" within biofilm (Hanke & Kielian, 2012).

6.3.1 Staphylococcus *spp.*

Staphylococcus spp. like *S. aureus* and *S. epidermidis* causes chronic infections like endocarditis, periodontitis, rhinosinusitis, and osteomyelitis (Archer et al., 2011). The mature biofilm of *S. aureus* contains nutrient channel through which human leukocytes penetrate but do not kill the bacteria (Leid, Shirtliff, Costerton, & Stoodley, 2002). The catheter-associated biofilm infection was studied using mouse model, where the immune cells like macrophages and neutrophils are attracted by chemokines CCL2 (C motif chemokine ligand 2).The tumor necrosis factor alpha, interleukin [IL]-1β, seen in sterile catheter sites (Thurlow et al., 2011). The development of the biofilm is reduced by IL-1β by D88-dependent pathway. The production of arginase-1 increases the collagen synthesis around biofilm helping the bacteria to escape immune cells attack including TLR2 (T cell Receptor 2) and TLR9 (T-cell Receptor 9) recognition (Thurlow et al., 2011). Innate immune system can recognize biofilm formed by *S. aureus*, but the activity depends on the strain, population of immune cells, opsonization by antibodies.

6.3.2 Candida *spp.*

The monocytes in the immune system influence the formation of biofilm by *Candida albicans* with increased release of proinflammatory cytokines like IL-1β and reduced IL-6, IL-10, MCP-1, I-309, and tumor necrosis factor alpha (Chandra, McCormick, Imamura, Mukherjee, & Ghannoum, 2007). In the mature biofilm, the fungal cells are present within the biofilm that makes it difficult for the monocytes to phagocyte them (Chandra et al., 2007). This is evident from the lower production of cytokine in biofilm-phagocyte culture in comparison with phagocyte cell from planktonic mix (Chandra et al., 2007). The more mature the biofilm, the more difficult to break it also it is not easy to activate the reactive oxygen species response (Xie et al., 2012). *C. albicans* mature biofilm consists of β-glucans in the extracellular matrix which helps in escaping from the innate immune attack (Chandra et al., 2007). Immune system factors like interferon-γ and granulocyte colony-stimulating factor enhance the activity of PMNs against *Candida* spp. (planktonic). When biofilm was exposed to PMNs that are cytokine activated did not enhance its activity against biofilm (Katragkou et al., 2011). Elimination of *C. albicans* from the host requires ligation of many receptors (mannoproteins, mannans, or β-glucans).The in vitro study shows the presence of (1→3)-β-D-glucans and *Hyr1* encoding cell surface glycosyl phosphatidylinositol anchored proteins on the surface showing resistance to neutrophil killing also in a biofilm model (oral mucosal tissue) (Dongari-Bagtzoglou, Kashleva, Dwivedi, Diaz, & Vasilakos, 2009).

7. Therapies against biofilms

The most common preventive measures to be taken to avoid the spread of biofilm is proper hygiene and antibiotic therapy. The implementation of antibiotic course must be done under strict considerations. New technologies are being used to prevent the bacterium getting attached to any surfaces (Camargo et al., 2009). In medical implants and catheters antimicrobial lock technique is used to resist biofilm development (Bordi & de Bentzmann, 2011). The biofilm-forming bacteria forms a protein layer before getting attached to any surfaces by applying antifouling compounds like polyethylene glycol or a zwitter ion compounds (Bordi & de Bentzmann, 2011). The compounds (polyethylene) and sulfonate exhibits resistance against *E. coli*, *S. epidermis*, *P. aeruginosa*, *C. tropicalis*, and *C. albicans* (Roosjen, van der Mei, Busscher, & Norde, 2004).

Recent advancements in Nanotechnology provides slow and prolonged release of drugs with less toxicity and improved stability. The penetration power is higher due to their smaller size. Nanoparticles has prolonged half-life and higher surface ratio making them better for drug encapsulation (Suci et al., 2007).

7.1 Nanoparticles in biofilm treatment

There are many nanoparticles that have antibiofilm, antibacterial properties that can be used as alternative for antibiotics. Some of the commonly known nanoparticles are silver, gold, titanium, and zinc. The metal composition in the nanoparticles has higher degree of antimicrobial activity when compared to its counter parts (Jones, Ray, Ranjit, & Manna, 2008). The nanomaterials are made of metal and metal oxides carbon-based nanomaterials, polymeric nanoparticles, nanocomposites, nano emulsions, lipid nanoparticles, etc. (Qayyum & Khan, 2016). The application of smart nano-materials shows some extent abilities like activation on external stimuli (pH, Temp), enzyme activation, and redox potential (Karimi et al., 2016).

7.1.1 Nanoparticle based on magnetic field

There are two common methods that are used in treating biofilm infection using magnetic fields are hyperthermia-induced drug release by magnetic field, drug targeting based on magnetic field (Schleich, Danhier, & Préat, 2015). The biofilm is disrupted by using friction (static) on magnetic iron oxide nanoparticle with the help of magnetic field (Li et al., 2013). The liquid metal compound-based droplets (Galinstan) form spheres under low magnetic field, but transform into rod or star-shaped particle under high magnetic field. They can penetrate cell wall also removes biofilm formed by the *S. aureus* and *P. aeruginosa* (Elbourne et al., 2020). In the presence of external magnetic field nanocomposites expressed higher antibiofilm property (*S. aureus*) and low cellular toxicity. The nanocomposite (400%) showed higher degree of bacterial death than selenium nanoparticles (51.6%). The silver ring-coated superparamagnetic iron oxide NPs (SPIONs) showed high activity against biofilm without harming the healthy cells (Mahmoudi & Serpooshan, 2012). The chitosan-coated iron oxide Fe_3O_4 when conjugated with PEL1 (polyvalent bacteriophage) helps in the penetration of biofilms formed by *P. aeruginosa* and *E. coli*. The PEL1 combines with CS-Fe_3O_4 (Colloidal nanoparticle cluster) kills the bacterial cells present inside the biofilm (40%) with efficacy of 88% (Kim et al., 2013).

7.1.2 Inorganic nanoparticles

Combinations of antibiotics and nanoparticles have a variety of antibacterial mechanisms for overcoming antibiotic resistance (Huh & Kwon, 2011). Nanoparticles are thus considered next-generation antibiotics. *S. aureus* MTCC 9760, *P. aeruginosa* MTCC 424, *Streptococcus pyogenes* MTCC 1926, *E. coli* MTCC 40, *Bacillus cereus* MTCC 430, and *Proteus mirabilis* MTCC 3310 were recently tested for antibacterial activity of Zn NPs. The antibacterial efficacy of the newly synthesized green Zn NPs against Gram-positive and Gram-negative bacteria was encouraging. These findings revealed a fast, low-cost, environmentally friendly, and convenient approach for making ZnO NPs, which could be employed as a potential antibacterial agent against drug-resistant bacteria (Gupta, Tomar, Kaushik, Mishra, & Sharma, 2018).

7.1.3 Silver nanoparticles

Silver is known for its antibacterial activity for long time. Silver is thought to be a powerful antibacterial agent. The most promising inorganic NPs used to treat a range of bacterial illnesses are silver nanoparticles (AgNPs) (Rather et al., 2021). By reducing its size to nanometers increases its surface area along with its antibacterial property (Baker, Pradhan, Paktis, Pochan, & Shah, 2005). Its activity is proven against super bugs like MRSA (methylene resistance *S. aureus*), by acting on the intermolecular forces in the biofilm (Ellis, 2007). Evidence shows that silver nanoparticle inhibits 95% of the biofilm formed by *P. aeruginosa*, *S. epidermidis*, *S. aureus* (MRSA), and methicillin-resistant *S. epidermidis* (MRSE) after 24h of treatment (Ansari, Khan, Khan, Cameotra, & Alzohairy, 2015). Silver ions interact with sulfhydryl groups that are present in the bacterial walls and disrupts the cell membrane, activities of the enzyme, and cellular proliferation (Loh, Percival, Woods, Williams, & Cochrane, 2009). Prolonged use of silver nanoparticle leads to resistance in MRSA and delayed wound recovery (Kostenko, Lyczak, Turner, & Martinuzzi, 2010).

7.1.4 Zinc oxide nanoparticle

Zinc oxide nanoparticles shows higher bactericidal activity and low toxicity on mammalian cells. They inhibit biofilm formation of *E. faecalis*, *S. aureus*, *S. epidermidis*, *B. subtilis*, and *E. coli* when compared to other metal nanoparticles (Lee, Kim, Cho, & Lee, 2014). When compared to titania nanophase, ZnO (nanotextured) has greater degree of bacterial resistance and bacteriostatic properties rather than *P. aeruginosa and Proteus*, which exhibits resistance to zinc nanoparticles. The chitosan-coated zinc nanoparticles have high antibiofilm activity against *M. luteus* and *S. aureus*, which was achieved by very low concentration at 0.625 to 0.156mg/mL (Dhillon, Kaur, & Brar, 2014). In case of *Streptococcus* spp., ZnO NPs (Zinc Oxide Nanoparticles) have shown activity after 96h of incubation confirmed by CLSM (confocal laser scanning microscopy) (Abdulkareem et al., 2015). To reduce the formation of biofilm in the wounds, ZnO NPs were combined with β-chitin in dressings resulted in reduced biofilm formation.

7.1.5 Titanium oxide nanoparticles

Titanium oxide nanoparticles are inert in nature, they inhibit biofilm by photocatalytic mode. They have antifungal biofilm activity against *C. albicans* (Dizaj, Lotfipour, Barzegar-Jalali, Zarrintan, & Adibkia, 2014). They form reactive oxygen species (ROS) in microbial cells, peroxidation of lipids, and finally oxidation of internal enzymes leading to cell death (Oku, Komatsu, Tajima,

Nakashimada, & Kato, 2012). When titanium oxide is combined with silicone (metal polymer composite) resulted in 90% reduction in attachment of *S. aureus* compared to normal silicone. When gallium and zinc nanoparticle are incorporated with polyether urethane showed reduction in *P. aeruginosa* infection with slow release of gallium nanoparticles (Ma, Darmawan, Zhang, Zhang, & Bryers, 2013).

8. Conclusion

Cell-signaling molecules, which are involved in QS and QQ, need to be investigated to see how and when they regulate biofilm formation. To create effective and broad-spectrum antibiofilm therapies, a better understanding of microbial interaction with biotic and abiotic surfaces, as well as how the host's unfavorable environment influences microbial survival, growth, and recalcitrance, is necessary. Understanding and developing antibiofilm agents requires the application of specific genetic factors essential for biofilm development in vitro and these findings in vivo, utilizing appropriate animal models that reflect the intricate interplay between biofilm and host. Intercellular interactions and communication are extremely important for biofilm growth and persistence, whether single-species or mixed-species biofilms. One of the future challenges in biofilm research is to dissect these relationships. Understanding the complexities of relationships within the biofilm community is particularly difficult.

Developing precise and realistic models of natural communities in the laboratory is one of the cornerstones to understanding complicated biological systems. Scientists have shown tremendous innovation in developing techniques for analyzing biofilm systems, and further work will undoubtedly be necessary in this field.

References

Abdulkareem, E. H., Memarzadeh, K., Allaker, R. P., Huang, J., Pratten, J., & Spratt, D. (2015). Anti-biofilm activity of zinc oxide and hydroxyapatite nanoparticles as dental implant coating materials. *Journal of Dentistry*, *43*(12), 1462–1469.

Absalon, C., Van Dellen, K., & Watnick, P. I. (2011). A communal bacterial adhesin anchors biofilm and bystander cells to surfaces. *PLoS Pathogens*, *7*(8), e1002210.

Aguilar, C., Vlamakis, H., Losick, R., & Kolter, R. (2007). Thinking about *Bacillus subtilis* as a multicellular organism. *Current Opinion in Microbiology*, *10*(6), 638–643.

Al-Ahmad, A., Wiedmann-Al-Ahmad, M., Faust, J., Bächle, M., Follo, M., Wolkewitz, M., et al. (2010). Biofilm formation and composition on different implant materials in vivo. *Journal of Biomedical Materials Research Part B: Applied Biomaterials*, *95*(1), 101–109.

Annous, B. A., Fratamico, P. M., & Smith, J. L. (2009). Scientific status summary. *Journal of Food Science*, *74*(1), R24–R37.

Ansari, M. A., Khan, H. M., Khan, A. A., Cameotra, S. S., & Alzohairy, M. A. (2015). Anti-biofilm efficacy of silver nanoparticles against MRSA and MRSE isolated from wounds in a tertiary care hospital. *Indian Journal of Medical Microbiology*, *33*(1), 101–109.

Archer, N. K., Mazaitis, M. J., Costerton, J. W., Leid, J. G., Powers, M. E., & Shirtliff, M. E. (2011). *Staphylococcus aureus* biofilms: Properties, regulation, and roles in human disease. *Virulence*, *2*(5), 445–459.

Baker, C., Pradhan, A., Paktis, L., Pochan, D. J., & Shah, S. I. (2005). Antimicrobial activity of silver nanoparticles. *Journal of Nanoscience and Nanotechnology*, *5*, 244–249.

Balestrino, D., Ghigo, J. M., Charbonnel, N., Haagensen, J. A., & Forestier, C. (2008). The characterization of functions involved in the establishment and maturation of *Klebsiella pneumoniae* in vitro biofilm reveals dual roles for surface exopolysaccharides. *Environmental Microbiology*, *10*(3), 685–701.

Beloin, C., Roux, A., & Ghigo, J. M. (2008). *Escherichia coli* biofilms. In T. Romeo (Ed.), *Bacterial biofilms. Current topics in microbiology and immunology (Vol. 322)*. Berlin, Heidelberg: Springer. https://doi.org/10.1007/978-3-540-75418-3_12.

Berne, C., Ducret, A., Hardy, G. G., & Brun, Y. V. (2015). Adhesins involved in attachment to abiotic surfaces by Gram-negative bacteria. *Microbiology Spectrum*, *3*(4). https://doi.org/10.1128/microbiolspec.MB-0018-2015. PMID: 26350310. PMCID: PMC4566860.

Bjarnsholt, T., Jensen, P.Ø., Jakobsen, T. H., Phipps, R., Nielsen, A. K., Rybtke, M. T., et al. (2010). Quorum sensing and virulence of *Pseudomonas aeruginosa* during lung infection of cystic fibrosis patients. *PLoS One*, *5*(4), e10115.

Bjarnsholt, T., Kirketerp-Møller, K., Jensen, P.Ø., Madsen, K. G., Phipps, R., Krogfelt, K., et al. (2008). Why chronic wounds will not heal: A novel hypothesis. *Wound Repair and Regeneration*, *16*(1), 2–10.

Bordi, C., & de Bentzmann, S. (2011). Hacking into bacterial biofilms: A new therapeutic challenge. *Annals of Intensive Care*, *1*(1), 1–8.

Bos, R., Van der Mei, H. C., & Busscher, H. J. (1999). Physico-chemistry of initial microbial adhesive interactions—Its mechanisms and methods for study. *FEMS Microbiology Reviews*, *23*(2), 179–230.

Bose, J. L., Lehman, M. K., Fey, P. D., & Bayles, K. W. (2012). Contribution of the *Staphylococcus aureus* Atl AM and GL murein hydrolase activities in cell division, autolysis, and biofilm formation. *PLoS ONE*, *7*(7), e42244. https://doi.org/10.1371/journal.pone.0042244.

Bouvet, M., Debarnot, C., Imbert, I., Selisko, B., Snijder, E. J., Canard, B., et al. (2010). In vitro reconstitution of SARS-coronavirus mRNA cap methylation. *PLoS Pathogens*, *6*(4), e1000863.

Boyd, A., & Chakrabarty, A. M. (1995). *Pseudomonas aeruginosa* biofilms: Role of the alginate exopolysaccharide. *Journal of Industrial Microbiology and Biotechnology*, *15*(3), 162–168.

Brackman, G., Cos, P., Maes, L., Nelis, H. J., & Coenye, T. (2011). Quorum sensing inhibitors increase the susceptibility of bacterial biofilms to antibiotics *in vitro* and *in vivo*. *Antimicrobial Agents and Chemotherapy*, *55*(6), 2655–2661.

Bramhachari, P. V., Ahmed, V. K., Selvin, J., & Hassan, S. (2019). Quorum sensing and biofilm formation by oral pathogenic microbes in the dental plaques: Implication for health and disease. In *Implication of quorum sensing and biofilm formation in medicine, agriculture and food industry* (pp. 129–140). Singapore: Springer.

Britigan, B. E., Hayek, M. B., Doebbeling, B. N., & Fick, R. B., Jr. (1993). Transferrin and lactoferrin undergo proteolytic cleavage in the *Pseudomonas aeruginosa*-infected lungs of patients with cystic fibrosis. *Infection and Immunity*, *61*(12), 5049–5055.

Bucior, I., Pielage, J. F., & Engel, J. N. (2012). *Pseudomonas aeruginosa* pili and flagella mediate distinct binding and signaling events at the apical and basolateral surface of airway epithelium. *PLoS Pathogens*, *8*(4), e1002616.

Cai, Q., Li, Z., Ouyang, Q., Luo, C., & Gordon, V. D. (2016). Singly flagellated *Pseudomonas aeruginosa* chemotaxes efficiently by unbiased motor regulation. *MBio*, *7*(2), e00013–e00016.

Cai, Z., Liu, Y., Chen, Y., Yam, J. K. H., Chew, S. C., Chua, S. L., et al. (2015). RpoN regulates virulence factors of *Pseudomonas aeruginosa* via modulating the PqsR quorum sensing regulator. *International Journal of Molecular Sciences*, *16*(12), 28311–28319.

Camargo, L. F. A., Marra, A. R., Büchele, G. L., Sogayar, A. M. C., Cal, R. G. R., de Sousa, J. M. A., et al. (2009). Double-lumen central venous catheters impregnated with chlorhexidine and silver sulfadiazine to prevent catheter colonisation in the intensive care unit setting: A prospective randomised study. *Journal of Hospital Infection*, *72*(3), 227–233.

Campisano, A., Schroeder, C., Schemionek, M., Overhage, J., & Rehm, B. H. (2006). PslD is a secreted protein required for biofilm formation by *Pseudomonas aeruginosa*. *Applied and Environmental Microbiology*, *72*(4), 3066–3068.

Campodónico, V. L., Llosa, N. J., Grout, M., Döring, G., Maira-Litrán, T., & Pier, G. B. (2010). Evaluation of flagella and flagellin of *Pseudomonas aeruginosa* as vaccines. *Infection and Immunity*, *78*(2), 746–755.

Carniello, V., Peterson, B. W., van der Mei, H. C., & Busscher, H. J. (2018). Physico-chemistry from initial bacterial adhesion to surface-programmed biofilm growth. *Advances in Colloid and Interface Science*, *261*, 1–14.

Chandra, J., McCormick, T. S., Imamura, Y., Mukherjee, P. K., & Ghannoum, M. A. (2007). Interaction of *Candida albicans* with adherent human peripheral blood mononuclear cells increases *C. albicans* biofilm formation and results in differential expression of pro-and anti-inflammatory cytokines. *Infection and Immunity*, *75*(5), 2612–2620.

Chapman, M. R., Robinson, L. S., Pinkner, J. S., Roth, R., Heuser, J., Hammar, M., et al. (2002). Role of *Escherichia coli* curli operons in directing amyloid fiber formation. *Science*, *295*(5556), 851–855.

Chen, K. M., Chiang, M. K., Wang, M., Ho, H. C., Lu, M. C., & Lai, Y. C. (2014). The role of pgaC in *Klebsiella pneumoniae* virulence and biofilm formation. *Microbial Pathogenesis*, *77*, 89–99.

Cheung, G. Y., Joo, H. S., Chatterjee, S. S., & Otto, M. (2014). Phenol-soluble modulins—Critical determinants of staphylococcal virulence. *FEMS Microbiology Reviews*, *38*(4), 698–719.

Chiang, W. C., Nilsson, M., Jensen, P.Ø., Høiby, N., Nielsen, T. E., Givskov, M., et al. (2013). Extracellular DNA shields against aminoglycosides in *Pseudomonas aeruginosa* biofilms. *Antimicrobial Agents and Chemotherapy*, *57*(5), 2352–2361.

Chrzanowski, Ł., Ławniczak, Ł., & Czaczyk, K. (2012). Why do microorganisms produce rhamnolipids? *World Journal of Microbiology and Biotechnology*, *28*(2), 401–419.

Chung, P. Y. (2016). The emerging problems of *Klebsiella pneumoniae* infections: Carbapenem resistance and biofilm formation. *FEMS Microbiology Letters*, *363*(20), fnw219.

Cohn, F. (1875). *Untersuchungen über Bacterien: I*. JU Kern.

Collinson, S. K., Emödy, L., Müller, K. H., Trust, T. J., & Kay, W. W. (1991). Purification and characterization of thin, aggregative fimbriae from *Salmonella enteritidis*. *Journal of Bacteriology*, *173*(15), 4773–4781.

Collinson, S. K., Parker, J. M. R., Hodges, R. S., & Kay, W. W. (1999). Structural predictions of AgfA, the insoluble fimbrial subunit of *Salmonella* thin aggregative fimbriae. *Journal of Molecular Biology*, *290*(3), 741–756.

Costa, O. Y. A., Raaijmakers, J. M., & Kuramae, E. E. (2018). Microbial extracellular polymeric substances: Ecological function and impact on soil aggregation. *Frontiers in Microbiology*, *9*, 1636. https://doi.org/10.3389/fmicb.2018.01636.

Costerton, J. W., Stewart, P. S., & Greenberg, E. P. (1999). Bacterial biofilms: A common cause of persistent infections. *Science*, *284*(5418), 1318–1322.

Dang, H., & Lovell, C. R. (2016). Microbial surface colonization and biofilm development in marine environments. *Microbiology and Molecular Biology Reviews*, *80*(1), 91–138.

Daniels, R., Vanderleyden, J., & Michiels, J. (2004). Quorum sensing and swarming migration in bacteria. *FEMS Microbiology Reviews*, *28*(3), 261–289.

Darouiche, R. O. (2004). Treatment of infections associated with surgical implants. *New England Journal of Medicine*, *350*(14), 1422–1429.

Das, T., Ibugo, A. I., Klare, W., & Manefield, M. (2016). Role of pyocyanin and extracellular DNA in facilitating *Pseudomonas aeruginosa* biofilm formation. *Microbial Biofilms-Importance and Applications*, *13*, 23–42.

Das, T., Kutty, S. K., Tavallaie, R., Ibugo, A. I., Panchompoo, J., Sehar, S., et al. (2015). Phenazine virulence factor binding to extracellular DNA is important for *Pseudomonas aeruginosa* biofilm formation. *Scientific Reports*, *5*(1), 1–9.

Das, T., & Manefield, M. (2012). Pyocyanin promotes extracellular DNA release in *Pseudomonas aeruginosa*. *PLoS ONE*, *7*(10), e46718. https://doi.org/10.1371/journal.pone.0046718.

Davey, M. E., & O'Toole, G. A. (2000). Microbial biofilms: From ecology to molecular genetics. *Microbiology and Molecular Biology Reviews*, *64*(4), 847–867.

Davies, D. (2003). Understanding biofilm resistance to antibacterial agents. *Nature Reviews Drug Discovery*, *2*(2), 114–122.

De Kievit, T. R. (2009). Quorum sensing in *Pseudomonas aeruginosa* biofilms. *Environmental Microbiology*, *11*(2), 279–288.

DePas, W. H., Syed, A. K., Sifuentes, M., Lee, J. S., Warshaw, D., Saggar, V., et al. (2014). Biofilm formation protects *Escherichia coli* against killing by *Caenorhabditis elegans* and *Myxococcus xanthus*. *Applied and Environmental Microbiology*, *80*(22), 7079–7087.

Dewanti, R., & Wong, A. C. (1995). Influence of culture conditions on biofilm formation by *Escherichia coli* O157: H7. *International Journal of Food Microbiology*, *26*(2), 147–164.

Dhillon, G. S., Kaur, S., & Brar, S. K. (2014). Facile fabrication and characterization of chitosan-based zinc oxide nanoparticles and evaluation of their antimicrobial and antibiofilm activity. *International Nano Letters*, *4*(2), 107.

Diago-Navarro, E., Chen, L., Passet, V., Burack, S., Ulacia-Hernando, A., Kodiyanplakkal, R. P., et al. (2014). Carbapenem-resistant *Klebsiella pneumoniae* exhibit variability in capsular polysaccharide and capsule associated virulence traits. *The Journal of Infectious Diseases*, *210*(5), 803–813.

Divya, M., Kiran, G. S., Hassan, S., & Selvin, J. (2019). Biogenic synthesis and effect of silver nanoparticles (AgNPs) to combat catheter-related urinary tract infections. *Biocatalysis and Agricultural Biotechnology*, *18*, 101037.

Dizaj, S. M., Lotfipour, F., Barzegar-Jalali, M., Zarrintan, M. H., & Adibkia, K. (2014). Antimicrobial activity of the metals and metal oxide nanoparticles. *Materials Science and Engineering: C*, *44*, 278–284.

Doggett, R. G. (1969). Incidence of mucoid *Pseudomonas aeruginosa* from clinical sources. *Applied Microbiology*, *18*(5), 936–937.

Dongari-Bagtzoglou, A., Kashleva, H., Dwivedi, P., Diaz, P., & Vasilakos, J. (2009). Characterization of mucosal *Candida albicans* biofilms. *PLoS One*, *4*(11), e7967.

Donlan, R. M. (2002). Biofilms: Microbial life on surfaces. *Emerging Infectious Diseases*, *8*(9), 881.

Donlan, R. M., & Costerton, J. W. (2002). Biofilms: Survival mechanisms of clinically relevant microorganisms. *Clinical Microbiology Reviews*, *15*, 167–193.

Dueholm, M. S., Albertsen, M., Otzen, D., & Nielsen, P. H. (2012). Curli functional amyloid systems are phylogenetically widespread and display large diversity in operon and protein structure. *PLoS One*, *7* (12), e51274.

Dufour, D., Leung, V., & Lévesque, C. M. (2010). Bacterial biofilm: Structure, function, and antimicrobial resistance. *Endodontic Topics*, *22*(1), 2–16.

Duperthuy, M., Sjöström, A. E., Sabharwal, D., Damghani, F., Uhlin, B. E., & Wai, S. N. (2013). Role of the *Vibrio cholerae* matrix protein Bap1 in cross-resistance to antimicrobial peptides. *PLoS Pathogens*, *9*(10), e1003620.

Ebrey, R., Hamilton, M. S., Cairns, G., & Lappin-Scott, H. M. (2004). Biofilms and hospital-acquired infections. In *Microbial biofilms* (pp. 294–313). American Society of Microbiology.

Elbourne, A., Cheeseman, S., Atkin, P., Truong, N. P., Syed, N., Zavabeti, A., et al. (2020). Antibacterial liquid metals: Biofilm treatment via magnetic activation. *ACS Nanotechnology*, *14*(1), 802–817.

Ellis, J. R. (2007). The many roles of silver in infection prevention. *American Journal of Infection Control*, *35*(5), E26.

Fahs, A., Quilès, F., Jamal, D., Humbert, F., & Francius, G. (2014). In situ analysis of bacterial extracellular polymeric substances from a *Pseudomonas fluorescens* biofilm by combined vibrational and single molecule force spectroscopies. *The Journal of Physical Chemistry B*, *118*(24), 6702–6713.

Fazeli, H., Akbari, R., Moghim, S., Narimani, T., Arabestani, M. R., & Ghoddousi, A. R. (2012). *Pseudomonas aeruginosa* infections in patients, hospital means, and personnel's specimens. *Journal of Research in Medical Sciences: The Official journal of Isfahan University of Medical Sciences*, *17*(4), 332.

Fazli, M., Bjarnsholt, T., Kirketerp-Møller, K., Jørgensen, B., Andersen, A. S., Krogfelt, K. A., et al. (2009). Nonrandom distribution of *Pseudomonas aeruginosa* and *Staphylococcus aureus* in chronic wounds. *Journal of Clinical Microbiology*, *47*(12), 4084–4089.

Flemming, H. C., & Wuertz, S. (2019). Bacteria and archaea on Earth and their abundance in biofilms. *Nature Reviews Microbiology*, *17*(4), 247–260.

Foreman, A., Singhal, D., Psaltis, A. J., & Wormald, P. J. (2010). Targeted imaging modality selection for bacterial biofilms in chronic rhinosinusitis. *The Laryngoscope*, *120*(2), 427–431.

Foster, T. J., Geoghegan, J. A., Ganesh, V. K., & Höök, M. (2014). Adhesion, invasion and evasion: The many functions of the surface proteins of *Staphylococcus aureus*. *Nature Reviews Microbiology*, *12*(1), 49–62.

Foulston, L., Elsholz, A. K., DeFrancesco, A. S., & Losick, R. (2014). The extracellular matrix of *Staphylococcus aureus* biofilms comprises cytoplasmic proteins that associate with the cell surface in response to decreasing pH. *MBio*, *5*(5), e01667-14.

Francolini, I., & Donelli, G. (2010). Prevention and control of biofilm-based medical-device-related infections. *FEMS Immunology & Medical Microbiology*, *59*(3), 227–238.

Friedman, L., & Kolter, R. (2004). Genes involved in matrix formation in *Pseudomonas aeruginosa* PA14 biofilms. *Molecular Microbiology*, *51* (3), 675–690.

Froeliger, E. H., & Fives-Taylor, P. (2001). *Streptococcus parasanguis* fimbria-associated adhesin Fap1 is required for biofilm formation. *Infection and Immunity*, *69*(4), 2512–2519.

Geoghegan, J. A., Corrigan, R. M., Gruszka, D. T., Speziale, P., O'Gara, J. P., Potts, J. R., et al. (2010). Role of surface protein SasG in biofilm formation by *Staphylococcus aureus*. *Journal of Bacteriology*, *192* (21), 5663–5673.

Ghafoor, A., Jordens, Z., & Rehm, B. H. (2013). Role of PelF in Pel polysaccharide biosynthesis in *Pseudomonas aeruginosa*. *Applied and Environmental Microbiology*, *79*(9), 2968–2978.

Gillet, Y., Issartel, B., Vanhems, P., Fournet, J. C., Lina, G., Bes, M., et al. (2002). Association between *Staphylococcus aureus* strains carrying gene for Panton-Valentine leukocidin and highly lethal necrotising pneumonia in young immunocompetent patients. *The Lancet*, *359* (9308), 753–759.

Giltner, C. L., Nguyen, Y., & Burrows, L. L. (2012). Type IV pilin proteins: Versatile molecular modules. *Microbiology and Molecular Biology Reviews*, *76*(4), 740–772.

Gjødsbøl, K., Christensen, J. J., Karlsmark, T., Jørgensen, B., Klein, B. M., & Krogfelt, K. A. (2006). Multiple bacterial species reside in chronic wounds: A longitudinal study. *International Wound Journal*, *3*(3), 225–231.

Gnanadhas, D. P., Elango, M., Janardhanraj, S., Srinandan, C. S., Datey, A., Strugnell, R. A., et al. (2015). Successful treatment of biofilm infections using shock waves combined with antibiotic therapy. *Scientific Reports*, *5*(1), 1–12.

González, J. E., & Keshavan, N. D. (2006). Messing with bacterial quorum sensing. *Microbiology and Molecular Biology Reviews*, *70*(4), 859–875.

Götz, F. (2002). *Staphylococcus* and biofilms. *Molecular Microbiology*, *43* (6), 1367–1378.

Gross, M., Cramton, S. E., Götz, F., & Peschel, A. (2001). Key role of teichoic acid net charge in *Staphylococcus aureus* colonization of artificial surfaces. *Infection and Immunity*, *69*(5), 3423–3426.

Gu, H., Hou, S., Yongyat, C., De Tore, S., & Ren, D. (2013). Patterned biofilm formation reveals a mechanism for structural heterogeneity in bacterial biofilms. *Langmuir*, *29*(35), 11145–11153.

Gualdi, L., Tagliabue, L., Bertagnoli, S., Ierano, T., De Castro, C., & Landini, P. (2008). Cellulose modulates biofilm formation by counteracting curli-mediated colonization of solid surfaces in *Escherichia coli*. *Microbiology*, *154*(7), 2017–2024.

Günther, F., Wabnitz, G. H., Stroh, P., Prior, B., Obst, U., Samstag, Y., et al. (2009). Host defence against *Staphylococcus aureus* biofilms infection: Phagocytosis of biofilms by polymorphonuclear neutrophils (PMN). *Molecular Immunology*, *46*(8–9), 1805–1813.

Gupta, P., Sarkar, S., Das, B., Bhattacharjee, S., & Tribedi, P. (2016). Biofilm, pathogenesis and prevention—A journey to break the wall: A review. *Archives of Microbiology*, *198*(1), 1–15.

Gupta, M., Tomar, R. S., Kaushik, S., Mishra, R. K., & Sharma, D. (2018). Effective antimicrobial activity of green ZnO nano particles of *Catharanthus roseus*. *Frontiers in Microbiology*, *9*, 2030.

Habash, M., & Reid, G. (1999). Ofloxacin for the treatment of urinary tract infections and biofilms in spinal cord injury. *Journal of Clinical Pharmacology*, *39*(9), 887.

Hammar, M. A. R. T. E. N., Bian, Z., & Normark, S. (1996). Nucleator-dependent intercellular assembly of adhesive curli organelles in *Escherichia coli*. *Proceedings of the National Academy of Sciences, 93*(13), 6562–6566.

Hammer, N. D., Schmidt, J. C., & Chapman, M. R. (2007). The curli nucleator protein, CsgB, contains an amyloidogenic domain that directs CsgA polymerization. *Proceedings of the National Academy of Sciences, 104*(30), 12494–12499.

Hanke, M. L., & Kielian, T. (2012). Deciphering mechanisms of staphylococcal biofilm evasion of host immunity. *Frontiers in Cellular and Infection Microbiology, 2*, 62.

Hannan, S., Ready, D., Jasni, A. S., Rogers, M., Pratten, J., & Roberts, A. P. (2010). Transfer of antibiotic resistance by transformation with eDNA within oral biofilms. *FEMS Immunology & Medical Microbiology, 59* (3), 345–349.

Hasman, H., Chakraborty, T., & Klemm, P. (1999). Antigen-43-mediated autoaggregation of *Escherichia coli* is blocked by fimbriation. *Journal of Bacteriology, 181*(16), 4834–4841.

Hassan, S., Ahmad, T., Bashir, M., Kiran, G. S., & Selvin, J. (2019). Novel perspectives on the quorum sensing inhibitors (QSIs)/quorum quenchers (QQs) in food preservation and spoilage. In *Implication of quorum sensing and biofilm formation in medicine, agriculture and food industry* (pp. 269–298). Singapore: Springer.

Hausner, M., & Wuertz, S. (1999). High rates of conjugation in bacterial biofilms as determined by quantitative in situ analysis. *Applied and Environmental Microbiology, 65*(8), 3710–3713.

Hay, I. D., Remminghorst, U., & Rehm, B. H. (2009). MucR, a novel membrane-associated regulator of alginate biosynthesis in *Pseudomonas aeruginosa*. *Applied and Environmental Microbiology, 75*(4), 1110–1120.

Hazes, B., Sastry, P. A., Hayakawa, K., Read, R. J., & Irvin, R. T. (2000). Crystal structure of *Pseudomonas aeruginosa* PAK pilin suggests a main-chain-dominated mode of receptor binding. *Journal of Molecular Biology, 299*(4), 1005–1017.

Henderson, I. R., Meehan, M., & Owen, P. (1997). Antigen 43, a phase-variable bipartite outer membrane protein, determines colony morphology and autoaggregation in *Escherichia coli* K-12. *FEMS Microbiology Letters, 149*(1), 115–120.

Hobley, L., Harkins, C., MacPhee, C. E., & Stanley-Wall, N. R. (2015). Giving structure to the biofilm matrix: An overview of individual strategies and emerging common themes. *FEMS Microbiology Reviews, 39* (5), 649–669.

Hobley, L., Ostrowski, A., Rao, F. V., Bromley, K. M., Porter, M., Prescott, A. R., et al. (2013). BslA is a self-assembling bacterial hydrophobin that coats the *Bacillus subtilis* biofilm. *Proceedings of the National Academy of Sciences, 110*(33), 13600–13605.

Høiby, N., Frederiksen, B., & Pressler, T. (2005). Eradication of early *Pseudomonas aeruginosa* infection. *Journal of Cystic Fibrosis, 4*, 49–54.

Huh, A. J., & Kwon, Y. J. (2011). "Nanoantibiotics": A new paradigm for treating infectious diseases using nanomaterials in the antibiotics resistant era. *Journal of Controlled Release, 156*(2), 128–145. https://doi.org/10.1016/j.jconrel.2011.07.002. Epub 2011 Jul 6 PMID: 21763369.

Jarraud, S., Mougel, C., Thioulouse, J., Lina, G., Meugnier, H., Forey, F., et al. (2002). Relationships between *Staphylococcus aureus* genetic background, virulence factors, agr groups (alleles), and human disease. *Infection and Immunity, 70*(2), 631–641.

Jennings, L. K., Storek, K. M., Ledvina, H. E., Coulon, C., Marmont, L. S., Sadovskaya, I., et al. (2015). Pel is a cationic exopolysaccharide that cross-links extracellular DNA in the *Pseudomonas aeruginosa* biofilm matrix. *Proceedings of the National Academy of Sciences, 112*(36), 11353–11358.

Jensen, P.Ø., Givskov, M., Bjarnsholt, T., & Moser, C. (2010). The immune system vs. *Pseudomonas aeruginosa* biofilms. *FEMS Immunology & Medical Microbiology, 59*(3), 292–305.

Jesaitis, A. J., Franklin, M. J., Berglund, D., Sasaki, M., Lord, C. I., Bleazard, J. B., et al. (2003). Compromised host defense on *Pseudomonas aeruginosa* biofilms: Characterization of neutrophil and biofilm interactions. *The Journal of Immunology, 171*(8), 4329–4339.

Jiang, Q., Jin, Z., & Sun, B. (2018). MgrA negatively regulates biofilm formation and detachment by repressing the expression of psm operons in *Staphylococcus aureus*. *Applied and Environmental Microbiology, 84* (16), e01008–e01018.

Johnson, T. L., Fong, J. C., Rule, C., Rogers, A., Yildiz, F. H., & Sandkvist, M. (2014). The Type II secretion system delivers matrix proteins for biofilm formation by *Vibrio cholerae*. *Journal of Bacteriology, 196* (24), 4245–4252.

Joly, V., Pangon, B., Vallois, J. M., Abel, L., Brion, N., Bure, A., et al. (1987). Value of antibiotic levels in serum and cardiac vegetations for predicting antibacterial effect of ceftriaxone in experimental *Escherichia coli* endocarditis. *Antimicrobial Agents and Chemotherapy, 31*(10), 1632–1639.

Jones, N., Ray, B., Ranjit, K. T., & Manna, A. C. (2008). Antibacterial activity of ZnO nanoparticle suspensions on a broad spectrum of microorganisms. *FEMS Microbiology Letters, 279*(1), 71–76.

Kaplan, J.Á. (2010). Biofilm dispersal: Mechanisms, clinical implications, and potential therapeutic uses. *Journal of Dental Research, 89*(3), 205–218.

Karimi, M., Ghasemi, A., Zangabad, P. S., Rahighi, R., Basri, S. M. M., Mirshekari, H., et al. (2016). Smart micro/nanoparticles in stimulus-responsive drug/gene delivery systems. *Chemical Society Reviews, 45*(5), 1457–1501.

Katragkou, A., Simitsopoulou, M., Chatzimoschou, A., Georgiadou, E., Walsh, T. J., & Roilides, E. (2011). Effects of interferon-γ and granulocyte colony-stimulating factor on antifungal activity of human polymorphonuclear neutrophils against *Candida albicans* grown as biofilms or planktonic cells. *Cytokine, 55*(3), 330–334.

Kennedy, C. A., & O'Gara, J. P. (2004). Contribution of culture media and chemical properties of polystyrene tissue culture plates to biofilm development by *Staphylococcus aureus*. *Journal of Medical Microbiology, 53*(11), 1171–1173.

Khatoon, Z., McTiernan, C. D., Suuronen, E. J., Mah, T. F., & Alarcon, E. I. (2018). Bacterial biofilm formation on implantable devices and approaches to its treatment and prevention. *Heliyon, 4*(12), e01067.

Kim, M. H., Yamayoshi, I., Mathew, S., Lin, H., Nayfach, J., & Simon, S. I. (2013). Magnetic nanoparticle targeted hyperthermia of cutaneous *Staphylococcus aureus* infection. *Annals of Biomedical Engineering, 41*(3), 598–609.

Kiran, G. S., Hassan, S., Sajayan, A., & Selvin, J. (2017). Quorum quenching compounds from natural sources. In *Bioresources and bioprocess in biotechnology* (pp. 351–364). Singapore: Springer.

Kiran, G. S., Sajayan, A., Priyadharshini, G., Balakrishnan, A., Prathiviraj, R., Sabu, A., et al. (2021). A novel anti-infective molecule nesfactin identified from sponge associated bacteria *Nesterenkonia* sp.

MSA31 against multidrug resistant *Pseudomonas aeruginosa*. *Microbial Pathogenesis*, *157*, 104923. https://doi.org/10.1016/j.micpath.2021.104923.

Kjærgaard, K., Schembri, M. A., Hasman, H., & Klemm, P. (2000). Antigen 43 from *Escherichia coli* induces inter- and intraspecies cell aggregation and changes in colony morphology of *Pseudomonas fluorescens*. *Journal of Bacteriology*, *182*(17), 4789–4796.

Koch, C., & Hoiby, N. (1993). Pathogenesis of cystic fibrosis. *The Lancet*, *341*(8852), 1065–1069.

Kokare, C. R., Chakraborty, S., Khopade, A. N., & Mahadik, K. R. (2009). *Biofilm: Importance and applications*.

Kokare, C. R., Kadam, S. S., Mahadik, K. R., & Chopade, B. A. (2007). Studies on bioemulsifier production from marine *Streptomyces sp.* S1. *Indian Journal of Biotechnology*, *6*, 78–84.

Kolenbrander, P. E., Palmer, R. J., Periasamy, S., & Jakubovics, N. S. (2010). Oral multispecies biofilm development and the key role of cell–cell distance. *Nature Reviews Microbiology*, *8*(7), 471–480.

Kostakioti, M., Hadjifrangiskou, M., & Hultgren, S. J. (2013). Bacterial biofilms: Development, dispersal, and therapeutic strategies in the dawn of the postantibiotic era. *Cold Spring Harbor Perspectives in Medicine*, *3*(4), a010306.

Kostenko, V., Lyczak, J., Turner, K., & Martinuzzi, R. J. (2010). Impact of silver-containing wound dressings on bacterial biofilm viability and susceptibility to antibiotics during prolonged treatment. *Antimicrobial Agents and Chemotherapy*, *54*(12), 5120–5131.

Kumar, V., Abbas Abul, K., Fausto, N., & Mitchell, R. (2007). *Robbins basic pathology* (8th ed.). Saunders Elsevier.

Kuypers, J. M., & Proctor, R. A. (1989). Reduced adherence to traumatized rat heart valves by a low-fibronectin-binding mutant of *Staphylococcus aureus*. *Infection and Immunity*, *57*(8), 2306–2312.

Lamont, R. J., & Jenkinson, H. F. (1998). Life below the gum line: Pathogenic mechanisms of *Porphyromonas gingivalis*. *Microbiology and Molecular Biology Reviews*, *62*(4), 1244–1263.

Larson, E. L., Gomez-Duarte, C., Lee, L. V., Della-Latta, P., Kain, D. J., & Keswick, B. H. (2003). Microbial flora of hands of homemakers. *American Journal of Infection Control*, *31*(2), 72–79.

Lasa, I., & Penadés, J. R. (2006). Bap: A family of surface proteins involved in biofilm formation. *Research in Microbiology*, *157*(2), 99–107.

Laverty, G., Gorman, S. P., & Gilmore, B. F. (2014). Biomolecular mechanisms of *Pseudomonas aeruginosa* and *Escherichia coli* biofilm formation. *Pathogens*, *3*(3), 596–632.

Lebeaux, D., Ghigo, J. M., & Beloin, C. (2014). Biofilm-related infections: Bridging the gap between clinical management and fundamental aspects of recalcitrance toward antibiotics. *Microbiology and Molecular Biology Reviews*, *78*(3), 510–543.

Lee, S. Y. (1996). High cell-density culture of *Escherichia coli*. *Trends in Biotechnology*, *14*(3), 98–105.

Lee, J. H., Kim, Y. G., Cho, M. H., & Lee, J. (2014). ZnO nanoparticles inhibit *Pseudomonas aeruginosa* biofilm formation and virulence factor production. *Microbiological Research*, *169*(12), 888–896.

Lee, J., & Zhang, L. (2015). The hierarchy quorum sensing network in *Pseudomonas aeruginosa*. *Protein & Cell*, *6*(1), 26–41.

Leid, J. G., Shirtliff, M. E., Costerton, J. W., & Stoodley, A. P. (2002). Human leukocytes adhere to, penetrate, and respond to *Staphylococcus aureus* biofilms. *Infection and Immunity*, *70*(11), 6339–6345.

Li, Y. H., & Tian, X. (2012). Quorum sensing and bacterial social interactions in biofilms. *Sensors*, *12*(3), 2519–2538.

Li, C., Zhang, X., Huang, X., Wang, X., Liao, G., & Chen, Z. (2013). Preparation and characterization of flexible nanoliposomes loaded with daptomycin, a novel antibiotic, for topical skin therapy. *International Journal of Nanomedicine*, *8*, 1285.

Li, B., Zhao, Y., Liu, C., Chen, Z., & Zhou, D. (2014). Molecular pathogenesis of *Klebsiella pneumoniae*. *Future Microbiology*, *9*(9), 1071–1081.

Limoli, D. H., Whitfield, G. B., Kitao, T., Ivey, M. L., Davis, M. R., Jr., Grahl, N., et al. (2017). *Pseudomonas aeruginosa* alginate overproduction promotes coexistence with *Staphylococcus aureus* in a model of cystic fibrosis respiratory infection. *MBio*, *8*(2), e00186-17.

Loh, J. V., Percival, S. L., Woods, E. J., Williams, N. J., & Cochrane, C. A. (2009). Silver resistance in MRSA isolated from wound and nasal sources in humans and animals. *International Wound Journal*, *6*(1), 32–38.

Lynch, S. V., Dixon, L., Benoit, M. R., Brodie, E. L., Keyhan, M., Hu, P., et al. (2007). Role of the rapA gene in controlling antibiotic resistance of *Escherichia coli* biofilms. *Antimicrobial Agents and Chemotherapy*, *51*(10), 3650–3658.

Ma, L., Conover, M., Lu, H., Parsek, M. R., Bayles, K., & Wozniak, D. J. (2009). Assembly and development of the *Pseudomonas aeruginosa* biofilm matrix. *PLoS Pathogens*, *5*(3), e1000354.

Ma, H., Darmawan, E. T., Zhang, M., Zhang, L., & Bryers, J. D. (2013). Development of a poly (ether urethane) system for the controlled release of two novel anti-biofilm agents based on gallium or zinc and its efficacy to prevent bacterial biofilm formation. *Journal of Controlled Release*, *172*(3), 1035–1044.

Ma, L. C., Fang, C. T., Lee, C. Z., Shun, C. T., & Wang, J. T. (2005). Genomic heterogeneity in *Klebsiella pneumoniae* strains is associated with primary pyogenic liver abscess and metastatic infection. *The Journal of Infectious Diseases*, *192*(1), 117–128.

Mahmoudi, M., & Serpooshan, V. (2012). Silver-coated engineered magnetic nanoparticles are promising for the success in the fight against antibacterial resistance threat. *ACS Nanotechnology*, *6*(3), 2656–2664.

Mai, G. T., Seow, W. K., Pier, G. B., McCormack, J. G., & Thong, Y. H. (1993). Suppression of lymphocyte and neutrophil functions by *Pseudomonas aeruginosa* mucoid exopolysaccharide (alginate): Reversal by physicochemical, alginase, and specific monoclonal antibody treatments. *Infection and Immunity*, *61*(2), 559–564.

Maki, D., & Mermel, M. (1998). Infections due to infusion therapy. In *Hospital infections* (4th ed., pp. 689–724). Philadelphia, PA: Lippincott-Raven.

Marsin, S., & Forterre, P. (1999). The active site of the rolling circle replication protein Rep75 is involved in site-specific nuclease, ligase and nucleotidyl transferase activities. *Molecular Microbiology*, *33*(3), 537–545.

Mathee, K., Ciofu, O., Sternberg, C., Lindum, P. W., Campbell, J. I., Jensen, P., et al. (1999). Mucoid conversion of *Pseudomonas aeruginosa* by hydrogen peroxide: A mechanism for virulence activation in the cystic fibrosis lung. *Microbiology*, *145*(6), 1349–1357.

McCallum, M., Tammam, S., Little, D. J., Robinson, H., Koo, J., Shah, M., et al. (2016). PilN binding modulates the structure and binding partners of the *Pseudomonas aeruginosa* type IVa pilus protein PilM. *Journal of Biological Chemistry*, *291*(21), 11003–11015.

McCrate, O. A., Zhou, X., Reichhardt, C., & Cegelski, L. (2013). Sum of the parts: Composition and architecture of the bacterial extracellular matrix. *Journal of Molecular Biology*, *425*(22), 4286–4294.

Mohamed, J. A., & Huang, D. B. (2007). Biofilm formation by enterococci. *Journal of Medical Microbiology*, *56*(12), 1581–1588.

Montanaro, L., Poggi, A., Visai, L., Ravaioli, S., Campoccia, D., Speziale, P., et al. (2011). Extracellular DNA in biofilms. *The International Journal of Artificial Organs*, *34*(9), 824–831.

Moormeier, D. E., Endres, J. L., Mann, E. E., Sadykov, M. R., Horswill, A. R., Rice, K. C., et al. (2013). Use of microfluidic technology to analyze gene expression during *Staphylococcus aureus* biofilm formation reveals distinct physiological niches. *Applied and Environmental Microbiology*, *79*(11), 3413–3424.

Moradali, M. F., Ghods, S., & Rehm, B. H. (2017). *Pseudomonas aeruginosa* lifestyle: A paradigm for adaptation, survival, and persistence. *Frontiers in Cellular and Infection Microbiology*, *7*, 39.

Moreillon, P., Entenza, J. M., Francioli, P., McDevitt, D., Foster, T. J., Francois, P., et al. (1995). Role of *Staphylococcus aureus* coagulase and clumping factor in pathogenesis of experimental endocarditis. *Infection and Immunity*, *63*(12), 4738–4743.

Morse, D. J., Wilson, M. J., Wei, X., Lewis, M. A. O., Bradshaw, D. J., Murdoch, C., et al. (2018). Denture-associated biofilm infection in three-dimensional oral mucosal tissue models. *Journal of Medical Microbiology*, *67*, 364–375.

Muhammad, M. H., Idris, A. L., Fan, X., Guo, Y., Yu, Y., Jin, X., et al. (2020). Beyond risk: Bacterial biofilms and their regulating approaches. *Frontiers in Microbiology*, *11*, 928.

Nadell, C. D., Xavier, J. B., Levin, S. A., & Foster, K. R. (2008). The evolution of quorum sensing in bacterial biofilms. *PLoS Biology*, *6*(1), e14.

Nelson, L. K., D'Amours, G. H., Sproule-Willoughby, K. M., Morck, D. W., & Ceri, H. (2009). *Pseudomonas aeruginosa* las and rhl quorum-sensing systems are important for infection and inflammation in a rat prostatitis model. *Microbiology*, *155*(8), 2612–2619.

Niveditha, S., Pramodhini, S., Umadevi, S., Kumar, S., & Stephen, S. (2012). The isolation and the biofilm formation of uropathogens in the patients with catheter associated urinary tract infections (UTIs). *Journal of Clinical and Diagnostic Research: JCDR*, *6*(9), 1478.

Novick, R. P., & Geisinger, E. (2008). Quorum sensing in *Staphylococci*. *Annual Review of Genetics*, *42*, 541–564.

Ofek, I., Hasty, D. L., Doyle, R. J., & Ofek, I. (2003). *Bacterial adhesion to animal cells and tissues. Vol. 1725*. Washington, DC: ASM Press.

Oku, S., Komatsu, A., Tajima, T., Nakashimada, Y., & Kato, J. (2012). Identification of chemotaxis sensory proteins for amino acids in *Pseudomonas fluorescens* Pf0-1 and their involvement in chemotaxis to tomato root exudate and root colonization. *Microbes and Environments*, ME12005.

Omadjela, O., Narahari, A., Strumillo, J., Mélida, H., Mazur, O., Bulone, V., et al. (2013). BcsA and BcsB form the catalytically active core of bacterial cellulose synthase sufficient for in vitro cellulose synthesis. *Proceedings of the National Academy of Sciences*, *110*(44), 17856–17861.

Oppenheimer-Shaanan, Y., Steinberg, N., & Kolodkin-Gal, I. (2013). Small molecules are natural triggers for the disassembly of biofilms. *Trends in Microbiology*, *21*(11), 594–601.

Ostrowski, A., Mehert, A., Prescott, A., Kiley, T. B., & Stanley-Wall, N. R. (2011). YuaB functions synergistically with the exopolysaccharide and TasA amyloid fibers to allow biofilm formation by *Bacillus subtilis*. *Journal of Bacteriology*, *193*(18), 4821–4831.

Otto, M. (2008). Staphylococcal biofilms. *Current Topics in Microbiology and Immunology*, *322*, 207–228. https://doi.org/10.1007/978-3-540-75418-3_10. PMID: 18453278. PMCID: PMC2777538.

Overhage, J., Campisano, A., Bains, M., Torfs, E. C., Rehm, B. H., & Hancock, R. E. (2008). Human host defense peptide LL-37 prevents bacterial biofilm formation. *Infection and Immunity*, *76*(9), 4176–4182.

Overman, P. R. (2000). Biofilm: A new view of plaque. *The Journal of Contemporary Dental Practice*, *1*(3), 18–29.

Palmer, J., Flint, S., & Brooks, J. (2007). Bacterial cell attachment, the beginning of a biofilm. *Journal of Industrial Microbiology and Biotechnology*, *34*(9), 577–588.

Parathon, H., Kuntaman, K., Widiastoety, T. H., Muliawan, B. T., Karuniawati, A., Qibtiyah, M., et al. (2017). Progress towards antimicrobial resistance containment and control in Indonesia. *BMJ*, *358*.

Paterson, D. L., & Bonomo, R. A. (2005). Extended-spectrum beta-lactamases: A clinical update. *Clinical Microbiology Reviews*, *18*(4), 657–686. https://doi.org/10.1128/CMR.18.4.657-686.2005. PMID: 16223952. PMC1265908.

Periasamy, S., Joo, H. S., Duong, A. C., Bach, T. H. L., Tan, V. Y., Chatterjee, S. S., et al. (2012). How *Staphylococcus aureus* biofilms develop their characteristic structure. *Proceedings of the National Academy of Sciences*, *109*(4), 1281–1286.

Pesavento, C., Becker, G., Sommerfeldt, N., Possling, A., Tschowri, N., Mehlis, A., et al. (2008). Inverse regulatory coordination of motility and curli-mediated adhesion in *Escherichia coli*. *Genes & Development*, *22*(17), 2434–2446.

Potts, S. B., Roggli, V. L., & Spock, A. (1995). Immunohistologic quantification of *Pseudomonas aeruginosa* in the tracheobronchial tree from patients with cystic fibrosis. *Pediatric Pathology & Laboratory Medicine*, *15*(5), 707–721.

Prathiviraj, R., Berchmans, S., & Chellapandi, P. (2019). Analysis of modularity in proteome-wide protein interaction networks of *Methanothermobacter thermautotrophicus* strain ΔH across metal-loving bacteria. *Journal of Proteins and Proteomics*, *10*, 179–190.

Prathiviraj, R., & Chellapandi, P. (2020). Modeling a global regulatory network of *Methanothermobacter thermautotrophicus* strain △H. *Network Modeling, Health Informatics and Bioinformatics*, *9*, 17.

Pratt, L. A., & Kolter, R. (1998). Genetic analysis of *Escherichia coli* biofilm formation: Roles of flagella, motility, chemotaxis and type I pili. *Molecular Microbiology*, *30*(2), 285–293.

Puiu, R. A., Dolete, G., Ene, A.-M., Nicoară, B., Vlăsceanu, G. M., Holban, A. M., et al. (2017). 2 – Properties of biofilms developed on medical devices. In Y. Deng, & W. Lv (Eds.), *Biofilms and implantable medical devices* (pp. 25–46). Woodhead Publishing, ISBN:9780081003824.

Qayyum, S., & Khan, A. U. (2016). Nanoparticles vs. biofilms: A battle against another paradigm of antibiotic resistance. *Medicinal Chemistry Communications*, *7*(8), 1479–1498.

Rather, G. A., Hassan, S., Pal, S., Khan, M. H., Rahman, H. S., & Khan, J. (2021). Antimicrobial efficacy of biogenic silver and zinc nanocrystals/nanoparticles to combat the drug resistance in human pathogens. In *Nanocrystals* IntechOpen.

Reisner, A., Maierl, M., Jörger, M., Krause, R., Berger, D., Haid, A., et al. (2014). Type 1 fimbriae contribute to catheter-associated urinary tract infections caused by *Escherichia coli*. *Journal of Bacteriology*, *196*(5), 931–939.

Røder, H. L., Herschend, J., Russel, J., Andersen, M. F., Madsen, J. S., Sørensen, S. J., et al. (2018). Enhanced bacterial mutualism through an evolved biofilm phenotype. *The ISME Journal*, *12*(11), 2608–2618.

Romero, D., Vlamakis, H., Losick, R., & Kolter, R. (2011). An accessory protein required for anchoring and assembly of amyloid fibers in *B. subtilis* biofilms. *Molecular Microbiology*, *80*(5), 1155–1168.

Römling, U., Bian, Z., Hammar, M., Sierralta, W. D., & Normark, S. (1998). Curli fibers are highly conserved between *Salmonella typhimurium* and *Escherichia coli* with respect to operon structure and regulation. *Journal of Bacteriology*, *180*(3), 722–731.

Roosjen, A., van der Mei, H. C., Busscher, H. J., & Norde, W. (2004). Microbial adhesion to poly (ethylene oxide) brushes: Influence of polymer chain length and temperature. *Langmuir*, *20*(25), 10949–10955.

Schachter, B. (2003). Slimy business—The biotechnology of biofilms. *Nature Biotechnology*, *21*(4), 361–365.

Schembri, M. A., Hjerrild, L., Gjermansen, M., & Klemm, P. (2003). Differential expression of the *Escherichia coli* autoaggregation factor antigen 43. *Journal of Bacteriology*, *185*(7), 2236–2242.

Schleich, N., Danhier, F., & Préat, V. (2015). Iron oxide-loaded nanotheranostics: Major obstacles to in vivo studies and clinical translation. *Journal of Controlled Release*, *198*, 35–54.

Seifi, K., Kazemian, H., Heidari, H., Rezagholizadeh, F., Saee, Y., Shirvani, F., et al. (2016). Evaluation of biofilm formation among *Klebsiella pneumoniae* isolates and molecular characterization by ERIC-PCR. *Jundishapur Journal of Microbiology*, *9*(1).

Serra, D. O., & Hengge, R. (2014). Stress responses go three dimensional—The spatial order of physiological differentiation in bacterial macro colony biofilms. *Environmental Microbiology*, *16*(6), 1455–1471.

Shannon, P., Markiel, A., Ozier, O., Baliga, N. S., Wang, J. T., Ramage, D., et al. (2003). Cytoscape: A software environment for integrated models of biomolecular interaction networks. *Genome Research*, *13*(11), 2498–2504.

Shunmugaperumal, T. (2010). *Biofilm eradication and prevention: A pharmaceutical approach to medical device infections*. John Wiley & Sons.

Singh, P. K., Parsek, M. R., Greenberg, E. P., & Welsh, M. J. (2002). A component of innate immunity prevents bacterial biofilm development. *Nature*, *417*(6888), 552–555.

Singh, P. K., Schaefer, A. L., Parsek, M. R., Moninger, T. O., Welsh, M. J., & Greenberg, E. P. (2000). Quorum-sensing signals indicate that cystic fibrosis lungs are infected with bacterial biofilms. *Nature*, *407*, 762–764.

Smith, R. S., Harris, S. G., Phipps, R., & Iglewski, B. (2002). The Pseudomonas aeruginosa quorum-sensing molecule N-(3-oxododecanoyl) homoserine lactone contributes to virulence and induces inflammation in vivo. *Journal of Bacteriology*, *184*(4), 1132–1139. https://doi.org/10.1128/jb.184.4.1132-1139.2002. PMID: 11807074. PMCID: PMC134808.

Solano, C., García, B., Valle, J., Berasain, C., Ghigo, J. M., Gamazo, C., et al. (2002). Genetic analysis of *Salmonella enteritidis* biofilm formation: Critical role of cellulose. *Molecular Microbiology*, *43*(3), 793–808.

Speziale, P., Pietrocola, G., Foster, T. J., & Geoghegan, J. A. (2014). Protein-based biofilm matrices in Staphylococci. *Frontiers in Cellular and Infection Microbiology*, *4*, 171.

Speziale, P., Pietrocola, G., Rindi, S., Provenzano, M., Provenza, G., Di Poto, A., et al. (2009). Structural and functional role of *Staphylococcus aureus* surface components recognizing adhesive matrix molecules of the host. *Future Microbiology*, *4*(10), 1337–1352.

Srey, S., Jahid, I. K., & Ha, S. D. (2013). Biofilm formation in food industries: A food safety concern. *Food Control*, *31*(2), 572–585.

Stahlhut, S. G., Struve, C., Krogfelt, K. A., & Reisner, A. (2012). Biofilm formation of *Klebsiella pneumoniae* on urethral catheters requires either type 1 or type 3 fimbriae. *FEMS Immunology & Medical Microbiology*, *65*(2), 350–359.

Stroh, P., Günther, F., Meyle, E., Prior, B., Wagner, C., & Hänsch, G. M. (2011). Host defence against *Staphylococcus aureus* biofilms by polymorphonuclear neutrophils: Oxygen radical production but not phagocytosis depends on opsonisation with immunoglobulin G. *Immunobiology*, *216*(3), 351–357.

Sturbelle, R. T., de Avila, L. F. D. C., Roos, T. B., Borchardt, J. L., Dellagostin, O. A., & Leite, F. P. L. (2015). The role of quorum sensing in *Escherichia coli* (ETEC) virulence factors. *Veterinary Microbiology*, *180*(3–4), 245–252.

Suci, P. A., Berglund, D. L., Liepold, L., Brumfield, S., Pitts, B., Davison, W., et al. (2007). High-density targeting of a viral multifunctional nanoplatform to a pathogenic, biofilm-forming bacterium. *Chemistry & Biology*, *14*(4), 387–398.

Szklarczyk, D., Gable, A. L., Lyon, D., Junge, A., Wyder, S., Huerta-Cepas, J., et al. (2019). STRING v11: Protein-protein association networks with increased coverage, supporting functional discovery in genome-wide experimental datasets. *Nucleic Acids Research*, *47*(D1), D607–D613.

Thurlow, L. R., Hanke, M. L., Fritz, T., Angle, A., Aldrich, A., Williams, S. H., et al. (2011). *Staphylococcus aureus* biofilms prevent macrophage phagocytosis and attenuate inflammation in vivo. *The Journal of Immunology*, *186*(11), 6585–6596.

Tielen, P., Kuhn, H., Rosenau, F., Jaeger, K. E., Flemming, H. C., & Wingender, J. (2013). Interaction between extracellular lipase LipA and the polysaccharide alginate of *Pseudomonas aeruginosa*. *BMC Microbiology*, *13*(1), 1–12.

Todar, K. (2008). *Textbook of bacteriology: Pathogenic E. coli* (pp. 1–4). University of Wisconsin: Madison.

Ton-That, H., Liu, G., Mazmanian, S. K., Faull, K. F., & Schneewind, O. (1999). Purification and characterization of sortase, the transpeptidase that cleaves surface proteins of *Staphylococcus aureus* at the LPXTG motif. *Proceedings of the National Academy of Sciences*, *96* (22), 12424–12429.

Toutain, C. M., Caiazza, N. C., & O'Toole, G. A. (2004). Molecular basis of biofilm development by pseudomonads. In *Microbial biofilms* (pp. 43–63). American Society of Microbiology.

Tremblay, Y. D., Lévesque, C., Segers, R. P., & Jacques, M. (2013). Method to grow *Actinobacillus pleuropneumoniae* biofilm on a biotic surface. *BMC Veterinary Research*, *9*(1), 1–8.

Vazquez, V., Liang, X., Horndahl, J. K., Ganesh, V. K., Smeds, E., Foster, T. J., et al. (2011). Fibrinogen is a ligand for the *Staphylococcus aureus* microbial surface components recognizing adhesive matrix molecules (MSCRAMM) bone sialoprotein-binding protein (Bbp). *Journal of Biological Chemistry*, *286*(34), 29797–29805.

Veening, J. W., Hamoen, L. W., & Kuipers, O. P. (2005). Phosphatases modulate the bistable sporulation gene expression pattern in *Bacillus subtilis*. *Molecular Microbiology*, *56*(6), 1481–1494.

Veerachamy, S., Yarlagadda, T., Manivasagam, G., & Yarlagadda, P. K. (2014). Bacterial adherence and biofilm formation on medical implants: A review. *Proceedings of the Institution of Mechanical Engineers, Part H: Journal of Engineering in Medicine*, *228*(10), 1083–1099.

Verderosa, A. D., Totsika, M., & Fairfull-Smith, K. E. (2019). Bacterial biofilm eradication agents: A current review. *Frontiers in Chemistry*, *7*, 824.

Vogeleer, P., Tremblay, Y. D., Mafu, A. A., Jacques, M., & Harel, J. (2014). Life on the outside: Role of biofilms in environmental persistence of Shiga-toxin producing *Escherichia coli*. *Frontiers in Microbiology*, *5*, 317.

Vuong, C., Saenz, H. L., Götz, F., & Otto, M. (2000). Impact of the agr quorum-sensing system on adherence to polystyrene in *Staphylococcus aureus*. *The Journal of Infectious Diseases*, *182*(6), 1688–1693.

Vuotto, C., Longo, F., Balice, M. P., Donelli, G., & Varaldo, P. E. (2014). Antibiotic resistance related to biofilm formation in *Klebsiella pneumoniae*. *Pathogens*, *3*(3), 743–758.

Wang, X., Hammer, N. D., & Chapman, M. R. (2008). The molecular basis of functional bacterial amyloid polymerization and nucleation. *Journal of Biological Chemistry*, *283*(31), 21530–21539.

Wang, S., Liu, X., Liu, H., Zhang, L., Guo, Y., Yu, S., et al. (2015). The exopolysaccharide Psl–eDNA interaction enables the formation of a biofilm skeleton in *Pseudomonas aeruginosa*. *Environmental Microbiology Reports*, *7*(2), 330–340.

Wiens, J. R., Vasil, A. I., Schurr, M. J., & Vasil, M. L. (2014). Iron-regulated expression of alginate production, mucoid phenotype, and biofilm formation by *Pseudomonas aeruginosa*. *MBio*, *5*(1), e01010–e01013.

Wilton, M., Charron-Mazenod, L., Moore, R., & Lewenza, S. (2016). Extracellular DNA acidifies biofilms and induces aminoglycoside resistance in *Pseudomonas aeruginosa*. *Antimicrobial Agents and Chemotherapy*, *60*(1), 544–553.

Wood, T. K. (2009). Insights on *Escherichia coli* biofilm formation and inhibition from whole-transcriptome profiling. *Environmental Microbiology*, *11*(1), 1–15.

Xie, Z., Thompson, A., Sobue, T., Kashleva, H., Xu, H., Vasilakos, J., et al. (2012). *Candida albicans* biofilms do not trigger reactive oxygen species and evade neutrophil killing. *The Journal of Infectious Diseases*, *206*(12), 1936–1945.

Yoon, S. S., Hennigan, R. F., Hilliard, G. M., Ochsner, U. A., Parvatiyar, K., Kamani, M. C., et al. (2002). *Pseudomonas aeruginosa* anaerobic respiration in biofilms: Relationships to cystic fibrosis pathogenesis. *Developmental Cell*, *3*(4), 593–603.

Zaph, C. (2010). Which species are in your feces? *The Journal of Clinical Investigation*, *120*(12), 4182–4185.

Zogaj, X., Nimtz, M., Rohde, M., Bokranz, W., & Römling, U. (2001). The multicellular morphotypes of *Salmonella typhimurium* and *Escherichia coli* produce cellulose as the second component of the extracellular matrix. *Molecular Microbiology*, *39*(6), 1452–1463.

Zorraquino, V., García, B., Latasa, C., Echeverz, M., Toledo-Arana, A., Valle, J., et al. (2013). Coordinated cyclic-di-GMP repression of Salmonella motility through YcgR and cellulose. *Journal of Bacteriology*, *195*(3), 417–428.

Chapter 19

Biofilm in antibiotic resistance and pathogenesis in relation to foodborne infection and control strategies

Javed Ahamad Khan[a], Fohad Mabood Husain[b], Gashahun Gezahegne[c], Samreen[d], and Iqbal Ahmad[d]

[a]*Department of Natural Resource Management, Mettu University, Bedele Campus, Mettu, Ethiopia,* [b]*Department of Food Science and Nutrition, King Saud University, Riyadh, Saudi Arabia,* [c]*Department of Agribusiness and Value Chain Management, College of Agriculture and Forestry, Mettu University, Bedelle, Ethiopia,* [d]*Department of Agricultural Microbiology, Faculty of Agricultural Sciences, Aligarh Muslim University, Aligarh, Uttar Pradesh, India*

1. Introduction

A "biofilm" is a thin layer of compactly packed microorganisms that is covered with self-produced exo-polymeric substances containing an aqueous matrix of proteins, nucleic acids, and polysaccharides. These microbes or microbial colonies in a biofilm are liable to flourish on moist surfaces, aggregating and forming colonies. Back in history, the first description of biofilm was suggested in the 17th century by the father of microbiology and the inventor of the microscope, Antonie Von Leeuwenhoek observed microbial aggregates (now known to be Biofilms) in scrapings of his teeth plaque. However later, the term "Biofilm" was coined by Bill Costerton in 1978. After almost two decades in 2002, Donlan and Costerton described the most prominently to the biofilm. According to them, biofilm is "a microbially derived sessile community characterized by cells that are irreversibly attached to a substratum or interface or to each other, embedded in a matrix of extracellular polymeric substances that they have produced, and exhibit an altered phenotype with respect to growth rate and gene transcription." Later in 2006, the European Society for Clinical Microbiology and Infectious Diseases (ESCMID) established a Biofilm Study Group (ESGB). In 2009, ESGB organized biannual Eurobiofilm meetings at different places in Europe such as Rome, Copenhagen and Gent (Hoibe, 2017).

Biofilm formation contributes an efficient adaptive strategy to microbial cells due to several advantages, which include tolerance to harmful environmental factors, increased nutrient availability for microbial growth, increased water-molecule binding, and proximity to progeny and other bacteria, allowing for faster DNA transfer rates. All of these conditions contribute to the survival of microbial cells; thus, inactivation of these microbial cells using standard means such as antibiotics and disinfectants is frequently found to be ineffective. The composition of biofilms varies depending on the system under investigation. Typically, biofilms consist of exopolymeric substances as the primary component. Aside from water and microbial cells, the biofilm matrix is made up of exopolysaccharides. In addition, other macromolecules such as proteins, DNA, and various products of microbial cell lysis are available in the biofilm matrix. The biofilms are made up of channels that allow water, nutrients, and oxygen to flow through the structure. Biofilms, which are inhabited by microbes, can be found in any environment, whether natural, industrial, or clinical. Microbial biofilm formation generally requires the occurrence of a hydrated environment as well as a minimal amount of nutrients. These biofilms can form on a wide range of surfaces, including those that are hydrophobic or hydrophilic, biotic, or abiotic. Most bacterial species are thought to be capable of forming biofilms in favorable environmental conditions (Hrubanova et al., 2018; Sutherland, 2001).

Microbial biofilms cause considerable economical losses in the food industry due to their ability to produce corrosion and biofouling on the various surfaces in food settings. This may entail the need for other equipment substitutes and higher energy consumption. Furthermore, the modification of organoleptic properties as a result of secretion of enzymes such as lipases or proteases by some microbes turns food unsuitable for consumption of humans. It again leads to economic losses to food manufacturers. Besides all these, food poisoning is identified as the main and most solemn problem due to the presence of human

Understanding Microbial Biofilms. https://doi.org/10.1016/B978-0-323-99977-9.00017-X

foodborne pathogens in biofilms. According to an estimate, biofilms are involved in 60% of foodborne illnesses and outbreaks, which are due to the ability of microbial cells to adhere to the surfaces in food settings. It leads to the formation of biofilm. The eradication of these biofilms is more complicated by mechanical washing and cleaning. These biofilms or microbial community structures are found more resistant to the process of disinfection. In fact, the increased natural resistance of microbial cells of biofilms against the biocides is the key factor that affects plant sanitation and product safety (Carrascosa, Raheem, Ramos, Saraiva, & Raposo, 2021). Therefore, it is urgent to extend our knowledge concerning biofilms focusing all panorama related to the food settings and environments, including biofilm formation, their structure, disinfection resistance, and new approaches for targeting them. The significant contribution role of biofilms in cross-contamination at the retail level is specifically important in the food industry and their relation to food contamination in food chain supply and consumption in outbreaks. The main objectives of this review were to recognize the important bacterial species in biofilms in the food settings and environment, the role of these biofilms to develop resistance against antibiotics, and the contribution in microbial pathogenesis. The role of quorum sensing (QS) in biofilm production and approaches to control and eradicate the biofilm are more precisely discussed.

2. Formation and development of biofilm

Mixed or polymicrobial biofilms are more common than mono-species biofilms in nature. However, mono-species biofilm forms at the surface of medical implants and is therefore considered as the reason for infections. The process of biofilm development is a complex process that includes a few distinct phases as depicted in Fig. 1 (Choudhary, Singh, & Agarwal, 2020). These phases are listed below.

I. Phase 1. Attachment of planktonic cells to the surface

During this phase, a conditioning layer is formed, which consists of loosely packed carbohydrates and proteins. These biological components further combine with minerals in hard water. This further attracts the microbial cells to get attached to the substratum.

II. Phase 2. Irreversible attachment of planktonic cells to the surface

As the conditioning layer forms at the surface, an electrical charge builds up at the surface, which attracts the bacterial cells of opposite charge. It results in the irreversible attachment of microbial cells nearby the surface. These charges are not much stronger, and so microbial cells can be easily removed by using the mild cleanser and sanitizers.

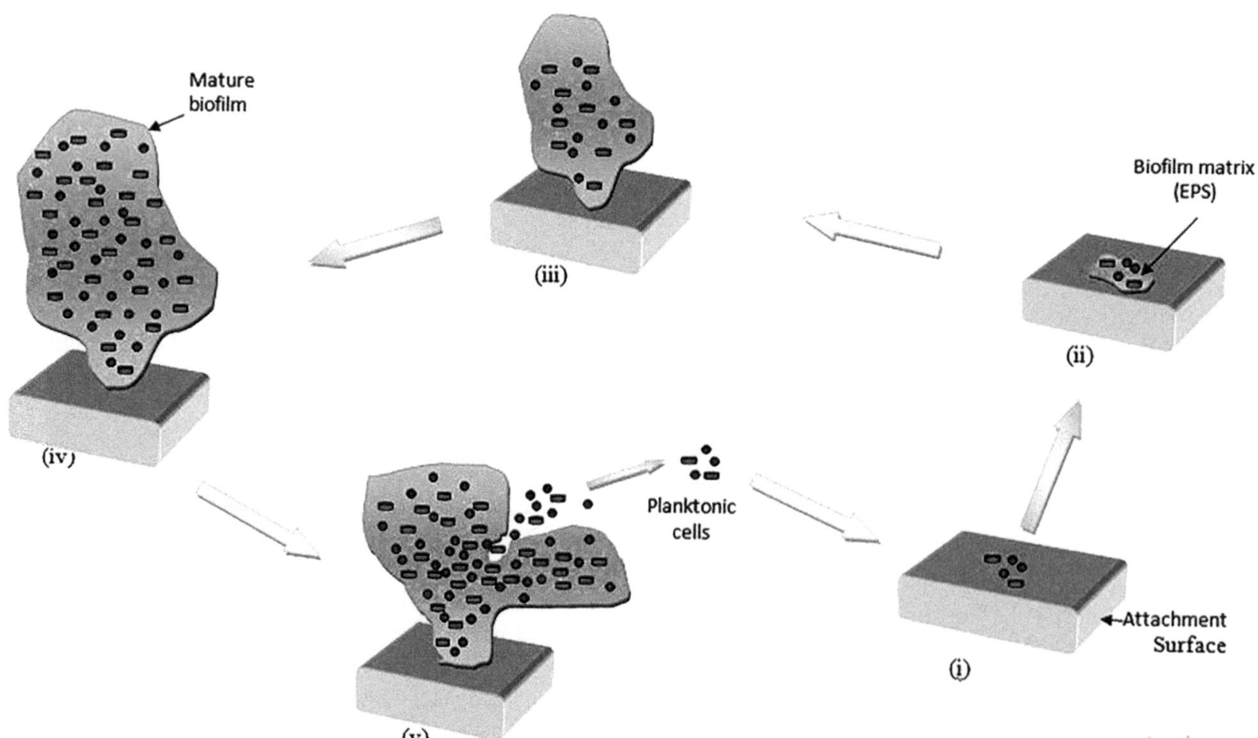

FIG. 1 Various phases of bacterial biofilm formation. (i) Attachment of planktonic bacterial cells to the surface. (ii) Irreversible attachment of planktonic cells to the surface. (iii) Proliferation and microcolony formation. (iv) Maturation of biofilm. (v) Dispersion of microbial cells.

III. Phase 3. Proliferation and microcolony formation

During this phase, microbial cells adhere to the surface and glue together in a matrix of exopolysaccharide (EPS) synthesized by microbial cells itself.

IV. Phase 4. Maturation of biofilm

The biofilm environment comprises the nutrient-rich layer that promotes the fast growth of microbial cells. In a mature well-developed biofilm, transport of nutrients, oxygen, and other components, which are essential for microbial growth, is done by complex diffusion channels. These channels also emit waste products and dead cells from biofilm structure.

V. Phase 5. Dispersion of microbial cells

During this phase, the dispersal of microbial cells from biofilm occurs actively growing in the structure. The biofilm continues to grow until supplied with the appropriate nutrients and returns to their planktonic mode by detaching itself from the surface under the absence or deficiency of nutrients (Choudhary et al., 2020).

3. Important foodborne bacteria and biofilm formation by them in food matrices

Microbial biofilm formation confronts many challenges in food processing settings and environments. For example, disinfectants or sanitizing agents commonly found in food processing settings can be selected to be antimicrobially effective that have established resistance to those chemicals. It's analogous to antibiotic selection for microorganisms that are resistant to antibiotics. Furthermore, because the diversity of commensal bacteria in those habitats is limited by chemicals, the resistant microbial strains will face less competition and have a better chance of growing. Microorganisms that commonly form biofilms differ in various food processing settings. The important bacterial pathogens may include *Staphylococcus* spp., *Micrococcus* spp., *Bacillus* spp., *Brochothrix thermosphacta Clostridium* spp., *Listeria monocytogenes*, *Lactobacillus* spp., *Salmonella enterica*, *Escherichia coli*, *Serratia* spp., *Campylobacter* spp., and *Pseudomonas* spp. (Bai, Nakatsu, & Bhunia, 2021).

L. monocytogenes proliferates at a variety of low temperatures ranging from −0.4°C to −18°C and forms pure culture biofilm or multispecies biofilm. In food processing environments, these biofilm-producing strains have good adhering nature contributed by bacterial structures such as pili, flagella, and membrane proteins. The composition of the biofilm formed by *L. monocytogenes* varies from the biofilms produced by other bacteria as it lacks alginate, a polysaccharide present in biofilm produced by *Pseudomonas* sp. Similarly, it also lacks poly-*N*-acetylglucosamine in *Staphylococcus* sp. A variety of genes play significant role in the regulation of *L. monocytogenes* biofilm formation. In a study, the microarray gene expression was analyzed at 4, 12, and 24h intervals to illustrate the gene expression during biofilm formation and found after 4h more than 150 genes were involved in biofilm regulation, whereas a total of 836 genes are found involved with the increase in time. The one or two genome approaches such as transposon insertion mutagenesis and transcriptome analysis are adopted for identifying the role of genes and regulatory pathways in biofilm formation. In transposons mutagenesis analysis, the identification of 70-week biofilm-producing mutants of *L. monocytogenes* was found possible with Himar 1 mariner transposon. A total of 38 genetic loci were identified, and from these four (*fli*D, *fli*Q, *fla*A, and *mot*A) of them were involved in bacterial motility, which is essential for initial surface attachment during biofilm formation. Another gene *prf*A, the listeriolysin positive regulatory factor A, was observed as necessary only at initial stage of biofilm formation and aggregation. However, it is not required later at colonization stage. In *L. monocytogenes* EGD-e, extracellular internalin A and surface BapL proteins are found involved in initial bacterial adhesion. Furthermore, mobility of this strain is assured by flagella, which are temperature-dependent influencing the formation of biofilm, and it is observed that transcription of flaA stopped above the temperature of 30°C. It is found that genes expressed during the entire biofilm formation process are also affected by biotic and abiotic factors. The *L. monocytogenes* strains dltABC were discovered to be deficient in biofilm formation. Transposon mutagenesis reveals that D-alanylation of teichoic acids plays an important role in biofilm formation. Due of the increased negative charge on the surface of teichoic acids in mutants lacking D-alanine, they develop a biofilm-negative phenotype (Galie, García-Gutiérrez, Miguélez, Villar, & Lombo, 2018; Grigore-Gurgu et al., 2019).

Staphylococcus aureus causes staphylococcal food poisoning and is a common foodborne infection (SFP). The bacterium can adhere on different food surfaces and matrices to develop biofilms, which results food quality and safety problem. In *S. aureus* biofilm, exopolysaccharide (PIA/PNAG) is commonly found in the extracellular matrix, but proteinaceous and extracellular DNA matrix can also be seen. Depending upon the environment, the biofilm matrix may also contain noncellular elements such as mineral corrosion particles, crystals, clay, or silt particles. In planktonic state, *S. aureus* is not much resistant toward the various disinfectants, as compared with other bacteria. However, it is observed that bacterium becomes much resistant when adhered to a surface. It also elevates the capability of biofilm formation with few variations in terms of their composition and structure particularly in wild-type biofilm-producing strains, in food settings. For instance, in *S. aureus* strains of food origin, protein-based sources are found accountable for the structure of biofilms similar to coagulase negative strains. Similarly, certain compounds such as carbohydrates, for example, lactose, are found involved in the modulation of biofilm formation by provoking the polysaccharide intercellular adhesins production. In *S. aureus*, *ica*ADBC operon contains the gene responsible for the cell adhesion to the food surfaces. This operon has been found involved in the biosynthesis of glucosamine polymer and polysaccharide intercellular adhesions. *S. aureus* adhesion genes such as *bap*, *bbp*, *clfA*, *clfB*, *cna*, *ebpS*, *fib*, *fnbA*, *fnbB*, *eno*, *icaAD*, *icaBC*, *sasG*, *sasC*, and *pls* were all involved in cellular agglomeration and bacterial gathering within the biofilm. The other set of genes *pro*P, *opu*D, *aap*A, and *dlt*A encode several transporter proteins (proP, opuD, aapA, and dltA). After 8h after the start of the biofilm, these genes are found to be elevated (Avila-Novoa et al., 2018; Grigore-Gurgu et al., 2019).

Micrococcus strains can be found in a wide range of terrestrial and aquatic environments, such as soil, fresh and saltwater, sand, and vegetation. *Micrococcus* sp. has been found in the skin of warm-blooded animals such as humans, and these organisms often contaminate foods of animal origin. Sometimes, *Micrococcus* spp. may cause invasive disease, mostly in immunocompromised patients. Numerous diseases caused by the genus *Micrococcus* have been reported, including pneumonia, meningitis linked with ventricular shunts, septic arthritis, bacteremia, peritonitis, endocarditis, endophthalmitis, and CR-BSI (Hatem, Rooijakkers, & Ekkelenkamp, 2017). In a study, biofilm formation by *Micrococcus luteus* isolated from cooling tower was evaluated. The authors evaluated the cell surface hydrophobicity, biofilm formation and growth rate using MATH and slide test assay with staining and numbering, respectively. In this study, a high degree of bacteria cell surface hydrophobicity and biofilm growth rate were noticed by the *M. luteus* (Shakeri, Kasra Kermanshahi, & Emtiazi, 2005). Few years back, the role of extracellular DNA (eDNA) in biofilm formation was evaluated. The role of hydrolytic enzymes DNaseI, proteas, cellulose, and mannanase, capable of biofilm dispersion, was revealed and found that DNaseI has significant ability to decrease biofilm components involved in adhesion and established the key structural function for eDNA (Blakeman et al., 2019).

Clostridium perfringens is a member of the genus *Clostridium* and is an opportunistic pathogen capable of causing food poisoning in humans and a range of enterotoxemia in species of animals. A terrible bit of knowledge about the *C. perfringens* biofilm and their exposure to subminimal inhibitory doses of antimicrobials. In a study, 230 *C. perfringens* from 277 human and animal origin isolates were found as biofilm producers. It has been noted that porcine clinical isolates were able to produce biofilm more significantly than the porcine commensal isolates. Additionally, it was discovered that the generated biofilm was capable of protecting bacterial cells from increased concentration of the antimicrobials studied. Further isolates were found producing more biofilm in the presence of low quantity of bacitracin and more area occupied by bacteria. It was noticed that biofilms produced without bacitracin were found lesser and thinner. The morphological examination revealed that *C. perfringens* forms flat biofilms, while only single isolate exhibited a mushroom-like architecture in biofilm. The examination of the biofilm matrix indicated the formation of proteins, beta-1,4-linked polysaccharides, butnopoly-beta-1,6-*N*-acetyl-D-glucosamine, and eDNA (Charlebois, Jacques, & Archambault, 2014). Infection with *Clostridium difficile* infection (CDI) is a severe infection that is frequently related with health care, has been reported all over the world. CDI is a serious health-care-associated disease that has been reported all over the world. Recurring infections and increased resistance against available antibiotic have created more complexity in the CDI treatment. Biofilm structure and complexity are determined by the microorganism and the surrounding environment. CDI has been connected to long-term antibiotic treatment, which has disrupted the native gut flora. *C. difficile* biofilms have been thought to be important for bacterial persistence in the gut and parroted infections in recent years (Frost, Cheng, & Unnikrishnan, 2021). Recently in a study, *C. perfringens* vegetative cells and spores were tested in planktonic and sessile settings for their disinfecting and environmental resistance capabilities. *C. perfringens* strains obtained from meat under sessile conditions were also studied for nucleotide polymorphisms and transcription. The rate of sporulation of sessile *C. perfringens* TYJAM-D-66 (cpe+) was around 19% on the fifth day, where it was only 0.26% and 0.67% on the seventh day for CMM-C-80 (cpe) and SDE-B-202 (cpe+), respectively. There was a marked increase in the vegetative cells of these strains after 36h of exposure to aerobic conditions showing reductions of 1.70 log, 5.36 log, and 5.67 log, respectively. The survival rates of TYJAM-D-66 vegetative cells (53.6%) and spores (82.3%) in biofilms were higher after treatment with sodium hypochlorite than those of planktonic cells. Biofilm- and spore-related genes expressed differently in TYJAM-D-66 (−4.66113.5), CMM-C-80 (−3.022.49), and SDE-B-202 (−5.072.73) (Hu, Woo, Kang, & Koo, 2021).

B. cereus is one of the pathogens commonly occurring in the environment and disseminates through different ways in the food settings such as shipping containers, stainless steel pipes, and storage tanks are a few examples. It can produce immersed as well as floating, both types of biofilms. They produce a variety of metabolites, surfactants, bacteriocins, and toxins through their secretions in their biofilms. They also secrete lipase and protease enzymes that can affect the quality of the foods. *B. cereus* is a spore-forming bacterium, which exists in vegetative state or sporal state in the biofilms. The sporal state has been observed with highly resistant and adhesive characteristics, and so they exhibited increased resistance characteristics toward the antimicrobials and disinfectants. In *B. cereus* biofilm formation, based on flagellar motility, four mechanisms are illustrated. According to first mechanism based on static state, bacteria produce biofilm, when they get their suitable place on an air-liquid surface in the environment. Second mechanism involves the channel formation in biofilm matrix for facilitating the admittance of nutrients from one end and diffusion of the toxic substance from other hand. Third mechanism involves the penetration of motile planktonic bacterial cells into the biofilm to increase the biofilm mass. The fourth mechanism involves the expansion of biofilm, depending on the capability of motile bacteria sited at the periphery of biofilm. It is observed that mature biofilms are developed from microcolonies of the bacterial cells as a result of EPS production. The extracellular biofilm matrix of genus *Bacillus* is similar in all species. However, in *B. cereus*, the *eps* genes encoding for EPS production are not necessarily required as in case of *B. subtilis*. The studies also showed that in *B. cereus*, CodY, a global regulator involved in nitrogen and carbon metabolism, and SpoOA, a master regulator of sporulation initiation, also play an important role in formation biofilm and its regulation. Furthermore, in *B. cereus*, *tap*A and *bsl*A genes encoding for structural proteins are found absent in the matrix of biofilms while present in *B. subtilis*. In case of *B. cereus*, the *tas*A gene encodes for the major matrix protein, encoded by the tapA-tasA-sipW operon facilitates structural support to floating biofilm and found essential for the development of biofilm. In floating biofilms of *Bacillus* spp., the production of "kurstakin," which is a lipopeptide biosurfactant regulated by means of quorum sensing (QS) signaling, is important. In *Bacillus subtilis*, the ComQXPA QS system supports production of surfactin production. This lipopeptide enhances transcription from several genes responsible for the synthesis of biofilm matrix via the involvement of master regulator Spo0A-P. It is hypothesized that inhibiting the QS system slows the formation of suspended biofilms. Mutants lacking the QS system form pellicles faster and produce more biofilm compounds than the wild type. It is also hypothesized that the ComQXPA-dependent signaling promotes sporulation, slowing the rate of growth in the wild-type *Bacillus subtilis* strain. Bacteria with the inactive QS were found to introduce endospore formation in biofilms subsequently and more continuously than those with the active PspoIIQ promoter (Lindback, Basset, Granum, Kuipers, & Kovacs, 2012; Spacapan, Danevčič, Štefanic, & Mandic-Mulec, 2019).

Lactobacillus spp. are identified as Gram-positive bacteria that generally occur under anaerobic environment and able to resist aerobic conditions. The members of genus *Lactobacillus* are considered as outstanding prebiotics. The advantages of biofilm-producing Lactobacillus strains over their planktonic counterparts include greater resistance to gastric pH, temperature, and mechanical forces. In addition, with the expansion of new encapsulation technologies where biofilm characteristics in the formation of double-coated capsules were exploited, which originated to fourth-generation probiotics including *Lactobacillus* spp. The information on biofilms of *Lactobacillus* strains used as probiotics, *specifically L. rhamnosus*, *L. plantarum*, *L. reuteri*, and *L. fermentum*, was depicted in a review. EPS generated by some biofilm-forming strains has been shown to retard biofilm development by certain pathogenic strains (Salas-Jara, Ilabaca, Vega, & Garcí, 2016). It is important for understanding the mechanism and process of interaction between host and microbes for the development of therapeutic interventions, and in this regard, assessment of time-dependent processes of biofilm development is highly significant. The time-dependent growth of *Lactobacillus* biofilms was investigated using two systems: a substantial biofilm output in a batch and continuous micro-fermenter and a real-time label-free cell analyzer based on electrical impedance. *L. plantarum* biofilm growth in a micro-fermenter system was videotaped and analyzed using the IMARIS software (Bitplane, Oxford Instrument Company, Concord, MA, United States). Furthermore, at time intervals of 24 h and 48 h, the assessment of tethered and distributed biofilm phases, as well as whole-genome expression, was performed. Except for dispersed phases, the dynamics of biofilm growth for bacterium *L. plantarum* were similar in both systems. The transcriptomes of attached and dispersed phases were compared, and it was discovered that 121 transcripts differed between the two phases at 24 h and 35 transcripts differed at 48 h in micro-fermenter growth. Furthermore, after 24 h, the main pathways such as regulation of transcription factors, biosynthesis of purine nucleotide, and L-aspartate were found to be substantially downregulated in attached phases compared with dispersed phases. Fatty acid and phospholipid metabolism, as well as ABC transporters and purine nucleotide biosynthesis, on the other hand, were upregulated. Four *Lactobacillus* spp. were found to be variable in terms of attachment phase duration and amplitude. However, the stages of growth were comparable (Martinez et al., 2020).

Anaerobic bacteria *Brochothrix thermosphacta* are Gram-positive facultative anaerobes found in beef, lamb, pig, fish, frozen vegetables, and in many dairy products bacteria. However, few detailed literature studies are available on characteristics of biofilm produced by the *B. thermosphacta*. In a recent study carried out in Austria, the characterization of biofilm-forming organisms isolated from meat-processing environment was performed. These organisms included 11 bacterial strains *Acinetobacter harbinensis* BF1, *Arthrobacter* sp. BF1, *Brochothrix thermosphacta* BF1, *Carnobacterium maltaromaticum* BF1, *Kocuria salsicia* BF1, *Lactococcus piscium* BF1, *Microbacterium* sp.

BF1, *Psychrobacter* sp. BF1, *Pseudomonas fragi* BF1, *Rhodococcus erythropolis* BF1, *Stenotrophomonas* sp. BF1. In order to illustrate genetic factors connected with biofilm behavior, whole-genome sequencing and subsequent genome analysis were implemented. An attempt was made to imitate the meat-processing environment by growing biofilms on stainless steel with a static mono-species biofilm model. Additional tests were carried out after 7 days at 100°C on the biofilm biomass and matrix elements such as proteins, carbohydrates, and extracellular DNA (eDNA). It has been found that all of the examined isolates except the *Kocuria salsicia* BF1 strain have genes linked with biofilm formation and motility. Among 82%, the cellulose metabolism genes were present, involved in cellulose metabolism. The differences of biofilm-forming abilities the screened organism also explored and observed the variations in the amount of the biofilm matrix constituents such as proteins, carbohydrates, and eDNA. In six strains, the detection of all three matrix components (carbohydrates, proteins, and eDNA) in their biofilm was observed at all phases. Besides these, two strains exhibited carbohydrates and eDNA, while in three strains, only the presence of carbohydrates was marked. The quantity of carbohydrate production was also assessed and found a range of 0.5–4.3 μg glucose equivalents/cm^2 among biofilms of all strains used. In contrast, *Microbacterium* sp. strain was observed as highest biofilm producer with a bacterial load of 8.7 log cfu/cm^2 and carbohydrates amount of 2.2 μg glucose equivalents/cm^2, proteins, and 549 ng/cm^2 eDNA. In conclusion, *B. thermosphacta* was observed as weak biofilm producer, depicting low bacterial count of 6.2 log cfu/cm^2 and low levels 0.5 μg glucose equivalents/$cm^2$of carbohydrates in EPS matrix (Wagner et al., 2021).

Salmonella species produce proteinaceous extracellular fibers called "Curli." Cell-to-cell and surface-to-surface interactions, as well as the stimulation of social behavior and host colonization, are all enhanced by the presence of these fibers. In addition to these curli, several serotype-specific fimbrial adhesions have been found that may have implications for biofilm formation. Biofilms that contain cellulose have been shown to be more resistant to mechanical pressures and ready to adhere to abiotic surfaces. A significant difference between different serovars has been observed regarding most persistent of the serovars in food processing environments those are capable of producing biofilms. In *S. enterica*, with the help of differential expression analysis, it has been determined that *ycfR* is highly conserved, as it is in many Gram-negative bacteria. It is discovered to be responsible for Salmonella's virulence and adhesion to substrates such as glass or polystyrene. Furthermore, in the *Salmonella* genus, biofilm formation is triggered by CsgD, a transcriptional regulatory protein that stimulates the expression of curli and cellulose. Additionally, the CsgBAC operon, which codes for "curli" structural subunits, indirectly stimulates the *adrA* gene, which is associated to cellulose formation. When the CsgD content is higher, the bacteria utilize a secondary messenger called cyclic-dimeric-guanosine monophosphate (c-di-GMP) (Grigore-Gurgu et al., 2019).

In *E. coli*, flagella, pili, and membrane proteins are utilized during biofilm formation to attach with nonliving surfaces. After the attachment with surface, the flagella are disappeared and organism starts to produce an extracellular polymeric substance (EPS). This EPS play a role in contributing resistance toward the disinfectants such as hypochlorites. Again the ability of bacterial pathogen to produce biofilm is also found variable according to serotypes or serovars. It is observed that in stress conditions, different serotypes of bacteria show different intensity of producing biofilm such as human isolates seropathotype A, O157: H7, and O157:NM associated with hemolytic uremic syndrome (HUS) exhibited higher ability of biofilm formation than the seropathotype B and C. The bacteria were also found to be involved in the production of mixed biofilms by exhibiting synergistic interaction with *Burkholderia caryophylli* and *Ralstonia insidiosa* in fresh-cut produce. Several genes involved in cell adhesion have been identified, including *fimA*, *yadK*, *yadN*, *yadM*, and *yadC*, which encode fimbriae-like proteins. Outer membrane proteins (htrE) and transcriptional regulators are also coexpressed with these genes (mngR and nhaR). This network of genes was also discovered to be strain-specific (Grigore-Gurgu et al., 2019).

Serratia marcescens is a Gram-negative bacterium, and until recently, it was thought to be a harmless bacterium. When comparing with typical Gram-negative pathogens, its presence in food is typically overlooked. The bacterium is now known to have clinical significance and can cause infections and diarrhea in infants and immunocompromised patients. *S. marcescens* can potentially be transferred to humans through food. The presence of *S. marcescens* and the production of prodigiosin pigment in foodstuff were a source of astonishment and joy in the 13th century. Biofilm development was investigated using the Congo red agar method in a study, and 85% of isolates were shown to be capable of producing biofilm. It is identified that *S. marcescens* utilizes QS, and by using QS it controls swarming motility through serrawettin, which is a surfactant. The bacteria produce biofilms through defined sequence of stages. A signal synthase mutant is shown to create a flat sort of biofilm, which lacks the typical cell clumps and chains and also shows poor attachment to inanimate surfaces. It is also noticed that *S. marcescens* formed a classical biofilm, which consisted of microcolonies. It was discovered that by changing the medium after biofilm formation, the filamentous biofilm has the potential to be transformed into a microcolony-type biofilm. By adjusting the nutrient composition, microcolony biofilm could be easily converted to filamentous biofilm. Under high-nutrient conditions, for example, an *N*-acyl homoserine lactone QS mutant of *S. marcescens* formed biofilm that was distinct from wild-type biofilms (Amorim et al., 2018; Ray, Shenoy, Orihuela, & Gonzalez-Juarbe, 2017; Rice et al., 2005).

Campylobacter jejuni is an anaerobic bacterium, which is found involved in producing biofilm under microaerophilic (5% O_2 and 10% CO_2) as well as aerobic conditions (20% O_2). To our knowledge, the formation and development of *C. jejuni* biofilms are not much explored. However, in a study, various factors involved in *C. jejuni* biofilm formation and development were investigated. The bacterial cells enriched in matrix of biofilm are found protected better from effect of oxygen and survive for many days in food processing setting or environment. It is believed that *C. jejuni* biofilm formation and directed motility or chemotaxis are correlated. The chemotactic pathway in *C. jejuni* shares similar feature with *E. coli*. The membrane-bound chemoreceptors called as transducer-like proteins (Tlps) contain a series of Che proteins. These Che proteins control *C. jejuni* motility. It has been discovered that non-flagellated *C. jejuni* cannot produce biofilms. *Campylobacter jejuni* is found to produce four major types of glycosylated compounds that play a significant role in the establishment of *C. jejuni*. Lipooligosaccharides (LOS), N-linked and O-linked glycosylated proteins are the names given to these compounds. The environment of *C. jejuni* is also important in stimulating biofilm formation. A few signals, such as oxygen saturation, capacity for energy production, metabolites, and mineral nutrition are all important factors to consider that assist in the transition from planktonic to biofilm states. Furthermore, it has been hypothesized that the ability to control intracellular phosphate levels is critical for the development of *C. jejuni* biofilm. The accumulation of inorganic polyphosphate acts as a phosphate pool for the downstream biosynthetic pathways of other molecules. It has been observed that *C. jejuni* strains lacking polyphosphate kinase genes PPK1 and PPK2 are unable to synthesize inorganic polyphosphate, resulting in a diminished ability to infect the host and withstand stress. Consequently, hyperbiofilm-forming phenotype of both PPK1 and PPK2 mutant strains was observed. Likewise, the mutant strains that are deficient in alkaline phosphatase (PhosX) show decreased infectivity and survivability while increasing biofilm formation. As a result, it has been suggested that upregulation of biofilms may operate as a protective mechanism for cells when phosphate metabolism is disturbed and energy generation is impaired. Oxygen saturation could also play a role in biofilm formation. Microaerophilic *C. jejuni* exhibits a significant enhancement in biofilm formation under aerobic conditions. It suggests that the formation of biofilms is a mechanism by which *C. jejuni* tolerates oxidative stress. Also, the surface adherence capability of *C. jejuni* has been observed to significantly elevate under aerobic conditions with increased biomass as compared with those formed under microaerobic condition. Surprisingly, *C. jejuni* mutant strains lacking oxidative stress factors such as alkyl hydroperoxide reductase (AhpC) and the Campylobacter oxidative stress regulator (CosR) show an increased capacity for biofilm formation. It is assumed that in response to increased oxidative stress, the accumulation of radical oxygen species may act as a trigger to increase the level of biofilm produced. This could suggest that *C. jejuni* prefers to live in biofilms in aerobic circumstances external of the host. Furthermore, nutrients and metabolites found in the environment have been shown to have a strong influence on the formation of *C. jejuni* biofilm. In one study, nutrient-poor media were found to be more suitable for the development of biofilms, whereas high osmolarity was found to be inhibitory. Few antibiotics increase biofilm formation in sensitive strains of *C. jejuni*, whereas this effect is reversed in resistant strains. Interestingly, "fucose" is an uncommon glycan metabolite that tends to inhibit the production of *C. jejuni* biofilms. Due to the large amount of fucose in mammalian intestinal mucins, it may have a role in *C. jejuni* pathogenicity in host cells (Tram, Day, & Korolik, 2020; Zhong et al., 2020).

Pseudomonas aeruginosa is appraised as an opportunistic bacterial pathogen and responsible for acute as well as chronic infections in human beings. The capability of biofilm formation by the bacterium on animate and inanimate surfaces makes this bacterium highly resistant toward the immune system of host and current antibiotic therapies. Furthermore, *P. aeruginosa* is an established biofilm producer and an ideal model for research on biofilm development. A resilient biofilm is a critical accoutrement for this bacterium to compete, survive, and dominate in the certain infections such as cystic fibrosis. The bacteria may spread through contaminated food spread in food settings, industry, or environment through mixing in food tanks, vats, and tubing. Polysaccharides, eDNA, proteins, and lipids make up the bulk of the bacterial biofilm matrix. It serves as a platform or scaffold for attachment to both living and nonliving surfaces and as a place of refuge for bacteria in the face of antibiotics and host immunological responses. The biofilm matrix is found accountable for more than 90% of biofilm biomass. The three EPSs, namely Psl, Pel, and alginate, are found actively involved in surface attachment, formation and the stability of biofilm architecture. "Psl" is a neutral pentasaccharide consisting of D-glucose, D-mannose, and L-rhamnose components. The cells attached to a surface are called as "sessile cells" and the EPS "Psl" is found essential for adhesion of sessile cells to the surfaces and cell-to-cell interactions during biofilm initiation of both non-mucoid and mucoid strains. "Psl" also acts as a signaling molecule to promote the formation of c-di-GMP (bis-(3′-5′)-cyclic dimeric guanosine monophosphate). The increased level of c-di-GMP results in thicker and robust biofilm formation. In addition, "Psl" protects bacterial cells from antimicrobial compounds and neutrophil phagocytosis. As a result, it adds to a strong defense against a long-term infection. Polysaccharide "Pel" is composed of *N*-acetyl-D-glucosamine and *N*-acetyl-D galactosamine units that have been largely deacetylated. Nonmucoid strains such as "Psl" and "Pel" have it as a critical biofilm matrix component. It helps to initiate surface adhesion and maintain the integrity of biofilms. In a static broth culture, the cationic polysaccharide is also responsible for the formation of the pellicle biofilm at the air-liquid interface. "Pel" also gives support to the idea that biofilm-embedded bacteria are resistant to aminoglycoside medicines. Other biofilms including "Pel" have been discovered to be persistent in their actions against last-resort colistin antibiotics such as polymyxin E, which are employed against multidrug-resistant Gram-negative infections, including pneumonia. Human HL-60 cell-derived neutrophils used to mediate killing are less effective against these biofilms. Because of a mutation in the mucA22 allele, many mucoid *Pseudomonas* strains create an acetylated polymer called "alginate," which is composed of mannuronic and guluronic acids. As a sign that an infection has progressed from the acute stage to the chronic stage, mucoid characteristics are most commonly found in cystic fibrosis isolates. Many significant roles of alginate have been identified including biofilm formation, protection against phagocytosis and opsonization, and reduced antibiotic diffusion through biofilm (Thi, Wibowo, & Rehm, 2020).

4. Role of biofilm in antibiotic resistance

Antibiotic resistance is a growing effect of pathogenic bacteria biofilm formation that results in chronic infections (Cepas et al., 2019). Biofilm formation is a bacterial resistance mechanism against antimicrobial drugs (Høiby, Bjarnsholt, Givskov, Molin, & Ciofu, 2010; Lu et al., 2019; Navon-Venezia, Kondratyeva, & Carattoli, 2017). Biofilms are resistant to the action of antibacterial drugs and are responsible for most of the persistent and recurring infections caused by pathogenic bacteria (Verderosa, Totsika, & Fairfull-Smith, 2019). Biofilms are difficult to control and/or eradicate due to presence of the complex matrix that encapsulates the bacteria residing in the biofilm mode making them recalcitrant to the action of antibiotics, disinfectants, and sterilizing agents (Bjarnsholt, 2013; Satpathy, Sen, Pattanaik, & Raut, 2016). The extrapolymeric substances (EPSs) matrix prevents the bacteria from the action of antibiotics by preventing its entry inside the EPS envelope, and thus, biofilm bacteria are several times more resistant to antibiotics than free-living bacteria (Rabin et al., 2015). Nature of the biofilm structure and physiological adaptations help the bacteria to resist the antibiotics. Certain factors that contribute to the antibiotic resistance of the biofilm are (a) EPS matrix that acts a barrier to the entry of the antibiotics, (b) interaction with the EPS envelope reduces the activity of the antibiotics activity by EPS envelope interaction, (c) resistance mediated by enzymes such as lactamases, (d) metabolic changes in biofilm-forming cells, (e) target site genetic manipulation or hiding, (f) antibiotic efflux-pump removal, and (g) in Gram-negative bacteria, there is an outer membrane. These mechanisms are critical in the development of resistance to antibiotics and pathogen survival in biofilms (Høiby et al., 2010; Singh, Singh, Chowdhury, & Singh, 2017). Reports have emerged that have demonstrated that the bacteria inside biofilm have evolved molecular mechanisms to counter the antibiotics such as interaction of EPS matrix with the antibacterial drugs to reduce their activity, slowed growth rates of bacteria to render the antibiotics ineffective, resistance associated with genetics and production of persister cells that are antibiotic-tolerant (Koo, Allan, Howlin, Stoodley, & Hall-Stoodley, 2017). High rate of mutations has been reported in biofilm cells that enables these cells to develop resistance by getting the genes to produce enzymes that deactivate the drugs or remove the drugs by the efflux transporters (Ciofu & Tolker-Nielsen, 2019; Dzianach, Dykes, Strachan, Forbes, & Pérez-Reche, 2019). Bacteria residing in biofilm produce metabolically inert "persister cells" that have the ability to survive high concentrations of antibiotics (Keren, Kaldalu, Spoering, Wang, & Lewis, 2004; Wood, Knabel, & Kwan, 2013). Further, due to the dense assortment of bacteria inside the biofilm, horizontal transfer of resistance and virulence genes is efficient. Biofilm assumes an important role in the transfer of the conjugative plasmid due to dense colonization of bacteria. Thus, due to the close proximity of the cells, whole community acquires the resistance genes (Balcázar, Subirats, & Borrego, 2015; Costerton, Montanaro, & Arciola, 2005). Therefore, it can be speculated that the genetic diversification in biofilm of bacteria helps in shaping the antibiotic resistance (Abebe, 2020).

5. Relation between quorum sensing and biofilm production

It is necessary for bacteria to coordinate their gene expression in order to form biofilms, and one method for using it is called "Quorum Sensing" (QS) (Rabin et al., 2015). QS is a cell-to-cell communication system that affects group behavior and is density-dependent (Lyon & Muir, 2003). In response to a range of intra and intercellular signaling molecules known as autoinducers, populations of bacteria coordinate their gene expression (West, Winzer, Gardner, & Diggle, 2012). QS in Gram-negative bacteria is mediated by acyl homoserine lactones (AHLs). Many AHLs have been categorized by their length and functionalization of the acyl side chain. Alternatively, Gram-positive bacteria's QS-mediated actions are regulated by short peptides, whereas autoinducer 2 (AI-2) controls gene expression in both Gram-positive and Gram-negative bacteria. QS has been proven to govern the production of biofilms in a wide range of organisms (Jakobsen, Tolker-Nielsen, & Givskov, 2017).

Biofilm formation and maturation are both regulated by genes and QS regulatory circuits that have been discovered in a wide variety of bacteria (O'Toole, Kaplan, & Kolter, 2000). The eDNA release and biofilm structure are both controlled by QS systems based on AHL in the model organism *P. aeruginosa* PAO1 (Allesen-Holm et al., 2006). QS modulates PEL exopolysaccharide synthesis in PA14 strain (Sakuragi & Kolter, 2007). *S. aureus* AgrC detects the peptide AIP, a product of the *agrD* gene, and triggers the regulator AgrA. Regulating the transcription of genes coding for biofilm-dispersing enzymes is AgrA's primary role (Rabin et al., 2015). Surfactin, a QS molecule found in *Bacillus subtilis*, is critical in the development of biofilms in this bacterium. Because of the potassium leakage, genes involved in the synthesis of extracellular matrix are activated (Branda, Chu, Kearns, Losick, & Kolter, 2006).

6. Approaches for biofilm control and its eradication

Pathogenic foodborne bacterial pathogens cause food poisoning, placing consumers' health at risk and causing substantial economic damage. These foodborne infectious biofilms can form on or in a wide range of biological and non-biological objects. Bacteria are found on surfaces of food or other contact surfaces, air, and pipelines. The presence of microbial biofilms in food settings and environments makes it more difficult to maintain high food hygiene standards. Thus, effective biofilm prevention and removal are required. The development and eradication of biofilms are dependent on the bacteria, QS, attachment surfaces, medium, and other environmental conditions such as moisture, temperature (Zhao, Zhao, Wang, & Zhong, 2017). Thus, biofilm controlling strategies include physical, chemical, herbal, and biological methods. Some of them are listed here:

6.1 Physical approaches for the control and eradication of microbial biofilms

Among the physical approaches of controlling the microbial biofilms in food settings mainly include ultrasound (US), electric (magnetic) field, plasmas, and irradiation applications. The summary of the various investigations has been highlighted here. A summary of various strategies used to prevent or eradicate biofilms is presented in Fig. 2.

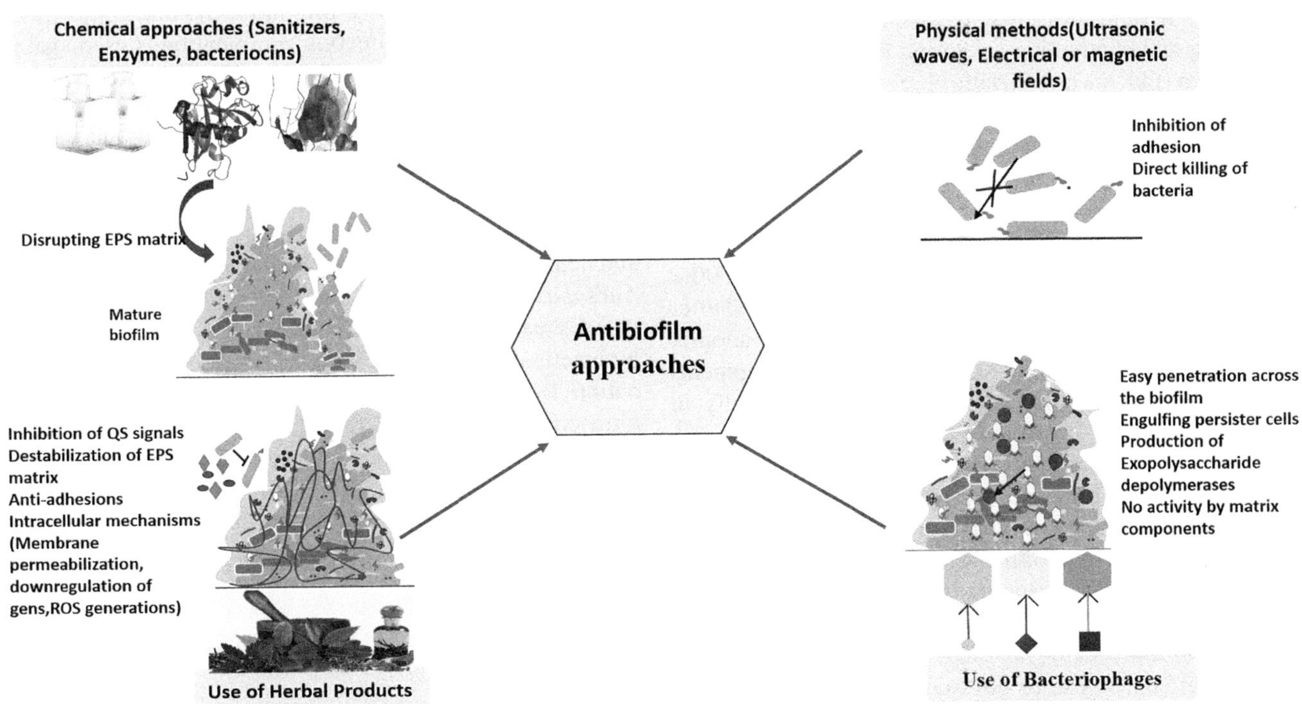

FIG. 2 Various strategies to control/eradicate biofilms.

6.1.1 Ultrasonic waves and their utilization as antibiofilm agent

Ultrasonic waves are a form of mechanical energy that disrupts biofilm from a surface by generating shear forces. The application of ultrasonic waves using ultrasound has been identified to eradicate pathogenic biofilms from food metrices and food settings (Bigelow, Northagen, Hill, & Sailer, 2008; Oulahal-Lagsir, Martial-Gros, Boistier, Blum, & Bonneau, 2000; Yu et al., 2020). The ultrasonic disruption of bacterial biofilm theorized as in bubble and fluid flow phenomena occur during sonication. Because ultrasonic waves propagate in liquid media, longitudinal waves are generated, alternating compression and expansion zones. Pressure changes cause cavitation and tiny bubble nuclei. It now oscillates, grows, and contracts in a series of dynamic processes. The affricate collapse of cavitating bubbles creates high temperatures, pressures, shear forces, and shock waves, which causes biofilm fragmentation on the surface (Verhaagen & Rivas, 2016; Vyas et al., 2019). The other principal methods for eradicating biofilm from surface include microstreaming, acoustic streaming, and microstreamers are all terms for the same thing. Acoustic streaming occurs when an acoustic wave's momentum is transferred to a liquid (Nowak, Cairos, Batyrshin, & Mettin, 2015). Sonic streaming can improve the permeability of bacterial membranes and reduce the thickness of the boundary layer for mass transfer. As a result, it helps to remove biofilm from the surface (Ananta, Voigt, Zenker, Heinz, & Knorr, 2005). Microstreaming, a fluid flow that occurs in conjunction with expanding and contracting cavitation bubbles, can lift biofilms from the surface (Halford et al., 2012). Microstreamers are ribbons of cavitating microbubbles that may aid in the removal of biofilm (Reuter, Lauterborn, Mettin, & Lauterborn, 2017). The effectiveness of ultrasound at a low-frequency level (35 kHz) for eliminating *L. monocytogenes* biofilm formed on a polystyrene surface was investigated in a study, and it was discovered that more than 90% of the *L. monocytogenes* in the biofilm was inactivated after the ultrasonic treatment (Torlak & Sert, 2013). In another study, ultrasonography (37 kHz, 1200 W for equipment) had no bactericidal effect on biofilm when especially in comparison to a control (Na-Young, Seok-Won, & Sang-Do, 2014). The variation in bactericidal activity in *L. monocytogenes* biofilm exposed to ultrasound could be attributed to variations in ultrasound frequency, intensity/power, and microbial strain resistance. Thus, ultrasonic cleaning at frequencies ranging from 10 to 200 kHz necessitates additional research to determine the relationship between biofilm removal and this frequency range (Fuchs, 2015). The use of ultrasound in conjunction with a biological or chemical sanitizer to remove biofilm has also been investigated (Bang, Park, Kim, Rahaman, & Ha, 2017; Baumann, Martin, & Feng, 2009; do Rosario et al., 2017; Duckhouse, Mason, Phull, & Lorimer, 2004; Fink, Oder, Strazar, & Filip, 2017; Jose & Vanetti, 2012; Shao, Dong, Chen, Xu, & Wang, 2020; Stanley, Golden, Williams, & Weiss, 2004; Zhao, Zhang, & Yang, 2017). Ultrasound combined with enzymes demonstrated substantially higher efficacy in eliminating biofilms by damaging heterogenic EPS matrix (Oulahal, Martial-Gros, Bonneau, & Blum, 2003, 2004, 2007). Using this approach, some investigators combined the use of ozonation and sonication for removing the *L. monocytogenes* biofilm from stainless steel food contact surfaces and observed a significant potential of such combination (Baumann et al., 2009). This decrease was significantly found higher than treatment used alone. The microbial cells are eliminated by the disrupting the cell envelope, which causes cell contents leakage. Some researchers looked at the effectiveness of using a combination of ultrasonication (45 kHz) and four disinfectants, including sodium dichloroisocyanurate, hydrogen peroxide, chlorine dioxide, and peracetic acid, to decontaminate natural contaminant microbiota and *Salmonella typhimurium* biofilm on marginally packaged cherry tomatoes. A high elevation in biofilm reduction was attained by combined ultrasonication and used disinfectants (Jose & Vanetti, 2012). Few year later, the combination of ultrasonic waves and Benzakolium chloride, 2.5% didecyldimethylammonium chloride, 10% alcohols (commercially available cleaning agent with the brand name 2% P3Tresolin STV-R) was used against *B. cereus* biofilm and found significant synergistic antibiofilm effect against the *B. cereus* (Fink et al., 2017). Similarly, the ultrasonic waves and peracetic acid disinfection combination is used against *Cronobacter sakazakii* and *Salmonella enterica* biofilms. Because of the oxidizing impact of peracetic acid, the combination boosted the penetration of peracetic acid into the biofilm matrix, which increased cell wall vulnerability and denaturation of proteins and enzymes (do Rosario et al., 2017). In another study, the efficacy of power ultrasound and neutralized electrolyzed water against *Escherichia coli*, *Aureobasidium pullulans*, and *Pichia pastoris* was investigated separately and in tandem. The results showed that by inactivating and detaching the bacterial cell, the combination technique had a greater antibacterial effect. The ultrasonic vibrations aided bacterial detachment, enhanced HClO dispersion in liquid media, and facilitated HClO penetration into the removed cells. Ultrasonic waves and acidic electrolyzed water (AEW) had a synergistic antibiofilm effect against *S. aureus* and *Salmonella* spp. (Shao et al., 2020).

6.1.2 Electrical field (EF) and pulse electric field (PEF)

The electric field effect of on bacterial cellular system has been marked also. In both prokaryotic and eukaryotic cells, the organization of biological membranes, metabolic processes, cell behavior, and the different cellular responses get affected by the EF and currents (Ravikumar, Basu, & Dubey, 2019). Repulsive electrostatic forces help in mediating the microbial cell adhesion to the surface (Jucker, Harms, & Zehnder, 1996; Rijnaarts, Norde, Bouwer, Lyklema, & Zehnder, 1993). The use of electrical current generates repulsive forces, allowing bacteria biofilms to separate from their surfaces. This is referred to as the "electrical effect" (Van der Borden, Van der Mei, & Busscher, 2005). Additionally, it has been shown that electrical current can function synergistically with certain biocides or antibacterial compounds, which is referred to as the "bioelectric effect." Bacterial cell detachment from surfaces such as stainless steel (Van der Borden, Van der Mei, & Busscher, 2004) and indium tin oxide surfaces (Poortinga, Bos, & Busscher, 2000) has been reported to have electrical discharges caused by low-intensity direct current (DC) that may be related to the electricidal effect. The staphylococcal antibiofilm effect was observed among 75% initially adhering bacterial cells on stainless steel. The electric current application causes detachment of bacterial cells from stainless steel and prevents the reunion of detached bacterial cells on the surface (Van der Borden, Van der Werf, Van der Mei, & Busscher, 2004). In biofilms of *S. epidermidis*, *S. aureus*, and *P. aeruginosa*, the application of low-ampere current or intermittent current for 4h per day reduced its formation. Therefore, the amount and time interval play an important role in biofilm reduction or removal (Schmidt-Malan et al., 2015).

In food processing technology, the term "pulsed electric field (PEF)" refers to the application of information in the form of brief, high-voltage pulses to a food substance implanted between two or more electrodes. By disrupting the cell membrane, the pulses enhance cell permeability. When the transmembrane potential is sufficiently high, electroporation occurs. Additionally, if pores are not resealed, cell death follows. In the majority of food packaging applications, structures are designed to allow liquid to flow via pipes, passing through a few regions where the fluid flows between both the electrodes that apply the PEFs. Thermosonication (TS) were investigated in combination with PEF to examine its influence on *P. fluorescens* and *E. coli* inactivation and sublethal damage. Thermosonication was used at low wave amplitudes (18.6mm) and high wave amplitudes (27.9mm), while pulsed electric field was used at low (29kV/cm) and high electrical field strengths (32kV/cm). For *E. coli*, TS/PEF inactivated up to 66% of the population while damaging roughly 26%. In entire liquid eggs, pulsed electric field revealed synergistic potential when used with additives such as EDTA or triethyl citrate to inhibit the growth *Salmonella* serovars. The combination of PEF and other obstacles demonstrates promise for eradicating persistent clones capable of forming biofilms.

6.1.3 Plasma

After solid, liquid, and gas, plasma is considered the fourth state of matter. Plasma is a primarily or fully ionized gas, according to physics principles. Photons, ions, ultraviolet rays, free electrons, and an element with a net neutral charge in the ground or excited state make up the majority of the gas mixture (Zhu, Li, Cui, & Lin, 2020). Plasma can be classified as either thermal plasma (TP) or nonthermal plasma (NTP) based on the thermodynamic balance among electrons and ions (NTP). Low-temperature plasma is another name for NTP (Gilmore et al., 2018). Thermal plasma has a temperature of many thousand Kelvins because it is in a condition of local equilibrium. In NTP, the electrons receive the majority of the linked energy and their temperature rises to a high level, while the neutral particles and ions remain cold (Mugica-Vidal et al., 2019). With its simple design and easy maintenance, NTP accurately retains the advantages of food color, texture, and nutritional composition. As a result, it's gaining a lot of interest in the food industry.

Reactive oxygen species (ROS) and reactive nitrogen species (RNS or RONS) may play a role in microbiocidal effect of plasma, resulting from complex plasma-chemical processes (Graves, 2012). Various active components such as H_2O_2, O_3, and NO_x (stable moiety), O_2^-, $^1O_2$, OH (unstable moiety), NO_2, and $ONOO^-$ may be developed based on the received gas composition and plasma discharge factors, but the existence of a specific particle tends to depend on the discharge character, surrounding gas, and other parameters. These reactive charged molecules are responsible for the physical harm caused by microbial cells (Liu, Yin, Weng, & Cai, 2016). The microbiocidal mechanism of plasma has been found based on the following oxidative stress, etching effect, and QS signal regulations. Cold plasma can disrupt cellular membranes in several ways. It can oxidize lipids and membrane constituents in biological membranes, causing biofilm substrate collapse, reduced adhesion, and biofilm breakup (Gupta, Karki, Fournier, & Ayan, 2018). Plasma active species such as ROS and RNS have a severe oxidizing impact on bacterial cell surface structure. It damages the cell membrane and intracellular components such as DNA, protein, and carbohydrates (Cheng, Liu, Lu, & Liu, 2016). The charged moieties produced by the plasma can break the chemical bonds in biofilm structures. It also damages the biofilm matrix and EPS structure, removing the biofilm's protective mechanism against pathogens. This is called "Etching" (Zhu et al., 2020). The etching of the cell membrane promotes the transfer of active chemicals created during plasma discharge. The substance piercing the cell membrane erodes and destroys bacterial cells. It damages the cell membrane or DNA.

It is known that QS systems govern a variety of physiological processes, including biofilm formation, microbial virulence expression, and diffusion. Plasma has the ability to disable QS signaling molecules and thus inhibit the expression of virulence genes. According to a study, cold plasma can inhibit the release of QS harmful components in Gram-negative bacteria. During cold plasma treatment for biofilm treatment, acyl homoserine lactones (AHL, a signaling molecule in Gram-negative bacteria) undergo fast modification, producing secondary derivatives, and finally, lose their activity with prolonged plasma treatment (Flynn et al., 2016).

6.1.4 Magnetic field (mf)

A magnetic field is a region of space in which a magnetic body has the ability to magnetize the surroundings around it. Experiments involving the inactivation of microbes through pulsed magnetic field have yielded limited results (Gu, Lee, Carnicelli, Zhang, & Ren, 2020). The antibacterial activity of gentamicin antibiotic against biofilm of *P. aeruginosa* has been observed to be enhanced by static magnetic fields (Benson, Grissom, Burns, & Mohammad, 1994). After almost two decades, few investigators designed a magnetic fields system and applied to biofilms. Thus, low-level magnetic fields in conjunction with an appropriate antibiotic may be useful in preventing and eliminating bacterial biofilms (Bandara, Nguyen, Mogarala, Osinski, & Smyth, 2015; McLeod & Sandvik, 2010). Additionally, it is discovered that the use of cyclic magnetic fields enhances the antibacterial activity of gentamycin, ciprofloxacin, chlorhexidine, ethacridine lactate octenidine, and polihexanidine, against *S. aureus* and *P. aeruginosa* biofilms (Junka et al., 2018). Recent research examined the influence of a constant magnetic field on biofilm development as well as other virulence factors in *P. aeruginosa* and isogenic sod mutants (Raouia, Hamida, Khadidja, Ahmed, & Abdelwaheb, 2020). The antibiofilm activity of microenvironment-responsive magnetic nanoparticles was also enhanced by magnetic field (Li et al., 2019; Quan et al., 2019; Wang et al., 2018). The magnetic nanoparticles allowed deeper penetration through the biofilm matrix and therefore created an artificial channel. Antibiotic penetration through these constructed conduits in the biofilm matrix boosted biofilm clearance by four to six times (Quan et al., 2019). This is another physical method for destroying biofilms. Because these irradiations can destroy microbial DNA, they are used to ensure quality and food safety (CFR, 2016; Diehl, 2002). The FDA has approved this ionizing radiation since it does not require heat or chemicals, ionizing irradiation offers numerous advantages over conventional sanitation technologies. However, it may cause minor nutritional alterations (Zehi, Afshari, Noori, Jannat, & Hashemi, 2020). Three *S. enterica* serovars were analyzed to check the responsiveness of planktonic cells and biofilms to ionizing radiation. The numbers of both planktonic and biofilm-associated *S. enterica* were decreased efficiently after treating with ionizing radiation and found biofilm phenotype more sensitive to the ionizing radiation as compared with their planktonic counterparts toward antimicrobial treatments (Niemira & Solomon, 2005). Ultraviolet-C (UV-C) is also found effective to reduce bacterial biofilm populations (Adhikari, Syamaladevi, Killinger, & Sablani, 2015; Jahid, Han, Srey, & Ha, 2014; Srey, Park, Jahid, & Ha, 2014). DNA damage, namely cross-linking of neighboring pyrimidine bases in DNA strands, causes its antibacterial effect on microbes (Lopez-Malo & Palou, 2005). The number of *L. monocytogenes* biofilms on the surfaces of pear, cabbage, strawberry, and cantaloupe were reduced by UV-C irradiation (Adhikari et al., 2015; Srey et al., 2014). *S. typhimurium* biofilms and cultivable indigenous bacteria were decreased after UV-C treatment (Jahid et al., 2014). On stainless steel and eggshell surfaces, UV-C irradiation and NaOCl treatment had a synergistic effect on *L. monocytogenes* biofilms (Jung, Park, & Ha, 2018; Kim, Park, & Ha, 2016). On *S. typhimurium*, a synergistic antibiofilm action of X-ray irradiation and NaOCl was identified. On biofilms, the X-ray/NaOCl therapy provided the most synergistic impact. Irradiation generates highly reactive species such as free radicals and hydrogen peroxide to form. Nucleic acid strands, cell membranes, and genetic material are all examples of this (microbial DNA). This has an unpredictable effect on biological activity and limits the replication and regeneration of microbial cells. Even if the DNA of an injured microbial cell is undamaged, damage to cellular proteins and membranes can prohibit the cell from surviving. While radical generation does not obstruct biofilm development, radical mobility in the EPS matrix may, resulting in self-quenching and decreased efficacy. The free radicals may also damage the structural features of the EPS matrix, aiding removal of biofilms (Ravindran & Jaiswal, 2019; Seifert et al., 2016; Sommers, 2003).

6.2 Chemical approaches for the control and eradication of microbial biofilms

6.2.1 Chemical or synthetic sanitizers and enzymes

As the knowledge toward the microbial biofilm occurrence and its deleterious effect is increasing, the demand of antibiofilm agents is increasing also. Although natural biofilm inhibitors have been found as being safer and more environmentally friendly, the demand for putative microbial biofilm inhibitors has surged, the chemically synthesized inhibitors are discovered rapidly. For example, a number of chemicals are discovered against *E. coli* biofilm. A thermoresponsive oligomer, such as *N*-vinylcaprolactam, has been identified, which is involved in *E. coli* O157:H7 biofilms inhibition through downregulation of "Curli gene" and "shiga-like toxin-1 and 2." The electrostatics spraying of 2% lactic acid and malic identified as potent inhibitor of *E. coli* O157:H7 biofilms via reduction in attachment and biofilm formation. Another chemical substance trisodium phosphate (1%) has found involved in the inactivation of stainless steel formed *E. coli* O157:H7 biofilm. Cetyl pyridinum chloride also showed the similar inactivation of biofilms. The application of lactic acid and steam treatment also found to reduce the *E. coli* O157:H7 biofilm formation. In a study it was observed that quaternary ammonium compounds and acid-based sanitizer inactivated 7-day-old biofilm of *E. coli* O157:H7 formed on stainless steel surface. In another study, sodium hypochlorite significantly reduced the *E. coli* O157:H7 biofilm on spinach and lettuce leaves. Furthermore, some investigators found that peroxyacetic acid is involved in decontamination of *E. coli* O157:H7 biofilm cells on fresh beef and stainless steel coupons (Oloketuyi & Khan, 2017). Some chemical sanitizers based on NaOCl, peracetic acid, NaOH, H_2O_2 found potent biofilm inhibitors in food settings. Their antibiofilm action is based on oxidation of cellular structure. Some enzymes such as proteases and cellulases are found as microbial biofilm inhibitors by disrupting on extracellular matrix (Mayton, Walker, & Berger, 2021).

Nowadays, the application of nanotechnology-based antimicrobials has come in highlight specifically in food industry settings, for controlling biofilm formation, and nanoemulsions and nano-particles (NPs) approaches are found to be important alternatives with more specific precise uses and advantages. In few studies carried out on model system using polystyrene well plates and real systems including lettuce, fresh pineapple, tofu, the antibiofilm activity of essential oils (EOs) is explored in the form of nanoemulsion also. The four various nanoemulsions of EOs, namely *Citrus medica L. var. sarcodactylis* (particle size 73 nm), *Cymbopogon flexuosus* (lemongrass) (particle size 78.46 ± 0.51 nm), Trans-CA (particle size >100 and <100 nm), and Linalool (particle size 10.9 ± 0.1 nm), were utilized for their antibiofilm activity against *S. aureus* (ATCC 29213), *P. aeruginosa* (PA01 and CMCC 10104), and *S. typhimurium* (ATCC 1331). The study revealed that nanoemulsions of *Citrus medica L. var. sarcodactylis* were exhibiting the significant antibiofilm activity against *S. aureus* through inhibiting the ability of bacteria to adhere on surfaces. These nanoparticles can be utilized to prevent biofilm formation as well as remove existing biofilms. Silver, gold, and metal oxide nanoparticles having antibacterial activity can be employed to develop nanocomposite materials that limit bacterial adhesion to food contact various substrates. *S. aureus* biofilm formation on polystyrene polymer was reduced by cysteine dithiothreitol and beta-mercaptoethanol. Furthermore, by breaking the bacterial cell membrane, silver salt 12-tungstophosphoric acid nanoparticles (AgWPA-NPs) can be used to create new food preservation materials. The polyethylene-glycol-coated Ag-ZnO-rGO (AZGP) nanocomposite showed extraordinary antibiofilm action against Gram-positive (*S. aureus* ATCC 25923) and Gram-negative (*P. aeruginosa* MTCC 2453) bacteria. Titanium-based nanoparticles were utilized to obstruct biofilm formation in *P. fluorescens* on the TiO_2/polystyrene nanocomposite film surfaces. It was observed that these nanostructured TiO_2 in combination with UVA irradiation can be utilized to annihilate the biofilm of *L. monocytogenes*, whereas silver nanoparticles with 15 μg/mL concentration inhibited the *S. aureus* and *E. coli* biofilms. Besides these, the ability of superparamagnetic iron oxide also has shown promising results against *B. subtilis* biofilm. Nanoparticles based on Ag^{2+}, Fe_3O_4, TiO_2, ZnO, CuO, and MgO are observed as potent microbial biofilm inhibitors by altering of bacterial membrane. Some biosurfactants, such as lichenysin and surfactin, work to prevent bacterial biofilm formation by preventing bacterial cell adhesion to surfaces. The combination of sodium hypochlorite and hydrogen peroxide was also found to be more successful at controlling and removing *P. aeruginosa* biofilms than either sanitizer used alone under the same concentrations (Galie, García-Gutiérrez, Miguélez, Villar, & Lombo, 2018; Grigore-Gurgu et al., 2019; Wang, 2019).

The several enzymes such as oxidoreductases (e.g., glucose oxidase, curvularia haloperoxidase), showed potential to control the *Pseudomonas aeruginosa* biofilms by direct or indirect retardation of bacterial growth through production of H_2O_2. Against similar organism, transferases such as transaminase showed the antibiofilm effect through the disruption of EPS matrix. The enzymes hydrolases such as AiiA, α-amylase, and proteinase K are found targeting QS molecules, exopolysaccharides, and exoproteins during their mode of action against *P. aeruginosa* biofilms. The lyases such as alginate lyase target exopolysaccharide matrix of biofilm for its degradation. Another enzyme named as "Dioxygenases," destroys a quinolone-based signaling molecule called as 2-heptyl-3-hydroxy-4(1*H*), and it is found that it inhibits the quinolone signals in the *P. aeruginosa* QS system. The investigators studied the mechanism of β-glucosidase and lyticase enzymes on the biofilms produced by various strains of *P. aeruginosa* and observed that enzymes degraded biofilms as well as altered the biofilm states (Vetrivel et al., 2021). A number of disruptive enzymes, including proteases, amylases, cellulases, glycosidases, and DNAses, have been identified as a green alternative to chemical treatments frequently employed to combat biofilm formation in foods and food environments. These enzymes are nontoxic and can be used alone or in combination with industrial detergents to improve cleaning efficiency. *Bacillus* spp. are observed as remarkable source of bacterial proteases. These *Bacillus* spp. produced proteases are equipped with enhanced properties including stability in varied range of pH and extreme temperature, detergents, organic solvents, and oxidizing agents. The purified alkaline proteases by *B. subtilis* were accounted to deteriorate biofilm development by *E. coli* and *P. mendocina* bacteria within 10 min. On the other side, proteinase K protease derived from mold was confirmed to be effective against *L. monocytogenes* biofilms. It was found effective alone as well as in combination with other biofilm inhibitors. DNases and proteinase K work synergistically to disperse *L. monocytogenes* biofilms. Polysaccharide-hydrolyzing enzymes can help remove *Staphylococcal* and *Pseudomonas* biofilms from steel and polypropylene surfaces. These enzymes had no bactericidal action; thus, they were coupled with oxido-reductases enzymes for better results. Experiments also demonstrated that cellulase combined with cetyl trimethyl ammonium bromide may completely remove (100%) *S. enterica* mature biofilms in meat processing facilities (Grigore-Gurgu et al., 2019).

6.2.2 Bacteriocins

Bacteriocins are antimicrobial peptides synthesized ribosomally by bacteria to hinder or kill competitive microorganisms. "Nisin," a lantibiotic, has been shown to be effective against numerous Gram-positive bacteria, including foodborne pathogens. These nisins are the only bacteriocin permitted as food-grade additive currently. Nisin has been shown to enter the *S. aureus* biofilm and penetrate the sessile bacterial cells within the biofilm. Furthermore, nisin and its bioconstructed derivatives have been discovered to enhance the capacity of several conventional antibiotics, such as chloramphenicol, to combat *S. aureus* biofilms. Nonetheless, the nisin bacteriocin was also found to be beneficial toward Gram-negative bacteria such as *Salmonella typhimurium* biofilms in combination with P22 phage and EDTA due to combinational effect, which decreased the mature biofilm by 70%. Adsorption of these bioactive chemicals on surfaces in contact with food or in food settings can also help in the prevention of biofilms. In this scenario, Nisaplin (a heat-stable antimicrobial containing nisin A) adhered to three typical food-contact surfaces: stainless steel, PET, and rubber. Food-borne *L. monocytogenes* strains had lowered its adhesion on solid surface. The ability of nisin to prevent *L. monocytogenes* surface colonization was tested over 30 years ago. Similarly, a bacteriocin called "Sonorensis," generated by *Bacillus sonorensis*, was found to prevent the biofilm formation in *S. aureus* MT93 (Grigore-Gurgu et al., 2019).

In vitro antibiofilm activity of bacteriocin from *Bacillus* sp. strain Sh10 against *Proteus mirabilis* was evaluated. These two biofilm-forming *P. mirabilis* UCa4 and UCe1 strains were tested for bacteriocin sensitivity in planktonic and sessile (biofilm) stages. This study used the cell viability and crystal violet assays. A study using scanning electron microscopy (SEM) found that bacteriocin has bactericidal efficacy against biofilm-forming isolates in the planktonic condition. The cell shape of biofilm-associated cells was altered by bacteriocin, according to SEM examination. The SMIC50 of 32 and 128 g/mL of this *Bacillus* sp. strain Sh10 produced bacteriocin inhibited biofilm formation in *P. mirabilis* UCa4 and *P. mirabilis* UCe1 isolates. Bacteriocin's ideal SMIC50 against *P. mirabilis* UCa4 and UCe1 strain for biofilm destruction was found to be 128 and 256 g/mL, respectively (Shayesteh, Ahmad, & Usup, 2020). Other bacteriocin groups outside the famous "lantibiotics" used to target biofilms are also studied. "Sonorensin" is a novel bacteriocin with strong antibiofilm properties. It is a heterocycloanthracin bacteriocin subclass. This new bacteriocin had outstanding antibiofilm activity in *S. aureus*, which was attributable to enhanced membrane permeability. Enterocin AS-48, a class IIc circular bacteriocin, has shown potential against *S. aureus* biofilms. *S. aureus* biofilms were successfully treated with Enterocin AS-48 (50 g/mL) in combination with benzalkonium chloride, triclosan, and polyhexamethylene guanidium chloride. *L. monocytogenes* planktonic cells were similarly found to be sensitive to enterocin AS48. However, the effect of these enterocin was found variable in case of planktonic and sessile state. This bacteriocin was found ineffective in 10 μg/mL concentration against sessile cells of *L. monocytogenes*. Furthermore, protein synthesis takes priority over carbohydrate metabolism during sessile state under Enterocin AS48 treatment. Additionally, few stress response proteins such as GroEL and DnaK were found overexpressed in sessile state when exposed to the Enterocin AS48. The significant inhibition of *L. monocytogenes* biofilms attached to polystyrene plates was observed with 50 μg/mL concentrations of Enterocin AS-48, whereas the combined treatment with biocides has more remarkably increased the biofilm inhibition except the P3 topax 66 and P3 oxonia. Enterocin AS48 at 0.5–25 μg/mL concentration on polystyrene plates resulted in reduction of adherence capability and subsequent *L. monocytogenes* biofilm formation. Similarly, the activity of class IIb bacteriocins enterocin DD93 and DD28 against the biofilms of MRSA has been also described. These enterocins were found effective with the combination of antibiotics such as kanamycin or erythromycin against MRSA-1 biofilms on stainless steel and glass surfaces. Hyicin 4244, a new circular sactibiotic generated by *Staphylococcus hyicus* 4244, has been shown to have antibiofilm efficacy toward 14 *Staphylococcus* strains. Recently, "Licheniocin," a class II unmodified bacteriocin, was evaluated for anti *L. monocytogenes* and coagulase-negative *Staphylococcus* (CoNS) biofilms activity. Low quantities of licheniocin were found to be efficient in inhibiting the production of CoNS biofilms, as was the crude extract by BGBUI-4 strain effectively inhibiting *L. monocytogenes* biofilm development. The *E. coli* strain CIP54127 has previously been reported to show resistance toward disinfectants as well as oxidizing agents, like phenolic derivatives and cationic or amphoteric surfactants, because of its biofilm-forming capability. Additionally, the enterocin DD14, colistin and nisin combinations were also found to show antibiofilm activity in *E. coli* 184 (mcr1$^+$), which are acknowledged to have resistance against colistin. Sub-MIC levels of bacteriocin subtilosin were associated with the production of *L. monocytogenes* biofilms. The absence of an accompanying reduction in AI-2 production showed that mechanisms other than QS inhibition may be responsible for subtilosins antibiofilm activity. The presence of *Lactobacillus sanfranciscensis* DPMMA174 and *Lactobacillus plantarum* DPPMA20 really helped *Lactobacillus plantarum* DC400 in forming biofilms. In some situations, a powerful "probiotic" biofilm former such as *Lactobacillus plantarum* can guard against biofilms including other potentially pathogenic bacterial strains (Mathur et al., 2018).

6.3 Herbal approaches for the control and eradication of microbial biofilms

In last two decades, the herbal or plant-based approaches have been utilized against the various microbes and their metabolic activities. These herbals or plant-based products are rich in EOs, which contained a number of phytochemical compounds. Plants synthesize these phytocompounds as a defense strategy against diseases. The EOs act on cellular membrane through permeabilization when used against microbes affecting cell processes including membrane transport, energy synthesis, and metabolic regulation are disrupted. The evaluation of EO potential as disinfectants has revealed that the EOs from *M. spicata* and *S. aromaticum* contained principal compounds known as "R-(−) carvone" and "R-(−) eugenol" suppressed Listerial biofilm growth. However, when these "R-(−) carvone" and "eugenol" compounds were used alone, surprisingly, the biofilm biomass was increased. Previously, researchers found that α-pinene, 1,8-cineole, (+)-limonene, linalool, and geranyl acetate inhibited metabolic activity of biofilm bacteria, making them more resistant to hazardous chemicals. These findings revealed that antimicrobial potential of EOs was because of synergistic effect of chemical components, rather than a single compound's activity. Another study found that EOs from *Cymbopogon citratus* and *Cymbopogon nardus* were effective in reducing *L. monocytogenes* population completely from stainless steel surface. The disinfectant exhibited action after 60 min of interaction with the bacterial cells adhered and residing in a 240-hold biofilm. Thyme EO has also proven antimicrobial properties and markedly obstructs the biofilm growth of foodborne pathogens including *B. cereus*, *S. aureus*, and *E. coli*. Thymol, compared with carvacrol or eugenol, significantly reduced the bulk of the *Salmonella typhimurium* biofilm on stainless steel surfaces. The resistance toward antibiotics such as nalidixic acid by *S. typhimurium*, which are used to treat its related illnesses, was improved by these chemicals. Aromatic plants are abundant sources of the phytochemical eugenol. In methicillin-resistant *S. aureus* biofilms, on the other hand, it is discovered to disrupt a signaling pathway known as "quorum sensing (QS)." A key role is played by QS in host colonization, biofilm formation, and the development of defensive strategies against harmful agents. Because of this, bacterial cells are able to form social groups. Pimento berry, bay, and clove EOs, as well as eugenol, a prominent component, all played an important effect in inhibiting the growth of the *E. coli* O157:H7 biofilm. The benzene ring in eugenol's chemical structure led to its antibiofilm potential (Grigore-Gurgu et al., 2019).

6.4 Biological approaches for the control and eradication of microbial biofilms

Bacteriophages are able to infect bacterial cells and proliferate by utilizing the genetic apparatus of the host cells. As a result, they destroy bacteria when a sufficient number is obtained and emerge following lysis of host bacterial cells. They are abundantly found wherever host bacteria are available, and hence their potential as a natural antimicrobial agent for suppressing pathogenic bacteria in a variety of foods, food items, and food settings is well established. The bacteriophages have been utilized for their ability of targeting biofilm-producing bacteria such as *S. aureus*, *E. coli*, *B. cereus*, *L. monocytogenes*, and *S. enterica*, in food settings. As a result, the success of this technique in combating microbial biofilms is dependent on a variety of variables, including the composition and structure of microbial biofilms, their developmental stage or maturity, and the physiological status of host bacteria that are present inside the biofilms. Bacteriophage resistance is also thought to be conferred by biofilms. To decimate bacterial populations, these predator bacilli devised a slew of ways. It needs to get to the bacterial host's extracellular polymeric compounds, for their replication. As a result, they are able to increase their population. As a result, these viruses can degrade biofilms completely and inhibit their regeneration. Bacteriophages can also promote the production of depolymerizing enzymes from the host DNA. Persister bacterial cells are latent variations of regular microbes that really are particularly resistant to antimicrobials. The persister bacteria reactivate the bacterial lysis procedure. The "Bred-phage" can diminish *L. monocytogenes* biofilms on stainless steel. Bred-phage was found to be as effective as lactic acid (130 ppm) in removing

L. monocytogenes biofilms. Regardless of serotype, the P100 bacteriophage reduced the quantity of *L. monocytogenes* cells in biofilm on stainless steel coupon surface. Three bacteriophages, LiMN4L, LiMN4p, LiMN17, were tested for their ability to inactivate *L. monocytogenes* inside 7-day biofilms adhering to clean or fish-broth-coated stainless-steel coupons and dispersed cells of biofilm in a food setting. When applied over short periods of time, these phages outperformed entire biofilms, notably in extricated cells. Short-term phage therapies in fish processing settings should target disrupting biofilms to maximize efficiency. In chicken slaughterhouses, capability of development of biofilms by *Salmonella* spp. has found dependent on the type of adhering surface that may be present in slaughterhouses. Glass and stainless-steel substrates favored *Salmonella* biofilm development. The polyvinyl chloride surface developed less biofilm. Bacteriophages isolated from hospitals and livestock wastewater were tested for antibiofilm activity for all types of surfaces concentrated at 3h. The glass-attached biofilms were resistant to bacteriophage treatment for 6h, but after 9h, the viruses were mostly able to disrupt the biofilms. An effective biological antibiofilm agent in food settings, bacteriophage BPECO 19, was shown to suppress biofilm formation on abiotic (stainless steel, rubber, and minimal biofilm eradication concentration device) as well as biotic (lettuce leaves) substrates (Grigore-Gurgu et al., 2019).

7. Conclusion and future recommendations

Microbial biofilms have been identified as a major concern in the variety of foods, food settings, and their environment in the last three decades. This biofilm formation is critical as they are involved in the contamination and spoilage of the foods. As much as 20% cases of food poisoning can be attributed to biofilms, which have been found resistant to antibiotics compared with planktonic bacteria. Several species of bacteria are found enable to form biofilms, when exposed to the hostilities such as antibiotics and disinfectants, in the environment. Therefore, the changes in cleaning and disinfecting strategies in the food industry must be required for the elimination of biofilms. Biofilm formation results in economic losses in management of food settings, environment and health of the infected individuals, as well as negatively affecting their brand in comparison to their competitor. Historically, food manufacturers used chemical (sodium hydroxide or sodium hypochlorite solutions) or physical (hot water steam, ozone, mechanical eradication) means to remove or prevent biofilm from factory pipes and surfaces. These cost-effective solutions cannot limit biofilm development on some packing materials or food setting structures and procedures. Detergents with proteases, glysosidases, or DNase and medication of steel surfaces through the use of nanoparticles (such as silver, copper or zinc) are used to disrupt biofilms, while innovative antibiofilm polymers with lysozyme or bacteriocins and lichenysin biosurfactants are added to industrial detergents, are commonly adopted methods to control biofilms at food settings. Along these, the diagnostic approaches lead to advancements in nondestructive technology such as laser dissection, which should be employed to detect biofilms in the food sector, as well as to truly comprehend the physiological anatomy of bacteria and biofilms, and potential uses in food and their environmental setting. As a result, advancements in nondestructive technology such as laser dissection should be employed to detect biofilms in the food sector, as well as to truly comprehend the physiological anatomy of bacteria and biofilms, and potential uses in food settings (Carrascosa et al., 2021). However, in recent years, a few unique techniques have been created; however, their deployment in culinary contexts has been limited requiring permission from corresponding health authorities. Furthermore, the application of innovative techniques is sometimes restricted to a few commercial instances, such as the employment of bacteriophages to target specific biofilm-forming bacterial species. For example, *L. monocytogenes* is restricted to meat manufacturers and related products in the case of the phage P100. Nowadays, few novel promising alternatives come up in highlights in food setting, their environments and products, and can be applied in future. As an alternative, QS inhibitors such as paraoxonases, which destroy QS signal molecules, or lactic acid, rosmarinic acid, citric acid, furocoumarins, and various plant flavonoids can be used. Another option is to use plant EOs, which contain bioactive chemicals such as citral, cinnamaldehyde, carvacrol, or thymol that have been shown to be capable of decreasing bacterial biofilms on a variety of solid surfaces. The high hydrostatic pressure up to 900MPa, in combination with thermal applications of 50–100°C, is identified as effective for endospores removal in food settings. At the other end, nonthermal plasma treatment method is highly expensive. So the innovation of cost-effective approaches for controlling the biofilms in food settings and their environments is observed as a major challenge also. However, these all-novel approaches promise a positive future indication in the management of biofilm in various food environments.

References

Abebe, G. M. (2020). The role of bacterial biofilm in antibiotic resistance and food contamination. *International Journal of Microbiology, 2020*, 1–10. https://doi.org/10.1155/2020/1705814.

Adhikari, A., Syamaladevi, R. M., Killinger, K., & Sablani, S. S. (2015). Ultraviolet-C light inactivation of *Escherichia coli* O157:H7 and *Listeria monocytogenes* on organic fruit surfaces. *International Journal of Food Microbiology*, *210*, 136–142.

Allesen-Holm, M., Barken, K. B., Yang, L., Klausen, M., Webb, J. S., Kjelleberg, S., et al. (2006). A characterization of DNA release in *Pseudomonas aeruginosa* cultures and biofilms. *Molecular Microbiology*, *59*, 1114–1128. https://doi.org/10.1111/j.1365-2958.2005.05008.x.

Amorim, A. M. B., Melo, D. H., Souza, B. V., Medeiros, L. M., Mattoso, J. M. V., & Nascimento, J. S. (2018). A reddish problem: Antibiotic-resistant *Serratia marcescens* in dairy food commercialized in Rio de Janeiro. *International Food Research Journal*, *25*(2), 880–883.

Ananta, E., Voigt, D., Zenker, M., Heinz, V., & Knorr, D. (2005). Cellular injuries upon exposure of *Escherichia coli* and *Lactobacillus rhamnosus* to high-intensity ultrasound. *Journal of Applied Microbiology*, *99*(2), 271–278.

Avila-Novoa, M. G., Iniguez-Moreno, M., Solis-Velazquez, O. A., Gonzalez-Gomez, J. P., Guerrero-Medina, P. J., & Gutierrez-Lomel, M. (2018). Biofilm formation by *Staphylococcus aureus* isolated from food contact surfaces in the dairy industry of Jalisco, Mexico. *Journal of Food Quality*, *2018*, 1746139. 1–8.

Bai, X., Nakatsu, C. H., & Bhunia, A. K. (2021). Bacterial biofilms and their implications in pathogenesis and food safety. *Food*, *10*, 2117.

Balcázar, J. L., Subirats, J., & Borrego, C. M. (2015). The role of biofilms as environmental reservoirs of antibiotic resistance. *Frontiers in Microbiology*, *6*, 1–9. https://doi.org/10.3389/fmicb.2015.01216.

Bandara, H., Nguyen, D., Mogarala, S., Osinski, M., & Smyth, H. D. C. (2015). Magnetic fields suppress *Pseudomonas aeruginosa* biofilms and enhance ciprofloxacin activity. *Biofouling*, *31*(5), 443–457.

Bang, H. J., Park, S. Y., Kim, S. E., Rahaman, M. M. F., & Ha, S. D. (2017). Synergistic effects of combined ultrasound and peroxyacetic acid treatments against *Cronobactersakazakii* biofilms on fresh cucumber. *LWT*, *84*, 91–98.

Baumann, A. R., Martin, S. E., & Feng, H. (2009). Removal of *listeria monocytogenes* biofilms from stainless steel by use of ultrasound and ozone. *Journal of Food Protection*, *72*(6), 1306–1309.

Benson, D. E., Grissom, C. B., Burns, G. L., & Mohammad, S. F. (1994). Magnetic field enhancement of antibiotic activity in biofilm forming *Pseudomonas aeruginosa*. *ASAIO Journal (American Society for Artificial Internal Organs: 1992)*, *40*(3), M371–M376.

Bigelow, T. A., Northagen, T., Hill, T. M., & Sailer, F. C. (2008). Ultrasound histotripsy and the destruction of *Escherichia coli* biofilms. In *30th annual international conference: IEEE engineering in medicine and biology society, Vancouver, BC, Canada* (pp. 4467–4470).

Bjarnsholt, T. (2013). The role of bacterial biofilms in chronic infections. *APMIS*, *121*, 1–58. https://doi.org/10.1111/apm.12099.

Blakeman, J. T., Morales-García, A. L., Mukherjee, J., Gori, K., Hayward, A. S., Lant, N. J., et al. (2019). Extracellular DNA provides structural integrity to a *micrococcus luteus* biofilm. *Langmuir*, *35*(19), 6468–6475.

Branda, S. S., Chu, F., Kearns, D. B., Losick, R., & Kolter, R. (2006). A major protein component of the Bacillus subtilis biofilm matrix. *Molecular Microbiology*, *59*, 1229–1238. https://doi.org/10.1111/j.1365-2958.2005.05020.x.

Carrascosa, C., Raheem, D., Ramos, F., Saraiva, A., & Raposo, A. (2021). Microbial biofilms in the food industry—A comprehensive review. *International Journal of Environmental Research and Public Health*, *18*, 1–31. 2014.

Cepas, V., López, Y., Muñoz, E., Rolo, D., Ardanuy, C., Martí, S., et al. (2019). Relationship between biofilm formation and antimicrobial resistance in Gram-negative bacteria. *Microbial Drug Resistance*, *25*, 72–79. https://doi.org/10.1089/mdr.2018.0027.

CFR. (2016). *Title 21: Food & drugs. Part 179—Irradiation in the production, processing and handling of food*. https://www.accessdata.fda.gov/scripts/cdrh/cfdocs/cfCFR/CFRSearch.cfm?fr=179.26.

Charlebois, A., Jacques, M., & Archambault, M. (2014). Biofilm formation of *Clostridium perfringens* and its exposure to low-dose antimicrobials. *Frontiers in Microbiology*, *5*, 1–11.

Cheng, H., Liu, X., Lu, X., & Liu, D. (2016). Active species delivered by dielectric barrier discharge filaments to bacteria biofilms on the surface of apple. *Physics of Plasmas*, *23*(7), 073517.

Choudhary, P., Singh, S., & Agarwal, V. (2020). Microbial biofilms. In S. Dincer, M. S. Ozdenefe, & A. Arkut (Eds.), *Bacterial biofilms* (pp. 1–14). UK: InTech Open, ISBN:978-1-78985-900-3.

Ciofu, O., & Tolker-Nielsen, T. (2019). Tolerance and resistance of *Pseudomonas aeruginosa* biofilms to antimicrobial agents—How *P. aeruginosa* can escape antibiotics. *Frontiers in Microbiology*, *10*. https://doi.org/10.3389/fmicb.2019.00913.

Costerton, J. W., Montanaro, L., & Arciola, C. R. (2005). Biofilm in implant infections: Its production and regulation. *The International Journal of Artificial Organs*, *28*, 1062–1068.

Diehl, J. (2002). Food irradiation—Past, present and future. *Radiation Physics and Chemistry*, *63*(3–6), 211–215.

do Rosario, D. K. A., da Silva Mutz, Y., Peixoto, J. M. C., Oliveira, S. B. S., de Carvalho, R. V., Carneiro, J. C. S., et al. (2017). Ultrasound improves chemical reduction of natural contaminant microbiota and *Salmonella enterica subsp. enterica* on strawberries. *International Journal of Food Microbiology*, *241*, 23–29.

Duckhouse, H., Mason, T. J., Phull, S. S., & Lorimer, J. P. (2004). The effect of sonication on microbial disinfection using hypochlorite. *Ultrasonics Sonochemistry*, *11*(3–4), 173–176.

Dzianach, P. A., Dykes, G. A., Strachan, N. J. C., Forbes, K. J., & Pérez-Reche, F. J. (2019). Challenges of biofilm control and utilization: Lessons from mathematical modelling. *Journal of the Royal Society, Interface*, *16*, 20190042. https://doi.org/10.1098/rsif.2019.0042.

Fink, R., Oder, M., Strazar, E., & Filip, S. (2017). Efficacy of cleaning methods for the removal of *Bacillus cereus* biofilm from polyurethane conveyor belts in bakeries. *Food Control*, *80*, 267–272.

Flynn, P. B., Busetti, A., Wielogorska, E., Chevallier, O. P., Elliott, C. T., Laverty, G., et al. (2016). Non-thermal plasma exposure rapidly attenuates bacterial AHL dependent quorum sensing and virulence. *Scientific Reports*, *6*, 26320.

Frost, L. R., Cheng, J. K. J., & Unnikrishnan, M. (2021). *Clostridioides difficile* biofilms: A mechanism of persistence in the gut? *PLoS Pathogens*, *17*(3), e1009348.

Fuchs, F. (2015). Ultrasonic cleaning and washing of surfaces. In J. A. Gallego-Juarez, & K. F. Graff (Eds.), *Sensing for power ultrasonics: Applications of high-intensity ultrasound* (pp. 577–609). UK: Proceeding of Elsevier Ltd.

Galie, S., García-Gutiérrez, C., Miguélez, E. M., Villar, C. J., & Lombo, F. (2018). Biofilms in the food industry: Health aspects and control methods. *Frontiers in Microbiology*, *9*, 898.

Gilmore, B. F., Flynn, P. B., O'Brien, S., Hickok, N., Freeman, T., & Bourke, P. (2018). Cold plasmas for biofilm control: Opportunities and challenges. *Trends in Biotechnology*, *36*(6), 627–638.

Graves, D. B. (2012). The emerging role of reactive oxygen and nitrogen species in redox biology and some implications for plasma applications to medicine and biology. *Journal of Physics D: Applied Physics*, *45* (26), 263001.

Grigore-Gurgu, L., Bucur, F. I., Borda, D., Alexa, E., Neagu, C., & Nicolau, A. I. (2019). Biofilms formed by pathogens in food and food processing environments. In S. Dincer, M. S. Özdenefe, & A. Arkut (Eds.), *Bacterial biofilms*. UK: Intech Open. https://doi.org/10.5772/intechopen.90176.

Gu, H., Lee, S. W., Carnicelli, J., Zhang, T., & Ren, D. C. (2020). Magnetically driven active topography for long-term biofilm control. *Nature Communications*, *11*(1), 1–11.

Gupta, T. T., Karki, S. B., Fournier, R., & Ayan, H. (2018). Mathematical modelling of the effects of plasma treatment on the diffusivity of biofilm. *Applied Sciences*, *8*(10), 1729.

Halford, A., Ohl, C. D., Azarpazhooh, A., Basrani, B., Friedman, S., & Kishen, A. (2012). Synergistic effect of microbubble emulsion and sonic or ultrasonic agitation on endodontic biofilm in vitro. *Journal of Endodontics*, *38*(11), 1530–1534.

Hatem, D. J., Rooijakkers, S. H. M., & Ekkelenkamp, M. B. (2017). *Staphylococci* and *Micrococci*. In J. Cohen, W. G. Powderly, & S. M. Opal (Eds.), *Infectious diseases* (4th ed., pp. 1509–1522). Elsevier. ISBN 9780702062858.

Hoibe, N. (2017). A short history of microbial biofilms and biofilm infections. *Acta Pathologica Microbiologica et Immunologica Scandinavica*, *125*, 272–275.

Høiby, N., Bjarnsholt, T., Givskov, M., Molin, S., & Ciofu, O. (2010). Antibiotic resistance of bacterial biofilms. *International Journal of Antimicrobial Agents*, *35*, 322–332. https://doi.org/10.1016/j.ijantimicag.2009.12.011.

Hrubanova, K., Krzyzanek, V., Nebesarova, J., Ruzicka, F., Pilat, Z., & Samek, O. (2018). Monitoring *Candida parapsilosis* and *Staphylococcus epidermidis* biofilms by a combination of scanning electron microscopy and Raman spectroscopy. *Sensors (Basel)*, *18*(12), 4089. https://doi.org/10.3390/s18124089.

Hu, W. S., Woo, D. U., Kang, Y. J., & Koo, O. K. (2021). Biofilm and spore formation of *Clostridium perfringens* and its resistance to disinfectant and oxidative stress. *Antibiotics*, *10*(396), 1–14.

Jahid, I. K., Han, N. R., Srey, S., & Ha, S. D. (2014). Competitive interactions inside mixed-culture biofilms of *Salmonella typhimurium* and cultivable indigenous microorganisms on lettuce enhance microbial resistance of their sessile cells to ultraviolet C (UV-C) irradiation. *Food Research International*, *55*, 445–454.

Jakobsen, T. H., Tolker-Nielsen, T., & Givskov, M. (2017). Bacterial biofilm control by perturbation of bacterial signaling processes. *International Journal of Molecular Sciences*, *18*. https://doi.org/10.3390/ijms18091970.

Jose, J. F. B. S., & Vanetti, M. C. D. (2012). Effect of ultrasound and commercial sanitizers in removing natural contaminants and *Salmonella enterica typhimurium* on cherry tomatoes. *Food Control*, *24*, 95–99.

Jucker, B. A., Harms, H., & Zehnder, A. (1996). Adhesion of the positively charged bacterium *Stenotrophomonas (Xanthomonas) maltophilia* 70401 to glass and Teflon. *Journal of Bacteriology*, *178*(18), 5472–5479.

Jung, S. J., Park, S. Y., & Ha, S. D. (2018). Synergistic effect of X-ray irradiation and sodium hypochlorite against *Salmonella enterica serovar typhimurium* biofilms on quail eggshells. *Food Research International (Ottawa, Ont.)*, *107*, 496–502.

Junka, A. F., Rakoczy, R., Szymczyk, P., Bartoszewicz, M., Sedghizadeh, P. P., & Fijalkowski, K. (2018). Application of rotating magnetic fields increase the activity of antimicrobials against wound biofilm pathogens. *Scientific Reports*, *8*(167), 1–12.

Keren, I., Kaldalu, N., Spoering, A., Wang, Y., & Lewis, K. (2004). Persister cells and tolerance to antimicrobials. *FEMS Microbiology Letters*, *230*, 13–18. https://doi.org/10.1016/S0378-1097(03)00856-5.

Kim, M., Park, S. Y., & Ha, S. D. (2016). Synergistic effect of a combination of ultraviolet–C irradiation and sodium hypochlorite to reduce *Listeria monocytogenes* biofilms on stainless steel and eggshell surfaces. *Food Control*, *70*, 103–109.

Koo, H., Allan, R. N., Howlin, R. P., Stoodley, P., & Hall-Stoodley, L. (2017). Targeting microbial biofilms: Current and prospective therapeutic strategies. *Nature Reviews Microbiology*, *15*, 740–755. https://doi.org/10.1038/nrmicro.2017.99.

Li, Y. L., Pan, J., Wu, D., Tian, Y., Zhang, J., & Fang, J. (2019). Regulation of *Enterococcus faecalis* biofilm formation and quorum sensing related virulence factors with ultra-low dose reactive species produced by plasma activated water. *Plasma Chemistry and Plasma Processing*, *39*(1), 35–49.

Lindback, T., Basset, M. M. C., Granum, P. E., Kuipers, O. P., & Kovacs, A. T. (2012). CodY, a pleiotropic regulator, influences multicellular behaviour and efficient production of virulence factors in *Bacillus cereus*. *Environmental Microbiology*, *14*(8), 2233–2246.

Liu, X., Yin, H., Weng, C.-X., & Cai, Y. (2016). Low-frequency ultrasound enhances antimicrobial activity of colistin-vancomycin combination against pan-resistant biofilm of *Acinetobacter baumannii*. *Ultrasound in Medicine & Biology*, *42*(8), 1968–1975.

Lopez-Malo, A., & Palou, E. (2005). Ultraviolet light and food preservation. In G. V. Barbosa-Canovas, M. S. Tapia, & M. P. Cano (Eds.), *Sensing for novel food processing technologies* (pp. 405–422). UK: Proceeding of CRC Press.

Lu, L., Hu, W., Tian, Z., Yuan, D., Yi, G., Zhou, Y., et al. (2019). Developing natural products as potential anti-biofilm agents. *Chinese Medicine*, *14*, 11. https://doi.org/10.1186/s13020-019-0232-2.

Lyon, G. J., & Muir, T. W. (2003). Chemical signaling among bacteria and its inhibition. *Chemistry & Biology*, *10*, 1007–1021. https://doi.org/10.1016/j.chembiol.2003.11.003.

Martinez, S., Gomez-Garcia, J., Williams, R., Elmassry, M., West, A., Hamood, A., et al. (2020). *Lactobacilli spp.*: Real-time evaluation of biofilm growth. *BMC Microbiology*, *20*(64), 1–9.

Mathur, H., Field, D., Rea, M. C., Cotter, P. D., Hill, C., & Ross, R. P. (2018). Fighting biofilms with lantibiotics and other groups of bacteriocins. *Biofilms and Microbiomes*, *9*, 1–13.

Mayton, H. M., Walker, S. L., & Berger, B. W. (2021). Disrupting irreversible bacterial adhesion and biofilm formation with an engineered enzyme. *Applied and Environmental Microbiology*, *87*(13), e00265-21.

McLeod, B. R., & Sandvik, E. L. (2010). A biofilm growth protocol and the design of a magnetic field exposure setup to be used in the study of magnetic fields as a means of controlling bacterial biofilms. *Bioelectromagnetics*, *31*(1), 56–63.

Mugica-Vidal, R., Sainz-Garcia, E., Alvarez-Ordonez, A., Prieto, M., Gonzalez-Raurich, M., Lopez, M., et al. (2019). Production of antibacterial coatings through atmospheric pressure plasma: A promising alternative for combatting biofilms in the food industry. *Food and Bioprocess Technology*, *12*(8), 1251–1263.

Navon-Venezia, S., Kondratyeva, K., & Carattoli, A. (2017). *Klebsiella pneumoniae*: A major worldwide source and shuttle for antibiotic resistance. *FEMS Microbiology Reviews*, *41*, 252–275. https://doi.org/10.1093/femsre/fux013.

Na-Young, L., Seok-Won, K., & Sang-Do, H. (2014). Synergistic effects of ultrasound and sodium hypochlorite (NaOCl) on reducing *Listeria monocytogenes* ATCC19118 in broth, stainless steel, and iceberg lettuce. *Foodborne Pathogens and Disease*, *11*, 581–587.

Niemira, B. A., & Solomon, E. B. (2005). Sensitivity of planktonic and biofilm-associated *salmonella spp.* to ionizing radiation. *Applied and Environmental Microbiology*, *71*(5), 2732–2736.

Nowak, T., Cairos, C., Batyrshin, E., & Mettin, R. (2015). Acoustic streaming and bubble translation at a cavitating ultrasonic horn. In *Vol. 1685 (1). 20th international symposium: Nonlinear acoustics including the 2nd international sonic boom forum, Ecully, France* (p. 020002).

O'Toole, G., Kaplan, H. B., & Kolter, R. (2000). Biofilm formation as microbial development. *Annual Review of Microbiology*, *54*, 49–79. https://doi.org/10.1146/annurev.micro.54.1.49.

Oloketuyi, S. F., & Khan, F. (2017). Strategies for biofilm inhibition and virulence attenuation of foodborne pathogen-*Escherichia coli* O157: H7. *Current Microbiology*, *74*, 1477–1489.

Oulahal, N., Martial-Gros, A., Bonneau, M., & Blum, L. (2003). "*Escherichia coli*-milk" biofilm removal from stainless steel surfaces: Synergism between ultrasonic waves and enzymes. *Biofueling*, *19*(3), 159–168.

Oulahal, N., Martial-Gros, A., Bonneau, M., & Blum, L. (2004). Combined effect of chelating agents and ultrasound on biofilm removal from stainless steel surfaces. Application to "*Escherichia coli milk*" and "*Staphylococcus aureus milk*" biofilms. *Biofilms*, *1*(1), 65–73.

Oulahal, N., Martial-Gros, A., Bonneau, M., & Blum, L. (2007). Removal of meat biofilms from surfaces by ultrasounds combined with enzymes and/or a chelating agent. *Innovative Food Science & Emerging Technologies*, *8*(2), 192–196.

Oulahal-Lagsir, N., Martial-Gros, A., Boistier, E., Blum, L. J., & Bonneau, M. (2000). The development of an ultrasonic apparatus for the non-invasive and repeatable removal of fouling in food processing equipment. *Letters in Applied Microbiology*, *30*(1), 47–52.

Poortinga, A. T., Bos, R., & Busscher, H. J. (2000). Controlled electrophoretic deposition of bacteria to surfaces for the design of biofilms. *Biotechnology and Bioengineering*, *67*(1), 117–120.

Quan, K. C., Zhang, Z. X., Chen, H., Ren, X. X., Ren, Y. J., Peterson, B. W., et al. (2019). Artificial channels in an infectious biofilm created by magnetic nanoparticles enhanced bacterial killing by antibiotics. *Small*, *15*(39), 1902313–1902319.

Rabin, N., Zheng, Y., Opoku-Temeng, C., Du, Y., Bonsu, E., & Sintim, H. O. (2015). Biofilm formation mechanisms and targets for developing antibiofilm agents. *Future Medicinal Chemistry*, *7*, 493–512. https://doi.org/10.4155/fmc.15.6.

Raouia, H., Hamida, B., Khadidja, A., Ahmed, L., & Abdelwaheb, C. (2020). Effect of static magnetic field (200mT) on biofilm formation in *Pseudomonas aeruginosa*. *Archives of Microbiology*, *202*(1), 77–83.

Ravikumar, K., Basu, B., & Dubey, A. K. (2019). Analysis of electrical analogue of a biological cell and its response to external electric field. *Regenerative Engineering and Translational Medicine*, *5*(1), 10–21.

Ravindran, R., & Jaiswal, A. K. (2019). Wholesomeness and safety aspects of irradiated foods. *Food Chemistry*, *285*, 363–368.

Ray, C., Shenoy, A. T., Orihuela, C. J., & Gonzalez-Juarbe, N. (2017). Killing of *Serratia marcescens* biofilms with chloramphenicol. *Annals of Clinical Microbiology and Antimicrobials*, *16*(19), 1–6.

Reuter, F., Lauterborn, S., Mettin, R., & Lauterborn, W. (2017). Membrane cleaning with ultrasonically driven bubbles. *Ultrasonics Sonochemistry*, *37*, 542–560.

Rice, S. A., Koh, K. S., Queck, S. Y., Labbate, M., Lam, K. W., & Kjelleberg, S. (2005). Biofilm formation and sloughing in *Serratia marcescens* are controlled by quorum sensing and nutrient cues. *Journal of Bacteriology*, *187*(10), 3477–3485.

Rijnaarts, H. H., Norde, W., Bouwer, E. J., Lyklema, J., & Zehnder, A. J. B. (1993). Bacterial adhesion under static and dynamic conditions. *Applied and Environmental Microbiology*, *59*(10), 3255–3265.

Sakuragi, Y., & Kolter, R. (2007). Quorum-sensing regulation of the biofilm matrix genes (pel) of *Pseudomonas aeruginosa*. *Journal of Bacteriology*, *189*, 5383–5386. https://doi.org/10.1128/JB.00137-07.

Salas-Jara, M. J., Ilabaca, A., Vega, M., & Garcí, A. (2016). Biofilm forming *Lactobacillus*: New challenges for the development of probiotics. *Microorganisms*, *4*(35), 1–14.

Satpathy, S., Sen, S. K., Pattanaik, S., & Raut, S. (2016). Review on bacterial biofilm: An universal cause of contamination. *Biocatalysis and Agricultural Biotechnology*, *7*, 56–66. https://doi.org/10.1016/j.bcab.2016.05.002.

Schmidt-Malan, S. M., Karau, M. J., Cede, J., Greenwood Quaintance, K. E., Brinkman, C. L., Mandrekar, J. N., et al. (2015). Antibiofilm activity of low-amperage continuous and intermittent direct electrical current. *Antimicrobial Agents and Chemotherapy*, *59*(8), 4610–4615.

Seifert, T., Jaiswal, S., Martens, U., Hannegan, J., Braun, L., Maldonado, P., et al. (2016). Efficient metallic spintronic emitters of ultrabroadband terahertz radiation. *Nature Photonics*, *10*(7), 483–488.

Shakeri, S., Kasra Kermanshahi, R., & Emtiazi, G. (2005). Study of *Micrococcus luteus* biofilm formation and evaluation of antimicrobial effects on its biofilm by microtiter plate test. *Iranian Journal of Biology*, *18* (1), 5–14.

Shao, L., Dong, Y., Chen, X., Xu, X., & Wang, H. (2020). Modeling the elimination of mature biofilms formed by *Staphylococcus aureus* and *Salmonella spp.* using combined ultrasound and disinfectants. *Ultrasonics Sonochemistry*, *69*, 105269–105276.

Shayesteh, F., Ahmad, A., & Usup, G. (2020). *In vitro* anti-biofilm activity of bacteriocin from a marine *Bacillus sp.* strain Sh10 against *Proteus mirabilis*. *Iranian Journal of Microbiology*, *12*(1), 52–61.

Singh, S., Singh, S. K., Chowdhury, I., & Singh, R. (2017). Understanding the mechanism of bacterial biofilms resistance to antimicrobial agents. *The Open Microbiology Journal*, *11*, 53–62. https://doi.org/10.2174/1874285801711010053.

Sommers, C. H. (2003). Irradiation of minimally processed meats. In *Microbial safety of minimally processed foods* (pp. 301–318). Boca Raton, FL: CRC Press.

Spacapan, M., Danevčič, T., Štefanic, P., & Mandic-Mulec, I. (2019). Quorum sensing in *Bacillus subtilis* slows down biofilm formation by enabling sporulation bet hedging. *BioRxiv*, 768671.

Srey, S., Park, S. Y., Jahid, I. K., & Ha, S. D. (2014). Reduction effect of the selected chemical and physical treatments to reduce *L. monocytogenes* biofilms formed on lettuce and cabbage. *Food Research International*, *62*, 484–491.

Stanley, K. D., Golden, D. A., Williams, R. C., & Weiss, J. (2004). Inactivation of *Escherichia coli* O157: H7 by high-intensity ultrasonication in the presence of salts. *Foodborne Pathogens and Disease*, *1*(4), 267–280.

Sutherland, I. W. (2001). The biofilm matrix—An immobilized but dynamic microbial environment. *Trends in Microbiology*, *9*, 222–227. https://doi.org/10.1016/S0966-842X(01)02012-1.

Thi, M. T. T., Wibowo, D., & Rehm, B. H. A. (2020). *Pseudomonas aeruginosa* biofilms. *International Journal of Molecular Sciences*, *21* (8671), 1–25.

Torlak, E., & Sert, D. (2013). Combined effect of benzalkonium chloride and ultrasound against *Listeria monocytogenes* biofilm on plastic surface. *Letters in Applied Microbiology*, *57*(3), 220–226.

Tram, G., Day, C. J., & Korolik, V. (2020). Bridging the gap: A role for *Campylobacter jejuni* biofilms. *Microorganisms*, *8*(452), 1–11.

Van der Borden, A. J., Van der Mei, H. C., & Busscher, H. J. (2004). Electric-current-induced detachment of *Staphylococcus epidermidis* strains from surgical stainless steel. *Journal of Biomedical Materials Research Part B*, *68*(2), 160–164.

Van der Borden, A. J., Van der Mei, H. C., & Busscher, H. J. (2005). Electric block current induced detachment from surgical stainless steel and decreased viability of *Staphylococcus epidermidis*. *Biomaterials*, *26*(33), 6731–6735.

Van der Borden, A. J., Van der Werf, H., Van der Mei, H. C., & Busscher, H. J. (2004). Electric current-induced detachment of *Staphylococcus epidermidis* biofilms from surgical stainless steel. *Applied and Environmental Microbiology*, *70*(11), 6871–6874.

Verderosa, A. D., Totsika, M., & Fairfull-Smith, K. E. (2019). Bacterial biofilm eradication agents: A current review. *Frontiers in Chemistry*, *7*. https://doi.org/10.3389/fchem.2019.00824.

Verhaagen, B., & Rivas, D. F. (2016). Measuring cavitation and its cleaning effect. *Ultrasonics Sonochemistry*, *29*, 619–628.

Vetrivel, A., Ramasamy, M., Vetrivel, P., Natchimuthu, S., Arunachalam, S., Kim, G. S., et al. (2021). *Pseudomonas aeruginosa* biofilm formation and its control. *Biologics*, *1*, 312–336.

Vyas, N., Manmi, K., Wang, Q., Jadhav, A. J., Barigou, M., Sammons, R. L., et al. (2019). Which parameters affect biofilm removal with acoustic cavitation? A review. *Ultrasound in Medicine & Biology*, *45*(5), 1044–1055.

Wagner, E. M., Fischel, K., Rammer, N., Beer, C., Palmetzhofer, A. L., Conrady, B., et al. (2021). Bacteria of eleven different species isolated from biofilms in a meat processing environment have diverse biofilm forming abilities. *International Journal of Food Microbiology*, *349*, 1–11.

Wang, R. (2019). Biofilms and meat safety: A mini-review. *Journal of Food Protection*, *82*(1), 120–127.

Wang, X., Deng, A. P., Cao, W. W., Li, Q., Wang, L. N., Zhou, J., et al. (2018). Synthesis of chitosan/poly (ethylene glycol)-modified magnetic nanoparticles for antibiotic delivery and their enhanced anti-biofilm activity in the presence of magnetic field. *Journal of Materials Science*, *53*(9), 6433–6449.

West, S. A., Winzer, K., Gardner, A., & Diggle, S. P. (2012). Quorum sensing and the confusion about diffusion. *Trends in Microbiology*, *20*, 586–594. https://doi.org/10.1016/j.tim.2012.09.004.

Wood, T. K., Knabel, S. J., & Kwan, B. W. (2013). Bacterial persister cell formation and dormancy. *Applied and Environmental Microbiology*, *79*, 7116–7121. https://doi.org/10.1128/AEM.02636-13.

Yu, H., Liu, Y., Li, L., Guo, Y., Xie, Y., Cheng, Y., et al. (2020). Ultrasound-involved emerging strategies for controlling foodborne microbial biofilms. *Trends in Food Science and Technology*, *96*, 91–101.

Zehi, Z. B., Afshari, A., Noori, S. M. A., Jannat, B., & Hashemi, M. (2020). The effects of X-ray irradiation on safety and nutritional value of food: A systematic review article. *Current Pharmaceutical Biotechnology*, *21*(10), 919–926.

Zhao, L., Zhang, Y., & Yang, H. (2017). Efficacy of low concentration neutralised electrolysed water and ultrasound combination for inactivating *Escherichia coli ATCC* 25922, *Pichia pastoris* GS115 and *Aureobasidium pullulans* 2012 on stainless steel coupons. *Food Control*, *73*, 889–899.

Zhao, X., Zhao, F., Wang, J., & Zhong, N. (2017). Biofilm formation and control strategies of foodborne pathogens: Food safety perspectives. *Royal Society of Chemistry Advances*, *7*, 36670–36683.

Zhong, X., Wu, Q., Zhang, J., Ma, Z., Wang, J., Nie, X., et al. (2020). *Campylobacter jejuni* biofilm formation under aerobic conditions and inhibition by ZnO nanoparticles. *Frontiers in Microbiology*, *11*(207), 1–6.

Zhu, Y. L., Li, C. Z., Cui, H. Y., & Lin, L. (2020). Feasibility of cold plasma for the control of biofilms in food industry. *Trends in Food Science and Technology*, *99*, 142–151.

Chapter 20

Biofilms associated with biomedical implants and combating therapies

Hari Prasath Nagaiah, Karthikeyan Kandaswamy, Arumugam Priya, Thirupathi Kasthuri, and Shunmugiah Karutha Pandian

Department of Biotechnology, Alagappa University, Karaikudi, Tamil Nadu, India

1. Introduction

Medical implants are artificially made organs/portions/fragments, which can replace the anatomical structures of human body. Implants are made of synthetic biomaterials, which can be placed permanently or removed after doing their functions. From 2020 to 2027, the worldwide medical implants market is predicted to grow at a CAGR of 7.2%, from $85, 389 million in 2019 to $147, 464 million in 2027 (Global Industry Analysts, Inc, 2021). The market for orthopedic, cardiac, spinal, dental, ophthalmic, and cosmetic implants grew by 71% in 2017 and continues to grow due to increased demand. Advanced implants are in existence for the past 10 years due to their stability in host systems for a prolonged time period. The development of technologies in recent years leads to the development of implants with better life span and performance (Prakasam et al., 2017).

Even though implants with a better life span are beneficial, complications start with the accumulation of microbes and host metabolites to the implants. There are two possible ways of microbial incidence to implants. One is the human symbiotic microbiota near the site of implantation (Malard, Dore, Gaugler, & Mohty, 2020). This microbiota initially adheres to the implant substrate feeding in the nutrients and starts the cell division. The second possible way is pathogenic invasion to implant sites of hosts through the surgical procedure. Most commonly bacteria and fungi are responsible for the host-implant-associated infections. Some of the microbes with high virulence factors include *Escherichia coli*, *Staphylococcus aureus*, *Pseudomonas aeruginosa*, *Streptococcus mutans*, *Staphylococcus epidermidis*, *Klebsiella pneumoniae*, *Acinetobacter baumannii*, *Enterococcus faecalis*, and *Candida albicans* (Swetha, Vikraman, Nithya, Hari Prasath, & Pandian, 2020). Also, the pathogens and their probability in affecting specific tissues of human are listed in Table 1.

The microbes under stressful environment produce biofilm, which comprises extra polymeric substances (EPSs), extracellular DNA, and protein transporters. The worst factor of biofilm was drug resistance and higher virulence factor expression in host systems (Valliammai et al., 2021). Biomedical devices developed with conventional materials will be prone to protein fouling until nonfouling biomaterials become accessible and extensively utilized. Biomedical devices are often coated with a layer of glycoproteins and proteins, including albumin, fibrinogen, fibronectin, and immunoglobulins, many of which act as binding ligands for receptors on colonizing bacteria or invading mammalian cells. The surface chemistry of any material (e.g., ceramic, glass, titanium, and polyether urethane) is immediately affected by macromolecule adsorption. The transfer of microorganisms to the substrate surface interface, which is mediated by a mixture of transport mechanisms, continues biofilm production (Xu, Bauer, & Siedlecki, 2014). Microbial cells stick to a substratum surface through quasi or specific receptor: ligand-binding pathways. Through a mechanism known as quorum sensing, adherent cells upregulate the release of particular cell signal molecules that coordinate community-wide phenotypic responses within minutes of attachment. Upregulation of virulence genes and release of extracellular polymers are examples of population responses. By absorbing soluble resources and attracting other species of bacteria or mammalian cells, biofilms continue to develop. Depending on certain quorum cell signaling, portions of the combined bacterial and platelet biofilms can detach and then be transported downstream (Jothi, Hari Prasath, Gowrishankar, & Pandian, 2021).

Premature biofilm in host tissues and implants is hard to treat with antimicrobials and common antibiotics used to treat nosocomial infections. Biofilms are proven to cause chronic infection in healthcare maintenance, particularly those implantable medical devices including cardiovascular defibrillators, intravascular catheters, neurosurgical implants, orthopedic implants, and urinary catheters. Based on the site of infection, the biofilm formers and polymicrobial biofilm

Understanding Microbial Biofilms. https://doi.org/10.1016/B978-0-323-99977-9.00030-2

TABLE 1 Infections-causing pathogens and associated diseases.

S. no.	Types of infections	Causative pathogens	Infection-related diseases	References
1	Chronic lung infections	*Burkholderia cepacian, Haemophiles influenzae, MRSA,* and *Pseudomonas aeruginosa*	Cystic fibrosis, chronic pneumonia, lung cancer, interstitial lung disease and chronic bronchitis	Blasi (2018)
2	Dental plagues	*Bacteroides melaninogenicus, C. albicans, Lactobacillus* spp., *Porphyromonas gingivalis, Staphylococcus aureus,* and *Streptococcus mutans*	Peri-implantitis, demineralization and root canal infection	Priya, Selvaraj, Divya, Karthik Raja, and Pandian (2021)
3	Ear infections	*Streptococcus pneumoniae* and *Haemophiles influenzae*	Acute otitis media and chronic suppurative otitis media	Hoberg et al. (2017)
4	Eye infections	*Borrelia burgdorferi, Chlamydia* spp., *Gonorrhea* spp., *Haemophiles influenzae, Herpes Simplex Virus, Klebsiella pneumonia, Mycobacterium tuberculosis, P. aeruginosa, Staphylococcus* spp., and *Treponema pallidum*	Conjunctivitis, keratitis, stye, and uveitis	Teweldemedhin, Gebreyesus, Atsbaha, Asgedom, and Saravanan (2017)
5	Gastrointestinal infections	*E. coli, Salmonella* spp., Norovirus, Rotavirus, *Giardiasis,* and *Cryptosporidiosis*	Abdominal cramps, bloody diarrhea, gastroenteritis, and cryptosporidiosis	Pawlowski, Warren, and Guerrant (2009)
6	Heart infections	*Staphylococcus* spp. and *P. aeruginosa*	Infective endocarditis, pericarditis, and myocarditis	Murillo et al. (2016)
7	Orthopedic infections	*Bacillus subtilis, Brucella* spp., *Escherichia coli, Mycobacterium tuberculosis, Proteus mirabilis, P. aeruginosa, S. aureus,* and *Streptococcus* spp.	Periprosthetic joint infection; chronic osteomyelitis; implant-associated osteomyelitis	Singh, Tan, and Ahmad (2021)
8	Skin infections	*Staphylococcus* spp., *Streptococcus* spp., *P. aeruginosa,* and anaerobes	Cellulitis, septicemia, Diabetic neuropathy, and necrotizing fasciitis	Falcone et al. (2021), Mala, Rani, Prasath, and Celsia (2019)
9	Urinary tract infections	Anaerobes, *Enterococcus* spp., *Escherichia* spp., *Lactobacillus* spp., *Proteus mirabilis, Corynebacterium* spp., *S. aureus,* and *Staphylococcus epidermidis*	Urethritis, cystitis, and pyelonephritis	Flores-Mireles, Walker, Caparon, and Hultgren (2015)

population vary. Center for Disease Control (CDC) conformed that polymicrobial biofilms were considered more chronic due to their antimicrobial resistance up to 1000 times the normal biocidal concentration. The bactericidal/fungicidal drugs were not allowed into the matured biofilm EPS, allowing only essential nutrients to pass through it. Also, the microbes in a biofilm are considered to share the antimicrobial resistance genes, which thread the researchers to find a novel lead for polymicrobial biofilms (Centers for Disease Control and Prevention, 2019)

A medical device is considered implantable if that is partially or completely implanted into the human body, either surgically or therapeutically, and is meant to stay there after the process. Implantable medical devices are used to detect physiological responses in real time or to control physiological organs. Many researchers have recently concentrated on the creation of implanted real-time vital monitoring devices that run continuously in subsecond periods, thanks to the super miniaturization of digital circuits and mechanical constructions. Medical systems may transition from reacting to accidents after they happen to self-managing events before they happen, thanks to implanted real-time vital monitoring devices (Joung, 2013). The Medical Devices Bureau of Health, Canada, has identified four categories of medical devices depending on the quantity of control required to ensure the device's safety and effectiveness (Khatoon, McTiernan, Suuronen, Mah, & Alarcon, 2018):

1. **Class IV:** Poses the greatest danger and therefore is subject to rigorous testing and regulatory clearance prior to release. Implantable cardioverter defibrillators, pacemakers, and cardiac assist devices are examples.
2. **Class III:** This class poses a higher danger to the patient. This category contains bone cementing and hip implants, as well as hemodialysis machines and surgical meshes.
3. **Class II:** Require the products manufacturer to provide a declaration of products efficacy and safety. Contact lenses, ultrasound equipment, and medical catheters are all in Class II.
4. **Class I:** Poses a modest danger to patients and does not necessitate a license or the strictest regulatory standards. Surgical devices and dental materials are included under Class I. In addition to the stability and lifetime criteria of implanted biodevices, pathogen-free manufacture and storage of bio-medical implants have a greater influence on pharmaceutical manufacturing.

Implanting a contaminated product in living systems leads to the formation of biofilm and related serious conditions due to chronic pathogenicity product. Though the microbial adhesion and biofilm formation on the surfaces of implants were unavoidable, therapies were available by inhibiting biofilm adhesion and microbial growth initiation. The use of antibiotics was suggested not to use by the World Health Organization (WHO) due to the drug resistance formation, which in case decreases the curability. Rather than administering antibiotics, antibiofilm/antimicrobial compounds were employed to inhibit microbe invasion. Also, surface modification of implants and replacing the implant material with antiadhesive features were employed to prevent adhesion (Costerton, Montanaro, & Arciola, 2005). The medical implants and their related aggressive microbial biofilm promoters were covered in this chapter. In addition, this chapter also describes the current research and therapies that can be used to fight biofilms in medical implants. Some of the medical devices and its biofilm infection characteristics are listed in Table 2.

2. Types of medical devices and their associated biofilm infections

2.1 Intravascular implants

2.1.1 *Central venous catheters*

A central venous catheter (CVC) is a device that is introduced into a major vein and used to administer parenteral nourishment, blood products. CVC can also be utilized for hemodialysis and central venous pressure measurement. Nontunneled catheters, peripherally placed central catheters, tunneled catheters, and totally implantable venous access ports are the most common forms of CVC. The use of CVC has a high risk to form bold stream colonization and biofilm-associated infections. In 1982, *S. epidermidis* was found to cause CVC-associated biofilm infections. After this founding, many studies were carried out to find the potential biofilm formers on CVC. Two organisms, coagulase-negative *S. aureus* and *Candida* spp., were found to form potent biofilms in CVC-associated blood stream infection lately. Many factors were reported to enhance the biofilm attachment to CVC (Gominet, Compain, Beloin, & Lebeaux, 2017).

1. The formation of a conditioning film prior to the formation of the conditioning film.
2. Surface finish, wettability, and electrostatic attraction are all factors to consider when implanting.
3. Presence of adhesins such as flagella and fimbriae.
4. Surface-associated proteins that support anchoring the pathogen onto the catheter surface.

2.1.2 *Midline catheters*

Midline catheters can be used to administer antimicrobial/antibiofilm drugs, fluid replacement, and analgesics via peripheral vein. It's a vascular support device that goes into the patient peripheral vein in the upper arm, moreover with the basilic, cephalic orbrachial veins, through the internal tip point near the axilla and distal toward the shoulder. Infusions and short-term intravenous therapy are common uses. Midlines are designed to be implanted using ultrasound guidance. Ultrasound guidance allows clinicians to view their anatomical target and accompanying arteries, nerves, and tissues, as the needle progresses to the procedure's endpoint, rather than operating blindly. Standard midline catheters can be anything between 10 and 20 cm in length, with a single or double lumen (Adams, Little, Vinsant, & Khandelwal, 2016). This type of catheter significantly decreases the microbial attachment and biofilm formation upon attachment.

2.1.3 *Pulmonary artery catheters*

The addition of a catheter into a pulmonary artery is referred to as pulmonary artery catheterization (PAC). The pulmonary artery catheter provides for direct, simultaneous assessment of pressure levels in the left atrium's filling pressure, pulmonary artery, right atrium, and the right ventricle. Its primary use is diagnostic resolutions, and also, it is used to identify sepsis, track therapy and to assess the medication effects. The PAC can induce life-threatening problems such as arrhythmias, bleeding, pneumothorax, pseudoaneurysm development, pulmonary artery rupture sepsis, and thrombosis. Among other intravenous catheters, PAC projects higher microbial attachment and biofilm adhesion surfaces. A broad class of pathogens was involved in biofilm formation in PAC,

TABLE 2 Infection and pathogenicity that inflict damage on implantable devices.

S. no.	Medical devices	Potential biofilm formers	Characteristics of biofilm virulence	References
1	Artificial breasts	*S. aureus* and *S. epidermidis*	Biofilm maturation causes irritation and prolonged inflammation, resulting in capsular contracture.	Wikström, Isacsson, Nilsson, and Ahnesjö (2018)
2	Cardiac implantable devices	*P. aeruginosa, S. aureus, S. epidermidis,* and *S. hominis*	1. Reduce sensors ability on sensing devices 2. Endocarditis, pericarditis, and myocarditis	Liang, Beekmann, Polgreen, and Warren (2016)
3	Venous catheters	*B. cereus, Candida* spp., *P. aeruginosa, S. aureus,* and *S. epidermidis*	1. Biofilm dispersion from catheters causes catheter-related bloodstream infections. 2. Biofilm cells will survive host immune system and antimicrobials by producing virulence factors.	Rodríguez-Calero et al. (2020)
4	Contact lenses	*C. albicans, P. aeruginosa,* and *Staphylococcus* spp.	1. Biofilm in lenses causes microbial keratitis and affects the corneal conditions. 2. Pathogens retain in the inserter hole for prolonged periods and forms pathogens may be retained in the inserter hole for extended periods and forms acute Conjunctivitis.	Bispo, Haas, and Gilmore (2015)
5	Dental prosthetics	*Candida* species, *Capnocytophaga, Actinomyces* spp., *Fusobacterium,* and *Streptococcus* spp.	Neighboring tooth decay and root canal infections are major consequences due to biofilm infections in titanium implants.	Yeh, Lu, Chang, and Ge (2019)
6	Hemodialyzers	*P. aeruginosa, S. marcescens,* and *Staphylococci* spp.	Dispersal of microbial endotoxins produced by matured biofilm throughout the host systems via dialyzers	Edens et al. (2017)
7	Orthopedic prosthetics	*E. faecalis, P. acnes, S. aureus,* and *S. epidermidis*	1. Acute and chronic osteomyelitis due to poly-species biofilm formation on titanium/stainless steel implants 2. Demineralization of bones near the implanted site due to high level of biofilm virulence secretion	Campoccia et al. (2016)
8	Urinary catheters	*Candida* spp., *E. cloacae, E. coli, Enterococcus* spp., *Klebsiella* spp., *P. aeruginosa,* and *P. mirabilis*	1. Contently biofilm in catheters eradicate during catheters in-out cycles and infect the urinary bladder and nearby tissues 2. Creates acidic environment due to ammonia production which is suitable for the symbiotes to secrete virulence factors	Dadi, Radochová, Vargová, and Bujdáková (2021)
9	Ventilator	*A. baumannii, Candida* spp., *Enterococcus* spp., *Klebsiella* species, *P. aeruginosa,* and *S. maltophilia*	Endotracheal intubation and provide a constant flow of pathogens in critically sick patients	Diaconu, Siriopol, Poloşanu, and Grigoraş (2018)

which includes *A. baumannii*, *C. albicans*, *E. coli*, *K. pneumoniae*, *P. aeruginosa*, *S. mutans*, *S. aureus*, and *S. epidermidis*. Also, there are numerous numbers of controversies related to PAC and its associated biofilm infections (Alisiddiq & Butler, 2021).

2.2 Cardiovascular implants

2.2.1 Coronary stents

A coronary artery stent is a thin metallic mesh tubular structure that expands within the artery. Angioplasty is frequently followed by the placement of a stent. It aids in the prevention of the artery shutting up again. A drug-eluting stent contains medication that prevents the artery from shutting in the long run. Despite antibiotic therapy and/or surgical removal, vascular stent infectious diseases are an uncommon complication with a significant death rate; according to current research, fatality rates can exceed 40% (Wojcik & Kang, 2022). The apparatus reuse such as inadequate aseptic practices, guidewires, catheters, and balloons during the treatment are the most common causes of stent infections. These bacteria are the most commonly discovered in stent infections, and they come from the healthcare environment and human skin: There have also been reports of *Staphylococcus* spp., *Streptococcus* spp., *Pseudomonas* spp., fungi, and in rare cases, fast-developing mycobacteria (das Neves, Mortari, Schwartz, Kipnis, & Junqueira-Kipnis, 2019).

2.2.2 Defibrillators and associated devices

An implantable cardioverter defibrillator (ICD) is a rechargeable battery pack device that is implanted in the chest to identify and hold back abnormal heart pulse rate (arrhythmias). An ICD constantly monitors the heartbeat and, when necessary, provides electric shocks to restore a normal cardiac rhythm. The clinical chronicity of infection in ICD leads to mortality with a higher rate. It's critical to get a clear diagnosis of coronary implantable device (CID) infections before starting antimicrobial therapy. For identifying CID infections, traditional pocket swab and tissue specimens have limited sensitivity and specificity. Moreover, conventional cultures' diagnostic yield may be hampered by past antibiotic/antimicrobial therapy (Tom, 2016). Microorganisms in CID are *S. epidermidis*, *Corynebacterium striatum*, coagulase-negative, which forms chronic biofilms, which directly cause life-threatening minutes.

2.2.3 Implantable cardiovascular monitors

ICMs are miniature electrophysiology devices that are used to monitor a person's cardiac rate activity over time in order to detect arrhythmias. The device might do away with the necessity for a bulky exterior Holter monitor that needs the patient to wear wire connections. ICMs are small devices that can be implanted underneath the skin throughout an office visit and store cardiac data lasting up to 4.5 years. The proper use of cardiovascular implanted electronic devices in selected participants has been linked to improved survival and quality of life. Permanent pacemakers and ICD are the newest "breed" to join an existing and increasing group of permanent pacemakers and ICD. The quantity of cardiovascular devices transplanted each year is increasing at a rapid rate (Serhani, elKassabi, Ismail, & Navaz, 2020). Unfortunately, the advantages of these devices might be overshadowed by infectious biofilm complications because of the invasiveness of the implant surgeries necessary for device installation and various comorbid illnesses in device patients. Infection is a devastating consequence that necessitates the complete removal of both the infected device and systemic antimicrobial therapy.

2.2.4 Mechanical heart valves

Materials such as titanium and carbon are used to make mechanical heart valves. They are typically made up of two leaflets and a steel ring encircled by a knitted fabric ring that is sewed onto the heart in lieu of the native valve. Arterial and mitral replacement operations are offered in a variety of models. Mechanical valves have the benefit of being extremely durable. These valves, on the other hand, create a substrate on which blood blockages can easily form. As a result, everyone who has had a mechanical valve installed must take blood-thinning medicine for the rest of their lives; drug such as warfarin, to avoid clots that can lead to stroke or heart attack. Microorganisms can adhere to and develop biofilm on components of heart valves and the heart's surrounding tissues, causing prosthetic valve endocarditis. *Streptococcus* spp., *S. epidermidis*, *S. aureus*, *Enterococci* spp., and *Candida* spp. are most prevalent organisms that promote this condition (Harris, Croce, & Cao, 2015).

The biofilm pathogenicity on coronary stents and other heart implants was illustrated in Fig. 1. The detached highly resistant pathogens from biofilms invade the other tissues via bloodstream flow from heart. The mentioned protein MSCRAMM in Fig. 1 was proven as a highly adhesive molecule produced by *S. aureus*. Tissue damage occurs when the mechanical heart valve is implanted, and flowing platelets and fibrin tend to accumulate where the valves are joined. Microbes also have a strong propensity to inhabit these locations. Microbial biofilms develop around the prosthesis that adheres to the valve to the tissue instead of on the valve itself (Harris et al., 2015).

2.3 Neurosurgical implants

2.3.1 Intracranial pressure devices

Intracranial pressure (ICP) monitoring is used to treat a variety of neurological diseases, including severe traumatic brain damage, stroke, and brain bleeding. ICP tracking is a diagnostic technique that helps the physician identify whether your symptoms are caused by high or low CSF fluid pressure. The test uses a tiny pressure-sensitive probe placed through the skull to directly assess the severity in the head pressure. ICP transducers (ICPTs) pass through the major brain-protecting layers (blood-brain barrier, meninges, the skull, and the scalp). This potential exposure paves a way for bacterial contamination to invade the brain and cause chronic illness. External ventricular drains are

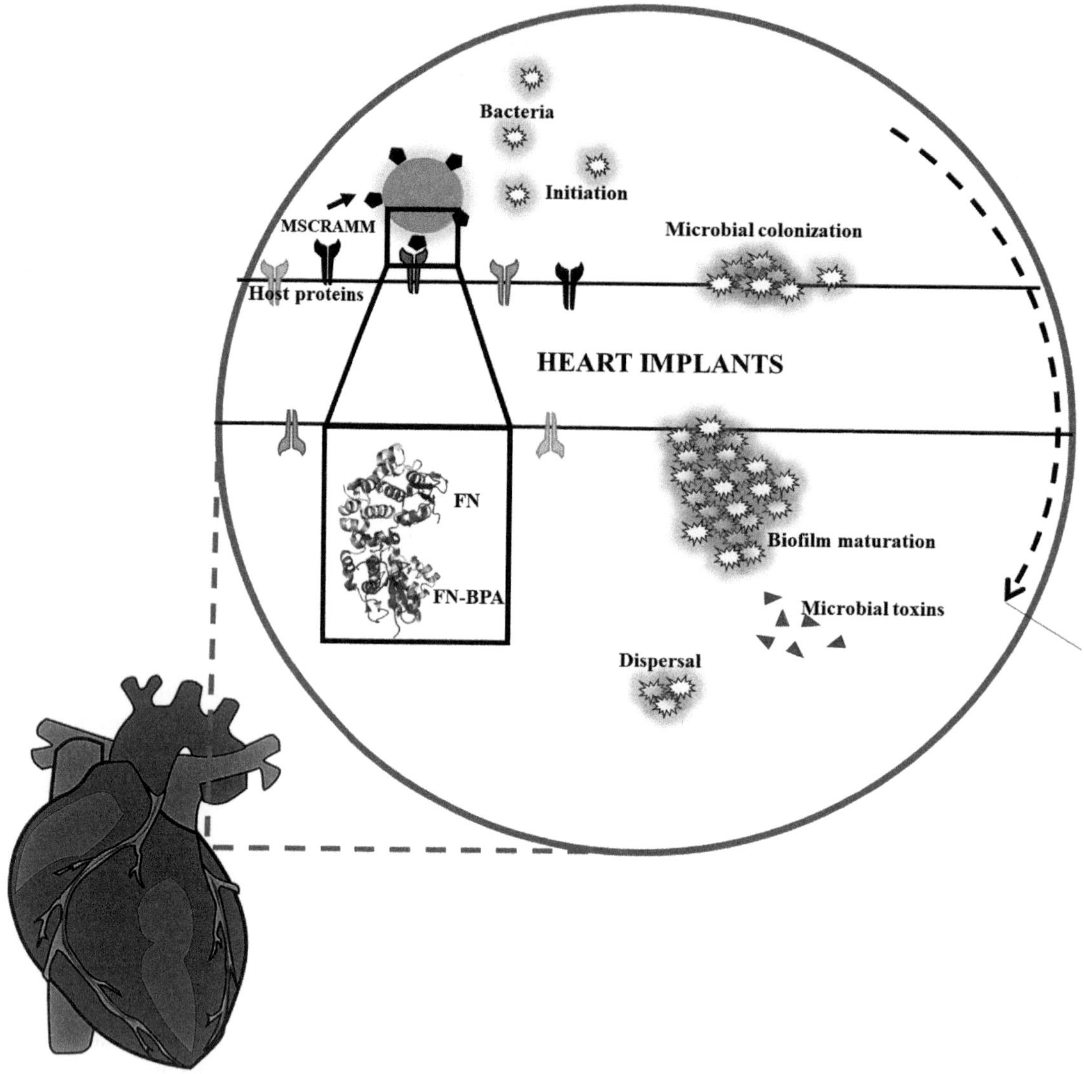

FIG. 1 Biofilm pathogenicity in heart implant and associated devices. Microbial biofilm model of *S. aureus* in metal substrate of heart implant. The biofilm forms after attachment of planktonic cells of *S. aureus* to implant substrate via adhesive molecule MSCRAMM. The pathogenicity starts after dispersion of chronic biofilms into the bloodstream.

associated with infections and microorganisms have previously been shown to travel down external ventricular drains, typically as biofilms, from the skin to the brain (Kawoos, McCarron, Auker, & Chavko, 2015).

2.3.2 Neurological stimulators

Spinal cord stimulators are composed of thin wires and a compact battery pack that resembles a pacemaker. The wires are put in the epidural space, which is the region between both the spine and the vertebrae, and the pacemaker is positioned beneath the skin, generally toward the buttocks or belly. Patients using spinal cord stimulators can use a remote control to emit electrical impulses when they are in pain (Sdrulla, Guan, & Raja, 2018). The antenna, as well as its remote control, is located on the exterior of the body. Biofilm in this device is associated to surgical infections caused by ESCAPE pathogens. There are three types of spinal cord stimulators:

1. A battery-operated neural code stimulator is also known as an implanted pulse generator (IPG). Because of its reduced electrical output, this device may be a useful alternative for persons who have discomfort in only one body area.
2. Rechargeable IPGs function similarly to traditional IPGs, with the exception that the battery may be recharged without the need for additional surgery. Because the electrical signal may reach deeper, this may be a better option for persons who have discomfort in their lower back or one or both legs.

3. The radiofrequency stimulator is powered by an external battery. Because of the device's strength, it may be beneficial for persons with lower back and leg discomfort (Garrigos, Farid, Bendel, & Rizwan Sohail, 2020).

2.3.3 *Ommaya reservoirs*

An Ommaya reservoir is a kind of intraventricular catheter that can be used to aspirate cerebrospinal fluid or inject medications (such as chemotherapy) further into cerebrospinal fluid. It is made up of a catheter implanted in one of the lateral ventricles and a reservoir inserted beneath the skull. Intrathecal medication administration is often used to treat brain tumors, leptomeningeal illness, and lymphoma. Ommaya reservoirs can lead to difficulties such as device bleeding, convulsions, and malfunction. The Ommaya reservoir, like all the other foreign medical components, is susceptible to microbial adhesion, colonization, and chronic biofilm-associated infection, followed by subsequent spread of the invasive infection into the CNS compartments, outer tissue, and soft tissues (Szvalb et al., 2014).

2.3.4 *Ventricular shunts*

Ventriculoperitoneal shunting is a procedure that removes excess CSF from the brain's cavities (ventricles). Hydrocephalus, which literally means "water on the brain," is a neurological condition that may be extremely severe. This CSF is routinely produced and reabsorbed at specific rates, and if these rates are out of balance, an excess in fluid can result in pressure and neurologic symptoms. CSF is generally found in the ventricles, which are chambers in the brain (Benachinmardi, Ravikumar, & Indiradevi, 2017). A ventricular perfusion system (VPS) is used to drain excess fluids from the ventricle. This is accomplished by inserting a catheter into the brain's ventricle, which flushes the fluid toward the abdomen. Ventricular blockage, infection, intraventricular bleeding, and overdrainage of CSF are all common consequences of ventriculoperitoneal shunt implantation. The most dangerous is infection, which frequently needs immediate treatment. The most frequent organism is coagulase-negative *Staphylococcus* spp., with an infection incidence ranging from 2% to 27% (Kumar et al., 2016).

2.4 Orthopedic implants

2.4.1 *Fracture fixation devices*

Fracture fixation devices are often evaluated in one of four ways: cadaveric specimens, finite element analysis, mechanical analogue bones, and synthetic materials or structures put up to imitate fundamental body stress conditions. These tests were designed to mimic the in situ situation of implantation. The number of difficulties and side effects associated with this type of surgery is also rising (Hayes & Richards, 2019). The emergence of a deep wound infection is one of the most serious consequences following orthopedic trauma surgery (SSI). SSIs are usually caused by perioperative surgical field contamination, which might manifest themselves early, late, or indefinitely (>2 years). It can also happen as a result of unrelated bacteremia causing secondary pathogen transport to the surgical site. Interestingly, a variety of risk variables have been investigated, along with the severity of the original soft tissue damage and bone injuries, patient comorbidities (for example, alcoholic, diabetes mellitus, and overweight), and the use of exterior fixation devices previous to internal fixation. Biofilm formation is primarily addressed about concerns in orthopedic surgery, wherein transplanting gadgets into the patient's psyche is the norm and biofilm creation on the surfaces of implants is so frequent. Biofilms have recently been discovered on fixation hardware as well as grafts/soft tissue material used in cartilage, ligamentous, and tendon restorations (Frey, Hügle, Jick, Meier, & Spoendlin, 2016). *A. baumannii*, *E. coli*, *E. faecalis*, *K. pneumoniae*, *P. aeruginosa*, *P. mirabilis*, *Streptococcus* spp., *Propionibacterium*, *S. epidermidis*, *S. viridans*, and *S. aureus* are among the bacteria. Certain fungal species (for instance, *Candida* spp.) are also frequently linked to fungal and multispecies biofilm communities.

2.4.2 *Joint prostheses*

Joint discomfort and impairment can be caused by a wide variety of conditions, leading patients to seek joint replacement surgery. Injury to the cartilage that covers the ends of the bones (articular cartilage) is a common source of joint pain, which can be caused by osteoarthritis, a fracture, or some other ailment. Prosthetic joints are more susceptible to infection than normal joints. They are commonly caused by perioperative inoculum of bacteria into the joint or postoperative bacteremia caused by invasive apparatus, pneumonia, falls, skin infection, and UTI. Two-thirds of patients develop joint biofilm infections within a year of surgery. *S. aureus*, broad class of Gram-negative organisms, anaerobes, and other polymicrobial biofilm flora are the most frequent cause within the first few months following surgery. Propionibacterium acne is particularly prevalent in contaminated prosthetic shoulder joints, and detection may take up to 2 weeks. In around 5% of instances, *Candida* spp. infect prosthetic joints (Vad, Adin, & Solomon, 2021).

2.4.3 *Spinal implants*

Spinal implants are utilized to stabilize a spine that has become unstable due to trauma or displaced vertebrae caused by degenerative intervertebral disks. The anteriorly technique might be used to provide surgical stability. A plates or rods with screws implanted into the vertebra join and fix neighboring vertebral bodies from the anterior approach. Rods attached to pedicle bolts are used to support the spines from the back. Titanium alloys and polyetheretherketone (PEEK) have been used to make spinal implants. In one type of spinal implant, cobalt-chromium alloys have also been used (disc replacements). The diversity of spinal implants is substantial when compared with other types of orthopedic implants, and design conception in this field is vast. The aim of therapy for spine implant-associated biofilm infections is for the patient to combat infection, maintain spinal stability, mobile, and pain-free (Danison, Lee, & Panchal, 2017). Most researchers concur that in early-onset infectious biofilm-associated diseases with well-fixed and operational cervical spine implants, the machine should then be debrided and maintained, despite a few other studies indicating complete elimination of the implants to accomplish cure of biofilm infection. High treatment failure rates have been recorded with implant retention in late-onset stent infection, which often presents as an indolent, slowly progressing illness. Installations should be removed in late-onset diseases with loosened implants and bone fusion. Given the significant morbidity associated with full removal, it is still debatable whether simply the loose portions or the entire implant should be replaced (Khatoon et al., 2018).

2.5 Urological—Inflatable penile implants

A urinary catheter is a tubular, semiflexible catheter that collects urine from the urinary bladder and transports it to a drainage bag. Urinary catheters are offered in a variety of sizes, dimensions, and styles. Rubber, plastic (PVC), and silicone are the substrate for building urinary catheters. Urinary catheters are typically needed when a person is unable to empty their bladder. Urine can accumulate in the bladder if it isn't drained, causing pressure on the kidneys. The pressure can cause renal failure, which is serious and can result in irreversible kidney damage. Urinary catheters may be required for a long time or permanently in the elderly and individuals with a chronic disability or severe illness. Indwelling, external, and short-term catheters are the three primary categories of catheters (Feneley, Hopley, & Wells, 2015).

2.5.1 *Indwelling catheters*

A catheter that is indwelled in the bladder is known as an indwelling catheter (Foley catheter). This kind is suitable for both short- and long-term use. An indwelling catheter is inserted further into bladder through the urethra. Also, the catheter is sometimes inserted into the bladder through a small hole in the abdomen. To restrict the catheter from falling out of the body, water is pumped into a smaller balloon at the end of the catheter. The balloon must be deflated by removing the water whenever the catheter needs to be removed from the patient.

The sensitivity of inert catheters material to bacteria colonization is linked to the etiology of CAUTI (Urinary Catheters, 2022). Biofilm pathogenicity of the polymicrobial biofilm deposited on the urinary catheters is illustrated in Fig. 2. The biofilm formed during the immobilized period of catheter was disintegrated during the urine outfall period and got inside the bladder during the catheter relaxation time for producing more virulence. The creation of a conditioning film of host urine components, such as electrolytes, proteins, and other organic compounds, is the initial stage in biofilm development on a urinary catheter. The surfaces of the indwelling catheter can be transformed using this conditioning coating, and any antiadhesive qualities can be neutralized. Free-swimming bacteria employ hydrophobic and electrostatic interactions, as well as flagella, to adhere to the surface. The creation of loosely packed three-dimensional structures with fluid channels between them is directed by cell-to-cell communication, allowing nutrients and wastes to be exchanged. The cycle is completed when individual organisms are detached from the biofilm (Niveditha, Pramodhini, Umadevi, Kumar, & Stephen, 2012).

2.5.2 *External catheters*

A condom catheter is a tube that is inserted into the body from the outside. It's usually required for males who have major functional or mental difficulties, such as dementia, but don't have urinary retention issues. The penis head is covered with a condom-like device. The condom device is connected to a drainage bag by a tube. In comparison to indwelling catheters, these catheters are often more comfortable and have a decreased risk of infection. Condom catheters are normally replaced every day; however, certain kinds are meant to last longer. These may cause less skin irritation than normal catheters, which must be removed and reapplied on a regular basis. Due to the reduced time period (>1 day) of exposure of this type of urinary catheters, there is no biofilm formation on the catheter implanted site (Anger et al., 2019).

2.5.3 *Short-term catheters (intermittent catheters)*

After surgery, a person may only require a catheter for a brief time until about the bladder empties. The short-term catheter must be removed after the bladder has emptied. This is referred to as in and out catheter by healthcare professionals. People are taught to deploy the catheter alone or with the help of a healthcare professional. This can be done either via urethra or a catheter's hole in the lower stomach. There is no biofilm formation on the catheter implanted site due to the shorter exposure duration of this kind of urine catheter (Saint & Chenoweth, 2003).

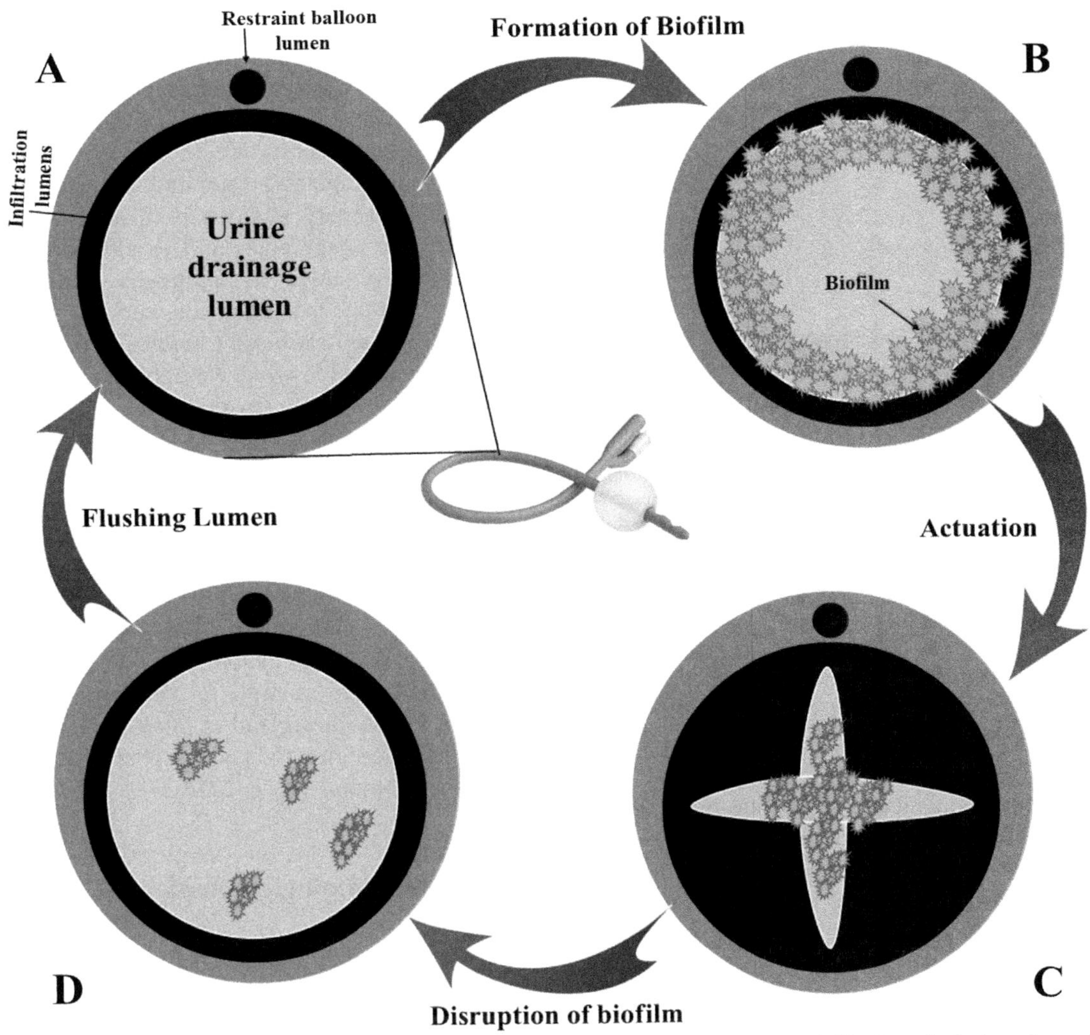

FIG. 2 Cross section of urinary catheter demonstrating the biofilm pathogenicity when at purpose. The biofilm forms in the inner layer of urinary catheters due to the contact of the inner layer with nutrient-rich sources. The microbial cells initially bind to the catheter and develop to matured biofilm. After development due to the compression and expansion of the catheter during work, the biofilm disrupts and pathogenesis occurs toward the inner organs such as bladder and kidney. (A) Cross section of urinary catheter. (B) Biofilm matures on the inner walls of catheter. (C) Actuation disrupts the formed biofilm in the inner walls. (D) Distrusted biofilm flushes onto the nearby orangs and causes infection.

2.6 Dental implants

The life quality and health of a patient can be considerably improved by replacing a missing tooth with a synthetic dental implant. Dental implant systems typically contain a dental implant body and abutment, as well as an abutment anchoring screw. In order to replace the tooth's root, the implant placement body is surgically implanted into the jawbone. The abutment anchoring screw attaches the dental implant abutment to the implant body and extends through the gum tissues into the mouth to maintain the connected artificial teeth (Singh, Kumar, Anwar, & Chand, 2015).

For each type of dental implant, prosthodontists can choose between a number of coatings, sizes, and connector options. The representation in Fig. 3 indicates the impact of dental implant biofilm contamination to the nearby natural teeth. The biofilm of the implant surfaces matures and protrudes toward the natural tooth. Due to the continuous exposure of biofilm bacterial colonization, deep tissue infection occurs forming natural tooth decay in a higher proposition. There are a variety of ways to place implants, they usually fall into two groups, namely Endosteal (Endosseous) Implants and Subperiosteal Implants (Singh, Kumar, et al., 2015).

2.6.1 *Endosteal (Endosseous) implants*

Endosteal implants are the most commonly used type of dental implant. They are something other than a linking or detachable denture based on circumstances. Screw (connected), cylinder (soft), and bladed endosteal implants are provided based on the seriality and locations. Endosteal

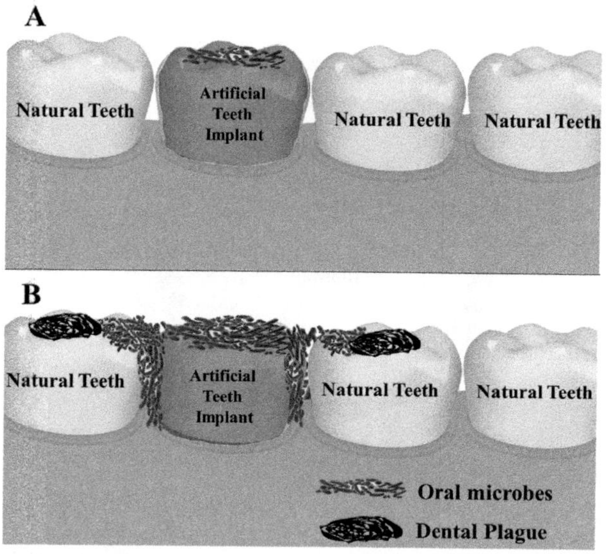

FIG. 3 The rough surfaces of titanium tooth implanted near the normal health teeth causes teeth decays due to biofilms and associated virulence factors production. (A) Biofilms from the contaminated tooth implant develop and affect the nearby health teeth. (B) Unlike dental metal implants, dementalization occurs in the healthy teeth due to deep tissue migration of biofilms and causes dental plagues.

implants have a reputation for being the most stable and natural-feeling options available. Endosteal implants start by inserting a titanium screw into the jaw bone, and it serves as a prosthetic root. The individual must wait for the tissue and bone around the root to recover before starting therapy. It's not unusual for this process will take many such weeks. The very first stage in implant placement is shafting the endosteal implant further into jawbone, which also requires adequate jawbone integrity and strength. Some patients may not have enough bone to sustain an endosteal implant if you have a thin jawbone ridge normally and one that is shorter or constricted and depleted due to the trauma or illness. A subperiosteal implant could be a possibility in this scenario (Kligman et al., 2021).

2.6.2 Subperiosteal implants

Subperiosteal implants were first used to secure dentures in people who didn't have enough skeletal height. This metal implant component protrudes through the gum tissue to support the dental implant. Subperiosteal implants were being embedded underneath the gums on the jaw bone, with the metal graft element elongating through the gum tissue to help the dental implant stability. The entire treatment process takes only two sessions and is usually significantly less time-intensive than an endosteal implant. Subperiosteal implants will not have the same degree of stability as standard implants because they do not penetrate the jaw bone but instead they lay on top of it and are fixed in place principally with the help of soft tissue (Abraham, 2014).

The microbes' major source of nutrition is the saliva and associated mucus from the patient. The acquired pellicle, a thin coating that coats the tooth and coats the enamel within minutes of brushing, is made up of salivary proteins. Proteins and glycoproteins are molecules that bind to the surface of the implants, restorations, tooth, and other surfaces. Their primary function is to increase oral bacteria adherence and coaggregation. The adhesion of bacteria to the pellicle is aided by particular surface molecules (adhesins) on the bacterial cell surface, mostly lectins. Microbes such as *S. mutans*, *C. albicans*, *S. sobrinus*, *S. viridans*, *S. mitis*, and *S. oralis* were found in virulent biofilms on dental caries. *S. aureus* exhibits a strong adherence to titanium surfaces and has been linked to suppuration and bleeding on probing. On the surface of *S. aureus*, many adhesins bind with a variety of host proteins, including collagen vitronectin, fibrinogen, fibronectin, and laminin. Socransky characterized the red and orange complex species as having a high prevalence in the microbiota of the implant in peri-implantitis. This microflora is dominated by red complex species including *T. forsythia*, *P. gingivalis*, and *T. denticola*, as well as orange complex species including *P. intermedia* and *F. nucleatum*. *C. albicans* has already been reported to have a higher level of attachment to titanium implants (Dhir, 2013).

2.7 Ophthalmological implants

2.7.1 Glaucoma tubes

Glaucoma sufferers typically have trouble draining aqueous humor through natural outflow channels (trabecular meshwork and Schlemm's canal). Elevated intraocular pressure is caused by an imbalance between the rate of aqueous humor production and outflow from the eye, which is a key risk factor for optic nerve injury and loss of vision. Glaucoma drainage devices (GDDs) are commonly implanted in individuals with refractory or inflamed glaucoma who are resistant to standard pharmacological or surgical treatments. These devices have a number of benefits over traditional trabeculectomy, including improved IOP control, less postoperative problems, and simplicity of operation. A small polymeric microscale tube links the forward chamber to a tiny silicone plate for draining of excess aqueous humor in traditional GDDs. GDDs are made of hydrophobic polymer materials including polydimethylsiloxane, polypropylene, and polymethylmethacrylate, which have a high affinity for interstitial proteins such as albumin, fibrinogen, and immunoglobulin, which adsorb onto the device surface minutes after implantation. Because most GDDs feature a drainage tube, various biofouling agents might easily obstruct the microscale channel (Raksha, Gangashettappa, Shantala, Nandan, & Sinha, 2020).

2.7.2 Lacrimal stents

Canalicular and bicanalicular implants are the two kinds of lacrimal devices. Mono-canalicular stents typically travel through a canaliculus to the lacrimal sac and nasolacrimal duct, whereas bicanalicular stents travel through both canaliculi to the duct and nasal cavity, where they are connected in a loop. The loop may occasionally only pass through the canaliculi through the common canaliculus. The operating extremities of bicanalicular silicon lacrimal intubation sets morphologically feature a metallic probe-like rod (bodkin) that may be taken out via the inferior meatus externally. It can be done under visualization with a hard nasal endoscope or metal to metal contact using a hook or Kelly's curved artery forceps. Mono-canalicular stents include Monoka stent, Monoka-Crawford stent, Masterka stent, and Disposable Punctum Dilator & Plug Inserter. Bicanalicular varieties include Crawford stents, Ritleng intubation device, O'Donoghue, Guboir&Bika (bi-canalicular) stents, and self-retaining bicanalicular stents. *Acanthamoeba*, fungi, *P. aeruginosa*, *Serratia* spp., and *S. aureus* (also other *Staphylococcal* spp.) cause device-associated biofilm infections such as endophthalmitis, keratitis, scleral buckle infection, lacrimal system infections, and periorbital infections and spread to nearby organs (Ricca, Kim, Chahal, & Shriver, 2018; Singh, Kamal, & Sowmya, 2015).

3. Various strategies to combat biofilms on medical devices

3.1 Biofilms in urinary catheters

As a result of the problems with CAUTI, various antimicrobial therapeutics have been developed to combat biofilm formation using various methods. When silver nanoparticles were used to formulate silver-coated catheters, many studies predicted a broad-spectrum antimicrobial effect, and all studies showed increased bacteriostatic activity (Wilks, Koerfer, Prieto, Fader, & Keevil, 2021). Silver interacts with the bacterial membrane to puncture it, then reacts with DNA to kill the microbes that form biofilms.

Fig. 4 illustrates the antimicrobial drug loading/surface coating on the catheters to improve antibiofilm efficiency (Belfield, Chen, Smith, Ashraf, & Bayston, 2019). Also, alternative ways such as touch killing strategies were employed in coating, where the pathogen upon interaction with the catheter will be killed due to the surface-functionalized antimicrobial agents. Microbes are less likely to be encrusted on a hydrogel-coated catheter due to the hydrophobic index of catheter upon modification. Because hydrogels act as a lubricant, bacteria have a hard time adhering to them, and hydrogels are insoluble (hydrophobic) in water. As a result, it will last for a long time. However, silver-coated catheters are more effective than

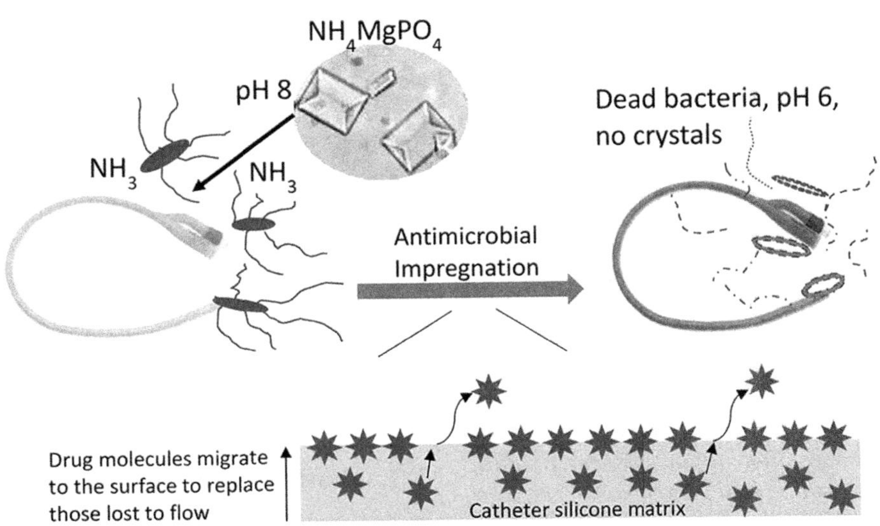

FIG. 4 Antimicrobial drug impregnated urinary catheter for enhanced antiadhesive/antibiofilm properties. Struvite, a versatile antimicrobial, antioxidant, and antibiofilm ammonium compound, is impregnated into the silicone matrix of the urinary catheters. Upon microbial attachment, the molecules on the surface of catheters interact with pathogens and stop initial attachment and also kill bacteria. The antimicrobial molecules utilized will be replaced by the underneath molecules in catheter. *(Data from Belfield, K., Chen, X., Smith, E. F., Ashraf, W., & Bayston, R. (2019). An antimicrobial impregnated urinary catheter that reduces mineral encrustation and prevents colonisation by multi-drug resistant organisms for up to 12 weeks.* Acta Biomaterialia, 90, *157–168. doi: https://doi.org/10.1016/J.ACTBIO.2019.03.042. Copyright (2019), Elsevier.)*

hydrogels against Gram-positive bacteria such as *E. coli* and *P. mirabilis* (uropathogenic bacteria). After a week, the hydrogel coating had a higher content of biomass than the silver coating catheter. CAUTI has been reduced by up to 40%–50% using a silver alloy-hydrogel coated catheter because silver halts the microbial colonization and eradicate the microbial genetic material rapidly. The application of yttrium fluoride (YF3) nanoparticle coatings to urinary catheters reduces biomass formation and reduces biofilm-associated virulence factors (Soto, 2014). Coatings of polyzwitterions are used to modify urinary catheters. Polyzwitterions are polymers that become neutral when they have an equal number of positive and negative charges. The hydrophilic layer of coatings repels the bacterial cell wall by preventing the microbial attachment in the catheters. Polymer coating made of 2-methacrylolo-xyethylphosphorylcholine copolymerized with long-chain alkyl methacrylate effectively repels *Proteus mirabilis* colonization and prevents catheter encrustation. Poly(sulfobetaine methacrylate) [PSBMA] polymers were used to modify a silicone catheter. PSBMA inhibited the formation of microbial biofilms in catheters and was most effective against *P. aeruginosa* and *S. aureus*. It repelled bacterial adhesion (hydrophilic layer) on urinary catheters due to its antiadhesion property. Cellobiose dehydrogenase (CDH) is an antimicrobial agent that is crucial on the silicone-coated catheter's surface. CDH uses an oxidation (electron donor) mechanism to create hydrogen peroxide, which is delivered continuously for 16 days to prevent biofilm development, particularly on *S. aureus*. Hydrogen peroxide is generated from the CDH during the oxidation process, which destroys the bacteria's cell wall and kills them (Andersen & Flores-Mireles, 2019).

3.2 Biofilms in dental caries

Because of the biofilm on dental surfaces, tooth decay is now extremely widespread in many from children to adults. Since it is difficult to avoid dental caries, it is necessary to utilize synthetic remedies such as tooth paste, mouthwash, and gel. A study involving tooth paste and mouthwash formulation of 20 traditional Chinese medicines (TCMs) showed that they have higher efficacy on killing common dental microbes implicated in dental caries such as *Streptococcus sanguis*, *S. mutans*, *Streptococcus mitis*, and *Porphyromonas gingivalis*. Out of 20 TCMs, 16 TCMs including *Cimicifugae rhizoma*, *Ramulus Cinnamomi*, *Gentianae radix*, *Radix et rhizomarhei*, *Pericarpium granati*, *Mentha haplocalyx*, *Forsythiae Fructus*, *Fructus Mume*, *Lonicerae japonicae flos*, *Caryophylli flos*, *Cortex phellodendri*, and *Cortex magnolia officinalis* (Wong et al., 2010) have enhanced antimicrobial activity upon dental care formulations. Biofilms formed by *Porphyromonas gingivalis*, *S. mutans*, *Streptococcus mitis*, and *Streptococcus sanguis* are inhibited *by Fructus Mume* in toothpaste formulation. External warts and corns are treated with *Fructus mume*. When compared with mouthwashes (chlorhexidine), it has a relatively modest toxicity level. The active components of *Fructus mume*, such as citric acid, malic acid, oxalic acid, succinic acid, fumaric acid, tartaric acid, and picric acid, provide the microbes with a very low pH environment. As a result, it may be able to combat microbes in high-acidity conditions (Wong et al., 2010). Flavonoids (apigenin, catechin, luteolin, morin, myricetin, naringin, quercetin, and rutin) have the potential to inhibit microbial attachment of nine dental complications causing microorganisms including *Streptococcus sanguinis*, *Streptococcus oralis*, *S. aureus*, *Lactobacillus casei*, *E. coli*, *E. faecalis*, *Aggregatibacter actinomycetemcomitans*, *Actinomyces viscosus*, and *Actinomyces naeslundii*, by high proton permeability. Rutin, quercetin, naringin, morin, luteolin have an effect against *S. aureus*, *Lactobacillus casei*, *E. coli*, *E. faecalis*, *Aggregatibacter actinomycetemcomitans*, *Actinomyces viscosus*, and *Actinomyces naeslundii*. Proanthocyanidin extracted from cranberries has inhibited the virulent organic acid synthesis by *S. mutans* and *Streptococcus sobrinus* as well as it inhibits the adhesion of *Streptococcus* species to hydroxyapatite pellets, which mimics oral calcium forms. As a function, flavonoids are an excellent alternative to antibiotics that inhibits organic acid synthesis by biofilms-forming pathogens in tooth, and they preserve dental enamel by covering it with tooth paste (Gutiérrez-Venegas, Gómez-Mora, Meraz-Rodríguez, Flores-Sánchez, & Ortiz-Miranda, 2019). Chlorhexidine is a common antiplaque and anti-gingivitis agent, as well as an anticaries agent that prevents *S. mutans* biofilm development. It has the highest rates of microorganism recolonization. When the negatively charged cell wall of *S. mutans* binds to chlorohexidine on oral surfaces, it exhibits bacteriostatic action (destroying the bacteria by disrupt its conversion of sucrose into lactic acid). In children, sodium fluorides has been used for the topical treatment in oral surfaces, which has the ability to decrease the production of *S. mutans* as a biofilm in artificial plaque source but not in human model. Comparatively, chlorhexidine inhibits the growth level of *S. mutans* in oral (salivary) surfaces by using as a mouthwash (Lobo et al., 2008).

3.3 Biofilms in breast implants

The formation of biofilms in breast implants resulted in capsular contracture, which was caused by one of the major causative agents, *S. epidermidis*.

The representation in Fig. 5 shows the biofilm-contaminated breast implant due to the symbiotic *S. epidermidis* population preexisting in the breast site. Other

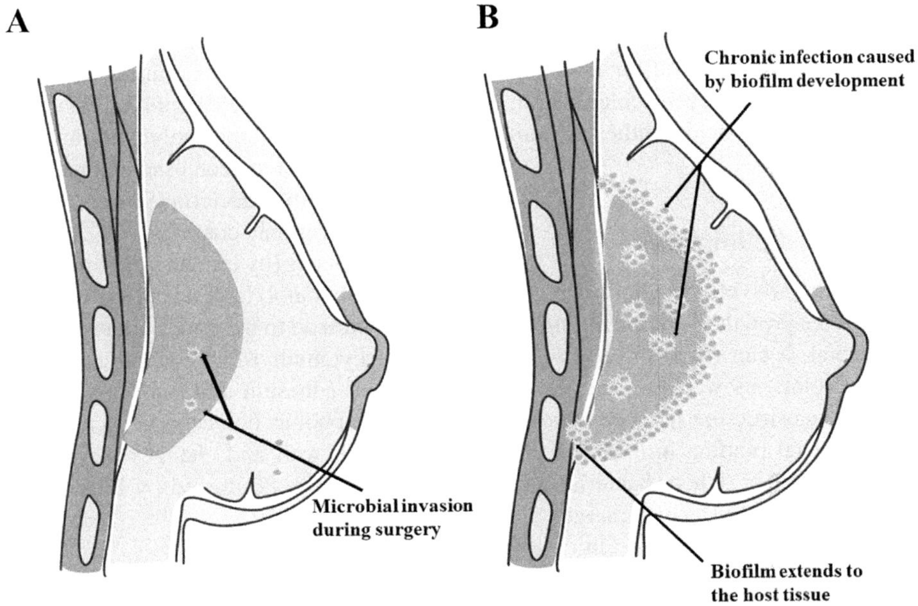

FIG. 5 Biofilm formation on the silicone breast implant. (A) Model of silicone implant on breast with initial microbial invasion. (B) Mature biofilm formation from pathogens colonized during surgery and symbiotic microbes, which will extend to the nearby host tissues forming pathogenicity.

common microorganisms present in the skin, which involve the formation of biofilms in breast implants, are *Propionibacterium acnes*, *S. aureus*, and other *Staphylococci* spp., *Streptococci* spp., *Bacillus* spp., *E. coli*, *Mycobacterium* species, *Corynebacterium, and Lactobacilli* spp. Due to biofilms of *S. epidermidis* and *P. acnes*, the majority of the breast implants were removed and replaced. The capsular contracture of silicone implants treated with *S. epidermidis* was studied using a rabbit as an animal model. Baker's grade III/IV capsular contracture, in the breast implants, was revealed by the effects of the inoculated pockets. Antibiotics and antibacterial agents that inhibit the higher propagation of biofilms in the skin/tissue may reduce capsular contracture. Betadine is an antiseptic with povidone-iodine as an active ingredient that prevents microbes from adhering to or growing on the skin, but it deflates breast implants (silicone). Gentamicin and cefazolin solution reduces the antibiotic effects of Baker's grade capsular contracture. In animal models, prophylactic intravenous and vancomycin were well-studied antibiotics for capsular contracture. As a result, concerned antibiotics have distinct characteristics against bacteria, such as vancomycin, which alters the bacterial membrane to eradicate it but not for a long time (Ajdic, Zoghbi, Gerth, Panthaki, & Thaller, 2016). Higher pathogenic colonization occurs to form capsular contracture sequel on breast implants, and those were Baker's grade IV capsular contracture in it. Antibiotics and capsulectomy helped to reduce contracture. The majority of research has shown that capsulectomy is the most effective treatment for capsular contracture on breast implants (Deva, Adams, & Vickery, 2013).

3.4 Biofilms in ophthalmological implants

Ophthalmological hitches are reduced when biomaterials such as intraocular lenses, orbital prosthesis, or lacrimal stents are used. Microbes such as *Streptococcus pneumoniae*, *S aureus*, *P. aeruginosa*, and *E. coli* colonize the surface of biomaterials. Intraocular lenses are replaced in the eyes during the cataract surgery. A study formulated porphyrin merged intraocular lenses, which induces photoinactivation by eradicating the bacterial cell wall up to 97%. Porphyrin used as a photosensitizer inhibits the growth of *S. epidermidis* due to the greater level of hydrophilicity. Silver particles at a rate of 350–700 ppm injected in the PMMA (polymethylmethacrylate) resin have been employed in the manufacture of ocular prostheses. Silver-coated resin has a better efficiency in killing pathogens such *S. pneumoniae*, *S. aureus*, *P. aeruginosa*, and *E. coli* than uncoated ocular prostheses (Anaya-Alaminos, Ibáñez-Flores, Aznar-Peña, & González-Andrades, 2015). Silicone lacrimal implants have threat on growth of microorganisms such as Gram-positive bacteria (*S. epidermidis* and *S. aureus*) and Gram-negative bacteria (*P. aeruginosa*). The single or numerous erythematous nodules are caused by nontuberculous mycobacterium, which is the causal agent for postoperative oculofacial infection and associated biofilms. Antibiotics including clarithromycin, doxycycline, ciprofloxacin, erythromycin, rifampin, and interferon gamma may possesses the effective measures against the microbes, and it may be done intracellularly by the process of phagocytosis, i.e., engulfing the microbes. To destroy pathogenic microorganisms, silicone lacrimal implants may also utilize mupirocin irrigation, gentian violet thiamphenicol glycinate, and acetylcysteine in larger concentrations (*S. epidermidis* and *S. aureus*). Antibiofilm medications may interact with pathogens and prevent colonization by damaging the pathogen's nucleic acid synthesis (Samimi et al., 2016).

3.5 Biofilms in orthopedic implants

Device-associated infection develops after orthopedic implantation as a result of the growth of colonization/or biofilms on the titanium metal. It can result in serious infections, particularly in hip joints, as well as relatively high trauma injuries. Surface nanostructure has been shown to reduce the amount of bacterial binding in implants and to rupture the cellular membranes when bacteria adhere to the implant surfaces. Surface charges and hydrophilicity-based coatings can help in biofilm attachment and rapid killing due to the hydrophilicity's repulsion effect (Ribeiro, Monteiro, & Ferraz, 2012). Selenium nanoparticles are emerging nanodelivery modes for antibiofilm effect by reducing the virulence factors. The therapeutic effectiveness of selenium nanoparticles coatings against methicillin-resistant *S. aureus* and methicillin-resistant *S. epidermidis* is unknown. Selenium nanoparticles, on the other hand, have a negligible concentration of cytotoxicity to RBC and reduce bacterial colonization (CFU)/or the bacterial viability on the plate. Manufacturers may also employ nanoparticle coating on gold and silver. Nanoparticles have the capacity to produce reactive oxygen species, which harm bacterial cell organelles, block transmembrane transport mechanisms, which halt the transfer of molecules (nutrients), and disrupt enzyme production by disrupting DNA/gene (Tran et al., 2019). In the case of hip and knee implants, biofilms produced infections such as osteomyelitis on the implant's surface. Chitosan was shown to reduce the negative effects of *S. aureus*–induced osteomyelitis in rabbits in the previous state of the art. The rate of propagation of pathogens such as *S. epidermidis* ATCC 35984, *S. aureus* 376, and *S. epidermidis* 389, which are extremely aggressive in colonization on metal orthopedic implants, has been lowered using Hydroxypropyltrimethyl Ammonium Chloride Chitosan (HACC) synthesized using chitosan polymer. To treat the preexisting biofilms on the implants, a larger proportion of HACC was necessary. HACC (chitosan) has been utilized in bone cement because it has the capacity to promote bone formation while also blocking bacterial colonization and enhancing antimicrobial and antibiofilm action. When positively charged chitosan and associated polymeric nanoparticles bind to the microbial cell membranes, they puncture and collapse the cellular structure, modify the transporter membranes, and eradicate the microbes by integrating with DNA (Peng et al., 2011). Stainless steel or titanium Locking Compression Plates (LCPs) were used as the orthopedic implants during mid-diaphyseal tibial osteotomy (incorporation of steel) by examining the plates coated with *N*, *N*-dodecyl,methyl-PEI (polyethylenimine) for biofilm formation. Hydrophobic polycationic coatings (*N*,*N* dodecyl, methyl-PEI) have been used against the adhesion properties of Gram-positive bacteria (*S. aureus*) to the implant. Because of the polycationic coatings, *N*,*N* dodecyl,methyl-PEI-coated titanium plates (by immersion of plated in hydrophobic polycation in butanol) have a bactericidal effect against *S. aureus*. In comparison to the control plate, bone plate treated with *N*, *N* dodecyl,methyl-PEI coating exhibits no bacterial cluster. Bacterial adhesion and biofilm formation can be inhibited by hydrophobic polycationic activity, which creates pores in the cell wall and changes the chemical structure of the microorganisms followed by killing (Schaer, Stewart, Hsu, & Klibanov, 2012).

3.6 Biofilms in cardiovascular implants

Infections are induced by biofilms of superbugs such as *S. aureus*, *E. coli*, and *S. epidermidis* in cardiac implanted electronic devices (CIED). The AIGISRX envelop has been used in cardiovascular devices (pacemakers) to prevent multidrug-resistant *S. aureus* strains from forming biofilms. The AIGISRX envelope features a bioresorbable polymer covering (a polypropylene mesh coating) that releases antibiotics (rifampin and minocycline) to prevent antimicrobial-resistant germs from growing. The combination of rifampin and minocycline has the potential to eliminate *Staphylococcal* biofilms while also reducing the activity of resistant species. The polypropylene mesh covered AIGISRX has the capacity to prevent *S. aureus*, *S. epidermidis*, and *E. coli* from adhering to the device and forming biofilms (Hansen, Berg, Johnson, Sanders, & Citron, 2010). Biofilm development in cardiac devices is influenced by catheter-associated bacteremia by *S. epidermidis* (pacemakers, implanted defibrillators, and prosthetic valves). *N*-Propanol, as well as propanol/ethanol/chlorhexidine and related mixtures, is effective against *S. epidermidis* biofilms. Hydrogen peroxide has also been shown to successfully impede microbial cells reproduction by causing cell shrinkage and obstructing *S. epidermidis* colony formation, as well as lowering biomass density (Presterl et al., 2007). Cardiovascular stents impart support to find out the vascular stenosis in a blood vessel at the place where lesions occur. Drug eluting stents are produced to elute antimicrobials to avoid the infections from occurring. Antiproliferative and antifungal characteristics are available with J&J's Cypher stents. The drug sirolimus (antiproliferative and immunosuppressant) or rapamycin (antifungal) was used in J&J's Cypher stents. The Cypher stents, which are made of polyethylene-*co*-vinyl acetate (PEVA) and poly(*n*-butylmethacrylate) (PBMA), have the ability to enlarge coronary arteries and clear blockages. The polymer coating on stents has antiproliferative, antibacterial, and antibiofilm properties, which reduces the number of biomasses collected by disrupting the cell cycle (mitosis) and aids in the healing process of the vascular system (Brooks, Brooks, & Grainger, 2013). The surface of cardiovascular equipment such as pacemakers, defibrillators, and related devices, as well as mechanical heart valves, had potential antibiofilm properties by inhibiting cell adhesion to values. The promotion of biofilm formation on the implantable devices results in a variety of negative consequences, including artery obstruction, gadget failure, and health problems. The colonization on the surface of the CIED was influenced by different microbial isolates (*S. aureus*, *S. epidermidis*, and *E. coli*). Hydrophobic surfaces and Zwitterionic charged coatings are two emerging strategies that may be used to prevent or hinder biofilm growth. For making hydrophobic surfaces, incorporating the functional domains prevents the hydrogen bonding on the surface of the device. As a result, the device's surface turns hydrophobic. Biocompatible fluorinated hydrophobic coatings with silica colloids or with slippery liquid infused porous surfaces (SLIPSs) are effective approach for decreasing the binding efficacy of *S. aureus* and *P. aeruginosa* on the implant surfaces. These polymer coverings may limit bacterial colony development, although overall biofilm inhibition is challenging. Because there is an equal amount of positively and negatively charged ions in zwitterionic charged coatings, they become neutral. Poly(sulfobetaine) and poly(carboxybetaine), which bond to the methacrylate or acrylamide backbones, are the most widely employed zwitterionic materials. Zwitterionic materials have the potential to stabilize the neutrally charged on the surface while also forming a hydration layer. As a result, the bacterial cell wall's adherence to the surface is disrupted. Polymer coatings on the device's surface destabilized the bacterial cell's osmolarity and potentially enhanced killing (Harding & Reynolds, 2014).

3.7 Combating biofilms in neurosurgical implants

Intracerebral hemorrhage caused by intracranial pressure can be treated using neurosurgical implants such as craniotomies, external ventricular and lumbar drainages, internal shunt systems, cranioplasty, and neurostimulators. Neurostimulators (spinal cord stimulators) have the potential to cure chronic refractory pain, while deep brain stimulators keep Parkinson's disease and dystonia in check. Biofilm deposition occurs on the surface of implants as a result of bacteria adhering to the device via extracellular polymeric substances (EPSs). Because of bacterial colonization, microorganisms in the stationary phase have the capacity to withstand antimicrobial treatment (Braxton et al., 2005). Biofilms have the potential to contaminate implants and disrupt the wound healing process, resulting in implant-related illnesses. Antibiofilm treatment helps to reduce the implant-associated infections. Rifampin is important as an antibacterial (when in stationary phase, rifampin is particularly active in eradicating microorganisms such as *Staphylococcus* spp.) and has a broad penetration in the cerebral fluid (CSF). The oral administration of rifampin in combination with cotrimoxazole, moxifloxacin/doxycycline, and levofloxacin is used to treat biofilm formed by methicillin-susceptible *S. aureus* or coagulase-negative *Staphylococci*, *Propionibacterium* spp., and also Gram-negative bacilli were inactivated by fluoroquinolones. The principal thread to the neurosurgical implants is *S. aureus*, *Streptococcus* spp., *Enterococcus* spp., and *P. acnes*, which causes biofilms and other severe diseases (meningitis) in the brain (Caceres & Goldstein, 2012). Biofilms in *P. aeruginosa* cause aggregation in spinal cord stimulators, which can be reduced by using Ceftazidime/or Cefepime intravenously. When antibiotics such as cotrimoxazole and doxycycline were combined with rifampin, it helps to combat biofilm and reduce antibiotic microbial resistance in bacteria such as quinolone-resistant Gram-negative bacilli and rifampin-resistant *Staphylococci*. Amoxicillin is used to treat Enterococcus species, *Streptococcus* species, and a combination of amoxicillin, Gentamicin, and Ceftriaxone is given intravenously to treat *Enterococcus* spp. and *Streptococcus* spp. Biofilm production will severely injure the soft tissue closest to the implant, hence it is vital to examine the implant periodically over a period of time. Because of its poor/insufficient penetration in CSF, antimicrobials such as vancomycin are not recommended for use with rifampin (Conen, Fux, Vajkoczy, & Trampuz, 2017). Rifampin/clindamycin-impregnated catheters have been used to avoid the adhesion of *Staphylococci* spp., and rifampin-loaded catheters may evade the *S. aureus*, *S. epidermidis*, which implanted into ventricular space. Other strategy to avoid ventricular drainage is Hydromer-coated CSF shunts, which are used to reduce the pathogenicity of biofilm formation by *Staphylococcus* sp., i.e., hydrophilic surface of CSF shunts (Hydromer coated in silicone rubber). It also exhibits the repulsion of the bacterial cell wall adhesion due to its negative charge. When bacterial adhesion occurs, hydromer layer of CSF shunts will destabilize its structure (osmolarity) to cause cell shrinkage. Because of significant risk concerns such as incorrect drainage, CSF shunts are rarely employed (Shunmugaperumal, 2010). Due to bacterial colonization, infections on the surface of the implants (known as ventriculitis) have caused issues with ventriculoperitoneal shunts. Antibiotic-coated (rifampin, amoxicillin) CSF shunts or ventriculostomy catheters change the chemical properties of polymers by incorporating the hydrophilic materials, which inhibit bacterial adhesion on the implant surface and long-lasting durability due to its insolubility. Infection-causing microorganisms linked with CSF shunts include biofilms of *S. aureus* and *S. epidermidis*, both of which were treated with antibiotic treatment (rifampin-cotrimoxazole combination). According to the majority of research investigations, antibiotics with antimicrobial combinations have the ability to evade bacterial adherence on the surface of neurosurgical implants while also protecting them from implant-related illnesses (Braxton et al., 2005).

4. Conclusion

The increase in the demand to biomedical devices/implants increases the rate of biofilm contaminations in products, which plays a dangerous problem in healthcare managements. The affected persons were subjected to prolonged motility sometimes morbidity based on the implant used and their function. The higher resistance of biofilm at non-toxic concentration of drugs makes research studies to find alternative therapies for antibiotics. Some of the alternative therapies comprise antimicrobial nanotechnologies, antimicrobial coatings, and modification of implant substrate with antiadhesive potential. In a cheaper way, the implants were surgically removed to stop biofilm pathogenicity on the implant-carrying host. However, not all biodevice/implant carriers were susceptible for removing the devices, considering life-saving measures. In recent two decades, many advancements target the biofilm virulence such as EPS production, virulence protein production, and antibiotic resistance. Many more target-specific studies were concluded in vitro and in vivo with a variety of potential drug actives, which can be an effective future drugs/device coatings/device and implant substrate. Also, antibiofilm drug targets with enhanced synergistic activity of drugs with antibiotics were discovered and used for maintaining biofilm in hospitals (amoxyclav). Concerning the antibiofilm drug toxicity and limited host tolerance, it is further recommended by the healthcare organizations to research more on alternative therapies that can be successfully used in clinical treatment protocols.

References

Abraham, C. M. (2014). A brief historical perspective on dental implants, their surface coatings and treatments. *The Open Dentistry Journal*, *8*(1), 50. https://doi.org/10.2174/1874210601408010050.

Adams, D. Z., Little, A., Vinsant, C., & Khandelwal, S. (2016). The midline catheter: A clinical review. *Journal of Emergency Medicine*, *51*(3), 252–258. https://doi.org/10.1016/j.jemermed.2016.05.029.

Ajdic, D., Zoghbi, Y., Gerth, D., Panthaki, Z. J., & Thaller, S. (2016). The relationship of bacterial biofilms and capsular contracture in breast implants. *Aesthetic Surgery Journal*, *36*(3), 297–309. https://doi.org/10.1093/ASJ/SJV177.

Alisiddiq, Z., & Butler, R. (2021). Pulmonary artery catheterization. *Medicine (United Kingdom)*, *46*(8), 496–498. https://doi.org/10.1016/j.mpmed.2018.05.002.

Anaya-Alaminos, R., Ibáñez-Flores, N., Aznar-Peña, I., & González-Andrades, M. (2015). Antimicrobial biomaterials and their potential application in ophthalmology. *Journal of Applied Biomaterials and Functional Materials*, *13*(4), e346–e350. https://doi.org/10.5301/jabfm.5000253.

Andersen, M. J., & Flores-Mireles, A. L. (2019). Urinary catheter coating modifications: The race against catheter-associated infections. *Coatings*, *10*(1), 23. https://doi.org/10.3390/COATINGS10010023. 2020.

Anger, H. A., Dabash, R., Durocher, J., Hassanein, N., Ononge, S., Frye, L. J., … Winikoff, B. (2019). The effectiveness and safety of introducing condom-catheter uterine balloon tamponade for postpartum haemorrhage at secondary level hospitals in Uganda, Egypt and Senegal: a stepped wedge, cluster-randomised trial. *BJOG: An International Journal of Obstetrics and Gynaecology*, *126*(13), 1612–1621. https://doi.org/10.1111/1471-0528.15903.

Belfield, K., Chen, X., Smith, E. F., Ashraf, W., & Bayston, R. (2019). An antimicrobial impregnated urinary catheter that reduces mineral encrustation and prevents colonisation by multi-drug resistant organisms for up to 12 weeks. *Acta Biomaterialia*, *90*, 157–168. https://doi.org/10.1016/J.ACTBIO.2019.03.042.

Benachinmardi, K. K., Ravikumar, R., & Indiradevi, B. (2017). Role of biofilm in cerebrospinal fluid shunt infections: A study at tertiary neurocare center from South India. *Journal of Neurosciences in Rural Practice*, *8*(3), 335. https://doi.org/10.4103/JNRP.JNRP_22_17.

Bispo, P. J., Haas, W., & Gilmore, M. S. (2015). Biofilms in infections of the eye. *Pathogens*, *4*(1), 111–136.

Blasi, F. (2018). Lung diseases: Chronic respiratory infections. *International Journal of Molecular Sciences*, *19*(10), 3051. https://doi.org/10.3390/ijms19103051.

Braxton, E. E., Ehrlich, G. D., Hall-Stoodley, L., Stoodley, P., Veeh, R., Fux, C., et al. (2005). Role of biofilms in neurosurgical device-related infections. *Neurosurgical Review*, *28*(4), 249–255. https://doi.org/10.1007/S10143-005-0403-8.

Brooks, B. D., Brooks, A. E., & Grainger, D. W. (2013). Antimicrobial medical devices in preclinical development and clinical use. In *Biomaterials associated infection: Immunological aspects and antimicrobial strategies* (pp. 307–354). https://doi.org/10.1007/978-1-4614-1031-7_13. 9781461410317.

Caceres, J. A., & Goldstein, J. N. (2012). Intracranial hemorrhage. *Emergency Medicine Clinics of North America*, *30*(3), 771–794. https://doi.org/10.1016/j.emc.2012.06.003.

Campoccia, D., Testoni, F., Ravaioli, S., Cangini, I., Maso, A., Speziale, P., … Arciola, C. R. (2016). Orthopedic implant infections: Incompetence of Staphylococcus epidermidis, Staphylococcus lugdunensis, and Enterococcus faecalis to invade osteoblasts. *Journal of Biomedical Materials Research. Part A*, *104*(3), 788–801. https://doi.org/10.1002/jbm.a.35564.

Centers for Disease Control and Prevention. (2019). *Antibiotic resistance threats in the United States, 2019*. https://doi.org/10.15620/CDC:82532.

Conen, A., Fux, C. A., Vajkoczy, P., & Trampuz, A. (2017). Management of infections associated with neurosurgical implanted devices. *Expert Review of Anti-Infective Therapy*, *15*(3), 241–255. https://doi.org/10.1080/14787210.2017.1267563.

Costerton, J. W., Montanaro, L., & Arciola, C. R. (2005). Biofilm in implant infections: Its production and regulation. *The International Journal of Artificial Organs*, *28*(11), 1062–1068. https://doi.org/10.1177/039139880502801103.

Dadi, N. C. T., Radochová, B., Vargová, J., & Bujdáková, H. (2021). Impact of healthcare-associated infections connected to medical devices—An update. *Microorganisms*, *9*(11), 2332.

Danison, A. P., Lee, D. J., & Panchal, R. R. (2017). Temporary stabilization of unstable spine fractures. *Current Reviews in Musculoskeletal Medicine*, *10*(2), 199. https://doi.org/10.1007/S12178-017-9402-Y.

das Neves, R. C., Mortari, M. R., Schwartz, E. F., Kipnis, A., & Junqueira-Kipnis, A. P. (2019). Antimicrobial and antibiofilm effects of peptides from venom of social wasp and scorpion on multidrug-resistant *Acinetobacter baumannii*. *Toxins*, *11*(4). https://doi.org/10.3390/TOXINS11040216.

Deva, A. K., Adams, W. P., & Vickery, K. (2013). The role of bacterial biofilms in device-associated infection. *Plastic and Reconstructive Surgery*, *132*(5), 1319–1328. https://doi.org/10.1097/PRS.0B013E3182A3C105.

Dhir, S. (2013). Biofilm and dental implant: The microbial link. *Journal of Indian Society of Periodontology*, *17*(1), 5. https://doi.org/10.4103/0972-124X.107466.

Diaconu, O., Siriopol, I., Poloşanu, L. I., & Grigoraş, I. (2018). Endotracheal tube biofilm and its impact on the pathogenesis of ventilator-associated pneumonia. *The Journal of Critical Care Medicine*, *4*(2), 50.

Edens, C., Wong, J., Lyman, M., Rizzo, K., Nguyen, D., Blain, M., … Patel, P. R. (2017). Hemodialyzer reuse and Gram-negative bloodstream infections. *American Journal of Kidney Diseases: The Official Journal of the National Kidney Foundation*, *69*(6), 726–733. https://doi.org/10.1053/j.ajkd.2016.09.022.

Falcone, M., Meier, J. J., Marini, M. G., Caccialanza, R., Aguado, J. M., Del Prato, S., & Menichetti, F. (2021). Diabetes and acute bacterial skin and skin structure infections. *Diabetes Research and Clinical Practice*, *174*, 108732. https://doi.org/10.1016/j.diabres.2021.108732.

Feneley, R. C. L., Hopley, I. B., & Wells, P. N. T. (2015). Urinary catheters: History, current status, adverse events and research agenda. *Journal of Medical Engineering & Technology*, *39*(8), 459. https://doi.org/10.3109/03091902.2015.1085600.

Flores-Mireles, A. L., Walker, J. N., Caparon, M., & Hultgren, S. J. (2015). Urinary tract infections: Epidemiology, mechanisms of infection and treatment options. *Nature Reviews Microbiology*, *13*(5), 269–284. https://doi.org/10.1038/nrmicro3432.

Frey, N., Hügle, T., Jick, S. S., Meier, C. R., & Spoendlin, J. (2016). Type II diabetes mellitus and incident osteoarthritis of the hand: A population-based case-control analysis. *Osteoarthritis and Cartilage*, *24*(9), 1535–1540. https://doi.org/10.1016/J.JOCA.2016.04.005.

Garrigos, Z. E., Farid, S., Bendel, M. A., & Rizwan Sohail, M. (2020). Spinal cord stimulator infection: Approach to diagnosis, management, and prevention. *Clinical Infectious Diseases: An Official Publication of the Infectious Diseases Society of America*, *70*(12), 2727–2735. https://doi.org/10.1093/CID/CIZ994.

Global Industry Analysts, Inc. (2021). *Implantable medical devices—Global market trajectory & analytics*. https://www.researchandmarkets.com/reports/4805500/implantable-medical-devices-global-market?utm_source=BW&utm_medium=PressRelease&utm_code=6t4k97&utm_campaign=1488019+-+Global+Orthopedic%2c+Dental+%26+Cardiovascular+Implantable+Medical+Devices+Market+Trajectory+%26+Analytics%2c+2012-2019+%26+2020-2027+with+COVID-19+Impacts+for+2020+%26+2021&utm_exec=chdo54prd.

Gominet, M., Compain, F., Beloin, C., & Lebeaux, D. (2017). Central venous catheters and biofilms: Where do we stand in 2017? *APMIS*, *125*(4), 365–375. https://doi.org/10.1111/APM.12665.

Gutiérrez-Venegas, G., Gómez-Mora, J. A., Meraz-Rodríguez, M. A., Flores-Sánchez, M. A., & Ortiz-Miranda, L. F. (2019). Effect of flavonoids on antimicrobial activity of microorganisms present in dental plaque. *Heliyon*, *5*(12). https://doi.org/10.1016/J. HELIYON.2019.E03013.

Hansen, L. K., Berg, K., Johnson, D., Sanders, M., & Citron, M. (2010). Efficacy of local rifampin/minocycline delivery (AIGIS (RX)®) to eliminate biofilm formation on implanted pacing devices in a rabbit model. *The International Journal of Artificial Organs*, *33*(9), 627–635. https://doi.org/10.1177/039139881003300908.

Harding, J. L., & Reynolds, M. M. (2014). Combating medical device fouling. *Trends in Biotechnology*, *32*(3), 140–146. https://doi.org/10.1016/J.TIBTECH.2013.12.004.

Harris, C., Croce, B., & Cao, C. (2015). Tissue and mechanical heart valves. *Annals of Cardiothoracic Surgery*, *4*(4), 399. https://doi.org/10.3978/J.ISSN. 2225-319X. 2015. 07. 01.

Hayes, J. S., & Richards, R. G. (2019). Osseointegration of permanent and temporary orthopedic implants. *Encyclopedia of Biomedical Engineering*, *1–3*, 257–269. https://doi.org/10.1016/B978-0-12-801238-3.00225-7.

Hoberg, S., Danstrup, C., Laursen, B., Petersen, N. K., Udholm, N., Kamarauskas, G. A., & Ovesen, T. (2017). Characteristics of CI children with complicated middle ear infections. *Cochlear Implants International*, *18*(3), 136–142. https://doi.org/10.1080/14670100.2017.1289298.

Jothi, R., Hari Prasath, N., Gowrishankar, S., & Pandian, S. K. (2021). Bacterial quorum-sensing molecules as promising natural inhibitors of candida albicans virulence dimorphism: An in silico and in vitro study. *Frontiers in Cellular and Infection Microbiology*, *11*, 1212. https://doi.org/10.3389/FCIMB.2021.781790/BIBTEX.

Joung, Y. H. (2013). Development of implantable medical devices: From an engineering perspective. *International Neurourology Journal*, *17*(3), 98. https://doi.org/10.5213/INJ.2013.17.3.98.

Kawoos, U., McCarron, R. M., Auker, C. R., & Chavko, M. (2015). Advances in intracranial pressure monitoring and its significance in managing traumatic brain injury. *International Journal of Molecular Sciences*, *16*(12), 28979. https://doi.org/10.3390/IJMS161226146.

Khatoon, Z., McTiernan, C. D., Suuronen, E. J., Mah, T. F., & Alarcon, E. I. (2018). Bacterial biofilm formation on implantable devices and approaches to its treatment and prevention. *Heliyon*, *4*(12), 1067. https://doi.org/10.1016/J.HELIYON.2018.E01067.

Kligman, S., Ren, Z., Chung, C. H., Perillo, M. A., Chang, Y. C., Koo, H., et al. (2021). The impact of dental implant surface modifications on osseointegration and biofilm formation. *Journal of Clinical Medicine*, *10*(8), 1641. https://doi.org/10.3390/JCM10081641.

Kumar, V., Shah, A. S., Singh, D., Loomba, P. S., Singh, H., & Jagetia, A. (2016). Ventriculoperitoneal shunt tube infection and changing pattern of antibiotic sensitivity in neurosurgery practice: Alarming trends. *Neurology India*, *64*(4), 671. https://doi.org/10.4103/0028-3886.185408.

Liang, S. Y., Beekmann, S. E., Polgreen, P. M., & Warren, D. K. (2016). Current management of cardiac implantable electronic device infections by infectious disease specialists. *Clinical Infectious Diseases*, *63*(8), 1072–1075.

Lobo, P. L. D., de Carvalho, C. B. M., Fonseca, S. G. C., de Castro, R. S. L., Monteiro, A. J., Fonteles, M. C., et al. (2008). Sodium fluoride and chlorhexidine effect in the inhibition of mutans streptococci in children with dental caries: A randomized, double-blind clinical trial. *Oral Microbiology and Immunology*, *23*(6), 486–491. https://doi.org/10.1111/J.1399-302X.2008.00458.X.

Mala, R., Rani, M. J., Prasath, N. H., & Celsia, A. R. (2019). A review on nano material based drug delivery for burn injury. In *vol. 2105. AIP Conference Proceedings* (1st edn., p. 020012). AIP Publishing LLC.

Malard, F., Dore, J., Gaugler, B., & Mohty, M. (2020). Introduction to host microbiome symbiosis in health and disease. *Mucosal Immunology, 14* (3), 547–554. https://doi.org/10.1038/s41385-020-00365-4.

Murillo, H., Restrepo, C. S., Marmol-Velez, J. A., Vargas, D., Ocazionez, D., Martinez-Jimenez, S., & Baxi, A. J. (2016). Infectious diseases of the heart: Pathophysiology, clinical and imaging overview. *Radiographics, 36*(4), 963–983.

Niveditha, S., Pramodhini, S., Umadevi, S., Kumar, S., & Stephen, S. (2012). The isolation and the biofilm formation of uropathogens in the patients with catheter associated urinary tract infections (UTIs). *Journal of Clinical and Diagnostic Research: JCDR, 6*(9), 1478–1482. https://doi.org/10.7860/JCDR/2012/4367.2537.

Pawlowski, S. W., Warren, C. A., & Guerrant, R. (2009). Diagnosis and treatment of acute or persistent diarrhea. *Gastroenterology, 136*(6), 1874–1886. https://doi.org/10.1053/j.gastro.2009.02.072.

Peng, Z. X., Tu, B., Shen, Y., Du, L., Wang, L., Guo, S. R., et al. (2011). Quaternized chitosan inhibits icaA transcription and biofilm formation by Staphylococcus on a titanium surface. *Antimicrobial Agents and Chemotherapy, 55*(2), 860–866. https://doi.org/10.1128/AAC.01005-10.

Prakasam, M., Locs, J., Salma-Ancane, K., Loca, D., Largeteau, A., & Berzina-Cimdina, L. (2017). Biodegradable materials and metallic implants—A review. *Journal of Functional Biomaterials, 8* (4). https://doi.org/10.3390/JFB8040044.

Presterl, E., Suchomel, M., Eder, M., Reichmann, S., Lassnigg, A., Graninger, W., et al. (2007). Effects of alcohols, povidone-iodine and hydrogen peroxide on biofilms of *Staphylococcus epidermidis. The Journal of Antimicrobial Chemotherapy, 60*(2), 417–420. https://doi.org/10.1093/JAC/DKM221.

Priya, A., Selvaraj, A., Divya, D., Karthik Raja, R., & Pandian, S. K. (2021). *In vitro* and *in vivo* anti-infective potential of thymol against early childhood caries causing dual species *Candida albicans* and *Streptococcus mutans. Frontiers in Pharmacology, 12*, 760768. https://doi.org/10.3389/fphar.2021.760768.

Raksha, L., Gangashettappa, N., Shantala, G., Nandan, B., & Sinha, D. (2020). Study of biofilm formation in bacterial isolates from contact lens wearers. *Indian Journal of Ophthalmology, 68*(1), 23. https://doi.org/10.4103/IJO.IJO_947_19.

Ribeiro, M., Monteiro, F. J., & Ferraz, M. P. (2012). Infection of orthopedic implants with emphasis on bacterial adhesion process and techniques used in studying bacterial-material interactions. *Biomatter, 2*(4), 176–194. https://doi.org/10.4161/biom.22905.

Ricca, A. M., Kim, A. K., Chahal, H. S., & Shriver, E. M. (2018, January). *Nasolacrimal stents: An introductory guide*. Eyerounds.Org. https://webeye.ophth.uiowa.edu/eyeforum/tutorials/Stents/index.htm.

Rodríguez-Calero, M. A., Blanco-Mavillard, I., Morales-Asencio, J. M., Fernandez-Fernandez, I., Castro-Sanchez, E., & de Pedro-Gómez, J. E. (2020). Defining risk factors associated with difficult peripheral venous Cannulation: A systematic review and meta-analysis. *Heart & Lung, 49*(3), 273–286.

Saint, S., & Chenoweth, C. E. (2003). Biofilms and catheter-associated urinary tract infections. *Infectious Disease Clinics, 17*(2), 411–432. https://doi.org/10.1016/S0891-5520(03)00011-4.

Samimi, D. B., Ediriwickrema, L. S., Bielory, B. P., Miller, D., Lee, W., & Johnson, T. E. (2016). Microbiology and biofilm trends of silicone lacrimal implants: Comparing infected versus routinely removed stents. *Ophthalmic Plastic and Reconstructive Surgery, 32*(6), 452–457. https://doi.org/10.1097/IOP.0000000000000590.

Schaer, T. P., Stewart, S., Hsu, B. B., & Klibanov, A. M. (2012). Hydrophobic polycationic coatings that inhibit biofilms and support bone healing during infection. *Biomaterials, 33*(5), 1245–1254. https://doi.org/10.1016/J.BIOMATERIALS.2011.10.038.

Sdrulla, A. D., Guan, Y., & Raja, S. N. (2018). Spinal cord stimulation: Clinical efficacy and potential mechanisms. *Pain Practice: The Official Journal of World Institute of Pain, 18*(8), 1048. https://doi.org/10.1111/PAPR.12692.

Serhani, M. A., elKassabi, H. T., Ismail, H., & Navaz, A. N. (2020). ECG monitoring systems: Review, architecture, processes, and key challenges. *Sensors (Basel, Switzerland), 20*(6). https://doi.org/10.3390/S20061796.

Shunmugaperumal, T. (2010). Biofilm eradication and prevention: A pharmaceutical approach to medical device infections. In *Biofilm eradication and prevention: A pharmaceutical approach to medical device infections*. https://doi.org/10.1002/9780470640463.

Singh, M., Kamal, S., & Sowmya, V. (2015). Lacrimal stents and intubation systems: An insight. *The Official Scientific Journal of Delhi Ophthalmological Society, 26*(1), 14–19. https://doi.org/10.7869/DJO.128.

Singh, M., Kumar, L., Anwar, M., & Chand, P. (2015). Immediate dental implant placement with immediate loading following extraction of natural teeth. *National Journal of Maxillofacial Surgery, 6*(2), 252. https://doi.org/10.4103/0975-5950.183864.

Singh, S., Tan, C. L., & Ahmad, A. R. (2021). Explaining osteomyelitis and prosthetic joint infections (PJI) in terms of biofilm - A review. *Malaysian Orthopaedic Journal, 15*(2), 1–8. https://doi.org/10.5704/MOJ.2107.001.

Soto, S. M. (2014). Importance of biofilms in urinary tract infections: New therapeutic approaches. *Advances in Biology, 2014*, 1–13. https://doi.org/10.1155/2014/543974.

Swetha, T. K., Vikraman, A., Nithya, C., Hari Prasath, N., & Pandian, S. K. (2020). Synergistic antimicrobial combination of carvacrol and thymol impairs single and mixed-species biofilms of *Candida albicans* and *Staphylococcus epidermidis. Biofouling, 36*(10), 1256–1271. https://doi.org/10.1080/08927014.2020.1869949.

Szvalb, A. D., Raad, I. I., Weinberg, J. S., Suki, D., Mayer, R., & Viola, G. M. (2014). Ommaya reservoir-related infections: Clinical manifestations and treatment outcomes. *Journal of Infection, 68*(3), 216–224. https://doi.org/10.1016/J.JINF.2013.12.002.

Teweldemedhin, M., Gebreyesus, H., Atsbaha, A. H., Asgedom, S. W., & Saravanan, M. (2017). Bacterial profile of ocular infections: A systematic review. *BMC Ophthalmology, 17*(1), 212. https://doi.org/10.1186/s12886-017-0612-2.

Tom, J. (2016). Management of patients with cardiovascular implantable electronic devices in dental, oral, and maxillofacial surgery. *Anesthesia Progress, 63*(2), 95. https://doi.org/10.2344/0003-3006-63.2.95.

Tran, P. A., O'Brien-Simpson, N., Palmer, J. A., Bock, N., Reynolds, E. C., Webster, T. J., et al. (2019). Selenium nanoparticles as anti-infective implant coatings for trauma orthopedics against methicillin-resistant *Staphylococcus aureus* and epidermidis: In vitro and in vivo assessment. *International Journal of Nanomedicine, 14*, 4613–4624. https://doi.org/10.2147/IJN.S197737.

Vad, V. B., Adin, D. R., & Solomon, J. (2021). Knee osteoarthritis. *Critical Reviews in Physical and Rehabilitation Medicine, 16*(3), 211–231. https://doi.org/10.1615/CritRevPhysRehabilMed.v16.i3.30.

Valliammai, A., Selvaraj, A., Muthuramalingam, P., Priya, A., Ramesh, M., & Pandian, S. K. (2021). Staphyloxanthin inhibitory potential of thymol impairs antioxidant fitness, enhances neutrophil mediated killing and alters membrane fluidity of methicillin resistant *Staphylococcus aureus*. *Biomedicine & Pharmacotherapy*, *141*, 111933. https://doi.org/10.1016/J.BIOPHA.2021.111933.

Wikström, K., Isacsson, U., Nilsson, K., & Ahnesjö, A. (2018). Reproducibility of heart and thoracic wall position in repeated deep inspiration breath holds for radiotherapy of left-sided breast cancer patients. *Acta Oncologica*, *57*(10), 1318–1324.

Wilks, S. A., Koerfer, V. V., Prieto, J. A., Fader, M., & Keevil, C. W. (2021). Biofilm development on urinary catheters promotes the appearance of viable but nonculturable bacteria. *MBio*, *12* (2). https://doi.org/10.1128/MBIO.03584-20/ASSET/2E9A2A8E-2564-42AA-BE6F-CDF89A3BA2D7/ASSETS/IMAGES/MEDIUM/MBIO.03584-20_F009.GIF.

Wojcik, S., & Kang, S. (2022). *Angioplasty and stent placement for the heart—Health encyclopedia*. University of Rochester Medical Center. Retrieved January 4, 2022, from https://www.urmc.rochester.edu/encyclopedia/content.aspx?contenttypeid=92&contentid=p07981.

Wong, R. W. K., Hägg, U., Samaranayake, L., Yuen, M. K. Z., Seneviratne, C. J., & Kao, R. (2010). Antimicrobial activity of Chinese medicine herbs against common bacteria in oral biofilm. A pilot study. *International Journal of Oral and Maxillofacial Surgery*, *39*(6), 599–605. https://doi.org/10.1016/J.IJOM.2010.02.024.

Xu, L.-C., Bauer, J., & Siedlecki, C. A. (2014). Proteins, platelets, and blood coagulation at biomaterial interfaces. *Colloids and Surfaces B: Biointerfaces*, *124*, 49. https://doi.org/10.1016/J.COLSURFB.2014.09.040.

Yeh, H. C., Lu, J. J., Chang, S. C., & Ge, M. C. (2019). Identification of microbiota in peri-implantitis pockets by matrix-assisted laser desorption/ionization time-of-flight mass spectrometry. *Scientific Reports*, *9*(1), 1–9.

Urinary Catheters. (2022). Retrieved 13 July 2022 from https://www.nhs.uk/conditions/urinary-catheters/types/.

Further reading

Anger, H. A., Durocher, J., Dabash, R., Hassanein, N., Ononge, S., Burkhardt, G., et al. (2021). Postpartum infection, pain and experiences with care among women treated for postpartum hemorrhage in three African countries: A cohort study of women managed with and without condom-catheter uterine balloon tamponade. *PLoS One*, *16*(2), e0245988. https://doi.org/10.1111/1471-0528.15948.

Caceres, J. A., & Goldstein, J. N. (2012). Intracranial hemorrhage. *Emergency Medicine Clinics of North America*, *30*(3), 771–794. https://doi.org/10.1016/j.emc.2012.06.003.

Section D

Molecular and biochemical aspect of microbial biofilms

Chapter 21

Influence of bacterial cell wall modulating genes and enzymes on biofilm formation with special emphasis on the role of DD-carboxypeptidases of bacteria

Sumit Kumar Rastogi and Anindya Sundar Ghosh
Molecular Microbiology Laboratory, Department of Biotechnology, Indian Institute of Technology Kharagpur, Kharagpur, West Bengal, India

1. Introduction

A biofilm is a complex microbial community where cells are held together by an extracellular matrix (ECM) produced by cells. The biofilm establishment involves a communication mechanism, and most research focuses on this area. However, the outer cell surface of bacteria, which presents an interface between bacterial cells and its surroundings, is crucial but primarily untouched. ECM with cell wall allows the individual biofilm cells to survive any physical and chemical stress. ECM and cell surface are the direct cell-to-cell contacts site. They are also crucial for adhering to the substratum (Branda, Vik, Friedman, & Kolter, 2005).

In the natural environment, bacteria experience many abiotic stresses such as temperature, nutrition, pH, osmolality, oxygen availability, etc. Cell wall and ECM are the key to overcoming these stresses, and in response to that, cell alters the expression of specific genes and the regulatory mechanism of the stress response. The cells in biofilms are resistant to most of the antimicrobials, allowing them to spread the infection. Thus, the cell envelope and ECM are crucial for the cell that serves as the first line of interaction with any external stimuli such as stress, antimicrobial, or while the cell-to-cell communication within a biofilm. The first and critical step in initiating biofilm formation is the adhesion to the substratum. It is affected by hydrophobicity, flagellation, motility, and roughness. Therefore, the nature of cell surface should affect the character of biofilms, and in reality, it does.

Cell envelopes of most of the bacteria are classified into two groups, primarily Gram-positive and Gram-negative. The cell envelope of Gram-negative bacteria consists of three layers: the outer membrane, the peptidoglycan (PG) or the cell wall, and the inner cytoplasmic membrane. They are often surrounded by the lipopolysaccharide layer linked to the outer membrane (Fig. 1). The cell envelopes of Gram-positive bacteria are composed of a comparably thicker peptidoglycan layer. Unlike Gram-negative bacteria, they lack an outer membrane. However, several long anionic polymers, namely teichoic acids, are embedded through the cell envelope or are covalently attached, which are mainly composed of repeating glycerol phosphate, glucosyl phosphate, or ribitol phosphate (Brauge et al., 2015). The peptidoglycan is a mesh-like structure comprising repeating units of *N*-acetlyglucosamine (GlcNAc) and *N*-acetlymuramic acid (MurNAc) cross-linked via peptide bridges attached to MurNAc of the adjacent chain (Vollmer, Blanot, & De Pedro, 2008). The biosynthesis of peptidoglycan takes place in the outer region of the cytoplasm. Though the PG provides rigidity to the cell wall, it is constantly remodeled to offer a certain level of flexibility (Dörr, Moynihan, & Mayer, 2019). A set of enzymes governs the remodeling of PG, often called autolysins that can cleave PG cross-links (endopeptidases), peptide side chains (carboxypeptidases and amidases), or the sugar backbone (muramidases) (Ghosh, Chowdhury, & Nelson, 2008; Scheurwater, Reid, & Clarke, 2008; Vollmer et al., 2008). Conversely, certain transferases such as PG-acetyltransferases impart the property of lysozyme resistance by transferring acetyl groups (Moynihan & Clarke, 2011) while L,D-transpeptidase can exchange terminal D-ala with some D-amino acids (Cava, de Pedro, Lam, Davis, & Waldor, 2011).

On the other hand, bacteria belonging to Genus *Mycobacterium* (a member of the order Corynebacteriales) display a cell envelope different from the usual bacterial envelopes. However, they are categorized as Gram-positive

Understanding Microbial Biofilms. https://doi.org/10.1016/B978-0-323-99977-9.00013-2

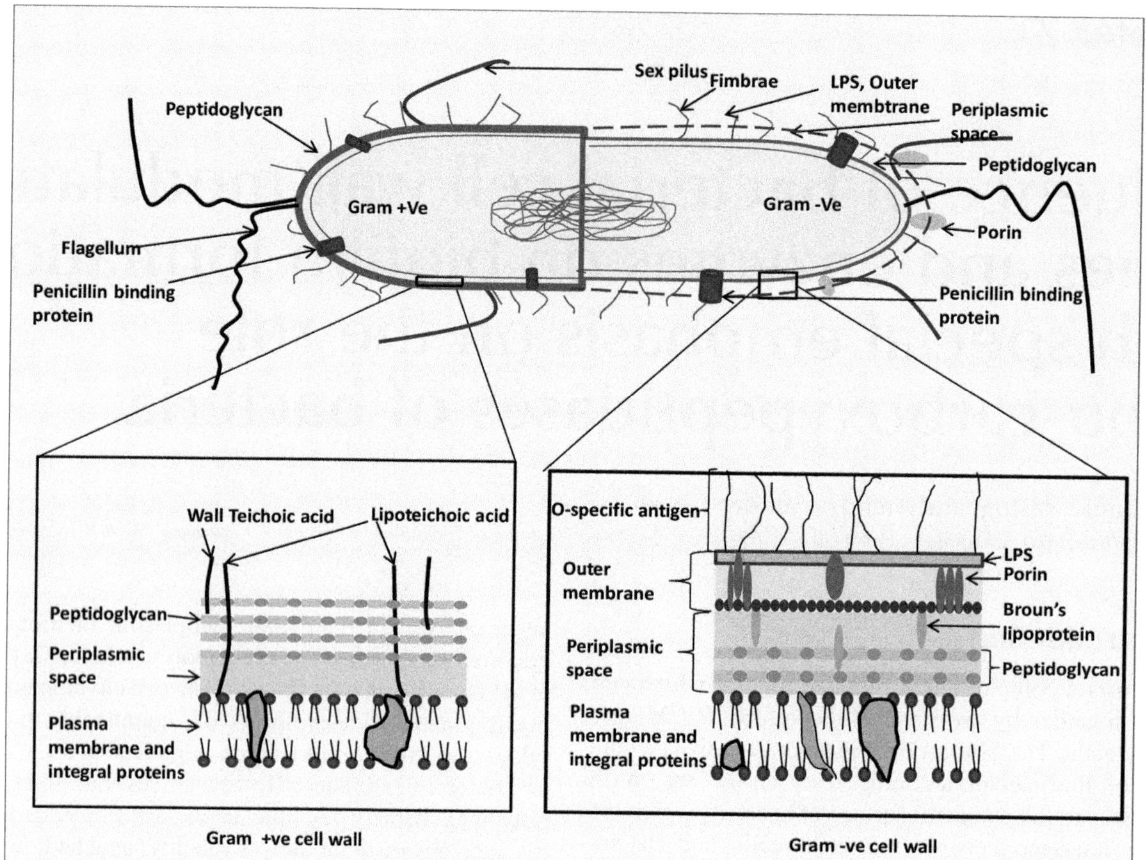

FIG. 1 Pictorial representation of the cell surface organization of Gram-positive and Gram-negative bacterium: Gram-positive and Gram-negative cell surfaces differ mainly in their peptidoglycan layer and thickness. Gram-positive cell wall presents an outer membrane present beyond the comparably thinner peptidoglycan layer. It also features O-antigens and lipopolysaccharide, which are not present in Gram-positive cell wall. The Gram-positive cell wall is particularly known to possess a thick peptidoglycan layer and wall teichoic acid. Cell surface of both the kinds of bacteria bear surface appendages such as flagella, fimbriae, and pili. They also have penicillin-binding proteins tethered to the outer fact of the cytoplasmic membrane. These surface features are important for cellular adhesion and biofilm formation.

bacteria. In addition to the cytoplasmic membrane and peptidoglycan layer, these bacteria possess an arabinogalactan layer that connects the peptidoglycan to the outer membrane, referred to as mycomembrane (Vincent et al., 2018). From inside to outside, the mycobacterial cell wall consists of (1) peptidoglycan; (2) a layer composed of arabinogalactan, lipoarabinomannan, and lipomannan; (3) another layer consists of mycolic acid and glycolipids; and (4) a capsular layer consisting of capsular protein and polysaccharides (Fig. 1). It is noteworthy that the basic PG structure is conserved in *Mycobacteria*, but it is modified to a certain extent. PG of both *M. tuberculosis* and *M. smegmatis* contains GlcNAc and MurNAc, whereas *M. leprae* only has MurNAc (Mahapatra, Crick, McNeil, & Brennan, 2008). The acetyl modifications of GlcNAc and MurNAc are not observed in *Mycobacterium* spp. (Vollmer et al., 2008). The *Mycobacterial* PG contains a high degree of cross-linked peptides than *E. coli*. Almost two-thirds of these cross-links are LD ($3 \rightarrow 3$) bonds linking two mesoDAP residues, while one-third of these cross-links are ($4 \rightarrow 3$) between D-Ala and mDAP. DD-Transpeptidases catalyze the 4-3 cross-link formation (Mainardi et al., 2005), and LD-transpeptidases catalyze the 3-3 cross-link formation (Sanders, Wright, & Pavelka, 2014). However, the switching between 4-3 and 3-3 cross-link or vice versa may lead to altered cell surface properties that may affect the physiology of the cell, including the surface attachment (Pandey et al., 2018) (Fig. 2).

Bacteria use several proteinaceous structures to adhere to a substratum. These are pili, fimbriae, flagella, curli, adhesins, intimins, and invasins. The filamentous appendages fimbriae are extended from cells, mainly like the bunch and are composed of a protein named pilin. Compared with flagella, fimbriae are the shorter, stiffer, and have smaller diameter. They are commonly found in Gram-negative bacteria and are usually involved in adhesion to surfaces and host tissues. Pili are very similar to fimbriae but are generally thin and present as a maximum number

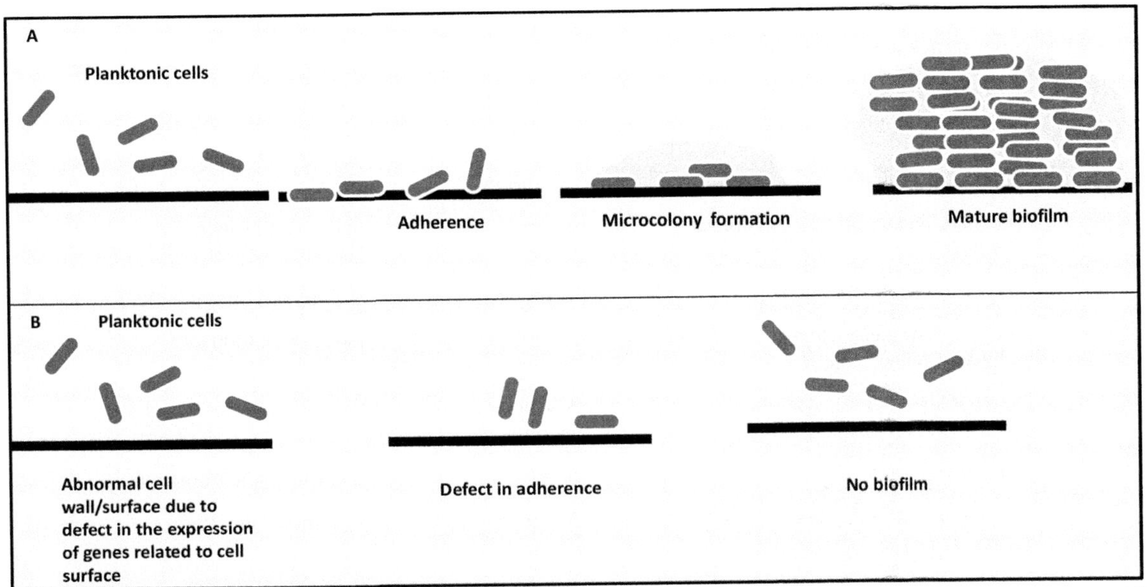

FIG. 2 Effect of cell surface in biofilm formation: (A) Planktonic bacterial cells with uniform cell surface adhere to the substratum to establish mature biofilm. (B) Abnormal cell surface due to defect in its structural components may lead to a defective adhesion that in turn hinders biofilm formation.

of 1–2 on the cell surface. Sometimes they are also involved in conjugation and twitching movement. Flagella provide motility to the bacterial cell, are very rigid, and act as a propeller. It has three main components: filament, hook, and motor. The motor provides the necessary torque required for rotation. Understanding the roles of these surface structures in biofilm formation and pathogenesis is still in an early phase, but considering their functions, their correlation with biofilm formation is likely.

The ECM surrounding the cell envelope also plays a crucial role in biofilm establishment. It consists of proteins, extracellular DNA, and polysaccharides. The polysaccharides of ECM provide adhesion, protection, and structural properties as they serve as molecular glue, allowing cell-to-cell and cell-to-surface contact (Limoli, Jones, & Wozniak, 2015). Bacteria can develop multiple polysaccharides depending on the exposure of bacterial strains to various environmental conditions, substrate, and nutrition (Kostakioti, Hadjifrangiskou, & Hultgren, 2013). Cells may produce a specific class of polysaccharides at a particular biofilm stage. Based on function, these polysaccharides may be categorized into three types: aggregate, protective, and architectural (Limoli et al., 2015). Considering their significance, it is important to look into their contribution in biofilm formation at genetic level.

The present chapter highlights the importance of cell wall constituents and their regulation in biofilm formation. Furthermore, the chapter deals with the genes and pathways involved in the maintenance of cell wall features and their possible role in biofilm formation, with a significant emphasis on the roles of DD-carboxypeptidases of bacteria.

2. Genes associated with cell surface are also involved in biofilm formation

Substantial impacts of bacterial biofilms on infections have led researchers to look for regulatory mechanisms behind their formation to identify novel targets for therapeutics. Global genomic and proteomic studies have pinpointed several genes associated with biofilm development. Many genomic and proteomic analyses were performed to look out for the genes differentially expressed in a biofilm compared with the planktonic cells for both laboratory and clinical isolates of bacteria. Significant proportions of cell-surface-related genes are involved in biofilm formation. A list of genes from various clinically relevant bacterial species involved in the biofilm mode of growth is provided in Tables 1 and 2. Table 1 enlists the genes required for biofilm formation, and Table 2 enlists differentially expressed cell-surface-associated genes in an established form of biofilm (Jefferson, 2004).

E. coli is commonly used as a model organism to study the mechanism of biofilm formation. A DNA microarray of genes expressed during the biofilm versus planktonic mode suggests that the genes responsible mainly for adhesion and autoaggregation were highly expressed in biofilms (Coli et al., 2003). The authors have compared the gene expression pattern in both exponential and stationary phases of planktonic cells with the cells in the biofilm. They observed that among 4290 genes, 233 genes are expressed differentially in biofilm compared with exponential. In comparison, 581 genes are differentially expressed in biofilm than the stationary phase cells. A total of six and nine genes are related to the cell structure in the respective

TABLE 1 Cell surface genes associated with biofilm.

Gene	Protein/function	Organism	Reference
Adhesion			
AbpA	Amylase binding	*S. gordonii*	Wilson and Devine (2003)
sspA/B	Human salivary protein and collagen binding	*S. gordonii*	Wilson and Devine (2003)
gbpA	Polysaccharide formation	*S. mutans*	Wilson and Devine (2003)
tarC	Regulator of glucosyltransferase S and glucan-binding protein	*S. mutans*	Idone et al. (2003)
icaADBC	Intercellular adhesin synthesis	*S. aureus, S. epidermidis*	Wilson and Devine (2003)
hla	Hemolytic toxin	*S. aureus*	Caiazza and O'Toole (2003)
clfA	Clumping factor A, fibrinogen-binding protein	*S. aureus*	Vaudaux et al. (1995)
dltA	D-Alanine esterification of teichoic acids	*S. aureus*	Gross, Cramton, Götz, and Peschel (2001)
atlE	Autolysin/adhesin	*S. epidermidis*	Wilson and Devine (2003)
aap	Accumulation-associated protein	*S. epidermidis*	Hussain, Herrmann, von Eiff, Perdreau-Remington, and Peters (1997)
bopABCD	Biofilm on plastic surfaces operon	*Enterococcus faecalis*	Hufnagel, Koch, Creti, Baldassarri, and Huebner (2004)
esp	Enterococcal surface protein	*E. faecalis*	Toledo-Arana et al. (2001)
agn43	Antigen protein involved aggregation	*E. coli*	Kjaergaard, Schembri, Ramos, Molin, and Klemm (2000)
Cell wall			
pbp2B	Peptidoglycan synthesis	*S. gordonii*	Wilson and Devine (2003)
pbpA	Peptidoglycan synthesis	*P. aeruginosa*	Wilson and Devine (2003)
pbp5	Peptidoglycan synthesis	*S. gordonii*	Wilson and Devine (2003)
dacB2	Peptidoglycan synthesis	*M. tuberculosis*	Wen and Burne (2002)
MSMEG_2432	Peptidoglycan remodeling	*M. smegmatis*	Pandey et al. (2018)
MSMEG_2433	Peptidoglycan remodeling	*M. smegmatis*	Pandey et al. (2018)
mrcB	Peptidoglycan synthesis	*E. coli*	Kumar, Sarkar, Ghosh, and Ghosh (2012)
dacA	Peptidoglycan remodeling	*E. coli*	Gallant et al. (2005)
dacB	Peptidoglycan remodeling	*E. coli*	Gallant et al. (2005)
pbpG	Peptidoglycan remodeling	*E. coli*	Gallant et al. (2005)
AmpG	Peptidoglycan remodeling	*E. coli*	Mallik, Pal, and Ghosh (2018)
glmM	Peptidoglycan synthesis	*S. gordonii*	Shimazu et al. (2008)
bacA	Peptidoglycan synthesis	*S. gordonii*	Loo, Corliss, and Ganeshkumar (2000)
brpA	Possible regulator of autolysis	*S. mutans*	Wen et al. (2018)

Updated list of genes (Jefferson, 2004) required for biofilm formation associated with adhesion and cell wall.

TABLE 2 Genes expressed in planktonic versus biofilm forms and their regulation in different bacterial species.

Gene	Function	Regulation	Species	Reference
Adhesion				
algC	Alginate synthesis	UP	*P. aeruginosa*	Davies, Chakrabarty, and Geesey (1993)
wcaB	Colanic acid synthesis	UP	*E. coli*	Prigent-Combaret and Lejeune (1999)
csgA	Curli	UP	*E. coli*	Prigent-Combaret et al. (2000)
clfA	Clumping factor/fibrinogen binding	DOWN	*S. aureus*	Wolz, Goerke, Landmann, Zimmerli, and Fluckiger (2002)
scaA	Coaggregation	DOWN	*S. gordonii*	Gilmore, Srinivas, Akins, Hatter, and Gilmore (2003)
abpA	Amylase binding	DOWN	*S. gordonii*	Gilmore et al. (2003)
rggD	Glucosyltransferase inducer	UP	*S. gordonii*	Gilmore et al. (2003)
int/CoA	Intrageneric coaggregation-relevant adhesin	DOWN	*S. gordonii*	Gilmore et al. (2003)
flpA	Fibronectin-binding	DOWN	*S. gordonii*	Gilmore et al. (2003)
has	Streptococcal hemagglutinin	DOWN	*S. gordonii*	Gilmore et al. (2003)
Cell wall				
mreC	Cell morphology	UP	*P. aeruginosa*	Gilmore et al. (2003)
dltA	D-Alanine-D-Alanyl carrier	DOWN	*S. gordonii*	Gilmore et al. (2003)
ddl	D-Alanine: D-Alanine	DOWN	*S. gordonii*	Gilmore et al. (2003)
Motility				
PA2128	Probable fimbrial protein	DOWN	*P. aeruginosa*	Whiteley et al. (2001)
pilA	Pilin protein	DOWN	*P. aeruginosa*	Whiteley et al. (2001)
flgD	Flagellar basal-body rod modification protein	DOWN	*P. aeruginosa*	Whiteley et al. (2001)
PA1092	Flagellin type B	DOWN	*P. aeruginosa*	Whiteley et al. (2001)
fliD	Flagellar capping protein	DOWN	*P. aeruginosa*	Whiteley et al. (2001)
flgE	Flagellar hook protein	DOWN	*P. aeruginosa*	Whiteley et al. (2001)
fliC	Flagellar synthesis	DOWN	*E. coli*	Whiteley et al. (2001)

List of differentially expressed genes for cell surface, when whole-genome transcription of biofilm and planktonic cells was compared for clinically relevant bacteria.

studies. A list of the cell surface genes with a statistically significant change in the fold of expression ratios is shown in Table 3. In another study, microarray analyses of *P. aeruginosa* biofilms with planktonic cells showed only 73 differentially expressed genes, and among these, 34 are upregulated while 39 genes are downregulated. They identified seven genes related to motility and attachment in that study (Whiteley et al., 2001). Another study on the whole-genome transcriptomic analysis assessed the genetic requirements during the planktonic, monolayer, and biofilm

TABLE 3 *E. coli* **genes related to cell surface and their differential expression in biofilm.**

Gene	Function	Differential expression
fimA	Major type 1 subunit fimbrin (pilin)	1.58
fimF	Fimbrial morphology	2.82
lpxC	Lipid A biosynthesis	4.03
murA	First step in murein biosynthesis	−1.15
nlpB	Lipoprotein-34	5.05
ompC	Outer membrane protein 1b (Ib; c)	10.90
lpxC	Lipid A biosynthesis	4.03
murA	First step in murein biosynthesis	−1.15
nlpB	Lipoprotein-34	−1.48
ompC	Outer membrane protein 1b (Ib; c)	2.61
ompF	Outer membrane protein 1a (Ia; b; F)	−2.80
ompT	Outer membrane protein 3b (a)	−1.65
pal	Peptidoglycan-associated lipoprotein	−1.28
slp	OMP induced after carbon starvation	6.28
yidC	60 kDa inner membrane protein	−1.30

stage for the pathogen *Vibrio cholerae*. A total of 128 genes are observed as differentially expressed in planktonic cells, 150 in monolayer while 383 genes are in the biofilm stage. Among these, 15 genes are differentially regulated for cell envelope and motility during the planktonic phase, 8 genes for the monolayer stage, and 23 genes are regulated during the biofilm stage (Moorthy & Watnick, 2005).

To study biofilm in various bacterial species such as *V. cholerae* (Fong, Karplus, Schoolnik, & Yildiz, 2006), *Staphylococcus aureus* (Resch et al., 2006), *Acinetobacter baumannii* (Eijkelkamp et al., 2013), and *M. tuberculosis* (Wang et al., 2019), proteomics tools are also found very useful and essential in the recent times. In a proteomic study involving *S. aureus*, a high-resolution 2-D PAGE and MS was used to compare the proteins synthesized by biofilm and planktonic cells. The protein pattern exhibited a significant difference in protein spots for 427 genes. It is observed that the levels of proteins related to cell attachment and peptidoglycan synthesis are significantly enhanced. It also showed the production of proteins for inducing intercellular adhesion operon, and cell wall metabolism is produced at significant levels (Resch et al., 2006). Similarly, in a comparative proteomics analysis between biofilm and planktonic cells of *M. tuberculosis*, a total of 47 different proteins are identified upon comparing with planktonic cells by 2-D PAGE and LC-MS/MS. Among these, seven proteins are associated with the cell wall and cell processes (Wang et al., 2019).

Thus, it is evident from numerous studies that cell-surface-related genes play a crucial role in biofilm formation. It is vital from initiation to detachment, valid for the Gram-positive, Gram-negative, and acid-fast bacterial strains. The involvement of different cell surface structures and associated genes in establishing a biofilm is addressed in the next section.

3. The genes with a role in peptidoglycan transpeptidation and transglycosylation processes are fundamental for biofilm formation and maturation

The peptidoglycan assembly needs the polymerization of NAG and NAM (catalyzed by transglycosylase), and they are cross-linked by muramyl pentapeptide moieties (catalyzed by transpeptidase). Two enzymes exhibit this phenomenon of peptidoglycan remodeling, DD-carboxypeptidase, which removes the terminal D-alanine from muramyl pentapeptide, thereby preventing unwanted cross-link formation and endopeptidases that hydrolyze the already formed unwanted cross-links between glycan units to regularize the PG structure (Ghosh et al., 2008). These phenomena of transpeptidation, transglycosylation, carboxypeptidation, and endopeptidation are catalyzed by a group of proteins named penicillin-binding proteins (PBPs) owing to their ability to bind with penicillins. Some hydrolases and LD-carboxypeptidases are also known to be involved in the process of remodeling (Ghosh et al., 2008). However, it is noteworthy to understand the involvement of the genes in peptidoglycan assembly and remodeling in influencing biofilm formation.

Beta-lactam antibiotics such as penicillins are substrate analogues of the PBPs as these enzymes are fooled by the muramyl pentapeptide mimicking the beta-lactam ring structure.

PBPs are termed acyl-serine transferases as they possess serine moiety at active sites and have three signature motifs: SXXK, SXN, and KTG responsible for catalysis (Ghosh et al., 2008; Sauvage, Kerff, Terrak, Ayala, & Charlier, 2008). Depending on the molecular mass, PBPs are classified as high molecular mass (HMM) or low molecular mass (LMM). There are 12 known PBPs of *E. coli*, and among them, there are 3 class A PBPs (PBP1a, PBP1b, and PBP1c), and 2 are class B PBPs (PBP2 and PBP3). PBP1a and PBPb are major transpeptidases and glycosylases. The two Class B PBPs are known to be monofunctional transpeptidases, and PBP2 is a known elongase while PBP3 is a part of the cell division complex (Den Blaauwen, de Pedro, Nguyen-Distèche, & Ayala, 2008). The rest seven LMM PBPs have roles in cellular separation, peptidoglycan maturation, and recycling. PBP4 and PBP7 have endopeptidase properties. The most abundant and significant carboxypeptidase in *E. coli* is PBP5, which cleaves the terminal D-Ala-D-Ala bonds (Vollmer et al., 2008). PBP6 and PBP6b are sequentially similar to PBP5 and also function as carboxypeptidases. It is reported in *E. coli* that most PBPs come under LMM compared with the HMM, though LMM-PBPs are nonessential for cell survival. Although, little is known about the exact mechanisms of the PBPs from different bacterial species influencing biofilm formation, we attempted to assort the available reports to provide a glimpse of the role of PBPs in the process.

PBPs are classified into class A and class B based on catalytic activity and structure of the N-terminal domain. The C-terminal penicillin-binding domain possesses the transpeptidase activity of both classes. Class A, N-terminal domain is responsible for catalyzing elongation of uncross-linked glycan chains. In Class B, the N-terminal domain is involved in cellular morphogenesis by cross talk with cell cycle proteins (Den Blaauwen et al., 2008). There are some monofunctional enzymes with glycosyltransferase activity of class A PBPs that are also known to exist in a few bacteria but are not functionally characterized (Spratt, Zhou, Taylor, & Merrick, 1996).

3.1 *Escherichia coli*

E. coli PBPs are predicted to be physiologically important for cells and, most importantly, are also involved in biofilm formation (Sauvage et al., 2008). It is observed that the bifunctional transpeptidase PBP1b displays a role in biofilm formation. The PBP1b deletion mutant significantly reduces biofilm formation and motility, while the PBP1a deletion mutant does not have any such effects. Complementation of the PBP1b mutants with ectopically expressed cloned PBP1b reverses the loss of biofilm-forming ability. Interestingly, PBP1b mutants show an enhanced concentration of extracellular indole, suggesting its possible role in biofilm assembly (Kumar et al., 2012).

A study by Gallant et al. showed that the deletion in PBPs such as PBP4, PBP5, and PBP7 individually or in a combination of two or three genes influences the biofilm-forming ability to harbor strains. The effect of deletion was cumulative upon biofilm formation as the triple mutant shows the highest reduction in biofilm formation. There was no growth defect in mutants compared with the parent strains, and the biofilm-forming property is found independent of the medium used (Gallant et al., 2005).

In another study, the deletion of PBP4 and PBP5 affects the adhesion property of the cells to abiotic surfaces and influences the curli formation, suggesting the possible roles of LMM PBPs in surface attachment and curli formation. The endopeptidase and DD-carboxypeptidase mutants have a role in maintaining or creating the 3D architecture of biofilm, and thus, their deletion affects the biofilm thickness and viability (Mallick, Kiran, Maiti, & Ghosh, 2021). The mutant's diminished EPS production is the likely reason for the reduced biofilm formation. The accumulation of xylose and production of xylitol in the EPS of the mutants is speculated as the possible reason for the diminished bacterial adhesion and biofilm formation (Mallick et al., 2021). It is also interesting to mention that a protein, AmpG, known for the peptidoglycan recycling and beta-lactamase induction, also influences biofilm formation as the *ampG* deletion mutants show significantly reduced appearance of biofilms (Mallik et al., 2018). However, there are ample scopes of future studies to understand the exact physiological roles of the *E. coli* PBPs related to biofilm formation and its underlying molecular basis.

3.2 *Pseudomonas aeruginosa*

P. aeruginosa genome carries eight PBPs (1a, 1b, 2, 3, 3a/3x, 4, 5/6, 7), and among them, PBP 1a, 1b, 3, and 3a/3x are categorized into HMM PBPs while PBPs 4, 5/6, and 7 are known to be LMM PBPs. PBP2 is a bifunctional Class B protein in *P. aeruginosa*, though not essential for cell survival. It is reported that the loss of PBP2 can alter biofilm formation in vitro, especially when coupled with the deletion of PBP1a or PBP1b. The very first study elucidating the link between LMM PBPs and biofilm in *P. aeruginosa* was demonstrated by Gallant et al., where they showed that overexpression of TEM1 beta-lactamases (shows a weak DD-CPase activity in vitro) impairs the interaction of *P. aeruginosa* with matrices, which is needed for planktonic to biofilm mode switching (Gallant et al., 2005).

3.3 *Bacillus subtilis*

Bacillus subtilis is the model organism for sporulating Gram-positive bacteria. A total of 16 PBPs are known to have a role in its development and sporulation. It has four class A PBPs, and among them, PBP1 is involved in cell division and sporulation. Among six class B PBPs, PBP2b is involved in cell division. Its PBP4 and PBP5 are equivalent to their counterpart in *E. coli*. PBP5 is required for spore cortex synthesis. PBP4 and PBP5 from *E. coli* and

B. subtilis have peptidase activity toward some noncanonical D-amino-acid-containing peptides. They impart a crucial role in removing D-Ala from muramyl pentapeptides leading to the cross-link formation. These events may affect bacterial growth and biofilm formation, suggesting their potential role in virulence as well (Miyamoto, Katane, Saitoh, Sekine, & Homma, 2020).

3.4 *Mycobacterium* species

While the PBPs from other microbes are studied concerning their respective roles in biofilm formation, little is known about the functional aspects of mycobacterial PBPs in biofilm formation. *M. tuberculosis* possesses two class A PBPs, two class B PBPs, and a lipoprotein that shares few motifs with the class B PBPs. *M. tuberculosis* H37Rv encodes three *E. coli* DD-CPase homologues, including Rv3330 (annotated as *dacB1*), Rv2911 (annotated as *dacB2*), and Rv3627c (direct homolog of *dacB* in *M. smegmatis*). It is reported that DacB2 plays some role in biofilm formation (Bourai, Jacobs, & Narayanan, 2012; Kandasamy & Narayanan, 2015), though the exact mechanism is unknown. PBPA of Mycobacteria is an ortholog of *E. coli* PBP2 required for cell septation/cell division, but its role in biofilm formation is not clear. The genome of *M. smegmatis* mc2155 carries four genes encoding putative DD-CPases-, namely MSMEG_2433, MSMEG_2432, MSMEG_1661, and MSMEG_6113. Both MSMEG_2432 and MSMEG_2433 are dual-function enzymes showing documented roles in maintaining cell shape. Though these two DD-carboxypeptidase homologues may not contribute to biofilm formation, the double deletion mutant of MSMEG_2432 and MSMEG_2433 (Δms33/32) shows the reduced biofilm-forming ability of the cells (Pandey et al., 2018, 2020). Considering the potential of PBPs in various physiological processes, research on the impact of other mycobacterial PBPs in biofilm formation would likely be worthwhile.

4. PBPs of other bacterial species and their role in biofilm formation

Whether it is a Gram-negative or Gram-positive bacterium, all of them possess PBPs in their genome, but very few of them are characterized. The genome of *S. pneumoniae* has six PBPs, namely PBP 1a, 1b, 2x, 2a, 2b, and PBP3. Except for PBP3, the rest are considered high-molecular-weight PBPs (Sauvage et al., 2008). But no evidence or study establishes these PBPs' role in biofilm formation. Instead, in a study involving *S. mutans* there is a report that its PBP1a is required in physiological processes and biofilm formation (Wen, Bitoun, & Liao, 2015). *Neisseria gonorrhoeae* possesses four PBPs, namely PBP1, PBP2, PBP3, and PBP4, with sequence similarity to its *E. coli* counterpart PBP1a, PBP3, PBP4, and PBP7, respectively (Sauvage et al., 2008). However, it is unclear whether these *Pseudomonas* PBPs mimic the already established roles of these *E. coli* PBPs, especially in biofilm formation. Group B *Streptococcus* biofilm regulatory protein A might require the activity of Pbp2 and Pbp2X as suggested by their differential expression in BrpA deletion mutant. Pgp1, a metal-dependent DL-carboxypeptidase, is reportedly involved in biofilm formation in *Campylobacter jejuni* (Frirdich et al., 2012). Few PBPs of *S. aureus*, *Listeria monocytogenes*, *Enterococci* are also known though we have little or no knowledge regarding their roles in biofilm formation. It is crucial to get information about the PBPs of most pathogenic microbes because of the multispecies involvement in biofilms. Learning about each individual is essential to implement better the idea for controlling the bacterial biofilms and the related infections.

5. Cell wall teichoic acid may play a vital role in biofilm initiation and pathogenesis

The cell wall peptidoglycans of Gram-positive bacteria display a modification of anionic polymers, which are carbohydrate-based and play a crucial role in membrane integrity (Neuhaus & Baddiley, 2003) and may be referred to as the outer membrane of bacteria. They are essential for membrane permeability, extracellular interaction and act as a scaffold for the enzymes needed for cell wall remodeling. These glycopolymers are teichoic acids present in both pathogenic and nonpathogenic bacteria. There are two types of teichoic acids, namely lipoteichoic acid and wall teichoic acid, based on their attachment to the plasma membrane and peptidoglycan, respectively (Swoboda, Campbell, Meredith, & Walker, 2010). Their role related to pathogenesis is still unexplored.

Nevertheless, as suggested by a few earlier studies, they seem to play a particular role in biofilm formation. In a study involving *S. aureus*, the mutants without wall teichoic acid (WTA) display reduced adherence to polystyrene plates. Although there was no change in the exopolysaccharide (EPS) production in the mutant, the ability to form biofilm was impaired, suggesting the dependence on teichoic acid. Another study indicates that the *S. aureus* WTA mutant was defective in adherence to nasal epithelial cells and could not colonize in the kidney and spleen endothelial tissues. As WTAs are required to initiate host infection, they play a vital role in pathogenesis. The enzymes responsible for its biosynthesis may serve as antimicrobial targets (Weidenmaier & Peschel, 2008). The inhibition of such cell surface components by tunicamycin can cause defects in biofilm formation, as revealed in a study involving *S. aureus* (Zhu et al., 2018). Teichoic acids are also a significant component in *L. monocytogenes* biofilm matrix that originates from the cell wall (Brauge et al., 2015). Even the abnormal glycosylation (Mistretta et al., 2019) and net charge (Gross et al., 2001) in teichoic acid can obstruct biofilm formation in *S. aureus*. In *S. epidermidis*, the deletion of the gene involved in the biosynthesis of teichoic acid also affected biofilm formation (Holland, Conlon, & O'Gara, 2011). Seeing the limited knowledge about the role of teichoic acid in biofilm formation, the above observations on its effect might prove to be a jackpot to target biofilm of various other Gram-positive bacterial species.

6. Genes and operons involved in exopolysaccharide synthesis are required for biofilm formation

The bacterial metabolic pathways involving the biosynthesis of numerous EPSs are also crucial for biofilm formation. The genes employed to synthesize them are usually clustered within the genome, and their regulation is critical for the successful construction of biofilm. Three major polysaccharides constitute biofilm at different stages: aggregative, protective, and architectural. There are commonly four mechanisms for the production of carbohydrate polymers for ECM, namely (a) Wzx/Wzy-dependent pathway, (b) ABC-transporter-dependent, (c) synthase-dependent, and (d) extracellular synthesis by a single sucrase protein (Schmid, Sieber, & Rehm, 2015).

The EPS synthesis using this pathway and its localization are mentioned in Table 4. The strains utilizing the Wzx/Wzy pathway possess flippase (Wzx) and polymerase (Wzy) genes in their extracellular polysaccharide operons. Pel (pellicle polysaccharide) and PIA (polysaccharide intracellular adhesin) are aggregative EPSs. The intercellular bacterial aggregation for *Staphylococci* biofilms requires PIA and is dependent on the expression of *ica* operon. Pel is produced by *P. aeruginosa* and is synthesized by *pel* operon. Pel operon (15 genes) expression leads to increased pellicle formation and allows aggregation in broth culture. Psl (polysaccharide synthesis locus) mutants form defective biofilms while overexpression displays thicker biofilms. The protective polysaccharides such as alginate allow cellular adhesion and are essential for providing a hydrated biofilm environment that enhances biofilm fluidity (Colvin et al., 2012). It protects from the innate immune response of the host and inhibits phagocytosis. Capsular polysaccharides are found on cell surfaces and are reportedly associated with phospholipids. Capsular polysaccharides inhibit adherence and subsequent biofilm formation. The genes related to its synthesis undergo downregulation upon contact with the substratum. Other known EPS Levan, confirs mucoid phenotype to the

TABLE 4 Exopolysaccharides of bacteria.

Exopolysaccharides	Localization	Pathway involved
Alginate	Cell-associated	Synthase-dependent
Cellulose	Secreted	Synthase-dependent
Colanic acid	Cell-associated	Wzx/Wzy-dependent
Pel	Secreted	Synthase-dependent
Psl	Secreted/cell-associated	Synthase-dependent
PIA	Secreted	Ica operon
Levan	Extracellular	Extracellular
Xanthan	Extracellular	Wzx/Wzy-dependent
Hyaluronic acid	Extracellular	Synthase-dependent

List of exopolysaccharides, their cellular or extracellular localization, and the pathways involved in their synthesis.

Bacillus subtilis and *Streptococcus mutans* biofilms (Limoli et al., 2015).

Further, colanic acid is an architectural polysaccharide, and 19 genes of *wca* cluster are responsible for its synthesis. The absence of *wcaJ* affects the biofilm formation phenomenon in *Klebsiella pneumoniae* by affecting colonic acid biosynthesis (Pal et al., 2019). In *E. coli*, the gene responsible for colonic acid production, *wcaF*, controls the biofilm formation (Zhang & Poh, 2018). In a study involving *B. cereus*, the genomic region required for capsular polysaccharide (eps2) production was overexpressed in the formed biofilm cells. It is thought to be needed for cellular adhesion and biofilm formation (Caro-Astorga et al., 2020). The EPS in *V. cholerae* is encoded by two *vpsl* operons (VpsI and VpsII). There are two major transcriptional regulators, VpsR and VpsT, which positively regulate the biofilm formation, while HapR is a negative regulator of EPS synthesis in *V. cholerae*. PBP4 and PBP5 in *E. coli* are also known to influence the EPS synthesis (Mallick et al., 2021). It is interesting to look for the role of other PBPs in EPS synthesis and regulation. The EPSs are thus an excellent entity to look for drugs to target bacterial biofilms. They are also known to have some biotechnological relevance. Knowledge of the genes responsible for their synthesis may provide a good target for the diminution of bacterial biofilms. Moreover, they may serve as a factory for producing complex polysaccharides with economic values. Extensive study is needed to develop these entities as an antimicrobial target.

7. Role of cell surface appendages and their encoding genes in biofilm formation

7.1 Flagella

The bacterial flagellum is often referred to as a motility organelle, but in a few years, the notion has changed and correlated with other biological functions. The whole flagellum is crucial from adhesion to invasion. Several bacterial species such as *E. coli*, *P. aeruginosa*, and *Clostridium difficile*, the flagellin or its cap proteins help in adhesion. The bacterial flagellar apparatus comprises 20 different proteins. The major subunit FliC plays a crucial role in immune response and is also involved in invasion via lipid rafts (Haiko & Westerlund-Wikström, 2013). In *V. cholerae*, the flagellum allows bacteria to detect a surface and is required for the initial step of biofilm formation. Both flagella and pili work synergistically for irreversible attachment with the substratum. This contact induces the expression of *vpsI* and *vpsII* operons that leads the bacterium to the third stage of biofilm development (Lauriano, Ghosh, Correa, & Klose, 2004). The flagellum of *V. cholera* comprises five flagellin subunits: FlaA, B, C, D and E. FlaA is the only flagellin required for filament synthesis and motility. Many adhesion genes are coregulated with the flagellar gene. It has also been shown that polymyxin B can interfere with biofilm formation in *V. cholera* during the initial step by affecting flagellar synthesis and shape (Giacomucci, Cros, Perron, Mathieu-Denoncourt, & Duperthuy, 2019).

P. aeruginosa does not require flagella or type IV pili for the initial attachment or to ensure the prominent biofilm architecture. It is supported by the fact that *ΔpilA*, *ΔfliM*, and *ΔpilAΔfliM* strains displayed different biofilm structures as compared with the wild-type strain. For initiation of biofilm formation, *P. aeruginosa* undergoes numerous genetic adaptations. First, it compromises its motility and forms a rugose small-colony variant (RSCV) phenotype. The significance of this type of phenotype is the overproduction of the EPSs Pel and Psl. Both the adaptations are linked as the flagellar mutation that leads to the overexpression of Pel and PsI in a contact-dependent manner (Harrison et al., 2020). In *E. coli*, it is also documented that bacterial flagellar class 3 genes undergo downregulation upon surface contact leading a protein *ycgR* to impair the rotation by interacting with FliG that acts as a brake. Most bacterial species have such brakes. So, it is interesting to play genetically with such brakes to control the biofilm initiation phenomenon (Pratt & Kolter, 1998).

7.2 Pili

The cell surface appendage pili have roles in forming a biofilm. Pili are essential for cellular adhesion to substratum. The model organism for biofilm formation, *P. aeruginosa* forms biofilms with spatial structure because of the presence of type IV pili at its surface. They play a significant role in microcolony formation, cellular aggregation, and biofilm differentiation. They also contribute to forming heterogeneous biofilm structures. In *Neisseria meningitidis*, it has been shown that pili and EPS allow cells to attach to the substratum and also help in creating biofilm architecture. It is also documented that bacterial pathogens such as *Enterococcus faecalis*, *Streptococcus pyogenes*, require pili to bind to the epithelial cells and biofilm formation. In nonpathogenic bacterium *Lactobacillus*, the pili allow firm adhesion to withstand the harsh natural environment (Drame et al., 2021). In *E. coli*, type I pili are essential for the initial attachment. The absence of functional genes responsible for the regulation and biogenesis of type I pili leads to defective adhesion as type I pili contain the adhesin. The role of flagella in adhesion to the abiotic surface also relies mainly on the functional type I pili. It is noteworthy that though *P. aeruginosa* strains lacking type IV pili can still form a biofilm monolayer, *E. coli* strain without a functional type I pili loses its ability to develop a sessile population (Pratt & Kolter, 1998). The role of pili in cell-cell interaction is studied, but its role in establishing the biomechanical property of biofilm is little known. The complex microstructure of biofilm needs to be revealed to control the biofilm process. Moreover, as the pili can shape the biofilm, it can be an essential handle in the biotechnological or probiotic industries. Thus, understanding the role of pili in generating sessile populations of cells can modulate biofilm's spatial structure.

7.3 Lipopolysaccharides (LPSs)

Lipopolysaccharides (LPSs) are the major cell surface component of the outer membrane of Gram-negative bacteria and are also involved in adhesion biofilm formation and colonization (Huszczynski, Lam, & Khursigara, 2020). It consists of Lipid A, core oligosaccharides, and O-antigen. Though the components primarily required to form biofilm are not fully understood, O-antigens are supposedly involved in *E. coli*, both static and dynamic growth conditions (Kumar et al., 2015; Lee, Lee, Hwang, Park, & Lee, 2016). It is also reported that the absence of O-antigen also reduces the biofilm formation of *Actinobacillus pleuropneumoniae* (Hathroubi et al., 2015). The *E. coli* mutants with defective Hep biosynthesis pathway that synthesizes its sugar components show abnormality in forming biofilm (Nakao, Ramstedt, Wai, & Uhlin, 2012). LPSs are also shown to be important for *P. aeruginosa* biofilm

(Huszczynski et al., 2020). In a study involving Gram-negative bacteria, it has been observed that deacylation of LPS rendered bacteria biofilm defective (Lee et al., 2016).

Therefore, for the above studies, it is evident that the cell surface components might play a key role in forming biofilms. However, large areas remain unexplored for correlating biofilms with bacterial cell surfaces.

8. Biotechnological implications of understanding surface genes related to biofilms

Bacterial biofilms may serve as factories for producing valuable compounds such as surfactants, ethanol, table olives, etc. Understanding the complex genetic machinery involved can facilitate us in developing useful products and their production. A difference between the expression pattern of planktonic and biofilm cells also is already discussed in the above sections. A clear idea of the genes involved in the planktonic and sessile growth forms can help us design the roadmap to obtain the desired product. It is widely observed that the genes mainly responsible for iron-sulfur metabolism, lipid metabolism, secondary metabolite production, carbohydrate and amino acid transport, stress response, and efflux pumps are induced in the cells forming sessile populations. Modulating these expressions can allow us to obtain the products in the ECM. ECM may serve as a reservoir of the cellular products possessing utility in cosmetic, food, and pharmaceutical industries. We can biotechnologically engineer the surface genes of the bacteria. For example, the production of acetic acid by acetic-acid-producing bacteria in the industries has widely accepted the use of biofilm where the productivity depends on the features of surface employed for biofilm formation. Thus, the knowledge of the cell surface genes that can affect the surface attachment is crucial. The surface proteins are also vital for surface repellence that may affect the microstructure of biofilms and productivity.

In industrial biotechnology, the routine method to produce valuable compounds is performed by cultivating suspended cells or using the immobilized enzyme instead of immobilized microbes. The concept of using biofilms as biofactories is in its naive stage and needs further study. The prospective approaches were examined primarily on a laboratory scale, and implementing such designs upon validation on a larger scale would benefit us in future.

9. Summary and possible future directions involving cell-surface components

Several genes and their products related to the cell wall/cell envelope involved in biofilm formation are reported. Nevertheless, the exact molecular mechanisms by which such gene products influence biofilm formation are unclear, from the establishment to dispersion. The proteins encoded by these genes might also be involved in other cellular processes and ultimately biofilm formation. Gathering more information about these genes will help us better understand biofilm formation's physiology in a better way, including the involvement of the interacting proteins in the process. Antibiotics are the primary choice for treating bacterial infections. However, sessile bacterial populations within biofilms are making them resistant, posing threats to antibiotic treatment. Whole-genome transcription profiling of biofilm-forming bacterial pathogens has revealed several uncharacterized genes. Among them, many cell-surface-related genes are not enough characterized and leave a scope to look for their promising roles in biofilm formation. Similarly, penicillin-binding proteins of all the clinically relevant bacterial species, especially low-molecular-mass PBPs that mainly express in the late-log to mid stationary cells, are not physiologically identified.

It is also apparent that the structural features of the cell-surface associated proteins warrant further study. These proteins might serve as the target to block the initial attachment to the surface and, consequently, biofilm formation. For example, a study by Svensson and coworkers reports new molecules mimicking the pilus protein that block pili assembly and reduce the expression of pili proteins, which in turn lowers biofilm formation in uropathogenic *E. coli* (Pinkner et al., 2006; Svensson et al., 2001). Likewise, searching for compounds that can block adhesion or block the glycosylation of adhesins might prove beneficial to control the initiation of the sessile community of cells. In addition, enhancement of biofilm dispersion may also be achieved through modeling EPS matrix by tweaking the pathways involved in its production. The polymicrobial biofilms present further threat in treatment; so, designing an effective strategy for dispersing microbial consortia, possibly by preventing the cross talk, through ECM and cell wall, between different species would be necessary. Though the challenges are unlimited, the cure should also be within. Therefore, it is the need of the hour to bridge the gap in our knowledge to accomplish the ultimate aim of targeting the reduction of bacterial biofilms.

References

Bourai, N., Jacobs, W. R., & Narayanan, S. (2012). Deletion and overexpression studies on DacB2, a putative low molecular mass penicillin binding protein from *Mycobacterium tuberculosis* H 37Rv. *Microbial Pathogenesis*, *52*(2), 109–116. https://doi.org/10.1016/j.micpath.2011.11.003.

Branda, S. S., Vik, Å., Friedman, L., & Kolter, R. (2005). Biofilms: The matrix revisited. *Trends in Microbiology*, *13*(1). https://doi.org/10.1016/j.tim.2004.11.006.

Brauge, T., Sadovskaya, I., Faille, C., Benezech, T., Maes, E., Guerardel, Y., et al. (2015). Teichoic acid is the major polysaccharide present in the *Listeria monocytogenes* biofilm matrix. *FEMS Microbiology Letters*, *363*(2). https://doi.org/10.1093/femsle/fnv229.

Caiazza, N. C., & O'Toole, G. A. (2003). Alpha-toxin is required for biofilm formation by *Staphylococcus aureus*. *Journal of Bacteriology*, *185*(10), 3214–3217. https://doi.org/10.1128/JB.185.10.3214-3217.2003.

Caro-Astorga, J., Álvarez-Mena, A., Hierrezuelo, J., Guadix, J. A., Heredia-Ponce, Z., Arboleda-Estudillo, Y., et al. (2020). Two genomic regions encoding exopolysaccharide production systems have complementary functions in *B. cereus* multicellularity and host interaction. *Scientific Reports*, *10*(1). https://doi.org/10.1038/s41598-020-57970-3.

Cava, F., de Pedro, M. A., Lam, H., Davis, B. M., & Waldor, M. K. (2011). Distinct pathways for modification of the bacterial cell wall by non-canonical D-amino acids. *The EMBO Journal*, *30*(16), 3442–3453. https://doi.org/10.1038/emboj.2011.246.

Coli, E., Schembri, B. A., Kjaergaard, K., Klemm, P., Schembri, M. A., Kjaergaard, K., et al. (2003). Global gene expression in *Escherichia coli* biofilms. *Molecular Microbiology*, *48*(1).

Colvin, K. M., Irie, Y., Tart, C. S., Urbano, R., Whitney, J. C., Ryder, C., et al. (2012). The Pel and Psl polysaccharides provide *Pseudomonas aeruginosa* structural redundancy within the biofilm matrix. *Environmental Microbiology*, *14*(8), 1913–1928. https://doi.org/10.1111/j.1462-2920.2011.02657.x.

Davies, D. G., Chakrabarty, A. M., & Geesey, G. G. (1993). Exopolysaccharide production in biofilms: Substratum activation of alginate gene expression by *Pseudomonas aeruginosa*. *Applied and Environmental Microbiology*, *59*(4), 1181–1186. https://doi.org/10.1128/aem.59.4.1181-1186.1993.

Den Blaauwen, T., de Pedro, M. A., Nguyen-Distèche, M., & Ayala, J. A. (2008). Morphogenesis of rod-shaped sacculi. *FEMS Microbiology Reviews*, *32*(2), 321–344. https://doi.org/10.1111/j.1574-6976.2007.00090.x.

Dörr, T., Moynihan, P. J., & Mayer, C. (2019). Editorial: Bacterial cell wall structure and dynamics. *Frontiers in Microbiology*, *10*. https://doi.org/10.3389/fmicb.2019.02051. Frontiers Media S.A.

Drame, I., Lafforgue, C., Formosa-Dague, C., Chapot-Chartier, M. P., Piard, J. C., Castelain, M., et al. (2021). Pili and other surface proteins influence the structure and the nanomechanical properties of *Lactococcus lactis* biofilms. *Scientific Reports*, *11*(1), 4846. https://doi.org/10.1038/s41598-021-84030-1.

Eijkelkamp, B. A., Stroeher, U. H., Hassan, K. A., Elbourne, L. D. H., Paulsen, I. T., & Brown, M. H. (2013). H-NS plays a role in expression of *Acinetobacter baumannii* virulence features. *Infection and Immunity*, *81*(7), 2574–2583. https://doi.org/10.1128/IAI.00065-13.

Fong, J. C. N., Karplus, K., Schoolnik, G. K., & Yildiz, F. H. (2006). Identification and characterization of RbmA, a novel protein required for the development of rugose colony morphology and biofilm structure in *Vibrio cholerae*. *Journal of Bacteriology*, *188*(3), 1049–1059. https://doi.org/10.1128/JB.188.3.1049-1059.2006.

Frirdich, E., Biboy, J., Adams, C., Lee, J., Ellermeier, J., Gielda, L. D., et al. (2012). Correction: Peptidoglycan-modifying enzyme Pgp1 is required for helical cell shape and pathogenicity traits in *Campylobacter jejuni*. *PLoS Pathogens*, *8*(10). https://doi.org/10.1371/annotation/aac1c73d-b014-4cee-8dbc-eb8472de211d.

Gallant, C. V., Daniels, C., Leung, J. M., Ghosh, A. S., Young, K. D., Kotra, L. P., et al. (2005). Common β-lactamases inhibit bacterial biofilm formation. *Molecular Microbiology*, *58*(4), 1012–1024. https://doi.org/10.1111/j.1365-2958.2005.04892.x.

Ghosh, A. S., Chowdhury, C., & Nelson, D. E. (2008). Physiological functions of D-alanine carboxypeptidases in *Escherichia coli*. *Trends in Microbiology*, *16*(7), 309–317. https://doi.org/10.1016/j.tim.2008.04.006.

Giacomucci, S., Cros, C. D. N., Perron, X., Mathieu-Denoncourt, A., & Duperthuy, M. (2019). Flagella-dependent inhibition of biofilm formation by sub-inhibitory concentration of polymyxin B in *Vibrio cholerae*. *PLoS One*, *14*(8). https://doi.org/10.1371/journal.pone.0221431.

Gilmore, K. S., Srinivas, P., Akins, D. R., Hatter, K. L., & Gilmore, M. S. (2003). Growth, development, and gene expression in a persistent *Streptococcus gordonii* biofilm. *Infection and Immunity*, *71*(8), 4759–4766. https://doi.org/10.1128/IAI.71.8.4759-4766.2003.

Gross, M., Cramton, S. E., Götz, F., & Peschel, A. (2001). Key role of teichoic acid net charge in *Staphylococcus aureus* colonization of artificial surfaces. *Infection and Immunity*, *69*(5), 3423–3426. https://doi.org/10.1128/IAI.69.5.3423-3426.2001.

Haiko, J., & Westerlund-Wikström, B. (2013). The role of the bacterial flagellum in adhesion and virulence. *Biology*, *2*(4), 1242–1267. MDPI AG https://doi.org/10.3390/biology2041242.

Harrison, J. J., Almblad, H., Irie, Y., Wolter, D. J., Eggleston, H. C., Randall, T. E., et al. (2020). Elevated exopolysaccharide levels in *Pseudomonas aeruginosa* flagellar mutants have implications for biofilm growth and chronic infections. *PLoS Genetics*, *16*(6). https://doi.org/10.1371/journal.pgen.1008848.

Hathroubi, S., Hancock, M. A., Bossé, J. T., Langford, P. R., Tremblay, Y. D. N., Labrie, J., et al. (2015). Surface polysaccharide mutants reveal that absence of O antigen reduces biofilm formation of *Actinobacillus pleuropneumoniae*. *Infection and Immunity*, *84*(1), 127–137. https://doi.org/10.1128/IAI.00912-15.

Holland, L. M., Conlon, B., & O'Gara, J. P. (2011). Mutation of tagO reveals an essential role for wall teichoic acids in *Staphylococcus epidermidis* biofilm development. *Microbiology*, *157*(2), 408–418. https://doi.org/10.1099/mic.0.042234-0.

Hufnagel, M., Koch, S., Creti, R., Baldassarri, L., & Huebner, J. (2004). A putative sugar-binding transcriptional regulator in a novel gene locus in *Enterococcus faecalis* contributes to production of biofilm and prolonged bacteremia in mice. *The Journal of Infectious Diseases*, *189*(3), 420–430. https://doi.org/10.1086/381150.

Hussain, M., Herrmann, M., von Eiff, C., Perdreau-Remington, F., & Peters, G. (1997). A 140-kilodalton extracellular protein is essential for the accumulation of *Staphylococcus epidermidis* strains on surfaces. *Infection and Immunity*, *65*(2), 519–524. https://doi.org/10.1128/iai.65.2.519-524.1997.

Huszczynski, S. M., Lam, J. S., & Khursigara, C. M. (2020). The role of *Pseudomonas aeruginosa* lipopolysaccharide in bacterial pathogenesis and physiology. *Pathogens*, *9*(1). https://doi.org/10.3390/pathogens9010006. MDPI AG.

Idone, V., Brendtro, S., Gillespie, R., Kocaj, S., Peterson, E., Rendi, M., et al. (2003). Effect of an orphan response regulator on *Streptococcus mutans* sucrose-dependent adherence and cariogenesis. *Infection and Immunity*, *71*(8), 4351–4360. https://doi.org/10.1128/IAI.71.8.4351-4360.2003.

Jefferson, K. K. (2004). What drives bacteria to produce a biofilm? *FEMS Microbiology Letters*, *236*(2), 163–173. https://doi.org/10.1016/j.femsle.2004.06.005.

Kandasamy, S., & Narayanan, S. (2015). Phenotypic characterization of a novel double knockout PknI/DacB2 from *Mycobacterium tuberculosis*. *Microbiological Research*, *170*, 255–262. https://doi.org/10.1016/j.micres.2014.10.002.

Kjaergaard, K., Schembri, M. A., Ramos, C., Molin, S., & Klemm, P. (2000). Antigen 43 facilitates formation of multispecies biofilms. *Environmental Microbiology*, *2*(6), 695–702. https://doi.org/10.1046/j.1462-2920.2000.00152.x.

Kostakioti, M., Hadjifrangiskou, M., & Hultgren, S. J. (2013). Bacterial biofilms: Development, dispersal, and therapeutic strategies in the dawn of the postantibiotic era. *Cold Spring Harbor Perspectives in Medicine*, *3*(4), a010306. https://doi.org/10.1101/cshperspect.a010306.

Kumar, A., Mallik, D., Pal, S., Mallick, S., Sarkar, S., Chanda, A., et al. (2015). *Escherichia coli* O8-antigen enhances biofilm formation under agitated conditions. *FEMS Microbiology Letters*, *362*(15), 1–8. https://doi.org/10.1093/femsle/fnv112.

Kumar, A., Sarkar, S. K., Ghosh, D., & Ghosh, A. S. (2012). Deletion of penicillin-binding protein 1b impairs biofilm formation and motility in *Escherichia coli*. *Research in Microbiology*, *163*(4), 254–257. https://doi.org/10.1016/j.resmic.2012.01.006.

Lauriano, C. M., Ghosh, C., Correa, N. E., & Klose, K. E. (2004). The sodium-driven flagellar motor controls exopolysaccharide expression in *Vibrio cholerae*. *Journal of Bacteriology*, *186*(15), 4864–4874. https://doi.org/10.1128/JB.186.15.4864-4874.2004.

Lee, K. J., Lee, M. A., Hwang, W., Park, H., & Lee, K. H. (2016). Deacylated lipopolysaccharides inhibit biofilm formation by Gram-negative bacteria. *Biofouling*, *32*(7), 711–723. https://doi.org/10.1080/08927014.2016.1193595.

Limoli, D. H., Jones, C. J., & Wozniak, D. J. (2015). Bacterial extracellular polysaccharides in biofilm formation and function. *Microbiology Spectrum*, *3*(3). https://doi.org/10.1128/microbiolspec.MB-0011-2014.

Loo, C. Y., Corliss, D. A., & Ganeshkumar, N. (2000). *Streptococcus gordonii* biofilm formation: Identification of genes that code for biofilm phenotypes. *Journal of Bacteriology*, *182*(5), 1374–1382. https://doi.org/10.1128/JB.182.5.1374-1382.2000.

Mahapatra, S., Crick, D. C., McNeil, M. R., & Brennan, P. J. (2008). Unique structural features of the peptidoglycan of *Mycobacterium leprae*. *Journal of Bacteriology*, *190*(2), 655–661. https://doi.org/10.1128/JB.00982-07.

Mainardi, J. L., Fourgeaud, M., Hugonnet, J. E., Dubost, L., Brouard, J. P., Ouazzani, J., et al. (2005). A novel peptidoglycan cross-linking enzyme for a β-lactam-resistant transpeptidation pathway. *Journal of Biological Chemistry*, *280*(46), 38146–38152. https://doi.org/10.1074/jbc.M507384200.

Mallick, S., Kiran, S., Maiti, T. K., & Ghosh, A. S. (2021). PBP4 and PBP5 are involved in regulating exopolysaccharide synthesis during *Escherichia coli* biofilm formation. *Microbiology (United Kingdom)*, *167*(3). https://doi.org/10.1099/mic.0.001031.

Mallik, D., Pal, S., & Ghosh, A. S. (2018). Involvement of AmpG in mediating a dynamic relationship between serine beta-lactamase induction and biofilm-forming ability of *Escherichia coli*. *FEMS Microbiology Letters*, *365*(8), fny065. https://doi.org/10.1093/femsle/fny065.

Mistretta, N., Brossaud, M., Telles, F., Sanchez, V., Talaga, P., & Rokbi, B. (2019). Glycosylation of *Staphylococcus aureus* cell wall teichoic acid is influenced by environmental conditions. *Scientific Reports*, *9*(1). https://doi.org/10.1038/s41598-019-39929-1.

Miyamoto, T., Katane, M., Saitoh, Y., Sekine, M., & Homma, H. (2020). Involvement of penicillin-binding proteins in the metabolism of a bacterial peptidoglycan containing a non-canonical D-amino acid. *Amino Acids*, *52*(3), 487–497. https://doi.org/10.1007/s00726-020-02830-7.

Moorthy, S., & Watnick, P. I. (2005). Identification of novel stage-specific genetic requirements through whole genome transcription profiling of *Vibrio cholerae* biofilm development. *Molecular Microbiology*, *57*(6), 1623–1635. https://doi.org/10.1111/j.1365-2958.2005.04797.x.

Moynihan, P. J., & Clarke, A. J. (2011). O-Acetylated peptidoglycan: Controlling the activity of bacterial autolysins and lytic enzymes of innate immune systems. *The International Journal of Biochemistry & Cell Biology*, *43*(12), 1655–1659. https://doi.org/10.1016/j.biocel.2011.08.007.

Nakao, R., Ramstedt, M., Wai, S. N., & Uhlin, B. E. (2012). Enhanced biofilm formation by *Escherichia coli* LPS mutants defective in Hep biosynthesis. *PLoS One*, *7*(12). https://doi.org/10.1371/journal.pone.0051241.

Neuhaus, F. C., & Baddiley, J. (2003). A continuum of anionic charge: Structures and functions of D-alanyl-teichoic acids in gram-positive bacteria. *Microbiology and Molecular Biology Reviews: MMBR*, *67*(4), 686–723. https://doi.org/10.1128/MMBR.67.4.686-723.2003.

Pal, S., Verma, J., Mallick, S., Rastogi, S. K., Kumar, A., & Ghosh, A. S. (2019). Absence of the glycosyltransferase WcaJ in Klebsiella pneumoniae ATCC13883 affects biofilm formation, increases polymyxin resistance and reduces murine macrophage activation. *Microbiology*, *165*(8), 891–904. https://doi.org/10.1099/mic.0.000827.

Pandey, S. D., Jain, D., Kumar, N., Adhikary, A., Ganesh, K. N., & Ghosh, A. S. (2020). MSMEG_2432 of *Mycobacterium smegmatis* mc2155 is a dual function enzyme that exhibits dd-carboxypeptidase and β-lactamase activities. *Microbiology (United Kingdom)*, *166*(6), 546–554. https://doi.org/10.1099/mic.0.000902.

Pandey, S. D., Pal, S., Ganesh Kumar, N., Bansal, A., Mallick, S., & Ghosh, A. S. (2018). Two DD-carboxypeptidases from *Mycobacterium smegmatis* affect cell surface properties through regulation of peptidoglycan cross-linking and glycopeptidolipids. *Journal of Bacteriology*, *200*(14). https://doi.org/10.1128/JB.00760-17.

Pinkner, J. S., Remaut, H., Buelens, F., Miller, E., Åberg, V., Pemberton, N., et al. (2006). Rationally designed small compounds inhibit pilus biogenesis in uropathogenic bacteria. *Proceedings of the National Academy of Sciences of the United States of America*, *103*(47). https://doi.org/10.1073/pnas.060679510.

Pratt, L. A., & Kolter, R. (1998). Genetic analysis of *Escherichia coli* biofilm formation: Roles of flagella, motility, chemotaxis and type I pili. *Molecular Microbiology*, *30*, 285–293.

Prigent-Combaret, C., & Lejeune, P. (1999). Monitoring gene expression in biofilms. *Methods in Enzymology*, *310*, 56–79. https://doi.org/10.1016/s0076-6879(99)10006-5.

Prigent-Combaret, C., Prensier, G., Le Thi, T. T., Vidal, O., Lejeune, P., & Dorel, C. (2000). Developmental pathway for biofilm formation in curli-producing Escherichia coli strains: Role of flagella, curli and colanic acid. *Environmental Microbiology*, *2*(4), 450–464. https://doi.org/10.1046/j.1462-2920.2000.00128.x.

Resch, A., Leicht, S., Saric, M., Pásztor, L., Jakob, A., Götz, F., et al. (2006). Comparative proteome analysis of *Staphylococcus aureus* biofilm and planktonic cells and correlation with transcriptome profiling. *Proteomics*, *6*(6), 1867–1877. https://doi.org/10.1002/pmic.200500531.

Sanders, A. N., Wright, L. F., & Pavelka, M. S. (2014). Genetic characterization of mycobacterial L,D-transpeptidases. *Microbiology (Reading, England)*, *160*(Pt 8), 1795–1806. https://doi.org/10.1099/mic.0.078980-0.

Sauvage, E., Kerff, F., Terrak, M., Ayala, J. A., & Charlier, P. (2008). The penicillin-binding proteins: Structure and role in peptidoglycan biosynthesis. *FEMS Microbiology Reviews*, *32*(2), 234–258. https://doi.org/10.1111/j.1574-6976.2008.00105.x.

Scheurwater, E., Reid, C. W., & Clarke, A. J. (2008). Lytic transglycosylases: Bacterial space-making autolysins. *The International Journal of Biochemistry & Cell Biology*, *40*(4), 586–591. https://doi.org/10.1016/j.biocel.2007.03.018.

Schmid, J., Sieber, V., & Rehm, B. (2015). Bacterial exopolysaccharides: Biosynthesis pathways and engineering strategies. *Frontiers in Microbiology*, *6*, 496. https://doi.org/10.3389/fmicb.2015.00496.

Shimazu, K., Takahashi, Y., Uchikawa, Y., Shimazu, Y., Yajima, A., Takashima, E., et al. (2008). Identification of the *Streptococcus gordonii* glmM gene encoding phosphoglucosamine mutase and its role in bacterial cell morphology, biofilm formation, and sensitivity to antibiotics. *FEMS Immunology and Medical Microbiology*, *53*(2), 166–177. https://doi.org/10.1111/j.1574-695X.2008.00410.x.

Spratt, B. G., Zhou, J., Taylor, M., & Merrick, M. J. (1996). Monofunctional biosynthetic peptidoglycan transglycosylases. *Molecular Microbiology*, *19*(3), 639–640. https://doi.org/10.1046/j.1365-2958.1996.442924.x.

Svensson, A., Larsson, A., Emtenas, H., Hedenstrom, M., Fex, T., Hultgren, S. J., et al. (2001). Design and evaluation of pilicides: Potential novel antibacterial agents directed against uropathogenic *Escherichia coli*. *Chembiochem*, *2*, 915–918. https://doi.org/10.1002/1439-7633(20011203)2:12<915::AID-CBIC915>3.0.CO;2-M.

Swoboda, J. G., Campbell, J., Meredith, T. C., & Walker, S. (2010). Wall teichoic acid function, biosynthesis, and inhibition. *Chembiochem: A European Journal of Chemical Biology*, *11*(1), 35–45. https://doi.org/10.1002/cbic.200900557.

Toledo-Arana, A., Valle, J., Solano, C., Arrizubieta, M. J., Cucarella, C., Lamata, M., et al. (2001). The enterococcal surface protein, Esp, is involved in *Enterococcus faecalis* biofilm formation. *Applied and Environmental Microbiology*, *67*(10), 4538–4545. https://doi.org/10.1128/AEM.67.10.4538-4545.2001.

Vaudaux, P. E., François, P., Proctor, R. A., McDevitt, D., Foster, T. J., Albrecht, R. M., et al. (1995). Use of adhesion-defective mutants of *Staphylococcus aureus* to define the role of specific plasma proteins in promoting bacterial adhesion to canine arteriovenous shunts. *Infection and Immunity*, *63*(2), 585–590. https://doi.org/10.1128/iai.63.2.585-590.1995.

Vincent, A. T., Nyongesa, S., Morneau, I., Reed, M. B., Tocheva, E. I., & Veyrier, F. J. (2018). The mycobacterial cell envelope: A relict from the past or the result of recent evolution? *Frontiers in Microbiology*, *9*(Oct). https://doi.org/10.3389/fmicb.2018.02341. Frontiers Media S.A.

Vollmer, W., Blanot, D., & De Pedro, M. A. (2008). Peptidoglycan structure and architecture. *FEMS Microbiology Reviews*, *32* (2). https://doi.org/10.1111/j.1574-6976.2007.00094.x.

Wang, C., Zhang, Q., Wang, Y., Tang, X., An, Y., Li, S., et al. (2019). Comparative proteomics analysis between biofilm and planktonic cells of *Mycobacterium tuberculosis*. *Electrophoresis*, *40*(20), 2736–2746. https://doi.org/10.1002/elps.201900030.

Weidenmaier, C., & Peschel, A. (2008). Teichoic acids and related cell-wall glycopolymers in Gram-positive physiology and host interactions. *Nature Reviews Microbiology*, *6*(4), 276–287. https://doi.org/10.1038/nrmicro1861.

Wen, Z. T., Bitoun, J. P., & Liao, S. (2015). PBP1a-deficiency causes major defects in cell division, growth and biofilm formation by *Streptococcus mutans*. *PLoS One*, *10*(4). https://doi.org/10.1371/journal.pone.0124319.

Wen, Z. T., & Burne, R. A. (2002). Functional genomics approach to identifying genes required for biofilm development by *Streptococcus mutans*. *Applied and Environmental Microbiology*, *68*(3), 1196–1203. https://doi.org/10.1128/AEM.68.3.1196-1203.2002.

Wen, Z. T., Scott-Anne, K., Liao, S., De, A., Luo, M., Kovacs, C., et al. (2018). Deficiency of BrpA in *Streptococcus mutans* reduces virulence in rat caries model. *Molecular Oral Microbiology*, *33*(5), 353–363. https://doi.org/10.1111/omi.12230.

Whiteley, M., Bangera, M. G., Bumgarner, R. E., Parsek, M. R., Teitzel, G. M., Lory, S., et al. (2001). Gene expression in *Pseudomonas aeruginosa* biofilms. *Nature*, *413*(6858), 860–864. https://doi.org/10.1038/35101627.

Wilson, M., & Devine, D. (Eds.). (2003). *Medical implications of biofilms* Cambridge University Press.

Wolz, C., Goerke, C., Landmann, R., Zimmerli, W., & Fluckiger, U. (2002). Transcription of clumping factor A in attached and unattached *Staphylococcus aureus* in vitro and during device-related infection. *Infection and Immunity*, *70*(6), 2758–2762. https://doi.org/10.1128/IAI.70.6.2758-2762.2002.

Zhang, J., & Poh, C. L. (2018). Regulating exopolysaccharide gene wcaF allows control of *Escherichia coli* biofilm formation. *Scientific Reports*, *8*(1). https://doi.org/10.1038/s41598-018-31161-7.

Zhu, X., Liu, D., Singh, A. K., Drolia, R., Bai, X., Tenguria, S., et al. (2018). Tunicamycin mediated inhibition of wall teichoic acid affects *Staphylococcus aureus* and *Listeria monocytogenes* cell morphology, biofilm formation and virulence. *Frontiers in Microbiology*, *9*(Jul). https://doi.org/10.3389/fmicb.2018.01352.

Further reading

Bhakta, S., & Basu, J. (2002). Overexpression, purification and biochemical characterization of a class A high-molecular-mass penicillin-binding protein (PBP), PBP1* and its soluble derivative from *Mycobacterium tuberculosis*. *The Biochemical Journal*, *361*(Pt 3), 635–639. https://doi.org/10.1042/0264-6021:3610635.

Dasgupta, A., Datta, P., Kundu, M., & Basu, J. (2006). The serine/threonine kinase PknB of *Mycobacterium tuberculosis* phosphorylates PBPA, a penicillin-binding protein required for cell division. *Microbiology (Reading, England)*, *152*(Pt 2), 493–504. https://doi.org/10.1099/mic.0.28630-0.

Ealand, C. S., Asmal, R., Mashigo, L., Campbell, L., & Kana, B. D. (2019). Characterization of putative DD-carboxypeptidase-encoding genes in *Mycobacterium smegmatis*. *Scientific Reports*, *9*(1). https://doi.org/10.1038/s41598-019-41001-x.

Goffin, C., & Ghuysen, J.-M. (2002). Biochemistry and comparative genomics of SxxK superfamily acyltransferases offer a clue to the mycobacterial paradox: Presence of penicillin-susceptible target proteins versus lack of efficiency of penicillin as therapeutic agent. *Microbiology and Molecular Biology Reviews: MMBR*, *66*(4), 702–738. table of contents https://doi.org/10.1128/MMBR.66.4.702-738.2002.

Kieser, K. J., Baranowski, C., Chao, M. C., Long, J. E., Sassetti, C. M., Waldor, M. K., et al. (2015). Peptidoglycan synthesis in *Mycobacterium tuberculosis* is organized into networks with varying drug susceptibility. *Proceedings of the National Academy of Sciences of the United States of America*, *112*(42), 13087–13092. https://doi.org/10.1073/pnas.1514135112.

Maitra, A., Munshi, T., Healy, J., Martin, L. T., Vollmer, W., Keep, N. H., et al. (2019). Cell wall peptidoglycan in *Mycobacterium tuberculosis*: An Achilles' heel for the TB-causing pathogen. *FEMS Microbiology Reviews*, *43*(5), 548–575. https://doi.org/10.1093/femsre/fuz016.

Chapter 22

Small regulatory RNAs in microbial pathogenesis and biofilm formation: An emerging role as potential drug targets

Thirupathi Kasthuri, Singh Dikshant, Singhal Hardik, Muruganandam Nandhakumar, Sivaraj Barath, and Shunmugiah Karutha Pandian

Department of Biotechnology, Alagappa University, Karaikudi, Tamil Nadu, India

1. Introduction

Small regulatory RNAs, a biological circuit, are the noticeable regulator and have been identified in a wide range of organisms and reported for their regulatory mechanisms at transcriptional and posttranscriptional level such as mRNA stability and modulation of translation initiation. Many small RNAs (sRNAs) are being reported for their virtual networks governing the expression of large gene pools related to the virulence factors, metabolic pathways, pathogenesis, and host-pathogen interactions. For instance, the noncoding RNA DsrA have proven their intrinsic regulatory functions on multidrug efflux pump system in *Escherichia coli* and lead to the emergence of antibiotic resistance (Nishino, Yamasaki, Hayashi-Nishino, & Yamaguchi, 2011). Small noncoding RNA *Rli60* regulates the transcription level of virulence genes *plcA*, *hly*, and *actA* in *Listeria monocytogenes* (Peng et al., 2016). During environmental adaptation, bacteria use sRNAs to establish a threshold to repress or activate the expression of a large fraction of their proteome, providing safety mechanism against random fluctuations and signals (Levine et al., 2008; Zhang et al., 2013). sRNAs are heterogeneous in size, ranging from 50 to 500 nucleotides, and regulate gene expression by direct base pairing with mRNA targets or by modifying the activity of concrete proteins. Many sRNAs regulate the gene expression through base pairing with mRNA molecules of their target (or) by sequestering the regulatory proteins involved in expression of target genes (Papenfort & Vogel, 2010). Based on the regulatory mechanism of gene expression, sRNAs in prokaryotes are categorized as *cis*-encoded sRNAs (antisense RNAs), *trans*-encoded sRNAs (resembles eukaryotic sRNAs) (Bloch, Węgrzyn, Węgrzyn, & Nejman-Faleńczyk, 2017; Ortega, Gonzalo-Asensio, & García-del Portillo, 2012). The intriguing sRNAs regulate the differential expression of most bacterial genes involved in cellular metabolism, virulence factors production including biofilm formation, sheltering the microbes from environmental stress, development of antimicrobial resistance, and host-pathogen interactions. Hence, understanding the entire small RNAome and their interaction networks in microbes and delineating their role in pathogenesis could be used for defining the regulatory circuits involved in pathogenesis, disease development, and emergence of antimicrobial resistance. Nevertheless, numerous sRNAs have been identified and functionally studied in prokaryotes, several research studies are aiming at understanding the sRNAs-mediated in-depth interaction networks in microbial pathophysiology. As identifying and analyzing the plethora of sRNAs from the entire RNA pool would take decades to conclude the interaction pathways, high-throughput RNA sequencing methods, biocomputational tools, and experimental approaches devoted for the identification of sRNAs and delineation of their interactions including RNA-RNA and RNA-protein interactions and their role in microbial pathogenesis. Understanding these sRNA-mRNA networks could open a new avenue for identifying a target specific potential drug for combating the antimicrobial resistance and also for treating the chronic infections.

2. Small regulatory RNAs (sRNAs)

The diversity in sRNAs structure is categorized based on their interaction with the RNA chaperone Hfq and direct mRNA base pairing. The sRNAs interact to the Hfq contain three functional domains, namely a highly conserved seed region for base pairing with target mRNA, Hfq-binding domain, and rho-independent terminator (Fig. 1A). The 5′ monophosphate of sRNA seed region allosterically activates the RNAse E, which degrades the cognate mRNA

Understanding Microbial Biofilms. https://doi.org/10.1016/B978-0-323-99977-9.00026-0

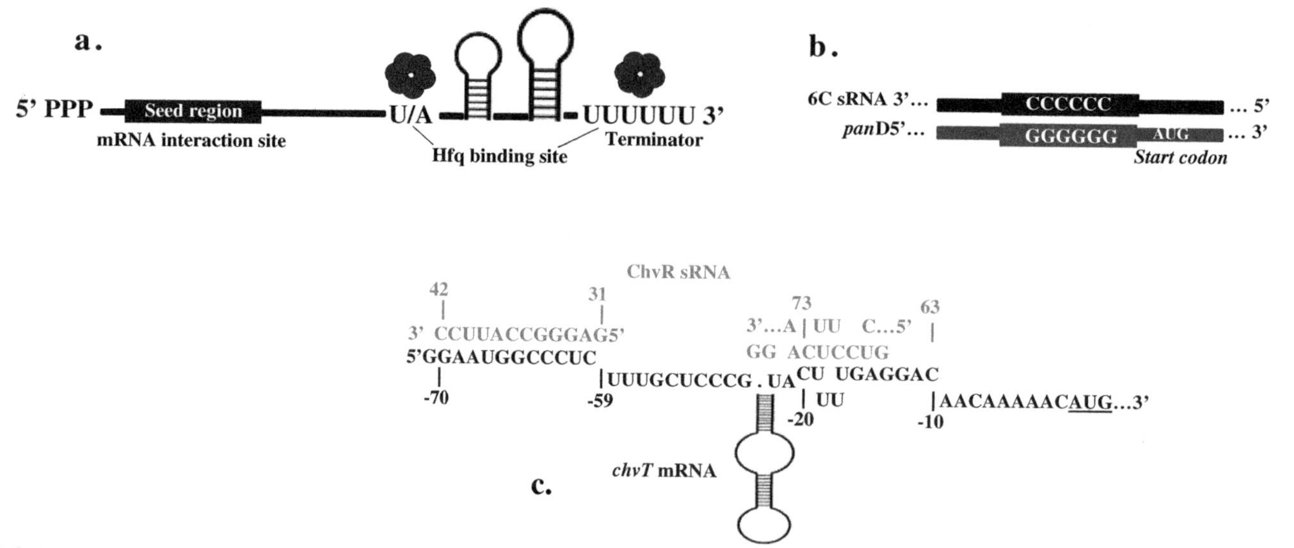

FIG. 1 Structure of sRNAs. Established structure of Hfq-dependent sRNA (A), interaction of Hfq-independent 6C sRNA with the *panD* target mRNA through C-rich region (B) and ChvR sRNA with *chvT* mRNA via direct base pairing.

from Hfq:sRNA:mRNA tenary complex. Moreover, presence of 5′ phosphate at the seed region protects the sRNA from RNase E cleavage (Bandyra et al., 2012; Miyakoshi, Matera, Maki, Sone, & Vogel, 2019). In Gram-positive bacteria, Hfq-independent sRNA, 6C having -C-rich loops pairs to the target mRNA, *panD* via complementary pairing with -G rich 5′ UTRs near the RBS (Fig. 1B) (Mai et al., 2019). In *Mycobacterium tuberculosis*, RNA chaperone-independent sRNA, ChvR negatively regulates the *ChvT* mRNA through direct base pairing at two distinct regions (Fig. 1C) (Fröhlich, Förstner, & Gitai, 2018).

3. Biogenesis of sRNAs

Biogenesis of sRNAs occurs through diverse mechanisms, which include (1) 5′ (or) 3′ UTRs derived, (2) independently transcribed from mRNA internal promoter and (3) processed from parental mRNA through RNase-E-mediated degradation, and (4) tRNA derived (Chao & Vogel, 2016; De Mets, Van Melderen, & Gottesman, 2019). Apart from this, sRNAs evolve more rapidly than the protein-coding counterparts through duplication events, horizontal gene transfer, and de novo emergence (Dutcher & Raghavan, 2018). A large number of sRNAs are encoded in intergenic regions of nucleic acids through their own promoters among the functionally characterized regulatory RNAs. Example, the OxyR-OxyS system, which controls the regulation of intracellular hydrogen peroxide level in *E. coli* (González-Flecha & Demple, 1999), and GcvA-GcvB system of *E. coli*, govern transportation of polar and branched amino acids and also modulate the glycine transport system (Stauffer & Stauffer, 2012).

In some of the RNA-mediated regulatory mechanisms, DNA strands opposite to the coding region encode the antisense RNAs (asRNAs), which complementarily pair to the target nucleic acids and regulate their expression through various modes. The asRNAs-mediated gene expression regulatory mechanisms include transcription interference, transcription attenuation, degradation of sense transcripts, stabilization of sense transcripts, protection of mRNAs from RNase E cleavage, and translation repression (Sesto, Wurtzel, Archambaud, Sorek, & Cossart, 2013).

sRNAs derived from 3′ UTR of mRNA belong to either type I or type II categories. The type I sRNAs are transcribed from an mRNA internal promoter in the upstream CDS, whereas type II sRNAs are generated by the processing of the Rho-independent mRNA in the 3′ region. For example, the *Salmonella enterica* serovar *T*yphimurium sRNA, DapZ, which represses the mRNAs of major ABC transporters during host cell invasion, is transcribed from a *dapB* gene-internal promoter (Chao, Papenfort, Reinhardt, Sharma, & Vogel, 2012).

The 5′ UTR region of mRNA also acts as a source for sRNA biogenesis despite the 5′ UTR region being well known for their function as riboswitches, thermosensors, and attenuators. The well-characterized examples for 5′ UTR transcribed *trans* acting sRNAs (*trans* acting riboswitches) are SreA and SreB. SreA and SreB are S-adenosylmethionine (SAM) riboswitches and function in *trans* by pairing with the 5′ UTR of *prfA* mRNA, which is associated with the virulence factors production in *L. monocytogenes* (Loh et al., 2009).

Another group of regulatory RNAs called intragenic sRNAs are transcribed from the already annotated protein coding functional genes. Hfq coimunoprecipitation followed by deep sequencing revealed the expression of 25 novel intragenic sRNAs in *E. coli* genome (Bilusic, Popitsch, Rescheneder, Schroeder, & Lybecker, 2014). Further, temperature-dependent differential regulation of intragenic sRNAs is reported in a spirochaete, *Borrelia burgdorferi* (Popitsch, Bilusic, Rescheneder, Schroeder, & Lybecker, 2017). Similarly, RIL-seq analysis of Hfq-bound RNAs in *E. coli* confirms the expression of putative intragenic sRNAs base pair with the 5′ mRNAs (Melamed et al., 2016).

Apart from the reported sRNAs sources, tRNAs also produce the regulatory RNAs termed as tRNA-derived small RNAs (tsRNAs) or tRNA-derived fragments (tRFs) or tRNA-derived, stress-induced RNAs (tiRNAs). tsRNAs are produced via cleavage at different sites of mature or pre-tRNAs. The tsRNAs derived from tRNAs through cleavage at specific site confirm the fact that the tsRNAs are not only the degradation products but are generated via precise regulatory mechanism (Kim, Yeom, & Kay, 2020).

4. Types of sRNAs

Small regulatory RNAs in prokaryotic origin are categorized into various types based on the method of biogenesis, mechanism of action, and various modes of gene regulations. Based on the encoding region and target site, sRNAs are classified into *cis*-encoded *cis*-acting, *cis*-encoded *trans*-acting, and *trans*-encoded *trans*-acting regulatory RNAs. *cis*-encoded *cis*-acting sRNAs are called riboswitches, riboregulators, and RNA thermometers, which are the integral part of 5′ UTR region of mRNAs otherwise called noncoding regulatory regions. Recently, *trans*-encoded *trans*-acting riboswitches that are regulating the gene expression are also reported (Loh et al., 2009). The *cis*-encoded *trans*-acting regulatory RNAs are known as antisense RNAs (asRNAs). Further, based on the interacting partners, sRNAs are categorized into three types as follows: (1) sRNAs interacting with proteins and other factors such as flavin mononucleotide (FMN), thiamine pyrophosphate (TPP), (2) sRNAs base pairing with target mRNA, and (3) bifunctional mRNAs with both protein coding and noncoding functions (called cncRNAs) regulate the gene expression through antisense mechanism (Kumari & Sampath, 2015). Additionally, based on the mode of gene regulation such as inhibition or activation of gene expression, sRNAs are categorized as positive regulators and negative regulators. The different categories of sRNAs based on their grouping systems are shown in Fig. 2.

5. Mechanism of action of sRNAs

Small regulatory RNAs are well known for their regulatory mechanism among three kingdoms of life, and they fall under various categories such as *cis*-encoded *cis*-acting regulatory RNAs, *cis*-encoded *trans*-acting regulatory RNAs, *trans*-encoded *trans*-acting regulatory RNAs, and bifunctional sRNAs that encode proteins as well as regulate the genes through complementary base pairing via antisense RNA mechanism (Fig. 2). The sRNAs exert various

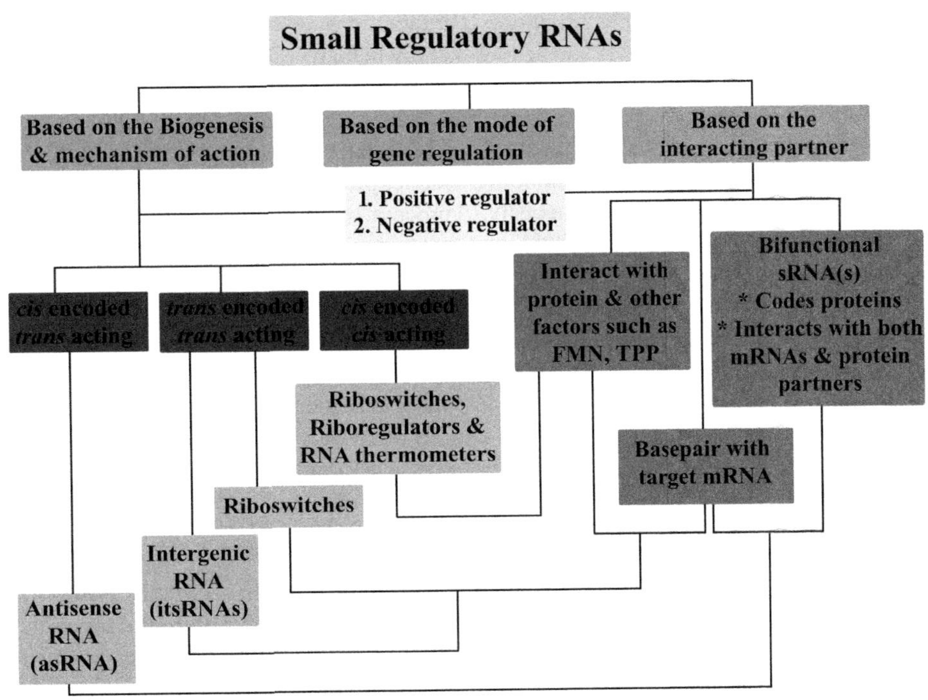

FIG. 2 Types and mechanism of action of sRNAs.

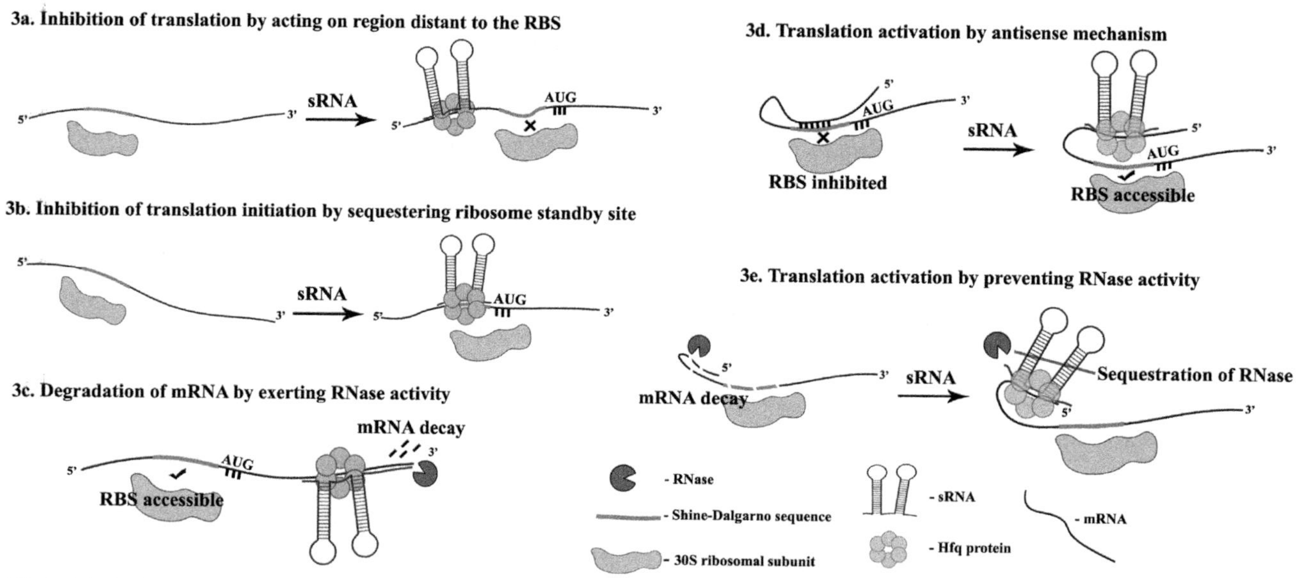

FIG. 3 Mechanism of action of sRNAs on regulation of target mRNAs (Azam & Vanderpool, 2020; Fröhlich & Papenfort, 2016).

mechanisms on target gene to activate or suppress the translational activity, which include inhibition of translation initiation by complementary base pairing at ribosome-binding site (RBS) (Fig. 3A), inhibition of translation by sequestering the RBS standby site of mRNA (Fig. 3B), activation of endonuclease-mediated sRNA-mRNA double-strand cleavage (Fig. 3C), activation of mRNA translation through antisense mechanism in which the sRNA removes the 5′ hairpin structure at the 5′ UTR of mRNAs similar to toehold switches mechanism (Fig. 3D), protection of mRNA from RNAse E cleavage by disabling the RNase E recognition site in mRNA (Fig. 3E) (Fröhlich & Papenfort, 2016).

5.1 Regulatory function of sRNAs in bacteria

Small RNA regulatory mechanism includes mRNA protection, transcription attenuation, and translational inhibition or activation (Watkins & Arya, 2019). Simultaneously, sRNA regulates the expression of mRNA and within an operon, alters the polarity pattern by complete or partial base pairing between the sRNA and the target mRNA (Massé, Majdalani, & Gottesman, 2003), it inhibits or changes the activity of protein by binding to it and also regulates the tRNA.

5.2 Base pairing of sRNA with target mRNA

sRNA molecules act as antisense to their target mRNAs. Interaction of sRNA with target mRNA promotes or inhibits the gene expression at the level of posttranscriptional and necessitates the particular degree of complementarity between sRNA and mRNA (Updegrove, Shabalina, & Storz, 2015). Base pairing of sRNA with target mRNA is a most considered sRNA regulation mechanism and divided in two classes: base pairing sRNA with high potential complementarity and base pairing sRNAs with low complementarity (Thomason & Storz, 2010). Inside the bacterial genome, the base pairing of sRNA with mRNA is classified into *cis* and *trans* based on the sRNA locations relative to mRNA.

RNA chaperone-independent sRNAs interact with mRNA to form a duplex sRNA-mRNA region through antisense mechanism via base pairing in the 5′ UTR leading to the mRNA destabilization and translational repression. For instance, 6C sRNA containing C-rich loops establishes stable interactions with mRNAs via multiple G-C pairings especially in high GC Gram-positive bacteria including *Mycobacterium* and *Streptomyces* spp. (Bouvier, Sharma, Mika, Nierhaus, & Vogel, 2008; Mai et al., 2019). Mai et al. (2019) have further reported that 6C sRNA directly regulates the expression of around 47 genes involved in cellular process of *M. tuberculosis* including DNA replication and protein secretion and experimentally demonstrated the differential expression of 15 genes including *panD*, *dnaB*, *gylR*, *espE*, *espF*, *mce1D*, and *mce1F* etc., upon overexpression of 6C sRNA. Another example for Hfq-independent sRNA is ChvR of *Caulobacter crescentus* expressed from its own promoter and is governed by the two-component system ChvI-ChvG. The ChvR expressed in response to acidic pH and DNA damage (Fröhlich et al., 2018).

5.2.1 *cis*-Encoded sRNA

Base pairing sRNAs directly transcribed from the DNA strand, which is opposite to the target mRNA, are described as *cis*-encoded sRNAs, and it shows high level of complementarity potential. The function is to maintain the copy number of mobile genetic elements in bacteria and bacteriophages.

5.2.2 trans-*Encoded sRNA*

Base pairing sRNAs transcribed from the DNA strand, which is located in the different place of the genome, are described as *trans*-encoded sRNAs and express low complementarity with target mRNA. The function is to change the mRNA stability, terminate or antiterminate the transcription, and inhibit or enhance the translation (Ghaz-Jahanian, Khodaparastan, Berenjian, & Jafarizadeh-Malmiri, 2013).

Base pairing of *trans*-encoded sRNA with mRNA downregulates the gene expression by blocking the ribosome-binding site (RBS) of target mRNA and degrades the target mRNA by RNase and upregulates the gene expression by uncovering the occluded RBS, which increases the translation process and prevents the degradation of target mRNA by RNase. Example, in *Salmonella*, the RybB RNA base pairing with the ompN mRNA, which prevents the translation process, the RybB RNA binds to the fifth codon of the open reading frame that inhibits the ribosome binding (Ghaz-Jahanian et al., 2013). In *Staphylococcus aureus*, the base pairing of RNA III with the target mRNA (hla) exposes the blocked RBS that enhances the translation (Nitzan, Rehani, & Margalit, 2017).

5.3 Interaction of sRNA with protein

The sRNA modulates the activity of proteins and regulates the RNA-binding protein. The sRNA–protein interactions fall into two categories,

1. sRNA Gives the specificity to the protein for RNA-protein interaction,
2. sRNA interacts and regulates the protein activity (Storz, Vogel, & Wassarman, 2011).

During metabolism, sRNAs interact with the proteins and form ribonucleoprotein complexes, used as a transporter in translational process. sRNA acts as a scaffold to attain a functional conformation of protein, controls the association and dissociation of proteins, and sustains the structure of proteins (Pichon & Felden, 2007). For example, the signal recognition particle (SRP) of *E. coli* contains 4.5S RNA and one protein (p48), the crystal structure of SRP shows the interaction of two internal loops in the 4.5S RNA with core domain [p48/54 homolog (Ffh)], the 4.5S RNA helps to maintain the domain structure (Batey, Rambo, Lucast, Rha, & Doudna, 2000). Some sRNA also regulates the activity of enzymes. For example, the *E. coli* 6S RNA acts as a global regulator, to regulate the transcription for cell survival. During stationary phase or stress condition, 6S RNA interacts with RNA polymerase holoenzyme and forms complex that suppresses the level of transcription (Cavanagh & Wassarman, 2014).

Majority of the sRNAs interact with cognate target transcripts through RNA chaperones such as Hfq, ProQ, and CsrA modulate the differential expression of interacting transcripts. Another group of RNA-binding proteins are ribonucleases, which include endonuclease RNase E, RNase III, and exonuclease PNPase. These ribonucleases are main players in both sRNA processing and transcripts degradation (Quendera et al., 2020). The sRNAs interact with RNA chaperones further categorized into Class I and Class II sRNAs based on the interacting region of sRNAs and interacting face to the chaperone proteins. Class I sRNAs interact with the proximal and rim face of the Hfq through U-rich tail and A/U-rich motifs of sRNAs, whereas class II sRNAs interact with the proximal and distal face of the Hfq through U-rich tail and (AAAN)n motifs (Jørgensen, Pettersen, & Kallipolitis, 2020).

5.3.1 *Hfq proteins*

The Hfq (host factor for Qb RNA replication) protein required when the level of complementarity is low and to make a best base pairing of *trans*-encoded RNA with target mRNA. Hfq protein from one bacterium varies from the other bacteria, and multiple Hfqs are present in a single bacterium, it shows the functional diversification. The main role of Hfq protein is a "RNA chaperone," enhances the stability of RNA, and prevents RNA misfolding. Hfq protein maintains the duplex of sRNA and mRNA by enhancing the rate of annealing (Ghaz-Jahanian et al., 2013). The Hfq binding to sRNA and target mRNA both contains the AU-rich single-stranded regions adjacent to the stem loop structure, provides the binding site for Hfq proteins, and also Hfq proteins recognize the polyU at the 3′ end of sRNA. Here Hfq also referred as loading site (Sauer & Weichenrieder, 2011). The result of Hfq binding alters the secondary structure of both transcripts that increase the amount of sRNA and mRNA base pairing. Hfq also binds to the ribosomes and RNase E endonuclease (Brennan & Link, 2007).

5.3.2 *CsrA proteins*

Carbon storage regulator protein A (CsrA) posttranscriptionally regulates the expression of mRNA especially by inhibiting the expression of most of the metabolites (Pourciau, Lai, Gorelik, Babitzke, & Romeo, 2020). RsmA is a structural and functional homolog of CsrA, which represses the stationary phase metabolites in *Pseudomonas aeruginosa*. In general, the CsrA proteins bind to the 5′ UTRs of mRNAs and prevent the RNA polymerase activity, whereas certain small RNAs sequester the CsrA proteins from the mRNA UTRs and lead to the expression of downstream gene cassettes (Ancona, Lee, & Zhao, 2016; Romeo, 1998). In *P. aeruginosa*, the RsmY and RsmZ sRNAs sequester the RsmA and RsmB from the mRNA and positively regulate the expression of virulence associated genes such as elastase, pyocyanin, and also the factors involved in the biofilm formation (Janssen et al., 2018).

5.3.3 ProQ proteins

The sRNA-binding ProQ is a FinO family member protein, which was initially identified as an osmoregulator, monitoring the optimal expression of the proline channel protein ProP. Recent studies establish the diverse regulatory functions of ProQ in divergent bacterial species including *E. coli*, *Legionella pneumophila*, *Salmonella* spp., and *Neisseria meningitidis* (Chaulk et al., 2011; Kerr, Culham, Marom, & Wood, 2014; Attaiech et al., 2016; Smirnov et al., 2016; Gonzalez et al., 2017). ProQ contains the characteristic N-terminal domain (NTD), linker region, and C-terminal domain (CTD). The positively charged residues such as R32, R69, R80, R100, K101, K107, and R114 of NTD of ProQ actively interact with the RNA strands and modulate their expression. The function of CTD is not clearly understood; nonetheless, it is also involved in RNA: ProQ interaction (Gonzalez et al., 2017).

6. Detection methods of sRNAs

The continually expanding interest in small regulatory RNA discovery is one of the main causes that have driven to exploit the recent development of novel high-throughput methods such as the cloning, NextGen sequencing, and computational/in silico approaches in sRNAs detection.

Most of the sRNAs in the past years have been discovered and detected using systematic screening that combined computational prediction of sRNA with experimental techniques such as microarray. Techniques such as 2D-PAGE are efficient methods for investigating high copy number sRNA. 2D-PAGE also allows us to immediately monitor changes that occur in RNA degradome under the effect of endogenous or exogenous stimuli (Nowacka et al., 2012). Microarray analysis has also been used to find new sRNAs that are generally transcribed from intergenic areas that lack protein-coding sequences, as well as to investigate their impact on gene expression (Ruiz-Larrabeiti et al., 2016). Due to its reliability, reasonable cost, and well-developed statistical techniques, microarrays are commonly employed in gene expression research to disclose datasets of differentially regulated transcripts that generally coincide well with those revealed by employing RNA sequencing (RNA-seq) methodologies.

In a variety of bacterial species, RNA sequencing (RNA-seq) technologies have been utilized to detect small RNAs (sRNAs) on a genome-wide scale. The processing of RNA-seq data to assemble and identify expressed transcripts is a crucial step in the detection of sRNAs from RNA-seq investigations. Many of the technical limitations of previous experimental approaches, such as low expression levels and small size and/or complex secondary structures of sRNAs, which make them poor substrates for microarrays, Northern blots, copurification, and cloning techniques, can be overcome with deep sequencing technology. This method does not require any prior knowledge of sequence or structural conservation and is an effective way to find new regulatory RNAs in bacteria (Sittka et al., 2008). In a variety of experimental and clinical research, next-generation sequencing (NGS) technologies are regarded as the most powerful and adaptable approaches for exploring short RNA (sRNA) transcriptomes.

Strand-specific RNA sequencing and methodologies to define the RNA boundaries have improved the transcriptome annotation studies. The unprocessed sRNA contains triphosphate at the 5′ UTR region, and hence, sequencing the strands containing triphosphate groups is one of the strategies to identify the sRNAs source from a pool of transcriptome. Terminator 5′—phosphate-dependent exonuclease (TEX), tobacco acid pyrophosphatase (TAP), or RNA 5′ pyrophosphohydrolase (RppH) enzymes remove the 5′ monophosphate and 5′ pyrophosphates respectively, leaving the 5′ triphosphate; these techniques are used to distinguish the 5′ transcriptional start sites (TSS) in transcripts. In another technology called Cappable-seq method, 5′ TPP end of the transcripts is coated with Desthiobiotin and enriched by binding to the Streptavidin beads followed by 5′ RNA sequencing (Ettwiller, Buswell, Yigit, & Schildkraut, 2016). Term sequencing employs the 3′ ends mediated enrichment of transcripts wherein, 3′ ends are ligated with adapters during library preparations, and it also can be combined with a Rho factor inhibitor Bicyclomycin to distinguish the Rho termination-dependent transcripts (Dar et al., 2016; Dar & Sorek, 2018).

Single-molecule, real-time (SMRT)-capable seq and simultaneous 5′ and 3′ end sequencing (SEnd-seq) employs the concurrent identification of both 5′ and 3′ ends of transcripts (Ju, Li, & Liu, 2019; Yan, Boitano, Clark, & Ettwiller, 2018). RNA-binding proteins mediated small transcripts enrichment techniques including cross-linking immune precipitation and sequencing (CLIP—seq) (Chihara et al., 2019), RNA interaction by ligation and sequencing (RIL—seq) (Melamed et al., 2016), cross-linking ligation, and sequencing of hybrids (Helwak & Tollervey, 2014) and another approach called gradient profiling by sequencing (Grad-seq) (Smirnov et al., 2016) are used as sRNAs detection methods. Furthermore, a high-throughput method such as cross-linking, ligation, and hybrid sequencing (CLASH) can also be used to experimentally identify the intramolecular and intermolecular RNA-RNA interactions (Kudla, Granneman, Hahn, Beggs, & Tollervey, 2011). RNA-seq data from a range of standard sequencing techniques, such as total RNA-seq, short RNA-seq, and high-throughput analysis of modified ribonucleotides (HAMR), may be used to discover and characterize the RNA modifications of the transcriptomes (Kuksa et al., 2017).

In the case of plants, the two primary approaches are green fluorescent protein (GFP)-based sensors and colorimetric in situ hybridization, which is commonly utilized. Small RNAs have been detected indirectly using a GFP-based sRNA sensor that detects the silencing of GFP by an sRNA (Huang, Baldrich, Meyers, & Caplan, 2019). Furthermore, locked nucleic acid (LNA) has been utilized in conjunction with nitro blue tetrazolium and 5-bromo-4-chloro-3-indolyl phosphate (NBT/BCIP) for colorimetric localization of sRNAs at the apex of the maize shoot (Javelle & Timmermans, 2012), mouse brain (Bak et al., 2008).

In silico approaches for detecting bacterial sRNAs have been developed in recent years. Among them, comparative genomics computational analysis of bacterial genomes predicted the existence of sRNAs, some of which were proven to be expressed in vivo. In most sRNA detection and analysis, three kinds of computational approaches are used: ab initio prediction based on sequence and structural characteristics, comparative genomic method based on evolutionary conservation, and integrated approach. The accuracy of these computational methods prediction is very varied for different methods (Zhou et al., 2009). The computational tools such as QRNA, RNA2, sRNA predict, sRNAN scanner, RNA surface, and other machine learning approaches including ANNogesic and APERO programs are used to detect the small transcript boundaries and also to predict the sRNA features (Adams & Storz, 2020).

Other tools and algorithms such as Manatee can also be employed to detect newly expressed noncoding loci and quantify sRNA classes. Manatee provides accurate transcriptome-wide sRNA expression measurement by combining previous sRNA identification with trustworthy alignment density information and thorough rescue of sometimes overlooked multimapped reads (Handzlik, Tastsoglou, Vlachos, & Hatzigeorgiou, 2020).

7. Small regulatory RNAs: A biological circuits in bacterial pathogenesis

Pathogenesis and disease persistence are primarily related to expression of virulence genes of pathogen, and they are governed by the many global regulators such as microbial two-component systems and quorum sensing (QS) molecules. During pathogenesis, the microbes enter into the new environment and adapt to the new habitat, proliferate, and then express the virulence genes and toxins into the host. In this adaptation process, numerous genes involved in metabolic function, cellular component, and biological process are differentially regulated by altering the gene expression profile. Recent advances in gene expression analysis and high-throughput sequencing (HTS) technologies have enabled to comprehend these complex regulatory mechanisms. Moreover, the emerging molecular studies on prokaryotic small regulatory RNAs in last two decades have evidenced that the small regulatory RNAs act as one of the global molecular regulators in modulating the gene expression profiles. These global regulators control the expression of many virulence-associated genes and also alter the bacterial physiology to the newer environmental cues leading to pathogenesis and disease development.

For example, the *P. aeruginosa* sRNAs, RmsY, and RmsZ control the expression of virulence-associated genes through positive regulation by sequestering the RsmA and RsmB proteins (belongs to the CsrA family proteins) from the 5′ mRNA region resulting in expression of genes involved in oxidative stress response, iron sequestration, components of type 6 secretion system (T6SS), elastase, LasA protease, hydrogen cyanide, and biofilm formation. The products of these genes play critical roles in persistent infection in cystic fibrosis patients (Janssen et al., 2018). String-based protein-protein interaction network also shows the interaction among the virulence-associated genes with RsmB and RsmA (KsgA) global regulators (Fig. 4).

Further, Toledo-Arana et al. (2009) have reported the role of *RliB*, which increases *L. monocytogenes* colonization in the spleen of mice. *Escherichia coli* noncoding RNAs, OxyS, and DsrA regulate the physiology of *Caenorhabditis elegans* (Liu et al., 2012). Regulatory RNAs that are controlling the expression of genes related to the normal microbial physiology and disease development are depicted in Fig. 5 and listed in Table 1. Further, the regulatory RNAs that are involved in host-pathogen interactions are discussed below.

7.1 Staphylococcus aureus

S. aureus causes severe respiratory syndrome by colonizing different parts of body surface. sRNA analysis provides an insight about the potential of the organism for adapting to different environments and establishing infection (González Plaza, 2020). But more than 600 regulatory sRNAs of this species remain largely uncharacterized (Tomasini et al., 2014). The protein encoded by sarA, which is a global regulator, directly or indirectly regulates the expression of genes responsible for virulence factors production. There are three promoters P2, P3, and P1, present 850 bp upstream to the coding sequence of sarA, which produces three overlapping transcripts including sarA ORF. Between these promoter regions, teg 49—a 196 nucleotide sRNA is located and probably generated at the promoter P3 mRNA. Several groups of genes involved in virulence, metabolism, and regulation were up- and downregulated by mutated teg 49 sRNA gene (Kim, Reyes, Beaume, Francois, & Cheung, 2014; Manna et al., 2018). Another sRNA involved in virulence factor

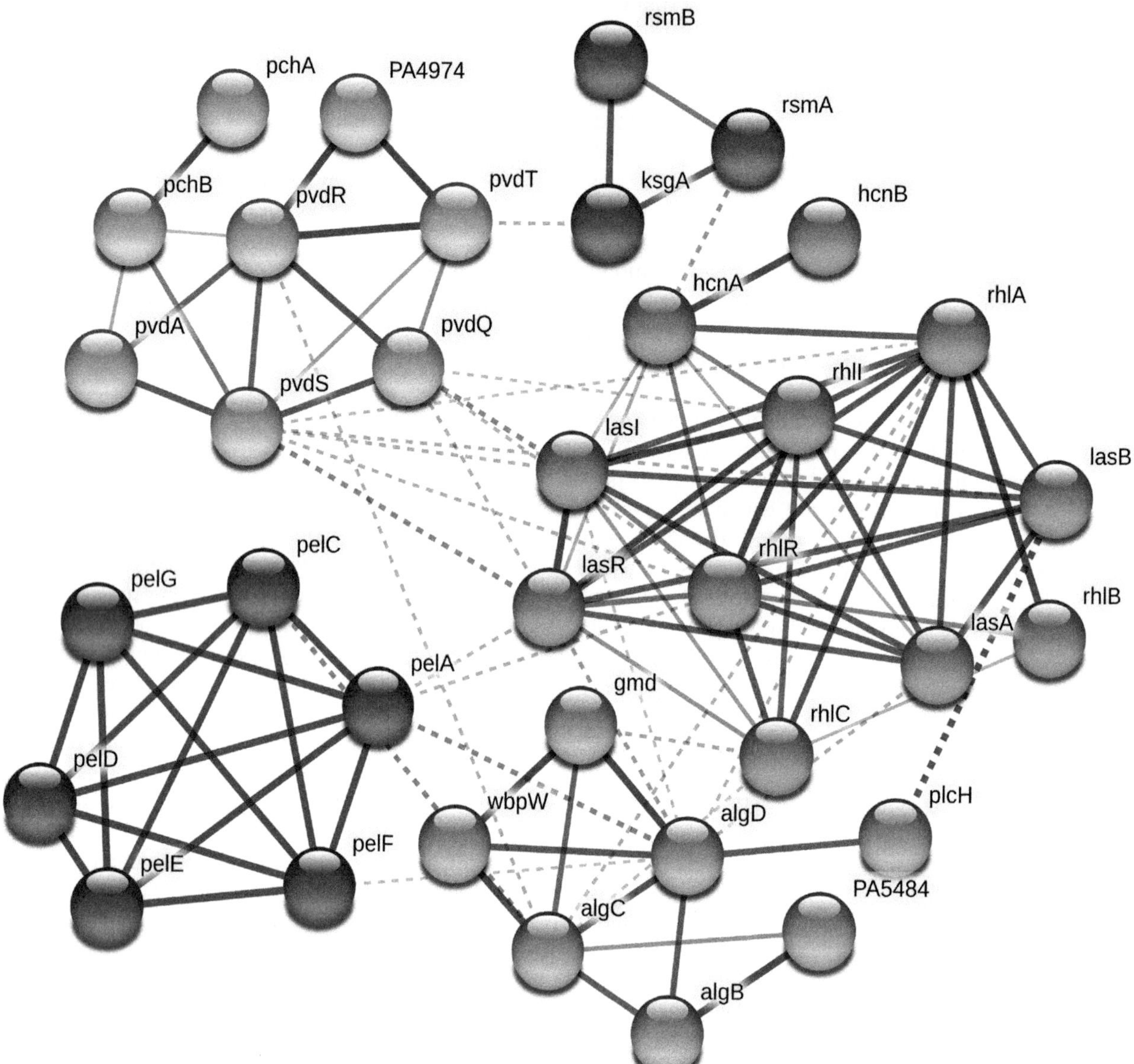

FIG. 4 Protein-protein interaction network of *P. aeruginosa* major virulence factors with RsmA and RsmB global regulators using string database.

regulation of *S. aureus* is teg 41. It is a 200nt RNA located downstream of the transcript region of the potent phenol soluble modulin (PSMs) toxin type α. It stabilizes the mRNA and regulates the production of toxin positively (Zapf et al., 2019).

7.2 Mycobacterium tuberculosis

Overexpression of sRNAs such as F6, B55, and B11 was observed in response to oxidative stress, which mimics oxidative stress caused by host macrophage. Hence, during early stage of infection, these RNAs could be associated in intracellular survival of pathogen. Overexpression of sRNA mcr11 and mrp4 during the transition from log to exponential phase suggests that they may involve in pathogenesis (Arnvig & Young, 2012).

7.3 *Escherichia coli* O157:H7

DicF is an sRNA-dependent on Hfq, which directly targets the Shine Dalgarno sequence of pchA (a transcriptional activator) and promotes its expression under low oxygen conditions (Zapf et al., 2019). In the presence of Hfq and lower oxygen condition, DicF increases the expression of the locus of enterocyte effacement (LEE) pathogenicity island indirectly and has an effect over virulence. Another Hfq-dependent sRNA Esr 41 also involved in regulation of LEE (Sudo et al., 2018).

7.4 Listeria monocytogenes

L. monocytogenes causes listeriosis. LhrC gene family of this organism containing seven sRNAs acts by regulating the target *tcs*A that encodes a virulence protein, CD4+ T

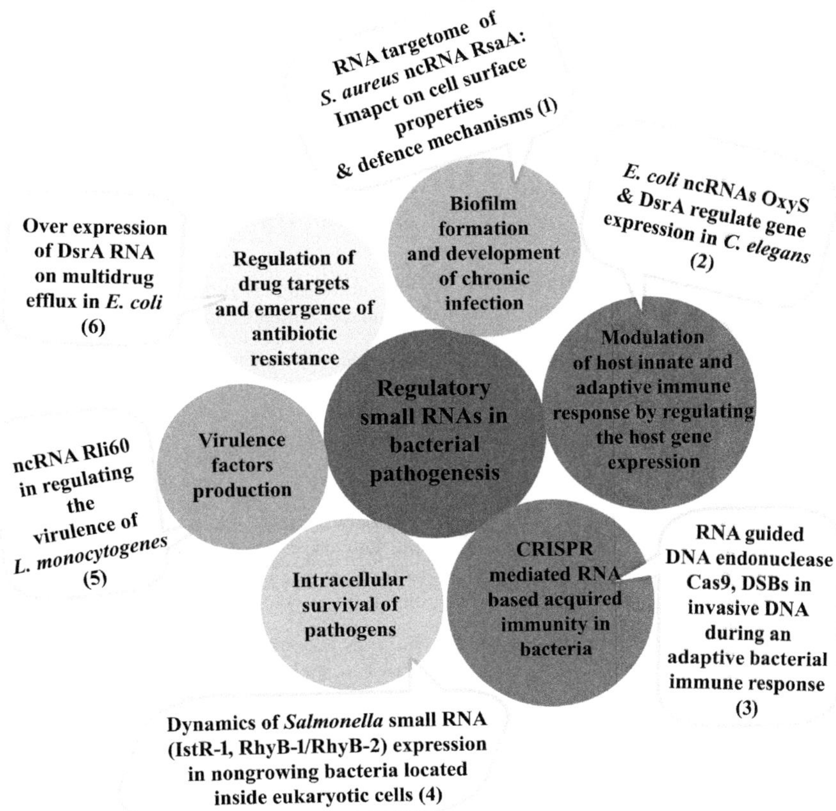

FIG. 5 Glimpses on sRNAs-mediated regulation of bacterial pathogenesis. (1) Tomasini et al. (2017); (2) Liu et al. (2012); (3) Jinek et al. (2014); (4) Ortega, Gonzalo-Asensio, and García-del Portillo (2012); (5) Peng et al. (2016); (6) Nishino, Yamasaki, Hayashi-Nishino, and Yamaguchi (2011).

cell-stimulating antigen. This family of sRNAs impairs the stability of target mRNA by binding far upstream of the RBS, through the participation of an unknown RNase decreasing the half-life of messenger RNA (Ross, Thorsing, Lillebæk, Teixeira Dos Santos, & Kallipolitis, 2019).

7.5 *Salmonella enterica* serovar Typhimurium

This organism causes typhoid fever with mortality rate of 10%–20%, if not treated properly. Two novel sRNAs are involved in positive (STnc1740) and negative (RssR) modulation of the growth of *S. enterica* serovar Typhimurium (Kröger et al., 2018). The RfrA and RfrB are two RybB homologous sRNAs in *S. enterica* that play an essential role in the intracellular replication in macrophage and regulate expression of virulence genes involved in the oxidative stress, iron homeostasis, and acid resistance within host cells. IsrM is SPI-encoding sRNA involved in direct regulation of HilE and SopA virulence genes, which are key regulators of *Salmonella* virulence (Ahmed, Zheng, & Liu, 2016).

7.6 *Legionella* spp.

Legionella uses effector proteins for modulating the function of host cell. It forms a membrane-bound vacuole designated the Legionella-containing vacuole (LCV) and establishes a replicative niche. Along with sRNA-binding protein, effector proteins such as VipA, RalF, and YlfA modulate the protein trafficking from endoplasmic reticulum to Golgi of host cell, aiding in the survival and replication of the organism as an intracellular pathogen. CsrA, which participates in the type IVB secretion system (T4SS), controls these effector proteins (Heidtman, Chen, Moy, & Isberg, 2009).

8. Effect of environmental cues on altered sRNA expression

Microbial systems encounter diverse environmental cues such as temperature, osmolarity, acidic/alkali environment, and changing nutrient gradients during infection establishment in the host cells. Microbes thrive in these complex environments by rapidly sensing the fluctuations in the host system via differentially regulating the genes associated with the cellular metabolism by selecting the Darwinian paradigm. Microbial two-component systems (TCSs), QS molecules, and sRNAs are connectively acting as a network and involved in rapid and stringent modulation of gene expression at various environmental cues. This part of the chapter discusses the differential expression of sRNAs and related gene regulatory networks in various environmental factors.

TABLE 1 Regulatory RNAs contributing to the microbial pathophysiology

Microbial strain	Regulatory RNA	Target	Phenotype/function	Reference
Bacteria				
S. Typhimurium	OxyS	Regulates about 40 genes; including *rpoS*	Under oxidative stress, blocks alternative stress adaption pathways	Hébrard et al. (2012)
S. Typhimurium	IsrM	*hilE, sopA* mRNA	hilE promotes SPI1 protein production and negatively regulates *SopA*	Hébrard et al. (2012)
S. Typhimurium	MicC	*ompD* mRNA	RNase E–dependent *ompD* mRNA degradation is promoted.	Pfeiffer, Papenfort, Lucchini, Hinton, and Vogel (2009)
Vibrio cholerae	IGR7	*mtlA*	Implicated in mannitol metabolism modulation	Liu et al. (2009)
V. cholerae	VrrA	*ompA, ompT,* and *RbmC*	Pairing with the 5′ end of the respective mRNAs—downregulate outer membrane complexes and biofilm matrix in host intestine	Song et al. (2008)
M. tuberculosis	6C sRNA	*panD* and DNA-B	Controls genes associated in several cellular activities, including DNA replication and protein synthesis	Mai et al. (2019)
Shigella flexneri	Ssr1	T3SS-related genes	Play a pivotal role in the tolerance and virulence to acidic stress	Wang et al. (2016)
S. flexneri	Ssr54	*treF*	Controls virulence and tolerance to high osmotic stress	Yang et al. (2020)
Vibrio anguillarum	Qrr1	*vanT*	Partially destabilized *vanT* mRNA-induced by VanO	Weber, Croxatto, Chen, and Milton (2008)
V. anguillarum	RNAβ	*fatDCBA*	Biosynthesis of siderophores is regulated	Stork, Di Lorenzo, Welch, and Crosa (2007)
Erwinia carotovora	CsrB	*csrA* antagonists	Takes part in global regulatory circuits that regulate central carbon flux, biofilm formation	Babitzke and Romeo (2007)
Legionella pneumophila	6S	RNAP	Controls the expression of genes that code for secretion system effectors of type IVB, as well as stress response genes	Faucher, Friedlander, Livny, Margalit, and Shuman (2010)
L. pneumophila	RsmY, Z	*rsmA, csrA*	Involved in the virulence regulation	Oliva, Sahr, and Buchrieser (2018)
P. aeruginosa	CrcZ	*crc*	Regulator of catabolite repression	Sonnleitner, Abdou, and Haas (2009)
P. aeruginosa	RsmY, Z	*rsmA*	Controls the RsmA and RsmF, which play a role in the transition from acute to chronic infection phenotype	Brencic et al. (2009)
Yersinia pseudotuberculosis	YsrH	*fabH2*	Required for the production of fatty acids-fatty acid biosynthesis	Lu, Zhang, Hu, Francis, and Chen (2014)
L. monocytogenes	RliB	*lmo2104*	In mice, the rliB mutation enhances spleen proliferation	Toledo-Arana et al. (2009)
S. aureus	RNAIII	*rot, spa*	Quorum-sensing and virulent transcriptional modulator	Chevalier et al. (2010)

TABLE 1 Regulatory RNAs contributing to the microbial pathophysiology—cont'd

Microbial strain	Regulatory RNA	Target	Phenotype/function	Reference
S. aureus	SprD	*sbi*	Sbi, an immunological evasion component, is suppressed	Chabelskaya, Gaillot, and Felden (2010)
Streptococcus pneumoniae	csRNA1–5	–	Autolysis in the stationary phase is controlled by csRNA 4 and 5	Halfmann, Kovacs, Hakenbeck, and Bruckner (2007)
Streptococcus pyogenes	FasX	*fpbA, mrp, ska, pel*	Enhance *S. pyogenes'* contact on epithelial cells	Ramirez-Peña, Treviño, Liu, Perez, and Sumby (2010)
S. pyogenes	Pel	*emm*, sic, *speB*	The SagA expression is regulated by a bifunctional RNA	Mangold et al. (2004)
S. Typhimurium	AmgR	*mgtBC*	Magnesium regulation and pathogenicity in mice are influenced	Lee et al. (2010)
Helicobacter. pylori	tmRNA	damaged mRNA	Involved in competence, response to antimicrobial compounds	Thibonnier, Thiberge, and De Reuse (2008)
H. pylori	NikS	*cagA, vacA*	In accordance to nickel availability, it coordinates posttranscriptional control of its targets	Eisenbart et al. (2020)
Fungi				
Botrytis cinerea	Bc-sRNAs	Argonaute 1 (AGO1) protein	Hijack the plant's own RNA interference machinery to produce an RNA-induced silencing complex	Weiberg et al. (2013)
Beauveria bassiana	bba-milR1	Argonaute 1 (AGO1)	By suppressing the expression of the mosquito Toll receptor, it lowers host immunity	Cui et al. (2019)
Verticillium dahliae	VdmilR1	VdHy1	Plays a regulatory role in pathogenicity and uncover an epigenetic mechanism for VdmilR1 in regulating a virulence target gene	Jin et al. (2019)
Puccinia striiformis f. sp. *tritici*	Pst-milR1	Bind to the wheat pathogenesis-related 2 (PR2) gene	Suppresses wheat defenses during wheat-Pst interactions. st-milR1 is an important pathogenicity factor in Pst, which acts as an effector to suppress host immunity	Wang et al. (2017)
Neurospora crassa	milR 1–4 milRNA	Repetitive transgene unique dsRNA	Posttranscriptional gene silencing of all homologous genes is triggered known as quelling	Lee et al. (2010)
N. craassa	disiRNAs	–	–	Lee et al. (2010)
Valsa mali	Vm-milR1	mdRlkT1 and mdRLkT2	Host receptor-like kinase genes are silenced, host defense responses are suppressed, and pathogen invasion is facilitated	Xu et al. (2022)
Mucor circinelloides	Ex-siRNA	Ago, RDP1	In response to environmental inputs, govern several physiological and developmental processes	Ruiz-Vázquez, Nicolás, Torres-Martínez, and Garre (2015)
Cryptococcus neoformans	miR1 and miR2	*RDP1, DCR1*, and *DCR2*	Transgene silencing via the canonical RNAi pathway	Jiang, Yang, Janbon, Pan, and Zhu (2012)

8.1 Temperature

Microbial adaptation to the vertebrate host requires extensive reprogramming of gene expression, especially adaptation to the temperature of the host system. For instance, *B. burgdorferi*, a causative agent of the zoonotic disease called Lyme disease, differentially expresses around 43% of sRNAs in temperature-dependent manner at 37°C when compared with the 23°C. These temperature-dependent sRNAs are expressed as antisense, intergenic, intragenic, and 5′ UTR-associated sRNAs (Popitsch et al., 2017).

8.2 Osmolarity

Major challenging factor for intestinal pathogens is exhibiting a tolerance to the high osmolarity of the intestinal lumen of the host system. In *E. coli* (a common intestinal pathogen), OmpF and OmpC are two major outer membrane porins (OMP) that facilitate the diffusion of small hydrophilic solutes across the OM and maintain the cellular integrity. The EnvZ/OmpR TCS of *E. coli* senses increased osmolarity of the surrounding environment and regulates the synthesis of OmpF and OmpC (Nikaido, 2003). The OmpR activates the MicF sRNA expression when the bacteria encounter the higher-osmolarity environment. MicF prevents the *omp*F mRNA translation by complementarily pairing to the ribosome binding site as well as start codon (Andersen & Delihas, 1990). Downregulation of OmpF porin, which is larger than the OmpC, lowers the intracellular uptake of solutes, which subsequently protects the cell from lysis during high-osmolarity conditions (Piattelli, Peltier, & Soutourina, 2020). Similarly, EnvZ/OmpR TCS of *V. cholerae* expresses the CoaR sRNA in high-osmolarity and acid stress conditions. The CoaR regulates the expression of *tcp*I, which negatively regulates the major pilin subunit, TcpA, and also contributes to the intestinal colonization of *V. cholera* against high-osmolarity condition (Xi et al., 2020).

8.3 Acidic/alkali environment

Acidic environment in the stomach is one of the first lines of defense mechanism of host. Pathogens will overcome this sudden exposure to acidic pH by enzymatic modification of already existing fatty acid in its membrane. This process is mediated by the enzyme cyclopropane fatty acid synthase, which incorporates methylene group into unsaturated fatty acids. Microbial system possesses noticeable signaling pathways such as TCSs, specialized sigma factors, and regulatory RNAs to enable crucial gene regulatory network to tackle the environmental threats in order to survive in vast hostile environments (Chao & Vogel, 2016). For example, among the Gram-negative bacteria especially in *E. coli* and *Salmonella* spp. the sRNA CpxQ is produced from 3′ UTR region of CpxP mRNA through RNase-E-mediated mRNA decay, which helps in maintaining the intracellular membrane homeostasis during alkaline stress conditions along with the CpxP protein, collectively called Cpx pathway (Chao and Vogel, 2016). In *E. coli*, two sRNAs, RydC and ArrS, stabilize and protect this enzyme's transcript from RNase activity (Bianco, Fröhlich, & Vanderpool, 2019). Similarly, ChvR (*trans*-acting sRNA) in *C. crescentus* expression is enriched in response to DNA damage, acidic pH, and minimal growth medium (Fröhlich et al., 2018).

8.4 Nutritional immunity and nutrient uptake

Microbial system must uptake the nutrients from extracellular environment through porins and passive diffusion except some of the nutrients such as CO_2 and NH_3, which can diffuse easily across the cell membrane (Simon & Gutknecht, 1980). During pathogenesis, microbial system has to equalize the essential elements concentration inside the cells by sequestering the host minerals and by increasing the nutrient uptake. In nutritional starvation conditions, multiple regulatory systems coordinate the energy production and cellular metabolism and also exhibit the differential expression of response regulators including sRNAs, which enable the microbial adaptation and nutrient uptake in the host environment. For example, *E. coli* sRNA, SgrS, regulates the glucose uptake through the membrane-bound $EIICB^{Glc}$ protein. In general, the *pts*G mRNA codes the $EIICB^{Glc}$ protein. In increased phosphogluocse stress condition, the sensor molecule SgsR activates the expression of SgrS, which impairs the *pts*G mRNA translation. This cascade process controls the cellular glucose uptake from the environment, which provides the nutritional immunity to the cells (Shimizu, 2014; Vanderpool & Gottesman, 2004). In *C. crescentus* species, CrfA sRNA induced upon carbon starvation. CrfA encodes TonB-dependent receptors on the cell surface, which facilitates the uptake of nutrients from the environment during carbon starvation (Landt, Lesley, Britos, & Shapiro, 2010). Another study explains the Fur regulated RyhB expression in iron starvation conditions, which represses the serine acetyltransferase and leads to increased production of siderophores (Salvail et al., 2010).

9. Cellular pathways that are differentially regulated by the sRNAs

Apart from the response to environmental cues, sRNAs also regulate the cellular pathways involved in both biological process and molecular functions such as cell division, oxidative stress response, QS, virulence, and antibiotic resistance development. This part exemplifies the some intrinsic mechanism of sRNAs involved in these regulatory processes.

9.1 Cell division

Distinct control on cell division is an important regulatory mechanism for continuous survival of an organism. Recent studies show that the sRNAs control the cell division and also influence the growth rate in prokaryotes. Stress-induced sRNA, StsR, exhibits StsR-UpsM (sRNA-sRNA) and StsR-dcw (sRNA-mRNA) interactions and leads to RNase-E-mediated cleavage of the targets, which subsequently limits the growth under starvation. UpsM is the most abundant sRNA in the exponential phase and derived from the 5′ UTR of the division and cell wall (*dcw*) gene cluster (Grützner et al., 2021).

9.2 Oxidative stress response

The extreme transient conditions such as oxidative stress caused by the host during oxidative burst largely induce the expression of RsaC sRNA in *S. aureus*. The sRNA binds to the start codon of the sodA mRNA, which gives protection against reactive oxygen species (ROS) by using manganese as cofactor. The expression of this gene is repressed and induces the expression of another enzyme called SodM, which uses iron as cofactor (Lalaouna et al., 2019). The sRNA OxyS is expressed in response to oxidative stress and represses the *rpo*S translation, whereas, the sRNAs DsrA and RprA enhance the *rpo*S expression (Kulesus, Diaz-Perez, Slechta, Eto, & Mulvey, 2008; Mulvey et al., 1998; Zhang et al., 1998; Zhang, Wassarman, Ortega, Steven, & Storz, 2002).

9.3 Quorum sensing

QS is a well-known microbial signaling system, which regulates the expression of various cellular pathways and is majorly controlled by the cell-density-dependent internal and external cues. The studies on sRNAs in the status quo have demonstrated the role of regulatory RNAs in inter- and intra-kingdom communications (Fu, Elena, & Marquez, 2019). sRNAs that are regulating the bacterial QS system are described in several studies. For instance, Qrr sRNAs function at the center of QS pathways in Vibrios. Qrr sRNAs modulate the expression of multiple mRNA targets including *lux*R, *lux*O, *lux*M, and *aph*A transcripts. Vibrios Qrr3 sRNA regulate the various QS-associated genes through four distinct mechanisms; (1) represses the *lux*R through catalytic degradation, (2) represses the *lux*M by sRNA-mRNA coupled degradation, (3) represses the *lux*O by sequestering a protein from 5′ region of mRNA transcript, and (4) activation of *aph*A expression via exposing the ribosome-binding site (Feng et al., 2015). Further, recent studies have demonstrated that microbes secrete extracellular vesicles (EVs) called outer membrane vesicles (OMVs) containing tRNA fragments (sRNAs) and secreted extracellularly to the host system. This OMV-mediated delivery of sRNAs confirms that the bacterial sRNAs are involved in the interkingdom communication during disease progression in the host system (Li & Stanton, 2021).

9.4 Virulence

Microbial virulence factors play a key role in establishing the infection in host environment and also increase the disease persistence in chronic infections. Secreted proteins, motility, chemotaxis, biofilm formation, and colonization are some of the noticeable virulence factors of microbial system. In *E. coli*, *flhDC* mRNA encodes the key regulator of flagellar synthesis, which is regulated by the transcription factors that respond to various environmental cues. Four sRNAs such as ArcZ, OmrA, OmrB, and OxyS negatively regulate and one sRNA, McaS, positively regulates the expression of *flh*DC and motility factors (De Lay & Gottesman, 2012).

9.5 Antibiotic resistance development

Some sRNAs impact the antibiotic resistance by modulating the expression of genes coding the efflux systems, peptidoglycan biosynthesis, and biofilm formation. Each antibiotic produces a large flux of sRNAs and creates a unique sRNA profile causing series of downstream effects that help to overcome the antibiotic challenge (Dersch, Khan, Mühlen, & Görke, 2017). The riboswitches such as *tetM* and *tetQ* are reported to be involved in tetracycline resistance development in *Enterococcus faecalis* (Su, He, & Clewell, 1992) and *Bacteroides* (Wang, Rotman, Shoemaker, & Salyers, 2005), respectively. Similarly, it is reported that the sRNA, DsrA, confers resistance trait to *E. coli* against oxacillin, novobiocin, and erythromycin antibiotics by overexpressing the efflux pump (Nishino et al., 2011). The sRNAs, MicF of *E. coli* and *Salmonella* spp. exert cephalosporin and norfloxacin resistance by differentially regulating the OmpF (Kim, Bak, Lee, & Kim, 2015). Dersch et al. (2017) have reviewed in detail about the list of riboswitches and sRNAs involved in the development of antibiotic resistance in various microbes.

10. sRNAs with functions relevant to microbial biofilm formation

The process of biofilm formation is interwined with many regulatory mechanisms and has cascade of events such as attachment, biofilm establishment, and dispersal. In the context of cellular physiology, biofilm formation reflects the transition from growth to stationary phase. In molecular level, biofilm formation is a hierarchical cascade and controlled by the various sigma factors and secondary messengers. Recent studies on understanding the molecular mechanism of biofilm formation unveil the involvement of sRNAs in controlling the biofilm growth phase of microbes. For example, the ArcZ sRNA indirectly stimulates the biofilm formation in *Salmonella* (Monteiro et al., 2012). Similarly, SdsR sRNA regulates the porin OmpD in *Salmonella* and regulates the biofilm formation by modulating the CsgD activity in *Salmonella* (Udekwu et al., 2005). MicA sRNA has reported for its role in *Salmonella* spp. biofilm formation (Kint, De Coster, Marchal, Vanderleyden, & De Keersmaecker, 2010). The homolog sRNAs OmrA and OmrB regulate the OmpF and OmpC and also influence the biofilm formation via negative regulation of *csgD* mRNA (Guillier & Gottesman, 2006; Holmqvist et al., 2010). The hydrogen peroxide stress induced sRNA, OxyS involved in cross-species regulation by modulating the host gene expression in eukaryotic system, *C. elegans* (Liu et al., 2012). The sRNA SrbA positively regulates the biofilm formation in *P. aeruginosa*. It is reported that the deletion mutant of SrbA showed 60% reduction in biofilm formation and also the mortality rate of *C. elegans* reduced to 39% when compared with the 78% of wild type in *C. elegans* at 3 days' interval (Taylor, Van Kessel, Colavita, Hancock, & Mah, 2017).

Apart from direct regulation, some of the sRNAs regulate the biofilm formation by altering the activity of transcription factors and signaling molecules. For instance, ArcZ also stimulates the accumulation of σ^s factor, which further activates the curli fiber and cellulose expressions in *Salmonella* spp. (Guillier & Gottesman, 2006). Further, RprA sRNA was reported to be involved in cyclic di-GMP regulation, which is responsible for transition from a motile to a sessile lifestyle (Crawford, Reeve, & Gunn, 2010), which in turn facilitates the biofilm dispersion and reinfection of the host system. The Hfq-binding sRNA, CsrB/C reversibly binds to the CsrA regulator, which regulates the expression of polysaccharide adhesin PGA (Wang et al., 2005). Nevertheless, the biofilm formation ability of an organism is directly regulated by the outer membrane integrity and homeostasis. As discussed earlier, some of the sRNAs modulate the OMP expression and lead to disintegration of OM, which influences the biofilm-forming ability of an organism (Andersen & Delihas, 1990; Mika & Hengge, 2013).

11. Regulatory small RNAs (sRNAs) among archaea

Along with bacteria and eukaryotes, sRNAs are also distributed in the third domain of life called the archaea, but their function in the third domain is not well understood and characterized (Babski et al., 2014; Gomes-Filho, Daume, & Randau, 2018; Prasse, Ehlers, Backofen, & Schmitz, 2013). The size of sRNA ranges from 50 to 500 nucleotides in archaea (Gelsinger & DiRuggiero, 2018). The first sRNA was found in the C/D box snoRNA (small nucleolar RNA) of an archaeon (Babski et al., 2014) and consists of a standard kink turn (k-turn) or kink loop structure in its structure (Gomes-Filho et al., 2018), guides ribose methylation at specific sites of the rRNAs and tRNAs, experimentally verified in *Sulfolobus acidocaldarius*. The H/ACA box snoRNA is also identified in archaea, which helps in conversion of uridine to pseudouridine at specific sites of rRNA (Babski et al., 2014; Dennis, Omer, & Lowe, 2001). Other sRNAs that are identified are *cis*- and *trans*-encoded sRNAs, tRNA-derived fragments, and CRISPR-RNAs (crRNAs) in different archaeon species such as *Methanocaldococcus jannaschii*, *Pyrococcus furiosus*, *Archaeoglobus fulgidus*, *Sulfolobus solfataricus*, *Methanosarcina mazei*, and *Haloferax volcanii*. (Babski et al., 2014; Marchfelder et al., 2012).

The identification of sRNA in archaea is achieved by bioinformatics, RNomics, HTS called RNA-Seq. The HTS has identified different sRNAs first in the *M. mazei* and confirmed differential expression of sRNA genes due to different nitrogen availability, differential expression of genes can also be studied by northern blotting and microarrays (Babski et al., 2014; Jäger et al., 2009). Tiling arrays analytic data when combined with the RNA-seq allowed for the identification of Transcription Start Site–associated RNAs (TSSaRNAs) in many archaea (Gomes-Filho et al., 2018).

The two major roles of sRNA in archaeon other than gene regulation, transposable element silencing, and defense against foreign elements are to guide the methyl group to 2′-O position of ribose, and the other is to act as the molecular chaperone to facilitate folding of rRNA and tRNA in the extreme conditions (Dennis et al., 2001; Gelsinger & DiRuggiero, 2018). sRNAs' functions in building the immunity against the foreign genetic elements are achieved by the presence of CRISPR/Cas systems in different archaeon species with exceptionally high number of fragmented tRNAs genes such as *P. furiosus* (Gomes-Filho et al., 2018; Hale, Kleppe, Terns, & Terns, 2008) generally using CRISPR/Cas system-III to cleave the foreign RNA instead of DNA as in prokaryotes, which uses system I or II (Marchfelder et al., 2012), *S. solfataricus* shows both CRISPR system I-A (DNA-specific endonuclease) and CRISPR system III (RNA-specific endonuclease) as the defense mechanism against the pathogenic genetic material (Marchfelder et al., 2012).

There are many intergenic sRNAs identified in archaeal species and are important regulators in many cellular function and gene expression (Gelsinger & DiRuggiero, 2018; Gomes-Filho et al., 2018). They are generally developed in specific environment conditions and target specific mRNAs in-*trans*, i.e., sRNAs are partially complimentary to mRNAs such that there is possibility of complementary base pairing between sRNA and multiple mRNAs resulting in the repression of mRNA translation due to the blockage of RBS within the 5′ untranslated region (UTR) of mRNA; however, the unmasking of the RBS relives the inhibition of translation bacteria (Babski et al., 2014; Gomes-Filho et al., 2018; Marchfelder et al., 2012). There are few understandable targets in few archaeal species, one of the first to be identified is *M. mazei* sRNA162—targets and binds in-trans to an internal RBS region within the bicistronic MM2440-MM2441 mRNA, which masks RBS of MM2441 mRNA, thereby suppresses the translation of second cistron leading to discoordinated operon expression. sRNA162 also targets and binds to 5′ leader region of MM2242 in-*cis*, which enables translational repression and subsequent mRNA degradation such that sRNA162 acts as antisense sRNA (asRNA) on *trans* and *cis*-encoded mRNA via two distinct domains (Babski et al., 2014; Gomes-Filho et al., 2018; Jäger et al., 2012), and other targets of different sRNAs are described in Table 2. 3′ UTRs are essential translational regulators as many species of archaeon do not have 5′ UTRs (Brenneis & Soppa, 2009). For example, *S. solfataricus* targets 3′ UTR of mRNA 1183, degrading the mRNA to regulate the phosphate levels (Buddeweg et al., 2018).

cis-Acting gene regulation can be achieved by riboswitches in archaea that can be predicted through in-silico analysis (Gomes-Filho et al., 2018). It is experimentally shown by incorporating bacterial tetracycline riboswitch into *M. acetivorans* to show the regulation (Demolli et al., 2014). Another example of this type of regulation was shown in *Thermococcus kodakarensis*, hyperthermophile uses fluoride responsive riboswitch for the regulation of genes involved in fluoride detoxification (Speed, Burkhart, Picking, & Santangelo, 2018).

tRNA-derived fragments (tRFs) have been found and reported in few species of archaea. These are usually coisolated with the ribosomes, for example—tRF of tRNA-Valine that binds to small subunit of ribosome, suppresses the translation, and is the most prominent coisolated tRF. It is most efficient under stressful environment. Therefore, tRFs are regulatory sRNAs in archaea (Babski et al., 2014).

In conclusion, there is a need for more understanding and research in finding the regulatory sRNAs in archaeal species other than mentioned above, to know the function of these in archaea at advance level. It is also required to develop the genomics and bioinformatics tools to identify and characterize different types of sRNAs in archaeon.

12. Conceptualization and implementation of methods for identification of sRNAs-based drug targets and therapeutic molecules

RNA has been considered as a drug target since the discovery of aminoglycosides that are targeting various regions of microbial ribosomal RNAs. Similarly, the selected sRNAs of any target pathogen could potentially be used as alternative drug targets to treat microbial infections, especially when the microbes elicit antimicrobial resistance for the conventional antibiotics. Small regulatory RNAs are well known for their global regulatory mechanisms in virulence gene expression. Recent developments such as genomic tiling arrays, high-throughput NGS technologies, and availability of many sRNA tools based on novel integrated platforms have provided the unparalleled opportunities for identification of target sRNAs from large RNA pool including their interactions with transcripts and metabolites.

The existing research on sRNAs and their regulatory mechanisms, indeed, confirms that a drug candidate targeting the specific sRNA can decrease the infection severity either through quenching the virulence properties or by killing the microbes via modulating the normal cellular metabolism. For implementing the idea of discovering the regulatory RNA-based targets and ligands followed by the successful development of the potential drugs, there are series of steps that have to be followed, which include: (1) analysis of sRNAs and their interaction with transcripts and proteins in microbial system using computational tools, (2) selection and functional analysis of sRNAs related to the virulence gene expression and pathogenesis, (3) identification of the role of bacterial sRNAs in modulation of host immune signaling, (4) in-depth identification of active sites in sRNAs using computational analysis, (5) discovery of sRNA specific ligand molecules using HTS in drug discovery, and (6) validation of the efficiency of drug molecules through in vitro and in vivo analyses. The idea behind the identification of sRNA-based drug molecules is represented as a schematic diagram in Fig. 6.

13. Conclusion and future perspectives

Small RNAs (sRNAs) are regulatory RNA molecules found in a variety of bacterial species that have been shown to regulate a variety of cellular functions through distinct regulatory mechanisms. Recent studies have unveiled the major role of sRNA posttranscriptional regulators as potent regulatory machineries involved in virulence factors production, biofilm formation, and infection establishment in the host system. sRNAs act as posttranscriptional regulators of protein synthesis by interacting with short stretches of base pairing target mRNAs to either promote or inhibit translation. Typically these sRNAs interact at (or) near the RBS of an mRNA transcript and might result in either negative regulatory effect by blocking the ribosome or positive regulatory effect by altering secondary structures through binding to the target mRNA and making the RBS available by altering the mRNA stability or by degrading the target mRNA through the recruitment of RNases. Therefore, a cumulative analysis on sRNA contributing to

TABLE 2 Regulatory sRNAs among archaea.

Archaeal strain	Regulatory RNA	Target	Phenotype/function	Reference
Sulfolobus acidocaldarius	C/D box snoRNA	Specific sites of rRNAs	Guide ribose methylation at specific sites of the rRNAs	Babski et al. (2014)
Pyrococcus sp.	C/D box snoRNA	Specific sites of rRNAs	Guide ribose methylation at specific sites of the rRNAs	Babski et al. (2014)
M. mazei	sRNA162	*In-trans* to an internal RBS region within the bicistronic MM2440-MM2441 mRNA	Masks RBS of MM2441 mRNA, thereby suppresses the translation of second cistron leading to dis-coordinated operon expression	Jäger et al. (2012), Gomes-Filho et al. (2018)
M. mazei	sRNA162	5′ leader region of MM2242 *in-cis*	Enables translational repression and subsequent mRNA degradation	Jäger et al. (2012), Gomes-Filho et al. (2018)
M. mazei	sRNA162	MM2441 mRNA	Regulates the metabolic switch between the carbon sources methanol and methylamine	Jäger et al. (2012)
M. mazei	sRNA154	nifH, nrpA, glnA2—mRNA	Expressed under N-deficiency to mask RBS, deletion of this causes decreased growth under N-limiting conditions, i.e., it has regulatory role	Prasse, Förstner, Jäger, Backofen, and Schmitz (2017)
M mazei	sRNA41	Operons encoding for acetyl-CoA-decarbonylase/synthase complexes (ACDS)	Highly expresses in N- sufficiency, masks 6 RBS region of targets repressing its expression. Downregulating sRNA41 provides sufficient amino acids for N-metabolism	Buddeweg, Sharma, Urlaub, and Schmitz (2018)
S. solfataricus	sRNA257	3′ UTR of 1183 mRNA encoding a putative phosphate transporter	Phosphate expression regulator. When phosphate levels decreased it also decreased and mRNA levels increased. Therefore, it inhibits translation and induces degradation of 1183 mRNA	Märtens, Manoharadas, Hasenöhrl, Manica, and Bläsi (2013)
Pyrobaculum sp.	asR3	3′ UTR of tpi mRNA	Compete against the formation of the intramolecular tpi element structure, and consequently, might modulate the function of the highly conserved tpi element	Babski et al. (2014)
H. volcanii	sRNA132	Operon for a phosphate-specific ABC transporter	Positive regulator of the target under low phosphate conditions	Kliemt, Jaschinski, and Soppa (2019)
P. aerophilum	PAEsR53	Etf operon at 3′ UTR	Enable a 20 -O-methyl modification of the central "A" nucleotide within the conserved TAA stop codon in all four species of *Pyrobaculum*	Bernick, Dennis, Lui, and Lowe (2012)
Pyrobaculum sp.	asR1	*Tfb1* mRNA	Unclear	Bernick et al. (2012)
Pyrobaculum sp.	asR2	Fur gene	Unclear	Bernick et al. (2012)

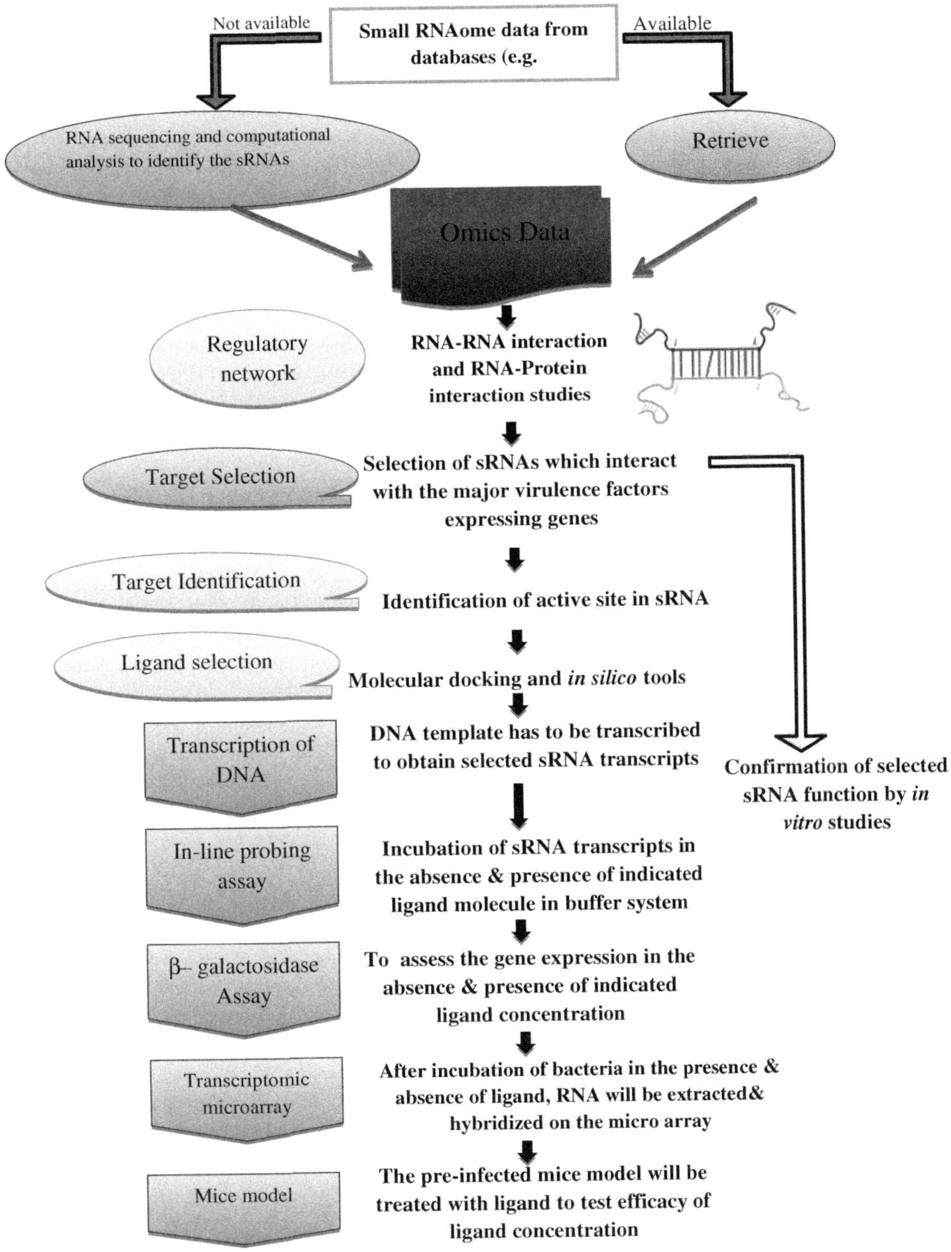

FIG. 6 Process in identification of sRNA-based drug molecules.

the biofilm formation and pathogenesis is of significant importance to unveil the cellular mechanisms and thereby aid in developing new treatments for bacterial infections.

References

Adams, P. P., & Storz, G. (2020). Prevalence of small base-pairing RNAs derived from diverse genomic loci. *Biochimica et Biophysica Acta (BBA) - Gene Regulatory Mechanisms*, *1863*(7), 194524.

Ahmed, W., Zheng, K., & Liu, Z. F. (2016). Small non-coding RNAs: New insights in modulation of host immune response by intracellular bacterial pathogens. *Frontiers in Immunology*, *7*, 431.

Ancona, V., Lee, J. H., & Zhao, Y. (2016). The RNA-binding protein CsrA plays a central role in positively regulating virulence factors in *Erwinia amylovora*. *Scientific Reports*, *6*(1), 1–12.

Andersen, J., & Delihas, N. (1990). micF RNA binds to the 5'end of *ompF* mRNA and to a protein from *Escherichia coli*. *Biochemistry*, *29*(39), 9249–9256.

Arnvig, K., & Young, D. (2012). Non-coding RNA and its potential role in *Mycobacterium tuberculosis* pathogenesis. *RNA Biology*, *9*(4), 427–436.

Attaiech, L., Boughammoura, A., Brochier-Armanet, C., Allatif, O., Peillard-Fiorente, F., Edwards, R. A., & Charpentier, X. (2016). Silencing of natural transformation by an RNA chaperone and a multi-target small RNA. *Proceedings of the National Academy of Sciences*, *113*(31), 8813–8818.

Azam, M. S., & Vanderpool, C. K. (2020). Translation inhibition from a distance: the small RNA SgrS silences a ribosomal protein S1-dependent enhancer. *Molecular Microbiology*, *114*(3), 391–408.

Babitzke, P., & Romeo, T. (2007). CsrB sRNA family: Sequestration of RNA-binding regulatory proteins. *Current Opinion in Microbiology*, *10*(2), 156–163.

Babski, J., Maier, L. K., Heyer, R., Jaschinski, K., Prasse, D., Jäger, D., et al. (2014). Small regulatory RNAs in archaea. *RNA Biology*, *11*(5), 484–493.

Bak, M., Silahtaroglu, A., Møller, M., Christensen, M., Rath, M. F., Skryabin, B., & Kauppinen, S. (2008). MicroRNA expression in the adult mouse central nervous system. *RNA*, *14*(3), 432–444.

Bandyra, K. J., Said, N., Pfeiffer, V., Górna, M. W., Vogel, J., & Luisi, B. F. (2012). The seed region of a small RNA drives the controlled destruction of the target mRNA by the endoribonuclease RNase E. *Molecular Cell*, *47*(6), 943–953.

Batey, R. T., Rambo, R. P., Lucast, L., Rha, B., & Doudna, J. A. (2000). Crystal structure of the ribonucleoprotein core of the signal recognition particle. *Science*, *287*(5456), 1232–1239.

Bernick, D. L., Dennis, P. P., Lui, L. M., & Lowe, T. M. (2012). Diversity of antisense and other non-coding RNAs in archaea revealed by comparative small RNA sequencing in four *Pyrobaculum* species. *Frontiers in Microbiology*, *3*, 231.

Bianco, C. M., Fröhlich, K. S., & Vanderpool, C. K. (2019). Bacterial cyclopropane fatty acid synthase mRNA is targeted by activating and repressing small RNAs. *bioRxiv*. https://doi.org/10.1128/jb.00461-419 (Preprint).

Bilusic, I., Popitsch, N., Rescheneder, P., Schroeder, R., & Lybecker, M. (2014). Revisiting the coding potential of the *E. coli* genome through Hfq co-immunoprecipitation. *RNA Biology*, *11*(5), 641–654.

Bloch, S., Węgrzyn, A., Węgrzyn, G., & Nejman-Faleńczyk, B. (2017). Small and smaller sRNAs and microRNAs in the regulation of toxin gene expression in prokaryotic cells: A mini-review. *Toxins*, *9* (6), 181.

Bouvier, M., Sharma, C. M., Mika, F., Nierhaus, K. H., & Vogel, J. (2008). Small RNA binding to 5′ mRNA coding region inhibits translational initiation. *Molecular Cell*, *32*(6), 827–837.

Brencic, A., McFarland, K. A., McManus, H. R., Castang, S., Mogno, I., Dove, S. L., et al. (2009). The GacS/GacA signal transduction system of *Pseudomonas aeruginosa* acts exclusively through its control over the transcription of the RsmY and RsmZ regulatory small RNAs. *Molecular Microbiology*, *73*(3), 434–445.

Brennan, R. G., & Link, T. M. (2007). Hfq structure, function and ligand binding. *Current Opinion in Microbiology*, *10*(2), 125–133.

Brenneis, M., & Soppa, J. (2009). Regulation of translation in haloarchaea: 5′-and 3′-UTRs are essential and have to functionally interact *in vivo*. *PLoS One*, *4*(2), e4484.

Buddeweg, A., Sharma, K., Urlaub, H., & Schmitz, R. A. (2018). sRNA41 affects ribosome binding sites within polycistronic mRNAs in *Methanosarcina mazei* Gö1. *Molecular Microbiology*, *107*(5), 595–609.

Cavanagh, A. T., & Wassarman, K. M. (2014). 6S RNA, a global regulator of transcription in *Escherichia coli, Bacillus subtilis*, and beyond. *Annual Review of Microbiology*, *68*, 45–60.

Chabelskaya, S., Gaillot, O., & Felden, B. (2010). A *Staphylococcus aureus* small RNA is required for bacterial virulence and regulates the expression of an immune-evasion molecule. *PLoS Pathogens*, *6*, e1000927. https://doi.org/10.1371/journal.ppat.1000927.

Chao, Y., Papenfort, K., Reinhardt, R., Sharma, C. M., & Vogel, J. (2012). An atlas of Hfq-bound transcripts reveals 3′ UTRs as a genomic reservoir of regulatory small RNAs. *The EMBO Journal*, *31*(20), 4005–4019.

Chao, Y., & Vogel, J. (2016). A 3′ UTR-derived small RNA provides the regulatory noncoding arm of the inner membrane stress response. *Molecular Cell*, *61*(3), 352–363.

Chaulk, S. G., Smith Frieday, M. N., Arthur, D. C., Culham, D. E., Edwards, R. A., Soo, P., et al. (2011). ProQ is an RNA chaperone that controls ProP levels in *Escherichia coli*. *Biochemistry*, *50*(15), 3095–3106.

Chevalier, C., Boisset, S., Romilly, C., Masquida, B., Fechter, P., Geissmann, T., et al. (2010). *Staphylococcus aureus* RNAIII binds to two distant regions of coa mRNA to arrest translation and promote mRNA degradation. *PLoS Pathogens*, *6*, e1000809. https://doi.org/10.1371/journal.ppat.1000809.

Chihara, K., Bischler, T., Barquist, L., Monzon, V. A., Noda, N., Vogel, J., et al. (2019). Conditional Hfq association with small noncoding RNAs in *Pseudomonas aeruginosa* revealed through comparative UV cross-linking immunoprecipitation followed by high-throughput sequencing. *mSystems*, *4*(6), e00590-19.

Crawford, R. W., Reeve, K. E., & Gunn, J. S. (2010). Flagellated but not hyperfimbriated *Salmonella enterica* serovar Typhimurium attaches to and forms biofilms on cholesterol-coated surfaces. *Journal of Bacteriology*, *192*, 2981–2990.

Cui, C., Wang, Y., Liu, J., Zhao, J., Sun, P., & Wang, S. (2019). A fungal pathogen deploys a small silencing RNA that attenuates mosquito immunity and facilitates infection. *Nature Communications*, *10*(1), 1–10.

Dar, D., Shamir, M., Mellin, J. R., Koutero, M., Stern-Ginossar, N., Cossart, P., et al. (2016). Term-seq reveals abundant ribo-regulation of antibiotics resistance in bacteria. *Science*, *352*(6282), aad9822.

Dar, D., & Sorek, R. (2018). High-resolution RNA 3′-ends mapping of bacterial Rho-dependent transcripts. *Nucleic Acids Research*, *46*(13), 6797–6805.

De Lay, N., & Gottesman, S. (2012). A complex network of small non-coding RNAs regulate motility in *Escherichia coli*. *Molecular Microbiology*, *86*(3), 524–538.

De Mets, F., Van Melderen, L., & Gottesman, S. (2019). Regulation of acetate metabolism and coordination with the TCA cycle via a processed small RNA. *Proceedings of the National Academy of Sciences*, *116*(3), 1043–1052.

Demolli, S., Geist, M. M., Weigand, J. E., Matschiavelli, N., Suess, B., & Rother, M. (2014). Development of β-lactamase as a tool for monitoring conditional gene expression by a tetracycline-riboswitch in *Methanosarcina acetivorans*. *Archaea, 2014*.

Dennis, P. P., Omer, A., & Lowe, T. (2001). A guided tour: Small RNA function in Archaea. *Molecular Microbiology*, *40*(3), 509–519.

Dersch, P., Khan, M. A., Mühlen, S., & Görke, B. (2017). Roles of regulatory RNAs for antibiotic resistance in bacteria and their potential value as novel drug targets. *Frontiers in Microbiology*, *8*, 803.

Dutcher, H. A., & Raghavan, R. (2018). Origin, evolution, and loss of bacterial small RNAs. *Microbiology Spectrum*, *6*(2). 6-2.

Eisenbart, S. K., Alzheimer, M., Pernitzsch, S. R., Dietrich, S., Stahl, S., & Sharma, C. M. (2020). A repeat-associated small RNA controls the major virulence factors of *Helicobacter pylori*. *Molecular Cell*, *80* (2), 210–226.

Ettwiller, L., Buswell, J., Yigit, E., & Schildkraut, I. (2016). A novel enrichment strategy reveals unprecedented number of novel transcription start sites at single base resolution in a model prokaryote and the gut microbiome. *BMC Genomics*, *17*(1), 1–14.

Faucher, S. P., Friedlander, G., Livny, J., Margalit, H., & Shuman, H. A. (2010). *Legionella pneumophila* 6S RNA optimizes intracellular multiplication. *Proceedings of the National Academy of Sciences of the United States of America*, *107*(16), 7533–7538.

Feng, L., Rutherford, S. T., Papenfort, K., Bagert, J. D., Van Kessel, J. C., Tirrell, D. A., et al. (2015). A qrr noncoding RNA deploys four different regulatory mechanisms to optimize quorum-sensing dynamics. *Cell*, *160*(1–2), 228–240.

Fröhlich, K. S., Förstner, K. U., & Gitai, Z. (2018). Post-transcriptional gene regulation by an Hfq-independent small RNA in *Caulobacter crescentus*. *Nucleic Acids Research*, *46*(20), 10969–10982.

Fröhlich, K. S., & Papenfort, K. (2016). Interplay of regulatory RNAs and mobile genetic elements in enteric pathogens. *Molecular Microbiology*, *101*(5), 701–713.

Fu, H., Elena, R. C., & Marquez, P. H. (2019). The roles of small RNAs: Insights from bacterial quorum sensing. *ExRNA*, *1*(1), 1–8.

Gelsinger, D. R., & DiRuggiero, J. (2018). The non-coding regulatory RNA revolution in archaea. *Genes*, *9*(3).

Ghaz-Jahanian, M. A., Khodaparastan, F., Berenjian, A., & Jafarizadeh-Malmiri, H. (2013). Influence of small RNAs on biofilm formation process in bacteria. *Molecular Biotechnology*, *55*(3), 288–297.

Gomes-Filho, J. V., Daume, M., & Randau, L. (2018). Unique archaeal small RNAs. *Annual Review of Genetics*, *52*, 465–487.

Gonzalez, G. M., Hardwick, S. W., Maslen, S. L., Skehel, J. M., Holmqvist, E., Vogel, J., et al. (2017). Structure of the *Escherichia coli* ProQ RNA-binding protein. *RNA*, *23*(5), 696–711.

González Plaza, J. J. (2020). Small RNAs as fundamental players in the transference of information during bacterial infectious diseases. *Frontiers in Molecular Biosciences*, *7*, 101.

González-Flecha, B., & Demple, B. (1999). Role for the *oxyS* gene in regulation of intracellular hydrogen peroxide in *Escherichia coli*. *Journal of Bacteriology*, *181*(12), 3833–3836.

Grützner, J., Remes, B., Eisenhardt, K. M., Scheller, D., Kretz, J., Madhugiri, R., et al. (2021). sRNA-mediated RNA processing regulates bacterial cell division. *Nucleic Acids Research*, *49*(12), 7035–7052.

Guillier, M., & Gottesman, S. (2006). Remodelling of the *Escherichia coli* outer membrane by two small regulatory RNAs. *Molecular Microbiology*, *59*(1), 231–247.

Hale, C., Kleppe, K., Terns, R. M., & Terns, M. P. (2008). Prokaryotic silencing (psi)RNAs in *Pyrococcus furiosus*. *RNA*, *14*(12), 2572–2579.

Halfmann, A., Kovacs, M., Hakenbeck, R., & Bruckner, R. (2007). Identification of the genes directly controlled by the response regulator CiaR in *Streptococcus pneumoniae*: Five out of 15 promoters drive expression of small non-coding RNAs. *Molecular Microbiology*, *66*, 110–126.

Handzlik, J. E., Tastsoglou, S., Vlachos, I. S., & Hatzigeorgiou, A. G. (2020). Manatee: Detection and quantification of small non-coding RNAs from next-generation sequencing data. *Scientific Reports*, *10*(1), 1–10.

Hébrard, M., Kröger, C., Srikumar, S., Colgan, A., Händler, K., & Hinton, J. C. (2012). sRNAs and the virulence of *Salmonella enterica* serovar Typhimurium. *RNA Biology*, *9*(4), 437–445.

Heidtman, M., Chen, E. J., Moy, M. Y., & Isberg, R. R. (2009). Large-scale identification of *Legionella pneumophila* Dot/Icm substrates that modulate host cell vesicle trafficking pathways. *Cellular Microbiology*, *11*(2), 230–248.

Helwak, A., & Tollervey, D. (2014). Mapping the miRNA interactome by cross-linking ligation and sequencing of hybrids (CLASH). *Nature Protocols*, *9*(3), 711–728.

Holmqvist, E., Reimegård, J., Sterk, M., Grantcharova, N., Römling, U., & Wagner, E. G. H. (2010). Two antisense RNAs target the transcriptional regulator CsgD to inhibit curli synthesis. *The EMBO Journal*, *29*(11), 1840–1850.

Huang, K., Baldrich, P., Meyers, B. C., & Caplan, J. L. (2019). sRNA-FISH: Versatile fluorescent *in situ* detection of small RNAs in plants. *The Plant Journal*, *98*(2), 359–369.

Jäger, D., Pernitzsch, S. R., Richter, A. S., Backofen, R., Sharma, C. M., & Schmitz, R. A. (2012). An archaeal sRNA targeting cis-and trans-encoded mRNAs via two distinct domains. *Nucleic Acids Research*, *40*(21), 10964–10979.

Jäger, D., Sharma, C. M., Thomsen, J., Ehlers, C., Vogel, J., & Schmitz, R. A. (2009). Deep sequencing analysis of the *Methanosarcina mazei* Gö1 transcriptome in response to nitrogen availability. *Proceedings of the National Academy of Sciences of the United States of America*, *106*(51), 21878–21882.

Janssen, K. H., Diaz, M. R., Golden, M., Graham, J. W., Sanders, W., Wolfgang, M. C., et al. (2018). Functional analyses of the RsmY and RsmZ small noncoding regulatory RNAs in *Pseudomonas aeruginosa*. *Journal of Bacteriology*, *200*(11), e00736-17.

Javelle, M., & Timmermans, M. C. (2012). In situ localization of small RNAs in plants by using LNA probes. *Nature Protocols*, *7*(3), 533–541.

Jiang, N., Yang, Y., Janbon, G., Pan, J., & Zhu, X. (2012). Identification and functional demonstration of miRNAs in the fungus *Cryptococcus neoformans*. *PLoS One*, *7*(12), e52734.

Jin, Y., Zhao, J. H., Zhao, P., Zhang, T., Wang, S., & Guo, H. S. (2019). A fungal milRNA mediates epigenetic repression of a virulence gene in *Verticillium dahliae*. *Philosophical Transactions of the Royal Society B*, *374*(1767), 20180309.

Jinek, M., Jiang, F., Taylor, D. W., Sternberg, S. H., Kaya, E., Ma, E., & Doudna, J. A. (2014). Structures of Cas9 endonucleases reveal RNA-mediated conformational activation. *Science*, *343*(6176), 1247997.

Jørgensen, M. G., Pettersen, J. S., & Kallipolitis, B. H. (2020). sRNA-mediated control in bacteria: An increasing diversity of regulatory mechanisms. *Biochimica et Biophysica Acta (BBA)-Gene Regulatory Mechanisms*, *1863*(5), 194504.

Ju, X., Li, D., & Liu, S. (2019). Full-length RNA profiling reveals pervasive bidirectional transcription terminators in bacteria. *Nature Microbiology*, *4*(11), 1907–1918.

Kim, S., Reyes, D., Beaume, M., Francois, P., & Cheung, A. (2014). Contribution of teg49 small RNA in the 50 upstream transcriptional region of sarA to virulence in *Staphylococcus aureus*. *Infections and Immunity*, *82*, 4369–4379.

Kerr, C. H., Culham, D. E., Marom, D., & Wood, J. M. (2014). Salinity-dependent impacts of ProQ, Prc, and Spr deficiencies on *Escherichia coli* cell structure. *Journal of Bacteriology*, *196*(6), 1286–1296.

Kim, H. K., Yeom, J. H., & Kay, M. A. (2020). Transfer RNA-derived small RNAs: Another layer of gene regulation and novel targets for disease therapeutics. *Molecular Therapy*, *28*(11), 2340–2357.

Kim, T., Bak, G., Lee, J., & Kim, K. S. (2015). Systematic analysis of the role of bacterial Hfq-interacting sRNAs in the response to antibiotics. *Journal of Antimicrobial Chemotherapy*, *70*(6), 1659–1668.

Kint, G., De Coster, D., Marchal, K., Vanderleyden, J., & De Keersmaecker, S. C. (2010). The small regulatory RNA molecule MicA is involved in *Salmonella enterica* serovar Typhimurium biofilm formation. *BMC Microbiology*, *10*(1), 1–8.

Kliemt, J., Jaschinski, K., & Soppa, J. (2019). A haloarchaeal small regulatory RNA (sRNA) is essential for rapid adaptation to phosphate starvation conditions. *Frontiers in Microbiology*, *10*, 1219.

Kröger, C., Rothhardt, J. E., Brokatzky, D., Felsl, A., Kary, S. C., Heermann, R., et al. (2018). The small RNA RssR regulates myo-inositol degradation by *Salmonella enterica*. *Scientific Reports*, *8*(1), 1–15.

Kudla, G., Granneman, S., Hahn, D., Beggs, J. D., & Tollervey, D. (2011). Cross-linking, ligation, and sequencing of hybrids reveals RNA–RNA interactions in yeast. *Proceedings of the National Academy of Sciences of the United States of America*, *108*(24), 10010–10015.

Kuksa, P. P., Leung, Y. Y., Vandivier, L. E., Anderson, Z., Gregory, B. D., & Wang, L. S. (2017). In silico identification of RNA modifications from high-throughput sequencing data using HAMR. In *RNA methylation* (pp. 211–229). New York, NY: Humana Press.

Kulesus, R. R., Diaz-Perez, K., Slechta, E. S., Eto, D. S., & Mulvey, M. A. (2008). Impact of the RNA chaperone Hfq on the fitness and virulence potential of uropathogenic *Escherichia coli*. *Infection and Immunity*, *76*(7), 3019–3026.

Kumari, P., & Sampath, K. (2015). cncRNAs: Bi-functional RNAs with protein coding and non-coding functions. In *Vol. 47. Seminars in cell & developmental biology* (pp. 40–51). Academic Press.

Lalaouna, D., Baude, J., Wu, Z., Tomasini, A., Chicher, J., Marzi, S., et al. (2019). RsaC sRNA modulates the oxidative stress response of *Staphylococcus aureus* during manganese starvation. *Nucleic Acids Research*, *47*(18), 9871–9887.

Landt, S. G., Lesley, J. A., Britos, L., & Shapiro, L. (2010). CrfA, a small noncoding RNA regulator of adaptation to carbon starvation in *Caulobacter crescentus*. *Journal of Bacteriology*, *192*(18), 4763–4775.

Lee, H. C., Li, L., Gu, W., Xue, Z., Crosthwaite, S. K., Pertsemlidis, A., et al. (2010). Diverse pathways generate microRNA-like RNAs and dicer-independent small interfering RNAs in fungi. *Molecular Cell*, *38*(6), 803–814.

Levine, E., Huang, M., Huang, Y., Kuhlman, T., Shi, H., Zhang, Z. G., & Hwa, T. *On noise and silence in small RNA regulation*. https://thefauve.hwa.ucsd.edu/~thwa/pub/sRNA-noise.pdf.

Li, Z., & Stanton, B. A. (2021). Transfer RNA-derived fragments, the underappreciated regulatory small RNAs in microbial pathogenesis. *Frontiers in Microbiology*, *12*, 1252.

Liu, J. M., Livny, J., Lawrence, M. S., Kimball, M. D., Waldor, M. K., & Camilli, A. (2009). Experimental discovery of sRNAs in *Vibrio cholerae* by direct cloning, 5S/tRNA depletion and parallel sequencing. *Nucleic Acids Research*, *37*(6), e46.

Liu, H., Wang, X., Wang, H. D., Wu, J., Ren, J., Meng, L., et al. (2012). *Escherichia coli* noncoding RNAs can affect gene expression and physiology of *Caenorhabditis elegans*. *Nature Communications*, *3* (1), 1–11.

Loh, E., Dussurget, O., Gripenland, J., Vaitkevicius, K., Tiensuu, T., Mandin, P., et al. (2009). A trans-acting riboswitch controls expression of the virulence regulator PrfA in *Listeria monocytogenes*. *Cell*, *139* (4), 770–779.

Lu, P., Zhang, Y., Hu, Y., Francis, M. S., & Chen, S. (2014). A *cis*-encoded sRNA controls the expression of fabH2 in Yersinia. *FEBS Letters*, *588* (10), 1961–1966.

Mai, J., Rao, C., Watt, J., Sun, X., Lin, C., Zhang, L., et al. (2019). *Mycobacterium tuberculosis* 6C sRNA binds multiple mRNA targets via C-rich loops independent of RNA chaperones. *Nucleic Acids Research*, *47*(8), 4292–4307.

Mangold, M., Siller, M., Roppenser, B., Vlaminckx, B. J., Penfound, T. A., Klein, R., et al. (2004). Synthesis of group A streptococcal virulence factors is controlled by a regulatory RNA molecule. *Molecular Microbiology*, *53*, 1515–1527.

Manna, A. C., Kim, S., Cengher, L., Corvaglia, A., Leo, S., Francois, P., et al. (2018). Small RNA teg49 is derived from a sarA transcript and regulates virulence genes independent of SarA in *Staphylococcus aureus*. *Infection and Immunity*, *86*(2), e00635-17.

Marchfelder, A., Fischer, S., Brendel, J., Stoll, B., Maier, L. K., Jäger, D., et al. (2012). Small RNAs for defence and regulation in archaea. *Extremophiles*, *16*(5), 685–696.

Märtens, B., Manoharadas, S., Hasenöhrl, D., Manica, A., & Bläsi, U. (2013). Antisense regulation by transposon-derived RNAs in the hyperthermophilic archaeon *Sulfolobus solfataricus*. *EMBO Reports*, *14*(6), 527–533.

Massé, E., Majdalani, N., & Gottesman, S. (2003). Regulatory roles for small RNAs in bacteria. *Current Opinion in Microbiology*, *6*(2), 120–124.

Melamed, S., Peer, A., Faigenbaum-Romm, R., Gatt, Y. E., Reiss, N., Bar, A., et al. (2016). Global mapping of small RNA-target interactions in bacteria. *Molecular Cell*, *63*(5), 884–897.

Mika, F., & Hengge, R. (2013). Small regulatory RNAs in the control of motility and biofilm formation in *E. coli* and *Salmonella*. *International Journal of Molecular Sciences*, *14*(3), 4560–4579.

Miyakoshi, M., Matera, G., Maki, K., Sone, Y., & Vogel, J. (2019). Functional expansion of a TCA cycle operon mRNA by a 3′ end-derived small RNA. *Nucleic Acids Research*, *47*(4), 2075–2088.

Monteiro, C., Papenfort, K., Hentrich, K., Ahmad, I., Le Guyon, S., Reimann, R., et al. (2012). Hfq and Hfq-dependent small RNAs are major contributors to multicellular development in *Salmonella enterica* serovar Typhimurium. *RNA Biology*, *9*(4), 489–502.

Mulvey, M. A., Lopez-Boado, Y. S., Wilson, C. L., Roth, R., Parks, W. C., Heuser, J., et al. (1998). Induction and evasion of host defenses by type 1-piliated uropathogenic *Escherichia coli*. *Science*, *282*(5393), 1494–1497.

Nikaido, H. (2003). Molecular basis of bacterial outer membrane permeability revisited. *Microbiology and Molecular Biology Reviews*, *67* (4), 593–656.

Nishino, K., Yamasaki, S., Hayashi-Nishino, M., & Yamaguchi, A. (2011). Effect of overexpression of small non-coding DsrA RNA on multidrug efflux in *Escherichia coli*. *Journal of Antimicrobial Chemotherapy*, *66* (2), 291–296.

Nitzan, M., Rehani, R., & Margalit, H. (2017). Integration of bacterial small RNAs in regulatory networks. *Annual Review of Biophysics*, *46*, 131–148.

Nowacka, M., Jackowiak, P., Rybarczyk, A., Magacz, T., Strozycki, P. M., Barciszewski, J., et al. (2012). 2D-PAGE as an effective method of RNA degradome analysis. *Molecular Biology Reports*, *39*(1), 139–146.

Oliva, G., Sahr, T., & Buchrieser, C. (2018). The life cycle of *L. pneumophila*: Cellular differentiation is linked to virulence and metabolism. *Frontiers in Cellular and Infection Microbiology*, *8*, 3.

Ortega, A. D., Gonzalo-Asensio, J., & García-del Portillo, F. (2012). Dynamics of *Salmonella* small RNA expression in non-growing bacteria located inside eukaryotic cells. *RNA Biology*, *9*(4), 469–488.

Papenfort, K., & Vogel, J. (2010). Regulatory RNA in bacterial pathogens. *Cell Host & Microbe*, *8*(1), 116–127.

Peng, Y. L., Meng, Q. L., Qiao, J., Xie, K., Chen, C., Liu, T. L., et al. (2016). The roles of noncoding RNA Rli60 in regulating the virulence of *Listeria monocytogenes*. *Journal of Microbiology, Immunology and Infection*, *49*(4), 502–508.

Pfeiffer, V., Papenfort, K., Lucchini, S., Hinton, J. C., & Vogel, J. (2009). Coding sequence targeting by MicC RNA reveals bacterial mRNA silencing downstream of translational initiation. *Nature Structural & Molecular Biology*, *16*(8), 840–846.

Piattelli, E., Peltier, J., & Soutourina, O. (2020). Interplay between regulatory RNAs and signal transduction systems during bacterial infection. *Genes*, *11*(10), 1209.

Pichon, C., & Felden, B. (2007). Proteins that interact with bacterial small RNA regulators. *FEMS Microbiology Reviews*, *31*(5), 614–625.

Popitsch, N., Bilusic, I., Rescheneder, P., Schroeder, R., & Lybecker, M. (2017). Temperature-dependent sRNA transcriptome of the Lyme disease spirochete. *BMC Genomics*, *18*(1), 1–14.

Pourciau, C., Lai, Y. J., Gorelik, M., Babitzke, P., & Romeo, T. (2020). Diverse mechanisms and circuitry for global regulation by the RNA-binding protein CsrA. *Frontiers in Microbiology*, *2709*.

Prasse, D., Ehlers, C., Backofen, R., & Schmitz, R. A. (2013). Regulatory RNAs in archaea: First target identification in Methanoarchaea. *Biochemical Society Transactions*, *41*(1), 344–349.

Prasse, D., Förstner, K. U., Jäger, D., Backofen, R., & Schmitz, R. A. (2017). sRNA154 a newly identified regulator of nitrogen fixation in *Methanosarcina mazei* strain Gö1. *RNA Biology*, *14*(11), 1544–1558.

Quendera, A. P., Seixas, A. F., Dos Santos, R. F., Santos, I., Silva, J. P., Arraiano, C. M., et al. (2020). RNA-binding proteins driving the regulatory activity of small non-coding RNAs in bacteria. *Frontiers in Molecular Biosciences*, *7*, 78.

Ramirez-Peña, E., Treviño, J., Liu, Z., Perez, N., & Sumby, P. (2010). The group A Streptococcus small regulatory RNA FasX enhances streptokinase activity by increasing the stability of the ska mRNA transcript. *Molecular Microbiology*, *78*(6), 1332–1347.

Romeo, T. (1998). Global regulation by the small RNA-binding protein CsrA and the non-coding RNA molecule CsrB. *Molecular Microbiology*, *29*(6), 1321–1330.

Ross, J. A., Thorsing, M., Lillebæk, E. M. S., Teixeira Dos Santos, P., & Kallipolitis, B. H. (2019). The LhrC sRNAs control expression of T cell-stimulating antigen TcsA in *Listeria monocytogenes* by decreasing *tcsA* mRNA stability. *RNA Biology*, *16*(3), 270–281.

Ruiz-Larrabeiti, O., Plágaro, A. H., Gracia, C., Sevillano, E., Gallego, L., Hajnsdorf, E., et al. (2016). A new custom microarray for sRNA profiling in *Escherichia coli*. *FEMS Microbiology Letters*, *363*(13).

Ruiz-Vázquez, R. M., Nicolás, F. E., Torres-Martínez, S., & Garre, V. (2015). Distinct RNAi pathways in the regulation of physiology and development in the fungus *Mucor circinelloides*. *Advances in Genetics*, *91*, 55–102.

Salvail, H., Lanthier-Bourbonnais, P., Sobota, J. M., Caza, M., Benjamin, J. A., Mendieta, M. E., et al. (2010). A small RNA promotes siderophore production through transcriptional and metabolic remodeling. *Proceedings of the National Academy of Sciences of the United States of America*, *107*, 15223–15228.

Sauer, E., & &Weichenrieder, O. (2011). Structural basis for RNA 3′-end recognition by Hfq. *The Proceedings of the National Academy of Sciences*, *108*(32), 13065–13070.

Sesto, N., Wurtzel, O., Archambaud, C., Sorek, R., & Cossart, P. (2013). The excludon: A new concept in bacterial antisense RNA-mediated gene regulation. *Nature Reviews Microbiology*, *11*(2), 75–82.

Shimizu, K. (2014). Regulation systems of bacteria such as *Escherichia coli* in response to nutrient limitation and environmental stresses. *Metabolites*, *4*(1), 1–35.

Simon, S. A., & Gutknecht, J. (1980). Solubility of carbon dioxide in lipid bilayer membranes and organic solvents. *Biochimica et Biophysica Acta (BBA) - Biomembranes*, *596*(3), 352–358.

Sittka, A., Lucchini, S., Papenfort, K., Sharma, C. M., Rolle, K., Binnewies, T. T., et al. (2008). Deep sequencing analysis of small noncoding RNA and mRNA targets of the global post-transcriptional regulator, Hfq. *PLoS Genetics*, *4*(8), e1000163.

Smirnov, A., Förstner, K. U., Holmqvist, E., Otto, A., Günster, R., Becher, D., et al. (2016). Grad-seq guides the discovery of ProQ as a major small RNA-binding protein. *Proceedings of the National Academy of Sciences of the United States of America*, *113*(41), 11591–11596.

Song, T., Mika, F., Lindmark, B., Liu, Z., Schild, S., Bishop, A., et al. (2008). A new *Vibrio cholerae* sRNA modulates colonization and affects release of outer membrane vesicles. *Molecular Microbiology*, *70*(1), 100–111.

Sonnleitner, E., Abdou, L., & Haas, D. (2009). Small RNA as global regulator of carbon catabolite repression in *Pseudomonas aeruginosa*. *Proceedings of the National Academy of Sciences of the United States of America*, *106*(51), 21866–21871.

Speed, M. C., Burkhart, B. W., Picking, J. W., & Santangelo, T. J. (2018). An archaeal fluoride-responsive riboswitch provides an inducible expression system for hyperthermophiles. *Applied and Environmental Microbiology*, *84*(7), e02306-17.

Stauffer, L. T., & Stauffer, G. V. (2012). The *Escherichia coli* GcvB sRNA uses genetic redundancy to control cycA expression. *International Scholarly Research Notices*, *2012*.

Stork, M., Di Lorenzo, M., Welch, T. J., & Crosa, J. H. (2007). Transcription termination within the iron transport-biosynthesis operon of *Vibrio anguillarum* requires an antisense RNA. *Journal of Bacteriology*, *189*(9), 3479–3488.

Storz, G., Vogel, J., & Wassarman, K. M. (2011). Regulation by small RNAs in bacteria: Expanding frontiers. *Molecular Cell*, *43*(6), 880–891.

Su, Y. A., He, P. I. N. G., & Clewell, D. B. (1992). Characterization of the tet (M) determinant of Tn916: Evidence for regulation by transcription attenuation. *Antimicrobial Agents and Chemotherapy*, *36*(4), 769–778.

Sudo, N., Soma, A., Iyoda, S., Oshima, T., Ohto, Y., Saito, K., et al. (2018). Small RNA Esr41 inversely regulates expression of LEE and flagellar genes in enterohaemorrhagic *Escherichia coli*. *Microbiology*, *164*, 821–834.

Taylor, P. K., Van Kessel, A. T., Colavita, A., Hancock, R. E., & Mah, T. F. (2017). A novel small RNA is important for biofilm formation and pathogenicity in *Pseudomonas aeruginosa*. *PLoS One*, *12*(8), e0182582.

Thibonnier, M., Thiberge, J. M., & De Reuse, H. (2008). Trans-translation in *Helicobacter pylori*: Essentiality of ribosome rescue and requirement of protein tagging for stress resistance and competence. *PLoS One*, *3*(11), e3810.

Thomason, M. K., & Storz, G. (2010). Bacterial antisense RNAs: How many are there and what are they doing? *Annual Review of Genetics*, *44*, 167–188.

Toledo-Arana, A., Dussurget, O., Nikitas, G., Sesto, N., Guet-Revillet, H., Balestrino, D., et al. (2009). The *Listeria* transcriptional landscape from saprophytism to virulence. *Nature*, *459*, 950–956.

Tomasini, A., François, P., Howden, B. P., Fechter, P., Romby, P., & Caldelari, I. (2014). The importance of regulatory RNAs in *Staphylococcus aureus*. *Infection, Genetics and Evolution*, *21*, 616–626.

Tomasini, A., Moreau, K., Chicher, J., Geissmann, T., Vandenesch, F., Romby, P., & Caldelari, I. (2017). The RNA targetome of *Staphylococcus aureus* non-coding RNA RsaA: Impact on cell surface properties and defense mechanisms. *Nucleic Acids Research*, *45*(11), 6746–6760.

Udekwu, K. I., Darfeuille, F., Vogel, J., Reimegård, J., Holmqvist, E., & Wagner, E. G. H. (2005). Hfq-dependent regulation of OmpA synthesis is mediated by an antisense RNA. *Genes & Development*, *19*(19), 2355–2366.

Updegrove, T. B., Shabalina, S. A., & Storz, G. (2015). How do base-pairing small RNAs evolve? *FEMS Microbiology Reviews*, *39*(3), 379–391.

Vanderpool, C. K., & Gottesman, S. (2004). Involvement of a novel transcriptional activator and small RNA in post-transcriptional regulation of the glucose phosphoenolpyruvate phosphotransferase system. *Molecular Microbiology*, *54*(4), 1076–1089.

Wang, X., Dubey, A. K., Suzuki, K., Baker, C. S., Babitzke, P., & Romeo, T. (2005). CsrA post-transcriptionally represses pgaABCD, responsible for synthesis of a biofilm polysaccharide adhesin of *Escherichia coli*. *Molecular Microbiology*, *56*(6), 1648–1663.

Wang, Y., Rotman, E. R., Shoemaker, N. B., & Salyers, A. A. (2005). Translational control of tetracycline resistance and conjugation in the Bacteroides conjugative transposon CTnDOT. *Journal of Bacteriology*, *187*(8), 2673–2680.

Wang, B., Sun, Y., Song, N., Zhao, M., Liu, R., Feng, H., et al. (2017). *Puccinia striiformis* f. sp. *tritici* microRNA-like RNA 1 (Pst-milR1), an important pathogenicity factor of Pst, impairs wheat resistance to Pst by suppressing the wheat pathogenesis-related 2 gene. *New Phytologist*, *215*(1), 338–350.

Wang, L., Yang, G., Qi, L., Li, X., Jia, L., Xie, J., et al. (2016). A novel small RNA regulates tolerance and virulence in *Shigella flexneri* by responding to acidic environmental changes. *Frontiers in Cellular and Infection Microbiology*, *6*, 24.

Watkins, D., & Arya, D. P. (2019). Regulatory roles of small RNAs in prokaryotes: Parallels and contrast with eukaryotic miRNA. *Non-coding RNA Investigation*, *3*, 28.

Weber, B., Croxatto, A., Chen, C., & Milton, D. L. (2008). RpoS induces expression of the *Vibrio anguillarum* quorum-sensing regulator VanT. *Microbiology*, *154*(3), 767–780.

Weiberg, A., Wang, M., Lin, F. M., Zhao, H., Zhang, Z., Kaloshian, I., et al. (2013). Fungal small RNAs suppress plant immunity by hijacking host RNA interference pathways. *Science*, *342*(6154), 118–123.

Xi, D., Li, Y., Yan, J., Li, Y., Wang, X., & Cao, B. (2020). Small RNA coaR contributes to intestinal colonization in *Vibrio cholerae* via the two-component system EnvZ/OmpR. *Environmental Microbiology*, *22*(10), 4231–4243.

Xu, M., Li, G., Guo, Y., Gao, Y., Zhu, L., Liu, Z., et al. (2022). A fungal microRNA-like RNA subverts host immunity and facilitates pathogen infection by silencing two host receptor-like kinase genes. *New Phytologist*, *233*, 2503–2519.

Yan, B., Boitano, M., Clark, T. A., & Ettwiller, L. (2018). SMRT-Cappable-seq reveals complex operon variants in bacteria. *Nature Communications*, *9*(1), 1–11.

Yang, G., Li, B., Jia, L., Qiu, H., Yang, M., Zhu, B., et al. (2020). A novel sRNA in *Shigella flexneri* that regulates tolerance and virulence under hyperosmotic pressure. *Frontiers in Cellular and Infection Microbiology*, *10*, 483.

Zapf, R. L., Wiemels, R. E., Keogh, R. A., Holzschu, D. L., Howell, K. M., Trzeciak, E., et al. (2019). The small RNA Teg41 regulates expression of the alpha phenol soluble modulins and is required for virulence in *Staphylococcus aureus*. *mBio*, *10*, e02484-18.

Zhang, A., Altuvia, S., Tiwari, A., Argaman, L., Hengge-Aronis, R., & Storz, G. (1998). The OxyS regulatory RNA represses rpoS translation and binds the Hfq (HF-I) protein. *The EMBO Journal*, *17*(20), 6061–6068.

Zhang, X., Lii, Y., Wu, Z., Polishko, A., Zhang, H., Chinnusamy, V., et al. (2013). Mechanisms of small RNA generation from cis-NATs in response to environmental and developmental cues. *Molecular Plant*, *6*(3), 704–715.

Zhang, A., Wassarman, K. M., Ortega, J., Steven, A. C., & Storz, G. (2002). The Sm-like Hfq protein increases OxyS RNA interaction with target mRNAs. *Molecular Cell*, *9*(1), 11–22.

Zhou, M., Wang, Q., Sun, J., Li, X., Xu, L., Yang, H., et al. (2009). *In silico* detection and characteristics of novel microRNA genes in the *Equus caballus* genome using an integrated *ab initio* and comparative genomic approach. *Genomics*, *94*(2), 125–131.

Chapter 23

Genetic basis of biofilm formation and their role in antibiotic resistance, adhesion, and persistent infections in ESKAPE pathogens

K. Prashanth, Ajit Ramesh Sawant, and Lipsa Panda
Laboratory No. 6, Nosocomial Pathogen Biology Laboratory, Department of Biotechnology, Pondicherry University, Kalapet, Puducherry, India

1. Introduction

Bacteria are capable of sensing their versatile surrounding environment, and they can modify or adapt their behavior based on their need and survival. Nosocomial pathogenic bacteria are no exception for such behavioral changes to the external environments such as hospital environs and in the infected host that adversely respond to their colonization. The infected host usually mobilizes strong immune response as their main antiinfectious strategies, which most of the ESKAPE group of pathogens easily ward off. The acronym ESKAPE includes six significant nosocomial pathogens, namely *Enterococcus faecium*, *Staphylococcus aureus*, *Klebsiella pneumoniae*, *Acinetobacter baumannii*, *Pseudomonas aeruginosa*, and *Enterobacter* spp. that display high levels of resistance to multiple drugs and increased virulence.

One of the most important adaptive processes of ESKAPE pathogens is their capability to change from the free, floating, or planktonic state to the adherent ones and to develop biofilms on alive and abiotic substrata. Biofilms can be defined as the communities of microorganisms wherein the cells glue to each other and often adhere to a surface. Such adherent cells are usually embedded within a self-produced matrix of extracellular polymeric substance (EPS). Biofilm community lifestyle is based on highly complex communication networks that is usually referred as quorum sensing (QS) and its response systems that usually confer them an increased phenotypic or behavioral resistance to different stress factors such as antibiotics and host defense mechanisms (Lazar, Holban, Curutiu, & Chifiriuc, 2021). As a consequence, infections caused by such biofilm phenotypes are difficult to diagnose and treat. Such infections are generally requiring a combination of multiple drug therapeutic regimens, which often fail to resolve the infection. Therefore, it is very important to devise novel and efficient antibiofilm strategies that disrupt the QS mechanisms that inhibit biofilm formation and target the individual cells, which can be easily eradicated. The main purpose of this review is to have a fleeting look into the progress made in investigating genetics of biofilms of ESKAPE pathogens and their role in persistent infections and antibiotic resistance. Further to identify the most promising potential leads that could be used in therapeutic management of highly resistant persistent infections caused by these pathogens. However, as large amount of research data are already available related to biofilms developed by *S. aureus* and *P. aeruginosa*, this review focuses more on the progress made on understanding of biofilms developed by less investigated ESKAPE pathogens such as *E. faecium*, *K. pneumoniae*, *A. baumannii*, and *Enterobacter* spp.

2. How biofilms are formed in bacterial pathogens

Biofilms exist on almost all kind of surfaces in the hospital environment; it could pose a serious problem in hospitals since it constitutes as a reservoir for many pathogens. The major characteristics of such biofilms are they are highly resistant to multiple antibiotics and also resistant to other various biotic and abiotic stress. The process of biofilm formation can be divided into five important stages, namely (i) initial reversible attachment, (ii) irreversible attachment by forming a monolayer, (iii) formation of microcolony, (iv) maturation of biofilm, and (v) dispersion (see Fig. 1). Biofilm matrix is mainly composed of EPSs that comprise polysaccharides, proteins, and extracellular DNA (eDNA), which forms a physical barrier to protect bacteria from antimicrobial agents or other harmful abiotic and biotic

Understanding Microbial Biofilms. https://doi.org/10.1016/B978-0-323-99977-9.00034-X

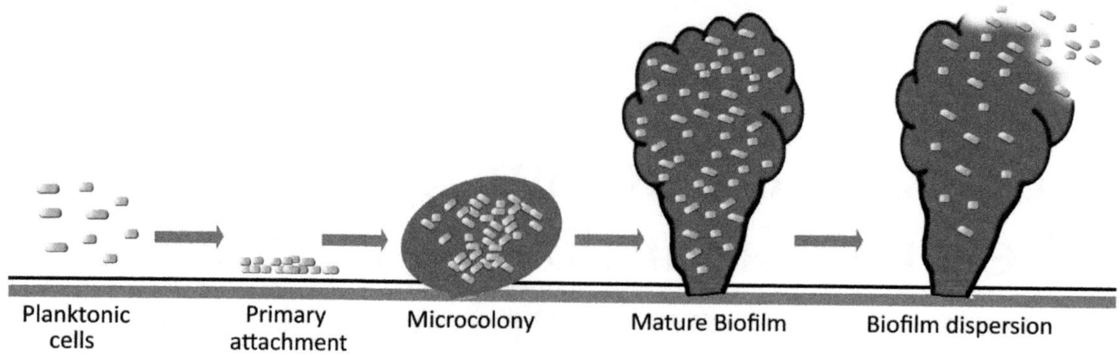

FIG. 1 Biofilm cycle—Different stages of biofilm development in bacteria.

stressors (Flemming et al., 2016; Høiby, Bjarnsholt, Givskov, Molin, & Ciofu, 2010). Biofilm microstructure based on its composition of EPS matrix assumes a variety of differentiated forms during the development such as filamentous streamers and mushroom-like microcolonies (Chew et al., 2014). Vertical structures in the biofilm are usually separated by interstitial spaces. The biofilms mainly acquire all the required nutrients easily and very quickly from the surrounding environment, and they usually move the toxic by-products away by the interstitial spaces (Secinti, Özalp, Attar, & Sargon, 2011). The mechanical stability and shapes of the biofilms are determined by the physical properties of the EPS matrix (Chew et al., 2014).

3. Microstructure of biofilms and their composition

Both Gram-positive and Gram-negative bacteria in the ESKAPE group of pathogens form biofilms, which comprise 90% of EPS material and a mere 10% microbial cell mass (Flemming & Wingender, 2010). This predominant matrix composition is secreted by the microbial cells themselves and is regulated by the strain of concern, nutrition availability, and environmental cues such as pH, temperature, metal ions (iron and zinc), etc. The components of matrix glue the microbial cells to the host surface through adhesion-cohesion forces through secretory polysaccharide, surface proteins, and eDNA (Foster, Geoghegan, Ganesh, & Höök, 2014). The EPS determines the porosity, density, water content, electrostatic charge, and hydrophobicity of the concerned biofilm (Flemming, Neu, & Wozniak, 2007). Depending on the genetic organization of a particular pathogen/strains, stress, temperature, and availability of nutrients in the surrounding environment, the EPS composition varies among biofilms. Bacterial extracellular appendages such as flagella, pili, and fimbriae assist the matrix in adherence and colonization. Biofilm EPS acts as a physical barrier by protecting the indwelling bacterial community from any physical damage or environmental stress caused by the host immune response and antimicrobial agents (Karygianni, Ren, Koo, & Thurnheer, 2020). The polysaccharide is the major component of the EPS whose composition varies among the ESKAPE group pathogens. The negative charge on eDNA facilitates its attachment to the matrix proteins and helps bacterial aggregation and stabilization of biofilm architecture. These electrostatic forces acting in the house of biofilm cells act as a chain to hold the microbial cells together in the stack of biofilm (Dengler, Foulston, DeFrancesco, & Losick, 2015). In coming paragraphs, biofilm composition and their genetic organization in each of the ESKAPE pathogens are discussed in detail.

3.1 Staphylococcus aureus

Infections caused by the *S. aureus* are generally of two types, namely acute and chronic. The adaptation to two different types of infections depends upon the ability of *S. aureus* to form biofilms on host surfaces. The acute infections caused by *S. aureus* are generally caused by planktonic cells while chronic infections are a result of the biofilm-forming ability of the bacterium. *S. aureus* produces biofilm composed of proteins having differential gene expression, glycocalyx, and some specific polysaccharide antigen such as polysaccharide intercellular antigen (PIA) comprising 80%–85% of biofilm. The (PIA)/poly-*N*-acetyl glucosamine (PNAG) is secreted by the expression of *icaADBC* (intercellular adhesion locus), and this locus is also essential for the structural integrity and virulence of *S. aureus* (Cramton, Gerke, Schnell, Nichols, & Götz, 1999). The biofilm matrix also contains phenol-soluble modulin (PSM) and eDNA, which are involved in the formation of amyloid fibers (Schwartz, Ganesan, Payne, Solomon, & Boles, 2016). The biofilm formation in *S. aureus* is regulated by negative regulator *IcaR* via binding at the promoter region of *icaADBC* locus. Besides, biofilm-associated proteins are also involved in biofilm formation in *S. aureus* independent of PIA (Lasa & Penadés, 2006). Since not all the *S. aureus* strains encode for *ica* locus, it is considered as a part of the "accessory genome" and not the core genome

(Arciola, Campoccia, Ravaioli, & Montanaro, 2015). Global regulators such as *sarA/agr* and *sigB* are also found to play important role in biofilm formation in a strain-specific manner. Cell surface adhesion protein such as protein A (Spa), fibronectin-binding proteins AB (fnbAB), clumping factor protein (*ClfA/B*), *S. aureus* surface protein (*SasC*) (Zapotoczna, O'Neill, & O'Gara, 2016), *SasG* (Geoghegan et al., 2010) are all the additional proteins that are shown to be involved in the initial attachment and biofilm formation. Proteins like the clumping factor (*ClfA/B*) play role in adhesion to fibrinogen and immune evasion, while fibronectin-binding proteins (*FnbpA/B*) are associated with the adhesion of *S. aureus* cells. Interestingly, one study by using an in vivo catheter infection model showed *MgrA*, an sarA family global transcriptional regulator that is present downstream of ArlRS two-component system (TCS), plays vital role in the biofilm formation of *S. aureus* on the surface of implanted catheters (Burgui, Gil, Solano, Lasa, & Valle, 2018). ArlRS-MgrA regulatory cascade controls the expression of several surface proteins. *MgrA* is also involved in the clumping of cells in the presence of soluble fibrinogen and controls the biofilm formation through repression of large cell-wall-associated surface proteins such as ebh, FmtB, FnbB, SasC, SasG, SdrD, Spa, and SraP (Crosby et al., 2016).

3.2 *Klebsiella pneumoniae*

In the case of *K. pneumoniae*, the initial attachment of cells to in vitro and in vivo surfaces is modulated by predominant Type I and Type III fimbriae (Schroll, Barken, Krogfelt, & Struve, 2010). It uses a chaperon usher pathway to produce type III fimbrial appendages. The major structural component of this appendage, MrkD protein polymerizes to form a fimbrial shaft, whereas the subsequent binding to host ECM collagen is mediated by the minor tip MrkA adhesion protein (Langstraat, Bohse, & Clegg, 2001). These two types of fimbriae work together in an augmented manner. Apart from the polysaccharides, proteins form major constituents of *K. pneumoniae* biofilms. One study identified the presence of 55 different proteins that include both the cytosolic and membrane proteins while deciphering the biofilm matrix of *K. pneumoniae* through matrix-assisted laser desorption/ionization mass spectroscopy/mass spectroscopy (MALDI MS/MS) (Singh et al., 2019).

3.3 *Pseudomonas aeruginosa*

The ability *P. aeruginosa* to form biofilm on surfaces of medical devices such as urinary catheters, central venous catheters (CVC), implants, etc., makes the treatment difficult. *Pseudomonas* biofilm matrix is encoded by at least three types of exopolysaccharides, namely mannose and galactose-rich type (Psl), glucose-rich type (Pel), and polyanionic alginate. Psl helps in the initial adherence of cells in a helical manner whose work is taken over by Pel during later stages of biofilm maturation (Ghafoor, Hay, & Rehm, 2011; Ryder, Byrd, & Wozniak, 2007). Strains lacking either of these two fail to produce biofilms. The third type, i.e., alginate exopolysaccharide traps live cells in the cap of the mushroom-shaped biofilm matrix. Any absence of the negatively charged alginate's activity is compensated by eDNA, which holds together the components of the biofilm matrix. Strains overproducing Pel and alginate form wrinkled and finger-like microcolonies, respectively. Apart from these secreted substances, bacteria's fimbriae, pili, and flagella are also known to bind adjacent cell types and ECM proteins of the host, thus initiating biofilm formation (O'Toole & Kolter, 1998; Wei & Ma, 2013). *P. aeruginosa* biofilm architecture depends on the type of carbon source present in the environment, i.e., if citrate is present, flat structure biofilm is formed (Heydorn et al., 2000), whereas if glucose is present as a carbon source mushroom-shaped biofilms are formed (Stewart, Peyton, Drury, & Murga, 1993). These mushroom-shaped biofilms are separated by distinct water channels supplying oxygen and other nutrients for the maintenance of biofilm. In response to environmental cues, twitching motility in bacteria is arrested leading to strong adherence among them. These nonmotile cells form the ECM stalk of the biofilm upon whom the rest of the cells use Type IV pili to climb up on the stalk forming a cap of biofilm (Klausen, Aaes-Jørgensen, Molin, & Tolker-Nielsen, 2003).

3.4 *Enterococcus faecium*

The Gram-positive bacteria *E. faecium* bind the host surface via microbial surface component adhesive matrix molecules (MSCRAMM). *Acm* (Nallapareddy, Weinstock, & Murray, 2003) and *Scm* (Sillanpää et al., 2008) proteins facilitate the bacterial interaction with Type I, IV collagen, and fibrinogen, respectively. The surface localization of Acm, a virulence factor, is mediated by autolysin $AtlA_{EFM}$ (Paganelli et al., 2013). Likewise, the binding to collagen Type V (Sillanpää et al., 2008) and nidogen 1 and nidogen 2 (Hendrickx et al., 2009) is facilitated by *Scm*, *Ecb*, and *SgrA* adhesion proteins, respectively. *SgrA* protein helps in binding abiotic surface in vitro; however, there is not much evidence about in vivo adhesion (Dunny, Hancock, & Shankar, 2014). The surface adhesion protein Esp helps in the initial attachment of bacterial cells to the abiotic surface, and it happens in a growth-condition-dependent manner (Heikens, Bonten, & Willems, 2007; Van Wamel et al., 2007). Following attachment, the bacteria is likely to acquire resistant genes by horizontal gene transfer of the secreted eDNAs present in the matrix. These extracellular DNAs are formed by the major autolysin $AtlA_{Efm}$, which glues the matrix components together (Paganelli et al., 2013). $AtlA_{Efm}$ is very important in biofilm stability and collagen binding, which is required for colonization and infection. Bacterial appendage such as the PilA and PilB helps in the adherence of cells (Hendrickx et al., 2008). Further expression of the PilA- and PilB-type pili is regulated in a temperature-dependent manner (only at 37°C and not lower temperatures), which contributes to *E. faecium* pathogenesis. With respect to second species of *Enterococcus*, which is of clinical importance, namely *E. faecalis*, one recent study revealed the involvement of *epa* (enterococcal polysaccharide antigen) operon, which plays a crucial role in biofilm-associated antibiotic resistance through modulation of polysaccharide content of biofilm, and it is also involved in the maintenance of hydrophobicity of cell surface (Korir, Dale, & Dunny, 2019). Deletion of *bph* gene that encodes for biofilm phosphatase protein (177 amino acid protein) showed a defect in biofilm formation; therefore, it was proposed that *bph* is involved in modulating global protein expression in *E. faecalis* (Willett, Ji, & Dunny, 2019).

3.5 *Enterobacter* spp.

Not much information is known about *Enterobacter* spp. biofilm structures and composition, but still some surface adhesion mechanisms have been elucidated off late. *Enterobacter* spp. binds various human tissue surfaces using type I and Type III fimbrial appendages and adhesion protein *ycfM* (Azevedo et al., 2018; Kim et al., 2012). Curli fimbriae mediate the adhesion and three-dimensional (3D) architectural layout of biofilms (Lazar et al., 2021; Mezzatesta, Gona, & Stefani, 2012; Zogaj, Bokranz, Nimtz, & Romling, 2003). An increase in mRNA expression levels of curli biogenesis genes such as csgA and csgD was observed in biofilm phenotype of *Enterobacter* spp. T6SS-2 secretion system also helps in *Enterobacter* spp. cell adherence and biofilm formation (Soria-Bustos et al., 2020). Host cell adhesion and invasion in *E. cloacae* species were assisted by protein extracellular fibers that control the biofilm formation and help in building the 3D architecture of biofilms. These protein extracellular fibers were also shown to modulate adherence on abiotic and biotic surfaces (Mezzatesta et al., 2012; Soria-Bustos et al., 2020).

4. Role of biofilms in adhesion, antibiotic resistance, and chronic infections

Having gained in-depth knowledge about the structure and the processes accompanying biofilm, the question arises why ESKAPE pathogens have to form a barrier coat in the first place. What survival advantages that the sessile microcolonies of bacterial cells possess so that planktonic bacteria indulge in sedimenting to form biofilms. Some of the forces driving the pathogen to form biofilm have been addressed in below paragraphs in response to the questions raised.

4.1 Survival in stress conditions

Pathogens residing in the biofilm community can withstand physical and chemical changes better than the planktonic both in vivo and in vitro conditions. Biofilm matrix lays down a physical barrier enabling them to withstand physiological challenges faced in the human host such as low oxygen concentration, osmotic stress, acidic environment, nutrient limitation, iron deprivation, and presence of antibiotics (Jefferson, 2004). Biofilms help in the survival of pathogens in all abovementioned adverse conditions, and we will describe some of these survival mechanisms here. In the case of cystic fibrosis, *P. aeruginosa* forms a slimy alginate polysaccharide coat making the antibiotics tough to penetrate the EPS, which allows their survival and difficult to eradicate through antibiotic treatment (Høiby et al., 2010). *S. aureus* and *A. baumannii* produce biofilm-associated proteins (BAPs), which have the capacity to aggregate to build up biofilm matrix in response to stressful conditions prevailing in the host during infection (Ayoub Moubareck & Hammoudi Halat, 2020). It was also recently shown that *A. baumannii* can directly regulate many stress-related pathways via the TCS regulatory system, BfmRS through which it directly controls the expression of many stress resistance genes (Palethorpe et al., 2022). With respect to *S. aureus*, the stress-tolerant persister cells remain metabolically active though they display an altered transcriptomic profile, which is consistent with activation of stress responses required for its survival. The survival response during the stress includes cell wall stress response and heat shock responses (Peyrusson et al., 2020). Besides, toxin-antitoxin (TA) systems play a crucial role in the persistence and survival of ESKAPE pathogens except for *Enterobacter* spp. where it is not identified till date. AsrR (antibiotic and stress response regulator) of *E. faecium* plays a vital role in the development of drug resistance and its survival when it is under oxidative stress within the host (Lebreton et al., 2012). It is said that bacterial cells in the biofilm are hundred times more resistant to antibiotics than the planktonic counterparts of the pathogens belonging to ESKAPE group, and such biofilm phenotypes are mainly responsible for causing persistent chronic infections (Esposito & De Simone, 2017).

4.2 Role of biofilms in adhesion and colonization

For the microbes, the human body is an alluring place to reside in unless seized by the patrolling immune cells of the host. In order to evade these patrolling immune cells, and to survive and flourish in nutrient-rich conditions of host body, bacteria had to give up their free-living mode and adapt to a sessile lifestyle. To aid bacterial attachment inside the host body, it requires pili, fimbriae, and surface adhesion proteins. In *Staphylococcus* species, adhesion of cells is facilitated by the irreversible binding to extracellular matrix (ECM) proteins that are collectively referred to as microbial surface component adhesive matrix molecule (MSCRAMM) (Foster et al., 2014). *K. pneumoniae* uses *kpc* and KPF-28 fimbriae appendages for adhesion while it is attaching initially on to in vivo surfaces. Fimbriae KPF-28 may account for the initial attachment to the mammalian intestine (Di Martino, Livrelli, Sirot, Joly, & Darfeuille-Michaud, 1996), and kpc fimbriae lead to hypermucoviscous forms of *K. pneumoniae*, which are highly mucoid and resistant (Wu, Huang, Fung, & Peng, 2010) and *K. pneumoniae* isolates that are demonstrating hypermucoviscosity have now become a distinguishing feature of many hypervirulent isolates that are known to cause complicated bacteraemia with multiple septic sites (central nervous system, liver, and muscles) according to the reports emerging across the world (Choby, Howard-Anderson, & Weiss, 2020; Fauvet et al., 2020; Wu et al., 2010). Hemolysin coregulated protein (Hcp) and pili of *A. baumannii* are involved in type VI secretion system and adhesion to the host tissues, which is subsequently followed by invasion and establishing infection (Massier et al., 2021; Pan et al., 2022). Surface protein of *E. faecium*, namely *Enterococcus* surface protein (esp), is the main virulence factor that also helps in the adhesion of cells to host intestine (Top et al., 2013). In case of *P. aeruginosa*, two lectins that are biofilm constituents, namely LecA and LecB, both of which are carbohydrate-binding proteins are very crucial for initial adhesion in the biofilm development and colonization in the host (Siebs et al., 2022).

4.3 Antibiotic resistance, bacterial persistence, and chronic infections

The main reason for the creation of mushroom and tower-shaped biofilm ultra structure by the bacteria is the availability of nutrients, water, and oxygen. Cells on the outer periphery of mushroom head and tower are exposed to the host's microenvironment where the availability of oxygen and water is not a problem. But as one descends across vertically, cells are exposed to dire environmental challenges wherein the availability of oxygen and nutrient, which are two metabolically important factors for bacterial growth in the biofilm matrix, is reduced to a minimum level. Moreover, it is also shown that low oxygen leads to an increase in resistance of biofilms toward aminoglycoside antibiotics (Høiby et al., 2010). On the other hand, penetration of positively charged antibiotics, namely tobramycin, through *P. aeruginosa* biofilm matrix was reduced due to the interaction of tobramycin with negatively charged eDNA leading to elevation of resistance in biofilm cells (Walters, Roe, Bugnicourt, Franklin, & Stewart, 2003). Besides, *P. aeruginosa* produces Psl, an exopolysaccharide composed of pentasaccharide subunits of D-mannose, D-glucose, and L-rhamnose that appears to be responsible for resistance toward colistin, polymyxin B, ciprofloxacin, and tobramycin in the initial stages of biofilm formation (Byrd et al., 2009). Decreased antibiotic penetration through the biofilm matrix is considered as a general mechanism of antibiotic resistance for pathogenic bacteria cells in the biofilm, apart from the presence of resistant genetic determinants. To sustain life in the anaerobic conditions prevailing in the inner matrix of biofilm, biofilms have water channels paving across deeper layers of EPS. The anionic EPS provides a physical barrier shutting off the penetration of cationic antimicrobial agents (for example, antibiotics belonging to class aminoglycosides).

Antibiotics resistance in biofilms is mainly due to: (i) transfer of resistant genes by horizontal gene transfer (Mah, 2012) and (ii) poor or less penetration of biofilms by different antimicrobial agents (Donlan et al., 2001). Besides, factors such as nutrients and oxygen in a gradient microenvironment account for the cell's adaptation to phenotypic heterogeneity (Lewis, 2010). The anaerobic condition and diminished nutrition availability in biofilm substratum impel the cells to have a dormant growth phase (Patel, 2005). These stochastically formed dormant cells have slower growth, can persist in limited nutritional conditions, and can greatly resist killing by antimicrobial agents (Lewis, 2010). These modifications and structure of the biofilm matrix prevent antimicrobials from reaching the bacterial cells, thus enabling the cells in the biofilm to tolerate 100–1000 times more of their determined MICs of different antibiotics.

Random transposon insertion mutation analysis of *P. aeruginosa* strain UCBPP-PA14 found that six loci (*ndvB*, PA1875–1877, *tssC1*, A0756–0757 (encoding a putative TCS), PA2070, and PA5033 (encoding hypothetical proteins having unknown functions)) do not affect biofilm formation but are involved in biofilm-specific antibiotic resistance wherein these genes are highly expressed in biofilm phenotype as compared with their planktonic counterpart. Deletion of these loci resulted in reduction in virulence animal model indicating their role in both biofilm-specific antibiotic resistance and persistence in vivo (Zhang et al., 2013). On the contrary, one recent study that investigated 104 clinical *A. baumannii* isolates for their biofilm-forming capacities revealed that nonmultidrug-resistant (MDR) isolates were strong biofilm formers, and biofilm-grown *A. baumannii* has induced antibiotic tolerance in non-MDR isolates, which is reversible but a higher level of irreversible resistance was observed in extensively drug-resistant (XDR) isolates (Shenkutie, Yao, Siu, Wong,

& Leung, 2020). Biofilm cells of *K. pneumoniae* produce β-lactamases, which digest the β-lactam class of antibiotics such as ampicillin and prevent cell death. While the absence of β-lactamases in biofilm cells due to deficient mutant of β-lactamase leads to higher penetration of ampicillin and higher killing activity (Anderl, Franklin, & Stewart, 2000). Therefore, antibiotic hydrolyzing enzymes are also important for bringing biofilm-specific antibiotic resistance (Anderl et al., 2000). Biofilms of *Staphylococcus* skew the host immune response away from a pro-inflammatory bactericidal kind of response toward an antiinflammatory and pro-fibrotic response, which favors its persistence (Hanke & Kielian, 2012). Here *S. aureus* biofilm cells persist as they were able to evade phagocytosis by directing the host to accumulate the alternatively activated M2 macrophages that show antiinflammatory and pro-fibrotic properties (Hanke & Kielian, 2012).

A subpopulation (1% of the total population) of cells in biofilm acquiring metabolically dormant state and exhibiting higher resistance to antimicrobial agents without change in genetic content are called as persister cells, and these persisters cause recalcitrant chronic infections (Wood, Knabel, & Kwan, 2013). TA model is the most accepted genetic basis of persister cell formation. Gain of function through mutation in hipA7 TA-locus leads to significantly higher persister formation in *E. coli*. Other TA systems such as *MqsR/MqsA* and *TisB/IstR-1* are responsible for persister cell formation since deletion of this TA system reduced the frequency of persister cell formation in *E. coli* (Dorr, Vulic, & Lewis, 2010; Kim & Wood, 2010). *Hfq*, a small RNA chaperon, regulates the persister formation in *E. coli* by modulating the outer membrane protein *YbfM* and transport protein OppA (Kim et al., 2010). In one of the recent studies, *Hfq* has been also shown to have a crucial role in physiology of *A. baumannii* wherein its deletion was affecting the vital cellular processes such as metabolism, stress tolerance, and virulence of the bacterium (Sharma et al., 2018). In the case of *S. aureus*, biofilm tolerance to amikacin is mainly due to the higher frequency of persister cells in biofilm than in the planktonic phase (Singh, Ray, Das, & Sharma, 2009). Bacterial growth kinetics significantly alter the frequency of persister cell formation wherein there exists a long lag phase due to antibiotic stress and acidic media that results in the formation of a small colony variant (SCV) that comprises the persister cells (Vulin, Leimer, Huemer, Ackermann, & Zinkernagel, 2018). A recent study showed that *S. aureus* cells within the host cells surviving the exposure of antibiotics are nothing but persister cells that show a characteristic biphasic killing curve (Peyrusson et al., 2020). These cells are metabolically active with altered transcriptome and had activated several stress response cascades in order to survive in the host system (Peyrusson et al., 2020).

Transcriptome analysis of persister cells generated after treating *A. baumannii* cells with 50-fold MIC revealed significantly higher expression of HigB/HigA and DUF1044/RelB TA systems along with the genes that are involved in the degradation of aromatic compounds (Alkasir et al., 2018). MerR-like transcriptional regulator BrlR (biofilm resistance locus regulator) contributes to the biofilm-specific antibiotic (cationic antimicrobial peptides) resistance of *P. aeruginosa*. *BrlR* is a transcriptional activator of mexAB-oprM and mexEF-oprN RND efflux pump, which are actively involved in antibiotic resistance of *P. aeruginosa* (Liao, Schurr, & Sauer, 2013). Alternatively, deletion of BrlR results in elevation of colistin resistance, while overexpression leads to colistin susceptible phenotype, and this is because BrlR represses the expression of *phoPQ* TCS involved in polymyxin resistance (Chambers & Sauer, 2013).

Biofilm-related infections account for more than 60% of all total microbial infections (Lewis, 2001). The slow growth rate of cells in biofilm affects the antibiotic activity especially the drugs that target the cell wall synthesis machinery leading to an increase in antibiotic tolerance among the cells within biofilms, which is the case in biofilm-related chronic infections. For instance, biofilms of *S. aureus* cause chronic wounds, eye, endocarditis, osteomyelitis (Parastan, Kargar, Solhjoo, & Kafilzadeh, 2020), periodontitis, chronic rhinosinusitis, and indwelling medical device infections (Archer et al., 2011). Biofilm formation also helps bacteria to escape from the immune system such as prevention of phagocytosis by releasing toxins. Biofilms formed by *P. aeruginosa* is majorly involved in inflammation and consequently cause cystic fibrosis in the lungs (Høiby et al., 2010), chronic wound infection, chronic rhinosinusitis, chronic otitis media (Römling & Balsalobre, 2012). Bacterial growth and biofilm formation mostly depend upon the nature of fluid flowing in the catheters, and *K. pneumoniae*, *P. aeruginosa*, and some *Enterobacter* species grow and colonize well in intravenous fluids (Donlan et al., 2001; Jamal et al., 2018). *A. baumannii* is referred as a critical pathogen for having strong capability to form biofilms, which has led to drug resistance and chronic infections. Ventilator-associated pneumonia, bloodstream infections, catheter-related urinary, and shunt-related meningitis were mainly caused by biofilm-forming isolates of *A. baumannii* (Rodríguez-Baño et al., 2008). *Enterococcus* biofilm also accounts for wound, urinary, gastrointestinal tract (GIT), and endocarditis infections (Ch'ng, Chong, Lam, Wong, & Kline, 2019). One of the recent studies showed that *E. faecium* was the predominant species isolated from trauma-related infections that are usually polymicrobial in nature, and most frequent cocolonizing bacteria were other ESKAPE pathogens (Heitkamp et al., 2018).

5. Major genetic elements and molecular pathways involved in biofilm formation

Biofilm-forming bacteria are highly adherent due to the expression of adhesion proteins and other structural proteins. Adherent cells are normally embedded within a self-produced matrix of EPS. From the initial attachment phase to bacterial release, the three-dimensional biofilm framework is highly modulated by various signaling pathways accompanied by structural proteins. In *P. aeruginosa*, *pel*, *psl*, and *alg* operons are shown to be responsible for the biosynthesis of EPS that plays a significant role in cell surface interactions during biofilm development. The three major types of polysaccharides in *P. aeruginosa* are pel, psl, and alginate, and each of these polysaccharides differs in their chemical structure. Pel is a cationic exopolysaccharide. However, the chemical structure of pel till now has not been resolved, though it is said that it is a glucose-abundant polysaccharide (Friedman & Kolter, 2004a, 2004b). Psl is mainly composed of D-glucose, D-mannose, and L-rhamnose repeats. The penta-polysaccharide is expressed in two forms, namely a soluble form and cell-bound high-molecular-weight form. Psl distributes itself helically around the cell surface, and it is involved in maintaining the biofilm organization, and it also facilitates the interactions between the cells and the cell surface (Ma et al., 2009). The production of Psl enhances the cross-linking and elasticity of the EPS matrix. Alginate that usually enhances surface adhesion is a high-molecular-weight linear anionic polysaccharide, which comprises β-1-4 glycosidic-linked α-L-guluronic acid and β-D-mannuronic acid (Hay, Rehman, & Rehm, 2010).

Recent studies have shown various pathways of *P. aeruginosa* such as cAMP/Vfr signaling pathway, QS pathway, Gac/Rsm pathway, and c-di-GMP signaling pathway are the main mechanisms involved in biofilm formation. In QS signaling pathway, when a threshold value of extracellular signaling molecule (auto-inducer) accumulates in the environment, it induces the release of eDNA into the biofilm matrix (Ibáñez de Aldecoa, Zafra, & González-Pastor, 2017; Spoering & Gilmore, 2006). The released anionic eDNA adheres onto the surface, glues the cells of biofilm together in a matrix while facilitating the exchange of various resistance determinants. Its multifaceted function also involves increasing antibiotic resistance by cationic antibiotic chelation whenever these antibiotics try to penetrate the biofilms, which was proved in biofilms of *P. aeruginosa* (Okshevsky & Meyer, 2015).

C-di-GMP regulates the production of ECM and components of biofilm-like proteins and polysaccharides in *P. aeruginosa*. C-di-GMP is an intracellular signaling molecule that plays important role in biofilm formation and dispersion, motility, and virulence of *P. aeruginosa* (Valentini & Filloux, 2016). The modulation in the intracellular concentration of this signaling molecule is correlated with motility and biofilm formation where a high intracellular level of C-di-GMP is associated with biofilm formation while a low level leads to their dispersion leading to generation of motile cells that are dispersed from biofilm. A similar role of C-di-GMP is also found in other clinically important bacteria such as *A. baumannii*, *E. coli*, *Salmonella enterica*, and *serovar typhimurium* (Ahmad, Nygren, Khalid, Myint, & Uhlin, 2020; Richter et al., 2020).

The key QS regulon of the Gram-positive *S. aureus* is the *agr* operon whose downregulation is associated with immature biofilm formation (Slany, Oppelt, & Cincarova, 2017). The two-component regulatory pathway, namely YycFG, is involved in the biofilm structural organization of *S. aureus* (Wu et al., 2019). Molecular signaling pathways contributing to biofilm formation in *S. aureus* (Gries et al., 2016) and *P. aeruginosa* (Liu, Wu, et al., 2020; Valentini & Filloux, 2016) have been extensively studied and discussed, and one can access this information from many exhaustive reviews (Otto, 2008; Paharik & Horswill, 2016; Valentini et al., 2016). So, in this review, we will be discussing more about the less addressed members of "ESKAPE group" of pathogens.

In *K. pneumoniae*, the initial attachment on ECM is primarily mediated by Type III fimbriae. This adhesion is coordinately modulated by CRP-cAMP and c-di GMP signaling pathways (Lin et al., 2016). In *K. pneumoniae*, there is an integration of both Type I and Type II QS systems for development of biofilms. In case of Type I QS as observed in *P. aeruginosa*, AHL signals are produced by Lux I synthase, which binds the corresponding LuxR protein. After reaching a threshold value, the AHL-Lux R regulates the gene expression based on the cell density (Balestrino, Haagensen, Rich, & Forestier, 2005). However, *K. pneumoniae* slightly differs from this pathway by being an orphan receptor. It lacks the Lux I synthase, and consequently, they are called non-AHL producers. This intraspecies communication among them is mediated by an orphan LuxR-type receptor, namely SdiA. The type II QS uses AI-2 signal produced by *LuxS* enzyme whose regulation depends on operons, Lsr and LsrRK, respectively (Chen et al., 2020). Both of these operons are upregulated in the mature biofilm of *K. pneumoniae*. AI-2 is involved in the exopolysaccharide synthesis of biofilms. Besides, CRP-cAMP regulates the expression of type III fimbriae via cyclic di-GMP in *K. pneumoniae*.

A. baumannii functions in similar fashion as *K. Pneumoniae* with respect to QS pathway. *A. baumannii* employs a variant of Type I (LuxR/LuxI) QS, namely AbaR/AbaI (Tang et al., 2020), which produces the acyl homo-serine lactone (A-HSL) molecule, *N*-(3-hydroxydodecanoyl)-L-HSL (3-hydroxy-C(12)-HSL) (Niu,

Clemmer, Bonomo, & Rather, 2008). Any mutations in AbaI and AbaR results in a reduction of biofilm in later stages of biofilm development in *A. baumannii* (Gaddy & Actis, 2009; Guo & Xiang, 2017). The initial twitching motility of *A. baumannii* facilitates adherence to the biotic and abiotic surface by the BfmRS TCS (Liou et al., 2014; Luo et al., 2015). The initial adhesion of cells to the abiotic surface is mediated by Csu/BABCDE chaperon usher operon (Gaddy et al., 2009), which is regulated by BfmRS TCS. Any aberration in the *BfmR* response regulator and *BfmS* sensor kinase impedes their synchronization leading to reduced adherence and biofilm formation of *A. baumannii* in eukaryotic host (Liou et al., 2014). Apart from this, two more TCSs are also associated with biofilm formation, AdeRS (Richmond et al., 2016) and GacSA (Cerqueira et al., 2014). Additionally, BAPs that are big surface-exposed proteins secreted via type I secretion system in *A. baumannii* are also shown to be essential for biofilm formation (Harding et al., 2017). BAP is a highly polymorphic protein that has repetitive, truncated, and disrupted coding gene sequence among different strains belonging to *Acinetobacter calcoaceticus-baumannii* (ACB) complex (De Gregorio et al., 2015). BAP is one of the important proteins required for *A. baumannii* in order to form biofilm and its maturation on both host cells and abiotic materials such as polystyrene and titanium, which are usually used for making medical devices. It helps the adhesion of *A. baumannii* by increasing bacterial surface hydrophobicity (Brossard & Campagnari, 2012; Goh et al., 2013).

With respect to *Enterobacter* species, it was found that it produces *N*-acyl C12-HSL in the human tongue surface when it causes infection (Yin, Purmal, Chin, Chan, & Chan, 2012). Biofilm formation in *Enterobacter* species is regulated by the expression level of curli Biogenesis genes, namely *csgA* and *csgD* (Soares, Costa, Melo, Mola, & Leal-Balbino, 2016) and through the production of a signaling molecule, C12-HSL in vivo (Yin et al., 2012). The ability to form biofilm is associated with emp_{fm} (*E. faecium*'s pilus) operon in *E. faecium* (Almohamad, Somarajan, Singh, Nallapareddy, & Murray, 2014). Expression of two wall-anchored surface proteins such as SgrA and EcbA that can bind to other ECM proteins and are involved in biofilm formation (Hendrickx et al., 2009). Additionally, AraC type of regulator, namely EbrB, positively regulates the *esp* operon and is implicated in biofilm formation as well as intestinal colonization of *E. faecium* (Top et al., 2013). Table 1 summarizes major signaling pathways and their mechanisms involved in biofilm formation in all the pathogens belonging to ESKAPE group. Fig. 2 illustrates the genetic organization of vital biofilm-related genes in ESKAPE pathogens.

TABLE 1 Signaling pathways and their mechanisms involved in biofilm formation in ESKAPE pathogens.

Signaling pathway	Role	References
Enterococcus faecium		
EbrB	AraC type of regulator regulating the *esp* operon that is involved in biofilm formation and intestinal colonization	Top et al. (2013)
$AtlA_{Efm}$	Major autolysin, involved in biofilm formation	Paganelli et al. (2013)
Staphylococcus aureus		
WalK/WalR TCS	Positively controls major autolysin AtlA and LytM and biofilm formation	Dubrac, Boneca, Poupel, and Msadek (2007)
agr QS	Has a repressive role in biofilm attachment and its reactivation leads to dispersal of cells	Boles and Horswill (2008)
SarA	Global Regulator of virulence determinants, positively controls biofilm formation through Bap protein	Trotonda, Manna, Cheung, Lasa, and Penadés (2005)
Klebsiella pneumoniae		
SdiA Type I QS Orphan LuxR receptor	Has a repressive role in fimbriae expression, cell adherences and consequently reduced biofilm formation	Pacheco et al. (2021)
ΔLuxS Type II QS	Play vital role in biofilm architecture with less surface coverage and decreased macro colony formation	Chen et al. (2020)
MrkJ	Inhibits intracellular cyclic-di-GMP accumulation resulting in decreased fimbriae and biofilm formation	Johnson and Clegg (2010)
MrkHI (TF)	Affects Type III fimbrial expression and biofilm formation by binding c-di-GMP	Johnson, Murphy, Sippy, Johnson, and Clegg (2011)

TABLE 1 Signaling pathways and their mechanisms involved in biofilm formation in ESKAPE pathogens—cont'd

Signaling pathway	Role	References
IscR (TF)	Directly represses mrkHIJ gene expression essential for c-di AMP mediated Type III fimbriae and biofilm formation	Lin et al. (2017)
SOS response	Helps in resistant biofilm formation in strains having DNA double stranded breaks	Liu, Wu, et al. (2020)
etcABC Putative EII complex	Positively regulate cAMP production leads to mrk operon activation and increased type III fimbriae	Panjaitan, Horng, Cheng, Chung, and Soo (2019)
Acinetobacter baumannii		
Cyclic di-GMP	Positively regulates biofilm formation and negatively regulates surface-associated motility	Ahmad et al. (2020)
AdeFGH	Overexpression is associated with biofilm formation	He et al. (2015)
adeABC	1. Mutant overexpressing the efflux pump underexpression of genes (*CsuA/B, CsuC,* and *FimA*) associated with pili formation proteins 2. Deletion of adeB leads to reduction in biofilm formation on abiotic and biotic surfaces as compare to WT strain	Yoon et al. (2015) Richmond et al. (2016)
BfmR/S TCS	Master regulator of biofilm initiation	Draughn et al. (2018)
adeIJK	Deletion OR disruption of *tetR* regulator, adeN leads to elevation in biofilm formation	Saranathan et al. (2017)
pgaABCD locus [poly-beta-(1-6)-*N*-acetylglucosamine (PNAG)]	Deletion of *pgaABCD* locus leads to loss of the strong biofilm phenotype	Choi, Slamti, Avci, Pier, and Maira-Litrán (2009)
Pseudomonas aeruginosa		
Cyclic AMP-Vfr	1. Negatively regulated by cyclic di-GMP produced by rugose small colony variants (RSCVs) 2. Deeper layers of biofilm have increased di-GMP and reduced cAMP levels	Almblad et al. (2015)
GacRS	Activates genes responsible for biofilm formation	Jimenez et al. (2012)
c-di-GMP	High c-di-GMP level decreases flagellar activity and stimulates cell adhesion and biofilm formation	Hengge (2009)
Enterobacter spp.		
pgaABCD locus [poly-beta-(1-6)-*N*-acetylglucosamine (PNAG)]	Involved in biofilm matrix formation	Skurnik et al. (2012)
csgA and *csgD*	Expression of these curli biogenesis genes regulates biofilm formation	Soares et al. (2016)

6. Biofilms and virulence

ESKAPE pathogens are mostly responsible for various nosocomial infections including UTI, QS-mediated virulence factors such as motility behavior, fimbriae and pili expression, and pyocyanin production appear to be important virulence determinants that are associated with biofilm formation that depends on QS, which is an intercellular communication mechanism that initiates biofilm development. *P. aeruginosa* produces two lectins, namely LecA and LecB, as part of its large arsenal of virulence factors (Siebs et al., 2022). Both are carbohydrate-binding proteins that are very essential for initial and later persistent infection processes, i.e., bacterial adhesion and biofilm formation. The biofilm structure is a major virulence determinant and protects the bacteria against the host immune response and antibiotic treatment. The poly-*N*-acetyl glucosamine (PNAG), a component of biofilm, is a key virulence factor in device-related infections caused both

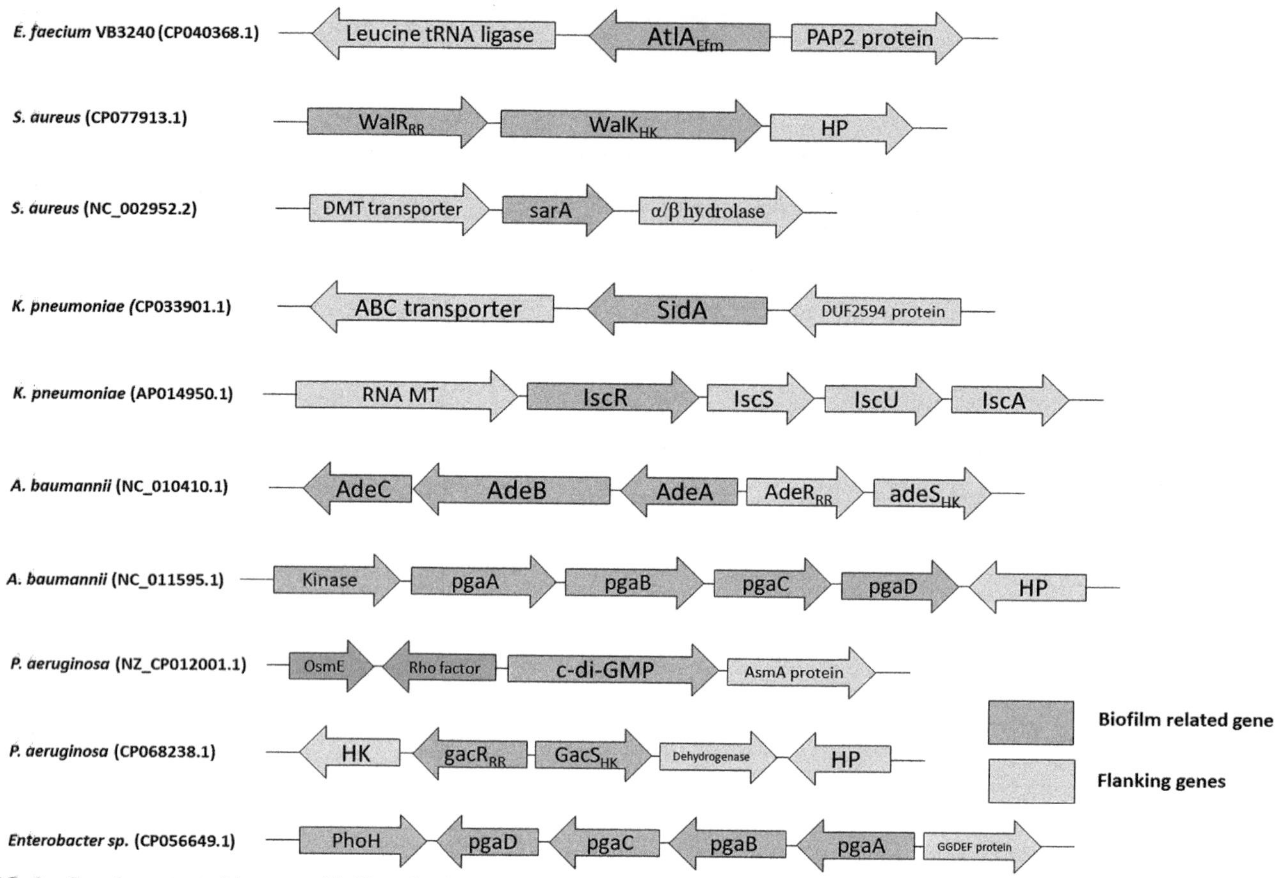

FIG. 2 Genetic context of important biofilm-related genes in ESKAPE pathogen, the data obtained from the NCBI genome database. *HK*, histidine kinase; *HP*, hypothetical protein; *RR*, response regulator.

A. baumannii and *S. aureus*. One recent study demonstrated through carbohydrate microarrays that there are similar specificities among bacteria such as *S. aureus* and *A. baumannii* in most of their intensely binding carbohydrates such as 3′ and 6′ sialyllactose. However, the differences are only in kind of binding ligands. An *N*-acetyl glucosamine-binding lectin function, which binds to PNAG identified on the *A. baumannii* cell surface, is likely to contribute to the biofilm structure and surface presentation of PNAG on cells of *A. baumannii*. Overall, the differences identified in surface presentation of PNAG and its accessibility for interactions on cell surfaces among these two bacteria might play vital role in biofilm-mediated pathogenesis (Flannery, Le Berre, Pier, O'Gara, & Kilcoyne, 2020). Other virulence traits such as ability to produce protease, lipase, hemolysin, and siderophores were additional factors in *A. baumannii*, which may contribute to its pathogenicity (Pan et al., 2022).

7. Antibiofilm strategies

An increase in antibiotic resistance in ESKAPE pathogens accelerated the need for diverse approaches in order to combat these ESKAPE pathogens. The slow growth of resistant cells in biofilm poses a great challenge in order to eradicate the biofilm since some of the antibiotics (penicillins and quniolones) require actively growing and dividing cells that can inhibit vital processes such as cell wall synthesis and DNA replication. Lower effectiveness of conventional or first-line antibiotics against the biofilm bacteria has forced the clinicians to use a combination of antibacterial agents for treating biofilm-related infections and to look for alternative treatment options. Antibiotics that are having different modes of action (colistin and tigecycline) can be used in combination as an effective strategy to cure biofilm-related infections caused by certain pathogens (e.g., *A. baumannii*). Similarly, *Staphylococcal* biofilm formed on orthopedic devices that cause implant infections can be cleared using a combination of clarithromycin and daptomycin (Fujimura et al., 2015).

7.1 EPS targeting

EPS is a critical and dynamic component of bacterial biofilm in which bacteria are embedded. It acts as a physical barrier that prevents penetration of antibiotics through it posing a real challenge in the eradication of biofilms that hold together the multicellular microbial communities and support microbial adhesion to a surface. Discovering the functions of EPS and genes involved in EPS production will be beneficial for devising strategies to prevent and control the biofilm formation. Secondary metabolites from diverse bacterial communities are employed to inhibit the QS, and various enzymes from diverse microbes were used to disrupt the architecture of biofilms in pathogenic bacteria (Khan, Oloketuyi, & Kim, 2019). Biofilm matrix-degrading enzymes such as alginate lyase produced by *B. circulans* are shown to degrade the EPS produced by *P. aeruginosa* and prevent biofilm formation (Alkawash, Soothill, & Schiller, 2006). Well-known flavonoid, quercetin, was shown to have antibiofilm activity against *Staphylococcus* by downregulating intracellular adhesion (*ica*) locus and inhibiting PIA production, which is very essential for cell adhesion (Mu, Zeng, & Chen, 2021). The use of proteases to degrade the proteinaceous matrix produced by *S. aureus* might be another way of disrupting and dispersing the biofilm (Lister & Horswill, 2014). Degradation of EPS will expose the bacterial cells in turn helping the antibiotics to access them better, and these bacteria can be effectively cleared even at lower doses. Phenyl lactic acid (PLA) inhibits the biofilm formation of *E. faecalis* by reducing EPS production through downregulation of *Epa* polysaccharide genes (*epaABE*) and *Ebp* pili genes (*ebpABC*) (Liu, Sun, Wang, et al., 2020). The use of glycoside degrading enzyme cellulase along with antibiotics such as ceftazidime has been shown to significantly inhibit the biofilm formation and aids in the effective eradication of *P. aeruginosa* in a concentration-dependent manner (Kamali, Jamali, Izanloo, & Ardebili, 2021). Alginate oligomer diffusion into *P. aeruginosa* biofilms leads to increased efficacy of antibiotics, disrupts the EPS matrix, and modifies the eDNA assembly in the biofilms (Powell et al., 2018). Such targeted disruption of the extracellular polymeric network is important for the treatment of biofilm-related infections. Alginate oligomer has been proven safe in inhalation therapy and is responsible for modulating the viscoelasticity of cystic fibrosis (CF) sputum in CF patients along with reducing bacterial burden in the biofilm. Flavonoids such as fisetin, phloretin, and curcumin showed antibiofilm activity against both laboratory reference strains and MDR clinical isolates of *A. baumannii* in a dose-dependent manner wherein curcumin showed the highest antibiofilm activity by inhibiting the pellicle formation and motility of bacterium (Raorane et al., 2019). In silico docking study of these flavonoids showed affinity toward response regulator *BfmR* that is involved in stress response and biofilm formation of *A. baumannii*.

7.2 Combination therapy

The combination of two antimicrobial agents is a widely used treatment method to treat biofilm-related infections, and it is found to be effective in treating ESKAPE pathogens. A combination of antibiotics is more effective against biofilm than the use of monotherapy, one of the studies using gentamicin/ciprofloxacin combination showed a 75% reduction in biofilm formation in *P. aeruginosa* (Wang, Di Luca, Tkhilaishvili, Trampuz, & Gonzalez Moreno, 2019). Ceftazidime in combination with colistin shows antibiofilm activity against *P. aeruginosa* biofilm suggesting the use of ceftazidime with colistin against biofilm-associated infections caused by *P. aeruginosa* (Gómez-Junyent et al., 2021). By screening of FDA-approved library of drugs for combinations with colistin, one investigation found that auranofin (analgesic) and clomiphene citrate (fertility drug) showed a significant reduction in preformed biofilm of both laboratory and clinical isolates of *P. aeruginosa* (Torres et al., 2018). Similarly, a combination of colistin and tigecycline is used as an effective strategy to clear *A. baumannii* biofilms. However, in case of *S. aureus*, *N*-acetylcysteine, the acetylated form of the amino acid L-cysteine in combination with ampicillin and vancomycin, was shown to be highly successful in lowering the levels of MIC and minimum biofilm eradication concentration (MBEC), which could be a newer alternative strategy of combined therapy against *S. aureus* (Kuyucuklu, Kaynak Onurdağ, & Eryıldız, 2021).

7.3 QS inhibition

QS is a highly conserved communication system encoded by bacteria in order to sense and respond to environmental signals. The general QS circuit in Gram-negative bacteria consists of auto-inducer synthases, signal molecules (auto-inducers), transcription regulators, and target genes. Cyclic-di-GMP is another important signaling system involved in regulating EPS production, enzyme synthesis, and bacterial adhesion in order to form biofilm. QS is the main system that assists bacteria in controlling the expression of specific genes involved in biofilm, swarming motility, EPS production, virulence, and cell aggregation based on bacterial density (Petrova & Sauer, 2016). Potential sites in QS circuit may be explored in order to inhibit the QS, which will affect the virulence and biofilm-forming potential of pathogens. Type VI secretion system (T6SS) is one such system involved in biofilm formation and expression of virulence factors in *P. aeruginosa*, ajoene, a sulfur-rich molecule from garlic found to affect the expression of T6SS along with genes involved in formation biofilm matrix (Jakobsen et al., 2017). Since QS plays a central role in modulating virulence characteristics of *A. baumannii* such as surface-associated motility and biofilm formation, it is also an attractive target for devising antimicrobial strategies associated with QS. The gene (*abaM*) located between *abaR* and *abaI* encoding *RsaM* family protein regulates the AHL-dependent QS, and disruption of this gene leads to an increase in AHL production, biofilm formation, and surface motility of *A. baumannii*; simultaneously it reduces the virulence potential (López-Martín, Dubern, Alexander, & Williams, 2021). In the case of *A. baumannii*, quorum quenching enzymes *Aii20J* and *abaI* (AHL synthesis) play a vital role in QS-mediated biofilm formation and surface-associated motility (Mayer et al., 2020). Using similar quorum quenching enzymes, enzymes degrading AHL molecules such as AHL lactonase (Chow, Yang, Tay, Chua, & Yew, 2014), or use of any analogs that can compete with the original QS molecule for binding to receptors, will help in preventing bacterial colonization.

7.4 Targeting the adhesion process

The initial step of bacterial colonization on host cells involves polysaccharides, proteins, and eDNA of the EPS. Antimicrobial peptides (AMPs) naturally occur in humans, animals, and plants that protect from infections. AMPs that are derived from humans are modified by synthesis, and using them therapeutically has shown a significant reduction of biofilms produced by human pathogens (de Breij et al., 2018). These peptides are able to disrupt the biofilm grown on medical devices formed by ESKAPE pathogens (Rajput & Kumar, 2018). *Hylarana guentheri* derived antimicrobial peptides modified by replacing histidine at both ends with lysine inhibit the initial adhesion and biofilm formation of *S. aureus* (Xie et al., 2019). One of the environmental strains of *A. baumannii* DSM30011 was found to have locus encoding contact-dependent growth inhibition (CDI) systems, and deletion of this system promotes adhesion of *A. baumannii* cells biotic and biotic surfaces (Roussin et al., 2019). The FhaB/FhaC 2-partner secretion system encoding large exoprotein and transport protein in *A. baumannii* is involved in adhesion to host epithelial cells (Pérez et al., 2017). Targeting such systems will result in preventing the initial and also very vital adhesion step of biofilm formation in the biofilm cycle, thereby preventing subsequent colonization in the host, and this could be an important strategy to combat biofilm-associated infections.

8. Diseases associated with biofilm

Bacterial colonization on indwelling medical devices such as central venous catheters, peritoneal dialysis catheters, heart valves, contact lenses, pacemakers, and urinary catheters implanted inside the host body is the major site of biofilm-related chronic infections. Colonization of bacterial cells on a variety of body surfaces, such as the skin or mucosal surfaces of the respiratory, urinary tract, and digestive tracts are also sites of biofilm formation. Most bacterial infections associated with implanted devices are due to nosocomial pathogens such as *S. aureus*, *S. epidermidis*, *E. faecium*, *Enterobacter* spp., *E. coli*, *K. pneumoniae*, *P. aeruginosa*, and *A. baumannii*. Biofilm-associated diseases are often difficult to treat due to prolonged hospital stays and immunocompromised conditions of patients. The use of conventional antibiotics for clearing the planktonic bacterial cells might not be effective in eradicating bacterial cells in the biofilm. Commonly known mechanisms of antibiotic resistance do not necessarily make biofilm cells resistant to antibiotics since bacterial cells after biofilm dispersal are susceptible to antibiotics. This suggests that the physical barrier created by biofilm does not allow antibiotics to penetrate into the biofilm allowing bacterial cells to survive in presence of antibiotics. In vitro analysis of biofilm may not give a clear picture of in vivo biofilm formation in the host since inside the host, the microenvironment is completely different, which probably has different implications on biofilm-forming abilities of bacteria. Table 2 summarizes some of the important potent antibiofilm agents discovered along with their respective targets in ESKAPE pathogens.

TABLE 2 List of potent antibiofilm agents with their targets in ESKAPE pathogens.

Name	Source	Organism(s)	Effect on biofilm	Reference
SAAP-148	Synthetic peptide	*A. baumannii*, *S. aureus*	Prevent the formation of predominantly polysaccharide, as well as proteinaceous biofilms	de Breij et al. (2018)
Octominin	AMP Derived from Octopus	*A. baumannii*	Alterations and damage to the bacterial surface, induction of membrane permeability	Jayathilaka, Rajapaksha, Nikapitiya, De Zoysa, and Whang (2021)
Curcumin	*Curcuma longa*	*A. baumannii*	Affect the motility and function of BfmRS TCS which regulate biofilm formation	Raorane et al. (2019)
Myrtenol	*Myrtus communis*	*A. baumannii*	Disrupting the mature biofilm without affecting bacterial fitness reduce expression of biofilm associated genes such as pgaA, pgaC, bfmR, bap, csuA/B, ompA, and katE	Selvaraj et al. (2020)

TABLE 2 List of potent antibiofilm agents with their targets in ESKAPE pathogens—cont'd

Name	Source	Organism(s)	Effect on biofilm	Reference
Honokiol and magnolol	Phenolic neolignans	*A. baumannii*	Inhibit pellicle formation and motility	Khadke, Lee, Woo, and Lee (2019)
Magainin 2	Clawed fogs *Xenopus laevis*	*A. baumannii*	Effective in clearing preformed biofilm	Kim et al. (2018)
Zerumbone	*Zingiber zerumbet* (L.) Smith	*A. baumannii*	Affect the biofilm formation and motility through down regulation of biofilm- and virulence-associated genes (bap, adeA, adeB, and adeC)	Kim, Shin, Jang, and Eom (2020)
Cec4	Synthetic peptide	*A. baumannii*	Eradication of biofilm formed by CRAB, affect the expression of membrane proteins, TCS, and pilus-related genes	Liu, Wu, et al. (2020)
Japonicin-2LF	Skin secretion of Fujian Large-headed Frog	MRSA	Eradication of biofilm matrix and killing of MRSA	Yuan et al. (2019)
Hordenine	Phenolic phytochemical from sprouting barley	*P. aeruginosa*	Suppressed production of alginate and loosed architecture of biofilm also suppress the expression lasR, lasI, rhlI, and rhlR genes	Zhou et al. (2018)
Biosurfactants	*L. plantarum*	*S. aureus*	Reduce the growth of biofilm by regulating the expression of biofilm-related genes, icaA, cidA, and dltB	Yan et al. (2019)
5-Hydroxymethyl furfural	*Musa acuminata*	*P. aeruginosa*	Inhibited the production of biofilm proteins, biofilm adherence, EPS, and downregulates the expression of QS-regulated virulence genes	Vijayakumar and Ramanathan (2020)
Epigallocatechin gallate (EGCG)	Synthetic	*E. coli*	Affect the curli production by suppressing the expression of curli-related proteins *csgA*, *csgB*, and *csgD* vital in biofilm formation	Arita-Morioka et al. (2018)
DJK-6	Synthetic variant of host defense peptide	*K. pneumoniae*	Increase in ability of meropenem to eradicate the biofilm by at least 16-fold	Ribeiro et al. (2015)
Paeoniflorin	Synthetic	*K. pneumoniae*	Strong inhibitory effect on the biofilm formation	Qian et al. (2020)
P5 and P16.2	Synthetic peptide	*P. aeruginosa* and *S. aureus*	P5 displayed antibiofilm activity on both bacterium while P6.2 only on inhibit biofilm of *S. aureus*	Martínez, Polizzotto, Flores, Semorile, and Maffía (2020)
Allicin	Garlic	*P. aeruginosa*	Inhibit virulence factor production and biofilm formation. Allicin attenuates the production of virulence factors, such as pyocyanin, pyoverdine, elastase, and rhamnolipids through inhibition of QS	Xu et al. (2019)
WLBU2 and LL37	Engineered cationic Peptides	ESKAPE pathogens	Affect the expression of biofilm associated genes	Lin, Deslouches, Montelaro, and Di (2018)
Zeaxanthin (Carotenoid)		*P. aeruginosa*	Reduce virulence potential and biofilm by inhibiting QS	Gökalsın, Aksoydan, Erman, and Sesal (2017)

9. Summary and conclusion

Understanding the exact nature and function of biofilms formed by ESKAPE pathogens is a key step in devising various strategies to eradicate these problematic pathogens. Biofilm phenotype is highly resistant for multiple antibiotics, and their developmental stages such as adhesion, formation of microcolonies, and formation of mature biofilm microstructure are vital in colonization of biotic surfaces within the host that enables these phenotypes to cause persistent and chronic infections. Biofilm phenotype lifestyle is regulated by highly complex communication networks involving various signaling (QS and TCS) and other regulatory systems (TA). These mechanisms usually confer ESKAPE pathogen an increased phenotypic or behavioral resistance to different stress factors such as host defense mechanisms and antibiotics and make them persistent and recalcitrant. All the factors influencing a planktonic pathogen to transform it into sessile one were discussed in this review. Understanding the molecular nature of response system leading to phenotype transformation of ESKAPE pathogen will allow us to identify potential targets that can help in killing them.

References

Ahmad, I., Nygren, E., Khalid, F., Myint, S. L., & Uhlin, B. E. (2020). A cyclic-di-GMP signalling network regulates biofilm formation and surface associated motility of *Acinetobacter baumannii* 17978. *Scientific Reports*, *10*(1), 1991. https://doi.org/10.1038/s41598-020-58522-5.

Alkasir, R., Ma, Y., Liu, F., Li, J., Lv, N., Xue, Y., et al. (2018). Characterization and transcriptome analysis of *Acinetobacter baumannii* persister cells. *Microbial Drug Resistance*, *24*(10), 1466–1474. https://doi.org/10.1089/mdr.2017.0341.

Alkawash, M. A., Soothill, J. S., & Schiller, N. L. (2006). Alginate lyase enhances antibiotic killing of mucoid *Pseudomonas aeruginosa* in biofilms. *APMIS (Acta Pathologica, Microbiologica et Immunologica Scandinavica)*, *114*(2), 131–138. https://doi.org/10.1111/j.1600-0463.2006.apm_356.x.

Almblad, H., Harrison, J. J., Rybtke, M., Groizeleau, J., Givskov, M., Parsek, M. R., et al. (2015). The cyclic AMP-Vfr signaling pathway in *Pseudomonas aeruginosa* is inhibited by cyclic Di-GMP. *Journal of Bacteriology*, *197*(13), 2190–2200. https://doi.org/10.1128/JB.00193-15.

Almohamad, S., Somarajan, S. R., Singh, K. V., Nallapareddy, S. R., & Murray, B. E. (2014). Influence of isolate origin and presence of various genes on biofilm formation by *Enterococcus faecium*. *FEMS Microbiology Letters*, *353*(2), 151–156. https://doi.org/10.1111/1574-6968.12418.

Anderl, J. N., Franklin, M. J., & Stewart, P. S. (2000). Role of antibiotic penetration limitation in *Klebsiella pneumoniae* biofilm resistance to ampicillin and ciprofloxacin. *Antimicrobial Agents and Chemotherapy*, *44*(7), 1818–1824. https://doi.org/10.1128/AAC.44.7.1818-1824.2000.

Archer, N. K., Mazaitis, M. J., Costerton, J. W., Leid, J. G., Powers, M. E., & Shirtliff, M. E. (2011). *Staphylococcus aureus* biofilms: Properties, regulation, and roles in human disease. *Virulence*, *2*(5), 445–459. https://doi.org/10.4161/viru.2.5.17724.

Arciola, C. R., Campoccia, D., Ravaioli, S., & Montanaro, L. (2015). Polysaccharide intercellular adhesin in biofilm: Structural and regulatory aspects. *Frontiers in Cellular and Infection Microbiology*, *5*, 7. https://doi.org/10.3389/fcimb.2015.00007.

Arita-Morioka, K. I., Yamanaka, K., Mizunoe, Y., Tanaka, Y., Ogura, T., & Sugimoto, S. (2018). Inhibitory effects of Myricetin derivatives on curli-dependent biofilm formation in *Escherichia coli*. *Scientific Reports*, *8*(1), 8452. https://doi.org/10.1038/s41598-018-26748-z.

Ayoub Moubareck, C., & Hammoudi Halat, D. (2020). Insights into *Acinetobacter baumannii*: A review of microbiological, virulence, and resistance traits in a threatening nosocomial pathogen. *Antibiotics*, *9*(3), 119. https://doi.org/10.3390/antibiotics9030119.

Azevedo, P., Furlan, J., Oliveira-Silva, M., Nakamura-Silva, R., Gomes, C. N., Costa, K., et al. (2018). Detection of virulence and β-lactamase encoding genes in *Enterobacter aerogenes* and *Enterobacter cloacae* clinical isolates from Brazil. *Brazilian Journal of Microbiology*, *49* (Suppl. 1), 224–228. https://doi.org/10.1016/j.bjm.2018.04.009.

Balestrino, D., Haagensen, J. A., Rich, C., & Forestier, C. (2005). Characterization of type 2 quorum sensing in *Klebsiella pneumoniae* and relationship with biofilm formation. *Journal of Bacteriology*, *187*(8), 2870–2880. https://doi.org/10.1128/JB.187.8.2870-2880.2005.

Boles, B. R., & Horswill, A. R. (2008). Agr-mediated dispersal of *Staphylococcus aureus* biofilms. *PLoS Pathogens*. https://doi.org/10.1371/journal.ppat.1000052.

Brossard, K. A., & Campagnari, A. A. (2012). The *Acinetobacter baumannii* biofilm-associated protein plays a role in adherence to human epithelial cells. *Infection and Immunity*, *80*(1), 228–233. https://doi.org/10.1128/IAI.05913-11.

Burgui, S, Gil, C, Solano, C, Lasa, I, & Valle, J. (2018). A systematic evaluation of the two-component systems network reveals that ArlRS is a key regulator of catheter colonization by *Staphylococcus aureus*. *Frontiers in Microbiology*, 9(342). https://doi.org/10.3389/fmicb.2018.00342.

Byrd, M. S., Sadovskaya, I., Vinogradov, E., Lu, H., Sprinkle, A. B., Richardson, S. H., et al. (2009). Genetic and biochemical analyses of the *Pseudomonas aeruginosa* Psl exopolysaccharide reveal overlapping roles for polysaccharide synthesis enzymes in Psl and LPS production. *Molecular Microbiology*, *73*(4), 622–638. https://doi.org/10.1111/j.1365-2958.2009.06795.x.

Cerqueira, G. M., Kostoulias, X., Khoo, C., Aibinu, I., Qu, Y., Traven, A., et al. (2014). A global virulence regulator in *Acinetobacter baumannii* and its control of the phenylacetic acid catabolic pathway. *The Journal of Infectious Diseases*, *210*(1), 46–55. https://doi.org/10.1093/infdis/jiu024.

Chambers, J. R., & Sauer, K. (2013). The MerR-like regulator BrlR impairs *Pseudomonas aeruginosa* biofilm tolerance to colistin by repressing PhoPQ. *Journal of Bacteriology*, *195*(20), 4678–4688. https://doi.org/10.1128/JB.00834-13.

Chen, L., Wilksch, J. J., Liu, H., Zhang, X., Torres, V., Bi, W., et al. (2020). Investigation of LuxS-mediated quorum sensing in *Klebsiella pneumoniae*. *Journal of Medical Microbiology*, *69*(3), 402–413. https://doi.org/10.1099/jmm.0.001148.

Chew, S. C., Kundukad, B., Seviour, T., van der Maarel, J. R., Yang, L., Rice, S. A., et al. (2014). Dynamic remodeling of microbial biofilms by functionally distinct exopolysaccharides. *mBio*, *5*(4), e01536-14.- https://doi.org/10.1128/mBio.01536-14.

Ch'ng, J. H., Chong, K., Lam, L. N., Wong, J. J., & Kline, K. A. (2019). Biofilm-associated infection by enterococci. *Nature Reviews Microbiology*, *17*(2), 82–94. https://doi.org/10.1038/s41579-018-0107-z.

Choby, J. E., Howard-Anderson, J., & Weiss, D. S. (2020). Hypervirulent *Klebsiella pneumoniae*—Clinical and molecular perspectives. *Journal of Internal Medicine*, *287*(3), 283–300. https://doi.org/10.1111/joim.13007.

Choi, A. H., Slamti, L., Avci, F. Y., Pier, G. B., & Maira-Litrán, T. (2009). The pgaABCD locus of *Acinetobacter baumannii* encodes the production of poly-beta-1-6-N-acetylglucosamine, which is critical for biofilm formation. *Journal of Bacteriology*, *191*(19), 5953–5963. https://doi.org/10.1128/JB.00647-09.

Chow, J. Y., Yang, Y., Tay, S. B., Chua, K. L., & Yew, W. S. (2014). Disruption of biofilm formation by the human pathogen *Acinetobacter baumannii* using engineered quorum-quenching lactonases. *Antimicrobial Agents and Chemotherapy*, *58*(3), 1802–1805. https://doi.org/10.1128/AAC.02410-13.

Cramton, S. E., Gerke, C., Schnell, N. F., Nichols, W. W., & Götz, F. (1999). The intercellular adhesion (ica) locus is present in *Staphylococcus aureus* and is required for biofilm formation. *Infection and Immunity*, *67*(10), 5427–5433. https://doi.org/10.1128/IAI.67.10.5427-5433.1999.

Crosby, H. A., Schlievert, P. M., Merriman, J. A., King, J. M., Salgado-Pabón, W., & Horswill, A. R. (2016). The *Staphylococcus aureus* global regulator MgrA modulates clumping and virulence by controlling surface protein expression. *PLoS Pathogens*, *12*(5), e1005604. https://doi.org/10.1371/journal.ppat.1005604.

de Breij, A., Riool, M., Cordfunke, R. A., Malanovic, N., de Boer, L., Koning, R. I., et al. (2018). The antimicrobial peptide SAAP-148 combats drug-resistant bacteria and biofilms. *Science Translational Medicine*, *10*(423), eaan4044. https://doi.org/10.1126/scitranslmed.aan4044.

De Gregorio, E., Del Franco, M., Martinucci, M., Roscetto, E., Zarrilli, R., & Di Nocera, P. P. (2015). Biofilm-associated proteins: News from *Acinetobacter*. *BMC Genomics*, *16*, 933. https://doi.org/10.1186/s12864-015-2136-6.

Dengler, V., Foulston, L., DeFrancesco, A. S., & Losick, R. (2015). An electrostatic net model for the role of extracellular DNA in biofilm formation by *Staphylococcus aureus*. *Journal of Bacteriology*, *197*(24), 3779–3787. https://doi.org/10.1128/JB.00726-15.

Di Martino, P., Livrelli, V., Sirot, D., Joly, B., & Darfeuille-Michaud, A. (1996). A new fimbrial antigen harbored by CAZ-5/SHV-4-producing *Klebsiella pneumoniae* strains involved in nosocomial infections. *Infection and Immunity*, *64*(6), 2266–2273. https://doi.org/10.1128/iai.64.6.2266-2273.1996.

Donlan, R. M., Murga, R., Bell, M., Toscano, C. M., Carr, J. H., Novicki, T. J., et al. (2001). Protocol for detection of biofilms on needleless connectors attached to central venous catheters. *Journal of Clinical Microbiology*, *39*(2), 750–753. https://doi.org/10.1128/JCM.39.2.750-753.2001.

Dorr, T, Vulic, M, & Lewis, K. (2010). Ciprofloxacin causes persister formation by inducing the TisB toxin in *Escherichia coli*. *PLoS Biology*, *8* (2). https://doi.org/10.1371/journal.pbio.1000317.

Draughn, G. L., Milton, M. E., Feldmann, E. A., Bobay, B. G., Roth, B. M., Olson, A. L., et al. (2018). The structure of the biofilm-controlling response regulator BfmR from *Acinetobacter baumannii* reveals details of its DNA-binding mechanism. *Journal of Molecular Biology*, *430*(6), 806–821. https://doi.org/10.1016/j.jmb.2018.02.002.

Dubrac, S., Boneca, I. G., Poupel, O., & Msadek, T. (2007). New insights into the WalK/WalR (YycG/YycF) essential signal transduction pathway reveal a major role in controlling cell wall metabolism and biofilm formation in *Staphylococcus aureus*. *Journal of Bacteriology*, *189*(22), 8257–8269. https://doi.org/10.1128/JB.00645-07.

Dunny, G. M., Hancock, L. E., & Shankar, N. (2014). Enterococcal biofilm structure and role in colonization and disease. In M. S. Gilmore, et al. (Eds.), *Enterococci: From commensals to leading causes of drug resistant infection* Massachusetts Eye and Ear Infirmary.

Esposito, S., & De Simone, G. (2017). Update on the main MDR pathogens: Prevalence and treatment options. *Le Infezioni in Medicina*, *25*(4), 301–310.

Fauvet, T., Tarantola, A., Colot, J., Merlet, A., Goarant, C., Marot, B., et al. (2020). Mucoviscous characteristics of *Klebsiella pneumoniae* strains: A factor of clinical severity? *Medecine et maladies infectieuses*, *50*(6), 500–506. https://doi.org/10.1016/j.medmal.2019.06.006.

Flannery, A., Le Berre, M., Pier, G. B., O'Gara, J. P., & Kilcoyne, M. (2020). Glycomics microarrays reveal differential in situ presentation of the biofilm polysaccharide poly-*N*-acetylglucosamine on *Acinetobacter baumannii* and *Staphylococcus aureus* cell surfaces. *International Journal of Molecular Sciences*, *21*(7), 2465. https://doi.org/10.3390/ijms21072465.

Flemming, H. C., & Wingender, J. (2010). The biofilm matrix. *Nature Reviews Microbiology*, *8*(9), 623–633. https://doi.org/10.1038/nrmicro2415.

Flemming, H. C., Neu, T. R., & Wozniak, D. J. (2007). The EPS matrix: The "house of biofilm cells". *Journal of Bacteriology*, *189*(22), 7945–7947. https://doi.org/10.1128/JB.00858-07.

Flemming, H. C., Wingender, J., Szewzyk, U., Steinberg, P., Rice, S. A., & Kjelleberg, S. (2016). Biofilms: An emergent form of bacterial life. *Nature Reviews Microbiology*, *14*(9), 563–575. https://doi.org/10.1038/nrmicro.2016.94.

Foster, T. J., Geoghegan, J. A., Ganesh, V. K., & Höök, M. (2014). Adhesion, invasion and evasion: The many functions of the surface proteins of *Staphylococcus aureus*. *Nature Reviews Microbiology*, *12*(1), 49–62. https://doi.org/10.1038/nrmicro3161.

Friedman, L., & Kolter, R. (2004a). Genes involved in matrix formation in *Pseudomonas aeruginosa* PA14 biofilms. *Molecular Microbiology*, *51* (3), 675–690. https://doi.org/10.1046/j.1365-2958.2003.03877.x.

Friedman, L., & Kolter, R. (2004b). Two genetic loci produce distinct carbohydrate-rich structural components of the *Pseudomonas aeruginosa* biofilm matrix. *Journal of Bacteriology*, *186*(14), 4457–4465. https://doi.org/10.1128/JB.186.14.4457-4465.2004.

Fujimura, S., Sato, T., Hayakawa, S., Kawamura, M., Furukawa, E., & Watanabe, A. (2015). Antimicrobial efficacy of combined clarithromycin plus daptomycin against biofilms-formed methicillin-resistant *Staphylococcus aureus* on titanium medical devices. *Journal of Infection and Chemotherapy: Official Journal of the Japan Society of Chemotherapy*, *21*(10), 756–759. https://doi.org/10.1016/j.jiac.2015.06.001.

Gaddy, J. A., & Actis, L. A. (2009). Regulation of *Acinetobacter baumannii* biofilm formation. *Future Microbiology*, *4*(3), 273–278. https://doi.org/10.2217/fmb.09.5.

Geoghegan, J. A., Corrigan, R. M., Gruszka, D. T., Speziale, P., O'Gara, J. P., Potts, J. R., et al. (2010). Role of surface protein SasG in biofilm formation by *Staphylococcus aureus*. *Journal of Bacteriology, 192*(21), 5663–5673. https://doi.org/10.1128/JB.00628-10.

Ghafoor, A., Hay, I. D., & Rehm, B. H. (2011). Role of exopolysaccharides in *Pseudomonas aeruginosa* biofilm formation and architecture. *Applied and Environmental Microbiology, 77*(15), 5238–5246. https://doi.org/10.1128/AEM.00637-11.

Goh, H. M., Beatson, S. A., Totsika, M., Moriel, D. G., Phan, M. D., Szubert, J., et al. (2013). Molecular analysis of the *Acinetobacter baumannii* biofilm-associated protein. *Applied and Environmental Microbiology, 79*(21), 6535–6543. https://doi.org/10.1128/AEM.01402-13.

Gökalsın, B., Aksoydan, B., Erman, B., & Sesal, N. C. (2017). Reducing virulence and biofilm of *Pseudomonas aeruginosa* by potential quorum sensing inhibitor carotenoid: Zeaxanthin. *Microbial Ecology, 74*(2), 466–473. https://doi.org/10.1007/s00248-017-0949-3.

Gómez-Junyent, J., Murillo, O., Yu, H. H., Azad, M., Wickremasinghe, H., Rigo-Bonnin, R., et al. (2021). In vitro pharmacokinetics/pharmacodynamics of continuous ceftazidime infusion alone and in combination with colistin against *Pseudomonas aeruginosa* biofilm. *International Journal of Antimicrobial Agents, 57*(2), 106246. https://doi.org/10.1016/j.ijantimicag.2020.106246.

Gries, C. M., Bruger, E. L., Moormeier, D. E., Scherr, T. D., Waters, C. M., & Kielian, T. (2016). Cyclic di-AMP released from *Staphylococcus aureus* biofilm induces a macrophage type I interferon response. *Infection and Immunity, 84*(12), 3564–3574. https://doi.org/10.1128/IAI.00447-16.

Guo, H. N., & Xiang, J. (2017). Influences of abaR gene on biofilm formation of *Acinetobacter baumannii*. *Zhonghua shao shang za zhi = Zhonghua shaoshang zazhi = Chinese Journal of Burns, 33*(4), 200–205. https://doi.org/10.3760/cma.j.issn.1009-2587.2017.04.003.

Hanke, M. L., & Kielian, T. (2012). Deciphering mechanisms of staphylococcal biofilm evasion of host immunity. *Frontiers in Cellular and Infection Microbiology, 2*, 62. https://doi.org/10.3389/fcimb.2012.00062.

Harding, C. M., Pulido, M. R., Di Venanzio, G., Kinsella, R. L., Webb, A. I., Scott, N. E., et al. (2017). Pathogenic *Acinetobacter* species have a functional type I secretion system and contact-dependent inhibition systems. *The Journal of Biological Chemistry, 292*(22), 9075–9087. https://doi.org/10.1074/jbc.M117.781575.

Hay, I. D., Rehman, Z. U., & Rehm, B. H. (2010). Membrane topology of outer membrane protein AlgE, which is required for alginate production in *Pseudomonas aeruginosa*. *Applied and Environmental Microbiology, 76*(6), 1806–1812. https://doi.org/10.1128/AEM.02945-09.

He, X., Lu, F., Yuan, F., Jiang, D., Zhao, P., Zhu, J., et al. (2015). Biofilm formation caused by clinical *Acinetobacter baumannii* isolates is associated with overexpression of the AdeFGH efflux pump. *Antimicrobial Agents and Chemotherapy, 59*(8), 4817–4825. https://doi.org/10.1128/AAC.00877-15.

Heikens, E., Bonten, M. J., & Willems, R. J. (2007). Enterococcal surface protein Esp is important for biofilm formation of *Enterococcus faecium* E1162. *Journal of Bacteriology, 189*(22), 8233–8240. https://doi.org/10.1128/JB.01205-07.

Heitkamp, R. A., Li, P., Mende, K., Demons, S. T., Tribble, D. R., & Tyner, S. D. (2018). Association of *Enterococcus* spp. with severe combat extremity injury, intensive care, and polymicrobial wound infection. *Surgical Infections, 19*(1), 95–103. https://doi.org/10.1089/sur.2017.157.

Hendrickx, A., Bonten, M., van Luit-Asbroek, M., Schapendonk, C., Kragten, A., & Willems, R. (2008). Expression of two distinct types of pili by a hospital-acquired *Enterococcus faecium* isolate. *Microbiology (Reading, England), 154*(Pt 10), 3212–3223. https://doi.org/10.1099/mic.0.2008/020891-0.

Hendrickx, A. P., van Luit-Asbroek, M., Schapendonk, C. M., van Wamel, W. J., Braat, J. C., Wijnands, L. M., et al. (2009). SgrA, a nidogen-binding LPXTG surface adhesin implicated in biofilm formation, and EcbA, a collagen binding MSCRAMM, are two novel adhesins of hospital-acquired *Enterococcus faecium*. *Infection and Immunity, 77*(11), 5097–5106. https://doi.org/10.1128/IAI.00275-09.

Hengge, R. (2009). Principles of c-di-GMP signalling in bacteria. *Nature Reviews Microbiology, 7*(4), 263–273. https://doi.org/10.1038/nrmicro2109.

Heydorn, A., Nielsen, A. T., Hentzer, M., Sternberg, C., Givskov, M., Ersbøll, B. K., et al. (2000). Quantification of biofilm structures by the novel computer program COMSTAT. *Microbiology (Reading, England), 146*(Pt. 10), 2395–2407. https://doi.org/10.1099/00221287-146-10-2395.

Høiby, N., Bjarnsholt, T., Givskov, M., Molin, S., & Ciofu, O. (2010). Antibiotic resistance of bacterial biofilms. *International Journal of Antimicrobial Agents, 35*(4), 322–332. https://doi.org/10.1016/j.ijantimicag.2009.12.011.

Ibáñez de Aldecoa, A. L., Zafra, O., & González-Pastor, J. E. (2017). Mechanisms and regulation of extracellular DNA release and its biological roles in microbial communities. *Frontiers in Microbiology, 8*, 1390. https://doi.org/10.3389/fmicb.2017.01390.

Jakobsen, T. H., Warming, A. N., Vejborg, R. M., Moscoso, J. A., Stegger, M., Lorenzen, F., et al. (2017). A broad range quorum sensing inhibitor working through sRNA inhibition. *Scientific Reports, 7*(1), 9857. https://doi.org/10.1038/s41598-017-09886-8.

Jamal, M., Ahmad, W., Andleeb, S., Jalil, F., Imran, M., Nawaz, M. A., et al. (2018). Bacterial biofilm and associated infections. *Journal of the Chinese Medical Association, 81*(1), 7–11. https://doi.org/10.1016/j.jcma.2017.07.012.

Jayathilaka, E., Rajapaksha, D. C., Nikapitiya, C., De Zoysa, M., & Whang, I. (2021). Antimicrobial and anti-biofilm peptide octominin for controlling multidrug-resistant *Acinetobacter baumannii*. *International Journal of Molecular Sciences, 22*(10), 5353. https://doi.org/10.3390/ijms22105353.

Jefferson, K. K. (2004). What drives bacteria to produce a biofilm? *FEMS Microbiology Letters, 236*(2), 163–173. https://doi.org/10.1016/j.femsle.2004.06.005.

Jimenez, P. N., Koch, G., Thompson, J. A., Xavier, K. B., Cool, R. H., & Quax, W. J. (2012). The multiple signaling systems regulating virulence in *Pseudomonas aeruginosa*. *Microbiology and Molecular Biology Reviews, 76*(1), 46–65. https://doi.org/10.1128/MMBR.05007-11.

Johnson, J. G., & Clegg, S. (2010). Role of MrkJ, a phosphodiesterase, in type 3 fimbrial expression and biofilm formation in *Klebsiella pneumoniae*. *Journal of Bacteriology, 192*(15), 3944–3950. https://doi.org/10.1128/JB.00304-10.

Johnson, J. G., Murphy, C. N., Sippy, J., Johnson, T. J., & Clegg, S. (2011). Type 3 fimbriae and biofilm formation are regulated by the transcriptional regulators MrkHI in *Klebsiella pneumoniae*. *Journal of Bacteriology, 193*(14), 3453–3460. https://doi.org/10.1128/JB.00286-11.

Kamali, E., Jamali, A., Izanloo, A., & Ardebili, A. (2021). In vitro activities of cellulase and ceftazidime, alone and in combination against *Pseudomonas aeruginosa* biofilms. *BMC Microbiology, 21*(1), 347. https://doi.org/10.1186/s12866-021-02411-y.

Karygianni, L., Ren, Z., Koo, H., & Thurnheer, T. (2020). Biofilm matrixome: Extracellular components in structured microbial communities. *Trends in Microbiology*, *28*(8), 668–681. https://doi.org/10.1016/j.tim.2020.03.016.

Khadke, S., Lee, J., Woo, J., & Lee, J. (2019). Inhibitory effects of honokiol and magnolol on biofilm formation by *Acinetobacter baumannii*. *Biotechnology and Bioprocess Engineering*, *24*, 359–365. https://doi.org/10.1007/s12257-019-0006-9.

Khan, F., Oloketuyi, S. F., & Kim, Y. M. (2019). Diversity of bacteria and bacterial products as antibiofilm and antiquorum sensing drugs against pathogenic bacteria. *Current Drug Targets*, *20*(11), 1156–1179. https://doi.org/10.2174/1389450120666190423161249.

Kim, Y., & Wood, T. K. (2010). Toxins Hha and CspD and small RNA regulator Hfq are involved in persister cell formation through MqsR in *Escherichia coli*. *Biochemical and Biophysical Research Communications*, *391*(1), 209–213. https://doi.org/10.1016/j.bbrc.2009.11.033.

Kim, M. K., Kang, N., Ko, S. J., Park, J., Park, E., Shin, D. W., et al. (2018). Antibacterial and antibiofilm activity and mode of action of magainin 2 against drug-resistant *Acinetobacter baumannii*. *International Journal of Molecular Sciences*, *19*(10), 3041. https://doi.org/10.3390/ijms19103041.

Kim, S. M., Lee, H. W., Choi, Y. W., Kim, S. H., Lee, J. C., Lee, Y. C., et al. (2012). Involvement of curli fimbriae in the biofilm formation of *Enterobacter cloacae*. *Journal of Microbiology*, *50*(1), 175–178. https://doi.org/10.1007/s12275-012-2044-2.

Kim, H. R., Shin, D. S., Jang, H. I., & Eom, Y. B. (2020). Anti-biofilm and anti-virulence effects of zerumbone against *Acinetobacter baumannii*. *Microbiology (Reading, England)*, *166*(8), 717–726. https://doi.org/10.1099/mic.0.000930.

Klausen, M., Aaes-Jørgensen, A., Molin, S., & Tolker-Nielsen, T. (2003). Involvement of bacterial migration in the development of complex multicellular structures in *Pseudomonas aeruginosa* biofilms. *Molecular Microbiology*, *50*(1), 61–68. https://doi.org/10.1046/j.1365-2958.2003.03677.x.

Korir, M. L., Dale, J. L., & Dunny, G. M. (2019). Role of *epaQ*, a previously uncharacterized *Enterococcus faecalis* gene, in biofilm development and antimicrobial resistance. *Journal of Bacteriology*, *201*(18), e00078-19. https://doi.org/10.1128/JB.00078-19.

Kuyucuklu, G., Kaynak Onurdağ, F., & Eryıldız, C. (2021). Stafilokok İzolatlarında Antibiyotiklerin Antibiyofilm Etkinliği Üzerine N-asetilsisteinin Etkisi [Effect of N-acetylcystein on antibiofilm efficiency of antibiotics in staphylococci isolates]. *Mikrobiyoloji bulteni*, *55*(2), 125–145. https://doi.org/10.5578/mb.20219902.

Langstraat, J., Bohse, M., & Clegg, S. (2001). Type 3 fimbrial shaft (MrkA) of *Klebsiella pneumoniae*, but not the fimbrial adhesin (MrkD), facilitates biofilm formation. *Infection and Immunity*, *69*(9), 5805–5812. https://doi.org/10.1128/IAI.69.9.5805-5812.2001.

Lasa, I., & Penadés, J. R. (2006). Bap: A family of surface proteins involved in biofilm formation. *Research in Microbiology*, *157*(2), 99–107. https://doi.org/10.1016/j.resmic.2005.11.003.

Lazar, V., Holban, A. M., Curutiu, C., & Chifiriuc, M. C. (2021). Modulation of quorum sensing and biofilms in less investigated gram-negative ESKAPE pathogens. *Frontiers in Microbiology*, *12*, 676510. https://doi.org/10.3389/fmicb.2021.676510.

Lebreton, F., van Schaik, W., Sanguinetti, M., Posteraro, B., Torelli, R., Le Bras, F., et al. (2012). AsrR is an oxidative stress sensing regulator modulating *Enterococcus faecium* opportunistic traits, antimicrobial resistance, and pathogenicity. *PLoS Pathogens*, *8*(8), e1002834. https://doi.org/10.1371/journal.ppat.1002834.

Lewis, K. (2001). Riddle of biofilm resistance. *Antimicrobial Agents and Chemotherapy*, *45*(4), 999–1007. https://doi.org/10.1128/AAC.45.4.999-1007.2001.

Lewis, K. (2010). Persister cells. *Annual Review of Microbiology*, *64*, 357–372. https://doi.org/10.1146/annurev.micro.112408.134306.

Liao, J., Schurr, M. J., & Sauer, K. (2013). The MerR-like regulator BrlR confers biofilm tolerance by activating multidrug efflux pumps in *Pseudomonas aeruginosa* biofilms. *Journal of Bacteriology*, *195*(15), 3352–3363. https://doi.org/10.1128/JB.00318-13.

Lin, Q., Deslouches, B., Montelaro, R. C., & Di, Y. P. (2018). Prevention of ESKAPE pathogen biofilm formation by antimicrobial peptides WLBU2 and LL37. *International Journal of Antimicrobial Agents*, *52*(5), 667–672. https://doi.org/10.1016/j.ijantimicag.2018.04.019.

Lin, C. T., Lin, T. H., Wu, C. C., Wan, L., Huang, C. F., & Peng, H. L. (2016). CRP-cyclic AMP regulates the expression of type 3 fimbriae via cyclic di-GMP in *Klebsiella pneumoniae*. *PLoS One*, *11*(9), e0162884. https://doi.org/10.1371/journal.pone.0162884.

Lin, T. H., Tseng, C. Y., Lai, Y. C., Wu, C. C., Huang, C. F., & Lin, C. T. (2017). IscR regulation of type 3 fimbriae expression in *Klebsiella pneumoniae* CG43. *Frontiers in Microbiology*, *8*, 1984. https://doi.org/10.3389/fmicb.2017.01984.

Liou, M. L., Soo, P. C., Ling, S. R., Kuo, H. Y., Tang, C. Y., & Chang, K. C. (2014). The sensor kinase BfmS mediates virulence in *Acinetobacter baumannii*. *Journal of Microbiology, Immunology, and Infection*, *47*(4), 275–281. https://doi.org/10.1016/j.jmii.2012.12.004.

Lister, J. L., & Horswill, A. R. (2014). *Staphylococcus aureus* biofilms: Recent developments in biofilm dispersal. *Frontiers in Cellular and Infection Microbiology*, *4*, 178. https://doi.org/10.3389/fcimb.2014.00178.

Liu, F., Sun, Z., Wang, F., Liu, Y., Zhu, Y., Du, L., et al. (2020). Inhibition of biofilm formation and exopolysaccharide synthesis of *Enterococcus faecalis* by phenyllactic acid. *Food Microbiology*, *86*, 103344. https://doi.org/10.1016/j.fm.2019.103344.

Liu, W., Wu, Z., Mao, C., Guo, G., Zeng, Z., Fei, Y., et al. (2020). Antimicrobial peptide Cec4 eradicates the bacteria of clinical carbapenem-resistant *Acinetobacter baumannii* biofilm. *Frontiers in Microbiology*, *11*, 1532. https://doi.org/10.3389/fmicb.2020.01532.

López-Martín, M., Dubern, J. F., Alexander, M. R., & Williams, P. (2021). AbaM regulates quorum sensing, biofilm formation, and virulence in *Acinetobacter baumannii*. *Journal of Bacteriology*, *203*(8), e00635-20. https://doi.org/10.1128/JB.00635-20.

Luo, L. M., Wu, L. J., Xiao, Y. L., Zhao, D., Chen, Z. X., Kang, M., et al. (2015). Enhancing pili assembly and biofilm formation in *Acinetobacter baumannii* ATCC19606 using non-native acyl-homoserine lactones. *BMC Microbiology*, *15*, 62. https://doi.org/10.1186/s12866-015-0397-5.

Ma, L., Conover, M., Lu, H., Parsek, M. R., Bayles, K., & Wozniak, D. J. (2009). Assembly and development of the *Pseudomonas aeruginosa* biofilm matrix. *PLoS Pathogens*, *5*(3), e1000354. https://doi.org/10.1371/journal.ppat.1000354.

Mah, T. F. (2012). Biofilm-specific antibiotic resistance. *Future Microbiology*, *7*(9), 1061–1072. https://doi.org/10.2217/fmb.12.76.

Martínez, M., Polizzotto, A., Flores, N., Semorile, L., & Maffía, P. C. (2020). Antibacterial, anti-biofilm and in vivo activities of the antimicrobial peptides P5 and P6.2. *Microbial Pathogenesis*, *139*, 103886. https://doi.org/10.1016/j.micpath.2019.103886.

Massier, S., Robin, B., Mégroz, M., Wright, A., Harper, M., Hayes, B., et al. (2021). Phosphorylation of extracellular proteins in *Acinetobacter baumannii* in sessile mode of growth. *Frontiers in Microbiology*, *12*, 738780. https://doi.org/10.3389/fmicb.2021.738780.

Mayer, C., Muras, A., Parga, A., Romero, M., Rumbo-Feal, S., Poza, M., et al. (2020). Quorum sensing as a target for controlling surface associated motility and biofilm formation in *Acinetobacter baumannii*

ATCC® 17978™. *Frontiers in Microbiology*, *11*, 565548. https://doi.org/10.3389/fmicb.2020.565548.

Mezzatesta, M. L., Gona, F., & Stefani, S. (2012). *Enterobacter cloacae* complex: Clinical impact and emerging antibiotic resistance. *Future Microbiology*, *7*(7), 887–902. https://doi.org/10.2217/fmb.12.61.

Mu, Y., Zeng, H., & Chen, W. (2021). Quercetin inhibits biofilm formation by decreasing the production of EPS and altering the composition of EPS in *Staphylococcus epidermidis*. *Frontiers in Microbiology*, *12*, 631058. https://doi.org/10.3389/fmicb.2021.631058.

Nallapareddy, S. R., Weinstock, G. M., & Murray, B. E. (2003). Clinical isolates of *Enterococcus faecium* exhibit strain-specific collagen binding mediated by Acm, a new member of the MSCRAMM family. *Molecular Microbiology*, *47*(6), 1733–1747. https://doi.org/10.1046/j.1365-2958.2003.03417.x.

Niu, C., Clemmer, K. M., Bonomo, R. A., & Rather, P. N. (2008). Isolation and characterization of an autoinducer synthase from *Acinetobacter baumannii*. *Journal of Bacteriology*, *190*(9), 3386–3392. https://doi.org/10.1128/JB.01929-07.

Okshevsky, M., & Meyer, R. L. (2015). The role of extracellular DNA in the establishment, maintenance and perpetuation of bacterial biofilms. *Critical Reviews in Microbiology*, *41*(3), 341–352. https://doi.org/10.3109/1040841X.2013.841639.

O'Toole, G. A., & Kolter, R. (1998). Flagellar and twitching motility are necessary for *Pseudomonas aeruginosa* biofilm development. *Molecular Microbiology*, *30*(2), 295–304. https://doi.org/10.1046/j.1365-2958.1998.01062.x.

Otto, M. (2008). Staphylococcal biofilms. *Current Topics in Microbiology and Immunology*, *322*, 207–228. https://doi.org/10.1007/978-3-540-75418-3_10.

Pacheco, T., Gomes, A., Siqueira, N., Assoni, L., Darrieux, M., Venter, H., et al. (2021). SdiA, a quorum-sensing regulator, suppresses fimbriae expression, biofilm formation, and quorum-sensing signaling molecules production in *Klebsiella pneumoniae*. *Frontiers in Microbiology*, *12*, 597735. https://doi.org/10.3389/fmicb.2021.597735.

Paganelli, F. L., Willems, R. J., Jansen, P., Hendrickx, A., Zhang, X., Bonten, M. J., et al. (2013). *Enterococcus faecium* biofilm formation: Identification of major autolysin AtlAEfm, associated Acm surface localization, and AtlAEfm-independent extracellular DNA release. *mBio*, *4*(2), e00154. https://doi.org/10.1128/mBio.00154-13.

Paharik, A. E., & Horswill, A. R. (2016). The staphylococcal biofilm: Adhesins, regulation, and host response. *Microbiology Spectrum*, *4* (2). https://doi.org/10.1128/microbiolspec.VMBF-0022-2015.

Palethorpe, S., Farrow, J. M., 3rd, Wells, G., Milton, M. E., Actis, L. A., Cavanagh, J., et al. (2022). *Acinetobacter baumannii* regulates its stress responses via the BfmRS two-component regulatory system. *Journal of Bacteriology*, *204*(2), e0049421. https://doi.org/10.1128/JB.00494-21.

Pan, P., Wang, X., Chen, Y., Chen, Q., Yang, Y., Wei, C., et al. (2022). Effect of Hcp iron ion regulation on the interaction between *Acinetobacter baumannii* with human pulmonary alveolar epithelial cells and biofilm formation. *Frontiers in Cellular and Infection Microbiology*, *12*, 761604. https://doi.org/10.3389/fcimb.2022.761604.

Panjaitan, N., Horng, Y. T., Cheng, S. W., Chung, W. T., & Soo, P. C. (2019). EtcABC, a putative EII complex, regulates type 3 fimbriae via CRP-cAMP signaling in *Klebsiella pneumoniae*. *Frontiers in Microbiology*, *10*, 1558. https://doi.org/10.3389/fmicb.2019.01558.

Parastan, R., Kargar, M., Solhjoo, K., & Kafilzadeh, F. (2020). *Staphylococcus aureus* biofilms: Structures, antibiotic resistance, inhibition, and vaccines. *Gene Reports*, *20*, 100739. https://doi.org/10.1016/j.genrep.2020.100739.

Patel, R. (2005). Biofilms and antimicrobial resistance. *Clinical Orthopaedics and Related Research*, *437*, 41–47. https://doi.org/10.1097/01.blo.0000175714.68624.74.

Pérez, A., Merino, M., Rumbo-Feal, S., Álvarez-Fraga, L., Vallejo, J. A., Beceiro, A., et al. (2017). The FhaB/FhaC two-partner secretion system is involved in adhesion of *Acinetobacter baumannii* AbH12O-A2 strain. *Virulence*, *8*(6), 959–974. https://doi.org/10.1080/21505594.2016.1262313.

Petrova, O. E., & Sauer, K. (2016). Escaping the biofilm in more than one way: Desorption, detachment or dispersion. *Current Opinion in Microbiology*, *30*, 67–78. https://doi.org/10.1016/j.mib.2016.01.004.

Peyrusson, F., Varet, H., Nguyen, T. K., Legendre, R., Sismeiro, O., Coppée, J. Y., et al. (2020). Intracellular *Staphylococcus aureus* persisters upon antibiotic exposure. *Nature Communications*, *11*(1), 2200. https://doi.org/10.1038/s41467-020-15966-7.

Powell, L. C., Pritchard, M. F., Ferguson, E. L., Powell, K. A., Patel, S. U., Rye, P. D., et al. (2018). Targeted disruption of the extracellular polymeric network of *Pseudomonas aeruginosa* biofilms by alginate oligosaccharides. *NPJ Biofilms and Microbiomes*, *4*, 13. https://doi.org/10.1038/s41522-018-0056-3.

Qian, W., Zhang, J., Wang, W., Wang, T., Liu, M., Yang, M., et al. (2020). Antimicrobial and antibiofilm activities of paeoniflorin against carbapenem-resistant *Klebsiella pneumoniae*. *Journal of Applied Microbiology*, *128*(2), 401–413. https://doi.org/10.1111/jam.14480.

Rajput, A., & Kumar, M. (2018). *Anti-biofilm peptides: A new class of quorum quenchers and their prospective therapeutic applications*. https://doi.org/10.1007/978-981-10-9026-4_5.

Raorane, C. J., Lee, J. H., Kim, Y. G., Rajasekharan, S. K., García-Contreras, R., & Lee, J. (2019). Antibiofilm and antivirulence efficacies of flavonoids and curcumin against *Acinetobacter baumannii*. *Frontiers in Microbiology*, *10*, 990. https://doi.org/10.3389/fmicb.2019.00990.

Ribeiro, S. M., de la Fuente-Núñez, C., Baquir, B., Faria-Junior, C., Franco, O. L., & Hancock, R. E. (2015). Antibiofilm peptides increase the susceptibility of carbapenemase-producing *Klebsiella pneumoniae* clinical isolates to β-lactam antibiotics. *Antimicrobial Agents and Chemotherapy*, *59*(7), 3906–3912. https://doi.org/10.1128/AAC.00092-15.

Richmond, G. E., Evans, L. P., Anderson, M. J., Wand, M. E., Bonney, L. C., Ivens, A., et al. (2016). The *Acinetobacter baumannii* two-component system AdeRS regulates genes required for multidrug efflux, biofilm formation, and virulence in a strain-specific manner. *mBio*, *7*(2), e00430-16. https://doi.org/10.1128/mBio.00430-16.

Richter, A. M., Possling, A., Malysheva, N., Yousef, K. P., Herbst, S., von Kleist, M., et al. (2020). Local c-di-GMP signaling in the control of synthesis of the *E. coli* biofilm exopolysaccharide pEtN-cellulose. *Journal of Molecular Biology*, *432*(16), 4576–4595. https://doi.org/10.1016/j.jmb.2020.06.006.

Rodríguez-Baño, J., Martí, S., Soto, S., Fernández-Cuenca, F., Cisneros, J. M., Pachón, J., et al. (2008). Biofilm formation in *Acinetobacter baumannii*: Associated features and clinical implications. *Clinical Microbiology and Infection: The Official Publication of the European Society of Clinical Microbiology and Infectious Diseases*, *14*(3), 276–278. https://doi.org/10.1111/j.1469-0691.2007.01916.x.

Römling, U., & Balsalobre, C. (2012). Biofilm infections, their resilience to therapy and innovative treatment strategies. *Journal of Internal Medicine*, *272*(6), 541–561. https://doi.org/10.1111/joim.12004.

Roussin, M., Rabarioelina, S., Cluzeau, L., Cayron, J., Lesterlin, C., Salcedo, S. P., et al. (2019). Identification of a contact-dependent growth inhibition (CDI) system that reduces biofilm formation and host cell adhesion of *Acinetobacter baumannii* DSM30011 strain. *Frontiers in Microbiology*, *10*, 2450. https://doi.org/10.3389/fmicb.2019.02450.

Ryder, C., Byrd, M., & Wozniak, D. J. (2007). Role of polysaccharides in *Pseudomonas aeruginosa* biofilm development. *Current Opinion in Microbiology*, *10*(6), 644–648. https://doi.org/10.1016/j.mib.2007.09.010.

Saranathan, R., Pagal, S., Sawant, A. R., Tomar, A., Madhangi, M., Sah, S., et al. (2017). Disruption of tetR type regulator adeN by mobile genetic element confers elevated virulence in *Acinetobacter baumannii*. *Virulence*, *8*(7), 1316–1334. https://doi.org/10.1080/21505594.2017.1322240.

Schroll, C., Barken, K. B., Krogfelt, K. A., & Struve, C. (2010). Role of type 1 and type 3 fimbriae in *Klebsiella pneumoniae* biofilm formation. *BMC Microbiology*, *10*, 179. https://doi.org/10.1186/1471-2180-10-179.

Schwartz, K., Ganesan, M., Payne, D. E., Solomon, M. J., & Boles, B. R. (2016). Extracellular DNA facilitates the formation of functional amyloids in *Staphylococcus aureus* biofilms. *Molecular Microbiology*, *99* (1), 123–134. https://doi.org/10.1111/mmi.13219.

Secinti, K. D., Özalp, H., Attar, A., & Sargon, M. F. (2011). Nanoparticle silver ion coatings inhibit biofilm formation on titanium implants. *Journal of Clinical Neuroscience: Official Journal of the Neurosurgical Society of Australasia*, *18*(3), 391–395. https://doi.org/10.1016/j.jocn.2010.06.022.

Selvaraj, A., Valliammai, A., Sivasankar, C., Suba, M., Sakthivel, G., & Pandian, S. K. (2020). Antibiofilm and antivirulence efficacy of myrtenol enhances the antibiotic susceptibility of *Acinetobacter baumannii*. *Scientific Reports*, *10*(1), 21975. https://doi.org/10.1038/s41598-020-79128-x.

Sharma, A., Dubey, V., Sharma, R., Devnath, K., Gupta, V. K., Akhter, J., et al. (2018). The unusual glycine-rich C terminus of the *Acinetobacter baumannii* RNA chaperone Hfq plays an important role in bacterial physiology. *The Journal of Biological Chemistry*, *293*(35), 13377–13388. https://doi.org/10.1074/jbc.RA118.002921.

Shenkutie, A. M., Yao, M. Z., Siu, G. K., Wong, B., & Leung, P. H. (2020). Biofilm-induced antibiotic resistance in clinical *Acinetobacter baumannii* isolates. *Antibiotics*, *9*(11), 817. https://doi.org/10.3390/antibiotics9110817.

Siebs, E., Shanina, E., Kuhaudomlarp, S., Da Silva Figueiredo Celestino Gomes, P., Fortin, C., Seeberger, P. H., et al. (2022). Targeting the central pocket of the *Pseudomonas aeruginosa* lectin LecA. *Chembiochem: A European Journal of Chemical Biology*, *23*(3), e202100563. https://doi.org/10.1002/cbic.202100563.

Sillanpää, J., Nallapareddy, S. R., Prakash, V. P., Qin, X., Höök, M., Weinstock, G. M., et al. (2008). Identification and phenotypic characterization of a second collagen adhesin, Scm, and genome-based identification and analysis of 13 other predicted MSCRAMMs, including four distinct pilus loci, in *Enterococcus faecium*. *Microbiology (Reading, England)*, *154*(Pt 10), 3199–3211. https://doi.org/10.1099/mic.0.2008/017319-0.

Singh, R., Ray, P., Das, A., & Sharma, M. (2009). Role of persisters and small-colony variants in antibiotic resistance of planktonic and biofilm-associated *Staphylococcus aureus*: An in vitro study. *Journal of Medical Microbiology*, *58*(Pt 8), 1067–1073. https://doi.org/10.1099/jmm.0.009720-0.

Singh, A. K., Yadav, S., Chauhan, B. S., Nandy, N., Singh, R., Neogi, K., et al. (2019). Classification of clinical isolates of *Klebsiella pneumoniae* based on their *in vitro* biofilm forming capabilities and elucidation of the biofilm matrix chemistry with special reference to the protein content. *Frontiers in Microbiology*, *10*, 669. https://doi.org/10.3389/fmicb.2019.00669.

Skurnik, D., Davis, M. R., Jr., Benedetti, D., Moravec, K. L., Cywes-Bentley, C., Roux, D., et al. (2012). Targeting pan-resistant bacteria with antibodies to a broadly conserved surface polysaccharide expressed during infection. *The Journal of Infectious Diseases*, *205* (11), 1709–1718. https://doi.org/10.1093/infdis/jis254.

Slany, M., Oppelt, J., & Cincarova, L. (2017). Formation of *Staphylococcus aureus* biofilm in the presence of sublethal concentrations of disinfectants studied via a transcriptomic analysis using transcriptome sequencing (RNA-seq). *Applied and Environmental Microbiology*, *83*(24), e01643-17. https://doi.org/10.1128/AEM.01643-17.

Soares, G., Costa, J., Melo, F., Mola, R., & Leal-Balbino, T. C. (2016). Biofilm production and resistance profile of Enterobacter sp. strains isolated from pressure ulcers in Petrolina, Pernambuco, Brazil. *Jornal Brasileiro de Patologia e Medicina Laboratorial*, *52*. https://doi.org/10.5935/1676-2444.20160045.

Soria-Bustos, J., Ares, M. A., Gómez-Aldapa, C. A., González-Y-Merchand, J. A., Girón, J. A., & De la Cruz, M. A. (2020). Two type VI secretion systems of *Enterobacter cloacae* are required for bacterial competition, cell adherence, and intestinal colonization. *Frontiers in Microbiology*, *11*, 560488. https://doi.org/10.3389/fmicb.2020.560488.

Spoering, A. L., & Gilmore, M. S. (2006). Quorum sensing and DNA release in bacterial biofilms. *Current Opinion in Microbiology*, *9*(2), 133–137. https://doi.org/10.1016/j.mib.2006.02.004.

Stewart, P. S., Peyton, B. M., Drury, W. J., & Murga, R. (1993). Quantitative observations of heterogeneities in *Pseudomonas aeruginosa* biofilms. *Applied and Environmental Microbiology*, *59*(1), 327–329. https://doi.org/10.1128/aem.59.1.327-329.1993.

Tang, J., Chen, Y., Wang, X., Ding, Y., Sun, X., & Ni, Z. (2020). Contribution of the AbaI/AbaR quorum sensing system to resistance and virulence of *Acinetobacter baumannii* clinical strains. *Infection and Drug Resistance*, *13*, 4273–4281. https://doi.org/10.2147/IDR.S276970.

Top, J., Paganelli, F. L., Zhang, X., van Schaik, W., Leavis, H. L., van Luit-Asbroek, M., et al. (2013). The *Enterococcus faecium* enterococcal biofilm regulator, EbrB, regulates the esp operon and is implicated in biofilm formation and intestinal colonization. *PLoS One*, *8*(5), e65224. https://doi.org/10.1371/journal.pone.0065224.

Torres, N. S., Montelongo-Jauregui, D., Abercrombie, J. J., Srinivasan, A., Lopez-Ribot, J. L., Ramasubramanian, A. K., et al. (2018). Antimicrobial and antibiofilm activity of synergistic combinations of a commercially available small compound library with colistin against *Pseudomonas aeruginosa*. *Frontiers in Microbiology*, *9*, 2541. https://doi.org/10.3389/fmicb.2018.02541.

Trotonda, M. P., Manna, A. C., Cheung, A. L., Lasa, I., & Penadés, J. R. (2005). SarA positively controls bap-dependent biofilm formation in *Staphylococcus aureus*. *Journal of Bacteriology*, *187*(16), 5790–5798. https://doi.org/10.1128/JB.187.16.5790-5798.2005.

Valentini, M., & Filloux, A. (2016). Biofilms and cyclic di-GMP (c-di-GMP) signaling: Lessons from *Pseudomonas aeruginosa* and other bacteria. *The Journal of Biological Chemistry*, *291*(24), 12547–12555. https://doi.org/10.1074/jbc.R115.711507.

Van Wamel, W. J., Hendrickx, A. P., Bonten, M. J., Top, J., Posthuma, G., & Willems, R. J. (2007). Growth condition-dependent Esp expression by *Enterococcus faecium* affects initial adherence and biofilm formation. *Infection and Immunity*, *75*(2), 924–931. https://doi.org/10.1128/IAI.00941-06.

Vijayakumar, K., & Ramanathan, T. (2020). Musa acuminata and its bioactive metabolite 5-hydroxymethylfurfural mitigates quorum sensing (las and rhl) mediated biofilm and virulence production of nosocomial pathogen *Pseudomonas aeruginosa* in vitro. *Journal of Ethnopharmacology, 246*, 112242. https://doi.org/10.1016/j.jep.2019.112242.

Vulin, C., Leimer, N., Huemer, M., Ackermann, M., & Zinkernagel, A. S. (2018). Prolonged bacterial lag time results in small colony variants that represent a sub-population of persisters. *Nature Communications, 9*(1), 4074. https://doi.org/10.1038/s41467-018-06527-0.

Walters, M. C., 3rd, Roe, F., Bugnicourt, A., Franklin, M. J., & Stewart, P. S. (2003). Contributions of antibiotic penetration, oxygen limitation, and low metabolic activity to tolerance of *Pseudomonas aeruginosa* biofilms to ciprofloxacin and tobramycin. *Antimicrobial Agents and Chemotherapy, 47*(1), 317–323. https://doi.org/10.1128/AAC.47.1.317-323.2003.

Wang, L., Di Luca, M., Tkhilaishvili, T., Trampuz, A., & Gonzalez Moreno, M. (2019). Synergistic activity of fosfomycin, ciprofloxacin, and gentamicin against *Escherichia coli* and *Pseudomonas aeruginosa* biofilms. *Frontiers in Microbiology, 10*, 2522. https://doi.org/10.3389/fmicb.2019.02522.

Wei, Q., & Ma, L. Z. (2013). Biofilm matrix and its regulation in *Pseudomonas aeruginosa*. *International Journal of Molecular Sciences, 14* (10), 20983–21005. https://doi.org/10.3390/ijms141020983.

Willett, J., Ji, M. M., & Dunny, G. M. (2019). Exploiting biofilm phenotypes for functional characterization of hypothetical genes in *Enterococcus faecalis*. *NPJ Biofilms and Microbiomes, 5*(1), 23. https://doi.org/10.1038/s41522-019-0099-0.

Wood, T. K., Knabel, S. J., & Kwan, B. W. (2013). Bacterial persister cell formation and dormancy. *Applied and Environmental Microbiology, 79*(23), 7116–7121. https://doi.org/10.1128/AEM.02636-13.

Wu, C. C., Huang, Y. J., Fung, C. P., & Peng, H. L. (2010). Regulation of the *Klebsiella pneumoniae* Kpc fimbriae by the site-specific recombinase KpcI. *Microbiology (Reading, England), 156*(Pt 7), 1983–1992. https://doi.org/10.1099/mic.0.038158-0.

Wu, M., Tong, X., Liu, S., Wang, D., Wang, L., & Fan, H. (2019). Prevalence of methicillin-resistant *Staphylococcus aureus* in healthy Chinese population: A system review and meta-analysis. *PloS One, 14*(10). https://doi.org/10.1371/journal.pone.0223599.

Xie, Z., Wei, H., Meng, J., Cheng, T., Song, Y., Wang, M., et al. (2019). The analogs of temporin-GHa exhibit a broader spectrum of antimicrobial activity and a stronger antibiofilm potential against *Staphylococcus aureus*. *Molecules, 24*(22), 4173. https://doi.org/10.3390/molecules24224173.

Xu, Z., Zhang, H., Yu, H., Dai, Q., Xiong, J., Sheng, H., et al. (2019). Allicin inhibits *Pseudomonas aeruginosa* virulence by suppressing the *rhl* and *pqs* quorum-sensing systems. *Canadian Journal of Microbiology, 65*(8), 563–574. https://doi.org/10.1139/cjm-2019-0055.

Yan, X., Gu, S., Cui, X., Shi, Y., Wen, S., Chen, H., et al. (2019). Antimicrobial, anti-adhesive and anti-biofilm potential of biosurfactants isolated from *Pediococcus acidilactici* and *Lactobacillus plantarum* against *Staphylococcus aureus* CMCC26003. *Microbial Pathogenesis, 127*, 12–20. https://doi.org/10.1016/j.micpath.2018.11.039.

Yin, W. F., Purmal, K., Chin, S., Chan, X. Y., & Chan, K. G. (2012). Long chain N-acyl homoserine lactone production by *Enterobacter sp.* isolated from human tongue surfaces. *Sensors, 12*(11), 14307–14314. https://doi.org/10.3390/s121114307.

Yoon, E. J., Chabane, Y. N., Goussard, S., Snesrud, E., Courvalin, P., Dé, E., et al. (2015). Contribution of resistance-nodulation-cell division efflux systems to antibiotic resistance and biofilm formation in *Acinetobacter baumannii*. *mBio, 6*(2), e00309–e00315. https://doi.org/10.1128/mBio.00309-15.

Yuan, Y., Zai, Y., Xi, X., Ma, C., Wang, L., Zhou, M., et al. (2019). A novel membrane-disruptive antimicrobial peptide from frog skin secretion against cystic fibrosis isolates and evaluation of anti-MRSA effect using *Galleria mellonella* model. *Biochimica et Biophysica Acta - General Subjects, 1863*(5), 849–856. https://doi.org/10.1016/j.bbagen.2019.02.013.

Zapotoczna, M., O'Neill, E., & O'Gara, J. P. (2016). Untangling the diverse and redundant mechanisms of *Staphylococcus aureus* biofilm formation. *PLoS Pathogens, 12*(7), e1005671. https://doi.org/10.1371/journal.ppat.1005671.

Zhang, L., Fritsch, M., Hammond, L., Landreville, R., Slatculescu, C., Colavita, A., et al. (2013). Identification of genes involved in *Pseudomonas aeruginosa* biofilm-specific resistance to antibiotics. *PLoS One, 8*(4), e61625. https://doi.org/10.1371/journal.pone.0061625.

Zhou, J. W., Luo, H. Z., Jiang, H., Jian, T. K., Chen, Z. Q., & Jia, A. Q. (2018). Hordenine: A novel quorum sensing inhibitor and antibiofilm agent against *Pseudomonas aeruginosa*. *Journal of Agricultural and Food Chemistry, 66*(7), 1620–1628. https://doi.org/10.1021/acs.jafc.7b05035.

Zogaj, X, Bokranz, W, Nimtz, M, & Romling, U. (2003). Production of cellulose and curli fimbriae by members of the family Enterobacteriaceae isolated from the human gastrointestinal tract. *Infection and immunity, 71*(7), 4151–4158. https://doi.org/10.1128/IAI.71.7.4151-4158.2003.

Further reading

Donlan, R. M. (2000). Role of biofilms in antimicrobial resistance. *ASAIO Journal, 46*(6), S47–S52. https://doi.org/10.1097/00002480-200011000-00037.

Hengge, R. (2016). Trigger phosphodiesterases as a novel class of c-di-GMP effector proteins. *Philosophical Transactions of the Royal Society of London. Series B, Biological Sciences, 371*(1707). https://doi.org/10.1098/rstb.2015.0498. 20150498.

Liu, Y., Pan, C., Ye, L., Si, Y., Bi, C., Hua, X., et al. (2020). Nonclassical biofilms induced by DNA breaks in *Klebsiella pneumoniae*. *mSphere, 5*(3), e00336-20. https://doi.org/10.1128/mSphere.00336-20.

Liu, C., Sun, D., Zhu, J., Liu, J., & Liu, W. (2020). The regulation of bacterial biofilm formation by cAMP-CRP: A mini-review. *Frontiers in Microbiology, 11*, 802. https://doi.org/10.3389/fmicb.2020.00802.

Lynch, S. V., Dixon, L., Benoit, M. R., Brodie, E. L., Keyhan, M., Hu, P., et al. (2007). Role of the rapA gene in controlling antibiotic resistance of *Escherichia coli* biofilms. *Antimicrobial Agents and Chemotherapy, 51*(10), 3650–3658. https://doi.org/10.1128/AAC.00601-07.

Mohamed, J. A., & Huang, D. B. (2007). Biofilm formation by enterococci. *Journal of Medical Microbiology, 56*(Pt. 12), 1581–1588. https://doi.org/10.1099/jmm.0.47331-0.

Sagar, K., Jin-Hyung, L., Woo, J.-T., & Lee, J. (2019). Inhibitory effects of honokiol and magnolol on biofilm formation by *Acinetobacter baumannii*. *Biotechnology and Bioprocess Engineering, 24*, 359–365.

Sauer, K., Camper, A. K., Ehrlich, G. D., Costerton, J. W., & Davies, D. G. (2002). *Pseudomonas aeruginosa* displays multiple phenotypes during development as a biofilm. *Journal of Bacteriology, 184*(4), 1140–1154. https://doi.org/10.1128/jb.184.4.1140-1154.2002.

Wu, S., Zhang, J., Peng, Q., Liu, Y., Lei, L., & Zhang, H. (2021). The role of *Staphylococcus aureus* YycFG in gene regulation, biofilm organization and drug resistance. *Antibiotics, 10*(12), 1555. https://doi.org/10.3390/antibiotics10121555.

Chapter 24

The emergence of predominance in the constitutive microflora of dairy membrane biofilms

Sanjeev Anand and Pratishtha Verma
Midwest Dairy Research Center, Department of Dairy and Food Science, South Dakota State University, Brookings, SD, United States

1. Introduction

The formation of membrane biofilms in dairy processing facilities is a critical issue that leads to irreversible fouling of the membranes, reducing the membrane flux, and finally, leading to frequent replacement of the membranes. In the dairy industry, the biofilms formed on equipment surfaces raise safety and quality concerns and may serve as a source of contamination of the final product with both spoilage and pathogenic bacteria, which are difficult to remove (Fadila, 2016). The bacterial attachment to the surface is different in biofilms than in their planktonic counterparts.

Multispecies biofilms are formed when the feed containing microorganisms comes in contact with the membrane surface. These biofilms are known to be more complex and tend to form more stable biofilm. It could be due to the production of extracellular polymeric substance (EPS) by different species. Certain species within the biofilm microflora acquire resilience over prolonged use of membrane and emerge predominantly over other strains (Nadell, Drescher and Foster, 2016a; Tang, Flint, Brooks, & Bennett, 2009).

This chapter will discuss the type of biofilm microflora found on filtration membranes, microbial interactions, the concept of emergence of predominance within the constitutive microflora of the membrane, and the release of antimicrobial substances. This chapter has been compiled under the key topic areas to understand the changes in biofilm microflora over prolonged use of membranes, its detrimental effect on membrane processes, and implementation of different techniques to remove biofilms from dairy processing plants.

2. The process of membrane fouling on dairy membrane

Filtration membranes have been extensively used in the dairy industry. During the filtration process as the product comes in contact with the membrane, it initiates bacterial colonization on the membrane surfaces (Marka & Anand, 2018). The attachment of the cell to the surface is called adhesion, and the cell to cell attachment is known as cohesion. The RO membranes used in dairy liquids provide a large surface area for the potential for the bacteria to colonize (Flint et al., 2020). These are also referred to as bacterial biofilms, which lead to biofouling. Biofouling is one of the biggest challenges faced by the dairy industry, which limits the membrane performance due to a severe decline in the flux rate.

An understanding of the process of biofilm formation has been generated over years. A 1980s study on microbial adhesion to surfaces described the process of microbial attachment to the surface as a three stage process: (1) adsorption, or the accumulation of an organism on a collector surface, i.e., substrate (deposition); (2) attachment, or the consolidation of the interface between an organism and a collector, often involving the formation of polymer bridges between the organism and collector; (3) colonization, or growth and division of organisms on the collector's surface (Garrett, Bhakoo, & Zhang, 2008), which was later described as an eight step process including the formation of initial conditioning layer, reversible and irreversible adhesion of bacteria, and the eventual detachment of cells from a mature biofilm for subsequent colonization (Garrett et al., 2008).

These biofilms are difficult to clean because of the multilayer spiral wound structure of the membranes. Many a times even the chemical cleaning protocols prove to be ineffective, leading to frequent premature replacement of membranes (Anand, Singh, Avadhanula, & Marka, 2014). Membrane replacement cost constitutes 25%–40% of the total cost of the membrane plants. That could vary from \$10 to \$15 million depending upon the type of membrane used (D'souza & Mawson, 2005).

Understanding Microbial Biofilms. https://doi.org/10.1016/B978-0-323-99977-9.00022-3

2.1 Factors influencing membrane fouling

Membrane fouling is affected by the hydrodynamics of the membrane filtration process, the interactions between the membrane and foulants in the feed stream and between the fouling layer and foulants (D'souza & Mawson, 2005). In a study on polyamide RO membranes, colonization was observed to be heaviest where a hydrophilic layer was applied to minimize the protein clogging of the membrane used for whey processing (Flint et al., 2020). The composition of the dairy feed is complex, which consists of different concentrations of proteins, carbohydrates (e.g., lactose), lipids (e.g., phospholipids), minerals, nonprotein N compounds, citric and lactic acids, and microorganisms (Marka & Anand, 2018). Depending upon the substrate deposition, the membrane fouling can be categorized into different types including particulate based, crystallization based, chemical binding based, and the most widespread is the adhesion of a variety of microorganisms to the membrane surface. This type of fouling is known as biofouling (James, Jing, & Chen, 2003). The presence of nutrients in the feed acts as a medium for microorganism to grow and attach to the membrane surface more firmly. The solid–liquid interfaces provide an ideal environment for the microbial attachment and formation of biofilm (Sharma & Anand, 2002). The temperature of the feed also plays a critical role in the formation and growth rate of bacterial communities to form firm biofilms on the membrane surface (Chamberland et al., 2019). Biofouling adversely affects the structural integrity of the membrane, leading to process failure, therefore increasing the operational and maintenance cost (Marka & Anand, 2018).

Studies have also reported the presence of foodborne pathogens and spoilage microorganisms as a part of biofilm microflora on a variety of surfaces used in food processing environments. There is a diverse range of factors that can influence bacterial colonization on food and dairy processing surfaces. Major factors include the type of microorganism and their growth stage, surface properties of the bacteria and their substrate, the microbial and nutritional quality of the feed stream, and pH of feed (Tang et al., 2009). Studies have also reported stronger attachment ability of the spores than vegetative cells, due to hydrophobicity and hair like structures on the cell surface (Husmark & Rönner, 1992; Kumar & Anand, 1998; Rönner, Husmark, & Henriksson, 1990). Spore formers make firmer biofilms, which also possess resistance to the cleaning regime (Avadhanula, 2011).

2.2 Bacterial fouling

The biofilms could be single or multispecies communities enclosed in a protected layer of EPS and attached to the substratum (Hassan, Anand, & Avadhanula, 2010). The biofilms on the surface show different stages of development including reversible, irreversible, formation of microcolonies, development of mature biofilm with three dimensional structure, cell detachment, and dispersion (Anand et al., 2014). The reversible biofilms can be removed with ease from the surface by rinsing. Whereas, the irreversible biofilm offers greater resistance to the clean in place (CIP) procedures. Once the irreversible biofilms are formed, strong forces are required for their removal from the surface of the membrane. The biofilm formation is a slow process; still, it accumulates a few millimeters thick layer in just a few days. As the biofilms mature and start to accumulate other components from the feed, these biofilms grow larger and finally get detached through the process of rinsing and shearing (Marka & Anand, 2018). The dislodged cells are distributed to new locations, where they start to develop into new biofilms. The biofilms have been observed to develop a hills and valley type of structure using scanning electron microscopy (SEM) (Fig. 1), where the hill represents a mushroom like appearance and the valley comprises dense matrices of EPS with embedded bacterial cells (Hassan et al., 2010). The mushroom looks like a three dimensional

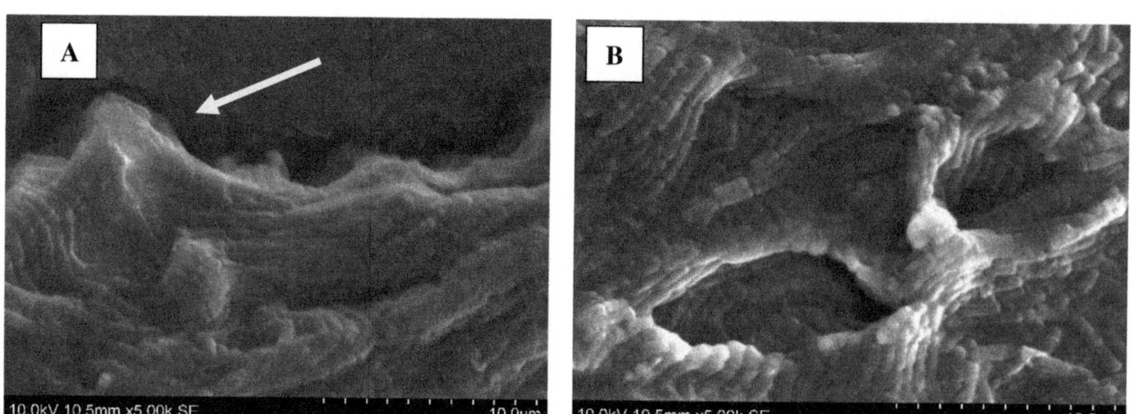

FIG. 1 A mushroom like structure on 72 h old in vitro biofilm forming using predominant *Bacillus subtilis* strain, isolated from an 18 month old reverse osmosis whey concentration membrane (A), a hill and valley structure of biofilm (B) (Unpublished work).

structure, possibly due to excessive secretion of EPS (Anand et al., 2014; Hassan et al., 2010). EPS may contain polysaccharides, proteins, phospholipids, teichoic and nucleic acids, and other polymers (Hassan et al., 2010). It is observed that the cells in biofilms are more resistant to the cleaning processes than their planktonic state (Anand & Singh, 2013). Biofilms are a complex form of microbial ecosystem that allows a high level of interaction among different organisms. Biofilms' temporal and spatial features can be influenced by these interactions. One of the outcomes of these interspecies interactions is the capacity to develop greater resistance to disinfectants (Alonso, Harada, & Kabuki, 2020).

3. Constitutive microflora reported on filtration membranes

The biofilm microflora can serve as a reservoir for different types of microflora causing contamination of the product (Avadhanula, 2011). The biofilm microflora may consist of viable or nonviable, single species, or multispecies communities of microorganisms embedded in EPS (Anand et al., 2014). Various studies have demonstrated the occurrence of multispecies biofilms on the filtration membranes (Anand, Hassan, & Avadhanula, 2012; Anand et al., 2014; Tang et al., 2009). Studies have also demonstrated the predominance of Gram positive microorganisms or thermoresistant species such as sporeforming *Bacillus* species on filtration membranes used to process whey (Friedrich & Lenke, 2006; Verma, Singh, & Anand, 2021). On the contrary, Tang et al. (2009) reported a higher proportion of Gram negative bacteria including coliforms as a part of biofilm microflora. Such results could be obtained in case of water contamination or general plant hygiene problems.

3.1 Microflora on RO membranes

Previous studies have reported the presence of *Enterococcus*, *Staphylococcus*, *Micrococcus*, *Lactococcus*, *Lactobacillus*, *Arthrobacter*, *Bacillus*, *Streptomyces*, *Corynebacterium*, *Klebsiella*, *Aeromonas*, *Methylobacterium*, *Cronobacter*, *Pseudomonas*, *Escherichia coli*, *Streptococcus*, *Chryseobacterium*, and bacteria from *Acinetobacter*, *Firmicutes*, and *Proteobacteria* phylum as constitutive microflora found on dairy separation membranes (Anand & Singh, 2013; Chamberland, Lessard, Doyen, Labrie, & Pouliot, 2017; Verma et al., 2021).

3.2 Microflora on UF membranes

Previous studies have isolated various kinds of bacteria from dairy UF membranes, which demonstrated the presence of *Klebsiella oxytoca*, *Citrobacter freundii*, *Bacillus licheniformis*, *Enterococcus faecalis*, *Rahnella aquatilis*, *Exiguobacterium* sp., *Enterobacter* sp., *Lactococcus lactis* ssp., *Bacillus* ssp., *Chryseobacterium* sp., and bacteria from *Acinetobactia*, *Firmicutes*, and *Proteobacteria* phylum were also found to be part of the constitutive microflora of UF separation membranes (Chamberland et al., 2017; Tang et al., 2009; Verma, 2020). It is important to note here some studies reported the presence of similar as well as different microflora on UF membranes, which could be due to the variations such as feed or operational parameters, as these factors are known to greatly affect the diversity of bacteria colonizing the membrane (Verma, 2020).

3.3 Resilient microflora on filtration membranes

Tang et al. (2009) have reported *Klebsiella* spp. and *Bacillus* spp. to be dominant on membranes used in the dairy industry. Several other studies have reported the persistence of *Bacillus* spp. on dairy filtration membrane and its resistance to the disinfectants and chemical cleaners used in the CIP procedures (Anand & Singh, 2013; Marka & Anand, 2018; Sharma & Anand, 2002). Some isolates within the biofilm microflora acquire resistance to the chemical cleaning protocols over prolonged use of membranes. Thus, constitutive bacteria are likely to compete, resulting in some species appearing to be dominant (Verma et al., 2021). Also, studies have reported the ability of persistent strains to produce significantly more biofilms than nonpersistent strains (Alvarez-Ordóñez, Coughlan, Briandet, & Cotter, 2019). Biofilm composing *Bacillus* species are often involved in sedulous contamination and spoilage of dairy products. They, therefore, present a major microbiological challenge in the field of dairy product quality and safety. Due to their substantial physiological multifariousness, *Bacillus* species can survive in various components of dairy manufacturing plants, leading to a high risk of product spoilage and potential dissemination of foodborne diseases. Furthermore, biofilm and heat resistant spore formation make these bacteria difficult to eliminate. Thus, some strategies have been employed to remove, prevent, or delay the formation of *Bacillus* biofilms in the dairy industry, but with limited success (Shemesh & Ostrov, 2020).

Studies on the change of biofilm microflora over prolonged use of membrane depicted the aerobic spore formers and lactic acid bacteria to be always a part of the constitutive microflora. Whereas, the occasional presence of other organisms such as coliforms, *Pseudomonas* spp., and *Staphylococcus aureus* was reported. This could be due to cross contamination during the whey concentration (Avadhanula, 2011).

3.4 Pathogens as part of biofilm microflora

The presence of pathogens such as *Listeria monocytogenes*, *Yersinia enterocolitica*, *Campylobacter jejuni*, *Salmonella* spp., *Staphylococcus* spp., *Bacillus cereus*, and even *E. coli* 0157:H7 (Avadhanula, 2011; Flint et al., 2020; Yuan et al., 2019) has been reported in the membrane biofilms. These foodborne pathogens could serve as a potential source of contamination and raise product quality and safety concerns. Studies have also reported an increase in the attachment and survival of pathogenic bacteria in the presence of other bacterial species (Flint et al., 2020). A previous study on biofilm microflora reported that *B. cereus* may constitute more than 12% of some of the biofilm microbial community (Sharma & Anand, 2002).

4. The emergence of predominance within the biofilm over prolonged use of membranes

Multispecies biofilms are considered to be more complex and appear to form thicker and more stable biofilms, which may be due to the release of EPSs by various bacterial species (Anand et al., 2014; Hassan et al., 2010). The EPS secreted by multiple species in a biofilm matrix could act as a shield and helps to develop resilient biofilms that are very difficult to clean (Anand et al., 2014). Even the cleaning and sanitization protocols prove to be ineffective for removing such resilient biofilms. Such biofouling not only results in a reduced flux rate but also serves as a constant source of contamination, hence, leading to premature replacement of separation membranes and economic losses faced by the dairy industry (Marchand et al., 2012). The membrane replacement cost is high and could constitute around 25%–40% cost of the membrane plant (Souza & Mawson, 2007).

Some organisms present within the biofilm microflora show predominance over others, over the prolonged use of membranes (Anand & Singh, 2013). Some of the other studies have also depicted the emergence of single species predominance in biofilms (Flint, Bremer, & Brooks, 2009; Nadell et al., 2016a, 2016b). Competition among microorganisms in a shared environment occurs in all but the simplest ecosystems (Fredrickson & Stephanopoulos, 1981). The populations of microorganisms within the common setting may outcompete one microbial population over the other when the competition is focused on a single resource, hence, emerging as predominant in a constitutive microflora.

5. Factors influencing the emergence of predominance

The emergence of predominance could be associated with several factors including competition for the nutrients, faster growing rate of one microorganism over the others, production of metabolites by the cells, production of bacteriocin, secretion of broad and narrow spectrum toxins with coupled privatized antitoxin, etc. (Amézquita & Brashears, 2002; Bowden & Hamilton, 1989; Nadell et al., 2016a, 2016b). Other decisive factors that can potentially lead to the prevalence of certain microbial species over the others within mixed species biofilms are the surface charge, cell chaining, inoculum composition, oxygen availability, or release of certain matrix protein like TasA, which provides structural integrity to *Bacillus subtilis* biofilms, etc. (Van Merode, Pothoven, Van Der Mei, Busscher, & Krom, 2007; Yannarell, Grandchamp, Chen, Daniels, & Shank, 2019). According to Van Merode et al. (2007), and the surface charge could also influence the prevalence of one microorganism within a mixed species biofilm. Besides, studies have demonstrated the importance of dual species biofilm in which one bacterial strain stimulates the attachment of other bacterial strains (Bradshaw, Marsh, Allison, & Schilling, 1996; Tait & Sutherland, 2002; Van Merode et al., 2007). This could protect other strains against the disinfectants, potentially leading to its predominance within a mixed species growth. Also, it is common to observe the occurrence of bacteriocin activity in the presence of closely related strains as a competitive trait (Tait & Sutherland, 2002).

The EPS is reported to be a major component in biofilm formation. It helps in cell to cell attachment and adhesion of bacterial biofilm to the membrane surface (Anand et al., 2014). The dairy biofilms are predominantly composed of EPS and milk residues, mostly proteins, and calcium phosphate (Simões, Simões, & Vieira, 2010). The multispecies biofilms are known to be more complex and tend to form thicker and more stable biofilms, which could be due to the release of EPSs by various bacterial species (Anand et al., 2014; Hassan et al., 2010). Also, relatively higher production of EPS has been associated with cells within the biofilm as compared with the planktonic cells. The EPS secreted by multiple species in a biofilm matrix may act as a shield and help develop resilient biofilms that are difficult to clean (Anand et al., 2014). It was also indicated that the colony and pellicle morphology were directly correlated with the enhanced production of an EPS. That is known to develop additional resistance to chlorine, hence, demonstrating a protective function of the EPS against biofilm embedded cells (Hobley, Harkins, MacPhee, & Stanley-Wall, 2015). Therefore, embedded biofilm microflora in a polysaccharide matrix may survive and serve as source of cross contamination (Anand et al., 2014).

The presence of mucoid colonies has been associated with the predominance of certain strains over others that do not produce mucoid colonies (Stratton, Gregory, & McNabb, 1986). Also, a colony is considered hypermucoviscosity with a string of $^{>}$5 mm. This could be one of the potential contributing factors for the predominance of one strain over the other in a biofilm matrix (Chinen, 2017; Verma, Singh, & Anand, 2021).

Another factor that could be responsible for the predominance of one species or the other is the action of natural selection. When various species and strains come in contact with each other, it is by default that the competitive phenotype will predominate. This competition has led to the development of numerous strategic strategies, ranging from rapid growth and resource acquisition to the use of adhesion and matrix processing to capture nutrient rich biofilm locations. One of the most common ways of intermicrobial competition is to secrete wide and narrow spectrum toxins in combination with privatized antitoxins that prevent self poisoning (Nadell et al., 2016a, 2016b).

The release of narrow or broad spectrum toxins by a strain is predicted to be most favored when the competition is confined, and the competing cell lineages are fairly well mixed in the space. The response of the toxin varies within different communities. The strength of each toxin secreting strain might be too low for an effective attack if the population mixture is too high. Whereas, the toxic secreting strain may not even function when these populations are isolated. Studies have therefore stated that if the cell lines are well mixed, the sensitive cells inside the same biofilm may become outcompeted by the toxin secreting species (Nadell et al., 2016a, 2016b; Schluter, Nadell, Bassler, & Foster, 2015).

Quorum sensing (QS) was also observed for controlling competitive traits such as bacteriocin release. Stress reactions to nutrient limitation and cell damage also upregulate the bacteriocins and antibiotics. The toxin secreting strains are known to inhibit the growth of the sensitive strains at or under localized conditions at a sufficiently high density of cells. Another mechanism for detecting the competitor's presence is because of the stresses they create when they are in close proximity. Such "competition sensing" may manifest as a response to nutrient limitation or cell damage, perhaps more reliably (Cornforth & Foster, 2013; Nadell et al., 2016a, 2016b).

Therefore, it is important to understand the emergence of predominance within membrane biofilms (Fig. 2). This can provide necessary information to bridge the gap between the development of mature biofilms on prolonged use of membranes and the potential of selective species to generate resistance, which leads to the predominance of one microorganism over the others in a biofilm microflora. This information would be vital to identify predominating species within biofilm microflora and to create new CIP strategies to eliminate resistant biofilm microflora to extend the service life of filtration membranes in the dairy industry.

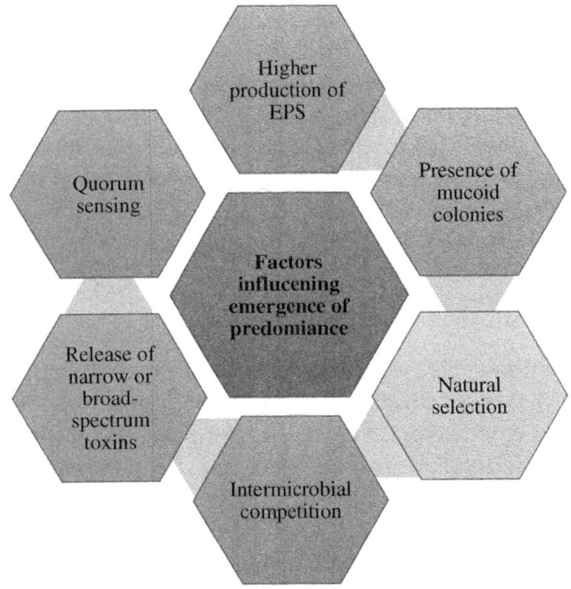

FIG. 2 A summary of factors influencing the emergence of predominance.

6. A case study to demonstrate factors influencing predominance

6.1 Competitive exclusion

Microorganisms found in the dairy industries can exist in the form of monospecies, dual species, or multispecies biofilms. Different species in a biofilm matrix can have cooperative, competitive, or neutral behavior. The bacterial species in a cooperative environment can be benefited by facilitating adhesion, growth, and protection against biocides. Whereas, the interactions among different bacterial species in a competitive environment may occur due to restricted growth environment, scarcity of nutrients, and energy sources (exploitative competition) or due to the production of secondary compounds such as bacteriocins, enzymes, hydrogen peroxide, and organic acids (interference competition), favoring competitive exclusion (Alonso et al., 2020; Nadell et al., 2016a, 2016b), therefore, creating a need to better understand microbial interactions in mixed biofilms.

A research study using biofilm microflora isolated from a used whey reverse osmosis membrane studied the microbial interactions in a controlled environment (Verma et al., 2021). The competitive exclusion was carried out using six isolates from a multispecies biofilm, recovered using the resuscitation step, using the coculture technique, and the process of natural selection. To establish predominance, the interaction within 6 isolates was investigated using 15 possible coculture combinations, where each coculture combination consisted of 2 isolates. From the competitive exclusion, one of the isolates named *B. subtilis* not only expressed itself in all the coculture combinations but also inhibited the growth of almost all other isolates. *B. subtilis* also emerged predominantly in the natural selection process. Competitive exclusion is a unique interaction between species that leads to predominance along with various degrees of coexistence.

Studies have discussed competitive exclusion within multiple strains and species that encounter each other and the predominance of competitive phenotypes, as the primary action of natural selection is to favor genetic lineages that benefit themselves over others (Nadell et al.,

2016a, 2016b). According to Nadell, the clearest incarnation of competitive strategies is the secretion of broad and narrow spectrum toxins, coupled with privatized antitoxins that prevent self poisoning. Another study suggests that *B. subtilis* produces YIT, a biofilm specific antibacterial toxin, which adds to the bacteria's competitiveness against sensitive strains and species (Arnaouteli, Bamford, Stanley-Wall, & Kovács, 2021).

6.2 Antimicrobial inhibition

Several strategies were developed to deal with the problems and challenges presented by the bacteria that exist in the hostile environment. Compared with planktonic cells, biofilms have high intrinsic resistance to antibiotics and sanitizers, which pose serious quality issues in the final products in the dairy and food industry (Flint et al., 2020; Shemesh & Ostrov, 2020). Bacterial strain can exist in the form of mono species, dual species, or multispecies biofilms, based on the previously acknowledged fact that bacterial strain interactions in a dual species biofilm can contribute to competition and cooperation. The cooperation among various bacterial strains in dual or multispecies biofilms can increase antimicrobial tolerance (Alonso et al., 2020).

The matured biofilms may result in a bacterial predominance with resilient strains on membranes with prolonged use. In a research study (Verma, 2020), the release of an antimicrobial compound by *B. subtilis* was established to be a potential cause of the predominance of that *B. subtilis* strain within the constitutive microflora of the biofilm. Initially, the antimicrobial activity of the predominant strain was screened using the cell free extract (CFE) of *B. subtilis* prepared at different propagation times, against the isolated constitutive microflora of the RO membrane (Verma & Anand, 2020), from which, CFE prepared at 12 h (propagation time) depicted the highest antimicrobial activity against other constitutive microflora such as *Exiguobacterium* sp., Exiguobacterium *auranticum*, and Acinetobacter*radioresistens*. Then, the proteinaceous nature of the CFE was established by treating the CFE with proteinase K. The proteinaceous nature was further validated by an amino acid profiling that revealed the total crude protein in CFE to be 51% (wt/wt).

Other studies have also demonstrated that the production of secondary or antimicrobial compounds as a response to interference competition by one microorganism against the other facilitates competitor dispersal (Alonso et al., 2020). A research study depicted a regulatory role of an antimicrobial compound release by *B. subtilis*, TasA, in cell physiology. The nearness of TasA strands inside the biofilm framework fortifies expression of motility qualities and downregulates framework expression in subpopulations of cells and hence contributes to colony spreading on surfaces. Cancellation of TasA has other pleiotropic impacts, counting the downregulation of qualities related to sporulation and an increment in the expression of framework and antimicrobial auxiliary metabolite related qualities (Arnaouteli et al., 2021).

7. Implications of the emergence of predominance

Several microorganisms tend to attach to the surface of the membrane and initiate bacterial colonization on the membrane surfaces (Anand et al., 2012). These are also referred to as biofilms, which consist of complex communities of microorganisms (Tait & Sutherland, 2002). These biofilms are difficult to clean and have been reported to develop resistance to chemical processes as compared with planktonic counterparts (Anand et al., 2014; Sayem, Chowdhury, Alam, & Sarker, 2018), proving the cleaning and sanitization protocols to be ineffective and result in the formation of resilient multispecies biofilms (Stoodley, Sauer, Davies, & Costerton, 2002). When the biofilm microflora is subjected to disinfectants, the sensitive cells die but resilient cells develop bacterial resistance against such chemical cleaner, thus, creating the demand to discover new ecofriendly natural antimicrobials for effective membrane cleaning. A study on the biofilm microflora has reported developing resistance within selective isolates over prolonged use of membranes and their potential to generate predominance in the biofilm matrix (Anand & Singh, 2013). The microbial resistance makes the CIP cleaning protocols unsuccessful by reducing the efficiency of almost all antimicrobial treatments. Besides, these strains rapidly generate additional resistance against new synthetic derivatives as these have already acquired resistance to the parent agents (Sumi, Yang, Yeo, & Hahm, 2015), hence, creating a great interest in the food industry on the development of new strategies and antibiofilm agents (Ripolles-Avila, Hascoët, Martínez-Suárez, Capita, & Rodríguez-Jerez, 2019). Under optimal conditions, the filtration membranes should be prevented from forming biofilms than later addressing biofilm issues. Unfortunately, currently, there is no suitable technique available for preventing biofilms to form on the membrane surfaces during processing. Therefore, dairy industries have adapted CIP regimes for the removal of biofilms. To ensure the efficient removal of foulants from membrane surfaces, proper cleaning regimes should be selected that do not promote microbial growth (Simões et al., 2010).

The standard CIP procedure used for cleaning RO membranes consists of five stages, which include an alkaline wash, an acid wash, an alkaline wash, an enzyme wash, and a final sanitizer wash (Tamime, 2009). The membranes are flushed with water in between different stages to remove adhered residues and biofilm debris (Garcia-Fernandez, 2016). Usually, not all stages of the CIP procedure are carried out at the end of each processing cycle. The most commonly used CIP procedure includes an alkaline wash, an acid wash, and a final alkaline wash to restore membrane pH with intermediate water flushing (Tang, Flint, Bennett, & Brooks, 2010). Sanitization wash after cleaning is very important to kill the remaining microorganisms. It is noteworthy that the CIP procedure can remove 90% of microorganisms from the surface but cannot be relied on to sterilize them, leading to the recolonization of bacteria on the membrane surfaces, which given time, water, and nutrients from stronger biofilms becomes more resilient to chemical cleaning protocols (Simões et al., 2010). Various studies have demonstrated the resilience of biofilm embedded bacteria to the cleaning protocols over their planktonic counterparts (Anand & Singh, 2013; Araújo, Lemos, Mergulhão, Melo, & Simões, 2011; Shi & Zhu, 2009), thus, proving that the cleaning and sanitization protocols tend to be ineffective in adequately removing resilient biofilms (Anand et al., 2012), which raises quality and safety concerns in the final product. Also, chemicals not being environmentally friendly contribute to the generation of nonbiodegradable waste and have a harsh effect on the membrane integrity (Regula et al., 2014). A research study demonstrated the development of resistance by selective isolates within the resilient biofilms over prolonged use of membranes and the potential of such isolates to acquire predominance over the biofilm microflora (Anand & Singh, 2013), further reducing the efficiency of the chemical cleaners and subsequently generating microbial resistance against almost all antimicrobial treatments. Also, these isolates generate additional resistance against the new synthetic derivatives as these have already developed resistance to the parent agents (Sumi et al., 2015), consequently, creating opportunities to develop novel alternative strategies to combat biofilm formation on membrane surfaces.

Biofilms growing on the membranes have been documented to be a problem, resulting in membrane blockage, product contamination, and reduced membrane life due to the microbial activity on the membrane material (Tang et al., 2010). Studies have stated that after cleaning, the thermophilic biofilms remain on the equipment surface, where the cells remain covered by foulant that persists after incomplete cleaning. It was also observed that binary species biofilms may survive sanitation regimes and may constitute reservoirs of contamination of the product resulting in spoilage and/or food safety hazards (Flint et al., 2020). Bacteriophages, lysozymes, bacteriocins, antimicrobial peptides, enzymes, extracellular vesicles, lytic agents, essential oils, QS inhibitors, and surface modifications are promising alternatives (Alonso et al., 2020). The antagonistic reaction depends on a few components, such as strain specific reaction, substrates, and the interaction between species. In common, the opposing measures can offer an elective to the aimless utilize of chemicals, and thus, it has pulled in expanding intrigued from analysts (Alonso et al., 2020).

7.1 Signaling molecules and biofilms (quorum sensing inhibition)

The biofilm uses several cell signaling pathways at high cell concentrations, which is referred to as quorum sensing. QS is a process in which a variety of autoinducers (high concentration chemical and peptide signals, such as homoserine lactones) are employed to activate the genetic expression of alginates' mechanical and enzymatic processors. Alginates are a key component of the extracellular matrix (Garrett et al., 2008).

Initiation, establishment, maintenance, and spread are all stages of biofilm formation. When free living bacterial cells reach a particular density, QS mechanisms enable them to live as multicellular entities, with cell to cell communication formed by the production of tiny signaling molecules. In the growth of biofilms, QS is critical. QS processes are used by pathogenic bacteria in biofilms to activate pathogenicity and establish antibacterial resistance (Saxena, Joshi, Rawat, & Bisht, 2019). These QS signaling molecules are involved in the communication at interspecies, intraspecies, and interkingdom levels. Therefore, impairing the capability of coordinated behavior of mixed species communities and target QS circuits have been evaluated as promising implications to eliminate persistent microorganisms in the form of biofilms (Khan, Pham, Tabassum, Oloketuyi, & Kim, 2020; Yuan et al., 2020).

The strategies to prevent biofilm formation include targeting the QS system that is based on inhibition of cell to cell communication, which can be executed in many ways, including the inhibition of pathway producing QS signaling molecules or the degradation of autoinducers, prevention of signaling peptide receptor binding, or inhibition of membrane transport system, and trapping of signaling molecules in the environment. Several QS signaling systems have been extracted from biological sources or chemically created because of this. However, because QS inhibitors are usually species specific, a mixture of inhibitors is necessary to remove mixed species biofilms (Khan et al., 2020; Mishra et al., 2020; Yuan et al., 2020).

7.2 Antimicrobial peptides

The use of antimicrobial peptides (AMPs) is a novel approach for the inhibition of resilient mixed species biofilms. These peptides are recognized as a promising alternative for the existing methodology as they are less prone to cause bacterial resistance (Mishra et al., 2020; Yuan et al., 2020). AMPs are oligopeptides with various numbers of amino acids obtained from microorganisms, animals, and plants. AMPs that are derived from microorganisms, also known as bacteriocin, have also been shown to be effective against broad spectrum of microorganisms (Singh, Ghosh, & Chauhan, 2021).

Some researchers have demonstrated or provided a proof of concept of the effectiveness of AMPs in the form of cell free extract obtained from one microorganism against resilient biofilms. Also, in vitro studies have demonstrated the elimination of biofilms on different surfaces. The potential of biocontrol of microorganisms producing AMPs has also been seen against the biofilms of pathogens (Flint et al., 2020; Koohestani, Moradi, Tajik, & Badali, 2018; Verma, 2020).

Bacteriocins from LAB have been used to prevent early cell adhesion and biofilm formation. They are considered "generally regarded as safe" (GRAS) and hence can be utilized in food manufacturing, including as food additives. The microorganisms that form biofilms generally communicate through QS signals. The ability of AMPs derived from natural sources to inhibit QS provides an antibiotic free alternative to treating biofilm related issues (Yuan et al., 2020).

7.3 Role of bacteriophages

The use of bacteriophages as an alternative to ineffective traditional treatment procedures for removing tenacious biofilms is another option. Bacteriophages (phages) are viruses that provide a natural, highly specific, and nontoxic approach to combat biofilms. The cell replication mechanism is used by phages to penetrate and multiply within bacterial cells, killing them in the process. In circumstances where antimicrobial compounds fail to destroy resilient cells due to the presence of various resistance mechanisms, phages can be used as bacteriolytic agents. Other viral proteins have been found and classed as antibacterial agents based on their activities in various stages of the viral life cycle, in addition to bacteriophages' overall cell killing effect (Khan et al., 2020; Yuan et al., 2020).

A review article reported some studies on the effectiveness of phages on the reduction of *E. coli* O157:H7 in commercial milk and biofilms. Also, some specific phages were proficient in reducing the biofilm numbers by 3–5 log CFU within 4 h of treatment (Flint et al., 2020). The efficacy of the phage in destroying resilient biofilms is dependent on the susceptibility of the matrix embedded microorganisms and the availability of receptor sites, potentially hidden by a physical barrier, i.e., biofilm matrix. Bacteriophages, particularly lytic phages, with significant host specificity, self propagation at the site of infection, rapid clearance, great diversity, relatively easy isolation for a range of pathogens, and the opportunity to make genetic modifications are some of the most promising biofilm control agents (Singh et al., 2021; Yuan et al., 2020).

7.4 Enzyme mediated disruption

The biofilms comprise the microbial cells and the secreted EPS that constitutes 90% of the biomass. The EPS matrix provides cohesion and adhesion sites for the bacterial cells to attach to the surface, inhibits antimicrobial agents from killing EPS embedded microorganisms, and enhances their resilience on the contact surfaces in the food and dairy industry (Saxena et al., 2019; Yuan et al., 2020). The enzyme mediated disruption solely or in combination with bactericides can be implemented as an alternative approach for effective removal of biofilms due to its potential to degrade the EPS network, hence making the chemicals and sanitizing agents easily penetrate, cell lysis, interference with the cell to cell QS signaling events responsible for biofilm formation and maintenance, catalyze antimicrobial substance formation, and would also help to reduce the use of chemical agents, water consumption, and energy cost (Singh et al., 2021; Yuan et al., 2020).

Anti QS enzymes, oxidative enzymes, polysaccharide degrading enzymes, and proteolytic enzymes are the four types of enzymes of considerable relevance for biofilm eradication. The EPS composition of biofilms, on the other hand, varies widely between bacterial species and is strongly influenced by growth circumstances. One of the components of EPS is eDNA, which acts as a factor in planktonic cell adhesion and aggregation. DNase I is an eDNA degrading enzyme. As a result, the presence of DNase I is linked to decreased bacterial adherence on surfaces, preventing biofilm formation (Das, Sharma, Busscher, Van Der Mei, & Krom, 2010; Kim, Kim, & Oh, 2021; Yuan et al., 2020). Given enzyme substrate specificity and the complex heterogeneous composition of the biofilm matrix, successful biofilm removal usually necessitates a combination of enzymes capable of degrading DNA, polysaccharides, proteins, and QS molecules, thereby broadening the range of actions of an enzymatic cleaner against mixed species biofilms. Because several environmental conditions in the food industry, such as temperature, pH, substrate, food residues, and a variety of food processing surfaces, might interfere with enzyme activity and efficiency, the use of enzymes in biofilm control is currently limited (Yuan et al., 2020).

8. Conclusions

The long filtration runs contribute to the deposition of different types of substrates coming from the feed stream onto the membrane surface leading to biofouling. This is one of the important challenges faced by the dairy industry. The biofilm microflora develops resistance to the cleaning procedures. The selective isolates from the biofilm microflora emerge predominantly over prolonged use of membranes. The emergence of predominance could be associated with several factors. However, the understanding of interaction with the mixed species biofilm and the ability to form resilient biofilm is still lacking. This is very important to implement optimal strategies to eliminate tenacious biofilms in the food and dairy industries.

The release of antimicrobial agents from a biofilm forming isolate due to the emergence of predominance could be used to develop an alternative, which could be natural, ecofriendly, and biodegradable. Being naturally produced from the microbial origin, it provides a promising alternative for a biofilm control strategy. Also, the release of antimicrobial by microbial origin could be linked with quorum signaling. This intervention strategy can be implemented to develop a biosanitizer exhibiting antimicrobial activity using quorum inhibitors for the prevention of biofilm formation on membranes along with other novel strategies to eliminate biofilm, such as bacteriophages, and enzyme mediated disruption.

References

Alonso, V. P. P., Harada, A. M. M., & Kabuki, D. Y. (2020). Competitive and/or cooperative interactions of *Listeria monocytogenes* with *Bacillus cereus* in dual-species biofilm formation. *Frontiers in Microbiology*, *11*(February), 1–10. https://doi.org/10.3389/fmicb.2020.00177.

Alvarez-Ordóñez, A., Coughlan, L. M., Briandet, R., & Cotter, P. D. (2019). Biofilms in food processing environments: Challenges and opportunities. *Annual Review of Food Science and Technology*, *10*(1), 173–195. https://doi.org/10.1146/annurev-food-032818-121805.

Amézquita, A., & Brashears, M. M. (2002). Competitive inhibition of *Listeria monocytogenes* in ready-to-eat meat products by lactic acid bacteria. *Journal of Food Protection*, *65*(2), 316–325. https://doi.org/10.4315/0362-028X-65.2.316.

Anand, S., Hassan, A., & Avadhanula, M. (2012). The effects of biofilms formed on whey reverse osmosis membranes on the microbial quality of the concentrated product. *International Journal of Dairy Technology*, *65*(3), 451–455. https://doi.org/10.1111/j.1471-0307.2012.00848.x.

Anand, S., & Singh, D. (2013). Resistance of the constitutive microflora of biofilms formed on whey reverse-osmosis membranes to individual cleaning steps of a typical clean-in-place protocol. *Journal of Dairy Science*, *96*(10), 6213–6222. https://doi.org/10.3168/jds.2013-7012.

Anand, S., Singh, D., Avadhanula, M., & Marka, S. (2014). Development and control of bacterial biofilms on dairy processing membranes. *Comprehensive Reviews in Food Science and Food Safety*, *13*(1), 18–33. https://doi.org/10.1111/1541-4337.12048.

Araújo, P., Lemos, M., Mergulhão, F., Melo, L., & Simões, M. (2011). Antimicrobial resistance to disinfectants in biofilms. In *2. Science against microbial pathogens: Communicating current research and technological advances* (pp. 826–834). Spain: Formatex Research Center Badajoz.

Arnaouteli, S., Bamford, N. C., Stanley-Wall, N. R., & Kovács, Á. T. (2021). *Bacillus subtilis* biofilm formation and social interactions. *Nature Reviews Microbiology*, *19*(9), 600–614. https://doi.org/10.1038/s41579-021-00540-9.

Avadhanula, M. (2011). *Formation of bacterial biofilms on spiral wound reverse osmosis whey concentration membranes* (MS thesis). Brookings, SD: South Dakota State Univ.

Bowden, G. H. W., & Hamilton, I. R. (1989). Competition between *Streptococcus mutans* and *Lactobacillus casei* in mixed continuous culture. *Oral Microbiology and Immunology*, *4*(2), 57–64. https://doi.org/10.1111/j.1399-302X.1989.tb00100.x.

Bradshaw, D. J., Marsh, P. D., Allison, C., & Schilling, K. M. (1996). Effect of oxygen, inoculum composition and flow rate on development of mixed-culture oral biofilms. *Microbiology*, *142*(3), 623–629. https://doi.org/10.1099/13500872-142-3-623.

Chamberland, J., Lessard, M. H., Doyen, A., Labrie, S., & Pouliot, Y. (2017). A sequencing approach targeting the 16S rRNA gene unravels the biofilm composition of spiral-wound membranes used in the dairy industry. *Dairy Science and Technology*, *96*(6), 827–843. https://doi.org/10.1007/s13594-016-0305-2.

Chamberland, J., Messier, T., Dugat-Bony, E., Lessard, M. H., Labrie, S., Doyen, A., et al. (2019). Influence of feed temperature to biofouling of ultrafiltration membrane during skim milk processing. *International Dairy Journal*, *93*, 99–105. https://doi.org/10.1016/j.idairyj.2019.02.005.

Chinen, K. (2017). *Klebsiella pneumoniae* liver abscess. *Journal of General and Family Medicine*, *18*(6), 466–467. https://doi.org/10.1002/jgf2.90.

Cornforth, D. M., & Foster, K. R. (2013). Competition sensing: The social side of bacterial stress responses. *Nature Reviews Microbiology*, *11*(4), 285–293. https://doi.org/10.1038/nrmicro2977.

D'souza, N. M., & Mawson, A. J. (2005). Membrane cleaning in the dairy industry: A review. *Critical Reviews in Food Science and Nutrition*, *45*(2), 125–134.

Das, T., Sharma, P. K., Busscher, H. J., Van Der Mei, H. C., & Krom, B. P. (2010). Role of extracellular DNA in initial bacterial adhesion and surface aggregation. *Applied and Environmental Microbiology*, *76*(10), 3405–3408. https://doi.org/10.1128/AEM.03119-09.

Fadila, M. (2016). Evaluation of a non-submerged cultivation assay combined to ESEM imaging for analysis of biofilms formed by dairy-associated sporeforming bacteria. *African Journal of Microbiology Research*, *10*(32), 1263–1273. https://doi.org/10.5897/ajmr2016.8203.

Flint, S. H., Bremer, P. J., & Brooks, J. D. (2009). Biofilms in dairy manufacturing plant—Description, current concerns and methods of control. *Biofouling*, *11*, 81–97. https://doi.org/10.1080/08927019709378321.

Flint, S., Bremer, P., Brooks, J., Palmer, J., Sadiq, F. A., Seale, B., et al. (2020). Bacterial fouling in dairy processing. *International Dairy Journal*, *101*, 104593. https://doi.org/10.1016/j.idairyj.2019.104593.

Fredrickson, A. G., & Stephanopoulos, G. (1981). Microbial competition. *Science*, *213*(4511), 972–979.

Friedrich, U., & Lenke, J. (2006). Improved enumeration of lactic acid bacteria in mesophilic dairy starter cultures by using multiplex quantitative real-time PCR and flow cytometry-fluorescence in situ hybridization. *Applied and Environmental Microbiology*, *72*(6), 4163–4171.

Garcia-Fernandez, N. (2016). *Role of exopolymeric substances in biofilm formation on dairy separation membranes*. South Dakota State University.

Garrett, T. R., Bhakoo, M., & Zhang, Z. (2008). Bacterial adhesion and biofilms on surfaces. *Progress in Natural Science*, *18*(9), 1049–1056. https://doi.org/10.1016/j.pnsc.2008.04.001.

Hassan, A. N., Anand, S., & Avadhanula, M. (2010). Microscopic observation of multispecies biofilm of various structures on whey concentration membranes. *Journal of Dairy Science*, *93*(6), 2321–2329. https://doi.org/10.3168/jds.2009-2800.

Hobley, L., Harkins, C., MacPhee, C. E., & Stanley-Wall, N. R. (2015). Giving structure to the biofilm matrix: An overview of individual strategies and emerging common themes. *FEMS Microbiology Reviews*, *39* (5), 649–669. https://doi.org/10.1093/femsre/fuv015.

Husmark, U., & Rönner, U. (1992). The influence of hydrophobic, electrostatic and morphologic properties on the adhesion of Bacillus spores. *Biofouling*, *5*(4), 335–344.

James, B. J., Jing, Y., & Chen, X. D. (2003). Membrane fouling during filtration of milk—a microstructural study. *Journal of Food Engineering*, *60*(4), 431–437.

Khan, F., Pham, D. T. N., Tabassum, N., Oloketuyi, S. F., & Kim, Y. M. (2020). Treatment strategies targeting persister cell formation in bacterial pathogens. *Critical Reviews in Microbiology*, *46*(6), 665–688. https://doi.org/10.1080/1040841X.2020.1822278.

Kim, U., Kim, J.-H., & Oh, S.-W. (2021). Review of multi-species biofilm formation from foodborne pathogens: Multi-species biofilms and removal methodology. *Critical Reviews in Food Science and Nutrition*, 1–11. https://doi.org/10.1080/10408398.2021.1892585.

Koohestani, M., Moradi, M., Tajik, H., & Badali, A. (2018). Effects of cell-free supernatant of *Lactobacillus acidophilus* LA5 and *Lactobacillus casei* 431 against planktonic form and biofilm of Staphylococcus aureus. *Veterinary Research Forum*, *9*(4), 301.

Kumar, C. G., & Anand, S. K. (1998). Significance of microbial biofilms in food industry: A review. *International Journal of Food Microbiology*, *42*(1–2), 9–27.

Marchand, S., De Block, J., De Jonghe, V., Coorevits, A., Heyndrickx, M., & Herman, L. (2012). Biofilm formation in milk production and processing environments; influence on milk quality and safety. *Comprehensive Reviews in Food Science and Food Safety*, *11*(2), 133–147. https://doi.org/10.1111/j.1541-4337.2011.00183.x.

Marka, S., & Anand, S. (2018). Feed substrates influence biofilm formation on reverse osmosis membranes and their cleaning efficiency. *Journal of Dairy Science*, *101*(1), 84–95. https://doi.org/10.3168/jds.2017-13249.

Mishra, R., Panda, A. K., De Mandal, S., Shakeel, M., Bisht, S. S., & Khan, J. (2020). Natural anti-biofilm agents: Strategies to control biofilm-forming pathogens. *Frontiers in Microbiology*, *11*(October). https://doi.org/10.3389/fmicb.2020.566325.

Nadell, C. D., Drescher, K., & Foster, K. R. (2016a). Spatial structure, cooperation and competition in biofilms. *Nature Reviews Microbiology*, *14*(9), 589–600. https://doi.org/10.1038/nrmicro.2016.84.

Nadell, C. D., Drescher, K., & Foster, K. R. (2016b). Spatial structure, cooperation and competition in biofilms. *Nature Reviews Microbiology*, *14*(9), 589.

Regula, C., Carretier, E., Wyart, Y., Gésan-Guiziou, G., Vincent, A., Boudot, D., et al. (2014). Chemical cleaning/disinfection and ageing of organic UF membranes: A review. *Water Research*, *56*, 325–365.

Ripolles-Avila, C., Hascoët, A. S., Martínez-Suárez, J. V., Capita, R., & Rodríguez-Jerez, J. J. (2019). Evaluation of the microbiological contamination of food processing environments through implementing surface sensors in an iberian pork processing plant: An approach towards the control of *Listeria monocytogenes*. *Food Control*, *99*(November 2018), 40–47. https://doi.org/10.1016/j.foodcont.2018.12.013.

Rönner, U., Husmark, U., & Henriksson, A. (1990). Adhesion of Bacillus spores in relation to hydrophobicity. *Journal of Applied Bacteriology*, *69*(4), 550–556.

Saxena, P., Joshi, Y., Rawat, K., & Bisht, R. (2019). Biofilms: Architecture, resistance, quorum sensing and control mechanisms. *Indian Journal of Microbiology*, *59*(1), 3–12. https://doi.org/10.1007/s12088-018-0757-6.

Sayem, S. M. A., Chowdhury, A., Alam, M. Z., & Sarker, P. K. (2018). Antibiofilm activity of crude cell free extract from *Bacillus subtilis* S01 against *E. coli*. *Journal of Scientific Research*, *10*(2), 211–221.

Schluter, J., Nadell, C. D., Bassler, B. L., & Foster, K. R. (2015). Adhesion as a weapon in microbial competition. *ISME Journal*, *9*(1), 139–149. https://doi.org/10.1038/ismej.2014.174.

Sharma, M., & Anand, S. K. (2002). Biofilms evaluation as an essential component of HACCP for food/dairy processing industry—A case. *Food Control*, *13*(6–7), 469–477. https://doi.org/10.1016/S0956-7135(01)00068-8.

Shemesh, M., & Ostrov, I. (2020). Role of *Bacillus* species in biofilm persistence and emerging antibiofilm strategies in the dairy industry. *Journal of the Science of Food and Agriculture*, *100*(6), 2327–2336.

Shi, X., & Zhu, X. (2009). Biofilm formation and food safety in food industries. *Trends in Food Science & Technology*, *20*(9), 407–413.

Simões, M., Simões, L. C., & Vieira, M. J. (2010). A review of current and emergent biofilm control strategies. *LWT - Food Science and Technology*, *43*(4), 573–583. https://doi.org/10.1016/j.lwt.2009.12.008.

Singh, B. P., Ghosh, S., & Chauhan, A. (2021). Development, dynamics and control of antimicrobial-resistant bacterial biofilms: A review. *Environmental Chemistry Letters*, *19*(3), 1983–1993. https://doi.org/10.1007/s10311-020-01169-5.

Souza, N. M. D., & Mawson, A. J. (2007). Membrane cleaning in the dairy industry: A review membrane cleaning in the dairy industry: A review. *Critical Reviews in Food Science and Nutrition*. https://doi.org/10.1080/10408690490911783. 8398.

Stoodley, P., Sauer, K., Davies, D. G., & Costerton, J. W. (2002). Biofilms as complex differentiated communities. *Annual Reviews in Microbiology*, *56*(1), 187–209.

Stratton, C. W., Gregory, W. J., & McNabb, P. C. (1986). Topics in clinical microbiology: *Pseudomonas cepacia*. *Infection Control*, *7*(5), 281–284. https://doi.org/10.1017/s0195941700064213.

Sumi, C. D., Yang, B. W., Yeo, I.-C., & Hahm, Y. T. (2015). Antimicrobial peptides of the genus *Bacillus*: A new era for antibiotics. *Canadian Journal of Microbiology*, *61*(2), 93–103.

Tait, K., & Sutherland, I. W. (2002). Antagonistic interactions amongst bacteriocin-producing enteric bacteria in dual species biofilms. *Journal of Applied Microbiology*, *93*(2), 345–352. https://doi.org/10.1046/j.1365-2672.2002.01692.x.

Tamime, A. Y. (2009). *Cleaning-in-place: Dairy, food and beverage operations. Vol. 13*. John Wiley & Sons.

Tang, X., Flint, S. H., Bennett, R. J., & Brooks, J. D. (2010). The efficacy of different cleaners and sanitisers in cleaning biofilms on UF membranes used in the dairy industry. *Journal of Membrane Science*, *352*(1–2), 71–75. https://doi.org/10.1016/j.memsci.2010.01.063.

Tang, X., Flint, S. H., Brooks, J. D., & Bennett, R. J. (2009). Factors affecting the attachment of micro-organisms isolated from ultrafiltration and reverse osmosis membranes in dairy processing plants. *Journal of Applied Microbiology*, *107*(2), 443–451. https://doi.org/10.1111/j.1365-2672.2009.04214.x.

Van Merode, A. E. J., Pothoven, D. C., Van Der Mei, H. C., Busscher, H. J., & Krom, B. P. (2007). Surface charge influences enterococcal prevalence in mixed-species biofilms. *Journal of Applied Microbiology*, *102* (5), 1254–1260. https://doi.org/10.1111/j.1365-2672.2006.03187.x.

Verma, P. (2020). *A natural antimicrobial from Bacillus subtilis as a biosanitizer for resilient membrane biofilms*. South Dakota State University.

Verma, P., & Anand, S. (2020). Antimicrobial activity as a potential factor influencing the predominance of *Bacillus subtilis* within the constitutive microflora of a whey reverse osmosis membrane biofilm. *Journal of Dairy Science*, *103*(11), 9992–10000. https://doi.org/10.3168/jds.2020-18576.

Verma, P., Singh, N., & Anand, S. (2021). Short communication: A competitive exclusion study reveals the emergence of *Bacillus subtilis* as a predominant constitutive microorganism of a whey reverse osmosis membrane biofilm matrix. *Journal of Dairy Science*, *104*(1), 221–227. https://doi.org/10.3168/jds.2020-18478.

Yannarell, S. M., Grandchamp, G. M., Chen, S. Y., Daniels, K. E., & Shank, E. A. (2019). A dual-species biofilm with emergent mechanical and protective properties. *Journal of Bacteriology*, *201*(18), e00670–e00718.

Yuan, L., Hansen, M. F., Røder, H. L., Wang, N., Burmølle, M., & He, G. (2019). Mixed-species biofilms in the food industry: Current knowledge and novel control strategies. *Critical Reviews in Food Science and Nutrition*, 1–17. https://doi.org/10.1080/10408398.2019.1632790.

Yuan, L., Hansen, M. F., Røder, H. L., Wang, N., Burmølle, M., & He, G. (2020). Mixed-species biofilms in the food industry: Current knowledge and novel control strategies. *Critical Reviews in Food Science and Nutrition*, *60*(13), 2277–2293. https://doi.org/10.1080/10408398.2019.1632790.

Chapter 25

Molecular basis of cariogenic biofilm and infections

Rosina Khan[a], Mohd Adil[a], and Asad U. Khan[b]

[a]*Department of Plant, Food and Environmental Sciences, Faculty of Agriculture, Dalhousie University, Truro, NS, Canada,* [b]*Interdisciplinary Biotechnology Unit, Aligarh Muslim University, Aligarh, Uttar Pradesh, India*

1. Introduction

Dental caries is a multifactorial biofilm-associated disorder, triggered by diet rich in sugar and microbiota-matrix interaction taking place on tooth surfaces, leading to deterioration of tooth tissue. Among the oral streptococci, *Streptococcus mutans* is the most vital cariogenic bacteria that contributes toward dental caries formation (Bowen, Burne, Wu, & Koo, 2018; Chen, Chakraborty, Zou, Burne, & Zeng, 2019; Maier, 2021). Biofilms consist of either single or multiple bacterial species that form structured communities that elicit many human infectious diseases, including dental caries (Table 1). The dental biofilm refers to a group of microorganisms dispersed in plethora of self-produced matrix rich in bacterial cell debris, extracellular DNA, secreted proteins, polysaccharides, and lipids (Lu & Collins, 2007). Biofilms can grow on biotic as well as abiotic surfaces, which may come across as either beneficial or detrimental, as is the case in dental caries. The existence of dental biofilm was first reported over 300 years ago when scientist Antonie Van Leeuwenhoek analyzed the plaque on his own teeth under the microscope (Jakubovics, 2015). As of today, dental biofilms have attracted immense attention globally.

Biofilm, with its unique composition, pose a formidable challenge to the immune system of the host and are highly resistant to treatments propounded by antibiotics. The cells residing within the biofilm are extremely tolerant (almost 1000-fold) to antibiotics than those residing as planktonic cells. The ability to exhibit such strong resistance may be attributed to increased expression of efflux pumps, transmission of antibiotic resistance markers, and acquired resistance (Soleimani, Mobarez, Olia, & Atyabi, 2015). The biofilm provides required nutrients to the bacteria as well as conditions necessary to promote its growth and metabolism. The requirement of energy and source of carbon by the bacteria for growth and reproduction is met through metabolism of organic and inorganic compounds available to it. Within the dental biofilm, fermentable carbohydrates act as the major source of energy for the bacteria. Under anaerobic conditions, *S. mutans* metabolizes glucose to give lactic acid as the final product. The acid production initiates the destruction and demineralization of the tooth enamel removing the mineral materials and thus inducing dental caries (Longbottom et al., 2021).

This chapter summarizes the role of *S. mutans* in the development of cariogenic biofilm, molecular basis of cariogenic biofilm formation by *S. mutans*, as well as the most recent approaches to control and eradicate biofilm-related infections.

2. Dental caries

There are up to 700 types of microbes present in the mouth, all of which thrive in this ideal ecosystem of nutrients, water, appropriate temperature, and a suitable pH. Many of these microbes are benign or even beneficial, but some are pathogenic too (Abranches et al., 2018; Kilian et al., 2016). The pathogenic microbes are known as dental pathogens. These pathogenic microbes are responsible for dental caries or tooth decay (House, Fry, & Brown, 2004; Nicolas & Lavoie, 2011). There are a few species from the genus *Streptococcus* that are known as Mutans Streptococci (MS). They were first discovered by scientist named Clarke who isolated single type of bacteria from tooth decay and named it *Streptococcus mutans* in 1924. *S. mutans* was linked to dental caries by Clark, but other researchers were unable to identify the bacteria, therefore it was not explored further. Later, it was discovered again and linked to tooth decay. It was also discovered that *S. mutans* composed of four genetic groupings and eight different serotypes, each of which was classified as a completely different species. The species that closely resembled Clarke's original isolation was given the name *S. mutans* (Loesche, 1986). About 70%–100% of *S. mutans* strains isolated from the oral cavity have serotype c *S. mutans*. *S. sobrinus* followed lead by

Understanding Microbial Biofilms. https://doi.org/10.1016/B978-0-323-99977-9.00033-8

TABLE 1 A list of common biofilm-forming microorganisms.

Microbe type	Species	References
Gram-positive rods	*Actinomyces odontolyticus, A. naeslundii, A. viscosus, Bifidobacterium dentium, A. israelii, Lactobacillus fermentum, Propionibacterium* spp.	Yadav and Prakash (2017)
Gram-positive cocci	*Streptococcus mutans, S. mitis, S. salivarius, S. sanguis, S. intermedius, S. vestibularis, Staphylococcus aureus, Atopobium* spp., *Peptostreptococcus* spp., *Enterococcus faecalis*	Kolenbrander et al. (2002)
Gram-negative cocci	*Neisseria* spp., *Veillonella parvula*	Marsh and Bowden (2000)
Gram-negative rods	*Escherichia coli, Haemophilus* spp., *Bacteroides denticola, Pseudomonas fluorescence, Fusobacterium necrophorum, F. mortiferum, Klebsiella pneumoniae, Enterobacter aerogenes, Citrobacter freundii, Prevotella* spp., *Leptotrichia* spp.	Aas et al. (2008)
Yeasts	*Candida* species (*C. albicans, C. glabrata, C. tropicalis*)	Koo, Andes, and Krysan (2018)

being the second most prevalent as well as the second most associated with tooth decay.

"The MS are not very good colonisers of the tooth surface," according to Loesche, "but *S. mutans* is one of the first MS to colonize infants immediately after their teeth emerge and was the only MS isolated from caries-affected infants in one research." Our bodies, however, have an inbuilt defense against microorganisms, for instance, the tooth surface, like most bacteria, has a net negative charge, which creates repulsion between the two. This natural barrier is compromised when plaque (biofilm) forms. Dental plaque attaches to the various surfaces of teeth with variable degrees of affinity depending on the morphology of the tooth, creating a sequestered environment in which the bacteria's acid can damage the tooth. When bacteria penetrate the hard enamel, infiltrate the pulp and dentin, and eventually eradicate/destroy the tooth, dental caries (tooth decay) occurs (Khan, Adil, Danishuddin, Verma, & Khan, 2012).

Caries is commonly considered to be preventable due to the relationship between caries and abundance of plaque formed. As a result, a significant portion of the therapeutic focus has been on plaque removal using physical and chemical methods (Urakov, Urakova, & Reshetnikov, 2019). Among the major etiological factors of dental caries, is the consumption of diet rich in fermentable carbohydrate partnered with the presence of cariogenic bacteria in the mouth. Over the time, if these conditions persist, development of carious lesions ensues. Demineralization of tooth surfaces results in the formation dental lesions associated with the indigenous oral flora. Once the balance of the indigenous oral flora shifts to support the growth of potentially pathogenic microorganisms, dental caries progresses, furthering the damage by the ability of these microbes to produce acid by fermenting the carbohydrates. Carbohydrate (such as sucrose) rich diet promotes growth of acidogenic (acid, particularly lactic acid, producing) species that are capable of fermenting carbohydrates (Rosier, De Jager, Zaura, & Krom, 2014). Acid metabolism, promoted by sugar intake, by oral cariogenic species leads to the demineralization of tooth surfaces. During demineralization, the carbonated hydroxyapatite crystal lattice of enamel is dissolved by hydrogen ions (from the acids) subsequently destroying the dentin and eventually penetrates the pulp (Rosier et al., 2014).

In its early stage, demineralization is reversible; however, continued if demineralization continues, it can lead to the development of cavitated lesions on the tooth surface (Beighton et al., 2010). Remineralization of tooth surfaces occurs naturally under healthy condition. Here, calcium and phosphate ions dissolved in saliva diffuse back into the porous tooth surface thereby balancing the damage caused by the demineralization. The cycle of demineralization/remineralization occurs following consumption of a diet rich in fermentable carbohydrates. A balance of these two processes determines whether a lesion would reverse, progress, or stops. Thus, the presence of fermentable carbohydrates promotes an environment that favors survival of acidogenic bacteria that actively participates in acid metabolism leading to a drop in pH of the oral cavity that further advances the demineralization process (Khan et al., 2012).

Plaque is a clump of bacteria and their products that aggregate on the surface of tooth. It is the principal etiological agent in most periodontal diseases (Katancik, Kumarswamy, Branch-Mays, & Califano, 2015). It develops chronologically within the host, from interactions within a specific bacterial species to intra species interactions, that permits the bacteria to first adhere and then colonize the tooth's surface followed by the formation of biofilm (Alejandra & Daniel, 2020). As long as there is balance between the demineralization and the remineralization of the tooth, this biofilm is often harmless, but once this equilibrium is disturbed due to environmental (unhygienic oral conditions and/or consumption of diet rich in fermentable carbohydrates), and bacterial conditions (constant low pH favoring acid producing and acid tolerating bacteria to thrive), caries and periodontal disease may occur. Biofilm formation is a sequential process involving different phases. Initially, bacteria are transported to a given tooth surface through Brownian movement and bacterial chemotaxis, which is promoted by either bacterial mobility or shearing effects of a fluid flowing over the surface (Muhammad et al., 2020). This is followed by interaction between that the bacteria and tooth surface (Carniello, Peterson, van der Mei, & Busscher, 2018). The Van der Waals forces and electrostatic attraction result in the initial adhesion. This is often reflected as the surface-free energy of the material (Kendall & Roberts, 2015). In the next step, specific extracellular proteins, through different types of bonds (hydrogen, ionic, and covalent), attach the bacteria firmly to the surface. In this phase, adhesion is influenced by bacterial affinity for salivary proteins, surface tension, and hydrophobicity (Huang, Li, & Gregory, 2011). During the last phase, different bacteria proliferate and colonize the surface by co-adhesion and coaggregation (Palmer et al., 2017). With the increase in the plaque levels, diversity among the species also evolves further strengthening the plaque and making more pathogenic in nature and much harder to remove (Lemos et al., 2019). The biofilm in this phase is reversible, as the initial attachments are mainly based on electrostatic or physical forces; however, it turns into a more established one when chemical forces become predominant. This enhanced attachment between bacteria and the tooth surface is now considered irreversible (Huang et al., 2011).

The saliva is crucial to the progression of dental caries (Dodds, Roland, Edgar, & Thornhill, 2015). Through the acquired pellicle, the salivary composition becomes the substrate for bacterial settlement (Cleaver, Moazzez, & Carpenter, 2021). Saliva is rich in proteins, carbohydrates, and lipids, which are variables dependent on the host's diet and oral hygiene (Bilbilova, 2020). There are various factors that influence a person's saliva composition. These include genetic, physiological, psychological, as well as environmental factors (Paula et al., 2012). The salivary components have protective nature that depends on the host's conditions (Vila, Rizk, Sultan, & Jabra-Rizk, 2019). Salivary pellicle develops almost immediately after brushing. It is rich in glycoproteins, exogenous DNA (eDNA), acid-rich proteins, mucins, sialic acid, and exoproducts. There is an increase in colonization between cells that is preceded by bacterial adhesion to acquired pellicle through primary colonizers and surface interactions (Bilbilova, 2020; Chawhuaveang et al., 2021; Ribet & Cossart, 2015). As the bacterial colonization and multiplication progresses, the conditions within the pellicle as well as the type of the microbes residing within this layer changes (Aruni, Dou, Mishra, & Fletcher, 2015). The role of the saliva is to provide buffering effect to decrease the dissolution of hydroxyapatite, which is the major component in the teeth, thereby preventing demineralization of teeth and neutralizing bacterial acid (Abou Neel et al., 2016; Lara-Carrillo, Montiel Bastida, Sánchez Pérez, & AlanísTavira, 2010). Salivary antibacterial components act by preventing attachment of bacteria. These include secretory IgA, lysozyme, lactoperoxidase, and lactoferrin (Pedersen & Belstrøm, 2019). Saliva also consists of histatins, Proline-rich proteins, and statherin that further inhibits bacterial adherence. A reduction in these proteins has been reported to exhibit increased caries because their absence promotes firmer attachment followed by bacterial growth (Chawhuaveang et al., 2021). There are also salivary agglutin that removes bacteria via deglutination and prevents bacterial binding and aggregation (Itzek, Chen, Merritt, & Kreth, 2017). Saliva's presence is indispensable for the maintenance of healthy teeth as well as the oral tissues and if there is a severe reduction in saliva production, there would be an expedited deterioration of oral health.

3. *Streptococcus mutans* and the cariogenic biofilm

Many studies have indicated a causative relationship between dental caries and the *S. mutans*. Substantial evidences have demonstrated the increased colonization with the MS is followed by development of caries (Childers et al., 2017; Sharma, Prabhakar, & Gaur, 2012). Serotypes c, e, and f of *S. mutans* and serotypes d and g of *S. sobrinus* have been identified as the primary etiological agents in dental caries initiation and progression (Aas et al., 2008). They are facultatively anaerobic, nonmotile, Gram-positive cocci (Patterson, 1996), isolated routinely from the mouth of adults and children, implying that dental caries is the most widespread infectious disease of humans (Taubman & Nash, 2006). Although other oral bacterial species also produce acids and are cariogenic too, MS possess unique biochemical features making them extremely efficient at developing caries (Chen et al., 2020). Exclusive properties include ability to produce ample amounts of lactic acid, and simultaneously tolerating extremes fluctuation in sugar concentration, ionic strength, and pH (Bowen et al., 2018). While *S. mutans* is associated with the initiation of carious lesions on tooth surfaces, *S. sobrinus* enhances the progression of these lesions (Lemos et al., 2019). Studies have revealed *S. mutans* comprise up to 50% of the plaque microbiota in carious lesions (Berkowitz, 2003). Studies on animal models have further confirmed the role of MS in the development of dental caries (Hamada, 2002; Lemos et al., 2019; Mattos-Graner, Klein, & Smith, 2014).

4. Virulence factors promoting oral biofilm formation

Streptococcus mutans has evolved several ways for colonizing tooth surfaces and establishing itself as a numerically significant member of the dental biofilm. As shown in Fig. 1, *S. mutans* initially adheres to the surfaces of the teeth where the salivary dental pellicle coats tooth surfaces with a fine layer of a mix of proteins, lipids, and glycoproteins from saliva and is the initial stage in biofilm formation (Abebe, 2021). Antigen I/II a well-studied adhesin, mediates attachment of *S. mutans* to the salivary pellicle while in *S. sobrinus* it is achieved by adhesin called Spa A. These adhesins are important for the colonization of the tooth surface by MS. This is followed by the accumulation of *S. mutans* (Conrads et al., 2014; Manzer, Nobbs, & Doran, 2020). Participating components for this process are sucrose, glucosyltransferases (GtfB, C, and D), and glucan-binding proteins (Gbps). Gtfs are enzymes that act on sucrose and synthesize extracellular glucan polymers. While GtfB produces water-insoluble glucans, GtfD produces a water-soluble glucan, and GtfC produces a mixture of both types of glucans (Ren, Chen, Li, & Li, 2016). Due to differences in their unique structures, the glucans play different role in the production of dental biofilm (Bowen & Koo, 2011). *S. mutans* produces large amounts of insoluble glucans, while *S. sobrinus* mainly produces water-soluble glucans. A variety of glucan binding proteins are produced by both *S. mutans* and *S. sobrinus*. These are receptor-like proteins capable of binding exclusively to glucans. Glucan-binding domains are also present on Gtfs that allows them to act as extracellular glucan receptors too. MS adherence is facilitated through Gtfs and Gbps that bind extracellular glucans and promotes sucrose-dependent adherence, which is crucial in biofilm development and the progress of dental caries (Lynch, Fountain, Mazurkiewicz, & Banas, 2007).

S. mutans biofilms form in a coherent way on tooth surfaces. First, surface antigen I/II, along with Gtfs and Gbps expression, mediates initial adhesion and accumulation. This is followed by production of soluble and insoluble glucans. During the concluding stage, considerable *S. mutans* accumulates and in the presence of sucrose they produce copious amounts of lactic acid which is the end-product of anaerobic respiration leading to the demineralization of tooth and the development of dental caries (Ahn, Wen, & Burne, 2007; Lemos et al., 2019). Through these steps, *S. mutans* form an extracellular matrix that provides them protection as well as to the other oral bacterial species from the immune response of the host, antimicrobial agents,

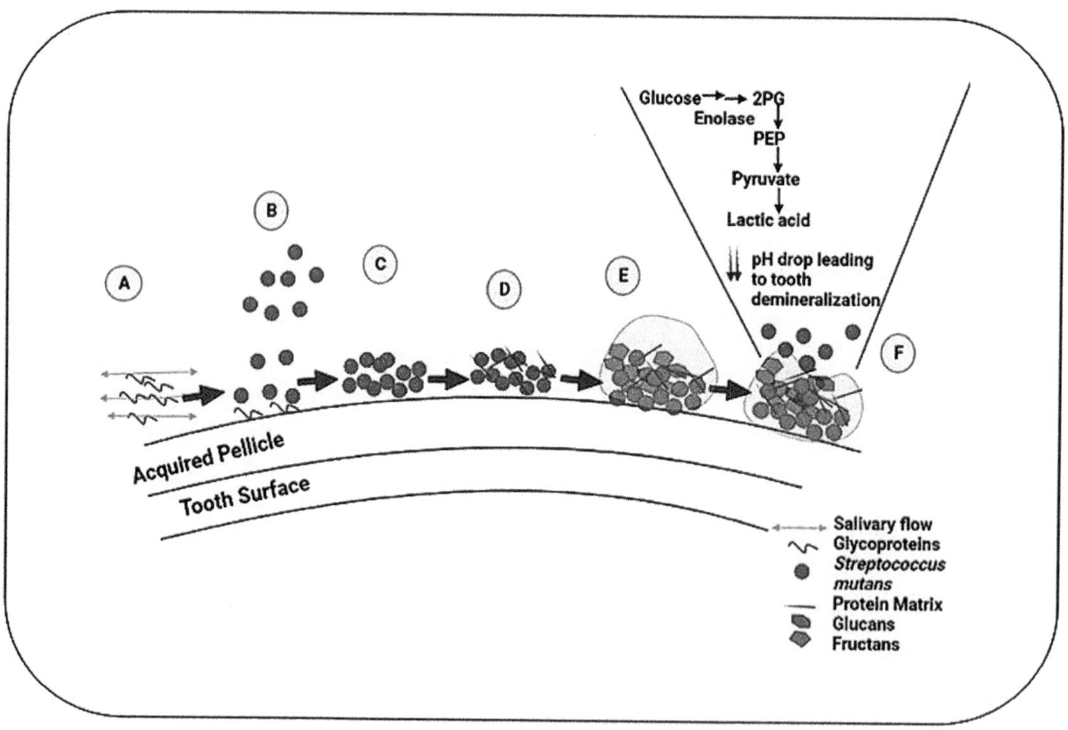

FIG. 1 Formation of *Streptococcus mutans* biofilm. **A**. Adsorption of an organic film, the pellicle, consisting of glycoproteins that acts as receptor ligand on the bacterial cell surface. **B**. Initial attachment on the acquired pellicle by the free floating planktonic bacterial cells. **C**. An irreversible process of bacterial adhesion. **D**. Formation of protein matrix. **E**. Formation of microcolonies within the exopolysaccharides comprising of water, glucans, fructans, proteins, minerals, lipids, etc. **F**. Formation of mature biofilm. Bacteria detach itself from the biofilm and disperse to colonize new sites in the oral cavity. Within the mature biofilm, sucrose is broken down into glucose and fructose. *PG*, phosphoglycerate, *PEP*, phosphoenol pyruvate. The illustration was created with BioRender.

and mechanical stress (Shemesh, Tam, Aharoni, & Steinberg, 2010).

Various studies have elucidated pivotal role of Gtfs and Gbps in caries progression. Introducing defect in the genes encoding Gtfs resulted in decreased amounts of glucans produced by *S. mutans*, diminished efficacy in adherence to tooth surfaces via sucrose-dependent pathway and reduced cariogenicity in animal models (Banas & Vickerman, 2003). Acidogenicity and aciduranc of *S. mutans* are key virulence features that allows these cariogenic bacteria to survive under low pH conditions created by acid formation in the oral cavity but also creating an environment that renders the survival of nonaciduric bacteria very difficult (Andreadis, Topitsoglou, & Kalfas, 2015). This ability is attributed to the presence of ATPase proton pumps that actively transports hydrogen ions (H^+) from the cytoplasm to the exteriors of the bacterial cell (Conrads et al., 2014; Matsui & Cvitkovitch, 2010). Overall, diet rich in sucrose shifts the microbial ecology toward conditions favoring acid-producing and acid-tolerant species that are associated with dental caries such as MS and Lactobacilli species (Bowen et al., 2018). As a result, the dental biofilm linked to caries differs from a healthy one in microbial ecology (Huang et al., 2011).

5. Survival of *Streptococcus mutans* in acidified cariogenic biofilm

In the presence of generous amounts of acids produced by *S. mutans*, there is a tremendous challenge that the bacteria must overcome to survive in the acidic environment created within the biofilm. This is where the aciduric nature of *S. mutans* comes into action that provides these bacteria, a significant competitive edge against other competing microbes present with in the oral cavity (Bowen & Koo, 2011; Lemos & Burne, 2008). The fall in the pH induces many transcriptional and physiological changes as an attempt to prevent any acid associated damage to *S. mutans*' sensitive molecules such as its DNA and enzymes (Matsui & Cvitkovitch, 2010). Acid tolerance response (ATR) is a mechanism paramount to *S. mutans*' sustenance under acidic stress conditions of the cariogenic biofilm and acts by altering the composition of the cell membrane and increasing protons extrusion from the cytoplasm to the exteriors of the bacterial cell (Baker, Faustoferri, & Quivey, 2017; Boisen, Davies, & Neilands, 2021). ATR helps the *S. mutans* achieve pH homeostasis by accomplishing the following:

- *Change in the proton permeability*. ATR triggered upon drop in pH due to the conversion of sugars to organic acid leads to the induction of F_1F_0-ATPases which hydrolyzes ATP to pump protons (H^+) actively to the exterior of the cytosol. It is one on the primary H^+ extrusion mechanism in *S. mutans* (MacGilvray, Lapek, Friedman, & Quivey, 2012; Sheng & Marquis, 2006).
- *Generation of neutralizing molecules*. In an attempt to maintain alkalinity of the cytoplasm, *S. mutans* utilizes the Agmatine deiminase system (AgDS) that catabolizes agmatine, a decarboxylated arginine, to Putrescine, ammonia (NH_3), carbon dioxide (CO_2), and ATP. While NH_3 enhances buffering in the cytoplasm, CO_2 is acted upon by carbonic anhydrase giving rise to bicarbonate ions (HCO_3^-) that further neutralizes the cytoplasm. ATP produced fuels the extrusion carried out by F_1F_0-ATPases (Liu & Burne, 2009).
- *Change in the cell membrane's fatty acid composition*. Upon detecting drop in the surrounding pH, the *S. mutans* cellular membrane shifts to a composition that consists of an increased percentage of monounsaturated fatty acids (MUFAs) (Quivey, Faustoferri, Monahan, & Marquis, 2000). Studies have shown acute defects in the ability by *S. mutans* to tolerate acid challenge and maintain pH homeostasis when the bacteria's biosynthetic machinery, that participates in transition to MUFA-dominant membrane was blocked (Fozo, Scott-Anne, Koo, & Quivey, 2007). The ability to shift its membrane fatty acid profile to long chained (for e.g., $C_{18:1}$ and $C_{20:1}$), monounsaturated profile (at pH 5) from a short-chained (for e.g., $C_{14:0}$ and $C_{16:0}$) saturated profile (at pH 7) implies that the modification is one of the constituents of the acid tolerance response in *S. mutans*. The concomitant reduction in the amount of saturated fatty acids and increase the proportion of as the environmental pH decreased is also survival component of the organism (Quivey, Kuhnert, & Hahn, 2001).
- *Activation of Malolactic fermentation (MLF)*. ATR induces cytoplasmic malic acid conversion to lactate by malolactic fermentation. This helps reduce the cytoplasmic pH as lactate is less acidic than malate. The reaction also generates ATP and CO_2 that further contributes toward attaining pH homeostasis as explained above (Griswold, Chen, & Burne, 2004).
- *Branched chain amino acid biosynthesis (BCAA)*. Through this pathway, *S. mutans* is capable of rerouting carbon away from pyruvate, thereby limiting its availability to fuel the carbon metabolism whose end products are acidic in nature. Furthermore, the reaction catalyzed BCAA aminotransferase produces NH_3, an alkaline molecule (Santiago, MacGilvray, Faustoferri, & Quivey, 2012; Santiago, Marek, Faustoferri, & Quivey, 2013).

With the advancement in molecular and other scientific techniques, increasing number of pathways are getting included among those influenced by induction of acid tolerance response. These include potassium transport (influx) and copper transport (efflux) machinery and alteration in sugar import and glycolytic end-products (Gong et al., 2009; Len, Harty, & Jacques, 2004; Zeng & Burne, 2013). In recent reports, it has been demonstrated that various other regions of the *S. mutans* genome play crucial role in acid tolerance. This includes several previously unidentified pathways as well as those that were not associated with acid tolerance in the past (Baker et al., 2015). A study based on genome-wide deletion library consisting of over thousand single-gene deletion mutants in *S. mutans* UA159, representing almost 60% of the genome revealed that ATR encompasses up to at least 20% of *S. mutans* genome (Quivey et al., 2015).

6. Biofilm properties contributing to antibiotic resistance

S. mutans within the biofilm differs from its planktonic counterpart in many ways that empowers it to become several folds more tolerant to antibiotics (Sharma, Misba, & Khan, 2019). The traditional use of fluoride as a means to prevent dental caries has led to development of resistance in bacterial strains. *S. mutans* has shown varied extent of reduced susceptibility toward antibiotics such as penicillin, clindamycin, erythromycin, and amoxicillin. The antimicrobial resistance in *S. mutans* can be attributed to biofilms that make these agents less successful at killing the bacteria present in established biofilm. The resistance contributing components include efflux pumps and modulation of expression of the cell membrane proteins. *S. mutans* remove antimicrobial agents and metabolites and other toxic substances by using efflux as a key mechanism of cellular responses to variations in their external environments in order to regulate their internal environments. These pumps play a vital role in multidrug resistance (Baugh, Phillips, Ekanayaka, Piddock, & Webber, 2014; Gupta, Singh, & Khan, 2017). They have also been described to be clinically important in exopolysaccharide (EPS) synthesis in several pathogens (Flemming & Wingender, 2010).

Researchers have also shown that an efflux pump, rcrRPQ, in *S. mutans* plays an indispensable role in stress tolerance and physiologic homeostasis by linking metabolism of highly phosphorylated guanosine, (p)ppGpp, genetic competence, and tolerance toward acid and oxidative stress (Ahn, Kaspar, Kim, Seaton, & Burne, 2014). A study conducted by Nilsson et al. (2016) studied the influence of dtaA mutant on biofilm formation and *Streptococcus mutans* susceptibility toward antibiotic gentamicin (positively charged aminoglycoside). The study revealed that the dltA mutant had intact biofilm; however, it was about 8 × less tolerant to gentamicin compared to wild type when growing as a biofilm. In many gram-positive bacteria, dltABCD operon is found to participate in D-alanylation of teichoic acid (Neuhaus & Baddiley, 2003). This operon is more highly expressed in *S. mutans* biofilms than in the planktonic cells (Mazda et al., 2012). The incapability of the mutant cells to add D-alanine to the cell membrane teichoic acid renders them more negatively charged than the wild type. The enhanced negative charge promotes the internalization of the positively charged gentamicin resulting in enhanced sensitivity toward this antibiotic (Nilsson et al., 2016).

7. Genetic regulation of *S. mutans* biofilm formation

7.1 Two-component signal transduction systems (TCS/TCSTS)

TCS is a signal transduction system that helps bacteria to adapt, survive, and exhibit virulence based on the variation in the environment that it senses and causes specific sets of genes to be expressed differentially thereby initiating coordinated responses to the environmental stimuli. There are 13 TCS reported in *S. mutans* (Hirakawa, Kurushima, Hashimoto, & Tomita, 2020). They are involved in the regulation of multiple cellular functions such as stress response, biofilm formation, competence, metabolism, other virulence traits, and antibiotic resistance. As depicted in Fig. 2, TCS generally comprises of a histidine kinase (HK), which is a transmembrane sensor and a cytoplasmic response regulator (RR) that links to the HK. They are encoded by genes located either adjacently within the same operon or as stand-alone genes coding for HKs or RRs.

HK and an RR can in some cases may exist as "hybrid" HK protein that may be found merged in the same polypeptide. When a specific environmental signal or stimulus is recognized by HK protein, its conserved histidine (H) residue gets auto phosphorylated, followed by aspartate residue of the corresponding RR getting phosphorylated. Proceeding to this step, RRs bind directly to the DNA and modulate specific genes expression resulting in a physiological response to the perceived signals (Hirakawa et al., 2020). Activation or repression of genes may occur as a result of the modulation, which leads to formation of products that are specifically utilized to respond to the given input signal. These TCS proteins have conserved functional and structural domains that are may be used for their classification. They present themselves as prospective targets for new antibacterial drugs development because such TCS proteins are absent in mammalian genomes (Galperin, Makarova, Wolf, & Koonin, 2018). We will be looking at TCS systems mainly associated with biofilm formation along with other virulence traits in *S. mutans* (Table 2).

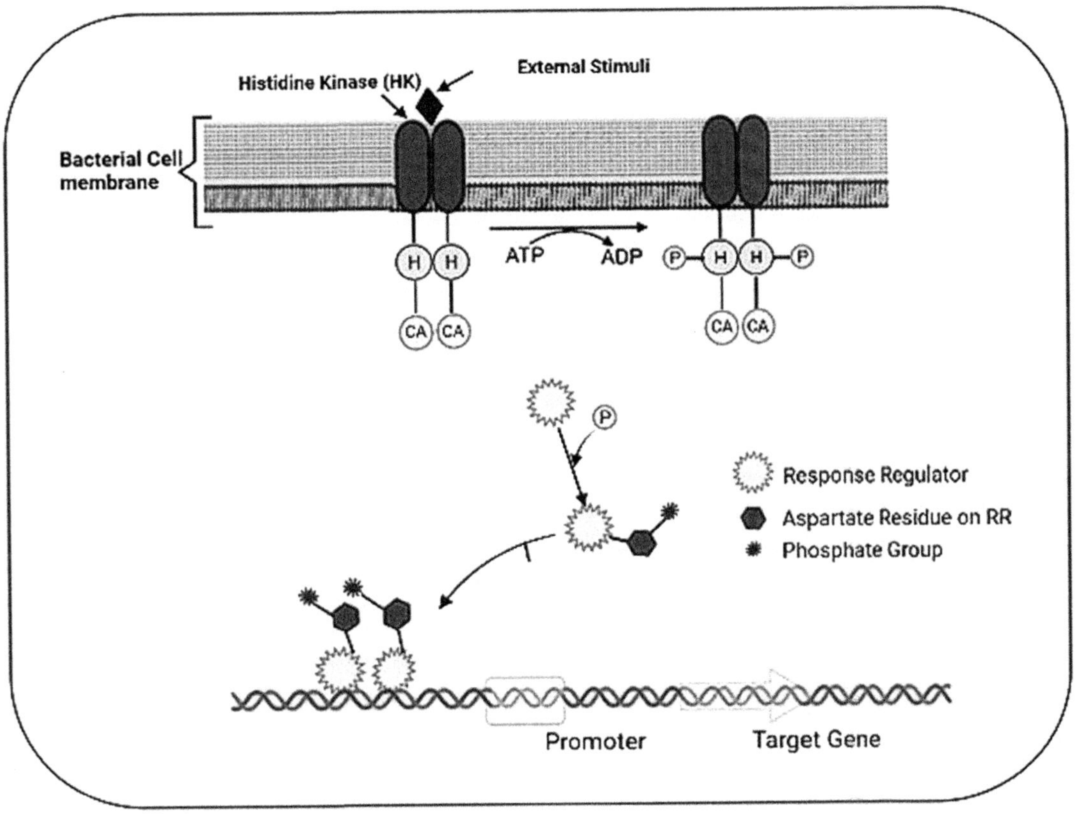

FIG. 2 Example of a typical two-component regulatory system transduction system (TCSTS) in *S. mutans*. Once an environmental stimulus is detected by the membrane-bound homo-dimeric histidine kinase (HK), its histidine residue (H) within the cytoplasmic region is auto-phosphorylated (P). It is followed by phosphate group transfer to the aspartate residue that is conserved in the receiving domain of the associated response regulator. The ATP-dependent kinase activity is associated with the catalytic domain (CA) of the HK. Homodimer consisting of phosphorylated response regulator pair is formed, which is followed by helix-turn-helix domain (HTH) of the RR binding to specific DNA sequences on or close to the target genes promoter. The illustration was created with BioRender.

7.1.1 VicRK

The VicRK TCS system believed to be the only TCS essential in *S. mutans* (Senadheera et al., 2009) has been discovered to regulate sucrose-dependent biofilm formation (Lee, Delaney, & Elkhateeb, 2004). Senadheera et al. (2005) demonstrated that VicK-deficient mutants showed defective *S. mutans* daughter cells separation as well as weak sucrose-dependent biofilm formation. It is also reported to influence acid and oxidative stress responses and competence (Senadheera et al., 2007, 2009). It has been projected in several studies that phosphotransfer crosstalk (a process where response regulators get phosphorylated by noncognate histidine kinases) between the TCS VicRK and the LiaFSR contributes to mutacin production, cell surface adhesion, and the ability to withstand environmental stresses such as damage to the cell envelope and tolerance toward heat shock (Shankar, Mohapatra, Biswas, & Biswas, 2015). In addition, VicRK in collaboration with the CovR is implicated in the synthesis of and interaction with extracellular polysaccharides by directly regulating the genes involved in these processes (Biswas & Biswas, 2006). For instance, *gbpC*, *gtfB*, and *gtfC* are repressed by CovR, while positive regulation of *gbpB* transcription is carried out by VicR (Duque et al., 2011; Stipp et al., 2013). There is substantial evidence accumulating over many that have clearly demonstrated that via its production of exopolysaccharides (EPS) and acids, *S. mutans* contributes significantly to dental caries thereby influencing the plaque microbiome.

7.1.2 CiaRH

He et al. (2008) reported that in *S. mutans*, the *ciaRH* system consists of three genes *ciaR*, *ciaH*, and *ciaX* where the CiaX contains a calcium-binding domain and the *ciaXRH* operon expression is repressed by calcium through CiaX. The *ciaH* is a histidine kinase sensor protein while *ciaR* gene includes its cognate response regulator. The inactivation of ciaH disturbs cellular functions ranging from absence of mutacin production, diminishing development of competence, sucrose-dependent biofilm formation being altered, to significantly retarding acid tolerance, but inactivation of *ciaR* did not reveal similar impact or any conspicuous change in

TABLE 2 TCS cluster and protein involved in biofilm formation and other functions in *Streptococcus mutans*.

TCS cluster/ protein	Functions	References
HK-VicK and RR-VicR (**TCS-1**)	Biofilm and competence development, acid and oxidative stress tolerance, autolysin production, glucan and fructan metabolism.	Stipp et al. (2013), Tremblay, Lo, Li, Halperin, and Lee (2009), Senadheera et al. (2005, 2007, 2009)
HK-CiaH and RR-CiaR (**TCS-2**)	Development of competence and sucrose-dependent biofilm formation, induction of multiple stress response, bacteriocin production.	Qi, Merritt, Lux, and Shi (2004), Mazda et al. (2012), He et al. (2008), Levesque et al. (2007)
HK-LytS and RR-LytT (**TCS-10**)	Increased biofilm formation, autolysis, oxidative stress tolerance, fructan metabolism.	Castillo Pedraza et al. (2017, 2019), Klein, Hwang, Santos, Campanella, and Koo (2015), Liao et al. (2014), Ahn, Rice, Oleas, Bayles, and Burne (2010)
HK-LiaS and RR-LiaR (**TCS-11**)	Enhance biofilm formation (sucrose-dependent adherence), induction of cell envelope stress response, acid tolerance, and bacteriocin resistance, bacteriocin production.	Suntharalingam, Senadheera, Mair, Lévesque, and Cvitkovitch (2009), Chong, Drake, and Biswas (2008), Banerjee and Biswas (2008), Lemos, Abranches, and Burne (2005)
HK-ComD and RR-ComE (**TCS-13**)	Development of biofilm formation and competence (quorum sensing), and bacteriocin production.	Khan et al. (2016), Merritt and Qi (2012), Li, Tian, Layton, Norgaard, and Sisson (2008), Havarstein (2010)

phenotype (Qi et al., 2004). However, it was observed that the *ciaR*, or *ciaX* inactivation resulted in diminished biofilm formation, while biofilm with very short chains and reduced biomass was obtained when there was deletion of CiaH, which implies that CiaH participated in the *S. mutans* cell growth and/or cell division regulation (He et al., 2008). In addition, ciaRH was found to impact teichoic acid biosynthesis operon *dlt* regulation that aids the bacteria to resist antimicrobial peptides (AMPs) in biofilm cells, as reported by Mazda et al. (2012). The Cia R/H system regulates genetic competence and stress response in coordination with Com CDE quorum-sensing system. Undoubtedly, CiaRH inactivation may lead to deterioration of the biofilm-forming capacity in streptococci; thus, presenting itself as a promising target against *S. mutans*' cariogenic biofilm-forming ability.

7.1.3 LytST

Sucrose and starch induce the expression of the TCS, *lytST* that increases extracellular DNA (eDNA) release within the extracellular matrix (ECM) during *S. mutans* biofilm formation (Castillo Pedraza et al., 2017; Klein et al., 2015). In *S. mutans* biofilms, *lytST* is associated with cell lysis that contributes toward the creation of an ECM that is bulky and sturdy (Florez Salamanca & Klein, 2018). The eDNA released interacts with GtfB thereby influencing EPS production and *S. mutans* colonization (Liao et al., 2014). A study revealed that a reduction in *lytST* gene expression results in a decrease in eDNA presence in the matrix during the development process and the pathogenicity of biofilms (Klein et al., 2010). However, in an in vitro study, contradicting results were obtained by Castillo Pedraza et al. (2017).They found that △*lytS*, △*lytT*, deletion strains for *lytST* contributed to extracellular eDNA abundance in the biofilm's matrix, compared to UA159, the parental *S. mutans* strain. This led to an increased amount of soluble and insoluble exopolysaccharides, implying that these biofilms could be more cariogenic and therefore possessed higher potential to cause more damage associated with caries. The study suggested that the presence of *lytT* in the Δ*lytS* or *lytS* in the Δ*lytT* strain, might have assisted *S. mutans* to successfully release eDNA or there might exist other factors influencing the amount of eDNA present in these deletion strains. On conducting in vivo experiment using *ΔlytS* and *ΔlytT* deletion strains, in a rodent model of dental caries (rats), to investigate their role in virulence, these strains that yielded more eDNA in the matrix were expected to cause more caries lesions; but the results were not in line with the expectations. The explanation presented was that the cell wall in *ΔlytS* and *ΔlytT* deletion strains might have modified the physiology to such an extent that it might have led to the changes in composition of the cell wall, the overall charge, and the cell wall turnover. Since the cell wall acts as an "anchor" that maintains cell to cell binding, cell to matrix, and also the cell-matrix-dental surface binding, changes in this anchor led to a fragile, and more permeable 3D structure, incapable of retaining acid within the biofilm and at the biofilm/dental surface interface, resulting in reduced caries lesions. Furthermore, fewer cells were found to aggregate with each other and to the matrix, i.e., a lower percentage of *S. mutans* recovered from rats' dental biofilms (Castillo Pedraza et al., 2019).

7.1.4 LiaSR

Indeed *S. mutans* has an advanced biofilm lifestyle for their survival in the dental plaque created on the surface of the tooth (Hamada & Slade, 1980). In order to maintain its presence in the oral cavity, this bacterium has developed several mechanisms to withstand drastic environmental changes (Ahn, Wen, & Burnem, 2006; Carlsson & Hamilton, 1994; Lemos et al., 2005). In *S. mutans*, cellular responses that the LiaSR two-component signal transduction system regulates include biofilm formation and those that induce cell envelope damages. The expression of *gbpC* gene, which encodes a glucan binding protein essential for biofilm-formation and cariogenicity, is negatively regulated by the *LiaS* from the LiaSR TCS (Biswas, Drake, Erkina, & Biswas, 2008; Chong et al., 2008). The regulation and secretion of the bacteriocin, mutacin, is also under the influence of *LiaS*. In the dental plaque, the growth of other competitor bacteria is suppressed and hindered by mutacin (Chong et al., 2008). *LiaS* mutations in *S. mutans* leads to abnormal biofilms production and a phenotype that is acid-sensitive, according to a study conducted by Li, Lau, et al. (2002). Expression profiling of multiple genes deletion strain of *S. mutans* UA159 under biofilm formation conditions was carried out by Banerjee and Biswas (2008), which revealed the involvement of over hundred and fifty genes that were affected and possibly regulated by *LiaR* either directly or indirectly. Thus, LiaSR appears to be critical for *S. mutans* pathogenesis in the oral cavity, as it impacts cell wall damaging stresses, tolerance against antibiotics, and acid during growth of this bacteria. They also found that *LiaR* inactivation did not affect any of the virulence factors' expression implying that either *LiaS* inactivates *LiaR* function or it might be involved in crosstalk with other TCS (Banerjee & Biswas, 2008).

7.2 Quorum sensing (QS)

Quorum sensing involves cell-cell biofilm community's communication to regulate nutrient competition and population. Several steps in the biofilm formation and maturation are believed to be mediated by bacterial communication (Kolenbrander, Palmer, Periasamy, & Jakubovics, 2010). The genetic components of the *S. mutans* quorum-sensing system were acknowledged due to realization of the fact that the genetic transformability of cells produced from biofilms was 10×–600× higher as compared to the cells derived from planktonic cultures. Through QS, various activities are coordinated including but not limited to bacterial aggregation, biofilm formation, motility, horizontal gene transfer, as well as antibiotic resistance and virulence (Whiteley, Diggle, & Greenberg, 2017). It also contributes to the regulation of bacterial virulence gene expression within biofilms (Bikash, Hamry, & Tal-Gan, 2018). Initially, it was discovered that the quorum sensing system responds to certain environmental pressures including cell density, via a competence stimulating peptide or CSP encoded by *comCDE* that functions as a regulatory system and that the peptide signal molecule is controlled by the *comC* gene. The *comCDE* genes work collectively to produce and respond to CSP and are located on the same locus (Li, Tang, et al., 2002; Perry, Cvitkovitch, & Lévesque, 2009). The precursor CSP is encoded by comC, while the histidine kinase (ComD) is encoded by *comDE* gene and its associated response regulator (ComE) (Li, Lau, Lee, Ellen, & Cvitkovitch, 2001). *cslAB* (*comAB*) and *comCDE* are upregulated by the CSP-mediated QS system in *S. mutans*, which is responsible for displaying the phenotypic traits (Fuqua, Parsek, & Greenberg, 2001). CSP-specific maturation and secretion complex components with an ATP-binding cassette (ABC) transporter (ComA) and an auxiliary factor are also encoded by the genes *cslA* and *cslB* (ComB). Upon reaching a threshold value of CSP concentration, the QS signaling network works at its best capacity. All this while, it relies completely and actively on the biofilm lifestyle. A study conducted by Li et al. in 2002 reported that mutant strains of *S. mutans* for *comC*, *comD*, and *comE* were capable of initiating biofilm formation, however, the biofilm formed, when as compared to wild-type biofilm, they lacked the architectural integrity of the biofilm. Furthermore, the mutants for *comD* and *ComE*-deficient mutants produced biofilms that had reduced biomass. Thus, it was apparent that *comCDE* was actively participating in biofilms formation in addition to competence development (Li et al., 2001).

S. mutans has evolved different mechanisms to cooperate or compete with other oral bacteria such as *S. mitis*, *S. oralis*, and *S. salivarius* that may reside within the dental biofilm. There are a number of bacteriocins produced by *S. mutans* and the major ones include mutacins I–IV, K8, and Smb. They exhibit antibacterial activity against several bacterial species. The expression of mutacin IV is a key factor for the quorum sensing system as well as responsible for genetic competence. Other mutacins regulated by ComDE are mutacins V and VI (Merritt & Qi, 2012). In 2010, Mashburn-Warren et al. discovered XIP in *S. mutans*, a comX/sigX inducing peptide that was encoded by comS gene. It was found to bind to an intracellular Rgg (regulator gene of glycosyltransferase) transcriptional regulator, ComR, to control competence (Mashburn-Warren, Morrison, & Federle, 2010). By activating transcription of ComX (an alternative sigma factor), competence is regulated in this species. The ComRS system is believed to participate in *S. mutans* competence induction (Havarstein, 2010). Transformation is tightly linked with

ComCDE and ComRS, and the alternative sigma factor controls a core transformation regulon. It is worth noticing that there are other genes that are regulated, despite not being directly engaged in the transformation process (Khan et al., 2016). Attempts to control biofilm formation by *S. mutans* could potentially be attenuated by targeting this cell-cell signaling mechanism controlled by quorum sensing.

The LuxS-mediated QS are well characterized to regulate multiple traits crucial to establish pathogenesis of *S. mutans*. The bacteria's ability to produce autoinducer-2 (AI-2) can be rendered inactive by targeting the *luxS* gene. In a diverse panel of bacteria, including oral streptococci, AI-2 has been shown to regulate various communicating activities among cells (Pereira, Thompson, & Xavier, 2013). Its role in interspecies communication within the dental plaque was reported by Wang, Li, and Ling (2017). The communication was shown to take place via AI-2 communication between *S. mutans* and *S. gordonii*, which was crucial for the formation of dual biofilm. It was also discovered that inactivation of the *luxS* gene (no AI-2 production) led to an increased sensitivity of the bacteria to chlorhexidine. Another study revealed that mutation of the *LuxS* gene led to an altered biofilm structure (Yoshida, Ansai, Takehara, & Kuramitsu, 2005). There is accumulating evidence indicating that *luxS* may operate as a global regulator, since *luxS* gene is highly conserved among the Gram-positive and Gram-negative bacteria and therefore presents itself as an essential factor for a drug target (Perry, Jones, Peterson, Cvitkovitch, & Levesque, 2009; Qi et al., 2005; Schauder & Bassler, 2001).

8. Treatment against dental biofilm

8.1 Phytochemical and bioactive compounds

Phytochemicals possess potential to be used as alternatives to the antibiotics used for the treatment of dental biofilm at present (Adil, Khan, & Rupasinghe, 2018). The phytochemicals present their action either by exhibiting bactericidal impact or through bacteriostatic effects on caries causing agents. Their mode of action includes prevention of bacterial adherence to the teeth surface, reduction of biofilm formation, inhibition of glycolytic enzymes, cell surface hydrophobicity, and pH drop. Table 3 presents some of the plants (or bioactive compounds) that have been reported to impact virulence in *S. mutans*, including their ability to form biofilm.

8.2 Nanoparticles as antibiofilm agents

Several metallic nanoparticles have been reported to have antibacterial and antibiofilm properties. These include gold, silver, and copper nanoparticles. They have reduced risk of developing resistance compared to their bulk counterpart (Jones, Ray, Ranjit, & Manna, 2008). These nanoparticles present notable antibiofilm effects of destroying biofilm matrices by disturbing the intermolecular forces (Ansari, Khan, Khan, Cameotra, & Alzohairy, 2015). A study on zinc oxide nanoparticles revealed that it proficiently inhibited the biofilm growth after incubating for 96 h. Nanoparticles composed of zinc doped copper oxide were developed sonochemically, generated ROS that were capable of inhibiting biofilm formation by *S. mutans* on artificial teeth (Eshed, Lellouche, Gedanken, & Banin, 2014). Silver nanoparticles (AgNPs) implied their use for the treatment of dental caries, after being reported to kill *S. mutans* isolated from clinical samples (Pérez-Díaz et al., 2015). In another study, it was reported that AgNPs were capable of inhibiting growth of established *S. mutans* biofilms (Martinez-Gutierrez et al., 2013).

8.3 Plant-based chemicals in conjunction with nanotechnology

Several plant-based compounds have been used in combination with nanotechnology approach to disrupt the biofilm or prevent its formation. Nano emulsion formulation for *Curcuma xanthorrhiza* oil (Xan) had shown prospective to be used as an antibiofilm agent by effectively inhibiting *S. mutans*' ability to form biofilm and demonstrated lower cell toxicity when compared to the commercially available mouthwash Listerine (Cho, Kang, Lee, & Kim, 2020). Qayyum, Oves, and Khan (2017) in their study utilized *Mangifera indica* inflorescence aqueous extract to obtain green silver nanoparticles (AgNPs). It was reported that the green AgNPs inhibited 75% biofilms of *S. mutans* while exhibiting no toxicity toward HeLa cell line viability. A study conducted by Araújo et al. (2020), reported effective bactericidal activity and the potential antibiofilm effect of zein nanoparticles comprising anacardic acid. The activity of anacardic acid to inhibit bacterial proliferation was attributed to its ability to cross the bilipid layer of cell membranes, resulting in its disruption. The biodegradable nanoparticles from corn protein zein provides stabilization of anacardic acid and enhances its antibacterial characteristic.

8.4 Vaccines

One of the caries prevention methods is the development of caries vaccine that has been recently explored for combating dental caries. These immunobiological substances may provide targeted immunity against caries. Vaccine act by boosting protective antibodies protection and activation other immune responses. They are prepared by using organisms (inactivated, live, or dead), isolated cellular fractions, toxoids, or their mixture (Jia, Yan, & Guo, 2020).

TABLE 3 Plants and plant-based compounds exhibiting *Streptococcus mutans* virulence inhibitory activity.

Plant/bioactive compound	Virulence property hindered	References
Cranberry (*Vaccinium macrocarpon* Aiton)	Inhibition of biofilm formation, reduction in hydrophobicity, F-ATPase activity inhibition and Inhibition of GTF B and C activities	Gregoire, Singh, Vorsa, and Koo (2007), Duarte et al. (2006)
Cocoa (*Theobroma cacao* L.)	Biofilm biomass reduction	Percival, Devine, Duggal, Chartron, and Marsh (2006)
Caraway seeds (*Trachyspermum ammi*)	Inhibition of genes involved in biofilm formation, reduction in hydrophobicity	Khan et al. (2012)
Red grape (*Vitis vinifera* L.)	Inhibition of adhesion	Furiga, Lonvaud-Funel, Dorignac, and Badet (2008)
Citral and Phloretin	Suppression of cell surface hydrophobicity, quorum sensing and biofilm formation genes	Adil, Baig, and Rupasinghe (2019)
Clove (*Syzygium aromaticum* L)	Inhibition of adhesion and reduction in hydrophobicity	Abd Rahim and Khan (2006)
Quercetin-3-*O*-alpha-L-arabinopyranoside (guaijaverin)	Reduction in hydrophobicity	Prabu, Gnanamani, and Sadulla (2006)
Ursolic acid	Adherence inhibition to tooth surface	Zhou et al. (2013)
Eugenol (*Syzygium aromaticum*)	Biofilm genes inhibition	Adil, Singh, Verma, and Khan (2014)
Methuselah's beard (*Usnea longissima* Ach.)	Glycolytic enzymes inhibition	Singh et al. (2015)
Green tea (*Camellia sinensis* L.)	Inhibition of glycolytic pH-drop	Xu, Zhou, and Wu (2011)
Indian gooseberry (*Emblica Officinalis* L.)	Inhibition of biofilm formation, adherence properties, reduction in glucan synthesis	Hasan et al. (2012)
Neem (*Azadirachta indica* A. Juss.)	Reduction in bacterial aggregation	Wolinsky, Mania, Nachnani, and Ling (1996)
Purple mangosteen (*Garcinia mangostana* L.)	Glycolytic enzymes inhibition	Nguyen and Marquis (2011)
Viburnum foetens	Inhibition of biofilm, glycolytic enzymes inhibition	Khan et al. (2012)

As is already well known, glucosyltransferases, AgI/II, and glucan-binding proteins from *S. mutans* have a role that is undeniable and crucial to the formation of cariogenic biofilm. In attempts to develop vaccine against caries, these constituents of *S. mutans* have gained major attention to be used as vaccine antigen candidates, although other components of the bacteria have also been studied (Russell, Childers, Michalek, Smith, & Taubman, 2004). pGJA-P/VAX a new fusion anticaries DNA vaccine, encoded two major antigenic domains of *S. mutans*, PAc and GLU, succeeded in repressing the ability of *S. mutans* to cause caries (Giasuddin, Huda, Jhuma, & Haq, 2017). Human experiments involving passive immunization against dental plaque that used mouth rinse containing egg yolk antibodies (IgY) specific to *Streptococcus mutans*, supported the effectiveness of IgY with specificity to *S. mutans* that was grown in the presence of sucrose, as an efficient method to control the bacterial colonization by *S. mutans*, in the oral cavity (Hatta et al., 1997). Other researchers have carried out studies where colonization of the *S. mutans* and its capability to induce dental caries was suppressed by using murine monoclonal antibodies and transgenic plant antibodies, by passive immunization. The antibodies used in the studies were raised against antigens of MS (Fukushima, Okada, & Ochiai, 1993; Larrick, Yu, Naftzger, Jaiswal, & Wycoff, 2001). Even if competently prepared, all vaccines appear to pose major health concerns and risks. For instance, it was reported that *S. mutans* immunized rabbits had antigens capable of eliciting antibodies that cross-reacted with tissues of the heart (Hughes, Machardy, Sheppard, & Woods, 1980). In another study, the sera of some rheumatic fever patients showed a serological cross-reactivity between antigens of heart tissue and certain antigens from *S. mutans*, which may induce myocarditis, representing an unacceptable risk to the health of the vaccine recipients Shanmugam et al., 2020.

9. Conclusion and future prospects

Cariogenic biofilm-associated diseases represent a global public health challenge. In this chapter, dental biofilms and their mechanisms are discussed in relation to their role in initiating and aggravating cariogenic ecosystems. Strategies to modulate and inhibit the damaging capacity of the biofilm are also outlined. It is a well-known fact that the bacteria residing within the biofilm are significantly less susceptible to antibacterial agents than are their counterparts residing as planktonic cells. Furthermore, a long tradition of using antibiotics for treating such bacterial infections has led to an enhanced resistance toward such drugs. Even though natural products have long been recognized as a valuable source of therapeutic medications, difficulties in their use arises from lack of standardized methods, inability to isolate pure chemical components, and determination of their mechanisms of action. With nanotechnology-based approaches being investigated to combat these bacterial biofilm infections, there exists crucial issues pertaining to nanoparticle-mediated resistance and their interaction with the exopolysaccharides of biofilm as well as those of the host. Other approaches for the treatment of biofilm associated oral diseases are metal ions, synthetic products, and antibodies, which have been acknowledged, but their thorough studies of mechanisms are not available and still under investigation. These approaches provide us with a hope of effective therapies that may succeed in eradicating the growing problem of biofilm infections. In view of the prevalence and cost related to prevention and treatment of biofilm-induced dental diseases, there is also a need for further the research to elucidate, in further detail, the complex relationship among different bacteria within the oral biofilms. Because of the high prevalence and high cost associated with prophylaxis as well as post infection treatment of biofilm-induced dental disorders, there is a growing need of the hour to better understand and escalate research elucidating the complex nature of oral biofilms. Future studies should aim for improved efficacy to identify agents, technologies, and methods that comply with health and are without adverse effects. Future research should focus on enhancing efficacy to develop drugs, technologies, and approaches that are safe for human use while presenting no or least severity of side effects.

References

Aas, J. A., Griffen, A. L., Dardis, S. R., Lee, A. M., Olsen, I., Dewhirst, F. E., et al. (2008). Bacteria of dental caries in primary and permanent teeth in children and young adults. *Journal of Clinical Microbiology*, *46*(4), 1407–1417.

Abd Rahim, Z. H., & Khan, H. B. S. G. (2006). Comparative studies on the effect of crude aqueous (CA) and solvent (CM) extracts of clove on the cariogenic properties of *Streptococcus mutans*. *Journal of Oral Science*, *48*(3), 117–123.

Abebe, G. M. (2021). Oral biofilm and its impact on oral health, psychological and social inter-action. *International Journal of Oral and Dental Health*, *7*, 127.

Abou Neel, E. A., Aljabo, A., Strange, A., Ibrahim, S., Coathup, M., Young, A. M., et al. (2016). Demineralization–remineralization dynamics in teeth and bone. *International Journal of Nanomedicine*, *11*, 4743.

Abranches, J., Zeng, L., Kajfasz, J. K., Palmer, S. R., Chakraborty, B., Wen, Z. T., et al. (2018). Biology of oral streptococci. *Microbiology Spectrum*, *6*(5). 6-5.

Adil, M, Baig, M. H., & Rupasinghe, H. V. (2019). Impact of citral and phloretin, alone and in combination, on major virulence traits of *Streptococcus pyogenes*. *Molecules*, *24*(23), 4237. https://doi.org/10.3390/molecules24234237.

Adil, M., Khan, R., & Rupasinghe, H. V. (2018). Application of medicinal plants as a source for therapeutic agents against *Streptococcus pyogenes* infections. *Current Drug Metabolism*, *19*(8), 695–703.

Adil, M., Singh, K., Verma, P. K., & Khan, A. U. (2014). Eugenol-induced suppression of biofilm-forming genes in *Streptococcus mutans*: An approach to inhibit biofilms. *Journal of Global Antimicrobial Resistance*, *2*(4), 286–292.

Ahn, S. J., Kaspar, J., Kim, J. N., Seaton, K., & Burne, R. A. (2014). Discovery of a novel peptides regulating competence development in *Streptococcus mutans*. *Journal of Bacteriology*, *196*(21), 3735–3745.

Ahn, S. J., Rice, K. C., Oleas, J., Bayles, K. W., & Burne, R. A. (2010). The *Streptococcus mutans* Cid and Lrg systems modulate virulence traits in response to multiple environmental signals. *Microbiology (Reading, England)*, *156*(Pt 10), 3136–3147.

Ahn, S. J., Wen, Z. T., & Burne, R. A. (2007). Effects of oxygen on virulence traits of *Streptococcus mutans*. *Journal of Bacteriology*, *189*(23), 8519–8527.

Ahn, S. J., Wen, Z. T., & Burnem, R. A. (2006). Multilevel control of competence development and stress tolerance in *Streptococcus mutans* UA159. *Infection and Immunity*, *74*, 1631–1642.

Alejandra, B. M., & Daniel, O. M. (2020). Virulence factors of *Streptococcus mutans* related to dental caries. In *Staphylococcus and Streptococcus* (p. 9). https://doi.org/10.5772/intechopen.85807.

Andreadis, G., Topitsoglou, V., & Kalfas, S. (2015). Acidogenicity and acidurance of dental plaque and saliva sediment from adults in relation to caries activity and chlorhexidine exposure. *Journal of Oral Microbiology*, *7*(1), 26197.

Ansari, M., Khan, H., Khan, A., Cameotra, S., & Alzohairy, M. (2015). Anti-biofilm efficacy of silver nanoparticles against MRSA and MRSE isolated from wounds in a tertiary care hospital. *Indian Journal of Medical Microbiology*, *33*(1), 101–109.

Araújo, J. T. C. D., Lima, L. A., Vale, E. P., Martin-Pastor, M., Lima, R. A., Silva, P. G. D. B., et al. (2020). Toxicological and genotoxic evaluation of anacardic acid loaded-zein nanoparticles in mice. *Toxicology Reports*, *7*, 1207–1215. https://doi.org/10.1016/j.toxrep.2020.08.024.

Aruni, A. W., Dou, Y., Mishra, A., & Fletcher, H. M. (2015). The biofilm community: Rebels with a cause. *Current Oral Health Reports*, *2*(1), 48–56.

Baker, J. L., Abranches, J., Faustoferri, R. C., Hubbard, C. J., Lemos, J. A., Courtney, M. A., et al. (2015). Transcriptional profile of glucose-shocked and acid-adapted strains of *Streptococcus mutans*. *Molecular Oral Microbiology*, *30*(6), 496–517.

Baker, J. L., Faustoferri, R. C., & Quivey, R. G., Jr. (2017). Acid-adaptive mechanisms of *Streptococcus mutans*—The more we know, the more we don't. *Molecular Oral Microbiology*, *32*(2), 107.

Banas, J. A., & Vickerman, M. M. (2003). Glucan-binding proteins of the oral streptococci. *Critical Reviews in Oral Biology and Medicine*, *14* (2), 89–99. https://doi.org/10.1177/154411130301400203.

Banerjee, A., & Biswas, I. (2008). Markerless multiple-gene-deletion system for *Streptococcus mutans*. *Applied and Environmental Microbiology*, *74*(7), 2037–2042.

Baugh, S., Phillips, C. R., Ekanayaka, A. S., Piddock, L. J., & Webber, M. A. (2014). Inhibition of multidrug efflux as a strategy to prevent biofilm formation. *Journal of Antimicrobial Chemotherapy*, *69*(3), 673–681. https://doi.org/10.1093/jac/dkt420.

Beighton, D., Al-Haboubi, M., Mantzourani, M., Gilbert, S. C., Clark, D., Zoitopoulos, L., et al. (2010). Oral Bifidobacteria: Caries-associated bacteria in older adults. *Journal of Dental Research*, *89*(9), 970–974.

Berkowitz, R. J. (2003). Causes, treatment and prevention of early childhood caries: A microbiologic perspective. *Journal-Canadian Dental Association*, *69*(5), 304–307.

Bikash, C. R., Hamry, S. R., & Tal-Gan, Y. (2018). Structure–activity relationships of the competence stimulating peptide in *Streptococcus mutans* reveal motifs critical for membrane protease sepm recognition and comd receptor activation. *ACS Infectious Diseases*, *4*, 1385–1394.

Bilbilova, E. Z. (2020). Dietary factors, salivary parameters, and dental caries. In *Dental caries* (pp. 1–18). IntechOpen.

Biswas, I., Drake, L., Erkina, D., & Biswas, S. (2008). Involvement of sensor kinases in the stress tolerance response of *Streptococcus mutans*. *Journal of Bacteriology*, *190*(1), 68–77. https://doi.org/10.1128/JB.00990-07.

Biswas, S., & Biswas, I. (2006). Regulation of the glucosyltransferase (gtfBC) operon by CovR in *Streptococcus mutans*. *Journal of Bacteriology*, *188*, 988–998.

Boisen, G., Davies, J. R., & Neilands, J. (2021). Acid tolerance in early colonizers of oral biofilms. *BMC Microbiology*, *21*(1), 1–9.

Bowen, W. H., Burne, R. A., Wu, H., & Koo, H. (2018). Oral biofilms: Pathogens, matrix, and polymicrobial interactions in microenvironments. *Trends in Microbiology*, *26*, 229–242.

Bowen, W. H., & Koo, H. J. C. R. (2011). Biology of *Streptococcus mutans*-derived glucosyltransferases: Role in extracellular matrix formation of cariogenic biofilms. *Caries Research*, *45*(1), 69–86.

Carlsson, J., & Hamilton, I. R. (1994). *Metabolic activity of oral bacteria* (2nd ed.). Copenhagen: Munksgaard.

Carniello, V., Peterson, B. W., van der Mei, H. C., & Busscher, H. J. (2018). Physico-chemistry from initial bacterial adhesion to surface-programmed biofilm growth. *Advances in Colloid and Interface Science*, *261*, 1–14.

Castillo Pedraza, M. C., Novais, T. F., Faustoferri, R. C., Quivey, R. G., Jr., Terekhov, A., Hamaker, B. R., et al. (2017). Extracellular DNA and lipoteichoic acids interact with exopolysaccharides in the extracellular matrix of *Streptococcus mutans* biofilms. *Biofouling*, *33*(9), 722–740.

Castillo Pedraza, M. C., Rosalen, P. L., Castilho, A. R. F. D., Freires, I. D. A., de Sales Leite, L., Faustoferri, R. C., et al. (2019). Inactivation of *Streptococcus mutans* genes lytST and dltAD impairs its pathogenicity in vivo. *Journal of Oral Microbiology*, *11*(1), 1607505.

Chawhuaveang, D. D., Yu, O. Y., Yin, I. X., Lam, W. Y. H., Mei, M. L., & Chu, C. H. (2021). Acquired salivary pellicle and oral diseases: A literature review. *Journal of Dental Sciences*, *16*(1), 523–529.

Chen, L., Chakraborty, B., Zou, J., Burne, R. A., & Zeng, L. (2019). Amino sugars modify antagonistic interactions between commensal oral streptococci and *Streptococcus mutans*. *Applied and Environmental Microbiology*, *85*(10). 00370-19.

Chen, X., Daliri, E. B. M., Kim, N., Kim, J. R., Yoo, D., & Oh, D. H. (2020). Microbial etiology and prevention of dental caries: Exploiting natural products to inhibit cariogenic biofilms. *Pathogens*, *9*(7), 569.

Childers, N. K., Momeni, S. S., Whiddon, J., Cheon, K., Cutter, G. R., Wiener, H. W., et al. (2017). Association of colonization with *Streptococcus mutans* genotypes from mothers and early childhood caries. *Pediatric Dentistry*, *39*(2), 130.

Cho, M. Y., Kang, S. M., Lee, E. S., & Kim, B. I. (2020). Antimicrobial activity of *Curcuma xanthorrhiza* nanoemulsions on *Streptococcus mutans* biofilms. *Biofouling*, *36*(7), 825–833.

Chong, P., Drake, L., & Biswas, I. (2008). LiaS regulates virulence factor expression in *Streptococcus mutans*. *Infection and Immunity*, *76*(7), 3093–3099.

Cleaver, L. M., Moazzez, R. V., & Carpenter, G. H. (2021). Evidence for proline utilization by oral bacterial biofilms grown in saliva. *Frontiers in Microbiology*, *11*, 3562.

Conrads, G., de Soet, J. J., Song, L., Henne, K., Sztajer, H., Wagner-Döbler, I., et al. (2014). Comparing the cariogenic species *Streptococcus sobrinus* and *S. mutans* on whole genome level. *Journal of Oral Microbiology*, *6*(1), 26189.

Dodds, M., Roland, S., Edgar, M., & Thornhill, M. (2015). Saliva A review of its role in maintaining oral health and preventing dental disease. *BDJ Team*, *2*, 15123.

Duarte, S., Gregoire, S., Singh, A. P., Vorsa, N., Schaich, K., Bowen, W. H., et al. (2006). Inhibitory effects of cranberry polyphenols on formation and acidogenicity of *Streptococcus mutans* biofilms. *FEMS Microbiology Letters*, *257*(1), 50–56.

Duque, C., Stipp, R. N., Wang, B., Smith, D. J., Höfling, J. F., Kuramitsu, H. K., et al. (2011). Downregulation of GbpB, a component of the VicRK regulon, affects biofilm formation and cell surface characteristics of *Streptococcus mutans*. *Infection and Immunity*, *79*(2), 786–796.

Eshed, M., Lellouche, J., Gedanken, A., & Banin, E. (2014). A Zn-doped CuO nanocomposite shows enhanced antibiofilm and antibacterial activities against *Streptococcus mutans* compared to nanosized CuO. *Advanced Functional Materials*, *24*(10), 1382–1390.

Flemming, H. C., & Wingender, J. (2010). The biofilm matrix. *Nature Reviews Microbiology*, *8*, 623–633.

Florez Salamanca, E. J., & Klein, M. I. (2018). Extracellular matrix influence in *Streptococcus mutans* gene expression in a cariogenic biofilm. *Molecular Oral Microbiology*, *33*(2), 181–193. https://doi.org/10.1111/omi.12212.

Fozo, E. M., Scott-Anne, K., Koo, H., & Quivey, R. G., Jr. (2007). Role of unsaturated fatty acid biosynthesis in virulence of *Streptococcus mutans*. *Infection and Immunity*, *75*(3), 1537–1539.

Fukushima, K., Okada, T., & Ochiai, K. (1993). Production, characterization, and application of monoclonal antibodies which distinguish three glucosyltransferases from *Streptococcus mutans*. *Infection and Immunity*, *61*(1), 323–328.

Fuqua, C., Parsek, M. R., & Greenberg, E. P. (2001). Regulation of gene expression by cell-to-cell communication: Acyl-homoserine lactone quorum sensing. *Annual Review of Genetics*, *35*(1), 439–468.

Furiga, A., Lonvaud-Funel, A., Dorignac, G., & Badet, C. (2008). In vitro anti-bacterial and anti-adherence effects of natural polyphenolic compounds on oral bacteria. *Journal of Applied Microbiology*, *105*(5), 1470–1476.

Galperin, M. Y., Makarova, K. S., Wolf, Y. I., & Koonin, E. V. (2018). Phyletic distribution and lineage-specific domain architectures of archaeal two-component signal transduction systems. *Journal of Bacteriology*, *200*. e00681–17.

Giasuddin, A. S. M., Huda, S. N., Jhuma, K. A., & Haq, A. M. M. (2017). Dental caries vaccine availability: Challenges for the 21st century. *Journal of Immunology and Immunotherapy*, *1*(002).

Gong, Y., Tian, X. L., Sutherland, T., Sisson, G., Mai, J., Ling, J., et al. (2009). Global transcriptional analysis of acid-inducible genes in *Streptococcus mutans*: Multiple two-component systems involved in acid adaptation. *Microbiology*, *155*(10), 3322–3332.

Gregoire, S., Singh, A. P., Vorsa, N., & Koo, H. (2007). Influence of cranberry phenolics on glucan synthesis by glucosyltransferases and *Streptococcus mutans* acidogenicity. *Journal of Applied Microbiology*, *103* (5), 960–1968.

Griswold, A. R., Chen, Y. Y. M., & Burne, R. A. (2004). Analysis of an agmatine deiminase gene cluster in *Streptococcus mutans* UA159. *Journal of Bacteriology*, *186*(6), 1902–1904.

Gupta, D., Singh, A., & Khan, A. U. (2017). Nanoparticles as efflux pump and biofilm inhibitor to rejuvenate bactericidal effect of conventional antibiotics. *Nanoscale Research Letters*, *12*(1), 1–6.

Hamada, S. (2002). Role of sweeteners in the etiology and prevention of dental caries. *Pure and Applied Chemistry*, *74*(7), 1293–1300. https://doi.org/10.1351/pac200274071293.

Hamada, S., & Slade, H. D. (1980). Biology, immunology, and cariogenicity of *Streptococcus mutans*. *Microbiological Reviews*, *44*(2), 331–384.

Hasan, S., Danishuddin, M., Adil, M., Singh, K., Verma, P. K., & Khan, A. U. (2012). Efficacy of *E. officinalis* on the cariogenic properties of *Streptococcus mutans*: A novel and alternative approach to suppress quorum-sensing mechanism. *PLoS One*, *7*(7), e40319.

Hatta, H., Tsuda, K., Ozeki, M., Kim, M., Yamamoto, T., Otake, S., et al. (1997). Passive immunization against dental plaque formation in humans: Effect of a mouth rinse containing egg yolk antibodies (IgY) specific to *Streptococcus mutans*. *Caries Research*, *31*(4), 268–274.

Havarstein, L. S. (2010). Increasing competence in the genus *Streptococcus*. *Molecular Microbiology*, *78*(3), 541–544.

He, X., Wu, C., Yarbrough, D., Sim, L., Niu, G., Merritt, J., et al. (2008). The cia operon of *Streptococcus mutans* encodes a unique component required for calcium-mediated autoregulation. *Molecular Microbiology*, *70*(1), 112–126.

Hirakawa, H., Kurushima, J., Hashimoto, Y., & Tomita, H. (2020). Progress overview of bacterial two-component regulatory systems as potential targets for antimicrobial chemotherapy. *Antibiotics*, *9* (10), 635.

House, D. R., Fry, C. L., & Brown, L. J. (2004). The economic impact of dentistry. *Journal of the American Dental Association*, *135*(3), 347–352.

Huang, R., Li, M., & Gregory, R. L. (2011). Bacterial interactions in dental biofilm. *Virulence*, *2*(5), 435–444.

Hughes, M., Machardy, S. M., Sheppard, A. J., & Woods, N. C. (1980). Evidence for an immunological relationship between *Streptococcus mutans* and human cardiac tissue. *Infection and Immunity*, *27*(2), 576–588. https://doi.org/10.1128/iai.27.2.576-588.1980.

Itzek, A., Chen, Z., Merritt, J., & Kreth, J. (2017). Effect of salivary agglutination on oral streptococcal clearance by human polymorphonuclear neutrophil granulocytes. *Molecular Oral Microbiology*, *32*(3), 197–210.

Jakubovics, N. S. (2015). Intermicrobial interactions as a driver for community composition and stratification of oral biofilms. *Journal of Molecular Biology*, *427*(23), 3662–3675. https://doi.org/10.1016/j.jmb.2015.09.022.

Jia, R., Yan, L., & Guo, J. (2020). Enhancing the immunogenicity of a DNA vaccine against *Streptococcus mutans* by attenuating the inhibition of endogenous miR-9. *Vaccine*, *38*(6), 1424–1430. https://doi.org/10.1016/j.vaccine.2019.11.083.

Jones, N., Ray, B., Ranjit, K. T., & Manna, A. C. (2008). Antibacterial activity of ZnO nanoparticle suspensions on a broad spectrum of microorganisms. *FEMS Microbiology Letters*, *279*(1), 71–76.

Katancik, J. A., Kumarswamy, A., Branch-Mays, G., & Califano, J. V. (2015). Infections of the periodontal apparatus. In *Head, neck, and orofacial infections. An interdisciplinary approach* (pp. 189–202). Elsevier.

Kendall, K., & Roberts, A. D. (2015). van der Waals forces influencing adhesion of cells. *Philosophical Transactions of the Royal Society B: Biological Sciences*, *370*(1661), 20140078. https://doi.org/10.1098/rstb.2014.0078.

Khan, R., Adil, M., Danishuddin, M., Verma, P. K., & Khan, A. U. (2012). In vitro and in vivo inhibition of *Streptococcus mutans* biofilm by *Trachyspermum ammi* seeds: An approach of alternative medicine. *Phytomedicine*, *19*(8–9), 747–755.

Khan, R., Rukke, H. V., Høvik, H., Åmdal, H. A., Chen, T., Morrison, D. A., et al. (2016). Comprehensive transcriptome profiles of *Streptococcus mutans* UA159 map core streptococcal competence genes. *Msystems*, *1*(2). e00038-15.

Kilian, M., Chapple, I. L. C., Hannig, M., Marsh, P. D., Meuric, V., Pedersen, A. M. L., et al. (2016). The oral microbiome—An update for oral healthcare professionals. *British Dental Journal*, *221*(10), 657–666.

Klein, M. I., DeBaz, L., Agidi, S., Lee, H., Xie, G., Lin, A. H. M., et al. (2010). Dynamics of *Streptococcus mutans* transcriptome in response to starch and sucrose during biofilm development. *PLoS One*, *5*(10), e13478.

Klein, M. I., Hwang, G., Santos, P. H., Campanella, O. H., & Koo, H. (2015). *Streptococcus mutans*-derived extracellular matrix in cariogenic oral biofilms. *Frontiers in Cellular and Infection Microbiology*, *5*, 10.

Kolenbrander, P. E., Andersen, R. N., Blehert, D. S., Egland, P. G., Foster, J. S., & Palmer, R. J., Jr. (2002). Communication among oral bacteria. *Microbiology and Molecular Biology Reviews*, *66*(3), 486–505.

Kolenbrander, P. E., Palmer, R. J., Periasamy, S., & Jakubovics, N. S. (2010). Oral multispecies biofilm development and the key role of cell–cell distance. *Nature Reviews Microbiology*, *8*(7), 471–480.

Koo, H., Andes, D. R., & Krysan, D. J. (2018). *Candida*–streptococcal interactions in biofilm-associated oral diseases. *PLoS Pathogens*, *14* (12), e1007342.

Lara-Carrillo, E., Montiel Bastida, N. M., Sánchez Pérez, L., & Alanís-Tavira, J. (2010). Effect of orthodontic treatment on saliva, plaque, and the levels of *Streptococcus mutans* and *Lactobacillus*. *Medicina oral, patología oral y cirugía bucal. Ed inglesa, 15*(6). https://roderic.uv.es/handle/10550/60233.

Larrick, J. W., Yu, L., Naftzger, C., Jaiswal, S., & Wycoff, K. (2001). Production of secretory IgA antibodies in plants. *Biomolecular Engineering, 18*(3), 87–94.

Lee, S. F., Delaney, G. D., & Elkhateeb, M. (2004). A two-component covRS regulatory system regulates expression of fructosyltransferase and a novel extracellular carbohydrate in *Streptococcus mutans*. *Infection and Immunity, 72*(7), 3968–3973.

Lemos, J. A., Abranches, J., & Burne, R. A. (2005). Responses of cariogenic streptococci to environmental stresses. *Current Issues in Molecular Biology, 7*(1), 95–107.

Lemos, J. A., & Burne, R. A. (2008). A model of efficiency: Stress tolerance by *Streptococcus mutans*. *Microbiology (Reading, England), 154*(Pt 11), 3247.

Lemos, J. A., Palmer, S. R., Zeng, L., Wen, Z. T., Kajfasz, J. K., Freires, I. A., et al. (2019). The biology of *Streptococcus mutans*. *Microbiology Spectrum, 7*(1). 7-1.

Len, A. C., Harty, D. W., & Jacques, N. A. (2004). Proteome analysis of *Streptococcus mutans* metabolic phenotype during acid tolerance. *Microbiology, 150*(5), 1353–1366.

Levesque, C. M., Mair, R. W., Perry, J. A., Lau, P. C., Li, Y. H., & Cvitkovitch, D. G. (2007). Systemic inactivation and phenotypic characterization of two-component systems in expression of *Streptococcus mutans* virulence properties. *Letters in Applied Microbiology, 45*(4), 398–404.

Li, Y. H., Lau, P. C. Y., Lee, L. H., Ellen, R. P., & Cvitkovitch, D. G. (2001). Natural genetic transformation of *Streptococcus mutans* growing in biofilm. *Journal of Bacteriology, 183*(3), 897–908.

Li, Y. H., Lau, P. C., Tang, N., Svensäter, G., Ellen, R. P., & Cvitkovitch, D. G. (2002). Novel two-component regulatory system involved in biofilm formation and acid resistance in *Streptococcus mutans*. *Journal of Bacteriology, 184*(2002), 6333–6342.

Li, Y. H., Tang, N., Aspiras, M. B., Lau, P. C. Y., Lee, L. H., & Ellen, R. P. (2002). A quorum sensing signaling system essential for genetic competence in *Streptococcus mutans* is involved in biofilm formation. *Journal of Bacteriology, 184*(10), 2699–2708.

Li, Y. H., Tian, X. L., Layton, G., Norgaard, C., & Sisson, G. (2008). Additive attenuation of virulence and cariogenic potential of *Streptococcus mutans* by simultaneous inactivation of the ComCDE quorum-sensing system and HK/RR11 two-component regulatory system. *Microbiology, 154*(11), 3256–3265.

Liao, S., Klein, M. I., Heim, K. P., Fan, Y., Bitoun, J. P., Ahn, S. J., et al. (2014). *Streptococcus mutans* extracellular DNA is upregulated during growth in biofilms, actively released via membrane vesicles, and influenced by components of the protein secretion machinery. *Journal of Bacteriology, 196*(13), 2355–2366.

Liu, Y., & Burne, R. A. (2009). Multiple two-component systems of *Streptococcus mutans* regulate agmatine deiminase gene expression and stress tolerance. *Journal of Bacteriology, 191*(23), 7363–7366.

Loesche, W. J. (1986). Role of *Streptococcus mutans* in human dental decay. *Microbiological Reviews, 50*(4), 353–380.

Longbottom, C., Vernon, B., Perfect, E., Haughey, A. M., Christie, A., & Pitts, N. (2021). Initial investigations of a novel bioluminescence method for imaging dental demineralization. *Clinical and Experimental Dental Research, 7*(5), 786–794.

Lu, T. K., & Collins, J. J. (2007). Dispersing biofilms with engineered enzymatic bacteriophage. *Proceedings of the National Academy of Sciences, 104*(27), 11197–11202. https://doi.org/10.1073/pnas.07046241.

Lynch, D. J., Fountain, T. L., Mazurkiewicz, J. E., & Banas, J. A. (2007). Glucan-binding proteins are essential for shaping *Streptococcus mutans* biofilm architecture. *FEMS Microbiology Letters, 268*(2), 158–165.

MacGilvray, M. E., Lapek, J. D., Jr., Friedman, A. E., & Quivey, R. G., Jr. (2012). Cardiolipin biosynthesis in *Streptococcus mutans* is regulated in response to external pH. *Microbiology, 158*(8), 2133.

Maier, B. (2021). How physical interactions shape bacterial biofilms. *Annual Review of Biophysics, 50*, 401–417.

Manzer, H. S., Nobbs, A. H., & Doran, K. S. (2020). The multifaceted nature of streptococcal antigen I/II proteins in colonization and disease pathogenesis. *Frontiers in Microbiology, 11*, 2948.

Marsh, P. D., & Bowden, G. H. (2000). Microbial community interactions in biofilms. In *Symposia-society for general microbiology* (pp. 167–198). Cambridge: Cambridge University Press. 1999.

Martinez-Gutierrez, F., Boegli, L., Agostinho, A., Sánchez, E. M., Bach, H., Ruiz, F., et al. (2013). Anti-biofilm activity of silver nanoparticles against different microorganisms. *Biofouling, 29*(6), 651–660.

Mashburn-Warren, L., Morrison, D. A., & Federle, M. J. (2010). A novel double-tryptophan peptide pheromone controls competence in *Streptococcus* spp. via an Rgg regulator. *Molecular Microbiology, 78*(3), 589–606. https://doi.org/10.1111/j.1365-2958.2010.07361.x.

Matsui, R., & Cvitkovitch, D. (2010). Acid tolerance mechanisms utilized by *Streptococcus mutans*. *Future Microbiology, 5*(3), 403–417.

Mattos-Graner, R. O., Klein, M. I., & Smith, D. J. (2014). Lessons learned from clinical studies: Roles of mutans streptococci in the pathogenesis of dental caries. *Current Oral Health Reports, 1*(1), 70–78.

Mazda, Y., Kawada-Matsuo, M., Kanbara, K., Oogai, Y., Shibata, Y., Yamashita, Y., et al. (2012). Association of CiaRH with resistance of *Streptococcus mutans* to antimicrobial peptides in biofilms. *Molecular Oral Microbiology, 27*(2), 124–135.

Merritt, J., & Qi, F. (2012). The mutacins of *Streptococcus mutans*: Regulation and ecology. *Molecular Oral Microbiology, 27*(2), 57–69.

Muhammad, M. H., Idris, A. L., Fan, X., Guo, Y., Yu, Y., Jin, X., et al. (2020). Beyond risk: Bacterial biofilms and their regulating approaches. *Frontiers in Microbiology, 11*, 928.

Neuhaus, F. C., & Baddiley, J. (2003). A continuum of anionic charge: Structures and functions of D-alanyl-teichoic acids in gram-positive bacteria. *Microbiology and Molecular Biology Reviews, 67*(4), 686–723.

Nguyen, P. T., & Marquis, R. E. (2011). Antimicrobial actions of α-mangostin against oral streptococci. *Canadian Journal of Microbiology, 57*(3), 217–225.

Nicolas, G. G., & Lavoie, M. C. (2011). *Streptococcus mutans* and oral streptococci in dental plaque. *Canadian Journal of Microbiology, 57* (1), 1–20.

Nilsson, M., Rybtke, M., Givskov, M., Høiby, N., Twetman, S., & Tolker-Nielsen, T. (2016). The dlt genes play a role in antimicrobial tolerance of *Streptococcus mutans* biofilms. *International Journal of Antimicrobial Agents, 48*(3), 298–304.

Palmer, R. J., Jr., Shah, N., Valm, A., Paster, B., Dewhirst, F., Inui, T., et al. (2017). Interbacterial adhesion networks within early oral biofilms of single human hosts. *Applied and Environmental Microbiology*, *83*(11), e00407–e00417.

Patterson, M. J. (1996). Streptococcus. In *Medical microbiology* (4th ed.). (Chapter 13) https://www.ncbi.nlm.nih.gov/books/NBK7611/.

Paula, J. S., Leite, I. C., Almeida, A. B., Ambrosano, G. M., Pereira, A. C., & Mialhe, F. L. (2012). The influence of oral health conditions, socioeconomic status, and home environment factors on schoolchildren's self-perception of quality of life. *Health and Quality of Life Outcomes*, *10*(1), 1–8.

Pedersen, A. M. L., & Belstrøm, D. (2019). The role of natural salivary defences in maintaining a healthy oral microbiota. *Journal of Dentistry*, *80*, S3–S12.

Percival, R. S., Devine, D. A., Duggal, M. S., Chartron, S., & Marsh, P. D. (2006). The effect of cocoa polyphenols on the growth, metabolism, and biofilm formation by *Streptococcus mutans* and *Streptococcus sanguinis*. *European Journal of Oral Sciences*, *114*(4), 343–348.

Pereira, C. S., Thompson, J. A., & Xavier, K. B. (2013). AI-2-mediated signalling in bacteria. *FEMS Microbiology Reviews*, *37*(2), 156–181.

Pérez-Díaz, M. A., Boegli, L., James, G., Velasquillo, C., Sánchez-Sánchez, R., Martínez-Martínez, R. E., et al. (2015). Silver nanoparticles with antimicrobial activities against *Streptococcus mutans* and their cytotoxic effect. *Materials Science & Engineering. C, Materials for Biological Applications*, *55*, 360–366. https://doi.org/10.1016/j.msec.2015.05.036.

Perry, J. A., Cvitkovitch, D. G., & Lévesque, C. M. (2009). Cell death in *Streptococcus mutans* biofilms: A link between CSP and extracellular DNA. *FEMS Microbiology Letters*, *299*(2), 261–266.

Perry, J. A., Jones, M. B., Peterson, S. N., Cvitkovitch, D. G., & Levesque, C. M. (2009). Peptide alarmone signalling triggers an auto-active bacteriocin necessary for genetic competence. *Molecular Microbiology*, *72*(4), 905–917.

Prabu, G. R., Gnanamani, A., & Sadulla, S. (2006). Guaijaverin—A plant flavonoid as potential antiplaque agent against *Streptococcus mutans*. *Journal of Applied Microbiology*, *101*(2), 487–495.

Qayyum, S., Oves, M., & Khan, A. U. (2017). Obliteration of bacterial growth and biofilm through ROS generation by facilely synthesized green silver nanoparticles. *PloS One*, *12*(8), e0181363. https://doi.org/10.1371/journal.pone.0181363.

Qi, F., Kreth, J., Lévesque, C. M., Kay, O., Mair, R. W., Shi, W., et al. (2005). Peptide pheromone induced cell death of *Streptococcus mutans*. *FEMS Microbiology Letters*, *251*(2), 321–326.

Qi, F., Merritt, J., Lux, R., & Shi, W. (2004). Inactivation of the ciaH gene in *Streptococcus mutans* diminishes mutacin production and competence development, alters sucrose-dependent biofilm formation, and reduces stress tolerance. *Infection and Immunity*, *72*(8), 4895–4899.

Quivey, R. G., Jr., Faustoferri, R., Monahan, K., & Marquis, R. (2000). Shifts in membrane fatty acid profiles associated with acid adaptation of *Streptococcus mutans*. *FEMS Microbiology Letters*, *189*(1), 89–92.

Quivey, R. G., Jr., Grayhack, E. J., Faustoferri, R. C., Hubbard, C. J., Baldeck, J. D., Wolf, A. S., et al. (2015). Functional profiling in *Streptococcus mutans*: Construction and examination of a genomic collection of gene deletion mutants. *Molecular Oral Microbiology*, *30* (6), 474–495.

Quivey, R. G., Jr., Kuhnert, W. L., & Hahn, K. (2001). Genetics of acid adaptation in oral streptococci. *Critical Reviews in Oral Biology and Medicine*, *12*(4), 301–314.

Ren, Z., Chen, L., Li, J., & Li, Y. (2016). Inhibition of *Streptococcus mutans* polysaccharide synthesis by molecules targeting glycosyltransferase activity. *Journal of Oral Microbiology*, *8*(1), 31095.

Ribet, D., & Cossart, P. (2015). How bacterial pathogens colonize their hosts and invade deeper tissues. *Microbes and Infection*, *17*(3), 173–183.

Rosier, B. T., De Jager, M., Zaura, E., & Krom, B. P. (2014). Historical and contemporary hypotheses on the development of oral diseases: Are we there yet? *Frontiers in Cellular and Infection Microbiology*, *4*, 92.

Russell, M. W., Childers, N. K., Michalek, S. M., Smith, D. J., & Taubman, M. A. (2004). A caries vaccine? *Caries Research*, *38*(3), 230–235. https://doi.org/10.1159/000077759.

Santiago, B., MacGilvray, M., Faustoferri, R. C., & Quivey, R. G., Jr. (2012). The branched-chain amino acid aminotransferase encoded by ilvE is involved in acid tolerance in *Streptococcus mutans*. *Journal of Bacteriology*, *194*(8), 2010–2019.

Santiago, B., Marek, M., Faustoferri, R. C., & Quivey, R. G., Jr. (2013). The *Streptococcus mutans* aminotransferase encoded by ilvE is regulated by CodY and CcpA. *Journal of Bacteriology*, *195*(16), 3552–3562.

Schauder, S., & Bassler, B. L. (2001). The languages of bacteria. *Genes & Development*, *15*(12), 1468–1480.

Senadheera, M. D., Guggenheim, B., Spatafora, G. A., Huang, Y. C., Choi, J., & Hung, D. C. (2005). A VicRK signal transduction system in *Streptococcus mutans* affects *gtfBCD*, *gbpB*, and *ftf* expression, biofilm formation, and genetic competence development. *Journal of Bacteriology*, *187*(12), 4064–4076.

Senadheera, D., Krastel, K., Mair, R., Persadmehr, A., Abranches, J., Burne, R. A., et al. (2009). Inactivation of VicK affects acid production and acid survival of *Streptococcus mutans*. *Journal of Bacteriology*, *191*(20), 6415–6424.

Senadheera, M. D., Lee, A. W., Hung, D. C., Spatafora, G. A., Goodman, S. D., & Cvitkovitch, D. G. (2007). The *Streptococcus mutans* vicX gene product modulates gtfB/C expression, biofilm formation, genetic competence, and oxidative stress tolerance. *Journal of Bacteriology*, *189*(4), 1451–1458.

Shankar, M., Mohapatra, S. S., Biswas, S., & Biswas, I. (2015). Gene regulation by the LiaSR two-component system in *Streptococcus mutans*. *PLoS One*, *10*(5), e0128083.

Shanmugam, K., Sarveswari, H. B., Udayashankar, A., Swamy, S. S., Pudipeddi, A., Shanmugam, T., et al. (2020). Guardian genes ensuring subsistence of oral *Streptococcus mutans*. *Critical Reviews in Microbiology*, *46*(4), 475–491. https://doi.org/10.1080/1040841X.2020.1796579.

Sharma, D., Misba, L., & Khan, A. U. (2019). Antibiotics versus biofilm: An emerging battleground in microbial communities. *Antimicrobial Resistance and Infection Control*, *8*(1), 1–10.

Sharma, R., Prabhakar, A. R., & Gaur, A. (2012). Mutans streptococci colonization in relation to feeding practices, age, and the number of teeth in 6 to 30-month-old children: An in vivo study. *International Journal of Clinical Pediatric Dentistry*, *5*(2), 124.

Shemesh, M., Tam, A., Aharoni, R., & Steinberg, D. (2010). Genetic adaptation of *Streptococcus mutans* during biofilm formation on different types of surfaces. *BMC Microbiology*, *10*(1), 1–10.

Sheng, J., & Marquis, R. E. (2006). Enhanced acid resistance of oral streptococci at lethal pH values associated with acid-tolerant catabolism and with ATP synthase activity. *FEMS Microbiology Letters*, *262*(1), 93–98.

Singh, B. N., Pandey, G., Jadaun, V., Singh, S., Bajpai, R., Nayaka, S., et al. (2015). Development and characterization of a novel Swarna-based herbo-metallic colloidal nano-formulation–inhibitor of *Streptococcus mutans* quorum sensing. *RSC Advances*, *5*(8), 5809–5822. https://doi.org/10.1039/C4RA11939H.

Soleimani, N., Mobarez, A. M., Olia, M. S. J., & Atyabi, F. (2015). Synthesis, characterization, and effect of the antibacterial activity of chitosan nanoparticles on vancomycin-resistant *Enterococcus* and other gram negative or gram-positive bacteria. *International Journal of Pure and Applied Sciences and Technology*, *26*(1), 14.

Stipp, R. N., Boisvert, H., Smith, D. J., Höfling, J. F., Duncan, M. J., & Mattos-Graner, R. O. (2013). CovR and VicRK regulate cell surface biogenesis genes required for biofilm formation in *Streptococcus mutans*. *PLoS One*, *8*(3), e58271.

Suntharalingam, P., Senadheera, M. D., Mair, R. W., Lévesque, C. M., & Cvitkovitch, D. G. (2009). The LiaFSR system regulates the cell envelope stress response in *Streptococcus mutans*. *Journal of Bacteriology*, *191*(9), 2973–2984.

Taubman, M. A., & Nash, D. A. (2006). The scientific and public-health imperative for a vaccine against dental caries. *Nature Reviews Immunology*, *6*(7), 555–563.

Tremblay, Y. D., Lo, H., Li, Y. H., Halperin, S. A., & Lee, S. F. (2009). Expression of the *Streptococcus mutans* essential two-component regulatory system VicRK is pH and growth-phase dependent and controlled by the LiaFSR three-component regulatory system. *Microbiology*, *155*(9), 2856–2865.

Urakov, A., Urakova, N., & Reshetnikov, A. (2019). Oxygen alkaline Dental's cleaners from tooth plaque, food debris, stains of blood, and pus: A narrative review of the history of inventions. *Journal of International Society of Preventive and Community Dentistry*, *9* (5), 427.

Vila, T., Rizk, A. M., Sultan, A. S., & Jabra-Rizk, M. A. (2019). The power of saliva: Antimicrobial and beyond. *PLoS Pathogens*, *15*(11), e1008058.

Wang, X., Li, X., & Ling, J. (2017). *Streptococcus gordonii* LuxS/autoinducer-2 quorum-sensing system modulates the dual-species biofilm formation with *Streptococcus mutans*. *Journal of Basic Microbiology*, *57*(7), 605–616.

Whiteley, M., Diggle, S. P., & Greenberg, E. P. (2017). Progress in and promise of bacterial quorum sensing research. *Nature*, *551*(7680), 313–320.

Wolinsky, L. E., Mania, S., Nachnani, S., & Ling, S. (1996). The inhibiting effect of aqueous *Azadirachta indica* (neem) extract upon bacterial properties influencing in vitro plaque formation. *Journal of Dental Research*, *75*(2), 816–822.

Xu, X., Zhou, X. D., & Wu, C. D. (2011). The tea catechin epigallocatechin gallate suppresses cariogenic virulence factors of *Streptococcus mutans*. *Antimicrobial Agents and Chemotherapy*, *55*(3), 1229–1236.

Yadav, K., & Prakash, S. (2017). Dental caries: A microbiological approach. *Journal of Clinical Infectious Diseases & Practice*, *2*(1), 1–15.

Yoshida, A., Ansai, T., Takehara, T., & Kuramitsu, H. K. (2005). LuxS-based signaling affects *Streptococcus mutans* biofilm formation. *Applied and Environmental Microbiology*, *71*(5), 2372–2380.

Zeng, L., & Burne, R. A. (2013). Comprehensive mutational analysis of sucrose-metabolizing pathways in *Streptococcus mutans* reveals novel roles for the sucrose phosphotransferase system permease. *Journal of Bacteriology*, *195*(4), 833–843.

Zhou, L., Ding, Y., Chen, W., Zhang, P., Chen, Y., & Lv, X. (2013). The in vitro study of ursolic acid and oleanolic acid inhibiting cariogenic microorganisms as well as biofilm. *Oral Diseases*, *19*(5), 494–500.

Section E

Biofilms and pathogenesis

Chapter 26

Salmonella biofilm and its importance in the pathogenesis

Jasmin Pradhan[a], Swarupa Mallick[a], Neha Mishra[a], Salina Patel[a,b], Jagannath Pradhan[a], and Vidya Devi Negi[a,*]

[a]*Laboratory of Infection Immunology, Department of Life Science, National Institute of Technology, Rourkela, Odisha, India*, [b]*Laboratory of Mycobacterial Immunology, Department of Life Science, National Institute of Technology, Rourkela, Odisha, India*

1. Introduction

Bacterial biofilms are surface-attached communities encased by extracellular polymeric substance (EPS), a self-produced matrix, which protects bacterial members from various environmental stress (Branda, Vik, Friedman, & Kolter, 2005; Flemming, Szewzyk, Steinberg, Rice, & Kjelleberg, 2016). Bacteria come in contact with the surface with the help of Brownian motion and gravitational force. Often bacterial cell surface composition affects surface attachment. The bacterial extracellular appendages such as pili or flagella help them overcome the hydrodynamic and repulsive force and contact the biotic or abiotic surfaces. After contacting the surfaces, bacteria secrete adhesive molecules that help to form microcolonies. This surface attachment is solely depending on the environmental stimuli because sensing nutrient availability or hydrodynamic force, bacteria can detach and rejoin their planktonic population (Kostakioti, Hadjifrangiskou, & Hultgren, 2013). This chapter will deal with the significant aspect of biofilm in Gram-negative bacteria *Salmonella enterica* serovar Typhimurium (*S.* Typhimurium). *Salmonella* is an anaerobic, flagellated, rod-shaped, intracellular Gram-negative foodborne pathogen that can infect various hosts (human, animal, and plant) and grow either as individual planktonic cells or in multicellular form within communities called biofilm (Flemming & Wuertz, 2019). There are more than 2500 serovars of *S. enterica*, and many serovars such as *Salmonella* Enteritidis, *S.* Typhimurium can infect a broad range of hosts, including humans and animals. In contrast, some serovars such as *Salmonella* Typhi (*S.* Typhi), *Salmonella* Paratyphi (*S.* Paratyphi), *Salmonella* Dublin (*S.* Dublin), *Salmonella* Gallinarum (*S.* Gallinarum) are host-specific (Foley, Lynne, & Nayak, 2008; Popoff, Bockemühl, & Gheesling, 2004; Uzzau et al., 2000). Salmonellosis is a foodborne disease caused by ingesting contaminated food and drink, increasing mortality and morbidity rates. Annually, around 93 million people get infected, and 155,000 people die with diarrhea worldwide (Majowicz et al., 2010). Microcolony formation of bacteria triggers quorum-sensing signals and helps to form mature biofilm. Here EPSs aid structural stability of biofilm and help in moisture retaining in a dry environment and prevent diffusion of harmful chemicals such as antibiotics or antimicrobial peptides. Often biofilm consists of channels that act as a circulatory system by allowing bacteria to exchange nutrients, water, signals, enzymes, etc., and show metabolic cooperation. Dispersal allows bacteria to attach to other surfaces and complete a sequential process (Bogino, Oliva, Sorroche, & Giordano, 2013). Biofilm plays a vital role in microbial survival in their natural environment. After sensing stressful environmental stimuli, such as shifting temperature, pH, oxygen and nutrient availability, individual bacterial cells are either attached to the biotic or abiotic surfaces and form biofilms. Bacteria can grow, mature, and get protected from various environmental challenges within this biofilm. Besides, this niche also provides bacteria with a stable environment to make a symbiotic relationship with other organisms, i.e., degrading harmful soil chemicals and promoting plant growth (Harrell et al., 2021). Biofilm protects the members from various environmental stresses (Fig. 1) such as pH variability, osmotic changes, UV light radiation, dehydration, disinfectants, and antimicrobial reagents, which help the bacteria such as *Salmonella* to survive and spread (Joseph, Otta, Karunasagar, & Karunasagar, 2001; Marin, Hernandiz, & Lainez, 2009).

S. Typhi follows the fecooral route for disease transmission via contaminated food or water. Typhoid is contagious until the bacteria are excreted in the infected person's stools. In the stool of chronic asymptomatic carriers, *S.* Typhi may be found for a more extended time, even more than a year (Gonzalez-Escobedo, Marshall, & Gunn, 2011).

* Current affiliation: Department of Biological Sciences, IISER Mohali, Mohali, Punjab, India.

Understanding Microbial Biofilms. https://doi.org/10.1016/B978-0-323-99977-9.00011-9

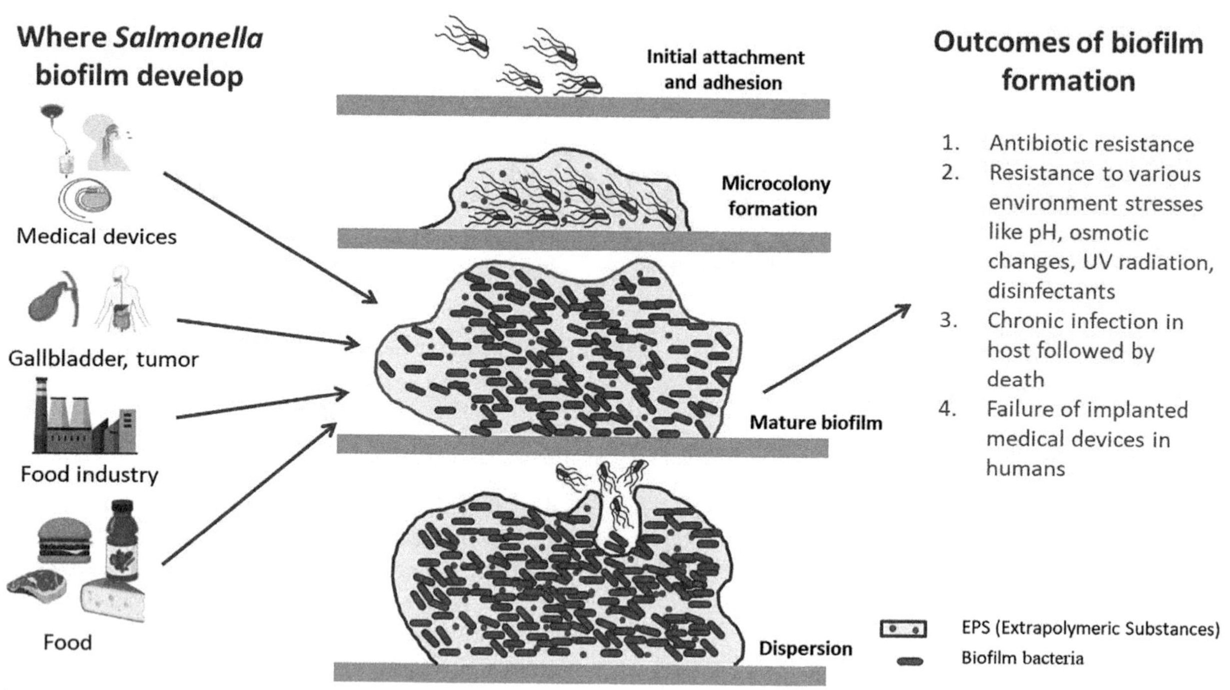

FIG. 1 Generalized idea about the surfaces where *Salmonella* develop biofilm and the outcome of biofilm formation indicating the impact of the biofilm forming capability of the microbes.

S. Typhi is adapted to induce human-specific disease and is the culprit for typhoid fever. Typhoid fever is an exhaustive illness marked by intestinal inflammation with symptoms appearing gradually and manifesting finally into bacterial dissemination to systemic locations (Arya, 2002). After entering the small intestine, *S.* Typhi crosses the epithelial barrier, mainly through the associated lymphoid tissues or invading the microfold (M) cells in the Peyer's patches. Then, the macrophages phagocytosed the bacteria and systemically spread them into various organs and blood vessels, thus causing acute illness. The most common bacterial dissemination sites are the bone marrow, ileum, liver, blood vessels and spleen, and liver ducts leading the infection to pass to the gall bladder. As per the Centers for Disease Control and Prevention (CDC) estimates, every year, typhoid fever affects approximately 22 million people globally, with many more unreported cases, due to insufficient diagnostic testing (Raza et al., 2011; Seth-Smith, 2008). The biofilm-forming capability of the bug further poses the risk of the disease and its treatment. The major components of *Salmonella* biofilm are proteinaceous curli fimbriae (Ledeboer, Frye, McClelland, & Jones, 2006), cellulose (Gerstel & Römling, 2003; Römling, 2005), colanic acid, biofilm-associated surface large proteins (BapA) (Latasa et al., 2005), O-antigen capsule (Barak, Jahn, Gibson, & Charkowski, 2007), flagella (Prouty & Gunn, 2003; Prouty, Schwesinger, & Gunn, 2002), extracellular DNA (eDNA) (Johnson et al., 2013). Different environmental signals influence the biofilm formation and composition of the extracellular matrix via a genetic regulatory network of different components such as CsgD, RpoS, OmpR, c-di-GMP, PhoPQ, Csr, RstA, BarA/SirA, Crl. Biofilms are essential for microbial survival in harsh environmental conditions and help the microbes to persist inside the host, leading to chronic infection. There are specific biofilm-related health complications where biofilm develops on medical devices such as catheters, stents, implants, dentures, and prosthetics (Fig. 1). Also, other life-threatening conditions developed such as respiratory diseases, chronic wound infections, chronic otitis media periodontitis, etc. (Percival, Suleman, Vuotto, & Donelli, 2015). Bacteria that survive after the treatment become persistent and form biofilm inside the host causing resistance to various antibiotics and hence become a challenge to antibiotic treatment. To avoid this, medical science has developed some medical devices coated with antibacterial material that prevents further biofilm development (Vasilev, Cavallaro, & Zilm, 2018). Infection due to biofilm becomes a significant health issue in medical science because it becomes chronic, antibiotic-resistant, and can result in implant failure. Therefore, further research is needed to know the genes and the genetic networks responsible and their regulations for the biofilm formation and improvement of novel preventative and therapeutic strategies needed for better treatment.

2. Biofilm formation in other Gram-positive and Gram-negative bacteria

Gram-positive bacteria such as lactic acid bacteria, *Listeria monocytogenes* (*L. monocytogenes*), *Bacillus* spp.,

Staphylococcus spp., and Gram-negative bacteria such as *Escherichia coli* (*E. coli*), *Pseudomonas aeruginosa* (*P. aeruginosa*), *Salmonella* are capable of forming biofilms. Gram-negative bacteria have a thicker cell wall and an outer membrane; these are the fundamental differences between the two categories of bacteria. The cell wall of Gram-positive bacteria is 20–80nm thick, while Gram-negative bacteria have a cell wall that is 1.5–10nm thick (Mai-Prochnow, Clauson, Hong, & Murphy, 2016). Biofilms are complex communities formed by enormous numbers of microorganisms living in close proximity in nature. Compared with their planktonic counterparts, members of these compact microbial particles displayed exceptional tolerance to antibiotics, biocides, and shear hydrodynamic pressures. Establishing these soil-associated bacterial communities is a highly organized and complex process requiring various signal transmission techniques to organize the many stages of development (Schembri, Givskov, & Klemm, 2002). The flagellar apparatus, which permits Gram-negative bacteria such as *Salmonella* and *E. coli* to swim in liquid or semiliquid environments, is required for active motility. In order to form a biofilm, bacterial movement or motility is crucial (Beloin, Roux, & Ghigo, 2008).

Both biotic and abiotic conditions allow *Salmonella* to form biofilms. Adhesin genes are among the most prevalent genes engaged in biofilm formation. Type 1 fimbriae is the best description of *Salmonella* fimbriae (Peng, 2016). The sturdiness in the *P. aeruginosa* biofilm structure is determined by at least three polysaccharides (alginate, Pel, and Psl) (Ryder, Byrd, & Wozniak, 2007). The qualitative structure of their polysaccharide in the biofilm matrix, particularly Pel/Psl or alginate, differs between mucoid and nonmucoid *P. aeruginosa* strains (Ma et al., 2009).

P. aeruginosa biofilms or aggregations are the most common cause of progressive lung infections during cystic fibrosis (CF). Biofilm can resist phagocytosis via polysaccharide synthesis locus (Psl) and other EPS components (Høiby, Bjarnsholt, Givskov, Molin, & Ciofu, 2010), which can cause lung tissue damage due to chronic inflammation (Mishra, Manent, Chabot, & Daneault, 2012).

An extracellular polymeric material matrix encircles the population of microorganisms, which may comprise live or dead cells, proteins, sugars or polysaccharides, metabolites, and quorum sensing signaling molecules. Lactic acid bacteria, *Bacillus* spp., *L. monocytogenes*, and *Staphylococcus* spp. are the most prevalent Gram-positive bacteria that create biofilms. Periodontal disease, catheters, implant surgery, open wounds, and other probable entrance points for pathogenic organisms.

Biofilms show a wide range of phenotypes depending on the environmental conditions and the microbe's nature. In the case of *L. monocytogenes*, biofilms formed under static circumstances typically have a planktonic form and are composed of microcolonies and a homogeneous layer of cells. The biofilms generated under a continuous flow environment are circular-shaped microcolonies encircled by a network of knitted chains of elongated cells (Schembri et al., 2002). The "Save our Souls" (SOS) response factor YneA is involved in cell elongation and is activated and required to develop this kind of *L. monocytogenes* biofilm. Multilayer biofilms occur in *L. monocytogenes* (Chavant, Martinie, Meylheuc, Bellon-Fontaine, & Hebraud, 2002), and recently, three-dimensional biofilm has been reported (Lee et al., 2019). Extracellular material anchors the cells in their places, and the biofilm is made up of cells piled on top of each other around and between pores (Garcia et al., 2003). Instead of being submerged in extracellular material as described for other biofilm-associated bacteria, the cells seemed to be connected by thread-like extracellular material. *L. monocytogenes* thin biofilms are less resistant to hypochlorite (Lee, Grodzinsky, Frank, Kamm, & Schoen, 1991), dodecylbenzenesulfonic acid, and benzalkonium chloride (Frank & Koffi, 1990) than thick biofilms. Furthermore, *L. monocytogenes* in multiple species biofilm is more resistant than *L. monocytogenes* in a biofilm that solely contains the *L. monocyte* gene.

In the case of *S. aureus*, the biofilms comprise a thick layer of cells surrounded by an intricate matrix containing several kinds of polymers. It is produced in both continuous and static growth conditions. The development and maturation of biofilm in *S. aureus*, *Lactobacillus plantarum*, and *Lactococcus lactis* require the interaction of multiple regulators encoded by *sarA*, *sigB*, *agr*, and *ica* described in Table 1. Biofilms of *Bacilli* are often found with a structured pellicle that floats on culture medium at the air-liquid interface or as architecturally sophisticated biofilms colonies on solid agar or culture media (Peng, 2016; Ryder et al., 2007; Wang et al., 2014). The latter form of biofilm produces aerial-like structures known as "fruiting bodies," the sporulation sites (Wang et al., 2014). The timing and spatial control of multiple pathways are required to develop these fruiting structures. *Bacilli* can also produce submerged biofilms in static and continuous flow environments (Peng, 2016; Ryder et al., 2007). *B. subtilis* may produce sturdy biofilms, which appear as highly structured floating pellicles on liquid culture surfaces and colonies on agar plates. The reinforcement of environmental signals, such as low oxygen levels, food depletion, or surface adhesion, converts planktonic *B. subtilis* into a nonmotile state by inhibiting flagellar genes and simultaneously upregulating genes needed for the extracellular matrix formation (Böttcher, Kolodkin-Gal, Kolter, Losick, & Clardy, 2013; Chaplan, Bach, Pogrel, Chung, & Yaksh, 1994; Sheu & Freese, 1972). During biofilm development, sessile cells form chains by suppressing cell-wall hydrolases, which are encased in a self-produced extracellular matrix and give rigidity to make a strong biofilm (Branda, Chu, Kearns, Losick, & Kolter, 2006). The activity of mobile cells and the production of surfactant molecules are responsible for biofilm growth (Evans

TABLE 1 Consolidated list of Gram-positive and Gram-negative bacteria producing biofilm and the major genes involved in the biofilm formation.

Bacterial strains	Major genes involved in biofilm formation
Gram-positive bacteria	
L. monocytogenes	*sigB* (van der Veen & Abee, 2010), *prfA* (Lemon, Freitag, & Kolter, 2010), *degU* (Gueriri et al., 2008), *relA* and *hpt* (Taylor et al., 2002), *luxS* (Sela, Frank, Belausov, & Pinto, 2006)
S. aureus	*icaABCD* (Mirzaee, Najar-Peerayeh, Behmanesh, & Moghadam, 2015), *fnbA, fnbB, sasG, bap, cifB* (Beenken, Blevins, & Smeltzer, 2003)
E. faecalis	*esp, ebpABC, ace, srtA, hyl* (Lim, Teh, & Thong, 2017; Sava, Heikens, & Huebner, 2010)
B. subtilis	*spo0A* (Branda, González-Pastor, Ben-Yehuda, Losick, & Kolter, 2001; Hamon & Lazazzera, 2001), *abbA, abrB* (Tucker et al., 2014), *sinR, tapA-sipW-tasA, epsA-O* (Chu, Kearns, Branda, Kolter, & Losick, 2006; Kearns, Chu, Rudner, & Losick, 2004), *slrR* (Chu et al., 2006)
Gram-negative bacteria	
S. Typhimurium	*csgD* (Liu, Niu, Wu, & Huang, 2014), *rpoS* (Römling, Bian, Hammar, Sierralta, & Normark, 1998), *bcsABZD* and *bcsEFG* (Solano et al., 2002), *adrA* (García et al., 2004), *csgBAC* and *csgDEFG* (Collinson, Clouthier, Doran, Banser, & Kay, 1996; Römling, Bian, et al., 1998)
P. aeruginosa	*pelA, pslA, algD, ppyR, algU, algL* (Billings et al., 2013)

et al., 2009). Extracellular matrix synthesis continues as the biofilm spreads and matures, and the biofilm acquires widespread wrinkles. The lateral compressive pressures created by the biofilm matrix's stiffness, followed by the matrix's mechanical buckling, are the consequence of spatially limited cell death (Asally et al., 2012). Overall, wrinkling offers several benefits for microcolonies. At first, it raises the surface-to-volume ratio, allowing cells to get more oxygen. Second, wrinkles increase the circulation of liquids by promoting the development of a complicated network of liquid channels inside the biofilms (Wilking et al., 2013). Finally, as the biofilm matures, aerial projections appear on the surface, which serves as preferred locations for sporulation and spore dissemination.

3. *Salmonella* and biofilm formation

In the natural environment, *Salmonella* forms biofilm on the plant, meat, meat processing environment, abiotic surfaces, e.g., glass, metals, plastics, animal epithelial cells, gallstones, etc. (Steenackers, Hermans, Vanderleyden, & De Keersmaecker, 2012). EPS composition of *Salmonella* depends on serovars and environmental conditions but primarily consists of biofilm-associated proteins (BapA), cellulose, curli fimbriae, O-antigen capsule, and extracellular DNA (eDNA). In the 1980s, for the first time in *E. coli*, a proteinaceous filamentous appendage with high adhesive properties allowing surface adhesions, cell aggregation, and biofilm formation that can activate host inflammation was discovered as curli fimbriae. The *csgBAC* and *csgDEFG* operons are involved in curli protein synthesis. In *csgBAC* operon, *csgA* synthesizes curlin, an essential subunit of curli, CsgB acts as a surface exposed nucleator for CsgA, and putative CsgC has an additional function in curli fimbriae development. *csgDEFG* encodes accessory protein for curli assembly. *CsgD* from *csgDEFG* operon is the master regulator of *Salmonella* biofilm formation, which regulates the synthesis of two crucial EPS components: curli and cellulose, as shown in Fig. 2. It also encodes a transcriptional regulator of LuxR super family. *csgD* induces curli fimbriae synthesis by directly binding to the promoter region of *csgBAC* operon.

Moreover, CsgD is also involved in BapA and O-antigen capsule production. In *Salmonella* Enteritidis, CsgE synthesizes extracellular thin aggregative fimbriae, whereas the function of CsgF and CsgG in *Salmonella* is still not clear. Cellulose is another crucial EPS found in *Salmonella* biofilm. This polysaccharide consists of β $(1\rightarrow4)$-linked D-glucose unit and aid matrix with tightly packed cells with hydrophobic bond. Bacterial cellulose synthesis operons *bcsABZD* and *bcsEFG* are involved in cellulose biosynthesis. The *bcsA*, *bcsZ*, and *bcsB* genes encode regulatory subunits, which bind to cyclic diguanylic acid c-di-GMP and endoglucanase, whereas *bcsG* also shows homology to an endoglucanase gene (García et al., 2004). AdrA, a putative inner membrane protein, positively regulates cellulose biosynthesis by directly binding to the *bcs* operon genes or indirectly with c-di-GMP (bis-3′-5′ cyclic dimeric guanosine monophosphate) at posttranscriptional level (García et al., 2004; Gerstel & Römling, 2003). Another *Salmonella* biofilm component is biofilm-associated protein BapA, a cell surface protein encoded by *bapBCD* operon downstream of *bapA* gene. This gene is highly conserved in *Salmonella* and regulated by *csgD*. Deleting the *bapA* gene leads to the loss of biofilm formation, and overexpression of this gene increases the biofilm biomass formation (García et al., 2004). *Salmonella* O-antigen capsule is also a vital component of the extracellular biofilm matrix, which protects against desiccation stress and is

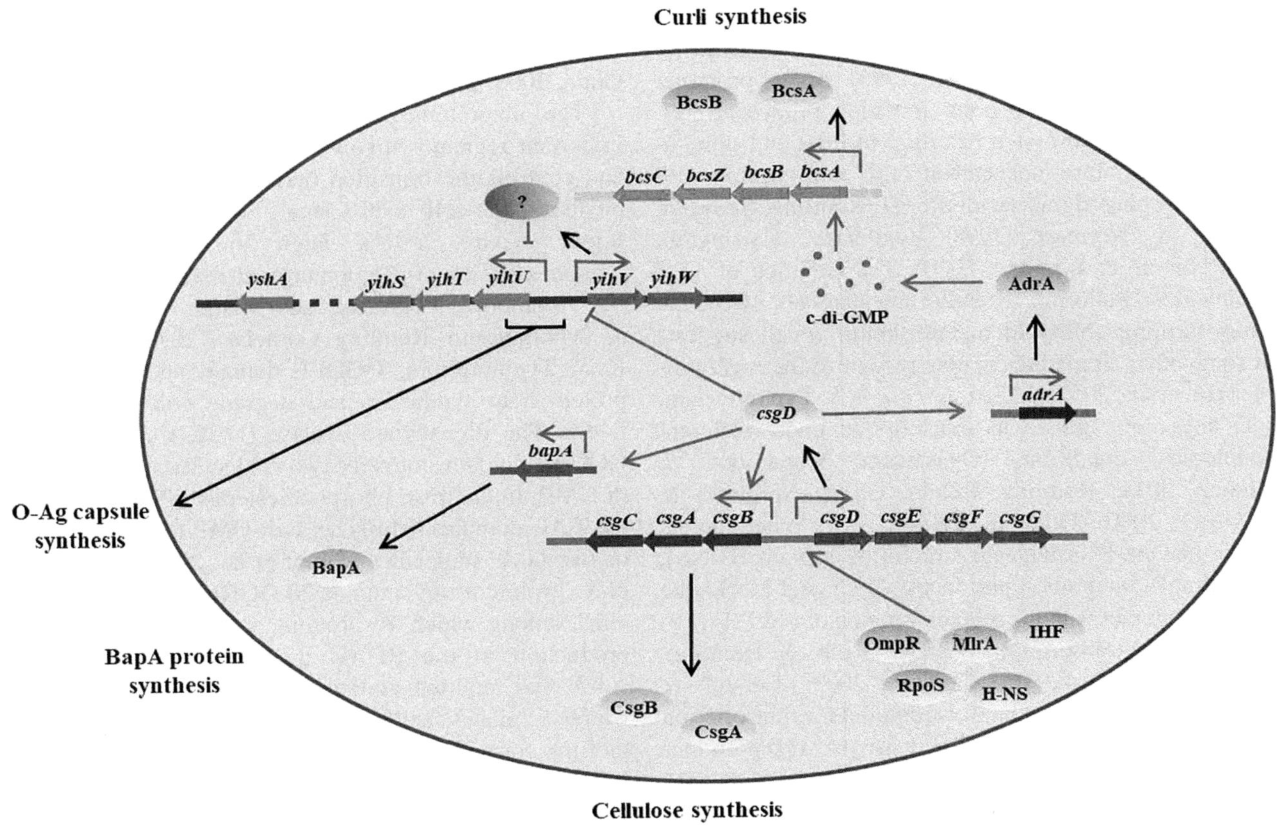

FIG. 2 Regulatory networks of *Salmonella* biofilm formation by *csgD*. The production of CsgD is regulated by many components like OmpR, MlrA, RpoS, H-NS, IHF. CsgD regulates cellulose synthesis by indirectly regulating AdrA, and AdrA further regulates the synthesis of BcsA and BcsB by modulating c-di-GMP activation. Curli synthesis is regulated directly by CsgD. CsgD also regulates the BapA protein and O-Ag capsule synthesis. In the presence of CsgD the *yihVW* operon gets repressed, and the *yihVW* operon produces a repressor which further represses *yihUTS…yshA*.

important for bacterial persistency in the environment. *yihU-yshA* and *yihVW* are the two operons involved in capsule surface assembly, translocation and differentially regulated with other matrix components by CsgD. This *yih* gene is highly conserved in *Salmonella* (Barak et al., 2007; Gibson et al., 2006; Maruzani, Sutton, Nocerino, & Marvasi, 2019). It shows destabilizing effect while forming a biofilm on abiotic surfaces by *S.* Typhi and *S.* Typhimurium (Özdemir, Akçelik, & Akçelik, 2018; Shatila, Yaşa, & Yalçın, 2021).

Upon sensing environmental stress, i.e., nutrient availability, oxygen tension, temperature, pH, etc., *S.* Typhimurium and *S.* Enteritidis show a change in their cellular morphology at 28°C. According to the Congo red binding ability on CR agar plate, cellular morphology of bacteria is grouped as follows: PDAR (pink, dry, and rough), RDAR (red, dry, and rough), SBAM (smooth, brown, and mucoid), BDAR (brown, dry, and rough), and SAW (smooth and white) (Jain & Chen, 2007). Each morphology indicates specific biofilm features depending on two essential matrix components: curli fimbriae and cellulose. Here, RDAR morphology suggests the presence of both cellulose and curli fimbriae, BDAR indicates lack of cellulose synthesis, PDAR implies a fault in curli expression, SBAM shows overproduction of capsular polysaccharide but the deficiency in cellulose production, and SAW morphology implies a deficiency in both curli and cellulose synthesis. The CsgD protein plays a vital role in RDAR morphotype expression. Mainly this multicellular behavior is observed during the stationary growth phase of bacteria when nutrients are rare and cell density increases and is the indicator of biofilm formation on the surface (Shatila et al., 2021; Simm, Ahmad, Rhen, Le Guyon, & Römling, 2014).

4. Regulation of *Salmonella* biofilm formation

The single motile cells attached to the surface become sessile during biofilm formation to form a multicellular community within EPS. This transition in bacterial physiology is accompanied by a sophisticated genetic network mainly regulated by CsgD. CsgD, known as the master regulator of biofilm, is an orphan response regulator with a C-terminal LuxR-like helix-turn-helix DNA binding motif and an N-terminal receiver domain with a conserved aspartate

(D59), a putative phosphorylation site. In the stationary phase, CsgD expression is regulated by the stress response gene *rpoS* (Römling, Bian, et al., 1998). At the promoter region of *csgD*, a single point mutation relieves RpoS-dependent *csgD* expression resulting in threefold upregulation of temperature-independent *csgD* transcription with the housekeeping sigma factor RpoD (Römling, Sierralta, Eriksson, & Normark, 1998; Zakikhany, Harrington, Nimtz, Hinton, & Römling, 2010). The mutation of *rpoS* in *Salmonella* Pullorum (*S.* Pullorum) does not affect the biofilm-forming ability of mutant strain, which suggests that some other sigma factors may recognize the *csgD* promoter (Lu et al., 2012). Unlike *E. coli*, in *S.* Typhimurium, CsgD does not regulate its transcription under different physiological conditions (Ogasawara, Yamamoto, & Ishihama, 2011; Römling, Rohde, Olsén, Normark, & Reinköster, 2000). Other than RpoS, some other factors such as histone-like nucleoid structuring protein (H-NS), RpoE, OmpR, integration host factor (IHF), and MerR-like MlrA also interact with the promoter region of *csgD* (Fig. 2) to regulate its transcription (Gerstel, Kolb, & Römling, 2006; Gerstel, Park, & Römling, 2003; Gerstel & Römling, 2003). Firstly, OmpR is required for transcription of *csgD* and has six binding sites (D1–D6) in *csgD* promoter region (Römling, Sierralta, et al., 1998). It was also reported that mutation of *ompR* in *S.* Pullorum and *S.* Enteritidis results in impaired cellulose, curli, and biofilm production (Dong et al., 2011; Lu et al., 2012). IHF and OmpR compete with each other to bind the overlapping region of IHF1/OmpR D3–D6 sites to activate transcription of *csgD* in response to oxygen tension (Gerstel et al., 2003). Mutation of IHF shows less RDAR morphotype (Gerstel et al., 2003). Upon deletion of *rpoE*, the mutant strain shows significantly reduced expression of *csgD* and biofilm, suggesting that RpoE is also a regulator that recognizes *csgD* promoter in *Salmonella* (Huang et al., 2015). The biofilm formation mechanisms in *S.* Typhimurium are similar to *E. coli*, but in *S.* Typhimurium, cAMP-CRP downregulates the transcription of *csgD*, unlike *E. coli*, which inhibits biofilm formation and promotes pellicle formation (Paytubi, Cansado, Madrid, & Balsalobre, 2017). CsgD is a regulator of curli synthesis operon (*csgBAC* and *csgDEFG*), which binds to the *csgB* and helps in the transcription of curli synthesis (Römling, Sierralta, et al., 1998; Zakikhany et al., 2010). CsgD also acts as a positive regulator of *adrA*, which ultimately positively regulates the transcription of cellulose synthase *bcsA* via an allosteric activator c-di-GMP and helps in cellulose synthesis (Römling et al., 2000; Simm, Morr, Kader, Nimtz, & Römling, 2004; Zakikhany et al., 2010). In vitro experiments suggest that, for the expression of *csgBA*, the master regulator CsgD interacts with the sigma factor RpoD and RpoS. In contrast, for *adrA* expression, it only interacts with RpoS (Zakikhany et al., 2010). In *Salmonella* Enteritidis, CsgD also regulates the expression of other extracellular components such as the O-antigen capsule and BapA (Crawford, Gibson, Kay, & Gunn, 2008; Latasa et al., 2005).

The ubiquitous second messenger c-di-GMP is an important regulator involved in many bacterial processes and controls the transition between motility and sessility on the single-cell level (Jenal, 2004; Ross et al., 1987; Simm et al., 2004). High intracellular c-di-GMP regulates biofilm production through master regulator CsgD by promoting extracellular matrix production in *S.* Typhimurium (Römling, Gomelsky, & Galperin, 2005). In *S.* Typhimurium, GGDEF domain and EAL domain protein help synthesize and degrade c-di-GMP, respectively. The diguanylate cyclase (DGCs) encoded in the GGDEF domain converts two molecules of GTP into c-di-GMP. In contrast, phosphodiesterase (PDE) encoded in the EAL domain hydrolyzes c-di-GMP to linear 5′-pGpG or two GMP molecules (Simm et al., 2004). The genome of *S.* Typhimurium contains 20 GGDEF/EAL domain proteins, among which 10 proteins are involved in cellulose production in the RDAR morphotype. In *Salmonella*, AdrA was reported as the first GGDEF domain protein, which regulates cellulose synthesis and has a role in biofilm formation (Zogaj, Nimtz, Rohde, Bokranz, & Römling, 2001). Posttranscriptionally, AdrA regulates the cellulose synthesis by modulating the *bcs* gene through c-di-GMP in *Salmonella*. Upon binding c-di-GMP with the C-terminal PilZ receptor domain, the cellulose synthase BcsA gets activated (Ryjenkov, Simm, Römling, & Gomelsky, 2006).

In *S.* Typhimurium, Hfq, a conserved RNA-binding protein, stabilizes sRNA-mRNA interactions that play an important role in biofilm formation. Bacterial sRNAs modulate gene expression in two ways: sRNA-protein interaction and sRNA-cognate mRNA interaction (Storz, Vogel, & Wassarman, 2011). In *S.* Typhimurium, it was also reported that Hfq partially regulates *csgD* expression via regulating the translation of *rpoS* (Mika & Hengge, 2014). In *E. coli*, some sRNAs such as GcvB, RprA, OmrA/B, and McaS downregulate CsgD translation (Boehm & Vogel, 2012). In *E. coli* and *Salmonella*, stimulation of RydC prevents initiation of CsgD translation, reducing biofilm formation due to impair curli synthesis (Bordeau & Felden, 2014). In *S.* Typhimurium, two Hfq-dependent sRNAs (SdsR and ArcZ) positively regulate rdar morphotype expression (Monteiro et al., 2012). Overexpression of sRNAs, OmrA/B inhibit the curli synthesis (Holmqvist et al., 2010). In *Salmonella*, six sRNAs are reported; *arcZ*, *sroC*, and *csrB* positively regulate biofilm formation while *dsrA*, *oxyS*, *rprA* significantly downregulate biofilm formation (Chambers & Sauer, 2013; Jennings et al., 2011).

Salmonella CsrA is a global regulator that controls carbon metabolism (Lawhon et al., 2003). Mutation of *csrA* in *S*. Typhimurium alters the expression of 8 among the 20 GGDEF and EAL domains and downregulates the RDAR morphotype expression. In *Salmonella*, FabR (a fatty acid biosynthesis repressor) targets FabB to regulate biofilm formation as the *fabR* mutant strain cannot form a biofilm (Hermans et al., 2016). Based upon the above stated information, we can understand how complex the regulation of biofilm formation in *Salmonella* is. This also indicates that the system has a significant role in bacterial physiology and virulence.

5. Role of biofilm in *Salmonella* pathogenesis

In the last five decades, the role of biofilm in chronic infection and pathogenesis has gotten considerable attention. According to the National Institutes of Health, over 80% of chronic infections are related to biofilm, suggesting biofilm's importance in the medical context (Davies, 2003). As biofilm members are resistant to multiple antibiotics, it is difficult to eradicate biofilms by standard antimicrobial treatment, which leads to chronic infection. Certain biofilm-associated diseases are chronic wound infections, otitis, respiratory diseases, etc. (Harrell et al., 2021; Rabin et al., 2015; Simm et al., 2014; Zeng, Zhang, Wang, & Meng, 2017). The trait for biofilm formation is quite common in most of the serotypes of *Salmonella enterica* enabling bacterial transmission, persistency, and survival inside the host. Various studies have shown that the development of *S*. Typhimurium biofilm formation on cultured epithelial cells, i.e., epithelial-like HEp-2 cells, mimics the early infection establishment where *S*. Typhimurium outcompetes with *E. coli* and displaces them by heterologous infection and biofilm formation (Steenackers et al., 2012). In HT-29 cell, biofilm produced by *S*. Typhimurium implied reduced virulence by producing pro-inflammatory cytokines IL-8 (Harrell et al., 2021). The biofilm formation of microorganisms that lead to chronic infections is mainly associated with increased cellular c-di-GMP concentrations (Römling, Galperin, & Gomelsky, 2013). The elevated level of c-di-GMP in *S*. Typhimurium inhibits major virulence determinants such as invasion and secretion of pro-inflammatory cytokine IL-8 in HT-29 cells by c-di-GMP-mediated inhibition of CsgD (Ahmad et al., 2011; Lamprokostopoulou, Monteiro, Rhen, & Römling, 2010). The c-di-GMP signaling system regulates the biofilm formation in bacteria and regulates the virulence gene expression of plant and animal pathogens (Tamayo, Pratt, & Camilli, 2007). CsgD regulates the expression of *csgA*, which is involved in the curli synthesis. However, AdrA is involved in cellulose synthesis by modulating the expression of *bcsA* operon at the posttranscriptional level. The single mutants Δ*bcsA* and Δ*csgA* of *S*. Pullorum strains are defective in the production of cellulose and curli, respectively, which results in less biofilm production in both the strains (El Hag et al., 2017). These two single-mutant strains of *E. coli* are not significantly affected in their adherence to HT-29 cells. However, the double-mutant Δ*bcsAcsgA* strain shows significantly reduced adherence to HT-29 cell compared with wild type, suggesting the synergistic role of curli and cellulose in cell adherence (Saldaña et al., 2009). Mutation of *csgA* in *S*. Pullorum showed reduced adhesion and invasion in Hela cell, whereas *bcsA* mutation strain has no effect (El Hag et al., 2017). In the chicken model, the *csgA* mutant strain shows attenuated virulence, which might be due to the reduced adherence and invasion or less proliferation inside macrophages. In contrast, the *bcsA* mutant shows virulence similar to the wild-type S6702 strain (El Hag et al., 2017). Inside the macrophages, MgtC represses cellulose synthesis, a major component of biofilm, which interferes with intracellular replication and virulence of bacteria in cell lines and mice (Pontes, Lee, Choi, & Groisman, 2015).

Gallstone is another well-studied biotic factor where *Salmonella* can form biofilm. The study has revealed that *S*. Typhimurium, *S*. Typhi, *S*. Enteritidis can form biofilm on gallstones surfaces within 14 days in vitro in bile-dependent way (Prouty et al., 2002). Besides bacterial virulence, biofilm enables the persistency of bacterium inside the host. Studies revealed that biofilm produced by *S*. Typhimurium acts as an important factor for causing asymptomatic infection in the cecum of new-born chicken (Harrell et al., 2021). In *Caenorhabditis elegans* gut, *S*. Typhimurium forms biofilm for persistency and survival inside the host (Desai, Padmanabhan, Harshe, Zaidel-Bar, & Kenney, 2019). Biofilm matrix components act as a virulence factor in acute infections. A recent study has shown that *S*. Typhimurium forms biofilm on the tumor cell upon intravenous injection of mice (Crull et al., 2011). Although plants are not considered a host for enteric pathogen *S*. Typhimurium, there was still evidence that *S*. Typhimurium forms biofilm within plant tissue, leading to significant damage to crops (Harrell et al., 2021; Yaron & Römling, 2014). Some evidences show both negative and positive correlations between biofilm and bacterial virulence. In *Salmonella*, *luxS* is a major biofilm regulator gene responsible for the synthesis of AI-2 synthase, an autoinducer, which regulates SPI-1 genes, hence bacterial entry and virulence (Choi, Shin, & Ryu, 2007). In some other bacteria, such as *Staphylococcus epidermidis*, *luxS* gene negatively regulates biofilm formation and virulence (Xu et al., 2006). There are reports describing the increased flagellar number and reduced biofilm production (van Ditmarsch et al., 2013). The laboratory-generated hypervirulent passaged strains of *S*. Typhimurium have increased flagella number, overexpression of *rpoS* and *luxS* in comparison to wild type but show reduced biofilm formation (Pradhan

& Negi, 2019). We can conclude that *Salmonella* biofilm and associated regulatory molecules have an essential role in enabling the bacteria to enter, persist, and disseminate in the host system and regulate virulence. The loss of any of the factors discussed above significantly impacts bacterial pathogenesis. The subsequent section discusses biofilm formation and its regulation in *S*. Typhi, a human-specific pathogen.

6. The role of biofilm in *Salmonella* Typhi pathogenesis

S. Typhi caused an acute illness, typhoid fever, with symptoms such as fever, nausea, vomiting, headache, and abdominal cramp. These microbes usually pass via the bowels around a week after the symptoms of the disease subside. A part of *S*. Typhi persists in an asymptomatic carrier by colonizing in gall bladders, generally linked by the presence of gallstones (Crawford, Rosales-Reyes, et al., 2010). Despite being extremely contagious, 25% of these carriers remain completely asymptomatic for decades during the chronic phase and make identifying the infected person difficult. Most people suffering from chronic infection recover after the requisite treatment, but 3%–5% of patients develop chronic infection (Tsai, Liang, Chen, & Chiu, 2020). The geographical distribution of endemic typhoid carriers and gallstones with biofilm was also burdened with high gall bladder carcinoma incidences (Dutta, Garg, Kumar, & Tandon, 2000). The capacity of *S*. Typhi to cause illness and elicit a protective immune response has traditionally been attributed to the presence of an exopolysaccharide capsule (Crawford et al., 2008). It is hypothesized that, in the gallbladder environment, the O-antigen capsule is induced by bile aid chronic carriage of *Salmonella* (Crawford et al., 2008). However, it is widely recognized that the bug can cause sickness even in the absence of O-antigen capsules (Arya, 2002; Wain et al., 2005). Because biofilm has a protective effect comparable to that of a capsule and acts as a shield, allowing for extended life in the body and justifies the carrier status of the bug (Raza et al., 2011). Another polysaccharide capsule attributing strong virulence to *S*. Typhi biofilm is the Vi antigen encoded on SPI-7 (Seth-Smith, 2008). This antigen resists complement deposition on the bacterium and reduces self-phagocytosis by reducing the PAMP expression (Atif, Winter, Winter, McSorley, & Bäumler, 2014; Sharma & Qadri, 2004; Wangdi et al., 2014). Also, Vi antigen and other exopolysaccharides involved in biofilm formation prevent TLR4-mediated recognition by macrophages and dendritic cells (DCs) by masking the lipopolysaccharide (LPS) (Wilson et al., 2008). An additional virulence factor contributing to the increase in the frequency of the carrier state pathogenesis of *S*. Typhi is the typhoid toxin. Typhoid toxin abrogates the intestinal inflammatory response by causing DNA damage in *S*. Typhi acute phase of infection while aiding in sustaining the chronic state of colonization by developing gallbladder carcinoma (Del Bel Belluz et al., 2016). The biofilm also helps in chronic persistence in the gallstones, leads to local inflammatory response, increases the epithelial exposure to toxins, and might lead to the onset of gallbladder cancer (Di Domenico, Cavallo, Pontone, Toma, & Ensoli, 2017).

S. Typhi is reported to form biofilm on gallstones, where the bacteria lead to its carrier state. Cholesterol is the main component of gall bladder stones, while bile duct stones are constituted majorly by calcium bilirubinate (Hofmann, 2005). Clinically administered antibiotics are often ineffective against dissolving of the chronic bacterial infection in patients with both *S*. Typhi and cholesterol gallstones in the gallbladder (Gonzalez-Escobedo et al., 2011) and pose a high risk for developing hepatobiliary carcinomas (Dutta et al., 2000; Kumar, Kumar, & Kumar, 2006). The primary bile acid degradation in the gall bladder by the bacterium and its chronic inflammation might affect the carcinoma development process. Cholecystectomy (gallbladder removal) has been the most successful treatment for chronic typhoid and gallstone carriers. *S*. Typhi initiates the attachment over the gallstone surfaces by FliC of the flagellar appendages (Crawford, Reeve, & Gunn, 2010).

S. Typhi biofilm can develop on live and dead tissues and is beneficial to the bug in terms of reduced bacterial clearance in association with the outer capsular polysaccharide. Biofilm also reduces the efficiency of antityphoidal first-line antibiotics, including ciprofloxacin, along with 60% of the isolated strains showing multidrug resistance (González, Alberts, Lee, Doolittle, & Gunn, 2018; Narasanna, 2017). Although these antibiotics successfully remove the planktonic *S*. Typhi, the *S*. Typhi exhibits slow, undistinguished, and intermittent symptoms by resisting the cellular and humoral immune response while persisting in the biofilm. *S*. Typhi also produces extended-spectrum ß-lactamase (ESBL), which provides resistance against cefotaxime at the MIC level (Narasanna, 2017).

Bile plays an essential physiological role in lipid emulsification along with its solubilization. Nevertheless, bile has detergent properties, and it shows effective antimicrobial activity by interacting with proteins and phospholipids of microbial membranes disturbing their homeostasis (Walawalkar, Vaidya, & Nayak, 2016). Since the primary site of *S*. Typhi carriage is the gallbladder, which is rich in bile, the bug is supposed to have a highly efficient bile resistance mechanism. Most *Salmonella* proteins being affected by bile are from PhoP-PhoQ (PhoPQ) regulon. The regulated genes required for LPS modification and those providing resistance from the antimicrobial peptides, and so, increase the bacterial persistence within macrophages during the acute infection (Groisman, Parra-Lopez, Salcedo, Lipps, & Heffron, 1992; Gunn et al., 1998; Guo et al., 1998). PhoPQ also plays a vital role in increased bile resistance as mutant bacteria constitutively expressing PhoPQ are highly resistant to bile. In contrast, *phoP* null mutant exhibits significantly lower bile resistance. However, several gene mutants regulated by PhoP were tested for alterations in transcription, and bile resistance patterns reveal that bile sensing by PhoPQ does not lead to enhanced resistance to bile or upregulated protein production in the presence of bile. Therefore, it is hypothesized that in *Salmonella*, both the PhoPQ regulon and the "bile regulon" contain overlapping genes, which provides enhanced bile resistance on activation (van Velkinburgh & Gunn, 1999).

7. Conclusions

The foodborne infection by *Salmonella* has become a global health problem, and biofilm-forming capability further increases the amount of effort we need to put into eradicating the bug. Biofilm is a complex multicellular community of microorganisms embedded in a self-produced extracellular matrix attached to biotic or abiotic surfaces. *Salmonella*'s biofilm formation ability helps them grow on different surfaces and persist in harsh environmental conditions for a more extended period. Microbes within biofilm become resistant to environmental stress compared with planktonic counterparts. However, biofilm involves 80% chronic infection, which increases hospitalization rates, morbidity rates, and mortality rates. Nowadays, infections caused by biofilm-associated bacteria have become a severe clinical issue globally. Eradication of biofilm-related infection becomes a challenge because of their antibiotic resistance properties and lack of understanding of the complex regulatory network of biofilm. The biofilm-forming capability of the pathogen plays an essential role in its persistence, dissemination and dispersal, and hence, overall pathogenesis. We still need further understanding to have better combat strategies and develop the therapeutics to deal with the pathogenesis of the *S.* Typhi/Typhimurium and any other biofilm-forming pathogens.

References

Ahmad, I., Lamprokostopoulou, A., Le Guyon, S., Streck, E., Barthel, M., Peters, V., et al. (2011). Complex c-di-GMP signaling networks mediate transition between virulence properties and biofilm formation in *Salmonella enterica* serovar Typhimurium. *PLoS One*, *6*(12), e28351.

Arya, S. C. (2002). Field effectiveness of Vi polysaccharide typhoid vaccine in the People's Republic of China. *The Journal of Infectious Diseases*, *185*(6), 845.

Asally, M., Kittisopikul, M., Rué, P., Du, Y., Hu, Z., Çağatay, T., et al. (2012). Localized cell death focuses mechanical forces during 3D patterning in a biofilm. *Proceedings of the National Academy of Sciences*, *109*(46), 18891–18896.

Atif, S. M., Winter, S. E., Winter, M. G., McSorley, S. J., & Bäumler, A. J. (2014). *Salmonella enterica* serovar Typhi impairs CD4 T cell responses by reducing antigen availability. *Infection and Immunity*, *82*(6), 2247–2254.

Barak, J. D., Jahn, C. E., Gibson, D. L., & Charkowski, A. O. (2007). The role of cellulose and O-antigen capsule in the colonization of plants by *Salmonella enterica*. *Molecular Plant-Microbe Interactions*, *20*(9), 1083–1091.

Beenken, K. E., Blevins, J. S., & Smeltzer, M. S. (2003). Mutation of sarA in *Staphylococcus aureus* limits biofilm formation. *Infection and Immunity*, *71*(7), 4206–4211.

Beloin, C., Roux, A., & Ghigo, J.-M. (2008). *Escherichia coli* biofilms. *Current Topics in Microbiology and Immunology*, *322*, 249–289.

Billings, N., Ramirez Millan, M., Caldara, M., Rusconi, R., Tarasova, Y., Stocker, R., et al. (2013). The extracellular matrix component Psl provides fast-acting antibiotic defense in *Pseudomonas aeruginosa* biofilms. *PLoS Pathogens*, *9*(8), e1003526.

Boehm, A., & Vogel, J. (2012). The csgD mRNA as a hub for signal integration via multiple small RNAs. *Molecular Microbiology*, *84*(1), 1–5.

Bogino, P. C., Oliva, M. D. L. M., Sorroche, F. G., & Giordano, W. (2013). The role of bacterial biofilms and surface components in plant-bacterial associations. *International Journal of Molecular Sciences*, *14*(8), 15838–15859.

Bordeau, V., & Felden, B. (2014). Curli synthesis and biofilm formation in enteric bacteria are controlled by a dynamic small RNA module made up of a pseudoknot assisted by an RNA chaperone. *Nucleic Acids Research*, *42*(7), 4682–4696.

Böttcher, T., Kolodkin-Gal, I., Kolter, R., Losick, R., & Clardy, J. (2013). Synthesis and activity of biomimetic biofilm disruptors. *Journal of the American Chemical Society*, *135*(8), 2927–2930.

Branda, S. S., Chu, F., Kearns, D. B., Losick, R., & Kolter, R. (2006). A major protein component of the *Bacillus subtilis* biofilm matrix. *Molecular Microbiology*, *59*(4), 1229–1238.

Branda, S. S., González-Pastor, J. E., Ben-Yehuda, S., Losick, R., & Kolter, R. (2001). Fruiting body formation by *Bacillus subtilis*. *Proceedings of the National Academy of Sciences*, *98*(20), 11621–11626.

Branda, S. S., Vik, Å., Friedman, L., & Kolter, R. (2005). Biofilms: The matrix revisited. *Trends in Microbiology*, *13*(1), 20–26.

Chambers, J. R., & Sauer, K. (2013). Small RNAs and their role in biofilm formation. *Trends in Microbiology*, *21*(1), 39–49.

Chaplan, S. R., Bach, F., Pogrel, J., Chung, J., & Yaksh, T. (1994). Quantitative assessment of tactile allodynia in the rat paw. *Journal of Neuroscience Methods*, *53*(1), 55–63.

Chavant, P., Martinie, B., Meylheuc, T., Bellon-Fontaine, M.-N., & Hebraud, M. (2002). *Listeria monocytogenes* LO28: Surface physicochemical properties and ability to form biofilms at different temperatures and growth phases. *Applied and Environmental Microbiology*, *68*(2), 728–737.

Choi, J., Shin, D., & Ryu, S. (2007). Implication of quorum sensing in *Salmonella enterica* serovar Typhimurium virulence: The luxS gene is necessary for expression of genes in pathogenicity island 1. *Infection and Immunity*, *75*(10), 4885–4890.

Chu, F., Kearns, D. B., Branda, S. S., Kolter, R., & Losick, R. (2006). Targets of the master regulator of biofilm formation in *Bacillus subtilis*. *Molecular Microbiology*, *59*(4), 1216–1228.

Collinson, S. K., Clouthier, S. C., Doran, J. L., Banser, P. A., & Kay, W. W. (1996). *Salmonella enteritidis* agfBAC operon encoding thin, aggregative fimbriae. *Journal of Bacteriology*, *178*(3), 662–667.

Crawford, R. W., Gibson, D. L., Kay, W. W., & Gunn, J. S. (2008). Identification of a bile-induced exopolysaccharide required for *Salmonella* biofilm formation on gallstone surfaces. *Infection and Immunity*, *76*(11), 5341–5349.

Crawford, R. W., Reeve, K. E., & Gunn, J. S. (2010). Flagellated but not hyperfimbriated *Salmonella enterica* serovar Typhimurium attaches to and forms biofilms on cholesterol-coated surfaces. *American Society for Microbiology*, *192*, 2981–2990.

Crawford, R. W., Rosales-Reyes, R., de la Luz Ramírez-Aguilar, M., Chapa-Azuela, O., Alpuche-Aranda, C., & Gunn, J. S. (2010). Gallstones play a significant role in *Salmonella* spp. gallbladder colonization and carriage. *Proceedings of the National Academy of Sciences*, *107*(9), 4353–4358.

Crull, K., Rohde, M., Westphal, K., Loessner, H., Wolf, K., Felipe-López, A., et al. (2011). Biofilm formation by *Salmonella enterica* serovar Typhimurium colonizing solid tumours. *Cellular Microbiology*, *13* (8), 1223–1233.

Davies, D. (2003). Understanding biofilm resistance to antibacterial agents. *Nature Reviews Drug Discovery*, *2*(2), 114–122.

Del Bel Belluz, L., Guidi, R., Pateras, I. S., Levi, L., Mihaljevic, B., Rouf, S. F., et al. (2016). The typhoid toxin promotes host survival and the establishment of a persistent asymptomatic infection. *PLoS Pathogens*, *12*(4), e1005528.

Desai, S. K., Padmanabhan, A., Harshe, S., Zaidel-Bar, R., & Kenney, L. J. (2019). *Salmonella* biofilms program innate immunity for persistence in *Caenorhabditis elegans*. *Proceedings of the National Academy of Sciences*, *116*(25), 12462–12467.

Di Domenico, E. G., Cavallo, I., Pontone, M., Toma, L., & Ensoli, F. (2017). Biofilm producing *Salmonella* typhi: Chronic colonization and development of gallbladder cancer. *International Journal of Molecular Sciences*, *18*(9), 1887.

Dong, H., Peng, D., Jiao, X., Zhang, X., Chen, S., Lu, Y., et al. (2011). Construction and characterization of an ompR gene deletion mutant from *Salmonella enteritidis*. *Wei sheng wu xue bao= Acta microbiologica Sinica*, *51*(9), 1256–1262.

Dutta, U., Garg, P. K., Kumar, R., & Tandon, R. K. (2000). Typhoid carriers among patients with gallstones are at increased risk for carcinoma of the gallbladder. *The American Journal of Gastroenterology*, *95*(3), 784–787.

El Hag, M., Feng, Z., Su, Y., Wang, X., Yassin, A., Chen, S., et al. (2017). Contribution of the csgA and bcsA genes to *Salmonella enterica* serovar Pullorum biofilm formation and virulence. *Avian Pathology*, *46*(5), 541–547.

Evans, P., Beardmore, A., Page, K., Osborne, J., O'Brien, P., Willingale, R., et al. (2009). Methods and results of an automatic analysis of a complete sample of Swift-XRT observations of GRBs. *Monthly Notices of the Royal Astronomical Society*, *397*(3), 1177–1201.

Flemming, H.-C., Wingender, J., Szewzyk, U., Steinberg, P., Rice, S. A., & Kjelleberg, S. (2016). Biofilms: An emergent form of bacterial life. *Nature Reviews Microbiology*, *14*(9), 563–575.

Flemming, H.-C., & Wuertz, S. (2019). Bacteria and archaea on Earth and their abundance in biofilms. *Nature Reviews Microbiology*, *17*(4), 247–260.

Foley, S., Lynne, A., & Nayak, R. (2008). *Salmonella* challenges: Prevalence in swine and poultry and potential pathogenicity of such isolates. *Journal of Animal Science*, *86*(Suppl. 14), E149–E162.

Frank, J. F., & Koffi, R. A. (1990). Surface-adherent growth of *Listeria monocytogenes* is associated with increased resistance to surfactant sanitizers and heat. *Journal of Food Protection*, *53*(7), 550–554.

García, B., Latasa, C., Solano, C., Portillo, F. G. D., Gamazo, C., & Lasa, I. (2004). Role of the GGDEF protein family in *Salmonella* cellulose biosynthesis and biofilm formation. *Molecular Microbiology*, *54*(1), 264–277.

Garcia, C., Robinson, J., Guethlein, L., Parham, P., Madrigal, J., & Marsh, S. (2003). Human KIR sequences 2003. *Immunogenetics*, *55*(4), 227–239.

Gerstel, U., Kolb, A., & Römling, U. (2006). Regulatory components at the csgD promoter–additional roles for OmpR and integration host factor and role of the 5′ untranslated region. *FEMS Microbiology Letters*, *261* (1), 109–117.

Gerstel, U., Park, C., & Römling, U. (2003). Complex regulation of csgD promoter activity by global regulatory proteins. *Molecular Microbiology*, *49*(3), 639–654.

Gerstel, U., & Römling, U. (2003). The csgD promoter, a control unit for biofilm formation in *Salmonella typhimurium*. *Research in Microbiology*, *154*(10), 659–667.

Gibson, D., White, A., Snyder, S., Martin, S., Heiss, C., Azadi, P., et al. (2006). *Salmonella* produces an O-antigen capsule regulated by AgfD and important for environmental persistence. *Journal of Bacteriology*, *188*(22), 7722–7730.

González, J. F., Alberts, H., Lee, J., Doolittle, L., & Gunn, J. S. (2018). Biofilm formation protects *Salmonella* from the antibiotic ciprofloxacin in vitro and in vivo in the mouse model of chronic carriage. *Scientific Reports*, *8*(1), 1–8.

Gonzalez-Escobedo, G., Marshall, J. M., & Gunn, J. S. (2011). Chronic and acute infection of the gall bladder by *Salmonella* Typhi: Understanding the carrier state. *Nature Reviews Microbiology*, *9*(1), 9–14.

Groisman, E. A., Parra-Lopez, C., Salcedo, M., Lipps, C. J., & Heffron, F. (1992). Resistance to host antimicrobial peptides is necessary for *Salmonella* virulence. *Proceedings of the National Academy of Sciences*, *89*(24), 11939–11943.

Gueriri, I., Cyncynatus, C., Dubrac, S., Arana, A. T., Dussurget, O., & Msadek, T. (2008). The DegU orphan response regulator of *Listeria monocytogenes* autorepresses its own synthesis and is required for bacterial motility, virulence and biofilm formation. *Microbiology*, *154*(8), 2251–2264.

Gunn, J. S., Lim, K. B., Krueger, J., Kim, K., Guo, L., Hackett, M., et al. (1998). PmrA–PmrB-regulated genes necessary for 4-aminoarabinose lipid A modification and polymyxin resistance. *Molecular Microbiology*, *27*(6), 1171–1182.

Guo, L., Lim, K. B., Poduje, C. M., Daniel, M., Gunn, J. S., Hackett, M., et al. (1998). Lipid A acylation and bacterial resistance against vertebrate antimicrobial peptides. *Cell*, *95*(2), 189–198.

Hamon, M. A., & Lazazzera, B. A. (2001). The sporulation transcription factor Spo0A is required for biofilm development in *Bacillus subtilis*. *Molecular Microbiology*, *42*(5), 1199–1209.

Harrell, J. E., Hahn, M. M., D'Souza, S. J., Vasicek, E. M., Sandala, J. L., Gunn, J. S., et al. (2021). *Salmonella* biofilm formation, chronic infection, and immunity within the intestine and hepatobiliary tract. *Frontiers in Cellular and Infection Microbiology*, *10*, 910.

Hermans, K., Roberfroid, S., Thijs, I. M., Kint, G., De Coster, D., Marchal, K., et al. (2016). FabR regulates *Salmonella* biofilm formation via its direct target FabB. *BMC Genomics*, *17*(1), 1–15.

Hofmann, A. F. (2005). *Helicobacter* and cholesterol gallstones: Do findings in the mouse apply to man? *Gastroenterology*, *128*(4), 1126–1129.

Høiby, N., Bjarnsholt, T., Givskov, M., Molin, S., & Ciofu, O. (2010). Antibiotic resistance of bacterial biofilms. *International Journal of Antimicrobial Agents*, *35*(4), 322–332.

Holmqvist, E., Reimegård, J., Sterk, M., Grantcharova, N., Römling, U., & Wagner, E. G. H. (2010). Two antisense RNAs target the transcriptional regulator CsgD to inhibit curli synthesis. *The EMBO Journal*, *29*(11), 1840–1850.

Huang, J., Chen, S., Huang, K., Yang, L., Wu, B., & Peng, D. (2015). Identification of rpoE gene associated with biofilm formation of *Salmonella* pullorum. *Wei sheng wu xue bao= Acta microbiologica Sinica*, *55*(2), 156–163.

Jain, S., & Chen, J. (2007). Attachment and biofilm formation by various serotypes of *Salmonella* as influenced by cellulose production and thin aggregative fimbriae biosynthesis. *Journal of Food Protection*, *70*(11), 2473–2479.

Jenal, U. (2004). Cyclic di-guanosine-monophosphate comes of age: A novel secondary messenger involved in modulating cell surface structures in bacteria? *Current Opinion in Microbiology*, *7*(2), 185–191.

Jennings, M. E., Quick, L. N., Soni, A., Davis, R. R., Crosby, K., Ott, C. M., et al. (2011). Characterization of the *Salmonella enterica* serovar Typhimurium ydcI gene, which encodes a conserved DNA binding protein required for full acid stress resistance. *Journal of Bacteriology*, *193*(9), 2208–2217.

Johnson, L., Horsman, S. R., Charron-Mazenod, L., Turnbull, A. L., Mulcahy, H., Surette, M. G., et al. (2013). Extracellular DNA-induced antimicrobial peptide resistance in *Salmonella enterica* serovar Typhimurium. *BMC Microbiology*, *13*(1), 1–8.

Joseph, B., Otta, S., Karunasagar, I., & Karunasagar, I. (2001). Biofilm formation by *Salmonella* spp. on food contact surfaces and their sensitivity to sanitizers. *International Journal of Food Microbiology*, *64* (3), 367–372.

Kearns, D. B., Chu, F., Rudner, R., & Losick, R. (2004). Genes governing swarming in *Bacillus subtilis* and evidence for a phase variation mechanism controlling surface motility. *Molecular Microbiology*, *52*(2), 357–369.

Kostakioti, M., Hadjifrangiskou, M., & Hultgren, S. J. (2013). Bacterial biofilms: Development, dispersal, and therapeutic strategies in the dawn of the postantibiotic era. *Cold Spring Harbor Perspectives in Medicine*, *3*(4), a010306.

Kumar, S., Kumar, S., & Kumar, S. (2006). Infection as a risk factor for gallbladder cancer. *Journal of Surgical Oncology*, *93*(8), 633–639.

Lamprokostopoulou, A., Monteiro, C., Rhen, M., & Römling, U. (2010). Cyclic di-GMP signalling controls virulence properties of *Salmonella enterica* serovar Typhimurium at the mucosal lining. *Environmental Microbiology*, *12*(1), 40–53.

Latasa, C., Roux, A., Toledo-Arana, A., Ghigo, J. M., Gamazo, C., Penadés, J. R., et al. (2005). BapA, a large secreted protein required for biofilm formation and host colonization of *Salmonella enterica* serovar Enteritidis. *Molecular Microbiology*, *58*(5), 1322–1339.

Lawhon, S. D., Frye, J. G., Suyemoto, M., Porwollik, S., McClelland, M., & Altier, C. (2003). Global regulation by CsrA in *Salmonella typhimurium*. *Molecular Microbiology*, *48*(6), 1633–1645.

Ledeboer, N. A., Frye, J. G., McClelland, M., & Jones, B. D. (2006). *Salmonella enterica* serovar Typhimurium requires the Lpf, Pef, and Tafi fimbriae for biofilm formation on HEp-2 tissue culture cells and chicken intestinal epithelium. *Infection and Immunity*, *74*(6), 3156–3169.

Lee, B.-H., Cole, S., Badel-Berchoux, S., Guillier, L., Felix, B., Krezdorn, N., et al. (2019). Biofilm formation of *Listeria monocytogenes* strains under food processing environments and pan-genome-wide association study. *Frontiers in Microbiology*, *10*, 2698.

Lee, R. T., Grodzinsky, A. J., Frank, E. H., Kamm, R. D., & Schoen, F. J. (1991). Structure-dependent dynamic mechanical behavior of fibrous caps from human atherosclerotic plaques. *Circulation*, *83*(5), 1764–1770.

Lemon, K. P., Freitag, N. E., & Kolter, R. (2010). The virulence regulator PrfA promotes biofilm formation by *Listeria monocytogenes*. *Journal of Bacteriology*, *192*(15), 3969–3976.

Lim, S. Y., Teh, C. S. J., & Thong, K. L. (2017). Biofilm-related diseases and omics: Global transcriptional profiling of *Enterococcus faecium* reveals different gene expression patterns in the biofilm and planktonic cells. *OMICS: A Journal of Integrative Biology*, *21*(10), 592–602.

Liu, Z., Niu, H., Wu, S., & Huang, R. (2014). CsgD regulatory network in a bacterial trait-altering biofilm formation. *Emerging Microbes & Infections*, *3*(1), 1–5.

Lu, Y., Chen, S., Dong, H., Sun, H., Peng, D., & Liu, X. (2012). Identification of genes responsible for biofilm formation or virulence in *Salmonella enterica* serovar Pullorum. *Avian Diseases*, *56*(1), 134–143.

Ma, L., Conover, M., Lu, H., Parsek, M. R., Bayles, K., & Wozniak, D. J. (2009). Assembly and development of the *Pseudomonas aeruginosa* biofilm matrix. *PLoS Pathogens*, *5*(3), e1000354.

Mai-Prochnow, A., Clauson, M., Hong, J., & Murphy, A. B. (2016). Gram positive and Gram negative bacteria differ in their sensitivity to cold plasma. *Scientific Reports*, *6*(1), 1–11.

Majowicz, S. E., Musto, J., Scallan, E., Angulo, F. J., Kirk, M., O'Brien, S. J., et al. (2010). The global burden of nontyphoidal *Salmonella* gastroenteritis. *Clinical Infectious Diseases*, *50*(6), 882–889.

Marin, C., Hernandiz, A., & Lainez, M. (2009). Biofilm development capacity of *Salmonella* strains isolated in poultry risk factors and their resistance against disinfectants. *Poultry Science*, *88*(2), 424–431.

Maruzani, R., Sutton, G., Nocerino, P., & Marvasi, M. (2019). Exopolymeric substances (EPS) from *Salmonella enterica*: Polymers, proteins and their interactions with plants and abiotic surfaces. *Journal of Microbiology*, *57*(1), 1–8.

Mika, F., & Hengge, R. (2014). Small RNAs in the control of RpoS, CsgD, and biofilm architecture of *Escherichia coli*. *RNA Biology*, *11*(5), 494–507.

Mirzaee, M., Najar-Peerayeh, S., Behmanesh, M., & Moghadam, M. F. (2015). Relationship between adhesin genes and biofilm formation in vancomycin-intermediate *Staphylococcus aureus* clinical isolates. *Current Microbiology*, *70*(5), 665–670.

Mishra, S. P., Manent, A.-S., Chabot, B., & Daneault, C. (2012). Production of nanocellulose from native cellulose–various options utilizing ultrasound. *BioResources*, *7*(1), 0422–0436.

Monteiro, C., Papenfort, K., Hentrich, K., Ahmad, I., Le Guyon, S., Reimann, R., et al. (2012). Hfq and Hfq-dependent small RNAs are major contributors to multicellular development in *Salmonella enterica* serovar Typhimurium. *RNA Biology*, *9*(4), 489–502.

Narasanna, R. (2017). Influence of subinhibitory-concentration (sub-MIC) Cefetoxime on biofilm formation. SEM study of ESBL-producing *Salmonella* typhi. *Journal of Microbiology and Infectious Diseases*, *7*(02), 67–74.

Ogasawara, H., Yamamoto, K., & Ishihama, A. (2011). Role of the biofilm master regulator CsgD in cross-regulation between biofilm formation and flagellar synthesis. *Journal of Bacteriology*, *193*(10), 2587–2597.

Özdemir, C., Akçelik, N., & Akçelik, M. (2018). The role of extracellular DNA in *Salmonella* biofilms. *Molecular Genetics, Microbiology and Virology*, *33*(1), 60–71.

Paytubi, S., Cansado, C., Madrid, C., & Balsalobre, C. (2017). Nutrient composition promotes switching between pellicle and bottom biofilm in *Salmonella*. *Frontiers in Microbiology*, *8*, 2160.

Peng, D. (2016). Biofilm formation of *Salmonella*. In D. Dhanesekaran (Ed.), *Microbial biofilms—Importance and applications* (pp. 231–242). Janeza Trdine, Croatia: InTech.

Percival, S. L., Suleman, L., Vuotto, C., & Donelli, G. (2015). Healthcare-associated infections, medical devices and biofilms: Risk, tolerance and control. *Journal of Medical Microbiology*, *64*(4), 323–334.

Pontes, M. H., Lee, E.-J., Choi, J., & Groisman, E. A. (2015). *Salmonella* promotes virulence by repressing cellulose production. *Proceedings of the National Academy of Sciences*, *112*(16), 5183–5188.

Popoff, M. Y., Bockemühl, J., & Gheesling, L. L. (2004). Supplement 2002 (no. 46) to the Kauffmann–White scheme. *Research in Microbiology*, *155*(7), 568–570.

Pradhan, D., & Negi, V. D. (2019). Repeated in-vitro and in-vivo exposure leads to genetic alteration, adaptations, and hypervirulence in *Salmonella*. *Microbial Pathogenesis*, *136*, 103654.

Prouty, A., & Gunn, J. S. (2003). Comparative analysis of *Salmonella enterica* serovar Typhimurium biofilm formation on gallstones and on glass. *Infection and Immunity*, *71*(12), 7154–7158.

Prouty, A., Schwesinger, W., & Gunn, J. (2002). Biofilm formation and interaction with the surfaces of gallstones by *Salmonella* spp. *Infection and Immunity*, *70*(5), 2640–2649.

Rabin, N., Zheng, Y., Opoku-Temeng, C., Du, Y., Bonsu, E., & Sintim, H. O. (2015). Biofilm formation mechanisms and targets for developing antibiofilm agents. *Future Medicinal Chemistry*, *7*(4), 493–512.

Raza, A., Sarwar, Y., Ali, A., Jamil, A., Haque, A., & Haque, A. (2011). Effect of biofilm formation on the excretion of *Salmonella enterica* serovar Typhi in feces. *International Journal of Infectious Diseases*, *15*(11), e747–e752.

Römling, U. (2005). Characterization of the rdar morphotype, a multicellular behaviour in Enterobacteriaceae. *Cellular and Molecular Life Sciences: CMLS*, *62*(11), 1234–1246.

Römling, U., Bian, Z., Hammar, M., Sierralta, W. D., & Normark, S. (1998). Curli fibers are highly conserved between *Salmonella typhimurium* and *Escherichia coli* with respect to operon structure and regulation. *Journal of Bacteriology*, *180*(3), 722–731.

Römling, U., Galperin, M. Y., & Gomelsky, M. (2013). Cyclic di-GMP: The first 25 years of a universal bacterial second messenger. *Microbiology and Molecular Biology Reviews*, *77*(1), 1–52.

Römling, U., Gomelsky, M., & Galperin, M. Y. (2005). C-di-GMP: The dawning of a novel bacterial signalling system. *Molecular Microbiology*, *57*(3), 629–639.

Römling, U., Rohde, M., Olsén, A., Normark, S., & Reinköster, J. (2000). AgfD, the checkpoint of multicellular and aggregative behaviour in *Salmonella typhimurium* regulates at least two independent pathways. *Molecular Microbiology*, *36*(1), 10–23.

Römling, U., Sierralta, W. D., Eriksson, K., & Normark, S. (1998). Multicellular and aggregative behaviour of *Salmonella typhimurium* strains is controlled by mutations in the agfD promoter. *Molecular Microbiology*, *28*(2), 249–264.

Ross, P., Weinhouse, H., Aloni, Y., Michaeli, D., Weinberger-Ohana, P., Mayer, R., et al. (1987). Regulation of cellulose synthesis in *Acetobacter xylinum* by cyclic diguanylic acid. *Nature*, *325*(6101), 279–281.

Ryder, C., Byrd, M., & Wozniak, D. J. (2007). Role of polysaccharides in *Pseudomonas aeruginosa* biofilm development. *Current Opinion in Microbiology*, *10*(6), 644–648.

Ryjenkov, D. A., Simm, R., Römling, U., & Gomelsky, M. (2006). The PilZ domain is a receptor for the second messenger c-di-GMP: The PilZ domain protein YcgR controls motility in enterobacteria. *Journal of Biological Chemistry*, *281*(41), 30310–30314.

Saldaña, Z., Xicohtencatl-Cortes, J., Avelino, F., Phillips, A. D., Kaper, J. B., Puente, J. L., et al. (2009). Synergistic role of curli and cellulose in cell adherence and biofilm formation of attaching and effacing *Escherichia coli* and identification of Fis as a negative regulator of curli. *Environmental Microbiology*, *11*(4), 992–1006.

Sava, I. G., Heikens, E., & Huebner, J. (2010). Pathogenesis and immunity in enterococcal infections. *Clinical Microbiology and Infection*, *16*(6), 533–540.

Schembri, M. A., Givskov, M., & Klemm, P. (2002). An attractive surface: Gram-negative bacterial biofilms. *Science's STKE*, *2002*(132), re6.

Sela, S., Frank, S., Belausov, E., & Pinto, R. (2006). A mutation in the luxS gene influences *Listeria monocytogenes* biofilm formation. *Applied and Environmental Microbiology*, *72*(8), 5653–5658.

Seth-Smith, H. M. (2008). SPI-7: *Salmonella*'s Vi-encoding pathogenicity island. *Journal of Infection in Developing Countries*, *2*(04), 267–271.

Sharma, A., & Qadri, A. (2004). Vi polysaccharide of *Salmonella typhi* targets the prohibitin family of molecules in intestinal epithelial cells and suppresses early inflammatory responses. *Proceedings of the National Academy of Sciences*, *101*(50), 17492–17497.

Shatila, F., Yaşa, İ., & Yalçın, H. T. (2021). Biofilm formation by *Salmonella enterica* strains. *Current Microbiology*, *78*(4), 1150–1158.

Sheu, C. W., & Freese, E. (1972). Effects of fatty acids on growth and envelope proteins of *Bacillus subtilis*. *Journal of Bacteriology*, *111* (2), 516–524.

Simm, R., Ahmad, I., Rhen, M., Le Guyon, S., & Römling, U. (2014). Regulation of biofilm formation in *Salmonella enterica* serovar Typhimurium. *Future Microbiology*, *9*(11), 1261–1282.

Simm, R., Morr, M., Kader, A., Nimtz, M., & Römling, U. (2004). GGDEF and EAL domains inversely regulate cyclic di-GMP levels and transition from sessility to motility. *Molecular Microbiology*, *53*(4), 1123–1134.

Solano, C., García, B., Valle, J., Berasain, C., Ghigo, J. M., Gamazo, C., et al. (2002). Genetic analysis of *Salmonella enteritidis* biofilm formation: Critical role of cellulose. *Molecular Microbiology*, *43*(3), 793–808.

Steenackers, H., Hermans, K., Vanderleyden, J., & De Keersmaecker, S. C. (2012). *Salmonella* biofilms: An overview on occurrence, structure, regulation and eradication. *Food Research International*, *45*(2), 502–531.

Storz, G., Vogel, J., & Wassarman, K. M. (2011). Regulation by small RNAs in bacteria: Expanding frontiers. *Molecular Cell*, *43*(6), 880–891.

Tamayo, R., Pratt, J. T., & Camilli, A. (2007). Roles of cyclic diguanylate in the regulation of bacterial pathogenesis. *Annual Review of Microbiology*, *61*, 131–148.

Taylor, C. M., Beresford, M., Epton, H. A., Sigee, D. C., Shama, G., Andrew, P. W., et al. (2002). *Listeria monocytogenes* relA and hpt mutants are impaired in surface-attached growth and virulence. *Journal of Bacteriology*, *184*(3), 621–628.

Tsai, M.-H., Liang, Y.-H., Chen, C.-L., & Chiu, C.-H. (2020). Characterization of *Salmonella* resistance to bile during biofilm formation. *Journal of Microbiology, Immunology and Infection*, *53*(4), 518–524.

Tucker, A. T., Bobay, B. G., Banse, A. V., Olson, A. L., Soderblom, E. J., Moseley, M. A., et al. (2014). A DNA mimic: The structure and mechanism of action for the anti-repressor protein AbbA. *Journal of Molecular Biology*, *426*(9), 1911–1924.

Uzzau, S., Brown, D. J., Wallis, T., Rubino, S., Leori, G., Bernard, S., et al. (2000). Host adapted serotypes of *Salmonella enterica*. *Epidemiology and Infection*, *125*(2), 229–255.

van der Veen, S., & Abee, T. (2010). Importance of SigB for *Listeria monocytogenes* static and continuous-flow biofilm formation and disinfectant resistance. *Applied and Environmental Microbiology*, *76*(23), 7854–7860.

van Ditmarsch, D., Boyle, K. E., Sakhtah, H., Oyler, J. E., Nadell, C. D., Déziel, É., et al. (2013). Convergent evolution of hyperswarming leads to impaired biofilm formation in pathogenic bacteria. *Cell Reports*, *4* (4), 697–708.

van Velkinburgh, J. C., & Gunn, J. S. (1999). PhoP-PhoQ-regulated loci are required for enhanced bile resistance in *Salmonella* spp. *Infection and Immunity*, *67*(4), 1614–1622.

Vasilev, K., Cavallaro, A., & Zilm, P. (2018). *Antibacterial materials and coatings*. Multidisciplinary Digital Publishing Institute.

Wain, J., House, D., Zafar, A., Baker, S., Nair, S., Kidgell, C., et al. (2005). Vi antigen expression in *Salmonella enterica* serovar Typhi clinical isolates from Pakistan. *Journal of Clinical Microbiology*, *43*(3), 1158–1165.

Walawalkar, Y. D., Vaidya, Y., & Nayak, V. (2016). Response of *Salmonella* Typhi to bile-generated oxidative stress: Implication of quorum sensing and persister cell populations. *FEMS Pathogens and Disease*, *74*(8), ftw090.

Wang, H., Huang, Y., Wu, S., Li, Y., Ye, Y., Zheng, Y., et al. (2014). Extracellular DNA inhibits *Salmonella enterica* serovar Typhimurium and *S. enterica* serovar Typhi biofilm development on abiotic surfaces. *Current Microbiology*, *68*(2), 262–268.

Wangdi, T., Lee, C.-Y., Spees, A. M., Yu, C., Kingsbury, D. D., Winter, S. E., et al. (2014). The Vi capsular polysaccharide enables *Salmonella enterica* serovar typhi to evade microbe-guided neutrophil chemotaxis. *PLoS Pathogens*, *10*(8), e1004306.

Wilking, J. N., Zaburdaev, V., De Volder, M., Losick, R., Brenner, M. P., & Weitz, D. A. (2013). Liquid transport facilitated by channels in *Bacillus subtilis* biofilms. *Proceedings of the National Academy of Sciences*, *110*(3), 848–852.

Wilson, R. P., Raffatellu, M., Chessa, D., Winter, S. E., Tükel, Ç., & Bäumler, A. J. (2008). The Vi-capsule prevents Toll-like receptor 4 recognition of *Salmonella*. *Cellular Microbiology*, *10*(4), 876–890.

Xu, L., Li, H., Vuong, C., Vadyvaloo, V., Wang, J., Yao, Y., et al. (2006). Role of the luxS quorum-sensing system in biofilm formation and virulence of *Staphylococcus epidermidis*. *Infection and Immunity*, *74*(1), 488–496.

Yaron, S., & Römling, U. (2014). Biofilm formation by enteric pathogens and its role in plant colonization and persistence. *Microbial Biotechnology*, *7*(6), 496–516.

Zakikhany, K., Harrington, C. R., Nimtz, M., Hinton, J. C., & Römling, U. (2010). Unphosphorylated CsgD controls biofilm formation in *Salmonella enterica* serovar Typhimurium. *Molecular Microbiology*, *77* (3), 771–786.

Zeng, L., Zhang, L., Wang, P., & Meng, G. (2017). Structural basis of host recognition and biofilm formation by *Salmonella* Saf pili. *eLife*, *6*, e28619.

Zogaj, X., Nimtz, M., Rohde, M., Bokranz, W., & Römling, U. (2001). The multicellular morphotypes of *Salmonella typhimurium* and *Escherichia coli* produce cellulose as the second component of the extracellular matrix. *Molecular Microbiology*, *39*(6), 1452–1463.

Chapter 27

Mycobacterial biofilm: Structure and its functional relevance in the pathogenesis

Assirbad Behura[a,*], Mousumi Das[a,*], Ashish Kumar[a], Lincoln Naik[a], Salina Patel[a,c], Dev Kiran Nayak[a], Abtar Mishra[a], Amit Mishra[b], and Rohan Dhiman[a]

[a]*Laboratory of Mycobacterial Immunology, Department of Life Science, National Institute of Technology, Rourkela, Odisha, India,* [b]*Cellular and Molecular Neurobiology Unit, Indian Institute of Technology Jodhpur, Jodhpur, Rajasthan, India,* [c]*Laboratory of Infection Immunology, Department of Life Science, National Institute of Technology, Rourkela, Odisha, India*

Abbreviations

CFP	culture filtrate protein
ECM	extra cellular matrix
eDNA	extracellular DNA
EPS	extracellular polymeric substances
ESAT 6	early secretory antigenic target 6
ESX	ESAT-6 secretion system
DTT	dithiothreitol
FAS II	fatty acid synthase II
GAG	glycosaminoglycans
GPLs	glycopeptidolipids
LAM	lipoarabinomannan
LM	laminin
LPS	lipopolysaccharide
M. tb	*Mycobacterium tuberculosis*
MMDAG	monomeromycolyl diacylglycerol
Mps	mycobacterial peptide synthase
NTMs	nontuberculous mycobacteria
PDIMs	phthiocerol dimycocerosates
PGL	phenolic glycolipids
ROS	reactive oxygen species
SEM	scanning electron microscope
TB	tuberculosis
TDM	trehalosedimycolate
TRS	thiol reductive stress

1. Introduction

Biofilm is a highly organized community of sessile bacteria embedded in a self-secreted matrix of extracellular polymeric substances composed of exopolysaccharides (EPSs), DNA, lipids, and amyloidogenic proteins. Donlan and Costerton elucidated the concept of biofilm as a "microbially derived sessile community characterized by cells that are irreversibly attached to a substratum or interface or each other, embedded in a matrix of extracellular polymeric substances that they have produced, and exhibit an altered phenotype with respect to growth rate and gene transcription" (Socransky, 2002). Biofilm formation imparts protection to bacteria from various environmental challenges such as antimicrobial agents, pH, salinity, dehydration, metal toxicity, mechanical forces, nutrient scarcity, oligo-elements, temperature, and UV exposure. They are a serious global health concern because of their ability to withstand host immune defense mechanisms and antibiotics. Bacterial biofilms are the primary source of all medical infections as they generally colonize the abiotic surfaces of all medical instruments and are responsible for nosocomial conditions (Bharti et al., 2021). In addition, microbes are also known to colonize on biotic surfaces and form biofilm inside the human body that gives rise to significant diseases. The biofilm thrives due to the coordinated transcriptional, proteomic, and metabolic profile of the heterogeneous population of bacteria having different growth and phenotypic properties. During biofilm formation, various bacterial communities communicate through quorum sensing (Jamal et al., 2018).

Many species from the genus *Mycobacterium* are known to produce biofilms. The concept of mycobacterial cells present in "aggregates" or "pellicles" was initially outlined by Robert Koch upon discovering mycobacteria. Robert Koch reported the presence of cells pressed with each other and organized in bundles in the etiology of TB (Koch, 1882). This genus harbors the highly successful human pathogen *Mycobacterium tuberculosis* (*M. tb*) and *M. leprae* that causes tuberculosis and leprosy, respectively. Another group of bacteria present under the genus is nontuberculous mycobacteria (NTMs), containing *M. ulcerans*, *M. avium*, *M. marinum*, *M. fortuitum*, *M. abscessus*, and *M. chelonae*, which are

*Contributed equally.

Understanding Microbial Biofilms. https://doi.org/10.1016/B978-0-323-99977-9.00018-1

environmental pathogens and cause a broad range of infections in humans other than TB and leprosy (Niño-Padilla, Velázquez, & Garibay-Escobar, 2021). The majority of mycobacteria known to form biofilms are NTMs and *M. tb.* Since NTMs are environmental pathogens, biofilms play a significant role in their ability to colonize and cause infection. In the environment, NTM biofilms are generally found in the water sources or medical instruments from where they infect the host. Inside the host, biofilm plays an essential role in virulence by increasing drug tolerance, evading host defense mechanisms, or easing the colonization of the host (Bernut et al., 2014; Fennelly et al., 2016; Yamazaki, Danelishvili, Wu, MacNab, & Bermudez, 2006). This explains the development of drug-resistant strains and the requirement of a more extended treatment regimen to cure tuberculosis and other NTM infections. Due to biofilm's multifaceted properties, understanding the role of mycobacterial biofilms in its virulence has garnered much interest in recent times. This chapter has discussed the current understandings of the mechanism and the characteristic feature of mycobacterial biofilms and their therapeutic potential.

2. Biofilm formation by bacteria

Globally, about 40%–80% of prokaryotes live in biofilm since it protects them from various environmental stresses, antibiotics and helps in resource capture (Penesyan, Paulsen, Kjelleberg, & Gillings, 2021). Further, the presence of biofilm is responsible for poor treatment outcomes in about 80% of bacterial infections (Oliva, Stefani, Venditti, & Di Domenico, 2021). Biofilm-associated diseases such as cystic fibrosis, diabetic foot infections, oral infections, etc., involve many Gram-positive and Gram-negative bacteria. In recent times, many groups have identified the presence of multibacterial and polymicrobial biofilms (Lopes, Azevedo, & Pereira, 2018; Ruhal & Kataria, 2021). The polymicrobial biofilm involves biofilm formation between multispecies microbes with either two different bacteria or a bacteria and fungi sp. (Lopes et al., 2018). Similarly, multibacterial biofilms involve biofilm formation by different Gram-type bacteria. It is reported that Gram-positive and Gram-negative bacteria coadhere to create biofilm in device-based infections (Arciola, Campoccia, & Montanaro, 2018). These complexes have an advantage of extra antibiotic resistance due to complex biochemical and physical characteristics and require antibiotics that act against all the microbes present in the biofilm (Beaudoin et al., 2017). Gram-positive and Gram-negative bacteria differ in their outer membrane and peptidoglycan layer. Despite differences in the cell wall, some biofilm formation patterns are similar in both types of bacteria.

The cell surface of Gram-negative and Gram-positive bacteria is negatively charged due to lipopolysaccharide (LPS) and teichoic acid, which aids in adhering to positively charged surfaces to form a biofilm (Ruhal & Kataria, 2021). The length and variation of different components of LPS determine the biofilm-forming capability of Gram-negative bacteria (Lee, Lee, Hwang, Park, & Lee, 2016). Mutations in various sections of LPS either impaired or enhanced the biofilm-forming ability of *Escherichia coli* and *Pseudomonas aeruginosa* (Chalabaev et al., 2014; Heacock-Kang et al., 2018; Ruhal & Kataria, 2021). In Gram-positive bacteria that lack the cell wall, teichoic acid is responsible for biofilm production regulation (Bentley et al., 2006). Deleting genes involved in the production of teichoic acid or enzymes accountable for modification of teichoic acid or changes in the glycolipid portion of teichoic acid resulted in defective biofilm production in *Bacillus subtilis*, *Staphylococcus aureus* (Bucher, Oppenheimer-Shaanan, Savidor, Bloom-Ackermann, & Kolodkin-Gal, 2015; Fedtke et al., 2007; Weidenmaier et al., 2004; Weidenmaier & Peschel, 2008). Recent discoveries have shown that several biofilm-associated proteins are phylogenetically similar in various bacterial matrix proteins (Foster, Geoghegan, Ganesh, & Höök, 2014; Guttenplan & Kearns, 2013; Jiang, Jin, & Sun, 2018). Several amyloid proteins and polysaccharides found in both Gram-type bacteria help in initiation, maturation, dispersal, and immune evasion (Beaudoin et al., 2017; Bleem et al., 2018; Chapman et al., 2002; Maher, Lim, Gunawan, & Cegelski, 2015; Soliman et al., 2018; Zheng, Joo, Nair, Le, & Otto, 2018). Biofilm has been regarded as the master regulator of different infectious diseases (Hall-Stoodley & Stoodley, 2009). Much evidence is present that proves biofilm's contribution to pathogenesis during chronic infections (Bjarnsholt, 2013). Bacteria residing in the biofilm have been shown to transfer genes horizontally due to their proximity (Casadevall & Pirofski, 2009). This is an important reason behind their survival that results in a high number of possible mutations accounting for antimicrobial resistance (Sousa, Bandeira, Carvalho, Duarte, & Jordao, 2015).

3. Structure and characteristics of mycobacterial biofilm

The sessile bacterial community differs significantly from their planktonic counterpart, having different growth rates, gene expression, transcription, and translation as they survive in a completely different microenvironment. The formation of three-dimensional bacterial biofilm is a well-organized process that requires cooperation, differentiation, capturing and sharing of nutrients, and equal division of labor between the colonized population of bacteria. The formation of mycobacterial biofilm is similar to other bacterial biofilms. Since mycobacteria lack pili or fimbriae, the only difference is that they produce specific proteins such as sulfated glycosaminoglycans (GAGs) that help in their aggregation and initial attachment to the substratum (Esteban & García-Coca, 2018; Menozzi et al., 1996). Mycobacteria can form biofilms on the surface and in the interphase of air and water. This peculiar characteristic is possible due to high lipid levels in their cell wall and the unique composition of their extracellular matrix (ECM). Biofilm formation is a cyclic process having five significant steps attachment, colonization, development, maturation, and dispersion (Fig. 1). Attachment involves the initial attachment of bacteria to the substratum through weak interactions such as van der Waal's force with biotic or abiotic components in a reversible manner (Bos, Van der Mei, & Busscher, 1999). During colonization, the bacteria adhere to the surface by a strong hydrophobic or hydrophilic interaction (Yin, Wang, Liu, & He, 2019). Similarly, the development includes the accumulation and proliferation of different layers of bacterial cells that generate and secrete extracellular polymeric substances (Flemming, Neu, & Wozniak, 2007; Flemming & Wingender, 2010). Maturation involves the formation of a 3D association that holds various channels to constructively disseminate the nutrients and signaling components inside the biofilm (Dufour, Leung, & Lévesque, 2010). Finally, dispersal involves segregating microorganisms in colonies or detachment because of several environmental factors. The uncoupled cells further colonize in different locations to start the cycle of biofilm formation (Srey, Jahid, & Ha, 2013).

Upon examination under the microscope, mycobacterial biofilms possess a plenteous aggregate of ECM. Biofilms are not monolayer structures. Instead, they are a diverse group of microcolonies embedded in the EPS matrix and separated from others by interstitial voids that allow the flow of oxygen, nutrients, and antimicrobial agents within the biofilm (Kumar et al., 2015). Scanning electron microscopy (SEM) analysis of different mycobacterial biofilms reveals that each species' biofilm ultrastructurally differs from others. The biofilms of *M. vaccae*, *M. fortuitum*, *M. obuense*, *M. gilvum*, and *M. chubuense* have uniformly arranged curved structures with apparent voids and regularly spaced long fiber and short fiber in the case of *M. fortuitum* with heterogeneous morphology and mycelia-like texture. On the other side, slow-growing mycobacteria such as *M. tb* and *M. avium* had globular and crystalline structures (Agustí, Astola, Rodríguez-Güell, Julián, & Luquin, 2008). Further, confocal analysis by Muñoz-Egea et al. found that different species form a biofilm at varying rates. It was observed that the biofilm of *M. fortuitum* and *M. chelonae* achieved maximum thickness in only 72h. In contrast, it took 92h for the rest of the rapidly growing NTMs to form thick biofilms (Muñoz-Egea, García-Pedrazuela, Mahillo, García, & Esteban, 2013). Further, it was also found that the fully formed biofilm of *M. fortuitum* covered much less area than *M. chelonae*. The biofilm development strictly depends on the availability of nutrients in the medium. Many factors such as humidity,

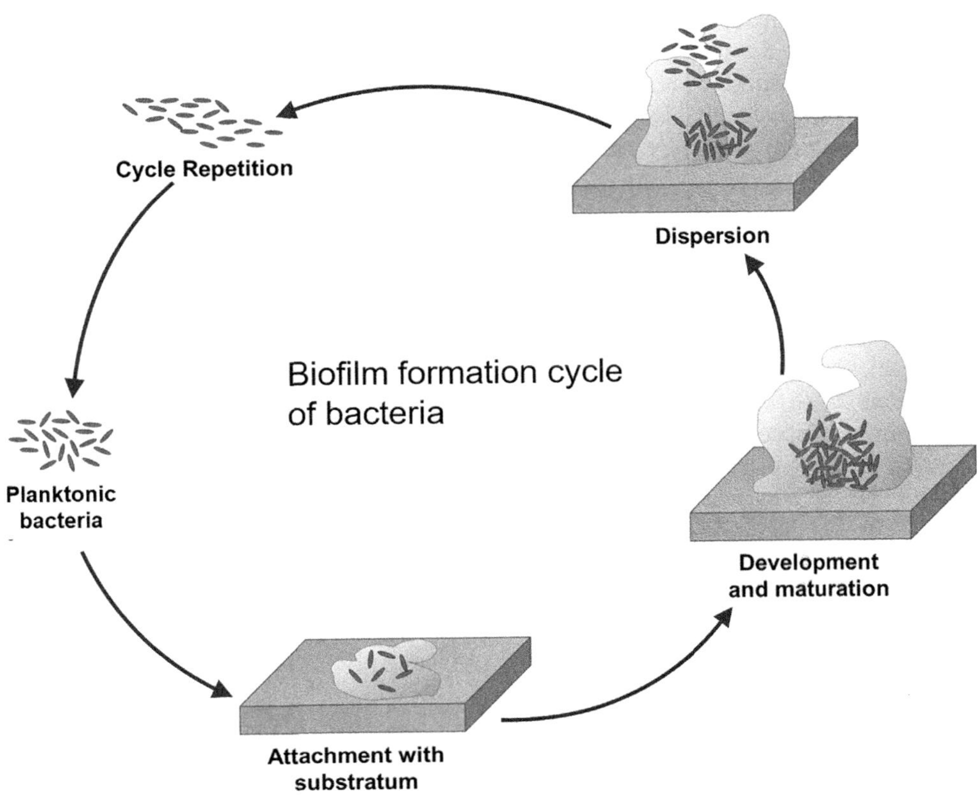

FIG. 1 Biofilm formation cycle of bacteria.

pH, temperature, contact surface, surface hydrophobicity, texture, and growth-stage regulate the strength of mycobacterial biofilm. Few mycobacterial species such as *M. obuense*, *M. gilvum*, *M. chubuense*, *M. vaccae*, *M. parafortuitum*, and *M. marinum* have rough colony morphology due to the presence of cording-like structure leading to the formation of spreading pellicles in vitro that foster biofilm formation and thereby making them more pathogenic than its smooth variants (Esteban et al., 2008).

Themycobacterial ECM differs in composition from species to species. For example, *M. smegmatis* forms a biofilm whose ECM is rich in lipid and fatty acid synthase (FAS)-II system (Ojha et al., 2005). FAS-II is known to regulate mycolic acid synthesis (Bhatt et al., 2007). In contrast, the biofilm of *M. avium* consists of mycolic acids and a unique type of glycopeptidolipid (GPL). GPLs play a significant role in pellicle biofilm formation (Chakraborty & Kumar, 2019). It is also demonstrated that in the case of *M. marinum*, phthiocerol dimycocerosates (PDIMs) and phenolic GPLs play a significant role in biofilm formation (Mohandas et al., 2016). The biofilm formed by *M. ulcerans* contains a high amount of carbohydrates, among which glucose is more elevated in proportion (Chakraborty & Kumar, 2019). Some lipid-like components are also found in ECM, such as phosphatidylinositol mannosides (PIM)-2, PIM-5, and PIM-6. These are mannosylated forms of a highly characterized and unusual phospholipid known as phosphatidylinositol in mycobacteria (Clary et al., 2018). The ECM also plays a significant role in attachment and invasion of tissue (Esteban & García-Coca, 2018; Menozzi et al., 1996). The ECM formed by biofilm-forming mycobacteria, when observed under ex-vivo conditions, usually shows intriguing nature of phenotypes in the case of solid and in liquid media (Zambrano & Kolter, 2005).

4. Biofilm formation by NTM

Many NTMs such as *M. avium*, *M. marinum*, *M. fortuitum*, *M. abscessus*, *M. ulcerans*, and *M. chelonae* are known to form biofilms. Each species has its own set of genes that regulate biofilm formation. *M. avium*, *M. marinum*, and *M. ulcerans* are slow-growing species that cause infection in birds, fish, and humans and are generally used as a model system to understand TB pathogenesis. *M. avium* and *M. marinum* form biofilm in 7 and 14 days, respectively, whereas the fast-growing *M. fortuitum* and *M. chelonae* form biofilm in about 48 h (Hall-Stoodley & Lappin-Scott, 1998; Johansen et al., 2009). The biofilm formed by NTMs is generally rich in GPLs and extracellular DNA (Aung et al., 2016; Rose, Babrak, & Bermudez, 2015). The extracellular DNA is an integral component of their biofilm matrix as biofilm exposure to DNase degraded the biofilm suggesting the crucial role of DNA in maintaining the structural integrity of biofilm (Rose et al., 2015). NTMs require the presence of different divalent cations such as Zn^{2+}, Mg^{2+}, and Ca^{2+} for biofilm formation, probably for stabilizing the extracellular DNA present in the biofilm (Carter, Wu, Drummond, & Bermudez, 2003). The biofilm of *M. ulcerans* is rich in glucose, PIM-2, PIM-5, and PIM-6, cardiolipin, phosphatidylinositol, phosphatidylethanolamine, phthiodiolone, and triacylglycerol (Chakraborty & Kumar, 2019). The presence of LPS is essential for the motility and biofilm formation of *M. marinum* (Mohandas et al., 2016). For the attachment of bacteria to the substratum, NTMs usually undergo sliding motility due to the presence of GPLs (Martínez, Torello, & Kolter, 1999; Recht, Martínez, Torello, & Kolter, 2000). In the case of *M. ulcerans*, an 18 kDa heat shock protein (Hsp18) enhances the attachment process (Pidot et al., 2010). The GPL synthesis in *M. abscessus* is controlled by specific genes such as *mmpL4b* or *mmpL4c*, whose deletions impair biofilm formation (Nessar, Reyrat, Davidson, & Byrd, 2011; Tsai et al., 2013). The biofilm formation in *M. avium* is strictly governed by enzymes of GPL synthesis, protein synthetase B (pstB), TCA cycle enzymes, and 6-oxodehydrogenase (Yamazaki, Danelishvili, Wu, MacNab, et al., 2006). *M. marinum* mutants defective in PGL or PDIM biogenesis pathway do not form functional biofilms and are more susceptible to antibiotics (Mohandas et al., 2016).

Aldecoa et al. have demonstrated that the release of DNA to the biofilm matrix in bacteria is mediated through membrane vesicles, active secretion, or autolysis (Ibanez de Aldecoa, Zafra, & González-Pastor, 2017). In *M. avium*, this process is strictly governed by quorum sensing, and in *M. abscessus*, it was found to be controlled by bicarbonate (Carter et al., 2003; Rose & Bermudez, 2016). Components such as peptone or glucose that act as carbon sources enhance *M. avium* to form a biofilm, whereas humic acid inhibits biofilm formation (Carter et al., 2003). The biofilm production is also dependent on oxidative stress induced by Autoinducer 2 (Geier, Mostowy, Cangelosi, Behr, & Ford, 2008). Similarly, it has been observed that the supernatants from *M. avium* biofilm culture, when added to planktonic cells, induce biofilm formation in those cells implying the role of quorum sensing in biofilm production (Carter et al., 2003). *M. abscessus* is known to form microcolonies in the alveolar walls. It forms rough and smooth colonies. The rough phenotype forms corded microcolonies and is incredibly virulent, whereas its smooth counterpart is more rounded and is less virulent (Byrd & Lyons, 1999). This difference in phenotype is due to the gene mab_3168c encoding acetyltransferase in the rough phenotypes and probably affects biofilm formation (Tsai et al., 2013).

5. Biofilm formation by *M. smegmatis*

M. smegmatis is a nonpathogenic and fast-growing mycobacterial species regularly used as a model system for studying physiological machinery (Chakraborty & Kumar, 2019; Reyrat & Kahn, 2001). It usually forms well-organized microcolonies encapsulated inside EPSs termed biofilms (Chakraborty & Kumar, 2019). The biofilm formation begins with colonization at the solid surface base through sliding motility (Martínez et al., 1999). The sliding motility is crucial for establishing colonies as it decreases the friction between the cell and substratum and enables the formation of a uniformly spaced single layer of bacteria. GPLs present on the mycobacterial cell wall initiate the bacteria's initial attachment with the substratum, and its acetylation is essential in determining the formation and structure of biofilm. The importance of GPL in biofilm formation was confirmed as bacteria lacking a functional GPL had impaired biofilm machinery (Recht & Kolter, 2001). Other than the solid surfaces, *M. smegmatis* is also reported to form a biofilm at the air-liquid interphase called a pellicle biofilm. The bacteria undergoing pellicle mode of biofilm formation are embedded in a self-produced EPS capsule rich in mycolic acids, which help in biofilm maturation (Marrakchi, Lanéelle, & Daffé, 2014; Sambandan et al., 2013). Ojha et al. have discovered that the mycolic acid present in the pellicle is far shorter than the one present in the cell wall. This was due to the activity of the GroEL1 chaperon that converts the membrane-bound mycolic acid into free mycolic acids (Ojha et al., 2005). The short-chain mycolic acids of biofilm are liberated from their membrane-bound form via hydrolysis of trehalosedimycolate (TDM) by serine esterase (Ojha, Trivelli, Guerardel, Kremer, & Hatfull, 2010).

Mathew et al. demonstrated that *M. smegmatis* lacking the rpoZ gene showed a different colony morphology, had altered sliding motility, and lacked biofilm-forming ability even though there was no change in overall GPL concentration (Mathew, Mukherjee, Balachandar, & Chatterji, 2006). Further studies revealed that the mutant had very few free short-chain mycolic acid, due to which it lacked an ECM in the pellicle mode of development (Mathew et al., 2006). Pacheco et al. further verified the role of mycolic acid in biofilm formation using *M. smegmatis* lacking mmpL1. This strain has a defective transfer of mycolate ester and monomeromycolyl diacylglycerol to the bacterial surface, which delayed biofilm formation (Pacheco, Hsu, Powers, & Purdy, 2013). Monomeromycolyl diacylglycerol plays a significant role in biofilm production as strains lacking monomeromycolyl diacylglycerol formed defective biofilms even with optimum levels of mycolic acids and GPLs in the case of Lsr2 mutants (Chen et al., 2006). Lsr2 also plays a significant role in defining the colony structure and morphology because the mutant bacteria had wet, smooth, and round colonies instead of dry, rough, and irregular shape (Chen et al., 2006). The role of Lsr2 in biofilm production was further confirmed by Arora et al. They discovered that the mutants have enhanced sliding motility and decreased surface hydrophobicity and pellicle mode of biofilm production. The reduced surface hydrophobicity is attributed to the lack of monomeromycolyl diacylglycerol on the cell surface that enhances hydrophobic interaction between the cells by holding them together for biofilm formation (Arora, Whiteford, Lau-Bonilla, Davitt, & Dahl, 2008). The cell surface composition also influences the biofilm formation process. The surface hydrophobicity is enhanced upon overexpression of peptidoglycan amidase that induces biofilm production and drug resistance in *M. smegmatis* (Padhi, Naik, Sengupta, Ganguli, & Sonawane, 2016). In contrast, inhibition of peptidoglycan synthesis in phosphoglucosamine mutase knockout *M. smegmatis* strains had significantly reduced ability to form biofilms and was sensitive to antibiotics (Kang et al., 2013).

The formation of a functional biofilm of *M. smegmatis* is strictly controlled by specific genes that are tightly regulated and are only expressed during particular stages of biofilm formation, starting from the aggregation of bacteria to maturation and dispersion. According to Yang et al., initial overexpression of Lsr2 in the planktonic *M. smegmatis* leads to accumulation of the cells, which escalates the expression of GroEL. Increased expression of GroEL enhances the production of free short mycolic acids that ultimately initiate biofilm formation (Yang et al., 2017). The biofilm formation is also actively regulated by the metabolic state of mycobacteria as increased intracellular NADH, or reductive thiol stress induces biofilm formation (Gupta, Kasetty, & Chatterji, 2015; Kumar et al., 2011; Vargas, Hageman, Gulati, Nobile, & Rawat, 2016).

6. Biofilm formation by *M. tuberculosis*

The discovery of the biofilm-forming ability of *M. tb* by Ojha et al. has opened many avenues of research. *M. tb* is also reported to form biofilm in vitro. This is predicted to be the reason behind the emergence of nonreplicating persistent bacterial cells that eventually develop into various drug-resistant strains. Three models of *M. tb* biofilm formation have been proposed, and they are pellicle biofilm formed at the interface of air and liquid, thiol reductive stress (TRS)-induced biofilm, and leukocyte lysate-induced biofilm (Ojha et al., 2008).

Pellicle: At the air-liquid interface, *M. tb* requires about 5–7 weeks to generate a pellicle biofilm (Ojha et al., 2008) (Fig. 2A). Branching filamentous fungus-like structure, presence of Zn^{2+}, and the water-repellent nature of mycobacteria give rise to this type of phenotype (Ojha et al., 2008). The presence of ketomycolic acid in ECM is necessary for *M. tb* to grow in the pellicle (Sambandan et al., 2013). Transformation of *M. tb* from planktonic stage to

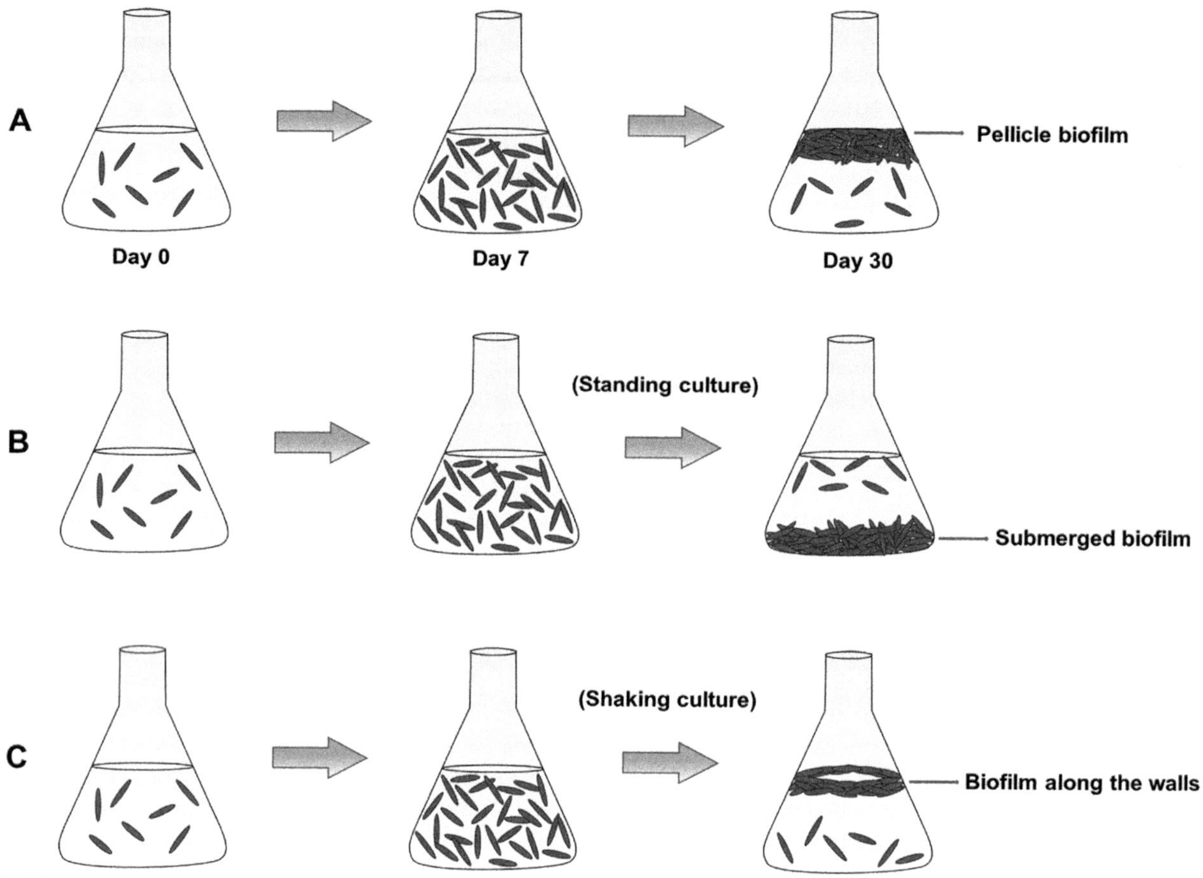

FIG. 2 Three model systems of *M. tb* biofilm growth.

pellicle mode of development imparts them with drug-tolerant activity, and mutants unable to grow in pellicles are found to be susceptible to antibiotics. This is a well-studied model of biofilm formation in *M. tb* as it allows the tracking of different developmental stages during maturation.

TRS-induced biofilm: Trivedi et al. discovered that exposure of actively growing *M. tb* in a flask to TRS such as dithiothreitol (DTT) leads to the attachment of the cells to the substratum within 30h of treatment (Trivedi, Mavi, Bhatt, & Kumar, 2016) (Fig. 2B). It was found that the architecture of the organized colonies dramatically depends on the culture conditions, such as biofilm forms at the air-liquid interface when the culture is under shaking conditions. In contrast, submerged biofilms form understanding culture conditions. Unlike the pellicle mode of biofilm formation, these biofilms are firmly attached to the substratum and do not disintegrate upon the addition of Tween80. This biofilm form is an intermediate between conventional biofilms and the pellicle mode of biofilm structure. It has been proved that many drugs-tolerant bacteria reside in TRS-induced biofilm. DNA and several proteins are the critical components of biofilm formation (Chakraborty & Kumar, 2019).

Leukocyte lysate-induced biofilm: Antibiotic-resistant *M. tb* was also found to form organized colonies attached to substrates upon addition of leukocyte lysate to the culture medium (Sambandan et al., 2013) (Fig. 2C). The resistance against antibiotics was lost upon disintegrating the biofilm with Tween 80 or DNase. Inside the necrotic granulomas of the host, *M. tb* is exposed to leukocyte lysate and might be forming these types of biofilm in vitro, resulting in the emergence of drug-resistant strains (Chakraborty & Kumar, 2019). This mode of biofilm formation is used to discover novel first-line anti-TB drugs that are antibiofilm agents and disperses the biofilm leading to the elimination of *M. tb* (Ackart et al., 2014).

7. Requirements for biofilm formation

For the formation of a functional biofilm of mycobacteria, the primary requirement is the presence of a substrate or air-liquid interface over which the biofilm forms. Several essential components required for biofilm formation are GPLs, short-chain mycolic acids, and GroEL1 chaperon. Apart from these requirements, nutritional supplements, different ions, and sources of carbon also play a significant role in influencing the behavior of mycobacteria inhabiting inside biofilm (Esteban & García-Coca, 2018). Pellicle mode of biofilm formed by mycobacteria requires a high amount of free mycolic acids formed by the breakdown of trehalosedimycolate (Ojha et al., 2010). Besides mycolic acids, meromycolyl diacylglycerol, glycopeptidolipids, extracellular DNA, and polysaccharides are also responsible for mycobacterial biofilm formation (Chakraborty & Kumar, 2019). Polysaccharides such as cellulose are generally found at the basal level of the microcolony stalk and between microcolonies (Chakraborty & Kumar, 2019). The function of cellulose as a fundamental component is supported by the fact that the *M. tb* genome encodes for cellulase, facilitating the dispersal of *M. tb* biofilms (Flemming & Wingender, 2010). Various mycobacterial cells were found to produce adhesins such as fibronectin-binding proteins, heparin-binding hemagglutinin adhesin, etc., that are crucial for biofilm formation and aggregation (Chakraborty & Kumar, 2019). Quorum sensing molecules enable communication between cells in biofilm, thus regarded as essential biofilm components. The overall structure of mycobacterial biofilm is shown in Fig. 3.

8. Mycobacterial genes involved in biofilm formation

Several genes play a crucial role in maintaining biofilm integrity. Around 115 genes of *M. tb* are associated with biofilm development and quorum sensing (Hegde, 2020). A master regulator of transcription, LuxR has a vital role in quorum sensing as it regulates the gene expression in case of the processes such as virulence, motility, and biofilm formation (Chen & Xie, 2011). Miller et al. have discovered that SecA secretory pathway helps export various proteins in *M. tb*, ultimately elevating their virulence (Miller, Zulauf, & Braunstein, 2017). Overexpression of Sec A is reported to induce biofilm formation in *Streptococcus mutans* and *M. avium* (Huang, Meng, Fan, Hu, & Bian, 2008; Limia, Sangari, Wagner, & Bermudez, 2001). Therefore, it is hypothesized that SecA might enhance *M. tb* virulence by improving biofilm formation.

Few genes are expressed only in Mycobacterial biofilms in a precise manner. The transposomal inactivation of GPL gene clusters and deletion of *tmtp* and *mps* in *M. smegmatis* curtailed the biofilm formation suggesting their importance in biofilm formation. Recently, Xu et al. have discovered MmbR, a master transcription regulator of fatty acid

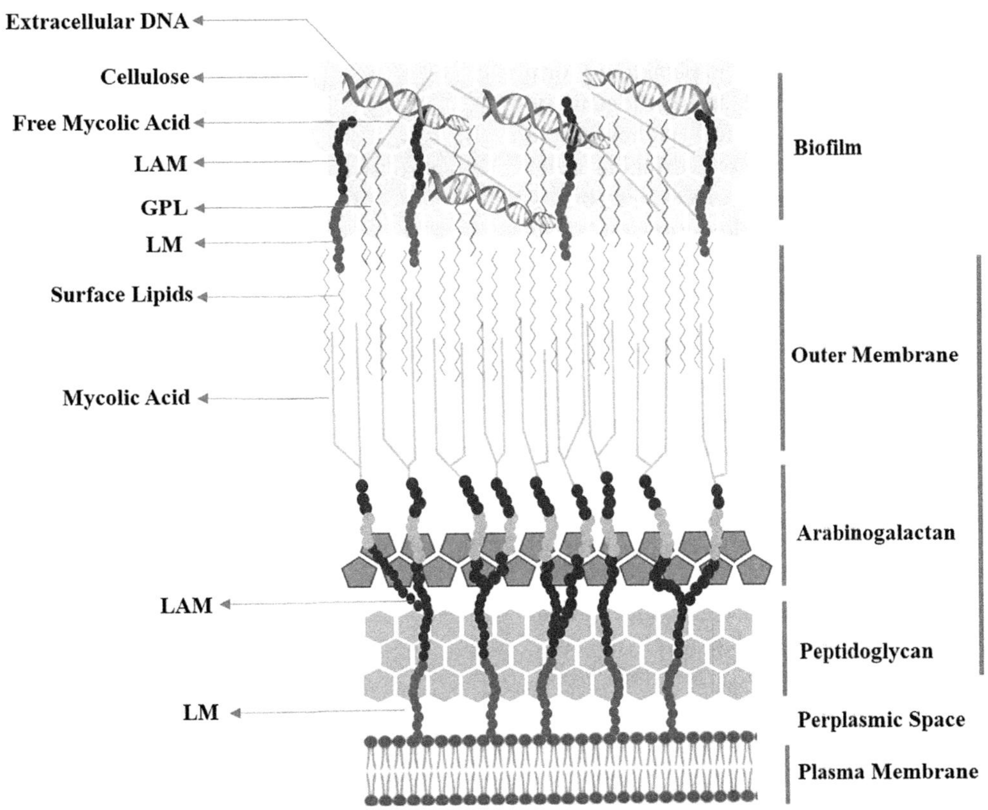

FIG. 3 Structural components of mycobacterial biofilm.

oxidation, which regulates biofilm formation in *M. smegmatis* (Xu, Su, Li, Deng, & He, 2021). They have observed that deletion of MmbR caused morphological changes in the biofilm by changing the lipid composition, and the bacteria were found more susceptible to anti-TB drugs (Xu et al., 2021). The initiation of *M. tb* biofilm is associated with the expression of *sigE*, *sigB*, and *whiB3*, whose functions are yet to be identified (Trivedi et al., 2016). The biofilm formation also coincides with the overexpression of type VII secretion system ESX-3 and *furA* that enhances iron uptake from the environment (Chakraborty & Kumar, 2019). The SenX3/RegX3 two-component system that senses the oxygen availability and regulates growth also is upregulated upon biofilm formation (Singh & Kumar, 2015; Trivedi et al., 2016). There are many genes involved in regulating mycobacterial biofilm, and discussing every gene in detail is beyond the scope of this review, so few of the essential genes associated with biofilm formation and maintenance of mycobacterial biofilm have been enlisted in Table 1.

TABLE 1 List of different mycobacterial genes involved in biofilm formation.

Gene	Functions
SecA2, SugA, SugB, SugC, HbhA, Rv3312A, Rv2136c, EsxA, EsxB, Rv1230c, ctpC, mmpS5, glf, Rv2326c	Cell wall formation and pathogenesis
luxR	Transcription regulator, quorum sensing
Rv1176c, Rv2216, Rv0021, Rv2102, Rv3519, Rv0574c, Rv1996, Rv0038, Rv2917, Rv3510c, Rv0433, Rv2323c, Rv1354c, Rv1357c, Rv3031	Conserved sequences
RplM, SigB, Hns, rplC, rplE, rplD, rpsC, rpsL, rpsO, rpsQ, rpsH, rpsF, rpsK, rpmD, rplY, gatA	Information pathways
Pks1, acpP, kasA, kasB, pks15, tgs1, fabD, Rv3790, Rv3234c, Rv0146, fbpA, accD5, inhA, fadE24, fadD9, fadE7, Rv0726c	Lipid metabolism
Rv0195, PknI, Rv1151c, Lsr2	Biofilm regulation
GroEL1, Rv0815c, cspB, Rv2005c, Rv2319c	Virulence
ltp3, pks10	Fatty acid biosynthesis
treX, treY, treZ	Mycolic acid synthesis
guaB2, pmmB, gtf, sucA, pstB	GPL biosynthesis
MoeW, RelA, NirB, Rv0097, Rv2465c, Rv2962c, Rv2993c, Rv0247c, gcvH, nuoM, purD, glnA4, ethA, atsA, rocD1, cydA, bfrB, lat, zwf1, purF, celA1	Intermediate metabolism
SigE, SigB, WhiB3	Formation of TRS-induced *M. tb* biofilm
Rv3312A	Formation of pellicle mode of *M. tb* biofilm
furA	Iron uptake in *M. tb* biofilms
PknA	Maintenance of bacterial cell growth in biofilm
BcsA, BcsB	Components of cellulose synthase

9. Functional relevance of biofilm formation by mycobacteria

Biofilm formation is the molecular arsenal of mycobacteria to adapt and survive in variable environmental conditions by establishing highly diverse microbial communities. Being intracellular pathogens, their ability to form biofilm has given them many advantages over the planktonic form of growth, increasing their virulence and pathogenicity. NTMs are opportunistic environmental pathogens to humans and are widely present in the soil, drinking water distribution system, and natural water sources (Esteban & García-Coca, 2018). Both rapidly and slow-growing NTMs are well reported to form biofilms responsible for their hydrophobicity in water distribution systems and household plumbing (Falkinham III, Iseman, de Haas, & van Soolingen, 2008). Increasing numbers of *M. avium*, *M. intracellulare*, and *M. phlei* found in water treatment plants are conferred to their biofilm-forming capability that imparts resistance to disinfectants and leads to outbreaks (El Helou, Viola, Hachem, Han, & Raad, 2013; Kennedy et al., 2012; Vijayaraghavan, Chandrashekhar, Sujatha, & Belagavi, 2006; Walker et al., 2017). Vijayraghavan et al. have shown that sessile mycobacteria are much more resistant to disinfectants than planktonic ones, as they found NTM in the bottom of the glutaraldehyde tray (Vijayaraghavan et al., 2006). Globally *M. chimaera* is reported to form biofilm on cardiac surgery equipment that requires special decontamination measures to make the instruments sterile (Kohler et al., 2015). NTMs actively form biofilm in biomaterials such as pacemakers, prosthetic joints and valves, abdominal mesh, and catheters, as antibiotic therapy is inefficient in eliminating the sessile mycobacteria; therefore, the infection requires removal of the implants (Al-Ghamdi et al., 2016; Bouchiat et al., 2015; Celdrán, Esteban, Manas, & Granizo, 2007; Eid et al., 2007). NTMs recovered from biofilms and grown in suspension show an increased resistance toward antimycobacterial agents. This resistance to enhance survival is adaptive and not genetic, as the resistance is lost after a day's growth in suspension (Falkinham III et al., 2008; Steed & Falkinham III, 2006). Further, NTMs are also reported to grow with free-living amoeba as endosymbionts in biofilms that enhance its resistance against disinfectants and act as a source of human infections (Marciano-Cabral, Jamerson, & Kaneshiro, 2010; Ovrutsky et al., 2013).

NTMs are also known to form biofilms in vivo. Fennelly et al. have successfully demonstrated the ability of *M. abscessus* to form biofilm in the lungs, suggesting why antibiotics do not cure *M. abscessus* infection (Fennelly et al., 2016). The virulent rough strain of *M. abscessus* is reported to form a biofilm-like extracellular structure when released from apoptotic macrophages that prevent its phagocytosis leading to uncontrolled growth and establishment of lethal infection in the zebrafish model (Bernut et al., 2014). Biofilm formation is also necessary to infect and translocate across bronchial epithelial cells by *M. avium* (Yamazaki, Danelishvili, Wu, Hidaka, et al., 2006). The presence of biofilm is responsible for developing resistance and partial activity of antibiotics against chronic lungs infection caused by *M. avium* (Griffith, Brown, Cegielski, Murphy, & Wallace, 2000). Carter et al. have discovered that Clarithromycin can effectively inhibit the growth of *M. avium* before biofilm formation, but it is ineffective once the infection is established with biofilm (Carter, Young, & Bermudez, 2004).

M. tb is also reported to form biofilms in biomaterials such as prosthetic joints, and biomaterial removal is the only possible way to manage the infection as the sessile bacteria are difficult to treat in biofilm (Esteban & García-Coca, 2018; Ha, Chung, & Ryoo, 2005). The role of biofilm in the pathogenesis of tuberculosis is unclear. Few groups have reported that *M. tb* biofilms participate in the formation of caseous necrosis and cavitation of lungs tissue leading to the successful establishment of disease (Basaraba & Ojha, 2017; Islam, Richards, & Ojha, 2012; Kulka, Hatfull, & Ojha, 2012). Recently, Chakraborty et al. have discovered that biofilm formation in lungs enhances virulence and drug tolerance of *M. tb*, leading to the establishment of chronic infections (Chakraborty, Bajeli, Kaushal, Radotra, & Kumar, 2021). Further, they have also shown that disruption of biofilms enhances the death of mycobacteria by antimycobacterial agents (Chakraborty et al., 2021). *M. tb* strains overexpressing cellulase do not form biofilms and are therefore eliminated in the late stages of infection. These findings effectively highlight that the biofilm production by *M. tb* at the early stages is essential for establishing infection and imparting protection against adaptive immunity in late stages (Chakraborty et al., 2021). Many mechanisms such as increased protection by polysaccharides acting as a physical barrier reducing the permeability of drugs, increased expression of efflux pumps, presence of ROS scavengers, and metabolic heterogeneity among the residents have been postulated to explain drug tolerance seen in biofilm-resident *M. tb* (Chakraborty & Kumar, 2019). *M. tb* is shown to form air-liquid pellicles within the lung cavities that are not the primary targets of the antimycobacterial drugs and therefore act as an ideal niche for their growth and proliferation, leading to the generation of persisters. The persistent *M. tb* is the primary source of drug-resistant strains and is responsible for TB reactivation (Adams et al., 2019; Bacon et al., 2014). Recently, Keating et al. have shown that *M. tb* residing in these pellicles tends to modify their cell wall carbohydrate content while their growth in biofilm, which leads to inactivation of complement system, ultimately avoiding the phagocytosis and exhibiting an extracellular fate in lungs (Keating et al., 2021).

10. Therapeutic implications of mycobacterial biofilm

Biofilm development imparts protection against many antibiotics and disinfectants that generally would have killed the bacteria in the planktonic state. It leads to increased virulence, treatment failure, development of drug-resistant strains, and disease reactivation. The emergence of drug-resistant *M. tb* and reactivation of TB have caused global insurgency to identify novel efficient drugs having fewer side effects (Esteban & García-Coca, 2018). Since several groups have demonstrated that physical eradication of biofilm makes them susceptible to antibiotics, and disinfectants such as Clarithromycin and Amikacin were found to be highly effective against planktonic *M. abscessus* but were minimally effective against *M. abscessus* in biofilm even at the highest concentrations tested (Greendyke & Byrd, 2008; Ortíz-Pérez et al., 2011). The drastic difference between minimum inhibitory concentration (MIC) and the minimum biofilm eradication concentration (MBEC) of antibiotics was not limited to *M. abscessus* alone; instead, Muñoz-Egea et al. found that there is a 100- and 100,000-fold difference between the MIC and MBEC of Ciprofloxacin in *M. mucogenicum* and *M. peregrinum*, respectively (Muñoz-Egea, García-Pedrazuela, Mahillo, & Esteban, 2015). Therefore, current research is focused on targeting biofilm to kill the bacteria.

Ojha et al. have proposed that since *M. tb* persists inside biofilms, identifying antibiofilm compounds could shorten the treatment of TB using standard antibiotics that will impact global health and drastically reduce the development of antibiotic resistance (Ojha et al., 2008). Muñoz-Egea et al. studied the effect of antibiotics on mycobacterial viability at different stages of biofilm development and found that antibiotics are most effective when added at the early stages of biofilm formation and kill the bacteria. Further, they have also shown that Ciprofloxacin is most potent against mycobacterial biofilm than Amikacin and Clarithromycin (Muñoz-Egea et al., 2015; Muñoz-Egea, García-Pedrazuela, Mahillo-Fernandez, & Esteban, 2016). Compounds such as Tween 80 that alter the structural integrity of proteins, lipids, and membranes and NAC that breaks the disulfide bond between polysaccharides also have antibiofilm properties (Olofsson, Hermansson, & Elwing, 2003; Teixeira et al., 2007). Cotreatment with these antibiofilm compounds and antibiotics has a synergistic effect as they effectively enhance the ability of Ciprofloxacin, Amikacin, and Clarithromycin to inhibit biofilm formation in *M. smegmatis*, *M. abscessus*, *M. chelonae*, and *M. fortuitum*, leading to their death (Muñoz-Egea, García-Pedrazuela, et al., 2016). Falkinham et al. discovered that *Methylobacterium* sp. inhibits the growth of *M. avium* in the water distribution system by inhibiting the ability of *M. avium* to form a biofilm (Falkinham III, Williams, Kwait, & Lande, 2016). The antibiofilm property of *Methylobacterium* sp. was further studied, and it was discovered that even the crude autoclaved extract of *Methylobacterium* sp. can inhibit biofilm formation in *M. abscessus* (Muñoz-Egea, Esteban, & Esteban, 2016). The antibiofilm effect of *Methylobacterium* sp. extract should also be checked on other pathogenic strains of mycobacteria such as *M. tb*. Chakraborty et al. have shown that administration of nebulized cellulose can disrupt *M. tb* biofilm in vivo, thereby enhancing mycobacterial killing by antibiotics (Chakraborty et al., 2021). But cellulose can elicit an immunological response in mammals as it is a foreign protein, and prolonged exposure has several side effects. So further studies need to be carried out to use cellulose in adjunct therapy, and small molecules capable of disrupting *M. tb* biofilm in vivo need to be discovered to be used in treatment. This novel approach is of great interest and can aid in finding other antibiofilm agents against mycobacteria and help treat patients in which biofilm cannot be surgically removed.

11. Conclusion

Mycobacterial biofilm is a unique field of research due to its clinical importance. Inhabitants of biofilm show variable features compared with free-living planktonic forms. This peculiar form of growth and colony formation endow the biofilm-related bacteria with high resistance capacity against extreme environments. Virulent and avirulent strains of mycobacteria form a fascinating array of biofilm structure and behavioral distinctions accordingly. Together, several agents contribute to the resistance of mycobacterial cells residing in the biofilm to antimicrobial compounds. Those are the distinct structural composition of the cell wall helping it act as a barrier to penetration by foreign materials, the capability of some virulent strains to thrive in phagocytic cells, mutation in specific genes, etc. To identify and find a solution to recurring TB infections, a thorough investigation of the bacterial endurance system is essential. An in-depth study about protective function adapted by clinically significant mycobacterial biofilms to survive even in critical environments would enlighten the path for the discovery of new treatment strategies and will also help in shortening the drug therapy duration.

References

Ackart, D. F., Lindsey, E. A., Podell, B. K., Melander, R. J., Basaraba, R. J., & Melander, C. (2014). Reversal of *Mycobacterium tuberculosis* phenotypic drug resistance by 2-aminoimidazole-based small molecules. *Pathogens and Disease*, *70*(3), 370–378.

Adams, K. N., Verma, A. K., Gopalaswamy, R., Adikesavalu, H., Singhal, D. K., Tripathy, S., et al. (2019). Diverse clinical isolates of *Mycobacterium tuberculosis* develop macrophage-induced rifampin tolerance. *The Journal of Infectious Diseases*, *219*(10), 1554–1558.

Agustí, G., Astola, O., Rodríguez-Güell, E., Julián, E., & Luquin, M. (2008). Surface spreading motility shown by a group of phylogenetically related, rapidly growing pigmented mycobacteria suggests that motility is a common property of mycobacterial species but is restricted to smooth colonies. *Journal of Bacteriology*, *190*(20), 6894–6902.

Al-Ghamdi, B., El Widaa, H., Al Shahid, M., Aladmawi, M., Alotaibi, J., Al Sanei, A., et al. (2016). Cardiac implantable electronic device infection due to *Mycobacterium* species: A case report and review of the literature. *BMC Research Notes*, *9*(1), 1–10.

Arciola, C. R., Campoccia, D., & Montanaro, L. (2018). Implant infections: Adhesion, biofilm formation and immune evasion. *Nature Reviews Microbiology*, *16*(7), 397–409.

Arora, K., Whiteford, D. C., Lau-Bonilla, D., Davitt, C. M., & Dahl, J. L. (2008). Inactivation of lsr2 results in a hypermotile phenotype in *Mycobacterium smegmatis*. *Journal of Bacteriology*, *190*(12), 4291–4300.

Aung, T. T., Yam, J. K. H., Lin, S., Salleh, S. M., Givskov, M., Liu, S., et al. (2016). Biofilms of pathogenic nontuberculous mycobacteria targeted by new therapeutic approaches. *Antimicrobial Agents and Chemotherapy*, *60*(1), 24–35.

Bacon, J., Alderwick, L. J., Allnutt, J. A., Gabasova, E., Watson, R., Hatch, K. A., et al. (2014). Non-replicating *Mycobacterium tuberculosis* elicits a reduced infectivity profile with corresponding modifications to the cell wall and extracellular matrix. *PLoS One*, *9*(2), e87329.

Basaraba, R. J., & Ojha, A. K. (2017). Mycobacterial biofilms: Revisiting tuberculosis bacilli in extracellular necrotizing lesions. *Microbiology Spectrum*, *5*(3). 5.3. 15.

Beaudoin, T., Yau, Y., Stapleton, P., Gong, Y., Wang, P., Guttman, D., et al. (2017). *Staphylococcus aureus* interaction with *Pseudomonas aeruginosa* biofilm enhances tobramycin resistance. *NPJ Biofilms and Microbiomes*, *3*(1), 1–9.

Bentley, S. D., Aanensen, D. M., Mavroidi, A., Saunders, D., Rabbinowitsch, E., Collins, M., et al. (2006). Genetic analysis of the capsular biosynthetic locus from all 90 pneumococcal serotypes. *PLoS Genetics*, *2*(3), e31.

Bernut, A., Herrmann, J.-L., Kissa, K., Dubremetz, J.-F., Gaillard, J.-L., Lutfalla, G., et al. (2014). *Mycobacterium abscessus* cording prevents phagocytosis and promotes abscess formation. *Proceedings of the National Academy of Sciences*, *111*(10), E943–E952.

Bharti, S., Maurya, R. K., Venugopal, U., Singh, R., Akhtar, M., & Krishnan, M. Y. (2021). Rv1717 Is a cell wall-associated β-galactosidase of *Mycobacterium tuberculosis* that is involved in biofilm dispersion. *Frontiers in Microbiology*, *3500*.

Bhatt, A., Fujiwara, N., Bhatt, K., Gurcha, S. S., Kremer, L., Chen, B., et al. (2007). Deletion of kasB in *Mycobacterium tuberculosis* causes loss of acid-fastness and subclinical latent tuberculosis in immunocompetent mice. *Proceedings of the National Academy of Sciences*, *104*(12), 5157–5162.

Bjarnsholt, T. (2013). The role of bacterial biofilms in chronic infections. *APMIS*, *121*, 1–58.

Bleem, A., Christiansen, G., Madsen, D. J., Maric, H., Strømgaard, K., Bryers, J. D., et al. (2018). Protein engineering reveals mechanisms of functional amyloid formation in *Pseudomonas aeruginosa* biofilms. *Journal of Molecular Biology*, *430*(20), 3751–3763.

Bos, R., Van der Mei, H. C., & Busscher, H. J. (1999). Physico-chemistry of initial microbial adhesive interactions—Its mechanisms and methods for study. *FEMS Microbiology Reviews*, *23*(2), 179–230.

Bouchiat, C., Saison, J., Boisset, S., Flandrois, J.-P., Issartel, B., Dauwalder, O., et al. (2015). Nontuberculous mycobacteria: An underestimated cause of bioprosthetic valve infective endocarditis. In *Open forum infectious diseases* Oxford University Press.

Bucher, T., Oppenheimer-Shaanan, Y., Savidor, A., Bloom-Ackermann, Z., & Kolodkin-Gal, I. (2015). Disturbance of the bacterial cell wall specifically interferes with biofilm formation. *Environmental Microbiology Reports*, *7*(6), 990–1004.

Byrd, T. F., & Lyons, C. R. (1999). Preliminary characterization of a *Mycobacterium abscessus* mutant in human and murine models of infection. *Infection and Immunity*, *67*(9), 4700–4707.

Carter, G., Wu, M., Drummond, D. C., & Bermudez, L. E. (2003). Characterization of biofilm formation by clinical isolates of *Mycobacterium avium*. *Journal of Medical Microbiology*, *52*(9), 747–752.

Carter, G., Young, L. S., & Bermudez, L. E. (2004). A subinhibitory concentration of clarithromycin inhibits *Mycobacterium avium* biofilm formation. *Antimicrobial Agents and Chemotherapy*, *48*(12), 4907–4910.

Casadevall, A., & Pirofski, L.-A. (2009). Virulence factors and their mechanisms of action: The view from a damage–response framework. *Journal of Water and Health*, *7*(S1), S2–S18.

Celdrán, A., Esteban, J., Manas, J., & Granizo, J.-J. (2007). Wound infections due to *Mycobacterium fortuitum* after polypropylene mesh inguinal hernia repair. *Journal of Hospital Infection*, *66*(4), 374–377.

Chakraborty, P., Bajeli, S., Kaushal, D., Radotra, B. D., & Kumar, A. (2021). Biofilm formation in the lung contributes to virulence and drug tolerance of *Mycobacterium tuberculosis*. *Nature Communications*, *12*(1), 1–17.

Chakraborty, P., & Kumar, A. (2019). The extracellular matrix of mycobacterial biofilms: Could we shorten the treatment of mycobacterial infections? *Microbial Cell*, *6*(2), 105.

Chalabaev, S., Chauhan, A., Novikov, A., Iyer, P., Szczesny, M., Beloin, C., et al. (2014). Biofilms formed by gram-negative bacteria undergo increased lipid a palmitoylation, enhancing in vivo survival. *MBio*, *5* (4). e01116-01114.

Chapman, M. R., Robinson, L. S., Pinkner, J. S., Roth, R., Heuser, J., Hammar, M., et al. (2002). Role of *Escherichia coli* curli operons in directing amyloid fiber formation. *Science*, *295*(5556), 851–855.

Chen, J. M., German, G. J., Alexander, D. C., Ren, H., Tan, T., & Liu, J. (2006). Roles of Lsr2 in colony morphology and biofilm formation of *Mycobacterium smegmatis*. *Journal of Bacteriology*, *188*(2), 633–641.

Chen, J., & Xie, J. (2011). Role and regulation of bacterial LuxR-like regulators. *Journal of Cellular Biochemistry*, *112*(10), 2694–2702.

Clary, G., Sasindran, S. J., Nesbitt, N., Mason, L., Cole, S., Azad, A., et al. (2018). *Mycobacterium abscessus* smooth and rough morphotypes form antimicrobial-tolerant biofilm phenotypes but are killed by acetic acid. *Antimicrobial Agents and Chemotherapy*, *62*(3). e01782-01717.

Dufour, D., Leung, V., & Lévesque, C. M. (2010). Bacterial biofilm: Structure, function, and antimicrobial resistance. *Endodontic Topics*, *22*(1), 2–16.

Eid, A. J., Berbari, E. F., Sia, I. G., Wengenack, N. L., Osmon, D. R., & Razonable, R. R. (2007). Prosthetic joint infection due to rapidly growing mycobacteria: Report of 8 cases and review of the literature. *Clinical Infectious Diseases*, *45*(6), 687–694.

El Helou, G., Viola, G. M., Hachem, R., Han, X. Y., & Raad, I. I. (2013). Rapidly growing mycobacterial bloodstream infections. *The Lancet Infectious Diseases*, *13*(2), 166–174.

Esteban, J., & García-Coca, M. (2018). *Mycobacterium* biofilms. *Frontiers in Microbiology*, *8*, 2651.

Esteban, J., Martín-de-Hijas, N. Z., Kinnari, T. J., Ayala, G., Fernández-Roblas, R., & Gadea, I. (2008). Biofilm development by potentially pathogenic non-pigmented rapidly growing mycobacteria. *BMC Microbiology*, *8*(1), 1–8.

Falkinham, J. O., III, Iseman, M. D., de Haas, P., & van Soolingen, D. (2008). *Mycobacterium avium* in a shower linked to pulmonary disease. *Journal of Water and Health*, *6*(2), 209–213.

Falkinham, J. O., III, Williams, M. D., Kwait, R., & Lande, L. (2016). Methylobacterium spp. as an indicator for the presence or absence of *Mycobacterium* spp. *International Journal of Mycobacteriology*, *5* (2), 240–243.

Fedtke, I., Mader, D., Kohler, T., Moll, H., Nicholson, G., Biswas, R., et al. (2007). A *Staphylococcus aureus* ypfP mutant with strongly reduced lipoteichoic acid (LTA) content: LTA governs bacterial surface properties and autolysin activity. *Molecular Microbiology*, *65*(4), 1078–1091.

Fennelly, K. P., Ojano-Dirain, C., Yang, Q., Liu, L., Lu, L., Progulske-Fox, A., et al. (2016). Biofilm formation by *Mycobacterium abscessus* in a lung cavity. *American Journal of Respiratory and Critical Care Medicine*, *193*(6), 692–693.

Flemming, H.-C., Neu, T. R., & Wozniak, D. J. (2007). The EPS matrix: The "house of biofilm cells". *Journal of Bacteriology*, *189*(22), 7945–7947.

Flemming, H.-C., & Wingender, J. (2010). The biofilm matrix. *Nature Reviews Microbiology*, *8*(9), 623–633.

Foster, T. J., Geoghegan, J. A., Ganesh, V. K., & Höök, M. (2014). Adhesion, invasion and evasion: The many functions of the surface proteins of *Staphylococcus aureus*. *Nature Reviews Microbiology*, *12*(1), 49–62.

Geier, H., Mostowy, S., Cangelosi, G. A., Behr, M. A., & Ford, T. E. (2008). Autoinducer-2 triggers the oxidative stress response in *Mycobacterium avium*, leading to biofilm formation. *Applied and Environmental Microbiology*, *74*(6), 1798–1804.

Greendyke, R., & Byrd, T. F. (2008). Differential antibiotic susceptibility of *Mycobacterium abscessus* variants in biofilms and macrophages compared to that of planktonic bacteria. *Antimicrobial Agents and Chemotherapy*, *52*(6), 2019–2026.

Griffith, D. E., Brown, B. A., Cegielski, P., Murphy, D. T., & Wallace, R. J., Jr. (2000). Early results (at 6 months) with intermittent clarithromycin-including regimens for lung disease due to *Mycobacterium avium* complex. *Clinical Infectious Diseases*, *30*(2), 288–292.

Gupta, K. R., Kasetty, S., & Chatterji, D. (2015). Novel functions of (p) ppGpp and cyclic di-GMP in mycobacterial physiology revealed by phenotype microarray analysis of wild-type and isogenic strains of *Mycobacterium smegmatis*. *Applied and Environmental Microbiology*, *81*(7), 2571–2578.

Guttenplan, S. B., & Kearns, D. B. (2013). Regulation of flagellar motility during biofilm formation. *FEMS Microbiology Reviews*, *37*(6), 849–871.

Ha, K.-Y., Chung, Y.-G., & Ryoo, S.-J. (2005). Adherence and biofilm formation of *Staphylococcus epidermidis* and *Mycobacterium tuberculosis* on various spinal implants. *Spine*, *30*(1), 38–43.

Hall-Stoodley, L., & Lappin-Scott, H. (1998). Biofilm formation by the rapidly growing mycobacterial species *Mycobacterium fortuitum*. *FEMS Microbiology Letters*, *168*(1), 77–84.

Hall-Stoodley, L., & Stoodley, P. (2009). Evolving concepts in biofilm infections. *Cellular Microbiology*, *11*(7), 1034–1043.

Heacock-Kang, Y., Zarzycki-Siek, J., Sun, Z., Poonsuk, K., Bluhm, A. P., Cabanas, D., et al. (2018). Novel dual regulators of *Pseudomonas aeruginosa* essential for productive biofilms and virulence. *Molecular Microbiology*, *109*(3), 401–414.

Hegde, S. R. (2020). Computational identification of the proteins associated with quorum sensing and biofilm formation in *Mycobacterium tuberculosis*. *Frontiers in Microbiology*, *10*, 3011.

Huang, M., Meng, L., Fan, M., Hu, P., & Bian, Z. (2008). Effect of biofilm formation on virulence factor secretion via the general secretory pathway in Streptococcus mutans. *Archives of Oral Biology*, *53*(12), 1179–1185.

Ibanez de Aldecoa, A. L., Zafra, O., & González-Pastor, J. E. (2017). Mechanisms and regulation of extracellular DNA release and its biological roles in microbial communities. *Frontiers in Microbiology*, *8*, 1390.

Islam, M. S., Richards, J. P., & Ojha, A. K. (2012). Targeting drug tolerance in mycobacteria: A perspective from mycobacterial biofilms. *Expert Review of Anti-Infective Therapy*, *10*(9), 1055–1066.

Jamal, M., Ahmad, W., Andleeb, S., Jalil, F., Imran, M., Nawaz, M. A., et al. (2018). Bacterial biofilm and associated infections. *Journal of the Chinese Medical Association*, *81*(1), 7–11.

Jiang, Q., Jin, Z., & Sun, B. (2018). MgrA negatively regulates biofilm formation and detachment by repressing the expression of psm operons in *Staphylococcus aureus*. *Applied and Environmental Microbiology*, *84* (16), e01008–e01018.

Johansen, T. B., Agdestein, A., Olsen, I., Nilsen, S. F., Holstad, G., & Djønne, B. (2009). Biofilm formation by *Mycobacterium avium* isolates originating from humans, swine and birds. *BMC Microbiology*, *9*(1), 1–10.

Kang, J., Xu, L., Yang, S., Yu, W., Liu, S., Xin, Y., et al. (2013). Effect of phosphoglucosamine mutase on biofilm formation and antimicrobial susceptibilities in *M. smegmatis* glmM gene knockdown strain. *PLoS One*, *8*(4), e61589.

Keating, T., Lethbridge, S., Allnutt, J. C., Hendon-Dunn, C. L., Thomas, S. R., Alderwick, L. J., et al. (2021). *Mycobacterium tuberculosis* modifies cell wall carbohydrates during biofilm growth with a concomitant reduction in complement activation. *The Cell Surface*, *7*, 100065.

Kennedy, B. S., Bedard, B., Younge, M., Tuttle, D., Ammerman, E., Ricci, J., et al. (2012). Outbreak of *Mycobacterium chelonae* infection associated with tattoo ink. *New England Journal of Medicine*, *367* (11), 1020–1024.

Koch, R. (1882). *Classifying infectious diseases: The etiology of tuberculosis*. Berlin: Robert Koch.

Kohler, P., Kuster, S. P., Bloemberg, G., Schulthess, B., Frank, M., Tanner, F. C., et al. (2015). Healthcare-associated prosthetic heart valve, aortic vascular graft, and disseminated *Mycobacterium chimaera* infections subsequent to open heart surgery. *European Heart Journal*, *36*(40), 2745–2753.

Kulka, K., Hatfull, G., & Ojha, A. K. (2012). Growth of *Mycobacterium tuberculosis* biofilms. *JoVE (Journal of Visualized Experiments)*, (60), e3820.

Kumar, A., Farhana, A., Guidry, L., Saini, V., Hondalus, M., & Steyn, A. J. (2011). Redox homeostasis in mycobacteria: The key to tuberculosis control? *Expert Reviews in Molecular Medicine*, *13*.

Kumar, T., Kumar, V., Shoorvir, S., Gupta, S., Chaubey, K., Jayaraman, S., et al. (2015). Chemical and ultrastructural characteristics of mycobacterial biofilms. *Asian Journal of Animal and Veterinary*, *10*(10), 592–622.

Lee, K.-J., Lee, M.-A., Hwang, W., Park, H., & Lee, K.-H. (2016). Deacylated lipopolysaccharides inhibit biofilm formation by Gram-negative bacteria. *Biofouling*, *32*(7), 711–723.

Limia, A., Sangari, F. J., Wagner, D., & Bermudez, L. E. (2001). Characterization and expression of secA in *Mycobacterium avium*. *FEMS Microbiology Letters*, *197*(2), 151–157.

Lopes, S. P., Azevedo, N. F., & Pereira, M. O. (2018). Quantitative assessment of individual populations within polymicrobial biofilms. *Scientific Reports*, *8*(1), 1–13.

Maher, M. C., Lim, J. Y., Gunawan, C., & Cegelski, L. (2015). Cell-based high-throughput screening identifies rifapentine as an inhibitor of amyloid and biofilm formation in *Escherichia coli*. *ACS Infectious Diseases*, *1*(10), 460–468.

Marciano-Cabral, F., Jamerson, M., & Kaneshiro, E. S. (2010). Free-living amoebae, Legionella and *Mycobacterium* in tap water supplied by a municipal drinking water utility in the USA. *Journal of Water and Health*, *8*(1), 71–82.

Marrakchi, H., Lanéelle, M.-A., & Daffé, M. (2014). Mycolic acids: Structures, biosynthesis, and beyond. *Chemistry & Biology*, *21*(1), 67–85.

Martínez, A., Torello, S., & Kolter, R. (1999). Sliding motility in mycobacteria. *Journal of Bacteriology*, *181*(23), 7331–7338.

Mathew, R., Mukherjee, R., Balachandar, R., & Chatterji, D. (2006). Deletion of the rpoZ gene, encoding the ω subunit of RNA polymerase, results in pleiotropic surface-related phenotypes in *Mycobacterium smegmatis*. *Microbiology*, *152*(6), 1741–1750.

Menozzi, F. D., Rouse, J. H., Alavi, M., Laude-Sharp, M., Muller, J., Bischoff, R., et al. (1996). Identification of a heparin-binding hemagglutinin present in mycobacteria. *The Journal of Experimental Medicine*, *184*(3), 993–1001.

Miller, B. K., Zulauf, K. E., & Braunstein, M. (2017). The Sec pathways and exportomes of *Mycobacterium tuberculosis*. *Microbiology Spectrum*, *5*(2), 5.2. 24.

Mohandas, P., Budell, W. C., Mueller, E., Au, A., Bythrow, G. V., & Quadri, L. E. (2016). Pleiotropic consequences of gene knockouts in the phthiocerol dimycocerosate and phenolic glycolipid biosynthetic gene cluster of the opportunistic human pathogen *Mycobacterium marinum*. *FEMS Microbiology Letters*, *363*(5).

Muñoz-Egea, M., Esteban, M., & Esteban, J. (2016). Inhibition of *Mycobacterium abscessus* biofilms by *Methylobacterium* sp. In *29th ECCMI (Amsterdam)*.

Muñoz-Egea, M.-C., García-Pedrazuela, M., Mahillo, I., & Esteban, J. (2015). Effect of ciprofloxacin in the ultrastructure and development of biofilms formed by rapidly growing mycobacteria. *BMC Microbiology*, *15*(1), 1–6.

Muñoz-Egea, M.-C., García-Pedrazuela, M., Mahillo, I., García, M. J., & Esteban, J. (2013). Autofluorescence as a tool for structural analysis of biofilms formed by nonpigmented rapidly growing mycobacteria. *Applied and Environmental Microbiology*, *79*(3), 1065–1067.

Muñoz-Egea, M.-C., García-Pedrazuela, M., Mahillo-Fernandez, I., & Esteban, J. (2016). Effect of antibiotics and antibiofilm agents in the ultrastructure and development of biofilms developed by nonpigmented rapidly growing mycobacteria. *Microbial Drug Resistance*, *22*(1), 1–6.

Nessar, R., Reyrat, J.-M., Davidson, L. B., & Byrd, T. F. (2011). Deletion of the mmpL4b gene in the *Mycobacterium abscessus* glycopeptidolipid biosynthetic pathway results in loss of surface colonization capability, but enhanced ability to replicate in human macrophages and stimulate their innate immune response. *Microbiology*, *157*(4), 1187–1195.

Niño-Padilla, E. I., Velazquez, C., & Garibay-Escobar, A. (2021). Mycobacterial biofilms as players in human infections: A review. *Biofouling*, *37*(4), 410–432.

Ojha, A., Anand, M., Bhatt, A., Kremer, L., Jacobs, W. R., Jr., & Hatfull, G. F. (2005). GroEL1: A dedicated chaperone involved in mycolic acid biosynthesis during biofilm formation in mycobacteria. *Cell*, *123*(5), 861–873.

Ojha, A. K., Baughn, A. D., Sambandan, D., Hsu, T., Trivelli, X., Guerardel, Y., et al. (2008). Growth of *Mycobacterium tuberculosis* biofilms containing free mycolic acids and harbouring drug-tolerant bacteria. *Molecular Microbiology*, *69*(1), 164–174.

Ojha, A. K., Trivelli, X., Guerardel, Y., Kremer, L., & Hatfull, G. F. (2010). Enzymatic hydrolysis of trehalose dimycolate releases free mycolic acids during mycobacterial growth in biofilms. *Journal of Biological Chemistry*, *285*(23), 17380–17389.

Oliva, A., Stefani, S., Venditti, M., & Di Domenico, E. G. (2021). Biofilm-related infections in gram-positive bacteria and the potential role of the long-acting agent dalbavancin. *Frontiers in Microbiology*, 3224.

Olofsson, A.-C., Hermansson, M., & Elwing, H. (2003). N-acetyl-L-cysteine affects growth, extracellular polysaccharide production, and bacterial biofilm formation on solid surfaces. *Applied and Environmental Microbiology*, *69*(8), 4814–4822.

Ortíz-Pérez, A., Martín-de-Hijas, N., Alonso-Rodríguez, N., Molina-Manso, D., Fernández-Roblas, R., & Esteban, J. (2011). Importance of antibiotic penetration in the antimicrobial resistance of biofilm formed by non-pigmented rapidly growing mycobacteria against amikacin, ciprofloxacin and clarithromycin. *Enfermedades Infecciosas y Microbiologia Clinica*, *29*(2), 79–84.

Ovrutsky, A. R., Chan, E. D., Kartalija, M., Bai, X., Jackson, M., Gibbs, S., et al. (2013). Cooccurrence of free-living amoebae and nontuberculous Mycobacteria in hospital water networks, and preferential growth of *Mycobacterium avium* in *Acanthamoeba lenticulata*. *Applied and Environmental Microbiology*, *79*(10), 3185–3192.

Pacheco, S. A., Hsu, F.-F., Powers, K. M., & Purdy, G. E. (2013). MmpL11 protein transports mycolic acid-containing lipids to the mycobacterial cell wall and contributes to biofilm formation in *Mycobacterium smegmatis*. *Journal of Biological Chemistry*, *288*(33), 24213–24222.

Padhi, A., Naik, S. K., Sengupta, S., Ganguli, G., & Sonawane, A. (2016). Expression of *Mycobacterium tuberculosis* NLPC/p60 family protein Rv0024 induce biofilm formation and resistance against cell wall acting anti-tuberculosis drugs in *Mycobacterium smegmatis*. *Microbes and Infection*, *18*(4), 224–236.

Penesyan, A., Paulsen, I. T., Kjelleberg, S., & Gillings, M. R. (2021). Three faces of biofilms: A microbial lifestyle, a nascent multicellular organism, and an incubator for diversity. *npj Biofilms and Microbiomes*, *7*(1), 1–9.

Pidot, S. J., Porter, J. L., Tobias, N. J., Anderson, J., Catmull, D., Seemann, T., et al. (2010). Regulation of the 18 kDa heat shock protein in *Mycobacterium ulcerans*: An alpha-crystallin orthologue that promotes biofilm formation. *Molecular Microbiology*, *78*(5), 1216–1231.

Recht, J., & Kolter, R. (2001). Glycopeptidolipid acetylation affects sliding motility and biofilm formation in *Mycobacterium smegmatis*. *Journal of Bacteriology*, *183*(19), 5718–5724.

Recht, J., Martínez, A., Torello, S., & Kolter, R. (2000). Genetic analysis of sliding motility in *Mycobacterium smegmatis*. *Journal of Bacteriology*, *182*(15), 4348–4351.

Reyrat, J.-M., & Kahn, D. (2001). *Mycobacterium smegmatis*: An absurd model for tuberculosis? *Trends in Microbiology*, *9*(10), 472–474.

Rose, S. J., Babrak, L. M., & Bermudez, L. E. (2015). *Mycobacterium avium* possesses extracellular DNA that contributes to biofilm formation, structural integrity, and tolerance to antibiotics. *PLoS One*, *10*(5), e0128772.

Rose, S. J., & Bermudez, L. E. (2016). Identification of bicarbonate as a trigger and genes involved with extracellular DNA export in mycobacterial biofilms. *MBio*, *7*(6). e01597-01516.

Ruhal, R., & Kataria, R. (2021). Biofilm patterns in gram-positive and gram-negative bacteria. *Microbiological Research*, *251*, 126829.

Sambandan, D., Dao, D. N., Weinrick, B. C., Vilchèze, C., Gurcha, S. S., Ojha, A., et al. (2013). Keto-mycolic acid-dependent pellicle formation confers tolerance to drug-sensitive *Mycobacterium tuberculosis*. *MBio*, *4*(3). e00222-00213.

Singh, N., & Kumar, A. (2015). Virulence factor SenX3 is the oxygen-controlled replication switch of *Mycobacterium tuberculosis*. *Antioxidants & Redox Signaling*, *22*(7), 603–613.

Socransky, S. S. (2002). Dental biofilms: Difficult therapeutic targets. *Periodontology 2000*, *28*, 12–55.

Soliman, C., Walduck, A. K., Yuriev, E., Richards, J. S., Cywes-Bentley, C., Pier, G. B., et al. (2018). Structural basis for antibody targeting of the broadly expressed microbial polysaccharide poly-N-acetylglucosamine. *Journal of Biological Chemistry*, *293*(14), 5079–5089.

Sousa, S., Bandeira, M., Carvalho, P. A., Duarte, A., & Jordao, L. (2015). Nontuberculous mycobacteria pathogenesis and biofilm assembly. *International Journal of Mycobacteriology*, *4*(1), 36–43.

Srey, S., Jahid, I. K., & Ha, S.-D. (2013). Biofilm formation in food industries: A food safety concern. *Food Control*, *31*(2), 572–585.

Steed, K. A., & Falkinham, J. O., III. (2006). Effect of growth in biofilms on chlorine susceptibility of *Mycobacterium avium* and *Mycobacterium intracellulare*. *Applied and Environmental Microbiology*, *72*(6), 4007–4011.

Teixeira, P. C., Leite, G. M., Domingues, R. J., Silva, J., Gibbs, P. A., & Ferreira, J. P. (2007). Antimicrobial effects of a microemulsion and a nanoemulsion on enteric and other pathogens and biofilms. *International Journal of Food Microbiology*, *118*(1), 15–19.

Trivedi, A., Mavi, P. S., Bhatt, D., & Kumar, A. (2016). Thiol reductive stress induces cellulose-anchored biofilm formation in *Mycobacterium tuberculosis*. *Nature Communications*, *7*(1), 1–15.

Tsai, S.-H., Shen, G.-H., Lin, C.-H., Liau, J.-R., Lai, H.-C., & Hu, S.-T. (2013). Mab_3168c, a putative acetyltransferase, enhances adherence, intracellular survival and antimicrobial resistance of *Mycobacterium abscessus*. *PLoS One*, *8*(6), e67563.

Vargas, D., Hageman, S., Gulati, M., Nobile, C. J., & Rawat, M. (2016). S-nitrosomycothiol reductase and mycothiol are required for survival under aldehyde stress and biofilm formation in *Mycobacterium smegmatis*. *IUBMB Life*, *68*(8), 621–628.

Vijayaraghavan, R., Chandrashekhar, R., Sujatha, Y., & Belagavi, C. (2006). Hospital outbreak of atypical mycobacterial infection of port sites after laparoscopic surgery. *Journal of Hospital Infection*, *64*(4), 344–347.

Walker, J., Moore, G., Collins, S., Parks, S., Garvey, M. I., Lamagni, T., et al. (2017). Microbiological problems and biofilms associated with *Mycobacterium chimaera* in heater–cooler units used for cardiopulmonary bypass. *Journal of Hospital Infection*, *96*(3), 209–220.

Weidenmaier, C., Kokai-Kun, J. F., Kristian, S. A., Chanturiya, T., Kalbacher, H., Gross, M., et al. (2004). Role of teichoic acids in *Staphylococcus aureus* nasal colonization, a major risk factor in nosocomial infections. *Nature Medicine*, *10*(3), 243–245.

Weidenmaier, C., & Peschel, A. (2008). Teichoic acids and related cell-wall glycopolymers in Gram-positive physiology and host interactions. *Nature Reviews Microbiology*, *6*(4), 276–287.

Xu, H., Su, Z., Li, W., Deng, Y., & He, Z. G. (2021). MmbR, a master transcription regulator that controls fatty acid β-oxidation genes in *Mycolicibacterium smegmatis*. *Environmental Microbiology*, *23*(2), 1096–1114.

Yamazaki, Y., Danelishvili, L., Wu, M., Hidaka, E., Katsuyama, T., Stang, B., et al. (2006). The ability to form biofilm influences *Mycobacterium avium* invasion and translocation of bronchial epithelial cells. *Cellular Microbiology*, *8*(5), 806–814.

Yamazaki, Y., Danelishvili, L., Wu, M., MacNab, M., & Bermudez, L. E. (2006). *Mycobacterium avium* genes associated with the ability to form a biofilm. *Applied and Environmental Microbiology*, *72*(1), 819–825.

Yang, Y., Thomas, J., Li, Y., Vilchèze, C., Derbyshire, K. M., Jacobs, W. R., Jr., et al. (2017). Defining a temporal order of genetic requirements for development of mycobacterial biofilms. *Molecular Microbiology*, *105*(5), 794–809.

Yin, W., Wang, Y., Liu, L., & He, J. (2019). Biofilms: The microbial "protective clothing" in extreme environments. *International Journal of Molecular Sciences*, *20*(14), 3423.

Zambrano, M. M., & Kolter, R. (2005). Mycobacterial biofilms: A greasy way to hold it together. *Cell*, *123*(5), 762–764.

Zheng, Y., Joo, H.-S., Nair, V., Le, K. Y., & Otto, M. (2018). Do amyloid structures formed by *Staphylococcus aureus* phenol-soluble modulins have a biological function? *International Journal of Medical Microbiology*, *308*(6), 675–682.

Chapter 28

Streptococcus pneumoniae biofilms and human infectious diseases: A review

Ruth Zomuansangi[a], Zothanpuia[b], Prashant Kumar Singh[b], Bhim Pratap Singh[d], Garima Singh[c], Purbajyoti Deka[e], Jae Jun Song[f], and Mukesh Kumar Yadav[b]

[a]*Department of Life-Science, Mizoram University (A Central University), Pachhunga University College Campus, Aizawl, Mizoram, India,* [b]*Department of Biotechnology (Life-Science), Mizoram University (A Central University), Pachhunga University College Campus, Aizawl, Mizoram, India,* [c]*Department of Botany (Life-Science), Mizoram University (A Central University), Pachhunga University College Campus, Aizawl, Mizoram, India,* [d]*Department of Agriculture and Environmental Sciences (AES), National Institute of Food Technology Entrepreneurship and Management (NIFTEM), Sonepat, Haryana, India,* [e]*Department of Biotechnology, Mizoram University, Aizawl, Mizoram, India,* [f]*Department of Otorhinolaryngology—Head and Neck Surgery, Korea University College of Medicine, Guro Hospital, Seoul, South Korea*

1. Introduction

Streptococcus pneumoniae (pneumococci) is a Gram-positive bacterium that occupies the nasal cavity within 3 months of birth and lives asymptotically and causes infection during immune-compressed conditions (Stoodley, Sauer, Davies, & Costerton, 2002). Pneumococci cause several infectious diseases in both children and elders such as otitis media (OM), pneumonia, sepsis, bacteremia. Pneumococci colonize in the nasopharynx in the form of adherent communities called biofilms, and the pneumococcal biofilms act as a reservoir for the bacteria (Munoz-Elias, Marcano, & Camilli, 2008). Pneumococcal biofilms are highly organized communities of bacteria that are adherent to the biotic or abiotic surface and produce apolymeric matrix consisting of protein, carbohydrate, and e-DNA (Costerton, Stewart, & Greenberg, 1999; Donlan & Costerton, 2002; Stoodley et al., 2002). The biofilm's mode of growth of pneumococci provides resistance to antibiotics and shields bacteria during stress and adverse environmental conditions (Donlan & Costerton, 2002). It is estimated that biofilm bacteria are more resistant (up to1000 times) to common antibiotics, such as erythromycin, clindamycin, amoxicillin, levofloxacin, gentamicin, tetracycline, penicillin when compared with the planktonic bacteria (del Prado et al., 2010; García-Castillo et al., 2007; Marks, Reddinger, & Hakansson, 2012). One of the reasons for elevated resistance of biofilm bacteria is due to the biofilm matrix that prevents the entry of antibiotics deep inside the biofilms and results in deep inside bacteria-related unaffected, which gives rise to persister cells and confers resistance to antibiotics (Costerton et al., 1999; Donlan & Costerton, 2002). In biofilms, pneumococci are in close proximity that increases the genetic transformation and results in the acquisition of resistant genes. In addition, biofilm bacteria follow the altered metabolic pathway, display different physiology and gene expression in comparison to the free-floating planktonic bacteria (Oggioni et al., 2006).

In the nasopharynx, pneumococci are carried in the form of biofilms without any disease or symptoms, and these biofilms act as a reservoir of the bacteria, and during immune-compressed conditions, it causes infections. From these nasopharynx biofilms, planktonic bacteria can transit to other sterile sites such as the lungs causing pneumonia, middle ear causing OM, in the brain causing sepsis, and in the blood causing bacteremia (Ash & Sheffield, 2013; Pichichero, 2013; Shak, Vidal, & Klugman, 2013). The role of biofilms in various human infections such as OM, pneumonia, sepsis, and bacteria is well established. In addition, the biofilms cause various infections related to indwelling medical devices (Yadav et al., 2019).

2. *Streptococcus pneumoniae*–related infections

The Gram-positive bacterium *Streptococcus pneumoniae* colonizes asymptomatically in the nasopharynx and causes infection under immune-suppressed individuals such as children and elderly. The bacteria become aggravated sometime after viral infection due to the decreased immune defense of the host and availability of favorable environment to infect host cells under the change of epidemiology of the host (Kim, Seon, & Rhee, 2017). The viral infection in the respiratory tract could be attained by the patients under mechanical ventilation. And the most

Understanding Microbial Biofilms. https://doi.org/10.1016/B978-0-323-99977-9.00002-8

common bacterial coinfection with viral infection is with individuals who are infected by the influenza virus (Mirzaei et al., 2020). The world's mortality and morbidity allied by the pneumococcal disease have created complications due to their resistance to the antimicrobials that are forced with excessive effort to develop the vaccines (Briles, Paton, Mukerji, Swiatlo, & Crain, 2019). High doses of antibiotic drug intake make it easy for the formation of persister cells, which results in antibiotic treatment failure (Mirzaei et al., 2020).

3. Role of *Streptococcus pneumoniae* biofilms in human infectious diseases

Streptococcus pneumonia is the global pathogenic bacterium causing disease in children and adults across the world (Berical, Harris, Dela Cruz, & Possick, 2016). The biofilm enables the bacterial cells to survive and protect themselves from the antibiotics and host immune factors. In addition, in the infections can be caused by devices, the biofilms can be formed on abiotic surface and cause infection (Del Pozo et al., 2018). Some of the *S. pneumonia* in which biofilms are involve are illustrated in Table 1 and briefly discussed below.

TABLE 1 **Infection caused by the *Streptococcus pneumoniae*; their location of infection and complications.**

Infection	Location of bacteria biofilms/ colonization infection	Complications
Otitis Media	• nasopharynx and eventually into the middle ear mucosa	• rupture eardrums • permanent hearing loss Infection that spread the fluid around the spinal cord and brain • infection that spread to the bones of the ears
Pneumonia	• nasopharynx and eventually to the respiratory system	• high risk of causing bacteremia • difficulty breathing • fluid accumulation around the lungs • lung abscess
Bronchitis	• nasopharynx and eventually to the bronchia	• it can lead to pneumonia • repeated bronchitis can be chronic obstructive pulmonary disease (COPD)
Medical device infection	• bladder catheters • central venous catheters • fracture fixation devices • dental implants • joint prostheses • vascular grafts • cardiac pacemaker • mammary implants • mechanical heart valves • heart assist devices	• depending on the loci of the infection, it can cause different chronic disease
Bacteremia	• into the bloodstream	• high risk of causing sepsis and septic shock

3.1 Otitis media

OM, a common infection, affects approx. 80% of children below age 5 years, is caused by upper respiratory tract complex microbial community including *Moraxella catarrhalis*, *Streptococcus pneumoniae*, Haemophilus influenza, etc. (Lieberthal, Carroll, & Chonmaitree, 2013; Revai, Mamidi, & Chonmaitree, 2008). And *S. pneumoniae* is an important bacterium that colonizes in the nasopharynx of 50% of children as the first step of infection (Bogaert, De Groot, & Hermans, 2004; Syrjanen, Kilpi, Kaijalainen, Herva, & Takala, 2001). *S. pneumoniae* after nasopharyngeal colonization gets the opportunity to invade the middle ear space and causes OM. Previously many studies focused on the planktonic mode of *S. pneumoniae* growth; however, biofilm mode of growth has been recognized as an important factor in the pathophysiology of OM (Dohar et al., 2005; Ehrlich et al., 2002; Hall-Stoodley et al., 2006; Post, 2001; Vermee, Cohen, Hays, et al., 2019). Indeed, Hall-Stoodley et al. (2006) directly detected biofilms on the tissue sample extracted from children suffering from OM (Hall-Stoodley et al., 2006). Thereafter, the role of biofilms in OM was studied using an animal model and reports suggest that the biofilm community of *S. pneumoniae* is very crucial for the outcome of OM (Yadav, Chae, & Song, 2012). Chronic OM with effusion, recurrent otitis media (rAOM), chronic tonsillitis, and cholesteatoma demonstrated microbial

TABLE 1 Infection caused by the *Streptococcus pneumoniae*; their location of infection and complications—cont'd

Infection	Location of bacteria biofilms/ colonization infection	Complications
Meningitis	• into the meninges; the membrane of the brain and spinal cord	• vulnerable to immune deficiency diseases such as HIV/AIDS, autoimmune disorder, chemotherapy, organ or bone marrow transplant

biofilms (Hall-Stoodley et al., 2006; Jensen, Johansen, Bjarnsholt, Eickhardt-Sørensen, & Homøe, 2017). In OM, the microbial biofilms formed on the surface of middle ear mucosa consisting of host factors also, example in middle ear effusion (MEE), the biofilms contain DNA of the host along with bacteria cells and biofilm matrix (Post, Stoodley, Hall-Stoodley, & Ehrlich, 2004). In addition, in MEE, the viscosity of effusion increases due to neutrophil extracellular traps (NETs) (Thornton et al., 2013). In addition, polyethylene (PE) tube is inserted as surgical intervention for the cure of middle ear abnormality, and pneumococcal biofilms have been detected on those PE tubes.

3.2 Pneumonia

Streptococcus pneumoniae causes pneumonia in children and immune-compressed individuals. Indeed, during COVID-19 pandemic, in more than 50 moderate-to-severe affected patients *S. pneumoniae* was detected as copathogen. It is suggested that the bacteria colonize in the nasopharynx and transit to the lungs. In addition, the viral infection created favorable condition for bacterial infection. Pneumonia is an infection in the lung, which causes dysregulation into the host respiratory system accumulating the fluid into the pulmonary tract because of the microorganisms present in the respiratory tract (Quinton, Walkey, & Mizgerd, 2018). Pneumonia is likely to be diagnosed in newborns as their lungs are susceptible to various viral and bacterial infections (Hooven & Polin, 2017). *Streptococcus pneumoniae* is one of the most common opportunistic bacteria pathogens reporting 25% of cases of community-acquired pneumonia worldwide, and the occurrence *Chlamydia pneumoniae*, *Mycoplasma pneumoniae*, and viruses is milder (Lanks, Musani, & Hsia, 2019). Pneumonia is likely to develop when the defense mechanisms of the respiratory tract are being attacked by the invasive pathogens. The entry of pathogen into the respiratory tract triggers the immune system of the host and leads to inflammation with an accumulation of fluids, cell debris, and possible white blood cells in the alveoli, which lead to further reduction in the pulmonary function (Leung, Wong, & Hon, 2018) (Fig. 1). Although vaccine against *S. pneumoniae* is available, in the developing countries, the pneumonia cases are more due to poor hygienic condition, circulating antibiotic strains or serotypes not covered in vaccine. *S. pneumoniae* consists of 96 serotypes based on capsule polysaccharides, and vaccine developed so far is for 23 types, i.e., 23-valent and that too has not reached the developing nations.

3.3 Bronchitis

Bronchitis is commonly caused by viral infection; among them, identified viruses are rhinovirus, influenza A and B,

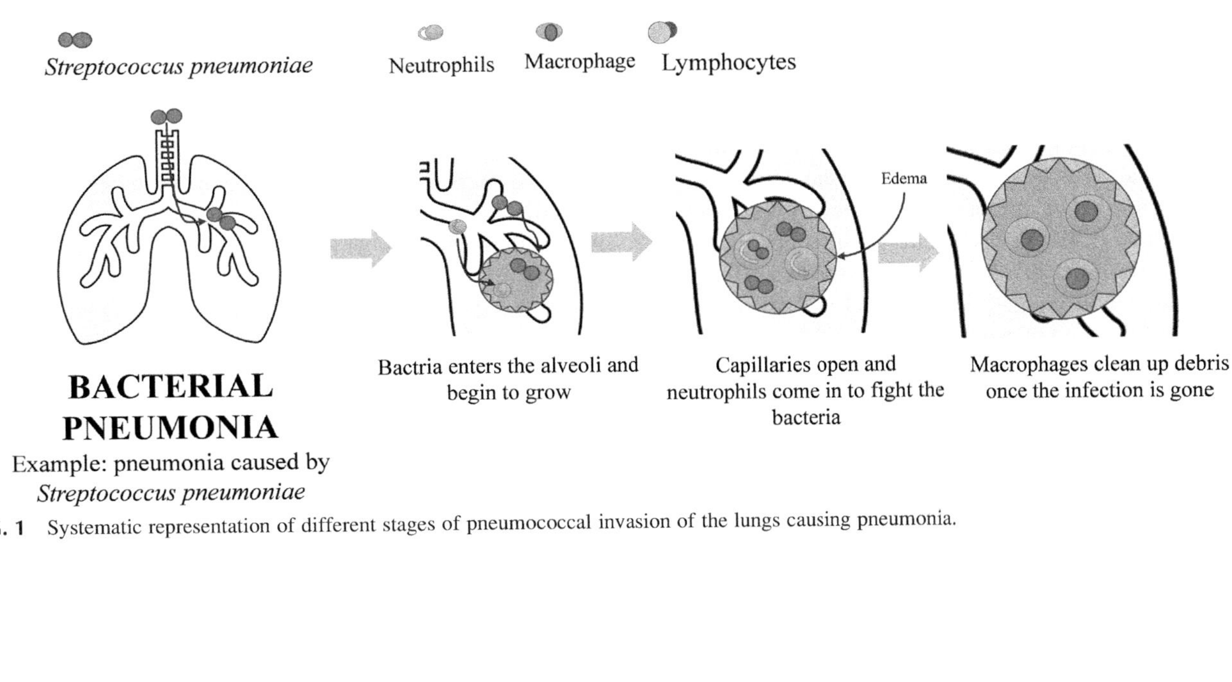

FIG. 1 Systematic representation of different stages of pneumococcal invasion of the lungs causing pneumonia.

coronavirus, parainfluenza, metapneumovirus, enterovirus, and respiratory syncytial viruses. In addition, a very less percentage of bacteria were detected to be the cause of bronchitis in which bacteria such as the *Chlamydia pneumoniae*, *Mycoplasma pneumoniae* are found to be causing acute bronchitis (Kinkade & Long, 2016). The hindrance in the airways mostly in the children having a history of congenital heart disease is likely to experience plastic bronchitis (PB) where the casts of PB are of mucin containing a small amount of fibrin (Rubin, 2016). Patients with PB attribute abnormality in their lymphatic system and central venous hypertension, which leads to the endobronchial outlet of lymph to precipitate forming airway cast (Chhabada & Khanna, 2020). In the human body, *S. pneumoniae* colonizes the upper respiratory tracts, and this colonization takes place in the upper mucosal surface of the nasopharynx during childhood and asymptomatically perseveres in healthy individuals into adolescence (Chao, Marks, Pettigrew, & Hakansson, 2015). *S. pneumoniae* possesses various virulence factors, which permit adherence to the host cells and thereby decrease the capability of the host's immune system to explicit the bacterium, and thus recommend invasion in the epithelial cells (Kadioglu, Weiser, Paton, & Andrew, 2008). In case the host is not able to explicit, *S. pneumoniae* can migrate to the tissues and organs instantly after colonization of the upper respiratory system and thus causes the infection (Steel, Cockeran, Anderson, & Feldman, 2013). Protracted bacterial infection is one of the major causes of persistent cough during childhood times, and *S. pneumoniae* and *Haemophilus influenza* are mostly responsible for chronic endobronchial infection such as acute pneumonia (Berndsen, Erlendsdóttir, & Gottfredsson, 2012).

3.4 Medical indwelling device-related infections

Certain bacteria can form a biofilm, which is an aggregation of the microorganisms living in a community forming a protective shield called extracellular protective substance (EPS), and certain strains of the *E. coli* are responsible for the mortality and morbidity as in the case of various medical-related infections such as the intravascular catheters, urethral, prosthetic grafts, joints, and shunts where a biofilm is formed in the area of contact by the microbes (Sharma et al., 2016). The mode of infection can occur majorly at the time of the surgery performed, contiguous and hematogenous spread at a site and nearby source and is commonly caused by bacteria such as *Propionibacterium acnes* and *Staphylococcus aureus* in the skin, and *S. pneumoniae* and *Haemophilus influenza* cause infection in the ventriculoatrial shunt (Seidelman & Lewis, 2018). The ventricular assist devices specifically at the site of the driveline pathogen such as *Staphylococcus aureus*, *Pseudomonas aeruginosa*, *Candida* spp., *Proteus* spp., *Corynebacterium* spp., *Enterococcus* spp., enteric Gram-negative bacteria, and coagulase-negative Staphylococci cause the infection (Koval et al., 2019). When the infection occurs, there is a rise of chances to uplevel in chronic infection as the bacteria aggregate into the surface of the implants or the devices; it forms a biofilm, and these biofilms often tolerate the antibiotic drugs (Stewart & Bjarnsholt, 2020).

3.5 Bacteremia

Bacteremia is the leading cause of the human pathogen bacteria *Staphylococcus aureus* (Tong et al., 2015). The infection is caused by the presence of bacteria in the bloodstream and is highly infected in younger children with immature immune systems (Pai, Enoch, & Aliyu, 2015) and according to the report of Centers for Disease Control Prevention (2015), it causes about 12,000 cases per year and usually accompanies other pneumococcal infection. Initially, Pneumococci colonize the human nasopharynx in the form of biofilms that act as bacterial reservoirs (Simell et al., 2012) and can also migrate to the bloodstream causing bacteremia or to the brain causing meningitis (Shak et al., 2013). Bacteremia can further cause systemic inflammation developing sepsis, the inflammation and the immunosuppression caused by the infection can cause damage in the host tissues, which can suppress the defense mechanisms to protect against the pathogen and secondary infections (Kwiecinski & Horswill, 2020). The infection of the *S. aureus* into the bloodstream has a higher risk of death of the patient with concomitant *Staphylococcus aureus* bacteremia (SAB) (van Hal et al., 2012). In both immunosuppressed and immunocompetent patients, *S. pneumoniae* (the pneumococcus) is importantly the well-known cause of bacteremia. The pneumococcal bacteremia can take place as a result of pneumococcal pneumonia or its absence. Secondary complications occur when bacteremia is present such as arthritis, meningitis, and/or endocarditis (Sexton, 2022).

3.6 Meningitis

Streptococcus pneumonia is the main bacterium causing bacterial meningitis, and it has been the leading cause of mortality and morbidity in younger children across the world (Ben-Shimol et al., 2018); thus, this condition is acute causing inflammation in the membranes of the brain and the spinal cord requiring a rapid treatment (Mańdziuk & Kuchar, 2021). The inflammation into the subarachnoid space and the meninges results to have symptoms such as headache, stiffness of the neck, fever, and an increase in leukocyte counts-pleocytosis (van de Beek et al., 2016). These pneumococcal diseases occur worldwide and are more common in young children, elderly people, and immunocompromised persons (Torres, Blasi, Dartois, & Akova, 2015).

For all the pneumococcal diseases, the migration of *S. pneumoniae* to sterile tissues and organs is the main cause, and likewise, when meninges occurs, the protective membranes surrounding the brain and the spinal cord become inflamed due to *S. pneumonia* infection, which is known as bacterial meningitis (Mayo Clinic Staff, 2016). According to the report of the Centers for Disease Control and Prevention (CDC), *S. pneumoniae* causes more than 50% of bacterial meningitis within the United States.

4. *Streptococcus pneumoniae* biofilms' structure

S. pneumoniae forms three-dimensional biofilms composed of polymeric matrix composed of lipids, polysaccharides, nucleic acids, and proteins (Chao, Bergenfelz, & Hakansson, 2019; Yadav et al., 2015) (Fig. 2). The extracellular polymeric matrix binds with the extracellular DNA for protection to the bacteria chemically and physically providing the cellular structural integrity to the biofilm (Domenech, Ruiz, Moscoso, & García, 2015). Domenech et al. (2015) described the structure of biofilms formed by *S. pneumoniae* observing the biofilm under the confocal laser scanning microscopy (CLSM) stained with fluorescent dyes. It was observed that the nonencapsulated pneumococcal biofilm was a three-dimensional complexed structure of thickness being 25–30 μm on the abiotic surface with small voids and channels that separate the colonies of the pneumococci (Domenech, García, & Moscoso, 2012). The structure observed by Domenech was called honey comb structure. Pneumococcal biofilms consist of both dead and live cells; however, the cell death or autolysis in biofilms is slow when compared with planktonic cell growth. It is reported that under in vitro condition and static model, pneumococcal biofilm reached at maximum growth at 12–18 h, after that cell death occurs (Vidal, 2011).

5. Pneumococci biofilms' properties

5.1 Antibiotic resistance

The formation of the biofilm by the microbes has caused over 60% of bacterial infections in humans and about 80% of the infection becomes chronic into the host system reducing the susceptibility to the antimicrobial agents (Domenech et al., 2015). The pneumococci become resistant to certain antibiotics due to the acquisition of an antibiotic resistance gene or mutation, whereas the horizontal gene transfer (HGT) of antibiotic resistance occurs by the mobile genetic elements (MGEs) or by the transformation (Lattar et al., 2018). Pneumococci resistance to the antibiotic penicillin was first reported in the 1980s, and the treatment of pneumococci infection with macrolides into the lower respiratory infection and to bacteremia was reported to be a failure (Schroeder & Stephens, 2016). The resistance toward macrolides can be due to two factors: (i) the target gene modification due to the ribosomal methylase by the *ermB* gene and (ii) the drug efflux by the *mefA* gene (Popovic et al., 2018). The pneumococci resistance to antibiotics such as penicillin, methicillin, ceftazidime has been the most concern worldwide since the 1970s (Cunha, 2000). It takes a whole species or genera to favor the resistance of the antibiotics (Livermore, 2000). When a patient is treated with a high dose of the vancomycin antibiotics such as in the case of bacterial meningitis, it results in side effects such as nephrotoxicity, hypersensitivity, and loss of hearing (Ben-Shimol et al., 2018).

5.2 Increases gene transfer frequency

One of the characteristics of microbial biofilms is the increased frequency of genetic exchange. Griffith was the first scientist who discovered natural genetic transformation in *S. pneumoniae* in the year 1928, and later on, Avery and coworkers discovered that DNA transfers in transformation (Avery, MacLeod, & McCarty, 1944; Griffith, 1928). The transformation process transfer result in horizontal gene transfers is critical for evolution that facilities emergence of unique and novel traits resulting in antibiotic resistance (Claverys, Martin, & Håvarstein, 2007; Hakenbeck et al., 2001; Majewski, Zawadzki, Pickerill, Cohan, & Dowson, 2000). The genetic transformation and acquisition of novel genes facilitate bacteria to survive in stress and adverse environmental condition (Johnsborg, Eldholm, & Håvarstein, 2007; Stewart & Carlson, 1986). In *S. pneumoniae*, natural transformation occurs and the antimicrobial-resistant genes are widely distributed among different serotypes.

FIG. 2 Scanning electron microscope images of *Streptococcus pneumoniae* in vitro biofilm grown on tissue culture microplate (Yadav et al., 2015).

Streptococcus pneumoniae horizontal gene transfer facilitates colonization, adaptation, fitness and infects the human host and causes invasive diseases. *S. pneumoniae* adapts to grow at a temperature close to human body temperature and under oxygen-limited conditions in the respiratory tract and is attached to mucosa epithelium form and organized in multicellular communities enclosed in a self-produced polymeric matrix called biofilms (Marks et al., 2012; Muñoz-Elías, Marcano, & Camilli, 2008). In pneumococci, in vitro biofilms competence-related gene upregulated and shielded bacteria from adverse environmental conditions (Oggioni et al., 2006; Trappetti et al., 2011). *S. pneumoniae* colonizes in the nasopharynx in the form of biofilms (Marks, Parameswaran, & Hakansson, 2012), and the tissue adherent bacteria demonstrated gene expression similar to biofilms and increase in transformation in comparison to planktonic bacteria (Oggioni et al., 2006). In addition, it has been reported that *Streptococcus pneumoniae* in sepsis and tissue adherent condition showed gene expressions similar to planktonic and biofilms respectively with less genetic transformation efficiency in sepsis (Marks, Parameswaran et al., 2012). Marks, Reddinger et al. (2012) tested the efficiency of *S. pneumoniae* genetic transformation, and concluded that the nontransformable or difficult to transformed strains of pneumococci can acquire genes when grown under the biofilm condition (Marks, Reddinger et al., 2012). In addition, they found that the lysed *S. pneumoniae* releases genetic material and acts as a donor for transformation, and the natural transformation occurs optimally in *S. pneumoniae* growing actively under biofilm conditions (Marks, Reddinger et al., 2012).

5.3 Persister cells

Persister composite plays a key role in the resistance of chronic infections to antibiotics (Lewis, 2010). Persister cells were known soon after the antibiotic was introduced, but it remained a health concern as these cells are persisting the infection and are ubiquitous among the bacteria and cause a failure toward the antibiotic treatment and support immunosuppression (Fisher, Gollan, & Helaine, 2017). The persister cells have the potential to enter the latent state and are immortal against the antibiotic drug by extremely reducing the metabolic activity (Wainwright, Hobbs, & Nakouti, 2021). Persister cells seem to form in response to stress, when stress like an antibiotic encountering the cell can create the persister cell, whereas reducing the production of protein by prohibiting the transcription with rifampicin, by discontinuing the transcription of tetracycline, and by hindering the translation by excluding the production of ATP with carbonyl cyanide *m*-chlorophenylhydrazone (Wood, Song, & Yamasaki, 2019). A study has suggested some ways to combat the persister cells: (i) to eradicate the persister cell while it is in its resting stage, (ii) to arouse the persister cells and then dose it with antibiotic drugs, and (iii) to inhibit the formation of persister cells, whereas an anticancer drug mitomycin C has successfully eradicated the persister cell while it was in its resting stage (Wood, 2016). The reservoir of the bacteria *S. pneumoniae* is present in the nasopharynx of humans asymptomatically (Ryan Gierke, Wodi, & Kobayashi, 2021) (Fig. 3). The pneumococci colonized into the nasopharynx disintegrating the mucus by exoglycosidases such as β-galactosidase, neuraminidase A and B, and

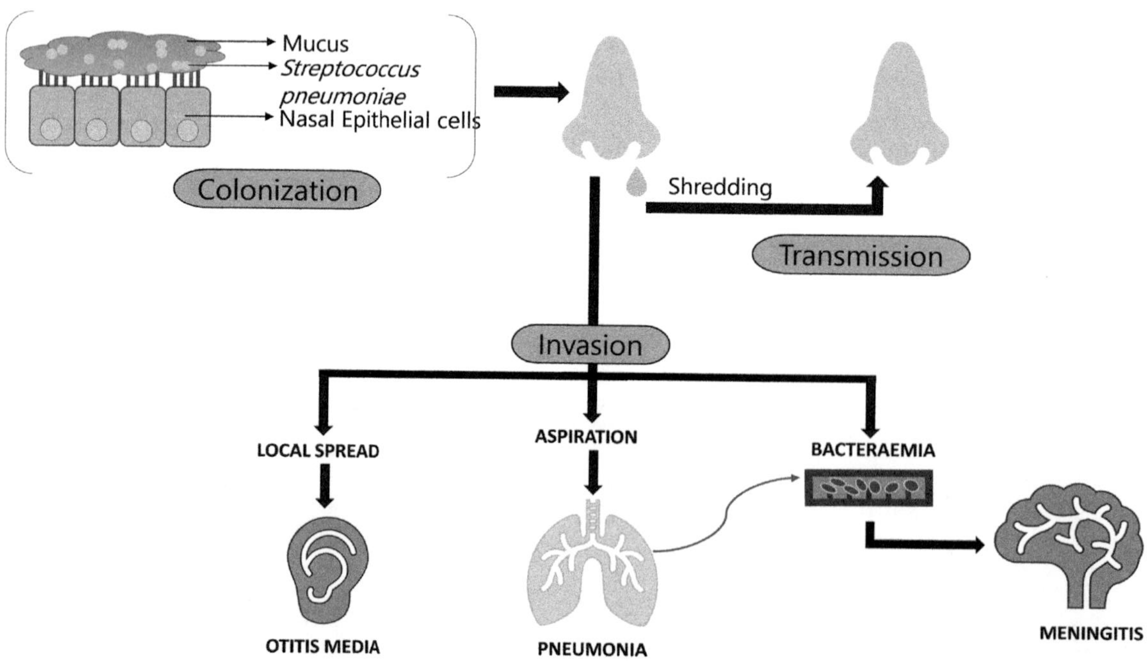

FIG. 3 Systematic representation of colonization/biofilms of *Streptococcus pneumoniae* in the nasopharynx and transition to other sterile sites.

β-*N*-acetylglycosamidase decreasing the viscosity of the mucus (Barichello, Generoso, Collodel, Moreira, & Almeida, 2012).

6. Conclusion

Streptococcus pneumoniae is opportunistic pathogen that occupies the nasopharynx within 3 months of birth and colonizes in the nasopharynx in the form of biofilms. Pneumococci cause infection in immune-compressed individuals when they transit to sterile sites from nasopharynx biofilms. These pneumococcal biofilms are carrier of bacteria and contribute in the spread of infectious diseases including pneumonia, OM, etc. The pneumococci biofilms are resistant to antibiotics, cause recurrent infection, give rise to persister cells, and increase the risk of antibiotic-resistant gene spread. Therefore, the study of pneumococci biofilms is vital to develop disease management strategy including finding alternative to antibiotics and biofilm decolonization. Nasopharynx is the prime location where pneumococci colonize, and decolonization of nasopharynx or mitigation of pneumococci in the nasopharynx could be a strategy to combat infectious diseases caused by *S. pneumoniae*.

Acknowledgment

This study is supported by funding to MKY from the Science and Engineering Research Board (SERB), Govt. of India, vide Project Sanction No: SRG/2021/000371.

References

Ash, S. Y., & Sheffield, J. V. L. (2013). Pneumococcus. *The Medical Clinics of North America*, *97*, 647–666.

Avery, O. T., MacLeod, C. M., & McCarty, M. (1944). Studies on the chemical nature of the substance inducing transformation of pneumococcal types: Induction of transformation by a desoxyribonucleic acid fraction isolated from pneumococcus type III. *The Journal of Experimental Medicine*, *79*, 137–158.

Barichello, T., Generoso, J. S., Collodel, A., Moreira, A. P., & Almeida, S. M. (2012). Pathophysiology of acute meningitis caused by *Streptococcus pneumoniae* and adjunctive therapy approaches. *Arquivos de Neuro-Psiquiatria*, *70*(5), 366–372. https://doi.org/10.1590/s0004-282x2012000500011. 22618789.

Ben-Shimol, S., Givon-Lavi, N., Greenberg, D., Stein, M., Megged, O., Bar-Yochai, A., et al. (2018). Impact of pneumococcal conjugate vaccines introduction on antibiotic resistance of *Streptococcus pneumoniae* meningitis in children aged 5 years or younger, Israel, 2004 to 2016. *Euro Surveillance*, *23*(47), 1800081. https://doi.org/10.2807/1560-7917.ES.2018.23.47.1800081. 30482264. PMCID: PMC6341944.

Berical, A. C., Harris, D., Dela Cruz, C. S., & Possick, J. D. (2016). Pneumococcal vaccination strategies. An update and perspective. *Annals of the. American Thoracic Society*, *13*(6), 933–944. https://doi.org/10.1513/AnnalsATS.201511-778FR.

Berndsen, M. R., Erlendsdóttir, H., & Gottfredsson, M. (2012). Evolving epidemiology of invasive Haemophilus infections in the post-vaccination era: Results from a long-term population-based study. *Clinical Microbiology and Infection*, *18*(9), 918–923.

Bogaert, D., De Groot, R., & Hermans, P. W. (2004). *Streptococcus pneumoniae* colonisation: The key to pneumococcal disease. *The Lancet Infectious Diseases*, *4*(3), 144–154.

Briles, D. E., Paton, J. C., Mukerji, R., Swiatlo, E., & Crain, M. J. (2019). Pneumococcal vaccines. *Microbiology Spectrum*, *7*(6). https://doi.org/10.1128/microbiolspec.GPP3-0028-2018. 31858954.

Centers for Disease Control Prevention. (2015). *Pneumococcal disease*. Available from: https://www.cdc.gov/pneumococcal/index.html (Accessed 20 March 2017).

Chao, Y., Bergenfelz, C., & Hakansson, A. P. (2019). Growing and characterizing biofilms formed by *Streptococcus pneumoniae*. In *Vol. 1968. Methods in molecular biology* (pp. 147–171). https://doi.org/10.1007/978-1-4939-9199-0_13. 30929213.

Chao, Y., Marks, L. R., Pettigrew, M. M., & Hakansson, A. P. (2015). *Streptococcus pneumoniae* biofilm formation and dispersion during colonization and disease. *Frontiers in Cellular and Infection Microbiology*, (4). https://doi.org/10.3389/fcimb.2014.00194.

Chhabada, S., & Khanna, S. (2020). Plastic bronchitis. *Anesthesiology*, *133*(2), 429. https://doi.org/10.1097/ALN.0000000000003381. 32433279.

Claverys, J. P., Martin, B., & Håvarstein, L. S. (2007). Competence-induced fratricide in streptococci. *Molecular Microbiology*, *64*, 1423–1433.

Costerton, J. W., Stewart, P. S., & Greenberg, E. P. (1999). Bacterial biofilms: A common cause of persistent infections. *Science*, *284*, 1318–1322.

Cunha, B. A. (2000). Antibiotic resistance. *The Medical Clinics of North America*, *84*(6), 1407–1429. https://doi.org/10.1016/s0025-7125(05)70295-9. 11155850.

Del Pozo, J. L. (2018). Biofilm-related disease. *Expert Review of Anti-infective Therapy*, *16*(1), 51–65. https://doi.org/10.1080/14787210.2018.1417036.

del Prado, G., Ruiz, V., Naves, P., Rodríguez-Cerrato, V., Soriano, F., del Carmen, P. M., et al. (2010). Biofilm formation by *Streptococcus pneumoniae* strains and effects of human serum albumin, ibuprofen, N-acetyl-l-cysteine, amoxicillin, erythromycin, and levofloxacin. *Diagnostic Microbiology and Infectious Disease*, *67*, 311–318.

Dohar, J. E., Hebda, P. A., Veeh, R., Awad, M., Costerton, J. W., Hayes, J., et al. (2005). Mucosal biofilm formation on middle-ear mucosa in a nonhuman primate model of chronic suppurative otitis media. *The Laryngoscope*, *115*(8), 1469–1472.

Domenech, M., García, E., & Moscoso, M. (2012). Biofilm formation in *Streptococcus pneumoniae*. *Microbial Biotechnology*, *5*(4), 455–465. https://doi.org/10.1111/j.1751-7915.2011.00294.x.

Domenech, M., Ruiz, S., Moscoso, M., & García, E. (2015). In vitro biofilm development of *Streptococcus pneumoniae* and formation of choline-binding protein-DNA complexes. *Environmental Microbiology Reports*, *7*(5), 715–727. https://doi.org/10.1111/1758-2229.12295. Epub 2015 Jun 18 25950767.

Donlan, R. M., & Costerton, J. W. (2002). Biofilms: Survival mechanisms of clinically relevant microorganisms. *Clinical Microbiology Reviews*, *15*, 167–193.

Ehrlich, G. D., Veeh, R., Wang, X., Costerton, J. W., Hayes, J. D., Hu, F. Z., et al. (2002). Mucosal biofilm formation on middle-ear mucosa in the chinchilla model of otitis media. *JAMA*, *287*(13), 1710–1715.

Fisher, R. A., Gollan, B., & Helaine, S. (2017). Persistent bacterial infections and persister cells. *Nature Reviews Microbiology*, *15*(8), 453–464. https://doi.org/10.1038/nrmicro.2017.42. Epub 2017 May 22 28529326.

García-Castillo, M., Morosini, M. I., Valverde, A., Almaraz, F., Baquero, F., Cantón, R., et al. (2007). Differences in biofilm development and antibiotic susceptibility among *Streptococcus pneumoniae* isolates from cystic fibrosis samples and blood cultures. *The Journal of Antimicrobial Chemotherapy, 59*, 301–304.

Griffith, F. (1928). The significance of pneumococcal types. *The Journal of Hygiene, 27*, 113–159.

Hakenbeck, R., et al. (2001). Mosaic genes and mosaic chromosomes: Intra- and interspecies genomic variation of *Streptococcus pneumoniae. Infection and Immunity, 69*, 2477–2486.

Hall-Stoodley, L., Hu, F. Z., Gieseke, A., Nistico, L., Nguyen, D., Hayes, J., et al. (2006). Direct detection of bacterial biofilms on the middle-ear mucosa of children with chronic otitis media. *JAMA, 296*(2), 202–211.

Hooven, T. A., & Polin, R. A. (2017). Pneumonia. *Seminars in Fetal & Neonatal Medicine, 22*(4), 206–213. https://doi.org/10.1016/j.siny.2017.03.002. Epub 2017 Mar 24 28343909. PMCID: PMC7270051.

Jensen, R. G., Johansen, H. K., Bjarnsholt, T., Eickhardt-Sørensen, S. R., & Homøe, P. (2017). Recurrent otorrhea in chronic suppurative otitis media: Is biofilm the missing link? *European Archives of Oto-Rhino-Laryngology, 274*, 1–7.

Johnsborg, O., Eldholm, V., & Håvarstein, L. S. (2007). Natural genetic transformation: Prevalence, mechanisms, and function. *Research in Microbiology, 158*, 767–778.

Kadioglu, A., Weiser, J. N., Paton, J. C., & Andrew, P. W. (2008). The role of *Streptococcus pneumoniae* virulence factors in host respiratory colonization and disease. *Nature Reviews Microbiology, 6*(4), 288–301. https://doi.org/10.1038/nrmicro1871.

Kim, G. L., Seon, S. H., & Rhee, D. K. (2017). Pneumonia and Streptococcus pneumoniae vaccine. *Archives of Pharmacal Research, 40* (8), 885–893. https://doi.org/10.1007/s12272-017-0933-y.

Kinkade, S., & Long, N. A. (2016). Acute bronchitis. *American Family Physician, 94*(7), 560–565. 27929206.

Koval, C. E., Stosor, V., & AST ID Community of Practice. (2019). Ventricular assist device-related infections and solid organ transplantation—Guidelines from the American Society of Transplantation Infectious Diseases Community of Practice. *Clinical Transplantation, 33*(9), e13552. https://doi.org/10.1111/ctr.13552. Epub 2019 Apr 29 30924952.

Kwiecinski, J. M., & Horswill, A. R. (2020). *Staphylococcus aureus* bloodstream infections: Pathogenesis and regulatory mechanisms. *Current Opinion in Microbiology, 53*, 51–60. https://doi.org/10.1016/j.mib.2020.02.005. Epub 2020 Mar 12 32172183. PMCID: PMC7244392.

Lanks, C. W., Musani, A. I., & Hsia, D. W. (2019). Community-acquired pneumonia and hospital-acquired pneumonia. *The Medical Clinics of North America, 103*(3), 487–501. https://doi.org/10.1016/j.mcna.2018.12.008. Epub 2019 Mar 7 30955516.

Lattar, S. M., Wu, X., Brophy, J., Sakai, F., Klugman, K. P., & Vidal, J. E. (2018). A mechanism of unidirectional transformation, leading to antibiotic resistance, occurs within nasopharyngeal pneumococcal biofilm consortia. *MBio, 9*(3), e00561-18. https://doi.org/10.1128/mBio.00561-18. 29764945. PMCID:PMC5954218.

Leung, A. K. C., Wong, A. H. C., & Hon, K. L. (2018). Community-acquired pneumonia in children. *Recent Patents on Inflammation & Allergy Drug Discovery, 12*(2), 136–144. https://doi.org/10.2174/1872213X12666180621163821. 29932038.

Lewis, K. (2010). Persister cells. *Annual Review of Microbiology, 64*, 357–372. https://doi.org/10.1146/annurev.micro.112408.134306. 20528688.

Lieberthal, A. S., Carroll, A. E., & Chonmaitree, T. (2013). The diagnosis and management of acute otitis media. *Pediatrics, 131*, 964–999.

Livermore, D. M. (2000). Epidemiology of antibiotic resistance. *Intensive Care Medicine, 26*(Suppl. 1), S14–S21. https://doi.org/10.1007/s001340051113. 10786953.

Majewski, J., Zawadzki, P., Pickerill, P., Cohan, F. M., & Dowson, C. G. (2000). Barriers to genetic exchange between bacterial species: *Streptococcus pneumoniae* transformation. *Journal of Bacteriology, 182*, 1016–1023.

Mańdziuk, J., & Kuchar, E. P. (2021). *Streptococcal meningitis [Updated 2021 Apr 19]*. Treasure Island, FL: StatPearls Publishing. https://www.ncbi.nlm.nih.gov/books/NBK554448/.

Marks, L. R., Parameswaran, G. I., & Hakansson, A. P. (2012). Pneumococcal interactions with epithelial cells are crucial for optimal biofilm formation and colonization in vitro and in vivo. *Infection and Immunity, 80*(8), 2744–2760. https://doi.org/10.1128/IAI.00488-12.

Marks, L. R., Reddinger, R. M., & Hakansson, A. P. (2012). High levels of genetic recombination during nasopharyngeal carriage and biofilm formation in *Streptococcus pneumoniae. MBio, 3*(5), e00200–12. https://doi.org/10.1128/mBio.00200-12.

Mayo Clinic Staff. (2016). *Meningitis*. Mayo Foundation for Medical Education and Research. https://www.mayoclinic.org/diseases-conditions/meningitis/symptoms-causes/syc-20350508. (Accessed 30 March 2017).

Mirzaei, R., Mohammadzadeh, R., Sholeh, M., Karampoor, S., Abdi, M., Dogan, E., et al. (2020). The importance of intracellular bacterial biofilm in infectious diseases. *Microbial Pathogenesis, 147*, 104393. https://doi.org/10.1016/j.micpath.2020.104393. Epub 2020 Jul 22 32711113.

Munoz-Elias, E. J., Marcano, J., & Camilli, A. (2008). Isolation of *Streptococcus pneumoniae* biofilm mutants and their characterization during nasopharyngeal colonization. *Infection and Immunity, 76*, 5049–5061.

Muñoz-Elías, E. J., Marcano, J., & Camilli, A. (2008). Isolation of *Streptococcus pneumoniae* biofilm mutants and their characterization during nasopharyngeal colonization. *Infection and Immunity, 76*, 5049–5061.

Oggioni, M. R., Trappetti, C., Kadioglu, A., Cassone, M., Iannelli, F., Ricci, S., et al. (2006). Switch from planktonic to sessile life: A major event in pneumococcal pathogenesis. *Molecular Microbiology, 61*, 1196–1210.

Pai, S., Enoch, D. A., & Aliyu, S. H. (2015). Bacteremia in children: Epidemiology, clinical diagnosis and antibiotic treatment. *Expert Review of Anti-Infective Therapy, 13*(9), 1073–1088. https://doi.org/10.1586/14787210.2015.1063418. Epub 2015 Jul 4 26143645.

Pichichero, M. E. (2013). Otitis media. *Pediatric Clinics of North America, 60*, 391–407.

Popovic, S., Hadnadjev, M., Gajic, I., Mijac, V., Kekic, D., Smitran, A., et al. (2018). Characterization of macrolide-resistant non-invasive pneumococci in the pre-vaccine era in Serbia. *Acta Microbiologica et Immunologica Hungarica, 65*(4), 477–488. https://doi.org/10.1556/030.65.2018.035. Epub 2018 Jul 16 30010393.

Post, J. C. (2001). Direct evidence of bacterial biofilms in otitis media. *The Laryngoscope, 111*(12), 2083–2094.

Post, J. C., Stoodley, P., Hall-Stoodley, L., & Ehrlich, G. D. (2004). The role of biofilms in otolaryngologic infections. *Current Opinion in Otolaryngology & Head and Neck Surgery, 12*, 185–190.

Quinton, L. J., Walkey, A. J., & Mizgerd, J. P. (2018). Integrative physiology of pneumonia. *Physiological Reviews, 98*(3), 1417–1464. https://doi.org/10.1152/physrev.00032.2017. 29767563. PMCID: PMC6088146.

Revai, K., Mamidi, D., & Chonmaitree, T. (2008). Association of nasopharyngeal bacterial colonization during upper respiratory tract infection and the development of acute otitis media. *Clinical Infectious Diseases*, *46*(4), e34–e37.

Rubin, B. K. (2016). Plastic bronchitis. *Clinics in Chest Medicine*, *37*(3), 405–408. https://doi.org/10.1016/j.ccm.2016.04.003. Epub 2016 Jun 15 27514587.

Ryan Gierke, A., Wodi, P., & Kobayashi, M. (2021). *Pneumococcal disease, epidemiology and prevention of vaccine-preventable disease*. Centers for Disease Control and Prevention (Chapter 17) https://www.cdc.gov/vaccines/pubs/pinkbook/downloads/pneumo.pdf.

Schroeder, M. R., & Stephens, D. S. (2016). Macrolide resistance in *Streptococcus pneumoniae*. *Frontiers in Cellular and Infection Microbiology*, *6*, 98. https://doi.org/10.3389/fcimb.2016.00098. 27709102. PMCID: PMC5030221.

Seidelman, J., & Lewis, S. S. (2018). Neurosurgical device-related infections. *Infectious Disease Clinics of North America*, *32*(4), 861–876. https://doi.org/10.1016/j.idc.2018.06.006. Epub 2018 Sep 18 30241709.

Sexton, D. J. (2022). *Invasive pneumococcal (Streptococcus pneumoniae) infections and bacteremia*. https://www.uptodate.com/contents/invasive-pneumococcal-streptococcus-pneumoniae-infections-and-bacteremia.

Shak, J. R., Vidal, J. E., & Klugman, K. P. (2013). Influence of bacterial interactions on pneumococcal colonization of the nasopharynx. *Trends in Microbiology*, *21*, 129–135. https://doi.org/10.1016/j.tim.2012.11.005.

Sharma, G., et al. (2016). Escherichia coli biofilm: Development and therapeutic strategies. *Journal of Applied Microbiology*, *121*(2), 309–319. https://doi.org/10.1111/jam.13078.

Simell, B., Auranen, K., Käyhty, H., Goldblatt, D., Dagan, R., O'Brien, K. L., et al. (2012). The fundamental link between pneumococcal carriage and disease. *Expert Review of Vaccines*, *11*, 841–855. https://doi.org/10.1586/erv.12.53.

Steel, H. C., Cockeran, R., Anderson, R., & Feldman, C. (2013). Overview of community-acquired pneumonia and the role of inflammatory mechanisms in the immunopathogenesis of severe pneumococcal disease. *Mediators of Inflammation*, *2013*, 490346. https://doi.org/10.1155/2013/490346.

Stewart, G. J., & Carlson, C. A. (1986). The biology of natural transformation. *Annual Review of Microbiology*, *40*, 211–235.

Stewart, P. S., & Bjarnsholt, T. (2020). Risk factors for chronic biofilm-related infection associated with implanted medical devices. *Clinical Microbiology and Infection*, *26*(8), 1034–1038. https://doi.org/10.1016/j.cmi.2020.02.027. Epub 2020 Feb 29 32120041.

Stoodley, P., Sauer, K., Davies, D. G., & Costerton, J. W. (2002). Biofilms as complex differentiated communities. *Annual Review of Microbiology*, *56*, 187–209.

Syrjanen, R. K., Kilpi, T. M., Kaijalainen, T. H., Herva, E. E., & Takala, A. K. (2001). Nasopharyngeal carriage of *Streptococcus pneumoniae* in Finnish children younger than 2 years old. *The Journal of Infectious Diseases*, *184*(4), 451–459.

Thornton, R. B., Wiertsema, S. P., Kirkham, L.-A. S., Rigby, P. J., Vijayasekaran, S., Coates, H. L., et al. (2013). Neutrophil extracellular traps and bacterial biofilms in middle ear effusion of children with recurrent acute otitis media—A potential treatment target. *PLoS One*, *8*(2), e53837.

Tong, S. Y. (2015). Staphylococcus aureus infections: epidemiology, pathophysiology, clinical manifestations, and management. *Clinical Microbiology Reviews*, *28*(3), 603–661. https://doi.org/10.1128/CMR.00134-14.

Torres, A., Blasi, F., Dartois, N., & Akova, M. (2015). Which individuals are at increased risk of pneumococcal disease and why? Impact of COPD, asthma, smoking, diabetes, and/or chronic heart disease on community-acquired pneumonia and invasive pneumococcal disease. *Thorax*, *70* (10), 984–989. https://doi.org/10.1136/thoraxjnl-2015-206780.

Trappetti, C., et al. (2011). The impact of the competence quorum sensing system on *Streptococcus pneumoniae* biofilms varies depending on the experimental model. *BMC Microbiology*, *11*, 75.

van de Beek, D., Brouwer, M., Hasbun, R., Koedel, U., Whitney, C. G., & Wijdicks, E. (2016). Community-acquired bacterial meningitis. *Nature Reviews Disease Primers*, *2*, 16074. https://doi.org/10.1038/nrdp.2016.74. 27808261.

van Hal, S. J., Jensen, S. O., Vaska, V. L., Espedido, B. A., Paterson, D. L., & Gosbell, I. B. (2012). Predictors of mortality in *Staphylococcus aureus* bacteremia. *Clinical Microbiology Reviews*, *25*(2), 362–386. https://doi.org/10.1128/CMR.05022-11. 22491776. PMCID: PMC3346297.

Vermee, Q., Cohen, R., Hays, C., et al. (2019). Biofilm production by *Haemophilus influenzae* and *Streptococcus pneumoniae* isolated from the nasopharynx of children with acute otitis media. *BMC Infectious Diseases*, *19*, 44. Published 2019 Jan 11.

Vidal, J. E. (2011). The LuxS-dependent quorum-sensing system regulates early biofilm formation by Streptococcus pneumoniae strain D39. *Infection and Immunity*, *79*(10), 4050–4060.

Wainwright, J., Hobbs, G., & Nakouti, I. (2021). Persister cells: Formation, resuscitation, and combative therapies. *Archives of Microbiology*, *203* (10), 5899–5906. https://doi.org/10.1007/s00203-021-02585-z. Epub 2021 Nov 5 34739553. PMCID: PMC8590677.

Wood, T. K. (2016). Combatting bacterial persister cells. *Biotechnology and Bioengineering*, *113*(3), 476–483. https://doi.org/10.1002/bit.25721. Epub 2015 Sep 3 26264116.

Wood, T. K., Song, S., & Yamasaki, R. (2019). Ribosome dependence of persister cell formation and resuscitation. *Journal of Microbiology*, *57* (3), 213–219. https://doi.org/10.1007/s12275-019-8629-2. Epub 2019 Feb 26 30806978.

Yadav, M. K., et al. (2015). The small molecule DAM inhibitor, pyrimidinedione, disrupts *Streptococcus pneumoniae* biofilm growth in vitro. *PLoS One*, *10*. https://doi.org/10.1371/journal.pone.0139238.

Yadav, M. K., et al. (2019). Ruthenium chloride-induced oxidative cyclization of trans-resveratrol to (±)-ε-viniferin and antimicrobial and antibiofilm activity against Streptococcus pneumoniae. *Frontier in Pharmacology*, (10), 890.

Yadav, M. K., Chae, S. W., & Song, J. J. (2012). In vitro *Streptococcus pneumoniae* biofilm formation and in vivo middle ear mucosal biofilm in a rat model of acute otitis induced by *S. pneumoniae*. *Clinical and Experimental Otorhinolaryngology*, *5*(3), 139–144.

Chapter 29

Oral biofilms: Architecture and control

Arumugam Priya, Hari Prasath Nagaiah, Nambiraman Malligarjunan, and Shunmugiah Karutha Pandian
Department of Biotechnology, Alagappa University, Karaikudi, Tamil Nadu, India

1. Introduction

"Mouth is a window into the overall health"

This may be a general statement but the scientific facts support that oral cavity can provide an overview and reflect comprehensive conditions of the heterogeneous tissues and organs of the human anatomy. World Health Organization (WHO) and Centers for Disease Control and Prevention (CDC) have also specified this fact as "oral health is a key indicator of overall health, well-being and quality of life." Thus, it is evidenced that there is a general connection between mouth and body, i.e., the oral health and systemic health of human being are interlinked. In addition to this, in the year 2000, the U.S. Surgeon General's report "oral health in America" emphasized numerous ways in which oral health and systemic health are linked (Public Health Service, United States). Fig. 1 provides comparative insights into the link between the oral health and systemic health in a pictorial form of oral manifestations of the systemic diseases and systemic diseases instigated by oral infections. Therefore, maintenance of oral health is extremely imperative beyond just for the perfect smile.

2. Human oral microbiome

Oral microbiota is the pivotal role player in this health association between mouth and the body. With respect to the microbial flora, oral cavity is one among the densely populated sites of the human body comprising more than 750 microbial species. Human Oral Microbiome Database (HOMD) catalogues comprehensive and curated information on these microbial species (http://www.homd.org/).

Microbial ecology of the oral cavity is particularly complex and a rich biological setting with unique niches that provide supportive environment for the colonization of microbes (Deo & Deshmukh, 2019). The distinctive niches of the oral cavity include saliva, teeth, tongue, cheek, hard and soft palates, gingival sulcus, floor of the mouth, and the throat (Benn, Heng, Broadbent, & Thomson, 2018; Dewhirst et al., 2010). Different niches of the mouth are colonized by oral organisms based on the adhesins present on their surface that interact with the complementary receptors of the oral surfaces (Aas, Paster, Stokes, Olsen, & Dewhirst, 2005). Normal microbiome of the oral cavity is encompassed with bacteria, fungi, viruses, archaea, and protozoa (Zaura, Nicu, Krom, & Keijser, 2014). In healthy mouth, there is a genus level conservation in the microbial community. In spite of similarities, individual and site-specific diversification is observed in the oral microbiome.

Rapid changes in the composition and activity of the oral microbiome are observed in temporal and spatial manner and developmentally dynamic with the host factors. Such dynamics are due to the result of several factors such as type and frequency of the diet, changes in pH, and microbial communications (McLean, 2014). On a mutual benefit basis, the microorganisms in the oral cavity maintain synergistic interactions and reside as normal flora without being virulent.

These symbionts prevent the adherence of pathogenic microbes to the niches of oral cavity and thereby regulate the oral health. Oral infections manifest when there is breach in the barriers of commensal organisms and the host immune system. Thus, advancement or decline in the health is directly proportional to the composition of microbiota. When there is alteration in the symbiotic oral ecosystem, there will be shift from the commensal flora to pathogenic microbiota, which can instigate the progression of various infectious diseases such as dental caries, periodontitis, mucosal diseases, implant-associated infections, etc. (Fig. 2).

For several decades, research had been focused on the planktonic part of the microorganisms, whereas in recent times the focus has been shifted to the sessile lifestyle as it was found that oral microorganisms are organized as well-structured biofilms. Microorganisms can settle on shedding surfaces of the oral cavity such as the lips, cheeks, tongue, and palate and also on nonshedding surfaces including the teeth, abiotic surfaces such as teeth fillings, fissure sealings, denture material, orthodontic implants, etc. (Marsh, Martin, Lewis, & Williams, 2009; Zaura, Keijser, Huse, & Crielaard, 2009). Of the various mucosal surfaces, multilayers of biofilm are formed on the tongue

Understanding Microbial Biofilms. https://doi.org/10.1016/B978-0-323-99977-9.00025-9

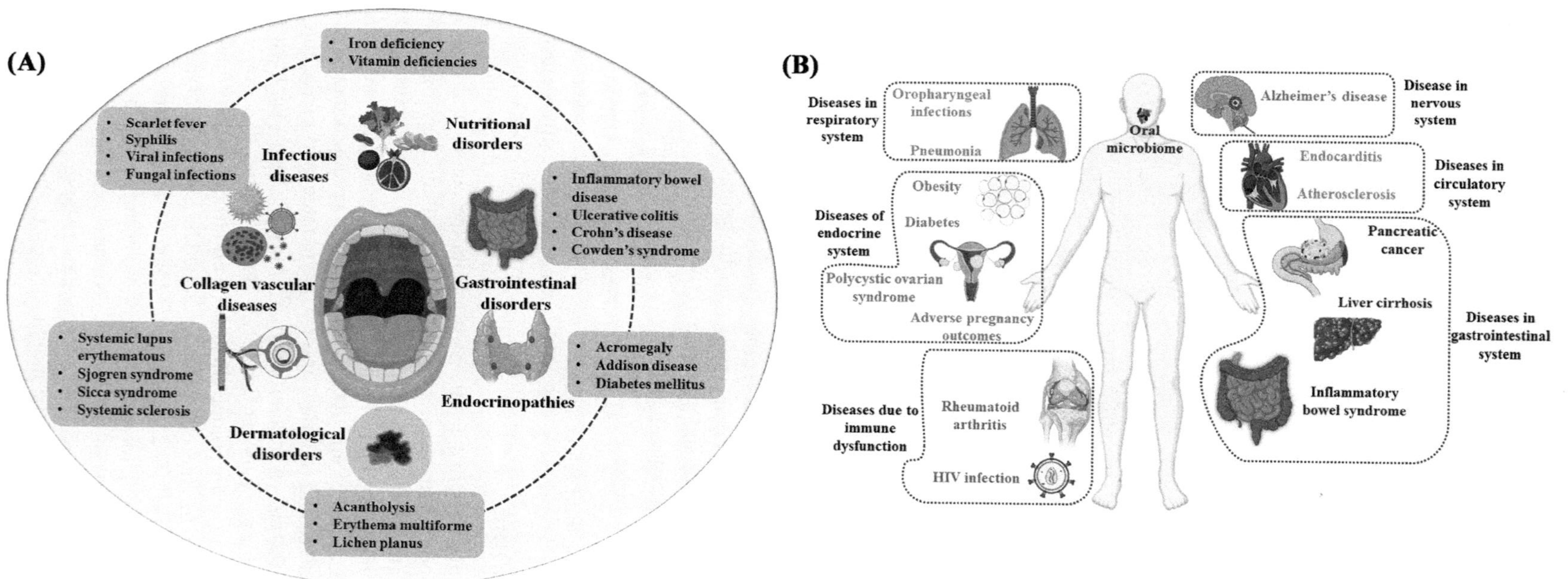

FIG. 1 Association between mouth and the body. (A) Oral manifestations of systemic diseases. (B) Systemic infections caused by the members of oral microbiome.

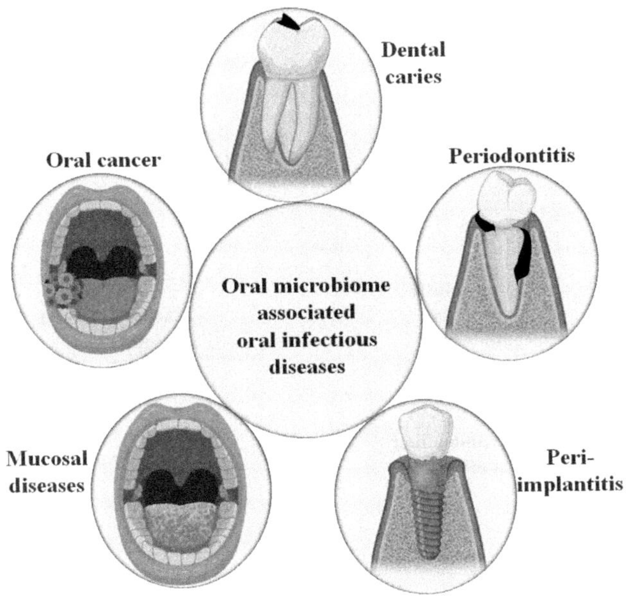

FIG. 2 Prevalent infectious diseases of the oral cavity.

as it embraces the major proportion of the oral microbial burden. Nonshedding surfaces of oral cavity are also heavily colonized by the microbial community (Lamont, Koo, & Hajishengallis, 2018).

3. Oral microbial biofilms

Contrary to the perspective of Koch's postulate, major oral infections are known to be caused by consortium of organisms embedded in biofilm rather than an individual pathogen (Jenkinson & Lamont, 2005) (Fig. 3).

Oral microbial biofilms are three-dimensional well-structured microbial communities that are adhered to the shedding and/or nonshedding and abiotic surfaces of the oral cavity (Socransky, 2002). Different oral surfaces are lodged by distinct microbial communities (Mager, Ximenez-Fyvie, Haffajee, & Socransky, 2003). Intricate nature and the architecture of the oral biofilms were spotlighted initially by the work of Listgarten, Mayo, and Tremblay (1975) through observation of epoxy resin crowns and extracted teeth under light and electron microscopes (Listgarten, 1976).

The formation of biofilm in dental plaque is a dynamic process and arbitrarily occurs in several stages. The biofilm formation in plaque is initiated with the acquired pellicle formation, where salivary proteins are adsorbed by the enamel forming a surface conditioning film. Pellicle formation is followed by the transport of microorganisms and reversible attachment by weak physicochemical interactions between the microbial cell surface and pellicle. In short span, the reversible attachments may become irreversible due to adhesin-receptor attachment that eventually leads to coaggregation by secondary colonizers, which further multiply and mature as biofilms (Marsh, 2006). Report from the Department of Health and Human Services, US NIH proclaimed that more than 80% of hard and soft tissue infections of human being are associated with the bacterial and fungal biofilms.

Stages of biofilm formation include initial adhesion of microorganisms to the surface, lag phase with formation of monolayer of microbial cells, and self-production of polymeric matrix material followed by microcolony formation with multilayered structure, which further matures and disperse (Fig. 3).

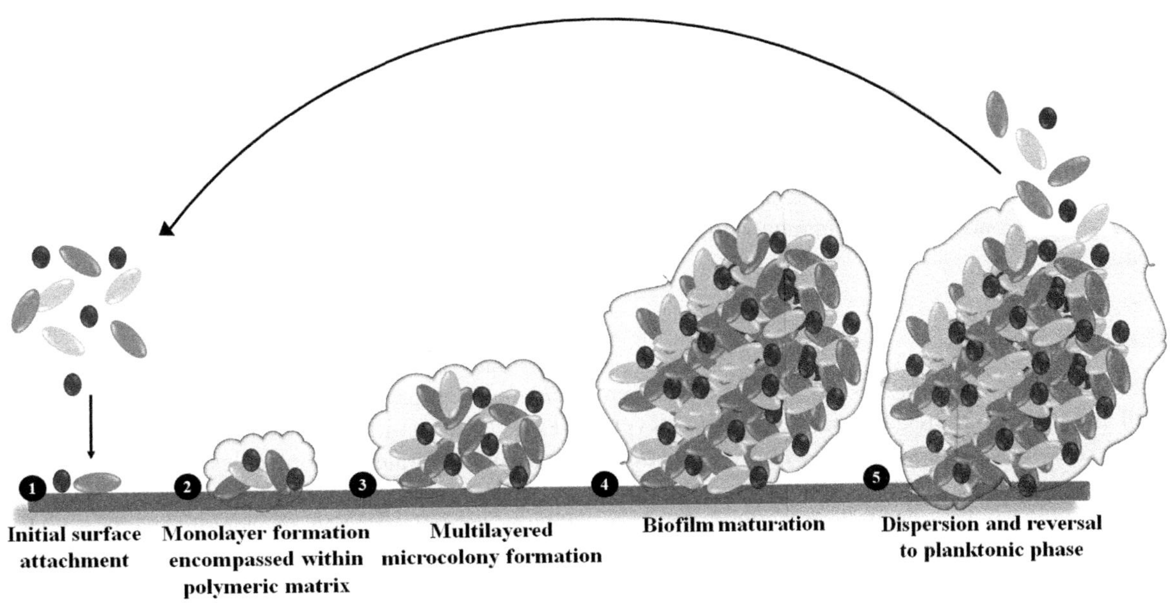

FIG. 3 Stages of biofilm formation.

In contrast to the planktonic phase, sessile phase contains 1000-fold more microbes per gram while on the other hand exhibits 1000-fold increased resistance against the antimicrobial agents (Arweiler & Netuschil, 2016). This is due to the fact that matrix of the biofilm restricts penetrance of the antibiotics while in the due course planktonic cells develop novel phenotypes and become less sensitive to antibiotics (Hall & Noverr, 2017). In addition to decreased penetrance, biofilm cells can acquire resistance to antibiotics based on other phenomenon, namely mutations, presence of drug efflux pumps, production of neutralizing enzymes, and prolonged usage leading to modification in the drug targets, etc. (Singh, Singh, Chowdhury, & Singh, 2017). Hence, biofilm cells are known to evade antimicrobial agents and host defense mechanisms (González, Hahn, & Gunn, 2018).

Every oral infectious disease is predominantly biofilm-associated, and this present chapter primarily focuses on diseases such as oral candidiasis, dental caries, and its severe form early childhood caries (ECC), which are instigated by fungal biofilm, bacterial biofilm, and the dual-species fungal-bacterial biofilm, respectively.

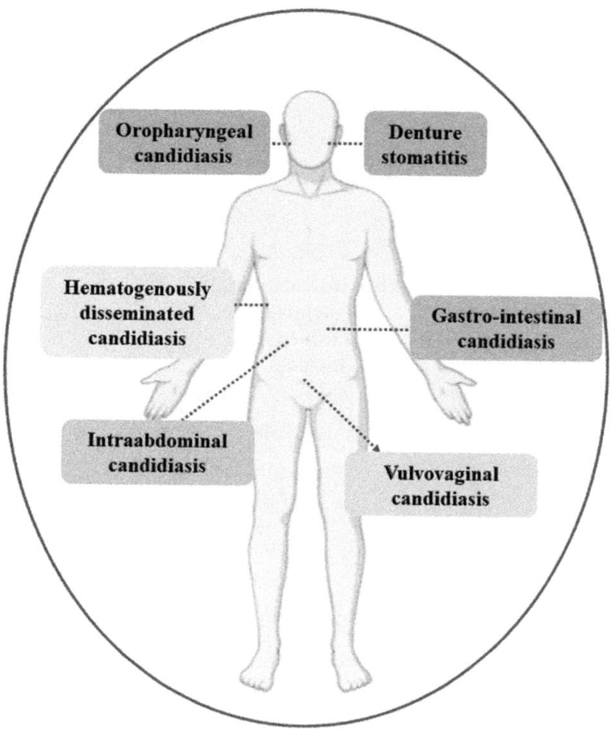

FIG. 4 Candidiasis of various anatomical sites.

4. Oral candidiasis

Infections elicited by the members of the *Candida* spp. are generally termed as candidiasis, which can affect various anatomical sites including oral cavity, skin, gastrointestinal tract, esophagus, vaginal tract, and vascular systems. Various forms of candidiasis are depicted in Fig. 4.

History of oral candidiasis dates back to the Hippocrates period (around 400 Before Christ), who in his book, "Of the epidemics," described oral candidiasis as "mouths affected with aphthous ulcerations" (Vila, Sultan, Montelongo-Jauregui, & Jabra-Rizk, 2020). Clinical presentation of candidiasis may appear as a mild disease with thrush on the mucosal membrane, but due to the recalcitrant nature can pose recurrent infections. Oral candidiasis is more frequently associated with the individuals of extreme ages such as newborn child and old-age adults. Primarily this infection can be more frequently observed in the individuals with severe underlying conditions and immunocompromised health status.

4.1 *Candida albicans*

Among the numerous *Candida* spp., *C. albicans* has been considered as an important fungal species, owing to its versatile nature and frequent association with the infectious diseases of humans. *C. albicans* can cause infections ranging from superficial mucosal infections to life-threatening systemic infections. *C. albicans* has the ability to reside as a symbiont in healthy individuals at various anatomical sites and can transit to pathobiont or dysbiont under unfavorable host conditions (Calderone & Fonzi, 2001; Hall & Noverr, 2017; Iliev & Leonardi, 2017; Jabra-Rizk et al., 2016).

In oral cavity, *C. albicans* is generally associated as a constituent of oral microbiota in health as well as disease (de Cássia Negrini, Koo, & Arthur, 2019; Xiao et al., 2018). Among the 150 other known species of *Candida*, 95% of oral candidiasis is primarily associated with *C. albicans* (Quindós et al., 2019). Few other species known to be involved in oral candidiasis include *C. glabrata*, *C. tropicalis*, *C. parapsilosis*, *C. krusei*, *C. dubliniensis*, etc. (Miranda-Cadena et al., 2018; Urizar, 2002). Colony morphology and biofilm architecture of *C. albicans* are displayed in Fig. 5.

Progression of candidiasis depends on virulence factors of the organism as well as the health status of the individual. Local and systemic risk factors that are associated with the oral candidiasis are listed in Fig. 6.

For the organism to survive under adverse niche conditions and to exert various pathological conditions, *C. albicans* expresses numerous virulence attributes (Table 1). Most important ones include expression of host recognition adhesin molecules, phenotypic switching between yeast and hyphal forms, secretion of proteases, lipases, hemolysin, and phospholipases (Sardi, Scorzoni, Bernardi, Fusco-Almeida, & Giannini, 2013). Virulence characteristics of *C. albicans* are depicted in Fig. 7.

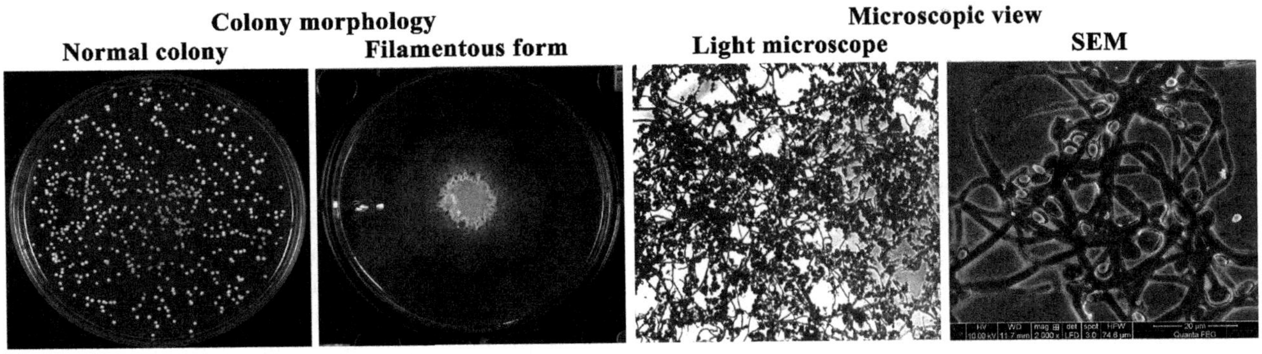

FIG. 5 Colony morphology and biofilm architecture of *C. albicans*.

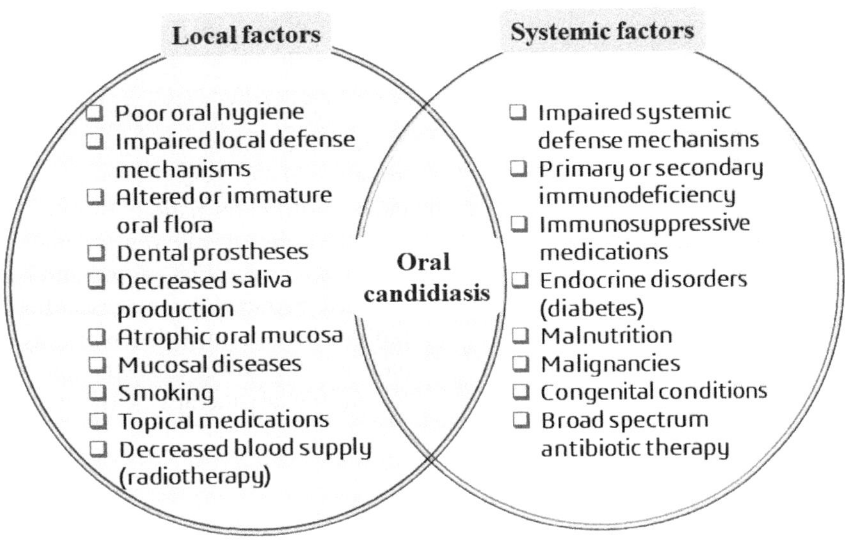

FIG. 6 Risk factors associated with development of oral candidiasis.

Host recognition factor of *C. albicans* enables the pathogen to adhere and bind to the host cell and proteins, which facilitate the primary colonization. In addition to biotic surfaces, *C. albicans* can adhere and form biofilm on abiotic implant surfaces. Specifically in denture stomatitis, *C. albicans* forms biofilm on the denture surface. Initial attachment of *C. albicans* is mediated through hydrophobic and electrostatic interactions between the adhesins present on the fungal cell surface to the ligands such as proteins, fibronectin, fibrinogen, etc., of the host cell surface (Li, Yan, & Xu, 2003; Verstrepen & Klis, 2006). In addition to this, the hydrolytic enzymes such as proteases, phospholipases, and hemolysins appear to play vital role in adherence, tissue penetration, destruction of host tissue, and invasion (Silva et al., 2012). Several reports have revealed the correlation of increased production of extracellular hydrolytic enzymes and pathogenic potential of yeast that leads to severe pathogenesis of candidiasis (Bramono, Yamazaki, Tsuboi, & Ogawa, 2006; Ingham et al., 2012).

Tremendous ability of *C. albicans* to form biofilm and transit between unicellular yeast cells to pseudohyphal or hyphal growth phase is tightly regulated by several transcription regulators. This phenotypic switching system enables the organism with increased virulence and pathogenicity by facilitating invasion to the host tissue and evasion from the immune attack (Cullen & Sprague, 2012; Jayatilake, Samaranayake, Cheung, & Samaranayake, 2006; Kumamoto & Vinces, 2005; Odds, 1988). Whereas the disseminated yeast cells are capable of initiating bloodstream infections (Lu, Su, Wang, & Liu, 2011).

4.2 Pathogenicity and virulence factors of *C. albicans*

Virulence is defined as the ability of pathogen to adhere, infect, and cause disease to the host cells. The pathogenicity of *C. albicans* is controlled by a number of virulence traits including adherence and biofilm formation on biotic (host tissues) and abiotic (medical devices), secretion of hydrolytic enzymes (e.g., proteinases, phospholipases, lipases, and hemolysins), morphogenesis, and phenotypic switching (yeast to hyphae or yeast to pseudohyphal forms) (Marak & Dhanashree, 2018).

TABLE 1 *C. albicans* **pathogenic attributes relevant to oral infection.**

Adherence to oral epithelial surfaces
• Cell surface hydrophobicity • Expression of cell surface adhesins
Biofilm formation
• Development of denture stomatitis • Failure of antifungal therapy
Evasion of host defenses
• Phenotypic switching • Binding to complement • Resistance to phagocytic stresses (oxidative and nitrosative stress response) • Proteolytic degradation of host immune factors (antibodies, antimicrobial peptides, etc.)
Invasion and destruction of host tissue
• Hyphal development and thigmotropism (tissue penetration) • Secretion of hydrolytic enzymes such as secreted aspartyl proteinases (SAPs), phospholipases, lipases (tissue degradation) • Secretion of hyphae specific toxin candidalysin • Degradation of E-cadherin • Induced endocytosis

4.2.1 Adhesion to epithelial cells

The very first event of *Candida* infections is the adherence to host surfaces for both colonization and establishment of infections. Several factors influence the adhesion, including cell wall proteins profile and physicochemical properties of host cell surface (Chaffin, 2008). In fact, *C. albicans* is more adherent than other non *albicans Candida* (NAC) species, wherein, adhesion is facilitated by a specialized set of proteins known as agglutinin-like sequence (ALS) proteins or adhesions; these *als* genes encode glycosylphosphatidylinositol (GPI)-linked cell surface glycoproteins. Among eight known Als proteins (Als 1–7 and Als9), Als3 codes for hyphae-associated proteins important for adhesion (Mayer, Wilson, & Hube, 2013).

4.2.2 Biofilm formation

Next step to initial adhesion of *Candida* spp. to biotic and abiotic surfaces is cell division, proliferation, and subsequent biofilm formation. Biofilm is the most ubiquitous growth form of microorganisms, described as surface-associated microbial communities embedded in self-secreted extracellular matrix. Microorganisms present in biofilm behave in a different way from planktonic cells and exhibit lower growth rate and higher resistance to antifungal therapy (Al-Fattani & Douglas, 2006).

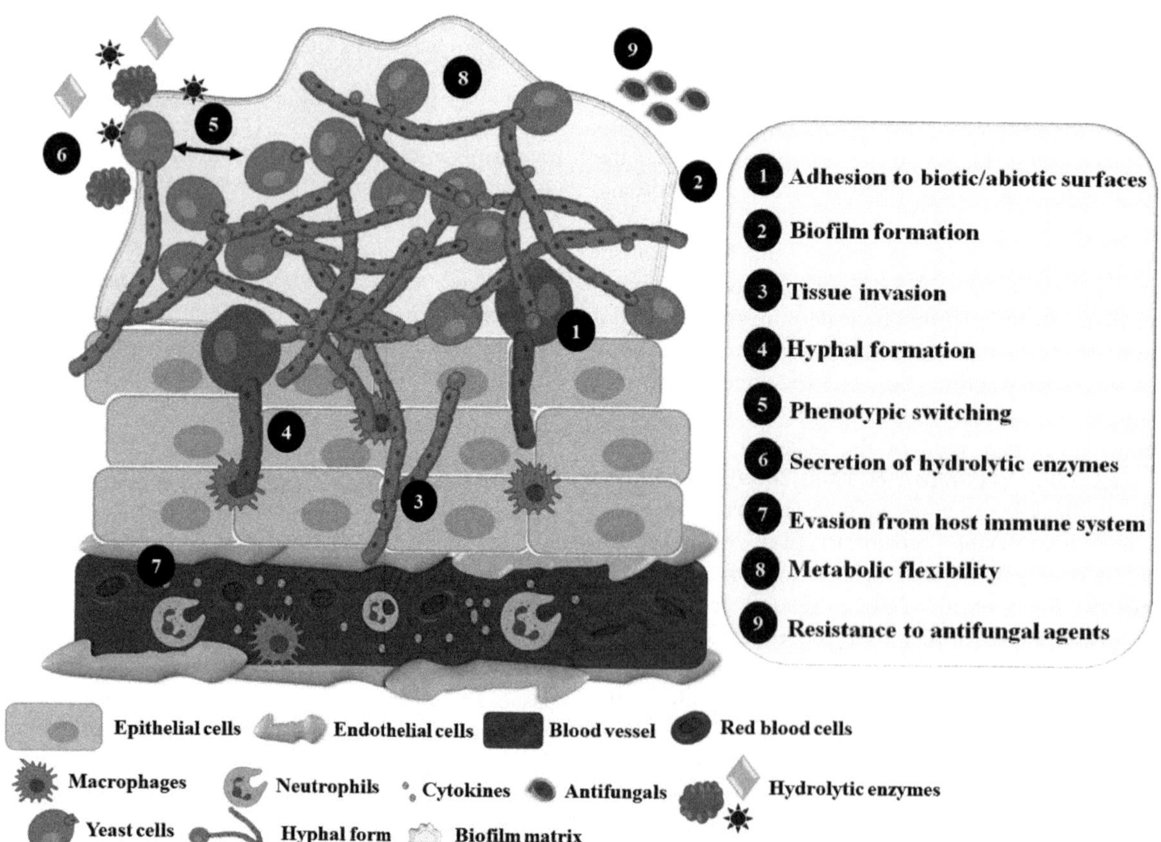

FIG. 7 Distinctive virulence attributes of *C. albicans*.

Candida species are identified as the fourth most common cause of nosocomial infections in United States, and biofilms are frequently found in medical devices such as urinary and vascular catheters, prostheses, cardioverter defibrillators (Cavalheiro & Teixeira, 2018). Typically, *Candida* biofilm is developed in four sequential steps, namely adhesion of candida cells to a surface, individual colonization and organization of cells, secretion of extrapolymeric substances (EPSs), and dissemination of progeny cells.

4.2.3 Yeast to hyphal transition

Among various fungal pathogens, *Candida* spp. are well known for their polymorphism, which facilitates to grow either as budding yeast cells or elongated cells with constriction at septa (pseudohyphae) or as parallel-walled long true hyphae. The transformation between yeast and hyphal growth forms is described as dimorphism and both forms have a unique role in pathogenicity. While hyphal cells are more invasive than yeast cells, yeast cells are mostly involved in dissemination (Subramenium, Swetha, Iyer, Balamurugan, & Pandian, 2018). In 2005, Kumamoto and Vinces had reported the following two major functions of hyphae, namely possibly involved in tissue penetration and escape from host cells following internalization/ endocytosis.

In *C. albicans*, several environmental factors affect morphology including pH, starvation, the presence of serum or *N*-acetylglucosamine, physiological temperature, and CO_2, which promote the formation of hyphae.

4.3 Denture stomatitis

In addition to oral candidiasis in normal individuals, *Candida*-associated denture stomatitis is the most common oral candidiasis that affects the denture wearers, and *C. albicans* has been reported as the primary etiological agent associated with this infection (Di Stasio et al., 2018; Gendreau & Loewy, 2011; Kulak, Arikan, & Kazazoglu, 1997; Webb, Thomas, Willcox, Harty, & Knox, 1998). Characteristics of the infection include inflamed mucosa, which has multifactorial epidemiology including denture trauma, continuous denture wearing, denture base material, denture cleanliness, denture age, salivary flow, and pH (Barbeau et al., 2003; Ramage, Tomsett, Wickes, López-Ribot, & Redding, 2004). *C. albicans* adheres to the denture acrylic surface such as polymethylmethacrylate surface either directly or via an intermediary layer of plaque-forming bacteria (Chandra et al., 2001).

4.4 Evolution of drug resistance in *C. albicans*

Drug resistance is one of the significant biological phenomena that pose enormous effect on the human health. Prevalence of drug resistance is increasing exponentially in recent decades due to frequent and overusage of the antimicrobial agents. Resistance to multiple antimicrobial agents limits the therapeutic options, which is an alarming issue. Resistance development to antimicrobials is not restricted to bacteria. Development of resistance by *C. albicans* against fluconazole emerged in 1990s (White, Marr, & Bowden, 1998).

For *C. albicans*, exposure to antifungal agents is one of the environmental stresses, which can stimulate an array of responses to alleviate dreadful effects of the drug. Genome plasticity permits the organism for continued survival under this selective pressure. Stress-responsive signaling systems of *C. albicans* are interrelated with the virulence and pathogenesis (Cannon et al., 2007). Major antifungal targets of *C. albicans* include pyrimidine biosynthesis, membrane integrity, ergosterol, and cell wall biosynthesis. However, the organism has developed increased resistance to all of these primary drug targets. Targets of the antifungal agents and mechanism of drug resistance are depicted in Fig. 8.

Expression of numerous virulence factors, ability to form biofilm on wide range of biotic and abiotic surfaces, and the advertent potential to resist the action of antifungal agents bestow *C. albicans* as a versatile fungal pathogen that is difficult to deal with.

5. Dental caries

Dental caries is one among the most common and major public oral health diseases that affects almost individuals of all the age groups and population worldwide (Petersen, Bourgeois, Ogawa, Estupinan-Day, & Ndiaye, 2005). The word caries had been derived from Latin meaning for rot or rotten. Dental caries is a contagious infection that arises due to complex interplay between the oral microorganisms in the plaque community, diet practices, and huge array of host factors (Yadav & Prakash, 2017). Risk factors associated with dental caries are illustrated in Fig. 9.

Dental caries is not an exclusive human dental infection. It has been found in numerous creatures of the Paleozoic, Mesozoic, Miocenic, and other eras (Kear, 2001; Kemp, 2003; Yadav & Prakash, 2017). Numerous concepts and theories have been proposed by various medical practitioners for decades. In 1890s, W.D. Miller proposed a concept named chemoparasitic caries theory through which it is explained that the bacteria colonized on the tooth surface produce acid by-products that dissolve the enamel in the presence of fermentable carbohydrates (Kleinberg, 2002). And it was Killian Clarke in 1924, who first described the existence of chains of spherical bacterium, which he named as *Streptococcus mutans* (Clarke, 1924).

This observation went unnoticed until Orland in 1959, who demonstrated that enterococci could cause dental caries in germ-free rats. Fitzgerald and colleagues

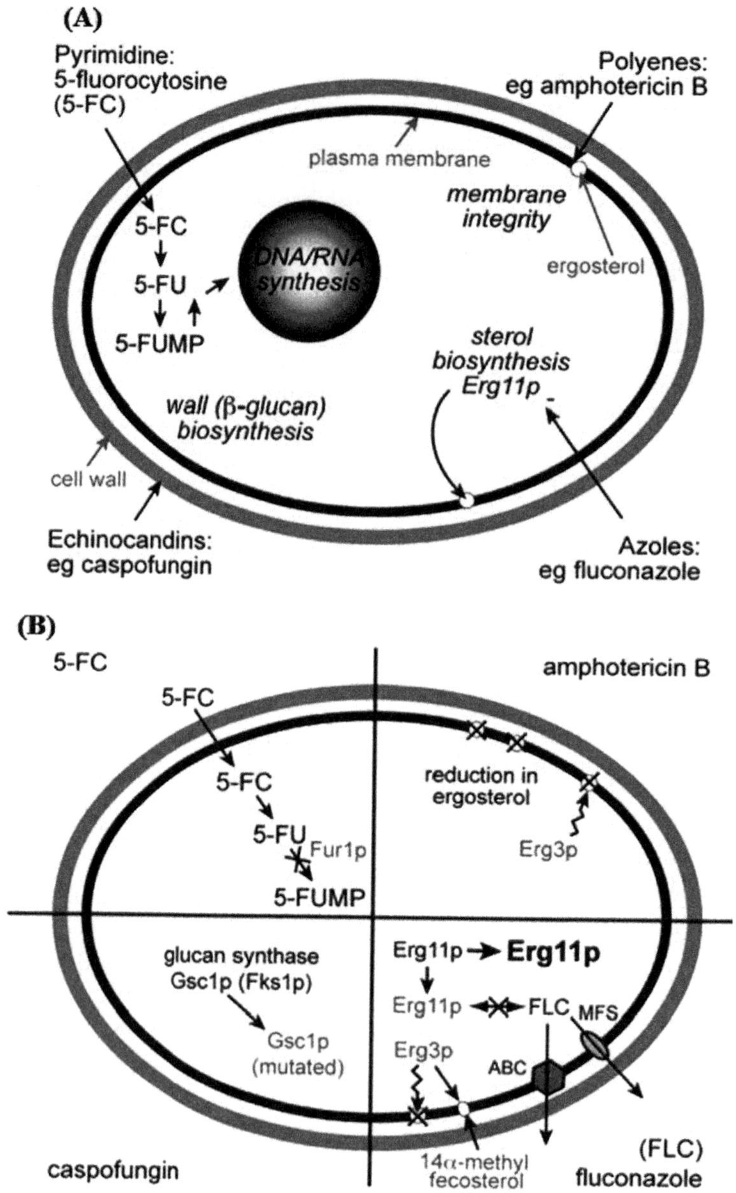

FIG. 8 Targets of antifungal drugs and resistance development by *C. albicans*. (A) Molecular targets of current antifungal drugs. (B) Resistance mechanisms developed by *C. albicans* against the drug targets.

(Fitzgerald & Keyes, 1960) found that the dental caries in germ-free animal models were caused by initial infection with the streptococci, which is of the type originally described by Clarke. Subsequent studies illustrated *S. mutans* as the principal etiological agent of dental caries in humans (Kuramitsu, 2003; Newbrun, 1978). Primary colonization of *S. mutans* initiates following tooth eruption, which occurs during the first year of human life. In most human beings, permanent colonization occurs during early childhood (Stiles, 1976). Dental caries induced by mutans streptococci group have been reported for the worldwide distribution since 1970s (Newbrun, 1978). *S. mutans* has adverse ability to survive in acidic environment and interact with other microorganisms in the ecosystem to form a complex diverse biofilm architecture (Krzyściak, Jurczak, Kościelniak, Bystrowska, & Skalniak, 2014). Peterson et al., in 2014, enumerated the numerical dominance of *Streptococcus* spp. in dental plaque by global gene expression analysis of plaque microbiota.

5.1 *S. mutans* in dental caries

Natural habitat of *S. mutans* is the oral cavity, thus, exclusively resides in the mouth and specifically in the niche of dental plaque (Lemos et al., 2019). *S. mutans* has been demonstrated as the primary etiological agent associated with the dental caries (Viszwapriya, Subramenium, Radhika, & Pandian, 2017). Fig. 10 depicts the colony and biofilm morphology of *S. mutans*.

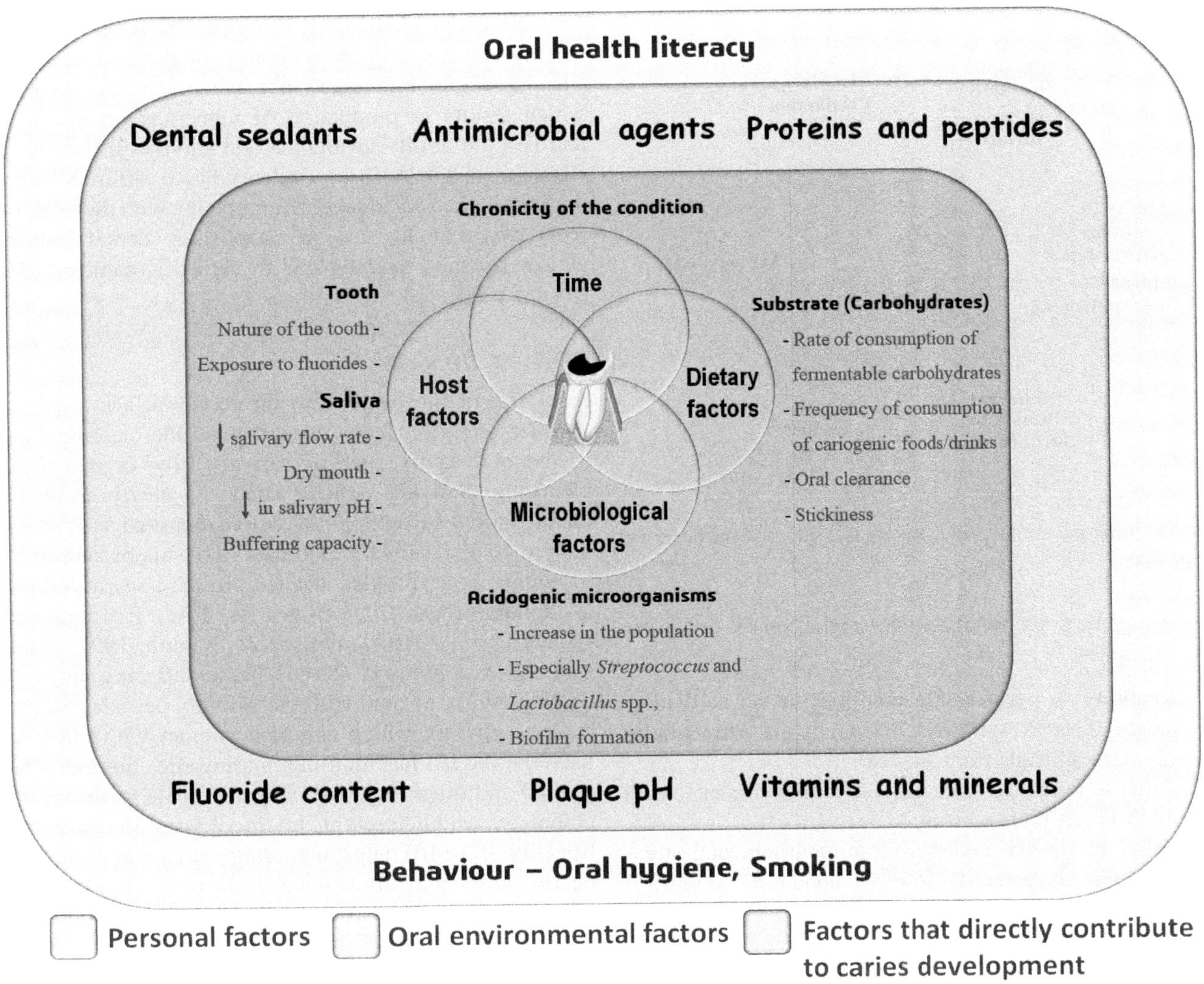

FIG. 9 Risk factors associated with the development and progression of dental caries.

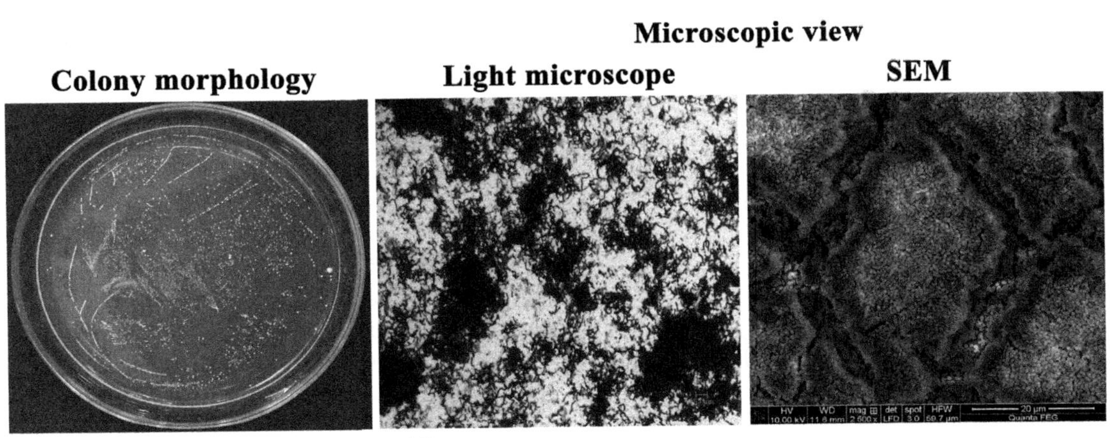

FIG. 10 Colony morphology and biofilm architecture of *S. mutans*.

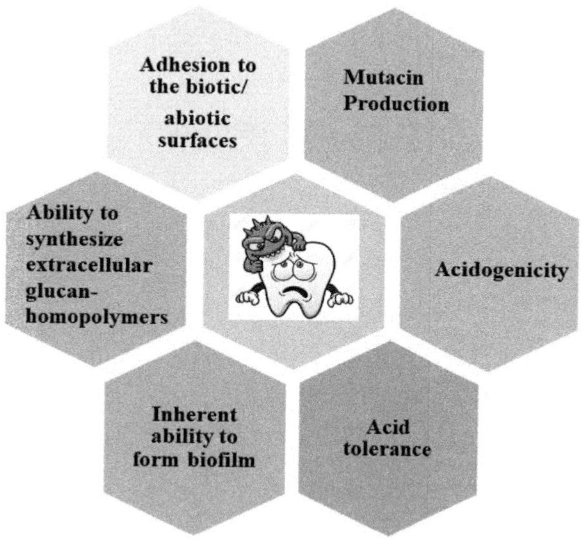

FIG. 11 Virulence attributes of *S. mutans* in dental caries initiation and progression.

Cariogenic potential of *S. mutans* relies on the following core attributes.

- Capability to permanently colonize on the hard teeth surface through synthesis of variety of extracellular polymeric glucans from sucrose
- Ability to ferment a wide range of carbohydrates to acid end products (acidogenicity)
- Ability to survive under adverse environmental stress conditions, importantly low pH condition (aciduricity) (Lemos & Burne, 2008).

Various virulence attributes of *S. mutans* in the implication of dental caries are illustrated in Fig. 11.

Whole-genome sequence of *S. mutans* strain UA159 has been published in the year 2002 (Ajdić et al., 2002). Through decades of research, it has been demonstrated that *S. mutans* is the principal cariogenic organism that contributes to the formation of dental biofilm matrix (Banas & Vickerman, 2003; Bowen & Koo, 2011; Gross et al., 2012; Lemos, Abranches, & Burne, 2005; Quivey, Kuhnert, & Hahn, 2001).

As the etiological agent of dental caries, *S. mutans* primarily survives in the biofilm form rather than free-living planktonic state (Lemos, Quivey, Koo, & Abranches, 2013).

5.2 Virulence factors of *S. mutans*

S. mutans follows an obligate biofilm lifestyle in its natural ecosystem, i.e., dental plaque for its survival and persistence (Ajdić et al., 2002; Burne, 1998; Shemesh, Tam, & Steinberg, 2007; Wen & Burne, 2002). The ability of *S. mutans* to initiate dental caries depends on several significant virulence traits, including: (i) adherence and accumulation on the tooth surface for initiation of biofilm formation; (ii) production of various bacteriocins, which can kill other species and outnumber in the dental biofilms; (iii) incredible efficiency in catabolizing carbohydrates, which results in production of numerous organic acids; and (iv) the ability to tolerate and survive in the low pH (Kuramitsu, 2003). These virulence factors afford *S. mutans* with ecological advantages in competing with the other oral pathogens and succeed in establishing dental biofilms, which can subsequently lead to caries (Kuramitsu, 2003; Marsh, 2000).

5.2.1 Adhesion

Adhesion of the bacteria to the tooth surface requires a specific retention factor to withstand the clearing forces in the oral cavity such as salivary flow or mechanical cleansing. Salivary pellicle coats the newly erupted or cleaned tooth surface. Initial colonizers such as *S. mutans* adhere to the salivary pellicles. The major adhesin of *S. mutans* is a 185-kDa surface protein known variously as antigen (Ag) I/II, antigen B, PAc, P1, and SpaP (Brady et al., 2010; Jenkinson & Demuth, 1997; Nobbs, Lamont, & Jenkinson, 2009). These antigenic molecules display ability to bind with the salivary pellicle receptors such as gp-340, which can also interact with other oral bacteria. Ag I/II has multiple binding sites through which gp-340 and other bacterial cells aggregate to form initial colonizers with heterotypic microcolonies. Versatile functionality of Ag I/II helps in binding with collage and fibronectin, which may allow the *S. mutans* to colonize on soft tissues of the oral cavity or other internal sites of the body if it gains systematic access.

5.2.2 Polysaccharide production

Followed by initial attachment mediated by Ag I/II, the retention and accumulation of *S. mutans* containing microcolonies are enhanced by the production of huge amounts of extracellular polysaccharide (Banas & Vickerman, 2003; Bowen & Koo, 2011). *S. mutans* produces glucosyltransferases (Gtf) and a fructosyltransferase (Ftf) on the cell wall. *S. mutans* can produce three Gtfs designated as GtfB, GtfC, and GtfD. GtfD and GtfB produce soluble linear glucan with α-1,6 glycosidic linkages and insoluble glucan rich in α-1,3 branch linkages, respectively. Whereas GtfC produces a mixture of low-molecular-mass insoluble and soluble glucans. The primary purpose of the insoluble glucans is to sequester the microcolonies within the matrix. It also serves as an extracellular long-term energy source. The extracellularly secreted Gtf enzymes are adsorbed to salivary pellicle to provide additional cohesive microenvironments. Glucose retained within the bacteria can be polymerized into a glycogen-like intracellular polysaccharide (IPS), which acts as a reserve energy source. In the deprivation of dietary carbohydrates, IPS will act as initiator for glycolysis, which will extend the duration of acid production (Hiratsuka, Wang, Sato, & Kuramitsu, 1998; Zeng & Burne, 2013).

5.2.3 Glucan-binding proteins

S. mutans expresses four different glucan-binding proteins (Gbp) A–D to facilitate the attachment of glucans produced by Gtf. Gbps are essential for *S. mutans* to accumulate into complex matrix, and they also determine the architecture of the biofilm community. GbpB may also function as a peptidoglycan hydrolase and contribute to cell wall cycling and synthesis (Banas & Vickerman, 2003).

5.2.4 Acid production

The virulence factors such as adhesion, polysaccharide production, and glucan-binding proteins are all required for the bacterium *S. mutans* to interact with the dental surface. While production of various organic acids is the major virulence factor to initiate damage in the tooth and for progression of dental caries. Sucrose is deliberated to be the most cariogenic sugar as it can be converted into polysaccharides and further fermented into organic acids. Among the organic acids produced by *S. mutans*, lactic acid is the strongest with the ionization constant of 3.5. Thus, lactic acid is considered the dominant organic acid associated with dental caries.

5.2.5 Acid tolerance

Many bacterial cells cannot tolerate and survive in low pH or acidic conditions. Hence, as carious lesion progress, selection pressure allows the existence of bacteria, which are resistant to low pH. *S. mutans* has developed various mechanisms to tolerate acidic conditions and prevents itself from acid damage (Lemos & Burne, 2008; Matsui & Cvitkovitch, 2010). The mechanisms include the following:

- **(i)** Pumping hydrogen ions out of the cell by F-ATPase activity
- **(ii)** Presence of long-chain and unsaturated fatty acids in the cell membrane
- **(iii)** Agmatine deiminase system (AgDS)
- **(iv)** Fermentation of malolactic acid

5.2.6 Host factors in caries

Several factors in the host oral environment modulate the susceptibility of the caries development and severity (Scannapieco, 1994). Saliva is the most important host factor in caries process, which affects in various ways such as

- **(i)** Salivary flow rate
- **(ii)** Salivary buffering capacity
- **(iii)** Supersaturation
- **(iv)** Direct antibacterial effectors in saliva

5.3 Biofilm formation by *S. mutans*

S. mutans is a robust biofilm forming organism, which can produce multilayered biofilm matrix. Metabolism of carbohydrates helps them to adhere to the tooth surface and form biofilm. Once stabilized within the biofilm matrix, *S. mutans* can tolerate and survive under rapid and frequent fluctuations in the oral environment such as deprived nutritional availability, transition from aerobic to anaerobic conditions, adverse pH, etc. (Lemos & Burne, 2008; Yoshida & Kuramitsu, 2002). Moreover, in response to the external physical and chemical stimuli, *S. mutans* can regulate diverse physiological processes in a cell-to-cell communication and cell-density-dependent manner known as quorum sensing (Bassler, 1999). Response through this system involves the two-component signal regulatory systems (TCRSs) (Barrett & Hoch, 1998; Beier & Gross, 2006). Numerous TCRSs are identified in *S. mutans*. Two most important TCRSs of *S. mutans* include comCDE and *vic*RK systems. *comCDE* system regulates the response to cell density and external stimuli (Li, Lau, Lee, Ellen, & Cvitkovitch, 2001), and *vicRK* system regulates sucrose-dependent biofilm formation in *S. mutans* (Senadheera et al., 2005). Through these available TCRSs, *S. mutans* establishes persistent dental biofilm with recalcitrant ability and lowers the pH of local environment that leads to erosion of teeth by imbalance between demineralization and remineralization process (Fig. 12).

5.4 Fluoride resistance in *S. mutans*

For decades, fluoride has been extensively used as a therapeutic agent against caries. Due to its ability to exert antimicrobial action and to protect the dental hard tissue, fluoride is widely used in oral care products (Anusavice, Zhang, & Shen, 2005). The antimicrobial action of fluoride against *S. mutans* is exhibited through regulation of three major virulence aspects such as acidogenicity, aciduránce, and adherence to the tooth surface (Hamilton & Ellwood, 1978; Meurman, 1988; Van Loveren, 2001). In addition to antimicrobial activity, fluoride in the saliva could be potentially absorbed on the surface of hydroxyapatite and can possibly arrest extensive demineralization process, and when pH of the environment transits from acidic to alkaline, the fluorohydroxyapatite enhances the remineralization process (Buzalaf, Pessan, Honório, & Ten Cate, 2011).

Resistance development against fluoride action has been observed in *S. mutans* during recent times (Liao et al., 2017). *S. mutans* has been reported to acquire fluoride resistance either in a transient or stable manner. Transient resistance was observed in patients with xerostomia, which is possibly related to the horizontal gene transfer (Brown, White, Horton, Dreizen, & Streckfuss, 1983; Streckfuss et al., 1980). Whereas the stable fluoride resistance is reported to arise due to chromosomal mutations (Liao et al., 2015).

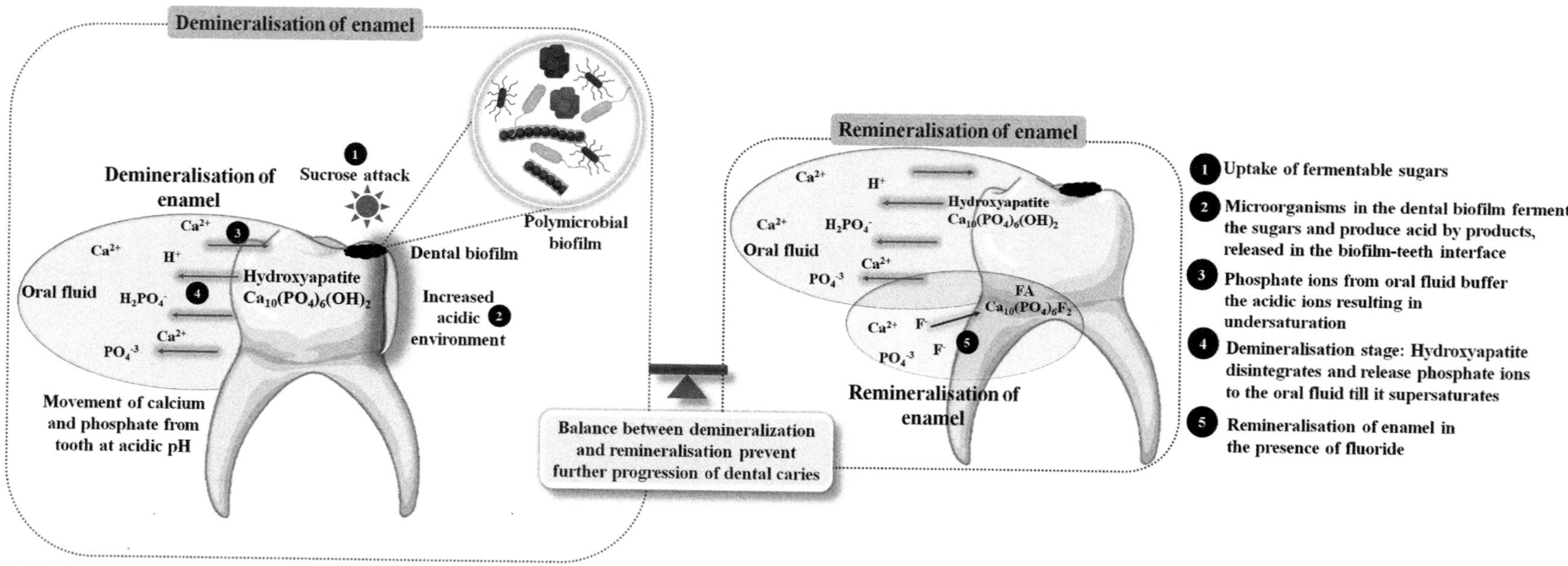

FIG. 12 Demineralization and remineralization in the teeth erosion process.

6. Early childhood caries (ECC)

ECC is one of the most prevalent infectious diseases of the children worldwide (Meyer & Enax, 2018). ECC is generally a more severe form of dental caries that affects the children of age group less than 6. Such types of caries that occur in the children have a special caries pattern that is different from the normal dental caries (Kawashita, Kitamura, & Saito, 2011). Caries in ECC significantly affects the maxillary primary incisors and first primary molars in similar way of the eruption of teeth (Kidd & Fejerskov, 2003). One of the primary reasons labeled for the development of ECC is inappropriate bottle feeding practices (Harris, Nicoll, Adair, & Pine, 2004; Narvey & Shwart, 2007).

Characteristics of ECC begin with white lesions on the upper primary incisors along the gingival margin and progression leads to complete destruction of crown (Kagihara, Niederhauser, & Stark, 2009). ECC not only affects primary or permanent dentitions of the children, but can impact overall health status.

Children with ECC are found to grow at slower pace, underweight, and subjected to iron deficiency (Acs, Lodolini, Kaminsky, & Cisneros, 1992; Clarke et al., 2006). ECC is primarily a familial disease, which is transmissible and often associated with poor dietary habits and increased sugar and carbohydrates consumption (Douglass & Clark, 2015). Risk factors associated with ECC are listed in Fig. 13.

Since 1980s, the studies on the microbial factors for ECC had begun (Van Houte, Gibbs, & Butera, 1982). Predominant microorganisms found to be involved are members of mutans *streptococci* and other acidogenic organisms (Matee, Mikx, Maselle, & Helderman, 1992; Starr et al., 2002). Several reports have evidenced the cariogenic potential of *C. albicans* and involvement of this fungal species in ECC (Nikkawa et al., 2003; Qiu, Li, Lin, Yu, & Zhao, 2015).

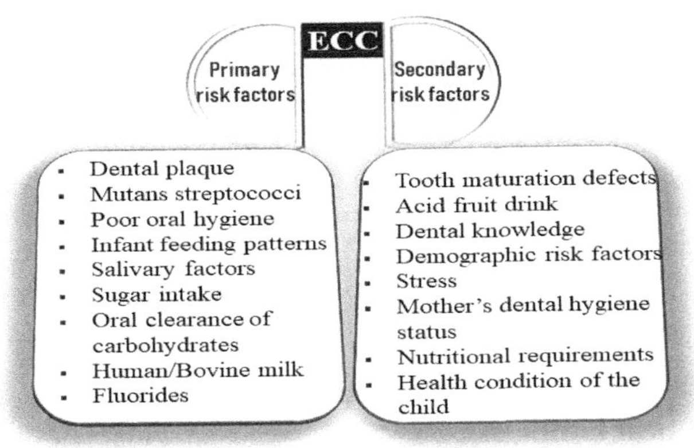

FIG. 13 Primary and secondary risk factors associated with ECC.

6.1 Synergistic interaction of *C. albicans* and *S. mutans* in development of ECC

For years, *S. mutans* has been known for its importance and role in ECC, whereas in recent times *C. albicans* has been frequently found in association with *S. mutans* from the carious lesions of ECC (Ellepola, Liu, Cao, Koo, & Seneviratne, 2017; Raja, Hannan, & Ali, 2010). Under normal circumstances, *C. albicans* neither synergizes with *S. mutans* nor colonizes on the tooth surface on its own (Falsetta et al., 2014; Klinke, Guggenheim, Klimm, & Thurnheer, 2011). Studies have shown enhanced adhesive interaction between these two organisms in the presence of sucrose, which leads to mixed species biofilm formation (Branting, Sund, & Linder, 1989; Gregoire et al., 2011; Metwalli, Khan, Krom, & Jabra-Rizk, 2013; Pereira-Cenci et al., 2008). This enhanced interaction is promoted by the secretion of exoenzyme glucosyltransferases (Gtfs) by *S. mutans* (Hwang, Marsh, Gao, Waugh, & Koo, 2015). Interaction between GtfB and *C. albicans* is facilitated by mannans that are located in the outermost cell wall of *C. albicans* (Hwang et al., 2017). Colony morphology and biofilm architecture of dual species of *C. albicans* and *S. mutans* are depicted in Fig. 14.

Within the dual species biofilm, the interaction between these two organisms dramatically modifies the environment. Thus, increased production of biofilm biomass and extracellular polysaccharides is observed. This synergistic alliance also displays increased virulence factors production by each organism in communication leading to rapid destruction of the teeth (de Carvalho, Silva, Hebling, Spolidorio, & Spolidorio, 2006; Joyner et al., 2010).

Though several studies have evidenced the interaction between *C. albicans* and *S. mutans* in the development and progression of ECC (Barbieri et al., 2007; Falsetta et al., 2014; Gregoire et al., 2011; Hwang et al., 2015, 2017; Metwalli et al., 2013; Yang et al., 2018), very few reports are available on the therapeutic regimens for prevention of the disease (Bombarda et al., 2019; Feldman, Ginsburg, Al-Quntar, & Steinberg, 2016; Li et al., 2019; Philip, Leishman, Bandara, & Walsh, 2019). Thus, identification of therapeutic molecule against the dual species of *C. albicans* and *S. mutans* is imperative to prevent and treat teeth destruction caused by ECC.

7. Biofilm-mediated antimicrobial resistance

Recalcitrant nature of the biofilm architecture remains as the basis for numerous persistent diseases owing to the diminished antimicrobial therapy (Mah, 2012). In particular, biofilm tolerance is a major clinical concern as more than 65%–80% of microbial infectious diseases involve

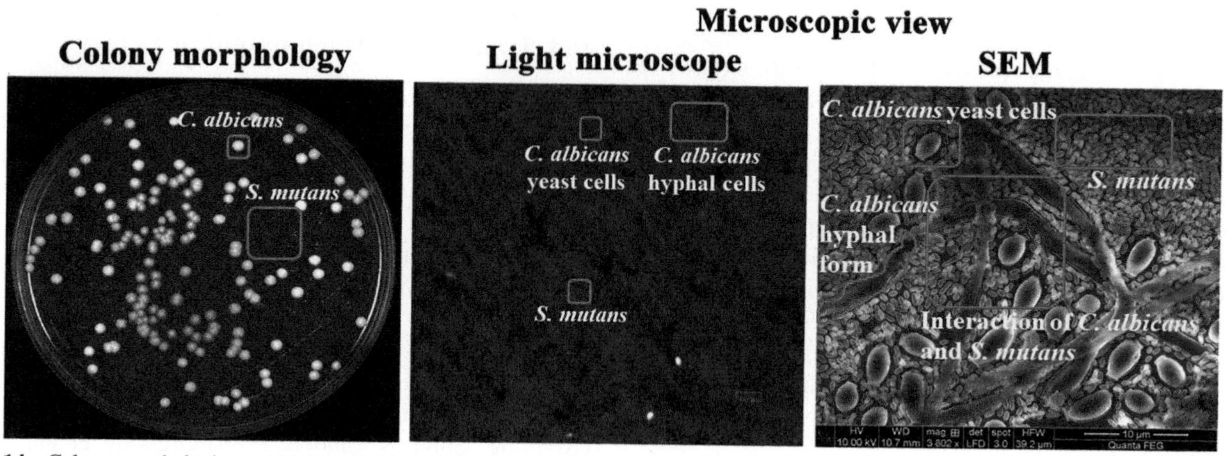

FIG. 14 Colony morphologies and biofilm architect of *C. albicans* and *S. mutans* dual species.

biofilm formation (Fux, Costerton, Stewart, & Stoodley, 2005; Van Acker, Van Dijck, & Coenye, 2014). Oral infectious diseases, primarily oral candidiasis, dental caries, and ECC, are biofilm-associated persistent infections. Existence within the biofilm provides these infection-causing microorganisms a basic survival strategy while being protected from the environmental influences such as antimicrobial therapy and/or host immune responses (Patel, 2005).

Though antibiotics can partially decrease the number of organisms residing in biofilms, complete eradication of sessile cells is impractical in in vivo conditions ensuing in recurrent infections. Further, concentration of antimicrobials required for antimicrobial activity against biofilm cells will be invariably higher than that required for planktonic cells. Alternatively, dispersed biofilm cells may attain susceptibility to antimicrobial agents, which suggests that biofilm-mediated resistance is not solely acquired through mutation or mobile element transformation. Thus, inherent resistance to the antimicrobials including antibiotics and other disinfectants is conferred through mechanisms that are apparently distinct from usual mechanisms of antimicrobial resistance such as efflux activation, target modifications, or modifications in the antimicrobial agents (Rodríguez-Martínez & Pascual, 2006). Single mechanism probably cannot be responsible for the biofilm-mediated antimicrobial resistance or tolerance. As biofilm encompasses physiologically different consortium of cells, various mechanisms act concurrently to establish a high level of resistance and/or tolerance.

8. Mechanism and risk factors associated with oral biofilm formation

Biofilm comprises communities of microbes aggregated inside an extracellular polysaccharide (EPS) layer produced by the microbial communities themselves, which aids to the biofilm architecture (Wingender & Flemming, 2010). The property of formation of biofilm is also a common trait in oral pathogens as it is one of the major virulence attributes that is exhibited by causative pathogens of oral diseases. Among all the other oral pathogens, *Streptococcus mutans*, *Porphyromonas gingivalis*, and *Candida albicans* are considered as prevalent pathogens of several severe oral conditions such as dental caries, gingivitis, and oral candidiasis (OC). Biofilms of these oral pathogens are a critical risk factor to deal with as they facilitate protection for these pathogens against external factors such as invading foreign microbes, and additionally, it aids in evasion of host immune response by acting as a barrier against host innate and adaptive immunity, which are much active against planktonic cells rather biofilm cells. For example, *P. gingivalis*, which is a member of the subgingival biofilm community, can facilitate downregulation of specific immune mediators such as proinflammatory interleukin-Iβ and NLRP3 inflammasome, which are essential for the bacterial elimination by the host. *P. gingivalis* inherits the mechanism to evade host immune response (Belibasakis, Guggenheim, & Bostanci, 2013). The enhancement of resistance against various antimicrobials such as antibiotics is facilitated by biofilm of oral pathogens. The constant increase in the usable concentration of the commercial antibiotics projects the ineffectiveness of antibiotics against microbial communities residing in the biofilms. The antimicrobial peptides (AMPs) are produced in oral cavity of host to act against pathogenic communities of oral microbiome. Salivary AMPs have decreased activity against biofilm-residing bacterial community in comparison with planktonic cells (Diamond, Beckloff, & Ryan, 2008; Wei, Campagna, & Bobek, 2006).

8.1 Risk factors associated with formation of oral biofilms

The formation of biofilm is itself a virulence characteristic that plays a vital role in the persistence of oral pathogens in oral cavity. In addition to its mere existence, which in itself a major drawback in the maintenance of oral care, it also enhances the severity of additional virulence characteristics of both *S. mutans* and *C. albicans*. *S. mutans* is known for its ability to produce lactic acid as a metabolic by-product through anaerobic respiration. For *S. mutans* cells, which reside in the biofilm, the production of lactic acid is enhanced, and it results in the inevitable enhancement of one of its virulence characters, i.e., acidogenicity. The communities of *S. mutans* in a biofilm formed by itself are highly durable in conditions such as high pH, low oxygen content, or presence of antimicrobial agents, when compared with the planktonic cells of the same species, as it is protected by an exopolysaccharide covering.

In case of *C. albicans*, following biofilm formation, the expression of other critical virulence characteristics is observed. The characteristics such as morphogenesis and secretion of hydrolytic enzymes are observed highly in cells of biofilm. The phenotypic switching capability of *C. albicans* helps in the maintenance of oral candidal population and also maintains the severity of OC. The emergence of oral biofilms among oral pathogens gives rise to the emergence of persister cells of the same oral pathogens, toward which many antimicrobials prove to be ineffective. The emerged persisting bacterial and fungal oral pathogens will eventually lead to the recurrence of the oral diseases. It also helps the pathogens to evade host immune response. The expression of immunoglobulins targeting *C. albicans* is the act of acquired immunity involved in the control of candidal population. In case of tissue penetration, the macrophages and polymorphonuclear leukocytes are involved in the abolition of colonization of *C. albicans* in tissue layers of the oral cavity (Challacombe, 1994). Secretory IgA (SIgA) is the primary immunoglobulin molecule found in the saliva. The evasion of *C. albicans* to these SigA is done by cell wall manno-proteins, which are considered as the receptor for SigA. The shedding of these receptors is done during progression of yeast cells to its hyphal form (Pontón et al., 1996). This results in the escape of clearance by *C. albicans*. The ability to form biofilm by the oral pathogens is a major virulent characteristic to make a note of as it is primarily responsible for both the induction and the enhancement of the oral diseases.

9. Alternative strategies to combat biofilm-associated drug resistance in pathogens

Besides increased antimicrobial resistance, rate of development of novel antibiotics has steeply declined where only very small pipeline of new therapeutics are available. For the rising pharmaceutical demand, it is prudent to search for alternative therapeutics. Efforts have been extensively made globally by numerous clinicians and researchers in developing novel conventional and nonconventional therapeutic strategies. Although certain alternatives exist already, major interrogation relies on its clinical efficacy. Thus, understanding and identifying novel alternatives to antibiotics for effective treatment of these oral infectious diseases will require experimental clinical evidence and not just the discovery of drug molecule (Czaplewski et al., 2016). Few emerging alternative approaches with promising activities include antibiofilm therapy, combinatorial strategy, use of natural bioactive molecules, phage therapy, bacteriocins, antibodies, lysins, probiotics, antimicrobial peptides, etc. In addition to the abovementioned naturally occurring alternatives, several synthetic and genetically engineered strategies such as synthetic mimics of antimicrobial peptides (SMAMP), innate defense regulatory peptides, antibacterial oligonucleotides, genetically modified bacteriophages, specifically targeted antimicrobial peptides (STAMP), CRISPR-Cas9, etc., are gradually emerging (Ghosh, Sarkar, Issa, & Haldar, 2019).

9.1 Targeting biofilm

As biofilm-associated cells are resistant against the antimicrobial agents and host immune system, traditional treatment of microbial infections fails often. In order to combat this dreadful situation, targeting biofilm either through inhibition of biofilm formation or through dispersal of preformed biofilm can provide a therapeutic advantage. Different stages of biofilm formation are well known. Each of the stages can be potentially targeted for control of infection. Initial stage of the biofilm formation can be interrupted by prohibiting or breaking interaction between the microorganisms and surface. Inhibition of extracellular polymeric substances can prevent the development of early-stage biofilm formation. Mature biofilm matrix can be potentially destroyed or dispersed by matrix degrading enzymes. Once the biofilm formation is inhibited or allowed to disperse, majority of the cells will turn into planktonic phenotypes, which can be taken care by the antimicrobial therapy and/or the host defense system and can effectively be cleared. Stage-dependent targeted therapy for biofilm inhibition is depicted in Fig. 15.

9.2 Bioactives to fight against biofilm and antimicrobial resistance

Natural resources are extensively used in ethnomedicine around the world from ancient times. Plants, as a part of their defense mechanisms, have limitless ability to synthesize substances that could fight against microorganisms. Several studies have shown that naturally occurring phytochemicals in the human diet and health care possess tremendous pharmacological potential

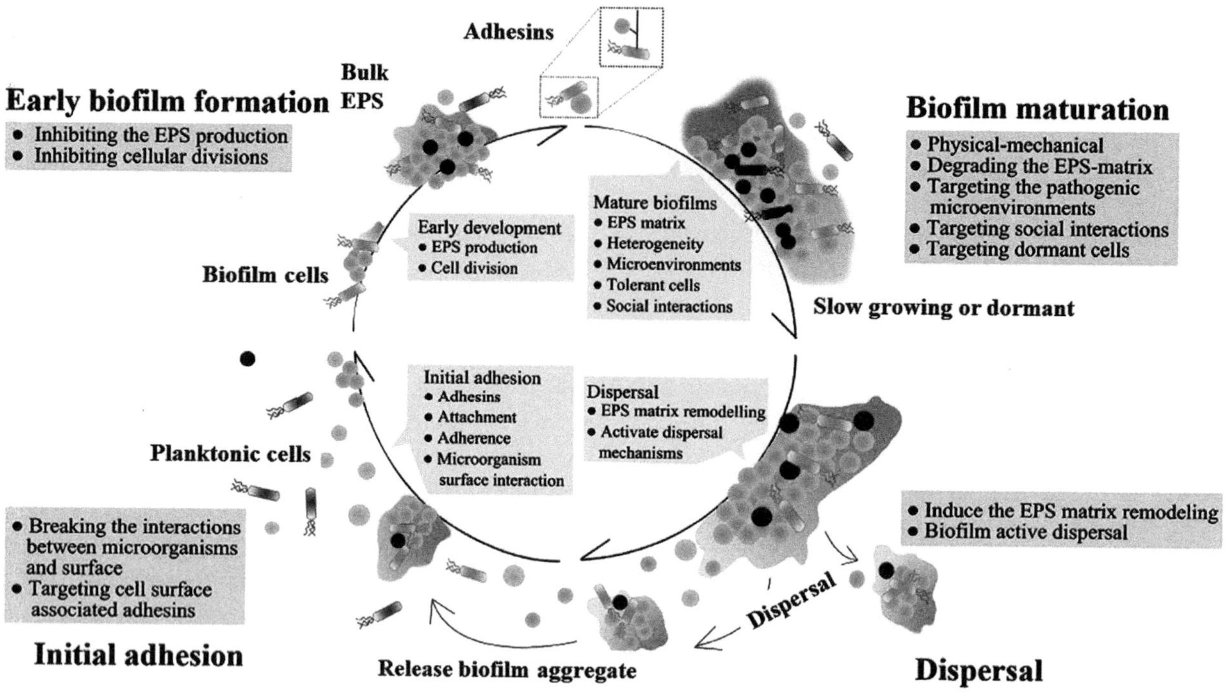

FIG. 15 Therapeutic interventions at stages of biofilm formation and development. *(Reproduced with permission from Jiang, Y., Geng, M., & Bai, L. (2020). Targeting biofilms therapy: Current research strategies and development hurdles.* Microorganisms, *8(8), 1222.)*

(Atanasov et al., 2015; Beutler, 2009). Besides plants, natural products from microbial sources have also been demonstrated to have therapeutic value in human health (Challinor & Bode, 2015). In addition to exerting direct antimicrobial efficacy, certain natural molecules have the prospect of synergizing either with antibiotics or with other natural bioactive molecules (Nitsch-Velasquez, 2020). Considering these facts, it could be stated that bioactives from natural sources can be effectively used for treatment of human infectious diseases.

9.3 Combinatorial drug therapy

Combinatorial drug therapy involves blending of more than one drug candidate at a specified ratio so as to attain improved or synergistic effect, wherein the activity of combination surpasses activity of individual drug candidate. Further, it also confers benefits such as less toxicity, less probability of drug resistance, and amplified antimicrobial spectrum and effectiveness (Chanda & Rakholiya, 2011). Success of combinatorial drug therapy relies on the right choice of drug candidates to be included in combination (Spitzer, Robbins, & Wright, 2017). With these benefits, combinatorial drug therapy becomes a part of emerging alternative strategies to combat microbial infections.

Combination therapy is mostly used in clinical practices because of the following reasons (Ahmed, Azim, Gurjar, & Baronia, 2014; Tamma, Cosgrove, & Maragakis, 2012):

- **Broad spectrum of activity**—Among the antimicrobial agents, at least one antimicrobial agent will sufficiently inhibit the pathogen(s) from causing infection(s).
- **Synergy—**A major advantage of addition of second antimicrobial agent is the synergistic effect, i.e., rapid killing of the pathogen (for example, combination of beta-lactams and aminoglycoside).
- **Polymicrobial infections**—Intra-abdominal infections contain two or more pathogens, which often require combination therapy.
- **Prevention of resistance—**Compared with single drug, chances of frequency of drug resistance against two drugs are very low.

The study of interaction between drugs has a long history; checkerboard method using fractional inhibitory concentration index (FICI) is used to measure the interaction between drugs, and FICI is calculated by the following equation:

$$\mathbf{FICI} = \mathrm{FIC_A} + \mathrm{FIC_B}$$

$$\mathbf{FIC_A} = \mathrm{MIC_A\ in\ combination/MIC_A\ tested\ alone}$$

$$\mathbf{FIC_B} = \mathrm{MIC_B\ in\ combination/MIC_B\ tested\ alone}$$

Previously, FICI <1.0 and FICI >1.0 are considered as synergy and antagonism, respectively. Later, in 2003, Odds defined synergy (FICI ≤ 0.5), antagonism (FICI >4.0), and "no interaction" (FICI >0.5–4.0). Additionally, intermediate categories such as "additive" (0.5–2.0), "subadditive," and "indifferent" are used to describe drug interaction (Mukherjee, Sheehan, Hitchcock, & Ghannoum, 2005).

9.4 Antimicrobial peptides (AMPs)

AMPs are evolutionarily conserved molecules found in all the organisms ranging from prokaryotes to humans. Lysozyme was the first described human antimicrobial protein, which was identified by Alexander Fleming in 1922 from nasal mucus. Followed by the discovery of lysozyme, Fleming identified the first antibiotic penicillin in the year 1928, and subsequently in 1940s, penicillin was exposed for its therapeutic use (Fleming, 1922; Johnson & Larson, 2005). Hence, the discovery of penicillin and the commencement of the "Golden Age of antibiotics" in 1940s overshadowed the effectuality and therapeutic potential of naturally occurring AMPs such as lysozyme. The emergence of multidrug resistance among various microbial pathogens prompted the search for alternate to antibiotics that have shed light on AMP research (Bodanszky, Izdebski, & Muramatsu, 1969; Meyers et al., 1970; Shemyakin et al., 1965; Tomasz, Albino, & Zanati, 1970).

AMPs have been identified in all the species that have been explored for this activity. Hans Boman in 1982 discovered the first major α-helical AMP, "Cecropin" from the Cecropia pupae (Hultmark, Engström, Bennich, Kapur, & Boman, 1982). The first cationic AMP to be characterized was "magainins" from the African clawed frog *Xenopus laevis* (Zasloff, 1987). Subsequently, numerous AMPs have been characterized from various sources. Presently, more than 3000 AMPs with known function from six kingdoms (338 bacteriocins/peptide antibiotics from bacteria, 5 from archaea, 8 from protists, 20 from fungi, 347 from plants, and 2278 from animals, including some synthetic peptides) are deposited in the Antimicrobial Peptide Database (APD) with the wide range of activities including antibacterial, antibiofilm, antifungal, antiviral, anticancer, insecticidal, antitoxin, etc. (http://aps.unmc.edu/AP/main.php).

9.4.1 Ideal properties of AMPs

Potential AMPs that are targeted to control or prevent the biofilm lifestyle should possess the following properties (Batoni, Maisetta, & Esin, 2016)

Should display rapid killing ability

- Should act in different microenvironments and target slow-growing or even nongrowing cells (persister cells)
- Must possess the ability to penetrate the extracellular matrix and/or interfere with its production
- Must interfere with bacterial cell communication machinery (quorum sensing system)
- Ability to modulate host response to biofilm
- Proficiency to synergize with other conventional and unconventional antimicrobial compounds

9.4.2 AMPs against cariogenic pathogen S. mutans

Few antibacterial peptides have been extensively studied for oral applications, and an even smaller subset of these has been taken for in vivo examination. Among them, few AMPs have been included as ingredients in commercially available oral care products, which are used to target and reduce oral microflora. Tong et al., in 2014, investigated the effect of free amino acids either alone or in combination with nisin on the biofilm and planktonic cells of major dental plaque forming pathogen *S. mutans* and reported the enhanced antimicrobial activity of nisin in combination with amino acids.

However, similar to chemical antimicrobial agents such as chlorohexidine, the active ingredients affect the plaque microflora by nonspecifically attacking either by enzymatic processes or by modulating the host innate immune response, and as a result, they provide no advantage for targeting specific pathogens. To address this issue, few groups have developed a new class of pathogen-selective molecules, called specifically (or selectively) targeted antimicrobial peptides (STAMPs). STAMPs are developed based on fusion of species- specific targeting peptide domain with an antibacterial domain. STAMPs have increased killing potency, selectivity, and kinetics against targeted bacteria (Eckert et al., 2006). Several studies reported the STAMP activity against the selective killing of cariogenic pathogen *S. mutans*, to leave rest of the oral microflora undisturbed. Eckert et al., in 2006, utilized the competence stimulating peptide (CSP), a pheromone produced by *S. mutans* as a STAMP targeting domain and succeeded in eliminating the *S. mutans* biofilm. Same group in 2010 went a step ahead to identify a potent and selective STAMP molecule against *S. mutans* via a "building block" approach, where they have used a combinatorial approach by utilizing number of AMPs, targeting and linker regions (He et al., 2010). Sullivan et al., in 2011, reported the in vitro efficacy of C16G2 STAMP oral rinse for specific elimination of *S. mutans* and found that the rinse solution is associated with reductions in plaque and salivary *S. mutans*, lactic acid production, and enamel demineralization. Li et al. (2010) reported a targeted antimicrobial therapy against *S. mutans*, which in addition to targeted killing presented a protective effect against subsequent infections. Mai et al., in 2011, reported a STAMP with the fusion of CSP as a targeting domain linked to killing domain containing a marine-derived, broad-spectrum AMP pleurocidin. Ahn, Kim, Kum, Yun, and Han (2017) investigated the effect of synthetic human beta-defensin-3 C15 peptide (HBD3-C15) alone and in combination with dental disinfectants calcium hydroxide and chlorhexidine and suggested HBD3-C15 as a potential alternate or additive disinfectant for the treatment of oral infectious diseases, including dental caries and endodontic infections.

9.4.3 Advantages of AMPs over antibiotics

Antimicrobial peptides possess several advantages over antibiotics, namely (Lazarev & Govorun, 2010):

- Limited ability for the development of resistance against AMP due to the ideal mechanism of antimicrobial action.
- Local application of the drug.
- Broad spectrum of antimicrobial action.
- Activity can be achieved in the nanomolar concentration range.
- Unlimited possibilities of chemical synthesis of AMP analogs with modified biological properties including decreased toxicity for the eukaryotic cell.
- Side effects during the use of AMPs are much lower than in the case of antibiotics.

10. Conclusion and prospects

Oral diseases are considered as one of the most predominant microbial infectious diseases among human population. Oral diseases have been reported to have impact on several systemic infections. Biofilm formation and quorum sensing are imparted as primary etiopathogenic mechanisms responsible for these infectious oral diseases. Thus, maintaining a proper oral health will provide a balance in the systemic health. Among the various bacterial and fungal infections of the oral cavity, dental caries and oral candidiasis are the most prevalent ones. As these infections are biofilm-associated, current therapeutic interventions could not provide complete cure. Also, the prevention strategies for oral infections are considerably more efficacious than the treatment. Though numerous oral care products are available in the market, which can possibly prevent these infections in early stage, usage of chemicals exerts certain adverse side effects. The existing commercial dentifrices either act as antibacterial or antiinflammatory agents, which can effectively act on the planktonic cells but fail to counteract the biofilm cells that are engrossed within polymeric matrix, and use of fluoride dentifrices can cause dental fluorosis. As biofilm is known to evade antimicrobial agents through increased resistance to antibiotics and host defense mechanisms, need for novel antiinfective agents is in surge. Natural products derived from various medicinal plants have proven to produce bioactive compounds. It is evident that phytochemicals and other plant-derived compounds exhibit possible preventive/therapeutic strategy against oral disease. In recent past, bioactive compounds from secondary metabolites of plants have proven to exhibit antibiofilm potential against oral pathogens. Thus, incorporation of alternative therapeutic strategies such as use of natural bioactives, antibiofilm, and combinatorial therapy could help in identification of potential antiinfective agents that could ably assist and/or replace the currently available prophylactic and medication regimens.

References

Aas, J. A., Paster, B. J., Stokes, L. N., Olsen, I., & Dewhirst, F. E. (2005). Defining the normal bacterial flora of the oral cavity. *Journal of Clinical Microbiology, 43*(11), 5721–5732.

Acs, G., Lodolini, G., Kaminsky, S., & Cisneros, G. J. (1992). Effect of nursing caries on body weight in a pediatric population. *Pediatric Dentistry, 14*(5), 303–305.

Ahmed, A., Azim, A., Gurjar, M., & Baronia, A. K. (2014). Current concepts in combination antibiotic therapy for critically ill patients. *Indian Journal of Critical Care Medicine: Peer-Reviewed, Official Publication of Indian Society of Critical Care Medicine, 18*(5), 310.

Ahn, K. B., Kim, A. R., Kum, K. Y., Yun, C. H., & Han, S. H. (2017). The synthetic human beta-defensin-3 C15 peptide exhibits antimicrobial activity against *Streptococcus mutans*, both alone and in combination with dental disinfectants. *Journal of Microbiology, 55*, 830–836.

Ajdić, D., McShan, W. M., McLaughlin, R. E., Savić, G., Chang, J., Carson, M. B., et al. (2002). Genome sequence of *Streptococcus mutans* UA159, a cariogenic dental pathogen. *Proceedings of the National Academy of Sciences, 99*(22), 14434–14439.

Al-Fattani, M. A., & Douglas, L. J. (2006). Biofilm matrix of *Candida albicans* and *Candida tropicalis*: Chemical composition and role in drug resistance. *Journal of Medical Microbiology, 55*(8), 999–1008.

Anusavice, K. J., Zhang, N. Z., & Shen, C. (2005). Effect of CaF2 content on rate of fluoride release from filled resins. *Journal of Dental Research, 84*(5), 440–444.

Arweiler, N. B., & Netuschil, L. (2016). The oral microbiota. *Microbiota of the Human Body*, 45–60.

Atanasov, A. G., Waltenberger, B., Pferschy-Wenzig, E. M., Linder, T., Wawrosch, C., Uhrin, P., et al. (2015). Discovery and resupply of pharmacologically active plant-derived natural products: A review. *Biotechnology Advances, 33*(8), 1582–1614.

Banas, J. A., & Vickerman, M. M. (2003). Glucan-binding proteins of the oral streptococci. *Critical Reviews in Oral Biology & Medicine, 14*(2), 89–99.

Barbeau, J., Séguin, J., Goulet, J. P., de Koninck, L., Avon, S. L., Lalonde, B., et al. (2003). Reassessing the presence of *Candida albicans* in denture-related stomatitis. *Oral Surgery, Oral Medicine, Oral Pathology, Oral Radiology, and Endodontology, 95*(1), 51–59.

Barbieri, D. D. S. A. V., Vicente, V. A., Fraiz, F. C., Lavoranti, O. J., Svidzinski, T. I. E., & Pinheiro, R. L. (2007). Analysis of the *in vitro* adherence of *Streptococcus mutans* and *Candida albicans*. *Brazilian Journal of Microbiology, 38*(4), 624–631.

Barrett, J. F., & Hoch, J. A. (1998). Two-component signal transduction as a target for microbial anti-infective therapy. *Antimicrobial Agents and Chemotherapy, 42*(7), 1529–1536.

Bassler, B. L. (1999). How bacteria talk to each other: Regulation of gene expression by quorum sensing. *Current Opinion in Microbiology, 2*(6), 582–587.

Batoni, G., Maisetta, G., & Esin, S. (2016). Antimicrobial peptides and their interaction with biofilms of medically relevant bacteria. *Biochimica et Biophysica Acta (BBA)-biomembranes, 1858*, 1044–1060.

Beier, D., & Gross, R. (2006). Regulation of bacterial virulence by two-component systems. *Current Opinion in Microbiology, 9*(2), 143–152.

Belibasakis, G. N., Guggenheim, B., & Bostanci, N. (2013). Down-regulation of NLRP3 inflammasome in gingival fibroblasts by subgingival biofilms: Involvement of *Porphyromonas gingivalis*. *Innate Immunity*, *19*(1), 3–9.

Benn, A. M. L., Heng, N. C. K., Broadbent, J. M., & Thomson, W. M. (2018). Studying the human oral microbiome: Challenges and the evolution of solutions. *Australian Dental Journal*, *63*(1), 14–24.

Beutler, J. A. (2009). Natural products as a foundation for drug discovery. *Current Protocols in Pharmacology*, *46*(1), 9–11.

Bodanszky, M., Izdebski, J., & Muramatsu, I. (1969). Structure of the peptide antibiotic stendomycin. *Journal of the American Chemical Society*, *91*, 2351–2358.

Bombarda, G. F., Rosalen, P. L., Paganini, E. R., Garcia, M. A., Silva, D. R., Lazarini, J. G., et al. (2019). Bioactive molecule optimized for biofilm reduction related to childhood caries. *Future Microbiology*, *14*(14), 1207–1220.

Bowen, W. H., & Koo, H. J. C. R. (2011). Biology of *Streptococcus mutans*-derived glucosyltransferases: Role in extracellular matrix formation of cariogenic biofilms. *Caries Research*, *45*(1), 69–86.

Brady, L. J., Maddocks, S. E., Larson, M. R., Forsgren, N., Persson, K., & Deivanayagam, C. C. (2010). The changing faces of *Streptococcus* antigen I/II polypeptide family adhesins. *Molecular Microbiology*, *77*, 27686.

Bramono, K., Yamazaki, M., Tsuboi, R., & Ogawa, H. (2006). Comparison of proteinase, lipase and alpha-glucosidase activities from the clinical isolates of *Candida* species. *Japanese Journal of Infectious Diseases*, *59*(2), 73–76.

Branting, C., Sund, M. L., & Linder, L. E. (1989). The influence of *Streptococcus mutans* on adhesion of *Candida albicans* to acrylic surfaces *in vitro*. *Archives of Oral Biology*, *34*(5), 347–353.

Brown, L. R., White, J. O., Horton, I. M., Dreizen, S., & Streckfuss, J. L. (1983). Effect of continuous fluoride gel use on plaque fluoride retention and microbial activity. *Journal of Dental Research*, *62*(6), 746–751.

Burne, R. A. (1998). Oral streptococci products of their environment. *Journal of Dental Research*, *77*, 445–452.

Buzalaf, M. A. R., Pessan, J. P., Honório, H. M., & Ten Cate, J. M. (2011). Mechanisms of action of fluoride for caries control. *Fluoride and the Oral Environment*, *22*, 97–114.

Calderone, R. A., & Fonzi, W. A. (2001). Virulence factors of *Candida albicans*. *Trends in Microbiology*, *9*(7), 327–335.

Cannon, R. D., Lamping, E., Holmes, A. R., Niimi, K., Tanabe, K., Niimi, M., et al. (2007). *Candida albicans* drug resistance—Another way to cope with stress. *Microbiology*, *153*(10), 3211–3217.

Cavalheiro, M., & Teixeira, M. C. (2018). *Candida* biofilms: Threats, challenges, and promising strategies. *Frontiers in Medicine*, *5*, 28.

Chaffin, W. L. (2008). *Candida albicans* cell wall proteins. *Microbiology and Molecular Biology Reviews*, *72*(3), 495–544.

Challacombe, S. J. (1994). Immunologic aspects of oral candidiasis. *Oral Surgery, Oral Medicine, Oral Pathology*, *78*(2), 202–210.

Challinor, V. L., & Bode, H. B. (2015). Bioactive natural products from novel microbial sources. *Annals of the New York Academy of Sciences*, *1354*(1), 82–97.

Chanda, S., & Rakholiya, K. (2011). Combination therapy: Synergism between natural plant extracts and antibiotics against infectious diseases. In *1. Microbiol book series* (pp. 520–529).

Chandra, J., Kuhn, D. M., Mukherjee, P. K., Hoyer, L. L., McCormick, T., & Ghannoum, M. A. (2001). Biofilm formation by the fungal pathogen *Candida albicans*: Development, architecture, and drug resistance. *Journal of Bacteriology*, *183*(18), 5385–5394.

Clarke, J. K. (1924). On the bacterial factor in the aetiology of dental caries. *British Journal of Experimental Pathology*, *5*(3), 141.

Clarke, M., Locker, D., Berall, G., Pencharz, P., Kenny, D. J., & Judd, P. (2006). Malnourishment in a population of young children with severe early childhood caries. *Pediatric Dentistry*, *28*(3), 254–259.

Cullen, P. J., & Sprague, G. F., Jr. (2012). The regulation of filamentous growth in yeast. *Genetics*, *190*(1), 23–49.

Czaplewski, L., Bax, R., Clokie, M., Dawson, M., Fairhead, H., Fischetti, V. A., et al. (2016). Alternatives to antibiotics—A pipeline portfolio review. *The Lancet Infectious Diseases*, *16*(2), 239–251.

de Carvalho, F. G., Silva, D. S., Hebling, J., Spolidorio, L. C., & Spolidorio, D. M. P. (2006). Presence of mutans streptococci and *Candida* spp. in dental plaque/dentine of carious teeth and early childhood caries. *Archives of Oral Biology*, *51*(11), 1024–1028.

de Cássia Negrini, T., Koo, H., & Arthur, R. A. (2019). *Candida*–bacterial biofilms and host–microbe interactions in oral diseases. *Oral Mucosal Immunity and Microbiome*, 119–141.

Deo, P. N., & Deshmukh, R. (2019). Oral microbiome: Unveiling the fundamentals. *Journal of Oral and Maxillofacial Pathology*, *23*(1), 122.

Dewhirst, F. E., Chen, T., Izard, J., Paster, B. J., Tanner, A. C., Yu, W. H., et al. (2010). The human oral microbiome. *Journal of Bacteriology*, *192*(19), 5002–5017.

Di Stasio, D., Lauritano, D., Minervini, G., Paparella, R. S., Petruzzi, M., Romano, A., et al. (2018). Management of denture stomatitis: A narrative review. *Journal of Biological Regulators and Homeostatic Agents*, *32*(2 Suppl. 1), 113–116.

Diamond, G., Beckloff, N., & Ryan, L. K. (2008). Host defense peptides in the oral cavity and the lung: Similarities and differences. *Journal of Dental Research*, *87*(10), 915–927.

Douglass, J. M., & Clark, M. B. (2015). Integrating oral health into overall health care to prevent early childhood caries: Need, evidence, and solutions. *Pediatric Dentistry*, *37*(3), 266–274.

Eckert, R., Qi, F., Yarbrough, D. K., He, J., Anderson, M. H., & Shi, W. (2006). Adding selectivity to antimicrobial peptides: Rational design of a multidomain peptide against *Pseudomonas* spp. *Antimicrobial Agents and Chemotherapy*, *50*, 1480–1488.

Ellepola, K., Liu, Y., Cao, T., Koo, H., & Seneviratne, C. J. (2017). Bacterial GtfB augments *Candida albicans* accumulation in cross-kingdom biofilms. *Journal of Dental Research*, *96*(10), 1129–1135.

Falsetta, M. L., Klein, M. I., Colonne, P. M., Scott-Anne, K., Gregoire, S., Pai, C. H., et al. (2014). Symbiotic relationship between *Streptococcus mutans* and *Candida albicans* synergizes virulence of plaque biofilms *in vivo*. *Infection and Immunity*, *82*(5), 1968–1981.

Feldman, M., Ginsburg, I., Al-Quntar, A., & Steinberg, D. (2016). Thiazolidinedione-8 alters symbiotic relationship in *C. albicans-S. mutans* dual species biofilm. *Frontiers in Microbiology*, *7*, 140.

Fitzgerald, R. J., & Keyes, P. H. (1960). Demonstration of the etiologic role of streptococci in experimental caries in the hamster. *The Journal of the American Dental Association*, *61*(1), 9–19.

Fleming, A. (1922). On a remarkable bacteriolytic element found in tissues and secretions. *Proceedings of the Royal Society of London. Series B, Containing Papers of a Biological Character*, *93*, 306–317.

Fux, C. A., Costerton, J. W., Stewart, P. S., & Stoodley, P. (2005). Survival strategies of infectious biofilms. *Trends in Microbiology*, *13*(1), 34–40.

Gendreau, L., & Loewy, Z. G. (2011). Epidemiology and etiology of denture stomatitis. *Journal of Prosthodontics: Implant, Esthetic and Reconstructive Dentistry*, *20*(4), 251–260.

Ghosh, C., Sarkar, P., Issa, R., & Haldar, J. (2019). Alternatives to conventional antibiotics in the era of antimicrobial resistance. *Trends in Microbiology*, *27*(4), 323–338.

González, J. F., Hahn, M. M., & Gunn, J. S. (2018). Chronic biofilm-based infections: Skewing of the immune response. *Pathogens and Disease, 76*(3), fty023.

Gregoire, S., Xiao, J., Silva, B. B., Gonzalez, I., Agidi, P. S., Klein, M. I., et al. (2011). Role of glucosyltransferase B in interactions of *Candida albicans* with *Streptococcus mutans* and with an experimental pellicle on hydroxyapatite surfaces. *Applied and Environmental Microbiology, 77*(18), 6357–6367.

Gross, E. L., Beall, C. J., Kutsch, S. R., Firestone, N. D., Leys, E. J., & Griffen, A. L. (2012). Beyond *Streptococcus mutans*: Dental caries onset linked to multiple species by 16S rRNA community analysis. *PLoS One, 7*(10), e47722.

Hall, R. A., & Noverr, M. C. (2017). Fungal interactions with the human host: Exploring the spectrum of symbiosis. *Current Opinion in Microbiology, 40*, 58–64.

Hamilton, I. R., & Ellwood, D. C. (1978). Effects of fluoride on carbohydrate metabolism by washed cells of *Streptococcus mutans* grown at various pH values in a chemostat. *Infection and Immunity, 19*(2), 434–442.

Harris, R., Nicoll, A. D., Adair, P. M., & Pine, C. M. (2004). Risk factors for dental caries in young children: A systematic review of the literature. *Community Dental Health, 21*(1), 71–85.

He, J., Yarbrough, D. K., Kreth, J., Anderson, M. H., Shi, W., & Eckert, R. (2010). Systematic approach to optimizing specifically targeted antimicrobial peptides against *Streptococcus mutans*. *Antimicrobial Agents and Chemotherapy, 54*, 2143–2151.

Hiratsuka, K., Wang, B., Sato, Y., & Kuramitsu, H. (1998). Regulation of sucrose-6-phosphate hydrolase activity in *Streptococcus mutans*: Characterization of the scrR gene. *Infection and Immunity, 66*, 373643.

Hultmark, D., Engström, A., Bennich, H., Kapur, R., & Boman, H. G. (1982). Insect immunity: Isolation and structure of cecropin D and four minor antibacterial components from Cecropia pupae. *European Journal of Biochemistry, 127*, 207–217.

Hwang, G., Liu, Y., Kim, D., Li, Y., Krysan, D. J., & Koo, H. (2017). *Candida albicans* mannans mediate *Streptococcus mutans* exoenzyme GtfB binding to modulate cross-kingdom biofilm development *in vivo*. *PLoS Pathogens, 13*(6), e1006407.

Hwang, G., Marsh, G., Gao, L., Waugh, R., & Koo, H. (2015). Binding force dynamics of *Streptococcus mutans*–glucosyltransferase B to *Candida albicans*. *Journal of Dental Research, 94*(9), 1310–1317.

Iliev, I. D., & Leonardi, I. (2017). Fungal dysbiosis: Immunity and interactions at mucosal barriers. *Nature Reviews Immunology, 17*(10), 635–646.

Ingham, C. J., Boonstra, S., Levels, S., De Lange, M., Meis, J. F., & Schneeberger, P. M. (2012). Rapid susceptibility testing and microcolony analysis of *Candida* spp. cultured and imaged on porous aluminum oxide. *PLoS One, 7*(3), e33818.

Jabra-Rizk, M. A., Kong, E. F., Tsui, C., Nguyen, M. H., Clancy, C. J., Fidel, P. L., et al. (2016). *Candida albicans* pathogenesis: Fitting within the host-microbe damage response framework. *Infection and Immunity, 84*(10), 2724–2739.

Jayatilake, J. A. M. S., Samaranayake, Y. H., Cheung, L. K., & Samaranayake, L. P. (2006). Quantitative evaluation of tissue invasion by wild type, hyphal and SAP mutants of *Candida albicans*, and non-albicans *Candida* species in reconstituted human oral epithelium. *Journal of Oral Pathology & Medicine, 35*(8), 484–491.

Jenkinson, H. F., & Demuth, D. R. (1997). Structure, function and immunogenicity of streptococcal antigen I/II polypeptides. *Molecular Microbiology, 23*, 18390.

Jenkinson, H. F., & Lamont, R. J. (2005). Oral microbial communities in sickness and in health. *Trends in Microbiology, 13*(12), 589–595.

Johnson, E. A., & Larson, A. E. (2005). Lysozyme. In *Antimicrobials in food* (pp. 369–395). CRC Press.

Joyner, P. M., Liu, J., Zhang, Z., Merritt, J., Qi, F., & Cichewicz, R. H. (2010). Mutanobactin A from the human oral pathogen *Streptococcus mutans* is a cross-kingdom regulator of the yeast-mycelium transition. *Organic & Biomolecular Chemistry, 8*(24), 5486–5489.

Kagihara, L. E., Niederhauser, V. P., & Stark, M. (2009). Assessment, management, and prevention of early childhood caries. *Journal of the American Academy of Nurse Practitioners, 21*(1), 1–10.

Kawashita, Y., Kitamura, M., & Saito, T. (2011). Early childhood caries. *International Journal of Dentistry, 2011*.

Kear, B. P. (2001). Dental caries in an Early Cretaceous ichthyosaur. *Alcheringa, 25*(4), 387–390.

Kemp, A. (2003). Dental and skeletal pathology in lungfish jaws and tooth plates. *Alcheringa, 27*(2), 155–170.

Kidd, E. A., & Fejerskov, O. (2003). *Dental caries: The disease and its clinical management* (pp. 245–250). Blackwell Munksgaard.

Kleinberg, I. (2002). A mixed-bacteria ecological approach to understanding the role of the oral bacteria in dental caries causation: An alternative to *Streptococcus mutans* and the specific-plaque hypothesis. *Critical Reviews in Oral Biology & Medicine, 13*(2), 108–125.

Klinke, T., Guggenheim, B., Klimm, W., & Thurnheer, T. (2011). Dental caries in rats associated with *Candida albicans*. *Caries Research, 45* (2), 100–106.

Krzyściak, W., Jurczak, A., Kościelniak, D., Bystrowska, B., & Skalniak, A. (2014). The virulence of *Streptococcus mutans* and the ability to form biofilms. *European Journal of Clinical Microbiology & Infectious Diseases, 33*(4), 499–515.

Kulak, Y., Arikan, A., & Kazazoglu, E. (1997). Existence of *Candida albicans* and microorganisms in denture stomatitis patients. *Journal of Oral Rehabilitation, 24*(10), 788–790.

Kumamoto, C. A., & Vinces, M. D. (2005). Contributions of hyphae and hypha-co-regulated genes to *Candida albicans* virulence. *Cellular Microbiology, 7*(11), 1546–1554.

Kuramitsu, H. K. (2003). Molecular genetic analysis of the virulence of oral bacterial pathogens: An historical perspective. *Critical Reviews in Oral Biology and Medicine, 14*, 331–344.

Lamont, R. J., Koo, H., & Hajishengallis, G. (2018). The oral microbiota: Dynamic communities and host interactions. *Nature Reviews Microbiology, 16*(12), 745–759.

Lazarev, V. N., & Govorun, V. M. (2010). Antimicrobial peptides and their use in medicine. *Applied Biochemistry and Microbiology, 46*, 803–814.

Lemos, J. A., Abranches, J., & Burne, R. A. (2005). Responses of cariogenic streptococci to environmental stresses. *Current Issues in Molecular Biology, 7*(1), 95–108.

Lemos, J. A., & Burne, R. A. (2008). A model of efficiency: Stress tolerance by *Streptococcus mutans*. *Microbiology (Reading, England), 154*(Pt 11), 3247–3255.

Lemos, J. A., Palmer, S. R., Zeng, L., Wen, Z. T., Kajfasz, J. K., Freires, I. A., et al. (2019). The biology of *Streptococcus mutans*. In *Gram-positive pathogens* (pp. 435–448). ASM.

Lemos, J. A., Quivey, R. G., Jr., Koo, H., & Abranches, J. (2013). *Streptococcus mutans*: A new Gram-positive paradigm? *Microbiology, 159* (Pt 3), 436–445.

Li, L. N., Guo, L. H., Lux, R., Eckert, R., Yarbrough, D., He, J., et al. (2010). Targeted antimicrobial therapy against *streptococcus mutans* establishes protective non-cariogenic oral biofilms and reduces subsequent infection. *International Journal of Oral Science*, *2*, 66.

Li, Y. H., Lau, P. C., Lee, J. H., Ellen, R. P., & Cvitkovitch, D. G. (2001). Natural genetic transformation of *Streptococcus mutans* growing in biofilms. *Journal of Bacteriology*, *183*(3), 897–908.

Li, X., Yan, Z., & Xu, J. (2003). Quantitative variation of biofilms among strains in natural populations of *Candida albicans*. *Microbiology*, *149* (2), 353–362.

Li, X., Yin, L., Ramage, G., Li, B., Tao, Y., Zhi, Q., et al. (2019). Assessing the impact of curcumin on dual-species biofilms formed by *Streptococcus mutans* and *Candida albicans*. *Microbiology Open*, *8*(12), e937.

Liao, Y., Brandt, B. W., Li, J., Crielaard, W., Van Loveren, C., & Deng, D. M. (2017). Fluoride resistance in *Streptococcus mutans*: A mini review. *Journal of Oral Microbiology*, *9*(1), 1344509.

Liao, Y., Chen, J., Brandt, B. W., Zhu, Y., Li, J., van Loveren, C., et al. (2015). Identification and functional analysis of genome mutations in a fluoride-resistant *Streptococcus mutans* strain. *PLoS One*, *10*(4), e0122630.

Listgarten, M. A. (1976). Structure of the microbial flora associated with periodontal health and disease in man: A light and electron microscopic study. *Journal of Periodontology*, *47*(1), 1–18.

Listgarten, M. A., Mayo, H. E., & Tremblay, R. (1975). Development of dental plaque on epoxy resin crowns in man: A light and electron microscopic study. *Journal of Periodontology*, *46*(1), 10–26.

Lu, Y., Su, C., Wang, A., & Liu, H. (2011). Hyphal development in *Candida albicans* requires two temporally linked changes in promoter chromatin for initiation and maintenance. *PLoS Biology*, *9*(7), e1001105.

Mager, D. L., Ximenez-Fyvie, L. A., Haffajee, A. D., & Socransky, S. S. (2003). Distribution of selected bacterial species on intraoral surfaces. *Journal of Clinical Periodontology*, *30*(7), 644–654.

Mah, T. F. (2012). Biofilm-specific antibiotic resistance. *Future Microbiology*, *7*(9), 1061–1072.

Marak, M. B., & Dhanashree, B. (2018). Antifungal susceptibility and biofilm production of *Candida* spp. isolated from clinical samples. *International Journal of Microbiology*, *2018*.

Marsh, P. D. (2000). Oral ecology and its impact on oral microbial diversity. In H. K. Kuramitsu, & R. P. Ellen (Eds.), *Oral bacterial ecology: The molecular basis* (pp. 11–65). Wymondham: Horizon Scientific Press.

Marsh, P. D. (2006). Dental plaque as a biofilm and a microbial community—Implications for health and disease. *BMC Oral Health*, *6* (1), 1–7.

Marsh, P. D., Martin, M. V., Lewis, M. A., & Williams, D. (2009). *Oral microbiology e-book*. Elsevier Health Sciences.

Matee, M. I. N., Mikx, F. H. M., Maselle, S. Y. M., & Helderman, W. V. P. (1992). Mutans streptococci and lactobacilli in breast-fed children with rampant caries. *Caries Research*, *26*(3), 183–187.

Matsui, R., & Cvitkovitch, D. (2010). Acid tolerance mechanisms utilized by *Streptococcus mutans*. *Future Microbiology*, *5*, 40317.

Mayer, F. L., Wilson, D., & Hube, B. (2013). *Candida albicans* pathogenicity mechanisms. *Virulence*, *4*(2), 119–128.

McLean, J. S. (2014). Advancements toward a systems level understanding of the human oral microbiome. *Frontiers in Cellular and Infection Microbiology*, *4*, 98.

Metwalli, K. H., Khan, S. A., Krom, B. P., & Jabra-Rizk, M. A. (2013). *Streptococcus mutans*, *Candida albicans*, and the human mouth: A sticky situation. *PLoS Pathogens*, *9*(10), e1003616.

Meurman, J. H. (1988). Ultrastructure, growth, and adherence of *Streptococcus mutans* after treatment with chlorhexidine and fluoride. *Caries Research*, *22*(5), 283–287.

Meyer, F., & Enax, J. (2018). Early childhood caries: Epidemiology, aetiology, and prevention. *International Journal of Dentistry*, *2018*.

Meyers, E., Weisenborn, F. L., Pansy, F. E., Slusarchyk, D. S., von Saltza, M. H., Rathnum, M. L., et al. (1970). Janiemycin, a new peptide antibiotic. *The Journal of Antibiotics*, *23*, 502–507.

Miranda-Cadena, K., Marcos-Arias, C., Mateo, E., Aguirre, J. M., Quindós, G., & Eraso, E. (2018). Prevalence and antifungal susceptibility profiles of *Candida glabrata*, *Candida parapsilosis* and their close-related species in oral candidiasis. *Archives of Oral Biology*, *95*, 100–107.

Mukherjee, P. K., Sheehan, D. J., Hitchcock, C. A., & Ghannoum, M. A. (2005). Combination treatment of invasive fungal infections. *Clinical Microbiology Reviews*, *18*(1), 163–194.

Narvey, A., & Shwart, L. (2007). Early childhood dental disease—What's in a name? *Journal of the Canadian Dental Association*, *73*(10), 929–930.

Newbrun, E. (1978). *Cariology*. Baltimore: Williams and Wilkins.

Nikkawa, H., Yamashiro, H., Makihira, S., Nishimura, M., Egusa, H., Furukawa, M., et al. (2003). *In vitro* cariogenic potential of *Candida albicans*. *Mycoses*, *46*(11–12), 471–478.

Nitsch-Velasquez, L. (2020). Synergistic effects of natural products and commercial antibiotics. *medRxiv*.

Nobbs, A. H., Lamont, R. J., & Jenkinson, H. F. (2009). *Streptococcus* adherence and colonization. *Microbiology and Molecular Biology Reviews*, *73*, 40750.

Odds, F. C. (1988). *Candida and candidosis* (2nd ed., p. 309). Philadelphia: Baillière Tindall.

Patel, R. (2005). Biofilms and antimicrobial resistance. *Clinical Orthopaedics and Related Research*, *437*, 41–47.

Pereira-Cenci, T., Deng, D. M., Kraneveld, E. A., Manders, E. M. M., Cury, A. A. D. B., Ten Cate, J. M., et al. (2008). The effect of *Streptococcus mutans* and *Candida glabrata* on *Candida albicans* biofilms formed on different surfaces. *Archives of Oral Biology*, *53*(8), 755–764.

Petersen, P. E., Bourgeois, D., Ogawa, H., Estupinan-Day, S., & Ndiaye, C. (2005). The global burden of oral diseases and risks to oral health. *Bulletin of the World Health Organization*, *83*, 661–669.

Philip, N., Leishman, S. J., Bandara, H., & Walsh, L. J. (2019). Polyphenol-rich cranberry extracts modulate virulence of *Streptococcus mutans*-*Candida albicans* biofilms implicated in the pathogenesis of early childhood caries. *Pediatric Dentistry*, *41*(1), 56–62.

Pontón, J., Bikandi, J., Moragues, M. D., Arilla, M. C., Elósegui, R., Quindós, G., et al. (1996). Reactivity of *Candida albicans* germ tubes with salivary secretory IgA. *Journal of Dental Research*, *75*(12), 1979–1985.

Qiu, R., Li, W., Lin, Y., Yu, D., & Zhao, W. (2015). Genotypic diversity and cariogenicity of *Candida albicans* from children with early childhood caries and caries-free children. *BMC Oral Health*, *15*(1), 1–6.

Quindós, G., Gil-Alonso, S., Marcos-Arias, C., Sevillano, E., Mateo, E., Jauregizar, N., et al. (2019). Therapeutic tools for oral candidiasis: Current and new antifungal drugs. *Medicina Oral, Patologia Oral y Cirugia Bucal*, *24*(2), e172–e180.

Quivey, R. G., Jr., Kuhnert, W. L., & Hahn, K. (2001). Genetics of acid adaptation in oral streptococci. *Critical Reviews in Oral Biology & Medicine*, *12*(4), 301–314.

Raja, M., Hannan, A., & Ali, K. (2010). Association of oral candidal carriage with dental caries in children. *Caries Research*, *44*(3), 272–276.

Ramage, G., Tomsett, K., Wickes, B. L., López-Ribot, J. L., & Redding, S. W. (2004). Denture stomatitis: A role for *Candida* biofilms. *Oral Surgery, Oral Medicine, Oral Pathology, Oral Radiology, and Endodontology*, *98*(1), 53–59.

Rodríguez-Martínez, J. M., & Pascual, A. (2006). Antimicrobial resistance in bacterial biofilms. *Reviews in Medical Microbiology*, *17*(3), 65–75.

Sardi, J. C. O., Scorzoni, L., Bernardi, T., Fusco-Almeida, A. M., & Giannini, M. M. (2013). *Candida* species: Current epidemiology, pathogenicity, biofilm formation, natural antifungal products and new therapeutic options. *Journal of Medical Microbiology*, *62*(1), 10–24.

Scannapieco, F. A. (1994). Saliva-bacterium interactions in oral microbial ecology. *Critical Reviews in Oral Biology and Medicine*, *5*, 20348.

Senadheera, M. D., Guggenheim, B., Spatafora, G. A., Huang, Y. C. C., Choi, J., Hung, D. C., et al. (2005). A VicRK signal transduction system in *Streptococcus mutans* affects gtfBCD, gbpB, and ftf expression, biofilm formation, and genetic competence development. *Journal of Bacteriology*, *187*(12), 4064–4076.

Shemesh, M., Tam, A., & Steinberg, D. (2007). Expression of biofilm-associated genes of *Streptococcus mutans* in response to glucose and sucrose. *Journal of Medical Microbiology*, *56*, 1528–1535.

Shemyakin, M. M., Vinogradova, E. I., Feigina, M. Y., Aldanova, N. A., Loginova, N. F., Ryabova, I. D., et al. (1965). The structure-antimicrobial relation for valinomycin depsipeptides. *Experientia*, *21*, 548–552.

Silva, S., Negri, M., Henriques, M., Oliveira, R., Williams, D. W., & Azeredo, J. (2012). *Candida glabrata*, *Candida parapsilosis* and *Candida tropicalis*: Biology, epidemiology, pathogenicity and antifungal resistance. *FEMS Microbiology Reviews*, *36*(2), 288–305.

Singh, S., Singh, S. K., Chowdhury, I., & Singh, R. (2017). Understanding the mechanism of bacterial biofilms resistance to antimicrobial agents. *The Open Microbiology Journal*, *11*, 53–62.

Socransky, S. S. (2002). Dental biofilms: Difficult therapeutic targets. *Periodontology 2000*, *28*, 12–55.

Spitzer, M., Robbins, N., & Wright, G. D. (2017). Combinatorial strategies for combating invasive fungal infections. *Virulence*, *8*(2), 169–185.

Starr, J. R., White, T. C., Leroux, B. G., Luis, H. S., Bernardo, M., Leitao, J., et al. (2002). Persistence of oral *Candida albicans* carriage in healthy Portuguese schoolchildren followed for 3 years. *Oral Microbiology and Immunology*, *17*(5), 304–310.

Stiles, H. M. (1976). Occurrence of *Streptococcus mutans* and *Streptococcus sanguis* in the oral cavity and feces of young children. *Microbial Aspects of Dental Caries*, 187–199.

Streckfuss, J. L., Perkins, D., Horton, I. M., Brown, L. R., Dreizen, S., & Graves, L. (1980). Fluoride resistance and adherence of selected strains of *Streptococcus mutans* to smooth surfaces after exposure to fluoride. *Journal of Dental Research*, *59*(2), 151–158.

Subramenium, G. A., Swetha, T. K., Iyer, P. M., Balamurugan, K., & Pandian, S. K. (2018). 5-hydroxymethyl-2-furaldehyde from marine bacterium *Bacillus subtilis* inhibits biofilm and virulence of *Candida albicans*. *Microbiological Research*, *207*, 19–32.

Tamma, P. D., Cosgrove, S. E., & Maragakis, L. L. (2012). Combination therapy for treatment of infections with gram-negative bacteria. *Clinical Microbiology Reviews*, *25*(3), 450–470.

Tomasz, A., Albino, A., & Zanati, E. V. E. (1970). Multiple antibiotic resistance in a bacterium with suppressed autolytic system. *Nature*, *227*, 138–140.

Urizar, J. M. A. (2002). Candidiasis orales. *Revista Iberoamericana de Micologı́a*, *19*, 17–21.

Van Acker, H., Van Dijck, P., & Coenye, T. (2014). Molecular mechanisms of antimicrobial tolerance and resistance in bacterial and fungal biofilms. *Trends in Microbiology*, *22*(6), 326–333.

Van Houte, J., Gibbs, G., & Butera, C. (1982). Oral flora of children with "nursing bottle caries". *Journal of Dental Research*, *61*(2), 382–385.

Van Loveren, C. (2001). Antimicrobial activity of fluoride and its *in vivo* importance: Identification of research questions. *Caries Research*, *35* (Suppl. 1), 65–70.

Verstrepen, K. J., & Klis, F. M. (2006). Flocculation, adhesion and biofilm formation in yeasts. *Molecular Microbiology*, *60*(1), 5–15.

Vila, T., Sultan, A. S., Montelongo-Jauregui, D., & Jabra-Rizk, M. A. (2020). Oral candidiasis: A disease of opportunity. *Journal of Fungi*, *6*(1), 15.

Viszwapriya, D., Subramenium, G. A., Radhika, S., & Pandian, S. K. (2017). Betulin inhibits cariogenic properties of *Streptococcus mutans* by targeting vicRK and gtf genes. *Antonie Van Leeuwenhoek*, *110*(1), 153–165.

Webb, B. C., Thomas, C. J., Willcox, M. D. P., Harty, D. W. S., & Knox, K. W. (1998). *Candida*-associated denture stomatitis. Aetiology and management: A review: Part1. Factors influencing distribution of *Candida* species in the oral cavity. *Australian Dental Journal*, *43*(1), 45–50.

Wei, G. X., Campagna, A. N., & Bobek, L. A. (2006). Effect of MUC7 peptides on the growth of bacteria and on *Streptococcus mutans* biofilm. *Journal of Antimicrobial Chemotherapy*, *57*(6), 1100–1109.

Wen, Z. T., & Burne, R. A. (2002). Functional genomics approach to identifying genes required for biofilm development by *Streptococcus mutans*. *Applied and Environmental Microbiology*, *68*, 1196–1203.

White, T. C., Marr, K. A., & Bowden, R. A. (1998). Clinical, cellular, and molecular factors that contribute to antifungal drug resistance. *Clinical Microbiology Reviews*, *11*(2), 382–402.

Wingender, J., & Flemming, H. C. (2010). The biofilm matrix. *Nature Reviews Microbiology*, *8*, 623–633.

Xiao, J., Huang, X., Alkhers, N., Alzamil, H., Alzoubi, S., Wu, T. T., et al. (2018). *Candida albicans* and early childhood caries: A systematic review and meta-analysis. *Caries Research*, *52*(1–2), 102–112.

Yadav, K., & Prakash, S. (2017). Dental caries: A microbiological approach. *Journal of Clinical Infectious Diseases & Practice*, *2*(1), 1–15.

Yang, C., Scoffield, J., Wu, R., Deivanayagam, C., Zou, J., & Wu, H. (2018). Antigen I/II mediates interactions between *Streptococcus mutans* and *Candida albicans*. *Molecular Oral Microbiology*, *33*(4), 283–291.

Yoshida, A., & Kuramitsu, H. K. (2002). Multiple *Streptococcus mutans* genes are involved in biofilm formation. *Applied and Environmental Microbiology*, *68*(12), 6283–6291.

Zasloff, M. (1987). Magainins, a class of antimicrobial peptides from *Xenopus* skin: Isolation, characterization of two active forms, and partial cDNA sequence of a precursor. *Proceedings of the National Academy of Sciences*, *84*, 5449–5453.

Zaura, E., Keijser, B. J., Huse, S. M., & Crielaard, W. (2009). Defining the healthy "core microbiome" of oral microbial communities. *BMC Microbiology*, *9*(1), 1–12.

Zaura, E., Nicu, E. A., Krom, B. P., & Keijser, B. J. (2014). Acquiring and maintaining a normal oral microbiome: Current perspective. *Frontiers in Cellular and Infection Microbiology*, *4*, 85.

Zeng, L., & Burne, R. A. (2013). Comprehensive mutational analysis of sucrose-metabolizing pathways in *Streptococcus mutans* reveals novel roles for the sucrose phosphotransferase system permease. *Journal of Bacteriology*, *195*, 83343.

Further reading

Mai, J., Tian, X. L., Gallant, J. W., Merkley, N., Biswas, Z., Syvitski, R., et al. (2011). A novel target-specific, salt-resistant antimicrobial peptide against the cariogenic pathogen *Streptococcus mutans*. *Antimicrobial Agents and Chemotherapy*, *55*, 5205–5213.

Mukherjee, P. K., Chandra, J., Kuhn, D. M., & Ghannoum, M. A. (2003). Mechanism of fluconazole resistance in *Candida albicans* biofilms: Phase-specific role of efflux pumps and membrane sterols. *Infection and Immunity*, *71*(8), 4333–4340.

Peterson, S. N., Meissner, T., Su, A. I., Snesrud, E., Ong, A. C., Schork, N. J., et al. (2014). Functional expression of dental plaque microbiota. *Frontiers in Cellular and Infection Microbiology*, *4*, 108.

Sullivan, R., Santarpia, P., Lavender, S., Gittins, E., Liu, Z., Anderson, M. H., et al. (2011). Clinical efficacy of a specifically targeted antimicrobial peptide mouth rinse: Targeted elimination of *Streptococcus mutans* and prevention of demineralization. *Caries Research*, *45*, 415–428.

Tong, Z., Zhang, L., Ling, J., Jian, Y., Huang, L., & Deng, D. (2014). An *in vitro* study on the effect of free amino acids alone or in combination with nisin on biofilms as well as on planktonic bacteria of *Streptococcus mutans*. *PLoS One*, *9*, e99513.

Chapter 30

Molecular mechanisms of *Acinetobacter baumannii* biofilm formation and its impact on virulence, persistence, and pathogenesis

K. Prashanth and Ajit Ramesh Sawant
Laboratory No. 6, Nosocomial Pathogen Biology Laboratory, Department of Biotechnology, Pondicherry University, Kalapet, Puducherry, India

1. Introduction

Several species of *Acinetobacter* are key nosocomial pathogens, with *Acinetobacter baumannii* being the most important species in a clinical setup having greatest worldwide concern because this bacterium has gained extremely high rates of multidrug resistance Ayoub Moubareck and Hammoudi Halat (2020). *A. baumannii* mainly causes hospital-associated infections such as ventilator-associated pneumonia (VAP), blood stream and urinary tract infections (UTIs), skin and wound infections in war-related injuries (Lee et al., 2017). Apart from pneumonia and bloodstream infections, almost one in five *A. baumannii* isolations in clinical settings is isolated from urinary sites, which makes it an important uropathogen (Di Venanzio et al., 2019). *A. baumannii* has a unique capacity to endure very stressful conditions, which is an important part of its abilities that allows it to spread successfully in the hospital environment and cause various kinds of nosocomial infections (Harding, Hennon, & Feldman, 2018). Colonization of hospital environment through formation of biofilms on abiotic surfaces promotes their environmental persistence and desiccation resistance. Both VAP and catheter-related infections due to *A. baumannii* are biofilm-associated infections that are highly resistant to antibiotic therapy (Gedefie et al., 2021). *A. baumannii* is armed with several resistant mechanisms such as enzyme-mediated degradation of antibiotics (β-lactamases), enzymes that can modify the antibiotics (aminoglycoside-modifying enzyme), efflux pumps effluxing out antibiotics out of the cell, target modifications, and permeability defects in porin proteins to overcome the toxicity and killing activity of antibiotics (Lee et al., 2017).

By definition, biofilms are referred as bacterial populations grown on biotic or abiotic surfaces that are encapsulated in the self-produced extracellular matrix (ECM). Biofilm-forming ability basically is dependent on pili formation, production of cell surface appendages, and ECM, which is abundantly produced by *A. baumannii*, and thus, *Acinetobacter* biofilms play important role in chronic recalcitrant infections. Besides, *Acinetobacter* biofilms play a crucial role in spread and establishment of infection among patients in hospitals, and such biofilms are capable of elaborating the pathogenesis making the treatment options more difficult (Ibrahim, Al-Saryi, Al-Kadmy, & Aziz, 2021; Pompilio et al., 2021). Different molecular mechanisms are involved in regulation and development of biofilms in *A. baumannii*. In this review, we will discuss in detail about different molecular mechanisms involved in biofilm development in *A. baumannii* and its impact on virulence, persistence, and pathogenesis of this bacteria.

2. Biofilms of *Acinetobacter baumannii*

Biofilm development on abiotic surfaces such as medical devices is considered as a major reservoir and source of infections in clinical settings across the globe. Biofilm is a reservoir of bacterial consortium embedded in self-produced ECM, which provides protection against antibiotics and host immune system by acting as a physical barrier. Biofilm formation is directly responsible for 80% infections in humans (Davies, 2003). *A. baumannii* biofilm formation on implanted medical devices such as catheters, prosthetic implants increase the clinical complications by making the treatment very difficult (Donlan, 2008). Different progressive steps involved in the development of biofilms are namely (i) initial attachment, (ii) subsequent strong adhesion, (iii) intercellular communication, (iv)

Understanding Microbial Biofilms. https://doi.org/10.1016/B978-0-323-99977-9.00027-2

colonization and growth, (v) production of exopolysaccharide, (vi) development of mature biofilm, and finally, the (vii) detachment and dispersal of cells from the mature biofilm. Fig. 1 illustrates different important steps involved in biofilm formation cycle and its maturation. Most of these steps are regulated and all controlled by cell-to-cell signaling mechanism that is called as quorum sensing (QS) (Jamal, Tasneem, Hussain, & Andleeb, 2015; Saipriya, Swathi, Ratnakar, & Sritharan, 2020). However, breach of skin and vasculature due to wound or device insertion is required for the invasion.

The nonspecific attachment of abiotic surface is aided by hydrophobicity of Gram-negative bacteria, and *A. baumannii* is no exception (Berne, Ducret, Hardy, & Brun, 2015; Skerniškytė et al., 2019). Once the bacteria enter through the breach caused by invasive medical device, the initial stage of colonization or infection cycle begins with the attempts of bacterial cells to interact with the host cells and primarily happens through specific receptors. Further, developing into microcolonies on living and no living surfaces leads to the creation of a cumulative structure network known as biofilm. Adherence of planktonic *A. baumannii* cells is the first step of biofilm formation on biotic and abiotic surfaces. Twitching mobility of *A. baumannii* on the moist surfaces uses type IV pili (TFP) to propel the cells forward and help in mobility of bacterial cells within the host. TFP also promote the adherence of *Acinetobacter* to host cell (Piepenbrink et al., 2016; Ronish, Lillehoj, Fields, Sundberg & Piepenbrink, 2019). After the initial attachment becomes stable, the EPS produced by the bacteria fixes the initial attachment into strong adhesion. Subsequently, the multiplication of cells starts, and they form a basic structural unit of biofilm, which is also referred as microcolony. The proximity of cells within the microcolony increases, and this plays an important role in the interchange of the nutrients, various substrates and distribution of major metabolites, and excretion of metabolic end-products (Garrett, Bhakoo, & Zhang, 2008; Jamal et al., 2018). The maturation of biofilm happens when all the cells irreversibly attached to the one another or to the surface it develops into a more organized and complex structure. The complexity and the shape of the mature biofilm are dependent on the source of available nutrients in the micro niche. *A. baumannii* cells in mature biofilm communicate with one another via QS that control their population density using signaling molecules known as auto-inducers (Niu, Clemmer, Bonomo, & Rather, 2008; Stewart & Franklin, 2008). QS also helps in accessibility of nutrient resources to all the cells wherein self-produced matrix generates interstitial spaces that act as channels, which are filled with water that aids in circulatory system (Hannan et al., 2010; Stewart & Franklin, 2008). Biofilm-encased cells thus developed have limited metabolic activity and are very well shielded by the ECM, making a solid physical barrier for antimicrobial agents, thus making these cells sessile and thousand times more resistant. The mature biofilm also prevents the entry of innate immune components of the host (Runci, Bonchi, Frangipani, Visaggio, & Visca, 2016). Finally, on the onset of favorable conditions for dissemination or when there is severe oxygen or nutrient starvation, the sessile cells

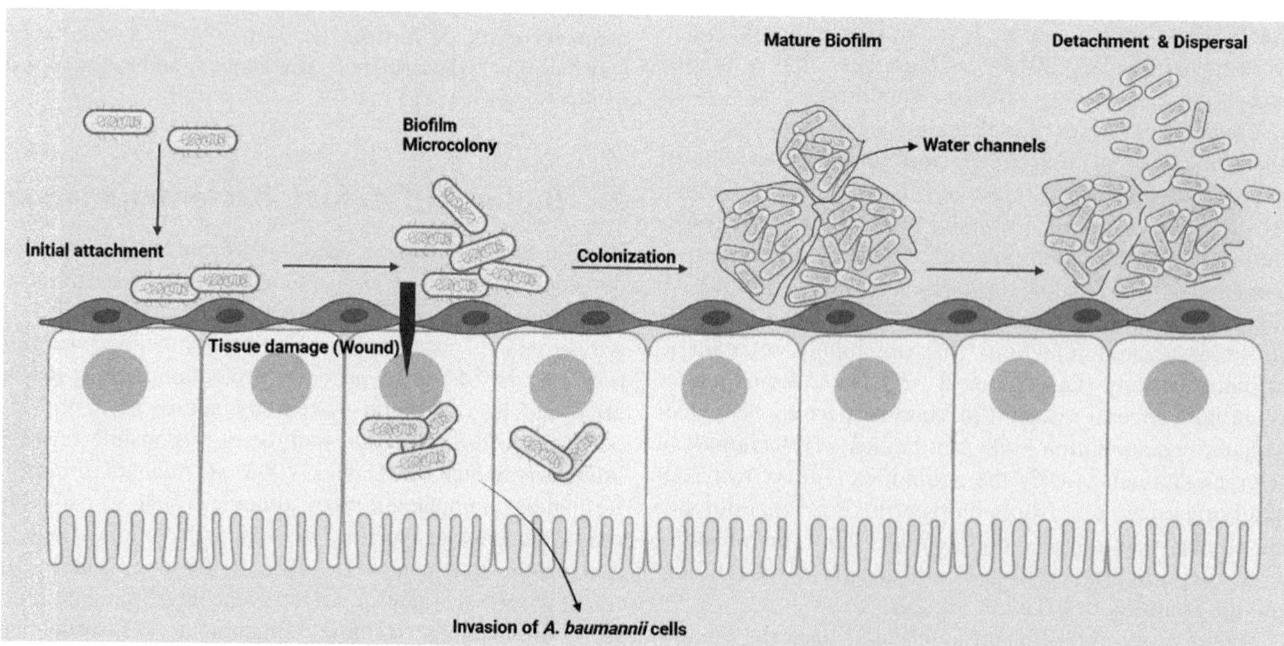

FIG. 1 Different stages of biofilm development and its maturation in *A. baumannii*.

transform into motile form, and it will trigger their dispersal. Detachment and dispersal of cells from the biofilm are aided by the dissolution of EPS by saccharolytic enzymes that allows the outermost layer of cells on the surface to be released first and rest of them may follow. These mobile cells may then find a new location for colonization (Parsek & Singh, 2003; Stewart & Franklin, 2008).

3. Microstructure of *A. baumannii* biofilms and its composition

The major constituents of this developing matrix of biofilms are carbohydrates, proteins, nucleic acids, and other macromolecules (Dufour, Leung, & Levesque, 2010). *A. baumannii* biofilm protects it from various damages caused by host response, disinfectants, and antimicrobial agents such as antibiotics. Biofilms contribute to prolonged long-lasting persistence of *A. baumannii* in the environment, and different factors such as bacterial microstructures, surface components, nutrients, extracellular DNA (e-DNA), QS, and other macromolecules will affect the biofilms (Dufour et al., 2010; Gaddy, Tomaras, & Actis, 2009). The developmental process of biofilm in *A. baumannii* has a positive correlation with expression of various virulence factors such as the extracellular polysaccharide poly-β-1,6-*N*-acetyl glucosamine (PNAG), OmpA, type I pili, BAP protein, RecA, and CarO outer membrane protein (Amala Reena, Subramaniyan, & Kanungo, 2018). Surface hydrophobicity and oxygen content in the environment influence the structure and development of biofilm in *A. baumannii* (Eze, Chenia, & El Zowalaty, 2018). Glucose seems to be the most important *A. baumannii* biofilm component, and its production is related to a range of additional biological and environmental elements (Gedefie et al., 2021). Many environmental factors influence the biofilm growth and maturation in a highly regulated manner at every stage of its development. Nutrients availability, ferrous iron concentration, osmolarity, temperature, moisture quality of surface material where biofilms are formed, light, and ambient acidic conditions are some of the main environmental elements that may affect the way the biofilm is formed (Toyofuku et al., 2016). In the below paragraphs, we describe some of the important components that contribute to biofilm development.

(i) Biofilm-associated proteins (BAP): These are large surface-exposed proteins secreted via type I secretion system in *A. baumannii* and are required for biofilm formation (Harding et al., 2017). It has a highly polymorphic protein having repetitive, truncated, and disrupted gene sequence in different strains of the *A. baumannii* and in strains of other species belonging to *Acinetobacter calcoaceticus-baumannii* (*Acb*) complex (De Gregorio et al., 2015). BAP is one of the important proteins required for *A. baumannii* in order to form biofilm and its maturation on both host cells and abiotic materials that are used to make medical devices such as polystyrene and titanium (Brossard & Campagnari, 2012). BAP helps the adhesion of *A. baumannii* by increasing the cell surface hydrophobicity (CSH) (Brossard & Campagnari, 2012; Goh et al., 2013). Interestingly, the capacity of modification of CSH depends on the given environmental conditions and growth phases in which the *A. baumannii* cells are in (Pour et al., 2011).

(ii) Csu operon: The development of biofilm in *A. baumannii* depends on synthesis and assembly of pili that enables the cell attachment to the surfaces, and it is coded through the *csuA/BABCDE* chaperon-usher system. *A. baumannii* produces a type I chaperone-usher pilus system that is simply called as *csu* pili. Csu pili expression is one of the important factors in *A. baumannii* required for production of pili, cell-to-cell attachment, and biofilm formation on abiotic surface, but is not very essential for attachment of cells on to biotic surfaces (de Breij et al., 2009; Tomaras, Dorsey, Edelmann, & Actis, 2003). The *csu* pili system consists of six genes, namely *csuA/B*, *csuA*, *csuB*, *csuC*, *csuD*, and *csuE* (Moon, Weber, & Feldman, 2017). CsuC and CsuA/B polymerize and form major subunit of pili. CsuD acts as an usher, and csuA/B forms a minor subunit of pili (Pakharukova, Tuittila, Paavilainen & Zavialov, 2015). The X-ray structure of the CsuC and CsuE chaperones reveals that CsuE is highly conserved gene responsible for three hydrophobic finger-like structures at the tip of pilus essential for bacterial attachment (Pakharukova, Tuittila, et al., 2015). Overall, the attachment of *A. baumannii* cells to surface in order to form biofilm is facilitated by csu operon system. The expression of *csu* operon is controlled and regulated by *BfmRS* TCS as shown by an investigation wherein the ability to form biofilm was abolished in those strains that are lacking *BfmRS* (Liou et al., 2014). However, a *csu* operon–independent short pilus is also produced in *A. baumannii*, which helps in bacterial attachment to biotic surfaces such as human respiratory cells (Pakharukova, Garnett, et al., 2015). It is important to know that such alternative mechanism is regulated by QS that is controlled by LuxI/LuxR TCS (Pakharukova, Garnett, et al., 2015; Sarshar, Behzadi, Scribano, Palamara, & Ambrosi, 2021).

(iii) Exopolysaccharides: An important element of exopolysaccharide is PNAG (Poly-β-1,6-*N*-acetyl-glucosamine), which constitutes *A. baumannii* biofilms (Flannery, Le Berre, Pier, O'Gara, & Kilcoyne, 2020). Desiccation resistance of *A. baumannii* was achieved by the production of capsular polysaccharide that covers bacterial cells grown in the biofilm under stressed conditions. Such polysaccharides retain the water content in the biofilm matrix. These polysaccharides constitute repeat moieties of carbohydrate units that make up these structures that act as glycan shields that surround the bacteria and protect it from external stress (Nguyen & Joshi, 2021).

The regulation of the biofilm-related genes and their expression by a range of regulatory networks together with two-component regulatory system (TCS) is the crux, and one has to understand all these at molecular level to devise antibiofilm strategies.

4. Regulation of biofilm formation

The TCS *BfmRS* appears to regulate the expression of *csu* operon since it was shown in one study that the biofilm-forming ability was abolished in those strains that are lacking *BfmRS* (Liou et al., 2014). The sensor kinase (A1S_0574) of TCS *GacSA* known to control the expression of phenylacetic acid catabolic pathway was also found to be regulating the pili synthesis in *A. baumannii*, which is required for ECM formation (Cerqueira et al., 2014). The regulation of biofilm by TCSs (*adeRS*, *BfmRS* and *GacSA*) directly or indirectly needs to be studied in detail in order to formulate the antibiofilm strategies to avoid colonization of *A. baumannii* in patients and on implanted devices. BfmR, response regulator of TCS in *A. baumannii*, is vital in order to form biofilm on abiotic surface and cell morphology (Tomaras, Flagler, Dorsey, Gaddy, & Actis, 2008). Through transcriptome studies of planktonic and biofilm *A. baumannii* cells, it was observed that the gene expression profile of cells in biofilm is distinct from free-living cells wherein more than 50 genes exclusively expressed in the cells present in biofilm (Rumbo-Feal et al., 2013).

In *A. baumannii* strain ATCC 17978, subinhibitory concentrations of sulfamethoxazole and trimethoprim abolish the expression of *csu* pili results in complete inhibition of biofilm formation (Moon et al., 2017). However, contrarily in another study, biofilm formation is found to be increased in sub-MIC concentration of colistin and of tigecycline (Sato, Ubagai, Tansho-Nagakawa, Yoshino, & Ono, 2021). Interestingly, one study showed that acquisition of colistin resistance is associated with loss in biofilm formation and decreased fitness of the *A. baumannii* (Dafopoulou et al., 2015), which shows that antibiotic resistance leads to compromise in fitness and virulence characters of *A. baumannii*. With respect to certain antibiotics such as colistin, less resistance is associated with isolates with strong biofilms indicating a certain trade-off between resistance and tolerance. Iron is one of the important factors required for biofilm formation in *A. baumannii*, and it was found that biofilm formation is stimulated by iron (Gentile et al., 2014). The carbon storage regulator CsrA plays a key role in the survival of *A. baumannii* wherein deletion mutant of *csrA* showed reduced ability to form biofilm and has improved the ability to survive in high-osmolarity condition (Farrow, Wells, Palethorpe, Adams, & Pesci, 2020).

Efflux pumps are membrane protein system required for bacteria to survive in the presence of toxic compounds by pumping out the toxic compounds. It is observed that RND efflux pump *adeFGH* plays a role in biofilm formation in clinical isolates of *A. baumannii* (He et al., 2015) where it was shown that overexpression of AdeFGH genes and AHL (acylated homoserine lactones) synthesis gene *abaI* is required for biofilm formation. On the contrary, isogenic mutants overexpressing *adeABC* and *adeIJK* showed decline in biofilm formation as compared with parent strain (Yoon et al., 2015). TCS such as *adeRS* modulating the expression of *adeABC* efflux pump is essential for biofilm formation in porcine mucosal model and on abiotic surface in strain-specific manner, which further showed different virulence phenotypes (Richmond et al., 2016).

Table 1 summarizes details of the major and important genetic regulatory components and their mechanisms, which are regulating the *A. baumannii* biofilm formation.

TABLE 1 Summary of regulatory components involved in *A. baumannii* biofilm development.

Gene name (description)	Role	References
RND efflux pump genes		
AdeFGH	Overexpression is associated with biofilm formation	He et al. (2015)
abaI (encode an AHL synthase)	Induction of biofilm formation via acyl homoserine lactone (AHL) production	
adeIJK	Deletion and disruption of *tetR* regulator, adeN leads to elevation in biofilm formation	Saranathan et al. (2017)
adeABC	i. Mutant over expressing the efflux pump underexpression of genes (*CsuA/B, CsuC,* and *FimA*) associated with pili formation proteins ii. Deletion of *adeB* in *A. baumannii* strain AYE leads to reduction in biofilm formation on abiotic and biotic surfaces as compared to wild type strain	Yoon et al. (2015) Richmond et al. (2016)

TABLE 1 Summary of regulatory components involved in *A. baumannii* biofilm development—cont'd

Gene name (description)	Role	References
A1S_1751 (*AdeA*, membrane fusion protein)	Exclusive expression in biofilm cells	Rumbo-Feal et al. (2013)
adeT	Exclusive expression in biofilm cells	Rumbo-Feal et al. (2013)
A1S_0009, A1S_0116, and *A1S_0538*	Upregulated in biofilms compared with stationary or planktonic cells	Rumbo-Feal et al. (2013)
MFS efflux pump genes		
A1S_1316	High expression in biofilm cells	Rumbo-Feal et al. (2013)
A1S_1117 (Sugar transporter protein)	Exclusive expression in biofilm cells	Rumbo-Feal et al. (2013)
Other genes		
A1S_0114 (Involved in the synthesis of acinetin 505)	Mutant lacking this gene showed reduction in mature biofilm formation, adhesion and virulence	Rumbo-Feal et al. (2013)
AvnR (*A1S_2006*)	Deletion of *avnR* affects the ability to form biofilm on abiotic surfaces	De Silva et al. (2020)
OmpA	Partly plays a role in the development of biofilm on plastic surface and essential for attachment of bacteria to human alveolar epithelial cells	Gaddy et al. (2009)
pglL (glycoprotein formation)	Deletion of pglL significantly decrease the biofilm production	Iwashkiw et al. (2012)
pgaABCD locus [poly-β(1-6)-*N*-acetylglucosamine (PNAG)]	Deletion of *pgaABCD* locus leads to loss of the strong biofilm phenotype	Choi, Slamti, Avci, Pier, and Maira-Litrán (2009)
bfmS (Sensor histidine kinase)	The *bfmS* knockout showed a reduction in biofilm formation, loss of adherence ability to eukaryotic cells	Liou et al. (2014)
BfmR/S TCS	Master regulator of biofilm initiation in *A. baumannii*	Draughn et al. (2018)
csrA	Biofilm formation and survival increases the tolerance to osmolarity	Farrow et al. (2020)
abaM	Regulate *N*-acyl homoserine lactone (AHL)-dependent quorum-sensing, its disruption due to T26 results in enhanced surface motility and biofilm formation	López-Martín, Dubern, Alexander, and Williams (2021)

5. Role of biofilms in virulence and pathogenesis

Around the world, *A. baumannii* infections are rapidly spread throughout hospitals, but the highest concentration of their infections occurs mostly in intensive care units (ICUs). Global spread of extensively drug-resistant *A. baumannii* and an array of its virulence factors makes it a very serious threat to public health worldwide. In spite of enormous research efforts performed to reveal its successes, this bacterium continues to confound researchers due to its highly adaptive nature, genome plasticity that helps it in mutating to meet the needs of a given environment (Mea, Yong, & Wong, 2021). Strong adherence and subsequent formation of biofilms, toxin secretions, and low activation of inflammatory responses are some of the mechanisms adopted by *A. baumannii* for manifestation of its pathogenesis (Ibrahim et al., 2021; Morris, Dexter, Kostoulias, Uddin, & Peleg, 2019). The most important factor associated with the initial stage of infection is host microenvironment adaptation that is facilitated by TCS and biofilm formation. *A. baumannii* cells in biofilms survive the harsh host condition and may develop dormancy and resist the environmental stress. Such biofilm cells in the host were to cause an array of chronic infections that are most difficult to eradicate (Greene, Vadlamudi, Newton, Foxman, & Xi, 2016; Richards, Abu Kwaik, & Lamont, 2015).

TABLE 2 Important virulence factors associated with biofilm formation and its stability in *A. baumannii*.

Gene	Description	Role in biofilm and pathogenesis	References
pglL	Involved in the synthesis of the O-pentasaccharide in glycoproteins and capsular polysaccharides of *A. baumannii*	Deletion leads to significant decrease in biofilm	Lees-Miller et al. (2013)
pglC		Deletion causes formation of rough and disordered biofilm	
$bla_{PER\text{-}1}$	Extended-spectrum-β-lactamase	Associated with cell adhesion to Caco2 cell lines	Sechi et al. (2004)
Ata	*Acinetobacter* trimeric autotransporter	Ata assist *A. baumannii* cells in formation of biofilm by binding to ECM like collagen type IV. It also vital for survival of *A. baumannii* in systemic infection in immunocompetent mice	Bentancor, Camacho-Peiro, Bozkurt-Guzel, Pier, and Maira-Litrán (2012)
pili	Essential for *A. baumannii* adherence	The chaperon usher system is essential for pili formation and attachment to plastic surface in order to form biofilm	Tomaras et al. (2003)
AbFhaB/FhaC	Involved in adhesion	Inactivation of this system decreases bacterial attachment to epithelial cells and also involved in virulence	Pérez et al. (2017)
plc1 and *plc2*	Phosphatidylcholine-specific phospholipase	Important for hemolytic and cytolytic activities when tested in a *Galleria mellonella* insect model	Fiester et al. (2016)
Ctp	Carboxy-terminal processing protease	Ctp negatively controls the biofilm formation and OMV biogenesis	Roy, You, Chang, Yang, and Lin (2021)
Pil (ABCFMNO)	Type IV pili system	Genes required for pili assembly and twitching motility	
ompA	OmpA is a β-barrel porin protein in *A. baumannii*	ompA involved in host cell death biofilm formation by interacting with host epithelial cells suing host fibronectin	Smani et al. (2014)
CarO	β barrel-shaped outer membrane channel protein required for influx of beta lactams antibiotics	Elevation in carO expression causes attenuation of proinflammatory response in lungs causing pneumonia	Sato, Unno, Kawakami, Ubagai, and Ono (2017)
OprD	Orthologous porin protein in *P. aeruginosa*	Deletion mutant did affect the antibiotic resistance in *A. baumannii* but in *A. baylyi* spp., it affected the MIC of carbapenem class antibiotics	Morán-Barrio et al. (2017)

Some of the significant virulence determinants associated with biofilm per se have been listed in Table 2, and among these, a few very important virulence attributes related to biofilms were discussed in below paragraphs.

We have relatively less information about the important factors that influence the pathogenesis of *A. baumannii*. Many previous investigations identified a number of important virulence factors such as capsular polysaccharides, lipopolysaccharide (LPS), outer membrane porins, proteases, phospholipases, diverse protein secretion systems, and iron-chelating systems that are involved in pathogenesis (Antunes, Imperi, Carattoli, & Visca, 2011; Lin & Lan, 2014; McConnell, Actis, & Pachón, 2013). However, multiple studies show components of cell wall and biofilm such as outer membrane protein A (OmpA), EPS, the K1 capsule, phospholipids, phospholipases play a significant role in pathogenicity of *A. baumannii* (Antunes et al., 2011; Cerqueira & Peleg, 2011; Lin & Lan, 2014; Richards et al., 2015; Sato et al., 2017). It was also shown that the capacity of cells to adhere to the surface and the expression level of OMPs mRNAs are the major determinants of virulence factors in case of *A. baumannii* pathogenesis (Sato et al., 2017). Besides, OmpA involved in biofilm development on epithelial cells assists the bacterial cell adherence to lung epithelial cells through its interaction with fibronectin on the surface of host cells (Gaddy & Actis, 2009). After the attachment process, OmpA also facilitates the host cell invasion and is involved in the commencement of apoptosis at the onset of infection (Guo, Xun, & Han, 2018; Nafarieh et al., 2017; Nie et al.,

2020). OmpA causes an increase in cell death mainly through mitochondrial and nuclear targeting (Smani et al., 2014). Further, outer membrane vesicles (OMVs) secreted by *A. baumannii* have abundant presence of OmpA in its membrane that confers OMVs their cytotoxicity (Jin et al., 2011). Another outer membrane protein, namely OMP38, induces apoptosis of human epithelial cells and monocytes in one of the study (Choi et al., 2005). Apoptosis of epithelial cells results in reduction of the mucosal surface that may provide easy access for bacteria to cause deep tissue infection (Choi et al., 2005). Furthermore, it was also previously shown that *A. baumannii* requires zinc for their growth, and it has to acquire the zinc from the host to survive and colonize the host. ZrlA of zinc acquisition system (ZnuABC, ZigA, and ZrlA) seems to be critical association among nutrient metal uptake and cell envelope homeostasis throughout pathogenesis of *A. baumannii*; therefore, Zn acquisition system is very essential for in vivo survival and persistence of *A. baumannii* (Lonergan et al., 2019). All the abovementioned components play multifaceted roles particularly in different stages of biofilm development and maturation that becomes an integral part of *A. baumannii*'s virulence and pathogenesis.

6. Biofilms and persistence of *Acinetobacter baumannii*

Persister cells are the antibiotic tolerant variants (cells that withstand antibiotic) in a bacterial population having similar genotype as that of susceptible cells, and they can switch between susceptible and persister cell based on antibiotic presence. Persister cells show biphasic killing curve when added with high concentration of antibiotic where susceptible cells are eradicated faster and a fraction of cells that survive were referred to as persister cells. Persister cell phenotype has been observed in most of the bacterial pathogens including *Mycobacterium tuberculosis*, *Staphylococcus aureus*, *Escherichia coli*, *Pseudomonas aeruginosa*, *Salmonella enterica* subsp. *enterica* serovar Typhimurium, and *A. baumannii* (Harms, Maisonneuve, & Gerdes, 2016; Helaine & Kugelberg, 2014).

Persister cells are usually formed in response to external stress such as the presence of antibiotics, extreme pH, DNA damage, nutrient limitation, etc. Persister cells left out at the end of antibiotic therapy are generally cleared by immune response; on the contrary, if those cells are embedded in self-produced ECM immune system does not have access to these persister cells. Biofilms developed out of stress usually grow very less, and they have substantial number of persister cells (Bhargava, Sharma, & Capalash, 2014; Lewis, 2010).

PNAG, which is the major component of ECM of *A. baumannii*, has a surface presentation and has an easy access to bacterial external proteins such as lectins (Amala Reena, Subramaniyan, & Kanungo, 2018). This PNAG not only contributes to the biofilm integrity, it also confers tolerance and persistence in hospital environments (Pompilio et al., 2021). Bacterial cells embedded in biofilm significantly affect the antibiotic therapy since biofilm acts as physical barrier between bacterial cells and antibiotics and is responsible for chronic infections. Persister cells often lead to treatment failure and cause recalcitrance of infections (Lewis, 2010). Along with antibiotic resistance, *A. baumannii* also reported to form persister cells in response to amikacin and carbenicillin (Bhargava et al., 2014). One recent investigation observed key elevation of components such as adhesion factors, acetate metabolic pathway, nutrient transporters, and proteins of secretion system among *A. baumannii* biofilm phenotypes studied through proteomic approach by using QExactive tandem mass spectrometry (Choudhary, Kaushik, Kapil, Shrivastava, & Vashistt, 2022). Such results imply that the presence of acetate metabolism and glyoxylate pathways might have potential role in persistence and biofilm formation in *A. baumannii*.

7. Novel antibiofilm treatment strategies

All the currently available antimicrobial agents, namely carbapenems, aminoglycosides, tigecycline, and colistin, have been found to be inadequate and ineffective due to emergence of multi and pan-drug-resistant strains, accentuating the dire necessity of alternative and novel therapeutic remedies. Increasing level of resistance is responsible for increase in mortality rates especially among patients admitted in ICUs and eradication of drug-resistant *A. baumannii* from either hospital or from the community has become herculean task for the physicians and healthcare workers. MDR pathogen *A. baumannii* has significantly threatened public health due to above reasons discussed in this review. Thus, there is an urgent need for better understanding molecular details of virulence and pathogenesis of this pathogen in order to devise novel therapeutics and control strategies to deal with this pathogen.

Biofilm disruption by quorum quenching strategy is performed through inhibition or degradation of QS signals, inhibition of signaling molecule synthesis, antagonizing the signaling molecules, signal transduction, and signal transport (Subhadra, Kim, Woo, Surendran, & Choi, 2018; Elkheloui et al., 2020). Flavonoids such as curcumin, fisetin, and phloretin showed dose-dependent inhibition of several MDR clinical isolates of *A. baumannii* in one of the investigations (Raorane et al., 2019). Further, it was shown that antibiofilm activity of curcumin was far greater than that of the known biofilm inhibitor gallium nitrate. In addition, in silico molecular docking studies revealed the increased binding efficacy of flavonoid with the biofilm response regulator BfmR that controls biofilm formation in *A. baumannii* (Raorane et al., 2019).

Octenidine dihydrochloride, a positively charged bispyridinamine, has been used in one of the studies to inhibit and disrupt the biofilms of *A. baumannii* present on various surfaces such as polystyrene, stainless steel, and urinary catheters (Narayanan et al., 2016). It has been also shown that bacterial growth and biofilm formation are inhibited by certain prebiotic metabolites, namely raffinose, riboflavin, inulin, citrate, sorbitol, and trehalose. All these prebiotic metabolites can be used to harness their effective antibiofilm activity. One such metabolite, namely inulin, produced by prebiotic *Lactobacillus rhamnosus* can inhibit around 80% of *A. baumannii* biofilm (Gedefie et al., 2021). Antimicrobial peptides AMP LL-37 of human origin and WAM-1 (marsupial AMP) were also able to successfully prevent the biofilm formation of *A. baumannii* (Asif, Alvi, & Rehman, 2018; Weinberg et al., 2020). Other compounds such as cinnamaldehyde are shown to possess a very high antibiofilm activity, which could be used in a clinical setup to clear *A. baumannii* (Mohamed, Salem, Azmy, & Fam, 2018).

In the presence of silver nanoparticles (AgNPs), *A. baumannii* is inhibited, and it has more pronounced antibiofilm activity, and it decreased the expression of virulence and biofilm-related genes such as *kpsMII*, *afa/draBC*, *BAP*, *OmpA*, and *csuA/B* genes (Hetta et al., 2021). Another possible novel alternative is the use of nonthermal plasma (NTP), a partially ionized gas with the energy stored particularly in the free electrons. It has antimicrobial and antibiofilm effects. NTP generates pores in the cell membrane, which in turn kills the bacterial cells (Scholtz et al., 2021).

8. Summary and conclusions

Biofilm-associated infections are threat to public health because they lead to persistent chronic infections that have direct economic and health impact. The ability of *A. baumannii* to grow as biofilm on the biotic and abiotic surfaces seems to be the most crucial feature that contributes their capacity to cause chronic and persistent infections. Biofilms of this bacterium also help them to build up their resistance to different antimicrobial agents and allow their long-lasting survival in the hospital environment. There have been considerable progresses made in understanding the way *A. baumannii* forms biofilm on different surfaces and what are their virulence characteristics, which are required for their pathogenesis. Biofilms turned out to be a key important feature behind their pathogenesis, which makes the available treatment options limited and complex. The production of biofilms in *A. baumannii* is influenced by several biological and environmental variables. Since high incidence of *A. baumannii* infections was linked to the invasive medical devices due to the biofilms formed by them, controlling those infections is very difficult. Targeting the biofilms through various antibiofilm molecules could be one of the attractive strategies for treating *A. baumannii* infections. Nevertheless, the need of the hour is more exhaustive, prospective, and in-depth studies that explore the magnitude of *A. baumannii* infections among patients and their pathogenesis in vivo and the possible aggravating factors of their infection that may help in finding possible therapeutics options.

References

Amala Reena, A. A., Subramaniyan, A., & Kanungo, R. (2018). Biofilm formation as a virulence factor of *Acinetobacter baumannii*: An emerging pathogen in critical care units. *Journal of Current Research in Scientific Medicine*, *3*, 74–78. [serial online] 2017 https://www.jcrsmed.org/text.asp?2017/3/2/74/222427.

Antunes, L. C., Imperi, F., Carattoli, A., & Visca, P. (2011). Deciphering the multifactorial nature of *Acinetobacter baumannii* pathogenicity. *PLoS One*, *6*(8), e22674. https://doi.org/10.1371/journal.pone.0022674.

Asif, M., Alvi, I. A., & Rehman, S. U. (2018). Insight into *Acinetobacter baumannii*: Pathogenesis, global resistance, mechanisms of resistance, treatment options, and alternative modalities. *Infection and Drug Resistance*, *11*, 1249–1260. https://doi.org/10.2147/IDR.S166750.

Ayoub Moubareck, C., & Hammoudi Halat, D. (2020). Insights into *Acinetobacter baumannii*: A review of microbiological, virulence, and resistance traits in a threatening nosocomial pathogen. *Antibiotics (Basel, Switzerland)*, *9*(3), 119. https://doi.org/10.3390/antibiotics9030119.

Bentancor, L. V., Camacho-Peiro, A., Bozkurt-Guzel, C., Pier, G. B., & Maira-Litrán, T. (2012). Identification of Ata, a multifunctional trimeric autotransporter of *Acinetobacter baumannii*. *Journal of Bacteriology*, *194*(15), 3950–3960. https://doi.org/10.1128/JB.06769-11.

Berne, C., Ducret, A., Hardy, G. G., & Brun, Y. V. (2015). Adhesins involved in attachment to abiotic surfaces by gram-negative bacteria. *Microbiology Spectrum*, *3*(4). https://doi.org/10.1128/microbiolspec.MB-0018-2015.

Bhargava, N., Sharma, P., & Capalash, N. (2014). Pyocyanin stimulates quorum sensing-mediated tolerance to oxidative stress and increases persister cell populations in *Acinetobacter baumannii*. *Infection and Immunity*, *82*(8), 3417–3425. https://doi.org/10.1128/IAI.01600-14.

Brossard, K. A., & Campagnari, A. A. (2012). The *Acinetobacter baumannii* biofilm-associated protein plays a role in adherence to human epithelial cells. *Infection and Immunity*, *80*(1), 228–233. https://doi.org/10.1128/IAI.05913-11.

Cerqueira, G. M., & Peleg, A. Y. (2011). Insights into *Acinetobacter baumannii* pathogenicity. *IUBMB Life*, *63*(12), 1055–1060. https://doi.org/10.1002/iub.533.

Cerqueira, G. M., Kostoulias, X., Khoo, C., Aibinu, I., Qu, Y., Traven, A., et al. (2014). A global virulence regulator in *Acinetobacter baumannii* and its control of the phenylacetic acid catabolic pathway. *The Journal of Infectious Diseases*, *210*(1), 46–55. https://doi.org/10.1093/infdis/jiu024.

Choi, C. H., Lee, E. Y., Lee, Y. C., Park, T. I., Kim, H. J., Hyun, S. H., et al. (2005). Outer membrane protein 38 of *Acinetobacter baumannii* localizes to the mitochondria and induces apoptosis of epithelial cells. *Cellular Microbiology*, *7*(8), 1127–1138. https://doi.org/10.1111/j.1462-5822.2005.00538.x.

Choi, A. H., Slamti, L., Avci, F. Y., Pier, G. B., & Maira-Litrán, T. (2009). The pgaABCD locus of Acinetobacter baumannii encodes the production of poly-beta-1-6-N-acetylglucosamine, which is critical for biofilm formation. *Journal of Bacteriology*, *191*(19), 5953–5963. https://doi.org/10.1128/JB.00647-09.

Choudhary, M., Kaushik, S., Kapil, A., Shrivastava, R., & Vashistt, J. (2022). Decoding Acinetobacter baumannii biofilm dynamics and associated protein markers: proteomic and bioinformatics approach. *Archives of microbiology*, *204*(4), 200. https://doi.org/10.1007/s00203-022-02807-y.

Dafopoulou, K., Xavier, B. B., Hotterbeekx, A., Janssens, L., Lammens, C., Dé, E., … Pournaras, S. (2015). Colistin-Resistant Acinetobacter baumannii Clinical Strains with Deficient Biofilm Formation. *Antimicrobial agents and chemotherapy*, *60*(3), 1892–1895. https://doi.org/10.1128/AAC.02518-15.

De Gregorio, E., Del Franco, M., Martinucci, M., Roscetto, E., Zarrilli, R., & Di Nocera, P. P. (2015). Biofilm-associated proteins: News from *Acinetobacter*. *BMC Genomics*, *16*, 933. https://doi.org/10.1186/s12864-015-2136-6.

Davies, D. (2003). Understanding biofilm resistance to antibacterial agents. *Nature Reviews Drug Discovery*, *2*(2), 114–122. https://doi.org/10.1038/nrd1008.

de Breij, A., Gaddy, J., van der Meer, J., Koning, R., Koster, A., van den Broek, P., et al. (2009). CsuA/BABCDE-dependent pili are not involved in the adherence of *Acinetobacter baumannii* ATCC19606(T) to human airway epithelial cells and their inflammatory response. *Research in Microbiology*, *160*(3), 213–218. https://doi.org/10.1016/j.resmic.2009.01.002.

De Silva, Patidar, R., Graham, C. I., Brassinga, A., & Kumar, A. (2020). A response regulator protein with antar domain, AvnR, in Acinetobacter baumannii ATCC 17978 impacts its virulence and amino acid metabolism. *Microbiology (Reading, England)*, *166*(6), 554–566. https://doi.org/10.1099/mic.0.000913.

Di Venanzio, G., Flores-Mireles, A. L., Calix, J. J., Haurat, M. F., Scott, N. E., Palmer, L. D., et al. (2019). Urinary tract colonization is enhanced by a plasmid that regulates uropathogenic *Acinetobacter baumannii* chromosomal genes. *Nature Communications*, *10*(1), 2763. https://doi.org/10.1038/s41467-019-10706-y.

Donlan, R. M. (2008). Biofilms on central venous catheters: Is eradication possible? *Current Topics in Microbiology and Immunology*, *322*, 133–161. https://doi.org/10.1007/978-3-540-75418-3_7.

Draughn, G. L., Milton, M. E., Feldmann, E. A., Bobay, B. G., Roth, B. M., Olson, A. L., et al. (2018). The Structure of the biofilm-controlling response regulator BfmR from *Acinetobacter baumannii* reveals details of Its DNA-binding mechanism. *Journal of Molecular Biology*, *430*(6), 806–821. https://doi.org/10.1016/j.jmb.2018.02.002.

Dufour, D., Leung, V., & Levesque, C. (2010). Bacterial biofilm: Structure, function, and antimicrobial resistance. *Endodontic Topics*, *22*. https://doi.org/10.1111/j.1601-1546.2012.00277.x.

Elkheloui, R., Laktib, A., Mimouni, R., Ait Alla, A., Hassi, M., Abdellah, E. B., Hamadi, F. (2020). *Acinetobacter baumannii* biofilm: Intervening factors, persistence, drug resistance and strategies of treatment. *Mediterranean Journal of Infection, Microbes and Antimicrobials*, *9*. https://doi.org/10.4274/mjima.galenos.2020.2020.7.

Eze, E. C., Chenia, H. Y., & El Zowalaty, M. E. (2018). *Acinetobacter baumannii* biofilms: Effects of physicochemical factors, virulence, antibiotic resistance determinants, gene regulation, and future antimicrobial treatments. *Infection and Drug Resistance*, *11*, 2277–2299. https://doi.org/10.2147/IDR.S169894.

Farrow, J. M., 3rd, Wells, G., Palethorpe, S., Adams, M. D., & Pesci, E. C. (2020). CsrA supports both environmental persistence and host-associated growth of *Acinetobacter baumannii*. *Infection and Immunity*, *88*(12), e00259-20. https://doi.org/10.1128/IAI.00259-20.

Fiester, S. E., Arivett, B. A., Schmidt, R. E., Beckett, A. C., Ticak, T., Carrier, M. V., et al. (2016). Iron-regulated phospholipase C activity contributes to the cytolytic activity and virulence of *Acinetobacter baumannii*. *PLoS One*, *11*(11), e0167068. https://doi.org/10.1371/journal.pone.0167068.

Flannery, A., Le Berre, M., Pier, G. B., O'Gara, J. P., & Kilcoyne, M. (2020). Glycomics microarrays reveal differential in situ presentation of the biofilm polysaccharide poly-*N*-acetylglucosamine on *Acinetobacter baumannii* and *Staphylococcus aureus* cell surfaces. *International Journal of Molecular Sciences*, *21*(7), 2465. https://doi.org/10.3390/ijms21072465.

Gaddy, J. A., & Actis, L. A. (2009). Regulation of *Acinetobacter baumannii* biofilm formation. *Future Microbiology*, *4*(3), 273–278. https://doi.org/10.2217/fmb.09.5.

Gaddy, J. A., Tomaras, A. P., & Actis, L. A. (2009). The *Acinetobacter baumannii* 19606 OmpA protein plays a role in biofilm formation on abiotic surfaces and in the interaction of this pathogen with eukaryotic cells. *Infection and Immunity*, *77*(8), 3150–3160. https://doi.org/10.1128/IAI.00096-09.

Garrett, T. R., Bhakoo, M., & Zhang, Z. (2008). Bacterial adhesion and biofilms on surfaces. *Progress in Natural Science*, *18*(9), 1049–1056. https://doi.org/10.1016/j.pnsc.2008.04.001.

Gedefie, A., Demsis, W., Ashagrie, M., Kassa, Y., Tesfaye, M., Tilahun, M., et al. (2021). *Acinetobacter baumannii* biofilm formation and its role in disease pathogenesis: A review. *Infection and Drug Resistance*, *14*, 3711–3719. https://doi.org/10.2147/IDR.S332051.

Gentile, V., Frangipani, E., Bonchi, C., Minandri, F., Runci, F., & Visca, P. (2014). Iron and Acinetobacter baumannii Biofilm Formation. *Pathogens (Basel, Switzerland)*, *3*(3), 704–719. https://doi.org/10.3390/pathogens3030704.

Goh, H. M., Beatson, S. A., Totsika, M., Moriel, D. G., Phan, M. D., Szubert, J., et al. (2013). Molecular analysis of the *Acinetobacter baumannii* biofilm-associated protein. *Applied and Environmental Microbiology*, *79*(21), 6535–6543. https://doi.org/10.1128/AEM.01402-13.

Greene, C., Vadlamudi, G., Newton, D., Foxman, B., & Xi, C. (2016). The influence of biofilm formation and multidrug resistance on environmental survival of clinical and environmental isolates of *Acinetobacter baumannii*. *American Journal of Infection Control*, *44*(5), e65–e71. https://doi.org/10.1016/j.ajic.2015.12.012.

Guo, Y., Xun, M., & Han, J. (2018). A bovine myeloid antimicrobial peptide (BMAP-28) and its analogs kill pan-drug-resistant *Acinetobacter baumannii* by interacting with outer membrane protein A (OmpA). *Medicine*, *97*(42), e12832. https://doi.org/10.1097/MD.0000000000012832.

Hannan, S., Ready, D., Jasni, A. S., Rogers, M., Pratten, J., & Roberts, A. P. (2010). Transfer of antibiotic resistance by transformation with eDNA within oral biofilms. *FEMS Immunology and Medical Microbiology*, *59*(3), 345–349. https://doi.org/10.1111/j.1574-695X.2010.00661.x.

Harding, C. M., Hennon, S. W., & Feldman, M. F. (2018). Uncovering the mechanisms of *Acinetobacter baumannii* virulence. *Nature Reviews Microbiology*, *16*(2), 91–102. https://doi.org/10.1038/nrmicro.2017.148.

Harding, C. M., Pulido, M. R., Di Venanzio, G., Kinsella, R. L., Webb, A. I., Scott, N. E., et al. (2017). Pathogenic *Acinetobacter* species have a functional type I secretion system and contact-dependent inhibition systems. *The Journal of Biological Chemistry, 292*(22), 9075–9087. https://doi.org/10.1074/jbc.M117.781575.

Harms, A., Maisonneuve, E., & Gerdes, K. (2016). Mechanisms of bacterial persistence during stress and antibiotic exposure. *Science (New York, N.Y.), 354*(6318), aaf4268. https://doi.org/10.1126/science.aaf4268.

He, X., Lu, F., Yuan, F., Jiang, D., Zhao, P., Zhu, J., et al. (2015). Biofilm formation caused by clinical *Acinetobacter baumannii* isolates is associated with overexpression of the AdeFGH efflux pump. *Antimicrobial Agents and Chemotherapy, 59*(8), 4817–4825. https://doi.org/10.1128/AAC.00877-15.

Helaine, S., & Kugelberg, E. (2014). Bacterial persisters: Formation, eradication, and experimental systems. *Trends in Microbiology, 22*(7), 417–424. https://doi.org/10.1016/j.tim.2014.03.008.

Hetta, H. F., Al-Kadmy, I., Khazaal, S. S., Abbas, S., Suhail, A., El-Mokhtar, M. A., et al. (2021). Antibiofilm and antivirulence potential of silver nanoparticles against multidrug-resistant *Acinetobacter baumannii. Scientific Reports, 11*(1), 10751. https://doi.org/10.1038/s41598-021-90208-4.

Ibrahim, S., Al-Saryi, N., Al-Kadmy, I., & Aziz, S. N. (2021). Multidrug-resistant *Acinetobacter baumannii* as an emerging concern in hospitals. *Molecular Biology Reports, 48*(10), 6987–6998. https://doi.org/10.1007/s11033-021-06690-6.

Iwashkiw, J. A., Seper, A., Weber, B. S., Scott, N. E., Vinogradov, E., Stratilo, C., … Feldman, M. F. (2012). Identification of a general O-linked protein glycosylation system in Acinetobacter baumannii and its role in virulence and biofilm formation. *PLoS Pathogens, 8*(6), e1002758. https://doi.org/10.1371/journal.ppat.1002758.

Jamal, M., Ahmad, W., Andleeb, S., Jalil, F., Imran, M., Nawaz, M. A., et al. (2018). Bacterial biofilm and associated infections. *Journal of the Chinese Medical Association, 81*(1), 7–11. https://doi.org/10.1016/j.jcma.2017.07.012.

Jamal, M., Tasneem, U., Hussain, T., & Andleeb, S. (2015). Bacterial biofilm: Its composition, formation and role in human infections. *Research & Reviews: Journal of Microbiology and Biotechnology, 4*, 1–14.

Jin, J. S., Kwon, S. O., Moon, D. C., Gurung, M., Lee, J. H., Kim, S. I., et al. (2011). *Acinetobacter baumannii* secretes cytotoxic outer membrane protein A via outer membrane vesicles. *PLoS One, 6*(2), e17027. https://doi.org/10.1371/journal.pone.0017027.

Lee, C. R., Lee, J. H., Park, M., Park, K. S., Bae, I. K., Kim, Y. B., et al. (2017). Biology of *Acinetobacter baumannii*: Pathogenesis, antibiotic resistance mechanisms, and prospective treatment options. *Frontiers in Cellular and Infection Microbiology, 7*, 55. https://doi.org/10.3389/fcimb.2017.00055.

Lees-Miller, R. G., Iwashkiw, J. A., Scott, N. E., Seper, A., Vinogradov, E., Schild, S., et al. (2013). A common pathway for O-linked protein-glycosylation and synthesis of capsule in *Acinetobacter baumannii. Molecular Microbiology, 89*(5), 816–830. https://doi.org/10.1111/mmi.12300.

Lewis, K. (2010). Persister cells. *Annual Review of Microbiology, 64*, 357–372. https://doi.org/10.1146/annurev.micro.112408.134306.

Lin, M. F., & Lan, C. Y. (2014). Antimicrobial resistance in *Acinetobacter baumannii*: From bench to bedside. *World Journal of Clinical Cases, 2* (12), 787–814. https://doi.org/10.12998/wjcc.v2.i12.787.

Liou, M. L., Soo, P. C., Ling, S. R., Kuo, H. Y., Tang, C. Y., & Chang, K. C. (2014). The sensor kinase BfmS mediates virulence in *Acinetobacter baumannii. Journal of Microbiology, Immunology, and Infection = Wei Mian Yu Gan Ran za Zhi, 47*(4), 275–281. https://doi.org/10.1016/j.jmii.2012.12.004.

Lonergan, Z. R., Nairn, B. L., Wang, J., Hsu, Y. P., Hesse, L. E., Beavers, W. N., et al. (2019). An *Acinetobacter baumannii*, Zinc-regulated peptidase maintains cell wall integrity during immune-mediated nutrient sequestration. *Cell Reports, 26*(8), 2009–2018.e6. https://doi.org/10.1016/j.celrep.2019.01.089.

López-Martín, M., Dubern, J. F., Alexander, M. R., & Williams, P. (2021). AbaM regulates quorum sensing, biofilm formation, and virulence in *Acinetobacter baumannii. Journal of Bacteriology, 203*(8), e00635-20. https://doi.org/10.1128/JB.00635-20.

McConnell, M. J., Actis, L., & Pachón, J. (2013). *Acinetobacter baumannii*: Human infections, factors contributing to pathogenesis and animal models. *FEMS Microbiology Reviews, 37*(2), 130–155. https://doi.org/10.1111/j.1574-6976.2012.00344.x.

Mea, H. J., Yong, P., & Wong, E. H. (2021). An overview of *Acinetobacter baumannii* pathogenesis: Motility, adherence and biofilm formation. *Microbiological Research, 247*, 126722. https://doi.org/10.1016/j.micres.2021.126722.

Mohamed, S., Salem, D., Azmy, M., & Fam, N. (2018). Antibacterial and antibiofilm activity of cinnamaldehyde against carbapenem-resistant *Acinetobacter baumannii* in Egypt: In vitro study. *Journal of Applied Pharmaceutical Science, 8*, 151–156. https://doi.org/10.7324/JAPS.2018.81121.

Moon, K. H., Weber, B. S., & Feldman, M. F. (2017). Subinhibitory concentrations of trimethoprim and sulfamethoxazole prevent biofilm formation by *Acinetobacter baumannii* through inhibition of Csu pilus expression. *Antimicrobial Agents and Chemotherapy, 61*(9), e00778-17. https://doi.org/10.1128/AAC.00778-17.

Morán-Barrio, J., Cameranesi, M. M., Relling, V., Limansky, A. S., Brambilla, L., & Viale, A. M. (2017). The *Acinetobacter* outer membrane contains multiple specific channels for carbapenem β-lactams as revealed by kinetic characterization analyses of imipenem permeation into *Acinetobacter baylyi* cells. *Antimicrobial Agents and Chemotherapy, 61*(3), e01737-16. https://doi.org/10.1128/AAC.01737-16.

Morris, F. C., Dexter, C., Kostoulias, X., Uddin, M. I., & Peleg, A. Y. (2019). The mechanisms of disease caused by *Acinetobacter baumannii. Frontiers in Microbiology, 10*, 1601. https://doi.org/10.3389/fmicb.2019.01601.

Nafarieh, T., Bandehpour, M., Hashemi, A., Taheri, S., Yardel, V., Jamaati, H., et al. (2017). Identification of antigens from nosocomial *Acinetobacter baumannii* clinical isolates in sera from ICU staff and infected patients using the antigenome technique. *World Journal of Microbiology and Biotechnology, 33*(10), 189. https://doi.org/10.1007/s11274-017-2355-x.

Narayanan, A., Nair, M. S., Karumathil, D. P., Baskaran, S. A., Venkitanarayanan, K., & Amalaradjou, M. A. (2016). Inactivation of *Acinetobacter baumannii* biofilms on polystyrene, stainless steel, and urinary catheters by octenidine dihydrochloride. *Frontiers in Microbiology, 7*, 847. https://doi.org/10.3389/fmicb.2016.00847.

Nguyen, M., & Joshi, S. G. (2021). Carbapenem resistance in *Acinetobacter baumannii*, and their importance in hospital-acquired infections: A scientific review. *Journal of Applied Microbiology, 131*(6), 2715–2738. https://doi.org/10.1111/jam.15130.

Nie, D., Hu, Y., Chen, Z., Li, M., Hou, Z., Luo, X., et al. (2020). Outer membrane protein A (OmpA) as a potential therapeutic target for *Acinetobacter baumannii* infection. *Journal of Biomedical Science, 27*(1), 26. https://doi.org/10.1186/s12929-020-0617-7.

Niu, C., Clemmer, K. M., Bonomo, R. A., & Rather, P. N. (2008). Isolation and characterization of an autoinducer synthase from Acinetobacter baumannii. *Journal of bacteriology, 190*(9), 3386–3392. https://doi.org/10.1128/JB.01929-07.

Pakharukova, N., Garnett, J. A., Tuittila, M., Paavilainen, S., Diallo, M., Xu, Y., et al. (2015). Structural insight into archaic and alternative chaperone-usher pathways reveals a novel mechanism of pilus biogenesis. *PLoS Pathogens*, *11*(11), e1005269. https://doi.org/10.1371/journal.ppat.1005269.

Pakharukova, N., Tuittila, M., Paavilainen, S., & Zavialov, A. (2015). Crystallization and preliminary X-ray diffraction analysis of the Csu pili CsuC-CsuA/B chaperone-major subunit pre-assembly complex from *Acinetobacter baumannii*. *Acta Crystallographica Section F: Structural Biology Communications*, *71*(Pt. 6), 770–774. https://doi.org/10.1107/S2053230X15007955.

Parsek, M. R., & Singh, P. K. (2003). Bacterial biofilms: An emerging link to disease pathogenesis. *Annual Review of Microbiology*, *57*, 677–701. https://doi.org/10.1146/annurev.micro.57.030502.090720.

Pérez, A., Merino, M., Rumbo-Feal, S., Álvarez-Fraga, L., Vallejo, J. A., Beceiro, A., et al. (2017). The FhaB/FhaC two-partner secretion system is involved in adhesion of *Acinetobacter baumannii* AbH12O-A2 strain. *Virulence*, *8*(6), 959–974. https://doi.org/10.1080/21505594.2016.1262313.

Piepenbrink, K. H., Lillehoj, E., Harding, C. M., Labonte, J. W., Zuo, X., Rapp, C. A., et al. (2016). Structural diversity in the type IV pili of multidrug-resistant *Acinetobacter*. *The Journal of Biological Chemistry*, *291*(44), 22924–22935. https://doi.org/10.1074/jbc.M116.751099.

Pompilio, A., Scribano, D., Sarshar, M., Di Bonaventura, G., Palamara, A. T., & Ambrosi, C. (2021). Gram-negative bacteria holding together in a biofilm: The *Acinetobacter baumannii* way. *Microorganisms*, *9*(7), 1353. https://doi.org/10.3390/microorganisms9071353.

Pour, N. K., Dusane, D. H., Dhakephalkar, P. K., Zamin, F. R., Zinjarde, S. S., & Chopade, B. A. (2011). Biofilm formation by *Acinetobacter baumannii* strains isolated from urinary tract infection and urinary catheters. *FEMS Immunology and Medical Microbiology*, *62*(3), 328–338. https://doi.org/10.1111/j.1574-695X.2011.00818.x.

Raorane, C. J., Lee, J. H., Kim, Y. G., Rajasekharan, S. K., García-Contreras, R., & Lee, J. (2019). Antibiofilm and antivirulence efficacies of flavonoids and curcumin against *Acinetobacter baumannii*. *Frontiers in Microbiology*, *10*, 990. https://doi.org/10.3389/fmicb.2019.00990.

Richards, A. M., Abu Kwaik, Y., & Lamont, R. J. (2015). Code blue: *Acinetobacter baumannii*, a nosocomial pathogen with a role in the oral cavity. *Molecular Oral Microbiology*, *30*(1), 2–15. https://doi.org/10.1111/omi.12072.

Richmond, G. E., Evans, L. P., Anderson, M. J., Wand, M. E., Bonney, L. C., Ivens, A., et al. (2016). The *Acinetobacter baumannii* two-component system AdeRS regulates genes required for multidrug efflux, biofilm formation, and virulence in a strain-specific manner. *mBio*, *7*(2), e00430-16. https://doi.org/10.1128/mBio.00430-16.

Ronish, L. A., Lillehoj, E., Fields, J. K., Sundberg, E. J., & Piepenbrink, K. H. (2019). The structure of PilA from *Acinetobacter baumannii* AB5075 suggests a mechanism for functional specialization in *Acinetobacter* type IV pili. *The Journal of Biological Chemistry*, *294*(1), 218–230. https://doi.org/10.1074/jbc.RA118.005814.

Roy, R., You, R. I., Chang, C. H., Yang, C. Y., & Lin, N. T. (2021). Carboxy-terminal processing protease controls production of outer membrane vesicles and biofilm in *Acinetobacter baumannii*. *Microorganisms*, *9*(6), 1336. https://doi.org/10.3390/microorganisms9061336.

Runci, F., Bonchi, C., Frangipani, E., Visaggio, D., & Visca, P. (2016). Acinetobacter baumannii Biofilm Formation in Human Serum and Disruption by Gallium. *Antimicrobial agents and chemotherapy*, *61*(1), e01563–16. https://doi.org/10.1128/AAC.01563-16.

Rumbo-Feal, S., Gómez, M. J., Gayoso, C., Álvarez-Fraga, L., Cabral, M. P., Aransay, A. M., et al. (2013). Whole transcriptome analysis of *Acinetobacter baumannii* assessed by RNA-sequencing reveals different mRNA expression profiles in biofilm compared to planktonic cells. *PLoS One*, *8*(8), e72968. https://doi.org/10.1371/journal.pone.0072968.

Saipriya, K., Swathi, C. H., Ratnakar, K. S., & Sritharan, V. (2020). Quorum-sensing system in *Acinetobacter baumannii*: A potential target for new drug development. *Journal of Applied Microbiology*, *128*(1), 15–27. https://doi.org/10.1111/jam.14330.

Saranathan, R., Pagal, S., Sawant, A. R., Tomar, A., Madhangi, M., Sah, S., et al. (2017). Disruption of tetR type regulator adeN by mobile genetic element confers elevated virulence in *Acinetobacter baumannii*. *Virulence*, *8*(7), 1316–1334. https://doi.org/10.1080/21505594.2017.1322240.

Sarshar, M., Behzadi, P., Scribano, D., Palamara, A. T., & Ambrosi, C. (2021). *Acinetobacter baumannii*: An ancient commensal with weapons of a pathogen. *Pathogens (Basel, Switzerland)*, *10*(4), 387. https://doi.org/10.3390/pathogens10040387.

Sato, Y., Unno, Y., Kawakami, S., Ubagai, T., & Ono, Y. (2017). Virulence characteristics of *Acinetobacter baumannii* clinical isolates vary with the expression levels of omps. *Journal of Medical Microbiology*, *66*(2), 203–212. https://doi.org/10.1099/jmm.0.000394.

Sato, Y., Ubagai, T., Tansho-Nagakawa, S., Yoshino, Y., & Ono, Y. (2021). Effects of colistin and tigecycline on multidrug-resistant Acinetobacter baumannii biofilms: advantages and disadvantages of their combination. *Scientific reports*, *11*(1), 11700. https://doi.org/10.1038/s41598-021-90732-3.

Scholtz, V., Vaňková, E., Kašparová, P., Premanath, R., Karunasagar, I., & Julák, J. (2021). Non-thermal plasma treatment of ESKAPE Pathogens: A review. *Frontiers in Microbiology*, *12*, 737635. https://doi.org/10.3389/fmicb.2021.737635.

Sechi, L. A., Karadenizli, A., Deriu, A., Zanetti, S., Kolayli, F., Balikci, E., et al. (2004). PER-1 type beta-lactamase production in *Acinetobacter baumannii* is related to cell adhesion. *Medical Science Monitor: International Medical Journal of Experimental and Clinical Research*, *10*(6), BR180–BR184.

Skerniškytė, J., Krasauskas, R., Péchoux, C., Kulakauskas, S., Armalytė, J., & Sužiedėlienė, E. (2019). Surface-related features and virulence among *Acinetobacter baumannii* clinical isolates belonging to international clones I and II. *Frontiers in Microbiology*, *9*, 3116. https://doi.org/10.3389/fmicb.2018.03116.

Smani, Y., Fàbrega, A., Roca, I., Sánchez-Encinales, V., Vila, J., & Pachón, J. (2014). Role of OmpA in the multidrug resistance phenotype of *Acinetobacter baumannii*. *Antimicrobial Agents and Chemotherapy*, *58*(3), 1806–1808. https://doi.org/10.1128/AAC.02101-13.

Stewart, P. S., & Franklin, M. J. (2008). Physiological heterogeneity in biofilms. *Nature Reviews Microbiology*, *6*(3), 199–210. https://doi.org/10.1038/nrmicro1838.

Subhadra, B., Kim, D. H., Woo, K., Surendran, S., & Choi, C. H. (2018). Control of biofilm formation in healthcare: Recent advances exploiting quorum-sensing interference strategies and multidrug efflux pump inhibitors. *Materials (Basel, Switzerland)*, *11*(9), 1676. https://doi.org/10.3390/ma11091676.

Tomaras, A. P., Dorsey, C. W., Edelmann, R. E., & Actis, L. A. (2003). Attachment to and biofilm formation on abiotic surfaces by *Acinetobacter baumannii*: Involvement of a novel chaperone-usher pili assembly system. *Microbiology (Reading, England)*, *149*(Pt. 12), 3473–3484. https://doi.org/10.1099/mic.0.26541-0.

Tomaras, A. P., Flagler, M. J., Dorsey, C. W., Gaddy, J. A., & Actis, L. A. (2008). Characterization of a two-component regulatory system from Acinetobacter baumannii that controls biofilm formation and cellular morphology. *Microbiology (Reading, England)*, *154*(Pt 11), 3398–3409. https://doi.org/10.1099/mic.0.2008/019471-0.

Toyofuku, M., Inaba, T., Kiyokawa, T., Obana, N., Yawata, Y., & Nomura, N. (2016). Environmental factors that shape biofilm formation. *Bioscience, Biotechnology, and Biochemistry*, *80*(1), 7–12. https://doi.org/10.1080/09168451.2015.1058701.

Weinberg, S. E., Villedieu, A., Bagdasarian, N., Karah, N., Teare, L., & Elamin, W. F. (2020). Control and management of multidrug resistant *Acinetobacter baumannii*: A review of the evidence and proposal of novel approaches. *Infection Prevention in Practice*, 2(3), 100077. https://doi.org/10.1016/j.infpip.2020.100077.

Yoon, E. J., Chabane, Y. N., Goussard, S., Snesrud, E., Courvalin, P., Dé, E., et al. (2015). Contribution of resistance-nodulation-cell division efflux systems to antibiotic resistance and biofilm formation in *Acinetobacter baumannii*. *mBio*, *6*(2), e00309–e00315. https://doi.org/10.1128/mBio.00309-15.

Further reading

Clemmer, K. M., Bonomo, R. A., & Rather, P. N. (2011). Genetic analysis of surface motility in *Acinetobacter baumannii*. *Microbiology (Reading, England)*, *157*(Pt. 9), 2534–2544. https://doi.org/10.1099/mic.0.049791-0.

Chapter 31

Polymicrobial biofilms: Impact on fungal pathogenesis

Caroline Barcelos Costa-Orlandi[a], Níura Madalena Bila[b], Carolina Orlando Vaso[b], Ana Carolina Moreira da Silva Pires[c], Samanta de Matos Silva[d], Kaila Petronila Medina Alarcón[a], Caroline Maria Marcos[e], Ana Marisa Fusco-Almeida[e], and Maria José Soares Mendes-Giannini[f]

[a]*Department of Clinical Analysis, School of Pharmaceutical Sciences, São Paulo State University (UNESP), Araraquara, SP, Brazil,* [b]*School of Pharmaceutical Science, Universidade Estadual Paulista, Araraquara, SP, Brazil,* [c]*Food and Nutrition, UNESP, Araraquara, São Paulo, Brazil* [d]*Sao Paulo State University, São Paulo, Brazil,* [e]*School of Pharmaceutical Sciences, São Paulo State University (UNESP), Araraquara, Brazil* [f]*Clinical Analysis, School of Pharmacy-UNESP, Araraquara, Sao Paulo, Brazil*

1. Introduction

Biofilms are sessile communities of microorganisms (MOs) that assemble in an organized three-dimensional (3D) microbial structure with cells enclosed within a self-produced extracellular matrix (ECM) that can be attached or not to a substrate (Arciola, Campoccia, & Montanaro, 2018; Costerton, Stewart, & Greenberg, 1999). MOs can also self-aggregate or coaggregate to form biofilms in pure or mixed cultures, respectively, if no solid substrate is available for their fixation (Alhede et al., 2011). These communities differ from free-living planktonic cells in growth, gene transcription, antimicrobials tolerance, and resistance by host defenses (Costerton, Geesey, & Cheng, 1978; Costerton, Lewandowski, Caldwell, Korber, & Lappin-Scott, 1995; Donlan, 2002; Donlan & Costerton, 2002).

Sessile communities can be formed on several surfaces, including living tissue, internal medical devices, industrial pipes, drinking water systems, or natural aquatic systems. Depending on the biofilm environment, mineral crystals, corrosion particles, clay or sludge particles, or even blood components may also be identified in the biofilm matrix (Burgos-Garay et al., 2021; Muhammad et al., 2020; Peters, Jabra-Rizk, O'May, Costerton, & Shirtliff, 2012).

MOs such as bacteria and fungi can form biofilms, which share many common attributes. Both can be found in the same microhabitats creating dynamic communities known as polymicrobial biofilms. Physical, chemical, and biological interactions between these MOs play a fundamental role in the performance of numerous ecosystems and impact the health and diseases of plants and animals (Deveau et al., 2018). Understanding the nature and the correct description of the bacterial-fungal interaction mechanism are essential tools for new treatment strategies in the case of diseases produced by these pathogens (Deveau et al., 2018). Polymicrobial interactions can be interspecies, intergenus, or even interkingdoms, with a spatial organization that depends on the functional regulation of the biofilm (Guo, Chen, Guo, & Chen, 2021). They can compete antagonistically for nutrients and adhesion sites, but they can also coexist in cooperative ways, supporting their mutual development neutrally or synergistically (Costa-Orlandi et al., 2017; Gabrilska & Rumbaugh, 2015; Garcia et al., 2020; Khoury et al., 2020).

Polymicrobial biofilms are significant from a human health point of view, prevalent throughout the human body, both during health and disease (Allison et al., 2016; Peters, Jabra-Rizk, O'May, Costerton, & Shirtliff, 2012; Stacy, McNally, Darch, Brown, & Whiteley, 2016). They are often worse than similar monomicrobial infections and exhibit more remarkable pathogen persistence at the site of infection, increased disease severity, and increased antimicrobial resistance (Peters et al., 2012). Moreover, biofilms can potentiate infection with chronic inflammatory induction that leads to collateral damage to host tissue while protecting pathogens against antimicrobial action, immune host factors, and conditions from the environment (Gabrilska & Rumbaugh, 2015). This condition is considered the primary cause of persistent nosocomial infections, of high virulence and recalcitrant to treatments, especially in immunocompromised patients who use devices such as catheters, contact lenses, prostheses, heart valves, dentures, among others (Ponde, Lortal, Ramage, Naglik, & Richardson, 2021; Roy, Tiwari, Donelli, & Tiwari, 2018). Fungal and bacterial biofilms account for approximately 65%–80% of all hospital-acquired infections (Harding, Marques, Howard, & Olson, 2009; Ramage, Rajendran, Gutierrez-Correa, Jones, & Williams, 2011). The bloodstream and urinary and respiratory tract are the significant locations of biofilm-associated nosocomial

Understanding Microbial Biofilms. https://doi.org/10.1016/B978-0-323-99977-9.00040-5

infections because those are common for plantation internal medical devices. The biofilm phenotype promotes advantages to MOs caused by physical and genetic factors and protects against the host's immune system through impaired activation of phagocytes and the complement system. In addition, for its eradication, it is necessary to increase the concentration of antimicrobials from 100 to 1000 times compared with planktonic cells (Roy et al., 2018). Persistent chronic infections in different parts of the human body such as the lungs, oral cavity, middle ear, genitourinary tract, nails, surgical or chronic wounds originate from polymicrobial biofilms commonly leading to worse prognoses (Garcia et al., 2020; Gupta & Foley, 2019; Orazi & O'Toole, 2019).

In this chapter, aspects related to polymicrobial biofilms will be addressed to unravel the exchange of functions in different interspecies or even in interkingdom conjunctions and their role in fungal pathogenesis.

2. Polymicrobial interactions

The assessment of biofilm phenotype during infection is still relatively recent, and minor interactions between MOs during polymicrobial growth and the consequences for infection mediated by them have recently improved (Elias & Banin, 2012; Peters, Jabra-Rizk, et al., 2012). Therefore, this section exposes some aspects of the polymicrobial interaction from bacterial-fungal to fungi-fungi associations, regarding the benefits or harms for each MO involved and relevance to human infections.

Bacteria, mainly *Staphylococcus* and *Streptococcus* species, can form mixed biofilms with fungal species (Ponde et al., 2021). The ability to colonize surfaces and form biofilms has been demonstrated and encompasses several fungi. Most studies on fungal biofilms have focused on *Candida albicans* and, more recently, on other yeast genera and *Candida* species, as well as on filamentous fungi. Recently, most fungi of clinical importance have produced biofilms in vivo. This group includes filamentous fungi (*Aspergillus*, *Fusarium*, zygomycetes, dermatophytes), yeasts (*Blastoschizomyces*, *Saccharomyces*, *Malassezia*, *Trichosporon*, *Cryptococcus*, *Rhodotorula*, and numerous *Candida* spp.), *Pneumocystis* species, and endemic fungi (*Histoplasma capsulatum*, *Paracoccidioides brasiliensis*, *Coccidioides immitis*) (Gupta & Foley, 2019; Nett & Andes, 2015; Sardi et al., 2014).

The polymicrobial biofilm offers numerous advantages to the MOs, including increased competitiveness, absent in planktonic cells (Lohse, Gulati, Johnson, & Nobile, 2018). Microbes housed in the ECM are more likely to resist mechanical stress and be better protected from the external environment and hostile host immune responses (Cavalheiro & Teixeira, 2018; Kernien, Snarr, Sheppard, & Nett, 2017; Mathé & Van Dijck, 2013). Mixed-species biofilms can produce and utilize a wide variety of nutrients through the formation of metabolic consortia in which waste products from one species can be metabolized by another (Elias & Banin, 2012). These communities exhibit enormous genomic diversity and, consequently, biofilms provide a suitable environment in which new genetic material can be acquired through the horizontal/lateral gene transfer process, particularly between bacterial species (Roberts & Kreth, 2014).

The polymicrobial relationships manifested in the mixed biofilm range from contact-mediated adhesion, cell-cell communication via quorum sensing (QS) cross talk, increased colonization, increased expression of virulence phenotypes, immunomodulation, and the combination of all these events (Krüger, Vielreicher, Kapitan, Jacobsen, & Niemiec, 2019).

Molecular techniques such as next-generation sequencing, transcriptomics analysis, and others evaluate the consequences of the polymicrobial interaction, such as the contribution to the colonization, survival, infection dynamics, and resistance to host factors or antimicrobials (Zhang, Hu, Andini, & Yang, 2019).

Fungi and bacteria are found on all epithelial mucosal surfaces in the human body, and the polymicrobial interactions can be classified as synergism when one MO creates a favorable niche for colonization or infection by another MO, or the MO can interact with the host and thus predispose to colonization by the second microbial. During this microbial interference, interaction with the host can also result in reduction or avoidance of colonization by the other MO, which is called antagonism, and in addition, when two nonpathogenic MOs start to cause disease when combined (Fig. 1) (Allison et al., 2016; Brogden, Guthmiller, & Taylor, 2005; Krüger et al., 2019).

Considering the infection by multiple MOs, the interaction between them is affected according to the host and its niche. The availability of nutrients, oxygen levels, and contact with the host's immune system differ according to this niche on the host. Such changes in homeostasis found by MOs can modify their metabolic status and their virulence potential, whether bacteria or fungus (Krüger et al., 2019).

Studying these polymicrobial interactions can contribute to identifying new targets to damage biofilm development.

2.1 Fungi-bacteria polymicrobial biofilms

Microbiome studies indicate that in the human body, there are 48 major microbial habitats (Bandara, Panduwawala, & Samaranayake, 2019; de Negrini, Koo, & Arthur, 2019; Ghannoum et al., 2010; Proal, Lindseth, & Marshall, 2017; Radaic & Kapila, 2021); 34% of all primary microbial habitats were associated with human skin, 25% were associated with the gastrointestinal tract, and 20% were associated with the head and neck region.

In the context of a biofilm, *C. albicans* may interact with several bacterial species (Gram-positive cocci and

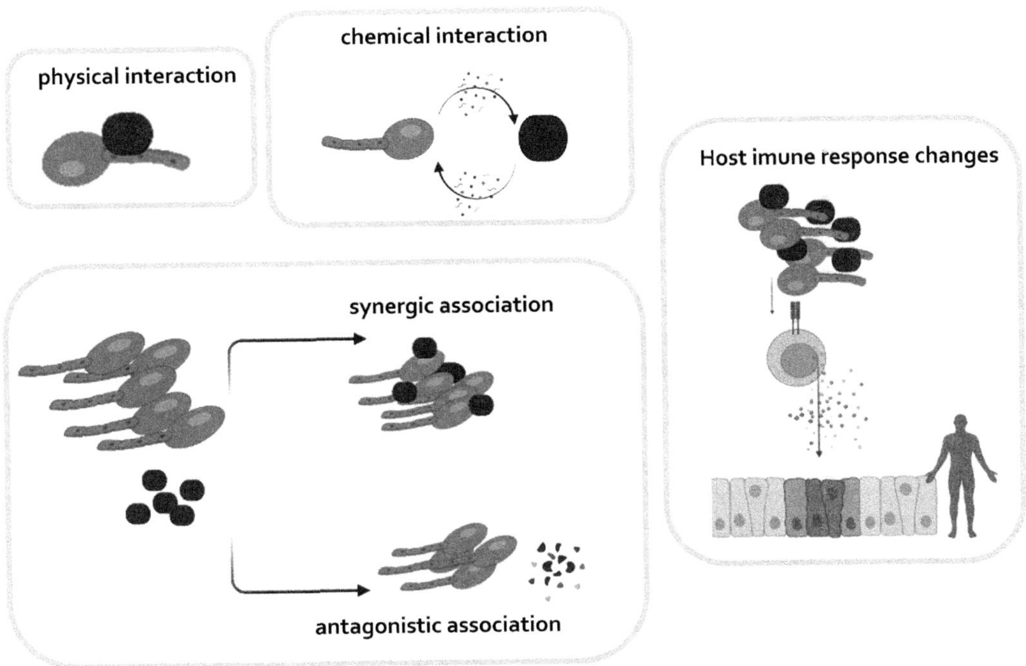

FIG. 1 Types of polymicrobial interactions. Mechanistic bases of polymicrobial interactions. The types of interaction described in this chapter are compiled here. Chemical interaction through signals, metabolites, proteins, or even host immune system or even direct contact represents a link allowing complex interaction between different MOs that can be synergistic or antagonistic (Created with BioRender.com).

bacilli, Gram-negative bacilli) and can directly bind to *C. albicans* hyphal cells.

Microbial populations represent the initial understanding and complexity of polymicrobial biofilm communities in the oral cavity (de Negrini et al., 2019; Peters, Jabra-Rizk, et al., 2012; Radaic & Kapila, 2021). The typical example of biofilm polymicrobial is related to dental plaques, and *Streptococcus gordonii* is one of the early-stage colonizers in the oral cavity and can bind directly to salivary proteins on the tooth surface, establishing a foundation for building the biofilm communities (Wang, Li, & Ling, 2017). MOs' communication plays an essential role in affecting the dental plaque community's composition and virulence properties (Kolenbrander, Egland, Diaz, & Palmer, 2005). Avenues of MO's communication include metabolic communication, genetic exchange, and QS (Hojo, Nagaoka, Ohshima, & Maeda, 2009; Wang et al., 2017). The interactions among MOs may occur by chemicals and QS, a microbial signaling method in which genes are regulated based on microbial cell density and the surrounding environment (Bassler, 1999; Wang et al., 2017). When the concentration of signaling molecules accumulates and reaches a maximum level, MOs can respond in diverse ways, including bacterial and fungal surface adhesion, extracellular matrix production, and virulence factor expression (Lohse et al., 2018; Ponde, Lortal, Ramage, Naglik, & Richardson, 2021; Wang, Li, & Ling, 2017).

As shown below, interactions between *C. albicans* and other microbes can be synergistic, mutually beneficial, or antagonistic, and these interactions can dictate the microbiota composition.

Regarding polymicrobial interactions, most studies describe the interaction of *Candida* species with other MOs, mainly *C. albicans*. This yeast can be a commensal in the host, occupying different niches (such as the gastrointestinal tract, reproductive tract, oral cavity, and skin) and, therefore, coexisting with numerous species of bacteria such as *Staphylococcus aureus*, *S. gordonii*, *Streptococcus mutans*, and others, on denture surfaces or oral mucosa of denture users (Allison et al., 2016; Satala et al., 2021; Baena-Monroy et al., 2005). *C. albicans* form biofilms composed of yeast cells, pseudohyphae, and hyphae. The transition from planktonic to biofilm growth is accompanied by a complex phenotypic remodeling followed by gene expression changes (Ponde et al., 2021). Hyphae are essential components that support the structural integrity of the biofilm and provide a framework for the attachment of additional yeast cells, pseudohyphae, other hyphae, and bacteria in multispecies biofilms (Chandra et al., 2001; Harriott & Noverr, 2009). *C. albicans* with other MOs can occur via coaggregation and coadhesion. *C. albicans* adhesins such as agglutinin-like sequence (Als) protein (Als1p, Als2p, Als3p) and hyphal wall protein 1 (Hwp1p) facilitate interaction with bacterial species such as *S. gordonii* and *S. aureus* (Allison et al., 2019; Hoyer, Oh, Jones, & Cota, 2014; Peters, Jabra-Rizk, et al., 2012; Ponde et al., 2021; Schlecht et al., 2015).

Dutton et al. (2016), through transcriptomic analysis of *C. albicans* and *S. gordonii* cocultivation, demonstrated the altered expression of essential genes as those related to the morphogenesis and arginine synthesis in *C. albicans*, the mixed biofilm of both favors the growth of *Candida* hyphae. This situation can occur since streptococci relieve the farnesol (produced by *Candida*) action, which consists of repression of hyphal development by *C. albicans* (Bamford et al., 2009). The presence of *S. mutans* and *C. albicans* in the oral microbiota substantially increases bacterial colonization and benefits the development of dental caries, suggesting that this interaction may be a crucial factor in improving the risk of disease (De Carvalho, Silva, Hebling, Spolidorio, & Spolidorio, 2006; Khoury et al., 2020). The Ssp and Csh polypeptides are crucial for *S. gordonii*, binding to various bacterial, human receptors, and in *C. albicans* (Demuth et al., 1996; Holmes, Mcnab, & Jenkinson, 1996; McNab, Jenkinson, Loach, & Tannock, 1994). *S. gordonii* binds directly to *C. albicans* hypha-specific adhesins Als3p, Eap1p, and Hwp1p through the streptococcal, belonging to the family of antigen I/II polypeptides (Bamford et al., 2009; Holmes et al., 1996; Silverman, Nobbs, Vickerman, Barbour, & Jenkinson, 2010). SspA and SspB mutants of *S. gordonii* show reduced expression of streptococcal adhesins for *C. albicans*, with less coaggregation of *C. albicans* with *S. gordonii* mutants (Klotz et al., 2007). There are two types of recognized QS systems in streptococci. The ComCDE system generates competence stimulating peptide (CSP), which enables intraspecies cell-to-cell communication and modulates the expression of many genes in the same bacterial species modulating the dual-species biofilm with *C. albicans* (Jack et al., 2015); the other is LuxS/autoinducer-2 (AI-2) system, which can mediate interspecies interactions in the multispecies plaque community. The *S. gordonii* QS factor AI-2 stimulates *C. albicans* filamentation, and *S. gordonii* can block the inhibitory effect of exogenous farnesol on *C. albicans* filamentation and biofilm formation (Bamford et al., 2009). *S. gordonii* expression of glucosyltransferase G (GtfG) also contributes to coaggregation with *C. albicans* (Ricker, Vickerman, & Dongari-Bagtzoglou, 2014).

This synergistic type of interaction between *C. albicans* and *S. gordonii* provides adhesion sites to streptococci that, on the other hand, excrete lactate employed in the yeast growth as a carbon source (Holmes, Cannon, & Jenkinson, 1995; Jenkinson, Lala, & Shepherd, 1990; Jenkinson & Lamont, 2005). Furthermore, chemical (diffusible) interaction with activation or inhibition of signaling pathways can occur during polymicrobial communities. Three mitogen-activated protein kinases of *Candida* important to the morphogenesis and cell wall biogenesis can be modulated by *S. gordonii*. As a result, the association of both leads to Cek1p (a) activation, inhibition of Mkc1p, and Hog1 activation by farnesol (Bamford et al., 2009; Silverman et al., 2010). These dates suggested that interactions between *C. albicans* and *S. gordonii* involve physical (adherence) and chemical signals that influence the development of biofilm communities. Thus, bacteria may play a significant role in modulating *Candida* carriage and infection processes in the oral cavity (Bamford et al., 2009). The formation of these mixed fungal/bacterial biofilms led to high levels of antibiotic resistance (Montelongo-Jauregui, Saville, & Lopez-Ribot, 2019).

The Gram-negative coccobacillus *Aggregatibacter actinomycetemcomitans* is a commensal MO in the oral niche but is responsible for periodontal disease with *C. albicans* coisolation in patients with severe conditions periodontitis (Brusca et al., 2010; Järvensivu, Hietanen, Rautemaa, Sorsa, & Richardson, 2004; Saarela, Doğan, Alaluusua, & Asikainen, 1999). As *S. gordonii* can secrete AI-2, synthetized by LuxS enzyme, this bacteria was able to repress *C. albicans* filamentation and reduce the biofilm formation (Bachtiar et al., 2014; Bamford et al., 2009), an observation that contrasts with founded to *Candida-S. gordonii* association in which AI-2 inhibits farnesol relieving its action and increasing the *C. albicans* biofilm formation (Bamford et al., 2009). These opposite responses related to the AI-2 of *S. gordonii* and *A. actinomycetemcomitans* can be due to differences in its chemical structure that *Candida* recognizes (Bachtiar et al., 2014), and the interspecies communication can be more complex than imagined.

S. mutans is the principal causative agent of dental caries in humans, presenting adhesion, acidogenicity, acid tolerance properties, and the ability to form biofilms (Esberg, Löfgren-Burström, Ohman, & Strömberg, 2012; Hwang, Klein, & Koo, 2014; Lemos, Abranches, & Burne, 2005; Quivey Jr., Faustoferri, Monahan, & Marquis, 2000; Wang & Ren, 2017). One of the best-characterized synergistic relationships is that of *C. albicans* and *S. mutans*. This association can be mediated by the glucosyltransferase B (GtfB), an exoenzyme of *S. mutans* that enables its tight binding to *C. albicans* and can be an explanation of the higher frequency of isolation of both in plaque biofilm from children affected with caries, demonstrating the effectiveness of these enzymes in modulating the cross-kingdom interaction (Hwang, Marsh, Gao, Waugh, & Koo, 2015). Glucosyltransferases from *S. mutans* bind directly to mannans on the surface of the yeast *C. albicans* and the cell walls of hyphae, promoting the development of an extensive ECM, accumulating more biomass, and the formation of a mixed-species biofilm (Ellepola, Liu, Cao, Koo, & Seneviratne, 2017; Hwang et al., 2015). The resulting three-dimensional biofilm architecture displays microcolonies of *S. mutans* surrounded by fungal cells, which are enmeshed in a dense matrix rich in extracellular polymeric substances (EPS). Coinfection synergistically increases biofilm virulence, leading to aggressive disease and supporting the development of dental caries, suggesting that this interaction may be a pivotal factor in increasing disease risk (De Carvalho et al., 2006; Khoury et al., 2020). *C. albicans* induces expression of virulence genes in *S. mutans* (gtfB, fabM), and *Candida*-derived β-1,3-glucans contribute to the structure of the EPS matrix, while fungal mannan and β-glucan provide sites for GtfB binding and activity. Mutants to glucosyltranferases in *S. mutans* demonstrate reduced ability to form mixed biofilms with *C. albicans* (Falsetta et al., 2014; Hwang et al., 2017). On the other hand, *S. mutans* excrete lactate, which acts as a carbon source for *C. albicans* (Metwalli, Khan, Krom, & Jabra-Rizk, 2013), and the yeast farnesol enhances *S. mutans* cell growth and microcolony development (Kim et al., 2017). *S. mutans* QS molecules (mutanobactin A, *trans*-2-decenoic acid, and competence-stimulating peptide) can inhibit *C. albicans* germ tube formation and the yeast-to-hyphae transition (Jarosz, Deng, van der Mei, Crielaard, & Krom, 2009; Joyner et al., 2010; Ponde et al., 2021; Vílchez et al., 2010). Overall, this demonstrates a mutually beneficial relationship between *C. albicans* and *S. mutans* with a synergistic partnership that can alter the oral microbiota and influence disease (Ponde et al., 2021).

The interaction between *S. aureus* and *C. albicans* is multifaceted, intrinsically linked to biofilm-phased development. Initial attachment of *S. aureus* to *C. albicans* hyphae appears to be mediated by *C. albicans* Als3 (Kean et al., 2017; Peters, Ovchinnikova, et al., 2012). In sequence, QS molecules have been shown to exhibit both synergistic and antagonistic effects, with either organism secreting reciprocal QS molecules to influence biofilm growth positively and negatively (Jabra-Rizk, Meiller, James, & Shirtliff, 2006; Lin, Alsad, Vogel, Koppar, et al., 2013). Additionally, *C. albicans* positively and negatively regulates cell division and morphogenesis from yeast to filamentous growth using the organic compounds tyrosol and farnesol, respectively (Chen, Fujita, Feng, Clardy, & Fink, 2004; Sato, Watanabe, Mikami, & Matsumoto, 2004).

Antagonistic effects can be seen in *Enterococcus faecalis-C. albicans* communication. *E. faecalis*, a Gram-positive bacterium, can colonize various niches of the host, including the gastrointestinal tract, the oral cavity, and is frequently isolated with *C. albicans* from endodontic infections (Dahlén, Samuelsson, Molander, & Reit, 2000; Gomes et al., 2008). *E. faecalis* genotype csp2 (capsular polysaccharide type 2) unencapsulated was able to inhibit the fungal-biofilm development in a time-dependent manner (48h, maturation phase) with reduction of transcription level of EFB1 mRNA, a biofilm maturity marker (Bachtiar, Dewiyani, Surono Akbar, & Bachtiar, 2016).

Another species found in the gastrointestinal tract is *Helicobacter pylori*, a Gram-negative bacterium responsible for causing gastric and extra-gastric pathologies. In addition to infecting eukaryotic cells, it also invades *Candida* yeasts. This last condition is favorable to bacteria, as yeasts are more resistant to environmental variations, such as nutrient deficiencies or changes in pH; in addition, *H. pylori* promotes changes in pH that facilitate the growth of *C. albicans* (García, Salas-Jara, Herrera, & González, 2014; Hathroubi, Servetas, Windham, Merrell, & Ottemann, 2018; Sánchez-Alonzo et al., 2021). More recently, Palencia, García, and Palencia (2022) demonstrated mechanisms that lead to the coaggregation and dual-species biofilm formation, *C. albicans* and *H. pylori*, contributing to the prevalence and pathogenicity of the bacteria through the following types, hydrophobic and van der Waals interactions between nonpolar peptides of *C. albicans* cell wall and lipid substance or nonpolar peptides from *H. pylori* membrane.

Among the Gram-negative bacteria that can interact with *Candida*, we can also highlight the *Salmonella enterica* serovar Typhimurium, which causes gastrointestinal infections. This polymicrobial infection shows antagonistic effects and *C. albicans* filamentation inhibition, and it is suggested that this occurs due to secreted molecules by *S. typhimurium* once the culture filtrate of the bacteria presents anticandidal activity (Tampakakis, Peleg, & Mylonakis, 2009). Posteriorly, (Kim & Mylonakis, 2011) showed that *C. albicans* supernatant could upregulate *S. typhimurium* genes (*sopB* and *sipB*) involved in the filaments killing and biofilm repression.

More recently, Bartnicka et al. (2019) evaluated the mutual colonization of *C. albicans* and *Porphyromonas gingivalis*, an anaerobic bacterium able to cause periodontal disease, which varies from curable gingivitis to irreversible severe periodontitis (Hajishengallis, 2015). The *C. albicans* biofilm comprises a hypoxic environment inside them that contributes to the growth of anaerobic bacteria (Fox et al., 2014).

A synergic interaction occurs between *C. albicans* and the periodontal pathogen *P. gingivalis*. This strict anaerobe bacterium benefits from fungal biofilm as it protects aerobic situations with increased viability of *P. gingivalis* (Karkowska-Kuleta et al., 2018). In *C. albicans-P. gingivalis* interaction, some factors can be involved, such as Als3, aspartic proteases (Sap6 and Sap9), enolase and Mp65 proteins of the fungi, and the extracellular enzyme peptidyl arginine deiminase used by the bacterium to adhere to fungi and proteolytic enzymes (arginine and lysine-specific cysteine proteases) called gingipains with hemagglutinin domains (Bartnicka et al., 2019; Karkowska-Kuleta et al., 2018).

C. albicans/S. aureus biofilm can be considered synergistic type since there is differential expression of proteins, and some are considered as virulence factors for the bacterium as the L-lactate dehydrogenase 1, critical to the resistance to host-derived oxidative stressors, indicating that *C. albicans* can increase the *S. aureus* pathogenesis (Peters et al., 2010). The coexistence of both species, like a consortium in a biofilm, can result in increased resistance of *S. aureus* to vancomycin, which is remarkable since *C. albicans* and *S. aureus* are often encountered in the bloodstream of hospitalized patients (O'Connell et al., 2006). Structurally, the ECM of *C. albicans* has been shown to advantageously support these dual-species biofilms through the prevention of diffusion of vancomycin to access *S. aureus*, with the bacteria potentially encasing itself within this polymeric material (Adam, Baillie, & Douglas, 2002; Harriott & Noverr, 2009). Therefore, drug diffusion seems to be reduced in mixed biofilm (Harriott & Noverr, 2009; Pammi, Liang, Hicks, Mistretta, & Versalovic, 2013). Kong et al. (2016) demonstrated that β-1,3-glucan produced by *C. albicans*, a vital component of the extracellular matrix, coats *S. aureus* and favors protection against antibiotics.

Schlecht et al. (2015) demonstrated using a murine model of oral candidiasis in which *S. aureus* was coinoculated with *C. albicans* that mortality and morbidity were higher than infection with the isolated pathogens. This result is consistent with systemic infection since *S. aureus* was detected in the analysis of renal load. Therefore, this double biofilm in the oral cavity can contribute to systemic diseases.

Dual-kingdom biofilm, fungi, and bacteria, as *C. albicans* and three pathogens of the respiratory system *Pseudomonas aeruginosa*, *Escherichia coli*, and *S. aureus* may be commonly founded on endotracheal tubes and favor the occurrence of a type of lung infection called ventilator-associated pneumonia (VAP) in hospitalized patients who are on mechanical breathing machines (Luo et al., 2021).

Biofilm formation is a significant concern for cystic fibrosis pneumonia caused by *P. aeruginosa* and urinary tract infections caused by *E. coli*, which are refractory to multiple antibiotics. *C. albicans* biofilm enhanced the number of *E. coli*, *S. aureus*, and the clinical isolate of methicillin-resistant *S. aureus* (Luo et al., 2021).

The interaction of *C. albicans* with *P. aeruginosa* results in an antagonism effect with the formation of a dense biofilm on *C. albicans* filaments that kills the fungus due to the action of important virulence factors of the bacteria as type IV pili, phospholipase C, and phenazines (a redox-active compound with antibiotic properties against bacterial and eukaryotic species). The destruction of hyphae but not *C. albicans* yeast cells suggests a potential survival mechanism developed by the fungus that allows yeast to evade destruction by *P. aeruginosa* (Gibson, Sood, & Hogan, 2009; Hogan & Kolter, 2002).

Additionally, *P. aeruginosa* produced 3-oxo-C12 homoserine lactone, a compound related to farnesol (a QS molecule known for inhibiting filamentation), also represses the fungal filamentation and kills hyphal cells (Gibson, Sood, & Hogan, 2009; Hogan, Vik, & Kolter, 2004). On the other hand, *C. albicans* secretes farnesol, a QS molecule that inhibits the Ras1-cyclic AMP (cAMP)-protein kinase A (PKA) (Ras1-cAMP-PKA) signaling pathway to repress hyphal development (Davis-Hanna, Piispanen, Stateva, & Hogan, 2008; Hornby et al., 2001), inhibits biofilm formation (Ramage, Saville, Wickes, & López-Ribot, 2002), and upregulates the expression of FCR1 and PDR16, which are involved in resistance to azole therapy (Cao et al., 2005). The inhibitory effect of *P. aeruginosa* also occurs in other species of *Candida* such as *Candida parapsilosis*, *Candida tropicalis*, *Candida dubliniensis*, and *Candida glabrata* (Bandara, Yau, Watt, Jin, & Samaranayake, 2010).

In addition to *P. aeruginosa*, other bacteria as *S. mutans* and *Burkholderia cenocepacia* were also able to produce homoserine lactone, as well as *trans*-2 decenoic and *cis*-2 decenoic acids, being these last two QS molecules are able to arrest *C. albicans* cells from undergoing germ-tube induction (Boon et al., 2008; Vílchez et al., 2010; Zhang, Cai, Yang, Weng, & Wang, 2011). *B. cenocepacia* can cause a severe pulmonary infection in patients with cystic fibrosis, frequently founded with *C. albicans*, and both present an antagonistic interaction (Boon et al., 2008; Nunvar, Capek, Fiser, Fila, & Drevinek, 2017). Besides the description of higher frequency of isolation of *Candida* and *P. aeruginosa* in the lungs, various authors mentioned the same for *Candida* and *Mycobacterium tuberculosis* in patients with tuberculosis, with this interaction showing synergic effects (Lohse et al., 2018; Mankiewicz & Liivak, 1960; Ochieng et al., 2005).

The association of *C. albicans* with *Klebsiella pneumoniae* in a biofilm model results in an antagonistic effect reducing the thickness of *C. albicans* biofilm (Fox et al., 2014). (Galdiero et al., 2021) evaluated integrated data on biomass, cell viability of polymicrobial biofilms with *C. albicans* and *K. pneumoniae* by a gas chromatography-mass spectrometry (GC-MS)-based metabolomics approach to explore changes that characterize the dynamics of biofilm development of the two genera. Metabolic data suggest that a prevalence of *C. albicans* in the early phase of bispecies biofilm culture may be due to the ability of the fungus to oppose the growth of *K. pneumoniae* by maintaining active glucose metabolism. There are also strict interactions between the two microbes in the fungal culture/biofilm stress-related metabolites produced by *C. albicans* (vitamin B6, trehalose) that are more abundant in the mature dual-species biofilm (Galdiero et al., 2021).

The *C. albicans-Clostridium perfringens* biofilm leads to an aggregation of *C. albicans* in a "mini-biofilm" with internal hypoxia that enables the anaerobic bacteria to survive in a usually toxic environment (Fox et al., 2014).

In the vaginal microbiota, lactobacilli act as a defense against vulvovaginal candidiasis. The defense employed by different strains of *Lactobacillus*, such as *Lactobacillus crispatus* and *Lactobacillus gasseri*, displays a better anti-*Candida* activity when present in a biofilm form, as demonstrated using metabolome assay. Antagonistic interactions between them may occur due to lactobacilli's nutrient competition and carbohydrate metabolism. Cell-free supernatants with higher anti-*Candida* activity showed increased amounts of pyruvate and acetate (Parolin et al., 2021), potentiated by the low pH triggered by lactic acid produced by lactobacilli by fermentation and acidification. The reduced pH can inhibit *Candida* adherence to the vaginal mucosa and hyphae formation (Boskey, Cone, Whaley, & Moench, 2001; O'Hanlon, Moench, & Cone, 2013; Parolin et al., 2021; Strus et al., 2005).

After coisolation of *E. coli* with *C. albicans* in a patient with symptomatic vulvovaginal candidiasis, Farrokhi, Al-Shibli, Al-Hameedawi, Neshati, and Makhdoumi (2021) verified that the bacteria contribute to the fungal pathogenicity potential. It increases the *C. albicans* filamentation, biofilm formation, expression of ALS3 (involved in the biofilm maintenance) and HWP1 (adhesin involved in the initial attachment of *C. albicans*), invasiveness, tissue damage, antibiotic resistance (to nystatin), and mortality in *Galleria mellonella* model of infection; drawing attention to the possibility of new therapy approaches even more considering the prevalence of vaginal candidias is and increased recurrence. Opposite findings have been described in the literature, with antagonistic effects of *C. albicans-E. coli* association, but explained by differences in experimental design, for example, culturing both strains simultaneously or forming the bacterial biofilm first and later the fungal one; furthermore, culture media composition as well the strains can influence (Bandara, Cheung, Watt, Jin, & Samaranayake, 2013; Cabral, Penumutchu, Norris, Morones-Ramirez, & Belenky, 2018; Farrokhi et al., 2021; Hall & Noverr, 2017).

Acinetobacter baumannii can be opportunistic in urinary tract infection, wound infections, and septicemia in immunocompromised patients, causing severe disease. In addition to the human alveolar epithelial cells attachment to forms biofilm, *A. baumannii* also can adhere to *C. albicans* hyphae (but not to yeast) and kill them by apoptosis (Gaddy & Actis, 2009; Gaddy, Tomaras, & Actis, 2009; Lee et al., 2008; Loehfelm, Luke, & Campagnari, 2008; Niu, Clemmer, Bonomo, & Rather, 2008; Tomaras, Dorsey, Edelmann, & Actis, 2003).

A polymicrobial biofilm can also occur in chronic wound infection, which causes significant mortality and morbidity, mainly in a diabetic and obese population. Kalan et al. (2016) isolated *C. albicans*, *Trichosporon asahii*, *Citrobacter freundii*, and *Staphylococcus simulans* from samples from patient's wounds and demonstrated their ability to form dual biofilms (*C. albicans* plus *C. freundii* and *T. asahii* plus *S. simulans*). *C. albicans* and *C. freundii* were analyzed in the human skin ex vivo model, showing that the bacteria can attach to the fungi mediating mannose-binding lectins that trigger fungal filamentation and increase the neutrophils death and inflammation (Cheong et al., 2021).

Succinctly, *C. albicans*-bacterial interactions can be synergistic when *C. albicans* biofilm can physically protect bacteria from antimicrobial agents or protect anaerobic bacteria from high oxygen concentrations by providing a low oxygen niche within the depths of the biofilm. On the other hand, signaling molecules produced by bacterial species and *C. albicans* enable interkingdom communication within mixed-species biofilms. *C. albicans*-bacterial interactions can be antagonistic when acid produced results in a lower local pH, inhibiting *C. albicans* hyphal formation. Cells within mixed-species biofilms can also exchange nutrients, cells can share nutrients, or one species can utilize available nutrients produced by the other.

Hoarau et al. (2016) evaluated the gut microbiota (bacteriome and mycobiome) of patients with Crohn's disease (CD), which occurs mediating an interplay between these microbial communities and host genetic factors that leads to an inflammatory bowel disease affecting several parts of the gastrointestinal tract. They identified microbial interactions that were subsequently confirmed using single and mixed biofilms. The most abundant fungus was *C. tropicalis*, and its presence showed correspondence with both bacteria, *Serratia marcescens*, and *E. coli*, suggesting an interaction between them.

The mass and thickness were greater than single or dual biofilms when considering triple biofilm *C. tropicalis-S. marcescens-E. coli*. The single biofilm of *C. tropicalis* is composed majority of blastopores. In contrast, there was an enrichment of hyphae in the mixed biofilms associated with pathogenic conditions. *E. coli* was closely apposed with the fungi, and *S. marcescens* interacted with the other two through its fimbriae (Hoarau et al., 2016). *E. coli* was also able to produce polysaccharides that contribute to *Candida* biofilm maturation (Bandara, Yau, Watt, Jin, & Samaranayake, 2009). Considering CD, the enteric bacteria, as well as components of the biofilm (lipopolysaccharides and β-glucan of bacteria and fungi, respectively), can influence the cytokines (Th17) production leading to oxidative damage and death of cells by apoptosis, and also the production of enzymes may compromise the barrier resulting in harmful consequences to tissues and lesions (Hoarau et al., 2016).

Margalit, Carolan, and Kavanagh (2021) reviewed the *Aspergillus fumigatus*-bacteria interactions; considering the lung niche in patients with immunocompromised airways, several pathogens can cause cystic fibrosis and chronic obstructive pulmonary infections disease and increase the probability of polymicrobial interactions. *A. fumigatus* and *P. aeruginosa* are the most frequently associated with infection in this patient, and both interact during the infection, influencing the pulmonary pathogenesis and leading to a worse prognosis (Margalit et al., 2021). The bacterium can outcompete the fungus through different antifungal mechanisms as QS molecules such as phenazines that inhibit the *A. fumigatus* biofilm development by targeting the mitochondria activity inducing the ROS production and consequently the fungal apoptosis (Briard et al., 2015; Price, Brown, Limoli, Phelan, & O'Toole, 2020; Price-Whelan, Dietrich, & Newman, 2006; Shirazi, Ferreira, Stevens, Clemons, & Kontoyiannis, 2016). *P. aeruginosa* also produces other QS molecules such as dirhamnolipids, a biosurfactant molecule, that change the ECM structure, enhancing the bacterium adhesion to the fungus, inducing melanin production (related to the immune evasion mechanisms), and decreasing the β-glucan synthase activity and thickening the fungal cell wall, compromising the fungal growth (Briard et al., 2017). On the other side, *A. fumigatus* induces the *P. aeruginosa* elastase synthesis, is able to inhibit the fungal growth, to be toxic to alveolar epithelial cells and related to the antimicrobial surfactant's degradation (Kuang et al., 2011; Mariencheck, Alcorn, Palmer, & Wright, 2003; Nomura et al., 2014; Smith et al., 2015). Another mechanism of fungal-biofilm development employed by the bacterium is related to its siderophores, able to chelate iron in the host niche, causing nutrient deprivation to the fungus (Briard et al., 2015; Sass et al., 2018). Moreover, *A. fumigatus* hyphae can have the ability to produce gliotoxin with anti-*P. aeruginosa* activity (Manavathu, Vager, & Vazquez, 2014; Margalit, Carolan, Sheehan, & Kavanagh, 2020; Reece, Doyle, Greally, Renwick, & McClean, 2018).

Among coinfections described in the literature, those with ocular diseases deserve attention, mainly by *A. fumigatus/S. aureus* biofilm. The description of dual biofilm in ocular infections is occasionally reported. However, Ramírez-Granillo et al. (2021), using cell culture of human-limbo corneal fibroblast, demonstrated the ability of both to form biofilm and that the mixed biofilm showed decreased density when compared with the single form; also, this antagonistic effect seems to affect mainly *A. fumigatus*. The same behavior was also observed by the in vitro polymicrobial biofilm formation isolated from a patient with infectious keratitis in which bacteria can induce the lysis of conidia impairing the hyphae formation (Ramírez Granillo et al., 2015).

Another antagonistic interaction of respiratory colonizers is the *A. fumigatus* and *S. pneumoniae* association, the latter pathogen involved in pneumonia and sepsis in an elderly patient. The bacterium can inhibit the fungal biofilm and disrupt preformed biofilm due to the mycelial fragmentation through products of its aerobic respiration such as hydrogen peroxide resulting in a biofilm with a predominance of pneumococcus (Iwahashi, Kamei, & Watanabe, 2020).

The behavior of the association between different species of *Aspergillus* (*Aspergillus. fumigatus*, *Aspergillus terreus*, *Aspergillus niger*, and *Aspergillus flavus*) and *K. pneumoniae* isolates presented variations in the biofilm formation ability (Nogueira, Pereira, Jenull, Kuchler, & Lion, 2019). The hyphae play an essential role in forming *Aspergillus* biofilm; they are related to tissue invasion and immune system evasion. *K. pneumoniae* acts inhibiting the spore germination that cannot transform in hyphae, as a consequence, the biofilm of *A. fumigatus* was thickened. These occur in a physically dependent manner and can be reversible after bacteria elimination. Additionally, in the mixed biofilm, *K. pneumoniae* provides a stress condition to the fungal cells, upregulating genes related to the cell wall, as a regulator of chitin synthase (CHS3), a conveyer of β-glucosidase (EXG1) activity, MPKC, a putative mitogen-activated protein kinase (MAPK), and downregulating genes related to the hyphae development as CRZA (transcription factor of calcineurin pathway) and PKAR (kinase of cAMP pathway).

P. brasiliensis, a dimorphic fungi causative agent of a systemic mycosis called paracoccidioidomycosis (PCM), can form biofilm in vitro and in vivo (Cattana et al., 2017; Sardi, Pitangui, Gullo, Fusco-Almeida, & Mendes-Giannini, 2015). Besides the oral involvement in the PCM, the lungs are frequently affected by *P. brasiliensis*, and in this host niche, this pathogen can interact with another one responsible for causing tuberculosis, *M. tuberculosis* (Pai et al., 2016; Queiroz-Telles & Escuissato, 2011). Tuberculosis (TB) and PCM are responsible for important systemic lung diseases and are distributed throughout the world and Latin America, respectively. Both diseases have been described in the literature, occurring between 5.5% and 28% in South American countries (Guery et al., 2015). Medina-Alarcón et al. (2021) demonstrated for the first time the ability of both species to form a mixed biofilm, highlighting the importance of this once both MOs are responsible for severe pulmonary infection and that can result in death. Fungal-bacteria interactions are summarized in Table 1.

2.2 Fungi-fungi polymicrobial biofilms

Most of the studies described in the literature focus on the polymicrobial interaction of mixed biofilms due to fungi-bacteria species, mainly *Candida* spp., as shown above, but the interaction between fungi species can also occur. *C. albicans is* frequently isolated with other non-*albicans Candida* species, but their communication remains unclear. The studies related to the relationship between *Candida* species comparing mono and mixed biofilm are

TABLE 1 Fungi-bacteria interaction.

Fungi	Bacteria	Type of interaction	Molecules/ compounds involved in the interaction	Host niche	References
C. albicans	*Streptococcus gordonii*	S	Bacteria relieves farnesol action; gene expression alteration: fungal morphogenesis and arginine synthesis Increase hypha development and biofilm by *Candida* Ssp and Csh polypeptides of *S. gordonii* binds to Als3, Eap1p, and Hw1p of *Candida* ComCDE system/LuxS-autoinducer-2 (AI-2) systems (*S. gordonii*) related to cell-to-cell communication AI-2 → stimulates *Candida* filamentation and biofilm formation Glucosyltransferase G (*S. gordonii*) → contribution to coaggregation with *Candida* Lactate production by *S. gordonii* → source of carbon to *Candida*	Oral cavity	Bamford et al. (2009), Dutton et al. (2016), Holmes, Mcnab, and Jenkinson (1996), Ricker, Vickerman, and Dongari-Bagtzoglou (2014)
	Streptococcus mutans	S and A	Increase bacteria colonization Glucosyltransferase B (*S. mutans*) binds to *Candida* mannans and β 1,3 glucan Increase mixed-biofilm formation Lactate produced by *S. mutans* → carbon source to *Candida* Farnesol produced by *Candida* → increase *S. mutans* growth and microcolony development Mutanobactin A, *trans*-2-decenoic acid, and competence-stimulating peptide (QS molecules produced by *S. mutans*) → inhibit the germ-tube formation and *Candida* morphogenesis	Oral cavity	Ellepola, Liu, Cao, Koo, and Seneviratne (2017), Falsetta et al. (2014), Hwang, Marsh, Gao, Waugh, and Koo (2015), Jarosz, Deng, van der Mei, Crielaard, and Krom (2009), Joyner et al. (2010), Kim et al. (2017), Ponde, Lortal, Ramage, Naglik, and Richardson (2021), Vílchez et al. (2010)

Continued

TABLE 1 Fungi-bacteria interaction—cont'd

Fungi	Bacteria	Type of interaction	Molecules/compounds involved in the interaction	Host niche	References
	Aggregatibacter actinomycetemcomitans	A	AI-2 repress *Candida* filamentation and reduce biofilm formation	Oral cavity	Bachtiar et al. (2014)
	Staphylococcus aureus	S and A	Prostaglandin E2 produced by *Candida* → stimulates growth and biofilm formation by *S. aureus* adhesion Als3 (*Candida*), proteinases (*S. aureus*) Farnesol (*Candida*) → reduces viability and biofilm formation by *S. aureus* Differential expression of virulence factors (lactate dehydrogenase) → increase resistance to host-oxidative stress ECM produced by *Candida* protects *S. aureus* against antibiotics *Candida* biofilm enhances the number of *S. aureus* cells	Oral cavity, wounds, skin	Harriott and Noverr (2009), Jabra-Rizk, Meiller, James, and Shirtliff (2006), Kong et al. (2016), Krause, Geginat, and Tammer (2015), Luo et al. (2021), Peters et al. (2010)
	Porphyromonas gingivalis	S	*Candida* biofilm provides a hypoxic environment that protects *P. gingivalis*	Oral cavity	Bartnicka et al. (2019)
	Enterococcus faecalis	S and A	QS molecules: EntV (bacteriocin) produced by *E. faecalis* → inhibits *Candida* filamentation and degrades its biofilm Adhesion, Sap9 of *Candida* to *E. faecalis*	Gastrointestinal tract	Bachtiar, Dewiyani, Surono Akbar, and Bachtiar (2016), Dutton et al. (2016), Graham, Cruz, Garsin, and Lorenz (2017)
	Helicobacter pylori		*Candida* protects *H. pylori* against nutrient and pH variations *H. pylori* change the environment pH → favors *Candida* growth	Gastrointestinal tract	García, Salas-Jara, Herrera, and González (2014), Hathroubi, Servetas, Windham, Merrell, and Ottemann (2018), Sánchez-Alonzo et al. (2021)

TABLE 1 Fungi-bacteria interaction—cont'd

Fungi	Bacteria	Type of interaction	Molecules/compounds involved in the interaction	Host niche	References
	Clostridium perfringens	A	Promotes *Candida* aggregation and protection the bacteria with internal hypoxia	Gastrointestinal tract	Fox et al. (2014)
	Escherichia coli	S and A	Coinfection → synergistic lethality (murine model) due to bacteria endotoxins Fungus suppresses the growth of bacteria directly or indirectly by dominant interaction during colonization	Gastrointestinal tract	Bandara, Cheung, Watt, Jin, and Samaranayake (2013), Cabral, Penumutchu, Norris, Morones-Ramirez, and Belenky (2018), Farrokhi, Al-Shibli, Al-Hameedawi, Neshati, and Makhdoumi (2021)
	Salmonella typhimurium	A	QS molecules from *S.* Typhimurium inhibits *Candida* filamentation *Candida* supernatant upregulates sopB and sipB of *S.* Typhimurium → filaments killing and biofilm repression.	Gastrointestinal tract	Kim and Mylonakis (2011), Tampakakis, Peleg, and Mylonakis (2009)
	P. aeruginosa	S and A	Elastase produced by bacteria → increase *Candida* virulence fermented products produced by fungi → increase phenazine secretion by bacteria farnesol produced by fungi → downregulates QS system of bacteria type IV pili, phospholipase C, and phenazines of *P. aeruginosa* → kills *Candida* filaments *P. aeruginosa* produces homoserine lactone → repress fungal filamentation, inhibits biofilm formation, and kill hyphae inhibitory effects of	Lungs, wounds	Bandara et al. (2010), Gibson et al. (2009), Hogan et al. (2004), Hogan and Kolter (2002), Peleg, Hogan, and Mylonakis (2010)

Continued

TABLE 1 Fungi-bacteria interaction—cont'd

Fungi	Bacteria	Type of interaction	Molecules/ compounds involved in the interaction	Host niche	References
			P. aeruginosa also occurs to *C. parapsilosis*, *C. tropicalis*, *C. dubliniensis*, and *C. glabrata*		
	Burkholderia cenocepacia	A	Homoserine lactone produced by *B. cenocepacia* → arrest *Candida* from undergoing germ-tube induction	Lungs	Boon et al. (2008)
	Mycobacterium tuberculosis	S	unknown	Lungs	Mankiewicz and Liivak (1960), Ochieng et al. (2005)
	Klebsiella pneumoniae	A	Bacterium adheres to *Candida* → inhibits the growth of the fungus; mechanism unknown	Lungs and gastrointestinal tract	Bandara, Yau, Watt, Jin, and Samaranayake (2010), Fox et al. (2014), Galdiero et al. (2021)
	Lactobacillus spp.	A	Nutrient competition *Lactobacillus* metabolites → anti-*Candida* activity	Vulvovaginal tract	Parolin et al. (2021)
	Acinetobacter baumannii	A	Bacteria adhere to *Candida* and kill hyphae by apoptosis via contact-dependent signals mediated by the outer-membrane protein A (OmpA)	Oral cavity, Urinary tract	Gaddy, Tomaras and Actis (2009)
	Citrobacter freundii	S	Bacterium attaches to *Candida* mediated by mannose-binding lectins → triggers fungal filamentation and increase the neutrophils death and inflammation	Wound	Cheong et al. (2021)
C. tropicalis	*E. coli*	S	Bacterium produces polysaccharides → *Candida* biofilm maturation	Oral cavity	Bandara, Yau, Watt, Jin, and Samaranayake (2009), Hoarau et al. (2016)
	Serratia marcescens	S	*Candida* apparently displays a stimulatory effect on the bacterium, enhancing its virulence; unknown mechanism	Lungs, urinary tract, wound	Hoarau et al. (2016)

TABLE 1 Fungi-bacteria interaction—cont'd

Fungi	Bacteria	Type of interaction	Molecules/ compounds involved in the interaction	Host niche	References
Aspergillus fumigatus	*S. aureus*	A	Bacteria can induce conidia lysis impairing the hyphae formation	Ocular niche	Ramírez Granillo et al. (2015), Ramírez-Granillo et al. (2021)
	P. aeruginosa	S and A	Phenazines (produced by bacterium) → inhibit the fungus biofilm development Dirhamnolipids (produced by bacterium) → increase bacterium adhesion to fungus, induces fungal melanin production, decreases β-glucan synthase activity (thickening cell wall) Elastase (bacterium) → inhibits fungal growth Bacterium siderophores chelates iron → promotes nutrient deprivation to the fungus Fungal gliotoxin → anti-*P. aeruginosa* activity	Lungs	Briard et al. (2015, 2017), Kuang et al. (2011), Mariencheck et al. (2003), Nomura et al. (2014), Price-Whelan et al. (2006), Sass et al. (2018), Shirazi et al. (2016), Margalit, Carolan, and Kavanagh (2021), Price, Brown, Limoli, Phelan, and O'Toole (2020), Reece, Doyle, Greally, Renwick, and McClean (2018), Smith et al. (2015)
	S. pneumoniae	A	Products of bacterium aerobic respiration → inhibits fungal biofilm development and disrupt preformed biofilm	Lungs	Iwahashi, Kamei and Watanabe (2020)
Aspergillus spp.	*K. pneumoniae*	A	*A. fumigatus, A. terreus,* and *A. flavus* Physical dependent manner: Bacterium inhibits spore germination → decreases fungal biofilm thickness Mixed-biofilm altered the gene expression of fungi	Lungs	Nogueira, Pereira, Jenull, Kuchler and Lion (2019)
Paracoccidioides brasiliensis	*M. tuberculosis*	–	Unknown mechanisms	Lungs	Medina-Alarcón et al. (2021)

contradictory, describing antagonistic and cooperative interactions.

Pathirana, McCall, Norris, and Edgerton (2019) evaluated the association of *C. albicans* and non-*albicans* species (*C. tropicalis*, *C. dubliniensis*, *C. parapsilosis*, *Candida lusitaniae*, and *Candida krusei*) in mixed biofilms, demonstrating that only *C. tropicalis* and *C. dubliniensis* (the only non-*albicans* species able to form true hyphae) adhere to *C. albicans*, which needs to be in the filamentous form. This fungi-fungi adhesion has no impact on the initial biofilm establishment but shows differences later, with speed to surface coverage to *C. albicans-C. dubliniensis* slower for *C. albicans-C. tropicalis* (maybe due to the inhibition by farnesol produced by *C. albicans*) but with a higher surface coverage than *C. albicans* alone and with a predominance of *C. albicans* at the maturation stage, indicating a benefit of interspecies communication through cell-cell adhesion that helps with the growth and maintenance of mixed biofilm. *C. glabrata* can also attach to *C. albicans*, and the precolonization with *C. albicans* can help *C. glabrata* proliferation in dual biofilms. Additionally, this polymicrobial interaction seems to contribute to pathogen's invasion in host tissue (Coco et al., 2008; Pathak, Sharma, & Shrivastva, 2012; Silva et al., 2013; Tati et al., 2016).

Mixed biofilm formed by *C. albicans/C. glabrata* results in a reduction of cell number; however, some virulence factors were altered, such as increased resistance to caspofungin, greater levels of secreted aspartic proteinase (SAP) and phospholipase (PL) proteins, higher expression of efflux-pump-related genes, and reduction of β-glucan unmasking, indicating a competitive interaction without prejudice to the virulence of them (Li et al., 2019).

The non-*albicans* species are often isolated from the oral niche and are also associated with biofilm formed on the surface of medical devices and tissues. Martins et al. (2016) showed in vitro that *C. albicans* could form a mixed biofilm with non-*albicans* species (*C. krusei*, *C. rugosa*, *C. tropicalis*, *C. parapsilosis*, and *C. glabrata*) and seems to present an antagonistic relationship once the hyphae presence is reduced in the mixed biofilm when compared with the monospecies biofilm.

Competitive interaction between *C. albicans* and *C. krusei* and *C. glabrata* was demonstrated by Rossoni et al. (2015). The mixed biofilm resulted in a reduced number of cells. The heterotypic suspension containing dual species of biofilm analyzed in *G. mellonella* and murine model resulted in increased survival of larvae and reduced fungal burden recovered from both infection models, indicating that the single infection with *C. albicans* is more deleterious to the animals than the mixed biofilms. A higher inhibitory condition in a mixed-biofilm was demonstrated for *C. albicans-C. krusei* species with a decrease in the metabolic activity of *C. albicans* (Santos, Piva, Vilela, Jorge, & Junqueira, 2016).

More recently, Oliveira et al. (2020) evaluated the polymicrobial biofilm of *P. brasiliensis/C. albicans*. These MOs cause disease depending on the host in different body parts, and they can be found in the oral cavity. The cocultivation of both species reduces the time needed for biofilm maturation, and this time is even shorter when the fungal is added to the culture after a preformed *C. albicans* biofilm. The interaction between both seems to occur by direct physical contact and through the ECM showing cooperation between both species. Additionally, the virulence of this dual biofilm in *G. mellonella* model is higher when compared with the monospecies suggesting that the coexistence of both MOs in the host can exacerbate the clinical and pathological manifestations (Oliveira et al., 2020).

In the scenario of superficial fungal infections, the most MOs associated with dermatomycosis are *Trichophyton rubrum*, *C. albicans*, and *C. parapsilosis*, in which all species form biofilms. Recently, Garcia et al. (2020) demonstrated that *T. rubrum* with both *Candida* species in a mixed biofilm shows antagonistic effects. The fungi biofilm seems to inhibit the *C. albicans* filamentation and *C. parapsilosis* development, but the exact mechanism by which this occurs has not yet been elucidated.

The knowledge about the mechanisms employed by pathogens to interact between them in a polymicrobial biofilm is essential to elucidate the impact on the host and clinical and disease manifestation. Moreover, understanding the physiology of this cross-kingdom interaction can explore combinations of treatment or new strategies to fight against these consortia of pathogens.

3. Factors affecting biofilm formation and development

The biofilm formation and maturation depend on several factors, such as the interaction between MOs, biotic or abiotic substrates for adhesion, including the ability to attach and aggregate, the physicochemical environment and nutrient levels surrounding the communities, and the host immune response (Costa-Orlandi et al., 2017; Garcia et al. (2020); Muhammad et al., 2020; Stacy et al., 2016).

Bacteria and fungi may form three-dimensional structures in traditional models in vitro biofilm. However, some of these MOs tend to clump together as multicellular aggregates in natural environments and during infection, and biofilms can also shed aggregates as part of the dispersal process (Kragh et al., 2016).

Interactions between organisms are dependent on the environment in which they develop, structure, and function (Burmølle, Ren, Bjarnsholt, & Sørensen, 2014; Elias & Banin, 2012; Rendueles & Ghigo, 2012). Such interactions can influence the growth and survival of biofilm community members and their potential virulence properties, influencing the overall pathogenicity of such structures (Peters et al., 2012). In the recent years, the research focused on the complexity and interactions in polymicrobial biofilms. The use of multiomic, imaging technologies, high-throughput, high-resolution methods facilitated this development by revealing the complex microbial spatial organization of polymicrobial biofilm infections interactions, including genetics, metabolite exchange, and signaling between MOs in biofilm communities (McLean & Kakirde, 2013; Møller et al., 1998; Stacy et al., 2016).

The biofilm formation by bacteria and fungi has some steps in common, such as adhesion mediated by adhesins and the secretion of EPSs, which make the cells permanently adhere to a surface or each other, allowing their aggregation. Adhesion is the first requirement for successful colonization. This attachment occurs through the specific interaction between cell surface adhesins and host receptors. However, MOs can also adhere nonspecifically to abiotic surfaces (Stacy et al., 2016). Oral cavity MOs can form a complex ecosystem that thrives in the dynamic oral environment in a symbiotic relationship with the human host. The oral cavity is an environment that exemplifies how attachment can influence colonization (Radaic & Kapila, 2021; Stacy et al., 2016). Polymicrobial biofilms of the oral cavity are formed after a process of dysbiosis of the indigenous microbiota, temporally, by the colonization of some microbial populations present in the microbiota, through a phenomenon called coaggregation (Peters et al., 2012; Rickard, Gilbert, High, Kolenbrander, & Handley, 2003). In this phenomenon, colonization by MOs is determined by the composition of the first colonizers, which serve as a substrate for the fixation of the others. After the transition from the planktonic to the sessile state and changes in gene regulation, the ECM is produced, involving the community, increasing adhesion and protection against the host's immune system and external factors, and the process evolves until maturation (Costerton et al., 1999; Peters et al., 2012; Rickard et al., 2003; Vuong et al., 2004).

The structures known as aggregates have been termed in vivo biofilms in which MOs grow and organize into small clusters (5–200 μm wide) and contrast with the large structures seen in vitro biofilm studies. Aggregation may involve collective behaviors or arise from single-cell events followed by clonal growth (Stacy et al., 2016). Many infections are colonized with aggregates, including wounds, abscesses, and cystic fibrosis lung. However, exceptions have been observed, such as thick biofilms often associated with middle ear infections, infection due to permanent tissue fillers both as to distribution (wound bed) and organization (monospecies or multispecies microcolonies) (Burmølle et al., 2014; Hall-Stoodley et al., 2006; Kirketerp-Møller et al., 2008; Stacy et al., 2016). The aggregation evolves to the formation of microcolonies covered or not by a polymeric ECM that ceases their development in the maturation stage and then disperses (Costa-Orlandi et al., 2017; Harding et al., 2009; Muhammad et al., 2020). The three-dimensional structure is established in the maturation stage, and cell-cell attachment depends on the ECM. When cell density is high, QS mechanisms are activated to regulate cell density and monitor the behavior of sessile cells. In the final stage, called dispersal, cells detach to colonize a new niche. Detachment occurs influenced by several factors, such as nutrient limitation, fluid volume dynamics, and shear effects, among others (Alotaibi & Bukhari, 2021). The regulation and gene expression at each step are diversified depending on the species of MO, and there is still a long way to go before gene regulation of mono- and multi-species biofilms formed by bacteria (Muhammad et al., 2020) and fungi to be fully understood (Mancera et al., 2021).

Another phenomenon that regulates MOs' biogeographic patterns is the biofilm's remodeling through their breakdown or synthesis. Remodeling occurs to increase the separation between MOs from the community instead of coaggregation (Stacy et al., 2016).

QS can be a mediator of intra and interspecies communication. Different bacteria use QS systems that produce, detect, and respond to extracellular signaling molecules called autoinducers. The population of bacteria has gene expression regulated through the production and response to these autoinducers (Alotaibi & Bukhari, 2021). Signaling mediated by autoinducers in QS allows the control of important processes such as virulence factors, biofilm formation, motility, sporulation, adaptation to stress, and production of secondary metabolites (Grainha, Jorge, Alves, Lopes, & Pereira, 2020). The role of QS molecules (QSm) has been extensively studied in some species of the fungi kingdom (Albuquerque & Casadevall, 2012; Chen et al., 2004; Costa-Orlandi et al., 2017; Hornby et al., 2001; Lorek, Poggeler, Weide, Breves, & Bockmuhl, 2008; Nickerson, Atkin, & Hornby, 2006; Sardi et al., 2014; Wongsuk, Pumeesat, & Luplertlop, 2016); however, much remains to be uncovered. QSms such as farnesol, tyrosol, phenylethanol, and tryptofhol can be produced during the developmental stages of biofilms and play important roles in morphogenesis, macroconidia germination, apoptosis, and fungal pathogenicity (Grainha et al., 2020).

Recent discoveries show the diverse roles of extracellular vesicles (EVs) in the formation of sessile communities, whether linked to ECM formation or as regulators of the functions of several events related to biofilms. (Zarnowski et al., 2021). The factors that interfere in the biofilm formation will be further detailed below.

The biofilm offers several favorable adaptations for the MO to grow. Some MOs secrete proteases, and they do not secrete in planktonic form. Furthermore, a cell that scatters from a biofilm generally has a more remarkable ability to adhere to new surfaces and appear more virulent (Uppuluri et al., 2010). When MOs form a biofilm, they shape their habitats by interacting with them, either by dissolving or precipitating minerals, changing the pH value, redox potential or oxygen concentration, or salinity. The different physiological interactions between the MOs themselves and the environment lead to steep gradients in these activities, providing habitat heterogeneity and supporting biodiversity within just a few micrometers of the distance between the cells (Flemming, 2020).

The attachment to surfaces alters the cellular configuration activating mechanosensitive channels and triggering programmed cell growth on surfaces (Flemming, 2020). In fungal biofilms, adhesion to abiotic surfaces, the interaction between cells, and in some cases, hyphae development are crucial for strong biofilm formation. Like bacterial biofilms, some fungal biofilms produce ECM, composed of EPSs that attach to cells, contributing to the complex three-dimensional architecture of the community (Mitchell, Zarnowski, & Andes, 2016b). ECM is considered a defining characteristic of biofilms, and its composition varies according to the species involved in the biofilm. Changes in the formation of this ECM can also negatively influence the biofilm formation of several microbial species (Borghi, Borgo, & Morace, 2016).

There are a variety of determinants that interfere with biofilm development (Kim, Park, & Chung, 2012). The conditions involved in the formation of biofilms by microbial cells are the concentration of nutrients, the temperature and pH of the environment in which the MO is found, and the ability to adhere to the surface, considering in this case the surface properties and the shear stress of the environment where it is inserted (Bajpai, 2015; Kim et al., 2012).

Environments with greater nutrient availability allow the formation of large quantities of thicker biofilms. Higher nutrient concentrations appear to produce an open structure in the biofilm, while the biofilm becomes more compact in low nutrient concentrations. This more open biofilm structure can provide a challenging environment for establishing nutrient gradients, genetic exchange, and quorum detection. The structural organization of biofilms and the presence of flowing water channels can allow the exchange of nutrients and gases while providing fungal cells with a sheltered niche for protection against environmental predators, immune cells, shear forces, and antifungals (Martinez & Casadevall, 2015).

The increase in nutrients in the medium is correlated with the increase in microbial cells, generating greater microbial multiplication, more adhered cells, and more biofilm formed. Low concentrations of nutrients do not prevent biofilm formation, as this characteristic will depend on the MO and the nutritional needs. However, if there are insufficient nutrients for the cells, there is no microbial multiplication, consequently less adhered cells and biofilm formation (Flemming, 2020). Nutrient intake is also related to an optimal temperature; when the MO is at its optimal growth temperature, it increases nutrient intake, resulting in rapid biofilm formation. Biofilm formation will depend on the temperature for enzyme activation, as these are often responsible for the metabolism of nutrients. Optimal temperatures result in the healthy growth of microbial populations in a sessile community, while distant temperatures result in reduced growth efficiency and reduced enzymatic reaction rates (Bajpai, 2015). The proper functioning of cells depends on maintaining their intracellular pH. In yeast, filamentous fungi, and higher plant cells, the proton-pumping activity of the plasma membrane P-ATPase has been recognized as the major factor responsible for pH cytoplasm homeostasis (Hesse, Ruijter, Dijkema, & Visser, 2002). Changes in pH can influence the growth of MOs. Significant variations in pH can be important inhibitors of metabolism, considering that some enzymes act in an optimal cytoplasmic pH (Garrett, Bhakoo, & Zhang, 2008; Hesse et al., 2002). Sudden changes in pH influence, for example, exopolymeric substances, generally involved in the ECM of microbial biofilms (Bajpai, 2015; Garrett et al., 2008).

In a study by Hesse et al. (2002), although the external pH ranged from 1.5 to 7, the fungal cells maintained the cytoplasmic pH, having suffered only a metabolic collapse when the pH was set to 0.7–0.8. The researchers also used some carbon sources, which could contribute to the maintenance of the cytoplasmic pH by the fungus during the variation of the external pH. In another study, Carmelo, Santos, and Sá-Correia (1997) demonstrated that glucose had little influence on the pH modification of the yeast *Saccharomyces cerevisiae*. However, acidification of the external environment with organic acids from pH 6.5 to 3.5 influenced the cytoplasmic pH and activated the plasma membrane ATPase to regulate hydrogen ions. When it comes to the pH of the human body, especially in fluids such as urine and blood, these pHs are influenced by diet, intermediate metabolites, and the level of final catabolic products; therefore, the growth of the MO will be favored according to the optimal pH of its growth. Generally, alkaline pH tends to favor the growth of MOs in catheters and medical devices (Di Bonaventura, Stepanović, Picciani, Pompilio, & Piccolomini, 2007).

The fixation of MOs on surfaces is a complex process that presents several variables (Martinez & Casadevall, 2015). Surface properties, including surface composition, morphology, structure, and material properties, affect biofilm development from the early stages of adhesion to the final stages (Kim et al., 2012). Certain surfaces tend to be easier for MOs to adhere to, such as rough or rough surfaces, which are more hydrophobic and coated with some conditioning (Garrett et al., 2008; Goller & Romeo, 2008;

Kim et al., 2012; Martinez & Casadevall, 2015). For example, pockets in rough surfaces provide a protective habitat with reduced shear stress. Shear stress is the force involved between the MO's adhesion to the surface and the movement of liquids in the medium or under the surface. This force exerted by the tension of a medium can inhibit the development of the biofilm as it hinders cell adhesion to the surface and increases their dispersion. This shear stress suppresses the development of a biofilm matrix but offers the possibility of cell dispersion. MOs' cells with high shear stress tend to form thinner monolayer biofilms, in addition to remaining in a "young" or early state. In contrast, the biofilm becomes thicker with a lower shear force, and the number of cells is higher (Kim et al., 2012).

Biofilm formation is complex and will depend on the necessary conditions of temperature and nutrients of each MO, while it will depend on where that MO is found. Fungi can grow on medical devices such as catheters and valves within the biological conditions of an individual's body and grow in the environment. Therefore, when studying the factors that affect biofilm formation, it is worth emphasizing that it is necessary to know which MO and especially which conditions favor its growth.

4. Composition of mono and polymicrobial biofilms

Microbial biofilms development occurs progressively and can generally be divided into four stages: (i) adsorption and adhesion of cells to a substrate, (ii) microcolony formation and ECM production, (iii) maturation, and (iv) dispersion of mature biofilm cells (Costa-Orlandi et al., 2017; Pammi, Zhong, Johnson, Revell, & Versalovic, 2014; Sardi et al., 2014; Weinstein et al., 1983).

The adhesion of MOs to surfaces, whether abiotic or biotic, is a prerequisite for initiating biofilm formation and establishing infection. At this stage, the participation of specific cell surface molecules called adhesins is essential (Demuyser, Jabra-Rizk, & Van Dijck, 2014; Sardi et al., 2014). The specificity of infecting a given host or tissue predilection is determined by interactions between adhesins and their specific receptors on the surface of the host tissue (Klemm & Schembri, 2000; Kumari et al., 2021). Several researchers have established the role of fungal adhesins in fungal colonization (Demuyser, Jabra-Rizk, & Van Dijck, 2014; Kumari et al., 2021; Nobile et al., 2008; Ponde et al., 2021; Sardi et al., 2014, 2015). In mycology, the most studied adhesins are those of *C. albicans*. New adhesins have been identified for most recurrent fungal infections caused by *A. fumigatus*, *C. neoformans*, *P. brasiliensis*, and *H. capsulatum*.

Three major families of *C. albicans* adhesins mediate adhesion during biofilm formation: (1) the Als family, (2) the hyphal wall protein (Hwp) family, and (3) an F/hyphally regulated family (Iff/Hyr) (Kumari et al., 2021; Ponde, Lortal, Ramage, Naglik, & Richardson, 2021). The ALS family is glycosylphosphatidylinositol (GPI) linked to b-1,6 glucans in the fungal cell wall and is composed of eight members (Als1p–Als7p and Als9p) (Ponde et al., 2021). These proteins have already been reported to play a key role in the adhesion of *C. albicans* (yeasts or hyphae) and *S. cerevisiae* to various substrates during biofilm formation and development (Nobile et al., 2008; Ponde et al., 2021). Although all are important adhesins for biofilm formation (whether they are mono and polymicrobial or with other fungi or bacteria) and involved in adhesion to surfaces, they exhibit functions such as cell-cell interactions and invasion of host tissue to iron acquisition (Demuyser, Jabra-Rizk, & Van Dijck, 2014; Garcia et al., 2011; Nobile et al., 2008; Ponde et al., 2021). ALS family proteins present an amyloid-forming region critical for cell aggregation (Garcia et al., 2011; Ponde, Lortal, Ramage, Naglik, & Richardson, 2021; Ramsook et al., 2010; Rauceo et al., 2004). In addition, these proteins have a binding affinity for ECM proteins and endothelial and epithelial cell components (Kumari et al., 2021; Sundstrom, 1999). The second family of adhesins consists of Hwp molecules (Hwp1, Hwp2, Rbt1, Eap1) containing the exposed extracellular N-terminal domain. Hwp is differentially expressed during germ tube and hyphal stages (Demuyser, Jabra-Rizk, & Van Dijck, 2014; Ponde, Lortal, Ramage, Naglik, & Richardson, 2021). Like Als proteins, Hwp proteins function in different processes such as biofilm formation and adhesion (De Groot, Bader, de Boer, Weig, & Chauhan, 2013; Demuyser, Jabra-Rizk, & Van Dijck, 2014). The specific hypha Hyr1p belongs to the Iff/Hyr family and is also associated with biofilm formation and development (Bailey, Feldmann, Bovey, Gow, & Brown, 1996).

Several adhesins have been described in the planktonic form of some fungi that form biofilms but have not been studied in mixed biofilms. RodA, RodB, RodD, and RodE are overexpressed in the early phase of *A. fumigatus* biofilms (Fanning & Mitchell, 2012; Liu et al., 2016; Morelli, Kerkaert, & Cramer, 2021). Cell floculin 1 (Cfl1) is an adhesin induced during *Cryptococcus* sexual development, and when overexpressed, the strains have increased flocculation and colonization (Wang, Zhai, & Lin, 2012).

Pitangui et al. (2012) demonstrated the ability to form a biofilm of *H. capsulatum* on abiotic surfaces and adhesion to lung cells. Heat shock protein 60 (Hsp60) has been shown to be important during the infection process, mediating the interaction with macrophages, a critical step in the infection process and interaction with host cells (Guimarães et al., 2011; Long, Gomez, Morris, & Newman, 2003). In addition, blocking Hsp60 in biofilms in vitro resulted in less dense and thinner biofilms, suggesting that this protein was crucial for biofilm adhesion and development (Fregonezi et al., 2021). Studies carried out with *P. brasiliensis* biofilm identified proteins such as GP43, GAPDH, and aspartyl proteinase with the function of cell adhesion and establishment of infection (Sardi et al., 2015).

Bacterial biofilm adhesion is also characterized by the production of bacterial adhesins that adhere to the surface, while cell-cell adhesion mechanisms mediate maturation. Many adhesins have been described as participating in mixed biofilms (fungi/bacteria), mainly with *C. albicans* (d'Enfert et al., 2021). Bacteria may adhere to an abiotic or biotic surface through surface charges and the production of pili, fimbriae, and exopolysaccharides (Schulze, Mitterer, Pombo, & Schild, 2021). Some of these adhesins were discussed in the polymicrobial section.

Cell wall proteins are essential for biofilm formation, as, in addition to having a role in adhesion, they may also have a sensory role that promotes adhesion-induced responses (Fanning & Mitchell, 2012).

The ECM is a 3D structure composed of a complex EPS set (polysaccharides, nucleic acids, proteins, lipids, and other biomolecules) that contributes to the biofilm's lifestyle and virulence. The EPS structure and composition may vary depending on MOs, local shear stress, nutrients/substrates, and the host environment. EPS synthesis and spatial organization also differ between monospecies and multispecies microbial communities (Karygianni, Ren, Koo, & Thurnheer, 2020).

The presence of an ECM between cells that form a community is one of the biofilm characteristics. ECM is a self-produced material that supports cell cohesion and adhesion to surfaces and provides nutritional resources, protection against xenobiotics, antimicrobials, and the host's immune system (Flemming & Wingender, 2010; Zarnowski et al., 2014). The composition of ECM has been extensively studied in bacterial biofilms. In fungi, studies are more concentrated on *S. cerevisiae*, *A. fumigatus*, *C. neoformans*, and *C. albicans* (Zarnowski et al., 2014). Although its composition varies according to the MOs that are part of the biofilm and the conditions of its formation, it is generally composed of macromolecules that include polysaccharides, lipids, nucleic acids, and proteins (Zarnowski et al., 2014).

C. albicans matrix is formed by β-1,3 glucan and more significant amounts of β-1,6 glucan and α-1,6 mannan with α-1,2 branches that constitute 25% of the ECM. The three polysaccharides were found to physically interact in a mannan-glucan (MGCx) complex, an undescribed structure in the cell wall. Then, cell wall and matrix polysaccharides are found in different proportions and with different bonds. Cell wall chitin was not detected in the ECM, suggesting that the cell wall and matrix are distinct (Mitchell et al., 2015). Seven genes have been identified that govern matrix mannan levels (ALG11, MNN9, VAN1, MNN4-4, PMR1, and VRG4), two genes govern matrix β-1,6 glucan levels (BIG1 and KRE5), and FKS1 encoding β-1,3 glucan synthase (Lohse et al., 2018; Mitchell et al., 2016b). Proteins represent 55% of a total matrix, with 458 distinct entities (Zarnowski et al., 2014), 15% include neutral and polar glycerolipids, sphingolipids (Zarnowski et al., 2014), 5% is DNA, mainly noncoding sequences (Zarnowski et al., 2014), and other components such as phosphorus, uronic acid, and hexosamine (Al-Fattani & Douglas, 2006). *C. neoformans* has a polysaccharide capsule composed primarily of glucuronoxylomannan (GXM), and this component was described surrounding the cells of the biofilm, besides other monosaccharides not found in GXM, suggesting that the ECM contains additional types of polysaccharides such as xylose, mannose, glucose, galactoxylomannan (Martinez & Casadevall, 2007). Some strains of *S. cerevisiae* exhibit flocculation, and others can form true surface-adherent biofilms (Bojsen, Andersen, & Regenberg, 2012; Verstrepen & Klis, 2006). *S. cerevisiae* can produce ECM in both the flocculating and biofilm forms. Both formations depend on the FLO family of cell surface adhesins, or flocculins, related to the ALS family and the Hwp1p adhesins in *C. albicans* (Ishigami, Nakagawa, Hayakawa, & Iimura, 2004; Nobile et al., 2008; Reynolds & Fink, 2001). ECM is formed by glucose, mannose in α-1,6 chains with α-1,2 and α-1,3 branches (Beauvais, Loussert, Prevost, Verstrepen, & Latgé, 2009), and galactose (Faria-Oliveira et al., 2015). Low protein levels were found (Beauvais et al., 2009) and several proteins present in the *S. cerevisiae* ECM, the glycolytic Tdh3 (glyceraldehyde-3-phosphate dehydrogenase), Tpi1 (triose phosphate isomerase), Pdc1 (pyruvate decarboxylase 1), Eno1/2 (enolase 1 and 2), Sod2 (superoxide dismutase 2), Hsp26 (heat-shock protein 26), and Cpr1 (a cytoplasmatic peptidyl-prolyl cis-trans isomerase) (Faria-Oliveira et al., 2014). The principal matrix components of *A. fumigatus* biofilms are polysaccharide (43%), lipid (14%), aromatic components such as melanin (3%) and protein (40%) including antigens and hydrophobins (Reichhardt et al., 2015). Three main secreted antigens: dipeptidyl peptidase V (DPPV), catalase B (CatB), and ribotoxin (ASPF1), besides hydrophobin proteins, have also been identified. In addition, the eDNA corresponding to the genomic DNA sequence was identified in the *A. fumigatus* biofilm matrix (Rajendran et al., 2013).

In vitro bacterial and fungal models have revealed important insights into biofilm physiology, tolerance to antimicrobials, phagocytic predators, and many other aspects of surface-associated growth. On the other hand, in vivo studies are much more challenging to demonstrate, and the biofilm aggregates are often numerous and small. However, there are important differences between in vitro biofilm structures and in vivo biofilms associated with chronic infections in humans and animal models. In most bacterial biofilm habitats (environmental and medical), grazers, phagocytes may be present, along with the presence of inflammation products. Many fungal and bacterial pathogens can form biofilms, establishing resilient communities on various clinical surfaces. Fungal biofilms can be found in various clinical niches and can vary significantly concerning available nutrients, flow conditions, immunological components, pH and substrate for cell adhesion, and initiation of biofilm growth (Nett & Andes, 2015). Animal models have been developed to simulate common infections associated with *Candida* devices, including vascular catheters, dentures, urinary catheters, and subcutaneous implants. The models also reproduced the most common mucosal biofilm infections, oropharyngeal and vaginal candidiasis (Dongari-Bagtzoglou, Kashleva, Dwivedi, Diaz, & Vasilakos, 2009; Kojic & Darouiche, 2004; Nett & Andes, 2015; Rodríguez-Cerdeira et al., 2019). Andes et al. (2004) reported differences in temporal development between studies of biofilms formed in vivo and in vitro. The duration of the initial phase and maturation of biofilm formation in vivo was shorter than that observed in vitro in the rat central venous catheter model (Andes et al., 2004).

C. albicans forms biofilms readily on abiotic substrates and the mucosal surfaces of the oral cavity. Nevertheless, conflicting observations have been made regarding the formation of biofilms on the vaginal mucosa (Harriott, Lilly, Rodriguez, Fidel, & Noverr, 2010; Paiva, Vidigal, Donatti, Svidzinski, & Consolaro, 2012; Swidsinski et al., 2019), suggesting that tissue/anatomical location and surrounding microbiota are likely to influence biofilm development. The most common polysaccharide found in the ECM of *C. albicans* biofilms is mannan, which is generally associated with beta-1,6 glucan (Pierce et al., 2017). The binding of *C. albicans* to receptors on host tissues is aided by the expression of ALSs, differentially regulated in biofilm-grown cells (Hoyer, 2001; Williams & Lewis, 2011). Als3 has been implicated in adherence to host tissue and bacterial species in mixed biofilm (Murciano et al., 2012; Peters et al., 2012; Phan et al., 2007). Several extracellularly released enzymes, such as lipases, esterases, and hemolysins, are crucial for host tissue invasion and nutrient acquisition of *C. albicans* (Naglik, Challacombe, & Hube, 2003). The secreted aspartyl proteinases (Saps) can directly induce host cell damage, facilitate the growth of hyphae for tissue invasion, increase adherence after exposure to receptor sites, and degrade antibodies and other host defense proteins (Schaller, Schackert, Korting, Januschke, & Hube, 2000; Williams & Lewis, 2011). Class B phospholipases (PLBs) are secreted and act by disrupting host cell membranes. Consequently, PLB1 and PLB5 deletions have been implicated in attenuating systemic *Candida* infection (Leidich et al., 1998; Theiss et al., 2006). Proteomic analysis of EPSs from *C. albicans* also reveals the presence of numerous host molecules, including those found in epithelial cells and neutrophils (Zarnowski et al., 2014).

Some studies have been made in clinical biofilms infections, as in aspergilloma, considered an *Aspergillus* biofilm. The same polysaccharides were found in ECM produced in vitro, α-1,3 glucan, galactomannan, and galactosamine. In the case of invasive infections, novel galactosaminogalactan (GAG) and galactomannan were the main ECM polysaccharides in vivo (Loussert et al., 2010).

Depending on the type of interaction between MOs, there may be an increase, decrease, or even no change in the amount of ECM produced in one or all of the species involved in the biofilm. The particular organization of MOs within the biofilm can be dependent on the production of ECM that will create the conditions for the growth of certain MOs, for example, creating environments with low oxygen concentration to promote the growth of anaerobic bacteria (Mitchell, Zarnowski, & Andes, 2016a).

Each component of the ECM plays a crucial role in biofilm formation and maintenance. Extracellular DNA (eDNA) from the biofilm of *C. albicans* and *A. fumigates* can be important in the early stages of biofilm formation in maintaining the structural integrity of mature biofilms (Martins et al., 2010; Rajendran et al., 2013) as well as in biofilm resistance to antifungal drugs (Martins, Henriques, Lopez-Ribot, & Oliveira, 2012; Rajendran et al., 2013). Removal of eDNA from mixed *S. mutans* and *C. albicans* biofilms has been reported to affect adhesion in the early phase of biofilm formation (Guo et al., 2021).

The polysaccharide elements of the extracellular matrix are distinct in each MO involved. These also differ in size, ramifications, or even combinations between monosaccharides compared with cell wall polysaccharides (Mitchell et al., 2016b; Pierce et al., 2017). The presence of β-1,3-glucans or GAG in the ECM is considered responsible for the resistance of biofilms to several antifungal agents (Loussert et al., 2010; Reichhardt, Stevens, & Cegelski, 2016). The analysis of proteins found in the biofilm matrix of *C. albicans* showed the presence mainly of enzymes related to carbohydrate and lipid metabolism and those involved in the degradation of the ECM to promote the biofilm dispersion (Pierce et al., 2017). Under in vivo biofilm conditions, the variability of proteins increases according to the host and the site of infection; 98% of the proteins may be related to the host's niche (Mitchell et al., 2016b). Few studies describe the composition of lipids in ECM. In *C. albicans* biofilms, the main ECM lipids are glycerolipids (99.5%), neutral or polar, and sphingolipids (0.05%). There are reports of the presence of ergosterol in the cell membrane (Mitchell et al., 2016b; Zarnowski et al., 2014).

Over the last decade, knowledge about the molecular mechanisms involved in the initiation and termination of biofilm formation has increased. It appears to occur in response to specific environmental stimuli, and transitions are mainly governed by adaptive response rather than specific genetic programs. Biofilm formation can occur through multiple pathways, and the spatial structure of biofilms is species-dependent and environmental conditions.

The regulation of matrix components is made in a coordinated and cooperative manner at the extracellular level (Mitchell et al., 2015). This fact may also explain the importance of eDNA for the formation and maintenance of biofilm integrity (Martins et al., 2010; Rajendran et al., 2013). Pierce et al. (2017) believe that eDNA and some proteins play a crucial role in interconnecting different ECM constituents. There are two groups of transcription factors that regulate ECM production: Rlm1 and Zap1 (also known as Csr1). Rlm1 is a positive regulator matrix formation through upregulation of the Gsc1 β-1,3-glucan synthase subunit (also known as Fks1), which catalyzes the β-1,3-glucan formation (Lohse et al., 2018; Ponde et al., 2021). The delivery of β-1,3 glucan from the cell membrane to the extracellular space is regulated by the glucan modifying proteins Bgl2p, Phr1p, and Xog1p (Mitchell et al., 2016b). The factor Zap1, unlike Rlm1, negatively regulates matrix formation through the downregulation of genes GCA1 and GCA2 that encode enzymes responsible for the hydrolysis of long-chain polysaccharides into smaller-chain polysaccharides (Lohse et al., 2018; Ponde et al., 2021). In addition, genes that encode the synthesis of matrix constituents, such as the synthesis of mannan, β-1,6-glucan, and β-1,3-glucan, have been shown to be crucial in matrix production and biofilm development (Lohse et al., 2018).

Many MOs, including bacteria, yeasts, and filamentous fungi, can produce persistent cells in planktonic cultures and biofilm (Galdiero et al., 2020). Persister cells are a small portion of the isogenic and dormant population (generally less than 1%) capable of surviving exposure to high doses of antimicrobial treatment (Galdiero et al., 2020; Khan, Pham, Tabassum, Oloketuyi, & Kim, 2020). There is a difference between persistent and resistant cells; in the case of resistance, a stable genetic modification occurs that gives rise to a completely resistant population, requiring doses far above minimum inhibitory concentrations to eliminate the population. In cases of persistence, they are triggered by physiological and nonhereditary factors. They can enter a metabolically inactive state as an antifungal stress response and return to an active growth and repopulation stage once the antimicrobial is removed ("Galdiero et al., 2020; Khan et al., 2020). So persistent cells can also survive exposure to very high doses of antimicrobials being correlated with the occurrence of recalcitrant infections (Galdiero et al., 2020).

Furthermore, if persistent cells are isolated, cultured again, and treated repeatedly with high concentrations of drugs, a biphasic response equal to the response of the original population (where the majority of the population is killed and only a small subpopulation survives) will be observed (Galdiero et al., 2020; Wuyts, Van Dijck, & Holtappels, 2018). The formation mechanisms of persister cells are not limited to the contact with antimicrobial agents; some factors that can induce their formation include environmental factors (such as lack of nutrients, oxidative stress, DNA damage), host-pathogen interaction, or even inter-/intraspecies through the QS signaling system (Khan et al., 2020). These mechanisms are well described in bacteria and poorly understood in fungi (Khan et al., 2020; Wuyts et al., 2018). Because persister cells are refractory to death even at high drug concentrations, they can only be detected using fungicidal drugs. The echinocandins, the polyenes, and one of the azoles (miconazole) have fungicidal activity against *C. albicans*, but persister cells are typically investigated using the polyene AmB (Galdiero et al., 2020; Borghi, Borgo and Morace, 2016; Wuyts et al., 2018). Due to its mechanism of action, which sequesters ergosterol from the fungal cell membrane without needing to be inside the cell, it hinders the development of resistant cells—increasing the chance that surviving colonies are persistent cells and not resistant isolates (Galdiero et al., 2020; Wuyts et al., 2018).

Glucose is the preferred carbon source for *C. albicans* and *Saccharomyces cerevisiae*. In the absence of glucose, fungal cells are forced to use alternative carbon sources. It has already been observed that all major glucose utilization pathways, such as the tricarboxylic acid cycle, glycolysis, and the pentose phosphate pathway, are downregulated in persister cells, suggesting that they utilize other energy sources (Wuyts et al., 2018).

Cells can resist high concentrations of reactive oxygen species, and to circumvent the mechanisms of death by apoptosis, persisters activate mechanisms of response to oxidative stress, such as increased expression of superoxide dismutases (SODs), which provide these cells with protection against ROS production (Wuyts et al., 2018).

In fungi, persisters were initially detected in mature biofilms of *C. albicans* (LaFleur, Kumamoto, & Lewis, 2006) and later in other *Candida* species (Galdiero et al., 2020; Wuyts et al., 2018). Galdiero et al. (2020) demonstrated the in vitro ability of persistent biofilms in *C. albicans* to recolonize, creating a new biofilm similar to the primary one. They also observed that both biofilms had similar amounts of biomass. However, the biofilms formed from the persistent ones became less compact, which could facilitate the dispersion of new cells and consequently contribute to the spread of infection (Galdiero et al., 2020). Persistent cells resulting from *S. cerevisiae* biofilms showed deletion of ergosterol synthesis genes (Bojsen, Regenberg, Gresham, & Folkesson, 2016); the same was observed in *C. albicans* biofilms (Wuyts et al., 2018), which may explain the low affinity to amphotericin B for membrane and cell persistence (Bojsen et al., 2016).

5. Biofilms and diseases

In the last decade, it has been estimated that 80% of human infections result from pathogenic biofilms contributing to high levels of morbidity and mortality in both adults and children because of their increased dissemination and the current lack of diagnostic, especially in a biofilm mode of growth.

Infection-associated biofilms can be found in tissue/mucus and nonforeign-associated and adherent to foreign bodies (Lebeaux, Chauhan, Rendueles, & Beloin, 2013). The first group includes cystic fibrosis (CF) patients with chronic lung infections and patients with chronic wound infections, where biofilms are found in the tissue and the sputum of CF patients (BBjarnsholt, Ciofu, Molin, Givskov, & Høiby, 2013; Kirketerp-Møller et al., 2008). The second group includes infections related to colonizing biofilms in devices. This group includes infections associated with orthopedic alloplastic devices (Gristina & Costerton, 1985), endotracheal tubes (Adair et al., 1999; Gil-Perotin et al., 2012; Inglis, Tit-Meng, Mah-Lee, Ee-Koon, & Kok-Pheng, 1995), intravenous catheters (Mermel et al., 2009), indwelling urinary catheters or urethral stents (Conway & Larson, 2012), and breast implants (Rieger et al., 2013). Other infections may be related, such as endocarditis, otitis media, chronic sinusitis, biliary stents, shunts, contact lenses, vocal prostheses, dialysis catheters, dentistry, intrauterine devices, artificial hearts, prosthetic valves (Høiby et al., 2015).

The overall burden of biofilm infections is considerable (Jamal et al., 2018; Römling et al., 2014; Zhang, Li, Yu, & Wang, 2020). Among >60,000 CF patients, almost 80% will develop a chronic lung (and sinus) biofilm infection. For patients with chronic wound infections (1%–2%), over 60% may involve biofilms. For all orthopedic alloplastic device patients, 0.5%–2% will develop an infection within the first two postoperative years (Stoodley et al., 2005). Furthermore, patients with IV catheters may develop catheter-related bloodstream infections in the intensive care unit (CR-BSI). The development of biofilm on the surface of endotracheal tubes (ETTs) is related to VAP, which occurs in 9%–27% of all intubated patients (Høiby et al., 2015). Rates for patients with indwelling urinary catheters are even higher because more than 50% of inserted catheters become colonized within the first 10–14 days of insertion (Hola & Ruzick, 2011). Polymicrobial infections often develop in individuals with immunocompromised or disrupted biological barrier function, in which a more favorable environment for coinfection may exist.

Biofilms are the higher resistance of their cells to chemical and physical treatments. Biofilm infections persist despite antimicrobials and the patient's innate and adaptive defense mechanisms (Rajendran et al., 2016). They are characterized by persistent and progressive pathology, mainly due to the inflammatory response related to the biofilm, and many can be challenging to diagnose and treat (Brogden et al., 2005; Harriott & Noverr, 2011; Lebeaux, Ghigo, & Beloin, 2014). Increased cancer risk is associated with persistent inflammation and chronic infections (Groeger & Meyle, 2019).

Most bacteria and fungi, such as *P aeruginosa*, *S. epidermidis*, *C. albicans*, *Acinetobacter baumannii*, *Helicobacter pylori*, *S. aureus*, *Listeria monocytogenes*, *Vibrio cholerae*, and *Salmonella enterica* can form biofilms (Chen et al., 2021; Tsui, Kong, & Jabra-Rizk, 2016; Zhang, Li, Yu, & Wang, 2020; Zhang et al., 2021). These MOs similarly form biofilms and share many common characteristics (Cavalheiro & Teixeira, 2018; Koo, Allan, Howlin, Stoodley, & Hall-Stoodley, 2017; Moormeier & Bayles, 2017).

C. albicans is the most polymicrobial biofilm reported in the scientific literature in association with other species of bacteria and fungi due to its opportunistic characteristic (Peters et al., 2012). Infections involving biofilm in tissues and organs, such as those on the skin, into systemic disease, CF with a chronic lung infection, in the oral cavity, in the gastrointestinal tract, in the vulvovaginal or devices (e.g., orthopedic alloplastic devices, endotracheal tubes, intravenous catheters, indwelling urinary catheters, tissue fillers) will be discussed below (Fig. 2). Biofilm growth on host tissues and medical devices has appeared as a key virulence mechanism for opportunistic pathogens such as *P. aeruginosa*, staphylococcal species, *C. albicans*, and *A. fumigatus* (Boisvert, Cheng, Sheppard, & Nguyen, 2016).

5.1 Systemic infections

Mixed biofilms of *C. albicans* and *S. epidermidis*, *Enterococcus* spp., and *S. aureus* have been found in systemic infections (Rodrigues & Černáková, 2020). *S. aureus* has been isolated in the high frequency from the blood of patients with candidemia, besides being associated with bloodstream infections of a neonatal population and infective endocarditis (Lebeaux et al., 2014; Marks et al., 2014; Pammi et al., 2014). Yeast adhesins facilitate the interaction between the *Candida* and *S. aureus* and are essential to transport *S. aureus* across mucosal barriers, leading to systemic infection in cocolonized patients (Schlecht et al., 2015). Therefore, the interaction between these two pathogens can be lethal to the host, causing candidemia and bacteremia (Klotz, Chasin, et al., 2007; Lebeaux, Ghigo, & Beloin, 2014; Sousa, Botelho, Rodrigues, Azeredo, & Oliveira, 2012).

5.2 Upper and low airways

Rhinosinusitis is a multifactorial condition related to mucosal inflammation and upper airway MOs such as *A.*

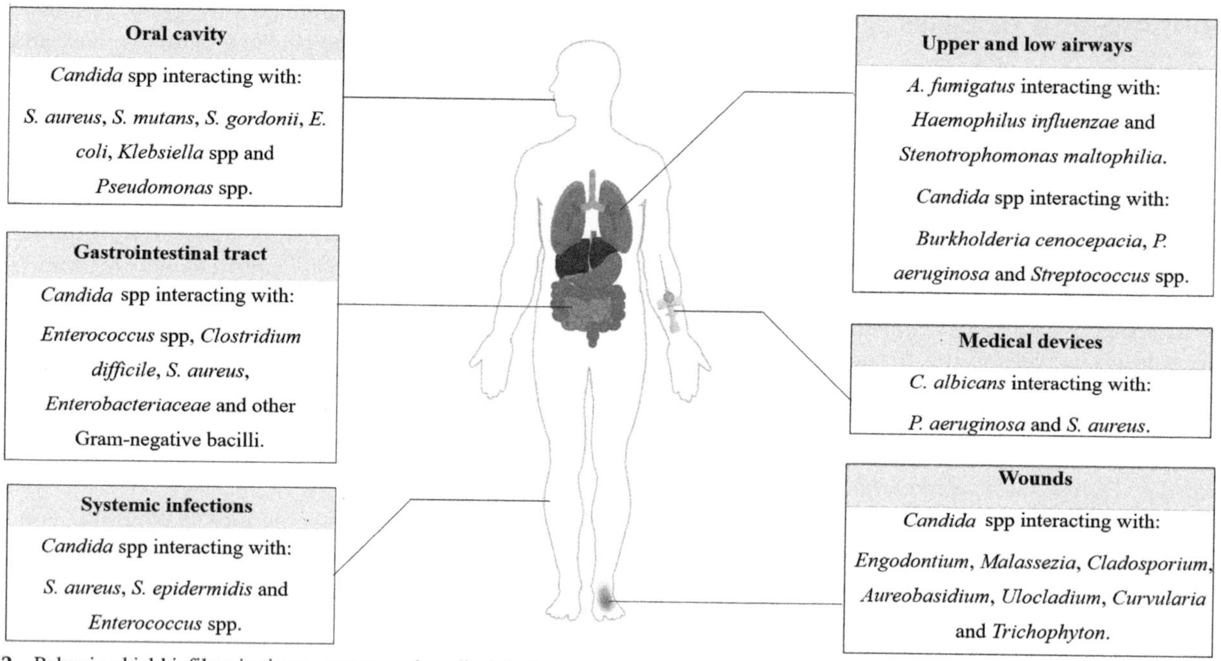

FIG. 2 Polymicrobial biofilms in tissue, organs, and medical devices.

fumigatus and *Haemophilus influenzae*. These fungal/bacteria can form dual-species biofilms, leading to enhancing antimicrobial drugs (Lee et al., 2015; Seidler, Salvenmoser, & Muller, 2008). Dellière et al. (2021) examined 45 samples of endoscopic surgery from patients with rhinosinusitis using metagenomics approaches and found coexistence without proven synergism or antagonism relationship interkingdom. Research performed by Healy, Leid, Sanderson and Hunsaker (2008) found evidence of polymicrobial biofilm with *A. fumigatus* and *H. influenzae* in 82% of sinus biopsies, demonstrating an interaction between them.

Pulmonary disease is the leading cause of morbidity and mortality in CF patients. CF is a genetic disease that usually results in polymicrobial infection resulting from the thick accumulation and dehydrated mucus in the lungs (Elborn, 2016). Many MOs are commonly coisolated from the airways of CF patients, including the filamentous fungus *A. fumigatus* and the Gram-negative bacillus *Stenotrophomonas maltophilia* (Roisin et al., 2020). These associations were proven to have poor clinical outcomes for patients diagnosed with coinfection resulting in more hospitalization and loss of lung function compared with single colonization by these pathogens (Reece et al., 2017). Additionally, *Candida* spp. has been reported to interact with *Burkholderia cenocepacia* in patients with CF (Lebeaux et al., 2014) and with *M. tuberculosis* in patients with tuberculosis (Ochieng et al., 2005).

Antagonistic interactions between *Candida* spp. and bacterial species have also been observed in the lungs, with *P. aeruginosa* killing yeast hyphae and biofilms of *C. albicans* (Lebeaux et al., 2014; Rodrigues & Černáková, 2020; Sardi, Scorzoni, Bernardi, Fusco-Almeida, & Mendes Giannini, 2013). Therefore, patients with CF may develop multispecies biofilm infection from lung-resident bacteria *P. aeruginosa* and *Streptococcus* spp. (Scott et al., 2019).

5.3 Oral cavity

The oral cavity houses diverse communities of MOs—viruses, mycoplasmas, bacteria, Archaea, fungi, and protozoa (Wade, 2013) that persist on all surfaces. The oral environment influences the composition of the microbiome, and changes in local conditions can affect microbial interactions and determine, in part, whether the relationship between the oral microbiome and the host is symbiotic or potentially harmful (dysbiotic), thus increasing the risk of diseases such as caries or periodontal diseases (Marsh, 2003; Roberts & Darveau, 2015). These diseases in the oral cavity are multispecies infections associated with biofilm. *Candida* spp. coexist with several bacterial species, including *S. aureus*, *S. mutans* (the main bacteria found in human caries), *S. gordonii*, *E. coli*, *Klebsiella* spp., and *Pseudomonas* spp. The formation of these polymicrobial biofilms is directly correlated with the use of prostheses, with the formation of biofilms on the surface of these prostheses or on the oral mucosa itself (Baena-Monroy et al., 2005; Lebeaux et al., 2014; Metwalli et al., 2013; Thein, Seneviratne, Samaranayake, & Samaranayake, 2009). Some bacterial species, such as *S. gordonii*, can potentiate the development of hyphae and biofilm formation by

C. albicans when in the presence of human saliva, thus contributing to the establishment of a difficult-to-treat polymicrobial biofilm. Cells from biofilms can spread throughout the body from the oral cavity, causing disseminated candidiasis (Miceli, Díaz, & Lee, 2011).

C. albicans, an opportunistic pathogen, was associated with several bacteria. Polymicrobial infections with *C. albicans* and *S. aureus* are common due to shared niches within the body, including coisolation from the skin, axillae, vagina, pharynx, nasal passages, and oral mucosa (Todd & Peters, 2019). These microbes are also commonly found together at various body sites and are implicated in a variety of diseases, including CF, VAP, urinary tract infections, superinfection of burn wounds, denture stomatitis, and keratitis (Garcia et al., 2020; Gupta & Foley, 2019; Gupta, Haque, Mukhopadhyay, Narayan, & Prasad, 2005; Orazi & O'Toole, 2019; Valenza et al., 2008). *Candida albicans* and *S. aureus* are two of the most commonly coisolated pathogens of mixed fungal-bacterial bloodstream infections, as well as biofilm-associated diseases such as CF, periodontitis, and catheter-associated infections indwelling medical devices (Harriott & Noverr, 2011; Todd & Peters, 2019). Harriott and Noverr (2011) demonstrated that *S. aureus* showed high tolerance to vancomycin during mixed biofilm growth and reduced susceptibility to doxycycline, nafcillin, and oxacillin (Kong et al., 2016).

5.4 Gastrointestinal tract

The human intestinal microbiome is diverse, and the coexistence of multiple MOs, among bacterial and fungi, is indispensable for its health functions. These MOs exhibit essential functions such as regulating host fat storage, protecting against cell injury, and stimulating intestinal angiogenesis (Eckburg et al., 2005). In some adverse conditions, fungi/bacteria can form polymicrobial biofilms in the human gut causing poor clinical outcomes. The intestinal tract may be the source for transmitting many nosocomial pathogens, including *Enterococcus* species, *Clostridium difficile*, *Candida* species, *Enterobacteriaceae*, and other Gram-negative bacilli. Recent data suggest that the intestinal tract of hospitalized patients may also be an important reservoir of *S. aureus* (Stiefel & Donskey, 2004). Although the clinical manifestations of these pathogens are diverse, common pathogenesis is involved in the colonization and spread of the intestinal tract (Stiefel & Donskey, 2004). The presence of *Candida* spp. in polymicrobial biofilms in the gastrointestinal tract is particularly problematic, as the associated infections have mortality rates relatively higher than those of solely bacterial polymicrobial biofilms (75% compared to 30%) (Dupont, 2002; Lebeaux et al., 2014).

5.5 Medical devices

Healthcare-related infections of polymicrobial biofilms can affect different human tissues through superficies of medical devices, for example, catheter-related infections, VAP, and titanium implants. Catheter-related infections due to *C. albicans* are an important problem to nosocomial infections; they can rapidly spread to the blood and cause serious clinical outcomes (Klotz, Chasin et al., 2007). VAP has a high mortality (10%–40%) and occurs rapidly after endotracheal intubation (Tablan et al., 2004). Both *C. albicans/P. aeruginosa* and *C. albicans/S. aureus* associations are known to form polymicrobial biofilms and are the leading cause of VAP in infections caused by medical devices. These associations are difficult to treat and need drug combinations to have effectiveness (Lou, Wang, Tang, & Chen, 2017; Rodrigues et al., 2017).

5.6 Wounds

Diabetic foot infection is a severe complication to many patients worldwide that leads to degeneration and amputation (Madmoli et al., 2019). Namgoong, Jung, Han, Kim, and Dhong (2019) found polymicrobial biofilms in 89% (67 of 75 patients) of cases with a diabetic foot infection; among 248 MOs isolated, 5% had *C. albicans* present and 2% other fungi not identified. Dowd et al. (2011) studied 915 chronic wound patients and assessed the prevalence of fungi species in polymicrobial infections. *Candida* species were the most prevalent fungi, even though *Engyodontium*, *Malassezia*, *Cladosporium*, *Aureobasidium*, *Ulocladium*, *Curvularia*, and *Trichophyton* were also present. Unlike most studies, fungi are shown to be important and diverse in a chronic wound; they can form polymicrobial biofilms and have a severe impact on effective treatment (Dowd et al., 2011).

Infections related to biofilms need the contribution of several research areas to develop more effective strategies to advance in better diagnoses, treatments, and more preclinical studies and multicenter clinical trials to evaluate these strategies.

6. Models of biofilm infection and interaction

In the study of polymicrobial interactions, understanding the mechanisms of biofilm formation can be challenging. In this regard, they use in vitro or in vivo methodologies (Fig. 3), which simulate the natural environment in which the MOs are inserted (Gabrilska & Rumbaugh, 2015; Khatoon, McTiernan, Suuronen, Mah, & Alarcon, 2018). These insertions can occur in two ways: simultaneously or staggered, to verify the interaction between different species/kingdoms. Among the main in vitro methodologies

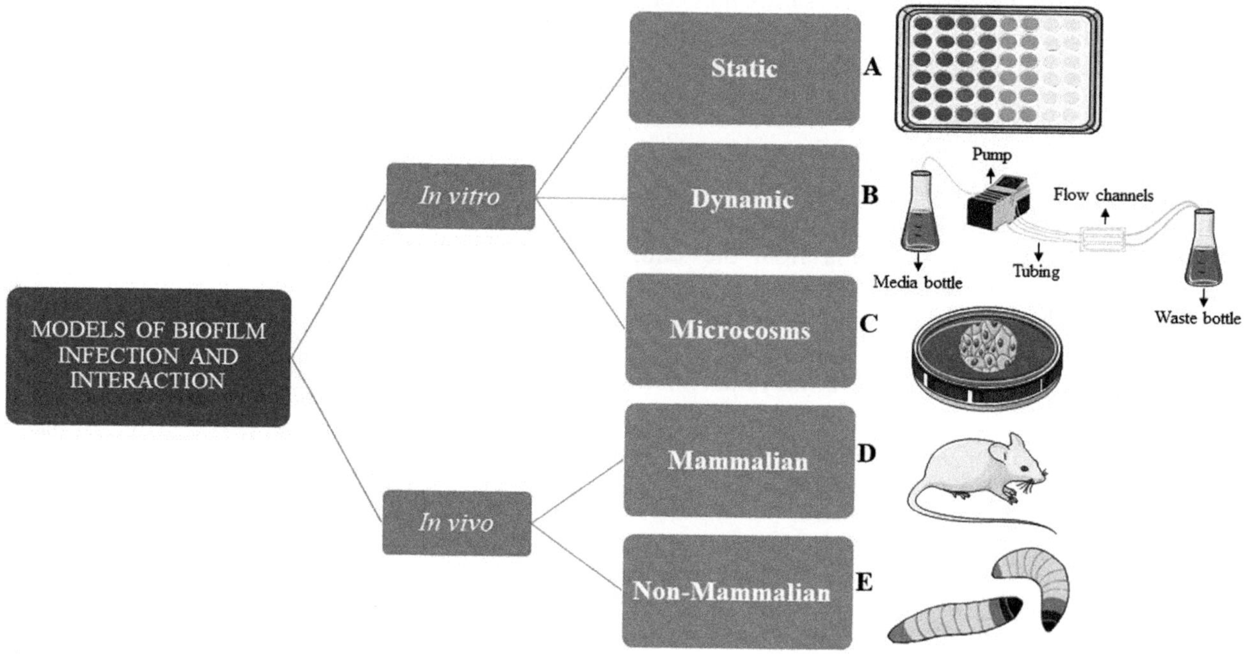

FIG. 3 Representative scheme of the principal methodologies used in vitro and in vivo to study the interaction and infection of polymicrobial biofilms. **A**—technique in microdilution plates; **B**—flow cell system setup technique; **C**—microcosms technique; **D**—mouse, and **E**—*Galleria mellonella. (Based on: Weiss Nielsen, M., Sternberg, C., Molin, S., & Regenberg, B. (2011).* Pseudomonas aeruginosa *and* Saccharomyces cerevisiae *biofilm in flow cells.* Journal of Visualized Experiments: JoVE, 47, *2383. https://doi.org/10.3791/2383.)*

are static, dynamic, and microcosms models (Lebeaux et al., 2013; Soares et al., 2015).

In the static model, the most used technique is microdilution plates (Fig. 3A). This methodology allows the study of some properties related to biofilms, such as metabolic activity, biomass, matrix quantification, antimicrobial tolerance, evaluation of the interaction of MOs with the environment, and tests with antimicrobials (Gabrilska & Rumbaugh, 2015; Garcia et al., 2020; Lebeaux et al., 2013; Medina-Alarcón et al., 2021; Peeters, Nelis, & Coenye, 2008). Most microplates are made of polystyrene, but other components can be included and/or excluded from the wells (Gabrilska & Rumbaugh, 2015; Peeters et al., 2008). One of the main disadvantages already observed is the limitation of nutrients and oxygenation that this technique provides (Almshawit, Macreadie, & Grando, 2014).

In the work by Peters, Ward, Rane, Lee, and Noverr (2013), silicone discs were added to the bottom of the microplates to simulate the microenvironment of the catheters. The mono- and polymicrobial biofilms of *C. albicans* with *S. aureus* were formed on these discs and treated with different percentages of ethanol to verify their action on the biofilms. With that, they proved the effectiveness of this compound in the decontamination of catheters. Other studies, such as the one conducted by Garcia et al. (2020), verified the interaction of *C. albicans*/*C. parapsilosis* with *Trichophyton rubrum*. Microplates and reagents were used to analyze metabolic activity, biomass, and ECM of polysaccharides. Medina-Alarcón et al. (2021) used the same parameters to analyze biofilms formed between *P. brasiliensis* and *M. tuberculosis*, demonstrating the importance of these techniques for understanding the interaction between different MOs.

Another static model methodology already described is the Calgary Biofilm Device, which uses 96-well microplates consisting of a lid with pegs that allow biofilms to adhere to this surface and be transported to other environments (Almshawit et al., 2014; Ceri et al., 1999). In the Ring Test biofilm model, MOs adhere to magnetic spheres, a fast and suitable methodology for high-throughput screening (Di Domenico et al., 2018; Olivares et al., 2020). Some agar-based models verify differences in phenotypes, colony size, growth, and other parameters (Abd El-Baky et al., 2020; Norouzi, Danesh, Mohseni, & Rabbani Khorasgani, 2018).

Dynamic models, also called open systems, can reproduce the environment in which the MOs are inserted continuously; they constantly supply nutrients and flow displacements to the biofilms. This type of model allows the study of physiological, morphological, growth dynamics, and antimicrobial sensitivity (Lebeaux, Chauhan, Rendueles and Beloin, 2013).

Among the dynamic models, the flow cell is one of the most used techniques, which consists of assembling a system consisting of a blade that has silicone lines, a microscope, syringes, and tubing for the entry of nutrients and the exit of waste (Fig. 3B) (Palmer, 1999; Weiss Nielsen, Sternberg, Molin, & Regenberg, 2011; Zhang, Sileika, et al., 2011). This methodology allows the control of growth, temperature, flow parameters and provides details on the development of biofilms in three dimensions (3D) (Pamp, Sternberg, & Tolker-Nielsen, 2009; Weiss Nielsen et al., 2011). For this, the authors interconnected the flow cell system in a confocal laser fluorescence microscope and made specific markings on the biofilms, allowing images to be captured at the desired times and in 3D (Crusz et al., 2012; Weiss Nielsen et al., 2011). In addition, this technique allows verifying the influence on growth between MOs existing in specific environments, such as oral, the presence of vesicles formed by the biofilm, and the release of metabolites (Mendez, Walker, Vipham, & Trinetta, 2020; Zhu et al., 2013).

The CDC biofilm reactor consists of a reactor connected to a flask of culture medium that distributes to eight racks, each housing three coupons. The MOs are incubated in the reactor, and biofilms are formed on the coupons (Galdiero et al., 2020; Lebeaux, Chauhan, Rendueles, & Beloin, 2013; Macià, Rojo-Molinero, & Oliver, 2014; Mendez et al., 2020). CDC scientists created this methodology to study biofilms in drinking water (Johnson, Petersen, & Goeres, 2021).

In the study conducted by Galdiero et al. (2020), the authors used coupons made of polycarbonate, polypropylene, and Teflon, which are standard components in medical devices. The authors tested the polymicrobial biofilm of *C. albicans* with *Klebsiella pneumoniae*; the MOs were able to form a biofilm (mono and polymicrobial) on the different materials used in the coupons. However, rough surfaces such as polypropylene had greater adherence than Teflon. Furthermore, they observed that components secreted by MOs could play fundamental roles in polymicrobial interactions.

BiofilmChip microfluidic systems provide significant control over nutrients, chemical gradients, real-time images of the development and growth of biofilms; in addition, they better mimic the environment in which these MOs are inserted, including concerning treatment, which can be a technique used for carrying out personalized treatments (Blanco-Cabra et al., 2021; Gabrilska & Rumbaugh, 2015; Hopke et al., 2021; Millet et al., 2019).

In addition to the models mentioned, another technique for studying biofilms in dynamic models is the modified Robbins device. It is a fluid system with six ports interconnected to the media containing the nutrients (Honraet & Nelis, 2006; Yassin, German, Rolland, Rickard, & Jakubovics, 2016). Yassin et al. (2016) observed that fluorine introduced into this system could interfere with interspecies and cell-cell interactions in forming mixed biofilms between *C. albicans*, *Lactobacillus casei*, and *Streptococcus mutans*.

Another in vitro technique used is microcosms models, which more reliably simulate the in vivo environment, as they use ex vivo tissues or fluids, such as saliva for the development of biofilms, or human cells to simulate the in vivo. They can be static or dynamic; however, they use surfaces and nutrients, which ensure the complexity and heterogeneity of the environments in which they are inserted (Fig. 3C) (Gabrilska & Rumbaugh, 2015; Lebeaux et al., 2013; McBain et al., 2005).

Filoche, Soma, van Bekkum, and Sissons (2008) used polystyrene plates coated with glass coverslips to investigate the microbial communities contained in the saliva of five different donors and performed the treatment with chlorhexidine and Listerine©. The authors concluded that the effective treatment of dental caries depended on the response of each microbiota. Other studies use reconstituted tissue from biopsy to verify the formation of biofilms in oral, urinary tract, gastrointestinal and vaginal microbiota (Kim, Hegde, & Jayaraman, 2010; Samaranayake et al., 2014).

In addition to this, other authors have shown that reconstituted tissues are excellent parameters for studying the interactions that occur within an organism. The work conducted by Schaller, Schäfer, Korting, and Hube (1998) used the culture of keratinocytes derived from an oral carcinoma, which were introduced into a controlled culture medium. The authors showed the capacity of reconstituted epithelium to be like that in vivo, including in terms of pathogenesis. The researchers infected these oral keratinocytes with *C. albicans* and found that the biofilm formed contained, in addition to yeast, bacteria that are naturally present in the oral cavity, forming a polymicrobial biofilm of bacteria and fungus.

The interaction and infection models seen so far can simulate what happens in an in vivo environment. However, some researchers bring approaches to studying polymicrobial biofilms in an in vivo environment, using mammals and nonmammals. Regarding nonmammals, the main models are *Galleria mellonella*, *Danio rerio*, *Drosophila melanogaster*, and *Caenorhabditis elegans*, also called alternative animals (Fig. 3D and E) (Coenye & Nelis, 2010; Fauvarque, 2014; Gabrilska & Rumbaugh, 2015; Galeane et al., 2019; Glavis-Bloom, Muhammed, & Mylonakis, 2012; Lebeaux et al., 2013; Scheunemann, Fortes, Lincopan, & Ishida, 2021; Singulani et al., 2017, 2018).

Despite the similarity between mammals, it is often difficult to reproduce human diseases. Most models have been tested on monospecies biofilms, and more recently, some research groups are expanding testing with polymicrobial biofilms (Benaducci et al., 2016; Gabrilska & Rumbaugh, 2015; Jensen, Johansen, & Jensen, 2017; Oliveira et al., 2020).

In Falsetta et al. (2014), the authors found that biofilms formed by *C. albicans* and *S. mutans* affect children's oral health with a high sugar diet. For this purpose, they used hyposalivation rats with a diet rich in sucrose and found that the interaction between MOs in biofilms in vivo had a hypervirulent relationship and produced severe damage to the animals' teeth, as had already been seen in children.

Other authors used mice to verify the formation of *C. albicans* biofilms with the bacterial microbiota present on pseudomembranous surfaces to characterize polymicrobial biofilms on the oral mucosa. The researchers proved that bacteria present in the mouth are present in the biofilms of *C. albicans*, contributing to the formation of pseudomembranous lesions in vivo in a murine model (Dongari-Bagtzoglou et al., 2009). In addition to these, another study verifies the interactions between fungi and bacteria in subcutaneous catheters inserted in murine models, demonstrating the interaction and colonization of these species in the formation of polymicrobial biofilms (Vila et al., 2021). Another aspect of using a murine model is researching intra-abdominal infections, which can become lethal due to the interaction between MOs (Ikeh, Fidel, & Noverr, 2018).

The use of alternative animals to mammals has been growing annually due to world legislation that recommends reducing the use of mammals in research. This recommendation is due to the principles of the 3RS created in 1959 by Russell and Burch, which mean replacement, reduction, and refinement (Arck, 2019; Sneddon, Halsey, & Bury, 2017). These alternative models to mammals are lower cost, less bureaucratic from the ethics committee's point of view, and high yield (Glavis-Bloom et al., 2012; Nett, 2018; Scheunemann et al., 2021).

Some studies address the interaction of polymicrobial biofilms and alternative animals. These proved to be reliable methods with good reproducibility, mainly to prove the pharmacological efficacy and interactions between MOs. The study conducted by Scheunemann et al. (2021) showed that caspofungin effectively treated *G. mellonella* infected with *C. albicans* and methicillin-resistant *S. aureus* biofilms increasing larvae survival and reducing fungal and bacterial load. In the study of interkingdom interaction, Sheehan, Tully, and Kavanagh (2020) showed that these larvae have a viable immune system, as proteomic analysis of larval hemolymph with coinfection of the species mentioned above revealed an abundance of proteins associated with the response to bacterial and fungal infections and very similar to the innate response of mammals to an infection.

Another alternative animal model is *Caenorhabditis elegans*. The study conducted by Holt, Houston, Adams, Edwards, and Kjellerup (2017) found that the combination of EPS-producing *S. epidermidis* with *C. albicans* did not result in a severe infection in this animal. Fourie, Albertyn, Sebolai, Gcilitshana, and Pohl (2021) demonstrated that combining the mutant strain of *C. albicans* (mutation that regulates morphogenesis) with *P. aeruginosa* could not reduce nematode survival. The authors found that the *C. albicans* strain could still form hyphae and pierce the nematode cuticle even with the mutation. These animals are known to share an innate immune response like humans, making viable models for understanding pathogenesis (Fourie et al., 2021; Pukkila-Worley, Ausubel, & Mylonakis, 2011).

Despite several approaches to polymicrobial biofilms and their interactions, much remains to be done on methods capable of predicting the behavior of interactions between species/kingdoms, especially in vivo models. Nowadays, many studies are being carried out to increase the knowledge of these relationships in both pathogenesis and treatment.

7. Treatment challenges

Tolerance to antimicrobials, biocides, and antiseptics is a common feature of microbial biofilms formation. Biofilm physiology contributes to antimicrobial tolerance by its physical structure, gene expression, different from planktonic cells. This tolerance is related to multidrug efflux pumps, changes in metabolism, response to stress regulation, and QS molecules. Microbial communities employ a specific response on regulatory mechanisms to antimicrobial agents. The presence of ECM provides a physical barrier to antifungal drugs, and cells within the biofilm exhibit increased cell density, increased stress response, and decreased metabolic activity contributing to antifungal drug resistance. Persister cells, nondividing cells with decreased metabolic activities, are highly resistant to antimicrobial drugs and likely to seed new biofilm infections after drug treatment. Efflux pumps export drug molecules from the inside the cell to the environment, and the expression of these is highly upregulated in biofilms, even in the absence of antifungal drugs, which contributes to the overall drug-resistant nature of biofilms (Cavalheiro & Teixeira, 2018; Costa-Orlandi et al., 2017; Ramage, Bachmann, Petterson, Wickes, & López-Ribot, 2002; Ramage et al., 2002; Santos et al., 2017; Sardi et al., 2014; Song et al., 2009).

The QS system regulates several cellular processes that enable microbial communication to regulate morphogenesis, pathogenesis, competence, biofilm formation, maturation, virulence, and synthesis of ECM compounds, through environmental stimuli, and plays an important regulatory role in the process of drug efflux pumps (Padder, Prasad, & Shah, 2018). In the past, QS has related a specialized system of certain bacteria, but new research on *C. albicans* filamentation using farnesol revealed the presence of QS also in fungi (Albuquerque & Casadevall, 2012; Bamford et al., 2009). QS plays an important interkingdom interaction role and is crucial for fungal/bacteria interaction, maintaining homeostasis and immunity response. The indirect role of QS in the emergence of multidrug resistance

(MDR) in microbial pathogens requires alternative antimicrobial therapies that target QS inhibition. A related phenomenon of quorum sensing inhibition (QSI) performed by small inhibitory molecules called quorum sensing inhibitors or quorum quenching (QQ) can efficiently reduce QS-regulated gene expression. For these reasons, QS mechanisms have been studied to target new antimicrobial therapies with significant potential against polymicrobial fungi/bacteria biofilms. Besides, in the context of MDR, they are an excellent strategy for alternative antimicrobial drugs discovery (Costa-Orlandi et al., 2017; Kovács & Majoros, 2020; Padder, Prasad, & Shah, 2018; Raina et al., 2009; Rasmussen & Givskov, 2006; Wu, Liu, Liu, Yang, & Qiao, 2020).

In clinical outcomes, these mechanisms of biofilms survival are related to failure in treatments, prolonged treatment regimens, hospitalization, and death, which has been considered a public health problem (Pires, Martinez, Mendes-Gianinni, & Stoianoff, 2021). Available antifungal agents' arsenal can treat invasive fungal infections (IFIs), which are limited to amphotericin B, echinocandins, isavuconazole, itraconazole, posaconazole, and voriconazole (Di Mambro, Guerriero, Magnani, & Marra, 2019; Perfect, 2017; Zhao, Khoo, & Tan, 2015; Zhang et al., 2021). Allylamines, azoles, and griseofulvin are the main drug classes used in dermatophytosis (de Pereira, 2021; Martinez-Rossi, Bitencourt, & Teixeira De Aguiar Peres, 2018). However, current antifungal options became more restricted due to toxicity and drug-resistant strains; cases of hepatotoxicity and nephrotoxicity are related to high doses of polyenes and azoles necessary to treat disease-associated biofilms (Olivares et al., 2019). Furthermore, biofilms of *Candida*, *Cryptococcus*, *Aspergillus*, dermatophytes have shown more resistance to all antifungal when compared with planktonic cells (Bila, Costa-Orlandi, & Vaso, 2021; B. C. Costa-Orlandi et al., 2017; C. Costa-Orlandi, Martinez, & Bila, 2021; C. Costa-Orlandi, Sardi, & Santos, 2014; Hawser & Douglas, 1995; Kuhn et al., 2004; de Pereira, 2021; Sardi et al., 2014; Scorzoni, Sangalli-Leite, & de Lacorte Singulani, 2016).

For this reason, Koo et al. (2017) argue that several approaches related to the interruption of biofilm formation should be encompassed. The first approach can prevent the attachment and interrupt the interactions between the MOs and the surface, directing the production of EPS and cell division. The rupture of the formed biofilms can be achieved by physical removal, degradation of the EPS matrix, aiming at the establishment of pathogenic microenvironments (low pH or hypoxia) and social interactions (in polymicrobial biofilms), as well as the elimination of dormant cells and the dispersion of the biofilm, can be induced by the remodeling of the EPS matrix or activation of mechanisms of dispersal.

Therefore, discovering new treatments as a strategy against biofilms is extremely important (Zhang et al., 2021). Some strategies available for treatments of biofilm-related infections are a concern to early diagnosis, vaccines, interference with biofilm regulators, antipersister strategies, QS inhibitors, ion chelators, photodynamic therapy, nonantimicrobial cell killing, ultrasounds, carrier systems to a specific target to biofilm surface, matrix degradation, detachment induction, and enhancement of drug penetration (Chen, Zeng, & Tian, 2013; Del Pozo, 2017; Zhang et al., 2021; Roy et al., 2018). There are several approaches to developing new antibiofilm treatments, from studying new chemical entities that act on new selective biofilm targets, modifying existing antifungal drugs, or repurposing existing drugs that are primarily used to treat conditions other than fungal/bacterial infections. This topic will cover new approaches against polymicrobial biofilms such as combined therapy, nanotechnology, photodynamic therapy, antimicrobial peptides, and QQ molecules, providing new insights into effective available polymicrobial biofilms treatments.

7.1 Combined therapy

Combined therapy is described as coadministration of multiple pharmacologically active compounds and has some advantages compared with monotherapy, such as minimization of resistance, synergism, broadening of the spectrum of action, and increase of drug's pharmacological effects (Gbejuade, Lovering, & Webb, 2015; Gill, Franco, & Hancock, 2015; Walvekar, Gannimani, & Govender, 2019; Worthington & Melander, 2013; Zhang et al., 2020). Combination therapy must consider the permeability to antimicrobials, MOs present in the polymicrobial biofilm, and targets involved. For that, the choice may include antibiotics, nonantibiotics, antifungals, natural compounds, antimicrobial peptides, and drug delivery systems (Khan, Pham, Oloketuyi, & Kim, 2019; Rose & Poppens, 2009; Worthington & Melander, 2013; Zhang et al., 2020).

Few studies have investigated drug intervention against polymicrobial biofilms. Rodrigues et al. (2017) observed that the combination of an antibiotic polymyxin B and an antifungal amphotericin B presented a potential synergy therapeutic effect against polymicrobial communities involving *P. aeruginosa* and *C. albicans* biofilm. (Li, Guo, Dawuti, and Aibai (2015) verified that fluconazole in combination with minocycline is highly effective against planktonic and biofilm dual-cultured species of *C. albicans* and *S. aureus*. Lown, Peters, Walraven, Noverr, and Lee (2016) assessed the association of micafungin, ethanol, and doxycycline to treating this fungi/bacteria biofilm present in implants. The results demonstrated that the association of these three drugs led to a 98% reduction of the mixed biofilm compared with a single antibacterial agent, which did not demonstrate significant results for *S. aureus*. Tan, Leonhard, and Moser (2019) evaluated the association of 2-aminobenzimidazole (2ABI) with curcumin (CUR) to inhibit the polymicrobial biofilm of *C. albicans* with *S. aureus*.

Rogiers, Holtappels, and Siala (2018) verified that anidulafungin acts synergistically when combined with tigecycline (a derivative of minocycline, which exhibits potent activity against a broad spectrum of bacteria including MRSA) against in vivo and in vitro mixed *C. albicans* and *S. aureus* biofilms.

7.2 Nanotechnology

Nanotechnology is a technology that aims to obtain nanoscale materials and particles with dimensions from 1 to 100nm are used as a therapeutic tool that can improve drug delivery, decreasing effective pharmacological dose of antifungal drugs and consequently their toxicity (Rout, Liu, & Wu, 2017; Voltan et al., 2017). For this purpose, nanoparticles (NPs) are widely used against biofilms for their ability to cross the biofilm and cause cell death (Galdiero, Lombardi, & Falanga, 2019; Medina-Alarcón et al., 2021). According to Pelgrift and Friedman (2013) and Pircalabioru and Chifiriuc (2020), drugs carried by NPs act on specific microbial cells, causing cell death by oxidative damage, disruption on the cell membrane, enzyme inhibition, DNA damage, protein inactivation, and changes in gene expression.

Besides these advantages, Ding, Songkiatisak, Cherukuri, Huang, and Xu (2018) demonstrated that NPs also increase the solubility, stability, and biocompatibility of drugs at sites of action and are an excellent choice for drug development. Among the nanocarriers against biofilms, nanocapsules, protein nanocomplexes, cyclodextrin nanocomplexes, dendrimers, core-shell, inorganic, metallic, and continuous lipid NPs, liposomes, nanoemulsion, and polymeric NPs have been used (Dos Santos Ramos, da Silva, & de Toledo, 2019; Zhang, Li, Yu, & Wang, 2020; Wang, Li, Liu, Chen, & Li, 2017).

Metallic NPs have an excellent antibacterial and antifungal potential due to their inert nature used as an antimicrobial carrier, decreasing the effective dose and side effects (Hemeg, 2017). Lara and Lopez-Ribot (2020) studied the activity of AgNPs (silver nanoparticles) against polymicrobial biofilm of *C. albicans* and *S. aureus* from contaminated catheters and observed a dose-dependent activity and inhibition of the biofilm growth. Ali, Morsy, El-Zawawy, Fareed, and Bedaiwy (2017) determined antimicrobial activity of ZnO_2-NPs against MDR *P. aeruginosa* and *A. niger* strains isolated from burn wound infections, demonstrating high antiinflammatory activity and healing of the patients.

Chitosan NPs are natural polymers, widely used to improve antibiofilm activity (Stuart, Schwartz, Beeson, & Owatz, 2006; Wu, Fan, Kishen, Gutmann, & Fan, 2014) and have shown high penetration into biofilm, allowing the disruption of multispecies biofilms and disinfection of root canals (Rodrigues et al., 2018). Ikono et al. (2019) demonstrated the ability of nano chitosan to reduce the polymicrobial biofilm of *C. albicans* and *S. mutans* from early childhood caries after 18h of treatment.

Liposomes are vesicles formed by physiologically compatible lipid bilayers of diameters <100–200nm, capable of penetrating and effective in the treatment of biofilms due to their versatility for antimicrobial agents and their ability to penetrate the EPS matrix (Forier et al., 2014; Rukavina & Vanić, 2016; Wang, van der Mei, Ren, Busscher, & Shi, 2020). According to Luo et al. (2021), liposomal amphotericin B (AmBisome) reduced polymicrobial biofilms of *C. albicans/S. aureus* and *C. albicans/S. aureus* methicillin-resistant. In contrast, liposomal amphotericin B did not demonstrate inhibition when *C. albicans* was associated with *E. coli*. Other studies were conducted to increase this system and result in effective treatment (Koo et al., 2017; Wang et al., 2020).

Finally, nanoemulsions (NEs) are a heterogeneous system formed by an oil phase dispersed in an aqueous phase, stabilized by a surfactant under high shear force (Medina-Alarcón et al., 2021; Voltan et al., 2017). These NEs are being studied in different MOs, such as bacteria, fungi, and virus demonstrating promising results (Pircalabioru & Chifiriuc, 2020; Ramalingam, Frohlich, & Lee, 2013). Medina-Alarcón et al. (2021) studied the action of 3'hydroxychalcone (NE3'chalc) against polymicrobial biofilm of *P. brasiliensis/M. tuberculosis* from hospitalized patients with comorbidities and found positive results that demonstrated in vitro NE3'chalc is a promising alternative treatment in this condition of fungi/bacteria biofilm.

7.3 Photodynamic therapy (PDT)

Photodynamic therapy (PDT) has been reported for over 100years, being used in different areas of microbiology, oncology, and ophthalmology. PDT was reported for the first time in treating herpes simplex through neutral red and colorless light (Warrier, Mazumder, Prabhu, Satyamoorthy, & Murali, 2021). Due to an increase of drug-resistant MOs, PDT emerges as an alternative treatment (Maisch, 2009; Ziental, Mlynarczyk, Czarczynska-Goslinska, Lewandowski, & Sobotta, 2021).

PDT, a technique that uses a photosensitizer (PS), consists of two steps. Initially, the PS is administered systemically or topically in a determined area, and then it is applied to lighting with a wavelength of light capable of exciting the PS (Cieplik et al., 2018). This method is based on chromophore absorption of light photons, causing an electron transfer that results in ROS production (Nikinmaa et al., 2020). Furthermore, PDT inhibits the adhesion capability of MOs to surfaces, with disruption of biofilm structures with damage to organelles and secretion of virulence factors, which inhibit efflux pump and QS molecules (Arciola, Montanaro, & Costerton, 2011; Mahdizade-ari, Pourhajibagher, & Bahador, 2019; Melo, Avci, de Nóbrega Oliveira, et al., 2013; Quishida et al., 2015).

Akhtar et al. (2021) investigated the efficacy of photodynamic therapy mediated by toluidine-blue-coated chitosan NPs with gold and silver incorporated (TBO-chit-Au-AgNPs). They demonstrated its use as a nontoxic antimicrobial therapy to fight ulcers of the diabetic foot (DFU) caused by drug-resistant multistrains in monomicrobial and polymicrobial states of infection. The in vitro efficacy of PDT mediated by TBO-chit-Au-AgNPs against polymicrobial biofilms was determined and compared with monomicrobial biofilms. The results showed a significant decrease in biofilm formation. Finally, its therapeutic potential was validated in vivo DFU model. The level of cytokines was reduced using a nanophototheranostic approach, indicating infection control. The growth factor expression profile confirmed the healing of DFU, leading to synergistic healing of the infection.

Diogo et al. (2017) compared the activity of PDT with the methyl ester of Zn(II) chlorin e6 (Zn(II) e6Me) activated by red light against monospecies and mixed biofilms of *C. albicans* and *Enterococcus faecalis*. Zn(II) e6Me, activated with light for 60 or 90 s, removed 60% of the biofilm biomass; the structure of the biofilms and the microbial cells were significantly damaged, in addition to being more efficient than Rose Bengal (RB) and Toluidine Blue-O (TBO).

The Carmello et al. (2017) group evaluated the efficacy of Photodithazine-mediated PDT in mono and duo-biofilms of *C. albicans*, *Candida glabrata*, and *Candida tropicalis*. Results showed biomass reduction of 24.4%, 39.2%, and 43.7% for *C. albicans*, *C. glabrata*, and *C. tropicalis* monospecies biofilm, respectively. PDT was more effective in monospecies than in duo-species biofilm.

7.4 Antimicrobial peptides (AMPs)

The treatment of bacteria-fungi polymicrobial biofilms requires a combination of antibiotics and antifungal agents. However, it has been reported that such a combination strategy shows poor efficacy, usually failing in 70% of infections (Harriott & Noverr, 2009), and in general, the patient is treated for monospecies bacterial infection. The cure is difficult because the fungus is not demonstrated at diagnosis. (Allison et al., 2016). AMPs are a growing strategy against polymicrobial biofilms, mainly for their multiple diversity and broad action spectrum against bacteria, fungi, and virus. They can be obtained from nature (plants, animals, and bacteria) or synthetic production. Synthetic AMPs have some advantages when compared with natural ones, concerning higher stability in pH changes, lower cost production, compatibility to delivery systems, and improvement of antimicrobial activity, and for these reasons, they can be used in association with other conventional strategies, such as combined therapy, or association to nanotechnology, to enhance more outstanding clinical outcomes (Batoni, Maisetta, & Esin, 2021; Huan, Kong, Mou, & Yi, 2020; Zhang, Li, Yu, & Wang, 2020).

AMPs can display several effects on cells, possessing different action mechanisms that operate simultaneously. Generally, AMPs' activity rests in cytoplasmic membranes disruption and perturbation via pore formation or targeting polypeptidoglycans, intracellular targets related to signaling pathways of cell permeability, inflammatory response, DNA/RNA biosynthesis, cell wall, and lipopolysaccharides synthesis and biofilm formation (Vanzolini, Bruschi, Rinaldi, Magnani, & Fraternale, 2022).

Gupta et al. (2019) studied 20 cholic acid-peptide conjugates (CAPs) in the face of *C. albicans*/*S. aureus* dual-species biofilm and found pieces of evidence that valine-glycine-derived CAP-3, in a concentration of 32 μM, reduced polymicrobial biofilm 4–5-log times in CFU for both MOs; in this concentration, CAP-3 also reduced silicon catheter-biofilm formation in vitro. The group showed CAP-3 efficacy in mice infected with *C. albicans*/*S. aureus* biofilm by reducing both species when compared with mice control without treatment. In addition, the researchers infected mice with bioluminescent *C. albicans* and *S. aureus* with a subcutaneous insertion catheter. CAP-3 significantly reduced the polymicrobial biofilms preformed on silicon catheters, while CAP-3-coated catheters prevented the formation of the same biofilms. Synthetic guanylate polymethacrylates, polymers that mimic AMP, were used against *C. albicans*/*S. aureus* demonstrating in vitro polymicrobial biofilm eradication with cell death rates above 80% for *C. albicans* and above 94% for *S. aureus* (Qu et al., 2015).

De Alteriis et al. (2018) used the membranotropic peptide gH625 and its derivative against dual-species biofilms of *C. tropicalis*/*S. aureus* and *C. tropicalis*/*Serratia marcescens* and demonstrated that the gH625-GCGKKK derivative increased the inhibition of biofilm formation by 80% compared with gH625. The derivate also decreases biomass, and both peptides could eliminate these MOs. gH625 and its derivate are promising molecules for further studies by preventing polymicrobial biofilm formation and treating resistant-treatment associations.

AMPs were also evaluated in *P. aeruginosa*/*S. aureus*/*C. albicans* polymicrobial biofilm with epsilon-poly-L-lysine, a natural peptide, incorporated into chitosan hydrogel using an ex vivo porcine biofilm model. Antibiofilm efficacy was

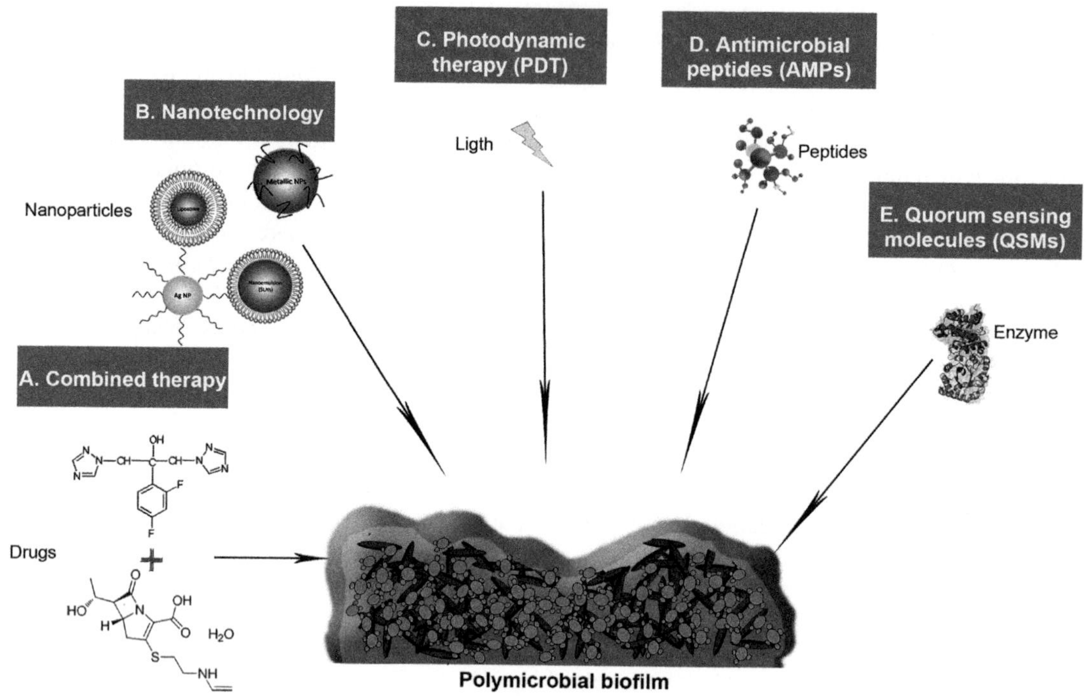

FIG. 4 Representative treatment scheme used to combat polymicrobial biofilms. A-Combination Therapy, B-Nanotechnology, C-Photodynamic Therapy (PDT), D-Antimicrobial Peptides (AMPs), and E-Quorum-Quenching Molecules (QSMs).

evaluated, showing effectiveness against *P. aeruginosa*, low activity against *S. aureus*, and no effective results in *C. albicans*; however, the hydrogel reduced 96% of early biofilm thickness and 70% of mature polymicrobial biofilm (Pati, Kurata, Horseman, & Pierce, 2021).

7.5 Quorum sensing molecules (QSMs)

Interference in the process of QS by inhibiting the synthesis or interaction is called quorum quenching. QQ promise novel therapeutic agents combating bacterial or fungal biofilms, but this approach needs further investigation. Studies in this field have evaluated two QSMs against *C. albicans/S. mutans* and *C. albicans/S. epidermidis* dual-species biofilms (Van Dyck, Pinto, Pully, & Van Dijck, 2021). Thiazolidinedione-8, a derivate Gram-negative bacteria QMS, could reduce *C. albicans/S. mutans* biomass, even though there was no effect against *S. mutans* (Feldman, Ginsburg, Al-Quntar, & Steinberg, 2016). Thiazolid inedione-8, a sustained-release membrane used in a model system with constant flow conditions, showed higher penetration on *C. albicans/S. mutans* mixed biofilm, changing morphology, reducing the metabolic activity and the production of EPS of both pathogens (Feldman, Shender ovich, Lavy, Friedman, & Steinberg, 2017).

Two molecules, QQ-5 and QQ-7, are promising as an antimixed biofilm. Initial studies demonstrated that both potentially interfere with the mono biofilm of *C. albicans* and *S. epidermidis* formation, respectively, by inhibiting the process of morphogenesis and the repressor for polysaccharide intracellular adhesin (PIA) (Weiland-Bräuer, Malek, & Schmitz, 2019) (Fig. 4).

Biofilm-related infections remain a severe concern in clinical settings. The high resistance of biofilms to current antimicrobial therapies is a significant challenge in this field, and biofilm eradication is demanding. New approaches such as AMPs and nanotechnology have been identified as promising, accompanied by a deeper understanding of biofilm resistance mechanisms. By associating these promising agents with antimicrobials, one can envision the eradication of biofilms in the future. However, to develop safe, effective, practical, and economically viable strategies against biofilm infections, more careful efforts are still needed in the field, including well-designed clinical trials.

References

Abd El-Baky, R. M., Mandour, S. A., Ahmed, E. F., Hashem, Z. S., Sandle, T., & Mohamed, D. S. (2020). Virulence profiles of some *Pseudomonas aeruginosa* clinical isolates and their association with the suppression of *Candida* growth in polymicrobial infections. *PLoS One*, *15*(12), e0243418. https://doi.org/10.1371/journal.pone.0243418.

Adair, C. G., Gorman, S. P., Feron, B. M., Byers, L. M., Jones, D. S., Goldsmith, C. E., et al. (1999). Implications of endotracheal tube biofilm for ventilator-associated pneumonia. *Intensive Care Medicine*, *25*(10), 1072–1076. https://doi.org/10.1007/s001340051014.

Adam, B., Baillie, G. S., & Douglas, L. J. (2002). Mixed species biofilms of *Candida albicans* and *Staphylococcus epidermidis*. *Journal of Medical*

Microbiology, *51*(4), 344–349. https://doi.org/10.1099/0022-1317-51-4-344.

Akhtar, F., Khan, A. U., Qazi, B., Kulanthaivel, S., Mishra, P., Akhtar, K., et al. (2021). A nano phototheranostic approach of toluidine blue conjugated gold silver core shells mediated photodynamic therapy to treat diabetic foot ulcer. *Scientific Reports*, *11*(1), 24464.

Albuquerque, P., & Casadevall, A. (2012). Quorum sensing in fungi—A review. *Medical Mycology*, *50*(4), 337–345. https://doi.org/10.3109/13693786.2011.652201.

Al-Fattani, M. A., & Douglas, L. J. (2006). Biofilm matrix of *Candida albicans* and *Candida tropicalis*: Chemical composition and role in drug resistance. *Journal of Medical Microbiology*, *55*(Pt 8), 999–1008. https://doi.org/10.1099/jmm.0.46569-0.

Alhede, M., Kragh, K. N., Qvortrup, K., Allesen-Holm, M., van Gennip, M., Christensen, L. D., et al. (2011). Phenotypes of non-attached *Pseudomonas aeruginosa* aggregates resemble surface attached biofilm. *PLoS One*, *6*(11), e27943. https://doi.org/10.1371/journal.pone.0027943.

Ali, S. S., Morsy, R., El-Zawawy, N. A., Fareed, M. F., & Bedaiwy, M. Y. (2017). Synthesized zinc peroxide nanoparticles (ZnO2-NPs): A novel antimicrobial, anti-elastase, anti-keratinase, and anti-inflammatory approach toward polymicrobial burn wounds. *International Journal of Nanomedicine*, *12*, 6059–6073. https://doi.org/10.2147/IJN.S141201.

Allison, D. L., Scheres, N., Willems, H. M. E., Bode, C. S., Krom, B. P., & Shirtliff, M. E. (2019). The host immune system facilitates disseminated *Staphylococcus aureus* disease due to phagocytic attraction to *Candida albicans* during coinfection: A case of bait and switch. *Infection and Immunity*, *87*(11), e00137-19. https://doi.org/10.1128/IAI.00137-19.

Allison, D. L., Willems, H. M. E., Jayatilake, J. A. M. S., Bruno, V. M., Peters, B. M., & Shirtliff, M. E. (2016). *Candida*-bacteria interactions: Their impact on human disease. *Microbiology Spectrum*, *4*(3). https://doi.org/10.1128/microbiolspec.VMBF-0030-2016.

Almshawit, H., Macreadie, I., & Grando, D. (2014). A simple and inexpensive device for biofilm analysis. *Journal of Microbiological Methods*, *98*, 59–63. https://doi.org/10.1016/j.mimet.2013.12.020.

Alotaibi, G. F., & Bukhari, M. A. (2021). Factors influencing bacterial biofilm formation and development. *American Journal of Biomedical Science and Research*, *12*(6), 617–626. https://doi.org/10.34297/AJBSR.2021.12.001820.

Andes, D., Nett, J., Oschel, P., Albrecht, R., Marchillo, K., & Pitula, A. (2004). Development and characterization of an in vivo central venous catheter *Candida albicans* biofilm model. *Infection and Immunity*, *72*(10), 6023–6031. https://doi.org/10.1128/IAI.72.10.6023-6031.2004.

Arciola, C. R., Campoccia, D., & Montanaro, L. (2018). Implant infections: Adhesion, biofilm formation and immune evasion. *Nature Reviews Microbiology*, *16*(7), 397–409. https://doi.org/10.1038/s41579-018-0019-y.

Arciola, C. R., Montanaro, L., & Costerton, J. W. (2011). New trends in diagnosis and control strategies for implant infections. *The International Journal of Artificial Organs*, *34*(9), 727–736. https://doi.org/10.5301/IJAO.2011.8784.

Arck, P. C. (2019). When 3 Rs meet a forth R: Replacement, reduction and refinement of animals in research on reproduction. *Journal of Reproductive Immunology*, *132*, 54–59. https://doi.org/10.1016/j.jri.2019.03.004.

Bachtiar, E. W., Bachtiar, B. M., Jarosz, L. M., Amir, L. R., Sunarto, H., Ganin, H., et al. (2014). AI-2 of *Aggregatibacter actinomycetemcomitans* inhibits *Candida albicans* biofilm formation. *Frontiers in Cellular and Infection Microbiology*, *4*, 94. https://doi.org/10.3389/fcimb.2014.00094.

Bachtiar, E. W., Dewiyani, S., Surono Akbar, S. M., & Bachtiar, B. M. (2016). Inhibition of *Candida albicans* biofilm development by unencapsulated *Enterococcus faecalis* cps2. *Journal of Dental Sciences*, *11* (3), 323–330. https://doi.org/10.1016/j.jds.2016.03.012.

Baena-Monroy, T., Moreno-Maldonado, V., Franco-Martínez, F., Aldape-Barrios, B., Quindós, G., & Sánchez-Vargas, L. O. (2005). *Candida albicans*, *Staphylococcus aureus* and *Streptococcus mutans* colonization in patients wearing dental prosthesis. *Medicina Oral, Patología Oral y Cirugía Bucal*, *10*(Suppl. 1), E27–E39.

Bailey, DA, Feldmann, PJ, Bovey, M, Gow, NA, & Brown, AJ. (1996). The Candida albicans HYR1 gene, which is activated in response to hyphal development, belongs to a gene family encoding yeast cell wall proteins. *Journal of Bacteriology*, *178*(18). https://doi.org/10.1128/jb.178.18.5353-5360.1996. 5353–60.

Bajpai, P. (2015). Pulp and paper chemicals**Some excerpts taken from Bajpai, P., 2004. Emerging Technologies in Sizing. PIRA Technology Report, Smithers PIRA, the worldwide authority on the Packaging, Print and Paper supply chains. In P. Bajpai (Ed.), *Pulp and paper industry* (pp. 25–273). Amsterdam: Elsevier (Chapter 3).

Bamford, C. V., d'Mello, A., Nobbs, A. H., Dutton, L. C., Vickerman, M. M., & Jenkinson, H. F. (2009). *Streptococcus gordonii* modulates *Candida albicans* biofilm formation through intergeneric communication. *Infection and Immunity*, *77*(9), 3696–3704. https://doi.org/10.1128/iai.00438-09.

Bandara, H. M. H. N., Cheung, B. P. K., Watt, R. M., Jin, L. J., & Samaranayake, L. P. (2013). Secretory products of *Escherichia coli* biofilm modulate *Candida* biofilm formation and hyphal development. *Journal of Investigative and Clinical Dentistry*, *4*(3), 186–199. https://doi.org/10.1111/jicd.12048.

Bandara, H. M. H. N., Panduwawala, C. P., & Samaranayake, L. P. (2019). Biodiversity of the human oral mycobiome in health and disease. *Oral Diseases*, *25*(2), 363–371. https://doi.org/10.1111/odi.12899.

Bandara, H. M. H. N., Yau, J. Y. Y., Watt, R. M., Jin, L. J., & Samaranayake, L. P. (2009). *Escherichia coli* and its lipopolysaccharide modulate in vitro *Candida* biofilm formation. *Journal of Medical Microbiology*, *58*(Pt 12), 1623–1631. https://doi.org/10.1099/jmm.0.012989-0.

Bandara, H. M. H. N., Yau, J. Y. Y., Watt, R. M., Jin, L. J., & Samaranayake, L. P. (2010). *Pseudomonas aeruginosa* inhibits in-vitro *Candida* biofilm development. *BMC Microbiology*, *10*, 125. https://doi.org/10.1186/1471-2180-10-125.

Bartnicka, D., Karkowska-Kuleta, J., Zawrotniak, M., Satała, D., Michalik, K., Zielińska, G., et al. (2019). Adhesive protein-mediated cross-talk between *Candida albicans* and *Porphyromonas gingivalis* in dual species biofilm protects the anaerobic bacterium in unfavorable oxic environment. *Scientific Reports*, *9*, 4376. https://doi.org/10.1038/s41598-019-40771-8.

Bassler, B. L. (1999). How bacteria talk to each other: Regulation of gene expression by quorum sensing. *Current Opinion in Microbiology*, *2*(6), 582–587. https://doi.org/10.1016/s1369-5274(99)00025-9.

Batoni, G., Maisetta, G., & Esin, S. (2021). Therapeutic potential of antimicrobial peptides in polymicrobial biofilm-associated infections.

International Journal of Molecular Sciences, *22*(2), 482. https://doi.org/10.3390/ijms22020482.

Beauvais, A., Loussert, C., Prevost, M. C., Verstrepen, K., & Latgé, J. P. (2009). Characterization of a biofilm-like extracellular matrix in FLO1-expressing *Saccharomyces cerevisiae* cells. *FEMS Yeast Research*, *9*(3), 411–419. https://doi.org/10.1111/j.1567-1364.2009.00482.x.

Benaducci, T., Sardi, J. D. C. O., Lourencetti, N. M. S., Scorzoni, L., Gullo, F. P., Rossi, S. A., et al. (2016). Virulence of *Cryptococcus* sp. biofilms in vitro and in vivo using *Galleria mellonella* as an alternative model. *Frontiers in Microbiology*, *7*, 290. https://doi.org/10.3389/fmicb.2016.00290.

Bila, N. M., Costa-Orlandi, C. B., Vaso, C. O., et al. (2021). 2-Hydroxychalcone as a potent compound and photosensitizer against dermatophyte biofilms. *Frontiers in Cellular and Infection Microbiology*, *13*, 679470. https://doi.org/10.3389/fcimb.2021.679470. 34055673.

Bjarnsholt, T., Ciofu, O., Molin, S., Givskov, M., & Høiby, N. (2013). Applying insights from biofilm biology to drug development—Can a new approach be developed? *Nature Reviews Drug Discovery*, *12* (10), 791–808. https://doi.org/10.1038/nrd4000.

Blanco-Cabra, N., López-Martínez, M. J., Arévalo-Jaimes, B. V., Martin-Gómez, M. T., Samitier, J., & Torrents, E. (2021). A new BiofilmChip device for testing biofilm formation and antibiotic susceptibility. *NPJ Biofilms and Microbiomes*, *7*(1), 62. https://doi.org/10.1038/s41522-021-00236-1.

Boisvert, A.-A., Cheng, M. P., Sheppard, D. C., & Nguyen, D. (2016). Microbial biofilms in pulmonary and critical care diseases. *Annals of the American Thoracic Society*, *13*(9), 1615–1623. https://doi.org/10.1513/annalsats.201603-194fr.

Bojsen, R. K., Andersen, K. S., & Regenberg, B. (2012). *Saccharomyces cerevisiae*—A model to uncover molecular mechanisms for yeast biofilm biology. *FEMS Immunology and Medical Microbiology*, *65* (2), 169–182. https://doi.org/10.1111/j.1574-695X.2012.00943.x.

Bojsen, R., Regenberg, B., Gresham, D., & Folkesson, A. (2016). A common mechanism involving the TORC1 pathway can lead to amphotericin B-persistence in biofilm and planktonic *Saccharomyces cerevisiae* populations. *Scientific Reports*, *6*, 21874. https://doi.org/10.1038/srep21874.

Boon, C., Deng, Y., Wang, L.-H., He, Y., Xu, J.-L., Fan, Y., et al. (2008). A novel DSF-like signal from *Burkholderia cenocepacia* interferes with *Candida albicans* morphological transition. *The ISME Journal*, *2*(1), 27–36. https://doi.org/10.1038/ismej.2007.76.

Borghi, E., Borgo, F., & Morace, G. (2016). Fungal biofilms: Update on resistance. *Advances in Experimental Medicine and Biology*, *931*, 37–47. https://doi.org/10.1007/5584_2016_7.

Boskey, E. R., Cone, R. A., Whaley, K. J., & Moench, T. R. (2001). Origins of vaginal acidity: High D/L lactate ratio is consistent with bacteria being the primary source. *Human Reproduction (Oxford, England)*, *16*(9), 1809–1813. https://doi.org/10.1093/humrep/16.9.1809.

Briard, B., Bomme, P., Lechner, B. E., Mislin, G. L. A., Lair, V., Prévost, M.-C., et al. (2015). *Pseudomonas aeruginosa* manipulates redox and iron homeostasis of its microbiota partner *Aspergillus fumigatus* via phenazines. *Scientific Reports*, *5*, 8220. https://doi.org/10.1038/srep08220.

Briard, B., Rasoldier, V., Bomme, P., ElAouad, N., Guerreiro, C., Chassagne, P., et al. (2017). Dirhamnolipids secreted from *Pseudomonas aeruginosa* modify anjpegungal susceptibility of *Aspergillus fumigatus* by inhibiting β1,3 glucan synthase activity. *The ISME Journal*, *11*(7), 1578–1591. https://doi.org/10.1038/ismej.2017.32.

Brogden, K. A., Guthmiller, J. M., & Taylor, C. E. (2005). Human polymicrobial infections. *Lancet (London, England)*, *365*(9455), 253–255. https://doi.org/10.1016/S0140-6736(05)17745-9.

Brusca, M. I., Rosa, A., Albaina, O., Moragues, M. D., Verdugo, F., & Pontón, J. (2010). The impact of oral contraceptives on women's periodontal health and the subgingival occurrence of aggressive periodontopathogens and *Candida* species. *Journal of Periodontology*, *81*(7), 1010–1018. https://doi.org/10.1902/jop.2010.090575.

Burgos-Garay, M., Ganim, C., de Man, T. J. B., Davy, T., Mathers, A. J., Kotay, S., et al. (2021). Colonization of carbapenem-resistant *Klebsiella pneumoniae* in a sink-drain model biofilm system. *Infection Control and Hospital Epidemiology*, *42*(6), 722–730. https://doi.org/10.1017/ice.2020.1287.

Burmølle, M., Ren, D., Bjarnsholt, T., & Sørensen, S. J. (2014). Interactions in multispecies biofilms: Do they actually matter? *Trends in Microbiology*, *22*(2), 84–91. https://doi.org/10.1016/j.tim.2013.12.004.

Cabral, D. J., Penumutchu, S., Norris, C., Morones-Ramirez, J. R., & Belenky, P. (2018). Microbial competition between *Escherichia coli* and *Candida albicans* reveals a soluble fungicidal factor. *Microbial Cell (Graz, Austria)*, *5*(5), 249–255. https://doi.org/10.15698/mic2018.05.631.

Cao, Y.-Y., Cao, Y.-B., Xu, Z., Ying, K., Li, Y., Xie, Y., et al. (2005). cDNA microarray analysis of differential gene expression in *Candida albicans* biofilm exposed to farnesol. *Antimicrobial Agents and Chemotherapy*, *49*(2), 584–589. https://doi.org/10.1128/AAC.49.2.584-589.2005.

Carmello, J. C., Alves, F., de Mima, E. G. O., Jorge, J. H., Bagnato, V. S., & Pavarina, A. C. (2017). Photoinactivation of single and mixed biofilms of *Candida albicans* and non-albicans *Candida* species using Photodithazine®. *Photodiagnosis and Photodynamic Therapy*, *17*, 194–199. https://doi.org/10.1016/J.PDPDT.2016.11.013.

Carmelo, V., Santos, H., & Sá-Correia, I. (1997). Effect of extracellular acidification on the activity of plasma membrane ATPase and on the cytosolic and vacuolar pH of *Saccharomyces cerevisiae*. *Biochimica et Biophysica Acta*, *1325*(1), 63–70. https://doi.org/10.1016/s0005-2736(96)00245-3.

Cattana, M. E., Tracogna, M. F., Marques, I., Rojas, F., Fernández, M., de los Ángeles Sosa, M., et al. (2017). In vivo *Paracoccidioides* sp. biofilm on vascular prosthesis. *Mycopathologia*, *182*(7), 747–749. https://doi.org/10.1007/s11046-017-0126-8.

Cavalheiro, M., & Teixeira, M. C. (2018). *Candida* biofilms: Threats, challenges, and promising strategies. *Frontiers in Medicine*, *5*. https://doi.org/10.3389/fmed.2018.00028.

Ceri, H., Olson, M. E., Stremick, C., Read, R. R., Morck, D., & Buret, A. (1999). The Calgary biofilm device: New technology for rapid determination of antibiotic susceptibilities of bacterial biofilms. *Journal of Clinical Microbiology*, *37*(6), 1771–1776. https://doi.org/10.1128/jcm.37.6.1771-1776.1999.

Chandra, J., Kuhn, D. M., Mukherjee, P. K., Hoyer, L. L., McCormick, T., & Ghannoum, M. A. (2001). Biofilm formation by the fungal pathogen *Candida albicans*: Development, architecture, and drug resistance. *Journal of Bacteriology*, *183*(18), 5385–5394. https://doi.org/10.1128/JB.183.18.5385-5394.2001.

Chen, H., Fujita, M., Feng, Q., Clardy, J., & Fink, G. R. (2004). Tyrosol is a quorum-sensing molecule in *Candida albicans*. *Proceedings of the*

National Academy of Sciences of the United States of America, *101* (14), 5048–5052. https://doi.org/10.1073/pnas.0401416101.

Chen, L., Yu, K., Chen, L., Zheng, X., Huang, N., Lin, Y., et al. (2021). Synergistic activity and biofilm formation effect of colistin combined with PFK-158 against colistin-resistant gram-negative bacteria. *Infection and Drug Resistance*, *14*, 2143–2154. https://doi.org/10.2147/idr.s309912.

Chen, Y., Zeng, H., Tian, J., et al. (2013). Antifungal mechanism of essential oil from Anethum graveolens seeds against Candida albicans. *Journal of Medical Microbiology*, *62*, 1175–1183. https://doi.org/10.1099/jmm.0.055467-0. 23657528.

Cheong, J. Z. A., Johnson, C. J., Wan, H., Liu, A., Kernien, J. F., Gibson, A. L. F., et al. (2021). Priority effects dictate community structure and alter virulence of fungal-bacterial biofilms. *The ISME Journal*, *15*(7), 2012–2027. https://doi.org/10.1038/s41396-021-00901-5.

Cieplik, F., Stanislaw Grinholc, M., Alves, E., Al-Ahmad, A., Hamblin, M. R., Y-y, H., et al. (2018). Antimicrobial photodynamic therapy to control clinically relevant biofilm infections. *Frontiers in Microbiology*, *1*, 1299. https://doi.org/10.3389/fmicb.2018.01299. www.frontiersin.org.

Coco, B. J., Bagg, J., Cross, L. J., Jose, A., Cross, J., & Ramage, G. (2008). Mixed *Candida albicans* and *Candida glabrata* populations associated with the pathogenesis of denture stomatitis. *Oral Microbiology and Immunology*, *23*(5), 377–383. https://doi.org/10.1111/j.1399-302X.2008.00439.x.

Coenye, T., & Nelis, H. J. (2010). In vitro and in vivo model systems to study microbial biofilm formation. *Journal of Microbiological Methods*, *83*(2), 89–105. https://doi.org/10.1016/j.mimet.2010.08.018.

Conway, L. J., & Larson, E. L. (2012). Guidelines to prevent catheter-associated urinary tract infection: 1980 to 2010. *Heart & Lung*, *41* (3), 271–283. https://doi.org/10.1016/j.hrtlng.2011.08.001.

Costa-Orlandi, B. C., Sardi, C. J., Pitangui, S. N., de Oliveira, C. H., Scorzoni, L., Galeane, C. M., et al. (2017). Fungal biofilms and polymicrobial diseases. *Journal of Fungi*, *3*(2). https://doi.org/10.3390/jof3020022.

Costa-Orlandi, C. B., Martinez, L. R., Bila, N. M., et al. (2021). Nitric oxide releasing nanoparticles are similar to efinaconazole in their capacity to eradicate *Trichophyton rubrum* biofilms. *Frontiers in Cellular and Infection Microbiology*, *15*, 684150. https://doi.org/10.3389/fcimb.2021.684150. 34336712.

Costa-Orlandi, C. B., Sardi,, J. C., Santos, C. T., et al. (2014). In vitro characterization of Trichophyton rubrum and T. mentagrophytes biofilms. *Biofouling*, *30*. https://doi.org/10.1080/08927014.2014.919282. 719–27 24856309.

Costerton, J. W., Geesey, G. G., & Cheng, K. J. (1978). How bacteria stick. *Scientific American*, *238*(1), 86–95. https://doi.org/10.1038/scientificamerican0178-86.

Costerton, J. W., Lewandowski, Z., Caldwell, D. E., Korber, D. R., & Lappin-Scott, H. M. (1995). Microbial biofilms. *Annual Review of Microbiology*, *49*, 711–745. https://doi.org/10.1146/annurev.mi.49.100195.003431.

Costerton, J. W., Stewart, P. S., & Greenberg, E. P. (1999). Bacterial biofilms: A common cause of persistent infections. *Science*, *284* (5418), 1318–1322.

Crusz, S. A., Popat, R., Rybtke, M. T., Cámara, M., Givskov, M., Tolker-Nielsen, T., et al. (2012). Bursting the bubble on bacterial biofilms: A flow cell methodology. *Biofouling*, *28*(8), 835–842. https://doi.org/10.1080/08927014.2012.716044.

d'Enfert, C., Kaune, A. K., Alaban, L. R., Chakraborty, S., Cole, N., Delavy, M., … Brown, A. (2021). The impact of the Fungus-Host-Microbiota interplay upon Candida albicans infections: current knowledge and new perspectives. *FEMS Microbiology Reviews*, *45* (3), fuaa060. https://doi.org/10.1093/femsre/fuaa060.

Dahlén, G., Samuelsson, W., Molander, A., & Reit, C. (2000). Identification and antimicrobial susceptibility of enterococci isolated from the root canal. *Oral Microbiology and Immunology*, *15*(5), 309–312. https://doi.org/10.1034/j.1399-302x.2000.150507.x.

Davis-Hanna, A., Piispanen, A. E., Stateva, L. I., & Hogan, D. A. (2008). Farnesol and dodecanol effects on the *Candida albicans* Ras1-cAMP signalling pathway and the regulation of morphogenesis. *Molecular Microbiology*, *67*(1), 47–62. https://doi.org/10.1111/j.1365-2958.2007.06013.x.

De Alteriis, E., Lombardi, L., Falanga, A., Napolano, M., Galdiero, S., Siciliano, A., et al. (2018). Polymicrobial antibiofilm activity of the membranotropic peptide gH625 and its analogue. *Microbial Pathogenesis*, *125*, 189–195. https://doi.org/10.1016/j.micpath.2018.09.027.

De Carvalho, F. G., Silva, D. S., Hebling, J., Spolidorio, L. C., & Spolidorio, D. M. P. (2006). Presence of mutans streptococci and *Candida* spp. in dental plaque/dentine of carious teeth and early childhood caries. *Archives of Oral Biology*, *51*(11), 1024–1028. https://doi.org/10.1016/j.archoralbio.2006.06.001.

De Groot, P. W., Bader, O., de Boer, A. D., Weig, M., & Chauhan, N. (2013). Adhesins in human fungal pathogens: Glue with plenty of stick. *Eukaryot Cell*, *12*(4), 470–481. https://doi.org/10.1128/EC.00364-12.

de Negrini, T. C., Koo, H., & Arthur, R. A. (2019). *Candida*-bacterial biofilms and host-microbe interactions in oral diseases. *Advances in Experimental Medicine and Biology*, *1197*, 119–141. https://doi.org/10.1007/978-3-030-28524-1_10.

de Pereira, F. O. (2021). A review of recent research on antifungal agents against dermatophyte biofilms. *Medical Micology*, *59*(4), 313–326. https://doi.org/10.1093/mmy/myaa114.

Del Pozo, J. L. (2017). Biofilm-related disease. *Expert Review of Anti-Infective Therapy*, *16*(1), 51–65. https://doi.org/10.1080/14787210.2018.1417036.

Dellière, S., Dannaoui, E., Fieux, M., Bonfils, P., Gricourt, G., Demontant, V., et al. (2021). Analysis of microbiota and mycobiota in fungal ball rhinosinusitis: Specific interaction between *Aspergillus fumigatus* and *Haemophilus influenza*? *Journal of Fungi*, *7*(7), 550. https://doi.org/10.3390/jof7070550.

Demuth, D. R., Duan, Y., Brooks, W., Holmes, A. R., McNab, R., & Jenkinson, H. F. (1996). Tandem genes encode cell-surface polypeptides SspA and SspB which mediate adhesion of the oral bacterium *Streptococcus gordonii* to human and bacterial receptors. *Molecular Microbiology*, *20*(2), 403–413. https://doi.org/10.1111/j.1365-2958.1996.tb02627.x.

Demuyser, L., Jabra-Rizk, M. A., & Van Dijck, P. (2014). Microbial cell surface proteins and secreted metabolites involved in multispecies biofilms. *Pathogens and Disease*, *70*(3). https://doi.org/10.1111/2049-632X.12123. 219–30.

Deveau, A., Bonito, G., Uehling, J., Paoletti, M., Becker, M., Bindschedler, S., et al. (2018). Bacterial-fungal interactions: Ecology, mechanisms and challenges. *FEMS Microbiology Reviews*, *42*(3), 335–352. https://doi.org/10.1093/femsre/fuy008.

Di Bonaventura, G., Stepanović, S., Picciani, C., Pompilio, A., & Piccolomini, R. (2007). Effect of environmental factors on biofilm formation by clinical *Stenotrophomonas maltophilia* isolates. *Folia Microbiologica, 52*(1), 86. https://doi.org/10.1007/BF02932144.

Di Domenico, E. G., Cavallo, I., Guembe, M., Prignano, G., Gallo, M. T., Bordignon, V., et al. (2018). The clinical Biofilm Ring Test: A promising tool for the clinical assessment of biofilm-producing *Candida* species. *FEMS Yeast Research, 18*(3). https://doi.org/10.1093/femsyr/foy025.

Di Mambro, T., Guerriero, I., Magnani, M., & Marra, E. (2019). The Yin and Yang of current antifungal therapeutic strategies: How can we harness our natural defenses? *Frontiers in Pharmacology, 10*, 80. https://doi.org/10.3389/fphar.2019.00080.

Ding, F., Songkiatisak, P., Cherukuri, P. K., Huang, T., & Xu, X. H. N. (2018). Size-dependent inhibitory effects of antibiotic drug nanocarriers against *Pseudomonas aeruginosa*. *ACS Omega, 3*(1), 1231–1243. https://doi.org/10.1021/ACSOMEGA.7B01956.

Diogo, P., Fernandes, C., Caramelo, F., Mota, M., Miranda, I. M., Faustino, M. A. F., et al. (2017). Antimicrobial photodynamic therapy against endodontic *Enterococcus faecalis* and *Candida albicans* mono and mixed biofilms in the presence of photosensitizers: A comparative study with classical endodontic irrigants. *Frontiers in Microbiology, 8*(MAR), 498. https://doi.org/10.3389/FMICB.2017.00498/bibTEX.

Dongari-Bagtzoglou, A., Kashleva, H., Dwivedi, P., Diaz, P., & Vasilakos, J. (2009). Characterization of mucosal *Candida albicans* biofilms. *PLoS One, 4*(11), e7967. https://doi.org/10.1371/journal.pone.0007967.

Donlan, R. M. (2002). Biofilms: Microbial life on surfaces. *Emerging Infectious Diseases, 8*(9), 881–890. https://doi.org/10.3201/eid0809.020063.

Donlan, R. M., & Costerton, J. W. (2002). Biofilms: Survival mechanisms of clinically relevant microorganisms. *Clinical Microbiology Reviews, 15*(2), 167–193.

Dos Santos Ramos, M. A., da Silva, P. B., de Toledo, L. G., et al. (2019). Intravaginal delivery of *Syngonanthus nitens* (Bong.) Ruhland fraction based on a nanoemulsion system applied to Vulvovaginal candidiasis treatment. *Journal of Biomedical Nanotechnology, 15*(5), 1072–1089. https://doi.org/10.1166/jbn.2019.2750. 30890237.

Dowd, S. E., Delton Hanson, J., Rees, E., Wolcott, R. D., Zischau, A. M., Sun, Y., et al. (2011). Survey of fungi and yeast in polymicrobial infections in chronic wounds. *Journal of Wound Care, 20*(1), 40–47. https://doi.org/10.12968/jowc.2011.20.1.40.

Dutton, L. C., Paszkiewicz, K. H., Silverman, R. J., Splatt, P. R., Shaw, S., Nobbs, A. H., et al. (2016). Transcriptional landscape of trans-kingdom communication between *Candida albicans* and *Streptococcus gordonii*. *Molecular Oral Microbiology, 31*(2), 136–161. https://doi.org/10.1111/omi.12111.

Dupont, H. (2002). Predictive factors of mortality due to polymicrobial peritonitis with *Candida* isolation in peritoneal fluid in critically ill patients. *Archives of Surgery, 137*(12), 1341. https://doi.org/10.1001/archsurg.137.12.1341.

Eckburg, P. B., Bik, E. M., Bernstein, C. N., Purdom, E., Dethlefsen, L., Sargent, M., et al. (2005). Diversity of the human intestinal microbial flora. *Science, 308*(5728), 1635–1638. https://doi.org/10.1126/science.1110591.

Elborn, J. S. (2016). Cystic fibrosis. *The Lancet, 388*(10059), 2519–2531. https://doi.org/10.1016/s0140-6736(16)00576-6.

Elias, S., & Banin, E. (2012). Multi-species biofilms: Living with friendly neighbors. *FEMS Microbiology Reviews, 36*(5), 990–1004. https://doi.org/10.1111/j.1574-6976.2012.00325.x.

Ellepola, K., Liu, Y., Cao, T., Koo, H., & Seneviratne, C. J. (2017). Bacterial GtfB augments *Candida albicans* accumulation in cross-kingdom biofilms. *Journal of Dental Research, 96*(10), 1129–1135. https://doi.org/10.1177/0022034517714414.

Esberg, A., Löfgren-Burström, A., Ohman, U., & Strömberg, N. (2012). Host and bacterial phenotype variation in adhesion of *Streptococcus mutans* to matched human hosts. *Infection and Immunity, 80*(11), 3869–3879. https://doi.org/10.1128/IAI.00435-12.

Falsetta, M. L., Klein, M. I., Colonne, P. M., Scott-Anne, K., Gregoire, S., Pai, C. H., et al. (2014). Symbiotic relationship between *Streptococcus mutans* and *Candida albicans* synergizes virulence of plaque biofilms in vivo. *Infection and Immunity, 82*(5), 1968–1981. https://doi.org/10.1128/iai.00087-14.

Fanning, S., & Mitchell, A. P. (2012). Fungal biofilms. *PLoS Pathogens, 8* (4), e1002585. https://doi.org/10.1371/journal.ppat.1002585.

Faria-Oliveira, F., Carvalho, J., Belmiro, C. L., Martinez-Gomariz, M., Hernaez, M. L., Pavão, M., … Ferreira, C. (2014). Methodologies to generate, extract, purify and fractionate yeast ECM for analytical use in proteomics and glycomics. *BMC Microbiology, 14*, 244. https://doi.org/10.1186/s12866-014-0244-0.

Faria-Oliveira, F., Carvalho, J., Ferreira, C., Hernáez, M. L., Gil, C., & Lucas, C. (2015). Quantitative differential proteomics of yeast extracellular matrix: There is more to it than meets the eye. *BMC Microbiology, 15*, 271. https://doi.org/10.1186/s12866-015-0550-1.

Farrokhi, Y., Al-Shibli, B., Al-Hameedawi, D. F.-J., Neshati, Z., & Makhdoumi, A. (2021). *Escherichia coli* enhances the virulence factors of *Candida albicans*, the cause of vulvovaginal candidiasis, in a dual bacterial/fungal biofilm. *Research in Microbiology, 172*(4–5), 103849. https://doi.org/10.1016/j.resmic.2021.103849.

Fauvarque, M.-O. (2014). Small flies to tackle big questions: Assaying complex bacterial virulence mechanisms using *Drosophila melanogaster*. *Cellular Microbiology, 16*(6), 824–833. https://doi.org/10.1111/cmi.12292.

Feldman, M., Ginsburg, I., Al-Quntar, A., & Steinberg, D. (2016). Thiazolidinedione-8 alters symbiotic relationship in *C. albicans-S. mutans* dual species biofilm. *Frontiers in Microbiology, 7*. https://doi.org/10.3389/fmicb.2016.00140.

Feldman, M., Shenderovich, J., Lavy, E., Friedman, M., & Steinberg, D. (2017). A sustained-release membrane of thiazolidinedione-8: Effect on formation of a *Candida*/bacteria mixed biofilm on hydroxyapatite in a continuous flow model. *BioMed Research International, 2017*, 1–9. https://doi.org/10.1155/2017/3510124.

Filoche, S. K., Soma, D., van Bekkum, M., & Sissons, C. H. (2008). Plaques from different individuals yield different microbiota responses to oral-antiseptic treatment. *FEMS Immunology and Medical Microbiology, 54*(1), 27–36. https://doi.org/10.1111/j.1574-695X.2008.00443.x.

Flemming, H. C. (2020). Biofouling and me: My Stockholm syndrome with biofilms. *Water Research, 173*, 115576. https://doi.org/10.1016/j.watres.2020.115576.

Flemming, H. C., & Wingender, J. (2010). The biofilm matrix. *Nature Reviews Microbiology, 8*(9), 623–633. https://doi.org/10.1038/nrmicro2415.

Forier, K., Raemdonck, K., De Smedt, S. C., Demeester, J., Coenye, T., & Braeckmans, K. (2014). Lipid and polymer nanoparticles for drug delivery to bacterial biofilms. *Journal of Controlled Release*. https://doi.org/10.1016/j.jconrel.2014.03.055.

Fourie, R., Albertyn, J., Sebolai, O., Gcilitshana, O., & Pohl, C. H. (2021). *Candida albicans* SET3 plays a role in early biofilm formation, interaction with *Pseudomonas aeruginosa* and virulence in *Caenorhabditis elegans. Frontiers in Cellular and Infection Microbiology, 11*, 680732. https://doi.org/10.3389/fcimb.2021.680732.

Fox, E. P., Cowley, E. S., Nobile, C. J., Hartooni, N., Newman, D. K., & Johnson, A. D. (2014). Anaerobic bacteria grow within *Candida albicans* biofilms and induce biofilm formation in suspension cultures. *Current Biology: CB, 24*(20), 2411–2416. https://doi.org/10.1016/j.cub.2014.08.057.

Fregonezi, N. F., Oliveira, L. T., Singulani, J.d. L., Marcos, C. M., dos Santos, & Fusco-Almeida, A. M. (2021). Heat shock protein 60, insights to its importance in histoplasma capsulatum: From biofilm formation to host-interaction. *Frontiers in Cellular and Infection Microbiology*. https://doi.org/10.3389/fcimb.2020.591950.

Gabrilska, R. A., & Rumbaugh, K. P. (2015). Biofilm models of polymicrobial infection. *Future Microbiology, 10*(12), 1997–2015. https://doi.org/10.2217/fmb.15.109.

Gaddy, J. A., & Actis, L. A. (2009). Regulation of *Acinetobacter baumannii* biofilm formation. *Future Microbiology, 4*(3), 273–278. https://doi.org/10.2217/fmb.09.5.

Gaddy, J. A., Tomaras, A. P., & Actis, L. A. (2009). The *Acinetobacter baumannii* 19606 OmpA protein plays a role in biofilm formation on abiotic surfaces and in the interaction of this pathogen with eukaryotic cells. *Infection and Immunity, 77*(8), 3150–3160. https://doi.org/10.1128/IAI.00096-09.

Galdiero, E., Di Onofrio, V., Maione, A., Gambino, E., Gesuele, R., Menale, B., … Guida, M. (2020). *Allium ursinum* and *Allium oschaninii* against *Klebsiella pneumoniae* and *Candida albicans* mono- and polymicrobic biofilms in in vitro static and dynamic models. *Microorganisms, 8*(3), 336. https://doi.org/10.3390/microorganisms8030336.

Galdiero, E., Lombardi, L., Falanga, A., et al. (2019). Biofilms: Novel strategies based on antimicrobial peptides. *Pharmaceutics*. https://doi.org/10.3390/pharmaceutics11070322. 31295834.

Galdiero, E., de Alteriis, E., De Natale, A., D'Alterio, A., Siciliano, A., Guida, M., et al. (2020). Eradication of *Candida albicans* persister cell biofilm by the membranotropic peptide gH625. *Scientific Reports, 10* (1), 5780. https://doi.org/10.1038/s41598-020-62746-w.

Galdiero, E., Salvatore, M. M., Maione, A., Carraturo, F., Galdiero, S., Falanga, A., et al. (2021). Impact of the peptide WMR-K on dual-species biofilm *Candida albicans/Klebsiella pneumoniae* and on the untargeted metabolomic profile. *Pathogens (Basel, Switzerland), 10* (2), 214. https://doi.org/10.3390/pathogens10020214.

Galdiero, E., Ricciardelli, A., D'Angelo, C., de Alteriis, E., Maione, A., Albarano, L., … Parrilli, E. (2021). Pentadecanoic acid against Candida albicans-Klebsiella pneumoniae biofilm: Towards the development of an anti-biofilm coating to prevent polymicrobial infections. *Research in Microbiology, 172*(7–8), 103880. https://doi.org/10.1016/j.resmic.2021.103880.

Galeane, M. C., Gomes, P. C., Singulani, J. D. L., Souza, B. M. D., Palma, M. S., Mendes-Giannini, M. J., et al. (2019). Study of mastoparan analog peptides against *Candida albicans* and safety in zebrafish embryos (*Danio rerio*). *Future Microbiology, 14*(13), 1087–1097. https://doi.org/10.2217/fmb-2019-0060.

Garcia, M. C., Lee, J. T., Ramsook, C. B., Alsteens, D., Dufrêne, Y. F., & Lipke, P. N. (2011). A role for amyloid in cell aggregation and biofilm formation. *PLoS One, 6*(3), e17632. https://doi.org/10.1371/journal.pone.0017632.

García, A., Salas-Jara, M. J., Herrera, C., & González, C. (2014). Biofilm and *Helicobacter pylori*: From environment to human host. *World Journal of Gastroenterology, 20*(19), 5632–5638. https://doi.org/10.3748/wjg.v20.i19.5632.

Garcia, L. M., Costa-Orlandi, C. B., Bila, N. M., Vaso, C. O., Gonçalves, L. N. C., Fusco-Almeida, A. M., et al. (2020). A two-way road: Antagonistic interaction between dual-species biofilms formed by *Candida albicans/Candida parapsilosis* and *Trichophyton rubrum. Frontiers in Microbiology, 11*, 1980. https://doi.org/10.3389/fmicb.2020.01980.

Garrett, T. R., Bhakoo, M., & Zhang, Z. (2008). Bacterial adhesion and biofilms on surfaces. *Progress in Natural Science, 18*(9), 1049–1056. https://doi.org/10.1016/j.pnsc.2008.04.001.

Gbejuade, H. O., Lovering, A. M., & Webb, J. C. (2015). The role of microbial biofilms in prosthetic joint infections: A review. *Acta Orthopaedica, 86*(2), 147. https://doi.org/10.3109/17453674.2014.966290.

Ghannoum, M. A., Jurevic, R. J., Mukherjee, P. K., Cui, F., Sikaroodi, M., Naqvi, A., et al. (2010). Characterization of the oral fungal microbiome (mycobiome) in healthy individuals. *PLoS Pathogens, 6*(1), e1000713. https://doi.org/10.1371/journal.ppat.1000713.

Gibson, J., Sood, A., & Hogan, D. A. (2009). *Pseudomonas aeruginosa-Candida albicans* interactions: Localization and fungal toxicity of a phenazine derivative. *Applied and Environmental Microbiology, 75* (2), 504–513. https://doi.org/10.1128/AEM.01037-08.

Gill, E. E., Franco, O. L., & Hancock, R. E. W. (2015). Antibiotic adjuvants: Diverse strategies for controlling drug-resistant pathogens. *Chemical Biology & Drug Design, 85*(1), 56–78. https://doi.org/10.1111/CBDD.12478.

Gil-Perotin, S., Ramirez, P., Marti, V., Sahuquillo, J. M., Gonzalez, E., Calleja, I., et al. (2012). Implications of endotracheal tube biofilm in ventilator-associated pneumonia response: A state of concept. *Critical Care, 16*(3), R93. https://doi.org/10.1186/cc11357.

Glavis-Bloom, J., Muhammed, M., & Mylonakis, E. (2012). Of model hosts and man: Using *Caenorhabditis elegans, Drosophila melanogaster* and *Galleria mellonella* as model hosts for infectious disease research. *Advances in Experimental Medicine and Biology, 710*, 11–17. https://doi.org/10.1007/978-1-4419-5638-5_2.

Goller, C. C., & Romeo, T. (2008). Environmental influences on biofilm development. In T. Romeo (Ed.), *Bacterial biofilms* (pp. 37–66). Berlin, Heidelberg: Springer Berlin Heidelberg.

Gomes, B. P. F. A., Montagner, F., Jacinto, R. C., Pinheiro, E. T., Zaia, A. A., Ferraz, C. C. R., et al. (2008). *Gemella morbillorum* in primary and secondary/persistent endodontic infections. *Oral Surgery, Oral Medicine, Oral Pathology, Oral Radiology, and Endodontics, 105*(4), 519–525. https://doi.org/10.1016/j.tripleo.2007.10.005.

Graham, C. E., Cruz, M. R., Garsin, D. A., & Lorenz, M. C. (2017). *Enterococcus faecalis* bacteriocin EntV inhibits hyphal morphogenesis, biofilm formation, and virulence of *Candida albicans. Proceedings of the National Academy of Sciences, 114*(17), 4507–4512. https://doi.org/10.1073/pnas.1620432114.

Grainha, T., Jorge, P., Alves, D., Lopes, S. P., & Pereira, M. O. (2020). Unraveling *Pseudomonas aeruginosa* and *Candida albicans* communication in coinfection scenarios: Insights through network analysis. *Frontiers in Cellular and Infection Microbiology, 10*, 550505. https://doi.org/10.3389/fcimb.2020.550505.

Gristina, A. G., & Costerton, J. W. (1985). Bacterial adherence to biomaterials and tissue. The significance of its role in clinical sepsis. *The Journal of Bone & Joint Surgery, 67*(2), 264–273. https://doi.org/10.2106/00004623-198567020-00014.

Groeger, S., & Meyle, J. (2019). Oral mucosal epithelial cells. *Frontiers in Immunology, 10*. https://doi.org/10.3389/fimmu.2019.00208.

Guery, R., Merat, C., Blanchet, D., Aznar, C., Labbe, S., & Djossou, F. (2015). Coinfection with tuberculosis and paracoccidioidomycosis in French Guiana: A common misdiagnosis. *International Journal of Dermatology, 54*(10), e436–e439. https://doi.org/10.1111/ijd.12907.

Guimarães, A. J., Nakayasu, E. S., Sobreira, T. J., Cordero, R. J., Nimrichter, L., Almeida, I. C., & Nosanchuk, J. D. (2011). Histoplasma capsulatum heat-shock 60 orchestrates the adaptation of the fungus to temperature stress. *PLoS One, 6*(2), e14660. https://doi.org/10.1371/journal.pone.0014660.

Guo, H., Chen, Y., Guo, W., & Chen, J. (2021). Effects of extracellular DNA on dual-species biofilm formed by *Streptococcus mutans* and *Candida albicans*. *Microbial Pathogenesis, 154*, 104838. https://doi.org/10.1016/j.micpath.2021.104838.

Gupta, A. K., & Foley, K. A. (2019). Evidence for biofilms in onychomycosis. *Giornale italiano di dermatologia e venereologia: organoufficiale, Societa italiana di dermatologia e sifilografia, 154*(1), 50–55. https://doi.org/10.23736/S0392-0488.18.06001-7.

Gupta, N., Haque, A., Mukhopadhyay, G., Narayan, R. P., & Prasad, R. (2005). Interactions between bacteria and *Candida* in the burn wound. *Burns, 31*(3), 375–378. https://doi.org/10.1016/j.burns.2004.11.012.

Gupta, S., Thakur, J., Pal, S., Gupta, R., Mishra, D., Kumar, S., et al. (2019). Cholic acid-peptide conjugates as potent antimicrobials against interkingdom polymicrobial biofilms. *Antimicrobial Agents and Chemotherapy, 63*(11). https://doi.org/10.1128/aac.00520-19.

Hajishengallis, G. (2015). Periodontitis: From microbial immune subversion to systemic inflammation. *Nature Reviews Immunology, 15* (1), 30–44. https://doi.org/10.1038/nri3785.

Hall, R. A., & Noverr, M. C. (2017). Fungal interactions with the human host: Exploring the spectrum of symbiosis. *Current Opinion in Microbiology, 40*, 58–64. https://doi.org/10.1016/j.mib.2017.10.020.

Hall-Stoodley, L., Hu, F. Z., Gieseke, A., Nistico, L., Nguyen, D., Hayes, J., et al. (2006). Direct detection of bacterial biofilms on the middle-ear mucosa of children with chronic otitis media. *JAMA, 296*(2), 202–211. https://doi.org/10.1001/jama.296.2.202.

Harding, M. W., Marques, L. L., Howard, R. J., & Olson, M. E. (2009). Can filamentous fungi form biofilms? *Trends in Microbiology, 17*(11), 475–480. https://doi.org/10.1016/j.tim.2009.08.007.

Harriott, M. M., Lilly, E. A., Rodriguez, T. E., Fidel, P. L., & Noverr, M. C. (2010). *Candida albicans* forms biofilms on the vaginal mucosa. *Microbiology, 156*(Pt 12), 3635–3644. https://doi.org/10.1099/mic.0.039354-0.

Harriott, M. M., & Noverr, M. C. (2009). *Candida albicans* and *Staphylococcus aureus* form polymicrobial biofilms: Effects on antimicrobial resistance. *Antimicrobial Agents and Chemotherapy, 53*(9), 3914–3922. https://doi.org/10.1128/AAC.00657-09.

Harriott, M. M., & Noverr, M. C. (2011). Importance of *Candida*–bacterial polymicrobial biofilms in disease. *Trends in Microbiology, 19*(11), 557–563. https://doi.org/10.1016/j.tim.2011.07.004.

Hathroubi, S., Servetas, S. L., Windham, I., Merrell, D. S., & Ottemann, K. M. (2018). *Helicobacter pylori* biofilm formation and its potential role in pathogenesis. *Microbiology and Molecular Biology Reviews: MMBR, 82*(2), e00001–e00018. https://doi.org/10.1128/MMBR.00001-18.

Hawser, S. P., & Douglas, L. J. (1995). Resistance of *Candida albicans* biofilms to antifungal agents in vitro. *Antimicrobial Agents and Chemotherapy, 39*(9), 2128–2131. https://doi.org/10.1128/aac.39.9.2128.

Healy, D. Y., Leid, J. G., Sanderson, A. R., & Hunsaker, D. H. (2008). Biofilms with fungi in chronic rhinosinusitis. *Otolaryngology–Head and Neck Surgery, 138*(5), 641–647. https://doi.org/10.1016/j.otohns.2008.02.002.

Hemeg, H. A. (2017). Nanomaterials for alternative antibacterial therapy. *International Journal of Nanomedicine, 12*, 8211–8225. https://doi.org/10.2147/IJN.S132163.

Hesse, S. J., Ruijter, G. J., Dijkema, C., & Visser, J. (2002). Intracellular pH homeostasis in the filamentous fungus *Aspergillus niger*. *European Journal of Biochemistry, 269*(14), 3485–3494. https://doi.org/10.1046/j.1432-1033.2002.03042.x.

Hoarau, G., Mukherjee, P. K., Gower-Rousseau, C., Hager, C., Chandra, J., Retuerto, M. A., et al. (2016). Bacteriome and mycobiome interactions underscore microbial dysbiosis in familial Crohn's disease. *MBio, 7* (5). https://doi.org/10.1128/mBio.01250-16.

Hogan, D. A., & Kolter, R. (2002). *Pseudomonas-Candida* interactions: An ecological role for virulence factors. *Science (New York, N.Y.), 296* (5576), 2229–2232. https://doi.org/10.1126/science.1070784.

Hogan, D. A., Vik, A., & Kolter, R. (2004). A *Pseudomonas aeruginosa* quorum-sensing molecule influences *Candida albicans* morphology. *Molecular Microbiology, 54*(5), 1212–1223. https://doi.org/10.1111/j.1365-2958.2004.04349.x.

Høiby, N., Bjarnsholt, T., Moser, C., Bassi, G. L., Coenye, T., Donelli, G., et al. (2015). ESCMID* guideline for the diagnosis and treatment of biofilm infections 2014. *Clinical Microbiology and Infection, 21*, S1–S25. https://doi.org/10.1016/j.cmi.2014.10.024.

Hojo, K., Nagaoka, S., Ohshima, T., & Maeda, N. (2009). Bacterial interactions in dental biofilm development. *Journal of Dental Research, 88* (11), 982–990. https://doi.org/10.1177/0022034509346811.

Hola, V., & Ruzick, F. (2011). The formation of poly-microbial biofilms on urinary catheters. In *Urinary tract infections* InTech. https://doi.org/10.5772/22680.

Holmes, A. R., Cannon, R. D., & Jenkinson, H. F. (1995). Interactions of *Candida albicans* with bacteria and salivary molecules in oral biofilms. *Journal of Industrial Microbiology, 15*(3), 208–213. https://doi.org/10.1007/BF01569827.

Holmes, A., Mcnab, R., & Jenkinson, H. (1996). *Candida albicans* binding to the oral bacterium *Streptococcus gordonii* involves multiple adhesin receptor interactions. *Infection and Immunity, 64*, 4680–4685. https://doi.org/10.1128/IAI.64.11.4680-4685.1996.

Holt, J. E., Houston, A., Adams, C., Edwards, S., & Kjellerup, B. V. (2017). Role of extracellular polymeric substances in polymicrobial biofilm infections of *Staphylococcus epidermidis* and *Candida albicans* modelled in the nematode *Caenorhabditis elegans*. *Pathogens and Disease, 75*(5), ftx052. https://doi.org/10.1093/femspd/ftx052.

Honraet, K., & Nelis, H. J. (2006). Use of the modified robbins device and fluorescent staining to screen plant extracts for the inhibition of *S. mutans* biofilm formation. *Journal of Microbiological Methods, 64* (2), 217–224. https://doi.org/10.1016/j.mimet.2005.05.005.

Hopke, A., Mela, A., Ellett, F., Carter-House, D., Peña, J. F., Stajich, J. E., et al. (2021). Crowdsourced analysis of fungal growth and branching on microfluidic platforms. *PLoS One, 16*(9), e0257823. https://doi.org/10.1371/journal.pone.0257823 (Accession No. 34587206).

Hornby, J. M., Jensen, E. C., Lisec, A. D., Tasto, J. J., Jahnke, B., Shoemaker, R., et al. (2001). Quorum sensing in the dimorphic fungus *Candida albicans* is mediated by farnesol. *Applied and Environmental Microbiology, 67*(7), 2982–2992. https://doi.org/10.1128/aem.67.7.2982-2992.2001.

Hoyer, L. L. (2001). The ALS gene family of *Candida albicans*. *Trends in Microbiology*, *9*(4), 176–180. https://doi.org/10.1016/s0966-842x(01)01984-9.

Hoyer, L. L., Oh, S.-H., Jones, R., & Cota, E. (2014). A proposed mechanism for the interaction between the *Candida albicans* Als3 adhesin and streptococcal cell wall proteins. *Frontiers in Microbiology*, *5*, 564. https://doi.org/10.3389/fmicb.2014.00564.

Huan, Y., Kong, Q., Mou, H., & Yi, H. (2020). Antimicrobial peptides: Classification, design, application and research progress in multiple fields. *Frontiers in Microbiology*, *11*, 582779. https://doi.org/10.3389/fmicb.2020.582779. 33178164.

Hwang, G., Klein, M. I., & Koo, H. (2014). Analysis of the mechanical stability and surface detachment of mature *Streptococcus mutans* biofilms by applying a range of external shear forces. *Biofouling*, *30*(9), 1079–1091. https://doi.org/10.1080/08927014.2014.969249.

Hwang, G., Liu, Y., Kim, D., Li, Y., Krysan, D. J., & Koo, H. (2017). *Candida albicans* mannans mediate *Streptococcus mutans* exoenzyme GtfB binding to modulate cross-kingdom biofilm development in vivo. *PLoS Pathogens*, *13*(6), e1006407. https://doi.org/10.1371/journal.ppat.1006407.

Hwang, G., Marsh, G., Gao, L., Waugh, R., & Koo, H. (2015). Binding force dynamics of *Streptococcus mutans*-glucosyltransferase B to *Candida albicans*. *Journal of Dental Research*, *94*(9), 1310–1317. https://doi.org/10.1177/0022034515592859.

Ikeh, M. A. C., Fidel, P. L., Jr., & Noverr, M. C. (2018). Prostaglandin E(2) receptor antagonist with antimicrobial activity against methicillin-resistant *Staphylococcus aureus*. *Antimicrobial Agents and Chemotherapy*, *62*(3). https://doi.org/10.1128/AAC.01920-17. e01920-01917.

Ikono, R., Vibriani, A., Wibowo, I., Saputro, K. E., Muliawan, W., Bachtiar, B. M., et al. (2019). Nanochitosan antimicrobial activity against *Streptococcus mutans* and *Candida albicans* dual-species biofilms. *BMC Research Notes*, *12*(1). https://doi.org/10.1186/S13104-019-4422-X.

Inglis, T. J. J., Tit-Meng, L., Mah-Lee, N., Ee-Koon, T., & Kok-Pheng, H. (1995). Structural features of tracheal tube biofilm formed during prolonged mechanical ventilation. *Chest*, *108*(4), 1049–1052. https://doi.org/10.1378/chest.108.4.1049.

Ishigami, M., Nakagawa, Y., Hayakawa, M., & Iimura, Y. (2004). FLO11 is essential for flor formation caused by the C-terminal deletion of NRG1 in *Saccharomyces cerevisiae*. *FEMS Microbiology Letters*, *237*(2), 425–430. https://doi.org/10.1016/j.femsle.2004.07.012.

Iwahashi, J., Kamei, K., & Watanabe, H. (2020). Disruption of *Aspergillus fumigatus* biofilm by *Streptococcus pneumoniae*: Mycelial fragmentation by hydrogen peroxide. *Journal of Infection and Chemotherapy*, *26*(8), 831–837. https://doi.org/10.1016/j.jiac.2020.03.015.

Jabra-Rizk, M. A., Meiller, T. F., James, C. E., & Shirtliff, M. E. (2006). Effect of farnesol on *Staphylococcus aureus* biofilm formation and antimicrobial susceptibility. *Antimicrobial Agents and Chemotherapy*, *50*(4), 1463–1469. https://doi.org/10.1128/AAC.50.4.1463-1469.2006.

Jack, A. A., Daniels, D. E., Jepson, M. A., Vickerman, M. M., Lamont, R. J., Jenkinson, H. F., et al. (2015). *Streptococcus gordonii* comCDE (competence) operon modulates biofilm formation with *Candida albicans*. *Microbiology (Reading, England)*, *161*(Pt 2), 411–421. https://doi.org/10.1099/mic.0.000010.

Jamal, M., Ahmad, W., Andleeb, S., Jalil, F., Imran, M., Nawaz, M. A., et al. (2018). Bacterial biofilm and associated infections. *Journal of the Chinese Medical Association*, *81*(1), 7–11. https://doi.org/10.1016/j.jcma.2017.07.012.

Jarosz, L. M., Deng, D. M., van der Mei, H. C., Crielaard, W., & Krom, B. P. (2009). *Streptococcus mutans* competence-stimulating peptide inhibits *Candida albicans* hypha formation. *Eukaryotic Cell*, *8*(11), 1658–1664. https://doi.org/10.1128/EC.00070-09.

Järvensivu, A., Hietanen, J., Rautemaa, R., Sorsa, T., & Richardson, M. (2004). *Candida* yeasts in chronic periodontitis tissues and subgingival microbial biofilms in vivo. *Oral Diseases*, *10*(2), 106–112. https://doi.org/10.1046/j.1354-523x.2003.00978.x.

Jenkinson, H. F., Lala, H. C., & Shepherd, M. G. (1990). Coaggregation of *Streptococcus sanguis* and other streptococci with *Candida albicans*. *Infection and Immunity*, *58*(5), 1429–1436. https://doi.org/10.1128/iai.58.5.1429-1436.1990.

Jenkinson, H. F., & Lamont, R. J. (2005). Oral microbial communities in sickness and in health. *Trends in Microbiology*, *13*(12), 589–595. https://doi.org/10.1016/j.tim.2005.09.006.

Jensen, L. K., Johansen, A. S. B., & Jensen, H. E. (2017). Porcine models of biofilm infections with focus on pathomorphology. *Frontiers in Microbiology*, *8*, 1961. https://doi.org/10.3389/fmicb.2017.01961.

Johnson, E., Petersen, T., & Goeres, D. M. (2021). Characterizing the shearing stresses within the CDC biofilm reactor using computational fluid dynamics. *Microorganisms*, *9*(8), 1709.

Joyner, P. M., Liu, J., Zhang, Z., Merritt, J., Qi, F., & Cichewicz, R. H. (2010). Mutanobactin A from the human oral pathogen *Streptococcus mutans* is a cross-kingdom regulator of the yeast-mycelium transition. *Organic & Biomolecular Chemistry*, *8*(24), 5486–5489. https://doi.org/10.1039/c0ob00579g.

Kalan, L., Loesche, M., Hodkinson, B. P., Heilmann, K., Ruthel, G., Gardner, S. E., et al. (2016). Redefining the chronic-wound microbiome: Fungal communities are prevalent, dynamic, and associated with delayed healing. *MBio*, *7*(5), e01058-16. https://doi.org/10.1128/mBio.01058-16.

Karkowska-Kuleta, J., Bartnicka, D., Zawrotniak, M., Zielinska, G., Kieronska, A., Bochenska, O., et al. (2018). The activity of bacterial peptidylarginine deiminase is important during formation of dual-species biofilm by periodontal pathogen *Porphyromonas gingivalis* and opportunistic fungus *Candida albicans*. *Pathogens and Disease*, *76*(4). https://doi.org/10.1093/femspd/fty033.

Karygianni, L., Ren, Z., Koo, H., & Thurnheer, T. (2020). Biofilm matrixome: Extracellular components in structured microbial communities. *Trends in Microbiology*, *28*(8), 668–681. https://doi.org/10.1016/j.tim.2020.03.016.

Kean, R., Rajendran, R., Haggarty, J., Townsend, E. M., Short, B., Burgess, K. E., et al. (2017). *Candida albicans* mycofilms support *Staphylococcus aureus* colonization and enhances miconazole resistance in dual-species interactions. *Frontiers in Microbiology*, *8*. https://www.frontiersin.org/article/10.3389/fmicb.2017.00258.

Kernien, J. F., Snarr, B. D., Sheppard, D. C., & Nett, J. E. (2017). The interface between fungal biofilms and innate immunity. *Frontiers in Immunology*, *8*, 1968. https://doi.org/10.3389/fimmu.2017.01968.

Khan, F., Pham, D. T. N., Oloketuyi, S. F., & Kim, Y.-M. (2019). Antibiotics application strategies to control biofilm formation in pathogenic bacteria. *Current Pharmaceutical Biotechnology*, *21*(4), 270–286. https://doi.org/10.2174/1389201020666191112155905.

Khan, F., Pham, D. T. N., Tabassum, N., Oloketuyi, S. F., & Kim, Y. M. (2020). Treatment strategies targeting persister cell formation in bacterial pathogens. *CRC Critical Reviews in Microbiology, 46*(6), 665–688. https://doi.org/10.1080/1040841x.2020.1822278.

Khatoon, Z., McTiernan, C. D., Suuronen, E. J., Mah, T.-F., & Alarcon, E. I. (2018). Bacterial biofilm formation on implantable devices and approaches to its treatment and prevention. *Heliyon, 4*(12), e01067. https://doi.org/10.1016/j.heliyon.2018.e01067.

Khoury, Z. H., Vila, T., Puthran, T. R., Sultan, A. S., Montelongo-Jauregui, D., Melo, M. A. S., et al. (2020). The Role of *Candida albicans* secreted polysaccharides in augmenting *Streptococcus mutans* adherence and mixed biofilm formation: In vitro and in vivo studies. *Frontiers in Microbiology, 11*, 307. https://doi.org/10.3389/fmicb.2020.00307.

Kim, J., Hegde, M., & Jayaraman, A. (2010). Microfluidic co-culture of epithelial cells and bacteria for investigating soluble signal-mediated interactions. *Journal of Visualized Experiments: JoVE*. https://doi.org/10.3791/1749.

Kim, J., Park, H. D., & Chung, S. (2012). Microfluidic approaches to bacterial biofilm formation. *Molecules, 17*(8), 9818–9834. https://doi.org/10.3390/molecules17089818.

Kim, D., Sengupta, A., Niepa, T. H. R., Lee, B.-H., Weljie, A., Freitas-Blanco, V. S., et al. (2017). *Candida albicans* stimulates *Streptococcus mutans* microcolony development via cross-kingdom biofilm-derived metabolites. *Scientific Reports, 7*, 41332. https://doi.org/10.1038/srep41332.

Kim, Y., & Mylonakis, E. (2011). Killing of Candida albicans filaments by Salmonella enterica serovar Typhimurium is mediated by sopB effectors, parts of a type III secretion system. *Eukaryot Cell, 10*(6). https://doi.org/10.1128/EC.00014-11. 782–90.

Kirketerp-Møller, K., Jensen, P., Fazli, M., Madsen, K. G., Pedersen, J., Moser, C., et al. (2008). Distribution, organization, and ecology of bacteria in chronic wounds. *Journal of Clinical Microbiology, 46*(8), 2717–2722. https://doi.org/10.1128/jcm.00501-08.

Klemm, P., & Schembri, M. A. (2000). Bacterial adhesins: Function and structure. *International Journal of Medical Microbiology, 290*(1), 27–35. https://doi.org/10.1016/S1438-4221(00)80102-2.

Klotz, S. A., Gaur, N. K., De Armond, R., Sheppard, D., Khardori, N., Edwards, J. E., Jr., … El-Azizi, M. (2007). Candida albicans Als proteins mediate aggregation with bacteria and yeasts. *Medical Mycology, 45*(4), 363–370. https://doi.org/10.1080/13693780701299333.

Klotz, S. A., Chasin, B. S., Powell, B., Gaur, N. K., & Lipke, P. N. (2007). Polymicrobial bloodstream infections involving *Candida* species: Analysis of patients and review of the literature. *Diagnostic Microbiology and Infectious Disease, 59*(4), 401–406. https://doi.org/10.1016/j.diagmicrobio.2007.07.001.

Kojic, E. M., & Darouiche, R. O. (2004). *Candida* infections of medical devices. *Clinical Microbiology Reviews, 17*(2), 255–267.

Kolenbrander, P. E., Egland, P. G., Diaz, P. I., & Palmer, R. J. J. (2005). Genome-genome interactions: Bacterial communities in initial dental plaque. *Trends in Microbiology, 13*(1), 11–15. https://doi.org/10.1016/j.tim.2004.11.005.

Kong, E. F., Tsui, C., Kucharíková, S., Andes, D., Van Dijck, P., & Jabra-Rizk, M. A. (2016). Commensal protection of *Staphylococcus aureus* against antimicrobials by *Candida albicans* biofilm matrix. *MBio, 7* (5). https://doi.org/10.1128/mBio.01365-16.

Koo, H., Allan, R. N., Howlin, R. P., Stoodley, P., & Hall-Stoodley, L. (2017). Targeting microbial biofilms: Current and prospective therapeutic strategies. *Nature Reviews Microbiology, 15*(12), 740–755. https://doi.org/10.1038/nrmicro.2017.99.

Kovács, R., & Majoros, L. (2020). Fungal quorum-sensing molecules: A review of their antifungal effect against *Candida* biofilms. *Fungi (Basel)*. https://doi.org/10.3390/jof6030099. 32630687.

Kragh, K. N., Hutchison, J. B., Melaugh, G., Rodesney, C., Roberts, A. E., Irie, Y., et al. (2016). Role of multicellular aggregates in biofilm formation. *MBio, 7*(2), e00237. https://doi.org/10.1128/mBio.00237-16.

Krause, J., Geginat, G., & Tammer, I. (2015). Prostaglandin E2 from Candida albicans stimulates the growth of Staphylococcus aureus in mixed biofilms. *PLoS One, 10*(8), e0135404. https://doi.org/10.1371/journal.pone.0135404.

Krüger, W., Vielreicher, S., Kapitan, M., Jacobsen, I. D., & Niemiec, M. J. (2019). Fungal-bacterial interactions in health and disease. *Pathogens (Basel, Switzerland), 8*(2), 70. https://doi.org/10.3390/pathogens8020070.

Kuang, Z., Hao, Y., Walling, B. E., Jeffries, J. L., Ohman, D. E., & Lau, G. W. (2011). *Pseudomonas aeruginosa* elastase provides an escape from phagocytosis by degrading the pulmonary surfactant protein-A. *PLoS One, 6*(11), e27091. https://doi.org/10.1371/journal.pone.0027091.

Kuhn, D. M., Mukherjee, P. K., Clark, T. A., Pujol, C., Chandra, J., Hajjeh, R. A., et al. (2004). *Candida parapsilosis* characterization in an outbreak setting. *Emerging Infectious Diseases, 10*(6), 1074–1081. https://doi.org/10.3201/eid1006.030873.

Kumari, A., Tripathi, A. H., Gautam, P., Gahtori, R., Pande, A., Singh, Y., … Upadhyay, S. K. (2021). Adhesins in the virulence of opportunistic fungal pathogens of human. *Mycology, 12*(4), 296–324. https://doi.org/10.1080/21501203.2021.1934176.

LaFleur, M. D., Kumamoto, C. A., & Lewis, K. (2006). *Candida albicans* biofilms produce antifungal-tolerant persister cells. *Antimicrobial Agents and Chemotherapy, 50*(11), 3839–3846. https://doi.org/10.1128/aac.00684-06.

Lara, H. H., & Lopez-Ribot, J. L. (2020). Inhibition of mixed biofilms of *Candida albicans* and methicillin-resistant *Staphylococcus aureus* by positively charged silver nanoparticles and functionalized silicone elastomers. *Pathogens, 9*(10), 784. https://doi.org/10.3390/pathogens9100784. 32992727.

Lebeaux, D., Chauhan, A., Rendueles, O., & Beloin, C. (2013). From in vitro to in vivo models of bacterial biofilm-related infections. *Pathogens, 2*(2), 288–356. https://doi.org/10.3390/pathogens2020288.

Lebeaux, D., Ghigo, J. M., & Beloin, C. (2014). Biofilm-related infections: Bridging the gap between clinical management and fundamental aspects of Recalcitrance toward antibiotics. *Microbiology and Molecular Biology Reviews, 78*(3), 510–543. https://doi.org/10.1128/mmbr.00013-14.

Lee, H.-W., Koh, Y. M., Kim, J., Lee, J.-C., Lee, Y.-C., Seol, S.-Y., et al. (2008). Capacity of multidrug-resistant clinical isolates of *Acinetobacter baumannii* to form biofilm and adhere to epithelial cell surfaces. *Clinical Microbiology and Infection: The Official Publication of the European Society of Clinical Microbiology and Infectious Diseases, 14*(1), 49–54. https://doi.org/10.1111/j.1469-0691.2007.01842.x.

Lee, M. J., Liu, H., Barker, B. M., Snarr, B. D., Gravelat, F. N., Al Abdallah, Q., et al. (2015). The fungal exopolysaccharide galactosaminogalactan mediates virulence by enhancing resistance to neutrophil extracellular traps. *PLoS Pathogens, 11*(10), e1005187. https://doi.org/10.1371/journal.ppat.1005187.

Leidich, S. D., Ibrahim, A. S., Fu, Y., Koul, A., Jessup, C., Vitullo, J., et al. (1998). Cloning and disruption of caPLB1, a phospholipase B gene involved in the pathogenicity of *Candida albicans*. *The Journal of*

Biological Chemistry, *273*(40), 26078–26086. https://doi.org/10.1074/jbc.273.40.26078.

Lemos, J. A. C., Abranches, J., & Burne, R. A. (2005). Responses of cariogenic streptococci to environmental stresses. *Current Issues in Molecular Biology*, *7*(1), 95–107.

Li, Q., Liu, J., Shao, J., Da, W., Shi, G., Wang, T., et al. (2019). Decreasing cell population of individual *Candida* species does not impair the virulence of *Candida albicans* and *Candida glabrata* mixed biofilms. *Frontiers in Microbiology*, *10*, 1600. https://doi.org/10.3389/fmicb.2019.01600.

Li, Z. J., Guo, X., Dawuti, G., & Aibai, S. (2015). Antifungal activity of ellagic acid in vitro and in vivo. *Phytotherapy Research*, *29*. https://doi.org/10.1002/ptr.5340. 25919446. 1019–25.

Lin, Y. J., Alsad, L., Vogel, F., Koppar, A., et al. (2013). Interactions between *Candida albicans* and *Staphylococcus aureus* within mixed species biofilms. *Bios*, *84*(1), 30–39.

Liu, H., Lee, M. J., Solis, N. V., Phan, Q. T., Swidergall, M., Ralph, B., & Filler, S. G. (2016). Aspergillus fumigatus CalA binds to integrin α5β1 and mediates host cell invasion. *Nature Microbiology*, *2*(2), 16211. https://doi.org/10.1038/nmicrobiol.2016.211.

Loehfelm, T. W., Luke, N. R., & Campagnari, A. A. (2008). Identification and characterization of an *Acinetobacter baumannii* biofilm-associated protein. *Journal of Bacteriology*, *190*(3), 1036–1044. https://doi.org/10.1128/JB.01416-07.

Lohse, M. B., Gulati, M., Johnson, A. D., & Nobile, C. J. (2018). Development and regulation of single- and multi-species *Candida albicans* biofilms. *Nature Reviews Microbiology*, *16*(1), 19–31. https://doi.org/10.1038/nrmicro.2017.107.

Long, K. H., Gomez, F. J., Morris, R. E., & Newman, S. L. (2003). Identification of heat shock protein 60 as the ligand on *Histoplasma capsulatum* that mediates binding to CD18 receptors on human macrophages. *Journal of Immunology*, *170*(1), 487–494. https://doi.org/10.4049/jimmunol.170.1.487.

Lorek, J., Poggeler, S., Weide, M. R., Breves, R., & Bockmuhl, D. P. (2008). Influence of farnesol on the morphogenesis of *Aspergillus niger*. *Journal of Basic Microbiology*, *48*(2), 99–103. https://doi.org/10.1002/jobm.200700292.

Lou, Z., Wang, H., Tang, Y., & Chen, X. (2017). The effect of burdock leaf fraction on adhesion, biofilm formation, quorum sensing and virulence factors of Pseudomonas aeruginosa. *Journal of Applied Microbiology*, *122*(3), 615–624. https://doi.org/10.1111/jam.13348.

Loussert, C., Schmitt, C., Prevost, M. C., Balloy, V., Fadel, E., Philippe, B., et al. (2010). In vivo biofilm composition of *Aspergillus fumigatus*. *Cellular Microbiology*, *12*(3), 405–410. https://doi.org/10.1111/j.1462-5822.2009.01409.x.

Lown, L., Peters, B. M., Walraven, C. J., Noverr, M. C., & Lee, S. A. (2016). An optimized lock solution containing micafungin, ethanol and doxycycline inhibits *candida albicans* and mixed *C. albicans - Staphylococcus aureus* biofilms. *PLoS One*, *11*(7). https://doi.org/10.1371/JOURNAL.PONE.0159225.

Luo, Y., McAuley, D. F., Fulton, C. R., Pessoa, J. S., McMullan, R., & Lundy, F. T. (2021). Targeting *Candida albicans* in dual-species biofilms with antifungal treatment reduces *Staphylococcus aureus* and MRSA in vitro. *PLoS One*, *16*(4), e0249547. https://doi.org/10.1371/JOURNAL.PONE.0249547.

Macià, M. D., Rojo-Molinero, E., & Oliver, A. (2014). Antimicrobial susceptibility testing in biofilm-growing bacteria. *Clinical Microbiology and Infection*, *20*(10), 981–990. https://doi.org/10.1111/1469-0691.12651.

Madmoli, M., Dehcheshmeh, Z. M., Rafi, A., Kord, Z., Mobarez, F., & Darabiyan, P. (2019). The rate of some complications and risk factors of diabetes in diabetic patients: Study on cases of 3218 diabetic patients. *Medical Science*, *23*(95), 63–68.

Mahdizade-ari, M., Pourhajibagher, M., & Bahador, A. (2019). Changes of microbial cell survival, metabolic activity, efflux capacity, and quorum sensing ability of *Aggregatibacter actinomycetemcomitans* due to antimicrobial photodynamic therapy-induced bystander effects. *Photodiagnosis and Photodynamic Therapy*, *26*, 287–294. https://doi.org/10.1016/J.PDPDT.2019.04.021.

Maisch, T. (2009). A new strategy to destroy antibiotic resistant microorganisms: Antimicrobial photodynamic treatment. *Mini Reviews in Medicinal Chemistry*, *9*(8), 974–983. https://doi.org/10.2174/138955709788681582.

Manavathu, E. K., Vager, D. L., & Vazquez, J. A. (2014). Development and antimicrobial susceptibility studies of in vitro monomicrobial and polymicrobial biofilm models with *Aspergillus fumigatus* and *Pseudomonas aeruginosa*. *BMC Microbiology*, *14*, 53. https://doi.org/10.1186/1471-2180-14-53.

Mancera, E., Nocedal, I., Hammel, S., Gulati, M., Mitchell, K. F., Andes, D. R., et al. (2021). Evolution of the complex transcription network controlling biofilm formation in *Candida* species. *eLife*, *10*. https://doi.org/10.7554/eLife.64682.

Mankiewicz, E., & Liivak, M. (1960). Effect of *Candida albicans* on the evolution of experimental tuberculosis. *Nature*, *187*(4733), 250–251. https://doi.org/10.1038/187250a0.

Margalit, A., Carolan, J. C., & Kavanagh, K. (2021). Bacterial interactions with *Aspergillus fumigatus* in the immunocompromised lung. *Microorganisms*, *9*(2), 435. https://doi.org/10.3390/microorganisms9020435.

Margalit, A., Carolan, J. C., Sheehan, D., & Kavanagh, K. (2020). The *Aspergillus fumigatus* secretome alters the proteome of *Pseudomonas aeruginosa* to stimulate bacterial growth: Implications for co-infection. *Molecular & Cellular Proteomics : MCP*, *19*(8), 1346–1359. https://doi.org/10.1074/mcp.RA120.002059.

Mariencheck, W. I., Alcorn, J. F., Palmer, S. M., & Wright, J. R. (2003). *Pseudomonas aeruginosa* elastase degrades surfactant proteins A and D. *American Journal of Respiratory Cell and Molecular Biology*, *28*(4), 528–537. https://doi.org/10.1165/rcmb.2002-0141OC.

Marks, D. J. B., Hyams, C., Koo, C. Y., Pavlou, M., Robbins, J., Koo, C. S., et al. (2014). Clinical features, microbiology and surgical outcomes of infective endocarditis: A 13-year study from a UK tertiary cardiothoracic referral centre. *QJM: An International Journal of Medicine*, *108*(3), 219–229. https://doi.org/10.1093/qjmed/hcu188.

Marsh, P. (2003). Plaque as a biofilm: Pharmacological principles of drug delivery and action in the sub- and supragingival environment. *Oral Diseases*, *9*, 16–22. https://doi.org/10.1034/j.1601-0825.9.s1.4.x.

Martinez, L. R., & Casadevall, A. (2007). *Cryptococcus neoformans* biofilm formation depends on surface support and carbon source and reduces fungal cell susceptibility to heat, cold, and UV light. *Applied and Environmental Microbiology*, *73*(14), 4592–4601. https://doi.org/10.1128/AEM.02506-06.

Martinez, L. R., & Casadevall, A. (2015). Biofilm formation by *Cryptococcus neoformans*. *Microbiology Spectrum*, *3*(3). https://doi.org/10.1128/microbiolspec.MB-0006-2014.

Martins, M., Henriques, M., Lopez-Ribot, J. L., & Oliveira, R. (2012). Addition of DNase improves the in vitro activity of antifungal drugs against *Candida albicans* biofilms. *Mycoses, 55*(1), 80–85. https://doi.org/10.1111/j.1439-0507.2011.02047.x.

Martinez-Rossi, N. M., Bitencourt, T. A., Teixeira De Aguiar Peres, N., et al. (2018). Dermatophyte resistance to antifungal drugs: Mechanisms and prospectus. *Frontiers in Microbiology, 9*, 1108. https://doi.org/10.3389/fmicb.2018.01108.

Martins, C. H. G., Pires, R. H., Cunha, A. O., Pereira, C. A. M., de Singulani, J. L., Abrão, F., et al. (2016). *Candida/Candida* biofilms. First description of dual-species *Candida albicans/C. rugosa* biofilm. *Fungal Biology, 120*(4), 530–537. https://doi.org/10.1016/j.funbio.2016.01.013.

Martins, M., Uppuluri, P., Thomas, D. P., Cleary, I. A., Henriques, M., Lopez-Ribot, J. L., et al. (2010). Presence of extracellular DNA in the *Candida albicans* biofilm matrix and its contribution to biofilms. *Mycopathologia, 169*(5), 323–331. https://doi.org/10.1007/s11046-009-9264-y.

Mathé, L., & Van Dijck, P. (2013). Recent insights into *Candida albicans* biofilm resistance mechanisms. *Current Genetics, 59*(4), 251–264. https://doi.org/10.1007/s00294-013-0400-3.

McBain, A. J., Sissons, C., Ledder, R. G., Sreenivasan, P. K., De Vizio, W., & Gilbert, P. (2005). Development and characterization of a simple perfused oral microcosm. *Journal of Applied Microbiology, 98*(3), 624–634. https://doi.org/10.1111/j.1365-2672.2004.02483.x.

McLean, R. J., & Kakirde, K. S. (2013). Enhancing metagenomics investigations of microbial interactions with biofilm technology. *International Journal of Molecular Sciences, 14*(11), 22246–22257. https://doi.org/10.3390/ijms141122246.

McNab, R., Jenkinson, H. F., Loach, D. M., & Tannock, G. W. (1994). Cell-surface-associated polypeptides CshA and CshB of high molecular mass are colonization determinants in the oral bacterium *Streptococcus gordonii*. *Molecular Microbiology, 14*(4), 743–754. https://doi.org/10.1111/j.1365-2958.1994.tb01311.x.

Medina-Alarcón, K. P., da Silva, I. P. T., Ferin, G. G., Pereira-da-Silva, M. A., Marcos, C. M., Dos Santos, M. B., et al. (2021). Mycobacterium tuberculosis and *Paracoccidioides brasiliensis* formation and treatment of mixed biofilm in vitro. *Frontiers in Cellular and Infection Microbiology, 11*, 681131. https://doi.org/10.3389/fcimb.2021.681131.

Melo, W., Avci, P., de Nóbrega Oliveira, M., et al. (2013). Photodynamic inactivation of biofilm: Taking a lightly colored approach to stubborn infection. *Expert Review of Anti-Infective Therapy, 11*(7), 669–693. https://doi.org/10.1586/14787210.2013.811861.

Mendez, E., Walker, D. K., Vipham, J., & Trinetta, V. (2020). The use of a CDC biofilm reactor to grow multi-strain *Listeria monocytogenes* biofilm. *Food Microbiology, 92*, 103592. https://doi.org/10.1016/j.fm.2020.103592.

Mermel, L. A., Allon, M., Bouza, E., Craven, D. E., Flynn, P., O'Grady, N. P., et al. (2009). Clinical practice guidelines for the diagnosis and management of intravascular catheter-related infection: 2009 update by the infectious diseases society of America. *Clinical Infectious Diseases, 49*(1), 1–45. https://doi.org/10.1086/599376.

Metwalli, K. H., Khan, S. A., Krom, B. P., & Jabra-Rizk, M. A. (2013). *Streptococcus mutans, Candida albicans*, and the human mouth: A sticky situation. *PLoS Pathogens, 9*(10), e1003616. https://doi.org/10.1371/journal.ppat.1003616.

Miceli, M. H., Díaz, J. A., & Lee, S. A. (2011). Emerging opportunistic yeast infections. *The Lancet Infectious Diseases, 11*(2), 142–151. https://doi.org/10.1016/s1473-3099(10)70218-8.

Millet, L. J., Aufrecht, J., Labbé, J., Uehling, J., Vilgalys, R., Estes, M. L., et al. (2019). Increasing access to microfluidics for studying fungi and other branched biological structures. *Fungal Biology and Biotechnology, 6*, 1. https://doi.org/10.1186/s40694-019-0071-z.

Mitchell, KF, Zarnowski, R, & Andes, DR. (2016a). Fungal super glue: The biofilm matrix and its composition, assembly, and functions. *PLoS Pathogens, 12*(9), e1005828. https://doi.org/10.1371/journal.ppat.1005828.

Mitchell, K. F., Zarnowski, R., & Andes, D. R. (2016b). The extracellular matrix of fungal biofilms. *Advances in Experimental Medicine and Biology, 931*, 21–35. https://doi.org/10.1007/5584_2016_6.

Mitchell, K. F., Zarnowski, R., Sanchez, H., Edward, J. A., Reinicke, E. L., Nett, J. E., et al. (2015). Community participation in biofilm matrix assembly and function. *Proceedings of the National Academy of Sciences of the United States of America, 112*(13), 4092–4097. https://doi.org/10.1073/pnas.1421437112.

Møller, S., Sternberg, C., Andersen, J. B., Christensen, B. B., Ramos, J. L., Givskov, M., et al. (1998). In situ gene expression in mixed-culture biofilms: Evidence of metabolic interactions between community members. *Applied and Environmental Microbiology, 64*(2), 721–732. https://doi.org/10.1128/aem.64.2.721-732.1998.

Montelongo-Jauregui, D., Saville, S. P., & Lopez-Ribot, J. L. (2019). Contributions of *Candida albicans* dimorphism, adhesive interactions, and extracellular matrix to the formation of dual-species biofilms with *Streptococcus gordonii*. *MBio, 10*(3). https://doi.org/10.1128/mBio.01179-19.

Moormeier, D. E., & Bayles, K. W. (2017). *Staphylococcus aureus* biofilm: A complex developmental organism. *Molecular Microbiology, 104*(3), 365–376. https://doi.org/10.1111/mmi.13634.

Morelli, K. A., Kerkaert, J. D., & Cramer, R. A. (2021). *Aspergillus fumigatus* biofilms: Toward understanding how growth as a multicellular network increases antifungal resistance and disease progression. *PLoS Pathogens, 17*(8), e1009794. https://doi.org/10.1371/journal.ppat.1009794.

Muhammad, M. H., Idris, A. L., Fan, X., Guo, Y., Yu, Y., Jin, X., et al. (2020). Beyond risk: Bacterial biofilms and their regulating approaches. *Frontiers in Microbiology, 11*, 928. https://doi.org/10.3389/fmicb.2020.00928.

Murciano, C., Moyes, D. L., Runglall, M., Tobouti, P., Islam, A., Hoyer, L. L., et al. (2012). Evaluation of the role of *Candida albicans* agglutinin-like sequence (Als) proteins in human oral epithelial cell interactions. *PLoS One, 7*(3), e33362. https://doi.org/10.1371/journal.pone.0033362.

Naglik, J. R., Challacombe, S. J., & Hube, B. (2003). *Candida albicans* secreted aspartyl proteinases in virulence and pathogenesis. *Microbiology and Molecular Biology Reviews, 67*(3), 400–428. table of contents https://doi.org/10.1128/mmbr.67.3.400-428.2003.

Namgoong, S., Jung, S.-Y., Han, S.-K., Kim, A.-R., & Dhong, E.-S. (2019). Clinical experience with surgical debridement and simultaneous meshed skin grafts in treating biofilm-associated infection: An exploratory retrospective pilot study. *Journal of Plastic Surgery and Hand Surgery, 54*(1), 47–54. https://doi.org/10.1080/2000656x.2019.1673170.

Nett, J. E. (2018). Special issue: *Candida* and candidiasis. *Journal of Fungi (Basel), 4*.

Nett, J. E., & Andes, D. R. (2015). Fungal biofilms: In vivo models for discovery of anti-biofilm drugs. *Microbiology Spectrum, 3*(3), E30. https://doi.org/10.1128/microbiolspec.MB-0008-2014.

Nickerson, K. W., Atkin, A. L., & Hornby, J. M. (2006). Quorum sensing in dimorphic fungi: Farnesol and beyond. *Applied and Environmental Microbiology*, *72*(6), 3805–3813. https://doi.org/10.1128/aem.02765-05.

Nikinmaa, S., Alapulli, H., Auvinen, P., Vaara, M., Rantala, J., Kankuri, E., et al. (2020). Dual-light photodynamic therapy administered daily provides a sustained antibacterial effect on biofilm and prevents *Streptococcus mutans* adaptation. *PLoS One*, *15*(5). https://doi.org/10.1371/JOURNAL.PONE.0232775.

Niu, C., Clemmer, K. M., Bonomo, R. A., & Rather, P. N. (2008). Isolation and characterization of an autoinducer synthase from *Acinetobacter baumannii*. *Journal of Bacteriology*, *190*(9), 3386–3392. https://doi.org/10.1128/JB.01929-07.

Nobile, C. J., Schneider, H. A., Nett, J. E., Sheppard, D. C., Filler, S. G., Andes, D. R., et al. (2008). Complementary adhesin function in *C. albicans* biofilm formation. *Current Biology*, *18*(14), 1017–1024. https://doi.org/10.1016/j.cub.2008.06.034.

Nogueira, M. F., Pereira, L., Jenull, S., Kuchler, K., & Lion, T. (2019). *Klebsiella pneumoniae* prevents spore germination and hyphal development of *Aspergillus* species. *Scientific Reports*, *9*(1), 218. https://doi.org/10.1038/s41598-018-36524-8.

Nomura, K., Obata, K., Keira, T., Miyata, R., Hirakawa, S., Takano, K., et al. (2014). *Pseudomonas aeruginosa* elastase causes transient disruption of tight junctions and downregulation of PAR-2 in human nasal epithelial cells. *Respiratory Research*, *15*(1), 21. https://doi.org/10.1186/1465-9921-15-21.

Norouzi, H., Danesh, A., Mohseni, M., & Rabbani Khorasgani, M. (2018). Marine actinomycetes with probiotic potential and bioactivity against multidrug-resistant bacteria. *International Journal of Molecular and Cellular Medicine*, *7*(1), 44–52. https://doi.org/10.22088/ijmcm.bums.7.1.44.

Nunvar, J., Capek, V., Fiser, K., Fila, L., & Drevinek, P. (2017). What matters in chronic *Burkholderia cenocepacia* infection in cystic fibrosis: Insights from comparative genomics. *PLoS Pathogens*, *13*(12), e1006762. https://doi.org/10.1371/journal.ppat.1006762.

O'Connell, H. A., Kottkamp, G. S., Eppelbaum, J. L., Stubblefield, B. A., Gilbert, S. E., & Gilbert, E. S. (2006). Influences of biofilm structure and antibiotic resistance mechanisms on indirect pathogenicity in a model polymicrobial biofilm. *Applied and Environmental Microbiology*, *72*(7), 5013–5019. https://doi.org/10.1128/AEM.02474-05.

Ochieng, W., Wanzala, P., Bii, C., Oishi, I., Ichimura, H., Lihana, R., et al. (2005). Tuberculosis and oral *Candida* species surveillance in HIV infected individuals in Northern Kenya, and the implications on tuberculin skin test screening for DOPT-P. *East African Medical Journal*, *82*(12), 609–613. https://doi.org/10.4314/eamj.v82i12.9365.

O'Hanlon, D. E., Moench, T. R., & Cone, R. A. (2013). Vaginal pH and microbicidal lactic acid when lactobacilli dominate the microbiota. *PLoS One*, *8*(11), e80074. https://doi.org/10.1371/journal.pone.0080074.

Olivares, E., Badel-Berchoux, S., Provot, C., Prévost, G., Bernardi, T., & Jehl, F. (2019). Clinical impact of antibiotics for the treatment of *Pseudomonas aeruginosa* biofilm infections. *Frontiers in Microbiology*, *10*. https://doi.org/10.3389/FMICB.2019.02894.

Olivares, E., Tasse, J., Badel-Berchoux, S., Provot, C., Prévost, G., & Bernardi, T. (2020). Clinical biofilm ring Test(®) reveals the potential role of β-lactams in the induction of biofilm formation by *P. aeruginosa* in cystic fibrosis patients. *Pathogens (Basel, Switzerland)*, *9*(12), 1065. https://doi.org/10.3390/pathogens9121065.

Oliveira, L. T., Medina-Alarcón, K. P., Singulani, J. D. L., Fregonezi, N. F., Pires, R. H., Arthur, R. A., et al. (2020). Dynamics of mono- and dual-species biofilm formation and interactions between *Paracoccidioides brasiliensis* and *Candida albicans*. *Frontiers in Microbiology*, *11*, 551256. https://doi.org/10.3389/fmicb.2020.551256.

Orazi, G., & O'Toole, G. A. (2019). "It takes a village": Mechanisms underlying antimicrobial recalcitrance of polymicrobial biofilms. *Journal of Bacteriology*, *202*(1). https://doi.org/10.1128/jb.00530-19.

Padder, S. A., Prasad, R., & Shah, A. H. (2018). Quorum sensing: A less known mode of communication among fungi. *Microbiological Research*, *210*, 51–58. https://doi.org/10.1016/j.micres.2018.03.007. 29625658. In this issue.

Pai, M., Behr, M. A., Dowdy, D., Dheda, K., Divangahi, M., Boehme, C. C., et al. (2016). Tuberculosis. *Nature Reviews Disease Primers*, *2*(1), 16076. https://doi.org/10.1038/nrdp.2016.76.

Paiva, L. C., Vidigal, P. G., Donatti, L., Svidzinski, T. I., & Consolaro, M. E. (2012). Assessment of in vitro biofilm formation by Candida species isolates from vulvovaginal candidiasis and ultrastructural characteristics. *Micron (Oxford, England : 1993)*, *43*(2–3). https://doi.org/10.1016/j.micron.2011.09.013. 497–202.

Palencia, S. L., García, A., & Palencia, M. (2022). Multiple surface interaction mechanisms direct the anchoring, co-aggregation and formation of dual-species biofilm between *Candida albicans* and *Helicobacter pylori*. *Journal of Advanced Research*, *35*, 169–185. https://doi.org/10.1016/j.jare.2021.03.013.

Palmer, R. J., Jr. (1999). Microscopy flowcells: Perfusion chambers for real-time study of biofilms. *Methods in Enzymology*, *310*, 160–166. https://doi.org/10.1016/s0076-6879(99)10014-4.

Pammi, M., Liang, R., Hicks, J., Mistretta, T.-A., & Versalovic, J. (2013). Biofilm extracellular DNA enhances mixed species biofilms of *Staphylococcus epidermidis* and *Candida albicans*. *BMC Microbiology*, *13*, 257. https://doi.org/10.1186/1471-2180-13-257.

Pammi, M., Zhong, D., Johnson, Y., Revell, P., & Versalovic, J. (2014). Polymicrobial bloodstream infections in the neonatal intensive care unit are associated with increased mortality: A case-control study. *BMC Infectious Diseases*, *14*(1). https://doi.org/10.1186/1471-2334-14-390.

Pamp, S. J., Sternberg, C., & Tolker-Nielsen, T. (2009). Insight into the microbial multicellular lifestyle via flow-cell technology and confocal microscopy. *Cytometry Part A*, *75*(2), 90–103. https://doi.org/10.1002/cyto.a.20685.

Parolin, C., Croatti, V., Laghi, L., Giordani, B., Tondi, M. R., De Gregorio, P. R., et al. (2021). *Lactobacillus* biofilms influence anti-*Candida* activity. *Frontiers in Microbiology*, *12*, 750368. https://doi.org/10.3389/fmicb.2021.750368.

Pathak, A. K., Sharma, S., & Shrivastva, P. (2012). Multi-species biofilm of *Candida albicans* and non-*Candida albicans Candida* species on acrylic substrate. *Journal of Applied Oral Science: Revista FOB*, *20*(1), 70–75. https://doi.org/10.1590/s1678-77572012000100013.

Pathirana, R. U., McCall, A. D., Norris, H. L., & Edgerton, M. (2019). Filamentous non-albicans *Candida* species adhere to *candida albicans* and benefit from dual biofilm growth. *Frontiers in Microbiology*, *10*, 1188. https://doi.org/10.3389/fmicb.2019.01188.

Pati, B. A., Kurata, W. E., Horseman, T. S., & Pierce, L. M. (2021). Antibiofilm activity of chitosan/epsilon-poly-L-lysine hydrogels in a porcine ex vivo skin wound polymicrobial biofilm model. *Wound Repair and Regeneration*, *29*(2), 316–326. https://doi.org/10.1111/wrr.12890.

Peeters, E., Nelis, H. J., & Coenye, T. (2008). Comparison of multiple methods for quantification of microbial biofilms grown in microtiter plates. *Journal of Microbiological Methods*, *72*(2), 157–165. https://doi.org/10.1016/j.mimet.2007.11.010.

Peleg, A. Y., Hogan, D. A., & Mylonakis, E. (2010). Medically important bacterial-fungal interactions. *Nat Rev Microbiol*, *8*(5). https://doi.org/10.1038/nrmicro2313. 340–9.

Pelgrift, R. Y., & Friedman, A. J. (2013). Nanotechnology as a therapeutic tool to combat microbial resistance. *Advanced Drug Delivery Reviews*. https://doi.org/10.1016/j.addr.2013.07.011.

Perfect, J. R. (2017). The antifungal pipeline: A reality check. *Nature Reviews Drug Discovery*, *16*(9), 603–616. 28496146.

Peters, B. M., Jabra-Rizk, M. A., O'May, G. A., Costerton, J. W., & Shirtliff, M. E. (2012). Polymicrobial interactions: Impact on pathogenesis and human disease. *Clinical Microbiology Reviews*, *25*(1), 193–213. https://doi.org/10.1128/CMR.00013-11.

Peters, B. M., Jabra-Rizk, M. A., Scheper, M. A., Leid, J. G., Costerton, J. W., & Shirtliff, M. E. (2010). Microbial interactions and differential protein expression in *Staphylococcus aureus-Candida albicans* dual-species biofilms. *FEMS Immunology and Medical Microbiology*, *59* (3), 493–503. https://doi.org/10.1111/j.1574-695X.2010.00710.x.

Peters, B. M., Ovchinnikova, E. S., Krom, B. P., Schlecht, L. M., Zhou, H., Hoyer, L. L., et al. (2012). *Staphylococcus aureus* adherence to *Candida albicans* hyphae is mediated by the hyphal adhesin Als3p. *Microbiology (Reading, England)*, *158*(Pt 12), 2975–2986. https://doi.org/10.1099/mic.0.062109-0.

Peters, B. M., Ward, R. M., Rane, H. S., Lee, S. A., & Noverr, M. C. (2013). Efficacy of ethanol against *Candida albicans* and *Staphylococcus aureus* polymicrobial biofilms. *Antimicrobial Agents and Chemotherapy*, *57*(1), 74–82. https://doi.org/10.1128/aac.01599-12.

Phan, Q. T., Myers, C. L., Fu, Y., Sheppard, D. C., Yeaman, M. R., Welch, W. H., et al. (2007). Als3 is a *Candida albicans* invasin that binds to cadherins and induces endocytosis by host cells. *PLoS Biology*, *5*(3), e64. https://doi.org/10.1371/journal.pbio.0050064.

Pierce, C. G., Vila, T., Romo, J. A., Montelongo-Jauregui, D., Wall, G., Ramasubramanian, A., et al. (2017). The *Candida albicans* biofilm matrix: Composition, structure and function. *Journal of Fungi (Basel)*, *3*(1). https://doi.org/10.3390/jof3010014.

Pircalabioru, G. G., & Chifiriuc, M. C. (2020). Nanoparticulate drug-delivery systems for fighting microbial biofilms: From bench to bedside. *Future Microbiology*, *15*(8), 679–698. https://doi.org/10.2217/FMB-2019-0251.

Pires, R. H., Martinez, L. R., Mendes-Gianinni, M. J. S., & Stoianoff, M. A. R. (2021). Editorial: Pathogenesis of fungal biofilms in different environmental conditions and clinical outcomes. *Frontiers in Cellular and Infection Microbiology*, *11*, 990. https://doi.org/10.3389/fcimb.2021.778458.

Pitangui, N. S., Sardi, J. C. O., Silva, J. F., Benaducci, T., Moraes da Silva, R. A., Rodríguez-Arellanes, G., et al. (2012). Adhesion of *Histoplasma capsulatum* to pneumocytes and biofilm formation on an abiotic surface. *Biofouling*, *28*(7), 711–718. https://doi.org/10.1080/08927014.2012.703659.

Ponde, N. O., Lortal, L., Ramage, G., Naglik, J. R., & Richardson, J. P. (2021). *Candida albicans* biofilms and polymicrobial interactions. *Critical Reviews in Microbiology*, *47*(1), 91–111. https://doi.org/10.1080/1040841X.2020.1843400.

Price, C. E., Brown, D. G., Limoli, D. H., Phelan, V. V., & O'Toole, G. A. (2020). Exogenous alginate protects *Staphylococcus aureus* from killing by *Pseudomonas aeruginosa*. *Journal of Bacteriology*, *202* (8). https://doi.org/10.1128/JB.00559-19.

Price-Whelan, A., Dietrich, L. E. P., & Newman, D. K. (2006). Rethinking "secondary" metabolism: Physiological roles for phenazine antibiotics. *Nature Chemical Biology*, *2*(2), 71–78. https://doi.org/10.1038/nchembio764.

Proal, A. D., Lindseth, I. A., & Marshall, T. G. (2017). Microbe-microbe and host-microbe interactions drive microbiome dysbiosis and inflammatory processes. *Discovery Medicine*, *23*(124), 51–60.

Pukkila-Worley, R., Ausubel, F. M., & Mylonakis, E. (2011). *Candida albicans* infection of *Caenorhabditis elegans* induces antifungal immune defenses. *PLoS Pathogens*, *7*(6), e1002074. https://doi.org/10.1371/journal.ppat.1002074.

Qu, Y., Locock, K., Verma-Gaur, J., Hay, I. D., Meagher, L., & Traven, A. (2015). Searching for new strategies against polymicrobial biofilm infections: Guanylated polymethacrylates kill mixed fungal/bacterial biofilms. *Journal of Antimicrobial Chemotherapy*, *71*(2), 413–421. https://doi.org/10.1093/jac/dkv334.

Queiroz-Telles, F., & Escuissato, D. L. (2011). Pulmonary paracoccidioidomycosis. *Seminars in Respiratory and Critical Care Medicine*, *32* (6), 764–774. https://doi.org/10.1055/s-0031-1295724.

Quishida, C. C., Carmello, J., Mima, E., Bagnato, V. S., Machado, A., & Pavarina, A. C. (2015). Susceptibility of multispecies biofilm to photodynamic therapy using Photodithazine®. *Lasers in Medical Science*, *30*(2), 685–694. https://doi.org/10.1007/s10103-013-1397-z.

Quivey, R. G., Jr., Faustoferri, R., Monahan, K., & Marquis, R. (2000). Shifts in membrane fatty acid profiles associated with acid adaptation of *Streptococcus mutans*. *FEMS Microbiology Letters*, *189*(1), 89–92. https://doi.org/10.1111/j.1574-6968.2000.tb09211.x.

Radaic, A., & Kapila, Y. L. (2021). The oralome and its dysbiosis: New insights into oral microbiome-host interactions. *Computational and Structural Biotechnology Journal*, *19*, 1335–1360. https://doi.org/10.1016/j.csbj.2021.02.010.

Raina, S., Vizio, D. D., Odell, M., Clements, M., Vanhulle, S., & Keshavarz, T. (2009). Microbial quorum sensing: A tool or a target for antimicrobial therapy? *Biotechnology and Applied Biochemistry*, *54*(2), 65–84. https://doi.org/10.1042/ba20090072.

Rajendran, R., Sherry, L., Nile, C. J., Sherriff, A., Johnson, E. M., Hanson, M. F., et al. (2016). Biofilm formation is a risk factor for mortality in patients with *Candida albicans* bloodstream infection—Scotland, 2012–2013. *Clinical Microbiology and Infection*, *22*(1), 87–93. https://doi.org/10.1016/j.cmi.2015.09.018.

Rajendran, R., Williams, C., Lappin, D. F., Millington, O., Martins, M., & Ramage, G. (2013). Extracellular DNA release acts as an antifungal resistance mechanism in mature *Aspergillus fumigatus* biofilms. *Eukaryotic Cell*, *12*(3), 420–429. https://doi.org/10.1128/ec.00287-12.

Ramage, G., Rajendran, R., Gutierrez-Correa, M., Jones, B., & Williams, C. (2011). *Aspergillus* biofilms: Clinical and industrial significance. *FEMS Microbiology Letters*, *324*(2), 89–97. https://doi.org/10.1111/j.1574-6968.2011.02381.x.

Ramage, G., Bachmann, S., Petterson, T. F., Wickes, B. L., & López-Ribot, J. L. (2002). Investigation of multidrug efflux pumps in relation to fluconazole resistance in *Candida albicans* biofilms. *Journal of Antimicrobial Chemotherapy*, *49*(6), 973–980. https://doi.org/10.1093/jac/dkf049.

Ramage, G., Saville, S. P., Wickes, B. L., & López-Ribot, J. L. (2002). Inhibition of Candida albicans biofilm formation by farnesol, a quorum-sensing molecule. *Applied and Environmental Microbiology*, *68*(11). https://doi.org/10.1128/AEM.68.11.5459-5463.2002. 5459–63.

Ramalingam, K., Frohlich, N. C., & Lee, V. A. (2013). Effect of nanoemulsion on dental unit waterline biofilm. *Journal of Dental Sciences*, *8*(3), 333–336. https://doi.org/10.1016/J.JDS.2013.02.035.

Ramírez Granillo, A., Canales, M. G. M., Espíndola, M. E. S., Martínez Rivera, M. A., de Lucio, V. M. B., & Tovar, A. V. R. (2015). Antibiosis interaction of *Staphylococcus aureus* on *Aspergillus fumigatus* assessed in vitro by mixed biofilm formation. *BMC Microbiology*, *15*(1), 33. https://doi.org/10.1186/s12866-015-0363-2.

Ramírez-Granillo, A., Bautista-Hernández, L. A., Bautista-De Lucío, V. M., Magaña-Guerrero, F. S., Domínguez-López, A., Córdova-Alcántara, I. M., et al. (2021). Microbial warfare on three fronts: Mixed biofilm of *Aspergillus fumigatus* and *Staphylococcus aureus* on primary cultures of human limbo-corneal fibroblasts. In *Vol. 11*. *Frontiers in cellular and infection microbiology* (p. 711). https://www.frontiersin.org/article/10.3389/fcimb.2021.646054.

Ramsook, C. B., Tan, C., Garcia, M. C., Fung, R., Soybelman, G., Henry, R., et al. (2010). Yeast cell adhesion molecules have functional amyloid-forming sequences. *Eukaryotic Cell*, *9*(3), 393–404. https://doi.org/10.1128/ec.00068-09.

Rasmussen, T. B., & Givskov, M. (2006). Quorum-sensing inhibitors as anti-pathogenic drugs. *International Journal of Medical Microbiology*, *296*(2–3). https://doi.org/10.1016/j.ijmm.2006.02.005. 16503194. 149–61.

Rauceo, J. M., Gaur, N. K., Lee, K. G., Edwards, J. E., Klotz, S. A., & Lipke, P. N. (2004). Global cell surface conformational shift mediated by a *Candida albicans* adhesin. *Infection and Immunity*, *72*(9), 4948–4955. https://doi.org/10.1128/iai.72.9.4948-4955.2004.

Reece, E., Doyle, S., Greally, P., Renwick, J., & McClean, S. (2018). *Aspergillus fumigatus* inhibits *Pseudomonas aeruginosa* in co-culture: Implications of a mutually antagonistic relationship on virulence and inflammation in the CF airway. *Frontiers in Microbiology*, *9*, 1205. https://doi.org/10.3389/fmicb.2018.01205.

Reece, E., Segurado, R., Jackson, A., McClean, S., Renwick, J., & Greally, P. (2017). Co-colonisation with *Aspergillus fumigatus* and *Pseudomonas aeruginosa* is associated with poorer health in cystic fibrosis patients: An Irish registry analysis. *BMC Pulmonary Medicine*, *17* (1). https://doi.org/10.1186/s12890-017-0416-4.

Reichhardt, C., Ferreira, J. A. G., Joubert, L.-M., Clemons, K. V., Stevens, D. A., & Cegelski, L. (2015). Analysis of the *Aspergillus fumigatus* biofilm extracellular matrix by solid-state nuclear magnetic resonance spectroscopy. *Eukaryotic Cell*, *14*(11), 1064–1072. https://doi.org/10.1128/EC.00050-15.

Reichhardt, C., Stevens, D. A., & Cegelski, L. (2016). Fungal biofilm composition and opportunities in drug discovery. *Future Medicinal Chemistry*, *8*(12), 1455–1468. https://doi.org/10.4155/fmc-2016-0049.

Rendueles, O., & Ghigo, J. M. (2012). Multi-species biofilms: How to avoid unfriendly neighbors. *FEMS Microbiology Reviews*, *36*(5), 972–989. https://doi.org/10.1111/j.1574-6976.2012.00328.x.

Reynolds, T. B., & Fink, G. R. (2001). Bakers' yeast, a model for fungal biofilm formation. *Science*, *291*(5505), 878–881. https://doi.org/10.1126/science.291.5505.878.

Rickard, A. H., Gilbert, P., High, N. J., Kolenbrander, P. E., & Handley, P. S. (2003). Bacterial coaggregation: An integral process in the development of multi-species biofilms. *Trends in Microbiology*, *11*(2), 94–100. https://doi.org/10.1016/s0966-842x(02)00034-3.

Ricker, A., Vickerman, M., & Dongari-Bagtzoglou, A. (2014). *Streptococcus gordonii* glucosyltransferase promotes biofilm interactions with *Candida albicans*. *Journal of Oral Microbiology*, *6*. https://doi.org/10.3402/jom.v6.23419.

Rieger, U. M., Mesina, J., Kalbermatten, D. F., Haug, M., Frey, H. P., Pico, R., et al. (2013). Bacterial biofilms and capsular contracture in patients with breast implants. *British Journal of Surgery*, *100*(6), 768–774. https://doi.org/10.1002/bjs.9084.

Roberts, F. A., & Darveau, R. P. (2015). Microbial protection and virulence in periodontal tissue as a function of polymicrobial communities: Symbiosis and dysbiosis. *Periodontology 2000*, *69*(1), 18–27. https://doi.org/10.1111/prd.12087.

Roberts, A. P., & Kreth, J. (2014). The impact of horizontal gene transfer on the adaptive ability of the human oral microbiome. *Frontiers in Cellular and Infection Microbiology*, *4*, 124. https://doi.org/10.3389/fcimb.2014.00124.

Rodrigues, C. F., & Černáková, L. (2020). Farnesol and tyrosol: Secondary metabolites with a crucial quorum-sensing role in Candida biofilm development. *Genes*, *11*(4), 444. https://doi.org/10.3390/genes11040444.

Rodrigues, C. T., de Andrade, F. B., de Vasconcelos, L. R. S. M., Midena, R. Z., Pereira, T. C., Kuga, M. C., et al. (2018). Antibacterial properties of silver nanoparticles as a root canal irrigant against *Enterococcus faecalis* biofilm and infected dentinal tubules. *International Endodontic Journal*, *51*(8), 901–911. https://doi.org/10.1111/IEJ.12904.

Rodrigues, M. E., Lopes, S. P., Pereira, C. R., Azevedo, N. F., Lourenço, A., Henriques, M., et al. (2017). Polymicrobial ventilator-associated pneumonia: Fighting in vitro *Candida albicans-Pseudomonas aeruginosa* biofilms with antifungal-antibacterial combination therapy. *PLoS One*, *12*(1), e0170433. https://doi.org/10.1371/journal.pone.0170433.

Rodríguez-Cerdeira, C., Gregorio, M. C., Molares-Vila, A., López-Barcenas, A., Fabbrocini, G., Bardhi, B., et al. (2019). Biofilms and vulvovaginal candidiasis. *Colloids and Surfaces B: Biointerfaces*, *174*, 110–125. https://doi.org/10.1016/j.colsurfb.2018.11.011.

Rogiers, O., Holtappels, M., Siala, W., et al. (2018). Anidulafungin increases the antibacterial activity of tigecycline in polymicrobial Candida albicans/Staphylococcus aureus biofilms on intraperitoneally implanted foreign bodies. *Journal of Antimicrobial Chemotherapy*, *73*, 2806–2814. https://doi.org/10.1093/jac/dky246. 30010876.

Roisin, L., Melloul, E., Woerther, P.-L., Royer, G., Decousser, J.-W., Guillot, J., et al. (2020). Modulated response of *Aspergillus fumigatus* and *Stenotrophomonas maltophilia* to antimicrobial agents in polymicrobial biofilm. *Frontiers in Cellular and Infection Microbiology*, *10*. https://doi.org/10.3389/fcimb.2020.574028.

Römling, U., Kjelleberg, S., Normark, S., Nyman, L., Uhlin, B. E., & Åkerlund, B. (2014). Microbial biofilm formation: A need to act. *Journal of Internal Medicine*, *276*(2), 98–110. https://doi.org/10.1111/joim.12242.

Rose, W. E., & Poppens, P. T. (2009). Impact of biofilm on the in vitro activity of vancomycin alone and in combination with tigecycline and rifampicin against *Staphylococcus aureus*. *The Journal of Antimicrobial Chemotherapy*, *63*(3), 485–488. https://doi.org/10.1093/JAC/DKN513.

Rossoni, R. D., Barbosa, J. O., Vilela, S. F. G., dos Santos, J. D., de Barros, P. P., de Prata, M. C. A., et al. (2015). Competitive Interactions between *C. albicans*, *C. glabrata* and *C. krusei* during biofilm formation and development of experimental candidiasis. *PLoS One*, *10* (7), e0131700. https://doi.org/10.1371/journal.pone.0131700.

Rout, B., Liu, C. H., & Wu, W. C. (2017). Photosensitizer in lipid nanoparticle: A nano-scaled approach to antibacterial function. *Scientific Reports*, *7*(1). https://doi.org/10.1038/S41598-017-07444-W.

Roy, R., Tiwari, M., Donelli, G., & Tiwari, V. (2018). Strategies for combating bacterial biofilms: A focus on anti-biofilm agents and their mechanisms of action. *Virulence*, *9*(1), 522–554. https://doi.org/10.1080/21505594.2017.1313372.

Rukavina, Z., & Vanić, Ž. (2016). Current trends in development of liposomes for targeting bacterial biofilms. *Pharmaceutics*. https://doi.org/10.3390/pharmaceutics8020018.

Saarela, M. H., Doğan, B., Alaluusua, S., & Asikainen, S. (1999). Persistence of oral colonization by the same *Actinobacillus actinomycetemcomitans* strain(s). *Journal of Periodontology*, *70*(5), 504–509. https://doi.org/10.1902/jop.1999.70.5.504.

Samaranayake, Y. H., Bandara, H. M., Cheung, B. P., Yau, J. Y., Yeung, S. K., & Samaranayake, L. P. (2014). Enteric Gram-negative bacilli suppress *Candida* biofilms on Foley urinary catheters. *APMIS*, *122* (1), 47–58. https://doi.org/10.1111/apm.12098.

Sánchez-Alonzo, K., Silva-Mieres, F., Arellano-Arriagada, L., Parra-Sepúlveda, C., Bernasconi, H., Smith, C. T., et al. (2021). Nutrient deficiency promotes the entry of *Helicobacter pylori* cells into *Candida* yeast cells. *Biology*, *10*(5), 426. https://doi.org/10.3390/biology10050426.

Santos, R., Costa, C., Mil-Homens, D., Romão, D., de Carvalho, C. C. C. R., Pais, P., et al. (2017). The multidrug resistance transporters CgTpo1_1 and CgTpo1_2 play a role in virulence and biofilm formation in the human pathogen *Candida glabrata*. *Cellular Microbiology*, *19*(5), e12686. https://doi.org/10.1111/cmi.12686.

Santos, J., Piva, E., Vilela, S., Jorge, A., & Junqueira, J. (2016). Mixed biofilms formed by *C. albicans* and non-albicans species: A study of microbial interactions. *Brazilian Oral Research*, *30*, e23. https://doi.org/10.1590/1807-3107BOR-2016.vol30.0023.

Sardi, J., Pitangui, N., Gullo, F., Fusco-Almeida, A., & Mendes-Giannini, A. (2015). Fungal biofilms: Formation, resistance and pathogenicity. In M. Shams-Ghahfarokhi, M. Razzaghi-Abyaneh, & M. Rai (Eds.), *Vol. 1. Medical mycology: Current trends and future prospects* (pp. 291–314). Oxford: Taylor & Francis Group.

Sardi, J. C. O., Scorzoni, L., Bernardi, T., Fusco-Almeida, A. M., & Mendes Giannini. (2013). Candida species: current epidemiology, pathogenicity, biofilm formation, natural antifungal products and new therapeutic options. *Journal of Medical Microbiology*, *62*(1), 10–24. https://doi.org/10.1099/jmm.0.045054-0.

Sardi, J. D. C. O., Pitangui, N. D. S., Rodríguez-Arellanes, G., Taylor, M. L., Fusco-Almeida, A. M., & Mendes-Giannini, M. J. S. (2014). Highlights in pathogenic fungal biofilms. *Revista Iberoamericana de Micología*, *31*(1), 22–29. https://doi.org/10.1016/j.riam.2013.09.014.

Sass, G., Nazik, H., Penner, J., Shah, H., Ansari, S. R., Clemons, K. V., et al. (2018). Studies of *Pseudomonas aeruginosa* mutants indicate pyoverdine as the central factor in inhibition of *Aspergillus fumigatus* biofilm. *Journal of Bacteriology*, *200*(1). https://doi.org/10.1128/JB.00345-17.

Satala, D., Gonzalez-Gonzalez, M., Smolarz, M., Surowiec, M., Kulig, K., Wronowska, E., … Karkowska-Kuleta, J. (2021). The role of *Candida albicans* virulence factors in the formation of multispecies biofilms with bacterial periodontal pathogens. *Frontiers in Cellular and Infection Microbiology*, *11*, 765942. https://doi.org/10.3389/fcimb.2021.765942.

Sato, T., Watanabe, T., Mikami, T., & Matsumoto, T. (2004). Farnesol, a morphogenetic autoregulatory substance in the dimorphic fungus *Candida albicans*, inhibits hyphae growth through suppression of a mitogen-activated protein kinase cascade. *Biological & Pharmaceutical Bulletin*, *27*(5), 751–752. https://doi.org/10.1248/bpb.27.751.

Schaller, M, Schackert, C, Korting, HC, Januschke, E, & Hube, B. (2000). Invasion of Candida albicans correlates with expression of secreted aspartic proteinases during experimental infection of human epidermis. *Journal of Investigative Dermatology*, *114*(4). https://doi.org/10.1046/j.1523-1747.2000.00935.x.0. 712–7.

Schaller, M., Schäfer, W., Korting, H. C., & Hube, B. (1998). Differential expression of secreted aspartyl proteinases in a model of human oral candidosis and in patient samples from the oral cavity. *Molecular Microbiology*, *29*(2), 605–615. https://doi.org/10.1046/j.1365-2958.1998.00957.x.

Scheunemann, G., Fortes, B. N., Lincopan, N., & Ishida, K. (2021). Caspofungin inhibits mixed biofilms of *Candida albicans* and methicillin-resistant *Staphylococcus aureus* and displays effectiveness in coinfected *Galleria mellonella* Larvae. *Microbiology Spectrum*, *9*(2), e0074421. https://doi.org/10.1128/Spectrum.00744-21.

Schlecht, L. M., Peters, B. M., Krom, B. P., Freiberg, J. A., Hänsch, G. M., Filler, S. G., et al. (2015). Systemic *Staphylococcus aureus* infection mediated by *Candida albicans* hyphal invasion of mucosal tissue. *Microbiology (Reading, England)*, *161*(Pt 1), 168–181. https://doi.org/10.1099/mic.0.083485-0.

Schulze, A., Mitterer, F., Pombo, J. P., & Schild, S. (2021). Biofilms by bacterial human pathogens: Clinical relevance—Development, composition and regulation—Therapeutical strategies. *Microbial Cell (Graz, Austria)*, *8*(2), 28–56. https://doi.org/10.15698/mic2021.02.741.

Scorzoni, L., Sangalli-Leite, F., de Lacorte Singulani, J., et al. (2016). Searching new antifungals: The use of in vitro and in vivo methods for evaluation of natural compounds. *Journal of Microbiological Methods*, *123*, 68–78. https://doi.org/10.1016/j.mimet.2016.02.005. 26853122.

Scott, J. E., Li, K., Filkins, L. M., Zhu, B., Kuchma, S. L., Schwartzman, J. D., et al. (2019). *Pseudomonas aeruginosa* can inhibit growth of streptococcal species via siderophore production. *Journal of Bacteriology*, *201*(8). https://doi.org/10.1128/jb.00014-19.

Seidler, M. J., Salvenmoser, S., & Muller, F.-M. C. (2008). *Aspergillus fumigatus* forms biofilms with reduced antifungal drug susceptibility on bronchial epithelial cells. *Antimicrobial Agents and Chemotherapy*, *52*(11), 4130–4136. https://doi.org/10.1128/aac.00234-08.

Sheehan, G., Tully, L., & Kavanagh, K. A. (2020). *Candida albicans* increases the pathogenicity of *Staphylococcus aureus* during polymicrobial infection of *Galleria mellonella* larvae. *Microbiology (Reading)*, *166*(4), 375–385. https://doi.org/10.1099/mic.0.000892.

Shirazi, F., Ferreira, J. A. G., Stevens, D. A., Clemons, K. V., & Kontoyiannis, D. P. (2016). Biofilm filtrates of *Pseudomonas aeruginosa* strains isolated from cystic fibrosis patients inhibit preformed *Aspergillus fumigatus* biofilms via apoptosis. *PLoS One*, *11*(3), e0150155. https://doi.org/10.1371/journal.pone.0150155.

Silva, S., Pires, P., Monteiro, D. R., Negri, M., Gorup, L. F., Camargo, E. R., et al. (2013). The effect of silver nanoparticles and nystatin on mixed biofilms of *Candida glabrata* and *Candida albicans* on acrylic. *Medical Mycology*, *51*(2), 178–184. https://doi.org/10.3109/13693786.2012.700492.

Silverman, R. J., Nobbs, A. H., Vickerman, M. M., Barbour, M. E., & Jenkinson, H. F. (2010). Interaction of *Candida albicans* cell wall Als3 protein with *Streptococcus gordonii* SspB adhesin promotes development of mixed-species communities. *Infection and Immunity, 78*(11), 4644–4652. https://doi.org/10.1128/IAI.00685-10.

Singulani, J. L., Scorzoni, L., de Oliveira, H. C., Marcos, C. M., Assato, P. A., Fusco-Almeida, A. M., et al. (2018). Applications of invertebrate animal models to dimorphic fungal infections. *Journal of Fungi (Basel), 4*(4). https://doi.org/10.3390/jof4040118.

Singulani, J. L., Scorzoni, L., Gomes, P. C., Nazaré, A. C., Polaquini, C. R., Regasini, L. O., et al. (2017). Activity of gallic acid and its ester derivatives in *Caenorhabditis elegans* and zebrafish (*Danio rerio*) models. *Future Medicinal Chemistry, 9*(16), 1863–1872. https://doi.org/10.4155/fmc-2017-0096.

Smith, K., Rajendran, R., Kerr, S., Lappin, D. F., Mackay, W. G., Williams, C., et al. (2015). *Aspergillus fumigatus* enhances elastase production in *Pseudomonas aeruginosa* co-cultures. *Medical Mycology, 53*(7), 645–655. https://doi.org/10.1093/mmy/myv048.

Sneddon, L. U., Halsey, L. G., & Bury, N. R. (2017). Considering aspects of the 3Rs principles within experimental animal biology. *Journal of Experimental Biology, 220*(17), 3007–3016. https://doi.org/10.1242/jeb.147058.

Soares, G. M., Teles, F., Starr, J. R., Feres, M., Patel, M., Martin, L., et al. (2015). Effects of azithromycin, metronidazole, amoxicillin, and metronidazole plus amoxicillin on an in vitro polymicrobial subgingival biofilm model. *Antimicrobial Agents and Chemotherapy, 59*(5), 2791–2798. https://doi.org/10.1128/aac.04974-14.

Song, J. W., Shin, J. H., Kee, S. J., Kim, S. H., Shin, M. G., Suh, S. P., et al. (2009). Expression of CgCDR1, CgCDR2, and CgERG11in *Candida glabrata* biofilms formed by bloodstream isolates. *Medical Mycology, 47*(5), 545–548. https://doi.org/10.1080/13693780802210726.

Sousa, C., Botelho, C., Rodrigues, D., Azeredo, J., & Oliveira, R. (2012). Infective endocarditis in intravenous drug abusers: An update. *European Journal of Clinical Microbiology & Infectious Diseases, 31*(11), 2905–2910. https://doi.org/10.1007/s10096-012-1675-x.

Stacy, A., McNally, L., Darch, S. E., Brown, S. P., & Whiteley, M. (2016). The biogeography of polymicrobial infection. *Nature Reviews Microbiology, 14*(2), 93–105. https://doi.org/10.1038/nrmicro.2015.8.

Stiefel, U., & Donskey, C. J. (2004). The role of the intestinal tract as a source for transmission of nosocomial pathogens. *Current Infectious Disease Reports, 6*(6), 420–425. https://doi.org/10.1007/s11908-004-0060-z.

Stoodley, P., Kathju, S., Hu, F. Z., Erdos, G., Levenson, J. E., Mehta, N., et al. (2005). Molecular and imaging techniques for bacterial biofilms in joint arthroplasty infections. *Clinical Orthopaedics and Related Research, 437*, 31–40. https://doi.org/10.1097/01.blo.0000175129.83084.d5.

Strus, M., Kucharska, A., Kukla, G., Brzychczy-Włoch, M., Maresz, K., & Heczko, P. B. (2005). The in vitro activity of vaginal *Lactobacillus* with probiotic properties against *Candida*. *Infectious Diseases in Obstetrics and Gynecology, 13*(2), 69–75. https://doi.org/10.1080/10647440400028136.

Stuart, C. H., Schwartz, S. A., Beeson, T. J., & Owatz, C. B. (2006). *Enterococcus faecalis*: Its role in root canal treatment failure and current concepts in retreatment. *Journal of Endodontics, 32*(2), 93–98. https://doi.org/10.1016/J.JOEN.2005.10.049.

Sundstrom, P. (1999). Adhesins in *Candida albicans*. *Current Opinion in Microbiology, 2*(4), 353–357. https://doi.org/10.1016/s1369-5274(99)80062-9.

Swidsinski, A., Guschin, A., Tang, Q., Dörffel, Y., Verstraelen, H., Tertychnyy, A., … Jiang, X. (2019). Vulvovaginal candidiasis: Histologic lesions are primarily polymicrobial and invasive and do not contain biofilms. *American Journal of Obstetrics and Gynecology, 220*(1), 91.e1–91.e8. https://doi.org/10.1016/j.ajog.2018.10.023.

Tablan, O. C., Anderson, L. J., Besser, R., Bridges, C., Hajjeh, R., CDC, et al. (2004). Guidelines for preventing health-care—Associated pneumonia, 2003: Recommendations of CDC and the Healthcare Infection Control Practices Advisory Committee. *MMWR. Recommendations and reports: Morbidity and mortality weekly report. Recommendations and reports, 53(RR-3)* (pp. 1–36).

Tampakakis, E., Peleg, A. Y., & Mylonakis, E. (2009). Interaction of *Candida albicans* with an intestinal pathogen, *Salmonella enterica* serovar Typhimurium. *Eukaryotic Cell, 8*(5), 732–737. https://doi.org/10.1128/EC.00016-09.

Tan, Y., Leonhard, M., Moser, D., et al. (2019). Antibiofilm efficacy of curcumin in combination with 2-aminobenzimidazole against single- and mixed-species biofilms of Candida albicans and Staphylococcus aureus. *Colloids Surf B Biointerfaces., 174*, 28–34. https://doi.org/10.1016/j.colsurfb.2018.10.079. 30412864.

Tati, S., Davidow, P., McCall, A., Hwang-Wong, E., Rojas, I. G., Cormack, B., et al. (2016). *Candida glabrata* binding to *Candida albicans* hyphae enables its development in oropharyngeal candidiasis. *PLoS Pathogens, 12*(3), e1005522. https://doi.org/10.1371/journal.ppat.1005522.

Thein, Z. M., Seneviratne, C. J., Samaranayake, Y. H., & Samaranayake, L. P. (2009). Community lifestyle of *Candida* in mixed biofilms: A mini review. *Mycoses, 52*(6), 467–475. https://doi.org/10.1111/j.1439-0507.2009.01719.x.

Theiss, S., Ishdorj, G., Brenot, A., Kretschmar, M., Lan, C. Y., Nichterlein, T., et al. (2006). Inactivation of the phospholipase B gene PLB5 in wild-type *Candida albicans* reduces cell-associated phospholipase A2 activity and attenuates virulence. *International Journal of Medical Microbiology, 296*(6), 405–420. https://doi.org/10.1016/j.ijmm.2006.03.003.

Todd, O. A., & Peters, B. M. (2019). *Candida albicans* and *Staphylococcus aureus* pathogenicity and polymicrobial interactions: Lessons beyond Koch's postulates. *Journal of Fungi, 5*(3), 81. https://doi.org/10.3390/jof5030081.

Tomaras, A. P., Dorsey, C. W., Edelmann, R. E., & Actis, L. A. (2003). Attachment to and biofilm formation on abiotic surfaces by *Acinetobacter baumannii*: Involvement of a novel chaperone-usher pili assembly system. *Microbiology (Reading, England), 149*(Pt 12), 3473–3484. https://doi.org/10.1099/mic.0.26541-0.

Tsui, C., Kong, E. F., & Jabra-Rizk, M. A. (2016). Pathogenesis of *Candida albicans* biofilm. *Pathogens and Disease, 74*(4), ftw018. https://doi.org/10.1093/femspd/ftw018.

Uppuluri, P., Chaturvedi, A. K., Srinivasan, A., Banerjee, M., Ramasubramaniam, A. K., Köhler, J. R., et al. (2010). Dispersion as an important step in the *Candida albicans* biofilm developmental cycle. *PLoS Pathogens, 6*(3), e1000828. https://doi.org/10.1371/journal.ppat.1000828.

Valenza, G., Tappe, D., Turnwald, D., Frosch, M., König, C., Hebestreit, H., et al. (2008). Prevalence and antimicrobial susceptibility of microorganisms isolated from sputa of patients with cystic fibrosis. *Journal of Cystic Fibrosis, 7*(2), 123–127. https://doi.org/10.1016/j.jcf.2007.06.006.

Van Dyck, K., Pinto, R. M., Pully, D., & Van Dijck, P. (2021). Microbial interkingdom biofilms and the quest for novel therapeutic strategies. *Microorganisms*, *9*(2), 412. https://doi.org/10.3390/microorganisms9020412.

Vanzolini, T., Bruschi, M., Rinaldi, A. C., Magnani, M., & Fraternale, A. (2022). Multitalented synthetic antimicrobial peptides and their antibacterial, antifungal and antiviral mechanisms. *International Journal of Molecular Sciences*, *23*(1), 545. https://doi.org/10.3390/ijms23010545.

Verstrepen, K. J., & Klis, F. M. (2006). Flocculation, adhesion and biofilm formation in yeasts. *Molecular Microbiology*, *60*(1), 5–15. https://doi.org/10.1111/j.1365-2958.2006.05072.x.

Vila, T., Kong, E. F., Montelongo-Jauregui, D., Van Dijck, P., Shetty, A. C., McCracken, C., et al. (2021). Therapeutic implications of *C. albicans-S. aureus* mixed biofilm in a murine subcutaneous catheter model of polymicrobial infection. *Virulence*, *12*(1), 835–851. https://doi.org/10.1080/21505594.2021.1894834.

Vílchez, R., Lemme, A., Ballhausen, B., Thiel, V., Schulz, S., Jansen, R., et al. (2010). *Streptococcus mutans* inhibits *Candida albicans* hyphal formation by the fatty acid signaling molecule trans-2-decenoic acid (SDSF). *Chembiochem: A European Journal of Chemical Biology*, *11*(11), 1552–1562. https://doi.org/10.1002/cbic.201000086.

Voltan, A. R., Alarcon, K. M., Fusco-Almeida, A., Soares, C. P., Mendes-Giannini, M. J., & Chorilli, M. (2017). Highlights in endocytosis of nanostructured systems. *Current Medicinal Chemistry*, *24*(18), 1909–1929. https://doi.org/10.2174/0929867324666170214111205.

Vuong, C., Voyich, J. M., Fischer, E. R., Braughton, K. R., Whitney, A. R., DeLeo, F. R., et al. (2004). Polysaccharide intercellular adhesin (PIA) protects *Staphylococcus epidermidis* against major components of the human innate immune system. *Cellular Microbiology*, *6*(3), 269–275. https://doi.org/10.1046/j.1462-5822.2004.00367.x.

Wade, W. G. (2013). The oral microbiome in health and disease. *Pharmacological Research*, *69*(1), 137–143. https://doi.org/10.1016/j.phrs.2012.11.006.

Walvekar, P., Gannimani, R., & Govender, T. (2019). Combination drug therapy via nanocarriers against infectious diseases. *European Journal of Pharmaceutical Sciences*, *127*, 121–141. https://doi.org/10.1016/J.EJPS.2018.10.017.

Wang, H., & Ren, D. (2017). Controlling *Streptococcus mutans* and *Staphylococcus aureus* biofilms with direct current and chlorhexidine. *AMB Express*, *7*(1), 204. https://doi.org/10.1186/s13568-017-0505-z.

Wang, D. Y., van der Mei, H. C., Ren, Y., Busscher, H. J., & Shi, L. (2020). Lipid-based antimicrobial delivery-systems for the treatment of bacterial infections. *Frontiers in Chemistry*, *7*, 872. https://doi.org/10.3389/FCHEM.2019.00872/bibTEX.

Wang, L., Zhai, B., & Lin, X. (2012). The link between morphotype transition and virulence in *Cryptococcus neoformans*. *PLoS Pathogens*, *8*(6), e1002765. https://doi.org/10.1371/journal.ppat.1002765.

Wang, X., Li, X., & Ling, J. (2017). Streptococcus gordonii LuxS/autoinducer-2 quorum-sensing system modulates the dual-species biofilm formation with Streptococcus mutans. *Journal of Basic Microbiology*, *57*(7), 605–616. https://doi.org/10.1002/jobm.201700010.

Wang, F., Li, L., Liu, B., Chen, Z., & Li, C. (2017). Hyaluronic acid decorated pluronic P85 solid lipid nanoparticles as a potential carrier to overcome multidrug resistance in cervical and breast cancer. *Biomedicine & Pharmacotherapy*, *86*, 595–604. https://doi.org/10.1016/J.BIOPHA.2016.12.041.

Warrier, A., Mazumder, N., Prabhu, S., Satyamoorthy, K., & Murali, T. S. (2021). Photodynamic therapy to control microbial biofilms. *Photodiagnosis and Photodynamic Therapy*, *33*. https://doi.org/10.1016/J.PDPDT.2020.102090.

Weiland-Bräuer, N., Malek, I., & Schmitz, R. A. (2019). Metagenomic quorum quenching enzymes affect biofilm formation of *Candida albicans* and *Staphylococcus epidermidis*. *PLoS One*, *14*(1), e0211366. https://doi.org/10.1371/journal.pone.0211366.

Weiss Nielsen, M., Sternberg, C., Molin, S., & Regenberg, B. (2011). *Pseudomonas aeruginosa* and *Saccharomyces cerevisiae* biofilm in flow cells. *Journal of Visualized Experiments: JoVE*, *47*, 2383. https://doi.org/10.3791/2383.

Williams, D., & Lewis, M. (2011). Pathogenesis and treatment of oral candidosis. *Journal of Oral Microbiology*, *3*. https://doi.org/10.3402/jom.v3i0.5771.

Wongsuk, T., Pumeesat, P., & Luplertlop, N. (2016). Fungal quorum sensing molecules: Role in fungal morphogenesis and pathogenicity. *Journal of Basic Microbiology*, *56*(5), 440–447. https://doi.org/10.1002/jobm.201500759.

Worthington, R. J., & Melander, C. (2013). Combination approaches to combat multidrug-resistant bacteria. *Trends in Biotechnology*, *31*(3), 177–184. https://doi.org/10.1016/J.TIBTECH.2012.12.006.

Wu, D., Fan, W., Kishen, A., Gutmann, J. L., & Fan, B. (2014). Evaluation of the antibacterial efficacy of silver nanoparticles against *Enterococcus faecalis* biofilm. *Journal of Endodontics*, *40*(2), 285–290. https://doi.org/10.1016/J.JOEN.2013.08.022.

Wu, S., Liu, J., Liu, C., Yang, A., & Qiao, J. (2020). Quorum sensing for population-level control of bacteria and potential therapeutic applications. *Cellular and Molecular Life Sciences: CMLS*, *77*(7), 1319–1343. https://doi.org/10.1007/S00018-019-03326-8.

Wuyts, J., Van Dijck, P., & Holtappels, M. (2018). Fungal persister cells: The basis for recalcitrant infections? *PLoS Pathogens*, *14*(10), e1007301. https://doi.org/10.1371/journal.ppat.1007301.

Yassin, S. A., German, M. J., Rolland, S. L., Rickard, A. H., & Jakubovics, N. S. (2016). Inhibition of multispecies biofilms by a fluoride-releasing dental prosthesis copolymer. *Journal of Dentistry*, *48*, 62–70. https://doi.org/10.1016/j.jdent.2016.03.001.

Zarnowski, R., Noll, A., Chevrette, M. G., Sanchez, H., Jones, R., Anhalt, H., et al. (2021). Coordination of fungal biofilm development by extracellular vesicle cargo. *Nature Communications*, *12*(1), 6235. https://doi.org/10.1038/s41467-021-26525-z.

Zarnowski, R., Westler, W. M., Lacmbouh, G. A., Marita, J. M., Bothe, J. R., Bernhardt, J., et al. (2014). Novel entries in a fungal biofilm matrix encyclopedia. *MBio*, *5*(4). https://doi.org/10.1128/mBio.01333-14. e01333-01314.

Zhang, Y., Hu, A., Andini, N., & Yang, S. (2019). A "culture" shift: Application of molecular techniques for diagnosing polymicrobial infections. *Biotechnology Advances*, *37*(3), 476–490. https://doi.org/10.1016/j.biotechadv.2019.02.013.

Zhang, K., Li, X., Yu, C., & Wang, Y. (2020). Promising therapeutic strategies against microbial biofilm challenges. *Frontiers in Cellular and Infection Microbiology*, *10*. https://doi.org/10.3389/fcimb.2020.00359.

Zhang, Y. Q., Cai, C., Yang, Y. X., Weng, L., & Wang, L. H. (2011). Blocking of Candida albicans biofilm formation by cis-2-dodecenoic acid and trans-2-dodecenoic acid. *Journal of Medical Microbiology*, *60*(Pt 11), 1643–1650. https://doi.org/10.1099/jmm.0.029058-0.

Zhang, W., Sileika, T. S., Chen, C., Liu, Y., Lee, J., & Packman, A. I. (2011). A novel planar flow cell for studies of biofilm heterogeneity and flow-biofilm interactions. *Biotechnology and Bioengineering*, *108*(11), 2571–2582. https://doi.org/10.1002/bit.23234.

Zhang, Q., Li, J., Nijjer, J., Lu, H., Kothari, M., Alert, R., Cohen, T., & Yan, J. (2021). Morphogenesis and cell ordering in confined bacterial biofilms. *Proceedings of the National Academy of Sciences*, *118*(31), Artigo e2107107118. https://doi.org/10.1073/pnas.2107107118.

Zhao, Y. J., Khoo, A. L., Tan, G., et al. (2015). Network meta-analysis and pharmacoeconomic evaluation of fluconazole, itraconazole, posaconazole, and voriconazole in invasive fungal infection prophylaxis. *Antimicrobial Agents and Chemotherapy*, *60*(1). https://doi.org/10.1128/AAC.01985-15. 376–86 26525782.

Zhu, Y., Dashper, S. G., Chen, Y. Y., Crawford, S., Slakeski, N., & Reynolds, E. C. (2013). *Porphyromonas gingivalis* and *Treponema denticola* synergistic polymicrobial biofilm development. *PLoS One*, *8*(8), e71727. https://doi.org/10.1371/journal.pone.0071727.

Ziental, D., Mlynarczyk, D., Czarczynska-Goslinska, B., Lewandowski, K., & Sobotta, L. (2021). Photosensitizers mediated photodynamic inactivation against fungi. *Nanomaterials (Basel)*, *11*(11).

Chapter 32

Molecular mechanism of biofilm formation of pathogenic microorganisms and their role in host pathogen interaction

Dhananjay Singh[a] **and Nishu Mittal**[b]
[a]*Department of Bioengineering and Biosciences, Integral University, Lucknow, Uttar Pradesh, India,* [b]*Institute of Biosciences and Technology, Shri Ramswaroop Memorial University, Barabanki, Uttar Pradesh, India*

1. Introduction

Biofilm, most dispersed and successful life form on earth, is assumed to be prominent microbial lifestyle in both natural and artificial atmosphere and found on all types of surfaces (Flemming et al., 2016; Lindsay & Von Holy, 2006; Płusa, 2019). Bacterial biofilm comes under their survival mechanism and is universal in nature. The hydrated biofilms create conductive surroundings for the microbial cells to stick to each other and onto different types of surfaces (Jun et al., 2010). Antoni van Leeuwenhoek in 1683 reported biofilm with the help of his primitive microscope from his own teeth. However, the biofilm lifestyle of microorganisms remains unnoticed to scientist until the early 1970s when Nils Høiby reported a connection in between etiology of continuous infection and bacterial aggregates in patients suffering from cystic fibrosis (CF) (Høiby, 2017). After that more focus was given to biofilm study that is involved in several clinical infections (Costerton, Stewart, & Greenberg, 1999; Hall-Stoodley & Stoodley, 2009), chiefly in dreadful infections (Bjarnsholt, 2013).

1.1 Microorganisms involved in biofilm formation

1.1.1 Gram-positive bacteria

Listeria monocytogenes, a foodborne pathogen, is found in soil, food, and water. Ingestion of *L. monocytogenes* can lead to fetus abortion in pregnant ladies and more complication to young children and elder persons. These bacteria grow and multiply at low temperature condition and can form either multispecies or pure culture biofilms (Chmielewski & Frank, 2003). *Staphylococcus aureus* and *S. epidermidis* bacteria are the most recurrent reason of hospital-borne infections on indwelling medical devices (Otto, 2008). The abundance of *S. epidermidis* in pure form on human skin shows its involvement in medical-device-related biofilm infections (Otto, 2009). *S. aureus* is nonmotile, facultative anaerobic, and nonspore-forming bacterium. It produces enterotoxin at wide range of temperatures from 10°C to 46°C. *S. aureus* bacteria can grow on the skin and mucous membrane of a person dealing with food and can be a major cause of infection in food industries (Giaouris et al., 2015). *Geobacillus stearothermophilus*, also known as *Bacillus stearothermophilus*, is a thermophilic, aerobic, or facultative anaerobic bacterium (Wu et al., 2019). This forms biofilm after the attachment to surfaces of stainless steel, processing lines in evaporators, and plate heat exchangers (Flint, Palmer, Bloemen, Brooks, & Crawford, 2001). *Anoxybacillus flavithermus* is a thermophilic, facultative anaerobic, and spore-forming nonpathogenic bacterium (Strejc et al., 2020). The spores of *A. flavithermus* are heat-resistant, and its vegetative cells can grow easily on stainless steel in the skimmed milk presence even at high temperatures (up to 65°C) (Sadiq et al., 2017).

1.1.2 Gram-negative bacteria

The Gram-negative bacteria are generally more involved in biofilm-associated infections. Among the Gram-negative bacteria, *Pseudomonas aeruginosa* causes serious long-term infection in patients suffering from CF (Høiby, Ciofu, & Bjarnsholt, 2010). *B. cereus* is an anaerobic or facultative anaerobic spore-forming bacterium that can multiply in wide range of temperatures from 4°C to 50°C. *B. cereus* is resistant to various chemicals, heat, and radiation (Bottone, 2010). It forms biofilm on the stainless-steel pipes, storage tanks, and conveyor belt that comes in direct contact of food material. *Campylobacter* spp. are anaerobic, spiral, rod-shaped, or curved flagella-bearing thermophilic and bipolar bacteria. *C. jejuni* forms biofilm in both microaerophilic (5% O_2 and 10% CO_2) and aerobic (20% O_2) environment (Taff et al., 2012). *Escherichia coli*, a rod-

Understanding Microbial Biofilms. https://doi.org/10.1016/B978-0-323-99977-9.00031-4

shaped bacterium, forms an important part of human intestinal flora and generally does not cause any threat to health. Among the several virulence strains of *E. coli*, O157:H7 EHEC is commonly linked with the enterohemorrhagic *E. coli* (EHEC) infections in human beings (Gould et al., 2013). To initiate attachment to the surface, *E. coli* uses flagella, pili, and membrane proteins. After the attachment over the surface, flagella are lost and bacteria start production of an extracellular polymeric substance (EPS) to protect itself from disinfectants (Lim, Koo, Kim, & Kim, 2019).

Salmonella enterica, a facultative aerobic, rod-shaped, and flagellated bacterium, causes gastroenteritis or septicemia (Giannella et al., 1996; Lamas et al., 2018). *Salmonella* spp. for the cell-to-cell contact and attachment to surfaces produces proteinaceous extracellular fibers (curli) (Ćwiek, Bugla-Płoskońska, & Wieliczko, 2019). *Pectinatus*, an anaerobic, nonspore-forming bacteria, is associated with formation of biofilms in breweries industries due to hygienic issues (Paradh, Mitchell, & Hill, 2011). *P. cerevisiiphilus* species is isolated from many breweries in many countries, namely Germany, Spain, Norway, Japan, the Netherlands, Sweden, and France (Paradh et al., 2011).

1.1.3 Fungal species in biofilm formation

Like bacterial species, fungal species also forms biofilms that are a collection of masses protected in an extracellular matrix (ECM) (Donlan, 2001; Hoyle, Jass, & Costerton, 1990). Several clinically related fungal species form biofilms, namely *Candida* spp. (Kojic & Darouiche, 2004), *Aspergillus* spp. (Beauvais et al., 2007), *Cryptococcus neoformans*, *Fusarium* spp. (Imamura et al., 2008), *Blastoschizomyces capitatus* (D'Antonio et al., 2004), *Malassezia pachydermatis* (Cannizzo et al., 2007), *Pneumocystis* spp. (Cushion, Collins, & Linke, 2009), *Trichosporon asahii* (Di Bonaventura et al., 2006), *Rhizopus* spp., and *Rhizomucor* spp. Fungus can grow and proliferate on nosocomial artificial medical devices including catheters, in addition to that on endothelial and epithelial surfaces (Escande et al., 2011; Ganguly & Mitchell, 2011; Harriott, Lilly, Rodriguez, Fidel, & Noverr, 2010; Seidler, Salvenmoser, & Müller, 2008). The *Aspergillus* spp. are saprophytic fungi. As they secrete several enzymes, proteins and their products cause severe infections in humans (González-Ramírez, Ramírez-Granillo, Medina-Canales, Rodríguez-Tovar, & Martínez-Rivera, 2016; Kaur & Singh, 2014; Müller, Seidler, & Beauvais, 2011). Aspergillosis infection is considered as the second serious cause of hospital-borne infection after *C. albicans* with high mortality rate (Kaur & Singh, 2014; Ramage, Rajendran, Gutierrez-Correa, Jones, & Williams, 2011). Among the fungal aggregates, Aspergilloma is a characteristic feature of biofilm, which is formed by *Aspergillus* spp. (Ramage et al., 2011). Like *C. albicans*, the biofilm synthesized by *Aspergillus* spp. can easily grow on nonliving surfaces (Ramage et al., 2011; Seidler et al., 2008). *Paracoccidioides brasiliensis*, a dimorphic fungus, can exist in both mold and yeast form and is responsible for paracoccidioidomycosis infection. In yeast, *P. brasiliensis* fungi synthesize biofilm with increase in level of expression of genes coding for adhesion and enzymes, namely GP43, enolase, GAPDH, and aspartyl proteinase while decreasing the action of phospholipase enzyme (Sardi et al., 2015). The first report of biofilm formation in *Histoplasma capsulatum* was given by Pitangui et al. (2012). In vitro formation of biofilm by *Trichophyton rubrum* and *T. mentagrophytes* was reported by Costa-Orlandi, Sardi, Santos, Fusco-Almeida, and Mendes-Giannini (2014). Paramonova, Krom, van der Mei, Busscher, and Sharma (2009) report the increase in compressive force of biofilm with the increase in filamentation. This makes biofilm tolerable to different unfavorable conditions such as sonication and vortex. In *C. albicans*, various genes play an important role in production of ECM, adhesion, development of biofilm, and quorum sensing (QS) (Blankenship & Mitchell, 2006; Bonhomme & d'Enfert, 2013; Finkel & Mitchell, 2011). *Candida* spp. are generally known as the causal organism of candidemia, hospital pneumonia and urinary tract infections (UTIs). These infections are commonly linked with the common use of medical devices having biofilm on their surface (McCarty & Pappas, 2015). The central venous catheter (CVC) used for fluid and medicine administration is generally colonized by *Candida* spp. (Seneviratne, Jin, & Samaranayake, 2008). Sometimes infusion fluid or catheter is contaminated, but generally yeast infection enters either through patients' skin or from infected hands of health professional (Seneviratne et al., 2008). The colonization of gastrointestinal track by *Candida* spp. can illicit its pathogenic behavior. Now it can easily penetrate intestinal mucosa and starts circulation inside blood stream where it can colonize the catheter endogenously. Patients undergoing chemotherapy as a part of treatment can easily develop this type of infection as chemotherapy causes damage to intestinal mucosa (Douglas, 2003). The in vivo and in vitro studies show the colonization and biofilm formation by *Candida* spp. including both *C. albicans* and *Candida* non-*albicans* (CAN) species causing candidemia (Lattif et al., 2010; Nett, Marchillo, Spiegel, & Andes, 2010; Ramage, Wickes, & López-Ribot, 2008; Silva et al., 2009).

Fusarium spp. are soil saprophyte, pathogenic to both plant and animal world. Among the several species of *Fusarium*, *F. solani*, *F. oxysporum*, *F. verticillioides*, and *F. moniliforme* are highly infectious to human beings (Nucci & Anaissie, 2007). Like *Aspergillus*, the impact of pathogenicity of *Fusarium* depends upon the immunity level of host organism. The *Fusarium* biofilm development on contact lenses, polystyrene surface, and fingernails of human has been explained both in vivo and in vitro (Sun et al., 2010; Vila, Quintanilha, & Rozental, 2015; Vila, Rozental, & Guimarães, 2015).

2. Biofilm-associated diseases

2.1 Otitis media

Otitis media (OM) infection is related to swelling of the middle ear cavity that is major reason behind the infection in preschool-going children (DeAntonio, Yarzabal, Cruz, Schmidt, & Kleijnen, 2016). OM can be further categorized into different classes, namely acute OM (AOM) and chronic supportive OM (CSOM) and OM with effusion (OME) (Schilder et al., 2016). This infection can cause impairment to hearing ability of ear (Monasta et al., 2012). Several bacteria such as *Streptococcus pneumoniae* and nontypeable *Haemophilus influenzae* (NT-Hi) colonize the nasopharynx area and can enhance the chances of frequent infection of AOM and OME in upcoming life (Chonmaitree et al., 2016; Faden et al., 1997). Current studies indicate the formation of biofilm in bacterial mass present in the mucosa of middle ear (Daniel, Imtiaz-Umer, Fergie, Birchall, & Bayston, 2012; Van Hoecke et al., 2016).

2.2 Cholesteatoma

In this type of infection, keratinizing squamous epithelium entrapped inside middle ear where presence of biofilm is already demonstrated (Galli et al., 2016; Kaya et al., 2013; Saunders, Murray, & Alleman, 2011). As biofilms are already indicated in the middle ear and mucosa, it can correlate to the presence of biofilm and causes supportive environment for cholesteatoma pathogenesis (Galli et al., 2016).

2.3 Infective endocarditis (IE)

The inflammation of endocardium is known as endocarditis. The biofilm in endocarditis consists of elements of both bacterial and host part present on the cardiac valve. The biofilm can lead to heart disease as given. The biofilm formation interferes with proper functioning of valve where closure of valve leads to leakage while opening of valve can cause turbulence and decrease in blood flow. Biofilm can also act as source of constant infection of the blood flow. Mixing of small part of biofilm with circulatory system can increase the vulnerability of the brain and kidney toward blood clotting. The *Staphylococci*, *Streptococci*, and *Enterococci* are generally isolated from the infective endocarditis and form a major part of infection (Elgharably, Hussain, Shrestha, Blackstone, & Pettersson, 2016).

2.4 Atherosclerosis

The fatty deposition and accumulation of calcium in the form of plaques in the arterial wall are known as atherosclerosis. It can cause decrease in lumen and elasticity of arteries and reduced flow of blood, which can lead to cardiovascular disease. Apart from this, breakage of a plaque can be deadly (Falk, Shah, & Fuster, 1995). In addition to this, the role of oral bacteria has been studied in the formation of atherosclerosis plaques. A group of researchers identified the presence of bacterial biofilm in atherosclerosis arteries with the help of fluorescence in situ hybridization (FISH) technique (Lanter, Sauer, & Davies, 2014; Snow et al., 2016).

2.5 Sialolithiasis

The formation of calcified masses inside salivary gland is known as sialolithiasis, which can be the reason behind the pain and swelling. Teymoortash, Wollstein, Lippert, Peldszus, and Werner (2002) showed the role of bacteria in the formation of sialolithiasis. (Kao, Chole, & Ogden, 2020) observed the presence of biofilm along with the immune cells, platelets, RBCs, WBCs, and calcium nanoparticles by using scanning electron microscopy (SEM).

2.6 Typhoid fever

Salmonella enterica serovar Typhi, a foodborne pathogen, causes systemic infection and typhoid fever in human beings. Symptoms of typhoid fever include high body temperature, sickness, headache, stomach ache, and constipation. If it is left untreated, severe health issues can occur such as intestinal bleeding, interruption of bowel movement, septicemia, meningitis, and fatality (Cohen, Bartlett, & Corey, 1987; Crum, 2003). *S. typhi* also causes chronic infections, which include hepatitis, cholecystitis, cholangitis, chronic diarrhea, and pancreatitis as well as hepatobiliary carcinoma (Crum, 2003; Vaishnavi et al., 2005). Acute infection can be treated by antibiotic treatment while chronic infection caused by colonization of *S. typhi* in gall bladder remains untreated (Dinbar, Altmann, & Tulcinsky, 1969; Lai, Chan, Cheng, Sung, & Leung, 1992). Colonization by *S. typhi* of gall bladder includes the adhesion and invasion of epithelial cell and further formation of biofilm on gall stones (Gonzalez-Escobedo, Marshall, & Gunn, 2011).

2.7 Inflammatory bowel disease and colorectal cancer

Inflammatory bowel disease (IBD) and colorectal cancer (CRC) (ulcerative colitis and Crohn's disease) are disorders of digestive system. The inflammation of digestive tract can lead to stomach pain, diarrhea, fatigue, and weight loss. In addition to that, a person suffering from IBD can develop CRC in future (Kim & Chang, 2014). The adhesion of biofilm to the epithelium tissue is connected to IBD and leads to activation of inflammatory response.

2.8 Chronic rhinosinusitis (CRS)

The inflammation of nose and paranasal sinuses is specified by blockage of nose, congestion, or discharge of mucus. Furthermore, symptoms are loss of smell sense, facial pain, and pressure. Most commonly viruses account for acute rhinosinusitis. The immune response of host to a viral infection leads to generation of dead epithelial cells, which can create favorable conditions for secondary bacterial infection and formation of biofilm. The presence of biofilm in persons suffering from CRS has been reported by using SEM and confocal laser microscopy (Foreman, Boase, Psaltis, & Wormald, 2012; Harvey & Lund, 2007; Ramakrishnan, Shields, Elbadawey, & Wilson, 2015). Apart from this, to some extent biofilm is also reported in healthy persons (Danielsen et al., 2014).

3. Mechanism of biofilm formation

In the microscopic globe, biofilms are symbolized as city of microbes (Fig. 1) (Kolter & Watnick, 2000; Nikolaev & Plakunov, 2007) where extracellular polysaccharides (EPSs) consist of 85% part of the total biofilm biomass, as "habitation of the biofilm cells" (Flemming, Neu, & Wozniak, 2007). The composition of EPS is biomolecules, extracellular DNA (eDNA), exopolysaccharides, and polypeptides, which synthesize a hugely hydrated polar blended mixture that leads to construction of structural organization of the biofilm (Branda, Vik, Friedman, & Kolter, 2005; Flemming, 2010; Sutherland, 2001a, 2001b).

3.1 In *Pseudomonas aeruginosa*

In *Pseudomonas aeruginosa* PAO1 life cycle of biofilm formation can be further classified into different five major steps. Step 1 begins with the reversible attachment of planktonic bacterium onto the appropriate surface of further growth and division. Now bacterium attaches irreversibly in the EPS matrix to form microcolonies (step 2). Gradually there will be expansion of bacterial microcolonies that can lead to better structured physical composition with noncolonized room (step 3). In next stage, noncolonized rooms are occupied with bacteria that will ultimately cover the whole surface (step 4). At the end, bacteria again enter planktonic stage where bacteria move freely from sessile stage, so that they spread and colonize nearby surfaces (step 5) (Davey, Caiazza, & O'Toole, 2003). Total three types of polysaccharides, such as alginate, Pel, and PsI, are produced by *P. aeruginosa*, which act as major determining factors for the firmness of the biofilm structure (Ghafoor, Hay, & Rehm, 2011; Ryder, Byrd, & Wozniak, 2007). A linear unbranched polymer alginate consists of D-mannuronic acid and L-guluronic acid that provide structural firmness and shielding to biofilm (Ertesvåg & Valla, 1998). Along with the stability of biofilm, alginate also helps in withholding of moisture and nutrients (Sutherland, 2001a, 2001b). Pel polysaccharide consists of glucose-rich matrix material (Franklin, Nivens, Weadge, & Howell, 2011; Friedman & Kolter, 2004) while PsI is composed of Penta-saccharides, namely D-mannose, L-rhamnose, and D-glucose (Byrd et al., 2009). Both Pel and PsI polysaccharides can act as a basic structural scaffold for the development of biofilm and are involved at early stages of biofilm formation (Colvin et al., 2011; Ma et al., 2012).

3.2 In *Candida albicans*

Development of biofilm in *Candida albicans* can be explained in four subsequent stages as described below:

Stage I: Attachment—In this commencing phase, yeast in suspension form and planktonic cells attach to surface in approximately 1–3 h.
Stage II: Intermediate phase—This phase lasts up to 11–14 h where development of biofilm occurs.
Stage III: Maturation phase—Here all the layers of cells attached to surface are submerged thoroughly in the polymetric matrix and take 20–48 h.
Stage IV: Dispersion —After 24 h of development, superficial cells start leaving the biofilm and colonize the surrounding area (Chandra et al., 2001).

In *C. albicans*, biofilm development rate is slower as compared with *Aspergillus*. In *C. albicans* lag phase, a phase between conidial attachment and development of first monolayer takes nearly for 10 h. After this stage, intense growth of hyphae takes place and secretion of ECM leads to enhanced structural entanglement within 48 h. In maturation phase after 72 h, biofilm becomes mature and dense (Mowat, Butcher, Lang, Williams, & Ramage, 2007; Seidler et al., 2008). In *C. albicans*, major adhesion families playing important role in regulating attachment during biofilm development are hyphal wall protein (Hwp) family, agglutinin-like sequence (Als) family, individual protein file family, and F/hyphally regulated (Iff/Hyr) family (de Groot, Bader, de Boer, Weig, & Chauhan, 2013). Among these key families, Als adhesins are studied extensively consisting of total eight members such as Als1p, Als2p, Als3p, Als4p, Als5p, Als6p, Als7p, and Als9p that are glycosylphosphatidylinositol (GPI)-linked to β-1,6 glucans in the cell wall of fungus (Hoyer & Cota, 2016).

In the adhesion of yeast and hyphae, the member of adhesins plays important role during biofilm development.

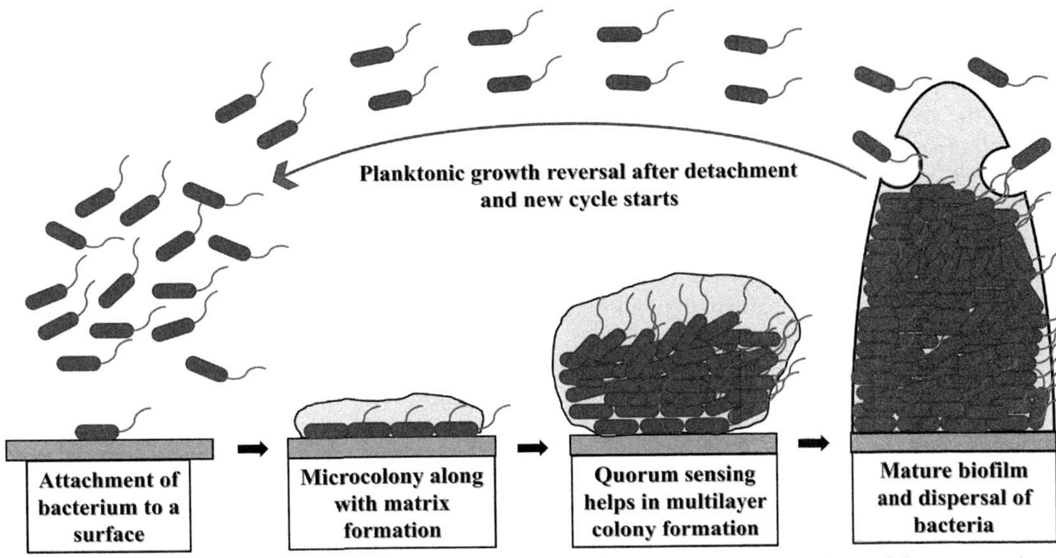

FIG. 1 The biofilm formation starts with the attachment of bacterium over a surface and after successful attachment, it forms a monolayer of bacteria. The extracellular matrix production also started with this monolayer and called microcolony. Soon, the bacteria in microcolony started producing quorum sensing signaling molecules, and it becomes multilayered colony. Post this process, a mature biofilm formed, and the bacterium started dispersing from this colony for initiating few more biofilm formation cycles.

At initial stage, Als1p and Als5p members of adhesin family help in attachment by using their amyloid synthesizing region, which is important in yeast for cell-cell accumulation and cell-substrate attachment (Garcia et al., 2011; Ramsook et al., 2010; Rauceo et al., 2004). Apart from Als1p and Als5p, Als1p and Als3p also have important roles in initiation of hyphal adhesion during development of biofilm (Nobile et al., 2008). The aggregates of yeast cell facilitate the founding of mixed species biofilm formation along with more fungal and bacterial species (Alsteens, Garcia, Lipke, & Dufrêne, 2010; Klotz et al., 2007; Lipke et al., 2012, 2014; Ramsook et al., 2010). Apart from the Als adhesin family, more adhesin families are crucial for the attachment of *C. albicans* yeast cell at the time of biofilm development. The presence of polystyrene protein Eap1p is important for regulating the attachment of *C. albicans* on the plastic surface of medical devices (Chong et al., 2018; McCall, Pathirana, Prabhakar, Cullen, & Edgerton, 2019). In vitro study performed on epithelial cell shows that deletion of EAP1 inhibits the attachment of *C. albicans* to plastic surfaces (Li, Svarovsky, et al., 2007).

3.3 Biofilm development

The development of biofilm is proceeded along with the change in cell number, structure, release of extracellular polymeric substance, and formation of extracellular matrix. After adherence, yeast cell grows and divides to develop microcolonies making the elementary layer of biofilm. The maturation of biofilm leads to enhancement in biomass of yeast cells, pseudo hyphae, hyphae, and ECM that contribute to major part of biofilm. The basic component of hyphae is the basic component of biofilm, which braces the structural integrity and provides a support system for adherence of pseudohyphae, other yeast and bacterial cells for the development of multispecies biofilm (Chandra et al., 2001; Harriott & Noverr, 2009). The ECM consists of a mixture of large molecules, having protein as major part (55%), carbohydrates (25%), nucleic acid (5%), and lipids (approximately 15%) (Zarnowski et al., 2014). Mannans are the major general polysaccharides present in ECM of biofilms of *C. albicans* (Pierce et al., 2017). The nucleic acid helps in maintenance of structural and protective ability of ECM in biofilm. *C. albicans* releases extracellular deoxyribonucleic acid (eDNA) within ECM, which is required for the stability of matured biofilm but not for the establishment (Al-Fattani & Douglas, 2006; Martins et al., 2010; Rajendran et al., 2014). The discharge of eDNA provides heterogeneity at the time of biofilm formation (Rajendran et al., 2014). Furthermore, presence of eDNA also regulates the response of immune system of host cell by the activation of myeloid dendritic cells (Kasai et al., 2006; Martins et al., 2010; Miyazato et al., 2009).

3.3.1 Biofilm dispersal

The dispersal of biofilm is the final step in biofilm formation where cells are discharged from a matured biofilm and move freely to colonize a surrounding surface area. The upper layer of biofilm contributes to dispersal of biofilm cells (Uppuluri et al., 2010). In *C. albicans*, several regulators such as Nrg1p, Pes1p, and Ume6p help in dispersion of biofilm where Nrg1p is a negative regulator to the switching of yeast to hypha (Braun, Kadosh, & Johnson,

2001; Murad et al., 2001). Uppuluri et al. showed in their study that overexpression of NRG1 gene enhances the rate of dispersal from biofilm (Uppuluri et al., 2010). In comparison to NRG1, the overexpression of Ume6 decreases the dispersal of cells from biofilm (Uppuluri et al., 2010). The morphology of dispersal cell is different from planktonic cell, including an increased ability to form biofilm, attachment to abiotic and biotic substrate, and reduction in responsiveness to antifungals. Dispersed cells show increased rate of pathogenicity in comparison to planktonic cells (Uppuluri et al., 2010).

3.4 Role of transcription factor in biofilm formation

The molecular mechanism of biofilm formation involves six transcription factors such as EFG1, TEC1, BCR1, NDT80, ROB1, and BRG1 that mediate the expression of nearly 1000 target genes. Among these six transcription factors, ROB1, BRG1, and NDT80 are newly identified while BCR1, TEC1, and EFG1 are formerly known to involve in development of biofilm. Banerjee et al. (2013) showed that the UME6 is involved in the regulation of hyphae in biofilm of *Candida albicans*. Taff et al. (2012) explained the role of genes coding BGL2, PHR2, and XOG1 enzymes in the synthesis of extracellular polysaccharides. Another transcription factor TEC1 mediates the expression of BCR1, which regulates the expression of genes involved in differentiation of hyphae (Nobile et al., 2006; Schweizer, Rupp, Taylor, Röllinghoff, & Schröppel, 2000). Another study shows that TEC1 mutant strain is inadequate to perform hyphal morphogenesis that ultimately led to development of defective biofilm. These enzymes are critical for the release of β-1,3-glucan for the biosynthesis of ECM. Although matrix biosynthesis is closely associated with biofilm development, still very less is known about its regulation mechanism. In *C. albicans*, ZAP1 transcription factor, which binds to zinc-responsive elements, regulates negatively β-1,3-glucan in biofilm formation (Nobile et al., 2009). A study shows that two glucoamylases, namely GCA1 and GCA2, have positive impact in matrix biosynthesis and enhance its functioning along with hydrolysis of insoluble β-1,3 glucan chains. Furthermore, enzyme alcohol dehydrogenases (ADH5, CSH1, and IFD6) also affect the matrix biosynthesis, where ADH5 positively regulates matrix biosynthesis while CSH1 and IFD6 negatively. Alcohol dehydrogenases also produce acyl and aryl alcohols involved in QS, which controls biofilm maturation at different levels (Nobile et al., 2009).

3.5 Quorum sensing mechanism in biofilm formation

QS, a communication mode, is explored by several bacteria to sense the size of their population. In this cell-to-cell communication method, bacteria release and sense signal molecules that are involved in the production of virulence factor, movement, and biofilm formation (Nadal Jimenez et al., 2012; Parsek & Greenberg, 2005). The QS regulatory connection network differs among different bacterial species. Therefore, QS can be classified into several categories based on the types of autoinducers (Ais) signaling molecules. The different types of QS are AHL system, AIP system (these two systems known as AI-1 systems, previously), AI-2 system, and AI-3 system. Formerly both AHL and AIP system were known as AI-1 system. Whereas in Gram-positive bacteria, AIP QS system exists, and autoinducing peptides (AIPs) are used as signaling molecules (Irrie & Parsek, 2008). Presently, there are many studies on these two systems.

3.5.1 Quorum sensing in Gram-negative bacteria

In Gram-negative bacteria, generally AHL system is reported where applied signaling molecules are *N*-acyl homoserine lactones (AHLs) (Slock, VanRiet, Kolibachuk, & Greenberg, 1990; Stevens, Dolan, & Greenberg, 1994). Nealson and Hastings in 1979 reported AHL system in *Vibrio fischeri* (Nealson & Hastings, 1979). In AHL system of Gram-negative bacteria, the signal molecule (*N*-(3-oxohexanoyl)-L-homoserine lactone) (OHHL) is synthesized by AI synthase LuxI gene. Then OHHL molecule diffuses out of the cell and binds to LuxR. The binding of signal to LuxR receptor occurs only when the concentration of OHHL crosses a threshold value. LuxR also acts as DNA-binding transcriptional activator and is involved in activation of expression of genes involved in biofilm synthesis (Engebrecht, Nealson, & Silverman, 1983; Engebrecht & Silverman, 1984). Presently, it is well-known regulatory process and a typical model for the regulation of biofilm formation by AHL systems in most Gram-negative bacteria.

In *P. aeruginosa*, there are two chief QS systems, namely *las* and *rhl*, which are involved in the production of LasI and RhlI synthases and the perception of *N*-(3-oxododecanoyl)-L-homoserine lactone (3dDHL) and *N*-butanoyl-L-homoserine lactone (BHL), respectively. The 3dDHL and BHL are autoinducer signaling molecules (Nadal Jimenez et al., 2012). The sequence of LasI and RhlI shows significant similarity with *luxI* gene of *V. fischeri* (Lee & Zhang, 2015). After reaching the threshold concentration, 3dDHL and BHL signaling molecules bind to their receptors LasR and RhlR, respectively, and activate the expression of genes involved in virulence factors and biofilm formation (Wade et al., 2005). The *rhl* QS system regulates the swarming motility, which is actively involved in the development of biofilm at initial stage (Khan, Pham, Oloketuyi, & Kim, 2020). Apart from this, *rhl* system biosynthesized rhamnolipid and pyocyanin virulence factors (Daniels, Vanderleyden, & Michiels, 2004; Dusane et al., 2010; Winzer et al., 2000). The elastase, endotoxin A, alkaline protease, and biofilm formation genes are controlled by *las* system (Wilder, Diggle, & Schuster, 2011). In addition to these two systems, *pqs* and *iqs* systems also exist in *P. aeruginosa*. The signaling molecules of *pqs* and *iqs* are PQS (2-heptyl-3-hydroxy-4-quinolone) and IQS (2-(2-hydroxyphenyl)-thiazole-4-carbaldehyde), respectively (Lee & Zhang, 2015). The *pqs* system is involved in the biosynthesis of extracellular DNA of bacterial cell, which plays an important role in biofilm formation (Allesen-Holm et al., 2006).

In *E. coli*, AHL system regulates the formation of biofilm (Walters & Sperandio, 2006). However, in *E. coli*, *sdiA* receptor gene is homologous to *luxR* and is present while gene encoding for AHL synthase, homologous to *luxI* is absent (Walters & Sperandio, 2006).

Apart from this study, the AI-2 system is part of biofilm formation in many more Gram-negative bacteria, such as *Helicobacter pylori*, *E. coli*, *V. parahemolyticus*, *P. aeruginosa*, and *V. cholerae* (Anderson et al., 2015; Guo et al., 2018; Hammer & Bassler, 2003; Li, Attila, et al., 2007; Li et al., 2015).

3.5.2 Quorum sensing in Gram-positive bacteria

In Gram-positive bacteria, AIPs are involved in the regulation of biofilm formation. A small-sized oligopeptide is synthesized inside a bacterial cell and further processed into mature form before its diffusion to outside of the cell (Sturme et al., 2002). The binding of AIP to the extracellular part of histidine kinase, a transmembrane receptor confined to the cell membrane. The binding of AIP to its receptor takes place only when the concentration of signaling molecules crosses its threshold value. The binding of AIP to its receptor leads to activation of kinase, which phosphorylates response regulatory factors and leads to regulation of genes involved in the formation of biofilm (Sturme et al., 2002).

In *S. aureus*, there are total five genes, namely *agrA*, *agrB*, *agrC*, *agrD*, and *hld*, in the *agr* operon comprising the agr AIP system (Painter, Krishna, Wigneshweraraj, & Edwards, 2014). *S. aureus* produces AgrD precursor oligopeptide, which is further processed into mature AgrD peptide by AgrB, a transmembrane protein. AgrB also plays a role in transportation of mature form AgrD outside the bacterial cell. Once the concentration of AgrD reaches the threshold value, it binds to AgrC, a histidine kinase, and activates kinase protein. Now the AgrC phosphorylates the AgrA, a response regulator, which binds to intergenic DNA found in between P2 and P3 promoters and finally activates transcription process. The *agr* system of QS plays an important role in dispersion phase (last stage) of biofilm formation (Yarwood, Bartels, Volper, & Greenberg, 2004).

3.5.3 Quorum sensing in fungal biofilm

QS is a common mode of communication among bacterial cells where it regulates secretion of virulence factors, antibiotic production, and biofilm formation. Current study now shows the presence of QS mechanism in fungal species also (Fox & Nobile, 2012; Miller & Bassler, 2001). The synchronized expression of genes coding virulence factors at the time of infection of a host contributes to survival rate of pathogen by increasing the chances of organizing infection and evading the immune response of host (Williams, 2007). Like bacteria, QS system in fungi also includes various signaling molecules, namely sphingolipids, farnesol, and oxylipins (Singh & Del Poeta, 2011). In *Saccharomyces cerevisiae*, aromatic alcohols phenylethanol and tryptophol act as signaling molecules in QS, whereas farnesol and tyrosol in *C. albicans*. In the initial stage of QS, autoinducers are secreted from fungi to external environment. Once the concentration of autoinducer reaches the threshold value, it binds with its receptor (Hense et al., 2007). Shirtliff and coworkers reported that 40 mM or 100 mM concentration of farnesol activates the regulation of genes intricated in defense mechanism against oxidative stress (Shirtliff et al., 2009). Apart from this, farnesol also regulates the action of drugs on planktonic cells of *C. albicans* (Sharma & Prasad, 2011). Ramage, Saville, Wickes, and López-Ribot (2002) studied the involvement of farnesol molecule in the development of biofilm and showed the inhibition of hyphae formation when farnesol is added to the stage 1 of biofilm formation. The microarray investigation of farnesol-treated biofilms shows its role in regulation of genes coding for maintenance of cell wall integrity (CHT2 and CHT3), hyphal development (TUP1 and CRK1), drug resistance (FCR1 and PDR16), and cell wall integrity (Cao et al., 2005). Another study shows the involvement of farnesol molecule in the suppression of genes coding for hypha while enhancing the level of expression of drug-resistance genes (Enjalbert & Whiteway, 2005). The CHK1 histidine kinase, an important member of a two-component signal transduction pathway, is also involved in farnesol signaling pathway. Genetic study on a mutant strain shows that *chk1* will synthesize biofilm in the presence of farnesol, recommending that CHK1 is a major regulator of the cellular response to QS (Kruppa et al., 2004). In *C. albicans*, nutrient deficiency at the time of biofilm maturation can lead to dispersion of loosely attached daughter cell to biofilm matrix (Blankenship & Mitchell, 2006). There are chances that yeast cells may contain an additional mechanism that helps them to modulate the functioning of adhesin proteins. Granger, Flenniken, Davis, Mitchell, and Cutler (2005) reported that YWP1, a yeast cell wall protein that can function as antiadhesin and deletion of YWP1, can increase the adherence property of cell to different type of surfaces.

4. Implications of biofilms during host-pathogen interaction

Microorganisms causing disease in their specific host are called pathogens. Whenever a pathogenic microorganism (bacterium, virus, fungus, and protozoal parasite) infects the human host, a tussle starts between the host's immune systems and the pathogen's virulence factors. But some pathogens can have diverse immune evasion tactics in the host, which is responsible for host cell invasion and colonization and access to target tissues (Table 1).

4.1 Bacterial

Biofilms have bacterial subpopulations, which are metabolically less active. Bacteria develop an intrinsic mechanism, i.e., production of enzymes that cause the inactivation of antibiotics or modify the cell permeability through efflux pumps. It may increase the survival of bacteria in the host (Parrino et al., 2019).

4.1.1 Pseudomonas aeruginosa

P. aeruginosa forms biofilms in human chronic wounds located in the subcutaneous fatty tissue, which prevents their detection through standard wound swabbing techniques (Fazli et al., 2009; Kirketerp-Møller et al., 2008). *P. aeruginosa* establishes chronic infection by the following mechanisms. One is that by which it secretes elastase and digests human thrombin, which prevents the toll-like receptor dimerization and ultimately kept away from host immune response (Van Der Plas et al., 2016). Other way to impair the host immune response is by rhamnolipid production, which impedes the calcium-regulated signaling pathways and activation of protein kinase C (PKC) (Dössel, Meyer-Hoffert, Schröder, & Gerstel, 2012). Complement system inhibition by biofilm matrix exopolysaccharides (Mishra et al., 2012) and bacterial lipopolysaccharides is another way of eliminating host immune response (Schiller & Joiner, 1986).

4.1.2 Enterococcus faecalis

Biofilm biosynthesis is the main mechanism used by *Enterococci* to eliminate the immune system. The reduced concentration of cytokine and chemokine induction during *E. faecalis* wound infection is defined that this bacterium could modulate the immune response in host and could promote its persistence in wounded tissues (Chong et al., 2017). Earlier studies stated that *E. faecalis* biofilms can lower the pro-inflammatory responses, which promotes their survival in phagocytes in comparison to planktonic cells in vitro (Baldassarri et al., 2001; Daw, Baghdayan, Awasthi, & Shankar, 2012; Mathew, Yaw-Chyn, & Kishen, 2010). *Enterococci* are also able to evade the immune system by showing resistance against the antimicrobial activity of lysozymes. Lysozyme is the main component of host's innate immune response and is found abundantly in the saliva, sweat, tears, blood, other body secretions, phagocytes, and at mucosal surfaces (Brouwer, van Leeuwen-Herberts, & Otting-van de Ruit, 1984; Faurschou & Borregaard, 2003; Ragland & Criss, 2017). *E. faecalis* has the mechanisms to counteract lysozymes; it can modify the peptidoglycan structure through *N*-acetylglucosamine (GlcNAc) deacetylation, *N*-acetylmuramic acid (MurNAc) *O*-acetylation, or reduce the net negative charge on cell via addition of positively charged D-alanine amino acid residues to wall teichoic acids and lipoteichoic acids (Benachour et al., 2012; Fabretti et al., 2006; Hébert et al., 2007; Le Jeune et al., 2010).

4.1.3 Escherichia coli

The complement system in the serum of the host rapidly kills *E. coli* or through the opsonization of bacterium by phagocytic cells. But the bacterium has developed the mechanism to alter the immune attacks over it and cause more invasive infections. Within biofilms, bacteria get the protection from adverse antimicrobial conditions, along with the endogenous host defense mechanisms (Fux, Costerton, Stewart, & Stoodley, 2005). The biofilm-like intracellular aggregates formation in epithelial cells of the urinary bladder plays a major role at the time of acute cystitis stage and during the recurrence of infections (Justice et al., 2004; Mysorekar & Hultgren, 2006). It is not necessary that in vitro biofilm formation reflects the same condition as in vivo situation. Exopolysaccharides also played an important role in bacterial persistence (Anderson et al., 2003). The *E. coli* K-antigen protects it from killing by neutrophils (Sarkar, Ulett, Totsika, Phan, & Schembri, 2014), which is an important defense mechanism of bacteria during their colonization in the bladder. Cellulose reduces the immune response to *E. coli* probably by covering some highly immunogenic structures present inside the host (Kai-Larsen et al., 2010).

TABLE 1 Biofilm forming pathogenic microorganisms and their interaction with host.

S. no.	Microorganism	Host	Disease(s) caused by microorganism	Role of biofilm in disease progression	References
1.	*Streptococcus pneumoniae*	Human	Pneumonia, otitis media, sinusitis, meningitis, and sepsis	Helps in adherence on the receptors of the respiratory epithelium and protection from the antimicrobial agents, immune responses. Persistence and dissemination are also improved through biofilm.	Bogaert, de Groot, and Hermans (2004) Bittaye and Cash (2015) Nobbs, Lamont, and Jenkinson (2009) Chao, Marks, Pettigrew, and Hakansson (2015) Weiser, Ferreira, and Paton (2018)
2.	*Pseudomonas aeruginosa*	Human	Hospital-acquired infections and colonization in lungs of cystic fibrosis patients	Guard the pathogen from host immune defenses.	Ramos, Somayaji, Nichols, and Goss (2018) Høiby et al. (2010)
3.	*Escherichia coli*	Human	Urinary catheter related infections and intestinal infections	Make bacteria less susceptible to antibiotics.	Høiby et al. (2010) Mittal, Sharma, and Chaudhary (2015)
4.	*Candida albicans*	Human	Vaginal and oral candidiasis	Protection from host immune system and from antifungal drugs.	Nett (2016) Cavalheiro and Teixeira (2018)
5.	*Mycobacterium ulcerans*	Human and snail	Buruli ulcer	Protection from host immune system and shows resistance against rifampicin.	Marsollier et al. (2007)
6.	*Mycobacterium tuberculosis*	Human	Tuberculosis	Protection from phagocytosis and increase drug tolerance.	Ojha et al. (2008) Basaraba and Ojha (2017)
7.	*Mycobacterium bovis*	Cattle and goat	Bovine tuberculosis	Defense mechanism against adverse environmental condition.	Srivastava et al. (2008) Costerton et al. (1999) Ishida, Arai, Niikawa, and Kobayashi (2011)
8.	*Staphylococcus aureus*	Human	Localized skin infections	Act as a reservoir to disseminate antimicrobial resistance.	Post et al. (2014) Yong, Dykes, and Choo (2019)
9.	*Aspergillus fumigatus*	Human	Bronchopulmonary aspergillosis	Resistance against potent antifungal drugs.	Skov, McKay, Koch, and Cooper (2005) Seidler et al. (2008)
10.	*Fusarium solani*	Human	Human fusariosis	Shows resistance against antifungal agents.	Guarro (2013) Zhang, Sun, Wang, Zhang, and Hou (2012)

4.1.4 Staphylococcus aureus

Staphylococcal biofilms show resistance against host immune clearing, and several significant efforts have characterized the mechanisms that contributed to this resistance (Costerton et al., 1999; Hanke & Kielian, 2012; Scherr, Heim, Morrison, & Kielian, 2014). Several studies have defined the effects of staphylococcal biofilms on immune cell function, host's response against the chronic staphylococcal infection, and the staphylococcal response against the host innate immune system. Polymorphonuclear leukocytes (PMNs) and macrophages are the first cells of innate immune system to respond against the staphylococcal infection (Cheng, DeDent, Schneewind, & Missiakas, 2011; Rigby & DeLeo, 2012; Verdrengh & Tarkowski, 1997, 2000). Several studies have been published on human PMNs in in vitro coculture and localization of *S. aureus* biofilm and phagocytosis of bacteria (Leid, Shirtliff, Costerton, & Stoodley, 2002; Scherr et al., 2013). The *agr* system in staphylococcal biofilm may be used to resist the PMN killing because a study on *S. aureus agr* mutant biofilm found it to be to be less cytotoxic to PMNs in coculture, in comparison to its wild type of counterpart and suggested that *agr* expressing biofilm cells could evade phagocytosis and killing by PMNs (Scherr et al., 2013). Neutrophil extracellular traps (NETs) showed that activated PMNs produced the thread-like projections, composing DNA and granular components in them (Brinkmann et al., 2004). *S. aureus* has the extracellular nuclease, which could degrade NETs, that is why it helps in resisting the NET-mediated killing of the bacteria (Berends et al., 2010; Thammavongsa, Missiakas, & Schneewind, 2013). The staphylococcal biofilms and host macrophages interactions have also been studied and based on that a model was developed, where staphylococci promote antiinflammatory, pro-fibrotic environment through alternative macrophage activation (Costerton et al., 1999; Scherr et al., 2014). Planktonic staphylococci are known to induce the classical activation of macrophages followed by phagocytosis (Thurlow et al., 2011). But *S. aureus* biofilm's coculturing with macrophages has shown little macrophagic phagocytosis of *S. aureus* cells (Scherr et al., 2013). *S. aureus* biofilms and macrophages-based study has shown that macrophages, which interacted closely with the biofilm, show little phagocytosis, and in the same study macrophage death also occurred in the cells that were in the proximity to the biofilm (Thurlow et al., 2011).

4.2 Fungi

Fungal biofilms are like bacterial biofilms where fungal adherent cells are surrounded by an ECM. These biofilms are made up during the infections caused by several pathogenic fungi. Eradication of the biofilm is difficult, and the formation of the biofilm can protect fungus severe defense of the innate immune system such as killing by neutrophils and monocytes.

4.2.1 Candida albicans

C. albicans biofilms can alter the monocytes defense mechanism and differ their cytokine profiling (Chandra, McCormick, Imamura, Mukherjee, & Ghannoum, 2007). Peripheral blood mononuclear cells (PBMCs) are usually phagocytosed planktonic fungus but fail to phagocytose *C. albicans* associated with biofilm (Chandra et al., 2007). Viability of these cells allows them to migrate within the biofilm, even provides them a stimulus for biofilm proliferation but mechanism is still unknown (Chandra et al., 2007; Katragkou et al., 2010). *C. albicans* biofilms are twofold to threefold more resistant for monocytes killing while comparing with others (Katragkou et al., 2010). *Candida* biofilms can influence the cytokine release by mononuclear cells through downregulation of TNF-α cytokine, which facilitates phagocyte activation. *C. albicans* biofilm exposure significantly reduces the production of TNF-α in monocytic cell line THP-1 (Katragkou et al., 2010). A recent work reported that *C. albicans* biofilm formation impairs the migratory capacity of macrophages (Alonso, Gow, Erwig, & Bain, 2017). Biofilm exposure reduced the migration of murine macrophages approximately twofold when comparing it with planktonic organisms.

4.2.2 Aspergillus fumigatus

Neutrophils are the major player of innate immune defense against *Aspergillus*. In vivo recruitment of the neutrophils to *Aspergillus* spores is critical for their engulfment from phagocytosis (Behnsen et al., 2007; Bonnett, Cornish, Harmsen, & Burritt, 2006; Sturtevant & Latgé, 1992). Neutrophils release NETs against the hyphal elements (Bianchi et al., 2009; Bruns et al., 2010; Jaillon et al., 2007; McCormick et al., 2010). NETs lack significant activity against conidia but show modest inhibitory activity against the hyphal forms of *A. fumigatus* (Bruns et al., 2010; McCormick et al., 2010). ECM production of biofilm protects *A. fumigatus* from neutrophil attack, and this protection is largely attributed to galactosaminogalactan (GAG) (Lee et al., 2015). Deficient GAG synthesis/treatment from recombinant glycoside hydrolases, which degrade GAG in *A. fumigatus*, reduces the growth of fungus in a murine model of aspergillosis (Snarr et al., 2017). The protection of GAG is dependent on NADPH oxidase enzyme and most probably involves defense against NETs, released via activation of this pathway. The immunomodulatory effects of GAG in biofilm formation are most probably multifactorial. A study says that hyphal-associated

GAG can mask β-glucans present on the cell wall of hyphae (Gravelat et al., 2010, 2013) so they alter the recognition from murine bone marrow–derived dendritic cells in vitro (Gravelat et al., 2013). Genetic inhibition of GAG synthesis leads to upregulation of pro-inflammatory cytokine via Dectin-1 signaling (Gravelat et al., 2013). The soluble GAG can modulate immunity via induction of apoptosis in neutrophils and stimulation of in vitro antiinflammatory IL-1Ra production from macrophages (Gresnigt et al., 2014; Robinet et al., 2014). Transcriptional analysis shows the high expression of RodA during *A. fumigatus* biofilm growth in comparison to planktonic conditions, and the role of RodA production in immune evasion during biofilm growth is also interesting (Gibbons et al., 2012).

5. Conclusion and future prospect

Biofilm formation is the most advanced and complicated mechanism shown by any microorganisms whether it is related to bacteria or fungi. Pathogenic microbes have used this mechanism for their best survival or to defend themselves against the host's immune system. Several microorganisms are involved in biofilm formation, and these are also known to cause several infectious diseases. Interestingly, these diseases are known to cause more complications, only because of biofilm formation inside their host(s). The biofilm is a complicated mixture of extracellular substances, and both the bacteria and fungi have unique molecular mechanism for their synthesis, and sometimes, it is found to be unique in particular species also. QS is also involved in the activation and induction of biofilm formation while bacteria are present in groups. Several studies have also explained that how these biofilm formations in pathogenic bacteria alter the host-pathogen interaction and implicated ways to protect themselves inside the host. So, based on the above studies and research outcomes, it can be concluded that biofilm is a kind of microbial groupism along with some extracellular substances, which improves their survival and defense strategies. Further studies on initialization of biofilm formation and the development of mechanism to restrict their growth at that stage will be a prominence in biofilm research.

References

Al-Fattani, M. A., & Douglas, L. J. (2006). Biofilm matrix of *Candida albicans* and *Candida tropicalis*: Chemical composition and role in drug resistance. *Journal of Medical Microbiology*, *55*(8), 999–1008.

Allesen-Holm, M., Barken, K. B., Yang, L., Klausen, M., Webb, J. S., Kjelleberg, S., et al. (2006). A characterization of DNA release in *Pseudomonas aeruginosa* cultures and biofilms. *Molecular Microbiology*, *59*(4), 1114–1128.

Alonso, M. F., Gow, N. A., Erwig, L. P., & Bain, J. M. (2017). Macrophage migration is impaired within *Candida albicans* biofilms. *Journal of Fungi*, *3*(3), 31.

Alsteens, D., Garcia, M. C., Lipke, P. N., & Dufrêne, Y. F. (2010). Force-induced formation and propagation of adhesion nanodomains in living fungal cells. *Proceedings of the National Academy of Sciences of the United States of America*, *107*(48), 20744–20749.

Anderson, J. K., Huang, J. Y., Wreden, C., Sweeney, E. G., Goers, J., Remington, S. J., et al. (2015). Chemorepulsion from the quorum signal autoinducer-2 promotes *Helicobacter pylori* biofilm dispersal. *MBio*, *6*(4), e00379-15.

Anderson, G. G., Palermo, J. J., Schilling, J. D., Roth, R., Heuser, J., & Hultgren, S. J. (2003). Intracellular bacterial biofilm-like pods in urinary tract infections. *Science*, *301*(5629), 105–107.

Baldassarri, L., Cecchini, R., Bertuccini, L., Ammendolia, M. G., Iosi, F., Arciola, C. R., et al. (2001). Enterococcus spp. produces slime and survives in rat peritoneal macrophages. *Medical Microbiology and Immunology*, *190*(3), 113–120.

Banerjee, M., Uppuluri, P., Zhao, X. R., Carlisle, P. L., Vipulanandan, G., Villar, C. C., et al. (2013). Expression of UME6, a key regulator of *Candida albicans* hyphal development, enhances biofilm formation via Hgc1-and Sun41-dependent mechanisms. *Eukaryotic Cell*, *12*(2), 224–232.

Basaraba, R. J., & Ojha, A. K. (2017). Mycobacterial biofilms: Revisiting tuberculosis bacilli in extracellular necrotizing lesions. *Microbiology Spectrum*, *5*(3). 5-3.

Beauvais, A., Schmidt, C., Guadagnini, S., Roux, P., Perret, E., Henry, C., et al. (2007). An extracellular matrix glues together the aerial-grown hyphae of *Aspergillus fumigatus*. *Cellular Microbiology*, *9*(6), 1588–1600.

Behnsen, J., Narang, P., Hasenberg, M., Gunzer, F., Bilitewski, U., Klippel, N., et al. (2007). Environmental dimensionality controls the interaction of phagocytes with the pathogenic fungi *Aspergillus fumigatus* and *Candida albicans*. *PLoS Pathogens*, *3*(2), e13.

Benachour, A., Ladjouzi, R., Le Jeune, A., Hébert, L., Thorpe, S., Courtin, P., et al. (2012). The lysozyme-induced peptidoglycan N-acetylglucosamine deacetylase PgdA (EF1843) is required for *Enterococcus faecalis* virulence. *Journal of Bacteriology*, *194*(22), 6066–6073.

Berends, E. T., Horswill, A. R., Haste, N. M., Monestier, M., Nizet, V., & von Köckritz-Blickwede, M. (2010). Nuclease expression by *Staphylococcus aureus* facilitates escape from neutrophil extracellular traps. *Journal of Innate Immunity*, *2*(6), 576–586.

Bianchi, M., Hakkim, A., Brinkmann, V., Siler, U., Seger, R. A., Zychlinsky, A., et al. (2009). Restoration of NET formation by gene therapy in CGD controls aspergillosis. *Blood, The Journal of the American Society of Hematology*, *114*(13), 2619–2622.

Bittaye, M., & Cash, P. (2015). *Streptococcus pneumoniae* proteomics: Determinants of pathogenesis and vaccine development. *Expert Review of Proteomics*, *12*(6), 607–621.

Bjarnsholt, T. (2013). The role of bacterial biofilms in chronic infections. *APMIS*, *121*, 1–58.

Blankenship, J. R., & Mitchell, A. P. (2006). How to build a biofilm: A fungal perspective. *Current Opinion in Microbiology*, *9*(6), 588–594.

Bogaert, D., de Groot, R., & Hermans, P. W. M. (2004). *Streptococcus pneumoniae* colonisation: The key to pneumococcal disease. *The Lancet Infectious Diseases*, *4*(3), 144–154.

Bonhomme, J., & d'Enfert, C. (2013). *Candida albicans* biofilms: Building a heterogeneous, drug-tolerant environment. *Current Opinion in Microbiology*, *16*(4), 398–403.

Bonnett, C. R., Cornish, E. J., Harmsen, A. G., & Burritt, J. B. (2006). Early neutrophil recruitment and aggregation in the murine lung inhibit germination of *Aspergillus fumigatus* conidia. *Infection and Immunity*, *74* (12), 6528–6539.

Bottone, E. J. (2010). *Bacillus cereus*, a volatile human pathogen. *Clinical Microbiology Reviews*, *23*(2), 382–398.

Branda, S. S., Vik, Å., Friedman, L., & Kolter, R. (2005). Biofilms: The matrix revisited. *Trends in Microbiology*, *13*(1), 20–26.

Braun, B. R., Kadosh, D., & Johnson, A. D. (2001). NRG1, a repressor of filamentous growth in C. albicans, is down-regulated during filament induction. *The EMBO Journal*, *20*(17), 4753–4761.

Brinkmann, V., Reichard, U., Goosmann, C., Fauler, B., Uhlemann, Y., Weiss, D. S., et al. (2004). Neutrophil extracellular traps kill bacteria. *Science*, *303*(5663), 1532–1535.

Brouwer, J., van Leeuwen-Herberts, T., & Otting-van de Ruit, M. (1984). Determination of lysozyme in serum, urine, cerebrospinal fluid and feces by enzyme immunoassay. *Clinica Chimica Acta*, *142*(1), 21–30.

Bruns, S., Kniemeyer, O., Hasenberg, M., Aimanianda, V., Nietzsche, S., Thywißen, A., … Gunzer, M. (2010). Production of extracellular traps against *Aspergillus fumigatus* in vitro and in infected lung tissue is dependent on invading neutrophils and influenced by hydrophobin RodA. *PLoS Pathogens*, *6*(4), e1000873.s.

Byrd, M. S., Sadovskaya, I., Vinogradov, E., Lu, H., Sprinkle, A. B., Richardson, S. H., et al. (2009). Genetic and biochemical analyses of the *Pseudomonas aeruginosa* Psl exopolysaccharide reveal overlapping roles for polysaccharide synthesis enzymes in Psl and LPS production. *Molecular Microbiology*, *73*(4), 622–638.

Cannizzo, F. T., Eraso, E., Ezkurra, P. A., Villar-Vidal, M., Bollo, E., Castellá, G., et al. (2007). Biofilm development by clinical isolates of *Malassezia pachydermatis*. *Medical Mycology*, *45*(4), 357–361.

Cao, Y. Y., Cao, Y. B., Xu, Z., Ying, K., Li, Y., Xie, Y., et al. (2005). cDNA microarray analysis of differential gene expression in *Candida albicans* biofilm exposed to farnesol. *Antimicrobial Agents and Chemotherapy*, *49*(2), 584–589.

Cavalheiro, M., & Teixeira, M. C. (2018). *Candida* biofilms: Threats, challenges, and promising strategies. *Frontiers in Medicine*, *5*, 28.

Chandra, J., Kuhn, D. M., Mukherjee, P. K., Hoyer, L. L., McCormick, T., & Ghannoum, M. A. (2001). Biofilm formation by the fungal pathogen *Candida albicans*: Development, architecture, and drug resistance. *Journal of Bacteriology*, *183*(18), 5385–5394.

Chandra, J., McCormick, T. S., Imamura, Y., Mukherjee, P. K., & Ghannoum, M. A. (2007). Interaction of *Candida albicans* with adherent human peripheral blood mononuclear cells increases *C. albicans* biofilm formation and results in differential expression of pro-and anti-inflammatory cytokines. *Infection and Immunity*, *75*(5), 2612–2620.

Chao, Y., Marks, L. R., Pettigrew, M. M., & Hakansson, A. P. (2015). *Streptococcus pneumoniae* biofilm formation and dispersion during colonization and disease. *Frontiers in Cellular and Infection Microbiology*, *4*, 194.

Cheng, A. G., DeDent, A. C., Schneewind, O., & Missiakas, D. (2011). A play in four acts: *Staphylococcus aureus* abscess formation. *Trends in Microbiology*, *19*(5), 225–232.

Chmielewski, R. A. N., & Frank, J. F. (2003). Biofilm formation and control in food processing facilities. *Comprehensive Reviews in Food Science and Food Safety*, *2*(1), 22–32.

Chong, P. P., Chin, V. K., Wong, W. F., Madhavan, P., Yong, V. C., & Looi, C. Y. (2018). Transcriptomic and genomic approaches for unravelling *Candida albicans* biofilm formation and drug resistance—An update. *Genes*, *9*(11), 540.

Chong, K. K. L., Tay, W. H., Janela, B., Yong, A. M. H., Liew, T. H., Madden, L., et al. (2017). *Enterococcus faecalis* modulates immune activation and slows healing during wound infection. *The Journal of Infectious Diseases*, *216*(12), 1644–1654.

Chonmaitree, T., Trujillo, R., Jennings, K., Alvarez-Fernandez, P., Patel, J. A., Loeffelholz, M. J., et al. (2016). Acute otitis media and other complications of viral respiratory infection. *Pediatrics*, *137*(4).

Cohen, J. I., Bartlett, J. A., & Corey, G. R. (1987). Extra-intestinal manifestations of salmonella infections. *Medicine*, *66*(5), 349–388.

Colvin, K. M., Gordon, V. D., Murakami, K., Borlee, B. R., Wozniak, D. J., Wong, G. C., et al. (2011). The pel polysaccharide can serve a structural and protective role in the biofilm matrix of *Pseudomonas aeruginosa*. *PLoS Pathogens*, *7*(1), e1001264.

Costa-Orlandi, C. B., Sardi, J. C. O., Santos, C. T., Fusco-Almeida, A. M., & Mendes-Giannini, M. J. S. (2014). In vitro characterization of *Trichophyton rubrum* and *T. mentagrophytes* biofilms. *Biofouling*, *30* (6), 719–727.

Costerton, J. W., Stewart, P. S., & Greenberg, E. P. (1999). Bacterial biofilms: A common cause of persistent infections. *Science*, *284* (5418), 1318–1322.

Crum, N. F. (2003). Current trends in typhoid fever. *Current Gastroenterology Reports*, *5*(4), 279–286.

Cushion, M. T., Collins, M. S., & Linke, M. J. (2009). Biofilm formation by *Pneumocystis* spp. *Eukaryotic Cell*, *8*(2), 197–206.

Ćwiek, K., Bugla-Płoskońska, G., & Wieliczko, A. (2019). *Salmonella* biofilm development: Structure and significance. *Advances in Hygiene & Experimental Medicine/ PostepyHigienyiMedycynyDoswiadczalnej*, *73*.

D'Antonio, D., Parruti, G., Pontieri, E., Di Bonaventura, G., Manzoli, L., Sferra, R., et al. (2004). Slime production by clinical isolates of *Blastoschizomyces capitatus* from patients with hematological malignancies and catheter-related fungemia. *European Journal of Clinical Microbiology and Infectious Diseases*, *23*(10), 787–789.

Daniel, M., Imtiaz-Umer, S., Fergie, N., Birchall, J. P., & Bayston, R. (2012). Bacterial involvement in otitis media with effusion. *International Journal of Pediatric Otorhinolaryngology*, *76*(10), 1416–1422.

Daniels, R., Vanderleyden, J., & Michiels, J. (2004). Quorum sensing and swarming migration in bacteria. *FEMS Microbiology Reviews*, *28*(3), 261–289.

Danielsen, K. A., Eskeland, O., Fridrich-Aas, K., Orszagh, V. C., Bachmann-Harildstad, G., & Burum-Auensen, E. (2014). Bacterial biofilms in patients with chronic rhinosinusitis: A confocal scanning laser microscopy study. *Rhinology*, *52*(2), 150–155.

Davey, M. E., Caiazza, N. C., & O'Toole, G. A. (2003). Rhamnolipid surfactant production affects biofilm architecture in *Pseudomonas aeruginosa* PAO1. *Journal of Bacteriology*, *185*(3), 1027–1036.

Daw, K., Baghdayan, A. S., Awasthi, S., & Shankar, N. (2012). Biofilm and planktonic *Enterococcus faecalis* elicit different responses from host phagocytes in vitro. *FEMS Immunology and Medical Microbiology*, *65*(2), 270–282.

de Groot, P. W., Bader, O., de Boer, A. D., Weig, M., & Chauhan, N. (2013). Adhesins in human fungal pathogens: Glue with plenty of stick. *Eukaryotic Cell*, *12*(4), 470–481.

DeAntonio, R., Yarzabal, J.-P., Cruz, J. P., Schmidt, J. E., & Kleijnen, J. (2016). Epidemiology of otitis media in children from developing countries: A systematic review. *International Journal of Pediatric Otorhinolaryngology*, *85*, 65–74.

Di Bonaventura, G., Pompilio, A., Picciani, C., Iezzi, M., D'Antonio, D., & Piccolomini, R. (2006). Biofilm formation by the emerging fungal pathogen *Trichosporon asahii*: Development, architecture, and antifungal resistance. *Antimicrobial Agents and Chemotherapy*, *50*(10), 3269–3276.

Dinbar, A., Altmann, G., & Tulcinsky, D. B. (1969). The treatment of chronic biliary *Salmonella* carriers. *The American Journal of Medicine*, *47*(2), 236–242.

Donlan, R. M. (2001). Biofilm formation: A clinically relevant microbiological process. *Clinical Infectious Diseases*, *33*(8), 1387–1392.

Dössel, J., Meyer-Hoffert, U., Schröder, J. M., & Gerstel, U. (2012). *Pseudomonas aeruginosa*-derived rhamnolipids subvert the host innate immune response through manipulation of the human beta-defensin-2 expression. *Cellular Microbiology*, *14*(9), 1364–1375.

Douglas, L. J. (2003). *Candida* biofilms and their role in infection. *Trends in Microbiology*, *11*(1), 30–36.

Dusane, D. H., Zinjarde, S. S., Venugopalan, V. P., Mclean, R. J., Weber, M. M., & Rahman, P. K. (2010). Quorum sensing: Implications on rhamnolipid biosurfactant production. *Biotechnology and Genetic Engineering Reviews*, *27*(1), 159–184.

Elgharably, H., Hussain, S. T., Shrestha, N. K., Blackstone, E. H., & Pettersson, G. B. (2016). Current hypotheses in cardiac surgery: Biofilm in infective endocarditis. *Seminars in Thoracic and Cardiovascular Surgery*, *28*(1), 56–59. WB Saunders.

Engebrecht, J., Nealson, K., & Silverman, M. (1983). Bacterial bioluminescence: Isolation and genetic analysis of functions from *Vibrio fischeri*. *Cell*, *32*(3), 773–781.

Engebrecht, J., & Silverman, M. (1984). Identification of genes and gene products necessary for bacterial bioluminescence. *Proceedings of the National Academy of Sciences of the United States of America*, *81* (13), 4154–4158.

Enjalbert, B., & Whiteway, M. (2005). Release from quorum-sensing molecules triggers hyphal formation during *Candida albicans* resumption of growth. *Eukaryotic Cell*, *4*(7), 1203–1210.

Ertesvåg, H., & Valla, S. (1998). Biosynthesis and applications of alginates. *Polymer Degradation and Stability*, *59*(1–3), 85–91.

Escande, W., Fayad, G., Modine, T., Verbrugge, E., Koussa, M., Senneville, E., et al. (2011). Culture of a prosthetic valve excised for streptococcal endocarditis positive for *Aspergillus fumigatus* 20 years after previous *A fumigatus* endocarditis. *The Annals of Thoracic Surgery*, *91* (6), e92–e93.

Fabretti, F., Theilacker, C., Baldassarri, L., Kaczynski, Z., Kropec, A., Holst, O., et al. (2006). Alanine esters of enterococcal lipoteichoic acid play a role in biofilm formation and resistance to antimicrobial peptides. *Infection and Immunity*, *74*(7), 4164–4171.

Faden, H., Duffy, L., Wasielewski, R., Wolf, J., Krystofik, D., & Tung, Y. (1997). Relationship between nasopharyngeal colonization and the development of otitis media in children. *The Journal of Infectious Diseases*, *175*(6), 1440–1445.

Falk, E., Shah, P. K., & Fuster, V. (1995). Coronary plaque disruption. *Circulation*, *92*(3), 657–671.

Faurschou, M., & Borregaard, N. (2003). Neutrophil granules and secretory vesicles in inflammation. *Microbes and Infection*, *5*(14), 1317–1327.

Fazli, M., Bjarnsholt, T., Kirketerp-Møller, K., Jørgensen, B., Andersen, A. S., Krogfelt, K. A., et al. (2009). Nonrandom distribution of *Pseudomonas aeruginosa* and *Staphylococcus aureus* in chronic wounds. *Journal of Clinical Microbiology*, *47*(12), 4084–4089.

Finkel, J. S., & Mitchell, A. P. (2011). Genetic control of *Candida albicans* biofilm development. *Nature Reviews Microbiology*, *9*(2), 109–118.

Flemming, H. C. (2010). Wingender: The biofilm matrix. *Nature Reviews Microbiology*, *8*, 623–633.

Flemming, H. C., Neu, T. R., & Wozniak, D. J. (2007). The EPS matrix: The "house of biofilm cells". *Journal of Bacteriology*, *189*(22), 7945–7947.

Flemming, H. C., Wingender, J., Szewzyk, U., Steinberg, P., Rice, S. A., & Kjelleberg, S. (2016). Biofilms: An emergent form of bacterial life. *Nature Reviews Microbiology*, *14*(9), 563–575.

Flint, S., Palmer, J., Bloemen, K., Brooks, J., & Crawford, R. (2001). The growth of Bacillus stearothermophilus on stainless steel. *Journal of Applied Microbiology*, *90*(2), 151–157.

Foreman, A., Boase, S., Psaltis, A., & Wormald, P. J. (2012). Role of bacterial and fungal biofilms in chronic rhinosinusitis. *Current Allergy and Asthma Reports*, *12*(2), 127–135.

Fox, E. P., & Nobile, C. J. (2012). A sticky situation: Untangling the transcriptional network controlling biofilm development in *Candida albicans*. *Transcription*, *3*(6), 315–322.

Franklin, M. J., Nivens, D. E., Weadge, J. T., & Howell, P. L. (2011). Biosynthesis of the *Pseudomonas aeruginosa* extracellular polysaccharides, alginate, Pel, and Psl. *Frontiers in Microbiology*, *2*, 167.

Friedman, L., & Kolter, R. (2004). Genes involved in matrix formation in *Pseudomonas aeruginosa* PA14 biofilms. *Molecular Microbiology*, *51* (3), 675–690.

Fux, C. A., Costerton, J. W., Stewart, P. S., & Stoodley, P. (2005). Survival strategies of infectious biofilms. *Trends in Microbiology*, *13*(1), 34–40.

Galli, J., Calò, L., Giuliani, M., Sergi, B., Lucidi, D., Meucci, D., et al. (2016). Biofilm's role in chronic cholesteatomatous otitis media: A pilot study. *Otolaryngology and Head and Neck Surgery*, *154*(5), 914–916.

Ganguly, S., & Mitchell, A. P. (2011). Mucosal biofilms of *Candida albicans*. *Current Opinion in Microbiology*, *14*(4), 380–385.

Garcia, M. C., Lee, J. T., Ramsook, C. B., Alsteens, D., Dufrêne, Y. F., & Lipke, P. N. (2011). A role for amyloid in cell aggregation and biofilm formation. *PLoS One*, *6*(3), e17632.

Ghafoor, A., Hay, I. D., & Rehm, B. H. (2011). Role of exopolysaccharides in *Pseudomonas aeruginosa* biofilm formation and architecture. *Applied and Environmental Microbiology*, *77*(15), 5238–5246.

Giannella, R. A., Baron, S., Albrecht, T., Castro, G., Couch, R. B., Davis, C. P., et al. (1996). *Baron's medical microbiology* (4th ed.). Galveston, TX: University of Texas Medical Branch. Salmonella.

Giaouris, E., Heir, E., Desvaux, M., Hébraud, M., Møretrø, T., Langsrud, S., … Simões, M. (2015). Intra-and inter-species interactions within biofilms of important foodborne bacterial pathogens. *Frontiers in Microbiology*, *6*, 841.

Gibbons, J. G., Beauvais, A., Beau, R., McGary, K. L., Latgé, J. P., & Rokas, A. (2012). Global transcriptome changes underlying colony growth in the opportunistic human pathogen *Aspergillus fumigatus*. *Eukaryotic Cell*, *11*(1), 68–78.

Gonzalez-Escobedo, G., Marshall, J. M., & Gunn, J. S. (2011). Chronic and acute infection of the gall bladder by *Salmonella* Typhi: Understanding the carrier state. *Nature Reviews Microbiology*, *9*(1), 9–14.

González-Ramírez, A. I., Ramírez-Granillo, A., Medina-Canales, M. G., Rodríguez-Tovar, A. V., & Martínez-Rivera, M. A. (2016). Analysis and description of the stages of *Aspergillus fumigatus* biofilm formation using scanning electron microscopy. *BMC Microbiology*, *16* (1), 1–13.

Gould, L. H., Mody, R. K., Ong, K. L., Clogher, P., Cronquist, A. B., Garman, K. N., et al. (2013). Increased recognition of non-O157 Shiga toxin–producing *Escherichia coli* infections in the United States during 2000–2010: Epidemiologic features and comparison with *E. coli* O157 infections. *Foodborne Pathogens and Disease*, *10*(5), 453–460.

Granger, B. L., Flenniken, M. L., Davis, D. A., Mitchell, A. P., & Cutler, J. E. (2005). Yeast wall protein 1 of *Candida albicans*. *Microbiology*, *151*(5), 1631–1644.

Gravelat, F. N., Beauvais, A., Liu, H., Lee, M. J., Snarr, B. D., Chen, D., et al. (2013). *Aspergillus* galactosaminogalactan mediates adherence to host constituents and conceals hyphal β-glucan from the immune system. *PLoS Pathogens*, *9*(8), e1003575.

Gravelat, F. N., Ejzykowicz, D. E., Chiang, L. Y., Chabot, J. C., Urb, M., Macdonald, K. D., et al. (2010). *Aspergillus fumigatus* MedA governs adherence, host cell interactions and virulence. *Cellular Microbiology*, *12*(4), 473–488.

Gresnigt, M. S., Bozza, S., Becker, K. L., Joosten, L. A., Abdollahi-Roodsaz, S., van der Berg, W. B., et al. (2014). A polysaccharide virulence factor from *Aspergillus fumigatus* elicits anti-inflammatory effects through induction of Interleukin-1 receptor antagonist. *PLoS Pathogens*, *10*(3), e1003936.

Guarro, J. (2013). Fusariosis, a complex infection caused by a high diversity of fungal species refractory to treatment. *European Journal of Clinical Microbiology & Infectious Diseases*, *32*(12), 1491–1500.

Guo, M., Fang, Z., Sun, L., Sun, D., Wang, Y., Li, C., et al. (2018). Regulation of thermostable direct hemolysin and biofilm formation of *Vibrio parahaemolyticus* by quorum-sensing genes luxM and luxS. *Current Microbiology*, *75*(9), 1190–1197.

Hall-Stoodley, L., & Stoodley, P. (2009). Evolving concepts in biofilm infections. *Cellular Microbiology*, *11*(7), 1034–1043.

Hammer, B. K., & Bassler, B. L. (2003). Quorum sensing controls biofilm formation in *Vibrio cholerae*. *Molecular Microbiology*, *50*(1), 101–104.

Hanke, M. L., & Kielian, T. (2012). Deciphering mechanisms of staphylococcal biofilm evasion of host immunity. *Frontiers in Cellular and Infection Microbiology*, *2*, 62.

Harriott, M. M., Lilly, E. A., Rodriguez, T. E., Fidel, P. L., Jr., & Noverr, M. C. (2010). *Candida albicans* forms biofilms on the vaginal mucosa. *Microbiology*, *156*(Pt 12), 3635.

Harriott, M. M., & Noverr, M. C. (2009). *Candida albicans* and *Staphylococcus aureus* form polymicrobial biofilms: Effects on antimicrobial resistance. *Antimicrobial Agents and Chemotherapy*, *53*(9), 3914–3922.

Harvey, R. J., & Lund, V. J. (2007). Biofilms and chronic rhinosinusitis: Systematic review of evidence, current concepts and directions for research. *Rhinology*, *45*(1), 3.

Hébert, L., Courtin, P., Torelli, R., Sanguinetti, M., Chapot-Chartier, M. P., Auffray, Y., et al. (2007). *Enterococcus faecalis* constitutes an unusual bacterial model in lysozyme resistance. *Infection and Immunity*, *75* (11), 5390–5398.

Hense, B. A., Kuttler, C., Müller, J., Rothballer, M., Hartmann, A., & Kreft, J. U. (2007). Does efficiency sensing unify diffusion and quorum sensing? *Nature Reviews Microbiology*, *5*(3), 230–239.

Høiby, N. (2017). A short history of microbial biofilms and biofilm infections. *APMIS*, *125*(4), 272–275.

Høiby, N., Ciofu, O., & Bjarnsholt, T. (2010). *Pseudomonas aeruginosa* biofilms in cystic fibrosis. *Future Microbiology*, *5*(11), 1663–1674.

Hoyer, L. L., & Cota, E. (2016). *Candida albicans* agglutinin-like sequence (Als) family vignettes: A review of Als protein structure and function. *Frontiers in Microbiology*, *7*, 280.

Hoyle, B. D., Jass, J., & Costerton, J. W. (1990). The biofilm glycocalyx as a resistance factor. *The Journal of Antimicrobial Chemotherapy*, *26*(1), 1–5.

Imamura, Y., Chandra, J., Mukherjee, P. K., Lattif, A. A., Szczotka-Flynn, L. B., Pearlman, E., et al. (2008). Fusarium and *Candida albicans* biofilms on soft contact lenses: Model development, influence of lens type, and susceptibility to lens care solutions. *Antimicrobial Agents and Chemotherapy*, *52*(1), 171–182.

Irrie, Y., & Parsek, M. R. (2008). Quorum sensing and microbial biofilms. *Bacterial Biofilms*, 67–84.

Ishida, S., Arai, M., Niikawa, H., & Kobayashi, M. (2011). Inhibitory effect of cyclic trihydroxamate siderophore, desferrioxamine E, on the biofilm formation of *Mycobacterium* species. *Biological and Pharmaceutical Bulletin*, *34*(6), 917–920.

Jaillon, S., Peri, G., Delneste, Y., Frémaux, I., Doni, A., Moalli, F., et al. (2007). The humoral pattern recognition receptor PTX3 is stored in neutrophil granules and localizes in extracellular traps. *Journal of Experimental Medicine*, *204*(4), 793–804.

Jun, W., Kim, M. S., Cho, B. K., Millner, P. D., Chao, K., & Chan, D. E. (2010). Microbial biofilm detection on food contact surfaces by macro-scale fluorescence imaging. *Journal of Food Engineering*, *99*(3), 314–322.

Justice, S. S., Hung, C., Theriot, J. A., Fletcher, D. A., Anderson, G. G., Footer, M. J., et al. (2004). Differentiation and developmental pathways of uropathogenic *Escherichia coli* in urinary tract pathogenesis. *Proceedings of the National Academy of Sciences of the United States of America*, *101*(5), 1333–1338.

Kai-Larsen, Y., Luthje, P., Chromek, M., Peters, V., Wang, X., Holm, A., et al. (2010). Uropathogenic *Escherichia coli* modulates immune responses and its curli fimbriae interact with the antimicrobial peptide LL-37. *PLoS Pathogens*, *6*(7), e1001010.

Kao, W. K., Chole, R. A., & Ogden, M. A. (2020). Evidence of a microbial etiology for sialoliths. *The Laryngoscope*, *130*(1), 69–74.

Kasai, M., Francesconi, A., Petraitiene, R., Petraitis, V., Kelaher, A. M., Kim, H. S., et al. (2006). Use of quantitative real-time PCR to study the kinetics of extracellular DNA released from *Candida albicans*, with implications for diagnosis of invasive candidiasis. *Journal of Clinical Microbiology*, *44*(1), 143–150.

Katragkou, A., Kruhlak, M. J., Simitsopoulou, M., Chatzimoschou, A., Taparkou, A., Cotten, C. J., et al. (2010). Interactions between human phagocytes and *Candida albicans* biofilms alone and in combination with antifungal agents. *The Journal of Infectious Diseases*, *201*(12), 1941–1949.

Kaur, S., & Singh, S. (2014). Biofilm formation by *Aspergillus fumigatus*. *Medical Mycology*, *52*(1), 2–9.

Kaya, E., Dag, I., Incesulu, A., Gurbuz, M. K., Acar, M., & Birdane, L. (2013). Investigation of the presence of biofilms in chronic suppurative otitis media, nonsuppurative otitis media, and chronic otitis media with cholesteatoma by scanning electron microscopy. *The Scientific World Journal*, *2013*, 638715.

Khan, F., Pham, D. T. N., Oloketuyi, S. F., & Kim, Y. M. (2020). Regulation and controlling the motility properties of *Pseudomonas aeruginosa*. *Applied Microbiology and Biotechnology*, *104*(1), 33–49.

Kim, E. R., & Chang, D. K. (2014). Colorectal cancer in inflammatory bowel disease: The risk, pathogenesis, prevention and diagnosis. *World Journal of Gastroenterology: WJG*, *20*(29), 9872.

Kirketerp-Møller, K., Jensen, P.Ø., Fazli, M., Madsen, K. G., Pedersen, J., Moser, C., et al. (2008). Distribution, organization, and ecology of bacteria in chronic wounds. *Journal of Clinical Microbiology*, *46*(8), 2717–2722.

Klotz, S. A., Gaur, N. K., De Armond, R., Sheppard, D., Khardori, N., Edwards, J. E., Jr., et al. (2007). *Candida albicans* Als proteins mediate aggregation with bacteria and yeasts. *Medical Mycology*, *45*(4), 363–370.

Kojic, E. M., & Darouiche, R. O. (2004). *Candida* infections of medical devices. *Clinical Microbiology Reviews*, *17*(2), 255–267.

Kolter, R., & Watnick, P. (2000). Biofilm, city of microbes. *Journal of Bacteriology*, *182*(10), 2675–2679.

Kruppa, M., Krom, B. P., Chauhan, N., Bambach, A. V., Cihlar, R. L., & Calderone, R. A. (2004). The two-component signal transduction protein Chk1p regulates quorum sensing in *Candida albicans*. *Eukaryotic Cell*, *3*(4), 1062–1065.

Lai, C. W., Chan, R. C. Y., Cheng, A. F. B., Sung, J. Y., & Leung, J. W. C. (1992). Common bile duct stones: A cause of chronic salmonellosis. *American Journal of Gastroenterology (Springer Nature)*, *87*(9).

Lamas, A., Miranda, J. M., Regal, P., Vázquez, B., Franco, C. M., & Cepeda, A. (2018). A comprehensive review of non-enterica subspecies of *Salmonella enterica*. *Microbiological Research*, *206*, 60–73.

Lanter, B. B., Sauer, K., & Davies, D. G. (2014). Bacteria present in carotid arterial plaques are found as biofilm deposits which may contribute to enhanced risk of plaque rupture. *MBio*, *5*(3), e01206–e01214.

Lattif, A. A., Mukherjee, P. K., Chandra, J., Swindell, K., Lockhart, S. R., Diekema, D. J., et al. (2010). Characterization of biofilms formed by *Candida parapsilosis*, *C. metapsilosis*, and *C. orthopsilosis*. *International Journal of Medical Microbiology*, *300*(4), 265–270.

Le Jeune, A., Torelli, R., Sanguinetti, M., Giard, J. C., Hartke, A., Auffray, Y., et al. (2010). The extracytoplasmic function sigma factor SigV plays a key role in the original model of lysozyme resistance and virulence of *Enterococcus faecalis*. *PLoS One*, *5*(3), e9658.

Lee, M. J., Liu, H., Barker, B. M., Snarr, B. D., Gravelat, F. N., Al Abdallah, Q., et al. (2015). The fungal exopolysaccharide galactosaminogalactan mediates virulence by enhancing resistance to neutrophil extracellular traps. *PLoS Pathogens*, *11*(10), e1005187.

Lee, J., & Zhang, L. (2015). The hierarchy quorum sensing network in *Pseudomonas aeruginosa*. *Protein & Cell*, *6*(1), 26–41.

Leid, J. G., Shirtliff, M. E., Costerton, J. W., & Stoodley, A. P. (2002). Human leukocytes adhere to, penetrate, and respond to *Staphylococcus aureus* biofilms. *Infection and Immunity*, *70*(11), 6339–6345.

Li, J., Attila, C., Wang, L., Wood, T. K., Valdes, J. J., & Bentley, W. E. (2007). Quorum sensing in *Escherichia coli* is signaled by AI-2/LsrR: Effects on small RNA and biofilm architecture. *Journal of Bacteriology*, *189*, 6011–6020.

Li, H., Li, X., Wang, Z., Fu, Y., Ai, Q., Dong, Y., et al. (2015). Autoinducer-2 regulates *Pseudomonas aeruginosa* PAO1 biofilm formation and virulence production in a dose-dependent manner. *BMC Microbiology*, *15*, 192.

Li, F., Svarovsky, M. J., Karlsson, A. J., Wagner, J. P., Marchillo, K., Oshel, P., et al. (2007). Eap1p, an adhesin that mediates *Candida albicans* biofilm formation in vitro and in vivo. *Eukaryotic Cell*, *6*(6), 931–939.

Lim, E. S., Koo, O. K., Kim, M. J., & Kim, J. S. (2019). Bio-enzymes for inhibition and elimination of *Escherichia coli* O157: H7 biofilm and their synergistic effect with sodium hypochlorite. *Scientific Reports*, *9*(1), 1–10.

Lindsay, D., & Von Holy, A. (2006). Bacterial biofilms within the clinical setting: What healthcare professionals should know. *Journal of Hospital Infection*, *64*(4), 313–325.

Lipke, P. N., Garcia, M. C., Alsteens, D., Ramsook, C. B., Klotz, S. A., & Dufrêne, Y. F. (2012). Strengthening relationships: Amyloids create adhesion nanodomains in yeasts. *Trends in Microbiology*, *20*(2), 59–65.

Lipke, P. N., Ramsook, C., Garcia-Sherman, M. C., Jackson, D. N., Chan, C. X., Bois, M., et al. (2014). Between amyloids and aggregation lies a connection with strength and adhesion. *New Journal of Science*, *2014*.

Ma, L., Wang, J., Wang, S., Anderson, E. M., Lam, J. S., Parsek, M. R., et al. (2012). Synthesis of multiple *Pseudomonas aeruginosa* biofilm matrix exopolysaccharides is post-transcriptionally regulated. *Environmental Microbiology*, *14*(8), 1995–2005.

Marsollier, L., Brodin, P., Jackson, M., Korduláková, J., Tafelmeyer, P., Carbonnelle, E., et al. (2007). Impact of *Mycobacterium ulcerans* biofilm on transmissibility to ecological niches and Buruli ulcer pathogenesis. *PLoS Pathogens*, *3*(5), e62.

Martins, M., Uppuluri, P., Thomas, D. P., Cleary, I. A., Henriques, M., Lopez-Ribot, J. L., et al. (2010). Presence of extracellular DNA in the *Candida albicans* biofilm matrix and its contribution to biofilms. *Mycopathologia*, *169*(5), 323–331.

Mathew, S., Yaw-Chyn, L., & Kishen, A. (2010). Immunogenic potential of *Enterococcus faecalis* biofilm under simulated growth conditions. *Journal of Endodontics*, *36*(5), 832–836.

McCall, A. D., Pathirana, R. U., Prabhakar, A., Cullen, P. J., & Edgerton, M. (2019). *Candida albicans* biofilm development is governed by cooperative attachment and adhesion maintenance proteins. *NPJ Biofilms and Microbiomes*, *5*(1), 1–12.

McCarty, T. P., & Pappas, P. G. (2015). Invasive candidiasis. *Infectious Disease Clinics of North America*, *15*(00114–2), 1445–1456.

McCormick, A., Heesemann, L., Wagener, J., Marcos, V., Hartl, D., Loeffler, J., et al. (2010). NETs formed by human neutrophils inhibit growth of the pathogenic mold *Aspergillus fumigatus*. *Microbes and Infection*, *12*(12–13), 928–936.

Miller, M. B., & Bassler, B. L. (2001). Quorum sensing in bacteria. *Annual Reviews in Microbiology*, *55*(1), 165–199.

Mishra, M., Byrd, M. S., Sergeant, S., Azad, A. K., Parsek, M. R., McPhail, L., et al. (2012). *Pseudomonas aeruginosa* Psl polysaccharide reduces neutrophil phagocytosis and the oxidative response by limiting complement-mediated opsonization. *Cellular Microbiology*, *14*(1), 95–106.

Mittal, S., Sharma, M., & Chaudhary, U. (2015). Biofilm and multidrug resistance in uropathogenic *Escherichia coli*. *Pathogens and Global Health*, *109*, 26–29.

Miyazato, A., Nakamura, K., Yamamoto, N., Mora-Montes, H. M., Tanaka, M., Abe, Y., et al. (2009). Toll-like receptor 9-dependent activation of myeloid dendritic cells by deoxynucleic acids from *Candida albicans*. *Infection and Immunity*, *77*(7), 3056–3064.

Monasta, L., Ronfani, L., Marchetti, F., Montico, M., VecchiBrumatti, L., Bavcar, A., et al. (2012). Burden of disease caused by otitis media: Systematic review and global estimates. *PLoS One*, *7*(4), e36226.

Mowat, E., Butcher, J., Lang, S., Williams, C., & Ramage, G. (2007). Development of a simple model for studying the effects of antifungal agents on multicellular communities of *Aspergillus fumigatus*. *Journal of Medical Microbiology*, *56*(9), 1205–1212.

Müller, F. M. C., Seidler, M., & Beauvais, A. (2011). *Aspergillus fumigatus* biofilms in the clinical setting. *Medical Mycology*, *49*(Suppl. 1), S96–S100.

Murad, A. M. A., Leng, P., Straffon, M., Wishart, J., Macaskill, S., MacCallum, D., … Brown, A. J. (2001). NRG1 represses yeast–hypha morphogenesis and hypha-specific gene expression in *Candida albicans*. *The EMBO Journal*, *20*(17), 4742–4752.

Mysorekar, I. U., & Hultgren, S. J. (2006). Mechanisms of uropathogenic *Escherichia coli* persistence and eradication from the urinary tract. *Proceedings of the National Academy of Sciences of the United States of America*, *103*(38), 14170–14175.

Nadal Jimenez, P., Koch, G., Thompson, J. A., Xavier, K. B., Cool, R. H., & Quax, W. J. (2012). The multiple signaling systems regulating virulence in *Pseudomonas aeruginosa*. *Microbiology and Molecular Biology Reviews, 76*(1), 46–65.

Nealson, K. H., & Hastings, J. W. (1979). Bacterial bioluminescence: Its control and ecological significance. *Microbiological Reviews, 43*(4), 496–518.

Nett, J. E. (2016). The host's reply to *Candida* biofilm. *Pathogens, 5*(1), 33.

Nett, J. E., Marchillo, K., Spiegel, C. A., & Andes, D. R. (2010). Development and validation of an in vivo *Candida albicans* biofilm denture model. *Infection and Immunity, 78*(9), 3650–3659.

Nikolaev, Y. A., & Plakunov, V. K. (2007). Biofilm—"City of microbes" or an analogue of multicellular organisms? *Microbiology, 76*(2), 25–138.

Nobbs, A. H., Lamont, R. J., & Jenkinson, H. F. (2009). *Streptococcus* adherence and colonization. *Microbiology and Molecular Biology Reviews, 73*(3), 407–450.

Nobile, C. J., Andes, D. R., Nett, J. E., Smith, F. J., Jr., Yue, F., Phan, Q. T., et al. (2006). Critical role of Bcr1-dependent adhesins in *C. albicans* biofilm formation in vitro and in vivo. *PLoS Pathogens, 2*(7), e63.

Nobile, C. J., Nett, J. E., Hernday, A. D., Homann, O. R., Deneault, J. S., Nantel, A., et al. (2009). Biofilm matrix regulation by *Candida albicans* Zap1. *PLoS Biology, 7*(6), e1000133.

Nobile, C. J., Schneider, H. A., Nett, J. E., Sheppard, D. C., Filler, S. G., Andes, D. R., et al. (2008). Complementary adhesin function in *C. albicans* biofilm formation. *Current Biology, 18*(14), 1017–1024.

Nucci, M., & Anaissie, E. (2007). Fusarium infections in immunocompromised patients. *Clinical Microbiology Reviews, 20*(4), 695–704.

Ojha, A. K., Baughn, A. D., Sambandan, D., Hsu, T., Trivelli, X., Guerardel, Y., et al. (2008). Growth of *Mycobacterium tuberculosis* biofilms containing free mycolic acids and harbouring drug-tolerant bacteria. *Molecular Microbiology, 69*(1), 164–174.

Otto, M. (2008). Staphylococcal biofilms. *Bacterial Biofilms*, 207–228.

Otto, M. (2009). *Staphylococcus epidermidis*—The 'accidental' pathogen. *Nature Reviews Microbiology, 7*(8), 555–567.

Painter, K. L., Krishna, A., Wigneshweraraj, S., & Edwards, A. M. (2014). What role does the quorum-sensing accessory gene regulator system play during *Staphylococcus aureus* bacteremia? *Trends in Microbiology, 22*(12), 676–685.

Paradh, A. D., Mitchell, W. J., & Hill, A. E. (2011). Occurrence of Pectinatus and Megasphaera in the major UK breweries. *Journal of the Institute of Brewing, 117*(4), 498–506.

Paramonova, E., Krom, B. P., van der Mei, H. C., Busscher, H. J., & Sharma, P. K. (2009). Hyphal content determines the compression strength of *Candida albicans* biofilms. *Microbiology, 155*(6), 1997–2003.

Parrino, B., Schillaci, D., Carnevale, I., Giovannetti, E., Diana, P., Cirrincione, G., & Cascioferro, S. (2019). Synthetic small molecules as anti-biofilm agents in the struggle against antibiotic resistance. *European Journal of Medicinal Chemistry, 161*, 154–178.

Parsek, M. R., & Greenberg, E. P. (2005). Sociomicrobiology: The connections between quorum sensing and biofilms. *Trends in Microbiology, 13*(1), 27–33.

Pierce, C. G., Vila, T., Romo, J. A., Montelongo-Jauregui, D., Wall, G., Ramasubramanian, A., et al. (2017). The *Candida albicans* biofilm matrix: Composition, structure and function. *Journal of Fungi, 3*(1), 14.

Pitangui, N. S., Sardi, J. C. O., Silva, J. F., Benaducci, T., Moraes da Silva, R. A., Rodríguez-Arellanes, G., et al. (2012). Adhesion of *Histoplasma capsulatum* to pneumocytes and biofilm formation on an abiotic surface. *Biofouling, 28*(7), 711–718.

Płusa, T. (2019). The importance of biofilm in the context of increasing bacterial resistance to antibiotics. *PolskiMerkuriuszLekarski: Organ PolskiegoTowarzystwaLekarskiego, 47*(281), 97–202.

Post, V., Wahl, P., Uckay, I., Zimmerli, W., Corvec, S., Loiez, C., et al. (2014). Biofilm formation and molecular characterisation of *Staphylococcus aureus* isolated from orthopaedic implant related infections depends on type of device. *Orthopaedic Proceedings, 96*(Suppl_11), 283. The British Editorial Society of Bone & Joint Surgery.

Ragland, S. A., & Criss, A. K. (2017). From bacterial killing to immune modulation: Recent insights into the functions of lysozyme. *PLoS Pathogens, 13*(9), e1006512.

Rajendran, R., Sherry, L., Lappin, D. F., Nile, C. J., Smith, K., Williams, C., et al. (2014). Extracellular DNA release confers heterogeneity in *Candida albicans* biofilm formation. *BMC Microbiology, 14*(1), 1–9.

Ramage, G., Rajendran, R., Gutierrez-Correa, M., Jones, B., & Williams, C. (2011). *Aspergillus* biofilms: Clinical and industrial significance. *FEMS Microbiology Letters, 324*(2), 89–97.

Ramage, G., Saville, S. P., Wickes, B. L., & López-Ribot, J. L. (2002). Inhibition of *Candida albicans* biofilm formation by farnesol, a quorum-sensing molecule. *Applied and Environmental Microbiology, 68*(11), 5459–5463.

Ramage, G., Wickes, B. L., & López-Ribot, J. L. (2008). A seed and feed model for the formation of *Candida albicans* biofilms under flow conditions using an improved modified Robbins device. *Revistaiberoamericana de micologia, 25*(1), 37.

Ramakrishnan, Y., Shields, R. C., Elbadawey, M. R., & Wilson, J. A. (2015). Biofilms in chronic rhinosinusitis: What is new and where next? *The Journal of Laryngology & Otology, 129*(8), 744–751.

Ramos, K. J., Somayaji, R., Nichols, D. P., & Goss, C. H. (2018). Comparative effectiveness research in pediatric respiratory disease: Promise and pitfalls. *Paediatric Drugs, 20*(1), 1–7. https://doi.org/10.1007/s40272-017-0256-y.

Ramsook, C. B., Tan, C., Garcia, M. C., Fung, R., Soybelman, G., Henry, R., et al. (2010). Yeast cell adhesion molecules have functional amyloid-forming sequences. *Eukaryotic Cell, 9*(3), 393–404.

Rauceo, J. M., Gaur, N. K., Lee, K. G., Edwards, J. E., Klotz, S. A., & Lipke, P. N. (2004). Global cell surface conformational shift mediated by a *Candida albicans* adhesin. *Infection and Immunity, 72*(9), 4948–4955.

Rigby, K. M., & DeLeo, F. R. (2012). Neutrophils in innate host defense against *Staphylococcus aureus* infections. *Seminars in Immunopathology, 34*(2), 237–259. Springer-Verlag.

Robinet, P., Baychelier, F., Fontaine, T., Picard, C., Debré, P., Vieillard, V., et al. (2014). A polysaccharide virulence factor of a human fungal pathogen induces neutrophil apoptosis via NK cells. *The Journal of Immunology, 192*(11), 5332–5342.

Ryder, C., Byrd, M., & Wozniak, D. J. (2007). Role of polysaccharides in *Pseudomonas aeruginosa* biofilm development. *Current Opinion in Microbiology, 10*(6), 644–648.

Sadiq, F. A., Flint, S., Yuan, L., Li, Y., Liu, T., & He, G. (2017). Propensity for biofilm formation by aerobic mesophilic and thermophilic spore forming bacteria isolated from Chinese milk powders. *International Journal of Food Microbiology, 262*, 89–98.

Sardi, J. D. C. O., Pitangui, N. D. S., Voltan, A. R., Braz, J. D., Machado, M. P., Fusco Almeida, A. M., et al. (2015). In vitro *Paracoccidioides brasiliensis* biofilm and gene expression of adhesins and hydrolytic enzymes. *Virulence, 6*(6), 642–651.

Sarkar, S., Ulett, G. C., Totsika, M., Phan, M. D., & Schembri, M. A. (2014). Role of capsule and O antigen in the virulence of uropathogenic *Escherichia coli*. *PLoS One, 9*(4), e94786.

Saunders, J., Murray, M., & Alleman, A. (2011). Biofilms in chronic suppurative otitis media and cholesteatoma: Scanning electron microscopy findings. *American Journal of Otolaryngology*, *32*(1), 32–37.

Scherr, T. D., Heim, C. E., Morrison, J. M., & Kielian, T. (2014). Hiding in plain sight: Interplay between staphylococcal biofilms and host immunity. *Frontiers in Immunology*, *5*, 37.

Scherr, T. D., Roux, C. M., Hanke, M. L., Angle, A., Dunman, P. M., & Kielian, T. (2013). Global transcriptome analysis of *Staphylococcus aureus* biofilms in response to innate immune cells. *Infection and Immunity*, *81*(12), 4363–4376.

Schilder, A. G., Chonmaitree, T., Cripps, A. W., Rosenfeld, R. M., Casselbrant, M. L., Haggard, M. P., et al. (2016). Otitis media. *Nature Reviews Disease Primers*, *2*(1), 1–18.

Schiller, N. L., & Joiner, K. A. (1986). Interaction of complement with serum-sensitive and serum-resistant strains of *Pseudomonas aeruginosa*. *Infection and Immunity*, *54*(3), 689–694.

Schweizer, A., Rupp, S., Taylor, B. N., Röllinghoff, M., & Schröppel, K. (2000). The TEA/ATTS transcription factor CaTec1p regulates hyphal development and virulence in *Candida albicans*. *Molecular Microbiology*, *38*(3), 435–445.

Seidler, M. J., Salvenmoser, S., & Müller, F. M. (2008). *Aspergillus fumigatus* forms biofilms with reduced antifungal drug susceptibility on bronchial epithelial cells. *Antimicrobial Agents and Chemotherapy*, *52*(11), 4130–4136.

Seneviratne, C. J., Jin, L., & Samaranayake, L. P. (2008). Biofilm lifestyle of *Candida*: A mini review. *Oral Diseases*, *14*(7), 582–590.

Sharma, M., & Prasad, R. (2011). The quorum-sensing molecule farnesol is a modulator of drug efflux mediated by ABC multidrug transporters and synergizes with drugs in *Candida albicans*. *Antimicrobial Agents and Chemotherapy*, *55*(10), 4834–4843.

Shirtliff, M. E., Krom, B. P., Meijering, R. A., Peters, B. M., Zhu, J., Scheper, M. A., et al. (2009). Farnesol-induced apoptosis in *Candida albicans*. *Antimicrobial Agents and Chemotherapy*, *53*(6), 2392–2401.

Silva, S., Henriques, M., Martins, A., Oliveira, R., Williams, D., & Azeredo, J. (2009). Biofilms of non-*Candida albicans Candida* species: Quantification, structure and matrix composition. *Sabouraudia*, *47*(7), 681–689.

Singh, A., & Del Poeta, M. (2011). Lipid signalling in pathogenic fungi. *Cellular Microbiology*, *13*(2), 177–185.

Skov, M., McKay, K., Koch, C., & Cooper, P. J. (2005). Prevalence of allergic bronchopulmonary aspergillosis in cystic fibrosis in an area with a high frequency of atopy. *Respiratory Medicine*, *99*(7), 887–893.

Slock, J., VanRiet, D., Kolibachuk, D., & Greenberg, E. P. (1990). Critical regions of the *Vibrio fischeri* LuxR protein defined by mutational analysis. *Journal of Bacteriology*, *172*(7), 3974–3979.

Snarr, B. D., Baker, P., Bamford, N. C., Sato, Y., Liu, H., Lehoux, M., et al. (2017). Microbial glycoside hydrolases as antibiofilm agents with cross-kingdom activity. *Proceedings of the National Academy of Sciences of the United States of America*, *114*(27), 7124–7129.

Snow, D. E., Everett, J., Mayer, G., Cox, S. B., Miller, B., Rumbaugh, K., et al. (2016). The presence of biofilm structures in atherosclerotic plaques of arteries from legs amputated as a complication of diabetic foot ulcers. *Journal of Wound Care*, *25*(Suppl. 2), S16–S22.

Srivastava, K., Chauhan, D. S., Gupta, P., Singh, H. B., Sharma, V. D., Yadav, V., et al. (2008). Isolation of *Mycobacterium bovis* & *M. tuberculosis* from cattle of some farms in north India—Possible relevance in human health. *Indian Journal of Medical Research*, *128*(1), 26–31.

Stevens, A. M., Dolan, K. M., & Greenberg, E. P. (1994). Synergistic binding of the *Vibrio fischeri* LuxR transcriptional activator domain and RNA polymerase to the lux promoter region. *Proceedings of the National Academy of Sciences of the United States of America*, *91*, 12619–12623.

Strejc, J., Kyselova, L., Cadkova, A., Matoulkova, D., Potocar, T., & Branyik, T. (2020). Experimental adhesion of *Geobacillus stearothermophilus* and *Anoxybacillus flavithermus* to stainless steel compared with predictions from interaction models. *Chemical Papers*, *74*(1), 297–304.

Sturme, M. H., Kleerebezem, M., Nakayama, J., Akkermans, A. D., Vaughan, E. E., & De Vos, W. M. (2002). Cell to cell communication by autoinducing peptides in gram-positive bacteria. *Antonie Van Leeuwenhoek*, *81*(1), 233–243.

Sturtevant, J., & Latgé, J. P. (1992). Participation of complement in the phagocytosis of the conidia of *Aspergillus fumigatus* by human polymorphonuclear cells. *Journal of Infectious Diseases*, *166*(3), 580–586.

Sun, Y., Chandra, J., Mukherjee, P., Szczotka-Flynn, L., Ghannoum, M. A., & Pearlman, E. (2010). A murine model of contact lens–associated *Fusarium* keratitis. *Investigative Ophthalmology & Visual Science*, *51*(3), 1511–1516.

Sutherland, I. W. (2001a). Biofilm exopolysaccharides: A strong and sticky framework. *Microbiology*, *147*(1), 3–9.

Sutherland, I. W. (2001b). The biofilm matrix—An immobilized but dynamic microbial environment. *Trends in Microbiology*, *9*(5), 222–227.

Taff, H. T., Nett, J. E., Zarnowski, R., Ross, K. M., Sanchez, H., Cain, M. T., et al. (2012). A *Candida* biofilm-induced pathway for matrix glucan delivery: Implications for drug resistance. *PLoS Pathogens*, e1002848.

Taff, H. T., Nett, J. E., Zarnowski, R., Ross, K. M., Sanchez, H., Cain, M. T., … Andes, D. R. (2012). A Candida biofilm-induced pathway for matrix glucan delivery: implications for drug resistance. *Plos Pathogens*, *8*, e1002848.

Teymoortash, A., Wollstein, A. C., Lippert, B. M., Peldszus, R., & Werner, J. A. (2002). Bacteria and pathogenesis of human salivary calculus. *Acta Oto-Laryngologica*, *122*(2), 210–214.

Thammavongsa, V., Missiakas, D. M., & Schneewind, O. (2013). *Staphylococcus aureus* degrades neutrophil extracellular traps to promote immune cell death. *Science*, *342*(6160), 863–866.

Thurlow, L. R., Hanke, M. L., Fritz, T., Angle, A., Aldrich, A., Williams, S. H., et al. (2011). *Staphylococcus aureus* biofilms prevent macrophage phagocytosis and attenuate inflammation in vivo. *The Journal of Immunology*, *186*(11), 6585–6596.

Uppuluri, P., Chaturvedi, A. K., Srinivasan, A., Banerjee, M., Ramasubramaniam, A. K., Köhler, J. R., et al. (2010). Dispersion as an important step in the *Candida albicans* biofilm developmental cycle. *PLoS Pathogens*, *6*(3), e1000828.

Vaishnavi, C., Kochhar, R., Singh, G., Kumar, S., Singh, S., & Singh, K. (2005). Epidemiology of typhoid carriers among blood donors and patients with biliary, gastrointestinal and other related diseases. *Microbiology and Immunology*, *49*(2), 107–112.

Van Der Plas, M. J., Bhongir, R. K., Kjellström, S., Siller, H., Kasetty, G., Mörgelin, M., et al. (2016). *Pseudomonas aeruginosa* elastase cleaves a C-terminal peptide from human thrombin that inhibits host inflammatory responses. *Nature Communications*, *7*(1), 1–13.

Van Hoecke, H., De Paepe, A. S., Lambert, E., Van Belleghem, J. D., Cools, P., Van Simaey, L., et al. (2016). *Haemophilus influenzae* biofilm formation in chronic otitis media with effusion. *European Archives of Oto-Rhino-Laryngology*, *273*(11), 3553–3560.

Verdrengh, M., & Tarkowski, A. (1997). Role of neutrophils in experimental septicemia and septic arthritis induced by *Staphylococcus aureus*. *Infection and Immunity*, *65*(7), 2517–2521.

Verdrengh, M., & Tarkowski, A. (2000). Role of macrophages in *Staphylococcus aureus*–induced arthritis and sepsis. *Arthritis & Rheumatism: Official Journal of the American College of Rheumatology*, *43*(10), 2276–2282.

Vila, T. V. M., Quintanilha, N. S., & Rozental, S. (2015). Miltefosine is effective against Candida albicans and Fusarium oxysporum nail biofilms in vitro. *Journal of Medical Microbiology*, *64*(11), 1–14.

Vila, T. V. M., Rozental, S., & Guimarães, C. M. D.de S. (2015). A new model of in vitro fungal biofilms formed on human nail fragments allows reliable testing of laser and light therapies against onychomycosis. *Lasers in Medical Science*, *30*(3), 1031–1039.

Wade, D. S., Calfee, M. W., Rocha, E. R., Ling, E. A., Engstrom, E., Coleman, J. P., et al. (2005). Regulation of *Pseudomonas* quinolone signal synthesis in *Pseudomonas aeruginosa*. *Journal of Bacteriology*, *187*, 4372–4380.

Walters, M., & Sperandio, V. (2006). Quorum sensing in *Escherichia coli* and *Salmonella*. *International Journal of Medical Microbiology*, *296*, 125–131.

Weiser, J. N., Ferreira, D. M., & Paton, J. C. (2018). *Streptococcus pneumoniae*: Transmission, colonization and invasion. *Nature Reviews Microbiology*, *16*(6), 355–367.

Wilder, C. N., Diggle, S. P., & Schuster, M. (2011). Cooperation and cheating in *Pseudomonas aeruginosa*: The roles of the las, rhl and pqs quorum-sensing systems. *The ISME Journal*, *5*(8), 1332–1343.

Williams, P. (2007). Quorum sensing, communication and cross-kingdom signalling in the bacterial world. *Microbiology*, *153*(12), 3923–3938.

Winzer, K., Falconer, C., Garber, N. C., Diggle, S. P., Camara, M., & Williams, P. (2000). The *Pseudomonas aeruginosa* lectins PA-IL and PA-IIL are controlled by quorum sensing and by RpoS. *Journal of Bacteriology*, *182*, 6401–6411.

Wu, P., Guo, Y., Golly, M. K., Ma, H., He, R., Luo, S., et al. (2019). Feasibility study on direct fermentation of soybean meal by *Bacillus stearothermophilus* under non-sterile conditions. *Journal of the Science of Food and Agriculture*, *99*(7), 3291–3298.

Yarwood, J. M., Bartels, D. J., Volper, E. M., & Greenberg, E. P. (2004). Quorum sensing in *Staphylococcus aureus* biofilms. *Journal of Bacteriology*, *186*(6), 1838–1850.

Yong, Y. Y., Dykes, G. A., & Choo, W. S. (2019). Biofilm formation by staphylococci in health-related environments and recent reports on their control using natural compounds. *Critical Reviews in Microbiology*, *45*(2), 201–222.

Zarnowski, R., Westler, W. M., Lacmbouh, G. A., Marita, J. M., Bothe, J. R., Bernhardt, J., … Andes, D. R. (2014). Novel entries in a fungal biofilm matrix encyclopedia. *MBio*, *5*(4), e01333–14.

Zhang, X., Sun, X., Wang, Z., Zhang, Y., & Hou, W. (2012). Keratitis-associated fungi form biofilms with reduced antifungal drug susceptibility. *Investigative Ophthalmology & Visual Science*, *53*(12), 7774–7778.

Further reading

Beauvais, A., & Latgé, J. P. (2015). *Aspergillus* biofilm in vitro and in vivo. *Microbial Biofilms*, 149–161.

Braun, D. K., Janssen, D. A., Marcus, J. R., & Kauffman, C. A. (1994). Cryptococcal infection of a prosthetic dialysis fistula. *American Journal of Kidney Diseases*, *24*(5), 864–867.

Costerton, J. W., Cheng, K. J., Geesey, G. G., Ladd, T. I., Nickel, J. C., Dasgupta, M., et al. (1987). Bacterial biofilms in nature and disease. *Annual Reviews in Microbiology*, *41*(1), 435–464.

Gonzalez-Escobedo, G., & Gunn, J. S. (2013). Gallbladder epithelium as a niche for chronic *Salmonella* carriage. *Infection and Immunity*, *81*(8), 2920–2930.

Hoveida, L., Halaji, M., Rostami, S., & Mobasherizadeh, S. (2019). Biofilm-producing ability of *Staphylococcus* spp isolated from different foodstuff products. *Annali di Igiene*, *31*(2), 140–147.

Kao, W. K., Chole, R. A., & Ogden, M. A. (2020). Evidence of a microbial etiology for sialoliths. *The Laryngoscope*, *130*(1), 69–74.

Klančnik, A., Šimunović, K., Sterniša, M., Ramić, D., Možina, S. S., & Bucar, F. (2021). Anti-adhesion activity of phytochemicals to prevent *Campylobacter jejuni* biofilm formation on abiotic surfaces. *Phytochemistry Reviews*, *20*(1), 55–84.

Langsrud, T., Doulgeraki, S., Nychas, A., Kaˇcániová, G.-J., Czaczyk, M., K., et al. (2015). Intra- and inter-species interactions within biofilms of important foodborne bacterial pathogens. *Frontiers in Microbiology*, *6*.

Lelouard, H., Henri, S., De Bovis, B., Mugnier, B., Chollat-Namy, A., Malissen, B., et al. (2010). Pathogenic bacteria and dead cells are internalized by a unique subset of Peyer's patch dendritic cells that express lysozyme. *Gastroenterology*, *138*(1), 173–184.

Ramos, G. P., Rocha, J. L., & Tuon, F. F. (2013). Seasonal humidity may influence *Pseudomonas aeruginosa* hospital-acquired infection rates. *International Journal of Infectious Diseases*, *17*(9), e757–e761.

Rovers, M. M., Schilder, A. G., Zielhuis, A. G., & Rosenfeld, R. M. (2004). Otitis media. *Lancet*, *363*, 465–473.

Tournu, H., & Van Dijck, P. (2012). *Candida* biofilms and the host: Models and new concepts for eradication. *International Journal of Microbiology*, *2012*.

Vila, T. V. M., Quintanilha, N. S., & Rozental, S. (2015). Miltefosine is effective against *Candida albicans* and *Fusarium oxysporum* nail biofilms in vitro. *Journal of Medical Microbiology*, *64*(11), 1436–1449.

Zarnowski, R., Sanchez, H., Covelli, A. S., Dominguez, E., Jaromin, A., Bernhardt, J., et al. (2018). *Candida albicans* biofilm–induced vesicles confer drug resistance through matrix biogenesis. *PLoS Biology*, *16*(10), e2006872.

Chapter 33

Pathogenic biofilms in environment and industrial setups and impact on human health

Samreen[a], Mohd Sajjad Ahmad Khan[b], and Iqbal Ahmad[a]

[a]Department of Agricultural Microbiology, Faculty of Agricultural Sciences, Aligarh Muslim University, Aligarh, Uttar Pradesh, India, [b]Department of Basic Sciences, Deanship of Preparatory Year and Supporting Studies, Imam Abdulrahman Bin Faisal University, Dammam, Saudi Arabia

1. Introduction

The formation of biofilm is a common phenomenon in nature, where microorganisms choose to survive on any suitable surface as community instead of planktonic cells (Flemming, 2020). The natural biofilm is the consortium made up of a single microbial species or a mix of bacteria, protozoa, archaea, filamentous fungi, algae, and yeast that form a strong bond with one other and with biotic or abiotic surfaces. All these cells are encased in a matrix of extracellular polymeric substances (EPSs) (Pinto, Langer, Hüffer, Hofmann, & Herndl, 2019; Tan, Lee, Burmølle, Kjelleberg, & Rice, 2017). In a protected environment such as biofilms, microorganisms have advantage in terms of cell proliferation and survival. Furthermore, biofilm facilitates easy access to the limited nutrition supply in the surrounding environment. In natural environmental settings, biofilms are developed by a variety of microbial species on food processing and medical equipment. The widespread usage of antiseptics, disinfectants such as quaternary ammonium compounds and other biocides has resulted in microbial resistance in the form of mixed biofilms (Pang, Wong, Chung, & Yuk, 2019). Biofilms can have a negative impact in natural and clinical settings. In environment, biofilm cause problems such as biocorrosion, decreased water quality, and foci for sanitary product contamination. On the other side, biofilm colonization can also occur on implanted biomedical devices and other surgical equipment, leading to higher infection rates and transmission of diseases (Abebe, 2020: Schulze, Mitterer, Pombo, & Schild, 2021). Biofilms have been found to develop on medical equipment surfaces and on human tissues in hospital environments, producing persistent infections (Schulze et al., 2021). Biofilms can form on medical implants, allowing infections to thrive and even facilitate human deaths as these hardy microbial colonies are resistant to drugs and immune system attack (Arciola, Campoccia, & Montanaro, 2018). Therefore, experts and general public at large have tended to concentrate on preventive and controlling strategies of biofilms.

Furthermore, groundwater flow biofilms are thought to be the excellent measures of the quality of the water and ecological condition of the system (i.e., ecosystem health) (Scharping & Garey, 2021). The evaluation of how biofilm communities adapt to anthropogenic pollution of aquatic ecosystems (e.g., rivers, lakes, and reservoirs) is of great interest. It is evident that numerous chemical compounds (personal care products, metals, and drugs used in veterinary and human medicine) are released into these water bodies primarily via effluents of wastewater treatment plants (WWTPs) and agrochemicals discharge (Calapez et al., 2020). According to the WHO, the fatality rate from water borne diseases surpasses 5 million people per year, with intestinal infections accounting for more than half of these. Cholera is a disease that has a high mortality rate (WHO, 2008). Furthermore, biofouling in industrial and drinking water system has negative consequences such as chemical and microbiological degradation of water quality, yield loss, reduction in heat transmission and exchange efficiency, and membrane technique failure (Yu et al., 2021). Microorganisms develop biofilms in response to a wide range of environmental stimuli, including nutrition availability, temperature, osmolarity, pH, iron, and oxygen concentrations (Primo et al., 2021; Rossi, Paroni, & Landini, 2018; Tallawi, Opitz, & Lieleg, 2017).

The ability of microorganisms to develop biofilm has been considered as an ancient survival system, which gives microorganisms more possibilities than planktonic cells, including a stronger way to thrive in oligotrophic environments (de Carvalho, 2018), greater access to nutritional supplies (Chen et al., 2018), enhanced resistance to biocides (Flemming et al., 2016), greater eutrophication (Dang &

Understanding Microbial Biofilms. https://doi.org/10.1016/B978-0-323-99977-9.00012-0

Lovell, 2016), as well as improved organism productivity and interconnections (Røder et al., 2018). The growth of biofilms proved harmful in a number of scenarios. Pathogenic bacteria, for example, can build biofilms inside food processing facilities, causing food deterioration and compromising consumer health (Galie, García-Gutiérrez, Miguélez, Villar, & Lombó, 2018; Mevo, Ashrafudoulla, Furkanur Rahaman Mizan, Park, & Ha, 2021). Therefore, we aimed to provide comprehensive aspect on environmental pathogenic biofilms and its impact on human health. A better understanding is expected to develop strategy to fight problematic biofilms in environment.

2. Development of environmental biofilms

Biofilms are surface-attached colonies of microbes bound together by a self-produced polymer matrix made primarily of polysaccharides, secreted proteins, and extracellular DNAs (Berne, Ellison, Ducret, & Brun, 2018; Rumbaugh & Sauer, 2020). The development of bacterial biofilms on to solid surfaces is a multistep process that begins with reversible adherence helped by hydrophobicity and intermolecular interactions, then progresses to the synthesis of EPS that allow the cells to stick to the substrate permanently (Caruso, 2020). The biofilm development process is divided into five phases: reversible adhesion, irreversible attachment, EPS generation, biofilm maturation, and dispersal/detachment (Toyofuku et al., 2016) (Fig. 1). The transcription and regulation processes of distinct bacteria species on various stages of biofilms, on the other hand, are highly variable.

The extracellular matrix is mostly made up of polysaccharides such as cellulose, proteins, and foreign DNA. This matrix can be attached to either hard surfaces (food or dairy industry equipment, transporting, dispensing, and storage surfaces, soil, and so on) or biological macromolecules (vegetables, meat, bones, fruits, etc.) (Carrascosa, Raheem, Ramos, Saraiva, & Raposo, 2021; Giaouris, Simões, & Dubois-Brissonnet, 2020; Neu & Lawrence, 2015). The extracellular polymeric matrix of biofilm plays a structural function in the biofilms' long-term persistence in many environments. In general, the production of bacterial biofilms is dependent on the interaction of bacterial cells, substrates, and the surrounding media (Fysun, Kern, Wilke, & Langowski, 2019; Stanley & Lazazzera, 2004).

Many microbial species can switch their mode of lifestyle (free-living vs attached) based on their physicochemical circumstances and the physiological state in their environment, providing advantage of the increased availability of organic materials in suspended matter and surfaces (Chatterjee, Samal, Singh, Pradhan, & Verma, 2020; Chua et al., 2014). Primarily, in the environmental settings, biofilms have an active role in the breakdown of organic matter, nutrient dynamics, and biogeochemical cycle of chemical substance, making them an important aspect of ecosystem function (Weigelhofer, Jirón, Yeh,

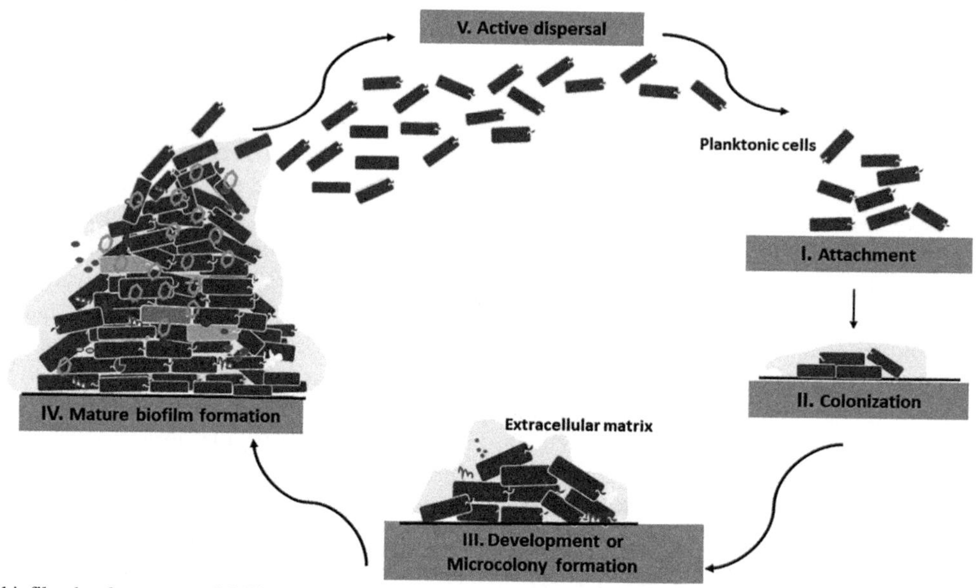

FIG. 1 Microbial biofilm development model. There are five phases to the formation of a biofilm: I. Attachment: Microbial cells reversibly adhere to the surface. II. Microbial cells attach to the surface irreversibly by flagella, pili, exopolysaccharides, and other mechanisms. III. Multilayered cells aggregate and generate extracellular polymeric compounds through development of EPS. IV. Maturation: the establishment of a stable three-dimensional microbial community. V. Active dispersal: microorganisms are dispersed and return to a planktonic form after being dispersed from the aggregate biofilm. *(Modified from Toyofuku, M., Inaba, T., Kiyokawa, T., Obana, N., Yawata, Y., & Nomura, N. (2016). Environmental factors that shape biofilm formation.* Bioscience, Biotechnology, and Biochemistry, 80*(1), 7–12.)*

Steniczka, & Pucher, 2020). Biofilms developed in mixed or polymicrobial state under natural environmental conditions or undifferentiated stress conditions performed abovementioned functions.

3. Biofilm under environmental stress

Bacteria choose biofilm development as a distinct growth mode in response to varied environmental challenges. Various studies have demonstrated that potential for biofilm development is critical for microorganisms to proliferate in a variety of extreme conditions. Biofilm formation, architecture, and function are dependent on a wide range and combination of surface morphologies, environmental factors, and various other microbiological factors. In a study conducted by Chu, Kilic, Cho, Groisman, and Levchenko (2018), uropathogenic *Escherichia coli* colonies showed increased and diverse biofilm formation under environmental stresses including mechanical and biological conditions. The mechanical stress initially emerges from the physical stress accompanying colony detention within microcavities or hydrogel environments similar to the cytosol of host cells. Biofilm formation can be enhanced by biofilm matrix deposition as it increases stress levels (Chu et al., 2018).

Biofilm architecture is a complex design of biofilm structure. Once the surface is conditioned with variety of organic pollutants or materials, sequential development of biofilm takes place, and its architecture is dependent upon various microbiological and environmental factors. In majority of environmental conditions, biofilm is not found as single species but as consortium. In biofilm, microbial cells come in association with each other either by direct contact or through signaling molecules, phenomenon called quorum sensing. Quorum sensing is widely recognized as an efficient mechanism to regulate expression of specific genes responsible for communal behaviors in bacteria. Exopolymeric substances (EPSs) are also an important constituent of biofilm formation at surface liquid interfaces. Bacteria and other microorganisms produce extracellular matrix components, which help them adhere to surfaces (Boo, Amaro, & Stan, 2021). However, the chemical composition of EPS matrix may differ, depending on the medium in which the biofilm is grown.

The age of a biofilm has also been suggested to influence both biofilm architecture and function, where an increase in *E. coli* O157:H7 population was observed as storage time progressed on surfaces encountered in meat processing plants and aged *Listeria monocytogenes* biofilms demonstrated resistance to desiccation. Temperature has an impact on bacteria, and biofilm can adequately describe the effects of microorganisms under severe temperatures. Cihan, Karaca, Ozel, and Kilic (2017) examined how thermophilic bacteria belonging from the Bacillaceae family formed biofilms at different temperatures and discovered that could form more efficient biofilm at 65°C incubation temperature than at 55°C for the thermophilic genera *Aeribacillus*, *Anoxybacillus*, *Thermolongibacillus*, and *Geobacillus*. Likewise, the cold-adapted *Bacteriovorax* bacteria revealed that at temperatures below 10°C, the quantity of bacteria in the water column is dramatically reduced, but not in superficial biofilms (Yin, Wang, Liu, & He, 2019). Williams, Turng, and Kelley (2009) found that *Bacteriovorax* lived 50% longer in biofilms than in suspension at 5°C. EPS in biofilms from Antarctic bacteria (cold-tolerant Winogradskyella CAL384 and CAL396, psychrophiles *Colwellia* GW185, and *Shewanella* CAL606) also exhibited the potential to create stable complexes to protect cells from freeze-thaw cycles, enhancing microbial cell adaptation to cold settings (Lo Giudice, Poli, Finore, & Rizzo, 2020). As a result, biofilm formation allows bacteria to be more tolerant to temperature stress in harsh settings. Biofilm communities can operate as a "protective clothing" for microbes, providing a suitable atmosphere for survival and metabolic activity when the environment changes (Yin et al., 2019). The impact of surrounding media has also been shown to influence the biofilm architecture whereby, elevated level of biofilm production was achieved in nutrient-limited conditions in case of *L. monocytogenes* (Russo et al., 2018). It has been shown in recent years that cells grown in different nutrient conditions produce diverse biofilm morphologies. For example, mature biofilms of meat *Salmonella* spp. produced in meat thawing loss broth had a "cloud-shaped" morphology, but biofilms grown in tryptic soy broth had a "reticular-shaped" morphology (Pradhan & Negi, 2019). Mature biofilms for a variety of food pathogens are often described as a collection of clusters or knitted chains may be ball-shaped (*L. monocytogenes*) (Cherifi, Jacques, Quessy, & Fravalo, 2017), mushroom-shaped (*Staphylococcus* spp.), or honeycomb-shaped *Vibrio cholerae* (Drescher et al., 2016). Microorganisms in biofilms are more resistant to high pressure than planktonic microorganisms, according to studies on high hydrostatic pressure (HHP) biofilms (Nogi, 2017). Hou, Veeregowda, van de Belt-Gritter, Busscher, and van der Mei (2018) tried to compare the mechanical pressure responses of biofilms from an EPS-producing *Staphylococcus aureus* ATCC 12600 and a non-EPS-producing *S. aureus* 5298 and discovered that the former bacterium's biofilm is more resistant to mechanical pressure since it can produce an immediate increase in polysaccharide composition. (Hou et al., 2018). Furthermore, biofilms also assist microorganisms to withstand the impacts of high pH, as biofilms typically contain both acidophiles and alkaliphiles (Hujslová, Bystrianský, Benada, & Gryndler, 2019; Li, Kappler, Jiang, & Bond, 2017). Charles et al. (2017) exhibited that *Alishewanella* and *Dietzia* can grow at pH values between 11.0 and 11.5 and maintain internal pH

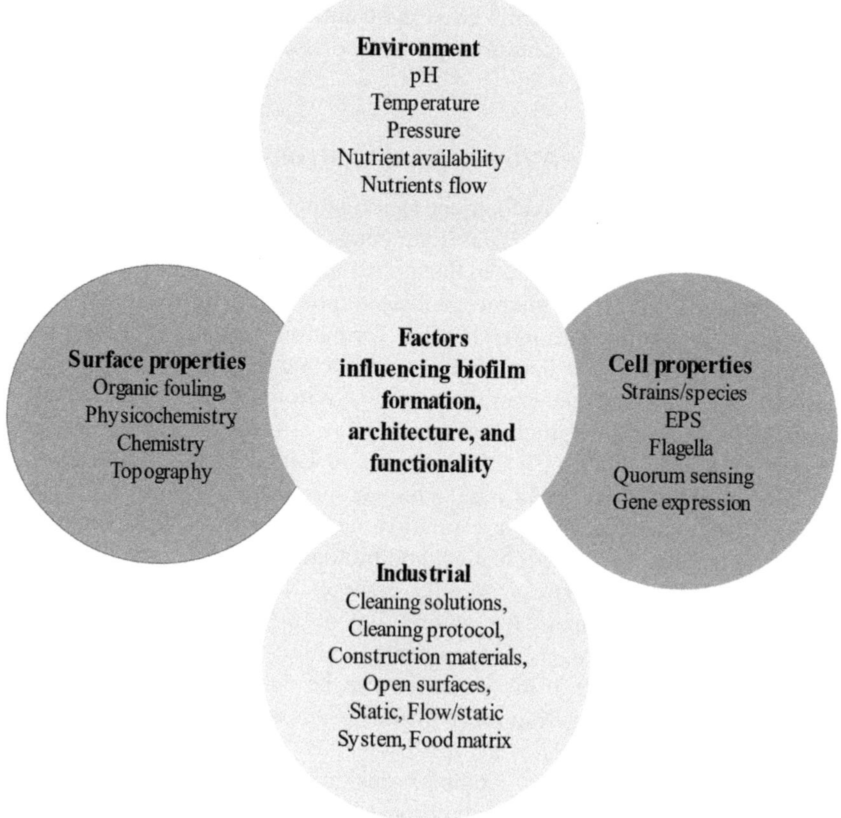

FIG. 2 Major factors affecting biofilm formation, architecture, and functionality. *(Modified from Whitehead, K.A., & Verran, J. (2015). Formation, architecture and functionality of microbial biofilms in the food industry.* Current Opinion in Food Science, 2, *84–91.)*

values of 10.4 and 10.7, respectively, under a thick EPS layer. They concluded that biofilm formation can provide a safe haven for alkaliphilic communities to survive hyperalkaline stress (Charles et al., 2017). Various factors influencing the formation, architecture, and functionality of biofilm is represented in Fig. 2.

4. Antibiotic resistance and tolerance in biofilm

The resistant bacteria can thrive at concentrations that would ordinarily have inhibitory effect on the growth, i.e., at bactericidal or bacteriostatic concentrations. Resistance has been most associated with mutations or the interchange of antibiotic resistance genetic elements (acquired resistance), but it can also be intrinsic and hence rely on wild-type genes and innate cell properties (Blair, Webber, Baylay, Ogbolu, & Piddock, 2015). Tolerance to an antimicrobial agent, on the other hand, is the ability of a bacterium to live in the presence of a bactericidal antimicrobial agent without growing or dying (Crabbé, Jensen, Bjarnsholt, & Coenye, 2019). Moreover, biofilms might give antibiotic tolerance to bacteria by acting as physical barriers, with their thickness and chemical composition preventing antibiotic penetration (Dincer, Uslu, & Delik, 2020). Uronic acids, glycoproteins, proteins, eDNA, glycolipids, and other anionic and cationic molecules can be found in the EPS of biofilms. They can also attach to charged antibiotics and provide microbes with a place of refuge (Simões, Pereira, Simões, Cagide, & Borges, 2021; Zhang, Dong, Hua, & Guo, 2018). In the *Pseudomonas aeruginosa* matrix of EPS, each exopolysaccharide is cross-linked with negatively charged eDNA (Jennings et al., 2015), which can also perform structural and protective roles in lowering its susceptibility to aminoglycoside antibiotics (Colvin et al., 2011). Physiological limitations including the growth rate, biofilm age, starvation, etc., can reduce biofilm susceptibility to the antibiotics. Williamson et al. (2012) found that a subset of inactive population of *P. aeruginosa* biofilms showed elevated level of resistance toward tobramycin and ciprofloxacin, but that the actively growing populations still showed susceptibility toward antibiotics.

5. Wastewater as hotspot of antibiotic resistance in environment

In the last decade, it has been revealed that drug resistance found in microorganisms that leads to clinical infection is mostly due to the environment (Pruden, 2014). In actuality, contaminants and drug-resistant bacteria reach the environment through wastewater released processed or untreated, as well as through animal farms, aquaculture, and industry (Berendonk et al., 2015; Gao et al., 2020). WWTPs are thought to be one of the most important reservoirs and sources of antibiotic resistance emergence and spread. Numerous studies have revealed that traditional biological wastewater treatments used in WWTPs are unable to entirely eradicate antimicrobial resistance genes (ARGs) (Korzeniewska & Harnisz, 2018; Osińska, Korzeniewska, & Harnisz, 2019). Antibiotic resistance is spread through wastewater treatment plants, which have been identified as hotspots (Karkman, Do, Walsh, & Virta, 2018). Inside the wastewater treatment plant, however, there could be unrecognized hotspots for antibiotic resistance, biofilm surfaces are one such place. EPSs form most of the biofilm matrix (Flemming, 2016). Antibiotic adsorption to EPS in activated sludge systems has been vastly studied (Wang, Yuan, Li, Zhang, & Sheng, 2021; Xu, Sheng, Ma, Wang, & Yu, 2013; Zhang et al., 2016). Human activities, such as continual discharges of waste from WWTP effluents, have an impact on many aquatic systems (lakes, rivers, and reservoirs) (Carey & Migliaccio, 2009). Under these conditions, macro- and microorganisms living in these bodies of water are exposed to a low but consistent concentration of a wide range of chemical pollutants (analgesics, antibiotics, antiinflammatory, and psychiatric pharmaceuticals, blockers, pesticides, etc.) that affect and alter their behavior at different levels, with repercussions that are not easy to understand (Bernier & Surette, 2013; Woegerbauer, Bellanger, & Merlin, 2020). When comparing antibiotic concentrations detected in contaminated aquatic ecosystems (ranging from ng/L to g/L) to those used to treat bacterial infections (therapeutic values, which are usually less than 1 mg/L), a severe problem emerges.

Antimicrobial chemicals in the environment are several folds lesser than the minimum inhibitory concentration (MIC) of most bacterial pathogens, therefore their antibiotic activity is questionable, if it exists at all (Blair, 2018; Davies & Davies, 2010; Waksman, 1961). At these sub-MIC concentrations, current evidence clearly suggests that antibiotics serve as signaling molecules, mediating a wide range of cell processes (gene transcription and expression, quorum sensing, inter- or intraspecies communication, biofilm formation, and so forth) (Andersson & Hughes, 2014; Ranieri, Whitchurch, & Burrows, 2018), rather than just inducing cell death or growth arrest. Furthermore, low antibiotic concentrations may trigger a variety of stress responses, which could hasten horizontal gene transfer (HGT) and the spread of ARGs in a variety of bacterial species (Sun, Jeannot, Xiao, & Knapp, 2019). Despite these considerations, it is now known that continuous antibiotic exposure, even at extremely low doses, generates and maintains a pool of antibiotic resistance in natural bacteria (Abe, Nomura, & Suzuki, 2020; Andersson & Hughes, 2014). It should be noted, however, that while many of these genetic elements when expressed confer a resistant phenotype, they are unlikely to be "real" resistance genes, with a function that is only distantly analogous compared with that in treatment settings (Allen et al., 2010; Martínez, Coque, & Baquero, 2015).

Wastewater treatment procedures have been found, in numerous studies, to be incapable of removing antimicrobial resistant bacteria (ARB) and ARGs from source water. Furthermore, WWTPs have been demonstrated to enhance the abundance of certain ARGs in plant effluent under certain instances (Kim et al., 2018; Mao et al., 2015). A metagenomic investigation found no significant differences in the ARGs observed in potable supply water and treated water at the extraction stage, indicating that ARGs were not eliminated during the drinking water treatment (Garner et al., 2018). Sewer pipes with inner surface biofilms, in addition to WWTPs, have been indicated as potential hotspots for ARG build-up and dispersal. According to Auguet et al. (2017), there are no significant variations in ARG concentrations in wastewater entering and exiting sewer systems (Auguet et al., 2017). Another potential hotspot in wastewater treatment plants is the gel/cake layer in tertiary treatment systems such as ultrafiltration and reverse osmosis. If the compound is a mutagen, high levels of the chemical in the biofilm create selection pressure for the bacteria in the gel/cake layer and in vicinity to evolve antibiotic resistance through mutation (Amarasiri, Sano, & Suzuki, 2020). ARB will be inactivated by disinfectants even if the membrane is breached. Environmental microorganisms can acquire ARGs and then become resistant if genetic material harboring ARGs that has been shown to be not totally removed by disinfection is released into aquatic settings (Manaia et al., 2018). The number and diversity of ARGs differ between Asia/Africa/South America and North America/Europe and Oceania, according to a recent metagenomics analysis conducted in 79 wastewater treatment facilities (WWTPs) in 60 nations. The North America/Europe/Oceania cluster had a small number of ARGs encoding macrolide resistance, whereas the Asia/Africa/South America cluster included ARGs related to sulfonamides and phenicols in significant abundance (Hendriksen et al., 2019). In Fig. 1, a representative list of antibiotics or antibiotic resistance genes (ARGs) found in various wastewater treatment plants is shown in Table 1.

6. Pathogenic biofilms as serious concern in environment and in humans

Biofilms formed by problematic microorganisms in medical and nonmedical settings can pose significant problems. Biofouling in potable water environments and food handling systems, and medical settings, for example, infections that

TABLE 1 Examples on various antibiotics and antibiotic resistance genes found in different wastewater environments.

Water environment	Antibiotics	Target ARGs	Major findings	References
Dewatered sludge	Sulfonamides, TC	*ermC*, *qnrB*, *sul1*, *sul2*, *tet*(A), *tet*(B), *tet*(E), *tet* (G), *tet*(H), *tet*(S), *tet*(T), *tet*(X)	Concentrations are higher than those found in the influent	Mao et al. (2015)
Fish intestine, Finland	–	*aadA1*, *aadA2*, *dfrA1*, *sul1*, *tet32*, *tet*(M), *tet* (O) *tet*(W)	In the fish farm, antibiotics aren't being used	Muziasari et al. (2017)
Seafood products, Japan		bla_{TEM-1}, bla_{CMY-2}, bla_{CMY-13}, bla_{CMY-39}, *qnrB2*, *qnrB6*, *qnrS1*	Products were either manufactured in Japan or imported from Chile, China, India, Indonesia, or the Pacific Northwest	Ahmed, Maruyama, Khalifa, Shimamoto, and Shimamoto (2015)
Active and closed down landfills	CEP, CIP, ERY, SMX, TMP		During the rainfall, antibiotics in leachate from active landfills were much higher than in closed landfills	Chung, Zheng, Burket, and Brooks (2018)
Landfill leachate	AMX, INN-d4, CEP, ENR, NOR, NOR-d5 OFX, PEF	*ampC*, bla_{CTX-M}, *qnrA*, *qnrB*, *qnrS*	–	You, Wu, Wei, Xie, and Lu (2018)
Hospital effluents		$bla_{SHV} > bla_{NDM-1} > bla_{CTX} > bla_{KPC}$	*Pseudomonas* spp., *Enterobacter* spp., *Klebsiella* spp., and *Citrobacter* spp. were the most common ARB taxa	Haller et al. (2018)
Hospital WWTP effluent	AMP, FEP, CAZ, TMP, VAN	*aacC2*, bla_{SHV}, *floR*, *mefA*, *sul1*, *sul2*, *tet*(A), *tet*(B), *tet*(C), *tet*(W)	–	Szekeres et al. (2017)
Activated sludge reactors	ERY, KAN, LIN, SMT, TC, TMP	*aadA*, *bacA*, chloramphenicol exporter, *sul1*, *tet* (C)	Minimum antibiotic concentration of 20 mg/L was used in this study	Zhao et al. (2019)
Sediments from 18 estuaries along over 4000 km of coastline in China	Macrolides, TC	*aacC*, *aadA5*, *mphA*, *oprD*, *oprJ*, *qacEΔ1*	–	Zhu et al. (2017)
Urban sewage pipes	SMX	FCA and MLSB resistance genes	–	Huang et al. (2019)
Maricultural site sediments	CIP, ENR, OTC, SDZ, SMX, TC	*bacA*, *mexF*, *mexB*, *sul1*, *tet*(C)	To represent the entire Chinese coastline, samples were collected from various locations	Gao et al. (2018)

AMP, ampicillin; *CIP*, ciprofloxacin; *ENR*, enrofloxacin; ERY, erythromycin; *FCA*, fluoroquinolone; FEP, cefepime; MLSB, Macrolide-Lincosamide-Streptogramin B resistance genes; NOR, norfloxacin; OFX, ofloxacin; SDZ, sulfadiazine; *SMX*, sulfamethoxazole; TC, tetracycline; TMP, trimethoprim; VAN, vancomycin; *WWTP*, wastewater treatment plant.

are persistent and recurring associated with medical devices have greater impact on human health. The growth of biofilms is a common phenomenon in industrial components and affects the environments that they grow on. Biofilms grow well in damp areas, spoiling products, reducing their efficiency, corroding metals, and reducing the lifetime of the equipment (Vishwakarma, 2020). Many bacterial species from the biofilm in the food industry spoiling food lead to serious hygienic problems and public health risks as well as great economic losses (Awad, Asker, & Hatton,

2018; George, Ramya, Ramachandran, & Mudali, 2013). In the dairy industry, the main source of nutrients for bacteria is the leftover milk products in all the equipment, which favor biofilm growth (Latorre et al., 2010). Bottling or wine industries are also facing undesirable effects of biofouling in storage tanks, pumps, valves, and fillers (Storgards et al. 2006). In the water industry, the blocking of pipes because of biofilms hinders the heat-transfer process and thus corrodes pipes. The water quality gets reduced, with a change of taste, color, and odor. This is due to the presence of many aerobic and anaerobic bacteria (Wingender & Flemming, 2011). In power plants, the interior of the condenser cooling tubes is affected by fouling. This increases the friction and energy needs, pipe pressure drops, (corrosion), blockage, and decreases heat-transfer efficiency (Nebot et al., 2010).

6.1 Biofilm in drinking water system (DWDS)

Waterborne diseases are the illnesses caused by drinking water that has been contaminated by human or animal feces, which contain pathogenic bacteria, or by chemical compounds. If pathogenic bacteria are not eradicated by disinfection and reach the consumer's tap, they may trigger illness outbreaks in the community (Han, An, Yang, & Zhang, 2020). Cholera, typhoid fever, meningitis, encephalitis, dysentery, hepatitis, legionellosis, pulmonary illness, poliomyelitis, leptospirosis, giardiasis, and salmonellosis are the most well-known diseases affiliated with waterborne outbreaks across the world (Andrade, O'Dwyer, O'Neill, & Hynds, 2018; Murphy, Pintar, McBean, & Thomas, 2014). Several incidents posing public health threats occurred because of faulty DW treatment plants and distribution networks that failed to maintain a suitable level of disinfectant to inhibit pathogen growth and/or harbored pathogens. Waterborne disease outbreaks caused by *E. coli* O157:H7, *Campylobacter* spp., *Helicobacter pylori*, *P. aeruginosa*, *Salmonella paratyphi*, *Cryptosporidium parvum*, *Giardia*, and viruses (e.g., norovirus, calcivirus, enterovirus) have been reported in Canada, France, Italy, England, Finland, Switzerland, Northern Ireland, Norway, Belarus, and New Zealand (Baum & Bartram, 2018; Marcheggiani et al., 2010; Reynolds, Checkley, Chui, Otto, & Neumann, 2020) (Fig. 3). Waterborne infections are one of the most serious water-related health issues (Forstinus, Ikechukwu, Emenike, & Christiana, 2015).

The most frequent type of microbial development in drinking water distribution systems is biofilms. (Liu et al., 2016; Mahapatra et al., 2015). Many problems in DWDS are microbial in nature, including biofilm growth, nitrification, microbially mediated corrosion, and the occurrence and persistence of pathogens. About 95% of the total biomass in water is estimated to attach to pipe walls, while only 5% is in the water phase. The creation of biofilms, also known as biofouling, is a well-known issue in drinking water industries. In general, biofouling refers to the undesired deposition of biotic materials on a surface. It has been shown to be of critical hygienic, operational, and economic benefit not just in DWDS but also in other purified water systems such as dentistry unit waterlines, dialysis units, reverse osmosis systems, the semiconductor sector, and so on (Flemming, 2020). Biofouling in commercial and drinking water has negative consequences such as chemical and microbiological degradation of water quality, yield loss, reduction in thermal transport and exchange efficiency, and failure of membrane techniques (Goh, Lau, Othman, & Ismail, 2018;Yu et al., 2021). Toxin-producing algae and fungi, as well as enteric bacteria such as *E. coli*, *Klebsiella*

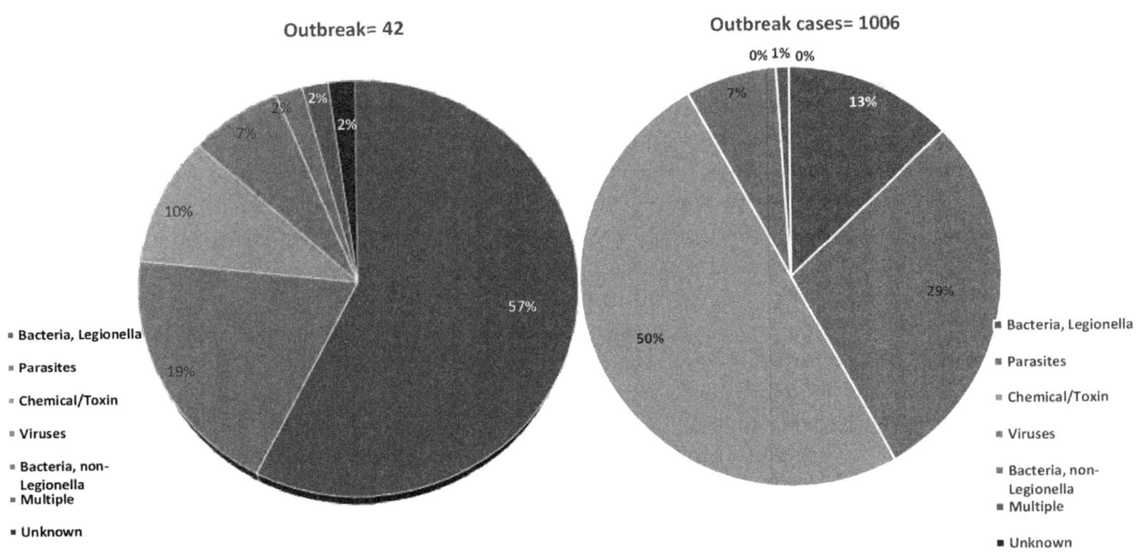

FIG. 3 Etiology of drinking water outbreaks and its related cases (Benedict et al., 2017).

pneumoniae, *K. oxytoca*, and *Enterobacter. cloacae*, *E. agglomerans*, *H. pylori*, *Shigella* spp., *Campylobacter* spp., *Salmonella* spp., *Clostridium perfringens*, *Enterococcus faecium*, *E faecalis*, and environmental pathogenic bacteria such as *L. pneumophila*, *P. aeruginosa*, *P. fluorescens*, and *Aeromonas hydrophila* are examples of opportunistic pathogens (Babič, Gostinčar, & Gunde-Cimerman, 2020; Falkinham, Pruden, & Edwards, 2015). These are often associated with biofilms present in drinking water.

Bacterial cells can adhere and develop to form biofilms on the internal surface of piping systems where pathogenic microorganisms are protected from stress conditions (chlorine, shear stress, temperature), which allows them to sustain. Furthermore, cells can be detached and released into the bulk water, resulting in pipe biocorrosion, unwanted water quality changes that impact color, taste, turbidity, and odors, and a reduction in heat exchange efficiency and lead to a possible outbreak of disease (Prest, Hammes, van Loosdrecht, & Vrouwenvelder, 2016; Simoes & Simões, 2013). The threat posed by microbial infections is nevertheless a daily challenge for water treatment engineers and scientists in the developed world.

Planktonic bacteria levels can be reduced by DWDS disinfection with chlorine dioxide and chlorine, while biofilm bacteria levels have little to no effect. Antimicrobial drugs are also delayed by biofilms, which act as a diffusion barrier. The deep embedded bacteria's natural resistance to antimicrobials can be mediated by very low metabolic levels and substantially reduced cell cycle rates (Zhu et al., 2021). Antimicrobial drugs interact with the components of EPS, lowering the concentration and efficacy of the agent (Karygianni, Ren, Koo, & Thurnheer, 2020). Another benefit of surviving in a biofilm is the potential for metabolic interactions between bacteria with differing physiological needs (Crabbé et al., 2019).

6.2 Biofilm in food industry

Biofilms can form on surfaces that encounter foods or sometimes not directly with food in the food industry (Phillips, 2016). Biofilms are to blame for over 60% of foodborne outbreaks (Han et al., 2017), and its presence in food processing environments constitutes a serious threat to food safety and the food premises (Ehling-Schulz, Frenzel, & Gohar, 2015; Galie et al., 2018). Contaminants in food processing settings mainly arise from the air, equipment, or from the food surfaces (Kumar & Anand, 1998). The growth of biofilms in food processing environments can therefore lead to food deterioration, posing a major public health risk to consumers as well as serious economic implications (Coughlan, Cotter, Hill, & Alvarez-Ordóñez, 2016; Galie et al., 2018). Foodborne diseases have always posed a concern to human health. They are regarded as an emerging public health threat all over the world. Many pathogenic bacteria build biofilms on food and food-contact surfaces under ideal conditions. Some of the representatives of foodborne pathogens in different food industry settings are listed in Table 2. According to the report of US Department of Agriculture, Economic Research Service (2014), major foodborne pathogens has resulted in approximately 17,571,792,712$ healthcare cost leading to significant economic loss. Biofilms are also responsible for the major technical issues in the food sector, as they can obstruct heat passage across metal substrates, raise fluid frictional resistance at the surfaces, and accelerate surface corrosion, resulting in the loss of manufacturing activities (Meesilp & Mesil, 2019). Pathogenicity, corrosion of metal surfaces, and alteration of organoleptic properties due to overexpression of microbially produced lipases or proteases are all important in some factories, where a variety of processes and structures (raw milk tanks, butter centrifuges, pipelines, cheese tanks, pasteurizers, and packing tools) act as surface precursors for biofilm development at various temperatures (Han, Mizan, Jahid, & Ha, 2016; Prest et al., 2016). *Pseudomonas* spp. and *Geobacillus stearothermophilus*, for example, are psychotropic bacteria. Pathogenic species (*A. hydrophila*, *L. monocytogenes*, *S. enterica*, or *Vibrio* spp.) may build biofilms on fresh fish products, resulting in serious health and economic consequences (Mizan, Jahid, & Ha, 2015; Rajkowski, 2009; Tasneem et al., 2018). Major biofilm-forming foodborne pathogens and their clinical impact are listed in Table 3.

Some biofilm-forming bacteria in food manufacturing settings include human pathogens, which is particularly important for the food safety concern in food industry. These bacteria can form biofilms on a variety of artificial substrates used in the food business, including stainless steel, polyethylene, wood, glass, polypropylene, and rubber, among others (Abdallah, Benoliel, Drider, Dhulster, & Chihib, 2014; Colagiorgi et al., 2017). Biofilms of *Enterobacter cloacae* have also been found in ready-to-eat (RTE) and raw foods, including seafood (Igbinosa, Beshiru, Egharevba, & Igbinosa, 2020). *V. cholera* is also a common cause of diarrhea in those who eat shellfish. During handling, production, and processing, *V. cholera* is a major source of seafood contamination, causing diarrhea (Elbashir et al., 2018), and is prevalent across many countries of South East Asia, Africa, and other developing countries (Jenson et al., 2011; Mahapatra et al., 2014; Marchello et al., 2021; Thomas et al., 2020). *Salmonella* spp. develop infections in poultry, shellfish, dairy products, pork, and beef (Augustin et al., 2020), as well as at high salt and temperature (Pradhan & Negi, 2019) and thus present a significant health threat. With 94,625 cases in 2016, *S. enterica* was the second most common agent (just after Campylobacter) of foodborne and waterborne outbreaks in the EU.

TABLE 2 Some of pathogenic bacteria of food industry.

S. no.	Pathogenic bacteria	Associated food industry					References
		PI	DI	REI	PI	FI	
1	*Listeria monocytogenes*	+	+	+	–	+	Rodríguez-López, Bernárdez, Rodríguez-Herrera, Comesaña, and Cabo (2019), Mazaheri, Cervantes-Huamán, Bermúdez-Capdevila, Ripolles-Avila, and Rodríguez-Jerez (2021)
2	*Escherichia coli*	+	+	+	–	–	(Zhang et al., 2016), Adator, Cheng, Holley, McAllister, and Narvaez-Bravo (2018), Liu et al. (2020)
3	*Staphylococcus aureus*	–	+	–	–	–	Karzis, Petzer, Naidoo, and Donkin (2021), Leijon, Atkins, Waller, and Artursson (2021)
4	*Pseudomonas* spp.	–	–	–	–	+	Saengsitthisak et al. (2020)
5	*Aeromonas* spp.					+	Chenia and Duma (2017)
6	*Salmonella typhimurium* and *Salmonella* spp.	–	–	–	+	+	Beshiru, Igbinosa, and Igbinosa (2018), Dhowlaghar et al. (2018)
7	*Campylobacter* spp. and *Campylobacter jejuni*	–	–	–	+	–	Würfel et al. (2019), Nohra et al. (2020)
8	*Vibrio alginolyticus*	–	–	–	–	+	Ashrafudoulla, Mizan, Park, and Ha (2021)

+, indicates predominant occurrence in associated industry; PI, produce industry; DI, dairy industry; REI, ready to eat food industry; PI, poultry industry; FI, fish industry.

TABLE 3 Some of the common representative foodborne pathogens forming biofilms in food industry.

Foodborne pathogens	Contact surface	Targeted spoiled food	Impact on humans	References
Listeria monocytogenes	Wastewater pipes, polypropylene, steel, rubber, or glass surfaces	Dairy products like soft cheeses, melons, meat and fish products that are ready to eat	Gastroenteritis and listeriosis	Galie et al. (2018), Wirtanen and Salo (2016)
Bacillus cereus	Stainless steel, plastic, glass wool, and soil	Sprouted seeds, fruit juices, fried rice, pasta dishes, meat products, cow meats, and dairy products vegetables, and milk products	Other strains generate the emetic toxin (heat-stable cereulide), which provokes vomiting symptoms like hemolytic enterotoxin NheA, cytotoxin K (CytK), hemolysin BL HblC, cell wall peptidase EntFM	Ribeiro, da Silva Fernandes, Kuaye, Jimenez-Flores, and Gigante (2017), Møretrø and Langsrud (2017), Galie et al. (2018)
Staphylococcus aureus	Stainless steel, polymers (such as polystyrene and polypropylene), and glass are some of the most common materials.	Dairy products, ready to eat meat products, ready to eat fish and seafood products	Endocarditis, osteomyelitis, skin and soft tissues infections, nasal colonization, urinary tract infection	Galie et al. (2018), Egege, Akani, and Nwankwo (2020)

Continued

TABLE 3 Some of the common representative foodborne pathogens forming biofilms in food industry—cont'd

Foodborne pathogens	Contact surface	Targeted spoiled food	Impact on humans	References
Escherichia coli	Food contact surfaces, stainless steel surfaces	Fermented dairy and meat products, sausage, meat, poultry, fish products, drinks, and vegetables	Hemolytic uremic syndrome, urinary tract infections (UTI), neonatal sepsis, and meningitis	Galie et al. (2018), Wirtanen and Salo (2016)
Pseudomonas spp.	Conveyor belts, floors, drains, slicing, and milk industry pipeline	Dairy products, red meat, and poultry	Urinary catheters, Contact lenses, pulmonary infections, Cystic fibrosis	Galie et al. (2018), Wirtanen and Salo (2016)
Clostridium spp.	Form mixed species biofilm with other bacteria	Fish, cattle meat, poultry, canned foods, and vegetables	Colitis and diarrhea	Wirtanen and Salo (2016), Anju et al. (2021)
Salmonella spp.	Stainless steel surfaces, concrete, and glass	Poultry and cow meat, pork and food surfaces (like lettuce and tomato)	Nausea, vomiting, fever, diarrhea, and stomach discomfort are all symptoms of gastroenteritis or septicemia (in the case of some serovars)	Wirtanen and Salo (2016), Galie et al. (2018), Dewey-Mattia, Manikonda, Hall, Wise, and Crowe (2018)
Cronobacter spp.	Powder service and powder packaging rooms, spray-drying areas, and evaporator rooms	Vegetables, dairy products, grains, bread, herbs, sausages, spices, and meat	Septicemia, meningitis, necrotizing enterocolitis, and brain abscesses	Wirtanen and Salo (2016), Ling et al. (2018)

S. enterica was responsible for 12,353 hospitalizations and 126 deaths in that year (50% of all outbreak-related deaths) (European Food Safety Authority & European Centre for Disease Prevention and Control, 2017). Another pathogen *L. monocytogenes* causes serious infection in freshwater fish, crabs, shrimp, and pacific oysters (Goudar & Prasad, 2020). This pathogen can proliferate at refrigeration temperatures after being contaminated in food (Salazar et al., 2020).

6.3 Biofilm in dairy industry

Due to changes in the worldwide economy, the dairy business has emerged as a major worldwide industry producing both perishable (yoghurt, butter, and cheese) and semiperishable food items (casein, milk powder). Foodborne pathogens have contributed to making biofilm in milk processing units by insufficient cleaning and sanitization (Ali, 2020; Gupta & Anand, 2018), resulting in both public health and commercial damage. Raw milk is an ideal culture medium for microorganisms to grow and may hold spoilage and/or health risks. On its journey from the farm to the consumer, milk encounters the walls of the equipment in which it is being processed and transported (Marchand et al., 2012). Microorganisms have been shown to linger on the surface of equipment despite sanitation and cleaning procedures (Wang et al., 2021). Biofilms are not just as source of pollution, but they can also limit heat transfer, speed up corrosion, and increase fluid friction resistance. Bacterial development in dairy products can have a negative impact on the functionality, quality, and safety of the product (Meesilp and Mesil, 2019). The most prevalent bacteria found in dairy businesses are *Enterobacter*, *Micrococcus*, *Listeria*, *Streptococcus*, *Bacillus*, and *Pseudomonas* (Bai, Nakatsu, & Bhunia, 2021; Cherif-Antar et al., 2016; Hassan, Ogbonna, & Obisike, 2020). The production of biofilms on the surfaces of milk transport pipes, milking containers, and accessories in the dairy industry is frequently linked to the main cause of contaminants of dairy products (Gupta & Anand, 2018). Biofilms form mostly in conditions that allow bacteria to easily stick to pipe walls, such as transport pipes that move milk from the farm to the processing plants (Skowron et al., 2019). As milk flows across surfaces, bacteria may dislodge from biofilms and contaminate it (Marchand et al., 2012). Furthermore, biofilm bacteria can cause metal pipes to corrode, impede transfer of heat, and elevate resistance toward fluid frictional (Gupta & Anand, 2018; Kumar & Anand, 1998). In dairy industries, bacterial biofilms can persist, potentially reducing the shelf life of pasteurized milk, cream, cheese, and other milk products (Galie et al., 2018). As a result, the majority of foodborne bacterial contaminations cause gastroenteritis, posing a substantial threat to public health and affecting livelihood of mankind.

7. Impact of biofilms in clinical settings

Biofilm formation in healthcare settings is an issue of great concern, as it is responsible for enhanced morbidity and mortality, and has led to a significant economic burden on the healthcare system (Subhadra, Kim, Woo, Surendran, & Choi, 2018). Moreover, it is observed that microbial growth is a common phenomenon in our daily-wear cotton/woolen clothes or clothes used in hospitals, wound dressings, sportswear, and results in wound infection, bad odor, dermatitis, and pruritus (Liu, Li, Cheng, Ren, & Huang, 2015; Wei, Tang, Yu, & Chen, 2017). Biofilms have been found on medical devices, including implants, catheters, endoscopes, and various environmental surfaces and held responsible in the transmission of pathogens. It is remarkable that biofilms can survive high-level disinfection and sterilization processes. Considering the significance of controlling microbial biofilms in healthcare settings, a considerable scientific and translational interest and research are emerging in understanding their formation, structure, and control. Also, various approaches are being adapted for identification and prevention of such a behavior of pathogens (Maillard & McBain, 2019).

7.1 Biofilm-associated diseases in humans

The clinical community has been challenged by the spread of a wide range of infections caused by bacterial biofilms. Biofilm-related organisms are thought to cause well over 65% of all microbial illnesses and have significant resistance toward drugs as well as innate and adaptive host defense components (Cepas et al., 2019; Vestby, Grønseth, Simm, & Nesse, 2020). Biofilms have a major influence on human healthcare industry. Device-related infections, persistent contaminations without an unfamiliar body, and even the breakdown of clinical devices are all examples of biofilm sickness. Consistent upper airways contamination, which is commonly found in patients with cystic fibrosis, chronic obstructive pulmonary diseases, persistent injury infections, endocarditis, periodontitis, inflammatory bowel diseases, and bacterial prostatitis are examples of biofilm infection not related to embedded devices (Del Pozo, 2018). In patients who can withstand routine treatment, pathogenic microbial populations can cause chronic infection (Ibberson & Whiteley, 2020). The embedded pathogens can withstand harsh environments because the microbial consortium of biofilm cells within the additional polymeric matrix is well shielded from a variety of factors. One of the reasons biofilm infections are rarely healed by an individual's immune system is because of matrix (Rybtke, Jensen, Nielsen, & Tolker-Nielsen, 2020). The pathogen's first line of defense is the additional polymeric matrix, and the presence of exopolysaccharide alginate shields microbial cells from phagocytosis, a condition in which macrophages and neutrophils fail to ingest the microbial cells (Rybtke et al., 2020).

P. aeruginosa biofilm, in particular, can cause serious lung infections in people with cystic fibrosis (Maurice, Bedi, & Sadikot, 2018). Otitis medium is caused by a biofilm of *Haemophilus influenza* (Akyıldız, Take, Uygur, Kızıl, & Aydil, 2013); biofilms formed by *Pseudomonas aerobicus* and *Fusobacterium nucleatum* are commonly responsible for periodontitis, an inflammation of the gums that affects the soft tissues as well as the bones that provide support to teeth. (Jamal et al., 2018). Scientific research and demonstrations have also established the function of biofilms in the transmission of gingival infections in adults, which account for about 40%–50% of all cases (Zawadzki et al., 2017). Moreover, biofilms were linked to 86% and 85% of pneumonias and bloodstream infections of clinical significance (Yousif, Jamal, & Raad, 2015). Polymicrobial communities are a diverse collection of organisms (fungi, bacteria, and viruses) that dwell at a phase or density interface and are coated in a self- and/or host-derived aqueous matrix, which is frequently polysaccharide-based forming biofilm.

Biofilm-forming bacteria have been reported to grow on medical device surfaces, dead tissues (such as bone sequestra), and inside living tissues (such as tooth surfaces and lung tissue) in hospital environments (Alav, Sutton, & Rahman, 2018). The conglomeration of biofilms formed on the surfaces of biomedical equipment such as pacemakers, contact lenses, catheters, artificial heart valves, breast implants, and cerebrospinal fluid shunts (Hall-Stoodley, Costerton, & Stoodley, 2004; Wu, Moser, Wang, Høiby, & Song, 2015). Microorganisms as single species or multispecies can irreversibly adhere on to the solid surface and establish biofilms on the surfaces of these devices; however, *S. epidermidis*, *S. aureus*, and *P. aeruginosa* are the most commonly reported biofilm-forming microorganisms (Ruhal & Kataria, 2021). *Staphylococcus* spp. are responsible for two-thirds of indwelling device-related illnesses (Khatoon, McTiernan, Suuronen, Mah, & Alarcon, 2018). Bacterial biofilms can also form in pipelines of hospitals. In the hospital water system, *P. aeruginosa* can build biofilms on the inner surfaces of metal pipes (Weinbren, Inkster, & Lafferty, 2021). Biofilm-associated indwelling medical devices are responsible for many UTIs and bloodstream infections (Yousif et al., 2015). Because the microorganisms that cause the infection are known for their ability to produce biofilms, biofilm infection was found in 95 of the patients with UTIs. Depending on circumstances, biofilms can be single or multispecies. Biofilms on urinary catheters can be made up of a single species at first, but over time they can become multispecies. Most common bacteria isolated from indwelling medical devices are *E. coli*, *E. faecalis*, *S. viridans*, *S. aureus*, *Proteus mirabilis*, *S. epidermidis*, *K. pneumoniae*, and *P. aeruginosa* (Bai, Cai, & Yang, 2019; Chatterjee, Maiti, Dey, Kundu, & Dey, 2014). Most prevalent bacteria found on cardiovascular devices are *S. epidermidis* and *S. aureus* Qu et al., 2020), causing 40%–50% of heart valve infections and 50%–70% of catheter infections. (Hoerr et al., 2018).

8. Conclusion and future prospects

Biofilms can form in both natural and man-made systems (such as implant materials, vessels, pipes, water bodies, food items, textiles surface, ship hulls, power plants) providing a suitable and optimal environment for the growth, activity, and interaction of different bacterial species. Biofilms offer struggling microbial cells to cope with transient or permanent stress conditions, promoting metabolic interactions and genetic interchange between different bacterial species. Therefore, biofilms are of major concern in almost every sector of environmental settings as they can contaminate industrial equipment and the surrounding environment. Simply a moist or damp area with limited nutrients is required by notorious microbes to slowly develop into a biofilm under any extreme condition. Biofilm formation by microorganisms provides protection to survive and thrive under extreme temperature and pH, high salinity, high pressure, poor nutrients, antibiotics, and under exposure to UV radiation etc. An in-depth study of the protective mechanism of biofilms in extreme environments is expected to resolve challenges faced by various industries such as dairy, water, food, and clinical settings. Increase in scientific research over the decade in biofilm formation has revealed possible targets to prevent or eradicate biofilm infections. Advances in the understanding of biofilm formation, coupled with emerging engineered biomaterials, provide many potential platforms and strategies to prevent or significantly reduce biofilm emergence in different environments. In addition, more research is needed to control problematic biofilms. By selecting suitable treatment techniques such as nanotechnology, industrial losses due to biofilm growth could be minimized.

Acknowledgment

The research was supported by Indo-UK joint Collaborative research project Respharm and SELECTAR from Department of biotechnology. The DST—INSPIRE fellowship (IF180419) support from Department of Science and Technology to Ms. Samreen is also thankfully acknowledged.

References

Abdallah, M., Benoliel, C., Drider, D., Dhulster, P., & Chihib, N. E. (2014). Biofilm formation and persistence on abiotic surfaces in the context of food and medical environments. *Archives of Microbiology*, *196*(7), 453–472.

Abe, K., Nomura, N., & Suzuki, S. (2020). Biofilms: Hot spots of horizontal gene transfer (HGT) in aquatic environments, with a focus on a new HGT mechanism. *FEMS Microbiology Ecology*, *96*(5), fiaa031.

Abebe, G. M. (2020). The role of bacterial biofilm in antibiotic resistance and food contamination. *International Journal of Microbiology*, *2020*.

Adator, E. H., Cheng, M., Holley, R., McAllister, T., & Narvaez-Bravo, C. (2018). Ability of Shiga toxigenic *Escherichia coli* to survive within dry-surface biofilms and transfer to fresh lettuce. *International Journal of Food Microbiology*, *269*, 52–59.

Ahmed, A. M., Maruyama, A., Khalifa, H. O., Shimamoto, T., & Shimamoto, T. (2015). Seafood as a reservoir of Gram-negative bacteria carrying integrons and antimicrobial resistance genes in Japan. *Biomedical and Environmental Sciences*, *28*(12), 924–926.

Akyıldız, I., Take, G., Uygur, K., Kızıl, Y., & Aydil, U. (2013). Bacterial biofilm formation in the middle-ear mucosa of chronic otitis media patients. *Indian Journal of Otolaryngology and Head & Neck Surgery*, *65*(3), 557–561.

Alav, I., Sutton, J. M., & Rahman, K. M. (2018). Role of bacterial efflux pumps in biofilm formation. *Journal of Antimicrobial Chemotherapy*, *73*(8), 2003–2020.

Ali, F. F. (2020). Detection of biofilm formation and antibiotics resistance for *Streptococcus* spp. isolated from some dairy products in Diwanyah City of Iraq. *Al-Mustansiriyah Journal of Science*, *31*(4), 28–35.

Allen, H. K., Donato, J., Wang, H. H., Cloud-Hansen, K. A., Davies, J., & Handelsman, J. (2010). Call of the wild: Antibiotic resistance genes in natural environments. *Nature Reviews Microbiology*, *8*(4), 251–259.

Amarasiri, M., Sano, D., & Suzuki, S. (2020). Understanding human health risks caused by antibiotic resistant bacteria (ARB) and antibiotic resistance genes (ARG) in water environments: Current knowledge and questions to be answered. *Critical Reviews in Environmental Science and Technology*, *50*(19), 2016–2059.

Andersson, D. I., & Hughes, D. (2014). Microbiological effects of sublethal levels of antibiotics. *Nature Reviews Microbiology*, *12*(7), 465–478.

Andrade, L., O'Dwyer, J., O'Neill, E., & Hynds, P. (2018). Surface water flooding, groundwater contamination, and enteric disease in developed countries: A scoping review of connections and consequences. *Environmental Pollution*, *236*, 540–549.

Anju, K., Karthik, K., Divya, V., Priyadharshini, M. L., Sharma, R. K., & Manoharan, S. (2021). Toxinotyping and molecular characterization of antimicrobial resistance in *Clostridium perfringens* isolated from different sources of livestock and poultry. *Anaerobe*, *67*, 102298.

Arciola, C. R., Campoccia, D., & Montanaro, L. (2018). Implant infections: Adhesion, biofilm formation and immune evasion. *Nature Reviews Microbiology*, *16*(7), 397–409.

Ashrafudoulla, M., Mizan, M. F., Park, S. H., & Ha, S. D. (2021). Current and future perspectives for controlling *Vibrio* biofilms in the seafood industry: A comprehensive review. *Critical Reviews in Food Science and Nutrition*, *61*(11), 1827–1851.

Auguet, O., Pijuan, M., Borrego, C. M., Rodriguez-Mozaz, S., Triadó-Margarit, X., Della Giustina, S. V., et al. (2017). Sewers as potential reservoirs of antibiotic resistance. *Science of the Total Environment*, *605*, 1047–1054.

Augustin, J. C., Kooh, P., Bayeux, T., Guillier, L., Meyer, T., Silva, J. D., et al. (2020). Contribution of foods and poor food-handling practices to the burden of foodborne infectious diseases in France. *Foods*, *9*(11), 1644.

Awad, T. S., Asker, D., & Hatton, B. D. (2018). Food-safe modification of stainless steel food-processing surfaces to reduce bacterial biofilms. *ACS Applied Materials & Interfaces*, *10*(27), 22902–22912.

Babič, M. N., Gostinčar, C., & Gunde-Cimerman, N. (2020). Microorganisms populating the water-related indoor biome. *Applied Microbiology and Biotechnology*, *104*(15), 6443–6462.

Bai, F., Cai, Z., & Yang, L. (2019). Recent progress in experimental and human disease-associated multi-species biofilms. *Computational and Structural Biotechnology Journal*, *17*, 1234–1244.

Bai, X., Nakatsu, C. H., & Bhunia, A. K. (2021). Bacterial biofilms and their implications in pathogenesis and food safety. *Foods*, *10*(9), 2117.

Baum, R., & Bartram, J. (2018). A systematic literature review of the enabling environment elements to improve implementation of water safety plans in high-income countries. *Journal of Water and Health*, *16*(1), 14–24.

Benedict, K. M., Reses, H., Vigar, M., Roth, D. M., Roberts, V. A., Mattioli, M., … Yoder, J. S. (2017). Surveillance for waterborne disease outbreaks associated with drinking water—United States, 2013–2014. *Morbidity and Mortality Weekly. Report*, *66*(44), 1216.

Berendonk, T. U., Manaia, C. M., Merlin, C., Fatta-Kassinos, D., Cytryn, E., Walsh, F., et al. (2015). Tackling antibiotic resistance: The environmental framework. *Nature Reviews Microbiology*, *13*(5), 310–370.

Berne, C., Ellison, C. K., Ducret, A., & Brun, Y. V. (2018). Bacterial adhesion at the single-cell level. *Nature Reviews Microbiology*, *16* (10), 616–627.

Bernier, S. P., & Surette, M. G. (2013). Concentration-dependent activity of antibiotics in natural environments. *Frontiers in Microbiology*, *4*, 20.

Beshiru, A., Igbinosa, I. H., & Igbinosa, E. O. (2018). Biofilm formation and potential virulence factors of *Salmonella* strains isolated from ready-to-eat shrimps. *PLoS One*, *13*(9), e0204345.

Blair, J., Webber, M. A., Baylay, A. J., Ogbolu, D. O., & Piddock, L. J. (2015). Molecular mechanisms of antibiotic resistance. *Nature Reviews Microbiology*, *13*(1), 42–51.

Blair, J. M. (2018). A climate for antibiotic resistance. *Nature Climate Change*, *8*(6), 460–461.

Boo, A., Amaro, R. L., & Stan, G. B. (2021). Quorum sensing in synthetic biology: A review. *Current Opinion in Systems Biology*, *28*, 100378.

Calapez, A. R., Elias, C. L., Alves, A., Almeida, S. F., Brito, A. G., & Feio, M. J. (2020). Shifts in biofilms' composition induced by flow stagnation, sewage contamination and grazing. *Ecological Indicators*, *111*, 106006.

Carey, R. O., & Migliaccio, K. W. (2009). Contribution of wastewater treatment plant effluents to nutrient dynamics in aquatic systems: A review. *Environmental Management*, *44*(2), 205–217.

Carrascosa, C., Raheem, D., Ramos, F., Saraiva, A., & Raposo, A. (2021). Microbial biofilms in the food industry—A comprehensive review. *International Journal of Environmental Research and Public Health*, *18*(4), 2014.

Caruso, G. (2020). Microbial colonization in marine environments: Overview of current knowledge and emerging research topics. *Journal of Marine Science and Engineering*, *8*(2), 78.

Cepas, V., López, Y., Muñoz, E., Rolo, D., Ardanuy, C., Martí, S., et al. (2019). Relationship between biofilm formation and antimicrobial resistance in gram-negative bacteria. *Microbial Drug Resistance*, *25* (1), 72–79.

Charles, C. J., Rout, S. P., Patel, K. A., Akbar, S., Laws, A. P., Jackson, B. R., et al. (2017). Floc formation reduces the pH stress experienced by microorganisms living in alkaline environments. *Applied and Environmental Microbiology*, *83*(6), e02985-16.

Chatterjee, S., Maiti, P. K., Dey, R., Kundu, A. K., & Dey, R. K. (2014). Biofilms on indwelling urologic devices: Microbes and antimicrobial management prospect. *Annals of Medical and Health Sciences Research*, *4*(1), 100–104.

Chatterjee, S., Samal, B., Singh, P., Pradhan, B. B., & Verma, R. K. (2020). Transition of a solitary to a biofilm community life style in bacteria: A survival strategy with division of labour. *International Journal of Developmental Biology*, *64*(4-5-6), 259–265.

Chen, X. P., Ali, L., Wu, L. Y., Liu, C., Gang, C. X., Huang, Q. F., et al. (2018). Biofilm formation plays a role in the formation of multidrug-resistant *Escherichia coli* toward nutrients in microcosm experiments. *Frontiers in Microbiology*, *9*, 367.

Chenia, H. Y., & Duma, S. (2017). Characterization of virulence, cell surface characteristics and biofilm-forming ability of *Aeromonas* spp. isolates from fish and sea water. *Journal of Fish Diseases*, *40* (3), 339–350.

Cherif-Antar, A., Moussa–Boudjemâa, B., Didouh, N., Medjahdi, K., Mayo, B., & Flórez, A. B. (2016). Diversity and biofilm-forming capability of bacteria recovered from stainless steel pipes of a milk-processing dairy plant. *Dairy Science & Technology*, *96*(1), 27–38.

Cherifi, T., Jacques, M., Quessy, S., & Fravalo, P. (2017). Impact of nutrient restriction on the structure of *Listeria monocytogenes* biofilm grown in a microfluidic system. *Frontiers in Microbiology*, *8*, 864.

Chu, E. K., Kilic, O., Cho, H., Groisman, A., & Levchenko, A. (2018). Self-induced mechanical stress can trigger biofilm formation in uropathogenic *Escherichia coli*. *Nature Communications*, *9*(1), 4087. https://doi.org/10.1038/s41467-018-06552-z.

Chua, S. L., Liu, Y., Yam, J. K., Chen, Y., Vejborg, R. M., Tan, B. G., et al. (2014). Dispersed cells represent a distinct stage in the transition from bacterial biofilm to planktonic lifestyles. *Nature Communications*, *5* (1), 1–2.

Chung, S. S., Zheng, J. S., Burket, S. R., & Brooks, B. W. (2018). Select antibiotics in leachate from closed and active landfills exceed thresholds for antibiotic resistance development. *Environment International*, *115*, 89–96.

Cihan, A. C., Karaca, B., Ozel, B. P., & Kilic, T. (2017). Determination of the biofilm production capacities and characteristics of members belonging to Bacillaceae family. *World Journal of Microbiology and Biotechnology*, *33*(6), 118.

Colagiorgi, A., Bruini, I., Di Ciccio, Zanardi, E., Ghidini, S., & Ianieri, A. (2017). Listeria monocytogenes biofilms in the wonderland of food industry. *Pathogens*, *6*(3), 41.

Colvin, K. M., Gordon, V. D., Murakami, K., Borlee, B. R., Wozniak, D. J., Wong, G. C., et al. (2011). The pel polysaccharide can serve a structural and protective role in the biofilm matrix of *Pseudomonas aeruginosa*. *PLoS Pathogens*, *7*(1), e1001264.

Coughlan, L. M., Cotter, P. D., Hill, C., & Alvarez-Ordóñez, A. (2016). New weapons to fight old enemies: Novel strategies for the (bio) control of bacterial biofilms in the food industry. *Frontiers in Microbiology*, *7*, 1641.

Crabbé, A., Jensen, P.Ø., Bjarnsholt, T., & Coenye, T. (2019). Antimicrobial tolerance and metabolic adaptations in microbial biofilms. *Trends in Microbiology*, *27*(10), 850–863.

Dang, H., & Lovell, C. R. (2016). Microbial surface colonization and biofilm development in marine environments. *Microbiology and Molecular Biology Reviews*, *80*(1), 91–138.

Davies, J., & Davies, D. (2010). Origins and evolution of antibiotic resistance. *Microbiology and Molecular Biology Reviews*, *74*(3), 417–433.

de Carvalho, C. C. (2018). Marine biofilms: A successful microbial strategy with economic implications. *Frontiers in Marine Science*, *5*, 126.

Del Pozo, J. L. (2018). Biofilm-related disease. *Expert Review of Anti-Infective Therapy*, *16*, 51–65. https://doi.org/10.1080/14787210.2018.1417036.

Dewey-Mattia, D., Manikonda, K., Hall, A. J., Wise, M. E., & Crowe, S. J. (2018). Surveillance for foodborne disease outbreaks—United States, 2009–2015. *MMWR Surveillance Summaries*, *67*(10), 1.

Dhowlaghar, N., Abeysundara, P. D., Nannapaneni, R., Schilling, M. W., Chang, S., Cheng, W. H., et al. (2018). Biofilm formation by *Salmonella* spp. in catfish mucus extract under industrial conditions. *Food Microbiology*, *70*, 172–180.

Dincer, S., Uslu, F. M., & Delik, A. (2020). Antibiotic resistance in biofilm. In *Bacterial biofilms* IntechOpen.

Drescher, K., Dunkel, J., Nadell, C. D., Van Teeffelen, S., Grnja, I., Wingreen, N. S., et al. (2016). Architectural transitions in *Vibrio cholerae* biofilms at single-cell resolution. *Proceedings of the National Academy of Sciences*, *113*, E2066–E2072.

Egege, S. R., Akani, N. P., & Nwankwo, C. E. (2020). Detection of methicillin-resistant *Staphylococcus aureus* in ready-to-eat shellfish (Corbiculid heterodont) in Bayelsa State, Nigeria. *Microbiology Research Journal International*, 22–35.

Ehling-Schulz, M., Frenzel, E., & Gohar, M. (2015). Food–bacteria interplay: Pathometabolism of emetic *Bacillus cereus*. *Frontiers in Microbiology*, *6*, 704.

Elbashir, S., Parveen, S., Schwarz, J., Rippen, T., Jahncke, M., & DePaola, A. (2018). Seafood pathogens and information on antimicrobial resistance: A review. *Food Microbiology*, *70*, 85–93.

European Food Safety Authority & European Centre for Disease Prevention and Control. (2017). The European Union summary report on trends and sources of zoonoses, zoonotic agents and food-borne outbreaks in 2016. *EFSA Journal*, *15*(12), e05077.

Falkinham, J. O., Pruden, A., & Edwards, M. (2015). Opportunistic premise plumbing pathogens: Increasingly important pathogens in drinking water. *Pathogens*, *4*(2), 373–386.

Flemming, H. C. (2016). EPS—Then and now. *Microorganisms*, *4*(4), 41.

Flemming, H. C. (2020). Biofouling and me: My Stockholm syndrome with biofilms. *Water Research*, *173*, 115576.

Flemming, H. C., Wingender, J., Szewzyk, U., Steinberg, P., Rice, S. A., & Kjelleberg, S. (2016). Biofilms: An emergent form of bacterial life. *Nature Reviews Microbiology*, *14*(9), 563–575.

Forstinus, N. O., Ikechukwu, N. E., Emenike, M. P., & Christiana, A. O. (2015). Water and waterborne diseases: A review. *International Journal of Tropical Disease & Health*, 1–4.

Fysun, O., Kern, H., Wilke, B., & Langowski, H. C. (2019). Evaluation of factors influencing dairy biofilm formation in filling hoses of food-processing equipment. *Food and Bioproducts Processing*, *113*, 39–48.

Galie, S., García-Gutiérrez, C., Miguélez, E. M., Villar, C. J., & Lombó, F. (2018). Biofilms in the food industry: Health aspects and control methods. *Frontiers in Microbiology*, *9*, 898.

Gao, Q., Dong, Q., Wu, L., Yang, Y., Hale, L., Qin, Z., et al. (2020). Environmental antibiotics drives the genetic functions of resistome dynamics. *Environment International*, *135*, 105398.

Gao, Q., Li, Y., Qi, Z., Yue, Y., Min, M., Peng, S., et al. (2018). Diverse and abundant antibiotic resistance genes from mariculture sites of China's coastline. *Science of the Total Environment*, *630*, 117–125.

Garner, E., Chen, C., Xia, K., Bowers, J., Engelthaler, D. M., McLain, J., et al. (2018). Metagenomic characterization of antibiotic resistance genes in full-scale reclaimed water distribution systems and corresponding potable systems. *Environmental Science & Technology*, *52*(11), 6113–6125.

George, R. P., Ramya, S., Ramachandran, D., & Mudali, U. K. (2013). Studies on biodegradation of normal concrete surfaces by fungus *Fusarium* sp. *Cement and Concrete Research*, *47*, 8–13.

Giaouris, E., Simões, M., & Dubois-Brissonnet, F. (2020). The role of biofilms in the development and dissemination of microbial resistance within the food industry. *Food*, *9*, 816.

Goh, P. S., Lau, W. J., Othman, M. H., & Ismail, A. F. (2018). Membrane fouling in desalination and its mitigation strategies. *Desalination*, *425*, 130–155.

Goudar, V., & Prasad, N. (2020). A critical review on *Listeria monocytogenes*. *International Journal of Innovations in Biological and Chemical Sciences*, *13*, 95–103.

Gupta, S., & Anand, S. (2018). Induction of pitting corrosion on stainless steel (grades 304 and 316) used in dairy industry by biofilms of common sporeformers. *International Journal of Dairy Technology*, *71*(2), 519–531.

Haller, L., Chen, H., Ng, C., Le, T. H., Koh, T. H., Barkham, T., et al. (2018). Occurrence and characteristics of extended-spectrum β-lactamase-and carbapenemase-producing bacteria from hospital effluents in Singapore. *Science of the Total Environment*, *615*, 1119–1125.

Han, Q., Song, X., Zhang, Z., Fu, J., Wang, X., Malakar, P. K., & Zhao, Y. (2017). Removal of foodborne pathogen biofilms by acidic electrolyzed water. *Frontiers in Microbiology*, *8*, 988.

Han, Z., An, W., Yang, M., & Zhang, Y. (2020). Assessing the impact of source water on tap water bacterial communities in 46 drinking water supply systems in China. *Water Research*, *172*, 115469.

Hall-Stoodley, L., Costerton, J. W., & Stoodley, P. (2004). Bacterial biofilms: From the natural environment to infectious diseases. *Nature Reviews Microbiology*, *2*(2), 95–108.

Han, N., Mizan, M. F., Jahid, I. K., & Ha, S. D. (2016). Biofilm formation by *Vibrio parahaemolyticus* on food and food contact surfaces increases with rise in temperature. *Food Control*, *70*, 161–166.

Hassan, A. O., Ogbonna, I. O., & Obisike, V. U. (2020). Biofilm forming ability and antibiotic susceptibility of food-borne pathogens isolated from common dairy products: Madara and Nono Vended in Makurdi Metropolis. *Microbiology Research Journal International*, 74–83.

Hendriksen, R. S., Munk, P., Njage, P., Van Bunnik, B., McNally, L., Lukjancenko, O., et al. (2019). Global monitoring of antimicrobial resistance based on metagenomics analyses of urban sewage. *Nature Communications*, *10*(1), 1–2.

Hoerr, V., Franz, M., Pletz, M. W., Diab, M., Niemann, S., Faber, C., et al. (2018). *S. aureus* endocarditis: Clinical aspects and experimental approaches. *International Journal of Medical Microbiology*, *308*(6), 640–652.

Hou, J., Veeregowda, D. H., van de Belt-Gritter, B., Busscher, H. J., & van der Mei, H. C. (2018). Extracellular polymeric matrix production and relaxation under fluid shear and mechanical pressure in *Staphylococcus aureus* biofilms. *Applied and Environmental Microbiology*, *84*(1), e01516-17.

Huang, H., Zeng, S., Dong, X., Li, D., Zhang, Y., He, M., et al. (2019). Diverse and abundant antibiotics and antibiotic resistance genes in an urban water system. *Journal of Environmental Management*, *231*, 494–503.

Hujslová, M., Bystrianský, L., Benada, O., & Gryndler, M. (2019). Fungi, a neglected component of acidophilic biofilms: Do they have a potential for biotechnology? *Extremophiles*, *23*(3), 267–275.

Ibberson, C. B., & Whiteley, M. (2020). The social life of microbes in chronic infection. *Current Opinion in Microbiology*, *53*, 44–50.

Igbinosa, I. H., Beshiru, A., Egharevba, N. E., & Igbinosa, E. O. (2020). Distribution of enterobacteria in ready-to-eat food in cafeterias and retail food outlets in Benin City: Public health implications. *Journal of Community Medicine and Primary Health Care*, *32*(2), 80–94.

Jamal, M., Ahmad, W., Andleeb, S., Jalil, F., Imran, M., Nawaz, M. A., et al. (2018). Bacterial biofilm and associated infections. *Journal of the Chinese Medical Association*, *81*(1), 7–11.

Jennings, L. K., Storek, K. M., Ledvina, H. E., Coulon, C., Marmont, L. S., Sadovskaya, I., et al. (2015). Pel is a cationic exopolysaccharide that cross-links extracellular DNA in the *Pseudomonas aeruginosa* biofilm matrix. *Proceedings of the National Academy of Sciences*, *112*(36), 11353–11358.

Jenson, D., Szabo, V., & Duke FHI Haiti Humanities Laboratory Student Research Team. (2011). Cholera in Haiti and other Caribbean regions, 19th century. *Emerging Infectious Diseases*, *17*(11), 2130.

Karkman, A., Do, T. T., Walsh, F., & Virta, M. P. (2018). Antibiotic-resistance genes in waste water. *Trends in Microbiology*, *26*(3), 220–228.

Karygianni, L., Ren, Z., Koo, H., & Thurnheer, T. (2020). Biofilm matrixome: Extracellular components in structured microbial communities. *Trends in Microbiology*, *28*(8), 668–681.

Karzis, J., Petzer, I. M., Naidoo, V., & Donkin, E. F. (2021). The spread and antimicrobial resistance of *Staphylococcus aureus* in South African dairy herds—A review. *The Onderstepoort Journal of Veterinary Research*, *88*(1).

Khatoon, Z., McTiernan, C. D., Suuronen, E. J., Mah, T. F., & Alarcon, E. I. (2018). Bacterial biofilm formation on implantable devices and approaches to its treatment and prevention. *Heliyon*, *4*(12), e01067.

Kim, S., Chu, K. H., Al-Hamadani, Y. A., Park, C. M., Jang, M., Kim, D. H., et al. (2018). Removal of contaminants of emerging concern by membranes in water and wastewater: A review. *Chemical Engineering Journal*, *335*, 896–914.

Korzeniewska, E., & Harnisz, M. (2018). Relationship between modification of activated sludge wastewater treatment and changes in antibiotic resistance of bacteria. *Science of the Total Environment*, *639*, 304–315.

Kumar, C. G., & Anand, S. K. (1998). Significance of microbial biofilms in food industry: A review. *International Journal of Food Microbiology*, *42*(1–2), 9–27.

Latorre, A. A., van Kessel, J. S., Karns, J. S., Zurakowski, M. J., Pradhan, A. K., Boor, K. J., et al. (2010). Biofilm in milking equipment on a dairy farm as a potential source of bulk tank milk contamination with *Listeria monocytogenes*. *Journal of Dairy Science*, *93*, 2792–2802.

Leijon, M., Atkins, E., Waller, K. P., & Artursson, K. (2021). Longitudinal study of *Staphylococcus aureus* genotypes isolated from bovine clinical mastitis. *Journal of Dairy Science*, *104*(11), 11945–11954.

Li, X., Kappler, U., Jiang, G., & Bond, P. L. (2017). The ecology of acidophilic microorganisms in the corroding concrete sewer environment. *Frontiers in Microbiology*, *8*, 683.

Ling, N., Li, C., Zhang, J., Wu, Q., Zeng, H., He, W., et al. (2018). Prevalence and molecular and antimicrobial characteristics of *Cronobacter* spp. isolated from raw vegetables in China. *Frontiers in Microbiology*, *9*, 1149.

Liu, G., Ali, T., Gao, J., ur Rahman, S., Yu, D., Barkema, H. W., et al. (2020). Co-occurrence of plasmid-mediated colistin resistance (mcr-1) and extended-spectrum β-lactamase encoding genes in *Escherichia coli* from bovine mastitic milk in China. *Microbial Drug Resistance*, *26*(6), 685–696.

Liu, S., Gunawan, C., Barraud, N., Rice, S. A., Harry, E. J., & Amal, R. (2016). Understanding, monitoring, and controlling biofilm growth in drinking water distribution systems. *Environmental Science & Technology*, *50*(17), 8954–8976.

Liu, Y., Li, J., Cheng, X., Ren, X., & Huang, T. S. (2015). Self-assembled antibacterial coating by N-halamine polyelectrolytes on a cellulose substrate. *Journal of Materials Chemistry B*, *3*(7), 1446–1454.

Mahapatra, T., Mahapatra, S., Babu, G. R., Tang, W., Banerjee, B., Mahapatra, U., et al. (2014). Cholera outbreaks in South and Southeast Asia: Descriptive analysis, 2003–2012. *Japanese Journal of Infectious Diseases*, *67*(3), 145–156.

Lo Giudice, A., Poli, A., Finore, I., & Rizzo, C. (2020). Peculiarities of extracellular polymeric substances produced by Antarctic bacteria and their possible applications. *Applied Microbiology and Biotechnology*, *104*(7), 2923–2934.

Mahapatra, A., Padhi, N., Mahapatra, D., Bhatt, M., Sahoo, D., Jena, S., et al. (2015). Study of biofilm in bacteria from water pipelines. *Journal of Clinical and Diagnostic Research: JCDR*, *9*(3), DC09.

Maillard, J. Y., & McBain, A. (2019). Biofilm in healthcare settings and their control. *Letters in Applied Microbiology*, *68*(4), 268.

Manaia, C. M., Rocha, J., Scaccia, N., Marano, R., Radu, E., Biancullo, F., et al. (2018). Antibiotic resistance in wastewater treatment plants: Tackling the black box. *Environment International*, *115*, 312–324.

Mao, D., Yu, S., Rysz, M., Luo, Y., Yang, F., Li, F., et al. (2015). Prevalence and proliferation of antibiotic resistance genes in two municipal wastewater treatment plants. *Water Research*, *85*, 458–466.

Marchand, S., De Block, J., De Jonghe, V., Coorevits, A., Heyndrickx, M., & Herman, L. (2012). Biofilm formation in milk production and processing environments; Influence on milk quality and safety. *Comprehensive Reviews in Food Science and Food Safety*, *11*(2), 133–147.

Marcheggiani, S., Puccinelli, C., Ciadamidaro, S., Della Bella, V., Carere, M., Francesca Blasi, M., et al. (2010). Risks of water-borne disease outbreaks after extreme events. *Toxicological and Environmental Chemistry*, *92*(3), 593–599.

Marchello, C. S., Fiorino, F., Pettini, E., Crump, J. A., & Vacc-iNTS Consortium Collaborators. (2021). Incidence of non-typhoidal *Salmonella* invasive disease: A systematic review and meta-analysis. *Journal of Infection*, *83*, 523–532.

Martínez, J. L., Coque, T. M., & Baquero, F. (2015). What is a resistance gene? Ranking risk in resistomes. *Nature Reviews Microbiology*, *13* (2), 116–123.

Maurice, N. M., Bedi, B., & Sadikot, R. T. (2018). Pseudomonas aeruginosa biofilms: Host response and clinical implications in lung infections. *American Journal of Respiratory Cell and Molecular Biology*, *58* (4), 428–439.

Mazaheri, T., Cervantes-Huamán, B. R., Bermúdez-Capdevila, M., Ripolles-Avila, C., & Rodríguez-Jerez, J. J. (2021). Listeria monocytogenes biofilms in the food industry: Is the current hygiene program sufficient to combat the persistence of the pathogen? *Microorganisms*, *9*(1), 181.

Meesilp, N., & Mesil, N. (2019). Effect of microbial sanitizers for reducing biofilm formation of *Staphylococcus aureus* and *Pseudomonas aeruginosa* on stainless steel by cultivation with UHT milk. *Food Science and Biotechnology*, *28*(1), 289–296.

Mevo, S. I., Ashrafudoulla, M., Furkanur Rahaman Mizan, M., Park, S. H., & Ha, S. D. (2021). Promising strategies to control persistent enemies: Some new technologies to combat biofilm in the food industry—A review. *Comprehensive Reviews in Food Science and Food Safety*, *20*, 5938–5964.

Mizan, M. F., Jahid, I. K., & Ha, S. D. (2015). Microbial biofilms in seafood: A food-hygiene challenge. *Food Microbiology*, *49*, 41–55.

Møretrø, T., & Langsrud, S. (2017). Residential bacteria on surfaces in the food industry and their implications for food safety and quality. *Comprehensive Reviews in Food Science and Food Safety*, *16*(5), 1022–1041.

Murphy, H. M., Pintar, K. D., McBean, E. A., & Thomas, M. K. (2014). A systematic review of waterborne disease burden methodologies from developed countries. *Journal of Water and Health*, *12*(4), 634–655.

Muziasari, W. I., Pitkänen, L. K., Sørum, H., Stedtfeld, R. D., Tiedje, J. M., & Virta, M. (2017). The resistome of farmed fish feces contributes to the enrichment of antibiotic resistance genes in sediments below Baltic Sea fish farms. *Frontiers in Microbiology*, *7*, 2137.

Nebot, E., Casanueva, J. F., Solera, R., Pendon, C., Taracido, L. J., Casanueva-Robles, T., et al. (2010). Marine biofouling in heat exchangers. In J. Chan, & S. Wong (Eds.), *Biofouling types, impact and anti-fouling* (pp. 65–104). New York, NY: Nova SciencePublishers.

Neu, T. R., & Lawrence, J. R. (2015). Innovative techniques, sensors, and approaches for imaging biofilms at different scales. *Trends in Microbiology*, *23*(4), 233–242.

Nogi, Y. (2017). Microbial life in the deep sea: Psychropiezophiles. In *Psychrophiles: From biodiversity to biotechnology* (pp. 133–152). Cham: Springer.

Nohra, A., Grinberg, A., Marshall, J. C., Midwinter, A. C., Collins-Emerson, J. M., & French, N. P. (2020). Shifts in the molecular epidemiology of *Campylobacter jejuni* infections in a sentinel region of New Zealand following implementation of food safety interventions by the poultry industry. *Applied and Environmental Microbiology*, *86*(5), e01753-19.

Osińska, E., Korzeniewska, M., & Harnisz, S. (2019). Niestępski quantitative occurrence of antibiotic resistance genes among bacterial populations from wastewater treatment plants using activated sludge. *Applied Sciences*, *9*(3), 387. https://doi.org/10.3390/app9030387.

Pang, X., Wong, C., Chung, H. J., & Yuk, H. G. (2019). Biofilm formation of *Listeria monocytogenes* and its resistance to quaternary ammonium compounds in a simulated salmon processing environment. *Food Control*, *98*, 200–208.

Phillips, C. A. (2016). Bacterial biofilms in food processing environments: A review of recent developments in chemical and biological control. *International Journal of Food Science and Technology*, *51*(8), 1731–1743.

Pinto, M., Langer, T. M., Hüffer, T., Hofmann, T., & Herndl, G. J. (2019). The composition of bacterial communities associated with plastic biofilms differs between different polymers and stages of biofilm succession. *PLoS One*, *14*(6), e0217165.

Pradhan, D., & Negi, V. D. (2019). Stress-induced adaptations in *Salmonella*: A ground for shaping its pathogenesis. *Microbiological Research*, *229*, 126311.

Prest, E. I., Hammes, F., van Loosdrecht, M., & Vrouwenvelder, J. S. (2016). Biological stability of drinking water: Controlling factors, methods, and challenges. *Frontiers in Microbiology*, *7*, 45.

Primo, M. G., Tipple, A. F., de Melo, C. D., Guadagnin, S. V., Azevedo, A. S., de Oliveira Leão, L. S., et al. (2021). Biofilm accumulation in new flexible gastroscope channels in clinical use. *Infection Control and Hospital Epidemiology*, 1–7.

Pruden, A. (2014). Balancing water sustainability and public health goals in the face of growing concerns about antibiotic resistance. *Environmental Science and Technology*, *48*(1), 5–14.

Qu, Y., McGiffin, D., Kure, C., Ozcelik, B., Fraser, J., Thissen, H., et al. (2020). Biofilm formation and migration on ventricular assist device drivelines. *The Journal of Thoracic and Cardiovascular Surgery*, *159*(2), 491–502.

Rajkowski, K. T. (2009). Biofilms in fish processing. In *Biofilms in the food and beverage industries* (pp. 499–516). Woodhead Publishing.

Ranieri, M. R., Whitchurch, C. B., & Burrows, L. L. (2018). Mechanisms of biofilm stimulation by subinhibitory concentrations of antimicrobials. *Current Opinion in Microbiology*, *45*, 164–169.

Reynolds, C., Checkley, S., Chui, L., Otto, S., & Neumann, N. F. (2020). Evaluating the risks associated with Shiga-toxin-producing *Escherichia coli* (STEC) in private well waters in Canada. *Canadian Journal of Microbiology*, *66*(5), 337–350.

Ribeiro, M. C., da Silva Fernandes, M., Kuaye, A. Y., Jimenez-Flores, R., & Gigante, M. (2017). Preconditioning of the stainless steel surface affects the adhesion of *Bacillus cereus* spores. *International Dairy Journal*, *66*, 108–114.

Røder, H. L., Herschend, J., Russel, J., Andersen, M. F., Madsen, J. S., Sørensen, S. J., et al. (2018). Enhanced bacterial mutualism through an evolved biofilm phenotype. *The ISME Journal*, *12*(11), 2608–2618.

Rodríguez-López, P., Bernárdez, M., Rodríguez-Herrera, J. J., Comesaña, Á. S., & Cabo, M. L. (2019). Identification and metagenetic characterisation of Listeria monocytogenes-harbouring communities present in food-related industrial environments. *Food Control*, *95*, 6–17.

Rossi, E., Paroni, M., & Landini, P. (2018). Biofilm and motility in response to environmental and host-related signals in Gram negative opportunistic pathogens. *Journal of Applied Microbiology*, *125*(6), 1587–1602.

Ruhal, R, & Kataria, R. (2021). Biofilm patterns in gram-positive and gram-negative bacteria. *Microbiological Research*, *251*, 126829.

Rumbaugh, K. P., & Sauer, K. (2020). Biofilm dispersion. *Nature Reviews Microbiology*, *18*(10), 571–586.

Russo, P., Hadjilouka, A., Beneduce, L., Capozzi, V., Paramithiotis, S., Drosinos, E. H., et al. (2018). Effect of different conditions on *Listeria monocytogenes* biofilm formation and removal. *Czech Journal of Food Sciences*, *36*(3), 208–214.

Rybtke, M., Jensen, P.Ø., Nielsen, C. H., & Tolker-Nielsen, T. (2020). The extracellular polysaccharide matrix of *Pseudomonas aeruginosa* biofilms is a determinant of polymorphonuclear leukocyte responses. *Infection and Immunity*, *89*(1), e00631-20.

Saengsitthisak, B., Chaisri, W., Punyapornwithaya, V., Mektrirat, R., Klayraung, S., Bernard, J. K., et al. (2020). Occurrence and antimicrobial susceptibility profiles of multidrug-resistant aeromonads isolated from freshwater ornamental fish in Chiang Mai province. *Pathogens*, *9*(11), 973.

Salazar, J. K., Natarajan, V., Stewart, D., Fay, M., Gonsalves, L. J., Mhetras, T., et al. (2020). Listeria monocytogenes growth kinetics in refrigerated ready-to-eat dips and dip components. *PLoS One*, *15*(6), e0235472.

Scharping, R. J., & Garey, J. R. (2021). Relationship between aquifer biofilms and unattached microbial indicators of urban groundwater contamination. *Molecular Ecology*, 324–342.

Schulze, A., Mitterer, F., Pombo, J. P., & Schild, S. (2021). Biofilms by bacterial human pathogens: Clinical relevance-development, composition and regulation-therapeutical strategies. *Microbial Cell*, *8*(2), 28.

Simões, M., Pereira, A. R., Simões, L. C., Cagide, F., & Borges, F. (2021). Biofilm control by ionic liquids. *Drug Discovery Today*, *26*, 1340–1346.

Simoes, L. C., & Simões, M. (2013). Biofilms in drinking water: Problems and solutions. *RSC Advances*, *3*(8), 2520–2533.

Skowron, K., Wiktorczyk, N., Grudlewska, K., Kwiecińska-Piróg, J., Wałecka-Zacharska, E., Paluszak, Z., et al. (2019). Drug-susceptibility, biofilm-forming ability and biofilm survival on stainless steel of Listeria spp. strains isolated from cheese. *International Journal of Food Microbiology*, *296*, 75–82.

Stanley, N. R., & Lazazzera, B. A. (2004). Environmental signals and regulatory pathways that influence biofilm formation. *Molecular Microbiology*, *52*(4), 917–924.

Subhadra, B., Kim, D. H., Woo, K., Surendran, S., & Choi, C. H. (2018). Control of biofilm formation in healthcare: Recent advances exploiting quorum-sensing interference strategies and multidrug efflux pump inhibitors. *Materials*, *11*(9), 1676.

Sun, D., Jeannot, K., Xiao, Y., & Knapp, C. W. (2019). Horizontal gene transfer mediated bacterial antibiotic resistance. *Frontiers in Microbiology*, *10*, 1933.

Szekeres, E., Baricz, A., Chiriac, C. M., Farkas, A., Opris, O., Soran, M. L., et al. (2017). Abundance of antibiotics, antibiotic resistance genes and bacterial community composition in wastewater effluents from different Romanian hospitals. *Environmental Pollution*, *225*, 304–315.

Tallawi, M., Opitz, M., & Lieleg, O. (2017). Modulation of the mechanical properties of bacterial biofilms in response to environmental challenges. *Biomaterials Science*, *5*(5), 887–900.

Tan, C. H., Lee, K. W., Burmølle, M., Kjelleberg, S., & Rice, S. A. (2017). All together now: Experimental multispecies biofilm model systems. *Environmental Microbiology*, *19*(1), 42–53.

Tasneem, U., Yasin, N., Nisa, I., Shah, F., Rasheed, U., Momin, F., et al. (2018). Biofilm producing bacteria: A serious threat to public health in developing countries. *Journal of Food Science and Nutrition*, *1* (2), 25–31.

Thomas, K. M., de Glanville, W. A., Barker, G. C., Benschop, J., Buza, J. J., Cleaveland, S., et al. (2020). Prevalence of Campylobacter and *Salmonella* in African food animals and meat: A systematic review and meta-analysis. *International Journal of Food Microbiology*, *315*, 108382.

Toyofuku, M., Inaba, T., Kiyokawa, T., Obana, N., Yawata, Y., & Nomura, N. (2016). Environmental factors that shape biofilm formation. *Bioscience, Biotechnology, and Biochemistry*, *80*(1), 7–12.

US Department of Agriculture, Economic Research Service (2014) Cost estimates of foodborne illnesses. 2014. Oct 7 [cited 2019 Dec 6]. https://www.ers.usda.gov/data-products/cost-estimates-of-foodborne-illnesses.

Vestby, L. K., Grønseth, T., Simm, R., & Nesse, L. L. (2020). Bacterial biofilm and its role in the pathogenesis of disease. *Antibiotics*, *9*(2), 59.

Vishwakarma, V. (2020). Impact of environmental biofilms: Industrial components and its remediation. *Journal of Basic Microbiology*, *60* (3), 198–206.

Waksman, S. A. (1961). The role of antibiotics in nature. *Perspectives in Biology and Medicine*, *4*(3), 271–287.

Wang, L., Yuan, L., Li, Z. H., Zhang, X., & Sheng, G. P. (2021). Quantifying the occurrence and transformation potential of extracellular polymeric substances (EPS)-associated antibiotic resistance genes in activated sludge. *Journal of Hazardous Materials*, *408*, 124428.

Wei, T., Tang, Z., Yu, Q., & Chen, H. (2017). Smart antibacterial surfaces with switchable bacteria killing and bacteria-releasing cap-abilities. *ACS Applied Materials & Interfaces*, *9*, 37511–37523.

Weigelhofer, G., Jirón, T. S., Yeh, T. C., Steniczka, G., & Pucher, M. (2020). Dissolved organic matter quality and biofilm composition affect microbial organic matter uptake in stream flumes. *Water*, *12*(11), 3246.

Weinbren, M., Inkster, T., & Lafferty, F. (2021). Drains and the periphery of the water system–what do you do when the guidance is outdated? *Infection Prevention in Practice*, *3*(4), 100179.

WHO (Ed.). (2008). *Recommendations: Vol. 1. Guidelines for drinking-water quality, incorporating 1st and 2nd addenda* (3rd ed.). Geneva: WHO.

Williams, H. N., Turng, B. F., & Kelley, J. I. (2009). Survival response of Bacteriovorax in surface biofilm versus suspension when stressed by extremes in environmental conditions. *Microbial Ecology*, *58*(3), 474–484.

Williamson, K. S., Richards, L. A., Perez-Osorio, A. C., Pitts, B., McInnerney, K., Stewart, P. S., et al. (2012). Heterogeneity in *Pseudomonas aeruginosa* biofilms includes expression of ribosome hibernation factors in the antibiotic-tolerant subpopulation and hypoxia-induced stress response in the metabolically active population. *Journal of Bacteriology*, *194*(8), 2062–2073.

Wingender, J., & Flemming, H. C. (2011). Biofilms in drinking water and their role as reservoir for pathogens. *International Journal of Hygiene and Environmental Health*, *214*(6), 417–423.

Wirtanen, G., & Salo, S. (2016). Biofilm risks. In *Handbook of hygiene control in the food industry* (pp. 55–79). Woodhead Publishing.

Woegerbauer, M., Bellanger, X., & Merlin, C. (2020). Cell-free DNA: An underestimated source of antibiotic resistance gene dissemination at the interface between human activities and downstream environments in the context of wastewater reuse. *Frontiers in Microbiology*, *11*, 671.

Wu, H., Moser, C., Wang, H. Z., Høiby, N., & Song, Z. J. (2015). Strategies for combating bacterial biofilm infections. *International Journal of Oral Science*, *7*(1), 1–7.

Würfel, S. F., Da Silva, W. P., de Oliveira, M. G., Kleinubing, N. R., Lopes, G. V., Gandra, E. A., et al. (2019). Genetic diversity of *Campylobacter jejuni* and *Campylobacter coli* isolated from poultry meat products sold on the retail market in Southern Brazil. *Poultry Science*, *98*(2), 932–939.

Xu, J., Sheng, G. P., Ma, Y., Wang, L. F., & Yu, H. Q. (2013). Roles of extracellular polymeric substances (EPS) in the migration and removal of sulfamethazine in activated sludge system. *Water Research*, *47*(14), 5298–5306.

Yin, W., Wang, Y., Liu, L., & He, J. (2019). Biofilms: The microbial "protective clothing" in extreme environments. *International Journal of Molecular Sciences*, *20*(14), 3423.

You, X., Wu, D., Wei, H., Xie, B., & Lu, J. (2018). Fluoroquinolones and β-lactam antibiotics and antibiotic resistance genes in autumn leachates of seven major municipal solid waste landfills in China. *Environment International*, *113*, 162–169.

Yousif, A., Jamal, M. A., & Raad, I. (2015). Biofilm-based central line-associated bloodstream infections Biofilm-based healthcare-associated infections. *Advances in Experimental Medicine and Biology*, *830*, 157–179.

Yu, H., Chang, H., Li, X., Zhou, Z., Song, W., Ji, H., et al. (2021). Long-term fouling evolution of polyvinyl chloride ultrafiltration membranes in a hybrid short-length sedimentation/ultrafiltration process for drinking water production. *Journal of Membrane Science*, *630*, 119320.

Zawadzki, P. J., Perkowski, K., Padzik, M., Mierzwińska-Nastalska, E., Szaflik, J. P., Conn, D. B., et al. (2017). Examination of oral microbiota diversity in adults and older adults as an approach to prevent spread of risk factors for human infections. *BioMed Research International*, *2017*.

Zhang, L., Dong, D., Hua, X., & Guo, Z. (2018). Inhibitory effects of extracellular polymeric substances on ofloxacin sorption by natural biofilms. *Science of the Total Environment*, *625*, 178–184.

Zhang, Y., Geng, J., Ma, H., Ren, H., Xu, K., & Ding, L. (2016). Characterization of microbial community and antibiotic resistance genes in activated sludge under tetracycline and sulfamethoxazole selection pressure. *Science of the Total Environment*, *571*, 479–486.

Zhao, R., Feng, J., Liu, J., Fu, W., Li, X., & Li, B. (2019). Deciphering of microbial community and antibiotic resistance genes in activated sludge reactors under high selective pressure of different antibiotics. *Water Research*, *151*, 388–402.

Zhu, Z., Shan, L., Zhang, X., Hu, F., Zhong, D., Yuan, Y., et al. (2021). Effects of bacterial community composition and structure in drinking water distribution systems on biofilm formation and chlorine resistance. *Chemosphere*, *264*, 128410.

Zhu, Y. G., Zhao, Y. I., Li, B., Huang, C. L., Zhang, S. Y., Yu, S., et al. (2017). Continental-scale pollution of estuaries with antibiotic resistance genes. *Nature Microbiology*, *2*(4), 1–7.

Chapter 34

Biofilm formation: A well-played game in bacterial pathogenesis

Vihang Thite[a,*], R. Kavi Bharathi[b,*], and C.S. Srinandan[b]

[a]*Department of Pharmacology, Sun Pharmaceutical Industries Limited, Vadodara, Gujarat, India,* [b]*Biofilm Biology Laboratory, Anusandhan Kendra-2, Centre for Research on Infectious Diseases (CRID), School of Chemical and Biotechnology, SASTRA Deemed to be University, Thanjavur, Tamil Nadu, India*

1. Introduction

Biofilm can be defined as a structured community of microorganisms that is firmly attached to a surface and proliferates by staying embedded in a self-fabricated matrix composed of extracellular polymeric substances, i.e., EPSs (Das, Singh, Matchado, Srivastava, & Bajpai, 2019; Flemming & Wingender, 2010; Karygianni, Ren, Koo, & Thurnheer, 2020). Anthony van Leeuwenhoek, father of microbiology, first time observed and described the presence of microorganisms in a community from scurf of his teeth and from particles scraped from his tongue using his primitive microscope during 1683–1708. Since then, till half of the 19th century, several studies regarding microbial biofilms present in diverse ecological habitat have been reported. In 1970–72, the first case of *Pseudomonas aeruginosa* infection in cystic fibrosis (CF) disease was reported. But it was 1982, when J. W. Costerton detected the *Staphylococcus aureus* infection from an implanted cardiac pacemaker surface and coined the term "Biofilm" (Costerton & Lewandowski, 1997; Del Pozo, 2018). Since the last couple of decades, healthcare-associated biofilm infections from medical devices and prosthetics have become a major issue of research because of certain outstanding features that appeared during biofilm development making the biofilm-forming pathogens wonderful players for the game.

2. Establishment and development of bacterial biofilm

Biofilm formation on any site is regulated by diverse physical, chemical, and biological processes. Initially, adhesion (cell to the surface attachment) followed by cohesion (cell-to-cell attachment) cumulatively determine the biofilm strength and properties. The sequential steps involved in the biofilm formation are commonly listed as (i) initial conditioning of the surface layer, (ii) accumulation/adsorption of microbes on the preconditioned surface, (iii) reversible and irreversible adhesion, (iv) initial microcolony establishment through cell proliferation and matrix production, (v) maturation of biofilm and interaction of microorganisms inside maturing biofilm, and (vi) dispersal of biofilm (Das et al., 2019; Justice et al., 2004). When it comes to infectious biofilms, the steps are more or less alike. Overall, the biofilm development process is outlined in Fig. 1.

2.1 Establishment of conditioning layer on substratum surface

The establishment site determines the structural complexity of the conditioning films. The more the unevenness of substratum surface, the more conditioning film develops (Wimpenny, 1996; Wimpenny, Manz, & Szewzyk, 2000). The conditioning layer is constituted when flow drifted organic/inorganic matter settles down on the biotic/abiotic substratum due to gravitational forces. Physicochemical interactions between settled matter and substratum surface alter the substratum surface charges and potentials that cumulatively intensify the frequency and magnitude of selective microbial adhesion to the preconditioned substratum surface, which was formerly unsupportive for bacterial attachment (Fletcher, 1991; Rutter & Vincent, 1984). Such conditioning layer from aquatic or terrestrial environments usually contains glycoproteins, complex polysaccharides, lipids, and humic compounds (Fletcher & Marshall, 1982). Whereas, when established in the upper respiratory, reproductive, or urinary tract of living animal/human hosts, the conditioning layers contain macromolecular components secreted by respective organs to maintain their functional normalcy. Host

* Dr. Vihang Thite and R. Kavi Bharathi would like to dedicate this chapter to late Dr. C.S. Srinandan, for his affection, interest, passion, and dedication towards biofilm related research.

Understanding Microbial Biofilms. https://doi.org/10.1016/B978-0-323-99977-9.00009-0

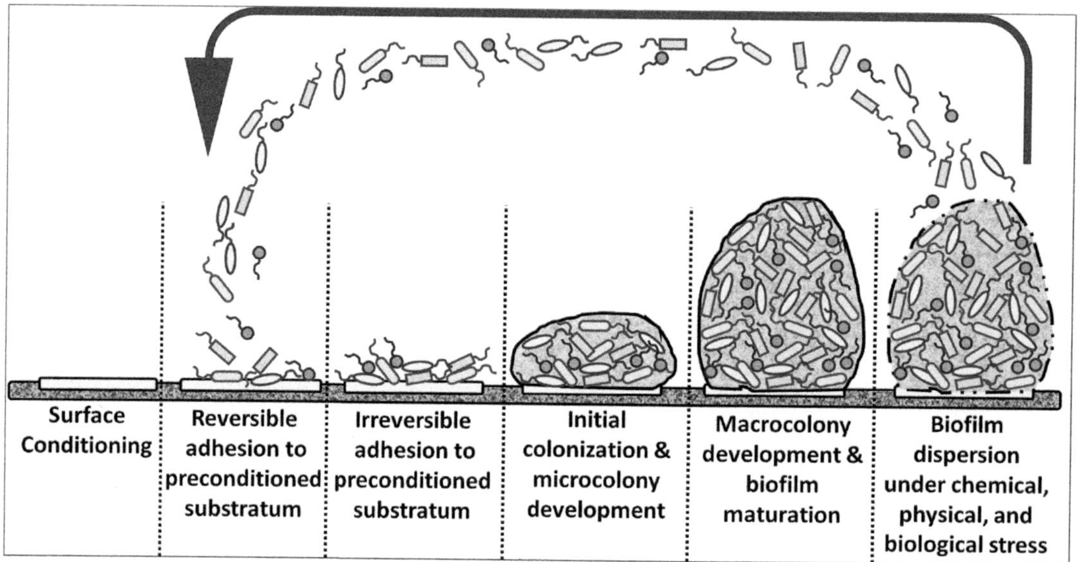

FIG. 1 The development of biofilm is categorized in different stages. The biofilm formation starts with attachment, progresses to microcolony formation, followed by maturation, and dispersion.

eukaryotic extracellular matrix (ECM) constituents such as the variety of proteins, glycoproteins, etc., often serve as conditioning layers targeted by microorganisms for adhesion (Koo & Yamada, 2016). For example, a pellicular proteinaceous film comprising glycoproteins, albumin, lipids, lysozyme, phosphoproteins, and salivary as well as gingival crevicular fluid components has been observed to condition the teeth enamel (Cugini, Shanmugam, Landge, & Ramasubbu, 2019). When established on abiotic surfaces of prosthetic implants, that are usually made up of metals, polyurethane, silicon, other polymers, or biomaterials, the conditioning films comprise host produced components, i.e., platelets, plasma and tissue proteins, erythrocytes, and fibrin derived from blood circulation that support and enhance biofilm formation by attracting contagion for colonization (O'Toole & Wong, 2016).

2.2 Adsorption and accumulation of microbes on the preconditioned surface

In a natural environment, the movement of microbial cells and their nutrients toward the preconditioned substratum surface is mostly governed by flow, gravitation, sedimentation, mass transport, thermal diffusion, and Brownian motion, etc. (Fletcher & Marshall, 1982). In the human body, fluids such as blood and urine move in a laminar manner with a rapid flow at the center and a smooth flow at the peripheries without any turbulent mixing that makes the farthest transportation of microbes and nutrients easy from the source site. Once the microbes enter from the central to peripheral flow, they tend to settle and adhere slowly fetching the nutrients (Percival & Walker, 1999). Whereas randomized turbulent flow leads to mixing of nutrients and bacteria that support minimal microbial movement required for adsorption of microbes on preconditioned surfaces before their adhesion (Fletcher, 1991; Percival & Walker, 1999).

2.3 Adhesion of microbes

Once the transported microbes accumulate in the proximity of the preconditioned substratum, the two-step adhesion process takes place. Initial week attachment of cells to the surface changes several cellular physicochemical properties, which gradually lead to strong and permanent irreversible adhesion. Van der Waals forces, steric interactions, electrostatic interactions, and molecule-specific interactions are the physicochemical forces involved in this process (Fletcher, 1991; Fletcher & Marshall, 1982; Rutter & Vincent, 1984). Hydrophobicity of cell surface, presence of fimbriae and flagella as well as the amount and composition of EPS matrix affect and control the degree of attachment. A complicated relationship between bacterial appendages and c-di-GMP governed system balances this reversible and irreversible adhesion process (Armbruster & Parsek, 2018; Collins, Smith, Sondermann, & O'Toole, 2020; Lee et al., 2018).

2.3.1 *Reversible adhesion*

"Surface naive" planktonic cells are the free-floating microbial cells without any exposure to the substratum surface. Once, they travel and reach the preconditioned substratum, continuous rotation and stroke of fimbriae and flagella-like appendages don't let these cells settle for longer time at one spot but move them to new ones making this momentarily adhesion reversible. In *P. aeruginosa*, the type IV pili in bacterial membrane detect such numerous short-term attachment-detachment cycles. Surface sensing Pil-Chp system cumulatively detects these and eventually triggers the upregulation of cAMP levels in the cytoplasm. Planktonic cells that have been associated with reversible adhesion and high cAMP levels are called "surface sentient." Orientation and surface properties of bacterial cells as well as temperature, pressure, and energy levels of the preconditioned substratum surface are the significant factors contributing to the adhesion process and let some of the bacteria adhere reversibly to the surface (Armbruster & Parsek, 2018; Collins et al., 2020; Das et al., 2019; Lee et al., 2018).

2.3.2 *Irreversible adhesion*

With several repeated reversible adhesions, many of the cells enter into irreversible adhesion to preconditioned substratum, which is facilitated by slowing down the bacterial movement through various physicochemical interactions occurring at molecular level in cytoplasm. This can be achieved through controlling flagellar regulation at two different levels. (i) Detection of several chemicals or molecules on the conditioning layer activates the molecular cascade altering the secondary signal molecule that triggers the motility responses by controlling speed and orientation of flagellar motor rotation. For chemoattractant molecules, the response usually slows down or stops the flagellar motor rotation restricting bacterial movement and allowing them to settle down and interact with the conditioning layer to form a strong association making the adhesion irreversible (de Weger et al., 1987; Guttenplan & Kearns, 2013). (ii) Activated Pil-Chp system during several reversible adhesions stimulates T-IV-P biogenesis, which helps in the secretion of the EPS matrix, increases cAMP levels, and initiates a cascade that ultimately activates signaling of the c-di-GMP, a secondary signal molecule. Increased c-di-GMP levels control the phosphorylation/dephosphorylation process of the flagellar switch protein FliM and promote flagellar shutdown as well as EPS matrix production. Both cumulatively result in the embedment of less motile bacterial cells in secreted EPS matrix initiating a biofilm establishment (Armbruster & Parsek, 2018; Collins et al., 2020; O'Toole & Wong, 2016).

2.4 Initial colonization, division of microbes, and development of microcolony

Once irreversible adhesion is established on the conditioning layer or host ECM, the bacterial cells receive the required nutrients for growth and exponentially proliferate. Thus, newly formed daughter cells keep on moving away from the substratum forming a mushroom-like microcolony structure. EPS, which is secreted by these bacterial cells, cements the intercellular space in biofilm. This Glycocalyx, i.e., the slime layer or capsular sheath, surrounds the bacterial cells, promotes irrevocable adhesion and cohesion in established microcolonies (Dragoš et al., 2018; Wood & Whitaker, 1999).

2.5 Microbial interactions and biofilm maturation

Incessant availability of nutrients through the EPS matrix makes microcolony-embedded bacterial cells continuously proliferate and produce more EPS, gradually developing in the macrocolonies and maturing further into the microbial biofilm. During this process, spatial and temporal heterogeneity occurring in microbial community as well as in nutrient resources from matrix can determine the *emergence of variations in intrinsic properties* of the biofilms. Intrinsic properties are the observed changes in the phenotypic properties of bacteria that emerged as an outcome of bacterial interactions in the multispecies community but were not visible from individual bacterial species when they were under isolation (D'Souza, 2020; Madsen, Sørensen, & Burmølle, 2018; Nasipuri et al., 2020). Gene expression studies on emergence of variation from multispecies biofilm unveiled altered expression patterns of several genes emphasizing that these community intrinsic properties in the biofilm system are dependent on up- and downregulation of various functions, which gradually develop in community based on spatial as well as temporal resource and microbial heterogeneity (D'Souza, 2020; Flemming et al., 2016).

2.5.1 *The spatial and temporal structure of EPS matrix and emergence of variation*

The different functions of EPS, which are attributed to its physicochemical characteristics, majorly contribute to the thriving of microbial communities and determine the three-dimensional architecture as well as the fate of biofilm. The resulting biofilm morphology can be smooth, flat, rough, fluffy, or filamentous, and the biofilm can also vary in its degree of porosity, having mushroom-like macrocolonies surrounded by water-filled spaces (Flemming & Wingender, 2010). This forms a continuous network of water channels throughout the macrocolonies facilitating the continuous nutrient supply from top to the basal layer (Costerton & Lewandowski, 1997; Costerton & Wilson, 2004; Nickel et al., 1990).

The EPS matrix comprises extracellular polysaccharides, proteins, glycoproteins, nucleic acids, lipids, surfactants, etc., serving as a major reservoir for the nutrients as it harbors more than half of the organic carbon content from the biofilm. These polymeric components impart thick slimy nature to the EPS matrix that provides greater cohesive forces to keep the bacterial cells together in each other's close proximity and helps in exchanging metabolites, signaling molecules, genetic materials, defensive compounds with each other. These cumulatively dictate the interactions between organisms of the microbial community (Flemming, Neu, & Wozniak, 2007; Karygianni et al., 2020; Xiao et al., 2012).

2.6 Maturation and dispersal

Once biofilm achieves the maturity, the cell concentration in biofilm is at its peak, and the rate of cell division slows down to balance the cell death rate, and daughter cells are at their utmost physiological activities. Maturation of the biofilm leads to certain environmental conditions that naturally initiate its dispersion. The gradient of chemicals, i.e., nutrients, oxygen, waste products, and signaling compounds such as c-di-GMP, etc., is the native and environmental dispersion-inducing condition. Instead of involving the entire biofilm, only cells from the selected area, which lies in the unfavorable region of the gradients, initiate a signaling cascade through a variety of the mechanisms and are dispersed from the mature biofilm (McDougald, Rice, Barraud, Steinberg, & Kjelleberg, 2012; Rumbaugh & Sauer, 2020; Solano, Echeverz, & Lasa, 2014). Detection of the nutrient gradient activates the molecular cascade activating the biopolymer lytic enzymes such as alginate lyase, *N*-acetyl-heparson lyase, and hyaluronidase, which breaks the polysaccharides that hold the biofilm together and releases the entrapped bacteria, allowing them to colonize on a fresh substrate (Choudhary & Cassaro, 2021; Flemming & Wingender, 2010). Similarly, alteration of the c-di-GMP molecules in the cytoplasm disturbs the balance of Palb-C-di-GMP system, which now releases the adhesion protein from the bacterial surface making the flock free from adhesion (Collins et al., 2020). On the other hand, in case of very drift and turbulent flow passing over the biofilm, the detachment is believed to be the main mechanism that limits the overall biofilm biomass accumulation due to mechanical or shear stress and refers to a passive release of biofilm particles. When biofilm particles from systemic or prosthetic infections in humans disperse in the blood flow, they create a chance of embolism as during the flow, the particle can get clotted at any point in the blood vessel putting the patient's life at risk (Das et al., 2019; Fletcher & Marshall, 1982).

3. Biofilm heterogeneity and pathogenesis

The heterogeneity or population variation in biofilms also leads to the development of tolerance to the antimicrobials and host immune system. The structural heterogeneity in biofilms creates a wide range of different populations of cells such as fast growers, slow growers, matrix producers, nonmatrix producers, or a combination of both in a single cell, persister cells, and cells with various acquired mutations. When compared with the free-living planktonic forms, the microbes existing as heterogeneous population in biofilm resiliently exhibit defensive features such as enhanced ability to tolerate and resist exposure to antimicrobials along with ability to invade host immune responses and ability to adhere firmly with biotic/abiotic surfaces, along with survival through extremely unfavorable conditions. Thus, the biofilm matrix components play a significant role in the evolution of the organism where the more fit ones are selected during evolution, which makes the bacterial strain more virulent and aids in pathogenesis cumulatively making such biofilms a fortress (Algburi, Comito, Kashtanov, Dicks, & Chikindas, 2017).

4. Biofilms: Making acute infections chronic

Mechanisms through which microorganisms trigger pathogenesis while establishing biofilms are somewhat alike to those discussed in Section 2. Such as (i) adherence to a solid substrate, (ii) increase in the metabolic efficiency, (iii) frustrated phagocytosis and achieving high-density population of pathogens, (iv) generation of relatively higher virulent strains, which are efficient toxin producers and protected against antimicrobial agents, (v) transmission of pathogens to different sites by detachment. For pathogens infecting the human host, the host ECM plays a pioneering role in establishing infection. Once established, the spatial and temporal changes occurring in biofilm dictate the heterogeneity within it. This gives rise to diverse intrinsic properties yielding stronger pathogenesis and making the acute infection chronic. Few of such intrinsic properties like tolerance to chemical and physical stresses such as, heavy metals, antimicrobial agents, drought, etc., have been briefly discussed in the context of the emergence of variation, which is of importance (Burmølle, Ren, Bjarnsholt, & Sørensen, 2014; Burmølle et al., 2006; Lee et al., 2014; Liu, Russel, Burmølle, Sørensen, & Madsen, 2018; Røder et al., 2018).

Microbial pathogens have evolved and adapted an alternative strategy of developing intrinsic properties for making acute infections chronic. The pathogen invades endothelium, reaches host ECM, initiates colonization, and releases metabolites that degrade and expose ECM components altering local environment. Such alterations in local environments make these niches susceptible for dense microbial biofilm growth. This leads to deeper invasion in host tissue where pathogens can hide and get protected in biofilm from the host immune responses and antimicrobials. Further on maturation, dispersal of such invaded biofilms disseminates the infectious pathogens to new sites. Thus, the initial acute infections, if left untreated, gradually penetrate deeper in tissues, establish a biofilm, and on dispersal make the infection persistent or chronic with prolonged time (Koo & Yamada, 2016).

4.1 Eukaryotic ECM—A conditioning layer for pathogens

Eukaryotic ECM is a three-dimensional network comprising extracellular macromolecules such as enzymes, glycoproteins, hydroxyapatite, collagens, as well as certain minerals and provides structural and biochemical support to surrounding tissue cells. This ECM composition varies according to the tissue organ to which it belongs. Laminins, collagens, and fibronectins are abundant host ECM components from basement membranes, interstitial spaces, as well as connective tissues, which are frequently targeted by microorganisms as a conditioning layer for adhesion. After implantation in hosts, surfaces of medical prostheses, which are usually made up of various biopolymers, abiotic and metallic materials, get surrounded by ECM enhancing the susceptibility of implants toward microbial adhesion. Dysbiosis of normal microbiota inhabiting the host organ systems triggers the imbalanced growth of opportunistic pathogens, which attach themselves with host ECM tissues through multiple surface proteins. The surface proteins vary in size and protein-binding domain structures, namely collagen-binding, fibronectin-binding, and laminin-binding adhesins or carbohydrate side chains binding adhesins (Koo & Yamada, 2016; von Eiff, Jansen, Kohnen, & Becker, 2005). This adherence initiates the microcolony formation creating an acute localized infection, and if left untreated, it matures into biofilm developing several intrinsic properties and critical responses in hosts, and all of which cumulatively make the infection chronic and facilitate microbial invasion to "metastasize" at other sites in the host body (Koo & Yamada, 2016).

4.2 Frustrated phagocytosis

Opportunistic pathogens express and possess a wide range of adhesins that target a variety of ECM components resulting in dynamic interactions between EPS and ECM components. This cumulatively modulates both bacterial and host cell responses. Biofilms activate lytic enzymes such as protease, polysaccharide hydrolase, etc., that clast the polymeric components, activate the host inflammatory and immune responses, recruiting neutrophils and phagocytes. Deeper invasion of pathogenic cells in ECM protects them from phagocytosis and triggers the rupture of phagocytes leading towards the mutilation of nearby healthy host tissue matrix. This damage further opens up the ECM from the surrounding region making the area infection-prone, a process called frustrated phagocytosis. Thus, biofilm infections become hardly detectable by the host defense system despite tremendous cellular and humoral immune reactions that damage the host tissues and ECM (Burmølle et al., 2006; Leid, 2009).

4.3 Matrix support to cooperative virulence

Proteins with enzymatic activities degrade the complex matrix components in EPS and break the biopolymers down to low-molecular-mass products that can be consumed as carbon and energy source by community microorganisms during biofilm growth. Moreover, the heterogeneity of cells with different metabolic capacities or physiological gradients provides opportunities for cooperation and the coexistence of microbes. In community, the inability of one genus to use a particular carbon or nitrogen product can be overcome as the neighboring microbes within the proximal region of the matrix can consume them and release their metabolites, which can be utilized by the former organisms as nutrient source (Dragoš et al., 2018; Flemming & Wingender, 2010; Karygianni et al., 2020).

4.4 EPS matrix giving rise to antimicrobial tolerance

Irrespective of their ionic state or charge density, EPS components act as molecular sieves, diffusion barriers, and adsorbents resulting in physicochemical phenomenon such as fluid frictional resistance and diffusion-reaction inhibition making pathogens tolerant to diverse antimicrobials (Flemming et al., 2007; Fletcher & Marshall, 1982; Percival & Walker, 1999; Sutherland, 2001). Thus, the term "tolerance" denotes a characteristic that is specific to biofilms, and it is lost following the dispersal to free-living bacterial cells. Antimicrobials with large sizes are sieved out and barred from crossing the EPS diffusion barrier. The smaller or hydrophilic ones, which diffuse through this barrier, are subjected to diffusion-reaction inhibition where they are quenched by ionic chelation through complex formation with polysaccharides and other polymers and subsequently subjected to enzymatic degradation. Thus, a very low concentration of active antimicrobial reaches the cells (Flemming et al., 2007). Thus, diffusion-reaction inhibition may promote the selection for antimicrobial tolerance against antimicrobial stress as the effective concentration of antimicrobials is reduced to its sublethal concentrations (Stewart, 2015).

4.5 EPS as a source for genetic alteration giving rise to antimicrobial resistance

The EPS also contributes in developing resistance in microbes against antimicrobial compounds. The term "resistance" denotes a genetic, heritable characteristic that is acquired either by mutation or by gene exchange and that remains even when cells in the biofilm are dispersed to a new place. Accumulation of higher number of mobile genetic elements such as e-DNA (secreted by living cells), as well as plasmids and gDNA fragments (released from dead or lysed cells) along with higher number of bacterial cells with increased genetic competencies in biofilm matrix cumulatively provides excellent conditions for efficient horizontal gene transfer. The plasmid and e-DNA, gDNA fragments possessing genes that confer resistance to several antibiotics can be horizontally transferred to the recipient cells through conjugation or transformation processes as the matrix provides mandatory stable cell to cell adhesion. Expression of these newly taken up genes can result in phenotype with resistance toward certain antimicrobials (Cepas et al., 2019; Patel, 2005).

4.6 EPS as a shield in unfavorable conditions and succession

Majority of the EPS components, which are spread throughout the matrix along with bacterial cells, are hydrophilic and can interact with each other, with bacterial cells, and retain the water molecules through hydrogen bond formation. Thus, the complex microporous network of the channel remains hydrated and water-filled in most cases. This protects the immobilized cells from drastic environmental changes, such as heat, cold, dryness, etc. The channels also distribute secreted microbial metabolites for inter and intracellular communications appropriately, i.e., social activities such as quorum sensing through autoinducer molecules, which can regulate several features of microbes such as the production of virulence factors, enzymes, etc. (Flemming et al., 2016). Thus, several microbial interactions cumulatively turn out to be the microbial succession, where each step represents a few common community features, which are under constant emergence of variations that gradually mold and lead the biofilm toward maturation.

4.7 Persister cells in biofilm

When bacterial cells in a matured biofilm reach the stationary phase, almost 1% of the organisms become metabolically inactive by reducing their growth rate. The sublethal concentration of antimicrobial, which has diffused from the surface of EPS to the cells, with the hope of targeting the metabolic machinery can't affect these 1% microbes as the target machinery is already switched off due to dormancy. These cells now can tolerate the antimicrobial concentrations and are called persister cells. This phenotype was first highlighted by Hobby in 1942 in *S. aureus* (Lewis, 2007). It should be noted that the persister cells do not undergo any genetic changes and hence, are not resistant mutants but are phenotypic variants that can tolerate the antimicrobials. When they relocate to the next place after dispersion, the cells relatively take a longer time to resurrect with delayed growth suggesting the persistence of cells in biofilms. On replenishing the nutrient conditions to favorable ones, these persister cells reestablish the biofilm making the biofilm infection persistent.

5. Relevance of biofilm disease in clinical practice

For ages, humans have suffered from certain life-threatening diseases caused by acute microbial infections. Despite diverse efforts for developing vaccines and antibiotics for more than two centuries, the evolution of microbial pathogens has barricaded human conquer on such infections. Adaptation to sessile or adhered lifestyle from free-floating planktonic style, upregulation or downregulation of diverse virulent factors and mutations therein, production and change in composition of EPS, resistance to host immune responses as well as antimicrobials, etc., are the features that contribute to complex pathogenic biofilm formation, a strategy that has been evolved with time (Fux, Costerton, Stewart, & Stoodley, 2005). Besides the normal microflora inhabiting host, which can elicit infection on dysbiosis or compromised host immune system, bacteria can also enter and infect through different pathways. Bacterial species from contaminated external environmental sources such as air, water and soil, or skin of healthcare workers and/or patients usually contaminate the indwelling devices and initiate biofilm formation (Malone et al., 2017; von Eiff et al., 2005). The mode of inoculation could be perioperative (during or after surgery through surgical equipment, use of contaminated washbasins, toilets, cots, beds, bedsheets, cloths, etc.), hematogenous (through the blood or lymph) or contiguous (through the spread of the pathogen from preexisting neighboring infections such as soft tissue lesions, penetrating trauma, or osteomyelitis) (Brady, Leid, Costerton, & Shirtliff, 2006; Cai et al., 2018; Lidgren, Knutson, & Stefánsdóttir, 2003; Wright & Nair, 2010). On successful entry and establishment of biofilm by planktonic infectious microorganisms in a host system, if left untreated, increased severity of the acute infection results in chronic ones threatening a life. Detection of such infection has not only been restricted to all major organ systems of human hosts such as respiratory, gastrointestinal, excretory, cardiac, circulatory, integumentary, reproductive, skeletal, auditory, olfactory, and ophthalmic systems, but also, from the indwelling implant or prosthetics, such as joint implants, stents, plugs, pacemakers, valves, etc. Thus, the biofilm diseases create great public health issues in underdeveloped, developing, and developed countries (del Pozo & Patel, 2007; Hall-Stoodley, Costerton, & Stoodley, 2004). Based on the establishment, biofilm diseases can be grouped as (i) nondevice-related chronic biofilm diseases and (ii) device-related biofilm diseases as well as biofilm-related device malfunction (Del Pozo, 2018; Vestby, Grønseth, Simm, & Nesse, 2020).

5.1 Nondevice-related chronic biofilm disease

When developed in any of the abovementioned systems, but not associated with any prosthetic implants, and left untreated, these biofilms become the responsible foreign entities that impede the normal functioning of organ systems creating a disease. Brief details on such biofilm-associated diverse systemic disease conditions and their establishments are described below.

5.1.1 *Respiratory-system-related infections*

Rhinosinusitis (RS)

Rhinosinusitis (RS) is an inflammation of paranasal sinuses, characterized by nasal congestion, obstruction, runny nose, and loss of smell, which is usually triggered by infection of viral agents, i.e., rhinovirus, influenza virus, adenovirus, respiratory syncytial virus (RSV), etc. Persistence of infection over 12 weeks converts acute rhinosinusitis (ARS) to chronic rhinosinusitis (CRS). In most chronic cases, viral infection triggers and activates nonspecific as well as specific host immune responses, which eventually invade the infectious agents (Vestby et al., 2020). This generates debris of dead epithelial and immune cells, creating an apt environment for opportunistic pathogens as well as anaerobic bacteria to establish secondary bacterial infections through biofilm formation (Maina, Patel, & Cohen, 2018). Once established, these pathogenic biofilms may persuade the epithelial layer destruction by altering the mucociliary layer and elicit local inflammatory responses in biofilm-associated CRS (Tan et al., 2013). The human nasopharynx is the main reservoir of bacterial pathogens. Pneumococcal colonization involving *Streptococcus pneumoniae* is believed to always precede bacterial infection (Hamilos, 2019; Wilson & Hamilos, 2014). The intracellular bacterial colonization of *S. aureus* takes place due to facilitated cellular invasion by pathogens and several superantigens, which helps the pathogens in adhesion and evade immune responses (Kadioglu, Weiser, Paton, & Andrew, 2008; Kumpitsch, Koskinen, Schöpf, & Moissl-Eichinger, 2019; Maina et al., 2018; Vestby et al., 2020).

Pharyngitis, laryngitis, and sore throat

The tonsils and adenoids are lymphoid structures respectively located at the back of the mouth on both sides of the throat and behind the nose. Any infection to tonsils, adenoids, throat, and larynx inflames these organs, and infection recurrence may result in tonsillar or adenoid hypertrophy (Zautner, 2012). Among observed cases, ~20%–25% cases are caused by bacterial infection, predominantly by *S. aureus*, *Haemophilus* spp., and *Streptococcus* spp. whereas 75%–80% of cases are caused by viral infections (Diaz et al., 2011). After tonsillectomy and adenoidectomy of severely inflamed hypertrophied tissues from chronic clinical cases, their electron and confocal microscopic studies unveiled the presence of bacterial biofilms. Biofilms on tonsils and adenoids can be blamed for the symptoms such as harsh vocal sound, apnea, and cervical adenopathy (Chole & Faddis, 2003; Roberts et al., 2012). There has been very sparse research on biofilm-associated with chronic laryngitis as the vocal fold biopsies had shown the presence of biofilm on the larynx in chronic laryngitis and the disease is mainly believed to be noninfectious (Kinnari, 2015).

Pertussis and other Bordetella infections

Pertussis or whooping cough is mostly caused by *Bordetella pertussis* and *Bordetella parapertussis* that might have evolved from a common ancestor *Bordetella bronchiseptica*, which infects several mammalian species. Despite no direct observation from the human host respiratory tract, mouse models for nasal and tracheal biofilm formation by *B. bronchiseptica* and *B. pertussis* have been established (Cattelan, Dubey, Arnal, Yantorno, & Deora, 2016). Distinct architectural features such as mats, towers, or pillars for *B. bronchiseptica* and clusters or macrocolonies for *B. pertussis* were found adherent to ciliated epithelium from mouse models (Cattelan, Jennings-Gee, Dubey, Yantorno, & Deora, 2017). Pertussis-positive patients have been detected with Bps (a mandatory polysaccharide expressed by pathogen at a higher-level during biofilm formation) and anti-Bps antibodies (expressed by the host). This cumulatively infers the involvement of either Bps or eDNA within the bacterial colonization on human respiratory tissues as biofilms (Conover, Sloan, Love, Sukumar, & Deora, 2010; Dorji et al., 2019; Paddock et al., 2008).

Cystic fibrosis (CF)

CF was probably the first and most thoroughly studied infection where biofilm was recognized as part of the etiology (Kumpitsch et al., 2019). CF is a hereditary life-limiting genetic disease caused by mutations in the cystic fibrosis transmembrane conductance regulator (CFTR) gene that primarily affects the respiratory and digestive systems by producing viscid mucus leading to chronic infections. In most cases, it becomes chronic, due to a defect in mucociliary clearance of mucopurulent secretion leading to morbidity and mortality (Kumpitsch et al., 2019; Vestby et al., 2020). Typical core bacterial CF pathogens such as *Rothia*, *Prevotella*, *Streptococcus*, *Actinomyces*, and *Veillonella* along with other CF-associated pathogens such as *P. aeruginosa*, *Haemophilus influenza*, *Burkholderia cepacia* complex, *Stenotrophomonas maltophilia*, and *S. aureus* inhabit the respiratory tract and can lead to chronic lung infection in CF (Boutin et al., 2015; Del Pozo, 2018; Khan & Oloketuyi, 2017). Microscopic analysis of lung tissue, lung abscess, and sputum of CF patients has shown the presence of *P. aeruginosa* microcolonies forming a primitive stage of biofilm adherent to samples. As a result of microcolonization, a large number of polymorphonuclear leukocytes (PMNLs) infiltrate the area, giving chronic inflammation with subsequent tissue damage, loss of lung function, and obstruction of the airways. The metabolic activity of bacteria and host cells consumes available oxygen-producing anaerobic conditions favoring the biofilm mode of *P. aeruginosa* (Boyle, Heilmann, van Ditmarsch, & Xavier, 2013). Thus, bacterial colonization starts with *P. aeruginosa* biofilm in the paranasal sinuses and nasopharynx, which serve as the reservoirs for repeated lung infections that finally become chronic (Inglis, Millar, Jones, & Robinson, 1989). Detection of CF-associated biofilms can be done by examining bacterial aggregates embedded in alginate including matrix in sputum or other tissue samples (Fodor et al., 2012). It has been reported that interactions in mixed eukaryotic-prokaryotic biofilms (polymicrobial infections) in CF lungs can lead to adverse clinical outcomes (Kumpitsch et al., 2019).

5.1.2 Infections related to digestive system

Periodontitis and gingivitis

The periodontal pockets harbor a dense microbiome, where a wide range of multispecies microorganisms interact with medically important opportunistic pathogenic species in many different ways contributing to periodontal health (Colombo et al., 2009; Del Pozo, 2018; Marsh & Zaura, 2017). Periodontitis and gingivitis are the most common infectious periodontal diseases associated with the establishment of a highly pathogenic biofilm that triggers host immune/inflammatory responses destructing the supporting periodontal tissue and eventually leading to tooth loss (Del Pozo, 2018; Vestby et al., 2020). Supragingival and subgingival plaques from healthy oral cavities have considerable biodiversity, i.e., the former one comprises mainly Gram-positive cocci and anaerobic Gram-positive rods with a predominance of *Streptococcus* spp. and *Actinomyces naeslundii*, respectively, whereas the latter one comprises anaerobic Gram-negative rods and facultative anaerobic Gram-positive cocci (Colombo et al., 2009; del Pozo & Patel, 2007). Networks of multiple synergistic and antagonistic interactions generate microbial interdependencies that alter the calcium flux, invade mucosal cells, release toxins. Due to the anatomical proximity of periodontal biofilm to the gingival bloodstream, periodontal pockets can act as reservoirs of microbial pathogens and their products, as well as inflammatory mediators and immune complexes that can disseminate to other sites of the human body (Han & Wang, 2013; Lamont & Jenkinson, 1998).

Sialadenitis and sialolithiasis

Sialolithiasis is a formation of calcified masses or calculi, i.e., sialoliths or salivary stones within a salivary gland leading to sialadenitis, i.e., swelling of the salivary glands. In literature, bacterial, viral, or fungal infections have been reported to cause sialadenitis (Kao, Chole, & Ogden, 2020; Vestby et al., 2020). Bacterial biofilms and host immune cells, platelets, and erythrocytes, as well as calcium nanoparticles, have been observed from the histological sections of the chronic obstructive sialadenitis patients through light and scanning electron microscopy. Many sialoliths exhibited the presence of bacterial biofilm structures at their center or at surfaces proving the hypothesis that local injury due to bacterial biofilms might have led to inflammation and calcium deposition gradually resulting in sialoliths. Patients with clinical postoperative infections, recurrent sialadenitis, or pus drainage indicate that the presence of bacterial biofilms may contribute to more severe and recurring cases of sialadenitis (Fusconi et al., 2016; Perez-Tanoira et al., 2019).

Recalcitrant typhoid fever

Typhoid fever is an acute foodborne illness, predominantly caused by *Salmonella enterica* serovar *Typhi*, often characterized by high fever, weakness, headache, abdominal pain, and constipation. Serious complications may arise, including intestinal bleeding, bowel perforation, septicemia, meningitis, and death if left untreated (Cohen, Bartlett, & Corey, 1987). After the acute phase of illness, 3%–5% of patients become chronic carriers, who are generally asymptomatic and constitute an important reservoir for *S. typhi* that can be shed in feces and urine, thereby spreading disease as exemplified by a famous historical case of typhoid carrier Mary Mallon (Marineli, Tsoucalas, Karamanou, & Androutsos, 2013). Chronic typhoid patients who are the carriers of *S. typhi*, suffer from complications such as hepatitis, cholecystitis, cholangitis, chronic diarrhea, and pancreatitis as well as hepatobiliary carcinomas. In chronic clinical cases, *S. typhi* is often detected to involve attachment and invasion of epithelial cells, whereas, in asymptomatic patients with cholelithiasis, *S. typhi* biofilm is formed on a large part of gall stones surface (Gonzalez-Escobedo, Marshall, & Gunn, 2011; Sinnott & Teall, 1987). Antibiotic treatment generally resolves the acute infection but fails to eradicate the chronic

S. typhi colonization in the gall bladder. Gradually host immune response changes from an early pro-inflammatory response to a later antiinflammatory response, whereas *S. typhi* increases persistence and biofilm-forming ability as well as reduces the ability to cause acute infections leading toward the host adaptions to the chronic infection (Crawford et al., 2010; Neiger et al., 2019; Vestby et al., 2020).

Inflammatory bowel disease

Inflammatory bowel disease (IBD, also known as ulcerative colitis and Crohn's disease) is characterized by chronic inflammation of the digestive tract. Common symptoms are pain, diarrhea, weight loss, and fatigue. The mucosa of the gastrointestinal tract is heavily laden with bacterial microbiota growing as healthy biofilm communities (Del Pozo, 2018; von Rosenvinge, O'May, Macfarlane, Macfarlane, & Shirtliff, 2013). Dysbiosis of mucosal microbiota disturbs the balanced microbial interactions breaching the protective mucosal layer that separates the colon epithelium from luminal microbiota and increases contact between microbes and epithelial cells providing the favorable conditions for opportunistic pathogens to invade, colonize, and hide in the epithelial layer by forming a biofilm. Though this hiding stimulates inflammatory responses, it protects biofilm infection from clearing and makes it persistent (Swidsinski, Weber, Loening-Baucke, Hale, & Lochs, 2005). These changes led to a pathogenic state that can trigger the development of IBD. In most of the IBD cases, biofilms are dominated by *Bacteroides fragilis* and *Escherichia coli* among the variety of commensal microbes, and if left untreated, IBD patients have a 10%–30% cumulative risk of developing colorectal cancer (CRC) within 30 years after IBD onset (Kim, 2014; Swidsinski, Weber, et al., 2005).

Colorectal cancer

Like IBD, studies also suggest an association between dysbiosis, colonic biofilm formation, and colorectal carcinogenesis (Kim, 2014; Swidsinski, Weber, et al., 2005). CRC is the most common intestinal cancer found in both males and females. CRC develops from the normal colorectal epithelium as a result of microbial dysbiosis, which is associated with an aberrant inflammatory response, augmenting genetic mutations and epigenetic changes in epithelial cells as described by the adenoma-carcinoma sequence model or driver-passenger model (Fearon & Vogelstein, 1990; Martin et al., 2004; Tjalsma, Boleij, Marchesi, & Dutilh, 2012). In this model, on dysbiosis, certain bacteria (bacterial drivers), that were part of the normal microbiota, can facilitate DNA damage in epithelial cells that may result in the initiation of CRC. Accumulation of pro-inflammatory agents disturbs regulation of inflammation, apoptosis, and cell proliferation. As toxigenic *B. fragilis* disturbs the epithelial homeostasis, and influences cell proliferation, it can initiate CRC. Whereas certain enteropathogenic *E. coli* produces toxins that induce chromosomal instability or mutations through double-stranded DNA breaks. Both of these have increased DNA damage, cell proliferation, faster initiation of tumorigenesis, and greater mortality (Dejea et al., 2018; Sears, Geis, & Housseau, 2014). Biological changes that occur during tumorigenesis alter the microenvironment that favors colonization by opportunistic bacteria (bacterial passengers). *Fusobacterium nucleatum* may not be involved in the early stages of carcinogenesis but may be involved in cancer progression as it gradually gets enriched in tumor tissue and can influence both inflammatory processes and cell proliferation (Hold & Allen-Vercoe, 2019; Vestby et al., 2020). Thus, the new microenvironment provided by the tumor may result in that the bacterial drivers being eventually outcompeted by bacterial passengers (Tjalsma et al., 2012). Biofilms associated with CRC, adenomas, and polyps-associated biofilms unveiled a significantly higher prevalence of *F. nucleatum* in biofilms from CRC than normal controls through SEM, FISH, and fluorescence microscopy (Tomkovich et al., 2019).

5.1.3 *Infections related to auditory system*

Otitis media

Otitis media (OM) is an inflammation of the middle ear cavity most commonly observed in young children (Auinger, Lanphear, Kalkwarf, & Mansour, 2003; DeAntonio, Yarzabal, Cruz, Schmidt, & Kleijnen, 2016). It can be subdivided into acute OM (AOM), chronic supportive OM (CSOM), and OM with effusion (OME) leading to temporary or permanent hearing loss (Schilder et al., 2016). During early childhood, the shorter, wider, and more horizontal course of the Eustachian tube promotes the transmission of *S. pneumoniae* and nontypeable *H. influenzae* (NT-Hi) and many other otopathogens from the nasopharynx to the middle ear cavity, increasing the consequences of AOM and OME in later life (Faden et al., 1997). With advancements in molecular microscopic techniques, plenty of bacteria *P. aeruginosa*, *Staphylococcus epidermidis*, etc., have been proven to contribute to biofilm formation during OM pathogenesis (Fergie, Bayston, Pearson, & Birchall, 2004).

5.1.4 *Infections related to the cardiovascular system*

Infective endocarditis

Infective endocarditis (IE) is a bacterial infection that leads to inflammation of the endocardium. *Staphylococci*, *Streptococci*, and *Enterococci* are the most commonly isolated microbes from the valve (Elgharably, Hussain, Shrestha, Blackstone, & Pettersson, 2016). Vigorous blood pumping and flow friction create an injury to the endothelial surface that triggers the sterile clot-like lesion at the injury site, from fibrin and platelets from the blood. Infectious bacteria traveling through blood gradually settle and adhere to the clot, develop microcolonies, and mature as a biofilm. The formation of biofilm at the cardiac valve disrupts the valve function, i.e., it disturbs the blood flow by creating turmoil when the valve is open, and it makes the blood leak as it interferes when the valve is closing. As the biofilm pieces are unceasingly shredded off in the blood flow, the biofilms continuously infect the blood, and this makes infection eradication difficult by antibiotics unless they are administered intravenously for a prolonged duration. A broken and released mature biofilm piece in blood flow can cause embolism at any point in circulation through vital organs. On failure of antibiotic treatments, the infected valve may be surgically excised and replaced (Long & Koyfman, 2018).

Atherosclerosis

In atherosclerosis, triglycerides and cholesterol forms plaques, i.e., fatty deposits along with calcium, at the luminal side of arterial endothelium. This narrows lumen, reduces arterial elasticity, and reduces blood flow causing cardiovascular diseases. Oral bacteria, which are responsible for periodontitis and gingivitis, have been reported from the infective endocarditis and atherosclerosis plaques. Sudden rupture of a plaque, like biofilm fractions in IE, maybe life-threatening (Lanter, Sauer, & Davies, 2014; Snow et al., 2016). The involvement of biofilms in pathogenesis of atherosclerosis and plaque rupturing has been proven through various microscopic techniques. But it is yet to be cleared, whether the bacteria initiate the arterial endothelial rupture and plaque formation, or they colonize the predeveloped plaques (Del Pozo, 2018; Falk, Shah, & Fuster, 1995; Vestby et al., 2020).

5.1.5 The integumentary system

Wound infections

The wound is damage to living tissue caused either by a trauma, i.e., cuts, abrasions, burns, and surgery, or as a consequence of underlying illnesses such as diabetes. Most wounds heal successfully despite bacterial contamination. However, sometimes bacterial proliferation, during severe infection, activates immune response, which damages the wound tissue and disrupts the healing (Fijan et al., 2019). Due to loss of primary innate barrier, i.e., skin and dermal appendages, both acute and chronic wounds are vulnerable to bacterial infections. Several bacterial genera such as *S. aureus*, *S. epidermidis*, and *P. aeruginosa*, and many more, can colonize chronic wounds, but most of the time, *S. aureus* dominates the crew at the surface. The high affinity of *S. aureus* to interact with fibronectin, collagen, and laminin from ECM with diverse receptors makes colonization and biofilm formation on the skin, soft tissue, and bone easy (Ciampolini, 2000). Whereas the deeper part of the wound was predominated by anaerobic species such as *Bacteroides* spp., *Fusobacterium* spp., *Peptostreptococcus* spp., as well as *Clostridium* spp. (Brackman, De Meyer, Nelis, & Coenye, 2013; Percival et al., 2012). As biofilm formation retards removal of microbes by evading the immune responses, supports microbial growth in damaged wound tissues, and resists the antimicrobial therapies cumulatively leading to longer persistence of microbial infections in chronic wounds (Cooper, Bjarnsholt, & Alhede, 2014; James et al., 2008). Thus, once established, chronic wound infections harbor different microorganisms, which could not be enlisted completely due to culturing technique limitations. But the involvement of advanced molecular identification techniques RT-PCR and visual techniques such as microscopies unveiled the presence of diverse viable but noncultivable varieties of bacteria from the wound infections (Del Pozo, 2018; Percival, 2017).

5.1.6 The urinary system

Urinary tract infections (UTIs)

Urinary tract infections (UTIs) are very common in humans, as the bacteria from the rectum, perineum (or vagina) often enter and colonize in the urethra, bladder, ureters, and/or kidneys. The major symptoms, i.e., urgent and frequent urination, dysuria, lower abdominal pain, etc., arise depending on the anatomical location of the infection. If the infection reaches the kidneys, sepsis and systemic symptoms may be triggered. Usually, UTI clears up on its own within a few days of onset, or if not, a short course of antibiotic treatment works but with higher recurrence chances due to the reintroduction of bacteria making it chronic. The persistent infection of uropathogenic *E. coli* (UPEC) in the bladder as intracellular bacterial communities (IBCs) contributes to more than 90% recurrence of UTI (Anderson et al., 2003; Conte et al., 2016; Scott, Haake, Churchill, Justice, & Kim, 2015). UPEC cells invade the urothelial cells, proliferate as IBC, and disperse into the bladder lumen following the same cycle of infection. The immune system fails to invade these hidden pathogens from IBC (Justice et al., 2004). Microscopic observations of urothelium with specialized bacterial staining techniques unveiled large lumps of bacterial populations on the urothelium surrounded by a matrix of polysaccharide and fimbriae proteins indicating the presence of bacterial biofilms inside the bladder epithelial cells (Anderson, Dodson, Hooton, & Hultgren, 2004; Anderson, Martin, & Hultgren, 2004; Robino et al., 2013).

Bacterial prostatitis

Bacterial prostatitis (BP) usually occurs in males when UTI causes bacteria to reach and inhabit the prostate gland. Reduced fertility, infection in the bladder, prostatic abscesses, urosepsis, and death are the complicated outcomes of acute BP. If left untreated and symptoms last for more than 3 months, the disease can develop into chronic bacterial prostatitis (CBP) (Yoon et al., 2012, 2013). *E. coli*, *Proteus mirabilis*, *P. aeruginosa*, *Enterococcus faecalis*, *Klebsiella* spp., and other *Enterobacteriaceae* family members are the most commonly associated with BP, and many of them produce biofilms (Wagenlehner, Diemer, Naber, & Weidner, 2008; Weidner et al., 1991). Involvement of biofilm-forming bacteria in CBP pathogenesis has been visualized microscopically as the presence of glycocalyx-enclosed bacterial microcolonies inside the prostate ducts, and their reminiscent on the surface of, and embedded in, the prostate calcifications from patients having chronic inflammation of gland due to CBP (Cai et al., 2018; Nickel et al., 1990). Just like UTI, the pathogenic bacteria execute the IBC formation as a self-protective strategy in CBP (Anderson et al., 2003; Justice et al., 2004).

5.1.7 The reproductive system

Bacterial vaginosis

Bacterial vaginosis (BV) is frequently relapsing and the most common genital tract infection in women during their reproductive years (Jung, Ehlers, Lombaard, Redelinghuys, & Kock, 2017). Any dysbiosis in the vaginal microflora increases anaerobic bacteria such as *Gardnerella vaginalis*, *Atopobium vaignae*, and others, reduces protective lactobacilli population leading to BV (Machado & Cerca, 2015). Detection of "clue cells," i.e., bacteria-covered vaginal epithelial cells, was one of the diagnostic criteria for BV. In more than 90% of cases, the desquamated epithelial cells were coated with multispecies bacterial biofilm dominated by *G. vaginalis* (Amsel et al., 1983; Machado, Castro, Palmeira-de-Oliveira, Martinez-de-Oliveira, & Cerca, 2016; Swidsinski, Weber, et al., 2005). Based on the cross talk with other bacterial species in the community, the commensal *G. vaginalis* alters the gene expression patterns and grabs a chance to exhibit its virulence traits, i.e., cytotoxicity, adherence, and unveils its opportunistic pathogenic nature to become probably the first species to adhere the vaginal epithelium creating a scaffold for other species to adhere (Castro et al., 2015; Castro, Machado, & Cerca, 2019; Swidsinski et al., 2005).

Chronic endometritis

Endometritis is the condition in which the peaceful coexistence between microbial community and host immunity breaks down. In normal physiological conditions, the presence of functional microbiome on the endometrial wall has been reported to comprise *Lactobacillus* as an abundant population, in homeostasis with *Gardnerella*, *Prevotella*, *Atopobium*, *Sneathia*, and many other common bacteria such as *E. faecalis*, *E. coli*, *G. vaginalis*, *Klebsiella pneumoniae*, *Proteus* spp., *P. aeruginosa*, *Staphylococcus* spp., and *Streptococcus* spp. in community. As this colonization is renewed after every menstrual cycle, and more or less the microbial homeostasis is persistent the endometritis is referred to as chronic endometritis (CE) (Cicinelli et al., 2009; Mitchell et al., 2015; Moreno & Franasiak, 2017). However, being asymptomatic, most of the time CE is less frequently suspected and diagnosed. In certain severe conditions, the dysbiosis in CE disturbs the immune responses altering the distribution of lymphocytes and resulting in inadequate secretion of various cytokines. Eventually, this turns out in the reduced endometrium receptivity leading to miscarriage (Moreno et al., 2016; Vestby et al., 2020).

Mastitis

Mastitis is an inflammation of the mammary gland primarily occurring during lactation that results in breast pain, swelling, warmth, and redness, sometimes fever, and chills. *S. aureus*, coagulase-negative *Staphylococci* (CNS), *E. coli*, and *Streptococci* are the etiological agents causing infectious mastitis (Vestby et al., 2020). Due to sparse cases in humans, mastitis has been majorly studied in dairy animals for their microbiological aspects and deserves more commercial importance (Halasa, Huijps, Østerås, & Hogeveen, 2007). Like in humans, dairy animals also, *S. aureus* is the main pathogen along with *Streptococcus agalactiae* (causing contagious mastitis), coagulase-negative *Staphylococci*, *E. coli*, *Klebsiella* spp., *Enterobacter* spp., *Citrobacter* spp., *Streptococcus dysgalactiae*, *Streptococcus uberis*, *Enterococci*, and *Pseudomonas* spp. frequently causing chronic and recurrent mastitis infections. *S. aureus*, isolated from clusters observed in the lumen of alveoli and lactiferous ducts from humans, has been shown to produce *Bap* surface protein involved in biofilm formation. Moreover, the presence of lactose in milk has promoted biofilm formations in vitro, suggesting biofilm formation by a pathogen. In veterinary medical cases, *S. aureus* has been shown to adhere and internalize into epithelial cells of the mammary gland, and this invasion as well as biofilm formation probably helped cocci to evade host defense in vivo (Hensen, Pavičić, Lohuis, de Hoog, & Poutrel, 2000; Seixas, Varanda, Bexiga, Tavares, & Oliveira, 2015).

5.1.8 Skeletal system

Septic arthritis and osteomyelitis

Septic arthritis and osteomyelitis are the two major bone infections caused by *Staphylococcus*. The infection can be through perioperative, hematogenous, or contiguous routes.

But due to lack of basal membrane, synovial cavity is more prone to get infection by hematogenous way. Joints with preexisting damage, such as in rheumatoid or osteoarthritis, are especially prone to septic arthritis. Once the infection initiated in the joint space, pathogens trigger intense immune response from host such as acute inflammation, which involves the surrounding tissues of bone. Endotoxins and superantigens produced by *Staphylococcus* yield a toxic shock syndrome, which damages the cartilaginous tissues from bones. This damages cartilages and opens up bone ECM leading to beginning of a microbial biofilm establishment (Harrison-Balestra, Cazzaniga, Davis, & Mertz, 2003; Teterycz et al., 2010; Wright & Nair, 2010).

5.2 Microbial contamination on medical implant-device-related biofilm disease

With advanced medical technology, medical device implantations have become an integral part to improve routine life quality. Many of such medical implants, which crucially help the patients in overcoming the diseased condition and/or the organ malfunction, are cardiac pacemakers and prosthetic heart valves, vascular prostheses, electrical dialyzers, prosthetic joints, alloplastic orthopedic devices, intravenous catheters, indwelling urinary catheters, urethral stents, endotracheal tubes, cerebrospinal fluid shunts, peritoneal dialysis catheters, biliary tract stents, breast implants, voice prostheses, intrauterine devices, abdominal drains, nephrostomy tubes, tissue fillers, and contact lenses, etc., and many more are under development. While these foreign devices try to establish normalcy of a vital organ within initial hours of implantation, host produced conditioning films, comprises platelets, plasma and tissue proteins on these implants to make theme highly vulnerable for the colonization of bacterial contagions. This can lead to pathogenic biofilm formation that gradually can result in recalcitrance and persistence of device-related infections, complicating and compromising the quality life of patients and sometimes can be fatal (del Pozo & Patel, 2007; Raad & Bodey, 1992). From 1972, till now, implanted device-related critical infectious cases are being reported (Hall-Stoodley et al., 2012; Marrie, Nelligan, & Costerton, 1982; Yadav, Vidal, & Song, 2020). Bacterial species from contaminated external environmental sources such as air, water and soil, or skin of healthcare workers and/or patients usually contaminate the indwelling devices and initiate biofilm formation (Jensen et al., 2017; von Eiff et al., 2005). The mode of inoculation could be any of the perioperative, hematogenous, or contiguous route. The synovium is highly vascularized and lacks a protective basement membrane; thus, the most common route of bacterial entry is via hematogenous spread. Nosocomial infections initiate with the bacterial colonization on the surface of medical device material followed by a complex metamorphosis of contaminant microorganisms forming biofilms. How these bacterial infections take place on diverse prosthetics and impair their function is described further (Trampuz & Zimmerli, 2005).

5.2.1 *Biofilms associated with catheters*

Among the most common hospital-acquired infections, catheter-associated urinary tract infections (CAUTIs) are the topmost in leading morbidity and mortality (Cortese, Wagner, Tierney, Devine, & Fogarty, 2018; Umscheid et al., 2011). Uropathogenic *E. coli*, *Enterococcus* spp., *K. pneumoniae*, *P. mirabilis*, and *P. aeruginosa* are the most frequent causative agents of CAUTI (Flores-Mireles, Walker, Caparon, & Hultgren, 2015). Furthermore, the urease producing *P. mirabilis* encrusts catheters through the precipitation of salts such as calcium and magnesium phosphate that significantly contribute to biofilm formation on catheters (Stickler, 2009). Latex and silicone surfaces of urinary catheters are prone to fetch bacterial contamination frequently, as the organisms adhere to the surface and form biofilm both intra- and extra-luminally. Tubes and catheters filled with liquid if not changed timely, the established biofilm can become the persistent source of uropathogens and the infecting bacteria can transit up to the upper urinary tract, and if left untreated, to the kidneys causing systemic damage (Mandakhalikar et al., 2018).

5.2.2 *Biofilms associated with ventilation-associated pneumonia*

Ventilator-associated pneumonia (VAP) is a critical nosocomial infection, leading to morbidity and mortality, acquired by young children or immunocompromised patients from intensive care units (Cernada, Brugada, Golombek, & Vento, 2014). During ventilation, the pathogens can enter through and adhere to the abiotic surface of the endotracheal tube (ETT) and start forming biofilms. These biofilms are complex multispecies communities, surrounded by self-produced EPS such as protein, nucleic acid, polysaccharides, and lipids, which yield antibiotic resistance to biofilms giving rise to some persister cells. Travel of these bacteria to the lower sterile respiratory tract and lung parenchyma can cause pneumonia (Adair et al., 1999; Feldman et al., 1999; Flemming & Wingender, 2010; Gil-Perotin et al., 2012; Pneumatikos, Dragoumanis, Bouros, Warner, & Warner, 2009). Among the pathogens forming biofilms on ETT, *Streptococcus* spp., *Staphylococcus* spp., *Candida* spp., and normal body flora are the dominant ones (Friedland, Rothschild, Delgado, Isenberg, & Holzman, 2001). Treatment of polymicrobial biofilm comprising a cooperative interaction between bacteria *Pseudomonas* spp. and fungi *Candida albicans* has been reported to need special attention as their combination is resistant to multiple drugs (Lindsay & Hogan, 2014; McAlester, O'Gara, & Morrissey, 2008; Rodrigues et al., 2017; Trejo-Hernández, Andrade-Domínguez, Hernández, & Encarnación, 2014).

5.2.3 Biofilms associated with orthopedic implants

Pins, screws, plates, external fixatives, and prosthetic joints, etc., are widely used orthopedic implantations in patients suffering from age-related abrasions and disorders of bones, accidentally fractured bones, or incurable bacterial infections to bones. Despite the involvement of antimicrobial nanomaterials in the manufacturing of metallic implants, the evolution of microflora has allegedly increased the infection chances for such orthopedic implants (Anderson, Martin, et al., 2004). Microbial contaminations in certain cases result in postoperative prosthesis failure, prolonged pain, immobility of limb or body part, recurrent infections with decreased recovery rate, and sometimes require surgical removal or replacement of implant. Although prosthetic joint infections (PJIs) are very less to occur, their treatment is more difficult leading to significantly higher morbidity and mortality (Esposito & Leone, 2008; Rohde et al., 2007). Despite belonging to normal flora, for patients with compromised immune systems, *S. epidermidis* and *S. aureus* become opportunistic pathogens that colonize on the surface of PJIs and form a multilayered biofilm (Campoccia, Montanaro, & Arciola, 2006; Esposito & Leone, 2008; Kalita & Verma, 2010).

5.2.4 Biofilms associated with cardiovascular implants

Cardiac implantable electronic devices (CIEDs) such as pacemakers and implantable cardioverter defibrillators (ICDs) are worldwide used to correct most life-threatening cardiac arrhythmias. To implant in the human body, a narrow pocket in the chest area must be prepared to fit the device. Microbes from the patient's skin penetrate these pockets, colonize and form biofilm resulting in primary CIED infections therein, which gradually cause secondary blood and endocarditis infections (Kirkfeldt, Johansen, Nohr, Jorgensen, & Nielsen, 2014; Klug et al., 2007). As a result of biofilm infections, the devices need to be removed, disinfected, and reimplemented giving rise to adverse reactions sometimes and if left untreated create a major health risk. *S. aureus* and *S. epidermidis*, *Streptococcus* spp., *Corynebacterium* spp., *Cutibacterium acnes*, Gram-negative *Bacillus* spp., and *Candida* spp. are frequently detected from pocket infections and endocarditis (Baddour et al., 2010; Kirkfeldt, Johansen, & Nielsen, 2016; Klug et al., 2007).

5.2.5 Biofilms associated with ocular implants

To overcome the vision impairments or optical abnormalities, a variety of ocular devices such as intraocular lenses, posterior contact lenses, lacrimal intubation devices, scleral buckles, conjunctival plugs, and orbital implants have been developed. These implants and devices when adjusted in the human body provide an abiotic surface that is more prone to microbial contamination and biofilm growth. Once established and if left untreated, this localized infection can lead to either occasional implant malfunction or sometimes to secondary bacterial endophthalmitis or ulceration resulting in permanent vision loss.

Biofilms on contact lenses

Refractive errors and cataracts are two of the major visual impairments that can be treated with contact lenses or the replacement of intraocular lenses respectively (Pascolini & Mariotti, 2012). Infectious keratitis, a microbial infection of the cornea, can advance as ulcerative keratitis resulting in vision loss if left untreated (Bourcier, Thomas, Borderie, Chaumeil, & Laroche, 2003; Vajpayee et al., 2000). The abiotic surface of lenses provides the substratum for adhesion and biofilm formation, and sometimes, various artificial tear solutions, which are used to keep lenses wet, can also support the contamination growth, leading to occasional microbial keratitis incidences (Guillon & Maissa, 2008; Teo et al., 2011; Yadav, Chuck, & Park, 2013). Depending upon the occupation and geographical location of patients, the causative agent of keratitis varies among the bacteria and fungi such as *S. aureus* and other *Staphylococcus* spp., *P. aeruginosa*, *Serratia* spp., *Fusarium* spp., *Candida* spp. (Cheng et al., 1999; Laverty, Gorman, & Gilmore, 2014; O'Callaghan, 2018; Shah, Sachdev, Coggon, & Hossain, 2011; Teo et al., 2011; Zegans, Becker, Budzik, & O'Toole, 2002).

Biofilms associated with scleral buckles

Rhegmatogenous retinal detachment (RRD), the most common type of retinal detachment, occurs when a retinal slit accumulates fluid and separates the neurosensory retina from the underlying retinal pigment epithelium. The situation can be treated through a common surgical process in which a permanent scleral buckle band is placed between the conjunctiva and sclera. The scleral bands are usually made up of silicone that provides a favorable surface for attachment and biofilm formation to pathogens. *S. aureus* and other coagulase-negative bacteria, along with nontuberculosis mycobacterium, have been detected from scleral buckle infections (Holland et al., 1991; Pathengay, Karosekar, Raju, Sharma, & Das, 2004).

Biofilms on conjunctival plugs and lacrimal intubation devices

The lacrimal system includes lacrimal gland that produces lacrimal secretion and nasolacrimal system comprising canaliculi, lacrimal sac, and nasolacrimal duct that drains it. Tears protect the conjunctival or corneal surface keeping it wet and lubricated. Malfunction of the system results in imbalanced production and drainage of tears causing irritated, itchy, dry, or watery eyes. Tropical medicinal treatment of dry eyes, caused by insufficient tear production, is accompanied by implanting punctal plugs at the lacrimal duct to block the tear draining. On the contrary, obstruction of the nasolacrimal duct results in itchy, irritated, and watery eyes, which can be treated by implanting a small, hollow tube (lacrimal stent) through the puncta and into the tear duct that allows the tears to drain appropriately. The punctal plugs are made of silicone, hydrogel, collagen, and hydrophobic acryle material, whereas the stents are made up of polyurethane or silicone material. All these provide abiotic surfaces for adherence and growth of bacterial as well as fungal pathogens. Once grown as a biofilm, they fail the device function. *Staphylococcus haemolyticus*, *Candida tropicalis*, *P. aeruginosa*, *S. epidermidis*, *S. viridans*, *Acanthamoeba*, *C. albicans*, etc., are the reported organisms to form multispecies biofilms on such implants (Osakabe et al., 2006; Samimi, Bielory, Miller, & Johnson, 2013; Sugita et al., 2001; Yokoi, Okada, Sugita, & Kinoshita, 2000).

5.3 Relevancy of Koch postulates in polymicrobial biofilm infections

As seen in Sections 5.1 and 5.2, most of the biofilm infections have been reported with more than one bacterial culture infecting the host and forming the biofilm. Earlier techniques to study the biofilm and detect the source microorganisms were completely dependent on culture and isolation-based techniques. Therefore, the count of biofilm-forming bacteria was just restricted to a few numbers of organisms as only those bacterial cultures that were cultivable on the particular media were selected, isolated, and identified for biofilm studies. With gradual advancement in bacterial visualization through various microscopic techniques and bacterial identification through molecular biology techniques, the uncultured bacteria were also identified suggesting the diversity of microorganisms from biofilms.

When it comes to the pathogenic microorganism or contagion, Robert Koch gave four criteria that should be fulfilled by pathogenic microorganisms. These criteria are called Koch's postulates and are defined as follows: (i) The microorganisms must be present in every case of diseases but absent from healthy organisms, (ii) the suspected microorganisms must be isolated and can be grown in as pure cultures, (iii) the same disease must result when the isolated microorganism is inoculated into a healthy host, and (iv) the same microorganisms must be isolated from this newly diseased host (Byrd & Segre, 2016).

But pathogens forming biofilms every time don't strictly adhere to these postulates due to several exceptions as listed: (i) Many opportunistic pathogens that establish biofilm infections in immunocompromised hosts are part of normal flora inhabiting the same organ system from a healthy host. (ii) Due to the lack of sufficient culturing techniques, many pathogenic bacterial species can't be isolated as pure culture from the pathogenic biofilms. (iii) The mature biofilm can give rise to certain intrinsic properties, which help a variety of pathogenic bacteria coexist (Antonelli & Cutler, 2016; Ehrlich & Arciola, 2012; Todd & Peters, 2019). Infecting a healthy host by isolated pathogenic microorganisms need not develop the same biofilm or pathogenic situation. This indicates a need to design new patterns and define new principles of Koch's postulates for confirming the role of microbes in biofilm pathogenesis (Hosainzadegan, Khalilov, & Gholizadeh, 2020).

6. Recent developments to overcome infectious biofilms and future prospects

Due to antimicrobial resistance and protection from adverse conditions, the treatment of infectious microbial biofilm has become a major challenge to medical world. Plenty of diverse aspects to overcome this issue have been explored by variety of groups and are summarized in literature, some of which are outlined over here (Fang, Park, & Hong, 2020; Gebreyohannes, Nyerere, Bii, & Sbhatu, 2019). (i) As microorganisms are communicating actively in EPS matrix, their signaling pathway can be modulated by artificially introducing agents that can quench the autoinducer signals during microbial quorum sensing. This can limit the growth and expression of virulence properties from pathogenic bacteria. (ii) Use of external enzymes, which can breakdown the EPS components to their smaller fractions decreasing the firmness of the biofilm and exposing the cells to harsh conditions or antimicrobials. (iii) Use of zwitterionic polymers, certain photoactive moieties, or temperature-responsive smart chemical materials, which on activation can trigger the detachment of bacteria from surface or kill them. (iv) Nanoparticles with metal ions such as Ag, Au, Cu, and Zn etc., are have been widely explored for their antimicrobial nature. Fullerene or carbon-based nanoparticles have also been developed for their use as antibacterial compounds. (v) Development of nano-texture material, i.e., material surfaces coated with variety of antibiotics, antimicrobials, nanoparticles, etc., at very fine nanoscale are also being used for development of variety of prosthesis and implant materials. (vi) Certain medical conditions where the microflora has undergone dysbiosis, replenishing of normal microbiota can help to overcome the damage done by the biofilms. Thus, development of antibiofilm treatments or process has very much vast efforts, which can't be concluded here. It's an open-end question, and medical world is still struggling to crack the fortress of pathogenic biofilm.

Acknowledgment

CSS thanks the Science and Engineering Research Board for currently supporting research in his lab by providing research grant (CRG/2018/001827).

References

Adair, C. G., Gorman, S. P., Feron, B. M., Byers, L. M., Jones, D. S., Goldsmith, C. E., et al. (1999). Implications of endotracheal tube biofilm for ventilator-associated pneumonia. *Intensive Care Medicine*, *25*(10), 1072–1076. https://doi.org/10.1007/s001340051014.

Algburi, A., Comito, N., Kashtanov, D., Dicks, L. M. T., & Chikindas, M. L. (2017). Control of biofilm formation: Antibiotics and beyond. *Applied and Environmental Microbiology*, *83*(3). https://doi.org/10.1128/AEM.02508-16.

Amsel, R., Totten, P. A., Spiegel, C. A., Chen, K. C. S., Eschenbach, D., & Holmes, K. K. (1983). Nonspecific vaginitis. *The American Journal of Medicine*, *74*(1), 14–22. https://doi.org/10.1016/0002-9343(83)91112-9.

Anderson, G., Dodson, K., Hooton, T., & Hultgren, S. (2004). Intracellular bacterial communities of uropathogenic in urinary tract pathogenesis. *Trends in Microbiology*, *12*(9), 424–430. https://doi.org/10.1016/j.tim.2004.07.005.

Anderson, G. G., Martin, S. M., & Hultgren, S. J. (2004). Host subversion by formation of intracellular bacterial communities in the urinary tract. *Microbes and Infection*, *6*(12), 1094–1101. https://doi.org/10.1016/j.micinf.2004.05.023.

Anderson, G. G., Palermo, J. J., Schilling, J. D., Roth, R., Heuser, J., & Hultgren, S. J. (2003). Intracellular bacterial biofilm-like pods in urinary tract infections. *Science*, *301*(5629), 105–107. https://doi.org/10.1126/science.1084550.

Antonelli, G., & Cutler, S. (2016). Evolution of the Koch postulates: Towards a 21st-century understanding of microbial infection. *Clinical Microbiology and Infection*, *22*(7), 583–584. https://doi.org/10.1016/j.cmi.2016.03.030.

Armbruster, C. R., & Parsek, M. R. (2018). New insight into the early stages of biofilm formation. *Proceedings of the National Academy of Sciences of the United States of America*, *115*(17), 4317–4319. https://doi.org/10.1073/pnas.1804084115.

Auinger, P., Lanphear, B. P., Kalkwarf, H. J., & Mansour, M. E. (2003). Trends in otitis media among children in the United States. *Pediatrics*, *112*(3), 514–520. https://doi.org/10.1542/peds.112.3.514.

Baddour, L. M., Epstein, A. E., Erickson, C. C., Knight, B. P., Levison, M. E., Lockhart, P. B., et al. (2010). Update on cardiovascular implantable electronic device infections and their management: A scientific statement from the American Heart Association. *Circulation*, *121*(3), 458–477. https://doi.org/10.1161/CIRCULATIONAHA.109.192665.

Bourcier, T., Thomas, F., Borderie, V., Chaumeil, C., & Laroche, L. (2003). Bacterial keratitis: Predisposing factors, clinical and microbiological review of 300 cases. *British Journal of Ophthalmology*, *87*(7), 834–838. https://doi.org/10.1136/bjo.87.7.834.

Boutin, S., Graeber, S. Y., Weitnauer, M., Panitz, J., Stahl, M., Clausznitzer, D., et al. (2015). Comparison of microbiomes from different niches of upper and lower airways in children and adolescents with cystic fibrosis. *PLoS One*, *10*(1), e0116029. https://doi.org/10.1371/JOURNAL.PONE.0116029.

Boyle, K. E., Heilmann, S., van Ditmarsch, D., & Xavier, J. B. (2013). Exploiting social evolution in biofilms. *Current Opinion in Microbiology*, *16*(2), 207–212. https://doi.org/10.1016/j.mib.2013.01.003.

Brackman, G., De Meyer, L., Nelis, H. J., & Coenye, T. (2013). Biofilm inhibitory and eradicating activity of wound care products against *Staphylococcus aureus* and *Staphylococcus epidermidis* biofilms in an in vitro chronic wound model. *Journal of Applied Microbiology*, *114*(6), 1833–1842. https://doi.org/10.1111/jam.12191.

Brady, R. A., Leid, J. G., Costerton, J. W., & Shirtliff, M. E. (2006). Osteomyelitis: Clinical overview and mechanisms of infection persistence. *Clinical Microbiology Newsletter*, *28*(9), 65–72. https://doi.org/10.1016/j.clinmicnews.2006.04.001.

Burmølle, M., Ren, D., Bjarnsholt, T., & Sørensen, S. J. (2014). Interactions in multispecies biofilms: Do they actually matter? *Trends in Microbiology*, *22*(2), 84–91. https://doi.org/10.1016/j.tim.2013.12.004.

Burmølle, M., Webb, J. S., Rao, D., Hansen, L. H., Sørensen, S. J., & Kjelleberg, S. (2006). Enhanced biofilm formation and increased resistance to antimicrobial agents and bacterial invasion are caused by synergistic interactions in multispecies biofilms. *Applied and Environmental Microbiology*, *72*(6), 3916–3923. https://doi.org/10.1128/AEM.03022-05.

Byrd, A. L., & Segre, J. A. (2016). Adapting Koch's postulates: Criteria for disease causation must take microbial interactions into account. *Science*, *351*(6270), 224–226. https://doi.org/10.1126/science.aad6753.

Cai, T., Tessarolo, F., Caola, I., Piccoli, F., Nollo, G., Caciagli, P., et al. (2018). Prostate calcifications: A case series supporting the microbial biofilm theory. *Investigative and Clinical Urology*, *59*(3), 187. https://doi.org/10.4111/icu.2018.59.3.187.

Campoccia, D., Montanaro, L., & Arciola, C. R. (2006). The significance of infection related to orthopedic devices and issues of antibiotic resistance. *Biomaterials*, *27*(11), 2331–2339. https://doi.org/10.1016/j.biomaterials.2005.11.044.

Castro, J., Alves, P., Sousa, C., Cereija, T., França, Â., Jefferson, K. K., et al. (2015). Using an in-vitro biofilm model to assess the virulence potential of Bacterial Vaginosis or non-Bacterial Vaginosis *Gardnerella vaginalis* isolates. *Scientific Reports*, *5*(1), 11640. https://doi.org/10.1038/srep11640.

Castro, J., Machado, D., & Cerca, N. (2019). Unveiling the role of *Gardnerella vaginalis* in polymicrobial Bacterial Vaginosis biofilms: The impact of other vaginal pathogens living as neighbors. *The ISME Journal*, *13*(5), 1306–1317. https://doi.org/10.1038/s41396-018-0337-0.

Cattelan, N., Dubey, P., Arnal, L., Yantorno, O. M., & Deora, R. (2016). Bordetella biofilms: A lifestyle leading to persistent infections. *Pathogens and Disease*, *74*(1), ftv108. https://doi.org/10.1093/femspd/ftv108.

Cattelan, N., Jennings-Gee, J., Dubey, P., Yantorno, O. M., & Deora, R. (2017). Hyperbiofilm formation by *Bordetella pertussis* strains correlates with enhanced virulence traits. *Infection and Immunity*, *85*(12). https://doi.org/10.1128/IAI.00373-17.

Cepas, V., López, Y., Muñoz, E., Rolo, D., Ardanuy, C., Martí, S., et al. (2019). Relationship between biofilm formation and antimicrobial resistance in Gram-negative bacteria. *Microbial Drug Resistance*, *25*(1), 72–79. https://doi.org/10.1089/mdr.2018.0027.

Cernada, M., Brugada, M., Golombek, S., & Vento, M. (2014). Ventilator-associated pneumonia in neonatal patients: An update. *Neonatology*, *105*(2), 98–107. https://doi.org/10.1159/000355539.

Cheng, K. H., Leung, S. L., Hoekman, H. W., Beekhuis, W. H., Mulder, P. G. H., Geerards, A. J. M., et al. (1999). Incidence of contact-lens-associated microbial keratitis and its related morbidity. *Lancet*, *354*(9174), 181–185. https://doi.org/10.1016/S0140-6736(98)09385-4.

Chole, R. A., & Faddis, B. T. (2003). Anatomical evidence of microbial biofilms in tonsillar tissues. *Archives of Otolaryngology–Head & Neck Surgery*, *129*(6), 634. https://doi.org/10.1001/archotol.129.6.634.

Choudhary, D., & Cassaro, C. J. (2021). Digging deeper in the biofilm. *Nature Reviews Microbiology*, *19*(8), 484. https://doi.org/10.1038/s41579-021-00584-x.

Ciampolini, J. (2000). Pathophysiology of chronic bacterial osteomyelitis. Why do antibiotics fail so often? *Postgraduate Medical Journal*, *76*(898), 479–483. https://doi.org/10.1136/pmj.76.898.479.

Cicinelli, E., De Ziegler, D., Nicoletti, R., Tinelli, R., Saliani, N., Resta, L., et al. (2009). Poor reliability of vaginal and endocervical cultures for evaluating microbiology of endometrial cavity in women with chronic endometritis. *Gynecologic and Obstetric Investigation*, *68*(2), 108–115. https://doi.org/10.1159/000223819.

Cohen, J. I., Bartlett, J. A., & Corey, G. R. (1987). Extra-intestinal manifestations of salmonella infections. *Medicine*, *66*(5), 349–388. https://doi.org/10.1097/00005792-198709000-00003.

Collins, A. J., Smith, T. J., Sondermann, H., & O'Toole, G. A. (2020). From input to output: The Lap/c-di-GMP biofilm regulatory circuit. *Annual Review of Microbiology*, *74*, 607–631. https://doi.org/10.1146/annurev-micro-011520-094214.

Colombo, A. P. V., Boches, S. K., Cotton, S. L., Goodson, J. M., Kent, R., Haffajee, A. D., et al. (2009). Comparisons of subgingival microbial profiles of refractory periodontitis, severe periodontitis, and periodontal health using the human oral microbe identification microarray. *Journal of Periodontology*, *80*(9), 1421–1432. https://doi.org/10.1902/jop.2009.090185.

Conover, M. S., Sloan, G. P., Love, C. F., Sukumar, N., & Deora, R. (2010). The Bps polysaccharide of *Bordetella pertussis* promotes colonization and biofilm formation in the nose by functioning as an adhesin. *Molecular Microbiology*, *77*(6), 1439–1455. https://doi.org/10.1111/j.1365-2958.2010.07297.x.

Conte, M. P., Aleandri, M., Marazzato, M., Conte, A. L., Ambrosi, C., Nicoletti, M., et al. (2016). The adherent/invasive *Escherichia coli* strain LF82 invades and persists in human prostate cell line rwpe-1, activating a strong inflammatory response. *Infection and Immunity*, *84*(11), 3105–3113. https://doi.org/10.1128/IAI.00438-16.

Cooper, R. A., Bjarnsholt, T., & Alhede, M. (2014). Biofilms in wounds: A review of present knowledge. *Journal of Wound Care*, *23*(11), 570–582. https://doi.org/10.12968/jowc.2014.23.11.570.

Cortese, Y. J., Wagner, V. E., Tierney, M., Devine, D., & Fogarty, A. (2018). Review of catheter-associated urinary tract infections and in vitro urinary tract models. *Journal of Healthcare Engineering*, *2018*. https://doi.org/10.1155/2018/2986742.

Costerton, J. W., & Lewandowski, Z. (1997). The biofilm lifestyle. *Advances in Dental Research*, *11*(1), 192–195. https://doi.org/10.1177/08959374970110011101.

Costerton, W. J., & Wilson, M. (2004). Introducing biofilms. *Biofilms*, *1*(1), 1–4. https://doi.org/10.1017/s1479050504001164.

Crawford, R. W., Rosales-Reyes, R., Ramírez-Aguilar, M. D. L. L., Chapa-Azuela, O., Alpuche-Aranda, C., & Gunn, J. S. (2010). Gallstones play a significant role in *Salmonella* spp. gallbladder colonization and carriage. *Proceedings of the National Academy of Sciences*, *107*(9), 4353–4358. https://doi.org/10.1073/pnas.1000862107.

Cugini, C., Shanmugam, M., Landge, N., & Ramasubbu, N. (2019). The role of exopolysaccharides in oral biofilms. *Journal of Dental Research*, *98*(7), 739–745. https://doi.org/10.1177/0022034519845001.

D'Souza, G. G. (2020). Phenotypic variation in spatially structured microbial communities: Ecological origins and consequences. *Current Opinion in Biotechnology*, *62*, 220–227. https://doi.org/10.1016/j.copbio.2019.12.013.

Das, S., Singh, S., Matchado, M. S., Srivastava, A., & Bajpai, A. (2019). Biofilms in human health. In S. Kumar, N. Chandra, L. Singh, M. Z. Hashmi, & A. Varma (Eds.), *Biofilms in human diseases: Treatment and control* (pp. 27–42). Springer International Publishing. https://doi.org/10.1007/978-3-030-30757-8_3.

de Weger, L. A., van Der Vlugt, C. I. M., Wijfjes, A. H. M., Bakker, P. A., Schippers, B., & Lugtenberg, B. (1987). Flagella of a plant-growth-stimulating *Pseudomonas fluorescens* strain are required for colonization of potato roots. *Journal of Bacteriology*, *169*(6), 2769–2773. https://doi.org/10.1128/jb.169.6.2769-2773.1987.

DeAntonio, R., Yarzabal, J.-P., Cruz, J. P., Schmidt, J. E., & Kleijnen, J. (2016). Epidemiology of otitis media in children from developing countries: A systematic review. *International Journal of Pediatric Otorhinolaryngology*, *85*, 65–74. https://doi.org/10.1016/j.ijporl.2016.03.032.

Dejea, C. M., Fathi, P., Craig, J. M., Boleij, A., Taddese, R., Geis, A. L., et al. (2018). Patients with familial adenomatous polyposis harbor colonic biofilms containing tumorigenic bacteria. *Science*, *359*(6375), 592–597. https://doi.org/10.1126/science.aah3648.

Del Pozo, J. L. (2018). Biofilm-related disease. *Expert Review of Anti-Infective Therapy*, *16*(1), 51–65. https://doi.org/10.1080/14787210.2018.1417036.

del Pozo, J. L., & Patel, R. (2007). The challenge of treating biofilm-associated bacterial infections. *Clinical Pharmacology & Therapeutics*, *82*(2), 204–209. https://doi.org/10.1038/sj.clpt.6100247.

Diaz, R. R., Picciafuoco, S., Paraje, M. G., Villegas, N. A., Miranda, J. A., Albesa, I., et al. (2011). Relevance of biofilms in pediatric tonsillar disease. *European Journal of Clinical Microbiology & Infectious Diseases*, *30*(12), 1503–1509. https://doi.org/10.1007/s10096-011-1249-3.

Dorji, D., Graham, R. M., Singh, A. K., Ramsay, J. P., Price, P., & Lee, S. (2019). Immunogenicity and protective potential of *Bordetella pertussis* biofilm and its associated antigens in a murine model. *Cellular Immunology*, *337*(December 2018), 42–47. https://doi.org/10.1016/j.cellimm.2019.01.006.

Dragoš, A., Kiesewalter, H., Martin, M., Hsu, C. Y., Hartmann, R., Wechsler, T., et al. (2018). Division of labor during biofilm matrix production. *Current Biology*, *28*(12), 1903–1913.e5. https://doi.org/10.1016/j.cub.2018.04.046.

Ehrlich, G. D., & Arciola, C. R. (2012). From Koch's postulates to biofilm theory. The lesson of Bill Costerton. *International Journal of Artificial Organs*, *35*(10), 695–699. https://doi.org/10.5301/ijao.5000169.

Elgharably, H., Hussain, S. T., Shrestha, N. K., Blackstone, E. H., & Pettersson, G. B. (2016). Current hypotheses in cardiac surgery: Biofilm in infective endocarditis. *Seminars in Thoracic and Cardiovascular Surgery*, *28*(1), 56–59. https://doi.org/10.1053/j.semtcvs.2015.12.005.

Esposito, S., & Leone, S. (2008). Prosthetic joint infections: Microbiology, diagnosis, management and prevention. *International Journal of*

Antimicrobial Agents, *32*(4), 287–293. https://doi.org/10.1016/j.ijantimicag.2008.03.010.

Faden, H., Duffy, L., Wasielewski, R., Wolf, J., Krystofik, D., & Tung, Y. (1997). Relationship between nasopharyngeal colonization and the development of otitis media in children. *The Journal of Infectious Diseases*, *175*(6), 1440–1445. https://doi.org/10.1086/516477.

Falk, E., Shah, P. K., & Fuster, V. (1995). Coronary plaque disruption. *Circulation*, *92*(3), 657–671. https://doi.org/10.1161/01.CIR.92.3.657.

Fang, K., Park, O. J., & Hong, S. H. (2020). Controlling biofilms using synthetic biology approaches. *Biotechnology Advances*, *40*(January), 107518. https://doi.org/10.1016/j.biotechadv.2020.107518.

Fearon, E. R., & Vogelstein, B. (1990). A genetic model for colorectal tumorigenesis. *Cell*, *61*(5), 759–767. https://doi.org/10.1016/0092-8674(90)90186-I.

Feldman, C., Kassel, M., Cantrell, J., Kaka, S., Morar, R., Goolam Mahomed, A., et al. (1999). The presence and sequence of endotracheal tube colonization in patients undergoing mechanical ventilation. *European Respiratory Journal*, *13*(3), 546–551. https://doi.org/10.1183/09031936.99.13354699.

Fergie, N., Bayston, R., Pearson, J. P., & Birchall, J. P. (2004). Is otitis media with effusion a biofilm infection? *Clinical Otolaryngology and Allied Sciences*, *29*(1), 38–46. https://doi.org/10.1111/j.1365-2273.2004.00767.x.

Fijan, S., Frauwallner, A., Langerholc, T., Krebs, B., ter Haar (née Younes), J. A., Heschl, A., et al. (2019). Efficacy of using probiotics with antagonistic activity against pathogens of wound infections: An integrative review of literature. *BioMed Research International*, *2019*, 1–21. https://doi.org/10.1155/2019/7585486.

Flemming, H. C., Neu, T. R., & Wozniak, D. J. (2007). The EPS matrix: The "house of biofilm cells". *Journal of Bacteriology*, *189*(22), 7945–7947. https://doi.org/10.1128/JB.00858-07.

Flemming, H. C., & Wingender, J. (2010). The biofilm matrix. *Nature Reviews Microbiology*, *8*(9), 623–633. https://doi.org/10.1038/nrmicro2415.

Flemming, H. C., Wingender, J., Szewzyk, U., Steinberg, P., Rice, S. A., & Kjelleberg, S. (2016). Biofilms: An emergent form of bacterial life. *Nature Reviews Microbiology*, *14*(9), 563–575. https://doi.org/10.1038/nrmicro.2016.94.

Fletcher, M. (1991). The physiological activity of bacteria attached to solid surfaces. *Advances in Microbial Physiology*, *32*(C), 53–85. https://doi.org/10.1016/S0065-2911(08)60005-3.

Fletcher, M., & Marshall, K. C. (1982). Are solid surfaces of ecological significance to aquatic bacteria? In *Advances in microbial ecology* (pp. 199–236). https://doi.org/10.1007/978-1-4615-8318-9_6.

Flores-Mireles, A. L., Walker, J. N., Caparon, M., & Hultgren, S. J. (2015). Urinary tract infections: Epidemiology, mechanisms of infection and treatment options. *Nature Reviews Microbiology*, *13*(5), 269–284. https://doi.org/10.1038/nrmicro3432.

Fodor, A. A., Klem, E. R., Gilpin, D. F., Elborn, J. S., Boucher, R. C., Tunney, M. M., et al. (2012). The adult cystic fibrosis airway microbiota is stable over time and infection type, and highly resilient to antibiotic treatment of exacerbations. *PLoS One*, *7*(9). https://doi.org/10.1371/JOURNAL.PONE.0045001.

Friedland, D. R., Rothschild, M. A., Delgado, M., Isenberg, H., & Holzman, L. (2001). Bacterial colonization of endotracheal tubes in intubated neonates. *Archives of Otolaryngology–Head and Neck Surgery*, *127* (5), 525–528. https://doi.org/10.1001/archotol.127.5.525.

Fusconi, M., Petrozza, V., Schippa, S., de Vincentiis, M., Familiari, G., Pantanella, F., et al. (2016). Bacterial biofilm in salivary gland stones. *Otolaryngology–Head and Neck Surgery*, *154*(3), 449–453. https://doi.org/10.1177/0194599815622425.

Fux, C. A., Costerton, J. W., Stewart, P. S., & Stoodley, P. (2005). Survival strategies of infectious biofilms. *Trends in Microbiology*, *13*(1), 34–40. https://doi.org/10.1016/j.tim.2004.11.010.

Gebreyohannes, G., Nyerere, A., Bii, C., & Sbhatu, D. B. (2019). Challenges of intervention, treatment, and antibiotic resistance of biofilm-forming microorganisms. *Heliyon*, *5*(8), e02192. https://doi.org/10.1016/j.heliyon.2019.e02192.

Gil-Perotin, S., Ramirez, P., Marti, V., Sahuquillo, J. M., Gonzalez, E., Calleja, I., et al. (2012). Implications of endotracheal tube biofilm in ventilator-associated pneumonia response: A state of concept. *Critical Care*, *16*(3), R93. https://doi.org/10.1186/cc11357.

Gonzalez-Escobedo, G., Marshall, J. M., & Gunn, J. S. (2011). Chronic and acute infection of the gall bladder by *Salmonella* Typhi: Understanding the carrier state. *Nature Reviews Microbiology*, *9*(1), 9–14. https://doi.org/10.1038/nrmicro2490.

Guillon, M., & Maissa, C. (2008). Contact lens wear affects tear film evaporation. *Eye & Contact Lens*, *34*(6), 326–330. https://doi.org/10.1097/icl.0b013e31818c5d00.

Guttenplan, S. B., & Kearns, D. B. (2013). Regulation of flagellar motility during biofilm formation. *FEMS Microbiology Reviews*, *37*(6), 849–871. https://doi.org/10.1111/1574-6976.12018.

Halasa, T., Huijps, K., Østerås, O., & Hogeveen, H. (2007). Economic effects of bovine mastitis and mastitis management: A review. *Veterinary Quarterly*, *29*(1), 18–31. https://doi.org/10.1080/01652176.2007.9695224.

Hall-Stoodley, L., Costerton, J. W., & Stoodley, P. (2004). Bacterial biofilms: From the natural environment to infectious diseases. *Nature Reviews Microbiology*, *2*(2), 95–108. https://doi.org/10.1038/nrmicro821.

Hall-Stoodley, L., Stoodley, P., Kathju, S., Høiby, N., Moser, C., William Costerton, J., et al. (2012). Towards diagnostic guidelines for biofilm-associated infections. *FEMS Immunology and Medical Microbiology*, *65*(2), 127–145. https://doi.org/10.1111/j.1574-695X.2012.00968.x.

Hamilos, D. L. (2019). Biofilm formations in pediatric respiratory tract infection. *Current Infectious Disease Reports*, *21*(2). https://doi.org/10.1007/s11908-019-0658-9.

Han, Y. W., & Wang, X. (2013). Mobile microbiome. *Journal of Dental Research*, *92*(6), 485–491. https://doi.org/10.1177/0022034513487559.

Harrison-Balestra, C., Cazzaniga, A. L., Davis, S. C., & Mertz, P. M. (2003). A wound-isolated *Pseudomonas aeruginosa* grows a biofilm in vitro within 10 hours and is visualized by light microscopy. *Dermatologic Surgery*, *29*(6), 631–635. https://doi.org/10.1046/j.1524-4725.2003.29146.x.

Hensen, S. M., Pavičić, M. J. A. M. P., Lohuis, J. A. C. M., de Hoog, J. A. M., & Poutrel, B. (2000). Location of *Staphylococcus aureus* within the experimentally infected bovine udder and the expression of capsular polysaccharide type 5 in situ. *Journal of Dairy Science*, *83*(9), 1966–1975. https://doi.org/10.3168/jds.S0022-0302(00)75073-9.

Hold, G. L., & Allen-Vercoe, E. (2019). Gut microbial biofilm composition and organisation holds the key to CRC. *Nature Reviews Gastroenterology & Hepatology*, *16*(6), 329–330. https://doi.org/10.1038/s41575-019-0148-4.

Holland, S. P., Pulido, J. S., Miller, D., Ellis, B., Alfonso, E., Scott, M., et al. (1991). Biofilm and scleral buckle-associated infections: A mechanism for persistence. *Ophthalmology*, *98*(6), 933–938. https://doi.org/10.1016/S0161-6420(91)32199-7.

Hosainzadegan, H., Khalilov, R., & Gholizadeh, P. (2020). The necessity to revise Koch's postulates and its application to infectious and non-infectious diseases: A mini-review. *European Journal of Clinical Microbiology and Infectious Diseases, 39*(2), 215–218. https://doi.org/10.1007/s10096-019-03681-1.

Inglis, T. J. J., Millar, M. R., Jones, J. G., & Robinson, D. A. (1989). Tracheal tube biofilm as a source of bacterial colonization of the lung. *Journal of Clinical Microbiology, 27*(9), 2014–2018. https://doi.org/10.1128/jcm.27.9.2014-2018.1989.

James, G. A., Swogger, E., Wolcott, R., Pulcini, E. D., Secor, P., Sestrich, J. , et al. (2008). Biofilms in chronic wounds. *Wound Repair and Regeneration, 16*(1), 37–44. https://doi.org/10.1111/j.1524-475X.2007.00321.x.

Jensen, L. K., Koch, J., Aalbaek, B., Moodley, A., Bjarnsholt, T., Kragh, K. N., et al. (2017). Early implant-associated osteomyelitis results in a peri-implanted bacterial reservoir. *APMIS, 125*(1), 38–45. https://doi.org/10.1111/apm.12597.

Jung, H.-S., Ehlers, M. M., Lombaard, H., Redelinghuys, M. J., & Kock, M. M. (2017). Etiology of bacterial vaginosis and polymicrobial biofilm formation. *Critical Reviews in Microbiology, 43*(6), 651–667. https://doi.org/10.1080/1040841X.2017.1291579.

Justice, S. S., Hung, C., Theriot, J. A., Fletcher, D. A., Anderson, G. G., Footer, M. J., et al. (2004). Differentiation and developmental pathways of uropathogenic *Escherichia coli* in urinary tract pathogenesis. *Proceedings of the National Academy of Sciences, 101*(5), 1333–1338. https://doi.org/10.1073/pnas.0308125100.

Kadioglu, A., Weiser, J. N., Paton, J. C., & Andrew, P. W. (2008). The role of *Streptococcus pneumoniae* virulence factors in host respiratory colonization and disease. *Nature Reviews Microbiology, 6*(4), 288–301. https://doi.org/10.1038/nrmicro1871.

Kalita, S. J., & Verma, S. (2010). Nanocrystalline hydroxyapatite bioceramic using microwave radiation: Synthesis and characterization. *Materials Science and Engineering C, 30*(2), 295–303. https://doi.org/10.1016/j.msec.2009.11.007.

Kao, W. K., Chole, R. A., & Ogden, M. A. (2020). Evidence of a microbial etiology for sialoliths. *The Laryngoscope, 130*(1), 69–74. https://doi.org/10.1002/lary.27860.

Karygianni, L., Ren, Z., Koo, H., & Thurnheer, T. (2020). Biofilm matrixome: Extracellular components in structured microbial communities. *Trends in Microbiology, 28*(8), 668–681. https://doi.org/10.1016/j.tim.2020.03.016.

Khan, F., & Oloketuyi, S. F. (2017). A future perspective on neurodegenerative diseases: Nasopharyngeal and gut microbiota. *Journal of Applied Microbiology, 122*(2), 306–320. https://doi.org/10.1111/JAM.13327.

Kim, E. R. (2014). Colorectal cancer in inflammatory bowel disease: The risk, pathogenesis, prevention and diagnosis. *World Journal of Gastroenterology, 20*(29), 9872. https://doi.org/10.3748/wjg.v20.i29.9872.

Kinnari, T. J. (2015). The role of biofilm in chronic laryngitis and in head and neck cancer. *Current Opinion in Otolaryngology & Head and Neck Surgery, 23*(6), 448–453. https://doi.org/10.1097/MOO.0000000000000200.

Kirkfeldt, R. E., Johansen, J. B., & Nielsen, J. C. (2016). Management of cardiac electronic device infections: Challenges and outcomes. *Arrhythmia & Electrophysiology Review, 5*(3), 183–187. https://doi.org/10.15420/aer.2016:21:2.

Kirkfeldt, R. E., Johansen, J. B., Nohr, E. A., Jorgensen, O. D., & Nielsen, J. C. (2014). Complications after cardiac implantable electronic device implantations: An analysis of a complete, nationwide cohort in Denmark. *European Heart Journal, 35*(18), 1186–1194. https://doi.org/10.1093/eurheartj/eht511.

Klug, D., Balde, M., Pavin, D., Hidden-Lucet, F., Clementy, J., Sadoul, N., et al. (2007). Risk factors related to infections of implanted pacemakers and cardioverter-defibrillators: Results of a large prospective study. *Circulation, 116*(12), 1349–1355. https://doi.org/10.1161/CIRCULATIONAHA.106.678664.

Koo, H., & Yamada, K. M. (2016). Dynamic cell-matrix interactions modulate microbial biofilm and tissue 3D microenvironments. *Current Opinion in Cell Biology, 42*(1), 102–112. https://doi.org/10.1016/j.ceb.2016.05.005.

Kumpitsch, C., Koskinen, K., Schöpf, V., & Moissl-Eichinger, C. (2019). The microbiome of the upper respiratory tract in health and disease. *BMC Biology, 17*(1), 1–20. https://doi.org/10.1186/S12915-019-0703-Z.

Lamont, R. J., & Jenkinson, H. F. (1998). Life below the gum line: Pathogenic mechanisms of *Porphyromonas gingivalis*. *Microbiology and Molecular Biology Reviews, 62*(4), 1244–1263. https://doi.org/10.1128/MMBR.62.4.1244-1263.1998.

Lanter, B. B., Sauer, K., & Davies, D. G. (2014). Bacteria present in carotid arterial plaques are found as biofilm deposits which may contribute to enhanced risk of plaque rupture. *MBio, 5*(3). https://doi.org/10.1128/mBio.01206-14.

Laverty, G., Gorman, S. P., & Gilmore, B. F. (2014). Biomolecular mechanisms of *Pseudomonas aeruginosa* and *Escherichia coli* biofilm formation. *Pathogens, 3*(3), 596–632. https://doi.org/10.3390/pathogens3030596.

Lee, C. K., De Anda, J., Baker, A. E., Bennett, R. R., Luo, Y., Lee, E. Y., et al. (2018). Multigenerational memory and adaptive adhesion in early bacterial biofilm communities. *Proceedings of the National Academy of Sciences of the United States of America, 115*(17), 4471–4476. https://doi.org/10.1073/pnas.1720071115.

Lee, K. W. K., Periasamy, S., Mukherjee, M., Xie, C., Kjelleberg, S., & Rice, S. A. (2014). Biofilm development and enhanced stress resistance of a model, mixed-species community biofilm. *ISME Journal, 8*(4), 894–907. https://doi.org/10.1038/ismej.2013.194.

Leid, J. G. (2009). Bacterial biofilms resist key host defenses. *Microbe, 4*(2), 66–70.

Lewis, K. (2007). Persister cells, dormancy and infectious disease. *Nature Reviews Microbiology, 5*(1), 48–56. https://doi.org/10.1038/nrmicro1557.

Lidgren, L., Knutson, K., & Stefánsdóttir, A. (2003). Infection of prosthetic joints. *Best Practice & Research: Clinical Rheumatology, 17*(2), 209–218. https://doi.org/10.1016/S1521-6942(03)00002-0.

Lindsay, A. K., & Hogan, D. A. (2014). *Candida albicans*: Molecular interactions with *Pseudomonas aeruginosa* and *Staphylococcus aureus*. *Fungal Biology Reviews, 28*(4), 85–96. https://doi.org/10.1016/j.fbr.2014.10.002.

Liu, W., Russel, J., Burmølle, M., Sørensen, S. J., & Madsen, J. S. (2018). Micro-scale intermixing: A requisite for stable and synergistic co-establishment in a four-species biofilm. *ISME Journal, 12*(8), 1940–1951. https://doi.org/10.1038/s41396-018-0112-2.

Long, B., & Koyfman, A. (2018). Infectious endocarditis: An update for emergency clinicians. *The American Journal of Emergency Medicine, 36*(9), 1686–1692. https://doi.org/10.1016/j.ajem.2018.06.074.

Machado, D., Castro, J., Palmeira-de-Oliveira, A., Martinez-de-Oliveira, J., & Cerca, N. (2016). Bacterial vaginosis biofilms: Challenges to current therapies and emerging solutions. *Frontiers in Microbiology, 6*(JAN). https://doi.org/10.3389/fmicb.2015.01528.

Machado, A., & Cerca, N. (2015). Influence of biofilm formation by *Gardnerella vaginalis* and other anaerobes on bacterial vaginosis. *Journal of Infectious Diseases*, *212*(12), 1856–1861. https://doi.org/10.1093/infdis/jiv338.

Madsen, J. S., Sørensen, S. J., & Burmølle, M. (2018). Bacterial social interactions and the emergence of community-intrinsic properties. *Current Opinion in Microbiology*, *42*, 104–109. https://doi.org/10.1016/j.mib.2017.11.018.

Maina, I. W., Patel, N. N., & Cohen, N. A. (2018). Understanding the role of biofilms and superantigens in chronic rhinosinusitis. *Current Otorhinolaryngology Reports*, *6*(3), 253–262. https://doi.org/10.1007/s40136-018-0212-6.

Malone, M., Goeres, D. M., Gosbell, I., Vickery, K., Jensen, S., & Stoodley, P. (2017). Approaches to biofilm-associated infections: The need for standardized and relevant biofilm methods for clinical applications. *Expert Review of Anti-Infective Therapy*, *15*(2), 147–156. https://doi.org/10.1080/14787210.2017.1262257.

Mandakhalikar, K. D., Rahmat, J. N., Chiong, E., Neoh, K. G., Shen, L., & Tambyah, P. A. (2018). Extraction and quantification of biofilm bacteria: Method optimized for urinary catheters. *Scientific Reports*, *8*(1), 1–9. https://doi.org/10.1038/s41598-018-26342-3.

Marineli, F., Tsoucalas, G., Karamanou, M., & Androutsos, G. (2013). Mary Mallon (1869-1938) and the history of typhoid fever. *Annals of Gastroenterology*, *26*(2), 132–134. http://www.ncbi.nlm.nih.gov/pubmed/24714738.

Marrie, T. J., Nelligan, J., & Costerton, J. W. (1982). A scanning and transmission electron microscopic study of an infected endocardial pacemaker lead. *Circulation*, *66*(6), 1339–1341. https://doi.org/10.1161/01.CIR.66.6.1339.

Marsh, P. D., & Zaura, E. (2017). Dental biofilm: Ecological interactions in health and disease. *Journal of Clinical Periodontology*, *44*, S12–S22. https://doi.org/10.1111/jcpe.12679.

Martin, H. M., Campbell, B. J., Hart, C. A., Mpofu, C., Nayar, M., Singh, R., et al. (2004). Enhanced *Escherichia coli* adherence and invasion in Crohn's disease and colon cancer 1 1The authors thank Professor T. K. Korhonen (Division of General Microbiology, University of Helsinki, Finland), who kindly donated *Escherichia coli* IH11165; Professor. *Gastroenterology*, *127*(1), 80–93. https://doi.org/10.1053/j.gastro.2004.03.054.

McAlester, G., O'Gara, F., & Morrissey, J. P. (2008). Signal-mediated interactions between *Pseudomonas aeruginosa* and *Candida albicans*. *Journal of Medical Microbiology*, *57*(5), 563–569. https://doi.org/10.1099/jmm.0.47705-0.

McDougald, D., Rice, S. A., Barraud, N., Steinberg, P. D., & Kjelleberg, S. (2012). Should we stay or should we go: Mechanisms and ecological consequences for biofilm dispersal. *Nature Reviews Microbiology*, *10* (1), 39–50. https://doi.org/10.1038/nrmicro2695.

Mitchell, C. M., Haick, A., Nkwopara, E., Garcia, R., Rendi, M., Agnew, K., et al. (2015). Colonization of the upper genital tract by vaginal bacterial species in nonpregnant women. *American Journal of Obstetrics and Gynecology*, *212*(5), 611.e1–611.e9. https://doi.org/10.1016/j.ajog.2014.11.043.

Moreno, I., Codoñer, F. M., Vilella, F., Valbuena, D., Martinez-Blanch, J. F., Jimenez-Almazán, J., et al. (2016). Evidence that the endometrial microbiota has an effect on implantation success or failure. *American Journal of Obstetrics and Gynecology*, *215*(6), 684–703. https://doi.org/10.1016/j.ajog.2016.09.075.

Moreno, I., & Franasiak, J. M. (2017). Endometrial microbiota—New player in town. *Fertility and Sterility*, *108*(1), 32–39. https://doi.org/10.1016/j.fertnstert.2017.05.034.

Nasipuri, P., Herschend, J., Brejnrod, A. D., Madsen, J. S., Espersen, R., Svensson, B., et al. (2020). Community-intrinsic properties enhance keratin degradation from bacterial consortia. *PLoS One*, *15*(1), 1–26. https://doi.org/10.1371/journal.pone.0228108.

Neiger, M. R., González, J. F., Gonzalez-Escobedo, G., Kuck, H., White, P., & Gunn, J. S. (2019). Pathoadaptive alteration of *Salmonella* biofilm formation in response to the gallbladder environment. *Journal of Bacteriology*, *201*(14). https://doi.org/10.1128/JB.00774-18.

Nickel, J. C., Olson, M. E., Barabas, A., Benediktsson, H., Dasgupta, M. K., & Costerton, J. W. (1990). Pathogenesis of chronic bacterial prostatitis in an animal model. *British Journal of Urology*, *66*(1), 47–54. https://doi.org/10.1111/j.1464-410X.1990.tb14864.x.

O'Callaghan, R. J. (2018). The pathogenesis of *Staphylococcus aureus* eye infections. *Pathogens*, *7*(1), 1–22. https://doi.org/10.3390/pathogens7010009.

O'Toole, G. A., & Wong, G. C. L. (2016). Sensational biofilms: Surface sensing in bacteria. *Current Opinion in Microbiology*, *30*(1999), 139–146. https://doi.org/10.1016/j.mib.2016.02.004.

Osakabe, Y., Yaguchi, C., Miyai, T., Miyata, K., Mineo, S., Nakamura, M., et al. (2006). Detection of Streptococcus species by polymerase chain reaction in infectious crystalline keratopathy. *Cornea*, *25*(10), 1227–1230. https://doi.org/10.1097/01.ico.0000240083.31432.70.

Paddock, C. D., Sanden, G. N., Cherry, J. D., Gal, A. A., Langston, C., Tatti, K. M., et al. (2008). Pathology and pathogenesis of fatal *Bordetella pertussis* infection in infants. *Clinical Infectious Diseases*, *47*(3), 328–338. https://doi.org/10.1086/589753.

Pascolini, D., & Mariotti, S. P. (2012). Global estimates of visual impairment: 2010. *British Journal of Ophthalmology*, *96*(5), 614–618. https://doi.org/10.1136/bjophthalmol-2011-300539.

Patel, R. (2005). Biofilms and antimicrobial resistance. *Clinical Orthopaedics and Related Research*, *437*(437), 41–47. https://doi.org/10.1097/01.blo.0000175714.68624.74.

Pathengay, A., Karosekar, S., Raju, B., Sharma, S., & Das, T. (2004). Microbiologic spectrum and susceptibility of isolates in scleral buckle infection in India. *American Journal of Ophthalmology*, *138*(4), 663–664. https://doi.org/10.1016/j.ajo.2004.04.056.

Percival, S. L. (2017). Importance of biofilm formation in surgical infection. *British Journal of Surgery*, *104*(2), e85–e94. https://doi.org/10.1002/bjs.10433.

Percival, S. L., Hill, K. E., Williams, D. W., Hooper, S. J., Thomas, D. W., & Costerton, J. W. (2012). A review of the scientific evidence for biofilms in wounds. *Wound Repair and Regeneration*, *20*(5), 647–657. https://doi.org/10.1111/j.1524-475X.2012.00836.x.

Percival, S. L., & Walker, J. T. (1999). Potable water and biofilms: A review of the public health implications. *Biofouling*, *14*(2), 99–115. https://doi.org/10.1080/08927019909378402.

Perez-Tanoira, R., Aarnisalo, A., Haapaniemi, A., Saarinen, R., Kuusela, P., & Kinnari, T. J. (2019). Bacterial biofilm in salivary stones. *European Archives of Oto-Rhino-Laryngology*, *276*(6), 1815–1822. https://doi.org/10.1007/s00405-019-05445-1.

Pneumatikos, I. A., Dragoumanis, C. K., Bouros, D. E., Warner, D. S., & Warner, M. A. (2009). Ventilator-associated pneumonia or endotracheal tube-associated pneumonia? *Anesthesiology*, *110*(3), 673–680. https://doi.org/10.1097/ALN.0b013e31819868e0.

Raad, I. I., & Bodey, G. P. (1992). Infectious complications of indwelling vascular catheters. *Clinical Infectious Diseases*, *15*(2), 197–210. https://doi.org/10.1093/clinids/15.2.197.

Roberts, A. L., Connolly, K. L., Kirse, D. J., Evans, A. K., Poehling, K. A., Peters, T. R., et al. (2012). Detection of group A *Streptococcus* in tonsils from pediatric patients reveals high rate of asymptomatic streptococcal carriage. *BMC Pediatrics*, *12*(1), 3. https://doi.org/10.1186/1471-2431-12-3.

Robino, L., Scavone, P., Araujo, L., Algorta, G., Zunino, P., & Vignoli, R. (2013). Detection of intracellular bacterial communities in a child with *Escherichia coli* recurrent urinary tract infections. *Pathogens and Disease*, *68*(3), 78–81. https://doi.org/10.1111/2049-632X.12047.

Røder, H. L., Herschend, J., Russel, J., Andersen, M. F., Madsen, J. S., Sørensen, S. J., et al. (2018). Enhanced bacterial mutualism through an evolved biofilm phenotype. *ISME Journal*, *12*(11), 2608–2618. https://doi.org/10.1038/s41396-018-0165-2.

Rodrigues, M. E., Lopes, S. P., Pereira, C. R., Azevedo, N. F., Lourenço, A., Henriques, M., et al. (2017). Polymicrobial ventilator-associated pneumonia: Fighting in vitro *Candida albicans-Pseudomonas aeruginosa* biofilms with antifungal-antibacterial combination therapy. *PLoS One*, *12*(1), 1–19. https://doi.org/10.1371/journal.pone.0170433.

Rohde, H., Burandt, E. C., Siemssen, N., Frommelt, L., Burdelski, C., Wurster, S., et al. (2007). Polysaccharide intercellular adhesin or protein factors in biofilm accumulation of *Staphylococcus epidermidis* and *Staphylococcus aureus* isolated from prosthetic hip and knee joint infections. *Biomaterials*, *28*(9), 1711–1720. https://doi.org/10.1016/j.biomaterials.2006.11.046.

Rumbaugh, K. P., & Sauer, K. (2020). Biofilm dispersion. *Nature Reviews Microbiology*, *18*(10), 571–586. https://doi.org/10.1038/s41579-020-0385-0.

Rutter, P. R., & Vincent, B. (1984). Physicochemical interactions of the substratum, microorganisms, and the fluid phase. *Microbial Adhesion and Aggregation*, *21–38*. https://doi.org/10.1007/978-3-642-70137-5_3.

Samimi, D. B., Bielory, B. P., Miller, D., & Johnson, T. E. (2013). Microbiologic trends and biofilm growth on explanted periorbital biomaterials: A 30-year review. *Ophthalmic Plastic and Reconstructive Surgery*, *29*(5), 376–381. https://doi.org/10.1097/IOP.0b013e31829a7313.

Schilder, A. G. M., Chonmaitree, T., Cripps, A. W., Rosenfeld, R. M., Casselbrant, M. L., Haggard, M. P., et al. (2016). Otitis media. *Nature Reviews Disease Primers*, *2*(1), 1–18. https://doi.org/10.1038/nrdp.2016.63.

Scott, V. C. S., Haake, D. A., Churchill, B. M., Justice, S. S., & Kim, J.-H. (2015). Intracellular bacterial communities: A potential etiology for chronic lower urinary tract symptoms. *Urology*, *86*(3), 425–431. https://doi.org/10.1016/j.urology.2015.04.002.

Sears, C. L., Geis, A. L., & Housseau, F. (2014). Bacteroides fragilis subverts mucosal biology: From symbiont to colon carcinogenesis. *Journal of Clinical Investigation*, *124*(10), 4166–4172. https://doi.org/10.1172/JCI72334.

Seixas, R., Varanda, D., Bexiga, R., Tavares, L., & Oliveira, M. (2015). Biofilm-formation by *Staphylococcus aureus* and *Staphylococcus epidermidis* isolates from subclinical mastitis in conditions mimicking the udder environment. *Polish Journal of Veterinary Sciences*, *18*(4), 787–792. https://doi.org/10.1515/pjvs-2015-0102.

Shah, A., Sachdev, A., Coggon, D., & Hossain, P. (2011). Geographic variations in microbial keratitis: An analysis of the peer-reviewed literature. *British Journal of Ophthalmology*, *95*(6), 762–767. https://doi.org/10.1136/bjo.2009.169607.

Sinnott, C. R., & Teall, A. J. (1987). Persistent gallbladder carriage of *Salmonella* Typhi. *The Lancet*, *329*(8539), 976. https://doi.org/10.1016/S0140-6736(87)90319-9.

Snow, D. E., Everett, J., Mayer, G., Cox, S. B., Miller, B., Rumbaugh, K., et al. (2016). The presence of biofilm structures in atherosclerotic plaques of arteries from legs amputated as a complication of diabetic foot ulcers. *Journal of Wound Care*, *25*(Suppl. 2), S16–S22. https://doi.org/10.12968/jowc.2016.25.Sup2.S16.

Solano, C., Echeverz, M., & Lasa, I. (2014). Biofilm dispersion and quorum sensing. *Current Opinion in Microbiology*, *18*(1), 96–104. https://doi.org/10.1016/j.mib.2014.02.008.

Stewart, P. S. (2015). Antimicrobial tolerance in biofilms examples of reduced biofilm susceptibility. *Microbiology Spectrum*, *3*(3), 1–30. https://doi.org/10.1128/microbiolspec.MB-0010-2014.Antimicrobial.

Stickler, D. J. (2009). Proteus mirabilis biofilm formation and catheter design. In *Biomaterials and tissue engineering in urology* (pp. 157–190). Elsevier. https://doi.org/10.1533/9781845696375.2.157.

Sugita, J., Yokoi, N., Fullwood, N. J., Quantock, A. J., Takada, Y., Nakamura, Y., et al. (2001). The detection of bacteria and bacterial biofilms in punctal plug holes. *Cornea*, *20*(4), 362–365. https://doi.org/10.1097/00003226-200105000-00005.

Sutherland, I. W. (2001). Biofilm exopolysaccharides: A strong and sticky framework. *Microbiology*, *147*(1), 3–9. https://doi.org/10.1099/00221287-147-1-3.

Swidsinski, A., Mendling, W., Loening-Baucke, V., Ladhoff, A., Swidsinski, S., Hale, L. P., et al. (2005). Adherent biofilms in bacterial vaginosis. *Obstetrics & Gynecology*, *106*(5, Part 1), 1013–1023. https://doi.org/10.1097/01.AOG.0000183594.45524.d2.

Swidsinski, A., Weber, J., Loening-Baucke, V., Hale, L. P., & Lochs, H. (2005). Spatial organization and composition of the mucosal flora in patients with inflammatory bowel disease. *Journal of Clinical Microbiology*, *43*(7), 3380–3389. https://doi.org/10.1128/JCM.43.7.3380-3389.2005.

Tan, N. C. W., Foreman, A., Jardeleza, C., Douglas, R., Vreugde, S., & Wormald, P. J. (2013). Intracellular *Staphylococcus aureus*: The Trojan horse of recalcitrant chronic rhinosinusitis? *International Forum of Allergy and Rhinology*, *3*(4), 261–266. https://doi.org/10.1002/alr.21154.

Teo, L., Lim, L., Tan, D. T. H., Chan, T. K., Jap, A., & Ming, L. H. (2011). A survey of contact lens complications in Singapore. *Eye & Contact Lens*, *37*(1), 16–19. https://doi.org/10.1097/ICL.0b013e3182048f99.

Teterycz, D., Ferry, T., Lew, D., Stern, R., Assal, M., Hoffmeyer, P., et al. (2010). Outcome of orthopedic implant infections due to different staphylococci. *International Journal of Infectious Diseases*, *14*(10), e913–e918. https://doi.org/10.1016/j.ijid.2010.05.014.

Tjalsma, H., Boleij, A., Marchesi, J. R., & Dutilh, B. E. (2012). A bacterial driver–passenger model for colorectal cancer: Beyond the usual suspects. *Nature Reviews Microbiology*, *10*(8), 575–582. https://doi.org/10.1038/nrmicro2819.

Todd, O. A., & Peters, B. M. (2019). *Candida albicans* and *Staphylococcus aureus* pathogenicity and polymicrobial interactions: Lessons beyond Koch's postulates. *Journal of Fungi*, *5*(3). https://doi.org/10.3390/jof5030081.

Tomkovich, S., Dejea, C. M., Winglee, K., Drewes, J. L., Chung, L., Housseau, F., et al. (2019). Human colon mucosal biofilms from healthy or colon cancer hosts are carcinogenic. *Journal of Clinical Investigation*, *129*(4), 1699–1712. https://doi.org/10.1172/JCI124196.

Trampuz, A., & Zimmerli, W. (2005). New strategies for the treatment of infections associated with prosthetic joints. *Current Opinion in Investigational Drugs*, *6*(2), 185–190.

Trejo-Hernández, A., Andrade-Domínguez, A., Hernández, M., & Encarnación, S. (2014). Interspecies competition triggers virulence and mutability in *Candida albicans-Pseudomonas aeruginosa* mixed biofilms. *ISME Journal*, *8*(10), 1974–1988. https://doi.org/10.1038/ismej.2014.53.

Umscheid, C. A., Mitchell, M. D., Doshi, J. A., Agarwal, R., Williams, K., & Brennan, P. J. (2011). Estimating the proportion of healthcare-associated infections that are reasonably preventable and the related mortality and costs. *Infection Control and Hospital Epidemiology*, *32*(2), 101–114. https://doi.org/10.1086/657912.

Vajpayee, R. B., Dada, T., Saxena, R., Vajpayee, M., Taylor, H. R., Venkatesh, P., et al. (2000). Study of the first contact management profile of cases of infectious keratitis: A hospital-based study. *Cornea*, *19*(1), 52–56. https://doi.org/10.1097/00003226-200001000-00011.

Vestby, L. K., Grønseth, T., Simm, R., & Nesse, L. L. (2020). Bacterial biofilm and its role in the pathogenesis of disease. *Antibiotics*, *9*(2), 59. https://doi.org/10.1016/0002-9343(83)91112-9.

von Eiff, C., Jansen, B., Kohnen, W., & Becker, K. (2005). Infections associated with medical devices. *Drugs*, *65*(2), 179–214. https://doi.org/10.2165/00003495-200565020-00003.

von Rosenvinge, E. C., O'May, G. A., Macfarlane, S., Macfarlane, G. T., & Shirtliff, M. E. (2013). Microbial biofilms and gastrointestinal diseases. *Pathogens and Disease*, *67*(1), 25–38. https://doi.org/10.1111/2049-632X.12020.

Wagenlehner, F. M. E., Diemer, T., Naber, K. G., & Weidner, W. (2008). Chronic bacterial prostatitis (NIH type II): Diagnosis, therapy and influence on the fertility status. *Andrologia*, *40*(2), 100–104. https://doi.org/10.1111/j.1439-0272.2007.00827.x.

Weidner, W., Schiefer, H. G., Jantos, C., Krauss, H., Friedrich, H. J., & Altmannsberger, M. (1991). Chronic prostatitis: A thorough search for etiologically involved microorganisms in 1,461 patients. *Infection*, *19* (S3), S119–S125. https://doi.org/10.1007/BF01643680.

Wilson, M. T., & Hamilos, D. L. (2014). The nasal and sinus microbiome in health and disease. *Current Allergy and Asthma Reports*, *14*(12), 485. https://doi.org/10.1007/s11882-014-0485-x.

Wimpenny, J. (1996). Ecological determinants of biofilm formation. *Biofouling*, *10*(1–3), 43–63. https://doi.org/10.1080/08927019609386270.

Wimpenny, J., Manz, W., & Szewzyk, U. (2000). Heterogeneity in biofilms. *FEMS Microbiology Reviews*, *24*(5), 661–671. https://doi.org/10.1111/j.1574-6976.2000.tb00565.x.

Wood, B. D., & Whitaker, S. (1999). Cellular growth in biofilms. *Biotechnology and Bioengineering*, *64*(6), 656–670. https://doi.org/10.1002/(SICI)1097-0290(19990920)64:6<656::AID-BIT4>3.0.CO;2-N.

Wright, J. A., & Nair, S. P. (2010). Interaction of staphylococci with bone. *International Journal of Medical Microbiology*, *300*(2–3), 193–204. https://doi.org/10.1016/j.ijmm.2009.10.003.

Xiao, J., Klein, M. I., Falsetta, M. L., Lu, B., Delahunty, C. M., Yates, J. R., et al. (2012). The exopolysaccharide matrix modulates the interaction between 3D architecture and virulence of a mixed-species oral biofilm. *PLoS Pathogens*, *8*(4), 7–9. https://doi.org/10.1371/journal.ppat.1002623.

Yadav, M. K., Chuck, R. S., & Park, C. Y. (2013). Composition of artificial tear solution affects in vitro *Pseudomonas aeruginosa* biofilm formation on silicone hydrogel lens. *Journal of Ocular Pharmacology and Therapeutics*, *29*(6), 591–594. https://doi.org/10.1089/jop.2012.0078.

Yadav, M. K., Vidal, J. E., & Song, J.-J. (2020). Microbial biofilms on medical indwelling devices. In *New and future developments in microbial biotechnology and bioengineering: Microbial biofilms* (pp. 15–28). Elsevier. https://doi.org/10.1016/B978-0-444-64279-0.00002-5.

Yokoi, N., Okada, K., Sugita, J., & Kinoshita, S. (2000). With biofilm formation on a punctal plug. *Japanese Journal of Ophthalmology*, *560*, 559–560.

Yoon, B. I., Han, D.-S., Ha, U.-S., Lee, S.-J., Sohn, D. W., Kim, H. W., et al. (2013). Clinical courses following acute bacterial prostatitis. *Prostate International*, *1*(2), 89–93. https://doi.org/10.12954/PI.12013.

Yoon, B. I., Kim, S., Han, D.-S., Ha, U.-S., Lee, S.-J., Kim, H. W., et al. (2012). Acute bacterial prostatitis: How to prevent and manage chronic infection? *Journal of Infection and Chemotherapy*, *18*(4), 444–450. https://doi.org/10.1007/s10156-011-0350-y.

Zautner, A. E. (2012). Adenotonsillar disease. *Recent Patents on Inflammation & Allergy Drug Discovery*, *6*(2), 121–129. https://doi.org/10.2174/187221312800166877.

Zegans, M. E., Becker, H. I., Budzik, J., & O'Toole, G. (2002). The role of bacterial biofilms in ocular infections. *DNA and Cell Biology*, *21*(5–6), 415–420. https://doi.org/10.1089/10445490260099700.

Section F

Application of microbial biofilms

Chapter 35

Plant growth promoting *Rhizobacteria* and their biofilms in promoting sustainable agriculture and soil health

Augustus Akshaya Rani, Samsudeen Malik Basha, Kannan Dheerka Darsha, Christudoss Abisha Christy, Hari Prasath Nagaiah, Thirupathi Kasthuri, and Shunmugiah Karutha Pandian

Department of Biotechnology, Alagappa University, Karaikudi, Tamil Nadu, India

1. Introduction

The rising demand for food at global level improved sustainable practices over traditional practices to witness the progress in agricultural productivity with substantially worsening soil and water quality in coming years. An efficient use of beneficial microorganisms to improve the availability of plant nutrients and plant health in an integrated manner is an attractive solution to meet out the global food needs (Koskey, Mburu, Awino, Njeru, & Maingi, 2021). There are abundant and diverse microbial populations that are present in the terrestrial environment, which compete with each other to sustain and modify the resource pools (Van Der Heijden, Bardgett, & Van Straalen, 2008). As plant tissues are rich in nutrients, they provide suitable protective oases for the microbes by creating complex organic compounds from inorganic matter and sunlight, releasing a portion of this material and offering a range of niches for a vast array of microorganisms. The 40% of the carbon fixed from the air by land plants is released into the soil as root exudates, including compounds rich in nitrogen and phosphorus. The release of root exudates with high nutrients (nitrogen and phosphorus) promotes bacterial growth in the rhizosphere, which in turn allows other organisms, most prominently protozoans, to feed on the bacteria. Therefore, the diversity of microorganisms residing around the plant profoundly depends upon the root exudates released through the plant roots (Andreote & Pereira e Silva, 2017).

Plants are colonized by diverse bacterial strains on their leaves, roots, seeds, and internal vasculature. All these tissues exhibit unique chemical and physical characteristics that possess a big challenge for microbes to colonize. The colonization of microbes onto different plant surfaces occurs passively or actively (Kandel, Joubert, & Doty, 2017). The passive deposition of microbes takes place when microbes are nonmotile, often requiring a transport agent for introduction of microbes onto the aerial plant surfaces. The transport agents are wind, rainwater, splatter, and aerosols that can transfer microbes onto leaves and flowers. Once deposited in the phyllosphere, motility may aid the bacteria in accessing a specific niche, but more often they congregate at their landing site (Andrews & Harris, 2000; Sivakumar, Sathishkumar, Selvakumar, Shyamkumar, & Arjunekumar, 2020). Water flow in the rhizosphere can act to inoculate cells and spores onto below ground tissues. Furthermore, physical wounding of plants will also introduce microbes from the external environment into internal tissues, including the vasculature. Insect and nematode activity can potentially inoculate microbes at any site on the plant, and sap-feeding insects are a common mode of transmission for vascular pathogens. The active deposition of microbes takes place when the microbes are motile and can undergo chemotaxis due to the presence of relatively nutrient-rich conditions generated by plant exudation. The motility of several different pseudomonads appears to be important for competitive root colonization and long-term survival in soils. Seed-associated bacteria may also colonize the developing rhizosphere via chemotaxis and motility (Andrews & Harris, 2000; Buée, de Boer, Martin, van Overbeek, & Jurkevitch, 2009; Bulgarelli, Schlaeppi, Spaepen, Van Themaat, & Schulze-Lefert, 2013; DeLeon-Rodriguez et al., 2013; Kandel et al., 2017; Sivakumar et al., 2020). In moist and nutrient-rich sites, the bacteria can grow into aggregates and form biofilms. Biofilms may form upon association or at later stages, with significant potential to direct or modulate the plant-microbe interaction. The terrestrial environment may vary constantly in terms of its water availability and saturation level. Hence, the bacteria that grow in that modifying environment experience different levels of hydration depending on the colonization site, prevailing climate conditions, and soil composition (Raaijmakers, Paulitz, Steinberg, Alabouvette,

Understanding Microbial Biofilms. https://doi.org/10.1016/B978-0-323-99977-9.00003-X

& Moënne-Loccoz, 2008). The increase in the branching of the roots occurs due to the release of nitrogen upon digestion of bacteria by protozoans. Bacteria have adapted to each of these microenvironments, and the biofilms thus formed reflect the nature of their colonization sites. For example, the surface characteristics vary along the length of the root, actively growing root tissues, leaf tissues with the different waxy cuticles in the upper and lower portions of the leaf, interspersed with veins and petioles, trichomes, and stomata. Phloem and xylem vessels are distinct tissue types within the vasculature that differ in fluid composition, architecture, and spatial arrangement within leaves, stems, and roots. The complexity in maintaining this system of ecology that actively modifies the colonized plant environment includes not only the organisms such as bacteria and protozoa but also the mutualistic and pathogenic fungi, nematodes, and arthropods. There are a lot of factors involved in plant-microbe associations, which are genotype of the plant and the soil composition, which account for selection and rejection of certain species and change their activity along the spectrum of mutualism and parasitism (Andreote & Pereira e Silva, 2017; Backer et al., 2018; Bais, Weir, Perry, Gilroy, & Vivanco, 2006; Berendsen, Pieterse, & Bakker, 2012; Raaijmakers et al., 2008).

Among all, the rhizosphere is a hotspot of plant-microbial interactions. The rhizosphere can be defined as the area of soil closely associated with plant roots, hairs, and their exudates. The three layers of rhizosphere are endorhizosphere, ectorhizosphere, and rhizoplane. The endorhizosphere constitutes the internal root area along with the cortical region that is significantly populated by high load of bacterial species with diverse function and metabolism. The ectorhizosphere and rhizoplane are the soil around the root and soil on the root, respectively. Together, the layers of rhizosphere are active and dynamic, harboring countless eukaryotic and prokaryotic life forms with the potential to modulate soil and plant health (Herlihy, 2021). A group of bacteria that aid in promoting plant growth and development and also to eliminate the attack of pathogens by suppressing their colonization on plants by associating themselves on and around plant roots are grouped as plant growth promoting rhizobacteria (PGPR). The biofilm formation of the colonized microorganisms can further serve as a fundamental strategy for the adaptation and survival under biotic and abiotic stress conditions of the important PGPR in the rhizosphere (Babalola, 2010; Backer et al., 2018).

2. Plant growth promoting rhizobacteria (PGPR)

PGPR are also known as nodule promoting rhizobacteria (NPR) or plant health promoting rhizobacteria (PHPR) as they exert beneficial role in maintaining plant health. PGPR are heterogeneous pool of microbial community found in the rhizosphere, are categorized into extracellular PGPR and intracellular PGPR based on their proximity to plant roots. Intracellular PGPR (iPGPR) are bacteria localized inside the plant cells and specialized structures that augment nodule formation where extracellular PGPR (ePGPR) reside outside the plant cells in the rhizosphere (Gray & Smith, 2005). The iPGPR group of organisms are mostly Gram-negative, rod-shaped, with a lesser proportion being Gram-positive rods, cocci, and pleomorphic forms. The iPGPR include bacteria belonging to the family Rhizobiaceae such as *Rhizobium*, *Bradyrhizobium*, *Sinorhizobium*, *Azorhizobium*, *Mesorhizobium*, and *Allorhizobium* that increase plant growth through nitrogen fixation. The PGPR class of bacterial genera including *Agrobacterium*, *Arthrobacter*, *Azotobacter*, *Azospirillum*, *Bacillus*, *Burkholderia*, *Caulobacter*, *Chromobacterium*, *Erwinia*, *Flavobacterium*, *Hyphomicrobium*, *Micrococcus*, *Pseudomonas*, *Serratia*, and free-living nitrogen-fixing bacteria do not facilitate nodule formation, instead stimulate plant growth directly by aiding in mobilization of nutrients and by conferring improved disease resistance to plants (Ahemad & Kibret, 2014; Backer et al., 2018; Bhattacharyya & Jha, 2012; Mustafa, Kabir, Shabbir, & Batool, 2019). The PGPR mechanisms of action are traditionally classified into direct and indirect means by which they have impact on plant growth as illustrated in Fig. 1.

2.1 Direct mechanisms

2.1.1 Nitrogen fixation

Plant growth and productivity are highly reliant on an abundant supply of nitrogen. However, nitrogen is the most limiting nutrient to plants due to the lack of potential mechanisms in plants to fix atmospheric nitrogen. PGPR fix nitrogen in the soil either symbiotically or nonsymbiotically. The symbiotic means of nitrogen fixation is a mutual association between plants and microbes that begins immediately after the colonization of rhizosphere and formation of root nodules with leguminous plants followed by the subsequent signal exchanges. The association of *Rhizobium*, *Bradyrhizobium*, *Sinorhizobium*, and *Mesorhizobium* with leguminous plants and *Frankia* bacteria with nonleguminous plants initiates the expression of rhizobial signaling molecules such as Nod factors upon secretion of flavonoids from roots, which are further recognized by LysM receptors of plant plasma membrane and activate the symbiotic signaling in the host. The nonsymbiotic nitrogen fixation is carried out by free-living associative endophytes (diazotrophs) including *Azoarcus*, *Azotobacter*, *Acetobacter*, *Azospirillum*, *Burkholderia*, *Diazotrophicus*, *Enterobacter*, *Gluconacetobacter*, *Pseudomonas*, and cyanobacteria

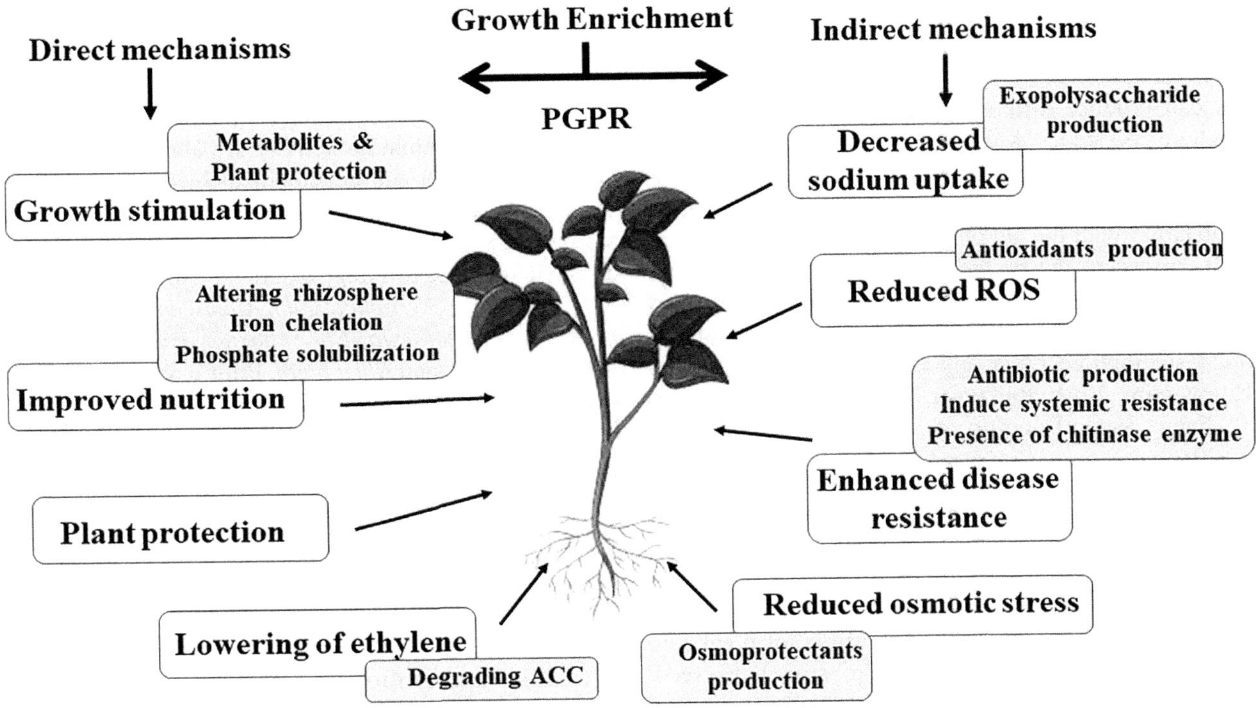

FIG. 1 PGPR mechanism of action in inducing plant growth.

(*Anabaena*, *Nostoc*) due to the presence of *nif* genes responsible for nitrogen fixation. The expression of *nif* genes is further responsible for activation of the Fe protein, iron/molybdenum cofactor biosynthesis, electron donation required for the synthesis and function of the molybdenum-nitrogenase complex and iron-nitrogenase complex. However, the nitrogen fixation is carried out by the activity of the molybdenum nitrogenase complex in all diazotrophs. The nonsymbiotic nitrogen fixing bacteria can fix smaller proportions of nitrogen than the plant's requirement (Kuan, Othman, Rahim, & Shamsuddin, 2016; Masclaux-Daubresse et al., 2010).

2.1.2 Phosphate solubilization

Phosphorus is the second most essential nutrient required for plant growth and development. The phosphate is profusely available in agricultural soils in insoluble, immobilized, and precipitated form. However, plants can utilize phosphate only in two soluble forms, the monobasic (H_2PO^{4-}) and the diabasic (HPO_4^{2-}) ions (Shen et al., 2011). The uses of phosphatic fertilizers are often observed in order to defeat the phosphate deficiency in agricultural soil irrespective of its cost and undesirable effects on environment. Phosphate solubilizing bacteria (PSB) among the PGPR are responsible for providing soluble forms of phosphate to plants. PSBs are gaining considerable attention in recent years due to their ability to serve as potent biofertilizers in place of chemical fertilizers (Zaidi, Khan, Ahemad, & Oves, 2009). The solubilization and uptake of phosphorus by plants can be induced by two different strategies including the production of chelating agents or mineral dissolving compounds including organic acid anions, hydroxyl or hydrogen ions, and CO_2 to solubilize inorganic phosphate compounds and production of extracellular enzymes (phosphatases/phytases) that mineralize organic forms of phosphate by the hydrolysis of phosphoric esters. *Azotobacter*, *Bacillus*, *Beijerinckia*, *Burkholderia*, *Enterobacter*, *Erwinia*, *Flavobacterium*, *Microbacterium*, *Pseudomonas*, *Rhizobium*, and *Serratia* are reported as the most significant phosphate solubilizing bacterial genera (Bianco & Defez, 2010; Cordell, Drangert, & White, 2009; Marra et al., 2012; Rodríguez & Fraga, 1999).

2.1.3 Siderophore production

Iron is an essential micronutrient required by all life forms. Iron, predominantly present in trivalent form, is sparsely soluble and is not readily assimilated by microorganisms and plants in spite of being fourth most abundant element on Earth. To combat this issue, a group of PGPR have evolved to acquire iron by the secretion of low-molecular-weight iron scavengers that chelate and transport iron to cells known as siderophores. Siderophores often exhibit high affinity to bind with iron. Based on the functional groups, siderophores can be classified into three distinct types: (1) hydroxamate-type siderophores (produced mainly by fungi); (2) catecholate-type siderophores (produced mainly by bacteria and have higher Fe binding affinity than hydroxamates); and (3) carboxylate-type siderophores (commonly produced by plants). Different PGPR strains can produce different types of siderophores. The siderophore associates with iron to form siderophore-Fe complex, which is then assimilated by reducing it into its respective divalent form or after siderophore cleavage upon internalization with the aid of specific receptors present in the cell surface of the bacteria. *Aeromonas*, *Azadirachta*, *Azotobacter*, *Bacillus*, *Burkholderia*, *Pseudomonas*, *Rhizobium*, *Serratia*, and *Streptomyces* sp. are some of the reported siderophore producing bacterial genera (Ahmed & Holmström, 2014; Rajkumar, Ae, Prasad, & Freitas, 2010; Saha, Saha, Donofrio, & Bestervelt, 2013; Saha et al., 2016).

2.1.4 Phytohormone production

Auxin

It is also known as indole acetic acid (IAA). IAA is a well-characterized signaling molecule responsible for plant cell division, extension, and differentiation; stimulates seed and tuber germination; increases the rate of xylem and root development; controls processes of vegetative growth; initiates lateral and adventitious root formation; mediates responses to light, gravity, and florescence; affects photosynthesis, pigment formation, biosynthesis of various metabolites, and resistance to stressful conditions. IAA is synthesized by five distinct pathways, which are named after their key intermediates as (1) indole-pyruvate pathway; (2) indole-acetamide pathway; (3) indole-acetonitrile pathway; (4) indole-acetaldehyde pathway; and (5) tryptamine pathway (Duca, Lorv, Patten, Rose, & Glick, 2014; Mohite, 2013). Around 80% of bacteria present in rhizosphere are able to synthesize and liberate IAA as their secondary metabolites. Tryptophan secreted in root exudates acts as a precursor for the synthesis of IAA. PGPR such as *Rhizobium*, *Bradyrhizobium*, *Pseudomonas*, *Enterobacter*, and *Klebsiella* use indole-pyruvate and indole-acetaldehyde pathways to synthesize IAA. Auxin synthesized by microorganisms is effective in stimulating plant growth owing to their slow and progressive release (Park, Kim, Hong, Shin, & Joo, 2021).

ACC (1-aminocyclopropane-1-carboxylate) deaminase

Ethylene is a vital hormone for plant growth and development as it plays a notable role in root initiation, fruit ripening, induction of seed germination, lowering of wilting, stimulation of leaf abscission, and activation of production of other plant hormones. Ethylene is otherwise known as stress hormone since stress conditions such as drought, water logging and salinity, heavy metal toxicity, and pathogenic infections trigger the release of ethylene in large proportions resulting in defoliation and diminished crop performance (Bleecker & Kende, 2000). The role of enzyme ACC deaminase is to counter balance the negative impacts of ethylene on plant physiology. Rhizobacterial strains displaying ACC deaminase activity are able to influence the host plant growth and development by lowering ethylene levels and thereby stimulating salt tolerance and drought resistance. PGPR with ACC deaminase activity are broadly distributed among different genera including *Acinetobacter*, *Achromobacter*, *Agrobacterium*, *Alcaligenes*, *Azospirillum*, *Bacillus*, *Burkholderia*, *Enterobacter*, *Pseudomonas*, *Ralstonia*, *Serratia*, and *Rhizobium*. Prior inoculation of PGPR with ACC deaminase activity in soil had demonstrated lower levels of ethylene production in response to a pathogenic infection and various positive effects such as root and shoot elongation, increased nodulation, and increased nutrient including sodium, phosphorus, and potassium (N, P, K) uptake (Ali, Charles, & Glick, 2014; Contesto et al., 2008; Glick, 2014).

Cytokinins and gibberellins

Cytokinins are N-6 aminopurines produced by algae, bacteria, and higher plants that play a key role in a wide range of physiological processes such as plant cell division, interruption of the quiescence of dormant buds, activation of seed germination, promotion of branching, root growth, accumulation of chlorophyll, leaf expansion, and delay of senescence. Cytokinin genes are evidently expressed in many PGPR, and their application to the growing plants can modify the plants' phytohormone composition. The phytopathogens induce greater production of cytokinin levels than PGPR, but their action is inhibitory in nature. However, lower proportions of cytokinin produced by PGPR exert stimulatory effects on plant growth. Several PGPR such as *Arthrobacter*, *Azospirillum*, *Paenibacillus polymyxa*, *Pseudomonas fluorescens*, and *Rhodospirillum rubrum* had been previously reported to synthesize cytokinins (Aloni, Aloni, Langhans, & Ullrich, 2006; Arkhipova et al., 2007; Liu, Xing, Ma, Du, & Ma, 2013). Gibberellin is another phytohormone that is synthesized by higher plants, fungi, and bacteria. They affect cell division and elongation and are involved in several plants' developmental processes, including seed germination, stem elongation, flowering, fruit setting, and delay of senescence in many organs of a range of plant species. Bacteria can produce four forms of gibberellins GAs (GA1, GA3, GA4, GA20) of which GA1 and GA4 being the most active. PGPR are capable of synthesizing GAs *Achromobacter xylosoxidans*, *Gluconobacter diazotrophicus*, *Acinetobacter calcoaceticus*, *Rhizobia*, *Azotobacter* spp., *Bacillus* spp., *Herbaspirillum seropedicae*, and *Azospirillum* spp. PGPR strains capable of producing either cytokinins or gibberellins or both of these include *Azotobacter* sp.,

Rhizobium sp., *Pantoea agglomerans*, *R. rubrum*, *P. fluorescens*, *Bacillus subtilis*, and *P. polymyxa* (Bottini, Cassán, & Piccoli, 2004; Kang et al., 2009, 2015).

2.2 Indirect mechanisms

2.2.1 Antagonistic role

Soil-borne phytopathogens tend to cause severe danger to global agricultural sustainability and ecosystem stability. They weaken soil fertility, undermine soil ecology, disrupt the environment, and subsequently show adverse consequences on human health, along with contaminating groundwater. Synthesis of antibiotics in response to phytopathogens is one of the most important indirect mechanisms of PGPR. Antibiotics are low-molecular-weight compounds derived from bacteria and fungi that suppress phytopathogens by disrupting 30s ribosomal RNA initiation complexes and inhibiting cell wall or cell membrane formation. Six distinct classes of antibiotics have been identified as efficient in reducing root infections, including phenazines, phloroglucinols, pyoluteorin, pyrrolnitrin, cyclic lipopeptides, and hydrogen cyanide (HCN). The antibiotic compounds produced by *Pseudomonas* species include Pyocyanin, phenazine-1-carboxamide, viscosinamide, Cepaciamide A, Rhamnolipids, Oomycin A, Phenazine-1-carboxyclic acid, Butyrolactones, Ecomycins, sulfonamide *N*-butylbenzene and 2,4 Diacetylphloroglucinol with antifungal activity; Karalicin with antiviral activity; Azomycin and Pseudomonic acid with antibacterial activity; Cepafungins with antitumor activity. *Bacillus*, on the other hand, produces number of metabolites including Kanosamine, Plipastatins A and B, Zwittermycin A, Bacillomycin, and Iturin A with antibiotic activity. Excessive reliance on antibiotic-producing bacteria as a biocontrol agent may be a drawback due to development of antibiotic resistance in coming years. HCN is the most potent volatile compound liberated by many PRGR. Antifungal activity of HCN released by antagonistic fluorescent pseudomonads against phytopathogens is exceptional (Fernando, Nakkeeran, & Zhang, 2005).

Antagonistic role of PGPR is mainly attributed to the production of distinct types of cell wall lysing enzymes, for instance, chitinases, cellulases, 1,3-glucanases, proteases, and lipases, which can lyse the cell wall and suppress the growth of pathogenic fungi through hyperparasitism. The cell wall of most of the phytopathogenic fungi (except oomycetes) is made up of chitin $(C_8H_{13}O_5N)_n$, which is an unbranched, long-chain polymer of glucose derivatives, composed of repeating units of β-1,4-linked units of the amino sugar *N*-acetyl-D-glucosamine (NAG). Hydrolytic enzymes are capable of breaking down glycosidic bonds in chitin. Microorganisms known for production of lytic enzymes include *Serratia marcescens*, *B. subtilis* NPU 001, *Paenibacillus* sp. strain 300 and *Streptomyces* sp. strain 385, *Paenibacillus* and *Streptomyces*, *P. fluorescens* LRB3W1 and *S. marcescens* B2 against *R. solani* and *F. Oxysporum* and *P. aeruginosa* PGPR2 against *Macrophomina* sp., *Rhizoctonia* sp., and *Fusarium* sp.

Most of the PGPR strains are involved in the production of bacteriocin. Bacteriocins are ribosomally produced antibacterial peptides, which are either bactericidal or bacteriostatic in mode of action against phytopathogens (Riley & Wertz, 2002). Mostly all bacterial species produce more than one type of bacteriocin at low concentrations in its inactive form. Specific immune proteins, encoded in the genome, are required for expression of bacteriocins, allowing the producer bacterial cell to resist the bacteriocin action (Cotter, Hill, & Ross, 2005; Cotter, Ross, & Hill, 2013). Bacteriocins are cationic peptides that bind to the negatively charged phospholipids of plasma membrane resulting in formation of pore, leakage of intracellular components consecutively leading to the cell death. *P. fluorescens* SF39a, isolated from the wheat rhizosphere, secretes bacteriocins that inhibit the growth of the phytopathogenic *Pseudomonas* and *Xanthomonas* strains. The small bacteriocin cerin 7, produced by *Bacillus cereus*, was the first reported bacteriocin-like compound from a *Bacillus* species. Interestingly, bacteriocins produced by *Bacillus* sp. have been found with broad-spectrum lethal activity. Bacteriocins produced by *Bacillus* species include Bac-GM17 from *Bacillus clausii*, GM17, H4, IH7, and Bac14B from *Bacillus subtilis*, thuricinS, thuricin17, entomocin110, morricin269, and tochicin from *B. thuringiensis*. *B. thuringiensis* is the most studied among *Bacillus* species due to its remarkable insecticidal property against diptera, coleoptera, and lepidoptera larvae (Cleveland, Montville, Nes, & Chikindas, 2001; Deegan, Cotter, Hill, & Ross, 2006; Gray et al., 2006; Palma et al., 2014).

2.2.2 Induced systemic response

Plants have wide range of defense mechanisms that are actively expressed only in response to biological stresses (pathogens and parasites) of various scales (ranging from microscopic viruses to phytophagous insect). The timing of this defense response is crucial for survival against biotic challenge caused by deadly pathogens/parasites. Induced systemic resistance (ISR) is a state of enhanced immune response developed by a plant against phytopathogens and systemic acquired resistance (SAR) is plant defense against subsequent attack of the same pathogen or parasite. ISR does not encounter the pathogen directly rather responds to prevent against the deleterious effects caused by the phytopathogens (Sticher, Mauch-Mani, & Métraux, 1997; Van Loon, Bakker, & Pieterse, 1998). Selected strains of PGPR suppress diseases by antagonism substances phytoalexins (proteinaceous antimicrobial compounds, polyphenolic and flavonoid) directly at the site of microbial infection and spread its synthesis to the adjacent cells resulting in rapid necrosis and death of the infected plant cells. Several rhizobacteria trigger the salicylic acid (SA)-dependent SAR pathway by producing SA at the root surface, whereas other rhizobacteria induce different signaling pathway independent of SA. The SA-independent ISR pathway is dependent on jasmonic acid (JA) and ethylene signaling. Combination of ISR and SAR can increase protection against pathogens that are resisted through both pathways besides extended protection to a broader spectrum of pathogens than ISR/SAR alone. The best characterized strains beside *Pseudomonas* species to initiate ISR are *Bacillus* spp. including *B. amyloliquefaciens*, *B. subtilis*, *B. pasteurii*, *B. cereus*, *B. pumilus*, *B. mycoides*, and *B. sphaericus* that elicits remarkable reduction in the incidence of diverse diseases of hosts. Some PGPR also use enzymes, for instance, phenylalanine ammonia lyase (PAL), chitinase, β-1,3-glucanase, peroxidase (PO), polyphenol oxidase (PPO), superoxide dismutase (SOD), catalase (CAT), lipoxygenase (LOX), ascorbate peroxidase (APX), and proteinase inhibitors to induce ISR (Alstrom, 1991; Annapurna et al., 2013; Dean & Kuć, 1985; Hammerschmidt, 1999; Iavicoli, Boutet, Buchala, & Métraux, 2003; Pieterse et al., 2000; Ross, 1961).

3. PGPR biofilms

Biofilms are defined as the well-organized cooperating communities of microorganisms enclosed in an extracellular matrix produced by them. The microbial community is associated with solid surfaces and in intimate contact with other microbial cells by producing a variety of extracellular polymeric substances (EPSs) including exopolysaccharides, proteins, and DNA in the complex matrix while being attached on the surface. The complexity in biofilm configuration ranges from simple to complex, flat to tower, featureless to clustered film, and making heterogeneous arrangements of microbes (Bacteria) (Davey & O'Toole, 2000; Kragh et al., 2016). Each cell within the biofilm will be heterogeneous and distinct in its response to nutrient and waste product diffusion gradients, modulating their metabolism as a function of their position within the biofilm, contacting adjacent cells, and engaging in cell-cell communication corresponding to their surroundings. Biofilms have enormous practical importance in industrial, medical, and agricultural settings, manifesting both beneficial and detrimental activities. The vigorous cellular colonization during the formation of biofilm helps the bacteria to survive the hostile condition. This cellular colonization confirms the nutrient utilization, expression of surface molecules, and virulence factors and furnishes bacteria with an arsenal of properties that facilitate their survival in unfavorable conditions (Cowan, Warren, & Fletcher, 1991; Donlan, 2000, 2002a; Fletcher, Lessmann, & Loeb, 1991; James, Beaudette, & Costerton, 1995; Tolker-Nielsen et al., 2000). The mechanisms involved in the process of biofilm formation differ depending on characteristics of bacterial strain and environmental conditions (Donlan, 2002a). There are numerous scientific evidences that are studied extensively in this regard for organisms such as *Pseudomonas aeruginosa*, *B. subtilis*, *Staphylococcus aureus*, *Escherichia. coli*, and other environmental bacteria. Similarly, biofilm formation under natural setting such as plant and associated soil is now the subject of intense investigation. Bacterial biofilms are widely distributed and play vital roles in many environments. The vigorous colonization of bacteria in various regions surrounding the plant roots ranges from rhizospheres rich in nutrients and root exudates to bulk soil deficient in nitrogen, phosphates, water, and other nutrients. Biofilm development contributes to the growth of the plant, for example, beneficial soil bacteria, including rhizobia, forming microcolonies or biofilms when they colonize roots. Similarly, it can also cause nonbeneficial virulence of phytopathogenic bacteria through various mechanisms, including blockage of xylem vessels, increased resistance to plant antimicrobial compounds, and/or enhanced colonization of specific habitats. The process of self-aggregation and the development of biofilm pertinent to both bacterial survival and host plant colonization. A variety of environmental, genetic, and structural factors affect bacterial adhesion, cell-cell interactions, and plant colonization, and ultimately plant-bacterial interactions in general (Costa, Raaijmakers, & Kuramae, 2018; Haque et al., 2020; Ude, Arnold, Moon, Timms-Wilson, & Spiers, 2006). Biofilm formation involves three steps as illustrated in Fig. 2.

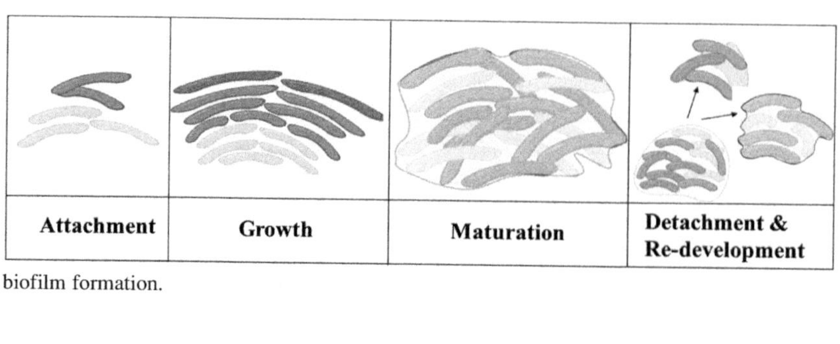

FIG. 2 Steps involved in biofilm formation.

3.1 Formation of biofilm

Rhizosphere microorganisms interact with plants and other soil microflora within the surroundings and have a great influence in promoting the growth of the plant. The complexity and fluctuating condition in the soil and the presence of various multiple numbers of microorganisms make it much harder to study the formation of biofilm in the rhizosphere. The rhizosphere forms a suitable environment for biofilm formation, which is referred to as the small area around the plant roots where convoluted biological and ecological processes take place that has sufficient moisture and inventory of nutrients that are contributed by the plant. The biofilm formation on the plant is due to the response of the associative, symbiotic, or pathogenic association of microbes, which constantly attempts to interact with plants and within each cell (Bais et al., 2006; Ramey, Koutsoudis, Bodman, & Fuqua, 2004; Rudrappa, Biedrzycki, & Bais, 2008). The biofilm formation process is a multistep process that involves the attachment of bacteria to the surface, growth and maturation of biofilm, and detachment and return to the planktonic growth. Fimbriae, pilli, flagella, and EPS attribute to the bacterial attachment to the conditioning film. Generally, many bacterial species are involved; fungi, protozoa, and algae may also contribute to the formation of biofilm, which can be of single species and many. Biofilms can be a single cell layer thick with the size of 6–8 cm while in most of the cases, it is 100 m thick on average. Bacterial attachment to the surface and biofilm formation involves the following steps:

3.1.1 Attachment of bacteria to the surface

The attachment of the microbial cells to the biotic or abiotic surfaces is the initial step in the formation of biofilm. This interaction reaction may be reversible or irreversible. Flagella, pili, and fimbriae of microbes help in promoting the cells to attach to the surface (Dang & Lovell, 2016; Koczan, Lenneman, McGrath, & Sundin, 2011; López, Vlamakis, & Kolter, 2010). While at the attachment stage, the rhizobacteria undergo physiological changes that lead to EPS production and fix the cells to the root surface and initiate the aggregation. The cells then start to multiply from microcolonies by clonal propagation. The maturation of biofilm of bacterial cells consists of many reversible and irreversible processes and also results in having diverse conserved and/or species-specific factors. At first, bacterial cells are introduced to a surface, operated through Brownian movement and gravitational forces, and simultaneously influenced by surrounding hydrodynamic forces. Then, inside the niche, the bacteria encounter attractive or repelling forces that differ, depending on nutrient levels, pH, ionic strength, and temperature. The velocity and direction toward or away from the contact surface are affected by the properties of the medium and the composition of the bacterial cell surface. The hydrodynamic and repulsive forces are counterbalanced by the motility of bacteria using its flagella and assist the competitive advantage in the formation of biofilm. The flagellar motility is important for initial attachment. The numbers of nutrients present in the surroundings also influence the attachment of the cells by chemotaxis in some bacterial species. Attachment of microbial cells to biotic surfaces requires more combined/complex interaction with plant root. The stimulation of exopolysaccharides production is a key factor in determining optimal cell-to-cell and cell-to-surface interaction and biofilm formation. Likewise, the stimulation of attachment of rhizobia by root exudates through the production of acid EPS and arabinogalactans proteins both in the leguminous and non-leguminous plants has also been reported. Some studies confirm that the surface attachment is regulated by the concentrations of cAMP and cyclic diGMP, which are secondary messengers, controlled by various environmental factors, such as carbon and oxygen. Similarly, rhizobial adhesion protein RapA1 was found to influence the colonization of biotic surfaces positively (Abdallah, Benoliel, Drider, Dhulster, & Chihib, 2014; Bacosa et al., 2018; Bogino, de las Oliva, Sorroche, & Giordano, 2013; Dang & Lovell, 2016; Hirakawa & Tomita, 2013; Jayathilake et al., 2017).

3.1.2 Maturation of the biofilm

After the attachment of bacteria to the surface, the maturation process begins. The maturation process requires two factors, and they are quorum sensing (QS) signal and EPS accumulation through continued cell division (Jayathilake et al., 2017). QS is the mechanism that mediates the communication among microbial cells (Nadell, Xavier, Levin, & Foster, 2008). The natural community of bacteria communicates by cell-to-cell signal and both genetically engineered and nonengineered bacteria can communicate using this method by forming bacterial products. The product that formed can diffuse from one cell to another and results in effective communication. The cell division while colonization increases the production of EPS and leads to its accumulation associated with the development of antibiotic resistance and other properties, including increased resistance to UV light, increased rates of genetic exchange, altered biodegradative capabilities, and increased secondary metabolite production. The variation in gene expression of the two bacterial growth states that are planktonic and sessile affects the adhesive capabilities of the population during surface colonization. For example, in sessile forms of bacteria, the surface appendages production is inhibited where motility is no longer needed. The expression of genes that are involved in the production of cell surface proteins and excretion products increases and consequently facilitates the transport of extracellular products in the cell and excretion of waste products out of the cell through surface proteins (porins) such as OprC and OprE and polysaccharides, respectively. *P. aeruginosa* is an organism that has been extensively studied for its extracellular matrix composition. It is shown to have variation in the composition of the extracellular matrix depending on the environmental conditions. In *P. aeruginosa*, as the biofilm matures, eDNA amount increases through lysis of a bacterial subpopulation in response to the *P. aeruginosa* quinolone signal (Pqs) QS system. The eDNA is organized in distinct patterns and is localized in the stalk portion of the mushroom-shaped biofilms. This localization may act as a scaffold for the formation of the mushroom structure, as type 4 pili show high eDNA binding affinity, inducing the accumulation of migrating bacteria toward the areas of high eDNA concentration (Ansari, Ahmad, Ansari, & Ahmad, 2018; Bogino et al., 2013; Kaplan, 2010; Ramey et al., 2004).

3.1.3 Detachment and return to the planktonic growth mode

Following the process of maturation of the biofilm, detachment of cells occurs in which a part or whole of the biofilm shears off from the surface causing a reduction of biofilm mass. The detached cells return to their planktonic mode of growth. The detachment of cells from biofilm will lead to the formation of a new cycle of biofilm development. Biofilm dispersal can be the result of several reasons, such as modifications in nutrient availability, fluctuations in oxygen demand, and an increase in the production of toxic substances or other stress-inducing conditions. Several sensory systems govern the levels of small molecules, as a proxy to environmental changes, and alter gene expression accordingly, promoting dispersal of the cells in the biofilm. The dispersion of cells is a mandatory process after maturation. The dispersion or detachment of cells is not a single-step process, and it can occur in various manners based on internal and environmental conditions. However, detachment is the term commonly referred, and it can also be classified into several different types: (I) sloughing, (II) erosion, and (III) seeding dispersal. Stripping off of biofilm as a lump from the attached surface, which in turn reduces the biofilm mass, is called sloughing. While erosion is a process in which a single cell or small portion of biofilm is gradually and continually detached from the outside. These two styles of detachment are considered to be shear-dependent and passive processes. The third style of the detachment of cells is named seeding dispersal, which is an active process, in this process, the release of single cells or a small portion of cells from the center part of the biofilm happens to leave the center part as a hallow cavity. The dispersion of cells in a biofilm can also be regulated by inducing the shift between sessility and motility in bacteria; typically, an increase in c-di-GMP favors sessility, whereas reduced c-di-GMP leads to upregulation of motility. Similarly, EPS-degrading enzymes, such as alginate lyase in *P. aeruginosa*, also contribute to bacterial detachment from the matrix, although a controlled rhamnolipid production takes place that attributes to channel formation within mature *P. aeruginosa* biofilm. An increase in the level of rhamnolipid aids bacterial dispersal (Bester, Wolfaardt, Joubert, Garny, & Saftic, 2005; Donlan, 2002b; Kostakioti, Hadjifrangiskou, & Hultgren, 2013; O'Toole, Kaplan, & Kolter, 2000; Rollet, Gal, & Guzzo, 2009).

3.2 Factors influencing biofilm formation

There are various factors governing in vitro and the soil environment that are known to promote plant growth, root colonization, and activities of microorganisms. Also, some other factors are contributed by the microbial cell structure and physiology, synthesis of exopolysaccharides, QS interference, interaction with another microorganism, the influence of plant root and root exudates, as well as physicochemical characteristics of soil and organic matter in soil. In the rhizosphere, plant roots interact with both pathogenic and beneficial bacteria. Some bacterial species produce exometabolites that can regulate root growth and other root-microbe interactions in the rhizosphere (Donlan, 2002b). Many primary and secondary microbial products can also influence the process of biofilm formation. Rudrappa et al. (2008) demonstrated the role of cyanide production in Pseudomonad affecting plant root growth and other rhizospheric processes. In this research, they have used the plant *Arabidopsis thaliana* Col-0 seedlings as a model, along with KCN, the plant was treated with direct and indirect forms of cyanide from different pseudomonad strains. The treatment causes significant inhibition of primary root growth due to suppression of an auxin-responsive gene, specifically at the root tip region by pseudomonad cyanogenesis. Pseudomonad cyanogenesis parallelly affects the other beneficial rhizospheric processes such as *Bacillus subtilis* colonization by biofilm formation on *A. thaliana* Col-0 roots. The effect of cyanogenesis on *B. subtilis* biofilm formation was further established by the downregulation of important *B. subtilis* biofilm operons *epsA* and *yqxM*. The authors demonstrated the functional significance of pseudomonad cyanogenesis in regulating the multitrophic rhizospheric interactions. An important advantage of the biofilm lifestyle for soil bacteria (rhizobacteria) is the protection against water deprivation (desiccation or osmotic effect). The composition and functions of bacterial biofilms in soil microniches are poorly understood. There are reported evidences on multibacterial communities established as biofilm-like structures in the rhizosphere of *Medicago sativa* (alfalfa) exposed to triple experimental conditions of water limitation. It was concluded that the whole biofilm-forming ability (WBFA) for rhizospheric communities exposed to desiccation is higher than that of communities exposed to saline or non-stressful conditions. The effect of various factors on biofilm formation/disruption is briefly summarized in the following subsections (Ansari, Jafri, Ahmad, & Abulreesh, 2017; Right, Alotaibi, & Bukhari, 2021).

3.2.1 Surfaces

Various factors influence the attachment of microbes that leads to biofilm formation under a rhizospheric environment. They range from soil environment, microbial characteristics, and finally, plant surfaces. Biofilm formation is dependent on the surrounding environmental conditions and substratum parameters such as biotic and abiotic surfaces. The physicochemical parameters of the surface are known to affect the initial colonization of the microbes. After the attachment of the cell, the surface chemistry and the topographic features allow maximum cell adhesion and cell-surface binding (Crawford, Webb, Truong, Hasan, & Ivanova, 2012). This process results in enhancing the strength of attachment and thus retention. Immediate upregulation of alginate synthesis is reported in a strain of *Pseudomonas aeruginosa* after its interaction with the solid surface. The binding of microorganisms to a surface can confer advantages to cell survival. The metal surfaces are highly susceptible to microbial attack, and it is majorly dependent on growth medium, characteristics of cell surface, and substratum. Metals such as aluminum, stainless steel, and copper are often suitable for cell attachment, and thus, biofilm formation can occur. Even after, some metals, such as aluminum or copper, are considered toxic to bacteria. It has been suggested that microbial resistance to some metals (e.g., lead acetate) can be attributed to the high lead content of disinfectants and antiseptics, while resistance to copper sulfate may be due to its use as an algicide (Crawford et al., 2012; Feng et al., 2015; Hori & Matsumoto, 2010; Li & Logan, 2004; Rather, Gupta, & Mandal, 2021; Right et al., 2021).

3.2.2 Temperature and moisture content

For the terrestrial bacterial communities that are exposed to various environmental stress factors, limited water availability implies a substantial effect on the survival and activity. The amount of dissolved solute through osmotic potential and the nature of the matrix determine the water availability of the soil (water potential, ψ). Water retention force on the ground and matric potential mainly contribute to the water availability in the surroundings. Different types of water deprivation affect bacterial physiology in different ways, which are constituted by the abovementioned factors. To reduce the occurrence of plant diseases, it is essential to understand the role of temperature and water stress concerning plants and beneficial rhizobacteria. The alteration in the temperature (high and low) and the water availability in the normal and reduced condition, the interaction of *Bacillus amyloliquefaciens* strain S499 and between the plant results in the induction of systemic resistance (ISR) were studied. ISR level was observed to be low when plants were subjected to stress before bacterization. Treatment with *B. amyloliquefaciens* strain S499 before the exposure to stress reduced this negative effect. Low temperature is found to be the favorable condition for *B. amyloliquefaciens* strain S499 to produce an increased amount of surfactin. This positively influences counterbalancing the negative effect on the traits associated with rhizosphere fitness (colonization, motility, and biofilm formation) (Pertot et al., 2013). The anaerobic bacteria such as *clostridium perfringens*'s morphology, thickness, and cell density reflect the temperature-dependent regulation of EPS production, and biofilm formation is radically influenced by temperature (Srivastava et al., 2008).

3.2.3 Salinity

Salinity in the soil influences the growth of plants and microbiome in the surroundings. The chemical and physical properties of soil can be altered by salinity, which creates an unsuitable environment for the growth of any plant and soil-bound organisms. Therefore, the tolerance toward salinity in plants has become a notable function in the survival mechanism. Microbes can reduce salinity in the soil area and promote plant growth (Dodd & Pérez-Alfocea, 2012; Shrivastava & Kumar, 2015; Shultana, Kee Zuan, Yusop, & Saud, 2020; Singh, Jha, & Jha, 2015; Soldan et al., 2019). The salt tolerance in a plant primarily depends on the ability of roots on factors such as controlled or restricted uptake of sodium and chloride and also continued uptake of essential elements, especially potassium and nitrate, respectively. In various halophytes and nonhalophytes, uptake of potassium over sodium contributes to the salt tolerance. The exopolysaccharides (EPSs) produced by bacteria are of highly potential enough to bind with cations such as sodium and promote salt tolerance in the plant. Excess production of EPSs and accumulation of endogenous osmolytes are the best approaches followed by bacteria for their successful survival in plant rhizosphere, especially during various stress conditions (Kasim, Gaafar, Abou-Ali, Omar, & Hewait, 2016; Morcillo & Manzanera, 2021a; Munns & Tester, 2008). The two native salt-tolerant strains such as *Oceanobacillus profundus* (Pmt2) and *Staphylococcus saprophyticus* (ST1) are examined to determine the osmoadaptation and their plant growth promoting abilities. These stains are used as inoculants of *Lens esculenta* Var. *masoor* 93 under salt stress in the laboratory. It is found that the bacterial growth, biofilm formation, EPS production, and endogenous osmolytes (proline, glycine, and betaine) accumulation increased at higher salt stress. The formation of biofilm, overproduction of EPSs, and accumulation of endogenous osmolytes were increased in higher salt stress conditions (Kasim et al., 2016; Morcillo & Manzanera, 2021a; Zörb, Geilfus, & Dietz, 2019).

3.2.4 Nutrient availability

Availability of nutrients is one of the influencing factors for the growth and other activities of microorganisms in soil and other habitats. Nutrient availability will be more in rhizosphere conditions due to nutrients released in root exudates by plants. The nutrients for bacteria to form biofilm are derived from concentrating trace organics on surfaces by the extracellular polymers, using the waste products from their neighbors and secondary colonizers, and using different enzymes to break down food supplies (Badri & Vivanco, 2008; Walker, Bais, Grotewold, & Vivanco, 2003). As the matrix of the biofilm is negatively charged many nutrients are attracted to its surface. However, nutrients with a negative charge can exchange with ions on the surface. Thus, the bacteria that are present within get more nutrient than that of surroundings (Garrett, Bhakoo, & Zhang, 2008). Various factors, including carbon source, amount of nitrate, calcium, phosphate, and magnesium as well as the effects of osmolarity, pH, and temperature, can influence the nutrition uptake of microbes. The biofilm formation is enhanced when there is an increased amount of nutrients present in the root zone. However, the extreme condition of temperature and pH negatively affect the biofilm formation of rhizobia (Bryers, 1987; Costerton, 1999; Costerton et al., 1987; Garrett et al., 2008).

3.2.5 Microbial products

There are various exometabolites of microorganisms that highly influence the biofilm formation, of which some are listed below.

QS signal molecules in biofilm formation

Biofilm formation completely relies on cell-to-cell communication, which is a widespread phenomenon controlling a broad range of activities. Thus, the mechanism of QS plays a major role in the communication between neighboring bacterial cells via signal molecules (Davies et al., 1998). The agent that helps in effective communication between the cells is called autoinducer, which is derived from the production of, secretion of, and response to small, diffusible signal molecules. These molecules are produced at the basal levels of bacterial growth. When there is an increase in the concentration of signal molecules, the bacterial population expands, finally when it reaches a quorum level (threshold level), it stimulates phenotypic effects by regulating QS-dependent target gene expression (Camilli & Bassler, 2006; Chan, Liu, & Chang, 2015; Fuqua, Winans, & Greenberg, 1994). All these processes take place in a well-organized manner through regulating the induction and repression of biofilm formation based on the species of bacteria and also the environmental conditions. In Gram-positive bacteria, the autoinducers are often peptides while in Gram-negative bacteria it is fatty acid (Camilli & Bassler, 2006; Davies et al., 1998; Fuqua et al., 1994).

Antimicrobial peptides

The soil rhizosphere is a suitable habitat for many functional microbial communities; these organisms compete with each other by producing antimicrobial metabolites for their survival. They are low-molecular-weight peptides ranging from 1 to 10 kDa and generally cationic. These antimicrobial compounds produced by bacteria are found in all major bacterial lineages and are produced by both Gram-negative and Gram-positive bacteria (Babasaki, Takao, Shimonishi, & Kurahashi, 1985). These peptides are cysteine-rich with varying numbers of amino acids (from five to over a hundred), most commonly L forms. They are also called host defense peptides. The lytic peptides bind with the LPS (lipopolysaccharide) moieties of the bacterial cell membrane and disrupt the membrane stability. They can destroy mature biofilms by encouraging microbial detachment or killing. Similarly, studies on *Staphylococcus aureus* have shown that the lytic peptide PTP-7 prevented in vitro biofilm formation and was also effective in diffusing into the deep layer of preformed biofilm that results in the killing of 99.9% of biofilm-forming bacteria. This peptide retains its activity under highly acidic environments and in the presence of an excess of metals, conditions that mimic the *S. aureus* biofilm environment (Ageitos, Sánchez-Pérez, Calo-Mata, & Villa, 2017; Batoni, Maisetta, Lisa Brancatisano, Esin, & Campa, 2011; Pfalzgraff, Brandenburg, & Weindl, 2018).

Exopolysaccharides

The association between EPSs production and biofilm formation is explored broadly and its synthesis by microbial cells varies greatly in their composition and hence in their chemical and physical properties. They play a critical role in cell-to-surface and cell-to-cell interactions that can influence biofilm formation and stabilization (Lembre, Lorentz, & Martino, 2012). Mutation in microbes may lead to failure in adherence and stabilization in the biofilm formation. These mutant microbes are highly susceptible to antibiotics and host immune defenses, which lead to the killing of these microbes. Similarly, the EPS produced by bacteria may inhibit or destabilize biofilm formation by other species. Polysaccharides with nonbiocidal antibiofilm properties have also been isolated from cell-free biofilm extracts of several species. The different characteristics approach of EPS against biofilm formation are achieved through altering the physical characteristics of bacterial cells or abiotic surfaces; acting as signaling molecules that impact the gene expression patterns of susceptible bacteria; or competitively inhibiting multivalent carbohydrate-protein interactions, thereby interfering with adhesion. Polymers are high-molecular-mass compounds formed by joining together a large number of repeating units of simple molecules, called monomers. Polymers can be natural polymers or synthetic polymers based on their origin. Synthetic polymers are made up of additive chemicals, such as antioxidants, light stabilizers, lubricants, pigments, and plasticizers, which are added to improve the desired physical and chemical properties of the material. However, these additives may strain into the surrounding environment and provide nutrients for microorganisms present. Phosphorus has been shown to increase the formation of biofilms on polyvinyl chloride in phosphorus-limited water. Several studies have shown that plastic materials can support the growth of biofilms, but it has been suggested that augmentation in plastic pipes is usually comparable with that on iron, steel, or cement (Maunders & Welch, 2017).

3.2.6 Soil enzymes

The soil ecosystem consists of many extracellular enzymes mainly contributed by microorganisms as well as plant roots. Thus the soil particles carry a persistent enzymatic catalytic capacity capable of responding to arriving substrates. These enzymes are also abundant in aquatic and terrestrial ecosystems and may be present in significant amounts in soil and water biofilm (Ansari et al., 2018; Bogino et al., 2013; Gupta, Snehi, & Singh, 2017). The humic matter from the soil and the enzyme together forms a complex that is highly resistant to various stress conditions such as high temperature, less water availability. These complexes are also resistant to proteolysis. *N*-Acetyl-D-glucosamine-1-phosphate acetyltransferase (GlmU) is an enzyme targeted for matrix disruption, which involves in the biosynthesis of activated UDP-GlcNAc, essential peptidoglycan, and LPS precursor in Gram-positive and Gram-negative bacteria, respectively. Inhibiting the GlmU enzyme using N-ethyl maleimide (NEM) and its analogs showed antibiofilm activity against *E. coli*, *P. aeruginosa*, *K. pneumoniae*, and *E. faecalis*. The enzymes DNase-I and Dispersin-B have also recently gained attention as potential antibiofilm agents, particularly against Gram-positive bacteria. The effects of DNase-I are linked with its ability to digest the eDNA found within the biofilm structure (Cai et al., 2019).

3.3 Constituents of biofilm matrix

The biofilm often constitutes less than 10% of dry mass and over 90% of matrix. Matrix is an extracellular material that is produced by the microorganisms onto which cells are adhered. The matrix serves as a reservoir for the lysed cell components. Stabilization of matrix is due to extracellular structures of the bacteria such as flagella, pili, and fimbriae. The production and quantity of EPSs rely on the growth and alimentation of structured microbial communities. Extracellular polymeric substances immobilize cells, enabling their intense interactions, including cell-to-cell communication, horizontal gene transfer, and the formation of synergistic microconsortia. EPS also serves as a nutrient source preventing the organisms in the biofilm from scarcity, biocides, antibiotics, heavy metals, ultraviolet radiation, host immune defenses, and many protozoan grazers. Due to the complexity of EPS, an extensive range of enzymes are required for its complete degradation. The cluster of diverse types of biopolymers, extracellular polymeric substances (EPSs) give rise to the three-dimensional structure of biofilm and are responsible for the attachment to surfaces (Flemming, Neu, & Wozniak, 2007; More, Yadav, Yan, Tyagi, & Surampalli, 2014). Acetyl groups of exopolysaccharides improve the adhesive and cohesive properties of EPS. Furthermore, the biofilm also consists of extracellular enzymes that help in the degradation of biopolymers. The biopolymers include water-soluble polymers such as polysaccharides, proteins, and nucleic acids and water-insoluble polymers such as cellulose, chitin, and lipids. The amount of proteins in the biofilm matrix can exceed the polysaccharide content. These enzymes serve as the external digestive system, which breaks down the biopolymers (Flemming & Wingender, 2010).

4. Role of rhizobacterial biofilm in agriculture

The rhizobacterial biofilms are the same as normal biofilms, enveloped by EPSs and attached to surfaces by extracellular appendages (flagella, slime layer, etc.). Biofilms formed by multiple species are known as mixed biofilms, which are greatly resistant to most of the antimicrobial substances because of higher EPS production, better nutrient utilization, and interspecific protection. The primary role of rhizobacterial biofilms is to enhance bacterial survival by protecting them against fluctuating environmental conditions, i.e., desiccation, low nutrient availability, high temperature and salt concentration, and extreme pH fluctuations as well as inducing plant growth by eliciting various PGPR-associated mechanisms as described above. Single and mixed biofilms of *Pseudomonas*, *Trichoderma*, *Bradyrhizobium*, and *Penicillium* had showed greater production of ammonia, auxin, increased phosphate solubilization, siderophore production, and/or nitrogenase activity than the planktonic cells. Moreover, the biofilms inoculated with cotton seeds induced its germination. The key advantage of biofilms over planktonic cells is their resistance to diverse antibiotics and improved survival in competitive terrestrial environment. The biofilms are also responsible in eliciting resistance mechanisms against pathogens. Apart from this, biofilms have potential applications in bioremediation of hazardous compounds, in maintenance of soil fertility, and formation of bio-barriers to protect soil and groundwater from contamination (Bogino et al., 2013; Gupta et al., 2017; Haque et al., 2020).

4.1 Importance of biofilm formation in plant growth promotion

Several studies have revealed the improvement in plant growth and development in response to seed or root inoculation with diverse PGPR biofilm forming inoculants proficient in producing plant growth regulators. The demonstrated plausible mode of action of PGPR biofilm in inducing plant growth by providing inorganic nutrients to plants by mineralizing organic and insoluble inorganic forms of nitrogen, phosphorous, and sulfur that plants cannot utilize directly has steered the development of considerable number of PGPR being tested as biofertilizers. Biofertilizers are the substances that can colonize the rhizosphere or the interior parts of the plants promoting plant growth and development. Biofertilizers are also responsible for providing essential micro and macronutrients to plants promoting the plant quality.

4.2 Role of PGPR biofilms in mitigating abiotic stress

Abiotic and biotic stress can be managed via eco-friendly, stress-tolerant, biofilm-forming rhizobacteria by ameliorating different stresses in plants, thereby promoting plant health and crop productivity. Plant stress is a reaction to altering environmental conditions. This greatly influences the rate of photosynthesis, further leading to the reduction of capability of the plants to convert energy into biomass. The plants may either be exposed to abiotic stress, for instance, low temperature, high temperature, salinity, and drought (Morcillo & Manzanera, 2021b).

4.2.1 Salt stress amelioration

Among abiotic stresses, soil salinization is most destructive and is considered as the most significant limiting factor for crop growth and productivity. In agricultural context, salinity is defined as the presence of soluble salts especially sodium chloride in high concentration in the rhizospheric soil or water. The salinity is caused due to three primary reasons: (1) primary salinization, due to rainfall and subsequent accumulation of salt or weathering of rocks; (2) secondary salinization (dry land salinity), due to rise of groundwater level and accumulation of salt to the surface; and (3) tertiary salinization, which is irrigated salinity due to reapplying water for several cycles (Flowers & Colmer, 2015; Flowers, Hajibagheri, & Clipson, 1986; Munns & Tester, 2008). Higher sodium concentration tends to limit mobilization of nutrients by acting upon enzymes affecting stomatal conductance and thereby decreasing the rate of photosynthesis in plants. Similarly, salinity also induces hormonal imbalance, programmed cell death inducing oxidative stress and hypertonic stress due to accumulation of higher amounts of sodium and chloride ions (Del Amor & Cuadra-Crespo, 2012; Flowers & Colmer, 2015; Flowers et al., 1986; Munns & Tester, 2008). Additionally, the elevated level of salinity also influences soil physiology by typically affecting its porosity, aeration, and water conductance. The biofilm-forming PGPR overcome salinity stress by producing acidic polysaccharides such as succinoglycan. However, inoculation of biofilm-forming bacteria wheat seedlings had exhibited the restriction in sodium ions uptake from the soil and uninterrupted plant growth even under high salt conditions. Similarly, inoculation of *Lens esculenta* Var. *masoor* with biofilm-forming bacterial strains *Oceanobacillus profundus* (Pmt2) and *Staphylococcus saprophyticus* (ST1) had exhibited enhanced plant growth and endogenous osmolytes accumulation in plants. Halophilic microorganism uses biofilm and EPS accumulation to withstand at higher salt levels by accumulating the salt ions in biofilm/EPS matrix. Co-cultivation of halophilic microorganisms including *Halomonas*, *Kushneria*, *Planococcus*, *Thalassobacillus*, and *Oceanobacillus* with plants had also demonstrated to alleviate the salinity levels and withstand the stress (Badri, Weir, van der Lelie, & Vivanco, 2009; Balsanelli, De Baura, De Oliveira Pedrosa, De Souza, & Monteiro, 2014; Barnawal et al., 2017; Chen et al., 2016; Dodd & Pérez-Alfocea, 2012; Duzan, Zhou, Souleimanov, & Smith, 2004; Gupta, Dey, & Gupta, 2013; Ilangumaran & Smith, 2017; Munns & Tester, 2008; Wallenstein & Hall, 2012).

4.2.2 Drought stress amelioration

Water scarcity occurring due to low annual precipitation is a major challenge in crop production globally. Several strains of PGPR overcome drought by undergoing several physiochemical modifications called rhizobacterial-induced drought endurance and resilience (RIDER) (Kang, Kim, Yun, & Chang, 2010; Kaushal & Wani, 2016; Rolli et al., 2015). The RIDER mechanisms are alterations of plant hormonal levels, metabolic variations, production of bacterial EPSs, biofilm formation, and antioxidant resistance, including the accumulation of many suitable organic solutes such as carbohydrates, amino acids, and polyamines. Drought stress decreases the supply of carbon dioxide required for photosynthesis, resulting in ROS production such as radical superoxide (O_2^-), radical hydroxyl (OH), and hydrogen peroxide, which induces damaging effect on membrane through lipid peroxidation. EPS-producing/biofilm-forming rhizobacteria such as *Bacillus amyloliquefaciens* improves soil structure by facilitating the formation of macropores and soil aggregation, contributing to increased supply of water and fertilizers. It also helps in evacuating micro aggregates responsible for causing poor aeration (Chaves, Flexas, & Pinheiro, 2009). Moreover, bacterial EPS is composed of 97% water, which plays a defensive role against desiccation by amplifying the water retention and by source regulation of nutrients. A study identified that *P. aeruginosa* (Pa2) to be the most potent strain in withstanding drought stress by increasing the uptake of water, protein, and sugar among the other EPS-producing and biofilm-forming bacterial strains *Proteus penneri* (Pp1) *and Alcaligenes faecalis* (AF3) when inoculated with *Zea Mays* (Praveen Kumar, Mir Hassan Ahmed, Desai, Leo Daniel Amalraj, & Rasul, 2014; Rehman & Nautiyal, 2002; Reynolds et al., 2007; Saleem et al., 2018; Sandhya, Shaik, Grover, Reddy, & Venkateswarlu, 2009; Susilowati, Puspita, & Yunus, 2018; Timmusk et al., 2014; Vurukonda, Vardharajula, Shrivastava, & SkZ, 2016).

4.3 Role of biofilm in biocontrol

Biofilm formation by PGPR strains not only aids in nutrient cycling but also in biocontrol of pests and diseases. Competition for nutrients, niche exclusion, induced systemic resistance and production of antimicrobials such as henazines, pyrrolnitrin, 2,4-diacetylphloroglucinol (DAPG), pyoluteorin, viscosinamide, and tensin are the probable means responsible for biocontrol activity of PGPRs. Among all the PGPR, *Pseudomonas*, a common soil bacterium that colonizes the plant root, forming biofilm-like structures and protects the plants from root pathogens, is the well-characterized biocontrol agent (Bais, Fall, & Vivanco, 2004; Gupta et al., 2017; Haque et al., 2020). Hydrolytic enzymes producing PGPR bacterial strains are also recognized as authentic mechanism to resist the attack of phytopathogens. Additionally, hydrogen cyanide production by Pseudomonads has excellent antagonistic antifungal activity against phytopathogens. *B. subtilis* can guard plants against attack of fungal pathogen, and it also plays an important function in the degradation process of organic polymers present in the soil. It has been demonstrated that *B. subtilis* could be utilized as a biocontrol agent (Barratt, Moran, Bigler, & van Lenteren, 2018; Barriuso, Ramos Solano, & Gutiérrez Mañero, 2008).

4.4 Role of biofilm in bioremediation

Increasing industrialization results in significant rise in accumulation of toxic wastes causing massive environmental contaminations. Bioremediation is an emerging strategy that involves the use of microbes or their consortia to decontaminate the environment. Bioremediation of toxic pollutants using microbial strains is a valuable strategy than physical or chemical means, in terms of their safety, cost, and efficiency. Biofilms possess enhanced probability to act upon common environmental pollutants including oil, pesticides, and herbicides due to their endurance and strong adaptation mechanisms (Cohen, 2002; Speight & El-Gendy, 2018). Cyanobacterial biofilms have been extensively studied for their effective accumulation of insecticides in water bodies and aquifers. Many cyanobacterial species including *Nostoc linckia* and *Nostoc muscorum* were reported to degrade methyl parathion, *Chlorella vulgaris* and *Scenedesmus obliquus* to degrade dichlorprop-methyl (2,4-DCPPM) and several number of cynobacterial species to degrade heavy metals by intracellular uptake through the EPS (Cain, Vannela, & Woo, 2008). Mucilaginous sheaths of *Aphanothece halophytica* and *Microcystis aeruginosa* have been known to possess strong affinity for zinc, lead, and copper ions. Additionally, nitrogenous and phosphorus compounds in wastewater along with harmful metal ions can be removed using phototrophic biofilms (Cain et al., 2008; Rajkumar, Prasad, Freitas, & Ae, 2009; Singh, Kumar, Rai, & Singh, 2016).

Biofilms are well-recognized strategy for degradation of polycylic hydrocarbon pollutants that pose severe threats to human health and environment. The biofilm forming and decomposing ability of marine bacterium, *Stenotrophomonas acidaminiphila* against complex hydrocarbons had already been reported. Moreover, the *Stenotrophomonas acidaminiphila* biofilm had shown increased degradation rate of around 71% of phenanthrene and 40% of pyrene than that of planktonic cells. Recently, biodegradation of chlorinated hydrocarbons has involved the usage of biofilm reactors. A membrane-aerated biofilm reactor (MBR) with a biofilm-*forming Pseudomonas* sp. was studied for removal of 1,2-DCA from water (Arulazhagan & Vasudevan, 2009; Mangwani, Kumari, & Das, 2015; Mangwani, Shukla, Kumari, Rao, & Das, 2014).

5. Future perspectives

PGPR strains are eco-friendly having promising applications in the field of agriculture. The application of PGPR to get rid of plant pathogens was pioneered in Soviet Union by 1958 using selective strains. The most crucial part is the selection of efficacious PGPR strain for agricultural development. The deeper understanding on mechanistic role of PGPR biofilms in sustainable agriculture is required in order to expand the exploitation of these biofertilizers for reducing the negative environmental effects. Furthermore, genetically engineered rhizobacterial strains can also be employed to remediate the contaminated soil and improvise the soil health. The genetic modification of diversely occurring PGPR strains with potential genes to alleviate the pest and pathogen attack on plants would further confer multiple benefits to the crop growers. The importance of PGPR and their biofilms in agriculture concised in this chapter demands the urgent need for the development of firm formulations concised of PGPR bacterial strains in place of chemical fertilizers. The successive commercialization of PGPR is based on collaborations between research organizations and industries. However, the future success of PGPR and their biofilms-producing industries is based on their business management, product marketing, and dedicated research.

6. Conclusion

The importance of PGPR and their biofilms in agriculture has been highlighted in this chapter. PGPR is crucial for soil health and agriculture due to their enhanced potential to acquire and assimilate nutrients, improvising soil porosity, in secreting growth promoting metabolites and other substances that elevate the plant growth. However, the understanding on mechanistic action PGPR-associated biofilms is still limited due to the complexity of microorganisms in the soil-root association and difficulties in studying the mixed biofilm under natural/simulated models. However, over the last decade, many researchers have explored and revealed the beneficial association of biofilm with plants, which can be exploited to enhance plant protection and promote growth even under stress conditions. Understanding the complex interactions between plants and microbes in the rhizosphere is still in its infancy. Recent metagenomic studies on root microbes still provide very limited information. Nevertheless, complex interactions among various groups of microorganisms inhabiting rhizospheres and their role in plant–microbe interactions in biofilms must be explored to gain a better understanding and for improved ecological niche engineering of plant and soil health in sustainable agriculture.

References

Abdallah, M., Benoliel, C., Drider, D., Dhulster, P., & Chihib, N. E. (2014). Biofilm formation and persistence on abiotic surfaces in the context of food and medical environments. *Archives of Microbiology, 196*(7), 453–472. Springer Verlag.

Ageitos, J. M., Sánchez-Pérez, A., Calo-Mata, P., & Villa, T. G. (2017). Antimicrobial peptides (AMPs): Ancient compounds that represent novel weapons in the fight against bacteria. *Biochemical Pharmacology, 133*, 117–138. Elsevier Inc.

Ahemad, M., & Kibret, M. (2014). Mechanisms and applications of plant growth promoting rhizobacteria: Current perspective. *Journal King Saudi University Science, 26*(1), 1–20. Elsevier B.V.

Ahmed, E., & Holmström, S. J. M. (2014). Siderophores in environmental research: Roles and applications. *Microbial Biotechnology, 7*(3), 196–208. John Wiley and Sons Ltd.

Ali, S., Charles, T. C., & Glick, B. R. (2014). Amelioration of high salinity stress damage by plant growth-promoting bacterial endophytes that contain ACC deaminase. *Plant Physiology and Biochemistry, 80*, 160–167. Elsevier Masson SAS.

Aloni, R., Aloni, E., Langhans, M., & Ullrich, C. I. (2006). Role of cytokinin and auxin in shaping root architecture: Regulating vascular differentiation, lateral root initiation, root apical dominance and root gravitropism. *Annals of Botany, 97*(5), 883–893.

Alstrom, S. (1991). Induction of disease resistance in common bean susceptible to halo blight bacterial pathogen after seed bacterization with rhizosphere pseudomonads. *Journal of General and Applied Microbiology, 37*(6), 495–501.

Andreote, F. D., & Pereira e Silva, M.d. C. (2017). Microbial communities associated with plants: Learning from nature to apply it in agriculture. *Current Opinion in Microbiology, 37*, 29–34. Elsevier Ltd.

Andrews, J. H., & Harris, R. F. (2000). The ecology and biogeography of microorganisms on plant surfaces. *Annual Review of Phytopathology, 38*, 145–180.

Annapurna, K., Kumar, A., Kumar, L. V., Govindasamy, V., Bose, P., & Ramadoss, D. (2013). PGPR-induced systemic resistance (ISR) in plant disease management. In *Bacteria in agrobiology: Disease management* (pp. 405–425). Berlin, Heidelberg: Springer. 9783642336393. Retrieved January 13, 2022, from https://link.springer.com/chapter/10.1007/978-3-642-33639-3_15.

Ansari, F. A., Ahmad, I., Ansari, F. A., & Ahmad, I. (2018). Biofilm development, plant growth promoting traits and rhizosphere colonization by *Pseudomonas entomophila* FAP1: A promising PGPR. *Advances in Microbiology, 8*(3), 235–251. Scientific Research Publishing. Retrieved January 13, 2022, from http://www.scirp.org/journal/PaperInformation.aspx?PaperID=83477.

Ansari, F. A., Jafri, H., Ahmad, I., & Abulreesh, H. H. (2017). Factors affecting biofilm formation in in vitro and in the rhizosphere. In *Biofilms in plant and soil health* (pp. 275–290). John Wiley & Sons, Ltd. Retrieved January 14, 2022, from https://www.researchgate.net/publication/318637545_Factors_Affecting_Biofilm_Formation_in_in_vitro_and_in_the_Rhizosphere.

Arkhipova, T. N., Prinsen, E., Veselov, S. U., Martinenko, E. V., Melentiev, A. I., & Kudoyarova, G. R. (2007). Cytokinin producing bacteria enhance plant growth in drying soil. *Plant and Soil, 292* (1–2), 305–315.

Arulazhagan, P., & Vasudevan, N. (2009). Role of a moderately halophilic bacterial consortium in the biodegradation of polyaromatic hydrocarbons. *Marine Pollution Bulletin, 58*(2), 256–262.

Babalola, O. O. (2010). Beneficial bacteria of agricultural importance. *Biotechnology Letters, 32*(11), 1559–1570.

Babasaki, K., Takao, T., Shimonishi, Y., & Kurahashi, K. (1985). Subtilosin a, a new antibiotic peptide produced by *Bacillus subtilis* 168: Isolation, structural analysis, and biogenesis. *Journal of Biochemistry, 98* (3), 585–603. Oxford University Press.

Backer, R., Rokem, J. S., Ilangumaran, G., Lamont, J., Praslickova, D., Ricci, E., et al. (2018). Plant growth-promoting rhizobacteria: Context, mechanisms of action, and roadmap to commercialization of biostimulants for sustainable agriculture. *Frontiers in Plant Science, 871*, 1473. Frontiers Media S.A.

Bacosa, H. P., Kamalanathan, M., Chiu, M. H., Tsai, S. M., Sun, L., Labonté, J. M., et al. (2018). Extracellular polymeric substances (EPS) producing and oil degrading bacteria isolated from the northern Gulf of Mexico. *PLoS One, 13*(12). Public Library of Science.

Badri, D. V., & Vivanco, J. M. (2008). Regulation and function of root exudates. In *Plant, cell & environment* Wiley-Blackwell.

Badri, D. V., Weir, T. L., van der Lelie, D., & Vivanco, J. M. (2009). Rhizosphere chemical dialogues: Plant-microbe interactions. *Current Opinion in Biotechnology*, *20*(6), 642–650.

Bais, H. P., Fall, R., & Vivanco, J. M. (2004). Biocontrol of *Bacillus subtilis* against infection of *Arabidopsis* roots by *Pseudomonas syringae* is facilitated by biofilm formation and surfactin production. *Plant Physiology*, *134*(1), 307–319. American Society of Plant Biologists.

Bais, H. P., Weir, T. L., Perry, L. G., Gilroy, S., & Vivanco, J. M. (2006). The role of root exudates in rhizosphere interactions with plants and other organisms. *Annual Review of Plant Biology*, *57*, 233–266.

Balsanelli, E., De Baura, V. A., De Oliveira Pedrosa, F., De Souza, E. M., & Monteiro, R. A. (2014). Exopolysaccharide biosynthesis enables mature biofilm formation on abiotic surfaces by *Herbaspirillum seropedicae*. *PLoS One*, *9*(10). Public Library of Science.

Barnawal, D., Bharti, N., Pandey, S. S., Pandey, A., Chanotiya, C. S., & Kalra, A. (2017). Plant growth-promoting rhizobacteria enhance wheat salt and drought stress tolerance by altering endogenous phytohormone levels and TaCTR1/TaDREB2 expression. *Physiologia Plantarum*, *161*(4), 502–514. Blackwell Publishing Ltd.

Barratt, B. I. P., Moran, V. C., Bigler, F., & van Lenteren, J. C. (2018). The status of biological control and recommendations for improving uptake for the future. *BioControl*, *63*(1), 155–167. Springer Netherlands.

Barriuso, J., Ramos Solano, B., & Gutiérrez Mañero, F. J. (2008). Protection against pathogen and salt stress by four plant growth-promoting rhizobacteria isolated from *Pinus* sp. on *Arabidopsis thaliana*. *Phytopathology*, *98*(6), 666–672.

Batoni, G., Maisetta, G., Lisa Brancatisano, F., Esin, S., & Campa, M. (2011). Use of antimicrobial peptides against microbial biofilms: Advantages and limits. *Current Medicinal Chemistry*, *18*(2), 256–279. Bentham Science Publishers Ltd.

Berendsen, R. L., Pieterse, C. M. J., & Bakker, P. A. H. M. (2012). The rhizosphere microbiome and plant health. *Trends in Plant Science*, *17*(8), 478–486.

Bester, E., Wolfaardt, G., Joubert, L., Garny, K., & Saftic, S. (2005). Planktonic-cell yield of a pseudomonad biofilm. *Applied and Environmental Microbiology*, *71*(12), 7792–7798.

Bhattacharyya, P. N., & Jha, D. K. (2012). Plant growth-promoting rhizobacteria (PGPR): Emergence in agriculture. *World Journal of Microbiology and Biotechnology*, *28*(4), 1327–1350.

Bianco, C., & Defez, R. (2010). Improvement of phosphate solubilization and Medicago plant yield by an indole-3-acetic acid-overproducing strain of *Sinorhizobium meliloti*. *Applied and Environmental Microbiology*, *76*(14), 4626–4632.

Bleecker, A. B., & Kende, H. (2000). Ethylene: A gaseous signal molecule in plant. *Annual Review of Cell and Developmental Biology*, *16*, 1–18.

Bogino, P. C., de las Oliva, M. M., Sorroche, F. G., & Giordano, W. (2013). The role of bacterial biofilms and surface components in plant-bacterial associations. *International Journal of Molecular Sciences*, *14*(8), 15838–15859.

Bottini, R., Cassán, F., & Piccoli, P. (2004). Gibberellin production by bacteria and its involvement in plant growth promotion and yield increase. *Applied Microbiology and Biotechnology*, *65*(5), 497–503.

Bryers, J. D. (1987). Biologically active surfaces: Processes governing the formation and persistence of biofilms. *Biotechnology Progress*, *3*(2), 57–68.

Buée, M., de Boer, W., Martin, F., van Overbeek, L., & Jurkevitch, E. (2009). The rhizosphere zoo: An overview of plant-associated communities of microorganisms, including phages, bacteria, archaea, and fungi, and of some of their structuring factors. *Plant and Soil*, *321*(1–2), 189–212.

Bulgarelli, D., Schlaeppi, K., Spaepen, S., Van Themaat, E. V. L., & Schulze-Lefert, P. (2013). Structure and functions of the bacterial microbiota of plants. *Annual Review of Plant Biology*, *64*, 807–838.

Cai, P., Sun, X., Wu, Y., Gao, C., Mortimer, M., Holden, P. A., et al. (2019). Soil biofilms: Microbial interactions, challenges, and advanced techniques for ex-situ characterization. *Soil Ecology Letters*, *1*(3), 85–93. Springer. Retrieved January 17, 2022, from https://link.springer.com/article/10.1007/s42832-019-0017-7.

Cain, A., Vannela, R., & Woo, L. K. (2008). Cyanobacteria as a biosorbent for mercuric ion. *Bioresource Technology*, *99*(14), 6578–6586.

Camilli, A., & Bassler, B. L. (2006). Bacterial small-molecule signaling pathways. *Science*, *311*(5764), 1113–1116.

Chan, K. G., Liu, Y. C., & Chang, C. Y. (2015). Inhibiting N-acyl-homoserine lactone synthesis and quenching *Pseudomonas* quinolone quorum sensing to attenuate virulence. *Frontiers in Microbiology*, *6* (OCT). Frontiers Media S.A.

Chaves, M. M., Flexas, J., & Pinheiro, C. (2009). Photosynthesis under drought and salt stress: Regulation mechanisms from whole plant to cell. *Annals of Botany*, *103*(4), 551–560.

Chen, L., Liu, Y., Wu, G., Veronican Njeri, K., Shen, Q., Zhang, N., et al. (2016). Induced maize salt tolerance by rhizosphere inoculation of *Bacillus amyloliquefaciens* SQR9. *Physiologia Plantarum*, *158*(1), 34–44. Blackwell Publishing Ltd.

Cleveland, J., Montville, T. J., Nes, I. F., & Chikindas, M. L. (2001). Bacteriocins: Safe, natural antimicrobials for food preservation. *International Journal of Food Microbiology*, *71*(1), 1–20.

Cohen, Y. (2002). Bioremediation of oil by marine microbial mats. *International Microbiology*, *5*(4), 189–193. Sociedad Espanola de Microbiologia.

Contesto, C., Desbrosses, G., Lefoulon, C., Béna, G., Borel, F., Galland, M., et al. (2008). Effects of rhizobacterial ACC deaminase activity on *Arabidopsis* indicate that ethylene mediates local root responses to plant growth-promoting rhizobacteria. *Plant Science*, *175*(1–2), 178–189.

Cordell, D., Drangert, J. O., & White, S. (2009). The story of phosphorus: Global food security and food for thought. *Global Environmental Change*, *19*(2), 292–305.

Costa, O. Y. A., Raaijmakers, J. M., & Kuramae, E. E. (2018). Microbial extracellular polymeric substances: Ecological function and impact on soil aggregation. *Frontiers in Microbiology*, *9*(JUL). Frontiers Media S.A.

Costerton, J. W. (1999). Introduction to biofilm. *International Journal of Antimicrobial Agents*, *11*(3–4), 217–221.

Costerton, J. W., Cheng, K. J., Geesey, G. G., Ladd, T. I., Nickel, J. C., Dasgupta, M., et al. (1987). Bacterial biofilms in nature and disease. *Annual Review of Microbiology*, *41*, 435–464.

Cotter, P. D., Hill, C., & Ross, R. P. (2005). Food microbiology: Bacteriocins: Developing innate immunity for food. *Nature Reviews Microbiology*, *3*(10), 777–788.

Cotter, P. D., Ross, R. P., & Hill, C. (2013). Bacteriocins-a viable alternative to antibiotics? *Nature Reviews Microbiology*, *11*(2), 95–105.

Cowan, M. M., Warren, T. M., & Fletcher, M. (1991). Mixed-species colonization of solid surfaces in laboratory biofilms. *Biofouling*, *3*(1), 23–34.

Crawford, R. J., Webb, H. K., Truong, V. K., Hasan, J., & Ivanova, E. P. (2012). Surface topographical factors influencing bacterial attachment. *Advances in Colloid and Interface Science*, *179–182*, 142–149.

Dang, H., & Lovell, C. R. (2016). Microbial surface colonization and biofilm development in marine environments. *Microbiology and Molecular Biology Reviews*, *80*(1), 91–138. American Society for Microbiology.

Davey, M. E., & O'Toole, G. A. (2000). Microbial biofilms: From ecology to molecular genetics. *Microbiology and Molecular Biology Reviews*, *64*(4), 847–867. American Society for Microbiology.

Davies, D. G., Parsek, M. R., Pearson, J. P., Iglewski, B. H., Costerton, J. W., & Greenberg, E. P. (1998). The involvement of cell-to-cell signals in the development of a bacterial biofilm. *Science*, *280* (5361), 295–298.

Dean, R. A., & Kuć, J. (1985). Induced systemic protection in plants. *Trends in Biotechnology*, *3*(5), 125–129.

Deegan, L. H., Cotter, P. D., Hill, C., & Ross, P. (2006). Bacteriocins: Biological tools for bio-preservation and shelf-life extension. *International Dairy Journal*, *16*(9), 1058–1071.

Del Amor, F. M., & Cuadra-Crespo, P. (2012). Plant growth-promoting bacteria as a tool to improve salinity tolerance in sweet pepper. *Functional Plant Biology*, *39*(1), 82–90.

DeLeon-Rodriguez, N., Lathem, T. L., Rodriguez-R, L. M., Barazesh, J. M., Anderson, B. E., Beyersdorf, A. J., et al. (2013). Microbiome of the upper troposphere: Species composition and prevalence, effects of tropical storms, and atmospheric implications. *Proceedings of the National Academy of Sciences of the United States of America*, *110* (7), 2575–2580.

Dodd, I. C., & Pérez-Alfocea, F. (2012). Microbial amelioration of crop salinity stress. *Journal of Experimental Botany*, *63*(9), 3415–3428.

Donlan, R. M. (2000). Role of biofilms antimicrobial resistance. *ASAIO Journal*, *46*(6). Lippincott-Raven Publ.

Donlan, R. M. (2002a). Biofilms: Microbial life on surfaces. *Emerging Infectious Diseases*, *8*(9), 881. Centers for Disease Control and Prevention. Retrieved January 13, 2022, from /pmc/articles/PMC2732559/.

Donlan, R. M. (2022b). Biofilms: Microbial life on surfaces. *Emerging Infectious Diseases*, *8*(9), 881–890.

Duca, D., Lorv, J., Patten, C. L., Rose, D., & Glick, B. R. (2014). Microbial indole-3-acetic acid and plant growth. *Antonie Van Leeuwenhoek*, *106* (1), 85–125. Kluwer Academic Publishers.

Duzan, H. M., Zhou, X., Souleimanov, A., & Smith, D. L. (2004). Perception of *Bradyrhizobium japonicum* Nod factor by soybean [*Glycine max* (L.) Merr.] root hairs under abiotic stress conditions. *Journal of Experimental Botany*, *55*(408), 2641–2646.

Feng, G., Cheng, Y., Wang, S. Y., Borca-Tasciuc, D. A., Worobo, R. W., & Moraru, C. I. (2015). Bacterial attachment and biofilm formation on surfaces are reduced by small-diameter nanoscale pores: How small is small enough? *Npj Biofilms and Microbiomes*, *1*(1), 1–9. Nature Publishing Group. Retrieved January 14, 2022, from https://www.nature.com/articles/npjbiofilms201522.

Fernando, W. G., Nakkeeran, S., & Zhang, Y. (2005). Biosynthesis of antibiotics by PGPR and its relation in biocontrol of plant diseases. In *PGPR: Biocontrol and biofertilization* (pp. 67–109). Dordrecht: Springer.

Flemming, H. C., Neu, T. R., & Wozniak, D. J. (2007). The EPS matrix: The "house of biofilm cells". *Journal of Bacteriology*, *189*(22), 7945–7947.

Flemming, H. C., & Wingender, J. (2010). The biofilm matrix. *Nature Reviews Microbiology*, *8*(9), 623–633.

Fletcher, M., Lessmann, J. M., & Loeb, G. I. (1991). Bacterial surface adhesives and biofilm matrix polymers of marine and freshwater bacteria. *Biofouling*, *4*(1–3), 129–140.

Flowers, T. J., & Colmer, T. D. (2015). Plant salt tolerance: Adaptations in halophytes. *Annals of Botany*, *115*(3), 327–331. Oxford University Press.

Flowers, T. J., Hajibagheri, M. A., & Clipson, N. J. W. (1986). Halophytes. *Quarterly Review of Biology*, *61*(3), 313–337.

Fuqua, W. C., Winans, S. C., & Greenberg, E. P. (1994). Quorum sensing in bacteria: The LuxR-LuxI family of cell density- responsive transcriptional regulators. *Journal of Bacteriology*, *176*(2), 269–275. American Society for Microbiology.

Garrett, T. R., Bhakoo, M., & Zhang, Z. (2008). Bacterial adhesion and biofilms on surfaces. *Progress in Natural Science*, *18*(9), 1049–1056. Elsevier.

Glick, B. R. (2014). Bacteria with ACC deaminase can promote plant growth and help to feed the world. *Microbiological Research*, *169* (1), 30–39.

Gray, E. J., Lee, K. D., Souleimanov, A. M., Di Falco, M. R., Zhou, X., Ly, A., et al. (2006). A novel bacteriocin, thuricin 17, produced by plant growth promoting rhizobacteria strain *Bacillus thuringiensis* NEB17: Isolation and classification. *Journal of Applied Microbiology*, *100* (3), 545–554.

Gray, E. J., & Smith, D. L. (2005). Intracellular and extracellular PGPR: Commonalities and distinctions in the plant-bacterium signaling processes. *Soil Biology and Biochemistry*, *37*(3), 395–412.

Gupta, K., Dey, A., & Gupta, B. (2013). Plant polyamines in abiotic stress responses. *Acta Physiologiae Plantarum*, *35*(7), 2015–2036.

Gupta, G., Snehi, S. K., & Singh, V. (2017). Role of PGPR in biofilm formations and its importance in plant health. In *Biofilms in plant and soil health* (pp. 27–42). John Wiley & Sons, Ltd.

Hammerschmidt, R. (1999). Induced disease resistance: How do induced plants stop pathogens? *Physiological and Molecular Plant Pathology*, *55*(2), 77–84. Academic Press.

Haque, M. M., Mosharaf, M. K., Khatun, M., Haque, M. A., Biswas, M. S., Islam, M. S., et al. (2020). Biofilm producing rhizobacteria with multiple plant growth-promoting traits promote growth of tomato under water-deficit stress. *Frontiers in Microbiology*, *11*, 2954. Frontiers Media S.A.

Herlihy, M. (2021). Rhizosphere. In *Encyclopedia of earth sciences series* (pp. 269–301). Elsevier.

Hirakawa, H., & Tomita, H. (2013). Interference of bacterial cell-to-cell communication: A new concept of antimicrobial chemotherapy breaks antibiotic resistance. *Frontiers in Microbiology*, *4*(MAY). Frontiers Research Foundation.

Hori, K., & Matsumoto, S. (2010). Bacterial adhesion: From mechanism to control. *Biochemical Engineering Journal*, *48*(3), 424–434.

Iavicoli, A., Boutet, E., Buchala, A., & Métraux, J. P. (2003). Induced systemic resistance in *Arabidopsis thaliana* in response to root inoculation with *Pseudomonas fluorescens* CHA0. *Molecular Plant-Microbe Interactions*, *16*(10), 851–858. American Phytopathological Society.

Ilangumaran, G., & Smith, D. L. (2017). Plant growth promoting rhizobacteria in amelioration of salinity stress: A systems biology perspective. *Frontiers in Plant Science*, *8*, 1768. Frontiers Media S.A.

James, G. A., Beaudette, L., & Costerton, J. W. (1995). Interspecies bacterial interactions in biofilms. *Journal of Industrial Microbiology*, *15* (4), 257–262. Springer-Verlag.

Jayathilake, P. G., Jana, S., Rushton, S., Swailes, D., Bridgens, B., Curtis, T., et al. (2017). Extracellular polymeric substance production and aggregated bacteria colonization influence the competition of microbes in biofilms. *Frontiers in Microbiology*, *8*(SEP), 1865. Frontiers Media S.A.

Kandel, S. L., Joubert, P. M., & Doty, S. L. (2017). Bacterial endophyte colonization and distribution within plants. *Microorganisms*, *5*(4). MDPI AG.

Kang, S. M., Joo, G. J., Hamayun, M., Na, C. I., Shin, D. H., Kim, H. Y., et al. (2009). Gibberellin production and phosphate solubilization by newly isolated strain of *Acinetobacter calcoaceticus* and its effect on plant growth. *Biotechnology Letters*, *31*(2), 277–281.

Kang, S. M., Khan, A. L., Waqas, M., You, Y. H., Hamayun, M., Joo, G. J., et al. (2015). Gibberellin-producing *Serratia nematodiphila* PEJ1011 ameliorates low temperature stress in *Capsicum annuum* L. *European Journal of Soil Biology*, *68*, 85–93. Elsevier Masson SAS.

Kang, B. G., Kim, W. T., Yun, H. S., & Chang, S. C. (2010). Use of plant growth-promoting rhizobacteria to control stress responses of plant roots. *Plant Biotechnology Reports*, *4*(3), 179–183.

Kaplan, J. B. (2010). Biofilm dispersal: Mechanisms, clinical implications, and potential therapeutic uses. *Journal of Dental Research*, *89*(3), 205–218.

Kasim, W. A., Gaafar, R. M., Abou-Ali, R. M., Omar, M. N., & Hewait, H. M. (2016). Effect of biofilm forming plant growth promoting rhizobacteria on salinity tolerance in barley. *Annals of Agricultural Science*, *61*(2), 217–227. Faculty of Agriculture, Ain-Shams University.

Kaushal, M., & Wani, S. P. (2016). Plant-growth-promoting rhizobacteria: Drought stress alleviators to ameliorate crop production in drylands. *Annals of Microbiology*, *66*(1), 35–42. Springer Verlag.

Koczan, J. M., Lenneman, B. R., McGrath, M. J., & Sundin, G. W. (2011). Cell surface attachment structures contribute to biofilm formation and xylem colonization by *Erwinia amylovora*. *Applied and Environmental Microbiology*, *77*(19), 7031–7039.

Koskey, G., Mburu, S. W., Awino, R., Njeru, E. M., & Maingi, J. M. (2021). Potential use of beneficial microorganisms for soil amelioration, phytopathogen biocontrol, and sustainable crop production in smallholder agroecosystems. *Frontiers in Sustainable Food Systems*, *5*, 130. Frontiers Media S.A.

Kostakioti, M., Hadjifrangiskou, M., & Hultgren, S. J. (2013). Bacterial biofilms: Development, dispersal, and therapeutic strategies in the dawn of the postantibiotic era. *Cold Spring Harbor Perspectives in Medicine*, *3*(4), a010306.

Kragh, K. N., Hutchison, J. B., Melaugh, G., Rodesney, C., Roberts, A. E. L., Irie, Y., et al. (2016). Role of multicellular aggregates in biofilm formation. *MBio*, *7*(2). American Society for Microbiology.

Kuan, K. B., Othman, R., Rahim, K. A., & Shamsuddin, Z. H. (2016). Plant growth-promoting rhizobacteria inoculation to enhance vegetative growth, nitrogen fixation and nitrogen remobilisation of maize under greenhouse conditions. *PLoS One*, *11*(3), e0152478.

Lembre, P., Lorentz, C., & Martino, P. D. (2012). Exopolysaccharides of the biofilm matrix: A complex biophysical world. In *The Complex World of Polysaccharides* IntechOpen. Retrieved January 17, 2022, from https://www.intechopen.com/chapters/40591.

Li, B., & Logan, B. E. (2004). Bacterial adhesion to glass and metal-oxide surfaces. *Colloids and Surfaces B: Biointerfaces*, *36*(2), 81–90.

Liu, F., Xing, S., Ma, H., Du, Z., & Ma, B. (2013). Cytokinin-producing, plant growth-promoting rhizobacteria that confer resistance to drought stress in *Platycladus orientalis* container seedlings. *Applied Microbiology and Biotechnology*, *97*(20), 9155–9164.

López, D., Vlamakis, H., & Kolter, R. (2010). Biofilms. *Cold Spring Harbor Perspectives in Biology*, *2*(7).

Mangwani, N., Kumari, S., & Das, S. (2015). Involvement of quorum sensing genes in biofilm development and degradation of polycyclic aromatic hydrocarbons by a marine bacterium *Pseudomonas aeruginosa* N6P6. *Applied Microbiology and Biotechnology*, *99*(23), 10283–10297. Springer Verlag.

Mangwani, N., Shukla, S. K., Kumari, S., Rao, T. S., & Das, S. (2014). Characterization of *Stenotrophomonas acidaminiphila* NCW-702 biofilm for implication in the degradation of polycyclic aromatic hydrocarbons. *Journal of Applied Microbiology*, *117*(4), 1012–1024. J Appl Microbiol. Retrieved January 16, 2022, from https://pubmed.ncbi.nlm.nih.gov/25040365/.

Marra, L. M., Sousa Soares, C. R. F., de Oliveira, S. M., Ferreira, P. A. A., Soares, B. L., de Carvalho, R. F., et al. (2012). Biological nitrogen fixation and phosphate solubilization by bacteria isolated from tropical soils. *Plant and Soil*, *357*(1), 289–307.

Masclaux-Daubresse, C., Daniel-Vedele, F., Dechorgnat, J., Chardon, F., Gaufichon, L., & Suzuki, A. (2010). Nitrogen uptake, assimilation and remobilization in plants: Challenges for sustainable and productive agriculture. *Annals of Botany*, *105*(7), 1141–1157.

Maunders, E., & Welch, M. (2017). Matrix exopolysaccharides; The sticky side of biofilm formation. *FEMS Microbiology Letters*, *364*(13), fnx120.

Mohite, B. (2013). Isolation and characterization of indole acetic acid (IAA) producing bacteria from rhizospheric soil and its effect on plant growth. *Journal of Soil Science and Plant Nutrition*, *13*(3), 638–649.

Morcillo, R. J. L., & Manzanera, M. (2021a). The effects of plant-associated bacterial exopolysaccharides on plant abiotic stress tolerance. *Metabolites*, *11*(6). Multidisciplinary Digital Publishing Institute (MDPI). Retrieved January 17, 2022, from /pmc/articles/PMC8225083/.

Morcillo, R. J. L., & Manzanera, M. (2021b). The effects of plant-associated bacterial exopolysaccharides on plant abiotic stress tolerance. *Metabolites*. https://doi.org/10.3390/metabo11060337. Retrieved January 14, 2022, from.

More, T. T., Yadav, J. S. S., Yan, S., Tyagi, R. D., & Surampalli, R. Y. (2014). Extracellular polymeric substances of bacteria and their potential environmental applications. *Journal of Environmental Management*, *144*, 1–25. Academic Press.

Munns, R., & Tester, M. (2008). Mechanisms of salinity tolerance. *Annual Review of Plant Biology*, *59*, 651–681.

Mustafa, S., Kabir, S., Shabbir, U., & Batool, R. (2019). Plant growth promoting rhizobacteria in sustainable agriculture: From theoretical to pragmatic approach. *Symbiosis*, *78*(2), 115–123. Springer Netherlands. Retrieved January 13, 2022, from https://link.springer.com/article/10.1007/s13199-019-00602-w.

Nadell, C. D., Xavier, J. B., Levin, S. A., & Foster, K. R. (2008). The evolution of quorum sensing in bacterial biofilms. *PLoS Biology*, *6*(1), 0171–0179.

O'Toole, G., Kaplan, H. B., & Kolter, R. (2000). Biofilm formation as microbial development. *Annual Review of Microbiology*, *54*, 49–79.

Palma, L., Muñoz, D., Berry, C., Murillo, J., Caballero, P., & Caballero, P. (2014). *Bacillus thuringiensis* toxins: An overview of their biocidal activity. *Toxins*, *6*(12), 3296–3325. MDPI AG.

Park, S., Kim, A. L., Hong, Y. K., Shin, J. H., & Joo, S. H. (2021). A highly efficient auxin-producing bacterial strain and its effect on plant growth. *Journal Genetic Engineering & Biotechnology*, *19*(1), 179.

Pertot, I., Puopolo, G., Hosni, T., Pedrotti, L., Jourdan, E., & Ongena, M. (2013). Limited impact of abiotic stress on surfactin production in planta and on disease resistance induced by *Bacillus amyloliquefaciens* S499 in tomato and bean. *FEMS Microbiology Ecology*, *86*(3), 505–519. Retrieved January 17, 2022, from https://www.researchgate.net/publication/247154819_Limited_impact_of_abiotic_stress_on_surfactin_production_in_planta_and_on_disease_resistance_induced_by_B_amyloliquefaciens_S499_in_tomato_and_bean.

Pfalzgraff, A., Brandenburg, K., & Weindl, G. (2018). Antimicrobial peptides and their therapeutic potential for bacterial skin infections and wounds. *Frontiers in Pharmacology*, *9*(MAR), 281. Frontiers Media S.A.

Pieterse, C. M. J., Van Pelt, J. A., Ton, J., Parchmann, S., Mueller, M. J., Buchala, A. J., et al. (2000). Rhizobacteria-mediated induced systemic resistance (ISR) in *Arabidopsis* requires sensitivity to jasmonate and ethylene but is not accompanied by an increase in their production. *Physiological and Molecular Plant Pathology*, *57*(3), 123–134. Academic Press.

Praveen Kumar, G., Mir Hassan Ahmed, S. K., Desai, S., Leo Daniel Amalraj, E., & Rasul, A. (2014). In vitro screening for abiotic stress tolerance in potent biocontrol and plant growth promoting strains of *Pseudomonas* and *Bacillus* spp. *International Journal of Bacteriology*, *2014*, 1–6 (Hindawi Limited).

Raaijmakers, J. M., Paulitz, T. C., Steinberg, C., Alabouvette, C., & Moënne-Loccoz, Y. (2008). The rhizosphere: A playground and battlefield for soilborne pathogens and beneficial microorganisms. *Plant and Soil*, *321*(1), 341–361. Springer. Retrieved March 4, 2022, from https://link.springer.com/article/10.1007/s11104-008-9568-6.

Rajkumar, M., Ae, N., Prasad, M. N. V., & Freitas, H. (2010). Potential of siderophore-producing bacteria for improving heavy metal phytoextraction. *Trends in Biotechnology*, *28*(3), 142–149. Elsevier Ltd.

Rajkumar, M., Prasad, M. N. V., Freitas, H., & Ae, N. (2009). Biotechnological applications of serpentine soil bacteria for phytoremediation of trace metals. *Critical Reviews in Biotechnology*, *29*(2), 120–130.

Ramey, B. E., Koutsoudis, M., Bodman, S. B. V., & Fuqua, C. (2004). Biofilm formation in plant-microbe associations. *Current Opinion in Microbiology*, *7*(6), 602–609. Elsevier Ltd.

Rather, M. A., Gupta, K., & Mandal, M. (2021). Microbial biofilm: Formation, architecture, antibiotic resistance, and control strategies. *Brazilian Journal of Microbiology*, *52*(4), 1701–1718. Springer Science and Business Media Deutschland GmbH.

Rehman, A., & Nautiyal, C. S. (2002). Effect of drought on the growth and survival of the stress-tolerant bacterium *Rhizobium* sp. NBRI2505 sesbania and its drought-sensitive transposon Tn5 mutant. *Current Microbiology*, *45*(5), 368–377.

Reynolds, J. F., Stafford Smith, D. M., Lambin, E. F., Turner, B. L., Mortimore, M., Batterbury, S. P. J., et al. (2007). Ecology: Global desertification: Building a science for dryland development. *Science*, *316* (5826), 847–851.

Right, C., Alotaibi, G. F., & Bukhari, M. A. (2021). Factors influencing bacterial biofilm formation and development. *The American Journal Biomedical Science Research*, (6), 2021–2033. Retrieved January 14, 2022, from www.biomedgrid.com.

Riley, M. A., & Wertz, J. E. (2002). Bacteriocins: Evolution, ecology, and application. *Annual Review of Microbiology*, *56*, 117–137.

Rodríguez, H., & Fraga, R. (1999). Phosphate solubilizing bacteria and their role in plant growth promotion. *Biotechnology Advances*, *17*(4–5), 319–339. Elsevier Science Ltd.

Rollet, C., Gal, L., & Guzzo, J. (2009). Biofilm-detached cells, a transition from a sessile to a planktonic phenotype: A comparative study of adhesion and physiological characteristics in *Pseudomonas aeruginosa*. *FEMS Microbiology Letters*, *290*(2), 135–142. Oxford Academic. Retrieved January 14, 2022, from https://academic.oup.com/femsle/article/290/2/135/490121.

Rolli, E., Marasco, R., Vigani, G., Ettoumi, B., Mapelli, F., Deangelis, M. L., et al. (2015). Improved plant resistance to drought is promoted by the root-associated microbiome as a water stress-dependent trait. *Environmental Microbiology*, *17*(2), 316–331.

Ross, A. F. (1961). Systemic acquired resistance induced by localized virus infections in plants. *Virology*, *14*(3), 340–358.

Rudrappa, T., Biedrzycki, M. L., & Bais, H. P. (2008). Causes and consequences of plant-associated biofilms. *FEMS Microbiology Ecology*, *64* (2), 153–166. Oxford Academic. Retrieved January 13, 2022, from https://academic.oup.com/femsec/article/64/2/153/643436.

Saha, R., Saha, N., Donofrio, R. S., & Bestervelt, L. L. (2013). Microbial siderophores: A mini review. *Journal of Basic Microbiology*, *53*(4), 303–317.

Saha, M., Sarkar, S., Sarkar, B., Sharma, B. K., Bhattacharjee, S., & Tribedi, P. (2016). Microbial siderophores and their potential applications: A review. *Environmental Science and Pollution Research*, *23*(5), 3984–3999. Springer Verlag.

Saleem, A. R., Brunetti, C., Khalid, A., Della, R. G., Raio, A., Emiliani, G., et al. (2018). Drought response of *Mucuna pruriens* (L.) DC. inoculated with ACC deaminase and IAA producing rhizobacteria. *PLoS One*, *13* (2). Public Library of Science.

Sandhya, V., Shaik, Z. A., Grover, M., Reddy, G., & Venkateswarlu, B. (2009). Alleviation of drought stress effects in sunflower seedlings by the exopolysaccharides producing *Pseudomonas putida* strain GAP-p45. *Biology and Fertility of Soils*, *46*(1), 17–26.

Shen, J., Yuan, L., Zhang, J., Li, H., Bai, Z., Chen, X., et al. (2011). Phosphorus dynamics: From soil to plant. *Plant Physiology*, *156*(3), 997–1005. American Society of Plant Biologists.

Shrivastava, P., & Kumar, R. (2015). Soil salinity: A serious environmental issue and plant growth promoting bacteria as one of the tools for its alleviation. *Saudi Journal of Biological Sciences*, *22*(2), 123–131. Elsevier. Retrieved January 17, 2022, from /pmc/articles/PMC4336437/.

Shultana, R., Kee Zuan, A. T., Yusop, M. R., & Saud, H. M. (2020). Characterization of salt-tolerant plant growth-promoting rhizobacteria and the effect on growth and yield of saline-affected rice. *PLoS One*, *15*(9), e0238537. Public Library of Science. Retrieved January 17, 2022, from https://journals.plos.org/plosone/article?id=10.1371/journal.pone.0238537.

Singh, R. P., Jha, P., & Jha, P. N. (2015). The plant-growth-promoting bacterium Klebsiella sp. SBP-8 confers induced systemic tolerance in wheat (*Triticum aestivum*) under salt stress. *Journal of Plant Physiology*, *184*, 57–67. Elsevier GmbH.

Singh, J. S., Kumar, A., Rai, A. N., & Singh, D. P. (2016). Cyanobacteria: A precious bio-resource in agriculture, ecosystem, and environmental sustainability. *Frontiers in Microbiology*, *7*(APR), 529. Frontiers Research Foundation.

Sivakumar, N., Sathishkumar, R., Selvakumar, G., Shyamkumar, R., & Arjunekumar, K. (2020). Phyllospheric microbiomes: Diversity, ecological significance, and biotechnological applications. *Plant Microbiomes for Sustainable Agriculture*, *25*, 113–172. Springer, Cham.

Soldan, R., Mapelli, F., Crotti, E., Schnell, S., Daffonchio, D., Marasco, R., et al. (2019). Bacterial endophytes of mangrove propagules elicit early establishment of the natural host and promote growth of cereal crops under salt stress. *Microbiological Research*, *223–225*, 33–43. Elsevier GmbH.

Speight, J. G., & El-Gendy, N. S. (2018). Bioremediation of marine oil spills. In *Introduction to petroleum biotechnology* (pp. 419–470). Elsevier.

Srivastava, S., Yadav, A., Seem, K., Mishra, S., Chaudhary, V., & Nautiyal, C. S. (2008). Effect of high temperature on *Pseudomonas putida* NBRI0987 biofilm formation and expression of stress sigma factor RpoS. *Current Microbiology*, *56*(5), 453–457. Retrieved January 14, 2022, from https://www.researchgate.net/publication/5631337_Effect_of_High_Temperature_on_Pseudomonas_putida_NBRI0987_Biofilm_Formation_and_Expression_of_Stress_Sigma_Factor_RpoS.

Sticher, L., Mauch-Mani, B., & Métraux, J. P. (1997). Systemic acquired resistance. *Annual Review of Phytopathology*, *35*, 235–270.

Susilowati, A., Puspita, A. A., & Yunus, A. (2018). Drought resistant of bacteria producing exopolysaccharide and IAA in rhizosphere of soybean plant (*Glycine max*) in Wonogiri Regency Central Java Indonesia. *IOP Conference Series: Earth and Environmental Science*, *142* (1). Institute of Physics Publishing.

Timmusk, S., Abd El-Daim, I. A., Copolovici, L., Tanilas, T., Kännaste, A., Behers, L., et al. (2014). Drought-tolerance of wheat improved by rhizosphere bacteria from harsh environments: Enhanced biomass production and reduced emissions of stress volatiles. *PLoS One*, *9*(5). Public Library of Science.

Tolker-Nielsen, T., Brinch, U. C., Ragas, P. C., Andersen, J. B., Jacobsen, C. S., & Molin, S. (2000). Development and dynamics of *Pseudomonas* sp. biofilms. *Journal of Bacteriology*, *182*(22), 6482–6489.

Ude, S., Arnold, D. L., Moon, C. D., Timms-Wilson, T., & Spiers, A. J. (2006). Biofilm formation and cellulose expression among diverse environmental *Pseudomonas* isolates. *Environmental Microbiology*, *8*(11), 1997–2011.

Van Der Heijden, M. G. A., Bardgett, R. D., & Van Straalen, N. M. (2008). The unseen majority: Soil microbes as drivers of plant diversity and productivity in terrestrial ecosystems. *Ecology Letters*, *11*(3), 296–310. John Wiley & Sons, Ltd. Retrieved January 13, 2022, from https://onlinelibrary.wiley.com/doi/full/10.1111/j.1461-0248.2007.01139.x.

Van Loon, L. C., Bakker, P. A. H. M., & Pieterse, C. M. J. (1998). Systemic resistance induced by rhizosphere bacteria. *Annual Review of Phytopathology*, *36*, 453–483.

Vurukonda, S. S. K. P., Vardharajula, S., Shrivastava, M., & SkZ, A. (2016). Enhancement of drought stress tolerance in crops by plant growth promoting rhizobacteria. *Microbiological Research*, *184*, 13–24. Elsevier GmbH.

Walker, T. S., Bais, H. P., Grotewold, E., & Vivanco, J. M. (2003). Root exudation and rhizosphere biology. *Plant Physiology*, *132*(1), 44–51. American Society of Plant Biologists.

Wallenstein, M. D., & Hall, E. K. (2012). A trait-based framework for predicting when and where microbial adaptation to climate change will affect ecosystem functioning. *Biogeochemistry*, *109*(1–3), 35–47.

Zaidi, A., Khan, M. S., Ahemad, M., & Oves, M. (2009). Plant growth promotion by phosphate solubilizing bacteria. *Acta Microbiologica et Immunologica Hungarica*, *56*(3), 263–284. Akademiai Kiado ZRt.

Zörb, C., Geilfus, C. M., & Dietz, K. J. (2019). Salinity and crop yield. *Plant Biology*, *21*, 31–38. Blackwell Publishing Ltd.

Further reading

Ali, S. Z., Sandhya, V., Grover, M., Kishore, N., Rao, L. V., & Venkateswarlu, B. (2009). *Pseudomonas* sp. strain AKM-P6 enhances tolerance of sorghum seedlings to elevated temperatures. *Biology and Fertility of Soils*, *46*(1), 45–55.

Bashan, Y., de-Bashan, L. E., Prabhu, S. R., & Hernandez, J. P. (2014). Advances in plant growth-promoting bacterial inoculant technology: Formulations and practical perspectives (1998-2013). *Plant and Soil*, *378*(1–2), 1–33. Kluwer Academic Publishers.

Beneduzi, A., Ambrosini, A., & Passaglia, L. M. P. (2012). Plant growth-promoting rhizobacteria (PGPR): Their potential as antagonists and biocontrol agents. *Genetics and Molecular Biology*, *35*(4 Suppl), 1044–1051.

Chaloner, T. M., Gurr, S. J., & Bebber, D. P. (2020). Plant pathogen infection risk tracks global crop yields under climate change. *bioRxiv*.

Chandler, D., Davidson, G., Grant, W. P., Greaves, J., & Tatchell, G. M. (2008). Microbial biopesticides for integrated crop management: An assessment of environmental and regulatory sustainability. *Trends in Food Science and Technology*, *19*(5), 275–283.

Chodak, M., Gołębiewski, M., Morawska-Płoskonka, J., Kuduk, K., & Niklińska, M. (2015). Soil chemical properties affect the reaction of forest soil bacteria to drought and rewetting stress. *Annals of Microbiology*, *65*(3), 1627–1637. Springer Verlag.

Compant, S., Duffy, B., Nowak, J., Clément, C., & Barka, E. A. (2005). Use of plant growth-promoting bacteria for biocontrol of plant diseases: Principles, mechanisms of action, and future prospects. *Applied and Environmental Microbiology*, *71*(9), 4951–4959.

de Clercq, P., Mason, P. G., & Babendreier, D. (2011). Benefits and risks of exotic biological control agents. *BioControl*, *56*(4), 681–698.

Dey, R., Pal, K. K., Bhatt, D. M., & Chauhan, S. M. (2004). Growth promotion and yield enhancement of peanut (*Arachis hypogaea* L.) by application of plant growth-promoting rhizobacteria. *Microbiological Research*, *159*(4), 371–394. Elsevier GmbH. Retrieved January 13, 2022, from http://www.ncbi.nlm.nih.gov/pubmed/15646384.

Getahun, A., Muleta, D., Assefa, F., & Kiros, S. (2020). Plant growth-promoting rhizobacteria isolated from degraded habitat enhance drought tolerance of Acacia (*Acacia abyssinica* Hochst. Ex benth.) seedlings. *International Journal of Microbiology*, *2020*. Hindawi limited.

Jiao, X., Takishita, Y., Zhou, G., & Smith, D. L. (2021). Plant associated rhizobacteria for biocontrol and plant growth enhancement. *Frontiers in Plant Science*, *12*, 420. Frontiers Media S.A.

Kahil, M. T., Dinar, A., & Albiac, J. (2015). Modeling water scarcity and droughts for policy adaptation to climate change in arid and semiarid regions. *Journal of Hydrology*, *522*, 95–109. Elsevier.

Kasim, W. A., Osman, M. E., Omar, M. N., Abd El-Daim, I. A., Bejai, S., & Meijer, J. (2013). Control of drought stress in wheat using plant-growth-promoting bacteria. *Journal of Plant Growth Regulation*, *32* (1), 122–130. Springer Science and Business Media, LLC.

Kavamura, V. N., Santos, S. N., da Silva, J. L., Parma, M. M., Ávila, L. A., Visconti, A., et al. (2013). Screening of Brazilian cacti rhizobacteria for plant growth promotion under drought. *Microbiological Research*, *168*(4), 183–191.

Park, S., Kim, A.-L., Hong, Y.-K., Shin, J.-H., & Joo, S.-H. (2021). A highly efficient auxin-producing bacterial strain and its effect on plant growth. *Journal, Genetic Engineering & Biotechnology*, *19*(1), 179. Springer Science and Business Media LLC. Retrieved January 13, 2022, from http://www.ncbi.nlm.nih.gov/pubmed/34859356.

Chapter 36

Antagonistic *Bacilli* as prospective probiotics against pathogenic biofilms

Satish Kumar Rajasekharan[a] and Moshe Shemesh[b,c]

[a]*School of Chemical Engineering, Yeungnam University, Gyeongsan, South Korea,* [b]*Volcani Institute, Food Science, Agricultural Research Organization, Rishon LeZion, Israel,* [c]*Biology Department, Northeastern University, Boston, MA, United States*

1. Introduction

1.1 Biofilms

Biofilms often comprise multicellular population of microbial species enclosed into self-produced exopolymeric substances, which may adhere to either living or nonliving surfaces (Donlan, 2002; Flemming et al., 2016). In general, microbes in biofilms are distinct from planktonic cells of the same organism in various physiological characteristics and gene expression patterns (Shemesh, Tam, & Steinberg, 2007). Within biofilms, the microbes generate copious extracellular polymeric substances (EPSs), which assist surface adhesion and matrix stability. A wide range of microorganisms are capable of forming biofilms, and as a result, biofilms exist in a variety of microenvironments (Flemming et al., 2016; Kostakioti, Hadjifrangiskou, & Hultgren, 2013; López, Vlamakis, & Kolter, 2010). Some biofilms play a beneficial role in nature by serving as nourishment for larger organisms in the food chain (Robertson & McLean, 2015). However, those formed by foodborne pathogens pose a significant threat to food safety and quality (Moser et al., 2017; Zhao, Zhao, Wang, & Zhong, 2017). Numerous topical outbreaks caused due to food spoilage are attributed to biofilms (Speranza & Corbo, 2017; Winkelströter, dos Reis Teixeira, Silva, Alves, & De Martinis, 2014). The capacity of biofilm-forming pathogen elevating resistence of microbial population to conventional cleaning and sanitation methods is intimidating (Galié, García-Gutiérrez, Miguélez, Villar, & Lombó, 2018). Further, a sophisticated and intricate network of cell-to-cell communication, for instance "quorum sensing (QS)" phenomenon (Parsek & Greenberg, 2005), also plays a key role in enhancing the virulence and biofilm-forming traits in foodborne bacterial pathogen. This type of communication helps the pathogenic biofilms to gain access to nutrients and favorable environmental conditions.

1.2 Biofilms in food industry

The food industry is particularly concerned about biofilm-forming species in food production environments because some of them are human pathogens. Food products and food contact surfaces (FCS) are the main sites for biofilm formation (Di Bonaventura et al., 2008; Han, Mizan, Jahid, & Ha, 2016; Joseph, Otta, Karunasagar, & Karunasagar, 2001). Apart from food products, biofilm-forming pathogens often grow in processing equipment, storage containers or vessels, and packaging materials (Chmielewski & Frank, 2003; Wong, 1998). Various artificial substrates commonly used in the food processing units are sites for biofilm-producing habitats for these pathogens. Stainless steel, polyethylene, wood, glass, polyethylene, and polypropylene are commonly collonized substrates by biofilm-forming bacteria in the food industry.

Considering biofilm-associated effects is especially vital in certain industries, such as dairy food manufacturing, where a variety of structural materials and milk composition act as a substrate for biofilm formation. Aside from inducing pathogenicity to humans or animals, microbial biofilms can also cause equipment surface corrosion and damage to the food processing units. For instance, in the dairy industries, tanks, pipelines, pasteurizers, and fermentation units are common sites that are often contaminated with microbial biofilms. Worldwide, milk and dairy products are considered one of the most promising sources of human nutrition. Therefore, spoilage of dairy products caused by pathogenic biofilms has been considered as critical microbiological problem (Galié et al., 2018; Ostrov, Sela, Belausov, Steinberg, & Shemesh, 2019). Milk-associated *Bacilli*, for example, adhere to surfaces of dairy equipment, circulate in pipelines, and deposit as sessile populations on milk filters (Ostrov, Harel, Bernstein, Steinberg, & Shemesh, 2016). Lately, certain *Bacilli* have been shown to form biofilm-like structures called biofilm bundles

Understanding Microbial Biofilms. https://doi.org/10.1016/B978-0-323-99977-9.00010-7

(Ben-Ishay et al., 2017; Pasvolsky, Zakin, Ostrova, & Shemesh, 2014). These bundles are thought to have undesirable effects on dairy products, causing spoilage or shortening shelf life. In addition, enzymes (such as lipases or proteases) secreted by the spoiling bacteria cause discoloration or off-flavor of dairy products. Various efforts are being made to produce safer, healthier, and high-quality dairy products by effectively eliminating the causative microbe (Shemesh & Ostrov, 2020). One of the most advanced antimicrobial strategies is the fortification of dairy products with metal ions to improve their microbial quality, which would be discussed in Section 3.1.

1.3 Biofilms onto food and food contact surfaces

Food safety problems and economic losses have been increasingly associated with biofilm formation throughout food production and processing. Perishable fresh produce can vastly become contaminated with pathogenic and spoilage microorganisms once they develop as biofilms. Food surfaces can be permanently contaminated when biofilms accumulate on them. Furthermore, biofilm-forming pathogens produce many enzymes and toxins that cause loss of quality and generate foul odor, off color, or a change in texture or flavor. Among the most common food-borne pathogens that affect human health include *Listeria monocytogenes*, *Salmonella* species, Enterohemorrhagic *E. coli*, *Staphylococcus aureus*, *Campylobacter jejuni*, and *Bacillus cereus*. All these pathogens may cause a variety of human illnesses when spread through food production chain, while the biofilm formation capabilities facilitate persistence of these pathogens within food producing environments, processes, and ready-to-eat food products (Fig. 1). Most importantly, biofilm mode of growth makes the pathogen resilient to most common food preservation tactics employed in food industries. For instance, most of the cleaning and sanitation processes are successful in eliminating the planktonic cells but are not so effective against biofilms. Biofilm bacteria often grow on hideout or dead-end sites, where the antimicrobial treatment barely reach to eliminate them. Meat, poultry, and fish products are often contaminated by biofilm-bacteria, such as of *E. coli*, *Pseudomonas* spp. *Salmonella* spp., *Vibrio* spp., or *L. monocytogenes*. Today's food processing industries and plants are concerned with the globalized production of food, and therefore, formation of bacterial biofilms represents growing threat of various FCS mainly due to cross-contamination processes. The bacteria in FCS are easily transferred to food during processing and packaging, a process that usually stays undetected. Bacterial biofilms are a common threat in various food-processing plants such as cheese processing, raw and cooked/fermented meats, raw and smoked fish, shellfish, shrimp, poultry and dairy products. Biofilms that form on FCS are very difficult to remove (Myszka & Czaczyk, 2011). They often result in persistent and endemic populations. Persistence of *Listeria* spp. is well known, and levels have been reported to range from 8 months to 10 years. This persistence in biofilms of individual species is due to the physical adaptation of attached cells and resistance to cleaning and disinfection measures (Ostrov et al., 2016). These species also have superior or enhanced adhesion to FCS and rapid growth within biofilms. certainly, biofilm-bacteria residing onto FCS may significantly shorten the shelf life of food products and might also lead to severe foodborne illness, which result in enormous economic losses.

2. Approaches to control pathogenic biofilms

Biofilms that form on abiotic surfaces such as FCS and industrial equipment may provide scaffold for other pathogens with access to food (Myszka & Czaczyk, 2011). Furthermore, surface-associated biofilms may consistently disseminate microbial resistance throughout food production chain (Giaouris & Simões, 2020). Chemical treatments with chlorine dioxide, sodium hydroxide, hydrogen peroxide, and paracetic acid are commonly used as measures to control biofilms on abiotic surfaces (Carrascosa, Raheem, Ramos, Saraiva, & Raposo, 2021). Most of these chemicals work by damaging and dissolving the biofilm structures. Other effective biofilm control agents include enzymatic disruption, the use of metal ions, biosurfactants, phytocompounds, and structurally similar synthetic analogs (Carrascosa et al., 2021). Details of biofilm inhibitory agents and their effects against selected foodborne pathogens are presented in Table 1. Although there are several strategies to control biofilms in the food industry, there is no effective measure that can be applied to food products. The control of biofilm formation in food products is rather challenging, as the process should be carried out without affecting the nutritional value of the food. A promising but less discussed approach is micobial biocontrol strategies to prevent establishment of pathogenic species within food. Furthermore, the use of synbiotics, a mixture of prebiotics (dietary fiber) and probiotics (beneficial bacteria) may also mitigate cross-contamination or *in vivo* colonization of pathogenic bacteria. The details and benefits of such approach are discussed in Section 2.2.

2.1 Metal ions and antibiofilm strategies

Prevention and elimination of biofilm-forming pathogens remain challenging in the dairy industry (Shemesh & Ostrov, 2020). Agents that prevent or mitigate biofilm formation have been being investigated as a measure to control biofilm-associated food spoilage. Recently, several metal ions have been investigated for their biofilm inhibitory

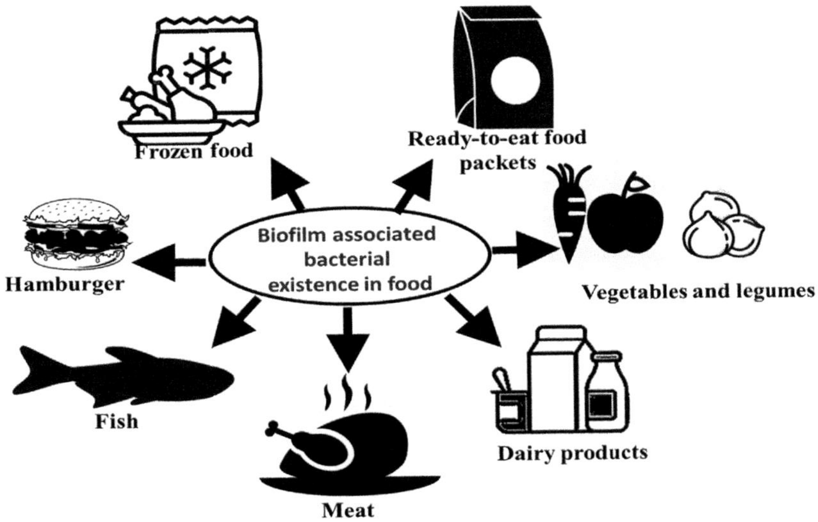

FIG. 1 Food varieties threatened by the biofilm-associated bacterial existence.

TABLE 1 Agents that exhibit biofilm inhibitory effects against food spoiling bacteria.

Phytochemicals or metal ions	Foodborne bacteria	Action	Reference
Mg^{2+} ions	*B. subtilis*	Biofilm inhibition	Oknin, Steinberg, and Shemesh (2015)
Zn^{2+} ions	*B. subtilis*	Inhibition of biofilm bundles	Hutchings, Rajasekharan, Reifen, and Shemesh (2020)
Zn^{2+} and Cu^{2+} ions	*B. subtilis*	Biofilm inhibition	García et al. (2020)
Ag^{+} ions	*S. typhimurium*	Antibacterial	Akhidime et al. (2019)
Metal coated surfaces (Silver, titanium, zinc, silicon, molybdenum, copper, iron)	*S. aureus* *E. coli* *L. monocytogenes*	Antibacterial	Akhidime et al. (2019)
Carvacrol, thymol, trans-cinnamaldehyde	*E. coli* O157:H7	Biofilm inhibition	Yuan and Yuk (2019)
Chalcone	*S. aureus*	Sortase A inhibitor	Zhang et al. (2017)
5-Iodoindole	*S. aureus* *E. coli* O157:H7	Biofilm inhibition	Lee, Kim, Gwon, Wood, and Lee (2016)
Chloroindoles	*V. parahaemolyticus*	Biofilm inhibition	Sathiyamoorthi, Faleye, Lee, Raj, and Lee (2021)
Small molecule inhibitors (T315)	*S. enterica*	Biofilm inhibition	Moshiri et al. (2018)
Resveratrol	*S. aureus* *L. monocytogenes* *E. coli* O157:H7 *C. jejuni*	Biofilm inhibition	Ma et al. (2018)
Sodium citrate	*E. coli* O157:H7 *S. aureus*	Biofilm inhibition	Liu et al. (2019)
Emodin	*S. aureus*	Biofilm inhibition	Yan et al. (2017)
Aleo-emodin	*S. aureus*	Biofilm inhibition	Xiang et al. (2017)

Continued

TABLE 1 Agents that exhibit biofilm inhibitory effects against food spoiling bacteria—cont'd

Phytochemicals or metal ions	Foodborne bacteria	Action	Reference
Green pepper essential oil	*P. aeruginosa* KM01	Antibacterial	Myszka et al. (2019), Ozogul, Kuley, Ucar, and Ozogul (2015)
Thyme and pine oil	*E. coli, S. paratyphi A, Y. enterocolitica, C. jejuni, E. faecalis, S. aureus*	Antibacterial	Ozogul et al. (2015)
Diphyllin	*Salmonella typhi*	Antibacterial, Biofilm inhibition	Thamburaj et al. (2020)
Cinnamic aldehyde	*S. aureus*	Biofilm inhibition	Liu et al. (2019)
Methylthioadenosine	*S. Typhimurium*	Antivirulent	
Small molecules F12 and F19	*S. aureus*	AgrA inhibitors	Greenberg et al. (2018)
β-Sitosterol glucoside	*E. coli* O157:H7	Biofilm inhibition	Vikram, Jayaprakasha, Uckoo, and Patil (2013)
Isolimonic acid	*E. coli* O157:H7	Biofilm inhibition	Vikram, Jesudhasan, Pillai, and Patil (2012)
ε-Polylysine	*S. aureus* *E. coli* *L. monocytogenes*	Antibacterial, Biofilm inhibition	Lee et al. (2021)

activities. Metal ions are essential for bacterial growth, although their excess can be harmful to the cells. Bacteria maintain tolerable metal ions by influx and efflux strategies based on their availability. Bacteria overcome metal limitations or respond to excess amounts, thus contributing to the maintenance of metal homeostasis in the cell. Metal ions such as silver, copper, titanium, gallium, nickel, aluminum, and zinc have been shown to destroy the biofilm and planktonic cells of foodborne pathogens such as *P. aeruginosa*, *S. aureus*, and *E. coli* (Gugala, Lemire, & Turner, 2017; Turner, 2017). Metal ions such as copper, gallium, and titanium also inhibited the preformed mature biofilms. Other divalent cations such as Mg^{2+} and Ca^{2+} as well as monovalent cations such as Na^{+} were studied against the biofilm formation of *B. subtilis*. Of all the metal ions studied, Mg^{2+} is becoming increasingly popular because of its use in dairy products (Demishtein, Reifen, & Shemesh, 2019). Microbes regularly require Mg^{2+} ions for their biological metabolism and function (Groisman et al., 2013). This divalent cation is the second most abundant cation in bacteria and is crucial for various cellular processes such as ribosome stabilization, DNA-protein interaction, replication, transcription, and RNA folding. In addition, Mg^{2+} is a vital ion for membrane stabilization and acts as a cofactor for various enzymatic reactions (Cromie, Shi, Latifi, & Groisman, 2006). Bacteria achieve Mg^{2+} homeostasis by regulating Mg^{2+} transporters and sensors that coordinate the influx and efflux of Mg^{2+} from the bacterial cell. The Gram-negative bacterium *Salmonella enterica serovar Typhimurium* is the best understood model for explaining Mg^{2+} homeostasis (Chandrangsu, Rensing, & Helmann, 2017). In *S. aureus*, Mg^{2+} has been shown to increase cell wall rigidity by binding to teichoic acids (García-Betancur et al., 2017); nonetheless, in *B. subtilis*, Mg^{2+} ions have been shown to repress extracellular matrix genes specified by the *epsH* and *tapA* operons (Oknin et al., 2015). An analogous study has also shown that enrichment of milk with Mg^{2+} ions inhibits biofilm bundles formation and may subsequently lead to the production of high-quality dairy products (Ben-Ishay et al., 2017). The observed results and reports suggest that Mg^{2+} has several molecular targets in bacterial cells that remain to be explored. Metal ions are also actively involved in regulating different physiological processes in bacterial cells (Table 1). For example, certain metal ions are known to inhibit electron transport chains in *B. subtilis* and *E. coli* by acting on cytochrome oxidase. Zn^{2+} was also shown to prevent biofilm formations of *S. epidermidis* and *S. aureus* (Conrady et al., 2008; Wu et al., 2013). Of late, Zn^{2+} ions were shown to mitigate the biofilm bundles of milk-associated pathogens (Hutchings et al., 2020). In addition, a combination of Zn^{2+} and Cu^{2+} was shown to inhibit *B. subtilis* biofilms (García et al., 2020). The Ag^{+} ions have been shown to decrease the activities of cytochrome C oxidase and succinate dehydrogenase, thus impairing respiratory functions in *E. coli* (Tambosi et al., 2018). Taken together, it is believed that metal ions play an important role in many biological processes, but excess metals can be toxic and detrimental (Chandrangsu et al., 2017). Elucidating the role of divalent metal ions in the inhibition of respiratory functions is another area that needs to be explored in detail.

2.2 Probiotics and postbiotics against pathogenic biofilms

An arm-race in the transmission and emergence of persistent pathogenic bacteria through the food chain emphasizes the need to explore alternative antimicrobial approaches. Therefore, there is an increasing interest in developing new strategies that use natural products to manage pathogen persistence on surfaces or equipment, especially in the food industry. One considerable approach to mitigate persistent pathogens could be using positive or beneficial biofilms formed by probiotic species. Probiotics are defined as live and beneficial microbes that may provide health benefits to humans or any other host organism (Williams, 2010). Probiotics play an essential role in restoring human gut flora and mitigating pathogenic bacteria (Gagliardi et al., 2018; Hemarajata & Versalovic, 2013; Szlufman & Shemesh, 2021). Most probiotics are known to secrete a wide range of metabolites or by-products (postbiotics) that have been shown to inhibit the growth of various bacterial and fungal pathogens (Barzegari et al., 2020). Moreover, the postbiotics are usually beneficial to the host as they could be used for various physiological processes. The most famous beneficial microbes that could potentially be used as probiotic supplements include *Bifidobacterium*, *Lactobacillus*, *Bacillus*, and *Saccharomyces* (Fijan, 2014). Other genera that have been shown to have probiotic potential include *Escherichia*, *Streptococcus*, and *Enterococcus*. The majority of probiotics are often associated with dairy products. Most of the probiotics were shown to inhibit the biofilms of prominent foodborne pathogens, which include *S. enterica*, *B. cereus*, *S. typhimurium*, *S. aureus*, *E. coli* O157:H7, and *C. jejuni* (Table 2). Probiotics also produce various metabolites that inhibit biofilms against various foodborne pathogens (Barzegari et al., 2020; Meroni et al., 2021). Exometabolites such as produced by several probiotic species have shown antibiofilm activity against foodborne pathogens such as *L. monocytogenes*, *S. aureus*, *P. aeruginosa*, and *S. typhimurium*. Certain antimicrobial peptides produced by probiotics, such as bacteriocins, subtilosin A, and sonorensin, showed significant antibiofilm potential against the above pathogens (Barzegari et al., 2020). For example, bacteriocins produced by lactic acid bacteria prevented the biofilm formation of *E. coli* and *S. typhimurium*. Subtilosin A synthesized by *B. subtilis* is effective against *L. monocytogenes* and its biofilms. Subtilosin A has also shown anti-QS activity against *E. coli* and *L. monocytogenes* (Algburi et al., 2017). Sonorensin, a bacteriocin produced by *B. sonorensis* MT93, has been shown to suppress biofilms of *S. aureus*, prevent adhesion of biofilm cells, and also act on mature biofilms (Chopra et al., 2015). The detailed list of probiotics, their derivatives, and their target is presented in Table 2.

3. Probiotic *Bacilli* supporting the health of host organism

Bacilli belong to a versatile class of bacteria that may provide abundant health benefits to the host organism (Cao et al., 2020; Cutting, 2011; Elshaghabee, Rokana, Gulhane, Sharma, & Panwar, 2017). *Bacilli* exhibit a tremendous ability to antagonize foodborne pathogens (Moore, Globa, Barbaree, Vodyanoy, & Sorokulova, 2013). Several species of *Bacilli* have been shown to produce a diverse spectrum of antimicrobial metabolites such as bacteriocins, iturinA, mycosubtilin surfactants, and plipastatin that account for its antagonistic activities against several bacterial and fungal pathogens (Urdaci & Pinchuk, 2004). Some of these metabolites are active against microbial biofilms and could mitigate the pathogens in human intestines. They also promote the growth of other probiotic bacteria (Piewngam et al., 2018; Ritter, Paula, Correa, Veras, & Brandelli, 2018). *Bacillus*-based probiotics are widely used worldwide, for instance they have been effectively used in poultry farming and other livestock productions (Mingmongkolchai & Panbangred, 2018). They have been shown to improve digestive health by minimizing human inflammatory response and augmenting intestinal barrier function (Cao et al., 2020; Rhayat et al., 2019). Probiotic *Bacilli* have also been involved in altering the microbiota in human GIT and supply health-promoting factors for the human or other hosts such as GABA and vitamins that elicit the host immune system to act against the pathogens. Other studies have shown that anticarcinogenic and antitumorigenic activities by suppressing the tumor inducing abilities of periodontal pathogens (Messora et al., 2016; Sourabh, Walia, & Kanwar, 2011). Specifically, it was shown that bacteriocins secreted by *Bacilli* play a critical role in suppressing cancer (Kaur & Kaur, 2015). In addition, the probiotic *Bacilli* also shown to repress and control influenza viral infectivity (Starosila, Rybalko, Varbanetz, Ivanskaya, & Sorokulova, 2017), which could be achieved by coordinated defense by the altered microbiota in the gut and also by secreting several antiviral substances. The coordination with the other microbiota is achieved by effective shielding or clothing of the bacterial community in the microbiota with the extracellular matrix produced by *Bacilli* (Yu et al., 2019). The matrix serves as a protective clothing against pathogens, osmotic stress, extreme cold or hot conditions, and desiccation stresses. That said, the probiotic candidature of *Bacilli* is rapidly escalating. A partial list of antagonistic *Bacilli* commercially available as probiotic supplements is provided in Table 3.

TABLE 2 Putative antagonistic action by probiotic bacteria against foodborne pathogens.

Probiotics	Models	Foodborne pathogens	Killing action	Reference
B. subtilis KATMIRA1933	In vitro model	*S. enterica*	Biofilm inhibition	Lin et al. (2021)
B. amyloliquefaciens B-1895	In vitro model	*S. enterica*	Biofilm inhibition	Lin et al. (2021)
Bacillus natto TK-1	In vitro model	*B. cereus* *S. typhimurium*	Antiadhesive effect, Lipopeptide production	Lin et al. (2021)
B. subtilis SN7	In vitro model	*S. aureus* *E. coli* O157:H7	Bacteriocin production	Lee and Chang (2017)
Bacillus species LM7	In vitro model	*B. cereus*	Bacillomycin D and surfactin production	Lee, Lee, et al. (2016)
B. subtilis SC-8	In vitro model	*B. cereus*	Surfactin, fengycin and iturin production	Yeo, Lee, and Hahm (2012)
B. subtilis NT-6	Shrimp	*Vibrio parahaemolyticus*	Amphipathic peptide AMPNT-6 production	Xu, Wang, Sun, Liu, and Li (2013)
B. subtilis SCK-2	In vitro model	*B. cereus*	Peptide production	Yeo, Lee, Cha, and Hahm (2011)
B. subtilis PS-216	In vitro model	*C. jejuni*	Bacillaene production and biofilm inhibition	Erega et al. (2021)
B. licheniformis B65-1	In vitro model	*S. aureus* *E. coli*	Phenylacetic acid and Lipopeptide production	Kim et al. (2004)
B. amyloliquefaciens EMD17	In vitro model	*B. cereus, L. monocytogenes*	Surfactin production	Lee, Shim, et al. (2016)
P. polymyxa NRRL B-30509	Chicken model	*C. jejuni*	Bacteriocin B602 production	Stern et al. (2005)
L. acidophilus LA 1	Intestinal epithelial cell	*E. coli* *Y. pseudotuberculosis* *S. typhimurium*	Prevention of cellular attachment and invasion	Bernet, Brassart, Neeser, and Servin (1994)
L. acidophilus A4	Intestinal epithelial cell	*E. coli* O157:H7	Activation of MUC2 mucin	Kim, Kim, Whang, Kim, and Oh (2008)
E. coli Nissle 1917	Gnotobiotic piglets	*S. enterica serovar Typhimurium*	Suppression of bacterial invasion	Altenhoefer et al. (2004)
E. coli Nissle 1917	In vitro model	*E. coli* (EHEC), *P. aeruginosa, S. aureus,* and *S. epidermidis*	Extracellular DegP activity	Fang, Jin, and Hong (2018)
Bifidobacteria species	Mice model	*E. coli* O157:H7	Antiinfectious	Valdés-Varela, Alonso-Guervos, García-Suárez, Gueimonde, and Ruas-Madiedo (2016)
B. longum and *B. breve*	Intestinal epithelial cell	*C. difficile*	Toxin neutralization	Valdés-Varela et al. (2016)

TABLE 2 Putative antagonistic action by probiotic bacteria against foodborne pathogens—cont'd

Probiotics	Models	Foodborne pathogens	Killing action	Reference
L. acidophilus La-5	Intestinal epithelial cell	*E. coli* O157:H7	QS signal inhibitor	Medellin-Peña, Wang, Johnson, Anand, and Griffiths (2007)
L. rhamnosus and *L. plantarum*	In vitro model	*P. aeruginosa, L. monocytogenes*	Antibacterial	Rezaei, Khanzadi, and Salari (2021)
P. pentosaceus E. faecium	In vitro model	*B. cereus, S. enterica, E. coli*	Positive biofilm, Nisin Production	Lin et al. (2021)
E. faecium NRRL B-30746	Broilers	*C. jejuni, S. enteritidis*	Bacteriocin E 50-52 production	Svetoch et al. (2008)
L. salivarius NRRL B-30514	Turkey	*C. jejuni*	Bacteriocin OR7 production	Cole et al. (2006)
E. faecium CCM 7420	Rabbit	Coagulase positive *Staphylococcus*	Bacteriocin PPB CCM 7420 production	Gopikrishna, Kumar, Perumal, and Elangovan (2021), Simonová et al. (2009)
E. faecium EK13	Japanese quails	*Salmonella*	Enterocin A production	Marciňáková, Strompfová, Boldižárová, Lauková, and Gancarčíková (2004)
Enterococcus sp. NRRL B-30745	Broilers	*C. jejuni*	E-760 production	Line et al. (2008)
B. subtilis	In vitro model	*L. monocytogenes*	Subtilosin A production	Algburi et al. (2017), Van Kuijk, Noll, and Chikindas (2012)
B. sonorensis MT93	In vitro model	*S. aureus*	Sonorensin production	Chopra, Singh, Jena, and Sahoo (2015)

TABLE 3 List of commercial probiotic supplement containing sporulating *Bacilli*.

Probiotic *Bacilli*	Commercial name and country
B. cereus	Bactisubtil (France), Cassella-Med (Germany) Biovicerin (Brazil), (Mekophar, Vietnam)
B. subtilis	LifeinU (Europe), Anaban (Europe), Bio-Kult (Protexin, UK), Lactipan Plus (Italy), Bibactyl (Vietnam), Biosubtyl DL (Vietnam), Bidisubtilis (Vietnam), Pastylbio (Vietnam), Medilac (Korea), Nature's First Food (US), Neolactoflorene (Italy), Primal Defense (US)
B. coagulans 15B and fructooligosaccharide	Nutrition essentials Probiotic (US)
B. subtilis, L. rhamnosus, L. casei, B. longum, L. acidophilus, L. plantarum, B. breve	Vital Probiotics (US)
B. indicus, B. subtilis, B. coagulans	MegaSporeBiotic (UK)
B. subtilis, B. coagulans	NutriCommit (US)
B. coagulans, S. boulardii	Flora3 (US)
B. coagulans	GanedenBC30 (US), THORNE (US) Sunny Green (US), Cleansing Green (US) Biosporin (Europe), Biofarma (Ukraine), Garars (Russia)
B. licheniformis and *B. subtilis*	
B. licheniformis and *B. subtilis*	BioPlus 2B (Denmark)

Continued

TABLE 3 List of commercial probiotic supplement containing sporulating *Bacilli*—cont'd

Probiotic *Bacilli*	Commercial name and country
B. polyfermenticus SCD	Bispan (South Korea)
B. subtilis, Lactobacillus, Bifidobacterium, and *S.* strains	Bio-Kult (UK)
B. clausii	Domuvar (Italy), Enterogermina (US), Lactopure (India), Lactospore (US), Neolactoflorene (Italy), Sustenex (US)
B. pumilus	Biosubtyl (Biophar Company, Vietnam)

3.1 *Bacilli* biofilms against pathogen biofilms

Bacilli are known for their stupendous antagonistic potentials (Urdaci & Pinchuk, 2004). Several different strains of *Bacilli* were found to be effective against food-borne pathogens (Table 2). *Bacilli* achieve pathogen elimination either through preventing biofilm formation by the pathogen or by producing secondary metabolites such as surfactin, fengycin, and iturin that elicit antimicrobial activities. Different *Bacilli* are readily available as a supplement to counteract food-borne pathogens. For example, one *B. subtilis* strain (PS-216) has displayed a tremendous ability to suppress *C. jejuni* growth, biofilm formation and colonization in poultry broilers, while augmenting positive benefits to the broilers such as their enhanced weight gain (Erega et al., 2021). *B. subtilis* also has remarkable ability to grow as biofilm (Shemesh & Chai, 2013). Lately, it was shown in in vitro conditions that the biofilms formed by *B. subtilis* protect *Lactobacilli*, another probiotic *Bacilli*. The exopolysaccharides produced by *B. subtilis* could provide the protective clothing for the *Lactobacilli* during dual-species growth (Kimelman & Shemesh, 2019). This strategy rescues *Lactobacilli* and confers enhanced survivability of *Lactobacilli* during exposure to desiccation stresses. *Lactobacilli* are prominent probiotics in food and nutrition. Therefore, they have tremendous potential to compete with pathogenic biofilms. Moreover, *Lactobacilli* are considered a barrier against pathogenic biofilms formed by food-spoiling bacteria. *Lactobacilli* can colonize surfaces and secrete postbiotics and other volatile molecules (Rajasekharan & Shemesh, 2022a). These secretions play a vital role in eliciting antibiofilm, bacteriostatic and fungistatic activities against various foodborne bacterial and fungal pathogens. Several species of *Lactobacilli* found in Brazilian foods were shown to arrest biofilms of *L. monocytogenes*, Enterohemorrhagic *E. coli* O157:H7, and *Salmonella enterica* subsp. *enterica serovar Typhimurium*. Several species of *Lactobacilli* were also demonstrated to secrete bacteriocins that effectively control *S. aureus*, *L. monocytogenes*, *P. aeruginosa*, *S. mutans*, and *E. faecalis* biofilm formation (Gómez et al., 2016). They are also effective in disrupting preformed mature biofilms. Some of the bacteriocins, for example, plantaricin A, a class IIb bacteriocin, secreted by *Lactobacilli* strains (*L. sanfranciscensis* DPMMA174 and *L. plantarum* DPPMA20) were shown to induce biofilm formation of other *L. plantarum* strain DC400 (Calasso et al., 2013). Such inducers can enrich the probiotic biofilms in the environment, thus providing a strong defense against pathogen invasion. In addition, most *Lactobacilli* have a stellar ability to autoaggregate within themselves (Rajasekharan & Shemesh, 2022b) as well as to coaggregate with other species. The mixed species coaggregation is a competition-exclusion approach the *Lactobacilli* implement to supersede the pathogen. By coaggregating the *Lactobacilli* successfully prevents the attachment of pathogens to surfaces. Lately, it has been shown that *Lactobacilli* biofilms successfully reduced *L. monocytogenes* in ready-to-eat poultry processing plants on abiotic surfaces. Overall, the approach using probiotic biofilms to counteract pathogenic biofilm is an emerging and promising strategy to fight foodborne pathogens. The approach relies on the competitive exclusion principle where probiotics compete with the pathogen and exclude them from either food or abiotic surfaces.

3.2 *Bacillus subtilis*, a potential probiotic model

B. subtilis is a beneficial Gram-positive, spore-forming bacterium that is being widely studied for use as a future probiotic species. This bacterium is found throughout the soil, GIT of ruminants/humans, and in various food processing facilities. *B. subtilis* exhibits diverse biofilm-forming abilities. Under laboratory conditions, *B. subtilis* forms colony-like biofilms at air-agar interfaces, pellicles at air-liquid interfaces, and biofilm bundles in liquid media (Rajasekharan et al., 2021). In addition, the bacterium is known to colonize plants, *C. elegans*, and resistant chickpea starch fibers (Beauregard, Chai, Vlamakis, Losick, & Kolter, 2013; Donato et al., 2017; Rajasekharan et al., 2021). Biofilm formation and colonization by *B. subtilis* are controlled by the Spo0A-SinI pathway under the influence of an external trigger(s) or environmental stimulus that activates histidine kinases (Cairns, Hobley, & Stanley-Wall, 2014). The perception of environmental signals is carried out by several histidine kinases. These kinases are responsible for activating the global master regulator Spo0A through phosphorylation process. The phosphorylated Spo0A triggers biofilm formation through a complex mechanism involving the SinR and AbrB regulators, which together repress biofilm matrix genes. Briefly, biofilm formation in *B. subtilis* is activated when a specific subset of founder cells activates the transcription of matrix operons (*epsA-O* and *tapA-sipW-tasA* operons) involved in extracellular matrix synthesis. A series of self-regulating histidine kinases (KinA-KinE) is activated by the external signals, which subsequently activates the i Spo0A through phosphorylation process. Upon reaching the threshold for phosphorylated Spo0A (Spo0A~P), SinI is actively transcribed, which blocks SinR and eventually leads to the de-repression or activation of the matrix operons. DegS-DegU, ComA-ComP, and LuxS are other pathways that regulate biofilms in an independent manner (Comella, 2004; Lombardía, Rovetto, Arabolaza, & Grau, 2006). In addition, *B. subtilis* produces other extracellular polymeric substances, such as poly-D-glutamine (PGA), which promotes surface attachment and development of submerged biofilms (Chettri, Bhutia, & Tamang, 2016). Moreover, the hydrophobic BslA protein has been shown to be abundant in biofilms of *B. subtilis*. It has been observed that BslA proteins are located on the surface of biofilms and confer strong hydrophobicity properties. The biofilm-forming potential of *B. subtilis* makes it an ideal candidate for widespread application in the food industry. One such application is to use *B. subtilis* biofilms adhered to edible dietary fibers, thus creating a probiotic-prebiotic complex (discussed in Section 4).

4. Probiotic-prebiotic complex

Prebiotics are dietary components , which may assist as a scaffold or substrate for bacterial or yeast colonization (Amoah, Rajasekharan, Reifen, & Shemesh, 2021; Rajasekharan et al., 2021). For instance, the nondigestible fiber portion of the food components that virtually pass intact through the GIT and stimulate the growth of specific gut microbiota (Markowiak & Śliżewska, 2017; Pandey, Naik, & Vakil, 2015). If pathogenic microbe colonizes food components, the effects may be detrimental to the food and the consumers. However, once the probiotic species colonize the dietary fibers, or when probiotics are artificially engineered to colonize the prebiotics, the outcome (synbiotic complex) becomes rewarding for the host's health and well-being (Olas, 2020; Pandey et al., 2015). The complex may address two major purposes: (1) fast and safe movement of probiotics through gastrointestinal tract, thus withstanding the altered pH conditions in GIT that are not favorable for the probiotics, and (2) once the probiotics reach the colon and initiate coordination with the gut flora, the residual prebiotics serve as a substrate for positive bacteria in the gut. These bacteria may degrade the prebiotic substances and produce by-products such as short-chain fatty acids (SCFAs that have tremendous antibacterial, antifungal, and antiviral activities. Also, they modulate the host immune system to fight pathogens. For instance, substances released from the degradation of certain prebiotics assist in fighting pathogenic bacteria by SCFA production (discussed in detail in Section 4.1).

4.1 Prebiotic degradation and SCFAs production as a measure to control pathogenic biofilms

Degradation of prebiotics (dietary fibers) generally results in the production of SCFAs, which have diverse biological functionalities (Davani-Davari et al., 2019; Markowiak-Kopeć & Śliżewska, 2020; McLoughlin, Berthon, Jensen, Baines, & Wood, 2017). The breakdown of SCFA yields by-products that benefitially affects the human . They regulate gut homeostasis, elicit an immune response by triggering anti-inflammatory cytokines, and are resource energy to the intestinal epithelial cells (Markowiak-Kopeć & Śliżewska, 2020). Butyrate, acetate, and propionate are the fundamental SCFAs produced by gut microbiota by the microbial degradation of fibers. Furthermore, SCFAs have the potential to inhibit several bacterial biofilms (Lamas, Regal, Vázquez, Cepeda, & Franco, 2019; Nakamura et al., 2020). They accomplish this by altering the bacterial membrane potential and by lowering the intracellular pH (Sun & O'Riordan, 2013). SCFAs also serve as a shield for the beneficial gut bacteria to defend or resist the invasion of several foodborne enteric pathogens. However, the biological activity of several SCFAs against enteric pathogens is nonlinear and concentration-dependent. For instance, it was shown that the composition of SCFAs in the distal ileum induced *Salmonella* virulence, while those in the large intestine suppressed it (Lawhon, Maurer, Suyemoto, & Altier, 2002). Lately, several studies have been conducted to understand the importance of SCFAs and the mode of action against several foodborne pathogens. Other benefits of SCFA are that they may serve as molecular signals, thus providing the ambient environment for establishment of the probiotic *Bacilli* community (Amoah et al., 2021; Pasvolsky et al., 2014). Due to their diverse traits, SCFAs are potentially additive to functional food and/or feeds. Fig. 2 depicts how SCFA are produced from a starch fiber and its putative beneficial roles to human.

4.2 SCFAs and histone deacetylase (HDAC) inhibition as a biofilm inhibition strategy

SCFAs can tremendously inhibit undesired bacterial and fungal biofilms (Machado, Sencio, & Trottein, 2021;

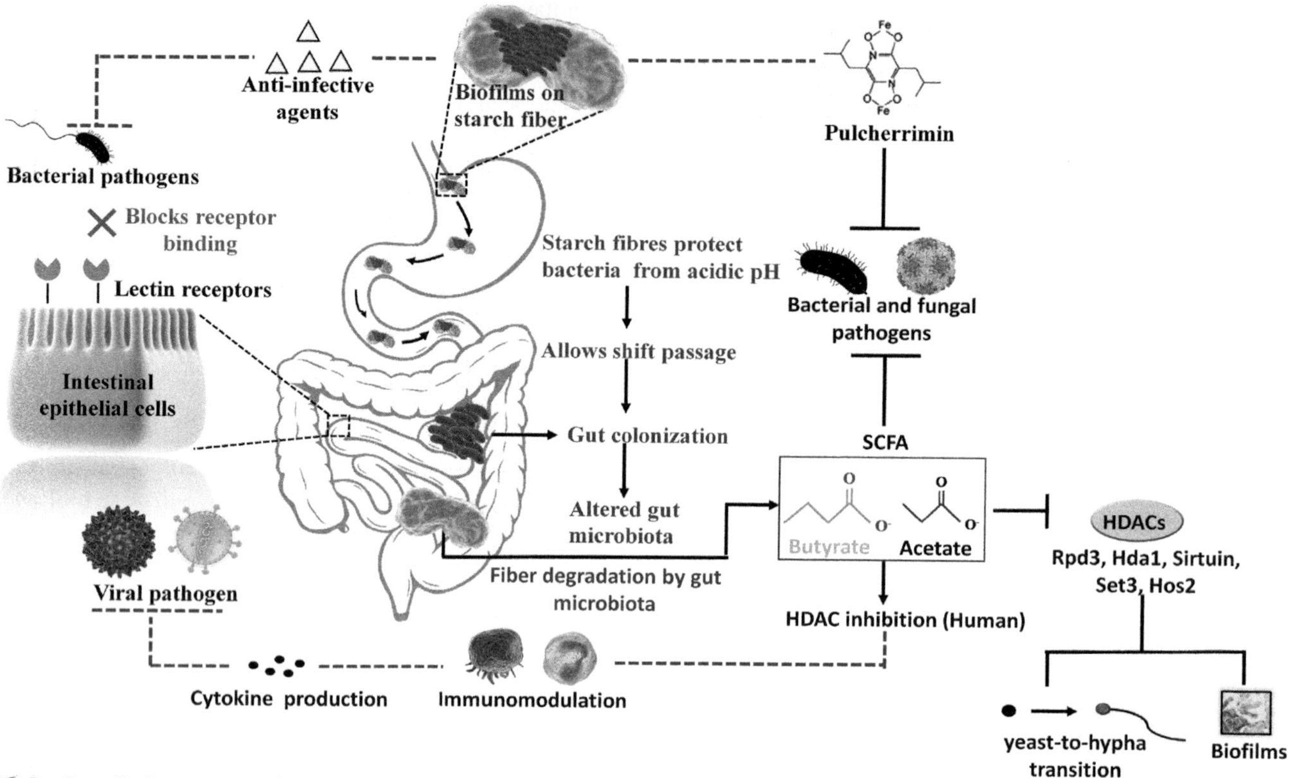

FIG. 2 Overall scheme representing how a probiotic-prebiotic complex (bacterial biofilms on starch fibers) survives and transits the gastrointestinal tract, reaches the gut, where the bacteria dislodge and colonize resulting in altered gut microbiota that may benefit the host. The residual fibers are degraded by the gut microbiota leading to the production of short-chain fatty acids (CSFAs) that exhibit diverse biological activities. SCFAs possess antibacterial, antifungal, and immunomodulatory activities leading to antiviral activities. SCFAs may also act as HDAC inhibitors, which control biofilm formations and yeast-to-hyphal switch in pathogenic yeasts

Nguyen, Lopes, Cordero, & Nosanchuk, 2011; Rajasekharan, Ramesh, & Bakkiyaraj, 2015). Analyzing the antagonistic effect of SCFAs against fungal pathogens has been challenging for researchers. Some of them functioned by regulating the cell's epigenetic machinery and not just at genomic or proteomic level. Lately, SCFAs such as sodium butyrate appeared to be potent inhibitors that may elicit stellar antagonistic effects against pathogenic yeast species (Nguyen et al., 2011). It was also shown to potentiate the activity of other antimicrobial agents or phytocompounds against pathogenic yeasts when used in combination. In fact, sodium butyrate is regarded as a eukaryotic histone deacetylase inhibitor (HDACi) that has remarkable potential to stall the functioning of both human and/or fungal histone deacetylase (HDACs) (Xu, Parmigiani, & Marks, 2007). Succinctly, histone acetylation and deacetylation are the epigenetic switching machinery that governs the expression, repression, and de-repression of several key genes. The lysine residues on H3 and H4 histone tails are the primary targets for acetylation. Histone deacetylases (HDACs) are the enzymes that remove the acetyl groups from these lysine residues (Xu et al., 2007). Removal of acetyl groups from lysine residues increases the electrostatic attraction between the positively charged histones and negatively charged DNA resulting in closed chromatin or heterochromatin structure. This structure is compressed and recalcitrant to transcription factors that need access to the DNA. Nevertheless, the process can be reversed by the use of histone deacetylase inhibitors (HDACi), which inhibit HDACs and facilitate euchromatin structure or transcriptional activation. HDACi is mechanism-based inhibitors, which have a long history of use in psychiatry and neurology as mood stabilizers, antiepileptics, anti-cancer, and anti-inflammatory agents. That said, SB is one such SCFA that functions as an HDACi in a fungal context. Synchronously, HDAC inhibition in humans by SCFAs results in positive health benefit (Xu et al., 2007). In humans, SB is known to uplift the digestive health, trigger the production of cytokines that fight bacteria and/or viruses, and is regarded as an effective antibiofilm drug. SIn humans, SB is known to uplift digestive health, trigger the production of cytokines that fight bacteria and/or viruses, and is regarded as an effective antibiofilm drug. Some HDAC inhibitors, such as Sulforaphane (SFN), are readily available in cruciferous vegetables, such as broccoli and broccoli sprouts (Dashwood & Ho, 2007). They primarily account for dietary HDAC inhibitors.

5. Promising functional foods

Legumes and grains in combination with probiotics represent promising synbiotic food systems. Legume-based milk products are rich in dietary fibers (Sethi, Tyagi, & Anurag, 2016); thus, they are ideal prebiotics. In addition, this plant-based milk has gained immense popularity among consumers. It has traditional values and is a famous cuisine in India, the Middle East, and the Mediterranean (Gupta et al., 2017; Rajasekharan et al., 2021). As such, they come with a rich nutritional package. For instance, chickpea (Cicer arietinum) legumes are rich in minerals, proteins, carbohydrates, and dietary fibers, including resistance starch, cellulose, hemicellulose, and lignin (Jukanti, Gaur, Gowda, & Chibbar, 2012). Using fiber-rich chickpea-based foods such as CPM also generates SCFA to fight infection (Fig. 2). It has also been shown that *B. subtilis* effectively utilizes CPM to establish probiotic biofilms (Rajasekharan et al., 2021). In CPM-based media, *B. subtilis* formed different biofilm phenotypes including the colony-type biofilms on air-agar interfaces, pellicles, biofilm bundles in liquid cultures. In addition, they displayed incredible potential to colonize the insoluble, autofluorescent resistant starch fibers. These biofilms on fibers further provided the bacteria the ability to survive under hostile conditions such as pasteurization and in vitro digestion. The additional exquisite discovery was the production of a reddish pink pigment later identified as pulcherrimin (will be discussed in Section 7). As depicted in Fig. 3 and from the earlier work (Rajasekharan et al., 2021), colonization of probiotic *B. subtilis* on starch fibers of CPM provides all the essential parameters to mobilize the probiotic and the diary fiber through the GIT to the gut, where the former can colonize. The latter may also serve as the feed for other beneficial gut bacteria.

6. Pulcherrimin as an antibiofilm agent

Pulcherrimin (IUPAC name: 1,4-dihydroxy-3,6-bis(2-methylpropyl) pyrazine-1,4-diium-2,5-diol; iron) is a reddish-pink pigment produced by several species of plant-pathogenic yeasts and other microbial species (*Bacillus* and *Streptomyces*) (Kántor, Hutková, Petrová, Hleba, & Kačániová, 2021). The pigment appears to be an iron chelator, which is formed when pulcherriminic acid, a cyclic dipeptide produced by certain microbes, chelates the ferric ions (Fe^{3+}). The pigment is recovered by growing microbial cells in Fe^{3+} containing growth media. The yeast *Metschnikowia pulcherrima* was first shown to synthesis this pigment (Pawlikowska, Kolesińska, Nowacka, & Kregiel, 2020), which were shown to possess strong antibacterial activities (Horváth et al., 2021; Oro, Ciani, & Comitini, 2014). The mutant defected in the pigment synthesis revealed almost no antagonistic abilities, clearly demonstrating the importance of the pigment in the observed antibacterial properties. Other yeasts that produce the pigment include the genera *Lipomyces*, *Kluyveromyces*, and *Dipodascopsis*. The pigment also possesses vigorous antifungal activities, especially against phytopathogenic molds (Sipiczki, 2020). It was first identified that the METSCH-Snf2 coding gene could be instrumental in antagonism. Later, the pathway regulating this pigment synthesis was identified. Succinctly, the pigment is produced by the enzymatic action of two catalytic enzymes, YvmC (Cyclo (L-leucyl-L-leucyl) synthase) and CypX (Pulcherrimin acid synthase) that convert two tRNA molecules of leucine to pulcherrimin acid (PA) in a two-step enzymatic process. PA is a yellow-colored pigment with a solid affinity for binding Fe3 +. PA is secreted extracellularly and binds with the iron in the surrounding environment to produce pulcherrimin. The pathway was confirmed by generating YvmC and CypX mutants that failed to produce the pigment in Fe^{3+}-rich media. The plausible mechanism of its antagonism could be linked to iron depletion the environments where the yeast grows, thus potentially limiting the access of iron acquisition by other microbes that require the metal for proper coordination and functioning of diverse cellular process (Horváth et al., 2021). Nonetheless, a comprehensive understanding of the antimicrobial mode of action remains to be accomplished.

Besides, since microbial cells that secrete pulcherrimin are considered ideal antagonists, the pigment is viewed as a prospective antimicrobial agent. Several studies have attempted to enhance the productivity or secretion of this pigment in culture media using various enhancers such as starch and glycerol (Li et al., 2017). Pigment production is of significant interest to biotechnologists as they have myriad functions. These include using fertilizers in agricultural fields, photoprotection against UV-induced cellular damage, and biofilm inhibitors in medicinal sectors (Li et al., 2017). Hence, microbes that produce large quantities of pulcherrimin greatly interest agricultural and medicinal biotechnology. *B. subtilis* displayed the ability to produce pulcherrimin on hard agars supplemented with Fe^{3+}. The pigment production was linked to biofilm formation in *B. subtilis* and was found to be a self-regulated to prevent biofilm outgrowth (Arnaouteli et al., 2019). Lately, the synthesis of pulcherrimin in chickpea milk (CPM) was revealed (Rajasekharan et al., 2021). CPM is a rich source of starch and iron Fe^{3+} and the growth of *B. subtilis* in CPM triggers the secretion of this pigment. It is hypothesized that the pigment may be instrumental in preventing food-spoiling pathogen growth in chickpea-based food products spiked with probiotic *B. subtilis*. Fig. 3 depicts a conceptual model to develop chickpea-based synbiotic food, for instance, plant-based hamburgers. Briefly, *B. subtilis* would blend with chickpea fibers in the food, thus producing the pigment

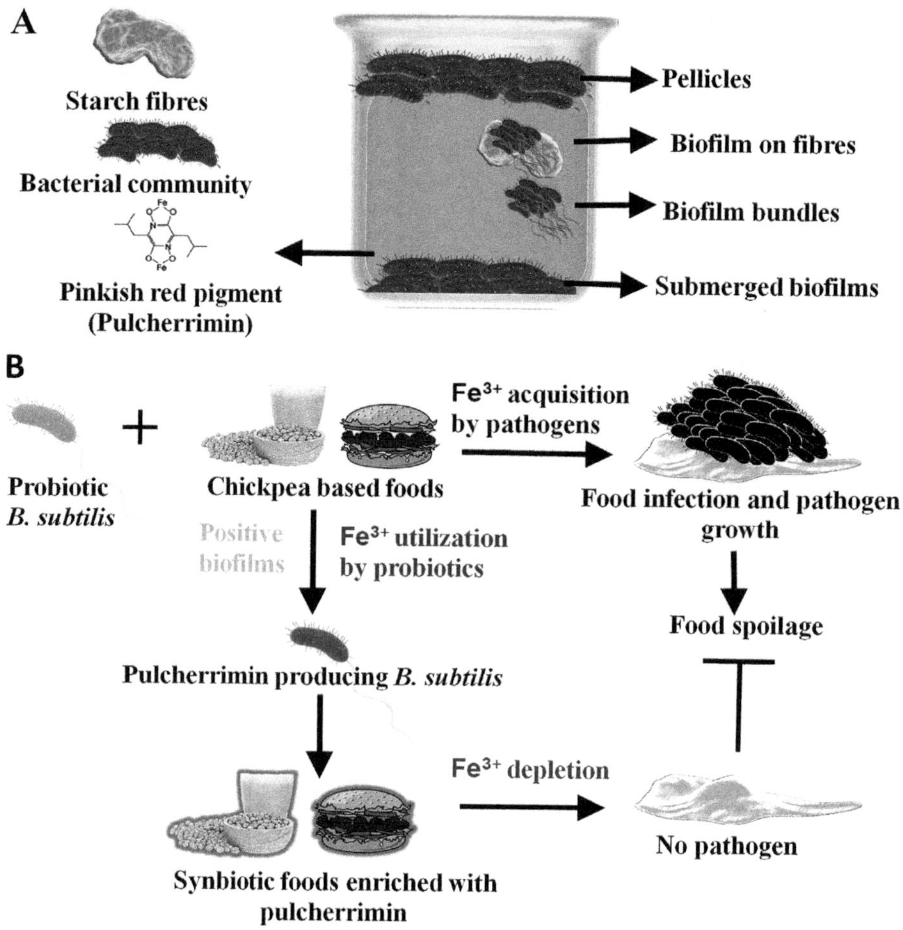

FIG. 3 (A) Different modes of biofilm formation by *B. subtilis* in chickpea milk (CPM) and (B) proposed scheme for developing *B. subtilis*-based synbiotic food.

by utilizing Fe^{3+} and starch from chickpea fibers. This iron acquisition by *B. subtilis* will lead to iron depletion, thus cutting down the iron supplement to the food-spoiling pathogens. In addition, it is believed that the pigment will play a significant role in preventing food collonisation of pathogens.

7. Concluding remarks and future prospects

This chapter shows how synbiotic foods might have a wide range of health benefits for humans. Probiotic microbiota is regarded as assisting in the digestion of dietary fibers by regulating the appropriate functioning of the innate and adaptive immune systems. Probiotic species, primarily *B. subtilis* could be used in clinical settings to boost the host's ability to defend itself against infectious diseases. Since tprobiotic *Bacilli* can effectively form biofilms on dietary fibers and secretes antimicrobial substances, they may play a significant role in health promotig processes of the host organism. Hence, probiotic *Bacilli* can be processed to promote human health in combinations with dietary fibers (which serve as scaffolds for bacteria to colonize). Fibers can protect probiotic bacteria by encapsulating them against various environmental and physical stresses that can typically kill the free-living bacteria. In addition, these fibers would serve as a prebiotic substance for the probiotic cells. The novel platform can be used for many applications, for instance, in developing functional and synbiotic foods.

References

Akhidime, I. D., Saubade, F., Benson, P. S., Butler, J. A., Olivier, S., Kelly, P., et al. (2019). The antimicrobial effect of metal substrates on food pathogens. *Food and Bioproducts Processing*, *113*, 68–76.

Algburi, A., Zehm, S., Netrebov, V., Bren, A. B., Chistyakov, V., & Chikindas, M. L. (2017). Subtilosin prevents biofilm formation by inhibiting bacterial quorum sensing. *Probiotics and Antimicrobial Proteins*, *9*, 81–90.

Altenhoefer, A., Oswald, S., Sonnenborn, U., Enders, C., Schulze, J., Hacker, J., et al. (2004). The probiotic *Escherichia coli* strain Nissle 1917 interferes with invasion of human intestinal epithelial cells by different enteroinvasive bacterial pathogens. *FEMS Immunology and Medical Microbiology*, *40*, 223–229.

Amoah, Y. S., Rajasekharan, S. K., Reifen, R., & Shemesh, M. (2021). Chickpea-derived prebiotic substances trigger biofilm formation by *Bacillus subtilis*. *Nutrients*, *13*, 4228.

Arnaouteli, S., Matoz-Fernandez, D., Porter, M., Kalamara, M., Abbott, J., MacPhee, C. E., et al. (2019). Pulcherrimin formation controls growth arrest of the *Bacillus subtilis* biofilm. *Proceedings of the National Academy of Sciences*, *116*, 13553–13562.

Barzegari, A., Kheyrolahzadeh, K., Khatibi, S. M. H., Sharifi, S., Memar, M. Y., & Vahed, S. Z. (2020). The battle of probiotics and their derivatives against biofilms. *Infection and Drug Resistance*, *13*, 659.

Beauregard, P. B., Chai, Y., Vlamakis, H., Losick, R., & Kolter, R. (2013). *Bacillus subtilis* biofilm induction by plant polysaccharides. *Proceedings of the National Academy of Sciences*, *110*, E1621–E1630.

Ben-Ishay, N., Oknin, H., Steinberg, D., Berkovich, Z., Reifen, R., & Shemesh, M. (2017). Enrichment of milk with magnesium provides healthier and safer dairy products. *npj Biofilms and Microbiomes*, *3*, 1–7.

Bernet, M.-F., Brassart, D., Neeser, J.-R., & Servin, A. (1994). *Lactobacillus acidophilus* LA 1 binds to cultured human intestinal cell lines and inhibits cell attachment and cell invasion by enterovirulent bacteria. *Gut*, *35*, 483–489.

Cairns, L. S., Hobley, L., & Stanley-Wall, N. R. (2014). Biofilm formation by *Bacillus subtilis*: New insights into regulatory strategies and assembly mechanisms. *Molecular Microbiology*, *93*, 587–598.

Calasso, M., et al. (2013). Effects of the peptide pheromone plantaricin A and cocultivation with Lactobacillus sanfranciscensis DPPMA174 on the exoproteome and the adhesion capacity of Lactobacillus plantarum DC400. *Applied and Environmental Microbiology*, *79*, 2657. https://doi.org/10.1128/AEM.03625-12.

Cao, J., Yu, Z., Liu, W., Zhao, J., Zhang, H., Zhai, Q., et al. (2020). Probiotic characteristics of *Bacillus coagulans* and associated implications for human health and diseases. *Journal of Functional Foods*, *64*, 103643.

Carrascosa, C., Raheem, D., Ramos, F., Saraiva, A., & Raposo, A. (2021). Microbial biofilms in the food industry—A comprehensive review. *International Journal of Environmental Research and Public Health*, *18*, 2014.

Chandrangsu, P., Rensing, C., & Helmann, J. D. (2017). Metal homeostasis and resistance in bacteria. *Nature Reviews Microbiology*, *15*, 338–350.

Chettri, R., Bhutia, M. O., & Tamang, J. P. (2016). Poly-γ-glutamic acid (PGA)-producing *Bacillus* species isolated from Kinema, Indian fermented soybean food. *Frontiers in Microbiology*, *7*, 971.

Chmielewski, R., & Frank, J. (2003). Biofilm formation and control in food processing facilities. *Comprehensive Reviews in Food Science and Food Safety*, *2*, 22–32.

Chopra, L., Singh, G., Jena, K. K., & Sahoo, D. K. (2015). Sonorensin: A new bacteriocin with potential of an anti-biofilm agent and a food biopreservative. *Scientific Reports*, *5*, 1–13.

Cole, K., Farnell, M., Donoghue, A., Stern, N., Svetoch, E., Eruslanov, B., et al. (2006). Bacteriocins reduce Campylobacter colonization and alter gut morphology in turkey poults. *Poultry Science*, *85*, 1570–1575.

Comella, N. (2004). *Characterization of a transcriptional response to population density in Bacillus subtilis*. Massachusetts Institute of Technology.

Conrady, D. G., Brescia, C. C., Horii, K., Weiss, A. A., Hassett, D. J., & Herr, A. B. (2008). A zinc-dependent adhesion module is responsible for intercellular adhesion in staphylococcal biofilms. *Proceedings of the National Academy of Sciences*, *105*, 19456–19461.

Cromie, M. J., Shi, Y., Latifi, T., & Groisman, E. A. (2006). An RNA sensor for intracellular Mg2+. *Cell*, *125*, 71–84.

Cutting, S. M. (2011). *Bacillus* probiotics. *Food Microbiology*, *28*, 214–220.

Dashwood, R. H., & Ho, E. (2007). Dietary histone deacetylase inhibitors: From cells to mice to man. In *Seminars in cancer biology* (pp. 363–369). Elsevier.

Davani-Davari, D., Negahdaripour, M., Karimzadeh, I., Seifan, M., Mohkam, M., Masoumi, S. J., et al. (2019). Prebiotics: Definition, types, sources, mechanisms, and clinical applications. *Foods*, *8*, 92.

Demishtein, K., Reifen, R., & Shemesh, M. (2019). Antimicrobial properties of magnesium open opportunities to develop healthier food. *Nutrients*, *11*, 2363.

Di Bonaventura, G., Piccolomini, R., Paludi, D., D'orio, V., Vergara, A., Conter, M., et al. (2008). Influence of temperature on biofilm formation by *Listeria monocytogenes* on various food-contact surfaces: Relationship with motility and cell surface hydrophobicity. *Journal of Applied Microbiology*, *104*, 1552–1561.

Donato, V., Ayala, F. R., Cogliati, S., Bauman, C., Costa, J. G., Lenini, C., et al. (2017). *Bacillus subtilis* biofilm extends *Caenorhabditis elegans* longevity through downregulation of the insulin-like signalling pathway. *Nature Communications*, *8*, 1–15.

Donlan, R. M. (2002). Biofilms: Microbial life on surfaces. *Emerging Infectious Diseases*, *8*, 881.

Elshaghabee, F. M., Rokana, N., Gulhane, R. D., Sharma, C., & Panwar, H. (2017). *Bacillus* as potential probiotics: Status, concerns, and future perspectives. *Frontiers in Microbiology*, *8*, 1490.

Erega, A., Stefanic, P., Dogsa, I., Danevčič, T., Simunovic, K., Klančnik, A., et al. (2021). Bacillaene mediates the inhibitory effect of *Bacillus subtilis* on *Campylobacter jejuni* biofilms. *Applied and Environmental Microbiology*, *87*. e02955-02920.

Fang, K., Jin, X., & Hong, S. H. (2018). Probiotic *Escherichia coli* inhibits biofilm formation of pathogenic *E. coli* via extracellular activity of DegP. *Scientific Reports*, *8*, 1–12.

Fijan, S. (2014). Microorganisms with claimed probiotic properties: An overview of recent literature. *International Journal of Environmental Research and Public Health*, *11*, 4745–4767.

Flemming, H. C., Wingender, J., Szewzyk, U., Steinberg, P., Rice, S. A., & Kjelleberg, S. (2016). Biofilms: An emergent form of bacterial life. *Nature Reviews Microbiology*, *14*, 563–575.

Gagliardi, A., Totino, V., Cacciotti, F., Iebba, V., Neroni, B., Bonfiglio, G., et al. (2018). Rebuilding the gut microbiota ecosystem. *International Journal of Environmental Research and Public Health*, *15*, 1679.

Galié, S., García-Gutiérrez, C., Miguélez, E. M., Villar, C. J., & Lombó, F. (2018). Biofilms in the food industry: Health aspects and control methods. *Frontiers in Microbiology*, *9*, 898.

García, C. F., Kretschmer, M., Lozano-Andrade, C. N., Schönleitner, M., Dragoŝ, A., Kovács, Á. T., et al. (2020). Metal ions weaken the hydrophobicity and antibiotic resistance of *Bacillus subtilis* NCIB 3610 biofilms. *npj Biofilms and Microbiomes*, *6*, 1–11.

García-Betancur, J.-C., Goñi-Moreno, A., Horger, T., Schott, M., Sharan, M., Eikmeier, J., et al. (2017). Cell differentiation defines acute and chronic infection cell types in *Staphylococcus aureus*. *eLife*, *6*, e28023.

Giaouris, E., & Simões, M. (2020). The role of biofilms in the development and dissemination of microbial resistance within the food industry. *Foods*, *9*(6), 816.

Gómez, N., et al. (2016). Use of potential probiotic lactic acid bacteria (LAB) biofilms for the control of Listeria monocytogenes, Salmonella Typhimurium, and Escherichia coli O157:H7 biofilms formation. *Frontiers in Microbiology*, *7*, 863.

Gopikrishna, T., Kumar, H. K. S., Perumal, K., & Elangovan, E. (2021). Impact of *Bacillus* in fermented soybean foods on human health. *Annals of Microbiology*, *71*, 1–16.

Greenberg, M., Kuo, D., Jankowsky, E., Long, L., Hager, C., Bandi, K., et al. (2018). Small-molecule AgrA inhibitors F12 and F19 act as antivirulence agents against Gram-positive pathogens. *Scientific Reports*, *8*, 1–12.

Groisman, E. A., Hollands, K., Kriner, M. A., Lee, E.-J., Park, S.-Y., & Pontes, M. H. (2013). Bacterial Mg2+ homeostasis, transport, and virulence. *Annual Review of Genetics*, *47*, 625–646.

Gugala, N., Lemire, J. A., & Turner, R. J. (2017). The efficacy of different anti-microbial metals at preventing the formation of, and eradicating bacterial biofilms of pathogenic indicator strains. *The Journal of Antibiotics*, *70*, 775–780.

Gupta, R. K., Gupta, K., Sharma, A., Das, M., Ansari, I. A., & Dwivedi, P. D. (2017). Health risks and benefits of chickpea (*Cicer arietinum*) consumption. *Journal of Agricultural and Food Chemistry*, *65*, 6–22.

Han, N., Mizan, M. F. R., Jahid, I. K., & Ha, S.-D. (2016). Biofilm formation by *Vibrio parahaemolyticus* on food and food contact surfaces increases with rise in temperature. *Food Control*, *70*, 161–166.

Hemarajata, P., & Versalovic, J. (2013). Effects of probiotics on gut microbiota: Mechanisms of intestinal immunomodulation and neuromodulation. *Therapeutic Advances in Gastroenterology*, *6*, 39–51.

Horváth, E., Dályai, L., Szabó, E., Barna, T., Kalmár, L., Posta, J., et al. (2021). The antagonistic *Metschnikowia andauensis* produces extracellular enzymes and pulcherrimin, whose production can be promoted by the culture factors. *Scientific Reports*, *11*, 1–14.

Hutchings, C., Rajasekharan, S. K., Reifen, R., & Shemesh, M. (2020). Mitigating milk-associated bacteria through inducing zinc ions antibiofilm activity. *Foods*, *9*, 1094.

Joseph, B., Otta, S., Karunasagar, I., & Karunasagar, I. (2001). Biofilm formation by *Salmonella* spp. on food contact surfaces and their sensitivity to sanitizers. *International Journal of Food Microbiology*, *64*, 367–372.

Jukanti, A. K., Gaur, P. M., Gowda, C., & Chibbar, R. N. (2012). Nutritional quality and health benefits of chickpea (*Cicer arietinum* L.): A review. *British Journal of Nutrition*, *108*, S11–S26.

Kántor, A., Hutková, J., Petrová, J., Hleba, L., & Kačániová, M. (2021). Antimicrobial activity of pulcherrimin pigment produced by *Metschnikowia pulcherrima* against various yeast species. *Journal of Microbiology, Biotechnology and Food Sciences*, *2021*, 282–285.

Kaur, S., & Kaur, S. (2015). Bacteriocins as potential anticancer agents. *Frontiers in Pharmacology*, *6*, 272.

Kim, Y., Cho, J.-Y., Kuk, J.-H., Moon, J.-H., Cho, J.-I., Kim, Y.-C., et al. (2004). Identification and antimicrobial activity of phenylacetic acid produced by *Bacillus licheniformis* isolated from fermented soybean, Chungkook-Jang. *Current Microbiology*, *48*, 312–317.

Kim, Y.-H., Kim, S.-H., Whang, K.-Y., Kim, Y.-J., & Oh, S.-J. (2008). Inhibition of *Escherichia coli* O157:H7 attachment by interactions between lactic acid bacteria and intestinal epithelial cells. *Journal of Microbiology and Biotechnology*, *18*, 1278–1285.

Kimelman, H., & Shemesh, M. (2019). Probiotic bifunctionality of *Bacillus subtilis*—Rescuing lactic acid bacteria from desiccation and antagonizing pathogenic *Staphylococcus aureus*. *Microorganisms*, *7*, 407.

Kostakioti, M., Hadjifrangiskou, M., & Hultgren, S. J. (2013). Bacterial biofilms: Development, dispersal, and therapeutic strategies in the dawn of the postantibiotic era. *Cold Spring Harbor Perspectives in Medicine*, *3*, a010306.

Lamas, A., Regal, P., Vázquez, B., Cepeda, A., & Franco, C. M. (2019). Short chain fatty acids commonly produced by gut microbiota influence *Salmonella enterica* motility, biofilm formation, and gene expression. *Antibiotics*, *8*, 265.

Lawhon, S. D., Maurer, R., Suyemoto, M., & Altier, C. (2002). Intestinal short-chain fatty acids alter *Salmonella typhimurium* invasion gene expression and virulence through BarA/SirA. *Molecular Microbiology*, *46*, 1451–1464.

Lee, S. G., & Chang, H. C. (2017). Assessment of *Bacillus subtilis* SN7 as a starter culture for Cheonggukjang, a Korean traditional fermented soybean food, and its capability to control *Bacillus cereus* in Cheonggukjang. *Food Control*, *73*, 946–953.

Lee, J.-H., Kim, Y.-G., Gwon, G., Wood, T. K., & Lee, J. (2016). Halogenated indoles eradicate bacterial persister cells and biofilms. *AMB Express*, *6*, 1–12.

Lee, M.-H., Lee, J., Nam, Y.-D., Lee, J. S., Seo, M.-J., & Yi, S.-H. (2016). Characterization of antimicrobial lipopeptides produced by *Bacillus* sp. LM7 isolated from chungkookjang, a Korean traditional fermented soybean food. *International Journal of Food Microbiology*, *221*, 12–18.

Lee, D.-U., Park, Y. J., Yu, H. H., Jung, S.-C., Park, J.-H., Lee, D.-H., et al. (2021). Antimicrobial and antibiofilm effect of ε-polylysine against *Salmonella Enteritidis*, *Listeria monocytogenes*, and *Escherichia coli* in tryptic soy broth and chicken juice. *Foods*, *10*, 2211.

Lee, J. Y., Shim, J. M., Yao, Z., Liu, X., Lee, K. W., Kim, H.-J., et al. (2016). Antimicrobial activity of *Bacillus amyloliquefaciens* EMD17 isolated from Cheonggukjang and potential use as a starter for fermented soy foods. *Food Science and Biotechnology*, *25*, 525–532.

Li, X., Wang, D., Cai, D., Zhan, Y., Wang, Q., & Chen, S. (2017). Identification and high-level production of pulcherrimin in *Bacillus licheniformis* DW2. *Applied Biochemistry and Biotechnology*, *183*, 1323–1335.

Lin, Y.-C., Wu, C.-Y., Huang, H.-T., Lu, M.-K., Hu, W.-S., & Lee, K.-T. (2021). *Bacillus subtilis* natto derivatives inhibit enterococcal biofilm formation via restructuring of the cell envelope. *Frontiers in Microbiology*, 3839.

Line, J., Svetoch, E., Eruslanov, B., Perelygin, V., Mitsevich, E., Mitsevich, I., et al. (2008). Isolation and purification of enterocin E-760 with broad antimicrobial activity against gram-positive and gram-negative bacteria. *Antimicrobial Agents and Chemotherapy*, *52*, 1094–1100.

Liu, L., Ye, C., Soteyome, T., Zhao, X., Xia, J., Xu, W., et al. (2019). Inhibitory effects of two types of food additives on biofilm formation by foodborne pathogens. *MicrobiologyOpen*, *8*, e00853.

Lombardía, E., Rovetto, A. J., Arabolaza, A. L., & Grau, R. R. (2006). A LuxS-dependent cell-to-cell language regulates social behavior and development in *Bacillus subtilis*. *Journal of Bacteriology*, *188*, 4442–4452.

López, D., Vlamakis, H., & Kolter, R. (2010). Biofilms. *Cold Spring Harbor Perspectives in Biology*, *2*, a000398.

Ma, D. S., Tan, L. T.-H., Chan, K.-G., Yap, W. H., Pusparajah, P., Chuah, L.-H., et al. (2018). Resveratrol—Potential antibacterial agent against foodborne pathogens. *Frontiers in Pharmacology*, *9*, 102.

Machado, M. G., Sencio, V., & Trottein, F. (2021). Short-chain fatty acids as a potential treatment for infections: A closer look at the lungs. *Infection and Immunity*. 00188-00121.

Marciňáková, M., Strompfová, V., Boldižárová, K., Lauková, A., & Gancarčíková, S. (2004). Effect of potential probiotic activity of

Enterococcus faecium EE3 strain against *Salmonella* infection in Japanese quails. *Bulletin of the Veterinary Institute in Pulawy*, *48*, 387–390.

Markowiak, P., & Śliżewska, K. (2017). Effects of probiotics, prebiotics, and synbiotics on human health. *Nutrients*, *9*, 1021.

Markowiak-Kopeć, P., & Śliżewska, K. (2020). The effect of probiotics on the production of short-chain fatty acids by human intestinal microbiome. *Nutrients*, *12*, 1107.

McLoughlin, R. F., Berthon, B. S., Jensen, M. E., Baines, K. J., & Wood, L. G. (2017). Short-chain fatty acids, prebiotics, synbiotics, and systemic inflammation: A systematic review and meta-analysis. *The American Journal of Clinical Nutrition*, *106*, 930–945.

Medellin-Peña, M. J., Wang, H., Johnson, R., Anand, S., & Griffiths, M. W. (2007). Probiotics affect virulence-related gene expression in *Escherichia coli* O157:H7. *Applied and Environmental Microbiology*, *73*, 4259–4267.

Meroni, G., Panelli, S., Zuccotti, G., Bandi, C., Drago, L., & Pistone, D. (2021). Probiotics as therapeutic tools against pathogenic biofilms: Have we found the perfect weapon? *Microbiology Research*, *12*, 916–937.

Messora, M. R., Pereira, L. J., Foureaux, R., Oliveira, L. F., Sordi, C. G., Alves, A. J., et al. (2016). Favourable effects of *Bacillus subtilis* and *Bacillus licheniformis* on experimental periodontitis in rats. *Archives of Oral Biology*, *66*, 108–119.

Mingmongkolchai, S., & Panbangred, W. (2018). *Bacillus* probiotics: An alternative to antibiotics for livestock production. *Journal of Applied Microbiology*, *124*, 1334–1346.

Moore, T., Globa, L., Barbaree, J., Vodyanoy, V., & Sorokulova, I. (2013). Antagonistic activity of *Bacillus* bacteria against food-borne pathogens. *Journal of Probiotics & Health*, *1*, 110.

Moser, C., Pedersen, H. T., Lerche, C. J., Kolpen, M., Line, L., Thomsen, K., et al. (2017). Biofilms and host response—Helpful or harmful. *APMIS*, *125*, 320–338.

Moshiri, J., Kaur, D., Hambira, C. M., Sandala, J. L., Koopman, J. A., Fuchs, J. R., et al. (2018). Identification of a small molecule antibiofilm agent against *Salmonella enterica*. *Frontiers in Microbiology*, *9*, 2804.

Myszka, K., & Czaczyk, K. (2011). Bacterial biofilms on food contact surfaces—A review. *Polish Journal of Food and Nutrition Sciences*, *61*.

Myszka, K., Olejnik, A., Majcher, M., Sobieszczańska, N., Grygier, A., Powierska-Czarny, J., et al. (2019). Green pepper essential oil as a biopreservative agent for fish-based products: Antimicrobial and antivirulence activities against *Pseudomonas aeruginosa* KM01. *LWT*, *108*, 6–13.

Nakamura, K., O'Neill, A. M., Williams, M. R., Cau, L., Nakatsuji, T., Horswill, A. R., et al. (2020). Short chain fatty acids produced by Cutibacterium acnes inhibit biofilm formation by *Staphylococcus epidermidis*. *Scientific Reports*, *10*, 1–12.

Nguyen, L. N., Lopes, L. C. L., Cordero, R. J., & Nosanchuk, J. D. (2011). Sodium butyrate inhibits pathogenic yeast growth and enhances the functions of macrophages. *Journal of Antimicrobial Chemotherapy*, *66*, 2573–2580.

Oknin, H., Steinberg, D., & Shemesh, M. (2015). Magnesium ions mitigate biofilm formation of *Bacillus* species via downregulation of matrix genes expression. *Frontiers in Microbiology*, *6*, 907.

Olas, B. (2020). Probiotics, prebiotics and synbiotics—A promising strategy in prevention and treatment of cardiovascular diseases? *International Journal of Molecular Sciences*, *21*, 9737.

Oro, L., Ciani, M., & Comitini, F. (2014). Antimicrobial activity of *Metschnikowia pulcherrima* on wine yeasts. *Journal of Applied Microbiology*, *116*, 1209–1217.

Ostrov, I., Harel, A., Bernstein, S., Steinberg, D., & Shemesh, M. (2016). Development of a method to determine the effectiveness of cleaning agents in removal of biofilm derived spores in milking system. *Frontiers in Microbiology*, *7*, 1498.

Ostrov, I., Sela, N., Belausov, E., Steinberg, D., & Shemesh, M. (2019). Adaptation of *Bacillus* species to dairy associated environment facilitates their biofilm forming ability. *Food Microbiology*, *82*, 316–324.

Ozogul, Y., Kuley, E., Ucar, Y., & Ozogul, F. (2015). Antimicrobial impacts of essential oils on food borne-pathogens. *Recent Patents on Food, Nutrition & Agriculture*, *7*, 53–61.

Pandey, K. R., Naik, S. R., & Vakil, B. V. (2015). Probiotics, prebiotics and synbiotics—A review. *Journal of Food Science and Technology*, *52*, 7577–7587.

Parsek, M. R., & Greenberg, E. (2005). Sociomicrobiology: The connections between quorum sensing and biofilms. *Trends in Microbiology*, *13*, 27–33.

Pasvolsky, R., Zakin, V., Ostrova, I., & Shemesh, M. (2014). Butyric acid released during milk lipolysis triggers biofilm formation of *Bacillus* species. *International Journal of Food Microbiology*, *181*, 19–27.

Pawlikowska, E., Kolesińska, B., Nowacka, M., & Kregiel, D. (2020). A new approach to producing high yields of pulcherrimin from *Metschnikowia* yeasts. *Fermentation*, *6*, 114.

Piewngam, P., Zheng, Y., Nguyen, T. H., Dickey, S. W., Joo, H.-S., Villaruz, A. E., et al. (2018). Pathogen elimination by probiotic *Bacillus* via signalling interference. *Nature*, *562*, 532–537.

Rajasekharan, S. K., & Shemesh, M. (2022a). The bacillary postbiotics, including 2-undecanone, suppress the virulence of pathogenic microorganisms. *Pharmaceutics*, *14*, 962.

Rajasekharan, S. K., & Shemesh, M. (2022b). Spatiotemporal bio-shielding of bacteria through consolidated geometrical structuring. *NPJ Biofilms and Microbiomes*, *8*, 1–10.

Rajasekharan, S. K., Paz-Aviram, T., Galili, S., Berkovich, Z., Reifen, R., & Shemesh, M. (2021). Biofilm formation onto starch fibres by *Bacillus subtilis* governs its successful adaptation to chickpea milk. *Microbial Biotechnology*, *14*, 1839–1846.

Rajasekharan, S. K., Ramesh, S., & Bakkiyaraj, D. (2015). Synergy of flavonoids with HDAC inhibitor: New approach to target *Candida tropicalis* biofilms. *Journal of Chemotherapy*, *27*, 246–249.

Rezaei, Z., Khanzadi, S., & Salari, A. (2021). Biofilm formation and antagonistic activity of *Lacticaseibacillus rhamnosus* (PTCC1712) and *Lactiplantibacillus plantarum* (PTCC1745). *AMB Express*, *11*, 1–7.

Rhayat, L., Maresca, M., Nicoletti, C., Perrier, J., Brinch, K. S., Christian, S., et al. (2019). Effect of *Bacillus subtilis* strains on intestinal barrier function and inflammatory response. *Frontiers in Immunology*, *10*, 564.

Ritter, A. C., Paula, A., Correa, F., Veras, F., & Brandelli, A. (2018). Characterization of *Bacillus subtilis* available as probiotics. *Journal of Microbiology Research*, *8*, 23–32.

Robertson, S. R., & McLean, R. J. (2015). Beneficial biofilms. *AIMS Bioengineering*, *2*, 437–448.

Sathiyamoorthi, E., Faleye, O. S., Lee, J.-H., Raj, V., & Lee, J. (2021). Antibacterial and antibiofilm activities of chloroindoles against *Vibrio parahaemolyticus*. *Frontiers in Microbiology*, 2206.

Sethi, S., Tyagi, S. K., & Anurag, R. K. (2016). Plant-based milk alternatives an emerging segment of functional beverages: A review. *Journal of Food Science and Technology*, *53*, 3408–3423.

Shemesh, M., & Chai, Y. (2013). A combination of glycerol and manganese promotes biofilm formation in Bacillus subtilis via histidine kinase KinD signaling. *Journal of Bacteriology*, *195*(12), 2747–2754.

Shemesh, M., & Ostrov, I. (2020). Role of *Bacillus* species in biofilm persistence and emerging antibiofilm strategies in the dairy industry. *Journal of the Science of Food and Agriculture*, *100*, 2327–2336.

Shemesh, M., Tam, A., & Steinberg, D. (2007). Differential gene expression profiling of *Streptococcus mutans* cultured under biofilm and planktonic conditions. *Microbiology*, *153*, 1307–1317.

Simonová, M. P., Lauková, A., Chrastinova, L., Strompfová, V., Faix, Š., Vasilková, Z., et al. (2009). *Enterococcus faecium* CCM7420, bacteriocin PPB CCM7420 and their effect in the digestive tract of rabbits. *Czech Journal of Animal Science*, *54*, 376–386.

Sipiczki, M. (2020). *Metschnikowia pulcherrima* and related pulcherriminproducing yeasts: Fuzzy species boundaries and complex antimicrobial antagonism. *Microorganisms*, *8*, 1029.

Sourabh, A., Walia, S., & Kanwar, S. (2011). Role of probiotics in colorectal cancer. *International Journal of Biomedical and Pharmaceutical Sciences*, *5*, 1–6.

Speranza, B., & Corbo, M. R. (2017). The impact of biofilms on food spoilage. In *The microbiological quality of food* (pp. 259–282). Elsevier.

Starosila, D., Rybalko, S., Varbanetz, L., Ivanskaya, N., & Sorokulova, I. (2017). Anti-influenza activity of a *Bacillus subtilis* probiotic strain. *Antimicrobial Agents and Chemotherapy*, *61*. e00539-00517.

Stern, N. J., Svetoch, E. A., Eruslanov, B. V., Kovalev, Y. N., Volodina, L. I., Perelygin, V. V., et al. (2005). *Paenibacillus polymyxa* purified bacteriocin to control *Campylobacter jejuni* in chickens. *Journal of Food Protection*, *68*, 1450–1453.

Sun, Y., & O'Riordan, M. X. (2013). Regulation of bacterial pathogenesis by intestinal short-chain fatty acids. *Advances in Applied Microbiology*, *85*, 93–118.

Svetoch, E. A., Levchuk, V. P., Pokhilenko, V. D., Eruslanov, B. V., Mitsevich, E. V., Mitsevich, I. P., et al. (2008). Inactivating methicillinresistant *Staphylococcus aureus* and other pathogens by use of bacteriocins OR-7 and E 50-52. *Journal of Clinical Microbiology*, *46*, 3863–3865.

Szlufman, C., & Shemesh, M. (2021). Role of probiotic *Bacilli* in developing synbiotic food: Challenges and opportunities. *Frontiers in Microbiology*, *12*, 638830.

Tambosi, R., Liotenberg, S., Bourbon, M.-L., Steunou, A.-S., Babot, M., Durand, A., et al. (2018). Silver and copper acute effects on membrane proteins and impact on photosynthetic and respiratory complexes in bacteria. *MBio*, *9*. e01535-01518.

Thamburaj, S., Ramaraj, E., Sethupathy, S., Kamalanathan, C., Raji, A., & Rajasekharan, S. K. (2020). Antibacterial and antibiofilm activities of diphyllin against fish pathogens. *Microbial Pathogenesis*, *145*, 104232.

Turner, R. J. (2017). Metal-based antimicrobial strategies. *Microbial Biotechnology*, *10*, 1062–1065.

Urdaci, M., & Pinchuk, I. (2004). Antimicrobial activity of *Bacillus* probiotics. In *Bacterial spore formers—Probiotics and emerging applications* (pp. 171–182). Norfolk: Horizon Bioscience.

Valdés-Varela, L., Alonso-Guervos, M., García-Suárez, O., Gueimonde, M., & Ruas-Madiedo, P. (2016). Screening of bifidobacteria and lactobacilli able to antagonize the cytotoxic effect of *Clostridium difficile* upon intestinal epithelial HT29 monolayer. *Frontiers in Microbiology*, *7*, 577.

Van Kuijk, S., Noll, K., & Chikindas, M. (2012). The species-specific mode of action of the antimicrobial peptide subtilosin against *Listeria monocytogenes* Scott A. *Letters in Applied Microbiology*, *54*, 52–58.

Vikram, A., Jayaprakasha, G., Uckoo, R. M., & Patil, B. S. (2013). Inhibition of *Escherichia coli* O157:H7 motility and biofilm by β-sitosterol glucoside. *Biochimica et Biophysica Acta (BBA) - General Subjects*, *1830*, 5219–5228.

Vikram, A., Jesudhasan, P. R., Pillai, S. D., & Patil, B. S. (2012). Isolimonic acid interferes with *Escherichia coli* O157:H7 biofilm and TTSS in QseBC and QseA dependent fashion. *BMC Microbiology*, *12*, 1–13.

Williams, N. T. (2010). Probiotics. *American Journal of Health-System Pharmacy*, *67*, 449–458.

Winkelströter, L. K., dos Reis Teixeira, F. B., Silva, E. P., Alves, V. F., & De Martinis, E. C. P. (2014). Unraveling microbial biofilms of importance for food microbiology. *Microbial Ecology*, *68*, 35–46.

Wong, A. C. L. (1998). Biofilms in food processing environments. *Journal of Dairy Science*, *81*, 2765–2770.

Wu, C., Labrie, J., Tremblay, Y., Haine, D., Mourez, M., & Jacques, M. (2013). Zinc as an agent for the prevention of biofilm formation by pathogenic bacteria. *Journal of Applied Microbiology*, *115*, 30–40.

Xiang, H., Cao, F., Ming, D., Zheng, Y., Dong, X., Zhong, X., et al. (2017). Aloe-emodin inhibits *Staphylococcus aureus* biofilms and extracellular protein production at the initial adhesion stage of biofilm development. *Applied Microbiology and Biotechnology*, *101*, 6671–6681.

Xu, W., Parmigiani, R., & Marks, P. (2007). Histone deacetylase inhibitors: Molecular mechanisms of action. *Oncogene*, *26*, 5541–5552.

Xu, D., Wang, Y., Sun, L., Liu, H., & Li, J. (2013). Inhibitory activity of a novel antibacterial peptide AMPNT-6 from *Bacillus subtilis* against *Vibrio parahaemolyticus* in shrimp. *Food Control*, *30*, 58–61.

Yan, X., Gu, S., Shi, Y., Cui, X., Wen, S., & Ge, J. (2017). The effect of emodin on *Staphylococcus aureus* strains in planktonic form and biofilm formation in vitro. *Archives of Microbiology*, *199*, 1267–1275.

Yeo, I.-C., Lee, N. K., Cha, C.-J., & Hahm, Y. T. (2011). Narrow antagonistic activity of antimicrobial peptide from *Bacillus subtilis* SCK-2 against *Bacillus cereus*. *Journal of Bioscience and Bioengineering*, *112*, 338–344.

Yeo, I.-C., Lee, N. K., & Hahm, Y. T. (2012). Genome sequencing of *Bacillus subtilis* SC-8, antagonistic to the *Bacillus cereus* group, isolated from traditional Korean fermented-soybean food. *Journal of Bacteriology*, *194*, 536–537.

Yu, T., Kong, J., Zhang, L., Gu, X., Wang, M., & Guo, T. (2019). New crosstalk between probiotics *Lactobacillus plantarum* and *Bacillus subtilis*. *Scientific Reports*, *9*, 1–9.

Yuan, W., & Yuk, H.-G. (2019). Effects of sublethal thymol, carvacrol, and trans-cinnamaldehyde adaptation on virulence properties of *Escherichia coli* O157:H7. *Applied and Environmental Microbiology*, *85*. e00271-00219.

Zhang, B., Teng, Z., Li, X., Lu, G., Deng, X., Niu, X., et al. (2017). Chalcone attenuates *Staphylococcus aureus* virulence by targeting sortase A and alpha-hemolysin. *Frontiers in Microbiology*, *8*, 1715.

Zhao, X., Zhao, F., Wang, J., & Zhong, N. (2017). Biofilm formation and control strategies of foodborne pathogens: Food safety perspectives. *RSC Advances*, *7*, 36670–36683.

Chapter 37

Use of bacterial biofilms to produce high added-value compounds

Ana Azevedo[a,b,*], Fábio M. Carvalho[a,b,*], Luciana C. Gomes[a,b], and Filipe J.M. Mergulhão[a,b]

[a]*LEPABE—Laboratory for Process Engineering, Environment, Biotechnology and Energy, Faculty of Engineering, University of Porto, Porto, Portugal,* [b]*ALiCE – Associate Laboratory in Chemical Engineering, Faculty of Engineering, University of Porto, Porto, Portugal*

1. Introduction

Research in microbial biofilms commonly addresses their detrimental effects in the health and industrial sectors (Carrascosa, Raheem, Ramos, Saraiva, & Raposo, 2021; Roberts, 2013). Biofilms can cause persistent infections on medical devices, as well as equipment corrosion and clogging, heat transfer reduction, and product degradation in industry (Moreira et al., 2016; Yan & Bassler, 2019). They are the prevailing lifestyle of microorganisms, conferring protection to harsh conditions (e.g., pH, temperature, antibiotics, hydraulic shock, contaminants, and/or toxic substances), tolerance to antibacterial treatments, and increased availability of nutrients in low-nutrient environments (Ercan & Demirci, 2015; Roberts, 2013; Roy, Tiwari, Donelli, & Tiwari, 2018). However, despite their detrimental impacts, biofilms can also be beneficial for health, as normal components of animal and human microflora (Yan & Bassler, 2019), and agricultural and industrial settings (Muhammad et al., 2020) as biocontrol agents to enhance crop production, bioremediation, and wastewater treatment (Berlanga & Guerrero, 2016; Edwards & Kjellerup, 2013; Lewandowski & Boltz, 2011; Timmusk, Behers, Muthoni, Muraya, & Aronsson, 2017). Besides these applications, the robust biocatalytic activity of biofilms turned increasing attention to their employment in chemical synthesis, in the production of recombinant heterologous proteins (Soares, Azevedo, Gomes, & Mergulhão, 2019; Soares, Gomes, & Mergulhão, 2018) and other high-value products (Ercan & Demirci, 2015).

Recombinant proteins (RPs) are proteins obtained by the isolation of the gene encoding the target protein, engineering their sequence, and introducing it into an expression host (Overton, 2014; Rosano & Ceccarelli, 2014). These proteins have captured the interest of the biotechnological industries as these proteins can be expressed at much higher levels than those obtained naturally (Overton, 2014).

The first step for RP production is the selection of a suitable expression system that should take into consideration the transcriptional and translational issues (Overton, 2014; Rosano & Ceccarelli, 2014). The next step is the selection of a host that can be chosen among different bacteria, yeast, filamentous fungi, mammalian cells, plants, and insect cells (Chen, 2012; Demain & Vaishnav, 2009). Despite the great variety of host cells available for RP production, the selection is often limited to a few options due to the protein characteristics and the processes to be developed (Gomes & Mergulhão, 2019; Mergulhão, Monteiro, Cabral, & Taipa, 2004). For research purposes, host selection should be made taking into consideration the intrinsic protein properties, level of production, cell growth, and scalability potential, to which are added the time for setup and development, safety component, regulatory issues, patenting issues, and production cost when moving toward industrial applications (Gomes & Mergulhão, 2019; Soares et al., 2019).

RPs have been mainly produced using suspended cultures, although some evidence has shown that biofilm cultures may be more appealing (Gomes & Mergulhão, 2017a; Gomes, Monteiro, & Mergulhão, 2020; O'Connell, Niu, & Gilbert, 2007). For instance, the insertion of a gene of interest into a multicopy plasmid imposes an extra metabolic burden to the host cell due to the energy and metabolites required for the replication of plasmid DNA and the production of RP (Hoffmann & Rinas, 2004; Overton, 2014). In planktonic cells, this event may decrease the cellular growth and biomass yields, but also decrease the target protein production level (Hoffmann & Rinas, 2004). On the other hand, the presence of plasmid vectors in biofilm cells was shown to benefit biofilm formation and lead to higher production of RP when compared with planktonic cells (Gomes & Mergulhão, 2017a; O'Connell et al., 2007; Soares et al., 2018).

*These authors contributed equally to this work.

Understanding Microbial Biofilms. https://doi.org/10.1016/B978-0-323-99977-9.00015-6

RPs and other added-value compounds are being produced resorting to biofilms reactors due to the many advantages of these platforms when compared with suspended growth systems. Products from biofilm reactors are easily recovered and biofilm reactors can retain 5–10 more biomass per unit volume, thus increasing production rates and yields, reducing the risk of washout, and eliminating the necessity of repeated inoculations during subsequent batch fermentations (Demirci, Pongtharangkul, & Pometto, 2007; Ercan & Demirci, 2015). Nevertheless, biofilm reactors have limited oxygen and substrate diffusion, which may increase population heterogeneity, are difficult to scale up, and the maintenance of a pure culture in consecutive batch operations is complicated (Demirci et al., 2007). Furthermore, the downstream processing of an intracellular product is more complicated in these systems. Fig. 1 summarizes the benefits and handicaps of biofilm reactors.

From the organisms that can be chosen for the production of added-value substances in biofilms, bacteria remain the favorite organism for recombinant protein production, contributing to approximately 30% of current biopharmaceuticals on the market (Huang, Lin, & Yang, 2012; Overton, 2014), and it has been successfully used in the synthesis of other high-value products such as ethanol, organic acids, and antimicrobial compounds (Demirci et al., 2007). Bacterial expression systems are the most attractive because of their biochemical, physiological, and process simplicity, low cost and shorter time processes, high productivity, and ability to achieve high-density cultures in inexpensive substrates (Overton, 2014; Terpe, 2006). *Escherichia coli* is one of the favorite hosts in the biotechnology field (Becker & Wittmann, 2016) because of its well-characterized genetics, fast growth on cheap medium, and its ability to accumulate recombinant products at a high level and secret them to the culture medium (Mergulhão, Gabriel, Cabral, & Taipa, 2003; Mergulhão, Summers, & Monteiro, 2005). In recent years, *E. coli* has been regularly used to produce biofuels, organic acids, alcohols, amino acids, and other valuable molecules (Chen et al., 2013).

Several studies suggested the application of biofilms as robust, self-immobilized, and self-regenerating systems in the production of added-value chemicals and specific proteins (Cheng, Demirci, & Catchmark, 2010; Mahdinia & Demirci, 2020; Muffler et al., 2014). Thus, this chapter intends to summarize the latest advances on the production of recombinant proteins using bacterial biofilms, as well as to give an overview of the main added-value compounds produced using bacterial biofilms as a biocatalytic system (Fig. 2).

2. Production of recombinant proteins using biofilms

During the past three decades, bacterial biofilms have been studied as platforms for the production of RPs (Table 1) such as β-galactosidase (Huang et al., 1992, 1994a; Huang, Peretti, & Bryers, 1994b), enhanced green fluorescent protein (eGFP) (Gomes, Carvalho, Briandet, & Mergulhão, 2016; Gomes, Monteiro, & Mergulhão, 2020; Gomes & Mergulhão, 2017b; O'Connell et al., 2007; Soares et al., 2018, 2019, 2021), and iturin A (Rahman et al., 2007).

The main platforms for producing recombinant proteins in biofilms are microplates, the parallel-plate flow chamber (PPFC), and the modified Robbins devices. Microplates are regularly used in biofilm development studies as they are easy to handle, have low volume, and can be used in both static and controlled shaken conditions (Gomes et al., 2014, 2021). PPFCs allow in situ and real-time visualization of cell adhesion and consequent biofilm formation, and require lower medium volume compared to modified Robbins devices. Nevertheless, PPFCs have lower throughput than microplates and modified Robbins devices (Alves, Gomes, Vorobii, Rodriguez-Emmenegger, & Mergulhão, 2020). The Robbins device was first developed to monitor biofilm formation in water systems, but then several modifications were introduced. Gomes et al. (2016) used a custom-made and semicircular flow cell with a set of hydrodynamic features completely characterized (Gomes & Mergulhão, 2021).

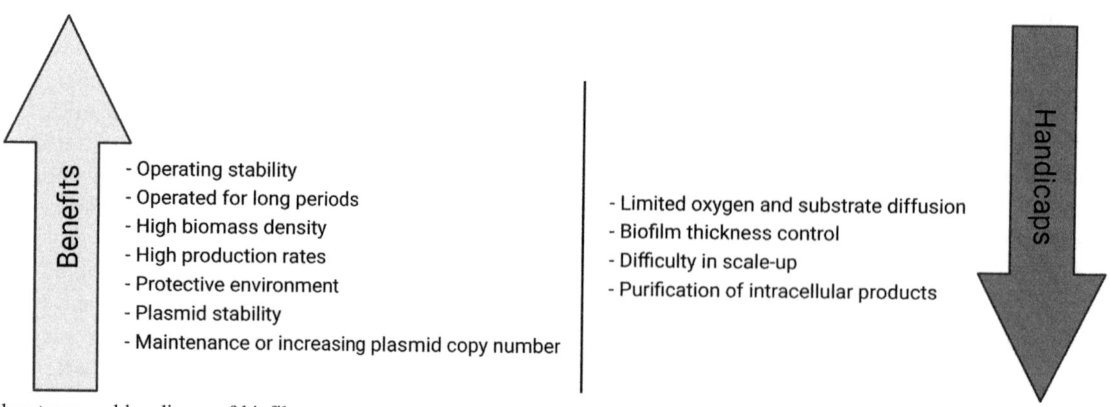

FIG. 1 Advantages and handicaps of biofilm reactors.

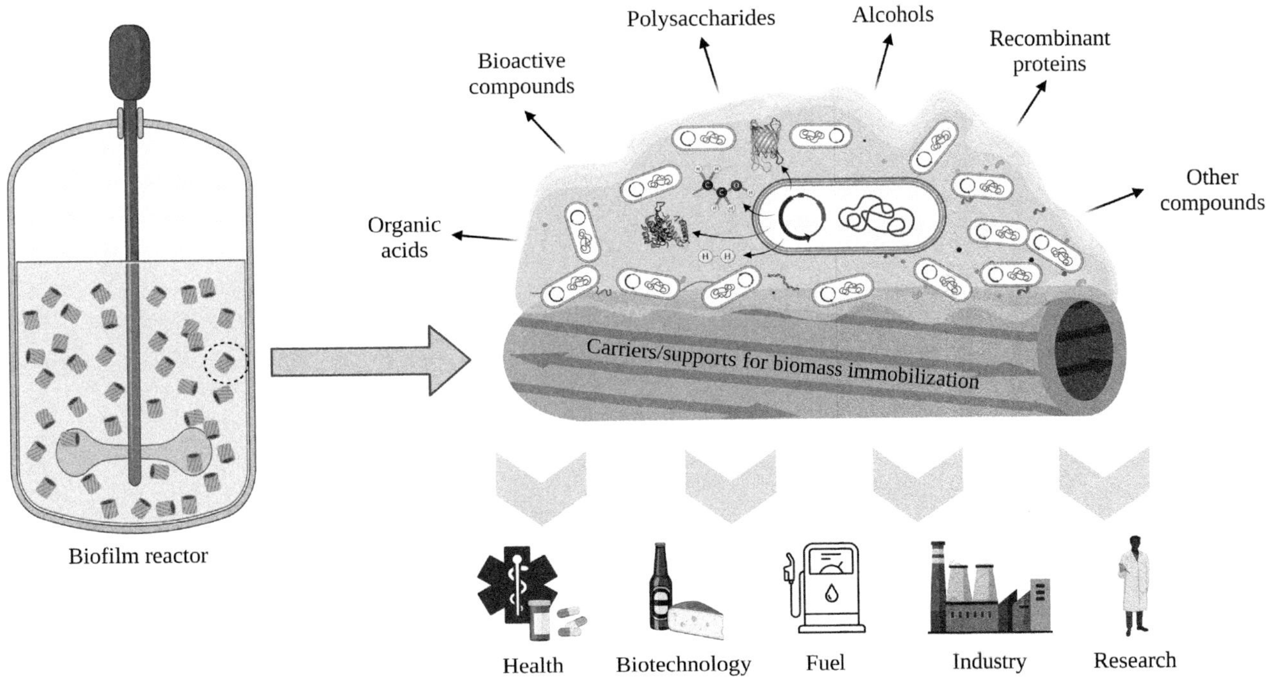

FIG. 2 Added-value compounds produced in biofilm reactors for different applications.

2.1 β-Galactosidase

The synthesis of recombinant proteins using biofilms was firstly described in 1992 by Huang et al. (1992), who produced β-galactosidase in *E. coli* DH5α cells using a PPFC system. During the experiment, the production of β-galactosidase increased with isopropyl β-D-1-thiogalactopyranoside (IPTG) induction, and no basal expression was detected. The maximum production was obtained 24 h after induction with yields of 0.08, 0.1, and 0.12 pg cell^{-1} for IPTG concentrations of 0.17, 0.34, and 0.51 mM, respectively (Huang et al., 1992).

Afterward, Huang et al. (1994a) studied the production of β-galactosidase with another plasmid but using the same host (*E. coli* DH5α) and the same cultivation and induction conditions. The results showed that β-galactosidase production reached its maximum after 36 h of induction for 0.17 and 0.34 mM IPTG with 0.027 and 0.036 pg cell^{-1}, respectively, and after 48 h of induction for 0.51 mM IPTG with 0.050 pg cell^{-1}. Further, the synthesis rates of β-galactosidase mRNA increased 4-fold under 0.17 mM IPTG and increased almost 12-fold under 0.34 and 0.51 mM IPTG after 36 h of induction. Still, the production of β-galactosidase did not increase at the same proportion of mRNA synthesis rate, which suggests that mRNA was less stable at higher expression levels (Huang et al., 1994a).

2.2 Green fluorescent protein (GFP)

In 2007, O'Connell et al. (2007) described a biofilm system with an *E. coli* strain harboring a pEGFP plasmid to investigate the production of enhanced green fluorescent protein (eGFP) in a PPFC reactor. The authors investigated the effect of ampicillin concentration on cell fluorescence, revealing that low antibiotic concentrations (between 33 and 100 ppm) lead to 60% of strongly eGFP-producing cells (O'Connell et al., 2007). The results of this study also showed that the biofilm environment enhanced plasmid maintenance and heterologous protein production when compared with planktonic cells, as opposed to previously found by Huang et al. (1992, 1994a).

Subsequently, Gomes and colleagues have studied the eGFP production in *E. coli* JM109(DE3) planktonic and biofilm cells (Gomes, Carvalho, Briandet, & Mergulhão, 2016; Gomes, Monteiro, & Mergulhão, 2020; Gomes & Mergulhão, 2017a, 2017b; Soares, Azevedo, Gomes, & Mergulhão, 2019; Soares, Gomes, & Mergulhão, 2018; Soares, Gomes, Monteiro, & Mergulhão, 2021). The studies were all carried out in a modified Robbins device with fixed temperature (37°C) and hydrodynamic conditions (turbulent flow, Reynolds number of 4600) throughout the assays. The first assays performed by Gomes et al. (2016) used Lysogeny broth (LB) and established a combination of techniques that could be used to monitor and quantify fluorescent recombinant protein expression in biofilm cells. This study revealed that the biofilm population would become increasingly heterogeneous during the experiment, which is consistent with the results obtained by O'Connell et al. (2007). Regarding the spatial distribution, it was shown that eGFP-expressing cells were mainly located at the external regions of biofilms (Gomes et al., 2016).

TABLE 1 Synopsis of the published work on the production of recombinant proteins in bacterial biofilms.

Recombinant protein	Bacterial host	Cultivation conditions							Specific production	Ref.
		Reactor	Material	Culture medium	Temp. (°C)	Hydrodynamics	Time (days)	Induction		
β-Galactosidase	*Escherichia coli* DH5α (pMJR1750)	PPFC	Glass slides	M9 minimal	37	Laminar flow ($Re=20$)	4–5	IPTG (0.17–0.51 mM)	0.08–0.12 pg cell^{-1}	Huang, Peretti, and Bryers (1992)
	Escherichia coli DH5α (pTKW106)						4–5		0.027–0.050 pg cell^{-1}	Huang, Peretti, and Bryers (1994a)
eGFP	*Escherichia coli* ATCC33456	PPFC	Glass	LB	37	Laminar flow ($Re=32$)	6	–	0.01–0.16 g L^{-1}	O'Connell et al. (2007)
	Escherichia coli JM109(DE3)	Flow cell	PVC surfaces	LB	30	Turbulent flow ($Re=4600$)	11	–	n.d.	Gomes, Carvalho, Briandet, and Mergulhão (2016)
				DM and LB			12	–	5.7–12 fg cell^{-1}	Gomes and Mergulhão (2017b)
				LB			11	–	n.d.	Soares et al. (2018)
				LB			11	IPTG (2 mM)	17 fg cell^{-1}	Gomes, Monteiro, and Mergulhão (2020)
				LB and M9ZB			10	–	1.51–15.96 fg cell^{-1}	Soares, Gomes, Monteiro, and Mergulhão (2021)
D-Amino acid oxidase	*Escherichia coli* TOP10	Static and shaken reactors	–	HSG4	30	Static conditions	7	IPTG (0.1 mM)	1.2 U g^{-1}	Setyawati, Chien, and Lee (2008)
			Cellulose nanofibers			170 rpm			2.1 U g^{-1}	
Iturin A	*Bacillus subtilis*	24-well plates	–	LB	28	Static conditions	6	–	0.6 g L^{-1}	Rahman, Ano, and Shoda (2007)
mCherry, EgTrp, and EgA31 (part of fusion proteins)	*Bacillus subtilis*	Well plates with 22 mm^2 surface area and agar plates	–	MSgg	30	Static conditions	3	–	n.d.	Vogt, Schraner, Aguilar, and Eichwald (2016)

Abbreviations: *IPTG*, isopropyl β-D-1-thiogalactopyranoside; *n.d.*, not described; *PPFC*, parallel-plate flow chamber; *PVC*, polyvinyl chloride; *Re*, Reynolds number; *Temp.*, temperature; *U*, activity unit.

Then, Gomes and Mergulhão (2017b) compared eGFP production between two different culture media (LB and Diluted Medium) and antibiotic concentrations (20 and 30 μg mL^{-1} kanamycin). LB (composed of 10 g L^{-1} tryptone and 5 g L^{-1} yeast extract) has a large amount of carbon and nitrogen and is frequently used for the expression of recombinant proteins (Donovan, Robinson, & Glick, 1996); the diluted medium (DM; made of 0.55 g L^{-1} glucose, 0.25 g L^{-1} peptone, and 0.125 g L^{-1} yeast extract) was described as a good medium for biofilm development (Teodósio, Simões, Melo, & Mergulhão, 2011). These experiments showed that eGFP expression was higher with LB supplemented with 20 μg mL^{-1} kanamycin, showing a production of 12 fg cell^{-1} against 5.7 and 6.2 fg cell^{-1} with DM containing 20 and 30 μg mL^{-1} kanamycin, respectively. Gomes and Mergulhão (2017b) concluded that eGFP production was higher when produced in LB, and that the increased antibiotic concentration had no effect on RP expression.

Using the same bacterial host, plasmid, and cultivation conditions as Gomes et al. (2016), Soares et al. (2018) showed that the average specific eGFP production in biofilms cells was 10-fold higher than in planktonic cells, and that the percentage of total expressing cells in biofilms was 4-fold higher than in suspended cultures, which could be associated with the potential of biofilms to retain plasmids.

In 2020, Gomes, Monteiro, and Mergulhão (2020) compared the eGFP protein expression in noninduced and induced biofilms, with chemical induction performed with IPTG at a final concentration of 2 mM, and revealed that eGFP production remained constant in the induced biofilm culture over the operation time, with a productivity of around 17 fg cell^{-1}, whereas in the noninduced biofilms the eGFP production decreased around 31%.

In 2021, Soares et al. (2021) investigated the influence of nutrient conditions on RP production in biofilms. The assay was done comparing LB and M9ZB medium, demonstrating that M9ZB favored biofilm development, but had an inhibitory effect on eGFP expression due to the glucose present in the medium. On the other hand, the number of eGFP-expressing cells and eGFP yield were higher in LB medium likely because of the higher nitrogen content.

2.3 Other proteins

D-amino acid oxidase (DAAO) is a protein from *Rhodosporidium toruloides* that was produced in *E. coli* TOP10 (Setyawati et al., 2008). This study was carried out comparing static and shaken cultivation with IPTG induction and revealed that DAAO production was almost 2-fold higher in shaken rather than in static conditions (Setyawati et al., 2008).

The production of recombinant proteins in biofilms has also been studied in Gram-positive bacterium as *Bacillus subtilis* (Rahman et al., 2007; Vogt et al., 2016). In 2007, Rahman et al. (2007) used the *B. subtilis* 168 strain for the production of a lipopeptide called iturin A. These authors studied the production of iturin A in biofilm at different temperatures, being the best production obtained at 28 °C with 0.6 g L^{-1} of iturin A. Later, in 2016, Vogt et al. (2016) used *B. subtilis* biofilms and engineered a fusion protein of TasA with the red fluorescent protein mCherry, revealing that this fusion was homogeneously distributed within the biofilm and abundant throughout. In the same study, Vogt et al. (2016) also produced fusion proteins of TasA with *Echinococcus granulosus* antigenic peptides, paramyosin and tropomyosin, indicating that antigens could be expressed and were efficiently located in the biofilm matrix.

3. Production of added-value compounds in biofilms

During the past decades, bacterial biofilms have demonstrated great potential for the sustainable production of bulk and fine chemicals, as well as other added-value compounds (Halan, Buehler, & Schmid, 2012; Rosche, Li, Hauer, Schmid, & Buehler, 2009). The compounds produced by productive bacterial biofilms include the following classes: organic acids, polysaccharides, bioactive compounds, alcohols, and other products (Table 2). An assortment of reactor configurations have been utilized for the cultivation of immobilized cells and biofilms, some of them operated continuously for months (Cho et al., 1996; Ferone et al., 2018; Liu et al., 2005). The packed-bed and stirred-tank reactors are the most commonly used reactor types in the production of added-value compounds (Cheng et al., 2009a; Horiuchi et al., 2000; Napoli et al., 2010; Pongtharangkul & Demirci, 2006). Other suitable configurations also include fluidized bed reactors (Dumsday, Zhou, Buhmann, Stanley, & Pamment, 1997), rotating-disc reactors (Amin & Doelle, 1990), and membrane biofilm reactors (Gross et al., 2013). The major advantages and disadvantages of the different biofilm reactor types have already been reviewed (Cheng et al., 2010; Mahdinia & Demirci, 2020; Muffler et al., 2014; Qureshi, Annous, Ezeji, Karcher, & Maddox, 2005). Additionally, it is of crucial importance to accumulate large biomass concentrations inside the reactor. This is achieved by immobilizing cells on various supports (Ercan & Demirci, 2015). Many studies demonstrated that the implementation of plastic composite supports (PCSs) as carriers/supports for biomass immobilization enhanced the production of several chemical compounds (Mahdinia & Demirci, 2020; Todhanakasem, 2017) such as ethanol (Kunduru

TABLE 2 Examples of added-value compounds produced in bacterial biofilm reactors.

Product	Organism	Culture medium	Reactor type (feeding strategy)	Material	Maximum productivity ($g\,L^{-1}\,h^{-1}$)	Process time (h)	Ref.
Organic acids							
Lactic acid	*Lactobacillus casei* subsp. *rhamnosus*	Complex medium with glucose as CS and yeast extract as NS	Stirred-tank reactor (C)	PCS tubes attached to the agitator shaft	9.88	n.d.	Cotton, Pometto, and Gvozdenovic-Jeremic (2001)
			Stirred-tank reactor (B, FB)[a]	PCS rings and discs	4.20	40–90[b]	Velázquez, Pometto, Ho, and Demirci (2001)
	Lactobacillus delbrueckii	MRS medium with molasses as CS	Dual reactor system: packed-bed biofilm reactor and stirred-tank reactor (C)	Polyurethane foam	5	1000	Rangaswamy and Ramakrishna (2008)
		Complex medium with glucose as CS and yeast extract as NS	Tubular biofilm reactor (C)	Glass	10	504	Cuny et al. (2019)
Succinic acid	*Actinobacillus succinogenes*	Complex medium with xylose as CS and yeast extract as NS	Stirred-tank reactor (C)	Porous polypropylene or silicone protruding arms	2.64	1550	Bradfield et al. (2015)
Acetic acid	*Acetobacter pasteurianus*	Complex medium with glucose as CS and ethanol as the BS	Packed-bed biofilm reactor (C)	Charcoal pellets	6.45	4320	Horiuchi, Tabata, Kanno, and Kobayashi (2000)
Glycolic acid	*Pseudomonas diminuta*	M12 medium with ethylene glycol as the BS	Aerated trickle-bed biofilm reactor (C)	Stainless steel structured packing	1.6	1536	Li, Hauer, and Rosche (2013)

Volatile fatty acids (acetate, propionate, and butyrate)	Acidogenic microbial consortium	Olive mill wastewater	Packed-bed biofilm reactor (C)	Ceramic cubes of Vukopor S10	2.76 g COD L^{-1} day^{-1}	984	Monti, Scoma, Martinez, Bertin, and Fava (2014)
Polysaccharides							
Bacterial cellulose	*Acetobacter xylinum*	Modified corn steep liquor with fructose as CS	Stirred-tank reactor (B)	PCS tubes attached to the agitator shaft	0.059	120	Cheng, Catchmark, and Demirci (2009a, 2009b)
Xanthan gum	*Xanthomonas campestris*	Complex medium with glucose as CS and yeast extract as NS	Centrifugal packed-bed reactor (B)[a]	Rotating fibrous matrix	1	24–26[b]	Yang, Lo, and Min (1996)
Bioactive compounds							
Nisin	*Lactococcus lactis* subsp. *lactis*	M17 medium with lactose as CS and whey permeate with casein hydrolysate	Packed-bed biofilm reactor (C)	Spiral wound fibrous matrix	5.2×10^7 AU L^{-1} h^{-1}	4320	Liu, Chung, Yang, and Yousef (2005)
		Complex medium with sucrose as CS	Stirred-tank reactor (B)[a]	PCS tubes attached to the agitator shaft	5.5×10^5 IU L^{-1} h^{-1}	24[b]	Pongtharangkul and Demirci (2006)
Pediocin	*Pediococcus acidilactici*	MRS medium	Packed-bed biofilm reactor (C)	Spiral wound fibrous matrix	4.2×10^5 AU L^{-1} h^{-1}	2160	Cho, Yousef, and Yang (1996)
Alcohols and solvents							
Ethanol	*Zymomonas mobilis*	Complex medium with glucose as CS and yeast extract as NS	Vertical rotating immobilized cell reactor (C, B)[a]	Polyurethane foam discs	63	1728	Amin and Doelle (1990)
			Packed-bed biofilm reactor (C)	Polypropylene and PCS chips	536	1440	Kunduru and Pometto (1996b)
	Zymomonas mobilis pure and mixed with *Streptomyces viridosporus*	Complex medium with glucose as CS and yeast extract as NS	Packed-bed biofilm reactor (C)	PCS chips	374	24[b]	Kunduru and Pometto (1996b)
	Escherichia coli	Modified Luria broth with glucose as CS, and tryptone and yeast extract as NS	Fluidized-bed reactor (C)	Macroporous glass beads	1.12	648	Dumsday, Zhou, Buhmann, Stanley, and Pamment (1997)

Continued

TABLE 2 Examples of added-value compounds produced in bacterial biofilm reactors—cont'd

Product	Organism	Culture medium	Reactor type (feeding strategy)	Material	Maximum productivity ($g L^{-1} h^{-1}$)	Process time (h)	Ref.
1-Octanol	Recombinant *Pseudomonas putida*	M9 minimal medium with octane as BS	Biofilm membrane reactor (C)	Silicone membrane	0.05	720	Gross, Buehler, and Schmid (2013)
Cyclohexanol	*Synechocystis* sp. and *Pseudomonas taiwanensis*	YBG 11 medium with cyclohexane as BS	Capillary reactor (C)	Glass	0.2	720	Hoschek et al. (2019)
Butanol	*Clostridium acetobutylicum*	Complex media with lactose as CS and yeast extract as NS	Packed-bed biofilm reactor (C)	Tygon rings	4.4	750	Napoli, Olivieri, Russo, Marzocchella, and Salatino (2010)
Acetone, butanol, and ethanol	*Clostridia beijerinckii*	Solventogenic media containing glucose as CS and yeast extract as NS	Biofilm reactor (C)	Corn stalk pieces	5.06	480	Zhang, Ma, Yang, and Zhang (2009)
Others compounds							
Hydrogen	Anaerobic sludge	Vegetable waste extract	Acidogenic sequencing batch biofilm reactor (B)[a]	Inert stone chips	14.1 mol $_{H2}$ m^{-3} day^{-1}	24^b	Mohanakrishna, Subhash, and Venkata Mohan (2011)
		Sucrose-based synthetic wastewater	Stirred anaerobic sequencing batch biofilm reactor (FB, B)[a]	Low-density polyethylene pellets	81.2 mol $_{H2}$ m^{-3} day^{-1}	2^b	Inoue, Lima, Rodrigues, Ratusznei, and Zaiat (2014), Manssouri, Rodrigues, Ratusznei, and Zaiat (2013)
	Acidogenic microbial consortium	Dephenolized olive mill wastewater	Packed-bed biofilm reactor (C)	Ceramic cubes of Vukopor S10	250 mL $_{H2}$ L^{-1} day^{-1}	1272	Monti et al. (2014)
Polyhydroxyalkanoates	*Bacillus* sp.	Mineral salt media with date syrup	Stirred-tank reactor (B)[a]	PCS tubes attached to the agitator shaft	0.195	30^b	Khiyami, Alfadul, and Bahkali (2011)

(*S*)-Styrene oxide	*Pseudomonas* sp. strain VLB120ΔC	M9 minimal medium with glucose as CS and styrene as BS	Tubular membrane reactor (C)	Silicone membrane	2.92	1200	Gross, Lang, Bühler, and Schmid (2010)
		M9 medium with styrene as BS	Biofilm membrane reactor (C)	Microporous ceramic aeration unit	1.17	720	Halan, Schmid, and Buehler (2010)
			Solid support membrane-aerated biofilm reactor (C)	Expanded polytetrafluoroethylene membrane	1	384	Halan, Letzel, Schmid, and Buehler (2014)
Dihydroxyacetone	*Gluconobacter oxydans*	Complex medium with glycerol as CS and yeast extract as NS	Packed-bed bubble column reactor (FB)[a]	Silicone-coated Ralu rings	5.9	432	Hekmat, Bauer, and Neff (2007)
Poly(3-hydroxybutyrate)	*Alcaligenes eutrophus*	Glucose as CS	Packed-bed biofilm reactor (C)	Anion exchange resin	0.038	74	Zhang, Norrlöw, Wawrzynczyk, and Dey (2004)

Abbreviations: *AU*, Anson unit; *B*, batch culture; *BS*, biotransformation substrate; *C*, continuous feeding; *CS*, carbon source; *FB*, fed-batch culture; *IU*, international unit; *n.d.*, not described; *NS*, nitrogen source; *PCS*, plastic composite support.

[a]*Repeated-batch or fed-batch mode.*

[b]*Batch duration.*

& Pometto, 1996a), lactic acid (Cotton et al., 2001), bacterial cellulose (Cheng et al., 2009b), and polyhydroxyalkanoates (Khiyami et al., 2011). PCSs are usually fixed to the agitator shaft of stirred-tank biofilm reactors or packed in bed or fluidized reactors (rings/discs), and are commonly produced by the extrusion of a mixture of polypropylene and nutritious compounds, such as agricultural materials (e.g., oat hulls, soybean flour and corn starch), microbial nutrients (e.g., yeast extract, bovine albumin, and red blood cells), and mineral salts (Cotton et al., 2001; Demirci & Pometto, 1995; Ho, Pometto, & Hinz, 1997; Ho, Pometto, Hinz, & Demirci, 1997; Velázquez et al., 2001). These supports were extensively studied due to their capacity to stimulate biofilm formation and provide additional nutrients to the attached and suspended cells. Other materials applied as biofilm supports included ceramics (Monti et al., 2014), silicone membranes (Gross et al., 2010), polyurethane foam (Rangaswamy & Ramakrishna, 2008), and fibrous matrix (Yang et al., 1996).

3.1 Organic acids

Biofilm reactors have been studied to produce a broad variety of organic acids, in particular, lactic, succinic, acetic, glycolic, and propionic acids. Since the production of organic acids is often inhibited by the pH decrease, biofilms might be favorable for their production due to the higher robustness toward changing environmental conditions (Muffler et al., 2014). Cotton et al. (2001) designed a novel immobilized-cell biofilm reactor for the continuous production of lactic acid by *Lactobacillus casei*. They have fixed PCS tubes composed of 50% (wt/wt) agricultural products and 50% (wt/wt) polypropylene to an agitator shaft in a stirred-tank bioreactor in order to stimulate biofilm formation and supply nutrients to the attached and suspended cells while producing lactic acid. The PCS tubes demonstrated a positive effect on cell density, production rates, and yields, achieving average productivity of $9.0\,g\,L^{-1}\,h^{-1}$ and yields of approximately 70% at a dilution rate of $0.4\,h^{-1}$ and 125-rpm agitation. More recently, Cuny et al. (2019) characterized the performance of *Lactobacillus delbrueckii* for lactic acid production using a horizontal tubular biofilm reactor made of glass. The productivity of the biofilm system was maintained stable for about 3 weeks in continuous mode, demonstrating its suitability for long-term cultivation. The biofilm performance was evaluated at three different flow velocities, whereby the productivity increased with the flow velocity. At low flow velocities, the higher retention times caused a strong pH drop generated by the production of lactic acid, which significantly inhibited growth rate and production. A maximum productivity of $7–10\,g\,L^{-1}\,h^{-1}$ and a product yield of $0.94\,g\,g^{-1}$were obtained at a flow velocity of $2.12\,mm\,s^{-1}$. The biofilm system significantly increased the cell density by a factor of 19 compared with planktonic cells, which led to increased lactic acid productivity (by a factor of 6–8). Similarly, Rangaswamy and Ramakrishna (2008) found that the volumetric productivity of a dual reactor system composed of a packed-bed biofilm reactor and a stirred-tank reactor with *L. delbrueckii* immobilized in polyurethane foam was about four times higher than the productivity obtained using a suspended system.

The continuous production of succinic acid using a pure xylose feed stream and a xylose-enriched, deacetylated, dilute acid pretreated corn stover hydrolysate (DDAP-H) was explored using immobilized *Actinobacillus succinogenes* (Bradfield et al., 2015). In order to increase the surface area and cell density, and thereby support biofilm development, a stirred-tank biofilm reactor outfitted with porous polypropylene or silicone arms was operated for 1550 h. The maximum succinic acid yield on xylose was similar for both substrates ($0.78\,g\,g^{-1}$) and the maximum productivity achieved for the pure xylose fermentation ($2.64\,g\,L^{-1}\,h^{-1}$) exceeded the productivity obtained using DDAP-H ($1.77\,g\,L^{-1}\,h^{-1}$). However, the possibility of employing an industrially relevant biomass-derived feed stream to produce added-value compounds might improve the economic landscape of the process.

Also, Horiuchi et al. (2000) conducted a continuous acetic acid fermentation for 180 days in a packed-bed bioreactor using charcoal pellets derived from the waste mushroom medium as solid support for biofilm formation. The productivity of acetic acid was limited by the oxygen transfer, where a supply of O_2-enriched air (40%) enhanced the conversion of ethanol in acetic acid compared with normal aeration, achieving a maximum productivity of $6.45\,g\,L^{-1}\,h^{-1}$ at a dilution rate of $0.375\,h^{-1}$. Besides, the porosity and high specific surface area of charcoal pellets demonstrated to be advantageous for biofilm reactors.

Li et al. (2013) evaluated the effect of dilution rate on biotransformation of ethylene glycol to glycolic acid by *Pseudomonas diminuta* biofilms in an aerated trickle-bed biofilm reactor with structured packing stainless steel. For over 2 months of continuous operation, the dilution rate was periodically varied between 0.14 and $0.33\,h^{-1}$, achieving maximum steady-state productivity of $1.6\,g\,L^{-1}\,h^{-1}$ at a dilution rate of $0.33\,h^{-1}$. The results showed a positive correlation between glycolic acid productivity and the experimental dilution rates, and this biofilm reactor design demonstrated excellent reproducibility and robustness. Also, the authors assessed the productivity of suspended cells, attaining a steady-state productivity of $0.29\,g\,L^{-1}\,h^{-1}$, which was approximately 20% of the productivities using biofilms.

3.2 Polysaccharides

Compared with other valuable substances, the application of bacterial biofilm reactors for the production of polysaccharides has been sparsely studied. Bacterial cellulose, a polysaccharide commonly used in the food, paper manufacturing, and electronics industries, was successfully produced in batch fermentation by *Acetobacter xylinum* using a PCS biofilm reactor (Cheng et al., 2009b). The implementation of PCS tubes on the bioreactor enhanced bacterial cellulose production ($7.05\,g\,L^{-1}$) by 2.5-fold compared with the conventional suspended-cell cultures ($2.82\,g\,L^{-1}$). This is a consequence of the higher biomass density accumulated on PCS and the dependent relation between cell density and cellulose production. Furthermore, improved mechanical properties and thermal stability were observed for the PCS-grown bacterial cellulose.

Additionally, the production of xanthan gum was evaluated in a biofilm centrifugal packed-bed reactor with *Xanthomonas campestris* immobilized in a rotating fibrous matrix (Yang et al., 1996). The immobilized cells enabled stable productions of xanthan gum in repeated batches, reaching a xanthan yield from glucose of 85% and a volumetric xanthan productivity of around $3\,g\,L^{-1}\,h^{-1}$, in contrast with the suspended cell reactor where the productivity was 6-fold lower. The production of xanthan gum in biofilms was demonstrated to overtake the problems found in conventional stirred-tank fermentation related to the high operation costs, and agitation and aeration limitations due to the high broth viscosity.

3.3 Bioactive compounds

Several studies have reported the successful application of biofilms in the synthesis of bioactive compounds such as bacteriocins, small peptides, or proteins produced by lactic acid bacteria with antagonistic bactericidal activity against some foodborne pathogenic and spoilage microorganisms (Cotter, Ross, & Hill, 2013). The continuous production of nisin, a bacteriocin widely used as biopreservative in the food industry (Benmechernene et al., 2013), was investigated by Liu et al. (2005) using a packed-bed bioreactor with *Lactobacillus lactis* immobilized on a spiral wound fibrous matrix. This matrix accomplished effective cell immobilization by natural attachment on the fibber surfaces and entrapment in the void volume within the fibrous matrix, with about 96% of the cells being immobilized. The bioreactor was operated for 6 months at different cultivation conditions without encountering any clogging or contamination problem. Commensurable nisin productivities were observed using laboratory media ($5.2\times10^{7}\,AU\,L^{-1}\,h^{-1}$) and whey permeate ($1.0\times10^{7}\,AU\,L^{-1}\,h^{-1}$). The use of whey permeate might be economically advantageous for bacteriocin production since it is an inexpensive by-product of the cheese industry.

Analogously, Cho et al. (1996) operated a continuous bioreactor packed with a spiral wound fibrous matrix for the production of pediocin PO2 using immobilized *Pediococcus acidilactici*. Pediocin is a bacteriocin with potential use as a food preservative due to its high inhibitory action against Gram-positive bacteria, including *Listeria monocytogenes*, commonly found in foodborne outbreaks (Klaenhammer, 1988). The maximum bacteriocin productivity was $1.0\times10^{7}\,AU\,L^{-1}\,day^{-1}$ at a dilution rate of $1.58\,day^{-1}$, and no clogging or contamination were observed during the 3 months of operation, demonstrating long-term stability for bacteriocin production.

3.4 Alcohols and solvents

The production of alcohols and solvents such as ethanol, butanol, and acetone is a very prominent example of biotechnologically manufactured compounds using biofilm reactors. Over the past decades, biofilm reactors have been studied extensively for ethanol production. Amin and Doelle (1990) investigated the continuous conversion of glucose into ethanol using a vertical rotating immobilized cell reactor with *Zymomonas mobilis* entrapped in polyurethane foam discs. A volumetric ethanol productivity of $63\,g\,L^{-1}\,h^{-1}$ and an ethanol concentration of $80\,g\,L^{-1}$ were maintained after more than 72 days. In addition, since the continuous exposure of immobilized biomass to high concentrations of ethanol in the presence of residual glucose may cause cell growth inhibition, they performed a repeated-batch fermentation, whereby an efficiency of glucose conversion into ethanol of 95% and a volumetric productivity of up to $13.3\,g\,L^{-1}\,h^{-1}$ were obtained. These productivity values were described to be 10 times higher than those of the corresponding traditional free-cell fermentation. On the other side, Kunduru and Pometto (1996b) operated a biofilm reactor packed with PCS chips containing polypropylene and agricultural materials in continuous ethanol fermentation. Fermentations were performed with the ethanol producer *Z. mobilis* in pure and mixed cultures with *Streptomyces viridosporus*, a biofilm former. The pure culture of *Z. mobilis* achieved a maximum ethanol productivity of $374\,g\,L^{-1}\,h^{-1}$, against $149.5\,g\,L^{-1}\,h^{-1}$ for a mixed culture with *S. viridosporus*. The lower productivities in mixed-culture fermentations could be caused by the aeration, needed for the growth of the aerobic *S. viridosporus*, but which have been reported to decrease the cell growth rate and ethanol productivity for *Z. mobilis*. Additionally, a 3-fold decrease in ethanol productivity was obtained with *Z. mobilis* in suspended culture. The same authors tested a long-term fermentation of 60 days with *Z. mobilis* using a packed-bed reactor (Kunduru & Pometto, 1996b). Maximum ethanol productivities of 536 and $499\,g\,L^{-1}\,h^{-1}$ were obtained on polypropylene supports and soybean hull-zein-PCS, respectively, being 100-fold higher than the productivity obtained with suspended culture in continuously stirred fermenters.

Acetone, butanol, and ethanol production was also investigated by Zhang et al. (2009) using corn stalks as a support to immobilize *Clostridia beijerinckii* in glucose fermentation. Corn stalks were chopped into small pieces and treated with hydrochloric acid to activate the surface and increase the porosity and surface area. The effect of the dilution rate on solvent production was continuously examined for 20 days. Increasing the dilution rate from 0.2 to $1.0\,h^{-1}$ increased solvent productivity until a maximum of $5.06\,g\,L^{-1}\,h^{-1}$. The productivity in a suspended cell bioreactor was reported as $0.22\,g\,L^{-1}\,h^{-1}$, a much lower value than the one achieved in the biofilm reactor. Between these solvents, butanol is gaining particular interest as a biofuel because of the higher energy content in comparison with ethanol (Cascone, 2007; Qureshi et al., 2005). Butanol can be produced from a number of microbial cultures. Napoli et al. (2010) characterized butanol production by immobilizing *Clostridium acetobutylicum* in a continuous packed-bed reactor using Tygon rings as a carrier. The maximum butanol productivity was $4.4\,g\,L^{-1}\,h^{-1}$, recorded at a dilution rate of $0.97\,h^{-1}$. Besides butanol, ethanol and acetone were produced at lower concentrations. The authors also investigated the effects of pH on the reactor performance, and the results confirmed the existence of pH gradients moving deep into the biofilm, suggesting that the pH in the bulk should be set at higher values than the optimal for suspended cell processes.

Later on, Gross et al. (2013) assessed the selective hydroxylation of octane to 1-octanol by two recombinant *Pseudomonas* sp. strains (with plasmids containing the alkane monooxygenase system alkBGT/alkST) in a biofilm membrane reactor with silicone tubes used as support material. *Pseudomonas* sp. strain VLB120 pBT10 and *Pseudomonas putida* PpS81 pBT10 biofilms were able to continuously produce 1-octanol from octane with a maximum productivity of $0.74\,g\,L^{-1}\,day^{-1}$ and $1.3\,g\,L^{-1}\,day^{-1}$ for about 30 and 7 days, respectively. The volumetric productivities of biofilm experiments with the two strains were similar to a suspended cell reactor ($1.0\,g\,L^{-1}\,day^{-1}$). More recently, a dual-species high-cell density biofilm composed of the cyanobacterium *Synechocystis* sp. and *Pseudomonas taiwanensis* was applied for cyclohexanol production in a capillary reactor. Both strains contained a recombinant cyclohexane monooxygenase, which was responsible for the oxyfunctionalization of cyclohexane to cyclohexanol. These two microbial species were cocultivated due to their complementary properties regarding O_2 metabolism, since the O_2 accumulation, and consequent inhibition of the photoautotrophic *Synechocystis* sp., was overcome by using the O_2 respiring *P. taiwanensis*. This cooperation enhanced biomass formation compared with single-species biofilms, achieving high cell densities. These biofilms enabled continuous cyclohexanol production for a month with a productivity of $0.2\,g\,L^{-1}\,h^{-1}$ and achieved remarkable cyclohexane conversion (98.9%).

3.5 Miscellaneous

So far, biofilm reactors have been successfully used to produce alcohols, acids, polysaccharides, and antimicrobials. Other examples of industrial chemicals produced in biofilm reactors include biohydrogen, (*S*)-styrene oxide, benzaldehyde, and dihydroxyacetone. Hydrogen is an interesting eco-friendly alternative energy source because of its high energy potential and combustion yields (Manssouri et al., 2013). The biological anaerobic fermentation in the production of hydrogen from different wastewaters has been applied in biofilm reactors (Venkata Mohan, Vijaya Bhaskar, & Sarma, 2007). Inoue et al. (2014) and Manssouri et al. (2013) developed similar studies operating stirred anaerobic sequencing batch biofilm reactor (AnSBBR) for biohydrogen production by an anaerobic sludge with diverse microbial population from sucrose-based synthetic wastewater. Biomass was immobilized on low-density polyethylene pellets. The maximum daily molar productivity of hydrogen obtained by Manssouri et al. (2013) was $39.9\,mol_{\,H2}\,m^{-3}\,day^{-1}$ using a batch strategy, whereas Inoue et al. (2014) found that the formation of hydrogen was higher for the fed-batch strategy than for the batch strategy, reaching a maximum of $81.2\,mol_{\,H2}\,m^{-3}\,day^{-1}$. Additionally, fill time in the fed-batch mode was shown to interfere with substrate consumption rates and hydrogen formation, minimizing the inhibition caused by substrate excess. Lower productivity values of $14.1\,mol_{\,H2}\,m^{-3}\,day^{-1}$ were obtained by Mohanakrishna et al. (2011) in an AnSBBR with inert stone chips used as fixed-bed packing material using vegetable-based solid waste extract as substrate.

Khiyami et al. (2011) optimized the production of polyhydroxyalkanoates (PHAs), a renewable and biodegradable group of biopolymers with physical characteristics close to oil-based plastics, using *Bacillus* sp. biofilms. Biofilms were developed on PCS tubes in repeated-batch cultures. The authors investigated the effect of different concentrations of date syrup, agitations, nitrogen sources, and batch periods on the accumulation of PHA in cells. PHA accumulation and cellular dry matter were significantly higher with a date syrup concentration of 15% (v/v), $0.5\,g\,L^{-1}$ urea as nitrogen source, 120-rpm agitation, and 30 h of fermentation. The maximum cellular dry matter was $8.3\,g\,L^{-1}$ and the PHA accumulation was $70.5\,g\,L^{-1}$. Comparing the accumulated PHA in suspended cells and PCS biofilms, PCS stimulated the development of biofilm and, consequently, PHA production was higher.

The robustness of biofilm reactors is the main argument for their application in producing foreign compounds for the producing strain. Moreover, biofilm membrane reactors were employed for the continuous epoxidation of styrene to (*S*)-styrene oxide, a toxic and unstable high-value chemical synthon, using the engineered *Pseudomonas* sp. strain VLB120DC as biocatalyst (Gross et al., 2010; Halan et al., 2014). Gross et al. (2010) used a tubular membrane reactor with a silicone membrane as biofilm support. A maximum (*S*)-styrene oxide volumetric productivity of $70\,g\,L^{-1}\,day^{-1}$ was achieved over more than 50 days. A critical aspect in the biofilm reactor that influenced the biotransformation rate was oxygen limitation, as a consequence of the high biomass concentrations obtained. Productivity was also linearly dependent on the specific membrane area and the tube wall thickness. On the other side, Halan et al. (2014) utilized a solid support membrane-aerated biofilm reactor with an expanded polytetrafluoroethylene membrane as biofilm substratum. The catalytic biofilms of *Pseudomonas* sp. strain VLB120ΔC produced (*S*)-styrene oxide with a productivity value of $24.4\,g\,L^{-1}\,day^{-1}$.

Finally, Hekmat et al. (2007) explored the effect of *Gluconobacter oxydans* immobilization on Ralu-rings coated with a porous silicone matrix for the production of dihydroxyacetone (DHA). A repeated-fed-batch feeding strategy was operated in a packed-bed bubble column reactor with and without cell immobilization. The DHA yield from glycerol for the experiments with and without cell immobilization was similar (0.87 and $0.85\,kg\,kg^{-1}$, respectively). However, the product formation rate increased from $3.7\,kg\,m^{-3}\,h^{-1}$ in suspended biomass to $5.9\,kg\,m^{-3}\,h^{-1}$ in immobilized cells.

4. Conclusions and future remarks

Biofilm reactors demonstrated high potential to be employed in biotechnology/biotransformation processes that involve various chemicals. However, the production of recombinant proteins in biofilms is taking its first steps. A comparison of biofilm reactors with suspended cell cultures demonstrated that biofilm reactors can be operated for a longer period, offering higher productivities. The good operational stability, increased tolerance of biofilms to toxic substrates and products, the robustness of the immobilized cells toward fluctuating process conditions, and high cell densities found in biofilm reactors demonstrated to enhance the volumetric productivity rates of several products, even on dilute feed streams. Although the production of chemicals has been consistently reported at a bench scale, the implementation of commercial biofilm processes has not been fully established. Thus, industrial biotechnology needs to take the next leap to the implementation of this laboratory pilot-scale biofilm reactors into large-scale production. This can be achieved with a deep knowledge of the fundamentals physical and chemical properties of a biofilm to run the biofilm reactor at optimum conditions, as well as the factors that improve biofilm formation, such as support materials and bioreactor configuration. Since biofilm systems can be unpredictable due to the dynamic and complex interactions with the liquid bulk, the lack of biofilm reproducibility can be an obstacle for industrial applications. Scale-up studies on several parameters, such as culture conditions, mass and heat transfer constraints, kinetics, and production modeling are needed beforehand.

Thus, one of the biggest challenges to be addressed in the upcoming years is to provide modeling tools to predict biofilm behavior and develop analytical tools to obtain real-time data from inside the bioreactors (Halan et al., 2012). One of the driving forces for the increased application of biofilm processes is the stress to produce bulk chemicals with a low ecological footprint. Some studies have been employing agro-wastes and biorefinery residues for their bioconversion into valuable chemical and pharmaceutic compounds in order to meet economic process sustainability (Bengtsson et al., 2017; Longanesi, Frascari, Spagni, DeWever, & Pinelli, 2018; Vecchiato et al., 2019). A potential application for biofilms in the pharmaceutic field is the use of enzyme cascades within biofilm cells to synthesize complex products (Rosche et al., 2009).

To conclude, it is believed that productive biofilms have a strong potential for application in several areas, particularly in the production of a range of chemicals, food additives, biofuels, and bioactive compounds.

Acknowledgments

This work was financially supported by Base Funding-UIDB/00511/2020 of the Laboratory for Process Engineering, Environment, Biotechnology and Energy-LEPABE—funded by national funds through the FCT/MCTES (PIDDAC), and by Project PTDC/BII-BIO/29589/2017-POCI-01-0145-FEDER-029589—funded by FEDER funds through COMPETE2020-Programa Operacional Competitividade e Internacionalização (POCI) and by national funds (PIDDAC) through FCT/MCTES. A.A. acknowledges the receipt of a Ph.D. Grant from the Portuguese Foundation from Science and Technology (FCT) (2020.07427.BD). L.C.G. thanks FCT for the financial support of her work contract through the Scientific Employment Stimulus-Individual Call—[CEECIND/01700/2017].

References

Alves, P., Gomes, L. C., Vorobii, M., Rodriguez-Emmenegger, C., & Mergulhão, F. J. (2020). The potential advantages of using a poly(HPMA) brush in urinary catheters: Effects on biofilm cells and architecture. *Colloids and Surfaces B: Biointerfaces*, *191*, 110976. https://doi.org/10.1016/j.colsurfb.2020.110976.

Amin, G., & Doelle, H. W. (1990). Production of high ethanol concentrations from glucose using a vertical rotating immobilized cell reactor of the bacterium *Zymomonas mobilis*. *Acta Biotechnologica*, *10*(1), 35–40. https://doi.org/10.1002/abio.370100110.

Becker, J., & Wittmann, C. (2016). Systems metabolic engineering of *Escherichia coli* for the heterologous production of high value molecules—A veteran at new shores. *Current Opinion in Biotechnology*, *42*, 178–188. https://doi.org/10.1016/j.copbio.2016.05.004.

Bengtsson, S., Karlsson, A., Alexandersson, T., Quadri, L., Hjort, M., Johansson, P., et al. (2017). A process for polyhydroxyalkanoate (PHA) production from municipal wastewater treatment with biological carbon and nitrogen removal demonstrated at pilot-scale. *New Biotechnology, 35*, 42–53. https://doi.org/10.1016/j.nbt.2016.11.005.

Benmechernene, Z., Fernandez-No, I., Kihal, M., Böhme, K., Calo-Mata, P., & Barros-Velazquez, J. (2013). Recent patents on bacteriocins: Food and biomedical applications. *Recent Patents on DNA & Gene Sequences, 7* (1), 66–73. https://doi.org/10.2174/187221561130701001O.

Berlanga, M., & Guerrero, R. (2016). Living together in biofilms: The microbial cell factory and its biotechnological implications. *Microbial Cell Factories, 15*(1), 165. https://doi.org/10.1186/s12934-016-0569-5.

Bradfield, M. F., Mohagheghi, A., Salvachúa, D., Smith, H., Black, B. A., Dowe, N., et al. (2015). Continuous succinic acid production by *Actinobacillus succinogenes* on xylose-enriched hydrolysate. *Biotechnology for Biofuels, 8*, 181. https://doi.org/10.1186/s13068-015-0363-3.

Carrascosa, C., Raheem, D., Ramos, F., Saraiva, A., & Raposo, A. (2021). Microbial biofilms in the food industry—A comprehensive review. *International Journal of Environmental Research and Public Health, 18*(4), 2014. https://doi.org/10.3390/ijerph18042014.

Cascone, R. (2007). Biobutanol—A replacement for bioethanol? *Chemical Engineering Progress, 104*, 4.

Chen, R. (2012). Bacterial expression systems for recombinant protein production: *E. coli* and beyond. *Biotechnology Advances, 30*(5), 1102–1107. https://doi.org/10.1016/j.biotechadv.2011.09.013.

Chen, X., Zhou, L., Tian, K., Kumar, A., Singh, S., Prior, B. A., et al. (2013). Metabolic engineering of *Escherichia coli*: A sustainable industrial platform for bio-based chemical production. *Biotechnology Advances, 31*(8), 1200–1223. https://doi.org/10.1016/j.biotechadv.2013.02.009.

Cheng, K.-C., Catchmark, J. M., & Demirci, A. (2009a). Effect of different additives on bacterial cellulose production by *Acetobacter xylinum* and analysis of material property. *Cellulose, 16*(6), 1033–1045. https://doi.org/10.1007/s10570-009-9346-5.

Cheng, K.-C., Catchmark, J. M., & Demirci, A. (2009b). Enhanced production of bacterial cellulose by using a biofilm reactor and its material property analysis. *Journal of Biological Engineering, 3*, 12. https://doi.org/10.1186/1754-1611-3-12.

Cheng, K. C., Demirci, A., & Catchmark, J. M. (2010). Advances in biofilm reactors for production of value-added products. *Applied Microbiology and Biotechnology, 87*(2), 445–456. https://doi.org/10.1007/s00253-010-2622-3.

Cho, H. Y., Yousef, A. E., & Yang, S. T. (1996). Continuous production of pediocin by immobilized *Pediococcus acidilactici* PO2 in a packed-bed bioreactor. *Applied Microbiology and Biotechnology, 45*(5), 589–594. https://doi.org/10.1007/s002530050734.

Cotter, P. D., Ross, R. P., & Hill, C. (2013). Bacteriocins—A viable alternative to antibiotics? *Nature Reviews Microbiology, 11*(2), 95–105. https://doi.org/10.1038/nrmicro2937.

Cotton, J. C., Pometto, A. L., 3rd, & Gvozdenovic-Jeremic, J. (2001). Continuous lactic acid fermentation using a plastic composite support biofilm reactor. *Applied Microbiology and Biotechnology, 57*(5–6), 626–630. https://doi.org/10.1007/s002530100820.

Cuny, L., Pfaff, D., Luther, J., Ranzinger, F., Ödman, P., Gescher, J., et al. (2019). Evaluation of productive biofilms for continuous lactic acid production. *Biotechnology and Bioengineering, 116*(10), 2687–2697. https://doi.org/10.1002/bit.27080.

Demain, A. L., & Vaishnav, P. (2009). Production of recombinant proteins by microbes and higher organisms. *Biotechnology Advances, 27*, 297–306.

Demirci, A., & Pometto, A. L. (1995). Repeated-batch fermentation in biofilm reactors with plastic-composite supports for lactic acid production. *Applied Microbiology and Biotechnology, 43*(4), 585–589. https://doi.org/10.1007/BF00164758.

Demirci, A., Pongtharangkul, T., & Pometto, A. L. (2007). Applications of biofilm reactors for production of value-added products by microbial fermentation. In A. L. Pometto, & A. Demirci (Eds.), *Biofilms in the food environment* (pp. 167–190). Wiley.

Donovan, R. S., Robinson, C. W., & Glick, B. R. (1996). Review: Optimizing inducer and culture conditions for expression of foreign proteins under the control of the lac promoter. *Journal of Industrial Microbiology, 16*(3), 145–154. https://doi.org/10.1007/BF01569997.

Dumsday, G. J., Zhou, B., Buhmann, S., Stanley, G. A., & Pamment, N. B. (1997). Continuous ethanol production by *Escherichia coli* KO11 in continuous stirred tank and fluidized bed fermenters. *Australasian Biotechnology, 7*(4), 300–303.

Edwards, S. J., & Kjellerup, B. V. (2013). Applications of biofilms in bioremediation and biotransformation of persistent organic pollutants, pharmaceuticals/personal care products, and heavy metals. *Applied Microbiology and Biotechnology, 97*(23), 9909–9921. https://doi.org/10.1007/s00253-013-5216-z.

Ercan, D., & Demirci, A. (2015). Current and future trends for biofilm reactors for fermentation processes. *Critical Reviews in Biotechnology, 35*(1), 1–14. https://doi.org/10.3109/07388551.2013.793170.

Ferone, M., Raganati, F., Ercole, A., Olivieri, G., Salatino, P., & Marzocchella, A. (2018). Continuous succinic acid fermentation by *Actinobacillus succinogenes* in a packed-bed biofilm reactor. *Biotechnology for Biofuels, 11*(1), 138. https://doi.org/10.1186/s13068-018-1143-7.

Gomes, L. C., Carvalho, D., Briandet, R., & Mergulhão, F. J. (2016). Temporal variation of recombinant protein expression in *Escherichia coli* biofilms analysed at single-cell level. *Process Biochemistry, 51*, 1155–1161.

Gomes, M., Gomes, L. C., Teixeira-Santos, R., Pereira, M. F. R., Soares, O. S. G. P., & Mergulhão, F. J. (2021). Optimizing CNT loading in antimicrobial composites for urinary tract application. *Applied Sciences, 11*(9). https://doi.org/10.3390/app11094038.

Gomes, L. C., & Mergulhão, F. J. (2017a). Heterologous protein production in *Escherichia coli* biofilms: A non-conventional form of high cell density cultivation. *Process Biochemistry, 57*. https://doi.org/10.1016/j.procbio.2017.03.018.

Gomes, L. C., & Mergulhão, F. J. (2017b). Effects of antibiotic concentration and nutrient medium composition on *Escherichia coli* biofilm formation and green fluorescent protein expression. *FEMS Microbiology Letters, 364*(5). https://doi.org/10.1093/femsle/fnx042.

Gomes, L. C., & Mergulhão, F. J. (2019). Production of recombinant proteins in *Escherichia coli* biofilms: Challenges and opportunities. In *Advances in medicine and biology* (p. 181). Nova Science Publishers.

Gomes, L. C., & Mergulhão, F. J. M. (2021). A selection of platforms to evaluate surface adhesion and biofilm formation in controlled hydrodynamic conditions. *Microorganisms, 9*(9), 1993. https://doi.org/10.3390/microorganisms9091993.

Gomes, L. C., Monteiro, G. A., & Mergulhão, F. J. (2020). The impact of IPTG induction on plasmid stability and heterologous protein expression by *Escherichia coli* biofilms. *International Journal of Molecular Sciences*, *21*(2). https://doi.org/10.3390/ijms21020576.

Gomes, L. C., Moreira, J. M., Teodósio, J. S., Araújo, J. D., Miranda, J. M., Simões, M., et al. (2014). 96-well microtiter plates for biofouling simulation in biomedical settings. *Biofouling*, *30*(5), 535–546. https://doi.org/10.1080/08927014.2014.890713.

Gross, R., Buehler, K., & Schmid, A. (2013). Engineered catalytic biofilms for continuous large scale production of n-octanol and (S)-styrene oxide. *Biotechnology and Bioengineering*, *110*(2), 424–436. https://doi.org/10.1002/bit.24629.

Gross, R., Lang, K., Bühler, K., & Schmid, A. (2010). Characterization of a biofilm membrane reactor and its prospects for fine chemical synthesis. *Biotechnology and Bioengineering*, *105*(4), 705–717. https://doi.org/10.1002/bit.22584.

Halan, B., Buehler, K., & Schmid, A. (2012). Biofilms as living catalysts in continuous chemical syntheses. *Trends in Biotechnology*, *30*(9), 453–465. https://doi.org/10.1016/j.tibtech.2012.05.003.

Halan, B., Letzel, T., Schmid, A., & Buehler, K. (2014). Solid support membrane-aerated catalytic biofilm reactor for the continuous synthesis of (S)-styrene oxide at gram scale. *Biotechnology Journal*, *9* (10), 1339–1349. https://doi.org/10.1002/biot.201400341.

Halan, B., Schmid, A., & Buehler, K. (2010). Maximizing the productivity of catalytic biofilms on solid supports in membrane aerated reactors. *Biotechnology and Bioengineering*, *106*(4), 516–527. https://doi.org/10.1002/bit.22732.

Hekmat, D., Bauer, R., & Neff, V. (2007). Optimization of the microbial synthesis of dihydroxyacetone in a semi-continuous repeated-fed-batch process by in situ immobilization of *Gluconobacter oxydans*. *Process Biochemistry*, *42*, 71–76. https://doi.org/10.1016/j.procbio.2006.07.026.

Ho, K. L., Pometto, A. L., 3rd, & Hinz, P. N. (1997). Optimization of L-(+)-lactic acid production by ring and disc plastic composite supports through repeated-batch biofilm fermentation. *Applied and Environmental Microbiology*, *63*(7), 2533–2542. https://doi.org/10.1128/aem.63.7.2533-2542.1997.

Ho, K. G., Pometto, A. I., Hinz, P. N., & Demirci, A. (1997). Nutrient leaching and end product accumulation in plastic composite supports for L-(+)-lactic acid biofilm fermentation. *Applied and Environmental Microbiology*, *63*(7), 2524–2532. https://doi.org/10.1128/aem.63.7.2524-2532.1997.

Hoffmann, F., & Rinas, U. (2004). Stress induced by recombinant protein production in *Escherichia coli*. *Advances in Biochemical Engineering/Biotechnology*, *89*, 73–92. https://doi.org/10.1007/b93994.

Horiuchi, J., Tabata, K., Kanno, T., & Kobayashi, M. (2000). Continuous acetic acid production by a packed bed bioreactor employing charcoal pellets derived from waste mushroom medium. *Journal of Bioscience and Bioengineering*, *89*(2), 126–130. https://doi.org/10.1016/s1389-1723(00)88725-3.

Hoschek, A., Heuschkel, I., Schmid, A., Bühler, B., Karande, R., & Bühler, K. (2019). Mixed-species biofilms for high-cell-density application of *Synechocystis* sp. PCC 6803 in capillary reactors for continuous cyclohexane oxidation to cyclohexanol. *Bioresource Technology*, *282*, 171–178. https://doi.org/10.1016/j.biortech.2019.02.093.

Huang, C. J., Lin, H., & Yang, X. (2012). Industrial production of recombinant therapeutics in *Escherichia coli* and its recent advancements. *Journal of Industrial Microbiology and Biotechnology*, *39*(3), 383–399. https://doi.org/10.1007/s10295-011-1082-9.

Huang, C. T., Peretti, S. W., & Bryers, J. D. (1992). Plasmid retention and gene expression in suspended and biofilm cultures of recombinant *Escherichia coli* DH5α(pMJR1750). *Biotechnology and Bioengineering*, *41*(2), 211–220. https://doi.org/10.1002/bit.260410207.

Huang, C. T., Peretti, S. W., & Bryers, J. D. (1994a). Effects of inducer levels on a recombinant bacterial biofilm formation and gene expression. *Biotechnology Letters*, *16*, 903–908.

Huang, C. T., Peretti, S. W., & Bryers, J. D. (1994b). Effects of medium carbon-to-nitrogen ratio on biofilm formation and plasmid stability. *Biotechnology and Bioengineering*, *44*(3), 329–336. https://doi.org/10.1002/bit.260440310.

Inoue, R. K., Lima, D. M., Rodrigues, J. A., Ratusznei, S. M., & Zaiat, M. (2014). Effect of organic loading rate and fill time on the biohydrogen production in a mechanically stirred AnSBBR treating synthetic sucrose-based wastewater. *Applied Biochemistry and Biotechnology*, *174*(6), 2326–2349. https://doi.org/10.1007/s12010-014-1205-7.

Khiyami, M., Alfadul, S., & Bahkali, A. (2011). Polyhydroxyalkanoates production via *Bacillus* PCS biofilm and date palm syrup. *Journal of Medicinal Plant Research*, *5*, 3312–3320.

Klaenhammer, T. R. (1988). Bacteriocins of lactic acid bacteria. *Biochimie*, *70*(3), 337–349. https://doi.org/10.1016/0300-9084(88)90206-4.

Kunduru, M. R., & Pometto, A. L. (1996a). Evaluation of plastic composite-supports for enhanced ethanol production in biofilm reactors. *Journal of Industrial Microbiology*, *16*(4), 241–248. https://doi.org/10.1007/BF01570028.

Kunduru, M. R., & Pometto, A. L. (1996b). Continuous ethanol production by *Zymomonas mobilis* and *Saccharomyces cerevisiae* in biofilm reactors. *Journal of Industrial Microbiology*, *16*(4), 249–256. https://doi.org/10.1007/bf01570029.

Lewandowski, Z., & Boltz, J. (2011). Biofilms in water and wastewater treatment. In *Vol. 4. Water-quality engineering* (pp. 529–570). Elsevier. https://doi.org/10.1016/B978-0-444-53199-5.00095-6.

Li, X. Z., Hauer, B., & Rosche, B. (2013). Catalytic biofilms on structured packing for the production of glycolic acid. *Journal of Microbiology and Biotechnology*, *23*(2), 195–204. https://doi.org/10.4014/jmb.1207.07057.

Liu, X., Chung, Y.-K., Yang, S.-T., & Yousef, A. E. (2005). Continuous nisin production in laboratory media and whey permeate by immobilized *Lactococcus lactis*. *Process Biochemistry*, *40*(1), 13–24. https://doi.org/10.1016/j.procbio.2003.11.032.

Longanesi, L., Frascari, D., Spagni, C., DeWever, H., & Pinelli, D. (2018). Succinic acid production from cheese whey by biofilms of *Actinobacillus succinogenes*: Packed bed bioreactor tests. *Journal of Chemical Technology & Biotechnology*, *93*(1), 246–256. https://doi.org/10.1002/jctb.5347.

Mahdinia, E., & Demirci, A. (2020). Biofilms in fermentation for the production of value-added products. In *Microbial biofilms* (pp. 73–108). CRC Press. https://doi.org/10.1201/9780367415075-6.

Manssouri, M., Rodrigues, J. A., Ratusznei, S. M., & Zaiat, M. (2013). Effects of organic loading, influent concentration, and feed time on biohydrogen production in a mechanically stirred AnSBBR treating sucrose-based wastewater. *Applied Biochemistry and Biotechnology*, *171*(7), 1832–1854. https://doi.org/10.1007/s12010-013-0457-y.

Mergulhão, F. M., Gabriel, A. M., Cabral, J. M. S., & Taipa, M. A. (2003). Design of bacterial vector systems for the production of recombinant proteins in *Escherichia coli*. *Journal of Microbiology and Biotechnology*, *14*(1), 1–14.

Mergulhão, F., Monteiro, G., Cabral, J., & Taipa, M. A. (2004). Design of bacterial vector systems for the production of recombinant proteins in *Escherichia coli*. *Journal of Microbiology and Biotechnology*, *14*, 1–14.

Mergulhão, F. J., Summers, D. K., & Monteiro, G. A. (2005). Recombinant protein secretion in *Escherichia coli*. *Biotechnology Advances*, *23*(3), 177–202. https://doi.org/10.1016/j.biotechadv.2004.11.003.

Mohanakrishna, G., Subhash, D. G. V., & Venkata Mohan, S. (2011). Adaptation of biohydrogen producing reactor to higher substrate load: Redox controlled process integration strategy to overcome limitations. *International Journal of Hydrogen Energy*, *36*. https://doi.org/10.1016/j.ijhydene.2011.04.138.

Monti, M., Scoma, A., Martinez, G., Bertin, L., & Fava, F. (2014). Uncoupled hydrogen and volatile fatty acids generation in a two-step biotechnological anaerobic process fed with actual site wastewater. *New Biotechnology*, 32. https://doi.org/10.1016/j.nbt.2014.08.002.

Moreira, J. M. R., Fulgêncio, R., Alves, P., Machado, I., Bialuch, I., Melo, L. F., et al. (2016). Evaluation of SICAN performance for biofouling mitigation in the food industry. *Food Control*, *62*, 201–207. https://doi.org/10.1016/j.foodcont.2015.10.023.

Muffler, K., Lakatos, M., Schlegel, C., Strieth, D., Kuhne, S., & Ulber, R. (2014). Application of biofilm bioreactors in white biotechnology. *Advances in Biochemical Engineering/Biotechnology*, *146*, 123–161. https://doi.org/10.1007/10_2013_267.

Muhammad, M. H., Idris, A. L., Fan, X., Guo, Y., Yu, Y., Jin, X., et al. (2020). Beyond risk: Bacterial biofilms and their regulating approaches. *Frontiers in Microbiology*, *11*, 928. https://doi.org/10.3389/fmicb.2020.00928.

Napoli, F., Olivieri, G., Russo, M. E., Marzocchella, A., & Salatino, P. (2010). Butanol production by *Clostridium acetobutylicum* in a continuous packed bed reactor. *Journal of Industrial Microbiology and Biotechnology*, *37*(6), 603–608. https://doi.org/10.1007/s10295-010-0707-8.

O'Connell, H. A., Niu, C., & Gilbert, E. S. (2007). Enhanced high copy number plasmid maintenance and heterologous protein production in an *Escherichia coli* biofilm. *Biotechnology and Bioengineering*, *97*(3), 439–446. https://doi.org/10.1002/bit.21240.

Overton, T. W. (2014). Recombinant protein production in bacterial hosts. *Drug Discovery Today*, *19*(5), 590–601. https://doi.org/10.1016/j.drudis.2013.11.008.

Pongtharangkul, T., & Demirci, A. (2006). Evaluation of culture medium for nisin production in a repeated-batch biofilm reactor. *Biotechnology Progress*, *22*(1), 217–224. https://doi.org/10.1021/bp050295q.

Qureshi, N., Annous, B. A., Ezeji, T. C., Karcher, P., & Maddox, I. S. (2005). Biofilm reactors for industrial bioconversion processes: Employing potential of enhanced reaction rates. *Microbial Cell Factories*, *4*(1), 24. https://doi.org/10.1186/1475-2859-4-24.

Rahman, M. S., Ano, T., & Shoda, M. (2007). Biofilm fermentation of iturin A by a recombinant strain of *Bacillus subtilis* 168. *Journal of Biotechnology*, *127*(3), 503–507. https://doi.org/10.1016/j.jbiotec.2006.07.013.

Rangaswamy, V., & Ramakrishna, S. V. (2008). Lactic acid production by *Lactobacillus delbrueckii* in a dual reactor system using packed bed biofilm reactor. *Letters of Applied Microbiology*, *46*(6), 661–666. https://doi.org/10.1111/j.1472-765X.2008.02362.x.

Roberts, C. G. (2013). The role of biofilms in reprocessing medical devices. *American Journal of Infection Control*, *41*(5 Suppl), S77–S80. https://doi.org/10.1016/j.ajic.2012.12.008.

Rosano, G. L., & Ceccarelli, E. A. (2014). Recombinant protein expression in *Escherichia coli*: Advances and challenges. *Frontiers in Microbiology*, *5*, 172. https://doi.org/10.3389/fmicb.2014.00172.

Rosche, B., Li, X. Z., Hauer, B., Schmid, A., & Buehler, K. (2009). Microbial biofilms: A concept for industrial catalysis? *Trends in Biotechnology*, *27*(11), 636–643. https://doi.org/10.1016/j.tibtech.2009.08.001.

Roy, R., Tiwari, M., Donelli, G., & Tiwari, V. (2018). Strategies for combating bacterial biofilms: A focus on anti-biofilm agents and their mechanisms of action. *Virulence*, *9*(1), 522–554. https://doi.org/10.1080/21505594.2017.1313372.

Setyawati, M. I., Chien, L. J., & Lee, C. K. (2008). Self-immobilized recombinant *Acetobacter xylinum* for biotransformation. *Biochemical Engineering Journal*, *43*, 78–84. https://doi.org/10.1016/j.bej.2008.09.002.

Soares, A., Azevedo, A., Gomes, L. C., & Mergulhão, F. J. (2019). Recombinant protein expression in biofilms. *AIMS Microbiology*, *5*(3), 232–250. https://doi.org/10.3934/microbiol.2019.3.232.

Soares, A., Gomes, L. C., & Mergulhão, F. J. (2018). Comparing the recombinant protein production potential of planktonic and biofilm cells. *Microorganisms*, *6*(2). https://doi.org/10.3390/microorganisms6020048.

Soares, A., Gomes, L., Monteiro, G., & Mergulhão, F. (2021). The influence of nutrient medium composition on *Escherichia coli* biofilm development and heterologous protein expression. *Applied Sciences*, *11*(8867). https://doi.org/10.3390/app11188667.

Teodósio, J. S., Simões, M., Melo, L. F., & Mergulhão, F. J. (2011). Flow cell hydrodynamics and their effects on *E. coli* biofilm formation under different nutrient conditions and turbulent flow. *Biofouling*, *27*(1), 1–11. https://doi.org/10.1080/08927014.2010.535206.

Terpe, K. (2006). Overview of bacterial expression systems for heterologous protein production: From molecular and biochemical fundamentals to commercial systems. *Applied Microbiology and Biotechnology*, *72*(2), 211–222. https://doi.org/10.1007/s00253-006-0465-8.

Timmusk, S., Behers, L., Muthoni, J., Muraya, A., & Aronsson, A.-C. (2017). Perspectives and challenges of microbial application for crop improvement. *Frontiers in Plant Science*, *8*(49). https://doi.org/10.3389/fpls.2017.00049.

Todhanakasem, T. (2017). Developing microbial biofilm as a robust biocatalyst and its challenges. *Biocatalysis and Biotransformation*, *35*(2), 86–95. https://doi.org/10.1080/10242422.2017.1295230.

Vecchiato, S., Skopek, L., Russmayer, H., Steiger, M. G., Aldrian, A., Beer, B., et al. (2019). Microbial production of high value molecules using rayon waste material as carbon-source. *New Biotechnology*, *51*, 8–13. https://doi.org/10.1016/j.nbt.2019.01.010.

Velázquez, A. C., Pometto, A. L., 3rd, Ho, K. L., & Demirci, A. (2001). Evaluation of plastic-composite supports in repeated fed-batch biofilm lactic acid fermentation by *Lactobacillus casei*. *Applied Microbiology and Biotechnology*, *55*(4), 434–441. https://doi.org/10.1007/s002530000530.

Venkata Mohan, S., Vijaya Bhaskar, Y., & Sarma, P. N. (2007). Biohydrogen production from chemical wastewater treatment in biofilm configured reactor operated in periodic discontinuous batch mode by selectively enriched anaerobic mixed consortia. *Water Research*, *41*(12), 2652–2664. https://doi.org/10.1016/j.watres.2007.02.015.

Vogt, C. M., Schraner, E. M., Aguilar, C., & Eichwald, C. (2016). Heterologous expression of antigenic peptides in *Bacillus subtilis* biofilms. *Microbial Cell Factories*, *15*(1), 137. https://doi.org/10.1186/s12934-016-0532-5.

Yan, J., & Bassler, B. L. (2019). Surviving as a community: Antibiotic tolerance and persistence in bacterial biofilms. *Cell Host & Microbe*, *26* (1), 15–21. https://doi.org/10.1016/j.chom.2019.06.002.

Yang, S.-T., Lo, Y.-M., & Min, D. B. (1996). Xanthan gum fermentation by *Xanthomonas campestris* immobilized in a novel centrifugal fibrous-bed bioreactor. *Biotechnology Progress*, *12*(5), 630–637. https://doi.org/10.1021/bp9600501.

Zhang, Y., Ma, Y., Yang, F., & Zhang, C. (2009). Continuous acetone-butanol-ethanol production by corn stalk immobilized cells. *Journal of Industrial Microbiology and Biotechnology*, *36*(8), 1117–1121. https://doi.org/10.1007/s10295-009-0582-3.

Zhang, S., Norrlöw, O., Wawrzynczyk, J., & Dey, E. S. (2004). Poly(3-hydroxybutyrate) biosynthesis in the biofilm of *Alcaligenes eutrophus*, using glucose enzymatically released from pulp fiber sludge. *Applied and Environmental Microbiology*, *70*(11), 6776–6782. https://doi.org/10.1128/aem.70.11.6776-6782.2004.

Chapter 38

Biofilms as sustainable tools for environmental biotechnologies: An interdisciplinary approach

Maria Pia Busnelli[a,*], Irene C. Lazzarini Behrmann[a,*], Andrea M. Monroy[a,b,*], Maria Alejandra Daniel[a], and Diana L. Vullo[a,b,*]

[a]*Chemistry Area, National University General Sarmiento, Buenos Aires, Argentina,* [b]*CONICET, Buenos Aires, Argentina*

1. Introduction to biofilms in environmental biotechnology

1.1 Beneficial properties of the biofilm state

Bacteria act according to changes in their environment and the formation of biofilms is one of the utmost fascinating adaptations, showing the ease in which the bacterial cells counteract the rough surroundings and interact with them. The impact of biofilm formation on ecology, medicine, and industry has led to a substantial effort by scientists to improve their understanding of microbial communities (Singh, Singh, Haibi, & Singh, 2020). Biofilms have been defined as aggregates of microorganisms. The term "aggregate" is used because most cells in multilayered biofilms experience cell-to-cell contact, either in surface-attached biofilms, in which only one layer is in direct contact with the substratum, or in flocs, which are mobile biofilms formed in the absence of any substratum. In all cases, they represent survival strategies in response to various environmental disturbances, especially the self-aggregation that can be generated by the shortage of oxygen, conditions of lack of nutrients, temperature changes, predation, or other situations of environmental stress (Trunk, Khalil, & Leo, 2018). The aggregation process is usually considered as a step prior to adhesion to a surface for the biofilm establishment described as the second microbial lifestyle. However, the formation of clusters that remain in suspension has already been described as a third microbial lifestyle (Cai, 2020). This ubiquitous distribution of bacterial clusters suspended in aqueous ecosystems, made up of multiple or single species, leads to further study of this phenomenon in terms of knowing the causes that trigger their formation dynamics or what is the function or role of this style of microbial life. From controlling the pathogenicity of certain self-aggregating strains to optimizing wastewater treatment processes through proper maintenance of flocs, center the main objectives of interest on which research is currently focused on. Intercellular interactions, both social and physical along with the matrix properties and changes in gene expression and regulation, make the biofilm cells differ both phenotypically and metabolically from the free-living bacterial cells. Therefore, biofilm communities have characteristics that are not predictable from the study of free-living bacterial cells (Ghiasian, 2020). After microorganisms have successfully attached to a surface, they form aggregates and produce an extracellular polymeric matrix that holds the cells together and provides protection from environmental threats. Extracellular bacterial structures such as pili, fimbria, and flagella can stabilize the matrix and create bridges between cells and surfaces. The biofilm matrix is a complex microenvironment that contains proteins, polysaccharides, DNA, RNA, lipids, dissolved nutrients, phages, and host-derived substances. High-molecular-weight matrix compounds such as adhesive pili, capsular polysaccharides, and extracellular DNA (eDNA) are known to mediate many basic biofilm phenotypes including surface attachment, intercellular adhesion, and biocide resistance.

Under this wide spectrum of beneficial properties exclusive of the biofilm state, multiple biotechnological applications, such as bioremediation of contaminated environments, biocatalysis, probiotics, biofertilizers based on plant-growth-promoting rhizobacteria (PGPR), development of new materials can be considered among others. This low-cost strategy contributes to reducing the potential chemical footprint when applying nonbiological technologies to the environment (Mohsin et al., 2021). Useful biofilms can further be engineered in vitro for different biotechnological applications. Observations of bacteria

*Authors who equally contributed to this chapter.

Understanding Microbial Biofilms. https://doi.org/10.1016/B978-0-323-99977-9.00028-4

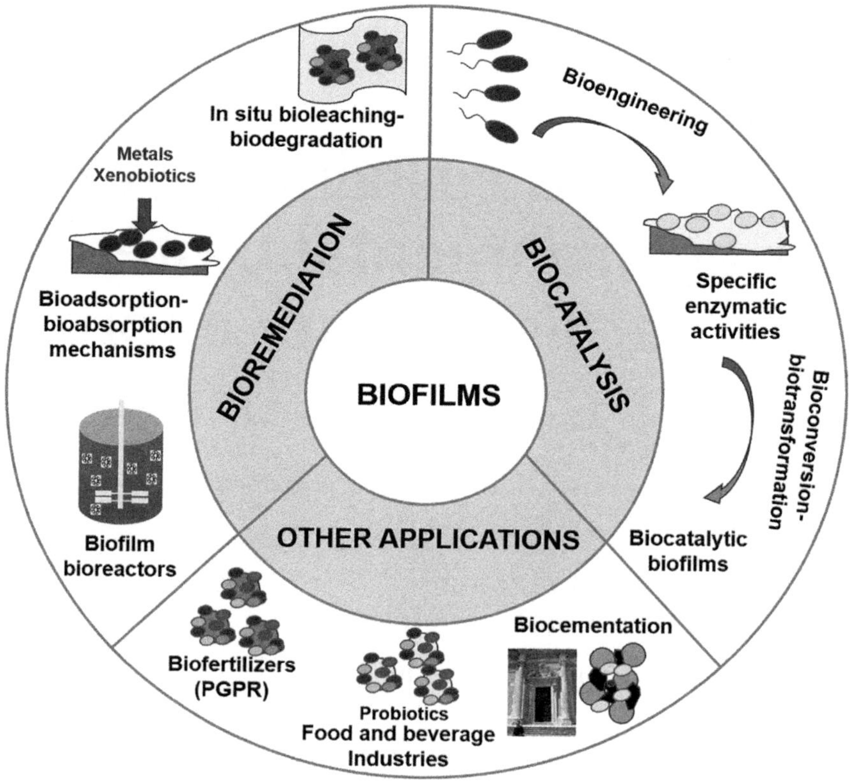

FIG. 1 Biotechnological application of the biofilm state. *(Based on Mohsin, M. Z., Omer, R., Huang, J., Mohsin, A., Guo, M., Qian, J., et al. (2021). Advances in engineered Bacillus subtilis biofilms and spores, and their applications in bioremediation, biocatalysis, and biomaterials.* Synthetic and Systems Biotechnology, *6(3), 180–191. https://doi.org/10.1016/j.synbio.2021.07.002.)*

associated with plants increasingly unfold biofilm-type structures that vary from small clusters of cells to wide biofilms. The plant-associated biofilms are highly capable of providing protection from external stress, decreasing microbial competition, and giving protecting effects to the host plant supporting growth, yield, and crop quality. Examples of useful biofilm applications are detailed below and schematized in Fig. 1.

1.2 Biofilms in industrial processes

According to literature, there are several biofilm-based reactors applied in industrial processes including wastewater treatments, such as the moving-bed bioreactor or the packed-bed bioreactor, both illustrated in Fig. 2. It was demonstrated that biofilm fermentations can potentially improve productivity and product yields by increasing biomass retention, allowing continuous operation at high dilution rates. Also, biofilms can tolerate hazardous environments, which enhance the conversion of crude biomass under substrate and product inhibitory conditions. Actually, examples for the improved conversion of pure and crude substrates into bulk chemicals by mixed microbial biofilms are proved, which can benefit from microenvironments in biofilms for synergistic multispecies reactions and improved resistance to contaminants. In addition, the use of mathematical models should not be disregarded as useful tools to supplement experimental insights related to the effects of physicochemical and biological phenomena on the process (Leonov, Flores-Alsina, Gernaey, & Sternberg, 2021). Nevertheless, major challenges for biofilm fermentations arise from inconsistent fermentation performance, slow reactor start-up, biofilm carrier costs and carrier clogging, insufficient biofilm monitoring and process control, reactor sterilization and scale-up, as well as issues in recovering dilute products. Several studies converge that the key to a successful process is likely to be an interdisciplinary approach. Major research areas might include genetic engineering in combination with the development of specialized biofilm reactors, biofilm carrier development, in situ biofilm monitoring, model-based process control, mixed microbial biofilm technology, development of suitable biofilm reactor scale-up criteria, and in situ product recovery (Leonov et al., 2021).

Recent studies have shown both the interaction of biofilms (commensal, mutualistic, antagonistic or parasitic/pathogenic) with different types of counterparts (living and nonliving) and their applications in several fields (Ghiasian, 2020). The differentiation of beneficial biofilms in vitro has shown many positive effects on agricultural and biotechnological applications. Regarding agriculture, phyllospheric or rhizospheric biofilms contribute positively for plant development. A direct interaction canalized through vegetal exudates, microbial N_2 fixation, mineral phosphate

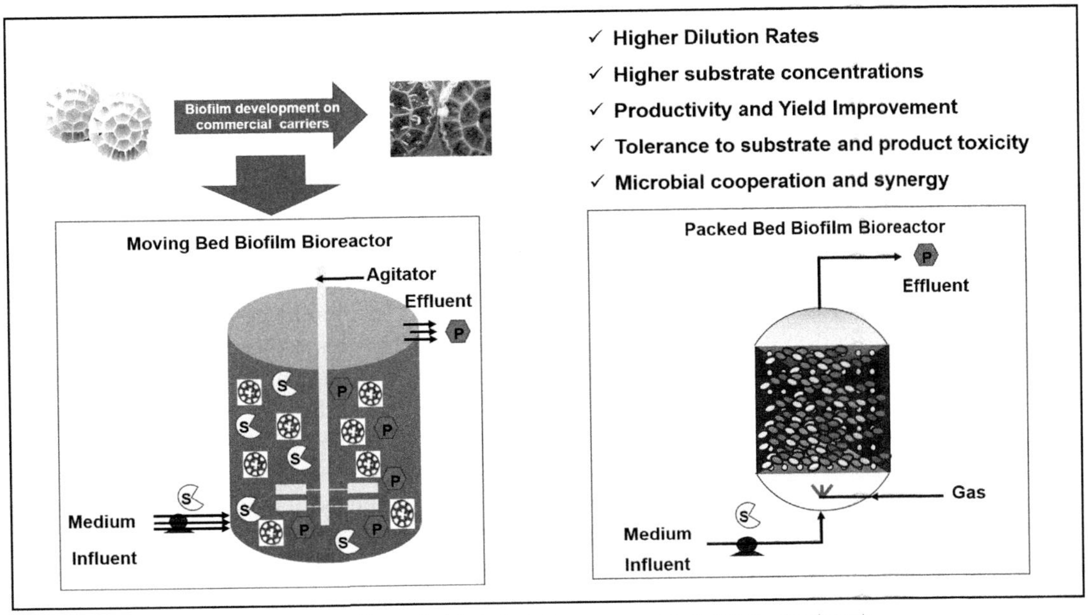

FIG. 2 Examples of biofilm-based bioreactors usually applied in industrial processes or wastewater treatments.

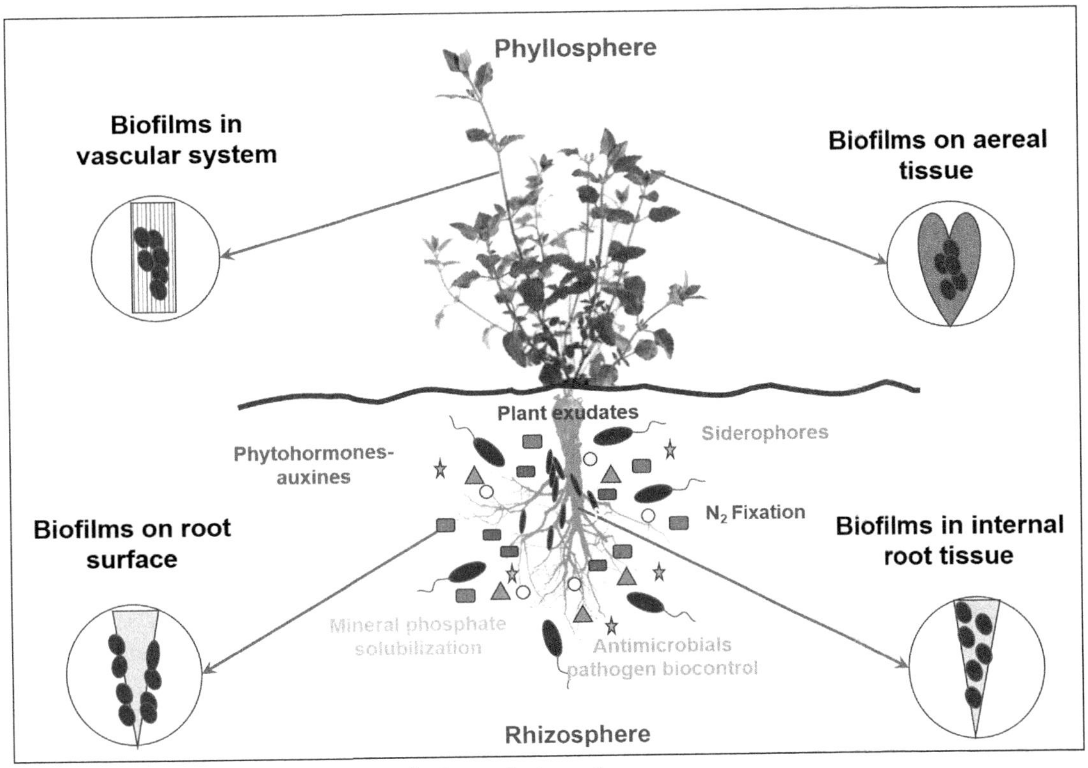

FIG. 3 Location of beneficial biofilm development contribution to plant growth.

solubilization, and secretion of siderophores and phytohormones, or even including the release of volatile compounds, can also influence on plant health exerting pathogenic species biocontrol (Fig. 3). By observational in vitro studies, the formation of fungal-bacterial biofilms/fungal-rhizobial biofilms (FBBs/FRBs) by colonizing bacteria on biotic fungal surfaces increased metabolic activities while compared with monocultures. Also, various compositions of the FBBs/FRBs inoculates have been shown to: amend nodulation and N_2 fixation in *Rhizobium*-legume symbiosis, colonize nonlegume plant roots, improve growth, increase carbon (C), nitrogen (N), and phosphorus (P) availabilities, biosolubilize rock phosphate, produce higher acidity and plant-growth-promoting hormones (auxins), and generate bioactive compounds, which also could increase the biodegradability of synthetic polymers (Fig. 3).

In the food industry, the biofilm usage is highly attractive. In fact, an increased tendency toward green food production was recently observed, such as plant-based diets. They have been widely recognized as crucial for achieving a healthy diet from sustainable food systems, but the short shelf-life narrows down the accessibility. Fermentation appears as an effective approach to extend storage and facilitate transportation, improving in this way the food supply chain. Plant-based fermented foods such as vinegar, kimchi, table olives, wine, and kombucha are produced or preserved by the metabolic activities of microorganisms, which deliver unique flavors and high nutrition. During the fermentation process, biofilm formation often occurs on the surface of the fermenting plant-based foods and their related liquids. So, a form of an inherent environmental adaptation from the planktonic state of microorganisms in harsh conditions such as acidic pH, high osmotic pressure, and insufficient nutrition can be considered in this case (Gong et al., 2021). Recent studies have shown that also yeast species such as *Saccharomyces cerevisiae* may have other biotechnological applications. A tryptophol and phenylethyl alcohol might be controlled by a signaling system in *S. cerevisiae* while these molecules are produced by both *Candida albicans* and *S. cerevisiae*. Lactic acid bacteria are also key organisms in fermented products. Probiotics are live microorganisms generally administered in food, which confer beneficial effects by helping prevent or treat diseases through improving the host's microbiota. The administration of probiotic strains in the biofilm forms allows a safe gut colonization because of greater resistance acquisition to high temperatures, or gastric pH stress, or osmotic stress and mechanical forces, while compared with their planktonic counterparts (Hayta, Ertelt, Kretschmer, & Lieleg, 2021). Lactic acid bacteria such as *Lactobacillus plantarum*, *L. rhamnosus*, and *L. fermentum* are able to join a biofilm matrix and have become industrially significant. Some other examples of probiotics that can be found in the literature are: *L. reuteri* biofilm on dextranomer microspheres (Navarro, Mashburn-Warren, Bakaletz, Bailey, & Goodman, 2017); *Lacticaseibacillus paracasei* in calcium pectinate beads (Heumann et al., 2020); *L. rhamnosus* in calcium alginate (Cheow & Hadinoto, 2013).

According to Singh et al. (2020), *Aspergillus niger* biofilm can be used for cellulase production, well-known enzyme with applications for food, textile, pulp, and paper processing.

Other applications of biofilms are in bioengineering and biotechnology fields. In this sense, different microorganisms such as *Bacillus subtilis*, commonly used as a commercial species, are capable of producing both biofilms and spores. The spore-forming microorganisms are *Bacillus*, *Clostridium*, *Anoxybacillus*, *Geobacillus*, and *Sporolactobacillus*. As mentioned above, biofilms are matrix-encased multicellular communities that comprise several components including exopolysaccharides, proteins, eDNA, and organic acids. These biofilms resist environmental conditions such as oxidative stress and hence have applications in bioremediation technologies. Furthermore, biofilms and spores can be engineered for environmentally friendly and safe production of bioproducts such as enzymes. The ability to withstand harsh conditions and produce spores makes *Bacillus* a suitable candidate for this technology. In recent years, the spores of such species are widely classified as Generally Recognized As Safe, usually known as GRAS (FDA, 2019). Advances in synthetic biology have enabled the reprogramming of biofilms to improve their functions and enhance the production of value-added products. Globally, there is increased interest in the production of engineered biosensors, biocatalysts, and biomaterials. The elastic modulus and gel properties of *B. subtilis* biofilms have been used to develop living materials. *B. subtilis* biofilm engineering for biotechnological processes, and their increasing application in bioremediation and biocatalysis, leads to produce functional living materials with self-regenerating, self-regulating, and environmentally responsive characteristics (Mohsin et al., 2021).

2. Relevance of interactions of biofilm components with contaminants

2.1 Biofilm matrix: Chemical structure

The abbreviation EPS (extracellular polymeric substance) is a general term to represent different types of macromolecules such as polysaccharides, proteins, nucleic acids, lipids, and other polymeric components present within microbial aggregates or in biofilm matrices where cells are embedded in. EPSs are mainly secretions of microorganisms with a high molar mass and products of cell lysis and hydrolysis of macromolecules, located both outside cells and inside microbial aggregates that constitute the way of life of most microorganisms, such as biofilms or bioflocs. Regardless of their origin, EPSs can be classified as bound EPS (attached to cell surface) or soluble EPS (suspended in the surrounding liquid medium). Related to this definition, the structure of the bound EPS is generally represented by a two-layer model. The inner layer, which consists of tightly bound EPS, has a certain shape and is firmly and stably attached to the cell surface (Sheng, Yu, & Li, 2010). On the other hand, the outer layer consists of lightly bound

EPS, a dispersible mucous layer. EPSs are composed of polysaccharides, proteins, humic substances, nucleic acids, etc. These exopolymers present various charged functional groups on the surface such as carboxyl, amino, phosphates, carbonyl, and hydroxyl groups, which could interact with dissolved contaminants through ion exchange and complex formation. Like most exopolymers produced by bacteria, they have a polarity, so they can exist in different configurations depending on pH, temperature, ionic strength, electrolytes in solutions, etc. These factors can change the complete conformation and orientation of functional groups affecting the effective charge and therefore, for example, the adsorption properties to metal cations (Mao, Hu, & Yan, 2011). Fig. 4 resumes EPS characteristics, types, and composition, which define certain microbial behaviors under different environmental conditions.

EPSs fulfill vital functions within the microbial agglomerates; among many of them, it can be mentioned as examples that they allow the formation of a three-dimensional matrix that keeps the microorganisms together, and they also intervene in the adhesion of the biofilm to the surface or help to retain water in cases of scarcity (Cordoba Velandia, 2011). Another important function is the accumulation of active enzymes, such as those involved in the digestion of exogenous macromolecules for the acquisition of nutrients when environmental conditions are hostile, increasing in this way the possibility of survival. Likewise, they stabilize the structure of the biofilm, as is the case of extracellular DNA, which proved to be a fundamental component since it mediates cell-cell interaction. Indeed, removal of extracellular DNA by early-stage DNase treatment has been shown to interfere with biofilm maturation. But, in mature biofilms, the same treatment showed little impact on biofilm architecture (Ghafoor, Hay Iain, & Rehm Bernd, 2011). EPS also forms a protective barrier that provides resistance to biocides and other harmful agents. It allows microorganisms to live continuously at high cell densities in stable communities of mixed populations. In other words, the EPS matrix is a medium that allows cooperation and communication between cells in microbial aggregates. Adhesion and cohesion occur between EPS and biomass, among other suspended solids. This formation of complexes is caused by different interactions such as London and Van der Waal's forces, electrostatic interactions, and hydrogen bonds, which lead to the stabilization of bioflocs. These properties of exopolymers make them suitable for many applications such as flocculation, sedimentation, dehydration, metal binding, and removal of toxic organic compounds. In addition, the biosynthesis of EPS promotes the attachment of cells to a solid support: this helps in the establishment, evolution into a mature biofilm and protecting from environmental stress (Nouha, Kumar, Balasubramanian, & Tyagi, 2018). Fig. 5 summarizes the principal functions and properties of the EPS matrix that confers to biofilms survival particularities, since 80% of the world's microbial biomass is found in this lifestyle.

Critical and metabolic pathways of EPS type-dependent biosynthesis, e.g., homopolysaccharides or heteropolysaccharides, are subjects of multiple studies. The three main steps involved in EPS synthesis are: carbon substrate assimilation, intracellular polysaccharide synthesis, and EPS exudation outside the cell. However, these steps depend among others on multiple factors such as the microbial species

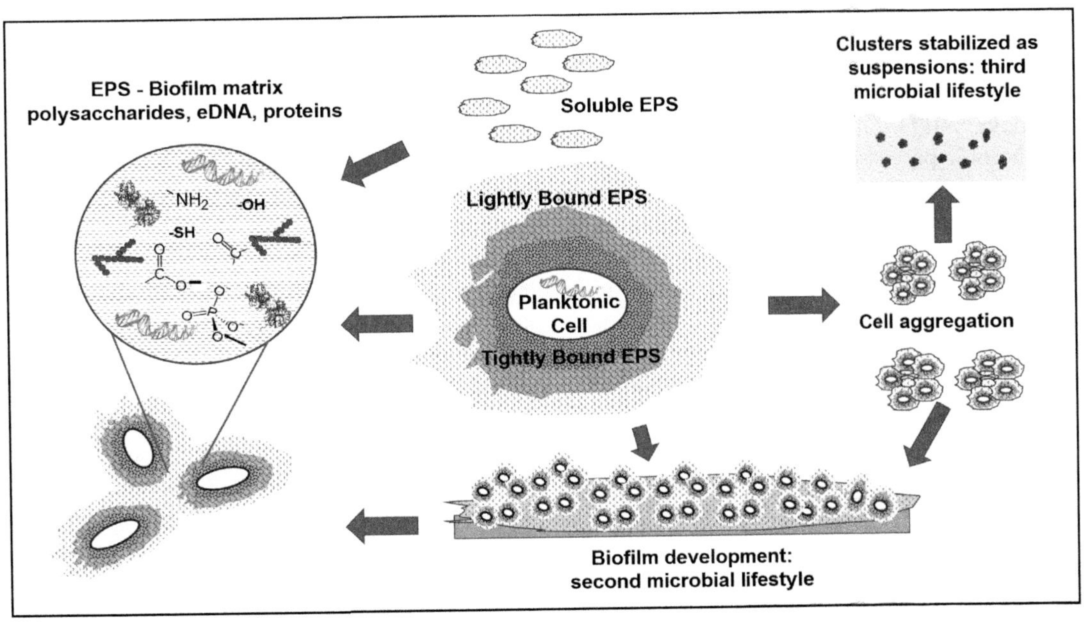

FIG. 4 Characteristics of microbial EPS and biofilm matrix.

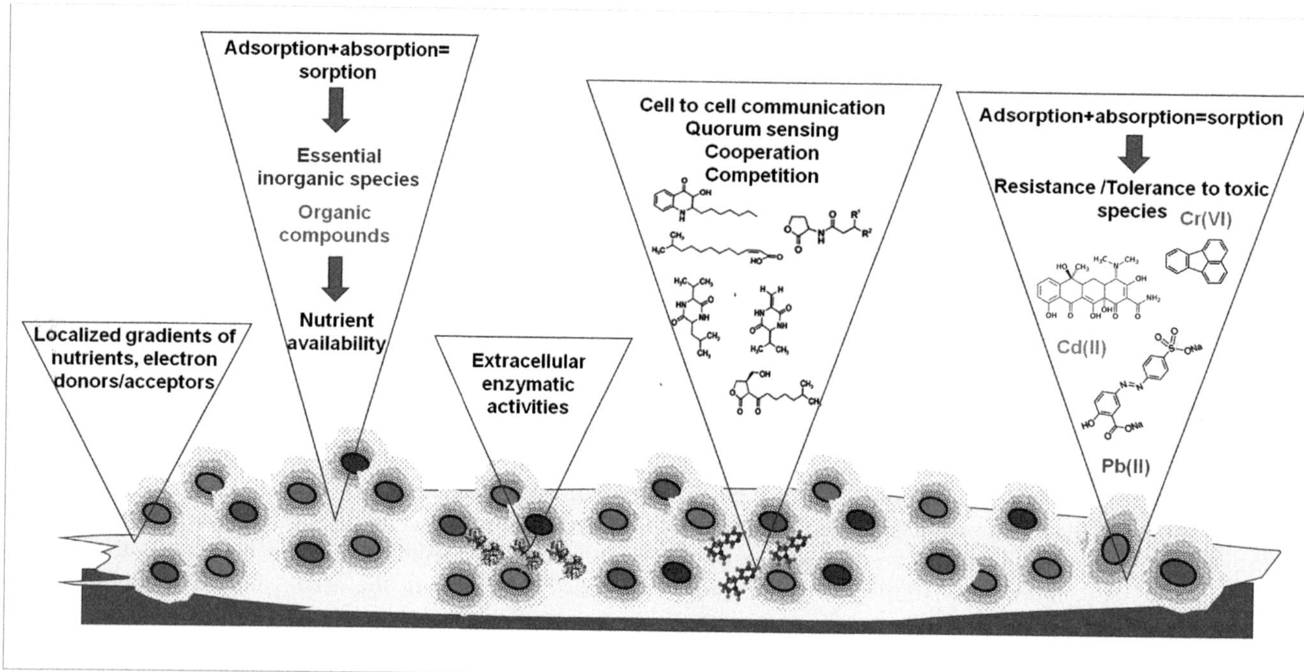

FIG. 5 Principal properties of EPS-biofilm matrix considered as a fortress. *(Based on Flemming, H.-C., Wingender, J., Szewzyk, U., Steinberg, P., Rice, S. A., & Kjelleberg, A. (2016). Biofilms: An emergent form of bacterial life.* Nature Reviews Microbiology, 14, *563–575. https://doi.org/10.1038/nrmicro.2016.94.)*

(genes involved), the medium composition (carbon and nitrogen source and C/N ratio), and the working conditions (pH, temperature, dissolved oxygen).

Some studies have reported that the amount of EPS released in anaerobic wastewater treatment processes is less than that released in an aerobic process. This may be because EPS production in an anaerobic process is restricted by methanogenic bacteria or because the biopolymers are degraded to CO_2 and CH_4. However, in terms of composition, the fraction of proteins produced in an anaerobic process is greater than the one for an aerobic process (Cordoba Velandia, 2011). Other studies show that EPS extraction in anaerobic processes is more difficult than that one in aerobic activated sludge processes, due to the mineral fraction present in granular sludge and the compact conformation of anaerobic granules. The importance of EPS in biological wastewater treatment systems lies largely in its relevance of different aspects for microorganism aggregates, as well as in its relationship with sludge sedimentation (Cordoba Velandia, 2011). The chemical structure of polymeric substances secreted by microbial cells depending on different environmental conditions is diverse. Polysaccharides and proteins are the main components of EPS (comprise between 75% and 90% of its mass), and to a lesser extent it can be found in nucleic acids, lipids, uronic acids, glycoproteins, and some inorganic compounds. In the EPS of wastewater treatment sludge, almost 20% by weight is made up of humic substances, while in cell cultures of *Pseudomonas* NCIMB 2021, the composition of soluble and capsular EPS is a copolymer of carbohydrates and proteins. For the exopolymer extracted from the biofilm of *P. putida*, proteins were the main component (75%) followed by polysaccharides and DNA. The protein fraction of the EPS matrix is not strongly affected by the type of carbon source in the culture medium. This corresponds to the observations of some works, where a persistent content of proteins (from the EPS of the activated sludge) was reported, when the microorganisms were supplemented with different types of carbon sources such as glucose or sodium acetate. On the other hand, the addition of different nitrogen sources affects the protein content. For example, it was reported with NH_4Cl, an increase in the *P. veronii* 2E-EPS protein proportion (Ferreira et al., 2017).

Another polymer that constitutes EPS is DNA, an intracellular component that is released when the cell is lysed, and which can be absorbed by the polymeric matrix. Humic substances are components that are also naturally present in activated sludge from the hydrolysis of organic waste. They can be adsorbed onto the polymeric matrix (biofilm) by different functional groups such as carboxylic and phenolic groups. DNA and humic substances have been reported to influence the properties and stability of biofilms. Extracellular DNA is required for the initial establishment of the biofilm in *P. aeruginosa* because it helps bacteria-surface adhesion by modulating the charges and hydrophobicity of the interactions between the microorganism and the abiotic surface (Ghafoor et al., 2011). Similarly, the biofilm formed by many other bacteria specifically releases DNA under conditions of stress or due to cell lysis. In conclusion, taking into account the data found in the specialized bibliography, the biochemical composition of EPS is affected by many factors such as the extraction method used, nutrient supplementation, the origin, the type of sludge analyzed, the phase of growth, the microorganism, etc. The molecular composition of the EPS is very important since it defines its suitability for various types of uses. Among the newer applications, one of the most important is the removal inorganic or organic contaminants with different chemical structures (Nouha et al., 2018).

2.2 Bioavailability of xenobiotics in biofilm structure

The bioavailability of hydrophobic compounds is usually defined as the fraction of the total amount present in a specific environmental compartment that is either available or potentially available for their biologically uptake (Ferreira et al., 2013). The anthropogenic chemicals, termed xenobiotics, can be accidentally or deliberately released by processing industries and cause acute or chronic toxicity depending on their bioavailability. Pesticides, hydrocarbon derivatives, halocarbons, azo-dyes, nitroaromatics, plastics, alkylbenzene sulfonates are the most common pollutants included in this category. Xenobiotics may be associated to soils and sediments by sorption (adsorption and absorption) processes that promote their uptake and transformation by cells. In this way, biodegradation rates may be dependent on restricted phase exchanges, acquiring persistence for long periods of time with the consequent environmental risks (Krell et al., 2013). Adsorption is a physicochemical process in which both inorganic and organic species are bound to surfaces as consequence of a combination of electrostatic, Van der Waals, induced dipole-dipole, dipole-dipole, and hydrogen bounding forces. This mechanism is restricted to the polarity of the involved compounds. Hydrophobic nonpolar molecules can only "dissolve" within the interior of nonaqueous systems, consequence of retentive forces that go beyond being surface forces, and the process is usually named as absorption. So, the generic term sorption, combining both phenomena, is more appropriately used for hydrophobic fractions of soil, sediments, and EPS-biofilm matrix when their interactions with xenobiotics or other nonpolar chemicals are involved. The prefix "bio" is often added when biological materials are in charge of adsorption, absorption, or sorption procedures. Additionally to sorption, relying on strong complexes and low bioavailability of xenobiotics, the proved cell resistance/tolerance within biofilms could be related to a generally less metabolic activity than planktonic cells. Therefore, it implies a significantly lower sensitivity to many antimicrobials' action mechanisms, which includes some metabolic pathways. Thus, apart from helping microorganisms to persist in the environment improving the capability to resist desiccation and UV light, biofilms are also an efficient barrier for treatments with sanitizers and antibiotics, constituting a real fortress (Rossi et al., 2018).

2.3 Biofilm matrix complexing capacity and metal immobilization

Taking into account all the metals present on Earth, they can be classified into three large groups (Duffus, 2002). Class A metals are essential for biological processes and nontoxic at high concentrations; borderline class metals are essential but toxic when exceeding μM concentrations; and finally, the fully toxic Class B metals. This classification is based on the Lewis acid behavior that each metal cation possesses, since its tendency toward the formation of complexes with different families of ligands depends on it. For example, Ca and K are nontoxic metals, belonging to Class A; Cr, Cu, and Mn are essential metals for certain biological processes, but with a toxicity level due to the ability to form complexes with biological ligands (borderline); and lastly Cd, Hg, and Pb that do not have biological utility but high affinity for functional groups with O, N, and S and hence high toxicity (Class B). Borderline and Class B metals can be considered as "environmentally relevant metals" (ERMs) since they can interfere with biological mechanisms and in fact with the biosphere equilibrium. The way that ERMs appear in the environment is only as consequence of anthropogenic activities as mining, electroplating, metallurgy, and tanning procedures. ERMs are not degradable species, but their toxicity is related to the metal chemical speciation: free cations and labile complexes are bioavailable because of their ability to cross biological membranes (Ferreira et al., 2013). In the case of strong complexes, metal-cell interactions are reduced.

EPS binding capacity is directly related to the presence of functional groups, which act as ligands, boosting metal removal from aquatic environments by promoting metal immobilization. As reported, metal binding constants for EPS extracted from activated sludges varied over a wide range: for Cd(II): $-3.7 < \log K' < 2.3$, and for Pb(II): $-1.4 < \log K' < 4.1$ (Guibaud, Bordas, Saaid, d'Abzac, & Van Hullebusch, 2008; Guibaud, van Hullebusch, & Bordas, 2006; Guibaud, van Hullebusch, Bordas, d'Abzac, & Joussein, 2009). Several analytical techniques are usually applied to visualize EPS complexing capacity and hence metal retention. For example, it can be determined by applying electroanalytical methods. The soluble EPS binding capacity and the number of binding sites could be calculated by anodic stripping voltammetry (ASV) titrations. Regarding Cd(II), for *P. veronii* 2E soluble EPS fractions, the log of conditional binding constants (Log K′) of 5.47 ± 0.01 and 5.48 ± 0.08, with 65 ± 3 and 66 ± 6 48 ± 3 μM ligand/g EPS, at pH = 7.5 and 8.1 respectively was effectively calculated applying ASV methodology. In this way, soluble EPS fractions were characterized as ligands of moderate strength (Ferreira et al., 2017). By analysis of Fourier transform infrared (FT-IR) spectroscopy, characteristic functional groups are associated to each absorbance peak, consequence of stretching or bending deformations. As the frequencies of the vibration modes are sensitive to changes in microenvironments, those derived from the presence of metals and complex formations can be specifically identified (Ferreira et al., 2020). In this study, attenuated total reflectance-Fourier transform infrared spectroscopy (ATR-FTIR) in combination with multivariate analysis (principal component analysis) was used to study the interaction of Cd(II), Cu(II), and Zn(II) and *P. veronii* 2E cells, including bound EPS and its different soluble EPS fractions. Most remarkable results indicated that amino groups present in both cells and soluble EPS exopolysaccharide fraction were responsible for Zn(II) and Cu(II) complexation, as well as carboxylates chelated Cd(II). Meanwhile in the soluble EPS lipopolysaccharide fraction, phosphoryl and carboxyl sites were involved in Cd(II) and Cu(II) binding and Zn(II) interacted with amino groups (Ferreira et al., 2020).

3. Biofilm-dependent microbial survival strategies in polluted environments

Microorganisms have multiple strategies to survive in the environment, including those that are more extreme or those that are highly contaminated. Through biofilm development, they are organized in such a way that they cooperate with each other, forming a consortium aimed at the survival and development of each of the members living in the community (Daniel, 2018). Its formation depends on various environmental factors and in particular its composition: it will not be the same biofilm found in an extreme saline marine environment than in a metal-contaminated soil.

As mentioned above, the biofilm represents a structure that protects microorganisms from harmful compounds, avoiding their diffusion inward to exert a negative action. As the same gradient also happens to nutrients necessary for the microbial viability and division maintenance, lower growth rates are usually observed within the biofilm matrix (Rittmann, 2018). Also, a wide spectrum of protection effects by the biofilm matrix to the microbial community from environmental stress factors such as water loss, nutrient limitation, and susceptibility to antimicrobial agents were described (Geng et al., 2019). In addition to these functional and structural abilities, sessile growth provides the proper environment for cell-cell communication. This is an essential step, for example, in the colonization by opportunistic pathogens since it allows them to adapt to the different infection areas through the differential expression of biofilm compositions (Paluch, Rewak-Soroczyńska, Jędrusik, Mazurkiewicz, & Jermakow, 2020). The first phases of adherence and beginning in the development of the biofilm are of great clinical importance. The better understanding in how to avoid the formation of this polymeric initial structure or how to eliminate it efficiently in this preliminary stage could be the bases of tools for the treatment of infections and cleaning of prostheses or catheters. The chronic development of infections in patients is very common as consequence of the remains of biofilm and the presence of viable cells within the matrix, immersed in a cooperative growth as an effective survival strategy for those cells. Similar cases have been reported in the food industry. *P.. fluorescens* has been found in milk and other dairy production machinery. This psychrophilic bacterium has the ability to produce biofilm and hence adheres to tanks, resisting even after the routines of cleaning and sanitizing processes that are carried out (Rossi et al., 2018).

Several hypotheses have been developed about the mechanism by which the biofilm protects cells within the structure, such as the impossibility of antibiotics to pass through water-filled channels or anoxic environments that modify the microenvironment where the antibiotic must act (Mahamuni-Badiger et al., 2020). Therefore, the development of bacterial growth in a biofilm style is undoubtedly a true protection of cells against the host's immune system, sanitization processes, and pharmacological therapies.

But microorganisms not only face the action of humans trying to eliminate them or to achieve aseptic conditions. In natural environments, adversities such as high-intensity ultraviolet (UV) radiation; high or low temperatures; high alkalinity, acidity, salinity and pressure; nutrient scarcity and antibiotics put microorganisms to work in redoubling their efforts to live and survive under those conditions, emerging the term extremophiles for them. Collaboration and cooperation seem to be, without question, the best strategy for what the development of biofilm is fundamental, in so much level as to be recognized as the "protective clothing" (Yin, Wang, Liu, & He, 2019). Some environmental stressors have the potential to damage microbial development, for which the biofilm style represents a way of adaptation to these hostile environments.

Temperature is an important factor for microbiological growth. Outside the optimal temperature, the development of microorganisms slows down, either by destruction of proteins and membranes by thermal shock at high temperatures or by a decrease in the fluidity of biological processes at low temperatures. Therefore, those microorganisms that live in thermal environments, geysers, etc., are adapted to these conditions, being the biofilm production a tool that helps them in these extreme climates. It has been reported that species of the genus *Thermolongibacillus*, *Geobacillus*, *Aeribacillus*, *Sulfolobus*, and *Anoxybacillus* have a greater biofilm development at a temperature above 60°C, which suggests the protective role of this matrix (Yin et al., 2019). At lower temperatures, this property has not been deeply studied yet. Zhang et al. (2020) research was focused on partial nitrification-anammox process (PNA) performance set up in a sequencing batch biofilm bioreactor (SBBR) at low temperatures. A steady nitrogen removal was achieved in domestic sewage treatment despite the low temperature range 12.8–16.3°C. The nitrogen removal obtained in the PNA-SBBR system was related to the establishment of a microbial community structure enriched in psychrotolerant microorganisms. Apparently, EPS secretion was increased at low temperatures, especially the tightly bound EPS with a higher protein content, conferring an improved protective effect under adverse conditions and facilitating in fact the stable operation of the designed bioreactor. Another stress factor is light, in particular ultraviolet light in its entire spectrum range (UV-A; UV-B; or UV-C), which can cause damage to DNA, proteins, and lipidic membranes and also generate an increase of the reactive oxygen species (ROS) production. ROS act nonspecifically by oxidizing all biological molecules and causing oxidative stress and cell damage. The biofilm formation protects the cells from UV radiation by the presence of secreted absorbing compounds and thus acting as a "protective umbrella" (Yin et al., 2019).

Regarding pH, habitats with very low pH (such as mining excavations) tend to have acidophilic microorganisms that oxidize inorganic compounds and metals as part of their metabolism. Among them, the chemolithotrophic *Acidithiobacillus*, *Leptospirillum*, and *Acidiferrobacter* species (Bellenberg, Huynh, Poetsch, Sand, & Vera, 2019) can be found. Apart from the bacterial consortia that tolerate these conditions, acidophilic fungi that participate in these communities have also recently been described. *Acidomyces acidophilus* lives in these heterogeneous biofilms and possesses interesting biotechnological properties. *Acidomyces*, *Acidiella*, and *Acidotrix* excrete enzymes and can participate in bioremediation processes such as wastewater recovery and xenobiotic degradation (Hujslová, Bystrianský, Benada, & Gryndler, 2019). Although many of these properties have only been tested in pure cultures, the versatility of these organisms and their promising activities well suggest that biotechnological benefits can be achieved by working with the right consortium composition. The lack of water availability can severely affect free cells, putting their viability at risk. In the presence of biofilms, their upper layer could help preventing water loss by the presence of hydrophilic polysaccharides and proteins (Flemming et al., 2016). Another phenomenon of cellular communication—not occurring in planktonic cells—is the secretion of other type of molecules able to retain water and, in this way, contributing to the biofilm hydration maintenance.

As mentioned above, the nutrient gradient is established in these structures but, even in oligotrophic environments, biofilms have the ability to capture nutrients from the surrounding water flow. This concentration process can be directly associated with adsorption and absorption mechanisms, representing one more advantage of this lifestyle. The interaction between the diversity of the positively and negatively charged groups on the biofilm matrix allows the entry and retention of various molecules that serve as a nutritional source (Flemming et al., 2016). Chelation promotes the capture of ions, preventing their precipitation and losing its bioavailability (Decho, Visscher, & Reid, 2005). Another way of nutrient entry is possible through canal-like structures that collaborate with the mass transport toward the interior (Billings, Birjiniuk, Samad, Doyle, & Ribbeck, 2015). But this passive sorption property is not exclusive to beneficial compounds for cells, since being a nonspecific transport it allows also the entry of harmful species for cell development such as some metals, xenobiotics, or even antibiotic molecules (Fig. 5). In the particular case of metals and their toxicity, the same cells have charged groups on their membranes that allow ionic-type interaction with metallic cations (Ferreira et al., 2020). Growth in biofilm can also be seen on particles, which can be used as substratum. The presence of macro and micro plastics in the sea provides a surface to be colonized, promoting the formation of aggregates. These aggregates modify their buoyancy and confer an additional survival strategy ensuring a permanency up to several days even in marine sediments (Michels, Stippkugel, Lenz, Wirtz, & Engel, 2018).

4. Application of biofilms in bioremediation processes

4.1 In situ bioremediation mediated by biofilms

The first step to consider in a bioremediation is to establish the optimal conditions, such as a suitable temperature, pH, nutrients, oxygen availability, among other variables, so that the microbial community can successfully remove the contaminant. As remarked above, the biofilm structure supports the development of microorganisms in such a way that allows a solid bioremediation process durable over time (Rittmann, 2018). Besides, pollutant bioavailability is absolutely relevant as a first issue to be considered since abiotic interactions such as sorption or nonmiscibility phenomena can interfere and limit microbial accessibility. Bioremediation as environmental biotechnology is beginning to be studied for pollutants such as metals, antibiotics, pesticides, or hydrocarbons among other xenobiotics. In situ bioremediation can be intrinsic when the indigenous microbiota is in charge of the whole process and does not require any amendment to efficiently remove pollutants. If necessary, some innovations can be considered to improve efficiency, as biostimulation—by N and P or electron acceptor/donor supply—or bioaugmentation by the inoculation of specific microorganisms to increase degraders' population. In some treatments, commercial surfactants can be applied to improve bioavailability in case of hydrophobic compounds, as in Deepwater Horizon Oil spill occurred in 2010. The dispersant effect on nonmiscible products in the environment leads to small droplets formation, increasing the specific surface area and promoting microbial colonization as biofilm state over them. The tensioactive agent spread—in water or soil ecosystems— may act as a double-edged sword, since the applied product may produce adverse effects on microbiota and, in fact, it could not be necessarily biodegraded as well. Therefore, microbial biosurfactants either in situ secreted or externally added remain being the best option when bioavailability improvement is required avoiding at the same time the introduction of another perturbation.

In situ biological metal removal is a widely applied strategy due to its greater sustainability and lower costs compared with physical or chemical remediation. In particular, the optimal methodology is phytoremediation, which includes the epiphytic microbiome that actively inhabits the leaves of aquatic plants. Leaves immersed in water contaminated with metals provide a surface covered by exudates, polymeric substances, and metabolites as part of the microenvironment within the biofilm matrix (Geng et al., 2019). The removal of metal cations is possible thanks to the polyanionic charge on the surface of the biofilm, allowing adsorption and transport phenomena. As a negative consequence of this interaction, the contamination produced by these metals or even by organic molecules can travel long distances, generating accumulation and reaching the food chain (Bradney et al., 2019). Phytoremediation can be also applied to metal contaminated soils. In order to ensure the bioremediation process success, a responsible monitoring of the metal accumulating plants is necessary. Their collection is mandatory when vegetal life cycle ends accompanied by an environmentally safe disposal of the plant remains, including a metal recovery that should not be disregarded (van der Ent et al., 2015). Fig. 6 resumes a brief scheme of in situ bioremediation processes often applied to aquatic or soil ecosystems for both metal and xenobiotic removal.

In addition to the natural properties of biofilm, modifications can be introduced to obtain special properties directed to a specific process. The application of genetically modified microorganisms could increase the production or improve the performance of a particular enzyme or molecule by changing, for example, pH range and temperature of the activity of interest. Additionally, the biofilm development in the presence of nanoparticles or other molecules leads to a mature stage with robust and different properties (Hayta et al., 2021).

Biocementation is a potential beneficial aspect of biofilms, which can be exploited to decrease or even avoid the natural degradation of stone buildings or monuments, based on the process by which microorganisms precipitate calcium carbonate as calcite (Robertson & McLean, 2015). In this case, a mixture of bacteria with a variety of calcium salts, urea, and a carbon source is used. By biological hydrolysis of urea, NH_3 and CO_2 are produced, increasing pH values. As consequence of CO_3^{2-} formation, precipitation of $CaCO_3$ occurs. The biofilm matrices promote mineral formation by metal binding as well, serving as a connection to calcite and other mineral components.

With regard to mining, biohydrometallurgy, including bioleaching and biooxidation, is an alternative technology with low costs, simplicity, less energy consumption, greater profitability at long terms while compared with traditional methods in obtaining metals from sulfides in low-grade mines (Moazzam et al., 2021). This process is used in different parts of the world, for example, bioleaching of copper mines in Chile, biooxidation to obtain gold in Fairview Mine in South Africa (Gu, Rastegar, Mousavi, Li, & Zhou, 2018). The bacterial chemolithotropic consortium forms a biofilm on the mineral particles. Sessile cells embedded in the EPS matrix are capable of interacting with the mineral and as a result the redox reaction happens, oxidizing and solubilizing the metal of interest (Giese, 2019). This process is known as "contact leaching"; however, other coupled mechanisms should not be disregarded (Mahmoud, Cézac, Hoadley, Contamine, & D'Hugues, 2017). Planktonic cell-dependent "cooperative leaching" contributes to complement colloids solubilization with acidic species release, promoting a "noncontact leaching," as summarized in Fig. 7.

4.2 Ex situ bioremediation: Biofilm-based bioreactors

Several reported examples of the use biofilm-based processes beneficial to ecosystem restoration can be found. As consequence of the increasing industrial activities, high pollutant amounts are discharged into water bodies causing huge damage to the environment. To prevent contamination, biotreatments should be implemented before their release. In the literature, hybrid biological treatment systems, such as the moving-bed biofilm reactor (MBBR) or fixed-bed biofilm reactor (both described in Fig. 2),

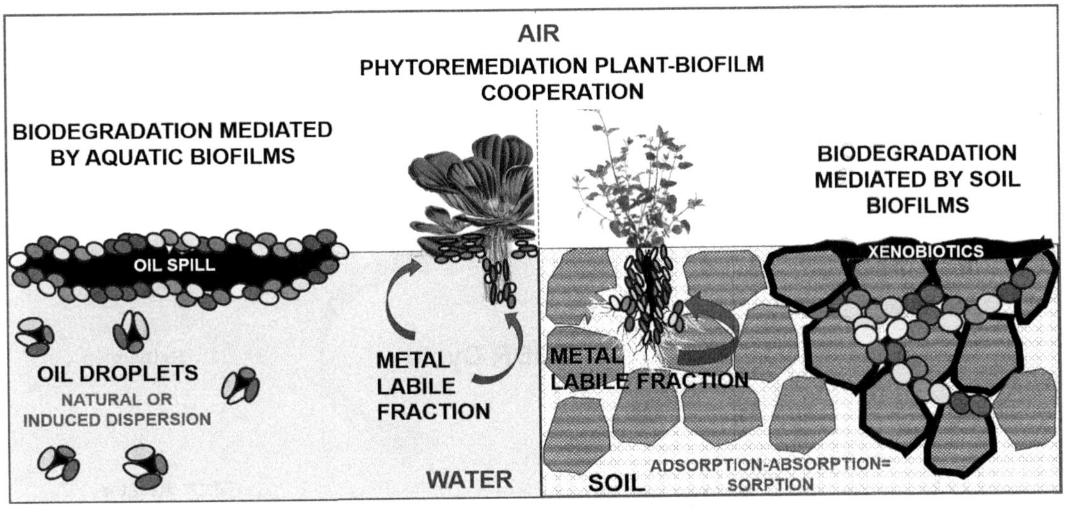

FIG. 6 Schematic representation of in situ bioremediation processes mediated by biofilms in nature.

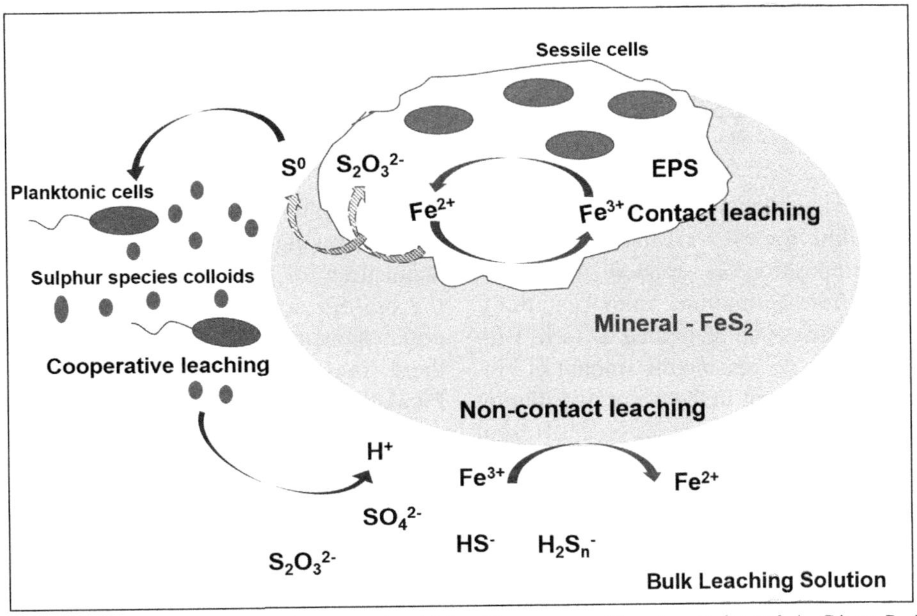

FIG. 7 Schematic representation of the coupled steps for a complete bioleaching process. *(Based on Mahmoud, A., Cézac, P., Hoadley, A. F. A., Contamine, F., & D'Hugues, P. (2017). A review of sulfide minerals microbially assisted leaching in stirred tank reactors.* International Biodeterioration and Biodegradation, 119, *118–146. https://doi.org/10.1016/j.ibiod.2016.09.015.)*

sequencing batch biofilm reactor (SBBR) and integrated fixed-film activated sludge are most commonly used to effectively remove including micropollutants (Yakamercan & Aygün, 2021). In this way, Yakamercan and Aygün (2021) proposed modifications to biological treatment systems into more effective hybrid biological processes using biocarriers. This innovative initiative showed high removal efficacy with low-cost efficiency for pollutants from textile industry wastewater, such as phthalic acid esters (PAEs) and chlorophenols, classified as priority pollutants in the European Union Water Framework Directive. The use of this treatment is desirable when, on the one side, physicochemical methods (e.g., advanced oxidation, membrane, adsorption, and coagulation-flocculation processes) are not convenient, either for its operational or investment costs, process difficulties (membrane fouling), or the generation of more toxic compounds as by-products, and, on the other side, conventional biological treatment processes can be insufficient for removing some micropollutants. The SBBR (Fig. 8) was selected then because of its easy operation, with low cost, does not require an additional settling tank, and achieves high efficiency when combined with biocarriers (Yakamercan & Aygün, 2021).

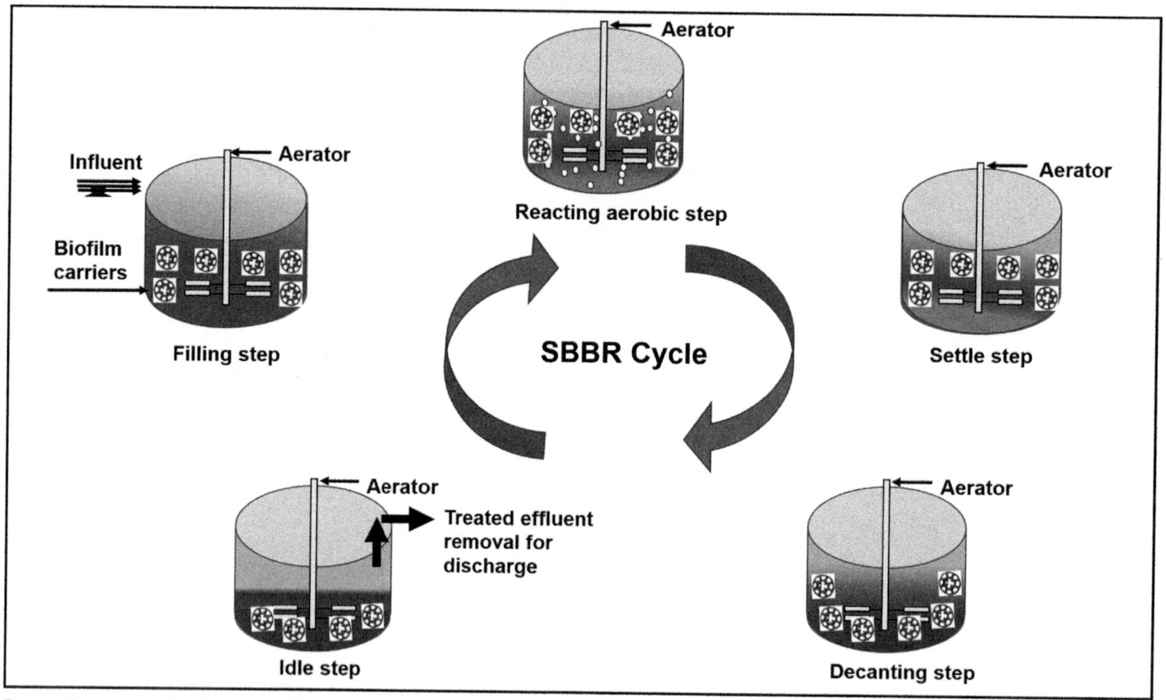

FIG. 8 Innovative sequencing batch biofilm bioreactor (SBBR) by commercial biocarrier utilization.

Water and wastewater treatments have successfully used the advantages of biofilms for over a century. Not only biofilm-mediated decomposition is applied to human wastes, but wastewater from agriculture, animal facilities, and food processing industries can be treated as well. Biological filters are engaged in decreasing the amount of biodegradable organic carbon present in the water distribution systems. In addition, extra nutrients such as N and P from secondary effluents can be treated by phototrophic cyanobacterial biofilms in wastewater plants. In fact, biofilm bioreactors eliminate organic carbon and nitrogen in large-scale aquaria. As increased levels of NH_4^+ from animal wastes are toxic to numerous forms of aquatic life, biofilters, including biofilms with nitrifying bacteria, transform NH_4^+ to the less toxic species NO_3^- (Singh et al., 2020). Other reported examples describe highly loaded MBBRs used to remove pollutants such as pulp and paper (P&P) from wastewater, one of the most polluting industrial sectors in the world. According to the European and US markets, the P&P segment generated approximately 2.5 billion m^3 of wastes in 2015, which represented around 42% of the total industrial wastewater production in these regions. Batch tests with suspended biomass and biofilms were performed to separately assess the activity of each biomass fraction. Data from the continuous reactors' operation were fitted to the Kincannon-Stover kinetic model for assessing both the maximum substrate utilization rate and the wastewater biodegradability. The results of this work point toward a better dimensioning of MBBRs treating P&P wastewater, either prioritizing: (i) a bigger reactor (higher HRT) mostly relying on the suspended biomass activity, saving in nutrients supplementation and carriers' acquisition; or (ii) a smaller MBBR mostly dependent on the biofilm activity, saving in construction and aeration costs. Nonetheless, further studies are needed regarding these issues (Matheus, Ekenberg, Bassin, Dezotti, & Piculell, 2021).

The use of *B. subtilis* biofilms in consortia with other strains, for example, *P. aeruginosa*, allows the elimination of volatile organic compounds known as BTX (benzene, toluene, and xylene) and other hydrocarbons (Mohsin et al., 2021). The diversity of microbial metabolic pathways allows the efficient use of these compounds as carbon source, making it an excellent tool for their removal. Furthermore, working with bacterial consortia favors bioremediation by promoting a collaborative task. As a complement, production of biosurfactants and extracellular enzymes plays a fundamental role to ensure the bioavailability of hydrophobic organic compounds. Several industrial effluents do not accomplish discharge guidelines for their safe disposal. In some cases, conventional (physicochemical) treatments are not efficient enough by the low pollutant concentration. Furthermore, the development and implementation of new technologies based on biological treatments look being the best option. Particularly, *B. subtilis* and *B. cereus* produce a mixed biofilm that allows the removal of chromium (III), achieving a reduction from an initial 1000 ppm to 2 ppm by applying a batch system to treat tannery industry effluents (Fathima, Rao, & Unni Nair, 2012). The same consortium immobilized in coarse sand in a continuous flow system (flow rate 20 mL/min) achieved a Cr(III) removal of 94% after 24 h after diluting 1:100 from an initial concentration of 3400 mg/L (Sundar, Sadiq, Mukherjee, & Chandrasekaran, 2011).

The use of biofilms in fluidized-bed reactors has been reported for the phosphorus removal from household waters in Norway. The flow passes through an anaerobic zone and then an aerobic zone, achieving an average phosphorus remotion efficiency of $94 \pm 0.5\%$ accompanied by a soluble chemical oxygen demand (SCOD) removal of $66 \pm 0.07\%$ (Rudi et al., 2019).

Biofilms are generally viewed negatively in the industry for being harmful and causing economic losses. Despite this, there are various processes that are catalyzed by microorganisms, and in turn, if these are immobilized as biofilm, they will evidence a better performance compared with planktonic cells, as the production of alcohols and organic acids (Jiang et al., 2021). In addition, using waste from other industries as raw materials should not be disregarded. An example is the biofilm of *Zymomonas mobilis* used in obtaining ethanol from wheat bran. Another advantage is that the biofilm confers the microorganisms a greater resistance to the solvent/alcohol produced, obtaining higher yields (Leonov et al., 2021).

5. Biofilm monitoring in nature: Image acquisition and processing

Because of the biofilm environmental relevance, different techniques of image capture accompanied by the corresponding software for processing have been growing in the last years of research. The idea of a direct 3D visualization, identifying matrix fractions and cells, discriminating the maturing steps in real time or even interactions with pollutants gains special interest for biofilm-mediated bioremediation studies (da Silva, Pennafirme, Lopes, Lima, & Crapez, 2017). Several alternatives are available to be applied depending on the study goal. Scanning electron microscopy (SEM) and transmission electron microscopy (TEM) are the traditional techniques, initially used as tools for biofilm evaluation thanks to their high resolution and real-time analysis in case of SEM and the allowance of the evaluation of the compositional characteristics of the sample in case of TEM. However, in both electron microscopies, the difficulty is focalized in sample preparation complexity, which includes an OsO_4 fixation and a coupled X-Ray microanalyzer in TEM, with the risk of losing some EPS elements during this process. An example of innovating development of microscopical methods is the scanning electrochemical microscopy (SECM). SECM can detect microbial interactions between biofilms, by a real-time spatial quantification of small molecules in 3D-microenvironments. For example, the electroactive signaling molecule pyocyanin produced in *P. aeruginosa* could be quantified by SECM, evidencing cell-cell communication and cooperative activities within the matrix (Connell, Kim, Shear, Bard, & Whiteley, 2014). Another innovative technique is the scanning transmission X-ray microscopy (STXM). Through an X-ray absorption mechanism, STXM collects sequences of images of biofilms performing an exhaustive mapping of the chemical species present.

To improve the image quality and reproducibility especially when environmental samples are processed, new advanced techniques were developed and are continuously evolving for bacterial biofilm evaluation, including confocal laser scanning microscopy (CLSM) and X-ray computed microtomography (μCT). In one hand, CLSM enables captures of natural living biofilm structure and morphology with a high resolution that provides qualitative and quantitative integrative 3D analyses. CLSM limitations are circumscribed to the image postprocessing, sample preparation, and depth of laser penetration caused by absorption and scattering phenomena. On the other side, μCT is a noninvasive useful tool for 2D and 3D image acquisition, which allows biofilm growth monitoring in situ with a temporal scale analysis, but with radiation damage and a complex post computing analysis required (da Silva et al., 2017).

In terms of cell quantification, epifluorescence microscopy allows direct counts of viable and nonviable microorganisms within biofilm matrix. This type analysis of viability and microbial metabolic conditions as well is a way to put in evidence the potential perturbations that pollutants exert on biofilm inhabitants.

Once images are captured, adequate software must be used for a correct processing guiding to a consistent interpretation. IMAGE J, COMSTAT, PHILIP, DAIME, ISA-3D, CONAN, and MATLAB are useful computational tools for an excellent image processing (Daniel, Barrionuevo, Doyle, & Vullo, 2016; Neu et al., 2010). The consideration of the advantages and disadvantages of the described techniques and an optimal combination of them are required to elucidate complex natural or engineered microbial biofilm structures, especially when the bacterial populations' resistance or tolerance to pollutants is subject of interest for a successful bioremediation process.

6. Concluding remarks

Developing a biofilm-mediated environmental technology requires an integrative approach to understand from different but converging perspectives the mechanisms involved. First, the biofilm formation, maturing, and maintenance are important factors to be explored in order to ensure efficiency in a process design. In addition, the EPS matrix composition and properties are relevant to infer the diversity of chemical interactions and hence widen the biotechnological application spectrum. Finally, the monitoring of cell viability and metabolic conditions, microbial cooperation, synergism and antagonism as social behaviors contributes to the biofilm dynamics and functionality. To conclude, this global knowledge represents the main issue to focus on to begin exploiting this second microbial biofilm lifestyle toward a sustainable restoration of the anthropogenically modified biosphere.

Acknowledgments

This work was financially supported by the Universidad Nacional de General Sarmiento and Consejo Nacional de Investigaciones Científicas y Tecnológicas (CONICET).

References

Bellenberg, S., Huynh, D., Poetsch, A., Sand, W., & Vera, M. (2019). Proteomics reveal enhanced oxidative stress responses and metabolic adaptation in *Acidithiobacillus ferrooxidans* biofilm cells on pyrite. *Frontiers in Microbiology*, *10*, 593. https://doi.org/10.3389/fmicb.2019.00592.

Billings, N., Birjiniuk, A., Samad, T. S., Doyle, P. S., & Ribbeck, K. (2015). Material properties of biofilms—A review of methods for understanding permeability and mechanics. *Reports on Progress in Physics*, *78*(3), 036601. https://doi.org/10.1088/0034-4885/78/3/036601.

Bradney, L., Wijesekara, H., Palansooriya, K. N., Obadamudalige, N., Bolan, N. S., Ok, Y. S., et al. (2019). Particulate plastics as a vector for toxic trace-element uptake by aquatic and terrestrial organisms and human health risk. *Environment International*, *131*, 104937. https://doi.org/10.1016/j.envint.2019.104937.

Cai, Y. M. (2020). Non-surface attached bacterial aggregates: A ubiquitous third lifestyle. *Frontiers in Microbiology*, *11*, 557035. https://doi.org/10.3389/fmicb.2020.557035.

Cheow, W. S., & Hadinoto, K. (2013). Biofilm-like *Lactobacillus rhamnosus* probiotics encapsulated in alginate and carrageenan microcapsules exhibiting enhanced thermotolerance and freeze-drying resistance. *Biomacromolecules*, *14*, 3214–3222. https://doi.org/10.1021/bm400853d.

Connell, J. L., Kim, J., Shear, J. B., Bard, A. J., & Whiteley, M. (2014). Real-time monitoring of quorum sensing in 3D-printed bacterial aggregates using scanning electrochemical microscopy. *Proceedings of the National Academy of Sciences of the United States of America*, *111* (51), 18255–18260. https://doi.org/10.1073/pnas.1421211111.

Cordoba Velandia, D. C. (2011). *Caracterización de EPS y SMP comoproteínas Y Carbohidratosdurante la operación de un BAnMI*. Colombia: Universidad de los Andes. http://hdl.handle.net/1992/24615.

da Silva, G. O. A., Pennafirme, S., Lopes, R. T., Lima, I., & Crapez, M. A. C. (2017). Imaging techniques for monitoring bacterial biofilms in environmental samples—An important tool for bioremediation studies. *BAOJ Microbiology*, *3*, 026.

Daniel, M. A. (2018). *Biotratamiento de efluentesindustriales que contenganmetales* (Ph.D. thesis). Universidad Nacional de General Sarmiento, Instituto de Ciencias. https://repositorio.ungs.edu.ar/handle/UNGS/398.

Daniel, M. A., Barrionuevo, M. R., Doyle, S. R., & Vullo, D. L. (2016). Kinetics of *Pseudomonas veronii* 2E biofilm development under different nutritional conditions for a proper bioreactor design. *Biochemical Engineering Journal*, *105*, 150–158. https://doi.org/10.1016/j.bej.2015.09.001.

Decho, A. W., Visscher, P. T., & Reid, R. P. (2005). Production and cycling of natural microbial exopolymers (EPS) within a marine stromatolite. *Palaeogeography Palaeoclimatology Palaeoecology*, *219*, 71–86. https://doi.org/10.1016/j.palaeo.2004.10.015.

Duffus, J. H. (2002). "Heavy metals"—A meaningless term? IUPAC technical report. *Pure and Applied Chemistry*, *74*(5), 793–807.

Fathima, A., Rao, J. R., & Unni Nair, B. (2012). Trivalent chromium removal from tannery effluent using kaolin-supported bacterial biofilm of *Bacillus* sp. isolated from chromium polluted soil. *Journal of Chemical Technology and Biotechnology*, *87*, 271–279. https://doi.org/10.1002/jctb.2710.

FDA. (2019). *Generally recognized as safe (GRAS)*. https://www.fda.gov/food/food-ingredients-packaging/generally-recognized-safe-gras. (accessed 05.01.22).

Ferreira, M. L., Casabuono, A. C., Stacchiotti, S. T., Couto, A. S., Ramirez, S. A., & Vullo, D. L. (2017). Chemical characterization of *Pseudomonas veronii* 2E soluble exopolymeras Cd(II) ligand for the biotreatment of electroplating wastes. *International Biodeterioration and Biodegradation*, *119*, 605–613. https://doi.org/10.1016/j.ibiod.2016.10.013.

Ferreira, M. L., Gerbino, E., Cavallero, G. J., Casabuono, A. C., Couto, A. S., Gomez-Zavaglia, A., et al. (2020). Infrared spectroscopy with multivariate analysis to interrogate the interaction of whole cells and secreted soluble exopolimeric substances of *Pseudomonas veronii* 2E with Cd(II), Cu(II) and Zn(II). *Spectrochimica Acta Part A, Molecular and Biomolecular Spectroscopy*, *228*, 117820. https://doi.org/10.1016/j.saa.2019.117820.

Ferreira, M. L., Lazzarini Behrmann, I. C., Ceretti, H. M., Zalts, A., Vullo, D. L., & Ramirez, S. A. (2013). Towards understanding the influence of cadmium on microbial behavior. In M. Hasanuzzaman, & M. Fujita (Eds.), *Cadmium: Characteristics, sources of exposure, health and environmental effects* (pp. 151–170). New York: Nova Science Publishers, Inc.

Flemming, H.-C., Wingender, J., Szewzyk, U., Steinberg, P., Rice, S. A., & Kjelleberg, A. (2016). Biofilms: An emergent form of bacterial life. *Nature Reviews Microbiology*, *14*, 563–575. https://doi.org/10.1038/nrmicro.2016.94.

Geng, N., Wu, Y., Zhang, M., Tsang, D. C. W., Rinklebe, J., Xia, Y., et al. (2019). Bioaccumulation of potentially toxic elements by submerged plants and biofilms: A critical review. *Environment International*, *131*, 105015. https://doi.org/10.1016/j.envint.2019.105015.

Ghafoor, A., Hay Iain, D., & Rehm Bernd, H. A. (2011). Role of exopolysaccharides in *Pseudomonas aeruginosa* biofilm formation and architecture. *Applied and Environmental Microbiology*, *77*(15), 5238–5246. https://doi.org/10.1128/AEM.00637-11.

Ghiasian, M. (2020). Microbial biofilms: Beneficial applications for sustainable agriculture. In A. A. Rastegari, A. N. Yadav, & N. Yadav (Eds.), *New and future developments in microbial biotechnology and bioengineering*. Amsterdam: Elsevier Inc. https://doi.org/10.1016/b978-0-12-820526-6.00009-9.

Giese, E. (2019). Evidences of EPS-iron (III) ions interactions on bioleaching process mini-review: The key to improve performance. *Orbital: The Electronic Journal of Chemistry*. https://doi.org/10.17807/orbital.v11i3.1389.

Gong, C., He, Y., Tang, Y., Hu, R., Lv, Y., Zhang, Q., et al. (2021). Biofilms in plant-based fermented foods: Formation mechanisms, benefits and drawbacks on quality and safety, and functionalization strategies.

Trends in Food Science and Technology, *116*, 940–953. https://doi.org/10.1016/j.tifs.2021.08.026.

Gu, T., Rastegar, S. O., Mousavi, S. M., Li, M., & Zhou, M. (2018). Advances in bioleaching for recovery of metals and bioremediation of fuel ash and sewage sludge. *Bioresource Technology*, *261*, 428–440. https://doi.org/10.1016/j.biortech.2018.04.033.

Guibaud, G., Bordas, F., Saaid, A., d'Abzac, P., & Van Hullebusch, E. (2008). Effect of pH on cadmium and lead binding by extracellular polymeric substances (EPS) extracted from environmental bacterial strains. *Colloids and Surfaces, B: Biointerfaces*, *63*, 48–54. https://doi.org/10.1016/j.colsurfb.2007.11.002.

Guibaud, G., van Hullebusch, E., & Bordas, F. (2006). Lead and cadmium biosorption by extracellular polymeric substances (EPS) extracted from activated sludges: pH-sorption edge tests and mathematical equilibrium modelling. *Chemosphere*, *64*, 1955–1962. https://doi.org/10.1016/j.chemosphere.2006.01.012.

Guibaud, G., van Hullebusch, E., Bordas, F., d'Abzac, P., & Joussein, E. (2009). Sorption of Cd(II) and Pb(II) by exopolymeric substances (EPS) extracted from activated sludges and pure bacterial strains: Modeling of the metal/ligand ratio effect an role of the mineral fraction. *Bioresource Technology*, *100*, 2959–2968. https://doi.org/10.1016/j.biortech.2009.01.040.

Hayta, E. N., Ertelt, M. J., Kretschmer, M., & Lieleg, O. (2021). Bacterial materials: Applications of natural and modified biofilms. *Advanced Materials Interfaces*, *8*, 2101024. https://doi.org/10.1002/admi.202101024.

Heumann, A., Assifaoui, A., Da Silva Barreira, D., Thomas, C., Briandet, R., Laurent, J., et al. (2020). Intestinal release of biofilm-like microcolonies encased in calcium-pectinate beads increases probiotic properties of *Lacticaseibacillus paracasei*. *npj Biofilms Microbiomes*, *6*, 44. https://doi.org/10.1038/s41522-020-00159-3.

Hujslová, M., Bystrianský, L., Benada, O., & Gryndler, M. (2019). Fungi, a neglected component of acidophilic biofilms: Do they have a potential for biotechnology? *Extremophiles*, *23*, 267–275. https://doi.org/10.1007/s00792-019-01085-9.

Jiang, Y., Liu, Y., Zhang, X., Gao, H., Mou, L., Wu, M., et al. (2021). Biofilm application in the microbial biochemicals production process. *Biotechnology Advances*, *48*, 107724. https://doi.org/10.1016/j.biotechadv.2021.107724.

Krell, T., Lacal, J., Reyes-Darias, J. A., Jimenez-Sanchez, C., Sungthong, R., & Ortega-Calvo, J. J. (2013). Bioavailability of pollutants and chemotaxis. *Current Opinion in Biotechnology*, *24*, 451–456. https://doi.org/10.1016/j.copbio.2012.08.011.

Leonov, P. S., Flores-Alsina, X., Gernaey, K. V., & Sternberg, C. (2021). Microbial biofilms in biorefinery—Towards a sustainable production of low-value bulk chemicals and fuels. *Biotechnology Advances*, *50*, 107766. https://doi.org/10.1016/j.biotechadv.2021.107766.

Mahamuni-Badiger, P. P., Patil, P. M., Badiger, M. V., Patel, P. R., Thorat-Gadgil, B. S., Pandit, A., et al. (2020). Biofilm formation to inhibition: Role of zinc oxide-based nanoparticles. *Materials Science and Engineering: C*, *108*, 110319. https://doi.org/10.1016/j.msec.2019.110319.

Mahmoud, A., Cézac, P., Hoadley, A. F. A., Contamine, F., & D'Hugues, P. (2017). A review of sulfide minerals microbially assisted leaching in stirred tank reactors. *International Biodeterioration and Biodegradation*, *119*, 118–146. https://doi.org/10.1016/j.ibiod.2016.09.015.

Mao, Y., Hu, H., & Yan, Y. (2011). Biosorption of cesium(I) from aqueous solution by a novel exopolymers secreted from *Pseudomonas fluorescens* C-2: Equilibrium and kinetic studies. *Journal of Environmental Sciences (China)*, *23*(7), 1104–1112. https://doi.org/10.1016/S1001-0742(10)60522-0.

Matheus, M. C., Ekenberg, M., Bassin, J. P., Dezotti, M. W. C., & Piculell, M. (2021). High loaded moving bed biofilm reactors treating pulp & paper industry wastewater: Effect of hydraulic retention time, filling degree and nutrients availability on performance, biomass fractions and nutrients utilization. *Journal of Environmental Chemical Engineering*, *9*(1), 940–953. https://doi.org/10.1016/j.jece.2020.104944.

Michels, J., Stippkugel, A., Lenz, M., Wirtz, K., & Engel, A. (2018). Rapid aggregation of biofilm-covered microplastics with marine biogenic particles. *Proceedings of the Royal Society B*, *285*, 20181203. https://doi.org/10.1098/rspb.2018.1203.

Moazzam, P., Boroumand, Y., Rabiei, P., Baghbaderani, S. S., Mokarian, P., Mohagheghian, F., et al. (2021). Lithium bioleaching: An emerging approach for the recovery of Li from spent lithium ion batteries. *Chemosphere*, *277*, 130196. https://doi.org/10.1016/j.chemosphere.2021.130196.

Mohsin, M. Z., Omer, R., Huang, J., Mohsin, A., Guo, M., Qian, J., et al. (2021). Advances in engineered *Bacillus subtilis* biofilms and spores, and their applications in bioremediation, biocatalysis, and biomaterials. *Synthetic and Systems Biotechnology*, *6*(3), 180–191. https://doi.org/10.1016/j.synbio.2021.07.002.

Navarro, J. B., Mashburn-Warren, L., Bakaletz, L. O., Bailey, M. T., & Goodman, S. D. (2017). Enhanced probiotic potential of *lactobacillus reuteri* when delivered as a biofilm on dextranomer microspheres that contain beneficial cargo. *Frontiers in Microbiology*, *8*. https://doi.org/10.3389/fmicb.2017.00489.

Neu, T. R., Manz, B., Volke, F., Dynes, J. J., Hitchcock, A. P., & Lawrence, J. R. (2010). Advanced imaging techniques for assessment of structure, composition and function in biofilm systems. *FEMS Microbiology Ecology*, *72*(1), 1–21. https://doi.org/10.1111/j.1574-6941.2010.00837.x.

Nouha, K., Kumar, R. S., Balasubramanian, S., & Tyagi, R. D. (2018). Critical review of EPS production, synthesis and composition for sludge flocculation. *Journal of Environmental Sciences (China)*, *66*, 225–245. https://doi.org/10.1016/j.jes.2017.05.020.

Paluch, E., Rewak-Soroczyńska, J., Jędrusik, I., Mazurkiewicz, E., & Jermakow, K. (2020). Prevention of biofilm formation by quorum quenching. *Applied Microbiology and Biotechnology*, *104*, 1871–1881. https://doi.org/10.1007/s00253-020-10349-w.

Rittmann, B. E. (2018). Biofilms, active substrata, and me. *Water Research*, *132*, 135–145. https://doi.org/10.1016/j.watres.2017.12.043.

Robertson, R. S., & McLean, R. J. C. (2015). Beneficial biofilms. *AIMS Bioengineering*, *2*, 437–448. https://doi.org/10.3934/bioeng.2015.4.437.

Rossi, C., Serio, A., Chaves-López, C., Anniballi, F., Auricchio, B., Goffredo, E., et al. (2018). Biofilm formation, pigment production and motility in *Pseudomonas* spp. isolated from the dairy industry. *Food Control*, *86*, 241–248. https://doi.org/10.1016/j.foodcont.2017.11.018.

Rudi, K., Goa, I. A., Saltnes, T., Sørensen, G., Angell, I. L., & Eikås, S. (2019). Microbial ecological processes in MBBR biofilms for biological phosphorus removal from wastewater. *Water Science and Technology*, *79*, 1467–1473. https://doi.org/10.2166/wst.2019.149.

Sheng, G. P., Yu, H. Q., & Li, X. Y. (2010). Extracellular Polymeric Substances (EPS) of microbial aggregates in biological wastewater treatment systems: A review. *Biotechnology Advances*, *28*(6), 882–894. https://doi.org/10.1016/j.biotechadv.2010.08.001.

Singh, M. P., Singh, P., Haibi, L., & Singh, R. K. (2020). Microbial biofilms: Development, structure, and their social assemblage for beneficial applications. In A. A. Rastegari, A. N. Yadav, & N. Yadav (Eds.), *New and future developments in microbial biotechnology and bioengineering*. Amsterdam: Elsevier Inc. https://doi.org/10.1016/B978-0-444-64279-0.00010-4.

Sundar, K., Sadiq, I. M., Mukherjee, A., & Chandrasekaran, N. (2011). Bioremoval of trivalent chromium using *Bacillus* biofilms through continuous flow reactor. *Journal of Hazardous Materials*, *196*, 44–51. https://doi.org/10.1016/j.jhazmat.2011.08.066.

Trunk, T., Khalil, H. S., & Leo, J. C. (2018). Bacterial autoaggregation. *AIMS Microbiology*, *4*(1), 140–164. https://doi.org/10.3934/microbiol.2018.1.140.

van der Ent, A., Baker, A. J., Reeves, R. D., Chaney, R. L., Anderson, C. W., Meech, J. A., et al. (2015). Agromining: Farming for metals in the future? *Environmental Science & Technology*, *49*(8), 4773–4780. https://doi.org/10.1021/es506031u.

Yakamercan, E., & Aygün, A. (2021). Fate and removal of pentachlorophenol and diethylhexyl phthalate from textile industry wastewater by sequencing batch biofilm reactor: Effects of hydraulic and solid retention times. *Journal of Environmental Chemical Engineering*, *9* (4), 105436. https://doi.org/10.1016/j.jece.2021.105436.

Yin, W., Wang, Y., Liu, L., & He, J. (2019). Biofilms: The microbial "protective clothing" in extreme environments. *International Journal of Molecular Sciences*, *20*, 3423. https://doi.org/10.3390/ijms20143423.

Zhang, J., Miao, Y., Zhang, Q., Sun, Y., Wu, L., & Peng, Y. (2020). Mechanism of stable sewage nitrogen removal in a partial nitrification-anammox biofilm system at low temperatures: Microbial community and EPS analysis. *Bioresource Technology*, *297*, 122459. https://doi.org/10.1016/j.biortech.2019.122459.

Chapter 39

Use of biofilm bacteria to enhance overall microbial fuel cell performance

Ankit Rao[a], Shipra Varshney[a], Sudipa Bhadra[b], Anubha Kaushik[a], Anshu Gupta[a], and Surajbhan Sevda[b]
[a]*University School of Environment Management, Guru Gobind Singh Indraprastha University, New Delhi, India,* [b]*Department of Biotechnology, National Institute of Technology Warangal, Warangal, Telangana, India*

1. Introduction

Energy demands in every sector have been consistently increasing for efficient working in present times. Continuous rise in industrialization and urbanization is causing energy consumption at high pace exerting pressure on the energy resource (Rahimnejad, Ghoreyshi, Najafpour, Younesi, & Shakeri, 2012). Along with the energy crisis, the industrialization and urbanization also result in generation of large amount of wastewater. Both the overexploitation of the resources for energy production and the generated wastewater are detrimental for the environmental health requiring attention and serious consideration (Nikhil, Chaitanya, Srikanth, Swamy, & Mohan, 2018). Therefore, there is a need to search for an option that not only generates energy following a sustainable route but also utilizes the waste and thereby reduces environmental burden. One such emerging area of interest for the above-stated conditions is microbial fuel cell (MFC) (He et al., 2017; Khan et al., 2020; Sevda, Dominguez-Benetton, Vanbroekhoven, Sreekrishnan, & Pant, 2013).

MFC has generated incredible interest as a framework reaction unit for producing bioenergy, treating wastewater, and biodegrading the organic matter (Rahimnejad, Adhami, Darvari, Zirepour, & Oh, 2015). MFCs have evident validated benefits; for instance, they are credited for converting the substrate into energy by using biocatalysts and are capable of operating in varied range of different temperatures (low or high), pH, and substrate sources (He et al., 2017; Palanisamy et al., 2019). MFCs employ unique microorganisms, which have the capability to transfer electrons to an outer electron acceptor, and these are termed as electroactive bacteria (EAB) or exoelectrogens (Logan, Rossi, Ragab, & Saikaly, 2019; Sydow, Krieg, Mayer, Schrader, & Holtmann, 2014). These electroactive microorganisms survive in the anodic chamber by feeding on available substrate. The wastewater generated from different sources can be used as substrate in this chamber for growth of microorganisms. Different wastewater, which are fed into anode chamber, contains variety of microorganisms, which forms an active biofilm mass on electrode surface (Liu et al., 2011) (Fig. 1). However, in most of the cases, microbial inoculum is added into anode chamber to accelerate the biofilm formation process (Hemdan, El-Taweel, Goswami, Pant, & Sevda, 2021; Sevda et al., 2016; Sevda, Dominguez-Benetton, Vanbroekhoven, Sreekrishnan, & Pant, 2013). In case of mixed bacterial inoculum source, many exoelectrogens and nonexoelectrogenic bacteria are found in the biofilm, nonetheless over the course of time, metabolically active EAB community is enriched on the anode electrode (Mathuriya, 2013).

Electrochemically active microorganisms play key role of biocatalysts in oxidizing the required substrates to carbon dioxide and hence producing electrons at the anode (Brutinel & Gralnick, 2012). EAB bacteria are confined in biofilm matrix framework formed by most of biota, which includes exopolysaccharides, eDNA, lipids, and extracellular proteins. Matrix encompasses and concretes cells together, and furthermore, it sorts out them into firm multicellular structure. Consequently, extracellular network provides a platform for cell association keeping the microbial strains in the biofilm matrix glued together, thus making intramatrix interactions effective. Biofilm matrix comprises extracellular polymeric substances (EPS), which is a fundamental requirement for the formation of biofilm. It also empowers the microbial life span, influencing the microenvironment of the biofilm (Garlapati, Naha, Sharma, Goswami, & Sevda, 2020; Hobley, Harkins, MacPhee, & Stanley-Wall, 2015; Teschler et al., 2015). EAB biofilm in MFC has been leveraged for vast number of applications including MFC-based biosensors (Sun, Zhang, Wu, Dong, & Angelidaki, 2019), wastewater treatment, metal removal (Singh & Kaushik, 2021a), and as mini power source (Zhang et al., 2019). Thus, formation of a biofilm with viable and metabolically active EAB is conducive for continuous operation of MFC. Accurate analysis and monitoring of biofilm formation and its regulation ascertain

Understanding Microbial Biofilms. https://doi.org/10.1016/B978-0-323-99977-9.00014-4

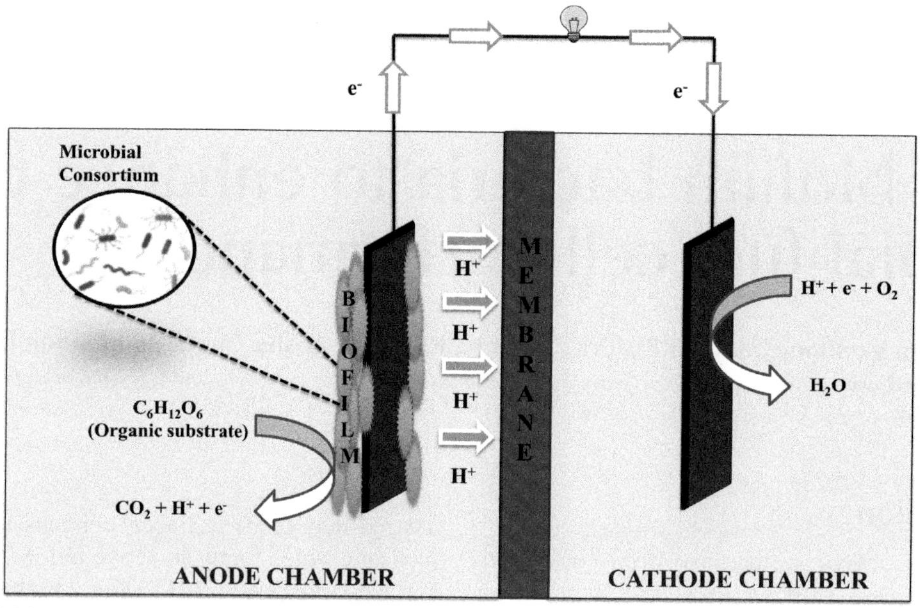

FIG. 1 Basic microbial fuel cell architecture with biofilm developed anode.

the smooth running of MFC. This chapter provides an overview of EAB biofilm formation and its dynamic role in affecting the electrochemical output of an MFC.

2. MFC design and operation

MFCs are bioelectrochemical systems, which rely on the metabolic activity of electroactive organisms to oxidize the substrate. MFC architecture combines the biological component (biofilm) and their associated oxidation-reduction reactions for generation of electric current and simultaneous substrate degradation. Wastewater as a source of substrate is fed into the MFC's anode chamber, and a final current output is harvested through electrical circuit (Munoz-Cupa, Hu, Xu, & Bassi, 2021).

Basic components of MFC include an anode chamber along with cathode chamber, which are separated by membrane and are connected via external electrical circuit. Anode chamber hosts the electroactive bacterial biofilm, whereas cathode chamber is utilized for reduction reactions to complete the electrical circuit (Logan et al., 2006; Singh & Kaushik, 2021). MFC size and reactor configuration have varied from a large-size reactor of 1000L (Liang et al., 2018) to small size of few microliters (Liu & Choi, 2017). Reducing the working volume of MFC unit shows positive effect on power generation (Singh & Kaushik, 2021b). To eliminate membrane requirement and reactor complexity, single-chambered MFCs are now widely used (Zhang, Zhu, et al., 2017), and to generate and harvest large power output with high treatment efficiency, stacking of various small-size reactors is also suggested (Ieropoulos, Greenman, & Melhuish, 2008). Various design configurations in MFCs are summarized below:

3. Double-chamber MFC (DCMFC)

Basic architecture of double-chambered MFC incorporates two distinct chambers, one is anaerobic anodic compartment, and other is cathode, which is usually aerobic. Proton exchange membrane separating both the chambers guarantees proper transfer of protons, which is essential for the high-power output. Presence of proton exchange membrane and two distinct chambers increases the electrode spacing and owing to the complicated design of the system, double-chambered MFC possesses high internal resistance, thereby reducing the power generation. However, power generation can be greatly enhanced in double-chamber MFC by employing high proton conductive membrane (Sirajudeen, Annuar, Ishak, Yusuf, & Subramaniam, 2021). Anode and cathode chamber in DCMFC can be operated in different shapes and size based on the intended use. Different shapes have been developed in DCMFC over the years ranging from conventional H-shaped design, cubic-shaped, and cylindrical-shaped (Do et al., 2018). A typical DCMFC is shown in Fig. 2A with two distinctive anode and cathode chambers separated by PEM. Double-chambered MFC provides an added benefit of utilizing the cathode chamber for heavy metal reduction (Mathuriya & Yakhmi, 2014) or cotreatment of wastewater in anode and cathode chamber simultaneously (Vélez-Pérez et al., 2020).

4. Single-chamber MFC (SCMFC)

Single-chambered MFCs comprise a basic framework, which has single anode chamber and open air cathode (Fig. 2B). Fundamental principle standardized for making single-chamber MFC is unique as it maintains both the connectivity between the anodic chamber and the open-air cathode along with keeping them isolated from one another by separation layer such as PEM restricting oxygen transfer to the anode (Choudhury et al., 2017). Researchers now prefer using single-chambered MFC, because of their relatively simple design and high power output. Single-chamber MFC also lowers the cost of reactor fabrication by eliminating one chamber and thus is considered for scale-up studies (Liu, Cheng, Huang, & Logan, 2008). Electrode distance in MFC reactors is considered to be major contributor to internal resistance (Rossi & Logan, 2020). Thus in a single-chambered MFC, cathode is placed adjoining the proton exchange membrane reducing the anode-cathode distance. However, certain single-chambered MFCs are also developed without the application of PEM (Liu et al., 2008) reducing the manufacturing cost further (Vicari, Albamonte, Galia, & Scialdone, 2018).

5. Stacked MFC

Effective technique adopted to arrange MFCs for enhancing the energy unit's performance is stacking various MFC units together either in series or in parallel circuit combination (Fig. 2C). Series combination increases voltage output, while parallel association upgrades electric current (Abdallah, Feroz, Alani, Sayed, & Shanableh, 2019). Furthermore, adequate designing of MFC unit depends on numerous factors such as the direction of stack either horizontal or vertical, type of the electrode, size and shape of the reactor. Mumford, Barringer, Benzel, Reilly, and Young (2012) reported that in 72 L MFC system, the stacking of units enhanced the COD removal to 97% and aided in attainment of maximum power density of 51 W/m^3. As documented by Liang et al. (2018), stacking of 50 20-L MFC units corresponding to 1000-L MFC module led to accomplish power density of 125 W/m^3 and up to 80% COD reduction.

Due consideration of conducted studies infers that use of stacked framework can be reasonable for wastewater treatment processes owing to easy modification, circulation of wastewater with no entanglements, maintaining capillary

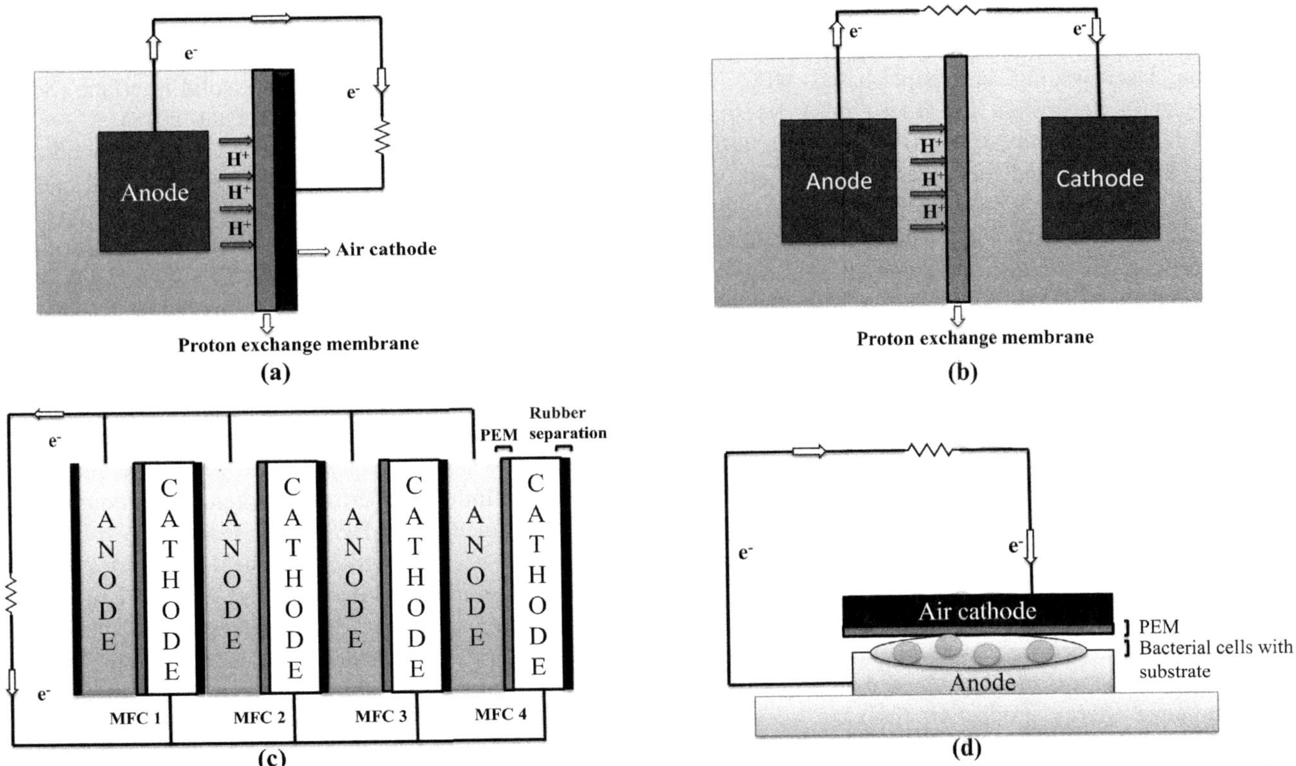

FIG. 2 Different designs of microbial fuel cell. (A) Double-chamber MFC, (B) single-chamber MFC, (C) stacked MFC (parallel connection), (D) micro-sized MFC.

action, and it is cost-effective to assemble such units (Happe et al., 2016). But, when operated for long duration, organic matter starts depleting, and this further leads to an outrageous loss of power as microorganisms in the biofilm die or disperse. Also, association of MFC units in series leads to voltage inversion owing to an imbalance in substrate loading (Choudhury et al., 2017; Flimban, Ismail, Kim, & Oh, 2019; Mukherjee et al., 2022).

6. Micro-sized MFC

With the continuous research to develop a high-energy MFC, it has been well established that smaller-sized reactors are better suited for high-power harvesting. Micro-sized MFCs are favored to have small start-up time and enhanced power output (Qian, He, Thelen, & Li, 2011). In a typical micro-sized MFC, proton exchange membrane layer is sandwiched between small-sized electrode attached with some solid support layer as depicted in Fig. 2D (Choi & Chae, 2013). Micro-sized MFC of 25 μL fabricated using graphene as anode and air cathode produced a high current density of 1190 A/m^3 while utilizing human saliva as substrate source (Mink, Qaisi, Logan, & Hussain, 2014). Moreover, microfluidics MFC could provide valuable push toward development of self-powered biosensor systems and implantable devices (Parkhey & Sahu, 2021). Few widely employed methods for microfabrication of MFC reactors involve lithographic fabrication, laser writing, and paper-based MFC (Goel, 2018). Recently, Jayapiriya and Goel (2021) developed a cellulose filter paper–based micro-MFC while utilizing *E. coli* as biocatalyst and metal-based ink as both anode and cathode electrodes. Cellulose filter worked as both separator between anode and cathode provided support for electrode placement generating a power output of 11.8 μW/cm^2. To increase the continuous power output from micro-MFC, an innovative method was adopted by Mohammadifar and Choi (2019) to solidify the microbial nutrients and ion exchanging mediums in MFC-based bio-battery for IoDT (Internet of Disposable things) application. The *Shewanella oneidensis* MR-1-based bio-battery generated a maximum power density of 0.4 μW/cm^2 with 96 h of operation, whereas liquid-based MFC sustained for only 4 h.

7. Electroactive biofilms in MFC

A plethora of microbes are present in every single niche of this planet, and the utilization of these microorganisms for the sustainable energy production from the waste is the recent field of innovation and exploration. One such novel technology serving dual benefits of electricity generation and wastewater treatment is MFC (Logan et al., 2019; Mohan, Velvizhi, Modestra, & Srikanth, 2014). Generation of electrical energy from the bio-catalyzed MFC is due to the presence of bacterial strains in the system, which have the potential of transforming the chemical energy to the electrical energy (Obata et al., 2020). At anode, microorganism acts as biocatalyst permitting the electron capture during the anaerobic oxidation of substrates (Patil, Harnisch, Kapadnis, & Schröder, 2010). The working compartment of the MFC, i.e., the anodic compartment upholds the formation of the biofilm. Design and fabrication of the anode influence the development, composition, thickness, and conductivity of the biofilm (Katuri et al., 2020).

Development of biofilm is a continual, steady, and sequent process involving the aggregation of the microbial cells to form the microcolonies. Biofilm formation is controlled and mediated through various pathways relying upon the physical, chemical, biological, and electrochemical parameters, namely type and amount of substrate, specific microbial species used in MFC, the electrode material, and the reaction conditions of the reactor including pH, temperature, etc. (Kumar, Singh, Wahid, & Din, 2015; Mohan et al., 2014) (Fig. 3). Microbial cells are supported on the solid matrix of the anode, which imparts them the mechanical stability and tolerance to physicochemical stress conditions. Biofilm may comprise pure strain or mixed microbial cultures (Koch & Harnisch, 2016); the diversity of the microbial cultures in the biofilms develops the synergistic effect, henceforth enhancing the efficiency of the energy unit (Kim et al., 2006). Electroactive biofilms can be characterized as the network of electroactive microbial cultures flourishing on a solid interface (Sevda, Sarma, Mohanty, Sreekrishnan, & Pant, 2018).

Biofilm is an organized colony of microbial cells appended to a solid support surface and implanted in extracellular polymeric substances (EPSs) network (Xiao & Zhao, 2017). Structure and functionality of the developed biofilm are highly influenced by the EPS. EPS mainly comprises polysaccharides, lipids, proteins, nucleic acids, and other biomolecules (Koo & Yamada, 2016). Framework and functionality of the developing biofilm are dictated by the biomolecules excreted extracellularly (Flemming et al., 2016). Once formed, EPS frames a network that encompasses and concretes cells together, keeping them in close proximity and permitting intercellular communications in confined space. EPS matrix plays a key role in providing the necessitated mechanical support, required adequate chemical environment, and structural stability to the generated biofilm. EPS impacts the dispersion of different particles inside the biofilm, subsequently making way to bridge the nutritional requirement of the developed microcolony (Nadell, Ricaurte, Yan, Drescher, & Bassler, 2017; Yahav, Berkovich, Ostrov, Reifen, & Shemesh, 2018).

Electroactive microorganisms can be classified broadly into two categories—exoelectrogens and electrotrophs. Exoelectrogens are known for performing the extracellular

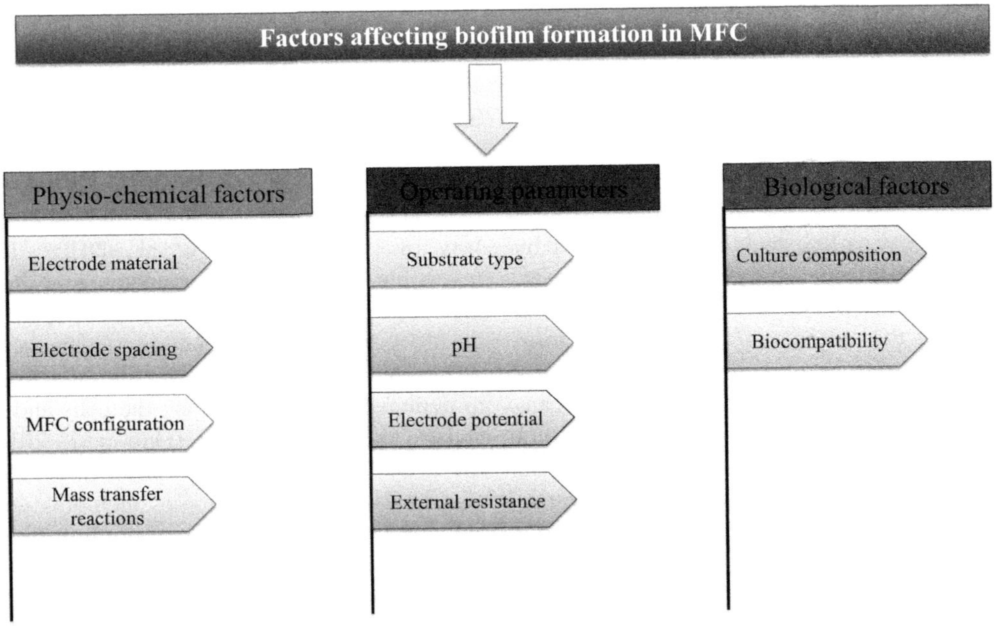

FIG. 3 Different factors affecting biofilm formation in microbial fuel cells.

electron transport generated by substrate oxidation to anode, whereas the electrotrophs are the acceptors of the electrons from the cathode electrode for their own metabolic activities. Former plays a vital role in MFC's anode chamber for the degradation of organic pollutants, wastewater treatment, electricity generation, and the latter are utilized for value-added products generation in cathodic compartment of the MFC unit (Liu, Li, & Yu, 2014; Logan et al., 2019; Rana, Singh, & bin ab Wahid, 2019).

Electron transfer process by exoelectrogens in the MFC can opt two different routes including mediated electron transport by utilizing the external mediators (electron shuttle molecules), and secondly, by direct electron transfer through conductive nanowires and cell-membrane-bound c-type cytochromes (Kumar et al., 2015) (Fig. 4). During

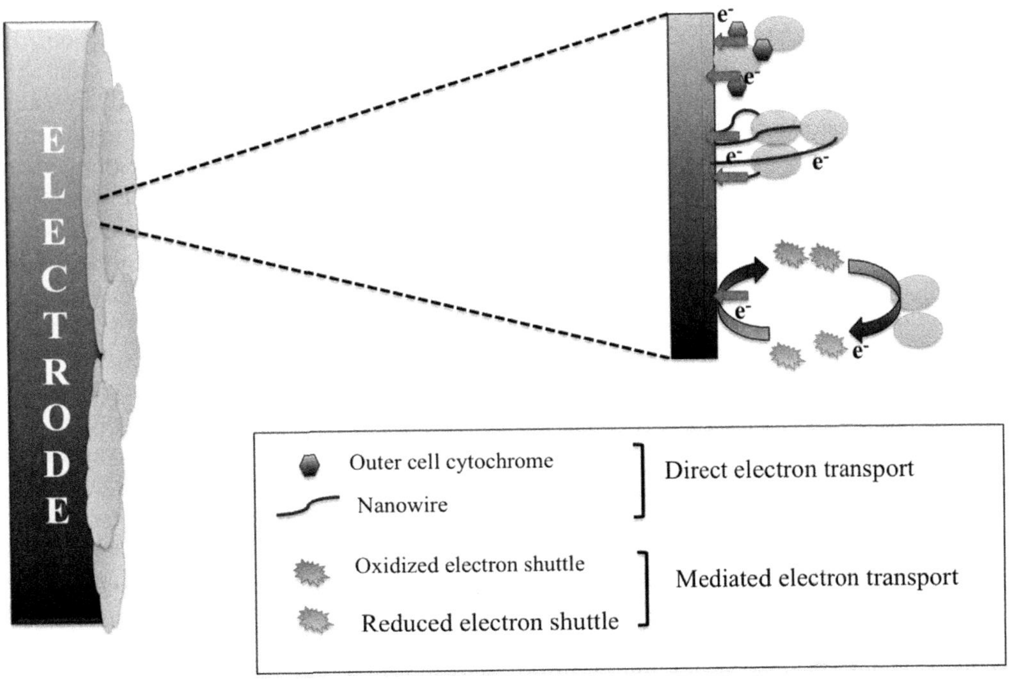

FIG. 4 Extracellular electron transport mechanisms.

extracellular electron transfer process, the electrons make their way across the cell membrane utilizing a network that is proteinaceous in nature (Korth & Harnisch, 2019). In particular, electron transport pathway is initiated by transfer of electron generated through oxidation of substrate to the intracellular electron carrier, followed by electron transport chain, and then finally to terminal electron acceptor. Ability of the microbial culture to participate in EET is examined by the presence of the multiheme c-type cytochrome (Shi, Squier, Zachara, & Fredrickson, 2007). Scrutiny of the genomic data reflects the presence of the genes for porin cytochrome conduit system in the Gram-negative bacteria entitling them as evident electroactive biofilm forming candidature strains. Gram-positive microorganisms do not possess the porin–cytochrome channels as detected in numerous electroactive Gram-negative microorganisms. However, among the few, a strain namely *Thermincola potens*, a Gram-positive thermophile, has been identified and reported to have the conduits for direct electron transfer (Lovley & Holmes, 2022). Extremophilic electroactive microbes application in MFC is the new area of research to widen application of microbial fuel cells in extreme environmental conditions (Chaudhary, Yadav, Singh, Sadhotra, & Patil, 2022). Extremophiles are defined as the microbes tolerant to a list of various physical-chemical parameters—pH, temperature, salinity, pressure, ionizing radiations, nutrition (Dhakar & Pandey, 2016; Gunjal Aparna, Waghmode Meghmala, & Patil Neha, 2021). Electroactive extremophiles are the microbes capable of transferring the electron in the extremities of the above-stated conditions. The electrogenic activity of the microbes in the thermophilic range has been shown by some extremophilic microbes such as *Thermincola ferriacetica*, *Calditerrivibrio nitroreducens* (Fu et al., 2013; Parameswaran et al., 2013). Exoelectrogenically capable microorganisms have also been reported in extreme pH conditions; profoundly saline; for instance, *G. ahangari* and *Acidiphilum* sp. (Kashefi et al., 2002; Malki, De Lacey, Rodríguez, Amils, & Fernandez, 2008). Outrageous conditions, such as high temperatures, low pH, high salinity, can have noted benefits for MFC application, for example, fast pace of tolerant microbial strain development and decreased possibilities of contamination by other microorganisms (Lovley & Holmes, 2021; Lusk et al., 2018).

8. Biofilm formation and regulation in MFC

Environmental conditions lead the microorganisms to develop a protective network for their dwelling and upliftment. Additionally, the microbes metabolize and flourish, owing to their ability to form biofilm. Biofilm matrix development initiates with planktonic cells adjoining to the surface. Surface attachment occurs in two steps, reversible attachment phase followed by irreversible attachment phase. Reversible attachment includes the attachment of pili, flagella to the surface. This phase acts as decision-making stage in which the microscopic organisms can either focus on forming the biofilm or can pass on a superficial level and return back to the planktonic way of life (Toyofuku et al., 2016). Transition from reversible to irreversible attachment is due to secondary messengers such as cAMP and c-Di-GMP as these regulates the EPS production (Römling, Galperin, & Gomelsky, 2013; Zhao et al., 2013). Irreversible attachment phase leads to permanent linkages among the cells and the linkages of cells to the anodic surface (Ono et al., 2014). This is followed by biofilm maturation phase in which bacterial cells start to develop microcolonies including cell division and cell aggregation. In the mature biofilm, EPS is responsible for attachment to surfaces, adjoining cells together, safeguarding from environmental stress, and maintaining the architecture of the biofilm (Fig. 5). EPS holds quorum signaling (QS) molecules, extracellular catalysts, and metabolic prerequisites enabling cell-to-cell communication (Drescher, Nadell, Stone, Wingreen, & Bassler, 2014). Few microbial cells move to planktonic stage from the mature biofilms. These free cells investigate new opportunities to append to another surface or move to new niches. Henceforth, dispersal isn't the last phase of the biofilm life cycle yet, in addition it is the beginning of another life cycle (McDougald, Rice, Barraud, Steinberg, & Kjelleberg, 2012).

Biofilm and microbial mat development is attributed and regulated by the EPS discharge cells. EPS acts as holding material empowering cell connection. Biofilm matrix structure is formed owing to the adhesive properties of the EPS. Enveloped by EPS, the microbial cells in biofilm matrix are shielded from the external environment enabling them to stay dynamic under external harsh conditions (Flemming et al., 2016; Ras, Lefebvre, Derlon, Paul, & Girbal-Neuhauser, 2011). Accordingly, it holds the biofilm matrix throughout a significant stretch of time, permitting the improvement of synergistic connections among species and empowering them to exhibit optimum performance (Blanco et al., 2019).

Thickness of biofilm plays pivotal role in regard to age of the biofilm but it doesn't remain electrochemically dynamic all through, due to the development of an internal layer of dead cells in the biofilm framework (Islam, Ehiraj, Cheng, Dubey, & Khan, 2019).

Despite the fact that EPS is an overpowering part in biofilm matrix formation and arrangement, the interaction is also affected by other set of variables, namely exopolysaccharides, lipopolysaccharides (LPS), and quorum sensing (QS) signaling molecules. Cellular interactions among the bacterial cells are coordinated owing to the

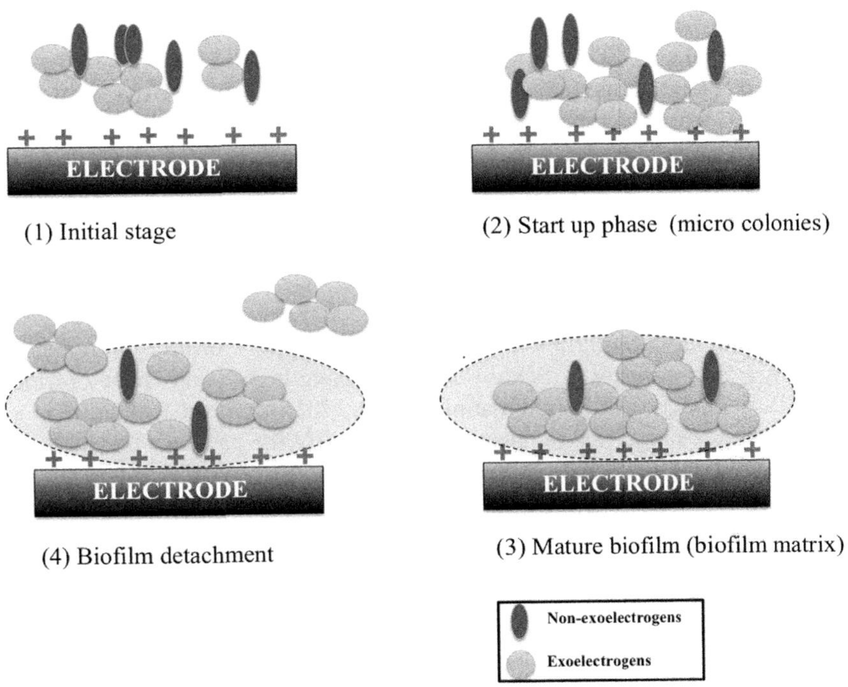

FIG. 5 Biofilm formation process at anode.

release of autoinducers from the microbial cells giving rise to the quorum sensing (Monzon, Yang, Li, & Alvarez, 2016). Microbes' physiological and biological pathways are also impacted by QSMs. QSM enables the microbes to adapt to their respective surroundings (Das, Das, & Ghangrekar, 2019). Therefore, the major parameters associated with biofilms development and regulation are QS signaling molecules and EPS (Angelaalincy et al., 2018).

9. Electrochemical characterization of biofilms

As only certain specific bacterial species are endowed with the ability of generating electricity in the biofilm matrix, i.e., electroactive bacteria, which have the mechanism of extracellular electron transport (EET) (Zhao et al., 2021). Biofilms developed at the anode electrode behave differently in various operating conditions. To analyze the biofilm formation, various methods used in MFC research include protein concentration determination, scanning electron microscopy (SEM) analysis, and microbial community analysis employing pyrosequencing (Zhi, Ge, He, & Zhang, 2014). However, these methods could not provide the exact measurement of electrocatalytic capabilities of bacterial biofilm, as various nonexoelectrogenic species are also present in biofilm such as methanogens and fermentative bacteria (Kim, Rossi, Regan, & Logan, 2021). These bacteria take part in substrate degradation but do not contribute to the electrons generation (Neto, Reginatto, & De Andrade, 2018). Electrochemical characterization of anode biofilms proves to be a robust method for studying the electrocatalytic activity of bacteria biofilms at anode. Different electrochemical methods employed for biofilm characterization include studying the polarization curve with open-circuit voltage (OCV), electrochemical impedance spectroscopy (EIS), and cyclic voltammetry (CV) studies. These methods provide useful insights into electron transfer capabilities of bacterial biofilm and their reaction kinetics (Martin, Savadogo, Guiot, & Tartakovsky, 2013). Most simplistic and widely used method to measure the electrochemical output of MFC is through polarization curve.

Cyclic voltammetry is a strong electrochemical technique, well exploited in studying bacteria-electrode interaction dynamics in MFCs. Cyclic voltammetry provides valuable insights about the electron transfer mechanisms followed by bacterial species, by sweeping the applied potential at a defined scan rate, the peak current observed at the defined potential scan rate in the cyclic voltagramms can be comprehended to observe the electrocatalytic activity of bacterial biofilms (Fricke, Harnisch, & Schröder, 2008). Moreover, CV could be utilized to differentiate and compare various operating conditions affecting the electrocatalytic output (López Zavala, Gonzalez Pena, Cabral Ruelas, Delgado Mena, & Guizani, 2019). Katuri, Kavanagh, Rengaraj, and Leech (2010) utilized cyclic voltammetry to study the electron transfer mechanism in *Geobacter sulfurreducens* biofilm subjected to changing applied potential and presence of electron acceptor. In another study, Sharma et al. (2016) observed electrochemical behavior of novel exoelectrogenic bacterial strain named as SCS5, which showed two redox peaks of -197 mV (Vs SHE) and 175 mV (Vs SHE) in cyclic voltagramms studies similar to redox peaks for known exoelectrogenic species, which ascertains the electrocatalytic and electron transfer capabilities of identified bacteria. EIS is another well-known nondestructive technique to characterize the internal resistance in a bioelectrochemical system. Ohmic losses along with mass transfer resistance and charge transfer resistance together contributes to total internal resistance in BES. Thus, EIS proves to be efficient in defining the internal resistance contributed by each factor (Sekar & Ramasamy, 2013). In EIS technique, AC (alternating current) is applied over a frequency range, and the resulted responding potential is measured and the spectra obtained are fitted to equivalent circuit to get the quantitative estimate (Kashyap et al., 2014). In a mediatorless MFC inoculated with *Escherichia coli* cells, internal resistance decreases from the initial levels of 5.6 MΩ to 3 MΩ after 5 h of biofilm development in MFC after that it increases further in operation resulting in lower output potential levels; however, with further addition of the substrate, internal resistance resulted to the same level. These EIS-based internal resistance values show that the resistance is affected with the bacterial growth phase, whereas on nutrients replenishment, internal resistance and output return to the initial values quickly (Herrero-Hernandez, Greenfield, Smith, & Akid, 2019). Thus, with the application of different electrochemical techniques, anodic biofilms characteristic and functions can be well studied and provide with an accurate measurement of electrocatalytic properties of developed biofilms.

10. Biofilm morphology and viability

The working of the MFC can be made more efficient and effective by taking a keynote of the physiology and morphology of the microbial biofilms utilized within the anodic chamber (Beyenal, Donovan, Lewandowski, & Harkin, 2004). MFC performance is dependent on type of biofilm formed, and henceforth, consideration of the optimized microbial strains for the biofilm generation is vital. The morphological identification and characterization of the biofilm can be done by using the analytical instruments, for instance, confocal laser scanning microscopy (CLSM), scanning electron microscopy (SEM) (Ishii, Shimoyama, Hotta, & Watanabe, 2008; Nandy et al., 2019).

As established and validated by the literature, type of microbes forming biofilm matrix, biofilm thickness, roughness coefficient, surface area, and volumetric parameters of the biofilm are considered as essential for describing the biofilm structure (Heydorn et al., 2000; Kierek & Watnick, 2003). Development of thicker biofilms may cause obstruction reducing the electrical productivity of the unit because of slow rate or inability of electron transfer by bacterial cells far away in the biofilm. Additionally, as width of the biofilm affirmed on anodic substrate increases, the diffusional resistance is added between the inner lying biofilm layer and the electron donor (Marcus, Torres, & Rittmann, 2007). Further the accessibility of deeper layers is highly reduced, and as microcolonies are away from the solution, they start to die and slough off (Islam et al., 2017). Excellent electrochemical performance of MFCs can be observed when the roughness coefficient of the concerned sample biofilm is minimum after polarization, demonstrating higher consistency of the biofilm (Artyushkova et al., 2015). CLSM technique is a well-suited technique to identify the presence of viable or dead cells in the bacterial biofilm, which is ignored in SEM images of the biofilm; however, the application of CLSM to study the biofilm morphology and viability is still very limited and needs more exploration (Saba et al., 2017).

Most dynamic part of biofilm, playing a significant job in the EET cycle, is within 10–20 μm of the electrode as the cells further away in biofilm find it difficult to transfer electrons to solid electron acceptor and become unviable after some time (Read, Dutta, Bond, Keller, & Rabaey, 2010). Thus, Islam et al. (2017) suggested the mild ultrasound treatment of anodic biofilm to slough off the dead cells in biofilm matrix. In their study, they observed the stable power output from the *K. pneumoniae* operated dual-chambered MFC with 30-min low ultrasound frequent exposure. More recently, same group has suggested flow-induced shear stress to remove the ineffective biofilm mass from anode chamber. Three different shear stress values of 4.38 mPa, 9.34 mPa, 14.32 mPa were applied to MFCs after 21 days when the current output values started to drop, Thus, after 5 min of moderated shear stress of 9.34 mPa, current output was increased and remained (Islam et al., 2019).

Ledezma, Greenman, and Ieropoulos (2012) through their experiments established that the growth of EAB biofilms can be controlled to get optimized power output from the biofilm-mediated substrate degradation. They studied specific growth of *S. oneidensis* MR-1 in small MFC reactor by varying the substrate flow rate, thus controlling the carbon/energy input. The bacterial cell showed a proportional increase in specific growth rate and subsequently increase in power output up to a limit of CFR (critical flow rate). Thus, knowing the growth and substrate requirement kinetic of specific EAB species could provide a framework to get optimized substrate requirement and higher power output.

11. Microbial community analysis in MFC

Anode chamber reactions are the foremost step in the working of MFC where it hosts the special kind of microorganisms capable of transferring electrons to an electron acceptor (Logan et al., 2019). *Geobacter* and *Shewanella* are two mostly commonly employed and extensively studied exoelectrogens in MFC (Kondaveeti, Lee, Park, & Min, 2020; Zhang, Liu, Qu, & Feng, 2017). Pure culture of bacteria (Bond & Lovley, 2003) or mixed culture of two or more microbes (Islam, Ong, Ethiraj, Cheng, & Khan, 2018; Saba et al., 2017) can be used as biocatalyst in MFC. Pure culture MFC has an advantage of fast start-up and high electrogenic activity as reported with pure culture of *Geobacter sulfurreducens* strain (Bond & Lovley, 2003). Pure culture MFC produces higher current output as mostly electroactive bacteria are selectively utilized (Jung & Pandit, 2019) but at the same time possess drawbacks with their sterile handling requirement and chances of contamination; moreover, utilizing pure culture is not feasible for the large-scale application in wastewater treatment; thus, there are limited studies on pure culture bacteria in MFC (Prathiba, Kumar, & Vo, 2022). To study the electron transport mechanism of specific microbes, pure culture inoculated MFCs are however often employed (Luo, Li, Zhou, & Hu, 2015). A synergistic relationship between two microorganisms toward efficient EET is well explored in MFCs (Islam et al., 2020; Liu & Choi, 2017). Culture of one phototrophic bacterium and heterotrophic exoelectrogens can overcome the requirement of continuous food supply to the bacteria as organic matter generated by photosynthetic bacteria can power the oxidation reaction of exoelectrogens for electricity generation. This coculture of two bacteria in micro-sized 90 μL MFC generated a self-sustained current output for 13 days without the external supply for organic food (Liu & Choi, 2017).

Pure culture and coculture inoculated MFC, being dependent on a single or two microbial species for electricity generation and waste degradation, makes the operation of MFC specific to operating condition and substrate used. Mixed culture microbial community at anode electrode is however more resilient and feasible for wide range of operating conditions and substrate. A very insightful study conducted by Jothinathan and Wilson (2017) to study the effect of pure culture, coculture, and mixed culture of microbes on MFC performance reported voltage generation of 180 mV and 300 mV for *Bacillus thuringiensis* DRR-1 and *Paracoccus homiensis* DRR-3 strains respectively. Whereas mixed culture inoculated MFC produced maximum voltage output of 500 mA confirming the suitability of utilizing mixed culture MFC for improved performance.

Also in mixed bacterial anode biofilm, microbial community is influenced by a variety of internal and external factors. Researchers have documented the changes in the microbial community structure in accordance with substrate used (Mei, Guo, Liu, Tang, & Xing, 2015), different electrode materials (Yu, Feng, He, Yang, & Xun, 2021), source of inoculation (Singh & Kaushik, 2021b), anode electrode potential (Hari, Katuri, Logan, & Saikaly, 2016) along with external factors such as pH and resistance applied (Jiang et al., 2021). Microbial community structure at anode can also be influenced by mode of operation. Two H-type reactors inoculated with 1:1 ratio of activated sludge and synthetic wastewater were operated in two modes of MFC and microbial electrolysis cell (MEC) (Almatouq, Babatunde, Khajah, Webster, & Alfodari, 2020). The MFC mode of operation leads to the development of more diverse bacterial community as compared with MEC mode. However, major bacterial species found in both reactors belonged to same classes of bacteria, i.e., Bacteroidetes, Proteobacteria, and Firmicutes.

In a mixed culture anodic biofilm, certain bacterial strains perform the function of EET dominantly and generate bioelectricity; these bacteria are gradually enriched in biofilm and soon take over the major proportion of MFC to deliver stable voltage output (Li et al., 2018). Researchers have opted for different methods to get exoelectrogens-rich biofilm in MFC, which includes physical, chemical, and biological approaches (Pham et al., 2008; Sharma et al., 2020; Varanasi, Sinha, & Das, 2017). Sui et al. (2021) studied the enrichment of exoelectrogenic species in dark and light exposure condition in three-electrode anaerobic reactor with indium tin oxide (ITO) as working electrode. Formation of electrogenically active biofilm in dark-operated reactors led to current generation of 4.5 A/m^2, which was many times more than the EAB formed under light condition. Dark exposed EAB majorly contained 85% *Geobacter*, whereas *Rhodopseudomonas* (75%) was dominant on 1200 lx light-operated biofilm. This study provided useful insights for developing highly enriched and electroactive biofilms for MFCs application. Singh and Kaushik (2020) selected wetland sediments as inoculum source to get sustained high power output with active biofilm formed anode while treating distillery wastewater. Nanotechnology-based approaches have also been well investigated and have proven to be effective methods in obtaining high electrocatalytic activity. In one study, gold nanoparticles and multiwalled carbon nanotube–based coating on anode electrode led to development of rich electroactive bacterial community mainly belonging to Gammaproteobacteria and Negativicutes class. Nano-modified anode MFC led to increase in power density from 114.24 ± 3.66 mW/m^2 by bare anode to 178.34 ± 4.79 mW/m^2 by Biogenic Au/MWCNT anode (Wu et al., 2018). Additionally, Li et al. (2021) explored the FeS nanoparticle and biochar composite–based modification on carbon cloth anode in two-chambered MFC. To investigate the microbial community development, anode electrodes were studied, and results indicated that the FeS/BC anode contained 75.16% *Geobacter* sp., whereas bare carbon cloth contained only 4.81% indicating a highly enriched EAB sp. at modified electrode.

12. Conclusion

The EAB formation process is discussed in detail for improving the performance of MFC. The wastewater treatment process is carried out in the various designs of MFC such as the single chamber, dual chamber, stacked chamber, and micro-sized MFC. In all the MFC designs, the basic active force is healthy EAB on anode electrode for higher efficiency of bioelectricity generation and wastewater treatment. The recent development in the electrochemical analysis of EAB and microbial community analysis for electroactive bacteria involved in the formation of biofilm give future insight for a robust MFC performance with complex wastewater treatments and its use in biosensor and other fuel generation from wastewater. The biofilm regulation strengthens the reuse of MFC for real wastewater treatment and bioenergy generation.

Acknowledgments

Dr. Surajbhan Sevda thanks NIT Warangal for the Research Seed grant (P1128) for a support of this work.

References

Abdallah, M., Feroz, S., Alani, S., Sayed, E. T., & Shanableh, A. (2019). Continuous and scalable applications of microbial fuel cells: A critical review. *Reviews in Environmental Science and Biotechnology, 18*(3), 543–578.

Almatouq, A., Babatunde, A. O., Khajah, M., Webster, G., & Alfodari, M. (2020). Microbial community structure of anode electrodes in microbial fuel cells and microbial electrolysis cells. *Journal of Water Process Engineering, 34*, 101140.

Angelaalincy, M. J., Navanietha Krishnaraj, R., Shakambari, G., Ashokkumar, B., Kathiresan, S., & Varalakshmi, P. (2018). Biofilm engineering approaches for improving the performance of microbial fuel cells and bioelectrochemical systems. *Frontiers in Energy Research, 6*, 63.

Artyushkova, K., Cornejo, J. A., Ista, L. K., Babanova, S., Santoro, C., Atanassov, P., et al. (2015). Relationship between surface chemistry, biofilm structure, and electron transfer in *Shewanella* anodes. *Biointerphases, 10*(1), 019013.

Beyenal, H., Donovan, C., Lewandowski, Z., & Harkin, G. (2004). Three-dimensional biofilm structure quantification. *Journal of Microbiological Methods, 59*(3), 395–413.

Blanco, Y., Rivas, L. A., González-Toril, E., Ruiz-Bermejo, M., Moreno-Paz, M., Parro, V., et al. (2019). Environmental parameters, and not phylogeny, determine the composition of extracellular polymeric substances in microbial mats from extreme environments. *Science of the Total Environment, 650*, 384–393.

Bond, D. R., & Lovley, D. R. (2003). Electricity production by *Geobacter sulfurreducens* attached to electrodes. *Applied and Environmental Microbiology, 69*(3), 1548–1555.

Brutinel, E. D., & Gralnick, J. A. (2012). Shuttling happens: Soluble flavin mediators of extracellular electron transfer in *Shewanella. Applied Microbiology and Biotechnology, 93*(1), 41–48.

Chaudhary, S., Yadav, S., Singh, R., Sadhotra, C., & Patil, S. A. (2022). Extremophilic electroactive microorganisms: Promising biocatalysts for bioprocessing applications. *Bioresource Technology*, 126663.

Choi, S., & Chae, J. (2013). Optimal biofilm formation and power generation in a micro-sized microbial fuel cell (MFC). *Sensors and Actuators A: Physical, 195*, 206–212.

Choudhury, P., Uday, U. S. P., Mahata, N., Tiwari, O. N., Ray, R. N., Bandyopadhyay, T. K., et al. (2017). Performance improvement of microbial fuel cells for waste water treatment along with value addition: A review on past achievements and recent perspectives. *Renewable and Sustainable Energy Reviews, 79*, 372–389.

Das, S., Das, S., & Ghangrekar, M. M. (2019). Quorum-sensing mediated signals: A promising multi-functional modulator for separately enhancing algal yield and power generation in microbial fuel cell. *Bioresource Technology, 294*, 122138.

Dhakar, K., & Pandey, A. (2016). Wide pH range tolerance in extremophiles: Towards understanding an important phenomenon for future biotechnology. *Applied Microbiology and Biotechnology, 100*(6), 2499–2510.

Do, M. H., Ngo, H. H., Guo, W. S., Liu, Y., Chang, S. W., Nguyen, D. D., et al. (2018). Challenges in the application of microbial fuel cells to wastewater treatment and energy production: A mini review. *Science of the Total Environment, 639*, 910–920.

Drescher, K., Nadell, C. D., Stone, H. A., Wingreen, N. S., & Bassler, B. L. (2014). Solutions to the public goods dilemma in bacterial biofilms. *Current Biology, 24*(1), 50–55.

Flemming, H. C., Wingender, J., Szewzyk, U., Steinberg, P., Rice, S. A., & Kjelleberg, S. (2016). Biofilms: An emergent form of bacterial life. *Nature Reviews Microbiology, 14*(9), 563–575.

Flimban, S. G., Ismail, I. M., Kim, T., & Oh, S. E. (2019). Overview of recent advancements in the microbial fuel cell from fundamentals to applications: Design, major elements, and scalability. *Energies, 12* (17), 3390.

Fricke, K., Harnisch, F., & Schröder, U. (2008). On the use of cyclic voltammetry for the study of anodic electron transfer in microbial fuel cells. *Energy & Environmental Science, 1*(1), 144–147.

Fu, Q., Kobayashi, H., Kawaguchi, H., Wakayama, T., Maeda, H., & Sato, K. (2013). A thermophilic gram-negative nitrate-reducing bacterium, *Calditerrivibrio nitroreducens*, exhibiting electricity generation capability. *Environmental Science & Technology, 47*(21), 12583–12590.

Garlapati, V. K., Naha, S., Sharma, S., Goswami, P., & Sevda, S. (2020). Electroactive biofilms (EAB): Role in a bioelectrochemical system for wastewater treatment and bioelectricity generation. In *Microbial biofilms* (pp. 207–226). CRC Press.

Goel, S. (2018). From waste to watts in micro-devices: Review on development of membraned and membraneless microfluidic microbial fuel cell. *Applied Materials Today, 11*, 270–279.

Gunjal Aparna, B., Waghmode Meghmala, S., & Patil Neha, N. (2021). Role of extremozymes in bioremediation. *Research Journal of Biotechnology, 16*(3), 240–252.

Happe, M., Sugnaux, M., Cachelin, C. P., Stauffer, M., Zufferey, G., Kahoun, T., et al. (2016). Scale-up of phosphate remobilization from sewage sludge in a microbial fuel cell. *Bioresource Technology, 200*, 435–443.

Hari, A. R., Katuri, K. P., Logan, B. E., & Saikaly, P. E. (2016). Set anode potentials affect the electron fluxes and microbial community structure in propionate-fed microbial electrolysis cells. *Scientific Reports, 6*(1), 1–11.

He, L., Du, P., Chen, Y., Lu, H., Cheng, X., Chang, B., et al. (2017). Advances in microbial fuel cells for wastewater treatment. *Renewable and Sustainable Energy Reviews*, *71*, 388–403. https://doi.org/10.1016/j.rser.2016.12.069.

Hemdan, B. A., El-Taweel, G. E., Goswami, P., Pant, D., & Sevda, S. (2021). The role of biofilm in the development and dissemination of ubiquitous pathogens in drinking water distribution systems: An overview of surveillance, outbreaks, and prevention. *World Journal of Microbiology and Biotechnology*, *37*(2). https://doi.org/10.1007/s11274-021-03008-3.

Herrero-Hernandez, E., Greenfield, D., Smith, T. J., & Akid, R. (2019). Evaluation of the performance of a mediatorless microbial fuel cell by electrochemical impedance spectroscopy. *Electroanalysis*, *31*(6), 1189–1194.

Heydorn, A., Nielsen, A. T., Hentzer, M., Sternberg, C., Givskov, M., Ersbøll, B. K., et al. (2000). Quantification of biofilm structures by the novel computer program COMSTAT. *Microbiology*, *146*(10), 2395–2407.

Hobley, L., Harkins, C., MacPhee, C. E., & Stanley-Wall, N. R. (2015). Giving structure to the biofilm matrix: An overview of individual strategies and emerging common themes. *FEMS Microbiology Reviews*, *39* (5), 649–669. https://doi.org/10.1093/femsre/fuv015.

Ieropoulos, I., Greenman, J., & Melhuish, C. (2008). Microbial fuel cells based on carbon veil electrodes: Stack configuration and scalability. *International Journal of Energy Research*, *32*(13), 1228–1240.

Ishii, S. I., Shimoyama, T., Hotta, Y., & Watanabe, K. (2008). Characterization of a filamentous biofilm community established in a cellulose-fed microbial fuel cell. *BMC Microbiology*, *8*(1), 1–12.

Islam, M. A., Ehiraj, B., Cheng, C. K., Dubey, B. N., & Khan, M. M. R. (2019). Biofilm re-vitalization using hydrodynamic shear stress for stable power generation in microbial fuel cell. *Journal of Electroanalytical Chemistry*, *844*, 14–22.

Islam, M. A., Karim, A., Mishra, P., Dubowski, J. J., Yousuf, A., Sarmin, S., et al. (2020). Microbial synergistic interactions enhanced power generation in co-culture driven microbial fuel cell. *Science of the Total Environment*, *738*, 140138.

Islam, M. A., Ong, H. R., Ethiraj, B., Cheng, C. K., & Khan, M. M. R. (2018). Optimization of co-culture inoculated microbial fuel cell performance using response surface methodology. *Journal of Environmental Management*, *225*, 242–251.

Islam, M. A., Woon, C. W., Ethiraj, B., Cheng, C. K., Yousuf, A., & Khan, M. M. R. (2017). Ultrasound driven biofilm removal for stable power generation in microbial fuel cell. *Energy & Fuels*, *31*(1), 968–976.

Jayapiriya, U. S., & Goel, S. (2021). Influence of cellulose separators in coin-sized 3D printed paper-based microbial fuel cells. *Sustainable Energy Technologies and Assessments*, *47*, 101535.

Jiang, J., Wang, H., Zhang, S., Li, S., Zeng, W., & Li, F. (2021). The influence of external resistance on the performance of microbial fuel cell and the removal of sulfamethoxazole wastewater. *Bioresource Technology*, 125308.

Jothinathan, D., & Wilson, R. T. (2017). Comparative analysis of power production of pure, coculture, and mixed culture in a microbial fuel cell. *Energy Sources, Part A: Recovery, Utilization, and Environmental Effects*, *39*(5), 520–527.

Jung, S. P., & Pandit, S. (2019). Important factors influencing microbial fuel cell performance. In *Microbial electrochemical technology* (pp. 377–406). Elsevier.

Kashefi, K., Tor, J. M., Holmes, D. E., Van Praagh, C. V. G., Reysenbach, A. L., & Lovley, D. R. (2002). *Geoglobus ahangari* gen. nov., sp. nov., a novel hyperthermophilic archaeon capable of oxidizing organic acids and growing autotrophically on hydrogen with Fe (III) serving as the sole electron acceptor. *International Journal of Systematic and Evolutionary Microbiology*, *52*(3), 719–728.

Kashyap, D., Dwivedi, P. K., Pandey, J. K., Kim, Y. H., Kim, G. M., Sharma, A., et al. (2014). Application of electrochemical impedance spectroscopy in bio-fuel cell characterization: A review. *International Journal of Hydrogen Energy*, *39*(35), 20159–20170.

Katuri, K. P., Kamireddy, S., Kavanagh, P., Muhammad, A., Conghaile, P., Kumar, A., et al. (2020). Electroactive biofilms on surface functionalized anodes: The anode respiring behavior of a novel electroactive bacterium, *Desulfuromonas acetexigens*. *Water Research*, *185*, 116284.

Katuri, K. P., Kavanagh, P., Rengaraj, S., & Leech, D. (2010). *Geobacter sulfurreducens* biofilms developed under different growth conditions on glassy carbon electrodes: Insights using cyclic voltammetry. *Chemical Communications*, *46*(26), 4758–4760.

Khan, N., Anwer, A. H., Ahmad, A., Sabir, S., Sevda, S., & Khan, M. Z. (2020). Investigation of CNT/PPy-modified carbon paper electrodes under anaerobic and aerobic conditions for phenol bioremediation in microbial fuel cells. *ACS Omega*, *5*(1). https://doi.org/10.1021/acsomega.9b02981.

Kierek, K., & Watnick, P. I. (2003). Environmental determinants of *Vibrio cholerae* biofilm development. *Applied and Environmental Microbiology*, *69*(9), 5079–5088.

Kim, K. Y., Rossi, R., Regan, J. M., & Logan, B. E. (2021). Enumeration of exoelectrogens in microbial fuel cell effluents fed acetate or wastewater substrates. *Biochemical Engineering Journal*, *165*, 107816.

Kim, G. T., Webster, G., Wimpenny, J. W. T., Kim, B. H., Kim, H. J., & Weightman, A. J. (2006). Bacterial community structure, compartmentalization and activity in a microbial fuel cell. *Journal of Applied Microbiology*, *101*(3), 698–710.

Koch, C., & Harnisch, F. (2016). Is there a specific ecological niche for electroactive microorganisms? *ChemElectroChem*, *3*(9), 1282–1295.

Kondaveeti, S., Lee, S. H., Park, H. D., & Min, B. (2020). Specific enrichment of different *Geobacter* sp. in anode biofilm by varying interspatial distance of electrodes in air-cathode microbial fuel cell (MFC). *Electrochimica Acta*, *331*, 135388.

Koo, H., & Yamada, K. M. (2016). Dynamic cell–matrix interactions modulate microbial biofilm and tissue 3D microenvironments. *Current Opinion in Cell Biology*, *42*, 102–112.

Korth, B., & Harnisch, F. (2019). Corrigendum: Spotlight on the energy harvest of electroactive microorganisms: The impact of the applied anode potential. *Frontiers in Microbiology*, *10*, 2744.

Kumar, R., Singh, L., Wahid, Z. A., & Din, M. F. M. (2015). Exoelectrogens in microbial fuel cells toward bioelectricity generation: A review. *International Journal of Energy Research*, *39*(8), 1048–1067.

Ledezma, P., Greenman, J., & Ieropoulos, I. (2012). Maximising electricity production by controlling the biofilm specific growth rate in microbial fuel cells. *Bioresource Technology*, *118*, 615–618.

Li, Z., Zhang, P., Qiu, Y., Zhang, Z., Wang, X., Yu, Y., et al. (2021). Biosynthetic FeS/BC hybrid particles enhanced the electroactive bacteria enrichment in microbial electrochemical systems. *Science of the Total Environment*, *762*, 143142.

Li, M., Zhou, M., Tian, X., Tan, C., McDaniel, C. T., Hassett, D. J., et al. (2018). Microbial fuel cell (MFC) power performance improvement through enhanced microbial electrogenicity. *Biotechnology Advances*, *36*(4), 1316–1327.

Liang, P., Duan, R., Jiang, Y., Zhang, X., Qiu, Y., & Huang, X. (2018). One-year operation of 1000-L modularized microbial fuel cell for municipal wastewater treatment. *Water Research*, *141*, 1–8.

Liu, H., Cheng, S., Huang, L., & Logan, B. E. (2008). Scale-up of membrane-free single-chamber microbial fuel cells. *Journal of Power Sources*, *179*(1), 274–279.

Liu, L., & Choi, S. (2017). Self-sustaining, solar-driven bioelectricity generation in micro-sized microbial fuel cell using co-culture of heterotrophic and photosynthetic bacteria. *Journal of Power Sources*, *348*, 138–144.

Liu, X. W., Li, W. W., & Yu, H. Q. (2014). Cathodic catalysts in bioelectochemical systems for energy recovery from wastewater. *Chemical Society Reviews*, *43*(22), 7718–7745.

Liu, G., Yates, M. D., Cheng, S., Call, D. F., Sun, D., & Logan, B. E. (2011). Examination of microbial fuel cell start-up times with domestic wastewater and additional amendments. *Bioresource Technology*, *102*(15), 7301–7306.

Logan, B. E., Hamelers, B., Rozendal, R., Schröder, U., Keller, J., Freguia, S., et al. (2006). Microbial fuel cells: Methodology and technology. *Environmental Science & Technology*, *40*(17), 5181–5192.

Logan, B. E., Rossi, R., Ragab, A., & Saikaly, P. E. (2019). Electroactive microorganisms in bioelectochemical systems. *Nature Reviews Microbiology*, *17*(5), 307–319. https://doi.org/10.1038/s41579-019-0173-x.

López Zavala, M.Á., Gonzalez Pena, O. I., Cabral Ruelas, H., Delgado Mena, C., & Guizani, M. (2019). Use of cyclic voltammetry to describe the electrochemical behavior of a dual-chamber microbial fuel cell. *Energies*, *12*(18), 3532.

Lovley, D. R., & Holmes, D. E. (2022). Electromicrobiology: The ecophysiology of phylogenetically diverse electroactive microorganisms. *Nature Reviews Microbiology*, *20*(1), 5–19.

Luo, J., Li, M., Zhou, M., & Hu, Y. (2015). Characterization of a novel strain phylogenetically related to *Kocuria rhizophila* and its chemical modification to improve performance of microbial fuel cells. *Biosensors and Bioelectronics*, *69*, 113–120.

Lusk, B. G., Peraza, I., Albal, G., Marcus, A. K., Popat, S. C., & Torres, C. I. (2018). pH dependency in anode biofilms of *Thermincola ferriacetica* suggests a proton-dependent electrochemical response. *Journal of the American Chemical Society*, *140*(16), 5527–5534.

Malki, M., De Lacey, A. L., Rodríguez, N., Amils, R., & Fernandez, V. M. (2008). Preferential use of an anode as an electron acceptor by an acidophilic bacterium in the presence of oxygen. *Applied and Environmental Microbiology*, *74*(14), 4472–4476.

Martin, E., Savadogo, O., Guiot, S. R., & Tartakovsky, B. (2013). Electrochemical characterization of anodic biofilm development in a microbial fuel cell. *Journal of Applied Electrochemistry*, *43*(5), 533–540.

Marcus, A. K., Torres, C. I., & Rittmann, B. E. (2007). Conduction-based modeling of the biofilm anode of a microbial fuel cell. *Biotechnology and Bioengineering*, *98*(6), 1171–1182.

Mathuriya, A. S. (2013). Inoculum selection to enhance performance of a microbial fuel cell for electricity generation during wastewater treatment. *Environmental Technology*, *34*(13–14), 1957–1964. https://doi.org/10.1080/09593330.2013.808674.

Mathuriya, A. S., & Yakhmi, J. V. (2014). Microbial fuel cells to recover heavy metals. *Environmental Chemistry Letters*, *12*(4), 483–494.

McDougald, D., Rice, S. A., Barraud, N., Steinberg, P. D., & Kjelleberg, S. (2012). Should we stay or should we go: Mechanisms and ecological consequences for biofilm dispersal. *Nature Reviews Microbiology*, *10* (1), 39–50.

Mei, X., Guo, C., Liu, B., Tang, Y., & Xing, D. (2015). Shaping of bacterial community structure in microbial fuel cells by different inocula. *RSC Advances*, *5*(95), 78136–78141.

Mink, J. E., Qaisi, R. M., Logan, B. E., & Hussain, M. M. (2014). Energy harvesting from organic liquids in micro-sized microbial fuel cells. *NPG Asia Materials*, *6*(3), e89.

Mohammadifar, M., & Choi, S. (2019). A solid phase bacteria-powered biobattery for low-power, low-cost, internet of Disposable Things. *Journal of Power Sources*, *429*, 105–110.

Mohan, S. V., Velvizhi, G., Modestra, J. A., & Srikanth, S. (2014). Microbial fuel cell: Critical factors regulating bio-catalyzed electrochemical process and recent advancements. *Renewable and Sustainable Energy Reviews*, *40*, 779–797.

Monzon, O., Yang, Y., Li, Q., & Alvarez, P. J. (2016). Quorum sensing autoinducers enhance biofilm formation and power production in a hypersaline microbial fuel cell. *Biochemical Engineering Journal*, *109*, 222–227.

Mukherjee, A., Patel, V., Shah, M. T., Jadhav, D. A., Munshi, N. S., Chendake, A. D., et al. (2022). Effective power management system in stacked microbial fuel cells for onsite applications. *Journal of Power Sources*, *517*, 230684.

Mumford, A. C., Barringer, J. L., Benzel, W. M., Reilly, P. A., & Young, L. Y. (2012). Microbial transformations of arsenic: Mobilization from glauconitic sediments to water. *Water Research*, *46*(9), 2859–2868.

Munoz-Cupa, C., Hu, Y., Xu, C., & Bassi, A. (2021). An overview of microbial fuel cell usage in wastewater treatment, resource recovery and energy production. *Science of the Total Environment*, *754*, 142429.

Nadell, C. D., Ricaurte, D., Yan, J., Drescher, K., & Bassler, B. L. (2017). Flow environment and matrix structure interact to determine spatial competition in *Pseudomonas aeruginosa* biofilms. *eLife*, *6*, e21855.

Nandy, A., Sharma, M., Venkatesan, S. V., Taylor, N., Gieg, L., & Thangadurai, V. (2019). Comparative evaluation of coated and non-coated carbon electrodes in a microbial fuel cell for treatment of municipal sludge. *Energies*, *12*(6), 1034.

Neto, S. A., Reginatto, V., & De Andrade, A. R. (2018). Microbial fuel cells and wastewater treatment. In *Electrochemical water and wastewater treatment* (pp. 305–331). Butterworth-Heinemann.

Nikhil, G. N., Chaitanya, D. K., Srikanth, S., Swamy, Y. V., & Mohan, S. V. (2018). Applied resistance for power generation and energy distribution in microbial fuel cells with rationale for maximum power point. *Chemical Engineering Journal*, *335*, 267–274.

Obata, O., Salar-Garcia, M. J., Greenman, J., Kurt, H., Chandran, K., & Ieropoulos, I. (2020). Development of efficient electroactive biofilm in urine-fed microbial fuel cell cascades for bioelectricity generation. *Journal of Environmental Management*, *258*, 109992.

Ono, K., Oka, R., Toyofuku, M., Sakaguchi, A., Hamada, M., Yoshida, S., et al. (2014). cAMP signaling affects irreversible attachment during biofilm formation by *Pseudomonas aeruginosa* PAO1. *Microbes and Environments*, *29*(1), 104–106.

Palanisamy, G., Jung, H. Y., Sadhasivam, T., Kurkuri, M. D., Kim, S. C., & Roh, S. H. (2019). A comprehensive review on microbial fuel cell technologies: Processes, utilization, and advanced developments in electrodes and membranes. *Journal of Cleaner Production*, *221*, 598–621.

Parameswaran, P., Bry, T., Popat, S. C., Lusk, B. G., Rittmann, B. E., & Torres, C. I. (2013). Kinetic, electrochemical, and microscopic characterization of the thermophilic, anode-respiring bacterium *Thermincola ferriacetica*. *Environmental Science and Technology*, *47*(9), 4934–4940.

Parkhey, P., & Sahu, R. (2021). Microfluidic microbial fuel cells: Recent advancements and future prospects. *International Journal of Hydrogen Energy*, *46*(4), 3105–3123.

Patil, S. A., Harnisch, F., Kapadnis, B., & Schröder, U. (2010). Electroactive mixed culture biofilms in microbial bioelectrochemical systems:

The role of temperature for biofilm formation and performance. *Biosensors and Bioelectronics*, *26*(2), 803–808.

Pham, H. T., Boon, N., Aelterman, P., Clauwaert, P., De Schamphelaire, L., Van Oostveldt, P., et al. (2008). High shear enrichment improves the performance of the anodophilic microbial consortium in a microbial fuel cell. *Microbial Biotechnology*, *1*(6), 487–496.

Prathiba, S., Kumar, P. S., & Vo, D. V. N. (2022). Recent advancements in microbial fuel cells: A review on its electron transfer mechanisms, microbial community, types of substrates and design for bioelectrochemical treatment. *Chemosphere*, *286*, 131856.

Qian, F., He, Z., Thelen, M. P., & Li, Y. (2011). A microfluidic microbial fuel cell fabricated by soft lithography. *Bioresource Technology*, *102* (10), 5836–5840.

Rahimnejad, M., Adhami, A., Darvari, S., Zirepour, A., & Oh, S. E. (2015). Microbial fuel cell as new technology for bioelectricity generation: A review. *Alexandria Engineering Journal*, *54*(3), 745–756.

Rahimnejad, M., Ghoreyshi, A. A., Najafpour, G. D., Younesi, H., & Shakeri, M. (2012). A novel microbial fuel cell stack for continuous production of clean energy. *International Journal of Hydrogen Energy*, *37*(7), 5992–6000.

Rana, S., Singh, L., & bin ab Wahid, Z. (2019). Electrotroph as an emerging biocommodity producer in a biocatalyzed bioelectrochemical system. In *Waste to sustainable energy* (pp. 67–84). CRC Press.

Ras, M., Lefebvre, D., Derlon, N., Paul, E., & Girbal-Neuhauser, E. (2011). Extracellular polymeric substances diversity of biofilms grown under contrasted environmental conditions. *Water Research*, *45*(4), 1529–1538.

Read, S. T., Dutta, P., Bond, P. L., Keller, J., & Rabaey, K. (2010). Initial development and structure of biofilms on microbial fuel cell anodes. *BMC Microbiology*, *10*(1), 1–10.

Römling, U., Galperin, M. Y., & Gomelsky, M. (2013). Cyclic di-GMP: The first 25 years of a universal bacterial second messenger. *Microbiology and Molecular Biology Reviews*, *77*(1), 1–52.

Rossi, R., & Logan, B. E. (2020). Unraveling the contributions of internal resistance components in two-chamber microbial fuel cells using the electrode potential slope analysis. *Electrochimica Acta*, *348*, 136291.

Saba, B., Christy, A. D., Yu, Z., Co, A. C., Islam, R., & Tuovinen, O. H. (2017). Characterization and performance of anodic mixed culture biofilms in submersed microbial fuel cells. *Bioelectrochemistry*, *113*, 79–84.

Sekar, N., & Ramasamy, R. P. (2013). Electrochemical impedance spectroscopy for microbial fuel cell characterization. *Journal of Microbial and Biochemical Technology*, *S6*(2).

Sevda, S., Dominguez-Benetton, X., Vanbroekhoven, K., Sreekrishnan, T. R., & Pant, D. (2013). Characterization and comparison of the performance of two different separator types in air-cathode microbial fuel cell treating synthetic wastewater. *Chemical Engineering Journal*, *228*. https://doi.org/10.1016/j.cej.2013.05.014.

Sevda, S., Dominguez-Benetton, X., Graichen, F. H. M., Vanbroekhoven, K., De Wever, H., Sreekrishnan, T. R., & Pant, D. (2016). Shift to continuous operation of an air-cathode microbial fuel cell long-running in fed-batch mode boosts power generation. *International Journal of Green Energy*, *13*(1), 71–79. https://doi.org/10.1080/15435075.2014.909363.

Sevda, S., Sarma, P. J., Mohanty, K., Sreekrishnan, T. R., & Pant, D. (2018). Microbial fuel cell technology for bioelectricity generation from wastewaters. In *Waste to wealth* (pp. 237–258). Springer.

Sharma, S. C. D., Feng, C., Li, J., Hu, A., Wang, H., Qin, D., et al. (2016). Electrochemical characterization of a novel exoelectrogenic bacterium strain SCS5, isolated from a mediator-less microbial fuel cell and phylogenetically related to *Aeromonas jandaei*. *Microbes and Environments*, ME15185.

Sharma, M., Nandy, A., Taylor, N., Venkatesan, S. V., Kollath, V. O., Karan, K., et al. (2020). Bioelectrochemical remediation of phenanthrene in a microbial fuel cell using an anaerobic consortium enriched from a hydrocarbon-contaminated site. *Journal of Hazardous Materials*, *389*, 121845.

Shi, L., Squier, T. C., Zachara, J. M., & Fredrickson, J. K. (2007). Respiration of metal (hydr) oxides by *Shewanella* and *Geobacter*: A key role for multihaem c-type cytochromes. *Molecular Microbiology*, *65*(1), 12–20.

Singh, A., & Kaushik, A. (2020). Suitability of wetland microbial consortium for enhanced and sustained power generation from distillery effluent in microbial fuel cell. *Energy Sources, Part A: Recovery, Utilization, and Environmental Effects*, 1–17.

Singh, A., & Kaushik, A. (2021a). Removal of Cd and Ni with enhanced energy generation using biocathode microbial fuel cell: Insights from molecular characterization of biofilm communities. *Journal of Cleaner Production*, 127940.

Singh, A., & Kaushik, A. (2021b). Sustained energy production from wastewater in microbial fuel cell: Effect of inoculum sources, electrode spacing and working volume. *3 Biotech*, *11*(7), 1–8.

Singh, A., & Kaushik, A. (2021). Sustained energy production from wastewater in microbial fuel cell: Effect of inoculum sources, electrode spacing and working volume. *3. Biotech*, *11*(7), 1–8.

Sirajudeen, A. A. O., Annuar, M. S. M., Ishak, K. A., Yusuf, H., & Subramaniam, R. (2021). Innovative application of biopolymer composite as proton exchange membrane in microbial fuel cell utilizing real wastewater for electricity generation. *Journal of Cleaner Production*, *278*, 123449.

Sui, M., Li, Y., Jiang, Y., Zhang, Y., Wang, L., Zhang, W., et al. (2021). Light exposure interferes with electroactive biofilm enrichment and reduces extracellular electron transfer efficiency. *Water Research*, *188*, 116512.

Sun, H., Zhang, Y., Wu, S., Dong, R., & Angelidaki, I. (2019). Innovative operation of microbial fuel cell-based biosensor for selective monitoring of acetate during anaerobic digestion. *Science of the Total Environment*, *655*, 1439–1447.

Sydow, A., Krieg, T., Mayer, F., Schrader, J., & Holtmann, D. (2014). Electroactive bacteria—Molecular mechanisms and genetic tools. *Applied Microbiology and Biotechnology*, *98*(20), 8481–8495.

Teschler, J. K., Zamorano-Sánchez, D., Utada, A. S., Warner, C. J., Wong, G. C., Linington, R. G., et al. (2015). Living in the matrix: Assembly and control of *Vibrio cholerae* biofilms. *Nature Reviews Microbiology*, *13*(5), 255–268.

Toyofuku, M., Inaba, T., Kiyokawa, T., Obana, N., Yawata, Y., & Nomura, N. (2016). Environmental factors that shape biofilm formation. *Bioscience, Biotechnology, and Biochemistry*, *80*(1), 7–12.

Varanasi, J. L., Sinha, P., & Das, D. (2017). Maximizing power generation from dark fermentation effluents in microbial fuel cell by selective enrichment of exoelectrogens and optimization of anodic operational parameters. *Biotechnology Letters*, *39*(5), 721–730.

Vélez-Pérez, L. S., Ramirez-Nava, J., Hernández-Flores, G., Talavera-Mendoza, O., Escamilla-Alvarado, C., Poggi-Varaldo, H. M., et al. (2020). Industrial acid mine drainage and municipal wastewater co-treatment by dual-chamber microbial fuel cells. *International Journal of Hydrogen Energy*, *45*(26), 13757–13766.

Vicari, F., Albamonte, M., Galia, A., & Scialdone, O. (2018). Effect of mode of operation, substrate and final electron acceptor on single-chamber membraneless microbial fuel cell operating with a mixed community. *Journal of Electroanalytical Chemistry*, *814*, 104–110.

Wu, X., Xiong, X., Owens, G., Brunetti, G., Zhou, J., Yong, X., et al. (2018). Anode modification by biogenic gold nanoparticles for the improved performance of microbial fuel cells and microbial community shift. *Bioresource Technology*, *270*, 11–19.

Xiao, Y., & Zhao, F. (2017). Electrochemical roles of extracellular polymeric substances in biofilms. *Current Opinion in Electrochemistry*, *4*(1), 206–211.

Yahav, S., Berkovich, Z., Ostrov, I., Reifen, R., & Shemesh, M. (2018). Encapsulation of beneficial probiotic bacteria in extracellular matrix from biofilm-forming *Bacillus subtilis*. *Artificial Cells, Nanomedicine, and Biotechnology*, *46*, 974–982.

Yu, B., Feng, L., He, Y., Yang, L., & Xun, Y. (2021). Effects of anode materials on the performance and anode microbial community of soil microbial fuel cell. *Journal of Hazardous Materials*, *401*, 123394.

Zhang, P., Liu, J., Qu, Y., & Feng, Y. (2017). Enhanced *Shewanella oneidensis* MR-1 anode performance by adding fumarate in microbial fuel cell. *Chemical Engineering Journal*, *328*, 697–702.

Zhang, Y., Liu, M., Zhou, M., Yang, H., Liang, L., & Gu, T. (2019). Microbial fuel cell hybrid systems for wastewater treatment and bioenergy production: Synergistic effects, mechanisms and challenges. *Renewable and Sustainable Energy Reviews*, *103*, 13–29.

Zhang, L., Zhu, X., Li, J., Kashima, H., Liao, Q., & Regan, J. M. (2017). Step-feed strategy enhances performance of unbuffered air-cathode microbial fuel cells. *RSC Advances*, *7*(54), 33961–33966.

Zhao, J., Li, F., Cao, Y., Zhang, X., Chen, T., Song, H., et al. (2021). Microbial extracellular electron transfer and strategies for engineering electroactive microorganisms. *Biotechnology Advances*, *53*, 107682.

Zhao, K., Tseng, B. S., Beckerman, B., Jin, F., Gibiansky, M. L., Harrison, J. J., et al. (2013). Psl trails guide exploration and microcolony formation in *Pseudomonas aeruginosa* biofilms. *Nature*, *497*(7449), 388–391.

Zhi, W., Ge, Z., He, Z., & Zhang, H. (2014). Methods for understanding microbial community structures and functions in microbial fuel cells: A review. *Bioresource Technology*, *171*, 461–468.

Chapter 40

Industrial applications and implications of biofilms

Toleti Subba Rao[a,b,c]

[a]School of Arts & Sciences, Sai University, Chennai, Tamil Nadu, India, [b]Biofouling and Biofilm Processes Section, Water & Steam Chemistry Division, Bhabha Atomic Research Centre Facilities, Kalpakkam, Tamil Nadu, India, [c]Homi Bhabha National Institute, Mumbai, Maharastra, India

1. Biofilms in perspective

Biofilms are transitory adherent microbial communities, which develop on immersed surfaces in aquatic environments. The biofilm develops as a heterogeneous unit composed of organic and inorganic entities as well as microorganisms, namely algae, bacteria, fungi, protozoans, and larvae of invertebrate organisms (Allison & Gilbert, 1992; Azam et al., 1983; Battin, Kaplan, Newbold, & Hansen, 2003; Bhinu, 2005; Bott, 1993). Biofilm formation at water-pipeline interface of cooling water distribution systems results in reduction of heat transfer efficiency and reduction of water flow, thus adversely affecting the operation of industrial units (Bhinu, 2005). The anisotropic nature of bacterial biofilms can result in spatial variations on pipeline surfaces of metal substratum resulting in the formation of electrochemical corrosion cells. This process subsequently propagates and manifests into microbially induced corrosion (MIC) (Characklis & Cooksey, 1983).

Generally, biofilms constitute a distinct growth phase of microorganisms that are markedly different from the planktonic microbiota. During the adhesion process, bacterial cells undergo phenotypic changes in response to surface proximity. In the course of biofilm formation, the sessile bacteria form microcolonies, which are in juxtaposition with cells of the same species. The microenvironment surrounding the microcolonies, along with the copious exopolymeric matrix, results in biofilm formation on a surface. The microenvironments are specific for each microbe, with different growth patterns, and with time a structurally complex biofilm develops (Costerton, Lewandowski, Caldwell, Korber, & Lappin-Scott, 1995). The typical structure of biofilm formation on surfaces is outlined into the following phases shown schematically in Fig. 1.

Phase I—Revocable Adhesion: It is a vital step in primary biofilm formation, wherein the bacteria approach the conditioned surface. This may be both active and passive process, depending on whether the bacteria are motile or transported by diffusion process of the surrounding aqueous phase. During the initial attachment phase, the physicochemical properties of the bacterial cell surface are important in determining whether the cell may attach or not. The primary forces that are involved in cell adhesion are superficial monomolecular polymers, which possess polar low energy force and are the main driving force for hydrophobic interactions and to a smaller degree a polar or Lifshitz-van der Waals forces (Cowan, Warren, & Fletcher, 1991; Fletcher, 1988). In primary bacterial adhesion process, Lawrence and Caldwell (1987) reported that the main body of the bacterium does not have direct contact with the substratum surface. The angle of alignment at solid-liquid interface may be variable and can be described using colloidal chemistry equations. However, when considering such theories, it is important to remember that most bacteria are not smooth round particles. Although the microbial cell surface is having the negative charge, which at times can be similar to that of the substratum, the bacterium can still move toward the conditioned surface by van der Waals force attraction (Allison & Gilbert, 1992; Lawrence & Caldwell, 1987). In this phase, bacterial cells can get desorbed due to liquid shear forces (Bhinu, 2005; Lawrence, Delaquis, Korber, & Caldwell, 1987).

Phase II – Irreversible Adsorption: This phase is an important step in biofilm formation, wherein the microorganisms attach to the surface irreversibly. Even though repulsive forces inhibit the bacterium from making direct contact on to the surface, bacteria adhere to the surface by producing surface appendages, such as fimbriae and exopolymers. Since electrostatic repulsion depends on radius of curvature, bacterial surface appendages such as flagella, fimbriae, and exo-polymeric fibers enter the primary minimum zone by overcoming the barrier energy (Rao, Kesavamoorthy, Babu, & Nair, 1997). Bacterial cells can hang at this phase and take part in short-range interactions such as dipole-dipole, ion-dipole, and hydrophobic interactions. Exopolymeric fibers form a bridge between the

Understanding Microbial Biofilms. https://doi.org/10.1016/B978-0-323-99977-9.00029-6

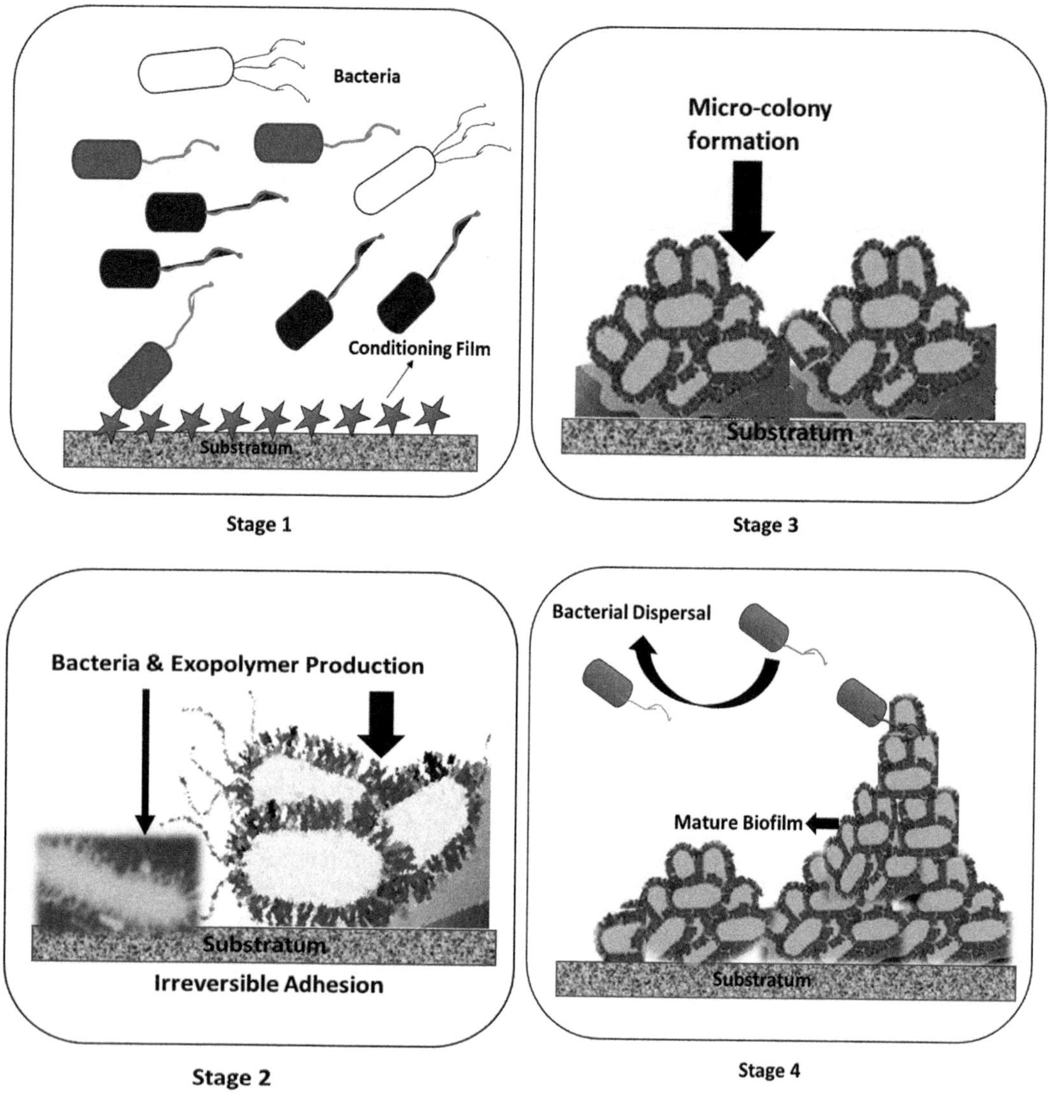

FIG. 1 Illustration of various stages (1–4) of primary biofilm formation.

bacterium and the surface, thus reinforcing irreversible attachment (Rao, 2020). During this phase, the attached bacteria begin to secrete more exopolymeric substances to further strengthen the cell adhesion to the surface and stabilize the microbial colony from surrounding environmental disturbances (Rao, 2003). The exopolymeric glue is not an inert matrix, and it exhibits physical, chemical, and electrical responses to environmental stimulus (Rao, Rani, Venugopalan, & Nair, 1997). The polymeric matrix secreted by the bacteria provides the biofilm structural stability (Rao, Kumar, & Chandra, 2020).

Phase III—Microbial Biofilm Formation: The accumulation of macromolecules on solid surfaces serves as carbon and nitrogen sources for bacteria present at the solid-liquid interface. Bacterial colonization of the surface may result in utilization of the macromolecules almost as rapidly as they adsorb at the surface, and it depends on the nutritional characteristics of the bacteria found at the surface. Biofilm populations predominate in virtually all nutrient-sufficient aquatic systems, which are independent of the cooling system geometry and the ecosystem. Further development of the biofilm involves microbial cell-cell interactions between the primary colonizing organisms and other microorganisms possessing different nutritional requirements. At this stage, biofilms demonstrate significant functional homogeneity. However, the individual cells and their activities are dependent on the structural integrity of the biofilm in toto. Costerton et al. (1995) suggested that biofilm could be considered as a "quasi tissue" with measurable rates of respiration and nutrient uptake. Biofilms resemble a tissue in their physiological cooperativity and protect themselves from variations in the bulk phase conditions by a primitive homeostasis. The spatial organization of microbial cells in the biofilm is also subject to variations. The growth characteristics and metabolic activities of the biofilm bring in structural heterogeneity. The physiological congruity is a typical feature of mixed microbial communities. Eventually, the biofilm reaches a plateau phase with a certain thickness and metabolic activity. The biofilm can now provide the basis for succession and development, wherein colonization by higher organisms is supported. This leads to the process of macro-fouling, a common problem described especially in marine aquatic systems (Rao, 2015a, 2021; Rao, Kumar, Balamurugan, & Vithal, 2017).

Phase IV – Biofilm Dispersal: In order to achieve propagation, the biofilm bacteria detach, disperse, and colonize new niches. This requirement is analogous to that of higher organisms such as the mycelial fungi that grow attached to surfaces. At different points in their life cycle, these organisms differentiate and produce spores, which are spread to new locations by dispersal. Bacterial growth in biofilms also brings with it the need to disperse. As the biofilm matures, there will be nutrient competition and some bacteria release enzymes, which will release the bacteria in to the water phase. The dispersed bacteria settle on a new virgin surface and initiate biofilm formation (Costerton et al., 1995; Rao et al., 2017; Shukla & Rao, 2013).

1.1 Exopolymers

The vast majority of microorganisms lives and grows in aggregated forms such as biofilms and flocs (also termed as planktonic biofilms). The common feature of all the biofilm types is that the microorganisms are embedded in a matrix of extracellular polymeric substances (EPSs). The production of EPS is a general property of microorganisms in natural environments and has been shown to occur both in prokaryotic (Bacteria, Archaea) and in eukaryotic (algae, fungi) organisms (Liu, Lau, Chau, & Pacepavicius, 1993). Commonly, biofilms contain mixed populations of these organisms, which are ubiquitously distributed in natural soil and aquatic environments, on tissues of plants, animals, and in humans as well as on engineering components such as filters and other porous materials, plumbing systems, pipelines, ship hulls, heat exchangers, separation membranes (Maki, Riftschof, Schmidt, Snyder, & Mitchell, 1989). Biofilms are accumulations of microorganisms (prokaryotic and eukaryotic unicellular organisms), EPS, multivalent cations, biogenic and inorganic particles as well as colloidal and dissolved compounds. EPSs are mainly responsible for the structural and functional integrity of biofilms and are considered as the key components that determine the physicochemical and biological properties of biofilms (Rao, 2018).

Pseudomonas aeruginosa synthesizes an exopolysaccharide called alginate in response to environmental conditions. Alginate serves to protect the bacteria from adversity in its surroundings and enhances adhesion to solid surfaces. Transcription of the alginate biosynthetic genes is induced upon attachment to the substratum, and this leads to increased alginate production. As a result, biofilms mode of development is an advantage for the survival and growth of the bacteria. In certain circumstances, *P. aeruginosa* produces an alginate lyase enzyme, which cleaves the polymer into short oligosaccharides. This negates the anchoring properties of the alginate and results in increased detachment of the bacteria from the surface, allowing them to spread and colonize new sites. Thus, both alginate biosynthetic and degradative enzymes are important for the development, maintenance, and dispersal of *P. aeruginosa* biofilms (O'Toole & Kolter, 1998).

2. Biofilm formation

Formation of nontransient adherent communities of microorganisms on submerged surfaces (biofilm) is ubiquitous. Over 90% of the total aquatic bacteria are found associated with interfaces such as the sediment water interface and surface microlayer (Pedersen, 1982). The overall biofilm formation process is schematically presented in Fig. 2.

Biofilm definition: *"A biofilm can be defined as a consortium of bacteria, cyanobacteria and algae attached to a substratum by the extracellular polymeric matrix which entraps soluble and particulate organic and inorganic matter, thereby acting as a sink for nutrients, extracellular enzymes and eDNA."*

The biofilm composition may vary, and both spatially and temporally and greatest differences are usually associated with shifts in the relative importance of autotrophic and heterotrophic microorganisms (Miller & Ahearn, 1987). Mutual interactions of microbial cells in biofilms result in the development of community niches, which are habitat-specific. The biofilm community interactions take place at scales of tens of microns (James, Korber, Caldwell, & Costerton, 1995). The presence of various species of bacteria and algae in successive layers of the biofilm provides clues about the energy and substrate transfer within the biofilm. Water quality parameters affect the biofilm formation because of their role on the regulation of bacterial metabolism (Korber, Lawrence, Lappin-Scott, & Costerton, 1995). Recently, a limited number of investigations have been carried out to understand the mechanisms of biofilm development and the physiological activities of the microorganisms associated with it (Mercedes & Blanch, 1994). Differences or changes in the structural or physicochemical properties of the biofilm are determined by the organic and inorganic constituents of the surrounding environment. Various physical, chemical, and biological factors govern the formation of biofilms; among them, nutrient concentrations attain much significance (Presson & Molin, 1987).

The adhesion of microorganisms to solid surfaces has practical implications in industrial units. Following adhesion to a surface, a bacterial cell undergoes a

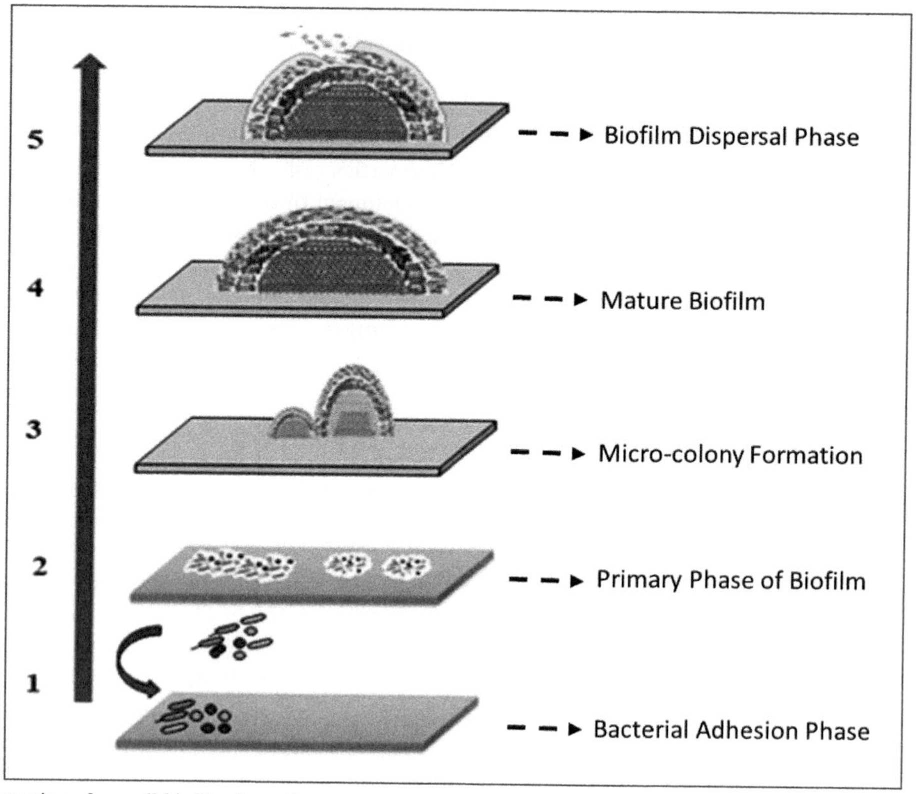

FIG. 2 Schematic illustration of overall biofilm formation process.

phenotypic change and starts cell division, which would lead to the development of microcolonies. The microcolony is the basic unit of biofilm, and each biofilm bacterium lives in a customized micro-niche and enjoys a measure of homeostasis, circulatory system, and metabolic cooperativity, in this complex microcommunity (Stewart & Franklin, 2008). Although 90% of the microbial population in nature occurs in biofilms (Stewart, Peyton, Drury, & Murga, 1993), most of the laboratory studies on microorganisms have been based on suspensions. Moreover, due to lack of convenient experimental systems for the study of surface-bound microbial growths, relatively few investigations have been carried out (Van der Kooij, Veenendaal, Baars-Lorist, Van der Klift, & Drost, 1995; Van Loosdrecht, Lyklema, Norde, & Zehnder, 1990). However, with the development of biofilm reactors and flow cells, studies on microcolony/biofilm formation under laboratory conditions got augmented. The outcome of the research efforts provided valuable information on the behavior of mono and mixed culture bacterial biofilms under various flow conditions, which mimicked industrial cooling circuits (Rao, Eswaran, Venugopalan, Nair, & Mathur, 1993), on ordering characteristics (Rao, Kesavamoorthy, & Rao, 1996) and behavior of bacteria in the hydrodynamic boundary layer (Rao, Webb, & Kjelleberg, 2005). Stewart (2012) used light microscopy to observe the depth variability in *Pseudomonas aeruginosa* biofilms, the data provided refined fluid frictional coefficients and models of molecular diffusion. Currently, with modern tools such as confocal scanning laser microscopy (CSLM), one can make direct observation of structure and function of living biofilms. Thus, examination of *Pseudomonas fluorescens* biofilms by CSLM has thrown light on spatial variability that exists within pure cultures. Costerton et al. (1995) have reported that adhesion of microbes to surfaces triggers the expression of a σ factor that derepresses many genes, so that biofilm cells are phenotypically distinct from their planktonic counterparts.

Furthermore, biofilms cause fouling of water distribution systems, industrial equipment such as heat exchangers, cooling towers, submerged sight glass and sensors, and ship hulls. Biofilms can significantly decrease conductive heat transfer (Bott, 1993; Gordon, Davis-Fields, Kovach, & Rodesney, 2017), since it has a thermal conductivity similar to that of water (about 0.6 W/m/K). Rheological measurements indicate that biofilms are viscoelastic in nature (Characklis & Cooksey, 1983; Pavlovsky, Younger, & Solomon, 2013). Therefore, biofilms can cause significant energy losses as they contribute to increased fluid frictional resistance. It is well accepted that the major problem concerning biofilm formation on heat transfer systems is the resultant technoeconomics in heat exchanger performance. It is reported that annual cost due to heat exchanger fouling in the United Kingdom is around 500 million pounds, in the United States, the cost is \$1.5 billion.

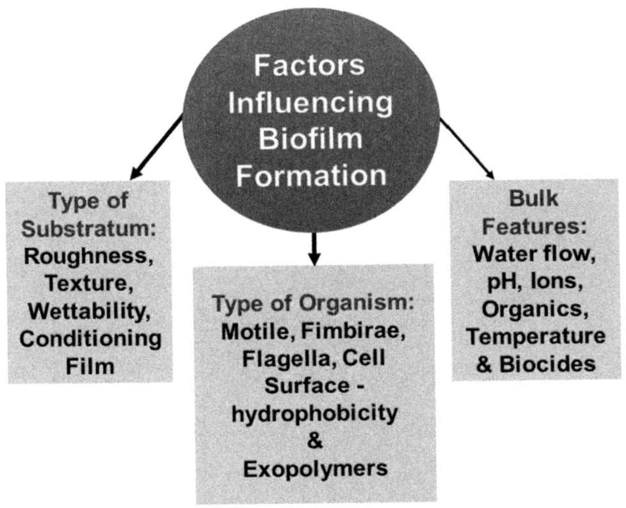

FIG. 3 Schematic showing the common factors that influence biofilm formation.

Biofilms can readily form the largest resistance to heat transfer in systems cooled by water within the temperature ranges conducive to microbial growth (Krieg & Holt, 1984). Biofilm control in industrial systems, pipelines, and ship hulls has an expensive proposition, which is usually based on mechanical treatment or use of antimicrobial agents. Although chlorine and other chemical formulations are in use for biofouling control, their use probably will be restricted or prohibited in future due to their toxic effects on the ecosystem (Rao, Nancharaiah, & Nair, 1998). Controlling bacterial adhesion alone per se inhibits the total biofilm accumulation on a surface, which is also unrealistic. However, it was also reported that controlling biofilm differentiation and integrity of quorum sensing molecules has important implications in biofilm control (Starr, Stolp, Truper, Balows, & Schlegel, 1986). Fig. 3 describes the various factors that play a role in biofilm formation, while Fig. 4 provides time series confocal microscopy images of biofilm formation.

3. Industrial applications of biofilms

Although biofilms are frequently recognized as potential destructive bacterial consortia, in clinical and other industrial fields, several biofilm types are beneficial, and there are a number of reports related to the positive applications of the biofilms. Beneficial biofilms can be used in food, agricultural, medical, environment, and other fields, with properties such as antibacterial agents, in food fermentation, as a biofertilizer, in filtration, in prevention of biofouling and corrosion, wastewater treatment as well as bioremediation and microbial fuel cells. Among all, food and agricultural industries are the most benefited from biofilms in terms of their biochemical, fermentative, antimicrobial, and biotechnological applications (Morikawa, 2006; Rao et al., 2017). Many published reports highlighted that certain strains of bacteria such as *Bacillus* species such as: (*B. subtilis*, *B. thuringiensis*, *B. brevis*, *B. licheniformis*, *B. polymyxa*, *B. liquefaciens*); *Lactobacillus* species such as *L. casei*, *L. paracasei*, *L. acidophilus*, *L. plantarum*, *L. reuteri*; *Enterococcus* species such as *E. casseliflavus*, *E. faecalis*, *E. faecium*; *Pseudomonas* bacteria such as *P. fluorescens*, *P. putida* and *P. chlororaphis*, *Acetobacter aceti*, some fungi, and *Pseudoalteromonas* species are involved in the formation of beneficial biofilms (Robertson & McLean, 2015). Biofilm bacteria enmeshed in the exopolymeric matrix produce many specific secondary metabolites and the cells become robust. The resident bacteria in the biofilm matrix will positively improve the quality characteristics of food products such as texture, biochemical composition, and sensorial properties *by* the production of a variety of secondary metabolites. In addition, biofilms also have an important role in water and soil safety of agricultural land (Toyofuku et al., 2016). Fig. 5 shows multiple images describing the biofilm formed by various microbes such as bacteria, fungi, and algae.

Biochemical-process-oriented bioreactors play a vital role in the bioprocess industry as the rate of reaction, ease, and length of reactor operation affect the productivity and hence play a role in process economics. The imperative to have a suitable bioreactor is that the reaction rates should be high enough and the bioreactor conformation should also be simple for an industrial operation. By increasing the biomass concentration in the bioreactor, the biochemical reaction rates can be increased for increasing the yield of the product. Likewise, by optimizing the reactor operation parameters such as pH, temperature, substrate loading, and nutrition components, the bioprocess can be standardized. Chiefly there are two techniques used for increasing cell biomass: (1) a permeable membrane to retain cells and (2) immobilized biomass procedure, which are commonly used to concentrate the end-product production in a bioreactor. Commonly, membrane-based bioreactors allow easy transfer of liquids and separation of substrates and the product out of the bioreactor while the cells/biomass is retained. In addition, relatively high cell biomass content can be obtained. Inappropriately, for some bioprocesses such as wastewater treatment, the membrane-based bioreactors are not ideal, because of their high capital and running costs, and there are chances of membrane fouling. Other bioprocess products such as vinegar or acetic acid, where the high cost of the bioreactors does not hamper their use since the products are popular, and they are produced in large volumes (Todhanakasem, 2013).

Immobilized cell bioreactors offer high reaction rates, and in these bioreactors, very high cell biomass concentrations are attained, since the biomass is fixed on to various support materials. However, it may be noted that some

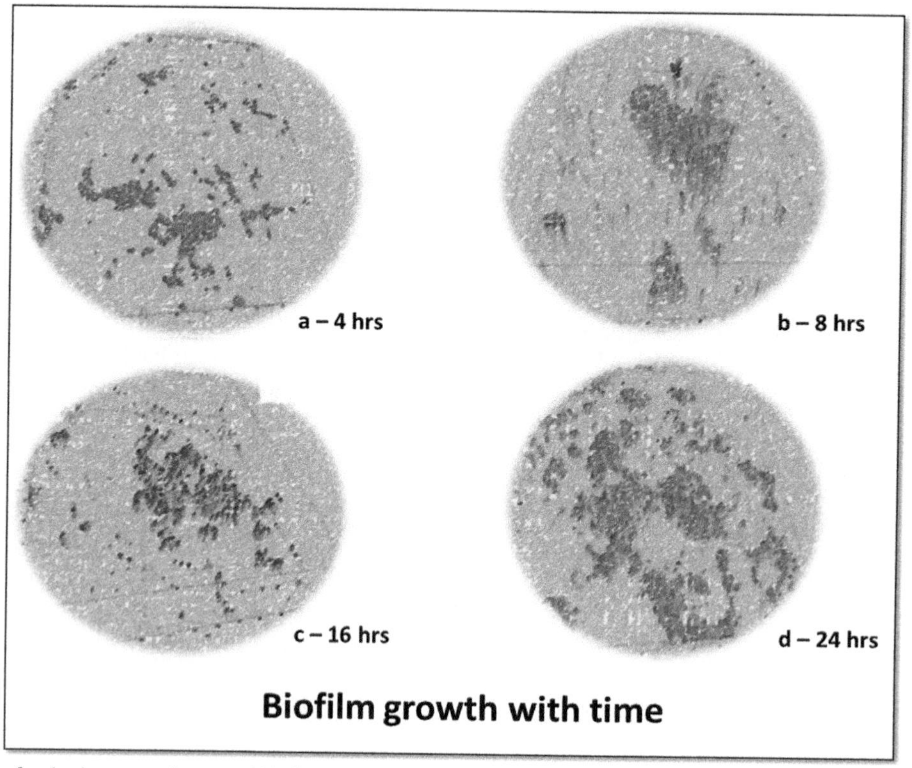

FIG. 4 Time series confocal microscopy images of biofilm formation.

bacterial cells get released or dispersed from immobilized cell bioreactors; this also requires separation of cells from the bioreactor effluent, before the product is removed possibly by centrifugation. Generally, bacterial cells can be immobilized by three different techniques, namely adsorption, entrapment, and by forming covalent bonds with the substrate. Entrapment and covalent bond formation require biochemicals that increase the cost of produced

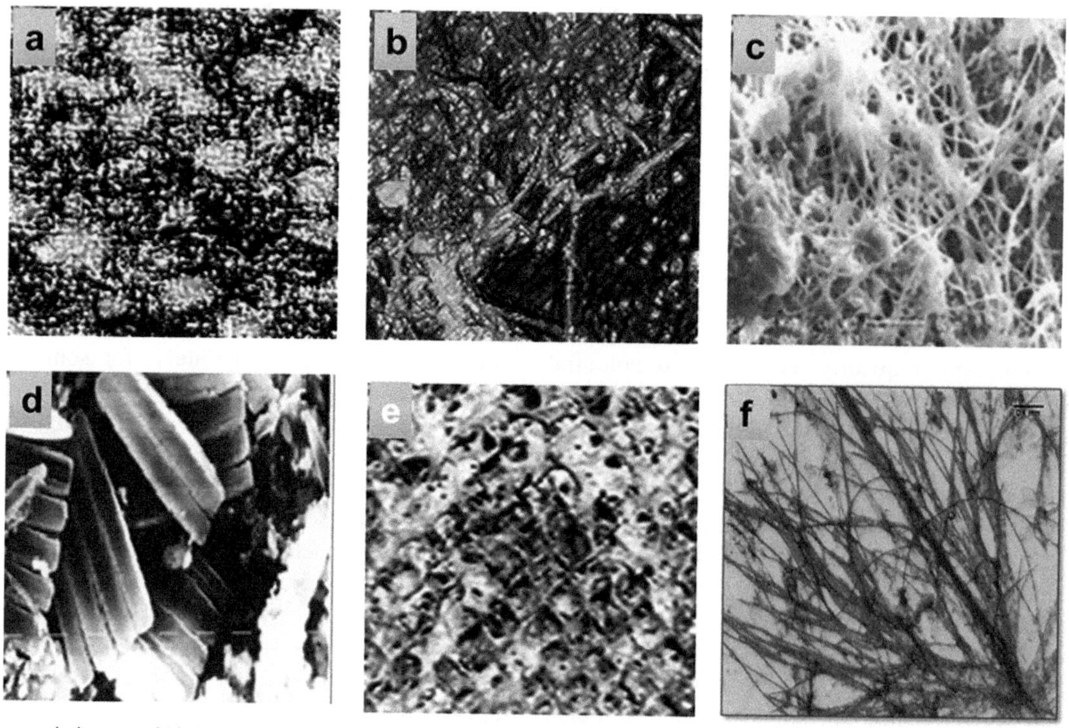

FIG. 5 Microscopic images of biofilm growth of bacteria, fungi, and algae (A–F).

product, which perhaps scuttle the propagation of biomass or increase the biomass concentration inside the bioreactor. While the third method is natural by origin wherein the bacterial cells adsorb or adhere to a substratum naturally and firmly. This biotechnical process is called cell adhesion and has been used widely in the literature to describe the adsorption of bacterial cells (Houdt & Michiels, 2010; Petrova & Sauer, 2012; Rao et al., 2017; Sarjit, Tan, & Dykes, 2015).

4. Biofilm bioreactors for beneficial applications

Major opinions that describe microbial biofilms are happening as a result of attentive research projects as well as their realistic applications. The practice of using bioreactors for the management of domestic wastewaters and industrial effluents is becoming an important aspect of beneficial uses of biofilms. Although, applied research has existed, it provided only a basis for systematic application of microbial biofilm formation on cooling water system materials. However, the experiential information consequent to such practical research was used to build biofilm reactors with proper design criteria. This approach still remains the origin for designing many biofilm bioreactors regardless of the novel mathematical models as reliable tools for both basic and applied research. Nevertheless, there is a wide knowledge gap on the existing perceptive of fundamental aspect of biofilms and bioreactor scale-up (Stavridou & Forzi, 2011; Wood et al., 2016).

Generally biofilm reactors were mainly represented for their application in wastewater treatment. Bacterial biofilms that are growing in these bioreactors assist as a primary means for transformation of environmental pollutants such as nutrients and other organic components such as biodegradable organic matter, nitrogen, and phosphorus through biological processes. In wastewater treatment, numerous biofilm reactors are used; however, considerable focus was made on moving-bed biofilm reactors (MBBRs) and other bioprocesses, for instance, membrane-supported biofilm reactors (MBfRs) and bacteria-supported granular biomass processes. Lewandowski and Boltz (2011) have highlighted the dichotomy by describing up-to-date basic research and the application-oriented useful biofilm systems for wastewater treatment.

MBBRs and MBfRs bioprocesses are well-established technologies, which are showing sustainable development, while state-of-the-art MBBRs and MBfRs processes use submerged free-moving biofilm carriers. These reactors are designed for application in organic carbon oxidation, denitrification, deammonification, and in nitrification processes (Michela et al., 2014). Nevertheless, the MBBR reactor system is an active platform for synchronized fractionation of nitrification and deammonification processes. The AnitaMOX is a patented process, which is commercially available having both MBBR/IFAS paraphernalia and augments a restricted anammox/nitritation process (Ziye, Tommaso, van Loosdrecht, & Boran, 2013). However, recent research developments have extended our understanding regarding the carriers of biofilms and their role on mass transfer and hydrodynamics, which are interrelated in biochemical transformation processes. World over, it was assessed that different types of MBBRs units are equally distributed in various industrial and municipal wastewater treatment facilities. Fig. 6 illustrates the advantages and disadvantages of different types of bioreactors.

It was proved that aerobic microbial granular biomasses are grown and utilized in sequencing batch reactors (SBRs). They are highly promising environmental biotechnology processes for the treatment of polluted water streams with sewage or in water bodies where organically rich food and chemical industry effluents are released. Although Beun et al. (1999) reported the potential of SBRs in stable granular biomass development, it took a decade to implement the bioprocess of aerobic granule formation for municipal waste treatment. Presently, more than 50 WWTPs utilizing aerobic granular bioprocesses are being operated or under construction at various places in Africa, Australia Europe, India, and South America. Various types of WWTPs are designed for removal of biologically formed nutrients from sewage. To date, the functional highest capacity aerobic microbial granular bioprocess-based SBR can treat a daily flow of $55{,}000\,m^3\,day^{-1}$of sewage, which is located in Rio de Janeiro, Brazil. NEREDA is the name given for commercially available aerobic granular sludge system that has been used for successful biological nutrient removal from selectively separated wastewaters. Pronk et al. (2015) have reported an operational full-scale NEREDA process at Garmerwolde, the Netherlands. In NEREDA bioprocess, a constant liquid/biomass volume ratio is maintained and the filling, settling, and decanting steps happen concurrently during one-third of the operational period, while the two-thirds remaining operational time was reserved for aeration (i.e., reaction time for bioprocess) (Jun, Li-Bin, Cai, Guo-Xian, & Hprn, 2014). Generally, 15 min time is required for starting the next cycle with influent feed water input from the bottom of the SBR. These distinctive operational parameters, along with proper influent wastewater characteristics, can result in discharges having $<5\,gm\,m^{-3}$ total nitrogen and $\sim 1\,gmm^{-3}$ of total phosphorus (Vasanthadevi, Satoshi, Kenji, & Kazuya, 1998). The SBRs are the simplest bioreactor systems principally having a hollow tank with aeration (fine-bubble) and a feed water line of wastewater for distribution from the bioreactor bottom. The treated wastewater effluent overflows the weir, which is situated at the top portion of the tank. The bioreactor has no mixers/impellers, but has an overflow

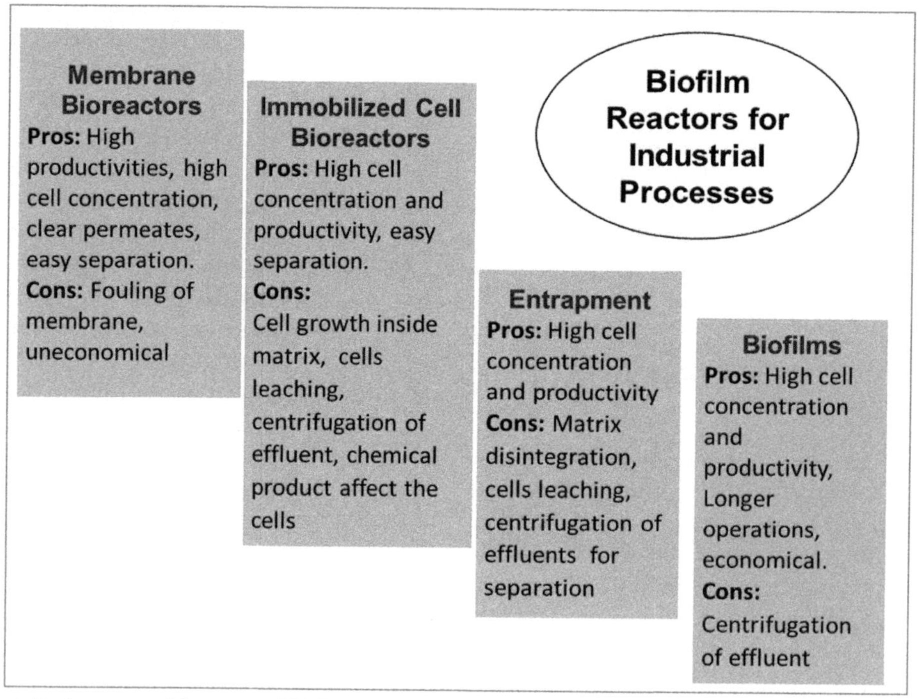

FIG. 6 Advantages and disadvantages of few types of biofilm reactors.

weir for effluent discharge and a biomass sludge collection system, which is situated near the sludge bed to promote slow settling of sludge (Isanta et al., 2012). An additional, useful benefit of the bioreactor is the presence of a cyclone for the selective retention of granular biomass by using screens. These granules can also be used for treating high ammonia-nitrogen containing wastewaters from digested sludge dewatering and anaerobically treated wastewater through a bioprocesses called ANAMMOX (Abma et al., 2007; Van der Star et al., 2007).

A couple of decades have observed an increased interest in membrane bioreactors for wastewater treatment. The membrane-aerated biofilm reactor (MABR), wherein the biomass is immobilized on membranes through which oxygen is supplied, seems to be the most promising concept design. Results from the studies with MABRs have been reported for the degradation of phenol, chlorophenols, xylene, and ammonia (Anjali & Sabumon, 2017; Chelliapan, Din, Anuar, & Selvam, 2017; Guglielmi & Andreottola, 2011). Stripping losses of volatile organic compounds are minimized, and the oxygen partial pressure in the gas compartment allows easy control of oxygen penetration into the biofilm. The dissolved oxygen gradient across the membrane and the biofilm offers an ideal environment for aerobic bacteria to grow, and foaming problem due to surfactants can be stopped (Shin, Lee, Jung, Chung, & Noh, 2005). For high organic strength wastewaters, the possibility of enhanced oxygen penetration into the depths makes MABRs an attractive bioprocess for pollutant degradation. Casey, Glennon, and Hamer (2000) have reported frequent observation of dense biofilm growth; therefore, biofilm growth has to be controlled during the operation of this bioreactor.

MBfR is an alternative type of biofilm reactor that displays great prospects for industrial applications. In this bioreactor, the microbial biofilm directly grows on a membrane, which is tubular, hollow fiber in structure. This bioreactor has a diverse potential and has widespread applications in many bioprocesses (Brindle & Stephenson, 1996). In MBfR bioreactor liquid phase, the gas delivery takes place by means of a membrane unit. In MBfR bioreactors, two types of systems are in practice: (1) the hydrogen-based MBfR wherein hydrogen is an electron donor to a biofilm, and (2) the oxygen-based MBfR, wherein oxygen is an electron acceptor in the biofilm. The latter is also termed as the aerated/aerobic membrane biofilm reactor (Dutta & Sarkar, 2015). Hydrogen-based MBfRs were demonstrated for the biochemical transformation of arsenate, bromate, chromate, nitrate, nitrite, perchlorate, and selenate/selenite. Since the MBfR allows the delivery and higher control of electron donor/acceptor in biofilms with defined or strong redox stratification. This type of process can be modified for a simultaneous aerobic/anaerobic process such as nitritation/anammox (Guglielmi & Andreottola, 2011).

The regular developments and enactment of new water quality guidelines, and the innovations in new bioprocesses, have resulted in development of mature biofilm bioreactor models that are relevant to up-to-date trends and challenges of wastewater treatment. For example, as per new regulation, the selenium concentration in drinking water has to be <0.05 mg L^{-1} as per the US Environmental Protection Agency. Biofilm bioreactors that are capable of operating under anaerobic conditions are to be designed and fine-tuned for effective operation. Therefore, particulate biofilm bioreactors, as described by Nicolella, van Loosdrecht, and Heijnen (2000a), have the capability and are versatile for operation. Likewise, biofilm bioreactors such as deep-bed filters and upflow anaerobic sludge mode can also do the reduction effectively. Consequently, the production of elemental selenium was preferred through biological transformation of selenate and selenite.

A couple of decades of research and development in biofilms have thoroughly studied to generate electricity production through microorganisms, and these were termed as microbial fuel cells (MFCs). The MFCs can generate electricity by biologically converting complex organic compounds present in municipal wastewaters. Although there are several methods reported for constructing MFCs, there are several ways of constructing MFCs and comparing the devices on an equivalent basis. They also reviewed a range of related scientific principles, from the disciplines of environmental engineering to microbiology and electrochemistry. The objective was to establish an MFC that can yield sufficient electrical output, which is economically viable in production and utilization. MFCs, which are operated in the electrolysis mode such as microbial electrolysis cells, can produce useful chemical products such as hydrogen and ethanol (Gude, 2016; Hoh, Watson, & Kan, 2016; Munoz-Cupa, Hu, Xu, & Bassi, 2021; Zhou, Wang, Hassett, & Gu, 2013).

5. Types of biofilm reactors

Biofilm bioreactors can be designed and fabricated into a number of configurations and operated in batch, continuous stirred-tank (CSTR) such as agitating and rotary continuous bioreactors, packed bed (PBR), trickling bed (TBR), fluidized bed (FBR), airlift bioreactors (ALR), upflow anaerobic sludge blanket (UASB), and expanded bed reactors. The operation of these bioreactors changes from reactor to reactor. In a batch mode biofilm bioreactor, the immobilized cells on to a matrix have to be utilized for repeated operations in batches. However, it is likely that during the late stationary phase of chemical production, the bacterial culture can experience growth inhibition, thereby reducing the production of the product. Also, in a batch bioreactor, production would be reduced due to downtime, which is the time taken to fill and empty the bioreactors. If the bioreactor is packed with biofilm particles, some cells may die or become inactive due to lack of feed during emptying and filling of the bioreactor. Therefore, it can be inferred that batch mode bioreactors are not practical for biofilm mode of reactors (Boltz et al., 2017; Malovanyy, Yang, Trela, & Plaza, 2015; McQuarrie & Boltz, 2011).

5.1 Continuous stirred-tank reactor (CSTR)

The CSTR is a stirred-type bioreactor with a mechanical impeller. In a CSTR, the medium/liquid is fed to the bioreactor and the end product is withdrawn at the same rate as feed is added. Since there is no agitation, CSTRs cannot be packed with the adsorbent support covered by biofilms. However, fibrous material can be used as bed support for adsorption of cells for development of biofilm. In such an event, cells can grow and form a biofilm on the fibrous bed wherein agitation can be provided. CSTR-type bioreactors with a fibrous bed are used for butanol and lactic acid production, in continuous mode operation with a constant feed supply and product removal from the bioreactor. In few instances due to operational exercise, there can be excessive biomass growth on the surface of the fibrous bed and the biofilm cell layers can be sheared off the bed support, thereby increasing the turbidity. These fibrous bed biofilm CSTRs types are also called as agitating continuous bioreactors. Another type of CSTRs called rotating CSTRs have similar length/diameter (L/D) ratio as in CSTRs. The rotating CSTRs are placed horizontally and are rotated along the horizontal axis (Arooj, Han, Kim, Kim, & Shin, 2008).

5.2 Packed-bed reactors

PBRs (packed-bed reactors) are a different type of bioreactors, they are packed with proper support material followed by inoculation of a log phase bacterial culture to form an active biofilm. In a PBR, the feed is supplied that is rich in nutrients. Depending on the culture, nutrients, and support material, biofilm formation may take from a few to several days. Such reactors are usually fed at the bottom and getting the product at the top of the reactor. However, these bioreactors are likely to get blocked due to dense bacterial growth. For example, in *C. acetobutylicum* biofilm mode PBR, biochemical reaction rates are 45 times more than that of batch bioreactors (Mueller, 2012).

In comparison, TBR bioreactors are different from PBRs as the feed to the reactor is from the top, and the end product is released from the bottom. Yet, in such bioreactors, the biofilm grown on the substrate may be heterogeneous and dissimilar; this in turn may result in feed deficiency for the biofilms, thereby adversely affecting the bioreactor efficiency/productivity. Besides, in gaseous fermentation process, gas may occupy significant space in the bioreactor resulting in stagnant pockets/voids formation. Overall this may also affect the performance of the bioreactor. These types of bioreactors have been used successfully in large-scale production of acetic acid and in the anaerobic wastewater treatment (Nemec & Levec, 2005).

5.3 Fixed-bed reactors

As per literature reports, FBRs (fixed-bed reactors) have played an effective role in the degradation of phenolic and other toxic chemicals as well as butanol production. In this type of bioreactors, cell growth occurs on the circumference of the adsorbent particles. Formation of dynamic biofilms on the adsorbent particles results in accumulation of adequate biomass in the bioreactor, which may take 14–28 days to generate. A significant advantage of these bioreactors is that they can be operated for longer time than PBRs or CSTRs. These bioreactors do not undergo blockage due to excessive growth. For example, in these bioreactors the rate of butanol production was increased by 40–50 times than that of the batch mode bioreactors. These bioreactors are operated successfully for longer time periods in continuous mode operation (Janecki, Burghardt, & Bartelmus, 2016; Noorman & Annaland, 2007).

Likewise, airlift-type bioreactors contain two concentric tubes, a riser tube (inner tube) and a downcomer tube (outer tube). In this type of bioreactors, mixing is achieved by circulating air from the bottom of the bioreactor; as a result, the force applied by the air (at the bottom of the inner tube), the liquid moves up and then overflows downward the inner tube, thereby creating eddies/flux to thoroughly mix the liquid medium. In other airlift bioreactors, downcomer tube is replaced with an external loop to circulate the fermented broth. Such bioreactors where air is replaced by an anoxic gas are called gas-lift bioreactors (Ensley & Kurisko, 1994).

5.4 Upflow anaerobic sludge blanket reactors

Upflow anaerobic sludge blanket reactors (UASBRs) were developed by Lettinga, van Velsen, Hobma, de Zeeuw, and Klapwijk (1980). Fig. 7 illustrates a schematic diagram of various bioreactor types, which are used in several bioprocesses. In contrast to airlift bioreactors, UASBRs contain granular biofilm particles that are used for anaerobic treatment of municipal wastewaters as well as industrial effluents. As the name suggests, the flow in these bioreactors is in upward direction, accordingly at the top, provision is made for the escape of gases and simultaneously sludge particles settle at the bottom of the bioreactor. The effluent after the bioprocess is removed from the top of the bioreactor. Prior to the development of UASB, interest in anaerobic treatment systems in wastewater treatment is scarce. Interestingly, the development of UASB saw a significant increase in anaerobic biodegradation of various chemicals from the industrial wastewaters. The examples include anaerobic removal of pentachlorophenol, nitrogen removal dechlorination using *Dehalospirillum multivorns* anaerobic treatment of municipal solid leachate and starch degradation. Veeresh, Kumar, and Mehrotra (2005) reviewed the anaerobic degradation of phenols and cresols in UASB reactors. The success of the UASB concept relies on the establishment of a dense sludge bed in the bottom of the bioreactor as a result of microbial growth and incoming feed with sludge particles.

6. Bioreactors and denitrification process

Denitrification is a microbially mediated process, where nitrate is used for generating energy by denitrifying bacteria. The complete denitrification process can be expressed as a redox reaction:

$$2\,NO^{3-} + 10\,e^- + 12\,H^+ \rightarrow N_2 + 6\,H_2O.$$

The protons that are generated are transported across the membrane by the initial NADH reductase enzyme, quinones, and nitrous oxide reductase to produce the electrochemical gradient that is critical for respiration of the biomass. In denitrification process with volatile solid concentrations between 30,000 and 40,000 mg L^{-1}, 99% of influent nitrate content was removed under hydraulic retention times as low as 6 min. In fluidized-bed biofilm reactors (FBBRs), an exceptionally high rate of denitrification (up to 99.8%) can be achieved with constant nitrate nitrogen concentration of 1000 mg (N) L^{-1}. The study also reported that higher denitrification rates can be achieved at relatively low superficial velocities because it is possible to maintain stable and high biomass concentration at lower velocities. However, when very low velocities are used, agglomeration of media particles would occur and the bioprocess may fail. The efficacy of the FBBR units can be up to 10 times greater than that of the activated sludge process and typically occupies very small footprint ~10% of the space required by stirred-tank bioreactors of similar capacities. The greater efficiency of FBBRs is due to their higher biomass concentration, which can be up to 40,000 mg L^{-1} when compared with 3000 mg L^{-1} in the activated sludge process (Burghate & Ingole, 2013; De Beer, Klein Gunnewiek, Veenhuis, Bock, & Laanbroek, 1991; Kowalski & Lewandowski, 1984; Zou, Papirio, Lakaniemi, Ahoranta, & Puhakka, 2016).

Commonly, nitrogen removal from domestic sewage and industrial wastewater effluents can be achieved in two steps bioprocess: (1) microbial nitrification of ammonium (aerobic process) to nitrate, followed by (2) denitrification (anoxic process) or reduction of nitrate to nitrogen. This conservative method employs a two-step sequence of aerobic and anaerobic processes, wherein two different environmental conditions are provided (Simone

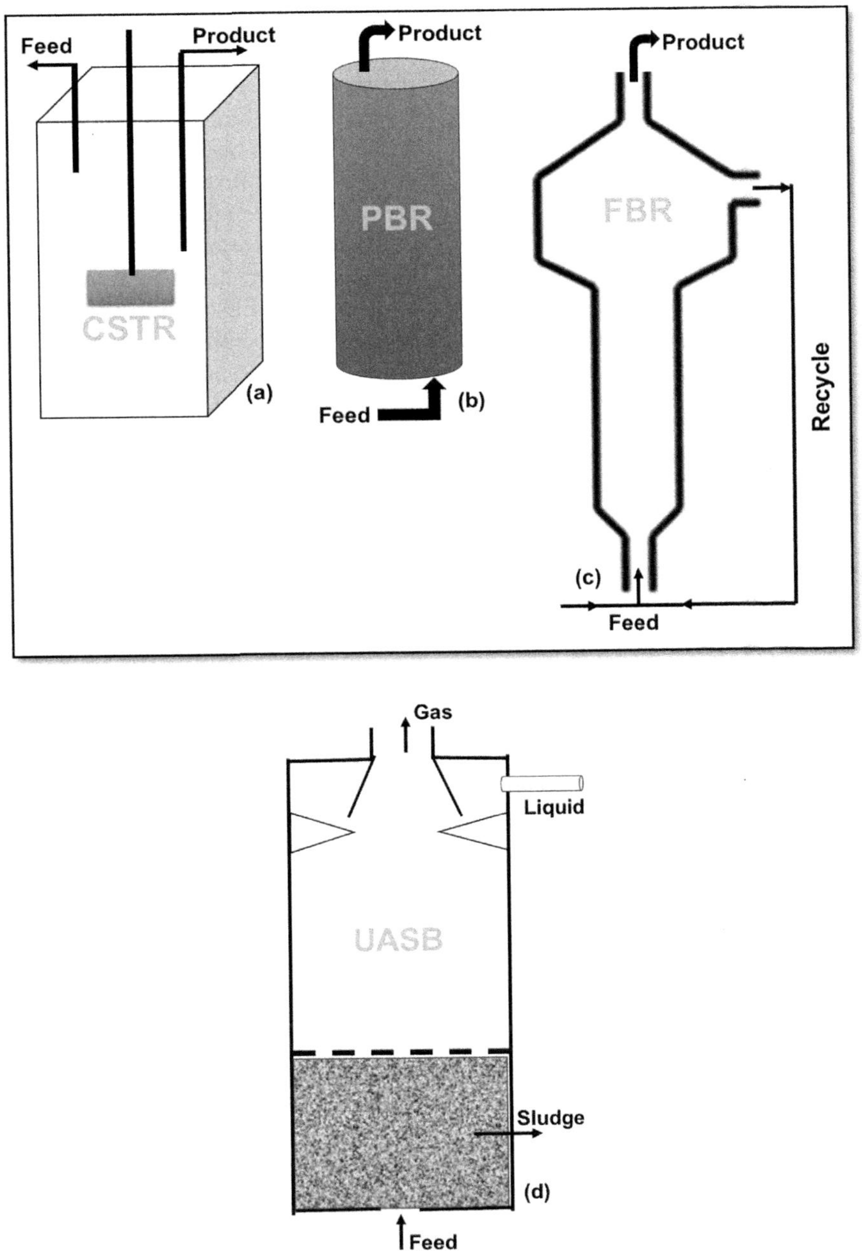

FIG. 7 Types of bioreactors commonly used in various bioprocesses and wastewater treatment.

et al., 2019). However, recent studies have showed that these two vital bioprocess steps can occur simultaneously in one single bioreactor. This bioprocess is also termed simultaneous nitrification and denitrification (SND) procedure. In SND bioprocess, nitrification is restricted to the outer aerobic zone where microbial flocs are in high concentration, whereas denitrification occurs principally in the inner anaerobic zone. To check the hypothesis that SND is a physical phenomenon, Pochana and Keller (1999) carried out batch experiments to determine the effect of floc size on SND. Typical floc sizes as measured in their studies were 50–100 μm, which is typically a large size. Such large floc sizes could create an anaerobic zone inside the flocs resulting in denitrification process. It was further concluded that a substantial anaerobic microbial biomass segment exists in the center of the floccular biomass. This is caused mainly due to limitation in oxygen diffusion into the floccular biomass. The sequencing batch and single activated sludge flocs bioreactors require significant O_2 concentration to initiate nitrification process (NO_3^-/NO_2^- generation), which is the precursor for denitrification process (Lei, Yu, & Joo-Hwa, 2004).

Robertson and Kuenen (1983) noticed that under fully aerobic condition, *Thiosphaera pantotropha* can carry out the following biochemical reaction sequentially and simultaneously, in the presence of a suitable electron donor such as acetate (see Eq. 1).

$$NH_4^+ \rightarrow NH_2OH \rightarrow NO_2^- \quad (1)$$

$$NO_2^- \rightarrow N_2O \rightarrow N_2$$

$$O_2 \rightarrow H_2O$$

This implies that the bacterium can convert ammonia into nitrogen gas without intermediary accumulation of nitrite. Further studies showed that this organism denitrifies under aerobic conditions, because of denitrifying enzymes, which are present even when the bacterium was growing aerobically without nitrate.

7. Biofilm reactors in wastewater treatment

In early 1900, the use of biofilm technology in wastewater treatment was initiated based on the operational experience of trickling filters in the city of Wales, Great Britain. The biofilm reactors with a number of full-scale application showed superior efficiency and are used throughout the world for industrial and wastewater treatment. The application of biofilm processes in wastewater treatment can be divided into two main categories: (a) the fixed-bed systems where the biofilm is static in the bioreactors and the biological reactions take place in the biofilm matrix that is developed on the static support, and (b) the moving-bed system where the biofilm matrix on the support is kept continually moving by means of mechanical, hydraulic system, and air is used to generate shear force. The moving-bed reactor includes rotating biological units, horizontally moving-bed biofilm reactors, vertically moving biofilm reactors, and fluidized-bed biofilm reactors, while the fixed-bed systems include trickling filters and aerated biological filters. Rotating biological contactors (RBCs) have been widely used in biological treatment of wastewater for reducing chemical oxygen demand (COD) and biological oxygen demand (BOD) as well as for nitrification and denitrification processes. RBCs treat wastewater streams using a thin biofilm matrix of aerobic microorganisms on rotating cylinders or bio-disc systems. The rate of rotation is selected to provide optimum contact of the wastewater with the biofilm for efficient oxygen transfer and biological activity (Singh & Mittal, 2012).

Anaerobic treatment of wastewater in fluidized-bed reactors is another area that has been studied extensively. The anaerobic fluidized-bed technology offers a number of advantages for treating wastewater including high concentration of biomass attached to the dense support matrix that makes it possible to operate them at high dilution rate without cell washout. In these reactors, no plugging, gas hold-up, or channeling occurs. The FBBRs have been used for more than two decades for treating industrial wastewaters. In this bioreactor, the particles move up and down within the bed in a well-defined zone of the bioreactor system. Immobilized bacterial systems that are configured as FBBRs offer many technical advantages. In this bioreactor, chemical wastes that are to be treated are injected into the recycle process where toxic chemicals are instantly diluted. This makes the microorganisms to adapt to the chemical concentrations and become more resistant to direct chemical toxicity than any other conventional treatment system. Since FBBRs are usually aerated by supplying air into the recycle loop, a high growth rate of microbes occurs with significant microbial activity, which is further supported with minimal air stripping of volatile chemicals (Ensley & Kurisko, 1994). Jesis and Owen (1977) studied FBBR, and they found that the use of small, fluidized media has enabled the FBBR to retain high biomass concentrations and can be operated at reduced hydraulic retention times.

Before considering the application of biofilm-based treatment systems for the wastewater treatment, it will be worthwhile to determine whether the naturally occurring microorganisms are able to form the biofilm. The biofilm reactor system should simultaneously reduce the COD and BOD of the wastewater. If there is a need to augment the bioprocess, new bacterial strains may be inoculated in the bioreactor. The most commonly used rotating biofilm reactor unit is the rotating bio-disc with various applications that are based on modifications. For the treatment of high-strength wastewaters, moderate aeration of the liquid phase has shown 40% improvement in the COD reduction (Meng et al., 2018). In studies using synthetic wastewater, Kargi and Eker (2001) reported that a rotating perforated-tube biofilm reactor is more effective in COD removal. In some cases, thermophilic aerobic systems are used to biodegrade high-strength wastewaters, wherein remarkable reduction in COD was noticed both in lab and pilot-scale experiments. However, the thermophilic systems showed poor bacterial flocculation property due to dispersal of the biofilm, which makes bacterial separation difficult from the treated effluent.

8. Biofilms for gas and odor treatment

Conventionally, physicochemical methods such as adsorption, scrubbing, condensation, and oxidation processes were used to treat industrial waste gases (Burgess, Parsons, & Stuetz, 2001). Industrial waste gas treatment using biological process is an attractive and environmentally friendly alternate to the popular physicochemical methods. A close look at the chemistry of industrial waste gases suggests that they can serve as carbon or energy source for growth as well as bacterial metabolism. Besides, gases such as H_2S and NH_3 can be directly treated by using autotrophic microbes that have the capability to consume CO_2 as a carbon source for anabolic processes (Stockar & Liu, 1999). To significantly reduce the impact of industrial air pollutants from wastewater treatment plant, Koe and Tan (1987) suggested that open sources of odorous emissions such as inlet works, primary sedimentation units, aeration tanks, clarifiers, sludge units, and wastewater channels must be closed and the released odorous gases have to be treated before releasing to the ambient atmosphere.

The biofilter, trickling filter, and bioscrubber are three major bioreactor designs that are normally used for treating waste gases. Generally, a biofilter consists of a filter-bed unit made of a carrier material such as sawdust, dry sludge, etc., as support material for the bacteria to adhere and use as carbon source. Fig. 8 describes the aerobic bacterial granules, phototrophic algal granules, and biofilm growth on support material.

8.1 Biofilters

A biofilter operates by enabling the obnoxious gas transfer from waste air blown through the biofilter unit, then into biofilm matrix, which is formed on the support particles of biofilter medium in which bacteria, fungi, and other microbes are immobilized. In contrast, waste gas treatment in trickling biofilter unit involves the use of a microbial filter that is continuously fed with a liquid medium and packed with a synthetic carrier on which biofilm is formed. Trickling biofiltration has been used for removal of odorous gases such as H_2S. Several bacterial species can oxidize H_2S to form sulfuric acid, for example, *Thiobacillus thiooxidans* is capable of oxidizing H_2S at pH as low as 1.5. Sulfide-oxidizing bacteria are an ideal habitat for effective removal of H_2S odor; however, care should be taken to avoid the growth of other competing microorganisms, which typically predominate in aerobic treatment processes (Yuan & Blackall, 2002). Earlier, DeBeer, Stoodley, Roe, and Lewandowski (1994) demonstrated that the water flow channels neighboring the biofilm cell clusters increase the supply of oxygen and other nutrients to microbial cells within the biofilm, which can be related to biofilm structure and function. Commonly, the microbial biofilm architecture seems to be largely influenced by the production of EPSs, which gives the structural support to the biofilm matrix. The structure of biofilms is largely determined by a number of biological factors such as bacterial growth rate, motility, quorum sensing, and the production of EPS. The biofilm growth and environment also play a significant role in the determination of the biofilm propagation and hence the efficacy of the process. Nevertheless, dense biofilm formation can lead to blockage of the bioreactor filter bed. However, microbial biomass growth and biofilm formation can be limited by reducing nutrient supply, albeit this may

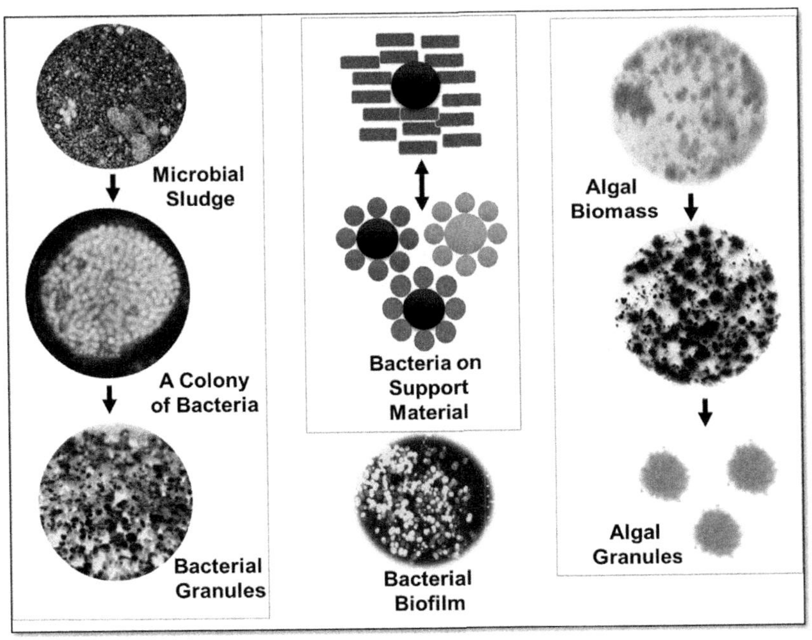

FIG. 8 Digital image description of aerobic bacterial granules, phototrophic algal granules, and biofilm growth on support material.

decrease bioreactor performance, since higher biomass growth shows higher substrate consumption. Hence, it is important to balance the excessive biomass growth to prevent blockage of biofilter unit and the efficiency of odorous gas removal (Venugopalan et al., 2005).

8.2 Bioscrubbers

Waste gases that are characterized by high concentrations of water-soluble pollutants can be treated using a bioscrubber. The bioscrubber commonly consists of two reactor subunits. The first reactor unit is an absorption column where pollutants are adsorbed and move to a liquid phase. Then the liquid phase moves to the second reactor, which consists of a filter bed with an activated carbon matrix that supports microbial growth. The high metabolic activity of the microbial biomass in the bioscrubber system augments the conversion of waste gases into nonhazardous and small compounds that have no odor. The effluent or mineralized gases leaving the bioscrubber can be recirculated in the first bioreactor unit (the absorption column); this know-how further allows good gas cleaning, particularly when the gaseous pollutants are highly water soluble (Hecht, Brebbermann, Bremer, & Deckwer, 1995).

9. Biofilm reactors in ethanol production

Ethanol production is commonly done with fermentation process using *Saccharomyces cerevisiae* and *Zymomonas mobilis*. The *Saccharomyces* species is the most common microorganism type that is behind the fermentation process and has been involved in winemaking, baking, and brewing processes since bygone era. Later on the cells of *Zymomonas mobilis* are being used in ethanol production, wherein the cells are adsorbed onto vermiculite support material resulting in an active biofilm formation for enhanced ethanol production (Demirci, Pometto, & Ho, 1997). The concept of ion exchange resins and biofilm formation was used by Daugulis, Krug, and Choma (1985) to produce ethanol in high production rate bioreactors using *Z. mobilis*. Ten ion exchange resin types, activated carbon, and ceramic chips were tested to find out a suitable adsorbent. A cationic macro-reticular resin has shown efficient immobilization/adsorption of cells of *Z. mobilis*. The immobilized cells were used in a continuous column and 100 g L^{-1} glucose was fed to the bioreactor. Since the biofilm formation in the bioreactor was excellent, ethanol production was high up to 135.8 g $L^{-1}h^{-1}$. However, the bioreactor could not be operated beyond 200h of operation and stopped working due to dense biofilm growth and blockage issues in the bioreactor (Shiva, 2012).

Qureshi, Annous, Ezeji, et al. (2005) reported ethanol production from xylose as carbon source in a bioreactor of genetically modified *Escherichia coli*. The *E. coli* biofilm was formed on baked clay particles, and the bioreactor was operated on continuous mode for 100 days. To optimize the bioprocess, the bioreactor was operated at various flow rates. The ethanol production showed significant increase when compared with the fermentation by planktonic cell batch process. In another study, Demirci et al. (1997) developed a new support material using a mixture of grounded hull of soybean and polypropylene, which is heated to high temperature and molded into disks and rings for the growth of *S. cerevisiae* biofilm. However, no data have been provided by the authors on the rate of biofilm formation as well as on the thickness of biofilm. Adsorbed cells of *Saccharomyces cerevisiae* were used in a packed-bed continuous bioreactor to produce ethanol from molasses. The cells were immobilized onto a support material of natural origin, such as sugarcane bagasse. Based on the volumetric analysis of the biofilm bioreactor operation, the ethanol production was up to 105 g L^{-1} h^{-1}, while typically in a batch or planktonic cell bioreactor, ethanol production was <4 g $L^{-1}h^{-1}$. The increase in ethanol production in biofilm bioreactor was due to the formation of active biofilm onto the adsorbent and effective utilization of the carbon source (Todhanakasem et al., 2019).

10. Biofilm bioreactors and production of biochemicals

Commercial production of acetic acid or vinegar using biofilm reactors has been practiced for many years. Large biofilm fermenters of size up to 60 m^3 have been used. Production of other industrial biochemicals (organic acids) using biofilm bioreactors includes acetic acid, fumaric acid, lactic acid, and succinic acid. Acetic acid is commonly produced in a trickling-bed biofilm reactor, which is a matured technology and is commonly applied at the commercial level. The acetic acid is commonly produced by bacteria that are grouped in the two genera, *Gluconobacter* and *Acetobacter*. The common species that are used commercially include *Acetobacter aceti*, *A. pasteurianus*, and *Gluconobacter oxydans* (Mamlouk & Gullo, 2013). In the trickling-bed biofilm reactor, beech wood shavings are packed and alcohol solution is used as starting material and is sprayed over the packing surface. Subsequently, growth nutrients and bacteria strains (inoculum) are added for biofilm formation on the beech wood shavings. After the fermentation, the liquid that trickles to the bottom of the bioreactor contains acetic acid. To enhance the concentration of acetic acid, the warm fermented liquid is cooled and recycled to the top of the bioreactor to increase acetic acid concentration. About 90% of the alcohol added is converted to acetic acid during the trickling process, a final concentration of 120 g L^{-1} acetic acid is produced in

72 h of bioreactor operation (Jun-Ichi, Kouji, Tohru, & Masayoshi, 2000).

Lactic acid, production in biofilm mode bioreactors, is another example of industrial biochemical production. Bastarrachea, Britt, and Demirci (2022) have studied several support systems for biofilm formation using lactic acid producing bacteria; they concluded that in case of polypropylene composite chips support material, the biofilm formation was significant. The lactic acid producing and biofilm-forming bacteria were *Pseudomonas fragi*, *Streptomyces viridosporus*, and *Thermoactinomyces vulgaris* species. The polypropylene composite chip support material contains polypropylene and 25% (w/w) and agricultural material. A mixture of several support material components was extruded to form chip-like designs of desired dimensions. Bastarrachea et al. (2022) also reported on synthesis, testing, and using various support materials for biofilm formation and lactic acid production. The authors further reported a lactic acid yield of $4.26\,g\,L^{-1}\,h^{-1}$ in repeated batch biofilm bioreactor. A maximum lactic acid concentration of $60\,g\,L^{-1}$ was produced in biofilm bioreactors wherein plastic composite support material was used for adsorption and biofilm formation. In another study on the production of lactic acid by *Rhizopus oryzae*, cell culture was immobilized on a fibrous bed consisting of a sheet of 100% cotton cloth onto which the bacterial culture was adsorbed and used as a bioreactor support material. In fed-batch bioreactor, a high yield of 90% lactic acid amounting to $2.5\,g\,L^{-1}\,h^{-1}$ was obtained with a high product concentration, when glucose was used as a substrate. However, when cornstarch was used instead of glucose, the lactic acid yield was improved to 100%. Alternatively, when starch was used as a substrate, a lactic acid concentration of $126\,g\,L^{-1}$ was obtained.

Succinic acid is a dicarboxylic acid, which can be used as a feedstock chemical for the production of high-value products such as 1,4-butanediol, tetrahydrofuran, adipic acid, γ-butyrolactone, and *n*-methylpyrrolidone, which have wide applications in agriculture, food, medicine, plastics, cosmetics, and textile industries. Succinic acid is a biochemical that can be synthesized or produced using biofilm bioreactors. In a study on succinic acid production using *Actinobacillus succinogenes*, Urbance, Pometto, Dispirito, and Demirci (2003) utilized a customized plastic composite support material and 20 other different blends with and without mineral salt additions and designed and carried out several simulated experiments in sequential-batch fermentations using $MgCO_3$ for pH adjustment and CO_2 supply. The tailored plastic composite support blends were screened for biofilm formation and succinic acid production simultaneously. Succinic acid concentration, percentage yield of succinic acid, and biofilm formation on each of plastic composite support blend were determined. The authors found that there was no correlation between biofilm formation and succinic acid production. Nevertheless, a customized plastic composite support blend with a bacterial strain of *A. succinogenes* showed 70% better yield for succinic acid when compared with suspended cell bioreactor for succinic acid production (Nghiem, Kleff, & Schwegmann, 2017).

Butanol production with superior productivity is achieved through batch, fed-batch, and planktonic cell continuous fermentation processes. In addition, biofilm reactors have been practiced in numerous models at laboratory scale. Among all the bioreactors, the most prominent bioreactors are the packed-bed and fluidized-bed bioreactors. In the biofilm-based bioreactors, butanol is produced in the concentration range of $4.5–15.8\,g\,L^{-1}\,h^{-1}$. As per the current scenario of increasing fuel prices, butanol production will be an advantageous proposition, in order to have a high butanol yield, it is suggested that fluidized-bed bioreactors can be upscaled to pilot plant level to further commercialize this fermentation process (Birgen, Dürre, Preisig, et al., 2019).

Production of other industrial chemicals such as lactic acid and 2,3-butanediol is of economic relevance; however, they are demonstrated at pilot plant level. Nicolella, van Loosdrecht, and Heijnen (2000b) reported that biofilm reactors are in operation at industrial scale throughout the world. Use of biofilm reactors is expected to be economical for the production of some industrial chemicals. A comparison of biofilm reactors with other bioreactor systems suggests that biofilm reactors are simple and result in higher production than other reactor systems. In biofilm reactors, cells can be adsorbed within the bioreactor structure without the use of chemicals, and the bioreactors can be operated for longer time. These bioprocesses generally cost less and reduce the process time. These bioreactors are already in use for wastewater treatment and in acetic acid/vinegar production by fermentation process. From the various studies, it is clear that the consistent use of batch-scale bioreactors has helped in increased production of various other chemicals. As productivities in these simple biofilm reactors are high, their full prospective should be employed for biotechnological/bioconversion processes (Sanchez & Demain, 2011).

11. Industrial ramifications of biofilms

The main consequences of fouling in marine environment are problems associated with marine structures such as pier pilings, oil platform, sea and land-based aquaculture units, and maritime industries. Biofilms formed on offshore oil platforms, seawater intake systems grow up to few millimeters and result in the propagation of biofouling phenomena. Macro-fouling can cause the fouling layer to increase to several centimeters in thickness, this will significantly reduce fuel efficiency of ships, and in addition, biofilms/macro-fouling also initiates corrosion of substratum metal. Biofilm/biofouling growth results in reduction of water intake volumes required for an industry, as a consequence, power stations where seawater is used to cool the condensers/heat exchangers, very dense biofouling hampers the power plant operation. Technoeconomics involved in the biofouling problem of power plants has been estimated elsewhere. A 5 mmHg condenser back pressure is equal to loss of 3 MW of power generating capacity. Barnacles, tubeworms, bryozoans, and mussels are the dominant biofouling species. Maximum fouling accumulation occurs at the turn of the bilge, below the keel, on the rudder surface, propeller, sea chests, and the water line. Types of ship and trade patterns influence docking time, bulkers hold on average 5 days at port compared with just 1 day for car carriers. Heavy fouling was also noticed on trawlers, which have an average speed of 7–11 knots. It has been estimated that the growth of marine fouling organisms costs the shipping industries over $6.5 billion per year (Hall, Costerton, & Stoodley, 2004; Rao, 2012, 2015a).

Biofouling refers to the undesirable deposition of biological materials on equipment surfaces, resulting in a drop in the performance and operational efficiency of the equipment. Biofouling affects almost every human activity connected with the sea including off-shore drilling platforms, hulls of ships, cooling conduits of power stations, mariculture equipment, heat-exchanger units of Ocean Thermal Energy Conversion (OTEC) plants, defense and surveillance technology (surface vessels, submarines, sonar, and optical instrumentation). Biofouling is mostly constituted by initial colonization by microorganisms followed by the settlement of larvae of macro-invertebrates, which generally inhabit the benthic habitats of aquatic systems. Survival of the larvae of fouling organisms depends on the amount of food available and therefore, larval releases by major fouling organisms (e.g., mussels, barnacles, etc.) are often timed to coincide with plankton blooms. Major conditions that enhance organism growth inside cooling conduits are a continuous supply of water rich in oxygen and food, reduction in density of predators, lack of competition from algal communities (due to the absence of light), and reduction in silt deposition. Macro-foulers are mostly represented by numerous species of algae, amphipods, ascidians, bryozoans, barnacles, mollusks, polychaetes, and sponges (Holmström & Kjelleberg, 1999; Holmström, Rittschof, & Kjelleberg, 1992; Murthy, Veeramani, Mohamed Ershath, & Venugopalan, 2011; Nair, Murugan, & Eswaran, 1988; Rajagopal & Jenner, 2012; Rao, Murthy, Veeramani, et al., 2021; Thiyagarajan, Stanley, Cheung, & Quian, 2006; Zardus, Nedved, Huang, Tran, & Hadfield, 2008).

12. Industrial biocorrosion problems

Microbial corrosion is a multifaceted process, the magnitude of this problem has slowly been realized during the present century. Iron-age civilization has brought in new varieties of iron types and use of iron has brought in the problem of iron corrosion. Albeit, some of the medieval iron structures are still corrosion-resistant. Interestingly, the iron produced in antiquity is free from corrosion than the iron products, which are manufactured in recent years. The best example is the most famous iron technology specimen, the Ashoka Pillar in Delhi, which is available even today without corrosion. Zajic briefly described corrosion in the following way:

> "Fundamentally, the theory of corrosion is simple but, it is very composite degradation process. In its broadest concept, corrosion is an electrochemical reaction. It is caused by any kind of heterogeneity at different points on a metal surface. Once a surface makes a contact with an electrolyte, however weak, different heterogeneous points show a variation in potential. They develop an electromotive force (emf) and form micro corrosion cells. The more reactive metal acts as the anode, and its atoms pass into the ionic state; at the same time, hydrogen ions are driven back to the less reactive metal which acts as the cathode. When hydrogen forms at the cathode, the corrosion cell becomes polarized. Thus, the metal seeks an equilibrium state and once this state is reached, metal dissolution at the anode starts and it continues until emf decreases and corrosion stops. Outside forces, can interfere and sustain or increase corrosion."

Although the above description is cursory, it directs attention to the outside factors, which contribute to the corrosion process. If the outside factors are microorganisms, the process is generally referred to as microbiologically influenced corrosion (MIC). The primary effect would be the formation of differential concentration cells on the metal surfaces by the respiring periphytic microcolony of organisms causing a difference in oxygen concentration between its microenvironment and the surrounding area. This will give rise to potential differences, with subsequent corrosion currents between the microbial colony sites and the surrounding metal surface. Direct removal of corrosion products, e.g., a respiring microbial colony removes the oxidized metal (at anodic sites) or processes that utilize hydrogen (at cathodic sites) will cause electrochemical corrosion reactions to be biased toward metal dissolution. The basic cause of corrosion is the thermodynamic instability of metals in their refined forms. It was the basis of the laws of electrochemistry laid down by Faraday that gave the relationship between chemical action and the generation of electric current, upon which the theory of electrolytic corrosion was based. The first evidence that corrosion was essentially electrochemical in nature was provided by the work of Whitney. Metals and alloys are subject to deterioration due to reaction with air, acids, bases, seawater, soils, etc., with which they come in contact. Considerable quantity of materials undergoes deterioration due to corrosion resulting in loss of several hundred crores of rupees with consequent adverse impact on the country's economic development. Sulfate-reducing bacteria (SRB) are the single most common causative organism in microbial corrosion, constituting 50% of all instances of corrosion (Iverson, 1987; Licina, 1989; Little, Wagner, & Mansfeld, 1991; Rao, ArunaJyothi, Anupkumar, Narasimhan, & Feser, 2005; Rao & Nair, 1998; Rao, Sairam, Viswanathan, & Nair, 2000).

The possibility that microorganisms exert an influence on the corrosion of metals was first mentioned by Garrett. He reported that the increase in corrosion of lead might be due to the ammonia, nitrites, and nitrates produced by bacterial action. Gaines suggested that the corrosion of iron in aqueous and soil environments might be caused by SRB, sulfur bacteria, and iron bacteria. The iron bacteria *Gallionella ferruginea* was isolated from the corrosion products and high concentration of sulfur and organic matter were noted. The formation of deposits of iron in water pipes was noted by Ellis and Harder (Rao & Nair, 1998). The above investigations on electrochemical corrosion of iron were concerned with corrosion in the presence of oxygen (aerobic conditions). Von Wolzogen Khur showed that iron could also be attacked in the absence of oxygen (anaerobic condition) as a result of the activities of SRB growth. Further studies by von Wolzogen Khur and van der Vlugt have provided convincing evidence on the corrosion of iron water pipes in anaerobic soils corroded due to graphitization. Later investigations into bacterial corrosion process were pursued by many workers and resulted in reviews (Little et al., 2020).

As early as 1954, Great-house and Wessel estimated 500–2000 million dollars loss per year due to biocorrosion, wherein they primarily estimated the deterioration of buried pipelines in the United States. Seed (1990) reported that in the United States, every year \$82 billion losses are due to corrosion. In the present scenario, the cost of corrosion is so enormous that it is measured as a fraction of a country's national gross product (GNP). In many countries, the cost of corrosion is higher than the total expenditure on R&D activities or the total production in significant industrial sectors (Seed, 1990). Significant corrosion losses worldwide are attributed to microorganisms. India incurs an estimated corrosion loss of Rs 24,000 crores (equivalent to \$28 billion) every year. This amounts to ~2% of India's gross national income. Rao (2015b) quoted a National Bureau of Standards, United States, statement that biologically induced corrosion accounts to \$167 billion worldwide. Conservative estimates place the direct cost of MIC at \$30–50 billion per year in United States (Cheung, Walsh, Campbell, Chao, & Beech, 1994). However, there are no specific reports quantifying biocorrosion losses in India. In a study reported by R. Bhaskaran of Lovely Professional University, Punjab, India, and N.S. Rengaswamy, Central Electrochemical Research Institute, Karaikudi, India, the cost of corrosion was estimated using the National Bureau Standards input/output economic model for 2011 to 2012. The Indian study gave one of the most detailed sector-wise breakdowns of the national costs of corrosion. The direct cost of corrosion for India was \$26.1 billion or 2.4% of India GDP. The avoidable cost of corrosion was \$9.3 billion or 35% of the direct cost of corrosion. The indirect cost of corrosion was \$39.8 billion or 3.6% of India GDP. Several of the indirect costs of corrosion in the 1998 US study (Hoar method), include loss of product, loss of efficiency, and production loss and were classified as indirect costs in the US study. If only the social costs are classified as indirect costs, the direct cost of corrosion in India is 4.5% of our GDP. Based on the impact India study carried out under the supervision of National Association of Corrosion Engineers (NACE), Houston, United States, by NACE International Gateway India Section, Mumbai, the global cost of corrosion (billion USD) for India, by sector wise: Agriculture = \$17.7, Industry = \$20.3, and Services = \$32.3; the total cost \$70.3, which is equal to ~4.2% of our GDP (DNV GL, 2015).

13. Microbial corrosion in industrial cooling water systems

Although there are many types of materials available for application in cooling water distribution systems, their usage has resulted in several failures in the industry. Copper and its alloys were the choice few decades ago; however, they have been replaced by stainless steel alloys. and in recent times, different grades of titanium are being used for condenser cooling. The performances of most of the condenser material used in the industry are affected by biofilms and biofouling. Thus, without antifouling treatment, no power plant can operate to its full capacity. Materials used in distribution systems generally fail due to corrosion. These are primarily responsible for the presence of metal ions in the domestic water as well as in industrial effluents. The results from various corrosion studies strongly suggest that there is a microbial contribution to the material corrosion. Corrosion (biological and chemical) has implications for the management of cooling water/distribution systems as the installation of new tubes into actively corroding systems may initiate similar corrosion in the new pipeline networks through contamination from the existing pipes. Besides, the presence of organic carbon and metabolic by-products such as hydrogen sulfide in the distribution system pipelines also plays a significant role in their deterioration (e.g., copper alloys) (Khan, Patil, & Rao, 2012; Rao & Bera, 2021; Rao, Murthy, & Harinath, 2014; Rao & Nair, 1998).

Of late, stainless steel varieties and titanium grades are being used in industrial cooling systems as heat exchangers. Both the materials form oxide layers on the surface as protective films. Any gaps in the oxide films or factors that disrupt the oxide layers result in localized corrosion of these materials. This may ultimately result in stress corrosion cracking and pitting corrosion particularly in stainless steels. Although titanium is known to be corrosion-resistant with an impervious TiO_2 oxide layer, the metal also undergoes corrosion due to loss of oxide layer, which can be damaged in anoxic conditions due to the growth of bacteria. Rao et al. (2005) have reported that titanium can undergo pitting corrosion due to sulfide production by SRB. There are many reports on the failures of industrial cooling system materials. At times, the constituents of the cooling water play a role in failure; microorganisms were identified as a major factor in material failures. The following are few failure studies of condenser tube materials in power plant cooling water systems.

14. Implications of biofilms and biocorrosion

The heterogeneous and anisotropic nature of biofilm growth results in spatial variations in surface coverage when microorganisms colonize a metal substratum. Industrial processes use a large area of surfaces, such as in cooling circuits. A heat exchanger surface encourages biofilm formation and concentrates nutrients to support their growth. This may result in optimization of a focused utilization of nutrients, which results in bacterial cell digging into the nutrient matrix, thereby liberating large amounts of soluble nutrients such as organic acids, which alter their immediate milieu by proton extrusion. Depending on the physical conditions prevailing and the primary nutrients available, microbial consortia may contain a wide range of aerobic, facultative, and anaerobic organisms. These microbes together are capable of hydrolytic and fermentative metabolism of primary nutrient sources such as carbohydrate, protein and other organic detritus, or hydrocarbons. Microbial biofilm development in natural and industrial water systems depends on the type and concentrations of electron acceptors (e.g., oxygen, nitrate, and sulfate) and electron donors (organic substrates), as well as environmental factors including temperature. pH, salinity, and fluid dynamics. Biofilms favor the growth and multiplication of microorganisms that find themselves in a micro-niche. By the stimulation of molecular or proton exchange, complex microbial consortia are produced. These consortia are associated with the corrosion phenomenon, when they constitute a locally focused and attached microbiota on metal substrata. The initial activities of the aerobic and facultative species will tend to utilize the available oxygen and result in anaerobic conditions. It has variously been estimated that this effect is evident with biofilms of between 25 µm and 100 µm thickness. This leads to both the correct physicochemical condition for the growth of anaerobes and the continuous supply of their required nutrients, independent of the primary nutrient(s) supporting the growth of the biofilm as a whole. Differences in the degree of colonization of metals by biofilm bacteria (presence vs absence) can generate corrosion potentials almost as large as that generated between incompatible metals. Biocorrosion or MIC is defined as "the deterioration of materials by the functional activities of microorganisms and their metabolic processes." The most interesting aspect of this concept is that it uses established perceptions of microbial ecology to explain how bacterial growth can result in the establishment of a classical corrosion cell (Balamurugan, Joshi, & Rao, 2011; Blinkinsop & Costerton, 1991; Satpathy, Rao, Rajamohan, & Nair, 2001; Shukla & Rao, 2013).

The impact of the distribution system corrosion can range from significant physical damage to pipes and other system components due to uniform surface removal, pitting, or tuberculation. As the corrosion process advances, the attached microbiota and their metabolic by-products get released into the cooling water, thereby altering the water quality. Microbial colonization may also enhance the formation of ion concentration cells, increasing the net corrosion rate. As the biofilm grows on a metal surface, it results in oxygen depletion and the area beneath biofilm becomes more anodic. In addition, microbial metabolic by-products such as inorganic and organic acids and hydrogen sulfide enhance metal dissolution and precipitation, which, in turn, depolarizes the anode and furthers the corrosion process (Rao et al., 2015).

Corrosion processes are always complex, involving possible multifarious component reactions. They often act in concert. The oxygen-depleted regions become electrochemically active and promote anodic dissolution. At the cathodic regions, which are open, oxygen reduction takes place. Thus, corrosion proceeds, leading to the formation of deep pits. Metals that rely on the presence of stable oxide film for their corrosion protection such as aluminum and its alloys, stainless steels, copper and its alloys are particularly susceptible to this type of corrosion. The ability of organic and inorganic compounds produced by microorganisms to extract metals from surface and the acidic conditions created by microbial activity results in solubilization of the metal from pipe surfaces. A mature biofilm prevents the diffusion of oxygen to cathodic sites and the diffusion of aggressive anions, e.g., chloride to anodic sites. Outward diffusion of microbial metabolites and corrosion products is also impeded. Because of the electrochemical character of the primary corrosion process, deterioration is accelerated in an environment of high electrical conductivity. Differences of potential caused by variations in the environment of the pipe may cause rapid, local corrosion. Corrosion in cooling water systems is aggravated by the presence of exopolymer producing bacteria, which act by a combination of several mechanisms in corroding iron and copper surfaces. MIC of wrought iron and steel occurs in the form of deep pits, wave-like grooves, craters, or complete perforation. Most confirmed cases of MIC are localized. For example, distinctive hemispherical or conical tubercles are a result of tuberculation of carbon steel and can develop as a result of microbial activities, specifically iron-oxidizing bacteria (IOB). SRB act as depolarizers for corrosion cells, by oxidizing the polarizing hydrogen film on the metal surface. The anaerobic conditions for the growth of the SRB can be produced inside of tubercles and other porous deposits, which result in depletion of dissolved oxygen through the activity of other organisms. *Desulfovibrio* species of SRB produce conical pits and pits containing concentric rings in various steels and in cupronickel alloys. SRB attack on cast iron typically produces graphitization, i.e., the corroded areas are filled with soft graphite (Rao et al., 2000).

Microorganisms that promote corrosion generally include not only bacteria but also algae and fungi. Five categories of bacteria are commonly recognized; they are the IOB, nitrate-reducing (NRB), SRB, sulfur-oxidizing (SOB), and the exopolymer producing bacteria. *Gallionella*, *Leptothrix*, and *Sphaerotilus* species of iron oxidizing bacteria obtain their energy for carbon fixation from oxidation of ferrous iron to ferric ions. They deposit ferric oxide on carbon steel pipeline surfaces and promote tubercle formation. SRB such as *Desulfovibrio* are involved in the reduction of sulfate to sulfide. IOB and SRB are typical examples of the aerobic and anaerobic synergistic interaction, which is most frequent in microbial corrosion of steel and its alloys used in industrial cooling systems. The common SOB *Thiobacillus thiooxidans* is a chemolithotroph utilizing thiosulfate and sulfide as sources of energy to produce sulfuric acid. The principal slime forming bacterial genera include *Pseudomonas*, *Bacillus*, *Flavobacterium*, and *Aerobacter* species. Most of the slime formers that colonize metal surfaces produce polymers and form a gel matrix on the metal. Characterization of copper complexing property of an exopolymer produced by a mesophilic bacterium isolated from a corroded coupon has been carried out. The exopolymers influence interfacial processes by immobilizing water at the biofilm-metal interface, trapping metal species and corrosion products at the interface, and thereby decreasing diffusion. These exopolymers were found to be acidic and contain functional groups that bind metal ions from the aqueous phase. Algae range from very simple single cell plants to multicellular species. Algae generally flourish on wetted, well-lit surfaces such as cooling towers and distribution systems. Due to their capability to generate oxygen, organic acids, and nutrients for other organisms, algae also play a role in biocorrosion. Among the fungi, the important ones are the molds and yeasts. Fungi deteriorate cooling tower wood and some such as *Cladosporium resinae* are involved in the corrosion of aluminum alloys. They produce organic acids by metabolizing fuel components in aircraft fuel tanks, leading to fuel breakdown and corrosion of fuel tank. The corrosion of distribution systems can have several implications for treated water quality as well as for the integrity of the actual distribution system. Microbial after growth or regrowth in distribution systems has been linked to sloughing of the pipe biofilm into the flowing water (Balamurugan, Chandramohan, & Rao, 2016).

15. Prevention and control of biofilms in industries

The biofilm formation and biofouling are absolutely an aquatic natural process occurring in the pelagic waters of seas and oceans as soon as a substrate is lowered or moored for long term. Biofouling is a measurement of biomass deposition on to a defined surface and varies from substratum to substratum. Therefore, first, the assessing person must clearly define the procedures required to get quality data for his system or industrial unit. Second, the tolerable level of noise data due to biofouling and finally to determine the most convenient antifouling strategies for each case, under a rigorous qualification procedure. Special or meticulous care must be taken while designing an antifouling strategy, since it can also be influenced by local environment. The recommendation for biofouling control should be made as per site specificity (Jenner, Whitehouse, Taylor, & Khalanski, 1998; Venkateswarlu, 1990; Yebra, Kiil, & Johansen, 2004).

Biofouling control by manual cleaning of power plant cooling circuits is difficult as well as uneconomical. Although methods such as Amertap, sponge ball, screens of various sizes, heat treatment, and biocides are in use for prevention and control of biofouling, chlorination stands out as the most efficient method owing to its proven effectiveness, easy availability, and low cost. Yet, the increasing concern over chlorine's ecological impact has led to new regulations in lowering the discharge limit for residual chloride at an industrial effluent outfall. Besides, it should be noted that any antifouling control methods may not work for ever (Fusetani, 2004).

Chlorination of cooling water was first tried in the United States in 1924 at the Commonwealth Edison Company's Northwest Station on an experimental basis to combat biofouling. These formal experiments have shown chlorination would prevent biofouling. Over the years, chlorination has been accepted as the most widely used biocide for water treatment. The three most important points that favor its use as an antifouling measures are (a) it is very effective as a biocide (b) it's easy availability, and (c) it's relatively low cost. In India, all power stations nuclear as well as thermal use seawater in once-through cooling systems. Industrially, chlorine is typically used in an intermittent mode or in a continuous mode to prevent biofilm growth in the condenser section and biofouling in the precondenser conduits. Chlorine as a gas is added before condenser section as well as at the beginning of the precondenser sections to obtain chlorine residual of about 1.5–2 $mg L^{-1}$. The type of fouling organisms should be a major consideration in deciding the chlorination schedule. If foulants are soft-bodied like bryozoans and hydroids, intermittent chlorination can be used but not to the schedule or mode used for slime control. To overcome the problem of macro-fouling, low-level continuous chlorination is usually used and such a practice is effective in the control of slime. The amount employed is insufficient to kill the macrofoulants but sufficient to create an environment to deter them from settling in the cooling system. This mode of chlorination is reported to be economical and environmentally acceptable (Raveendran & LimnaMol, 2009).

In the case of slime control, chlorine is used on the basis of two principles: (i) let the slime accumulate for a predetermined length of time, then start chlorination for a fixed time during which the adhered slime forming organisms will be killed and then washed away by the velocity of water; (ii) the interval between successive treatment should not be long enough for the accumulated slime to attain a thickness, which will interfere with heat transfer. Such a practice involving operation of chlorination at predetermined intervals is known as intermittent chlorination. The frequency of chlorine dosing is determined by the above two factors. Duration is decided by the turnover time for the cooling system, while magnitude and type of residual chlorine (free or total) are determined by the type of slime. The point of chlorine application should be located as close as possible to the inlet of the heat exchanger to take care of the free chlorine component of the residual. Intermittent chlorination does not effectively deal with macro-fouling problems (Sweta, Kumar, Harinath, & Rao, 2013).

Presently, chlorination forms the main method of controlling the growth of fouling organisms in power plant cooling circuits. Other halogen-based formulations are also being used, but these are often supplementary in nature. Chlorine is procured as chlorine gas in cylinders, evaporated and dissolved in seawater, and dosed at various locations in the cooling water circuit. At the power station of Kalpakkam, India, the dosing is such that it produces low level of residuals (about 0.2–0.5 $mg L^{-1}$ in the forebay). At such low concentrations, chlorine is not expected to kill the attached organisms in any reasonable length of time; in fact, significant mortality using such low residuals would take weeks rather than days. But these low residuals discourage fresh fouling settlement by creating an unfavorable environment for the fouling larvae/juveniles, which enter the system along with the incoming water. In other words, continuous low-dose chlorination is more of a preventive measure, rather than a technique to dislodge established fouling. Our studies have shown that in Kalpakkam coastal waters, the breeding seasons of barnacles and mussels exhibit peaks in May–June and October–November. During such breeding seasons, chlorination should ideally be done uninterrupted. It has been pointed out that periodic heat treatment of the intake system would be a very effective means of controlling fouling growth. This would be prohibitively costly to retrofit on existing plants. Nevertheless, for new plants this is an option that can be evaluated for cost-effectiveness and environmental compatibility (Rajagopal, Nair, Azariah, Van der Velde, & Jenner, 1996).

The success of a cooling water treatment program depends on the properties of the available (feed) water supply. Despite a careful design of a cooling water circuit, the quality of the feed water can create serious problems for the various materials and equipment. Generally, the properties of water vary depending on site location and the main source (river, lake, or impoundment). Subsoil water and groundwater sources contain considerable amount of minerals and salts and result in relative hardness of water. Rivers, streams, and lakes usually have better water quality since their main source is rainfall. Although the technology of cooling water (mostly chemistry of cooling water) has made major advances in the recent times. The major objectives of successful cooling water treatment program are still the same: (1) maintaining heat transfer in the condenser unit; (2) life extension of cooling water equipment. Fig. 9 provides a comparative account of different biocides for biofilm/biofouling control (Petrucci, 2005).

Cooling water treatment should be mandatory for maintaining the efficient performance of heat exchangers; however, the mode of treatment has to be decided on a case-to-case basis. Online monitoring of cooling water system parameters provides vital inputs for sustained performance of the operating plants. Recently, new devices have come onto the market with improved performance for monitoring, thereby reducing or eliminating the need for reagents-based systems and offline analyses. An electromechanical device can record losses due to a reduction in performance of heat exchange due to microbial fouling. In one such device, cooling water flow and discharge temperature signals are sent to a microprocessor to continuously calculate and display the performance of individual condenser tube heat transfer coefficient. This provides valuable information on the performance of the condenser unit. The

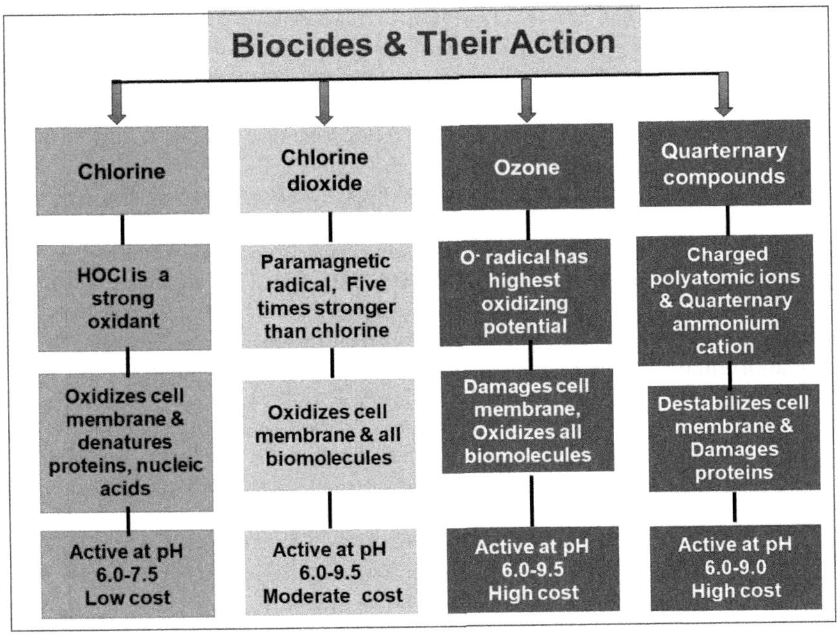

FIG. 9 Schematic illustration of different biocides and their mode of action.

new-generation membrane probes provide better management of fouling process without the need for electrode cleaning systems. These membrane cells are a closed-cell amperometric device with pH correction. This novel technology allows for continuous monitoring of, e.g., chlorine without use of chemical buffers (Khalanski & Jenner, 2012).

In modern water treatment, chromate has been discontinued due to its toxicity and environmental impact considerations. Currently, almost all the major power and process companies use alkaline cooling water technology and take advantage of the lower corrosivity nature of high pH waters saturated with respect to calcium carbonate index. Alkaline waters are generally less aggressive at near neutral pH water, because of their higher buffering capacity. It is encouraging to note that most of the cooling towers operating worldwide had already changed to alkaline treatment. Most industrial cooling water systems are operated at pH >7 in order to take advantage of the alkalinity of the water for corrosion control. Several different combination programs are available, offering corrosion and scale control in various combinations depending on the type of cooling system required. Besides, cooling water treatment package should also include biofouling control as an integral part of the operation. Alkaline cooling water treatments generally fall in the scaling range and will precipitate calcium carbonate quickly if scale inhibitors are not present. Complacency by operators at times can be disastrous and will be difficult and expensive to treat (Cloete, Jacobs, & Brozel, 1998).

A comprehensive plant audit inspection and monitoring program should be mandatory to increase the reliability of the cooling water system. A design evaluation to determine the components most likely to become fouled and the areas where increased monitoring and maintenance are required should be obligatory. The overall heat transfer coefficient of heat exchangers should be monitored as part of the surveillance procedure to assess the performance. Side-stream fouling monitors can be used in the condenser cooling system to monitor biofouling. Preoperational evaluation of the biofouling potential of cooling water source can be evaluated by exposure of panels to monitor the settlement and growth of fouling organism. Further laboratory studies with such dominant microorganisms can provide useful data in designing an effective cooling water treatment program for controlling biofouling and microbial corrosion. To activate future growth prospects in automation and online monitoring of vital cooling water parameters, there is a need to invest in advanced research and develop wireless sensors/monitors, which can detect the parameter of choice with high accuracy (Mazumder & Mohammad, 2020).

16. Control strategies for industrial cooling water systems

An effective cooling water treatment program depends on the water quality parameters of the feed water that is supplied to the industry. In a cooling water system, notwithstanding the careful design, the quality of the source water can deteriorate plant material and equipment. Water quality parameters vary depending on the location. Well waters that originate from groundwater where significant concentrations of minerals and salts are available give relatively a hard water property. Surface flowing waters such as rivers or streams typically have better water quality since the source is from rainfall.

Cooling water technology (i.e., chemistry of cooling water) particularly in industrial water treatment has made major developments since two decades. Though, the objectives of successful cooling water treatment program have remained the same since ages:

(1) Condenser or heat exchanger unit has to maintain effective heat transfer
(2) Extending the life of cooling water system equipment

The supposed objectives to some extend consummate in low corrosion rates of cooling system materials as well as reduce deposition of products on the metal surfaces. The aboriginal cooling water treatment programs started in the 19th century have addressed the concerns of corrosion, scaling, and deposit control. However, major changes or innovations in the cooling water treatment have happened since three decades particularly in terms of developing specialty chemicals for water treatment (Adu-Manu, Tapparello, Heinzelman, Katsriku, & Abdulai, 2017). Currently, to prevent bacterial biofilm growth and control corrosion in cooling water systems, the following treatment programs are predominantly in practice: (1) alkaline cooling water treatment program; (2) online dosing of biocides, corrosion inhibitors, and dispersants; (3) environmentally safe green chemicals technology.

17. Conclusion

Bacteria are pioneering microorganisms that are involved in biofilm formation, the biofilm growth is a sequential process wherein the microbes, exopolymeric matrix form a structurally complex system, which propagates on natural surfaces and industrial distribution systems. The bacterial growth furthers biofouling process and subsequently results in biocorrosion of metals. Various industrial metals, carbon steel, stainless steel (SS-304 and 316), copper alloys, and titanium undergo localized or pitting corrosion. Apart from material deterioration in industries, biofilms have beneficial applications wherein the bioreactors are used in production of vinegar, alcohol, organic acids such as lactic and succinic acids. Besides, various types of bioreactors are used in domestic and industrial effluents with membranes, biofilters, bioscrubbers being useful support materials. Finally, a brief description was given on the current cooling water treatment procedures, which are oriented toward environmentally benign green technologies.

References

Abma, W. R., Schultz, C. E., Mulder, J. W., van der Star, W., Strous, M., Tokutomi, T., et al. (2007). Full-scale granular sludge Anammox process. *Water Science and Technology: A Journal of the International Association on Water Pollution Research*, *55*, 27–33.

Adu-Manu, K., Tapparello, C., Heinzelman, W., Katsriku, F., & Abdulai, J. (2017). Water quality monitoring using wireless sensor networks: Current trends and future research directions. *ACM Transactions on Sensor Networks*, *13*, 1–41.

Allison, D. G., & Gilbert, P. (1992). Bacterial biofilms. *Science Progress*, *76*, 305–321.

Anjali, G., & Sabumon, P. C. (2017). Development of simultaneous partial nitrification, ANAMMOX and denitrification (SNAD) in a non-aerated SBR. *International Biodeterioration & Biodegradation*, *119*, 43–55.

Arooj, M. F., Han, S. K., Kim, S. H., Kim, D. H., & Shin, H. S. (2008). Continuous biohydrogen production in a CSTR using starch as a substrate. *International Journal of Hydrogen Energy*, *33*, 3289–3294.

Azam, F., Fenchel, T., Field, J. C., Grey, J. S., Meyer-Reil, L. A., & Thingstad, F. (1983). The ecological role of water-column microbes in the sea. *Marine Ecology Progress Series*, *10*, 257–263.

Balamurugan, P., Chandramohan, P. S., & Rao, T. (2016). Corrosion management of carbon steel material: Operational modes influence corrosion rate—An in vitro study. *RSC Advances*, *6*, 41122–41129.

Balamurugan, P., Joshi, M. H., & Rao, T. S. (2011). Microbial fouling community analysis of cooling water system of a nuclear test reactor: With emphasis on sulfate reducing bacteria. *Biofouling*, *27*, 967–978.

Bastarrachea, L. J., Britt, D. W., & Demirci, A. (2022). Development of bioactive solid support for immobilized *Lactobacillus lactis* biofilms in bioreactors for production of nisin. *Food and Bioprocess Technology*. https://doi.org/10.1007/s11947-021-02743-7.

Battin, T. J., Kaplan, L. A., Newbold, J. D., & Hansen, C. M. E. (2003). Contributions of microbial biofilms to ecosystem processes in stream mesocosms. *Nature*, *426*, 439–442.

Beun, J. J., Hendriks, A., Loosdrecht van, M. C. M., Morgenroth, E., Wilderer, P. A., & Heijnen, J. J. (1999). Aerobic granulation in a sequencing batch reactor. *Water Research*, *33*, 2283–2290.

Bhinu, V. S. (2005). Insight into biofilm associated microbial life. *Journal of Molecular Microbiology and Biotechnology*, *10*, 15–21.

Birgen, C., Dürre, P., Preisig, H. A., et al. (2019). Butanol production from lingo-cellulosic biomass: Revisiting fermentation performance indicators with exploratory data analysis. *Biotechnology for Biofuels*, *12*, 167. 1–15.

Blinkinsop, S. A., & Costerton, J. W. (1991). Understanding bacterial biofilms. *Trends in Biotechnology*, *9*, 138–143.

Boltz, J. P., Smets, B. F., Rittmann, B. E., van Loosdrecht, M. C. M., EberhardMorgenroth, G. T., & Daigger. (2017). From biofilm ecology to reactors: A focused review. *Water Science and Technology*, *75*, 1753–1760.

Bott, T. R. (1993). Aspects of biofilm formation and destruction. *Corrosion Reviews*, *11*, 1–24.

Brindle, K., & Stephenson, T. (1996). The application of membrane biological reactors for the treatment of wastewaters. *Biotechnology and Bioengineering*, *49*, 601–610.

Burgess, J. E., Parsons, S. A., & Stuetz, R. M. (2001). Developments in odour control and waste gas treatment biotechnology: A review. *Biotechnology Advances*, *19*, 35–63.

Burghate, S. P., & Ingole, N. W. (2013). Performance of fluidized bed biofilm reactor for nitrate removal. *International Journal of Research in Engineering and Science*, *7*, 38–47.

Casey, E., Glennon, B., & Hamer, G. (2000). Biofilm development in a membrane-aerated biofilm reactor: Effect of flow velocity on performance. *Biotechnology and Bioengineering*, *67*, 476–486.

Characklis, W. G., & Cooksey, K. E. (1983). Biofilms and microbial fouling. *Advances in Applied Microbiology*, *29*, 93–138.

Chelliapan, S., Din, M. F. M., Anuar, N. A., & Selvam, B. S. (2017). Effect of organic loading rate (OLR) on the performance of modified anaerobic baffled reactor (MABR) supported by slanted baffles. *Desalination and Water Treatment, 79*, 56–63.

Cheung, C. W., Walsh, F. C., Campbell, S. A., Chao, W. T., & Beech, I. B. (1994). Microbial contributions to the marine corrosion of steel piling. *International Biodeterioration & Biodegradation, 34*, 259–274.

Cloete, T. E., Jacobs, L., & Brozel, V. S. (1998). The chemical control of biofouling in industrial water systems. *Biodegradation, 9*, 23–37.

Costerton, J. W., Lewandowski, Z., Caldwell, D. E., Korber, D. R., & Lappin-Scott, H. M. (1995). Microbial biofilms. *Annual Review of Microbiology, 49*, 711–745.

Cowan, M., Warren, T. M., & Fletcher, M. (1991). Mixed species colonization of solid surfaces in laboratory biofilms. *Biofouling, 3*, 23–39.

Daugulis, A. J., Krug, T. A., & Choma, C. E. T. (1985). Filament formation and ethanol production by *Zymomonas mobilis* in adsorbed cell bioreactors. *Biotechnology and Bioengineering, 27*, 626–631.

De Beer, W., Klein Gunnewiek, P. J. A., Veenhuis, M., Bock, E., & Laanbroek, H. J. (1991). Nitrification at low pH by aggregated chemolithotrophic bacteria. *Applied and Environmental Microbiology, 57*, 3600–3604.

DeBeer, D., Stoodley, P., Roe, F., & Lewandowski, Z. (1994). Effect of biofilm structures on oxygen distribution and mass transfer. *Biotechnology and Bioengineering, 43*, 1131–1138.

Demirci, A., Pometto, A. L., & Ho, K. L. G. (1997). Ethanol production by *Saccharomyces cerevisiae* in biofilm reactors. *The Journal of Industrial Microbiology and Biotechnology, 19*, 299–304.

DNV GL. (2015). Assessment of global cost of corrosion. *Report No. OAPUS310GKOCH (PP110272)-1, Rev. 3*. www.dnvgl.com. 23 December 2015.

Dutta, A., & Sarkar, S. S. (2015). Sequencing batch reactor for wastewater treatment: Recent advances. *Current Pollution Reports, 1*, 177–190.

Ensley, B. D., & Kurisko, P. R. (1994). A gas lift bioreactor for removal of contaminants from the vapor phase. *Applied and Environmental Microbiology, 60*(1), 285–290.

Fletcher, M. (1988). Attachment of *Pseudomonas fluorescens* to glass and influence of electrolytes on bacterium-substratum separation distance. *Journal of Bacteriology, 170*, 2027–2030.

Fusetani, N. (2004). Biofouling and antifouling. *Natural Product Reports, 21*, 94–104.

Gordon, V. D., Davis-Fields, M., Kovach, K., & Rodesney, C. A. (2017). Biofilms and mechanics: A review of experimental techniques and findings. *Journal of Physics D: Applied Physics, 50*, 223002.

Gude, V. G. (2016). Microbial fuel cells for wastewater treatment and energy generation. In K. Scott, & E. Yu (Eds.), *Microbial electrochemical and fuel cells* (pp. 247–285). Woodhead Publishing, ISBN:9781782423751.

Guglielmi, G., & Andreottola, G. (2011). Alternate anoxic/aerobic operation for nitrogen removal in a membrane bioreactor for municipal wastewater treatment. *Water Science and Technology, 64*, 1730–1735.

Hall, S. L., Costerton, J. W., & Stoodley, P. (2004). Bacterial biofilms: From the natural environment to infectious diseases. *Nature Reviews Microbiology, 2*, 95–108.

Hecht, V., Brebbermann, D., Bremer, P., & Deckwer, W. D. (1995). Cometabolic degradation of trichloroethylene in a bubble column bioscrubber. *Biotechnology and Bioengineering, 47*, 461–469.

Hoh, D., Watson, S., & Kan, E. (2016). Algal biofilm reactors for integrated wastewater treatment and biofuel production: A review. *Chemical Engineering Journal, 287*, 466–473.

Holmström, C., & Kjelleberg, S. (1999). Marine *Pseudoalteromonas* species are associated with higher organisms and produce biologically active extracellular agents. *FEMS Microbial Ecology, 30*, 285–293.

Holmström, C., Rittschof, D., & Kjelleberg, S. (1992). Inhibition of settlement by larvae of *Balanus amphitrite* and *Ciona intestinalis* by a surface colonizing marine bacterium. *Applied and Environmental Microbiology, 58*, 2111–2115.

Houdt, R. V., & Michiels, C. W. (2010). Biofilm formation and the food industry, a focus on the bacterial outer surface. *Journal of Applied Microbiology, 109*, 1117–1131.

Isanta, E., María, S., Ángeles, V., Morales, N., Perez, J., & Carrera, J. (2012). Long term operation of a granular sequencing batch reactor at pilot scale treating a low-strength wastewater. *Chemical Engineering Journal, 198–199*, 163–170.

Iverson, W. P. (1987). Microbial corrosion of metals. *Advances in Applied Microbiology, 32*, 1–36.

James, G. A., Korber, D. R., Caldwell, D. E., & Costerton, J. W. (1995). Digital image analysis of growth and starvation responses of a surface-colonizing *Acinetobacter*. *Journal of Bacteriology, 177*, 907–915.

Janecki, D., Burghardt, A., & Bartelmus, G. (2016). Parametric sensitivity of a CFD model concerning the hydrodynamics of trickle-bed reactor (tbr). *Chemical and Process Engineering, 37*, 97–107.

Jenner, H. A., Whitehouse, J. W., Taylor, C. J. L., & Khalanski, M. (1998). Cooling water management in European power stations: Biology and control of fouling. *Hydroécologie Appliquée, 10*, 1–255.

Jesis, J. S., & Owen, R. W. (1977). Biological fluid-bed treatment for BOD and nitrogen removal. *Journal (Water Pollution Control Federation), 49*, 816–821.

Jun, L., Li-Bin, D., Cai, A., Guo-Xian, H., & Hprn, H. (2014). Aerobic sludge granulation in a full-scale sequencing batch reactor. *BioMed Research International*, 268789. https://doi.org/10.1155/2014/268789.

Jun-Ichi, H., Kouji, T., Tohru, K., & Masayoshi, K. (2000). Continuous acetic acid production by a packed bed bioreactor employing charcoal pellets derived from waste mushroom medium. *Journal of Bioscience and Bioengineering, 89*, 126–130.

Kargi, F., & Eker, S. (2001). Rotating-perforated-tubes biofilm reactor for high-strength wastewater treatment. *Journal of Environmental Engineering, 127*, 959–963.

Khalanski, M., & Jenner, H. A. (2012). Chlorination chemistry and ecotoxicology of the marine cooling water systems. In S. Rajagopal, H. A. Jenner, & V. P. Venugopalan (Eds.), *Operational and environmental consequences of large industrial cooling water systems* (pp. 183–226). New York: Springer.

Khan, A. F., Patil, A. P., & Rao, T. S. (2012). Corrosion resistance of a new Cu-alloy (low Ni) vis-à-vis Cu-10Ni alloy in sulfide polluted synthetic seawater. *Advanced Materials Research, 585*, 488–492.

Koe, L. C. C., & Tan, Y. G. (1987). GC-MS analysis of odorous emissions from the dissolved air flotation units treating surplus activated sludge at a wastewater treatment works. *International Journal of Environmental Studies, 30*(1), 37–44.

Korber, D. R., Lawrence, J. R., Lappin-Scott, H. M., & Costerton, J. W. (1995). Growth of microorganisms on surfaces. In H. M. Lappin-Scott, & J. W. Costerton (Eds.), *Microbial biofilms* (pp. 15–45). Cambridge: Cambridge University Press.

Kowalski, E., & Lewandowski, Z. (1984). Nitrification process in a packed bed reactor with a chemically active bed. *Water Research, 17*, 157–160.

Krieg, N. R., & Holt, J. G. (1984). *Bergey's manual of systematic bacteriology. Vol. 1* (pp. 140–407). London: Williams and Wilkins.

Lawrence, J. R., & Caldwell, D. E. (1987). Behavior of bacterial stream populations within the hydrodynamic boundary layers of surface microenvironments. *Microbial Ecology, 14*, 15–27.

Lawrence, J. R., Delaquis, P. J., Korber, D. R., & Caldwell, D. E. (1987). Behaviour of *Pseudomonas fluorescens* within the boundary layers of surface microenvironments. *Microbial Ecology, 14*, 1–14.

Lei, Q., Yu, L., & Joo-Hwa, T. (2004). Effect of settling time on aerobic granulation in SBR. *Biochemical Engineering Journal, 21*, 47–52.

Lettinga, G., van Velsen, A. F. M., Hobma, S. W., de Zeeuw, W., & Klapwijk, A. (1980). Use of the upflow sludge blanket (USB) reactor concept for biological wastewater treatment, especially for anaerobic treatment. *Biotechnology and Bioengineering, 22*, 699–734.

Lewandowski, Z., & Boltz, J. P. (2011). Biofilms in water and wastewater treatment. In *Vol. 4. Water-quality engineering* (pp. 529–570). The Netherland: Elsevier.

Licina, G. J. (1989). An overview of microbiologically influenced corrosion in nuclear power plant systems. *Materials Performance, 20*, 55–60.

Little, B. J., Blackwood, D. J., Hinks, J., Lauro, F. M., Marsili, E., Okamoto, A., et al. (2020). Microbially influenced corrosion—Any progress? *Corrosion Science, 170*, 108641. https://doi.org/10.1016/j.corsci.2020.108641.

Little, B., Wagner, P., & Mansfeld, F. (1991). Microbiologically influenced corrosion of metals and alloys. *International Materials Review, 36*, 253–271.

Liu, D., Lau, Y. L., Chau, Y. K., & Pacepavicius, G. J. (1993). Characterization of biofilm development on artificial substratum in natural water. *Water Research, 27*, 361–367.

Maki, J. S., Riftschof, D., Schmidt, A. R., Snyder, A. S., & Mitchell, R. (1989). Factors controlling attachment of bryozoan larvae. A comparison of bacterial films and un-filmed surfaces. *The Biological Bulletin, 117*, 295–302.

Malovanyy, A., Yang, J., Trela, J., & Plaza, E. (2015). Combination of upflow anaerobic sludge blanket (UASB) reactor and partial nitritation/anammox moving bed biofilm reactor (MBBR) for municipal wastewater treatment. *Bioresource Technology, 180*, 144–153.

Mamlouk, D., & Gullo, M. (2013). Acetic acid bacteria: Physiology and carbon sources oxidation. *Indian Journal of Microbiology, 53*, 377–384.

Mazumder, J., & Mohammad, A. (2020). A review of green scale inhibitors: Process, types, mechanism and properties. *Coatings, 10*(928), 1–29.

McQuarrie, J. P., & Boltz, J. P. (2011). Moving bed biofilm reactor technology: Process applications, design, and performance. *Water Environment Research, 83*, 560–575.

Meng, X., Wu, J., Kang, J., Gao, J., Liu, R., Gao, Y., et al. (2018). Comparison of the reduction of chemical oxygen demand in wastewater from mineral processing using the coagulation–flocculation, adsorption and Fenton processes. *Minerals Engineering, 128*, 275–283.

Mercedes, A., & Blanch, A. R. (1994). A set of keys for biochemical identification of environmental *Vibrio* species. *Journal of Applied Microbiology, 76*, 79–85.

Michela, L., Jia, Y., Suzanne, H., Huub, O. D. C., Mike, J., & Gianni, A. (2014). Coexistence of nitrifying, anammox and denitrifying bacteria in a sequencing batch reactor. *Frontiers in Microbiology, 5*. https://www.frontiersin.org/article/10.3389/fmicb.2014.00028.

Miller, M. J., & Ahearn, D. G. (1987). Adherence of *Pseudomonas aeruginosa* to hydrophilic contact lenses and other substrata. *Journal of Clinical Microbiology, 25*, 1392–1397.

Morikawa, M. (2006). Beneficial biofilm formation by industrial bacteria *Bacillus subtilis* and related species. *Journal of Bioscience and Bioengineering, 101*(1), 1–8.

Mueller, G. E. (2012). A simple method for determining sphere packed bed radial porosity. *Powder Technology, 229*, 90–96.

Munoz-Cupa, C., Hu, Y., Xu, C., & Bassi, A. (2021). An overview of microbial fuel cell usage in wastewater treatment, resource recovery and energy production. *Science of the Total Environment, 754*, 142429.

Murthy, P. S., Veeramani, P., Mohamed Ershath, M. I., & Venugopalan, V. P. (2011). Biofouling evaluation in the seawater cooling circuit of an operating coastal power plant. *Power Plant Chemistry, 13*, 314–319.

Nair, K. V. K., Murugan, P., & Eswaran, M. S. (1988). Macro-foulants in Kalpakkam coastal waters east coast of India. *Indian Journal of Marine Sciences, 17*, 341–343.

Nemec, D., & Levec, J. (2005). Flow through packed bed reactors: 2. Two-phase concurrent down-flow. *Chemical Engineering Science, 60*, 6958–6970.

Nghiem, N. P., Kleff, S., & Schwegmann, S. (2017). Succinic acid: Technology development and commercialization. *Fermentation, 3*, 26. https://doi.org/10.3390/fermentation3020026.

Nicolella, C., van Loosdrecht, M., & Heijnen, S. (2000a). Particle-based biofilm reactor technology. *Trends in Biotechnology, 18*, 312–320.

Nicolella, C., van Loosdrecht, M., & Heijnen, S. (2000b). Particle-based biofilm reactor technology. *Trends in Biotechnology, 18*(7), 312.

Noorman, S., & Annaland, M. (2007). Packed bed reactor technology for chemical-looping combustion. *Industrial & Engineering Chemistry Research, 46*. https://doi.org/10.1021/ie061178i.

O'Toole, G. A., & Kolter, R. (1998). The initiation of biofilm formation in *Pseudomonas fluorescens* WCS365 proceeds via multiple, convergent signaling pathways: A genetic analysis. *Molecular Microbiology, 28*, 449–461.

Pavlovsky, L., Younger, J. G., & Solomon, M. J. (2013). In situ rheology of *Staphylococcus epidermidis* bacterial biofilms. *Soft Matter, 9*, 122–131.

Pedersen, K. (1982). Factors regulating microbial biofilm development in a system with slowly flowing seawater. *Applied and Environmental Microbiology, 44*, 1196–1204.

Petrova, O. E., & Sauer, K. (2012). Sticky situations: Key components that control bacterial surface attachment. *Journal of Bacteriology, 194*, 2413–2425.

Petrucci, G. R. (2005). Chlorine dioxide in seawater for fouling control and cost of disinfection in potable water works. *Desalination, 182*, 283–291.

Pochana, K., & Keller, J. (1999). Study of factors affecting simultaneous nitrification and denitrification (SND). *Water Science and Technology, 39*, 61–68.

Presson, A., & Molin, G. (1987). Capacity for biosurfactant production of environmental *Pseudomonas* and *Vibrionaceae* growing on carbohydrates. *Applied Microbiology and Biotechnology, 26*, 439–442.

Pronk, M., de Kreuk, M., Bruin, B., Kamminga, P., Robbert, K., & van Loosdrecht, M. (2015). Full scale performance of the aerobic granular sludge process for sewage treatment. *Water Research, 84*, 207–217.

Qureshi, N., Annous, B. A., Ezeji, T. C., et al. (2005). Biofilm reactors for industrial bioconversion processes: Employing potential of enhanced reaction rates. *Microbial Cell Factories, 4*, 24.

Rajagopal, S., & Jenner, H. A. (2012). Biofouling in cooling water intake systems: Ecological aspects. In S. Rajagopal, H. A. Jenner, & V. P. Venugopalan (Eds.), *Operational and environmental consequences of large industrial cooling water systems* (pp. 13–32). New York: Springer.

Rajagopal, S., Nair, K. V. K., Azariah, J., Van der Velde, G., & Jenner, H. A. (1996). Chlorination and mussel control in the cooling conduits of a tropical coastal power station. *Marine Environmental Research, 41*, 201–220.

Rao, T. S. (2003). Temporal variations in an estuarine biofilm: With emphasis on nitrate reduction. *Estuarine, Coastal and Shelf Science, 56*, 1–9.

Rao, T. S. (2012). Microbial fouling and corrosion: Fundamentals and mechanisms. In S. Rajagopal, H. A. Jenner, & V. P. Venugopalan (Eds.), *Operational and environmental consequences of large industrial cooling water systems* (pp. 95–126). New York: Springer.

Rao, T. S. (2015a). Biofouling in industrial water systems. In Z. Amjad, & K. Demadis (Eds.), *Mineral scales and deposits: Scientific and technological approaches* (pp. 123–140). Amsterdam: Elsevier. Hardcover ISBN: 9780444632289; eBook ISBN: 9780444627520.

Rao, T. S. (2015b). Fundamentals of microbial corrosion and industrial failures. In *Internal corrosion of pipelines* (pp. 1–40). New Delhi: Narosa Publishers.

Rao, T. S. (2018). Biofilms: An enigmatic biochemistry of microbial life. *Journal of Biochemistry, 3*. 004 (1-5).

Rao, T. S. (2020). Mini review: Bacterial biofilms and implant infections: A perspective. *Archives of Orthopaedics, 1*, 98–105.

Rao, T. S. (2021). Microbial biofilms and biotechnology—Some perceptions. *Current Biotechnology, 9*, 263–268.

Rao, T. S., ArunaJyothi, K., Anupkumar, B., Narasimhan, S. V., & Feser, R. (2005). Pitting corrosion of titanium by a fresh-water strain of sulfate reducing bacteria (*Desulfovibrio vulgaris*). *Corrosion Science, 47*, 1071–1084.

Rao, T. S., & Bera, S. (2021). Protective layer *dissolution* by chlorine and corrosion of aluminum brass condenser tubes of a nuclear power plant. *Engineering Failure Analysis, 123*, 105307.

Rao, T. S., Eswaran, M. S., Venugopalan, V. P., Nair, K. V. K., & Mathur, P. K. (1993). Fouling and corrosion in an open recirculation cooling system. *Biofouling, 6*, 245–259.

Rao, T. S., Kesavamoorthy, R., Babu, R. C., & Nair, K. V. K. (1997). Influence of flow on ordering characteristics of a bacterial biofilm. *Current Science, 73*, 69–74.

Rao, T. S., Kesavamoorthy, R., & Rao, C. B. (1996). Orientational order in *Pseudomonas* bacterial colony formation. *Acta Stereologica, 15*, 187–192.

Rao, T. S., Kumar, R., Balamurugan, P., & Vithal, G. K. (2017). Microbial fouling problem in a water treatment plant and its control using biocides. *Biocontrol Science, 22*, 105–119.

Rao, T. S., Kumar, C. S., & Chandra, S. (2020). Comparative assessment of exopolymer production and chemical characteristics of two environmental biofilm isolates of *Pseudomonas aeruginosa*. *Chemical Reports, 2*, 132–143.

Rao, T. S., Kumar, R., More, N. S., Mandal, S. K., Prasad, N., Kakade, C. R., et al. (2015). Microbial corrosion of type 316 stainless steel of a process water cooling system in a nuclear power station. *Innovations in Corrosion and Materials Science, 5*, 53–61.

Rao, T. S., Murthy, P. S., & Harinath, Y. V. (2014). Microbiologically influenced pitting corrosion of stainless-steel type 316 in a seawater flow through system. *Innovations in Corrosion and Materials Science, 4*, 63–72.

Rao, T. S., Murthy, P. S., Veeramani, P., et al. (2021). Assessment of biogrowth assemblages with depth in a seawater intake system of a coastal power station. *Biofouling, 37*. https://doi.org/10.1080/08927014.2021.1933457.

Rao, T. S., & Nair, K. V. K. (1998). Microbiologically influenced stress corrosion cracking failure of admiralty brass condenser tubes in a nuclear power plant cooled by freshwater. *Corrosion Science, 40*, 1821–1836.

Rao, T. S., Nancharaiah, Y. V., & Nair, K. V. K. (1998). Biocidal efficacy of monochloramine against biofilm bacteria. *Biofouling, 12*, 321–332.

Rao, T. S., Rani, P. G., Venugopalan, V. P., & Nair, K. V. K. (1997). Biofilm formation in a fresh water impoundment under experimental photic and aphotic conditions. *Biofouling, 11*, 265–282.

Rao, T. S., Sairam, T. N., Viswanathan, B., & Nair, K. V. K. (2000). Carbon steel corrosion by iron oxidizing and sulfate bacteria in a freshwater cooling system. *Corrosion Science, 42*, 1417–1431.

Rao, D., Webb, J. S., & Kjelleberg, S. (2005). Competitive interactions in mixed-species biofilms containing the marine bacterium *Pseudoalteromonas tunicata*. *Applied and Environmental Microbiology, 71*, 1729–1736.

Raveendran, T. V., & LimnaMol, V. P. (2009). Natural product antifoulants. *Current Science, 97*, 509–520.

Robertson, L. A., & Kuenen, J. G. (1983). *Thiosphaera pantotropha* gen. nov. sp. a facultatively anaerobic, facultatively autotrophic sulphur bacterium. *Journal of General Microbiology, 129*, 2847–2855.

Robertson, S. R., & McLean, R. J. C. (2015). Beneficial biofilms. *Bioengineering, 2*(4), 437–448.

Sanchez, S., & Demain, A. L. (2011). Enzymes and bioconversions of industrial, pharmaceutical, and biotechnological significance. *Organic Process Research and Development, 15*, 224–230.

Sarjit, A., Tan, S. M., & Dykes, G. A. (2015). Surface modification of materials to encourage beneficial biofilm formation. *Bioengineering, 2*(4), 404–422.

Satpathy, K. K., Rao, T. S., Rajamohan, R., & Nair, K. V. K. (2001). Some investigations of condenser slime and biocorrosion at Rajasthan atomic power station. *Transactions of the Indian Institute of Metals, 54*, 125–130.

Seed, L. J. (1990). The significance of organisms in corrosion. *Corrosion Reviews, 9*, 1–102.

Shin, J. H., Lee, S. M., Jung, Y. J., Chung, C. Y., & Noh, H. S. (2005). Enhanced COD and nitrogen removals for the treatment of swine wastewater by combining submerged membrane bioreactor (MBR) and anaerobic upflow bed filter (AUBF) reactor. *Process Biochemistry, 40*, 3769–3776.

Shiva, S. (2012). Ethanol production by *Zymomonas mobilis* PTCC 1718 using low cost substrates. *African Journal of Microbiology Research, 6*. https://doi.org/10.5897/AJMR11.576.

Shukla, S. K., & Rao, T. S. (2013). Effect of calcium on *Staphylococcus aureus* biofilm architecture: A confocal laser scanning microscopic study. *Colloids and Surfaces B: Biointerfaces, 103*, 448–454.

Simone, M., Daniele, D., Elisabeth, L., Barbara, B., Gerhard, J., & Eva, S. (2019). Niche differentiation of aerobic and anaerobic ammonia oxidizers in a high latitude deep oxygen minimum zone. *Frontiers in Microbiology, 10*. https://doi.org/10.3389/fmicb.2019.02141.

Singh, V., & Mittal, A. (2012). Characterization of biofilm of a rotating biological contactor treating synthetic wastewater. *Water Science and Technology, 66*, 429–437.

Starr, M. P., Stolp, H., Truper, H. G., Balows, A., & Schlegel, H. G. (1986). *The prokaryotes: A handbook on habitat, isolation and identification of bacteria* (Vols. I and II). Springer Verlag, Berlin.

Stavridou, I., & Forzi, L. (2011). Biofilms: Friend or foe? *Virulence*, *2*(5), 475–476.

Stewart, P. (2012). Mini-review: Convection around biofilms. *Biofouling*, *28*, 187–198.

Stewart, P. S., & Franklin, M. J. (2008). Physiological heterogeneity in biofilms. *Nature Reviews Microbiology*, *6*, 199–210.

Stewart, P. S., Peyton, B. M., Drury, W. J., & Murga, R. (1993). Quantitative observation of heterogeneities in *Pseudomonas aeruginosa* biofilm. *Applied and Environmental Microbiology*, *59*, 327–329.

Stockar, U., & Liu, J. S. (1999). Does microbial life always feed on negative entropy? Thermodynamic analysis of microbial growth. *Biochimica et Biophysica Acta (BBA) - Bioenergetics*, *1412*, 191–211.

Sweta, S., Kumar, R., Harinath, Y. V., & Rao, T. S. (2013). Biocidal efficacy of ozone and chlorine on planktonic and biofilm cells of two marine bacteria species. *Ozone: Science & Engineering*, *35*, 90–100.

Thiyagarajan, V., Stanley, C. K. L., Cheung, S. C. K., & Quian, P. Y. (2006). Cypris habitat selection facilitated by microbial films influences the vertical distribution of subtidal barnacle *Balanus trigonus*. *Microbial Ecology*, *51*, 431–440.

Todhanakasem, T. (2013). Microbial biofilm in the industry. *African Journal of Microbiology Research*, *7*(17), 1625–1634.

Todhanakasem, T., O-Lan, S., Piyawit, K., Sanya, K., Pattanop, K., & Verawat, C. (2019). *Zymomonas mobilis* biofilm reactor for ethanol production using rice straw hydrolysate under continuous and repeated batch processes. *Frontiers in Microbiology*, *10*. https://www.frontiersin.org/article/10.3389/fmicb.2019.01777.

Toyofuku, M., Inaba, T., Kiyokawa, T., Obana, N., Yawata, Y., & Nomura, N. (2016). Environmental factors that shape biofilm formation. *Bioscience, Biotechnology, and Biochemistry*, *80*(1), 7–12.

Urbance, S., Pometto, A., III, Dispirito, A., & Demirci, A. (2003). Medium evaluation and plastic composite support ingredient selection for biofilm formation and succinic acid production by *Actinobacillus succinogenes*. *Food Biotechnology*, *17*, 53–65.

Van der Kooij, D., Veenendaal, H. R., Baars-Lorist, C., Van der Klift, D. W., & Drost, Y. C. (1995). Biofilm formation on surfaces of glass and teflon exposed to treated water. *Water Research*, *29*, 1655–1662.

Van der Star, W. R., Abma, W. R., Blommers, D., Mulder, J. W., Tokutomi, T., Strous, M., et al. (2007). Startup of reactors for anoxic ammonium oxidation: Experiences from the first full-scale anammox reactor in Rotterdam. *Water Research*, *41*(18), 4149–4163.

Van Loosdrecht, M. M., Lyklema, J., Norde, W., & Zehnder, A. B. (1990). Influence of interfaces on microbial activity. *Microbiological Reviews*, *54*, 75–87.

Vasanthadevi, A., Satoshi, T., Kenji, F., & Kazuya, K. (1998). Factors affecting nitrogen removal from domestic wastewater using immobilized bacteria. *Water Science and Technology*, *38*, 193–202.

Veeresh, G. S., Kumar, P., & Mehrotra, I. (2005). Treatment of phenol and cresols in upflow anaerobic sludge blanket (UASB) process: A review. *Water Research*, *39*, 154–170.

Venkateswarlu, K. S. (1990). Marine biofouling and its control in power industry. *Chemical Business*, 39–41.

Venugopalan, V. P., Kuehn, M., Hausner, M., Springael, D., Wilderer, P. A., & Wuertz, S. (2005). Architecture of a nascent *Sphingomonas* sp. biofilm under varied hydrodynamic conditions. *Applied and Environmental Microbiology*, *71*(5), 2677–2686.

Wood, T. L., Guha, R., Tang, L., Geitner, M., Kumar, M., & Wood, T. K. (2016). Living biofouling-resistant membranes as a model for the beneficial use of engineered biofilms. *Proceedings of the National Academy of Sciences of the United States of America*, *113*, E2802–E2811. 1–10.

Yebra, D. M., Kiil, S., & Johansen, K. D. (2004). Antifouling technology—Past, present and future steps towards efficient and environmentally friendly antifouling coatings. *Progress in Organic Coatings*, *50*, 75–104.

Yuan, Z., & Blackall, L. L. (2002). Sludge population optimization: A new dimension for the control of biological wastewater treatment systems. *Water Research*, *36*, 482–490.

Zardus, J. D., Nedved, B. T., Huang, Y., Tran, C., & Hadfield, M. G. (2008). Microbial biofilms facilitate adhesion in biofouling invertebrates. *The Biological Bulletin*, *214*, 91–98.

Zhou, M., Wang, H., Hassett, D. J., & Gu, T. (2013). Recent advances in microbial fuel cells (MFCs) and microbial electrolysis cells (MECs) for wastewater treatment, bioenergy and bioproducts. *Journal of Chemical Technology and Biotechnology*, *88*, 508–518.

Ziye, H., Tommaso, L., van Loosdrecht, M., & Boran, K. (2013). Nitrogen removal with the anaerobic ammonium oxidation process. *Biotechnology Letters*, *35*. https://doi.org/10.1007/s10529-013-1196-4.

Zou, G., Papirio, S., Lakaniemi, A. M., Ahoranta, S. H., & Puhakka, J. A. (2016). High rate autotrophic denitrification in fluidized-bed biofilm reactors. *Chemical Engineering Journal*, *284*, 1287–1294.

Index

Note: Page numbers followed by *f* indicate figures and *t* indicate tables.

N

O

P

T

U

V

W

X

Y

Z

9780323999779